ERRATA

VOLUME 19

Page	Line	For:	Read:
402	Fig. 12	Na_2S, Na_2CO_3, CaO	Na_2S, NaOH, $CaCO_3$
422	Fig. 3 Fig. 5	Transpose legends	
691	8 from bottom	Table 1	Table 5

The following are corrections for Table 6 of Radioisotopes; A = Avg. col., H = Half-life col., I = Intensity col., Is = Isotope col., RN = Registry number col., T = Type of radiation col. Number following letter indicates line in that column.

Page	Line	For:	Read:
692	H, 6	2.602 yr	2.602 yr 2
	I, 18	16.4 19	16.21 19
699	A, 2	9.2248 7	9.22482 7
	A, 3	9.25 14 7	9.25174 7
705	I, 34	1.14	1.1 4
717	H, 1	6.5 yr	6.5×10^6 yr 3
	H, 2	6.49 h	6.49 h 5
	I, 30	79.7 7	79 7
	A, 33	361.9	361.0 9
	A, 46	21.9403 3	21.9903 3
718	A, 27	23.1736 20	23.17360 20
720	H, 7	32.1 mo	32.1 min 3
721	I, 27	2.09	2.09 24
724	I, 2 from bottom	6.322	6.3 22
725	I, 43	4.24	4.2 4
	I, 44	0.304	0.30
	I, 50	(blank)	2.85 22
	I, 51	2.85 22	0.75
726	I, 7 from bottom	100.0 33	100.03 3
731	A, 8	19 3	193
	T, 26	max γ'	max $\gamma^{\pm}$
733	T, 18	weak βs	3 weak βs
	H, 4	13.16 d	13.16 d 3
734	I, 14	100.05	100.0 5
	H, 3	14.17 mo	14.17 min 7
	H, 4	32.2 mo	32.2 min 1
735	H, 2	1.35×10^6 yr 16	1.35×10^{11} yr 16
	I, 2	22.3 3	22 3
	I, 10	32.9 8	32.9 18

ERRATA

VOLUME 19

Page	Line	For:	Read:
736	Is, 7	^{145}Pm	^{145}Pr
	RN, 7	[15706-44-2]	[15765-23-8]
	RN, 8	(blank)	[15706-44-2]
	H, 8	17.7 yr 2	17.7 yr 4
738	I, 1	1.68 10	1.68 8
742	I, 52	7.59 18	4.59 18
743	A, 40	44.4816	44.4816 3
749	A, 35	97.286 9	97.2869 5
	A, 51	845.6808 3	84.6808 3
755	H, 4	4.02 d	4.02 d 1
756	I, 58	7.5 11	7.5 1
758	T, 7	x ray $K_{\alpha1}$	x ray $K_{\alpha2}$
	T, 8	x ray $K_{\alpha2}$	x ray $K_{\alpha1}$
	H, 1	94.4 min	94.4 min 8
759	H, 3	48.4 mo	48.4 min 3
761	I, 14	91.4 2	91.4 10
	I, 19	1.01 71	1.01 11
765	I, 34	20 5	20.3
766	Is, 1, 3, 4	(A decay)	(α decay)
767	H, 4	(blank)	21.773 yr 3
771	A, 22	415.76	415.76 4
776	H, 6	8.26 × 107 yr 8	8.26 × 10^7 yr 9
777	A, 20	20 27	2027

KIRK-OTHMER

ENCYCLOPEDIA OF CHEMICAL TECHNOLOGY

Third Edition

VOLUME 19

Powder Coatings
to
Recycling

EDITORIAL BOARD

HERMAN F. MARK
Polytechnic Institute of New York

DONALD F. OTHMER
Polytechnic Institute of New York

CHARLES G. OVERBERGER
University of Michigan

GLENN T. SEABORG
University of California, Berkeley

EXECUTIVE EDITOR

MARTIN GRAYSON

ASSOCIATE EDITOR

DAVID ECKROTH

KIRK-OTHMER

ENCYCLOPEDIA OF CHEMICAL TECHNOLOGY

THIRD EDITION

VOLUME 19

POWDER COATINGS
TO
RECYCLING

A WILEY-INTERSCIENCE PUBLICATION

John Wiley & Sons

NEW YORK • CHICHESTER • BRISBANE • TORONTO • SINGAPORE

Copyright © 1982 by John Wiley & Sons, Inc.

All rights reserved. Published simultaneously in Canada.

Reproduction or translation of any part of this work beyond that permitted by Sections 107 or 108 of the 1976 United States Copyright Act without the permission of the copyright owner is unlawful. Requests for permission or further information should be addressed to the Permissions Department, John Wiley & Sons, Inc.

Library of Congress Cataloging in Publication Data:

Main entry under title:
   Encyclopedia of chemical technology.

   At head of title: Kirk-Othmer.
   "A Wiley-Interscience publication."
   Includes bibliographies.
   1. Chemistry, Technical—Dictionaries. I. Kirk, Raymond Eller, 1890–1957. II. Othmer, Donald Frederick, 1904—
     III. Grayson, Martin. IV. Eckroth, David. V. Title: Kirk-Othmer encyclopedia of chemical technology.

TP9.E685   1978   660'.03   77-15820
ISBN 0-471-02072-9

Printed in the United States of America

CONTENTS

Powder coatings, 1
Powder metallurgy, 28
Power generation, 62
Pressure measurement, 95
Printing processes, 110
Process research and
    development, 164
Programmable pocket
    computers, 177
Propyl alcohols, 198
Propylene, 228
Propylene oxide, 246
Prosthetic and biomedical
    devices, 275
Proteins, 314
Psychopharmacological agents, 342
Pulp, 379
Pulp, synthetic, 420
Pyrazoles, pyrazolines,
    pyrazolones, 436
Pyridine and pyridine
    derivatives, 454
Pyrotechnics, 484

Pyrrole and pyrrole derivatives, 499

Quaternary ammonium
    compounds, 521
Quinoline and isoquinolines, 532
Quinones, 572

Radiation curing, 607
Radioactive drugs, 625
Radioactive tracers, 633
Radioactivity, natural, 639
Radioisotopes, 682
Radiopaques, 786
Radioprotective agents, 801
Rare-earth elements, 833
Rayon, 855
Reactor technology, 880
Recording disks, 915
Recreational surfaces, 922
Recycling, 936

EDITORIAL STAFF FOR VOLUME 19

Executive Editor: **Martin Grayson**
Associate Editor: **David Eckroth**
Production Supervisor: **Michalina Bickford**
Editors: **Caroline I. Eastman**   **Carolyn Golojuch**   **Anna Klingsberg**
         Goldie Wachsman   **Mimi Wainwright**

CONTRIBUTORS TO VOLUME 19

James Ackerman, *Sterling-Winthrop Research Institute, Rensselaer, New York,* Radiopaques

Harvey Alter, *Chamber of Commerce of the United States, Washington, D.C.,* Introduction and Plastics both under Recycling

L. R. Anderson, *GAF Corporation, Wayne, New Jersey,* Pyrrole and pyrrole derivatives

Sushil K. Batra, *GIMRET International, Shrewsbury, Massachusetts,* Power generation

Donald A. Becker, *National Bureau of Standards, Washington, D.C.,* Oil under Recycling

R. F. Benenati, *Polytechnic Institute of New York, Brooklyn, New York,* Programmable pocket computers

John W. Blieszner, *Amoco Chemicals Corporation, Naperville, Illinois,* Propylene

Gilbert Bourcier, *Reynolds Metal Co., Richmond, Virginia,* Nonferrous metals under Recycling

C. Raymond Brandt, *Honeywell Inc., Fort Washington, Pennsylvania,* Pressure measurement

Michael H. Bruno, *Consultant, Nashua, New Hampshire,* Printing processes

R. A. Budenholzer, *Illinois Institute of Technology, Chicago, Illinois,* Power generation

CONTRIBUTORS TO VOLUME 19

Edmund S. Copeland, *National Institutes of Health, Bethesda, Maryland,* Radioprotective agents

T. John Dempsey, *The Dow Chemical Company, Freeport, Texas,* Propylene oxide

James Early, *National Bureau of Standards, Washington, D.C.,* Ferrous metals under Recycling

K. Thomas Finley, *State University College at Brockport, Brockport, New York,* Quinones

Charles G. Gebelein, *Chemistry Department, Youngstown State University, and Pharmacology Department, Northeastern Ohio Universities, College of Medicine, Youngstown, Ohio,* Prosthetic and biomedical devices

Gerald L. Goe, *Reilly Tar & Chemical Corporation, Indianapolis, Indiana,* Pyridine and pyridine derivatives

James R. Grant, *Mytec, Inc., Battle Creek, Michigan,* Paper under Recycling

Allan M. Green, *New England Nuclear Corporation, Boston, Massachusetts,* Radioactive drugs

Irwin Gruverman, *New England Nuclear Corporation, Boston, Massachusetts,* Radioactive drugs

W. F. Hamner, *Monsanto Company, Research Triangle Park, North Carolina,* Recreational surfaces

Alexander P. Hardt, *Lockheed Missiles and Space Company, Inc., Palo Alto, California,* Pyrotechnics

Samuel N. Holter, *Koppers Company, Inc., Monroeville, Pennsylvania,* Quinolines and isoquinolines

W. D. Horst, *Hoffmann-LaRoche & Co., Inc., Nutley, New Jersey,* Psychopharmacological agents

Eugene V. Hort, *GAF Corporation, Wayne, New Jersey,* Pyrrole and pyrrole derivatives

Richard O. Kirk, *The Dow Chemical Company, Freeport, Texas,* Propylene oxide

Daniel L. Klayman, *Walter Reed Army Institute of Research, Washington, D.C.,* Radioprotective agents

Truman P. Kohman, *Carnegie-Mellon University, Pittsburgh, Pennsylvania,* Radioactivity, natural

Ernest I. Korchak, *Scientific Design Company, Inc., a division of The Halcon SD Group, Inc., New York, New York,* Process research and development

Gabriel Kornis, *The Upjohn Company, Kalamazoo, Michigan,* Pyrazoles, pyrazolines, pyrazolones

Leonard Laub, *Vision Three Inc., Pasadena, California,* Recording disks

John Lundberg, *Georgia Institute of Technology, Atlanta, Georgia,* Rayon

Paul Marsh, *Marsh-Eco-Service Co., Inc., Hamilton, Ohio,* Glass under Recycling

Murray J. Martin, *Oak Ridge National Laboratory, Oak Ridge, Tennessee,* Radioisotopes

Vincent D. McGinnis, *Battelle Columbus Laboratory, Columbus, Ohio,* Radiation curing

James Minor, *U.S. Forest Products Laboratory, Madison, Wisconsin,* Pulp

Robert E. O'Brien, *New England Nuclear Corporation, Boston, Massachusetts,* Radioactive tracers

T. A. Orofino, *Monsanto Company, Research Triangle Park, North Carolina,* Recreational surfaces

Anthony J. Papa, *Union Carbide Corporation, South Charleston, West Virginia,* Isopropyl alcohol under Propyl alcohols

Christos G. Papadopoulos, *Amoco Chemicals Corporation, Naperville, Illinois,* Propylene

John Paul, *Pedco Environmental, Inc., Arlington, Texas,* Rubber under Recycling

Peter L. Pellett, *University of Massachusetts, Amherst, Massachusetts,* Proteins

Terence Rave, *Hercules, Incorporated, Wilmington, Delaware,* Pulp, synthetic

Richard A. Reck, *Armak Company, Chicago, Illinois,* Quaternary ammonium compounds

D. S. Richart, *The Polymer Corporation, Reading, Pennsylvania,* Powder coatings

James H. Robins, *Mytec, Inc., Battle Creek, Michigan,* Paper under Recycling

Kempton H. Roll, *Metal Powder Industries Federation, American Powder Metallurgy Institute, Princeton, New Jersey,* Powder metallurgy

Morris R. Schoenberg, *Amoco Chemicals Corporation, Naperville, Illinois,* Propylene

F. H. Spedding, *Ames Laboratory, Iowa State University, Ames, Iowa,* Rare-earth elements

L. Spinicelli, *Celanese Chemical Co., Summit, New Jersey,* n-Propyl alcohol under Propyl alcohols

L. H. Sternbach, *Hoffmann-LaRoche & Co., Inc., Nutley, New Jersey,* Psychopharmacological agents

Barry L. Tarmy, *Exxon Research and Engineering Company, Florham Park, New Jersey,* Reactor technology

Albin Turbak, *Georgia Institute of Technology, Atlanta, Georgia,* Rayon

J. D. Unruh, *Celanese Chemical Company, Corpus Christi, Texas,* n-Propyl alcohol under Propyl alcohols

NOTE ON CHEMICAL ABSTRACTS SERVICE REGISTRY NUMBERS AND NOMENCLATURE

Chemical Abstracts Service (CAS) Registry Numbers are unique numerical identifiers assigned to substances recorded in the CAS Registry System. They appear in brackets in the *Chemical Abstracts* (CA) substance and formula indexes following the names of compounds. A single compound may have many synonyms in the chemical literature. A simple compound like phenethylamine can be named β-phenylethylamine or, as in *Chemical Abstracts*, benzeneethanamine. The usefulness of the *Encyclopedia* depends on accessibility through the most common correct name of a substance. Because of this diversity in nomenclature careful attention has been given the problem in order to assist the reader as much as possible, especially in locating the systematic CA index name by means of the Registry Number. For this purpose, the reader may refer to the CAS Registry Handbook-Number Section which lists in numerical order the Registry Number with the *Chemical Abstracts* index name and the molecular formula; eg, **458-88-8,** Piperidine, 2-propyl-, (*S*)-, $C_8H_{17}N$; in the *Encyclopedia* this compound would be found under its common name, coniine [*458-88-8*]. The Registry Number is a valuable link for the reader in retrieving additional published information on substances and also as a point of access for such on-line data bases as Chemline, Medline, and Toxline.

In all cases, the CAS Registry Numbers have been given for title compounds in articles and for all compounds in the index. All specific substances indexed in *Chemical Abstracts* since 1965 are included in the CAS Registry System as are a large number of substances derived from a variety of reference works. The CAS Registry System identifies a substance on the basis of an unambiguous computer-language description of its molecular structure including stereochemical detail. The Registry Number is a machine-checkable number (like a Social Security number) assigned in sequential order to each substance as it enters the registry system. The value of the number lies in the fact that it is a concise and unique means of substance identification, which is independent of, and therefore bridges, many systems of chemical nomenclature. For polymers, one Registry Number is used for the entire family; eg, polyoxyethylene (20) sorbitan monolaurate has the same number as all of its polyoxyethylene homologues.

Registry numbers for each substance will be provided in the third edition cumulative index and appear as well in the annual indexes (eg, Alkaloids shows the Registry Number of all alkaloids (title compounds) in a table in the article as well, but the intermediates have their Registry Numbers shown only in the index). Articles such as Analytical methods, Batteries and electric cells, Chemurgy, Distillation, Economic evaluation, and Fluid mechanics have no Registry Numbers in the text.

Cross-references are inserted in the index for many common names and for some systematic names. Trademark names appear in the index. Names that are incorrect, misleading or ambiguous are avoided. Formulas are given very frequently in the text to help in identifying compounds. The spelling and form used, even for industrial names, follow American chemical usage, but not always the usage of *Chemical Abstracts* (eg, *coniine* is used instead of *(S)-2-propylpiperidine*, *aniline* instead of *benzenamine*, and *acrylic acid* instead of *2-propenoic acid*).

There are variations in representation of rings in different disciplines. The dye industry does not designate aromaticity or double bonds in rings. All double bonds and aromaticity are shown in the *Encyclopedia* as a matter of course. For example, tetralin has an aromatic ring and a saturated ring and its structure appears in the

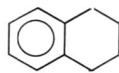

Encyclopedia with its common name, Registry Number enclosed in brackets, and parenthetical CA index name, ie, tetralin, [*119-64-2*] (1,2,3,4-tetrahydronaphthalene). With names and structural formulas, and especially with CAS Registry Numbers the aim is to help the reader have a concise means of substance identification.

CONVERSION FACTORS, ABBREVIATIONS, AND UNIT SYMBOLS

SI Units (Adopted 1960)

A new system of measurement, the International System of Units (abbreviated SI), is being implemented throughout the world. This system is a modernized version of the MKSA (meter, kilogram, second, ampere) system, and its details are published and controlled by an international treaty organization (The International Bureau of Weights and Measures) (1).

SI units are divided into three classes:

BASE UNITS

length	meter[†] (m)
mass[‡]	kilogram (kg)
time	second (s)
electric current	ampere (A)
thermodynamic temperature[§]	kelvin (K)
amount of substance	mole (mol)
luminous intensity	candela (cd)

[†] The spellings "metre" and "litre" are preferred by ASTM; however "-er" are used in the Encyclopedia.
[‡] "Weight" is the commonly used term for "mass."
[§] Wide use is made of "Celsius temperature" (t) defined by

$$t = T - T_0$$

where T is the thermodynamic temperature, expressed in kelvins, and $T_0 = 273.15$ K by definition. A temperature interval may be expressed in degrees Celsius as well as in kelvins.

FACTORS, ABBREVIATIONS, AND SYMBOLS

SUPPLEMENTARY UNITS

plane angle — radian (rad)
solid angle — steradian (sr)

DERIVED UNITS AND OTHER ACCEPTABLE UNITS

These units are formed by combining base units, supplementary units, and other derived units (2–4). Those derived units having special names and symbols are marked with an asterisk in the list below:

Quantity	Unit	Symbol	Acceptable equivalent
*absorbed dose	gray	Gy	J/kg
acceleration	meter per second squared	m/s^2	
*activity (of ionizing radiation source)	becquerel	Bq	1/s
area	square kilometer	km^2	
	square hectometer	hm^2	ha (hectare)
	square meter	m^2	
*capacitance	farad	F	C/V
concentration (of amount of substance)	mole per cubic meter	mol/m^3	
*conductance	siemens	S	A/V
current density	ampere per square meter	A/m^2	
density, mass density	kilogram per cubic meter	kg/m^3	g/L; mg/cm^3
dipole moment (quantity)	coulomb meter	C·m	
*electric charge, quantity of electricity	coulomb	C	A·s
electric charge density	coulomb per cubic meter	C/m^3	
electric field strength	volt per meter	V/m	
electric flux density	coulomb per square meter	C/m^2	
*electric potential, potential difference, electromotive force	volt	V	W/A
*electric resistance	ohm	Ω	V/A
*energy, work, quantity of heat	megajoule	MJ	
	kilojoule	kJ	
	joule	J	N·m
	electron volt[†]	eV[†]	
	kilowatt-hour[†]	kW·h[†]	

[†] This non-SI unit is recognized by the CIPM as having to be retained because of practical importance or use in specialized fields (1).

Quantity	Unit	Symbol	Acceptable equivalent
energy density	joule per cubic meter	J/m³	
*force	kilonewton	kN	
	newton	N	kg·m/s²
*frequency	megahertz	MHz	
	hertz	Hz	1/s
heat capacity, entropy	joule per kelvin	J/K	
heat capacity (specific), specific entropy	joule per kilogram kelvin	J/(kg·K)	
heat transfer coefficient	watt per square meter kelvin	W/(m²·K)	
*illuminance	lux	lx	lm/m²
*inductance	henry	H	Wb/A
linear density	kilogram per meter	kg/m	
luminance	candela per square meter	cd/m²	
*luminous flux	lumen	lm	cd·sr
magnetic field strength	ampere per meter	A/m	
*magnetic flux	weber	Wb	V·s
*magnetic flux density	tesla	T	Wb/m²
molar energy	joule per mole	J/mol	
molar entropy, molar heat capacity	joule per mole kelvin	J/(mol·K)	
moment of force, torque	newton meter	N·m	
momentum	kilogram meter per second	kg·m/s	
permeability	henry per meter	H/m	
permittivity	farad per meter	F/m	
*power, heat flow rate, radiant flux	kilowatt	kW	
	watt	W	J/s
power density, heat flux density, irradiance	watt per square meter	W/m²	
*pressure, stress	megapascal	MPa	
	kilopascal	kPa	
	pascal	Pa	N/m²
sound level	decibel	dB	
specific energy	joule per kilogram	J/kg	
specific volume	cubic meter per kilogram	m³/kg	
surface tension	newton per meter	N/m	
thermal conductivity	watt per meter kelvin	W/(m·K)	
velocity	meter per second	m/s	
	kilometer per hour	km/h	
viscosity, dynamic	pascal second	Pa·s	
	millipascal second	mPa·s	
viscosity, kinematic	square meter per second	m²/s	

FACTORS, ABBREVIATIONS, AND SYMBOLS

Quantity	Unit	Symbol	Acceptable equivalent
	square millimeter per second	mm²/s	
volume	cubic meter	m³	
	cubic decimeter	dm³	L(liter) (5)
	cubic centimeter	cm³	mL
wave number	1 per meter	m⁻¹	
	1 per centimeter	cm⁻¹	

In addition, there are 16 prefixes used to indicate order of magnitude, as follows:

Multiplication factor	Prefix	Symbol	Note
10^{18}	exa	E	
10^{15}	peta	P	
10^{12}	tera	T	
10^{9}	giga	G	
10^{6}	mega	M	
10^{3}	kilo	k	
10^{2}	hecto	h^a	^{a} Although hecto, deka, deci, and centi are SI prefixes, their use should be avoided except for SI unit-multiples for area and volume and nontechnical use of centimeter, as for body and clothing measurement.
10	deka	da^a	
10^{-1}	deci	d^a	
10^{-2}	centi	c^a	
10^{-3}	milli	m	
10^{-6}	micro	μ	
10^{-9}	nano	n	
10^{-12}	pico	p	
10^{-15}	femto	f	
10^{-18}	atto	a	

For a complete description of SI and its use the reader is referred to ASTM E 380 (4) and the article Units and Conversion Factors which will appear in a later volume of the *Encyclopedia*.

A representative list of conversion factors from non-SI to SI units is presented herewith. Factors are given to four significant figures. Exact relationships are followed by a dagger. A more complete list is given in ASTM E 380-79(4) and ANSI Z210.1-1976 (6).

Conversion Factors to SI Units

To convert from	To	Multiply by
acre	square meter (m²)	4.047×10^{3}
angstrom	meter (m)	1.0×10^{-10}†
are	square meter (m²)	1.0×10^{2}†
astronomical unit	meter (m)	1.496×10^{11}
atmosphere	pascal (Pa)	1.013×10^{5}
bar	pascal (Pa)	1.0×10^{5}†
barn	square meter (m²)	1.0×10^{-28}†

† Exact.

To convert from	To	Multiply by
barrel (42 U.S. liquid gallons)	cubic meter (m^3)	0.1590
Bohr magneton (μ_β)	J/T	9.274×10^{-24}
Btu (International Table)	joule (J)	1.055×10^3
Btu (mean)	joule (J)	1.056×10^3
Btu (thermochemical)	joule (J)	1.054×10^3
bushel	cubic meter (m^3)	3.524×10^{-2}
calorie (International Table)	joule (J)	4.187
calorie (mean)	joule (J)	4.190
calorie (thermochemical)	joule (J)	4.184†
centipoise	pascal second (Pa·s)	1.0×10^{-3}†
centistoke	square millimeter per second (mm^2/s)	1.0†
cfm (cubic foot per minute)	cubic meter per second (m^3/s)	4.72×10^{-4}
cubic inch	cubic meter (m^3)	1.639×10^{-5}
cubic foot	cubic meter (m^3)	2.832×10^{-2}
cubic yard	cubic meter (m^3)	0.7646
curie	becquerel (Bq)	3.70×10^{10}†
debye	coulomb·meter (C·m)	3.336×10^{-30}
degree (angle)	radian (rad)	1.745×10^{-2}
denier (international)	kilogram per meter (kg/m)	1.111×10^{-7}
	tex‡	0.1111
dram (apothecaries')	kilogram (kg)	3.888×10^{-3}
dram (avoirdupois)	kilogram (kg)	1.772×10^{-3}
dram (U.S. fluid)	cubic meter (m^3)	3.697×10^{-6}
dyne	newton (N)	1.0×10^{-5}†
dyne/cm	newton per meter (N/m)	1.0×10^{-3}†
electron volt	joule (J)	1.602×10^{-19}
erg	joule (J)	1.0×10^{-7}†
fathom	meter (m)	1.829
fluid ounce (U.S.)	cubic meter (m^3)	2.957×10^{-5}
foot	meter (m)	0.3048†
footcandle	lux (lx)	10.76
furlong	meter (m)	2.012×10^{-2}
gal	meter per second squared (m/s^2)	1.0×10^{-2}†
gallon (U.S. dry)	cubic meter (m^3)	4.405×10^{-3}
gallon (U.S. liquid)	cubic meter (m^3)	3.785×10^{-3}
gallon per minute (gpm)	cubic meter per second (m^3/s)	6.308×10^{-5}
	cubic meter per hour (m^3/h)	0.2271
gauss	tesla (T)	1.0×10^{-4}
gilbert	ampere (A)	0.7958
gill (U.S.)	cubic meter (m^3)	1.183×10^{-4}
grad	radian	1.571×10^{-2}
grain	kilogram (kg)	6.480×10^{-5}
gram force per denier	newton per tex (N/tex)	8.826×10^{-2}

† Exact.
‡ See footnote on p. xiv.

xviii FACTORS, ABBREVIATIONS, AND SYMBOLS

To convert from	To	Multiply by
hectare	square meter (m^2)	$1.0 \times 10^{4\dagger}$
horsepower (550 ft·lbf/s)	watt (W)	7.457×10^2
horsepower (boiler)	watt (W)	9.810×10^3
horsepower (electric)	watt (W)	$7.46 \times 10^{2\dagger}$
hundredweight (long)	kilogram (kg)	50.80
hundredweight (short)	kilogram (kg)	45.36
inch	meter (m)	$2.54 \times 10^{-2\dagger}$
inch of mercury (32°F)	pascal (Pa)	3.386×10^3
inch of water (39.2°F)	pascal (Pa)	2.491×10^2
kilogram force	newton (N)	9.807
kilowatt hour	megajoule (MJ)	$3.6^\dagger$
kip	newton (N)	4.48×10^3
knot (international)	meter per second (m/s)	0.5144
lambert	candela per square meter (cd/m^2)	3.183×10^3
league (British nautical)	meter (m)	5.559×10^3
league (statute)	meter (m)	4.828×10^3
light year	meter (m)	9.461×10^{15}
liter (for fluids only)	cubic meter (m^3)	$1.0 \times 10^{-3\dagger}$
maxwell	weber (Wb)	$1.0 \times 10^{-8\dagger}$
micron	meter (m)	$1.0 \times 10^{-6\dagger}$
mil	meter (m)	$2.54 \times 10^{-5\dagger}$
mile (statute)	meter (m)	1.609×10^3
mile (U.S. nautical)	meter (m)	$1.852 \times 10^{3\dagger}$
mile per hour	meter per second (m/s)	0.4470
millibar	pascal (Pa)	1.0×10^2
millimeter of mercury (0°C)	pascal (Pa)	$1.333 \times 10^{2\dagger}$
minute (angular)	radian	2.909×10^{-4}
myriagram	kilogram (kg)	10
myriameter	kilometer (km)	10
oersted	ampere per meter (A/m)	79.58
ounce (avoirdupois)	kilogram (kg)	2.835×10^{-2}
ounce (troy)	kilogram (kg)	3.110×10^{-2}
ounce (U.S. fluid)	cubic meter (m^3)	2.957×10^{-5}
ounce-force	newton (N)	0.2780
peck (U.S.)	cubic meter (m^3)	8.810×10^{-3}
pennyweight	kilogram (kg)	1.555×10^{-3}
pint (U.S. dry)	cubic meter (m^3)	5.506×10^{-4}
pint (U.S. liquid)	cubic meter (m^3)	4.732×10^{-4}
poise (absolute viscosity)	pascal second (Pa·s)	$0.10^\dagger$
pound (avoirdupois)	kilogram (kg)	0.4536
pound (troy)	kilogram (kg)	0.3732
poundal	newton (N)	0.1383
pound-force	newton (N)	4.448
pound per square inch (psi)	pascal (Pa)	6.895×10^3
quart (U.S. dry)	cubic meter (m^3)	1.101×10^{-3}

† Exact.

To convert from	To	Multiply by
quart (U.S. liquid)	cubic meter (m³)	9.464×10^{-4}
quintal	kilogram (kg)	$1.0 \times 10^{2\dagger}$
rad	gray (Gy)	$1.0 \times 10^{-2\dagger}$
rod	meter (m)	5.029
roentgen	coulomb per kilogram (C/kg)	2.58×10^{-4}
second (angle)	radian (rad)	4.848×10^{-6}
section	square meter (m²)	2.590×10^{6}
slug	kilogram (kg)	14.59
spherical candle power	lumen (lm)	12.57
square inch	square meter (m²)	6.452×10^{-4}
square foot	square meter (m²)	9.290×10^{-2}
square mile	square meter (m²)	2.590×10^{6}
square yard	square meter (m²)	0.8361
stere	cubic meter (m³)	$1.0^{\dagger}$
stokes (kinematic viscosity)	square meter per second (m²/s)	$1.0 \times 10^{-4\dagger}$
tex	kilogram per meter (kg/m)	$1.0 \times 10^{-6\dagger}$
ton (long, 2240 pounds)	kilogram (kg)	1.016×10^{3}
ton (metric)	kilogram (kg)	$1.0 \times 10^{3\dagger}$
ton (short, 2000 pounds)	kilogram (kg)	9.072×10^{2}
torr	pascal (Pa)	1.333×10^{2}
unit pole	weber (Wb)	1.257×10^{-7}
yard	meter (m)	$0.9144^{\dagger}$

Abbreviations and Unit Symbols

Following is a list of commonly used abbreviations and unit symbols appropriate for use in the *Encyclopedia*. In general they agree with those listed in *American National Standard Abbreviations for Use on Drawings and in Text (ANSI Y1.1)* (6) and *American National Standard Letter Symbols for Units in Science and Technology (ANSI Y10)* (6). Also included is a list of acronyms for a number of private and government organizations as well as common industrial solvents, polymers, and other chemicals.

Rules for Writing Unit Symbols (4):

1. Unit symbols should be printed in upright letters (roman) regardless of the type style used in the surrounding text.

2. Unit symbols are unaltered in the plural.

3. Unit symbols are not followed by a period except when used as the end of a sentence.

4. Letter unit symbols are generally written in lower-case (eg, cd for candela) unless the unit name has been derived from a proper name, in which case the first letter of the symbol is capitalized (W,Pa). Prefix and unit symbols retain their prescribed form regardless of the surrounding typography.

5. In the complete expression for a quantity, a space should be left between the numerical value and the unit symbol. For example, write 2.37 lm, *not* 2.37lm, and 35 mm, *not* 35mm. When the quantity is used in an adjectival sense, a hyphen is often used, for example, 35-mm film. *Exception:* No space is left between the numerical value and the symbols for degree, minute, and second of plane angle, and degree Celsius.

6. No space is used between the prefix and unit symbols (eg, kg).

7. Symbols, not abbreviations, should be used for units. For example, use "A," not "amp," for ampere.

8. When multiplying unit symbols, use a raised dot:

$$\text{N·m for newton meter}$$

In the case of W·h, the dot may be omitted, thus:

$$\text{Wh}$$

An exception to this practice is made for computer printouts, automatic typewriter work, etc, where the raised dot is not possible, and a dot on the line may be used.

9. When dividing unit symbols use one of the following forms:

$$\text{m/s } or \text{ m·s}^{-1} or \frac{\text{m}}{\text{s}}$$

In no case should more than one slash be used in the same expression unless parentheses are inserted to avoid ambiguity. For example, write:

$$\text{J/(mol·K) } or \text{ J·mol}^{-1} \cdot \text{K}^{-1} or \text{ (J/mol)/K}$$

but *not*

$$\text{J/mol/K}$$

10. Do not mix symbols and unit names in the same expression. Write:

$$\text{joules per kilogram } or \text{ J/kg } or \text{ J·kg}^{-1}$$

but *not*

$$\text{joules/kilogram } nor \text{ joules/kg } nor \text{ joules·kg}^{-1}$$

ABBREVIATIONS AND UNITS

A	ampere	AIME	American Institute of Mining, Metallurgical, and Petroleum Engineers
A	anion (eg, H*A*); mass number		
a	atto (prefix for 10^{-18})		
AATCC	American Association of Textile Chemists and Colorists	AIP	American Institute of Physics
		alc	alcohol(ic)
ABS	acrylonitrile–butadiene–styrene	Alk	alkyl
		alk	alkaline (not alkali)
abs	absolute	amt	amount
ac	alternating current, *n*.	amu	atomic mass unit
a-c	alternating current, *adj*.	ANSI	American National Standards Institute
ac-	alicyclic		
acac	acetylacetonate	AO	atomic orbital
ACGIH	American Conference of Governmental Industrial Hygienists	AOAC	Association of Official Analytical Chemists
		AOCS	American Oil Chemist's Society
ACS	American Chemical Society		
AGA	American Gas Association	APHA	American Public Health Association
Ah	ampere hour		
AIChE	American Institute of Chemical Engineers	API	American Petroleum Institute

aq	aqueous	cmpd	compound
Ar	aryl	CNS	central nervous system
ar-	aromatic	CoA	coenzyme A
as-	asymmetric(al)	COD	chemical oxygen demand
ASHRAE	American Society of Heating, Refrigerating, and Air Conditioning Engineers	coml	commercial(ly)
		cp	chemically pure
		cph	close-packed hexagonal
		CPSC	Consumer Product Safety Commission
ASM	American Society for Metals		
ASME	American Society of Mechanical Engineers	cryst	crystalline
		cub	cubic
ASTM	American Society for Testing and Materials	D	Debye
		D-	denoting configurational relationship
at no.	atomic number		
at wt	atomic weight	d	differential operator
av(g)	average	d-	dextro-, dextrorotatory
AWS	American Welding Society	da	deka (prefix for 10^1)
b	bonding orbital	dB	decibel
bbl	barrel	dc	direct current, n.
bcc	body-centered cubic	d-c	direct current, adj.
BCT	body-centered tetragonal	dec	decompose
Bé	Baumé	detd	determined
BET	Brunauer-Emmett-Teller (adsorption equation)	detn	determination
		Di	didymium, a mixture of all lanthanons
bid	twice daily		
Boc	t-butyloxycarbonyl	dia	diameter
BOD	biochemical (biological) oxygen demand	dil	dilute
		DIN	Deutsche Industrie Normen
bp	boiling point	dl-; DL-	racemic
Bq	becquerel	DMA	dimethylacetamide
C	coulomb	DMF	dimethylformamide
°C	degree Celsius	DMG	dimethyl glyoxime
C-	denoting attachment to carbon	DMSO	dimethyl sulfoxide
		DOD	Department of Defense
c	centi (prefix for 10^{-2})	DOE	Department of Energy
c	critical	DOT	Department of Transportation
ca	circa (approximately)		
cd	candela; current density; circular dichroism	DP	degree of polymerization
		dp	dew point
CFR	Code of Federal Regulations	DPH	diamond pyramid hardness
cgs	centimeter–gram–second	dstl(d)	distill(ed)
CI	Color Index	dta	differential thermal analysis
cis-	isomer in which substituted groups are on same side of double bond between C atoms		
		(E)-	entgegen; opposed
		ε	dielectric constant (unitless number)
cl	carload	e	electron
cm	centimeter	ECU	electrochemical unit
cmil	circular mil	ed.	edited, edition, editor

ED	effective dose	Gy	gray
EDTA	ethylenediaminetetraacetic acid	H	henry
		h	hour; hecto (prefix for 10^2)
emf	electromotive force	ha	hectare
emu	electromagnetic unit	HB	Brinell hardness number
en	ethylene diamine	Hb	hemoglobin
eng	engineering	hcp	hexagonal close-packed
EPA	Environmental Protection Agency	hex	hexagonal
		HK	Knoop hardness number
epr	electron paramagnetic resonance	HRC	Rockwell hardness (C scale)
		HV	Vickers hardness number
eq.	equation	hyd	hydrated, hydrous
esp	especially	hyg	hygroscopic
esr	electron-spin resonance	Hz	hertz
est(d)	estimate(d)	^{i}(eg, Pr^i)	iso (eg, isopropyl)
estn	estimation	i-	inactive (eg, i-methionine)
esu	electrostatic unit	IACS	International Annealed Copper Standard
exp	experiment, experimental		
ext(d)	extract(ed)	ibp	initial boiling point
F	farad (capacitance)	IC	inhibitory concentration
F	faraday (96,487 C)	ICC	Interstate Commerce Commission
f	femto (prefix for 10^{-15})		
FAO	Food and Agriculture Organization (United Nations)	ICT	International Critical Table
		ID	inside diameter; infective dose
		ip	intraperitoneal
fcc	face-centered cubic	IPS	iron pipe size
FDA	Food and Drug Administration	IPT	Institute of Petroleum Technologists
FEA	Federal Energy Administration		
		IPTS	International Practical Temperature Scale (NBS)
fob	free on board		
fp	freezing point	ir	infrared
FPC	Federal Power Commission	IRLG	Interagency Regulatory Liaison Group
FRB	Federal Reserve Board		
frz	freezing	ISO	International Organization for Standardization
G	giga (prefix for 10^9)		
G	gravitational constant = 6.67×10^{11} N·m^2/kg^2	IU	International Unit
		IUPAC	International Union of Pure and Applied Chemistry
g	gram		
(g)	gas, only as in H$_2$O(g)	IV	iodine value
g	gravitational acceleration	iv	intravenous
gem-	geminal	J	joule
glc	gas-liquid chromatography	K	kelvin
g-mol wt; gmw	gram-molecular weight	k	kilo (prefix for 10^3)
		kg	kilogram
GNP	gross national product	L	denoting configurational relationship
gpc	gel-permeation chromatography		
		L	liter (for fluids only)(5)
GRAS	Generally Recognized as Safe	l-	*levo*-, levorotatory
grd	ground	(l)	liquid, only as in NH$_3$(l)

LC_{50}	conc lethal to 50% of the animals tested	μ	micro (prefix for 10^{-6})
		N	newton (force)
LCAO	linear combination of atomic orbitals	N	normal (concentration); neutron number
LCD	liquid crystal display	N-	denoting attachment to nitrogen
lcl	less than carload lots		
LD_{50}	dose lethal to 50% of the animals tested	n (as n_D^{20})	index of refraction (for 20°C and sodium light)
LED	light-emitting diode	^{n} (as Bu^n), n-	normal (straight-chain structure)
liq	liquid		
lm	lumen	n	neutron
ln	logarithm (natural)	n	nano (prefix for 10^9)
LNG	liquefied natural gas	na	not available
log	logarithm (common)	NAS	National Academy of Sciences
LPG	liquefied petroleum gas		
ltl	less than truckload lots	NASA	National Aeronautics and Space Administration
lx	lux		
M	mega (prefix for 10^6); metal (as in MA)	nat	natural
		NBS	National Bureau of Standards
M	molar; actual mass		
$\overline{M}_w$	weight-average mol wt	neg	negative
$\overline{M}_n$	number-average mol wt	NF	*National Formulary*
m	meter; milli (prefix for 10^{-3})	NIH	National Institutes of Health
m	molal		
m-	meta	NIOSH	National Institute of Occupational Safety and Health
max	maximum		
MCA	Chemical Manufacturers' Association (was Manufacturing Chemists Association)		
		nmr	nuclear magnetic resonance
		NND	New and Nonofficial Drugs (AMA)
MEK	methyl ethyl ketone		
meq	milliequivalent	no.	number
mfd	manufactured	NOI-(BN)	not otherwise indexed (by name)
mfg	manufacturing		
mfr	manufacturer	NOS	not otherwise specified
MIBC	methyl isobutyl carbinol	nqr	nuclear quadruple resonance
MIBK	methyl isobutyl ketone	NRC	Nuclear Regulatory Commission; National Research Council
MIC	minimum inhibiting concentration		
		NRI	New Ring Index
min	minute; minimum	NSF	National Science Foundation
mL	milliliter	NTA	nitrilotriacetic acid
MLD	minimum lethal dose	NTP	normal temperature and pressure (25°C and 101.3 kPa or 1 atm)
MO	molecular orbital		
mo	month		
mol	mole	NTSB	National Transportation Safety Board
mol wt	molecular weight		
mp	melting point	O-	denoting attachment to oxygen
MR	molar refraction		
ms	mass spectrum	o-	ortho
mxt	mixture	OD	outside diameter

OPEC	Organization of Petroleum Exporting Countries	r-f	radio frequency, *adj.*
		rh	relative humidity
		RI	Ring Index
o-phen	*o*-phenanthridine	rms	root-mean square
OSHA	Occupational Safety and Health Administration	rpm	rotations per minute
		rps	revolutions per second
owf	on weight of fiber	RT	room temperature
Ω	ohm	^{s} (eg, Bu^s); *sec*-	secondary (eg, secondary butyl)
P	peta (prefix for 10^{15})		
p	pico (prefix for 10^{-12})		
p-	para	S	siemens
p	proton	(*S*)-	sinister (counterclockwise configuration)
p.	page		
Pa	pascal (pressure)	*S*-	denoting attachment to sulfur
pd	potential difference		
pH	negative logarithm of the effective hydrogen ion concentration	*s*-	symmetric(al)
		s	second
		(s)	solid, only as in H$_2$O(s)
phr	parts per hundred of resin (rubber)	SAE	Society of Automotive Engineers
p-i-n	positive-intrinsic-negative	SAN	styrene–acrylonitrile
pmr	proton magnetic resonance	sat(d)	saturate(d)
p-n	positive-negative	satn	saturation
po	per os (oral)	SBS	styrene–butadiene–styrene
POP	polyoxypropylene	sc	subcutaneous
pos	positive	SCF	self-consistent field; standard cubic feet
pp.	pages		
ppb	parts per billion (10^9)	Sch	Schultz number
ppm	parts per million (10^6)	SFs	Saybolt Furol seconds
ppmv	parts per million by volume	SI	Le Système International d'Unités (International System of Units)
ppmwt	parts per million by weight		
PPO	poly(phenyl oxide)		
ppt(d)	precipitate(d)	sl sol	slightly soluble
pptn	precipitation	sol	soluble
Pr (no.)	foreign prototype (number)	soln	solution
pt	point; part	soly	solubility
PVC	poly(vinyl chloride)	sp	specific; species
pwd	powder	sp gr	specific gravity
py	pyridine	sr	steradian
qv	quod vide (which see)	std	standard
R	univalent hydrocarbon radical	STP	standard temperature and pressure (0°C and 101.3 kPa)
(*R*)-	rectus (clockwise configuration)		
		sub	sublime(s)
r	precision of data	SUs	Saybolt Universal seconds
rad	radian; radius	syn	synthetic
rds	rate determining step	^{t} (eg, Bu^t), *t*-, *tert*-	tertiary (eg, tertiary butyl)
ref.	reference		
rf	radio frequency, *n.*		

T	tera (prefix for 10^{12}); tesla (magnetic flux density)	USDA	United States Department of Agriculture
t	metric ton (tonne); temperature	USP	*United States Pharmacopeia*
		uv	ultraviolet
TAPPI	Technical Association of the Pulp and Paper Industry	V	volt (emf)
		var	variable
tex	tex (linear density)	*vic-*	vicinal
T_g	glass-transition temperature	vol	volume (not volatile)
tga	thermogravimetric analysis	vs	versus
THF	tetrahydrofuran	v sol	very soluble
tlc	thin layer chromatography	W	watt
TLV	threshold limit value	Wb	Weber
trans-	isomer in which substituted groups are on opposite sides of double bond between C atoms	Wh	watt hour
		WHO	World Health Organization (United Nations)
TSCA	Toxic Substance Control Act	wk	week
TWA	time-weighted average	yr	year
Twad	Twaddell	*(Z)-*	zusammen; together; atomic number
UL	Underwriters' Laboratory		

Non-SI (Unacceptable and Obsolete) Units

		Use
Å	angstrom	nm
at	atmosphere, technical	Pa
atm	atmosphere, standard	Pa
b	barn	cm^2
bar†	bar	Pa
bbl	barrel	m^3
bhp	brake horsepower	W
Btu	British thermal unit	J
bu	bushel	m^3; L
cal	calorie	J
cfm	cubic foot per minute	m^3/s
Ci	curie	Bq
cSt	centistokes	mm^2/s
c/s	cycle per second	Hz
cu	cubic	exponential form
D	debye	C·m
den	denier	tex
dr	dram	kg
dyn	dyne	N
dyn/cm	dyne per centimeter	mN/m
erg	erg	J
eu	entropy unit	J/K
°F	degree Fahrenheit	°C; K
fc	footcandle	lx
fl	footlambert	lx
fl oz	fluid ounce	m^3; L
ft	foot	m
ft·lbf	foot pound-force	J

† Do not use bar (10^5Pa) or millibar (10^2Pa) because they are not SI units, and are accepted internationally only for a limited time in special fields because of existing usage.

Non-SI (Unacceptable and Obsolete) Units		Use
gf den	gram-force per denier	N/tex
G	gauss	T
Gal	gal	m/s^2
gal	gallon	m^3; L
Gb	gilbert	A
gpm	gallon per minute	(m^3/s); (m^3/h)
gr	grain	kg
hp	horsepower	W
ihp	indicated horsepower	W
in.	inch	m
in. Hg	inch of mercury	Pa
in. H$_2$O	inch of water	Pa
in.-lbf	inch pound-force	J
kcal	kilogram-calorie	J
kgf	kilogram-force	N
kilo	for kilogram	kg
L	lambert	lx
lb	pound	kg
lbf	pound-force	N
mho	mho	S
mi	mile	m
MM	million	M
mm Hg	millimeter of mercury	Pa
mμ	millimicron	nm
mph	miles per hour	km/h
μ	micron	μm
Oe	oersted	A/m
oz	ounce	kg
ozf	ounce-force	N
η	poise	Pa·s
P	poise	Pa·s
ph	phot	lx
psi	pounds-force per square inch	Pa
psia	pounds-force per square inch absolute	Pa
psig	pounds-force per square inch gauge	Pa
qt	quart	m^3; L
°R	degree Rankine	K
rd	rad	Gy
sb	stilb	lx
SCF	standard cubic foot	m^3
sq	square	exponential form
thm	therm	J
yd	yard	m

BIBLIOGRAPHY

1. The International Bureau of Weights and Measures, BIPM (Parc de Saint-Cloud, France) is described on page 22 of Ref. 4. This bureau operates under the exclusive supervision of the International Committee of Weights and Measures (CIPM).
2. *Metric Editorial Guide* (*ANMC-78-1*) 3rd ed., American National Metric Council, 1625 Massachusetts Ave. N.W., Washington, D.C. 20036, 1978.
3. *SI Units and Recommendations for the Use of Their Multiples and of Certain Other Units* (*ISO 1000-1981*), American National Standards Institute, 1430 Broadway, New York, N. Y. 10018, 1981.
4. Based on *ASTM E 380-79* (*Standard for Metric Practice*), American Society for Testing and Materials, 1916 Race Street, Philadelphia, Pa. 19103, 1979.
5. *Fed. Regist.*, Dec. 10, 1976 (41 FR 36414).
6. For ANSI address, see Ref. 3.

R. P. LUKENS
American Society for Testing and Materials

P *continued*

POWDER COATINGS

Fluidized-bed coating processes were discovered and developed in the mid-1950s (1). The first basic U.S. patent issued in 1958 (2) and exclusive rights to the process in North America were licensed to The Polymer Corporation of Reading, Pennsylvania. Acceptance of the new coating process was rather slow. In 1960, the sales of coating powders in the United States were below 450 metric tons per year. There are a number of reasons why the fluidized-bed process was slow in making significant inroads into the metal-coatings industry. Few people possessed the knowledge required to engineer, design, and build a suitable coating line. In addition, the available powder-coating materials were expensive since efficient production techniques had not been worked out and volume of production was low (see Coating processes; Coatings; Fluidization).

In the fluidized-bed coating process, the coating powder is placed in a container with a porous plate as its base. Air is passed through the porous plate which causes the powder to expand in volume and fluidize. In this state, the powder possesses some of the characteristics of a fluid. The part to be coated, usually metallic, is heated in an oven to a temperature above the melting point of the powder and dipped into the fluidized bed where the particles melt on the surface of the hot metal to form a coating. Using this process, it is possible to apply coatings ranging in thickness from ca 250 to 2500 μm. It is difficult to obtain coatings thinner than 250 μm and, therefore, fluidized-bed-applied coatings are generally referred to as thick-film coatings, differentiating them from most conventional thin-film coatings applied from solution at thicknesses of 25–75 μm (see also Film deposition techniques).

In the mid-1960s, electrostatic powder-spray-coating equipment became available, first in France and then in the United States. In the electrostatic spray process, the coating powder is dispersed in an air stream and passed through a high voltage field where the particles pick up an electrostatic charge. The charged particles are attracted to and deposited on the object to be coated which is usually at room temperature. The article is then placed in an oven where the powder melts and forms a coating. With this process it is possible to apply thin-film coatings comparable to conventional solution coatings. The electrostatic spray process gained rapid acceptance since by this time the general concept of coating with powders and the spray application of coatings had been well established.

Powder-coating processes are considered fusion-coating processes, that is, at some time in the coating process the powder particles must be fused or melted. This is usually carried out in an oven, although infrared and resistance- and induction-heating methods have also been used. Therefore, with minor exceptions, fixed installations are always used. This excludes the use of powders in maintenance coatings where they would perform with excellent results, but there is no practical way to apply them in the field. Another limitation of powder-coating processes is that the substrate must be able to withstand the temperatures required for fusion and curing. This limits powder coating to metal and glass substrates for the most part, although some plastics have been powder-coated successfully. However, compared with other coating methods, coating powders and powder-coating processes offer a number of significant advantages: They are essentially 100% nonvolatile; no solvents or other pollutants are given off during application or curing. They are ready to use; no thinning or dilution is required. They are easily applied by unskilled operators and automatic systems. The reject rate is low since a high degree of operator skill is not required for application and the finish is tougher and more abrasion resistant than that of most paints. Thicker films provide electrical insulation, corrosion protection, and other functional properties. Powder coatings cover sharp edges for better corrosion protection. The material is well utilized; overspray can be collected and reapplied. No solvent storage, solvent dry-off oven, or mixing room are required. Air from spray booths is filtered and returned to the room rather than exhausted to the outside; less air from the baking oven is exhausted to the outside; thus, energy is saved. Finally, no sludge from the spray-booth wash system has to be disposed of.

Coating powders are frequently separated into decorative and functional grades. Decorative grades are generally finer in particle size, and their color and appearance are important. They are applied to a cold substrate with electrostatic techniques at a relatively low film thickness, eg, 25–75 μm. Functional grades are usually applied in heavier films, eg, 250–2500 μm with fluidized-bed, flocking, or electrostatic spray-coating techniques to preheated parts. Corrosion resistance and electrical, mechanical, and other functional properties are important. The terms powder coating and coating powders are sometimes used interchangeably. To avoid confusion, here the term coating powder refers to the material and powder coating to the process and the applied film.

Coating powders are based on both thermoplastic and thermosetting resins. The volume of thermosetting powders sold exceeds the sales of thermoplastic powders by a factor of five, and this trend is expected to continue. Furthermore, although the variety of thermosetting powders is increasing each year as new resins and curing agents are developed, the thermoplastic resins available for use in coating powders are limited and tend to decrease.

For use as a powder coating, a resin should possess the following characteristics:

low melt viscosity, which affords a smooth continuous film; good adhesion to the substrate; good physical properties, eg, high toughness and impact resistance; light color, which permits pigmentation in white and pastel shades; good heat and chemical resistance; and good weathering resistance. It should remain stable on storage at 25°C for at least 6 mo.

Thermoplastic resins have a melt viscosity at least an order of magnitude higher than thermosetting resins at normal application temperatures. It is therefore difficult to pigment thermoplastic resins sufficiently to obtain complete hiding in thin films, ie, 25–75 μm, and still obtain sufficient flow to give a smooth coating. In addition, thermoplastic resins are much more difficult to grind than thermosetting resins and grinding is usually carried out under cryogenic conditions. Since powders designed for electrostatic spraying generally have a maximum particle size of about 75–100 μm (140–200 mesh), it is apparent why thermoplastic powders are predominant in the fluidized-bed coating process where heavier coatings are applied and the particle size of the powder is much coarser. Most thermoplastic coating powders require a primer to obtain good adhesion. This is undesirable because priming is a separate operation that requires time, manpower, and equipment. Furthermore, most primers are solvent-based, and all the problems associated with solvent-based coatings are introduced into an operation which, in many cases, has been installed to eliminate these problems. In some automotive applications, priming by electrocoating has been evaluated. Primers are not required for thermosetting-powder coatings.

Thermoplastic resins have one advantage over thermosetting resins, namely, they do not require a cure. The only postheating necessary is that required to complete melting or fusion of the powder particles. Thermoplastic resins have applications in coating wire, fencing, and other applications where the process involves continuous coating at high line speeds.

Thermoplastic Coating Powders

The physical properties of a polymeric material improve with increasing molecular weight. However, as molecular weight increases, melt viscosity also increases. As a coating powder, a thermoplastic resin must melt and flow at the application temperature without any significant degradation. If attempts are made to improve the melt-flow characteristics of a polymer by lowering its molecular weight and plasticizing or blending it with a compatible resin of lower molecular weight, physical properties suffer. In the applied coating, this results in poor impact resistance or a soft cheesy film. Attempts to improve the melt flow by increasing the application temperature are limited by the heat stability of the polymer. If the application temperature is too high, the coating shows a significant color change or evidence of heat degradation. As it is, most thermoplastic powder coatings are applied between 200 and 300°C, well above what is generally considered to be their upper temperature limits. The mitigating factor, of course, is the application time, usually 5 min or less. It is apparent that a thermoplastic resin must possess a unique combination of physical properties, melt viscosity, and heat stability to be useful in powder-coating processes. The principal polymer types in use today are based on plasticized PVC, polyamides, polyesters, and other specialty thermoplastics. Typical properties of coating powders based on these resins are given in Table 1.

4 POWDER COATINGS

Table 1. Physical and Coating Properties of Thermoplastic Powders[a]

Property	Vinyls	Polyamides	Polyester	Polyethylene	Polypropylene	Cellulosics
primer required	yes	yes	no	yes	yes	yes
melting point, °C	130–150	186	160–170	120–130	165–170	160–170
typical preheat/postheat, °C	290–230	310–250	300–250	230–200	250–220	280–230
specific gravity, g/cm^3	1.20–1.35	1.01–1.15	1.30–1.40	0.91–1.00	0.90–1.02	1.15–1.35
adhesion[b]	G–E	E	E	G	G–E	G–E
surface appearance	smooth	smooth	sl OP	OP	smooth	sl OP
gloss, Gardner 60° meter	40–90	20–95	60–95	60–80	60–80	80–90
hardness, Shore D	30–55	70–80	75–85	30–50	40–60	65–75
flexibility[c]	pass	pass	pass	pass	pass	pass
Resistance						
impact	E	E	G–E	G–E	G	G
salt spray	G	E	G	F–G	G	G
weathering	G	G	E	P	P	G
humidity	E	E	G	G	E	E
acid[d]	E	F	G	E	E	F
alkali[d]	E	E	G	E	E	P
solvent[d]	F	E	F	G	E	F

[a] E = Excellent, G = good, F = fair, P = poor, OP = orange-peel effect, sl OP = slight orange-peel effect.
[b] With primer where indicated.
[c] No cracking, 3-mm-dia mandrel bend.
[d] Inorganic; dilute.

PVC Coatings. All PVC powder coatings are plasticized formulations. Without plasticizers (qv), PVC resin is too high in melt viscosity and does not flow sufficiently under the influence of heat to form a continuous film. Suspension- and bulk-polymerized PVC homopolymer resins are used almost exclusively because vinyl chloride–vinyl acetate copolymer resins have poor heat stability. A typical melt-mixed PVC powder-coating formulation is given in Table 2.

The dispersion-grade PVC resin is added in a postblending operation. This step is essential to give a material with good fluidizing characteristics (4). Most PVC powder coatings are made by the dry-blend process (see under Manufacture). Melt-mixed

Table 2. Typical Melt-Mixed PVC Powder-Coating Formulation[a]

Ingredient	Parts by weight	%
PVC homopolymer resin	100	47.8
DNODP[b]	65	31.1
epoxidized soya oil	5	2.4
TiO$_2$/CaCO$_3$[c]	25	12.0
stabilizers		
barium–cadmium	6	2.9
organic phosphite	3	1.4
dispersion-grade PVC	5	2.4
Total	209	100.0

[a] Ref. 3.
[b] *n*-Octyl *n*-decyl phthalate.
[c] Pigment/extender.

formulations are used where superior performance is required such as in outdoor weathering applications and electrical insulation. Almost all PVC powder coatings are applied by the fluidized-bed coating process. Although some electrostatic spray-grade formulations are available, they are very erratic in their application characteristics. The resistivity of plasticized PVC powders is low compared with other powder-coating materials and the applied powder quickly loses its electrostatic charge (5). For the same reason, PVC powders show poor application characteristics in an electrostatic fluidized bed and are seldom used in this process.

Dishwasher baskets are usually coated with fluidized-bed PVC powder. Other applications are washing-machine retainer rings and various types of wire-mesh and chain-link fencing. PVC coatings have a cost/performance balance that is difficult to match with any of the other thermoplastic materials. Properly formulated PVC powders have good outdoor weathering resistance and have been used for many years to coat pole-line transformers. They are used in many applications where good corrosion resistance is required and are resistant to attack by most dilute chemicals except solvents. In addition, PVC coatings possess excellent edge coverage (see Fig. 1). Powder coatings as a class are superior to liquid coatings in their ability to coat sharp edges and isolate the substrate from contact with corrosive environments.

PVC coatings are softer and more flexible than any of the other powder-coating materials, either thermoplastic or thermosetting. As with other vinyl coatings, a primer is required to obtain adhesion to the substrate. Primers used for PVC plastisols have been found generally suitable for powder coatings as well (6).

Polyamides. Coating powders based on polyamide resins have been used in fusion-coating processes for a long time. Several examples are given in the original patent (1). Nylon-11 has been used almost exclusively; however, more recently, coating powders also have been sold based on nylon-12. The properties of these two resins are almost identical. Nylon-6 or nylon-6,6 are not used because their melt viscosity is too high and satisfactory coatings cannot be obtained (see Polyamides).

Polyamide powders are prepared by both the melt-mixed and dry-blend process. In the latter, the resin is ground to a fine powder and the pigments are mixed in with a high intensity mixer. Melt-mixed powders have a higher gloss, eg, 70–90% on the Gardner 60° glossmeter, whereas dry-blended powders have a gloss in the range of

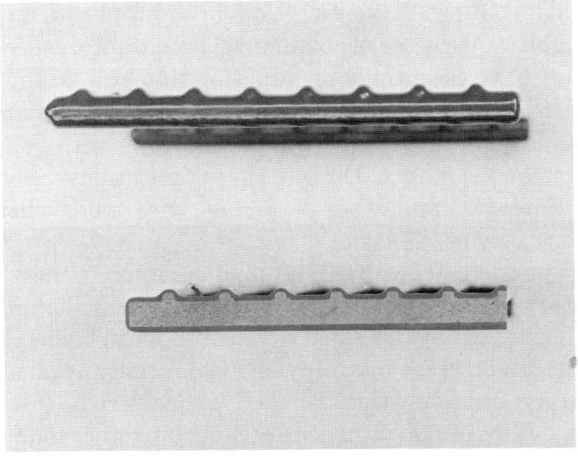

Figure 1. Edge coverage characteristic of PVC coatings. Courtesy of The Polymer Corporation.

40–70%. Because the pigment is not wetted out by the resin in the dry-blend process, it must be used at very low concentrations, usually less than 5%. Even in melt-mixed formulations, the concentration of pigment and fillers seldom exceeds about 30%.

Nylon coating powders are available for electrostatic-spray and fluidized-bed application. Nylon coatings are very tough, resistant to scratching and marring, and have a pleasing appearance. Because of these characteristics, they are used for chair bases, hospital furniture, office equipment, knobs, handles, and other hardware. Since nylon is a premium-priced powder coating, it is generally applied only to top-of-the-line items. Nylon coatings have good solvent and chemical resistance and are used for dishwasher baskets, hot-water heaters, plating and chemical-etching racks, and large-diameter water pipes in power-generating stations.

For maximum performance, a primer is used. Most nylon primers are unpigmented resin solutions applied at only ca 10% solids. This gives a dry-film thickness of about 3–5 μm. In the early days of fluidized-bed coating, it was discovered that a thin primer film gives better preheat resistance than the same primer applied at a more conventional thickness of about 25 μm (7). Preheat resistance relates to the length of time a primed part can remain in the preheat oven before being dipped in the fluidized bed and still give good adhesion. Since most primers are based on thermosetting resins, they continue to cross-link and cure during the preheating operation. Eventually, the time is reached where the topcoat no longer adheres to the primer. For most primers this time span is about 20–40 min, depending on the mass of the part. Heavier substrates, of course, take longer to come up to temperature than thinner- or lower-mass substrates and have a correspondingly longer preheat resistance. When applying nylon by electrostatic spray techniques, it is also necessary to prime the parts. If they are not primed, the electrostatically deposited layer of nylon loses adhesion and falls off the substrate at random locations as fusion starts to occur. With primed parts this electrostatic fall-off does not occur. Although the exact reason is not known, it is speculated that either the primer partially insulates the coating from the substrate and prevents the charge from leaking off too quickly, or the primer is tacky at the fusion temperature and causes the coating to adhere to the substrate. Nylon coating powders are discussed in more detail in ref. 8.

Polyethylene and Polypropylene. Polyethylene resins have a number of properties, including good chemical resistance and inertness, which should make them excellent candidates for powder coatings. However, powder coatings based on polyethylene have never achieved a significant degree of commercial acceptance, especially in the United States, because they have essentially no adhesion to metal. Many efforts have been made to obtain adhesion through resin modification, surface preparation of the metal, primers, and combinations of these (9). Furthermore, polyethylene coatings are soft and have poor abrasion resistance. Despite these limitations, polyethylene coatings are used for dishdrain racks and other decorative wire goods where adhesion is not a problem and a long service life is not expected.

Polypropylene powders likewise exhibit poor adhesion to metal without a primer. However, satisfactory primers for some applications are available. Polypropylene is used successfully for the coating of welded side seams of can bodies, but no details were ever reported in the literature. Polypropylene is normally applied by fluidized-bed coating methods to give relatively thick films. It is considered a functional rather than a decorative coating. Films of polypropylene are much tougher than polyethylene films and it is mainly used in applications where corrosion resistance is required. Excellent corrosion resistance can be obtained with certain resin grades modified to give good adhesion (10). Nevertheless, polypropylene coating powders have not captured any significant markets (see Olefin polymers).

Cellulose Esters. Although coating powders based on cellulose acetate butyrate resins enjoyed commercial success along with nylon powders in the early days of fluidized-bed coating, they were eventually replaced by PVC powder coating, which possessed many similar properties and cost less. In the early 1970s, cellulosic-based coatings were extensively modified with a polyester based on cyclohexanedimethanol–terephthalate–isophthalate (PCDT) (11). These coatings had good toughness and exterior durability and achieved some success in the exterior fencing and furniture market (see Cellulose derivatives, esters). However, their higher cost along with the general deficiencies of thermoplastic coating powders limited their growth and the resins were withdrawn from the market.

Polyester. Powder coatings based on polyester resins have been available since the early 1960s, but without achieving a high degree of commercial success. Polyester powders differ from powders based on other thermoplastic resins in one important respect: they possess good adhesion to metals without a primer. A principal application for these polyesters was in coating the outside of pipe used for gas distribution (12). However, because of marginal adhesion under tests simulating cathodic protection, it was necessary to treat the pipe with phosphate before coating (see Metal surface treatments). This proved to be too costly compared to epoxy-based coatings, which do not require chemical cleaning, and this application was eventually abandoned. Even though polyester coatings possess generally good chemical resistance, their main applications today are in decorative and exterior weathering applications. They possess a high gloss and have a handsome appearance although the surface is pockmarked like the skin of citrus fruit. This is called the orange-peel effect and is somewhat greater in polyester coatings than in cellulose ester coatings. Polyester coatings have been approved in some states for guardrails and interstate-highway signs, which attests to their excellent exterior durability.

The backbone for this polyester is based on a terephthalic acid–1,4-butanediol polymer. The terephthalic acid backbone provides a high T_g and, therefore, the resin does not sinter on storage. The backbone is also the basis for the excellent flexibility shown by these resins. Modification with other diols improves adhesion (11). Because of the unique properties of powders based on these polyester resins, thermoplastic polyester coatings will probably always be used for special applications (see Polyesters, thermoplastic).

Other Thermoplastic Coating Powders. Several other powder coatings based on specialty resins are available. These powders sell in the range of $11–22/kg and have very limited application. They are based on polymers such as ethylene–chlorotrifluoroethylene (E–CTFE), poly(vinylidene fluoride) (PVF_2), and poly(phenylene sulfide) (PPS) (see Fluorine compounds, organic; Polymers containing sulfur). These powders are used in functional applications where resistance to chemicals and elevated temperatures are required (13). They can be applied at very high temperatures by fluidized-bed or electrostatic spray processes (14). The recommended preheat and postheat temperature for PPS, for example, is 370°C (15). These powders will probably continue to be sold, although in very small quantities, because of their exceptional properties.

Thermosetting Coating Powders

Thermosetting coating powders, with minor exceptions, are based on resins that are cured by addition reactions rather than condensation reactions. Thus, for example, there are no coating powders based on unmodified phenolic resins of the resole type. The condensation products become trapped in the film and lead to voids or discontinuities. Thermosetting resins are more versatile than thermoplastic resins in the following respects: many types are available in varying molecular weight ranges and with different functional groups; a variety of cross-linking agents is available, and physical, electrical, and chemical properties of the applied film can be modified; they possess a low melt viscosity during application and thin films can be applied; because of lower melt viscosity, the quantities of pigments and fillers required to achieve opacity in thin films can be incorporated without adversely affecting flow; gloss, textures, and special effects can be produced by modifying the curing mechanism or through the use of additives; and manufacturing costs are lower since compounding is carried out at lower temperatures and the resins are friable and can be ground to a fine powder without using cryogenic techniques.

The properties of thermosetting coating powders are given in Table 3. The molecular weight, or more correctly the glass-transition temperature T_g of coating-powder thermosetting resins must be high enough to prevent the individual particles from sintering or fusing during transportation and storage. The minimum T_g required is in the range of 40–50°C and preferably above 50°C (16). Epoxy resins, because of their aromatic backbone, have the required T_g at a relatively low molecular weight and corresponding low melt viscosity. However, other thermosetting resins require some linear comonomers to achieve flexibility which results in a lower T_g. To maintain the necessary T_g, higher molecular weight resins must be used. Thus, at an equivalent range of T_g, an acrylic resin has a melt viscosity of about 3–4 times that of an epoxy resin (see Acrylic ester polymers). As pigment and filler loadings are increased, the

Table 3. Physical and Coating Properties of Thermosetting Powders[a]

Property	Epoxy	Polyurethane[b]	Polyester[c]	Hybrid	Acrylic[b]
fusion range, °C	120–200	160–220	160–220	140–210	120–200
cure time at °C, min	1–30 at 240–135	10 at 200	10 at 200	8 at 190	10 at 200
storage temp, °C max	30	30	30	30	30
adhesion	E	G–E	G–E	G–E	G
gloss, Gardner 60° meter	5–95	20–95	40–95	20–95	80–95
hardness	H-4H	H-2H	H-2H	H-2H	H-2H
flexibility	E	E	E	E	F
Resistance					
impact	E	G–E	G–E	G–E	F
overbake	F–P	G–E	E	G–E	G
weathering	P	G–E	E	P–F	G–E
acid[d]	G	F	G	G	F
alkali[d]	G	P	F	G	P
solvent	G	F	F–G	F	F

[a] E = Excellent, G = good, F = fair, P = poor.
[b] Hydroxy functional-blocked isocyanate cure.
[c] TGIC (triglycidyl isocyanurate) cure.
[d] Inorganic; dilute.

difference in flow becomes even more pronounced. Therefore, considerable efforts have been made to develop acrylic resins that give a smooth, glossy finish and good storage stability (17). The same problem exists with polyester resins but for somewhat different reasons.

In addition to the melt viscosity of the resin, the reaction rate of the resin and curing agent significantly affects the smoothness of the coating. During the heating operation, the individual particles must melt, fuse, and flow out to wet the substrate, form a continuous film, and cure or cross-link. The relative reaction rate of thermosetting powders can be determined easily by the gel time or stroke-cure test. The relationship between gel time and temperature for several thermoset powders is given in Figure 2. The effect of resin viscosity, reaction rates, and rate of heating on film properties such as gloss, adhesion, and surface finish have been evaluated (18–19). Thermosetting powders can be applied by all the fusion-coating processes to give coatings varying in thickness from ca 20 to 2500 µm. Surface appearance ranges from smooth to a slight orange-peel effect.

Epoxy Powder Coatings. Thermosetting coating powders based on epoxy resins have been used longest in fusion-coating processes and are sold in greater volume than any other class of resins (see Epoxy resins). The earliest epoxy powders were based on latent curing agents such as dicyandiamide and were at that time, of course, applied exclusively by the fluidized-bed coating process. But by today's standards these early powders, developed during the late 1950s and into the mid-1960s, are considered slow curing. Typical cure times are 20–30 min at 200°C (see Fig. 3). With minor modifications, some of these powders are still in use today in such applications as bus-bar and motor-core insulation and corrosion-resistant coatings where their excellent long-term performance has been well documented. With the advent of the electrostatic spray process and the ability to apply thin films, efforts were made to develop epoxy powders having faster cure rates and better appearance. In order to obtain a smooth coating

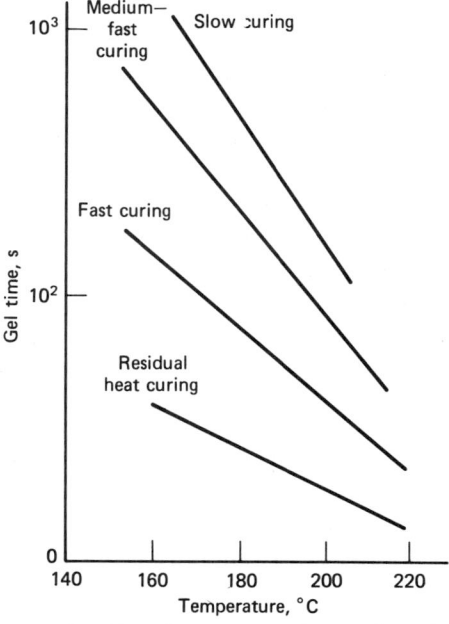

Figure 2. Gel time as a function of temperature for powders of various cure rates.

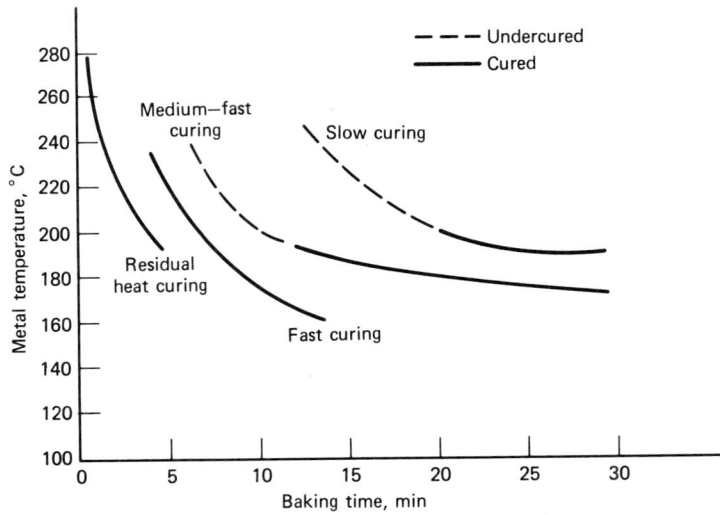
Figure 3. Baking time as a function of temperature for powders of various cure rates.

and at the same time allow titanium dioxide and other pigments required for opacity at this thickness to be incorporated, the melt flow of electrostatic-grade powders had to be improved significantly. This was eventually accomplished through improvements in raw materials, formulation, and processing techniques, which led to the development of the medium-fast curing powders. During the early 1970s, increasing amounts of epoxy resins were used for powder coatings and resins were developed specifically for this market. These resins are characterized by very good color, low viscosity, and uniform properties; they are very clean, since gels and other insoluble materials formed during the resin-making process are filtered. These improvements, the availability of new curing agents and the efforts of many powder manufacturers, led to the development of the fast-cure electrostatic powders in widespread use today. A special class of fast-cure powders is available that is cured by the residual heat of the part being coated. Such powders are used to coat the outside of gas pipes and oil-transmission lines (see Pipelines). Pipe is heated to 210–240°C, and powder is applied, usually by multiple electrostatic guns, while the pipe rotates and moves past the guns. The powder melts, fuses, and cures in 60 s or less. A comparison of the cure time required for the powders of various cure speeds is given in Figure 3.

A variety of curing agents is used in epoxy powder coatings. Modified and substituted dicyandiamide types have better compatibility than dicyandiamide alone and give coatings with higher gloss and improved physical properties (see Cyanamides). Anhydrides give coatings with better resistance to yellowing on overbake and better performance at elevated temperatures (20). Most epoxy powders contain a flow-control additive to reduce the surface tension and eliminate crater formation (21). Gloss in powder coating is not as easy to control as in solution coating, and special techniques are needed. Multiple curing agents lead to two distinct curing reactions and a controlled incompatibility in the final coating (22). Decorative epoxy powders are used in a wide variety of applications ranging from light fixtures, garden equipment, floor-cleaning machines, motor-control cabinets, pole-mounted transformers, to automotive items and many others. In outdoor applications, epoxy coatings have the disadvantage that they chalk readily; however, they protect the substrate for many years. Figure 4

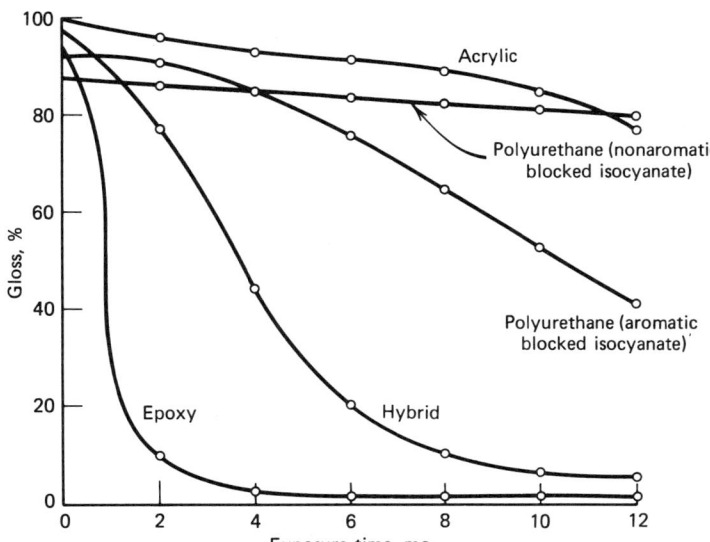
Figure 4. Gloss retention—outdoor exposure in Florida.

compares the gloss retention of epoxy coatings with other thermoset types. Many different epoxy coating-powder types are available with a wide range of properties. The most important properties of typical epoxy powders are compared with other thermosetting powders in Table 3.

Polyester and Polyurethane. By the early 1970s, the electrostatic spray process was widely accepted and many new installations were coming on-stream. Nevertheless, the market was still dominated by epoxy powders. Polyester and acrylic resin coating powders appeared on the market. Not surprisingly, some of these resins were based on chemistry similar to that employed in conventional polyester solution coatings. They were saturated, oil-free, polyester resins based primarily on isophthalic or terephthalic acid reactions with excess diols to give a hydroxy-functional polymer. This was then cured with a melamine–formaldehyde resin such as hexa(methoxymethyl)-melamine (HMMM) (see Amino resins). The proper balance of melt viscosity, T_g, molecular weight, and degree of branching is difficult to achieve (23). The HMMM lowered the T_g of the resin/hardener system, which created a severe problem of powder blocking during storage. In addition, the cured film had a surface haze due to microscopic imperfections caused by the evolution of methanol during curing. These problems were never successfully resolved and polyester resins based on this chemistry are no longer marketed.

Another system was introduced in the early 1970s; it is also based on hydroxy-functional polyester resin but uses a recently developed blocked isocyanate for curing. The isocyanate adduct has a sufficiently high T_g that sintering in storage is no problem. In addition, this system is extremely stable since essentially no reaction occurs until the unblocking temperature of the adduct is reached at ca 130°C and free isocyanate is generated. The isocyanate curing agent is isophorone diisocyanate (IPDI) blocked with caprolactam (24). Although the early resins varied considerably in hydroxyl number and degree of branching, the resins made today are uniform in structure and reactivity. Coating powders based on this technology are in widespread use today and, in the United States, are second only to epoxies in volume. Although powders based on this chemistry are technically polyurethanes (see Urethane polymers), they are sold by many manufacturers as polyesters (see also Isocyanates).

Polyurethane powders have excellent outdoor weathering characteristics, high gloss and film appearance, and good physical properties. The gloss retention of polyurethane powders cured with an aromatic and nonaromatic isocyanate adduct is shown in Figure 4. Polyurethane powders are widely used in coating lawn and garden furniture, garden tractors, exterior lights, telephone coin boxes, and many other items where exterior durability is important. They are usually applied by electrostatic spraying since the caprolactam evolved during curing causes pockmarks in films thicker than ca 75–100 μm. Cure times are in the range of 10–15 min at 180–200°C.

Another type of polyester with carboxyl functionality was introduced in Europe in 1973; it is cured with a nonyellowing polyepoxide, triglycidyl isocyanurate (TGIC). This system has excellent exterior durability and is gradually gaining acceptance in the United States (25). Sometimes referred to as a pure polyester, this system has a high degree of orange-peel effect, similar in appearance to a refrigerator finish. This lack of surface smoothness along with a premium price is the reason for the slow acceptance of this polyester system. Because no volatile materials are given off during curing, powder coatings based on this system can be applied in relatively thick films and cured at higher temperatures (or shorter times) than normally used for polyesters. According to the designation used in Figure 3, this system is in the fast-cure category.

Epoxy–Polyester Hybrids. Another series of acid-function polyester resins is designed specifically for coreaction with epoxy resins. A series of resins is available with an equivalent weight of 550–1100. They are generally produced by reaction with epoxy resins having an equivalent weight of 600–1000. Since stoichiometric ratios should be used, it is apparent that the ratio of epoxy to polyester can vary widely. In Europe, epoxy resins are more expensive than polyester resins, which led to the development of this class of resin. In the United States, polyester resins are slightly more expensive than epoxy resins, mainly because there are no domestic suppliers. However, epoxy resins still require curing agents that are generally more expensive than the resin itself, and therefore, a polyester resin as a curing agent is attractive economically.

Properties of the epoxy–polyester hybrids are similar to those of a straight epoxy resin. The hybrids are more resistant than amine-cured epoxies to color change on overbake and exposure to ultraviolet light but are only marginally superior to weathering, ie, resistance to chalking (see Fig. 4). Epoxy–polyester hybrids require a longer cure time than the fast-cure epoxies and do not give as smooth a coating. Cured films are also somewhat softer than a straight epoxy. Zinc oxide is an effective catalyst for the epoxy–polyester reaction and the use of 1–5 parts per hundred parts of resin improves the cure rate and physical properties (26). Combinations of various epoxy and polyester resins have been evaluated extensively as automotive primer and surfacer finishes (27). Coating powders based on epoxy–polyester systems are starting to find use in the United States and although the volume is still small, it could grow significantly as this system becomes more widely recognized and accepted.

Acrylic Resins. Coating powders based on acrylic resins have been available in both Europe and the United States since the early 1970s. However, powders based on a variety of different monomers and chemistry have come and gone without the combination of economics, storage stability, film properties, and exterior durability required to compete favorably with the currently available polyurethanes and polyesters. The main impetus behind developing acrylic powders has been the automotive top-coat market where the excellent performance of acrylic paints has been well documented. Most of the work today appears to be based on copolymerizing glycidyl methacrylate (GMA) into the acrylic backbone to provide the reactive group. A solid acid such as a C_{10}–C_{12} dibasic acid can be used for cross-linking or the acid cross-linking agent can be a carboxy-terminated polymer (28). However, even if suitable powders can be developed, the automotive industry presents many problems for powder coatings (29). There are a number of commercial coating installations currently in the United States, primarily in the appliance industry, utilizing acrylic powders based on hydroxy-functional resins cross-linked with a blocked isocyanate. However, the market for this type system appears limited.

Manufacture

Coating powders are either melt-mixed or dry-blended (see Fig. 5). Production methods based on spray drying have been investigated (30) but were never developed commercially. The ingredients given in Table 4 are dissolved in a volatile solvent such as methylene chloride and the pigments are dispersed in the vehicle in conventional paint-making equipment such as a ball mill or sand mill. The solution is then spray-

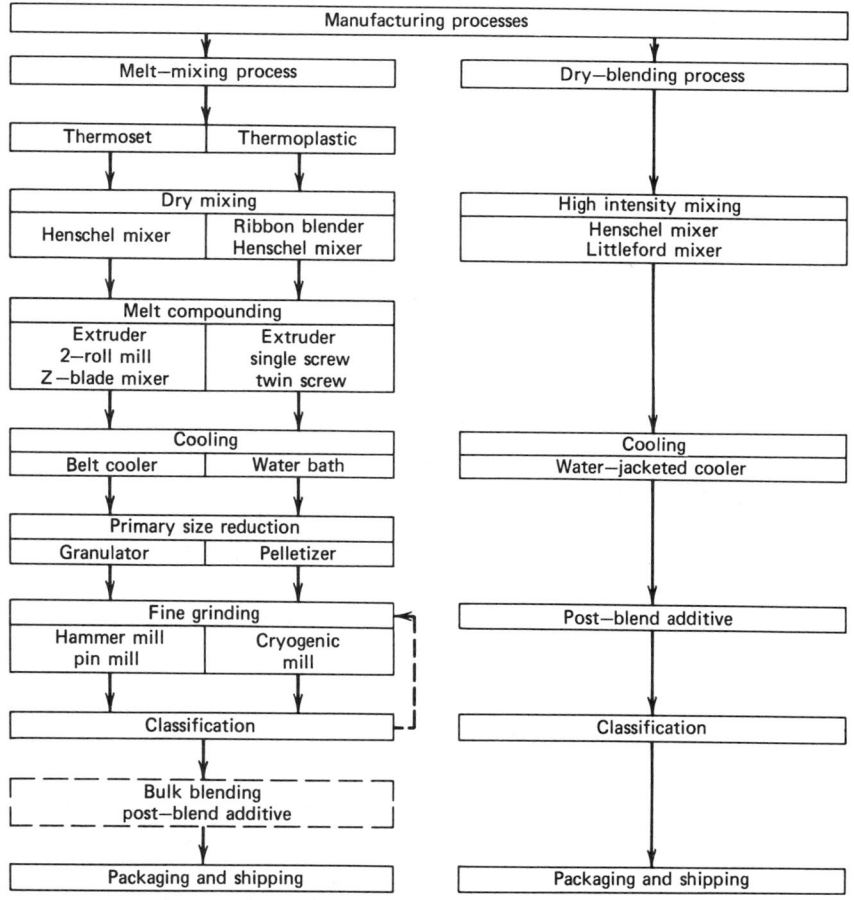

Figure 5. Flow diagram for powder-coating manufacture.

Table 4. Raw-Materials Selection[a]

Ingredient	Thermoset	Thermoplastic
resin	+	+
curing agent	+	−
plasticizer(s)	*	+
stabilizer(s)	*	+
flow additive	+	*
pigment(s)	+	+
extenders	+	+

[a] + = Normally used. − = Not normally used. * = Sometimes used.

dried to give a powder. This method has the disadvantage that it is extremely difficult to remove the last traces of solvent. During application, the solvent volatilizes and imperfections in the film result. In addition, a change of colors entails cleaning the equipment, a long and tedious procedure. Other wet methods, such as precipitating the powder from solution (31), have also been evaluated. These processes are seldom used commercially because it is difficult to dry the powder and because of the time-consuming cleaning of the equipment.

Equipment cleaning presents a problem in powder-coating manufacture. If in conventional paints a small amount of one color gets mixed with another, the only result is a slight shift in hue. However, in the case of coating powders, the result is a coating of one color contaminated with specks of another color. Therefore, most equipment used in powder manufacturing is stainless steel which can be washed with detergents and water.

Melt-Mixing. The dry ingredients given in Table 4 are weighed into a batch mixer. The most widely used type is a high speed, high intensity impeller mixer, such as a Henschel, Wellex, or Littleford mixer. The mixer imparts so much shear to the blend that mixing times are usually only 2–4 min. Longer mixing times would cause overheating and agglomeration (32). Other mixers, such as a ribbon blender or drum tumbler, can be used but the mixing is not as thorough, even after 15–30 min, and the compounded product varies in uniformity (see Mixing and blending).

The premix is then melt-compounded in a high-shear mixer. In this operation, the ingredients are compacted, the resin is melted, and the individual components are thoroughly dispersed in the molten resin. These compounding machines generate sufficient heat through the mechanical shear imparted to the resin that, after start-up, no external heat needs to be supplied (33). A single-screw machine with an interrupted flight is widely used; the screw reciprocates as well as rotates (34). Twin-screw compounders are also used, either the split-barrel or the modular type (35) (see Plastics processing).

These compounders all have a self-wiping action that prevents retention of material in the machine. Furthermore, residence time of the material is only 0.5–1.0 min, which accounts for the ability of these machines to compound thermosetting materials (36). Thermoplastic resins must be compounded on an extruder, ie, a machine in which the screw generates sufficient pressure to force the material through a die, usually in a continuous strand. The strand is water cooled and pelletized before grinding.

After compounding, thermosetting materials must be cooled quickly to ambient temperature to prevent reaction between the resin and hardener. Studies on epoxy

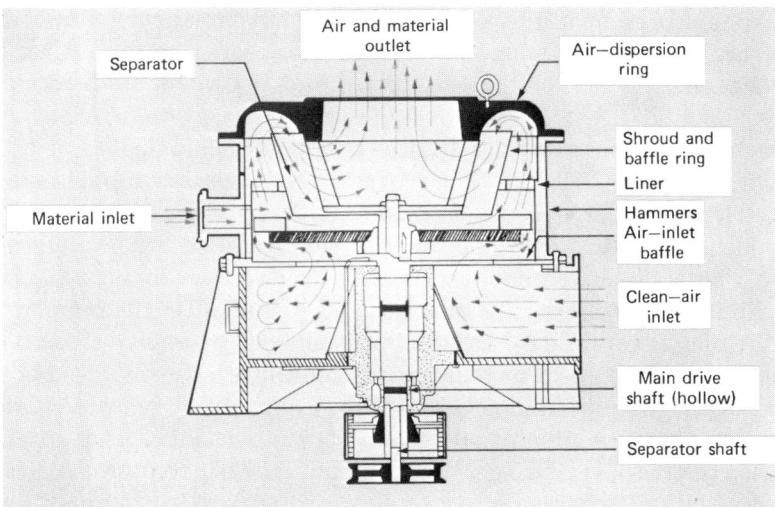

Figure 6. Air-classifying mill. Courtesy of Pulverizing Machinery Division of Mikropul Corporation.

resins have shown, for example, that only about 8–10% of the epoxy groups react during the compounding operation (37). The molten compound is cooled rapidly by passing it through the nip of a water-cooled roll and subsequently onto a continuous stainless-steel cooling belt. At the end of the belt, the thoroughly cooled ribbon is broken into small chips suitable for fine grinding.

Thermoplastic materials are normally ground on hammer mills under cryogenic conditions with liquid nitrogen as the coolant (38). Thermosetting resins do not require cryogenic temperatures for grinding since they are very brittle in the uncured state. The air-classifying mill grinds and classifies in the same operation and is widely used for grinding thermosetting resins (39) (see Fig. 6). A blower generates the air stream in which the product is entrained. The high air volume relative to the power of the mill minimizes the temperature rise in the material being ground. The product is separated from the air stream by a cyclone or dust collector. In most grinding operations, however, particles larger than desired are generated and must be separated (see Table 5). If the powder is used for fluidized-bed coating, particles above 210 µm (70 mesh) must be removed by screening. However, powders for application by electrostatic spray methods are considerably finer and an air classifier must be used. These are reasonably efficient in the separation of the mill product into the desired fractions when the cut point is relatively fine; eg, 74 µm (200 mesh). However, they are less efficient, ie, too

Table 5. Typical Particle-Size Range for Coating Powders

	Percent on indicated screen	
µm (mesh)	Fluidized bed	Electrostatic spray
250 (60)	0–1	0
210 (70)	0–5	0
149 (100)	20–60	0–1
74 (200)	40–80	3–20
44 (325)	60–90	2–60

many fines are present in the oversize fraction at coarser cut points. The oversize particles are returned to the mill for further size reduction (qv). After the desired particle-size distribution has been obtained, most powders are packaged for shipment.

Functional powders contain additives such as colloidal silica or other high oil absorption pigments to improve edge coverage and free-flowing characteristics (40). They are blended in large mixers in 1000–2000-kg batches.

Dry-Blending. Most PVC powders of fluidized-bed grade are made by dry-blending. This is a relatively simple process in which all the dry ingredients are loaded into a high intensity mixer. The mixer is started and, after the temperature begins to rise, the plasticizers and other liquids are added. The liquids are added slowly as the temperature increases. The temperature continues to rise as a result of the energy imparted by the high speed impeller until the dry point is reached. At this point, the resin absorbs the remaining plasticizer and the mix changes from a damp to a dry free-flowing powder (41). Mixing is continued until the temperature reaches 110–130°C and the powder is discharged into a cooling mixer where the temperature is rapidly reduced to below 37°C. Dispersion-grade PVC resin is added at this point and, after several minutes of mixing, the material is discharged across a coarse screen to remove any agglomerates. Then it is packaged for shipment.

Application Methods

Fluidized-Bed Coating. Fluidized-bed coating was the first commercial process for applying 100% nonvolatile polymeric materials to a substrate to form a uniform coating (42) (see Fig. 7). The process has gained widespread acceptance for reliability and is the method of choice for many applications where a heavy functional coating is required. The U.S. market for fluidized-bed coating powders is estimated to be in the range of 9,000–10,000 metric tons per year.

The process is relatively simple. The main variables are the temperature of the part as it enters the bed, the mass of the part being coated, dip time, and postheat temperature. Many other variables, such as motion of the part in the bed, and density and temperature of the powder in the bed, affect the quality of the coating (43–45). On a continuous system, it is possible to coat 300–500 parts per hour with few rejects.

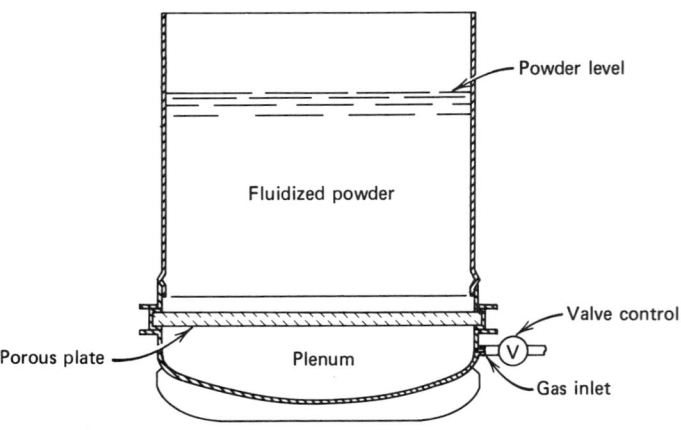

Figure 7. Fluidized bed.

The process is especially useful in coating objects with a high surface-to-mass ratio and fabricated wire goods with sharp edges and intersections such as dishwasher baskets and refrigerator shelves. However, the size of the parts that can be coated is obviously limited. Although there are a number of installations where relatively large parts are coated, such as bus bars 3–3.5-m long and pipes 6 m long, the cost of material to fill such a large bed becomes a significant consideration and a large installation is economically unattractive unless it can be utilized continually with the same coating material.

Electrostatic Fluidized-Bed Coating. In an electrostatic fluidized bed, the fluidizing container and the porous plate must be constructed of a dielectric material. The early models had a series of needle-shaped charging electrodes that projected through the porous plate and were in direct contact with the powder. This design, however, was too hazardous and in current models the fluidizing air is ionized by a high voltage source before passing through the porous plate, as illustrated in Figure 8. Electrostatic fluidized beds are relatively shallow. Parts to be coated in an electrostatic fluidized bed are passed over the bed and the charged powder is attracted onto the grounded part. If the parts to be coated are symmetrical, they are rotated as they pass over the charged cloud of powder to provide uniform deposition of powder.

Electrostatic fluidized-bed coating is an ideal method for the continuous coating of webs, wires, fencing, and other articles that are normally fabricated in continuous lengths and are essentially two-dimensional. In coating such products, two electrostatic fluid beds are used and the material is passed between them. In this fashion, both sides are coated simultaneously. A typical example is the coating of window screen (46). An electrostatic fluidized-bed coating device has also been developed to coat magnet wire (47).

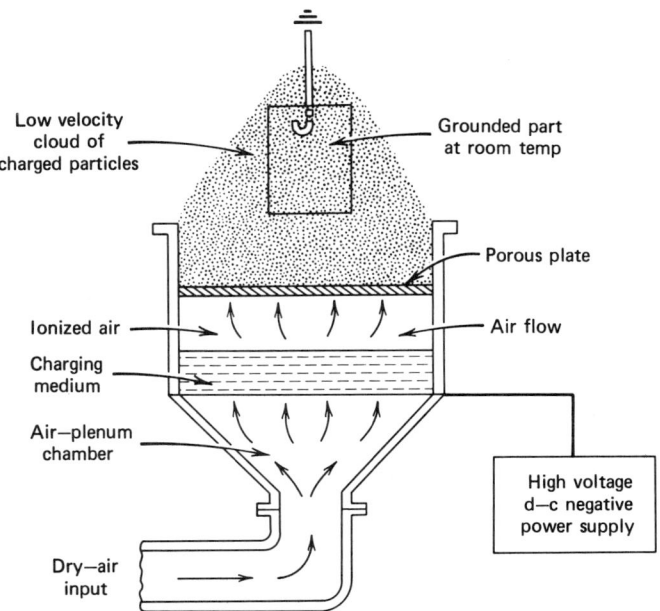

Figure 8. Electrostatic fluidized bed. Courtesy of Electrostatic Equipment Corporation.

Electrostatic Spray Coating. Electrostatic spray coating, or a variation of it, is the most widely utilized method of applying coating powders. The powder is maintained in a fluidized-bed reservoir, injected into an air stream, and carried to the gun where the powder is charged by a high voltage source. The charged powder is transported to the grounded part to be coated through a combination of electrostatic and aerodynamic forces, as illustrated in Figure 9. The powder is held by electrostatic forces to the surface of the substrate, which is subsequently heated in an oven where the powder particles fuse to form a continuous film. The processes involved are powder charging, powder transport, adhesion mechanisms, back ionization, and self-limitation (48–51). The thickness that can be applied is self-limiting. As charged powder particles and free ions generated by the high voltage source of the gun approach the layer already deposited, the point is eventually reached where the charge on the layer exceeds its dielectric strength and back-ionization occurs. At this point, any oncoming powder is rejected and loosely adhering powder on the surface falls off. The thickness of the self-limiting powder layer is quite consistent for a given powder and an electrostatic spray gun (52). This has significant implications. It makes possible, for example, the practical design of automatic-spray installations. Multiple electrostatic guns mounted on reciprocators are positioned in opposition to each other in an enclosed spray booth. Parts to be coated are moved between the two banks of guns where a uniform coating of powder is applied. Since the applied layer is self-limiting, more than the amount of powder required to provide a continuous coating can be sprayed on the parts to be sure all areas are coated. Any oversprayed powder is captured in the reclaim system and reused. Although automatic electrostatic spraying is carried out with conventional

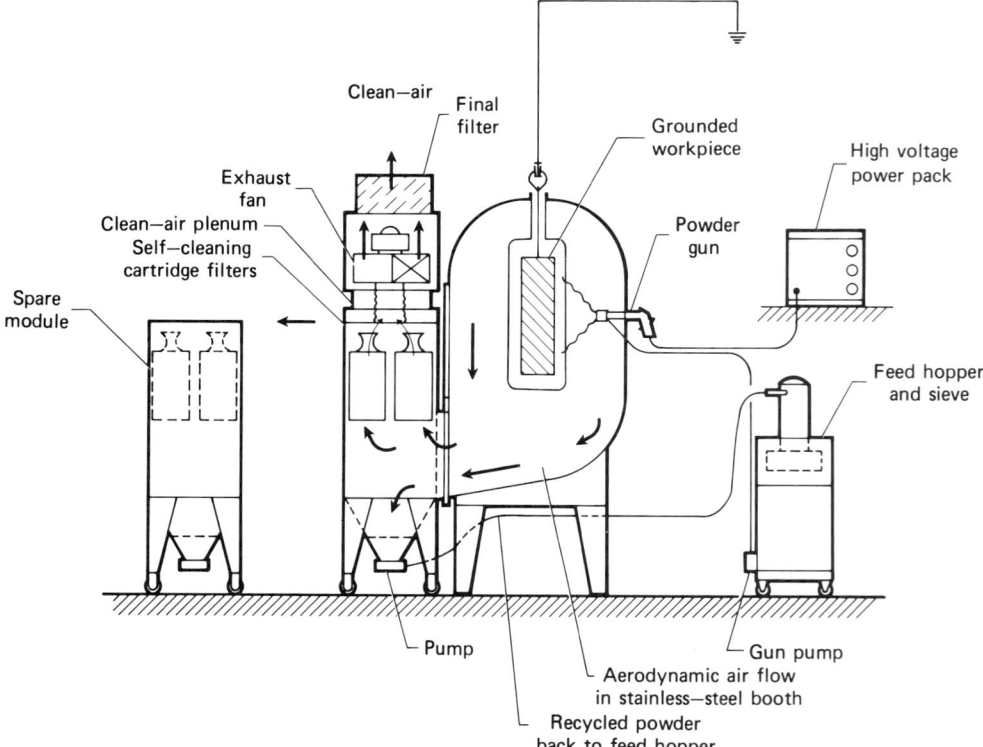

Figure 9. Schematic diagram of electrostatic powder-spray system. Courtesy of Nordson Corp.

paints, it is much more difficult to achieve the balance between providing sufficient coating in inaccessible sections of the part being coated while not getting too much on flat sections facing the spray gun or disk. In one case, uncoated sections result, whereas in the other runs, drips, or sags are obtained. With electrostatic powder coating, too much material cannot be applied and a fully coated part is obtained, even where the parts are complex in shape. For many years coating powders have been rated qualitatively as to their efficacy in various electrostatic spray systems, including the ability of the powder to penetrate corners and recesses, the degree of wrap-around, and the maximum coating thickness that can be applied. Recently, field equipment has become available that allows simple measurement of important electrical characteristics of the powder and equipment (53). This should make it possible to diagnose problems more effectively and make corrections based on sound technical principles rather than empirical judgements.

The cleaning of automatic powder-spray systems comprising guns, spray booth, dust collector, and reclaim system is a difficult and time-consuming task. Equipment manufacturers are trying to improve spray-booth and powder-collection design to reduce the difficulties involved in changing colors. For example, a continuous filter at the bottom of the spray booth has been proposed. The air flow through the spray booth is designed in such a manner that the overspray powder that is not deposited on the part is collected on the continuous filter. The powder is picked up by a vacuum system and returned directly to the feed reservoir of the electrostatic guns (54).

Another approach to the problem is illustrated in Figure 9. The booth is aerodynamically designed and fabricated from a very smooth stainless steel in order to keep the upper section virtually free of powder. The lower section, which accumulates only a thin layer of powder, is cleaned with a wide squeegee in a matter of minutes. The collector module is removed quickly from the back of the booth and replaced with another module. A separate module is required for each color. Other methods under evaluation are based on improving the efficiency of powder deposition in order to minimize or eliminate overspray (55–56). If the problem of quick color change can be solved, electrostatic powder-spray systems should gain even more widespread acceptance.

Hot Flocking. Several nonfluidized-bed coating methods are based on contacting a preheated substrate with powder to form a film. Although these techniques are not used widely, for certain parts they are the preferred method. For example, the coating of motor stators with a thermosetting powder, usually an epoxy, provides the primary insulation between the core and the windings. The motor core is preheated to about 200°C and placed on a water-cooled mandrel which prevents the inside diameter of the core, which communicates electrically with the rotor in the finished motor, from being coated. Powder is injected into an air stream from a fluidized-bed reservoir and directed through a series of tubes onto the heated motor core. The nozzles that terminate the transport tubes are set up in opposition to each other in such a way that the powder streams meet in the center of the slots in the stator core and cause a turbulent flow of powder which uniformly coats the slots while the stator rotates on the mandrel (see Fig. 10). If a residual heat-curing powder is used, sufficient heat remains in the stator core and no postheat is required. Insulation applied by this process is sometimes referred to as integral insulation since it becomes an integral part of the motor core, in contrast to conventionally used slot liners that mechanically separate the windings from the core (57).

Figure 10. Powder-spray machine. Courtesy of Possis Machine Corporation.

In a similar fashion, the inside of pipe can be coated by entraining powder in an air stream and blowing it through a preheated section of pipe. The pipe is first grit-blasted to remove any mill scale present and give a good anchor pattern, and sometimes primed with an anticorrosive primer barrier coating, and put in an oven where it is preheated to 180–200°C. Very uniform coatings in the range of 250–300 μm can be obtained using this process. Although mostly smaller-diameter pipe (5–10 cm) is coated by this process, pipe up to 45 cm can be coated with some process modifications.

Metal Cleaning and Preparation. As in any finishing operation, the surface of the object to be coated must be clean, dry, and free from rust, mill scale, grease, oil, drawing compounds, rust inhibitors, or any soil that might prevent good wetting of the surface by the coating powder (58). Steel should be sandblasted or centrifugally blast-cleaned to give an almost white metal finish, according to the Steel Structures Painting Council specification SSPC-SP10 (59), or chemically cleaned in accordance with well-established procedures. Phosphate and chromate conversion coatings improve the performance of powder coatings in harsh environments (60).

Economic Aspects

The market for coating powders has started to grow at a significant rate since about 1975. At the First North American Conference on Powder Coatings and the first U.S. conference in Cincinnati, both held in 1972, a market growth from the then estimated 22,000 metric tons per year to 97,000 t/yr was predicted for 1980 (61). However, a recent study indicated that the market for powder coatings in 1979 was only about 27,200 t (62). The same study projects an annual growth rate of over 9% for 1980–1985, whereas the growth rate for industrial finishes as a whole over the same period will be 1.8%. The projections for powder demand by use distribution are given in Table 6 and by resin type in Table 7. Another report estimates the 1979 market for electrostatic spray powders at 20,412 t with epoxy-based powders accounting for 68% of the total (63). A comprehensive report of the coating-powder market in Europe by country and use market for 1978 with projections for 1983 and 1988 was prepared by the Eu-

Table 6. Distribution of Powder-Coatings Consumption, Metric Tons [a]

	1981	1983[b]	1985[b]
Use			
metal decoration	19,504	24,041	28,577
electrical	4,990	6,350	7,258
wire goods	2,722	2,772	3,629
pipe	4,990	5,897	6,804
Total	*32,206*	*39,060*	*46,268*
constant $ value, 10^6 $	115	145	185
share of industrial finishing market, %	4.5	5.5	8.3

[a] Ref. 62.
[b] Projected.

Table 7. Resin Distribution in Powder-Coatings Consumption [a]

Resin	1981		1983[b]		1985	
	t	%	t	%	t	%
epoxy	21,092	65.5	24,041	61.5	28,577	62
polyester[c], thermoset	4,990	15.5	6,804	17.5	8,165	18
vinyl	3,175	10	3,175	8	2,948	6
acrylic, thermoset	2,268	7	3,629	9.5	4,536	10
other	680	2	1,361	3.5	1,814	4
Total	*32,205*	*100*	*39,010*	*100*	*46,040*	*100*

[a] Ref. 62.
[b] Projected.
[c] Includes polyurethanes.

ropean Federation of Paint Makers Associations (64) and is summarized in Table 8. Market share by resin type is given in Table 9. The market sizes for Europe and the United States are roughly comparable, but the distribution by resin type is different. The epoxy–polyester powders, for example, have a significant market share in Europe but not in the United States. On the other hand, the polyurethane powders are second

Table 8. Distribution and Consumption of Coating Powders, European Market [a,b]

	1978		1983[c]		1988[c]	
	t	%	t	%	t	%
furniture	8,430	33.5	13,110	31	16,500	28
appliances	2,845	11.5	3,950	9.5	4,870	8.5
building industry	5,114	21	8,485	19.5	11,490	19.5
automotive	3,445	13.5	7,435	17.5	12,075	20.5
machinery	1,445	5.5	2,360	5.5	3,695	6.5
protective industry	1,813	7.5	3,605	8.5	5,305	9
others	2,008	8	3,205	8	4,565	8
Total	*25,100*	*100*	*42,150*	*100*	*58,500*	*100*

[a] Ref. 64.
[b] Exclusive of United Kingdom.
[c] Projected.

Table 9. Consumption of Powder-Coating Type in 1978[a]

Resin	t	%
epoxy	13,810	55
epoxy–polyester	7,165	28.5
polyester	2,247	9
polyurethane	1,698	6.8
acrylic	180	0.7
Total	*25,100*	*100*

[a] Ref. 64.

only to epoxies in the U.S. market, whereas in Europe they rank lowest in volume except for acrylic resins. After getting off to a slow start, it appears that the market for coating powders is moving in a solid growth pattern. In 1981, General Motors became the first auto maker using powder, instead of solvent-base, for protective coatings and the newly formed Powder Coating Institute has projected increased growth (65).

PVC coatings have the lowest selling price and account for the highest volume of sales of all the thermoplastic powder-coating materials. Sales volume in 1979 was in the order of 4540–5440 metric tons, mostly to coat wire goods.

Test Methods

Test methods for evaluating the performance of powder coatings are the same as those used for conventional coatings and usually include application tests by fluidized bed, electrostatic fluidized bed, or electrostatic spray, as the case may be, in which test coatings are compared with standards. Specifications and such properties as particle size, reactivity, melt viscosity, cure time, and color are checked. Test methods for powder coatings have been reviewed in detail and reported in the literature (66–67). In addition, the American Society for Testing and Materials has now issued ASTM D 3451 "Standard Recommended Practices for Testing Polymeric Powders and Powder Coatings," a comprehensive standard that covers the most important testing methods for the characterization and evaluation of powder coatings (68).

Environmental and Energy Considerations

Certainly one of the main factors contributing to the growth of powder-coating processes has been the significant increase in environmental regulations and the legislative time limit for compliance. In August 1977, Congress passed the Clean Air Act Amendments which require that each state meet the requirements of the act by December 31, 1982. The EPA has issued a series of Control Technique Guidelines that set limits for various industries on the amounts of volatile organic material that can be emitted to the atmosphere. Furthermore, each state must submit plans to the EPA for the regulation of organic emissions. Users of Rule 66-type paints, for example, are required to reduce the hydrocarbon content from the current level of about 660–720 g/L (5.5–6 lb/gal) to 300–360 g/L (2.5–3 lb/gal). Powder coatings have been favorably evaluated by the EPA and recognized as one of several acceptable technologies to achieve the required level of volatile organic-compound emissions. The importance of this factor is emphasized in a recent market survey (62) which states: "The elimination of polluting emissions has been a vital factor in the justification of powders and now the nonpolluting characteristics are dramatically re-emphasized in the new EPA action" (see also Air pollution).

During the fusion or curing of a coating powder, some volatile organic compounds are liberated. With the exception of the caprolactam used as a blocking agent in the polyurethane powder coatings, however, most are only in trace amounts. Of the volatile materials emitted during the bake cycle of an epoxy, polyester–epoxy hybrid, and a polyester–TGIC powder, up to 99% was found to be water from absorbed moisture or condensed products (69). Volatile loss during baking of a polyurethane powder ranges from 5–10%, most of which is caprolactam, depending on the amount of blocked-isocyanate curing agent in the powder, film thickness, and baking temperature.

In addition to the environmental advantages, the low volatile emissions of powder coatings during the baking operation has economic and energy-saving advantages. Fewer air changes per hour in the baking oven are required for powder coatings than for solvent-based coatings, which saves fuel. For example, a manufacturer of electrical control panels reported using one half the amount of natural gas on a powder-coating line than was used on the paint system it replaced to finish the same amount of metal. The increased efficiency was the result of a number of factors. Line loading was increased, allowing a one-shift operation rather than two shifts, exhaust from the baking oven was considerably reduced over that previously required, and coating-booth exhaust was filtered and returned to the room rather than exhausted to the outside (70). Another study compared the energy requirements for 10 types of coatings. With a conventional solvent-based coating requiring a 20-min bake at 163°C as the control, an electrocoating showed the greatest energy savings at 71%, whereas several powder coatings requiring cure temperatures of 135–190°C showed energy savings of 18–31% (71). Several other studies have compared total coating costs for powder coating with other methods considering material costs, direct labor, maintenance, energy, ease of application, capital requirements, and other factors and concluded that powder coatings are definitely less expensive (72–73). Other advantages claimed for powder coating that reduce energy consumption are fewer rejects and fewer second operations (74), less stringent packaging requirements, and the ability to change from a two-coat wet system to one coat of powder on rough castings (75). Ample evidence shows powder coatings to be superior to most other coating technologies in reducing energy consumption.

Health and Safety Factors

Any finely divided organic material can form explosive mixtures with air at certain concentrations. The most significant hazard in the manufacture and application of coating powders is the potential of a dust explosion. A dust explosion is simple combustion, but at such a high rate of burning that pressure builds up rapidly if it occurs in an enclosed space. As the ratio of surface area to volume increases, ie, as particle size decreases, the burning rate increases and the rate of heat release is intensified. The severity of a dust explosion is primarily related to the material involved, and its particle size and concentration in air at time of ignition. The lower explosive limit (LEL) is the lowest concentration of a material dispersed in air that explodes when ignited (76). This is essentially the same as the minimum explosive concentration

(MEC) defined in Code No. 33 of the National Fire Prevention Association (NFPA) (77). The LEL values for a number of coating powders are given in Table 10. In powder-coating installations, the design of the spray booth and duct work should be such that the powder concentration is always kept below the LEL, with a safety factor of 50% (76). If the LEL of the powder is unknown, a value of 10 g/m^3 is suggested for design purposes (79). Air flow across all booth openings should be a minimum of ca 18 mm/min (60 ft/min); design should be based on ca 22–24 m/min (75–80 ft/min) (78). In a dust collector where the powder concentration cannot be kept below the LEL, either a pressure-relief or explosion-suppression system should be considered, if it cannot be located outside the building. Ducts from the explosion vents should be directed to the outside through short runs. Required explosion-vent areas and other design considerations can be found in Code No. 68 of the National Fire Protection Association (NFPA) (81). In an explosion-suppression system, the pressure generated by an explosion is detected within about 10–20 ms by sensors that activate a flame suppressant, usually a halogenated gas, stored under pressure; eg, 2480 kPa (360 psi) in a high discharge-rate extinguisher. Both pressure sensors and ultraviolet photoelectric sensors that detect the flame are used (82). Design requirements and other information regarding installation of an explosion-suppression system are given in NFPA Code No. 69 (83). Gun, spray booth, duct work, dust collection, and powder-reclaim system as well as the workpiece must be properly grounded (84). There are on the order of 800 electrostatic powder-spray lines, and a similar number of fluidized-bed, flocking, and other powder-coating installations in the United States today. Thus, powder coating is a safe operation if the system is properly designed. For additional general information on the safety aspects of powder coating, see refs. 85 and 86.

In considering the physiological hazards to personnel working with coating powders, particular attention must be paid to exposure by inhalation and skin contact. Skin and eye irritation as well as oral toxicity have to be considered also. Even though it is unlikely that any significant quantity of a coating powder would be ingested, acute oral toxicity provides a measure of toxicity compared with other materials. Inhalation toxicity tests are carried out by exposing the test animals, usually albino rats, for about one hour to the test material in the form of a dust, mist, or vapor. Skin irritation is also tested on animals by the usual procedures (88).

The results with five epoxy and one acrylic coating powders tested in animal studies showed an LD_{50} >15 g/kg, ie, negligible toxicity (87). A similar evaluation was carried out on a number of polyester resins for acute oral toxicity, skin irritation, and inhalation effect. In all cases the LD_{50} was >10 g/kg, and they were classified as practically nontoxic; skin irritation in all cases was negative (89). The following tests

Table 10. Lower-Explosive-Limits Range for Coating Powders

Resin type	g/m^3	Reference
epoxy	45–78	76
polyurethane	65–71	76
polyester/TGIC[a]	46	80
polypropylene	32	76

[a] Triglycidyl isocyanurate.

were also carried out on a clear TGIC powder-coating formulation: acute oral toxicity in rats, irritant effects on rabbit skin and rabbit-eye mucosa, and skin irritation effects on humans. Acute oral toxicity was >16 g/kg and irritant effects on rabbit skin and eyes were negative. The irritant effect on 50 human subjects was also negative after both 24 and 48 h (89).

Although these tests indicate that coating powders do not appear to pose any significant threats to the health of personnel working with them, worker exposure to them should nevertheless be minimized. They should be treated as nuisance dusts and the concentration in air kept below the threshold-limit value (TLV) of 10 mg/m^3, primarily through environmental controls (90). Hoods and proper ventilation should be provided, and ovens should be properly vented and provided with hoods. When environmental control is not practical, protective equipment such as dust and fume masks, protective clothing, gloves, etc, should be used (91–92). With proper procedures, the health hazard to personnel working with coating powders appears to be of a relatively low degree.

BIBLIOGRAPHY

1. Ger. Pat. 933,019 (Sept. 15, 1955), E. Gemmer (to Knapsack-Griesheim, A.G.).
2. U.S. Pat. 2,844,489 (July 22, 1958), E. Gemmer (to Knapsack-Griesheim, A.G.).
3. U.S. Pat. 3,640,747 (Feb. 8, 1972), D. S. Richart (to The Polymer Corporation).
4. U.S. Pat. 3,264,371 (Aug. 2, 1966), H. M. Gruber and L. Haag (to The Polymer Corporation).
5. A. Golovoy, "Charge Decay From Electrostatically Charged Powders," *Technical Paper FC-74-591*, Society of Manufacturing Engineers, Dearborn, Mich., 1974.
6. U.S. Pat. 3,008,848 (Nov. 14, 1961), R. W. Annonio (to Union Carbide Corp.).
7. U.S. Pat. 3,264,131 (Aug. 2, 1966), F. J. Nagel (to Polymer Processes Inc.).
8. D. S. Richart in M. I. Kohan, ed., *Nylon Plastics*, John Wiley & Sons, Inc., New York, 1973, Chapt. 14.
9. U.S. Pat. 4,007,298 (Feb. 8, 1977), C. E. Feehan and E. F. Wagner (to United States Pipe and Foundry Co.).
10. R. H. Mumma and C. E. Maag, "Polypropylene Powder Coatings," *Technical Paper FC-73-516*, Society of Manufacturing Engineers, Dearborn, Mich., 1973.
11. D. D. Taft, *Powder Finish. World*, 8 (1975).
12. J. K. Christensen, "Polyester Powder Coatings," *Technical Paper FC-72-941*, Society of Manufacturing Engineers, Dearborn, Mich., 1972.
13. E. B. Ewan, *Mater. Eng.* 38 (Apr. 1979).
14. J. P. Blackwell, D. G. Brady, and H. W. Hill, Jr., *J. Coat. Technol.* **50,** 62 (Aug. 1978).
15. *Technical Service Bulletin TSM-278*, Phillips Chemical Co., Bartlesville, Okla., Feb. 1978.
16. U.S. Pat. 4,002,699 (Jan. 11, 1977), S. S. Labana and Y. Chang (to Ford Motor Co.).
17. U.S. Pat. 3,998,768 (Dec. 21, 1976), P. J. Pettit, Jr. (to E. I. du Pont de Nemours & Co., Inc.).
18. T. Nakamichi, *Prog. Org. Coat.* **8,** 9 (1980).
19. R. R. Eley, *Org. Coat. Plast. Chem. Am. Chem. Soc.* **42,** 417 (1980).
20. M. Hoppe, *Paint Manuf. (UK)*, 11 (Oct. 1975).
21. L. Feyt and H. Bauwin, *J. Coat. Technol.* **52,** 87 (May 1980).
22. U.S. Pat. 4,007,299 (Feb. 8, 1977), F. Schulde and co-workers (to Veba-Chemie, A.G.).
23. E. M. Holda, *J. Paint Technol.* **44,** 75 (July 1972).
24. D. Obendorf, *Paint Manuf. (UK)*, 11 (Nov. 1974).
25. W. Marquardt and H. Gempler, *Polym. Paint Colour J. (UK)* **170,** 630 (Aug. 6–20, 1980).
26. J. E. Sreeves and L. Whitfield, *J. Oil Colour Chem. Assoc.* **62,** 293 (1979).
27. A. van de Werff, "Influence of Resin Composition on Automobile Powder Surfacer Properties," *FATIPEC Congress*, Amsterdam, The Netherlands, June 1980.
28. U.S. Pat. 3,781,380 (Dec. 25, 1973), S. Labana and Y. Chang (to Ford Motor Co.).
29. D. Schowiak, "Powder Coating Automobiles—The State of the Art," *Tech Paper FC-74-576m*, Society of Mechanical Engineers, Dearborn, Mich., 1974.

30. U.S. Pat. 3,561,003 (Feb. 2, 1971), B. J. Lanham and V. G. Hykel (to the Magnavox Co.).
31. U.S. Pat. 3,737,401 (June 5, 1973), I. H. Tsou and J. W. Garner (to Grow Chemical Corporation).
32. G. A. Campbell, "New Developments in High Intensity Mixing and Cooling of Powder Blend Formulations," *Tech Paper FC-72-947*, Society of Mechanical Engineers, Dearborn, Mich., 1972.
33. R. E. Smorodin, "The Practical Application of Modern Melt Mix Technology," *Tech Paper FC-73-551*, Society of Mechanical Engineers, Dearborn, Mich., 1973.
34. *Continuous Production of Powder Coating Materials With Buss PLK Lines*, Technical Bulletin #3041, Buss Limited, Basle, Switzerland.
35. S. Jakobin, "The Continuous Production of Powder Coating Resins," *Tech Paper FC-74-585*, Society of Mechanical Engineers, Dearborn, Mich., 1974.
36. R. E. Smorodin and J. W. Hunt, "Continuous Melt Mix Processing for Powder Coatings," *Tech Paper FC-72-948*, Society of Mechanical Engineers, Dearborn, Mich., 1972.
37. B. Dreher, *FATIPEC Congress Paper*, Budapest, June 1978, pp. 201–207.
38. *Powder. Finish. World*, 26 (1975).
39. C. J. Gero, "A Guide to Crushing and Grinding Practice," *Tech Paper FC-72-955*, Society of Mechanical Engineers, Dearborn, Mich., 1972.
40. U.S. Pat. 3,338,863 (Aug. 29, 1967), L. H. Haag (to The Polymer Corporation).
41. F. Roesler and H. M. Metz in L. I. Nass, ed., *Encyclopedia of PVC*, Vol. 2, Marcel Dekker, Inc., New York, 1977, Chapt. 20.
42. L. O. Gilbert, *Organic Coating Using the Fluidized Bed Technique*, Rock Island Arsenal Tech Report No. 62-3183, Sept. 20, 1962.
43. R. R. Sharetts and D. S. Richart, *Plast. Des. Proc.*, 10 (June 1962); 26 (July 1962).
44. C.K. Pettigrew, *Mod. Plast.* **43**(12), 111 (Aug. 1966); **44**(1), 150 (Sept. 1966).
45. A. H. Landrock, *Chem. Eng. Prog.* **63**(2), 86 (1967).
46. G. T. Robison, *Prod. Finish.*, 16 (Sept. 1976).
47. U.S. Pat. 4,188,413 (Feb. 12, 1980), J. H. Lupinski and B. Gorowitz (to General Electric Co.).
48. A. W. Bright, *J. Oil Colour Chem. Assoc.* **60**, 23 (1977).
49. G. D. Cheever, *J. Appl. Polym. Sci.* **19**, 147 (1975).
50. K. M. Osterle and I. Szasz, *J. Oil Colour Chem. Assoc.* **48**, 956 (1965).
51. S. Wu, *Polym. Plast. Technol. Eng.* **7**(2), 119 (1976).
52. G. F. Hardy, *J. Paint. Technol.* **46**, 73 (Dec. 1974).
53. J. F. Hughes, *Finish. Ind. (UK)*, 20, 52 (May 1980).
54. U.S. Pat. 4,153,008 (May 8, 1979), F. P. Marion and E. P. Light (to Interrad Corp.).
55. U.S. Pat. 4,204,497 (May 27, 1980), R. C. Lever (to Volstatic Holding Ltd.).
56. *Electrogasdynamic Powder Coating System*, Estey Dynamics Corp., Red Bank, N.J.
57. G. H. Baillie, *Insulation*, 78 (Sept. 1966).
58. "Heat Treating, Cleaning and Finishing," *Metals Handbook*, 8th ed., Vol. 2, American Society for Metals, Metals Park, Ohio, 1964.
59. *Near White Blast Cleaning*, Specification SSPC-SP10, Steel Structures Painting Council, Pittsburgh, Pa., 1963.
60. R. A. Kelly, "Treatments for Powder Coatings," *Tech Paper FC-74-588*, Society of Mechanical Engineers, Dearborn, Mich., 1974.
61. *Chem. Week*, 30 (Mar. 29, 1973).
62. *Plast. World* **38**, 35 (June 1980).
63. G. Cole, *Prod. Finish.*, 58 (Jan. 1980).
64. G. Venlet, *Powder Coat.* **2**, 2 (Dec. 1979).
65. *Chem. Week*, 16 (Sept. 2, 1981).
66. A. A. Worley, "Test Procedures and Specifications for Powders and Powder Coatings," *Tech Paper FC-73-535*, Society of Mechanical Engineers, Dearborn, Mich., 1973.
67. C. H. J. Klaren, *J. Oil. Colour Chem. Assoc.* **60**, 205 (1977).
68. "Paint-Tests for Formulated Products and Applied Coatings," *1979 Annual Book of ASTM Standards*, Part 27, American Society for Testing and Materials, Philadelphia, Pa.
69. B. D. Meyer and D. Kordes, *Farb + Lacke* **83**, 997 (1977).
70. *Ind. Finish.*, 32 (Oct. 1977).
71. *Prod. Finish.*, 55 (Sept. 1977).
72. P. G. deLange, *Finish. Ind.*, 32, 51 (Mar. 1978).
73. *Powder Finish. World* **2**(4), 14 (1975).
74. B. Haywood, *Ind. Finish.*, 28 (Dec. 1977).

75. *Powder Finish. World*, 16, 20 (Sept. 1974).
76. P. H. Dobson, *Ind. Finish.*, 77 (Sept. 1974).
77. *Spray Finishing Using Flammable and Combustible Materials*, Code No. 33, National Fire Protection Association, Boston, Mass., 1969.
78. J. L. Miller, "New Recovery System Concepts," *Tech Paper FC73-559*, Society of Mechanical Engineers, Dearborn, Mich., 1973.
79. W. D. Vork, "Powder, OSHA and EPA—An Equipment Manufacturers Viewpoint: How to Design for Compliance—How to Install," *SME Tech Paper FC73-554*, Society of Mechanical Engineers, Dearborn, Mich., 1973.
80. G. E. Cole, Jr. and D. Scarborough, *Finish. Int.* **11,** 63 (July 1979).
81. *Guide for Explosion Venting*, Code No. 68, National Fire Protection Association, Boston, Mass., 1954.
82. A. J. Martin, *Powder Finish. World* **III**(2), 17 (1976).
83. *Standard on Explosion Prevention Systems*, Code No. 69, National Fire Protection Association, Boston, Mass., 1970.
84. *Static Electricity*, Code No. 70, National Fire Prevention Society, Boston, Mass., 1971.
85. *Dust Explosion Prevention—Plastics Industry*, Code No. 654, National Fire Protection Association, Boston, Mass., 1970.
86. *OSHA Regulations*, *Code of Federal Regulations*, Title 29, Chapt. XVII, Part 1910, Sections 1910.93, 1910.94, and 1910.107, 1974.
87. R. F. Strobel, "Safety Considerations with Powder Coating," *SME Tech Paper FC73-555*, Society of Manufacturing Engineers, Dearborn, Mich., 1973.
88. J. H. Draize and co-workers, *J. Pharm. Exp. Ther.* **82,** 4 (Dec. 1944).
89. *Toxicological Studies of Uralac Powder Coating Resin and Powders*, report prepared by Scado B.V., Zwolle, The Netherlands, 1979.
90. *Threshold Limit Values for Chemical Substances and Physical Agents in the Workroom Environment*, American Conference of Governmental Industrial Hygienists, Cincinnati, Ohio, 1979.
91. *OSHA Regulations*, *Code of Federal Regulations*, Title 29, Chapt. XVII, Section 1910.34.
92. *American National Standard Practices for Respiratory Protection*, ANSI Z88.2-1969, American National Standards Institute, Inc., New York, 1969.

General References

S. T. Harris, *The Technology of Powder Coatings*, Portcullis Press, London, Eng., 1976.
M. W. Ranney, *Powder Coatings Technology*, Noyes Data Corporation, Park Ridge, N.J., 1975.
E. P. Miller and D. D. Taft, *Powder Coating*, Society of Manufacturing Engineers, Dearborn, Mich., 1976.
Powder Coatings Bibliography, 1972, Paint Research Association, Teddington, Middlesex, Eng., 1972.
M. W. Ranney, *Epoxy Resins and Products—Recent Advances*, Noyes Data Corporation, Park Ridge, N.J., 1977.
R. H. Chandler, *Epoxy Powder Coatings*, Bibliographies in Paint Technology No. 20, R. H. Chandler Ltd., Braintree, Essex, Eng., 1973.
R. H. Chandler, *New Epoxy Powder Coatings, 1973–79*, Bibliographies in Paint Technology No. 33, R. H. Chandler Ltd., Braintree, Essex, Eng., 1979.
Epikote Resin Powder Coatings—Formulation Manufacture and Testing, Technical Bulletin EP 2.10, Shell Int. Chem. Co., Ltd., London, Eng.

D. S. RICHART
The Polymer Corporation

POWDER METALLURGY

The production of metal powders and their utilization for the manufacture of massive materials and shaped objects is the formal definition of powder metallurgy. Powder metallurgy or P/M encompasses both the total technology and the application of metals in powder form from the production of the powders themselves to the manufacture of finished products. It is at once a science and a technology as well as an industry based on the metallurgical phenomenon called sintering, ie, the bonding of particles in a mass of metal powder by molecular (or atomic) attraction in the solid state through the application of heat below the melting point of the metal. Sintering causes the strengthening of the powder mass and normally results in densification and often recrystallization because of material transport (see Metal treatments).

In general practice metal powders or mixtures of various powders are fashioned into some form, a P/M gear for example, by first flowing them at room temperature into a die cavity shaped to that of the finished gear and then applying pressure, usually from above and below, to form a compact (see Fig. 1). After ejection from the die, the compact is heated (sintered) to form a coherent mass with the configuration of the original die and properties related to the metals used. A low density, highly porous compact such as a metal filter may be formed by pouring the powder into a die and then heating it without first applying pressure. The shape and size of the die cavity may be exactly as required by the finished part or it may be some intermediate configuration.

The pressing operation is usually carried out at room temperature, though warm or even hot pressing is sometimes used. The pressing operation consolidates the powder into a mass with sufficient strength to be handled without breaking after ejection from the die. During pressing, a coherent mass is formed through the process of interparticle binding and interlocking.

The application of heat during pressing, or as a separate step, does not cause melting of the powder except in special instances involving liquid-phase sintering. The temperature to which the pressed compact is then heated is usually slightly less than the melting point of the elemental metal or the solidus temperature of any alloy that may form. During heating, the bonding that was initiated in pressing is carried farther by the solid-state movement of atoms.

Liquid-phase sintering combines powder metallurgy and fusion metallurgy. In this procedure, part of the structure is liquid for part or all of the sintering time. The sintering temperature may be high enough to melt one constituent, which occasionally solidifies during the heating time, or the metal matrix may be infiltrated with another metal that melts at a lower temperature, such as the infiltration of copper in a sintered iron matrix.

After sintering, supplementary operations may be required. A sintered component may be worked by any number of metal-working procedures, such as swaging, rolling, or drawing. Shaped items, such as machine parts and structural components, may require some additional machining, plating, or other finishing operations. The result is a P/M product: a metal-shape equivalent in function, though usually of lower density, and often equivalent in physical and mechanical properties to a wrought metal product but produced faster, automatically, and normally at a lower cost in terms of labor, material, and energy.

In prehistoric times, humans learned that particles of metal such as gold could

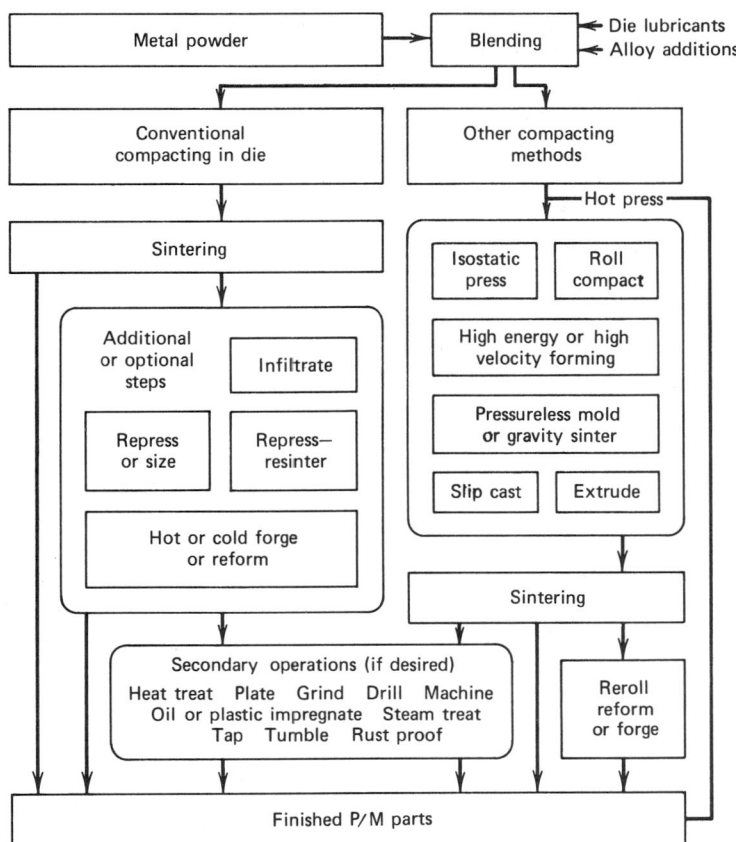

Figure 1. Basic steps in the powder-metallurgy process. Courtesy of Metal Powder Industries Foundation.

be joined together by hammering and, in the case of iron, by heating and hammering; the result was a solid metallic structure. This laborious technique was abandoned when furnaces developed temperatures high enough to melt these metals and to form alloys such as copper and tin to make bronze.

The ancient Egyptians used a sponge iron for the fashioning of implements. The iron was made by reducing iron oxide with charcoal and crushed shells and forging the soft, spongy mass into shape. This was a crude application of powder metallurgy which ultimately became one of the commercial methods for producing iron powder, known today as *sponge iron powder.*

Modern powder metallurgy was developed in the late 18th century when considerable attention was given to the manufacture of platinum powder. Platinum shapes were manufactured by forming the powder into a cylindrical mass which was then heated and hot-forged.

The first commercial application of powder metallurgy occurred when carbon and osmium were used for incandescent-lamp filaments. These materials were mixed with a binder, extruded into wire, and sintered. Tantalum and zirconium were also used for this application. The Coolidge process for pressing, sintering, and working tungsten was developed in 1900 and tungsten has been used for incandescent filaments since that time. Even today, essentially all tungsten products are made from powder, like other refractory metal products made from rhenium, molybdenum, and tantalum.

In the early 20th century, iron powder technology was developed commercially. During World War II, paraffin-impregnated sintered iron-driving bands for military projectiles were extensively used. Production reached a peak of ca 3200 metric tons per month.

With the advent of mass production in the U.S. automotive industry, iron and copper powders were used in large tonnages and provided an incentive for the technological advances of the modern P/M industry. Most industrial applications for powder metallurgy were developed in the automotive industry. The first commercial application of a powder-metallurgy product was a self-lubricating bearing made from a combination of copper and tin powders to produce a porous-bronze bearing capable of retaining oil within its pores by capillary attraction (see Bearing materials). Later, bearings were made from copper powders alone, followed by iron powder and steel P/M mechanical components such as gears, cams, and other structural shapes. Although copper powder remains an important P/M material, consumed on the order of 27,000 t/yr, it is overshadowed by iron and iron-base powders with markets of ca 180,000 t/yr.

Since the end of World War II and especially since the advent of aerospace and nuclear technology, new developments have been taking place in the powder metallurgy of refractory and reactive metals, eg, tungsten, molybdenum, niobium, titanium, and tantalum and the nuclear energy metals such as beryllium, uranium, zirconium, and thorium.

The latest P/M-wrought products are fully dense metal systems that are made from powders. Hot isostatically pressed superalloys, P/M forgings, P/M tool steels, roll-compacted strip, and dispersion-strengthened copper are examples of these products.

Powder-based high performance metals are a significant breakthrough in metal-working technology and new markets are opened up through superior performance coupled with cost-effectiveness and longer operational life.

P/M-wrought technology has been concerned mostly with superalloys, resulting principally from U.S. government-supported research and the desire to produce jet-engine components at a lower cost and with higher performance. In the 1980s, P/M-wrought products will be also investigated in the industrial sector principally for P/M tool steels and P/M forgings.

A number of societies guard the P/M interests of industry and professionals, including the Metal Powder Industries Federation (MPIF), the American Powder Metallurgy Institute (APMI), the American Society for Metals (ASM), and the Metallurgical Society of AIME (TMS–AIME).

The techniques of powder metallurgy are used to produce a wide range of products. Among these are all the common metals and alloys, high melting metals, composite metals, metal–nonmetal combinations, porous metals, alloys of unusual composition, metals of extremely high purity, mechanical components and bearings, and wear-surfacing coatings and decorative coatings such as gold and silver used in the graphic arts (see Table 1). Besides the special properties that can be attained through the use of metallurgy, powder P/M products are generally considered equivalent in properties to their cast counterparts and can be worked with equal ease. More-recent

Table 1. The Use of Metal Powders to Manufacture a Variety of Items[a]

Products	Constituents
Products manufactured only by P/M	
high melting metals	
incandescent-lamp filaments, nonconsumable welding electrodes, resistance wire for high temperature furnaces	W, Mo
electronic-tube components	W, Mo, Ta, Nb
heavy alloys for high mass and radiation-shielding	W–Ni–Cu
chemical-resistant containers	Ta
materials in which certain components of a composite must retain properties which would be lost if melted	
hard-metal alloys such as cemented carbides	carbides of W, Cr, Mo, Va, Nb, Ta, Ti, Zr; nitrides, borides, silicides
products with properties impossible to obtain by conventional methods	
catalysts	Pt, Ni, Fe, Cu
porous electrodes for alkaline batteries	Ni, Fe
porous bearings, machine components, filters, and diaphragms	Cu, Fe, Sn, stainless steel
friction materials	Cu–Sn, Fe, Pb, Sn, Al; oxides, silicides
cermets[b]	Al; oxides, borides, carbides, silicides with Fe, Ni, Ca, W, Mo, etc
magnetic cores of insulated particles	Fe
structures in which certain components do not alloy electrical contacts	W–Ag, Mo–Ag, Mo–Cu, W–Cu, W + Mo with oxides
materials in which ablation is provided by the volatilization of one component	W–Ag
Products best manufactured by P/M	
mass production cheaper by powder metallurgy	
machine and structural parts	Fe, Cu, Pb, Sn, Al, Ti
materials in which high purity, exact composition, or a high degree of structural homogeneity is essential	
permanent magnets	Fe–Ni–Co–Al
electronic metals with high permeability	Fe, Ni
semiconductor heat sinks	W, Mo, W–Ag
coins, medals, and jewelry	Ag, Au, Pt, Ni
dental alloys	Au, Pt
glass-sealing alloys	Fe–Ni–Co
nuclear and space applications	Be, Th, U, Zr

[a] Ref. 1.
[b] Combinations of metal and ceramic materials.

advances, such as P/M forging, often result in products possessing better properties than conventionally produced metals.

Powder metallurgy is employed to fabricate products that cannot be manufactured by other methods, eg, a metal with interconnected pores to provide self-lubrication. The products usually exhibit superior or special properties, eg, extremely high purity nickel strip roll compacted from nickel powder. In addition, P/M offers economic advantages, such as lower energy input and lower labor costs, because the P/M process is well-suited to automatic mass production. Furthermore, several design configurations can be combined into one component, such as a combination ratchet pinion or a gear with a cam surface that might otherwise be an assembly of components. Since powder metallurgy is a net or near-net process, ie, with little or no waste, it offers maximum material conservation with cost-saving advantages.

Thus, powder metallurgy competes with casting, stamping, screw-machine production, forging, and permanent mold casting. Although most metals in powdered form are more expensive than in other forms, the powders are cheaper to fabricate. The cost of the compacting tools is a principal factor in determining the competitiveness of the process.

Characteristics

Individual Particles. *Size.* The precise determination of particle size, usually referred to as the particle diameter, can actually be made only for spherical particles. For any other particle shape, a precise determination is practically impossible and such particle size represents an approximation only, based on an agreement between producer and consumer with respect to the testing methods.

Particles that are smaller than 44 μm (-325 mesh) are called fines; 44 μm is the finest sieve used on a large-volume basis (U.S. Standard). Size determination of fines is described in refs. 2 and 3.

Shape. Metal powder particles are produced in a greater variety of shapes, as shown in Figure 2. The shape usually depends to a large extent on the method of fabrication. Shape can be expressed as a deviation from a sphere of identical volume, or as the ratio between length, width, and thickness of a particle, as well as in terms of some shape factors.

Density. The density of a metal powder particle is not necessarily identical to the density of the material from which it is produced because of the particles' internal porosity. Therefore, it is difficult to determine the actual particle size and surface area.

Surface. Any reaction between two powder particles starts on the surface. The amount of surface area compared to the volume of the particle is, therefore, an important factor in powder technology. The particle–surface configuration, whether it is smooth or contains sharp angles, is another. Table 2 shows that the particle surface area depends strongly on the method of production, which usually determines the particle shape.

Microstructure. Powder particles may consist of a single crystal or many crystal grains of various sizes. The microstructure, ie, the crystal grain size, shape, and orientation, depend also on the method of powder fabrication. However, in many cases a correlation exists between particle size and grain size. In metal powders made by

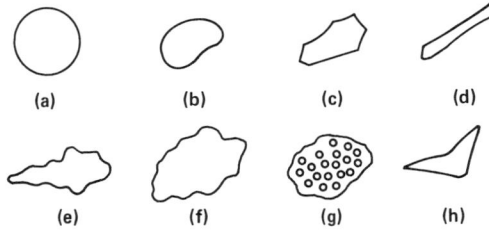

Figure 2. Various shapes of metal powder. (**a**) Spherical; (**b**) rounded; (**c**) angular; (**d**) acicular; (**e**) dendritic; (**f**) irregular; (**g**) porous; (**h**) fragmented.

Table 2. Approximate Surface Area of Powder Particles Fabricated by Various Methods[a]

Process	Particle shape	Approximate surface area[b]
carbonyl process	uniform spherical	πD^2
atomization	round irregular spheroids	1.5–2 πD^2
reduction of oxides	irregular spongy	7–12 πD^2
electrolytic process	dendritic	
mechanical comminution		
crushing	angular	3–4 πD^2
ball milling	flakes or leaves	varies over wide range

[a] Ref. 4.
[b] D = diameter.

atomization, the smaller atomized particles solidify faster than the larger ones, and therefore in small-size atomized powder particles there is usually less grain growth, producing smaller grains rather than coarse ones. The grain structure affects the activity of the powder particle, and to a certain extent, the amount of material transported by grain-boundary diffusion.

A faster cooling rate increases dendrite nuclei formation, resulting in smaller dendrites. Small dendrites produce a microstructure that is easier to homogenize during sintering. The finer the constituents, the more uniform are the properties of the powder.

Surface Oxide Layer. The possibility of reducing surface oxide layers in a reducing gas atmosphere (Cu, Fe, Ni, W, Mo, etc) depends on the type of metal. For a given thickness of the oxide layer, the amount of oxide in a powder changes with the particle size (see Table 3). The thickness of the oxide layer on an individual particle depends on the conditions under which oxidation occurs. For metals whose oxide layers can be reduced during sintering, the type and amount of oxide crystals that are converted to metal crystals greatly affect the activity of the particle surface. This is due to the increased mobility of atoms during the conversion of the crystal structure from the oxide to the metallic state. In many cases, an optimum amount of oxide determines the sinterability of the respective powders and their physical properties after sintering.

Particle Activity. The term activity refers to the reaction of a powder particle with its environment. The activity determines the type and rate of the reaction.

The total activity of a powder particle consists of the activity of the bulk and that of the surface. Both depend on the type and number of defects in the crystal structure. These defects are usually present in large numbers on the surface. Surface activity

Table 3. Particle Thickness and Oxide Content of Aluminum Flake Powder

Particle thickness, μm	Oxide content, vol %	Particle thickness, μm	Oxide content, vol %
0.1	30	100	0.03
1.0	3	1000	0.003
10	0.3		

increases with increasing ratio of surface area to volume. Small particles, therefore, usually show greater activity than larger ones. The shape and surface configuration of the particle also influence the activity. Particles with sharp-angled corners are more active than particles with rounded smooth surfaces.

Because grain boundaries contribute to the activity of a particle, particles with small crystal grains are more active than particles consisting of larger grains. With respect to diffusion, the activity of particles consisting of single grains is much lower than that of multicrystal particles.

The activity of a powder particle determines the rate of material transport by bulk and surface diffusion, the rate of adsorption and absorption, and other reactions with the environment.

Powder Mass. A mass of powder consists of a large quantity of particles. The characteristics of some iron powders are shown in Table 4. The most important properties of a good molding-grade powder are flow rate, particle size and size distribution, apparent density, green strength, compressibility, and dimensional stability during sintering. A powder must flow well in order to fill all parts of the die cavity evenly and move through the automatic equipment. The particle size and size distribution must maximize the compact density. There is a close relationship between particle size distribution and such factors as powder flow, apparent density and compressibility. Since the amount of powder needed for each compact is charged to a cavity of constant volume, the apparent density becomes extremely important. Although changes in the depth of a die cavity can be made with ease, it is most desirable that the powder has uniform apparent density batch-to-batch and hour-to-hour.

In addition to the properties discussed above, high compressibility is always desired. Compressibility is the density to which a powder may be pressed at any given pressure. As the compressibility of a powder increases, the pressure needed to obtain any given density decreases. Lower pressures result in lower tool and machine wear. Under the same compacting conditions, higher compressibility powders compact to a higher green strength. This is desirable since the compact must have enough strength to be transported either mechanically or by hand. Green strength of a pressed compact, like density, is generally considered a property of the powder.

Table 4. Characteristics of Iron Powders[a]

Method of fabrication	Average particle size, μm	Particle size distribution, %				Specific surface area, cm^2/g	Apparent density, g/cm^3
		>100 μm[b] (+150)	>74 to <100 μm (+200 to −150)	>44 to <74 μm (+325 to −200)	<44 μm (−325)		
electrolytic	78	29.26	16.39	54.16	10.0	265×10^4	3.32
	63	21.98	16.00	50.10	3.0	452×10^4	2.56
	53	2.38	21.0	74.0	1.0	115×10^5	2.05
reduced	68	28.5	15.5	54.5	12.0	516×10^4	3.03
	51		6.5	81.5	78.5	945×10^4	2.19
	6	3.5	2.0	13.5	95.5	516×10^5	0.97
carbonyl	7	2.5	0.1	1.0		346×10^5	3.40

[a] Ref. 4.
[b] Numbers in parentheses represent mesh sizes.

Average Particle Size. Average particle size refers to a statistical diameter, the value of which depends to a certain extent on the method of determination. The average particle size can be calculated from the particle-size distribution.

Particle-Size Distribution. For many P/M processes, the average particle size is not necessarily a decisive factor, whereas the distribution of the particles of various sizes in the powder mass is. The distribution curve can be irregular or show a rather regular distribution with one maximum. It may also have more than one maximum or can be perfectly uniform.

Specific Surface. The total surface area of 1 g of powder in cm^2/g is called its specific surface. The specific surface area is an excellent indicator for the conditions under which a reaction is initiated and also for the rate of the reaction. It correlates in general with the average particle size. The great difference in surface area between 6-μm reduced iron powder and 7-μm carbonyl iron powder (Table 4) cannot be explained in terms of particle size, but mainly by the difference between the very irregular-shaped reduced and the spherical carbonyl iron powders.

Determination of the specific surface area can be made by a variety of adsorption measurements or by air-permeability determinations. It is customary to calculate average particle size from the values of specific surface by making assumptions regarding particle-size distribution and particle shape, ie, assume it is spherical.

Apparent Density. This term refers to the weight of a unit volume of loose powder, usually expressed in g/cm^3 (5). The apparent density of a powder depends on the friction conditions between the powder particles, which are a function of the relative surface area of the particles and the surface conditions. It depends, furthermore, on the packing arrangement of the particles, which again depends on particle size, but mainly on particle-size distribution and the shape of the particles. The characteristics of a powder that determine its apparent density are rather complex, but some general statements with respect to powder variables and their effect on the density of the loose powder can be made.

The smaller the particles, the greater is the specific surface area of the powder. This increases the friction between the particles and lowers the apparent density but enhances the rate of sintering. This is shown in Table 4 for irregular-shaped electrolytic iron powder.

Powders with very irregular-shaped particles are usually characterized by a lower apparent density than more regular or spherical ones. This is shown in Table 5 for three different types of copper powders with identical particle-size distribution but different particle shape. Data in this table indicate that particle shape has a decisive influence on apparent density.

In any mixture of coarse and fine powder particles, an optimum mixture results in maximum apparent density. This optimum mixture is reached when the fine particles fill the voids between the coarse particles.

Table 5. Effect of Particle Shape on Apparent and Tap Density[a]

Particle shape	Density, g/cm^3		Increase, %
	Apparent	Tap	
spherical	4.5	5.3	18
irregular	2.3	3.14	36
flake	0.4	0.7	75

[a] Ref. 4.

Tap Density. Tapping a mass of loose powder, or more specifically, the application of vibration to the powder mass, separates the powder particles intermittently, and thus overcomes friction. This short-time lowering of friction results in an improved powder packing between particles and in a higher apparent density of the powder mass. Tap density is always higher than apparent density. The amount of increase from apparent to tap density depends mainly on particle size and shape as illustrated in Table 5.

Flow. The free flow of a powder through an orifice depends upon the orifice which is standardized for the test of the powder (6). Flow, therefore, depends not only on friction between powder particles, but also on friction between the particles and the wall of the orifice. Flow is usually expressed by the time necessary for a specific amount of powder (usually 50 g) to flow through the orifice.

Inasmuch as friction conditions determine the flow characteristics of a powder, coarser powder particles of spherical shape flow fastest and powder particles of identical diameter but irregular shape flow more slowly. Finer particles may start to flow, but stop after a short time and need tapping in order to start again. Very fine powders of less than approximately 20 μm do not flow at all. Addition of fine powder particles to coarser ones may increase the apparent density and usually decreases the flow quality.

Metal powders with a thin oxide film may flow well but may flow poorly when the oxide film is removed and the friction between the particles therefore increases.

The free flow of a powder is necessary for automatically filled compacting dies. Powders with low flow rates need vibratory filling in order to overcome friction, and powders that do not flow at all can be used only for manual filling of the die cavity.

Manufacture

The manufacture of metal in powder form is a complex and highly engineered operation. It is dominated by the variables of the powder, namely those that are closely connected with an individual powder particle, those that refer to the mass of particles which form the powder, and those that refer to the voids in the particles themselves. It must be remembered that in a mass of loosely piled powder, $\geq 60\%$ of the volume consists of voids.

The primary methods for the manufacture of metal powders are atomization, the reduction of metal oxides, and electrolytic deposition. An overview of production methods is given below (7–8).

Atomization. A stream of molten metal is struck with air or water jets; the particles formed are collected, sieved, and annealed. This is the most common commercial method in use for all powders.

Reduction. Reduction of iron oxides or other compounds in solid or gaseous media gives sponge iron or hydrogen-reduced mill scale.

Decomposition. Decomposition of liquid or gaseous metal carbonyls (iron or nickel) gives a fine powder.

Electrolytic Deposition. Electrolytic deposition from molten salts or solutions either gives powder directly, or an adherent mass that has to be mechanically comminuted.

Precipitation. For example, nickel ammonium carbonate gives nickel powder when subjected to hydrogen in an autoclave.

Spinning or Rotating Electrode. Molten metal droplets are centrifuged from a spinning electrode in a closed chamber.

High Energy Impaction. Brittle coarse shapes are impinged against a tungsten carbide target at high velocity and ambient or lower temperatures.

Mechanical Comminution. Machining: relatively coarse particles are produced. Milling: ball mills, impact mills, gyratory crushers, and eddy mills give fine powders of brittle materials.

Condensation. Condensation of metal vapors and deposition on cooler surfaces.

Decomposition of Metal Hydrides. Vacuum treatment of metal hydrides gives powders of fine particle size.

Reaction of a Metal Halide with Molten Magnesium. The Kroll process for titanium and zirconium results in a spongelike product.

Rapid Solidification Technology (RST). Molten metal is quench cast at a cooling rate up to $10^6 °C/s$ as a continuous ribbon which is subsequently pulverized to amorphous powder.

Intergranular Corrosion. Selective corrosion of carbide-rich grain boundaries in stainless steel.

Processing

Consolidation. Metal powders are consolidated by heat or by pressure followed by heat, or by heating during the application of pressure (9). Consolidation produces a coherent mass of definitive size and shape for further working, heat treating, or use as is.

During pressure application, the powder particles are first rearranged or packed. Then elastic and plastic deformation takes place, and third, the particles are cold-worked. Many attempts have been made to develop a mathematical relationship between compacting pressure and density and strength of the compact and strength of the sintered compact (10–13). Such a relationship has not been found to be of great practical significance. Particles bond and form a coherent mass under pressure because of liquid surface cementation; interatomic forces such as surface adhesion, cold welding, and surface tension; and mechanical interlocking of particles (14).

In the cementation process, the surface atoms are bonded by the heat generated during compaction. Although the temperature rise within the compact can be measured, this theory is not generally accepted.

Plasticity of the metal crystals plays a dominant role in interparticle bonding. This is a characteristic of each metal. It is affected to a large extent by the condition of the individual powder particles. Gold, silver, lead, and iron are highly ductile (plastic) metals, whereas chromium and tungsten deform with great difficulty. Plasticity also depends upon nature and history of the powder, impurities present (especially on the surface), and friction conditions between the particles. Powders that exhibit a high degree of plastic deformation form many areas of metal contact and therefore many interparticle bonds, whereas powders composed of harder particles form relatively few such bonds.

Probably the most important powder property governing the formation of atomic bonds is the surface condition of the particles, especially with respect to the presence of oxide films. If heavy oxide layers are present, they must be penetrated by projections on the particles. This results in only local rather than widespread bonding. A ductile metal such as iron with a heavy oxide layer may not form as strong or as many bonds as a less ductile metal that is relatively free of oxide.

Mechanical Interlocking. The fact that irregularly shaped powder particles form denser and stronger compacts under pressure than regularly shaped powder particles leads to the theory of interlocking particles. Interlocking is probably the principal strengthening mechanism and is probably also responsible for providing increased surface contact for cold welding. This mechanism plays an important role in the strengthening of compacts made from metals such as tungsten and chromium, which normally are not plastic at room temperature.

The characteristics of a pressed compact are influenced by the characteristics of the powder: rate and manner of pressure application, maximum pressure applied and for what period of time, shape of die cavity, temperature during compaction, additives such as lubricants and alloy agents, and die material and surface condition. The effect of various compaction variables on the pressed compact are shown in Figure 3.

Consolidation Techniques. *Unidirectional Compaction.* Uniaxial pressing of metal powders in a die of specific dimensions and configuration is the most frequently used technique for the consolidation of powders in the manufacture of P/M products. Powder flows automatically into a die cavity; the bottom punch or punches act as its bottom. An upper punch seals the top and pressure can be applied parallel to the direction of powder flow into the cavity by forcing the punches together. In single action either the bottom or the top punch is held rigid whereas the corresponding punch moves to press the powder. In double action both punches move. Double-acting punches produce a more uniformly dense compact.

After pressure application, the top punch is removed and the compact is ejected

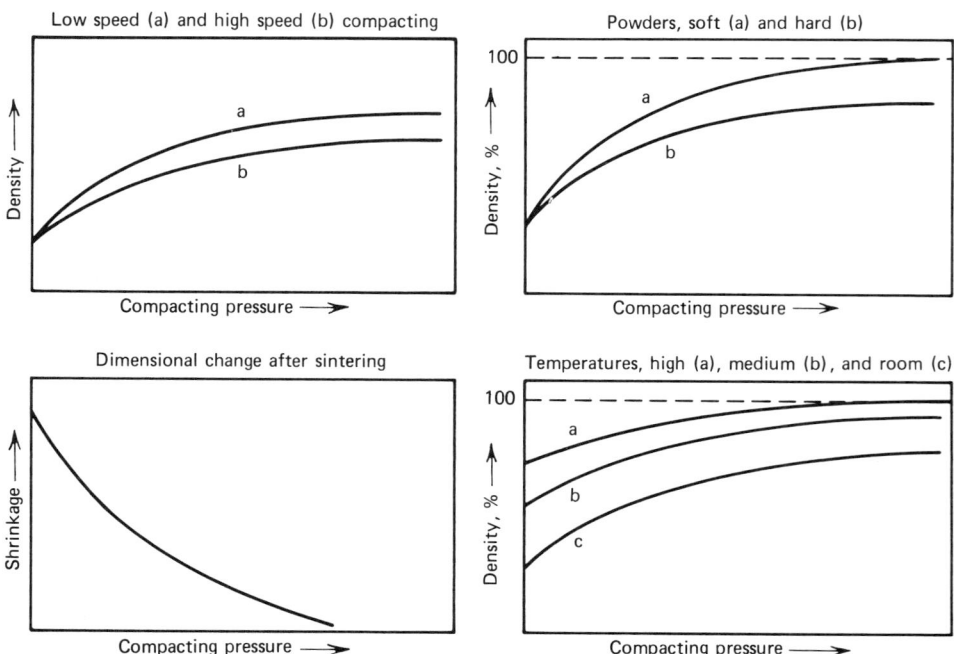

Figure 3. Effects on the pressed compact.

from the cavity by the bottom punch. The cavity is then reformed and is ready for another charge.

This cycle is repeated automatically at a rate that varies with the part and size and the complexity and flowability of the powder. The cycle may take 5–25 min. Pressing equipment producing relatively small, simple parts can operate at up to 200 parts/min. Rotary presses with multiple die sets are even faster. Table 6 gives the ranges of pressures that are used for various materials during die compaction.

Compacting tools must be properly designed, constructed, and fitted to the press. They may be made of heat-treated steel or cemented carbide, depending on the economics and the number of parts to be produced. Carbide tools are more expensive; however, they can be used much longer than steel tools.

Cold Isostatic Pressing. In the cold isostatic process, pressure is transmitted through a liquid medium, either oil or water, to a flexible container holding the powder (15). The liquid transmits the pressure uniformly throughout the chamber. Density distribution within the pressed compact, therefore, is very uniform.

Cold isostatic pressing may be either by the free-mold or wet-bag method, or the fixed-mold or dry-bag method. They differ only in the configuration and placement of the tooling. Both methods utilize a pressure vessel with a closure device designed to contain a fluid under high pressure, along with a pressurization system and controls. More advanced production-scale isostatic presses are equipped with various automated systems for powder fill, pressure vessel operation, and part ejection.

The free-mold process is used in batch-scale production, applications requiring complex shapes, and for most research and prototype studies. This technique provides flexibility in pressing a variety of sizes and shapes, limited only by the inside dimensions of the vessel.

The fixed-mold process is characterized by the flexible mold being positioned directly inside the pressure chamber and supported by a perforated outer shell with a lip on the mold, sealing it against the side wall of the chamber. In fixed-mold pressing, one part is ordinarily pressed per cycle per pressure chamber. The process lends itself to bars, rods, or tubes.

Table 6. Pressure Requirements and Compression Ratios[a] for Various Powder Products

Type of compact	MPa[b]	Compression ratio
brass parts	413–690	2.4–2.6:1
bronze bearings	207–276	2.5–2.7:1
carbon products	138–165	3.0:1
carbides	138–414	2.0–3.0:1
iron bearings	207–345	2.2:1
iron parts		
low density	345–483	2.0–2.4:1
medium density	483–552	2.1–2.5:1
high density	483–828	2.4–2.8:1
iron powder cores	138–690	1.5–3.5:1
tungsten	69–138	2.5:1
tantalum	69–138	2.5:1

[a] Compression ratio is the dimensional relationship between the loose and the compacted powder at a given compacting pressure.

[b] To convert MPa to psi, multiply by 145.

Isostatic compaction is used mainly for the formation of large compacts where the number of parts is limited. These parts, which may vary from a few cm to 2 m in length, and up to 30 cm dia are then sintered and used as extrusion, swaging, or drawing billets, or are machined into more intricate shapes. The isostatic process is used when a limited number of items with complicated shapes is desired, or perfectly uniform density is required. In general, the close tolerances achieved by rigid die compaction are not possible by isostatic methods, mainly because the mold dimensions change during pressure application. The process can be automated for the mass manufacture of small precision parts.

Hot Isostatic Pressing. Hot isostatic pressing (HIP) combines the application of pressure with sintering (16–17). During hot consolidation, the powder must be separated from its gaseous environment which increases the cost of the operation. The powder, therefore, is usually canned, encapsulated in a container to exclude air. The powder in the can is heated in an autoclave with pressure applied isostatically by a gas, commonly argon. Pressures are ca 100 MPa (15,000 psi). At temperatures below 1250°C, which includes the hot consolidation of high speed steel and superalloy powders, the temperature in the pressure vessel is usually maintained more or less constant not only during the actual pressing but also between cycles. This is possible because Kanthal heating elements may be exposed to air while hot whereas Mo, W, and C must be protected from oxidation when hot.

Hot isostatic pressing offers the advantage of producing directly from a powder a more complex shape than a simple billet. When this shape approaches that of the end product, it is called a near-net shape. Jet-engine turbine disks from nickel-base superalloy powders are an example. Hot isostatic pressing improves the homogeneity of the microstructure and the properties of the component. In addition, less raw material is required than for a casting which has to be forged and machined.

Substantial savings can be made also in tooling. Producing a shaped part from powder by the HIP process requires only a sheet steel can; this takes far less tooling than in die forging (18).

Powder Rolling. Powder conveyed either horizontally or vertically through a set of steel rolls is compacted in the roll gap and emerges as a porous sheet (19). This sheet is then sintered, rerolled (warm or cold, depending upon the material), annealed, and rolled again in a finishing operation. The rolling mills and furnaces are arranged for continuous production. Green strength of the sheet, as it emerges from the roll gap, is the limiting factor for both the thickness and width of the sheet. Roll size, roll gap, roll speed, and rate of powder feed have to be controlled. Compaction occurs essentially in only one direction since very little pressure is transmitted laterally.

Assuming a cost-effective starting material, powder rolling eliminates much of the equipment needed for the usual melting, casting, and rolling to produce thin sheet. Sheet can be rolled closely to finished size with a minimum loss of material. Most scrap generated by this process can be reclaimed as powder. This technique also permits the production of clad materials by using a metal strip as the carrier for the powder through the roll gap. Roll compacting of nickel powder produces strips from which blanks are made for coins. In the United States, large quantities of nickel and nickel–iron–cobalt powders are rolled into strip for glass-sealing alloys and various electronic applications. Multiple-ply strips may also be rolled directly by metallurgical sintering of the layered components in the roll gap. This process is used commercially to make tri-ply strip box composite bearings.

P/M Forging. Even after conventional repressing of a P/M component, it is still difficult to increase density above 95%. However, full density in a P/M part improves its properties. Hot isostatic pressing in autoclaves works well, especially for titanium and superalloy components, but the capital equipment is expensive and production rates are slow.

For ordinary materials and higher production rates, P/M forging can be used (18,20). After parts are compacted and sintered to medium density, they are reheated, lubricated, and fed into a hot-forming or P/M-forging press. The part is formed by one stroke of the press in a closed precision die. A typical hot-forming press set-up includes die sets, automatic die cooling and lubrication, transfer mechanism, an induction heating unit for preforms, and controls.

P/M forged parts with dimensions up to 65 cm^2 and weighing up to 2.3 kg are produced on hot-forming presses in the 725-t range. Usually the process is used for parts weighing 0.35–4.5 kg. Properties are generally equal or superior to those of conventional forgings. P/M forgings usually have a minimum density of 99% of theoretical and depending on alloy composition, can exhibit tensile strengths >1.38 GPa (200,000 psi) after heat treatment. Oil quenching and tempering at 230°C, for example, produces tensile strengths of ca 1.5 GPa (215,000 psi) (ultimate) and ca 1.25 GPa (185,000 psi) (yield) with 7% elongation and RC 43 hardness (see Hardness). Heat treating at 650°C reduces strength and hardness values by 50% but triples elongation. The cups and cones for tapered roller bearings, are forged from grade 4600 powder (2 wt % nickel, 1/2 wt % molybdenum alloy), which corresponds to AISI 4600 grade steel. They perform at least equal to and up to eight times longer than the same part produced from wrought steel.

The automotive industry is the principal user of P/M forgings, primarily for transmission and differential components, but also for engine parts. Other markets are in power tools and farm machinery. Cost effectiveness is generally the reason for substituting P/M forgings for conventionally forged, cast, or machined parts.

High Energy-Rate Compaction. Metal powders can be rapidly compacted in rigid dies by the application of high pressures at high speed (21). Special presses employ a pressure-upsetting system for rapid movement of the compacting tools. Explosives are also used. Single- and double-acting presses are capable of generating pressures on the order of several GPa (10^6 psi) for a time. The rams in presses employing explosives move at the rate of more than 1000 m/s, causing compaction to occur in microseconds. Shock fronts from explosives with pressures close to 14 GPa (2×10^6 psi) are generated.

High energy-rate-forming techniques have been used mainly for laboratory studies or to produce compacts with special properties, but these techniques are not of commercial interest.

Slip Casting. The slip casting of metal powders into useful articles is an interesting process but has only limited industrial application (22–23). It is sometimes used to produce large, very complicated parts from refractory metals.

Slip casting of metal powders closely follows ceramic slip casting techniques. Slip, which is a viscous liquid containing finely divided metal particles in a stable suspension, is poured into a plaster-of-Paris mold of the shape desired. As the liquid is absorbed by the mold, the metal particles are carried to the wall and deposited there. This occurs equally in all directions and equally for metal particles of all sizes which gives a uniformly thick layer of powder deposited at the mold wall.

Vibratory Consolidation. Powders are vibrated in a mold or other container in which they will be sintered, or in a metal container that will be used for extrusion or other metalworking process (23). Vibratory consolidation produces packings of UO_2 particles up to 95% of theoretical density.

Hot Pressing. Hot pressing may be used either to consolidate a powder that has poor compactability at room temperature, or to combine compaction and sintering in one operation. The technique is essentially the same as described for unidirectional die compacting. The powder is heated by either heating the entire die assembly in a furnace or by induction heating. In most instances, a protective atmosphere must be supplied.

Hot pressing produces compacts that have superior properties, mainly because of higher density and finer grain size. Closer dimensional tolerances than can be obtained with pressing at room temperature are also possible. Hot pressing is used only where the higher cost can be justified and has recently been useful in producing reactive materials. An interesting use is the combination of P/M and composites to produce hot-pressed parts that are fiber-reinforced. However, the technique is mainly employed in the laboratory.

Extrusion, Swaging, or Rolling. Metal powders first formed into ingots by isostatic pressing, or metal powders encased in a suitable container may be subjected to any number of operations such as extrusion, swaging, or rolling. In the absence of a container, a protective atmosphere is essential. This type of consolidation is usually performed at elevated temperature. For contained powder, the case becomes a sheath during working, which is subsequently removed by either machining or chemical methods. Canned extrusion is used commercially for tool steels, superalloys, and beryllium.

Injection Molding. P/M injection molding combines the characteristics of slip casting, investment casting, and powder metallurgy (18). It produces complex, high density (typically 96%) P/M parts. Tolerances can be close and virtually any material can be formed, including mild and nickel steels, stainless steels, superalloys, refractory metals, and carbides. High levels of ductility (20–30% elongation for mild steel) and toughness can be obtained with subsequent heat treatment.

The metal powders are blended with a thermoplastic resin, and the plastisol molded to an oversize replica of the part required. After solvent extraction of the resin, the mass is sintered to high density and net shape. The powders are extremely fine spheroidal particles, less than 10-μm dia. They may be prealloyed but usually are premixed and alloyed by solid-state diffusion during sintering. Molding pressures are low, ca 7 MPa (1000 psi) at 150°C.

The final density is not substantially affected by the molding operation, which is governed primarily by the powders, blending sequence, and subsequent sintering. The green or as-molded stage is an appropriate time for secondary operations such as drilling or machining.

The critical sintering operation is performed in oxidizing, reducing, or inert atmosphere, or vacuum, depending on the material. Shrinkage, on the order of 20–25% and very nearly isotropic, transforms the preform to its dense net shape. Shrinkage is reproducible from part to part and controlled by the plastic-to-metal ratio in the initial formulation and sintering parameters. Sintering time of ca 24 h is much longer than that required for conventional P/M parts.

Compacting Lubricants. The surface area of most moldable metal powders is in the range of 500–700 cm^2/g; finer powders can have a surface area as high as 1500 cm^2/g (24–25). A very large number of individual particles is involved. For example, 1 cm^3 uniformly filled with 2 μm spherical particles having a surface area of 1200 cm^2/g contains ca 1.2×10^9 particles. Because of this large surface area, a considerable amount of friction has to be overcome during powder consolidation. To one degree or another, friction is present in all consolidation methods.

Dry lubricants are usually added to the powder in order to decrease the friction effects. The more common lubricants include zinc stearate, lithium stearate, calcium stearate, stearic acid, paraffin, graphite, and molybdenum disulfide. Lubricants are generally added to the powder in a dry state in amounts of 0.25 to 1.0 wt % of the metal powder. Some lubricants are added by drying and screening a slurry of powder and lubricant. In some instances, lubricants are applied in liquid form to the die wall.

Lubricants protect die and punch surfaces from wear and burn out of the compact during sintering without objectionable effects or residues. They must have small particle size, and overcome the main share of the friction that is generated between the tool surfaces and powder particles during compaction and ejection. They must mix easily with the powder, and must not excessively impede powder flow (see Lubrication and lubricants).

Sintering. Sintering can be defined as the bonding of particles in a mass of powder by molecular (or atomic) attraction in the solid state through the application of heat, which produces strengthening of the powder mass and normally results in densification due to material transport (15,26–30).

During the sintering treatment, which usually occurs below the melting point of the metal powder except for liquid-phase sintering of some powder mixtures, material transport takes place in the solid state which results in some changes of the properties of the compacted powder, as shown in Figure 4. With increasing temperature and time, the strength of the powder mass increases, electrical resistivity and porosity decrease, and density, therefore, increases. The grain structure also undergoes some changes, and recrystallization and grain growth occur. In order to avoid oxidation of the metal-powder mass during the high temperature treatment, either a neutral or a reducing atmosphere is provided. The movement or transport of material during sintering is caused by surface diffusion, volume or lattice diffusion, grain-boundary diffusion, evaporation and condensation, and plastic or viscous flow (29,31). Probably several of these mechanisms act simultaneously, depending on the type of powder, its particle size and shape, and especially the temperature.

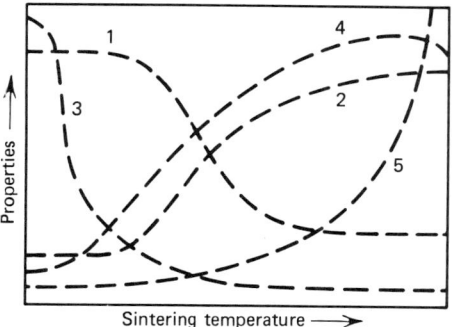

Figure 4. Effect on sintering on 1, porosity; 2, density; 3, electrical resistivity; 4, strength; and 5, grain size.

Diffusion is based mainly on the diffusion of vacancies; grain boundaries may act as sinks for these vacancies. This vacancy movement and annihilation cause the porosity of the powder compact to decrease during sintering.

Pressed powder (green) compacts are characterized by a porosity, or total pore volume, of approximately 10–40%. The number and size of the pores can be correlated with the size and shape of the powder particles from which the compact has been prepared, and the pressure applied during compacting. During sintering, the porosity undergoes a number of changes: with increasing sintering temperature or time, the total porosity decreases, the pores that originally are irregular or angular in shape become spherical; the average pore size becomes larger; the total number of pores decreases; the smaller pores disappear first; and the number of larger pores increases slightly.

The decrease of porosity during sintering results in shrinkage and, accordingly, in the densification of the powder compact. Density and densification rate during sintering are strongly affected by the particle size, the pressure applied during compacting (Fig. 5), and the sintering temperature and time. They depend further on the type of metal from which the powder has been prepared. The average sintering temperatures for various types of metals depend on their melting points (see Table 7).

Liquid-Phase Sintering. Sintering in the liquid state refers to the sintering of a powder mixture of two or more components, of which at least one has a melting temperature lower than the others. The sintering temperature is then selected in such a manner that a liquid phase is formed in which the solid powder particles of the other components rearrange. A high density powder compact is the result.

The properties of the sintered product depend to a large extent on the surrounding atmosphere. There are reducing atmospheres (hydrogen, dissociated ammonia, carbon monoxide, nitrogen plus hydrocarbons, or hydrocarbons) and neutral atmospheres (vacuum, argon, helium, or nitrogen). Metal powders susceptible to surface oxidation must be sintered in a reducing atmosphere in order to remove the oxide films. The selection of the atmosphere depends on the type of material to be sintered, whether a reaction between the metal and the atmosphere is desired, and the type of reaction desired. Cost is also a factor.

Both batch and continuous furnaces may be employed. The maximum temperature to be reached in a sintering furnace depends on the furnace and the heating methods (see Table 8) (15).

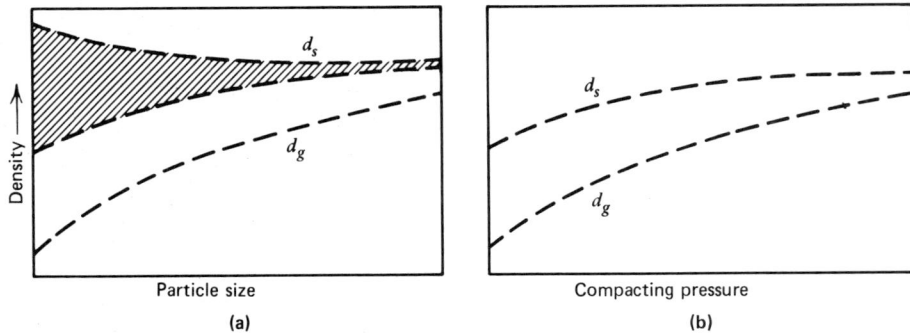

Figure 5. Effect of density of (**a**) particle size and (**b**) compacting pressure. d_g = density as compacted. d_s = density as sintered.

Table 7. Sintering Temperature and Time[a,b]

Material	Temp, °C	Time, min
bronze	760–871	10–20
copper	843–899	12–45
brass	843–899	10–45
iron, iron–graphite, etc	1010–1149	8–45
nickel	1010–1149	30–45
stainless steel	1093–1288	30–60
alnico magnets	1204–1302	121–150
90% tungsten–6% nickel–4% copper	1343–1593	10–120
tungsten carbide	1426–1482	20–30
molybdenum	2054	120[c]
tungsten	2343	480[c]
tantalum	2400[c]	480[c]

[a] Ref. 15.
[b] In high-heat chamber.
[c] Approximate.

Table 8. Sintering Methods and Temperatures

Method	Temp, °C
gas	1100
resistance-wire heating	
Nichrome	1150
Kanthal A	1300
molybdenum	1800
tungsten	2500
silicon carbide	1350
carbon short-circuit tube[a]	2750
direct resistance of the compact	3200
induction heating	3000

[a] Current goes directly through the furnace tube and no other heating elements are necessary.

Infiltration. In infiltration, the pores of a sintered solid are filled with a liquid metal or alloy (17). The most common application is the infiltration of a porous sintered steel matrix with copper, to give a copper-infiltrated P/M part. The aim of infiltration is to obtain a pore-free structure. The liquid and the solid must not react to form a solid compound or alloy having a specific volume as great as or greater than their combined preinfiltration specific volumes. The porosity in the porous matrix should be interconnected. Ideally, the matrix material should be insoluble in the liquid infiltrant. Simple liquid-phase sintering does not give pore-free structures; infiltration does, provided the pores are interconnected.

The process is used for ferrous P/M structural parts that have densities of at least 7.4 g/cm^3 and mechanical properties superior than those of parts that have been only compacted and sintered. Depending upon the application, the porous matrix may be infiltrated only partially or almost completely. Copper-base alloy infiltrants have been developed to minimize erosion of the iron matrix.

In one method, the matrix component has already been sintered and the green compact of infiltrant powders is positioned on or underneath it. The assembly is then heated to the infiltration temperature. Alternatively, sintering of the matrix and infiltration may be combined into one operation by positioning the green compact of the infiltrant on the green compact of the iron or steel matrix. This operation is called sintrating. By controlling the rate of heating, the matrix is adequately sintered by the time the melting point of the infiltrant is reached. If only part of the matrix is to be infiltrated, eg, the teeth of an infiltrated gear, the matrix may be positioned in a graphite container in which space is provided adjacent to the gear teeth to be preferentially infiltrated. The space is then filled with the appropriate amount of infiltrant in powder form. Tungsten for electrical contacts is often infiltrated in this manner with copper or silver.

Postsintering Treatments. The sintering process concludes the powder metallurgy processes of production and consolidation. However, some P/M parts may require a number of further operations.

Working Treatments. *Restricted plastic deformation* takes place entirely within the confines of a closed die cavity. A sintered part may be replaced in the die cavity and pressure applied to the part. This pressure generally is of the same magnitude as the original compaction pressure. This second application of pressure can be categorized as follows.

Sizing. The desired size is obtained by a final pressing of a sintered compact. During sintering, the compact may have expanded, shrunk, or changed dimensions slightly, which is corrected by the sizing operation.

Repressing. Pressure is applied to a previously pressed and sintered compact, usually for the purpose of improving a physical property, such as tensile strength, hardness, or density.

Coining. The sintered compact is pressed to obtain a definite surface configuration which changes the shape of the article. In some instances, the sintered piece is used as a blank with most of the surface configuration produced by coining such as in striking coins or medallions.

In general, sizing, repressing, and coining are performed at room temperature. With elevated temperatures, a protective atmosphere must be provided.

Unrestricted plastic deformation includes all of the metalworking procedures generally applied to cast metals. P/M ingots made by die pressing and sintering, or isostatic pressing and sintering may be forged, swaged, drawn, rolled, or extruded. These processes may also be applied to a mass of loose or loosely sintered powder. In most instances, the P/M ingot being worked has lower density and lower workability than cast ingots; temperatures are generally elevated. In many cases a neutral atmosphere is provided by enclosing the operation in a protective gas or in a sheath and providing the proper atmosphere within the sheath.

Heat Treatments. If the inherent porosity is taken into consideration, heat treatments performed on P/M parts do not differ substantially from the same treatments performed on cast or wrought metal with similar results.

The most common heating operation is resintering, usually under conditions similar to those of the first sintering operation, but at higher temperature. Resintering relieves stress or removes the effects of cold work imparted during coining or repressing.

Resintering is also undertaken for further densification. Cold-working, as in repressing or coining, reduces porosity and ruptures oxide films within the compact, which created new metal-contact areas. A resumption of sintering enables diffusion to proceed with fewer obstructions at a lower temperature. In general, it increases ductility and density with some loss in strength; however, hardness, tensile strength, and ductility are usually much improved.

Sintered metals may be softened or hardened by a number of procedures that are common in treating cast or wrought metals. Pure copper or silver may be annealed by heating above the recrystallization temperature. Certain copper-base alloys are precipitation-hardened, such as copper–beryllium, copper–chromium, and copper–zirconium by solution-treatment quenching and coining. Precipitation hardening is practiced on aluminum alloys that contain up to 5 wt % copper and 1.5 wt % silicon, or 0.5 wt % magnesium and 0.5 wt % manganese.

P/M steels can be heat-treated in the same manner as cast or wrought steels. They may be austenitized, quenched, and tempered. Surface hardening includes pack or gas carburization or nitriding, ie, heating in a nitrogen-containing atmosphere. Because of the greater amount of exposed surface area in the form of porosity, a protective atmosphere is needed (see Metal surface treatments).

Finishing. Finishing treatments include shaping operations such as machining, broaching, sizing, burnishing, grinding, straightening, deburring, and abrading, as well as surface treatments, such as steam oxidizing, coloring, plating, impregnating, dipping, or spraying.

Steam treatment imparts increased corrosion resistance for ferrous P/M parts. The parts are heated to 400–600°C and then exposed to superheated steam. After cooling, the parts are usually oil-dipped to further increase corrosion and wear resistance, and to enhance appearance. Heat-treated parts are seldom steam-treated, because of annealment.

The blue–black iron oxide formed in this process fills some of the interconnecting porosity and much of the surface. Hence the density is increased, resulting in higher compressive strength. Furthermore, the oxide coating increases hardness and wear resistance.

Bluing (Blackening). Ferrous P/M parts can be colored by several methods. To enhance corrosion resistance, the parts are blackened by heating in a furnace to the bluing temperature and then cooled. Oil dipping deepens the color and improves corrosion resistance. A dry-to-touch oil may leave a dry film on the parts.

Ferrous P/M parts can also be blackened chemically, using one of several commercial liquid salt baths. If the parts are below a density of 7.3 g/cm^3, entrapment of salt is avoided by impregnating the parts before blackening with a resin that does not break down in the bath. Nickel-or copper-bearing parts tend to adversely affect most blackening baths. Furthermore these materials seriously affect color. As is the case with furnace blackening, an oil dip improves appearance and corrosion resistance. However, the process causes a slight change in size and makes the parts more brittle and more difficult to machine.

Plating. All types of plating in general use, including copper, nickel, chromium, cadmium, and zinc, can be applied on P/M parts. High density (7.2 g/cm^3) and infiltrated parts can be plated by the same methods as wrought parts. To avoid entrapment of plating solutions in the pores, lower density parts should be sealed with resin. Before resin impregnation, the oil must be removed from all pores and surfaces. Electroless nickel plating can also be used, and peen (mechanical) plating is applicable to nonimpregnated ferrous parts with a density of 6.6–7.2 g/cm^3 (see Electroless plating; Electroplating).

Economic Aspects

Approximately 145,000 metric tons iron and iron-base P/M parts and 21,700 t copper and copper-base P/M parts were shipped in 1979. The annual growth rate for P/M parts has been about 10% per year.

About 80% of iron-powder production is used in the manufacture of P/M parts with the balance distributed among welding electrodes, cutting, scarfing and lancing applications, electronic and magnetic applications, and a host of other uses including iron-powder additives in pharmaceuticals and iron-enriched foodstuffs. The latter represents a market of about 1000 t/yr. Detailed statistical data are compiled by the Metal Powder Industries Federation.

About 87% of the copper and copper-base powder produced is used in the manufacture of P/M parts, more than half of this in the production of self-lubricating bearings and as elemental copper powder for additions to iron-base P/M parts.

Health and Safety

Metal powders possess an immensely high ratio of specific surface area to volume. This characteristic contributes to several potentially hazardous properties such as pyrophoricity, explosiveness, and toxicity (32–34). The problems associated with the fine particles can be minimized or eliminated with proper handling and good housekeeping procedures, such as storage in appropriate containers, processing in sparkproof equipment, avoiding exposure to open flames, and minimizing airborne particulates. A report on the *Explosibility of Metal Powders* (35) gives the results of investigations of various commercial metal powders with respect to ignition temperature, minimum explosive concentration, minimum ignition temperature for dust clouds, maximum pressure, and rate of pressure rise (32). The prevention of dust explosions is covered in codes issued by the National Fire Protection Association (34).

There is no difference for metal in powder form in terms of toxicity other than that fine particles can be ingested or inhaled more readily than massive metals (33). Conformance to OSHA requirements has not presented problems to the metal-powder producing or consuming industries. The limits of airborne particulates are being investigated by NIOSH.

There are no fumes or effluents generated in the processing of powders and the requirements of state and Federal environmental-protection agencies are met without difficulty. However, the production of powders has been subjected to the same concerns as most other metal refining and smelting operations.

The only aspect of the P/M industry with respect to safety in the workplace concerns the operation of compacting presses. Here, guarding devices are required by OSHA to prevent injuries. Those applying specifically to metal-powder compacting presses are described in a standard issued by the Metal Powder Industries Federation.

Applications

Structural Parts. Structural P/M parts are made from iron- and copper-base powders, and more recently, also from aluminum (see Fig. 6). Copper powder is used mainly for the manufacture of self-lubricating bearings whereas iron is used for structural components.

Figure 6. Examples of structures made from metal powder.

The tensile properties of sintered (not heat-treated) conventional P/M parts are generally comparable to cast metals of the same composition. Fatigue strength is about 38% of tensile values, whereas in wrought metals it is about 55%. Toughness, as a measure of impact strength, determined by Izod or Charpy methods, is density-dependent. The higher density, the greater the toughness. Properties of typical P/M parts are reported in ref. 36.

Because of the wide variety of raw materials and process techniques available, the properties of P/M parts can be tailored to meet the demands of most applications. Physical properties can match those of machined, cast, or even forged parts. Properties range from low density, porous self-lubricating bearings, to high density structural parts with tensile strengths >1.25 GPa (180,000 psi). Parts can be made hard and dense in one section and porous for self-lubrication in another.

Most P/M parts are manufactured on a custom-engineered basis, and orders are filled according to customer's blueprints. In addition, most manufacturers also offer operations such as plating, machining, heat-treating, steam-treating, plastic impregnation, sizing or coining, or grinding. Some companies specialize in certain areas of P/M techniques, including short-production runs, high density parts, helical gearing, split-die forming, infiltrating parts, parts with special electrical or magnetic properties, P/M forging, etc.

Although the P/M process is advantageous for the production of large quantities of parts in order to amortize the costs of tooling, short runs, ie, quantities of 50–50,000, are possible.

P/M parts may be made in great variety of materials, including iron, steels, low and high brass, bronze, nickel and nickel-base alloys, copper, aluminum, and various alloys including refractory metals. Copper–steel PM parts may contain up to 7 wt % copper to improve properties and dimensional control during the sintering operation. PM parts can be made smaller than a ball point on a pen or as large as a bearing weighing more than 50 kg. Because the parts are formed in precision dies under high pressure at room temperature, reproducibility is a great advantage. Production costs are low, and no environmental problems are presented.

Processing. The metal powders are blended or mixed with each other and with a lubricating agent. They are then fed into automatic molding presses where they are shaped into a "green" or molded part called a compact. The compact is heated at a specific temperature under a protective gas atmosphere, generally 1120°C for iron and steel, 1260°C for stainless steel, and 982°C for copper and copper alloys. In some instances, a separate presintering treatment (burn off) at a lower temperature is applied in order to volatilize the pressing lubricant and to sinter the compact to a specific degree for subsequent repressing. In general, the part is completed after a single sintering. Occasionally, coining or various heat treatments are necessary.

Changes in configuration may occur during sintering. Dimensional changes are generally a function of the powder, whereas shape changes are a function of density inequalities. Configuration changes are controlled either by a combination of powder properties or postsintering treatments such as sizing or machining; the latter, however, increase costs.

Porous Materials. In porous materials, the void space that determines the porosity is controlled as to amount, type, and degree of interconnection (17,37–39). Porous parts include self-lubricating bearings, bushings, certain types of P/M parts, and metallic filters. Their manufacture represents an important aspect of the P/M industry. The main applications for porous metals are in filters for separating combinations of solids, liquids, and gases; surge dampeners and flame arrestors for use with gases and liquids; metering devices and distribution manifolds; and storage reservoirs for liquids. The latter includes the self-lubricating bearings described below.

In addition, porous metals are used for the diffusion of air for aeration of liquids, and the physical separation of immiscible liquids such as gasoline and water, pressure-gauge equalizers in instrumentation, flow control, flame arrestors, and sound deadeners.

Porous parts and bearings are made by both the press and sinter techniques, whereas filters are made by loose-powder sintering. The metals most commonly used for P/M porous products are bronze, stainless steel (type 316), nickel-base alloys (Monel, Inconel, nickel), titanium, and aluminum.

Processing. Porous metal products are made by compacting and sintering, isostatic compacting, gravity sintering, sheet forming, and metal spraying.

Gravity (loose-powder sintering) is employed to make porous metal parts from powders that bond easily by diffusion, eg, bronze. No outside pressure is applied to shape the part. The appropriate material, graded for size, is poured into a steel or graphite mold that is heated to the sintering temperature of the metal; at this point bonding takes place. The part tolerances are necessarily liberal, although the inside diameters tend to be predictable because the material usually shrinks to the core during sintering. Design tolerances of ±2% for the outside diameter and ±3% for the part length are typical.

In the sheet-forming process, stainless steel, bronze, nickel-base alloys, or titanium powders are mixed with a thermosetting plastic and presintered to polymerize the plastic. Sintering takes place in wide, shallow trays. The specified porosity is achieved by selecting the proper particle size of the powder. Sheet is available in a variety of thicknesses between 16- × 30-mm and as much as 60- × 150-cm. A sheet can be sheared, rolled, and welded into different configurations.

Porous metal structures can also be created by spraying molten metal onto a base. Porosity is controlled by spraying conditions or by an additive that may be removed later.

Porous P/M products can be sinter-bonded to solid metals. They can also be welded, brazed, or soldered. Filling the voids with flux or molten metal has to be avoided. P/M porous products can be machined, but again the blocking of the porous passages has to be avoided. Press fitting and epoxy bonding are commonly used.

Self-Lubricating Parts. Self-lubrication depends upon the presence of oil within the pores of the bearing or bushing. A built-in oil reservoir provides a protective oil film that separates the bearing from the shaft, for example in a motor, and prevents metal-to-metal contact. During operation, the rotating shaft draws the oil in the bearing to the surface through capillary action. When not under load, most of the oil is drawn back into the bearing, leaving a layer of oil between the two metal parts.

Most bearing materials are made from bronze powders (see Bearing materials). Compositions may be 5–12 wt % tin and up to 6 wt % graphite. Standard compositions are 90.5 wt % Cu–7 wt % Sn–2.5 wt % graphite; 89 wt % Cu–7 wt % Sn–4 wt % graphite; 93 wt % Cu–7 wt % Sn; 96 wt % Cu–4 wt % graphite; and 100 wt % Cu. Compositions containing copper with up to 8 wt % Pb or up to 40 wt % Zn are also available.

Self-lubricating bearings are also made from iron usually containing 36 wt % Cu and 4 wt % Sn. Iron–lead alloys contain 2–6 wt % Pb and 2–4 wt % graphite. Iron-base materials offer increased hardness and strength; in addition, their coefficient of thermal expansion is close to that of the steel shaft. However, iron-base materials are generally not rated as high as copper-base materials as self-lubricating bearing materials.

Light-weight aluminum bearing compositions, commercialized in the 1960s, have excellent corrosion resistance against oxidized oils. Compositions generally include tin or lead.

Processing. Simple self-lubricating parts are produced on an automatic, rapid, high volume basis. The powder mixtures usually contain a few percent of organic lubricants. They are molded unidirectionally in a closed die cavity at pressures of 276–552 MPa (40,000–80,000 psi). During sintering, the lubricant volatilizes which facilitates some control over interconnection of the voids. Porosity is increased by volatile pore-forming agents that are added during sintering.

The sintering of bronze bearings is an example of liquid-phase sintering in which the liquid is present only during part of the sintering time. Alloying causes the liquid to solidify.

Bronze bearings are sometimes presintered at 370–480°C in order to control the degassing lubricants. During this treatment, the bronze alloy is formed by the diffusion of liquid tin into the copper, and it solidifies in the furnace. When the temperature is raised the tin-rich phase liquefies and is absorbed by the copper-rich matrix.

After sintering, P/M bearings are usually sized for dimensional accuracy on the same type of equipment used for molding. This produces a smooth surface and involves forcing the part back into a cavity of the dimensions desired. Lubricating oil is impregnated into the part either before or after sizing. The resulting part is an oil-impregnated bearing containing approximately 20 vol % oil.

A broad range of ferrous and nonferrous compositions is utilized in self-lubricating finished P/M parts. Among these materials are iron, brass, low carbon and low nickel steel, copper, nickel–silver, and stainless steel. Processing is essentially the same as for conventional bearings. P/M parts differ from conventional bearings only in shape and in the capability to withstand shear and compressive stresses considerably greater than those that bearings must withstand.

An important application of self-lubricating P/M parts is the oil-pump gear used in automotive engines. This part is made from a steel of eutectic composition containing 0.3 wt % carbon and 2 wt % copper; it is hardened after sintering. The gear teeth are resistant to wear and plastic deformation. The oil flowing through the interconnected pores lubricates the gear and increases corrosion resistance. High strength, nonporous sections can be combined with other sections having lower density and self-lubricating characteristics such as ratchet-and-cam combinations.

Filters. Compared to screens, metallic filters offer a tortuous path through which the fluid must flow and thus they provide depth filtration. Compared to glass or ceramic filters, P/M filters are strong and particles of the filtered material cannot break away in service and enter the filtrate. They are ductile and do not fracture under mechanical or thermal shocks. They can be fitted into housings by rolling, machining, press-fitting, or welding. In vibrating applications, the limited fatigue resistance of P/M filters must be taken into consideration (17). They also offer a wide range of design possibilities, which permits filtration at 0.5–200 μm and flow rates from a trickle to high volumes.

Porosity Measurement. Porosity is a quality that is difficult to define precisely because the total interconnected porosity reflects not only a range of hole diameters but also passages of varying lengths and tortuosity. Maximum pore size is determined by the *bubble-point* test. It measures the pressure at which a bubble forms at the part's surface when submerged in alcohol. The micrometer rating is roughly related to the ability of the filter to retain or remove particles. The lower the rating, the finer the powder used to make the part and the higher the pressure drop. For filter design, the maximum size particles in micrometers is specified that is acceptable in the filtrant. The minimum pressure drop at that micrometer rating should be attained.

P/M Tool Steels. In conventionally produced high alloy tool steels (slowly cooled cast ingots), carbide tends to segregate (42). Segregated clusters of carbide persist even after hot working, and cause undesirable effects on tool fabrication and tool performance. P/M tool steels, on the other hand, provide very fine and uniform carbides in the compact, the final bar stock, and the tools. Several tool-steel suppliers consolidate gas-atomized tool-steel powder by HIP to intermediate shapes, which are then hot-worked to final mill shapes. Water-atomized tool-steel powder is also available.

Small complex tool-steel parts are being made by conventional compaction and sintering in vacuum to near theoretical density. Applications include spade drills, knife blades, slotting cutters, insert blades for gear cutters, reamer blades, and cutting-tool inserts.

Friction Materials. Sintered friction materials are classified as metal–nonmetal combinations (41–42). They are manufactured best by the P/M process. Clutch plates, brake bands, brake blocks, and packing compositions are examples of friction materials (see Brake linings and clutch facings).

A sintered friction material is composed of a metal matrix, generally mainly copper, to which a number of other metals such as tin, zinc, lead, and iron are added. Important constituents include graphite and friction-producing components such as silica, emery, or asbestos.

Copper, with its high heat conductivity, resists frictional heat during service and is readily moldable. It is generally used as a base metal, at 60–75 wt %, whereas tin or zinc powders are present at 5–10 wt %. Tin and zinc are soluble in the copper, and they strengthen the matrix through the formation of a solid solution during sintering.

Iron or other higher melting metals insoluble in copper are added in amounts of 5–10 wt %. Harder particles, such as iron imbedded in the soft copper base, increase the coefficient of friction and also exert a scouring action on the surface. Lead, which is also insoluble in copper, is added in amounts of 5–15 wt %. Dispersed in the matrix, it acts as a lubricant during molding and as a lubricating film during operation of the friction material if the surface temperature is such that the lead liquefies. In addition, lead enhances the smooth engagement of sliding surfaces, which prevents erratic brake or clutch action. Graphite, 5–10 wt %, has effects similar to lead. Silica, emery, or other similar friction-producing materials, are added at 2–7 wt %. Minute quantities affect the coefficient of friction. A typical composition of friction–P/M material is given in Table 9.

Friction materials have to be able to wear the surface away in a uniform manner in order to provide new friction-producing conditions at the surface.

Processing. Friction materials are manufactured by cold pressing a mixture of the ingredients at relatively low pressures 138–276 MPa (20,000–40,000 psi) under a reducing atmosphere in steel dies. Large parts are sintered in bell-type furnaces, whereas small parts are sintered in conventional furnaces. The pressed compact is bonded to a backing plate during sintering. A number of compact-steel plate assemblies are stacked on top of each other and pressure is applied to the stacked assemblies during sintering. Postsintering treatments such as bending to shape, drilling holes, or machining to dimensions are usually necessary.

Electrical and Electronic Applications. *Contact Materials.* Electrical contact materials are produced by either slicing rod made from metal powder, infiltrating a porous refractory skeleton, or compaction and sintering of powders (43–45).

Tungsten contacts, cut from rod, are resistant to deformation under a large number of cycles at relatively high forces, have high hardness and strength, the ability to switch high currents without detrimental arcing or welding, and vaporize very little in an arc (if one should form). These contacts meet the requirements of many automotive, aviation, and appliance applications; however, they are limited in their use because an insulating oxide film forms during switching.

Copper and silver combined with refractory metals, such as tungsten, tungsten carbide, and molybdenum, are the principal materials for electrical contacts. A mixture of the powders is pressed and sintered, or a previously pressed and sintered refractory matrix is infiltrated with molten copper or silver in a separate heating operation. The composition is controlled by the porosity of the refractory matrix. Copper–tungsten contacts are used primarily in power-circuit breakers and transformer-tap charges. They are confined to an oil bath because of the rapid oxidation of copper in air. Copper–tungsten carbide compositions are used where greater mechanical wear resistance is necessary.

Table 9. Composition of Typical Friction Material

Constituent	Wt %
copper	62
lead	12
iron	8
graphite	7
silica	4

Tungsten–silver contacts are made similarly, but can be operated in air because of the greater stability of silver. The three standard compositions of this class include tungsten–silver, tungsten carbide–silver, and molybdenum–silver.

A third group includes silver–nickel, silver–cadmium oxide, and silver–graphite combinations. These materials are characterized by low contact resistance, some resistance to arc erosion, and excellent nonsticking characteristics. They can be considered intermediate in overall properties between silver alloys and silver or copper–refractory compositions. Silver–cadmium oxide compositions, the most popular of this class, have wide application in aircraft relays, motor controllers, and line starters and controls (see Electrical connectors).

Figure 7 illustrates the superior conductivity of P/M silver–nickel or silver–cadmium oxide contacts when compared with contacts made by standard melting techniques and formed from solid-solution alloys.

So-called sliding contacts transfer current in motors and generators. Sliding contacts are often called brushes. High electrical conductivity, wear resistance, lubricating qualities, and some arc-erosion resistance are the characteristics required by high current applications. Compositions are usually of copper and 5–7% graphite, silver and graphite, or bronze and graphite.

Magnets. Permanent magnets known as Alnico magnets 2, 4, and 5 are made by pressing and sintering powder mixtures (43). These materials are alloys of aluminum, nickel, and iron with additives such as cobalt, copper, and titanium. Sintered Alnico as compared to cast Alnico offers greater mechanical strength and closer tolerances without costly finishing operations.

Soft magnetic parts include iron pole pieces for small d-c motors or generators, armatures for generators, and sintered and rolled iron–nickel alloys for radio transformers, measuring instruments, and similar applications (see Magnetic materials).

Rare-Earth Magnets. The combination of rare earths with cobalt exhibits magnetic flux 4–19 times greater than conventional Alnico magnets (46). Because it is necessary to press the metal powders in a magnetic field to align the fine particles, ≤10 micrometers in size, powder metallurgy is the only practical way to obtain such superalloys. The rare earth-cobalt alloy powders are compacted in a mechanical press and sintered in argon or nitrogen at 1100–1200°C (see Cobalt and cobalt alloys). Samarium, mischmetal, and praseodymium are mostly used.

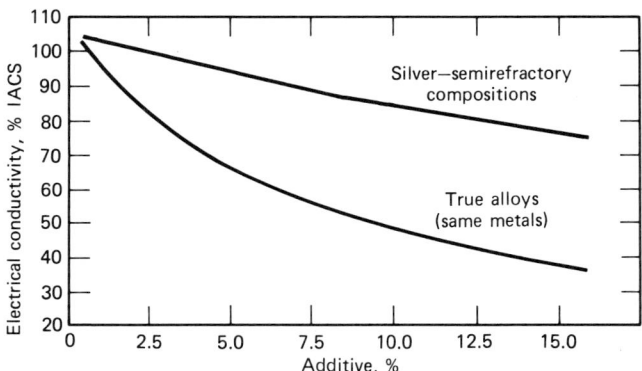

Figure 7. Conductivity of silver semirefractory compositions and solid-solution alloys (43).

Cores. Iron powder cores are manufactured for a-c self-inductance coils for high frequency applications in telephone, radio, and television systems. In the production of radio and television cores, iron powder is coated with an electrically insulating material, compacted in conventional P/M presses, ejected, and baked to fuse the coated particles together. Such a core can afford a large change of inductance by a simple movement in one direction in or out of a wire-wound coil. Fine iron powders of the electrolytic or carbonyl types are employed. These cores exhibit minimum eddy-current and hysteresis losses and the magnetic permeability returns to its original value after application of large magnetizing forces.

Electrolytic Capacitors. Tantalum, because of its high melting point of 2850°C, is produced as a metal powder. As such, it is molded, sintered, and worked to wire and foil, and used to build certain types of tantalum capacitors (43). Other capacitors are made by compacting and sintering the tantalum powder.

Batteries. In the nickel–cadmium rechargeable storage battery, the nickel powder forms a porous sheet around a woven wire grid (see Batteries). The sheet, with up to 90% porosity, is used as a reservoir for nickel and cadmium compounds and as an electron take-off. It offers a large surface area per unit weight or volume. The porous sheet is made either by sifting nickel powder into a ceramic mold of the correct size and containing the woven screen, or by applying a slurry to a thin perforated nickel-plated iron sheet. The composite is then sintered at a temperature that effects strengthening but not densification.

Dry-battery mercury anodes are pressed compacts of zinc–mercury amalgam. They were first developed and produced during World War II for walkie-talkie communication systems. Today, practically all hearing aids employ this type of battery.

Incandescent Lamps, Electronic Tubes, and Resistance Elements. Articles fashioned in any form from molybdenum and tungsten usually fall within the bounds of powder metallurgy since these metals normally are first produced as a powder. Both molybdenum and tungsten are used as targets in x-ray tubes, for structural shapes such as lead and grid wires in electron tubes, and as resistance elements in furnaces.

Iron–nickel and iron–nickel–cobalt alloys are prepared in wire and sheet form by either rolling an isostatic-pressed ingot or by directly rolling and sintering the metal powders. These materials have numerous applications in the electronic field as glass-sealing alloys since their coefficient of thermal expansion is close to that of certain glasses. Moreover, their coefficients can be tailored to the application by varying the composition, an easy accomplishment for the powder metallurgist.

Refractory Materials. Extremely high melting metals and those that are more resistant to deformation when hot are considered refractory metals (47). This generally refers to melting above the melting point of iron (1536°C). Tungsten, rhenium, tantalum, molybdenum, and niobium, belong to this group and also osmium, boron, titanium, thorium, zirconium, and vanadium. Refractory materials also include alloys of these metals (see High temperature alloys).

Refractory metals are associated with powder metallurgy because these metals are not easily melted; therefore in smelting their ores, the metal is recovered in powder form rather than melted.

Today, refractory metals are used mainly to produce filament wire for incandescent lamps. The manufacture of tungsten products from tungsten powder is given below as an example.

Tungsten powder of approximately 5 μm average particle size, derived from the hydrogen reduction of tungsten oxides, is compacted into ingots 1.25–1.87 cm high and wide, and 45–60 cm long. This bar is extremely weak and must be heated in hydrogen in a standard muffle furnace in order to strengthen it before sintering in a bell-shaped furnace under a hydrogen atmosphere. Heating is by resistance and is carried on until a certain density is reached. In this instance, the bar undergoes considerable linear shrinkage of 12–16%. The sintered bar is dense enough to be heated and worked in air. The bar is then worked first at 1550 to 1650°C and then at gradually decreasing temperatures. Metallurgically, this is cold working since recrystallization does not take place. It is swaged and drawn with intermediate annealings to wire filaments. During working, the sintered bar increases in tensile strength and ductility and becomes 100% dense.

Metalworking, such as swaging, drawing, rolling, etc, may be performed also on slabs or ingots of other metals prepared by any of the consolidation and sintering techniques previously described.

Refractory Metal Alloys. Alloys with refractory or nonrefractory metals are made by direct mixing of the elemental metal powders or by the incorporation of compounds of the solute in the processing step for subsequent reduction to the metal (48). In most instances, the metals are alloyed by solid-state diffusion during sintering. Tungsten and molybdenum form a continuous series of solid solutions which offer considerably greater workability compared to pure molybdenum and pure tungsten.

Tungsten with addition of as much as 5% thoria is used for thermionic emission cathode wires and as filaments for vibration-resistant incandescent lamps. Tungsten–rhenium alloys are employed as heating elements and thermocouples. Tantalum and niobium form continuous solid solutions with tungsten. Iron and nickel are used as alloy agents for specialized applications.

Heavy Metal. A comparison of the properties of the refractory metals is given in Table 10. The five principal refractory metals serve as main constituents in many important metal and alloy compositions. Their main areas of use as metals are electronics, alloying, nuclear power, aerospace, chemical catalysts, metal cutting and forming, mechanical parts, and mining/drilling (48). The tungsten–nickel–iron alloy nominally containing 93 wt % W–5 wt % Ni–2 wt % Fe is described as a heavy-metal composition, because its sintered density may exceed 17 g/cm^3. It has sufficient ductility to be worked and can be sintered to full density at 1400°C in contrast to pure tungsten. It is utilized when a high mass with good impact strength, ductility, and tensile strength is needed. Pure tungsten, for example, has more mass but is brittle. Heavy-metal composition is widely used for containers for shielding radioactive materials, counter-balances in airplane controls, and as cores in armor-piercing projectiles.

Cemented Carbides. Cemented carbides contain mostly tungsten carbide and lesser amounts of other hard-metal components, embedded in a matrix of cobalt (see Carbides).

The principal application of cemented carbides is as cutting tools for metals. They are also used in armor-piercing projectile cores and tips, and for carbide shot. Cemented carbides are used for parts requiring corrosion resistance, including burnishing tools and dies, pump valves, sandblast nozzles, gauges, rock and masonry drills, and guides of many types. Carbide compositions containing only tungsten carbide and cobalt are unsuitable for machining highly resistant materials. However, tungsten carbide can

Table 10. Properties of Refractory Metals[a]

Metals	Density, g/cm³	Mp, °C	Electrical conductivity, % IACS	Electrical resistivity, μΩ·cm	Thermal conductivity, W/(m·K)	Tensile strength, MPa[b,c] at RT	Tensile strength, MPa[b,c] at 500°C	Tensile strength, MPa[b,c] at 1000°C	Modulus of elasticity, MPa[b] at 500°C	Modulus of elasticity, MPa[b] at 1000°C
tungsten	19.3	3410	31.0	5.5	166.105	690–3448	690–2069	345–527	3.8	3.4
rhenium	21.04	3180	9.3	19.1	71.128	1931	924	454	3.8	
tantalum	16.6	2996	13.9	13.5	54.392	241–483	172–310	90–117	1.7	1.5
molybdenum	10.22	2610	34.0	5.7	146.44	828–1379	241–448	138–207	2.8	2.7
niobium	8.57	2468	13.2	14.1	52.30	207–414	138–276	55–103	0.9	0.8
titanium	4.5	1668	5.5	42	21.76	262–690	138–324		0.8	
iron	7.87	1536	19.1	10–12	83.68	172–283				
nickel	8.90	1453	25.2	6.8	90.048	345–690	131–221			
cobalt	8.85	1495	27.6	6.2	71.128	234–945				
copper	8.96	1083	100	1.7	393.296	221–379				
gold	19.32	1063	73.4	2.4	297.064	131–221				
aluminum	2.702	660	64.9	2.7	238.458	69–186				
zinc	7.13	420	28.3	5.9	122.958	283–324				
lead	11.36	327	8.3	20.7	33.472	14–18				

[a] Ref. 48.
[b] To convert MPa to psi, multiply by 145.
[c] Given in ranges because the tensile strengths vary considerably with form and processing.

be replaced with solid solutions of tungsten, molybdenum, and titanium carbides to machine such materials. Various rare-earth carbides have been added to tungsten carbide for special-purpose cutting tools. Tungsten-free compositions have also been developed as well as compositions containing diamonds.

Cemented carbides exhibit extreme hardness and toughness which are retained at the elevated temperatures that may occur between tool and work during cutting.

Processing. Tungsten carbide is made by heating a mixture of lampblack with tungsten powder in such proportions that a compound with a combined carbon of 6.25 wt % is obtained. The ratio of free to combined carbon is of extreme importance. Tantalum and titanium carbides are made by heating a mixture of carbon with the metal oxide. Multicarbide powders, such as Mo_2C–WC, TaC–NbC, and TiC–TaC–WC, are made by a variety of methods, the most important of which is carburization of powder mixtures.

These carbides are mixed with fine cobalt powder in a wet ball-milling operation. Water, benzene, or other organic materials may be used. Milling produces an intimate mixture of cobalt attached to the carbide particles. This is followed by mixing with a lubricant or other binder, such as paraffin, camphor, or stearic acid, in a separate operation. Dried and screened, the mass results in a lubricated, flowable powder that can be cold-molded in conventional presses and tooling. A low temperature, presinter treatment is carried out in order to strengthen the part through sintering of the binder metal and to evaporate the pressing lubricant. The presintered part is then formed by machining into the desired shape and dimensions. This is followed by high temperature sintering between 1350 and 1550°C for 1–2 h. Sintering is characterized by solution of the carbides in the liquid phase, and precipitation at areas of low energy. Recrystallized grains of carbide form and are embedded in a matrix of a solution of carbides and matrix metal. Linear dimensional shrinkage of 21–25% occurs.

Cermets. High temperature applications in space and nuclear technology created a need for materials to fill the gap between cobalt- and nickel-base superalloys and refractory metals such as tungsten and molybdenum. Pure ceramics are strong at elevated temperatures but lack sufficient ductility to be worked at room temperature, whereas metal alloys, although ductile at room temperature, are not strong enough at the higher temperatures. Mixtures of ceramics and metals have both the high temperature strength of ceramics, and sufficient ductility and thermal conductivity contributed by the metal to provide resistance from thermal shock at high temperatures and workability at room temperature. These compositions are termed ceramals and cermets (see Ceramics; High temperature composites). Actually, the cemented carbides can be considered cermets.

The cermet class of materials contains a large number of compositions.

Titanium carbides are mixtures of a hard and a soft phase containing TiC or TiN and Ni–Cr, Co–Cr, Ni–Cr–Co, Ni–Al, or Ni–Mo.

Borides are generally compounds of zirconium, titanium, and chromium or other metals (see Boron compounds).

Dispersions of flake aluminum powders with surface oxide up to 14% Al_2O_3 have been pressed, sintered, and worked to a material known as SAP (sintered aluminum powder). This product exhibits high strength at elevated temperatures. Nickel with small additions of thoria, known as TD-nickel, is also a high temperature cermet.

Thoria-dispersed nickel products are obtained by precipitating basic nickel compounds, whereby thoria particles of ca 100 nm are coated with layers of nickel to the extent that the product has a 2% thoria dispersion.

Dispersion-strengthened copper is made by dispersing a thoria or alumina phase through copper powder. The resulting P/M product retains its strength at elevated temperatures. It is used, for example, as the conductor or lead wire that supports the hot filament inside incandescent lamps.

Mechanical alloying is another method of producing dispersion-strengthened metals. In this process, the powdered constituents of the alloy are treated in an attrition mill. A finely distributed layer of the dispersed phase is distributed on particles of the base metal. Subsequent pressing and sintering strengthens the dispersion (17).

Space Applications. The growth of powder metallurgy in space technology is due to the difficulty of handling many materials with conventional fusion-metallurgy techniques, the need for controlled porosity, and the requirement of many special and unique properties (49–50). Powder metallurgy is applied in low density components with emphasis on porous tungsten for W–Ag structures, beryllium compounds, and dispersion-hardening systems.

Plates of beryllium metal were used as heat-shield shingles on the cylindrical section of the Mercury spacecraft used in some of the first unmanned suborbital flights. A fairing for the Agena-B space-vehicle interstage structure used beryllium machined from a pressed compact 180-cm dia and 70-cm high. Beryllium has also been used in the nose cap of the ascent shroud in several of the Ranger-Mariner space-probe series.

Beryllium has a favorable stiffness-to-density ratio and good thermal properties which make it acceptable as a shielding material against micrometeorites.

Special tungsten powders have been developed for space applications. Spherical tungsten particles are used to form porous tungsten bodies in the ionizing surface in ion-propulsion engines.

Silver-infiltrated tungsten parts have a density before infiltration of 70–80% of theoretical; 90% of the porosity is infiltrated with silver. These parts are used where the ablative cooling provided by the silver during solid–vapor change of state and transpiration cooling through the resulting porosity are desirable, as in reentry vehicles (see Ablative materials).

Space technology has always demanded materials that can operate at temperatures between those of superalloys and refractory metals and that have high temperature strength during operation and room-temperature ductility for fabrication. Part of the problem has been answered by the development of dispersion-strengthened and oxide alloy systems.

Space technology also utilizes sintered magnetic materials such as Alnico 2 and 6, sintered bronze bushings, electrical contacts and tantalum capacitors, and a variety of P/M porous filters. Aluminum and magnesium powders are blended with solid fuels for enhancing rocket-propulsion systems. The space shuttle Columbia is launched with 90,720 kg of aluminum powder plus ammonium perchlorate and oxidizers in each of its two booster rockets.

High vacuum is one of the main characteristics of space. Bearings with liquid lubricants would lose the lubricant through evaporation. Bearings produced by powder metallurgy techniques with imbedded MoS_2 give good service.

Nuclear Applications. Powder metallurgy is used in the fabrication of fuel elements as well as control, shielding, moderator, and other components of nuclear-power reactors (51) (see Nuclear reactors). The materials for fuel, moderator, and control parts of a reactor are thermodynamically unstable if heated to melting temperatures. These same materials are stable under P/M process conditions. It is possible, for example, to incorporate uranium or ceramic compounds in a metallic matrix, or to produce parts that are similar in the size and shape desired without effecting drastic changes in either the structure or surface conditions since only little postsintering treatment is necessary.

In nuclear technology, the P/M process is applied to beryllium, zirconium, uranium, and thorium. Uranium and thorium are used as fuel materials, beryllium for moderating purposes, and zirconium as construction material. These metals are used in elemental form, as alloys, or in metal–ceramic combinations.

These particular metal powders must be handled somewhat differently than ordinary metal powders. Beryllium powder is extremely toxic, zirconium is highly pyrophoric, and uranium and thorium powders are both toxic and pyrophoric. The vapors eminating from burning uranium are also extremely toxic. These powders are generally handled in closed containers, so-called dry boxes, usually filled with a protective atmosphere such as argon or helium.

Beryllium powders have extremely poor compactability at room temperature and are therefore generally processed by sintering in vacuum or HIP.

Both zirconium hydride and zirconium metal powders compact to fairly high densities at conventional pressures. During sintering the zirconium hydride decomposes and at the temperature of decomposition, zirconium particles start to bond. Sintered zirconium is ductile and can be worked without difficulty. Pure zirconium is seldom applied in reactor engineering, but the powder is used in conjunction with uranium powder to form uranium–zirconium alloys by solid-state diffusion. These alloys are important in reactor design since they change less under irradiation and are more resistant to corrosion.

BIBLIOGRAPHY

"Powder Metallurgy" in *ECT* 1st ed., Vol. 11, pp. 43–64, by Werner Leszynski, American Electro Metal Corporation; "Powder Metallurgy" in *ECT* 2nd ed., Vol. 16, pp. 401–435, by Arnold R. Poster, Metals Sintering Corp. and Henry H. Hausner, Consulting Engineering.

1. G. R. Bell, *J. Chem. Met. Min. Soc. S. Afr.* **56,** 260 (Jan. 1956).
2. R. R. Irani and C. F. Callis, *Particle Size: Measurement, Interpretation and Application*, John Wiley & Sons, Inc., New York, 1963.
3. H. E. Rose, *The Measurement of Particle Size in Very Fine Powders*, Chemical Publishing Co., Inc., New York, 1954.
4. H. H. Hausner "Basic Characteristics of Metal Powders" in A. R. Poster, *Handbook of Metal Powders*, Reinhold Publishing Corp., New York, 1966.
5. *MPIF Standards No. 4* (*Determination of apparent density of metal powders*), and *No. 28* (*Determination of apparent density of non-free flowing powders*), Metal Powder Industries Federation, Princeton, N.J., 1980.
6. *MPIF Standard No. 3* (*determination of flow rate of metal powders*), Metal Powder Industries Federation, Princeton, N.J., 1980.
7. J. S. Hirschhorn, *Introduction to Powder Metallurgy*, American Powder Metallurgy Institute, Princeton, N.J., 1976.
8. F. V. Lenel, *Powder Metallurgy—Principles and Applications*, Metal Powder Industries Federation, Princeton, N.J., 1980.
9. H. H. Hausner, *Planseeber. Pulvermet.* **12,** 172 (Dec. 1964).
10. D. Train and C. J. Lewis, *Trans. Inst. Chem. Eng.* (*London*) **40,** 235 (1962).
11. M. J. Donachie, Jr. and M. F. Burr, *J. Met.* **15,** 849 (Nov. 1963).
12. R. W. Heckel, *Trans. AIME* **221,** 671 (1961).

13. R. W. Heckel, *Trans. AIME* **222,** 1073 (1962).
14. C. G. Goetzel, *Treatise on Powder Metallurgy*, Vols. I (1949), II (1950), III (1952), IV (1963), Interscience Publishers, Inc., New York, 1949, pp. 259–312.
15. *Powder Metallurgy Equipment Manual*, Powder Metallurgy Equipment Association, Metal Powder Industries Federation, Princeton, N.J., 1977.
16. F. V. Lenel, *Powder Metallurgy—Principles and Applications*, Metal Powder Industries Federation, Princeton, N.J., 1980, Chapts. 12–13, 15, 21.
17. R. R. Irving, *Iron Age*, (July 28, 1980).
18. *Prod. Eng.*, 45 (Aug. 1979).
19. S. Storchheim, *Met. Prog.* **69,** 120 (Sept. 1956).
20. *Product Bulletin #182*, Hoeganaes Corp., Riverton, N.J., 1976.
21. P. D. Peckner, *Mater. Design Eng.* **51,** 89 (July 1960).
22. H. H. Hausner and A. R. Poster in W. Leszynski, ed., *Powder Metallurgy*, Interscience Publishers, Inc., New York, 1961, p. 461.
23. H. H. Hausner, "Compacting and Sintering of Metal Powders Without the Application of Pressure," *International Symposium Agglomeration*, Apr. 12–14, 1961.
24. I. Ljengberg and P. G. Arbstedt, *Proc. Ann. Meeting Metal Powder Assoc.* **12,** 78 (1956).
25. H. H. Hausner and I. Sheinhartz, *Proc. Ann. Meeting Metal Powder Assoc.* **10,** 6 (1954).
26. H. H. Hausner, ed., *Modern Developments in Powder Metallurgy: Proceedings International Powder Metallurgy Conference New York*, 1966, Plenum Press, New York, 1966.
27. E. V. Lenel and G. S. Ansell in Ref. 26, Vol. 1, pp. 281–296.
28. M. H. Tikkanen and S. Ylasaari in Ref. 26, Vol. 1, pp. 297–309.
29. F. Thummler and W. Thomma, *Met. Rev.* (115).
30. *Definitions of Terms Used in Powder Metallurgy*, MPIF Standard 9-62, Metal Powder Industries Federation, Princeton, N.J., 1982.
31. H. H. Hausner "Grain Growth During Sintering" in *Special Report No. 58 of the Iron and Steel Institute*, London, Eng., 1954, pp. 102–112.
32. H. Hausner, *Powder Metall.* 1(2), (Apr. 1965).
33. I. T. Brakhnova, *Environmental Hazards of Metals*, Consultant's Bureau, New York, 1975.
34. Codes 651 and 652, National Fire Protection Association, Boston, Mass., 1952.
35. *Explosibility of Metal Powders*, R.I. 6516, U.S. Bureau of Mines, Washington, D.C., 1965.
36. *MPIF Standard No. 35*, Metal Powder Industries Federation, Princeton, N.J., 1981.
37. Ref. 14, Vol. II, pp. 503–542.
38. *Products of Powder Metallurgy*, Engineering Manual E-64, Amplex Division, Chrysler Corp.
39. *Porous Metals Guidebook*, Porous Metals Council, Metal Powder Industries Federation, Princeton, N.J., 1980.
40. *Met. Prog.*, 100 (Jan. 1980).
41. Ref. 14, Vol. II, pp. 543–558.
42. B. J. Collins and C. P. Schneider in Ref. 26, Vol. 3, pp. 160–165.
43. A. S. Doty, *Proc. Ann. Meeting Metal Powder Assoc.* **12,** 46 (1956).
44. G. A. Meyer, *Met. Prog.* **86,** 95 (June 1965).
45. *Ibid.*, **87,** 92 (July 1965).
46. *PM Technology Newsletter*, American Powder Metallurgy Institute, Princeton, N.J., Oct./Nov. 1978.
47. Ref. 14, Vol. II, pp. 3–73.
48. *What are Refractory Metals?...*, Refractory Metals Association, MPIF, Princeton, N.J., 1980, pp. 4–5.
49. C. G. Goetzel and J. B. Rittenhouse, *J. Met.* **17,** 876 (Aug. 1965).
50. H. H. Hausner, *J. Met.* **16,** 894 (Nov. 1964).
51. H. H. Hausner, *Proc. Ann. Meeting Metal Powder Assoc.* **12,** 27 (1956).

General References

W. D. Jones, *Fundamental Principles of Powder Metallurgy*, Edward Arnold, London, 1960.
F. V. Lenel, *Powder Metallurgy—Principles and Applications*, Metal Powder Industries Federation, Princeton, N.J., 1980.
J. S. Hirschhorn, *Introduction to Powder Metallurgy*, American Powder Metallurgy Institute, Princeton, N.J., 1969.

H. H. Hausner, *Handbook of Powder Metallurgy*, Chemical Publishing, New York, 1973.
International Journal of Powder Metallurgy and Powder Technology (Quarterly), American Powder Metallurgy Institute, Princeton, N.J.
PM Technology Newsletter (monthly).

KEMPTON H. ROLL
Metal Powder Industries Federation
American Powder Metallurgy Institute

POWER GENERATION

The term power generation in the engineering sense implies the production of mechanical or electrical power from some other source of energy, eg, thermal, hydroelectric, or electrochemical energy. Conventional means of power generation have used air, steam, or water as a working medium for internal combustion engines, reciprocating steam engines, and steam and hydraulic turbines. Newer techniques such as magnetohydrodynamic, fuel-cell, wind, and solar power are under investigation and may eventually prove important (see Steam; Solar energy; Coal, magnetohydrodynamics; Batteries and electric cells, primary).

Geothermal Power

Geothermal energy (qv) is a vast potential source of power (1). This energy is thought to result from natural decay of radioactive materials, a virtually inexhaustible source of power. It raises the earth's temperature an average of 25°C with each kilometer of depth (see Radioactivity, natural).

Theoretically, such energy could be obtained anywhere by deep drilling. But in most parts of the world, the hot interior mass is too deep to reach with existing technology. However, in some areas, eg, many locations in the western United States, large heat sources are close enough to the surface to reach. Experts estimate that 33.7×10^6 EJ (32×10^{21} Btu) is within 10 km of the surface of the United States. Some 9.3×10^8 ha (23,000,000 acres) of Federal land has been leased for exploitation and development, and in 1979 drilling increased 25% over 1978.

It is estimated that by 2020 geothermal energy could be adding 19.5 EJ (18.5×10^{15} Btu) annually to the national energy pool. The commercial application of geothermal energy for generating electricity began in 1904, in Italy. Today, power is produced from geothermal sources in New Zealand, Japan, the U.S., Iceland, Italy, and the USSR. In the United States, the principal geothermal power plants are located at a natural-steam field, The Geysers, in California.

In the United States, three broad types of geothermal sources are under consideration for development in the near future: hydrothermal reservoirs, geopressured zones, and hot dry rock (see Table 1). Each type and each site poses technological and environmental problems. Hydrothermal reservoirs producing dry steam have been used mainly to generate electricity. Hot water, with or without steam, has been employed for heating purposes in a few locations, but the application to power generation

Table 1. Geothermal Resources[a]

Resource	Temperature, °C	Salinity, %	Commercial operation, yr	kW·h
vapor-dominated	171–196		1913	2.0–3.0
liquid-dominated	148–315	0.1–26	1958	2.8–7.5
geopressured	148–204	4–10	1986	
hot dry rock	148–287		1990	

[a] From ref. 2.

is now seriously considered. Development of geopressured zones is currently at the exploratory, resource-assessment stage. Dry hot rock, the largest heat resource, has not yet been exploited, but several new methods of utilization are under investigation.

Hydrothermal Energy. Hydrothermal systems consist of steam (vapor-dominated) or hot water (liquid-dominated) reservoirs trapped in fractured rocks or sediments below impermeable surface layers of earth. These systems are a significant commercial source of electricity (3).

Dry steam is the preferred, but rarest hydrothermal resource. Existing sources that emit dry and slightly superheated steam are located in Lardarello, Italy, and The Geysers. To harness energy from a dry-steam reservoir, a hole is drilled into the reservoir, the released steam is filtered to eliminate entrained solid material, and the steam is then piped directly to a turbine to generate electricity.

Power Cycles. In vapor-dominated systems, dry steam is conducted directly to a turbine, where the steam is expanded. Steam pressure and temperature are in the range of 7 MPa (ca 1000 psi) and 200°C, respectively.

Liquid-dominated systems may be direct or binary. In the direct process, water (brine) and steam pass through a separator, where additional steam is produced by flashing the hot liquid at a reduced pressure. The steam is then piped to a low pressure turbine; a single- or double-flash loop may be used. The separate brine fraction is discarded.

In another direct process, the total mixture of steam and water (brine) is delivered directly to a mixed-phase expander (1). Since all the available energy in the mixture is utilized, this method can achieve higher thermal efficiency than other processes.

In the binary cycle shown in Figure 1 (4), hot water (brine) is circulated in a heat exchanger and then transfers its heat to a secondary working fluid such as isobutane or Freon. The heated fluid is then expanded through a binary turbine, and the exhaust fluid is condensed. The geothermal effluent does not come in contact with the turbine or condenser, and problems of corrosion and scaling are confined to the heat exchanger (5). This cycle is particularly applicable for a moderate-temperature resource (100–200°C).

The geothermal steam-power plant differs from conventional fuel-fired steam units in the steam impurities. In addition, the geothermal plant requires no boiler, and the steam is at moderate pressure and temperature. Consequently, the specific volume and heat rate of the turbine are high, 23.2 MJ/(kW·h) (22,000 Btu/(kW·h)), and very large turbines are required relative to electric generation capacity.

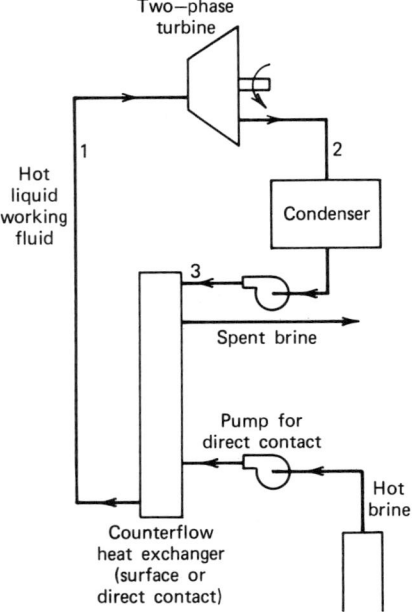

Figure 1. Geothermal binary cycle.

Environmental Aspects. Geothermal power generation has the advantage that almost all the activities related to the production of power are in the immediate vicinity of the plant. However, certain undesirable environmental effects must be considered.

Air. During the extraction of heat from hydrothermal reservoirs, a variety of gases may be released into the atmosphere, mainly carbon dioxide with varying amounts of hydrogen sulfide, methane, hydrogen, and ammonia. The greatest nuisance and a potential hazard is hydrogen sulfide, which is toxic to humans and has a very disagreeable rotten-egg odor at the very low concentrations associated with geothermal areas. The hydrogen sulfide concentrations at The Geysers have at times exceeded California air-quality standards. New improved hydrogen sulfide removal systems are under investigation at this facility. The emission levels vary substantially from site to site, depending on the composition of the geothermal fluid.

Geopressured systems present fewer air pollution problems than hydrothermal systems. Although fluids in geopressured reservoirs contain large amounts of methane, this gas can be recovered for use as a fuel. Moreover, air emissions from hot dry-rock geothermal resources do not present significant problems.

Water. Hydrothermal systems may pollute surface and groundwater because of the disposal of excess steam condensate and water. However, these spent fluids can be recycled into the geothermal reservoirs, thus minimizing the pollution problem (see Water pollution).

At geopressured sites, the great depths and pressures involved may make it economically unfeasible to inject fluids back into the reservoir (3), and the expected high solids content of the fluids would render them useless for most industrial purposes. Possible means of wastewater disposal include discharge into existing saline bodies of water, injection into subsurface saline aquifers, or desalinization.

Land Stability. The withdrawal of geothermal fluids from the earth may cause the sinking of the land (subsidence), an environmental problem long associated with groundwater wells, oil-field development, and mining operations.

Land subsidence has not occurred during the development of two existing vapor-dominated geothermal fields at The Geysers and Lardarello because of the sturdy geological structures under which such systems form. Hot-water systems appear more likely to cause subsidence. For example, at a power plant in Wairakei, New Zealand, a subsidence rate of up to 0.3–1.0 m/yr has been measured. However, it is expected that fluid-reinjection systems will minimize subsidence.

At hot dry rock sites, the great depths (about 3.5 km), small fractured areas, and the use of externally introduced fluids make subsidence unlikely. Fluid injection and withdrawal could increase seismic activity, although none has been detected to date at the Fenton Hill site in New Mexico.

Noise. The noise generated by the venting of geothermal wells can be effectively reduced by muffling systems, currently in use at The Geysers. At hot-water hydrothermal sites, noise is not a problem, whereas the high pressures of geopressured reservoirs may generate considerable noise. Significant noise from hot dry-rock sites is not anticipated (see Noise pollution).

Outlook. Efforts to develop geothermal power are being accelerated. By conservative estimates, about 100 GW of generating capacity could, with vigorous efforts, be developed by the end of the century. Despite the optimistic outlook, geothermal energy utilization is in its infancy, and substantial technical problems remain to be solved. Exploration and prospecting techniques are in the early stages of development and need to be improved. Control of corrosion caused by mineral-laden hot water and the designing of turbines that can operate efficiently at low temperatures may be crucial to the exploitation of this energy source on a large scale. The environmental problems are not regarded as unsolvable.

Solar Power

Solar energy utilization has recently received increased attention, and the research in this field has expanded tremendously since 1970. Solar energy is not expected to adversely affect the environment, but it is a diffuse source and only available part of the year.

The sun emits energy at the rate of 3.86×10^{17} GW (3.86×10^{33} erg/s) (6). In passing through the earth's atmosphere, the sun's radiation is partially and selectively absorbed, scattered, and reflected by water vapor, ozone, air molecules, natural dust, clouds, and pollutants. The supply of solar energy, however, is so great that it can be considered virtually inexhaustible (see Solar energy).

Although it is relatively easy to define a standard solar spectrum outside the earth's atmosphere, the spectrum at sea level is difficult to determine because of atmospheric effects, reflection, absorption, and scattering. The solar energy arriving at sea level also varies with the time of day and season. The situation is complicated and can only be approached empirically with the use of site-specific data (7). Climate, cloud conditions, atmospheric haze, the geographical and annual distribution of sunshine, and seasonal distribution of rainfall and diurnal cloudiness must be considered. Finally, ground-level insolation is not predictable by theory, and empirical site-specific data must be used (8–14).

Photovoltaic Systems. Photovoltaic cells (qv) directly convert solar-radiation energy into electricity without the intervention of thermal cycles. Several materials have been utilized, including p–n junctions of various semiconductors. The efficiency of the cells is wave-length dependent. Photovoltaic cells made from silicon are receiving most of the attention, whereas cadmium sulfide and gallium arsenide cells are considered second-generation materials that will come into use between 1980 and 1985. Although employed in unusual small-scale applications, large-scale development of photovoltaic systems has not taken place because of extremely high cost and relatively low efficiency. Of primary importance today are the silicon batteries that provide power for space laboratories and communication satellites. Table 2 compares various photovoltaic devices (15).

Attempts to lower the cost of photovoltaic systems include the production of silicon solar-cell modules and the development of new materials for concentrating systems that use relatively inexpensive reflective components to reduce the area of expensive solar cells. With a present conversion efficiency of 10%, a collection area of approximately 9.5 m²/kW is required for silicon cells. The price of one watt of electricity generated by photovoltaic cells has been cut considerably in the past few years. The DOE aims to lower the cost in 1980 dollars of solar-cell modules or arrays to $2.80/W in 1982, 70¢/W by 1986, and 15–50¢/W by 1990.

Wind Power

Wind has been used for centuries as an energy source; the applications range from the pumping of water and grinding of grain to the generation of electricity and the propulsion of ships. In this century, however, its use has not been extensive in developed countries. Nevertheless, because of the recent oil and gas shortages, interest in this technology has been revived.

Since 1890, windmills have been used in conjunction with generators to produce electricity for day-to-day needs. In the 1930s, a USSR system employed a Balaclava generator for several years, which supplied about 250,000 kW·h per year to the Yalta electric grid. The rotor of this machine was 30-m dia, and it was situated on a tower 30 m high.

The design of wind-energy conversion systems (WECS) is based on the fact that wind power is proportional to the cube of the wind velocity. The annual average wind

Table 2. Comparison of Photovoltaic Devices[a]

Device	Status	Efficiency, % Predicted	Attained
single-crystal silicon	commercial	20	18
polycrystalline silicon	laboratory		12
gallium arsenide	laboratory	29	26
CdS/CdTe heterojunction	laboratory	16	8
CdS/Cu$_2$S heterojunction	commercial	16	8
amorphous silicon	laboratory		5.5
multilayered cell stack	laboratory	35	28.5
cells with spectral splitting	proposed	35	
thermophotovoltaic conversion	laboratory	50	26

[a] Ref. 15.

speed at any site depends upon its geographical position, altitude, distance from the sea, its exposure to prevailing wind, and the shape of the land in the immediate vicinity. Favorable sites for wind-power installations include mountain tops and ridges, the western Great Plains, and the sea (16).

The frequency of wind speed or its variability is important to determine the peak generation capacity of the installation and gives an assessment of the variation in power generation (17–18). Wind data showing the time-duration curve for Plum Brook, Ohio, are given in Figure 2 (19).

Wind turbines are categorized in terms of orientation of the axis of rotation relative to the wind speed (20). The horizontal-axis wind-turbine generator (HA-WTG) has its axis of rotation parallel to the direction of the wind and the earth's surface. Upwind rotors are designed with blades that rotate in front of the towers with respect to the direction of the wind stream; downwind rotors are designed with blades that rotate in back of the tower.

The vertical-axis wind-turbine generator (VA-WTG) has its axis of rotation perpendicular to both the horizontal earth's surface and the wind stream. An example of this type is the Darrieus rotor, which looks somewhat like an eggbeater. The blades are high performance symmetric airfoils formed into a curve to reduce the bending stresses.

A Savonius rotor is composed of two semicylindrical offset cups rotating about a vertical axis. It is a slow-speed turbine with characteristically high starting torque accompanied by low overall aerodynamic efficiency. Performance curves of various types of rotors are shown in ref. 16.

Efficient conversion of wind energy to mechanical power promises to contribute significantly to the world's energy requirements. The main problems in wind-energy utilization are the intermittent nature and relatively low energy level of wind compared to conventional fuels. The low energy density of wind and the cost of site operation and power collection imply that WECS for utility systems must be made as large as possible. However, there are constraints on the size and, therefore, on the power output of conventional WECS that can be built and operated reliably. Research and development efforts are currently directed to improve the control systems and the strength of the materials used in the blade construction (see Solar energy).

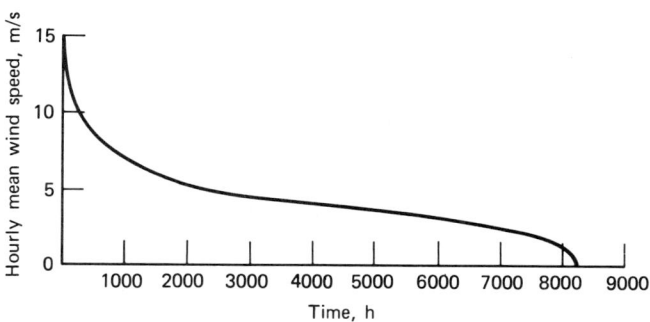

Figure 2. Wind velocity duration curve, Plum Brook, Ohio, 1972.

Nuclear Power

The first commercial nuclear power station in the United States was the Shippingport, Pa., plant of the Duquesne Light Company. Operation of the plant began in December 1957. Its success provided the necessary impetus for the rapid growth of this new form of power generation (21).

A nuclear power plant uses a different source of heat supply than a conventional power plant. In the latter, heat is obtained from the combustion of fossil fuel in a boiler furnace, combustion chamber, or internal-combustion engine cylinder, as the case may be (22). In a nuclear power plant, heat is obtained from the release of energy associated with the fission or splitting of ^{235}U into lighter elements (see Nuclear reactors).

Hydroelectric Power

Hydroelectric power plants may be classified roughly as high-head, medium-head, and low-head plants, depending upon the height of the water level in the reservoir above the plant. High-head plants are usually considered to be those operating on heads of ca 160 m or more, and low-head plants are those operating on heads of <16 m. High-head plants require much less flow than low-head plants of the same capacity. Therefore, turbines and generators may be designed for higher speeds, which permits smaller units and housing. This advantage is offset by the fact that long conduits or penstocks are necessary between dams and power plant.

Because of the great volumes of water that have to be handled, turbine generators for low-head plants are necessarily large low speed reaction-type machines designed for axial or radial flow.

Dams vary greatly in design. In the arid regions of the West, the primary purpose is water storage. These dams may also have facilities for hydroelectric power generation, but unless a given flow is assured throughout the entire year, such facilities are usually built for limited service only.

Flood-control dams provide storage reservoirs capable of holding large volumes of water when flood conditions arise. Such dams may or may not be compatible with the demands of electric power generation. If, for example, the reservoir is maintained at a high level for power generation, it has little reserve capacity to absorb flood water, and its primary purpose is defeated. On the other hand, if the reservoir is allowed to fall to a low level in order to provide storage space for flood conditions, power generation may at times have to be curtailed altogether should the anticipated flood conditions fail to occur.

The pumped-storage plant is a facility that can be operated in conjunction with either a steam system or a conventional hydroelectric system. The purpose of such a plant is to level out the peaks and valleys in the load curve, thereby increasing the capacity and efficiency of the overall system. The pumped-storage plant consists of one reservoir at high elevation and another at a lower elevation. They are connected through a reversible pump-turbine unit. When the demand for electrical energy is light, additional power can be generated in the steam or hydroplants of the system for pumping water from the lower to the higher reservoir. Conversely, when demand is high, the water in the upper reservoir can be used to generate additional power upon its return to the lower reservoir through the pump-turbine unit. A number of pumped-storage systems have been in operation in the United States since the mid-1960s, but the total capacity from this source is not expected to exceed 17 GW by the year 1990.

Capacity and Growth Potential. Total U.S. hydroelectric capacity in 1979 was about 64 GW, with a heavy concentration in the Columbia River Basin of the Pacific Northwest, where the Northwest power pool has a total capacity of ca 29 GW. Of this, approximately 17 GW is provided by 11 hydroelectric projects on the main branch of the Columbia River, including the largest in the United States, the Grand Coulee Dam, with a capacity of 5.463 GW and a projected capacity of 10.363 GW.

Because of a lack of new hydroelectric sites as well as environmental and esthetic considerations, hydroelectric capacity is not expected to increase greatly in the future. In 1980, the North American Electric Reliability Council estimated that by 1989 the total capacity would increase by only 3 GW to a total of about 67 GW.

There is, however, a trend toward redevelopment of previously retired hydroelectric sites. These sites have become economically attractive because of the high cost of generating electricity by other means. The Federal Energy Regulatory Commission identified 3,009 small-plant sites, most of which are identified as low-head hydro, that had discontinued power production from 1930–1970 (23). All but ca 300 of these are estimated to have redevelopment potentials comparable to their original installed capacity of 1.3 GW.

Fuel Cells

A fuel cell converts the chemical energy of a fuel directly to d-c power without going through a thermal or combustion cycle in the process (24–26). A fuel cell, somewhat similar to the cell in a conventional lead-acid battery, is comprised of two electrodes and an electrolyte. The principal reactions create an electrical current at the electrodes. However, unlike a conventional lead-acid battery cell, the reactants in a fuel cell can be replenished externally, thereby enabling the fuel cell to operate continuously at a constant power output (see Batteries and electric cells, primary—fuel cells).

Fuel cells are classified according to the type of electrolyte or ion-conducting media used and the temperature of operation (see Fig. 3) (24).

Internal-Combustion Engines

Internal-combustion engines are used principally in automobiles, aircraft, trucks, locomotives, ships, excavating machinery, and all types of apparatus where relatively small amounts of power are required. They also are used for the production of electrical energy in small stationary power plants.

The principal fuels for this type of engine are gasoline, diesel oil, and natural gas (see Gasoline and other motor fuels; Gas, natural). More recently, gasohol, methanol, and ethanol have received a great deal of attention as substitutes for gasoline. The power is produced by the expansion of the working substance against a piston that reciprocates inside a cylinder. The reciprocating motion of the piston is converted into rotary motion by means of a connecting rod, crankshaft, and flywheel; the latter serves as a device for smoothing out variable forces exerted against the crankshaft.

Internal-combustion engines can be built with several cylinders. For small ma-

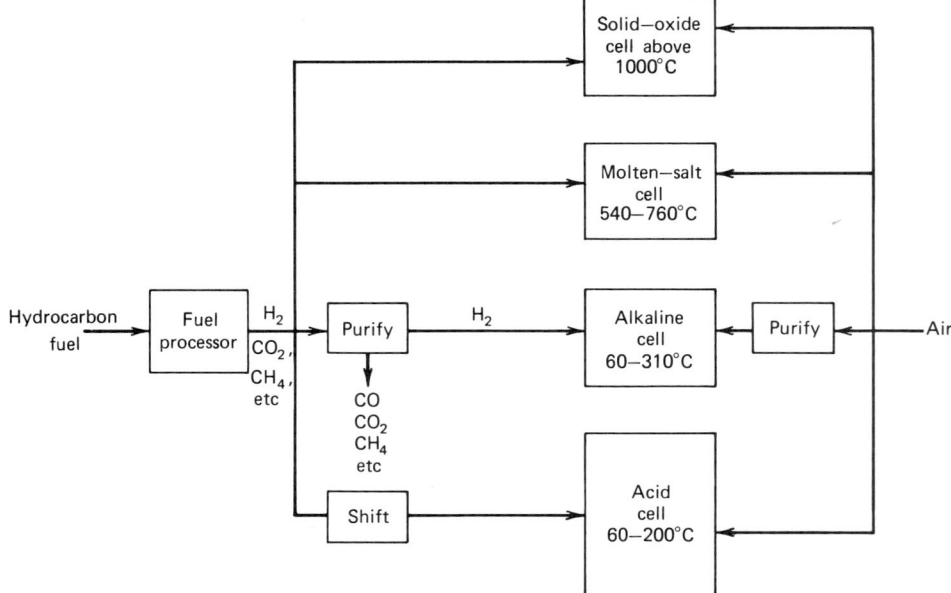

Figure 3. Fuel-cell types.

chines, one cylinder may suffice, whereas for large diesel power plants, automobiles, or aircraft engines, many cylinders may be required. Cylinders may be arranged in a variety of ways, but the most popular are the in-line and V arrangements used in automobiles and stationary diesel-engine power plants. Other arrangements are the opposed-piston and radial type. The former is used principally in diesel engines, the latter in aircraft engines.

Internal-combustion engines range in size from small-model airplane engines to large units capable of delivering several megawatts. Rotating speeds are 50–5000 rpm, depending upon size and design. Thermal efficiencies are usually high in well-designed units, ranging from ca 20% for automotive and aircraft engines to as high as 40% for diesel-locomotive and large stationary diesel power-plant engines.

Engine Cycles. The two most important cycles employed for internal-combustion engines are represented in Figure 4 on pressure–volume and temperature–entropy diagrams. The cycles shown are ideal cycles, in which all processes are assumed to take place in a thermodynamically reversible manner. In the case of the Otto-cycle engine, intake begins at point *0* in Figure 4a. The piston moves toward the crank end under the impetus of the flywheel, and intake air accompanied by a charge of gasoline vapor or gaseous fuel is drawn into the cylinder along the line *0–1*. In the case of the diesel cycle, exactly the same processes are followed except that fuel is injected into the cylinder in atomized form at point *2* (see Fig. 4b).

They are called four-stroke cycles because four strokes of the piston (intake, compression, expansion, and exhaust) are required to complete the cycle. By arranging exhaust and intake ports around the cylinder near the end of the expansion stroke of the piston, the exhaust gases are displaced with a new charge of fresh air before the piston returns on its compression stroke. With such an arrangement, the piston itself uncovers the exhaust ports near the end of the power stroke. Shortly thereafter, the intake ports are also uncovered by the piston, and a fresh supply of air, under a slight

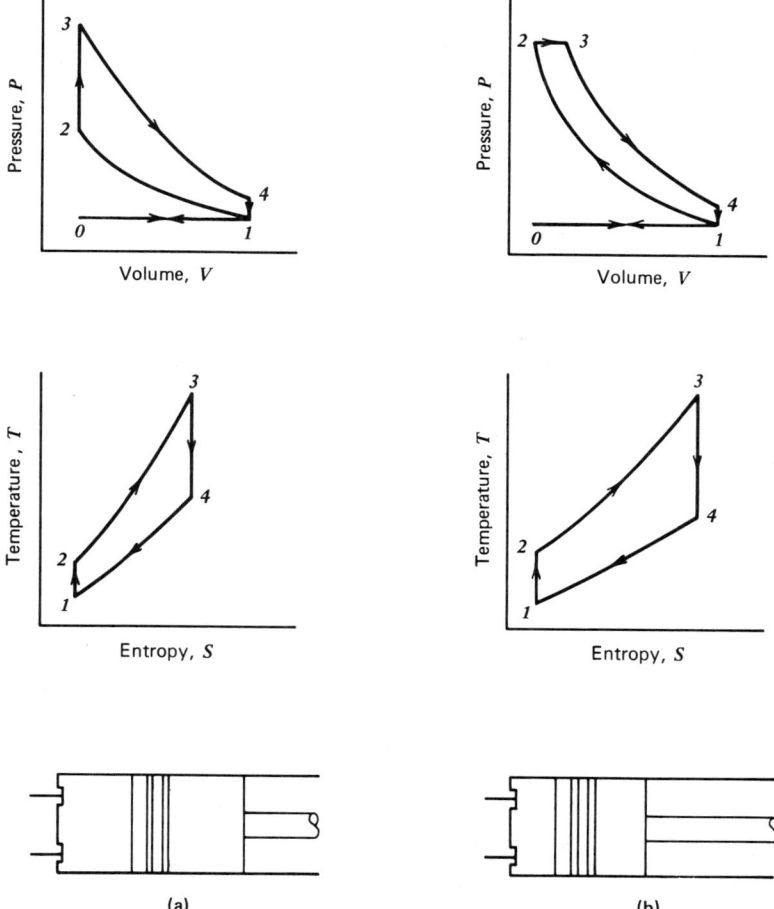

Figure 4. Pressure–volume and temperature–entropy diagrams for ideal internal-combustion engine cycles. (**a**) Four-stroke Otto cycle. (**b**) Four-stroke diesel cycle. *1*, intake valve closes and piston returns to head end; *2*, air and fuel are compressed isentropically; *2* → *3*, combustion takes place at constant volume (**a**) or pressure (**b**); *3* → *4*, expansion continues, delivering a powerful turning moment to the crankshaft and flywheel; *4*, exhaust valve opens.

pressure generated in the crankcase by the moving piston, is swept into the cylinder, thus driving the exhaust gases out. This procedure eliminates the necessity of a separate intake and exhaust stroke and permits the cycle to be completed with only two strokes, compression and expansion. Engines that operate in this manner are called two-stroke-cycle engines. They may operate on either the Otto or the diesel cycle, depending upon the manner in which the fuel is admitted and burned. Such two-stroke-cycle engines are employed widely for powering small devices such as lawn mowers, snow blowers, outboard motors, hand tools, and similar articles.

In all actual engines, the processes depart rather widely from those just described because of frictional effects, the necessity of cooling the cylinder to prevent overheating, the imperfect control of the fuel-injection rate, leakage of air and gases past the piston, imperfect combustion, variation of the properties of air and combustion gases with temperature and pressure, and many other less evident factors. For these reasons, actual thermal efficiencies computed from the ideal cycles must be calculated with a correction factor.

Cycle Efficiencies. Because the thermal efficiency of the ideal cycle is a measure of performance and can be computed easily, it provides a basis for calculating thermal efficiencies of actual engines. The ideal-cycle efficiency η_T for the Otto cycles may be computed readily from the pressure–volume or temperature–entropy diagram.

$$\eta_T = 1 - \frac{1}{r_c^{k-1}} \tag{1}$$

where r_c = ratio of compression V_1/V_2 (see Fig. 4a), and k = isentropic exponent for air C_p/C_v (C_p = specific heat at constant pressure, C_v = specific heat at constant volume).

If it is assumed that throughout the cycle air has a value of k equal to 1.4, corresponding to cold air, the ideal cycle is said to be computed on the cold-air standard. If the value of k is taken to be 1.3, corresponding to hot air, it is said to be computed on the hot-air standard. Efficiencies computed on the hot-air standard are less than those computed by the cold-air standard, but they are somewhat closer to those actually encountered in practice because of the high average temperature of combustion gases in engines.

For the ideal diesel-cycle engine, the thermal efficiency is given by

$$\eta_T = 1 - \frac{1}{r_c^{k-1}} \left[\frac{r_f^k - 1}{k(r_f - 1)} \right] \tag{2}$$

where r_f = fuel cutoff ratio V_3/V_2 (see Fig. 4b).

Actual thermal efficiencies are somewhat lower than those computed by equations 1 and 2. The factor by which the ideal-cycle efficiency must be multiplied in order to give the brake-thermal efficiency is called the brake-engine efficiency. For well-designed engines at rated load, this factor is usually ca 50–60%.

Applications and Performance Characteristics. Internal-combustion engines operating on the Otto cycle are well adapted to applications where speed and power requirements vary over a wide range. This is particularly true of automobile and aircraft engines, which must be capable of adjusting rapidly to sudden variations in load and speed. With modern methods of carburetion, the correct fuel–air ratio is automatically supplied at all speeds and, thus, it is possible to operate the engines at near-peak efficiency regardless of operating conditions. Otto-cycle engines employed in automobiles have brake-thermal efficiencies of approximately 25% which have increased little over the years. Improvement in fuel economy in recent years is largely due to better engineering.

Diesel engines are used widely for stationary and marine power plants, and for railroad locomotives. For stationary power-plant application, the frequency of the electric current generated must remain constant. This permits a constant-speed design that yields optimum efficiency throughout the normal load range. In locomotive applications, diesel engines usually drive electric generators which, in turn, furnish electric power to traction motors. With this arrangement, the speed of the diesel engine can be held at values conducive to highest efficiency, with variations in train speed and load handled electrically.

Diesel engines are also used for trucks and heavy-duty excavating and road-building equipment. For these applications, the engines must be designed to maintain reasonably high efficiencies throughout wide ranges of speed and load. Compression ratios for diesel engines range from about 11–22 and are typically about twice those for Otto-cycle engines. These higher compression ratios result in higher efficiencies, which are ca 40%.

Because of heightened concern over environmental problems, a great deal of effort has recently gone into designing automobile engines that are more fuel efficient and free from harmful exhaust gases (see Exhaust control, automotive).

Gas Turbines

Simple Open-Cycle Gas Turbine Without Regeneration. The simplest gas-turbine cycle is essentially the Brayton cycle illustrated in Figure 5. Its efficiency is computed easily from the temperature–entropy diagram assuming that air follows the ideal gas laws, that the specific heat of air at constant pressure is a constant, and that the weight of fuel supplied contributes a negligible additional weight to the heated air flowing through the turbine.

$$\eta_T = 1 - \frac{1}{r_p^{(k-1)/k}} \qquad (3)$$

where r_p = pressure ratio P_2/P_1.

In an actual gas turbine, both compressor and turbine have internal efficiencies of less than 100%, and the net energy delivered to the generator shaft of the turbine is substantially less than values computed by equation 3.

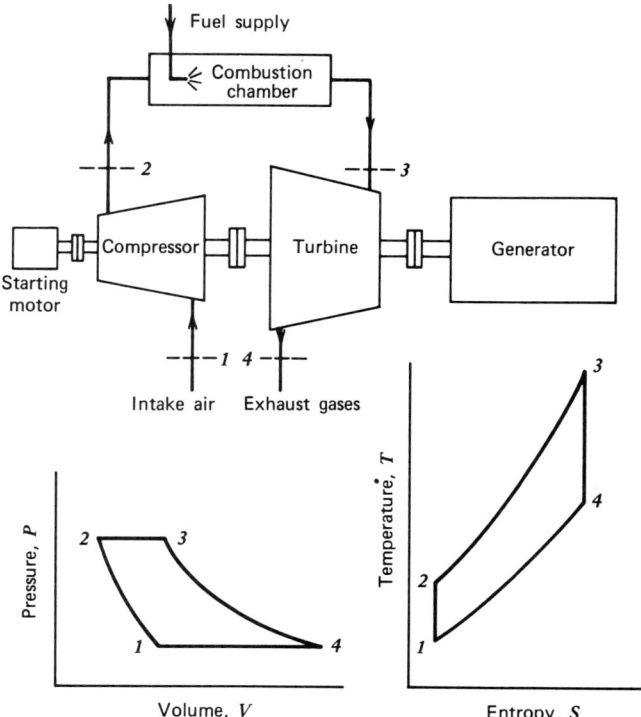

Figure 5. Simple open-cycle gas turbine; pressure–volume and temperature–entropy diagrams for ideal cycle.

Open-Cycle Gas Turbine with Regeneration. The efficiency of the simple open-cycle gas turbine can be improved considerably by employing a regenerator in which the exhaust gases are passed countercurrent to the air leaving the compressor. In this manner, air is heated by the exhaust gases before entering the combustion chamber, thereby reducing the amount of heat that must be supplied by the fuel.

If a source of cooling water is available, the compression process can be divided into two stages with intercooling. The work of compression can thus be reduced and the efficiency of the entire cycle improved. In larger units, overall efficiency can be improved if the expansion process is divided into two parts with reheating to the initial temperature before the second expansion. This is usually accomplished by the use of both a high and a low pressure turbine with a second combustion chamber placed between them.

A cycle employing regeneration, intercooling, and reheating is illustrated in Figure 6, including pressure–volume and temperature–entropy diagrams.

Working fluids other than air can be accommodated if the cycle is closed so that the gas leaving the turbine is cooled and returned to the compressor inlet for reuse. This cycle requires a separate furnace for supplying heat to the working fluid. It permits maintaining the gas at a high pressure level, thus reducing the size of the rotating

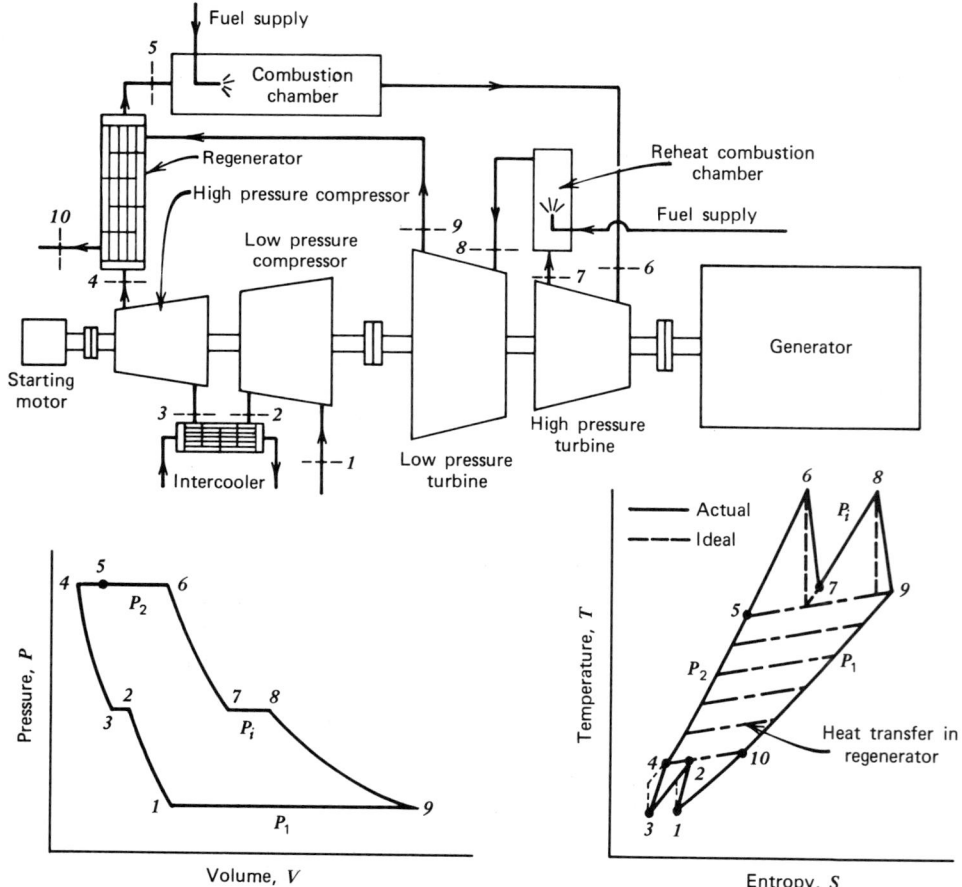

Figure 6. Open-cycle gas turbine with regeneration, reheat, and intercooling; pressure–volume and temperature–entropy diagrams.

machinery for a given power output. Although the closed-cycle gas turbine is not employed in the United States, several closed-cycle plants have been built in Europe that operate on a wide variety of fuels, including coal, oil, mine gas, natural gas, and nuclear energy. Because of the attractiveness of burning coal, renewed interest in the closed-cycle gas turbine has been expressed by the DOE and the electric-utility industry, and research directed toward electric-power production and cogeneration applications is now underway. Of particular interest is the fluidized-bed combustion of coal for supplying primary heat to the working fluid. The objective is the twofold goal of raising the turbine inlet temperature to ca 815°C by using ceramic tubing in the hotter portions of the furnace, and reducing harmful emissions. Success in this direction would provide a competitive alternative to the regenerative-cycle coal-burning steam-power plant, with its associated scrubber and other air-quality-control equipment (see Air pollution control methods).

Characteristics. The curves in Figure 7 represent the efficiency of various gas-turbine cycles as a function of pressure ratio and turbine inlet temperature. Curve A is for the ideal cycle as computed for air, using $k = 1.4$. With this cycle, the efficiency depends only on the pressure ratio. Curves C, D, and E represent cycle efficiencies computed for an open-cycle gas turbine without regeneration, but with internal efficiencies of the compressor and turbine less than 100% and with three maximum air temperatures of 815, 650, and 540°C. For these curves, the internal efficiency of the compressor was assumed to be 85%, whereas that of the turbine was assumed to be 87%. The intake air temperature was assumed to be 27°C, with a value of k equal to 1.4. The curves indicate that, for a given turbine and compressor efficiency, the cycle efficiency reaches a maximum value at a definite pressure ratio that depends upon the maximum cycle temperature T_3 in Figure 5. They also show the importance of a high turbine-inlet temperature in achieving high thermal efficiency. For a maximum temperature of ca 540°C, for example, a peak cycle efficiency of only ca 18% can be

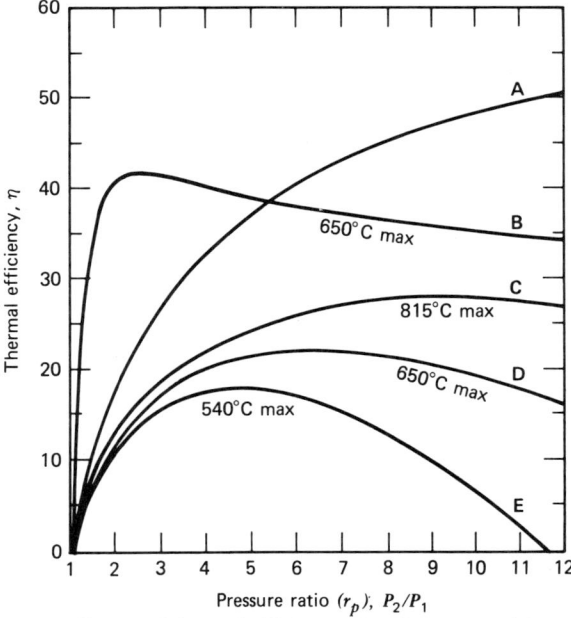

Figure 7. Curves of thermal efficiency of various gas-turbine cycles.

expected, whereas with 650 and 815°C, peak cycle efficiencies of 22 and 28%, respectively, can be reached. Because of mechanical and generator losses, actual efficiencies would be about 90–95% of those indicated by the curves.

A concerted effort is now underway in the area of high temperature turbine technology (HTTT) to raise the maximum cycle temperature of open-cycle units as high as ca 1425°C and eventually to 1650°C by water cooling or transpiration air cooling in the hot parts of the gas turbine. Success in this endeavor would result in cycle efficiencies far above those indicated in Figure 7.

Improvement can be realized by combining regeneration with intercooling and reheating as illustrated by curve B in Figure 7. A maximum cycle temperature of 650°C was assumed, and the specific heat for air at constant pressure was taken as 1.0 J/(g·K) (0.24 cal/(g·°C)) for the compressor and 1.08 J/(g·°C) (0.26 cal/(g·°C)) for the turbine in order to simulate actual conditions more closely. Values of k computed from these specific heats were also used. The efficiency of each portion of the compression process was assumed to be 85%, whereas that for each portion of the expansion process was assumed to be 87%. The regenerator was assumed to heat the air to within 28°C of the temperature of the exhaust gases leaving the turbine. The division between high and low pressure processes in both the compressor and turbine was chosen to ensure that maximum efficiency would be obtained. This occurs when the intermediate pressure P_i is equal to $\sqrt{P_1 P_2}$. Curve B shows that, for this particular cycle, maximum efficiency occurs at a very low pressure ratio, but that the efficiency is always much higher than for the corresponding open cycle without regeneration, reheating, or intercooling (curve D). With conditions similar to those indicated for this more complicated cycle, actual thermal efficiencies greater than 30% should be realized for all loads carried except the very lightest. A maximum output can be obtained at some particular pressure ratio, depending on the maximum temperature assumed and the efficiency of turbine and compressor. Maximum output does not necessarily occur at the same pressure ratio as maximum efficiency.

Applications. The most important gas-turbine application is for jet-propulsion aircraft (see also Aviation and other gas turbine fuels). Air flows into the compressor at the front end of the engine and is compressed to several hundred kPa (several atm) before entering a ring of small combustion chambers located between the compressor outlet and turbine inlet. In these chambers, a special grade of kerosene fuel is burned, which heats the air to 650–815°C. From the combustion chambers the hot gases expand through a turbine, thereby furnishing the motive power for driving the compressor attached to the turbine shaft. The hot gases continue their expansion through the exhaust nozzle of the engine, and create a high velocity jet that drives the plane forward by the force of reaction. Jet-propulsion engines are most efficient at very high speeds and at high altitudes where the ambient air temperature is low.

A modification of the conventional jet engine is the bypass or fan-jet engine illustrated in Figure 8. In this version, the blading of the first few stages of the compressor is lengthened, and a portion of the air flowing into the compressor is discharged without passing through the combustion chamber or turbine. Fuel consumption is reduced significantly, and the range and economy of operation are increased.

Gas turbines are also used for locomotive and marine applications, but only to a limited extent, partly because of the low efficiency of the simple open-cycle gas turbine compared to the diesel engine. Development of an automotive gas turbine has so far not been successful. A few gas-turbine-powered trucks and automobiles have

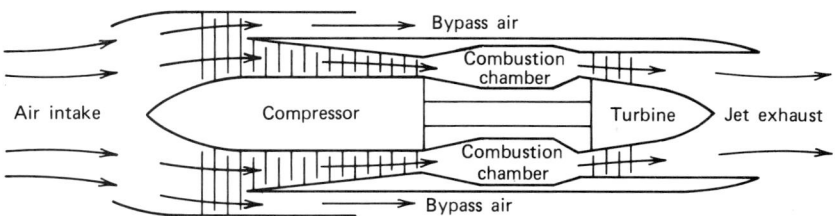

Figure 8. Diagram illustrating principle of operation of the bypass or fan-jet engine.

been built and tested, but many problems remain to be solved before they can be marketed.

Stationary gas turbines are employed in the oil and gas industries to drive axial or centrifugal compressors. Such units are used for compressing hydrocarbon gases and numerous other purposes, eg, the transport of gas from the field to metropolitan areas through large high pressure pipelines (see High pressure technology; Pipelines).

Gas turbines for pipeline-compressor applications vary in complexity from simple open-cycle units of ca 25% efficiency to combined gas-turbine/steam-turbine installations with efficiencies of ca 40%. In the combined-cycle units, the high temperature turbine exhaust gas is utilized for heating feedwater and to vaporize and superheat steam for powering a steam turbine that can be directly connected to the gas-turbine-compressor drive shaft. In a typical installation, the steam turbine contributes ca 35% of the total shaft power and the gas turbine supplies the remainder (27). The U.S. pipeline system alone amounts to ca 12 GW (16×10^6 hp) as of 1980.

Gas turbines have innumerable applications in the petrochemical industry and other processing industries. They can be used with a wide variety of fuels and are particularly suitable for cogeneration purposes. The large amount of high temperature turbine exhaust gas can be used to perform numerous heating tasks, or for regeneration purposes to improve cycle efficiency. Combined gas-turbine/steam-turbine cycles can be tailored to suit almost any energy need and can compete on an economic basis with alternative systems.

Another application in the oil industry is the Houdry process for refining gasoline (see Petroleum, refinery processes). In this process, air leaving the compressor passes over a hot catalyst, where it supplies oxygen for burning off carbon that has collected. The resulting hot-combustion gases are passed through a turbine, and enough work is obtained to drive the compressor and still supply a small amount of by-product power. A similar application exists in the steel industry, where blast-furnace gas is used as fuel in a gas turbine that compresses the air required for operation of the blast furnaces.

In the electric-utility industry, gas turbines are employed for peak-load generation or for standby service. They also supply power for fully automatic booster stations at the end of long transmission lines. Combined gas-turbine/steam-turbine cycles offer better station heat rate than conventional cycles.

Steam Engines

The reciprocating steam engine operates on the principle of the expansion of steam against a piston that moves back and forth inside a cylinder. For many years, this type of engine carried the burden of power generation, but the advent of inter-

78   POWER GENERATION

nal-combustion engines and steam turbines has greatly reduced this application. The steam engine is definitely limited in size, speed, and capacity and, except in the smallest sizes, it cannot possibly compete with the steam turbine on an economic basis. Moreover, even in the smaller sizes, cheaper power-generating devices that do not require steam generation are available.

Nevertheless, occasionally a steam engine is the proper choice. This may be the case when steam is already available, or will be needed for other purposes, and where the exhaust steam from the engine can be used for heating or process work.

Figure 9 is an illustration of a typical steam-engine cylinder. The ideal-indicator diagram represents graphically the pressure inside the cylinder versus the piston position. Steam produced in a separate steam generator is admitted under high pressure, forcing the piston toward the crank end to cutoff point *1*. At this point, the intake valve closes, and steam is allowed to continue its expansion until the piston reaches the end of its power stroke at point *2*. The exhaust valve then opens and the pressure drops at constant volume to exhaust pressure. The steam is then forced out as the piston moves toward the head end under the impetus of the flywheel.

In an actual engine, the exhaust valve is closed near the end of the piston's forward movement, thus trapping a small amount of steam in the head end of the cylinder. This

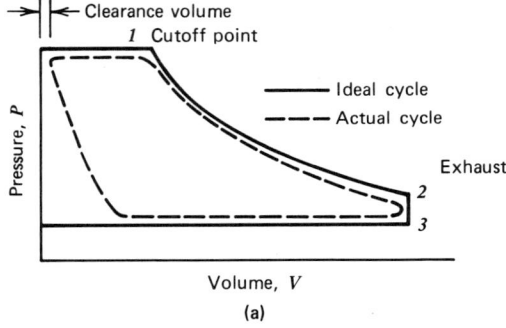

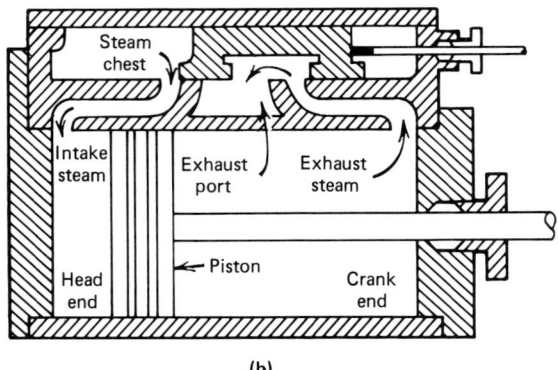

Figure 9. (a) Ideal and actual indicator diagrams for head end and (b) diagram of a D-slide valve steam-engine cylinder.

steam is compressed as the piston moves further forward, and it serves as a cushion against impact when new steam is admitted at the beginning of the power stroke.

Most early steam engines used a slow-acting D-slide valve, which resulted in considerable pressure drop as the steam passed in and out of the cylinder. A more rapid valve action, as exemplified by the Corliss valve, eliminated this effect.

In the uniflow engine, one of the most widely used steam engines, steam is admitted at either end of the cylinder and exhausts near the center, where the piston itself uncovers the exhaust ports. The problem of initial steam condensation is reduced to a minimum, since the hottest portions of the cylinder where intake occurs are not cooled by exhaust steam as in other types of engines.

The approximate power obtainable from a steam engine is easily estimated by first determining from the ideal-indicator diagram the mean effective pressure acting on the piston. This is multiplied by the piston area, length of stroke, number of power strokes per minute, mechanical efficiency, and a diagram factor F_D to correct for failure to achieve ideal conditions.

For D-slide valve and uniflow engines, the value of F_D at rated load is about 0.75, whereas for Corliss-type engines it is about 0.80. The mechanical efficiency is in the neighborhood of 95%. The expansion ratio V_2/V_1 varies between 4 and 6. Because of limitations imposed by the valves, steam temperatures should not exceed 315°C, and steam pressures should not be above ca 1.75 MPa (265 psi).

Steam Turbines

The steam turbine is the prime mover for the production of large quantities of power. All steam turbines operate on the principle of expanding steam through a series of nozzle and blade elements designed to convert the energy of expansion directly into rotational motion.

In general, a large rotor is suspended in a casing or cylinder to which rows of blading are attached that extend radially outward. The steam flows between the blades to the turbine exhaust. Suspended radially inward from the casing are also rows of blading forming nozzle passages that fit between the rows of blading attached to the rotor. The velocity of steam expanding through the turbine is increased in the stationary nozzle passages, and the jet created is then directed against the blades on the rotor, causing it to spin at high speed. In a typical moderate capacity central-station steam turbine employing reheat, steam enters the turbine through control valves which automatically regulate the flow rate in accordance with the load to be carried. After flowing through the first stages of the high pressure section, the steam leaves the turbine and is returned to the steam generator for reheating before returning to the turbine through another valve. After passing through additional stages of blading, it leaves the high pressure section of the turbine and is directed into the low pressure section. Here, the steam is divided into two paths, each flowing in opposite directions through several additional stages of blading before it is exhausted into a condenser located beneath the low pressure section. At various points along the flow path, openings are provided from which steam may be extracted for the purpose of heating boiler feedwater.

Steam turbines are either condensing or noncondensing. In condensing turbines, the steam, after expanding through the blading, is exhausted into a condenser operating under low pressure, usually between 3.4–6.8 kPa (25–50 mm Hg). In the condenser, latent heat contained in the exhaust steam is removed by the circulation of large quantities of cooling water. The collected condensate is pumped through a series of feedwater heaters back to the boiler.

In noncondensing turbines, the steam is exhausted either to the atmosphere or to some other apparatus at ≥101 kPa (1 atm). These turbines are used in plants where power is generated as a by-product and where exhaust steam is used for process work or heating.

Because of the relatively large loss in thermal efficiency that results from high exhaust pressure, all central-station steam-power plants employ condensing turbines. These are attached to large high voltage electric generators. Power is distributed to the customer through complex electrical networks that include substations and transformers.

Both condensing and noncondensing turbines may be classified as straight-through turbines, bleeder turbines, or automatic extraction turbines. In straight-through turbines, the steam flows through the turbine from inlet to exhaust without any portion being added or abstracted. This type is employed mainly for driving pumps, fans, or other high speed equipment, or for generating relatively small amounts of power.

Bleeder turbines are those from which steam is extracted at various points along the path of flow for the purpose of heating condensate feedwater on its way to the boiler. Since no attempt is made to maintain the pressure of the extracted steam at a constant level, it varies almost directly in proportion to the load being carried. This type is used widely in modern central-station power generation because a much higher thermal efficiency can be achieved. The total quantity of steam bled from the turbine is ca 30% of that supplied to the throttle. The overall reduction in heat input required to produce the same turbine output with a bleeder turbine compared to that obtained from a straight-through turbine under the same conditions is about 10%.

Automatic extraction turbines permit the extraction of steam at constant pressure at one or more points along the flow path. Steam extracted in this way may be used for process work or heating. Special valve arrangements are needed since both the speed and extraction pressure are held constant regardless of the load carried. Automatic extraction turbines are used in industrial plants where electric power is generated and, at the same time, steam is supplied for other processes. The steam can be exhausted to a condenser at low pressure or to some other apparatus, eg, a heating system, at moderate pressure.

Characteristics. Figure 10 is a set of characteristic curves for a typical 20-MW 3600-rpm, ASME-preferred standard turbine operating at ca 6 MPa (865 psi) and 482°C with four extraction openings for feedwater heating. For either case, the throttle flow varies almost linearly with load. This is typical of all steam turbines, regardless of size. For extraction operation, the heat rate is at a minimum in the neighborhood of rated load where maximum economy is usually desired.

The curves in Figure 11 show throttle flow versus load for a typical 3600-rpm single automatic-extraction turbine. Throttle pressure and temperature are 2.86 MPa (415 psi) and 315°C, respectively. Exhaust and extraction pressures are 6.8 kPa (50 mm Hg) and 343 kPa (50 psi), respectively. The curves indicate that, for any constant load, an increase in extraction flow of 1.25 kg/s requires an increase in throttle flow of ca 0.7 kg/s. The line of maximum extraction on the left represents the limiting amount of steam that can be safely extracted at any given load without reducing the exhaust

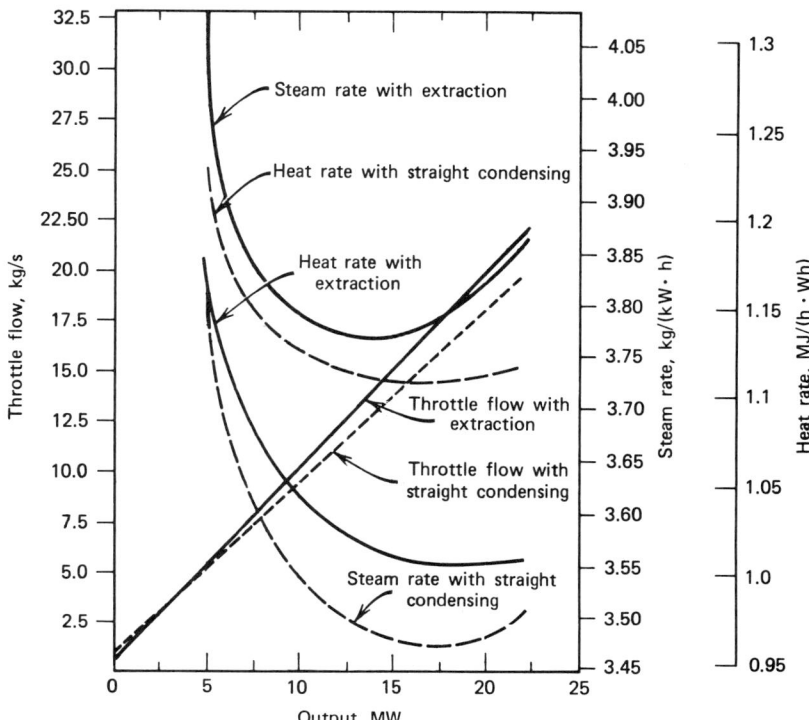

Figure 10. Characteristic curves for a 20-MW ASME-preferred standard turbine generator. Steam conditions: 5.96 MPa (865 psi), 482°C. Solid lines are for four-heater performance. To convert MJ to Btu, divide by 1.054×10^{-3}.

flow to a value too small to carry away the frictional heat generated in the turbine stages beyond the point of extraction. The line on the lower right represents the limiting load that can be carried without overloading the condenser. Maximum generator capacity is 2.75 MW and maximum throttle flow is 7.56 kg/s.

Central-Station Power Plants. Most central-station power-plant turbines are for capacities exceeding 100 MW. In fact, at the end of 1978, 95% of the new fossil-fuel generating capacity scheduled for operation by 1986 was for units exceeding 100 MW and 60% was for units of 500–884 MW. Today, turbine generator units as large as 1.3 GW are available for nuclear plants. Most of these turbines are individually designed to meet the particular conditions of a given plant. However, by making use of the same components in different combinations, costs can be reduced considerably.

Turbines for fossil-fuel power plants use highly superheated steam with reheat. Typical steam conditions at inlet for the newer units of 500 MW and above are pressures of 17–24 MPa (2400–3500 psi) and temperatures of ca 540°C. Most units are reheated once to 540–565°C, but some are reheated twice, each to ca 540–565°C.

Because of economical and technical limitations imposed by water-cooled nuclear reactors, turbines designed for pressurized-water and boiling-water nuclear power plants are limited to a maximum inlet pressure of ca 6.5 MPa (950 psi). Since a successful nuclear superheater has not yet been developed, steam is supplied to these turbines in the dry saturated state at ca 280°C. Because of these disadvantages, turbines employed in nuclear power plants are considerably larger and less efficient than

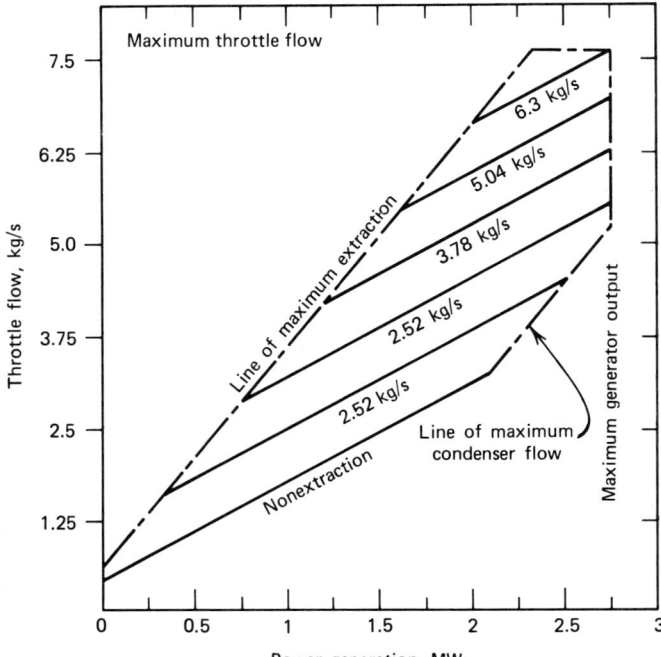

Figure 11. Curves of throttle flow vs power generation for various extraction rates for a 2-MW-automatic extraction turbine.

those employed in fossil-fuel plants. In addition, they require complicated devices for extracting the moisture that is released from the steam as it expands through the turbine. However, the lower efficiency is not as important for a nuclear plant as for a fossil-fuel plant because of the relatively low cost of nuclear fuel.

Steam Power Plants

Rankine Cycle. The Rankine cycle is the basis from which all modern steam-power-plant cycles have evolved. Figure 12 is a sketch of the equipment required together with the corresponding temperature–entropy and enthalpy–entropy diagrams.

The work of the cycle is the difference between the work of the turbine and the work of the feed pump. From the steady-flow energy equation, this may be shown to be $(h_1 - h_2) - (h_4 - h_3)$. The heat added is $(h_1 - h_4)$. The thermal efficiency of the cycle is therefore given by

$$\eta_T = \frac{(h_1 - h_2) - (h_4 - h_3)}{(h_1 - h_4)} = \frac{(h_1 - h_2) - PW}{(h_1 - h_3) - PW} \tag{4}$$

where h = enthalpy of working substance in J/kg (Btu/lb) of working substance, and PW = feed-pump work $(h_4 - h_3)$ in J/kg (Btu/lb) of working substance.

The pump work PW is negligibly small compared to other quantities, except for pressures above ca 2.86 MPa (415 psi). Graphically, the work of the ideal cycle is represented by the enclosed area W on the temperature–entropy chart, and the heat rejected is represented by the area Q_R. The heat added is represented by the sum of these two areas. In the actual cycle, expansion through the turbine occurs inefficiently,

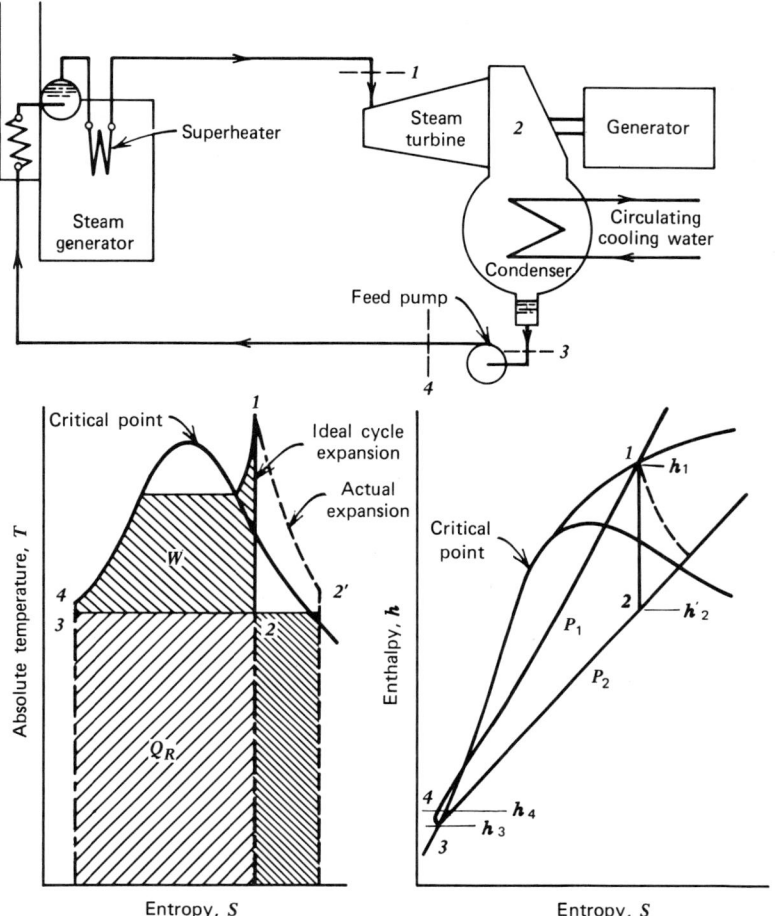

Figure 12. Rankine-cycle power plant with corresponding temperature–entropy and enthalpy–entropy diagrams.

which causes the steam to leave the turbine at point 2′ instead of 2, ie, a greater heat rejection takes place in the condenser. This additional heat is represented by the darker-shaded area on the temperature–entropy chart or by the enthalpy difference $(h_{2'} - h_2)$ on the enthalpy–entropy chart.

Because of the greater heat rejection, the cycle efficiency is reduced. The ratio of the actual energy released by the steam during its expansion $(h_1 - h_{2'})$ compared to the ideal release $(h_1 - h_2)$ is called the turbine internal efficiency. In modern central station steam turbines, it is ca 85%. The actual output of the generator is less than the energy release $(h_1 - h_{2'})$, ie, it is diminished by the exhaust loss and the mechanical and generator losses. Inasmuch as these losses constitute about 5% of the generator output in modern units, the overall output of the turbogenerator is equal to about 80% of that represented by the ideal isentropic expansion $(h_1 - h_2)$. This percentage is called the overall engine efficiency of the turbogenerator unit.

From the area represented on the temperature–entropy chart, it is obvious that the higher the initial pressure and temperature, the higher the cycle efficiency. This fact has led to a gradual trend toward higher pressures and temperatures in power-plant design.

Regenerative Cycle. The regenerative cycle is similar to the Rankine cycle except that certain modifications increase the overall thermal efficiency. Steam is bled from the turbine at several points for the purpose of heating the condensate feedwater before returning it to the boiler. In the reheat-regenerative cycle, steam that has already expanded partially through the turbine is withdrawn and reheated at constant pressure to a higher temperature (usually equal or close to the initial steam temperature) before it is returned to the turbine for completion of expansion. Reheating increases thermal efficiency, a result of the addition of heat to the cycle at a higher average temperature than that used for the straight regenerative cycle. Furthermore, the exhausted steam contains less moisture. Thus, frictional losses and water erosion are reduced in the last stages of the turbine, and the efficiency is increased even further.

A typical modern central-station reheat-regenerative power-plant cycle designed for a typical steam turbine is shown schematically in Figure 13, which includes temperature–entropy and enthalpy–entropy diagrams (see also Thermodynamics).

The temperature–entropy and enthalpy–entropy diagrams show the weight of the working substance involved in each portion of the cycle. It can be seen that none of the extracted steam flows to the condenser and, consequently, less heat is rejected than in the case of the corresponding Rankine cycle. Hence, a larger portion of the heat added is converted into work, and the cycle is more efficient. The entropy increase during expansion through the turbine results from internal friction that is converted into heat. This raises the enthalpy of the expanded steam by the amount indicated on the enthalpy–entropy diagram.

The regenerative cycle can be modified with a gas turbine. For example, a gas turbine is added to the standard steam cycle in such a manner that its exhaust gases are used to supply oxygen for fuel combustion in the steam generator. This is feasible because only ca 25% of the oxygen in the compressed air is used chemically in the combustion chamber of the gas turbine. The remaining 75% remains available in the exhaust gases and can be used to burn fuel in the boiler furnace. Such an arrangement permits a gain of 4 or 5% in station efficiency (28). Alternatively, a steam generator is equipped with a pressurized boiler furnace in which air from the gas-turbine compressor supplies oxygen for burning the fuel in the furnace under several hundred kPa (several atm) pressure. The boiler furnace serves as the combustion chamber for the gas turbine as well as the heat source for the generation of steam. After passing through the boiler furnace, the hot gases are returned to the gas turbine for the production of power, and then exhausted through feedwater heaters of the steam portion of the cycle before being directed to the stack. The gain in efficiency of this system is expected to average 5–6%. However, this system can be used only for plants burning gas or oil, because the fly ash associated with the burning of pulverized coal could not be tolerated by a gas turbine. Many other combinations devised to improve the steam cycle make use of exhaust heat from gas turbines. However, because of the high cost of oil and gas, combined cycles capable of utilizing coal in one form or another as fuel are preferred.

Cogeneration. The term power plant is often used loosely to designate any plant in which steam is generated, regardless of whether power is produced (29). In a more exact sense, an industrial steam power plant is one in which power is generated from steam, either for use *in situ* or for sale as by-product power. In the oil, steel, and chemical industries, large quantities of steam are often required. Whenever this is the case, a careful economic analysis will reveal whether or not it is an advantage to generate power and, if so, how much (30).

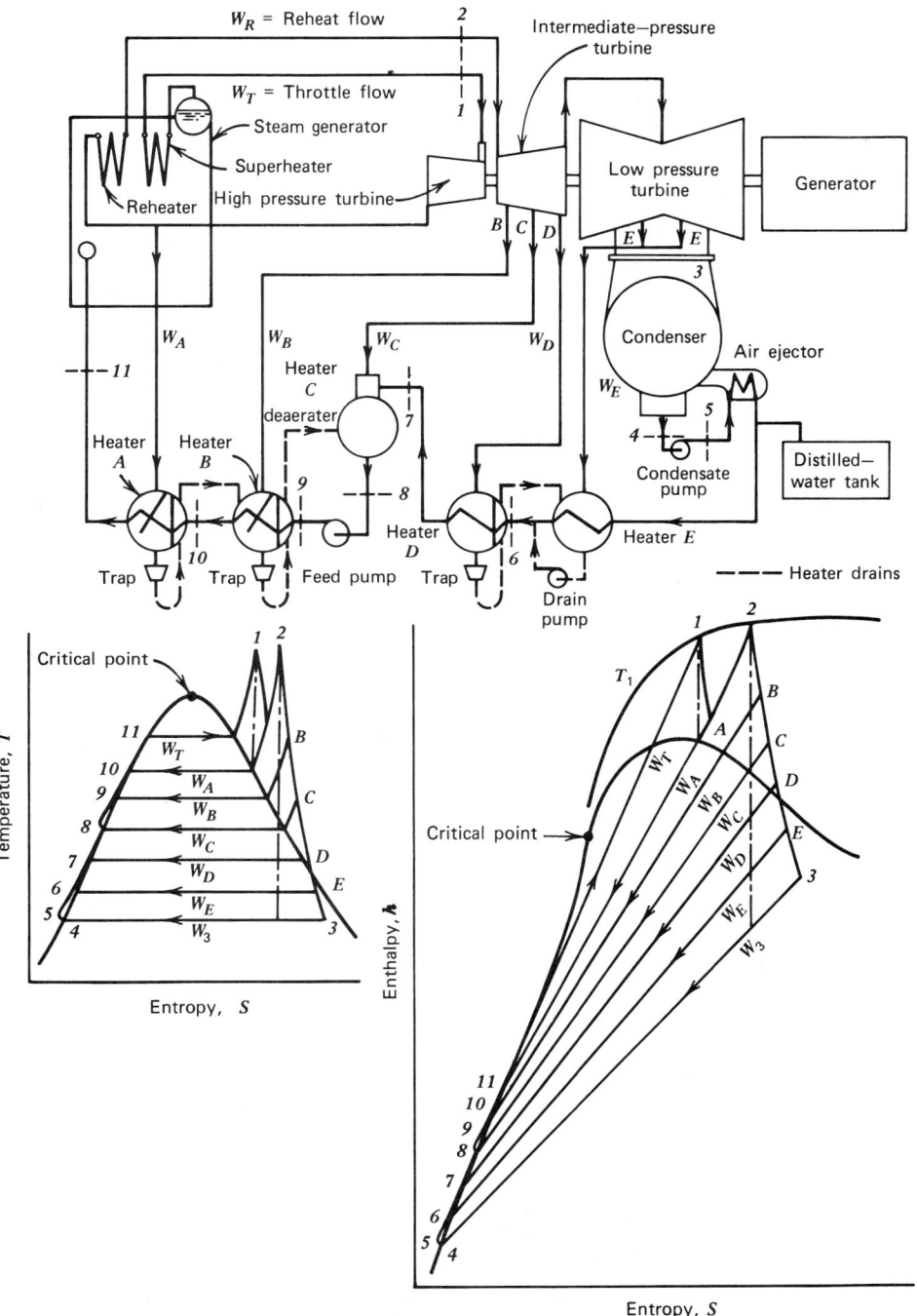

Figure 13. Flow diagram for a reheat-regenerative cycle with temperature–entropy and enthalpy–entropy diagrams. Dashed lines in entropy diagrams represent ideal cycle; solid lines, actual cycle.

86 POWER GENERATION

Because of the necessity and economic advantage of conserving fuel, governmental authorities have encouraged the generation of electric power as a by-product in plants built primarily for supplying process and other heat requirements. The term cogeneration has been coined to designate this type of operation.

Since it costs only a little more to build a high pressure steam generator than a low pressure steam generator, cogeneration may offer a decided economic advantage. By installing a turbogenerator with a high pressure steam generator, the necessary power and process steam requirements are supplied. Such an installation might prove especially attractive in cases where steam could be extracted at two pressure levels, ie, a high pressure for process work and a low pressure for heating during the winter. During the summer, the low pressure extraction could be curtailed and additional generated power could be employed for the operation of an air-conditioning system. During the seasons of the year when excess power is generated, this power can usually be sold to a utility at a mutually satisfactory rate.

The real advantage of cogeneration is its more efficient use of fuel (28). Since ca 44% of the heat supplied to a conventional central station power plant is lost to the cooling water in the condenser, this loss should be prevented by any means available (31). With a cogeneration plant with MW electric power output, the efficiency of fuel utilization may be expressed as

$$\text{Fuel utilization efficiency} = \frac{Q_P + \text{MW}}{Q_{HHV}}$$

where Q_P = heat use for process and other requirements, MW (949 Btu/s) and Q_{HHV} = higher heating value of fuel supplied, MW.

For a central-station plant, the fuel utilization efficiency is equal to the thermal efficiency, namely

$$\eta = \frac{\text{MW}}{Q_{HHV}}$$

Clearly, from the standpoint of energy utilization alone, cogeneration is highly desirable. However, to be economically attractive, it requires a reasonably satisfactory balance of energy requirements for process or space heating in comparison to electric needs, and each case must be judged on its own merits. Figure 14 illustrates the distribution of fuel utilization for a cogeneration plant as compared to a typical power-generation plant (32).

Because of the wide variety of needs, there is no one preferred industrial or cogeneration cycle. Any cycle that is adopted is essentially either the Rankine or the

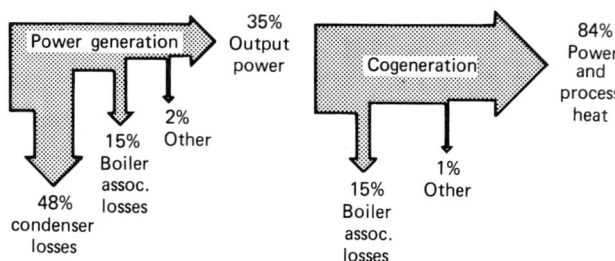

Figure 14. Fuel utilization efficiency for a typical power-generation plant as compared to a cogeneration plant. Courtesy of American Power Conference.

regenerative cycle with modifications for extracting steam at various pressures and in various amounts of process or other heating requirements.

Other Methods

Magnetohydrodynamics. In magnetohydrodynamic (MHD) power generation, electric power is produced by the movement of an electrically conducting gas through a magnetic field (33–34). In a simple open-cycle MHD generator (see Fig. 15) (35), hot partially ionized gas is expanded in a duct whose walls are electrically insulated from each other. When a strong magnetic field is applied across the duct, the interaction between the conducting gas and the magnetic field induces potential in the electrodes, and a current flows through the gas, electrodes, and external load. The output per unit volume of an MHD generator is proportional to the electrical conductivity, the square of the flow velocity, and the square of the magnetic flux density. Temperatures in excess of 3000 K are necessary for the required gas ionization. In the presence of seeding materials such as potassium or cesium, the temperature can be reduced to ca 2750 K. Most of the technical problems in open-cycle MHD are caused by the high temperature, the alkali-metal additive, the coal slag, and the d-c electric currents, which impose strict material requirements (see Coal, magnetohydrodynamics).

An MHD generator could be used in conjunction with a conventional power plant by passing the high temperature ionized gas through the MHD generator and using the still hot exhaust gas for generating steam in a boiler.

Advantages of MHD are the lack of moving parts and the direct generation of power from high temperature gases. Studies carried out for a group of electric utility companies (36) indicate that a 500-MW plant in which 70% of the power is generated by MHD and the balance by conventional means would have an efficiency of about 53%. This compares to about 40% for the best steam power plants.

A disadvantage of MHD generation is that the dc produced has to be converted to ac for suitable transmission and distribution. Other formidable problems caused mostly by the high temperature required have to be solved before this type of power generation can become a practical reality.

Electrogasdynamics (EGD). Like magnetohydrodynamics, electrogasdynamics is a method that directly converts the enthalpy of a working fluid to electrical energy.

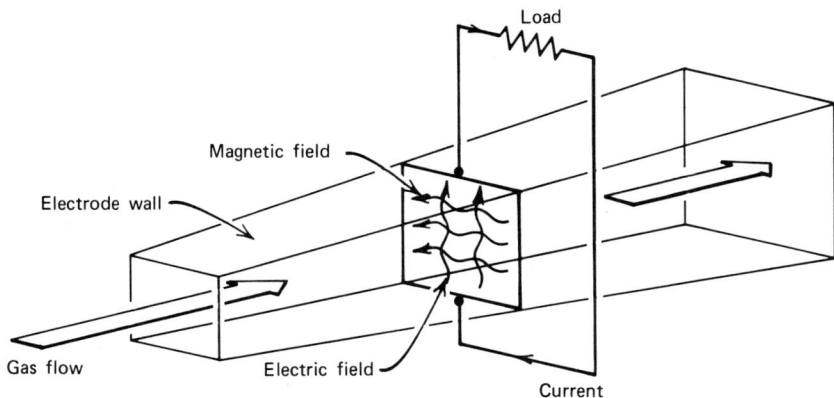

Figure 15. Principle of magnetohydrodynamics.

The generator consists of a duct subjected to an electrical potential imposed in the direction of the flow. The working fluid is seeded with unipolar ions as it enters the duct. The energy in the working medium transports the ions against the imposed potential. The ions are then collected at the duct outlet at high potential. When the inlet ion source and outlet collector are connected with an external load, a flow of electrical energy results.

This process does not require the extremely high temperatures of MHD; it can be incorporated into a power-generation scheme without the need for combining it with a steam plant of overall efficiencies comparable to that achievable with MHD. However, only very low current, high voltage electrical power is produced. Because of electrostatic-field effects, the maximum channel size is severely limited, and thousands of channels in parallel and series are required to give power levels of commercial interest.

Intensive work on the EGD power generation system was started in the United States in 1966.

Fusion Power. A great deal of interest has been focused in recent years on the possibility of generating power by thermonuclear fusion. It has long been known that the heat generated in the sun comes from the fusion of four hydrogen nuclei into a helium nucleus. Controlled fusion could open up an inexhaustable supply of fuel (see Fusion energy; Plasma technology).

Thermoelectric Generators. Thermoelectric generators utilize the Seebeck effect which operates on the same principle as a thermocouple. If two wires of dissimilar metal are joined to form a loop, an electric current circulates through the wires if one of the junctions is held at a higher temperature than the other. The current obtainable with metal junctions is very small, but with junctions of p- and n-semiconductors, it becomes much larger (see Semiconductors). By using several hot and cold junctions connected in series by semiconducting materials, the voltage that causes the current to flow can be increased in proportion to the number of junctions. Many different thermoelectric generators have been built or are being developed for use in space exploration (36). The capacity of these units varies from ca 2.5 to 5000 W at efficiencies of 0.7–6%. Fuels for keeping the hot junctions warm vary from hydrocarbons to radioisotopes such as strontium-90. Solar radiation supplied directly or through parabolic collectors may also be used for heating the junctions (see Thermoelectric energy conversion).

Because of their low efficiency and power output, thermoelectric generators are not expected to have a significant effect on present methods of large-scale power generation.

Thermionic Generators. If a metal is heated to a sufficiently high temperature, electrons are emitted from its surface. This principle may be used to generate electric power in a device called a thermionic generator. If two parallel metal plates are separated by an ionized gas, and if one plate is heated and the other kept cool, electrons flow through the ionized gas from the hot plate to the cold plate. By connecting the two plates through an external circuit, a path is provided for the return of the electrons to the hot plate, and a continuous electric current flows.

Like the thermoelectric generator, the thermionic generator has the advantage of no moving parts, but has the disadvantage of the same Carnot-cycle efficiency limitations of other heat engines. Tests on a large number of laboratory models show efficiencies of 5–16% with power densities of 1–22 W/cm^2. Emitter temperatures in these tests are 1200–1800°C.

Applications for the thermionic generator are mostly in space vehicles where compactness, low weight, and a high sink temperature are needed. They are not expected to become commercially significant.

Energy Storage

With the increase in oil prices since the 1970s and the uncertainty of the future supply of oil and gas, utilities and consumers are seriously considering energy management and energy-storage systems to meet peak requirements. Several research and development as well as demonstration programs are under way. The total savings for the United States at the turn of the century could be as high as $(24-48) \times 10^6$ m^3 $((1.5-3.0) \times 10^8$ bbl) oil per year. For the homeowner, energy storage could save 5–8 m^3 (1300–2100 gal) oil per year (37).

Energy-storage systems provide economies for both utilities and customers. Storage systems have lower capital costs than cycling-coal units. In addition, utilities benefit from the reduction of reserve generating capacity. The installation of energy-storage systems at the customers' end would reduce the energy cost considerably.

Principal energy-storage media are water, ie, pumped hydroelectric energy (pumped-hydro), compressed air, heat, and chemical batteries. Until recently, the only economic storage option available to electric utilities was the conventional pumped-hydro in which, during low demand periods, water is pumped to large storage areas at higher elevations. During peak demand periods, the water is released to fall to the original elevation and, in the process, it operates a hydraulic turbine that drives an electric generator. This method has been in use since 1929. About 35 such systems, with a generating capacity of more than 25 MW, are either in operation or under construction in the United States. Because the sites appropriate for such systems are limited, research and development is currently directed to other options, such as underground pumped-hydro, compressed-air storage, and batteries.

Compressed-air energy storage (CAES) has several advantages over pumped-hydro, including a wider choice of geologic formations and smaller minimum capacity. In this system, air is compressed during off-peak demand periods, stored in underground caverns, and then withdrawn and used with turbine systems to generate electric power during peak demand periods. The world's first commercial compressed-air storage facility at Huntorf, FRG (see Fig. 16) (38) has a capacity of 290 MW. This plant began operation in December 1978, with two 140 km^3 salt caverns (38). The main advantage of the CAES system is that the two principal operations, ie, compressing the air and driving the turbine, take place at different times of day. In a conventional gas-combustion turbine, about two-thirds of the output is needed to drive the compressor and only one-third can be converted into electricity. At the Huntorf plant, the entire output of the plant (290 MW) can be used for power generation.

Another potentially attractive form of energy storage for peak use by electric utilities is thermal-energy storage. In this system, thermal energy is extracted as steam from a variety of electric generating plants, or water is extracted from the boiler–feedwater return path. Storage media, for example, can be hot rock, oil, or hot water. The latter can be stored under pressure. The stored thermal energy can then be converted to steam by heat exchanges or flash evaporators for conversion to electricity.

Yet another advanced technology, batteries (qv), can be utilized for daily cycling.

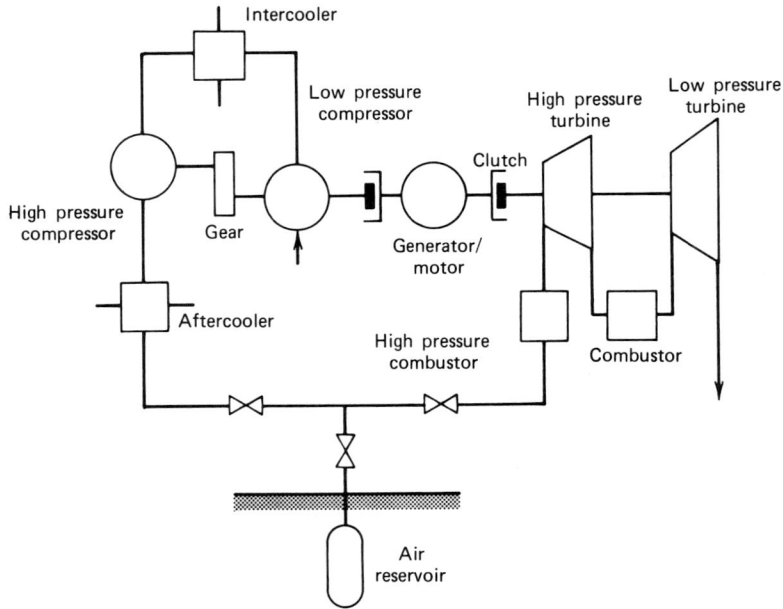

Figure 16. Schematic of the Huntorf CAES system.

During periods of low electricity demand, high voltage ac is converted into lower voltage d-c power for storage in the batteries. Batteries have the potential advantage of ease of siting, operating flexibility, and short installation lead time. Although lead–acid batteries have been in use for several years, the conventional lead–acid battery is too expensive for large-scale application to modern power systems. A generation of advanced batteries, with potentially lower initial costs and longer service life, are being developed.

In view of the need to find substitutes for oil and gas to meet peak electric loads, most utilities eventually will adopt an integrated energy management system that incorporates both system-level and user energy storage. Technologically, energy-storage development is proceeding rapidly. Plans will be announced soon by the utilities for the construction of underground pumped-hydro or compressed-air storage systems. Utility-level battery storage systems could come on-line in the late 1980s. Equipment to provide storage for residential- and commercial-water and -space heating is already in widespread use in Europe. Regulatory strategies, pricing, and tax policies could increase the attractiveness and importance of energy storage and help to promote its application.

Current Electric Power Generation and Projections

In 1981, the North American Electric Reliability Council (see Fig. 17) completed its "11th Annual Review of Overall Reliability and Adequacy of the North American Bulk Power Systems" (39). The total electric-utility power production of North America was reviewed by principal energy sources for the years 1971 and 1980, and was projected to the year 1989 (40). The trend toward greatly increased consumption of coal and nuclear power is shown in Figure 18 (40). Power generation rose by 43% between 1971 and 1980, and a similar increase is projected for the next decade.

Figure 17. The North American Electric Reliability Council, NERC. ECAR, East Central Area Reliability Coordination Agreement; ERCOT, Electric Reliability Council of Texas; MAAC, Mid-Atlantic Area Council; MAIN, Mid-America Interpool Network; MARCA, Mid-Continent Area Reliability Coordination Agreement; NPPC, Northeast Power Coordinating Council; SERC, Southeastern Electric Reliability Council; SPP, Southwest Power Pool; WSCC, Western Systems Coordinating Council.

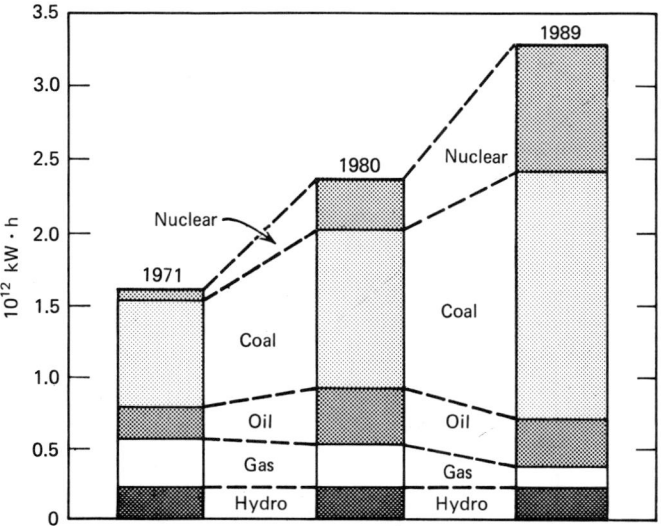

Figure 18. Electric power production by principal energy sources, contiguous U.S., 1971–1989. Courtesy of North American Electric Reliability Council.

Table 3. Electric Power Generation, % by Principal Energy Sources[a]

Area[b]	Coal		Nuclear		Oil		Gas		Hydro		Other	
	1981	1990	1981	1990	1981	1990	1981	1990	1981	1990	1981	1990
NERC	51.2	52.8	13.2	25.9	10.7	8.1	14.0	4.9	10.8[c]	8.8[c]		
ECAR	90.4	80.2	5.3	16.3	4.2[c]	3.5[c]						
ERCOT	34.2	47.8		15.0	0.6[c]	8.1[c]	65.2	29.0				
MAAC	55.7	47.0	24.0	42.0	12.7	8.7	5.4	1.1	2.1[c]	1.3[c]		
MAIN	70.9	55.3	21.8	41.1	4.9	2.5						
MARCA	65.6	73.8	21.5	16.2	0.7	0.9			12.3[c]	9.1[c]	2.4	1.1
NPCC	13.1	29.8	24.9	37.6	43.0	20.7	4.7		14.3[c]	11.9[c]		
SERC	56.9	49.8	21.8	37.2	11.7	8.0			9.7[c]	4.9[c]		
SPP	35.4	56.6	4.3	13.7	1.7	8.8	56.3	19.3	2.2[c]	1.6[c]		
WSCC	29.7	37.6	5.7	16.7	13.6	10.6	14.1	3.2	33.5	25.3	3.4	6.6

[a] Compiled from ref. 39.
[b] See Figure 17 for explanation of abbreviations.
[c] Includes other sources.

The North American Electric Reliability Council (NERC) was formed by the electric-utility industry in 1968 to promote the reliability and adequacy of bulk power supply in the electric-utility systems of North America. It consists of nine Regional Reliability Councils encompassing essentially all power systems in the United States and most Canadian systems.

The principal energy sources for electric power generation for the nine NERC regions are given in Table 3.

With the worldwide shortage of oil and gas, the outlook for near-term power production would be bleak indeed were it not for the abundant sources of coal and uranium which could satisfy U.S. energy needs for centuries to come. Should these ever dwindle, solar and fusion power would offer a limitless supply (see Fuels, survey).

BIBLIOGRAPHY

Power Generation" in *ECT* 1st ed., Vol. 11, pp. 65–87, by R. A. Budenholzer, Illinois Institute of Technology; "Power Generation" in *ECT* 2nd ed., Vol. 16, pp. 436–469, by R. A. Budenholzer, American Power Conference, Illinois Institute of Technology.

1. *Geothermal Energy*, U.S. Energy Research and Development Administration, Washington, D.C., April 1975.
2. L. M. Pruce, *Power* **123,** 37 (Oct. 1979).
3. *Geothermal Energy and Our Environment*, U.S. DOE, Washington, D.C., 1980.
4. D. R. Butler, *Proc. 3rd Annual Geothermal Conference and Workshop*, Electric Power Research Institute, Palo Alto, Calif., Oct. 1979, pp. 5-1 to 5-2.
5. D. G. Elliott, *Proc. 3rd Annual Geothermal Conference and Workshop*, Electric Power Research Institute, Palo Alto, Calif., Oct. 1979, pp 6-6 to 6-10.
6. G. P. Kuiper, *The Sun*, Vol. I. of *The Solar System*, University of Chicago Press, Chicago, Ill., 1953.
7. J. C. Denton, *Energy Convers.* **16,** 181 (1977).
8. M. K. Selcuk, *Survey of Several Central Receiver Solar Thermal Power Plant Design Concepts*, Internal Report No. 900-714, Jet Propulsion Laboratory, Pasadena, Calif., Aug. 1975.
9. *Highlights of the Central Receiver Pilot Plant and Test Facility Projects*, The Aerospace Corporation, El Segundo, Calif., Nov. 1975.
10. R. Manvi and T. Fujita, *Proc. 13th Intersociety Energy Conversion Engineering Conference*, San Diego, Calif., Vol. II, Aug. 1978, pp. 1535–1540.
11. V. A. Burns, *Sol. Energy* **1,** 2 (1957).
12. *Solar Thermal Conversion Workshop Proceedings*, University of Maryland, College Park, Md., 1974.
13. *Solar Thermal Conversion Mission Analysis*, Vol. 1, The Aerospace Corporation, El Segundo, Calif., 1975.
14. C. Starr, *paper presented at AAAS Annual Meeting*, Denver, Colo., 1977.
15. L. M. Pruce, *Power* **123**(8), 84 (Aug. 1979).
16. D. K. McLaughlin and W. L. Hughes, "Wind Power" in *Marks' Standard Handbook for Mechanical Engineers*, McGraw-Hill, New York, 1978.
17. C. Glasgow and G. Birchenough in ref. 10, pp. 2052–2059.
18. T. R. Richards and H. E. Noustadter in ref. 10, pp. 2060–2063.
19. R. Thomas, R. Puthoff, and J. Savino, *Paper presented at IEEE/ASME Meeting*, Portland, Ore., Sept. 1975.
20. L. A. Kilar, *Power* **123**(5), 40 (May 1979).
21. *Coal and Nuclear Generating Costs*, EPRI PS-455, Palo Alto, Calif., April 1977.
22. W. Marshall, *Chem. Br.* **17**(10), 466 (1981).
23. *Staff Report on Retired Hydroelectric Plants in the United States*, FERC, Washington, D.C., Dec. 1980.
24. T. G. Benjamin, E. H. Camara, and L. G. Marianowski, *Handbook of Fuel Cell Performance*, Institute of Gas Technology, Chicago, Ill., May 1980.
25. L. K. Kilar, *Power* **123**(5), 37 (May 1979).

26. W. J. Lueckel, L. G. Eklund, and S. H. Law, *Paper presented at IEEE Winter Meeting*, New York, Jan. 1972.
27. M. Axford, *Turbomachinery International* **21**(6), 17 (July/Aug. 1980).
28. R. C. Sheldon and T. D. McKone, *Proc. Am. Power Conf.* **24,** 350 (1962).
29. *Chem. Week*, 23 (Sept. 30, 1981).
30. *Electr. World* **192**(10), 29 (Nov. 1979).
31. P. E. Graybeal and A. H. Manchester, *Chemtech*, 48 (Jan. 1982).
32. J. M. Koracik, *Am. Power Conf.* **41,** 764 (1979).
33. *Evaluation of Phase 2 Conceptional Designs and Implementation Assessment Resulting from the Energy Conversion Alternatives Study (ECAS)*, NASA TM X-73515, April 1977.
34. R. V. Shanklin and E. Levi in ref. 10, pp. 1242–1247.
35. *EPRI Journal* **5,** 21 (Apr. 1980).
36. A. B. Cambel and co-workers, *Energy R&D and National Progress*, U.S. Govt. Printing Office, Washington, D.C., 1964.
37. Ref. 35, pp. 6–13.
38. *Compressed Air Mag.* **85,** 5 (Aug. 1980).
39. *11th Annual Review of Overall Reliability and Adequacy of The North American Bulk Power Systems*, NERC, Princeton, N.J., Aug. 1981.
40. *10th Annual Review of Overall Reliability and Adequacy of the North American Bulk Power Systems*, NERC, Princeton, N.J., Aug. 1980.

General References

P. Kruger and C. Otte, *Geothermal Energy*, Stanford University Press, Stanford, Calif., 1973.
A. B. Minel and M. P. Minel, *Applied Solar Energy*, Addison-Wesley Publishing Co., Inc., Reading, Mass., 1976.
Solar Cells, National Academy of Science, Washington, D.C., 1972.
P. C. Putnam, *Power From the Wind*, D. Van Nostrand Rheinhold Co., New York, 1948.
A. McDougall, *Fuel Cells*, John Wiley & Sons, Inc., New York, 1976.
A. L. Hammond, W. D. Metz, and T. H. Maugh II, *Energy and the Future*, AAAS, Washington, D.C., 1973.
J. A. Booth and D. H. Hall, *Turbomachinery International* **22**(4), 13 (Apr. 1981).
A. W. Culp, Jr., *Principles of Energy Conversion*, McGraw-Hill Inc., New York, 1979.
Energy Conversion, American Nuclear Society, La Grange, Ill., 1978.
R. A. Knief, *Nuclear Energy Technology*, Hemisphere Publishing Corp., Washington, D.C., 1981.
R. L. Foster and R. L. Wright, Jr., *Basic Nuclear Engineering*, 2nd ed., Allyn and Bacon, Inc., Boston, Mass., 1973.
J. W. Sawyer, *Gas Turbine Engineering Handbook*, 2nd ed., Vol. 1, *Theory and Design;* Vol. II, *Applications;* Vol. III, *Maintenance-Basic Fundamentals*, Gas Turbine Publications, Stamford, Conn., 1976.
U.S. Department of Energy, *Proceedings of the Department of Energy Advanced Gas Turbine Central Power Systems Workshop*, CONF 8004103, Washington, D.C., April 30–May 2, 1980.

SUSHIL K. BATRA
GIMRET International, Inc.

R. A. BUDENHOLZER
Illinois Institute of Technology

PRASEODYMIUM. See Rare-earth elements.

PRESERVATIVES. See Antioxidants and antiozonants; Coatings; Food additives; Paint; Wood.

PRESSURE MEASUREMENT

Measurement of pressure in the chemical industries and laboratories is of interest for a variety of reasons: differential pressure is the driving force in fluid dynamics; pressure is one of the fundamental terms in the Ideal Gas law and its corollaries and is one of the determining factors in vapor–liquid equilibria; and pressure is a safety consideration in the operation of process equipment (see High pressure technology; Vacuum technology).

Units of Measurement

Pressure is defined as force per unit of area. It can be expressed in a wide variety of units. Conversion factors from non-SI units to the SI pressure unit, the pascal (Pa), are given in Table 1. The pascal is defined as 1 N/m^2 (see also Units and conversion factors; Front matter to this volume). Relationships that are exact in terms of base units are followed by an asterisk. Relationships that are not followed by an asterisk are either the results of physical measurements or are only approximate. The factors are written as numbers greater than one and less than 10 with six or less decimal places. Each number is followed by the letter E (exponent), a plus or minus symbol, and two digits that indicate the power of 10 by which the number must be multiplied to obtain the correct value. For example,

$$3.523\ 907\ \text{E}{-}02 = 3.532\ 907 \times 10^{-2} = 0.035\ 239\ 07$$

Table 1. Conversion of Pressure Units to the SI Unit[a]

To convert from	To	Multiply by[b]
atmosphere (normal = 760 torr)	pascal (Pa)	1.013 25 E+05
atmosphere (technical = 1 kgf/cm^2)	pascal (Pa)	9.806 650*E+04
bar	pascal (Pa)	1.000 000*E+05
centimeter of mercury (0°C)	pascal (Pa)	1.333 22 E+03
centimeter of water (4°C)	pascal (Pa)	9.806 38 E+01
decibar	pascal (Pa)	1.000 000*E+04
dyne/$centimeter^2$	pascal (Pa)	1.000 000*E−01
foot of water (39.2°F)	pascal (Pa)	2.988 98 E+03
gram-force/$centimeter^2$	pascal (Pa)	9.806 650*E+01
inch of mercury (32°F)	pascal (Pa)	3.386 389 E+03
inch of mercury (60°F)	pascal (Pa)	3.376 85 E+03
inch of water (39.2°F)	pascal (Pa)	2.490 82 E+02
inch of water (60°F)	pascal (Pa)	2.488 4 E+02
kilogram-force/$centimeter^2$	pascal (Pa)	9.806 650*E+04
kilogram-force/$meter^2$	pascal (Pa)	9.806 650*E+00
kilogram-force/$millimeter^2$	pascal (Pa)	9.806 650*E+06
kip/$inch^2$ (ksi)	pascal (Pa)	6.894 757 E+06
millibar	pascal (Pa)	1.000 000*E+02
millimeter of mercury (0°C)	pascal (Pa)	1.333 224 E+02
poundal/$foot^2$	pascal (Pa)	1.488 164 E+00
pound-force/$foot^2$	pascal (Pa)	4.788 026 E+01
pound-force/$inch^2$ (psi)	pascal (Pa)	6.894 757 E+03
torr (mm Hg, 0°C)	pascal (Pa)	1.333 22 E+02

[a] Ref. 1.
[b] E = exponent, base ten (eg, E + 02 = 10^2). Asterisk means conversion is exact.

96 PRESSURE MEASUREMENT

An asterisk (*) after the sixth decimal place indicates that the conversion factor is exact and that all subsequent digits are zero. All other conversion factors have been rounded to the figures given in accordance with standard practice for rounding numbers (1).

Terminology

Atmospheric or barometric pressure is the pressure exerted by the column of air on the earth's surface. It varies with place, time, elevation, and weather conditions. These variables have been eliminated by establishing a normal atmosphere of 101,325 Pa. This is equal to the pressure exerted by a column of mercury 760 mm high at a temperature of 0°C, or 29.921 in. Hg, or 14.696 psi.

Absolute pressure is pressure measured from zero pressure. However, pressure gauges frequently measure from atmospheric pressure. Positive pressure measured in this manner is called gauge pressure. Gauge pressure is equal to the absolute pressure minus the atmospheric pressure. It has often been reported specifically, ie, $lbf/in.^2$ gauge, or psig. The expression psia has often been used to emphasize that the measurement is absolute pressure. In SI units, no official provision has been made to differentiate between gauge or absolute pressure. If the context does not clarify which pressure is meant, then a statement should be included to identify the pressure. A gauge that measures vacuum pressure reads the amount by which the pressure is less than atmospheric pressure. A compound gauge measures a pressure range that is above and below atmospheric pressure, ie, both gauge pressure and vacuum on the same scale.

Differential pressure is the difference between two pressures, neither one of which may be zero absolute or zero gauge pressure. Differential pressure often is measured to determine the quantity of flow in a pipe into which an artificial restriction has been placed.

Pressure-Measuring Devices

Pressure or vacuum are generally measured by directly actuated mechanical elements, particularly where they are monitored as opposed to being controlled. However, various electronic measuring devices continue to be introduced and are useful in certain areas. The mechanical elements are reliable and inexpensive. Liquid-filled indicators are used where vibration, pulsation, or atmospheric corrosion are a problem. Designs with blow-out backs are utilized for safety (2). Plastic mechanical movements on indicators exhibit good wear and are light weight and corrosion resistant. The electronic devices are available in a large variety of forms. A broad choice of options, costs, and operating characteristics is offered. There is to date little standardization or interchangeability. In plants where economies can be effected by extensive automatic instrumentation, and where precision of control, automatic data processing, or quick analysis of operations are important, electronic instrumentation meets these needs (see Instrumentation and control).

Diaphragm. A diaphragm element is a thin flexible disk upon which pressure acts to create a force. The force causes a deflection of the diaphragm which, in turn, serves to move an indicator pointer, recorder pen, or other mechanism. A diaphragm can be either metallic or nonmetallic. A nonmetallic diaphragm is sometimes referred to as a slack diaphragm, since it has little rigidity. Because a slack diaphragm does not have a suitable pressure-deflection relationship by itself, it must be provided with a calibrated spring whose gradient determines the deflection for an applied pressure. Metallic diaphragms usually provide their own suitable pressure-deflection characteristics. Most diaphragms have circular convolutions. These make possible deflections with a proportional pressure-to-deflection relationship many times greater than those of flat plates (3). Diaphragm elements are sensitive to small pressure changes and, therefore, are particularly useful in the measurement of low pressures.

Inverted Bell. An inverted-bell pressure element consists of two inverted bells that are partly immersed in oil, which provides a liquid seal. The bells are suspended from the opposite ends of a balance beam and are arranged so that a pressure can be introduced under each bell. This arrangement weighs the minutest difference between the two pressure lines. One of these lines is open to atmospheric pressure and the other to the pressure to be measured. The bell with the higher pressure rises in the oil, and the beam tilts and moves a pointer on a scale and linkage to a controller (4–5).

This element is sensitive to within 0.1 Pa (0.0005 in. H_2O). It is used for measuring relatively low pressure and vacuum, eg, in furnace drafts. The ranges available are 0–0.5 kPa (0–2 in. H_2O) to 0–0.05 kPa (0–0.2 in. H_2O) pressure or vacuum.

Diaphragm-Capsule. The diaphragm-capsule pressure element is made up of two or more circular metal diaphragms that are welded at the inner and outer edges. Figure 1 shows a typical assembly. The fabricated unit becomes a flexible sac or container, which is sealed at one end and open to a connecting tube at the other end for the pressure connection. The diameter and number of diaphragms used to make the complete unit depend on the pressure range and amount of desired deflection. The process pressure is applied to the inside of the capsule through the connecting tube; it expands the capsule with a resulting movement at the closed end. The actuating

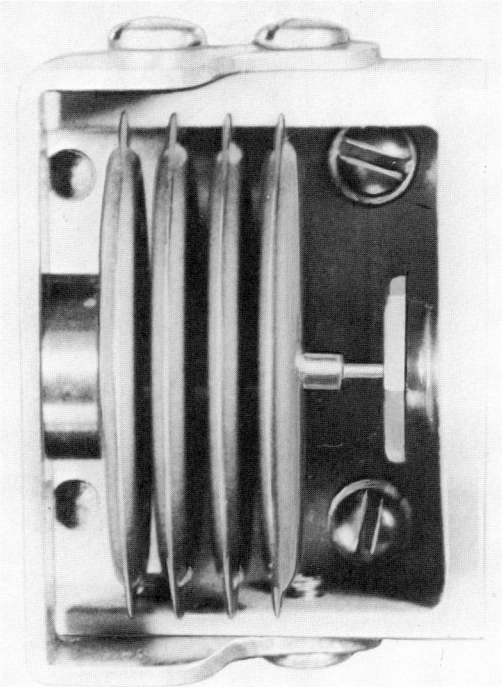

Figure 1. Diaphragm-capsule pressure element.

98 PRESSURE MEASUREMENT

linkage for the pointer or pen arm is attached to this closed end. Diaphragm capsules are used for pressure ranges of 0–2.5 kPa (0–10 in. H$_2$O) to 0–690 kPa (0–100 psi). The materials may be phosphor bronze, stainless steel, or alloys of any type.

Bourdon Tube. A Bourdon tube is made from a flattened or elliptical tube, with one end sealed and the other open to the process pressure through connecting tubing. The final shape of the tube and the amount of flatness determine the trade name of the element and identify the overall shape and form. There are spiral-, helix-, and C-type Bourdon tubes.

A spiral and a helix are shown in Figures 2 and 3, respectively. These elements are made from a thin-wall tube, which is flattened to produce a long, narrow, elliptical cross section. It is then formed into a spiral or helix as illustrated. When the process pressure is applied through the connecting tube, the resulting force tends to uncoil or straighten the tubing. The rotating motion of the spiral or helix through a suitable linkage arrangement can be used to actuate a pointer or pen arm. The spiral is normally used for pressure ranges of 0–138 kPa to 0–27.6 MPa (0–20 to 0–4000 psi) and the helix

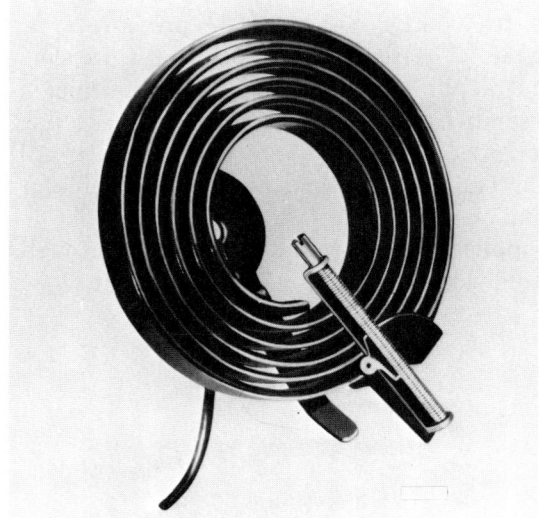

Figure 2. Spiral pressure element.

Figure 3. Helix pressure element.

for ranges of 0–690 kPa to 0–690 MPa (0–100 to 0–100,000 psi). The element material can be bronze, steel, stainless steel, or a special alloy.

Figure 4 shows a C-type Bourdon tube. This also is made from a thin-wall tube, which can be flattened a small or a large amount, depending on the material and the pressure range. The tubing is formed into a C shape, with one end closed and free to move, and the other end fixed and open to a connecting tube to the pressure. The force from the applied pressure tends to straighten the tube and, thus, produces tip travel. A suitable linkage transfers this tip travel to a pointer or pen arm. These elements are used for pressures of 0–103 kPa to 0–69 MPa (0–15 to 0–10,000 psi). Bronze, steel, stainless steel, or special alloys can be used.

Spring-and-Bellows. Figure 5 shows a cross section of a spring-and-bellows pressure element. The bellows is formed from a length of thin-wall tubing by hydraulic

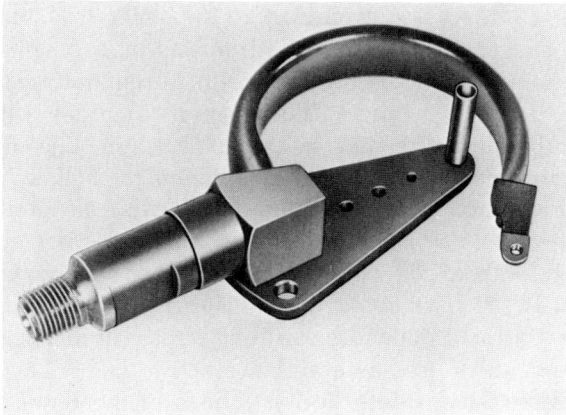

Figure 4. C-type Bourdon-tube pressure element.

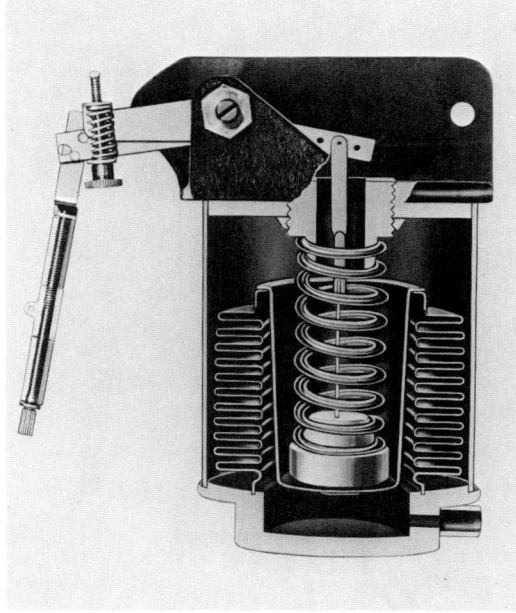

Figure 5. Spring-and-bellows pressure element.

extrusion in a die. The bellows is enclosed in a metal shell connected by tubing to the source of pressure. A compression spring is mounted inside and rests against the bottom of the bellows. It is restrained at the top by a formfitted nut. A rod resting on the bottom of the bellows transmits any vertical motion of the bellows through a suitable linkage arrangement to a pointer or pen readout. As the pressure inside the metal shell increases, the bellows moves vertically and compresses the spring. The bellows spring gradient is small compared to the gradient of the spring itself and, therefore, the pressure range is a function primarily of the spring gradient. A spring-and-bellows pressure element can be used at pressures from ca 0–1.25 kPa (0–5 in. H_2O) to 0–350 kPa (0–50 psi). The lower pressures require bellows of larger diameters than the higher pressures. The bellows is usually made of phosphor bronze or brass but it also can be supplied in many other metals.

Absolute-Pressure Gauge. When industrial-process, low vacuum measurements are required from 0–13.3 kPa (0–100 mm Hg) to 101.3 kPa (0–30 in. Hg abs), it is often necessary to compensate for the normal variations in atmospheric or barometric pressure. Figure 6 shows a spring-and-bellows element that automatically compensates for the barometric pressure changes. The element includes a double bellows arrangement with both bellows fixed at the top and bottom. The adjacent end of each bellows is attached to a movable plate, which transmits the bellows movement through a suitable linkage to a pointer or pen. The upper bellows is evacuated to a nearly perfect vacuum, ie, absolute zero, and is sealed. The process vacuum is applied to the lower bellows, which then tends to collapse or close the lower bellows, thereby moving the center plate down. If the barometric or atmospheric pressure changes, the upper bellows expands or contracts, depending on any decrease or increase in the barometer. The bellows can be made of brass or stainless steel.

Differential-Pressure Gauge (Meter Body). The formed bellows, diaphragm capsule, and single diaphragm are used in differential-pressure meter bodies. These units can be used to measure differences in pressure between two pipes, two stills, etc, from 25

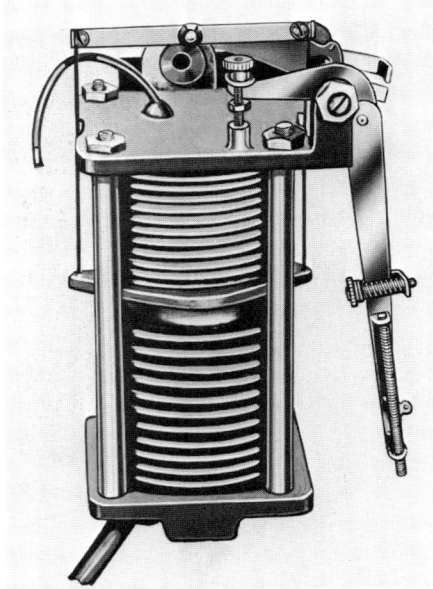

Figure 6. Absolute-pressure-gauge element, bellows type.

Pa (0.1 in. H₂O) to 0–4825 kPa (0–700 psi) and with operating pressures as great as 69 MPa (10,000 psi). Figure 7 illustrates a bellows-actuated meter body. The high pressure and low pressure bellows are joined by the center-stem assembly. The entire volume inside the bellows is filled with liquid, and the bellows is sealed. When the process pressure at the high pressure tap is greater than the process pressure at the low pressure tap, the high pressure bellows moves to the right and, by means of the center stem and liquid filling, moves the low pressure bellows to the right. Motion stops when the force on the range spring equals the force of the differential pressure, ie, the difference between the high and low process pressures. The cable and motion take-off arm translate the center-stem movement to the torque tube, and this is connected to a linkage mechanism for positioning of the pointer or pen arm.

Piezoelectric. Designs of piezoelectric pressure elements are based on the principle that a piece of quartz, when properly cut and oriented with respect to its crystallographic axes, generates a small electric charge on certain surfaces when stressed. In practice, a stack of properly cut quartz plates is mounted in a housing, which has a thin diaphragm at one end. The housing is usually designed to be mounted in the wall of a pressure vessel. The diaphragm is exposed to the pressure and deflects, thereby applying a compressive force to the quartz stack which, in turn, generates a charge directly proportional to the force. Such devices are available in pressure ranges between 0–350 kPa (0–50 psi) and 0–69 MPa (0–10,000 psi), although some are available in ranges as low as 0–69 Pa (0–.01 psi) and as high as 0–830 MPa (0–120,000 psi). Response time is generally very fast, with frequency response as high as 500 kHz. A typical design is shown in Figure 8.

Linear-Variable-Differential-Transformer (LVDT) and Reluctive Pressure Transducers. In a LVDT pressure transducer, the pressure to be measured is fed to a Bourdon tube or diaphragm. The motion of this element is transferred to the magnetic core of a transformer. With a-c excitation of the primary coil of the transformer, a varying voltage is produced in the secondary coil as the core is moved. The two secondary coils are wound in opposite directions. Therefore, when the core is centered between them, the voltages in them cancel each other. When the core is moved from center, a differential voltage is produced. This voltage represents the amount of motion of the Bourdon tube or diaphragm, which is proportional to the pressure (6).

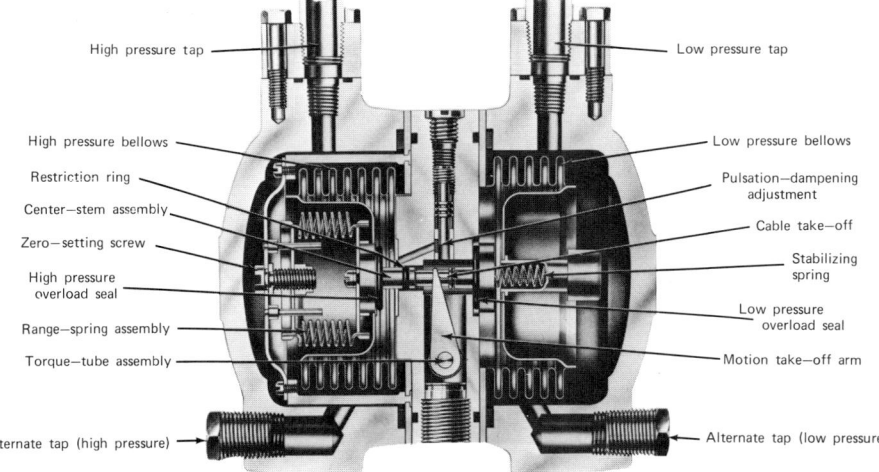

Figure 7. Differential-pressure-gauge element, bellows type.

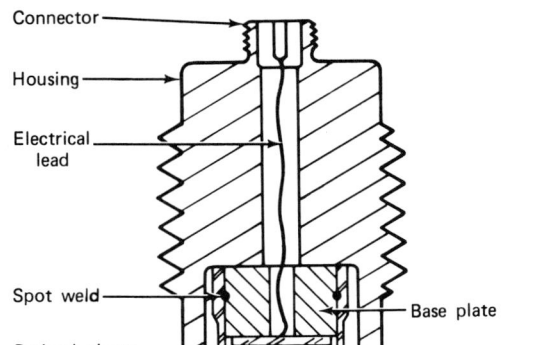

Figure 8. Piezoelectric pressure sensor.

Measurement by a reluctive transducer is based on the ratio of the reluctance of the magnetic flux path of two coils. In the diaphragm type shown in Figure 9, a diaphragm of magnetically permeable material is supported between two symmetrical E-core inductance assemblies and completes a magnetic circuit with each core. Application of pressure to the diaphragm causes it to deflect, thereby increasing the gap in the magnetic flux path of one core and decreasing the gap in the other an equal amount. Since the magnetic reluctance varies with the gap, the inductance ratio changes with the position of the diaphragm. This can be measured in a bridge circuit, which produces an output voltage proportional to the pressure. The pressure range can be changed easily in the field by replacing the diaphragm. Pressure ranges from 0–86 Pa (0–0.0125 psi) to 0–860 MPa (0–125,000 psi) are available (7).

Piezoresistive Sensors (Integrated-Circuit Sensors). Piezoresistive transducers convert a change in pressure into a change in resistance caused by strain. Pressure applied to one side of a silicon wafer strains resistors diffused into the wafer. The change in applied pressure causes a linear change in the resistance value, which can be converted and amplified to a usable output signal. In some designs, the transducer consists of a single silicon-crystal sensor and its supporting electronics mounted in one pressure-tight housing on a ceramic substrate. If the transducer is to be used for

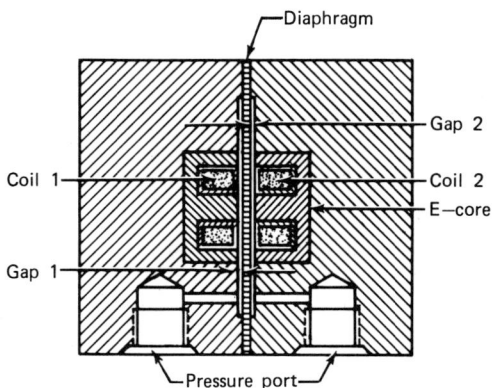

Figure 9. Reluctance-type pressure sensor. Courtesy of Validyne Engineering Corp.

a corrosive or conductive process fluid, the housing is filled with a silicone oil. An elastic sock isolates the oil from the process fluid while allowing pressure to be applied to the oil.

The piezoresistive sensor is very reliable for use under conditions of severe shock and vibration. It has high natural frequencies of greater than 50 kHz (8). Pressure can be measured from 70–35,000 kPa (10–5,000 psi). Operating temperatures are −1 to 85°C.

In other designs, a diffused silicon sensor, such as is pictured in Figure 10, is mounted in a meter body which is designed to permit calibration, convenient installation in pressure systems and electrical circuits, protection against overload, protection from weather, and isolation from corrosive or conductive process fluids and, in some cases, to meet standard requirements, eg, of Underwriters' Laboratory. Figures 11 and 12 show a typical process-pressure meter body and transmitter. Pressure measurement from 0–750 Pa (0–3 in. H$_2$O) to 0–70 MPa (0–10,000 psi) is available for process temperatures of −40 to 107°C. Differential pressure- and absolute pressure-measuring meter bodies are also available. As transmitters, the output of these devices typically is 4–20 mA dc with 25 V d-c supply voltage.

Strain-Gauge Sensors. The distinction between strain-gauge sensors and the piezoresistive sensors is minor, since both function by measuring the strain on a flexible member as it is subjected to pressure. Primarily, strain-gauge sensors are transducers with separate pressure elements and gauges. These include unbonded metal-wire

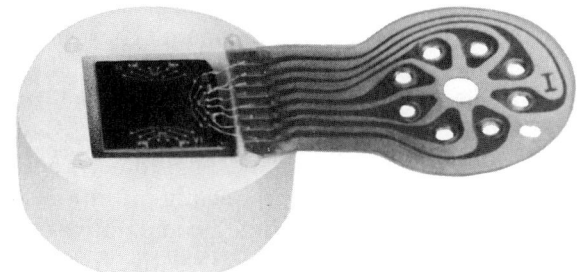

Figure 10. Diffused-silicon pressure sensor (piezoresistive).

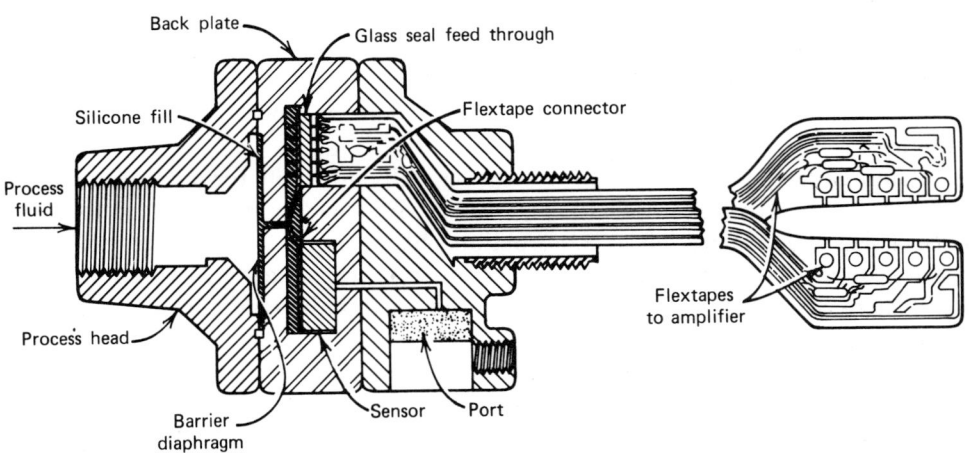

Figure 11. Cross-section of meter body with diffused-silicon sensor for pressure measurement.

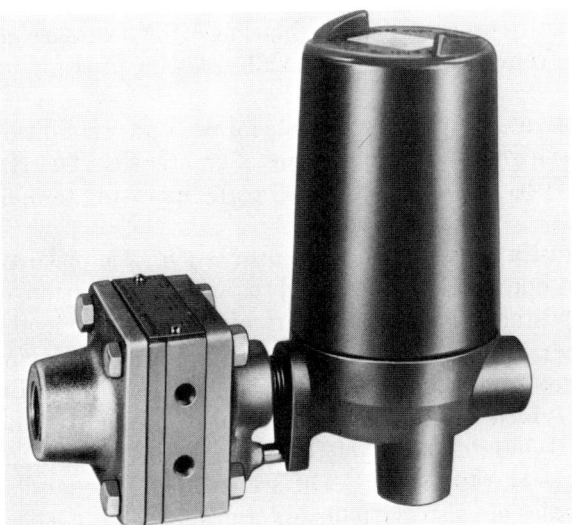

Figure 12. Pressure transmitter with diffused-silicon sensor.

gauges, bonded metal-wire gauges, bonded metal-foil gauges, and bonded semiconductor gauges. Unbonded metal-wire elements consist of wire that is stretched unsupported between a fixed and a moving or stressed end. The wire may be looped several times between end posts. This type has high sensitivity, but is also sensitive to environmental vibration. Bonded strain gauges are attached by a permanent bond for their entire length and width to the elastic element. These are less sensitive to vibration. Bonded semiconductor gauges have larger outputs than the wire types (9).

All of the preceding gauges convert a change in pressure into a change in resistance. The resistors are usually arrayed in four arms of a Wheatstone bridge. The most common wire materials used are platinum and Nichrome. The advantages of the strain-gauge transducer are fast response, practically infinite resolution, minimum motion of internal mechanical parts, high accuracy, comparative ease of compensation for temperature effects, low source impedance, and relative freedom from acceleration effects. The disadvantages include the difficulty of obtaining zero output at zero pressure, ie, bridge unbalance, relatively high vibration error at low ranges (<100 kPa), low output levels, necessity for isolating excitation ground from output ground, and signal conditioning requirements, which include zero nulling and calibration. With application of 10 V and with compensating and adjusting resistors in place, the transducer output is ca 20–30 mV if it is a bonded gauge, and 30–40 mV if it is unbonded.

Strain-gauge pressure transducers are typically available for measurements of 0–35 kPa (0–5 psi) to 0–350 MPa (0–50,000 psi) with a wide variety of options in the choice of size, shape, materials of construction, output range, temperature limits, and pressure connections. A typical strain-gauge pressure transducer is pictured in Figure 13, but there is almost no standardization in the designs and configurations available from the various manufacturers.

Potentiometric Pressure Transducers. Pressure is fed into a mechanical element, eg, a Bourdon tube or diaphragm capsule, which deflects with applied pressure and moves a slider or wiper across a resistance element. The resistor is either wire-wound, carbon ribbon, or deposited conductive film. The motion of the slider results in a

Figure 13. Strain-gauge pressure sensors.

change in resistance, which may be either linear or one of a variety of characteristic curves, depending on the design of the resistance element. Certain available designs isolate the process fluid that is being measured from the pressure element by means of an inert isolation fluid and a diaphragm.

The advantageous features of these devices are that they are inexpensive, a relatively high output is available, they can be used with either a-c or d-c excitation, often no amplification or impedance matching is necessary, a very wide variety of functions is available, and they lend themselves to rugged construction. The potential disadvantages are the finite resolution of the wire-wound types, mechanical friction, limited life due to wear of wire or slider, development of noise with wear, large displacement, and a relatively low frequency response. Pressure ranges of 0–35 kPa (0–5 psi) to 0–70 MPa (0–10,000 psi) are generally available, and specialized devices for measurements of 0–250 Pa (0–1 in. water) also are available.

Capacitive Pressure Transducers. Capacitive sensors generally have a deflectable metal diaphragm positioned between two cavities, each of which contains a stationary metal plate that is positioned parallel to the diaphragm. Pressure deflects the diaphragm, which changes the capacitive coupling between it and the two plates. The change can be sensed either by passing an a-c signal across the plates or by using the change to alter the frequency of an oscillator. These devices are generally small, and have excellent frequency response, good linearity, and good resolution. They also are extremely sensitive and are capable of a wide range of measurement in a single sensor/readout combination. Capacitance manometers can be used as secondary standards for calibrating other gauges. Ranges of 10^{-5} Pa (10^{-7} mm Hg) to 35 MPa (5000 psi) are available. Capacitive sensors are also available that are mounted in meter bodies designed to operate as transmitters of 4–20 mA dc or 10–50 mA dc.

Vacuum Measurement. Weak vacuums or negative pressures can be measured by most of the pressure sensors described above. These negative pressures typically occur where the process itself produces a vacuum, such as in fractionating-column evaporators and the exhaust of a turbine. In cases where this vacuum must be known, sensors, eg, the diaphragm capsule and Bourdon-tube gauges, semiconductor devices, and piezoresistive, LVDT, and potentiometeric sensors can be used to measure it (10).

For measurement of high [$(0.13$–$1.3) \times 10^{-4}$ Pa (10^{-3}–10^{-6} mm Hg)], very high [1.3×10^{-4}–1.3×10^{-7} Pa (10^{-6}–10^{-9} mm Hg)], and ultra-high [$<1.3 \times 10^{-7}$ Pa ($<10^{-9}$ mm Hg)] vacuums, thermoelectric and ionization sensors can be used. The thermoelectric sensor operates on the principle that the heat loss from a hot wire varies as the pressure of the gas or vapor surrounding the hot wire varies. This variation in heat loss with pressure is relatively large in the high vacuum ranges for which it is used.

Figure 14 illustrates a resistance-bridge type of thermoelectric sensor where the heat lost by the coil of resistance wire in the measuring cell is indicated directly by a resistance change in the leg of the bridge circuit. The compensating cell contains a second coil of resistance wire that is sealed at a pressure well below 100 Pa (0.75 mm Hg). This coil is designed so that changes in resistance with changes in temperature balance those changes in the measuring cell resistance and, thus, automatically compensate for temperature variations.

Figure 15 shows the diagram of a thermocouple type of thermoelectric sensor. The filaments are continuously and uniformly heated by means of the constant-voltage regulator and transformer. There are two sections, one sealed under high vacuum and a second connected to the process pressure. A small, sensitive thermocouple is located on each of the filaments, and each pair of couples is connected in series to increase the generated emf. The two thermocouples in the reference chamber are connected in opposition to the two in the measuring chamber. Thus, their emfs oppose each other. This difference in emf is a measure of the difference in pressure between the measuring chamber and the reference chamber. This type of sensor is used in the high and medium vacuum ranges.

The ionization sensor is illustrated in Figure 16. The measurement is based on the ability of electrons emitted from a hot filament to bombard the molecules of the residual gas in an evacuated system, thereby generating a current from the resulting ions. The magnitude of the current flow is directly proportional to the number of ions

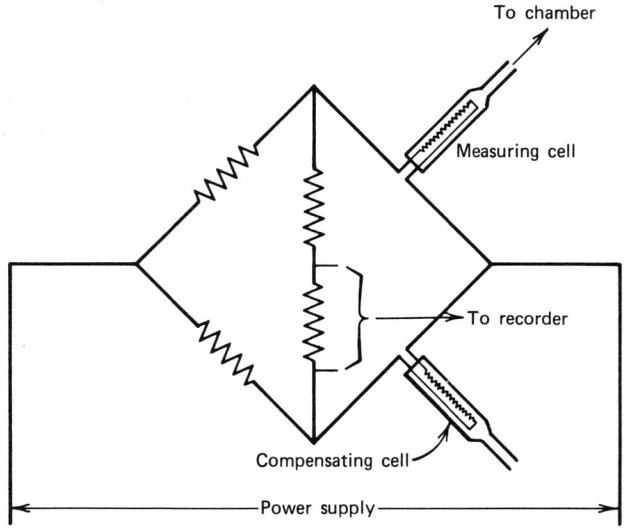

Figure 14. Circuit of resistance-type thermal gauge.

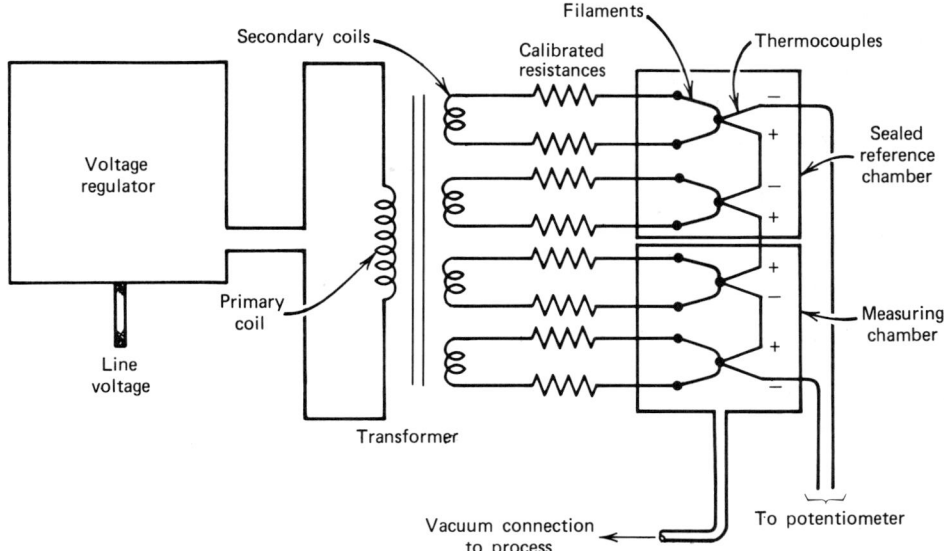

Figure 15. Circuit of thermocouple-type thermal gauge.

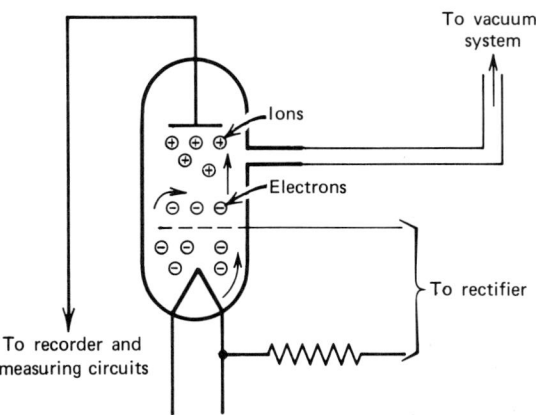

Figure 16. Ionization gauge.

formed. This is an indication of the amount of gas present, which is a measure of the degree of vacuum. The sensor is essentially a triode tube, and the electron emission from the cathode is held constant by a precision bridge circuit. The electrons are attracted to the grid, which is at a high positive potential with respect to the cathode, and the momentum of the electrons carries them past the grid to the plate. The plate is held at a negative potential with respect to the grid and repels the electrons, driving them among the molecules of the gas. This bombardment of the gas causes ions to form. Since a prior potential difference has been imposed, the ions are attracted to the plate, and the current flow is proportional to the number of molecules present, and the magnitude of the current flow is a measure of the vacuum. Ionization sensors are used in ultra-high and very high vacuums and must be protected against too high pressures, which would burn out the filaments.

For measurement to 1.3×10^{-5} Pa (10^{-7} mm Hg), a cold-cathode ionization gauge, sometimes called a Penning or Philips gauge, can be used. The principle of operation is the same as with the ionization gauges described earlier: positive ions are produced by loss of electrons and the current produced by these ions gives a measure of the pressure. However, in this case, the electrons originate at the cathodes instead of being produced by a hot filament. A magnetic field forces the electrons to move in spiral paths instead of traveling directly between cathode and anode, thereby causing formation of more ions and, thus, larger currents. Advantages of these devices are rugged construction and simple electronic circuitry. Their accuracy, however, is not comparable to the hot-cathode sensor, nor can they measure as high a vacuum; they are limited to 1.3×10^{-5} Pa (10^{-7} mm Hg) (11). Also, a high voltage is used, which can be hazardous.

Gauges to measure very low pressures have been developed in recent years, particularly in connection with space exploration, development of various electronic devices, research in solid-state physics, etc. They include modified Bayard-Alpert gauges, the cold-cathode inverted magnetron gauge, the hot-cathode magnetron ionization gauge, and mass spectrometers (12).

Modifications of the Bayard-Alpert gauge increase the length of the path of the electrons and result in more ionization and, therefore, in greater sensitivity. It is possible to measure pressures down to ca 0.13 nPa (ca 10^{-12} mm Hg) with these modifications. The cold-cathode magnetron gauge is a common cold-cathode gauge and is modified to increase the length of path of the electrons so as to obtain more ionization at low pressures. This gauge is pictured in Figure 17**a**. It is sometimes called the Redhead gauge and can measure pressures of 0.013–0.13 nPa (1–10^{-12} mm Hg). The hot-cathode magnetron ionization gauge is also modified from the basic thermionic ionization gauge to increase the path length of the electrons. This gauge is illustrated in Fig. 17**b**. In theory, this gauge should be able to measure to 1.3 pPa (10^{-14} mm Hg), but the inability of the external circuitry to measure the ion currents at this level restrict its use to somewhat higher pressures, ca 13 pPa (ca 10^{-13} mm Hg) (13). Certain

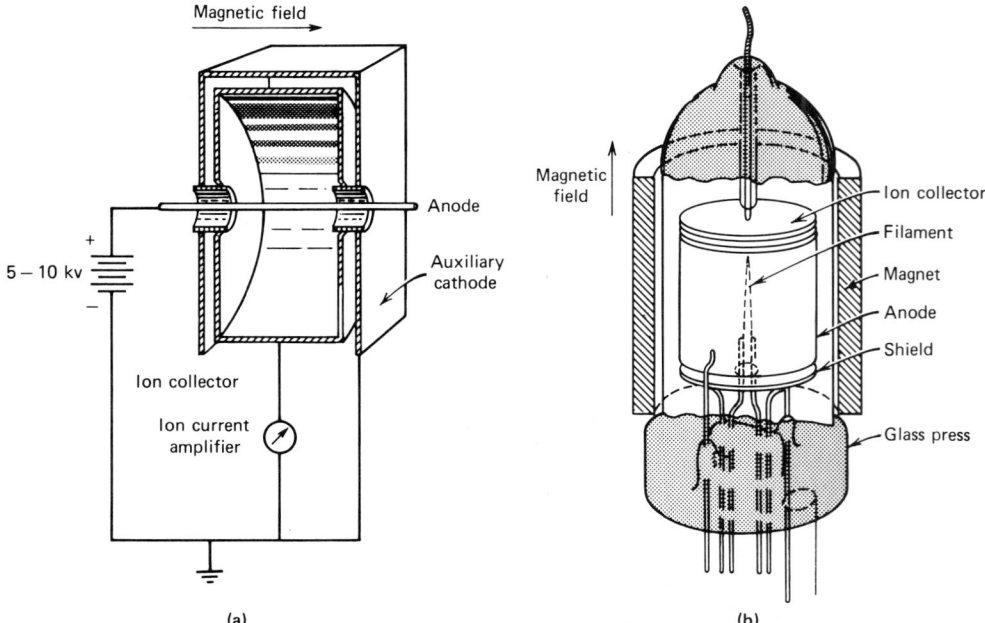

Figure 17. (**a**) Inverted magnetron gauge (Hobson and Redhead). (**b**) Hot-cathode magnetron ionization gauge (Lafferty).

types of mass spectrometers can be used to measure pressures as low as 13 pPa (10^{-13} mm Hg). With a secondary-emission electron multiplier cooled in liquid nitrogen, measurements to 0.13 pPa (10^{-15} mm Hg) are possible.

Other types of low pressure measuring devices coming into use are time-of-flight instruments and ion-resonance devices with designs based on r-f fields and strong-focusing quadrupole lenses (14).

Installation and Maintenance of Pressure Gauges

Industrial pressure gauges must be protected from excessive overload pressures, high process temperatures, and corrosive or solid-entrained fluids, which deteriorate or clog the pressure element. The instruction data supplied with the gauge should specify the precautions that must be taken for the given unit. Diaphragm seals isolate the pressure element from the process fluid, if necessary (15). Porous-plug snubbers in the line dampen pulsating pressures to prevent wear of the mechanical parts of some elements (16). Actual application, installation, and maintenance should be carried out in accordance with the manufacturer's recommendations. A detailed discussion of the design of pressure-gauge connections is given in ref. 17.

BIBLIOGRAPHY

"Pressure Measurement" in *ECT* 2nd ed., Vol. 16, pp. 470–481, by Charles F. Cusick, Honeywell, Inc.

1. *Metric Practice Guide*, ASTM Bulletin E 380-79, American Society for Testing and Materials, Philadelphia, Pa., 1979.
2. B. G. Liptak in B. G. Liptak, ed., *Instrument Engineers Handbook*, Vol. 1, Chilton Book Company, Philadelphia, Pa., 1969, p. 249.
3. H. E. Soisson, *Instrumentation in Industry*, John Wiley & Sons, Inc., New York, 1975, p. 72.
4. Ref. 3, p. 66.
5. Ref. 2, p. 222.
6. R. P. Benedict, *Fundamentals of Temperature, Pressure and Flow Measurements*, John Wiley & Sons, Inc., New York, 1969.
7. *Short Form Catalogue No. VI*, Validyne Engineering Corporation, Northridge, Calif., 1980.
8. J. Hall, *Instrum. Control Syst.*, 28 (Apr. 1979).
9. Ref. 2, p. 233.
10. J. Hall, *Instrum. Control Syst.*, 42 (Apr. 1980).
11. Ref. 10, p. 43.
12. A. Guthrie, *Vacuum Technology*, John Wiley & Sons, Inc., New York, 1963, p. 190.
13. Ref. 12, pp. 190–193.
14. *Measurements and Data.*, 104 (Jan.–Feb. 1975).
15. Ref. 2, pp. 256–262.
16. Ref. 3, p. 87.
17. J. A. Masek, *Chem. Eng.*, 71 (May 4, 1981).

General References

References 2, 3, 6, and 12 are general references.
G. C. Carroll, *Industrial Process Measuring Instruments*, McGraw-Hill Book Company, Inc., New York, 1962, pp. 117–145.
S. Dushman in J. M. Lafferty, ed., *Scientific Foundations of Vacuum Technique*, 2nd ed., John Wiley & Sons, Inc., New York, 1962, pp. 258–370.
D. P. Eckman, *Industrial Instrumentation*, John Wiley & Sons, Inc., New York, 1950, pp. 216–246.
S. M. Elonka and A. R. Parsons, "Measuring Systems," *Standard Instrumentation Questions and Answers*, Vol. I, McGraw-Hill Book Company, Inc., 1962, pp. 87–116.

110   PRESSURE MEASUREMENT

A. E. Fribance, *Industrial Instrumentation Fundamentals*, McGraw-Hill Book Company, Inc., New York, 1962, pp. 187–297.

J. H. Morrison in D. M. Considine and S. D. Ross, eds., *Handbook of Applied Instrumentation*, McGraw-Hill Book Company, Inc., New York, 1964, pp. 4-10 to 4-23.

Note: In addition, all of the main instrument manufacturers have available descriptive and instructive data on their pressure gauges and transducers.

C. RAYMOND BRANDT
Honeywell Inc.

PRESSURE VESSELS.   See High pressure technology.

PRIMING COMPOSITIONS.   See Explosives and propellants.

PRINTING INK.   See Inks.

PRINTING PROCESSES

Prepress: the preparatory stages,   112
Process photography,   113
Photochemical methods of platemaking,   128
Printing,   153
Finishing and binding,   159
Comparison of the three major printing processes,   160
Economic aspects,   160
Environmental considerations,   161

The invention of printing in the western world is credited to Johannes Gutenberg in ca 1450 AD; however, movable type was used to print on a press as early as 1041 AD by the Chinese and later by the Japanese and Koreans. What Gutenberg contributed to printing was the realization of its cultural and commercial possibilities and the introduction to western civilization of the integrated concept of using movable type to print with ink on paper that is on a press. This elevated the publishing of books from the slow, laborious operation of handwriting by monastic scribes to a practical, fast, and economical production process. Printing is considered to be one of the most important inventions in the history of mankind and one of the main factors advancing civilization from the Dark and Middle Ages, when knowledge was restricted to the privileged few, to the Renaissance, when education became available to all. From the hand-set type of Gutenberg we have advanced through the Industrial Revolution to machine composition and, in the Electronic Age, to computer-set type.

There are four main printing processes: relief or letterpress, intaglio or gravure, planographic or lithography, and stencil or porous printing. These are illustrated in Figure 1. Letterpress printing traditionally excelled in the reproduction of text and pictorial matter. However, improvements in the other processes have made it possible to produce reproductions equivalent in quality to letterpress. In the production of all printing there are two physical areas: the printing or image areas, and the nonprinting or nonimage areas. In relief printing, the image or printing area is raised above the nonprinting area. Ink is applied to the raised surface, which is brought into direct contact with the paper or other surface to be printed on. Relief printing is the method by which typewriters, rubber stamps, and letterpress printing operate. The letterpress printing process is used to print magazines, newspapers, advertising brochures, business forms, labels, packages, invitations, etc. Handset or machine-set cast-metal type can be used for direct printing but, for long printing runs, it is common to prepare printing plates or engravings from the type. Plates are made of all illustration materials to be printed. The printing plates can be on zinc, magnesium, or copper and can be plated with nickel or chromium. They also can be on plastic. When rubber or other elastomeric plates and water-base or solvent inks are used in printing, the process is called flexography. Letterset describes the use of relatively thin relief plates for printing by the offset principle.

In the intaglio process, the nonprinting area is at a common surface level and the printing area is recessed and consists of wells etched or engraved, usually to different depths. The most typical method of intaglio printing is the gravure process. Solvent inks with the consistency of light cream are transferred to the whole surface and a metal doctor blade is used to remove the excess ink from the nonprinting surface. Gravure printing is used to print long-run magazines, mail-order catalogues, newspaper supplements, preprints for newspapers, plastic laminates, floor coverings, etc. Paper currency, stock and bond certificates, stamps, some greeting cards, letterheads, business cards, and other specialties are printed by the intaglio process, which is also called steel-die and copperplate engraving. These methods involve the use of a heavy ink, and a special wiping paper is used instead of a doctor blade to clean the plate.

In the planographic or lithographic process, the image and nonimage areas are on the same plane, and the difference between image and nonimage areas is maintained by the physicochemical principle that grease and water do not mix. The image area is grease-receptive and water-repellent; the nonimage area is water-receptive and grease-repellent. Therefore, the ink adheres only to the image areas, from which it is transferred to the surface to be printed, usually by the offset method. This process is used for printing general commercial literature, books, catalogues, greeting cards, letterheads, business forms, checks, maps, art reproductions, labels, packages, etc.

In the stencil or screen printing process, a stencil representing the nonprinting areas is applied to a silk, nylon or stainless-steel fine-mesh screen to which ink with

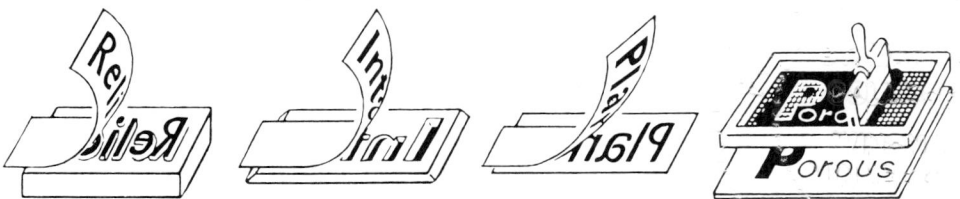

Figure 1. The four major printing processes: relief (letterpress), intaglio (gravure), planography (lithography), and porous (screen process) (1). Courtesy of The Printing Industries of America.

the consistency of paint is applied and transferred to the surface to be printed by scraping with a rubber squeegee. This process is used for printing displays, posters, signs, instrument dials, wallpaper, textiles, etc.

Direct printing is the transfer of the image directly from the image carrier to the paper. Most letterpress and gravure and all screen printing is done by this method. In indirect or offset printing, the image is transferred from the image carrier to an intermediate rubber-covered blanket cylinder, from which it is transferred to the paper (see Fig. 2). Most lithography is printed in this way and lithography is usually referred to as offset printing. Letterpress and gravure also can be printed by the offset method (see Lithography).

Image carriers are made in a number of different ways, depending on the printing process. Letterpress printing is the only process that involves cast-metal type for printing. All other processes must produce image carriers by manual, chemical, or mechanical means. By far the greatest number of plates and image carriers are made by photomechanical means. These systems are characterized by photographic images and light-sensitive coatings which, with chemical etching or other treatments, lead to the formation of a printing surface.

Prepress: The Preparatory Stages

Prepress consists of the operations involved in converting the original to be reproduced into a printing plate or image carrier. The starting point in all printing processes is the original or copy. This can be in many and varied forms, such as type matter or text and pictorial matter consisting of line drawings, art sketches, black-and-white and color photographs, paintings, etc. In some cases the original can be produced directly on the plate surface, as in stone lithography, linoleum blocks, steel-die and copperplate engraving, and screen printing. In all other cases, the original is converted to the form for printing by photomechanical means.

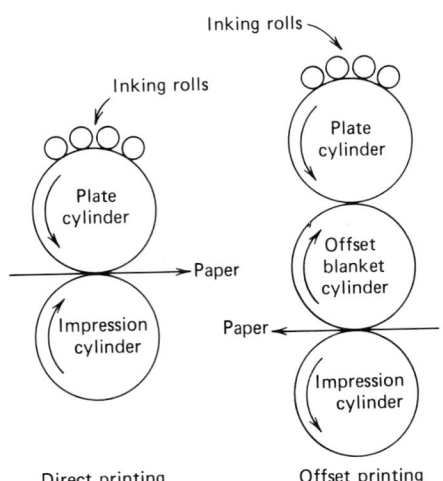

Figure 2. Printing cycle for direct and offset printing.

TYPESETTING

A vast amount of printing begins with typesetting, which can be done by hand, machine, photography, or electronics. The second milestone in the printing industry was reached with the invention of the linotype machine in 1886. This is also known as a line-casting machine, because it casts a slug or line of type at a time. The intertype, which was invented in 1911, is a similar machine. The monotype machine, which was invented in 1887, casts single characters at a time from tape, which is punched on a special keyboard by machine. Line-casting machines can also be operated by punched tape. The tape can be punched on special machines or on a computer. These hot-metal processes result in raised cast-metal type consisting of an alloy of lead, tin, and antimony and known as type metal.

As distinguished from hot-metal type, there are two other forms of type composition, which were called cold type. One is strike-on composition, which includes typesetting for reproduction on special typewriters with carbon-paper ribbons. Some of these include electronic memory devices and are called word-processing machines. The other form of composition is electronic typesetting, which consists of phototypesetting and computer typesetting. The third milestone in the printing industry was reached in 1952 with the invention of the Photon phototypesetter. Photo- and computer typesetting have displaced more than two thirds of cast-metal typesetting and the degree of complexity extends to the electronic setting of complete pages without any photographic film or handwork in the assembly of the images on the pages.

COPY ASSEMBLY

All originals, consisting of text, pictures, and illustrations, must be photographed to convert them into the proper positive or negative films for the photomechanical plate processes by which they will be reproduced. For the production of a satisfactory printed result, the copy must be clean, sharp, in focus, square, in the right position, and the correct size. If the printed job consists of combinations of type and pictorial matter, it can be pasted up before photography, provided everything is in the correct size proportion. On complicated jobs, eg, color advertising, magazines, books, etc, in which the copy can come from a number of sources, the separate pieces of copy are photographed at the correct size and assembled in film form. This completed form is called a flat and the operation of film assembly is image assembly or stripping.

Process Photography

Process photography is the term used for the photographic techniques employed in the graphic-arts processes. Until the introduction of electronic page-makeup systems, all the printing processes involving photomechanical methods for making printing plates required photography for the production of the negatives or positives. Some platemaking methods require negatives and others require positives. Some are reverse-reading on the emulsion side and some are straight-reading, depending on the printing process. The photographic materials can be continuous-tone, line, or halftone. Many of the materials and techniques used in process photography are similar

114 PRINTING PROCESSES

to those in photography (qv). Graphic arts involves highly specialized photographic equipment, methods, and materials. The photographic requirements of the various printing processes are diverse and exacting. The successful execution of any printing process involving photomechanical methods for making the image carriers depends to a great extent on the quality of its photographic components.

Letterpress, lithography, screen printing, and lateral-dot gravure are binary processes, ie, at one time or in one impression on the press, they can only print a solid color in the image areas on the press and no color in the nonimage areas. They cannot print intermediate tones or gradations of tone, ie, continuous tone. Most pictures or scenes to be reproduced have many intermediate tones between the shadows and the highlights. Such a picture is reproduced by a process involving a simulation of continuous tone called halftone. This is an optical illusion in which the tones are represented by solid dots, which are spaced equally but which vary in area (see Figs. 3 and 4). There are a number of screen rulings or spacings, from 24 to 120 lines per centimeter, depending on the type and quality of printing. Letterpress, lithography, screen printing, and some gravure printing involve both line and halftone photography.

In some types of gravure printing, varying thicknesses of ink are printed to produce pictures with a wide range of tones. In these processes, continuous-tone positives are used and an overall screen or halftone positive is used only to produce well walls with a constant height for the doctor blade to wipe. For the overall screen, the individual image wells are the same size and shape but vary in depth; this is known as conventional gravure printing. In the process involving the combination of continu-

Figure 3. Halftone illustration and enlarged portion showing halftone dots. Courtesy of *Graphic Arts Manual*.

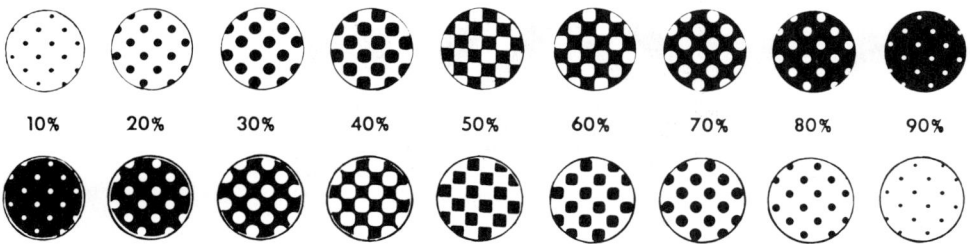

Figure 4. Diagram of halftone dots. The dots in the upper row are positive, those in the lower row are negative. Courtesy of Graphic Arts Technical Foundation.

ous-tone and halftone positives, both the area and the depth of the wells vary. In lateral-dot gravure printing, only a halftone positive is used and all the wells are of the same depth but vary in area.

Collotype or photogelatin is a continuous-tone printing process. Plates are coated with bichromated gelatin and are exposed through continuous-tone negatives. The gelatin is hardened in proportion to the amount of exposure received and, after processing, prints ink density in proportion to the exposure. Techniques have been developed for making lithographic plates from continuous-tone negatives or positives, and the process is known as screenless lithography (see Screenless Printing). The advantages of screenless printing are high resolution, no screen moiré (see Screen Angles), and greater purity of color, especially in the highlights and middletones.

In the direct printing processes, eg, letterpress and gravure, the image must be reverse-reading on the plate, ie, from right to left, in order that it be read correctly from left to right on the printed surface. The negatives or positives, which are used to make the plates, must be straight-reading, ie, from left to right, on the emulsion side. For the printing processes based on the offset principle, the reverse is true. The plate must be straight-reading, so the negative or positives that are which used to make them, must be reverse-reading on the emulsion side.

EQUIPMENT

Cameras. The most commonly used piece of equipment in process photography is the process camera, which is like an oversized enlarger. These cameras have either a shock- or vibration-resistant bed or overhead suspension on which are mounted a movable glass-covered copyboard, a stationary vacuum film back, and a movable lens board, which is connected to the film back with a bellows. Most process cameras are of the darkroom type, for which the film back is mounted in a darkroom wall. Vertical cameras and enlargers are also used (Fig. 5). Cameras usually vary in size of film from 41 × 51 cm to 102 × 102 cm. Larger and smaller cameras have been built for special purposes.

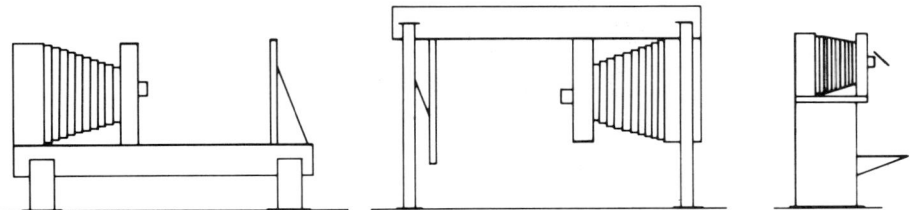

Figure 5. Diagram of horizontal (bed and overhead types) and vertical cameras (1). Courtesy of The Printing Industries of America.

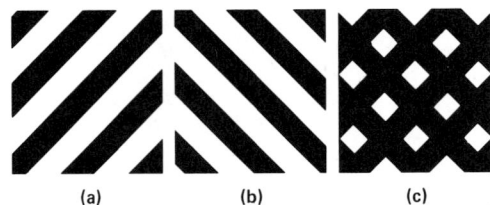

Figure 6. Construction of the crossline halftone screen. (**a**) and (**b**) show the opaque rules on each of two glass plates. (**c**) shows the crossline pattern formed after the two plates are cemented (1). Courtesy of The Printing Industries of America.

Lenses. Special lenses are used on process cameras. They are color-corrected, ie, apochromatic, and are relatively free of spherical and chromatic aberrations and distortion. For this reason, most process lenses are symmetrical in design and their maximum apertures are between F/8 and F/11. Focal lengths vary from 200 mm for wide-angle lenses used on enlargers and compact horizontal and vertical cameras to 1220 mm for lenses used on some of the larger cameras. Since the regular photographic process produces negatives that are reverse-reading (right to left) on the emulsion side, prisms or mirrors are used on cameras to produce negatives that are straight-reading (left to right) on the emulsion side.

Halftone Screens. Halftone screens are used to produce the halftone images, and there are both glass and contact screens. The former are usually crossline screens and consist of two pieces of glass, each with inked rulings, which are cemented together at right angles to each other (Fig. 6). The rulings are approximately the same width as the clear spaces between them and their spacings are 24–120 lines/cm. Screens of 24–33 lines/cm are used for letterpress newspapers, 33–40-line/cm screens for offset newspapers, 47–52-line/cm screens for letterpress magazines, and 52–60-line/cm screens for lithography. Finer-screen rulings are used for special and high quality color printing. Contact screens are on film and are usually made from glass screens. Most contact screens consist of variable-density vignetted dots in a pattern corresponding to the screen ruling of the original screen from which they are made (Fig. 7).

Glass screens are used at a fixed separation distance from the film so they require special precision holders in the back of the camera. The screens can be either square or circular. If the screen is circular, the holder is arranged so that it can be rotated to provide the different screen angles needed for color reproduction (see Color Reproduction). Contact screens are used in contact with the film on which exposure is to be made; therefore, their use requires vacuum backs on process cameras or enlargers. Contact-screen halftones can also be made in a vacuum printing frame.

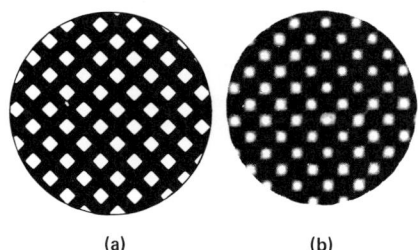

Figure 7. (**a**) Enlarged glass screen. (**b**) Enlarged contact screen.

Processing Equipment. Developing and fixing or processing can be done in trays in temperature-controlled sinks. Since automatic processors were introduced in ca 1960, much of the industry has converted to their use. Advantages associated with their use are increased productivity, economies in time and cost, and products that are more consistent in quality. They are used for processing contact and camera-line, halftone, and continuous-tone negative and positive films and papers.

Lights and Exposure Controls. Arc lights for exposure have been replaced by pulsed xenon and quartz iodine lamps and ensure better consistency of product. Light-integrating devices are usually used to control exposures on cameras, enlargers, and printing frames. These are especially important when automatic processors are used.

MATERIALS

Photographic Emulsions. Three types of photographic emulsions are used in photomechanical processes. For line and halftone reproductions, high contrast orthochromatic or panchromatic emulsions of the lith type are needed. These emulsions are slow, thin, have a high silver content and fine grain, and their characteristic curves show little or no toe, very high gamma (ca 6–10), and maximum density of over 4.0 (see Photography). For single-color gravure and other continuous-tone photography, a low contrast, orthochromatic commercial emulsion is used. The emulsion is characterized by a long tone scale with a long straight-line portion and a maximum density of ca 2.0. For color-separation photography, low contrast panchromatic emulsions with characteristics similar to those of the commercial film are used.

Film Substrate. Glass is seldom used as a substrate or support for photographic emulsions. Cellulose nitrate and cellulose acetate, which were the first film bases used, likewise are seldom used; this is also true for the mixed-ester bases, which consisted of cellulose acetate and butyrate. The most commonly used cellulose ester is cellulose triacetate (see Cellulose derivatives, esters). It has good mechanical strength and fair resistance to hygroexpansion, which is stretch or shrinkage resulting from the absorption or desorption of moisture. The copolymer of vinyl acetate and vinyl chloride was one of the first dimensionally stable film bases, but its use is almost completely discontinued because it has a low softening temperature (ca 60°C). Polycarbonate and polystyrene bases are used to some extent but by far the most popular film base is polyester (see Film and sheeting materials; Polyesters). Of the three, the latter has the best mechanical properties and, with its low temperature coefficient and low hygroexpansion, it gives the film good dimensional stability.

Processing. Regular continuous-tone developers (see Photography) are used for the commercial and color-separation films. Fixing of all films with regular "hypo" (sodium thiosulfate) and high speed x-ray ammonium thiosulfate is the same as in photography. Development of the high contrast lith-type films, however, is different. The contrast of these emulsions is increased with a special type of developer (2). A typical formula for such a developer (Kodak D-85 Developer) is

water (not over 32°C), mL	750
sodium sulfite (anhydrous), g	30
paraformaldehyde, g	7.5
potassium metabisulfite, g	2.6

boric acid crystals, g	7.5
hydroquinone, g	22.5
water	to make 1 L

Note that Kodak D-85 developer does not contain an alkali, which is required for development. The alkali forms by reaction of paraformaldehyde, which is converted to formaldehyde in solution, with sodium sulfite; sodium hydroxide and a formosulfite addition product form according to the following equation:

$$HCHO + Na_2SO_3 + H_2O \rightarrow NaOH + HCHO \cdot NaHSO_3$$

Paraformaldehyde–hydroquinone developers owe their extreme contrast and density properties to the so-called infectious nature of the development (3). In this type of process, density and contrast continue to increase on exposed areas because alkali is formed while sodium sulfite is removed from the reaction site. This happens at the expense of fine detail. Fine lines and letters tend to fill in. This detail can be preserved and resolution can be increased by stopping agitation during development so that the developer becomes exhausted at the development sites. This is known as still development.

Fixing and washing are as important to proper processing as is development, but they are no different in graphic arts than in conventional photography. The automatic processor is only effective if it is used properly and its operation is controlled. Photographic processing is a function of four variables: time, temperature, chemical activity of processing solutions, and agitation. Automatic processors have means for keeping these variables under control but the variables must be checked frequently with the use of control strips to make sure the controls are working effectively.

Dot Etching. After processing, corrections are often required to compensate for minor errors in photography or to make deliberate changes in the reproduction. This can be done by reducing the amount of silver in the image areas. There are a number of chemical reducers that can be used to dissolve and remove silver from negatives or positives. These are often used to remove stains or fog. By far their greatest use is in dot etching, which is used in color reproduction to reduce the size of dots on halftone positives without reducing their density and, thereby, reducing the amount of color in the treated area (4). This is a most important means of manual color correction, which is necessary when color changes in the original are required. A typical reducer is Farmer's reducer; its formula is as follows:

stock solution A	water, L	1
	potassium ferricyanide, g	80
stock solution B	water, L	1
	sodium thiosulfate (hypo), g	255

One part of A, four parts of B, and thirty-two parts of water are used. The chemical reactions involved in the use of this reducer on a silver halide photographic emulsion are as follows:

$$4\,Ag + 4\,K_3Fe(CN)_6 \rightarrow Ag_4Fe(CN)_6 + 3\,K_4Fe(CN)_6$$

$$3\,Ag_4Fe(CN)_6 + 16\,Na_2S_2O_3 \rightarrow 4\,Na_5Ag_3(S_2O_3)_4 + 3\,Na_4Fe(CN)_6$$

LINE PHOTOGRAPHY

Line copy consists entirely of solids, lines, figures, or text matter. In photography, the copy is usually placed on the copyboard of the camera; illuminated by high intensity lights, eg, pulsed xenon or quartz iodine; and focused to the correct size on a ground glass in the film plane. The film is placed on the vacuum back of the camera, which is placed in the image plane in place of the ground glass, and the exposure is made through a shutter operated by an automatic timer or a light-integrating meter. In all photography and particularly in line photography, it is important that the image is the correct size and in sharp focus and that the exposure is correct. If the focus is not sharp or the exposure not right, fine lines or serifs on the type blur or disappear or the type becomes too broad or too fine. After exposure and processing, the negative should be clear and transparent in the areas corresponding to the image on the copy and opaque in the areas corresponding to the white paper on the copy.

Contact Negatives and Positives. A camera is not used to make contact positives from negatives. These are made by placing an unexposed film and the negative in a vacuum frame, in which contact is maintained between the two films by exhausting the air between a plate covered with a corrugated rubber blanket and a sheet of flawless plate glass. When the films are placed emulsion to emulsion, the image on the positive is straight-reading. If it is desired that the positive be reverse-reading, as is needed in the deep-etch and some of the bimetal platemaking methods of lithography, the negative can be placed so that its emulsion side is away from the emulsion side of the unexposed film. If a good point source of light is used at a moderate distance so that the light rays are almost parallel and the exposure is timed accurately, the positive image does not show much sign of light spread, which results in image sharpening. Controls like the GATF (Graphic Arts Technical Foundation) sensitivity guide, star target, and dot-gain scale should be used to control the exposure and the processing (5–7).

Multiple negatives or positives can be made by alternating exposure and development of the negative and positive. Exposure and development are extremely critical to prevent image sharpening or spread. Duplicating films are available and allow the making of a duplicate negative from a negative or a duplicate positive from a positive without involvement of an intermediate film. Sometimes, as in color reproduction, considerable handwork, eg, dot etching, is done on positives to produce the correct tone values. After such handiwork, it is safer to make duplicate positives than to risk the chance of a tone change in making an intermediate negative and then a positive. The duplicating films require a special light source for exposure as well as a special developer for processing.

HALFTONE PHOTOGRAPHY

Tone Reproduction. Tone reproduction and contrast of reproduction are two conditions that halftone photography attempts to satisfy. If a stepped gray scale is considered (as in Fig. 3), normal tone reproduction in halftone photography is the result achieved when the darkest area of the subject prints as a solid tone and the lightest area prints as a white with no evidence of a screen in either tone. The intermediate tones contain varying sizes of dots in a regular progression from ca 3% dot area in the highlight end to ca 95% dot area in the shadow end of the scale.

120 PRINTING PROCESSES

High contrast is evident when two or three steps of the gray scale in the shadow end print solid and/or several steps in the highlight end print white, with a corresponding increase in density increment, ie, high gamma, between the intermediate steps of the gray scale. Low contrast results when the solid contains 80–90% dot area and/or the white end of the scale contains 10–20% dot area with a corresponding decrease in density increment, ie, low gamma, between gray-scale steps in the rest of the scale. The contrast can be increased or decreased at either end of the scale without appreciably affecting the rest of the scale. When the contrast of the light end of the scale is increased, the effect is highlighting or drop-out. Separation of contrast between tones in the shadow end of the gray scale, especially in dark subjects, is improved by flashing, which is an overall exposure of the film to light with the halftone screen in place. This exposure effectively overcomes the inertia of the film so that the rest of the exposure is off the toe and on the straight-line portion of the sensitimetric curve, thereby causing evening of the increment between gray-scale steps in the dark or shadow end of the scale without appreciably affecting the other tones.

Glass Screens. In photography through a glass screen, the screen is accurately positioned in the camera so that it is at a fixed distance from the emulsion of the film to be exposed. During exposure, each of the square openings in the screen grid acts as a pinhole, which produces an image or dot element proportional in size to the amount of light reflected from the corresponding area of the copy (Fig. 8). If the area of the copy being photographed is dark, there is little exposure and the dot is small. If the area is light so that it reflects much light, the dot size is large. In a middle-tone area, the dot pattern appears as a checkerboard. The distance between the centers of the dot elements is the same and corresponds to the spacing between the line rulings in the screen.

The dot areas change according to the tone values they represent. The tone reproduction or the range of tones that can be reproduced and the resolution of the image depend on the lens aperture and the screen distance or the separation between the screen and the film emulsion. The contrast can be lowered by reducing the lens aperture and by flashing the emulsion directly with white light. The larger the aperture and the shorter the screen distance, the higher the contrast and the sharper the reproduction. There are a number of formulas and aperture systems for making halftone negatives or positives with glass screens. Photography with glass screens is an intricate, tedious process and requires considerable skill and experience to produce good results.

Contact Screens. Photography with contact screens is different and requires less skill for acceptable results than with glass screens. The photography is carried out with the screen in contact with the emulsion of the film to be exposed. The contact screen

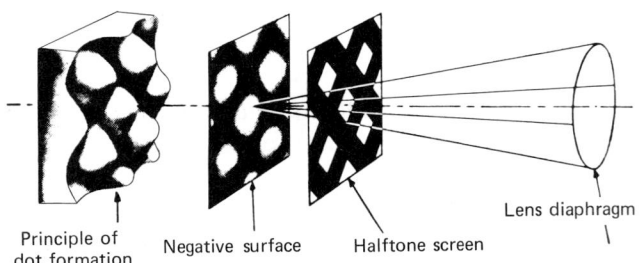

Figure 8. Principle of halftone-dot formation. Courtesy of Graphic Arts Technical Foundation.

consists of a grid of vignetted dot elements. The contact screen is made from a glass screen by adjusting the screen distance, the lens aperture, processing, and other conditions so that each dot element on the contact screen has a predetermined density distribution, which varies depending on whether the screens are designed for making positives or negatives. Some screens have silver images and are gray. In others, the silver is converted to a magenta dye image. Because of the fixed-density distribution in the dots, contact screens have fairly fixed tone-reproduction characteristics with limited variability.

Contrast of reproduction can be varied in several ways. It can be lowered by flashing the film to yellow light with the screen in place. The overall exposure lowers contrast by making the dots larger in the dark tones and shadows and decreasing the difference in dot sizes between gray-scale steps. Contrast can be increased by turning the screen so that its emulsion side is away from the emulsion of the film to be exposed. The separation of the thickness of the film base of the screen increases the contrast of reproduction by increasing the difference in dot sizes between gray-scale steps, but it reduces the resolution or sharpness slightly. The contrast of the highlights can be increased by removing the screen and continuing the exposure without the screen for a short time. This removes the dots from the white portions of the copy and increases the differences between dot sizes in the light end of the gray scale.

With dyed screens, contrast also can be varied with the use of filters: a yellow filter reduces contrast; a magenta filter increases it. Most dyed screens are magenta. Use of a yellow filter over the lens during photography increases the contrast or density distribution across the individual dots in the screen. With such high contrast in the dots in the screen, a large difference in intensity of the light, which is reflected from different areas of the original, produces a small difference in dot sizes in the reproduction so the contrast or density difference between neighboring steps on the gray scale is reduced. When a magenta filter is used, the opposite is true: the contrast of the individual dot elements in the screen is reduced. With low contrast in the screen dots, a small difference in light intensity from different areas of the original produces a large difference in dot sizes of the reproduction, thus contrast is increased.

Dot Fringe. Since maximum contrast is desired in the individual dot elements, the same type of lith film and infectious development is used as for line and contact photography. In camera halftones made through a glass screen, the dots have a variable density gradient, especially at the edges, which is called dot fringe. The amount of this fringe differs depending on the exposure conditions, film, and processing. It can be detected by dark-field illumination. Because of this fringe, glass-screen halftones can be dot-etched considerably; ie, a 40% tone can be reduced chemically to a 10% tone (see Fig. 3). Contact-screen halftones do not have as much fringe and, therefore, cannot be dot-etched as much without losing density over the entire halftone. Although the fringe is desirable for dot-etching, it is very undesirable for platemaking, since the dot size on the plates can vary depending on the amount of exposure to the plates. This disadvantage is eliminated on critical work by making contact or duplicate negatives or positives from halftones. On these, the density distribution across the dots is more uniform and the amount of fringe is reduced appreciably.

Gravure. The type of screen used in conventional gravure printing is different from the previously described screens. The only purpose of the screen is to maintain a grid of constant height over which the doctor blade can wipe. The conventional gravure screen consists of clear lines and black squares with the clear lines having approximately one third the width of the black squares (3:1 screen). Screen rulings of 60–120 lines/cm are used, and these are still used for printing text matter by the gravure process. Halftones for the variable-area gravure processes can be made with special contact screens or by using specially shaped apertures in the lens in conjunction with the glass screen.

COLOR REPRODUCTION

Until after the middle of the nineteenth century, all color printing was like painting. A separate plate or printing was used for each different color in the original to be reproduced. For example, if there were a red, an orange, and a yellow, each of the three would be printed separately. In 1869, the first three-color, ie, yellow, red, and blue, print was produced on paper. Today, practically all color reproduction is done by four-color process printing (8). In this method, the color original is separated into four separate images, each of which is printed from a separate printing plate or image carrier with a different ink to recreate a visual impression of the color original. The four generally used colors are yellow, magenta (blue–red), cyan (blue–green), and black. The different colors in the reproduction are produced by combinations of the yellow, magenta, and cyan inks. The black is used mainly to make the shadows, which are neutral in color. If the color inks are transparent, combinations of them produce most of the colors in the spectrum. Variations of the four-color process method are two-color process printing, in which orange and cyan are used as the printing colors, and fake-color process printing, in which the regular four-color inks are used but the color separations for each of the colors is made manually, usually from a black-and-white image and not from a color original.

Color Theory. Like color photography (qv), process color printing is based on the Young-Helmholtz theory of three-color vision. According to this theory, white light, which is a combination of all the wavelengths of light, consists of three primary colors: blue, green, and red. These are broad bands of color as distinguished from the physical concept of color in which each wavelength of light varies in color from every other wavelength (see Color).

In 1861, it was demonstrated that white light can be produced from broad bands of blue, green, and red light and that any other color can be produced by appropriate mixtures of these colors (9). Illuminating an area or spinning a disk with approximately equal areas or sectors of blue, green, and red light produces white light. Covering the red area with a black sector and spinning produces a blue–green, which is called cyan. Cyan reflects blue and green light and absorbs red light. Covering the green area with the black sector produces a blue-red, which is called magenta. Magenta reflects blue and red light and absorbs green light. Covering the blue area produces a yellow. Yellow reflects red and green light and absorbs blue. These are recognized as the colors of the printing inks used in four-color process reproductions. They are sometimes called complementary colors but are usually referred to as subtractive primaries. Each is a combination of the two colors that are left when one primary color is subtracted from white light. Blue, green, and red light are additive primaries, as these colors of light add to form white light. When two subtractive primaries are printed over each other, they produce an additive primary. Overprinting yellow and cyan, yellow and magenta, and magenta and cyan forms green, red, and blue, respectively.

When all three subtractive primaries are overprinted, they should produce black but, actually, a brown is produced. This is because the colors, as in color photography, are not ideal. An ideal magenta should reflect all the red and blue light and absorb all the green. Actual magentas are good in red reflectance and green absorption but are poor in blue reflectance; they behave as though they have yellow in them. An ideal cyan should reflect all the blue and green light and absorb red light. Actual cyans are poor in blue and green reflectance and fair in red absorption. Cyan colors are dirty and behave as though they have red in them. Yellows are quite good; most yellows reflect green and red light well and absorb blue. Color pigments are poorer in their spectral characteristics than the dyes used in color photography. A typical good set of process color inks consists of dispersions of benzidine yellow, rhodamine Y (magenta), and phthalocyanine blue (cyan) pigments in the proper vehicles. For greater permanence or resistance to fading, rubine is often substituted for or mixed with the rhodamine. Rubine is a poorer pigment in blue reflectance than rhodamine so its use further affects color balance.

Color Separation. Color reproduction is based on the three-color theory of vision in terms of duplicating the operations that the eye and brain perform when a color scene is viewed. The scene is usually photographed on color film and the transparency or a color print made from it is used as the original for reproduction. The copy can also consist of the original object or paintings, drawings, or other colored originals. These are mounted on the copy holder of the camera or enlarger and are illuminated properly. Color filters corresponding in color transmission to the additive primary colors of light, ie, blue, green, and red, are used on the lights or in the lens during exposure to divide the original into three separate color records or separations, each representing one of the primary colors of light (10). A negative is produced in the camera or enlarger. Color transparencies can also be separated by contact in a vacuum frame. The separations are usually continuous-tone negatives but they can also be halftone negatives, as in direct screening by which the color separations are made through a contact screen (11).

A negative made with the blue filter is a recording of all the blue light reflected from or transmitted, in the case of transparencies, through the copy. When a positive is made from this negative, it becomes a recording of the red and green colors in the original. In effect, the negative serves to subtract the blue from the original. The color that reflects red and green light is yellow. In color reproduction, the positive made from the blue filter or blue separation negative is printed with yellow ink. The negative made with the green filter is a recording of the green light, which is reflected or transmitted, and the positive made from this negative is printed in magenta ink. Likewise, the negative made with the red filter records the red light from the original and the positive made from it is printed in cyan ink. Thus, in photography, the negatives separate the color original into three records, each representing a primary color of light. In printing, each negative is converted into a positive record, which is printed in the subtractive-primary ink that is complementary to the color of the filter used to make the negative.

Color Correction. Because the printing-ink colors are not ideal, a reproduction made according to the preceding simple principles lacks crispness, cleanness, and color purity and saturation (12). Most colors, except yellows and reds, are dirty and muddied. There is too much yellow in the reds and greens and too much red in the blues and purples. Even the best magenta ink does not reflect enough blue light and acts as though it has yellow in it. Wherever magenta ink is printed, there is an excess of yellow.

The reds are orange and the blues are dirty and purplish. Because the reflectances of blue and green light are weak in cyan inks, they act as though they contain yellow and magenta. Wherever cyan ink is printed, there is an excess of yellow and magenta, which makes the greens yellowish and dirty and further contributes to the blues being dirty and purplish.

Color correction of the negatives, positives, or printing plates removes the excess color resulting from the improper spectral reflectances of the printing inks. The amount of correction needed depends on the inks and their departure from ideality. Because of the excess yellow in the magenta and cyan inks, the amount of yellow in the yellow printer, the positive made from the blue separation negative, must be reduced in the areas where magenta and cyan print with yellow. Because of the excess magenta in the cyan inks, the amount of magenta in the magenta printer, the positive made from the green separation negative, must be reduced in proportion to the amount of cyan printing with magenta. This color correction can be accomplished by hand, photographic masking, or electronic scanning.

Hand correction can be by retouching the continuous-tone separation negatives or continuous-tone positives made from them, as used in conventional gravure printing. In lithography, in the powderless-etching photoengraving processes, and in the variable-area gravure processes, correction on halftone positives usually is by chemical reduction of the silver in the dots, which reduces their size (4). This is called dot etching. In conventional photoengraving and gravure printing, the final corrections are made directly on the engraved plates or cylinders by hand-tooling or local chemical etching and it is called re-etching or fine etching.

Hand correction is an art that requires considerable skill and is very time consuming and costly. It has been largely replaced by photographic masking and electronic scanning. It is useful, however, in instances where the colors in the original are not correct and must be changed or where the client is not completely satisfied with the original and wants to make changes. In such instances, hand correction is needed to supplement correction by masking or scanning. Except in the most complex page-composition and color-correction systems, hand retouching is required when the original is not correct and changes must be made in the colors in it.

Photographic masking is done by making supplementary images on film that is used in contact with the original when the film is a color transparency, in a special holder in the back of the camera, or in contact with the separation negative. The photographic masks in the first two cases are used while the color-separation negatives are being made; those in the third case are used while the halftone positives are being made from the separation negatives. The masks on film that is used in contact with the original are either silver or colored, eg, Kodak Trimask or Gevaert Multimask. The masks in the second case can be silver halide, dyed magenta, or other colored film. The masks in the third case are always silver halide. Although the first two types of masking are used to some extent, by far the most versatile system of masking is that on a film that is in contact with the separation negative (13).

The color-separation negatives are made with the appropriate filters. The blue, green, and red separation negatives are made with a Wratten No. 47 blue filter, Wratten No. 58 green filter, and Wratten No. 25 red filter, respectively. A fourth separation negative for the black printer is made with a Wratten No. 8 yellow filter (see Filters, optical). With the exception of the black printer, the color-correction masks are made from the color-separation negatives. The color errors in one separation negative are corrected by making a positive from another separation negative to a predetermined density range depending on the set of inks to be used for printing. With a well-balanced set of process color inks consisting of benzidine yellow, rhodamine Y, and phthalocyanine blue pigments, a positive is made from the green separation negative with a density range corresponding to ca 40% of the full range of the negative. This low range positive mask is placed in register with the image on the blue separation negative. A halftone positive representing a color-corrected yellow printer is made from this combination. The positive mask serves to subtract yellow from the magenta and cyan printing areas, where these colors print with yellow.

The 40% value is determined from the color matrix for a good set of inks. The color matrix is a diagram of the optical-reflection densities of the three printing inks as measured through the three color-separation filters. A matrix for a well-balanced set of process color inks is as follows:

	Filters		
Printed inks	Red 25	Green 58	Blue 47
yellow	0.01	0.06	1.00
magenta	0.10	1.20	0.48
cyan	1.25	0.40	0.16

An ideal set of inks would have the following matrix (14):

	Filters		
Printed inks	Red 25	Green 58	Blue 47
yellow	0.00	0.00	1.50
magenta	0.00	1.50	0.00
cyan	1.50	0.00	0.00

Optical density is a measure of light absorption. The figures that appear in the spaces in the first matrix where there is 0.00 in the second matrix are measures of the color errors in the inks. In the first matrix, the ratios of blue to green densities for the magenta and cyan inks are the same. In both cases, they are 0.40 (0.48:1.20 = 0.16:0.40 = 0.40). This indicates a set of balanced inks with respect to the yellow errors in the magenta and cyan inks. Thus, a single 40% mask made from the green separation negative that is placed on the blue separation negative should correct the errors in blue reflectances of both the magenta and the cyan inks. If the ratios were different, then two separate masks, ie, two-stage masking, would be needed or the masking would need to be supplemented by hand correction, since the yellow in one of the colors would be either over- or undercorrected.

The magenta error in the cyan ink is corrected by making a positive from the red separation negative in a percentage equivalent to the ratio of the green-to-red densities or 32%. This positive is placed in register with the image on the green separation negative from which a color-corrected halftone magenta printer is made.

The red-filter negative or cyan printer does not require any color correction, but it is usually masked to increase the color saturation of cyan printing in solids. The mask is usually made from the yellow separation negative or black printer. Its strength depends on the strength of the four-color inks and their gray balance in four-color process printing.

Black is prevented from printing in all the colors and, thereby, dirtying them by making a special mask for the black printer or yellow filter negative. The mask is made from the original by means of a combination of filters that transmits a narrow band of spectral yellow light. The use of this mask ensures a minimum of black printing in the colors and just enough to make the shadows and deep tones neutral in color. This is called a skeleton black.

Undercolor Removal (UCR). In magazine printing, especially by letterpress, it is economical to print a full black and reduce the amount of color printing in the darker colors in proportion to the increased black. Use of undercolor removal (UCR), not only saves on the costs of inks, since black inks are less expensive than color inks, it also helps to alleviate improper transfer of ink or trapping. It is used extensively in high speed rotary letterpress magazine and catalogue printing where trapping is usually affected if more than a total of 240% in screen values is printed, eg, yellow 60%, magenta 60%, cyan 80%, black 40%. Because of the thinner ink films used, 280–300% is the limit in web offset and much lithography is without any UCR. Undercolor removal can be accomplished in masking by making a fuller black and low density positive masks, which are applied to each of the color negatives to reduce the color in proportion to the amount of black added.

Electronic scanning can be used to produce the equivalent of color correction by photographic masking. Depending on the system employed, a light beam is used to scan a color transparency, a color print, a set of separation negatives, a painting, or the original colored art (Fig. 9). A photocell evaluates each minute area electronically in terms of the proportions of each of the three printing colors that are to be used. It translates these color values into electric currents, which are fed into four separate analogue computers, one for each color and one for the black printer, which is computed from the other three signals. The computers modify the currents, depending on the inks, paper, tonal range, and other conditions to be met in the color reproduction. The modified currents are then fed into an exposing light. This light varies in intensity in proportion to the corrected values of each element in the scanned area as it exposes the corrected color separation on film.

Scanners can produce continuous-tone negative or positive color separations or halftones. Halftones are produced by using contact screens during exposure of the separations or by electronic digitizers with lasers or quartz–halogen lamps supplying light for exposure. As in masking, scanners cannot normally introduce local color changes, such as changing the color of a background or dress from red to blue, etc. Such

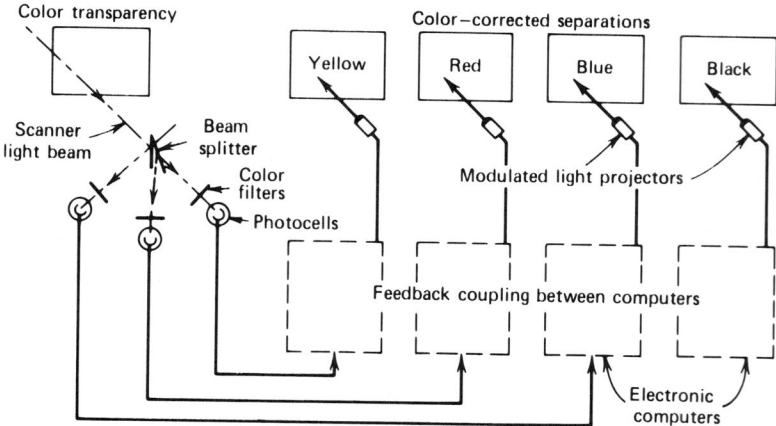

Figure 9. Diagram of typical electronic scanner for color separation and correction. Courtesy of Graphic Arts Technical Foundation.

changes can be made by hand-retouching on the separations, by dot-etching on the halftone positives, or by initiating special digital computer programs in the advanced electronic page-composition systems.

Screen Angles. In printing processes, halftones are made in the usual way using either glass screens or contact screens, as described in Halftone Photography. In color reproduction, however, the angles of the halftone for each color must be changed in order to avoid objectionable patterns or moiré. When two screen patterns of the same ruling are placed over each other, an interference or beat pattern forms and varies in spacing, depending on the angle between the screens. There are first-, second-, and third-order patterns, depending on the number of overprinting colors. The pattern is minimized when the angle between colors is 30°. With conventional crossline screens, in which the patterns repeat every 90°, there are only three possible positions at which 30° angles can be maintained between colors. This is ideal for three-color printing. In four-color printing, however, either two colors must be printed at the same angle or one color is printed at an intermediate angle. Because the most objectionable patterns are generally caused when there are slight errors (as small as 0.10°) between colors, the practice has been adopted of printing the yellow, which is a light and less noticeable color, at 15° between two other colors, usually the magenta and cyan (Fig. 10). The closer the colors are, the more severe is the moiré pattern. For this reason, screen angles are often changed, depending on which colors are the predominating or critical ones in a reproduction. If red is the critical color, the yellow is usually left at 90° or 0°, the cyan at 15° or 105°, and the magenta is placed at 45°. If green is the important color, the cyan is placed at 45°.

Trapping. There always is some pattern and this is aggravated in printing by slight misregister between printing colors on the press and by improper transfer of ink, which is known as poor- or undertrapping. Proper trapping is the condition in printing when the same amount of ink transfers to a previously inked area as to unprinted blank paper. Undertrapping is the condition when less ink transfers to an inked area than to blank paper. This is a serious defect in printing on high speed, multicolor presses, on which ink must transfer to wet ink films. If the ink is not formulated to trap prop-

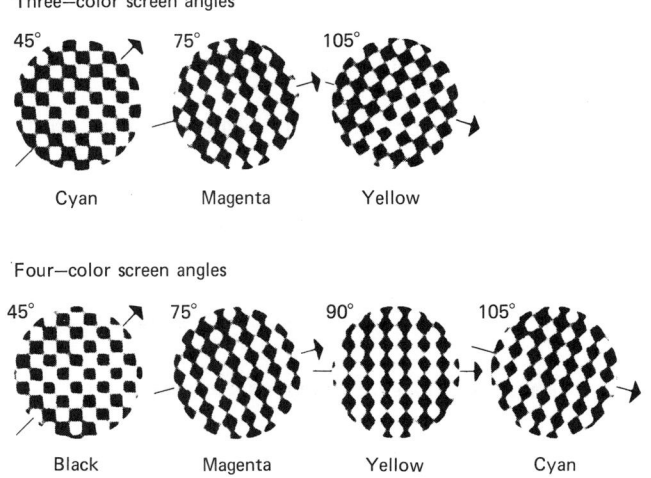

Figure 10. Typical screen angles for minimum moiré patterns in three- and four-color process printing. Courtesy of Graphic Arts Technical Foundation.

erly, poor trapping results and shows as weak overprint colors and accentuated moiré patterns. Undercolor removal is often used to alleviate poor trapping. In the gravure process, because inks are dried between printings, overtrapping usually results. The dried ink seals the paper and, therefore, succeeding colors dry with higher gloss and apparent ink density on previously printed inks than on the unprinted paper.

ELECTRONIC PAGE-COMPOSITION SYSTEMS

The main thrust of new developments in the preparatory areas is the design of complete prepress systems, which print directly from copy to plate, thereby eliminating time- and labor-intensive manual operations and cost of materials in photography and image assembly. The systems will integrate photo- and computer typesetting with electronic color scanners, digitizing tablets, video displays, central processing units, and laser scanners to set type, lay it out on a page, scan pictures, size, crop, color-correct them and position them on the page, assemble the pages in their proper imposition for platemaking, and make the plates on the laser scanner, all without the use of any photography, film, or handwork in the assembly of the images. A number of systems accomplishing many of these functions are in use, eg, the Information International Inc. AIDS, the Sci-Tex Response 300, Hell Chromacom, and Crosfield Magnascan 570. Others, eg, Dainippon Screen and the KC-Digital Litho System are in development.

In addition, a number of new systems have been developed to speed up image or film assembly into the layouts used for plate exposure and to eliminate much of the time-consuming and expensive handwork. There are new systems for automated step-and-repeat processes that use computer programs to move films of the images and fix their position on the plate. There are optical imposition cameras which project images of pages onto a film mounted on an image back whose movements are controlled by a computer program. There is also equipment for microstripping, ie, pages are reduced to 70-mm film and reprojected onto plates in the proper position for printing.

Photomechanical Methods of Platemaking

Printing-image carriers, eg, plates or cylinders, can be produced in any or a combination of six different ways.

Manually: hand tools, engravers, knives, etc, can be used to produce image carriers for the relief, intaglio, and stencil processes; greasy crayon or tusche can be used for making hand-drawn lithographic plates.

Mechanically: engraving and geometric lathes, ruling machines, pantographs, etc, can be used for the relief and intaglio processes; hand transferring and Benday machines can be used for lithographic plates.

Electrochemically: electrodeposited metals can be used to produce long-wearing images.

Electronically: relief and intaglio-image carriers can be produced by electronic engravers.

Electrostatically: relief and lithographic plates can be produced based on xerographic and/or electrofax principles (see Electrophotography).

Photomechanically: the printing images are produced from photographic images.

The photomechanical method is by far the most widely used and most important, as it is a vital part of all the other methods except manual and mechanical ones. This method has been responsible for accelerating the reproduction of pictorial and text matter and for improving their quality.

Each of the different types of printing involves platemaking or image-forming methods that are characteristic of the type of printing. Letterpress plates are quite thick, up to 23.3 mm, and are usually made for individual subjects or pages of metal plates, which are mounted on wooden blocks or metal bases. They are composed into larger forms for printing by locking or arranging the individual plates and cast-metal type, if it is used, into the correct position, called an imposition, in a frame or chase for printing on a flat-bed or platen press. Cast-metal type cannot be used for printing on a rotary press; it must be converted to plates. Plates for printing on a rotary press must be thinner and curved for mounting on the press cylinder. In lithography and letterset printing, the images are composed into the proper position as a unit on a single plate, 0.15–0.76-mm thick, which is mounted on the plate cylinder of the press. These are called wraparound plates. Wraparound plates can also be used in gravure printing but, most often, the printing is from copper-plated cylinders that have been etched and chromium-plated after the images have been transferred to them.

In addition to printing from cast-metal type, letterpress printing can be done from original, duplicate, or wraparound plates. Original letterpress plates are made on zinc, magnesium, or copper by photoengraving. Duplicate plates for printing are plastic, rubber, stereotypes, or electrotypes, depending on the type of mold from which they are made. Wraparound plates can be either plastic or metal. Lithographic plates are usually wraparound, made on aluminum, and can be surface, deep-etch, or bimetal plates. They can be plant-coated, precoated, or presensitized. Some sheet-fed gravure printing is done from copper wraparound plates, but the majority of gravure printing is from copper-plated cylinders. The printing image is produced on these cylinders with carbon tissue, photographic transfer film, or direct coatings. Screen-process stencils are made on a silk, nylon, or stainless-steel screen and can be made by means of carbon tissue or other transfer film or by coating directly on the screen for the printing image. All these processes of platemaking involve photomechanical methods.

CHEMISTRY

Photomechanical methods involve exposure of photographic images on light-sensitive coatings that are either directly on the printing member or can be transferred to it. The distinguishing feature of the light-sensitive coatings is that, on exposure to light, they undergo changes in physical characteristics, usually solubility in water or in other chemicals, so that they can be developed to produce images that serve either as the printing images or as resists for producing the printing images. Where the light-exposed or hardened coatings are used as resists, eg, in deep-etch and bimetal plates, they must remain soluble in other chemicals so that they can be removed after the images have been produced (see Photoreactive polymers).

Means must be provided for applying the light-sensitive coating to the printing member in the processes where the printing plates are coated directly. A whirler is usually used for this purpose. It consists of a rotating table on which the plate is placed for coating. It is in a housing that can be heated and be made light-tight. The table can be either horizontal, vertical, or at an angle of about 15° from the vertical. During coating, the sensitizer solution is poured near the center of the plate as it rotates. The rotation of the plate causes the solution to spread out to the edges of the plate by centrifugal force on a horizontal whirler and by a combination of centrifugal force and gravity on the vertical type. A fairly even coating is produced, the thickness of which depends on the speed of rotation, density and viscosity of the solution, amount of moisture on the plate at the time of coating, roughness of the surface of the plate, relative humidity, and temperature. Coating of wipe-on plates does not require a whirler; they can be coated by hand or on a simple roller-coating machine.

For exposure of the coated plate to the negative or positive, a vacuum frame, such as is used for making contact exposures in photography, is usually used. Since most photomechanical coatings are of the print-out type, much light is required. Therefore, pulsed-xenon or metal-halide lights are used. Light-integrating meters are often employed for controlling exposures. For exposing larger plates in lithography and wraparound letterpress printing where multiple exposures of the same subjects are often required, a step-and-repeat machine is used. Other equipment needed in photomechanical platemaking includes developing sinks, tables, pads, wipes, squeegees, etc. Automatic processing machines are available for making lithographic plates. Special apparatus is needed in gravure and screen-process printing for coating directly on or transferring carbon tissue or transfer film to the cylinders or screens.

Until recently, the majority of the coatings used in photomechanical processes were bichromated colloids (15). Although most of these have been replaced by diazo compounds, photopolymers, silver halide photographic emulsions, and electrostatics, a knowledge of their chemistry and uses is important to understanding photomechanical platemaking processes.

Bichromated Colloids. Bichromated colloid coatings have been used for all four main printing processes. Bichromated albumin, casein, and poly(vinyl alcohol) were used for surface plates in lithography but this type of plate is obsolete. Bichromated gum arabic is still used for some deep-etch plates for lithography. Bichromated gelatin is used in collotype or photogelatin and it is the active ingredient in carbon tissue, which is used in gravure and screen printing. Ammonium bichromate is usually used as the sensitizer. Potassium bichromate is sometimes used in collotype coatings for special long-scale or low contrast reproduction effects.

Solutions of bichromated colloid coatings are fairly stable but, once they are dry, light makes them insoluble by a process similar to leather tanning. Part of the hexavalent chromium (Cr^{6+}) is reduced to trivalent (Cr^{3+}), and a colloidal complex $(Cr_2O_3)_x(CrO_3)_y$ is formed. At the same time, the coating undergoes a sol–gel transformation. As the light exposure proceeds, the gel becomes stronger and less capable of hydration. The isoelectric point also changes. For example, the isoelectric point of egg albumin is ca 4.9, whereas that of a suitably exposed bichromated albumin coating is 3.8–4.0. The resulting protein gel is oleophilic and capable of retaining greasy ink, even under conditions of maximum hydration at or near its isoelectric point.

The light sensitivity of bichromated colloid coatings is governed by a number of factors, of which the following are the most important: *bichromate:colloid ratio*—sensitivity increases with the bichromate concentration up to the point where recrystallization of the bichromate takes place; *coating thickness*—sensitivity decreases with increased coating thickness; only wavelengths of light below 450 nm are very effective for exposure and the yellow coating acts as a light filter; *pH value*—sensitivity of the dried coating increases as its pH value decreases; *moisture content*—sensitivity increases with relative humidity and hygroscopic moisture content; and *temperature*—sensitivity increases with temperature.

The dried bichromated colloid coating gradually becomes insoluble in the absence of light. This deterioration or dark reaction is more rapid with increased temperature and relative humidity. Therefore, plates must be exposed and developed within a time limit established by experience. For example, an albumin-coated plate may be usable for three or four days at 24°C and 30% rh; but at 32°C and 75% rh, it does not develop after ca 2 hours.

Since insolubilization of the coating is a chemical reaction, its rate is the product of all the factors. If any one factor is low, the result is low. This accounts for the fact that bichromated coatings can be stored in a refrigerator, even though the relative humidity is very high, because the temperature is low.

Consistent success in platemaking requires control of the bichromate:colloid ratio, coating thickness, pH, temperature, relative humidity, and light exposure. With an exposing light of constant intensity, exposure can be timed. With a fluctuating light source, an integrating light meter is necessary. Controls are available to indicate variations in sensitivity and to help establish controls to compensate for them (5). The GATF sensitivity guide is a continuous-tone gray scale with an increment of ca 0.15 density units between numbered steps. The number of steps that print solid on the plate are a function of the light sensitivity of the coating and the light intensity of the light used for exposure. If the light intensity is constant, the number of solid steps on the sensitivity guide can be used as a measure of coating sensitivity and as a means of controlling or compensating for it when it changes. A difference of two steps in the sensitivity guide is equivalent to either doubling or halving the sensitivity, depending on whether it is increased or reduced.

Diazo Coatings. Diazo resins are used primarily as coatings for lithographic presensitized and wipe-on plates because they are not affected as much by temperature and relative humidity as are the bichromated colloids (16). The diazo resin most commonly used for negative plates, ie, plates made from negatives, is the condensation product of 4-diazodiphenylamine salt with formaldehyde (1); this product is usually used as the zinc chloride salt. Diazo oxides, eg, pyrido[1,2-a]benzimidazol-8-yl 3(4H)-diazo-4(3H)-oxo-1-naphthalenesulfonate (2) are used for both negative and positive plates.

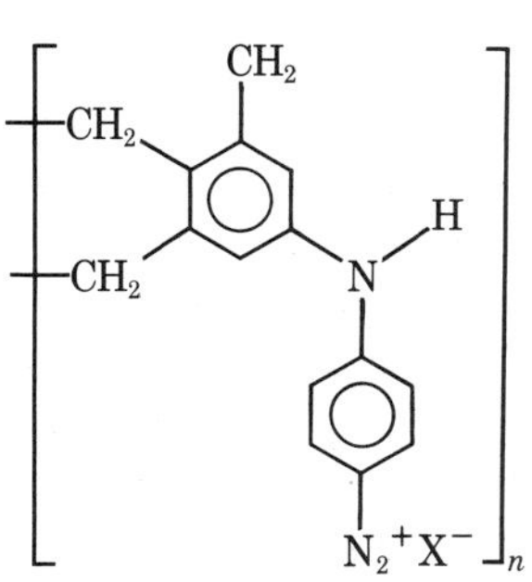

(1) 4-diazodiphenylamine–formaldehyde condensate

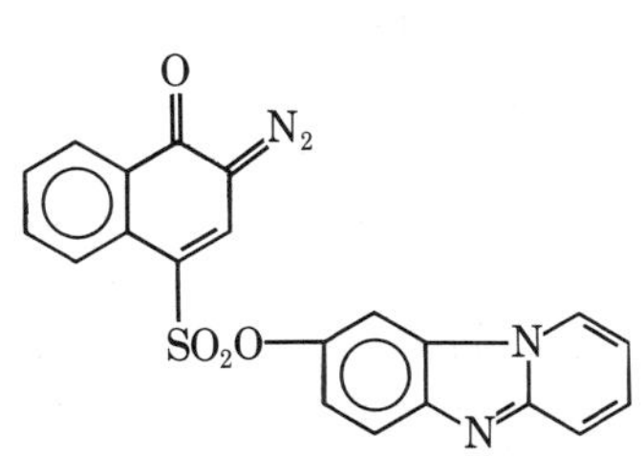

(2) pyrido[1,2-*a*]benzimidazol-8-yl 3(4*H*)-diazo-4(3*H*)-oxo-1-naphthalenesulfonate

132　PRINTING PROCESSES

The mechanisms of the photochemical reactions of diazonium and diazo oxide compounds are not known exactly. They are suspected to be like the photochemical reactions of similar but less-complex diazo materials. The following equations show the photochemical decomposition of diazonium salt, which is comparable to that of diazo resin (eq. 1), and 2-diazo-1(2H)-naphthalenone, which is comparable to diazo oxide (eq. 2).

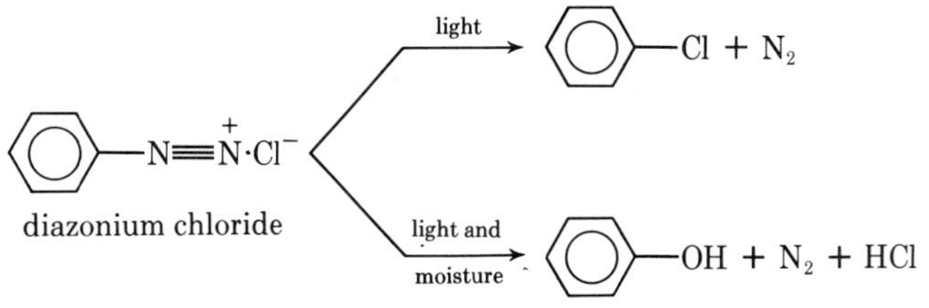

$$\text{diazonium chloride} \xrightarrow{\text{light}} \text{PhCl} + N_2$$
$$\text{diazonium chloride} \xrightarrow{\text{light and moisture}} \text{PhOH} + N_2 + HCl \tag{1}$$

$$\text{2-diazo-1-(2H)-naphthalenone} \xrightarrow{\text{light}} \text{ketene} + N_2$$
$$\text{2-diazo-1-(2H)-naphthalenone} \xrightarrow{\text{light and moisture}} \text{CO}_2\text{H derivative} + N_2 \tag{2}$$

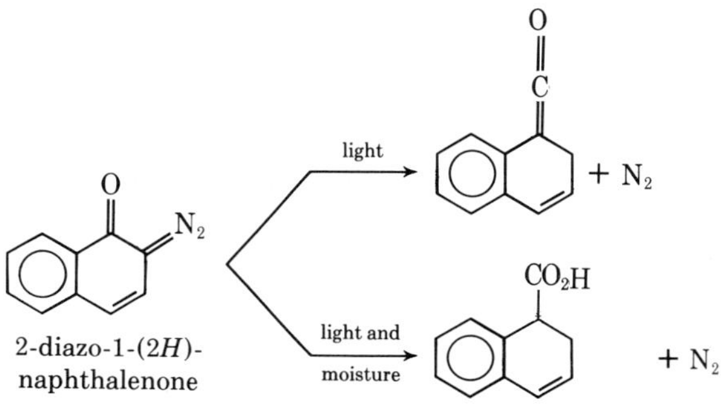

The solubility of the complex diazo polymer (1) results from the presence of ionic diazo groups. The destruction of these groups by light renders the decomposed polymer insoluble in water. On exposure to a negative, the exposed and photochemically decomposed image becomes the insoluble ink-receptive printing image. The diazo oxide (eq. 2) can be used for either negative or positive plates, depending on the solubility differential of the photochemically decomposed product in acid and alkaline solutions. The light-exposed material is less soluble in acid than in alkaline solutions. If acid solutions are used, the exposed material remains as the printing image; therefore, it is a negative plate system. If alkaline solutions are used, the unexposed material remains as the printing image and it is a positive plate system. Although theoretically the same plate could be used for both systems, in practice the particular resin is selected that works best for one or the other system.

Extended exposure to temperatures above 52°C can cause decomposition of the diazonium compound, thereby resulting in scum, ie, unwanted ink-receptive material in the nonprinting areas, on presensitized plates. Also, solutions of diazo resin (1) deteriorate on standing (17). This is not much of a problem with presensitized plates that have been coated by the manufacturer, but it is particularly troublesome on wipe-on plates. If solutions of diazo resin (1) are allowed to stand too long before the plates are coated, the plates may not develop properly and may take ink or scum on press. In any case, the age of the coating is a serious source of variability with wipe-on plates. The best practice is to mix only the amount that is needed for a 24-h period. Not much deterioration takes place during this short period, especially if the solution is refrigerated when not in use.

Surface Treatments. The diazo resin (1) is very reactive, therefore, metals must be specially pretreated prior to coating; otherwise short storage life results. This is especially true for presensitized plates. Usual storage life for diazo presensitized plates is one year. The base metal of most wipe-on and presensitized plates is aluminum. The surface treatments (18) that are used to protect the metal from reaction with the diazo resin include silicates, anodizing, potassium zirconium fluoride (potassium fluorozirconate) (19), organic phosphates (20), phosphates (21), acrylic monomers (22), titanium tetrachloride (23), tetraisopropyl titanate (24), boehmite ($Al_2O_3 \cdot H_2O$) (25), phosphomolybdates (26), and a cycloaliphatic polyphosphoric acid, eg, phytic acid (27). Plates coated with diazo oxides do not require these special treatments. The plate surface must be clean and usually some form of mechanical roughening or graining is used to increase the surface area of the plate and improve its wettability for coating. Surface treatments are described in detail in ref. 18 (see Metal surface treatments—cleaning, pickling, and related processes).

Photopolymer Coatings. Photopolymer coatings are either insolubilized by polymerization or cross-linking during exposure or solubilized in developers by photodegradation after exposure. These coatings are very inert and have excellent abrasion resistance. Photopolymer coatings are used for making letterpress, flexographic, and lithographic plates from negatives or positives. They are also used for making printed circuits. There are a number of types of photopolymer coatings in use: those that are developed in organic solvents, those that are developed in water or alcohol solutions, liquid polymers of the vinyl urethane type, and photosensitive synthetic rubbers (see Photoreactive polymers). Coatings of the first type are mainly cinnamic ester resins and are used extensively as etching resists for printed circuits, engravings, and lateral dot gravure and for making lithographic plates. Coatings of the second type made with polymethacrylates and polyamides are used mainly for letterpress plates. Coatings of the third and fourth types are used mainly for newspaper and flexographic plates. Combinations of diazo resins and photopolymers are used for lithographic plates. The main feature of photopolymer coatings is that they are affected very little by temperature and relative humidity and, therefore, have a very long shelf life as precoated materials.

Cinnamic Ester Resins. Cinnamic resins of poly(vinyl alcohol), cellulose, and starch are known, but the preferred lithographic and printed-circuit and acid-resist coating materials are based on epoxy resins. The structural formula for the cinnamic ester of an epoxy resin of epichlorhydrin and bisphenol A (4,4′-isopropylidenediphenol) is given below (16) (see Epoxy resins).

$$\left[-O-\underset{CH_3}{\overset{CH_3}{\underset{|}{\overset{|}{C}}}}-\text{\char 0}-OCH_2\underset{OCCH=CHC_6H_5}{\overset{O}{\underset{\|}{\overset{\|}{C}}}}HCH_2- \right]_n$$

In the exposure of cinnamic ester resins, the photochemical reaction most probably involves a cross-linking mechanism between cinnamic ester units in the polymer chains, which produces a rigid, insoluble structure. The principle of the cross-linking mechanism for the classical dimerization reaction of cinnamic acid to form truxillic acid (*trans,trans*-2,4-diphenyl-1,3-cyclobutanedicarboxylic acid) is as follows:

The number of cross-linkings per unit of light or time of exposure is increased by a number of compounds, including hydrocarbons, amines, nitro compounds, ketones, quinones, etc. Commercial coatings based on cinnamic resins usually include one of these organic sensitizers (see also Dyes, sensitizing).

The cinnamic resins are water-insoluble and the solubility differential is established by developing the coating in a suitable solvent system. In practice, this is done either in a vapor degreaser or by using an emulsion of the solvent as dispersed in an aqueous phase of gum and phosphoric acid. The photochemical reaction renders the exposed areas insoluble and these form the printing image. In addition to being little affected by temperature and relative humidity, these coatings have excellent water-, acid-, and abrasion-resistance. On a lithographic press, they are capable of long runs. Some plates have run satisfactorily for editions in excess of 10^6 impressions.

Polymethacrylate. Polymethacrylate and polyamide coatings are two types of photopolymerizable coatings used to produce relief plates for letterpress or letterset printing. The polymethacrylate plate consists of a base metal sheet coated with an antihalation layer, which prevents light from reflecting into the coating, and a coating containing monomeric and polymeric methyl methacrylates, poly(ethylene glycol dimethacrylate), ethyl acrylates, poly(vinyl acetate), polystyrene, etc, and a photoinitiator or addition-polymerization catalyst, eg, benzoin, benzoin methyl ether, α-methylbenzoin, α-allylbenzoin, diacetyl, or 1,1'-azodicyclohexanecarbonitrile (28) (see also Initiators). The coating may also contain polymerization inhibitors, such as antioxidants like hydroquinone, *tert*-butylcatechol, etc, to prevent spontaneous polymerization before it is desired. Plates are usually conditioned in a carbon dioxide atmosphere for this reason. Light exposure activates the photoinitiator which causes polymerization and cross-linking of the various monomers and polymers in the coating. Maximum differentiation in solubility between exposed and unexposed areas is produced when an appreciable proportion of the monomers and polymers consists of cross-linking materials. Unexposed coatings containing a high proportion of monomers and a low proportion of polymers are soluble in alkaline solutions. For unexposed coatings containing higher proportions of polymers, organic solutions are required for development (see Methacrylic polymers).

Polyamide. Polyamide coatings are not photosensitive in themselves; the polyamides must be mixed with photosensitive polymerizable unsaturated compounds, which are capable of cross-linking with the polyamides on exposure to light (29). Materials useful for this purpose are N,N'-methylenebisacrylamide, N,N'-hexamethylenebismethacrylamide, and related compounds. As with the polymethacrylate coatings, a polymerization inhibitor and photoinitiator are needed in the coating composition. Polymerization inhibitors are antioxidant materials, eg, pyrogallol, quinone, hydroquinone, methylene blue, etc. The photoinitiators used are benzophenone, benzoin, benzaldehyde, acetophenone, and similar compounds. It is believed that during exposure, cross-linkings are formed between the polyamide resins and the photosensitive, polymerizable unsaturated compounds. The photoinitiator serves to accelerate the reaction so that the exposure time is reduced. Since these coatings consist mainly of polymers and only a small percentage of monomers, they must be developed in alcohol solutions (see Acrylamide polymers; Polyamides).

Vinyl Urethane Systems. A number of photosensitive systems have been developed for the production of photoelastomer plates, such as are used in flexographic plates and, to some extent, in newspaper plates, which do not require elastomers but similar photosensitive systems are used to produce them. Vinyl urethane systems are based on liquid polymers and were among the first photopolymer systems used for newspaper plates, eg, the W.R. Grace Letterflex and Hercules Merigraph. Modifications of these systems are used for flexographic plates.

The first stage in the preparation of the vinyl urethane prepolymer involves the principal reaction of polyurethane chemistry: the reaction between a hydroxyl group and an isocyanate group (30).

$$\underset{\text{hydroxyl group}}{\text{R—OH}} + \underset{\text{isocyanate group}}{\text{O=C=N—R}'} \longrightarrow \underset{\text{urethane linkage}}{\text{R—O}\overset{\overset{\text{O}}{\|}}{\text{C}}\text{NH—R}'}$$

Use of di- or polyfunctional hydroxyl compounds and di- or polyfunctional isocyanates makes possible the preparation of a wide range of polyurethanes (see Urethane polymers).

$$\text{HOROH} + \text{OCNR}'\text{NCO} \longrightarrow \text{—(ORO}\overset{\overset{\text{O}}{\|}}{\text{C}}\text{NHR}'\text{NH}\overset{\overset{\text{O}}{\|}}{\text{C}}\text{)}_n\text{—}$$

In the preparation of flexographic plates, hydroxyl-terminated polyethers and polyesters are used.

Photosensitive Synthetic Rubber. Photosensitive synthetic-rubber coatings consist of synthetic rubber materials compounded with vinyl monomers and standard photoinitiators in such a way that exposure to uv radiation yields lightly cross-linked elastomers. The vinyl cross-linking agents consist mainly of standard acrylic and methacrylic monomers but may also include low molecular weight vinyl urethanes. There are many flexographic plates based on these photosensitive systems. One of the most popular plates of this type is the DuPont Cyrel plate, which is the product of cross-linking poly(2-chloro-1,3-butadiene) and the vinyl monomer 1,1,1-trimethylolpropane triacrylate (31).

Silver Halide Photographic Systems. Silver halide photographic systems are used in the four main printing processes (see Photography). Their main advantage is exposure speed, so that in most cases exposure can be made by projection or contact. The colloid in the photographic emulsion, usually gelatin, is hardened or tanned corresponding to the exposed and developed silver image. In the systems used for letterpress, gravure, and screen printing, the unhardened gelatin in the nonprinting areas is removed by soaking in hot water. The hardened gelatin serves as a resist or stencil for etching in the letterpress and gravure processes and for printing in the case of screen printing (1). In the system used for lithography, the photographic material becomes the printing plate; the tanned gelatin is ink-receptive and the unexposed and unhardened gelatin remains water receptive (32–33).

Diffusion transfer (DTR) is another photographic system used for making lithographic plates, and it is the principle involved in the production of Polaroid prints (see Photography; Color photography, instant). Diffusion transfer is a positive–positive process consisting of two specially coated parts: a negative material and a receiver sheet, which becomes the final image or printing plate. The image is exposed on the negative material and, after special processing, is brought into contact with the receiver sheet where the unexposed area or positive image is transferred by diffusion and produces the printing image. These plates are used mainly for reprography on small presses.

A high speed photopolymer plate was introduced in 1980 and is a combination photopolymer–silver halide plate. It consists of a photopolymer coating on anodized aluminum that is overcoated with a silver halide emulsion. After exposure and development of the silver halide coating, an overall exposure produces an image on the photopolymer coating which is converted to the printing image.

PHOTOENGRAVING

Photoengraving is the photomechanical production of relief plates for letterpress, letterset, and flexographic printing in line, halftone, and color. Although the use of the letterpress process is declining, flexography is gaining and methods for making relief plates will continue to expand. There are two general types of engravings: line and halftone. These are divided into five groups: line engravings in one color, halftone engravings in one color, combination line and halftone engravings in one color, sets of engravings for multicolor printing either in line or halftone or in line and halftone combinations, and four-color process engravings for full-color printing.

Zinc, magnesium, and copper with small amounts of other alloying metals are the three commonly used metals for photoengraving original plates. Zinc and magnesium are used almost exclusively in the United States for line engravings, and copper is used for halftones and for line engravings where fine lines and many molds are involved. In Europe, zinc and magnesium are used for halftones as well as for line engravings. In the United States, copper is seldom used as a printing plate; it is used as the original to make the mold from which electrotype duplicate plates are made. The actual printing is done from the electrotypes, thereby preserving the original in case of damage to the printing element.

Etching. In photoengraving, the metal in the nonprinting areas is removed by chemical etching or mechanical routing. Nitric acid is used as the etchant for zinc and magnesium and ferric chloride is used for copper. There is a very wide range in the depth of etching. On a 52-line/cm halftone engraving, it varies from 0.036 mm in the shadows to 0.051 mm in the middletones and 0.074 mm in the highlights. It can vary from ca 0.5 mm on line engravings for printing on high finish papers to >1 mm for flexographic printing on rough papers, eg, the liner for corrugated board (34).

The main problem in etching is to maintain the correct dot or line area at the proper etch depth. Classically, this has been done by either scale compression or successive four-way powdering. Scale compression is achieved when photographing the negative. Much larger white dots are left in the highlights to compensate for the lateral etching, which reduces the area of the highlight dots. It is difficult to control middletone and shadow dot size by this method. Four-way powdering provides protection of the sides of etched image elements with acid-resistant materials, which prevent attack on the sides of the images as the proper depth is achieved. The acid-resistant materials consist of fusible natural resins, eg, dragon's blood (a dark-red resinous substance), asphaltum, colophony, shellac, mastic, damar, copal, and synthetic resins of the phenol–formaldehyde type. Four-way powdering is a manual operation that is time consuming and requires considerable skill and judgment.

Powderless Etching. The powderless etching system was developed in the early 1950s (35). A special etching machine is required and the etching bath consists of an emulsion, which continuously applies an acid-resistant coating on the sides of the image elements to control sidewise etching and depth of etch. The etch bath consists of nitric acid, dioctyl sodium sulfosuccinate, diethylbenzene (mixed isomers), gelatin, and a wetting agent. The mechanism of the reaction is not known, but it is suspected that the diethylbenzene and succinate ester form an acid-resistant film on the etched areas and this film protects the sides of the image elements while the bottom continues to be etched. It has been claimed that the resist-coated areas are heat sinks in which the temperature is cool enough for the film to remain on the slopes (35). In the bottom, active etching produces locally high temperatures and the oil–surfactant layer is soluble or nonadherent. This system is usable on magnesium and zinc.

A mechanically similar system has been developed for etching copper (36). In this system, ferric chloride is the etchant and one of the film-forming additives is formamidine disulfide, $NH=C(NH_2)SSC(NH_2)=NH$. Mechanically, this mixture works in a manner similar to that of the powderless etching system used on magnesium and zinc; chemically, it is quite different. The protective banking agent that prevents side etching is the reaction product of the organic compound and the copper dissolved from the plate. This agent forms a gelatinous deposit on the sides of image elements being etched.

Line Engravings. Line engravings are made from photographic line negatives of text matter, type, line drawings, graphs, block diagrams, etc. A typical process for making these plates in the conventional way without powderless etching includes the following steps. (*1*) Clean the zinc surface by scrubbing it with powdered pumice and water, followed by a weak acid wash. (*2*) Coat the plate in a whirler with bichromated albumin solution and dry (bichromated shellac can also be used). (*3*) Expose the coated plate to light while it is in contact with a line negative. (*4*) For albumin coatings: coat the exposed plate with etching ink, develop it under running water by light rubbing with cotton to remove the unexposed albumin coating, and dry. Shellac coatings are developed in alcohol and water. (*5*) Powder the ink image with powdered resin, and heat to sinter the resin. (*6*) Paint the margins of the plate with an acid-resistant paint or lacquer. The backs of the plates are usually factory-coated. (*7*) Etch the bare metal with nitric acid so as to leave the protected lines and solids in relief. Undercutting the image is prevented by etching in several stages. After each etch or bite, the plate is reinked and powdered, and the resin is brushed in four directions to protect the sides of the lines; then the plate is reheated to sinter the resin. (*8*) When the plate has been etched to the desired depth, the resist is removed with a solvent. Large nonimage areas are routed to a depth of 0.5 mm or more, and the edges are trimmed and beveled.

The powderless etching steps are as follows. Steps (*1*) and (*3*) are the same as in the preceding process. (*2*) Coat the plate with a bichromated poly(vinyl alcohol) (PVA) sensitizer, as other sensitizers are attacked by the acid–organic solvent mixture used in the etching machine. (*4*) Develop the exposed plate in water. (*5*) Bake the PVA coating at 163°C to complete the polymerization and to convert it to an acid-resistant enamel. (*6*) Apply a descumming agent to remove any bichromate remaining in the coating. (7) Wash the plate thoroughly to remove all traces of descumming agent. (8) Mount the plate in the etching machine, turn the machine on, and let the etching proceed for 5–12 min.

Halftone Engravings. A typical U.S. process for making halftone engravings in the conventional manner without powderless etching involves the following steps. (1) Clean the copper plate with a detergent to remove grease and sand with FFF pumice. (2) Coat the plate with a solution of bichromated fish glue in a whirler and dry it. (3) Expose the coated plate, which is in contact with a halftone negative, to light. (4) Place the exposed plate in a solution of methyl violet dye to color the coating; then develop it under water to remove any unexposed coating. Dry the plate. (5) Heat the plate to convert the coating on the image areas to an acid-resistant enamel. (6) Paint the margins with asphalt paint or lacquer. The backs are factory-coated. (7) Etch the bare metal to leave the halftone image in relief. The metal is etched with ferric chloride solution (sp gr 1.41 (42° Bé)) in several stages. After each bite, the plate is dusted four ways with powdered dragon's blood to prevent undercutting of the dots. Since the depth of etching varies with the tone of the subject, the shadow areas are painted or staged with acidproof varnish after the first or second bite and etching is continued. Staging is repeated for each successively lighter tone. The final etch determines the depth in the highlights. (8) The etched plate is cleaned and proved with ink to determine if the tone values are correct. If they are not, the plate is re-etched locally to reduce the dots to the desired size. (9) When the proof is satisfactory, the plate is trimmed and beveled, and any large blank areas are routed to increase their depth.

The etching can be done in a tray but is usually carried out in an etching machine. There are three types of etching machines: one splashes the etch against the plate, another type sprays the etch, and another type etches electrolytically by means of a sodium chloride bath. Uniform, controlled etching requires controlled renewal of the etching bath to maintain a nearly constant composition and concentration.

The procedure for the powderless etching of halftone engravings on copper is the same as for powderless etching of zinc or magnesium line engravings. Either bichromated glue or PVA can be used as the sensitized coating. The chemicals that are used are different but the operations are similar. In the etching machine, a gelatinous film forms on the etched areas. The etchant spray breaks up the film so that etching takes place in the bottoms or open areas and flushes the material sideways, thereby causing it to accumulate as a protective bank along the sides of the relief-image elements. Because the etchant strikes the banking agent on the sidewalls at a large angle, it does not have enough force to dislodge the protective material, and sidewall etching is inhibited.

Photopolymer Plates. Original photoengravings can also be made on photopolymer materials. There are four steps in making the polymethacrylate plate. (1) Conditioning of the plates in a carbon dioxide atmosphere to remove oxygen from the coating, which retards the action of the photoinitiator. (2) Exposure of the coated plate to negatives. (3) Washing the exposed plate in an alkaline solution, which dissolves the unexposed areas. (4) Drying. In the making of sensitized polyamide or nylon plates, the plates are exposed, developed, and dried. Development is carried out in an alcohol solution. Processing of the vinyl urethane and photosensitive synthetic rubber plates is equally simple.

A number of photopolymer relief plates are in use mainly by newspapers and flexographic printers, with the exceptions of DuPont Dycril, which is a polymethacrylate plate, and BASF Nyloprint, which is a polyamide plate. These two plates are used mainly in letterpress magazine and commercial printing. The other photopolymer plates are W.R. Grace Letterflex, which is a liquid urethane system used for newspapers, Cameron belt presses, and flexography; Dynaflex, which is used by newspapers and is a dry prepolymer system that must be refrigerated until exposed; NAPP, which is a denatured poly(vinyl alcohol) plate used for newspapers; Hercules Merigraph, which is a liquid urethane system used for newspapers, belt presses, and flexography; and a number of Japanese plates, which have not been used in the United States. These include the following plates: Sonne KPM 2000, Toplon, Tevista, Torelief, and Toyobo Printight, which is a water-developable nylon plate.

Wraparound Plates. Wraparound plates are relief plates on thin metal shells, 0.43–0.8 mm thick. The plates are the full size of the plate cylinder and they are flexible enough to be mounted on the plate cylinder for printing. These plates are used on special wraparound presses on which the plates print directly on the paper. They are used even more extensively for printing letterset on offset presses. All lithographic plates are wraparound plates.

Electromechanical Engraving. Engravings have also been produced electromechanically on special equipment, which scans an original photoelectrically and simultaneously etches a plastic printing plate with a heated stylus. One was used successfully to produce engravings for small newspapers, printers, and engravers for a number of years after its introduction in 1947 (37). A more advanced electromechanical engraver made photoengravings on plastic sheets directly from colored copy (38). The plastic engravings could be used as printing plates or they could be contacted with film to make negatives. The engravers also made four-color corrected plates, which either were enlarged or reduced, directly from a color original, which was either a transparency or flat art. It was the only one of the small scanners that could enlarge or reduce the image and produce screened separations from the original. Its main disadvantage was the time required for the process: 4–6 hours for a set of separations at an enlargement ratio of 6× to 8×. Also, repairs and local corrections could not be made on the plates, and molds for electrotypes were difficult to make.

Duplicate Plates. Line and halftone photoengravings can be used directly in letterpress printing. Generally, however, the original engravings are used to make molds from which duplicate plates are made for the actual printing.

Stereotypes. Stereotypes have been used almost exclusively in letterpress newspaper printing, but their use is declining as photopolymer plates are replacing them (39). Stereotypes are used to some extent in the printing of books and short-run trade magazines. In the making of stereotypes, a matrix or mat is molded from the original engraving or type with heat and pressure in a matrix material consisting of a sheet of cellulose fibers and with a smooth, coated, malleable surface. Plastics can also be used for molds. After proper treatment and mounting of the mat, a stereotype is made by pouring molten metal in the mat. After casting and cooling, the stereotype is trimmed and curved into individual plate units for mounting on the press. Nickel- or chromium-plated plates can be used for long runs.

Electrotypes. Electrotypes are duplicates used for high quality commercial, book, and magazine letterpress printing (40). All electrotypes are made essentially by taking an impression of the original engraving in hot vinyl plastic, which is sprayed with silver to conduct current. Then a thin shell of copper or nickel is deposited by an electrolytic process, and the shell is backed with molten metal. As with stereotypes, electrotypes are plated with nickel or chromium for long runs.

Plastic and Rubber Plates. Plastic and rubber plates have the advantages over metal plates of light weight and facile curving for mounting on printing cylinders (41). They are quite durable, but they give relatively poor rendition of fine halftone images. The matrix material for these plates is a rigid board, which is impregnated with a phenolic thermosetting resin. Plastic plates are molded from thermoplastic vinyl resins. Rubber plates are molded from either natural or synthetic rubber or combinations of them, depending on what solvents are used in the printing inks. The rubber is vulcanized during the molding process to increase its hardness and resistance to solvents. The hardness is varied, depending on the materials to be printed. In flexographic printing, for which rubber plates are used extensively, the hardness of the plates is 45–55 on the Durometer A scale.

GRAVURE PLATEMAKING AND CYLINDERMAKING

The adaptation of the crossline screen for photogravure printing initiated the modern commercial process. Modifications for multicolor printing have been made (42–43).

Printing Surfaces. Modern gravure printing is done principally from images etched in cylinders on web presses and is generally referred to as rotogravure. On sheet-fed presses, the printing element is a thin copper plate, which is wrapped around the cylinder. Preparation of the printing surface is essentially the same for cylinders and for plates.

For monochrome printing by conventional gravure, bichromate sensitized carbon tissue or transfer film is contact-printed through a continuous-tone positive and then exposed a second time while in contact with a screen consisting of transparent lines and opaque square dots, 24 or 28 per cm (60 or 70 per in.). The ratio of line to dot width usually is 1:3. The exposed carbon tissue or transfer film is moistened and squeegeed in contact with the clean copper surface. Warm water is applied and the paper of the carbon tissue or backing of the film is peeled off. The gelatin thus transferred to the copper surface is further developed with warm water to produce a gelatin relief resist. Etching is done with 35–43 wt % ferric chloride solution (sp gr 1.34–1.45 (37–45° Bé)). This solution etches the copper to different depths, depending on the thickness of the gelatin resist in the different tone areas. Usually several solutions of different densities are used. The areas corresponding to the screen lines remain unetched and provide lands, which support the doctor blade in printing. For long runs and to minimize wear, the etched cylinder or plate is chromium-plated.

For multicolor printing, the principal process involves a combination of halftone and continuous-tone positives for each color. The halftone positive has a lateral dot formation, which is similar to conventional gravure in the shadows but with varying dot sizes in the middletone and highlights. The two positives are contact-printed successively, in register, onto a sheet of carbon tissue and the gelatin is transferred to the copper cylinder (42). Development and etching are essentially the same as for conventional gravure. The printing surface thus consists of disconnected ink cells of varying size and depth corresponding to the desired tone values. In the direct-transfer process, only the special halftone positive is made (43). This is contact-printed directly onto the copper cylinder, which has been sensitized with a photopolymer of the cinnamic ester type. After being etched, the printing surface consists of disconnected ink cells of varying size but approximately the same depth. A number of modifications of these two processes are in commercial use. For long runs the cylinders are chromium-plated.

Gravure plates can be used only for one image, which can be rerun. The image cannot be removed nor the plate be reused. On gravure cylinders, however, the electroplated copper that contains the old image can be stripped off and the cylinder can be replated for reuse (44).

The photochemistry of bichromated gelatin is similar to that of other bichromated colloids. Light sensitivity is governed by the same factors, namely, the bichromate: protein ratio, thickness of the layer, pH value, moisture content, and temperature. The thickness of the layer has an added importance, since the purpose of exposure of the carbon tissue is to insolubilize the gelatin layer to different depths, depending on the intensity of the light that is transmitted by different tones of the positives. The bichromate is yellow and absorbs actinic light, but its effect does not produce sufficient differentiation in depth of hardening within the layer. This differentiation is effected by pigmenting the gelatin layer with a reddish-brown, semitransparent pigment.

After the gelatin layer has been transferred face down to the copper surface, treatment with warm water dissolves the unhardened parts, leaving a gelatin relief resist. The thinnest areas are penetrated first by the ferric chloride solution, are etched deepest, and print the shadow tones of the picture. The areas where the resist is thickest are penetrated last by the etchant, are etched the least, and print the lightest tones.

Other Methods. A serious problem with gravure cylinder making has been the difficulty of reproducing identical cylinders from the same films. This is so because of differences in materials, etching solution, impurities in the copper, environmental conditions, and other factors that affect bichromated colloids. As a result, prepress proofing has not been successful and corrections must be made on the cylinders, which is expensive. Other methods have been tried for making cylinders, including the use of controlled etching systems, powderless etching as is used for copper engravings, electromechanical engraving, laser engraving, and offset positives. Electromechanical engraving, as is done on the Hell Helioklischograph, has been very promising. Also promising is the new Crosfield Lasergravure system. Another novel means of producing cylinders is the Toppan TH process, by which positives that are made for lithographic printing are used in making the cylinders. Gravure printing is one of the first processes involving no intermediate steps from the original copy to processing of the cylinder. This is made possible by the use of the advanced, expensive Hell HDP Direct Engraving system, which consists of the Hell scanner with a Chromacom page-composition system and a Helioklischograph.

LITHOGRAPHIC PLATEMAKING

Lithography (literally stone writing) was invented by Alois Senefelder, a Bavarian, in ca 1798 (45). He discovered that if he drew characters on smooth Bavarian Solnhofen limestone with a greasy crayon and then dampened the surface with gum water, he could repeatedly ink the greasy design and pull impressions on paper. In 1906, the offset principle was introduced. Until then, all lithographic printing on paper involved a direct transfer of ink from the stone or plate to the paper. The rotary offset press embodied an additional cylinder covered with a rubber blanket between the plate and impression cylinders. Thus, the ink was transferred first from the plate to the rubber blanket, then from the blanket to the paper. Until 1906, with the invention of the rotary offset press, no one had made use of the offset principle in printing on paper.

The offset press is characterized by the following important advantages over other printing methods.

(1) The rubber printing surface conforms to irregularities in the paper surface. Less printing pressure is needed, and print quality is improved. Halftones of high quality can be printed on rough papers.

(2) Paper does not come into contact with the metal printing plate. Therefore, the plate is less subject to abrasive wear and can run much longer editions.

(3) Speed of printing is increased. The effect of press improvements on printing speeds is shown by the following statistics for offset presses:

Press	Printing speed, impressions/h
litho hand-proving press	10–20
litho power-proving press (hand-fed)	1,200
direct rotary press (hand-fed)	2,000
rotary offset press (stream-fed)	up to 15,000
web offset press (roll-fed)	up to 60,000

(4) The image on an offset printing plate reads right instead of in reverse or wrong reading. This facilitates both hand and photomechanical preparation.

(5) Less ink is required for equal coverage. This reduces trapping problems and decreases the tendency of the printed sheets to smudge and set off, ie, produce a mirror image on the back, in the delivery pile. Also, drying is quicker.

The first offset presses were single-color, but soon two-, three-, and four-color presses were developed. The proportion of multicolor offset presses has increased. These include two-, three-, four-, five-, and six-color presses. Web- or roll-fed offset presses were developed and their number is steadily increasing. Commercial sheet-fed offset presses range in size from 43 × 56 cm to 140 × 198 cm. Sizes above 96 × 127 cm are used mainly for specialties, eg, packaging, greeting cards, labels, wraps, etc. Web presses range in web width from 19–193 cm. In addition, there are a large number of offset duplicating presses in sizes of 25 × 36 cm to 36 × 51 cm. These are mostly used in quick or instant-printing installations and in offices and plants for noncommercial, in-plant printing.

The developments in lithographic printing equipment have resulted mainly from improvements in the quality of printing plates and the efficiency of platemaking methods. Shortly after the advent of the offset press in 1906, methods were developed for making printing plates photomechanically, and the term photolithography was coined. Practically all printing plates are of this type. Despite these advancements, original stone plates are still in use, mainly by artists.

The first photolithographic printing plates were sensitized to light with bichromated egg albumin. Other light-sensitive materials are used in the same way, and the term surface plate applies to this group. Deep-etch plates were developed in ca 1930. Bimetal plates, although developed during the late 1930s, were not in general use until after World War II. This period is marked by the development of paper and plastic plates, primarily for the small duplicating presses. In ca 1950, presensitized diazo plates on metal were introduced. These and wipe-on plates, which were developed in ca 1957, have completely replaced albumin for surface plates. The newest developments are long run photopolymer, electrostatic, and laser plates.

Plate Materials. Most lithographic printing is from aluminum plates. These are manufactured in thicknesses of 0.15–0.51 mm, depending on the size and type of press, with thickness tolerances of ±0.013 mm for the smaller sizes and ±0.025 mm for the larger sizes. The plates must be flat with no buckles. Other metals, eg, zinc, mild steel, stainless steel, and brass are used as bases for bimetal plates and they must be made to the same tolerances as aluminum plates.

The plates are grained or roughened prior to coating or processing. This can be done in a flat, circularly oscillating tub by means of steel balls or marbles or abrasive grit and water, or by dry grit blasting or brush graining with nylon or steel brushes and abrasive. Grain depth usually is 3–8 μm. The plate grain provides protrusions for anchorage of the coating and ink, and recesses which help the surface carry moisture. Surface treatments that lower the surface energy and increase wettability have made possible the use of finer grains; even grainless plates are being employed. Paper, plastic, and foil-laminated plates are also used for the smaller sizes of presensitized, direct-image, projection, and electrostatic plates.

Chemistry of Photolithography. The chemistry of the photolithographic principle is described in ref. 46. The image areas on a lithographic plate must be ink-receptive and water-repellant, and the nonimage areas must be water-receptive and not be wet by ink. The wider the difference in ink and water receptivities of these two areas, the better the plate is and the easier the printing.

Roughening of the plate surface by ball or brush graining or by sand or grit blasting improves wettability. Surface treatments also have been used, especially on diazo-sensitized plates. Before diazo plates were developed, chemical surface treatments, eg, Cronak for zinc and Brunak for aluminum plates, were used (47–48). Surface treatments were especially effective on albumin- or casein-surface plates, for which they were used as a posttreatment after development.

Besides surface treatments, a most important way to maintain wettability is by the use of hydrophilic materials, eg, gum arabic, or arabogalactan. Gum arabic is the most widely used hydrophilic agent in lithography (see Resins, water-soluble; Gums). A mixture of calcium and magnesium salts of arabic acid, it has enough free carboxyl groups in the molecule to give good adhesion to metal surfaces. It is usually used in combination with phosphoric acid and salts, eg, ammonium bichromate, diammonium monohydrogen phosphate, and zinc or magnesium nitrate, to produce a hydrophilic layer on a lithographic plate. This is called a desensitizing etch or simply an etch. The gum arabic provides the water-receptive surface; the acid and salts serve as buffers and conditioners to passivate the metal surface and promote better adhesion of the gum to it. The gum arabic is used as a dilute, ca 15 wt % solution (sp gr 1.06 (8° Bé)) and dries on the plate as a protective layer to keep dirt, dust, fingerprints, and other grease smudges from printing. Care must be used to keep the gum off the image areas. If gum arabic dries on image areas, the areas become hydrophilic and refuse to wet with ink on the press, which results in blind spots.

The chemistry of plate desensitization is not well understood. The fact that desensitizing etches deposit an insoluble but hydrophilic gum film on the nonprinting areas has been established by dye absorption, contact-angle measurements, and radioactive tracer techniques. A suitable gum in the etch seems to be essential. Without it, the only desensitizing effect results from the subsequent gumming, which is the

144 PRINTING PROCESSES

final step in plate preparation. Only gums that contain free carboxyl groups seem to have appreciable desensitizing action. Other gums, eg, mesquite gum, do not desensitize the image nor the nonimage areas as well as gum arabic.

The effect of negative ions on the desensitizing film seems to be important, but not much is known about it. Dichromate, phosphate, nitrate, tannate, and gallate ions increase desensitization, whereas chloride and sulfate ions generally decrease it. Any particular desensitizing etch is characterized by an optimum pH value; most are 2.0–2.5.

Platemaking Processes. Lithographic plates are of many types and can be made by many processes. An excellent treatise on lithographic platemaking is given in ref. 49. Many types of plates and the processes used or planned for making them are illustrated in Figure 11. In the figure, the thick lines indicate the conventional processes in common use. The thin solid lines show the processes whose feasibility has been demonstrated and which are being used to a limited extent. The dotted lines are for processes that are in various stages of development.

The many different processes by which lithographic printing plates are made are distinguished by the ways the images are produced. Surface plates are those in which the light-sensitive coating eventually becomes the image that accepts ink during the printing cycle on the press. There are two types of surface plates: additive plates, on which an oleophilic lacquer or ink is applied to the image during development, and subtractive plates, which are precoated with oleophilic lacquer, which is removed in the nonimage areas during development. In deep-etch plates, the coating is removed from the image areas during development and these areas are chemically coppered

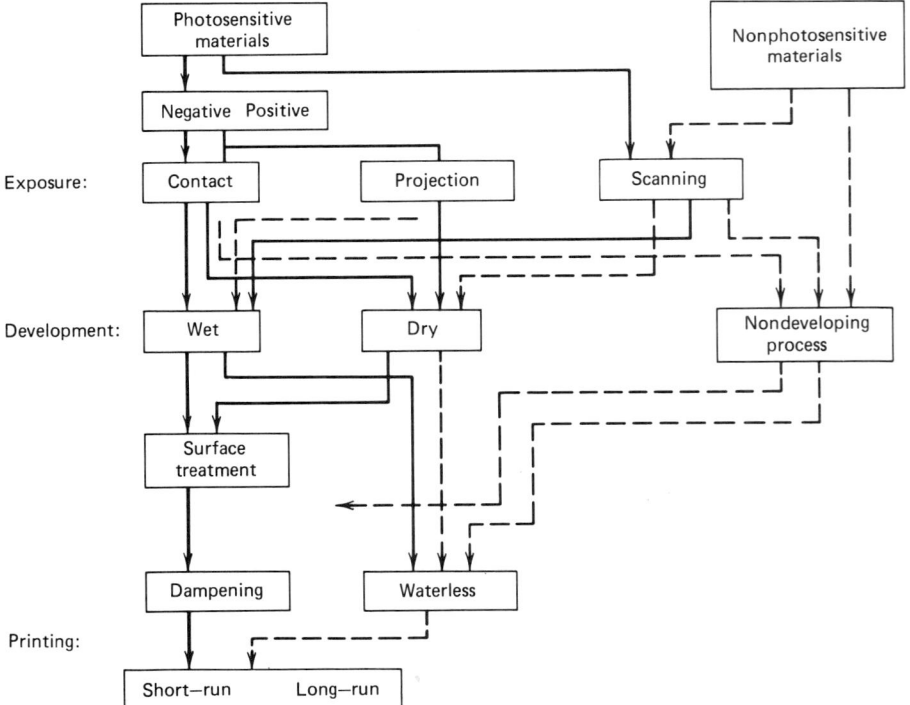

Figure 11. Lithographic plate processes: (━) in general use; (—) in limited use; (- - -) in development.

and/or lacquered and inked so that they become the oleophilic image on the press. Bimetal plates are similar to deep-etch plates in that the coating is removed from the image areas in development, but the areas consist of brass or copper either as the base metal or electroplated on another base metal. Laser imaging refers to plates that are exposed by computerized lasers or by real-time scanning of paste-ups. Other types of plates, which are used mostly in duplicating or printing on small presses, are direct-image plates, on which the image is drawn or typed directly on the plate; projection and diffusion-transfer plates, on which the image consists of a specially hardened photographic emulsion which is ink-receptive on the press; and electrostatic plates, on which the images are produced by fused ink-receptive toners (48).

Surface Plates. *Albumin Plates.* Until the introduction of presensitized and wipe-on plates, the most popular surface plates were those made with bichromated albumin or casein. Albumin plates are practically obsolete now but a description of them is useful to understand the lithographic principle and appreciate the improvements introduced with presensitized and wipe-on plates. The following steps are involved in making an albumin plate: (*1*) counteretch the metal plate with a weak acid solution, eg, 0.75 wt % HCl, to make sure it is chemically clean; (*2*) coat with bichromated albumin or casein coating on a whirler; (*3*) expose to a negative; (*4*) apply a developing ink; (*5*) develop in water or a very dilute ammonia solution, eg, 29 mL NH_4OH to 3.8 L solution; (*6*) apply a surface treatment, ie, Cronak for zinc, Brunak for aluminum, to remove residual protein from the nonprinting areas; (*7*) desensitize the plate with an etch; and (*8*) gum with a dilute solution of gum arabic (sp gr 1.06 (8° Bé)). Unless a surface treatment is used to remove residual coating from the nonprinting areas, these plates are problematic during printing: the nonprinting areas are not well desensitized and do not pick up ink uniformly or they pick up scum.

Diazo Presensitized and Wipe-On Plates. Diazo presensitized and wipe-on plates, which are additive plates, are much simpler to make and generally print cleaner than albumin plates, because the surface treatments used on them prior to coating are very water-receptive and inhibit any tendency to scum. Presensitized plates are already coated. Wipe-on plates are coated with a solution of the diazo resin by wiping the coating on with a sponge and drying it with a cloth or by applying the coating with a simple roller coater. After exposure to a negative, the presensitized and wipe-on plates can be handled in the same manner. An emulsion lacquer consisting of a pigmented lacquer, acidified gum arabic solution, and a suitable emulsifier is applied over the whole plate. The acidified gum dissolves the coating from the unexposed nonimage areas and the lacquer deposits on the exposed image areas, rendering them visible and ink-receptive. After washing, the plate is gummed and it is ready for use. Once mixed, wipe-on coatings should be used within 24 h as they deteriorate on standing. The diazo resins should be refrigerated as they decompose on standing at room temperatures.

Subtractive plates are presensitized plates that are precoated with a lacquer. After exposure and during development, the lacquer and coating are removed from the nonprinting areas. These plates are more consistent in their running characteristics and generally are capable of longer runs than the diazo presensitized or wipe-on plates.

There are some presensitized plates that can be made from positives. Diazo oxides are used, and the developing solutions are alkaline and remove the exposed areas and leave the unexposed ones as the printing areas. These are inked to protect them from light degradation and the plate is gummed. Most positive plates are subtractive. Few, if any, satisfactory subtractive or positive wipe-on processes have been developed.

Photopolymer Plates. Photopolymer plates of the cinnamic ester type have been used for lithography and they are available as wipe-on or presensitized plates. All are exposed to negatives. Since the coatings are water-insoluble, they must be developed in organic solvents. This can be done by one of three ways: wiping the solvents over the plate, using a vapor-degreasing machine, or using an emulsion of the solvent with the desensitizing agent for rendering the nonimage areas water-receptive. The plates are desensitized with an etch and are gummed.

Photopolymer plates that are made like polymethacrylate plates have been used. The presensitized plate is first exposed to a negative, then developed in an aqueous solution, desensitized, gummed, and reexposed to light, which hardens the image areas for long runs (46). Other photopolymer plates are made with combinations of polymers of the polyacrylate or poly(vinyl acetate) type and a diazo resin as a sensitizer (49). An important feature of some photopolymer plates is that run life on the press can be increased considerably by baking the plates in an oven at 250–300°C for ca 4.5–5.5 min. Runs of over 10^6 on web presses have been reported.

Printing Characteristics. As with all bichromated coatings, albumin and casein plates were affected by temperature and relative humidity. Also, if not made properly or if a surface treatment was not used after development, they had a tendency to scum on the press. This is one of the main reasons why presensitized plates have replaced them. These plates are not affected much by relative humidity but can deteriorate if exposed for any length of time to temperatures exceeding 52°C. The additive-presensitized and wipe-on plates are not very resistant to abrasion on the press, and they are not practical for runs over 5×10^4 impressions on sheet-fed presses or over ca 10^5 impressions on newspaper and other web presses. The prelacquered, presensitized plates are capable of longer runs. Photopolymer plates show good resistance to temperature and relative humidity and to abrasion on the press. They are capable of runs in excess of 2.5×10^5 and baked plates have run over 10^6 impressions.

Deep-Etch Plates. With the introduction of positive photopolymer plates and baking, deep-etch plates are gradually becoming obsolete. There are several presensitized deep-etch plates, but by far the majority of deep-etch plates are of grained aluminum with bichromated gum arabic coatings. Most plates in Europe are made on anodized aluminum. The most popular deep-etch plate in the United States is the copperized aluminum plate, in which copper is deposited chemically on the image areas. The following steps are needed to make a deep-etch plate.

(1) Counteretch the grained aluminum plate with a solution containing ca 12 mL 85 wt % phosphoric acid and water to make 3.8 L solution.

(2) Coat with bichromated gum arabic solution on a whirler. A typical coating formula is 2.8 L gum arabic solution (sp gr 1.11 (14° Bé)), 200 mL ammonium bichromate, 141 mL 28 wt % ammonium hydroxide, water to make 3.8 L, pH of final solution = 8.5–9.5, sp gr 1.11 (ca 14° Bé), average thickness of dried coating = 12–13 μm.

(3) Expose through positives.

(4) Stage out, ie, paint, all areas that did not expose but are not to print, such as dust specks, film edges, crop marks, etc, with a pigmented cellulose nitrate lacquer, which is alcohol-soluble.

(5) Develop in a nearly saturated salt solution. Water cannot be used for development because the exposed coating is too soluble in water. The solubility differential between exposed and unexposed bichromated gum arabic is maximized in almost saturated salt solutions containing an organic acid. A typical formula for such a de-

veloper is 3.6 L 39 wt % calcium chloride solution (sp gr 1.38 (40° Bé)), 192 mL 85 wt % lactic acid. The solution is applied over the plate with a plush pad until metal is bared in the printing area or until step (9) (see below) is reached on the GATF sensitivity guide. The developer is squeegeed off and two more fresh applications of developer are made for the same length of time or until step 6 is reached on the GATF sensitivity guide.

(6) Deep etching is also done with a nearly saturated salt solution. A typical formula is: 2.6 L 39 wt % calcium chloride (sp gr 1.38 (40° Bé)), 1.04 L zinc chloride, 0.75 L 48 wt % ferric chloride (sp gr 1.53 (50° Bé)), 37 mL 37 wt % hydrochloric acid. This solution is applied to the plate with a plush pad for 1–1.5 min and squeegeed off.

(7) The alcohol wash removes the residue from deep-etching. Approximately four washes of anhydrous denatured ethyl alcohol or anhydrous isopropyl alcohol are used. The first wash removes the staging lacquer and vigorous rubbing with embossed paper wipes is recommended to remove iron that was deposited in the image area during deep-etching. In some platemaking procedures, developer is applied for 2–3 min to remove the iron.

(8) While the plate is wet with the last alcohol wash, copperizing solution is poured on the plate. The formula for such a solution follows: 950 mL 99 wt % isopropyl alcohol, 30 mL cuprous chloride, 30 mL 37 wt % hydrochloric acid. Copper deposition is complete in ca 5–7 min. The plate then is washed with two more applications of anhydrous alcohol.

(9) When the plate is thoroughly dry, it is coated with a thin layer of ink-receptive, vinyl-type lacquer or epoxy, either of which has excellent nonblinding characteristics, ie, neither wets with acid and gum on the press. After the lacquer is thoroughly dry, developing ink is applied, rubbed well, and dried.

(10) The dried, lacquered, and inked plate is soaked for ca 10 min in warm water (ca 38°C) to soften the gum stencil, after which it is scrubbed under running water to remove all visible traces of the stencil and its lacquer and ink coating.

(11) An etch is applied to the plate and rubbed over it for ca 1.5 min.

(12) The etch is rinsed off the plate; it is squeegeed to remove most of the water and is placed on a flat table. A solution of gum arabic (sp gr 1.06 (8° Bé)) is applied over the whole plate and the plate is rubbed smooth until it is dry. Drying with a fan is continued to ensure removal of all moisture.

Deep-etch plates are very dependable and durable and have been used for the majority of medium to long run, quality lithographic jobs. They have also been used extensively for color printing, for web offset, and for packaging printing. Their principal disadvantage, in addition to the total time required to make the plates, is the time required for staging or painting unwanted work on the plates. If any spots are missed in the platemaking operation, the press must be stopped while the spots are polished from the plates. With the introduction of pin-register systems and long-running baked photopolymer plates, much printing that was done from positives on deep-etch plates has been converted to negative plates. Also, much of the work that is in positive form is being printed from long-run positive photopolymer plates, so that use of deep-etch plates is declining.

Bimetal Plates. On bimetal plates the ink-receptive image and the water-receptive nonimage areas consist of different metals (50). The image metal is usually copper but can be brass; the nonimage metal can be chromium, aluminum, or stainless steel. The feature of these metals is that copper is ink-receptive in the presence of ions, such as phosphate and nitrate, which render chromium, aluminum, and stainless steel water-receptive.

Type I bimetal plates consist of 2–2.5 µm of copper, which is electroplated on aluminum or stainless steel. They are usually made from negatives. The regular platemaking procedure for these plates is as follows. (*1*) Clean with pumice and 2 wt % sulfuric acid. (*2*) Coat with a deep-etch light-sensitive coating. (*3*) Expose to light through a negative. (*4*) Develop with a deep-etch developer as in deep-etch platemaking. (*5*) Wash with alcohol and dry as in deep-etch platemaking. (*6*) Etch the bared copper in the nonimage areas with 40 wt % ferric chloride solution (sp gr 1.41 (42° Bé)) if the base metal is stainless steel or ferric nitrate solution if the base metal is aluminum. (*7*) Remove the stencil from the image areas with pumice and 2 wt % sulfuric acid. (*8*) Ink the copper image with a rub-up ink, which is like a developing ink but with a higher consistency. (*9*) Etch and gum the plate as in deep-etch platemaking.

There are some presensitized plates of this type that can be made from positives. These have diazo oxide coatings. After exposure through a positive and development in an alkaline developer, which dissolves the exposed coatings, the unexposed image areas are protected with a fixing solution, which prevents the acid in the copper etch from attacking the image.

Type II bimetal plates consist of 1–2 µm of chromium electroplated on 3–8 µm of copper, which is plated on a base metal of either zinc, aluminum, mild steel, or stainless steel. Some of the plates that are used on web presses have also been made by plating the chromium directly on sheets of copper or brass. Type II plates are always made from positives. The platemaking procedure for these plates is as follows. (*1*) Remove the gum. (After plating, gum arabic is applied to the chromium to protect it and keep it water-receptive. Wash the gum off with water. Sometimes the plates are counteretched with dilute sulfuric acid or phosphoric acid.) (*2*) Coat with a deep-etch coating. (*3*) Expose to light through a positive. The positives should have slightly smaller dots than for deep-etch plates because there is some lateral etching as the chromium is removed which enlarges the printing image elements slightly. (*4*) Stage as in deep-etch platemaking. (*5*) Develop as in deep-etch platemaking. (*6*) Etch through the chromium layer in the image areas. A suitable chromium etch formula is: 2.8 L 32 wt % aluminum chloride solution (sp gr 1.28 (32° Bé)), 2.5 kg technical granular zinc chloride, 148 mL 85 wt % phosphoric acid (51). (*7*) Alcohol-wash as in deep-etch platemaking. (*8*) Sensitize copper to ink by covering the image areas with a thin solution of asphaltum and oleic acid. This protects the copper from corrosion and renders it ink-receptive. (*9*) Remove the gum stencil as in deep-etch platemaking. (*10*) Etch and gum as in deep-etch platemaking.

Bimetal plates are the most rugged plates used in lithography; they are capable of runs in excess of 10^6 impressions. Since they have very good abrasion resistance, they are used under the most difficult conditions of printing, eg, on poorer grades of board and paper. Their use requires a minimum of water to keep the nonprinting areas clean and, because they are relatively smooth, the images are solid, clear, and sharp. They are the easiest plates to handle on the press because a single treatment is all that is needed to restore them to printing condition when something happens on the press. A treatment with dilute nitric or phosphoric acid renders the copper or brass ink-receptive and the chromium, aluminum, or stainless steel remain water-receptive. With other types of plates, this type of treatment can damage the image areas while improving the nonimage areas, and much more careful treatment is needed on the press.

The main disadvantages of bimetal plates are their high cost, the length of time to make them, and the presence of heavy metals in the developing and etching baths which can cause water pollution if the baths are dumped in municipal sewers. This is also a problem with all processes involving bichromated colloids, eg, photoengraving, deep-etch, collotype, and gravure.

Waterless Plates. A number of attempts have been made to eliminate dampening in the lithographic process. Most have involved emulsifying the dampening fluid with the ink (52). In 1970, 3M announced the Driographic process, which involves the use of a special planographic plate which need not be dampened for printing (53). The plate consisted of an anodized aluminum base coated with a diazo sensitizer, which was overcoated with a silicone rubber. After exposure through a negative and development, the silicone and diazo were removed from the image areas, which were inked, leaving the silicone rubber on the nonprinting areas. Silicones have such a low surface energy that they refuse to wet with ink; therefore, no water is needed to print the plates on the press. In practice, the plates did pick up some ink causing deposition of scum over the nonprinting areas. Modifying the ink with silicones helped, but as the ink heated up on the press, it started to scum. Temperature control of the inking rollers reduced the scumming but the problem was so persistent that 3M withdrew the plates from the market in 1977.

Others experimented with the principle and a number of patents have been issued. The Japanese have been reasonably successful with the Toray plate (54). The plates have a disadvantage in that they must be made from positives; consequently, they are less desirable for the U.S. market. Negative plates are being developed. If the success of these plates continues, this new plate method can be expected to command a good share of the lithographic plate market, as it eliminates one of lithography's most serious disadvantages, ie, the necessity for an ink–water balance which causes variations in color balance during printing which, in turn, increases waste.

Electrostatic Plates. Electrostatically produced images can be used for photomechanical platemaking. Two electrophotographic processes are in common use for producing images; these are xerography and electrofax (see Electrophotography). In these processes, a photoconductive surface accepts and holds a uniform electrostatic charge in the absence of light but loses the charge when exposed to light. When such a surface is exposed to a light image, the charge is retained only in the unexposed or image areas. The charged image can be developed by attracting toners which are finely divided, oppositely charged pigmented particles. These are positive–positive processes.

In xerography, the photoconductor consists of selenium in a suitable binder, which is coated permanently on a drum or plate (55). The powder image is electrostatically transferred to paper for final fixing, and the image on paper can then be used as a copy or can be used as a plate for an offset duplicating press from which a number of copies can be printed. It can also be transferred to a metal plate which, after proper fixing of the image and treatment of the nonprinting areas, becomes a lithographic plate for appreciable runs. In the electrofax process, the photoconductor is zinc oxide in a suitable binder (56). This photoconducting material is coated directly on paper or metal for duplicating or reprography. It has been used for making relief plates and modifications of the process are used for newspapers. The main advantage of these systems is the absence of an intermediate silver film. With the escalating silver and film-base costs, this results in appreciable cost savings.

An application of the electrofax process is the 3M Pyrofax platemaking system, which is used extensively by the newspaper industry (57). The equipment consists of an imager/camera and a plate processor (Fig. 12). An image is made from a camera-ready paste-up of the newspaper page with the special imager/camera on an intermediate film coated with zinc oxide photoconductor. After toning, the image is transferred to a clean blanket mounted on the drum pin-bar in the imager. An uncoated litho plate is registered in the plate processor, the imaged blanket is mounted on the plate-processor register pins, and the toner image is transferred from the blanket to the plate after which it is fused and treated with a prepress solution to assure fast roll-up on the press. The photoconductor-coated film is reused and multiple images can be transferred to plates. The fused toner on the plate is usable for runs in excess of 10^5 impressions.

Another electrostatic plate process is the Kalle Elfasol plate system. It is similar to electrofax but an organic photoconductor and liquid toner are used for higher resolution. Also, the photoconductor is coated directly on anodized aluminum, and therefore, no transfer is involved. The image can be produced in a camera from a paste-up or it can be made by laser exposure (see Laser Platemaking). After exposure, toning, and fusing, the photoconductor must be chemically removed from the nonimage areas of the plate, which are then etched and thereby made hydrophilic.

Another plate process involving electrostatic principles is the KC-Plate system (58). This process is based on the use of vacuum-deposited anisotropic crystallites of cadmium sulfide on a metal base. The system is completely inorganic, and it is relatively unaffected by changes in temperature and relative humidity. The coating is extremely thin, ca 0.33 μm, and has very high resolution. In use, the coating is charged, exposed, and toned in a liquid toner. The toner can either be fused on the plate and the plate converted in a chemical bath to render the nonprinting areas hydrophilic, or the toner image can be transferred to a less expensive lithographic plate, as in the Pyrofax process.

Laser Platemaking. In several processes, lasers (qv) are used for exposing images from paste-ups or computer memories onto film or plates. One system scans or reads a paste-up with a helium/neon (He/Ne) laser and transmits the data to the write laser,

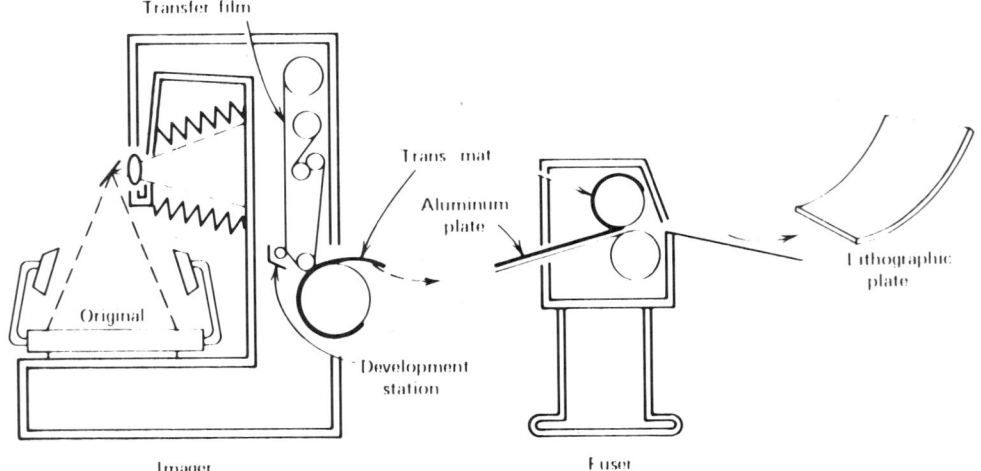

Figure 12. 3M Pyrofax system. Courtesy of 3M Printing Products Division.

which is an argon-ion laser, to produce a reproduction of the paste-up on film or on a lithographic plate (59). The operation can be transmitted to one or more other locations where the write operation is performed on film or on lithographic plates. In another system, a paste-up is scanned with a He/Ne laser and an yttrium aluminum garnet (YAG) laser is used to produce an image on an intermediate Lasermask, which consists of a plastic film coated with a black-pigmented resin (60). The Lasermask can be placed in contact with an uncoated lithographic plate or a receiver sheet. During exposure, the YAG laser softens the coating, which transfers to the substrate in contact with it, thereby producing a negative on the Lasermask and imaged plate or proof simultaneously. The main advantage of laser platemaking is that the paste-up can be eliminated entirely and the plate can be made from scans that are stored in a computer memory. This makes the computer-based, electronic page-composition and -imposition systems complete prepress systems, which do not involve any intermediate films or manual handling of such films.

Other Processes. A number of lithographic plate-making processes that are in development or testing stages are described in ref. 46.

Collotype Printing. Although collotype printing is not strictly lithography, it is similar in principle (61). Collotype printing is done both from glass plates on flat-bed presses and metal plates on direct rotary presses on which the image is transferred directly from the plate to the paper. The plates are made with a coating of bichromated gelatin. Potassium bichromate is generally used as the sensitizer as it has a longer scale of reproduction than ammonium bichromate; ie, it reproduces more steps on a gray scale. A continuous-tone negative with a density range of ca 1.2 and a gamma close to 1.0 produces reproductions of good quality. The gelatin is differentially hardened inversely to the density of silver in the negative. The plate is developed by washing in water after exposure. The temperature at which the coating dries and the temperature of the washwater control the degree of reticulation of the gelatin. The reproductions are not truly continuous in tone but show the irregular pattern of the reticulation. Before printing on the press, the plates are soaked in a glycerol or ethylene glycol bath. The gelatin swells in inverse proportion to the amount of light that is received during exposure and, consequently, accepts ink in inverse proportion to the swelling or direct proportion to the light exposure.

Screenless Printing. Screenless printing is done from continuous-tone films on presensitized plates and involves regular lithographic presses and techniques (62). Positive systems seem to work best. The random grain pattern of the images is almost indistinguishable from the reticulation pattern of the gelatin, which is characteristic of collotype. The process has not been used much because of variations in plate coatings, which affect the consistency of printing results. The tone reproduction of plate coatings depends on the coating thickness and grain of the plate, each of which is dependent on the other and neither can be controlled accurately (63–64). Research is improving plate consistency and predictability and many plants in the United States and the United Kingdom are using the process with moderate success.

Direct-Image Plates. Direct-image plates usually have a paper, resin-impregnated, or plastic base. They are not characterized by quality capabilities suitable for commercial advertising printing, but the quality is sufficient for in-plant and quick printing operations. Direct-image plates are like the original lithographic plates drawn on stone. They are called masters and the image can be drawn, lettered, painted, ruled, traced, typed, or written on them using pencil, crayon, ink, carbon paper, fabric or carbon-paper ribbon, rubber stamp, numbering machine, brush, or air brush. Preprinted masters can be made by printing the image by the letterpress process or offset with oil-base inks, which are ink-receptive on the press. Guide lines and instructions can be printed with water-color inks, as these do not print. These plates are ideal for systems printing and are used extensively in personalized check printing for magnetic ink character recognition (MICR) and for other in-plant uses.

Making direct-image plates is very simple and only two steps are involved. (1) The image is applied to the plate by any of the methods or materials mentioned. The plate must be protected from grease spots, such as fingerprints, grease, or oil from the machine, etc, as these print. (2) The plate is mounted on the press, the etch designed for the plate is passed over the plate, the ink and dampening rollers are dropped on the plate, and printing can begin.

With paper-base plates, the fountain solution used on the press must be the one designed for this type of plate. Most of these fountain solutions contain glycols or glycerol. The inks should be designed to be compatible with these polyalcohols, as some inks tend to emulsify with these materials.

IMAGE CARRIERS FOR THE STENCIL PROCESSES

There are two stencil processes in general use: stencil duplicating and screen printing. (See Reprography for description of stencil duplicating.) Screen-printing image carriers can be produced manually or by photomechanical means (65). The screens can consist of silk bolting cloth with taffeta weave and mesh counts of 40–80 openings per lineal centimeter. Nylon screens are used for textile printing and metal screens of phosphor bronze and stainless steel are used for fine detail printing in meshes as fine as 120 openings/cm. The screen material is attached to a rigid frame and is stretched tightly so that it is level and smooth. The stencil is applied to the bottom side of the screen, ie, the side in contact with the surface to be printed. Ink with a consistency similar to thick paint is used in the screen and the ink is transferred by rubbing on the screen surface with a rubber squeegee. The screens can be used for ca 10^5 impressions.

Manual stencils are made by knife-cutting special film stencil materials. These consist of two plastic layers. The image to be printed is cut through one layer, and this part of the stencil is placed in contact with the underside of the screen. A solvent, which is insoluble in the ink but attaches the cut stencil to the screen, is applied and then the backing layer is removed. Manual stencils can also be produced by drawing directly on the screens with special materials. When screen printing is used for art reproduction it is called seriography.

Photomechanical stencils are of two types, direct coatings and transfer films. Direct coatings are either bichromated gelatin or bichromated PVA. The coated screens are exposed through a positive, washed, and inspected. These screens are used for printing electronic components. They are not very practical for short-run commercial work because of the difficulty of reclaiming the screen after use.

There are four transfer-film methods for making screens: carbon tissue as used in gravure printing, unsensitized film, presensitized film, and photographic transfer film. The carbon-tissue method is almost the same as the gravure process, except that the tissue is transferred to a temporary vinyl support for development before application to the screen. The unsensitized and presensitized films are similar in their use in that there is no double transfer, which is necessary with carbon tissue. These materials are mounted on plastic films. The unsensitized film must be sensitized before use with an alcoholic solution of ammonium bichromate.

The latest method of producing images for screen printing is with rotary screens, which are made by plating a metal cylinder electrolytically on a steel cylinder, removing the cylinder after plating, applying a photopolymer coating to the cylinder, exposing it through a positive and a screen, developing the image, and etching it. The result is a cylinder with solid metal in the nonimage areas and pores in the image area. On rotary screen presses, the ink is pumped into the cylinder, and the squeegee, which is inside the cylinder, controls the flow of ink through the image pores to the substrate.

Printing

In general, printing is done by feeding the paper into the press in sheets, ie, sheet-fed, or from rolls, ie, web-fed. Newspapers and magazines are printed on presses that are web-fed. Greeting cards, maps, and most packaging are printed on sheet-fed presses. Many presses are multicolor: they can print a number of colors in succession. Usually each color requires a separate complete unit, ie, inking, plate, and impression mechanism, on the press. A two-color press consists of two such units, a four-color press consists of four. Perfecting presses print both sides of the sheet in one pass through the press. They can be either blanket-to-blanket, as in most web-offset presses and some sheet-fed offset presses for printing books, or the units can be in line and the sheets are turned over between printings.

LETTERPRESS

Letterpress presses are of three types: platen, flat-bed cylinder, and rotary (Fig. 13). Platen presses have two flat surfaces, the bed and the platen. Type and plates can be mounted on the bed. These are inked by inking rollers and the impression is made or pulled on sheets, which are fed manually or automatically on the platen. Maximum size of platen presses is 45.7 × 61 cm. They are used for many different purposes, eg, printing of paper and paperboard, envelopes, embossing, steel-rule die-cutting, and gold-leaf stamping.

Flat-bed cylinder presses are available with horizontal and vertical beds and they can print type and plates. The horizontals print larger sheets (up to 106.7 × 142.2 cm) than the platen presses and were designed in three types: single color, two color, and perfecting presses. Their manufacture was discontinued in the United States in 1962, but some are still being built in Europe. The vertical presses are popular job presses and their associated maximum sheet size is 35.6 × 50.8 cm. Their rated speed is 5000 impressions per hour (iph).

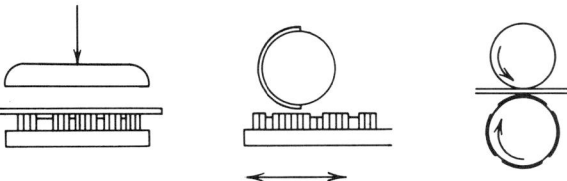

Figure 13. Diagram of platen, flat-bed, and rotary presses for letterpress printing (1). Courtesy of The Printing Industries of America.

154 PRINTING PROCESSES

By far the greatest amount of letterpress is printed on rotary presses. This is the type on which long-run commercial work, packaging, newspapers, and magazines are run. The sheet-fed presses were made in sizes up to 138.4 × 195.6 cm with rated speeds as high as 152 m/h (6000 iph), but the medium and large sizes are no longer made in the United States. Plates must be curved for mounting on these presses. Stereos, electros, photopolymer, and wraparound plates can be used on them. One type of wraparound press was built like a sheet-fed offset press with three cylinders. The plate was mounted on what would be the blanket cylinder, and what would normally be the plate cylinder was used as a large ink drum, which distributed the ink evenly.

Web-fed rotary letterpress presses are of many sizes and styles. In magazine printing, presses of the unit type are used, which include a separate complete unit for each color printed, and the common-impression cylinder types, often called satellite type, on which the printing units are situated around one large cylinder which serves as the impression cylinder for all the printing units. Newspaper presses are built in couples with each unit printing both sides of the paper in succession, usually 16 pages (8 pages on each side), at a time. Flexographic presses are of three types: stack, central-impression cylinder, which is similar to the common-impression-cylinder magazine press; and in-line, which is similar to the unit-type press (Fig. 14).

Makeready. Because pressure is needed for ink wetting and transfer and because of the variable size of the image elements in letterpress, the same amount of printing pressure or squeeze exerts more pressure on highlight dots than on shadow dots. This necessitates considerable makeready to even the impression so that the highlights print correctly and do not puncture the paper. Precision electros, wraparound plates, and premakeready systems have helped reduce makeready time, but it is still appreciable for quality printing and is a reason letterpress has been replaced by other processes for some types of work.

GRAVURE

Not much gravure printing is from sheet-fed presses, which involve plates. By far the greatest amount of gravure printing is from cylinders on rolls of paper or film. The cylinders are removable and cylinders of different diameters can be used so that different print lengths can be accommodated from job to job, if necessary. This is desirable in packaging but not as necessary in magazine printing. The gravure printing unit consists of a printing cylinder, an impression cylinder, and an inking system, as indicated in Figure 15. Ink is applied to the printing cylinder by an ink roll or spray

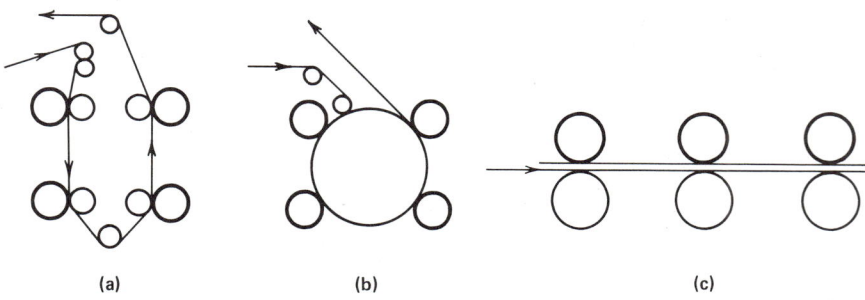

Figure 14. Diagram of (**a**) stack-type, (**b**) central-impression cylinder-type, and (**c**) in-line type of flexographic presses (30). Courtesy of The Printing Industries of America.

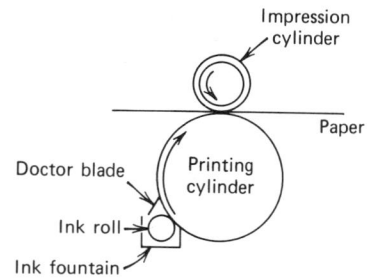

Figure 15. Diagram of a gravure printing unit.

and the excess is removed by a doctor blade and returned to the ink fountain. The impression cylinder is covered with a resilient rubber composition, which presses the paper into contact with the ink in the tiny cells of the printing surface.

Gravure ink consists of pigment, a resin binder, and a volatile solvent. It is quite fluid and dries entirely by evaporation. For high speed printing, the solvents are quite volatile and the inking system must be enclosed. In multicolor printing, where two or more gravure units operate in tandem, each color dries before the next is printed. The web, therefore, is passed through a heated dryer after each impression. In magazine printing, where the same substrate and ink solvent is used, the dryers are usually connected to a solvent-recovery system to conserve solvents and to eliminate air pollution.

Single-color rotogravure printing yields excellent pictorial reproduction on a wide range of papers. Its reproduction of type matter and line drawings leaves something to be desired, however, because the screen somewhat reduces sharpness. Color reproduction is done in three or four colors on multicolor presses. Gravure printing is widely used for newspaper preprints and magazine supplements, magazines, mail-order catalogs, cartons, labels, and in the printing of cellophane, plastic films, foils, and plastic laminates. It is the most practical process for the printing of gold, ie, bronze, silver, ie, aluminum, and opaque whites.

A serious problem in gravure printing has been the necessity for very smooth papers, otherwise there are skips in the printing (see Paper). The introduction of trailing-blade-coated papers in ca 1957 was a benefit to gravure printing but it did not benefit printing on newsprint or on rough boards. The development of the electrostatic assist has helped to solve this problem and to raise the general level of quality of gravure printing on all paper and paperboard stocks (66).

LITHOGRAPHY

The lithographic press is based on the principle that grease and water do not mix. Design of a sheet-fed, offset printing unit is shown in Figure 16. Multicolor presses consist of 2–6 of the units shown in Figure 16 in tandem, with a single feeder and delivery. There are some presses that include a common-impression cylinder for printing two colors. For web presses, paper is fed from a roll and delivery is in the form of a roll, cut sheets, or folded signatures. Most web-offset presses are of the blanket-to-blanket design, in which two units print simultaneously on both sides of the sheet and one blanket acts as the impression cylinder for the other blanket and vice versa. There are also satellite or common-impression cylinder presses and some newspaper presses include combinations of blanket-to-blanket units interspersed with common-impression cylinder units on which three colors are printed in succession.

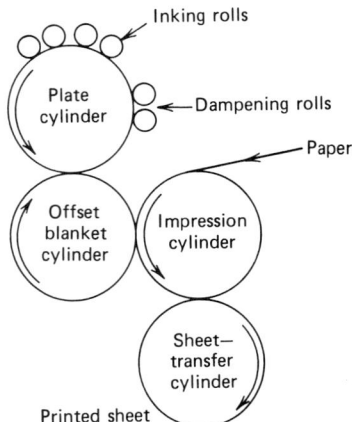

Figure 16. Diagram of a sheet-fed, offset lithographic printing unit.

Direct Lithography. Some newspapers have been faced with the problem that their letterpress equipment is not old enough to be replaced and yet the economics and quality of the lithographic process is desired. In ca 1974, the American Newspaper Publishers Association Research Institute developed the concept of di-litho or direct lithography (67). Letterpress plate cylinders were converted so that lithographic plates and dampening systems could be attached to them. A number of presses were converted to this process. In recent years, offset conversion systems have been developed to convert letterpress presses to print by offset by the addition of blanket cylinders to the di-litho cylinders.

Ink. Lithographic ink is basically a concentrated dispersion of pigment in a viscous oil vehicle with various additives, which give it suitable working properties. One type contains a drier, which accelerates hardening of the vehicle after printing (see Driers and metallic soaps). In another type, the oil vehicle consists of a resin in a volatile solvent. This type dries by evaporation of the solvent and penetration of the solvent into the paper. Various combinations of drying oils, resins, and solvents are used. There are also solventless inks which can be cured by uv and electron-beam radiation. For lithographic ink, careful selection of ingredients is essential. Since the ink comes into intimate and continuous contact with water during printing, it must be free from any tendency to bleed or to form an ink-in-water emulsion. The formation of water-in-ink emulsion is unavoidable, but this does no harm unless the working consistency of the ink is affected. During normal printing, the ink takes up 5–30 wt % water as a water-in-ink emulsion. The surface chemistry of this ink–water relationship is little known (see Inks).

Dampening. In the printing cycle, water or fountain solution is fed to the plate just before it contacts the inking or form rollers. This is done by means of rollers, which are covered with cotton flannel and are called molleton rollers, or with parchment paper, to which the water is metered from its fountain. Rubber composition and plastic rollers also are used to transfer the fountain solution to the plate. Very little moisture is required. The moisture film, which is transferred on the plate, is continuous on the nonimage areas of the plate and acts as a barrier thereby preventing adhesion of ink. Any moisture on the greasy image areas is discontinuous and does not prevent transfer of ink to them. To keep the nonprinting areas clean during long runs, the dampening water should contain some acid and a desensitizing gum. It is assumed that their function is to maintain the desensitizing film, which was produced when the plate was made, but little is known of the surface chemistry involved. With well-desensitized plates, the pH value of the fountain solution or dampening water may be as high as 5–6. With poorly desensitized plates, it may be as low as 3–4. Up to 30 wt % isopropyl alcohol is used in new dampening systems as a dampening-solution ingredient. In some systems, the water–alcohol mixture is fed directly to the first ink-form roller. These methods increase the efficiency of the dampening action but produce no deleterious effect on the paper, ink, or plate.

Blankets. The standard offset blanket has a three-ply fabric base, on one surface of which is a skim coat of a rubber compound 0.305–0.381 mm thick. The total thickness is ca 1.63 mm. This blanket is wrapped around the blanket cylinder and is held under a tension of ca 890 kg/m (ca 50 lb/in.) of width. The rubber surface receives ink from the image areas of the plate and transfers it to the surface of the substrate being printed. The life of the blanket is limited, since it is susceptible to mechanical damage, to swelling resulting from absorption of ink vehicle or solvent, and to development of tackiness and glazing, which are caused by oxidation that is stimulated by absorbed ink driers, principally cobalt and manganese soaps. Great improvements in resistance to ink vehicles, solvents, and driers have been made through the development of synthetic-rubber compositions and suitable antioxidants (see Antioxidants and antiozonants). Also, compressible blankets, which improve sharpness of printing, have been developed.

Paper. Although lithographic printing can be done on practically all types of paper, certain special requirements are necessary for best results in long runs. These include freedom from loose surface fibers and unbound mineral filler, sufficient bond or pick strength to resist the pull of tacky lithographic inks, minimum tendency to curl, freedom from chemicals that could sensitize the plate or cause the formation of an ink-in-water emulsion, and in coated papers, resistance of the coating adhesive to water. Offset lithography produces quality printing on both rough and smooth papers, but the trend in recent years has been toward coated papers because they require less ink and give more brilliance to colors.

COLLOTYPE

In collotype or photogelatin printing, no dampening is used during printing as on a lithographic press. The presses are direct printing and the plates are mounted on them after they have been soaked in a glycerol or ethylene glycol bath. The gelatin attracts ink in proportion to the amount of exposure it has received. The shadow areas, with much exposure, take the most ink; the highlight areas, with only a small amount of exposure, take very little ink. The hygroscopic polyalcohols absorb moisture from the air, thereby maintaining proper moisture content in the gelatin which controls the tone reproduction. The relative humidity of the pressroom is gradually increased to compensate for the loss of moisture from the gelatin at each impression. As the moisture content of the gelatin decreases, its tone reproduction flattens so that more moisture is needed to increase contrast. Eventually a point is reached where the flattening can no longer be compensated by raising the relative humidity in the air and the plate must be resoaked. The relative humidity must be lowered and the cycle between moisture content of the gelatin and relative humidity is repeated. From 500 to 1000 good prints are obtained between soakings. Although the collotype process is capable of beautiful printing, no two prints are exactly alike as the contrast of the printing varies with the moisture content of the gelatin. It is a short-run process, is slow, and is expensive. It could eventually be replaced by screenless lithographic, if the variables affecting tone reproduction of lithography plates can be controlled.

SCREEN

Screen printing equipment can be very simple, consisting only of a table, screen frame, and squeegee. However, power-screen presses with mechanical feed and delivery are in common use and produce 450–3500 prints per hour, depending on the sheet size. Special presses are designed to print objects of irregular shape, eg, milk and soft-drink bottles, and use ceramic colors, plastic and metal containers, etc. Rotary screen presses not only increase printing speed but improve quality and extend the range of products produced by screen printing. Screen-printing inks are usually of the drying-oil type and have the consistency of thick paints. The amount of ink applied in screen printing is far greater than in letterpress printing or lithography, and the prints must be racked separately until dry or be passed through a heated tunnel before they can be piled. Uv-curing inks are also used.

Screen printing is in common use for the production of art prints, posters, decalcomania transfers, greeting cards, menus, program covers, and wallpapers. It is particularly adapted to the printing of fabrics, felt, leather, metals, glass, ceramic materials, and plastics, both flat and in finished form. Printing an adhesive size and then dusting with cotton, silk, or rayon flock, makes the appearance of the finished design like felt or suede leather.

Screen printing has distinct advantages for short runs because of the simplicity of the equipment needed. For longer runs, the advantage is soon lost, since other printing methods are much faster. For many of the applications listed, however, screen printing is the only practical process.

OTHER

Micropublishing. Micropublishing is an important process for the storage and retrieval of scientific and technical reports, directories, catalogues, medical and business records, engineering and architectural drawings, and rare books in libraries. Aperture cards, roll film, and microfiche are the systems in use, of which microfiche seems to be gaining because of the ease of designing readers and retrieving the information. Ordinary microfiche has ca 30 21.6- $\times$ 27.9-cm pages on a 10.2- $\times$ 15.2-cm film. Ultramicrofiche involving 75–200$\times$ reductions and magnifications has been used for special applications, eg, large catalogues, but its use is not growing because of the extreme requirements for cleanliness in the photography.

Micropublishing offers duplicating speeds of 650 pages per second and savings in mailing costs. It is developing a place in the printing and publishing industries in the recording of information from computers (COM), and in the storage and retrieval of records in offices and reference materials in libraries.

Ink-Jet Printing. Ink-jet is a pressureless printing process, which produces an image with jets of colored ink similar to fountain-pen ink. The ink is ejected through a nozzle under pressure, vibrated into uniform droplets which are charged electrostatically, and deflected by a computer or other image-generating device. The process can be done in two ways. A single nozzle activated by a computer can oscillate across a sheet much as an electron beam produces an image on a television tube. This is the principle of the AB Dick Videojet system, which is used in coding equipment for containers and packaging materials and for addressing printed materials while they are being printed.

The second method of ink-jet printing involves a bank of nozzles, each of which is digitally controlled by a computer program. As many as 118 jets per centimeter are used and can produce droplets at speeds as high as 8×10^4 per second to image as many as 1.5×10^5 characters per second or 7×10^4 newspaper lines of type a minute. This is the basis of the Mead Dijit System, which is used in many applications of variable information printing, eg, computer letters, and with improving quality, it can become a book publishing process.

Electronic Printing. Electronic printing is a process utilizing an electroconductor drum, like a copier, which can be imaged or erased sequentially during printing. New thin-film photoconductors have been developed and can be exposed by a laser beam activated by digital signals from a computer program to produce charged images on the drum. The images can be developed by toner and be transferred to paper. This is the principle of the Xerox 9700, IBM 6670, and Wang Image Printer. These systems are nearing printing and publishing capabilities.

The world's first completely electronic publishing system has been assembled by Mitel Corp. in Canada for technical documentation. It consists of a Xerox 9700, with software, supplied by Krontek Corp., which inputs graphics digitized on an AM/ECRM Autokon 8400 laser camera, to the 9700 and connects the scanned information directly into a PDP 11/34 computer and a Kurzweil Omni-Font OCR (optical character reader) to convert existing documentation produced in a Compugraphic 7700 phototypesetter to a computer-compatible magnetic-tape format. The data are transferred directly from keyboard entry to the printed page without any phototypesetting paper, platemaking, printing press, or ink. Such systems should be considerably useful in the short-run book market and business areas, such as billing, reporting, bank statements, etc.

Finishing and Binding

The printed material is subjected to a number of finishing operations. Most printing is done in large sheets with a number of the same or different subjects on the sheet. Labels are varnished and cut to size after printing. Calendars are cut to size and stapled. In check printing, some are cut to size and stapled, and some are perforated, cut, stapled, or drilled for insertion into a binder. Some greeting cards are embossed, dusted with gold bronze, and folded. Some printed material is folded into pamphlets for mailing. Some is folded into signatures of usually 16 or 32 pages, which later are bound together in various ways to make books or magazines. Some printing, like letterheads, is just cut and packaged into reams. Most package printing is scored and die-cut for shaping into packages. Printed metal is formed into cans, boxes, trays, and into odd shapes like toys and globes. There are other finishing operations, eg, pasting, mounting, laminating, and collating, depending on the use of the product. These operations are mostly mechanical and, with the exception of those involving glue, none are affected or influenced by advances in chemical technology.

There are many ways to bind a book, including edition binding, also known as hardcover or case binding; perfect binding or adhesive binding, which is used on paperback books; and mechanical binding for manuals and notebooks. Perfect binding, which was developed to reduce the expense of sewing and casebinding books, is being used increasingly. Many developments are being made in adhesives (qv), binding-machine automation, and materials handling to quicken operations and reduce costs.

Comparison of the Printing Processes

Lithography is an ideal process for text and pictorial reproduction for short and medium runs that are sheet-fed or web-fed up to ca 10^6 impressions; letterpress printing is good for text matter, flexibility, broad lettering, solid backgrounds and large expanses of color from short to long runs; gravure printing is best for long-run pictorial reproduction, such as magazines and mail-order catalogues and packaging. Research in these and other processes, eg, flexography and waterless lithography, could result in improvements that could change these use categories radically. Other printing processes, eg, ink-jet and electronic printing, could be competitive with lithography, letterpress, flexography, or gravure printing.

PROCESS DISTRIBUTION

An estimate of the distribution by printing process within the printing, publishing, and packaging industries, not including quick and in-plant printing, is given in Table 1 (68).

Letterpress printing will decrease in use. By 1990, about the only uses of letterpress printing will be a few large daily newspapers, some label and packaging printing, and some business forms. Flexography will increase and be the dominant relief process as it is used increasingly in newspaper-, book-, and magazine-publishing areas. Lithography is the dominant process; when its use will level depends on whether the problems of solvent inks and high investment costs related to gravure printing are solved. The increase in screen printing and other processes by 1990 will result from increased use of ink-jet and electronic printing.

Economic Aspects

The graphic arts, as the printing processes are known, is growing rapidly as an industry. In many of the large cities and states of the United States, eg, New York,

Table 1. Distribution by Process Printing, Publishing, and Packaging[a,b]

	1981	1985	1990
letterpress	20%	12%	5%
flexography	14	15	18
lithography	44	47	42–47
gravure	17	20	20–25
screen printing and other processes	5	6	10

[a] Ref. 68.
[b] Does not include instant and in-plant printing.

Illinois, and California, printing ranks among the leading industries. Among all U.S. manufacturing industries, it ranks high in the number of industrial establishments, payroll paid, and value added by manufacture, which is a value exclusive of the cost of materials used. In 1981, the printing and publishing industry was estimated to produce products valued at almost 75×10^9 (Table 2) (69). The amounts listed in Table 2 do not include printing for packaging or for copying and office duplicating (see Reprography).

Environmental Considerations

New government regulations may severely restrict all solvents that are commonly used in printing inks from being emitted to the atmosphere without some control method, eg, incineration, solvent recovery, or electrostatic precipitation (70). Enforcement of such regulations may stimulate development and use of radiation-cured, chemically reactive, or water-base inks (see Radiation curing). Increasing costs and possible shortages of energy sources from fossil fuels, eg, natural gas and oil, will increase the development of alternative, more energy-efficient methods of ink drying or curing.

In addition to air-quality regulations, printers will also be faced with new water-pollution, solid-waste, and health and safety regulations. Printers using bichromated coatings in plate and cylinder making and electroplating and etching of heavy metals will be particularly affected by water-pollution regulations. Noise levels of press and bindery equipment have been the subject of serious study and new regulations are expected to reduce acceptable levels from 90 to 85 dB (see Noise pollution). Video display terminals (VDT) are also being studied for any possible deleterious effects on the eyes. Thus, the printing and publishing industry will have to contend with many more factors in addition to the impact of new technology.

Table 2. Printing and Publishing Industries, 1979–1985 Value of Shipments (10^6 1980 Dollars)

	1979	% Change 1978–1979	1980	% Change 1979–1980	1981	% Change 1980–1981	1985[a]	% Annual change 1981–1985
printing and publishing	62,248	11.4	68,063	9.3	74,865	10.0	109,900	10.0
newspapers	16,125	11.2	17,500	8.5	19,250	10.0	27,400	9.5
magazine publishing	8,052	12.5	8,937	11.0	9,920	11.0	15,000	11.0
book publishing and printing	7,975	10.0	8,815	10.6	9,790	11.0	14,300	10.0
business forms	4,001	21.0	4,441	11.0	4,707	6.0	7,700	11.8
commercial printing	18,200	10.3	19,700	8.2	21,800	10.7	31,500	10.5
other printing and publishing[b]	7,895	10.9	8,670	9.8	9,575	10.4	14,000	10.0

[a] 1985 projections based on information in ref. 69.
[b] Includes engraving and plate printing, photoengraving, greeting-card publishing, bankbooks, bookbinding, typesetting, electrotyping and stereotyping, lithographic-plate services, and miscellaneous printing.

BIBLIOGRAPHY

"Printing and Reproducing Processes" in *ECT* 1st ed., Vol. 11, pp. 126–149, by R. F. Reed and M. H. Bruno, Lithographic Technical Foundation; "Printing Processes" in *ECT* 2nd ed., Vol. 16, pp. 494–546, by Michael H. Bruno, International Paper Company.

1. V. Strauss, *The Printing Industry*, Printing Industries of America, Washington, D.C., 1967, p. 221.
2. J. M. Sturges, ed., *Nebletle'e Handbook of Photography and Reprography*, Van Nostrand Reinhold Co., New York, 1977, p. 32.
3. C. E. K. Mees and T. N. James, *Theory of the Photographic Process*, 3rd ed., The Macmillan Co., New York, 1966.
4. B. R. Halpern, *GATF Bull. No. 510–511*, 1956.
5. C. Shapiro, ed., *Lithographers Manual*, Graphic Arts Technical Foundation, Pittsburgh, Pa., 1980, p. 9:11.
6. G. W. Jorgensen, *GATF Res. Prog. Rep. No. 52*, 1 (1961).
7. Z. Elyjiw, *GATF Res. Prog. Rep. No. 69*, 1 (1965).
8. Ref. 5, p. 7:1.
9. R. M. Evans, *An Introduction to Color*, John Wiley & Sons, Inc., New York, 1948.
10. J. N. Field, ed., *Graphic Arts Manual*, Arno Press, New York, 1980, Chapt. 9.
11. F. R. Clapper, *J. Photogr. Sci.* **12**, 28 (1964).
12. J. A. C. Yule, *Principles of Color Reproduction*, John Wiley & Sons, Inc., New York, 1967, Chapt. 4.
13. Ref. 5, Chapt. 7.
14. Ref. 5, p. 7:14.
15. G. W. Jorgensen and M. H. Bruno, *GATF Publ. No. 218*, (1954).
16. A. H. Smith, *Print. Technol.* **11**, 19 (Apr. 1967).
17. A. R. Materazzi, *TAGA (Technical Association of the Graphic Arts) Proceedings*, Rochester, N.Y., 1967, p. 229.
18. J. Kosar, *Light-Sensitive Systems: Chemistry and Application of Nonsilver Halide Photographic Processes*, John Wiley & Sons, Inc., New York, 1965, Chapt. 7.
19. U.S. Pat. 3,160,506 (Dec. 8, 1964), G. F. O'Connor and S. L. Chin (to Polychrome Corp.).
20. U.S. Pat. 3,220,832 (Nov. 30, 1965), F. Uhlig (to Cyoplate Corp.).
21. G. W. Jorgensen in *TAGA Proceedings*, TAGA, Rochester, N.Y., 1952, pp. 97–104.
22. U.S. Pat. 3,064,562 (Nov. 20, 1962), E. Deal (to Lithoplate, Inc.).
23. U.S. Pat. 3,196,785 (July 27, 1965), R. L. Eissler (to Bull Brothers Co.).
24. U.S. Pat. 3,281,243 (Oct. 25, 1966), J. L. Sorkin and D. C. Thomas (to Harris-Intertype Corp.).
25. U.S. Pat. 3,210,184 (Oct. 5, 1965), F. Uhlig (to Azoplate Corp.).
26. U.S. Pat. 3,247,791 (Mar. 26, 1966), R. F. Leonard (to Litho Chemical & Supply Co., Inc.).
27. U.S. Pat. 3,307,951 (Mar. 7, 1967), D. N. Adams and D. C. Thomas (to Lithoplate, Inc.).
28. U.S. Pat. 2,760,863 (Aug. 28, 1956), L. Plambeck, Jr. (to E. I. du Pont de Nemours & Co., Inc.).
29. U.S. Pat. 3,081,168 (Mar. 12, 1963), R. M. Leekley and R. L. Sorenson (to Time, Inc.).
30. N. L. Moore, *Prof. Printer* **22**(3), (1978).
31. U.S. Pat. 3,024,180 (Mar. 6, 1962), W. J. McGraw (to E. I. du Pont de Nemours & Co., Inc.).
32. U.S. Pat. 2,273,740 (Feb. 17, 1942), B. F. Terry.
33. "The Photo Direct Process—How It Works," *Multigraph Sales Training Bull.*, Vol. XII, No. 23, Addressograph Multigraph Co., Cleveland, Ohio.
34. J. S. Mertle and G. L. Monsen, *Photomechanics and Printing*, Mertle Publishing Co., Chicago, Ill., 1957.
35. J. A. Easley, *Penrose Ann.* **49**, 87 (1955).
36. P. Borth and M. C. Rogers, *TAGA Proceedings*, TAGA, Rochester, N.Y., 1961, p. 1; U.S. Pats. 3,033,725 (May 8, 1962), P. M. Dougherty and H. C. Vaughn (to Photoengravers Research, Inc.); 3,033,793 (May 8, 1962), J. W. Bradley, L. W. Eltson, and W. H. Burrows (to Photoengravers Research, Inc.).
37. S. Whinne, R. N. Hotchkiss, and F. P. Willcox, *TAGA Proceedings*, TAGA, Rochester, N.Y., 1955, p. 48.
38. O. Eisenschmid, *Photoengravers Bull.*, 195 (Nov. 1960).
39. Ref. 38, p. 205.
40. Ref. 38, p. 206.
41. Ref. 38, p. 207.

42. U.S. Pat. 2,040,247 (May 12, 1936), A. Dultgen.
43. U.S. Pat. 2,182,559 (Dec. 12, 1939), C. L. Henderson (to Paper Patents Co.).
44. U.S. Pat. 1,831,645 (Nov. 10, 1931), E. S. Ballard (to Ballard Process Co.).
45. Ref. 5, p. 1:1.
46. P. J. Hartsuch, *Chemistry of Lithography*, 2nd ed., GATF, Pittsburgh, Pa., 1980.
47. R. F. Reed, *Offset Platemaking*, GATF, Pittsburgh, Pa., 1967, pp. 23–25.
48. R. F. Reed, *Offset Lithographic Platemaking*, GATF, Pittsburgh, Pa., 1967, p. 68.
49. T. Yamaoka, *TAGA Proceedings*, TAGA, Rochester, N.Y., 1978, pp. 1–44.
50. U.S. Pat. 2,291,854 (Aug.8, 1942), P. Whyzmusis (to Interchemical Corp.).
51. U.S. Pat. 2,599,914 (June 10, 1952), P. J. Hartsuch and C. Wachtl.
52. M. H. Bruno, *What's New(s) in Graphic Communications*, Nashua, N.H., No. 11, Dec. 1977.
53. Ref. 10, pp. 395–397; U.S. Pat. 3,511,178 (May 12, 1970), J. F. Curtin (to Minnesota Mining & Manufacturing Co.).
54. Ref. 52, No. 10, Oct. 1977.
55. R. M. Schaffert, *Electrophotography*, Pitman Publishing Corp., New York, 1965.
56. M. L. Sugarman, "Electrofax—A New Tool for the Graphic Arts," *TAGA Proceedings*, 1955, p. 59.
57. Ref. 10, p. 396.
58. M. H. Bruno, *TAGA Proceedings*, 1977, TAGA, Rochester, N.Y., pp. 112–122.
59. L. G. Larson, *TAGA Proceedings*, TAGA, Rochester, N.Y., 1975, pp. 109–119.
60. J. R. Werner, *TAGA Proceedings*, TAGA, Rochester, N.Y., 1979, pp. 278–294.
61. G. B. Mayer, *Inland Printer*, (Dec. 1932–May 1933).
62. U.S. Pat. 3,282,208 (Nov. 1, 1966), M. Ruderman.
63. I. Pobboravsky and M. Pearson, *TAGA Proceedings*, TAGA, Rochester, N.Y., 1967, p. 229.
64. W.H. Banks, *Advances in Printing Science and Technology*, 1979 IARIGAI Conference, Pentech Press, London, Eng., 1980, pp. 104–115.
65. Ref. 1, p. 270.
66. J. F. Hutchinson, *TAGA Proceedings*, TAGA, Rochester, N.Y., 1968, pp. 175–197.
67. K. Bardin, *McGraw-Hill Yearbook of Science and Technology*, 1978, McGraw-Hill Book Co., New York, pp. 305–306.
68. M. H. Bruno, *Status of Printing in the U.S.A.—1981*, New England Printer and Publisher, Salem, N.H., July 1981.
69. *1981 U.S. Industrial Outlook*, U.S. Government Printing Office, Washington, D.C., 1981.
70. *Forecast of Graphic Arts in 1981*, GATF, Pittsburgh, Pa., 1981, p. 8.

MICHAEL H. BRUNO
Consultant

PROCESS RESEARCH AND DEVELOPMENT

The aim of process research and development is to adapt a laboratory-scale procedure to a commercial process (see also Pilot plants and microplants; Operations planning). Further input may be needed to resolve problems that arise on start-up and for optimization of performance. During process development, a stage may be reached where further research and development is transferred to a plant technical staff; the latter phase is called process improvement.

Process research and development combines experimental work with technical and economic calculations, which are guided largely by chemical and chemical engineering principles. Because the ultimate aim is an operating plant, the technical aspects are inseparable from the economic and legal (patent) ones. If these activities are not properly integrated, the process design will not be satisfactory (see Research management).

The focus of process research is on experimental work of a chemical engineering nature. Included are reaction studies in sufficient detail to obtain a conceptual flow sheet and economic data; preliminary process design and mathematical modeling; the determination of phase-equilibrium data; and the experimental simulation of separation schemes that can be calculated from, eg, vapor–liquid equilibrium data. Process research is undertaken by commercial as well as noncommercial organizations, eg, universities or research institutes, and is strictly a technical activity. An up-to-date discussion of design-oriented process development is given in an excellent monograph (1).

The commercialization of a process may differ greatly, depending on the purposes of the organization carrying out the work (2). Frequently, a company develops new technology for its own use which, however, does not preclude licensing such technology to others. Some research institutes and companies emphasize licensing.

The process-design basis is the documentation that contains all the information necessary for the engineering and design of a plant. The extent of this information varies according to the requirements and experience of the engineering group responsible for the project. Experimental data may be included, but more important is the correlation of such data in the form of mathematical models, ie, the presentation of reaction kinetics and constants for the appropriate phase-equilibrium equations. The process-design basis usually contains specifications for the optimum operating conditions and resulting selectivities and conversion for reactions, and an overall processing scheme with a flow sheet and material and energy balances. A competent engineering group can develop and optimize a process design from such information.

The process-design basis includes considerable detail regarding the operability of various steps, from the point of view of the process characteristics. This aspect of operability differs from that discussed below, where the convenience and flexibility of the operation is weighed against capital expenditures and maintenance (qv) requirements. Operability data from the process point of view include, for example, the imposition of temperature limitations in distillation columns to prevent product decomposition that could foul the equipment or lead to gas evolution. Another example is a description of the variable effects in reactions and how they influence turn-down conditions and start-up and shutdown procedures.

Certain areas of process uncertainty and an analysis of possible problems are included in process design, such as scale-up difficulties, tray efficiencies, or fouling factors. A risk analysis is a study not only of steps where some risk is involved, but also of the overall risk in commercializing the process. Process and operability risk are weighed against market requirements and economics of scale. After all factors have been considered, a smaller first-of-a-kind plant might be favored, followed by larger, improved versions (3).

The process research and development activity for a new chemical process assumes that the demand for the product warrants the efforts and expenditures involved, and that previously developed laboratory or commercial procedures are technically inadequate or less economical than the new technology. Market studies must indicate a strong demand for a new product and the availability of raw material.

An exploratory research program, usually on a small scale, is carried out by chemists or chemical engineers and experts in economic evaluation in order to determine if it is possible to develop a technically feasible process with economic advantages over existing processes.

A patent search discovers prior art or conflicting patents. The organization's patent position depends on whether the technology is intended for internal use, for licensing, or both. For licensing, a strong patent position is often desirable if not essential, whereas a manufacturing company with a strong market position and little desire to license may give less consideration to the patent situation and rely on proprietary technical information. The risk of this approach is that a competitor may later patent the invention and thus interfere with its practice (see Patents, management).

If the technical, economic, and legal prognosis for the new process is optimistic, the decision must be made whether to proceed with the program. Such a decision should be backed up by capital cost and production cost estimates. The degree of uncertainty that can be tolerated in the total project cost determines the amount of technical and economic detail required to proceed.

For the project to be successful, the plant design must be sufficiently flexible to allow smooth start-up and shutdown as well as operation at intermediate capacities, all of which must be under acceptable control by the instrumentation system. To achieve an operable plant the program manager must consult representatives of start-up operation, production, and maintenance staffs.

Personnel and Resources

The process research and development group must cover an extremely broad range of expertise. Specialists are included, eg, patent attorneys, cost estimators, and analytical chemists. Most of the work, however, whether experimental, computational, or economic, requires a familiarity with chemical engineering principles as well as a thorough understanding of the process chemistry. A strong background in the principles and applications of mass and energy transfer, applied thermodynamics, kinetics, and chemistry, and some plant experience are needed by the members of this group.

Some essential resources for successful process development include engineering and workshop facilities to construct and design experimental equipment rapidly within the research and development departments (the ability of a competent and experienced research staff to make rapid modifications and ensure that equipment is operable can shorten development time considerably); an analytical laboratory with suitable equipment; and computer facilities, preferably with the capacity to handle data acquisition and control problems, and complicated mathematical models. The data base should be readily accessible to both research and development engineers. Such resources are expensive and require a high level professional and technician staff.

Early Stage

Research frequently provides the concept for the process, ie, the new chemistry, and always generates the necessary data. Research begins with an exploratory phase during which chemistry is at the center of the activities. It is followed by the process research stage, which involves the experimental development of a processing scheme based on chemical engineering principles, and leads to start-up and operation.

Analysis and subsequent synthesis of data involving suitable physical property correlations, reaction kinetics, and thermodynamic models is followed by a synthesis into mathematical models of the operations. The synthesis is generally rudimentary in the exploratory phase of the project, but should be expanded and refined greatly during the research and development phase. This function covers process research, process development, and conceptual and preliminary process design. It is frequently executed by the development department.

Economic evaluation with the use of conceptual estimates is based on conceptual process flow sheets. Conceptual process flow sheets are produced either by the development department or by a separate group with a technical background and expertise in economic evaluations. A representative sequence of events is shown in Table 1.

Critical Areas. Most of the main advances in chemical processes during the past 20–30 years have been based on new or improved catalyst systems. Catalysis (qv) will no doubt continue to be the central factor in new or improved process technology. Obvious benefits are to be gained from high selectivity, high activity, and better utilization of feed materials.

Energy has become a critical parameter because of the steep rise in costs. Better energy usage necessitates greater complexity of design which, in turn, requires increased capital investment and more complicated control systems. However, the availability of computers and microprocessors facilitates such operations (see Energy management).

Various U.S. government agencies have published a number of reports to encourage economic optimization by energy conservation for various operations (4–6).

Economic Evaluation

The economic evaluation group begins its work long before the process research and development phase. Economic evaluation may be the impetus for technical personnel to investigate a certain reaction system. An important aspect is the preparation of conceptual process flow sheets based on reaction data, by-products and separation schemes, recycle streams, product purification, and materials of construction. Such flow sheets are the basis for a cursory plant design, followed by estimates of capital

Table 1. Elements of Process R&D Program

Research	Development and economic evaluation		
Exploratory phase			
exploratory and catalysis research	conversion and selectivity, basic model	conceptual process flow sheets	conceptual estimates and economics[a]
Process R&D phase			
basic process research; reaction system	simple kinetic model, variable effects in reactions	reactor type, effluent determination	preliminary process flow sheets
primary separations, VLE[b], LLE[c] data	VLE, LLE models	reactor sizing	revised estimates[a]
recycle simulation product purity equipment tests	revised kinetic model	revised process flow sheets separation scheme	
	equipment specification	energy balance, control schemes, safety analysis, environmental considerations, detailed mathematical simulation	
	process-design basis	equipment sizing, preliminary process design	appropriation grade estimate; typically, −15% to +20% accuracy[a]

[a] Decision point.
[b] Vapor–liquid equilibrium.
[c] Liquid–liquid equilibrium.

investment, utilities, plot plan, and environmental safeguards. Working with the flow sheet, the economic evaluations group comments on criticalities in a proposed process and recommends technical solutions (see Economic evaluation).

It is difficult to estimate capital costs during the exploratory research and the early parts of the process research and development phases before the process is well defined. Process design is on a preliminary basis because of the sketchy data available, and some pieces of equipment either are not included in the process flow sheet or are not sized accurately. Clearly this lack of definition limits the accuracy of estimates of the "major equipment" category, which comprises all discrete equipment items such as reactors, vessels, columns, heat exchangers, pumps, compressors, etc. The estimation of the "other equipment" or "materials of construction" category (these items are used interchangeably) can be less accurate than major equipment by a factor of two or more. Other equipment generally includes piping, foundations, buildings, structural steel, insulation and fireproofing, instrumentation, electrical equipment, equipment erection, and painting. Specialized correlations between major equipment and other equipment categories as a function of materials of construction and plant size are available (7–8) (see Plant layout).

Alternative Process Schemes. The economic evaluation group must be able to draw up conceptual flow sheets of alternative proposals and guide the experimental and computational development work to the best economics. New concepts formulated by research and development engineers must be rapidly evaluated for their economic effect and incorporated in the flow sheets.

Reaction conditions also need economic evaluation, specifically the effects of temperature, pressure, and concentrations on selectivity and conversion. From these data, recycle requirements and by-product formation can be estimated.

Energy requirements are important cost items that are expected to increase in the foreseeable future.

Detailed Estimates. After preliminary estimates have indicated a favorable prognosis for the new process, a detailed and consistent process flow sheet is drawn in order to obtain price quotations for large items of equipment. This more detailed flow sheet can be designed after the main process difficulties have been resolved and results in a much more accurate estimate. They provide a firm basis for management decisions on the viability of the project. Presentations backed up by defensible flow sheets and economics can be made at this point to potential licensees, partners, or production divisions.

Project Management

The management of a process research and development project requires that interaction between process research, economic evaluation, development, legal, and other supporting or advisory groups be swift and effective. Adequate resources and support services are required within the constraints of budget, relative importance, and priority, and high professional standards must be maintained among the professionals.

Activity in the process research and development phase implies that management has decided to proceed beyond the exploratory phase, after having considered all alternatives.

A general outline of a research and development (R&D) organization is given in Figure 1. Some companies have centralized research and development organizations, in others responsibilities are decentralized and distributed among various operating groups. Some companies emphasize a strict line organization, others use the matrix system, with line functions and project functions interacting, as shown in Figure 1.

In a small company, one manager is sufficient to direct the research, development, and economic evaluation sections. In large organizations, a matrix organization may be appropriate, in which a project manager has authority across the line organization of research, development, and economic evaluations groups. Intermediate management schemes have also been devised. Harmonious relations between the managers of the various groups working on the project must be maintained, priorities set, manpower allocated, and a link must exist to higher management within the company and outside organizations (see also Research management).

Decision making in research programs depends mostly on information, and new data must be passed quickly up and down the line. Delays may have a direct bearing on the rate of progress.

Objective decisions in research are arrived at by a management technique called decision analysis, a combination of operations research, systems analysis, and statistical decision theory that provides a logical thought process for complicated situations with uncertain and incomplete data. In its formal applications, logical and mathematical procedures are used that may provide a measure of the uncertainty of any one decision. A logical sequence of decision points is assumed, with probabilities assigned to alternative events. This assignment entails considerable judgment and has to be carried out by senior management personnel (see Operations planning).

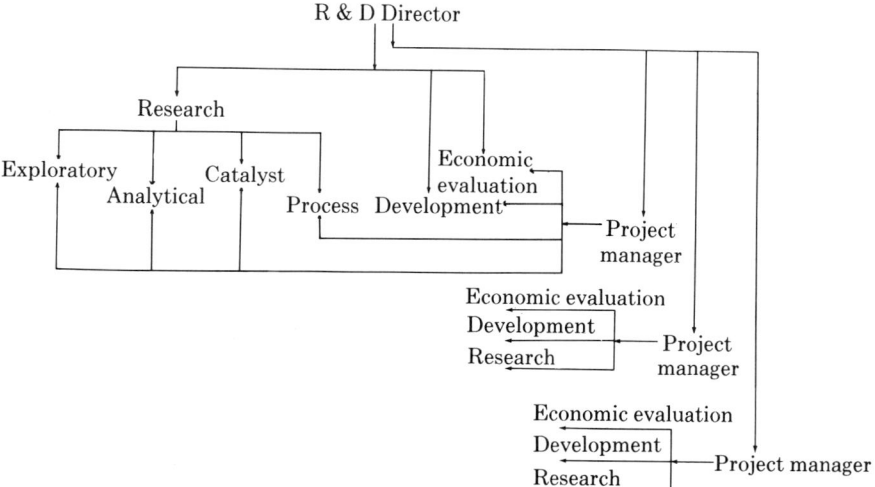

Figure 1. Typical matrix R&D organization.

Case studies from the process industry are described in a decision analysis approach in ref. 9, and an application to synthetic fuels strategy is described in ref. 10.

Process Research

Reaction Engineering. The reactors are the central concern in most chemical processes. Their performance determines the development of the process.

The selection of a suitable construction material is based on reaction selectivity and rate, the heat of reaction, the catalyst system, and the phase of reactants and products. It must be tested for corrosion resistance at an early stage (see Corrosion).

Equilibrium Limitations: Reversible Reactions. Determination of the reaction regime is vital for a meaningful interpretation of reaction studies. In the case of reversible reactions, the equilibrium constant permits the investigator to outline the region of desirable reaction conditions. A summary of heterogeneous and homogeneous reversible reactions is given in ref. 11 and in refs. 12–14, respectively.

Mass Transfer and Chemical-Rate-Controlled Reactions. In industry, mass-transfer or chemical-rate-limited reactions are more common than equilibrium-limited reactions. The energy of activation is an indication of rate limitation, since mass-transfer operations such as diffusion through catalyst pores or interfacial films are not highly temperature dependent, ie, they have low energy of activation (see Mass transfer).

With a heterogeneous catalyst, two mass-transfer-limiting regimes can occur as well as two chemical-rate-limiting regimes (11) (see Fig. 2). The rate-limiting mechanism is important in scale-up to commercial reactors.

Reactor Type. The reactor design is based on preliminary process development and makes use of exploratory information on the reaction characteristics, including the type of catalyst to be used (powdered, granular, pelletized, or homogeneous liquid phase); the phase of catalyst system, reactants, and products; reaction conditions, temperature, and pressure; the capacity of reactors and feedstock availability; residence time required for desired conversion; and mass-transfer limitations.

Important system properties are the volatility of the reactants, products, and catalyst system, and the heat of reaction (11,15) (see Reactor technology).

Catalyst Systems. The process research engineers explore the catalyst system to the point where a model with a sufficiently wide range of variables has been developed and proved in a steady-state mode, including recycle streams (16) (see Catalysis). This work includes the modeling of the rate equation. At this time, sufficient data must be obtained to establish the design basis for the control system, specifically a suitable mathematical model. Control considerations have a fundamental impact on the design of the reaction system, its instrumentation, and operation. Dynamic studies may also indicate that a cycled reactor may outperform a steady-state one because of adsorption–desorption kinetics.

The development of a heterogeneous catalyst system is closely connected to catalyst performance. The variables that affect the performance of a solid catalyst are usually not understood sufficiently to allow extrapolation without experimental verification. The process researchers must therefore carry out their investigation with a catalyst of the same size and composition as the one to be used in production. This requirement determines the size of the experimental unit.

If the process research program appears successful and is expected to result in

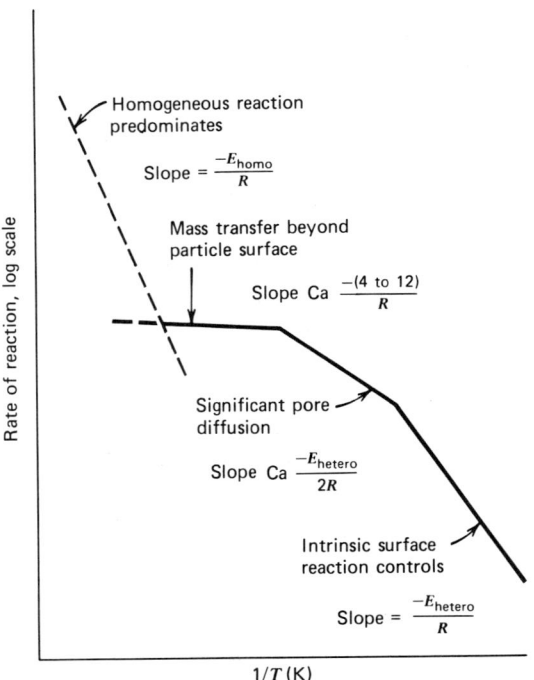

Figure 2. Possible kinetic regimes in a gas-phase reaction occurring on a porous solid catalyst. E = energy of activation, R = gas constant.

commercialization of the process, a development program for catalyst manufacture is called for.

Performance tests and quality control have to be established for the catalyst development program, including tests for chemical composition and physical properties, eg, density, particle crush strength, abrasion resistance, packing density, particle size, pore volume, surface area, pore size distribution, and other variables (16).

Reactor Control. The process control and safety system is based on process dynamics (17–19). The following data are needed for its design: a reaction rate model, physical properties of all components, a residence time-distribution model, a heat-transfer model, response models for measuring elements such as analytical instruments, temperature-, flow-, and pressure-control elements, and response models for process controllers, eg, control valves. The reaction rate model is also required for scale-up and includes start-up and shutdown. This is one reason why reactor control must be considered fairly early in the process-research phase. Dynamic kinetics may differ from steady-state kinetics, because of, for instance, the nature of the adsorption–desorption kinetics (see Instrumentation and control).

Physical and thermodynamic properties data are required for the mathematical modeling of reaction, and mass and heat transfer (20).

Residence time distributions (RTD) are accurately determined in experimental and commercial equipment by a variety of tracer methods (21–22). With this information and basic reaction-rate kinetics, the conversion and hence effluent composition could be computed. In practice, however, reactors tend to be designed in such a way that either plug flow or well-stirred, completely backmixed conditions are closely approached in experimental and commercial units. In most cases, similar conditions are assumed in experimental and full-scale units, and experimental rate data are used for scale-up and as a basis for a mathematical rate model for the full-scale unit. The heat-transfer model can be relatively simple, eg, in heat exchangers external to the actual reactor and in internal heat exchangers used in well-stirred reactors (see Heat exchange technology). Tubular reactors with a significant heat of reaction are more complicated. The reactor tubes must be designed in such a way that the hot-spot temperature can be controlled at an acceptable value, eg, by means of a differential model (23).

Instrument response times should, if possible, be an order of magnitude faster than the reactor modes. Measurement and response delays should, therefore, be a negligible factor in the control model. Control requirements frequently demand analytical instrumentation for on-line analysis. These instruments may be integrated in a control loop or used for fast manual response by operators. The pilot plant is the best place for testing such instruments.

Selectivity and By-Products. Selectivity is the percentage of reactant actually transformed that gives the desired product. Stated another way, selectivity is the yield of product divided by the conversion per pass. Rarely can the determination of the selectivity be limited to the desired product. The result of a careful by-product study as a function of reactor variables is then a selectivity profile of all products of the reaction. Changes in the by-product selectivities may point to potential separation problems downstream when the reactor is operated in a start-up or shutdown mode, or under other than design conditions.

By-Product Identification and Separation. After investigation of the reactions, the biggest problem in process research is the separation of by-products. By-products must be purged from the system. Economics usually determines the composition of the purge streams. An optimum design must be chosen that would weigh capital and utility requirements for the separation of undesirable by-products against the expense of less clean separations; the latter might purge the product or precursor along with the by-product. The buildup of by-products in recycle streams has a detrimental effect on process economics.

Separation Schemes. After the principal components in the reactor effluent are identified, their relative volatilities are determined, and their separation is evaluated (20,22,24–25). A processing scheme can be designed for the most economical separation by simple distillation of the various categories of compounds that may be present in the reactor effluent (see Distillation). Frequently, azeotropes or compounds with close boiling points are present. More complicated, and generally more expensive, separation techniques are then required, eg, azeotropic distillation, extractive distillation, solvent extraction (see Azeotropic and extractive distillation), chromatographic separations, adsorption, and membrane separations (22,26–30) (see Membrane technology).

Process design of the separation scheme follows basically the same principles, regardless of the particular separation operation used for any one step. Computation may determine which schemes promise the best results. The separation scheme is first tested in the laboratory because nonidealities in the equilibrium data may have been missed when certain data were measured or estimated; contacting efficiency of separation equipment, ie, tray efficiency or height of a transfer unit, is often difficult to estimate; potential problems may come to light during a laboratory simulation, eg, polymerization, decomposition, fouling of equipment, and local corrosion; and simulation of the reactor effluent in continuous equipment allows processing of real effluent rather than synthetic mixtures. Buildup of unexpected impurities may be detected.

For some unit operations, scale-up and mathematical modeling procedures are not adequate, eg, for filtration, centrifugation, crystallization, drying, and other operations involving solids. For these operations, it is advisable to work closely with knowledgeable equipment manufacturers who have the appropriate test facilities.

Product Purification. Purification is an extension of by-product separation. Product specifications are determined by specifications of products from competing processes and by purity requirements set by applications of the product in further processing. Specifications must be checked by a range of tests, and test procedures that can be used in plant control as well as in the research laboratory must be developed.

Pure standard samples must also be provided, and it might be necessary to build a pilot plant of a size sufficient to supply a test product. Much larger quantities of a test product may be required for market testing than for the acquisition of technical information needed for a satisfactory process-design basis.

Materials of Construction. Materials of construction must be evaluated throughout the process research and development. Like catalyst life-testing, corrosion tests, preferably under conditions of continuous operation, should be started as soon as stream compositions and temperatures are reasonably well known (31) (see Corrosion and corrosion inhibitors). Potentiometric corrosion testing indicates the extent of corrosion by instantaneous potentiometric measurement (32).

Process Development

Flow Sheet and Mathematical Simulation. The performance of the reaction systems is the core of the entire process research and development effort. The specification of conversion and selectivity determines the reaction system flow sheet, and the physical properties of reactants and products determine the downstream process scheme. The latter can be calculated as long as the required data can be computed, estimated, or obtained from the laboratory.

The earlier a comprehensive mathematical model of the process is developed, the more time is saved on the subsequent recalculations and optimizations. The development of section models is useful, and, as the process develops, they can be expanded to other sections (33). Mathematical modeling has been greatly assisted by the remarkable lowering of computer hardware costs.

Microprocessors permit either local or integrated computer control of single items of equipment or sections of a plant. Mathematical modeling is necessary in order to utilize such a plant-control system.

Software for mathematical modeling is also readily available commercially, eg, the widely used and successful system Flowtran (34). A highly complex system, Aspen, was made available to the public in 1981 (35) (see Information retrieval). Integrated mathematical models contain subroutines for processing unit designs, eg, distillation columns and heat exchangers, and material and energy balances are thus integrated with the sizing of the processing units. The Aspen program also has provisions for estimating physical properties and equipment cost.

Process Control. Process control cannot be separated from mathematical modeling. Process control imposes limits on the variables as correlated by the mathematical model. These limits may be the allowed variation around a desired control point, which may be a reactor temperature, reboiler temperature, distillation column overhead pressure, etc. At this stage of the process, ie, when a model and variable ranges are determined, even if in a preliminary form, the overall process-control strategy should be developed.

Equipment and Process Operability. Scale-up and mathematical modeling techniques are not adequate for some unit operations, mostly those involving solids separation or size change. The manufacturer's equipment must be thoroughly tested before scale-up.

The following list gives certain essential aspects of an operable plant: convenience in start-up and shutdown operations; safety during start-up and shutdown mode, particularly in emergencies; hazard analysis through statistical methods, ie, consideration of any conceivable sequence or coincidence of events that could cause dangerous conditions; rapid response to equipment failures, which determines policy on spare units, rapid replacement with warehouse spares, and accessibility for maintenance; effect of breakdowns on overall plant operation, including considerations of requirements for intermediate product-storage capacity; effects of breakdowns on on-stream time, and reliability of the overall plant; effect of operating upsets on product quality, including the ability to recycle material that does not have the correct specifications; and the ability of the plant to operate safely at low capacities and at other conditions deviating from design conditions, eg, abnormally high and low ambient temperatures.

Health and Safety Factors; Environmental Considerations

The process-research phase offers the opportunity to investigate any potentially dangerous features of the process, eg, potentially explosive gas mixtures (see Plant safety). Determination of explosive regions of the gas mixtures encountered in the process is necessary for safe equipment, process and control-system design (36–37).

Toxicity of all compounds must be checked and measures taken accordingly. If monitoring procedures are required, eg, in carbonylation processes where carbon monoxide could escape, they should be tested before incorporation in the plant design.

Surveys of toxic properties of chemicals are provided in the United States by the National Library of Medicine (NLM) and by the data base Toxline, NLM's computerized service with constant access. Other sources of toxicity information are available.

The Ames test for mutagenicity should be performed by a competent laboratory if there is any question about carcinogenic effects.

The process development engineer is usually responsible for technical representation to the plant operators and the environmental agencies. In many locations, governmental regulations require a detailed description of composition and flow rate of any process stream or emission leaving the plant. Emissions may include leaks from stuffing boxes and valve-stem packings.

Provisions must be made for containment of accidental plant emissions such as emergency dumping, blowing of pressure relief valves, and disposal of contaminated storm water or cooling water.

Any stream that could leave the process, either as a purge stream or as a safety measure, must be analyzed and evaluated for potential pollution hazard. If there is a potential pollution hazard, safe disposal methods must be devised. Every case and every stream must be evaluated individually. These pollution and disposal considerations should be incorporated into the process design during the flow sheet development phase (see Air pollution control methods).

Patents

The patent work should be well under way by the time a project has entered the process research and development phase (38). Patent searches are made early in the exploratory phase, with reasonably firm indications that the chemistry of the process is free of prior patents or that accommodation can be reached with prior patent holders (see Patents). Patents are filed if there is no prior art, and a patent position is thus established. Sometimes it may be preferable not to apply for a patent and thereby maintain the invention as a trade secret. This decision should be based on possible difficulties in enforcing the patent, or difficulties in definition which would make it possible to avoid patent claims. It may also be necessary to provide unpatentable information with the examples or claims, the value of which would exceed the value of the patent claims.

Patent applications must fulfill the following basic criteria in order to be considered favorably by the Patent Office: the patented idea must be new; it must not be obvious from prior art to one skilled in the art; and the invention must be useful.

Process Patents. Regardless of whether the chemistry of a new process is known and patented or not, there is generally room for further inventive improvement. Criticalities of certain variables or specific limits of the ranges of such variables may be discovered once a thorough study of the kinetics and thermodynamics of the reaction takes place. Certain solvents may have unexpected beneficial effects, or there may be a concentration range in which a catalyst promoter has a positive effect. A particular separation scheme may have decided advantages over others, or the suitability of certain special equipment might be discovered. There is an endless variety of aspects in a process where inventions can be made and patented. Such patents may not establish a fundamental position in the particular technology; nevertheless, they can prevent other practitioners of similar technology from applying certain advantages. Such patents can therefore contribute to a stronger licensing position or a more competitive manufacturing position.

Close cooperation between patent attorneys and technical personnel can substantially shorten the time between invention and the filing of a patent application. This can mean the difference between obtaining or losing a patent position in countries where the date of filing decides which of several competing applicants will be granted the patent. In the United States, however, the patent is granted to the first inventor, if he has diligently pursued its filing. Therefore, a later inventor can be successfully challenged even if he has filed first.

BIBLIOGRAPHY

1. J. R. Fair, *AIChE Monogr. Ser.* **76**(13), (1980).
2. R. Landau, ed., *The Chemical Plant—From Process Selection to Commercial Operation*, Reinhold Publishing Corp., New York, 1966.
3. R. Malpas, *Chem. Ind. (London)*, 826 (Dec. 1, 1979).
4. U.S. DOE, *Energy Conservation: A Route to Improved Distillation Profitability*, DOE/CS4431-TI, N.T.I.S., U.S. Dept. of Commerce, Washington, D.C., 1980.
5. U.S. DOE, *Computer Technology Can Enhance Industrial Energy Efficiency*, DOE/CS/2123-TI, N.T.I.S., Dept. of Commerce, Washington, D.C., Oct. 1979.
6. U.S. DOE, *Computer Technology: Its Potential for Industrial Energy Conservation*, DOE/CS/2123-T2, N.T.I.S., Dept. of Commerce, Washington, D.C., 1979.
7. K. M. Guthrie, *Process Plant Estimating Evaluation and Control*, Craftsman Book Co., Solana Beach, Calif., 1974.
8. F. C. Jelen, *Cost and Optimization Engineering*, McGraw-Hill Inc., New York, 1970.
9. H. U. Balthasar, R. A. A. Boschi, and M. M. Menke, *Harvard Bus. Rev.* **56**(3), 151 (1978).
10. E. G. Cazalet, "SRI-Gulf Energy Model: Overview of Methodology" from *Readings in Decision Analysis*, 2nd ed., SRI International, Menlo Park, Calif., 1977.
11. C. N. Satterfield, *Mass Transfer in Heterogeneous Catalysis*, The M.I.T. Press, Cambridge, Mass., 1970.
12. M. Walas, *Reaction Kinetics for Chemical Engineers*, McGraw-Hill Inc., New York, 1959.
13. A. A. Frost and R. G. Pearson, *Kinetics and Mechanism*, John Wiley & Sons, Inc., New York, 1953.
14. S. L. Friess and A. Weissberger, *Investigation of Rates and Mechanisms of Reactions*, Interscience Publishers, Inc., New York, 1953.
15. W.-D. Deckwer and E. Alper, *Chem. Ing. Tech.* **52**(3), 219 (1980).
16. C. N. Satterfield, *Heterogeneous Catalysis in Practice*, McGraw-Hill Inc., New York, 1980.
17. P. S. Buckley, *Techniques of Process Control*, John Wiley & Sons, Inc., New York, 1964.
18. J. O. Hougen, *Measurements and Control Applications*, 2nd ed., Instrument Society of America, Pittsburgh, Pa., 1979, Chapts. 1 and 8.
19. W. L. Luyben, *Process Modeling, Simulation and Control for Chemical Engineers*, McGraw-Hill Inc., New York, 1973.

20. R. C. Reid, J. M. Prausnitz, and T. K. Sherwood, *The Properties of Gases and Liquids*, 3rd ed., McGraw-Hill Inc., New York, 1977.
21. P. V. Danckwerts, *Chem. Eng. Sci.* **2**(1), 1 (1953).
22. G. Jordan, *Chemical Process Development*, Interscience Publishers, a division of John Wiley & Sons, Inc., New York, 1968.
23. J. J. Carberry, *Chemical and Catalytic Reaction Engineering*, McGraw-Hill Inc., New York, 1976.
24. E. Hala, J. Pick, V. Fried, and O. Vilim, *Vapor-Liquid Equilibrium*, trans. by G. Standard, 2nd ed., Pergamon Press, Inc., 1958.
25. H. Renon and J. M. Prausnitz, *AIChEJ* **14**(1), 135 (1968).
26. E. J. Hoffman, *Azeotropic and Extractive Distillation*, Interscience Publishers, a division of John Wiley & Sons, Inc., New York, 1964.
27. R. E. Treybal, *Mass Transfer Operations*, 3rd ed., McGraw-Hill Inc., New York, 1980.
28. R. E. Treybal, *Liquid Extraction*, 2nd ed., McGraw-Hill Inc., New York, 1963.
29. A. W. Francis, *Liquid-Liquid Equilibriums*, John Wiley & Sons, Inc., Krieger Publishing Co., New York, 1963.
30. J. H. Perry, *Chemical Engineers' Handbook*, 5th ed., McGraw-Hill Inc., New York, 1973.
31. M. G. Fontana and N. D. Greene, *Corrosion Engineering*, 2nd ed., McGraw-Hill Inc., New York, 1978.
32. N. D. Greene, *Corrosion* **18**, 136 (1961).
33. A. W. Westerberg, H. P. Hutchison, R. L. Motard, and P. Winter, *Process Flowsheeting*, Cambridge University Press, Cambridge, Mass., 1979.
34. J. D. Seader, W. D. Seider, and A. C. Pauls, *Flowtran Simulation—An Introduction*, 2nd ed., CACHE (Computers and Chemical Engineering Education), Cambridge, Mass., 1974.
35. L. B. Evans, *paper presented at First International Conference on Foundations of Computer-Aided Chemical Process Design*, Henniker, N.H., July 10, 1980, pp. 425–460.
36. M. G. Zabetakis, *Flammability Characteristics of Combustible Gases and Vapors*, Bulletin 627, U.S. Bureau of Mines, Washington, D.C., 1965.
37. B. Lewis and G. von Elbe, *Combustion Flames and Explosions of Gases*, 2nd ed., Academic Press, Inc., New York, 1961.
38. J. T. Maynard, *Understanding Chemical Patents*, American Chemical Society, Washington, D.C., 1978.

ERNEST I. KORCHAK
Scientific Design Company
a division of The Halcon SD Group, Inc.

PRO DRUGS. See Pharmaceuticals, controlled release; Pharmacodynamics.

PRODUCER GAS. See Fuels, synthetic.

PROGRAMMABLE POCKET COMPUTERS

There are two firms that market programmable computers, the Hewlett-Packard Corp. (HP) and the Texas Instrument Co. (TI), each of which has several models, which vary in cost and capability. (As this article went to press, several new computers were announced by Panasonic, Quasar, Radio Shack, and Sharp. These models are ca 2.5 cm wider and ca 7.6 cm longer than the HP and TI models described herein, and they are programmable in a higher level language, such as BASIC, rather than being step-programmable. There are other differences too numerous to be mentioned here.) Programmable computers are distinguished from programmable calculators since the latter have no logic or branching capability and are, therefore, quite limited in their applications (see also Computers). They are all pocket size, weigh less than 0.5 kg, and are battery-powered. Alternating-current adapter cords are available and most models have rechargeable battery packs. The face of the computer has a single line-display area and a collection of 35–45 keys. The display area consists of light-emitting diodes (LEDs) (qv) or liquid crystals (LCs) (qv) and can display up to 8 or 10 decimal digits plus a two-digit exponent (one model can display up to 12 alpha characters). The key area is for data entry and to invoke the many function capabilities designed into the computer. Most keys have two or more functions associated with each of them, thus making it possible for a relatively small keyboard to access the computer's more than one hundred functions.

These computers contain a program memory, data-storage area, and some form of operational stack. The program memory can vary from a few hundred to several thousand program steps, depending on the model and manufacturer. The data-storage registers also vary in number and, in some models, the data-storage area can be increased or decreased as required by interchanging the data-storage and program-storage area, usually on the basis that eight program steps equals one data-storage register. Some models have a continuous-memory feature, which permits the computer to retain data and program information even when the computer is turned off. Also contained within some computers is a small electric motor, which drives a magnetic-card reader, thus permitting program and/or data to be both read from or written to small magnetic cards, which measure 12.7 × 71.4 mm for the HP kind and 19.1 × 76.2 mm for those of TI. It is this ability to store programs and data on these tiny cards that contributes so much to the power, versatility, and general usefulness of these small machines. Some of these pocket computers are capable of being connected to a small thermal printer, which permits hard copy output to be obtained at the expense, however, of no longer being pocket size. Another feature available on some of the newer models is plug-in modules of read-only memory. These are preprogrammed to solve specific problems and permit very substantial increases in total memory capacity.

The programmable pocket computer measures ca 7.6 × 12.7 × 2.5 cm, consumes only a few milliwatts of power for its operation, requires no special cooling whatsoever, and can perform common arithmetic functions in ca 10^{-4} s. It is estimated that to date several hundred thousand of these computers have been purchased and that new users are being added at a rate in excess of several hundred per day. The purchase price of these pocket computers is ca 0.5% of the annual rental fee of the IBM 650, which was a first-generation computer with roughly the same capacity of the current pocket machines. Thus, computing power, which at the onset of the computer age was available to relatively few, is readily available to anyone in the fields of science and technology.

Computer Languages

The HP and TI pocket computers, although basically competitive with each other, are based on fundamentally different concepts insofar as programming is concerned. The latter is based on algebraic notation, and the former is based on Polish notation so-called because of the difficulty in pronouncing the name of its inventor, Jan Lukasiewicz. When one writes a formula, for example, that includes the addition of two variables x and y, one generally writes $x + y$; that is to say, one places the addition operator between the two operands. This is generally called infix notation and is the essence of algebraic programming language. If instead of writing about the addition of two values, one performed the addition with pencil and paper, one would first write down the value of the first operand, then write down the value of the second operand, and then perform the addition. Thus, the operands precede the operator. This is postfix notation and in computer parlance is called Polish notation.

As applied to programmable pocket calculators, algebraic notation requires that the operator be entered between the operands, at least for diadic operators, eg, $+$, $-$, $*$, $\div$. Thus, if one were to add $A + B$, one would first key in the value of A, then press the addition key, then key in the value of B, and press the equal key, which causes the pending addition operation to be performed and the sum to be shown in the calculator's display. Polish notation requires that the operands always precede the operators. Thus, to add $A + B$, one would first key in the value of A, then key in the value of B, and then press the addition key. The operation of addition is performed and the value of the sum of $A + B$ shows in the calculator's display. If two operands are entered one after the other, it is necessary to indicate that the value of the first one has been entered by pressing the key marked ENTER before starting to enter B. Thus, in this example, the total number of key strokes involved is exactly the same as for algebraic notation. This example is, however, atypical, and in general, economy in key strokes is experienced with postfix notation. However, economy of key strokes is not the most important difference in these two languages. Algebraic notation involves infix notation for diadic operators, ie, those requiring two operands, and postfix notation for monadic operators, ie, those requiring one operand, involves a hierarchy of operator precedences and the use of parentheses to alter the hierarchical precedence. The usual hierarchical precedence places monadic operators at the highest level, followed by exponentiation, followed by multiplication and division, which occupy the same hierarchical level, and finally by addition and subtraction which also occupy the same level. Inner parentheses take precedence over outer parentheses and, at the same hierarchical level, operators are executed from left to right in the expression. Thus, to evaluate the expression:

$$A + B \div C*(D + E)$$

one keys in the value of A, then presses the addition key, then keys in the value of B, then presses the divide key, then keys in the value of C, etc. Actually no computation takes place until the multiply key is pressed, at which time B, which was pending, is divided by C, which was also pending. On continuing to key in the above expression, no further arithmetic is performed until the closing parenthesis is pressed, at which point the sum of D and E would be formed and multiplied by the quotient of B/C and the product is added to A. Thus, to implement algebraic notation, the computer must be able to store or stack the operands in the order they were keyed in. In addition, it must be able to stack the pending operators as well as all open parentheses. The size of these operand and operator stacks determines the degree of complexity of the algebraic statement that can be handled as well as the depth of nesting of parentheses that is permitted. Evaluation of a fifth-order polynomial expressed in the nested parenthetical form:

$$a_0 + x(a_1 + x(a_2 + x(a_3 + x(a_4 + x*a_5))))$$

requires stacks as follows:

Operand stack	Operator stack
a_0	+
x	*
a_1	(
x	+
a_2	*
x	(
a_3	+
x	*
a_4	(
x	+
a_5	*
	(
	+
	*
	=

Thus, before the first arithmetic operation takes place, a stack of eleven pending operands and a stack of fifteen pending operators and parentheses have been constructed. With pressing of the equal key, all open parentheses close and statement evaluation proceeds from the innermost parentheses to the outermost ones.

Polish notation involves no parentheses nor rules of hierarchical precedence but involves postfix placement of the operator for both diadic and monadic operators. It does require a stack for pending operands and the number of pending operands is limited by the size of that stack. Using the same example:

$$A + B \div C*(D + E)$$

and assuming a four-position stack, one would key as follows:

A, ENTER, B ENTER, C, $\div$, D, ENTER, E, +, *, +

Each arithmetic operation occurs when its operator key is pressed, ie, there is never any pending operator. At one point, the operand holds four pending operands. To evaluate the fifth-order polynomial

$$a_0 + x(a_1 + x(a_2 + x(a_3 + x(a_4 + x*a_5))))$$

one starts at the inner parenthesis and keys as follows:

a_5, ENTER, x, *, a_4, +, x, *, a_3, +, x, *, a_2, +, x, *, a_1, +, x, *, a_0, +

The stack never holds more than two operands and no operators are ever pending.

Programming

Programming the pocket computer simply means recording each keystroke in solving the problem. Virtually none of the problems of learning a programming language, eg, FORTRAN or PL1, are associated with the programmable pocket computer. For example, to calculate the vapor–liquid equilibrium curve for a binary mixture with constant relative volatility using the Smoker (1) equation:

$$y = \frac{\alpha x}{1 + (\alpha - 1)x} \tag{1}$$

Suppose that the value of the relative volatility α was already stored in register 0 and that the value of the liquid-phase composition x was stored in register 1. The key sequence to solve for y is

Polish

RCL 0, RCL 1, *, RCL 0, 1, −, RCL 1, *, 1, +, ÷

Algebraic

RCL 0, * RCL 1, ÷, (, 1, +, (, RCL 0, −, 1,), *, RCL 1, =

where RCL = recall. These sequences of key strokes are used either to solve the problem or to program its solution. Thus, to solve the problem is to program the problem. Once a particular problem has been programmed, it can be solved again for new values of the input variables with only a minimal effort being required by the user. Furthermore, once a program has been written, it can be used by many other persons and the total amount of time saved can be substantial.

Both HP and TI operate clearinghouses of user-developed programs for their respective machines. These clearinghouses and their related newsletters provide a vast store of programs covering a wide variety of interests from computer games to programs that solve practical problems in diverse areas of human activity (see also ref. 2). Technical publications are also a fruitful source of program listings, eg, refs. 3–5. The *Oil and Gas Journal* publishes program listings related to petroleum technology (6). There also is a Professional Program Exchange (PPX) for the TI-59 with its own newsletter (7).

Several authors have assembled large numbers of programs that are related in a single subject area and have published them in book form. For example, ref. 8 contains ca 40 different programs related to hydrologic and hydraulic calculations and that are intended for 10 different models of HP and TI machines. A number of programs representative of problems encountered in biochemistry, with each program being in Polish and algebraic notation, have been published (9). Programs involving numerical methods for the solution of differential equations in physics are given in ref. 10.

Data Correlation. Data correlation, which encompasses all aspects of fitting a suitable equation to a set of data points, is of almost universal interest (see Engineering and chemical data correlation). The two programs presented in this category show the wide range of possibilities that exist.

Linear Forms. The program in Listing 1 below is based on an expression of the form:

$$y = a + b * f(x) \qquad (2)$$

and is a modification of a program in ref. 3. As coded, the program determines the coefficients a and b of equation 1 for seven different functions of x, as described below, and permits the user to change any or all of these functions to other functions of x of the user's choice. If y increases with x and the data follow a trend that is concave upward, appropriate forms of $f(x)$ might be $f(x) = 10^x, f(x) = e^x, f(x) = x^2$, or $f(x) = x^n$ where $n > 1.0$. If, however, the trend of the data is concave downward, appropriate forms of $f(x)$ might be $f(x) = -1/x, f(x) = \ln(x), f(x) = \log(x), f(x) = \sqrt{x}$, or $f(x) = x^n$ where $0 < n < 1$. If y decreases when x increases, the above functions result in concavity in the direction opposite to that mentioned above.

Listing 1. Data correlation—linear forms

LBLA, CF1, STOB, Σ+, R↓, STOA, GSB1, LBLØ, DSPØ, P⇌S, ST+3, X², ST+2, RCLA, RCLC, ×, ST+1, RCLA, RCLD, Σ+, R↓, RCLE, P⇌S, ST+3, ×, ST+1, RCLE, X², ST+2, RCL9, R/S, F3?, GTO5, GTOØ, LBL5, F1?, GTOC, GTOA, LBLB, CLRG, P⇌S, CLRG, CLX, R/S, LBLC, SF1, STOB, Σ+, R↓, STOA, GSB2, GTOØ, LBLØ, DSP4, P⇌S, RCL7, RCL6, RCL9, ÷, STOI, RCL6, ×, −, STOC, GSB3, P⇌S, GSB3, R/S, LBLD, CF1, CF3, STOB, Σ−, R↓, STOA, GSB1, LBL4, P⇌S, ST−3, X², ST−2, RCLA, RCLC, ×, ST−1, RCLA, RCLD, Σ−, R↓, RCLE, P⇌S, ST−3, ×, ST−1, RCLE, X², ST−2, R/S, F3?, GTO6, GTOØ, LBL6, F1?, GTOE, GTOD, R/S, LBLE, SF1, F3?, STOB, Σ−, R↓, STOA, GSB2, GTO4, LBL1, RCLB, 1/X, STOC, RCLB, LOG, STOD, RCLB, √X, STOE, RCLC, RTN, LBL2, RCLB, e^x, STOC, RCLB, 10^x, STOD, RCLB, X², STOE, RCLC, RTN, LBL3, RCL8, RCLI, RCL4, ×, −, STOØ, RCL5, RCL4, X², RCL9, ÷, −, STOE, ÷, PRTX, RCL4, ×, RCL9, ÷, CHS, RCLI, +, PRTX, RCLØ, X², RCLE, ÷, RCLC, ÷, √X, R/S, RCL1, RCLI, RCL3, ×, −, STOØ, RCL2, RCL3, X², RCL9, ÷, −, STOE, ÷, PRTX, RCL3, ×, RCL9, ÷, CHS, RCLI, +, PRTX, RCLØ, X², RCLE, ÷, RCLC, ÷, √X, R/S, RTN, R/S

Program 1, as it is coded, is capable of determining the coefficients a and b of equation 1 as well as the correlation coefficient for all of the above functional relationships and, with minor modification, can compute many other functions. To use the program, start by pressing key B for BEGIN, which clears all the data registers to zero. Next key in the value of y_1, press ENTER, key in the value of x_1, and press key A if the functions $1/x$, $\log x$ or $\sqrt{x}$ are wanted or press key C if the functions e^x, 10^x, or x^2 are wanted. Shortly the computer stops showing 1. in the display. For each additional data pair, key in Y_i, ENTER, x_i, R/S. Each time the computer stops, the display shows the total number of data pairs already entered and is ready to receive the next data pair. When there are no more data pairs to be entered, press R/S and the computer proceeds with the least-squares fitting of the selected functions. The program displays the slope b for the equation $y = a + bx$ for 5 s, after which it continues execution, then displays the intercept a for 5 s, and finally displays the correlation coefficient and stops. Pressing R/S causes this display sequence to be repeated for the second function chosen. Pressing R/S again displays results for the third function chosen, and so on for the fourth function chosen. Comparison of the values of the correlation coefficient for each of the four equations permits selection of the one that best fits the data.

Suppose one or two of the data pairs appears to be erratic and the user wants to determine what the correlation would look like if that data pair had not been included. The user can simply key in the y and x values for the maverick and press key D. (If key C had been used instead of key A at the start of data entry, key E now should be used instead of key D.) When the computer stops, other erratic data pairs can be deleted by repeating this procedure or else to reevaluate the new slopes, intercepts, and correlation coefficients for all four functions, press R/S. Keys D or E can be used during data entry to delete a set of y and x values that may have been entered in error.

To modify any one or more of the functions tested by this program, first delete the unwanted step, then replace it with the desired function. For example, to replace $f(x) = x^2$ with $f(x) = x^{1.65}$, proceed as follows.

(1) Move slide switch to W/PROG.

(2) Press the following keys in sequence: (a) GO TO .131; (b) h dEL; (c) 1; (d) .; (e) 6; (f) 5; (g) h y^x.

(3) Move slide switch to RUN.

To modify any of the other functions, change step (2a) to GO TO .nnn and replace steps (2c) through (2g) with any sequence of key strokes needed to evaluate the specific functions. The step addresses nnn, which are associated with each of the other programmed functions, are

$f(x)$	$1/x$	$\log x$	$\sqrt{x}$	e^x	10^x	x^2
nnn	118	121	124	130	133	136

Example. Suppose one collected the data shown in Figure 1 and wanted to fit an equation to the data. Noting that the curvature appears to be concave downward, one would start to enter the data with key A, as described above. A few seconds after the last data point is entered, the computer displays the slope, then the intercept, and then the correlation coefficient for each of the four equation forms represented by key A. The results of the example are as follows and are shown in Figure 2.

Form	Equation	Correlation coefficient
linear	$y = 0.329 + 0.496\, x$	0.856
reciprocal	$y = 2.317 - 0.6781$	0.756
logarithmic	$y = 1.005 - 12.586 \log(x)$	0.931
square root	$y = -0.907 + 1.715 \sqrt{x}$	0.920

The correlation coefficient indicates that a better fit is obtained with logarithmic and square-root forms in this instance. Figure 2, which is a plot of these four equations and the original data points, shows that the square-root function fits the data better

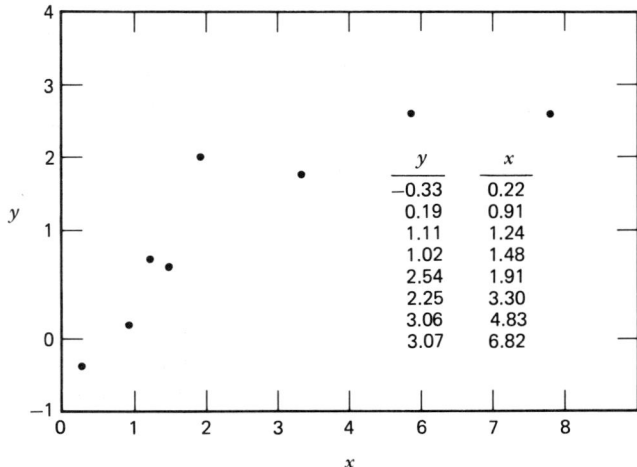

Figure 1. Raw data, y vs x.

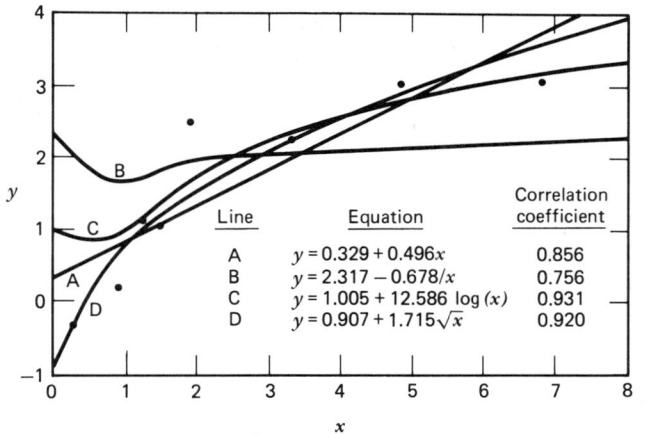

Figure 2. Correlations developed by program 1 for data of Figure 1.

at low values of x but that the logarithmic function fits the data better at high values of x. It also appears from Figure 2 that the fifth data point ($y = 2.54$, $x = 1.91$) might be a maverick. To determine what the correlations are with this single point omitted, one keys in the y and x values for this point, then presses key D and, when the computer stops, presses R/S, after which the following results are displayed:

Form	Equation	Correlation coefficient
linear	$y = 0.073 + 0.525\ x$	0.938
reciprocal	$y = 2.209 - 0.654/x$	0.754
logarithmic	$y = 0.896 + 2.559 \log(x)$	0.961
square root	$y = -1.108 + 1.747 \sqrt{x}$	0.977

There has been an obvious improvement in the correlation coefficient for all but the reciprocal form, and the logarithmic and square-root forms have switched their relative positions, although both still show very high correlation.

Polynomial Forms. The program in Listing 2 below is based on an expression of the form:

$$y = a_0 + a_1 x + a_2 x^2 + \ldots + a_n x^n$$

Listing 2. Data correlation—polynomial forms
LBL1, RCLB, GSBE, STO1, RCL5, RCLB, 3, Y^x, ×, 4, ×, ST+2, RCLB, X^2, RCL4, ×, 3, ×, ST+2, RCL3, 2, ×, RCLB, ×, ST+2, RCL5, ENT↑, +, RCLB, ×, RCL4, +, 3, ×, RCLB, ×, ST+3, RCL5, 4, ×, RCLB, ×, ST+4, RCLC, STOI, LBL3, RCLI, PRTX, DSZI, GTO3, R/S, LBLE, STOD, RCLC, STOI, CLX, LBL4, RCLD, ×, RCLI, +, DSZI, GTO4, RTN, LBLa, RCLØ, 7, ×, RCL7, −, 2, 1, ÷, STO1, RTN, LBLb, RCL6, 2, 8, ÷, RCLA, ÷, STO2, RTN, LBLc, RCL7, RCLØ, 4, ×, −, 8, 4, ÷, RCLA, X^2, ÷, STO3, RTN, LBLd, RCL8, RCL6, 7, ×, −, 2, 1, 6, ÷, RCLA, 3, Y^x, ÷, STO4, RTN, LBL9, RCL8, 4, 9, ×, CHS, RCL6, P⇄S, RCL7, ×, +, RCL5, ÷, P⇄S, RCLA, ÷, STO2, RTN, LBLA, 2, STOC, GSBb, RCLØ, 7, ÷, STO1, GTO1, LBLe, CLRG, STOA, 3, STO1, ×, +, CHS, STOB, 7, STOI, LBLØ, RCLI, R/S, ST+Ø, RCL1, ×, ST+6, LSTX, ×, ST+7, LSTX, ×, ST+8, LSTX, ×, ST+9, 1, ST−1, DSZI, GTOØ, CLX, R/S, LBLB, 3, STOC, GSBc, GSBb, GSBa, GTO1, LBLC, 4, STOC, GSBd, GSBc, GSB9, GSBa, GTO1, LBLD, 5, STOC, GSB9, GSBd, RCL9, P⇄S, RCL8, ×, RCL9, P⇄S, RCL7, ×, −, RCLØ, P⇄S, RCLØ, ×, +, RCL1, ÷, P⇄S, STO1, 4, PRTX, R/S, LBLD, 6, 7, 9, RCL9, 6, 7, ×, −, RCLØ, P⇄S, RCL2, X, −, RCL3, ÷, P⇄S, RCLA, X^2, ÷, STO3, RCL9, 7, ×, RCL7, 6, 7, ×, −, RCLØ, P⇄S, RCL4, ×, +, RCL3, ÷, P⇄S, RCLA, 4, Y^x, ÷, STO5, GTO1

This program requires that the following constants be stored in the indicated registers

Reg. #	0	1	2	3	4	5	6	7	8	9
Constant	524	924	840	3168	72	1512	49	397	21	245

Fitting high order polynomials generally overtaxes the data storage capabilities of pocket computers; however, much simplification is possible if the values of the independent variable are equidistant. Such a condition does not usually occur with experimental data but is easily achievable if it is necessary to fit some points from a smooth curve to an algebraic expression. Particularly useful are the equations representing a least-squares fit of a fourth-order polynomial involving seven data points that are equally spaced in x and with the coordinates chosen so that $x = 0$ for the central data point. The following solution for this special case is as follows (11):

$$7a + 28h^2c + 196h^4e = \Sigma y$$
$$28h^2b + 196h^4d = h\Sigma ky$$
$$28h^2a + 196h^4c + 1{,}588h^6e = h^2\Sigma k^2y$$
$$196h^4b + 1{,}577h^6d = h^3\Sigma k^3y$$
$$196h^4a + 1{,}588h^6c + 13{,}636h^8e = h^4\Sigma k^4y$$

Here h is the size of the x interval and k is the coefficient in the equation $x = kh$ in the transposed coordinate system, ie, $x_1 = -3h$, $x_2 = -2h$, $x_3 = -h$, $x_4 = 0$, $x_5 = h$, $x_6 = 2h$, and $x_7 = 3h$. With the recognition that the constants of all polynomials up through the fourth order can be obtained from subsets of these equations, direct solutions for these constants can be developed and are presented in Table 1.

To use this program, key in the value of x_1 and h, then press key E. The computer stops showing 1 in the display at which time, key in y_1 and press R/S. Repeat these steps for all seven values of y. Finally, press key A for linear fit, B for a quadratic fit, C for a cubic fit, or D for a fourth-order fit. Each case results in a display of the appropriate coefficients a, b, c, and d.

Iterative Solutions. Every branch of engineering has unique problems that require iterative techniques for their solution, ie, trial-and-error solutions. Typical of such problems is the equilibrium flash calculation of a multicomponent mixture.

Table 1. Summary of Solutions for Polynomial Forms[a]

	Linear	Quadratic	Cubic	4th Order
Transposed coordinate system				
a'	$\Sigma y/7$	$\frac{1}{21}(7\Sigma y - \Sigma k^2 y)$	$\frac{1}{21}(7\Sigma y - \Sigma k^2 y)$	$\frac{1}{924}(524\Sigma y - 245\Sigma k^2 y + 21\Sigma k^4 y)$
b'	$\frac{1}{28}h(\Sigma ky)$	$\frac{1}{28}h(\Sigma ky)$	$\frac{1}{1512}h(397\Sigma ky - 49\Sigma k^3 y)$	$\frac{1}{1512}h(397\Sigma ky - 49\Sigma k^3 y)$
c'		$\frac{1}{84}h^2(\Sigma k^2 - 4\Sigma y)$	$\frac{1}{84}h^2(\Sigma k^2 y - 4\Sigma y)$	$\frac{1}{3168}h^2(679\Sigma k^2 y - 840\Sigma y - 67\Sigma k^4 y)$
d'			$\frac{1}{216}h^3(\Sigma k^3 y - 7\Sigma ky)$	$\frac{1}{216}h^3(\Sigma k^3 y - 7\Sigma ky)$
e'				$\frac{1}{3168}h^4(7\Sigma k^4 y - 67\Sigma k^2 y + 72\Sigma y)$
Original coordinate system				
a	$a' + b'\delta$	$a' + b'\delta + c'\delta^2$	$a' + b'\delta + c'\delta^2 + d'\delta^3$	$a' + b'\delta + c'\delta^2 + d'\delta^3 + e'\delta^4$
b	b'	$b' + 2c'\delta$	$b' + 2c'\delta + 3d'\delta^2$	$b' + 2c'\delta + 3d'\delta^2 + 4e'\delta^3$
c		c'	$c' + 3d'\delta$	$c' + 3d'\delta + 6e'\delta^2$
d			d'	$d' + 4e'\delta$
e				e'

[a] $\delta = -x(4)$.

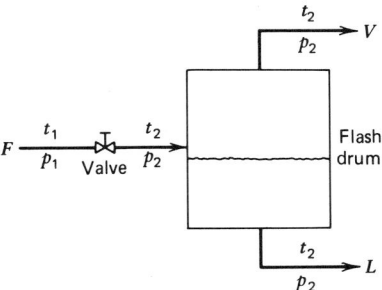

Figure 3. Equilibrium-flash separator.

A typical situation is as shown in Figure 3. The process involves the regulation of a binary or multicomponent mixture to some pressure and temperature such that when the stream is released into the flash drum, a fraction V is withdrawn as vapor in equilibrium with the residue liquid fraction L. If the flash drum temperature t_2 is known, the process is an isothermal flash and its solution involves iteration for the quantities V and L. If only the feed temperature t_1 is known, the process is an adiabatic flash and involves a double iteration for the temperature t_2 as well as the quantities V and L.

For the isothermal flash, the feed composition and the equilibrium constants K_i for each component in the feed stream are presumed known. The mass-balance constraint requires that for each component:

$$z_i F = V y_i + L x_i \tag{3}$$

where z_i, y_i, and x_i are mole fractions in the feed, vapor fraction, and liquid fraction, respectively. The equilibrium constraint requires that for each component:

$$y_i = K_i x_i \tag{4}$$

Solving equations 3 and 4 simultaneously results in

$$v_i = V y_i = \frac{z_i F}{1 + \dfrac{L}{V K_i}} \tag{5}$$

$$\Sigma v_i = V \tag{6}$$

$$L = F - V \tag{7}$$

The classical solution involves a guess of L/V followed by a calculation of all v_i by equation 5, after which the assumed L/V can be checked and the whole process is repeated until satisfactory agreement has been reached.

Consider a mixture containing 45.1 mol % propane, 18.3 mol % isobutane, and 36.6 mol % n-butane flashed at 93°C and 2.4 MPa (350 psia), at which conditions the K values are 1.42, 0.86, and 0.72, respectively. A number of iterations, based on the classical approach, are given in Table 2. Based on slide-rule calculations, any one of the guesses, each of which agrees to the second decimal place, might be accepted as the converged solution, whereas only the last one is accurate enough to be correct. A meaningful solution to the problem requires more significant digits in the calculation than is possible using a slide rule. Furthermore, the calculation can be quite tedious when the number of components is large.

Table 2. Classical Approach to Multicomponent Flash Calculations, Using Slide-Rule Solutions

Component	X_f	K	Guess $L/V = 0.2$ v	Guess $L/V = 0.4$ v	Guess $L/V = 0.5$ v	Guess $L/V = 0.65$ v
C_3	0.451	1.42	0.3953	0.34188	0.33355	0.3094
iso-C_4	0.183	0.86	0.1485	0.1249	0.11572	0.1042
nC_4	0.366	0.72	0.2864	0.23528	0.21600	0.1924
			$\Sigma = 0.8302$	$\Sigma = 0.71207$	$\Sigma = 0.65527$	$\Sigma = 0.60596$
$v = \dfrac{X_f}{1 + \dfrac{L}{VK}}$			$L/V = 0.2045$	$L/V = 0.404$	$L/V = 0.503$	$L/V = 0.65028$

The program shown in Listing 3 below solves a modification of equation 5, namely:

$$y_i = \frac{z_i k_i}{1 + (V/F)(K_i - 1)} \qquad (8)$$

which is an improvement over equation 5 in two respects: the iteration variable V/F can only vary between 0 and 1, whereas L/V can vary between 0 and infinity; and the program converges more rapidly. This program, as written, can operate with a feed stream of up to 20 components. This can be increased if the total number of data registers is increased beyond 25. Data packing, ie, the storing of two pieces of data in one data register, has been used to increase the effective available storage capacity. This is possible in the above case, since the number of significant digits in the feed mole fractions and in the component K values is unlikely ever to exceed four each, and there is a total of 10 significant digits available in each data-storage register.

Listing 3. Multicomponent equilibrium-flash calculation
LBLA, CLRG, P⇌S, CLRG, CLX, LBLØ, R/S, F3?, GTO1, RCLI, STOC, RCLØ, STOA, CLX, STOØ, LBL9, GSBb, ÷, ST+Ø, DSZI, GTO9, RCLØ, 1, X>Y?, R/S, RCLC, STOI, CLX, STOØ, LBL8, GSBb, ×, ST+Ø, DSZI, GTO8, RCLØ, 1, X>Y?, GTO7, RCLC, STOI, RCLA, STOØ, CLX, GSBa, STOB, RCLØ, STOD, R↓, X>Ø?, SF2, =, Ø, 5, F2?, CHS, ST+Ø, LBL3, RCLC, STOI, CLX, GSBa, PSE, STOA, ABS, EEX, 5, CHS, X>Y?, GTO2, RCLD, CHS, RCLØ, STOD, +, RCLA, RCLB, −, ÷, RCLA, ×, CHS, RCLD, +, STOØ, RCLA, STOB, GTO3, LBL1, ISZI, 1, X>Y?, SF2, R↓, EEX, 4, ×, +, STOI, F2?, GTOØ, FRC, ST+Ø, GTOØ, LBL4, GSBb, GSBc, RCLE, ×, RCLØ, ×, ENT↑, CHS, RCLI, FRC, +, R/S, DSZI, GTO4, GTOA, LBLa, GSBb, GSBc, 1, RCLE, −, ×, +, DSZI, GOTa, RTN, LBLb, RCLI, FRC, LSTX, INT, EEX, 4, ÷, RTN, LBLc, STOE, 1, −, RCLØ, ×, 1, +, ÷, RTN, LBL2, RCLC, STOI, GTO4

To use this program, press key C to clear all registers after which key in x_f, ENTER↑, K, and R/S for each component. After the data for the last component has been entered, press key D to begin the flash calculation. When computation stops, the value of l_n is displayed and the value of v_n is in the y register. Repeated pressing of the R/S key results in the display of l_{n-1}, v_{n-1}, then l_{n-2}, etc, until the distribution between liquid and vapor of all components has been displayed. There is neither a need to specify a starting trial value of L/V nor a need in any way to intervene in the computation until the final results are displayed. For a starting value of V/F, the program assumes that all components with $k \geq 1.0$ enter the vapor phase. The component material balance, equation 8, is solved for each component, and the error is taken as $\Sigma y_i - 1.0$. For the second trial, V/F is changed by 0.05 and, for all subsequent trials, a Newton-Raphson technique is used to a convergence tolerance of 10^{-5}.

Graphical Solutions. Another area, upon which programmable pocket computers can have a significant impact, involves problems that used to be solved graphically. Two such situations arise in distillation, namely the McCabe-Thiele solution of a binary-distillation problem and the graphical integration required in the Rayleigh equation for batch distillation (12–13).

McCabe-Thiele. The process involved in a binary distillation is shown in Figure 4. In a typical design problem, the following question is asked: for a given feed composition and for given product compositions, how many separation stages are required? Usual input specifications include other parameters describing the feed conditions, reflux conditions, stage efficiency, etc. The McCabe-Thiele solution technique results from material balance equations, which are developed for both the top section and the bottom section of the column:

$$y_{n+1} = \frac{L}{V} x_n + \frac{Dx_D}{V} \tag{9}$$

$$y_{m+1} = \frac{\overline{L}}{\overline{V}} x_m - \frac{Bx_b}{\overline{V}} \tag{10}$$

where n and m refer to the stage indexes in the top and bottom sections of the column, respectively. Based on certain simplifying assumptions, which lead to L/V being constant in the upper or rectifying section of the column and $\overline{V}/\overline{L}$ being constant in the lower or stripping section of the column, these equations plot as straight lines on the equilibrium diagram and the number of ideal stages can be stepped off, ie, determined by drawing a series of horizontal and vertical lines, as shown in Figure 4. This graphical procedure is a little more complicated when a Murphree tray efficiency has been specified and when the count of real rather than ideal stages is wanted (14). When

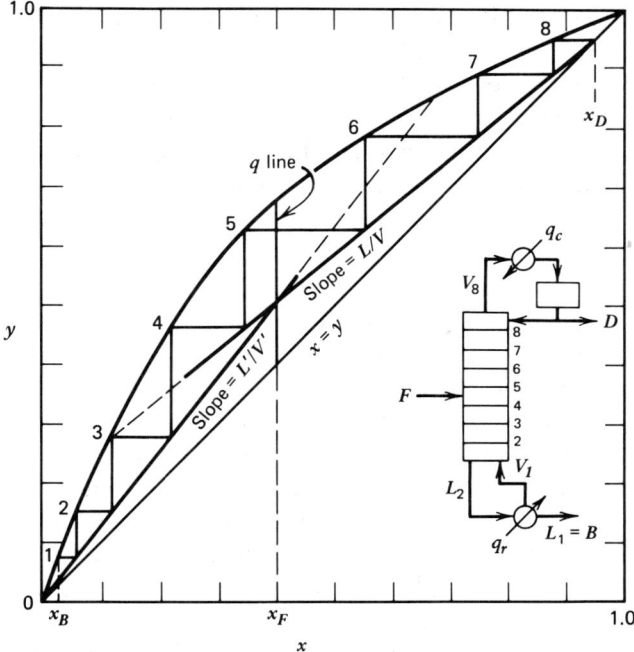

Figure 4. Continuous distillation: McCabe-Thiele.

very pure products are required, ie, $x_B \to 0.0$, $x_D \to 1.0$, the graphical stepping-off process becomes very imprecise unless very large graph paper and very sharp pencils are used (see Distillation).

The program presented in Listing 4 below solves this problem in terms of the various ways in which the problem might be specified. One starts by keying in the values of q, α, the reflux multiplier, and the stage efficiency, followed by key E. The reflux multiplier is to be used with the minimum reflux ratio to obtain the actual reflux ratio L/D. If the actual reflux ratio is to be specified instead of the reflux multiplier, the former can be keyed in instead of the reflux multiplier; however, it should be entered as a negative number. The programmed response is to ignore the minimum reflux ratio and to take the absolute value of the number that was entered as the actual reflux ratio. Stage efficiencies are sometimes specified as overall efficiencies and sometimes as Murphree efficiencies. Here too, entering a positive number signals the former interpretation and a negative number signals the latter interpretation.

Listing 4. Graphical solutions: McCabe-Thiele distillation
LBLE, CFØ, X<Ø?, SFØ, ABS, STO3, R↓, X<Ø7, SF2, ABS, STOC, R↓, STO1, 1, −, X⇌Y, X≠Ø?, GSBd, EEX, 4, CHS, +, LBLd, STOØ, ×, STOA, R/S, LBLC, STO4, R↓, STO5, X⇌Y, STO6, −, RCL4, RCL6, −, ÷, STO8, GTOa, LBLB, STO8, 1, R↑, STO5, R↑, STO4, R↑, ×, −, X⇌Y, RCL8, −, ÷, STO6, GTOa, LBLD, 1, X⇌Y, −, STO8, R↓, STO5, X⇌Y, STO6, LSTX, ×, −, RCL8, ÷, STO4, LBLa, RCLC, F2?, GTOØ, 1, RCL1, −, RCLØ, RCL5, +, ×, RCL1, +, STOB, X^2, 4, RCLA, ×, RCL5, ×, +, $\sqrt{X}$, STOD, RCLB, −, 2, ÷, RCLA, ÷, STO9, RCL1, ×, RCL9, RCL1, 1, −, ×, 1, +, ÷, RCL6, −, RCL9, LSTX, −, ÷, STO9, 1, RCL9, −, ÷, RCLC, ×, LBLØ, STO2, STOE, 1, +, STOB, RCL5, RCL2, ×, RCLØ, RCL6, ×, +, RCLØ, RCL2, +, ÷, STO9, CLX, STOI, RCL6, STO7, LBLØ, ISZI, RCL7, RCL1, RCL1, 1, −, RCL7, ×, −, ÷, RCL2, ×, RCL6, +, RCLB, ÷, FØ?, GTOA, GTOb, LBLA, RCL7, −, RCL3, ×, RCL7, +, LBLb, PRTX, STO7, RCL9, X≤Y?, GTOØ, F2?, GTOe, SF2, RCL6, STOA, RCL2, 1, RCL8, −, STOD, ×, RCLØ, +, STO2, RCL4, STO9, RCL8, ×, CHS, STO6, RCLB, 1, RCL8, −, ×, RCLØ, 1, −, +, STOB, GTOc, LBLe, RCLI, FØ?, GTOd, RCL3, ÷, INT, LBLd, RCLE, RCLD, RCLA, R↑, PRST, R/S

What is required next is to designate the degree of separation desired, and this can be expressed in several alternative forms: (*1*) specify compositions of overhead, feed, and bottoms, in that order; (*2*) specify compositions of feed and bottoms and quantity of bottom product per mole of feed; and (*3*) specify compositions of overhead and feed and the quantity of overhead product per mole of feed. The above specification forms are related through the overall material-balance equation

$$Fx_f = Bx_B + Dx_D \tag{11}$$

Thus, any of the three specifications is sufficient to uniquely specify the separation desired. To complete the problem, key in the three compositions and/or flow ratios required for any one of the above specifications and then press key C for specification (*1*), or press key D for specification (*2*), or press key B for specification (*3*). The computer then steps-off plates from the top of the column to the bottom of the column. As each plate is calculated, there is a 3–4-s pause, with the vapor composition showing in the display. This pause is long enough to write down the displayed value. When all the plates have been calculated, the computer displays the actual reflux ratio, then the overhead product rate per mole of feed, and stops with the total number of real plates showing in the display. The total time of the calculation is the sum of those 3–4-s pauses. All of the usual factors associated with graphical computations are eliminated and are substituted with speed and accuracy.

Rayleigh Distillation. A differential batch distillation, which sometimes is referred to as a Rayleigh distillation, is represented by the diagram in Figure 5 and by the equation below (13):

$$\int_{L_0}^{L} \frac{dL}{L} = \int_{x_0}^{x} \frac{dx}{y^* - x} \qquad (12)$$

The equation results from the unsteady-state material balance on the more volatile ingredient. Since the equation of the equilibrium line y^* vs x is generally unknown, ie, most data is presented in tabular form, the integral on the right is generally evaluated graphically. In a typical Rayleigh distillation, the feed quantity L_0 and the feed composition x_0 are known. The distillation is to be carried out for some time, after which a quantity of residue L with composition x remains in the still. Meanwhile, a quantity of vapor $L_0 - L$ has been collected from the condenser and its average composition is $\bar{y}$. The overall material balance is

$$L_0 x_0 = Lx + (L_0 - L)\bar{y} \qquad (13)$$

The problem is determining how long the distillation should be carried out. Again, there are three common ways for specifying the conditions that should prevail at the conclusion of the distillation: *(1)* the still temperature or still composition can be specified; (2) the quantity of residue left in the still or the quantity of distillate collected can be specified; and (3) the average composition of the distillate can be specified. In case (*1*), the limits on the integral on the right side of equation 12 are known and, therefore, the graphical integration can be performed directly. In case (2), the limits on the integral on the left side are known and, since this integrates to $\ln L_0/L$, the numerical value of the area required from the graphical integration is known and the value of x, ie, the upper limit on the integral, must be determined by trial and error. In case (3), neither of the integration limits L or x is known. Thus, one must choose an x, solve for L in equation 12, then solve for $\bar{y}$ in equation 13 for the arbitrarily chosen x and compare this calculated $\bar{y}$ with the specified value. Then another x must be chosen, etc, and the process is repeated until satisfactory convergence is reached.

The program presented in Listing 5 below does all of this and substitutes speed and accuracy for the tedium and general annoyance of graphical integration. To use this program, start by pressing key E and, each time the computer stops, enter the equilibrium vapor composition corresponding to the liquid composition shown in the display. After entering nine such values, enter either the quantity of residue in the still w and press key A, or the composition of the final residue in the still x_2 and press

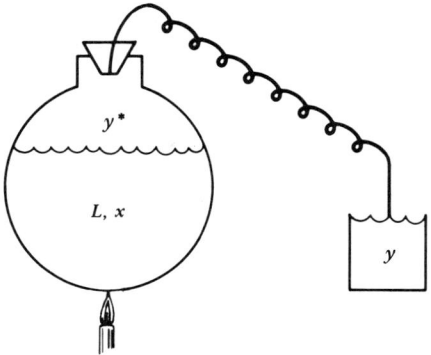

Figure 5. Rayleigh distillation.

key B, or the average composition of the distillate $\bar{y}$ and press key C, or the quantity of the distillate desired D and press key D. For keys A and C, the program stops with a display of x_2 and $\bar{y}$. For key B, it stops with a display of D and $\bar{y}$. For key C, it stops with a display of D and x_2.

Listing. 5. Graphical solutions: Rayleigh distillation
LBLA, STO9, 1/X, LN, STO2, GSB5, GSB6, PRTX, RCL3, R/S, LBLB, STO5, RCL4, X⇌Y, −, 2, ∅, ÷, STO6, SF∅, GSB5, 1, 9, STOI, LBL4, GSB1, DSZI, GTO4, GSB7, PRTX, GSB6, R/S, LBLC, STO5, CLX, STO2, SF∅, GSB5, LBL∅, GSB7, GSB6, RCL5, X>Y?, GTOc, F2?, GTOd, GSB5, LSTX, 2, ×, ST+∅, RCL6, 2, ×, ST+3, 1, ∅, ST÷6, SF2, GSB1, GTO∅, LBLd, RCL9, PRTX, RCL3, R/S, LBLD, CHS, 1, +, GTOA, LBL5, RCL4, GSB∅, STO7, LBL1, RCL6, ST−3, RCL3, GSB∅, STO8, RCL∅, RCL2, X≤Y?, GTO3, RCL8, STO7, GTO1, LBL3, F∅?, RTN, F2?, RTN, SF2, RCLD, ST−∅, RCL6, ST+3, 1, ∅, ST÷6, GTO1, LBL6, RCL4, RCL9, RCL3, ×, −, 1, RCL9, −, ÷, RTN, LBL7, e^x, 1/X, STO9, RTN, LBL2, RCL7, RCL8, +, 2, ÷, RCL6, ×, STOD, ST+∅, RTN, LBL∅, STOD, GSBb, RCLD, −, 1/X, RTN, LBLa, STOD, RCL1, ×, RCL1, 1, −, RCLD, ×, 1, +, ÷, RTN, LBLE, 1, CHS, STOI, ∅, STO∅, LBL9, RCLI, 1, +, STOI, RCL∅, −, 1, +, STO∅, R/S, GSB8, GSB8, ÷, STO1, 8, RCLI, X≠Y?, GTO9, RCL8, STO9, P⇌S, R/S, LBLb, 1, ∅, ×, −, 5, −, STOI, RCLI, STO1, X⇌Y, −, 5, +, 1, ∅, ÷, ×, LSTX, RCL1, 1, −, ×, 1, +, ÷, RTN, R/S

Computer-Aided Design. Computer-aided design is an area generally not thought of as being within the realm of programmable pocket computers because of their limited memory capacity and because of the transient nature of the display. This example is presented to show that these disadvantages can be overcome and that the programmable pocket computer can make a significant contribution. The example chosen involves a rather detailed design of a heat exchanger, in which two fluids are respectively heated and cooled without phase change (see Heat exchange technology). A program is shown in Listing 6.

Listing 6. Computer-aided design: heat exchangers
Card 1
LBLA, STO3, R↓, STO2, R↓, STO1, RCL2, −, RCL3, ×, X⇌Y, STO∅, ×, P⇌S, F2?, GTO∅, SF2, STOB, R/S, GTOA, LBL∅, 1, PSE, X⇌Y, PRTX, RCLB, X≤Y?, X⇌Y, STOA, RCL1, P⇌S, RCL1, X≤Y?, P⇌S, RCL1, RCL2, P⇌S, RCL1, RCL2, STO8, −, STOB, LSTX, +, X⇌Y, STOC, RCLB, −, STOD, R↓, −, STOE, LSTX, +, RCL8, −, STO8, LSTX, +, RCLC, −, STOC, RCLB, X≤Y?, X⇌Y, ÷, STOI, X^2, 1, +, $\sqrt{X}$, STO9, RCLD, RCLE, X=Y?, GTO2, −, RCLD, RCLE, ÷, LN, GTO3, LBL2, RCLD, RCLE, +, 2, LBL3, ÷, 2, PSE, X⇌Y, PRTX, 1/X, RCLA, ×, STOA, P⇌S, RCL1, RCL2, +, 2, ÷, STO1, P⇌S, RCL1, RCL2, +, 2, ÷, STO1, RCLC, RCLB, X≤Y?, X⇌Y, RCL8, ÷, STO7, CHS, 1, +, 1, RCL7, RCLI, ×, −, X=Y?, GTO5, ÷, LH, RCL9, ×, RCLI, 1, −, ÷, LBL4, 2, RCL7, ÷, 1, −, RCLI, −, STO6, RCL9, +, RCL6, RCL9, −, ÷, LN, ÷, 3, PSE, X⇌Y, PSE, 1/X, RCLA, ×, STOA, RCL9, X=∅?, P⇌S, 4, R/S, LBL5, CLX, RCL7, X⇌Y, +, 2, $\sqrt{X}$, ×, GTO4, LBLB, STO6, R↓, STO5, R↓, STO4, R/S, P⇌S, STO6, R↓, STO5, R↓, STO4, 5, R/S, LBLC, STOC, R↓, STOB, R↓, 1/X, RCLA, ×, 1, 2, ×, PI, ÷, RCLB, INT, EEX, CHS, 4, ×, ÷, 6, PSE, X⇌Y, R/S, STOD, ÷, R/S, STOE, R↓, STOI, 7, R/S

Card 2
LBLA, P⇌S, STO7, =, ∅, 4, 5, RCL∅, RCL6, RCL7, ÷, $\sqrt{X}$, ÷, $\sqrt{X}$, ×, LBL∅, 1, PSE, X⇌Y, R/S, STO9, F2?, GTO1, R/S, SF2, X⇌Y, =, 4, 5, 3, ×, RCLI, RCLE, −, 2, ÷, −, RCL9, ×, RCLC, X^2, ÷, RND, ×, 2, PSE, X⇌Y, PRTX, RCL9, GTO2, LBL1, PRTX, LBL2, X^2, RCL6, ×, 1/X, =, ∅, 5, ∅, 7, ×, RCL∅, ×, 3, PSE, X⇌Y, PRTX, X^2, RCL6, ×, 1, 6, 2, EEX, CHS, 6, ×, 4, PSE, X⇌Y, R/S, LBLB, P⇌S, STO7, RCL∅, =, 3, 7, 5, Y^x, 2, ∅, ÷, GTO∅, LBLE, STO9, R↓, P⇌S, STO9, P⇌S, R↑, ÷, RCL∅, ×, STO8, =, ∅, 5, ∅, 7, ×, RCL6, ÷, RCLB, FRC, 1, ∅, ×, STO2, X^2, ÷, 5, PSE, X⇌Y, PRTX, 5, =, 3, 7, EEX, CHS, 7, RCL5, =, 2, Y^x, ×, RCL8, 1, =, 8, Y^x, ×, RCL2, 6, ×, RCLD, +, ×, P⇌S, RCL9, P⇌S, ×, RCL6, ÷, RCL2, 4, =, 8, Y^x, ÷, 6, PSE, X⇌Y, R/S, 1, =, 9, RCL8, RCL2, ÷, =, 8, Y^x, ×, RCL3, 3, 1/X, Y^x, ×, RCL4, 2, ENT↑, 3, ÷, Y^x, ×, RCLB, INT, EEX, 4, ÷, STO8, ÷, RCL5, =, 4, 6, 7, Y^x, ÷, STO2, 7, PSE, X⇌Y, PRTX, RCL8, =, 7, 5, Y^x, 3, 7, =, 2, ×, P⇌S, R/S

Card 3
LBLA, STO2, GSB2, RCL2, ×, STO8, GSB4, RCL8, ÷, RCLD, ×, 1, 2, ×, INT, 1/X, 1, 2, ×, RCLD, ×,
1, PSE, R↓, R/S, LBLB, STO9, GSB5, GSBØ, RCL8, ÷, STO1, RCL6, ÷, X≤Y?, GTOa, 2, PSE, R↓, PRTX,
R↓, R/S, LBLa, GSB4, RCL9, ×, 1/X, GSBØ, ×, STO7, RCL6, 4, GSB5, X>Y?, GTOb, 3, PSE, R↓, X⇌Y,
PRTX, R↓, R/S, LBLb, RCL1, RCL7, ÷, 4, X>Y?, GTOc, PSE, X⇌Y, R/S, LBLc, GSB3, 1, +, 1, RCL2,
−, ×, RCLD, ×, RCL9, ÷, RCL7, 1, =, 8, Y^x, ×, 7, =, 4, 4, EEX, CHS, 4, ×, RCL5, =, 2, Y^x, ×, RCLC, GSB1,
−, =, 2, Y^x, ÷, RCL6, ÷, 5, PSE, X⇌Y, PRTX, RCL1, X^2, RCL6, ÷, 1, =, 1, 7, EEX, CHS, 4, ×, 1, 2, RCLD,
mult, RCL9, ÷, 1, −, ×, 6, PSE, R↓, PRTX, +, R/S, ÷, =, 3, 5, 7, Y^x, RCL9, ×, R/S, LBLC, RCLI, RCL7,
×, √X, STO1, 9, R/S, LBL5, GSB1, RCL9, ÷, X^2, 1, 7, 2, 2, 5, ×, RTN, LBL4, RCLC, GSB1, −, GSB3,
×, LSTX, RCLC, ×, GSB1, +, −, RCLI, +, RTN, LBL3, RCLE, GSB1, −, RCLC, ÷, INT, RTN, LBL2,
RCLI, X^2, P⇌S, RCL9, P⇌S, GSB1, X^2, ×, −, Pi, ×, 4, ÷, RTN, LBL1, RCLB, INT, EEX, 4, ÷, RTN, LBLØ,
=, Ø, 4, RCLØ, ×, RTN

Card 4
LBLA, RCL1, =, 6, Y^x, GSB1, =, 4, Y^x, ÷, 6, 4, ×, RCL3, 3, 1/X, Y^x, ×, RCL4, 2, ENT↑, 3, ÷, Y^x, ×, RCL5,
=, 2, 6, 7, Y^x, ÷, 1, PSE, X⇌Y, STO2, R/S, LBLB, +, STO8, R↓, GSB1, GSBØ, ÷, ×, RCL8, +, RCL2, 1/X,
+, P⇌S, RCL2, 1/X, +, 1/X, 2, PSE, X⇌Y, PRTX, STO2, RCLA, Pi, 1, 2, ÷, GSB1, ×, RCLD, ×, RCL9,
×, STO8, 1/X, RCLA, ×, 3, PSE, X⇌Y, PRTX, RCL2, X⇌Y, ÷, PRTX, 4, PSE, RCL8, R/S, LBL1, RCLB,
INT, EEX, 4, ÷, RTN, LBLØ, RCLB, FRC, 1, Ø, ×, RTN, LBLC, STOD, R↓, STOE, R↓, STOI, 2, R/S

A typical design situation involves knowing the flow rates and terminal temperatures of both fluid streams and the allowable pressure drops for each fluid. The important design details include tube diameter, tube length, number of tubes, shell diameter, nozzle sizes on both the shell and tube sides of the exchanger, and shell baffle details, eg, the number of baffles and the baffle window area. All of these physical parameters influence exchanger performance and must be consistent with the desired heat transfer and allowed pressure drop.

A special form was devised to provide a record of the input data and calculated results and to provide instructions for operating the program. This form is shown in Figure 6. The input data are unboxed and all computed results are boxed.

Thus, starting at the upper left and after passing a program card through a card reader or reading in of the first program card, data is entered for the shell-side flow, temperatures, and fluid heat capacity. The computer displays the shell-side duty or heat-transfer requirement and stops, awaiting the entry of similar data for the tube side. Program continuation is effected by pressing the key indicated at each point in the instructions. Following entry of the tube-side conditions, there is displayed in succession the tube-side duty (the greater of the tube-side and shell-side duties is automatically preserved as the design duty), the counterflow LMTD, and the multipass correction factor for a 1–2 exchanger. Program execution can be halted at this point to alter F_{1-2}, for example, or it can be aborted if, for example, the shell and tube-side duties are in disagreement. Next the thermal conductivity, viscosity, and density of the shell fluid are entered and are followed by entry of similar data for the tube-side fluid, after which the computer stops, awaiting additional data. A rough guess of the overall coefficient is entered with the tube dimensions and the tube pitch; equilateral triangular pitch is assumed. The computer responds with the total length of tubing that is required. The individual tube length is entered and the computer responds with the required tube count, which permits the user to respond with the shell diameter and the outer-tube limit.

At this point, the user reads in the second program card, which alters the stored program but leaves all the stored data quantities untouched. The allowed shell-side pressure drop is entered. The computer responds with a calculated shell-size nozzle diameter and waits, affording the opportunity for the user to alter this diameter to that of the closest commercial pipe size, for example. After this, the user enters a zero,

Read in Card 1
Shell Side
flow, kg/h or lb/h = _____
hot temp, °C or °F = _____
cold temp, °C or °F = _____
C_p, kJ/(kg·K) or Btu/(lb·°F) = _____
KEY A

Q_s, W or Btu/h = ☐

STOP
Tube Side
flow, kg/h or lb/h = _____
hot temp, °C or °F = _____
cold temp, °C or °F = _____
C_p, kJ/(kg·K) or Btu/(lb·°F) = _____
KEY R/S

1—Q_T, W or Btu/h = ☐

2—LMTD, °C or °F = ☐

3—F_{1-2} = ☐

During pause showing F_{1-2},
R/S can be used to stop
computer to change F_{1-2},
after which, R/S to restart
4—(STOP)
shell side
  k, W/(m²·K) or (Btu·ft)/(°F·ft²·h) = _____
  μ, cP = _____
  p, kg/m³ or lb/ft³ = _____
KEY B
STOP (showing k)
tube side
  k, W/(m²·K) or (Btu·ft)/(°F·ft²·h) = _____
  μm, Pa·s or cP = _____
  p, kg/m³ or lb/ft³ = _____
KEY R/S
5—(STOP)
U_{guess}, W/(m²·K) or Btu/(h·ft²·°F) = _____
$10^4 D_o + 10^{-1} D_i$, cm or in. = _____
pitch, cm or in. = _____
KEY C

6—total lineal meter or feet = ☐

tube length, m or ft = _____
KEY R/S

tube count required = ☐

STOP
D_s, cm or in. = _____
KEY R/S
7—(STOP)

Read in Card 2
shell ΔP, MPa or psi = _____
KEY A FOR GAS, B FOR LIQUID

1—DNS (calc), cm or in. = ☐

STOP
DNS (specified), cm or in. = _____
KEY R/S
STOP [showing DNS (specified)]
no. of internal domes = 0/1/2
KEY R/S

2—TLTD = ☐

3—VNS, m/s or ft/s = ☐

4—DPNS, MPa or psi = ☐

STOP
tube DP, MPa or psi = _____
KEY A FOR GAS, B FOR LIQUID

1—DNC (calc), cm or in. = ☐

STOP
DNC (specified), cm or in. = _____
KEY R/S

3—VNC, m/s or ft/s = ☐

4—DPNC, MPa or psi = ☐

STOP
NTPAS = _____
NTA = _____
KEY E

5—VT, m/s or ft/s = ☐

6—ΔP_T, MPa or psi = ☐

REPEAT UNTIL Δ IS OK. CORRECT
NTA FOR NO. OF PASSAGES, TIE RODS,
NTLD, ETC
KEY R/S

7—h_i, W/(m²·K) or Btu/(h·ft²·°F) = ☐

BP$_{max}$, cm or in. = ☐

STOP

Read in Card 3

Trials	1	2	3	4	5	6
F_1 =	____	____	____	____	____	____

Figure 6. Heat-exchanger design: data sheet and operating instructions.

KEY A

1—BP (calc), cm or in. = ☐☐☐☐☐☐

STOP

BP (specified), cm or in. = _____ _____ _____ _____ _____ _____

KEY B

2—$VB_{cut\,out}$, m/s or ft/s = ☐☐☐☐☐☐

$V_{critical}$, m/s or ft/s = ☐☐☐☐☐☐

STOP (see 1)

3—V_{cross}, m/s or ft/s = ☐☐☐☐☐☐

$V_{critical}$, m/s or ft/s = ☐☐☐☐☐☐

STOP (see 2)

4—GB/GC = ☐☐☐☐☐☐

STOP (see 3)

5—DP_{cross}, MPa or psi = ☐☐☐☐☐☐

6—DP_B, MPa or psi = ☐☐☐☐☐☐

DP_S, MPa or psi = ☐☐☐☐☐☐

┌─STOP
│ DP_{SALL}, MPa or psi = _____ _____ _____ _____ _____ _____
│ KEY R/S
ELSE
│ NEW BP, cm or in. = ☐☐☐☐☐☐
│
│ STOP
└→ **KEY C**
   9 (STOP)

Read in Card 4
KEY A

1—h_o, W/(m²·K) or Btu/(h·ft²·°F) = ☐

STOP
RD_I = _____
RD_O = _____
RD_M = _____
KEY B

2—u_{calc}, J or Btu = ☐

3—$u_{required}$, = ☐

u_{ratio} = ☐

4—area, m² or ft² = ☐

STOP

For new L and/or DS
L, ft = _____
DS, in. = _____
OTL, in. = _____
KEY C
2 (STOP)
Read in Card 2 and continue from Card 2 instructions

Figure 6. (*continued*)

one, or two, indicating the number of internal domes needed and the computer responds with the total count of tubes to be removed from the bundle to provide this dome space and quickly follows this with the nozzle velocity and the nozzle pressure drop. Essentially the same procedure is followed for the tube side, except for the detail of the internal domes.

The number of tube passes and the actual tube count can now be entered. The computer responds with the tube velocity and total pressure drop. If these values are satisfactory, the user can proceed; otherwise the number of tube passes or the number of tubes, or both, must be altered and this step repeated until a satisfactory pressure drop is obtained, after which the computer displays the tube-side film coefficient and the maximum baffle pitch permitted and stops, awaiting the third program card.

The third program card determines the shell-side design based on the use of segmental baffles. A trial baffle cut is entered, and the computer responds with a suggested baffle pitch, which may be accepted by the user or changed. The computer then calculates the velocity through the bundle and through the baffle window. If either of these velocities exceeds a calculated critical velocity, an appropriate stop and display occurs. Mass velocities through the bundle and through the baffle window are computed and compared with each other. Another stop and display is provided if the ratio of these mass velocities is greater than 4.0. If all of these tests are passed satisfactorily, the pressure drop through the bundle, the pressure drop through the baffle window, and the total shell-side pressure drop are displayed. If these values are unsatisfactory, the user can either reenter the allowed pressure drop, from which the computer provides a new estimate of baffle pitch, or the user can arbitrarily change the baffle cut or the baffle pitch, or both, and repeat this phase of the calculation as often as desired.

With the shell-side completed, the fourth and last program card is read in and the shell-side coefficient is displayed. Dirt factors and the tube metal resistance are entered next. The computer responds with a display of the calculated overall coefficient, the required overall coefficient, the ratio of one to the other, and the total amount of outside surface area in the unit. If the unit is excessively over- or undersize, new values for tube length, shell diameter, and OTL can be entered, after which a return to program card 2 is required.

The entire procedure of reading in the several program cards, copying the displayed data into the form shown in Figure 6 and going through the several iteration steps provided takes only about five minutes, if first guesses are reasonable. The procedure can take as long as 15 min if the guesses are unreasonable.

Another computer-aided design program involves distillation. The design of a distillation column for separating a multicomponent mixture is so involved that various short-cut procedures have been developed over the years to speed up preliminary sizing calculations. Three such developments are the Fenske equation for total reflux (15), the Underwood equation for minimum reflux ratio (16), and the Gilliland correlation for number of stages versus reflux (17).

The Fenske equation:

$$\left(\frac{X_i}{X_r}\right)_D = \alpha^N \left(\frac{X_i}{X_r}\right)_B$$

is used with an average value of the relative volatility α to determine the number of stages at total reflux. Once the number of stages N_m is determined, this same equation

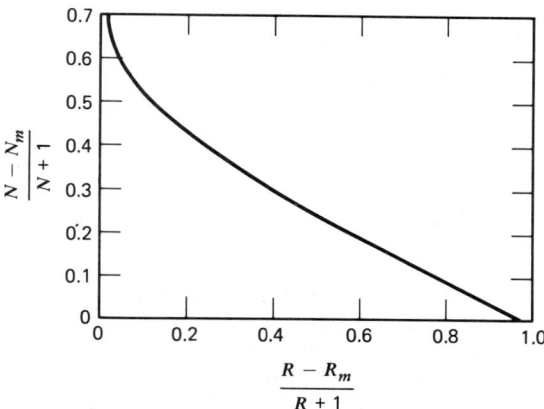

Figure 7. Gilliland correlation.

can be used to determine the distribution of all components between the overhead and bottoms products.

The Underwood equation:

$$\sum_{i=1}^{n} \frac{\alpha_i D_i}{\alpha_1 - \theta} = 1 - q$$

is first used to determine θ, after which another form of the Underwood equation:

$$R_m + 1 = \sum_{i=1}^{n} \frac{\alpha_i D_i}{\alpha_i - \theta}$$

can be solved for the minimum reflux ratio R_m.

The Gilliland correlation shown in Figure 7 graphically relates N vs R based on the knowns N_m and R_m. This correlation has been expressed as (18)

$$\frac{N - N_m}{N + 1} = 0.75 - 0.75 \left(\frac{R - R_m}{R + 1}\right)^{0.5668}$$

The above equations have been programmed using the algebraic notation of the TI-59 computer for a feed mixture containing up to eight components (19). This program is reproduced as Listing 7 below. To use this program, one must first store the input-feed composition in registers 11–18, the relative volatilities in registers 21–28, the bottoms-product compositions in registers 41–48, the overhead compositions in registers 51–58, and the feed condition q in register 50. The value of R_m can then be determined by pressing 2nd B' and the value of N_m can be determined by pressing 2nd C'. At this point, the desired R, which is greater than R_m, is entered and 2nd D' is pressed to determine the actual number of plates N.

Listing 7. Fenske—Underwood—Gilliland distillation (Courtesy of *Chemical Engineering*, McGraw-Hill, Inc.).
Find heavy key
LBL, B, RCL, 4Ø, STO, Ø1, STO, 19, STO, 2Ø, LBL, EE, 2, Ø, +, RCL, Ø1, =, STO, 35, 1, X⇌T, RCL, Ø1, −, 1, =, STO, Ø1, RC*, 35, INV, GE, EE, RCL, Ø1, +, 2, Ø, =, STO, 35, RC*, 35, STO, 33, Ø, STO, Ø1, GTO, Ø1, 26,

Calculate θ
LBL, A↑, STO, 1Ø, RCL, 19, STO, 4Ø, Ø, STO, 32, CP, LBL, LNX, (, 2, Ø, +, RCL, 4Ø,), STO, 35, (, 1, Ø, +, RCL, 4Ø,), STO, 31, (, ST*, 35, RC*, 31, ÷, (, RC*, 35, −, RCL, 1Ø,),), SUM, 32, (, RCL, 4Ø, −, 1,), STO, 4Ø, INV, EQ, LNX, (, RCL, 32, +, RCL, 5Ø, −, 1,), STO, 34, RCL, 34, RTN, 1, ., Ø, Ø, Ø, 5, PGM, Ø8, A, RCL, 33, PGM, Ø8, B, ., 1, PGM, Ø8, Ø, ., Ø, Ø, 1, PGM, Ø8, D, PGM, Ø8, E, Ø,

Calculate R_{min}
STO, 37, LBL, X², CP, 5, Ø, +, RCL, 19, =, STO, 35, 2, Ø, +, RCL, 19, =, STO, 36, RCL*, 35, ×, RC*, 36, ÷, (, RC*, 36, −, RCL, 1Ø,), =, SUM, 37, RCL, 19, −, 1, =, STO, 19, INV, EQ, X², RCL, 37, −, 1, =, STO, 29, R/S

Calculate N_{min}
LBL, C↑, 2, Ø, +, RCL, 2Ø, =, STO, 35, 1, X⇄T, RCL, 2Ø, −, 1, =, STO, 20, RCL*, 35, INV, GE, C↑, RCL, 2Ø, +, 5, 1, =, STO, 35, −, 1, =, STO, 31, RCL, 2Ø, +, 4, Ø, =, STO, 38, +, 1, =, STO, 39, RC*, 31, ×, RC*, 39, ÷, RC*, 38, =, LOG, ÷, RCL, 33, LOG, =, STO, 49, R/S

Calculate one of N, R, N_{min}, R_{min}
LBL, D', CP, RCL, 29, STO, Ø1, RCL, 49, STO, Ø3, LBL, SUM, RCL, Ø1, INV, EQ, $\sqrt{x}$, SBR, LOG, GTO, COS, LBL, LOG, RCL, Ø4, −, RCL, Ø3, =, ÷, (, RCL, Ø4, +, 1,), =, STO, Ø5, (, 1, −, 1, ., 3, 3, 3, ×, RCL, Ø5,), Y^X, 1, ., 7, 6, 4, 3, =, STO, Ø6, RTN, LBL, COS, RCL, Ø6, ×, (, RCL, Ø2, +, 1,), =, +/−, +, RCL, Ø2, =, R/S, LBL, $\sqrt{X}$, RCL, Ø2, INV, EQ, 1/X, SBR, LOG, (, RCL, Ø1, +, RCL, Ø6,), ÷, (, 1, −, RCL, Ø6,), =, R/S, LBL, 1/X, RCL, Ø3, INV, EQ, CE, SBR, CLR, GTO, SIN, LBL, CLR, RCL, Ø2, −, RCL, Ø1, =, ÷, (, RCL, Ø2, +, 1,), =, STO, Ø7, Y^X, −, 5, 6, 6, 8, =, x, ., 7, 5, +/−, +, ., 7, 5, =, STO, Ø8, RTN, LBL, SIN, RCL, Ø8, X, (, RCL, Ø4, +, 1,), =, +/−, +, RCL, Ø4, =, R/S, LBL, CE, SBR, CLR, RCL, Ø3, +, RCL, Ø8, =, +/−, ÷, (, RCL, Ø8, −, 1,), =, R/S

Nomenclature

B	= bottoms product flow rate
D	= distillate product flow rate
F	= feed flow rate
K	= equilibrium constant = y/x
L	= liquid flow rate
LMTD	= log mean temperature difference
N	= total number of stages in distillation
OTL	= outer tube limit
q	= function defining thermal condition of distillation-column feed
R	= reflux ratio = L/D
V	= vapor flow rate
x	= liquid composition
y	= vapor composition
y*	= equilibrium vapor composition
α	= relative volatility = $y/(1-y)/x/(1-x)$

Subscripts

0	= designates initial condition
b	= bottoms product
D	= distillate product
f	= designates feed stream
i	= designates any component
m	= designates minimum reflux
m	= plate number in stripping column
n	= plate number in rectifying section of column

BIBLIOGRAPHY

1. E. H. Smoker, *Trans. AIChE* **34,** 165 (1938).
2. Texas Instruments Learning Center and Rensselaer Polytechnic Institute, *Sourcebook for Programmable Calculators*, McGraw-Hill, New York, 1979.
3. R. F. Benenati, *Chem. Eng.* **84**(5), 201 (1977); **84**(6), 129 (1977).
4. W. Volk, **86**(9), 128 (1979); **86**(12), 133 (1979); **86**(19), 131 (1979); **86**(27), 93 (1979).
5. J. H. Weber, *Chem. Eng.* **86**(7), 173 (1979); **86**(10), 95 (1979); **86**(13), 111 (1979); **86**(16), 79 (1979); **86**(24), 111 (1979); **87**(1), 105 (1980); **87**(4), 93 (1980); **87**(10), 151 (1980); **87**(19), 155 (1980); **88**(1), 127 (1981); **88**(5), 91 (1981), **88**(9), 67 (1981).

6. *Oil Gas J.* **76**(2), 106 (1978); **76**(11), 70 (1978); **78**(12), 168 (1978); **78**(13), 130 (1978); **78**(30), 183 (1978); **78**(32), 107 (1980); **79**(18), 247 (1981); **79**(22), 133 (1981).
7. *Professional Program Exchange*, PPX, Box 53, Lubbock, Texas.
8. T. E. Croley, *Hydrologic and Hydraulic Computations on Small Programmable Computers*, Iowa Institute of Hydraulic Research, 1979.
9. J. E. Barnes and A. J. Waring, *Pocket Programmable Calculators in Biochemistry*, John Wiley & Sons, Inc., New York, 1980.
10. R. M. Eisberg, *Applied Mathematical Physics with Programmable Pocket Calculators*, McGraw-Hill, New York, 1976.
11. T. K. Sherwood and C. E. Reed, *Applied Mathematics in Chemical Engineering*, McGraw-Hill, New York, 1939, p. 266.
12. W. L. McCabe and E. W. Thiele, *Ind. Eng. Chem.* **17**, 605 (1925).
13. Lord J. W. Rayleigh, *Phil. Mag.* **4**, 521 (1902).
14. E. V. Murphree, *Ind. Eng. Chem.* **17**, 747 (1925).
15. M. R. Fenske, *Ind. Eng. Chem.* **24**, 482 (1932).
16. A. J. U. Underwood, *Chem. Eng. Prog.* **44**, 603 (1948).
17. E. R. Gilliland, *Ind. Eng. Chem.* **32**, 1101 (1940).
18. H. E. Eduljee, *Hydro Proc.* **54**(9), 120 (1975).
19. M. Kesler, *Chem. Eng.* **88**(15), 85 (1981).

R. F. BENENATI
Polytechnic Institute of New York

PROLINE (2-PYRROLIDINECARBOXYLIC ACID), $C_5H_9NO_2$. See Amino acids.

PROMETHIUM. See Rare-earth elements.

PROPANE. See Hydrocarbons, C_1–C_6.

PROPANOLAMINES. See Alkanolamines.

PROPANOLS. See Propyl alcohols.

PROPELLANTS. See Explosives and propellants.

PROPENE. See Propylene.

PROPIONALDEHYDE. See Aldehydes.

PROPIONIC ACID. See Carboxylic acids, survey.

PROPYL ALCOHOLS

Isopropyl alcohol, 198
n-Propyl alcohol, 221

ISOPROPYL ALCOHOL

Isopropyl alcohol [67-63-0] (2-propanol, dimethylcarbinol, sec-propyl alcohol) is the lowest member of the class of secondary alcohols. As one of the lower (C_1–C_5) alcohols, isopropyl alcohol is second in commercial production to methanol (qv) (see also Alcohols, higher). U.S. production of isopropyl alcohol in 1978 was 7.9×10^5 metric tons (1). Higher demand for isopropyl alcohol over ethyl alcohol can be accounted for by the former's lower price and freedom from government regulations. Changing technology has also influenced the relative production of these alcohols in recent years, eg, loss of isopropyl alcohol production to acetone is somewhat less influential than that of ethyl alcohol for use in the acetaldehyde-based manufacture of n-butanol and 2-ethyl-1-hexanol. The properties, preparation, and uses of isopropyl alcohol are described in ref. 2.

Physical Properties

Isopropyl alcohol is a colorless, volatile, flammable liquid. Its odor is slight, resembling a mixture of ethyl alcohol and acetone. Unlike ethyl alcohol, it has a bitter, unpotable taste. The physical and chemical properties of isopropyl alcohol reflect its secondary hydroxyl functionality. For example, its boiling and flash points are lower than n-propyl alcohol, whereas its vapor pressure and freezing point are significantly higher. Thus, isopropyl alcohol boils only 4°C higher than ethyl alcohol and possesses similar solubility properties, which accounts for the competition between these two products in many solvent applications.

Three grades of isopropyl alcohol, which differ mainly in water content, are marketed in the United States: anhydrous, 95 vol %, and 91 vol %. The 91 vol % is the azeotrope (with water) and usually is referred to as CBM (constant-boiling-mixture) isopropyl alcohol. A listing of some important physical constants of anhydrous and CBM isopropyl alcohol are given in Table 1. Because of its tendency to associate in solution, isopropyl alcohol forms azeotropes with compounds from a variety of classes, including hydrocarbons, esters, halocarbons, amines, ketones, and aromatics. Examples of some binary azeotropes are shown in Table 2. It does not form binary compounds with acetone, ethyl alcohol, ethylbenzene, hexylamine, or methyl isobutyl ketone. However, it forms ternary systems with many of these and other compounds (2) (see also Azeotropic and extractive distillation).

Chemical Properties

Most of isopropyl alcohol chemistry involves the introduction of the isopropyl or isopropoxy group into other organic molecules. The use of isopropyl alcohol for this purpose accounts for ca 60% of its production (9). Much of the production is for the manufacture of agricultural chemicals, pharmaceuticals, process catalysts, and solvents.

Table 1. Physical Properties of Isopropyl Alcohol[a]

Property	Grade Anhydrous	91 vol %
molecular weight	60.10	60.10
boiling point (at 101.3 kPa), °C[b]	82.3	80.4
freezing point, °C	−88.5	−50
sp gr, 20/20°C	0.7861	0.8179
density (at 20°C), g/cm^3	0.7849	
surface tension (at 20°C), mN/m (= dyn/cm)	0.0213	0.0214[c]
specific heat (liquid at 20°C), J/(kg·K)[d]	2510.4	
refractive index, n_D^{20}	1.3772	1.3769
heat of combustion (at 25°C), kJ/mol[d]	2005.8	
latent heat of vaporization (at 101.3 kPa[b]), kJ/mol[d]	39.8	
vapor pressure (at 20°C), kPa[b]	4.4	4.5
critical temperature, °C	235.2	
critical pressure (at 20°C), kPa[b]	4760	
viscosity, mPa·s (= cP)		
at 0°C	4.6	
at 20°C	2.4	2.1[c]
at 40°C	1.4	
solubility (at 20°C)		
in water	complete	complete
water in	complete	complete
coefficient of expansion[e]	$V_t = V_0 [1 + (1.0743 \times 10^3) t + (3.28 \times 10^{-7}) t^2]$	
flammability limit in air, vol %[f]		
lower	2.02	
upper	7.99	
flash point, °C		
Tag open cup	17.2	21.7
closed cup	11.7	18.3

[a] Refs. 3–5 except where noted.
[b] To convert kPa to mm Hg, multiply by 7.50.
[c] At 25°C.
[d] To convert J to cal, divide by 4.184.
[e] Ref. 6.
[f] Ref. 7.

Isopropyl alcohol undergoes reactions typical of an active secondary alcohol. It can be dehydrogenated, oxidized, esterified, etherified, aminated, halogenated, or otherwise modified at this site more readily than primary alcohols, eg, n-propyl alcohol or ethyl alcohol (see Propyl alcohols, n-propyl alcohol; Ethanol). Manufacture of the commercially important aluminum isopropoxide and isopropyl halides illustrates this reactivity. The former reaction involves replacement of the hydrogen atom of the hydroxyl group with concomitant hydrogen evolution and, in the latter, the hydroxyl group is displaced. Thus, aluminum isopropoxide is produced in quantitative yield by refluxing isopropyl alcohol with aluminum turnings (10–11) (see Alkoxides, metal).

$$6 (CH_3)_2CHOH + 2 Al \rightarrow 2 [(CH_3)_2CHO]_3Al + 3 H_2$$

Catalytic amounts of mercuric chloride are usually employed in this preparation. Aluminum isopropoxide is a useful Meerwein-Ponndorf-Verley reducing agent in

Table 2. Azeotropes of Isopropyl Alcohol[a]

Second components	Boiling point at 101.3 kPa (= 1 atm), °C	Azeotrope boiling point[b] at 101.3 kPa, °C	Azeotrope composition, wt %
water	100.0	80.3	87.4
			12.6
water		120.5[c]	88.3
			11.7
toluene	110.6	80.6	69
			31
methyl propionate	79.6	77	28
			72
methyl ethyl ketone	79.6	77.9	32
			68
ethyl acetate	77.05	75.9	25
			75
2-chlorobutane	68.25	64	18
			82
hexane	68.9	62.7	23
			77
cyclohexane	80.8	68.6	33
			67
butylamine	77.8	84.7	60
			40
isopropyl ether	69	66.2	16.3
			83.7

[a] Ref. 8.
[b] Azeotropes determined at 101.3 kPa (1 atm), except where noted.
[c] At 411.5 kPa (4.06 atm).

certain ester exchange reactions and is a precursor for aluminum glycinate, a buffering agent (12).

Displacement of the hydroxyl group is exemplified by the production of isopropyl halides, eg, isopropyl bromide, by refluxing isopropyl alcohol with a halogen acid, eg, hydrobromic acid (13).

$$(CH_3)_2CHOH + HBr \rightarrow (CH_3)_2CHBr + H_2O$$

The order of reactivity with acid is HI > HBr > HCl. Reaction with hydrochloric acid to form isopropyl chloride is facilitated by a zinc chloride catalyst.

Esterification. Isopropyl alcohol is esterified readily by treatment with carboxylic acids in the presence of an acidic catalyst, eg, p-toluenesulfonic acid. An equilibrium is established in the reaction.

$$RCO_2H + (CH_3)_2CHOH \underset{}{\overset{acid}{\rightleftarrows}} RCO_2CH(CH_3)_2 + H_2O$$

The equilibrium reaction is typically carried out at 100–160°C, 101.3 kPa (1 atm) (depending on the carboxylic acid employed) and with an excess of alcohol. Energy is supplied to remove the water as an azeotrope, thus forcing the reaction in the desired direction. Excess alcohol is distilled and recycled, and yields of ester are nearly quantitative. For example, isopropyl acetate can be prepared by the reaction of isopropyl alcohol with acetic acid in the presence of sulfuric acid catalyst and toluene as the azeotroping agent. Esterification of isopropyl alcohol with myristic acid forms isopropyl myristate, which is an emollient and lubricant in various cosmetic products and topical medicinals (12,14). A jellied product is marketed as Estergel (14).

Xanthate esters are prepared by reaction of isopropyl alcohol with carbon disulfide. Isopropyl xanthates have wide use in mineral flotation processes, and sodium isopropyl xanthate, is a useful herbicide for bean and pea fields (14) (see Flotation; Herbicides).

$$(CH_3)_2CHOCSNa \text{ with } S=$$

$$(CH_3)_2CHO\overset{\overset{S}{\|}}{C}SNa$$

Phosphite esters are formed readily by the reaction of phosphorus halides with isopropyl alcohol. For example, triisopropyl phosphite is prepared from phosphorus trichloride and isopropyl alcohol at low temperatures in the presence of an acid scavenger, eg, pyridine.

$$3 (CH_3)_2CHOH + PCl_3 \rightarrow [(CH_3)_2CHO]_3P + 3 HCl$$

Similarly, another important esterification reaction of isopropyl alcohol involves the production of tetraisopropyl titanate, a commercial polymerization catalyst, from titanium tetrachloride and isopropyl alcohol.

Isopropyl nitrate can be prepared by the reaction of isopropyl alcohol with nitric acid.

$$(CH_3)_2CHOH + HNO_3 \rightarrow (CH_3)_2CHONO_2 + H_2O$$

The reactants are fed separately into a still, from which the product is removed continuously by distillation (15) (see also Nitration). Isopropyl nitrate is a valuable engine-starter fuel and can be used in explosives (16) (see Explosives and propellants). The nitrite ester, isopropyl nitrite, can be prepared from the reaction of isopropyl alcohol with either nitrosyl chloride or nitrous acid at ambient temperature (17). The ester is used as a jet-engine propellant (14).

Etherification. Glycol ethers can be prepared from isopropyl alcohol by reaction with olefin oxides, eg, ethylene or propylene oxide (see Glycols). Reaction is generally catalyzed by an alkali hydroxide.

$$(CH_3)_2CHOH + \overset{O}{\underset{CH_2-CH_2}{\triangle}} \xrightarrow{KOH} (CH_3)_2CHOCH_2CH_2OH$$

2-isoproxyethanol
(isopropyl Cellosolve)

Higher alkoxylated products (oligomers) are formed by secondary reaction of oxide with the hydroxy group of the product.

$$(CH_3)_2CHOCH_2CH_2OH + \overset{O}{\underset{CH_2-CH_2}{\triangle}} \xrightarrow{KOH} (CH_3)_2CHO(CH_2CH_2O)_xH$$

This is a particularly troublesome competing reaction when the olefin oxide, eg, ethylene oxide, produces the more reactive terminal primary hydroxy group. Glycol ethers are used as solvents in lacquers, enamels, and waterborne coatings to improve gloss and flow.

Isopropyl alcohol can be dehydrated in either the liquid phase over acidic catalysts, eg, sulfuric acid, or in the vapor phase over acidic aluminas to give diisopropyl ether and propylene (qv).

$$(CH_3)_2CHOH \rightarrow CH_3CH{=}CH_2 + H_2O$$

$$2\,(CH_3)_2CHOH \rightarrow [(CH_3)_2CH]_2O + H_2O$$

Either product can be favored over the other by proper selection of catalyst and reaction conditions. However, the principal source of diisopropyl ether is as a by-product from isopropyl alcohol production. Typically, excess diisopropyl ether is recycled over acidic catalysts in the alcohol process where it is hydrated to isopropyl alcohol. Diisopropyl ether is used to a minor extent in industrial extraction and as a solvent, but it presents an explosive hazard because of the associated ease of peroxide formation.

Dehydrogenation. Isopropyl alcohol can be catalytically dehydrogenated by a wide variety of catalysts in high conversions (75–95 mol %) in an endothermic [66.5 kJ/mol (15.9 kcal/mol) at 327°C], vapor-phase process. Operation at 300–500°C and moderate pressures [207 kPa (2.04 atm)] provides acetone in yields up to 90 mol %. The most useful catalysts contain Cu, Cr, Zn, and Ni either alone, as oxides, or in combinations on inert supports (18–21).

$$(CH_3)_2CHOH \xrightarrow[\text{ZnO catalyst}]{380°C,\ 100-200\ kPa\ (1-2\ atm)} CH_3COCH_3 + H_2$$

Although the selectivity is high, minor amounts of by-products, eg, propylene, diisopropyl ether, mesityl oxide, acetaldehyde, and propionaldehyde, form by dehydration, condensation, and oxidation (see Acetone).

Oxidation. Isopropyl alcohol can be catalytically oxidized with air or oxygen at high temperatures to give acetone and water.

$$(CH_3)_2CHOH + \tfrac{1}{2}O_2 \xrightarrow[\text{catalyst}]{400-600°C} CH_3COCH_3 + H_2O$$

The catalysts are of the same general type as those used for dehydrogenation processes. In contrast to dehydrogenation, oxidation is highly exothermic [180 kJ/mol (43 kcal/mol) at 295°C]. Therefore, careful control of processing conditions is critical in order to minimize formation of by-products, especially those of dehydration (22–23). It is possible to run this oxidation and the dehydrogenation reactions simultaneously by proper choice of catalysts and conditions. Conversion of isopropyl alcohol to acetone by this technology is minimal compared to the dehydrogenation route.

Isopropyl alcohol can be partially oxidized by a noncatalytic, liquid-phase process at low temperatures and pressure to produce hydrogen peroxide (qv) and acetone (24–26).

$$(CH_3)_2CHOH + O_2 \xrightarrow[\text{253 kPa (2.5 atm)}]{\text{peroxide 90-140°C}} CH_3COCH_3 + H_2O_2$$

Oxygen or air can be used with a peroxide, eg, hydrogen peroxide, initiator. The oxidation rate is sensitive to the quantity of by-product acetic acid generated. The theoretical yield ratio of acetone to hydrogen peroxide produced by weight is 1.7. The process is normally employed where hydrogen peroxide is desired, in which case acetone and unreacted isopropyl alcohol are recycled. This process is used by Shell, where hydrogen peroxide is used for oxidation of alkyl alcohol to acrolein, and Burmah Oil Company, where hydrogen peroxide is converted to peracetic acid (26–27) (see Peroxides and peroxy compounds, organic).

Isopropyl alcohol can be oxidized by reaction with an α,β-unsaturated aldehyde or ketone at high temperature over metal oxide catalysts (28). In a Shell process for the manufacture of allyl alcohol, a vapor mixture of isopropyl alcohol and acrolein, which contains two to three moles of alcohol per mole of aldehyde, is passed over a bed of uncalcined magnesium oxide and zinc oxide at ca 400°C.

$$(CH_3)_2CHOH + CH_2=CHCHO \rightarrow CH_3COCH_3 + CH_2=CHCH_2OH$$

The process yields ca 77% allyl alcohol based on acrolein.

Amination. Isopropyl alcohol can be aminated by either ammonolysis in the presence of dehydration catalysts or reductive ammonolysis with hydrogenation catalysts. Both methods produce two amines: isopropylamine and diisopropylamine. Virtually no trisubstituted amine, ie, triisopropylamine, is produced. The ratio of mono- to diisopropylamine produced depends on the molar ratio of isopropyl alcohol and ammonia employed; molar ratios of ammonia and hydrogen to alcohol are 2:1–5:1 (29–30).

$$(CH_3)_2CHOH + NH_3 \xrightarrow[\Delta,\ pressure]{catalyst} (CH_3)_2CHNH_2 + H_2O$$

$$(CH_3)_2CHOH + (CH_3)_2CHNH_2 \xrightarrow[\Delta,\ pressure]{catalyst} [(CH_3)_2CH]_2NH + H_2O$$

In the reductive ammonolysis process, there is virtually no consumption of added hydrogen. One main function of the hydrogen is to increase catalyst life by inhibiting coking and tar formation (see Amines, lower aliphatic). In this reaction, a gaseous mixture of the alcohol, ammonia, and hydrogen is fed to a fixed-bed reactor that contains a dehydrogenation catalyst, eg, Cu, Cr, or Ni, supported on alumina; operating conditions are 150–250°C and 790–2860 kPa (100–400 psig) (29–34). The liquid hourly space velocity (LHSV) of isopropyl alcohol is ca 0.5/h. Isopropyl alcohol conversions per pass are in excess of 85% and yields are >90%. By-products, eg, nitrile (from dehydrogenation of the amine) and amide (from hydrolysis of the nitrile) are produced, but they are recycled to increase productivity.

Direct ammonolysis involving dehydration catalysts is generally run at higher temperatures (300–500°C) and at about the same pressure as reductive ammonolysis. Many catalysts are active, including aluminas, silica, titanium dioxide, and aluminum phosphate (35–37). Yields are acceptable (>80%), and coking and nitrile formation are negligible. However, little control is possible over the composition of the mixture of primary and secondary amines that can be obtained.

Isopropylamine is the most widely used of the propylamines. Most of it is consumed for herbicide manufacture (largely 2-chloro-4-ethyl-6-isopropylamino-*sym*-triazine) and a smaller quantity for pesticide manufacture (34,38) (see Herbicides). Diisopropylamine is used chiefly in pesticides and as a corrosion inhibitor, eg, diisopropylammonium nitrate (38–39) (see Corrosion and corrosion inhibitors).

Halogenation. Normally, 2-halopropane derivatives are prepared from isopropyl alcohol most economically by reaction with the corresponding acid halide. However, under the appropriate conditions, other reagents, eg, phosphorus halides and elemental halogen, also react with replacement of the hydroxyl group to give the halide (40).

$$3 \; (CH_3)_2CHOH + PBr_3 \rightarrow 3 \; (CH_3)_2CHBr + H_3PO_4$$

Halogenation of isopropyl alcohol in aqueous solution results in concomitant oxidation. Thus, chlorination at 65°C produces a mixture of chloroacetone derivatives; chiefly 1,3-dichloroacetone and 1,1,3-trichloroacetone (41–42). Further chlorination at 70–100°C provides nearly complete conversion of lower chloroacetones into 1,1,1,3,3-pentachloroacetone and hexachloroacetone (42,43–46). Chlorination of isopropyl alcohol reportedly can be conducted to give 1,1,1,3-tetrachloroacetone, which is converted to 1,1,1-trichloro-2,3-epoxypropane, a useful intermediate for synthesis of agricultural and pharmaceutical chemicals and for the preparation of plastics with low flammability (47–48). However, commercial processes to chloroacetones are believed to be based on chlorination of acetone rather than isopropyl alcohol, because of undesired by-products produced in the latter route. In addition to their use as chemical intermediates, chloroacetones have excellent solvent properties, especially for plastics; however, their toxicity may limit their use.

Halogenated 2-propanol derivatives, eg, 1,3-dichloro-2-propanol, are generally prepared from glycerol (see Chlorohydrins). These materials are used in the preparation of halogen-containing phosphates to plasticize and to lower the flammability of plastics, eg, polyurethanes and cellulosics (see Flame retardants).

Miscellaneous Reactions. Several reactions of potential commercial significance include acylation by ketene

$$(CH_3)_2CHOH + CH_2\!\!=\!\!C\!\!=\!\!O \rightarrow CH_3CO_2CH(CH_3)_2$$

and the Ritter reaction to prepare N-isopropylacrylamide from acrylonitrile and isopropyl alcohol:

$$CH_2\!\!=\!\!CHCN + (CH_3)_2CHOH \rightarrow CH_2\!\!=\!\!CHCONHCH(CH_3)_2$$

Manufacture

There are two basic processes for the commercial manufacture of isopropyl alcohol, and each involves synthesis from propylene. The standard method since ca 1920 is the indirect-hydration process (esterification–hydrolysis or the sulfuric acid method), which involves reaction of propylene with sulfuric acid. Although indirect hydration is the only process used in the United States, it is being replaced rapidly in Europe and Japan by newer processes involving catalytic hydration of propylene, ie, direct hydration, with the use of superheated steam and high pressures. High corrosion, high energy costs, and air pollution by the indirect process led to alternative processes in Europe. However, high purity propylene feedstock is not required in the indirect hydration process. Thus, a C_3-feedstock stream from refinery off-gases containing 40–60 wt % propylene is used in the United States.

Other potential synthetic methods include fermentation (qv) of certain carbohydrates, oxidation of propane, hydrogenation of acetone, and hydrolysis of isopropyl acetate (see Hydrocarbon oxidation). None of these methods is practiced commercially.

Indirect Hydration. *Chemistry.* Indirect hydration is based on a two-step reaction of propylene with sulfuric acid. In the first step, mixed sulfate esters, primarily isopropyl hydrogen sulfate, form. They are then hydrolyzed, forming the alcohol and sulfuric acid.

Step 1. Formation of mono- and diisopropyl sulfates

$$CH_3CH{=}CH_2 + H_2SO_4 \rightarrow (CH_3)_2CHOSO_3H$$
$$\text{isopropyl hydrogen sulfate}$$
$$(CH_3)_2CHOSO_3H + CH_3CH{=}CH_2 \rightarrow [(CH_3)_2CHO]_2SO_2$$
$$\text{diisopropyl sulfate}$$

Step 2. Hydrolysis of the sulfates to isopropyl alcohol

$$(CH_3)_2CHOSO_3H + H_2O \rightleftharpoons (CH_3)_2CHOH + H_2SO_4$$
$$[(CH_3)_2CHO]_2SO_2 + 2\,H_2O \rightleftharpoons 2\,(CH_3)_2CHOH + H_2SO_4$$

By-Products. Diisopropyl ether is the principal by-product formed by reaction of the intermediate sulfate esters with isopropyl alcohol.

$$(CH_3)_2CHOSO_3H + (CH_3)_2CHOH \rightleftharpoons [(CH_3)_2CH]_2O + H_2SO_4$$
$$\text{diisopropyl ether}$$
$$[(CH_3)_2CHO]_2SO_2 + (CH_3)_2CHOH \rightleftharpoons (CH_3)_2CHOSO_2OH + [(CH_3)_2CH]_2O$$

Other significant by-products include polymers, acetone, propionaldehyde, hydrocarbons, and carbonaceous material; many arise from acid-catalyzed polymerization and oxidation of propylene or isopropyl alcohol. Minor contaminants arise from impurities in the feed. Acetone, an oxidation product, also forms from thermal decomposition of the intermediate sulfate esters, eg,

$$(CH_3)_2CHOSO_3H \xrightarrow{\text{heat}} CH_3COCH_3 + SO_2 + H_2O$$

In addition to generating malodorous sulfur dioxide, the acetone formed can undergo further condensation in the acidic medium to generate mesityl oxide and higher products.

$$CH_3COCH_3 \xrightarrow{H^+} (CH_3)_2C{=}CHCOCH_3 + H_2O$$
$$\text{mesityl oxide}$$

High propylene concentrations in the presence of acids can form dimers, trimers, and higher homologues, which can polymerize or hydrate to C_6, C_9, and higher alcohols, and olefins. These derivatives can emit musty, woody, and camphoraceous odors, and their reaction products with sulfur-containing compounds can give a cat odor to the product (49). Odor can be improved by employing appropriate reaction conditions and by contacting isopropyl alcohol with various metals, eg, copper and nickel, or certain partially reduced metal oxides (50–51).

Process. A typical indirect-hydration process is presented schematically in Figure 1 (52–53).

In the process, crude liquid propylene reacts with sulfuric acid (>60 wt %) in agitated reactors at moderate pressure [2.2–2.9 MPa (300–400 psig)]. The isopropyl sulfate esters form and are maintained in the liquid state at 20–70°C. Low propylene concentrations, ie, 50 wt %, can be tolerated, but concentrations of 65 wt % or higher are preferred to achieve high alcohol yields. Since the reaction is exothermic, cooling helps minimize corrosion.

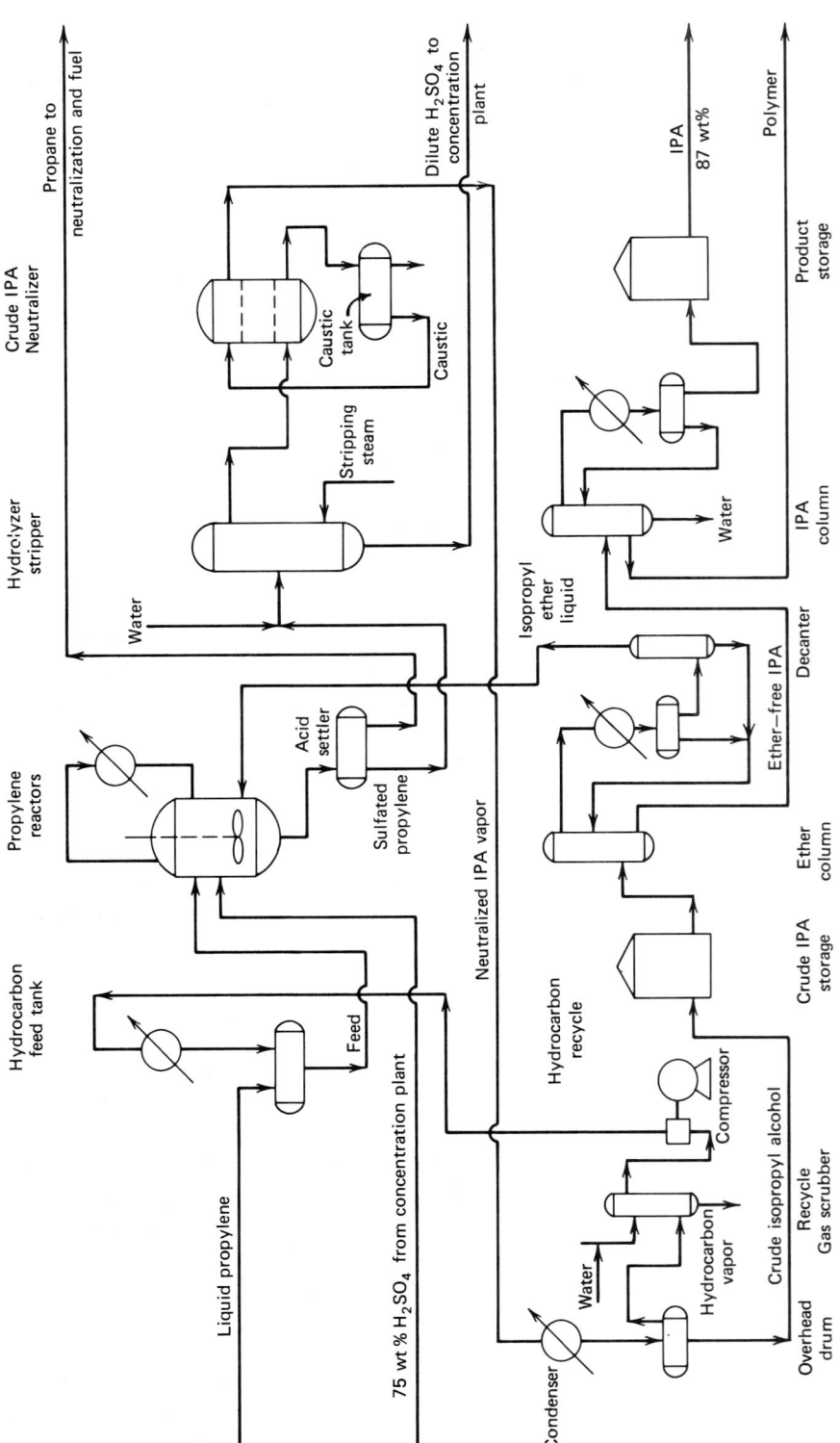

Figure 1. Indirect hydration for manufacture of isopropyl alcohol (IPA). Courtesy of Gulf Publishing Company (52).

There are two general operational modes practiced for conducting the reaction. In the two-step strong-acid process, separate reactors are used for the propylene absorption and sulfate ester hydrolysis stages. The reaction occurs at high sulfuric acid concentration (>80 wt %) and at 1–1.2 MPa (130–160 psig) and low temperature, eg, 20–30°C. The weak-acid process is conducted in a single stage at low acid concentration (60–80 wt %) and at higher pressure and temperature, ie, 2.5 MPa (350 psig) and 60–65°C, respectively. Isopropyl alcohol selectivities in excess of 90 wt % are obtained from both acid processes.

The sulfate ester hydrolysate is separated in a stripper to give a mixture of isopropyl alcohol, isopropyl ether, and water overhead and dilute sulfuric acid bottoms. The overhead is neutralized in a scrubbing tower containing sodium hydroxide and is refined in a two-column distillation system. Diisopropyl ether is taken overhead in the first, ie, ether, column. This stream is generally recycled to the reactors to produce additional isopropyl alcohol by the following equilibrium reaction:

$$[(CH_3)_2CH]_2O + H_2SO_4 \rightleftharpoons (CH_3)_2CHOSO_3H + (CH_3)_2CHOH$$

Wet isopropyl alcohol (87 wt %; 91 vol %) is taken overhead in the second still. More than 93 wt % of the charged propylene is converted to isopropyl alcohol in this system.

The bottoms from the stripper (40–60 wt % acid) are sent to the acid reconcentration unit for upgrading to the proper acid strength for recycle to the reactor. Because of the associated high energy requirements, reconcentration of the diluted sulfuric acid is a costly operation. However, a propylene-gas stripping process, which utilizes only a small amount of added water for hydrolysis, has been recently described (54). In this modification, the equilibrium quantity of isopropyl alcohol is stripped so that acid is recycled without reconcentration. Equilibrium is attained rapidly at 50°C as isopropyl alcohol is removed from the hydrolysis mixture. Similarly, the weak sulfuric acid process minimizes reconcentrating the acid and its associated corrosion and pollution problems.

The 91 vol % alcohol is sold as such or is dehydrated by extractive distillation with diisopropyl ether or cyclohexane to produce an anhydrous product (55). The wet isopropyl alcohol is fed at about the center of a dehydrating column, and the azeotroping agent is fed near the top. As the ternary azeotrope forms, it is taken overhead, condensed, and the layers are separated. The upper layer, which is mainly azeotroping agent and alcohol, is returned to the top of the column as reflux. The lower layer is mostly water. Anhydrous isopropyl alcohol is removed from the base of the column.

Acid corrosion presents a problem in isopropyl alcohol factories. Steel is a satisfactory material of construction for tanks, lines, and columns where concentrated acid (>65 wt %) and moderate temperature (<60°C) are employed. For dilute acid and higher temperatures, however, stainless steel, tantalum, Hastelloy, and the like are required for corrosion resistance and to ensure product purity (56).

The extent of purification depends on the use requirements which can be from 91 vol % azeotrope to essence grade. Generally, either intense aqueous extractive distillation or posttreatment by fixed-bed absorption with the use of activated carbon, molecular sieves (qv), and certain metals on carriers are employed to improve odor and to remove minor impurities. Essence grade is produced by final distillation in nonferrous, eg, copper, equipment (57).

Manufacturing plants in the United States are believed to use solely indirect propylene hydration. Several European companies, eg, British Petroleum, Shell, and Deutsche Texaco, also employ this older technology in plants in Europe and Japan (58).

Direct Hydration. The acid-catalyzed direct hydration of propylene is exothermic and resembles the preparation of ethanol from ethylene.

$$CH_3CH=CH_2 + H_2O \underset{}{\overset{catalyst}{\rightleftharpoons}} (CH_3)_2CHOH \qquad \Delta H = -50 \text{ kJ/mol} (-12 \text{ kcal/mol})$$

The equilibrium can be controlled to favor product alcohol if high pressures and low temperatures are applied. However, the advantage of low temperature cannot be utilized, because all known catalysts require moderate temperatures to be effective.

Process. There are three basic processes in commercial operation:

(1) Vapor-phase hydration over a fixed-bed catalyst of supported phosphoric acid (Veba-Chemie (59–62), or silica-supported tungsten oxide with zinc oxide promoter (ICI) (57–58,63).

(2) Mixed vapor-liquid phase hydration at low temperature (150°C) and high pressure [10.13 MPa (100 atm)] with a strongly acidic cation-exchange resin catalyst (Deutsche Texaco AG) (64–68).

(3) Liquid-phase hydration at high temperature and high pressure [270°C, 20.3 MPa (200 atm)] in the presence of a soluble tungsten catalyst (Tokuyama Soda) (69–71).

Process schemes for direct hydration of propylene are shown in Figures 2–4. Plants employing direct hydration are listed in Table 3. Turn-key plants based on this technology are available (62,76).

The manufacture of isopropyl alcohol by the direct catalytic hydration of propylene was begun in 1951 by ICI. The plant used a WO_3–ZnO catalyst supported on SiO_2, high temperature (230–290°C), and high pressure [20.3–25.3 MPa (200–250 atm)]. Similarly, in the Veba-Chemie process (see Fig. 2), a vaporized stream of propylene and water is passed through an acidic catalyst bed (H_3PO_4 supported on SiO_2) at 240–260°C and 2.5–6.6 MPa (25–65 atm) (60). The gas stream from the reactor is cooled and fed to a scrubber where the remaining isopropyl alcohol is removed. Isopropyl alcohol selectivity is ca 96% for the gas-phase process. Owing to equilibrium limitations in the gas phase at high temperature and low pressure, a low propylene

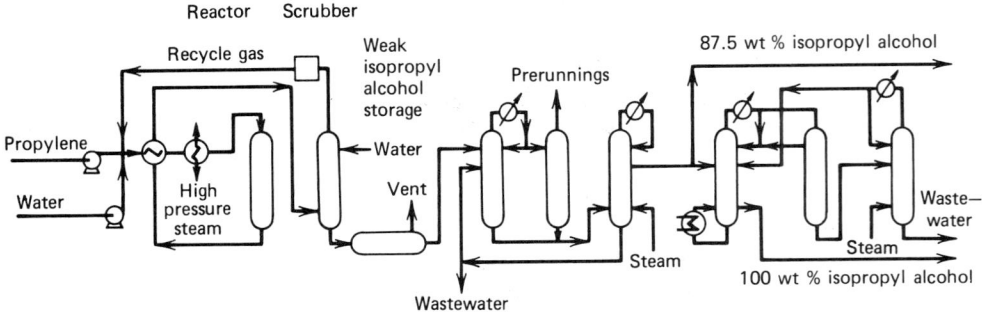

Figure 2. Veba-Chemie's direct hydration process for the manufacture of isopropyl alcohol. Courtesy of Gulf Publishing Company (60).

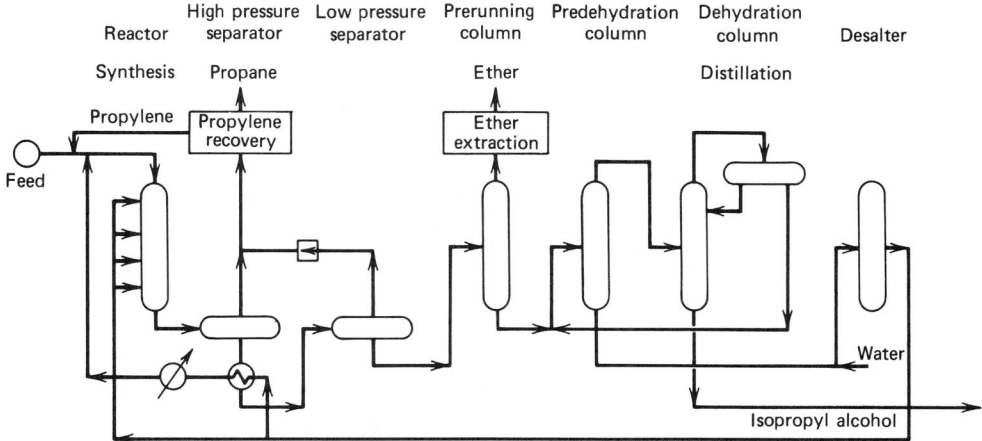

Figure 3. Deutsche Texaco's AG direct hydration process for the manufacture of isopropyl alcohol. Courtesy of Gulf Publishing Company (65).

conversion (5–6%) results and, thus, a large amount of unreacted propylene is recycled. Both processes involve high plant costs owing to high pressure requirements, gas recycles, and the requirement for high purity propylene (ca 99 wt %).

Deutsche Texaco developed a trickle-bed process to avoid the disadvantages of the gas-phase process (see Fig. 3). In this process, a mixture of liquid water and propylene gas in a molar ratio of (12–15):1 is introduced at the top of a fixed-bed reactor and allowed to trickle down over a sulfonic acid ion-exchange resin. Reaction between the liquid and gas phases takes place at 130–160°C and high pressure [8–10 MPa (80–100 atm)]; thus, aqueous isopropyl alcohol is formed. Propylene conversions per pass are greater than 75%, and isopropyl alcohol selectivity is 93%. Only 92 wt % propylene purity is needed for this process. Approximately 5% diisopropyl ether and some alcohols of the higher oligomers form as by-products. The life of the cation-exchange resin is at least eight months.

A liquid-phase variation of the direct hydration was developed by Tokuyama Soda (see Fig. 4). The disadvantages of the gas-phase processes are largely avoided by employing a weakly acidic aqueous catalyst solution of a silicotungstate (89). Preheated propylene, water, and recycled aqueous catalyst solution are pressurized and fed into a reaction chamber. They react in the liquid state at 270°C and 20.3 MPa (200 atm), and form aqueous isopropyl alcohol. Propylene conversions of 60–70% per pass are obtained, and selectivity to isopropyl alcohol is 98–99 mol % of converted propylene. The catalyst is recycled and requires very little replenishment compared to other processes. Corrosion and environmental problems are also minimized because the catalyst is a weak acid and because the system is completely closed. Because of the low gas-recycle ratio, regular commercial propylene of 95% purity can be used as feedstock.

After flashing the propylene, the aqueous solution from the separator is sent to the purification section where the catalyst is separated by azeotropic distillation; 88 wt % isopropyl alcohol is obtained overhead. The bottoms aqueous catalyst solution is recycled to the reactor, and the light ends are stripped of low boiling impurities, eg, diisopropyl ether and acetone. Azeotropic distillation yields dry isopropyl alcohol, and the final distillation column yields a product of more than 99.99% purity.

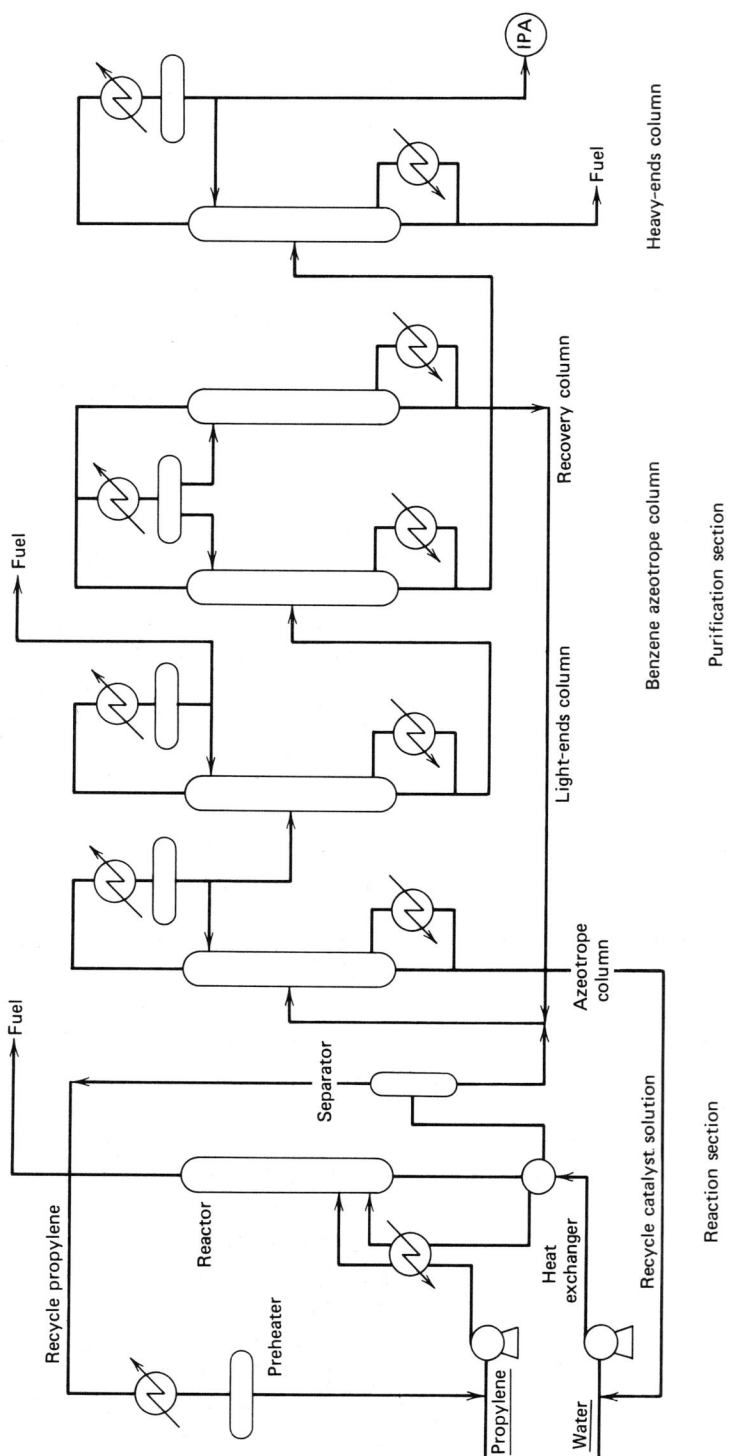

Figure 4. Tokuyama Soda's direct hydration process for the manufacture of isopropyl alcohol. Courtesy of *Chemtech* (69).

Table 3. Isopropyl Alcohol Plants Employing Direct Hydration Processes

Company	Location	Capacity, 10^3 t/yr	Technology	Completion date	Refs.
British Petrochemical	UK	50	Veba-Chemie	1974	72–73
Deutsche Texaco	FRG	80	own	1972	65, 67
Imperial Chemical Industries	UK	45	own	1951	58, 63
Nippon Petrochemical	Japan	60	Veba-Chemie	1973, 1974[a]	74–75
Petroleos del Peru	Peru	10	Deutsche Texaco	1977	76
Petroleos Mexicanos	Mexico	75	Tokuyama Soda	1983	77–78
Petrosul CA	Venezuela	30	Tokuyama Soda	1980	79–82
Tokuyama Soda	Japan	30	own	1972	83–84
Veba-Chemie	FRG	110	own	1965	85–88

[a] Capacity expanded from 40,000 to 60,000 t/yr in 1974.

Catalysts. Because low temperature and high pressure favor product formation in propylene hydration, one of the earliest efforts in developing this technology was a search for a catalyst that maximizes alcohol productivity at low temperatures within a reasonable time. Various tungsten compounds were claimed in patents issued in the early 1950s (90–94). Studies of acidic supports and of acids adsorbed on various porous supports were conducted at about the same time (95–106). Since then, many other acidic materials have been claimed, including cation-exchange resins (107–112), molybdophosphoric acid (113), titanium and zinc oxides (114), tungsten and zirconium oxides (115), silicotungstates (89,116), molybdenum oxalate (117), and zeolites (118) (see Catalysis).

Reaction Mechanism. Propylene hydration in dilute acid solution probably proceeds according to the rate-determining formation of propyl carbonium ion (119).

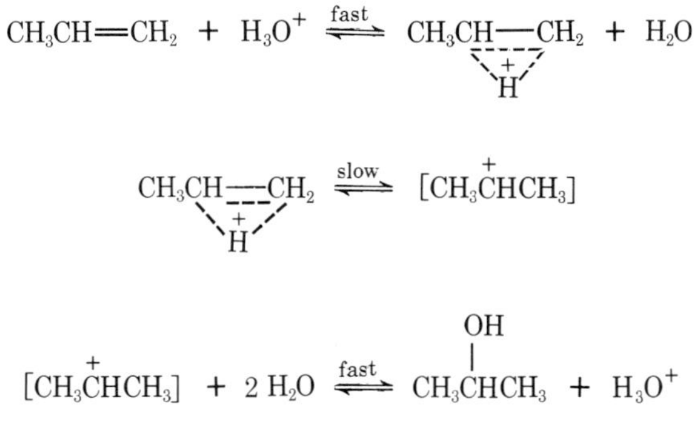

$$CH_3CH{=}CH_2 + H_3O^+ \underset{}{\overset{fast}{\rightleftharpoons}} CH_3CH\cdots CH_2\cdots H^+ + H_2O$$

$$CH_3CH\cdots CH_2\cdots H^+ \underset{}{\overset{slow}{\rightleftharpoons}} [CH_3\overset{+}{C}HCH_3]$$

$$[CH_3\overset{+}{C}HCH_3] + 2\,H_2O \underset{}{\overset{fast}{\rightleftharpoons}} CH_3CH(OH)CH_3 + H_3O^+$$

$$\text{rate} = K[C_3H_6][H_3O^+]$$

According to this mechanism, the reaction rate is proportional to the concentration of hydronium ion and is independent of the associated anion. However, evidence suggests that the anion may play a marked role in hydration rate. Thus, phosphomolydate and phosphotungstate anions exhibit hydration rates two or three times that of sulfate or phosphate (69). Association of the polyacid anion with the propyl carbonium ion is suggested. Because protonization of propylene occurs more readily than that of ethylene as a result of formation of a more stable secondary carbonium ion, higher conversions are achieved in propylene hydration.

Thermochemical Data. Equilibrium considerations significantly limit alcohol yield at low pressures in the vapor-phase process (120). Consequently, conditions controlling equilibrium constants have been determined with the resulting relation (120–121)

$$\log K = 2624/T - 7.584$$

Likewise, hydration in the liquid phase can be expressed by the free-energy equation

$$\Delta F° = 23.25\,T - 9352$$

The effect of pressure and temperature on the equilibrium concentration of alcohol in both phases of hydration of propylene with both liquid and vapor phases present has been calculated and is presented in Table 4 (120). It is significant that low temperature reduces by-product diisopropyl ether.

Other Processes. Isopropyl alcohol can be prepared by the liquid-phase oxidation of propane with the use of soluble vanadium catalysts (122). It is produced incidentally by the reductive condensation of acetone, and partly it is recovered from fermentation (123).

Large-scale, commercial biological production of isopropyl alcohol from carbohydrate raw materials is being studied. Approximately 50 wt % of C_3 chemicals probably will be supplied by fermentation (qv) when carbohydrates become competitive with petroleum feedstocks in ca 2000 (124–127) (see Chemurgy).

Table 4. Calculated Vapor–Liquid Equilibrium Composition for Propylene Hydration[a]

Pressure, MPa[b]	Alcohol concentration in liquid phase, mol %	Concentrations in vapor phase, mol %			
		Isopropyl alcohol	Water	Propylene	Isopropyl ether
at 200°C					
10.13	3.8	17.3	23.3	54.8	4.6
20.26	5.5	24.0	18.9	48.8	8.3
30.39	6.2	28.3	18.2	43.6	9.9
40.52	6.3	31.2	18.8	37.6	12.4
50.65	5.7	35.3	19.8	31.9	13.0
at 250°C					
10.13	1.2	7.9	49.3	42.5	0.3
20.26	2.1	13.0	37.6	48.6	0.8
30.39	2.6	14.5	36.1	48.5	0.9
40.52	3.2	15.3	36.1	47.5	1.1
50.65	3.7	15.2	37.8	45.7	1.3
at 275°C					
10.13	0.9	3.9	69.7	26.3	0.1
20.26	1.6	8.1	52.0	39.7	0.2
30.39	2.0	9.5	49.2	41.0	0.3
40.52	2.3	9.8	48.5	41.3	0.4
50.65	2.4	9.7	51.0	38.9	0.4

[a] Ref. 120.
[b] To convert MPa to atm, divide by 0.1013.

Storage and Shipping

Typically, anhydrous isopropyl alcohol is shipped in plain steel railroad tank cars, tank trucks, 3.8-L (1-gal) glass jugs, and 19- and 208-L (5- and 55-gal) DOT 17E phenolic-lined steel pails and drums, respectively. Plain steel is not suitable for isopropyl alcohol containing water, since rusting can result. Instead, baked phenolic-lined steel tanks are used. Aluminum is unsuitable since it is attacked by isopropyl alcohol, especially the anhydrous grade, which results in the formation of aluminum isopropoxide. Containers must comply with specifications of the DOT. Tanks, piping, and equipment can be made of similar material. Anhydrous isopropyl alcohol as well as water solutions to 91 vol % alcohol are considered red label (flammable) materials by the DOT, since both have flash points below 37.8°C by the Tag closed-cup method.

Economic Aspects

Economic comparisons of indirect vs direct hydration processes show a cost savings for use of the latter technology (70,128). The largest savings are in capital investment and processing and maintenance necessitated by the troublesome sulfuric acid reconcentration and the corrosion and pollution problems associated with indirect hydration. However, the drawback of direct hydration is its high energy usage and the need for highly concentrated propylene feedstock. Economics of the various direct hydration processes are roughly comparable (70).

Price and Demand. In terms of production volume, isopropyl alcohol is about the fourth largest chemical produced from propylene (57). Total 1981 U.S. and western Europe nameplate capacities for isopropyl alcohol production are 1285×10^6 and 8.52×10^5 metric tons, respectively (see Table 5) (129–130). The U.S. prices (February 1982) were $0.52/L ($1.96/gal) for refined 91 vol %, $0.54/L ($2.04/gal) for refined 95 vol %, and $0.59/L ($2.15/gal) for anhydrous alcohol. The average price has increased from ca $0.18/L ($0.70/gal) in 1977 and $0.25/L ($0.95/gal) in 1979.

Traditionally, isopropyl alcohol production has been dependent on demand for acetone, which accounts for roughly half the alcohol consumed (131). Since acetone is also produced as a coproduct in the cumene-based process for phenol production (see Phenol; Cumene), the future growth of isopropyl alcohol depends on the demand for phenol, since it is a cheaper source of the ketone. However, it is expected that demand for isopropyl alcohol will be high, since vast amounts of by-product phenol may become available before ca 2000 from various plants employing coal gasification for fuel and from synthesis gas production (132). When the cumene process is no longer required for phenol, substantial capacity for acetone will be required.

The near-term demand for isopropyl alcohol is expected to grow at 3%/yr to ca 9.9×10^5 t/yr by 1982 because of the demand for solvents and miscellaneous uses (133). However, isopropyl alcohol demand, ie, acetone demand, may decline somewhat by the mid-1980s if the new technology for methyl methacrylate is commercialized (see Methacrylic acid and derivatives).

Specifications

Typical specifications for the three basic grades of isopropyl alcohol are shown in Table 6. Other grades that are marketed include a cosmetic-grade (91 vol % and

Table 5. Isopropyl Alcohol Capacities

Producer	Plant location	1981 Capacity, 10^3 t
United States[a]		
Arco Chem. Co.	Channelview, Tex.	23
Exxon Corp.	Baton Rouge, La.	425
Shell Chem. Co.	Deer Park, Tex.	273
	Wilmington, Calif.	91
	Wood River, Ill.	155
Union Carbide Corp.	Texas City, Tex.	318
	Total	1285
Western Europe[b]		
France		
Shell Chimie SA	Berre-L'Etang	108
Federal Republic of Germany		
Chemische Werke Huls AG-CWH	Herne	110
Deutsche Texaco AG-DTA	Meerbeck	140
Henkel & Cie GmbH	Dusseldorf	na
The Netherlands		
Shell Nederland Chemie BV	Rotterdam-Pernis	250
Spain		
Industrias Quimicas Asociadas, SA-IQA	Tarragona	36
United Kingdom		
BP Chemicals, Ltd.	Park Talbot	85
Imperial Chemical Industries, Ltd.	Billingham	18
Shell Chemicals UK, Ltd.	Ellesmere Park	105
	Total	852

[a] Ref. 129. Courtesy of SRI International.
[b] Ref. 130. Courtesy of SRI International.

anhydrous) containing perfume and an electronic grade of low conductivity. Other, more restrictive specifications for special grades are shown in Table 7.

Analytical and Test Methods

Purity of commercial aqueous isopropyl alcohol mixtures is most simply and accurately determined by specific gravity measurement. Gas chromatography is an excellent technique for determining isopropyl alcohol in the presence of other organic substances, eg, ethyl alcohol, methyl alcohol, acetone, etc (134). Colorimetric methods can be used to determine trace amounts of ethyl alcohol upon the addition of specific compounds known to form complexes with alcohols. In the case of isopropyl alcohol, some of the colorimetric methods make use of its facile oxidation to acetone, which can provide highly colored complexes. Colorimetric methods have been reviewed (135). Trace amounts of isopropyl alcohol can also be determined photometrically in the presence of acetone and acetic acid (136). A conductometric method has been developed for isopropyl alcohol–water mixtures (137).

Table 6. Typical Isopropyl Alcohol Specifications

Requirement	Grade		
	91 vol %	95 vol %	Anhydrous
purity, vol %, min	91.0	95.0	99.85
sp gr (20/20°C)			
min	0.8169	0.8035	0.7861
max	0.8193	0.8055	0.7866
acidity, wt % as acetic acid, max	0.0024	0.002	0.002
acidity, mg KOH/g sample	0.023		0.019
distillation, °C			
initial, min	79.9	80.0	82.05
dry point, max	80.9	83.0	82.55
range, max	1.0		0.50
color, APHA, max	10	10	10
water, wt %, max			0.1
nonvolatile matter, g/100 mL, max	0.002		0.001
permanganate time test, minutes, min	30		
odor		no foreign or residual	
water solubility	complete	complete	complete

Health and Safety Factors (Toxicology)

Although alcohols as a class have a low order of toxicity, isopropyl alcohol is about twice as toxic as ethyl alcohol. There is no known systematic investigation of the effects of inhalation, eg, from aerosols, of isopropyl alcohol in humans. The known human toxicity is based on numerous cases of accidental ingestion or topical application. Toxic doses of ingested isopropyl alcohol, usually as rubbing alcohol, may produce narcosis, anesthesia, coma, and death. The fatal dose for man is ca 166 mL; death occurs from paralysis of the central nervous system. Approximately 70–90% of ingested isopropyl alcohol is oxidized to acetone in the body; thus, it is not a cumulative poison (138). Acetone appears on the breath within 15 min. There is no fixed relationship between blood and urine isopropyl alcohol concentrations.

Use of isopropyl alcohol in industrial applications does not present a health hazard. It produces anesthetic effects in very high vapor concentration. Consequently, the TLV for isopropyl alcohol has been established at 400 ppm (0.098 mg/L) for an 8-h exposure (TWA) (139). This level causes a mild irritation of the eyes, nose, and throat (140). However, the TLV level does not produce symptoms of anesthesia (141). Further indications of the toxicity of isopropyl alcohol is provided by tests with animals (see Table 8).

Uses

The uses of isopropyl alcohol are chemical, solvent, and medical. Estimated U.S. uses in 1977 were acetone, 38%; other chemicals, 7%; coatings and other solvents, 30%; drugs and cosmetics, 10%; miscellaneous, 8%; and export, 7% (9).

Chemical. Growth of the use of isopropyl alcohol as a feedstock for the production of acetone will be influenced by alternative routes to, and markets for, the production of acetone. In addition, isopropyl alcohol is consumed in the production of other chemicals (see Chemical Properties).

Table 7. Special Specifications for Isopropyl Alcohol

Property	ACS reagent 1975	1980 USP XX and NF XV	ASTM D 770-80	Federal TT-1-735, grade A	ASTM D 1310-81, 91 vol % IPA
acidity, wt % acetic acid, max	0.0008	0.0024	0.002	0.002	
alkalinity, wt % NaOH, max	0.0005				
appearance				CFSM[a]	
color (Pt–Co), max	10	colorless	10	10	
corrosion on copper				no pitting or black stain	
distillation, initial bp, min–dry point, max, °C				81.3–83.0	
range, °C, max	1.0 (must include 82.3°C)		1.5 (must include 82.3°C)		1.0 (must include 80.4°C)
nonvolatile matter, g/100 mL, max	0.0008	0.005	0.005	0.002	
odor initial			characteristic	characteristic	
residual			none	none	
refractive index, n_D^{20}		1.376–1.378			
sp gr (20/20°C)	0.7883	0.783–0.787[b]	0.785–0.787	0.7862–0.7870	0.8175–0.8185
uv spectrum	smooth curve[c]				
max absorbance at 210 nm	1.00				
220	0.40				
230	0.20				
245	0.08				
260	0.04				
275	0.03				
300	0.02				
330	0.01				
water content, wt % max	0.2		0.2	0.10	
water solubility	clear		no turbidity	no turbidity	

[a] Clear and free of suspended matter.
[b] At 25°C.
[c] For spectrometric grade only.

Solvent. Because of the balance between alcohol, water, and hydrocarbonlike characteristics, isopropyl alcohol is an excellent, low cost solvent which is free from the government regulations and taxes that apply to ethyl alcohol. The lower toxicity of isopropyl alcohol favors its use over methyl alcohol, even though the former is somewhat higher in cost. Consequently, isopropyl alcohol is used as a solvent in many consumer products as well as industrial products and procedures, eg, gasification and extractions. It is a good solvent for a variety of oils, gums, waxes, resins, and alkaloids and, consequently, it is used for preparing cements, primers, varnishes, paints, printing inks, etc.

Table 8. Toxicological Properties of Isopropyl Alcohol[a]

Alcohol	Single oral LD_{50}[b], rats	Single skin penetration[c] LD_{50}, rabbits, mL/kg	Single inhalation[d] concentrated vapors in ppm, rats	Primary skin irritation[e], rabbits	Eye injury[f], rabbits
isopropyl alcohol, 91 vol %	10.7 g/kg	>20		none	moderate
isopropyl alcohol, anhydrous	6.48 mL/kg	8.0	8 h at 8,000 ppm killed none of 12	none	moderate

[a] Ref. 4.
[b] The term LD_{50} refers to that quantity of chemical which kills 50% of dosed animals within 14 days.
[c] Single skin penetration refers to a 24-h covered skin contact with the liquid chemical.
[d] Single inhalation refers to the continuous breathing of certain concentrations of chemical for the stated period.
[e] Primary irritation refers to the skin response 24 h following application of 0.01 mL to uncovered skin.
[f] Eye injury refers to surface damage produced by the liquid chemical.

Isopropyl alcohol is also employed widely as a solvent for cosmetics (qv), eg, lotions, perfumes, shampoos, skin cleansers, nail polishes, make-up removers, deodorants, body oils, and skin lotions. In cosmetic applications, the acetonelike odor of isopropyl alcohol is masked by the addition of fragrance (142).

Over 68 aerosol products containing isopropyl alcohol solvent have been reported (143). Aerosol formulations include hair sprays (144), floor detergents (145), shoe polishes (146), insecticides (147–148), burn ointments (149), window cleaners, waxes and polishes, paints, automotive products, eg, windshield deicer, insect repellents, flea and tick spray, air refreshers, disinfectants, veterinary wound and pinkeye spray, first-aid spray, foot fungicide, and fabric-wrinkle remover (150) (see Aerosols).

Medical. Isopropyl alcohol also is used as an antiseptic and disinfectant for home, hospital, and industry (see Disinfectants and antiseptics). It is about twice as effective as ethyl alcohol in these applications (151–152). Rubbing alcohol, a popular 70 vol % isopropyl alcohol-in-water mixture, exemplifies its medicinal use. Other examples include 30 vol % isopropyl alcohol solutions for medicinal liniments, tinctures of green soap, scalp tonics, and tincture of mercurophen. It is contained in pharmaceuticals, eg, local anesthetics, tincture of iodine, and bathing solutions for surgical sutures and dressings. Over 200 uses of isopropyl alcohol have been tabulated (2).

BIBLIOGRAPHY

"Isopropyl Alcohol" in *ECT* 1st ed., Vol. 11, pp. 182–190, by J. G. Park and C. M. Beamer, Enjay Company, Inc.; "Isopropyl Alcohol" in *ECT* 2nd ed., Vol. 16, pp. 564–578, by E. J. Wickson, Enjay Chemical Laboratory.

1. *Synthetic Organic Chemicals*, U.S. Production and Sales, U.S. Tariff Commission, U.S. Govt. Printing Office, Washington, D.C., 1978.
2. L. F. Hatch and W. R. Fenwick, *Isopropyl Alcohol*, Enjay Chemical Company, New York, 1966.
3. Unpublished physical property data, Union Carbide Corp., New York.
4. *Alcohols*, Brochure F-42379C, Union Carbide Corp., New York, Oct. 1979.
5. *Solvent Selector*, Bulletin F-7465T, Union Carbide Corp., New York, Apr. 1979.
6. R. F. Brunel, J. L. Crenshaw, and E. Tobin, *J. Am. Chem. Soc.* **43,** 561 (1921).
7. Louis and Entezam, *Ann. Combustible Liquids* **14,** 21 (1939).

8. L. H. Horsley, "Azeotropic Data-III," *Advances in Chemistry Series 116*, American Chemical Society, Washington, D.C., 1973.
9. *Chem. Mark. Rep.*, 9 (June 5, 1978).
10. U.S. Pat. 2,394,848 (Feb. 12, 1946), T. F. Doumani (to Union Oil Co.).
11. C. F. Brown, *paper presented at 116th Meeting of the American Chemical Society*, Atlantic City, N.J., Sept. 1949.
12. *Isopropanol*, Brochure F-42437, Union Carbide Corp., New York, Apr. 1969.
13. J. F. Norris, W. S. Johnson, H. D. Hirsch, and C. R. McCullough, *Rec. Trav. Chim.* **48,** 885 (1929).
14. M. Windholz, ed., *The Merck Index*, 9th ed., Merck and Co., Inc., Rahway, N.J., 1976.
15. U.S. Pat. 2,647,914 (Aug. 4, 1953), W. G. Allan and T. J. Tobin (to Imperial Chemical Industries, Ltd.).
16. Ger. Pat. 2,019,808 (Nov. 5, 1970), W. A. Craig and O. A. Gurten (to Imperial Chemical Industries, Ltd.).
17. M. Arvis and L. Gilles, *J. Chim. Phys. Phys-Chim. Biol.* **67**(9), 1538 (1970).
18. Jap. Pat. 42,11351 (June 26, 1967), (to Mitsubishi Chemical).
19. Brit. Pat. 868,023 (May 17, 1961), L. E. Addy (to British Hydrocarbon Chemicals).
20. Brit. Pat. 1,097,819 (Jan. 3, 1968), (to Usines de Melle).
21. U.S. Pat. 2,586,694 (Feb. 19, 1952), H. O. Mottern (to Standard Oil Development).
22. P. W. Sherwood, *Pet. Refiner* **33**(12), 147 (1954).
23. S. S. Lokras, P. K. Deshpande, and N. R. Kuloor, *Ind. Eng. Chem. Prod. Des. Devel.* **9**(2), 293 (1970).
24. U.S. Pat. 2,871,104 (Jan. 27, 1959), F. F. Rust (to Shell Development Co.).
25. U.S. Pat. 2,871,101 (Jan. 27, 1959), F. F. Rust and M. L. Porter (to Shell Development Co.).
26. *Hydrocarbon Process. Pet. Refiner* **40**(11), 249 (1961).
27. *Eur. Chem. News* **15**(400), 30 (1969).
28. Brit. Pat. 619,014 (Mar. 2, 1949), S. A. Ballard, H. de V. Finch, and E. A. Peterson (to N. V. Bataafsche Petroleum Maatschappy).
29. U.S. Pat. 2,636,902 (Apr. 28, 1953), A. W. C. Taylor, P. Davies, and P. W. Reynolds (to Imperial Chemical Industries, Ltd.).
30. U.S. Pat. 2,609,394 (Sept. 2, 1952), P. Davies, P. W. Reynolds, R. R. Coats, and W. C. Taylor (to Imperial Chemical Industries, Ltd.).
31. P. H. Groggins, *Unit Processes in Organic Synthesis*, McGraw-Hill Book Company, Inc., New York, 1958, pp. 407, 434.
32. U.S. Pat. 2,349,222 (May 16, 1944), R. H. Goshorn (to Sharples Chemicals).
33. U.S. Pat. 2,365,721 (Dec. 26, 1944), J. Olin and J. McKenna (to Sharples Chemicals).
34. P. Richter and J. Pasek, *Chemicky Prumysl* **17**(7), 353 (1967).
35. Brit. Pat. 649,980 (Feb. 7, 1951), W. Whitehead (to Imperial Chemical Industries, Ltd.).
36. S. Coffey, ed., *Rodd's Chemistry of Carbon Compounds*, 2nd ed., American Elsevier Publishers, Inc., New York, 1965, pp. 114–115.
37. M. R. A. Rao, *J. Indian Inst. Sci.* **39,** 138 (1957).
38. C. Matasa and E. Tonca, *Basic Nitrogen Compounds*, 3rd ed., trans. from Romanian by S. Marcus, Chemical Publishing Co., Inc., New York, 1973, p. 282.
39. G. T. Austin, *Chem. Eng.*, 101 (May 27, 1974).
40. Z. E. Zolles, *Bromine and its Compounds*, Academic Press, Inc., New York, 1966, pp. 65, 387.
41. E. H. Huntress, *Organic Chlorine Compounds*, John Wiley & Sons, Inc., New York, 1948, p. 774.
42. U.S. Pat. 1,391,757 (Sept. 27, 1921), H. E. Buc (to Standard Oil Company).
43. USSR Pat. 211,528 (Feb. 18, 1968), E. V. Sergeev, T. V. Uvarova, and A. N. Smirnova; *Chem Abstr.* **69,** 43452 (1968).
44. Fr. Pat. 816,956 (Aug. 21, 1937), (to I. G. Farbenind. A.G.).
45. Fr. Pat. 818,131 (May 26, 1937), (to I. G. Farbenind. A.G.).
46. U.S. Pat. 3,325,545 (June 13, 1967), W. W. Levis, Jr. (to BASF Wyandotte Chemicals Corp.).
47. Ref. 36, page 25.
48. U.S. Pat. 3,361,657 (Jan. 2, 1968), M. Kokorudz (to BASF Wyandotte Chemicals Corp.).
49. H. Maarse and M. C. Ten Noever de Brauw, *Chem. Ind. (London)* **1,** 36 (1974).
50. Brit. Pat. 2,004,538 (Apr. 4, 1979), C. Savini (to Exxon Research Engineering Co.).
51. U.S. Pat. 4,219,685 (Aug. 26, 1980), C. Savini (to Exxon Research Engineering Co.).
52. *Pet. Refiner* **38**(11), 264 (1959).
53. *Hydrocarbon Process. Pet. Refiner* **40**(11), 260 (1961).

54. T. Horie, M. Imaizumi, and Y. Fujiwara, *Hydrocarbon Process.* **49**(3), 119 (1970).
55. Jap. Pat. 7 7012-166 (Apr. 5, 1977), T. Sato, R. Ohuji, and H. Yamanovchi (to Tokuyama Soda).
56. F. C. Fetter, *Chem. Eng.* **55,** 235 (Oct. 1948).
57. J. C. Fielding in E. G. Hancock, ed., *Propylene and its Industrial Derivatives*, John Wiley & Sons, Inc., New York, 1973.
58. K. Weissermel and H. J. Arpe, *Industrial Organic Chemistry*, Springer-Verlag Chemie, Weinheim, Austria, 1978.
59. Belg. Pat. 683,923 (Dec. 16, 1966), (to Hibernia-Chemie GmbH).
60. *Hydrocarbon Process.* **46**(11), 195 (1967).
61. U.S. Pat. 3,955,939 (May 11, 1976), A. Sommer and M. Urban (to Veba-Chemie A.G.).
62. *Eur. Chem. News*, 32 (July 24, 1970).
63. *Petroleum (London)* **16,** 19 (1953).
64. W. Neier and J. Woellner, *Chemtech*, 95 (Feb. 1973).
65. *Hydrocarbon Process.* **58**(11), 181 (1979).
66. W. Neier and J. Woellner, *Erdoel Kohle* **28**(1), 19 (1975).
67. W. Neier and J. Woellner, *Hydrocarbon Process.* **5**(11), 113 (1972).
68. *Hydrocarbon Process.* **52**(11), 141 (1973).
69. Y. Onoue, Y. Mizutani, S. Akiyama and Y. Izumi, *Chemtech*, 432 (July 1978).
70. Y. Onoue and Y. Izumi, *Chem. Econ. Eng. Rev.* **6**(7), 48 (1974).
71. U.S. Pat. 3,758,615 (Sept. 11, 1973), Y. Izumi, Y. Kawasaki, and M. Tani (to Tokuyama Soda).
72. *Eur. Chem. News*, 8 (June 1, 1973).
73. *Chem. Ind. (London)*, 911 (Oct. 6, 1973).
74. *Jpn. Chem. Week*, 2 (July 25, 1974).
75. *Jpn. Chem. Week*, 2 (Nov. 8, 1973).
76. *Eur. Chem. News*, 14 (July 25, 1975).
77. *Eur. Chem. News*, 32 (Mar. 3, 1980).
78. *Chem. Mark. Rep.*, 7 (Feb. 25, 1980).
79. *Jpn. Chem. Week*, 2 (Dec. 7, 1978).
80. *Chem. Age*, 1 (Sept. 9, 1976).
81. *Eur. Chem. News*, 51 (Sept. 10, 1976).
82. *Japan Econ. J.*, 8 (Aug. 8, 1976).
83. *Eur. Chem. News*, 4 (July 28, 1972).
84. *Chem. Mark. Rep.*, 3 (Dec. 18, 1972).
85. *Chem. Age London* **99**(6), 2599 (1969).
86. *Chem. Age London* **99**(10), 2593 (1969).
87. *Eur. Chem. News*, 24 (Feb. 23, 1973).
88. *Chem. Age* **95,** 132 (1966).
89. U.S. Pat. 3,758,615 (Sept. 11, 1973), Y. Izumi, Y. Kawasaki, and M. Tani (to Tokuyama Soda).
90. Brit. Pat. 622,937 (May 10, 1949), P. W. Reynolds and co-workers (to ICI).
91. Brit. Pat. 718,723 (Nov. 7, 1954), D. A. Dowden (to ICI).
92. U.S. Pat. 2,683,753 (July 13, 1954), N. Levy and co-workers (to ICI).
93. U.S. Pat. 2,725,403 (Nov. 29, 1955), M. A. E. Hodgson (to ICI).
94. U.S. Pat. 2,755,309 (July 17, 1956), P. W. Reynolds and co-workers (to ICI).
95. U.S. Pat. 2,504,618 (Apr. 18, 1950), R. C. Archibald and co-workers (to Shell Development Co.).
96. U.S. Pat. 2,579,601 (Dec. 25, 1951), R. C. Nelson and co-workers (to Shell Development Co.).
97. U.S. Pat. 2,658,924 (Nov. 10, 1953), S. J. Lukasiewicz and co-workers (to Socony-Vacuum Oil).
98. U.S. Pat. 2,663,744 (Dec. 22, 1953), S. J. Lukasiewicz (to Socony-Vacuum Oil).
99. Ger. Pat. 963,238 (May 2, 1957), C. Wagner (to Bayer).
100. U.S. Pat. 2,825,704 (Mar. 4, 1958), H. R. Arnold and co-workers (to E. I. du Pont de Nemours & Co., Inc.).
101. Brit. Pat. 750,176 (June 13, 1956), H. Newby (to Huels).
102. Brit. Pat. 996,917 (June 30, 1965), (to Gulf Research & Development).
103. Brit. Pat. 1,159,666 (July 30, 1969), (to Hibernia Chemie).
104. Brit. Pat. 1,159,667 (July 30, 1969), (to Hibernia Chemie).
105. Fr. Pat. 1,531,086 (July 17, 1968), (to Scholven-Chemie).
106. Jap. Pat. 47-23524 (June 30, 1972), R. Ono, T. Sugirua, and K. Takemori (to Mitsui Toatsu Chemicals).
107. U.S. Patent 3,256,250 (June 14, 1966), V. J. Frilette (to Socony Mobil Oil).

108. Brit. Pat. 1,238,556 (July 7, 1971), R. H. Scott and D. L. Gaulding (to Celanese Corp.).
109. Ger. Pat. 2,147,737 (Mar. 29, 1973), G. Brands and co-workers (to Deutsche Texaco).
110. Ger. Pat. 2,147,739 (Apr. 5, 1973), G. Brands and co-workers (to Deutsche Texaco).
111. Ger. Pat. 2,147,740 (Apr. 5, 1973), G. Brands and co-workers (to Deutsche Texaco).
112. Ger. Pat. 2,147,738 (Mar. 29, 1973), G. Brands and co-workers (to Deutsche Texaco).
113. U.S. Pat. 3,644,497 (Feb. 22, 1972), F. G. Mesick (to Celanese Corp.).
114. Jap. Pat. 47-23523 (June 30, 1972), K. Tabe and I. Matsuzaki (to Mitsui Toatsu Chemicals).
115. U.S. Pat. 3,450,777 (June 17, 1969), Y. Mizutani (to Tokuyama Soda).
116. Brit. Pat. 1,281,120 (July 12, 1972), Y. Izumi, M. Tani, and Y. Kawasaki (to Tokuyama Soda).
117. U.S. Pat. 3,705,912 (Dec. 12, 1972), S. N. Massie (to Universal Oil Products).
118. U.S. Pat. 4,214,107 (July 22, 1980), C. D. Chang and N. J. Morgan (to Mobil Oil Corp.).
119. R. W. Taft, Jr., *J. Am. Chem. Soc.* **74**, 5372 (1952).
120. C. S. Cope, *J. Chem. Eng. Data* **11**(3), 379 (1966).
121. F. M. Majewski and L. F. Marek, *Ind. Eng. Chem.* **30**, 203 (1938).
122. B. W. Kiff, Union Carbide Corporation, patent to be issued.
123. G. T. Austin, *Chem. Eng.*, 101 (May 27, 1974).
124. *Chem. Age*, 15 (July 21, 1978).
125. *Eur. Chem. News*, 30 (July 21, 1978).
126. *Chem. Eng. News*, 28 (July 24, 1978).
127. *Chem. Eng. Prog.* 70 (Apr. 1978).
128. *Eur. Chem. News*, (March 10, 1972).
129. *Directory of Chemical Producers—United States*, SRI International, Menlo Park, Calif., 1981, p. 678.
130. *Directory of Chemical Producers—Western Europe*, SRI International, Menlo Park, Calif., 1981, pp. 1317–1318.
131. *Synthetic Organic Chemicals*, U.S. Production and Sales, U.S. Tariff Commission, U.S. Govt. Printing Office, Washington, D.C., 1978, p. 311.
132. Brit. Pat. 1,289,158 (Sept. 13, 1972), P. C. Keith, E. S. Johanson, R. H. Walk, S. B. Alpert, and S. C. Schuman (to Hydrocarbon Research, Inc.).
133. *Chem. Mark. Rep.*, 9 (June 5, 1978).
134. D. W. Hessel and F. R. Modglin, *J. Forensic Sci.* **9**, 255 (1964).
135. W. H. Simmons, *Perfum. Essen. Oil Rec.* **18**, 168 (1927).
136. M. Mantel and M. Anbar, *Anal. Chem.* **36**(4), 936 (1964).
137. A. M. Arjuna, *Anales Real Soc. Espan. Fis. Quim. (Madrid) Ser. B* **61**(3), 591 (1965).
138. S. Zakhori, P. Levy, M. Leibowitz, and D. M. Aviado in L. Golberg, ed., *Isopropanol and Ketones in the Environment*, CRC Press, Inc., Cleveland, Oh., 1977.
139. *American Conference of Governmental Industrial Hygienists, Documentation of the Threshold Limit Values*, 4th ed., Cincinnati, Oh., 1980, p. 238.
140. K. W. Nelson and co-workers, *J. Ind. Hyg. Toxicol.* **25**, 282 (1943).
141. G. D. Clayton and F. E. Clayton, *Patty's Industrial Hygiene and Toxicology*, Vol. II, Interscience Publishers, Inc., a division of John Wiley & Sons, Inc., New York, 1979, p. 853.
142. V. Lechnitz, *Kosmet. Aerosole* **44**, 65 (1971).
143. M. N. Gleason and co-workers, *Clinical Toxicology of Commercial Products*, Williams and Wilkins, Baltimore, Md., 1969.
144. Ger. Pat. 2,239,690 (Feb. 22, 1973), D. Y. Hsiung (to Gillette Co.).
145. U.S. Pat. 3,650,956 (Mar. 21, 1972), D. L. Strand and R. L. Abler (to Minnesota Mining and Manufacturing Co.).
146. U.S. Pat. 3,231,397 (Jan. 25, 1966), A. Kessler, G. L. Layne, and C. L. Spector (to Proctor and Gamble Co.).
147. V. M. Tsetlin, I. V. Bessanova, and E. B. Zhuk, *Khim. Promst. (Moscow)* **47**, 31 (1971).
148. U.S. Pat. 3,244,502 (Apr. 5, 1966), S. M. Woogerd (to Hercules Glue Co.).
149. Ger. Pat. 1,935,939 (Feb. 4, 1971), H. Augart (to Goedecke).
150. U.S. Pat. 3,600,325 (Aug. 17, 1971), K. L. Kaufman, D. N. Martin, and W. J. Brown (to CPC International Inc.).
151. Sister M. John, *Hosp. Manage.* **57**, 86 (1944).
152. W. R. Straughn, *Mod. Hosp.* **66**, 90 (1946).

ANTHONY J. PAPA
Union Carbide Corporation

n-PROPYL ALCOHOL

n-Propyl alcohol [71-23-8], 1-propanol, $CH_3CH_2CH_2OH$ (mol wt, 60.09), is a clear, colorless liquid with a typical alcohol odor, and it is miscible in water, ethyl ether, and alcohols. 1-Propanol occurs in nature in fusel oils and forms from fermentation and spoilage of vegetable matter (1).

Properties

A number of physical and chemical properties of 1-propanol are listed in Table 1 (2–3). The chemistry of 1-propanol is typical of low molecular weight primary alcohols (see also Alcohols, higher). Biologically, 1-propanol is easily degraded by activated sludge and is the easiest alcohol to degrade (4).

Manufacture

1-Propanol has been manufactured by hydroformylation of ethylene followed by hydrogenation of propionaldehyde (see Oxo process) and as a by-product of vapor-phase oxidation of propane (see Hydrocarbon oxidation). Celanese operated the only commercial vapor-phase oxidation facility at Bishop, Texas. After this facility was shut down in 1973 (5–6), hydroformylation or oxo technology has been the only basis for commercial manufacture of 1-propanol. Some attempts have been made to hydrate propylene in an anti-Markownikoff fashion to produce 1-propanol (7–9). However, these attempts have not been commercially successful.

The production of 1-propanol by hydroformylation or oxo technology is a two-step process in which ethylene is first hydroformylated to produce propanal. The resulting propanal is hydrogenated to 1-propanol (eqs. 1 and 2).

$$CH_2{=}CH_2 + CO + H_2 \xrightarrow[\Delta,\ \text{pressure}]{\text{catalyst}} CH_3CH_2CHO \quad (1)$$

$$CH_3CH_2CHO + H_2 \xrightarrow[\Delta,\ \text{pressure}]{\text{catalyst}} CH_3CH_2CH_2OH \quad (2)$$

Propane, 1-propanol, and heavy ends (the latter are made by aldol condensation) are minor by-products of the hydroformylation step. A number of transition-metal carbonyls, eg, Co, Fe, Ni, Rh, and Ir, have been used to catalyze the oxo reaction, but cobalt and rhodium are the only economically practical choices. In the United States, Texas Eastman (10), Union Carbide (11), and Celanese (12) make 1-propanol by oxo technology. Texas Eastman apparently uses conventional cobalt oxo technology with an $HCo(CO)_4$ catalyst.

Conditions of the conventional cobalt oxo process are 150–200°C, 20–30 MPa (200–300 atm), 1:1 H_2:CO, and 0.1–1.0 wt % cobalt (13). Under these conditions, the ethylene conversion to propanal is ca 90%. Approximately 2–3 wt % of the aldehyde is converted to high boilers by aldol condensation and Tischenko ester formation. The substantial high boiler efficiency is effected by $HCo(CO)_4$ since it is rather acidic (K_a = 1) (14). The concentration of cobalt in solution is a function of temperature and CO partial pressure, since $HCo(CO)_4$ is unstable in the absence of CO (15). The cobalt catalyst must be efficiently removed from the oxo reaction product, as it may de-

Table 1. Physical and Chemical Constants of 1-Propanol[a]

Property	Value
freezing point, °C	−126.2
boiling point, °C	97.20
vapor pressure, kPa[b]	
at 20°C	1.987
at 40°C	6.986
at 60°C	20.292
at 80°C	50.756
Antoine eq. (2–120°C): $\log P_{kPa}{}^{c} = 6.97257 - \dfrac{1499.21}{(204.64 - t)}$, $t =$ °C	
vapor density (air = 1)[d]	2.07
density (at 20°C), g/cm^3	0.80375
Francis eq. (−21 to 180°C): dens $= 0.8813 + (5.448\ t \times 10^{-4}) - \left[\dfrac{21.536}{(313.09 - t)}\right]$, $t =$ °C	
refractive index, n_D^{20}	1.38556
viscosity (at 20°C), mPa·s (= cP)[d]	2.256
surface tension (at 20°C), mN/m (= dyn/cm)	23.75
critical temperature, °C	263.56
critical pressure, kPa[b]	5169.60
critical density, g/cm^3	0.275
heat capacity (liquid at 25°C), J/(mol·K)[e]	141
heat of vaporization, kJ/mol[e]	
at 25°C	47.53
at 97.20°C	41.78
heat of combustion (liquid at 25°C), kJ/mol[e]	−201.98
heat of formation (liquid at 25°C), kJ/mol[e]	−304.01
flash point (Tag open cup), °C[d]	29
autoignition temperature, °C[d]	371.1
explosive limit (in air), vol %[d]	
lower	2.1
upper	13.5
electrical conductivity (at 25°C), S (= mho)[d]	2×10^{-8}

[a] Ref. 2 except where noted.
[b] To convert kPa to mm Hg, multiply by 7.5.
[c] To convert $\log P_{kPa}$ to $\log P_{mm\ Hg}$, add 0.8751 to the constant.
[d] Ref. 3.
[e] To convert J to cal, divide by 4.184.

compose (plug equipment) or poison the aldehyde hydrogenation catalyst (16). Further loss of aldehyde downstream from the reactor by condensation reactions can also occur (16). Many of the more recent improvements in the process have been in catalyst-handling techniques (17–19).

The generally accepted catalytic cycle is illustrated in Figure 1 with ethylene (20–21). Figure 1 shows that many sources of cobalt yield the active species. Although neither $HCo(CO)_3$ nor the π-alkyl intermediate have been isolated, they are probable intermediates.

The rhodium–triphenylphosphine catalyst system is being used instead of cobalt in oxo processes because it has higher reaction rates, greater stability, lower operation pressure, and higher linear aldehyde selectivity, ie, lower by-product production.

$$2\text{ Co(metal, oxide, salt, etc)} + 8\text{ CO} \rightarrow \text{Co}_2(\text{CO})_8 \underset{-H_2}{\overset{+H_2}{\rightleftharpoons}} 2\text{ HCo(CO)}_4 \underset{+CO}{\overset{-CO}{\rightleftharpoons}} \text{HCo(CO)}_3 \overset{CH_2=CH_2}{\rightleftharpoons}$$

$$\text{HCo(CO)}_3 \overset{CH_2=CH_2}{\rightleftharpoons} \text{C}_2\text{H}_5\text{Co(CO)}_3 \overset{CO}{\rightleftharpoons} \text{C}_2\text{H}_5\text{Co(CO)}_4 \rightleftharpoons \text{C}_2\text{H}_5\text{COCo(CO)}_3 \overset{H_2}{\rightleftharpoons} \text{C}_2\text{H}_5\text{COCo(H)}_2(\text{CO})_3 \rightarrow$$
$$\text{C}_2\text{H}_5\text{CHO} + \text{HCo(CO)}_3$$

Figure 1. Hydroformylation of ethylene by a cobalt-catalyzed oxo process.

Simple rhodium carbonyls are ca 100–1000 times more reactive than cobalt carbonyls (22). When triphenylphosphine is added, the reaction rate increases monotonically until the triphenylphosphine:rhodium mole ratio reaches 20:1 to 50:1 (23). Above this ratio, the rate of reaction decreases. A substantial amount of research on the Rh oxo process has been directed toward improving the linear aldehyde selectivity and linear-to-branched aldehyde ratio. In general, increasing triphenylphosphine concentration or decreasing CO partial pressure leads to higher selectivity to linear aldehyde, ie, the selectivities to branched aldehyde and internal olefin decrease. The rate of hydroformylation decreases as the selectivity to linear aldehyde increases. Since ethylene can lead neither to internal olefin nor to branched aldehyde, it is best to use the triphenylphosphine-to-rhodium ratio that maximizes the rate. However, catalyst stability problems may dictate operation at higher triphenylphosphine-to-rhodium ratios. A rhodium-catalyzed, ethylene hydroformylation process operates at 90–120°C, 2.17–3.55 MPa (300–500 psig), 1:1–10:1 H_2:CO, 1–10 mM rhodium, and 0.02–0.4 M triphenylphosphine. The chemical efficiency to propanal under these conditions is 98–99% (0.5–1.0% efficiency to ethane and 0.5–1.0% efficiency to heavy ends). The rhodium can be obtained from almost any source except those that contain halogen, since halogen forms complexes that have very low activity (24).

Rhodium-catalyzed hydroformylation has been studied extensively (24–37). The most active catalyst source is a carbonylhydrotris(triphenylphosphine)rhodium (HRhCO[P(C$_6$H$_5$)$_3$]$_3$) (29). However, a molecule of triphenylphosphine is presumed to dissociate to form the active species (29,36). Further dissociation could occur as shown in equation 3.

$$\text{HRhCO[P(C}_6\text{H}_5\text{)}_3\text{]}_3 \underset{P(C_6H_5)_3}{\overset{CO}{\rightleftharpoons}} \text{HRh(CO)}_2[\text{P(C}_6\text{H}_5\text{)}_3]_2 \underset{+P(C_6H_5)_3}{\overset{-P(C_6H_5)_3}{\rightleftharpoons}} \text{HRh(CO)}_2\text{P(C}_6\text{H}_5\text{)}_3 \qquad (3)$$

For each active species, (HRh[CO]$_2$[P(C$_6$H$_5$)$_3$]$_2$) or [HRh(CO)$_2$P(C$_6$H$_5$)$_3$], there exists a catalytic cycle for hydroformylation which is very similar to the mechanism shown in Figure 1 for HCo(CO)$_4$-catalyzed hydroformylation. This is illustrated for the HRh[CO]$_2$[P(C$_6$H$_5$)$_3$]$_2$ species in Figure 2.

$$\text{HRh(CO)}_2[\text{P(C}_6\text{H}_5\text{)}_3]_2 \overset{CH_2=CH_2}{\rightleftharpoons} \text{HRh(CO)}_2[\text{P(C}_6\text{H}_5\text{)}_3]_2 \rightleftharpoons$$

$$\text{C}_2\text{H}_5\text{Rh(CO)}_2[\text{P(C}_6\text{H}_5\text{)}_3]_2 \rightleftharpoons \text{C}_2\text{H}_5\text{CORhCO}[\text{P(C}_6\text{H}_5\text{)}_3]_2 \overset{H_2}{\rightleftharpoons}$$

$$\text{C}_2\text{H}_5\text{CORh(H)}_2\text{CO}[\text{P(C}_6\text{H}_5\text{)}]_2 \rightarrow \text{C}_2\text{H}_5\text{CHO} + \text{HRhCO}[\text{P(C}_6\text{H}_5\text{)}_3]_2 \overset{CO}{\rightleftharpoons} \text{HRh(CO)}_2[\text{P(C}_6\text{H}_5\text{)}_3]_2$$

Figure 2. Hydroformylation of ethylene by a rhodium-catalyzed oxo process.

It has been suggested that rhodium with three phosphines bound to it, eg, $HRhCO[P(C_6H_5)_3]_3$, can also catalyze hydroformylation (38–41). It also has been suggested that CO rather than phosphine might dissociate prior to coordination of olefin (37,38). Whatever the case, the greater the number of phosphine molecules bound to rhodium, the slower the rate of hydroformylation, but the greater the selectivity to linear aldehyde.

The technology for hydrogenating the propanal to 1-propanol is well established. Commercially, nickel-based catalysts, eg, Raney nickel or supported nickel, and copper chromium oxide catalysts are used (42). Vapor-phase or liquid-phase hydrogenation can be carried out. The reactor can be a fixed-bed, slurry-bed, or trickle-bed system (43–44). Conditions for liquid-phase hydrogenation are 2.17–4.24 MPa (300–600 psig) and 100–170°C. Vapor-phase reactors generally operate below 790 kPa (100 psig) (45). Efficiencies to 1-propanol of >95% can be expected, and the principal by-products are acetals, ethers, esters, and diols (44). Both CO and triphenylphosphine poison the hydrogenation catalysts and should be removed prior to hydrogenation of propanal.

Figure 3 is a composite flow diagram for the production of 1-propanol by a two-step oxo hydrogenation process (23,46–47). The oxo catalyst is rhodium–triphenylphosphine and the propanal is removed by vapor stripping with excess synthesis gas $(CO + H_2)$ (23,47). After condensation, the propanal is sent to a CO stripping column to remove traces of CO prior to hydrogenation. The crude 1-propanol is purified in a standard two-tower purification system. If $HCo(CO)_4$ is the oxo catalyst, the propanal must be removed as a liquid stream because of the higher reactor pressure, and liquid recycle would be necessary. Also, provisions for catalyst removal and handling would have to be installed.

Economic Aspects

1-Propanol is produced by Celanese, Texas Eastman, and Union Carbide in the United States. BASF AG and Shell Chemical produce 1-propanol in Western Europe, and Diacel, Ltd., is the only 1-propanol producer in Japan. No propanol was manufactured in Eastern Europe or in the Far East in 1979. Worldwide capacity for 1-pro-

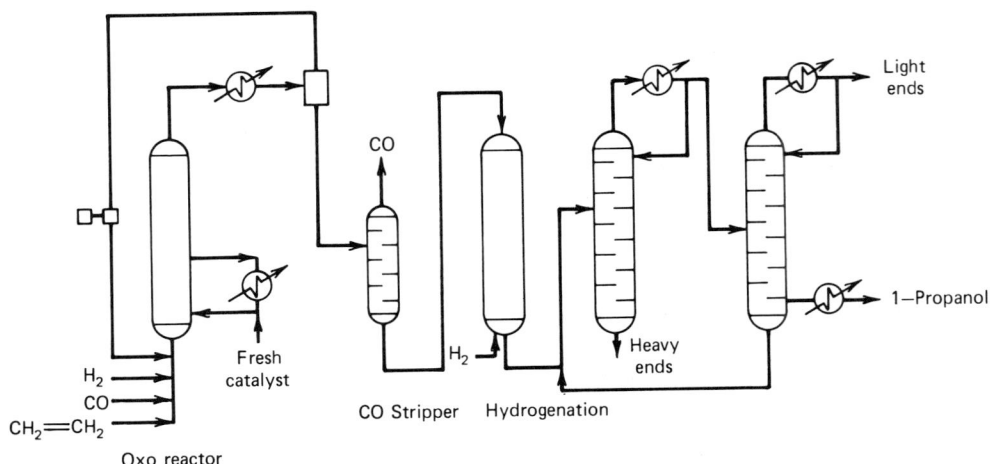

Figure 3. Oxo/hydrogenation process for 1-propanol.

panol in 1979 was in excess of 1.3×10^5 metric tons, with most of the capacity in the United States. Approximately 85,000 t of 1-propanol was produced in the United States in 1979. The February 1981 delivered price for 1-propanol was 88¢/kg (48).

1-Propanol economics are sensitive to the raw material costs of ethylene and the feedstock for synthesis gas (see Feedstocks). Although most of the 1-propanol produced in the United States is based on natural gas as a source of synthesis gas, less expensive liquid feedstocks could play a more important role in the future as the United States decontrols the price of natural gas (see Fuels, survey).

Analytical and Test Methods

The separation and analysis of 1-propanol are straightforward, with gas chromatography being the principal mode. Other instrumental techniques, eg, nmr, ir, and classical organic qualitative analysis are useful. Molecular sieves have been used to separate 1-propanol from ethanol and methanol. Commercial purification is accomplished by distillation.

Health and Safety Factors; Toxicity

Eye contact can cause irritation or burns. Repeated skin contact can result in dermatitis. Exposure to excessive vapor concentrations irritates the eyes and respiratory tract. Very high concentrations have a narcotic effect. 1-Propanol gives negative results in the Ames Test and in the Mouse Lymphoma Forward Mutation Assay (49–50). Toxicity data are given in Table 2.

Uses

1-Propanol is used mainly as a solvent and as a chemical intermediate. Historically, its chief use has been as a specialty solvent in flexographic printing inks, particularly for printing on polyolefin and polyamide film (53) (see Printing processes). n-Propyl acetate, a principal derivative of 1-propanol, is a powerful solvent and is used in nitrocellulose lacquers, cellulose esters and ethers, waxes (qv), and insecticide formulations (see Cellulose derivatives; Insect control technology) (54–55). 1-Propanol improves the drying rate, reduces foaming, and controls viscosity at constants solids level of water-based printing inks (56). It is also used as a ruminant feed supplement and reportedly improves the water tolerance of motor fuels (57).

Table 2. 1-Propanol Toxicity

Type	LD_{50}
acute single oral dose (rats), g/kg[a]	5.4
acute single skin dose (rabbits), g/kg[a]	6.7
eye irritation (rabbits)[b]	43/110 (moderate)
TLV[b,c]	200 ppm

[a] Ref. 51.
[b] Ref. 52.
[c] 8-h time-weighted average.

PROPYL ALCOHOLS (n-PROPYL)

As a chemical intermediate, 1-propanol has been used increasingly in the production of n-propylamines. n-Propylamines are used to manufacture preemergence herbicides (qv) for use on cotton (qv), vegetables, beans, rice, and other foods (58–59). Other uses include the production of n-propyl halides, specialty esters, and ethers for applications in the pharmaceutical, coatings, and dyestuff industries.

BIBLIOGRAPHY

"n-Propyl Alcohol" in *ECT* 1st ed., Vol. 11, pp. 178–182, by M. J. Curry, Celanese Corporation of America; n-Propyl Alcohol" in *ECT* 2nd ed., Vol. 16, pp. 559–564, by Joseph J. Wocasek, Celanese Chemical Co.

1. J. A. Monick, *Alcohols, Their Chemistry, Properties and Manufacture*, Reinhold, New York, 1968, pp. 117–119.
2. R. C. Wilhoit and B. J. Zwolinski, *J. Phys. Chem. Ref. Data* **2**, Suppl. 1, 1–66 (1973).
3. *Normal Propyl Alcohol*, Celanese Chemical Company, New York, Feb. 1977.
4. J. C. Buzzell, Jr. and co-workers, *Behavior of Organic Chemicals in the Aquatic Environment, Part III—Behavior in Aerobic Treatment Systems (Activated Sludge)*, Association of Manufacturing Chemists, Washington, D.C., 1969, pp. 26–31.
5. *Chemical Products Synopsis*, Mannsville Chemical Products, Mannsville, N.Y., Oct. 1977.
6. *Chem. Mark. Rep.*, 3, 15 (Oct. 6, 1976).
7. U.S. Pat. 2,830,091 (Apr. 8, 1958), B. Freedman and F. Marritz (to Sinclair Refining Co.).
8. Ger. Pat. 1,041,938 (Oct. 30, 1958), O. Bankowski and G. Hoffman (to VEB Leuna-Werke "Walter Ulbricht").
9. U.S. Pat. 2,873,290 (Feb. 10, 1959), D. Esmay and C. Johnson (to Standard Oil Co.).
10. *Mod. Paint*, 64 (Feb. 1980).
11. *Chem. Mark. Rep.*, 3 (June 24, 1974); *Chem. Eng.* **84**(26), 109 (1977).
12. *Chem. Purch.*, 13 (Dec. 1979).
13. J. Falbe in E. G. Hancock, ed., *Propylene and Its Industrial Derivatives*, John Wiley & Sons, Inc., New York, pp. 333ff.
14. W. Hieber and E. Linder, *Chem. Ber.* **94**, 1417 (1961).
15. J. Falbe, *Carbon Monoxide in Organic Synthesis*, trans. by C. R. Adams, Springer-Verlag, New York, Inc., 1970, pp. 14–15.
16. R. G. Denney, *Stanford Research Institute No. 21*, Menlo Park, Calif., 1966, pp. 38ff.
17. H. Lemke, *Pet. Refiner* **45**, 149 (1966).
18. Ger. Pat. 1,206,419 (Dec. 9, 1965), H. Lemke (to Kuhlman).
19. U.S. Pat. 3,763,247 (Oct. 2, 1973), H. Lemke and R. Duval (to Société Anonyme: Ugine Kuhlmann).
20. R. F. Heck and D. S. Breslow, *J. Am. Chem. Soc.* **83**, 1097, 4023 (1961); *Chem. Ind. London*, 467 (1960).
21. M. Orchin and W. Rupilius, *Catal. Rev.* **6**, 85 (1972).
22. H. Wakamotsu, *Nippon Kagaku Zasshi* **85**, 257 (1964).
23. R. L. Olivier and F. B. Booth, *Hydrocarbon Process.*, 112 (1970).
24. J. A. Osborn, G. Wilkinson, and J. F. Young, *Chem. Commun.*, 17 (1965).
25. F. H. Jardine, J. A. Osborn, G. Wilkinson, and J. F. Young, *Chem. Ind. London*, 560 (1965).
26. J. A. Osborn, F. H. Jardine, J. F. Young, and G. Wilkinson, *J. Chem. Soc. A*, 1711 (1966).
27. P. S. Hallman, D. Evans, J. A. Osborn, and G. Wilkinson, *Chem Commun.*, 305 (1967).
28. D. Evans, G. Yagupsky, and G. Wilkinson, *J. Chem. Soc. A*, 2660 (1968).
29. D. Evans, J. A. Osborn, and G. Wilkinson, *J. Chem. Soc. A*, 3133 (1968).
30. M. C. Baird, J. T. Mague, J. A. Osborn, and G. Wilkinson, *J. Chem. Soc. A*, 1347 (1967).
31. M. C. Baird, C. J. Nyman, and G. Wilkinson, *J. Chem. Soc. A*, 348 (1968).
32. G. Wilkinson, *Bull. Soc. Chim Fr.* **12**, 5055 (1968).
33. C. K. Brown and G. Wilkinson, *Tetrahedron Lett.* **22**, 1725 (1969).
34. M. Yagupsky, C. K. Brown, G. Yugupsky, and G. Wilkinson, *J. Chem. Soc. A*, 937 (1970).
35. G. Yugupsky, C. K. Brown, and G. Wilkinson, *J. Chem. Soc. A*, 1392 (1970).
36. C. K. Brown and G. Wilkinson, *J. Chem. Soc. A*, 2753 (1970).
37. Ger. Pat. 1,939,322 (Feb. 12, 1970), G. Wilkinson (to Johnson, Matthey and Co.).
38. J. D. Unruh and J. R. Christenson, *J. Molecular Catalysis* **14**, 1 (1982).

39. O. R. Hughes and J. D. Unruh, *J. Molecular Catalysis* **13,** 71 (1981).
40. O. R. Hughes and D. Young, *J. Am. Chem. Soc.* **103,** 6636 (1981).
41. R. L. Pruett, *Science* **211,** (1981).
42. H. Adkins, *Reactions of Hydrogen with Organic Compounds over Chromium Oxide and Nickel Catalysts*, University of Wisconsin Press, Madison, Wisc., 1946.
43. U.S. Pat. 3,491,159 (Jan. 20, 1970), M. Reich and K. Schneider (to Chemische Werke Huls).
44. Brit. Pat. 1,182,797 (March 4, 1970), M. W. Fewlass and T. M. B. Wilson (to B. P. Chemicals Ltd.).
45. J. J. McKetta, *Encyclopedia of Chemical Processing and Design*, Vol. 5, Marcel Dekker, Inc., New York, 1977, pp. 393–394.
46. O. A. Hershamann and co-workers, *Ind. Eng. Chem. Prod. Res. Dev.* **8,** 372 (1969).
47. U.S. Pat. 4,247,486 (Jan. 27, 1981), E. A. V. Brewster and R. L. Pruett (to Union Carbide).
48. *Chem. Mark. Rep.*, 45 (Feb. 9, 1981).
49. Celanese Chemical Co., Inc., *Mutagenicity Evaluation of n-Propyl Alcohol In the Ames Salmonella/Microsome Plate Test*, by Litton Bionetics, Inc., Kensington, Md., March 1978, unpublished.
50. Celanese Chemical Co., Inc., *Mutagenicity Evaluation of n-Propyl Alcohol In the Mouse Lymphoma Forward Mutilation Assay*, by Litton Bionetics, Inc., Kensington, Md., Oct. 1978, unpublished.
51. Celanese Corp., *Range Finding Toxicity Test on n-Propanol*, Industrial Hygiene Foundation of America, Inc., Pittsburgh, Pa., July 1962, unpublished.
52. OSHA, S1910.1000, Chapt. XVII, Table Z-1, Washington, D.C.
53. Brit. Pat. 1,002,039 (Aug. 18, 1965), G. Schmidt and L. J. van Vlerken.
54. *Chemical Products Synopsis*, Mannsville Chemical Products, Mannsville, N.Y., Oct. 1977.
55. *n-Propyl Alcohol*, Celanese Chemical Company, Inc., Dallas, Texas, 1978.
56. G. H. Hutchinson, *Am. Inkmaker* **58**(2), 74 (Feb. 1980).
57. J. L. Keller, *Hydrocarbon Process.* **58**(5), 133 (May 1979).
58. U.S. Pat. 3,257,190 (June 21, 1966), Q. F. Soper (to Eli Lilly and Co.).
59. U.S. Pat. 2,913,327 (Nov. 17, 1959), H. Tiller and J. Antognini (to Stauffer Chemical Co.).

J. D. UNRUH
L. SPINICELLI
Celanese Chemical Company, Inc.

PROPYLAMINES. See Amines, lower aliphatic.

PROPYLENE

Propylene [115-07-1] (propene, $CH_3CH{=}CH_2$) is perhaps the oldest petrochemical feedstock and is one of the main light olefins (1) (see Feedstocks). It is used widely as an alkylation or polymer-gasoline feedstock for octane improvement (see Alkylation; Gasoline). In addition, large quantities of propylene are used in plastics as polypropylene, and in chemicals, eg, acrylonitrile (qv), propylene oxide (qv), 2-propanol, and cumene (qv) (see Olefin polymers, polypropylene; Propyl alcohols). It is produced primarily as a by-product of petroleum refining and of ethylene production by steam pyrolysis (see Ethylene).

Physical Properties

Physical properties of propylene are listed in Tables 1–4. Principal values are presented in Table 1 (2). Parameters for the Van der Waals equation of state and for the virial equation are given in Table 2. Other pressure-volume-temperature (PVT) relationships are referenced.

Table 1. Propylene Physical Properties[a]

Property	Value
mol wt	42.081
fp, K	87.9
bp, K	225.4
critical temperature, K	365.0
critical pressure, MPa[b]	4.6
critical volume, cm^3/mol	181.0
critical compressibility	0.275
Pitzer's acentric factor	0.148
liquid density (at 223 K), g/cm^3	0.612
dipole moment, 10^{-30} C·m[c]	1.3
std enthalpy of formation, kJ/mol[d]	20.42
std Gibbs energy of formation for ideal gas [at 101.3 kPa (= 1 atm)], kJ/mol[d]	62.72
heat of vaporization at bp, kJ/mol[d]	18.41
Lennard Jones potential[e]	
T, nm	0.4678
ϵ_0/K, K	298.9
solubility (at 20°C and 101.3 kPa), mL gas/100 mL solvent	
in water	44.6
in ethanol	1250
in acetic acid	524.5
refractive index, n_D	1.3567

[a] Ref. 2 except where noted.
[b] To convert MPa to atm, divide by 0.1013.
[c] To convert C·m to debye, divide by 3.336×10^{-30}.
[d] To convert J to cal, divide by 4.184.
[e] Ref. 3.

Table 2. Pressure–Volume–Temperature Relationship for Propylene

Relationship	Refs.
Van der Waals equation (per mole) $(P + a/V^2)(V − b) = RT^a$ $a = 6.373^b$ $b = −0.08272$	
Benedict, Webb, Rubin equations of state	4–7
Benedict, Webb, Rubin, Starling equation of state	8
Redlich equation of state	9
Redlich–Kwong equation of state	10
virial equation of state $\frac{PV}{RT} = 1 + \frac{B(T)}{V}$	11

T, K	B(T)
280	−392 ± 5
300	−342 ± 5
320	−297 ± 5
340	−260 ± 5
380	−204 ± 5
420	−162 ± 5
460	−132 ± 5
500	−105 ± 5

^{a} P = kPa, V = L, R = 8.314 J/(mol·K).
^{b} For P = atm, a = 8.379.

The following relationship exists for liquid density ρ_l on the saturation line and is valid at 87.85 to 365.05 K:

$$\rho_l = AB^{-(1-T/T_c)^{2/7}} \text{ g/cm}^3 \quad T = \text{K} \tag{1}$$

where $A = 0.2252$, $B = 0.2686$, and T_c = critical temperature, 365.05 K (12).

For the vapor pressure P_v of propylene, equations 2 and 3 apply:
Antoine equation

$$\log_{10} P_v = A + \frac{B}{T + C} \tag{2}$$

where P_v = vapor pressure in kPa, T = K, and $A = 5.94327$, $B = 784.86$, and $C = -26.15$. If P_v is given in mm Hg, $A = 6.81837$. The Antoine equation is valid for $T = 160–240$ K. For temperatures of 123–365 K, equation 3 should be used:

$$\log_{10} P_v = A + \frac{B}{T} + C \log T + DT \tag{3}$$

where P_v = kPa, T = K, $A = 34.752$, $B = -1725.5$, $C = -12.057$, and $D = 8.9948 \times 10^{-3}$ (12). If P_r is given in mm Hg, $A = 36.877$.

Ideal gas properties and other useful thermal properties are reported in Table 3. Experimental solubility data are evaluated in refs. 13 and 14. Extensive data on propylene solubility in water have been measured (15). Vapor-liquid-equilibrium (VLE) data for propylene are given in refs. 16–30. Correlations of VLE data are discussed in refs. 31–37. Henry's Law constants are given in refs. 38–41. Equations for the transport properties of propylene are given in Table 4.

Table 3. Thermal Properties of Propylene[a]

Property	Value
ideal gas, heat capacity, J/mol[b] $C_p^\circ = A + BT + CT^2 + DT^3$	$A = 2.85$ $B = 0.238$ $C = -1.2 \times 10^{-4}$ $D = 2.3 \times 10^{-8}$
ideal gas, heat of formation, kJ/mol[c] $\Delta H_f^\circ = A + BT + CT^2$	$A = 35.3$ $B = -5.77 \times 10^{-2}$ $C = 2.22 \times 10^{-5}$
ideal gas, free energy of formation, kJ/mol[d] $\Delta G_f^\circ = A + BT$	$A = 75.3$ $B = 1.75$
heat of vaporization, kJ/kg[e] $\Delta H_v = \Delta H_{v_i} \left(\dfrac{T_c - T}{T_c - T_i} \right)^n$	$\Delta H_{v_i} = 437.6$ kJ/kg $T_i = 225.45$ K $T_c = 365.05$ K $n = 0.38$
liquid heat capacity, J/(kg·K)[f] $C_p = A + BT + CT^2 + DT^3$	$A = 1969$ $B = 7.04$ $C = -7.04 \times 10^{-2}$ $D = 1.84 \times 10^{-4}$

[a] Ref. 12.
[b] Valid for range 298–1500 K. To convert J to cal, divide by 4.184.
[c] Valid for range 298–1500 K.
[d] Valid for range 298–1500 K.
[e] Valid for range 87.85–365.05 K.
[f] Valid for range 88–373 K.

Chemical Properties

The chemistry of propylene is characterized both by its double bond and by its allylic hydrogen atoms.

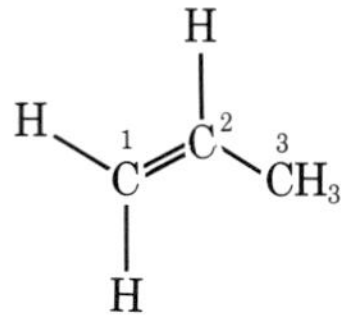

$$\begin{array}{c} H \\ | \\ H\diagdown \underset{1}{C} = \underset{2}{C} \diagdown \underset{3}{CH_3} \\ | \\ H \end{array}$$

Carbon atoms 1 and 2 have a trigonal planar geometry identical to that of ethylene. They generally are not free to rotate because of the double bond. Carbon atom 3 is tetrahedral, like methane. Hydrogen atoms attached to this carbon atom are allylic hydrogens.

The propylene double bond consists of a σ bond formed by two overlapping sp^2 orbitals, and a π bond formed above and below the plane between the two carbon atoms by the side overlap of two p orbitals. The π bond is responsible for many of the reac-

Table 4. Transport Properties of Propylene[a]

Property	Value
gas thermal conductivity[b], W/(m·K)[c] $$k_G = A + BT + CT^2 + DT^3$$	k_G (25°C) = 1.84 × 10^{-2} $A = -7.577 \times 10^{-3}$ $B = 6.096$ $C = 9.96 \times 10^{-8}$ $D = 3.84 \times 10^{-11}$
gas viscosity, μPa·s[d] $$\mu_G = A + BT + CT^2$$	μ_G (25°C) = 8.39 × 10^{-6} $A = -5.601 \times 10^{-7}$ $B = 31.88 \times 10^{-9}$ $C = -62.91 \times 10^{-13}$
liquid thermal conductivity[e], W/(m·K)[c] $$k_L = A + BT + CT$$	k_L (20°C) = 0.115 $A = 2.9 \times 10^5$ $B = -6.05 \times 10^{-2}$ $C = 1.26 \times 10^{-2}$
liquid viscosity, mPa·s (= cP) $$\log \mu_L = A + \frac{B}{T} + CT + DT^2$$	A B $C \times 10^2$ $D \times 10^6$ −27.84[f] 1.096[f] 26.02[f] −863.5[f] −5.009[g] 413.2[g] 1.771[g] −30.92[g]
surface tension, mN/m (= dyn/cm)[h] $$\sigma = n \cdot \sigma_l \frac{T_c T}{T_c - T_l}$$	$\sigma_l = 19.98$ mN/m $T_l = 203$ K $T_c = 365.05$ K $n = 1.1797$

[a] Ref. 11.
[b] Range, −100 to 1000°C.
[c] To convert W/(m·K) to cal/(s·cm·K), divide by 418.4.
[d] Range, −10 to 1000°C; to convert Pa·s to centipoise, multiply by 10^3.
[e] Range, −185 to 70°C.
[f] Range, −185.3 to 160°C.
[g] Range, −160 to 91.9°C.
[h] Range, −185.3 to 91.9°C.

tions that are characteristic of alkenes. It serves as a source of electrons for electrophilic reactions; addition reactions typically are in this category. Simple examples are the addition of hydrogen or a halogen, eg, chlorine:

$$CH_3CH{=}CH_2 + H_2 \xrightarrow{catalyst} CH_3CH_2CH_3 \quad (4)$$

$$CH_3CH{=}CH_2 + Cl_2 \xrightarrow{catalyst} CH_3CHClCH_2Cl \quad (5)$$

Reactions of alkenes are described in refs. 42 and 43 (see also Olefins, higher).
The allylic hydrogens in propylene often distinguish its chemistry from that of ethylene. For example, their presence causes cross-linked, gummy materials to form when propylene polymerizes with peroxide initiators (44). The effect of the allyl hydrogens on propylene reactions can be explained by the stability of allyl radicals and allyl carbocations. When an allylic hydrogen is abstracted from propylene, the sp^3

hybridized carbon of the methyl group changes to sp^2. The p orbital of this carbon can then overlap with the p orbitals that formed the π bond. This forms a new π bond, which overlaps all three carbon atoms. The electrons from the alkene π bond and the free-radical electron are delocalized over the entire molecule.

Resonance theory can also account for the stability of the allyl radical. The two equivalent structures for the allyl radical are

$$CH_2=CHCH_2\cdot \leftrightarrow \cdot CH_2CH=CH_2$$

The radical is much more stable if both structures exist. Moreover, both molecular orbital theory and resonance theory show that the allyl carbocation is relatively stable.

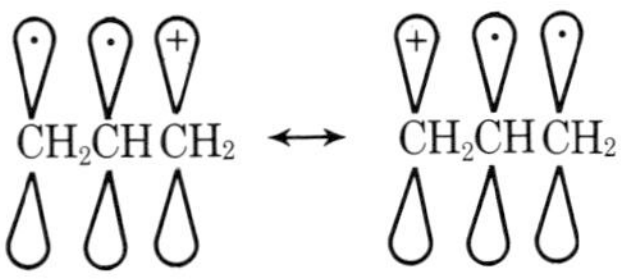

Manufacture

Steam Cracking. In steam cracking, a mixture of hydrocarbon and steam is preheated to ca 870 K in the convective section of a pyrolysis furnace. Then it is further heated in the radiant section to as much as 1170 K (see Ethylene; Petroleum, refinery processes). Steam reduces the hydrocarbon partial pressure in the reactor. The steam-to-hydrocarbon weight ratio is generally a function of the feedstock and ranges from ca 0.2 for ethane to ≥ 2.0 for gas oils (45). The amount of steam used is probably a compromise between yield structure (olefin selectivity), energy consumption, and furnace run length, which is limited by coking. The residence time in the radiant section varies from ca 1 s in older plants to as low as 0.1 s in some newer furnaces. The residence time influences olefin selectivity. Generally, selectivity to ethylene improves as residence time decreases. However, a given furnace is relatively inflexible to gross residence time changes for a specific feedstock because of hydrodynamic and heat-flux limitations.

In the radiant section, the hydrocarbon mixture undergoes reactions involving free radicals (46). More recently, these mechanisms have been generalized to include the molecular reactions shown below:

Chain initiation reactions: R—R' $\rightarrow$ R$\cdot$ + R'$\cdot$
Hydrogen-abstraction reactions: R$\cdot$ + R'H $\rightarrow$ RH + R'$\cdot$
Radical-decomposition reactions: R$\cdot$ $\rightarrow$ RH + R'$\cdot$
Radical-addition reactions to unsaturated molecules: RH + R'$\cdot$ $\rightarrow$ R''$\cdot$
Chain-termination reactions: R$\cdot$ + R'$\cdot$ $\rightarrow$ R—R'

Molecular reactions: RH + R'H → R''H + R'''H
Radical-isomerization reactions: R'· → R''·

The total number of reactions depends on the number of constituents present in the hydrocarbon feedstock. As many as 2000 reactions can occur simultaneously.

The constituents in the furnace effluent are the same for all hydrocarbon feedstocks. These include all hydrocarbons lighter than pentane plus heavier material, eg, gasoline and fuel oil. The portion of these components in the effluent depends on the feedstock. For example, the furnace effluent of a plant designed for pure ethane cracking contains large amounts of hydrogen, methane, unconverted ethane, and ethylene. Only small amounts of other constituents are produced. Conversely, a furnace cracking-gas oil produces large amounts of gasoline and fuel oil in addition to significant quantities of useful olefins, eg, propylene.

The separation train of the plant is designed to recover important constituents present in the furnace effluent. The modern olefin plant must be designed to accommodate various feedstocks, ie, it usually is designed for feedstock flexibility in both the pyrolysis furnaces and the separation system (47). For example, a plant may crack feedstocks ranging from ethane to naphtha or naphtha to gas oils.

The yield of propylene produced in a pyrolysis furnace is a function of the feedstock and the operating severity of the pyrolysis. (For typical yields of propylene for various feedstock, see Ethylene.) Under practical operating conditions, ethylene yield increases with increasing severity of feedstock conversion. Propylene yield passes through a maximum, as shown in Figure 1 (48). The economic optimum effluent composition for a furnace usually is beyond the propylene maximum. The furnace operation usually is dictated by computer optimization, ie, linear programming, of the whole plant, where an economic optimum for the plant is based on feedstock price, yield structures, energy considerations, and market conditions for the multitude of products obtained from the furnace. Thus, propylene produced by steam cracking varies according to economic conditions.

In an olefins-plant separation train, propylene is obtained by distillation of a mixed C_3 stream, ie, propane, propylene, and minor components, in a C_3-splitter tower. Propylene is produced as the overhead distillation product, and the bottoms are a propane-rich stream. The size of the C_3-splitter depends on the purity of the propylene product. Two grades of propylene are commonly produced: a chemical grade, which consists of 92–94 wt % propylene, and a >99 wt % polymer-grade propylene. Specifi-

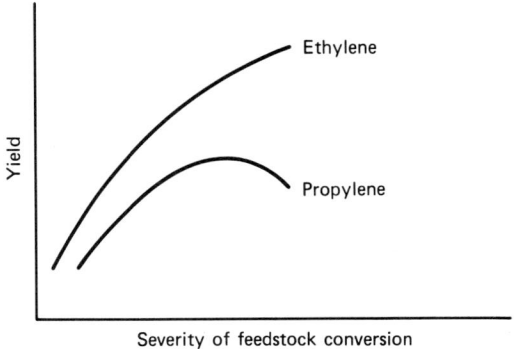

Figure 1. Ethylene and propylene yields.

cations for these two grades are listed in Tables 5–6 (49–50). A larger number of theoretical distillation trays is required to produce polymer-grade propylene than to produce the chemical grade because of the close relative volatilities of propane and propylene.

Refinery Production. Refinery propylene is formed as a by-product of fluid catalytic cracking of gas oils and, to a far lesser extent, of thermal processes, eg, coking. The total amount of propylene that is produced depends on the mix of these processes and the refinery product slate. For example, in the United States, refiners have maximized gasoline production. This results in more propylene production than in Europe, where proportionally more heating oil is produced.

In fluid catalytic cracking, a partially vaporized gas oil is contacted with zeolite catalyst. Contact time is 5 s–2 min and pressure is 250–400 kPa (2.5–4 atm), depending on the design of the unit. Reaction temperatures are 720–850 K (see Butylenes).

Table 5. Product Specification for Chemical-Grade Propylene

Specification	ASTM test method
Component, wt %	
propylene, 92–94	D 2163
ethane and lighter, <0.4	D 2163
ethylene, <0.02	D 2723
C_4 and heavier hydrocarbons, <0.2	D 2712
C_5 and heavier hydrocarbons, <0.005	D 2162
propane, <8.0	D 2163
Impurities, ppmwt	
acetylene + methyl acetylene + propadiene, <100	D 2712
total H_2, O_2, CO, CO_2, and N_2, <100	D 2504
sulfur, <10	D 3120
water, <50	D 2713
halides, <10	
alcohols, <50	
amines, <5	
butadiene, <20	D 2712
butenes, <125	
dimethyl formamide, 0.3	

Table 6. Product Specifications for Polymer-Grade Propylene

Specification	ASTM test method
Component	
propylene, wt %, 99.5	D 2163
ethane, ppm wt, <1000	D 2163
total acetylenes, dienes, and other unsaturates, ppm wt, <10	D 2712
hydrogen, ppm wt, <2	D 2504
oxygen, ppm wt, <8	D 2504
carbon monoxide, ppm wt, <4	D 2504
water, ppm wt, <10	
sulfur, ppm wt, <5	D 3120
total nitrogen, aldehydes, ketones, and alcohols, ppm wt, <10	D 2504

Converted feedstock forms gasoline-boiling-range hydrocarbons, C_4 and lighter gas, and coke. Propylene yield varies, depending on reaction conditions, but yields of 2–5% based on feedstock are common (51–52).

Two thermal-cracking processes, ie, delayed coking (53) and Flexicoking or fluid coking (54), are used to convert residuum into more valuable products. In delayed coking, residuum and steam are heated in a furnace and then fed into an insulated drum where the free-radical decomposition of the feedstock takes place. Coke eventually fills the drum and must be removed. In fluid coking, a residuum feed is injected into a reactor, where it cracks thermally. Coke formed during the process deposits on other fluidized coke particles and is removed by fluid coking or is gasified by Flexicoking. Both fluid and delayed coking occur at 300–600 kPa (3–6 atm). Delayed coking is a lower temperature process (720 vs 820 K) and, thus, should have lower total olefin yields than fluid coking.

Refinery propylene is recovered at the vapor-recovery unit. Refinery wet gas is passed through an absorber, where it contacts a hydrocarbon liquid, usually a heavy naphtha. The heavier molecular weight material dissolves in the liquid, and the lighter material, eg, hydrogen and methane, passes through. The absorbent or rich oil is then passed to a stripper, where dissolved hydrocarbons are removed. The lean oil is recycled to the absorber, and the absorbate is passed to a depropanizer where a propane–propylene stream is taken as the overhead. This refinery-grade stream may require further treatment to remove acid gases, such as hydrogen sulfide, carbonyl sulfide, and carbon dioxide. A chemical- or polymer-grade propylene can be made by further distillation in a propylene-concentration unit.

Future Processes. *Advanced Cracking Techniques.* Several groups are studying techniques to pyrolyze whole crude oil or various heavy petroleum fractions. These methods have in common very high temperatures, ultrashort residence times in the reactor zone, and rapid quench of the reaction products to minimize undesirable by-products. Among these techniques are the advanced cracking reactor from Union Carbide Corporation and thermal regenerative cracking from Gulf Oil, and Stone and Webster Engineers (55).

Synthetic Fuels. Hydrocarbon liquids made from nonpetroleum sources can be used in steam crackers to produce olefins. Fischer-Tropsch liquids, oil-shale liquids, and coal-liquefaction products are examples (56) (see Fuels, synthetic). Recent work with Fischer-Tropsch catalysts indicates that olefins can be made directly from synthesis gas–carbon monoxide and hydrogen (57–58). Shape-selective molecular sieves (qv) also are being evaluated (59).

Catalytic Processes. Advances in catalysis (qv) may allow the direct catalytic dehydrogenation of alkanes to form alkenes. At least one process is claimed to make propylene with high selectivity and good yields from pure propane or propane mixed with C_4-hydrocarbon feedstock (60).

Economic Aspects

Estimated 1972–1979 worldwide propylene production is listed in Table 7 (61–62). Production figures can only be approximated, since refinery propylene may be diverted captively to fuel or gasoline uses when recovery is uneconomic, and steam-cracker propylene production varies with feedstock and operating conditions. Moreover, since propylene is a by-product, production rates depend on gasoline or ethylene demand.

236 PROPYLENE

Table 7. Propylene Capacity and Production

Year	1000 metric tons/yr, nameplate capacity	Nameplate capacity used, %	Ref.
1972	5,170		61
1973	5,440		61
1974	5,710		61
1975	6,120		61
1976	7,170	72	62
1977	7,940	72	62
1978	8,620	68	62
1979	9,070	72	62
1980	9,072	69	62
1981	9,767	68	62
1982 (estd)	10,470	68	62

Propylene prices for 1970–1980 are shown in Figure 2 (63–64). Rapid increases in prices are related to similar feedstock-cost increases, which occurred in 1973 and 1979. Although the actual price of propylene affects supply and demand, the propylene-to-ethylene price ratio is also important. If ethylene and propylene prices are similar, operators of steam crackers tend to choose feedstocks that produce more propylene. When propylene prices are significantly lower than ethylene prices, propylene-derived products are substituted for ethylene derivatives in areas where applications overlap. Since producers have the option of diverting propylene to fuel use or cracking feed,

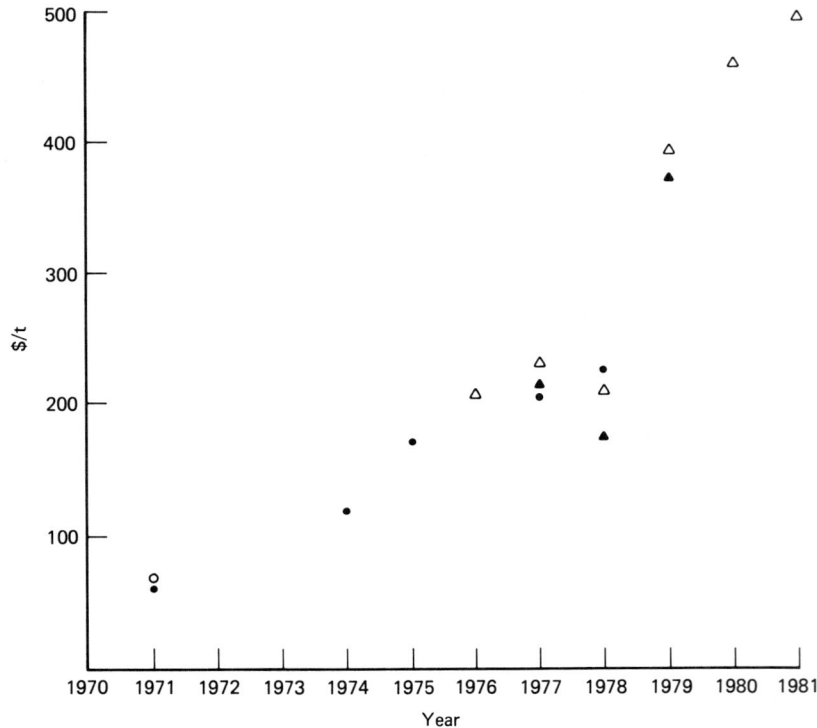

Figure 2. Propylene and ethylene prices. ●, See Ethylene. ▲ = chemical grade propylene (63). △ = polymer grade propylene (63). ○ = ref. 64.

the market price of propylene for chemical uses and gasoline blending must exceed its value as fuel or feedstock for steam cracking.

Propylene production in the 1980s is expected to increase at ca 4%/yr (65). The actual growth rate probably will depend on gasoline demand during this period and the dominant feedstock for steam cracking. If gas oil becomes a principal cracking feedstock, propylene production may increase at a greater rate. If LPG is preferred, considerably less by-product propylene will be produced relative to ethylene. The position of propylene as a basic raw material in a mature, heavy organic chemical industry favors steady, slow growth of production. Two future events that could significantly affect propylene production would be a substantial decline in gasoline demand and the development of advanced cracking techniques, which would enable production of ethylene with very little by-product propylene. If these traditional propylene sources declined, attention would turn to production of propylene as a primary product. Direct production of propylene has been the subject of speculation during previous periods of shortage (62). At present, there are no significantly new applications of propylene. However, new uses in liquid fuels have been reviewed (66).

Storage and Handling

Precautions must be taken to avoid health and fire hazards wherever propylene is handled (67). Equipment capable of causing ignition should be shut down while connecting, disconnecting, loading, and unloading equipment (68). Electrical installations in unloading areas should be classified under Division II requirements of the National Electrical Code (69). No part of any cylinder containing propylene should be subjected to temperatures above 325 K.

Propylene is very volatile and is usually stored as a liquid under pressure. However, it can be stored safely at ambient temperature in approved containers. Storage tanks should be of welded-steel construction in accordance with the ASME Code for Unfired Pressure Vessels (70). All piping and related equipment should be steel and conform to the piping codes of ANSI (71). Waste mixtures containing propylene should not be allowed to enter drains or sewers where the danger of vapor ignition exists. Commercial quantities of propylene are shipped by tanker or pipeline and are stored above ground in pressure vessels or underground in brine caverns.

Health and Safety Factors

Propylene is a colorless gas under normal conditions, has anesthetic properties at high concentrations, and can cause asphyxiation. It does not irritate the eyes, and its odor is characteristic of olefins. Propylene is a flammable gas under normal atmospheric conditions. Vapor-cloud formation from liquid or vapor leaks is the main hazard that can lead to explosion. The autoignition temperature is 731 K in air and 696 K in oxygen (72). Evaporation of liquid propylene can cause skin burns. Propylene also reacts vigorously with oxidizing materials. Under unusual conditions, eg, 96.8 MPa (995 atm) and 600 K, it explodes. It reacts violently with NO_2, N_2O_4, and N_2O (73). Explosions have been reported when liquid propylene contacts water at 315–348 K (74). Table 8 shows the ratio T_w/T_{sl}, where T_w is the initial water temperature, and T_{sl} is the superheat limit temperature of the hydrocarbon.

Table 8. T_w/T_{sl} Ratios for Liquid Propylene Spills in Water[a]

Initial water temperature, T_w, K	T_w/T_{sl}	Result
311–314	0.95–0.96	ice formation
315–348	0.97–1.07	explosion
353–358	1.08–1.10	rapid pops

[a] Ref. 74. T_w = initial water temperature; T_{sl} = superheat limit temperature of propylene.

Uses

Propylene has many commercial and potential uses. The actual utilization of a particular propylene supply depends not only on the relative economics of other petrochemicals and the value of propylene in various uses, but also on the location of the supply and the form in which the propylene is available. For example, economics dictate that recovery of high purity propylene for polymerization from a small-volume, dilute off-gas stream is not feasible; whereas polymer-grade propylene is routinely recovered from large refineries and olefins steam crackers. A synthetic-fuels project located in the western United States might use propylene as fuel rather than recover it for petrochemical use, as would occur in a plant on the Gulf Coast (see Fuels, synthetic).

The uses of propylene may be loosely categorized as refinery or chemical purpose. In the refinery, propylene occurs in varying concentrations in fuel-gas streams. As a refinery feedstock, propylene is alkylated by isobutane or dimerized to produce polymer gasoline for gasoline blending. Commercial chemical derivatives include polypropylene, acrylonitrile, propylene oxide, isopropyl alcohol and others. In 1979, ca 58% of U.S. propylene supplies were consumed in the production of chemicals (75). The relative amounts of propylene in various chemical applications are shown in Table 9 (76–77). Polypropylene has been the largest consumer of propylene since the early 1970s and is likely to dominate propylene utilization for some time to come.

Refinery. *Fuel.* Propylene has a net heating value of 45.8 MJ/kg (19,700 Btu/lb) and is often contained in refinery fuel-gas streams. However, propylene is diverted from streams for refinery fuel use in large quantities only when economics for other uses are unfavorable, or equipment for propylene recovery does not exist or is limited in capacity. Propylene is also contained in LPG (liquid petroleum gas), but is limited

Table 9. Relative Amounts of Propylene Feedstock Consumed in the U.S. Chemical Industry, %

Derivative	Year					
	1974[a]	1975[b]	1976[b]	1977[b]	1978[b]	1979[b]
polymer	22.4	22.5	23.0	25.0	25.0	25.0
acrylonitrile	15.6		16.0	15.0	15.0	15.0
isopropyl alcohol	14.3	15.0	14.0	12.0	10.0	10.0
propylene oxide	14.3		13.0	13.0	10.0	10.0
oligomers				12.0	10.0	
cumene	10.7		10.0	10.0		

[a] Ref. 76.
[b] Ref. 77.

to a maximum concentration of 5 vol % in certain grades (78) (see Liquefied petroleum gas).

Alkylate. In petroleum refining, alkylation is the reaction of light olefins with isoparaffins (principally isobutane) to produce isoparaffins of higher molecular weight for gasoline blending. Alkylation produces material of higher octane rating than polymerization does, and alkylation predominated as gasoline-pool octane requirements increased after World War II. Although butenes are the preferred olefins for alkylation, propylene is also used, depending on the availability of butenes and alkylate demand. The basic alkylation reactions are acid-catalyzed (79). Propylene alkylation requires higher temperatures and acid strengths than alkylation of butenes, and butenes are good promoters of propylene alkylation (80). The principal isomer of propylene alkylation by isobutane is 2,3-dimethylpentane as compared to the higher octane quality 2,2,4-, 2,3,3-, and 2,3,4-trimethylpentanes obtained from butenes (see Alkylation). Commercial alkylation units are described in ref. 81.

Polymer Gasoline. Polymerization of butenes and propylene in the refinery, first through thermal, ie, free-radical, reactions and later by acid-catalyzed mechanisms, was the first successful process for upgrading light olefinic gases, which are coproduced in petroleum-cracking processes. Although alkylation gained in importance as gasoline-pool octane requirements increased, polymerization has been the subject of new interest as isobutane supplies decrease (82) (see Gasoline and other motor fuels).

One of the newest forms of polymerization for gasoline is the Dimersol process (66). Refinery propylene, ie, 67 wt % propylene, 33 wt % propane, reacts in the liquid phase with a nickel coordination complex and aluminum alkyl catalyst at ca 330 K and 1725 kPa (250 psi) (83). Propylene conversions are 90–97%. The heat of reaction is removed by circulation through an external cooler. Ammonia and water are injected into the reactor effluent to neutralize the catalyst; hydrocarbon and aqueous phases are separated; and the hydrocarbon is fractionated to produce LPG, which contains unreacted propane and propylene, and dimerization products consisting mainly of isohexenes. The process can also be based on butane/butene feeds. As a gasoline blending component, the Dimersol product may have clear octane ratings of $(R + M)/2 = 89$, which is somewhat higher than that of traditional propylene polymer gasoline, ie, $(R + M)/2 = 87$ (66,79).

Chemical. *Propylene Oligomers.* Through acid-catalyzed Friedel-Crafts processes similar to those in refinery alkylation and polymerization, propylene forms oligomers, eg, nonenes, dodecenes, and higher molecular weight olefins and viscous polypropenes (84). These materials are used in the production of alkyl-aryl sulfonate detergents and motor-oil additives (85–86).

Polypropylene. One of the most important applications of propylene is as a monomer for the production of polypropylene. Propylene is polymerized by Ziegler-Natta coordination catalysts (87–88). Polymerization is carried out either in the liquid phase with the polymer forming a slurry of particles or in the gas phase with the polymer forming dry solid particles. Propylene polymerization is an exothermic reaction (89).

$$C_3H_6 \text{ (gaseous)} \rightarrow [C_3H_6]_n \qquad \Delta H^\circ_{298} = -104.1 \text{ kJ/mol } (-24.89 \text{ kcal/mol}) \qquad (6)$$

$$C_3H_6 \text{ (liquid)} \rightarrow [C_3H_6]_n \qquad \Delta H^\circ_{298} = -89.1 \text{ kJ/mol } (-21.30 \text{ kcal/mol}) \qquad (7)$$

The main objective of commercial research is to develop catalysts that would make it unnecessary to remove catalyst residues and undesirable atactic polypropylene from the polymer product. The most desirable catalyst is one that polymerizes propylene to the crystalline polymer.

Most commercial processes produce polypropylene by a liquid-phase slurry process. Hexane or heptane are the most commonly used diluents. However, there are a few examples in which liquid propylene is used as the diluent. The leading companies involved in propylene processes are Amoco Chemicals (Standard Oil, Indiana), El Paso (formerly Dart Industries), Exxon Chemical, Hercules, Hoechst, ICI, Mitsubishi Chemical Industries, Mitsubishi Petrochemical, Mitsui Petrochemical, Mitsui Toatsu, Montedison, Phillips Petroleum, Shell, Solvay, and Sumimoto Chemical. Eastman Kodak has developed and commercialized a liquid-phase solution process. BASF has developed and commercialized a gas-phase process, and Amoco has developed a vapor-phase polymerization process that has been in commercial operation since early 1980.

Polypropylene is used in battery cases and in the replacement of metal parts in automobiles. It is also used widely in consumer products, eg, kitchen wares, trays, toys, and packaging materials. Its future applications are expected to include an increased portion of the fibers and filaments markets with continued growth in carpet backing and carpet face yarns. Film, both oriented and unoriented, is also expected to be a significant growth market for polypropylene.

Acrylonitrile. Catalytic oxidation of propylene in the presence of ammonia yields acrylonitrile (90).

$$CH_2=CHCH_3 + NH_3 + 3/2\ O_2 \rightarrow CH_2=CHCN + 3\ H_2O \qquad (8)$$

Yields based on propylene are 50–75%, and the main by-products are acetonitrile and hydrogen cyanide. Recently, claims have been made for catalysts that give yields of over 80% acrylonitrile (91).

Propylene Oxide. Propylene oxide is produced from propylene by two main processes. The first is chlorohydrination of propylene at ca 310 K,

$$CH_3CH=CH_2 + HOCl \rightarrow CH_3CHOHCH_2Cl \qquad (9)$$

followed by epoxidation (qv) of the chlorohydrin by calcium hydroxide.

$$2\ CH_3CHOHCH_2Cl + Ca(OH)_2 \rightarrow 2\ CH_3CH\overset{O}{-}CH_2 + CaCl_2 + 2\ H_2O \qquad (10)$$

The second process involves reaction of propylene with peroxides, as in the Oxirane process (92) in which either isobutane or ethylbenzene is oxidized to form a hydroperoxide.

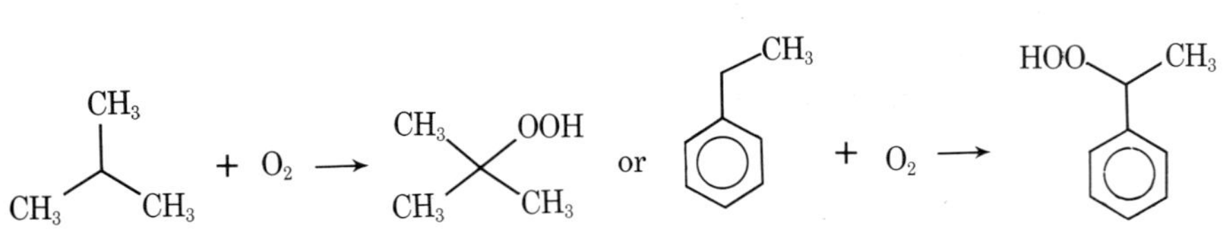

$$(CH_3)_2CHCH_3 + O_2 \rightarrow (CH_3)_2C(OOH)CH_3 \quad \text{or} \quad C_6H_5CH_2CH_3 + O_2 \rightarrow C_6H_5CH(OOH)CH_3 \qquad (11)$$

The hydroperoxide reacts with propylene, forming propylene oxide and an alcohol. *tert*-Butyl alcohol can be dehydrated to form isobutylene, and α-methylbenzyl

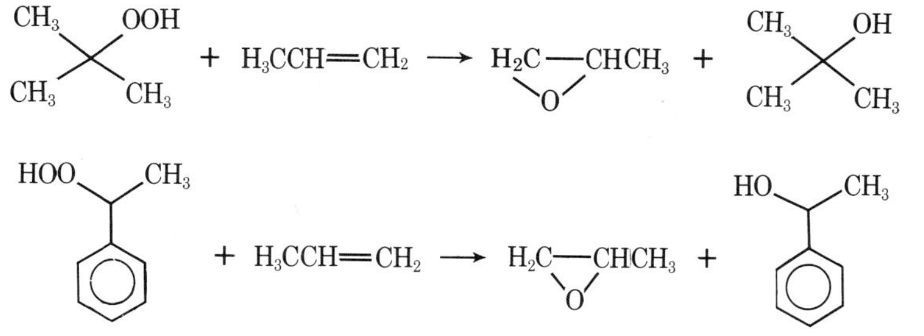

$$\underset{CH_3}{\overset{CH_3}{>}}\!\!\!\!\!\!\!\!\!\!\!\!\!\!\!\underset{CH_3}{\overset{OOH}{<}} + H_3CCH{=}CH_2 \longrightarrow H_2C\underset{O}{\overset{}{-}}CHCH_3 + \underset{CH_3}{\overset{CH_3}{>}}\!\!\!\!\!\!\!\!\!\!\!\!\!\!\!\underset{CH_3}{\overset{OH}{<}}$$

$$\underset{C_6H_5}{\overset{HOO}{>}}\!\!\!\!\!\!\!\!\!\!\!\!\!\underset{H}{\overset{CH_3}{<}} + H_3CCH{=}CH_2 \longrightarrow H_2C\underset{O}{\overset{}{-}}CHCH_3 + \underset{C_6H_5}{\overset{HO}{>}}\!\!\!\!\!\!\!\!\!\!\!\!\underset{H}{\overset{CH_3}{<}} \qquad (12)$$

alcohol can be dehydrated to form styrene (qv). Thus, in either reaction sequence, a by-product of significant value is obtained. Hydroperoxide formation occurs under mild conditions [400 K, 440–3500 kPa (4.3–34.5 atm)] to minimize decomposition to the alcohol (93). Epoxidation of the propylene occurs in the liquid phase under similar conditions in the presence of a catalyst containing Mo, V, or Ti. Conversions are ca 10–20% for propylene and greater than 95% for the hydroperoxide. One other commercialized process involving hydroperoxides is the Daicel process (94). Peracetic acid is produced by oxidation of acetaldehyde in solution in the presence of metal ion catalyst at 300–320 K and 2.53–4.05 MPa (25–40 atm) (see Peroxides, organic).

$$CH_3\overset{O}{\overset{\|}{C}}HH + O_2 \longrightarrow CH_3\overset{O}{\overset{\|}{C}}OOH \qquad (13)$$

Peracetic acid then reacts with propylene at 320–350 K and 1.00–1.34 MPa (10–13.2 atm), forming propylene oxide and acetic acid.

$$CH_3\overset{O}{\overset{\|}{C}}OOH + H_3CCH{=}CH_2 \longrightarrow H_2C\underset{O}{\overset{}{-}}CHCH_3 + CH_3\overset{O}{\overset{\|}{C}}OH \qquad (14)$$

Propylene oxide uses include manufacture of polyurethanes, unsaturated polyesters, propylene glycols and polyethers, and propanolamines (see Urethane polymers; Polyesters, unsaturated; Glycols; Polyethers; Alkanolamines).

Isopropyl Alcohol. Propylene may be easily hydrolyzed to isopropyl alcohol. Early commercial processes involved the use of sulfuric acid in an indirect process (95). The disadvantage of this process was the need to reconcentrate the sulfuric acid after hydrolysis. Direct catalytic hydration of propylene to 2-propanol followed commercialization of the sulfuric acid process and eliminated the need for acid reconcentration, thus reducing corrosion problems, energy use, and air pollution by SO_2 and organic sulfur compounds. Gas-phase hydration takes place over supported oxides of tungsten at 540 K and 25 MPa (247 atm) or over supported phosphoric acid at 450–540 K and 2.5–6.5 MPa (25–64 atm) (95).

$$CH_3CH{=}CH_2 + H_2O \underset{}{\overset{catalyst}{\rightleftharpoons}} CH_3\overset{OH}{\overset{|}{C}}HCH_3 \qquad (15)$$

At conditions of high temperature and low pressure, for sufficient catalyst activity and acceptable reaction rates, equilibrium conversions may be as low as 5%, necessitating recycle of large amounts of unreacted propylene (96).

Conversions of ca 75% are obtained for propylene hydration over cation-exchange resins in a trickle-bed reactor (97). Excess liquid water and gaseous propylene are fed concurrently into a downflow, fixed-bed reactor at 400 K and 3.0–10.0 MPa (30–100 atm). Selectivity to isopropanol is ca 92%, and the product alcohol is recovered by azeotropic distillation with benzene.

A third catalytic route to isopropyl alcohol from propylene involves the use of polytungsten compounds in solution in a liquid-phase reactor (96). Propylene is hydrated at ca 540 K under pressure. Conversions are 60–70% and selectivity to 2-propanol is 99%.

At one time, one of the largest sources of acetone was isopropyl alcohol. However, cumene oxidation to produce phenol and acetone (qv) has become the chief new acetone source (98). Isopropyl alcohol is used as a solvent; it is also used in paint, pharmaceuticals, and cosmetics, and as a gasoline additive to prevent carburator icing, and as an antifreeze (see Antifreezes and deicing fluids). Isopropyl acetate also can be obtained from 2-propanol by esterification with acetic acid (see Propyl alcohols).

Cumene. Cumene is produced by Friedel-Crafts alkylation of benzene by propylene (99–100). The main application of cumene is the production of phenol and by-product acetone; minor amounts are used in gasoline blending (101) (see Cumene).

Butyraldehydes. Normal and iso-butyraldehydes are produced from propylene by the oxo or hydroformylation process (see Oxo process).

$$CH_3CH{=}CH_2 \xrightarrow[\text{catalyst}]{H_2/CO} CH_3CH_2CH_2CHO + CH_3CH(CH_3)CHO \quad (16)$$

The two main industrial processes that are employed are described in ref. 102 (see Butyraldehyde). Normal butyraldehyde is the product of primary interest (99,103).

Other. Ethylene can be produced by steam co-cracking of propylene with ethane and propane. Ethylene and butenes can also be produced by catalytic disproportionation of propylene (104).

$$CH_2{=}CHCH_3 \rightleftarrows CH_2{=}CH_2 + CH_3CH{=}CHCH_3 \quad (17)$$

2-butene is the main C_4 olefin isomer.

Acrolein can be obtained by propylene oxidation in a process similar to ammoxidation (105) (see Acrolein and derivatives).

$$CH_2{=}CHCH_3 \xrightarrow{O_2} CH_2{=}CHCHO + H_2O \quad (18)$$

High temperature chlorination of propylene yields allyl chloride, which is used in glycerol (qv) production (106):

$$CH_2{=}CHCH_3 + Cl_2 \xrightarrow{770\ K} CH_2{=}CHCH_2Cl + HCl \quad (19)$$

Allyl acetate can be obtained by the vapor-phase reaction of propylene and acetic acid

$$CH_2{=}CHCH_3 + CH_3COOH \xrightarrow{O_2} CH_2{=}CHCH_2OCOCH_3 \quad (20)$$

over a supported Pd catalyst (106). Reaction of acrylic acid and propylene yields isopropyl acrylate,

$$CH_2=CHCH_3 + CH_2=CHCOOH \xrightarrow{H^+} CH_2=CHCOCH(CH_3)_2 \quad \text{(with C=O)} \tag{21}$$

and catalytic reaction with acetic acid produces isopropyl acetate (106).

$$CH_2=CHCH_3 + CH_3COOH \xrightarrow{catalyst} CH_3COCH(CH_3)_2 \quad \text{(with C=O)} \tag{22}$$

Cresols can be made from propylene by reaction with toluene to produce cymene (107):

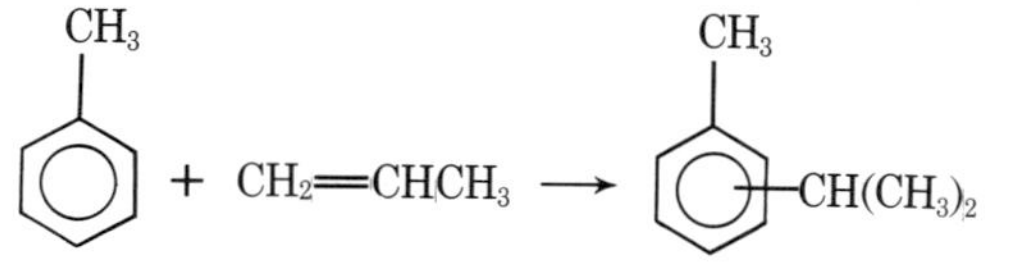

$$\text{toluene} + CH_2=CHCH_3 \longrightarrow \text{p-cymene} \tag{23}$$

The cymene is oxidized to cymene hydroperoxide, which decomposes to cresols and acetone.

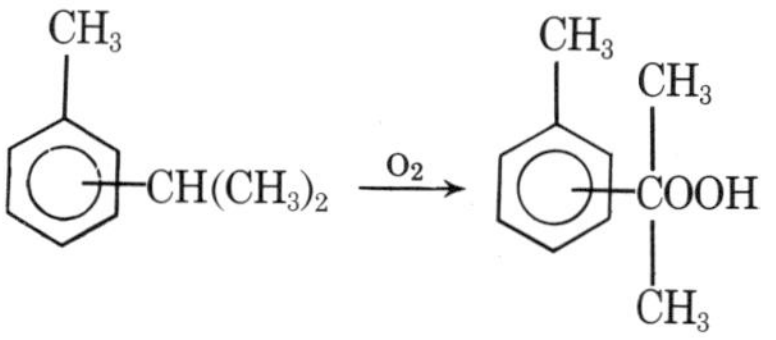

$$\text{cymene} \xrightarrow{O_2} \text{cymene hydroperoxide} \tag{24}$$

$$\text{cymene hydroperoxide} \xrightarrow{H^+} \text{cresol} + (CH_3)_2C=O \tag{25}$$

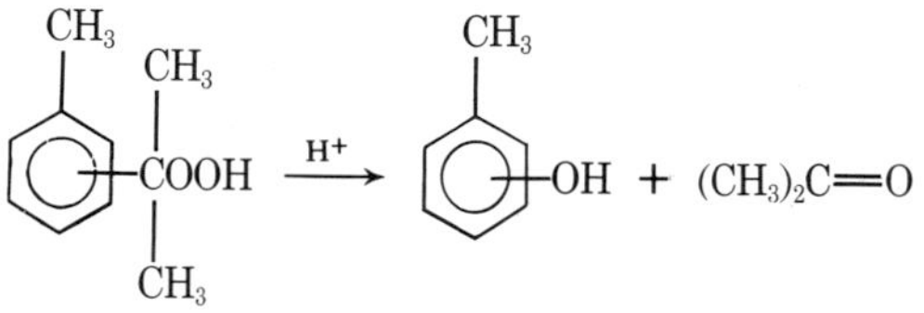

The process is similar to phenol (qv) production from cumene.

BIBLIOGRAPHY

"Propylene" in *ECT* 1st ed., Vol. 11, pp. 193–197, by John Happel and W. H. Kapfer, New York University; "Propylene" in *ECT* 2nd ed., Vol. 16, pp. 579–584, by William H. Davis, Texas National Bank of Commerce, and Leland K. Beach, Enjay Chemical Laboratory.

1. D. S. Sanders, D. M. Allen, and W. T. Sappenfield, *Chem. Eng. Prog.* **73**(7), 40 (1977).
2. R. C. Reid, T. M. Prausnitz, and T. K. Sherwood, *The Properties of Gases and Liquids*, 3rd ed., McGraw Hill, Inc., New York, 1977.
3. R. A. Svehla, *NASA Tech. Dept. R-132*, Lewis Research Center, Cleveland, Oh., 1962.
4. M. Benedict, G. B. Webb, and L. C. Rubin, *Chem. Eng. Prog.* **17**(8), 419 (1951).
5. H. W. Cooper, and T. C. Goldfrank, *Hydrocarbon Process.* **46**(12), 141 (1967).
6. S. K. Sood and G. A. Haselden, *AIChE J.* **16**, 891 (1970).
7. E. Bender, *Cryogenics* **15**, 667 (Nov. 1975).

8. K. E. Starling and Y. C. Kwok, *Hydrocarbon Process.* **50**(10), 90 (1971).
9. Otto Redlich, *Ind. Eng. Chem. Fundam.* **14**(3), 257 (1975).
10. B. D. Djordjevic and co-workers, *Chem. Eng.* **32**, 1103 (1977).
11. J. H. Dymond and E. B. Smith, *The Virial Coefficients of Gases, A Critical Compilation*, Clarendon Press, London, UK, 1969, p. 72.
12. C. L. Yaws, *Physical Properties*, McGraw-Hill, Inc., New York, 1977, pp. 167, 197, 217.
13. R. Battino, *Chem. Rev.* **66**, 395 (1966).
14. E. Wilhelan, R. Battino, and R. T. Wilcock, *Chem. Rev.* **77**(2), 219 (1977).
15. A. Azarnoosh and J. J. McKetta, *J. Chem. Eng. Data* **4**, 211 (1957).
16. I. Wichterle, J. Linek, and E. Hale, *Vapor-Liquid Equilibrium Data Biography*, Elsevier North Holland, Inc., New York, 1973, p. 221; *Supplement*, 1977.
17. H. H. Reamer and B. H. Sage, *Ind. Eng. Chem.* **43**, 1628 (1951).
18. K. Ishii, S. Hayami, T. Shirai, and K. Ishida, *J. Chem. Eng. Data* **11**(3), 288 (1966).
19. D. D. Li and J. J. McKetta, *J. Chem. Eng. Data* **8**, 271 (1963).
20. R. B. Williams and D. L. Katz, *Ind. Eng. Chem.* **46**, 2512 (1954).
21. S. L. McCurdy and D. L. Katz, *Oil Gas J.* **43**(45), 102 (1945).
22. W. G. Schneider and O. Maass, *Can. J. Res.* **19**(10), 231 (1941).
23. G. G. Haselden, F. A. Holland, M. B. King, and R. F. Strickland-Constable, *Proc. Royal Soc. London Ser. A* **240**, 1 (1957).
24. H. Lee, D. M. Newitt, and M. Rubemann, *Proc. Royal Soc. London Ser. A* **178**, 506 (1941).
25. R. A. McKay, H. H. Reamer, B. H. Sage, and W. N. Lacey, *Ind. Eng. Chem.* **43**, 2112 (1951).
26. G. H. Hanson and co-workers, *Ind. Eng. Chem.* **44**, 604 (1952).
27. A. N. Mann and co-workers, *J. Chem. Eng. Data* **8**(4), 499 (1963).
28. D. B. Manley and G. W. Swift, *J. Chem. Eng. Data* **16**(3), 301 (1971).
29. D. R. Laurence and G. W. Swift, *J. Chem. Eng. Data* **17**(3), 333 (1972).
30. G. H. Goff and co-workers, *Ind. Eng. Chem.* **42**, 735 (1950).
31. M. Hirata, S. Ohe, and K. Nagahama, *Computer Aided Data Book of Vapor-Liquid Equilibria*, Kadausha Ltd., Elsevier, Tokyo, 1975.
32. M. Sagara, Y. Arai, and S. Saito, *J. Chem. Eng. Jpn.* **5**(4), 418 (1972).
33. H. K. Bae, K. Nagahama, and M. Hirata, *J. Jpn. Pet. Inst.* **21**(4), 249 (1978).
34. M. L. McWilliams, *Chem. Eng.* **80**(25), 138 (Oct. 29, 1973).
35. C. S. Howat and G. W. Swift, *Ind. Eng. Chem. Process Des. Dev.* **19**, 318 (1980).
36. E. W. Funk and J. M. Prausnitz, *AIChE J.* **17**(1), 254 (1971).
37. S. E. M. Haman, W. K. Chung, I. M. Epshayal, and B. C. Y. Lu, *Ind. Eng. Chem. Process Des. Dev.* **16**(1), (1977).
38. J. Y. Lenoir and co-workers, *J. Chem. Eng. Data* **16**(3), 340 (1971).
39. S. Ng, H. G. Harris, and J. M. Prausnitz, *J. Chem. Eng. Data* **14**(4), 482 (1969).
40. H. Sayara and co-workers, *J. Chem. Eng. Jpn.* **8**(2), 98 (1975).
41. G. T. Preston and J. M. Prausnitz, *Ind. Eng. Chem. Fundam.* **10**(3), 384 (1971).
42. R. T. Morrison and R. N. Boyd, *Organic Chemistry*, 3rd ed., Allyn and Bacon, Inc., Boston, Mass., 1977, p. 143.
43. T. G. W. Solomons, *Organic Chemistry*, John Wiley & Sons, Inc., New York, 1977.
44. H. Wittcoff, *J. Chem. Educ.* **57**, 707 (1980).
45. S. B. Zdonik, E. J. Green, and L. P. Hallee, *Manufacturing Ethylene*, The Petroleum Publishing Co., Tulsa, Okla., 1970.
46. A. G. Goosens, M. Dente, and E. Ranzi, *Hydrocarbon Process.* **57**(9), 227 (1978).
47. S. B. Zdonik, E. J. Bassler, and L. P. Hallee, *Hydrocarbon Process.* **53**, 73 (Feb. 1974).
48. L. E. Chambers and W. S. Potter, *Hydrocarbon Process.* **53**(1), 121 (1974).
49. Amoco Chemicals Manufacturing Specifications.
50. A. Hahn, A. Chaptal, and J. Sialelli, *Hydrocarbon Process.* **54**(2), 89 (1975).
51. E. G. Wollaston, W. J. Haflin, and W. D. Ford, *Hydrocarbon Process.* **54**, 93 (Sept. 1975).
52. R. B. Ewell and G. Gadmer, *Hydrocarbon Process.* **57**(4), 125 (1978).
53. J. H. Gary and G. E. Handwerk, *Petroleum Refining*, Marcel Dekker Inc., New York, 1975, p. 52.
54. *Oil Gas J.*, 53 (Mar. 10, 1975).
55. H. G. Davis and R. G. Keister, *ACS Symposium Ser.*, 32 (1976); C. Bowen, paper presented at National Meeting of the American Institute of Chemical Engineers, Series 32, Houston, Tex., Apr. 5-9, 1981, p. 158.
56. H. J. Glidden and C. F. King, *Chem. Eng. Prog.* **76**(12), 47 (1980).

57. V. U. S. Rao and R. J. Gormley, *Hydrocarbon Process*, **59**(11), 139 (1980).
58. B. Bussemeir, C. D. Frohning, and B. Cornils, *Hydrocarbon Process.* **55**(11), 105 (1976).
59. P. D. Caesar, J. A. Brennan, W. E. Garwood, and J. Cirik, *J. Catal.* **56**, 274 (1979).
60. S. Gussow, D. C. Spence, and E. A. White, *Oil Gas J.*, (Dec. 8, 1980).
61. T. C. Ponder, *Hydrocarbon Process.* **53**(7), 104 (1974).
62. *Chem. Eng. News*, 15 (Oct. 10, 1977); 9 (Mar. 6, 1978); 8 (Oct. 9, 1978); 10 (Apr. 2, 1979).
63. *Chem. Eng. News*, 11 (Nov. 15, 1976); 8 (Nov. 7, 1977); 12 (Nov. 20, 1978); 16 (Nov. 17, 1980); 12 (Jan. 5, 1981).
64. *Oil Gas J.*, 68 (June 28, 1971).
65. J. M. Winton, *Chem. Week*, 44 (Apr. 16, 1980).
66. J. W. Andrews, P. Bonnifay, B. Cha, D. Douillet, and J. Raimbault, *paper presented at NPRA Annual Meeting*, Mar. 28, 1976, San Antonio, Tex., AM-76-25.
67. *Chemical Safety Data Sheet SD-59, Propylene*, Manufacturing Chemists Association, Washington, D.C., revised 1974.
68. Standards for Storage Handling of Liquefied Petroleum Gases, National Fire Code, Sect. 5, No. 58, National Fire Protection Assoc., Boston, Mass., 1977.
69. National Electric Code, No. 70, National Fire Protection Assoc., Boston, Mass., 1971.
70. ASME Boiler and Pressure-Vessel Code, Section 8, ASME, New York, 1980.
71. ANSI B 31 Series, B31.3., ASME, New York, 1980.
72. H. F. Coward and co-workers, *Limits of Flammability of Gases and Vapors*, Bull. 503, U.S. Bureau of Mines, Washington, D.C., 1952.
73. N. I. Sax, *Dangerous Properties of Industrial Materials*, 5th ed., Van Nostrand Reinhold Co., New York, 1979.
74. W. M. Porteous and R. C. Reid, *Chem. Eng. Prog.* **72**, 83 (1976).
75. *Chemical Economics Handbook*, SRI International, Menlo Park, Calif., Dec. 1980.
76. P. H. Spitz, *Hydrocarbon Process.* **55**(7), 131 (1976).
77. *Chem. Eng. News*, 8 (Mar. 10, 1975); 14 (Nov. 15, 1976); 12 (Nov. 7, 1977); 14 (Nov. 20, 1978); 16 (Nov. 17, 1980).
78. *Publication 2140-77*, Gas Processors Association, Tulsa, Okla., 1977.
79. R. J. Hengstebeck, *Petroleum Processing, Principles and Applications*, McGraw-Hill, Inc., New York, 1959.
80. W. S. Knoble and F. E. Herbert, *Petr. Refiner*, 101 (Dec. 1959).
81. C. R. Cupit, J. E. Cwyu, and E. C. Jernigan, *Petro/Chem. Eng.*, 203 (Dec. 1961); *Petro/Chem. Eng.*, 207 (Jan. 1962).
82. G. E. Weisingntel, *Chem. Eng.*, 77 (June 16, 1980).
83. P. M. Kohn, *Chem. Eng.*, 114 (May 23, 1977).
84. E. K. Jones, *Adv. Catal.* **8**, 219 (1956).
85. A. Schrieshein and I. Kirschenbaum, *Chemtech.*, 310 (May 1978).
86. C. M. Fontana, R. J. Herold, E. J. Kinney, and R. C. Miller, *Ind. Eng. Chem.* **44**, 2955 (1952).
87. P. Pin and R. Mülhempt, *Angew. Chem. Int. Ed. Engl.* **19**, 857 (1980).
88. J. Boor, Jr., *Ziegler-Natta Catalysts and Polymerizations*, Academic Press, Inc., New York, 1979.
89. G. S. Parks and H. P. Mosher, *J. Polym. Sci. Part A-1*, 1979 (1963).
90. D. J. Hadley, R. E. Saunders, and P. T. Mapp in E. G. Hancock, ed., *Propylene and its Industrial Derivatives*, Earnest Benn Ltd., London, UK, 1973, pp. 416 ff.
91. P. R. Pujado, B. V. Vora, and A. P. Krueding, *Hydrocarbon Process.* **56**(5), 169 (1977).
92. R. Landau, G. A. Sullican, and D. Brown, *Chemtech.*, 602 (Oct. 1979).
93. K. Weissermel and H. J. Arpe, *Industrial Organic Chemistry*, Verlag-Chemie, Weinheim, Austria, 1978, p. 239.
94. *1979 Petrochemical Handbook*, *Hydrocarbon Process.* **59**(11), 240 (1979).
95. Ref. 14, p. 175.
96. Y. Onove, Y. Mizutani, S. Akiyama, and Y. Izumi, *Chemtech*, 432 (July 1978).
97. W. Nier and J. Woellner, *Hydrocarbon Process.* **51**(11), 113 (1972).
98. A. M. Brownstein, *U. S. Petrochemicals*, The Petroleum Publishing Co., Tulsa, Okla., 1972, p. 324.
99. W. C. Fernelius, H. Wittcoff, and R. E. Varneria, *J. Chem. Educ.* **57**, 707 (1980).
100. J. C. Fielding in ref. 90, p. 244 ff.
101. R. A. Persale, E. L. Pollitzer, D. J. Ward, and P. R. Pujado, *Chem. Econ. Eng. Rev.* **10**(7), 25 (1978).
102. R. Fowler, H. Conner, and R. A. Gaehl, *Hydrocarbon Process.* **55**(9), 247 (1976).

103. H. Weber, W. Demmeling, and A. M. Desai, *Hydrocarbon Process.* **55**(4), 129 (1976).
104. Ref. 14, pp. 77–80.
105. G. E. Schaal, *Hydrocarbon Process.* **52**(9), 218 (1973).
106. L. F. Hatch and S. Matar, *Hydrocarbon Process.* **57**(6), 149 (1978).
107. K. Sto, *Hydrocarbon Process.* **52**(8), 89 (1973).

MORRIS R. SCHOENBERG
JOHN W. BLIESZNER
CHRISTOS G. PAPADOPOULOS
Amoco Chemicals Corporation

PROPYLENE OXIDE

Propylene oxide [75-56-9] (methyloxirane, 1,2-epoxypropane) was first prepared in 1860 by Oser by the classical chlorohydrin method (1) (see Chlorohydrins). This synthesis is still one of the two principal processes for propylene oxide manufacture and accounts for about half of U.S. production (2). Considerable research effort has failed to devise a direct-oxidation process comparable to the excellent commercial ethylene oxide process (see Ethylene oxide). An indirect-oxidation process with the use of an organic hydroperoxide accounts for the balance of propylene oxide production capacity. The total U.S. sales value of propylene oxide for 1980 of ca 6.5×10^8 makes it a significant organic chemical product (3–4).

Physical Properties

Propylene oxide is a colorless, low-boiling liquid. It is miscible with most organic solvents but forms a two-layer system with water. Propylene oxide can exist as two optical isomers; the racemic mixture is the commercial product. Except as noted, properties are of the racemic mixture. Physical properties are summarized in Tables 1–3 (5–14); the propylene oxide–water system is shown in Figure 1 (7,15–16). Propylene oxide forms azeotropes with methylene chloride, ether, and several hydrocarbons (17).

The specific rotation $[\alpha]_D^{18}$ of the two optical isomers has been reported as +12.72 and −8.26° for (R) and (S)-propylene oxide, respectively (18). Values for (R)-propylene oxide in various solvents are given in ref. 19. Infrared and Raman spectra have been reported (20–22).

Chemical Properties

Propylene oxide exhibits a high degree of reactivity, a result of the presence of the strained, three-membered oxirane ring. Reactions such as those with hydrogen halides and ammonia proceed at a satisfactory rate without a catalyst. Most reactions,

Table 1. Some Physical Constants of Propylene Oxide, C_3H_6O

Property	Value	Refs.
mol wt	58.08	
bp at 101.3 kPa[a], °C	34.2	5
Δbp/pressure from 98.66–101.3 kPa[a], K/kPa	0.28	6
freezing point, °C	−112	5
coefficient of cubical expansion at 20°C, per °C	0.00151	7
critical pressure, MPa[b]	4.92	8, 9
critical temperature, °C	209.15	8, 9
critical density, g/cm^3	0.312	8, 9
critical compressibility factor	0.2284	10
dipole moment, C·m[c]	6.61×10^{-30}	11
explosive limits in air, vol %		
upper	37	5
lower	2.3	
flash point, calculated, °C	<−20	5
heat of fusion, kJ/mol[d]	6.54	12
index of refraction, n_D^{25}	1.36322	10

[a] To convert kPa to mm Hg, multiply by 7.5.
[b] To convert MPa to atm, divide by 0.1013.
[c] To convert C·m to debye, divide by 3.336×10^{-30}.
[d] To convert J to cal, divide by 4.184.

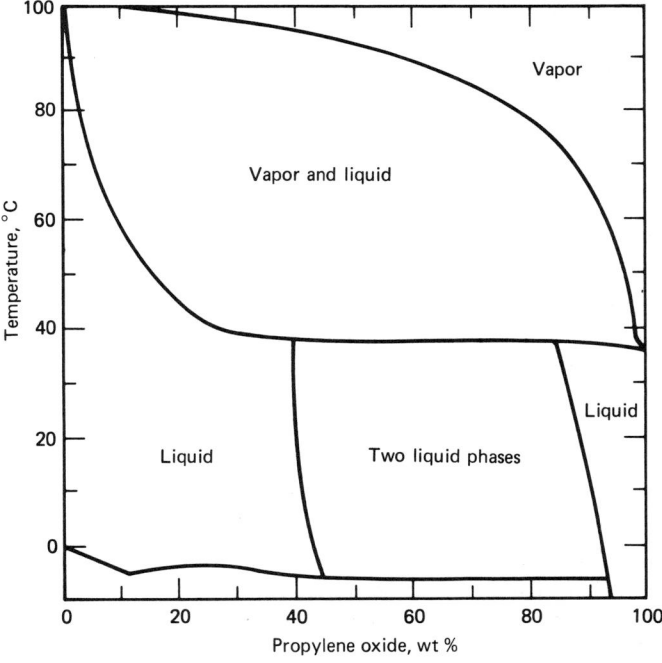

Figure 1. Propylene oxide–water system (7,15–16).

however, require an acidic or basic catalyst; in both cases, the reaction proceeds through a nucleophilic substitution (S_N2) mechanism.

In base-catalyzed reactions, the nucleophilic reagent attacks the least-substituted oxirane carbon, and the primary product is the secondary alcohol.

Table 2. Physical Properties of Propylene Oxide Liquid from −20 to +209°C

Temperature, °C	Vapor pressure[a], kPa[b]	Enthalpy of vaporization, J/g[c]	Heat capacity[d], J/(kg·K)[c]	Density[e], g/cm³	Surface tension, mN/m (= dyn/cm)	Viscosity[f], mPa·s (= cP)	Thermal conductivity, W/(m·K)
−20	8.62	531.7	1949	0.878	27.99	0.48	0.17
−10	14.20		1973				
0	24.13	506.6	2004	0.854	25.03	0.38	0.15
10	39.30		2038				
20	58.61	485.7[g]	2059	0.830	22.14	0.31	0.14
25		480.6	2074			0.28	
30	86.19						
40	124.1	464.7		0.805	19.31	0.25	
50	157.2						
60	222.7	443.8		0.783	16.55	0.22	
70	317.2						
80	420.6	418.7		0.758	13.88	0.20	
90	530.9						
100	689.5	393.6		0.730	11.30		0.18
120	1069	364.3		0.695	8.82		
140	1482	330.8		0.660	6.46		
160	2151	288.9		0.62	4.3		
180	2910	238.6		0.55	2.2		
200	4275	150.7		0.44			
209	4923						

[a] Refs. 8–9.
[b] To convert kPa to mm Hg, multiply by 7.5.
[c] To convert J to cal, divide by 4.184.
[d] Estimated from ref. 12.
[e] Refs. 5–6, 8–9.
[f] Ref. 9.
[g] Ref. 13.

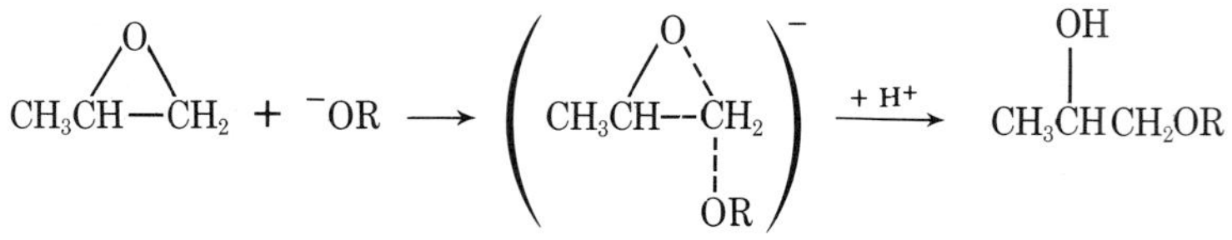

$$\text{CH}_3\text{CH}-\text{CH}_2 + {}^-\text{OR} \longrightarrow \left(\text{CH}_3\text{CH}-\text{CH}_2 \atop \text{OR}\right)^- \xrightarrow{+\text{H}^+} \text{CH}_3\text{CHCH}_2\text{OR} \text{ (OH)}$$

In the transition state of the acid-catalyzed reaction, the oxygen is protonated, and bond cleavage is more complete than bond formation. Both partial bonds are longer than usual, and a partial positive charge develops on the central carbon. The methyl group stabilizes the partial positive charge by electron release; however, this mechanism is also subject to steric hindrance from the methyl group. Thus, a mixture of products is obtained (23):

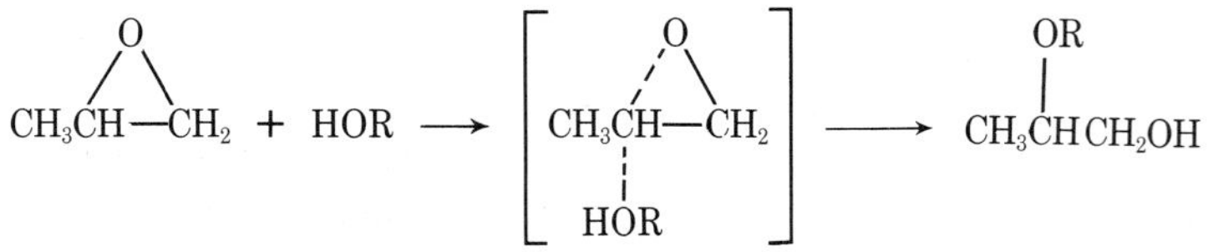

$$\text{CH}_3\text{CH}-\text{CH}_2 + \text{HOR} \longrightarrow \left[\text{CH}_3\text{CH}-\text{CH}_2 \atop \text{HOR}\right] \longrightarrow \text{CH}_3\text{CHCH}_2\text{OH} \text{ (OR)}$$

Table 3. Physical Properties of Propylene Oxide Vapor from 298 to 1000 K (Ideal Gas State)

Temperature, K	Entropy[a], J/(mol·K)[b]	Heat of formation[a], kJ/mol[b]	Free energy of formation[a], kJ/mol[b]	Heat capacity[a], J/mol·K	Viscosity, μPa·s[c]	Thermal conductivity, W/(m·K)	log Kp[a]
298	286.9	−92.82	−25.79	72.39	8.8	0.011	4.517
300	288.2	−92.95	−25.41	72.77	8.9	0.011	4.418
400	311.1	−97.93	−2.09	92.78	11.9	0.021	0.271
500	333.8	−102.03	22.35	110.7	14.7	0.032	−2.335
600	355.4	−105.30	47.56	125.9	17.4	0.043	−4.137
700	375.8	−107.81	73.27	138.6	20.0	0.055	−5.462
800	395.0	−109.74	99.22	149.3	22.5	0.066	−6.475
900	413.2	−111.08	125.4	158.6	25.0	0.078	−7.276
1000	430.3	−111.87	151.8	166.5	27.3	0.089	−7.924

[a] Ref. 14.
[b] To convert J to cal, divide by 4.184.
[c] 1 μPa·s = 10^{-5} P.

Polymerization. *Polyether Polyols.* The formation of polyether polyols is commercially the most important reaction of propylene oxide (24). A polyol is the product of reaction of an epoxide and compounds or initiators (eg, glycols, amines, acids, or water) that contain active hydrogens.

Poly(propylene glycol), the simplest propylene oxide-based polyol, is prepared by the base-catalyzed polymerization of propylene oxide with propylene glycol as the initiator (25):

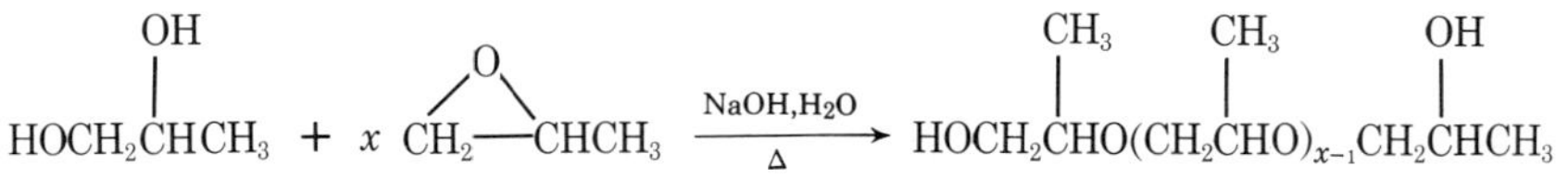

$$\text{HOCH}_2\overset{\overset{\displaystyle\text{OH}}{|}}{\text{CH}}\text{CH}_3 + x\ \text{CH}_2\overset{\displaystyle\diagup\text{O}\diagdown}{\text{—}}\text{CHCH}_3 \xrightarrow[\Delta]{\text{NaOH,H}_2\text{O}} \text{HOCH}_2\overset{\overset{\displaystyle\text{CH}_3}{|}}{\text{CH}}\text{O}(\text{CH}_2\overset{\overset{\displaystyle\text{CH}_3}{|}}{\text{CH}}\text{O})_{x-1}\text{CH}_2\overset{\overset{\displaystyle\text{OH}}{|}}{\text{CH}}\text{CH}_3$$

Such a polyol is commonly known as a polyol diol; a polyol triol results from the polymerization of propylene oxide initiated with glycerol (25–26):

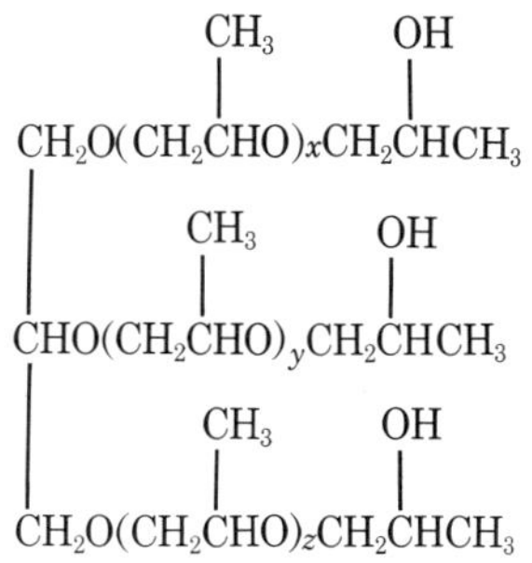

$$\begin{array}{c}
\overset{\displaystyle\text{CH}_3\ \ \ \ \ \ \text{OH}}{|\ \ \ \ \ \ \ \ \ \ \ \ \ \ \ |}\\
\text{CH}_2\text{O}(\text{CH}_2\text{CHO})_x\text{CH}_2\text{CHCH}_3\\
|\\
\overset{\displaystyle\text{CH}_3\ \ \ \ \ \ \text{OH}}{|\ \ \ \ \ \ \ \ \ \ \ \ \ \ \ |}\\
\text{CHO}(\text{CH}_2\text{CHO})_y\text{CH}_2\text{CHCH}_3\\
|\\
\overset{\displaystyle\text{CH}_3\ \ \ \ \ \ \text{OH}}{|\ \ \ \ \ \ \ \ \ \ \ \ \ \ \ |}\\
\text{CH}_2\text{O}(\text{CH}_2\text{CHO})_z\text{CH}_2\text{CHCH}_3
\end{array}$$

Other polyol triols may be obtained by initiating the reaction with trimethylolpropane, triethanolamine, and hexanetriols (27–28).

Polyols with a larger number of terminal hydroxyl groups result when the reaction is initiated with compounds such as pentaerythritol, 2,2,6,6-tetrakis(hydroxymethyl)cyclohexanol, or with natural products, eg, sucrose, raffinose, or D-mannitol. Amines such as ethylenediamine, diethylenetriamine, and 2,4-diaminotoluene may also be used as the active-hydrogen-containing initiator (27–28).

Propylene oxide also reacts with active hydroxyl hydrogen derived from the ring opening of other compounds such as ethylene oxide (25,29) and tetrahydrofuran (30–31); thus, a copolymer polyol is obtained.

Typically, polyols are obtained from base-catalyzed reactions with aqueous ammonia, sodium or potassium hydroxide, or lower alkyl tertiary amines such as trimethyl- and triethylamine (25,32–33). The reaction of propylene oxide with tetrahydrofuran is catalyzed by boron trifluoride etherate (31).

The molecular weights of polyols prepared according to the reactions described above range from ca 200 to 7000 (25) (see Glycols).

High Polymers. Poly(propylene oxide) polymers with molecular weights of 100,000 or more can be prepared with a catalyst that consists of $FeCl_3$ and approximately five equivalents of propylene oxide (34–35). The addition of small amounts of toluene 2,4- and 2,6-diisocyanates greatly increases the molecular weights of the polymers obtained (36). Propylene oxide homopolymers can also be prepared with catalysts such as diethylzinc (37) and trialkylaluminum compounds (38) (see Polyethers).

Propylene oxide is copolymerized with CO_2 and anhydrides to form the polycarbonate and polyesters, respectively (39–40). When epoxides such as butadiene monoxide or allyl glycidyl ether, which contain carbon–carbon double bonds, are copolymerized with propylene oxide, the resulting product has unsaturated, reactive side chains (41–42).

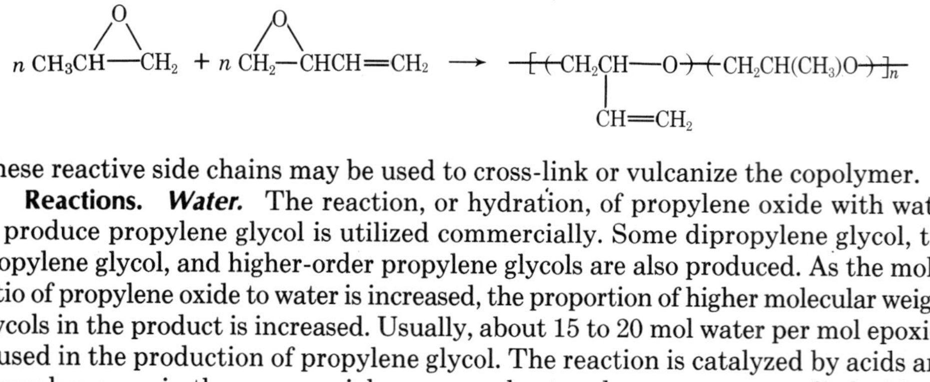

$$n\ CH_3CH{-}CH_2 + n\ CH_2{-}CHCH{=}CH_2 \longrightarrow {-}[{-}(CH_2CH{-}O){-}(CH_2CH(CH_3)O){-}]_n$$
$$\qquad\qquad\qquad\qquad\qquad\qquad\qquad\qquad\qquad\qquad\qquad\qquad |$$
$$\qquad\qquad\qquad\qquad\qquad\qquad\qquad\qquad\qquad\qquad\qquad\qquad CH{=}CH_2$$

These reactive side chains may be used to cross-link or vulcanize the copolymer.

Reactions. Water. The reaction, or hydration, of propylene oxide with water to produce propylene glycol is utilized commercially. Some dipropylene glycol, tripropylene glycol, and higher-order propylene glycols are also produced. As the molar ratio of propylene oxide to water is increased, the proportion of higher molecular weight glycols in the product is increased. Usually, about 15 to 20 mol water per mol epoxide is used in the production of propylene glycol. The reaction is catalyzed by acids and bases; however, in the commercial processes, heat and pressure are applied without catalyst (5).

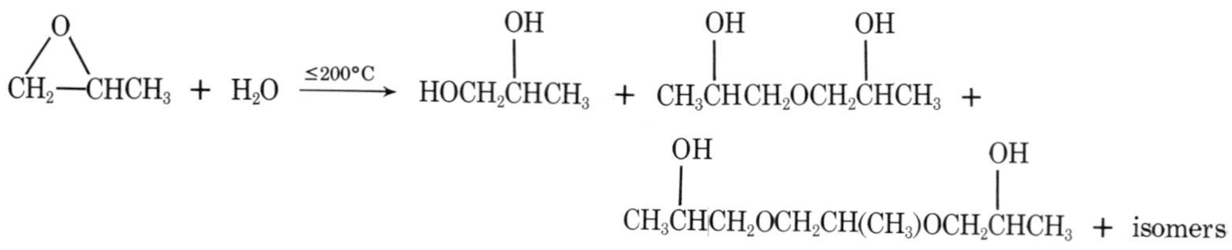

$$CH_2{-}CHCH_3 + H_2O \xrightarrow{\leq 200°C} HOCH_2\underset{OH}{CHCH_3} + CH_3\underset{OH}{CHCH_2}OCH_2\underset{OH}{CHCH_3} +$$

$$CH_3\underset{OH}{CHCH_2}OCH_2CH(CH_3)OCH_2\underset{OH}{CHCH_3} + \text{isomers}$$

Carbon dioxide in admixture with compounds such as potassium hydroxide, tetraethylammonium bromide, potassium carbonate (43), and tributylmethylphosphonium iodide (44) has been reported to be an effective catalyst for the hydrolysis of propylene oxide.

Ammonia and Amines. Propylene oxide and ammonia with a small amount of water give isopropanolamine. Further reaction yields the di- and triisopropanolamines. The ratio of primary, secondary, and tertiary amines obtained depends upon the molar ratios (see Alkanolamines).

Aryl and alkyl primary and secondary amines give the corresponding N- or N,N-disubstituted isopropanolamines. Once the amino hydrogens are replaced, the isopropanolamines may react through the hydroxyl group with additional epoxide to form poly(propylene glycol) derivatives of amines (45).

Carbon Dioxide, Carbon Disulfide. Propylene oxide reacts with carbon dioxide to yield propylene carbonate which, in turn, can be hydrolyzed to propylene glycol. The reaction is catalyzed by potassium iodide, tetraalkylammonium bromides, calcium bromide, or magnesium bromide (46–48).

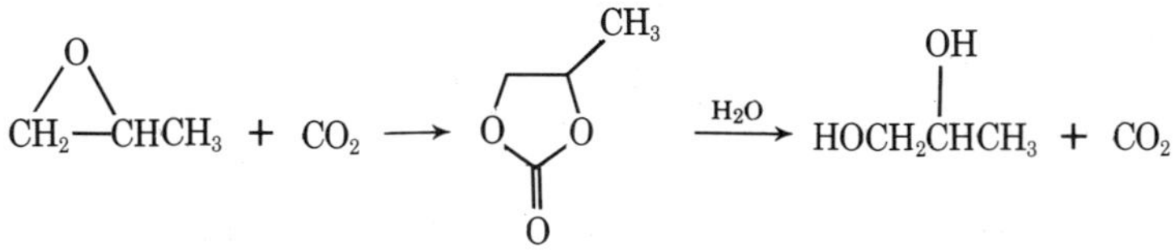

$$\underset{\text{CH}_2\text{—CHCH}_3}{\triangle\!\!\!\!\text{O}} + \text{CO}_2 \longrightarrow \text{propylene carbonate} \xrightarrow{\text{H}_2\text{O}} \text{HOCH}_2\text{CHCH}_3\text{(OH)} + \text{CO}_2$$

In the presence of catalytic quantities of diethylzinc and hexamethylphosphoric triamide, carbon disulfide and propylene oxide form a low molecular weight copolymer (49). In the presence of magnesium oxide, the episulfide, methylthiirane, is formed (50), whereas trimethylamine catalyzes the formation of propylene trithiocarbonate (51).

Organic Acids and Anhydrides. Carboxylic acids and propylene oxide give a mixture that contains the monoesters of the primary and the secondary alcohol groups of propylene glycol. The monoesters may then react with additional acid to form the glycol diester. With a sufficient concentration of epoxide, the monoester may add propylene oxide to yield the ester of dipropylene glycol or of one of the higher-order glycols. Transesterification may give propylene glycol, higher poly(propylene glycols) and their esters, and the glycol diester. The esterification is catalyzed with sodium or potassium hydroxide, and anhydrous chromium(III) tricarboxylate salts (52–55).

Cyclic carboxylic acid anhydrides, eg, phthalic anhydride, give polyesters (qv) with propylene oxide. The reaction is catalyzed by diethylzinc, lithium chloride, tertiary amines, and quaternary ammonium halides (16,56).

Alcohols and Phenols. Alcohols or phenols and propylene oxide give monoethers of propylene glycol. These glycol ethers may then react further to produce di-, tri-, and poly(propylene glycol) ethers. As the ratio of alcohol to epoxide in the reaction mixture is increased, the molecular weight of the products tends to decrease, ie, the yield of propylene glycol ether increases relative to that of the di-, tri-, and poly(propylene glycol) ethers. A basic catalyst favors formation of secondary alcohols, whereas an acidic catalyst leads to a mixture of the primary and secondary alcohols (57–59):

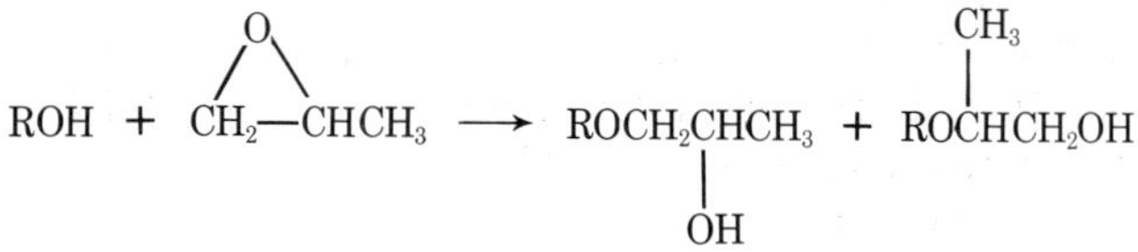

$$\text{ROH} + \underset{\text{CH}_2\text{—CHCH}_3}{\triangle\!\!\!\!\text{O}} \longrightarrow \text{ROCH}_2\text{CHCH}_3\text{(OH)} + \text{ROCH(CH}_3\text{)CH}_2\text{OH}$$

Catalysts include stannic and aluminum halides, zinc and nickel sulfates, potassium hydroxide, boron trifluoride, and a combination of palladium chloride, copper chloride, and oxygen (59–61) (see also Ethers).

Hydrogen Sulfide and Mercaptans. Propylene oxide and hydrogen sulfide form 1-mercapto-2-propanol, which may then react further to form the thiodiglycol, bis(2-hydroxypropyl) sulfide (62). Reaction of the epoxide with an alkyl- or aryl mercaptan in a basic medium affords the 1-alkylthio- or the 1-arylthio-2-propanol, respectively (63–64). Sodium or lithium hydroxide or the sodium salt of the mercaptan catalyze the reaction (65–67).

Natural Products. Starches, sugars, cellulose, and glycerol react with propylene oxide in the presence of alkaline catalysts to produce propylene glycol monoethers and poly(propylene glycol) ethers (68–70). Reaction with fatty acids gives mono- and diesters (71). Cotton, castor oil, and wood resins have been modified with propylene oxide (72–74).

Friedel-Crafts. In the presence of catalytic amounts of aluminum chloride, propylene oxide and aromatic compounds such as benzene or toluene give 2-arylpropanols (see Friedel-Crafts reactions). The reaction is conducted at low temperatures in an anhydrous medium. With monosubstituted aromatic compounds, a mixture of ortho, meta, and para isomers is obtained (75–76).

Grignard Reagents. Reaction with a Grignard reagent, RMgX, yields a secondary alcohol, $RCH_2CHOHCH_3$, as the principal product; a common by-product is the halohydrin $CH_3CHOHCH_2X$ (77–79). Rearrangements are avoided with dialkylmagnesiums (80). Yields of desired alcohols may be increased by the use of a copper(I) halide catalyst (81) (see Grignard reaction).

Isomerization. Propylene oxide isomerizes to allyl alcohol, propionaldehyde, and acetone (82–84). A 95% yield of allyl alcohol from epoxide may be obtained with a supported Li_3PO_4 catalyst (85). Moderate to high yields of propionaldehyde are obtained by heating propylene oxide over catalysts such as silica gel, alumina, CdO–$CdCl_3$, or WO_3–Fe_2O_3. Acetone may be a by-product (86–87).

Carbonyl Compounds. Ketones and aldehydes give cyclic ketals and acetals (dioxolanes), respectively. Stannic chloride, quaternary ammonium halides, and molybdenum acetylacetonate catalyze the reaction. Alkali-stable solvents and protecting groups for organic synthesis are prepared by this reaction (88–90).

Reduction. Hydrogenation of propylene oxide over nickel affords 1-propanol. With sodium amalgam, sodium in ammonia (91), lithium aluminum hydride (92), or lithium dimethylcuprate (93), the product is isopropyl alcohol. In the presence of (cyclopentadienyl)dicarbonyl ferrate (94) or metallocenes of the formula $[M(C_5H_5)Cl_2]$ (M = W, Mo, Ti, or Zr) and sodium amalgam (95), propylene is obtained.

Other Inorganic Reagents. Reaction with hydrogen halides yields a mixture of 1-halo-2-propanol and 2-halo-1-propanol (96). With nitric acid, a mixture of the isomeric nitrate monoesters of propylene glycol is obtained (97). Sodium sulfite gives the sodium salt of 2-hydroxypropanesulfonic acid (98–99).

Boric acid and propylene oxide form a mixture of isomeric biborates (100). Boron trichloride and propylene oxide give substituted boranes, eg, dichloro(2-chloroisopropoxy)borane and chlorobis(2-chloroisopropoxy)borane, and esters of boric acid, eg, tris(2-chloroisopropoxy)borate; the number of chloropropoxy groups per boron atom depends upon the initial molar ratio of epoxide to boron trichloride (101–102). Diisopropoxyborane and an oligomer, $H{+}CH(CH_3)CH_2O{+}_6BH_2$, result from the reaction with diborane (103–104).

In the presence of a basic ion-exchange resin, hydrogen cyanide and propylene oxide in an aqueous medium give propylene cyanohydrin (105).

Metallic halides of the form MCl_x, where M = P, Bi, As, Ti, or Be, condense with propylene oxide to form compounds of the form $(C_3H_6ClO)_{(x-y)}MCl_y$; the degree of substitution of chloropropoxy groups for the chlorine on the metal atom depends upon the molar ratio of the reactants (106–107).

Other Reactions. Propylene oxide (PO) and water form a clathrate hydrate, $PO \cdot 17H_2O$, mp $-3.8°C$ (108) (see Clathration).

Propylene oxide is used for the preparation of crown ethers. The cyclic tetramer, tetramethyl-12-crown-4, is obtained by heading the epoxide in the presence of boron trifluoride etherate (109–110). An optically active crown ether and its isomers are prepared from propylene oxide, propylene glycol, and catechol; the product resulting from this multistep synthesis is tetramethyldibenzo-18-crown-6 and other isomers (111) (see Chelating agents).

Trialkylaluminums and propylene oxide form 2-methyl-substituted alcohols; with trimethylaluminum, 2-methyl-1-butanol is obtained (112–113).

In the presence of a mercury(II) salt HgZ where Z = Cl, SCN, or NCS, and an alkyl iodide, RI, propylene oxide yields $CH_3CHORCH_2Z$ (114).

Propylene oxide undergoes a number of other reactions; however, they are not of industrial significance. Nucleophilic reactions occur with reagents such as potassium thiocyanate, *tert*-butyl hydroperoxide, sodium azide, diethyl sodiomalonate, and diethyl phosphite (115). Electrophilic additions with nitrosyl chloride, dinitrogen tetroxide, diethoxychlorophosphine (diethyl phosphorochloridite), trichloromethylsulfenyl chloride, triphenylmethyl bromide, and sulfur dioxide have also been reported (116).

Manufacture

Propylene oxide is produced either by the chlorohydrin or the hydroperoxide process. The former is older and involves the reaction of propylene with chlorine and water to produce propylene chlorohydrin, followed by dehydrochlorination with lime or caustic to give propylene oxide and a salt. This process was first used in Germany during World War I by BASF and others (117). At present, the Dow Chemical Company plants at Freeport, Texas, and Plaquemine, Louisiana, are the only active chlorohydrin units in the United States (4) (see Chlorohydrins). In the hydroperoxide process, an organic hydroperoxide is used to epoxidize propylene; an organic alcohol is a coproduct. Currently, *tert*-butyl alcohol and α-methylbenzyl alcohol are produced this way. The former can be sold as a gasoline octane booster or be dehydrated to isobutylene, whereas α-methylbenzyl alcohol is converted to styrene (qv). Oxirane, a division of Arco Chemicals, has units in Bayport, Texas, where *tert*-butyl alcohol is a coproduct, and a unit at Channelview, Texas, where styrene is a coproduct (118).

Published information on commercial propylene oxide processes is scarce and of a general nature. Patent examples offer much of the insight into various process conditions and procedures; however, they do not provide state-of-the-art technology information. Therefore, the process-data flow sheets presented here are estimates of current commercial activity.

Chlorohydrin Process. The chlorohydrin process has been used for both ethylene and propylene oxide. However, ethylene is now produced by direct oxidation, and ethylene oxide chlorohydrin units have been converted to propylene oxide. The present commercial process is based on mixing propylene and chlorine in ca molar amounts with an excess of water to form a dilute solution of propylene chlorohydrin and minor amounts of chlorinated hydrocarbon by-products. This solution is then treated with a base, eg, lime or caustic soda, to form crude propylene oxide and an effluent water stream of calcium chloride or sodium chloride.

The main steps are chlorohydrination, epoxidation, wastewater effluent treatment, and propylene oxide purification. The most frequently proposed mechanism for the chlorohydrination is as follows (119):

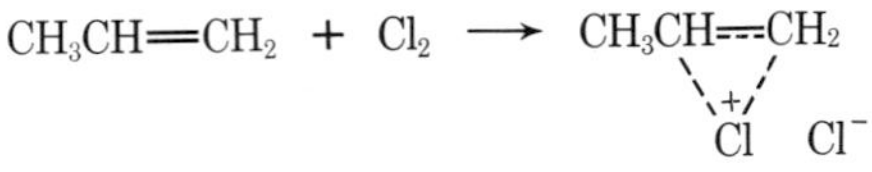

$$CH_3CH=CH_2 + Cl_2 \rightarrow \underset{\underset{Cl^+ \quad Cl^-}{\diagdown+\diagup}}{CH_3CH\text{---}CH_2}$$

propylene–chloronium complex

This ion complex reacts with water to form the propylene chlorohydrin isomers 1-chloro-2-propanol and 2-chloro-1-propanol plus hydrogen chloride:

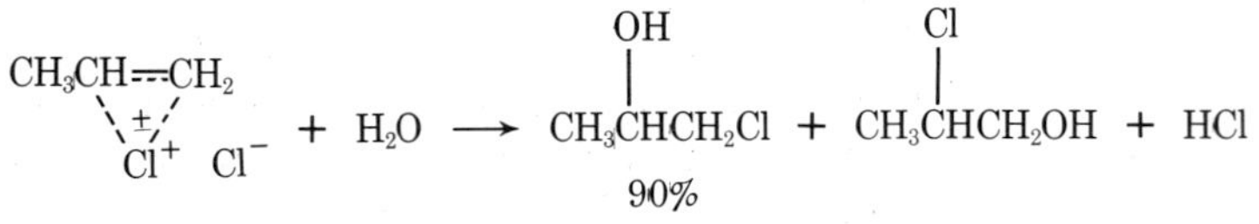

$$\underset{\underset{Cl^+ \quad Cl^-}{\diagdown+\diagup}}{CH_3CH\text{---}CH_2} + H_2O \rightarrow \underset{90\%}{CH_3\underset{OH}{\overset{|}{C}}HCH_2Cl} + CH_3\underset{Cl}{\overset{|}{C}}HCH_2OH + HCl$$

The chlorohydrin mixture formed during reaction is referred to as propylene chlorohydrin. The chloronium ion can undergo other reactions to form propylene dichloride and bis(1-chloroisopropyl) ether, the main by-products:

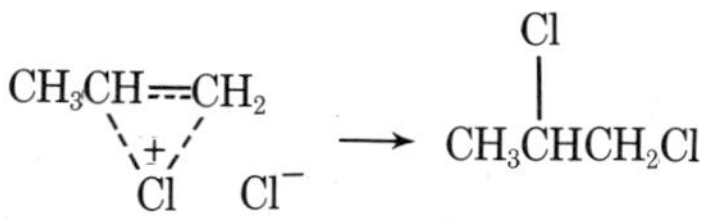

$$\underset{\underset{Cl^+ \quad Cl^-}{\diagdown+\diagup}}{CH_3CH\text{---}CH_2} \rightarrow CH_3\underset{Cl}{\overset{|}{C}}HCH_2Cl$$

and

$$\underset{\underset{Cl^+ \quad Cl^-}{\diagdown+\diagup}}{CH_3CH\text{---}CH_2} + CH_3\underset{OH}{\overset{|}{C}}HCH_2Cl \rightarrow (ClCH_2\underset{CH_3}{\overset{|}{C}}H)_2O$$

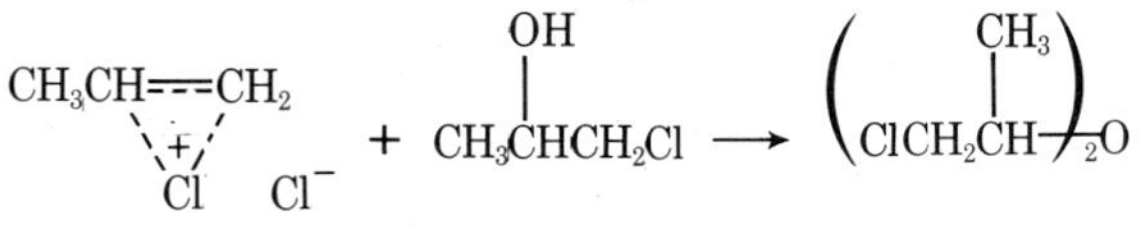

Monochloroacetone, 1,2-dichloro-3-propanol, and other chlorinated compounds are produced in small amounts. Commercial success depends on minimizing these by-products. The main by-product, propylene dichloride, represents a 4–8% loss of propylene and chlorine feed. Because of the limited solubility of the dichloride and the dependence of ether formation on propylene chlorohydrin concentration, water is added to maintain the propylene chlorohydrin at 4–5.5 wt % in the reaction product (119–121).

Chlorohydrination. A flow sheet of the chlorohydrin process is shown in Figure 2 (118–121). Propylene and chlorine in ca equal molar amounts are mixed with an excess of water, 4.8–7.3 mol water per kg of propylene (120). The excess water reduces the propylene chlorohydrin and chloride ion concentration in the reactor, thereby minimizing the formation of by-product propylene dichloride and ether formation.

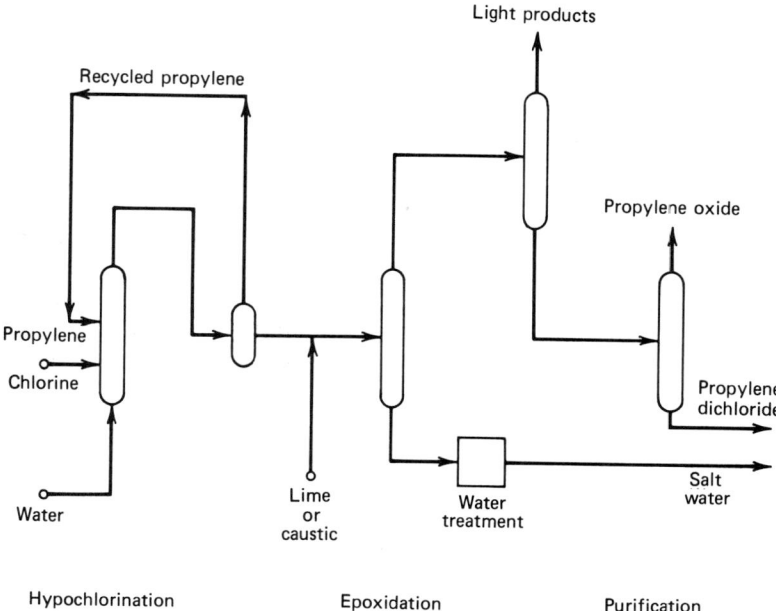

Figure 2. Chlorohydrin process (118–121).

Excess water also prevents formation of an organic phase of propylene dichloride in which propylene and chlorine would react rapidly.

The organic phase can be avoided by stripping the propylene dichloride from the reaction mixture with heat or excess propylene (120); the unreacted propylene is water-washed and recycled. The chlorohydrin reaction must be controlled in order to minimize by-product formation and to maintain safe operation. The propylene-to-chlorine ratio must be closely controlled to maintain the proper amount of recycle and avoid an explosive reaction between the hydrocarbon and chlorine in the vent system. Because of the corrosive nature of the reaction mixture, an aqueous mixture of organic compounds and hydrogen chloride, the preferred materials of construction are brick, rubber, and plastic-lined equipment (119–120). This limits the reaction pressure to about atmospheric or slightly above. Reaction temperature is in the range of 40–90°C. Depending on the water-to-chlorine ratio, the exothermic chlorohydrination reaction raises the temperature 10–40°C.

Epoxidation. The chlorohydrin is treated with aqueous base, eg, lime or caustic soda (possibly as chlorine-cell liquor). Approximately half of the base is consumed in neutralizing the by-product hydrogen chloride. The balance reacts with the propylene chlorohydrin:

$$CH_3\overset{\overset{\displaystyle OH}{|}}{C}HCH_2Cl + MOH \underset{K'}{\overset{K}{\rightleftharpoons}} CH_3CH\!\!\overset{\diagdown\;\;\diagup}{\underset{O}{-}}\!\!CH_2 + H_2O + MCl$$

or

$$CH_3\overset{\overset{\displaystyle Cl}{|}}{C}HCH_2OH$$

The equilibrium is far to the right. Removal or reaction of the propylene oxide drives the reaction further to the right. Excess alkalinity, ca 10% based on the amount of chlorohydrin and hydrogen chloride, is used (122–123). Intimate mixing of the chlorohydrin solution and the base reduces further reaction of propylene oxide to propylene glycol (124). After being mixed, the propylene oxide is removed from the alkaline solution in a stripping column. The reaction kinetics of dehydrochlorination of propylene chlorohydrin and a model of the propylene oxide stripping are developed in refs. 125–126. Propylene oxide is fed to the purification section and the water effluent to treatment.

Effluent Wastewater Treatment. Although not a process step, treatment and disposal of the large water effluent from the epoxidation section contributes to the commercial success of the chlorohydrin process (127). This stream contains 5–6 wt % calcium chloride or 5–10 wt % sodium chloride, several hundred ppm propylene glycol, and trace amounts of other organic compounds at a pH of about 11 (128). The costs associated with this stream are a serious drawback of the conventional chlorohydrin process. Several methods for removal of the organic contaminants have been proposed (129). Biological oxidation effectively reduces the organic contents to a concentration acceptable for discharge (130). Disposal of calcium or sodium chloride presents a problem and may have been a factor in the shutdown of two production units on inland waterways (131). Concentration of the sodium chloride for recycle to a chlorine plant or calcium chloride for sale is possible, but energy cost for recovery is high. Therefore, future conventional-technology chlorohydrin propylene oxide production will undoubtedly be located at sites where a treated brine can be discharged into the environment (118,132).

Purification. Several steps are needed to obtain propylene oxide of sufficient purity to be used in the production of polyurethane polyols. The overhead product from the propylene oxide stripper in the epoxidation section contains approximately 70% water, 26% propylene oxide, 3% propylene dichloride, and 1% other compounds (133). After condensation, the crude propylene oxide is sent to a distillation tower where propylene, inert components, and low-boiling organic compounds are removed. The underflow from this tower is sent to the finishing column where propylene oxide product is taken off overhead. Propylene dichloride, water, and others are removed in the bottoms stream. Propylene dichloride can be recovered for sales or fed to a chlorinated-solvents plant. To reduce hydrolysis of propylene oxide to propylene glycol, the distillation takes place under vacuum (133).

Modifications. Several modified chlorohydrin processes have been proposed (134–135), including the production of propylene oxide in a conventional chloralkali electrolysis cell (136–137). Such a cell eliminates the large dilute brine effluent and creates a closed-loop system. Another process is based on cation-exchange-membrane electrolysis cells (127). Commercialization of these processes is not known.

A closed-loop chlorohydrin process has been developed that would eliminate the dilute brine effluent and be useful in locations where salt, lime, or water is not readily available (134) (see Fig. 3) (138–139). *tert*-Butyl alcohol and chlorine, caustic, and water give *tert*-butyl hypochlorite and sodium chloride. Propylene chlorohydrin and *tert*-butyl alcohol are obtained in the second reaction step where the hypochlorite, propylene, and water react. The chlorohydrin is epoxidized to propylene oxide, much as in the conventional chlorohydrin process, and the alcohol is recycled. Sodium chloride, the coproduct in the hypochlorite formation and epoxidation reactions, is

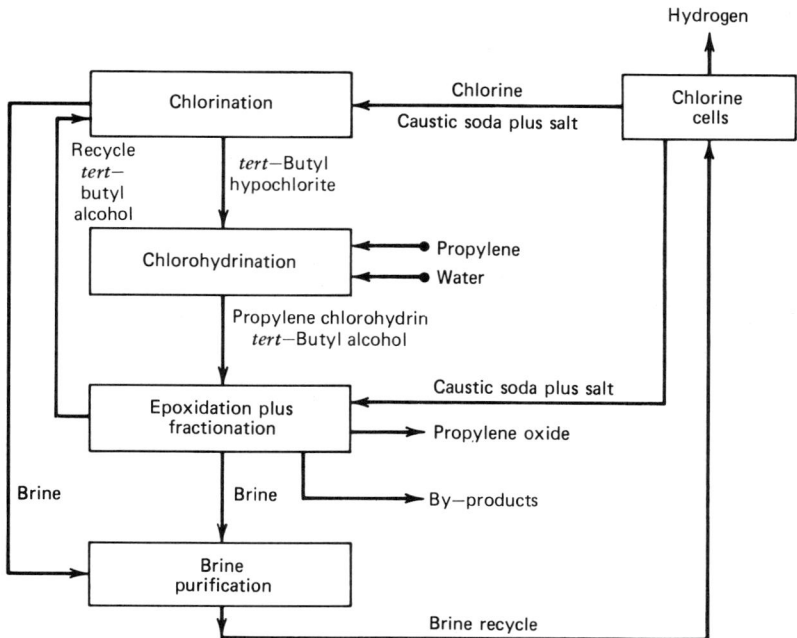

Figure 3. Hypochlorite chlorohydrin process developed by Lummus Industries (138–139).

recycled to the electrolysis chlorine cells after cleanup, without the need for evaporation (138). Yields are reported to be equal to those of the currently operating chlorohydrin processes (139).

Hydroperoxide Process. The other process for propylene oxide is based on the use of an organic hydroperoxide as an oxygen carrier to epoxidize propylene. It was developed by Halcon International and the Atlantic Richfield Corporation. They formed a third company, Oxirane, which carried out the commercial development of this process. Shell Chemical has developed a similar process that is now in operation in the Netherlands. Chemical syntheses with the use of organic peroxides are not novel (140). The epoxidation of propylene by an organic hydroperoxide is based on the following reactions (141):

$$RH + O_2 \longrightarrow ROOH$$

$$ROOH + CH_3CH{=}CH_2 \longrightarrow CH_3CH\underset{O}{\overset{}{-\!\!-}}CH_2 + ROH$$

At present, ethylbenzene and isobutane are being used industrially as the starting materials; thus, isobutane is oxidized to *tert*-butyl hydroperoxide:

$$(CH_3)_3CH + O_2 \rightarrow (CH_3)_3COOH$$

Some *tert*-butyl alcohol is also formed. The next step is the epoxidation of propylene in the presence of a metal catalyst:

$$(CH_3)_3COOH + CH_3CH{=}CH_2 \xrightarrow{M} CH_3CH\underset{O}{\overset{}{-\!\!-}}CH_2 + (CH_3)_3COH$$

tert-Butyl alcohol can be used as is or dehydrated to isobutylene.

With ethylbenzene, styrene is the coproduct. Ethylbenzene is first oxidized to ethylbenzene hydroperoxide: (1-phenylethane 1-hydroperoxide)

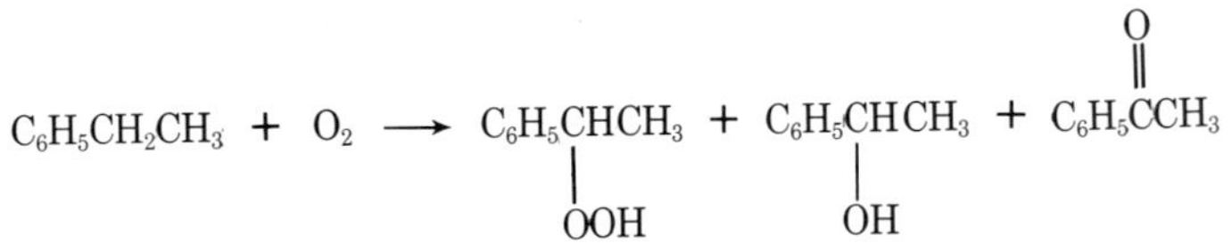

$$C_6H_5CH_2CH_3 + O_2 \rightarrow C_6H_5CH(OOH)CH_3 + C_6H_5CH(OH)CH_3 + C_6H_5C(O)CH_3$$

During this oxidation step, some by-product α-methylbenzyl alcohol and acetophenone may be formed (142). Propylene is then epoxidized with the hydroperoxide in the presence of a metal catalyst,

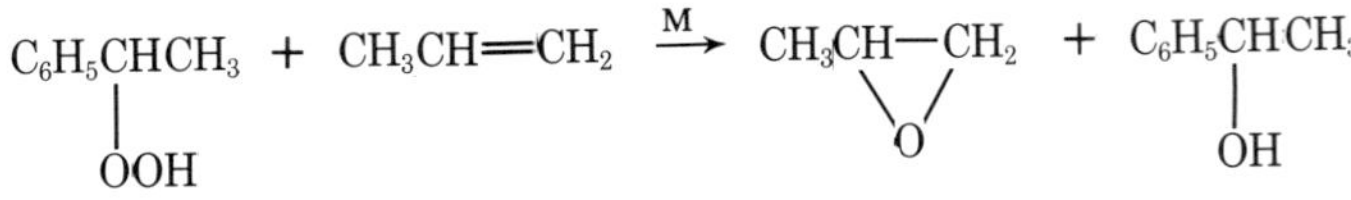

$$C_6H_5CH(OOH)CH_3 + CH_3CH{=}CH_2 \xrightarrow{M} CH_3CH{-}CH_2\text{(O)} + C_6H_5CH(OH)CH_3$$

and α-methylbenzyl alcohol is dehydrated to styrene:

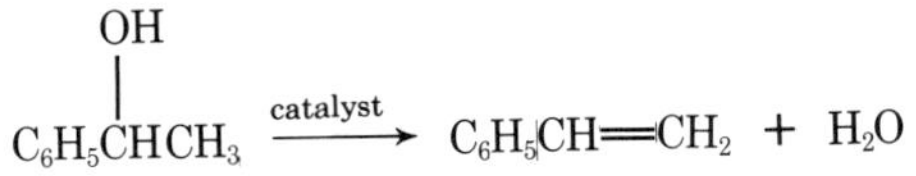

$$C_6H_5CH(OH)CH_3 \xrightarrow{catalyst} C_6H_5CH{=}CH_2 + H_2O$$

By-product acetophenone is hydrogenated to α-methylbenzyl alcohol and recycled.

Theoretically, one mole of coproduct is produced for each mole of propylene oxide. In practice, about three kg *tert*-butyl alcohol per kg propylene oxide is produced when starting with isobutane, whereas 2.5 kg of styrene is obtained with ethylbenzene (142). Both *tert*-butyl alcohol and styrene can be converted to the starting materials. This is economically attractive only when the value of propylene oxide less the cost of recycle is greater than values obtained by alternative means to produce isobutylene and styrene (127,132).

tert-Butyl Alcohol as Coproduct. A flow sheet of the propylene oxide–*tert*-butyl alcohol process is shown in Figure 4 (118,142–143). The various reaction parameters are covered by numerous patents. The first step, the liquid-phase air oxidation of isobutane, is carried out between 2.2 and 5.6 MPa (300 and 800 psig) at 110–150°C (144–146). The oxidation is done by either oxygen or air. In both cases, an inert recycle stream is needed to maintain nonexplosive conditions in the vapor space of the reactor. Patents suggest reaction times of 4–8 h (146). It could be assumed that, with the proper combination of temperature, pressure, and oxygen concentration, reaction time can be maintained in the 0.3–2 h range. Conversion of the isobutane is kept fairly low, 10–25%, to improve the selectivity to *tert*-butyl hydroperoxide. Further reaction of the hydroperoxide to *tert*-butyl alcohol increases with higher conversion. Isobutane from the reaction product is recovered by distillation and is recycled. The bottoms product, a mixture of *tert*-butyl hydroperoxide and *tert*-butyl alcohol, is fed to the epoxidation reactor.

Propylene epoxidation is carried out in a liquid-phase mixture of *tert*-butyl hydroperoxide, *tert*-butyl alcohol, propylene, and propylene oxide over a soluble metal catalyst, eg, molybdenum, titanium, vanadium, and others (147–148). Molybdenum naphthenate is often suggested. Reaction conditions can vary over a broad range: temperature, 80–130°C; contact time, 0.3–2 h; pressure, 1.8–7 MPa (250–1000 psig); and catalyst concentration, 0.001–0.006 mol per mole of hydroperoxide (149). Pro-

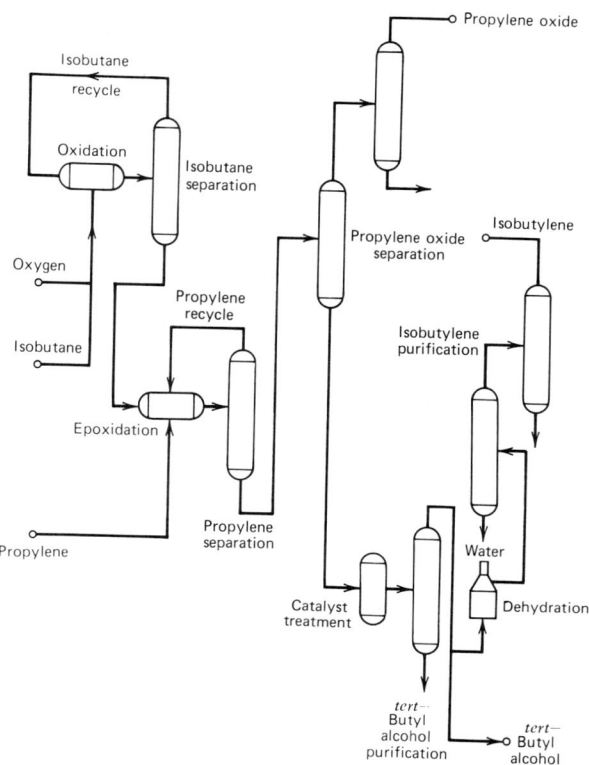

Figure 4. Propylene oxide–*tert*-butyl hydroperoxide process (118,142–143). Production of isobutylene is optional.

pylene comprised of make-up and recycle is fed in a ratio of 2–6 mol per mol hydroperoxide. Essentially 100% of the hydroperoxide is converted.

Unreacted propylene is first separated by distillation and then recycled. The product propylene oxide and *tert*-butyl alcohol along with high molecular weight material and catalyst are fed to a separation and purification train. Commercial-specification propylene oxide and gasoline-grade *tert*-butyl alcohol are taken off as overhead streams of the final finishing towers. *tert*-Butyl alcohol can be dehydrated to produce isobutylene, which has a substantial market (150). Hydrogenation of isobutylene gives the starting material isobutane. The metal catalyst from the epoxidation is contained, along with any high molecular weight material produced, in the bottoms from the *tert*-butyl alcohol purification. The molybdenum content of this stream represents most of the catalyst cost, and its recovery is a necessary part of the process.

Styrene as Coproduct. A flow sheet of the propylene oxide–styrene process is shown in Figure 5. Oxidation of ethylbenzene is similar to the isobutane/propylene oxidation. Ethylbenzene and oxygen are fed to a pressurized, liquid- and vapor-phase reactor where 15–25% ethylbenzene is converted per pass (151). High yields of hydroperoxide are obtained by maintaining a low ethylbenzene conversion. Reaction parameters are much the same as for the isobutane process. Temperature is controlled by evaporation of ethylbenzene and condensation in heat exchangers external to the oxidation reactor (152). Air or pure oxygen can be used. In either case, a nonexplosive condition in the reactor is maintained, and the oxygen concentration in the reactor is controlled at a level to maximize the yield of hydroperoxide. Ethylbenzene is distilled

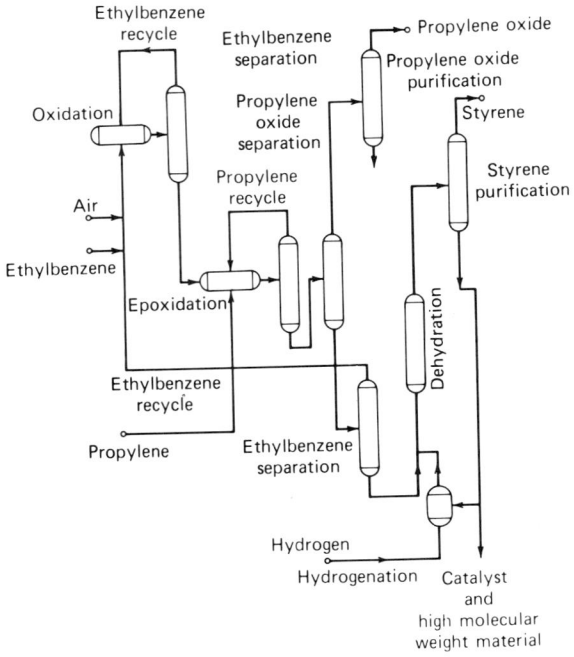

Figure 5. Propylene oxide–styrene hydroperoxide process (118,121,142).

from the reactor product and recycled. Ethylbenzene hydroperoxide concentration in the distillation tower is closely controlled, because this compound is highly reactive at concentrations over approximately 30 wt %.

After concentration of the ethylbenzene hydroperoxide, the crude mixture is fed together with propylene to the epoxidation reactor. A variety of catalysts, eg, compounds of molybdenum, titanium, tungsten, and similar materials, have been proposed for this step. Patent examples suggest that molybdenum and titanium are preferred (153–154). The catalyst form is the main difference between the Oxirane and the Shell processes. The former uses a soluble liquid-phase catalyst system, whereas Shell has numerous patents on a solid heterogeneous catalyst (155–157). Both systems operate at similar reaction conditions and obtain nearly equal results. Selectivity to the epoxide for the Shell system is improved by treatment of the catalyst with chlorotrimethylsilane (158). This heterogeneous system eliminates the catalyst separation and recovery procedures required in the Oxirane process. The various reactor parameters are given in Table 4. The epoxidation reactor product, which contains propylene oxide, propylene, ethylbenzene, α-methylbenzyl alcohol, acetophenone, and miscellaneous by-products, is separated in a distillation train. Propylene is recycled without further cleanup, followed by propylene oxide purification. Small amounts of acetaldehyde and propionaldehyde formed during the oxidation reaction are concentrated in the crude propylene oxide stream. A total aldehyde content of 20 ppm or less is desirable in propylene oxide for polyurethane production. To obtain such high purity material, a multitower distillation sequence is needed (159–160).

In the Oxirane process, the catalyst remains mixed with ethylbenzene and must be deactivated before further processing. Its acidic nature causes dehydration of the α-methylbenzyl alcohol to styrene, followed by polymerization. References 161–163 suggest hydrogenation and various organic and inorganic treatments of this stream, eg, with an aqueous base, to reduce loss of the α-methylbenzyl alcohol. A basic, aqueous solution of an alkali or alkaline earth metal hydroxide, carbonate, bicarbonate or oxide,

Table 4. Epoxidation of Propylene with Ethylbenzene Hydroperoxide

Reactor parameter	Oxirane[a]	Shell[b]
catalyst	molybdenum-liquid phase	titanium on silica gel
ethylbenzene hydroperoxide conversion, %	99	99
propylene conversion, %	10–20	10–20
pressure, MPa[c]	1.7–5.5	3.1
temperature, °C	100–130	80
contact time, h	1–3	0.4
liquid hourly space velocity per h		2
catalyst concentration, $\dfrac{\text{mol catalyst}}{\text{mol ethylbenzene hydroperoxide}}$	0.001–0.006	
propylene yield to propylene oxide, %	95	92–95

[a] Ref. 149.
[b] Ref. 158.
[c] To convert MPa to psi, multiply by 145.

ie, sodium hydroxide, would seem to be the preferred method. After catalyst removal, ethylbenzene is separated and recycled. The α-methylbenzyl alcohol–acetophenone mixture is then dehydrated to styrene in the vapor phase over a silica gel (164) or titanium dioxide catalyst (165) packed in tubes at 180–400°C. Styrene is recovered from the unreacted α-methylbenzyl alcohol and acetophenone by distillation. The acetophenone is hydrogenated and recycled. Reaction conditions of 80–130°C at ca 8.4 MPa (1200 psig) with the use of a copper-containing catalyst give >95% yield and conversion of acetophenone to α-methylbenzyl alcohol (166–167). Both Oxirane's and Shell's hydroperoxide processes are reported to produce about 2.8 kg styrene per kg of propylene oxide with yields ranging from ca 88 to ca 95%. These yields are somewhat above those reported in patents, but current operations should have benefited from 12 years' commercial experience (168).

Other Processes. *Peracetic Acid.* At present, the Japanese Daicel Ltd. peracetic acid route is the only other commercial propylene oxide process. Acetaldehyde, ethyl acetate, a metal ion catalyst, and air are mixed in a gas-sparged reactor to give peracetic acid (169). The product is concentrated to approximately 30% and fed to the epoxidation reactors.

$$CH_2{=}CHCH_3 + CH_3COOH \longrightarrow CH_3CH{-}CH_2\!\!\underset{O}{\diagdown\diagup} + CH_3COH$$

Propylene oxide and acetic acid are formed in a series of gas-sparged, tray-tower reactors. Propylene, acetic acid, and ethyl acetate are purified by distillation. Daicel has operated the process on a scale of 12,000 metric tons per year since 1969; the process seems economically feasible only under special circumstances (170).

Hydrogen Peroxide. The oxidation of propylene to propylene oxide with hydrogen peroxide has been studied for many years (171–173).

$$CH_2{=}CHCH_3 + H_2O_2 \xrightarrow{\text{catalyst}} CH_3CH{-}CH_2\!\!\underset{O}{\diagdown\diagup} + H_2O$$

Because of an increase of activity in this area, commercialization seems likely in the near future. From a chemical standpoint, hydrogen peroxide is an almost ideal oxidizer; few organic coproducts are produced, and the aqueous effluent is free of inorganic chlorides. However, relatively low selectivities of hydrogen peroxide to propylene oxide have hindered commercial development. Bayer and Degussa jointly developed a propylene oxide process based on hydrogen peroxide and propionic acid to the stage of consideration of a 100,000–150,000 t/yr plant. Propylox, the Interox–Carbochimique joint venture, is also working on this process. The principal reactions are formation of the peracid (peroxypropionic shown), followed by epoxidation of propylene:

$$CH_3CH_2\overset{O}{\overset{\|}{C}}OH + H_2O_2 \xrightarrow[\text{solvent}]{H^+} CH_3CH_2\overset{O}{\overset{\|}{C}}OOH + H_2O$$

$$CH_3CH_2\overset{O}{\overset{\|}{C}}OOH + CH_3CH{=}CH_2 \longrightarrow CH_3CH\underset{\underset{O}{\diagdown\diagup}}{\text{---}}CH_2 + CH_3CH_2\overset{O}{\overset{\|}{C}}OH$$

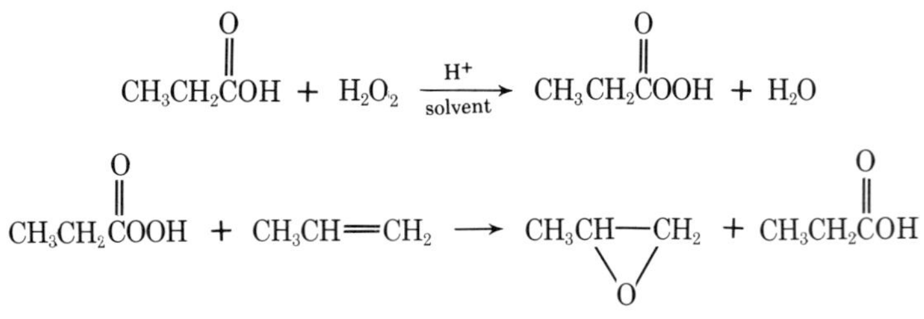

Oxidation of propionic acid is carried out in an inert solvent over an acid catalyst (174), followed by concentration of the peracid and epoxidation of propylene in a separate reactor. Propylene oxide and the acid are recovered and purified by distillation; the acid is recycled to the first step. Because of its high cost, maximum utilization of hydrogen peroxide is mandatory (45–75 ¢/kg, 100% basis, 1981). A patent to Interox (174) suggests a hydrogen peroxide to propylene oxide yield in the high 80% range.

Direct Oxidation. Liquid- and vapor-phase direct oxidation, catalytic and noncatalytic, of propylene to propylene oxide has been the subject of great interest (175–177), and numerous patents have been issued (178–182). However, these methods have not been commercialized because of low yields, low raw-material conversion, low productivities (kg propylene oxide/unit of catalyst or reactor volume), numerous byproducts, and complex catalyst systems. It seems that the formation of absorbed intermediates on the catalyst surface is the basic reason why vapor oxidation of propylene to propylene oxide does not proceed to high selectivities as does ethylene oxidation (183–184). Infrared spectra of absorbed species on a silver-on-silica catalyst established the presence of acrolein, which probably polymerizes and then oxidizes to carbon dioxide and water (183).

A process developed by Chem Systems (185–186) involves the reaction of propylene, oxygen, and acetic acid to form propylene glycol monoacetate. The monoacetate is cracked to yield propylene oxide and acetic acid for recycle. Minor amounts of propionaldehyde and acetone are formed. This process is in the pilot-plant stage.

Storage and Transportation

Storage, Materials of Construction. Carbon steel or stainless steel should be used for equipment in propylene oxide service. Teflon-filled spiral-wound stainless steel or glass-filled Teflon can be used for gaskets and seals. For pressures above 200 kPa (15 psig), storage tanks should be designed in accordance with the ASME code for unfired pressure vessels. All-welded construction is suggested. Propylene oxide tanks should be insulated, protected by sprinklers, diked, and electrically grounded. New equipment should be absolutely dry and free of air, ammonia, acetylene, hydrogen sulfide, rust, or other contaminants (187). A gas inert to the reactants, preferably nitrogen or methane, should be kept over the oxide during storage and transfer. Enough inert gas pressure should be maintained to keep the vapor phase out of the flammable range. The inert gas must be dry and free of impurities. Tanks and equipment should be inspected frequently for leaks (188).

Transportation. Propylene oxide is classified as a flammable liquid by the DOT. Surface and air shipments must show the red label, Flammable Liquid (189).

Large amounts may be transported by ship (190) or barge (191). In each case, the cargo must be under an inert-gas pad. Ships must be equipped with a cooling system in order to maintain the cargo at or below 40°C; barge cargos are shipped at ambient temperature.

Several types of rail cars are certified by the DOT for propylene oxide (192), eg, tank cars such as the DOT-111A100W4, DOT-105A100W, and DOT-111A100W1. Type DOT-MC330 and DOT-MC331 cargo tanks are examples of those currently in use for shipment by motor vehicle (192).

A 4-L package is the maximum amount of propylene oxide, in one package, that may be shipped on a cargo aircraft. Shipment in any quantity on a passenger aircraft is forbidden (193–194).

Small amounts may be shipped in cylinders that are prescribed for compressed gases other than acetylene, eg, drums of volumes from 22.7–45.4 L that conform to DOT Specifications 5B, 5C, and 42B.

Detailed information on the shipping and packaging of propylene oxide is given in references 189–194. Familiarity with applicable U.S. Government regulations is advised.

Economic Aspects

In the 1970s, several companies produced propylene oxide, including Jefferson, BASF, and Olin. However, after the introduction of the Oxirane process, only Dow and Oxirane remained in this market (see Table 5) (118,195).

Total U.S. production capacity for 1980 was ca 1.17×10^6 metric tons. Production is located along the Texas Gulf Coast near propylene supplies. Because of current excess capacity, no new producers are foreseen for the early 1980s.

Table 5. U.S. Propylene Oxide Capacities and Technology [a]

Producer	Location	Capacities, 1000 t/yr		Technology
		1979	Planned increase	
Dow	Freeport, Texas	436	364 by 1985	chlorohydrin
	Plaquemine, La.	155	45 by 1982	chlorohydrin
Oxirane	Bayport, Texas	398	45 by 1981	hydroperoxide, tert-butyl alcohol coproduct
	Channelview, Texas	180		hydroperoxide, styrene coproduct

[a] From refs. 118, 195.

United States production history is given in Table 6 (196–198). Propylene oxide manufacture consumes ca 13% of U.S. propylene production. Total world production outside the United States in 1979 was approximately 1.0×10^6 t with a capacity of approximately 1.9×10^6 t (196). More than 95% of the total production of propylene oxide is converted to derivatives. About 60% of the propylene oxide is used in the manufacture of urethane polyether polyols (197). Overall growth for propylene oxide is expected to be about 4% per year (197). This decline in growth from the 6–8% per year (198) in the late 1970s has been caused by steep cutbacks in the automobile and home markets. Propylene glycol accounts for 25% of propylene oxide consumption and is used mainly for unsaturated thermoset polyester resins with extended use in glass-fiber-reinforced plastics (see Polyesters, thermoplastic; Laminated and reinforced plastics). These plastics have been hard hit by the reduced demand for automobiles, boats, and appliances (197). Growth in propylene oxide will closely parallel the economic recovery in the United States and other industrial nations. As of 1981, propylene oxide plants are producing at 50–60% of capacity. Announced capacity increases for Dow and Oxirane may well be delayed to await improvement of the economy (131). The bulk price for propylene oxide has ranged from 50 to 62 ¢/kg over the past several years (197), with increases being tied to the cost advances of propylene.

Table 6. U.S. Propylene Oxide Production 1960–1980 [a]

Year	Production, 1000 metric tons
1960	141
1965	276
1970	536
1972	691
1974	798
1976	829
1978	907
1979	1040
1980	857

[a] Refs. 196–198.

Specifications and Storage

Propylene oxide is used mostly as an intermediate, thus its specifications vary slightly with the application. Typical sales specifications are listed in Table 7 (5).

Table 7. Typical Specifications for Propylene Oxide [a]

Assay	Value
density at 25/25°C, g/cm^3	0.829–0.831
acidity, as acetic acid, ppm, max	20
water, ppm, max	500
chlorides, as Cl, ppm, max	40
color, APHA, max	5
total aldehydes, ppm, max	100

[a] Ref. 5.

Commercial propylene oxide is of such high purity that no direct epoxide analysis is performed, and only the analyses of the impurities are reported.

Analysis. An 8-h time-weighted average (TWA) exposure to propylene oxide in air is determined by gas chromatography. Air is drawn through a charcoal-filled cartridge. The sample is then eluted with carbon disulfide, and an aliquot of the resulting solution is injected into the gas chromatograph for analysis. The method is validated in the range of 121–482 mg/m^3 (199).

In a similar gas chromatographic method, a porous polymer is used for the adsorption of the sample. The sample is eluted by placing the tube with the polymer directly into the carrier gas stream of the chromatograph, and the tube is heated to 200°C (200).

Colorimetric methods include the use of sodium periodate (143,201) and 4-(4-nitrobenzyl)pyridine (202). Direct (203) and indirect (204) titrations with hydrogen halides can also be employed.

Infrared (205–208), ^{1}H nmr (209), ^{13}C nmr (210), and chemical ionization and electron impact mass spectra have been recorded (211). A multigas analyzer that utilizes microwave spectroscopy has been developed in order to monitor propylene oxide and other gases in air (212).

Health and Safety Factors

The lower flammable (explosive) limit of propylene oxide is 2.3 vol % in air; the upper limit is 37 vol % (5). There is no difference in the terms flammable and explosive when they pertain to the upper and lower limits of flammability (213). The vapor density is 2.0; thus, the vapor may travel a long distance to a source of ignition and, flash back (214).

Dry chemical, alcohol foam, or carbon dioxide may be used to extinguish propylene oxide fires, whereas water is ineffective (213–214).

A leaking container of propylene oxide should be considered a total loss and should be immersed, if possible, in running water. If vapor cannot accumulate, the hazard of fire or explosion may be eliminated by adding a minimum of 36 parts water to 1 part epoxide at 10°C, or 144 parts water to 1 part epoxide at 38°C. Where there is a risk of exposure to propylene oxide, protective equipment should be worn, including a respirator, goggles, gloves, and boots (215).

Propylene oxide should not be stored in the presence of acids, bases, chlorides of iron, aluminum, and tin, or peroxides of iron and aluminum; any of these may cause violent polymerization. Even an uncontrolled reaction with water may become violent (213,215).

In processes that employ propylene oxide, the epoxide reservoir, piping, and valves should be constructed of steel or stainless steel. Explosion-proof electric motors should be used. Other equipment should be vapor-tight, sparkproof, or both (215).

A conscious victim of ingestion should be induced to vomit. Vomiting should be induced at least three times if possible. A tablespoonful of epsom salt in a glass of water should be given. The victim should receive medical attention promptly.

Physiological Effects. Propylene oxide is recognized as a toxic substance by NIOSH; its NIOSH number is TZ2975000 (216).

The permissible exposure limit (PEL) or the acceptable 8-h TWA exposure established by OSHA for propylene oxide in air is 100 ppmv or 240 mg/m^3 at 25°C and

101.3 kPa (760 mm Hg). Exposure to propylene oxide during any 8-h shift of a 40-h week should not exceed the PEL (217). The ACGIH has proposed a TLV of 20 ppmv, which will probably go into effect in 1982 (218).

Exposure to concentrations of propylene oxide significantly greater than the TLV may cause irritation of the eyes and upper respiratory tract (219), and drowsiness and weakness (220); oxygen may be administered (5). Early symptoms of acute exposure in laboratory animals include gasping and labored breathing, lachrymation, salivation, and nasal discharge (221). Chronic exposure to 457 ppm propylene oxide leads to moderate alveolar hemorrhage and edema, interstitial edema, and congestion of the lungs (222) in rats and guinea pigs.

The median detectable concentration of propylene oxide vapor is 200 ppm. The odor may indicate the danger of acute exposure (223).

Inhalation studies designed to determine the teratogenic risk (224), neurotoxicity (225–226), acute pulmonary response, and the long-term toxic dose-response of propylene oxide are in progress (227–228).

The lowest lethal concentration (LC_{Lo}) is 4000 ppm for four hours in mice and guinea pigs, and 2005 ppm for four hours in dogs (216). For other LC_{50} data, see refs. 229–230.

Generally, neat propylene oxide is not injurious to the skin if the compound is not restricted and can evaporate readily (223). Close contact with even dilute aqueous solutions can result in edema, blistering, and burns in six minutes (222). The LD_{50} for ingestion of propylene oxide through the skin is 1500 mg/kg in rabbits (216).

Gaseous propylene oxide irritates the eyes; the liquid causes severe injury to the eyes of laboratory rabbits (231). Eyes should be washed with water immediately after contact.

Laboratory rats survived after single oral doses of 0.3 g/kg and showed a normal weight gain thereafter (222). The oral LD_{50} is 930 mg/kg in rats (216).

Propylene oxide is mutagenic (232–235) in the Ames test (232,236). It is believed that mutagenesis arises as a result of alkylation of DNA by propylene oxide (237–239).

Propylene oxide is a suspected animal carcinogen (216). Tumors developed in laboratory rats (240) and mice (241) after subcutaneous injection in an oil base.

Uses

Propylene oxide is certified for use as a package fumigant for dried fruit and as a bulk fumigant for foodstuffs such as cocoa, spices, processed nutmeats, starch, and gums. It may also be used in admixture with carbon dioxide (242–243).

In admixture with a variety of other compounds, propylene oxide is a stabilizer for methylene chloride. Chlorinated hydrocarbons are stabilized with respect to color, pH, and solvent degradation (244–248). Fuel and heating oils are stabilized with respect to sludge and color formation by direct treatment with propylene oxide (250). Wood is made resistant to attack by termites (249).

Cotton fibers treated with propylene oxide exhibit improved moisture sorption and dye uptake (251) (see Textiles).

Slurries of crude polyolefins are freed of residual polymerization catalyst by treatment with propylene oxide or propylene oxide in combination with alcohols (252–255).

Propylene oxide can be used as an acid scavenger and pH control agent because of its reaction with hydrogen chloride to produce propylene chlorohydrin (256–257).

In conjunction with ethylene oxide, propylene oxide is a fuel–air explosive in munitions. The epoxide mixture is preferable to ethylene oxide alone with respect to storage, vapor pressure, and toxicity (258).

The use of propylene oxide in a mixture with propylene carbonate has been investigated as a solvent for lithium perchlorate in batteries (259).

Derivatives

Polyether Polyols. Polyether polyols are based on tri- and polyhydric alcohols (260). Polyols with mol wt of about 3000 or more are used to produce flexible polyurethanes, and polyols of about 300 to 1200 mol wt are used for rigid polyurethanes (261) (see Urethane polymers).

Propylene Glycol. Propylene glycol is used primarily in the production of unsaturated thermoset polyester resins (260–261). Unlike ethylene glycol, propylene glycol is nontoxic and is employed as a solvent in food, drug, and cosmetic preparations. It is included in the GRAS list published by the FDA and is considered to be a multipurpose food substance (260,262) (see Food additives). Propylene glycol is also used in antifreezes and hydraulic fluids (132).

Di- and Tripropylene Glycol. Commercial dipropylene glycol is a mixture of three isomers obtained in the production of propylene glycol; tripropylene glycol is also an isomeric mixture. Dipropylene glycol is used in hydraulic fluids (qv), cutting oils, textile lubricants, industrial soaps, as a solvent (263), and as an indirect food additive (264). Tripropylene glycol is used in cleansing creams, textile soaps, lubricants, and cutting-oil concentrates (263) (see Glycols).

Poly(Propylene Glycol)s. Poly(propylene glycol)s are low molecular weight liquids obtained from propylene oxide and water or propylene glycol; they range in mol wt from 400 to ca 4000. Their viscosity increases with molecular weight, and they are immiscible with water at mol wt of ca 500–1000. They are used as rubber lubricants and mold-release agents (see Abherents); lubricants in metal rolling, drawing and machining; antifoam agents; in hydraulic fluids (5); and deicing formulas for gasoline (265) (see Polyethers).

Surfactants. Propylene oxide is homopolymerized, or copolymerized, with ethylene oxide; water, alcohols, glycols, polyhydroxyl alcohols, amines, and acids are used as initiators to give surface-active agents. The hydrophobic/hydrophilic nature of the copolymer surfactants can be altered by changing the ratio of propylene oxide to ethylene oxide. An increase results in a more hydrophobic surfactant. These surfactants are used in a variety of ways, eg, in detergents, textiles, defoamers, hairdressings, and lubricants (266–273) (see Surfactants and detersive systems).

Glycol Ethers. Mono-, di-, and tripropylene glycol alkyl ethers are obtained from the reactions of the epoxide and an alcohol. Usually the alcohol is methanol or ethanol. These compounds are used as solvents for paints, resins, and inks. Owing to their miscibility in water, they can be used in aqueous solvent systems. They are also used in heat-transfer fluids and as chemical intermediates (274–275).

Isopropanolamines. Mono-, di-, and triisopropanolamines are formed from the reaction of ammonia and propylene oxide. The isomers are soluble in water, but become less soluble with increasing molecular weight. The isopropanolamines are used in conjunction with fatty acids as emulsifiers in cosmetics and as detergents and soaps. In the rubber industry, they are used as secondary vulcanization accelerators. They are also used in cyanide-free electroplating systems (276) (see Alkanolamines).

Other Derivatives. Propylene oxide is polymerized to obtain rubbers of very high molecular weight that have resistance to oil, hydrocarbons, and degradation by oxygen (277). Lower molecular weight block copolymers of propylene oxide and ethylene oxide serve as plasticizers in elastomers (278).

Derivatives have been used in the following: cyclic acetals from propylene oxide and low molecular weight aldehydes are employed as alkali-stable solvents (279); propylene carbonate is a solvent for organic and some inorganic compounds and also is used in gas conditioning for the removal of H_2S, CO_2, COS, and mercaptans (280); adducts of propylene oxide and boric acid are used as wood preservatives (281); propoxylation of photographic gelatin in films increases its silver capacity (282); an inhibitor consisting of the adduct of propylene oxide and thiourea significantly reduces the corrosion of ferrous metals by acids (283); and propylene oxide derivatives are also used in conjunction with organosilicon compounds to increase the bonding of glass fibers to polymers (284), and as scale inhibitors (285), chelating compounds (286), and catalysts for aldol reactions (287).

BIBLIOGRAPHY

"Propylene Oxide" in *ECT* 1st ed., under "Ethylene Oxide," Vol. 5, pp. 922–923, by R. S. Aries, Consulting Chemical Engineer, and Henry Schneider, R. S. Aries & Associates; "Propylene Oxide" in *ECT* 2nd ed., Vol. 16, pp. 595–609, by L. H. Horsley, The Dow Chemical Company.

1. B. Oser, *Bull. Soc. Chim. Fr.*, 235 (1860).
2. *Chem. Purchasing* **16**(7), 47 (1980).
3. *Chem. Eng. News* **58**(27), 11 (1980).
4. *Chem. Week* **127**(14), 47 (1980).
5. *Alkylene Oxides, Product Bulletin, Form 110-551-77R*, The Dow Chemical Co., Midland, Mich., 1977.
6. T. R. Bott and H. N. Sadler, *J. Chem. Eng. Data* **11**, 25 (1966).
7. *Propylene Oxide, Product Bulletin F-41180*, Union Carbide Corp., New York, Jan. 1965.
8. R. W. Gallant, *Hydrocarbon Process.* **46**(3), 143 (1967).
9. C. L. Yaws and M. P. Rackley, *Chem. Eng.* **83**, 429 (1976).
10. K. A. Kobe, A. E. Ravicz, and S. P. Vohra, *Ind. Eng. Chem.* **1**, 50 (1955).
11. M. T. Rogers, *J. Am. Chem. Soc.* **69**, 2544 (1947).
12. F. L. Oetting, *J. Chem. Phys.* **41**(1), 149 (1964).
13. G. C. Sinke and D. L. Hildenbrand, *J. Chem. Eng. Data* **1**(1), 74 (1962).
14. D. R. Stull, E. F. Westrum, Jr., and G. C. Sinke, *The Chemical Thermodynamics of Organic Compounds*, John Wiley & Sons, Inc., New York, 1969, p. 420.
15. J. N. Wickert, W. S. Tomplin, and R. L. Shank, *Chem. Eng. Prog. Symp. Ser.* **48**(2), 92 (1952).
16. G. O. Curme and F. Johnston, *Glycols, ACS Monograph No. 114*, Reinhold Publishing Corp., New York, 1952, Chapt. 11.
17. L. H. Horsley, ed., *Azeotropic Data, Advances in Chemistry Series*, No. 6, American Chemical Society, Washington, D.C., 1952; No. 35, 1962.
18. E. Abderhalden and E. Eichwald, *Chem. Ber.* **51**, 1312 (1918).
19. Y. Kumata, J. Furukawa, and T. Fueno, *Bull. Chem. Soc. Jpn.* **43**(12), 3920 (1970).
20. H. H. Kirchner, *Z. Phys. Chem. (Frankfurt am Main)* **39**, 273 (1963).
21. O. D. Shreeve and M. R. Heether, *Anal. Chem.* **23**, 271 (1951).

22. J. R. Villarreal and J. Laane, *J. Chem. Phys.* **62,** 303 (1975).
23. R. E. Parker and N. S. Isaacs, *Chem. Rev.* **59,** 737 (1959).
24. *Chem. Eng. News* **58,** 11 (1980).
25. J. H. Saunders and K. C. Frisch, *Polyurethanes: Chemistry and Technology, Part 1, Chemistry,* Vol. XVI of *High Polymers,* Interscience Publishers, a division of John Wiley & Sons, Inc., New York, 1962, pp. 32–34.
26. *Voranol Polyether Polyols, Product Bulletin, Form 194-45-75,* The Dow Chemical Co., Midland, Mich., 1975.
27. U.S. Pat. 3,370,056 (Feb. 20, 1968), M. Yotsuzuka and co-workers (to Takeda Chemical Industries, Ltd.).
28. U.S. Pat. 3,433,751 (Mar. 18, 1969), M. Yotsuzuka and co-workers (to Takeda Chemical Industries, Ltd.).
29. U.S. Pat. 4,026,941 (May 31, 1977), R. B. Login and co-workers (to BASF Wyandotte Corp.).
30. L. P. Blanchard, J. Singh, and M. D. Baijal, *Can. J. Chem.* **44,** 2679 (1966).
31. L. A. Dickinson, *J. Polym. Sci.* **58,** 857 (1962).
32. U.S. Pat. 4,166,172 (Aug. 28, 1979), H. P. Klein (to Texas Development Corp.).
33. U.S. Pat. 3,865,806 (Feb. 11, 1975), L. R. Knodel (to The Dow Chemical Co.).
34. U.S. Pat. 2,706,181 (Apr. 12, 1955), M. E. Pruitt and J. M. Baggett (to The Dow Chemical Co.).
35. G. Gee, W. C. E. Higginson, and J. B. Jackson, *Polymer* **3,** 231 (1962).
36. U.S. Pat. 3,338,873 (Aug. 29, 1967), A. E. Gurgiolo (to The Dow Chemical Co.).
37. M. Nakaniwa, K. Ozaki, and J. Furukawa, *Makromol. Chem.* **138,** 197 (1970).
38. U.S. Pat. 3,468,817 (Sept. 23, 1969), H. L. Hsieh (to Phillips Petroleum Co.).
39. K. Soga, H. Hyakkoku, and S. Ikeda, *J. Polym. Sci. Poly. Chem. Ed.* **17,** 2173 (1979).
40. S. Inoue, K. Kitamura, and T. Tsuruta, *Makromol. Chem.* **126,** 250 (1969).
41. U.S. Pat. 3,957,697 (May 18, 1976), R. K. Schlatzer (to The B. F. Goodrich Co.).
42. U.S. Pat. 3,398,126 (Aug. 20, 1968), M. J. Shrype (to Allied Chemical Corp.).
43. Brit. Pat. Application G.B. 2,023,601 A (Jan. 3, 1980), G. A. Taylor (to Union Carbide Corp.).
44. U.S. Pat. 4,160,116 (July 3, 1979), M. Mieno and co-workers (to Showa Denko, K.K.).
45. K. Slipko and J. Chlebicki, *Pol. J. Chem.* **53,** 2231 (1979).
46. Brit. Pat. Application GB 2,011,402 A (July 11, 1979), C. H. McMullen and co-workers (to Union Carbide Corp.).
47. W. J. Peppel, *Ind. Eng. Chem.* **50,** 767 (1958).
48. A. J. Gait in E. G. Hancock, ed., *Propylene and Its Industrial Derivatives,* Halsted Press, a division of John Wiley & Sons, Inc., New York, 1973, p. 285.
49. N. Adachi, Y. Kida, and K. Shikada, *J. Polym. Sci. Polym. Chem. Ed.* **15,** 937 (1977).
50. U.S. Pat. 3,542,808 (Nov. 24, 1970), R. C. Vander Linden and co-workers (to Esso Research and Engineering Co.).
51. U. A. Durden, Jr., H. A. Stansbury, Jr., and W. H. Catlette, *J. Am. Chem. Soc.* **82,** 3082 (1960).
52. F. Scholnick, H. A. Monroe, Jr., E. J. Saggese, and A. N. Wrigley, *J. Am. Oil Chem. Soc.* **44,** 40 (1967).
53. J. D. Malkemus, *J. Am. Oil Chem. Soc.* **33,** 571 (1956).
54. J. B. Lewis and G. W. Hedrick, *Ind. Eng. Chem. Prod. Res. Dev.* **9,** 304 (1970).
55. U.S. Pat. 4,017,929 (Apr. 22, 1977), R. B. Steele and A. Katzakian, Jr. (to Aerojet General Corp.).
56. U.S. Pat. 3,355,434 (Nov. 28, 1967), J. G. Milligan (to Jefferson Chemical Company, Inc.).
57. J. Chlebicki, *Roczniki Chemii* **48,** 1241 (1974).
58. *Ibid.,* **49,** 207 (1975).
59. U.S. Pat. 3,843,706 (Oct. 22, 1974), J. K. Weil and A. J. Stirton (to The United States of America as represented by the Secretary of Agriculture).
60. A. Rosowsky in A. Weissberger, ed., *Heterocyclic Compounds,* Vol. 19, Pt. 1, Wiley-Interscience, New York, 1964, pp. 289, 308.
61. U.S. Pat. 4,118,426 (Oct. 3, 1978), N. C. Holy and T. E. Nalesnik (to Western Kentucky University).
62. Ref. 60, p. 328.
63. Ref. 60, p. 331.
64. Ref. 60, p. 337.
65. M. Prochazka and P. Durdovic, *Coll. Czech. Chem. Comm.* **42,** 2401 (1976).
66. U.S. Pat. 3,440,287 (Apr. 22, 1969), L. H. Horsley (to The Dow Chemical Co.).
67. U.S. Pat. 4,031,023 (June 21, 1977), J. L. Musser and F. W. Koch (to The Lubrizol Corp.).

68. U.S. Pat. 3,317,508 (May 2, 1967), A. D. Winquist and L. F. Theiling (to Union Carbide Corp.).
69. U.S. Pat. 3,476,598 (Nov. 4, 1969), H. L. Sanders (to Varney Chemical Corp.).
70. U.S. Pat. 3,890,300 (June 17, 1975), M. Huchette and G. Fleche (to Roquette Freres).
71. A. N. Wrigley, F. D. Smith, and A. J. Stirton, *J. Am. Oil Chem. Soc.* **36,** 34 (1959).
72. M. A. Rousselle, M. L. Nelson, H. H. Ramey, Jr., and G. L. Barker, *Ind. Eng. Chem. Prod. Res. Dev.* **19,** 654 (1980).
73. U.S. Pat. 3,661,782 (May 9, 1972), M. K. Smith (to The Baker Castor Oil Co.).
74. U.S. Pat. 3,591,573 (July 6, 1971), J. B. Class (to Hercules Inc.).
75. M. Inoue, K. Chano, O. Itoh, T. Sugita, and K. Ichikawa, *Bull. Chem. Soc. Jpn.* **53,** 458 (1980).
76. Ref. 60, p. 433.
77. Ref. 60, pp. 399–400.
78. N. G. Gaylord and E. I. Becker, *Chem. Rev.* **49,** 413 (1951).
79. R. E. Parker and N. S. Isaacs, *Chem. Rev.* **59,** 737 (1959).
80. S. Winstein and R. B. Henderson in R. C. Elderfield, ed., *Heterocyclic Compounds*, Vol. 1, John Wiley & Sons, Inc., New York, 1950, p. 57.
81. C. Huynh, F. Derquini-Boumechal, and G. Linstrumelle, *Tetrahedron Lett.* (17), 1503 (1979).
82. Ref. 48, p. 285.
83. T. Imanaka, Y. Okamoto, and S. Teranishi, *Bull. Chem. Soc. Jpn.* **45,** 3251 (1972).
84. *Ibid.*, 1353 (1972).
85. U.S. Pat. 4,065,510 (Dec. 27, 1977), G. Schreyer and co-workers (to Deutsche Gold- und Silber Scheideanstalt Vormals Roessler).
86. M. J. Astle, *The Chemistry of Petrochemicals*, Reinhold Publishing Corp., New York, 1956, pp. 184–185.
87. Ref. 60, p. 320.
88. J. F. W. McOmie, *Protective Groups in Organic Chemistry*, Plenum Publishing Corp., London, UK, 1973, p. 326.
89. U.S. Pat. 3,725,438 (Apr. 3, 1973), B. J. Barone and W. F. Brill (to PetroTex Chemical Corp.).
90. J. L. E. Erickson and F. E. Collins, Jr., *J. Org. Chem.* **30,** 1050 (1965).
91. Ref. 60, p. 181.
92. E. L. Eliel and M. N. Rerick, *J. Am. Chem. Soc.* **82,** 1362 (1960).
93. R. W. Herr, D. M. Wieland, and C. R. Johnson, *J. Am. Chem. Soc.* **92,** 3813 (1970).
94. W. P. Giering, M. Rosenblum, and J. Tancrede, *J. Am. Chem. Soc.* **94,** 7170 (1972).
95. M. Berry, S. G. Davies, and M. L. H. Green, *J. Chem. Soc. Chem. Comm.*, 99 (1978).
96. Ref. 60, pp. 350–351.
97. Ref. 60, p. 365.
98. Ref. 60, p. 346.
99. G. S. Yoneda, M. T. Griffin, and D. W. Carlyle, *J. Org. Chem.* **40,** 375 (1975).
100. Brit. Pat. 1,560,884 (Feb. 13, 1980), W. Reuther and co-workers (to BASF Aktiengesellschaft).
101. H. Steinberg, *Organoboron Chemistry*, Vol. 1, Interscience Publishers, a division of John Wiley & Sons, Inc., New York, 1964, pp. 82, 520, 549.
102. J. D. Edwards, W. Gerrard, and M. F. Lappert, *J. Chem. Soc.*, 348 (1957).
103. F. G. A. Stone and H. J. Emeleus, *J. Chem. Soc.*, 2755 (1950).
104. Ref. 101, p. 487.
105. Ref. 60, p. 384.
106. N. I. Shuikin and I. F. Bel'skii, *J. Gen. Chem. USSR* **29,** 2936 (1959).
107. Ref. 60, p. 450.
108. L. Carbonnel and J.-C. Rosso, *J. Solid State Chem.* **8,** 304 (1973).
109. J. L. Down, J. Lewis, B. Moore, and G. Wilkinson, *J. Chem. Soc.*, 3767 (1959).
110. J. A. Orvik, *J. Am. Chem. Soc.* **98,** 3322 (1976).
111. D. G. Parsons, *J. Chem. Soc. Perkin Trans. I*, 245 (1975).
112. W. Kuran, S. Pasynkiewicz, and J. Serzyko, *J. Organomet. Chem.* **73,** 187 (1974).
113. A. J. Lundeen and A. C. Oehlschlager, *J. Organomet. Chem.* **25,** 337 (1970).
114. N. Watanabe, S. Uemura, and M. Okano, *Bull. Chem. Soc. Jpn.* **52,** 3611 (1979).
115. Ref. 60, pp. 340–342, 421, 428, 430, and 432.
116. Ref. 60, pp. 440, 442–443, 451, and 456.
117. J. F. Norris, *J. Ind. Eng. Chem.* **11,** 817 (1919).
118. R. Landau and co-workers, *Chemtech* **9**(10), 602 (1979).
119. A. C. Fyvie, *Chem. Ind. London* (10), 384 (1964).

120. Ref. 48, pp. 276–277.
121. R. B. Stobaugh and co-workers, *Hydrocarbon Process.* **52**(1), 102 (1973).
122. Jpn. Pat. 73-34,809 (May 22, 1973), K. Suzuki and co-workers (to Asahi Glass Co., Ltd.).
123. Ref. 119, p. 385.
124. Ger. Pat. 2,056,198 (May 25, 1972), E. Bartholome and co-workers (to Badische Anilin- und Soda-Fabrik A-G.).
125. S. Carr and co-workers, *Chem. Eng. Sci.* **34**(9), 1123 (1979).
126. Ref. 125, pp. 1133–1140.
127. K. H. Simmrock, *Hydrocarbon Process.* **57**(11), 110 (1978).
128. M. A. Zeitoun and W. F. McIlhenny, *Treatment of Wastewater from the Production of Polyhydric Organics*, EPA Project, 12020-EEQ, 1971, pp. 11–24.
129. *Ibid.*, pp. 7–9.
130. *Ibid.*, pp. 1–3.
131. *Chem. Week* **127**(14), 47 (1980).
132. K. Weissermel and H.-J. Arpe, *Industrial Organic Chemistry*, trans. by Alexander Mullen, Verlag, Chemie, New York, 1978, p. 244.
133. Ref. 121, p. 100.
134. U.S. Pat. 4,008,133 (Feb. 15, 1977), J. T. Kwon and A. P. Gelbein (to The Lummus Co.).
135. Ger. Pat. 2,658,189 (May 24, 1978), K. H. Simmrock and G. Hellemanns (to Metallgesellschaft A.-G.).
136. U.S. Pat. 3,427,235 (Feb. 11, 1969), J. A. M. LeDue (to Pullman Inc.).
137. Ger. Pat. 1,290,926 (Mar. 20, 1969), W. Kroenig and P. Konard (to Fabenfabriken Bayer A.-G.).
138. U.S. Pat. 4,126,526 (Nov. 21, 1978), J. T. Kwan and A. P. Gelbein (to The Lummus Co.).
139. *Hydrocarbon Process.* **59**(11), 239 (1979).
140. Ref. 118, p. 606.
141. Ref. 118, p. 604.
142. J. Poloczek and J. Bobinski, *Int. Chem. Eng.* **11**(1), 87 (1971).
143. F. E. Critchfield and J. B. Johnson, *Anal. Chem.* **29**, 797 (1957).
144. U.S. Pat. 3,445,523 (Sept. 23, 1975), H. R. Grane (to Atlantic Richfield Co.).
145. U.S. Pat. 4,128,587 (Dec. 5, 1978), J. C. Jubin (to Atlantic Richfield Co.).
146. U.S. Pat. 3,478,108 (Nov. 11, 1969), H. R. Grane (to Atlantic Richfield Co.).
147. U.S. Pat. 3,362,972 (Jan. 8, 1968), J. Kollar (to Halcon International, Inc.).
148. U.S. Pat. 3,507,809 (Apr. 21, 1970), J. Kollar (to Halcon International, Inc.).
149. Ref. 142, p. 85.
150. Ref. 118, p. 605.
151. U.S. Pat. 3,459,810 (Aug. 5, 1969), C. Y. Choo and R. L. Golden (to Halcon International, Inc.).
152. U.S. Pat. 4,066,706 (Jan. 3, 1978), J. P. Schmidt (to Halcon International, Inc.).
153. U.S. Pat. 3,351,635 (Nov. 7, 1967), J. Kollar (to Halcon International, Inc.).
154. U.S. Pat. 3,849,451 (Nov. 19, 1974), T. W. Stein and co-workers (to Halcon International, Inc.).
155. U.S. Pat. 3,634,464 (Jan. 11, 1972), H. P. Wulff and F. Wattlemenn (to Shell Oil Co.).
156. U.S. Pat. 3,829,392 (Aug. 13, 1974), H. P. Wulff (to Shell Oil Co.).
157. U.S. Pat. 3,702,855 (Nov. 14, 1972), C. S. Bell and H. P. Wulff (to Shell Oil Co.).
158. U.S. Pat. 3,642,833 (Feb. 15, 1972), H. Wulff and P. Haynes (to Shell Oil Co.).
159. U.S. Pat. 3,632,482 (Jan. 4, 1972), S. E. Heary and S. F. Newmann (to Shell Oil Co.).
160. U.S. Pat. 3,881,996 (May 6, 1975), J. P. Schmidt (to Oxirane Corp.).
161. U.S. Pat. 3,860,662 (Jan. 14, 1975), J. Kollar (to Halcon International, Inc.).
162. U.S. Pat. 3,947,500 (Mar. 30, 1976), J. Kollar (to Halcon International, Inc.).
163. U.S. Pat. 3,988,363 (Oct. 26, 1976), J. P. Schmidt (to Halcon International, Inc.).
164. U.S. Pat. 4,049,736 (Sept. 20, 1977), J. J. Lamson and co-workers (to The Dow Chemical Co.).
165. U.S. Pat. 3,442,963 (May 6, 1969), E. I. Korchak (to Halcon International, Inc.).
166. U.S. Pat. 3,927,720 (Dec. 16, 1975), H. R. Grane and T. S. Zak (to Atlantic Richfield Co.).
167. U.S. Pat. 3,927,121 (Dec. 16, 1975), H. R. Grane and T. S. Zak (to Atlantic Richfield Co.).
168. Ref. 118, p. 607.
169. K. Yamagishi and co-workers, *Hydrocarbon Process.* **55**(11), 102 (1976).
170. Ref. 169, p. 103.
171. Jpn. Pat. 78,137,903 (Dec. 1, 1978), H. Miyamori and H. Mosai (to Mitsubishi Gas Chem. Co., Inc.).
172. U.S. Pat. 4,024,165 (May 17, 1977), T. M. Shryne and L. Kim (to Shell Oil Co.).

173. Ger. Pat. 2,605,041 (Aug. 26, 1976), M. Pralus and co-workers (to Ugine Kohlmann).
174. U.S. Pat. 4,177,196 (Dec. 4, 1979), A. M. Hildon and P. F. Greenholgh (to Interox Chemicals Limited).
175. *Chem. Eng. News* **50**(47), 22 (1972).
176. C. Daniel, J. R. Mannier, and G. W. Keulks, *J. Catal.* **31**, 360 (1973).
177. N. W. Cant and W. K. Hall, *J. Catal.* **52**, 84 (1978).
178. Brit. Pat. 1,373,489 (Nov. 13, 1974), P. Hayden and co-workers (to Imperial Chemical Industry, Ltd.).
179. U.S. Pat. 4,046,783 (Sept. 6, 1977), S. B. Cavitt (to Texaco Development Corp.).
180. Ger. Pat. 2,332,285 (Jan. 17, 1974), A. J. Kolombas and co-workers (to British Petroleum International Ltd.).
181. Ger. Pat. 2,336,396 (Feb. 7, 1974), D. Bryce-Smith and E. T. Blues.
182. U.S. Pat. 3,316,279 (Apr. 25, 1967), D. M. Fenton (to Union Oil of Calif.).
183. I. L. C. Freriks, R. Bowman, and P. V. Gennen, *J. Catal.* **65**, 311 (1980).
184. C. Daniel and G. W. Kevlkr, *J. Catal.* **24**, 529 (1972).
185. U.S. Pat. 4,012,424 (Mar. 15, 1977), M. B. Sherwin and J. Perers (to Chem Systems Inc.).
186. Ger. Pat. 2,632,158 (Jan. 20, 1977), M. B. Sherwin and I. Der Hvang (to Chem Systems Inc.).
187. Ref. 5, p. 22.
188. Ref. 5, p. 19.
189. *U.S. Code of Federal Regulations*, Title 49, paragraphs 172.101, 172.407, 172.419, 173.115, U.S. Government Printing Office, Washington, D.C., 1979.
190. *U.S. Code of Federal Regulations*, Title 46, paragraph 153, U.S. Government Printing Office, Washington, D.C., 1979.
191. Ref. 190, paragraph 151.
192. Ref. 190, Title 49, paragraph 173.119.
193. Ref. 192, paragraph 172.
194. *Restricted Articles Regulations*, 23 ed., International Air Transport Association, Montreal, Quebec, Canada, 1980.
195. *Chemical Profiles*, *Propylene Oxide*, Schnell Publishing Co. Inc., New York, 1979.
196. Selected Issues of *Chem. Mark. Rep.* (1980).
197. *Chem. Eng. News* **58**(27), 11 (1980).
198. *Chem. Eng. News* **58**(18), 35 (1980); **58**(27), 11 (1980).
199. *NIOSH Manual of Analytical Methods*, Vol. 2, 2nd ed., U.S. Department of Health, Education and Welfare, Washington, D.C., Apr. 1977, Method No. S-75.
200. J. W. Russel, *Environmental Sci. Technol.* **9**, 1175 (1975).
201. H. E. Mishmash and C. E. Meloan, *Anal. Chem.* **44**, 835 (1972).
202. S. C. Agarwal, B. L. Van Duuren, and T. J. Kneip, *Bull. Environm. Contam. Toxicol.* **23**, 825 (1979).
203. R. Dijkstra and E. A. M. F. Dahmen, *Anal. Chim. Acta* **31**, 38 (1964).
204. J. L. Jungnickel, E. D. Peters, A. Polgar, and F. T. Weiss in J. Mitchell, Jr., I. M. Kolthoff, E. S. Prokaver, and A. Weissberger, eds., *Organic Analysis*, Vol. I, Interscience Publishers, Inc., New York, 1953, p. 127.
205. O. D. Shreve and M. R. Heether, *Anal. Chem.* **23**, 277 (1951).
206. J. R. Villarreal and J. Laane, *J. Chem. Phys.* **62**, 303 (1975).
207. Sadtler Research Laboratories, 3316 Spring Garden St., Philadelphia, Pa., IR Spectrum No. 15270.
208. Ref. 207, Vapor-phase IR Spectrum No. 49.
209. Ref. 207, NMR Spectrum No. 10797.
210. Ref. 207, Carbon-13 NMR Spectrum No. 1824.
211. S. Suzuki, Y. Hori, R. C. Das, and O. Koga, *Bull. Chem. Soc. Jpn.* **53**, 1451 (1980).
212. L. H. Hrubesh, A. S. Maddux, D. C. Johnson, R. L. Morrison, J. N. Nielsen, and M. Malachosky, *Operational Manual for Microwave Multi-gas Analyzer*, UCID-17751, Lawrence Livermore Laboratory, prepared for NIOSH under contract No. IA-77-17, available from National Technical Information Service, Springfield, Va., 1978.
213. *National Fire Codes, A Compilation of NFPA Codes, Standards, Recommended Practices, and Manuals*, Volume 12, National Fire Protection Association, Boston, Mass., 1979, p. 325M-6.
214. Ref. 213, Volume 13, pp. 49–225.
215. Ref. 5, p. 18.

216. *Registry of Toxic Effects of Chemical Substances* (*RTECS*), U.S. Department of Health and Human Services, NIOSH, Cincinnati, Oh., 1980.
217. *U.S. Code of Federal Regulations*, Title 29, paragraph 1910.1000, U.S. Government Printing Office, Washington, D.C., 1980, p. 589.
218. *TLVs® Threshold Limit Values for Chemical Substances and Physical Agents in the Workroom Environment with Intended Changes for 1980*, ACGIH, Cincinnati, Oh., 1980, p. 38.
219. N. H. Proctor and J. P. Hughes, *Chemical Hazards of The Workplace*, J. B. Lippincott Company, Philadelphia, Pa., 1978, p. 430.
220. *Encyclopedia of Occupational Health and Safety*, International Labour Office, McGraw-Hill Book Company, New York, 1971, p. 468.
221. K. H. Jacobson, E. B. Hackley, L. Feinsilver, *AMA Arch. Ind. Health* **13,** 237 (1958).
222. V. K. Rowe, R. L. Hollingsworth, F. Oyen, D. D. McCollister, and H. C. Spencer, *AMA Arch. Ind. Health* **13,** 228 (1956).
223. C. H. Hine and V. K. Rowe in F. A. Patty, ed., *Industrial Hygiene and Toxicology*, 2nd Revised Edition, Vol. II, Interscience Publishers, New York, 1958, p. 1642.
224. R. Niemeir, *Toxicology Research Projects Directory*, 4(2), 1–46 (1979).
225. B. Johnson, *Toxicology Research Projects Directory*, 5(3), 1–14 (1980).
226. *Ibid.*, 1–48 (1980).
227. T. Lewis, *Toxicology Research Projects Directory*, 4(3), 1–48 (1979).
228. W. Moorman in ref. 225.
229. A. L. Bridie, C. J. M. Wolff, and M. Winter, *Water Res.* **13,** 623 (1979).
230. R. C. Crews, *Effects of Propylene Oxide on Selected Species of Fishes*, AFATL-TR-74-183, Air Force Armament Laboratory, 1974, available from the National Technical Information Service, Springfield, Va.
231. C. P. Carpenter and H. P. Smyth, *Am. J. Opthalmol.* **29,** 1363 (1946).
232. D. R. Wade, S. C. Airy, and J. E. Sinsheimer, *Mutat. Res.* **58,** 217 (1978).
233. R. E. McMahon, J. C. Cline, and C. Z. Thompson, *Cancer Res.* **39,** 682 (1979).
234. G. Kolmark and N. H. Giles, *Genetics* **40,** 890 (1955).
235. J. Bootman, D. C. Lodge, and H. E. Whalley, *Mutat. Res.* **67,** 101 (1979).
236. E. H. Pfeiffer and H. Dunkelberg, *Food Cosmet. Toxicol.* **18,** 115 (1980).
237. K. Hemminki, J. Paasivirta, T. Kurkirinne, and L. Virkki, *Chem.-Biol. Interactions* **30,** 259 (1980).
238. B. Singer, *J. Toxicol. Environm. Health* **2,** 1279 (1977).
239. K. Hemminki, *Chem.-Biol. Interactions* **28,** 269 (1979).
240. A. L. Walpole, *Ann. N.Y. Acad. Sci.* **68,** 750 (1958).
241. H. Dunkelberg, *Br. J. Cancer* **39,** 588 (1979).
242. *U.S. Code of Federal Regulations*, Title 21, paragraph 193.380, U.S. Government Printing Office, Washington, D.C., 1980.
243. U.S. Pat. 3,919,189 (Nov. 11, 1975), R. A. Empey and D. J. Pettitt (to Kelco Company).
244. U.S. Pat. 3,670,036 (June 13, 1972), T. A. Vivian (to The Dow Chemical Co.).
245. U.S. Pat. 3,887,628 (June 3, 1975), N. L. Beckers (to Diamond Shamrock Corp.).
246. U.S. Pat. 3,900,524 (Aug. 19, 1975), N. L. Beckers (to Diamond Shamrock Corp.).
247. U.S. Pat. 4,108,910 (Aug. 22, 1978), M. Godfroid and R. Gerkens (to Solvay and Cie).
248. U.S. Pat. 4,032,584 (June 28, 1977), M. R. Irani (to Stauffer Chemical Co.).
249. R. M. Rowell, S. V. Hart, and G. R. Esenther, *Wood Science* **11,** 271 (1979).
250. U.S. Pat. 3,425,814 (Feb. 4, 1969), W. J. Mattox (to Esso Research and Engineering Co.).
251. M.-A. Rouselle, M. L. Nelson, H. H. Ramey, Jr., and G. L. Barker, *Ind. Eng. Chem. Prod. Res. Dev.* **19,** 654 (1980).
252. Jpn. Kokai 77 32,089 (Mar. 10, 1977), G. Kakokawa and co-workers (to Mitsubishi Chemical Industries Co., Ltd.).
253. Jpn. Kokai 80 23,171 (Feb. 19, 1980), Y. Takemoto and co-workers (to Sumitomo Chemical Co., Ltd.).
254. Jpn. Kokai 80 07,845 (Jan. 21, 1980), H. Katada and co-workers (to Sumitomo Chemical Co., Ltd.).
255. Jpn. Kokai 80 09,653 (Jan. 23, 1980), H. Katada and co-workers (to Sumitomo Chemical Co., Ltd.).
256. U.S. Pat. 3,901,316 (Aug. 26, 1975), R. H. Knapp (to Shell Oil Co.).
257. U.S. Pat. 3,925,437 (Dec. 9, 1975), R. L. Rowton (to Jefferson Chemical Co., Inc.).
258. U.S. Pat. 4,157,928 (June 12, 1979), C. W. Falterman and co-workers (to U.S.A. as represented by the Secretary of the Navy).

259. U.S. Pat. 3,960,595 (June 1, 1976), G. Lehmann and T. Rassinoux (to Saft-Société des Accumulateurs Fixés et de Traction).
260. *Chem. Eng. News* **58**(27), 11 (1980).
261. Ref. 48, p. 273.
262. *U.S. Code of Federal Regulations*, Title 21, paragraphs 170.20–170.38, 177.2600, 182.1, 182.90, 182.1666, 182.4666, U.S. Government Printing Office, Washington, D.C., 1980.
263. *Properties and Uses of Glycols*, Product Bulletin, Form No. 110-285-74, The Dow Chemical Co., Midland, Mich., 1974.
264. *U.S. Code of Federal Regulations*, Title 21, paragraphs 175.105, 175.320, 176.170, 176.200, 177.2420, U.S. Government Printing Office, Washington, D.C., 1980.
265. Brit. Pat. 1,418,526 (Dec. 24, 1974), B. H. Garth (to E. I. du Pont de Nemours & Co., Inc.).
266. L. E. St. Pierre and A. S. Kastens in N. G. Gaylord, ed., *High Polymers, Vol. XIII, Polyethers, Part I*, Interscience Publishers, a division of John Wiley & Sons, Inc., New York, 1963, Chapts. 3–4.
267. U.S. Pat. 3,456,013 (July 15, 1969), R. R. Egan and L. D. Smiens (to Ashland Oil and Refining Company).
268. U.S. Pat. 3,956,401 (May 11, 1976), M. Scardera and R. N. Scott (to Olin Corp.).
269. U.S. Pat. 3,830,627 (Aug. 20, 1974), M. Daeuble and co-workers (to Badische Anilin-und Soda Fabrik Aktiengesellschaft).
270. U.S. Pat. 4,151,269 (Apr. 24, 1979), K. Torii and K. Tomita (to Shieseido Co., Ltd.).
271. U.S. Pat. 4,060,501 (Nov. 29, 1977), C. G. Naylor and E. L. Yeaky (to Texaco Development Corp.).
272. U.S. Pat. 3,892,522 (July 1, 1975), W. Schade and co-workers (to Henkel and Cie, GmbH).
273. U.S. Pat. 3,927,104 (Dec. 16, 1975), E. F. Miller and W. W. Hellmuth (to Texaco, Inc.).
274. Ref. 261, p. 288.
275. *Dowanol Solvents*, Product Bulletin, Form No. 125-154-68, The Dow Chemical Co., Midland, Mich., 1968.
276. *Alkanolamines From Dow*, Product Bulletin, Form No. 118-428-75, The Dow Chemical Co., Midland, Mich., 1975.
277. E. W. Duck and B. J. Ridgewell, *Chem. Ind. London*, 254 (1969).
278. U.S. Pat. 3,919,448 (Nov. 11, 1975), E. R. Dufresne (to Chicago Rawhide Manufacturing Co.).
279. U.S. Pat. 3,725,438 (Apr. 3, 1973), B. J. Barone and W. F. Brill (to Petro-Tex Chemical Corp.).
280. *Hydrocarbon Process.* **58**(4), 112 (1979).
281. Brit. Pat. 1,560,884 (Feb. 13, 1980), W. Reuther and co-workers (to BASF Aktiengesellschaft).
282. U.S. Pat. 3,436,220 (Apr. 1, 1969), F. Dersch and S. L. Paniccia (to G.A.F. Corp.).
283. U.S. Pat. 3,440,095 (Apr. 22, 1969), L. C. Larsonneur (to Nalco Chemical Co.).
284. U.S. Pat. 3,993,837 (Nov. 23, 1976), K. M. Foley and F. M. Vigo (to Owens-Corning Fiberglass Corp.).
285. U.S. Pat. 3,617,578 (Nov. 2, 1971), J. R. Stanford and P. G. Vogelsang, Jr. (to Nalco Chemical Co.).
286. U.S. Pat. 3,515,673 (June 2, 1970), P. W. Kersnar and S. Taormina (to Progressive Products Co.).
287. U.S. Pat. 4,102,930 (July 25, 1978), M. L. Deem (to Union Carbide Corp.).

RICHARD O. KIRK
T. JOHN DEMPSEY
The Dow Chemical Company

PROSTAGLANDINS. See Supplement Volume.

PROSTHETIC AND BIOMEDICAL DEVICES

A prosthetic or biomedical device is an artificial part or device that replaces or augments a part of the human body. The area of prosthetic and biomedical devices is extremely broad and encompasses hundreds of specific devices (1–10) (see also Biomedical automated instrumentation).

Historically, the first use of prosthetic and biomedical devices goes back to antiquity. Sutures were first used about 4000 BC and gold plates for skull repair before 1000 BC. Artificial limbs are mentioned in Roman writing and in the Middle Ages; later on prosthetic devices were used as replacement dentures. Substitution vascular grafts, first reported early in this century, have been used clinically during the past 25–30 years. Experimentation with an artificial kidney was started ca 70 years ago and the first successful clinical application occurred in 1943. A heart–lung machine was investigated toward the end of the 19th century but was actually used in the 1940s. The membrane oxygenator, which is incorporated in heart–lung machines, was developed in 1955.

Heart pacemakers and artificial heart valves were first developed in the early 1950s. In 1969, an artificial heart was implanted for the first time in a dying patient who needed a heart transplant but lacked a suitable donor. The patient remained on the artificial heart for 64 hours and then a heart transplant was made. Unfortunately, the patient died 32 hours later of an overwhelming pneumonia. In calves, artificial hearts have functioned as long as 268 days, and the prospects for human use are considered very good.

In the early 1940s, acrylate and methacrylate materials were adapted to replace skull bones and other hard tissue and even corneas. At the same time plates of nylon, polytetrafluoroethylene, polyethylene, Orlon, and Dacron were introduced in surgery. Acrylates were used for hip prostheses, and poly(vinyl chloride) copolymers for implants.

An important development of the 1960s was the discovery of hemoperfusion using a loose bed of granular charcoal. This technique, now much improved, is used frequently to treat drug overdose, acute poisoning, and other severe toxic conditions. The American Society for Artificial Internal Organs was founded in 1955 and the International Society for Artificial Organs in 1977.

The materials of construction are metals, ceramics (including glass), and polymers including natural polymers such as collagen (see also Biopolymers). Synthetic polymers are the principal materials (see Table 1). The glass-transition temperature T_g corresponds to the temperature above which a polymer is flexible. Thus, a polymer with a T_g below the normal body temperature of 37°C is flexible, whereas a polymer with a T_g above this value is rigid. For example, polydimethylsiloxane with a T_g of -123°C can be used as a soft-tissue replacement whereas poly(methyl methacrylate) with a T_g of 105°C is used in dentures. Obviously no single material is suitable for all biomedical devices, because the specific requirements vary greatly. The range of requirements is so broad that it is necessary to delineate the specific requirements for each type of biomedical device and then find or develop a suitable material or combination of materials. This aim is difficult to achieve and many materials are not completely satisfactory.

Table 1. Synthetic Polymers Commonly Used in Prosthetic Devices

Name	Structure	T_g, °C	Abbreviation or trade name
polyethylene	 $-(CH_2CH_2)_n-$	−120	PE
polypropylene	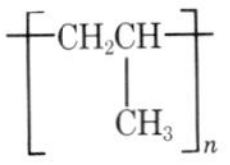 $-(CH_2CH(CH_3))_n-$	−18	PP
poly(vinyl chloride)	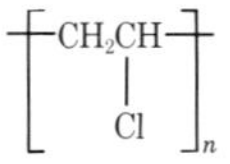 $-(CH_2CH(Cl))_n-$	87	PVC
polytetrafluoroethylene	 $-(CF_2CF_2)_n-$	126	Teflon, PTFE
polyacrylonitrile	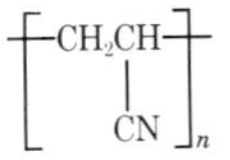 $-(CH_2CH(CN))_n-$	104	Orlon
poly(vinyl acetate)	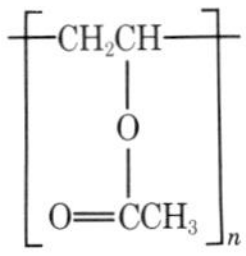 $-(CH_2CH(O-COCH_3))_n-$	29	PVAc
poly(vinyl alcohol)	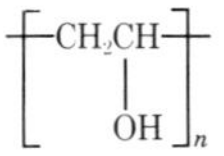 $-(CH_2CH(OH))_n-$	85	PVAl
poly(methyl methacrylate)	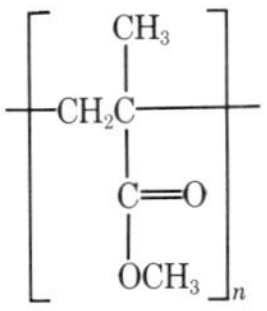 $-(CH_2C(CH_3)(C(=O)OCH_3))_n-$	105	Plexiglas, Lucite
poly(cis-1,4-isoprene)	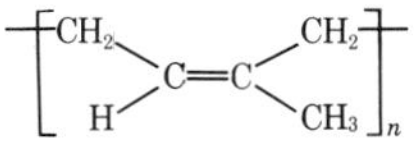 $-(CH_2C(H)=C(CH_3)CH_2)_n-$	−73	natural rubber
polydimethylsiloxane	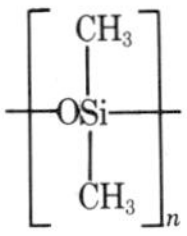 $-(OSi(CH_3)_2)_n-$	−123	Silastic
poly(ethylene terephthalate)	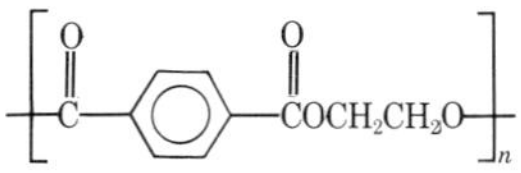 $-(C(=O)-C_6H_4-C(=O)OCH_2CH_2O)_n-$	69	Dacron
poly(hexamethylenediamine adipate)	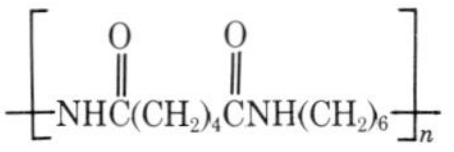 $-(NHC(=O)(CH_2)_4C(=O)NH(CH_2)_6)_n-$	50	nylon-6,6

The main requirements for biomaterials are biocompatibility with the body tissues and fluids, blood compatibility, external-environment compatibility (wear, discoloration, etc), and physical properties such as strength, flexibility, permeability, and degradation.

Biocompatibility

Biocompatibility refers to the manner in which a prosthetic or biomedical device interacts with body fluids and tissues. It depends to a great extent on where and how it contacts the body. A device could be completely external, completely implanted under the skin (subcutaneous), protrude through the skin (percutaneous), or be deeply implanted in the body. Essentially, any foreign body contact with body tissues or fluids elicits some type of response. These reactions could be due to the shape, design, or movement of the device, its chemical nature, extractable materials (usually plasticizers, stabilizers, catalysts, or degradation products), and infections caused by improper sterilization. The reactions range from outright rejection to relatively benign tolerance. Biological rejection means that the body does not accept the implant which is manifested in several ways including extrusion of the implant from the body, destruction of the implant by enzymatic, phagocytic, or other action, or encapsulation of the implant by a fibrous tissue. The term biocompatibility can be defined in a variety of ways. If the definition (11) is accepted that a biocompatible material has no effect on the surrounding tissues and has no adverse effect on the normal healing process in the body, then no material currently in use is completely biocompatible. Usually, a fibrous capsule grows around implanted materials which is a definite response by the body (12–13). Even a completely external device may produce a rash or callous formation on the skin. The materials used today might better be termed biotolerable, but this term is not often used. In practice, materials have various levels or degrees of biocompatibility and an attempt is made to employ only those materials that most nearly approach the ideal specified in ref. 11. Although many suture materials, such as catgut, silk, and multifilament nylon, produce a considerable amount of adverse tissue response, they are considered acceptable (14). In this article, the term biocompatibility describes various degrees of interaction of materials with the body. Thus, a completely biocompatible material shows no effect or interaction with the body, a satisfactory biocompatible material shows only slight interaction, and an incompatible material elicits a severe reaction.

The importance of a satisfactory biocompatible material can hardly be overemphasized. The most perfectly designed or engineered prosthesis or biomedical device could be rejected if the materials used are not tolerated reasonably well. Unfortunately, each application has its own specific set of requirements and potential interactions, and broad generalizations could lead to a poor choice of biomedical material. Even though the ultimate test criteria for a prosthesis is the actual performance in the human body, a number of screening tests have been devised, including: infrared spectroscopy on the pseudo-extracellular fluid (PECF) extracts of the material; tissue-culture methods such as agar overlay, direct contact, and suspension culture; animal implant studies; and utilization tests. In the case of blood contact, further tests involving blood are essential.

Infrared spectroscopy of PECF extracts is a simple screening test that gives reasonably good correlation to tissue-culture tests. The material is extracted with a simulated body fluid (PECF) under carefully controlled conditions and the amount estimated by infrared analysis. The result is usually specified as the total amount of CH bonds equivalent to the moles per ppm of n-hexanol, the calibration standard. Obviously, polymers such as polytetrafluoroethylene would register zero in this test, as would metals and ceramics, yet they can cause adverse tissue response. The method is rapid, relatively inexpensive, and eliminates unsuitable materials early in a research program (15–16).

In the widely used tissue-culture methods, human or animal cells are grown *in vitro* and their interaction with the biomaterial is tested. In the agar-overlay method, a layer of these cells is formed in a Petri dish and then covered with a thin layer of agar. The biomaterials are placed directly on the agar and their effect on the cells is ascertained after incubation for 24 hours at 37°C in an atmosphere containing 5% carbon dioxide (17–19). In the direct-contact tissue method, a film or coating of the biomaterial is placed in a Petri dish under a suspension of tissue cells. The extent of cell attachment to this biomaterial, the growth of the cells, and the time required to form a complete monolayer (if this occurs) determines the degree of biocompatibility (19). However, this test is valid only for a specific type of cell. It is, however, sometimes more sensitive than the agar-overlay method.

In the suspension-culture method, a cell suspension is prepared in contact with the biomaterial. The ATP (adenosine triphosphate) concentration in the cells is determined enzymatically with a DuPont Luminescence Biometer. The results of this test correlate well with the agar-overlay method (19).

In implant studies, the material to be tested is implanted subcutaneously, percutaneously, or into deep muscle in humans or laboratory animals (18,20–22). However, different results obtained with humans and animals interfere with the interpretation of data. In general, a material exhibiting good biocompatibility in several different types of animals is considered a good prospect for human studies. Although subcutaneous and percutaneous implantations give useful information, deep muscle implantation is generally preferred for determining the potential toxicity of a biomaterial. The most widely used method is a modified USP rabbit-muscle implantation (18,20).

Another screening program includes tissue-culture agar-overlay, rabbit-muscle implantation, rabbit blood hemolysis, and tests on biomaterial extracts (18). The extraction media are saline, polyethylene glycol 400, and cottonseed oil. The sample is extracted for one hour in an autoclave at 121°C. Each extract is tested by the tissue-culture agar-overlay method and intracutaneous injection into rabbits, and for systemic toxicity in mice. Finally, a distilled-water extract is tested for cell-growth inhibition in a standardized procedure (18). Only the materials that survive this regimen are tested on humans.

A few polymers appear to have adequate biocompatibility for limited use including silicone-rubber polymers, polyethylene, polytetrafluoroethylene, some polyether polyurethanes, acrylates, methacrylates, poly(ethylene terephthalate), and nylon. Pyrolytic carbon (23) and certain types of hydrogels show few adverse interactions with body tissues and are considered fairly biocompatible. The polymers must be free from plasticizers, fillers, stabilizers, catalysts, and other additives that may cause tissue reaction. Similarly, powdered polymers usually evoke a severe reaction (14,24). It is not known why certain polymers give toxic reactions whereas others are nearly inert, although some theories have been proposed (24–25).

Ceramic materials, such as TiO_2 and Al_2O_3, usually show good biocompatibility, although most become encapsulated in a fibrous capsule. Pure metals, on the other hand, cause a severe tissue reaction. Rabbit-muscle implantation studies show that the metallic components of Vitallium, an alloy of cobalt, nickel, chromium, and molybdenum, or 316 stainless steel migrate and appear in many body organs (14). A plastic coating can prevent migration. Some metals or metal compounds, such as those of cadmium, hexavalent chromium, and nickel cause tumors in rats, whereas aluminum, copper, gold, iron, lead, and silver do not (26). Both chromium and nickel occur in Vitallium and stainless steel alloys and, although these metals are widely used in implants, no cases of implants causing tumor formation in humans have been reported. In a similar way, a number of studies have shown tumor formation in animals, presumably caused by polymer implants, but no cases of human cancer caused by such implants have been observed to date (27–28). This observation may be species-related phenomena or owing to the length of time needed for tumors to appear in humans, but most observers feel that the former is the explanation. Thus, the ideal biocompatible material for implanted prostheses or biomedical devices has not yet been developed.

Blood Compatibility

Blood compatibility is much more complex than the compatibility of a biomaterial with other body fluids or tissues. The extent of the compatibility of blood with a specific biomaterial depends on whether the blood is moving (as in a heart device or blood vessel) or static (as in a storage bag or bottle); whether the blood is arterial or venous; flow patterns and especially changes in flow patterns; and interactions with red cells, white cells, platelets, plasma proteins and other blood components. Blood is a heterogeneous, non-Newtonian fluid consisting of ca 45% solids (red cells, white cells, platelets) and 55% plasma. The plasma contains a variety of inorganic ions and a series of soluble proteins which can be classified as albumins, fibrinogens, and globulins (see Blood, coagulants and anticoagulants). Blood forms a clot or thrombus when injury occurs or when it is contacted by a foreign substance. This basic defense mechanism prevents fatal bleeding. Almost all biomaterials set off this clot-formation process and soon become coated with an irreversible clot of varying size that could have an adverse effect on the utility of the biomedical device and even be fatal to the patient. Biomedical and prosthetic devices that contact blood are either extracorporal devices (tubing and membranes used in heart–lung machines, dialysis units, etc) or internal devices (replacement blood vessels, heart-assist devices, total artificial hearts, etc). Their main difference with regard to blood compatibility is the length of time of contact with the blood. In addition, anticoagulants and clot filters can be used with external but not with internal devices.

Several theories have been advanced to explain the differences in compatibility of various biomaterials; for example, blood compatibility is inversely related to the wettability of the polymers. Although this idea holds for certain hydrophobic polymers, such as polydimethylsiloxane and the polyether polyurethane ureas (PEUU), it obviously fails for certain hydrophilic (wettable) polymers, such as hydrogels. In addition, the natural blood-contacting surfaces in blood vessels and the heart are hydrophilic and blood compatible (24,29). A second theory claims that the zeta potential of the compatible surface must be highly negative since the natural blood-contacting tissues have this property. Although this approach succeeds for certain ionomer-type polymers and electrets (charged polymers), it does not work well for hydrophobic materials. In addition, glass has a high negative zeta potential and is thrombogenic (30–32). A third theory is that the differences in the nature of the proteins that adsorb on the biomaterial surface cause the differences in blood compatibility. Whenever the blood contacts a foreign surface (eg, a polymer) various plasma proteins adsorb on this surface. This fact has been known for many years and has been the subject of many studies (33–38), not all of which agree with each other. It does appear, however, that surfaces that adsorb mostly γ-globulin and fibrinogen show thrombogenicity, whereas those that adsorb albumin are relatively thromboresistant. This generalization appears to work for such different materials as the hydrophobic polyether polyurethane ureas and the hydrophilic hydrogels. However, protein adsorption does not reach equilibrium for many hours or longer whereas thrombus formation begins in a matter of minutes. Many biomaterials show initial thromboresistance but develop clots after several weeks, possibly because of changes in the adsorbed protein layer. Finally, this protein adsorption appears to be related primarily to the fact that the biomaterial is a foreign substance and there is no evidence to suggest that the uninjured, natural blood vessels adsorb proteins from the plasma. No theory, at the present time, is completely adequate to explain all the variations in blood compatibility for natural and synthetic materials (39–44).

A number of biomaterials have limited utility in various extracorporal devices. Polydimethylsiloxane and other types of tubing can be used with heart-lung machines, dialyzers, and related devices if a suitable anticoagulant, such as heparin, is added to the blood. However, administration of heparin reduces or prevents the natural clotting of the blood. Materials that have shown promise for use as membranes in blood oxygenators, usually with an anticoagulant, include polydimethylsiloxane, perfluoroacyl ethyl cellulose, polyalkylsulfones, and various heparinized polymers (see Fig. 1). Heparin, a naturally occurring polyanionic mucosaccharide with a molecular weight of ca 12,000–16,000 that is synthesized in the liver, has been attached to various surfaces by a variety of techniques (40,43). Although ionically bound heparin confers a significant degree of thromboresistance to the surface, the heparin desorbs with time and the basic thrombogenic nature of the surface prevails. This does not, however, pose a problem with short-term, extracorporal use and would eliminate the need for administering heparin to the patient. Covalently bonded heparin maintains its thromboresistance longer, although the heparin is usually somewhat less active than the natural material. It appears likely that many experimental nonthrombogenic amido–amine polymers are thromboresistant because heparin is adsorbed at the amido–amine sites since many of the heparinization techniques involve a quaternary ammonium compound and heparin does form complexes with amino groups.

Several experimental polymer systems have shown promise for extracorporeal (or internal) use. These include the Ioplex materials and other hydrogels such as those based on 2-hydroxyethyl methacrylate or acrylamide (see Fig. 1). These materials may contain 50–80% water and it was claimed formerly that this was the basis of thromboresistance of hydrogels. More recent studies have shown that blood compatibility does not depend on the water content of hydrogels (45). Hydrogels normally lack physical or mechanical strength, a problem that has been partially solved by grafting hydrogels onto other substrates or by making a composite material with the hydrogel surface contacting the blood (46–47). Various lysing agents, such as urokinase or streptokinase, have been bonded to polymers with the intention of lysing any clotted material that might form on the surface (48–49).

Certain polyether polyurethane ureas (PEUU) show good thromboresistance and are generally regarded as promising materials for internal use (blood vessels, artificial hearts). The PEUU system (see Fig. 1), can be made with a wide variety of alkyl and/or

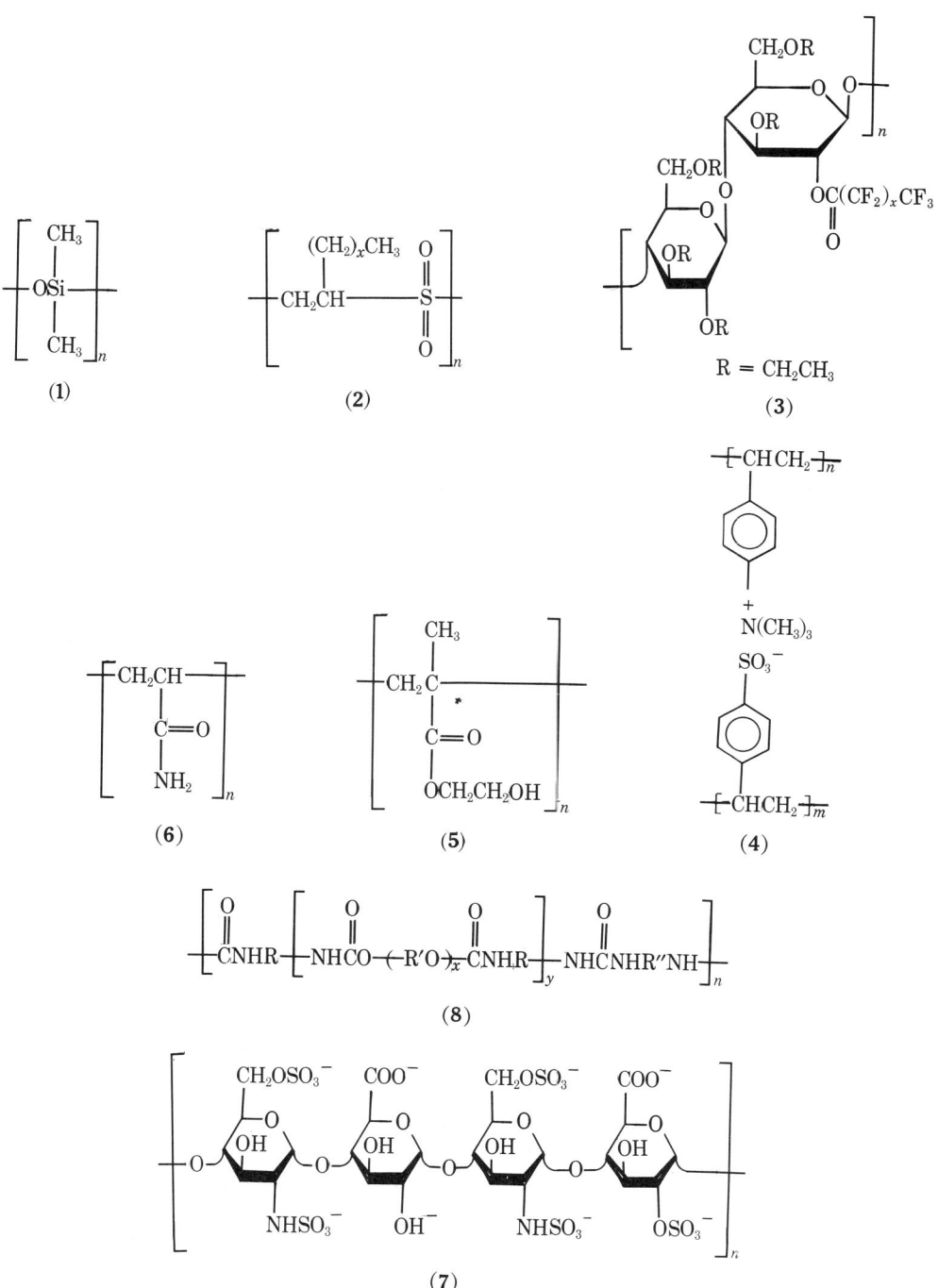

Figure 1. Structures of some thromboresistant polymers. (1) polydimethylsiloxane, (2) polyalkylsulfone, (3) perfluoroacyl ethyl cellulose, (4) Ioplex 101, (5) poly(2-hydroxyethyl methacrylate), (6) polyacrylamide, (7) heparin, (8) polyether polyurethane urea (PEUU).

aryl polymer groups and is often referred to as a segmented polyurethane. Such materials as Biomer (Ethicon Corp.) and Avcothane (a copolymer with a silicone; Avco-Everrett Laboratories) have widespread use in heart-assist devices (especially intraaortic balloons), artificial hearts, artificial blood vessels, and related areas. Devices made from these hydrophobic polymers often show no evidence of thrombus formation after having been in contact with circulating blood for a substantial period of time (ie, they are clean), but in many cases, emboli (a dislodged clot that is found removed from the original formation site) are noted in other parts of the test animal's body. Furthermore, surface and bulk morphology of PEUU materials are not the same and depend on the formation of the surface (ie, in contact with air, glass, etc). In spite of these drawbacks, PEUU materials are highly promising for use in blood-contacting applications because of their excellent strength and flexibility, important characteristics for blood-pumps (50–51).

Biolized materials often show good blood compatibility. Biolization is a term coined to describe a process in which natural tissue (eg, collagen, gelatin, or albumin) is made biologically inactive by chemical or thermal treatment (52). Biolized material can be used as coating or as the sole material for a biomedical device. Biolized materials show enhanced thromboresistance compared to untreated natural tissue or uncoated substrate. The natural protein is coated on a polymeric substrate such as Hexsyn (a 1,4-hexadiene polymer) and then biolized to obtain a material with greater strength and flexibility. The biolization process involves treatment with an aldehyde, eg, formaldehyde or glutaraldehyde, and heating. Treatment of synthetic or natural rubber in this manner, in the absence of the natural polymers, does not improve thromboresistance. Cardiac prostheses fabricated from such biolized materials have shown blood compatibility when implanted in calves for up to ten months without the use of heparin or other anticoagulants (52–53).

Prostheses coated with low temperature isotropic pyrolytic carbon (LTI carbons) show excellent thromboresistance and good overall biocompatibility. These LTI carbon prostheses have excellent mechanical strength and wear resistance and have been used in heart valves since 1969 without failure of the biomaterial. Whether this material will prove as useful in other blood-contact applications, such as artificial hearts or blood vessels, remains to be seen (23) (see Carbon and artificial graphite).

A number of tests have been developed to ascertain the blood compatibility of biomaterials. In the Lee-White *in vitro* test the blood is placed in a standard test tube that is coated with the plastic to be tested; three such tubes are used for each evaluation. The first is inverted every 30 seconds until blood flow no longer occurs. The second tube is then inverted in the same way, followed by the third after flow ceases in number two. The time at which blood flow no longer occurs in the third tube is the clotting time of the material. The test is simple and inexpensive, and is a fairly reproducible screening method but it does not correlate well with *in vivo* experience (29). The lack of correlation may be partly caused by the air–blood interface in this test. A simple flow-through cell has been developed consisting of a small glass chamber in which the polymer film is clamped and the cell is filled with saline solution. Blood is then drawn into the cell to displace the saline solution and thus a blood–air interface is avoided. This cell can be used directly with human blood and appears to give some ranking of the relative thrombogenicity of different polymers when short time intervals are used. At longer times, all the materials showed about the same amount of clotting (33). A somewhat similar kinetic clotting test uses a small closed cell in which two disks of the biomaterial to be tested are placed and then covered with saline solution. The blood, previously withdrawn from fasting, anesthetized dogs, is then drawn into this cell between the two disks by removal of 2 cm³ saline solution; eight cells of the test material and eight control cells lined with Silastic are run in each test and pairs are opened periodically during the two hour test interval. The cells are rotated slowly to prevent sedimentation of the blood. The amount of clotting is determined by direct weight and by determining the hemoglobin in the unclotted blood by colorimetric optical density analysis determinations at 540 nm. The results are expressed graphically and the difference in area under the curve between the test material and the Silastic control is the kinetic clotting index. Platelet adhesion can also be determined on the same type cell after ten minutes contact time. The test can also be run with human blood. The test results are reproducible and show good general agreement with *in vivo* results (54–55).

In the *in vivo* test, the device is implanted in an experimental animal, eg, dog, calf, pig, and primates; each offers various advantages and disadvantages. Species-related effects do sometimes occur. A relatively commonly used test is the Gott ring test in which rings of a specific size are implanted in the vena cavae of dogs. Assuming survival, the animals are sacrificed at various times and the extent of clotting determined. Such *in vivo* tests are essential for any material before it could be considered for use in humans.

Tubular Prostheses, Plastic Surgery, and Related Devices

Tubular Prostheses and Devices. A wide variety of tubular prostheses and devices is used in medicine including drains, catheters, cannula, shunts, and reconstructions or replacements of natural tubular-type organs (eg, windpipe or intestines). Some of these devices are lifetime implants; others are used only briefly. Some implants are internal; others protrude through the skin. Some are exposed to the blood and must be nonthrombogenic (see also Sutures). The materials used most frequently for tubular applications are polyethylene, polypropylene, polytetrafluoroethylene, poly(vinyl chloride), polydimethylsiloxane, and natural rubber (56–57).

A drain prosthesis removes liquid from one region of the body to the outside or to some other region. Drains are normally inserted following surgery to alleviate fluid build-up at the operation site; they are generally for short-term use. Other types are used for longer periods; these include drains for mucus secretions in sinus conditions, eustachian-tube drains, and the hydrocephalus shunt. In the latter case, a tube is run from the subarachnoid region of the head to some lower part of the body (often the heart, a blood vessel, or the peritoneal cavity) to remove excess cerebrospinal fluid and thereby correct congenital hydrocephalus; this device is also used for the pressure relief of inoperable tumors and for postoperative surgery drainage. A typical hydrocephalus shunt, made from silicone rubber, is shown in Figure 2a. In hydrocephalus treatment, the device is implanted for long-term use and must not clog at any point. Several one-way valves prevent fluid from rising in the tube and thereby increasing pressure on the brain. Most of these devices are implanted in infancy and must be modified with growth. For this reason, the tubing is fitted with male and female joints, and a section can be replaced with a longer one (56–57).

A catheter is a tubular device for introduction into canals, passages, or tubes (including blood vessels). Such devices can be used for drainage, fluid sampling, or

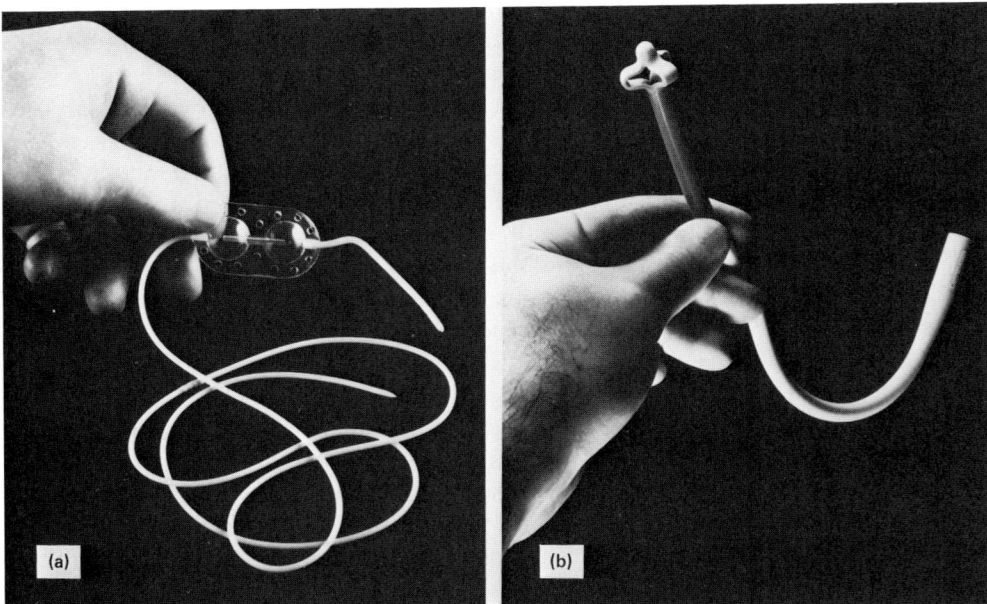

Figure 2. Some tubular devices. (a) Silastic hydrocephalus shunt. (b) Silastic malecot catheter. Courtesy of Dow Corning Corp.

the introduction of fluids, medication, or devices such as an intraaortic balloon. A typical catheter, made of silicone rubber, is shown in Figure 2b. Catheters are usually temporary devices, but are sometimes used for many months. Blood compatibility, essential for vascular applications, frequently presents difficulties in long-term catheters because of phlebitis, thrombosis, or septicemia. Among experimental materials, an ethylene–acrylic acid copolymer has shown promise (58).

A cannula is a tube inserted into a cavity or another tube in the body for various purposes, usually by percutaneous insertion into a blood vessel and for long periods of time. Apart from blood compatibility, the main problem associated with such devices is to prevent motion, irritation, infection, and the formation of a sinus tract with weeping of fluid. These objectives can be achieved by anchoring the devices in various ways. For example, the attachment of material containing holes or fenestrations for tissue ingrowth (21) or a textured surface on the implanted tube made of foamed material or nylon or Dacron velour. The texturized surfaces interact with the surrounding tissues and a bond is formed. Unfortunately, however, these texturized surfaces may be extruded gradually as a natural consequence of cell maturation which results in cells migrating toward the surface. Since the texturized implant is bonded to these cells, it is also carried toward the surface and extruded (59). It may be possible to solve this problem by special surgical techniques (60). In the case of prolonged renal dialysis, cannulas are usually placed on the nondominant forearm of the patient to allow ready access to the blood vessels. These cannulae often cause infection, hemorrhaging, or irritation; new surgical techniques may alleviate these problems (61).

In a tracheotomy, a circular opening is cut through the skin into the trachea or windpipe, and a tube, usually made of silicone rubber or polyethylene, inserted to permit ready passage of air. Replacement or reconstruction of the trachea, when necessary, is effected with polytetrafluoroethylene tubes or polypropylene mesh. Replacement of the larynx with plastic tubing results in loss of the voice. Certain vocal-cord impairments have been treated experimentally by injecting polymer solutions or suspensions into the larynx (62) but this technique is not widely used. The esophagus (food pipe) has been replaced surgically with a polypropylene mesh or with Dacron tubes containing nylon rings. The ureter has been replaced with various types of tubing; silicone seems best. In ureter replacement, stone formation can obstruct the passageway; a similar problem occurs in bile-duct replacement. The bladder can be replaced by a silicone rubber device, with loss of muscle contraction. The gastrointestinal tract is sometimes replaced partially with plastic tubing, with corresponding loss of physiological function. Most tubular parts of the body have been replaced surgically, at least on an experimental basis, and polysilicone rubber tubes have been placed in the body as nerve cuffs to prevent damage during neuronal regeneration (56–57).

Ear Prostheses. The ear consists of an external part (auricula), a middle portion containing the tympanic membrane (eardrum) and the ossicular bones, and the inner ear which contains the nerve connections. The semicircular canals or labryinth, which is involved in balance control, is also situated in the ear. The auricula directs sound into the middle ear. At times it is necessary to replace this external ear because of congenital defects, disease, or accidents. The replacement may be a device that the patient can remove, or it may be a permanent implant. The replacement may be modeled from pictures of the ear before trauma or from the other ear. In the case of an externally attached ear, the pigmentation is closely matched to the patient. The materials most commonly used include natural rubber, plasticized poly(vinyl chloride), and polysilicone rubber. In the implant technique (otoplasty prosthesis), a silicone elastomer framework replaces the natural cartilage of the ear and is covered with Dacron mesh. The patient's skin is then grafted onto this mesh creating a new ear with a natural appearance. The implanted device, which is shown in Figure 3, is easy to keep clean and does not discolor by exposure to the environment or light since the outer layer is actual skin (57,63–64).

Myringoplasty is concerned with replacement or repair of the eardrum. This is usually done with sheets of collagen, or related material, cemented to the remaining parts of the eardrum. Tympanoplasty is the repair or reconstruction of the middle ear which consists of three small, articulated bones. Replacement materials include 316 stainless steel, platinum, tantalum, alumina ceramic, polyethylene (plain or porous), and polytetrafluoroethylene (plain or porous). The polymeric materials are the most widely used. It is estimated that the likelihood of hearing improvement, under optimum circumstances, is at least 85% (64–66).

Some research is in progress to develop an artificial ear that permits a deaf person to hear. These devices stimulate the branches of the 7th nerve in the cochlea and show much promise (10,67). Ménière's disease, which afflicts the semicircular canal system, is treated by placing a plastic tube (silicone rubber, polyethylene, or polytetrafluoroethylene) in the ear to drain the excess endolymphatic fluid into the subarachnoid part of the head (64).

Eye Prostheses. When an eye must be replaced because of disease or an accident, much of the surrounding orbital area usually needs replacement as well (see also Contact lenses). The eyeball is generally replaced with a plastic globe, made from acrylic or silicone plastics, although glass had been used in the past. The patient

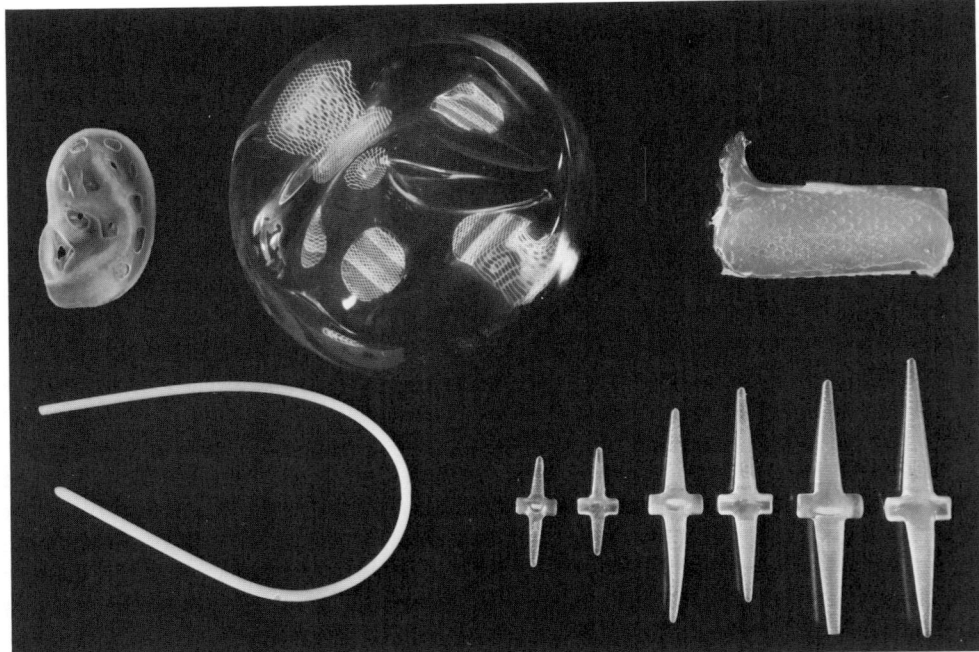

Figure 3. Various soft-tissue devices made from silicone rubber. Clockwise from the upper left: ear prosthesis; mammary (breast) prosthesis; chin prosthesis; finger joint prostheses in various sizes shown top and bottom views; tendon prosthesis. Courtesy of Dow Corning Corp.

normally does not have control of the eyelids, or eyeball movement, although some experimental work has been done to connect the eye muscles to the prosthesis for better cosmetic effect (63,68). Some research has been done on a visual prosthesis to enable the blind to see. This involves implanting electrodes on the occipital cortex of the brain to produce a visual sensation of light (phosphenes) when stimulated. Coupling these electrodes to a miniature television camera, via a microcomputer, could permit the recipient to see to a degree. Such prostheses are still highly experimental but do hold promise for the future (10,69–70).

The cornea, which is the anterior portion of the eye, sometimes becomes opaque and must be replaced to restore vision (keratoprosthesis). This is accomplished by a cornea transplant or by an implant usually made of poly(methyl methacrylate). The prosthesis often had a pronounced tendency to be extruded and various anchoring methods have been used to solve this problem, including Dacron mesh plates and various fenestrated disks made of collagen or aluminum oxide (68,71).

Cataract formation often necessitates removal of the opacified lens to restore vision. Corrective spectacle lenses can be used to compensate for the loss of focusing, but they cause a 25% magnification, high distortion, and often intolerable double vision. The implantation of an intraocular lens alleviates these problems. Such lenses are made from poly(methyl methacrylate); other materials have been tried. With such a device, 20/20 vision is achieved ca 28% of the time and at least 20/40 vision 73% of the time. Although these results appear to be slightly better than those obtained with the corrective spectacle lens, the intraocular lens causes dislocation and cornea endothelial cell destruction. Dislocation of the implanted lens occurs in about 4% of the cases and requires another operation to reposition the lens. Conventional cataract removal causes loss of about 7–8% of the corneal endothelial cells whereas the intraocular implantation technique causes losses of 35–50% (72). These losses may be due to the adhesion of these cells to the hydrophobic surface of the implant or to the surgical gloves used. The use of a hydrophilic surface or of a hydrophilic polymer solution, eg, poly(vinylpyrrolidinone), appears to reduce endothelial-cell adhesion (73). Monomeric methyl methacrylate, which can be cytotoxic at moderate concentrations, does not appear to present a problem in intraocular lenses since the residual monomer concentrations are below those that cause cell damage (74).

A detached retina is normally fused by laser when it is in contact with the choroid layer of the eye. When these layers are separated, they are pushed together from the outside using a variety of plastic strips, bands, sponges, or buckles. Silicone rubber sponges are considered the best for this application since they result in the least erosion and abscessing of the sclera (68).

Craniofacial Reconstruction Prostheses. Reconstruction of the head is necessary to correct deformations due to heredity, disease, or accident. This reconstruction is complex because the head contains skin, soft tissue, cartilage and bone with very different physical characteristics that must be matched by the reconstruction biomaterials. When skin must be replaced, as in burn damage, it is usually grafted from another part of the body, if possible. Facial regions can be augmented or structurally modified by implantation or by affixing an external prosthesis. In the latter case almost any stable material can be used that does not cause an adverse reaction in the surrounding or underlying surfaces and that is well matched to the surrounding facial material in texture, feel, and color. The materials most commonly used include silicone rubber, polyurethane, poly(vinyl chloride), and natural rubber. These external prostheses are constructed from a model of the part to be replaced. A good prosthesis is nearly impossible to detect except by very close examination.

Implants are covered with grafted skin for cosmetic effect. The most satisfactory soft tissue implant is silicone rubber which can be wired in place or fixed via tissue ingrowth through fenestrations in the implant or onto a fabric backing (usually Dacron). This material has been widely used in nose reconstruction (rhinoplasty), chin augmentation, and in the cheek or forehead (57). Silicone rubber, reinforced with wire and sometimes cancellous bone autographs, has been used in mandible replacement (75), as has poly(methyl methacrylate), polyethylene, and polytetrafluoroethylene. A silicone rubber chin prosthesis is shown in Figure 3. The alveolar ridge has been reconstructed with autogeneous bone, bone derivatives, or poly(methyl methacrylate) in order to make this ridge more effective in supporting a lower denture, but these materials are not considered satisfactory. A ceramic made from calcium aluminate has been found satisfactory for alveolar-ridge augmentation in dog experiments; trials have not yet been tried on humans (76). Other alveolar augmentation materials include a polytetrafluoroethylene–carbon composite (Proplast), porous alumina, and Bioglass. Poly(methyl methacrylate), nylon, and metal plates have been used for cranial bone replacement for some time. These are usually fixed in place by screws in order to prevent movement which could erode the nearby bone or cause other damage (eg, to the brain or skin). The metal plates are heavy and conduct heat. In a more recent technique, Dacron mesh is impregnated with a polyester polyurethane. The mesh prosthesis can be draped over a solid model and cured thermally to give a material with sufficient rigidity for bone replacement in the mandibular or the cranial regions. In addition, an autogenous bone graft aids in new bone formation (77–78).

Soft-Tissue Prostheses. A soft-tissue replacement or augmentation prosthesis takes the place of relatively soft muscle, fatty tissue, or connective tissue within the body. The skin, which may be considered as soft tissue, is discussed in the following section and some soft-tissue replacements were covered in the preceding section.

Wall defect and hernia repairs are internal prostheses and must exhibit good biocompatibility and remain soft indefinitely. Certain soft tissues, such as a breast, may be replaced by an external plastic device made from almost any material, eg, poly(vinyl chloride), polyurethanes, silicone rubbers, and natural rubber. These external devices are attached by friction, adhesives, or straps. Until fairly recently, the usable soft-tissue materials were transplanted from another part in the patient's body; cadaver transplants cause rejection problems. Many synthetic materials have been tried with limited success. In the older literature, it is not clear what plastic was used, nor is it always clear that sufficient care had been taken to avoid contamination when infection or inflammation occurred. The materials were only scantily described, and details such as molecular weight averages or distributions were very rarely known. Gradually, this problem was recognized and later studies were reported in more detail. Specific requirements for soft-tissue prostheses as suggested by several researchers (79–80) are suitable physical and mechanical properties (eg, soft and/or rubbery); long-term retention of properties in a physiological environment; ease of sterilization; chemical inertness; lack of toxicity and immunogenic or allergenic properties; freedom of carcinogenicity; lack of infection, inflammation, and erosion of surrounding tissue; painlessness; inertness to surrounding tissue; minimal adverse tissue reaction; lack of fibrous-tissue ingrowth; and nonthrombogenicity. Very few materials meet these requirements and most current applications use polysilicone rubbers, fabrics, or meshes made from polytetrafluoroethylene, Dacron, nylon, some polyurethanes, and some natural materials such as collagen. Fibrous tissue can form around most implants and ingrowth may harden the implant. For these reasons, textile materials are normally used only for anchoring a soft tissue prosthesis or for hernias or wall repairs.

Probably the most widely known type of soft tissue prosthesis is the implanted breast or mammary prosthesis of which the best designs consist of a preshaped, flexible silicone rubber bag filled with a silicone rubber gel. In some cases, these have patches of Dacron mesh attached to the back to aid in tissue fixation via fibrous ingrowth into the mesh, which anchors the prosthesis. The entire prosthesis is implanted under the skin and gives a good simulation of a natural breast in shape, weight, and feel (see Fig. 3). Breast prostheses are widely used after mastectomy and also for cosmetic breast augmentation. These prostheses can cause diagnostic difficulties, since they are opaque to x-rays. The use of silicone or other injectable fluids, paraffin, or any mobile material for breast augmentation is dangerous and illegal. Such material can migrate and cause death which would not occur with the prosthesis described above.

Almost any soft tissue can be replaced by a silicon rubber bag containing a silicone rubber gel; the rigidity of this system can be increased by increasing the gel cross-linking. Testicles have been replaced by silicone-rubber spheres and by polyethylene. Similarly, a missing penis can be replaced with plastic devices usually made from silicone rubber but, like the testicle prosthesis, these are nonfunctional. Several implantable penis prostheses have been developed that are either permanently rigid or inflatable to provide erection. Hernias and wall defects are normally repaired by a plastic-mesh prosthesis which bridges the defect and permits ready tissue regrowth. These meshes are usually made from polyethylene, polypropylene, or silicone rubber; textile fabrics (eg, Dacron) have been used as well.

Artificial Skin. The skin, the largest organ in the body from the standpoint of volume and weight, serves a variety of important functions. It contains the epidermis (outer skin), dermis, hair, sebaceous glands, sweat glands, blood vessels, and nerves. Any attempt to duplicate this complex organ exactly would probably be unsuccessful with today's technology. Nevertheless, over 100,000 people are hospitalized each year with severe skin damage requiring immediate treatment to prevent gross bacterial contamination and the loss of essential body fluids and electrolytes. Such damage can result from disease, accident, or burns. The preferred mode of treatment is to transplant, or graft, skin from another part of the body (autografts). If the damage amounts to ca 50% or more of the skin area, another technique must be used, namely, homografts, xenografts, or artificial skin. A homograft is a transplant from another person or more usually a cadaver. Homograft tissue is difficult to store and often not available. Usually it is rejected in three weeks or less. Xenografts are transplants from an animal, usually a pig; these seldom last more than a week before rejection. At one time pig xenografts were believed to have antibacterial properties but these were caused by impregnation of the material with neomycin.

Obviously, the need for an artificial skin is great and has led to much research during the past twenty years. Most of this research is designed to develop an artificial skin that could function as a temporary wound dressing. A promising method, tested clinically with success, utilizes a velour fabric (usually nylon or Dacron), backed by a polymeric film (81). The regenerating skin ingrows the velour and makes a good seal to control bacterial growth and rejection. The outer polymeric film helps to control moisture transmission. The most effective polymeric barriers for this artificial skin are silicone rubber and a peptide (82). A variety of peptides can be synthesized by the anionic polymerization of N-carboxyanhydrides (NCAs or Leuch anhydrides) and these may be used as artificial skin (83). It is not known how these would interact with the human body on a long-term basis or whether they would elicit an immunological response. Collagen, in the form of sponges or films, has been much studied as a temporary skin replacement (84–87). The sponge adheres well to burn wounds and aids in healing but the collagen is attacked by the body enzymes and is liquefied readily, leading to rejection and infection. Recently, a composite system consisting of an inner layer of a collagen–glycosaminoglycan membrane and an outer layer of silicone rubber has been developed which shows considerable clinical promise (85–87). Dextran hydrogels also show promise as temporary artificial skin (88). These biomedical devices are all designed for short term use. The need remains for long-lasting artificial skin where severe damage renders the natural regenerative process inoperative.

Bone, Joint, and Limb Prostheses

The human skeleton contains 206 bones; most are connected by movable joints to other bones. Several types of movable joints exist in the body but most are either a hinge type (eg, fingers, elbow) or a ball-in-socket type (eg, hip, shoulder). Problems that require a prosthesis can arise in bones, the joints, ligaments, or tendons; sometimes an entire limb or section thereof must be replaced. Many different techniques of restoration have been attempted, not always with success. The principal materials used internally are cobalt alloys, titanium alloys, stainless steel, aluminum oxide ceramic, Bioglass, methyl methacrylate cements, polyethylene, polytetrafluoroethylene, and silicone rubber; numerous others are used externally.

290   PROSTHETIC AND BIOMEDICAL DEVICES

Casts, Braces, and Splints. Most bone fractures can be repaired by resetting the break externally and restraining the limb, etc, in a cast, brace, or splint (5). Casts are generally made from plaster but plastics are much lighter and work at least as well. These consist of a polypropylene or polyethylene sheet molded thermally to match the patient's contours and normally lined with a polyester or polyurethane foam for comfort (see Fig. 4). These devices are rigid enough to replace braces used for additional support (eg, on a leg); their light weight is an advantage. In addition, these bracing casts can often be worn with any type shoe, rather than with cumbersome orthopedic shoes. In a similar way, plastic rods and sheets have been used for finger splints, in place of metallic or wooden devices.

Internal Bone Fixation. In cases of severe bone fracture or destruction by disease, the bone must be repaired internally. Usually a metal plate is placed on the bone to bridge the fracture and give rigidity and strength during healing. Such plates are normally made from cobalt or titanium alloys or stainless steel. Unfortunately, the desired purpose may be defeated by body resorption of the bone when insufficient stress is placed on it. In addition, most metals readily undergo fatigue fracture in physiological environments if the bone does not heal. Consequently, this technique sometimes fails. Studies have shown that a more flexible bone plate promotes better bone regrowth. A material more resistant to biological degradation is needed (89–90).

Attempts at bone repair using polymeric foams or cements have not been very successful. These attempts included poly(vinyl formal) sponges and a polyurethane cure-in-place foam (called Ostamer). In clinical tests, the latter had a success rate of only ca 18% and this approach has been discontinued (91). Stronger materials, however,

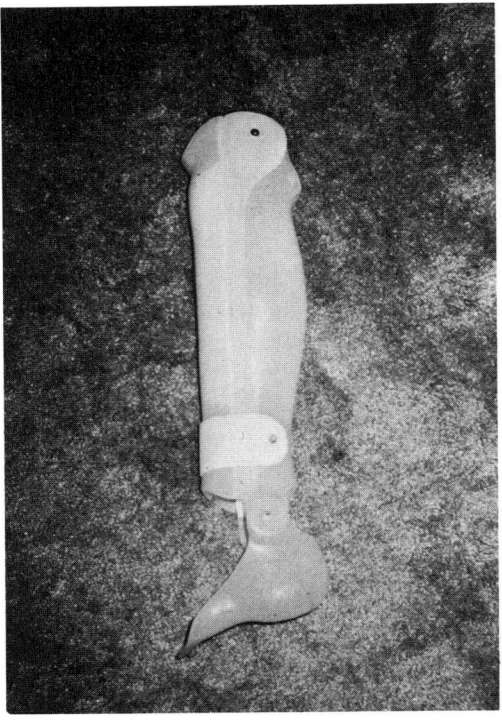

Figure 4.   A polypropylene bracing cast. Courtesy of Shamp Prosthetic Center, Inc.

might succeed. Bioglass, a ceramic consisting of SiO_2, Na_2O, CaO, P_2O_5, and sometimes CaF_2 or B_2O_3 in various proportions, has been used for bone repair, cementing, or replacement. Strong interfacial bonds develop between Bioglass and bone, but failure may occur because the Bioglass is not sufficiently strong (92). Methyl methacrylate, polymerized *in situ*, is sometimes used for bone repair or replacement, although its primary use has been in the fixation of joint prostheses. Aside from the necrosis of nearby tissues from residual monomer, polymeric methyl methacrylate is not strong enough to replace a bone that is placed under much stress (eg, in a leg), although it is used to replace cranial bones, etc. The use of metal rods often leads to bone resorption. Various ceramic materials, such as aluminum oxide and silicon nitride, are being studied. Bone transplants can sometimes be made but rejection problems usually occur except for autografts. At present, the best internal fixation technique is still the use of metal bone plates, in spite of their shortcomings.

Tendon and Ligament Prostheses. Ligaments connect bones and help maintain joint articulation, whereas tendons connect the muscles to the bone. These are normally repaired by direct anastomosis, where possible, or by replacement with an autograft transplant. Prostheses have been made of a fabric, eg, polytetrafluoroethylene, nylon, or Dacron, that is sometimes impregnated with silicone rubber and often used inside a tube or sheath of silicone rubber or polyethylene (see Fig. 3). At the ends an open mesh permits ready ingrowth by muscle or other tissue to give a permanent fixation. Some of these have shown clinical success (93).

Joint Prostheses. Although the human body has many different types of joints, prostheses are mostly applied to the hinge-type (finger, elbow, or knee) or the ball-and-socket type (hip or shoulder). Freely movable joints (diarthoses) have several features in common. In these joints, also called synovial joints, the contacting surfaces are covered with cartilage and the entire joint is enclosed in a fibrous capsule. The viscous fluid, called the synovial fluid, lubricating the joint is secreted from a synovial membrane within the fibrous capsule. The motion of the joint is normally restricted by ligaments or tendons as well as the design. For example, the motion of the hip joint is constrained by the positioning of the hip socket in the pelvic bone. The shoulder, which is also a ball-and-socket joint, is not restricted in the same way and is capable of greater circular motion. Any component of these joints may be damaged by accident or disease and require medical treatment for restoration of function. Some conditions, such as a sprain, heal spontaneously if the joint is allowed to rest. A joint dislocation can usually be corrected externally, without prosthesis. A joint prosthesis must meet the following requirements in order to function properly: maintenance of normal joint space; good, steady, natural joint motion; durable fixation; self-lubrication or possibility to be readily lubricated; stress resistance; resistance to deterioration or erosion; simple, efficient design; ease of fabrication, implantation, and sterilization; biocompatibility with surrounding tissues; lack of leachable materials, and of carcinogenicity.

These characteristics would result in a prosthesis that closely duplicates a normal healthy joint. However, this is seldom possible at present, although some prosthetic joints are fairly good. Most are subject to wear and erosion, resulting in unsteady motion with loosening of the joint fixation. Such joints are now being used in active young adults and problems with wear, etc, are to be expected in the future. The ideal prosthetic joint remains to be developed and will probably require new biomaterials as well as improved design and surgical implantation techniques. The two problems whose severity is not always recognized are carcinogenicity and biocompatibility. As noted above, certain metals and metallic compounds cause tumor formation in rats (26). Chromium and nickel are present in most alloys used in the metal parts of joint prostheses. Although no human tumor formation has been shown to be due to these implants, caution does appear warranted here, especially since joint implants are becoming more common in younger patients.

Biocompatibility problems can arise in several ways. The presence of leachable material usually causes inflammation and tissue necrosis. If this material is a plasticizer or a lubricant, leaching would affect the properties of the joint; therefore, mobile substances are highly undesirable in the joint or other implant. Many joints are fixed with methyl methacrylate cement polymerized *in situ*. Residual monomer would be toxic to nearby tissues but is usually metabolized. As a prosthetic device wears, particles of polymer or metal are released into the surrounding area and cause inflammation. These may be enclosed in a fibrous capsule by the body, which can alleviate the problem, but inflammation often requires surgery. Replacement of the prosthesis is sometimes required. Autopsy studies have shown that a fibrous capsule grows around most joint prostheses which may impede blood flow to remote areas. For example, the fibrous tissue from a hip prosthesis has been found to block the ureter (12).

Finger Joint Prostheses. Many attempts have been made to replace finger joints and restore use to hands disabled by accidents, rheumatism, or arthritis. These have included the use of metal-hinge prostheses and a number of polymeric devices. In the United States, over 400,000 such replacements are made annually. The most successful device is the Swanson design prosthesis and related designs made from silicone rubber. As can be seen in Figure 3, this finger (metacarpal) prosthesis consists of two triangular rods of silicone rubber joined in the center by a concave hinge; these are available in several sizes. In some designs (eg, the Niebauer-Cutter), the ends are covered with a Dacron mesh to aid in fixation. In the surgery, a space is excavated in each part of the finger and the sterilized prosthetic joint inserted. Many variations of this technique are known. It has been effective in tens of thousands of implants, relieving pain and restoring much of the normal hand function (57,94). Silastic has also been used to replace the scaphoid and lunate bones in the carpal region of the hand. Silicone-rubber prostheses have been very successful because of their simple, effective design, the inertness of the biomaterial, the ability of the prosthesis to survive over 10×10^7 flexings, and the fact that the soft biomaterial does not promote nearby bone resorption. Recently, experimental finger joint prostheses have been developed where the polymeric portion is poly(1,4-hexadiene) with exceptional flexural strength and durability.

Hip Prostheses. The natural hip joint consists of a ball on the femoral leg bone inserted into a socket in the acetabulum. Hip prostheses attempt to duplicate this basic design (see Fig. 5), using different acetabular–femoral combinations including metal–metal, plastic–plastic, plastic–metal, and ceramic–metal. In the United States, approximately 250,000 full or partial hip prostheses are implanted each year. The metal–metal prostheses exhibit relatively high coefficients of friction and tend to corrode, especially if different types of metal are used (95). Plastic–plastic prostheses do not hold up well since the plastic materials, such as acrylics, nylons, polytetrafluoroethylene, and polyethylene, tend to wear or fragment or have too little strength to maintain adequate joint stability (93). The most successful designs are the plastic–metal and ceramic–metal systems; the former are much more widely used.

The preferred type of hip prosthesis is the Charnley prosthesis. It consists of a

Figure 5. Hip joint prostheses. Left and back: metal femoral ball and shaft portion; front and center: plastic acetabular socket.

plastic acetabular socket and a metal femoral ball. The socket material is a high molecular weight, high-density (linear) polyethylene (HDPE) that wears about 15 times better than polytetrafluoroethylene in clinical tests. Other materials tried include various polyamides, polyesters, and acrylics, but HDPE has proved the most reliable to date. Several types of metal have been studied for the femoral ball but most prostheses are made of 316-L stainless steel or Vitallium.

Surgically, a region is excavated in the acetabulum and femor and the devices are implanted. They are attached with a methyl methacrylate bone cement that anchors the device in place; threaded prostheses have also been used. Methyl methacrylate cement causes necrosis in nearby tissue, including bone, temporarily but the anchoring is considered adequate in most cases. Solid metal shafts are preferred for the femoral portion since fenestrated shafts are too difficult to remove from the bone if the prosthesis needs replacement (96–97). More recently, Bioglass has been used to anchor both parts of a hip prosthesis (ceramic–metal type) in animals; fixation is good. Bioglass usually forms a bond with the bone which involves a SiO_2-rich and a Ca–P-rich layer. Such bonds develop quickly in experimental animals (98). The HDPE–metal (Vitallium or 316 stainless steel) hip prostheses show a wear depth of only 1.3 to 1.5 mm in ten years. They show a low coefficient of friction and are easy to sterilize, fabricate, and implant. The coefficients of friction for some combinations of Vitallium alloy with plastic, metal, or ceramic using distilled water lubrication are 0.044, HDPE; 0.12, vitreous carbon; 0.17, poly(ethylene terephthalate); and 0.377, Vitallium. Decreased wear correlates well with the lower coefficients of friction but lubrication is also important; synovial fluid usually gives a lower coefficient of friction than distilled water (97). Polytetrafluoroethylene, however, appears to wear more rapidly with synovial fluid (96).

The wear of the HDPE socket-metal ball prosthesis does sometimes present a problem but at autopsy most prostheses show smooth, polished plastic surfaces with no metallic staining. When wear does occur, particulate matter leaves the socket and lodges in the surrounding tissue, where it normally becomes enclosed in a fibrous capsule. Such particulate matter can cause inflammation and pain sometimes so severe that the implant must be replaced. In the past, most hip prostheses were implanted into less active, older patients. How well they will survive in more active, younger patients remains to be seen. Stresses at the femoral head in vigorous motions, such as those involved in tennis, are different from those involved in simple walking. The most common source of failure is loosening of the joint from its setting in the acetabular or femoral bone. This problem will probably become more frequent with younger recipients, and better prostheses will be needed requiring new biomaterials and design improvements. Other metals being used, although less often than Vitallium or stainless steel, include titanium metal and alloys, which are very corrosion resistant, and tantalum metal. Some ceramics, such as alumina, are being explored for the socket portion as are various other types of polymers (eg, polyacetals). Fatigue of the metallic femoral component causes hip-prosthesis failure more often than wear at the acetabular socket, although the latter could give rise to unstable motion in the joint. Coating of the metal component with pyrolytic carbon may aid in reducing corrosion and possibly fatigue fracture. Nevertheless, a more durable metal and a better design are definitely needed.

Knee-Joint Prostheses. The knee-joint is primarily a hinge with some lateral and rotational (axial) motion in addition to the flexing or bending motion. The primary joint articulation consists of two condyles on the lower end of the femur that fit on two oval surfaces of the tibia covered with cartilage sections called the medial and lateral meniscus. A second joint articulation occurs between the femur and the patella (kneecap); this has the effect of restraining forward motion in the knee. The joint is lubricated with synovial fluid. Each year about 100,000 full or partial knee prostheses are implanted in the United States. These prostheses are either nonarticulated devices, such as the Gunston and geometric devices, or articulated, such as the Walldius and Spherocentric devices. Over 80 different types of knee prosthetic joints are in current use but many have similar designs. Unfortunately, the knee prosthesis is not nearly as reliable as the hip prosthesis and failure rates as high as 20% have been reported. However, an actual failure rate of 8–10% appears more typical. These failures, which require reoperation, normally occur in less than two years even though the patients are almost always older, less active people. Failure almost always occurs at the tibular portion of the prosthesis and appears to occur more frequently with heavier patients, whereas the failure in hip prostheses appears independent of body weight. The total force on the tibular portion of the prosthesis has been estimated to be between 4 and 11 times the body weight of the usual patient (55–100 kg) (99). This is much higher than the force of 2–6 times the body weight that might be expected in the hip prosthesis. This problem, coupled with the more complex motion of the knee, has made this prosthesis more difficult to design. Nevertheless, knee prostheses have successfully relieved pain and restored mobility in thousands of patients.

The Walldius knee prosthesis was developed in the early 1950s and was first made from an acrylic plastic which wore poorly. Present models are made from a cobalt–chromium–molybdenum alloy. The Walldius knee consists of a metal hinge with long metal rods that are inserted into the intramedullary spaces of the tibia and the femur, thereby connecting these bonds by a mechanical joint. Failure usually occurs by tissue inflammation caused by debris from the metal–metal hinge or loosening of the prosthesis from the bone with attendant bone damage. The Walldius joint does, however, provide relief from arthritic pain and restores much of the normal knee function. It is still widely used after about three decades.

Several designs for nonarticulated joints rely primarily upon replacing the damaged load-bearing sections of the knee without a mechanical femur-tibia linkage.

The classical example is the Gunston knee prosthesis, introduced in 1971, which utilizes the same type of metal–plastic combination used in the Charnley hip prosthesis. It consists of metal runners cemented to the femur that ride on high density polyethylene (HDPE) tracks cemented to the tibia. Methyl methacrylate cement is used for fixation. The geometric knee prosthesis is similar, except that it consists of two rather than four parts which makes surgical alignment simpler. It restores up to 130° flexion (bending) motion to the knee (normal value is about 135°). The femoral component, a spherical runner, is made from a metal alloy (usually cobalt–chromium–molybdenum) and cemented to the condyle section of the femur with methyl methacrylate polymerized *in situ*. The tibial track component has a matching surface made from HDPE which is also cemented in place using methyl methacrylate. The stability of the geometric device depends largely on the ligaments and muscles of the patient in a manner somewhat similar to a natural knee joint. Geometric knee prostheses usually fail at the tibial section, because inadequate skeletal fixation permits a positional change that leads to the crushing of the underlying bone. This bone collapse then permits more positional shifting with even greater bone fragmentation and crushing. In some cases, the angle of shifting is as much as 10° from the originial angle that occurred at arthroplasty. Much pain results, along with erratic joint motion, and surgical correction is required. The problem can be alleviated by using thicker HDPE tracks (10 vs 5 mm), by making certain that the tibial portion is fixed more firmly, and by correct surgical positioning of each component in the prosthesis. The geometric knee prosthesis, in its various models, is probably the most widely used design. It restores much knee motion while reducing pain. Failure, which can be as high as 20%, usually occurs within two years of implantation (99–100).

A third type is the spherocentric knee prosthesis developed in the early 1970s. This articulated knee combines features of the geometric knee and the hip prosthesis. The femur section consists of a metal casing that is cemented in place and contains a socket of HDPE. The metal forms a runner at the edges that rides on a HDPE track on the tibia. The tibial section contains a metal ball and shaft that is cemented in place and fits into the HDPE plastic socket in the femur. Unlike the Walldius articulated joint, there is no metal–metal contact in the spherocentric joint and the ball-in-socket connection consists of a metal ball in the plastic socket (like the hip prosthesis). The spherocentric knee, which is still under clinic evaluation, can give about 120° flexional motion, 30° axial rotation, and 5° lateral motion. All plastic-bearing surfaces are supported by metal (a cobalt–chromium–molybdenum alloy) in the device. Although long-term data are not yet available, it appears that failure occurs in the tibial section, as with other replacement knees (100).

At present, the knee prosthesis has a fairly high failure rate, mostly because of inadequate joint fixation. This difficulty might be alleviated by design modifications that permit bone ingrowth into the prosthesis or by better cementing methods and materials. A design using several materials in both the femoral and tibial portions (as in the Spherocentric knee) might help solve the problems.

Another area of knee surgery is kneecap replacement. In the past, the kneecap (patella) has been replaced with celluloid, but polydimethylsiloxane rubber has been preferred since the early 1960s. A polyurethane rubber also shows promise because of its greater strength.

Joint Lubrication. No movable joint would function well for long without lubrication. This is true both for mechanical or biological joints. The lubrication of natural joints is not completely understood but appears to depend on both the synovial fluid and the articular cartilage that lines each joint. The synovial fluid is a yellow liquid with about 1000 times the viscosity of water. It is usually considered to be a dialysate of the blood and contains the mucosaccharide hyaluronic acid, and some proteins and enzymes in a medium with a pH of about 7.3. It is non-Newtonian in its viscosity characteristics and shows lower viscosity at higher shear rates. Diseases such as rheumatoid arthritis can cause a breakdown of the synovial fluid. It seems to function more as a wear inhibitor than as an actual lubricant like oil in a mechanical joint. Natural joints can be lubricated nearly as well with a low viscosity, buffered solution. When the synovial fluid is compressed, it forms a gel that prevents cartilage–cartilage contact at the joint. Such contact would cause excessive wear and joint damage. The cartilage structure is quite complex and it appears that the articular cartilage, which is kept moist and nourished by the synovial fluid, serves as the lubrication site for the joints. Breakdown of this lubrication system can occur from disease affecting the cartilage or the synovial fluid, or from abnormal stress on the joint. Such problems can be treated by medication, a synthetic lubricant, or an artificial joint (101–102).

Silicone fluids were tried clinically as synovial-fluid replacements with some success in 1968. However, a double-blind clinical test showed saline solution to be as good as or better than silicone fluid. It was suggested that the saline solution might have stimulated an increase in synovial-fluid production, which improves joint lubrication (102). Since normal joint lubrication depends on both fluid and cartilage, the saline solution probably functioned like a normal lubricant, whereas the silicone fluid operated differently (possibly by boundary lubrication) since it could not penetrate the cartilage. Several other fluids have been examined as possible joint lubricants, including aqueous solutions of methyl cellulose or polyvinylpyrrolidinone but no completely successful replacement has been found.

Lubrication is essential in artificial replacement joints. As noted above, most of these joints contain metal-plastic interfaces. These joints show low coefficients of friction when lubricated with water or synovial fluid but wear badly when dry. Any joint replacement must provide lubrication. Normally, a small space between joint interfaces permits synovial fluid to enter and serve as a lubricant.

Artificial Limbs. When injury or disease disables part or all of a limb, a replacement becomes necessary. This may be an artificial foot, leg, hand, or arm. The replacement becomes more difficult as the total portion replaced increases. Thus, a hand or a foot is easier to replace than an entire arm or leg. Artificial limbs can be attached externally by means of buckles or straps, or they can be attached directly to the skeletal system. Obviously, the latter procedure is more difficult, since the device has to pass through the skin with all the accompanying biocompatibility problems. Nevertheless, direct attachment would make an artificial limb a better replacement, if this could be done safely. At present, the direct-attachment approach is considered experimental, and artificial limbs are attached externally in almost all cases. It is important that amputation be performed in such a manner that the artificial limb can be attached readily by either external or internal devices. Generally, this means preserving as much of the natural limb as possible. In the past, this was not always done and good limb attachment was not always possible.

The total number of patients who might benefit from an artificial limb is difficult to estimate but there appear to be over 25,000 amputations per year in the United States. War tends to increase the rate greatly. In addition, there are over 500,000 quadriplegics in the United States and an unknown number with birth defects (103). Although as many as 5×10^6 people could be candidates for an artificial limb, they represent only 2% of the population of the United States and, therefore, incentive for extensive research funding has been lacking. Nevertheless, many advances have been made in the design of artificial limbs. They can be cosmetic (63) or functional (103–106). Only the latter are considered here. Such devices are powered or controlled by the patient's muscles; some designs have used small motors for additional power. Myoelectric control, via the electrical impulses of other muscles in the patient, has been employed with some degree of success (107). The primary purpose of these limbs is to permit the patient to lead a fairly normal life, including moving about without attracting undue attention, and being accepted by others.

Lower-Limb Replacements. Lower-limb replacements include replacements for the foot or part or all of the leg, have been made from wood, leather, metals, and plastics. Plastics are increasingly popular because they are light in weight and can be shaped readily. They must, however, sometimes be reinforced by interior metal rods. Although the simple peg-leg-type design is still occasionally used, modern techniques have created devices that function and look better. Several types of foot-ankle units are available, but the most widely used devices are the solid-ankle cushion-heel (SACH) foot and the single-axis foot. The SACH type has no moving or rotating parts, whereas the single-axis foot permits motion of the foot relative to the leg. Either device permits the patient to function fairly well. Both can be made from wood or plastic with rubber bumpers to reduce impact shock (5,104).

For amputations above the knee, a knee unit must be added to the prosthetic device. This unit must permit bending for kneeling or sitting. In addition, it must be capable of staying rigid for standing and must permit walking and climbing stairs. Above-knee units are usually attached by a suction socket into which the leg stump is strapped. Obviously, such a knee prosthesis must be hinged, and several types have been designed that are mechanial or hydraulic in operation. The bending motion is controlled by the motion of the upper leg (swing-phase control), whereas the vertical-standing control is normally achieved by a friction-braking action (5,104,106). When an entire leg is amputated, the prosthesis must then be attached to the hip and is strapped in place. An entire leg prosthesis combines the problems of the above-knee prosthesis with the additional difficulty that there are no leg muscles available to power the unit. Myoelectric control by small motors is possible but has not yet been developed into an effective unit. Most present units are powered by shifting the body weight and propelling the prosthetic leg from the hip (5). The two basic designs for these artificial legs include a hollow device constructed from a rigid plastic or wood with metal reinforcements at the hinges, and rubber or sponge cushions where needed to absorb impact. Alternatively, the device can be constructed with an articulated metal rod in the center that is surrounded by foam sponge and covered with a plastic film with a feel similar to skin. Either type can be color matched to the patient's skin, but the second type has a more natural feel. In addition, the articulated joints can be made better and more durable.

Upper-Limb Prostheses. Upper-limb prostheses may replace the hand, the elbow or the entire arm. Like the lower limb devices, these may be cosmetic, functional, or both. They are held in place by a suction cup into which the remaining stump is strapped. They are powered by the patient through a strap-and-cable device or by motors controlled myoelectrically. The key portion of the upper limb is, of course, the hand which is involved in various types of prehension, such as: palmar pinch, tip prehension, lateral pinch, cylindrical grasp, spherical grasp, and hook grasp. Unfortunately, no artificial hand can duplicate all these functions. Essentially, all artificial hands provide palmar prehension (pinching object between the thumb and the index and middle fingers). These devices can resemble a hand, but a hook is usually more functional. For this reason, many patients use a hand for esthetic purposes in public, but a hook device for performing various tasks in private. These hands-hooks can be voluntary-opening or voluntary-closing devices, depending on how they are constructed. (The voluntary-opening device more closely resembles the function of a normal hand.) Prehension of small objects is usually difficult with most of these devices and, naturally, they lack a tactile sense and must be guided visually. A hand-type prosthesis can be covered with a plastic colored to match the skin; hair can be implanted for a more natural appearance. Even the best devices, however, are not nearly as effective as a natural hand and patients with an intact natural hand perform most functions with it and use the artificial hand or limb only for tasks requiring both hands (5,104).

Replacing the hand only has the advantage that many muscles and points of control remain to power the device. When amputation occurs above the elbow, these power and control points are reduced and more functions must be performed with fewer sources. This is even true when the entire arm is replaced. In such cases, much of the control and powering is obtained from the muscles of the back and shoulders and is transmitted via straps, cables, and pulleys or myoelectrically. Although many modern devices utilize battery-powered motors to move these artificial limbs, it is possible to effect motion by the back and shoulders muscles with springs being used to return the device to the original position. A typical myoelectric arm is shown in Figure 6. Sensing devices are placed on the skin or implanted through the skin to detect the myoelectric currents in muscles and these are used to control the motion of the arm. Normally, only one type of motion is possible at a time, eg, the arm can be rotated or bent, but not both at the same time. However, most devices provide a multiplicity of motions in sequence, eg, rotation after bending. Such myoelectric devices can terminate in a hand-type unit (as shown in Fig. 6) or in a hook-type unit. The hook unit is normally preferred for manual tasks, and a hand is substituted for social activities (5,104).

Organ Replacement or Augmentation Prostheses

The proper function of several organs is essential for the maintenance of good health and even life itself. When such an organ (eg, heart, lungs, liver, pancreas, kidneys) begins to fail, the patient soon develops characteristic symptoms that lead to death, if not treated. Treatment can consist of medication; modification of life style; transplantation; or artificial organs including assist devices. Medication and life-style modification are the preferred approaches. Thus, a diabetic controls sugar intake and injects insulin, whereas a person with a badly defective heart avoids unusually strenuous activities and takes the appropriate medication. Unfortunately, these approaches are limited and alternatives such as transplantation or artificial organs must be considered. Organ transplant from a living donor is limited to the kidney. The donated

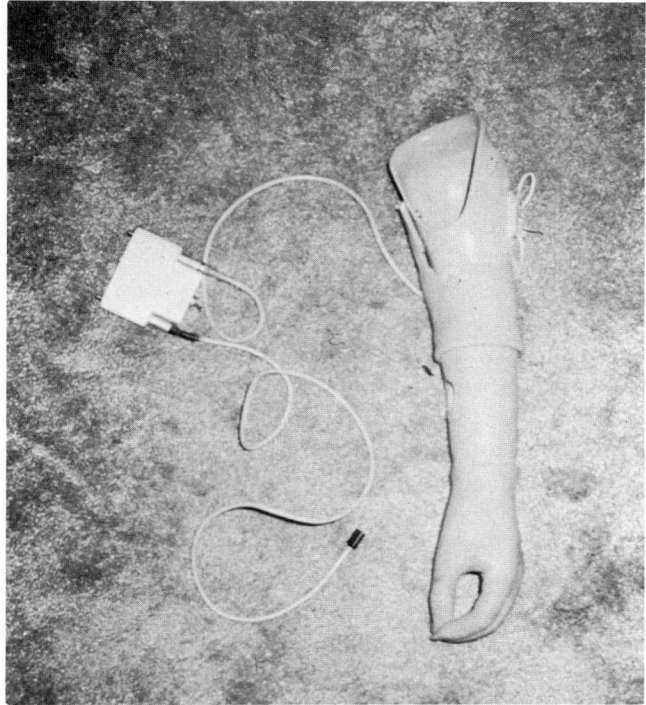

Figure 6. A myoelectric arm. Courtesy of Shamp Prosthetic Center, Inc.

organ is normally recognized by the body as foreign tissue and a series of immunological responses is initiated by which the recipient's body attempts to destroy or reject the transplant. This is a serious drawback and only those organs from donors with closely similar physiological characteristics have any chance of successful transplantation. The greatest success rate occurs with kidney transplants between close relatives although nonrelatives can sometimes be used. Under these conditions, the success rate is about 75%. With other organs, such as heart transplants, the rate is lower. Certain drugs suppress the immunological response of the body but these leave the patient highly susceptible to disease. Furthermore, the donor organ is not always available when needed. The time an organ can be stored before implantation is fairly short and varies with the organ. Today, the short storage life of a donated organ does not present a problem because demand surpasses supplies. Finally, the problems that caused the original organ to fail could cause failure of the transplanted organ as well.

An artificial organ could eliminate some of these difficulties since it can be built in advance and stored until needed. The problem of rejection is not necessarily eliminated since the body views an artificial organ as an intrusion that must be attacked and destroyed. However, the materials of construction can be designed or selected to resist such an attack. Each artificial organ is designed according to specific material and functional requirements: it must be biocompatible in order to avert rejection; since it has direct contact with the blood, it must be blood compatible; it must be capable of replacing or assisting the natural organ's function in a manner similar to the natural method, although some variation may be possible, eg, a nonpulsatile heart; long-term operation (10 years or more) is desirable; and an appropriate power source might be included and possibly some means of renewing a chemical (eg, insulin release in an artificial pancreas) or a physical activity (eg, dialysis in an artificial kidney). Ideally, these requirements should be met without restricting the patient's mobility (eg, by an attached electrical cord, etc) or impeding the pursuit of a fairly normal life. Although expense might seem trivial when the quality or maintenance of life is being considered, it cannot be completely ignored. A person might develop hostility to a device that consumed all of his or her resources, and this may reduce the actual quality of life. The ideal artificial organ must also be capable of implantation in about the same space as the natural organ. These requirements are very difficult to achieve in actual practice.

Many different artificial organs are in use at the present time, but all fall short of the ideal. Some are implanted (eg, heart valves), whereas others are solely extracorporal devices (eg, the artificial kidney). Much research is expanded to develop a total artificial heart, whereas little attention is given to the development of an implantable artificial kidney since a patient can be maintained on an extracorporal dialysis unit. However, maintenance of life with a heart–lung machine does not permit the patient to approach a normal life style and is therefore a highly unsatisfactory approach. Ideally, any artificial organ is more satisfactory as an implanted device but some extracorporal systems, such as the present artificial kidney, can be tolerated whereas others cannot.

Heart Pacemakers. Cardiac pacemakers regulate the heart beat and are employed to treat Stokes-Adams disease (heart block) in which the heart slows down to 20–30 beats per minute resulting in weakness, fainting spells, and sometimes sudden death. Pacemakers operate by electrical stimulation of the heart to beat faster. The first experimental heart pacemakers were implanted in animals in 1932; Stokes-Adams disease was treated with electrodes placed on the patient's chest in 1952, whereas implanted pacemakers were first developed in 1958. Since then, pacemakers have been developed that can vary the beat rates and over 30,000 have been installed. The plastic-coated electrodes are usually sewn into the myocardium of the heart and powered by mercury or lithium cells (108–109).

Heart Valves. The heart has four valves, ie, aortic, mitral, tricuspid, and pulmonary. Most replacements are made on the aortic (58.6%) or mitral (41.2%) valves (110–112). Many different replacement heart valves have been developed (113–114). Valve replacements are made either from polymers and metals or natural materials (homografts or modified heterografts). Figure 7 is reproduction of a Starr-Edwards mitral valve prosthesis consisting of a metal ring and cage (struts) made from Stellite (cobalt–chromium–molybdenum alloy) with a polymeric ball (silicone rubber) in the cage. The ring is covered with Teflon or Dacron mesh to anchor the prosthesis by endothelial ingrowth. In some models the struts are covered with the same mesh to reduce noise and thrombus formation. Other heart valves contain polymeric (Teflon, Delrin, or Silastic) or pyrolytic carbon disks, balls, or leaflets. An anticoagulant, eg, sodium warfarin, prevents excessive thrombus formation.

Natural tissue valves are constructed from pig (porcine) heart valves, bovine pericardium, or human dura mater which had been treated with glutaraldehyde or glycerol (110,112,115–116) to reduce antigenicity. The porcine valves taken from freshly killed pigs are reconstructed on a flexible metal frame, usually Elgiloy (a cobalt–nickel alloy) (110). The dura-mater valves are made from material obtained from accident victims within 14 hours of death and treated with 98% glycerol for 2–3 weeks. The modified dura-mater homografts are then cut and fitted on either a rigid (stainless

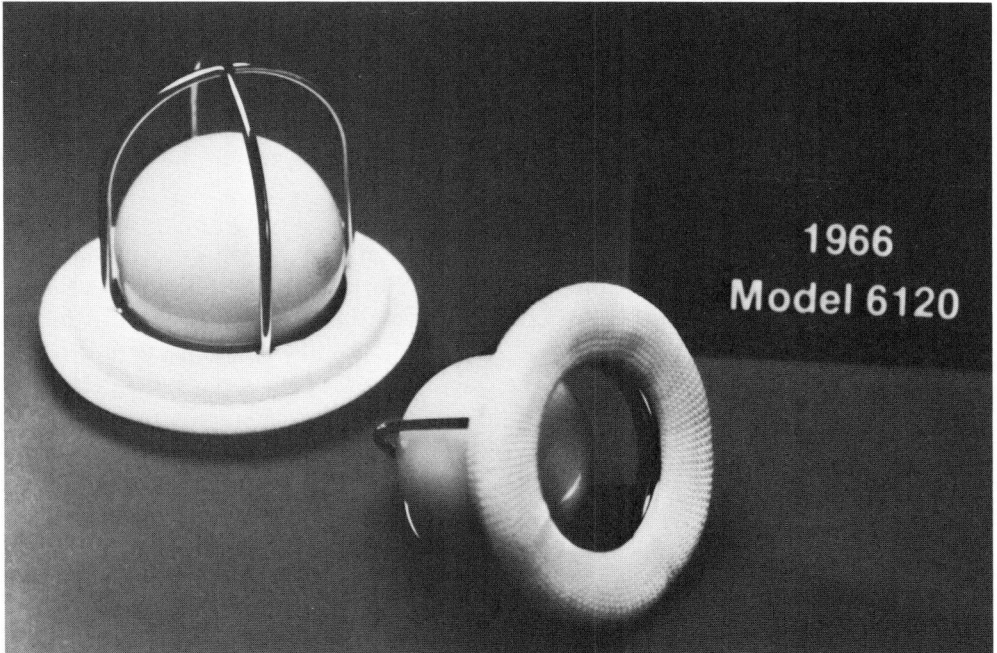

Figure 7. The Starr-Edwards mitral valve prosthesis. Courtesy of American Edwards Laboratories.

steel) or flexible (polypropylene) frame (struts) (115–116). Naturally derived biomaterial prosthetic valves show a lower rate of thromboembolism, even without anticoagulants, than the valves derived from completely synthetic materials. With either type, long-term success can be obtained, provided the valves are properly sterilized and free of microdirt on the prosthetic surface (111).

Blood-Vessel Replacement. Various diseases, accidents, and other sources of trauma can necessitate the replacement of all or part of a blood vessel. Approximately 110,000 coronary-artery by-pass operations were performed in the United States in 1980 and this number seems to be increasing. Some injuries can be repaired by suturing the blood vessels together after removing the injured portion, but this is only possible when short segments of the vessel are removed. Many materials have been investigated as blood vessel replacement. Experiments were made as early as the turn of the century. Metals were tried first, but rapidly caused thrombosis. Polyethylene, nylon, and Orlon were tried in the 1950s, but clotting that blocked the tube occurred. Silicone rubber also cause blood clotting in spite of the fact that it can be used successfully as the ball in heart valves. Various other plastic materials in the form of solid tubes, sponges, and fabrics were completely unsuccessful. Currently the knitted or woven Dacron prosthesis and the expanded polytetrafluoroethylene graft (Gore-Tex) are widely used. Natural materials have also been tried. In coronary by-pass surgery and other blood vessel replacement the saphenous vein from the leg is transplanted to the appropriate location in the body. This technique, first introduced in 1951, is generally considered the best method available, although some surgeons prefer the Dacron prosthesis for larger arteries such as the aorta (117). Unfortunately, in many cases of advanced disease, blood vessels, including these saphenous veins, are not in sufficiently good condition to use as a transplant. Glutaraldehyde-stabilized human umbilical-cord veins have been used successfully for blood vessel replacement and have shown patency in humans for more than four years (118). Obviously, the umbilical cord vein must be clean and fresh. Umbilical cord veins can be longer than 50 cm and are useful for many types of replacement operations. However, they are not as readily available as polymeric material. Various heterografts, such as bovine blood vessels, have been tried but mostly without success.

The preferred polymeric blood-vessel prosthesis is a knitted or woven Dacron tube (see Figure 8). It is ribbed and flexible. In implantation, a clot forms in the pores of the Dacron mesh and seals the prosthesis against further blood leakage. Eventually, this clot is replaced with a neointima which has essentially the same structure as the inside surface of the natural blood vessel. As a result, the blood actually contacts this natural neointima surface rather than the Dacron itself. Since poly(ethylene terephthalate) is thrombogenic, this is an important advantage. Experimental work continues on Dacron prostheses and some current research suggests that a double-velour prosthesis implants easier and develops a more adherent, better formed neointima than the knitted or woven Dacron prosthesis (119). However, the former can be used only for relatively large blood vessels. With vessels below ca 6 mm, the neointima ingrowth blocks the tube and renders it nonpatent. The newer, expanded polytetrafluoroethylene (PTFE) grafts (Gore-Tex), are composed of a microporous PTFE with a surface consisting of small nodes interconnected with many thin, flexible fibrils. The neointima is formed rapidly into a smooth thin layer that transverses the suture lines. These PTFE grafts are considered very promising and have been used in hundreds of cases with excellent patency (usually above 70%). The material is easily sterilized and stored, and is readily available for surgery at any time (unlike natural prostheses). Tubes as

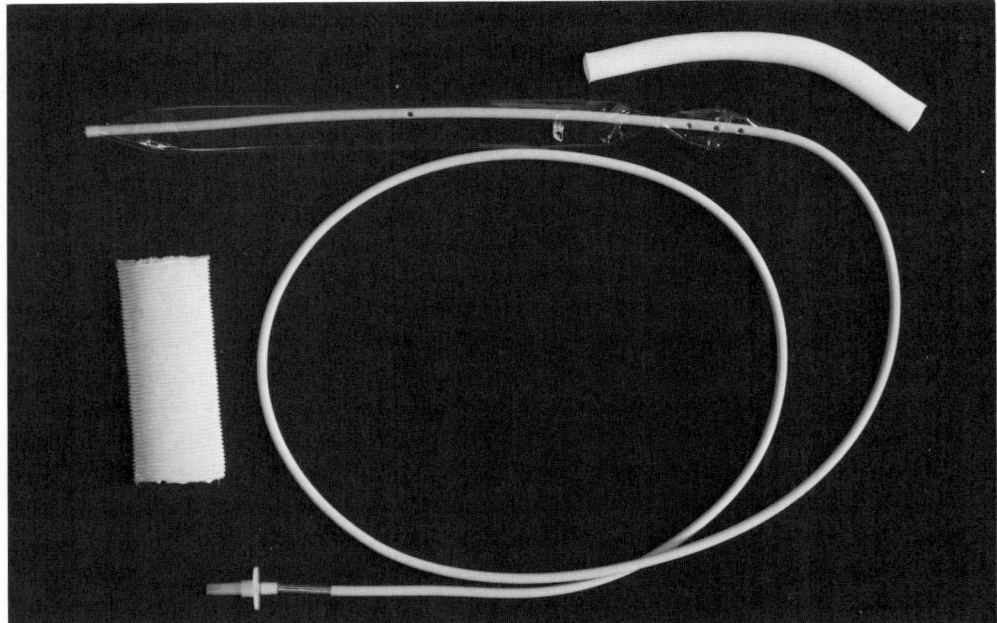

Figure 8. Some cardiovascular devices. Right: Dacron aorta prosthesis; bottom: expanded PTFE arterial prosthesis. Courtesy of Gore-Tex Corp.; center: intra-aortic balloon heart assist device. Courtesy of Datascope Corp.

narrow as 4 mm have been successfully implanted into humans with patency greater than 70%. In dog experiments, vessels as narrow as 3 mm were used successfully (120–123). An example of the Gore-Tex PTFE graft is shown in Figure 8.

Many blood vessels are much narrower than 4 mm and at the present time there are no prostheses available to replace such vessels. The most promising lines of current research seem to be the polyether polyurethane ureas (PEUU) (33,51,124) and the hydrogel systems (45–47). The PEUU systems can be made in small diameters, and some relatively high patency rates (above 75%) have been reported for 4-mm implants in dogs (124). Long-term data (several years) are not yet available but this procedure shows promise. As noted earlier, it is generally believed that the blood compatibility of these PEUU systems is derived from preferential adsorption of albumin. The hydrogel systems also show much promise in the form of hydrogels grafted to another polymer in order to achieve sufficient strength. Some vessels made from these grafted hydrogels have shown good results in 2-h vena-cava ring tests in dogs (46) but this technique is still highly experimental and will probably not be available for human use for several years.

Heart-Assist Devices. In the United States, more than 500,000 persons are hospitalized each year with some form of heart attack (myocardial infarction). Several devices have been developed to aid a failing heart such as an intraaortic balloon pump or an implanted mechanical left-ventricular assist device. In open-heart surgery a heart-lung machine takes over. The simplest heart-assist device is the intraaortic balloon-pumping (IABP) system shown in Figure 8 (125–131). The IABP is inserted into the aorta, usually through the femoral artery, and the balloon is alternately inflated and deflated at a rate to match the heart beat. The IABP consists of a PEUU balloon set mounted on a hollow catheter which is connected to the pumping device. The IABP is usually 30 cm long and has an 18 mm distal-occluding balloon and a 14 mm proximal-pumping balloon. The IABP system is generally considered to be the best available heart-assist device, mainly because of its ease of insertion and removal and the simplicity of operation. Circulation is improved in about 75% of patients with refractory cardiogenic shock, but the mortality rate is still in the range of 65–90%.

In many cases of advanced heart disease or damage, the IABP cannot maintain an adequate blood flow and some other type of device is necessary. A number of left-ventricular assist devices (LVAD) have been developed over the past two decades (132–137). Both implantable and extracorporal devices are available, but since they are all externally powered, they do not permit much patient mobility; battery-powered units are being developed. In general, LVADs are considered temporary replacements rather than permanent-assist devices (see Fig. 9). Animal experiments have shown that the devices can function safely for longer than 8–10 months which is beyond the normally expected time of use (138). The blood is withdrawn from the left atrium into the pump and then returned to the ascending aorta. In effect, LVADs are half a mechanical heart and artificial hearts are made by connecting a pair. LVADs vary greatly in design and may be a tubular- or sac-type device made from many different types of materials. The most commonly used material appears to be PEUU for the blood-contacting surface although Dacron-flocked and biolized surfaces are also available. LVADs can be implanted or used in a paracorporal manner and, although most devices are planned for short-term use (two weeks or less), systems are under investigation that could be used for months or even years. It has been estimated that the need for such devices is in the range of up to 50,000 annually.

304 PROSTHETIC AND BIOMEDICAL DEVICES

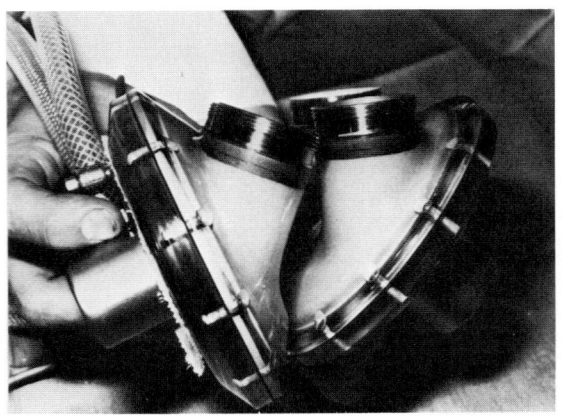

(a)

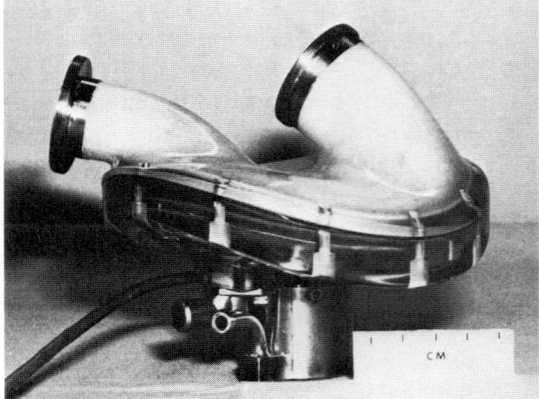

(b)

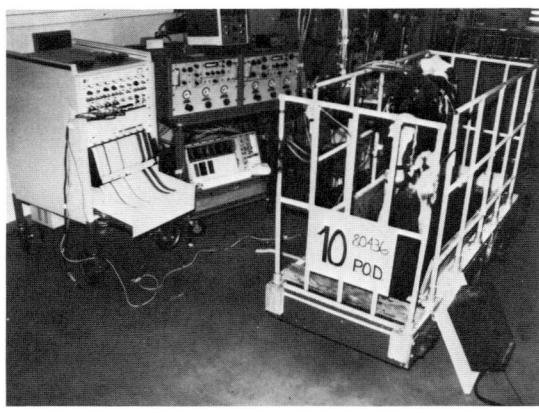

(c)

Figure 9. Total artificial heart (TAH) prosthesis. (**a**) Complete artificial heart. (**b**) One ventricle. (**c**) Calf with implanted TAH showing maintenance equipment. Courtesy of Artificial Organs Dept., Cleveland Clinic.

The heart–lung machine is an extracorporal blood-pumping system as well as an oxygenation device. In the United States, more than 75,000 annually are used in various cardiac and other operations.

Total Artificial Heart. Although modern medical techniques, including pacemakers, prosthetic valves and drugs, and changes in life style and diet, greatly reduce the stress on an ailing heart, heart diseases account for approximately 10^6 deaths in the United States each year. At the present time, transplantation of another heart is the only alternative to death when this organ ceases to function. Heart-transplantation techniques have improved greatly since the initial work of Christiaan Barnard in 1967 and about 65% of the current recipients survive at least a year, some even for over a decade. There are, however, obvious limitations to this technique. First and foremost is the fact that heart transplantation requires a heart from another person. Second, there are always problems of rejection by the recipient's body. Although rejection can be controlled by various drugs, it leaves the patient in a weakened condition and highly susceptible to infection. These problems are accentuated when a transplant is attempted using a heart from an animal, such as a chimpanzee; such transplants are short-lived. The need for a totally artificial heart (TAH) is readily apparent although this device could give rise to some difficult ethical and moral considerations as could any artificial vital organ. An artificial heart was implanted in a human in 1969 to sustain life for 64 hours until a natural heart became available for a transplant (139–140). No further attempts of this type have been reported since then until 1981 when another total artificial heart was used to sustain a patient's life until a transplant became available (50).

Among the specific requirements for a TAH are biocompatibility, blood compatibility, size, pumping capacity, ability to vary the pumping rate, and durability. A typical TAH and the support equipment are shown in Figure 9. The main problems in maintaining life in animals are caused by a weight increase after implantation (eg, from ca 90 kg to over 160 kg during a seven-month interval) and infection at the percutaneous sites where the life-support tubes enter the animal's body. Most TAHs are constructed with PEUU as the blood-contacting surface, although biolized poly(1,4-hexadiene) is also used. In earlier work poly(vinyl chloride) or poly(dimethylsiloxane) was used, but these materials are fairly thrombogenic and not widely used today. Most current experiments are performed on calves but also on dogs, sheep, pigs, and goats since the mature animals are more readily handled in the laboratory. Life in calves has been maintained up to 268 days. The power source is normally a pneumatic system using compressed air but several electric-motor systems have been developed and a nuclear engine is being investigated. The latter devices permit mobility to the recipient. Nuclear-powered motors could potentially operate uninterruptedly for over ten years whereas electric devices could be powered by rechargeable, portable battery packs. Research on the TAH is being conducted at over fifteen centers in at least twelve countries and approval for possible long-term human use is being sought (50,52,139,141–145).

Artificial Kidney. Although the human body is equipped with two kidneys, it can function adequately with one. Consequently, the preferred medical treatment of a severely defective kidney is transplantation. The first kidney transplant was made in 1954 and these operations are fairly common now. Immunological rejections do occur unless the donor and recipient are closely matched and, therefore, close relatives are often the donors. When transplantation is not possible, the patient is connected to a hemodialyzer or dialysis unit. These extracorporal devices are normally connected to cannulae in the patient's nondominant arm and the blood is passed through the dialysis unit where the various metabolic waste products are removed by passage through a semipermeable membrane into a dialysis fluid. The first hemodialyzers were developed in the early 1940s. The key unit is the membrane which may be in the form of a coil, a plate, or hollow fibers (see Dialysis; Hollow-fiber membranes). The membranes are made of cellulosic derivatives, although polyacrylonitrile is also used. Blood clotting is controlled by addition of heparin or other anticoagulants, or by heparinization of the tubing and membranes (146–149). The membranes are discarded after use.

In 1982, approximately 40,000 people in the United States and over 100,000 people worldwide are maintained on dialysis units. Many others undergo hemodialysis treatment for brief treatment times to correct a temporary problem. The present artificial kidney is not a perfect solution to kidney problems and can cause secondary disorders, such as anemia, hypertension, and hemolysis. Psychiatric problems may occur owing to the fact that the patient depends on the machine and feels restricted in mobility and activity. A more effective membrane unit reduces the frequency of dialysis, and home dialysis units are also available. A portable artificial kidney would permit the patient even greater mobility (10). An implantable device may be the best solution but little research is being done in this area since the present hemodialysis system is considered adequate until a transplant can be made.

Artificial Lung. Implantable, artificial lungs do not exist at present but various extracorporal oxygenators have been developed. These devices circulate the patient's blood through a system in which the blood becomes oxygenated. Oxygenators can be used as part of a heart–lung machine or as a separate unit when the heart is functioning adequately. Different types include bubble oxygenators, film oxygenators, disk oxygenators, and membrane oxygenators (see Table 2). The system with the highest oxygenation rate causes the greatest blood damage and requires the largest amount of blood for priming. The most widely used system, the membrane and the bubble oxygenators, show the best combination of characteristics. In the bubble oxygenator, oxygen (or air) is bubbled through the blood, whereas in the membrane oxygenator the blood is passed through a coiled membrane where the oxygen exchange occurs. The surface of a membrane oxygenator, 2–4 m^2, is much larger than that of an artificial kidney (0.5–1.5 m^2), and the blood flow rate is much faster (5 L/min vs 0.2 L/min). This makes it much more difficult to achieve blood compatibility and avoid blood damage (150–151). Most recent work has concentrated on improving membranes and membrane polymers. The most widely used material is silicone rubber but other materials, such as a silicone rubber–polycarbonate copolymer, polyalkylsulfones (PAS), and ethyl cellulose perfluorobutyrate polymers (ECFB) show much promise (152–153). A strong, thin, pinhole-free film is needed that transmits oxygen and carbon dioxide across the membrane without damaging the blood. Artificial lung devices are used in about 100,000 cases a year, usually in conjunction with a heart–lung machine; the membranes are discarded after each use to prevent cross-contamination of the blood.

Artificial Liver. The liver is the main detoxification organ in the body and can be damaged by poisoning, drug overdoses, allergies, and acute hepatitis. No true artificial liver has yet been developed and transplantation is very difficult and seldom attempted. Fortunately, the liver has an enormous regenerative ability and short-term substitution devices often enable the patient to survive a temporary crisis. The most common treatment is hemoperfusion in which the blood is passed through a column or bed of some sorbent material to remove the toxic substances. The sorbent material can be charcoal, ion-exchange resins, affinity-chromatography materials, immobilized

Table 2. Characteristics of Different Blood Oxygenators

Characteristic	Oxygenator type			
	Membrane	Bubble	Film	Disk
priming blood volume	small	medium	large	large
blood damage	lowest	moderate	moderate	high
oxygenation capacity	medium	medium	high	high

hepatic microsomes, or liver material in the form of slices or enclosed in artificial cells. Extracorporal perfusion can also be performed via the liver of another person or an animal (usually a primate). The main problems encountered are blood incompatibility, blood damage, low efficiency, and reactor-particle carry-over into the blood stream. Nevertheless, hemoperfusion is widely used to treat drug overdose and poisoning as well as hepatic failure due to disease (154–157).

Artificial Pancreas. The pancreas produces enzymes that promote digestion in the small intestine, and hormones, including insulin, that are secreted by the islets of Langerhans (see Insulin and other antidiabetic agents). The device termed an artificial pancreas is actually a pump that releases insulin at a controlled rate. The NIH estimates that in the United States about 10.2×10^6 people have diabetes but only about half of these cases are detected. Although less than 10% of these cases have insulin-dependent diabetes, many use insulin to control the disease symptoms. The NIH estimates that diabetes is controlled by insulin in 1.1×10^6, by oral antidiabetic drugs in 2.1×10^6, and by diet and exercise alone in 2.0×10^6 people. Insulin controls the metabolism of glucose in the blood and in a normal person it is released from the beta cells of the pancreas at a rate sufficient to maintain the glucose concentration in the normal range. In the case of a diabetic, the glucose levels can greatly exceed the normal range when the insulin is injected only once or twice daily. For about the past fifteen years, various research centers have experimented with various types of infusion pumps to deliver controlled amounts of insulin. Most of these devices have been external, wearable, portable pumps but one type was implanted. The purpose of the implant is to obtain better control of glucose concentration by controlling the rate of insulin release. Infusion pumps do not precisely match the behavior of a healthy, normal pancreas, although most have some means to increase insulin release after a meal. Ideally, these devices could be controlled by a glucose sensor but present devices are far too large to be portable. It is hoped that implantable glucose sensors, complete with a microprocessor, will be developed by the late 1980s (158–159).

Medication Administration Devices

Better methods to administer medication are constantly needed and a number of devices and approaches have been developed in recent years, including controlled-release systems and polymeric drugs. These systems control the medication dose rate, transport the medication to the proper site, reduce toxic side effects, and prolong drug activity (see Pharmaceuticals, controlled release).

Controlled-Release Systems. All medicines are most effective in an optimum concentration range. Below this concentration, little therapeutic activity occurs whereas above this range, the toxic side effects may be severe. Normal medication procedures, by oral or injection routes, result in varying concentrations of the drug in the body. The therapeutic range is difficult to maintain unless several small doses are taken daily. Controlled-release systems achieve the desired drug release level in a simple manner. The drug is usually contained within a polymeric matrix or membrane and is released at a fairly constant rate. The simplest device is the reservoir type in which the drug is encapsulated within the polymeric matrix, whereas in the monolithic type the drug is dispersed throughout the polymeric matrix. Either system could be employed for the controlled release of a specific drug, although each system follows different release kinetics. The polymers are nondegradable, eg, silicone rubber or poly(vinyl acetate), or biodegradable, eg, polylactic acid or polyglycolic acid. The drug release is also controlled by the type of coating. The reservoir or monolithic type systems control the release of a wide variety of drugs including salicylates, steroids, antifertility drugs, pilocarpine (for glaucoma), and addiction-control drugs (160–161). The osmotic-pump system regulates the drug delivery by water penetrating through a semipermeable membrane into the unit and this, in turn, forces a solution of the drug out into the body (161). Either type is implanted in the body near the diseased site. In the case of glaucoma, the device is placed under the eyelid, whereas for fertility control the device is implanted in the uterus. In each case the device can be retrieved after the drug is released. For surgically implanted controlled-release systems, a biodegradable polymer is preferred.

Polymeric Drugs. In a polymeric drug, the drug unit is attached to a polymer molecule. Several functions are controlled by different parts or units. These systems contain the chemotherapeutic or drug unit, units or groups to control the solubility of the copolymer, and groups or units that direct the copolymer to some specific organ or diseased tissue. In principle, these functions remain independent in a high molecular weight polymer, because they are located in different regions. On the other hand, they interact in low molecular weight systems and often cancel out the desired combination of effects. At the present time, polymeric drug systems are in the experimental stage

Table 3. Annual U.S. Consumption and Cost Estimates of Biomedical and Prosthetic Devices

Device	Quantity, in thousands	Approximate unit cost, $	Total expenditure, 10^6 $
dental fillings[a]	400,000	15	6,000
dentures[a]	1,100	250	2,750
dental implants[b]	300		
contact lenses[a]	6,000	150	900
artificial lenses or corneas[b]	600		
joint replacement	350	3,000	1,050
finger joints[b]	400	1,000	400
total or partial hip replacement[b]	250		
knee prostheses[b]	100		
artificial limbs	30	1,500	45
hernia repair[c]	35		
breast prostheses and augmentation[d]	200	2,000	400
facial plastic surgery	200	3,000	600
artificial kidney, dialysis[e]	40	20,000	800
hydrocephalus shunts[f]	>10		
heart–lung machines	100	10,000	1,000
coronary bypass[e]	90	15,000	1,350
intra-aortic balloons[f]	15		
heart valves[f]	25	10,000	250
pacemakers	60	4,000	240

[a] Ref. 165.
[b] Ref. 166.
[c] Ref. 7.
[d] Ref. 13.
[e] Ref. 167.
[f] Estimated.

but simple systems, such as a divinyl ether–maleic anhydride copolymer, are being tested clinically. Although many solubilizing groups and drug units are available, truly effective directing units or groups remain to be developed (162–164).

Economic Aspects

The total number of people with biomedical implants or devices is not known for certain but is in the millions (10^6). A recent study at the Implant Retrieval Center of the Hospital of Case Western Reserve University revealed that 9–15% of their autopsies had some type of implant, exclusive of contact lenses or dental repairs (12). This figure indicates that there are between 20 and 34×10^6 people in the United States with some type of implanted biomedical device. Many 10^6 more people have dental fillings, implants, or dentures (see Dental materials), or wear contact lenses. The total cost of these devices readily exceeds 10^9 dollars. Data on the use and costs of various prosthetic or biomedical devices in the United States are given in Table 3. The total U.S. annual cost for prosthetic and biomedical devices is nearly 16×10^9 dollars. The overall expenditures are probably more than double this amount when research costs are included. The actual cost of the devices would be less than the total expenditures given in Table 3, but are increased by multiple usage.

BIBLIOGRAPHY

1. C. G. Gebelein and F. F. Koblitz, eds., *Biomedical and Dental Applications of Polymers*, Plenum Publishing Corp., New York, 1981.
2. J. B. Park, *Biomaterials, An Introduction*, Plenum Publishing Corp., New York, 1979.
3. H. P. Gregor, ed., *Biomedical Applications of Polymers*, Plenum Publishing Corp., New York, 1975.
4. R. L. Kronenthal, Z. Oser, and E. Martin, *Polymers in Medicine and Surgery*, Plenum Publishing Corp., New York, 1975.
5. A. Tohen Z. in R. W. Milam and E. Lopez, translators, *Manual of Mechanical Orthopaedics*, Charles C Thomas, Publisher, Springfield, Il., 1973.
6. D. F. Williams and R. Roaf, *Implants in Surgery*, W. B. Saunders, London, 1973.
7. H. Lee and K. Neville, *Handbook of Biomedical Plastics*, Pasadena Technology Press, Pasadena, Calif., 1971.
8. A. Rembaum and M. Shen, *Biomedical Polymers*, Marcel Dekker, Inc., New York, 1971.
9. R. L. White and J. D. Meindl, *Science* **195**, 1119 (March 1977).
10. W. J. Kolff, *Artificial Organs* **1**(1), 8 (1977).
11. R. I. Leininger, *CRC Crit. Rev. Bioeng.* **1**, 333 (1972).
12. J. M. Anderson in Ref. 1, pp. 11–20.
13. M. B. Habel, *Biomater., Med. Devices Artif. Organs* **7**(2), 229 (1979).
14. Ref. 2, Chapt. 8.
15. C. A. Homsy, R. Hodge, N. Gordon, E. E. Braggs, and M. Estrella, *J. Biomed. Mater. Res.* **3**, 235 (1969).
16. C. A. Homsy, K. D. Ansevin, W. O'Bannon, S. A. Thompson, R. Hodge, and M. E. Estrella in Ref. 8, pp. 121–140.
17. W. L. Guess, S. A. Rosenbluth, B. Schmidt, and J. Autian, *J. Pharm. Sci.* **54**, 1545 (1965).
18. J. Autian, *Artificial Organs* **1**(1), 53 (1977).
19. Z. Oser, R. A. Abodeely, and R. G. McGunnigle, *Int. J. Polym. Mater.* **5**, 177 (1977).
20. J. E. Turner, W. H. Lawrence, and J. Autian, *J. Biomed. Mater. Res.* **7**, 39 (1973).
21. D. E. Ocumpaugh and H. L. Lee, *J. Macromol. Sci. Chem.* **A4**(3), 595 (1970).
22. J. Autian in Ref. 4, pp. 181–203.
23. J. C. Bokros, R. J. Akins, H. S. Shim, A. D. Haubold, and N. K. Agarwar, *Chemtech* **7**, 40 (1977).
24. R. Bagnell, *Chem. Br.* **14**, 598 (1978).

25. R. Lefaux, *Practical Toxicology of Plastics*, CRC Press, Inc., Cleveland, Ohio, 1968.
26. *Chem. Eng. News*, 20 (Sept. 8, 1975).
27. J. Autian, A. R. Singh, J. E. Turner, G. W. C. Hung, L. J. Nunez, and W. H. Lawrence, *Cancer Res.* **35,** 1591 (1975).
28. M. B. Habal and R. D. Powell, *J. Biomed. Mater. Res.* **14,** 447 (1980).
29. Ref. 7, Chapt. 3, pp. 3-1 to 3-42.
30. S. Srinivasan and P. N. Sawyer in Ref. 8, pp. 51–66.
31. P. N. Sawyer, S. Srinivasan, B. Stanczewski, N. Ramasamy, and W. Ramsey, *J. Electrochem. Soc.* **121**(7), 221C (1974).
32. P. Murphy, A. Lacroix, and S. Merchant in Ref. 8, pp. 67–86.
33. D. J. Lyman, *Angew. Chem. Int. Ed. Eng.* **13,** 108 (1974).
34. R. E. Baier, *Ann. N.Y. Acad. Sci.* **283,** 17 (1977).
35. S. W. Kim and E. S. Lee, *J. Polym. Sci. Symp.* **66,** 429 (1979).
36. E. S. Lee and S. W. Kim, *ASAIO J.* **3**(2), 50 (1980).
37. H. Y. K. Chuang, W. F. King, and R. G. Mason, *J. Lab. Clin. Med.* **92**(3), 483 (1978).
38. H. Y. K. Chuang, S. F. Mohammad, N. C. Sharma, and R. G. Mason, *J. Biomed. Mater. Res.* **14,** 467 (1980).
39. S. D. Bruck, *Blood Compatible Synthetic Polymers–An Introduction*, Charles C Thomas, Publisher, Springfield, Il., 1974.
40. R. I. Leininger in Ref. 1, pp. 99–109.
41. S. D. Bruck, *Chemtech* **7,** 240 (1977).
42. S. D. Bruck, *Pure Appl. Chem.* **46,** 221 (1976).
43. R. I. Leininger, *Chemtech* **5,** 172 (1975).
44. S. D. Bruck, *Polymer* **16,** 409 (1975).
45. B. D. Ratner, A. S. Hoffman, S. R. Hanson, L. A. Harker, and J. D. Whiffen, *J. Polym. Sci. Symp.* **66,** 363 (1979).
46. J. D. Andrade, ed., *Hydrogels for Medical and Related Applications*, American Chemical Society, Washington, D.C., 1976.
47. B. D. Ratner, A. S. Hoffman, and J. D. Whiffen, *J. Bioeng.* **2,** 313 (1978).
48. B. K. Kusserow, R. Larrow, and J. E. Nichols, *Trans. Amer. Soc. Art. Intern. Organs* **19,** 8 (1973).
49. T. Ohshiro and G. Kosaki, *Artificial Organs* **4,** 58 (1980).
50. T. Akutsu, N. Yamamoto, M. A. Serrato, J. Denning, and M. A. Drummond in Ref. 1, pp. 119–142.
51. K. Knutson and D. J. Lyman in Ref. 1, pp. 173–188.
52. S. Murabayashi and Y. Nose in Ref. 1, pp. 111–118.
53. H. Kambic, S. Barenburg, H. Harasaki, D. Gibbons, R. Kiraly, and Y. Nose, *Trans. Amer. Soc. Artif. Intern. Organs* **24,** 426 (1978).
54. Y. Nose, H. E. Kambic, R. J. Kiraly, T. Komai, and J. U. Urzua in P. Didisheim, T. Shimamoto, and H. Yamazaki, eds., *Platelets, Thrombosis and Inhibitors*, Springer-Verlag, Inc., New York, 1974, pp. 89–95.
55. H. E. Kambic, R. J. Kiraly, and Y. Nose, *J. Biomed. Mater. Res. Symp.* **7,** 561 (1976).
56. Ref. 7, Chapts. 5 and 8.
57. Dow Corning Bulletins, Midland, Mich., L080-0010, *Silastic Gel Filled Mammary Implant* (May 1980); 51-314, *Gel Filled Testicular Implant* (Dec. 1976); 51-286, *Silastic Carpal Lunate Implant H. P.* (April 1977); 51-287, *Silastic Carpal Scaphoid Implant H. P.* (Dec. 1977); 80-5, *Silastic Chin Implant* (May, 1979); 51-238, *Silastic Finger Joint Implant H. P.* (Feb. 1975); 51-048D, *Silastic Foley Catheter* (June 1976); 51-051B, *Silastic Hydrocephalus Shunt* (Dec. 1972); 51-270, *Silastic Malecot Catheter* (May 1976); 51-173, *Silastic Otoplasty Prothesis* (May 1978); 51-175A, *Silastic Rhinoplasty Implant* (March 1975); 51-246A, *Silastic Tracheostomy Tube* (May 1979).
58. P. N. Sawyer, W. Ramsey, B. Stanczewski, R. Turner, W. Liebig, G. W. Kammlott, and B. Braun, *Med. Instrum. Baltimore* **11,** 221 (1977).
59. C. W. Hall, L. M. Adams, and J. J. Ghidoni, *Trans. Amer. Soc. Artif. Intern. Organs* **21,** 281 (1975).
60. C. W. Hall and J. J. Ghidoni, *J. Surg. Res.* **25,** 122 (1978).
61. J. L. Giacchino, W. P. Geis, J. M. Buckingham, L. L. Vertuno, and V. K. Bansal, *Arch. Surg. Chicago* **114,** 403 (1979).
62. N. A. Peppas and R. E. Benner, Jr., *Biomaterials* **1,** 158 (1980).
63. C. B. Edwards, *Beyond Plastic Surgery*, Wayne State University Press, Detroit, Mich., 1972.
64. Ref. 7, Chapt. 9.

65. K. Jahnke, D. Plester, and G. Heimke, *Arch. Oto Rhino Laryngology* **223**, 373 (1979).
66. E. A. Ramirez-Garcia, J. M. Courtney, and W. Lang, *RAPRA Biomed. Appl. Polymers* **11**(1), 1 (1981).
67. S. E. Kinney, *Artificial Organs* **3**, 379 (1979).
68. Ref. 7, Chapt. 10.
69. W. Kolff in Ref. 4, pp. 1–28.
70. T. D. Sperling, E. A. Bering, S. V. Pollack, and H. G. Vaughan, eds., *Visual Prosthesis, The Interdisciplinary Dialogue*, Academic Press, New York, 1971.
71. F. M. Polack and G. Heimke, *Opthalmology* **87**, 693 (1980).
72. R. H. S. Langston, *Artificial Organs* **2**(1), 55 (1978).
73. H. E. Kaufman, J. Katz, J. Valenti, J. W. Sheets, and E. P. Goldberg, *Science* **198**, 525 (1977).
74. L. Turkish and M. A. Galin, *Arch. Ophthalmol.* **98**, 120 (1980).
75. M. B. Habal and V. A. Chalian, *J. Prosthet. Dent.* **32**(2), 292 (1974).
76. W. B. Hammer, R. G. Topazian, R. V. McKinney, Jr., and S. F. Hulbert, *J. Dent. Res.* **52**, 356 (1973).
77. D. L. Leake and M. B. Habal, *J. Trauma* **17**, 299 (1977).
78. M. B. Habal, D. L. Leake, J. E. Maniscalco, and J. Kim, *Plast. Reconstr. Surg.* **61**, 394 (1978).
79. S. A. Braley in A. L. Bement, Jr., ed., *Biomaterials*, U. Washington Press, Seattle, 1971, pp. 277–283.
80. R. Johnsson-Hegyeli in Ref. 79, pp. 207–233; S. Braley in L. Stark and G. Agarwal, eds., *Biomaterials*, Plenum Publishing Corp., New York, 1969, pp. 67–89; Ref. 7, Chapt. 13.
81. M. Spira and C. W. Hall, *Symposium on Burns*, Vol. 5, C. V. Mosby Co., St. Louis, 1973, 182–188.
82. M. Spira, J. Fissette, C. W. Hall, S. B. Hardy, and F. J. Gerow, *J. Biomed. Mater. Res.* **3**, 213 (1969).
83. P. D. May in A. L. Bement, Jr., ed., *Biomaterials*, University of Washington Press, Seattle, Wash., 1971, pp. 257–268.
84. M. Chvapil, *J. Biomed. Mater. Res.* **11**, 721 (1977).
85. I. V. Yannis and J. F. Burke, *J. Biomed. Mater. Res.* **14**, 65 (1980).
86. I. V. Yannas, J. F. Burke, P. L. Gordon, C. Huang, and R. H. Rubenstein, *J. Biomed. Mater. Res.* **14**, 107 (1980).
87. N. Dagalakis, J. Flink, P. Stasikelis, J. F. Burke, and I. V. Yannas, *J. Biomed. Mater. Res.* **14**, 511 (1980).
88. P. W. Wang and N. A. Samji in Ref. 1, pp. 29–37.
89. G. B. McKenna, G. W. Bradley, H. K. Dunn, and W. O. Statton, *J. Biomed. Mater. Res.* **13**, 783 (1979).
90. G. W. Bradley, G. B. McKenna, H. K. Dunn, A. V. Daniels, and W. O. Statton, *J. Bone Jt. Surg.* **61-A**, 866 (1979).
91. S. F. Hulbert and L. S. Bowman in Ref. 4, pp. 161–166.
92. L. L. Hench, *Biomat. Med. Dev. Art. Org.* **7**, 339 (1979).
93. Ref. 7, Chapt. 12.
94. J. L. Goldner and J. R. Urbaniak in S. F. Hulbert, S. N. Levine and D. D. Moyle, eds., *Materials and Design Considerations for the Attachment of Prostheses to the Musculo-Skeletal System*, Wiley-Interscience, New York, 1973, pp. 137–163.
95. R. M. Rose in H. S. Stanley, ed., *Biomedical Physics and Biomaterials Science*, MIT Press, Cambridge, Mass., 1972, pp. 169–189.
96. J. Charnley, *Plast. Rubber* **1**(2), 59 (1976).
97. P. S. Walker and E. Salvati, *J. Biomed. Mater. Res. Symp.* **4**, 327 (1973).
98. L. L. Hench, C. G. Pantano, Jr., P. J. Buscemi, and D. C. Greenspan, *J. Biomed. Mater. Res.* **11**, 267 (1977).
99. P. Ducheyne, A. Kagan, II, and J. A. Lacey, *J. Bone Jt. Surg.* **60-A**, 384 (1978).
100. D. A. Sonstegard, L. S. Matthews, and H. Kaufer, *Sci. Am.* **238**(1), 44 (1978).
101. P. A. Torzilli in D. G. Fleming and B. N. Feinberg, eds., *Handbook of Engineering in Medicine and Biology*, CRC Press, Cleveland, Ohio, 1976, pp. 225–251.
102. D. Dowson and V. Wright in R. M. Kenedi, ed., *Perspectives in Biomedical Engineering*, Macmillan Press Ltd., London, 1973, pp. 103–107.
103. J. B. Reswick and L. Vodovnik in J. F. Dickson III and J. H. U. Brown, eds., *Future Goals of Engineering in Biology and Medicine*, Academic Press, New York, 1969, pp. 147–166.

104. E. F. Murphy and A. B. Wilson, Jr. in M. Clynes and J. H. Milsum, eds., *Biomedical Engineering Systems*, McGraw-Hill, Inc., New York, 1970, pp. 489–549.
105. E. F. Murphy, *J. Biomed. Mater. Res. Symp.* **4**, 275 (1973).
106. G. Murdoch and J. Hughes in R. M. Kenedi, ed., *Perspectives in Biomedical Engineering*, Macmillan Press Ltd., London, 1973, pp. 67–72.
107. R. N. Scott in S. N. Levine, ed., *Advances in Biomedical Engineering and Medical Physics*, Vol. 2, Wiley-Interscience, New York, 1968, pp. 45–72.
108. G. H. Myers and V. Parsonnet, *Engineering in the Heart and Blood Vessels*, Wiley-Interscience, New York, 1969, Chapts. 2–7.
109. G. H. Myers and V. Parsonnet in Y. Nose and S. N. Levine, eds., *Cardiac Engineering*, Vol. 3 of *Advances in Biomedical Engineering and Medical Physics*, Wiley-Interscience, New York, 1970, pp. 335–368.
110. Product Bulletin, American Edwards Laboratories, *Starr-Edwards and Carpentier-Edwards Heart Valve Prostheses*, Santa Ana, Calif., Feb. 1980.
111. P. N. Sawyer, B. Stanczewski, J. G. Stempak, and G. W. Kammlott in J. C. Davila, ed., *Second Henry Ford Hospital International Symposium on Cardiac Surgery*, Appleton-Century-Crofts, New York, 1977, pp. 408–432.
112. E. J. Zerbini, K. Nakiri, and L. B. Puig, *Artificial Organs* **2**(Suppl.), 26 (1978).
113. Ref. 108, pp. 160–165.
114. M. D. Silver, B. N. Datta, and V. F. Bowes, *Arch. Pathol.* **99**, 132 (March 1975).
115. E. A. Lefrak and A. Starr, *Cardiac Valve Prostheses*, Appleton-Century-Crofts, New York, 1979, pp. 341–352.
116. H. Harasaki, J. Snow, R. Cloesmeyer, and Y. Nose, *Inter. J. Artificial Organs* **2**(2), 73 (1979).
117. D. L. Bricker, A. C. Beall, Jr., and M. E. DeBakey, *Chest* **58**, 566 (1970).
118. R. E. Baier, C. K. Akers, J. R. Natiella, M. A. Meenagham, and J. Wirth, *Vasc. Surg.* **14**(3), 145 (1980).
119. J. G. Bennett, R. Trono, J. C. Norman, and D. A. Cooley, *Cardiovascular Diseases* **4**(1), 18 (1977).
120. C. D. Campbell, D. Goldfarb, and R. Roe, *Ann. Surg.* **182**(2), 138 (1975).
121. G. D. Vaughan, K. L. Mattox, D. V. Feliciano, A. C. Beall, Jr., and M. E. DeBakey, *J. Trauma* **19**, 403 (1979).
122. H. Haimov, F. Giron, and J. H. Jacobson, *Arch. Surg. Chicago* **114**, 673 (1979).
123. D. Raithel and H. Groitl, *World J. Surg.* **4**, 223 (1980).
124. D. J. Lyman, K. B. Seifert, H. Knowlton, and D. Albo, Jr. in Ref. 1, pp. 163–171.
125. A. Kantrowitz, S. Tjønneland, P. S. Freed, S. J. Phillips, A. F. Butner, and J. L. Sherman, Jr., *J. Am. Med. Assoc.* **203**(2), 135 (1968).
126. A. Kantrowitz, S. Tjønneland, J. Krakauer, A. N. Butner, S. J. Phillips, W. Z. Yahr, M. Shapiro, P. S. Freed, D. Jaron, and J. L. Sherman, Jr., *Trans. Am. Soc. Artif. Int. Organs* **14**, 344 (1968).
127. A. Kantrowitz, J. S. Krakauer, G. Zorzi, M. Rubenfire, P. S. Freed, S. Phillips, M. Lipsius, C. Titone, P. Cascade, and D. Jaron, *Transplant. Proc.* **3**, 1459 (1971).
128. D. Bregman, E. N. Parodi, K. Reemtsm, and J. R. Malm in J. C. Norman, ed., *Coronary Artery Medicine and Surgery: Concepts and Controversies*, Appleton-Century-Crofts, New York, 1975, pp. 413–420.
129. D. Bregman, E. N. Parodi, S. M. Haubert, R. Szarnicki, R. N. Edie, H. M. Spotnitz, F. O. Bowman, Jr., K. Reemtsma, and J. R. Malm, *Med. Instrum. Baltimore* **10**, 232 (1976).
130. D. Bregman, M. Bailin, F. O. Bowman, Jr., E. N. Parodi, S. M. Haubert, R. N. Edie, H. M. Spotnitz, K. Reemtsma, and J. R. Malm, *Ann. Thoracic Surg.* **24**, 574 (1977).
131. D. Bregman, A. B. Nichols, M. B. Weiss, E. R. Powers, E. C. Martin, and W. J. Casarella, *Am. J. Cardiol.* **46**, 261 (1980).
132. D. Liotta, C. W. Hall, W. S. Henly, D. A. Cooley, E. S. Crawford, and M. E. DeBakey, *Am. J. Cardiol.* **12**, 399 (1963).
133. M. DeBakey, D. Liotta, and C. W. Hall in *Mechanical Devices to Assist the Failing Heart*, NAS Publ. No. 1283, Washington, D.C., 1966, p. 223.
134. A. Kantrowitz, J. L. Sherman, Jr., and J. Krakauer, *Prog. Cardiovasc. Dis.* **10**(2), 134 (1967).
135. A. Kantrowitz, T. Akutsu, P. A. Chaptal, J. Krakauer, A. R. Kantrowitz, and R. T. Jones, *J. Am. Med. Assoc.* **197**, 525 (1966).
136. E. K. Olsen, W. S. Pierce, J. H. Donachy, D. L. Landis, G. Rosenberg, W. M. Phillips, G. A. Prophet, M. J. O'Neill, and J. A. Waldhausen, *Inter. J. Artif. Organs* **2**, 197 (1979).

137. W. E. Pae, Jr. and W. S. Pierce in J. M. Moran and L. L. Michaelis, eds., *Surgery for the Complications of Myocardial Infarction*, Grune & Stratton, Inc., New York, 1980, pp. 411–425.
138. Y. Nose, L. R. Golding, and F. D. Loop, *Transplant. Proc.* **11,** 313 (1979).
139. R. K. Jarvik, *Sci. Am.* **244**(1), 74 (1981).
140. D. A. Cooley, D. Liotta, G. L. Hallman, R. D. Bloodwell, R. D. Leachman, and J. D. Milam, *Am. J. Cardiol.* **24,** 723 (1969).
141. J. I. Wright and Y. Nose in Y. Nose and S. N. Levine, eds., *Advances in Biomedical Engineering and Medical Physics*, Vol. 3, Wiley-Interscience, New York, 1970, pp. 295–317.
142. G. Jacobs, N. Tsushima, R. Kiraly, and Y. Nose, *Proceedings of the International Conference on Cybernetics and Society*, IEEE, New York, Sept. 19–21, 1977, pp. 567–571.
143. W. S. Pierce, J. A. Brighton, J. H. Donachy, D. L. Landis, G. Rosenberg, G. A. Prophet, W. J. White, and J. A. Waldhausen, *Arch. Surg. Chicago* **112,** 1430 (1977).
144. K. Cheng, J. W. Meador, M. A. Serrato, and T. Akutsu, *Cardiovasc. Dis.* **4**(1), 9 (1977).
145. J. Lawson and W. J. Kolff, *Practical Cardiology* **5**(7), 95 (1979).
146. M. Gutcho, *Artificial Kidney Systems*, Noyes Data Corp., Park Ridge, N.J., 1970.
147. Y. Nose in S. N. Levine, ed., *Advances in Biomedical Engineering and Medical Physics*, Vol. 4, Wiley-Interscience, New York, 1971, pp. 163–224.
148. R. J. Wineman in F. Villarroel and R. L. Dedrick, eds., *AIChE Symposium Series No. 187*, AIChE, New York, 1979, pp. 1–9.
149. P. C. Farrell in Ref. 148, pp. 10–23.
150. P. M. Galletti in Ref. 107, pp. 121–167.
151. J. Ketteringham and T. J. Driscoll, *Ann. N.Y. Acad. Sci.* **283,** 410 (1977).
152. W. M. Zapol and J. Ketteringham in Ref. 4, pp. 287–312.
153. D. N. Gray in Ref. 1, pp. 21–27.
154. T. M. S. Chang, *Artificial Kidney, Artificial Liver and Artificial Cells*, Plenum Publishing Corp., New York, 1978.
155. Y. Nose, P. S. Malchesky, I. Koshino, F. Castino, and K. Scheucher in R. M. Kenedi, J. M. Courtney, J. D. S. Gaylor, and T. Gilchrist, eds., *Artificial Organs*, University Park Press, Baltimore, 1977, pp. 372–377, 378–387.
156. K. D. Kulbe, *Artificial Organs* **3,** 143 (1979).
157. J. L. Rosenbaum, *Kidney Int.* **18**(Suppl. 10), 5106 (1980).
158. H. J. Sanders, *Chem. Eng.* 30 (March 2, 1981).
159. J. V. Santiago, A. H. Clemens, W. L. Clarke, and D. M. Kipnis, *Diabetes* **28**(1), 71 (1979).
160. M. Nakano, *J. Membrane Sci.* **5,** 355 (1979).
161. A. Zaffaroni in Ref. 1, pp. 293–313.
162. L. G. Donaruma and O. Vogl, *Polymeric Drugs*, Academic Press, New York, 1977.
163. C. G. Gebelein in Ref. 1, pp. 191–201.
164. C. G. Gebelein, *Polym. News* **4,** 163 (1978).
165. C. G. Gebelein, *Org. Coatings Plast. Chem.* **42,** 70 (1980).
166. L. L. Hench, *J. Biomed. Mater. Res.* **14,** 803 (1980).

CHARLES G. GEBELEIN
Department of Chemistry
Youngstown State University

Department of Pharmacology
Northeast Ohio Universities College of Medicine

PROTACTINIUM. See Actinides and transactinides.

PROTEINS

Proteins are complex macromolecules that are fundamental to life. Much of the cellular content of plants and animals is protein, and metabolism is dependent on protein enzymes. Metabolic processes regulated by protein hormones and bioelectric processes at the cellular level are dependent on protein–ion interactions.

Plants can synthesize all their needs, including amino acids, but animals require certain (essential) amino acids. Human diets in developed countries contain much protein from animal sources (meat, milk, eggs, and fish), but poorer countries are heavily dependent on cereals and legumes.

Much of traditional technology is dependent on the properties of proteins. Baking of bread requires the properties of wheat gluten in forming an elastic network and thus holding moisture and gases. Brewing requires yeast protein enzymes to transform sugars to alcohol (see Beer), and the enzymes from lactobacilli are the basis of cheesemaking (see Milk and milk products). Flavor formation often depends on protein behavior during cooking.

Proteins are widely used in clothing (wool, silk, and leather). Adhesives are obtained from casein, blood albumin, collagen, and its breakdown product gelatin (see Adhesives). Casein, soybean, and egg albumin have also been used for fiber production. The production of the protein fractions from blood for use in medicine is a further example of protein technology (see Blood). Food, however, constitutes the largest use of protein by far, and accordingly is emphasized in this article.

Early ideas of protein metabolism are linked to the discovery of nitrogen and its distribution in nature (1–3). In the early 19th century, the importance of nitrogenous compounds was recognized, and the name protein was first used in 1838. Because of the complexity of protein structure, chemical studies were initially unrewarding. Although the essential amino acid threonine was not identified until the 1930s (4), techniques for the biological evaluation of proteins had been developed much earlier and were based on nitrogen balance and on rat growth. In the 1940s, analytical techniques permitted the correlation of amino acid data with protein quality (5). During this period, the Krebs ornithine cycle (6) and other studies in protein metabolism had opened a new era in biochemical thought. Following World War II, a great resurgence in protein research climaxed in epoch-making advances in molecular biology and biochemistry (7–8). During this period, protein quality was a prime concern and many assay techniques were developed (see Analytical methods). With increasing concern over protein malnutrition, a worldwide crisis was expected. Subsequently, however, evidence accumulated that protein requirements were lower than previously believed and that when energy intake was limited, protein could be deaminated and used for energy purposes (9–14).

Structure

Although several hundred amino acids are known, only about 20 occur in proteins (see Amino acids). The possible number of the different proteins that could be produced from only 20 amino acids, even with a moderate chain length of 100 residues (20^{100}), is so enormous that all protein molecules could be unique (15). However, the actual number of different proteins in an individual, although large (10^4–10^5), is only a tiny fraction of the theoretical possibilities.

The molecular weights of the residues range from 57 to 186 with the weighted average in a typical protein being about 110. Thus, a protein with a molecular weight of 33,000 contains ca 300 residues.

The amino acid composition of proteins may be expressed in many different ways, depending on the context, eg, nutritional considerations or composition and sequence data. Although composition can vary greatly, certain amino acids, eg, tryptophan, are generally present in much smaller amounts than others, eg, alanine.

Proteins, like nucleic acids, are derived from straight-chain molecules with one standard linkage. In peptides, all amino acid residues are of the α-type with L-configuration at the α-carbon atom. Differences and therefore information are restricted to the rather short side chains of the amino acids. The relation of amino acid properties to protein structure is shown in Table 1.

After or even during the synthesis on the ribosome, the polypeptide chain folds spontaneously to a globular protein dependent on the amino acid sequence by adopting a state of lower free energy. The resulting chain fold determines the specificity of the protein (8,16). Exceptions are some very elongated ribosomal proteins (17). Spontaneous folding of the chain is also a probable reason for the evolutionary selection of amino acids. Chains can also be formed from β amino acids but because of the extra free rotation possible on the α–β bond, spontaneous chain folding would be prevented.

Peptides are synthesized on ribosomes by the formation of peptide bonds. The chain direction is from the amino end (N terminus) to the carboxyl end (C terminus), and is defined as coinciding with the chain synthesis (Fig. 1) (18). The planarity and rigidity of the bond are due to the fact that the α amino proton is shared and the N—C and C—O bonds are neither single nor double bonds. Other features of protein structure, such as the helix and pleated sheets, are possible because of the geometry of the peptide bond in the protein backbone. The rigidity of the peptide chain is further increased by steric hindrance within the main chain and between main- and side-chain atoms. Although a cis bond is possible, the trans bond is geometrically superior and is the usual form of the bond.

Protein structure, formation, and stability also depend heavily on noncovalent forces, such as the repulsion between nonbonded electron shells; coupling between oscillating dipoles giving rise to an attraction; electrostatic attraction and repulsion of partial charges as well as attraction of full charges as found in salt bridges; and hydrogen bonds. Since functional proteins exist only in aqueous environments, the combined peptide chain and solvent system must be taken into account. Protein stability is mainly due to hydrophobic forces. The protein interior is as tightly packed as that of a crystal which proves that the noncovalent forces are efficiently used.

In the patterns of folding and association of peptide chains, four levels of organization have been recognized (19). Although these concepts cannot be defined with precision, the IUPAC–IUB Commission on Biochemical Nomenclature (20) has recommended the following definitions.

(1) The primary structure of a segment is the amino acid sequence without regard to spatial arrangement (except for configuration at the α-carbon atom). Since this definition does not include the positions of disulfide bonds, it is not identical with covalent structure.

(2) The secondary structure of a segment is the local spatial arrangement of its main-chain atoms without regard to the conformation of its side chains or to its relationship with other segments.

Table 1. Properties of Side Chains in Relation to Protein Structure [a]

Amino acid	Abbreviation	Observation
glycine	Gly	Increases main-chain flexibility. No asymmetry at $C\text{-}\alpha$.
alanine	Ala	Small nonpolar residue. Abundant, little preference for inside or surface of protein.
leucine	Leu	Branching nonpolar side chain allows for large stiffened side chains with limited flexibility which facilitate chain folding. Mainly inside molecules.
isoleucine	Ileu	
valine	Val	
phenylalanine	Phe	The presence of a $C\text{-}\beta$ methylene between $C\text{-}\alpha$ and the aromatic ring allows moderate side-chain flexibility. Without the methylene group there would be severe steric hindrance at $C\text{-}\alpha$. Mainly inside molecules.
tyrosine	Tyr	
tryptophan	Trp	
proline	Pro	Completely nonpolar, side chain curls back to main chain, produces rigid side chain. Fixes dihedral angle between $C\text{-}\alpha$ and the peptide nitrogen to a small range of $\pm 20°$C.
methionine	Met	Flexible side chain with one sulfur atom in a thioether bond, sulfur atom introduces an electric dipole.
cysteine	Cys	All polar and form hydrogen bonds. Cysteine has special role because of formation of cystine cross bridges. Hydroxyl groups (Ser, Thr) and amido groups (Gln, Asn) facilitate hydrogen-bond formation.
serine	Ser	
threonine	Thr	
asparagine	Asn	
glutamine	Gln	
tyrosine	Tyr	
histidine	His	Imidazole ring can be charged or uncharged in the physiological pH range. When uncharged, accepts H^+, thus often found in active center of enzymes.
aspartic acid	Asp	Most charged side chains are at the molecular surface. Long and flexible Lys and Arg residues increase solubility. Lys residues often react at the exposed $\epsilon\text{-}NH_2$.
glutamic acid	Glu	
lysine	Lys	
arginine	Arg	
hydrophobic amino acids		Most amino acids except the hydrophilic-charged Asp, Glu, Lys, and Arg are hydrophobic. The formation of a hydrophobic core strongly affects folding. Hydrophobisity can be evaluated by measuring free energy when amino acids are transferred from water to an organic solvent.
size of side chain		Amino acids can be arranged to reveal exchange groups wherein amino acids exchange preferentially with each other. In general, large and small residues find themselves in different groups; thus, it appears that size of side chain may be almost as important as the chemical nature.

[a] Ref. 15.

(3) The tertiary structure of a protein molecule or of a subunit of a protein molecule, is the arrangement of all its atoms in space, without regard to its relationship with neighboring molecules or subunits.

(4) The quaternary structure of a protein molecule is the arrangement of its subunits in space and the ensemble of its intersubunit contacts and interactions, without regard to the internal geometry of the subunits (see Biopolymers).

It should be noted that a protein molecule that is not made up of at least poten-

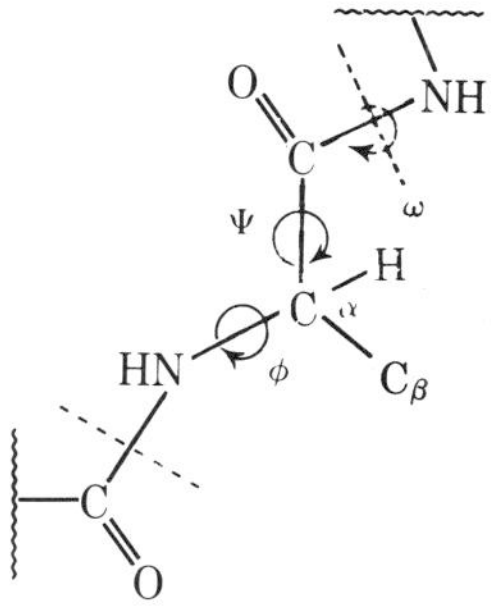

Figure 1. Section of a polypeptide chain representing two peptide units in a fully extended conformation. Sum of angles ϕ, ψ, and ω all equal 180°. Dotted lines define the limits of a residue.

tially separable subunits (not connected by covalent bonds) possesses no quaternary structure, eg, ribonuclease (1 chain) and chymotrypsin (3 chains).

Supersecondary structures denote physically preferred aggregates of secondary structure, and domains refer to those parts of the protein that form well-separated globular regions (15).

Secondary structures are stabilized by hydrogen bonds between peptide, amide, and carboxyl groups. The backbone of a peptide, like any long-repeating polymer tends to form a coil, usually an α helix. Here the NH of each amino acid residue is hydrogen-bonded to the C=O groups of the fourth following residue which is now on an adjacent turn of the helix (Fig. 2). There are actually 3.6 amino acid residues in each complete turn of the helix and hence 100° of turn per residue. Other helix types are less common (see Table 2).

The fact that the α helix is the most abundant secondary structure in proteins indicates that it is a stable conformation (18). Collagen is a superhelix formed by three parallel left-handed helices. Collagen is one of the most abundant proteins in nature and its extended peptide chain results in resistance to strong mechanical tension along its axis. Hence it acts as a force transmitter in tendons.

Extended peptide chains can also associate by hydrogen bonding to form folded sheetlike structures. Pleated sheets are either parallel or antiparallel (see Fig. 3 and Table 2). In the former, carboxyl and amino terminal ends are in the same direction, whereas in the latter they are in opposite directions. Each peptide bond is involved in hydrogen-bond formation with an adjacent peptide bond from another chain. Such a conformation places the R groups on both sides of the sheet.

Because of the known relationships between the properties of side chains and protein structure, the secondary structure of proteins can be predicted in some degree from amino acid sequence data. About two thirds of all residues can be assigned to the correct secondary structure. For the upper levels, the structural considerations become much more complex and the predictions are generally less accurate.

Symmetry, however, is present at all levels above the primary. In secondary structures and aggregates, linear-group symmetry is found. Point-group symmetry is found in all levels above secondary structures, and space-group symmetry is found in aggregates (15). In tertiary structures, the folded chains (secondary structure) are arranged into a specific shape that is maintained by salt bonds, hydrogen bonds, —S—S— bridges, Van der Waals' forces, and hydrophobic interactions. Some of these are covalent linkages, others are noncovalent. Van der Waals potentials are noncovalent and combine electron-shell repulsion, dispersion forces, and electrostatic interactions.

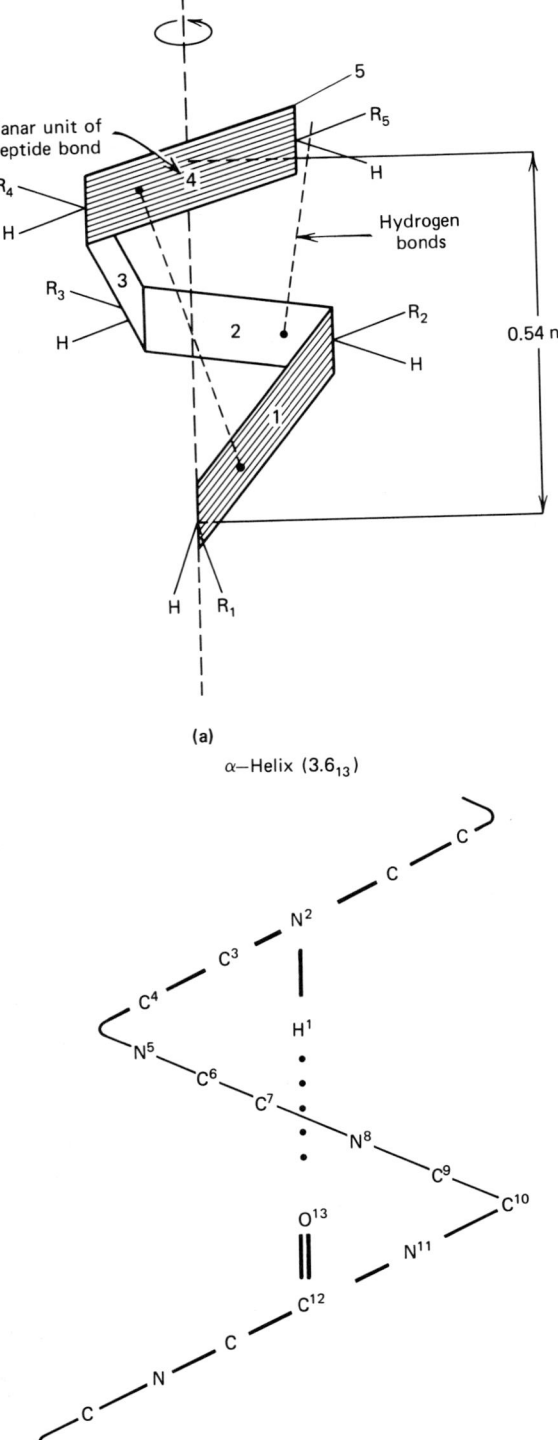

Figure 2. Peptide α-helix. (**a**) Conformation. Courtesy of C. V. Mosby Company (21). (**b**) H bonds. Courtesy of W. B. Saunders Company (22).

Table 2. Linear Groups Formed by Peptide Chains[a]

Linear group	Observed	Residues per turn n and chirality[b]	Rise per residue, nm	Radius of helix, nm
planar parallel sheet	rarely	±2.0	0.32	0.11
planar antiparallel sheet	rarely	±2.0	0.34	0.09
twisted parallel or antiparallel sheet	abundantly	−2.3	0.33	0.10
3_{10}-helix	small pieces	+3.0	0.20	0.19
α-helix (right-handed)	abundantly	+3.6	0.15	0.23
α_L-helix (left-handed)	hypothetical	−3.6	0.15	0.23
π-helix	hypothetical	+4.3	0.11	0.28
collagen-helix	in fibers	−3.3	0.29	0.16

[a] Ref. 15. Courtesy of Springer-Verlag.
[b] Plus and minus correspond to right- and left-handed helices, respectively.

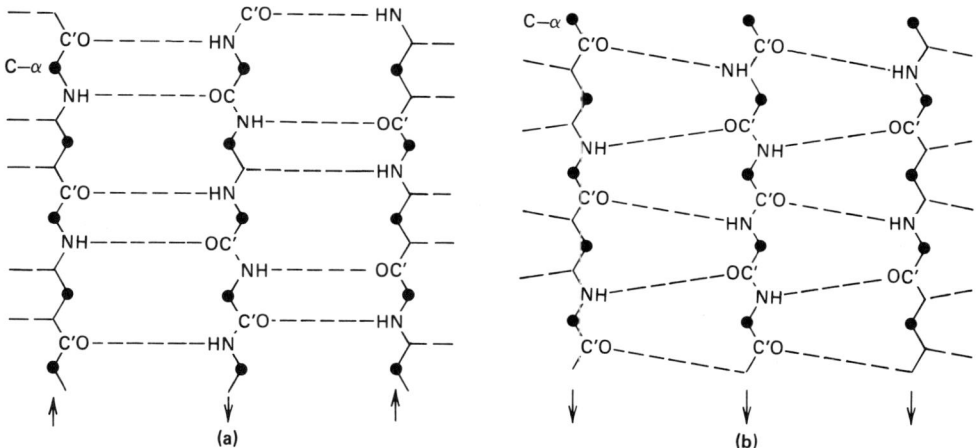

Figure 3. β-Pleated sheet. Hydrogen bonds are indicated by dashed lines and chain directions by arrows. C-α-atoms are marked by dots. (a) Hydrogen-bond pattern of antiparallel three-stranded β-sheet. (b) Hydrogen-bond pattern of a parallel three-stranded β-sheet (15).

Although these forces contribute to protein stability, they are difficult to calculate and measure. The surrounding liquid medium also contributes to stability.

Disulfide bonds are covalent linkages that appear to form spontaneously *in vitro* and can occur both between and within peptide chains (see Fig. 4). Disulfide bonds give extra stability to proteins, but they can be sometimes broken without loss of the protein's function. They tend to be more common in extracellular proteins and are especially important in bread technology and protein fibers (23–24).

Quaternary structures are the association of protein subunits into oligomers with some degree of symmetry. Since weaker noncovalent forces are holding the subunits together they can be dissociated under relatively mild conditions. Such a combination of subunits is essential for the aggregate structures to function (eg, hemoglobin, actinomycin). Isozymes (isoenzymes) usually result from varying proportion of several subunits thus conveying different kinetic properties to the enzyme.

The ultimate goal of the elucidation of protein structure is the explanation of biological functions. Proteins operate in environments of high complexity and thus

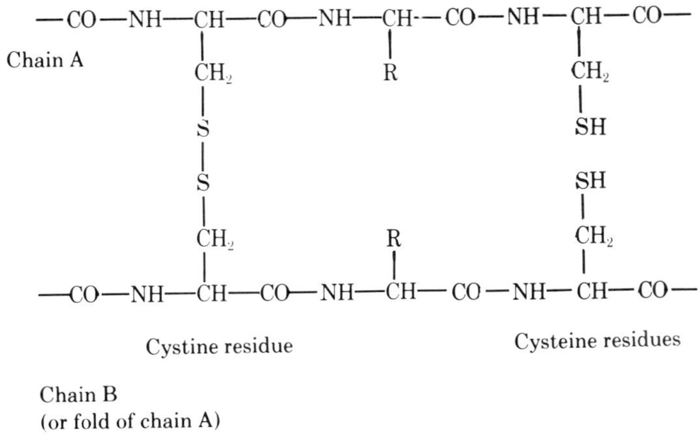

```
                —CO—NH—CH—CO—NH—CH—CO—NH—CH—CO—
Chain A              |              |              |
                    CH₂             R             CH₂
                     |                             |
                     S                             SH
                     |
                     S                             SH
                     |                             |
                    CH₂             R             CH₂
                     |              |              |
                —CO—NH—CH—CO—NH—CH—CO—NH—CH—CO—
```

           Cystine residue   Cysteine residues

Chain B
(or fold of chain A)

Figure 4. Disulfide bonding.

natural proteins are generally selectively limited in their binding capacities. These properties are determined by their folding characteristics. The mechanisms of hemoglobin and the peptide-splitting enzyme, chymotrypsin, are well-known; muscle protein is also studied (25–27). The molecular basis of muscle contraction is mediated by calcium-binding proteins and consists of interaction between Mg–ATP (adenosine triphosphate) and the proteins myosin and actin. The calcium-binding protein calmodulin (28) appears to have a fundamental role in regulating a large variety of cellular activities. Although in general a protein has a special structure and hence function, multifunctional proteins exist (29). They combine several autonomous functions on one peptide chain. In these cases, the various functions exist at different domains on the overall peptide chain.

Properties, Separation, and Classification

    A classification of proteins, based on solubility and composition, is given in Table 3.

    Proteins are amphoteric and function as acids or bases, depending on the conditions. The isoelectric point (pI) is the pH at which a protein does not migrate in an electric field. The minimum solubility of many proteins is at this point. Electrophoretic mobility is zero at the isoelectric point and thus electrophoresis at different pH can be used for protein separation and identification. These techniques are used especially for the separation of the blood serum protein fractions into various components.

    Solubility varies greatly and the earliest systems of classification were based on solubility. Some proteins are water-soluble, others, such as keratin, are completely insoluble.

    Changes of pH affect solubility and proteins are separated by isoelectric focusing. In this procedure, the proteins are mixed with synthetic low molecular weight (300–600) polyampholytes and an electric field is applied. Proteins migrate to the position of their respective isoelectric point and can thus be separated. Alternatively, the proteins are precipitated with aqueous solutions of organic solvents such as methanol, ethanol, and acetone in varying concentrations, resulting often in some degree of denaturation.

    The molecular weights of proteins range from 10^4 to 10^6. At the lower end, the distinction between peptides and proteins in conventional usage is far from clear-cut;

Table 3. Protein Classification Based on Solubility and on Composition

Protein	Characteristics	Examples	Nonprotein components	Ref.
Solubility				
globular				
albumin	sol water, dil acid, base, and salt solutions; precipitated by saturation with $(NH_4)_2SO_4$	lactalbumin ovalbumin serum albumin		
globulin	sol salt solutions, insoluble in water	immunoglobulin fibrinogen myosin	carbohydrate	
histones	sol water, dil acids and alkalies, insol dil ammonia	thymus histone nucleoproteins	nucleic acids	
protamines	sol water, dil acids, alkalies, and ammonia; high arginine content	sperm protamine	nucleic acids	
fibrary sclero				
collagens	insol, resistant to digestive enzymes, high in hydroxyproline, low in SAA[a]	skin, tendon, bone proteins		24, 30–31
elastins	insol, partly resistant to digestive enzymes	artery, tendon, and elastic tissue protein		
keratins	highly insol, resistant to digestive enzymes; S—S bridges	skin, hair, nail protein		30
Composition				
nucleoproteins	often histones present that stabilize DNA in a compact form	viruses, chromosomes	nucleic acids	32–33
glycoproteins, mucoproteins	carbohydrate chains, aldehyde groups present in short, highly branched chains; high concentration of hexosamine	immunoglobulin serum, globulins, blood-group specificity proteins	carbohydrates, eg, mucopoly-saccharides, hexosamines	34–35
lipoproteins, proteolipids	may have solubility characteristics of proteins or lipids, depending on proportions; present in cell and organelle membranes and nerve sheaths	chylomicrons, HDL, LDL, and VLDLs Myelin	lipids, eg, triglycerides, cholesterol, cholesterol esters	36–37
phosphoproteins	present in proteins of eggs and milk; calcium present by chelation	phosphovitin in eggs, casein	ester-linked phosphoric acid	38–40
metalloproteins	metals attached directly to protein, ie, not part of a nonprotein prosthetic group	cerruloplasmin ferritin carboxypeptidase transferrin xanthine oxidase (metalloenzyme)	copper iron zinc iron, manganese, copper molybdenum, iron, FAD	41–42
chromoproteins	proteins in association with a nonprotein pigment;	flavoproteins myoglobin	riboflavin (FAD), heme, (iron-	

321

Table 3. (continued)

Protein	Characteristics	Examples	Nonprotein components	Ref.
	frequently involved in oxidation and reduction reactions	hemaglobin cytochromes	containing porphyrin), hemin–heme	

a Sulfur amino acids.

strictly, all proteins are peptides. Macromolecules (large peptides) can be separated in a centrifuge by the speed with which they move through a solution. Ultracentrifuges develop forces up to 50,000 g which permits the observation of movement of peptides of mol wt as low as 10^4. The shape of the molecule also affects separation.

In gel-permeation chromatography, a protein mixture is passed through a porous gel that separates the components according to their molecular size and shape. Molecular weight may be estimated from the rate of passage through such a gel.

Vigorous shaking or stirring, uv radiation, ultrasonic vibration, and significant changes in pH or temperature result in denaturation of the protein solution (43). Urea, guanidine, or organic solvents have the same effect. It can, however, be minimized at low temperature. Denaturation is usually irreversible, although there are exceptions, such as the denaturation of hemoglobin or pancreatic ribonuclease.

A wide variety of techniques is available for the purification and separation of proteins, including molecular-sieve analysis (44–45), electrofocusing (46), electrophoresis (47–48), gel filtration (49–50), chromatography (51), sedimentation (52), and fractionation at low temperatures (53).

After separation and purification, other procedures are available to determine the amount of protein present (see Table 4).

Protein Sequence and Biosynthesis

As indicated above, the nature and function of a protein are determined by its primary structure (also called sequence), ie, the amino acids present and the order in which they are linked. The first protein to be sequenced was bovine insulin (mol wt 6000) in the classic work of Sanger (65). Protein sequences have since been reported in great numbers (20,66) and can now be determined by commercially available automatic equipment. The general technique (67–68) is outlined in Table 5. Protein sequencing is crucially important to an understanding of the molecular basis of heredity (21–22,69). The genetic code for translation of nucleotide triplets to amino acids in protein synthesis entails five principal steps.

(1) The amino acids are activated.

(2) The peptide chain is initiated, starting at the amino end.

(3) Elongation of the chain occurs because of codon-directed binding. (Amino acid–tRNA + polysome → polypeptide–polysome complex.)

(4) Termination requires guanosine triphosphate (GTP) and protein release factors that recognize the termination codons UAA, UAG, and UGA (U = uracil, A = adenine, and G = guanine). (Peptide–polysome complex → polypeptide + polysome.)

(5) Postribosomal modification (Fig. 5) occurs after synthesis on the mRNA–ribosome complex and involves many types of modification.

Table 4. Analytical Procedures for the Determination of Protein

Analysis	Method	Remarks	Refs.
nitrogen			
Dumas	Sample heated with catalyst in a stream of CO_2; direct volumetric measurement of gaseous nitrogen	Basis of comparison for other methods	
Kjeldahl	Digestion of sample in H_2SO_4 with a catalyst. Separation of ammonia. Determination of ammonia.	Standard procedure for food and feeds. Has been successfully automated and can be used for very small samples. Protein = N × 6.25.	54–55 54–55
other elements	Iodine in thyroglobulin, ferritin, hemoglobin, etc, direct determination of element can be used to determine protein present.	Only suitable when composition of protein is already known.	55
amino acids			
total	Total amino acid composition determined by ion-exchange chromatography. Summation of amino acids can allow estimation of total protein.	Allowance must be made for destruction of amino acids during hydrolysis.	5, 56
specific	Color reactions for specific amino acids can be used to estimate total protein.	Must be standardized for each protein since proteins differ in their amino acid composition.	55
peptide bonds	Lowry modification uses biuret reagent followed by phosphomolybdotungstic acid.	Interference by purines, salicylates, and Tris buffer. Can be used for less than 1 μg protein. Suitable for serum proteins.	55, 58
dye binding	Dependent on binding of dye to basic groups of proteins and measurement of color before and after addition of the protein.	Procedures are empirical and results vary depending on the amounts of dye-binding sites present in the protein.	59–60
whole macromolecule activity	Biological protein activity of enzymes and hormones. Activity per unit protein can indicate degree of purity. Analytical techniques can overlap with separation techniques.	Specific for individual proteins but if reaction chosen is too general can be misleading. Immunoelectrophoresis, radioimmunoassay, enzyme activity	61–62
ultraviolet absorption	Maximum absorption at 280 nm mainly due to the aromatic amino acids. Additional absorption maxima below 230 nm, specific for peptide bonds and amino acids. Nucleic acids may interfere.	Metalloproteins such as ferritin can have intense uv absorption due to the metal and thus lead to erroneous results.	55, 63
near-infrared-reflectance spectroscopy, NIRS	Response proportional to the content of protein.	Measurement of protein content in cereals and legumes in plant-breeding programs. Must be standardized.	64

324 PROTEINS

Table 5. Determination of Protein Sequence

Step	Remarks
preliminary information	Purity is absolute requirement and homogeneity must be proved. Mol wt also necessary together with identification of subunits, if any. Each chain is a separate sequencing problem.
amino acid composition	Ion-exchange chromatography, glc, and hplc. Special techniques for sulfur AA and tryptophan. Number of each residue calculated by use of mol wt and compositional data.
determination of terminal amino acids (AA)	N-terminal AA with 2,4-dinitrofluorobenzene or dansyl chloride[b]. C-terminal AA by hydrazine treatment or carboxypeptidases. Presence of —S—S— linkages may give erroneous conclusions. Special problems with cyclic peptides and with N-acetylated terminal groups.
fragmentation and separation of peptides	Cyanogen bromide cleavage. Mild acid hydrolysis is less specific. Enzymic procedures can preserve asparagine and glutamine. Phenylisothiocyanate[c] couples with N-terminal AA and is followed by cleavage. Basis of automated sequencers.
structure reconstruction	Computer techniques extensively used.
confirmation by synthesis	Peptide chains resynthesis. Merrifield solid-state synthesis (C-terminal linked to ion-exchange resin) for small peptides.

[a] Refs. 67–68.
[b] 1-Dimethylamino-5-naphthalenesulfonyl chloride.
[c] Edmans reagent.

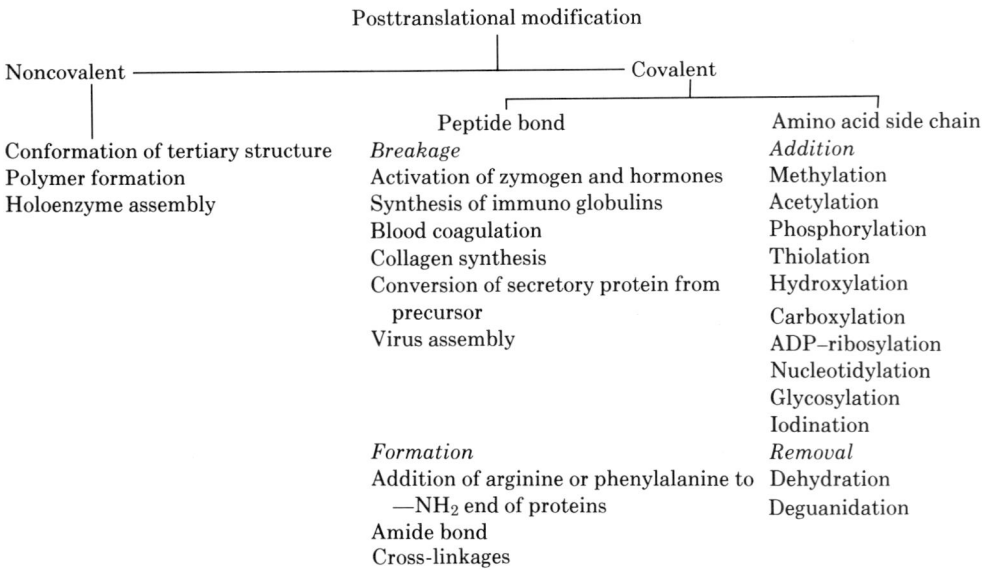

Figure 5. Posttranslational modification reactions of proteins (70).

Nutrition

Since 1972, the North American grain surpluses can no longer be relied upon to meet world food requirements, and supplemental protein from unconventional sources may be needed (71–74). However, the principal means of increasing future food production will be the use of more energy.

According to recent estimates, 70% of world protein production is derived from plants (75–76), and the remaining 30% from animal sources. In the developed countries, animal sources provide most of the dietary protein. The less-developed countries have argued that scarce grain reserves should be diverted from livestock to human consumption, but it is not clear how such a change would produce any long-term benefit.

Protein and food-energy data for various regions are given in Table 6; comparative costs are given in Table 7.

The trends in U.S. domestic food production and disposition, and projected world protein supply and demand have been extensively reviewed (75). In Figure 6, the flow of basic inputs into the production of edible protein is shown.

Protein takes many forms and is fundamental to life; its biological relationships

Table 6. Protein and Food-Energy Consumption Per Capita Per Day for Various Countries and Regions of the World[a]

	Protein, g			kJ[b]	Protein–energy ratio
	Total	Vegetable	Animal		
Africa	58.7	46.7	12.0	9,657	0.102
North America (including Central America)	92.7	36.2	56.5	13,452	0.115
U.S.	106.2	33.5	72.7	14,799	0.120
South America	66.2	36.8	29.3	10,732	0.103
Asia	58.3	46.2	12.1	9,523	0.102
USSR	103.2	52.1	51.1	14,406	0.120
Europe	96.0	43.2	52.8	14,267	0.113
Oceania	95.8	33.3	62.5	13,406	0.120
world	69.3	44.8	24.4	10,418	0.107

[a] Ref. 77.
[b] To convert kJ to kcal, food calorie, divide by 4.184.

Table 7. Costs of Protein from Various Sources[a,b]

Protein source	Product, cents/kg	Protein	
		Content, g/kg	Cost, $/kg
SCP			
(algae, sewage substrate)	7–20	500	0.14–0.40
(yeast, carbohydrate substrate)	15–20	500	0.30–0.40
(bacteria, methanol substrate)	25.0	600	0.42
soybean meal (feed grade)	8.9	440	0.20
FPC	40.0	800	0.50
peanut flour	55.4	600	0.92
casein, food grade	88.6	950	0.93
skim-milk powder	44.4	370	1.20
beans	46.6	220	4.50
beef	79.0	150	8.25
eggs	98.0	110	8.25
lamb	132.0	120	10.00

[a] Refs. 78–79.
[b] Wholesale mid-1970s prices.

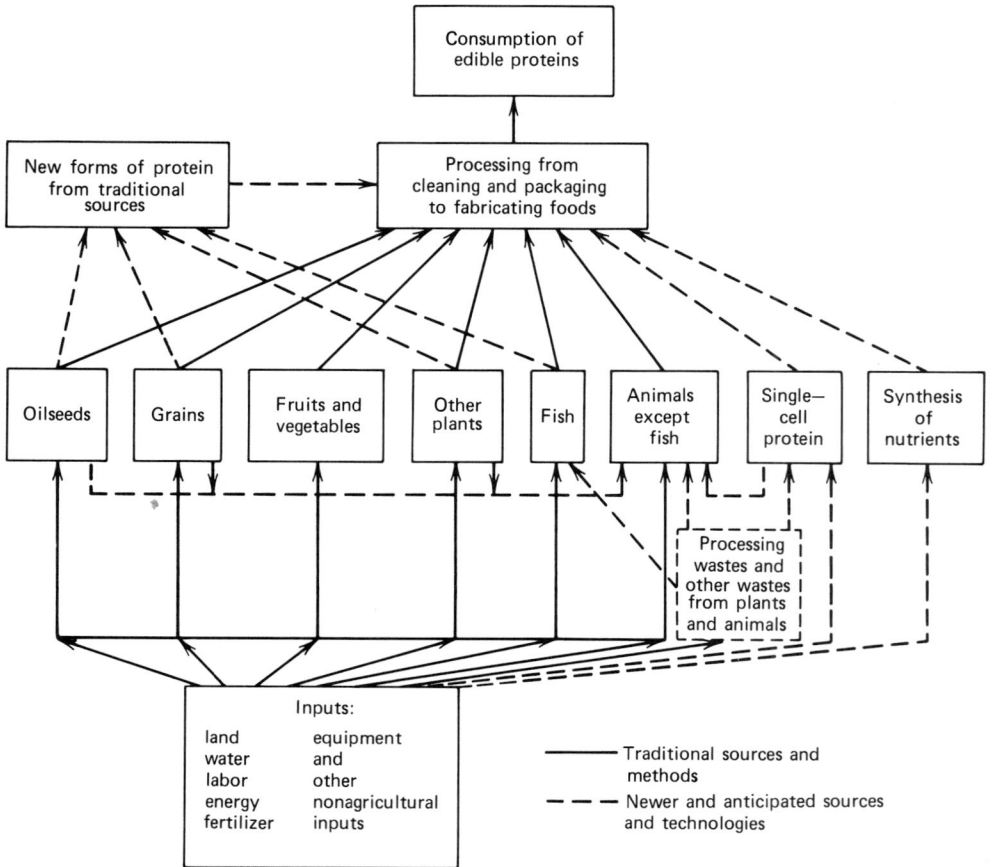

Figure 6. Flow of inputs in producing and processing edible proteins. Courtesy of Avi Press (80).

encompass all of biochemistry, and the elucidation of the biosynthesis of proteins has been a significant accomplishment. Mammalian protein metabolism overlaps with carbohydrate and lipid metabolism and is an important aspect of nutritional studies (81–83). The principal metabolic pathways are shown in Figure 7.

Quality of Dietary Proteins. The capacity of a protein to meet the amino acid and nitrogen requirements of an organism depends not only on the amino acid composition and digestibility of the protein itself but also upon the composition and adequacy of the diet and the health status of the consumer. These factors interact in a complex manner to modify the utilization of dietary protein. Furthermore, all animals, including humans, have some ability to reuse amino acids for the synthesis of proteins.

Secondary and tertiary protein structures are of some nutritional importance, since they can affect utilization of such indigestible proteins as hair and feathers. Digestibility and availability is an inherent part of the protein quality (56,84). In practice, digestibility is included in the determination of protein quality by any animal procedure based on growth or nitrogen retention that does not include fecal collection and analysis. It is, however, not always considered in chemical predictive procedures such as protein scoring.

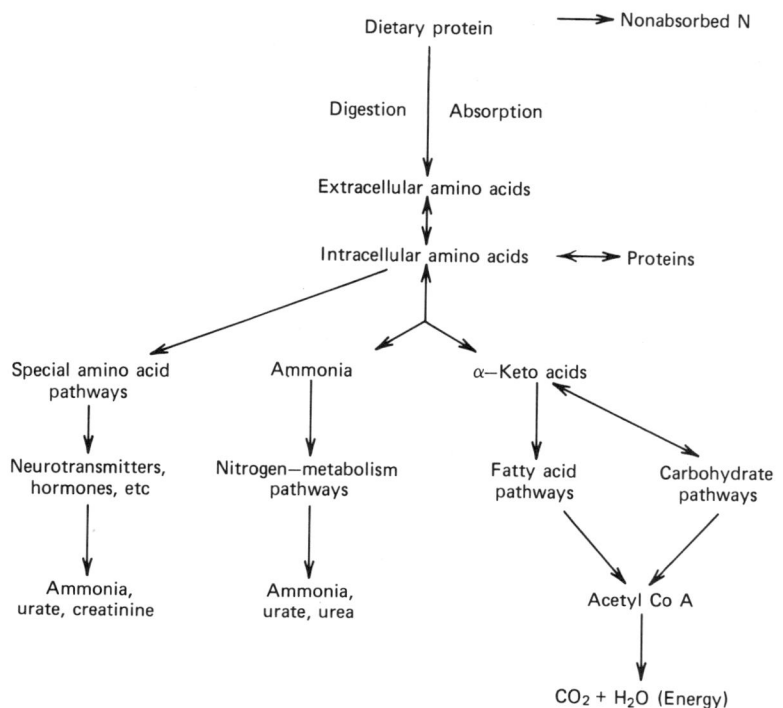

Figure 7. Outline of major pathways in the metabolism of dietary protein.

Animal Assays. A full description of animal-assay procedures including many references is given in the recent United Nations publication on Nutritional Evaluation of Protein Foods (56). Most bioassay procedures are rat assays (see Table 8). The best-known procedures and the official procedure for regulatory purposes within the

Table 8. Rat Assays for Protein Quality [a]

Type	Name	Abbreviation	Introduced in
Two-point assays			
retention of nitrogen			
intake–output	net protein utilization, balance	NPU	1924
carcass analysis	net protein utilization, carcass	NPU	1955
weight gain			
related to intake	protein-efficiency ratio	PER	1919
related to intake and to nonprotein diet	net protein ratio	NPR	1957
related to intake and to nonprotein diet and compared to a standard protein	relative net protein ratio	RNPR	1975
Multipoint assays			
N balance vs N absorbed	nitrogen-balance index	NBI	1945
growth vs intake	nitrogen-growth index	NGI	1945
response vs intake	relative nutritive value	RNV	1965
response vs intake	relative protein value	RPV	1978

[a] Ref. 56.

United States is the protein-efficiency ratio (PER). It is defined as the weight gain of a rat per gram of protein consumed under standardized experimental conditions. In spite of its long history, wide usage, and official status, PER is not a good assay procedure. Some improvement can be made by including a group of animals consuming a nonprotein diet for a similar period. This modification is termed the net protein ratio (NPR) and is more precise and reproducible. In the relative NPR (RNPR), the results are expressed relative to the value obtained with standard protein. For a more accurate assay, slope-ratio assay procedures are recommended, but they are more complex and expensive. Well-known procedures such as PER or NPU (net protein utilization) also measure slope but use only two points and are thus less accurate.

Methods for the evaluation of protein quality in humans have been discussed (85) and simple nitrogen-balance techniques are no longer adequate for comparison of protein quality. Multiple-level studies are now recommended. Routine testing on human subjects, however, is impractical because of the enormous biological variability of subjects, the expense, and the difficulties involved in such studies. Thus, rat assays and chemical procedures remain the routine techniques employed. It must, however, be demonstrated at some point that these techniques give results comparable to results obtained with humans.

Chemical and Microbiological Assays. Total protein is always assayed chemically, usually by the Kjeldahl procedure. However, some nitrogen in foods may be in the form of free amino acids or peptides that can be used for protein synthesis; thus, the distinctions are not clear-cut. In practice, however, assay of protein is in fact evaluation of nitrogen expressed as crude protein (N × 6.25). For labeling regulations and food-table data, other factors may be used for the expression of protein. Products such as single-cell protein (SCP) require special calculations because of the high content of nucleic acids (56).

Dietary protein needs can be met by nitrogen and sulfur in metabolically active forms together with the so-called essential amino acids which the body is not capable of synthesizing; these are histidine, isoleucine, leucine, lysine, methonine (plus cystine), phenylalanine (plus tyrosine), threonine, tryptophan, and valine. The status of histidine as an essential amino acid for all ages has not been unequivocally demonstrated and requirements include usually only the other ten amino acids. The other amino acids are termed nonessential or dispensable amino acids, since they can be synthesized from amino nitrogen and metabolites. The nutritive value of a protein depends primarily on its capacity to satisfy the needs for nitrogen and essential amino acids. Thus, after determination of the amino acid composition, usually by ion-exchange chromatography, protein quality can be estimated by comparison with some standard and the value expressed as an amino acid score.

Reference protein-scoring patterns were introduced by FAO in 1957 and again by the FAO/WHO expert group in 1965. The term amino acid score was used in the FAO/WHO 1973 (86) report which suggested a new pattern based on the more recent evaluation of human amino acid requirements. The NAS/NRC 1980 (87) amino acid pattern for high quality proteins is similar to the FAO/WHO 1973 pattern, except that histidine levels are included and the recommended level for methionine and cystine is 25% lower than that recommended by FAO/WHO (see Table 9). In spite of certain disadvantages, scoring in combination with assay for digestibility and availability can give very accurate predictions of protein quality (88). Another rapid procedure involves a microbiological assay in association with an *in vitro* determination for digestibility

Table 9. FAO/WHO Amino Acid Scoring Patterns[a]

	Amino acid		
	mg/g Protein	mg/g N	millimol/mol N
histidine			
isoleucine	40	250	26.7
leucine	70	440	47.0
lysine	55	340	32.6
methionine and cystine	35	220	15.8
phenylalanine and tyrosine	60	380	30.8
threonine	40	250	29.4
tryptophan	10	60	4.1
valine	50	310	37.1
Total	*360*	*2250*	*223.5*

[a] Ref. 86.

(89). It gives a satisfactory, quick prediction of the PER value. Other techniques for the estimation of protein quality are based on the formation of an insoluble dye complex which is measured colorimetrically (59–60).

Problems still exist in protein quality assay in both methodology and interpretation. Nevertheless, most procedures whether chemical or biological will rank proteins in a similar order. However, it has been argued that an absolute value for protein quality, true under all circumstances, is an unattainable goal (90). Thus, it has been proposed that the defineable characteristics of the protein, ie, total nitrogen, total and available amino acids, and digestibility are the best criteria of quality. These should be modified in practice by the biological and dietary considerations of the conditions under which that protein is likely to be consumed.

Protein Requirements. A number of national and international committees have made estimates of human protein requirements and dietary protein allowances (1, 86–87). Considerable changes in the recommended allowances, both for adult and child, have occurred over the years though they have remained constant for the past decade. The estimates for protein requirements have been the basis for food and agricultural policies and programs and for the planning and evaluation of diets. The proposed standards have a profound effect on the interpretation of the possible nutritional status of a population.

In general, recommended dietary allowances (RDA) for protein are established as follows. The minimum requirement of high-quality protein for maintenance of nitrogen equilibrium is estimated. It is adjusted to allow for the poorer utilization of proteins from normal mixed diets, as compared to diets containing only high quality proteins. The protein allowance is adjusted to allow for the extra needs of growth, pregnancy, and lactation.

The approved procedures for each of these three steps are controversial. The most common technique for the establishment of minimum requirements uses the factorial method given in ref. 86. This approach assumes that nitrogen losses associated with a protein-free intake can be used to predict the nitrogen losses occurring when intakes just meet the minimum physiological requirement for high quality protein. There is, however, no physiological basis for the assumption that these losses are equivalent to maintenance nitrogen needs (13).

The procedure for the determination of protein and amino acid requirements depends upon nitrogen balance, ie, differences between nitrogen intake and excretion are calculated. This method has certain limitations: body nitrogen equilibrium does not necessarily reflect a steady state of organ protein metabolism or nutritional status because it fails to reveal alterations in the distribution of tissue metabolism and protein content within the body. Furthermore, it is difficult to measure quantitatively all the routes of nitrogen loss from the body; nitrogen loss from one route may be compensated by changes in another. It is thus possible that erroneous conclusions can be drawn from apparent nitrogen-balance values. Such errors are larger when the diet is high in nitrogen and some studies on essential amino acid requirements have provided relatively high nitrogen intakes. Other factors influencing the sensitivity and absolute value of the nitrogen balance include the level of energy intake, composition of diet, and the prior nutritional status of the subject (91).

Despite these limitations, current estimates of protein requirements are based upon nitrogen-balance techniques and the factorial approach. In Table 10, the safe levels of protein intake recommended by FAO/WHO, are shown. These recommended levels are increased when the protein consumed is assumed to have a quality lower than that of milk or eggs. Both the 1974 and 1980 NAS recommended dietary allow-

Table 10. Levels of Protein Intake Recommended by FAO/WHO[a]

Age group	Body weight, kg	Per kg per day		Per day		Adjusted level of protein of different quality, g per person per day		
		Protein, g	Nitrogen[b], millimoles	Protein, g	Nitrogen[b], millimoles	Score 80	Score 70	Score 60
infants[c], months								
6–11	9.0	1.53	17.5	14	160	17	20	23
children, yr								
1–3	13.4	1.19	13.6	16	183	20	23	27
4–6	20.2	1.01	11.5	20	229	26	29	34
7–9	28.1	0.88	10.1	25	286	31	35	41
adolescents, yr								
male								
10–13	36.9	0.81	9.3	30	343	37	43	50
13–15	51.3	0.72	8.2	37	423	46	53	62
16–19	62.9	0.60	6.9	38	434	47	54	63
female								
10–12	38.0	0.76	8.7	29	331	36	41	48
13–15	49.9	0.63	7.2	31	354	39	45	52
16–19	54.4	0.55	6.3	30	343	37	43	50
adults								
male	65.0	0.57	6.5	37	423	46	53	62
female	55.0	0.52	5.9	29	331	36	41	48
pregnancy, latter half				add 9	add 103	add 11	add 13	add 15
lactation, first 6 months				add 17	add 194	add 21	add 24	add 28

[a] Ref. 86.
[b] Assumes protein = N × 6.25.
[c] Infants <3 months 2.40 and 3–6 months 1.85 g protein/kg/d.

ances are based upon the FAO/WHO conclusions (86) (see Table 11). The values assume a 75% efficiency of utilization of the protein in a mixed diet.

A further problem relating to protein requirements or allowances is caused by the interaction between protein and other nutrients; for example, the amount of dietary protein significantly enhances urinary calcium excretion without proportionately increasing calcium absorption, and a decrease in calcium balance results (92). Thus, 500 mg calcium is sufficient at a protein intake of RDA level, whereas 800 mg calcium may be marginal at protein intakes of 100 g protein per day or more. Since such intakes are not uncommon in the United States, the resulting precarious calcium balance in many individuals may be partly responsible for the high incidence of osteoporosis in the elderly.

Malnutrition. There are two forms of protein-energy malnutrition (PEM) (9,11). In the kwashiorkor syndrome, first described in the 1930s, edema, skin changes, and many biochemical abnormalities are a predominant feature. In nutritional marasmus the main features are those of starvation.

In general, there is cause for optimism in the decreasing prevalence of many nutritional-deficiency diseases. Curative and preventative programs based on purely nutritional measures such as increasing the supply of the missing nutrient by fortification or direct administration are continuing to produce results. Unfortunately, in spite of enormous research involvement, protein-energy malnutrition, especially of the nutritional marasmus type, remains a worldwide problem.

Since in the kwashiorkor syndrome, low protein staples predominated in the diet resulting in low serum proteins, more protein was the advocated cure and a widespread worldwide deficiency of protein was assumed (93). A United Nations Protein Advisory

Table 11. U.S. Daily Protein Allowances Recommended by NAS[a]

	Age, yr	Body weight, kg	Protein[b], g	Nitrogen millimoles[c]
infants	0.0–0.5	6	2.2[d]	25[d]
	0.5–1.0	9	2.0[d]	23[d]
children	1–3	13	23	263
	4–6	20	30	343
males	11–14	45	45	514
	15–18	66	56	640
	19–22	70	56	640
	23–50	70	56	640
	>51	70	56	650
females	11–14	46	46	526
	15–18	55	46	526
	19–22	55	44	526
	23–50	55	44	526
	>51	55	44	526
pregnant			add 30	add 343
lactating			add 20	add 229

[a] Ref. 87.
[b] Protein Quality (NPU) assumed to be 75%.
[c] Assuming protein = N × 6.25.
[d] Per kg.

Group (PAG) was established to study protein needs. Nutritionists, on the basis of the African kwashiorkor experience, advised that more protein was needed to remedy malnutrition, and efforts were made by various means to increase the production of utilizable protein. It is now realized, however, that protein requirements may have been set too high and that the relationship between energy and protein utilization may have been overlooked. For example, when total food-energy intake is low, protein may be used as a source of energy and is thus less utilized for nutritional purposes. Under these circumstances, the adaptive mechanisms of the body assume that survival is more important than growth. Similar conclusions can be drawn from food-balance-sheet data that indicate that the world is barely meeting its needs for food energy but may be exceeding the requirements for protein. Challenges to the theory of the impending world protein crisis have been growing for the last decade (9–11). Others argue that the 1973 FAO/WHO recommended allowances may be too low (12–14). The conclusion as to whether protein or energy is the limiting feature within a diet can lead to very different recommendations for action (11,94).

The lesser role of protein as a causative feature of world malnutrition is illustrated by the various names of the Protein Advisory Group (PAG) of the United Nations. After its formation during the era in which a world protein crisis was expected, it was renamed the Protein–Calorie Advisory group as the interaction between protein and energy became apparent. Its replacement is now called the Subcommittee on Nutrition of the Administrative Committee on Coordination.

Food Technology

Extraction and Processing. Since cereals often contain more than 12% protein and are grown worldwide, they are a vast potential protein resource. Grain processing (wet and dry milling), with or without fractionation to improve protein quality, has been widely advocated as a new protein source (71,95–97). The protein content of legumes, especially soybeans, is much higher than that of cereals and the preparation of soy protein is an important industry in the United States. Soybeans are processed for both oil and protein. The residue from oil production is used as animal feed, but for human protein food the residue is extracted with hexane. An outline of the procedures for soy-protein production is shown in Figure 8. The undenatured, dehulled, and defatted flakes containing 50% protein may be upgraded to protein concentrates. Commercial yields are ca 60% of a product containing 70% protein. Soy-protein isolates are made commercially by treating the 50% protein flakes with alkali to dissolve the protein. It is then coagulated into a curd by shifting the pH to 4.5; yields are ca 33% (101). The approximate compositions are summarized in Table 12. Both concentrates and isolates are of granular, powdery nature and are transformed by fiber spinning or thermoplastic extrusion into products with fibrous texture to be appealing as food products. Fiber-spun products are truly fabricated foods but are relatively expensive. In thermoplastic extrusion, a cooker–extruder forces the thermoplastic protein material through a die that controls the textured material size and shape. After drying, the unflavored product consists of 50% protein and has excellent storage stability (see Food processing). Nutritionally, soy-protein products, unless supplemented with methionine or other proteins are of poorer quality than animal proteins. Soy proteins are unlikely to displace animal proteins in consumer preference but will continue to supplement animal proteins (see Soybeans and other seed proteins).

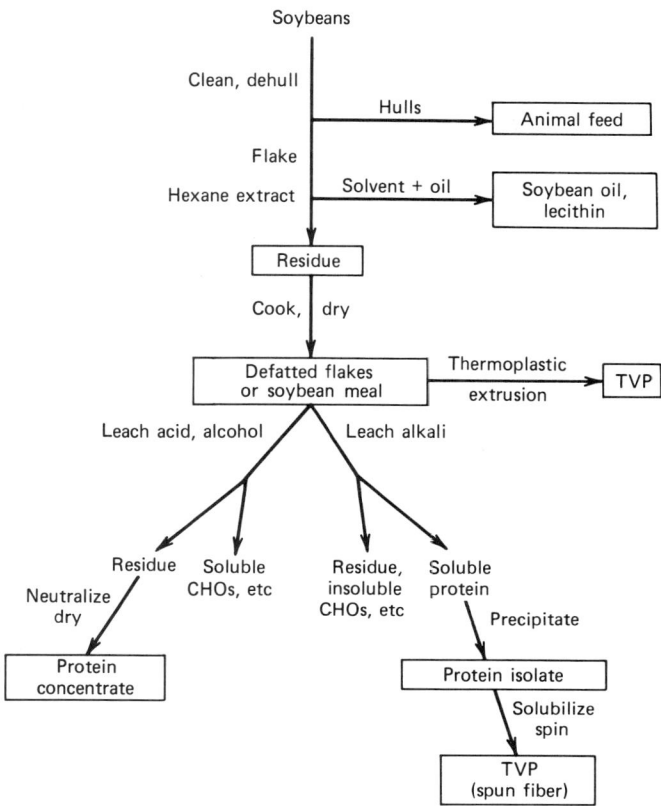

Figure 8. Outline of processes for the preparation of soybean protein products. TVP = textured vegetable protein.

Table 12. Composition of Soy Products, Moisture-Free Basis[a], per 100 g

Component	Flours	Concentrates	Isolates
protein	56.0	72.0	96.0
fat	1.0	1.0	0.1
fiber	3.5	4.5	0.1
ash	6.0	5.0	3.5
carbohydrates			
soluble	14.0	2.5	0
insoluble	19.5	15.0	0.3

[a] Ref. 98.

Modification. Alterations of the chemical structure of food proteins may be caused by deteriorative reactions arising during processing and storage (Table 13) or by modification with chemical reagents in order to change the protein properties (84,99–101). Deteriorative reactions include the Maillard reaction (Fig. 9) and alkaline degradations leading to the formation of compounds such as lysinoalanine. Lysinoalanine formation can be lessened by acylation (102). Although chemical modification of food proteins is currently applied only to a limited extent, it offers opportunities for improving food proteins and extending their availability from nonconventional sources. Chemical modification of protein side chains most frequently in-

334 PROTEINS

Table 13. Processing Damage to Proteins

Condition	Amino acid affected	Remarks
mild to moderate heat	Lys	common in milk products
elevated temperatures	all	lowering of nutritional value
alkaline treatment	Lys, Cys, Arg, Ileu, Thr, Ser	change detectable in acid hydrolysate
oxidation	Met, Cys, cystine	methionine sulfoxide detectable in product

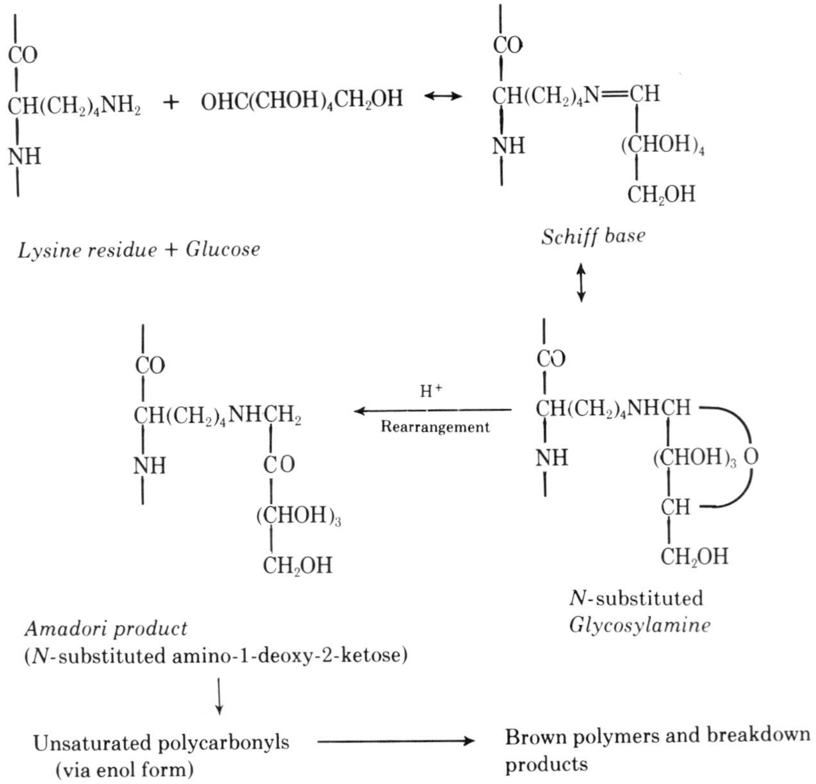

$$\underset{\text{Lysine residue + Glucose}}{\overset{|}{\underset{|}{\text{CO}}}\text{CH(CH}_2)_4\text{NH}_2 \atop \overset{|}{\underset{|}{\text{NH}}}} + \text{OHC(CHOH)}_4\text{CH}_2\text{OH} \longleftrightarrow \underset{\text{Schiff base}}{\overset{|}{\underset{|}{\text{CO}}}\text{CH(CH}_2)_4\text{N}=\text{CH} \atop \overset{|}{\underset{|}{\text{NH}}} \quad \overset{|}{\underset{|}{\text{(CHOH)}_4}} \atop \text{CH}_2\text{OH}}$$

$$\underset{\text{Amadori product}\atop(N\text{-substituted amino-1-deoxy-2-ketose})}{\text{CO}\atop\text{CH(CH}_2)_4\text{NHCH}_2\atop\text{NH}\atop\text{CO}\atop(\text{CHOH})_3\atop\text{CH}_2\text{OH}} \xleftarrow{\text{H}^+ \atop \text{Rearrangement}} \underset{N\text{-substituted}\atop Glycosylamine}{\text{CO}\atop\text{CH(CH}_2)_4\text{NHCH}\atop\text{NH}\atop(\text{CHOH})_3\atop\text{CH}\atop\text{CH}_2\text{OH}}$$

$$\downarrow$$

Unsaturated polycarbonyls (via enol form) $\longrightarrow$ Brown polymers and breakdown products

Figure 9. Early stages of the Maillard reaction. Note: glucose is shown as an example but nonenzymatic browning occurs also with pentoses and other hexoses.

volves the ϵ amino group of lysine and the sulfhydryl or disulfide group of cystine. Side-chain modification can improve physical state or nutritional quality by blocking deteriorations. However, safety and consumer acceptability have to be considered. Chemical modification of soy proteins includes treatment with acids and alkalies, acylation, alkylation and esterification, and oxidation and reduction. In most instances, these reactions have been applied to heterogenous protein mixtures that contain nonprotein impurities. Nevertheless, it is evident (103) that protein functional properties can be altered significantly by such reactions (see Table 14).

Modification also includes enzymatic action (see Table 15), by which the peptide bonds are hydrolyzed (104) (see Enzymes).

Table 14. Functional Properties of Proteins in Food Systems

Property	Function	Remarks
adhesion, cohesion	binding in meats and meat analogues	soy proteins enhance ability
coagulation and curding	gel structure in milk products: matrix in holding other components	rennin coagulates milk proteins in presence of Ca^{2+}
emulsification	stability and capacity of emulsions in sausages, mayonnaise, coffee whiteners	affected by proteases on meat and protein concentrates
elasticity and dough formation	bread doughs: proteins affect dough strength	soybean lipoxygenase oxidizes —SH groups and increases cross-linkage; ascorbic acid oxidase also affects dough quality
fat binding	sausages, meat products, doughnuts	soy proteins enhance ability
gel formation	gel structures act as a matrix for holding moisture, lipids and polysaccharides; cheese and yogurt	rennin and pepsin function in cheesemaking in association with coagulation and curding; proteases can reduce
solubility	allows selective precipitation to form structured products; meat analogues, cheese, yogurt	proteases and alkaline treatment increase solubility
viscosity	proteins increase viscosity in protein containing foodstuffs, soups, gravies, etc	proteases reduce viscosity by increasing solubility of proteins
water binding and absorption	prevents moisture loss in breads, cakes, etc	protein polar groups bind water by hydrogen bonding
whippability	foaming and air-holding capacity enhanced by soluble proteins (albumins); reduction of surface tension	proteases can increase foam volume but also decrease stability

Followed by resynthesis, enzymatic degradation can improve both the nutritive value of food proteins and their functional quality (99,105–106). The mechanism of resynthesis (plastein reaction) is complicated but both condensation and transpeptidation reactions are known to be involved. Since it is possible to add amino acids, plasteins of improved essential amino acid composition can be formed. Amino acid esters can also be incorporated directly into proteins (106). Similar enzymatic processes improve solubility and remove impurities that may contribute undesirable flavor, odor, color, or toxicity (107). Although such reactions have been demonstrated under laboratory conditions, the commercial feasibility of the processes remains to be demonstrated.

Proteolytic enzymes decrease gel formation and viscosity and modify foaming and emulsification properties.

New Protein Products. *Single-Cell Protein.* Single-cell protein (SCP) is the generic term for protein produced by unicellular organisms such as bacteria, yeast, and single-cell algae (108–111). The distinction into photosynthetic and nonphotosynthetic single-cell protein is obvious and depends on the presence of chlorophyll in the organism producing the proteins (see Foods, nonconventional).

In the production of SCP, the growth medium consists of a carbon source (sugar

Table 15. Proteolytic Enzymes Action in Protein Modification[a]

Food	Purpose
baked goods	softening action, reduction of mixing time, increased extensibility of dough, improvement in texture, grain and loaf volume; liberation of β-amylase
brewing	body, flavor, and nutrient development during fermentation; aid in filtration and clarification; chill-proofing
cereals	increased drying rate, improved product-handling characteristics
cheese	casein coagulation; characteristic flavor development during aging
chocolate, cocoa	action on beans during fermentation
egg, egg products	improved drying properties
feeds	conversion of waste product to feeds; digestive aids, particularly for pigs
fish	solubilization of FPC; recovery of oil and proteins from inedible parts
legumes	hydrolyzed protein products, removal of flavor, plastein formation
meats	tenderization, recovery of protein from bones
milk	coagulation, preparation of soybean milk
protein hydrolysates	condiments such as soy sauce and tamar sauce; bouillon, dehydrated soups, gravy powders, processed meats, special diets
antinutrient factor removal	specific protein inhibitors of proteolytic enzymes and amylases
wines	clarification[b]
in vivo processing[c]	conversion of zymogens to enzymes, fibrinogen to fibrin, proinsulin to insulin; macromolecular assembly; collagen biosynthesis

[a] Ref. 104. Courtesy of the American Chemical Society.
[b] In large part caused by other than proteolytic enzymes.
[c] Representative examples given.

or hydrocarbon), a nitrogen source (usually ammonium salts), and minerals dissolved in water. The organisms (most often yeasts) are grown in a fermenter that requires oxygen transfer and heat removal together with continuous stirring (see Fig. 10). The cells and the spent medium are removed on a continuous basis and the organisms are harvested, washed, and dried. Substrates include natural gas, paraffins, alcohols, and cellulose.

As with other proteins, nutritional qualities are primarily determined by amino acid composition, but acceptability is determined by the level of nucleic acids. SCP often contains too much RNA (ribonucleic acid) for human consumption. However, recent reports (111–112) conclude that there is no reason for delaying the production and sale of such thoroughly tested SCP products as Toprina (derived from *Candida*

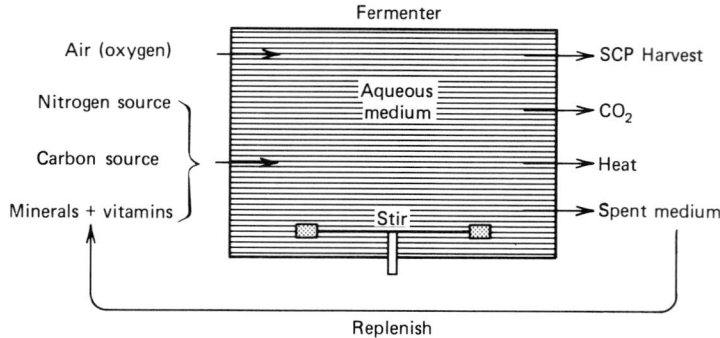

Figure 10. Schematic for nonphotosynthetic single-cell protein (SCP) production (98).

lipolytica, grown on *n*-paraffins), Liquipron (*Candida maltosa*, grown on *n*-paraffins), and Pruteen, a bacterial product derived from *Methylophilus methylotrophus*, produced on a methanol substrate.

Leaf-Protein Concentrate (LPC). Green leaves constitute the largest source of protein, but leafy plants also contain high percentages of fibrous structural material (eg, cellulose) that cannot be digested by humans because of the lack of enzymes capable of splitting β-linked glycosides. Ruminant animals (cattle, sheep) and animals with functional caeca can utilize large amounts of cellulose because of the action of symbiotic microorganisms in the rumen or caecum. Leafy plants have a great potential as a protein source if the protein can be separated from other leaf components and recovered in forms suitable as feed for poultry and swine and as food ingredients for humans (113–114). All procedures are based on macerating the leafy material followed by recovery of protein from the juice. There are several grades of LPC: whole green-leaf protein, decolorized whole green-leaf protein, protein fractionated to yield a white and a green fraction, and protein fractionated to yield pure crystalline protein plus other protein fractions. Extraction procedures are complicated by the fact that, although fresh leaves contain 90% water, only 20% of the total water contains the chloroplasts and other cytoplasmic bodies where most of the protein is concentrated. If the plant is grown directly for LPC production, little fertilizer is needed. Crops grown for LPC production include alfalfa, grasses, clovers, vegetable wastes, and a large number of tropical and subtropical plants.

The original Pirie process (113) and the USDA Pro-Xan process (115) have different objectives. The former attempted to develop a procedure for developing countries. In this process the freshly chopped forage is pulped and then treated in a belt press where the juice is forced out through a perforated pulley. The juice is purified by passing through a coarse cloth filter and the protein is coagulated by the direct injection of steam. The protein is skimmed from the surface and concentrated. Further washing, drying (60°C), and grinding produces a pale green powder (116).

In the USDA Pro-Xan process, research has been directed toward a large-scale commercial process suitable for use in developed countries to increase the efficiency of utilization of forages and leafy vegetable wastes (114–115). A more recent modification allows the preparation of a bland colorless protein fraction suitable for human consumption (Pro-Xan 2 process). Sodium metabisulfite added before the juice extraction improves the the whiteness of LPC and protects the sulfur amino acids from oxidation; yields, however, are very low and current research activity is directed toward increased yields. Nutritional evaluation of LPC products have shown generally high levels of essential amino acids; the white product has a higher concentration of sulfur amino acid than the green. On the other hand, total yields of green protein are considerably higher than the white.

Fish-Protein Concentrate. Fish-protein concentrate (FPC), originally known as fish flour, was developed as a source of protein at a time when protein deficiency was seen as a worldwide problem. It is prepared from edible-grade whole fish in accordance with food processing standards, but was opposed by the FDA for sale within the United States because of the presence of viscera, scales, etc. It now appears that FPC was vastly overrated as a solution to the problem of protein-energy malnutrition (see Aquaculture).

Protein-Rich Food Mixtures. Early discontinuance of breast feeding in the urban areas of most developing countries has been a serious concern for many years, and has often proved disastrous to infants, especially where hygenic practices are inadequate for safe use of breast-milk replacements. The food industry, in close cooperation with governments and physicians, has attempted to develop low cost protein foods to supplement breast feeding (117).

Food mixtures may be manufactured products or they may be prepared at home. True complementation occurs only when the quality of the mixture is better than that of the components. Many mixtures contain cereal–legume combinations ranging from 70–30 to 60–40. Much emphasis has been placed by nutrition scientists, UN agencies, and governments on these low cost foods to solve worldwide malnutrition. However, the commercial mixtures are not very effective. Protein complementation is now more a home activity (118).

Synthesis of Protein and Amino Acids. Traditional methods of increasing the quantity and quality of food protein production include modifications of plants and animals, leaf-protein concentrate, and some varieties of single-cell protein. These are all extensions of traditional agriculture in the sense that they utilize biological transformation of solar energy. Chemical synthesis, in conjunction with an efficient energy-producing system, by allowing development on nonagricultural land, could greatly increase food production and permit enormous increases in population density. Such changes remain far in the future because of current energy constraints. Present techniques, however, suggest the direction for future research (119). Synthesis of essential amino acids is quite feasible, but the costs remain excessively high.

Outlook. As a result of an NSF/Rann grant, a broad, multidisciplinary inquiry was initiated in the early 1970s to survey the needs, evaluate the issues, and establish priorities for research on new protein sources. Although the final document (71) was only published in 1978, the summary and research recommendations were published in 1975 by NSF/Rann.

BIBLIOGRAPHY

"Proteins" in *ECT* 1st ed., Vol. 11, pp. 226–248, by H. B. Vickery, The Connecticut Agricultural Station; "Proteins" in *ECT* 2nd ed., Vol. 16, pp. 610–640, by Michael V. Tracey, CSIRO (Commonwealth Scientific and Industrial Research Organization) Division of Food Preservation, Australia.

1. H. N. Munro in H. N. Munro and J. B. Allison, eds., *Mammalian Protein Metabolism*, Vol. I, Academic Press, New York, 1964, Chapt. 1.
2. E. V. McCollum, *A History of Nutrition*, Houghton Mifflin Co., Boston, Mass., 1957, p. 451.
3. P. R. Srinavasan, J. S. Fruton, and J. T. Edsall, eds., *Ann. N.Y. Acad. Sci.* **325,** 1 (1979).
4. H. E. Carter, *Fed. Proc.* **38,** 2684 (1979).
5. R. J. Block and M. M. Mitchell, *Nutr. Abst. Rev.* **16,** 249 (1946).
6. F. L. Holmes, *Fed. Proc.* **39,** 216 (1980).
7. S. Moore and W. H. Stein, *Science* **180,** 458 (1973).
8. C. B. Anfinsen, *Science* **181,** 223 (1973).
9. D. S. McLaren, *Lancet* **2,** 93 (1974).
10. J. C. Waterlow and P. R. Payne, *Nature* **258,** 113 (1975).
11. P. L. Pellett, *Ecol. Food Nutr.* **5,** 205 (1976).
12. N. S. Scrimshaw, *Nutr. Rev.* **35,** 321 (1977).
13. V. R. Young and N. S. Scrimshaw in M. Milner, N. S. Scrimshaw, and D. I. C. Wang, eds., *Protein Resources and Technology: Status and Research Needs*, Avi Publishing Co., Westport, Conn., 1978, p. 136.
14. C. Garza, N. S. Scrimshaw, and V. R. Young, *J. Nutr.* **107,** 335 (1977).

15. G. E. Schulz and R. H. Schirmer, *Principles of Protein Structure*, Springer-Verlag, New York, 1978, p. 314.
16. C. B. Anfinsen and H. Scheraga in C. B. Anfinsen, J. T. Edsall, and F. M. Richards, eds., *Advances in Protein Chemistry*, Vol. 29, Academic Press, New York, 1975, p. 205.
17. G. W. Tischendorf, H. Zeichardt, and G. Stöffler, *Proc. Natl. Acad. Sci. USA* **72,** 4820 (1975).
18. L. Pauling, R. B. Corey, and H. R. Branson, *Proc. Natl. Acad. Sci. USA* **37,** 205 (1951).
19. K. U. Linderström-Lang and J. A. Schellman in P. D. Boyer, ed., *The Enzymes*, Vol. I, 2nd ed., Academic Press, New York, 1959, p. 443.
20. G. D. Fasman, ed., *Handbook of Biochemistry and Molecular Biology, Protein*, Vols. I–III, 3rd ed., CRC Press, Cleveland, Ohio, 1976.
21. R. Montgomery, R. L. Dryer, T. W. Conway, and A. A. Spector, *Biochemistry: A Case Oriented Approach*, 2nd ed., C. V. Mosby Co., St. Louis, Mo., 1977.
22. R. W. McGilvery, *Biochemistry—A Functional Approach*, 2nd ed., W. B. Saunders, Philadelphia, Pa., 1979.
23. J. S. Wall, *J. Agr. Food Chem.* **19,** 619 (1971).
24. R. S. Asquith, ed., *Chemistry of Natural Protein Fibers*, Plenum Press, New York, 1977.
25. H. G. Mannherz and R. S. Goody, *Ann. Rev. Biochem.* **45,** 427 (1976).
26. D. E. Goll, R. M. Robson, and M. H. Stromer in J. R. Whitaker and S. R. Tannenbaum, *Food Proteins*, Avi Publishing Co., Inc., Westport, Conn., 1977.
27. W. F. Harrington in H. Neurath, R. L. Hill, and C. L. Boeder, eds., *The Proteins*, Vol. IV, 3rd ed., Academic Press, New York, 1979, p. 246.
28. J. L. Marx, *Science* **208,** 274 (1980).
29. H. Bisswanger and E. Schmincke-Ott, eds., *Multifunctional Proteins*, Wiley-Interscience, New York, 1980.
30. J. H. Bradbury in Ref. 16, Vol. 27, 1973, p. 111.
31. P. Bornstein and W. Traub in Ref. 27, Vol. IV, 1979, p. 412.
32. S. C. R. Elgin and H. Weintraub, *Ann. Rev. Biochem.* **44,** 725 (1975).
33. R. J. Delange and E. L. Smith in Ref. 27, Vol. IV, 1979, p. 119.
34. R. G. Spiro in Ref. 16, Vol. 27, 1973, p. 349.
35. E. F. Walborg, Jr., ed., *Glycoproteins and Glycolipids in Disease Processes*, ACS Symposium Series 80, Washington, D.C., 1978.
36. J. C. Osborne, Jr. and H. B. Brewer, Jr. in Ref. 16, Vol. 31, 1977, p. 253.
37. W. A. Bradley and A. M. Gotto, Jr. in J. M. Dietschy, A. M. Gotto, Jr., and J. A. Ontko, eds., *Disturbances in Lipid and Lipoprotein Metabolism*, American Physiological Society, Bethesda, Md., 1978, p. 111.
38. G. Taborsky in Ref. 16, 1974, p. 1.
39. C. R. Rubin and O. M. Rosen, *Ann. Rev. Biochem.* **44,** 831 (1975).
40. M. Weller, *Protein Phosphorylation*, Pion, Ltd., London, Eng., 1979.
41. K. T. Yasunobu, H. F. Mower, and O. Hayaishi, *Adv. Exp. Biol. Med.* **5,** 74 (1976).
42. E. B. Brown, P. Aisen, J. Fielding, and R. R. Crichton, eds., *Proteins of Iron Metabolism*, Grune and Stratton, New York, 1977.
43. J. R. Whitaker in Ref. 26, 1977, p. 14.
44. G. K. Ackers in Ref. 27, 1975, p. 1.
45. S. Hierten in A. Niederwieser and G. Pataki, *New Techniques in Amino Acid Peptide and Protein Analysis*, Ann Arbor Science Publishers, Inc., Ann Arbor, Mich., 1971, p. 227.
46. C. W. Wrigley in Ref. 45, 1971, p. 291.
47. H. Hyden and P. W. Lange in Ref. 45, 1971, p. 271.
48. N. Catsimpoolas in Ref. 26, 1977, p. 106.
49. B. G. Johansson in Ref. 45, 1971, p. 249.
50. K. Weber and M. Osborn in Ref. 27, Vol. I, 1975, p. 180.
51. J. Porath and T. Kritiansen in Ref. 27, Vol. I, 1975, p. 95.
52. K. E. Van Holde in Ref. 27, Vol. I, 1975, p. 226.
53. P. Douzou and C. Balney in Ref. 16, Vol. XXXII, 1978, p. 77.
54. S. Jacobs, *Methods Biochem. Anal.* **13,** 241 (1965).
55. H. N. Munro and A. Fleck in Ref. 1, Vol. III, 1960, p. 423.
56. P. L. Pellett and V. R. Young, eds., *Nutritional Evaluation of Protein Foods*, United Nations University, Tokyo, Jpn., WHTR-3/UNUP-129, 1980.

57. P. E. Hare in S. B. Needleman, ed., *Protein Sequence Determination: A Source Book of Methods and Techniques*, Springer-Verlag, New York, 1975, p. 204.
58. O. H. Lowry, N. J. Rosebrough, A. L. Farr, and R. J. Randall, *J. Biol. Chem.* **193,** 265 (1951).
59. A. L. Lakin in J. W. G. Porter and B. A. Rolls, eds., *Proteins in Human Nutrition*, Academic Press, New York, 1973, p. 179.
60. R. F. Hurrel, P. Lerman, and K. J. Carpenter, *J. Food. Sci.* **44,** 1221 (1979).
61. R. S. Yalow and S. A. Berson, *J. Clin. Invest.* **39,** 1157 (1960).
62. E. Afonso, *Clin. Chim. Acta* **14,** 63 (1966).
63. W. E. Groves, F. C. Davis, Jr., and B. H. Sels, *Anal. Biochem.* **22,** 195 (1968).
64. P. C. Williams, S. G. Stevenson, P. M. Starkey, and G. C. Hawtin, *J. Sci. Food. Agric.* **29,** 285 (1978).
65. F. Sanger and H. Tuppy, *Biochem. J.* **49,** 463 (1961); F. Sanger and E. O. P. Thompson, *Biochem. J.* **53,** 353 (1963).
66. M. O. Dayhoff, ed., *Atlas of Protein Sequence and Structure*, Vol. 5, Supplement No. 3, National Biomedical Research Foundation, Georgetown University Medical Center, Washington, D.C., 1979.
67. Ref. 57, entire book.
68. S. B. Needleman, ed., *Advanced Methods in Protein Sequence Determination*, Springer-Verlag, New York, 1977.
69. J. D. Watson, *Molecular Biology of the Gene*, 3rd ed., Benjamin-Cummings; Addison Wesley, Menlo Park, Calif., 1975.
70. W. K. Paik and S. Kim, *Protein Methylation*, Wiley-Interscience, New York, 1980, p. 282.
71. Ref. 13, entire book.
72. D. W. Stanley, E. D. Murray, and D. H. Lees, *Utilization of Protein Resources*, Food and Nutrition Press, Inc., Westport, Conn., 1981.
73. M. W. Adams, M. Milner, G. M. Montfort, and L. M. Rockland in Ref. 13, 1978, p. 302.
74. N. W. R. Daniels in Ref. 72, 1981, p. 188.
75. W. W. Gallimore, A. Dasgupta, and C. Chee-Khoon in Ref. 13, 1978, p. 47.
76. G. E. Inglett in Ref. 26, 1977, p. 267.
77. *Food Balance Sheets 1975–1977 Average*, FAO, Rome, Italy, 1980.
78. J. C. Abbott in R. Ferrando, M. Ganzin, and P. R. Payne, *Conventional and Non Conventional Proteins*, Folia Veterinaria Latina, Vol. VI, Suppl. 1, Casa Editrice 'Il Ponte', Milan, Italy, 1976, p. 75.
79. R. P. Oullette, N. W. Lord, and P. N. Cheremisinoff, *Food Industry Energy Alternatives*, Food and Nutrition Press, Inc., Westport, Conn., 1980.
80. Ref. 13, p. 72.
81. D. W. Ribbons and H. Brew, eds., *Proteolysis and Physiological Regulation*, Academic Press, Inc., New York, 1976.
82. J. S. Garrow, *Energy Balance and Obesity in Man*, American Elsevier Publishing Co., New York, 1974.
83. B. Schepartz, *Regulation of Amino Acid Metabolism in Mammals*, W. B. Saunders Co., Philadelphia, Pa., 1973.
84. J. Mauron in *Proceedings 10th Congress Food. Science and Technology*, Vol. I, Madrid, Spain, 1974, p. 564.
85. D. E. Bodwell, ed., *Evaluation of Proteins for Humans*, Avi Publishing Co., Inc., Westport, Conn., 1977.
86. FAO/WHO, *Energy and Protein Requirements*, Report of a FAO/WHO Ad Hoc Expert Committee, Food and Agricultural Organization of the United Nations, Rome, Italy, 1973.
87. Committee on Dietary Allowances, Food and Nutrition Board, National Research Council, *Recommended Dietary Allowances*, 9th rev. ed., National Academy of Sciences, Washington, D.C., 1980.
88. L. D. Satterlee, H. F. Marshall, and J. M. Tennyson, *J. Am. Oil Chem. Soc.* **56,** 103 (1979).
89. H. W. Hsu, N. E. Sutton, M. O. Banjo, L. D. Satterlee, and J. G. Kendrick, *Food. Technol.* **32**(12), 69 (1978).
90. P. L. Pellett, *Food. Technol.* **32**(5), 60 (1978).
91. D. M. Hegsted, *J. Nutr.* **106,** 307 (1976).
92. W. Mertz, *J. Am. Diet. Assoc.* **77,** 258 (1980).
93. *International Action to Avert the Impending Protein Crisis*, E 68 XIII 2, United Nations, New York, 1968.
94. R. G. Whitehead in Ref. 59, 1973, p. 103.

95. R. A. Anderson in G. E. Inglett, ed., *Wheat Production and Utilization*, Avi Publishing Co., Westport, Conn., 1974, p. 355.
96. J. E. Cluskey, D. A. Fellers, G. E. Inglett, H. C. Neilsen, Y. Pomeranz, R. L. Roberts, R. M. Saunders, A. D. Shepherd, J. S. Wall, and Y. V. Wu in Ref. 13, 1978, p. 256.
97. W. Bushuk in Ref. 72, 1981, p. 208.
98. F. E. Horan, *J. Am. Oil Chem. Soc.* **51**, 67A (1974).
99. R. E. Feeney and J. R. Whitaker, eds., *Food Proteins: Improvement Through Chemical and Enzymic Modification*, Advances in Chemistry Series No. 160, American Chemical Society, Washington, D.C., 1977.
100. J. C. Cheftel in Ref. 26, 1977, p. 401.
101. W. D. Powrie and S. Nakai in Ref. 72, 1981, p. 73.
102. M. Friedman in M. Friedman, ed., *Nutritional Improvement of Food and Feed Proteins*, Plenum Press, New York, 1978, p. 613.
103. A. Pour-El, *Functionality and Protein Structure*, ACS Symposium Series, No. 92, American Chemical Soc., Washington, D.C., 1979.
104. J. R. Whitaker in Ref. 99, 1977, p. 95.
105. T. Richardson in Ref. 99, 1977, p. 185.
106. S. Arai, M. Yamashitam, and M. Fujimaki in Ref. 102, 1978, p. 663.
107. I. E. Liener in Ref. 99, 1977, p. 283.
108. S. R. Tannenbaum and I. C. Wang, eds., *Single Cell Protein II*, M.I.T. Press, Cambridge, Mass. and London, 1975.
109. J. E. Kinsella and K. J. Shetty in Ref. 102, 1978, p. 797.
110. S. R. Tannenbaum, C. L. Cooney, A. M. Demain, and L. Haverberg in Ref. 13, 1978, p. 502.
111. S. Garattini, S. Paglialunga, and N. S. Scrimshaw, eds., *Single Cell Protein—Safety for Animal and Human Feeding*, Pergamon Press, New York, 1979.
112. Protein Advisory Group, *PAG Bulletin*, Guidelines Nos. 6 and 12, Vol. II, 1972, and Vol. IV, 1974.
113. N. W. Pirie in N. W. Pirie, ed., *Leaf Protein—Its Agronomy, Preparation Quality and Use*, Blackwell, Oxford, 1971, p. 53.
114. G. O. Kohler, S. G. Wildman, N. A. Jorgensen, R. V. Enochian and W. J. Bray in Ref. 13, 1978, p. 543.
115. *Twelfth Technical Alfalfa Conference Proceedings*, USDA, Western Regional Reserach Center and American Dehydrators Association, Berkeley, Calif., USDA Western Regional Research Service, 1975.
116. W. J. Bray, *Cost of producing leaf protein concentrate in an Indian Village*, Meals for Millions, Freedom from Hunger Foundation, Santa Monica, Calif., 1980.
117. P. L. Pellett in D. S. McLaren, ed., *Nutrition in the Community*, John Wiley & Sons, Inc., London and New York, 1976, p. 185.
118. M. Cameron and Y. Hofvander, *Manual on Feeding Infants and Young Children*, 2nd ed., Protein–Calorie Advisory Group of the United Nations System, United Nations, New York, 1976.
119. S. W. Fox, J. W. Frankenfeld, D. Romsos, D. M. Robinson, and S. A. Miller in Ref. 13, 1978, p. 569.

PETER L. PELLETT
University of Massachusetts

PROTEINS FROM PETROLEUM. See Foods, nonconventional.

PROTOZOAL INFECTIONS, CHEMOTHERAPY. See Chemotherapeutics, protozoal.

PRUSSIAN BLUE. See Iron compounds; Pigments, inorganic.

PSYCHOPHARMACOLOGICAL AGENTS

For unknown centuries humans have employed various mind-altering, herbacious concoctions for recreational and mystical purposes, but the earliest recorded use of drugs for psychotherapeutic reasons dates to the third century BC. At that time cathartic herbs, particularly hellebore, were commonly used in the treatment of melancholy (1). Melancholy was believed to result from an excess of black bile in the blood and the atomists reasoned that cathartics opened pores in the intestinal wall. These pores would permit the passage of the black bile into the intestinal lumen and thus restore the body's humoral composition to normal (2). Despite their presumed ineffectiveness, cathartics were employed in the treatment of emotional disorders for several centuries (see Gastrointestinal agents). Hippocrates postulated the connection between mental and emotional disorders and brain function ca 2400 years ago (3). The tranquilizing properties of *Rauwolfia* extracts have been known for many centuries by some eastern cultures. Modern research has verified this property of *Rauwolfia* by identifying reserpine [50-55-5] as an active constituent (4). Serious research regarding the use of drugs in the treatment of mental disorders began during the 19th century when clinical investigations of the use of *cannabis* in various disorders, particularly depression, were conducted (5).

Cannabis has a brief but stimulating and euphoric influence on depressed patients. Although the effects of *cannabis* appeared to be both dramatic and desirable, the action was short-lived and the preparations available for use by the practicing physician were so unreliable that *cannabis* never achieved widespread clinical use. Clinical experimentation with extracts of cocoa leaves and cocaine also occurred during the latter half of the 19th century. These preparations had therapeutic use in the treatment of alcoholism and morphinism (6).

The development of drugs that are useful in the treatment of specific emotional and mental disorders occurred during the 1950s. The therapeutic effect of chlorpromazine [69-09-0] in psychotic patients was demonstrated in 1952 (6). The specific tranquilizing effect, with no general sedation, of chlorpromazine represented a new and exciting step in the development of psychopharmacology. Iproniazid [54-92-2], a drug used in the treatment of tuberculosis, was observed to have antidepressant effects in 1952 and became available to practicing physicians in 1957 (6). The antidepressant actions of the tricyclic compound imipramine [50-49-7] was also described in 1957 (7).

In 1954, meprobamate [57-53-4] (1) was identified as a specific anxiolytic and is useful in the treatment of neuroses (6). Clinical trials of another new class of psychoactive structures, the benzodiazepines, began in 1958 and resulted in the introduction of the anxiolytic drug chlordiazepoxide [58-25-3] in 1960 (8). The benzodiazepines have become the most widely prescribed drugs in the history of medicine.

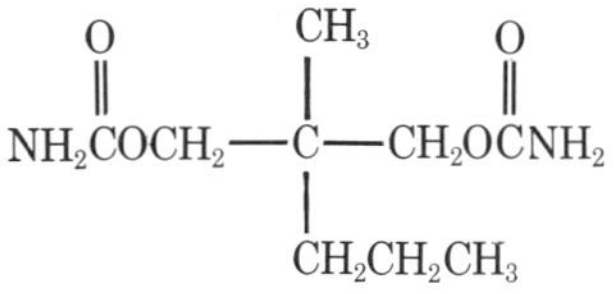

$$NH_2COCH_2-\underset{\underset{CH_2CH_2CH_3}{|}}{\overset{\overset{CH_3}{|}}{C}}-CH_2OCNH_2$$

(1)

Although the number of marketed psychotherapeutic drugs has expanded tremendously since the 1950s, there remains considerable need for improved agents in this area of pharmacology. Most of today's drugs are structural analogues or are believed to have identical mechanisms of action to those pioneer drugs developed in the 1950s and early 1960s and, therefore, possess many of their therapeutic limitations. Although no two compounds are exactly alike with regard to their pharmacological and pharmacokinetic properties, none of the psychotherapeutic agents introduced onto the market since 1965 represents a significant advance over those drugs described prior to that time.

The need for psychotherapeutic agents is clear. It is estimated that in the United States, ca 1% of the population is schizophrenic, 1% is deeply depressed, and another 25% suffer from mild to moderate depression and anxiety. Mental disorders still account for more hospital admissions than any other single illness (9).

Recent purchase trends in the United States for the three main classes of psychotherapeutic drugs are illustrated in Table 1. Although total sales have remained relatively constant during the 1970s, the number of prescriptions has decreased significantly since 1975. The reasons for this decline are not entirely understood but are believed to result, in part, from a trend toward writing prescriptions for larger amounts of drugs. This shift in prescription size apparently is caused by an increase in government-supported health programs.

Psychotherapeutic agents have received broad acceptance and have played an important role in reducing the population of resident patients of our mental institutions. From 1956 to 1975, resident mental hospital patient populations decreased by more than 50% (10). In addition, the lives of countless nonhospitalized patients afflicted with emotional disorders have been improved by the use of drugs.

The discussion in this article is limited to the most used groups of psychotherapeutic agents, ie, the anxiolytics, the antipsychotics, and the antidepressants, and concerns itself mainly with the drugs marketed in the United States.

Table 1. 1970–1979 Psychotherapeutic Purchase Trends in the United States[a]

	1970	1975	1979	% change, 1970–1975	% change, 1975–1979	% change, 1970–1979[b]
total sales, 10^6 $	574	741	805	+29	+9	−4
total prescriptions, 10^6	132	160	122	+21	−24	
anxiolytics						
sales, 10^6 $	277	418	433	+51	+4	+7
prescriptions, 10^6	83	103	75	+24	−27	
antipsychotics						
sales, 10^6 $	125	138	153	+10	+11	−16
prescriptions, 10^6	22	24	18	+9	−25	
antidepressants						
sales, 10^6 $	172	185	219	+8	+18	−13
prescriptions, 10^6	27	33	29	+22	−12	

[a] Courtesy of IMS America, Ltd., 1971–1980.
[b] Corrected for rate of monetary inflation.

Anxiolytics

Anxiety is an emotion experienced to some degree by nearly everyone. Generally, anxiety is an unpleasant sensation that may play an important and necessary role in motivation and drive. It is well known, however, that excessive anxiety levels can interfere with normal function. Clinical anxiety consists of a pervasive, subjective feeling of apprehension and foreboding not related to a specific external threat. Clinical anxiety may also manifest itself through somatic signs, eg, palpitations, hyperventilation, and tics or twitching. Apprehension and somatic symptoms may appear together or separately (11).

Anxiolytic drugs, such as meprobamate and various benzodiazepines, reduce both the feelings of apprehension and somatic symptoms not only in cases of clinical anxiety, where no definable external threat can be identified, but also in those cases where anxiety accompanies some threatening event, such as a physical illness (12). The majority of prescriptions are written for patients with a primary diagnosis of a physical disorder as opposed to emotional disorders. Benzodiazepines, particularly chlordiazepoxide and diazepam, are the most widely used anxiolytics (13).

Preclinical Pharmacology. In addition to their specific effects on anxiety, anxiolytics also possess muscle relaxant, sedative, and anticonvulsant properties. It is not known whether or not all of these responses are elicited through a common mechanism. However, these properties have proven useful for the preclinical identification of new anxiolytics. Anxiolytics inhibit the clonic convulsions induced in mice or rats by pentylenetetrazole, induce muscle relaxation in cats, and have taming effects on cynomolgus monkeys (14). Although the potencies of the benzodiazepines in these procedures correlate well with clinical anxiolytic activity, their relevance to anxiety is in question. Nevertheless, antipentylenetetrazole and muscle relaxation determinations are frequently employed for the characterization of anxiolyticlike drugs.

Perhaps a more definitive but more complex test for anxiolytic activity is provided by the conflict behavioral test. In this procedure, animals are taught to press a lever to obtain a reward of food; however, in the presence of a conditioning signal, the animals are simultaneously rewarded with food and punished by an electric shock. The punished and unpunished trials are interspersed. Anxiolytics, in a wide dose range, increase the rate of response in the presence of shock punishment but have no effect on the unpunished response rate (15). The sedative properties of the anxiolytics become evident at high doses, at which the unpunished responses are depressed. No other class of psychoactive drug exhibits this profile in the conflict procedure. Sedative compounds, such as pentobarbital [76-74-4] or chlorpromazine, influence both punished and unpunished behavior at the same doses. Because drug potencies in the conflict procedure correlate well with clinical anxiolytic potencies, conflict has become an important tool in defining preclinical, anxiolyticlike activity (16).

Recently, a specific, high affinity, benzodiazepine binding site has been described in brain tissue (17–18). The affinity of various benzodiazepines for this site also correlates with their pharmacological and clinical potencies, suggesting that this binding site may be the site of benzodiazepine pharmacological activity. On this basis, a number of laboratories have begun to employ in vitro ^{3}H-diazepam binding procedures as a primary screening test for new compounds having anxiolytic activity. Although the validity of this approach for the identification of new anxiolytics has not been confirmed, the procedure is advantageous in that it makes possible evaluation of large numbers of compounds in a short time.

Mechanisms of Action. The precise mechanisms whereby drugs exert their anxiolytic properties are not known. The problem is compounded by their multiple actions of anxiolytics, ie, muscle relaxation, sedation, and anticonvulsant effects, and by the lack of understanding of the anatomical and biochemical substrates of anxiety (see Hypnotics, sedatives, anticonvulsants). Hypotheses of the mechanisms of action of benzodiazepines have recently been reviewed (19). The benzodiazepines decrease the rates of utilization of a number of important brain neurotransmitters, including norepinephrine, dopamine, serotonin, and acetylcholine (see Neuroregulators). Whether these influences are important for their anxiolytic action is not known. The fact that the benzodiazepines do not appear to have direct effects on the metabolism and disposition of these neurotransmitters suggests that they may mediate their influences indirectly, perhaps by altering the activity of other inhibitory neurohumors, such as glycine or gamma aminobutyric acid (GABA).

Benzodiazepines interfere with the *in vitro* binding of ^{3}H-strychnine to glycine receptors in brain tissue (20). Other workers, however, have determined that diazepam does not inhibit strychnine-induced convulsions, nor does strychnine reverse flurazepam-induced neural depression (21-22). In addition, the diazepam concentration required to inhibit ^{3}H-strychnine binding is ca 200 times higher than that required to produce pharmacologic responses (23).

Studies of the influences of the benzodiazepines on GABA-related functions have provided more promising results. Diazepam influences on presynaptic inhibition in the spinal cord and on the suppression of electrical activity in the cerebral cortex depend on GABA concentrations in those tissues (24-25). Furthermore, diazepam responses are blocked by the GABA receptor antagonist bicuculline [485-49-4]. Diazepam has GABA-like influences on cyclic guanosine monophosphate [85-32-5] levels in the cerebellum of rats (26). Diazepam does not mimic GABA at GABA receptor sites, nor does it appear to influence the release or inactivation of GABA; rather, it seems to increase the sensitivity of GABA receptors.

Research on the benzodiazepine binding sites described in references 17 and 18 has provided some additional insights regarding the mechanisms of benzodiazepine action. These binding sites are distributed heterogenously in the brain and exhibit a high affinity for the benzodiazepines. Although meprobamate has a very low affinity for these binding sites, the anxiolytic benzodiazepines have affinities that correlate very well with their activities in muscle relaxation and anticonflict procedures (20,27). At least some of the benzodiazepine binding sites are closely associated with GABA receptors and the physiological states of these two sites are interdependent (28-29). Based on recent findings, a model for this receptor interaction has been proposed (29). This model suggests that the occupation of benzodiazepine binding sites increases the passage of chloride ions through membrane ion channels that are regulated by GABA receptors. Since GABA is believed to exert its inhibitory influences on neuronal firing by increasing intraneuronal chloride ion concentrations, then the benzodiazepine-induced increase in chloride ion transport is expected to potentiate the physiological response to GABA.

Although there is no question that the benzodiazepines potentiate postsynaptic responses to GABA, there is still considerable question as to whether this GABA potentiation is responsible for their anxiolytic effects. It is apparent that not all benzodiazepine's binding sites are associated with GABA receptors and that alterations in GABA sensitivity do not account for all of its pharmacological activity (30-32). Several workers have attempted to alter conflict behavior by altering the availability of GABA at postsynaptic receptor sites. The evidence to date is inconclusive. In some experiments, the GABA antagonist picrotoxin [124-87-8] has no effect on benzodiazepine anticonflict activity; whereas in others, picrotoxin selectively blocks the anticonflict but not the sedation activity of benzodiazepines (33–36). Increasing GABA availability by inhibiting its metabolism also is reported to have no influence on conflict behavior nor to potentiate the anticonflict effect of diazepam (37). The GABA agonist muscimol is reported to have weak and inconsistant influences on anticonflict activity (27). The final resolution of the GABA-antianxiety controversy may depend on the discovery of more specific pharmacological tools.

The discovery of specific benzodiazepine binding sites in the brain has led to the suggestion that the brain may contain an unidentified substance that normally interacts with that site (17,19). The isolation and identification of such a substance would have important implications in understanding more precisely the mechanisms of action of the benzodiazepines as well as furthering the understanding of the basic mechanisms involved in the anxiety response.

Chemistry. The first synthetic drugs to be used as tranquilizers were the barbiturates, which in small doses possessed the desired properties. Despite the advent of other more specific drugs, some barbiturates are still used as anxiolytics, eg, phenobarbital [50-06-6] (2), butabarbital [125-40-6] (3), and amobarbital [57-43-2] (4). The most important currently used anxiolytics belong to two classes of compounds: propanediols and benzodiazepines.

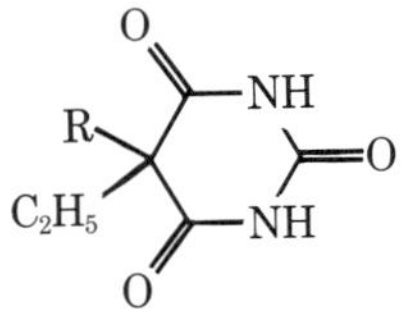

(2) R = C_6H_5
(3) R = $CHCH_2CH_3$
       |
       CH_3
(4) R = CH_2CH_2CH $\diagup CH_3$ $\diagdown CH_3$

Propanediols. Propanediols were studied in connection with the search for compounds that would be superior to mephenesin [59-47-2] (5), a short-acting muscle relaxant and anticonvulsant (38). These studies culminated in the synthesis and introduction of the first tranquilizer, meprobamate [57-53-4] (9), which became widely used under the trade names Miltown (1955, Wallace Laboratories) and Equanil (1957, Wyeth) (39). It is marketed domestically and abroad by many companies under its generic name and a number of trade names.

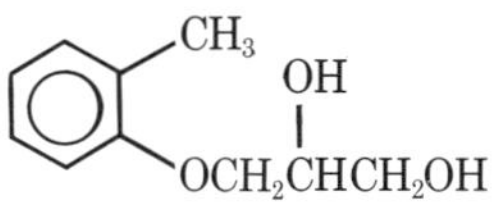

Two synthetic methods that are particularly useful for the synthesis of the product and related compounds are outlined in Figure 1. The propanediol (**6**) is used as starting material and converted in two simple steps into (**9**). Related compounds were introduced but never achieved the general acceptance accorded to meprobamate. Its sales in the United States were highest ($50 × 10^6) in 1963. However, its use fell gradually with the appearance of the 1,4-benzodiazepines, which are the most widely used anxiolytic agents (40–42).

1,4-Benzodiazepines and Related Compounds. 1,4-Benzodiazepines have pronounced anxiolytic properties, combined with very low toxicity and a minimal effect on the autonomic nervous system (43). Anxiolytic 1,4-benzodiazepines that are marketed in the United States and those that are marketed abroad are listed with their CAS Registry No. in Tables 2 and 3, respectively. The first compound of this series was prepared as shown in Figure 2 (44). Three routine steps lead from (**27**) to (**30**). Upon treatment with methylamine, compound (**30**) undergoes a ring enlargement in which the pyrimidine ring is converted into a 1,4-diazepine ring. This unusual transformation yields the pharmacologically active product (**31**). Its hydrochloride was introduced in 1960 under the trade name Librium (**10**) (see Table 2) by Roche Laboratories in the United States and worldwide by F. Hoffman-La Roche & Co., Ltd. Librium achieved broadest acceptance and was the most frequently used prescription drug in the United States until 1969 when sales of Valium (**11**) surpassed its sales.

The synthesis of Valium is simpler than that of chlordiazepoxide and can be realized by various routes (54). One of the most practical methods is depicted in Figure 3. The starting material is again the aminobenzophenone (**27**) which is transformed stepwise into diazepam (**11**).

Oxazepam (**12**) was introduced by Wyeth Laboratories under the trade name Serax in the United States and under various other names in other countries. Its synthesis starts with the chlormethylquinazoline 3-oxide (**30**) which, on treatment with alkali, undergoes a ring enlargement forming benzodiazepinone 4-oxide (**34**) (54)

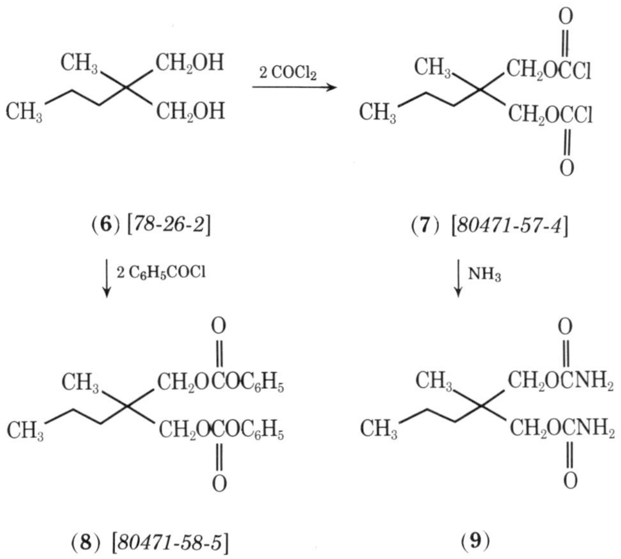

Figure 1. Two synthetic routes to meprobamate.

Table 2. Anxiolytic 1,4-Benzodiazepines Marketed in the United States[a]

Structure no.	Generic name	CAS Registry No.	Structural formula	Trade name[b]	Pharmaceutical co. (year introduced)	Ref.
(10)	chlordiazepoxide hydrochloride	[438-41-5]	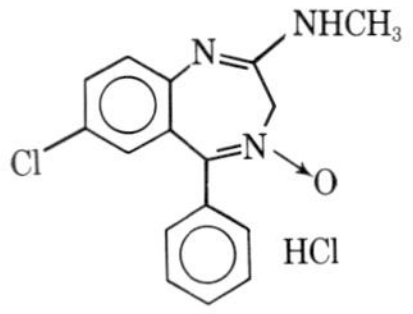	Librium	Roche Laboratories (1960)	42
(11)	diazepam	[439-14-5]	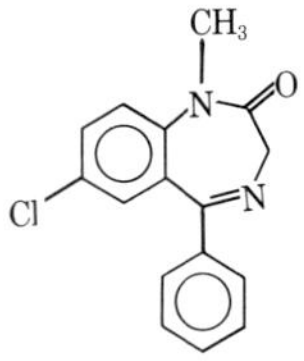	Valium	Roche Laboratories (1963)	45
(12)	oxazepam	[604-75-1]	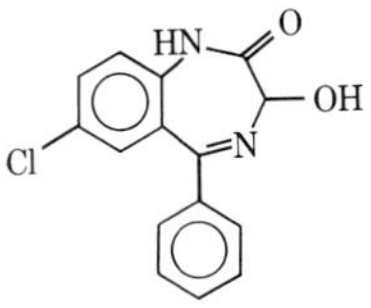	Serax	Wyeth Laboratories (1965)	46
(13)	clorazepate dipotassium	[57109-90-7]	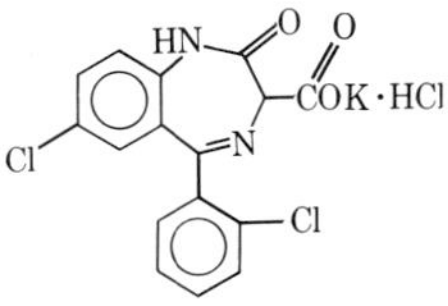	Tranxene	Abbott Laboratories (1972)	47
(14)	prazepam	[2955-38-6]	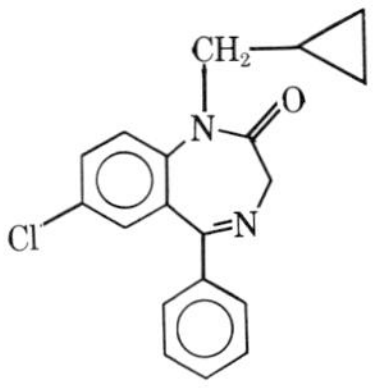	Verstran	Parke-Davis (1977)	48
(15)	lorazepam	[846-49-1]	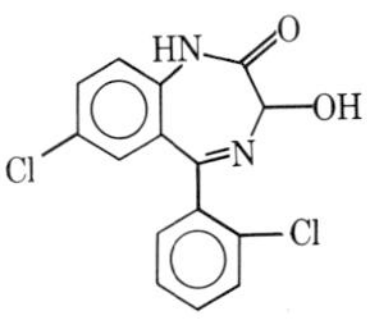	Ativan	Wyeth Laboratories (1977)	49
(16)	halazepam	[23092-17-3]	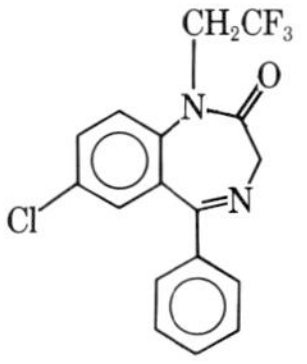	Paxipam	Schering-Plough (1981)	50

Table 2. (continued)

Structure no.	Generic name	CAS Registry No.	Structural formula	Trade name[b]	Pharmaceutical co. (year introduced)	Ref.
(17)	alprazolam	[28981-97-7]	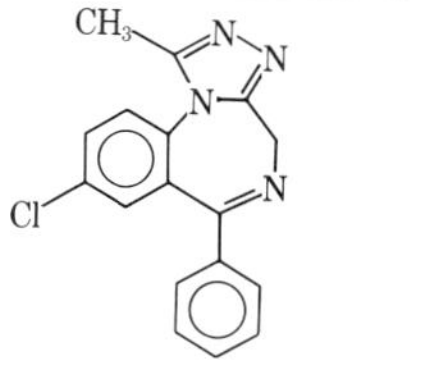	Xanax	Upjohn (1981)	51

[a] Only anxiolytics are listed; closely related benzodiazepines, which are mainly used as hypnotics or anticonvulsants, are not included.

[b] Since many compounds are sold under various proprietary names, only the first or the most generally used name is indicated.

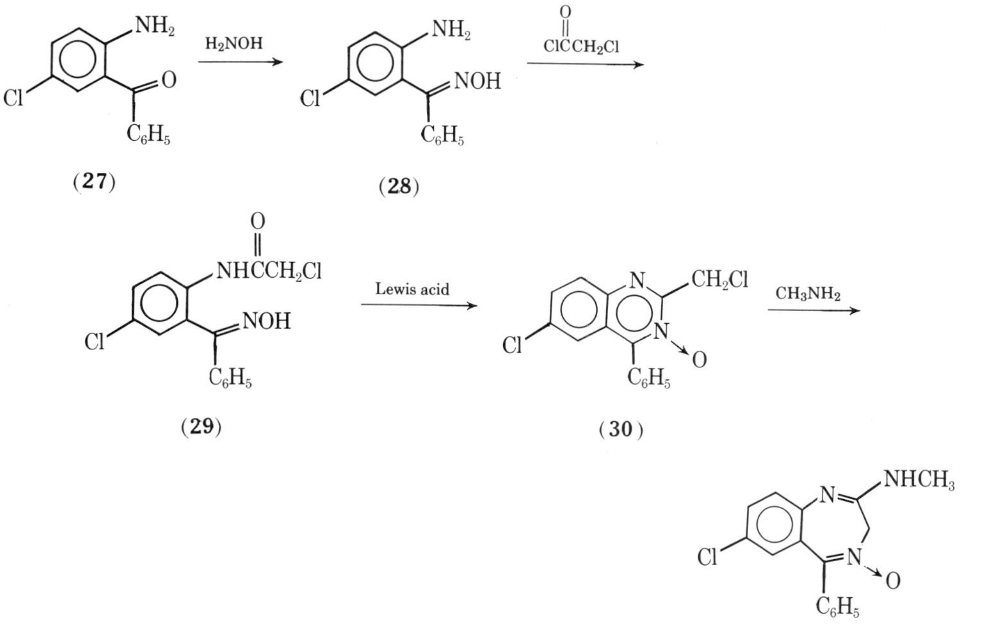

Figure 2. Synthesis of chlordiazepoxide.

(see Fig. 4). This, on treatment with acetic anhydride at elevated temperature, undergoes the Polonovski rearrangement to yield compound (35), which in turn is hydrolyzed to give the 3-hydroxy derivative oxazepam (12) (46).

Introduction of compounds (10), (11), and (12) was followed by the introduction of a number of related 1,4-benzodiazepine derivatives, which are listed in Tables 2 and 3. The compounds resemble each other in their pharmacological activity but differ in some cases in their potency, duration of action, and in their spectra of biological activity, ie, sedation, muscle relaxation, etc.

The preparation of chlorazepate dipotassium (13) is shown in Figure 5 (47).

Table 3. Anxiolytic 1,4-Benzodiazepines Marketed Abroad[a]

Structure no.	Generic name	CAS Registry No.	Structural formula	Trade name[b]	Pharmaceutical co. (year introduced)	Ref.
(18)	medazepam	[2898-12-6]	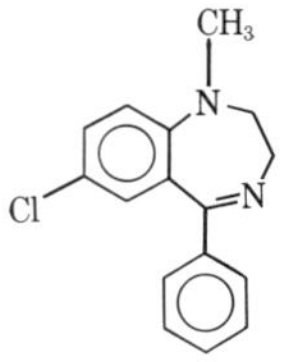	Nobrium	F. Hoffmann-La Roche & Co., Ltd., Basle (1968)	52
(19)	temazepam	[846-50-4]	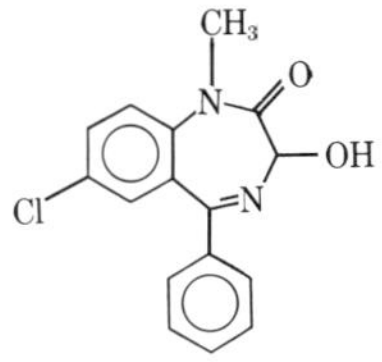	Levanxol[c]	Erba (1970)	46
(20)	oxazolam	[24143-17-7]	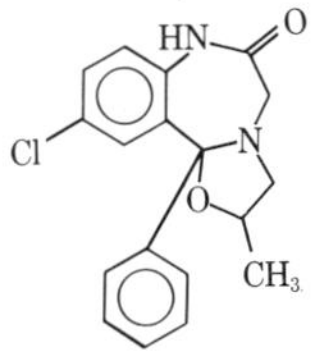	Serenal	Sankyo (1971)	53
(21)	desmethyldiazepam	[1088-11-5]	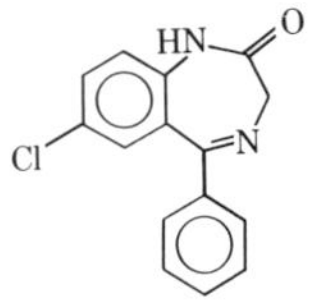	Madar	Ravizza (1973)	54
(22)	cloxazolam	[24166-13-0]	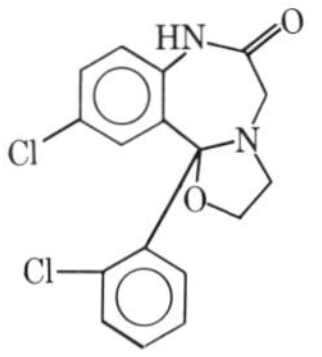	Sepazon	Sankyo (1974)	53
(23)	bromazepam	[1812-30-2]	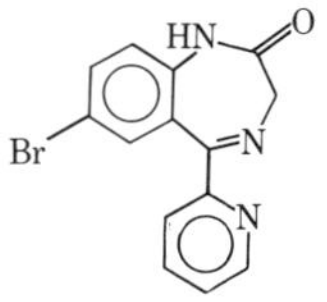	Lexotanil	F. Hoffmann-La Roche & Co., Ltd., Basle (1974)	55
(24)	pinazepam	[52463-83-9]	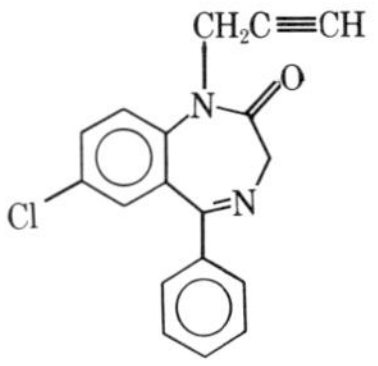	Domar	Zambeletti (1975)	56
(25)	nimetazepam	[2011-67-8]	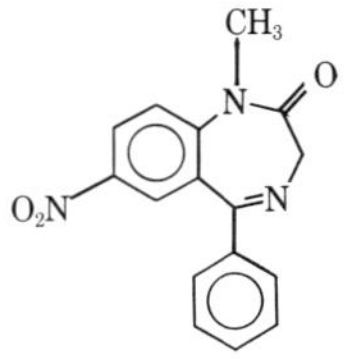	Erimin	Sumitomo, Kaganaku (1977)	57

Table 3 (*continued*)

Structure no.	Generic name	CAS Registry No.	Structural formula	Trade name[b]	Pharmaceutical co. (year introduced)	Ref.
(26)	camazepam	[36104-80-0]		Albego	Simes, Milano (1977)	58

[a] Only anxiolytics are listed; closely related benzodiazepines, which are used mainly as hypnotics or anticonvulsants, are not included.
[b] Since many compounds are sold under various proprietary names, only the first or the most generally used name is indicated.
[c] This compound was introduced by Sandoz Pharmaceuticals in the U.S. in 1981 as a hypnotic under the proprietary name Restoril.

Compounds (14) and (24) are prepared in analogy to diazepam, except that a reagent other than a methyl halide or sulfate is used for the alkylation of (21). Compounds (23) and (25) are prepared according to Figure 3 using as starting materials aminoketones that differ from 2-amino-5-chlorobenzophenone (27). Compound (19) is prepared according to Figure 4 by first methylating compound (34). Lorazepam (15) is prepared according to the sequence shown for (12) in Figure 4, by using the quinazoline derivative (30) with a chlorine substituent in the 2′-position. Camazepam (26) is prepared from temazepam (19) in a two-step reaction (58).

Halazepam (16) is prepared by the sequence of reactions outlined in Figure 3, with the substituted benzophenone (**Q**) as the starting material (50). The triazolo-1,4-benzodiazepine alprazolam (17) was synthesized from the thiolactone (**R**) by treatment with acetic hydrazide (51). Medazepam (18) is prepared as shown in Figure 6 (52).

The oxazolodiazepines (20) and (22) can be prepared by three methods as shown in Figure 7 (56). Two additional compounds that are similar to 1,4-benzodiazepines

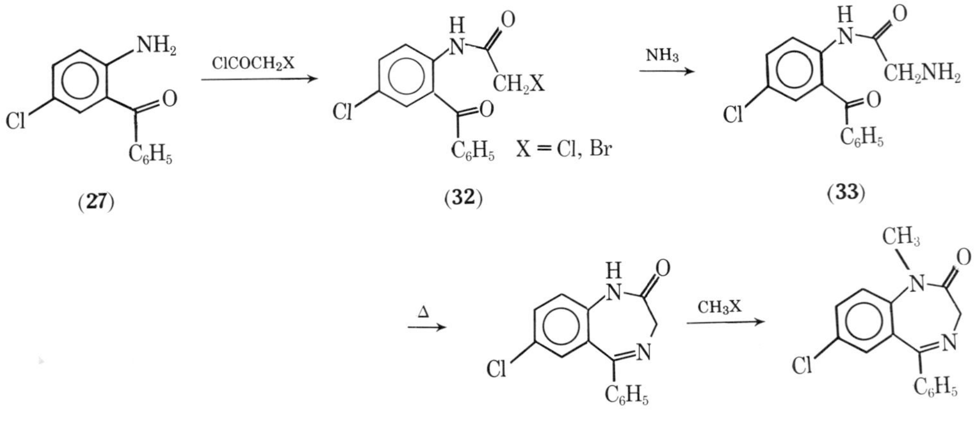

352 PSYCHOPHARMACOLOGICAL AGENTS

Figure 3. Synthesis of diazepam.

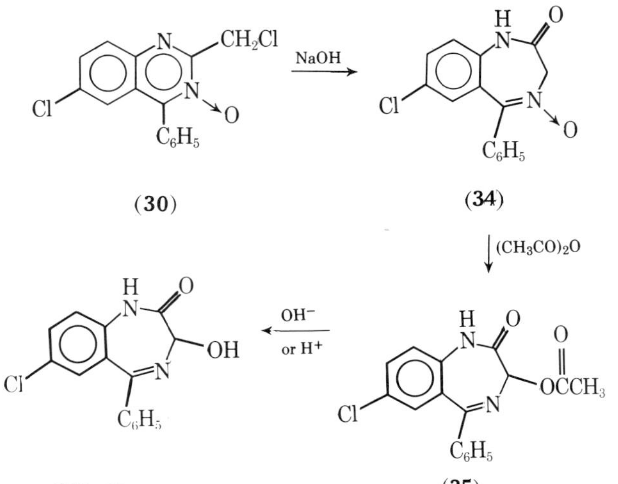

Figure 4. Synthesis of oxazepam.

in structure have anxiolytic properties and are marketed under the trade names Urbanyl [clobazam (**50**)] and Rize [clotiazepam (**52**)], respectively. The preparation of clobazam (**50**), which was introduced by Boehringer, is shown in Figure 8 (59). Clotiazepam (**52**) is prepared from the appropriate aminothienophenylketone (**51**) (60).

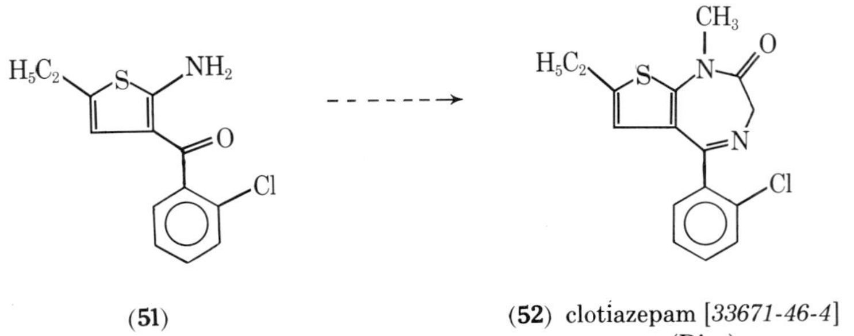

(**52**) clotiazepam [33671-46-4]
(Rize)

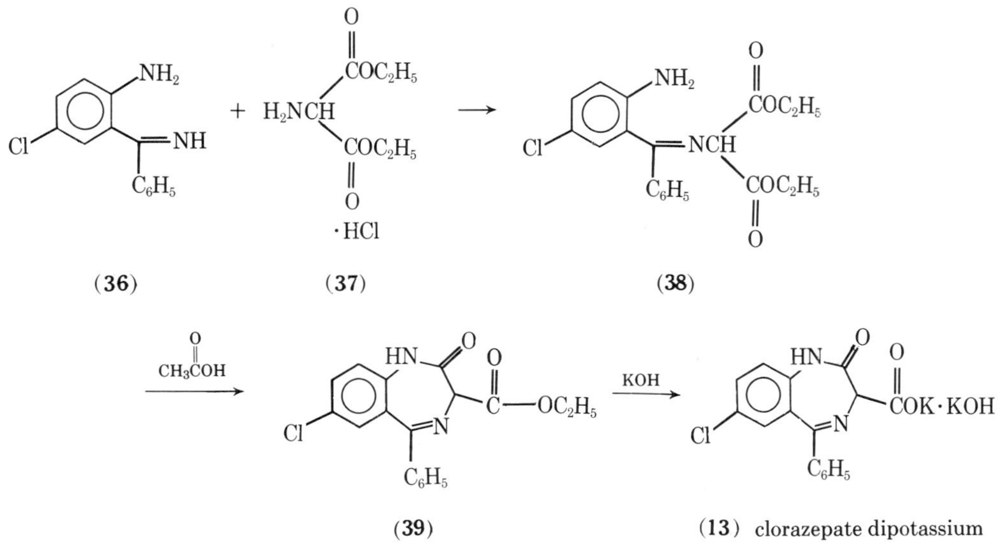

Figure 5. Synthesis of clorazepate dipotassium.

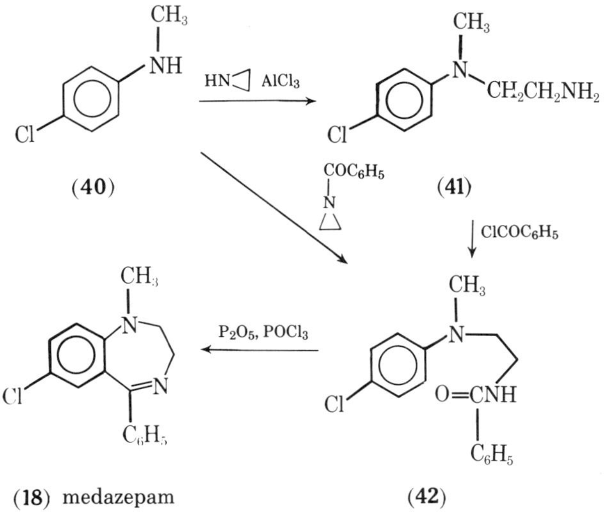

Figure 6. Synthesis of medazepam.

Clotiazepam is marketed in Japan by Yoshitomi.

Other Anxiolytics. Less-used anxiolytics that are not related structurally to meprobamate or the benzodiazepines are hydroxyzine (**55**) and chlormezanone (**57**). The syntheses of these products are shown in Figures 9 and 10, respectively. The structure of hydroxyzine resembles that of the antihistamines; whereas chlormezanone is chemically unrelated to other classes of biologically active agents (61–62).

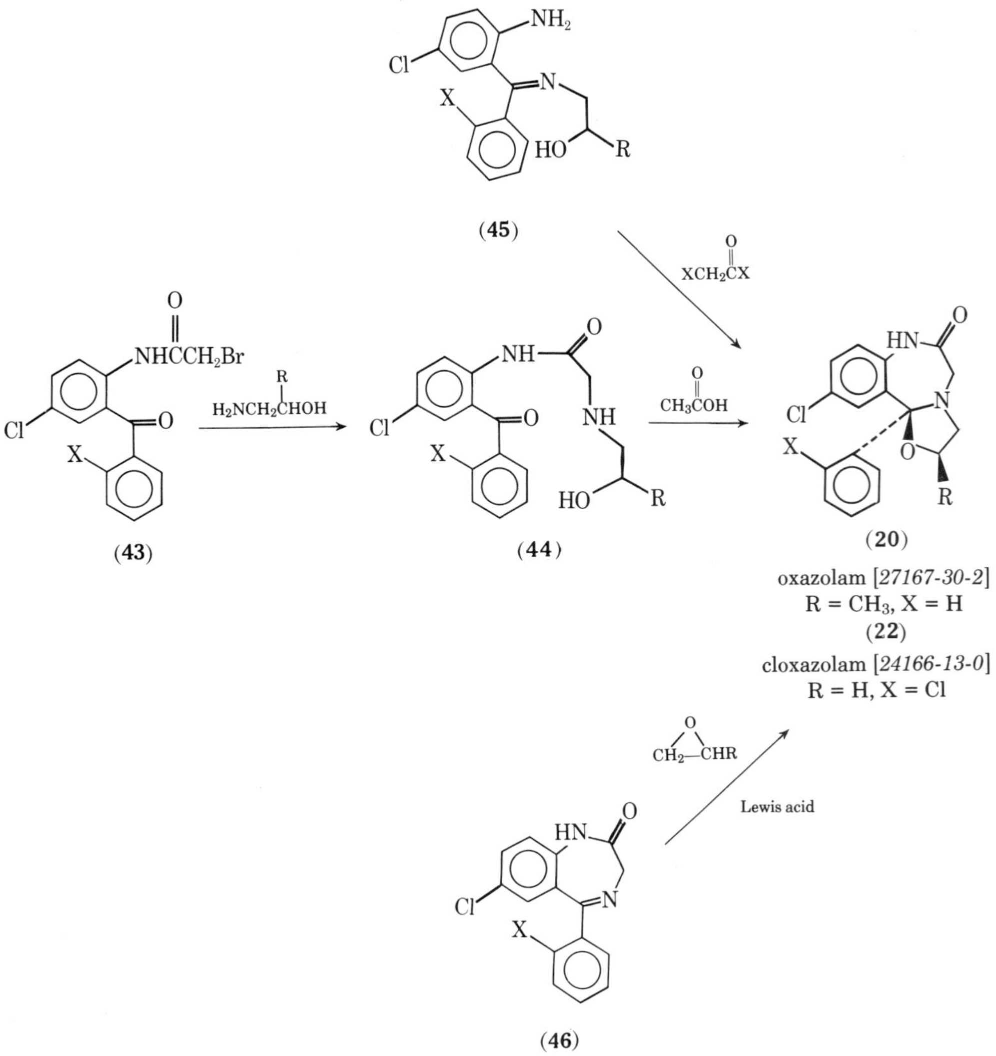

Figure 7. Three synthetic routes to oxazolams.

(45)

(43) → (44) → (20) oxazolam [27167-30-2] R = CH₃, X = H
(22) cloxazolam [24166-13-0] R = H, X = Cl

(46)

The dihydrochloride of hydroxyzine is marketed under the trade name Atarax by Roerig, and the pamoate is marketed as Vistaril by Pfizer Laboratories. Chlormezanone is prepared as shown in Figure 10 and is marketed as Trancopal by Breon Laboratories.

Side Effects and Toxicity. Anxiolytic drugs are relatively free of toxicity, addiction potential, and unpleasant side effects. The most common side effects experienced are ataxia, drowsiness, and dizziness and, at higher doses, some impairment of cognitive function may occur (63). In most cases, satisfactory therapeutic efficacy can be

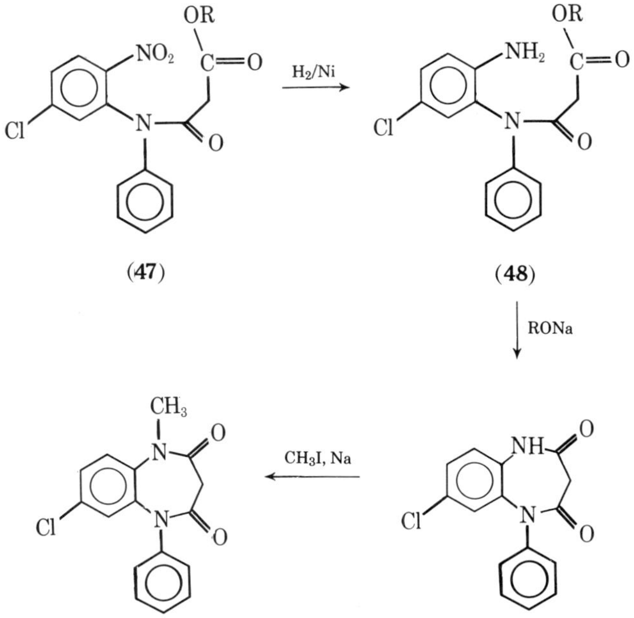

(**47**) (**48**) (**49**)

(**50**) clobazam [*22316-47-8*]
(Urbanyl)

Figure 8. Synthesis of clobazam.

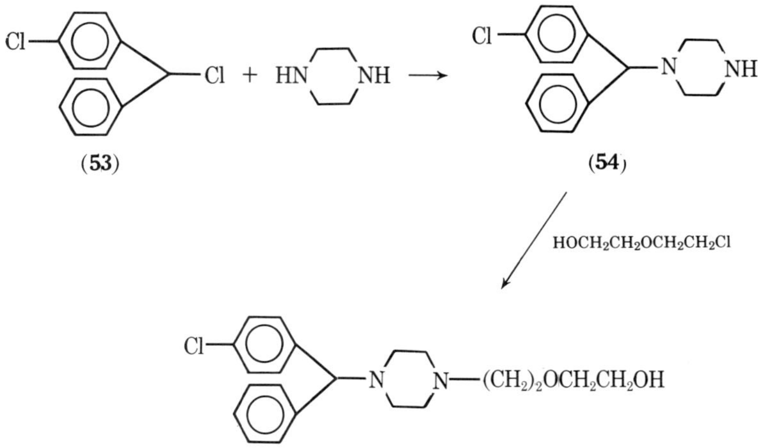

(**53**) (**54**)

(**55**) hydroxyzine [*68-88-2*]

Figure 9. Synthesis of hydroxyzine.

achieved at doses that produce minimal side effects. The sedative properties of all anxiolytics are potentiated substantially by alcohol.

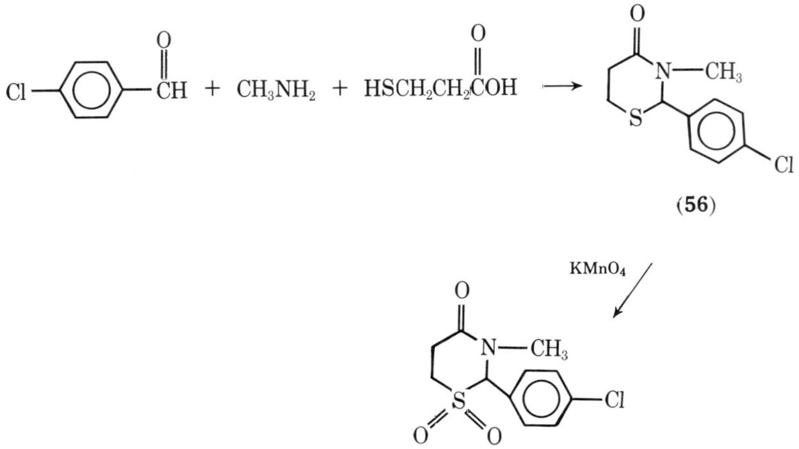

$$Cl-\underset{}{\bigcirc}-\overset{O}{\underset{\parallel}{C}}H + CH_3NH_2 + HSCH_2CH_2COH \longrightarrow$$

(56)

(57) chlormezanone [80-77-3]

Figure 10. Synthesis of chlormezanone.

Antipsychotics (Neuroleptics)

Although the incidence of schizophrenia is much lower than that of neurosis or depression, its impact on society is tremendous. Schizophrenia disables its victims during the prime years of their lives and treatment requires substantial institutional and professional support, often for long periods of time (64). The discovery of drugs with a specific influence on psychoses has been a milestone in the history of psychological medicine. Therapy that includes the use of antipsychotic drugs has been shown to be the most effective means of treating psychoses (65–66).

Antipsychotic drugs reduce or reverse the fundamental symptoms of schizophrenia, such as thought disorder, blunted affect, and autistic withdrawal (66). Secondary symptoms, including hallucinations, paranoia, and belligerence, are also attenuated by antipsychotic therapy (66). Although the antipsychotics are often referred to as tranquilizers, their effects are not caused by simple sedation or tranquilization, but they appear specifically to inhibit or reverse psychotic symptoms. For example, they normalize the psychomotor behavior of both the retarded and the hyperactive patient (66). The antipsychotics have been of substantial benefit in 75% of treated patients; another 20% of treated patients improve to a lesser degree. In comparable groups of untreated patients, only 25% show a moderate to marked improvement, and 50% fail to improve or become worse.

Preclinical Pharmacology. The preclinical behavioral and biochemical actions of the antipsychotics have recently been reviewed (67–69). The clinically active antipsychotics are well known for their ability to block postsynaptic dopaminergic receptors in the brain; thus the various *in vivo* and *in vitro* tests employed for identifying new antipsychotic drugs are based on this property. Preliminary *in vivo* tests include the inhibition of apomorphine, which is a dopamine receptor agonist, or amphetamine-, which is a dopamine releasing agent, induced stereotypic behavior, the induction of jumping behavior in mice induced by coadministration of amphetamine and L-DOPA [L-3-(3,4-dihydroxyphenyl)alanine)], and apomorphine-induced emesis in dogs. Activity in all of these tests correlates well with clinical antipsychotic activity; however, the jumping-mouse test is perhaps the best since it involves a small inexpensive laboratory animal and drug effects can be measured automatically by appropriate instrumentation.

The administration of apomorphine or amphetamine to rats, in which the nigro-striatal dopamine pathway has been unilaterally lesioned, causes circling behavior. This behavior pattern is inhibited by antipsychotic drugs. The potency of drugs in blocking circling behavior correlates with their propensity for inducing extrapyramidal side effects in man. This correlation may be accounted for by the restriction of pharmacological activity to striatal dopaminergic influences in this model and the role of striatal dopamine in extrapyramidal function (70).

Perhaps the most reliable and sensitive *in vivo* test for antipsychotics is the shock-avoidance behavioral model. This procedure consists of training rats or monkeys to press a lever within a given time interval in order to avoid receiving a painful electric shock. If the animal fails to avoid the shock, it can terminate the stimulus by either pressing a lever or escaping to a safe area of the enclosure. Antipsychotics possess a unique property of inhibiting shock-avoidance behavior at doses that do not prevent the escape response. Sedative or central depressant drugs inhibit both the avoidance and escape behavior.

Recent investigations of the *in vitro* binding of antagonists to neurotransmitter receptors has provided new techniques for identifying potential antipsychotic activity in new drugs. ^{3}H-Haloperidol and ^{3}H-spiroperidol bind with high affinity to dopamine receptors in several brain areas (69). The binding of these radiolabeled butyrophenones is inhibited by drugs with antipsychotic activity. The potencies of various antipsychotics in inhibiting butyrophenone binding correlates very well with their ability to inhibit apomorphine- and amphetamine-induced stereotypic behavior and at the average daily clinical dose required to achieve antipsychotic therapy.

In addition to identifying potential antipsychotic activity, binding procedures are also useful for predicting autonomic and extrapyramidal side effects. Autonomic side effects result from the blocking of postsynaptic alpha-adrenergic receptors. The potency of drugs in eliciting this effect may be determined through the use of ^{3}H-WB4101 (a specific postsynaptic alpha antagonist) binding procedure (69).

The potent anticholinergic, ie, muscarinic, property of some antipsychotics is believed to prevent the appearance of extrapyramidal symptoms in man (71). Use of ^{3}H-QNB (quinuclidinyl benzilate), which binds to muscarinic cholinergic receptors in the brain, is a convenient procedure for evaluating a drug's anticholinergic potential.

In vitro binding procedures provide rapid and economical methods for predicting the antipsychotic and side effect potential of new compounds. However, like all *in vitro* tests, they reveal no information concerning drug absorption or transport into the brain. The latter properties must be verified by *in vivo* procedures.

Mechanism of Action. Drugs useful in the treatment of psychoses have many pharmacological properties, but the one property that they share is their ability to block dopaminergic transmission in brain. Reserpine [50-55-5] (**58**) accomplishes this by depleting stores of dopamine in presynaptic nerve endings (72), whereas the phenothiazines and butyrophenones inhibit postsynaptic dopamine receptors (69). It is widely believed that these antidopaminergic properties may account for both the antipsychotic effects and extrapyramidal influences of these drugs (69–70,73).

However, the ratios of antipsychotic to extrapyramidal activities vary among the antipsychotic agents. For example, thioridazine [50-52- 2] produces very few Parkinsonian effects at therapeutic doses (74). Thioridazine's unusual profile may be explained by its apparent selectivity for a specific dopaminergic pathway. There are two main dopaminergic pathways, which project forward from the midbrain. One originates in the substantia nigra and terminates in the striatum. The second originates in the ventral tegmental area and projects forward to the limbic forebrain and cortex. Blockade of the nigro–striatal pathway mediates extrapyramidal symptoms and inhibition of the tegmental–limbic pathway is believed to mediate antipsychotic actions (70,75).

Investigations in laboratory animals suggests that thioridazine selectively inhibits the tegmental–limbic pathway but has little influence on the nigro–striatal dopaminergic neurons (73). This in turn suggests that the dopamine receptors are slightly different in the two pathways. The possible difference is supported by the recent discovery of multiple dopamine receptors (76). An alternative hypothesis to explain thioridazine's profile is suggested by the observation that thioridazine is a more potent anticholinergic agent than are other antipsychotics. The anticholinergic activity may attenuate the antidopaminergic action in the nigro–striatal tract but not in the tegmental–limbic pathway (see also Cholinesterase inhibitors).

Although there is abundant evidence to suggest that the antipsychotics exert their therapeutic effects by disrupting dopaminergic transmission, the role of dopamine in schizophrenia is not understood. Available evidence suggests that schizophrenia is not accompanied by excessive dopaminergic activity in the brain (73).

Chemistry. *Reserpine.* The first drug used in the treatment of psychotic patients was reserpine (58), which is an alkaloid obtained from the roots of the plant *Rauwolfia serpentina* (see Alkaloids) (4). Its use as an antipsychotic drug decreased considerably after the development of the phenothiazine-type antipsychotics. Reserpine preparations are mostly used in the treatment of high blood pressure.

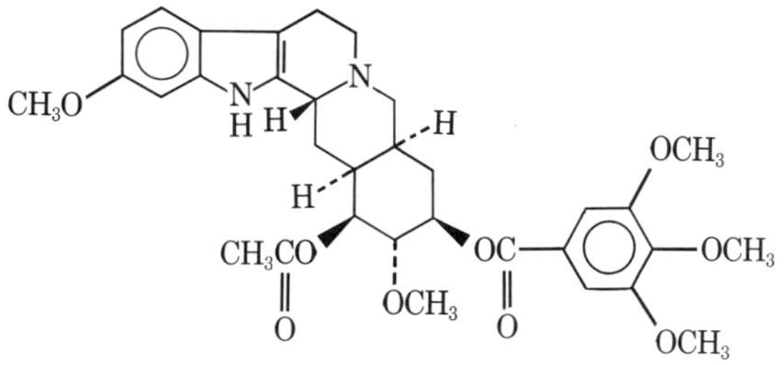

(58) Reserpine [50-55-5]

Phenothiazines. The synthesis of chlorpromazine and the discovery of its psychopharmacological properties in the early 1950s introduced the period of the most extensive use of antipsychotic agents (77–80). The valuable properties of this compound led to broad research programs which resulted in the synthesis and pharmacological evaluation of over 3000 products and in the final selection of the 20–30 compounds in worldwide use under a multitude of trade names. These products differ only by the substituents in the 2- and 10-positions of the phenothiazine ring system. Differences between compounds are in potency, duration of action, and the character of side effects, which in many instances appear after prolonged use. The neuroleptic phenothiazine derivatives which are marketed in the United States are listed in Table

4. The references in the table refer to the first published syntheses of the compounds. Not included are phenothiazines that are used mainly because of their antihistaminic or antiemetic properties.

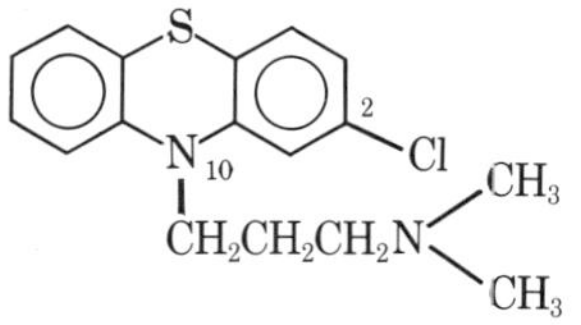

(**59**) chlorpromazine [50-53-3]

The starting materials for phenothiazine neuroleptics are the appropriately 2-substituted phenothiazines (**75**). Phenothiazines with Cl, CF_3, or SCH_3 substituents are prepared by heating the corresponding diphenylamines (**74**) with sulfur (95). In most cases, mixtures of the 2- and 3-isomers are obtained. These can be separated quite readily by crystallization.

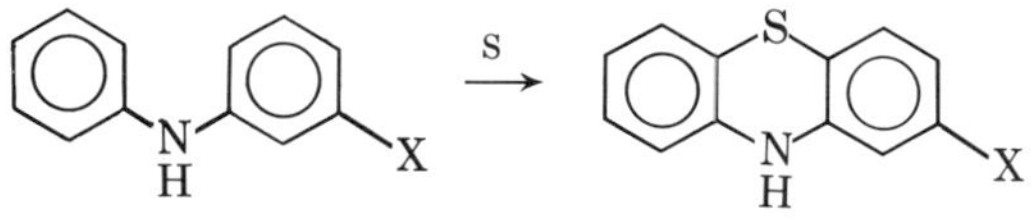

(**74**)   (**75**)

Those phenothiazines bearing an alkanoyl group in the 2-position (**77**) are prepared from the unsubstituted phenothiazines by Friedel-Crafts acylation (96).

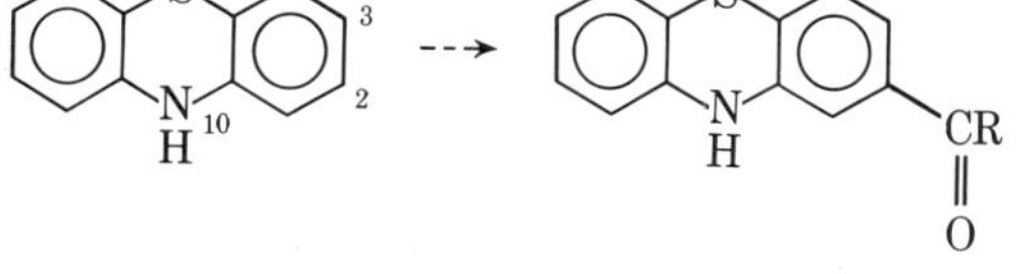

(**76**)   (**77**)

R = CH_3, 2-acetylphenothiazine [6631-94-3]
C_2H_5, 2-propionylphenothiazine [92-33-1]
C_3H_7, 2-butyrylphenothiazine [25244-91-1]

The sulfinyl derivative (**81**), which is the starting material for mesoridazine besylate (**69**), is prepared by oxidation of (**79**), followed by removal of the acetyl group, as shown in Figure 11 (80,97).

Table 4. Phenothiazine Neuroleptics

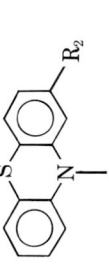

Structure no.	Generic name	CAS Registry No.	R_1	R_2	Salt	Trade name (pharmaceutical co.)	Ref.
(60)	chlorpromazine hydrochloride	[69-09-0]	—(CH$_2$)$_3$—N(CH$_3$)$_2$	Cl	HCl	Thorazine (SKF)	81–82
(61)	prochlorperazine dimaleate	[84-02-6]	—(CH$_2$)$_3$—N(piperazine)—CH$_3$	Cl	2 $\genfrac{}{}{0pt}{}{CHCO_2H}{\parallel\ CHCO_2H}$	Compazine (SKF)	83
(62)	perphenazine	[58-39-9]	—(CH$_2$)$_3$—N(piperazine)—CH$_2$CH$_2$OH	Cl		Trilafon (Schering)	84
(63)	trifluopromazine hydrochloride	[1098-60-8]	—(CH$_2$)$_3$—N(CH$_3$)$_2$	CF$_3$	HCl	Vesprin (Squibb)	85
(64)	trifluoperazine dihydrochloride	[440-17-5]	—(CH$_2$)$_3$—N(piperazine)—CH$_3$	CF$_3$	2 HCl	Stelazine (SKF)	86
(65)	fluphenazine dihydrochloride	[146-56-5]	—(CH$_2$)$_3$—N(piperazine)—CH$_2$CH$_2$OH	CF$_3$	2 HCl	Prolixin (Squibb)	87
(66)	fluphenazine enanthate	[2746-81-8]	—(CH$_2$)$_3$—N(piperazine)—CH$_2$CH$_2$—O—CO—CH$_3$(CH$_2$)$_5$	CF$_3$		Prolixin enanthate (Squibb)	88

(67)	fluphenazine decanoate	[5002-47-1]	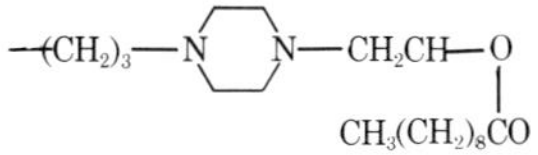 $-(CH_2)_3-N\!\!\bigcirc\!\!N-CH_2CH-O-CO(CH_2)_8CH_3$	CF_3		Prolixin decanoate (Squibb)	88
(68)	thioridazine hydrochloride	[130-61-0]	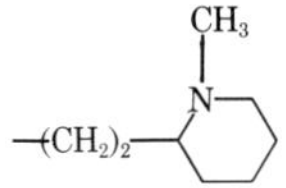 $-(CH_2)_2-\!\!\bigcirc\!\!{N-CH_3}$	SCH_3	HCl	Mellaril (Sandoz)	89
(69)	mesoridazine besylate	[32672-69-8]	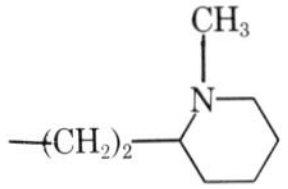 $-(CH_2)_2-\!\!\bigcirc\!\!{N-CH_3}$	$\overset{O}{=}SCH_3$	$C_6H_5SO_3H$	Serentil (Boehringer-Ingelheim)	90
(70)	piperacetazine	[3819-00-9]	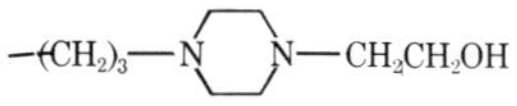 $-(CH_2)_3-N\!\!\bigcirc\!\!N-CH_2CH_2OH$	$\overset{O}{=}CCH_3$		Quide (Dow Pharmaceutical Co.)	91
(71)	acetophenazine dimaleate	[5714-00-1]	$-(CH_2)_3-N\!\!\bigcirc\!\!N-CH_2CH_2OH$	$\overset{O}{=}CCH_3$	$2\;\|\!\!\underset{CHCO_2H}{CHCO_2H}$	Tindal (Schering)	92
(72)	carphenazine dimaleate	[2975-34-0]	$-(CH_2)_3-N\!\!\bigcirc\!\!N-CH_2CH_2OH$	$\overset{O}{=}CC_2H_5$	$2\;\|\!\!\underset{CHCO_2H}{CHCO_2H}$	Proketazine (Wyeth)	93
(73)	butaperazine dimaleate	[10213-91-9]	$-(CH_2)_3-N\!\!\bigcirc\!\!N-CH_3$	$\overset{O}{=}CC_3H_7$	$2\;\|\!\!\underset{CHCO_2H}{CHCO_2H}$	Repoise (Robins)	94

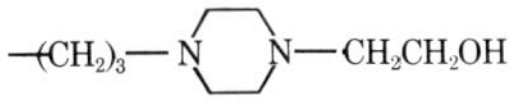

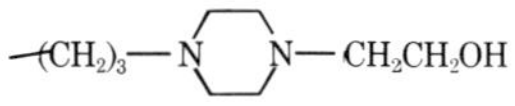

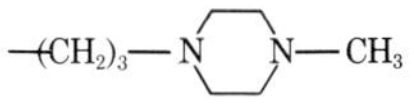

The substituent in the 10-position is introduced by treatment of the phenothiazines in the presence of a base, eg, NaNH$_2$ or LiH, with an appropriately substituted haloalkylamine:

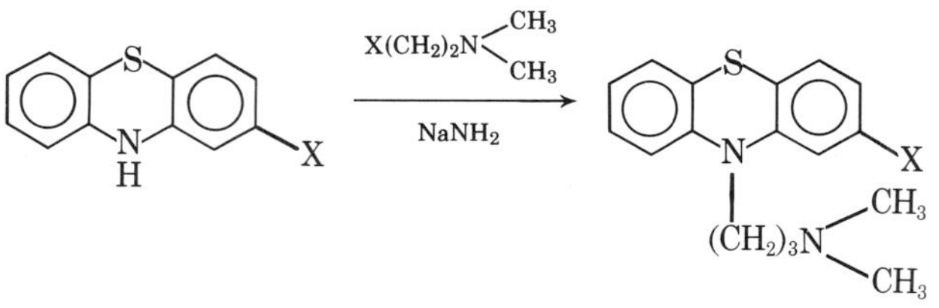

(75)

(59) X = Cl: chlorpromazine
(63) X = CF$_3$; salt = HCl: trifluopromazine hydrochloride

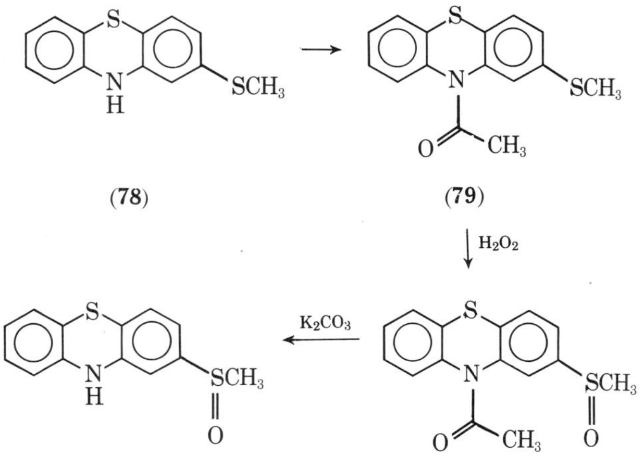

(78)   (79)

(81) 2-(methylsulfinyl)phenothiazine   (80)
[27612-10-8]

Figure 11. Synthesis of 2-(methylsulfinyl)phenothiazine.

Heavier substituents can be introduced in several steps, as shown in Figure 12. Compounds bearing a hydroxyl group on the basic substituent, eg, fluphenazine (84), can be esterified with fatty acid chlorides to yield esters, eg, fluphenazine enanthate (66) and fluphenazine decanoate (67) (88). These esters provide prolonged action on parenteral administration.

The large number of phenothiazine derivatives available to the clinicians and their value in the treatment of psychotic patients has made them the most widely used class of antipsychotic drugs. Their valuable clinical properties has prompted increasing research related to tricyclic ring systems that show a formal similarity to phenothiazines, eg, the thioxanthenes.

Thioxanthenes. The investigation of thioxanthenes led to the syntheses of many representative compounds and to the introduction of thioxanthene derivatives that bear substituents corresponding to those present in the clinically used phenothiazines (80,97). Their antipsychotic properties are similar to those of the phenothiazines, but the former never achieved the widespread popularity of the latter compounds.

In the United States, only two thioxanthene derivatives, chlorprothixene (**88**) and thiothixene (**93**), are marketed, but other derivatives are available to the clinicians in other countries. Chlorprothixene (98–100) can be synthesized as shown in Figure 13. The starting material is the appropriately substituted thioxanthenone (**85**), and the basic side chain is introduced via a Grignard reaction. The intermediate carbinol (**86**) is dehydrated and the cis and trans isomers are separated by fractional crystallization of the bases or of their oxalate salts. In the case of chlorprothixene (**88**), the α or trans isomer is the biologically more active compound, whereas in the case of thiothixene (**93**), the cis compound shows higher activity. Thiothixene was first described in 1964 and can be prepared by several methods (101). The best yields are obtained by the sequence of reactions shown in Figure 14 (102). The cis and trans isomers are separated by fractional crystallization of their salts and, in this case, the cis isomer is more potent and is marketed under the trade name Navane (Roerig) (**93**).

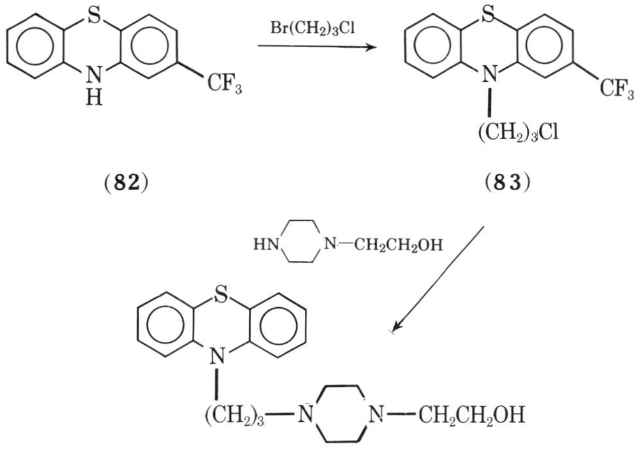

(**84**) fluphenazine [69-23-8]
(**65**) salt = 2 HCl: fluphenazine dihydrochloride

Figure 12. Synthesis of fluphenazine.

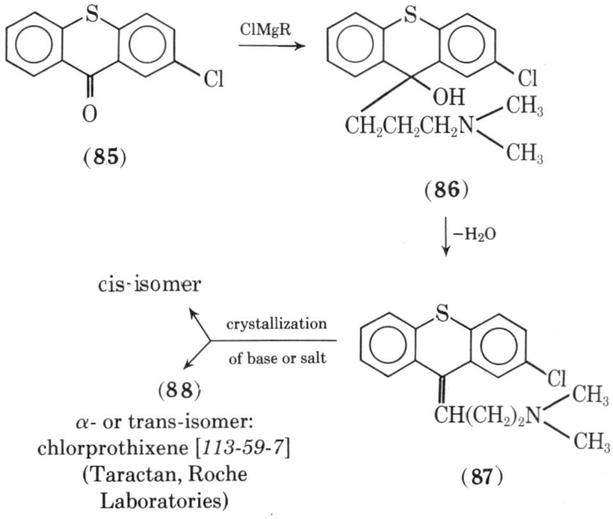

(**88**) α- or trans-isomer: chlorprothixene [113-59-7] (Taractan, Roche Laboratories)

Figure 13. Synthesis of chlorprothixene.

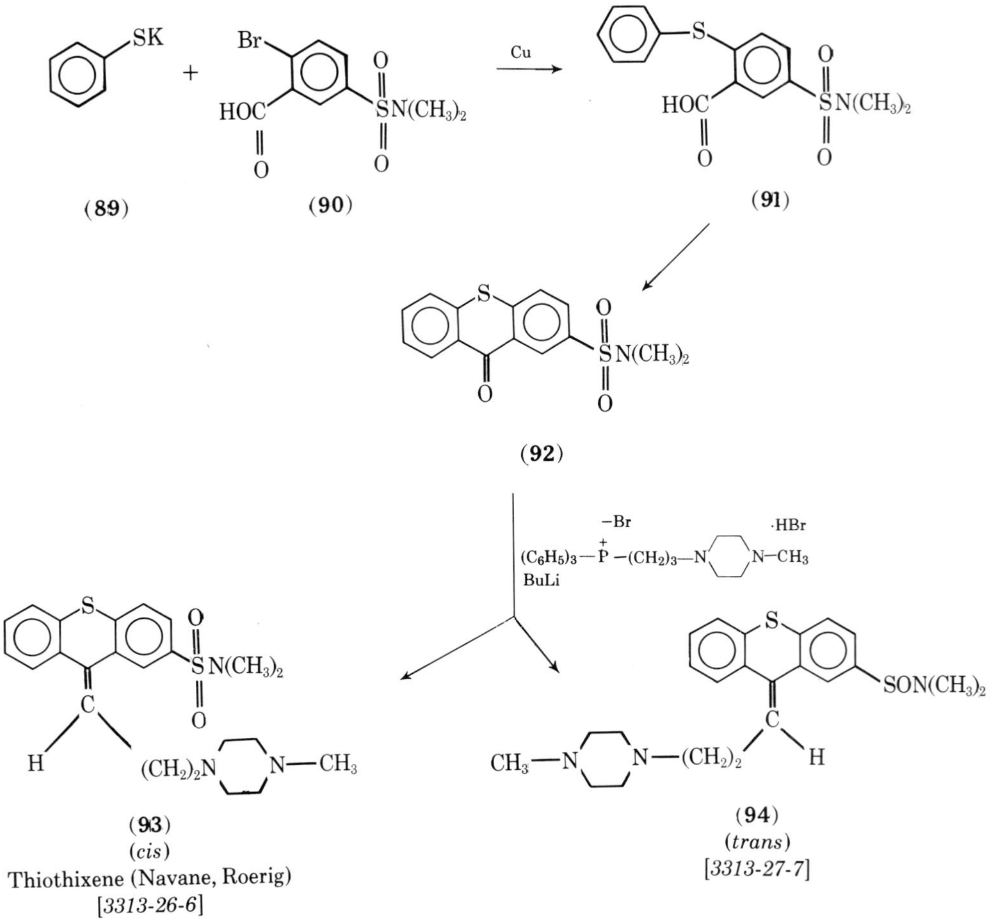

Figure 14. Synthesis of thiothixene.

Butyrophenones. The butyrophenones have pronounced antipsychotic properties. An excellent review of their discovery is given in reference 103. The first compound of this series to be clinically tested was haloperidol (**98**). It was synthesized as shown in Figure 15 and is marketed in the United States as Haldol by McNeil Laboratories and worldwide under various other trade names (104). The valuable properties of this compound and the ease of preparation of analogous butyrophenones resulted in the syntheses and biological evaluation of over 5000 analogues and related products. Ultimately, approximately ten different butyrophenone antipsychotics were introduced worldwide. In the United States in addition to Haldol, one other butyrophenone, droperidol (**99**), is marketed (105). It is used only as an injectable preparation under the trade name Inapsine (Janssen Pharmaceutical Inc.). Because of its rapid onset and relatively short duration of action, it is used only as an adjunct in general anesthesia (see Anesthetics). It is synthesized in a manner similar to haloperidol by condensation of a butyrophenone and a basic moiety. Some of the butyrophenones that are marketed abroad are compounds (**100**)–(**104**). These compounds are effective but they differ in potency, duration of action, and side effects.

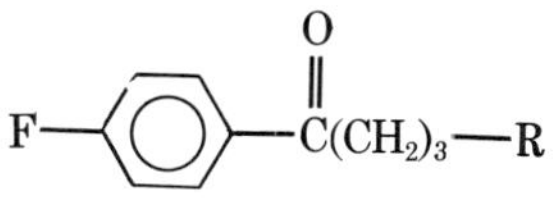

F—⟨phenyl⟩—C(=O)(CH$_2$)$_3$—R

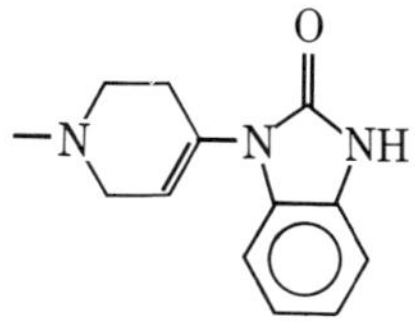

R =

(99), —N⟨piperidine⟩—N⟨benzimidazolone⟩NH   droperidol [548-73-2]

(100), —N⟨piperazine⟩N—⟨phenyl-OCH$_3$⟩   fluanisone [1480-19-9]

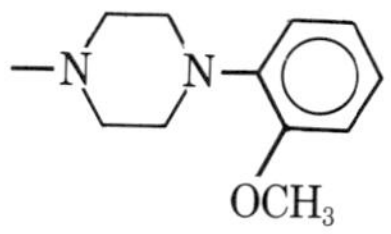

(101), —N⟨piperidine⟩(OH)—⟨phenyl⟩—CF$_3$ trifluperidol [749-13-3]

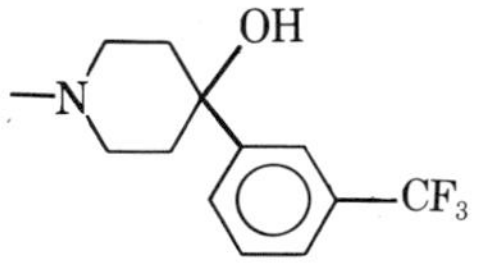

(102), —N⟨piperidine⟩(C(=O)NH$_2$)(N⟨piperidine⟩)   pipamperone [1893-33-0]

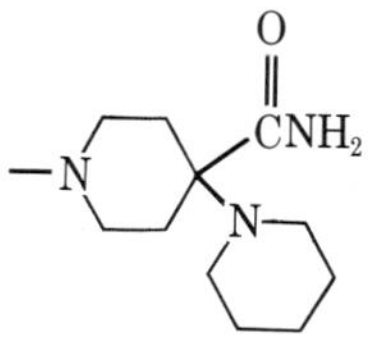

(103) 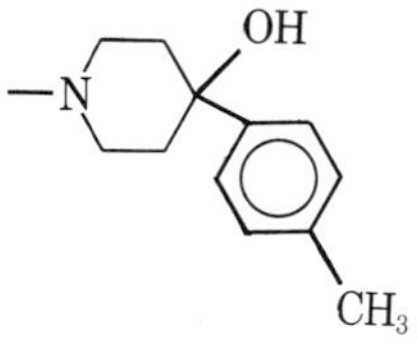 —N⟨piperidine⟩(OH)—⟨phenyl⟩—CH$_3$ moperone [1050-79-9]

(104) 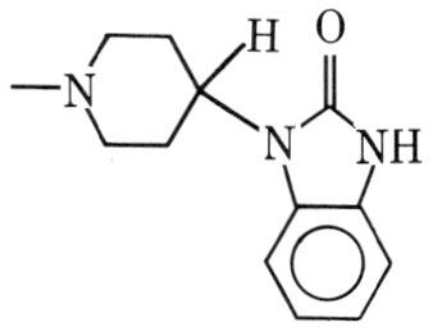 —N⟨piperidine⟩(H)—N⟨benzimidazolone⟩NH   benperidol [2062-84-2]

Other Antipsychotics. Loxapine (**107**) differs structurally from the previously discussed classes of antipsychotics. It shows a formal resemblance to the tricyclic antidepressants (see Antidepressants), but its properties and side effects resemble those of the other antipsychotic agents. The synthesis proceeds via (**105**) and (**106**) (106). Its hydrochloride is marketed by the Dome Division of Miles Laboratories, Inc. as Daxoline C and by Lederle Laboratories as Loxitane C. Loxapine succinate [27833-64-3] is marketed as Daxoline and Loxitane, respectively, by the same two companies.

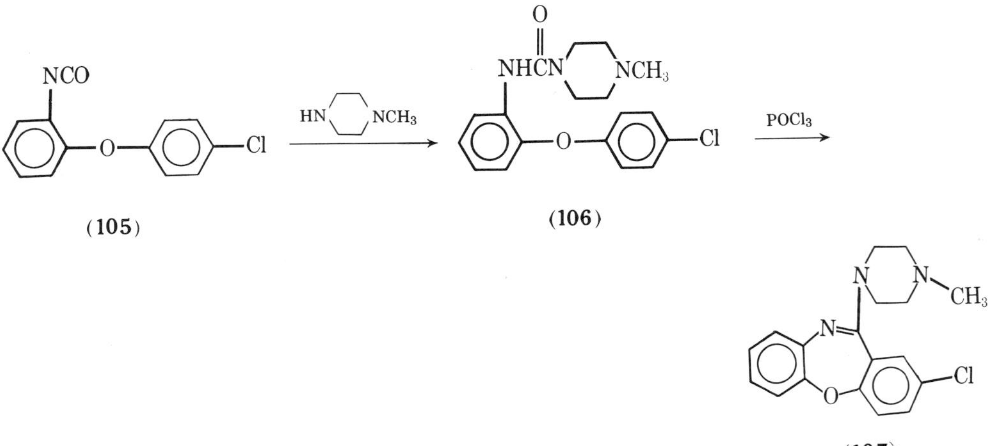

(105) (106) (107)

Another neuroleptic is molindone (111) (107). It is synthesized as shown in Figure 16 from the tetrahydro-oxoindole (110), which is prepared by a reductive condensation of the dione (108) and the oxime (109). A Mannich reaction converts (110) into mol-

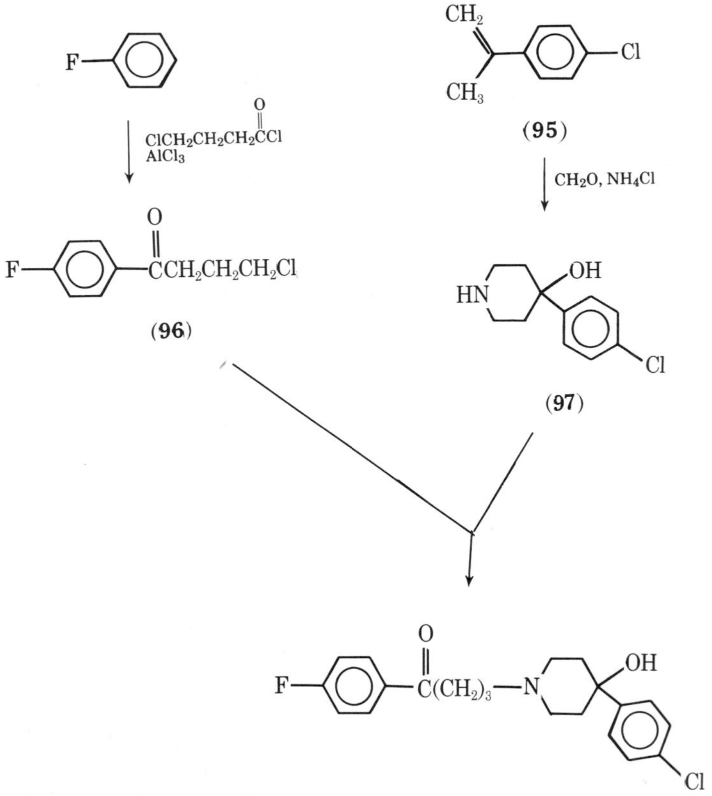

(98) haloperidol (Haldol) [52-86-8]

Figure 15. Synthesis of haloperidol.

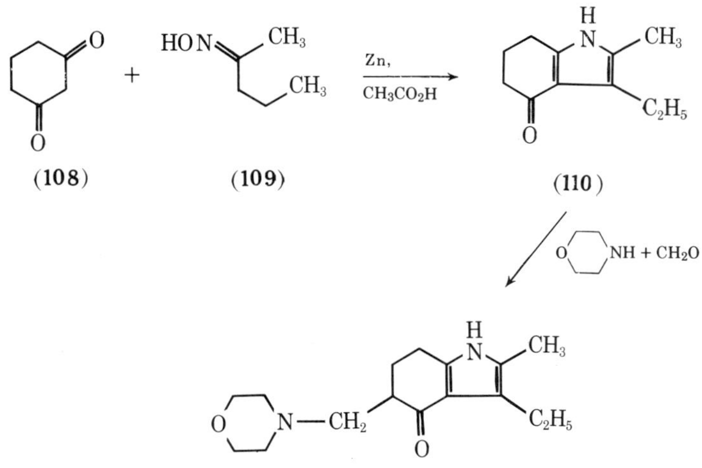

(111) molindone [*7416-34-4*]

Figure 16. Synthesis of molindone.

indone which is marketed as molindone hydrochloride [*15622-65-8*] by Abbott and Endo Laboratories under the trade names Lidone and Mobane, respectively.

A product that is not marketed in the United States but is marketed widely abroad is sulpiride (**114**) (108). It is synthesized by condensation of (**112**) with (**114**). It was

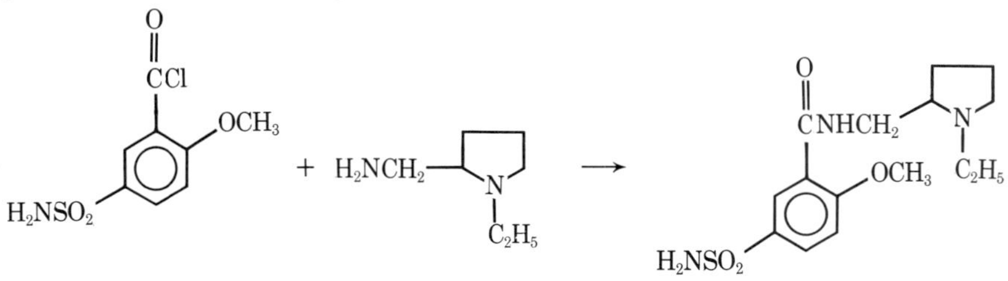

(**112**) (**113**) (**114**) sulpiride [*15676-16-1*]

first introduced in France by Delagrange under the trade name Dogmatil, and has shown valuable properties as an neuroleptic and an antiemetic.

Side Effects and Toxicity. The phenothiazine antipsychotics influence a number of important neurotransmitters and, consequently, produce several undesirable side effects (109) (see Neuroregulators). These include anticholinergic and antiadrenergic effects, eg, dry mouth, flushing, blurred vision, nasal congestion, constipation, and postural hypotension. Most of these effects tend to diminish as therapy continues, or they may be controlled by adjusting drug doses. The butyrophenone antipsychotics are less prone to produce autonomic effects than are the phenothiazines. The antidopaminergic property of the antipsychotics increases the release of prolactin from the pituitary gland, which frequently results in galactorrhea.

The most serious side effects of the antipsychotics are the extrapyramidal reactions, which cause disturbances in motor function (109–110). Patients on antipsychotic therapy may develop dyskinesias, such as bizarre movements of the face, tongue, neck or eyes; Parkinson's syndrome; or akathisia. Dyskinesia and Parkinson's syndrome may be controlled by coadministering anti-Parkinson drugs and akathisia usually responds to a decreased dose of the antipsychotic (110). Dyskinesia and Parkinson's syndrome occur in a high percentage of patients and are frequently sufficiently disconcerting to the patient that many physicians routinely administer anti-Parkinson drugs to their outpatients prior to the onset of these motor symptoms.

Prolonged exposure to antipsychotic drugs sometimes causes abnormal movements of the mouth, face, and tongue as well as choreic movements of the trunk and extremities. This phenomenon has been termed tardive dyskinesia (111). Tardive dyskinesias are estimated to occur in 10–20% of patients who are exposed to antipsychotic drugs for longer than a year (112). The cause of this reaction is not known but increased dopamine receptor sensitivity is believed to be a contributing factor (112). Tardive dyskinesias become more severe if antipsychotic therapy is suddenly stopped and are attenuated by increased doses of antipsychotics. Although there is no known treatment for tardive dyskinesia, motor control is generally improved by drugs that block dopaminergic transmission or enhance cholinergic or GABA-ergic transmission (113). Upon gradual withdrawal of antipsychotic therapy, some tardive dyskinesia may remit spontaneously.

The development of tardive dyskinesia after extended exposure to antipsychotics is a severe and disturbing phenomenon for which there is no known treatment. Nevertheless, a recent consensus suggests that discontinuing therapy with currently available drugs is not warranted and recommends the continued judicious use of these agents in the treatment of psychoses (112).

Antidepressants

Depression, like anxiety, is an emotion experienced by everyone to some degree. Estimates suggest that 10–12% of the population in the United States will be sufficiently depressed sometime in their lives to require treatment. Depression is a normal reaction to certain life events, eg, loss, physical illness, chronic stress, old age, and starvation. Generally, the depression spontaneously remits when the causative factors are removed or dealt with. In some cases, depression appears when no apparent external cause exists. This state is endogenous depression. Depressive states are often difficult to diagnose but elaborate schemes have been developed to diagnose and categorize them (114). In its simplest form, depression is characterized by a diminished capacity to experience and express emotions, a sense of despair, decreased appetite and libido, impaired sleep, and retarded body movements and thought processes. The symptomatology of depression is frequently complicated by the presence of anxiety or psychosis.

Drugs employed in the treatment of depression belong to either of two classes: the tricyclic antidepressants or the monoamine oxidase inhibitors (MAOIs). The tricyclic antidepressants are by far the most frequently used agents in the treatment of depression and are effective in ca 70% of treated patients (115). The MAOIs are usually reserved for those patients who do not respond to treatment with tricyclics. Although the overall efficacy rate of the MAOIs is considerably less than for the tricyclics, there appears to be a subpopulation of depressed patients who respond particularly well to MAOI treatment. In addition to these two main classes of antidepressants there are a few unrelated compounds that have shown valuable antidepressant properties (see Monoamine Oxidase Inhibitors).

Preclinical Pharmacology. Tricyclic antidepressants influence as many as 20 different biochemical events in the brain and alter the activity of a number of important neurotransmitters, including acetylcholine, norepinephrine, dopamine, serotonin, and histamine. Which of these actions, if any, is responsible for their antidepressant properties is not known.

Prominent among the tricyclic actions is their ability to potentiate central adrenergic and serotonergic functions by blocking the re-uptake of these amines into presynaptic neurons (116–117). The re-uptake mechanism is believed to be the principal means of terminating the synaptic action of these neurotransmitters. Thus, many test procedures for identifying new antidepressants have been designed to quantitate adrenergic or serotonergic potentiation (117). These procedures have included the prevention or reversal of reserpine and tetrabenazine [58-46-8] effects, the inhibition of muricide activity in rats, the potentiation of L-DOPA and L-tryptophan and the direct evaluation of drugs on amine uptake. Although these procedures may be useful for identifying tricycliclike antidepressants, there is considerable doubt about their specificity and relevance to depression. Examples of drugs that show activity in these various procedures but that are avoid of clinical antidepressant activity and those that are clinically effective but are inactive in these laboratory tests have been reported (118–120).

Some recent laboratory studies of antidepressants may suggest new approaches to identifying antidepressants with novel pharmacological profiles. Studies with imipramine [50-49-7] (**116**) or desipramine [50-47-5] (**121**) having a tritium label reveal binding sites in brain tissue for these antidepressants (121–122). A variety of structures reported to have antidepressant activity interact at these sites. None of the known endogenous brain neurotransmitters or modulators recognize the tricyclic binding sites. The correlation between antidepressant activities and affinities for the tricyclic binding site has prompted further investigations into the possible relevance of these binding sites to depression and the antidepressant properties of drugs.

The chronic administration of antidepressant drugs to rats produces a hyposensitization of β-adrenergic receptors (123). Drug-induced down regulation of β-receptors is of particular interest, because the treatment time required for the onset of β-receptor desensitization in rats is approximately the same as the length of time of antidepressant therapy required for the onset of clinical efficacy.

Some behavioral procedures believed to involve components of despair are also being studied as possible indicators of antidepressant activity. These include swimming-induced immobility in rodents and learned helplessness induced in rats and dogs by subjecting them to unavoidable shocks (124–125). Pharmacological investigations are also being conducted with infant monkeys that are depressed by isolating them from their mothers (126).

None of the above biochemical or behavioral models has been demonstrated to reliably predict clinical antidepressant efficacy. The absence of a reliable test for antidepressant activity is a serious deterrent to progress in this area of psychopharmacology.

The limited use of MAOIs in antidepressant therapy has discouraged developmental work with this type of drug; however, two types of monoamine oxidases have been described (117). One type appears to be localized predominantly in neuronal tissues, and the other exists in nonneuronal tissues. Because a selective inhibitor of the neuronal MAO may produce desired antidepressant effects but not the undesirable potential for hypertensive crisis following tyramine ingestion, some attempts have been made to identify such a selective inhibitor. No such selectivity has yet been demonstrated in humans, so this hypothesis has not been put to trial.

Mechanism of Action. The mechanisms by which antidepressants exert their therapeutic effects are a matter for conjecture. The multiple and diverse pharmacological actions of the tricyclic antidepressants have spawned an equally diverse array of hypotheses to account for their action. The most prominent among these has been the monoamine hypotheses. The various evidences supporting these theories have been extensively reviewed (127). The monoamine hypotheses suggest that depression results from a functional decrease in the activity of noradrenergic or serotonergic pathways in the brain, presumably within the limbic system, and that antidepressants restore the activity of these aminergic neurotransmitters to normal functional levels. These hypotheses are supported by observations that reserpine, which is an amine depletor, can induce a depressive syndrome in man. In some studies, depressed patients have reduced concentrations of monoamine metabolites. Intraneuronal monoamine concentrations are believed to be regulated by the activity of monoamine oxidase, and the synaptic activity of these neurotransmitters is thought to be terminated by their re-uptake into presynaptic terminals. Thus, it has been suggested that the MAOIs reverse depression by increasing the concentrations of intraneuronal amines available for release and that the potent inhibition of amine re-uptake by the tricyclic antidepressants prolongs the activity of synaptic norepinephrine or serotonin. The serotonin hypothesis is further supported by the observations that the therapeutic effect of imipramine depends upon serotonin concentrations in the brain (128).

Not all observations of depression and the antidepressants are consistent with the amine hypotheses. For example, the effects of the tricyclics on amine re-uptake occur immediately, whereas the onset of clinical efficacy of these drugs requires one to three weeks of treatment. In addition, the potencies of the tricyclics for blocking the re-uptake of amines does not correlate with their therapeutic potencies. The potent antiserotonergic properties of some antidepressants is particularly difficult to reconcile with the serotonin potentiation hypothesis. Although the onset and extent of clinical efficacy for the MAOIs does appear to coincide with the onset and degree of enzyme inhibition, these drugs may have stimulant effects independent on their influences on MAOI (129). Finally, investigations of the functional status of amines in depressed individuals have provided inconsistent results, and considerable controversy surrounds these investigations.

The recent observations that the chronic administration of antidepressants causes a decrease in β-adrenergic receptor sensitivity has suggested the novel hypothesis that depression results from the hyperactivity of noradrenergic pathways (130).

Chemistry. *Tricyclics.* Tricyclic antidepressants have a structural similarity to the tricyclic antipsychotics of the phenothiazine and thioxanthene group and, like them, were the result of the search for new antihistaminics and anticholinergics (80,131–132). The sulfur atom joining the two benzenoid rings in the antipsychotics is replaced by a two-carbon chain. Thus, instead of the (six-membered) thiazine or thiopyrane ring, there is a (seven-membered) azepine or cycloheptadiene ring system. The basic substituent attached in the 5-position is usually a mono- or dimethylaminopropyl group.

The first tricyclic antidepressant to be used clinically was imipramine (**116**), which is synthesized by alkylation of the dibenzazepine (**115**). Imipramine hydrochloride was introduced in 1959 by Geigy under the trade name Tofranil and is widely used (133). Imipramine pamoate [*10246-75-0*] is marketed as Tofranil PM.

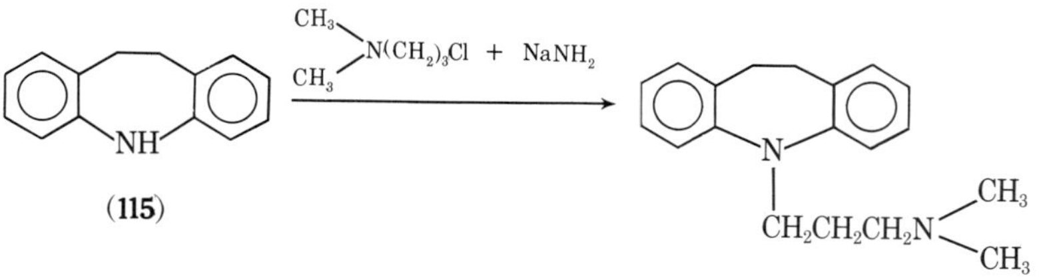

[Structure of compound (115) + CH₃N(CH₃)(CH₂)₃Cl + NaNH₂ → (116) Imipramine]

The next tricyclic antidepressant to be marketed was amitriptyline (119), which is a dibenzocycloheptane derivative (134). It was synthesized and patented by Hoffmann-La Roche & Co. and Merck & Co. Its valuable antidepressant properties led to the introduction in 1958 of amitriptyline hydrochloride [549-18-8] under the trade name Elavil, by Merck in the United States and later in the United States as Endep by Hoffmann-La Roche. A combination of the hydrochloride with the anxiolytic chlordiazepoxide is marketed by Roche under the trade name Limbitrol.

One of the synthetic routes proceeds via a Grignard reaction followed by dehydration, as shown in Figure 17.

Five other tricyclic antidepressants are marketed in the United States; two are analogues of amitriptyline, two are related to imipramine, and the fifth is a dibenzoxepine. These compounds differ in duration of action; side effects, eg, sedation, anticholinergic properties, etc; and potencies.

Desipramine is a monomethylamino derivative and a biologically active metabolite of imipramine. Its sedative and anticholinergic side effects are, however, less pronounced than those of imipramine. It is synthesized as shown in Figure 18 and is marketed as desipramine hydrochloride [58-28-6] by Merrell-National under the trade name Norpramine and by U.S.V. Laboratories as Pertofrane (135).

Nortriptyline (124) is a pharmacologically active metabolite of imipramine and can be prepared by several methods, eg, by the sequence of reactions outlined in Figure 19 (136). It is marketed as nortriptyline hydrochloride [894-71-3] by Lilly under the trade name Aventyl hydrochloride and by Sandoz Pharmaceuticals as Pamelor.

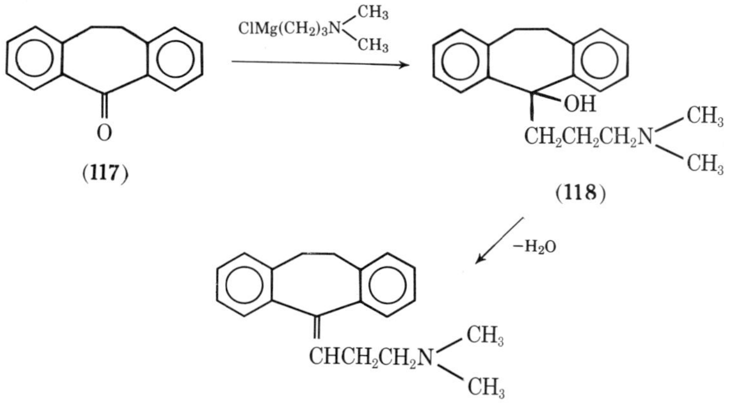

(117) → [ClMg(CH₂)₃N(CH₃)₂] → (118) → −H₂O → (119) amitriptyline [50-48-6]

Figure 17. Synthesis of amitriptyline.

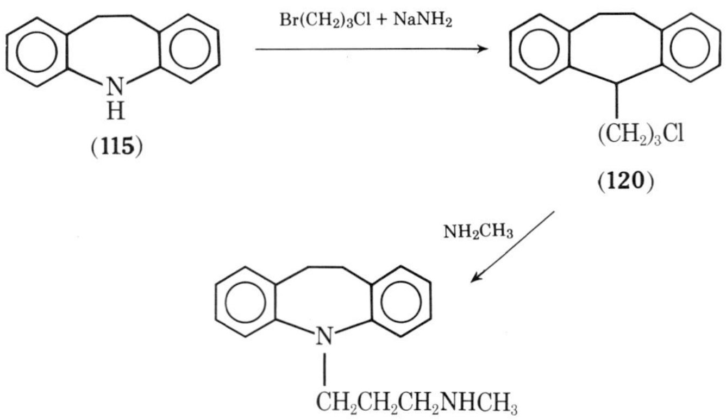

(121) desipramine

Figure 18. Synthesis of desipramine.

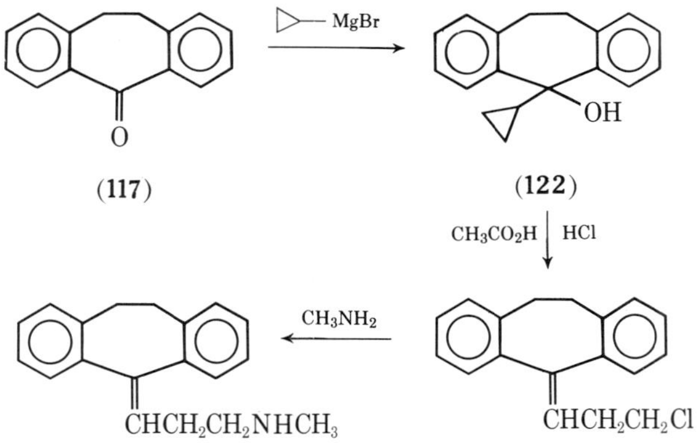

(124) nortriptyline [*72-69-5*] (123)

Figure 19. Synthesis of nortriptyline.

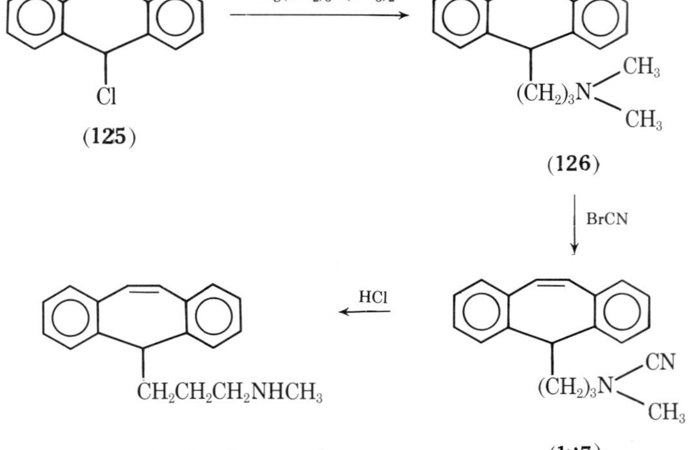

(128) protriptyline [*438-60-8*] (127)

Figure 20. Synthesis of protriptyline.

Protriptyline is an isomer of nortriptyline and differs in the position of a double bond. It can be prepared by several methods, eg, by the reaction sequence shown in Figure 20 (137–138). It is marketed as protriptyline hydrochloride [1225-55-4] by Merck, Sharp, and Dohme under the trade name Vivactil.

Trimipramine (**129**) is a homologue of imipramine (**116**) and is synthesized from (**115**) (139). Trimipramine maleate [521-78-8] is marketed as Surmontil by Ives Laboratories Inc.

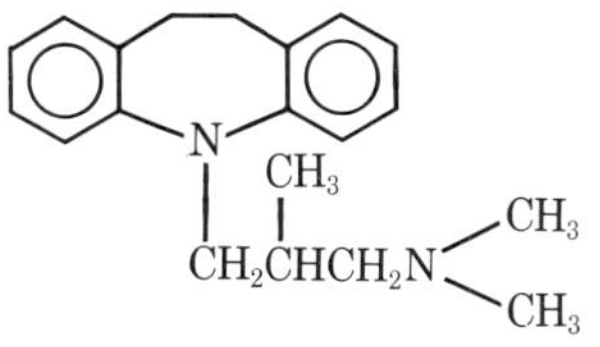

(**129**) trimipramine [739-71-9]

Doxepin (**131**) is an oxa analogue of imipramine (**116**) and has similar side effects. It is prepared in the same manner as amitryptyline and is marketed as doxepin hydrochloride [1229-29-4], which is the hydrochloride of a mixture of the cis- and trans isomers (ca 1:5) by Pfizer Laboratories under the trade name Sinequan and by Pennwalt under the trade name Adapin (140).

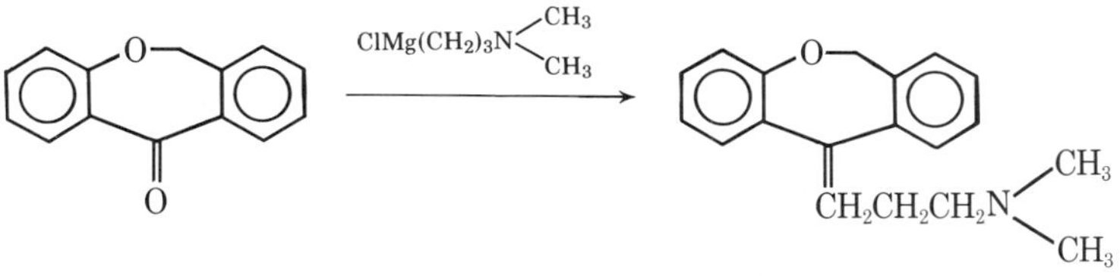

(**130**)   (**131**) doxepin [1668-19-5]

Monoamine Oxidase Inhibitors (MAOIs). The antidepressant properties of the MAOIs were discovered in the early 1950s in connection with a study of the antitubercular drug, iproniazid [54-92-2] (1-isonicotinyl-2-isopropylhydrazide). The euphoria and elevated mood of the patients was first thought to be the natural reaction to their improved health. On closer investigation, it was realized that this mood elevation is a specific characteristic of the drug. The mood-elevating effects of a number of other MAOIs were studied and new compounds were synthesized and investigated. Three MAOIs are marketed in the United States and a few others are marketed abroad. The simplest MAOI to be used as an antidepressant is phenethylzine (**132**), which is prepared by treatment of phenethyl bromide with hydrazine (141). Phenethylzine sulfate [156-51-4] is marketed by Parke-Davis as Nardil.

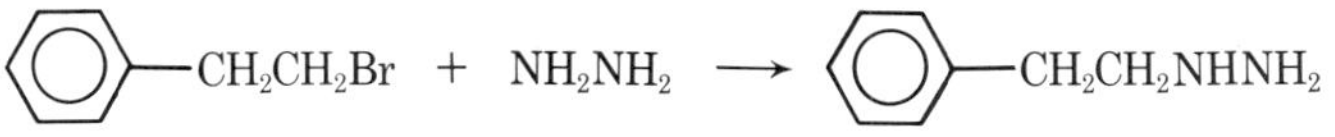

(**132**) phenethylzine [51-71-8]

Another hydrazine—MAOI is isocarbazid (135); it is marketed by Hoffmann-La Roche Inc. as Marplan. It is synthesized as shown by condensation of the hydrazine (133) with the ester (134) (142).

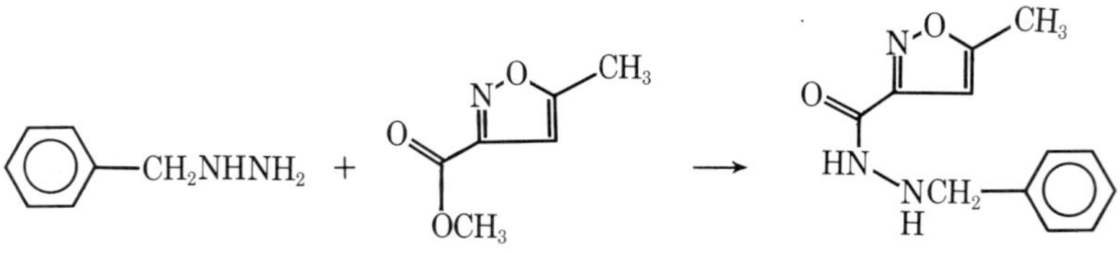

(133)   (134)   (135) isocarbazid [59-63-2]

A third MAO inhibitor, tranylcypromine (140), is not a hydrazine derivative. It is synthesized as shown in Figure 21 (143). The sulfate is marketed by Smith, Kline & French Laboratories as Parnate.

An Antidepressant Not Belonging to the Two Main Classes. Recently, a new antidepressant, trazodone monohydrochloride [25332-39-2], was introduced in the United States. It is a triazolopyridine (141) and can be prepared as shown below.

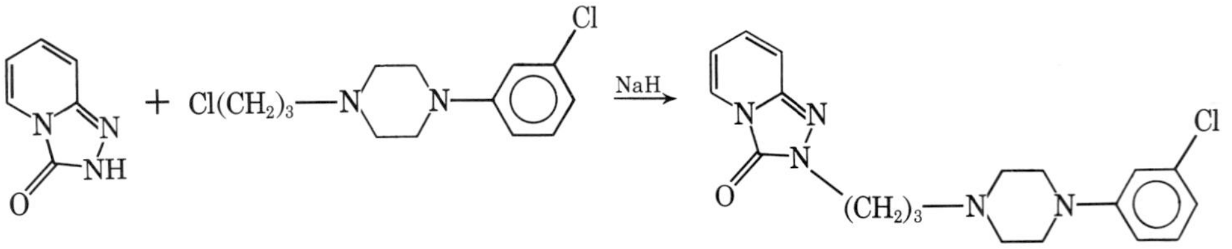

(141) [19794-93-5]

Trazodone (141) was introduced in the United States by Bristol-Myers in the form

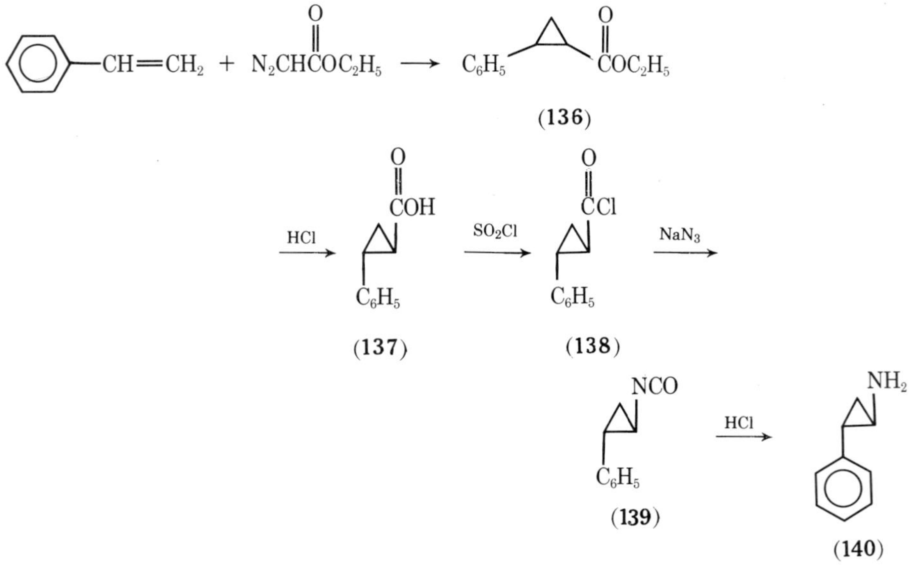

(136)

(137)   (138)

(139)   (140)

tranylcypromine [155-09-9]

Figure 21. Synthesis of tranylcypromine.

of its hydrochloride under the trade name Desyrel (144–145). It causes fewer anticholinergic side effects than the widely used tricyclic compounds.

Side Effects and Toxicity. Antidepressant compounds exhibit a number of undesirable side effects and at high doses can be very toxic (115). The tricyclic compounds are well known for their potent inhibition of cholinergic, adrenergic (alpha), serotonergic, and histaminic receptors. In therapy, anticholinergic effects are particularly troublesome, causing dizziness, postural hypotension, tachycardia, dry mouth, and constipation. Frequently, the severity of these side effects diminishes during the first few weeks of treatment. In some patients, normal therapeutic doses of tricyclic antidepressants may supress cardiac function with life-threatening consequences. Patients with existing cardiovascular disease are particularly susceptible to tricyclic cardiotoxicity (63).

The MAOIs produce many autonomic side effects like those of the tricyclics, ie, dizziness, dry mouth, postural hypotension, and constipation. In addition, the MAOIs are more prone than are the tricyclics to induce CNS (central nervous system) side effects, such as mania, psychosis, and disorientation (63). Perhaps the most serious side effect of MAOI therapy is the hypertensive crisis brought on by the ingestion of dietary and medicinal sympathomimetic amines (63,115). Many foods, eg, cheese, wine, chicken livers, herring, and broad beans, contain large concentrations of tyramine. Tyramine is normally converted to inactive metabolites by MAO in the gut, but in the presence of MAOI, this amine is absorbed intact. Once absorbed, tyramine causes a massive release of norepinephrine in heart and vascular tissue, which results in marked elevations of blood pressure and heart rate. Such hypertensive crises can be of sufficient magnitude, especially if untreated, so as to result in death. In general, the MAOIs are less efficacious and more toxic than are the tricyclic antidepressants.

BIBLIOGRAPHY

"Psychopharmacological Agents" in *ECT* 1st ed., Suppl. 1, pp. 720–743, by Maxwell Gordon and G. E. Ullyot, Smith, Kline & French Laboratories; "Stimulants and Depressants of the Nervous System" in *ECT* 1st ed., Vol. 13, pp. 1–45, by Mark Nickerson and E. F. Domino, University of Michigan; "Psychopharmacological Agents" in *ECT* 2nd ed., Vol. 16, pp. 640–679, by Maxwell Gordon and Glenn E. Ullyot, Smith, Kline & French Laboratories.

1. R. Burton, *The Anatomy of Melancholy*, 1st ed., Henry Cripps, Oxford, 1621, p. 444.
2. G. Zilboorg, *A History of Medical Psychology*, W. W. Norton and Co., New York, 1941, p. 72.
3. *Ibid.*, p. 48.
4. E. Schlitter and A. J. Plummer in M. Gordon, ed., *Psychopharmacological Agents*, Vol. 1, Academic Press, New York, 1964, p. 9.
5. H. Ey and H. Mignot, *Ann. Med. Psychol.* **105,** 226 (1947).
6. A. E. Caldwell in W. G. Clark and J. Delgiudice, eds., *Principles of Psychopharmacology*, 2nd ed., Academic Press, New York, 1978, p. 9.
7. R. Kuhn, *Schweiz. Med. Wochenschr.* **87,** 1135 (1957).
8. L. H. Sternbach in S. Garattini, E. Mussini, and L. O. Randall, eds., *The Benzodiazepines*, Raven Press, New York, 1973, p. 1.
9. Mental Health Association, Arlington, Va., 1980.
10. *Prescription Drug Industry*, The Pharmaceutical Manufacturers Association, Washington, D.C., 1980.
11. M. Lader and I. Marks, *Clinical Anxiety*, Grune and Stratton Inc., New York, 1971, p. 29.
12. B. J. Blackwell, *J. Am. Med. Assoc.* **225,** 1637 (1973).
13. H. J. Perry and co-workers, *Arch. Gen. Psychiatry* **28,** 769 (1973).

14. W. Schallek in ref. 6, p. 325.
15. I. Geller and co-workers, *Psychopharmacologia* **3,** 374 (1962).
16. L. Cook and A. B. Davidson in ref. 8, p. 327.
17. R. Squires and C. Braestrup, *Nature (London)* **266,** 132 (1977).
18. H. Möhler and T. Okada, *Science* **198,** 849 (1977).
19. W. Schallek and co-workers, *Adv. Pharmacol. Chemother.* **16,** 45 (1977).
20. A. B. Young and co-workers, *Proc. Natl. Acad. Sci. U.S.A.* **71,** 2246 (1974).
21. D. R. Curtis and co-workers, *Br. J. Pharmacol.* **56,** 307 (1976).
22. A. Dray and D. W. Straughan, *J. Pharm. Pharmacol.* **28,** 314 (1976).
23. E. Costa and co-workers, *Life Sci.* **17,** 167 (1975).
24. P. Polc and co-workers, *Naunyn-Schmiedeberg's Arch. Pharmacol.* **284,** 319 (1974).
25. V. V. Zakusov and co-workers, *Arch. Int. Pharmacodyn. Ther.* **214,** 188 (1975).
26. C. C. Mao and co-workers, *Naunyn-Schmiedebergs Arch. Pharmacol.* **289,** 369 (1975).
27. J. Sepinwall and L. Cook, *Fed. Proc. Fed. Am. Soc. Exp. Biol.* **39,** 3024 (1980).
28. A. Guidotti and co-workers in ref. 27, p. 3039.
29. D. W. Gallager and co-workers in ref. 27, p. 3043.
30. R. S. L. Chang and co-workers, *Brain Res.* **190,** 95 (1980).
31. A. S. Lippa and co-workers, *Fed. Proc. Fed. Am. Soc. Exp. Biol.* **36,** 1044 (1977).
32. L. Juhasz and W. Dairman in ref. 31, p. 377.
33. A. S. Lippa and co-workers in T. Hanin and E. Usdin, eds., *Animal Models in Psychiatry and Neurology*, Pergamon Press, New York, 1977, p. 279.
34. M. L. Billingsley and R. K. Kudena, *Life Sci.* **22,** 897 (1978).
35. J. Sepinwall and L. Cook in L. L. Iversen, S. D. Iversen, and S. H. Snyder, eds., *Handbook of Psychopharmacology*, Vol. 13, Plenum Press, New York, 1978, p. 345.
36. L. Stein and co-workers, *Am. J. Psychiatry* **134,** 665 (1977).
37. N. C. Tye and co-workers, *Neuropharmacology* **18,** 689 (1979).
38. B. J. Ludwig and E. C. Piech, *J. Am. Chem. Soc.* **73,** 5779 (1951).
39. F. M. Berger and B. J. Ludwig in ref. 4, pp. 103–135.
40. L. H. Sternbach, L. O. Randall, and S. Gustafson in ref. 39, p. 137.
41. L. H. Sternbach, L. O. Randall, R. Banziger, and H. Lehr in A. Burger, eds., *Medicinal Research Series*, Vol. 2, Marcel Dekker, Inc., New York, 1968, p. 237.
42. L. H. Sternbach in R. Jucker, ed., *Progress in Drug Research*, Vol. 22, Birkhaüser Verlag, Basel and Stuttgart, 1978, p. 229.
43. L. O. Randall, *Dis. Nerv. Syst.* **22,** 7, 11 (1961).
44. L. H. Sternbach and E. Reeder, *J. Org. Chem.* **26,** 1111 (1961).
45. *Ibid.*, p. 4936.
46. S. C. Bell and S. J. Childress, *J. Org. Chem.* **27,** 1691 (1962).
47. J. Schmitt, P. Comoy, M. Suquet, J. Boitard, J. LeMeur, J.-J. Basselier, M. Brunaud, and J. Salle, *Chim. Ther.* **2,** 254 (1967).
48. U.S. Pat. 3,192,200 (June 29, 1965), H. M. Wuest.
49. S. C. Bell, R. J. McCaully, C. Gochman, S. J. Childress, and M. I. Gluckman, *J. Med. Chem.* **11,** 457 (1968).
50. M. Steinman, J. G. Topliss, R. Alekel, Y-S. Wong, and E. E. York, *J. Med. Chem.* **16,** 1354 (1973).
51. J. M. Hester, Jr., A. D. Rudzik, and B. V. Kamdar, *J. Med. Chem.*, **14,** 1078 (1971).
52. H. H. Kaegi, *J. Label. Congr.* **4,** 363 (1968).
53. T. Miyadera, A. Terada, M. Fukunaga, Y. Kawano, T. Kamioa, C. Tamura, H. Takagi, and R. Tachikawa, *J. Med. Chem.* **14,** 520 (1971); T. L. Lemke and A. R. Hanze, *J. Heterocycl. Chem.* **8,** 125 (1971); M. E. Derieg, J. V. Earley, R. I. Fryer, R. J. Lopresti, R. M. Schweininger, L. H. Sternbach, and H. Wharton, *Tetrahedron* **27,** 2591 (1971).
54. L. H. Sternbach, R. I. Fryer, W. Metlesics, G. Sach, and A. Stempel, *J. Org. Chem.* **27,** 3781 (1962).
55. R. I. Fryer, R. A. Schmidt, and L. H. Sternbach, *J. Pharm. Sci.* **53,** 264 (1964).
56. Belg. Pat. 803,315 (Aug. 7, 1973), F. Benconi, R. Tagliabue, and L. Molteni (to Dr. L. Zambeletti S.p.A.).
57. L. H. Sternbach, R. I. Fryer, O. Keller, W. Metlesics, G. Sach, and N. Steiger, *J. Med. Chem.* **6,** 261 (1963).
58. U.S. Pat. 3,799,920 (March 26, 1974), G. Ferrari and C. Casagrande (to Siphar S.A.).
59. Belg. Pat. 707,667 (July 1968), (to Roussel Uclaf); S. African Pat. 6,800,803 (July 1968), K. H.

Hauptmann, K. H. Weber, K. Zeile, P. Danneberg, and K. Giesemann (to Boehringer Ingelheim GmbH); *Chem Abstr.* **70,** 106579 (1969); S. Rossi, O. Pirola, and R. Maggi, *Chim. Ind.* **51,** 479 (1969).
60. Ger. Offen. 2,107,356 (Aug. 26, 1971), M. Nakanishi and co-workers (to Yoshitomi Pharmaceutical Ind. Ltd.); M. Nakanishi, T. Tsumagari, Y. Takigawa, S. Shuto, T. Kenjo, and T. Kukuda, *Arzneimittel-Forsch* **22,** 1905 (1972).
61. H. Morren, R. Denayer, S. Trolin, E. Grivsky, R. Linz, H. Strubbe, G. Dony, and J. Marico, *Ind. Chim. Belg.* **19,** 1176 (1954); *Chem. Week* **79**(5), 78 (1956).
62. A. R. Surrey, W. G. Webb, and R. M. Gesler, *J. Am. Chem. Soc.* **80,** 3469 (1958).
63. J. M. Davis and R. C. Casper in ref. 6, p. 479.
64. D. M. Turns in H. C. B. Denber, ed., *Schizophrenia*, Marcel Dekker, Inc., New York, 1978, p. 17.
65. B. Pasamanick and co-workers, *Schizophrenics in the Community: An Experimental Study in the Prevention of Hospitalization*, Appleton-Century-Crofts, New York, 1967, p. 29.
66. J. M. Davis and D. L. Garver in ref. 35, Vol. 10, p. 129.
67. S. Fielding and H. Lal in ref. 35, Vol. 10, p. 91.
68. P. A. Janssen and W. F. M. Venbever in ref. 6, p. 279.
69. I. Creese and co-workers in ref. 35, Vol. 10, p. 37.
70. O. Horneykiewicz in D. B. Calne, T. H. Chase, and A. Barbeau, eds., *Advances in Neurology*, Vol. 9, Raven Press, New York, 1975, p. 155.
71. S. H. Snyder, *Science* **184,** 1243 (1974).
72. P. A. Shore and A. Giachetti in ref. 35, Vol. 10, p. 197.
73. S. Matthysse and J. Sugarman in ref. 35, Vol. 10, p. 221.
74. J. O. Cole and D. J. Clyde, *Rev. Can. Biol.* **20,** 565 (1961).
75. M. LeMoal and co-workers in E. Costa and G. L. Gessa, eds., *Advances in Biochemical Psychopharmacology*, Vol. 16, Raven Press, New York, 1977, p. 237.
76. A. R. Cools and J. M. Van Rossum, *Life Sci.* **27,** 1237 (1980).
77. E. Schenker and H. Herbst in ref. 42, Vol. 5, p. 269.
78. S. Massie, *Chem. Rev.* **54,** 797 (1954).
79. M. Gordon in M. Gordon, ed., *Psychopharmacological Agents*, Vol. 2, Academic Press, New York, 1967, pp. 1, 305.
80. C. L. Zirkle and C. Kaiser in M. Gordon, ed., *Psychopharmacological Agents*, Vol. 3, Academic Press, New York, 1974, p. 39.
81. P. Charpentier, P. Gailliot, R. Jacob, J. Gaudechon, and P. Buisson, *Compt. Rend.* **235,** 59 (1952).
82. U.S. Pat. 2,645,640 (July 4, 1953), P. Charpentier (to Rhône-Poulenc, Paris).
83. Brit. Pat. 780,193 (July 31, 1957), R. J. Horclois (to Société des Usines Chimiques Rhône-Poulenc, Paris).
84. U.S. Pat. 2,838,507 (June 10, 1958), J. W. Cusic and R. W. Hamilton (to G. D. Searle & Co.); U.S. Pat. 2,860,138 (Nov. 11, 1958), M. H. Sherlock and N. Sperber (to Schering Corporation).
85. H. L. Yale, F. Sowinski, and J. Bernstein, *J. Am. Chem. Soc.* **79,** 4375 (1957).
86. P. N. Craig, E. A. Nodiff, J. J. Lafferty, and G. E. Ullyot, *J. Org. Chem.* **22,** 709 (1957).
87. H. L. Yale and F. Sowinski, *J. Am. Chem. Soc.* **82,** 2039 (1960).
88. U.S. Pat. 3,194,733 (July 13, 1965), H. L. Yale and R. C. Merrill (to Olin Mathieson Chemical Corporation); U.S. Pat. 3,394,131 (July 23, 1968), H. L. Yale and R. C. Merrill (to E. R. Squibb & Sons).
89. J.-P. Bourquin, G. Schwarb, G. Bamboni, R. Fischer, L. Ruesch, S. Guldimann, V. Theus, E. Schenker, and J. Renz, *Helv. Chim. Acta* **41,** 1072 (1958).
90. U.S. Pat. 3,084,161 (Apr. 2, 1963), J. Kenz and J. P. Bourquin (to Sandoz Ltd.).
91. Brit. Pat. 861,807 (March 1, 1961), (to G. D. Searle & Co.).
92. U.S. Pat. 2,985,654 (May 23, 1961), M. H. Sherlock and N. Sperber (to Schering Corporation).
93. U.S. Pat. 3,023,146 (Feb. 27, 1962), R. Tislow, W. F. Bruce, and J. A. Page (to American Home Products).
94. Ger. Pat. 1,120,451 (Dec. 28, 1961), U. Hoerlein, K. H. Risse, and W. Wirth (to Farbenfabriken Bayer A. G.).
95. A. Bernthsen, *Ber.* **16,** 2896 (1883).
96. J. Schmitt, J. Boitard, P. Comoy, A. Hallot, and M. Suquet, *Bull. Soc. Chim. Fr.*, 938 (1957).
97. P. V. Petersen and I. M. Nielsen in ref. 4, p. 301.
98. P. V. Petersen, N. Lassen, T. Holm, P. Kopf, and I. M. Nielsen, *Arzneim. Forsch.* **8,** 395 (1958).
99. Brit. Pat. 829,763 (Mar. 9, 1960), (to Merck & Co.); Belg. Pat. 558,171 (Dec. 6, 1957), K. Doebel, G.

Rey-Bellet, R. Schlapfer, and H. Spiegelberg (to Hoffmann-La Roche, A.-G.); Ger. Pat. 1,048,589 (Jan. 15, 1959), (to Hoffmann-La Roche & Co., Akt.-Ges.).
100. G. E. Bonvicino, H. G. Arlt, Jr., K. M. Pearson, and R. B. Hardy, Jr., *J. Org. Chem.* **26**, 2383 (1961).
101. Belg. Pat. 647,006 (Oct. 26, 1964), (to Chas. Pfizer & Co., Inc.).
102. J. F. Muren and B. M. Bloom, *J. Med. Chem.* **13**, 17 (1970).
103. P. A. J. Janssen in ref. 77, p. 199.
104. P. A. J. Janssen, C. Van De Westeringh, A. H. M. Jageneau, P. J. A. Demoen, B. K. F. Hermans, G. H. P. Van Daele, K. H. L. Schellekens, C. A. M. Van Der Eycken, and C. J. E. Niemegeers, *J. Med. Pharm. Chem.* **1**, 281 (1959).
105. U.S. Pat. 3,141,823 (July 21, 1964), P. A. J. Janssen and J. F. Gardocki (to N. V. Research Laboratorium).
106. J. Schmutz, F. Künzel, F. Hunziker, and R. Gauch, *Helv. Chim. Acta* **50**, 245 (1967).
107. Belg. Pat. 670,798 (Jan. 31, 1966), (to Endo Laboratories, Inc.).
108. U.S. Pat. 3,342,826 (Sept. 19, 1967), C. S. Miller, E. L. Engelhardt, and M. Tominet (to Societe d'Etudes Scientifiques et Industrielles de l'Ile-de-France).
109. O. Hornykiewicz and H. L. Klawans, Jr. in ref. 4, p. 297.
110. J. M. Davis and R. C. Casper in ref. 6, p. 479.
111. R. J. Baldessarini and D. Tarsy in W. E. Fann, R. C. Smith, J. M. Davis, and E. F. Domino, eds., *Tardive Dyskinesia*, Spectrum Publications, Inc., New York, 1980, p. 181.
112. *Am. J. Psychiatry* **137**, 1163 (1980).
113. T. N. Chase and C. A. Tamminga in F. Cattabeni, P. F. Spano, G. Racagni, and E. Costa, eds., *Advances in Biochemical Psychopharmacology*, Vol. 24, Raven Press, New York, 1980, p. 457.
114. W. W. K. Zung in W. E. Fann, I. Karacan, A. D. Pokorny, and R. L. Williams, eds., *Phenomenology and Treatment of Depression*, Spectrum Publications, Inc., New York, 1977, p. 217.
115. D. J. Kupfer and T. P. Detre in ref. 35, Vol. 14, p. 199.
116. R. A. Maxwell and H. L. White in ref. 35, Vol. 14, p. 83.
117. F. Sulser in ref. 35, Vol. 14, p. 157.
118. L. Hekimian and co-workers, *Arzneim. Forsch.* **19**, 955 (1969).
119. R. M. Itil and co-workers, *Curr. Ther. Res. Clin. Exp.* **14**, 395 (1972).
120. P. F. Fell and co-workers, *Eur. J. Clin. Pharmacol.* **5**, 116 (1973).
121. R. A. O'Brien and co-workers, *Soc. for Neuroscience Abstr.* **4**, 430 (1978).
122. A. Biegon and D. Samuel, *Biochem. Pharmacol.* **28**, 3361 (1979).
123. J. Vetulani, *Naunyn-Schmiedebergs Arch. Pharmacol.* **293**, 109 (1976).
124. R. D. Porsolt and co-workers, *Arch. Int. Pharmacodyn. Ther.* **229**, 327 (1978).
125. W. R. Miller and co-workers in J. D. Maser and M. E. P. Seligman, eds., *Psychopathology: Experimental Models*, W. H. Freeman & Co., San Francisco, Calif., 1977, p. 104.
126. S. J. Suomi and H. F. Harlow in ref. 125, p. 131.
127. R. J. Baldessarini in F. F. Flach and S. C. Draghi, eds., *The Nature and Treatment of Depression*, John Wiley & Sons, Inc., New York, 1975, p. 347.
128. B. Shopsin and co-workers, *Psychopharmacol. Commun.* **1**, 239 (1975).
129. M. R. Mandel and G. Klerman in ref. 6, p. 537.
130. F. Sulser, *Trends Pharmacol. Sci.* **1**, 92 (1979).
131. F. Häfliger and V. Burckhardt in ref. 4, p. 35.
132. C. Kaiser and C. L. Zirkle in A. Burger, ed., *Medicinal Chemistry*, Part 2, Wiley-Interscience, New York, 1970, p. 1470.
133. W. Schindler and F. Häfliger, *Helv. Chim. Acta* **37**, 472 (1954); U.S. Pat. 2,554,736 (May 29, 1951), F. Häfliger and W. Schindler (to J. R. Geigy A.-G.).
134. U.S. Pat. 3,384,663 (May 21, 1968, filed March 27, 1959), G. Rey-Bellet and H. Spiegelberg (to Hoffmann-La Roche & Co., Inc.); Can. Pat. 744,730 (Oct. 18, 1966, filed Aug. 24, 1959), E. L. Engelhardt (to Merck & Co., Inc.).
135. Brit. Pat. 908,788 (Oct. 24, 1962), (to J. R. Geigy, A.-G.); Belg. Pat. 614,616 (Sept. 3, 1962), (to J. R. Geigy, A.-G.).
136. R. D. Hoffsommer, D. Taub, and N. L. Wendler, *J. Org. Chem.* **27**, 4134 (1962).
137. U.S. Pat. 3,922,305 (Nov. 25, 1975), E. L. Engelhardt (to Merck & Co., Inc.).
138. E. L. Engelhardt, M. E. Christy, C. D. Colton, M. B. Freedman, C. C. Boland, L. M. Halpern, V. G. Vernier, and C. A. Stone, *J. Med. Chem.* **11**, 325 (1968); Belg. Pat. 617,967 (Nov. 22, 1962), E. L. Engelhardt and M. E. Christy (to Merck & Co., Inc.).

139. R.-M. Jacob and M. Messer, *Compt. Rend.* **252,** 2117 (1961).
140. K. Stach and F. Bickelhaupt, *Monatsh.* **93,** 896 (1962).
141. U.S. Pat. 3,000,903 (Sept. 15, 1959), J. H. Biel (to Lakeside Laboratories, Inc.).
142. T. S. Gardner, E. Wenis, and J. Lee, *J. Med. Pharm. Chem.* **2,** 133 (1960).
143. A. Burger and W. L. Yost, *J. Am. Chem. Soc.* **70,** 2198 (1948); U.S. Pat. 2,997,422 (Jan. 9, 1959), R. E. Tedeschi (to Smith Kline & French Laboratories).
144. U.S. Pat. 3,381,000 (April 30, 1968), G. Palazzo and B. Silvestrini (to Aziende Chemishe Riunite Francesco Angelini).
145. G. Palazzo, *Curr. Ther. Res.* **15,** 745 (1973).

L. H. STERNBACH
W. D. HORST
Hoffmann-La Roche & Co., Inc.

PULP

Pulp is the raw material for the production of paper (qv), paperboard, fiberboard, and similar manufactured products. In purified form, it is a source of cellulose (qv) for rayon (qv), cellulose esters (qv), and other cellulose-derived products. Pulp is obtained from plant fiber and is, therefore, a renewable source (see also Chemurgy). Fibrous plants have been used as a source for writing materials, eg, papyrus, since the earliest Babylonian and Egyptian civilizations. The origin of papermaking, which is the formation of a cohesive sheet from the rebonding of separated fibers, has been attributed to Ts'ai-Lun in China in 105 AD, who used bamboo, mulberry bark, and rags. At the same time in Europe, parchment was used and, during the Middle Ages, the use of rags and rope supplanted the above material. The use of wood (qv) as a source of papermaking fiber was not commercially applied until the mid-1800s. The principal wood-pulping processes in use today, eg, the groundwood, soda, SO_2 or acid sulfite, and the sulfate or kraft processes were developed in 1844, 1853, 1866, and 1870, respectively. Since their development, the basic processes have been modified and adapted and the technology has been highly refined. However, the scientific base for this technology is considerably slower in its development, largely because of the physical and chemical heterogeneity of wood, and the complexity of its component polymers and their interactions.

As with most industries, the environmental and energy concerns of the 1970s effected large changes in the operation of pulp and paper mills as well as much research effort to develop the most energy-efficient and cleanest methods of production. In most cases, the practical result for the short term has been add-on methods, eg, scrubbers, precipitators, holding ponds, etc, which minimize the discharge of effluents. For the future, dramatic modifications of existing procedures or totally new technologies are being considered. Recent trends have been the increasing use of high yield pulps by modifying the groundwood process to improve pulp quality, the use of more of the tree in harvesting and chipping, and elimination or minimization of malodorous sulfur compounds in pulping and of the toxic and corrosive chlorine compounds from bleaching.

Wood is the original source of 99% of the pulp fiber produced in the United States. Although virtually any wood can be pulped by some process, there are certain species commonly used for pulp because of desirability of fiber, ease of pulping, availability, competition with other wood products, etc. The common pulpwoods in the United States are listed in Table 1.

Wood

In terms of abundance and suitability for pulping, there are two chief botanical classifications of trees: the softwoods or evergreens, which are gymnosperms, and the hardwoods or broad-leaved deciduous trees, which are dicotyledon angiosperms. The chemistry and anatomy of wood varies somewhat with the species of tree, but there are gross similarities within the two classifications. The softwoods, which are preferred for most pulp products because of their longer fibers, generally contain a higher percentage of lignin (26–32% on an extractives-free basis) and a lower percentage of hemicellulose (14–17%) than the hardwoods, which contain 17–26% lignin and 18–27% hemicellulose (see Lignin).

Anatomy and Morphology. A cross section of pine is shown in Figure 1 as a representation of the anatomy of softwoods. The main cell type is the axially aligned tracheid (TR). Although in botanical terminology tracheids are not considered to be true fibers, they are the papermaking fibers from softwoods and are referred to as fibers throughout this article as is common practice in the industry. Other cell types in softwoods are the ray cells, ie, the fusiform wood ray (FWR) and wood ray (WR) cells, and the longitudinal and epithelial parenchyma, which are the cells surrounding the horizontal and vertical resin ducts (HRD and VRD, respectively).

As a tree grows, the cells are produced in concentric lamella in the cambium layer, which is between the bark and the wood. In the spring, when moisture is plentiful and the tree is growing rapidly, the tracheid cell wall is thin (3–4 μm) and the hollow center or lumen is relatively large (26–43 μm). This portion is called springwood (Sp). During the summer or later in the growing season, the cell wall thickness increases to 8–12 μm and the outside diameter decreases from 47–29 μm in short-leaf pine. These cells

Table 1. Pulpwood Species by Main U.S. Pulp-Producing Regions

Region	Softwoods		Hardwoods	
	Dominant	Secondary	Dominant	Secondary
Northeast	spruce	hemlock	oak	aspen
	fir	tamarack	hickory	poplar
		white pine	maple	
South	yellow pines	cypress	oaks	
			gums	
Northwest	douglas fir	true firs	red alder	
	hemlock	spruce		
Lake States	jack pine	white pine	red oak	birch
	red pine	tamarack	aspen	
			maple	

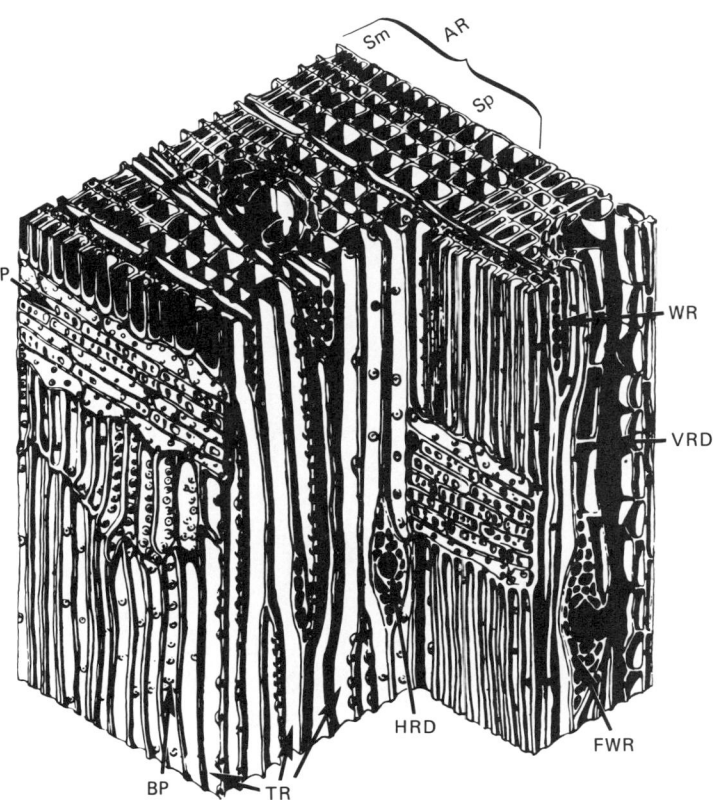

Figure 1. Schematic section of softwood. AR = annual ring, BP = bordered pits, FWR = fusiform wood ray, HRD = horizontal resin ducts, P = primary wall, Sm = summerwood, Sp = springwood, TR = tracheid, VRD = vertical resins ducts, WR = wood ray.

form the summerwood (Sm). The sequential combination of seasonal cell types leads to the characteristic annual ring (AR) of trees, which is more or less distinct in softwoods, depending on the species.

In the living tree, nutrients flow through the cells of the sapwood. The pattern of liquid conductance is important for the penetration of chemicals in the initial stage of chemical pulping. In softwoods, liquid is transferred from rays to tracheids and between tracheids through tiny voids or pits (P) in the cell wall. Usually, pits in adjacent cells are aligned so that a passage between the lumens of the two cells is formed that is blocked only by a thin pit membrane of intercellular substance. In bordered pits (BP), this membrane contains a thickened circular portion called a torus, which functions as a check valve to seal the passage against a return flow of liquid.

Figure 2 is a cross section of the structure of yellow poplar as a typical hardwood. Hardwoods have a more varied and more complex arrangement of cells than softwoods. The main structural element of hardwoods is the wood fiber (F), which is significantly shorter than the softwood tracheid (1–2 mm vs 3–6 mm), and generally is thinner, ie, ca 20 μm in dia. The true fibers are uniform throughout the annual ring. Hardwoods also contain a sizable proportion of short, large-diameter cells or vessels (V) through which sap is transported. Vessels may be larger in springwood, eg, oaks, hickories, etc, or uniform throughout the annual ring, eg, yellow poplar, aspen, etc. Vessels have open ends or a connecting gratelike tissue called a scalariform plate (SC). Hardwood fibers

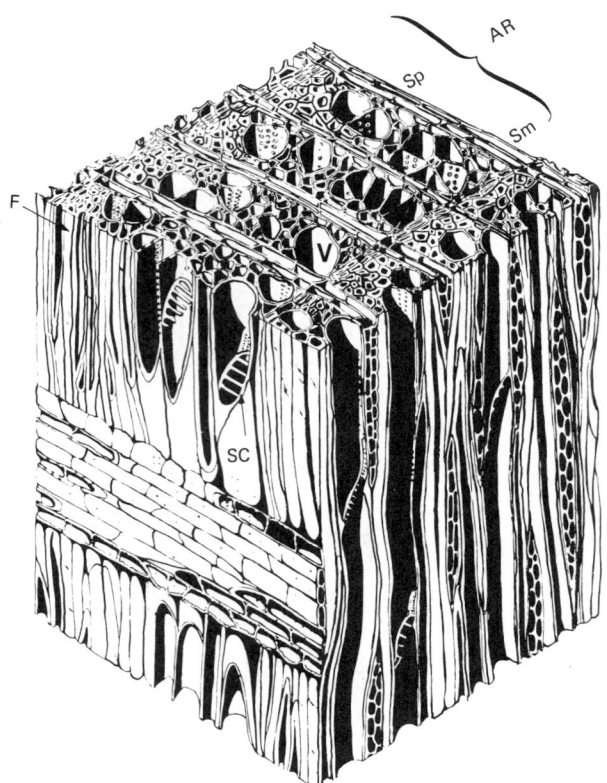

Figure 2. Schematic section of hardwood. F = wood fiber, SC = scalariform plate, Sm = summerwood, Sp = springwood, AR = annual ring.

have simple pits, which are smaller, and do not contain the torus system of the softwood bordered pits. Apparently, this is related to the fact that the vessels perform the primary liquid transport function in hardwoods. Thus, the cross-fiber liquid flow is greatly restricted in hardwoods in contrast with softwoods. In general, there is less differentiation in springwood (Sp) and summerwood (Sm) fibers than in those of softwoods. The AR is shown by the one or two layers of terminal parenchyma cells which form at the end of the growing season.

With maturation of softwoods and hardwoods, the parenchyma cells at the core die. This portion of the wood is called heartwood and often contains polyphenols, flavones, and other colored compounds that do not occur in the contrasting sapwood. This usually provides a clear, visual distinction between heartwood and sapwood, depending on the species. Heartwood compounds, eg, dihydroquercetin [taxifolin, 2-(3,4-dihydroxyphenyl)-2,3-dihydro-3,5,7-trihydroxy-4H-1-benzopyran-4-one] in douglas fir, may cause problems in pulping or bleaching.

Other distinct classes of wood in a tree include the portion formed in the first 10–12 yr of a tree's growth, ie, juvenile wood, and the reaction wood formed when a tree's growth is distorted by external forces. Juvenile fibers from softwoods are slightly shorter and the cell walls thinner than mature wood fibers. Reaction wood is of two types because the two classes of trees react differently to externally applied stresses. Tension wood forms in hardwoods and compression wood forms in softwoods. Compression wood forms on the side of the tree subjected to compression, eg, the underside of a leaning trunk or branch. Tension wood forms on the upper or tension side. In compression wood, the tracheid cell wall is thickened until the lumen essentially disappears; in tension wood, true fiber lumens are filled with a gel layer of hemicellulose. The chemical compositions of the reaction woods are also different. Lignin, which contributes significantly to the compressive strength of wood, is present to a greater extent in compression wood. Cellulose and hemicelluloses, which are largely responsible for tensile strength, are present in greater quantity in tension wood. Normally, these types of wood are a minor portion of the total amount being pulped, and their influence on the average pulp property is insignificant. However, in certain tree stands, such as short-rotation coppice or high elevation regions, the juvenile wood or the reaction wood becomes a significant portion of the total, and allowance must be made for the different pulping and fiber characteristics.

Chemical Composition. The basic structural element of the cell wall is cellulose. Lignin and hemicelluloses are also distributed throughout the cell wall in an incompletely understood manner. The intercellular substance, which is primarily lignin, must be softened or dissolved to free individual fibers.

Wood also contains 3–10% of extracellular, low molecular weight constituents, many of which can be extracted from the wood with neutral solvents and, therefore, are commonly called extractives. These include the food reserves, the fats and their esters in parenchyma cells, the terpenes and resin acids in epithelial cells and resin ducts, and phenolic materials in the heartwood. Resin materials occur in the vessels of some hardwood heartwood. Tannins usually occur in bark but may be in the wood of some species such as redwood and oak.

Many of these chemicals are recovered as by-products of the pulping operation, eg, tall oil (qv) and turpentine (see Terpenoids). Some of them cause problems in pulping or bleaching, eg, the heartwood phenols react with lignin during acid sulfite pulping more rapidly than the pulping chemicals do and inhibit the lignin solubilization reaction. Condensed lignins and phenolics are dark or form colored salts with metal ions. These compounds are not easily bleached and consume excess bleaching chemical. In alkaline pulping, the fatty acids and their glycerides form sodium soaps, thereby causing foam problems that reduce washing and evaporator efficiency (see Soap). Western red cedar contains a group of tropolones, eg, thujic acid (5,5-dimethyl-1,3,6-cycloheptatriene-1-carboxylic acid), which necessitates the use of special corrosion-resistant alloys in the digester. High rosin-containing species, eg, pine, cause problems in groundwood production because of the tacky nature of the rosin.

All trees contain inorganic minerals as nutrients which are essential for growth. Generally, these amount to less than 0.5% of the weight of the wood, and a larger amount occurs in the bark. The principal constituents are calcium, magnesium, and phosphorus. Heavy metals and many other elements are present in smaller amounts. The metals are present as carbonates, silicates, oxalates, and phosphates and usually occur in the intercellular region. They also may be associated with the carboxylic acids of lignin and carbohydrates in the cell wall (see Mineral nutrients).

Other Fiber Sources. A wide variety of plants can be used as a source of papermaking fibers. The only requirement is the ability of the fibers to bond to one another with sufficient strength so that a cohesive sheet is formed. However, there are several considerations that determine whether or not pulp from a particular plant source is suitable for the commercial production of paper or other fiber products. These include the characteristics of the fiber, supply, ease of storage, yield of desirable fibers, and wastes generated.

384 PULP

In countries where the wood supply is scarce, plants such as bamboo, rice, esparto, and sugarcane residues or bagasse are used to produce pulp. Because of the increasing demand for paper and other wood products, alternative fiber sources, including less desirable wood species, annual plants specifically for fiber use, tropical woods, and agricultural residues, are being sought. Approximately 22% of the pulp produced in the United States in 1978 was secondary, eg, recycled fibers from newspapers, used corrugated boxes, computer printouts, etc (see Recycling).

Pulp Fibers

A photomicrograph of douglas fir kraft pulp is shown in Figure 3. Even though most of the cell-wall lignin has been removed from the tracheids, the physical structure remains virtually unaffected. The long, narrow shape and tapered ends are characteristic of the structure in wood. The bordered pits are visible on the cell wall. When made into paper, the fiber is collapsed into a ribbonlike structure, as shown in the electron micrograph in Figure 4. With collapse, the surface area for bonding between fibers is increased and, very probably, there is new intrafiber bonding within the cell wall and between the collapsed lumen surfaces. The extent of collapsibility and potential bonding is largely determined by the pulping process. Mechanical pulps retain the stiffness and rigidity of wood, whereas chemical pulp fibers, from which nearly all of the lignin is removed, are very flexible and collapse easily.

A hardwood kraft pulp of white oak is shown in Figure 5. The fibers are finer and shorter than the softwood tracheid and many more cell types are evident. Very noticeable in a porous hardwood pulp are the short and very wide vessels. These elements are not long enough to bond across numerous fibers and, therefore, are easily lost from a paper sheet. In printing, this leads to picking, ie, the ink-coated vessel element ad-

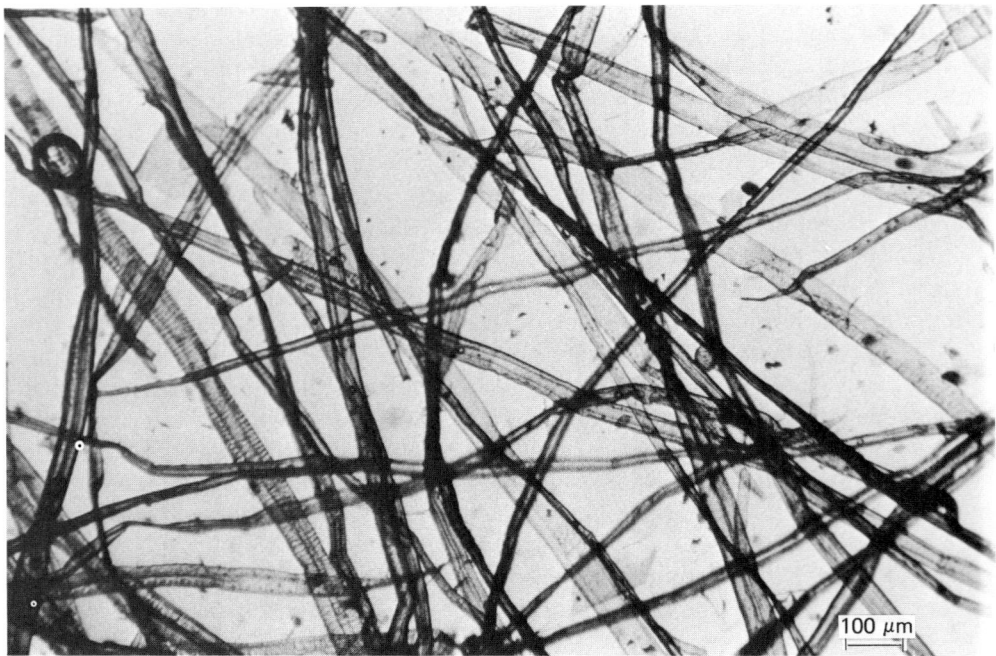

Figure 3. Photomicrograph of Douglas fir kraft pulp.

can have crystalline and fibrillar structure as well (2). The three-dimensional lignin polymer shows no tendency to crystallize. It is not known precisely how lignin and hemicelluloses are distributed throughout the cell wall. Although they cannot be seen, it has been demonstrated by gel permeation and absorption techniques that the cell wall is filled with microcapillary pores with openings of up to ca 2 nm in diameter in greenwood (3–4). These pore openings essentially close upon drying of the wood. The pore volumes increase in the presence of water and certain chemicals, eg, alkalies. During chemical pulping, the dissolved chemicals diffuse into the submicroscopic capillary water and react with the lignin polymer. Polymer fragments, which become solubilized, diffuse out of the cell wall if they are small enough. Frequently, and perhaps unavoidably, lignin removal is accompanied by removal of hemicelluloses.

Chemical Constituents of the Cell Wall. The variation in chemical composition across the cell wall is also shown in Figure 6. The principal constituents of cellulose, hemicellulose, and lignin are present throughout the cell wall, but in different proportions. Cellulose is not present in the interfiber middle lamella.

Cellulose. Cellulose (qv) is a straight chain β-1→4-linked D-glucan of high molecular weight, the value of which varies with the source and method of isolation but is ca 8000–10,000 $\overline{DP}_w$ (weight-average degree of polymerization) for native wood cellulose. The supramolecular structure of cellulose, ie, the alignment of the cellulose chains to form the cell-wall fibrils, has long been an enigma because various physical and chemical properties have been difficult to reconcile into a universally acceptable, coherent theory. X-ray diffraction patterns show a distinct crystallinity for various cellulose preparations. There are, however, at least four different polymorphic forms that are identified as cellulose I, II, III, or IV. Different preparations of cellulose or different lignocellulosic samples also have varying degrees of crystallinity. These phenomena are ultimately related to chemical accessibility and the physical properties of the pulp fiber.

Hemicelluloses. The hemicelluloses are lower molecular weight ($\overline{DP}_w$ = 100–200), mixed-sugar polysaccharides. Hemicelluloses are of various types and their exact nature varies with tree species and location in the tree. However, there are types that are common to hardwoods and to softwoods, as listed in Table 2. In some instances, the hemicelluloses are mixtures of closely related polymers, but the general structural characteristics of the most common hemicelluloses are given on page 388.

Hemicelluloses are largely responsible for hydration and development of bonding during beating of chemical pulps. Pulps containing high percentages of hemicellulose are typically weak in tensile strength but develop high bonding strengths. It is important to learn how to modify these properties if yields of chemical pulps are to be maximized.

Table 2. Composition of the Carbohydrate Fraction of Wood, wt %[a]

Component	Hardwoods	Softwoods
cellulose	40–45	39–42
O-acetyl-4-O-methylglucuronoxylan	20–30	
O-acetylgalactoglucomannan		16–20
glucomannan	2–3	
4-O-methylglucuronoarabinoxylan		8–11
pectic materials	1	1
arabinogalactan	trace	2–3
galactoglucomannan	trace	
starch	trace	trace

[a] Wt % except where noted.

Softwood Hemicelluloses: O-acetylgalactoglucomannan

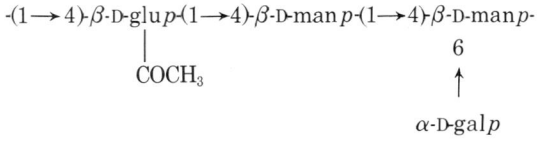

mannose/glucose/galactose/acetyl
3 : 1 : 0.15 : 0.24

arabino(4-O-methylglucurono)xylan

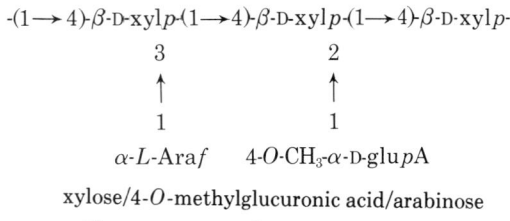

xylose/4-O-methylglucuronic acid/arabinose
10 : 2 : 1.3

Hardwood Hemicelluloses: O-Acetyl(4-O-methylglucurono)xylan

(1→4)-β-D-xyl$p$-(1→4)-β-D-xyl$p$-(1→4)-β-
2
↑
1
4-O-CH$_3$-β-D-glu$p$A

xylose/4-O-methylglucuronic acid/acetyl
10 : 1 : 7

Minor Constituents. Other hemicellulosic materials occur in extracellular regions of wood; in softwoods, such materials include a 1→3,1→6-linked arabinogalactan. The compound middle lamella, which is composed of the middle lamella and the primary cell wall, contains three hemicelluloses called the pectic-group substances. These consist of pectin, which is a polygalacturonic acid, a 1→5 α-L-linked arabinan, and an arabinogalactan with α-L-arabinofuranose units attached to the C-6 position of a β-D-1→4 linked galactan. Compression wood, which is highly lignified, also contains a high proportion of galactan. Larch is unusual among the softwoods in that it contains very significant amounts of the water-soluble 1→3,1→6 linked arabinogalactan.

Lignin. Lignin is a highly branched alkylaromatic thermoplastic polymer. It is incompletely characterized, but contains the structural elements and linkages as shown in Figure 4 in the article, Lignin (5). The experimental difficulties associated with the characterization of lignin as it exists in wood are extensive because of the variety of bond types and possible condensation or degradation during attempted isolation and purification from the polysaccharides with which lignin is intimately associated. In softwoods, the aromatic skeletal unit is the guaicyl type, as shown in Figure 7. Hardwood lignin has a higher methoxyl content, and units of the syringyl type have been isolated from degradative studies. The ratio of OCH$_3$ to C$_9$ units in hardwood lignins is ca 1.3 to 1.7 (see Lignin).

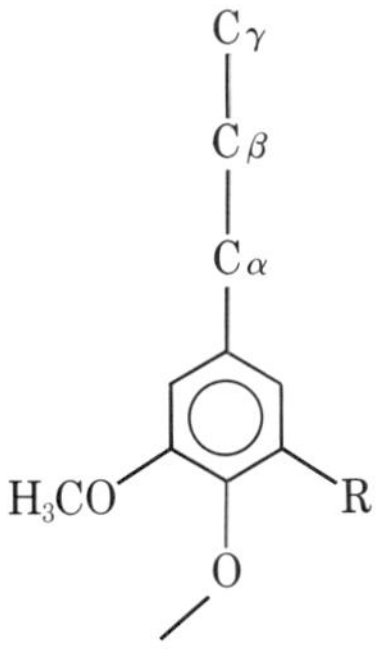

R = H:guaicyl, propyl

or

R = OCH$_3$:syringyl, propyl

Figure 7. Building units of wood lignins.

The aromatic nature of lignin contrasts with the aliphatic structure of the carbohydrates and permits the selective use of electrophilic substitution reactions, eg, chlorination or nitration. A portion of the phenolic hydroxyl units, which are estimated to comprise 30 wt % of softwood lignin, are unsubstituted, and in alkaline systems the ionized hydroxyl group is highly susceptible to oxidative reactions.

Reductive processes involving intermediate quinonemethides are important in alkaline pulping. Highly colored quinones and other conjugated chromophores are readily formed during alkaline pulping. These constituents impart the brown color to kraft pulps and necessitate subsequent bleaching operations. The benzylic alpha-carbon position is reactive and can be substituted by nucleophilic reagents. A common depolymerization site is the β-aryl ether link, the cleavage of which can be facilitated by neighboring-group participation of substituents on either the α- or γ-carbons.

The molecular weight of lignin in the wood, ie, of protolignin, is unknown. In addition to the difficulties of isolation and purification, the polymer exhibits strong solvent, ionic, and associative effects in solution, and an unequivocal method of measurement has not been developed.

The physical association of lignin with carbohydrates is very strong, even when purified samples of the separated polymers are remixed. Thus, the existence of covalent bonds between lignin and carbohydrates has been difficult to demonstrate. The accumulated evidence is strong that such bonds exist, but bonded compounds have not been isolated and characterized. Lignin–carbohydrate complexes can be obtained in a variety of ways. The usual starting material is vibratory ball-milled wood. Approximately 20% of the lignin can be selectively dissolved in solvents, eg, aqueous dioxane. Most carbohydrates can be removed from the lignin by enzymatic hydrolysis. The remaining complex contains sugars and lignins, which cannot be separated from one another and presumably are chemically bonded. The most likely linkages to lignin are to the side chain at the α- or β-carbons or glycosidic links to phenolic hydroxyls. With the carbohydrates, the evidence is fairly good that uronic acids form esters with lignin and that ether bonding occurs at the C-6 position of hemicellulose hexoses (6). There is also evidence that arabinose and xylose are bonded to softwood lignin (6). Approximately one phenylpropane unit in 30 in softwoods is calculated to be bonded to carbohydrates (7). If one assumes that there is no bonding to cellulose, there would be 3–6 bonds to lignin in each hemicellulose chain.

Wood Preparation

Harvesting. The wood-processing operations of harvesting, topping and delimbing, barking, and chipping are combined into as few individual steps as possible, and all of them can take place on site in the woods. There are several factors that influence how much mechanization or which combination of procedures are used in a particular operation. These include land ownership, tree size, terrain, climate, required chip quality, and other uses for the wood. In an optimized situation, such as a company-owned pine plantation on level ground in the southeastern United States where pulp chipping is the only objective, the entire operation can take place on site with trees being felled at 30 per minute and delimbing at over 75 m^3/h. Delimbing of over 200 m^3/h is achievable if poor quality can be tolerated. Unbarked trees can be chipped at 12 m in 30 s yielding 230–540 metric tons per day of whole tree chips, or trees can be barked at 24 m/min and chipped separately yielding 140–180 t/d of clean chips. A common procedure, where road access is limited, is to cut and stack the trees with a feller-buncher, eg, the Koehring feller-forwarder, or units produced by Drott, Forano, or OSA. The bunches of 30–40 trees are picked up by a separate vehicle and taken to the road. The trees can then be delimbed, barked, and chipped at a stationary roadside location with equipment such as a Forano multiple-tree delimbing unit and a Nicholson utilizer (Nicholson Manufacturing Co.). Delimbed trees can be delivered to the mill in random lengths or cut into 1.2-, 2.4-, or 5-m lengths in the woods or at the road.

In eastern and central Canada, much of the forested land is swampy; therefore, harvesting is done in the winter when the ground is frozen. On the west coast of North America, the large trees are used for plywood and sawtimber, and sawmill wastes provide most of the pulp-chip supply.

Where the woods are located in small individually owned farms or lots, or where the terrain is mountainous or otherwise difficult to reach, or where thinning as opposed to clear-cutting is the preferred silvicultural practice, advanced mechanization is not as suitable. Some novel approaches include the use of helicopters or helium balloons. Although the wood is usually supplied to the mills through a dealer, the harvesting operation can be done by the owner or by small crews, with the use of chain saws and tractors. It has been proposed that the early method of horse-drawn logs may still be the most suitable from the standpoint of energy requirements and minimal damage to the forest (8).

Barking. Although the bark of some tree species, eg, mulberry, contains bast fibers which may be used in papermaking, the outer bark of trees usually does not contain fibrous material and, therefore, is a contaminant in the pulping and papermaking process. Under some circumstances, varying amounts of bark can be tolerated, as with thin-barked species and where the final product, eg, fiberboard or corrugated medium, can accept a dirtier, lower quality pulp than can be used for fine papers. When the demand for fibers is great, whole-tree chipping is sometimes employed to increase the fiber yield. In this process, the entire tree, including bark, limbs, and leaves, is passed through the chipper. The higher fiber yield is bought at a price, because more extensive cleaning is necessary. The presence of bark necessitates higher chemical consumption in pulping and bleaching and a higher load on the recovery system. Bark has a higher ash content than wood, and more silicates from sand and dirt are entrapped, particularly if the tree has been dragged on the ground. The silicates cause severe erosion problems in processing and cleaning equipment, mud settling problems in the lime cycle, and evaporator scaling. Systems that upgrade whole-tree chips either on site or at the mill are being developed (9).

Logs are barked in the woods where humus and nutrients are returned to the soil, or they are barked at the mill where the waste bark is used as fuel. Several types of barkers are in use in pulp mills. Drum barkers are large open-ended cylinders that are rotated on a sloped axis; this enables logs entering at one end to move towards and out the other. The bark, which is broken off by rubbing and pummelling between the logs themselves and between the logs and the cylinders, drops out through slots in the cylinder. These units have a relatively high capacity. Inline barking units, which remove the bark from one log at a time, are either of the hydraulic or mechanical-friction type. With the former, the bark is removed by jets of water at pressure in excess of 6800 kPa (1000 psi). This system is used only with very large logs such as those available on the west coast of North America. With friction barkers, eg, the Cambio barker, special tools, which are pressed against the log, are rotated around it as it is passed through. This unit processes logs at a linear throughput of ca 45–55 m/min.

Chipping. The purpose of chipping for pulping is to reduce the wood to a size that allows penetration and diffusion of the processing chemicals without excessive cutting or damage to the fibers. The chips, which are ca 20 mm long, are fairly free-flowing and can be transported pneumatically or on belts and then stored in piles or bins. The cost of moving chips in and out of storage is normally less than that of logs; consequently, a chip pile has become a common method of storage for pulp mills.

In the Norman (Carthage Machine Co.) chipper, the cutting knives are mounted on the face of a large disk that is rotated on a horizontal axis. The log enters from a chute so arranged that the knives sequentially cut across the log at 37° to the fiber axis. The chips that form pass through slots in the disk. The knife severs the fibers as it enters the wood and, because of its wedgelike shape, places an increasing edge compression on the cut fibers. This pressure is relieved by shear along the grain. Thus, the thickness of the chip is related to the shearing forces at which the wood yields and is related to the length of the chip, which is normally specified to be 15–25 mm and is set by arrangement of the knives. In the Anglo drum chipper, logs of fixed length are fed from a magazine to a large rotating drum where the wafers are cut while passing through the drum. Another machine chips random-length logs and, thus, dispenses with the usual slicer. Drop-feed chippers can be modified to accept logs up to 5 m long. In the spiral H. P. chipper, logs are fed endwise and are sequentially engaged from both sides by a series of winged knives, which are mounted on two cones with their apexes facing each other and their common axis perpendicular to the log. The main knife edge peels, and the wing knife cuts across the fibers. This method results in chips of uniform length and thickness with less compression damage. The uniform thickness results in more even penetration of pulping liquors and, therefore, less screenings in the resultant pulp.

Screening. The chipped wood is screened to remove large knots and oversized chips, which are separately reduced in size by mechanical means, and to remove fines, which are burned or pulped separately. Conventionally, accepted chips are screened by length and width, but disk screens have been developed to give chips of uniform thickness.

Mechanical Pulps

Groundwood. Early production of groundwood pulp involved pressing wet wood against a wetted rotating grindstone, with the axis of the wood parallel to the axis of the wheel. Improvements have been made in the composition and speed of the grinding wheel, in methods of feeding the wood and pressing it against the stone, and in the size and capacity of the units. The original sandstone wheel has been replaced by mechanically stronger, synthetic composite stones produced from fine grits of silicone carbide or alumina embedded in a softer ceramic matrix; thus, the harder grit particles project from the surface of the wheel (see Abrasives). The synthetic stones have the mechanical strength to operate at peripheral surface speeds of ca 1200–1400 m/min under conditions that consume 0.37–3.7 MJ/s (500–5000 hp) per stone.

In the hydraulic-magazine grinder, the logs are fed from above and are pressed against the stone by a hydraulically operated pressure foot. The logs caught between the foot and the stone wear away and, when the foot reaches the end of its travel, it rapidly moves back, allowing more logs to drop down. These logs are then pressed against the stone, and the cycle is repeated. Two magazines are provided on either side of the stone, and pulp from the first magazine is removed by showers ahead of the second magazine. The period of interrupted pressure, ie, when the foot is reset, is ca 20 s. Grinders that eliminate this interruption have been developed.

The feeding of logs to the individual grinders has traditionally involved considerable labor. The Kone automated feeding line, developed in Finland requires one operator to handle a complete line of grinders. The logs are delivered by a series of belts and pushing devices and are available in bunches of appropriate size at each grinder; they are dropped to the magazine when required.

The quality of pulp depends on a number of variables. Grinder variables include peripheral stone speed, grit size and number per unit area, and stone surface. Wood variables include species and moisture content. Process variables include grinding pressure, pit consistency, and temperature. The combination of moisture and raised temperature tends to soften the lignin.

The grit diameter is ca 0.2 mm, ie, ca ten times the fiber diameter. The repeated compression–decompression action generates heat and softens the wood matrix before it reaches the grinding surface, where the loosened fibers are carried from the wood. Temperatures in the immediate grinding zone can be 180–190°C. The movement of the water and the removal of pulp controls and dissipates the heat, thus preventing charring of the wood. It is advantageous to allow the water temperature in the pit to rise to 70–90°C so as to achieve faster grinding rate, lower energy consumption, and higher pulp strength than are possible at lower temperatures.

Groundwood pulp contains a considerable proportion (70–80 wt %) of fiber bundles, broken fibers, and fines in addition to the individual fibers. The fibers are essentially wood with the original cell-wall lignin intact. They are, therefore, very stiff and bulky and do not collapse like the chemical-pulp fibers.

Since groundwood pulps are obtained in yields of ca 95%, their cost is relatively low. The main direct cost other than wood is power, which is ca 49–75 kJ (11.7–17.9 kcal)/t for normal paper grades. Somewhat less power is used in producing pulps for hardboard and low density wallboards, which contain coarser fiber than paper-grade pulps. The principal uses of paper-grade pulps are in newsprint, magazine papers including coated publication grades, board for folding and molded cartons, wallpapers, tissue, and similar products. The paper has high bulk, excellent opacity, but relatively low mechanical strength. The pulp can be partially bleached with peroxide or hydrosulfite without significant loss in yields. However, the paper, whether bleached or unbleached, has poor brightness stability, particularly in the presence of uv radiation, which makes it unsuitable for fine papers. Where light stability is significant, the more expensive bleached chemical pulps are normally used.

Modifications. There have been attempts to improve the pulp strength and decrease the energy requirements of the groundwood process by the addition of chemicals, eg, sodium carbonate or sodium sulfite. Although some benefits can be obtained, the additional costs and disposal problems have not been justified. Presteaming of wood is practiced in Europe to add moisture and to soften the lignin. The procedure is especially advantageous with some hardwoods. In the late 1970s and early 1980s, it was demonstrated that pressurizing the region surrounding a stone grinder to 100–300 kPa (1–3 atm) markedly improved the strength and fiber integrity of the pulp. This process retains the advantages of low energy consumption and high opacity (10). Figure 8 shows the modification of a Tampella-Great Northern type of pocket grinder to a pressurized system.

Refiner Mechanical Pulping. The conventional groundwood process requires bolts of roundwood as raw material. In the 1950s, the refiner mechanical pulping (RMP) process was developed, which produced a stronger pulp and utilized various supplies of wood chips, sawmill residuals, and sawdust. However, the energy requirement of RMP is higher, and the pulp does not have the opacity of stone-groundwood fibers.

The refiners are rotating-disk attrition mills. The disk plates have a construction of the type shown in Figure 9. The plates are paired face-to-face with a small interval between them. One disk rotates against a stationary disk or they both move in a counterrotating manner. The chips are fed through channels near the shaft in one of the disks and they move toward the periphery while undergoing attrition. The chips are first broken down into matchsticklike fragments by the action of the breaker bars, then into progressively smaller bundles as they move through the intermediate and

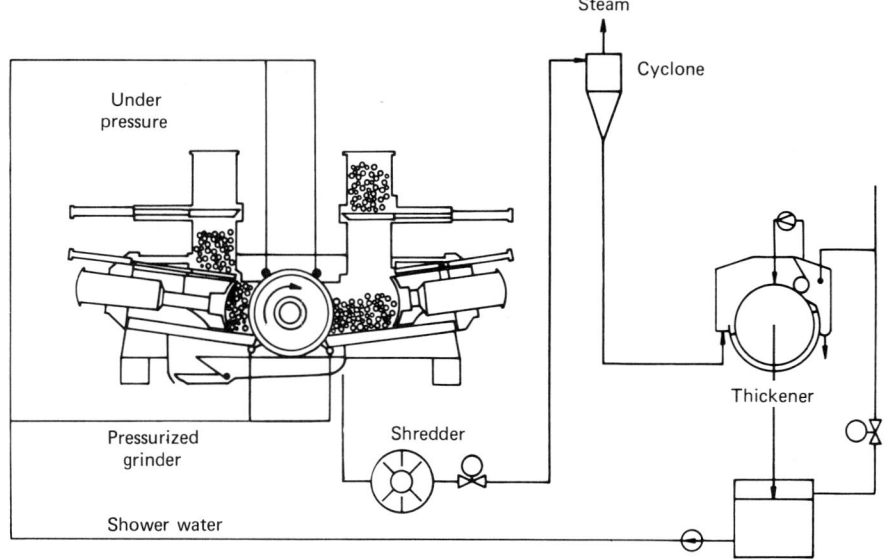

Figure 8. Modified Tampella-Great Northern grinder.

394 PULP

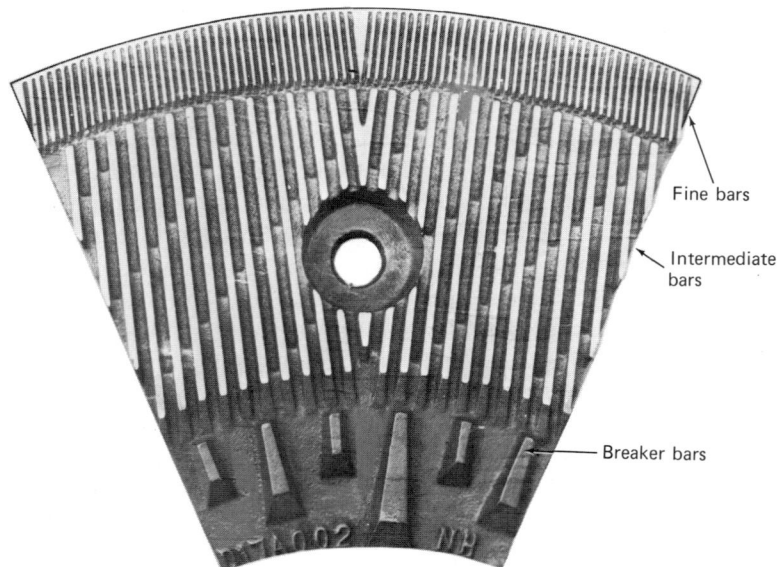

Figure 9. Refiner plate designs. Courtesy of Sprout-Waldron Division, Koppers Company, Inc.

fine-bar sections. They emerge from the periphery as single fibers or fiber fragments, including ribbons and fibrils that were formed by the unraveling of the spiral fiber walls of individual fibers. This process is termed fibrillation. These thin, flexible materials considerably improve the bonding properties of the mechanical pulps. Although it is possible to make refiner mechanical pulp in a single stage, normally two or three refining units are used in series.

The chips enter at a moisture content of ca 50 wt % and water is added at the eye of the disks to give a consistency or solids content of 20–30 wt %. At this consistency, considerable water evaporates by the heat generated, and the steam exits with the pulp, thus assisting the centrifugal movement. Some steam also vents out of the inlet, which helps soften the incoming chips. Untreated chips, however, are brittle, and considerable fiber fracture and debris formation occurs.

Thermomechanical Pulping. If chips are presteamed to 110–150°C, they become malleable and do not fracture readily under the impact of the refiner bars. This modification is called thermomechanical pulping (TMP). A thermoplasticization of the wood occurs when it is heated above the glass transition point of wet lignin. When these chips are fiberized in a refiner at high consistency, whole individual fibers are released; separation occurs at the middle lamella, and the same ribbonlike material described for RMP is produced from the S_1 layer of the cell wall. The amount of fibrillization depends on the refining conditions and is critical to the properties of the pulp. This material has a high light-scattering coefficient, although it is lower than that of groundwood, and is highly flexible, which gives good bonding and surface smoothness to the paper. The increased proportion of long fibers improves the tearing properties of TMP pulps, but the fibers in this fraction are stiff and contribute little to bonding. There is much less fiber fragmentation than in groundwood pulps or in those produced by RMP.

Two special conditions of thermochemical and refiner mechanical pulps must be recognized and treated or controlled by processing. The pulps tend to lint severely.

This problem can be minimized by increasing refiner energy or by pulp cleaning. Moreover, the pulps that are refined at high consistency become kinked and curled. When this occurs above the lignin glass-transition point and the pulp is subsequently cooled before the fibers have time to relax, the deformed state is retained. This condition, known as latency, is relieved by heating to above 80°C at low consistency.

The production of thermomechanical pulps increased dramatically after its introduction in the early 1970s because they could be substituted for conventional groundwood pulps in newsprint blends to give a stronger paper. Therefore, lower quantities of the more expensive, lower yield chemical pulps were required.

Chemithermomechanical Pulping. The strength properties of thermomechanical pulps can be increased further by a mild pretreatment with sodium sulfite at pH 9–10. The chips are impregnated with chemicals, steamed to 130–170°C, then refined. The yield is 90–92%, which is 2–3% lower than in TMP. A range of properties can be obtained by adjusting processing variables but, in general, chemithermomechanical pulping (CTMP) pulps have a greater long-fiber fraction and lower-fines fraction than a comparable thermomechanical pulp. The intact fibers are more flexible than TMP fibers and, consequently, better sheet-forming and bonding properties are obtained. The mild chemical treatment also allows the production of a long-fibered pulp with fast drainage, good absorption properties, and sufficient softness for tissue-grade pulp. Chemithermomechanical pulping is particularly suitable for pulping high density hardwoods.

Chemical Pulps

In chemical pulps, sufficient lignin is dissolved from the middle lamella to allow the fibers to separate with little, if any, mechanical action. However, a portion of the cell-wall lignin is retained in the fiber, and an attempt to remove this during digestion would result in excessive degradation of the pulp. For this reason, ca 3–4 wt % lignin is normally left in hardwood chemical pulps and 4–10 wt % lignin is left in softwood chemical pulps. The lignin is subsequently removed by bleaching in separate processing if completely delignified pulps are to be produced.

The concentration of the cooking liquor in contact with the wood influences the rate of delignification. Because of the time required for diffusion of the chemical through the wood structure and the depletion of the reagent concentration as it penetrates the chip, delignification proceeds more slowly at the center of the chip. This is particularly apparent in the case of oversize chips. In order to prevent overcooking of the principal portion of the pulp, digestion is normally halted before the centers of these larger chips are adequately delignified. The resultant pulp thus contains a portion of nondefibered wood fragments, which are separated by screening and returned to the digester or are fiberized mechanically.

Kraft. The dominant chemical wood-pulping process is the kraft or sulfate process. The alkaline pulping liquor or digesting solution contains about a 3-to-1 ratio of sodium hydroxide and sodium sulfide. The name kraft, which means strength in German, characterizes the stronger pulp produced when sodium sulfide is included in the pulping liquor, compared with the pulp obtained if sodium hydroxide alone is used, as in the original soda process. The alternative term, ie, the sulfate process, is derived from the use of sodium sulfate as a makeup chemical in the recovery process. Sodium sulfate is reduced to sodium sulfide in the recovery furnace by organic-derived carbon.

Chemistry of Delignification. As can be imagined from examining a partial structure of lignin, the chemistry of delignification is very complex. A variety of lignin model compounds have been studied. The results of the model studies are compared with the observed behavior of lignin during pulping. Although there is an extensive literature in this area, the chemistry of delignification is not well understood (11).

In the presence of alkali, the acidic phenolic units in lignin are ionized. Above 120°C, quinonemethides form from the phenolic units, as shown in equation 1.

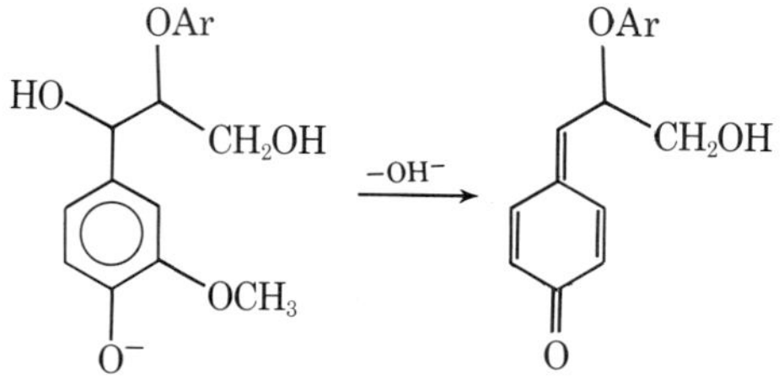

$$\text{[Structure with HO, OAr, CH}_2\text{OH, OCH}_3\text{, O}^-] \xrightarrow{-OH^-} \text{[Quinonemethide structure with OAr, CH}_2\text{OH, OCH}_3\text{, =O]} \quad (1)$$

In the presence of sulfide or sulfhydryl anions, the quinonemethide is attacked with the formation of a benzyl thiol. The β-aryl ether linkage to the next phenylpropane unit is broken by neighboring-group attack by the sulfur with elimination of the aryloxy group as a newly reactive phenolate ion (see eq. 2). If sulfide is not present, a principal reaction is formation of the very stable aryl enol ether, ArCH=CHOAr. A smaller amount of this product also forms in the presence of sulfhydryl anion.

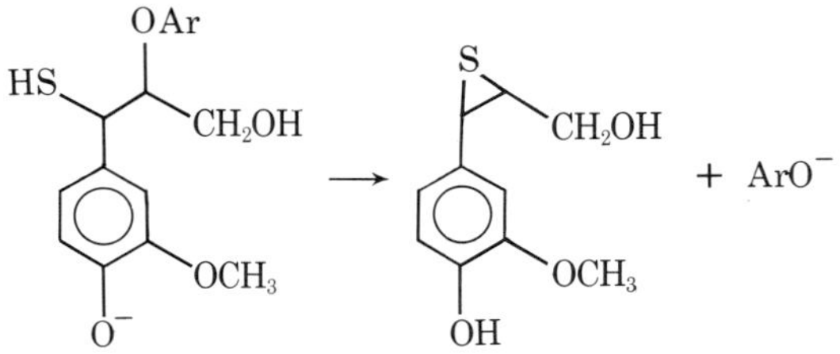

$$\text{[HS, OAr, CH}_2\text{OH, OCH}_3\text{, O}^-] \rightarrow \text{[episulfide structure with CH}_2\text{OH, OCH}_3\text{, OH]} + \text{ArO}^- \quad (2)$$

Some ether cleavage does take place, even if the phenolic hydroxyl is blocked as an ether link to another phenylpropane unit and quinonemethide formation is prevented. If the α- or γ-carbon hydroxyl is free, alkali-catalyzed neighboring-group attack can take place with epoxide formation and β-aryloxide elimination. In other reactions, blocked phenolic units are degraded if an α-carbonyl group is present.

Under acid or alkaline catalysis, condensation reactions take place homolytically within the lignin polymer and very likely between lignin and carbohydrates or extraneous components. These reactions are undesirable in delignification, and prevention of such condensation accelerates pulping. Formaldehyde is generated in the formation of the arylenol ether, as shown in equation 3, and it also causes condensation and cross-linking of phenylpropane units.

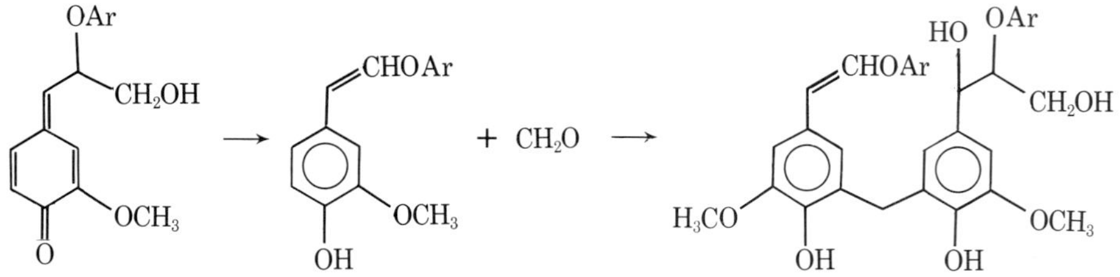

$$\text{[Quinonemethide with OAr, CH}_2\text{OH, OCH}_3\text{, =O]} \rightarrow \text{[aryl enol ether =CHOAr, OCH}_3\text{, OH]} + CH_2O \rightarrow \text{[cross-linked bis-phenylpropane structure]} \quad (3)$$

The very reactive quinonemethide intermediates are susceptible to attack by any carbanion. In kraft pulping, this type of condensation is minimized by the effective competition for the quinonemethide by the sulfhydryl anion.

Reactions of Polysaccharides. Although the goal of chemical pulping is to degrade the lignin polymer selectively, reactions with the polysaccharides are unavoidable and not always undesirable. Arabinogalactan is soluble in hot water and is therefore rapidly lost in any chemical pulping operation. In the alkaline kraft process, there is very little reaction of carbohydrates with the nucleophilic sulfide or sulfhydryl ions, as in the case of delignification. However, most of the alkali consumption results from reaction with carbohydrates and resinous materials. The acetyl groups on glucomannan in softwoods and xylan in hardwoods are easily saponified by alkali, and the uronic acid ester linkages to lignin are apparently cleaved at room temperature. Some hemicelluloses are extracted from wood by aqueous alkaline solutions. In addition, there are types of alkaline degradative reactions of polysaccharides at elevated temperatures.

The reducing-end units are very labile in alkaline solutions. After an initial attack by hydroxide ions at the hemiacetal function of C-1, a series of enolizations and rearrangements leads to deoxy acids, ie, saccharinic acids, and fragmentations. Substituents on one or more hydroxyl groups influence the direction, rate, and products of reaction.

The mechanism of the initial reaction is as shown:

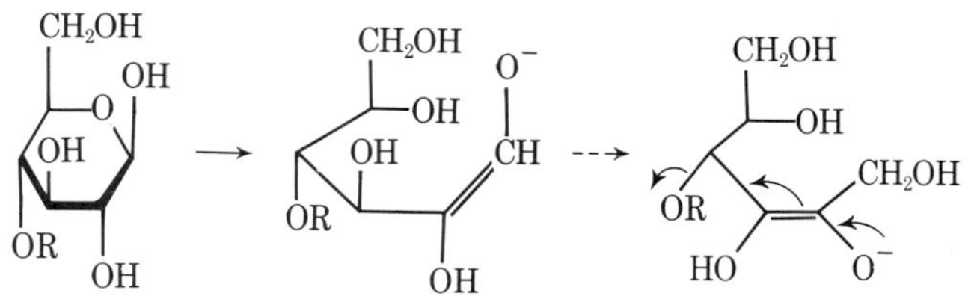

$$ (4) $$

Enolizations and tautomerizations take place easily because of the contiguous hydroxyl groups. The hydroxyl or substituted hydroxyl on the second, ie, β, carbon from a carbonyl group is released from the molecule by β elimination.

Glycosidic substituents are released more rapidly than are free hydroxyls. Thus, the remainder of the chain in a 1→3- or 1→4-linked polysaccharide is eliminated from the reducing-end unit undergoing this reaction. A new reducing end is created on the chain and the reactions may be repeated. This leads to a phenomenon termed peeling, in which polysaccharide chains are progressively shortened by the sequential reaction and loss of the reducing-end units. If there are no interfering structures on the chain, such as C-2 branching or substitution, peeling continues until the chain dissolves and is destroyed or until a hydroxyl group is eliminated in a competing reaction and an alkali-stable saccharinic acid forms by a benzylic acid rearrangement (see eq. 5).

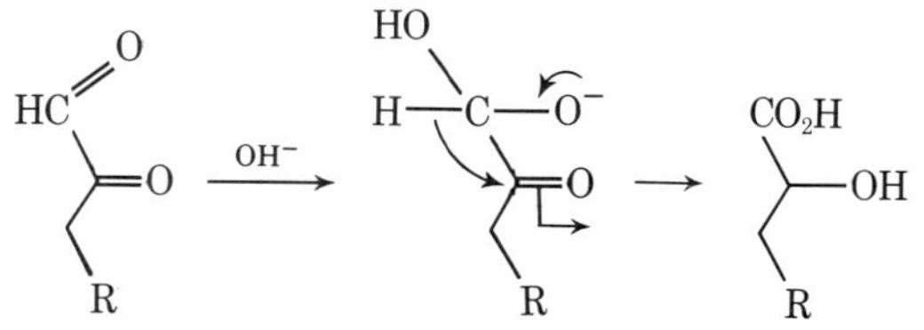

$$ (5) $$

The 4-O-methylglucuronoarabinoxylans are substituted at C-2 in the xylan backbone by the uronic acid group. This inhibits the tautomeric formation of a keto group at that position and, therefore, inhibits peeling. If the xylan unit is not substituted at C-2, but does carry an arabinose substituent at C-3, the arabinose is eliminated with formation of a xylometasaccharinic acid end unit, and peeling is inhibited. Galactoglucomannan, however, includes unsubstituted C-2 and C-3 positions and is highly susceptible to the alkaline peeling sequence. This is reflected in the much higher relative loss of galactoglucomannan during kraft pulping, as shown in Table 3. Cellulose is unsubstituted at C-2 and -3, but its higher molecular weight, crystallinity, and inaccessibility minimize the extent of degradation.

Polysaccharides are also susceptible to alkali-catalyzed glycosidic hydrolysis. Although much slower than acid-catalyzed hydrolysis, the rate of alkaline cleavage becomes significant at 170°C, which is the commonly used temperature in kraft pulping. When a chain is cleaved, a new reducing end group is formed and the peeling sequence may be reinstituted in a previously stabilized chain. Approximately 65 glucose units are degraded by peeling with each hydrolytic cleavage (12).

Diffusion and Topochemistry. Topochemical and diffusion processes are important in all chemical and semichemical pulping operations because of the solid nature of the wood components and the liquid or gaseous state of the pulping reagents. The initial liquid transport through wood chips is through the bordered pits of softwoods and the vessels of hardwoods. Liquid penetration of sapwood is higher than that of heartwood, especially in the case of some hardwood species where the heartwood vessels are plugged with tyloses. Penetration is also affected by the density of the wood. In general, penetration into softwoods is faster than into hardwoods, especially in the cross-fiber direction. With the strongly alkaline cooking liquors of the kraft process, diffusion takes place at nearly equal rates in all directions.

To make pulping as homogeneous as possible, an effort is made to achieve uniform penetration before pulping temperatures are reached. This may be done by presteaming chips at or above atmospheric pressure, applying hydrostatic pressure to chips submerged in pulping liquors, and allowing sufficient time during heating to process temperature.

When the chips are saturated with pulping liquors, the fiber lumens are filled with liquid. However, for fiber separation, the lignin in the intercellular middle lamella must be dissolved. Pulping is accomplished by transport of reagents through the submicroscopic pores of the fiber wall, contact and reaction with lignin in the cell wall and middle lamella, and transport of dissolved fragments back to the bulk solution in the lumen and that surrounding the chip. In the digester, the chips and liquor are kept in constant motion relative to one another to facilitate this transport. The exact topochemical sequence depends on the process (13). In kraft pulping, very little

Table 3. Yields of Carbohydrate Component from Loblolly Pine

Component	Wood	Chlorite holocellulose	Kraft
4-O-methylglucuronoarabinoxylan, wt %	11	5.8	3.9
galactoglucomannan, wt %	17	15.4	4.4
cellulose, wt %	41	40	36.8

middle-lamella lignin is lost until about half of the cell-wall lignin has been removed; then, middle-lamella delignification is very rapid. At the end of the cooking phase, the residual lignin is almost entirely in the cell wall. In the sulfite processes, particularly those involving neutral sulfite, the dissolution of lignin begins in the cell wall, but middle-lamella lignin removal begins earlier than in the kraft process. There is much less difference in rate of delignification in the two regions.

Pulping. Solutions of sodium sulfide and sodium hydroxide are in equilibrium: $H_2O + Na_2S \rightleftarrows NaHS + NaOH$. Aqueous sodium sulfide is therefore a source of hydroxide ions and this must be considered in adjusting the chemical charge. A system has been developed in the North American industry to put sodium hydroxide and sodium sulfide on an equivalent basis by expressing them both as their equivalent weight to sodium oxide, Na_2O. The percent of sodium sulfide in the mixture, when both Na_2S and NaOH are expressed as Na_2O, is known as the sulfidity. The chemicals are charged as a percent of the wood on an oven-dry (o.d.) basis, and the sum of Na_2S and NaOH as their Na_2O equivalent is called the active alkali. The actual concentration of pulping chemicals relative to lignin is determined by the liquor-to-wood ratio (L/W) as the total liquid volume, which includes moisture originally in the chips, divided by the dry wood weight. Since one-half a formula weight of Na_2S is equivalent to a formula weight of NaOH, the effective alkali is the sum of sodium hydroxide and one-half the sodium sulfide; both are expressed as percent of sodium oxide equivalent based on the wood. In practice, with recycled chemicals ca 15% of the sodium is present as sodium carbonate from incomplete causticizing, and other sodium sulfur compounds are present in small amounts as thiosulfate, sulfate, and sulfite.

The chemical charge, liquor composition, time of heat-up, and time and temperature of reaction are functions of the wood species or species mix being digested and the intended use of the pulp. A typical set of conditions for southern pine chips in the production of bleachable-grade pulp for fine papers is active alkali, 18%; sulfidity, 25%; liquor-to-wood ratio, 4/1; 90 min to 170°C; and 90 min at 170°C. Hardwoods require less vigorous conditions primarily because of the lower initial lignin content.

Both batch and continuous digesting systems are in operation. Continuous digesters usually are installed in new mills, but they have not completely replaced the batch systems, partly because of the high capital cost of replacement, and also because the batch digestion method has advantages, eg, flexibility to process and product changes, lower maintenance costs, and higher turpentine yields. The advantages of continuous digesters in addition to uninterrupted process flow include higher pulp yields and heat recovery, relative ease of automation, and in-line processing, eg, partial washing and blowline refining.

A schematic of the widely used Kamyr continuous kraft pulping system is shown in Figure 10. The chips are first continuously steamed at low pressure and the turpentine and gases are vented to the condenser. The chips are then brought to digester pressure of ca 1000 kPa (150 psi). They are picked up in a stream of recycled liquor to which white liquor, ie, fresh pulping solution, makeup has been added. This stream carries through to the top of the digester, where the recycling feed liquor is extracted from the chips. The chips with the balance of the liquor flow continuously down through the digester, their temperature being raised to ca 170°C in a top heating zone; this step requires ca 1.5 h. They are then held at 170°C for 1.5 h while passing through a second zone. At this point, the digestion is essentially complete; they next pass

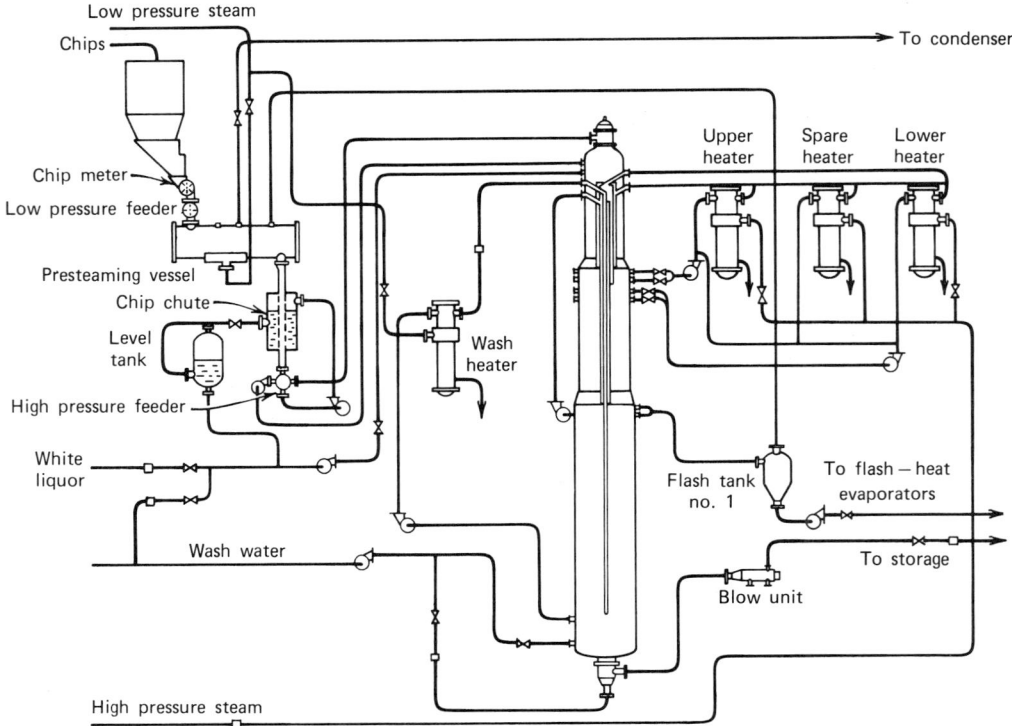

Figure 10. Kamyr continuous kraft pulping system.

continuously through a countercurrent washing and cooling zone; the black liquor or spent pulping solution leaves the digester from the top of this zone while the digested and partially washed chips leave the bottom of the digester through a blow valve to the blow tank. The force of the release is sufficient to fiberize the cooked chips. Cooling the chips prior to fiberization is necessary to assure adequate pulp properties.

Process Control. The rate of delignification depends on the concentrations of hydroxide and hydrosulfide ions as well as the lignin content of the wood. However, if values of lignin remaining in the wood are plotted as kinetically first order, they are described by three straight lines, as shown in Figure 11. The first and most rapid rate is initial delignification. During this phase, there is a rapid consumption of alkali by easily soluble carbohydrates and reactive functional groups, eg, esters. The initially removed lignin is probably associated with the solubilized fraction of hemicellulose. In commercial practice, this phase is usually completed during the extended heat-up period. The second and slower rate is the bulk delignification. In this phase, the decrease in effective alkali is much slower. Most of the lignin is removed during this time at the final pulping temperature. The amount of lignin removed relative to carbohydrates is greater in this phase than in initial delignification. Some readsorption of deesterified polysaccharides does occur. The lignin that remains, ie, residual lignin, is more difficult to remove than either initial or bulk-phase lignin. Pulping is usually terminated as this stage is approached. If desired, the residual lignin is removed in a bleaching operation where more selective, but more expensive, chemicals are employed.

For control purposes, the delignification rate is treated as a homogeneous reaction,

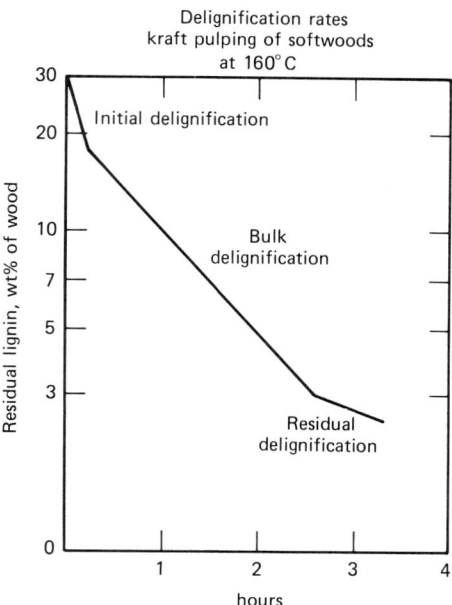

Figure 11. Rate of kraft pulping.

which is first order with respect to the lignin remaining in the wood. The influence of time and temperature has been incorporated into one term, called the H factor, by utilizing an Arrhenius-type of equation relating the change in the first-order rate constant to changes in absolute temperature. An H factor value of unity is assigned to the effect of pulping for 1 h at 100°C. Relative rates at other temperatures can then be calculated, and allowance can be made for the effect of varying heat-up times and maximum temperatures. For example, the kraft cooking of softwoods to a bleachable-grade pulp requires an H factor of ca 2000. The H factor does not take into account the effect of chemical concentrations on pulping rates, and modifications that include additional factors have been proposed. These modifications are designed to be programmed into computer control systems (see Instrumentation and control).

Computers (qv) are used to continuously perform calculations and adjust or recommend adjustments of process parameters. Developing analytical methods suitable for automation and, preferably, continuous monitoring has been a significant effort of the 1970s. The main control problem in pulping is the variability of the raw material, including the moisture content of the wood, the species mix, and the amount of bark, decay, knots, dirt, and other extraneous material.

A direct determination of the lignin remaining in the wood or pulp during the cooking phase requires sampling and analysis of large heterogeneous particles. This cannot be automated directly, but continuous monitoring of moisture, weight, and density as well as inline determination of the effective alkali can be accomplished. This information can be combined with computed H factors to give reproducible, automated control of both batch and continuous digesters.

Recovery System. Kraft pulping depends on its associated recovery process for producing the digestion liquor. This recovery is illustrated schematically in Figure 12. The chemical recovery cycle involves the separation of the black liquor from the pulp, its evaporation, and its combustion for the recovery of the inorganic chemicals as a smelt, which is subsequently converted to the white liquor for the next digestion.

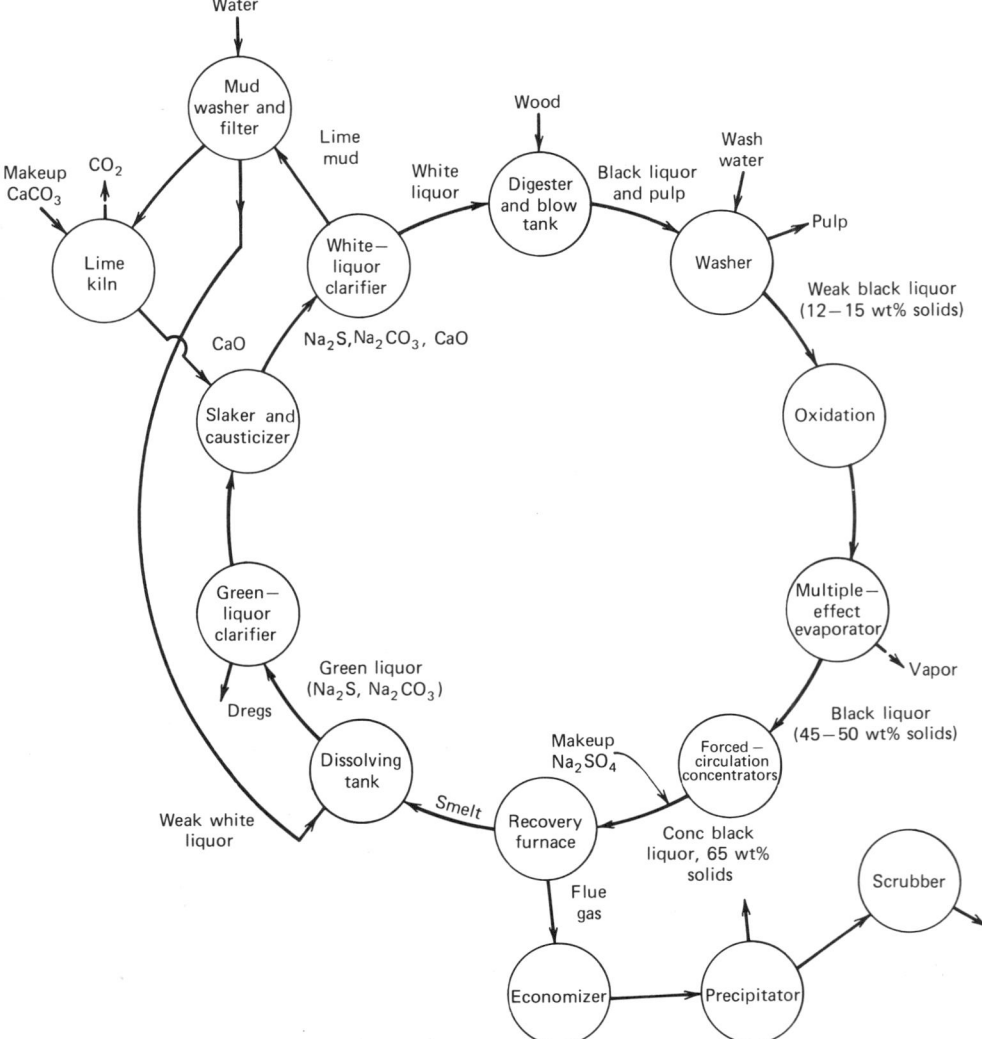

Figure 12. Schematic of kraft recovery system.

The recovery system, together with its heat and material balance, has been described in some detail in ref. 14.

Much of the equipment used in the recovery system is identical with, or closely related to, similar equipment used in other chemical industries. This includes the rotary filters used for brown-stock washing, the multiple-effect evaporators and forced-circulation concentrators, the causticizing equipment, and the lime kiln. The following discussion, therefore, concerns the function and nature of equipment essentially unique to the kraft recovery system.

Black-Liquor Processing. The black liquor is normally separated from the pulp by continuous diffusion washing, and two or three countercurrent stages are used. After this separation, it is subjected to evaporation, the last stage of which can be by contact with the recovery-furnace flue gas. Under these conditions, residual sodium hydrosulfide partially hydrolyzes to hydrogen sulfide and sodium hydroxide, whereas sodium methanethiolate, CH_3SNa, partially hydrolyzes to methyl mercaptan and sodium hydroxide. In order to prevent the escape of the volatile hydrogen sulfide and methyl mercaptan with the flue gas, the black liquor is oxidized with air or oxygen, which gives sodium thiosulfate and dimethyl disulfide, respectively, according to the following reactions:

$$2\,NaHS + 2\,O_2 \rightarrow Na_2S_2O_3 + H_2O \qquad (6)$$

$$2\,CH_3SNa + \tfrac{1}{2}\,O_2 + H_2O \rightarrow CH_3SSCH_3 + NaOH \qquad (7)$$

Black-liquor oxidation is normally carried out by passing a stream of liquor and a stream of air or oxygen to a tower fitted with perforated plates or with suitable packing to allow repeated, intimate contact of liquor and gas. Because of the presence of fatty acid sodium salts, ie, soaps, foaming may occur, particularly when unseasoned pine is used, and a number of designs have been used to rectify this problem (see also Defoamers). Less foaming is obtained at a higher solids concentration, and this can be obtained by recycling a portion of the evaporated liquor through the oxidation tower. Black-liquor oxidation must achieve 99% efficiency to meet emission standards where direct-contact evaporation is employed. The pH of the black liquor must also be kept high to minimize the hydrolysis to volatile compounds.

As the solids concentration of black liquor is increased as a result of evaporation, the liquor becomes increasingly viscous. For this reason, the evaporation is conducted in two separate operations. Multiple-effect evaporation is used to bring the liquor to a concentration of ca 45–50 wt % solids and forced-circulation evaporation or additional effects are used to bring it to ca 65–70 wt % solids, ie, the concentration at which it is fired in the recovery furnace (see Evaporation).

Evaporation of the black liquor by direct contact with the hot recovery-furnace flue gases was a common procedure and involved either cascade evaporators or venturi scrubbers. However, this method was a principal source of air pollution and, although various schemes were proposed to minimize the emissions problems, these methods of liquor concentration have generally been eliminated (see also Air pollution control methods).

Recovery Furnace. The function of the recovery furnace is to burn the organic matter, recover the chemical in the form of sodium carbonate and sodium sulfide, and generate steam through utilization of the heat that is liberated. The steam is generated at ca 1360 kPa (200 psi) for direct use as process steam or as superheated steam at up to 8160 kPa (1200 psi) for driving a turbine for electric-power generation; the turbine exhaust steam is used as process steam.

In the recovery furnace and after addition of makeup sodium sulfate, the evaporated liquor is heated to ca 120°C and is fired through one or more special oscillating nozzles or spray oscillators in the front of the furnace. These spray oscillators throw the droplets to the far and side walls. The remaining water quickly evaporates. The dried liquor increases to a certain depth on the wall and then drops off from its own weight in chunks and falls towards the hearth, which is covered with char. Primary air is admitted through a series of ports; thus, the depth of the charred mass is controlled by burning. Reducing conditions are maintained in this zone so that the sodium- and sulfur-containing compounds are largely reduced to sodium sulfide, as indicated by the following reaction:

$$Na_2SO_4 + 2\,C \rightarrow Na_2S + 2\,CO_2 \qquad (8)$$

The inorganic residue melts at ca 990–1020°C, depending on its exact composition, and runs continuously from the furnace through smelt spouts at the lower end of a sloping hearth. Secondary air is admitted at a higher level to limit the height of the bed and also to effect gas-phase reactions, eg, conversion of CO to CO_2 and H_2S to SO_2. Tertiary air is added at an even higher point and at relatively high velocity to ensure mixing of the gases and to complete their oxidation. A small excess of oxygen is required to ensure complete combustion of the gases; a large excess is avoided because it favors conversion of SO_2 to SO_3. The furnace temperature is ca 1250°C. Approximately 10% of the sodium salts is volatilized in the form of Na_2O, which quickly reacts with CO_2 to form Na_2CO_3; the latter reacts in the upper portion of the furnace with SO_3 and SO_2 to form Na_2SO_3 and Na_2SO_4. The walls of the furnace are water-cooled by an arrangement of fintubes connected to the water circulation of the boiler. The unit is designed so that the combustion gases are cooled to a temperature below the melting point of the inorganic constituents during the upward passage of the gases through the furnace. When the salts condense on the closely spaced boiler tubes at the top and back of the furnace, they are present as a dustlike solid which can be blown off, rather than as a slag which would block the gas passage. Automatically timed soot blowers, which impinge jets of steam or compressed air on the tubes, traverse the boiler banks, thereby cleaning them. The released ash falls either towards the hearth or to a special hopper, where it is collected and added to the black liquor for recycle to the furnace.

Proper control of the reactions in the furnace is important. Reducing conditions are necessary at the base of the furnace if the smelt is to contain sodium sulfide rather than the inert sodium sulfate. Slightly oxidizing conditions are needed above the bed, which causes the conversion of volatile sulfur compounds to sulfur dioxide; the latter not only reacts with the fume but is absorbed in the direct-contact evaporator, if one is used. Strongly oxidizing conditions in the upper portion of the furnace are undesirable, since sulfur trioxide forms in large amounts. Sulfur trioxide reacts with the sodium sulfate fume to form sodium pyrosulfate, which melts at 400°C and forms a glasslike, corrosive covering on the boiler tubes.

$$Na_2SO_4 + SO_3 \rightarrow Na_2S_2O_7 \qquad (9)$$

Accidental introduction of water into the smelt can result in serious explosions. Furnace explosions occur by mistake when weak rather than strong black liquor is fired in the furnace and reaches the pool of smelt on the hearth. Explosions also occur when one or more of the tubes in the wall rupture, thus allowing water to flow into the unit. This can occur if the furnace is relighted after a blackout that has resulted from an interruption of fuel or air and if the furnace is not first purged of the explosive gas mixture which may be present. If the primary explosion ruptures a tube, a large secondary explosion results.

The industry has considerable experience in the safe operation of recovery furnaces, and various safeguards are employed, eg, computer control. However, the explosion danger persists, and various proposals have been made to eliminate the smelt bed, improve the heat recovery, and minimize emissions. These proposals include the Swedish NSP furnace, St. Regis' hydropyrolysis system, Weyerhaeuser's dry pyrolysis, and adaptation of the SCA-Billerud NSSC recovery process to kraft black liquors.

Preparation of White Liquor. The smelt is continuously run from the furnace to the dissolving tank. The liquor is strongly agitated to dissipate the heat. The temperature in the tank is readily controlled by dispersion of the hot smelt with a steam jet before the smelt is added to the water, which prevents the explosive mixture that results from addition of water to a pool of hot smelt. Weak white liquor obtained from washing the lime mud in the causticizing plant is used for dissolving the smelt.

The green liquor, ie, the solution obtained on dissolving the smelt, contains an insoluble residue called dregs, which gives it a dark green appearance. The dregs contain a small amount of carbon plus a number of inorganic constituents, including iron sulfide, manganese dioxide, calcium carbonate, magnesium aluminum silicate, etc. These constituents originate in the wood and the process water or result from corrosion of equipment. The liquor is separated from the dregs by continuous decantation, the separated dregs are mixed with water, and the decanted solution is used for washing the lime mud at a subsequent stage.

The clarified green liquor is used for slaking the lime. For one to two hours, the mixture is passed through a series of tanks equipped with agitators. The causticizing reaction is as follows:

$$Na_2CO_3 + Ca(OH)_2 \rightarrow 2\ NaOH + CaCO_3 \qquad (10)$$

The white liquor is separated from the calcium carbonate by decantation in a clarifier and then is available for a new cycle. The underflow from the clarifier, which contains the calcium carbonate and is referred to as lime mud, is diluted with water and is passed to a second clarifier known as the lime-mud washer. The clarified weak white liquor then enters the dissolving tank, and the residue is passed to a rotary filter and then to the lime kiln where calcium carbonate is converted back to calcium oxide, thus completing the lime cycle.

Uses. The kraft process provides wide adaptability to the pulping of virtually any wood species and yields pulps that are suitable for a broad range of products. Packaging products have stringent strength requirements, and the use of kraft pulps in linerboard has enabled the replacement of wooden cases by corrugated cartons. Similarly, bags and multiwalled sacks of kraft pulp have replaced those made of cotton or jute. After partial or complete bleaching, kraft pulps are often blended with less expensive but weaker pulps, eg, groundwood, to add enough strength to meet product requirements as, for example, in newsprint, which must have sufficient strength to be run on modern, high speed printing presses. The brown color of kraft pulps is undesirable for applications such as writing and printing papers, but they can be bleached to remove the residual lignin and other chromophores. The bleached white product does not darken on aging.

By-products. There are three stages within the pulping operation at which wood-derived chemicals can be recovered as by-products. Turpentine is obtained from the relief gases after an initial steaming of chips in the digester. Better yields of turpentine are obtained from batch digesters than from continuous systems, and pines and firs are the best species. Turpentine is composed principally of unsaturated bicyclic hydrocarbons of which ca 90% are α- and β-pinenes and 5–12% other terpenes.

In the initial black-liquor concentration, saponified fatty and resin acid salts separate as tall oil soaps (see Tall oil). These soaps can be skimmed from the aqueous spent liquor, acidified, and refined to give a crude tall oil composed of resin acids, chiefly abietic and neoabietic; fatty acids, chiefly oleic and linoleic; and an unsaponifiable fraction composed of phytosterols, alcohols, and hydrocarbons. Tall oil is fractionated primarily into fatty acids (see Carboxylic acids).

The final source of by-products is the spent liquor. Possible utilization of kraft black liquors as a source of chemicals has been evaluated, but the mixture is too complex and the market value of the individual components too low to warrant the extensive separation and purification procedures that would be required. The organic constituents are mainly the various fragments and degradation products of lignin and of the carbohydrates, eg, saccharinic acids and their low molecular weight analogues such as lactic acid. The dissolved alkali lignin, which is still polymeric, can be separated from black liquor by acidification and can be further purified or chemically modified to give useful materials. The production from only a few mills is sufficient to meet the present world demand.

Another product from kraft black liquor is dimethyl sulfide. This chemical, which is added to natural gas to give it odor, is also oxidized to produce the versatile solvent, dimethyl sulfoxide (see Sulfoxides).

Modifications. Although the kraft process is a highly developed, adaptable, and efficient process, there are some problems and disadvantages to its use. Efforts are being made in individual mills to minimize energy, water, and chemical requirements, but there are two problems inherent in the chemistry of the process: low carbohydrate yield and the formation of malodorous organic sulfur compounds.

The low carbohydrate yield results largely from the alkaline peeling reactions. Since peeling is initiated at the hemiacetal or reducing end of the polysaccharide chains, modification of this functional group inhibits peeling and, thereby, increases the yield of carbohydrate material in alkaline pulping. A number of methods have proven effective in the laboratory, including glycosidation, reduction, reductive thiolation, and oxidation. To be effective as a process modification, however, the method must be compatible with kraft pulping and chemical recovery systems and be sufficiently inexpensive so that the cost is more than offset by the increased yield less fuel loss and other factors.

Polysulfide Process. One modification to the kraft process that is being applied commercially is the polysulfide process. When elemental sulfur is added to a solution of sodium sulfide and sodium hydroxide, the sulfur dissolves and forms a mixture of complexes with the general formula Na_2S_x (where x is 2–5, depending on equilibrium conditions and how much sulfur is added). Sulfur as Na_2S_x is an oxidizing agent which, under the conditions of kraft pulping, converts the hemiacetal function to a relatively alkali-stable aldonic acid. The reaction intermediate involves oxidation at C-2; thus, the epimeric acids are obtained. The increase in yield in the polysulfide process is proportional to the amount of added sulfur up to ca 10% based on wood. As expected from the sensitivity of the glucomannans to the peeling sequence, the greatest increase comes from their stabilization in softwoods (15) (see Fig. 13).

The additional sulfur for polysulfide pulping can upset the sodium–sulfur balance in the kraft recovery cycle, and sulfur-emission problems increase. In the MOXY (Mead Corp.) process, polysulfide is formed from kraft white liquor by catalytic oxygen oxidation of the sodium sulfide normally present. The need for additional sulfur is eliminated, but the sulfur level is limited to Na_2S_2. Under these conditions, yield gains of 2–2.5% based on wood are achieved and sulfur emissions are lower because of the reduced amount of Na_2S (16).

Anthraquinone. In the early 1970s, it was reported that the addition of sodium anthraquinone-2-sulfonate to sodium hydroxide solutions accelerates pulping to a rate similar to that obtained with sodium sulfide (17). The effect is partially additive, since improved yields and pulp properties are obtained from the kraft process when small amounts of the quinone salt are added to kraft pulping liquors. The sulfonate

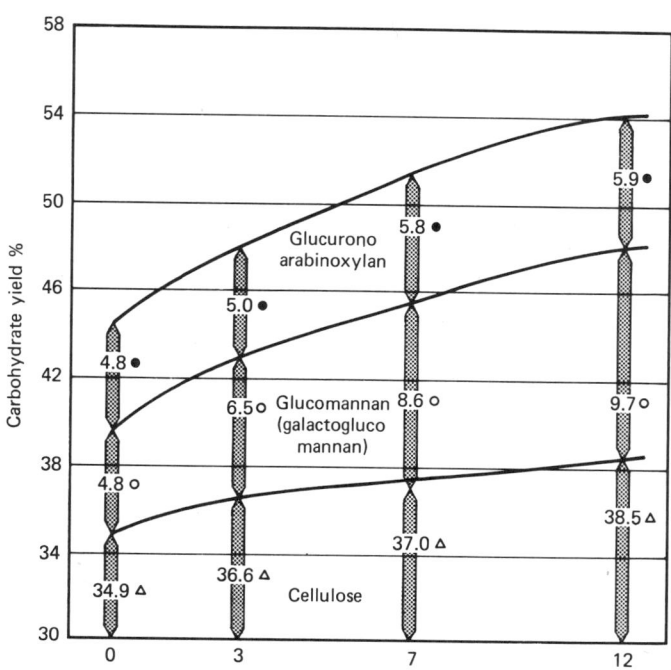

Figure 13. Polysulfide yield increase with added sulfur. △ = % carbohydrate yield attributed to cellulose; ○ = % carbohydrate yield attributed to glucomannan; ● = % carbohydrate yield attributed to glucuronoarabinoxylan.

salt is soluble in the alkaline solutions. A few years later it was discovered that the underivatized anthraquinone (AQ), although less soluble in alkaline solutions, is much more effective in accelerating pulping (18). Anthraquinone (qv) is being used as a kraft pulping additive in commercial production but, because it cannot be recovered, the economic benefits are marginal. The real potential for anthraquinone is in eliminating the use of sodium sulfide.

Modified Soda Pulping. Wood pulping with caustic soda solutions was the first chemical pulping process, but the beneficial effect of including sodium sulfide in the liquors was soon discovered. The soda process is used to advantage in pulping some hardwood species and nonwood plants, but wood pulping with sodium hydroxide was never widely used. The volatile malodorous sulfur compounds hydrogen sulfide, H_2S, methyl mercaptan, dimethyl sulfide, CH_3SCH_3, and dimethyl disulfide are produced as undesirable by-products of kraft pulping. These compounds are not easily contained in the large-volume pulping process, and research efforts have been aimed at eliminating the use of sulfur. Two possibilities have been tested on a pilot-plant scale: use of soda–oxygen or soda–anthraquinone.

Soda–Oxygen. Oxygen is an effective, readily available, and innocuous delignifying agent in aqueous alkaline solutions. However, it is a gas of low solubility and it degrades pulp polysaccharides. The diffusion of oxygen into the chips is the rate-determining step, and it is exacerbated by the rapid surface reaction with lignin, which depletes the oxygen supply. This problem can be minimized by increasing the oxygen pressure, maximizing gas contact, chipping the wood into thin wafers 1–2 mm thick, or decreasing the rate of the delignification reaction by lower temperatures (125–

140°C) and lower pH (pH 7–9). The latter steps increase the reaction time to 8 h or more.

A very high yield chemical pulp (52–53% bleached yield from softwoods) can be obtained but its strength properties are significantly inferior to those obtained from the kraft process. If a protector, eg, potassium iodide, is added, an additional 2–3% yield is obtained with an improvement in all strength properties. Kraft strengths can be achieved with some sacrifice in yield by postheating the oxygen-cooked chips in the absence of oxygen (19). The Middox process (a medium-consistency oxygen bleaching process) for pulp delignification with oxygen has recently been announced (20).

The gas-penetration problem can be minimized if fiberization is accomplished before treatment with oxygen (see also under Bleaching). Oxygen treatment of virtually all types of semichemical and mechanical pulps has been explored. In all cases, the replacement of the kraft process by such two-stage processes has not been justified over the alternative of pollution-abatement procedures.

Soda–Anthraquinone. A few mills worldwide use soda pulping of hardwoods. In such cases, the addition of anthraquinone is immediately justifiable in terms of increased yield and upgraded pulp quality. The conversion of existing kraft mills is not as simple, because AQ contributes no alkalinity to the process as sulfide does, and most kraft causticizing systems would have to be expanded by about 33%. This conversion is probably not justifiable in terms of the yield gain. The greatest benefit from AQ is for new mills in which air-pollution abatement devices could be eliminated.

Mechanism of Anthraquinone Acceleration. Anthraquinone is an effective pulping accelerator in very small quantities and functions as a catalyst in the process. Anthraquinone has limited solubility in alkaline solutions, but anthrahydroquinone (AHQ) (9,10-anthracenediol) is highly soluble in aqueous alkali. Alkaline solutions of sugars are strongly reducing because of the enhanced reactivity of the carbonyl-containing intermediates and degradation products. These solutions readily reduce AQ to AHQ. Anthrahydroquinone reacts with the lignin quinonemethide intermediates of alkaline pulping in reactions analogous to those of the sulfhydryl anion in kraft pulping (21–22). This addition compound is rapidly destroyed under pulping conditions to release the aryloxy group from the β position; thus, a new free phenolic unit is generated and, simultaneously, anthraquinone is produced, which starts the cycle again.

Anthraquinone reacts with the polysaccharide reducing end units to form an alkali-stabilizing aldonic acid (23). However, it accelerates glycosidic hydrolysis, at least in dissolved polysaccharides (24). Some minor side reactions remove anthraquinone from the catalytic cycle; these include the formation of stable ring compounds with lignin intermediates (25) and the further reduction of AHQ to anthracene and dihydroanthracene (26).

Sulfite Pulping. In the original sulfite pulping process, wood was pulped with an aqueous solution of SO_2 and lime. Calcium sulfite has very limited solubility above pH 2, and an excess of SO_2 gas was maintained in the digester to keep the pH below this level. Thus, the process was contrasted with the kraft or soda processes as being an acid process. Currently, bases other than calcium are used with SO_2 solutions, and sulfite pulping refers to a variety of processes in which the full pH range is utilized for all or part of the pulping. Magnesium, sodium, and ammonia are used as alternatives to calcium. Magnesium sulfite has decreasing solubility above pH 5, but sodium and ammonium sulfites are soluble at pH 1–14.

The chemical mechanism of sulfite delignification is not fully understood, but the chemistry of model compounds has been studied extensively, and attempts have been made to correlate the results with observations on the rates and conditions of delignification (27). The initial reaction is sulfonation of the aliphatic side chain, which occurs almost exclusively at the α-carbon by a nucleophilic substitution. The substitution displaces either a hydroxy or alkoxy group (see eq. 11).

$$\text{[Ar-CHOR with OCH}_3\text{, O-]} + HSO_3^- \longrightarrow \text{[Ar-CHSO}_3^-\text{ with OCH}_3\text{, O-]} + ROH \quad (11)$$

Sulfonation can take place at the γ-carbon, if that carbon is activated by an α-keto group or by extended conjugation through a double bond between the α- and β-carbons. The reaction does not occur through a quinonemethide under acidic conditions, because the p-phenolic function can be free or blocked. However, quinonemethide intermediates become important at neutral or alkaline pHs. The addition of the ionic sulfonic acid group imparts water solubility to fairly large lignin fragments, but there must be some depolymerization to permit the physical extraction from the cell wall.

The depolymerization mechanism is less well-understood. The sulfonated lignin polymer is subject to both acid- and alkali-catalyzed hydrolysis as well as sulfitolysis, which occurs concurrently with sulfonation. Under neutral conditions, β-ether cleavage occurs at 180°C, but little depolymerization takes place. It appears that lower molecular weight lignosulfonates are dissolved first while the higher molecular weight fraction is being slowly degraded (28). Acid-catalyzed condensations of the benzylic carbonium ion compete with sulfite and bisulfite delignification.

Acid-catalyzed glycosidic hydrolysis of polysaccharides is much more rapid than alkaline hydrolysis. Sulfite pulping at low pH must be carried out at a relatively low temperature (135°C) to avoid excessive polysaccharide depolymerization. The lower pulp strength of acid sulfite pulps compared with kraft pulps reflects this reaction. Other factors, such as accessibility and influence of the hemicellulose, are involved since the average $\overline{DP}_w$ in acid sulfite pulps can be higher than that in a comparable kraft pulp. The carbohydrate yield is higher in sulfite pulping (46% for a bleached softwood pulp compared with ca 44% for bleached kraft), because there is no peeling. The reducing-end functions of polysaccharide chains are oxidized to aldonic acid end units by bisulfite. However, a low hemicellulose-pulp that is suitable for producing cellulose derivatives, eg, rayon (qv) or cellulose acetate (qv), can be made by a two-stage cooking phase, eg, the Sivola process. The Sivola process is acidic in the first stage, which reduces the $\overline{DP}$ of the hemicelluloses, which are then dissolved and removed in an alkaline second stage. A more important stabilization mechanism in sulfite pulping is deacetylation of the softwood galactoglucomannan, in which the dissolved hemicellulose tends to crystallize or be redeposited on the cellulose microfibrils if the acetate substituents are removed. This may be promoted if a mildly alkaline first stage is followed by a more acidic stage, as in the Stora process.

Sulfite pulps have properties that are desirable for tissues and top quality, fine papers. Because sulfite pulping is not as versatile as kraft pulping, various options have been developed, and the choice of a specific process is very dependent on individual mill situations. There are pulping advantages and disadvantages to each of the base systems employed in sulfite pulping, and each has its own potential recovery systems except the calcium system, which is rapidly becoming obsolete. Calcium-base liquors can be burned, but scaling problems are severe, and conversion of the calcium sulfate to CaO is not economical. A comparative evaluation of the four bases used in sulfite pulping is given in Table 4.

In all sulfite pulping systems, any excess SO_2 in the digester can be collected easily from the blow gas by adsorption in fresh liquors. Ammonia-base liquors can be incinerated under oxidizing conditions to recover SO_2, but the ammonia is lost as nitrogen gas and water. Magnesium recovery is relatively simple because of the ready conversion of magnesium salts to MgO during combustion. Two types of systems are in use: a furnace and boiler, eg, the Babcock and Wilcox recovery unit, or a fluidized bed, eg, the Copeland process.

Chemical recovery in sodium-based sulfite pulping is more complicated, and a large number of processes have been proposed. The most common process, to which there are several variations, involves liquor incineration under reducing conditions to give a smelt, which is dissolved to produce a kraft-type green liquor. Sulfide is stripped from the liquor as H_2S after lowering of the pH with CO_2. The H_2S is oxidized to sulfur in a separate stream by reaction with SO_2, and the sulfur is subsequently burned to re-form SO_2. Alternatively, in a pyrolysis process such as SCA-Billerud, the H_2S gas is burned directly to SO_2. A rather novel approach is the Sonoco process, in which alumina is added to the spent liquors which are then burned in a kiln to form sodium aluminate. In another method, particularly in neutral sulfite semichemical processes, fluidized-bed combustion is used to give a mixture of sodium carbonate and sodium sulfate, which can be sold to kraft mills as makeup chemical.

Many other recovery alternatives have been proposed that include ion exchange (qv), pyrolysis, and wet combustion. However, these have not gained general acceptance. A limited number of calcium-based mills are able to utilize their spent pulping liquors to produce by-products, such as lignosulfonates for oil-well drilling muds, vanilla, yeast, and ethyl alcohol (see Petroleum; Vanillin).

Table 4. Comparison of Bases for Sulfite Pulping [a]

Property	Calcium	Magnesium	Sodium	Ammonium
SO_2 absorption system	complex	relatively simple	simple	simple
pH range for digestion	<2	<2	0–14	0–14
rate of pulping	intermediate	intermediate	slowest	fastest
level of screenings	moderate	moderate	low	low
scaling tendency	high	moderate	low	low
ease of liquor incineration	difficult, no base or SO_2 recovery	simple, base and SO_2 both recovered	complex, base and SO_2 both recovered	simple, no base recovery

[a] Ref. 29.

Modifications and Outlook. As a result of the discovery of the catalytic effect of anthraquinone in kraft and soda pulping, addition of AQ to the various sulfite pulping methods has also been investigated. In the neutral to strongly alkaline range, a catalytic effect is observed. The most significant observation is the marked increase in carbohydrate yield obtainable from AQ addition to neutral or slightly alkaline sulfite liquors. This produces a pulp with strength properties very similar to kraft except for decreased tear, which is typical of high hemicellulose pulps. Such a process would eliminate the air pollution problems of the kraft process and permit direct contact evaporation of the liquors. However, somewhat stronger cooking conditions, eg, a higher H factor, are required. Thus, the potential exists for sulfite pulp to regain some of the competitive market that has been lost to kraft pulp.

Uses. Full chemical sulfite pulps are used in tissue and sanitary paper products. Low hemicellulose pulps, ie, with high cellulose content, are marketed as dissolving pulp for cellulosic derivatives. High hemicellulose sulfite pulps are used to make glassine and greaseproof papers. Some sulfite chemical pulp is blended in newsprint and, to an increased extent, in printing and fine papers.

Miscellaneous Acid Pulping Processes. Delignification may be accomplished by electrophilic substitution of the aromatic nuclei by nitric acid. The pulp produced is similar in quality to acid sulfite pulp, and the spent nitrated liquors have good fertilizer value (see Fertilizers). The process is not used in the United States but may find application in favorable situations. Oxidative delignification by chlorine dioxide is conducted at acidic pHs. It is very selective and gives bleached pulps with very high hemicellulose contents. The process was evaluated briefly as holopulping in the United States and Japan (30).

Semichemical Pulping

The distinctions between semichemical and high yield chemical processes are very small and are more a matter of gradation between the mechanical and full chemical procedures. A semichemical process is essentially a chemical delignification process in which the chemical reactions are stopped at a point where mechanical treatment is necessary to separate fibers from the partially cooked chips. Any chemical pulping process can be used to produce semichemical pulp. The pulps, although less flexible, resemble chemical pulps more than mechanical pulps because they are not as dependent on rupture of the fiber wall for bonding (see Fig. 14). The yield is 60–85% with a lignin content of 15–20%. The lignin is concentrated on the fiber surface.

With shortages in fiber supply and consequent escalation of wood costs, high yield processes are used whenever possible. The high residual lignin is, however, responsible for the stiffness of the fibers and sensitivity to light aging. Semichemical pulping is particularly suited to the more abundant hardwood species. Softwoods require much more chemical treatment and refining.

Neutral Sulfite Semichemical (NSSC) Pulping. The characteristics of semichemical pulps are especially suited to the production of corrugated medium, which is the raw material for the fluted center ply of corrugated boxes. Neutral sulfite semichemical pulping was developed specifically as a semichemical process for corrugating medium and lends itself to small mills with minimal capital investment. For many years, this was the only semichemical pulping process.

The chemistry of NSSC pulping is similar to that of sulfite chemical pulping. It

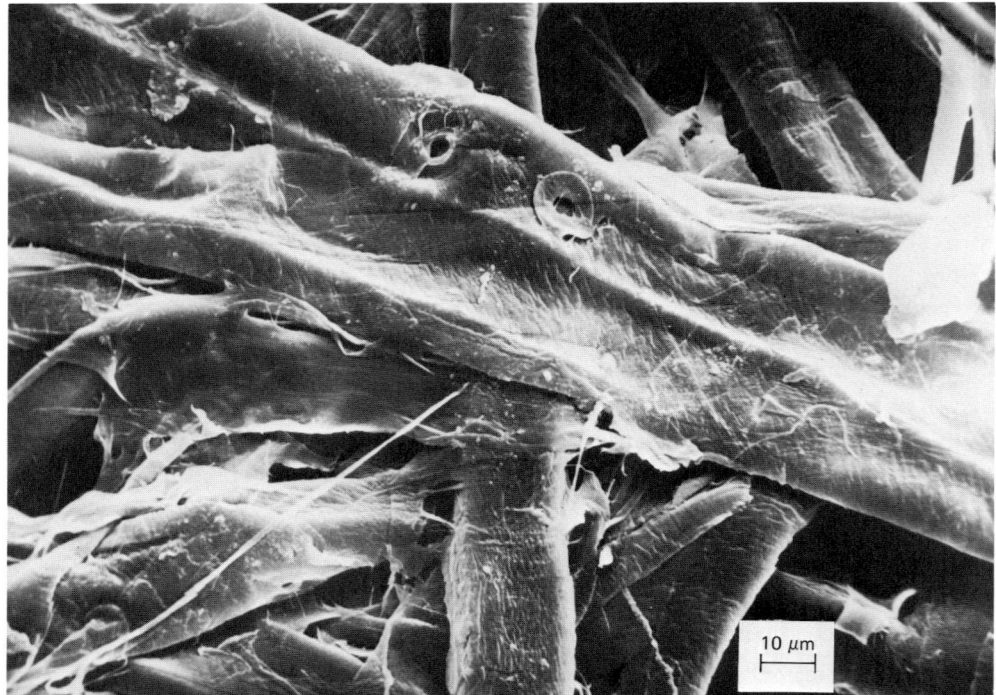

Figure 14. Electron micrograph of neutral sulfite semichemical (NSSC) pulp handsheet.

involves selective sulfonation and partial hydrolysis of the lignin. The selectivity is high, and corrosion is minimized if the pH is maintained at ca pH 7. The sodium sulfite solution is usually buffered with sodium bicarbonate. Additional buffer forms from the wood as pulping progresses, because the carbohydrate acetyl groups hydrolyze, wood acids dissolve, and lignosulfonates form.

As with other chemical pulping, continuous and batch digesters are used. In continuous pulping, the temperature is higher (190°C) and the time is shorter (10–12 min) than in batch systems (170–180°C for ca 20–60 min). At the higher temperatures, complete penetration of pulping chemicals into the chips is especially important. A screw feeder is used to compress the chips in the presence of pulping liquors. When the pressure is released, the liquid is rapidly absorbed by the wood. After being cooked, the chips are fiberized or screw-pressed up to 60% consistency before being fiberized in two or more refining stages before washing.

A computer control system has been developed in which the pulping rate is mathematically related to time, temperature, and chemical charge. The pH is monitored for fast feedback, and refiner energy is monitored as a delayed feedback (31).

Demands for pollution abatement forced the development of chemical-recovery methods for all sulfite pulping processes, including NSSC. One response to environmental concerns was the development of other semichemical pulping methods. Semichemical pulping can be performed using kraft green liquor. This procedure is especially advantageous if the semichemical pulp mill has access to an existing kraft recovery system. Neutral sulfite semichemical liquors can also be fed into a kraft recovery system as makeup for the kraft chemicals that are consumed, but the balance is unfavorable and the relative amount of NSSC liquors that could be utilized is small.

Semichemical pulping can be accomplished by a sulfur-free system of sodium hydroxide and sodium carbonate. The sodium carbonate is recovered by simple incineration, and sodium hydroxide is added as makeup. Advantages in recovery operation are obtained if potassium hydroxide is added occasionally to maintain ca 20 mol % potassium carbonate (32).

Bleaching

Pulps vary considerably in their color after pulping, depending on the wood species, method of processing, and extraneous components. For many paper types, particularly printing grades, bleaching of the raw pulp is required. The brightness standard is measured as the reflectance of light in the blue range (457 nm) compared with magnesium oxide as 100% white. Two scales are used, depending on the commercial meter. In the United States, the General Electric meter is the standard. In other countries, the Zeiss Elrepho is standard. In general, the GE brightness is 0.5–1% lower than the Elrepho value.

There are basically two types of bleaching operations: those that chemically modify the chromophoric groups by oxidation or reduction but remove very little lignin or other substances from the fibers, and those that complete the delignification and remove pitch and some carbohydrate material. In the special case of dissolving-pulp production, bleaching is a final purification of cellulose, and most of the residual hemicellulose is removed (see also Bleaching agents).

Mechanical Pulps. The lignin-retaining type of bleaching is used with high yield mechanical and chemimechanical pulps in paper grades, eg, newsprint, where brightness stability is not critical. The initial brightness values of these pulps usually are 50–65% GE. If sodium bisulfite is added in a chemimechanical process, the pulps are a few points brighter.

The most effective bleaching agent for most groundwoods is hydrogen peroxide (qv). Bleaching is performed in alkaline solutions; thus, sodium peroxide is also used. Sodium silicate and magnesium sulfate usually are added to buffer the solutions and to sequester metal ions, which would otherwise wastefully accelerate the decomposition of the peroxide (see Chelating agents). Typical conditions are 1–3 wt % hydrogen peroxide, 1.1 wt % sodium hydroxide, 5 wt % sodium silicate, and 0.05 wt % magnesium sulfate. The pH should be 10.5–11 and the consistency ca 12%, although higher consistencies have been used. The reaction requires 3 h at 40°C and is followed by a neutralization and destruction of excess peroxide with SO_2.

Brightening of high yield pulps can also be achieved reductively with sodium or zinc hydrosulfite. Sodium tripolyphosphate is added with the hydrosulfite at pH 6–7. Bleaching is performed with 0.5–1 wt % hydrosulfite for ca 2 hours at 55°C. Hydrosulfite bleaching is also performed in the refiner and latency-removal chest at pH 5.5 for zinc hydrosulfite.

Chemical Pulps. If all of the lignin, pitch, carbohydrate degradation products, and other chromophores and uv-absorbing materials are removed, a very white (over 90% GE), highly color-stable pulp can be obtained. This condition is limited to full chemical pulps, because it is much less expensive and more efficient to remove most of the lignin with pulping chemicals. The reagents for full bleaching are mostly oxidative. Since the carbohydrates are also susceptible to oxidation, bleaching is accomplished under the mildest conditions possible. It is not clear whether residual lignin is structurally different from the bulk lignin or whether the observed decrease in delignification rate during pulping is caused by other factors. In either case, the remaining lignin is intimately associated with the cell-wall polysaccharides. The final depolymerization and removal is performed gently in several stages.

Because of the variety of staging combinations that have been developed, a set of commonly accepted abbreviations are in use (33) (see Table 5). For example, the five-stage sequence CEDED refers to an initial chlorination of the lignin under acidic conditions followed by alkaline hydrolysis and extraction of the chlorinated lignin, mild oxidation with chlorine dioxide followed by another alkaline extraction, and final brightening with chlorine dioxide. Other common sequences are C/DEDED, CEHDED, and OCEDED where H and O refer to hypochlorite and oxygen, respectively. Washing is performed between stages when necessary.

Chlorination involves substitution of the aromatic ring and side chains and oxidative depolymerization through ether cleavages and ring openings. Chlorine also reacts by both ionic and free-radical mechanisms. It is desirable to minimize substitution and free-radical reactions for most effective bleaching and protection of carbohydrates. Inclusion of chlorine dioxide in the chlorination stage to function partly as a radical scavenger is common. The formation of mutagenic organochlorine compounds and discharge in the effluent is undesirable, and many attempts have been made to eliminate this stage. The reasons for the effectiveness of chlorine are not clear but, to date, a generally applicable substitute has not been found.

Chlorination can be carried out at 25°C or below; however, the reaction is exothermic and, as mills have closed their bleaching cycles, operating temperatures have unavoidably risen. Retention times are 30–60 min but decrease as temperature increases. The normal pulp consistency has been 3–4%, but the trend is toward higher consistency (ca 10%) or gas-phase chlorination. In the subsequent alkaline extraction, ca 1.5–3.5% NaOH on a pulp basis is used. Since sulfonated lignins are more water-soluble, the extraction is carried out at 20–40°C; with kraft pulps, 50–60°C is required.

Table 5. Symbols Representing Bleaching Stages[a]

Name of stage	Symbol	Chemical used
chlorination	C	chlorine gas or chlorine water
caustic extraction	E	sodium hydroxide solution
hypochlorite	H	sodium or calcium hypochlorite
chlorine dioxide	D	water solution of chlorine dioxide
oxygen	O	oxygen gas and alkali
peroxide	P	hydrogen peroxide (50 wt % soln)
ozone	Z	gaseous ozone (2 wt % in oxygen)
mixtures of chlorine and chlorine dioxide	C/D	chlorine–chlorine dioxide mixture
sequential dioxide and chlorine	(D–C)	chlorine dioxide followed by chlorine
oxidative extraction	E/H	inclusion of sodium hypochlorite in caustic extraction stage
peroxide extraction	E/P	inclusion of peroxide in an extraction stage
hydrosulfite	H	solution or solid sodium or zinc hydrosulfite
acid treatment	A	sulfuric acid
acid souring	S	sulfur dioxide gas

[a] Ref. 33.

Chlorine dioxide gas is essentially nondegrading to cellulose and is capable of producing very bright pulps, and the formation of chlorinated organic compounds is minimal. However, the gas is explosive at high concentration, toxic, and highly corrosive. For safety, it is generated at the point of use in low concentration. The use of proper vents and corrosion-resistant materials of construction, eg, titanium, is required. The reaction is conducted at 60–80°C for 2–6 h at pH 4 and consistency 12–15%. With sulfite and bisulfite pulps, a brightness of 90–92 can be reached at the end of this stage. With kraft pulps, the brightness is typically ca 85, and a second caustic extraction followed by a second chlorine dioxide stage is required to bring the kraft pulp brightness to the same level.

Hypochlorite is sometimes used following the first extraction stage or is included in an oxidative extraction stage (E/H). The pH must be maintained at ca pH 11 to minimize degradation to the carbohydrates.

Oxygen bleaching is a relatively new development but is fairly common in the industry. The reaction is conducted in a solution of sodium hydroxide. However, because the solubility of oxygen in aqueous solutions is low and decreases with increasing temperature, special attention must be given to maintaining adequate supplies of oxygen in solution. The original oxygen bleaching systems Sapoxal and MoDoCIL involve high consistency bleaching where the pulp, which is wet with alkaline solution, is pressed to 25–30% consistency, fluffed, and then bleached in a gas-phase oxygen reactor. Autoxidation and combustion, especially of pulp impurities such as pitch, are problems under these conditions, and systems that operate at lower consistencies have been proposed (34).

Oxygen bleaching is usually a prebleach stage before chlorination; one of the main advantages of oxygen bleaching is the compatibility of the spent liquors with the kraft recovery system. Under these conditions, it is used with 2–3 wt % NaOH for 1–2 h at 85–110°C to reduce the lignin content of softwood kraft pulp from 6 wt % to ca 3 wt %. Because the pulp polysaccharides are susceptible to oxygen degradation and the reaction is catalyzed by trace heavy metals, magnesium salts are usually added as protectors. The catalytic metal concentration can be minimized by incorporating a preliminary acid wash or A stage. Alternatively, oxygen bleaching is used in place of, or as part of, the caustic extraction stage following chlorination.

Pollution Control. The pulp-and-paper industry has had three very difficult pollution problems to solve: the development of suitable systems for sulfite waste liquor recovery; elimination of the volatile, malodorous, and toxic reduced sulfur compounds from kraft air emissions; and elimination of the toxic elements, bacterial oxygen demand (BOD), chemical oxygen demand (COD), and color from bleach-plant effluents. Of these, the bleach-plant problems have been the least amenable to solution. The energy requirements to evaporate the huge volumes of water from the dilute bleaching and extracting solutions are prohibitive, and the chlorination waste liquors are not readily compatible with the kraft recovery system because of corrosion, the volatility of NaCl, and the changes sodium chloride produces in the properties of the smelt.

The quantity of water necessary for bleaching has been decreased significantly by various schemes for recycle of liquors, eg, pulp washing with dilute spent liquors, countercurrent flow, etc. Highly reduced water use and reduced capital, chemical, and energy costs is achieved by displacement bleaching. In this system, the CEDED sequence occurs in a single upflow tower, and washing takes place only after the last stage. The Kamyr continuous-displacement bleaching system with medium consistency chlorination is illustrated in Figure 15.

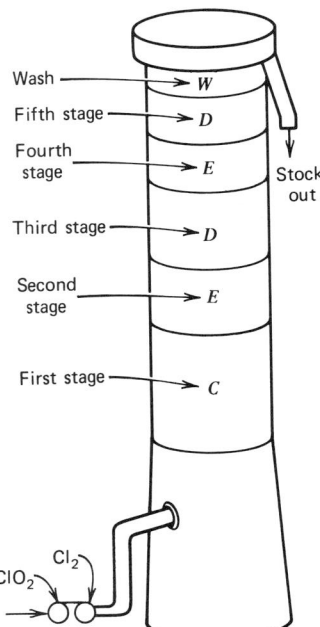

Figure 15. Kamyr continuous-displacement bleaching system. (For explanation of symbols, see Table 5.)

Some mills are attempting to minimize the amount of elemental chlorine by incorporating an oxygen stage or by substituting chlorine dioxide for most of the chlorine, either in the chlorination stage or sequentially. Two systems that involve cycling of chlorine-containing effluent into the kraft recovery system are being tested. In the Hooker APS system, the chlorination effluent is neutralized with lime and then is treated in standard effluent-treatment facilities. The extraction liquors, which contain some sodium chloride, are cycled into the recovery system. The sodium chloride is removed from the precipitator dust (35). In the Rapson-Reeve closed-cycle kraft mill, the chlorination liquors, after being used for brown stock washing, are combined with pulping black liquors, are evaporated, and are burned in the recovery furnace. The sodium chloride is removed from the causticized liquor by concentration and selective crystallization. The salt is used to regenerate chlorine and chlorine dioxide (36).

The other option is to eliminate the use of elemental chlorine entirely. This has not been possible with oxygen bleaching, but various combinations of ozone (Z) with peroxide or chlorine dioxide, such as ZED, ZEP, or OZEP, have been tested and seem promising (37).

Effluent Treatments

It has been reported that the tainting of fish taste by pulp-mill effluent compounds can be detected at levels below 0.1 ppb in fish flesh (38) and that the volatile reduced sulfur compounds, specifically dimethyl sulfide and hydrogen sulfide, of the kraft process can be detected by smell at ca 1 ppb in air (39) (see Water, water pollution; Air pollution). Most mills worldwide have the capability of treating their liquid effluent biologically, either in the plant or in cooperation with a municipality. Frequently, secondary and tertiary treatment is employed. The main sources of aqueous pulp-mill effluents are from barking operations, weak black liquor from final washings or leaks and spills, scrubbing operations, bleach plants, and paper machines. The capital expenditures of the pulp-and-paper industry for environmental protection through 1979 were ca 4.7×10^9: 2.5×10^9 for water, 2.0×10^9 for air, and 160×10^6 for solid wastes.

Pulp and paper-mill effluent contains 8 of the 129 priority water pollutants listed by the EPA: trichlorophenol, pentachlorophenol, chloroform, PCBs (polychlorobiphenyls), cyanides, phenol, zinc, and lead (40). In addition, the various effluents contain substances that are mutagenic (41), toxic to fish (42), and that taint the taste of fish (38). The principal identified hazardous components are low molecular weight chlorinated compounds from the chlorine bleaching stage, chlorinated phenols from the extraction stage, chloroform from the hypochlorite and the chlorine stages, and resin acids, fatty acids, diterpenes, and phenols from barking and pulping operations and their chlorinated analogs from bleaching. Efficient removal of environmental hazards by biological treatment has been demonstrated, but there are reports that some toxicity persists through aeration ponds (42–44).

Many questions remain about the magnitude and severity of environmental problems. Rapid progress is being made in the identification of chemicals present in the various effluent streams, but an enormous amount of testing remains to be done on linking specific chemicals and concentrations to mutagenic, toxic, and esthetic effects. In addition, there are questions of the indirect problems of biological and chemical degradability, effects on BOD, accumulation in the food chain, etc.

Economic Aspects

Pulp production and per capita consumption of paper and board for 1978 is shown in Table 6 (45). The United States, Canada, Sweden, and Finland are the principal exporters. In 1978, they supplied 77% of all pulp exports. The United States is, however, a net importer of pulp and paper. Worldwide demand for pulp products is expected to increase for many years.

The production of wood pulp for selected years through the 1970s is shown in Table 7 (46). In the recession year of 1975, overall demand was down, but the kraft process continued to increase in dominance at the expense of soda and sulfite. Groundwood production remained stable, but the refiner mechanical and semichemical pulp production increased. These trends may change because of recent developments, eg, pressurized groundwood and soda–anthraquinone and neutral sulfite–anthraquinone pulping.

Table 6. Pulp Production and Per Capita Consumption of Paper and Board for 1978[a]

Region	Production, 10^3 t	Per capita consumption, kg
North America	71,401	282
Western Europe	40,574	110
Asia and Oceania	32,140	14
Eastern Europe	15,569	40
Latin America	6,354	24
Africa	1,472	5

[a] Ref. 45.

Table 7. Estimated U.S. Production of Wood Pulp by Type, Metric Tons[a]

Pulp type	1969	1972	1975	1978
kraft				
unbleached	14,704	16,141	13,941	16,783
bleached	9,688	11,304	11,293	15,534
semibleached	1,561	1,428	1,266	
Subtotal	*25,953*	*28,873*	*26,500*	*32,317*
groundwood	4,006	4,208	3,956	3,963
semichemical	3,284	3,435	2,904	3,661
defibrated or exploded	1,666	2,159	2,294	2,762
sulfite	2,072	1,970	1,761	1,596
special alpha and dissolving	1,520	1,502	1,436	1,284
soda	219	167	136 (estd)	73 (estd)
other	118	111	98	100
Total	*38,838*	*42,425*	*39,085*	*45,756*

[a] Ref. 46.

Energy

The pulp-and-paper industry is a significant industrial consumer of energy, but much of the energy used can be generated from combustion of spent liquors, bark, and nonprocessed wood. The total energy usage of the industry in 1978 was 2.3×10^{18} J [2.2 quads (2.2×10^{24} Btu)] and of this, 47% was supplied from plant-generated wastes. The energy self-sufficiency of the pulp-and-paper industry probably will continue to increase for a number of years as energy efficiency and conservation is improved. Many companies are large woodlands owners and can utilize wood as process fuel. Many can generate their own electric power through back-pressure steam or hydroelectric turbines.

BIBLIOGRAPHY

"Pulp" in *ECT* 1st ed., Vol. 11, pp. 250–277, by R. S. Harch, Hudson Pulp and Paper Corp.; "Pulp" in *ECT* 2nd ed., Vol. 16, pp. 680–727, by G. H. Tomlinson II, Domtar Ltd.

1. C. E. Dunning, *Tappi* **52,** 1326 (1969).
2. T. F. Bobbitt, J. H. Nordin, M. Roux, J. F. Revol, and R. H. Marchessault, *J. Bacteriol.* **132**(2), 691 (1977).
3. J. E. Stone and A. M. Scallan, *J. Polym. Sci. Part C* **11,** 13 (1965).
4. H. Tarkow, W. C. Feist, and C. F. Southerland, *For. Prod. J.* **16**(10), 61 (1966).
5. J. M. Harkin, "A Natural Polymeric Product of Phenol Oxidation" in W. I. Taylor, and A. R. Battersby, eds., *Oxidative Coupling of Phenols*, Marcel Dekker, Inc., New York, 1967, p. 224.
6. J. L. Minor, *Wood Chem. Technol.*, in press, 1982.
7. J. R. Obst, *Tappi*, in press, 1982.
8. A. B. Berg, *A Commercial Thinning in Douglas-fir with a Horse*, Research Paper 6, Forest Research Laboratory, School of Forestry, Oregon State University, July 1966.
9. C. F. Mills, *Tappi* **63**(9), 39 (1980).
10. J. C. W. Evans, *Pulp Pap.* **54**(6), 76 (1980).
11. J. Gierer and S. Ljunggren, *Sven. Papperstid.* **82**(17), 503 (1979).
12. O. Franzon and O. Samuelson, *Sven. Papperstid.* **60**(23), 872 (1957).
13. A. R. Procter, W. Q. Yean, and D. A. I. Goring, *Pulp Pap. Can.* **68,** T445 (1967).

14. R. G. MacDonald and J. N. Franklin, eds., *Pulp and Paper Manufacture*, Vol. 1 of *The Pulping of Wood*, 2nd ed., McGraw-Hill Book Co., New York, 1969.
15. N. Sanyer and J. F. Laundrie, *Tappi* **47**(10), 640 (1964).
16. R. P. Green and Z. C. Prusas, *Pulp Pap. Can.* **76**(9), T272 (1975).
17. B. Bach and G. Fein, *Zellst. Pap.* **21**(1), 3 (1972).
18. H. H. Holton, *Pulp Pap. Can.* **78**(10), T218 (1977); U.S. Pat. 4,012,280 (March 15, 1977), (to Canadian Industries Ltd.).
19. J. L. Minor and N. Sanyer, *Tappi* **58**(3), 116 (1975).
20. *Chem. Week*, 17 (Oct. 21, 1981).
21. L. L. Landucci, *Tappi* **63**(7), 95 (1980).
22. J. Gierer, O. Lindeberg, and I. Noren, *Holzforschung* **33**, 213 (1979).
23. L. Lowendahl and O. Samuelson, *Tappi* **61**(2), 19 (1978).
24. F. L. A. Arbin, L. R. Schroeder, N. S. Thompson, and E. W. Malcolm, *Tappi* **63**(4), 152 (1980).
25. T. J. Fullerton and S. P. Ahern, *J. Chem. Soc. Chem. Comm.*, 457 (1979).
26. I. Gourang, R. Cassidy, and C. W. Dence, *Tappi* **62**(7), 43 (1979).
27. G. Gellerstedt, *Sven. Papperstid.* **79**(16), 537 (1976) and references therein.
28. W. Q. Yean and D. A. I. Goring, *Sven. Papperstidn.* **71**(20), 739 (1968).
29. James P. Casey, ed., *Pulp and Paper Chemistry and Chemical Technology*, 3rd ed., Vol. 1, John Wiley & Sons, Inc., New York, 1980, p. 301.
30. R. P. Whitney, N. S. Thompson, G. A. Nicholls, and S. T. Han, *Pulp Pap.* **43**(8), 68 (1969).
31. E. Jutila, M. Perron, H. Ahola, and R. Hipeli, *Tappi* **63**(11), 69 (1980).
32. P. E. Shick, *paper presented at TAPPI Conference on Alkaline Pulping and Secondary Fibers*, Washington, D.C., Nov. 7–10, 1977.
33. R. P. Singh, ed., *The Bleaching of Pulp*, 3rd ed., TAPPI Press, Atlanta, Ga., 1979, p. 6.
34. E. F. Elton, V. L. Magnotta, L. D. Markham, and C. E. Courchene, *Tappi* **63**(11), 79 (1980); P. J. Kleppe, P. C. Knutsen, and F. Jacobsen, *Tappi* **64**(6), 87 (1981).
35. Ref. 33, p. 571.
36. Ref. 33, p. 570.
37. Ref. 33, p. 23.
38. V. Naish and R. J. P. Bronzes, *Pulp Pap. Can.* **81**(10), T292 (1980).
39. P. Grennfelt and T. Lindvall, *Sven. Papperstid.* **79**(12), 389 (1976) and references therein.
40. *Pulp Pap.* **55**(1), 86 (1981).
41. E. G-H. Lee, J. C. Mueller, C. C. Walden, and H. Stich, *Pulp Pap. Can.* **82**(5), T149 (1981) and references therein.
42. C. C. Walden and T. E. Howard, *Pulp Pap. Can.* **82**(4), T143 (1981).
43. R. R. Claeys, L. E. LaFleur, and D. L. Borton, *NCASI Stream Improvement Technical Bulletin No. 332*, New York, May 1980.
44. B. Holmbom and K-J. Lehtinen, *Pap. ja Puu* **62**(11), 673 (1980).
45. J. E. Huber, ed., *Kline Guide to the Paper Industry*, 4th ed., C. H. Kline & Co. Inc., Fairfield, N.J., 1980.
46. *Current Industrial Reports*, *M26A*, from Department of Commerce and American Paper Institute, C. H. Kline & Co., Fairfield, N.J.

General References

Refs. 14, 29, 33, and 45 are general references.
D. Hunter, *Papermaking, the History and Technique of an Ancient Craft*, Dover Publications Inc., New York, 1978.
B. L. Browning, ed., *The Chemistry of Wood*, John Wiley & Sons, Inc., New York, 1963.
Statistics of Paper and Paperboard, American Paper Institute, New York, Annual.

JAMES MINOR
U.S. Forest Products Laboratory

PULP, SYNTHETIC

Synthetic pulp generally defines very fine, highly branched, discontinuous, water-dispersible fibers made from plastics. A photograph of polyolefin pulp in wet lap and fluff form is shown in Figure 1. The visual appearance and dimensions of synthetic pulps closely resemble those of cellulose pulps and asbestos (qv) (see Pulp). Synthetic pulps are not to be confused with extruded staple fibers, which are smooth rods of solid polymer (Fig. 2). Nearly all current pilot, semicommercial, and commercial production units produce pulps based on either high density polyethylene or polypropylene, with or without inorganic fillers (see Olefin polymers). The exceptions to this are pulps based on aromatic polyamides, ie, aramids (see Aramid fibers). A production unit for poly(p-phenyleneterephthalamide) [24938-64-5] (Kevlar) pulp was announced in 1980, and a sheet made from poly(m-phenyleneisophthalamide) [24938-60-1] (Nomex) pulp is commercially available, although the pulp is not (1–2). Pulps based on polystyrene [9003-53-6] were produced in pilot-plant equipment in Japan for several years, but their production apparently has been discontinued (3). Pulps prepared from many other polymers have been described, eg, aliphatic polyamides, poly(vinyl chloride) [9002-86-2], acrylonitrile homopolymer [25014-41-9] and copolymers with halogenated monomers, styrene copolymers, and mixtures of polymers, eg, polypropylene with low density polyethylene and high density polyethylene with polystyrene (3–9). However, these types of synthetic pulps have only been produced in the laboratory; little property or application data on them are available.

Polyolefin synthetic pulps are designed to be blended in all proportions with wood pulp and glass fibers and made into papers and boards using conventional papermaking equipment (see Paper). The sheets so produced may or may not be heated to melt the

Figure 1. Synthetic pulp in sheet and fluff form.

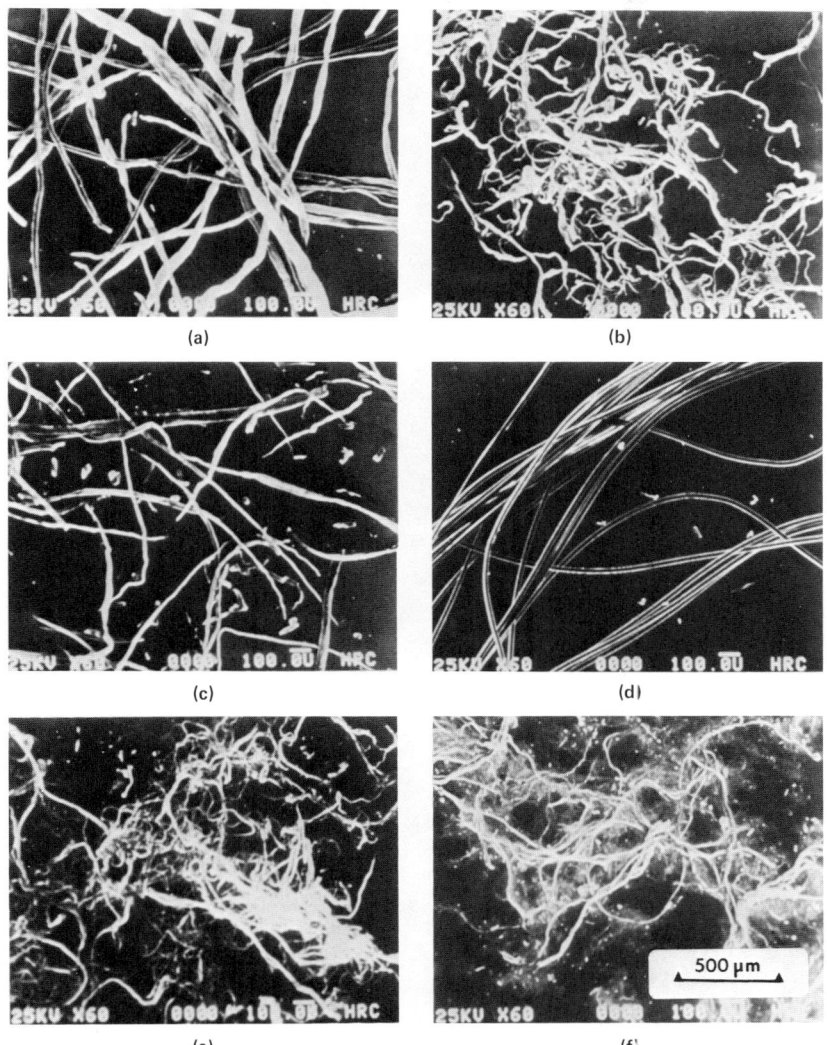

Figure 2. Photomicrographs of various fibers. (**a**), bleached softwood kraft. (**b**), polypropylene synthetic pulp. (**c**), bleached hardwood kraft. (**d**), polypropylene staple. (**e**), polyethylene synthetic pulp. (**f**), asbestos (type 40T).

polyolefin, depending on the intended application. In either event, the resulting materials differ significantly from the 100% plastic sheets or mats of continuous fibers made by spunbonding, extrusion, melt-blowing, and other processes (see Plastics processing; Nonwoven textiles, spunbonded) (3,10–12).

Properties

Comparisons of polyolefin synthetic pulps, cellulose pulps, and staple fiber are given in Figures 3–5. Whereas staple is a smooth rod of solid polymer, synthetic pulps have very irregular surfaces with many crevices and an almost filmlike nature. Their surface area is quite large, typically 5–20 m^2/g (150–590 yd^2/oz), compared with either staple or cellulose [<1 m^2/g (<30 yd^2/oz) by nitrogen absorption], which results in high

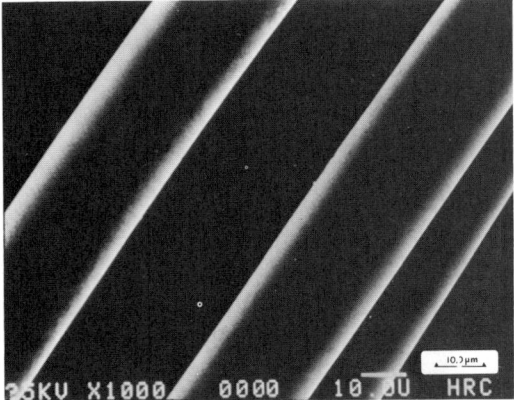

Figure 3. Magnification of polypropylene synthetic pulp.

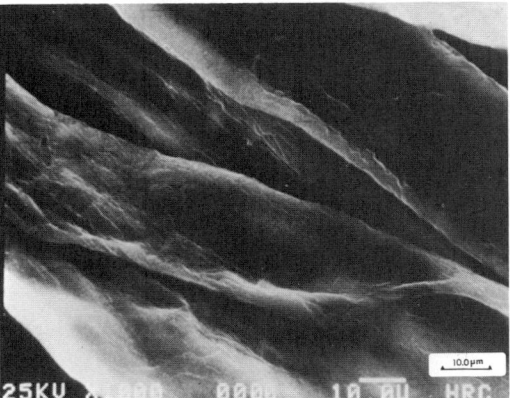

Figure 4. Magnification of bleached softwood kraft.

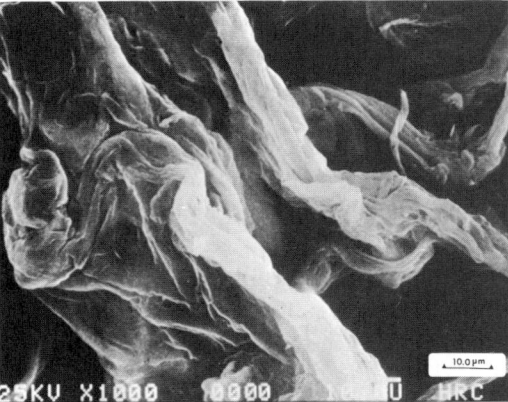

Figure 5. Magnification of polypropylene staple.

scattering coefficients, eg, 0.15–0.30. Synthetic pulps also are characterized by high brightness (>90%). Fiber densities of unfilled pulps are low because of high void volumes. Polypropylene (PP) pulps tend to have lower densities (ca 0.2 g/cm^3 measured by mercury porosimetry) than polyethylene pulps (ca 0.4 g/cm^3); and both are much less dense than the bulk polymers [0.91 g/cm^3 for PP and 0.94–0.97 g/cm^3 for high density polyethylene (HDPE)] or cellulose (1.5–1.6 g/cm^3 for crystalline cellulose and

0.9–1.1 g/cm^3 for many pulp fibers) (13). Average lengths of synthetic pulp fibers usually are 1 mm with a maximum length of 2.5–3 mm. For comparison, a very short staple fiber length is 3–5 mm. Fiber length distributions of synthetic pulps typically are broad and generally are measured using the Bauer-McNett classification procedure (14). Average synthetic pulp fiber diameters are 5–40 μm. Pulps dispersed with poly(vinyl alcohol) have a negative electrical charge, which is similar in magnitude to that of a bleached kraft softwood pulp.

An important property of polyolefin synthetic pulps is drainage rate (15). This usually is controlled for unfilled pulps by the amount of wetting agent applied during manufacture. Several procedures have been employed for determining drainage rate. One compares the drainage rate of a 70:30 blend of synthetic pulp and cellulose pulp refined to 500 CSF (Canadian standard freeness) with that of the cellulose pulp alone. Drainage rates either greater than or less than 100% can be obtained. A dispersibility index is used to measure the degree of water dispersibility in terms of the tendency of a synthetic pulp to rise to the surface from a standard quantity of aqueous slurry (16). Because of their low densities, unfilled polyolefin pulps float at slurry concentrations of ca 0.1 wt % or less. Floating problems are not enountered with blends of synthetic and cellulose pulps, however.

Another key property of polyolefin synthetic pulps is their thermoplasticity. The melting points for high density polyethylene and polypropylene are ca 132°C and 165°C, respectively, but softening occurs at lower temperatures (see Uses). Synthetic pulps also are inert to most chemicals and have outstanding dielectric properties.

The oxidative stabilities of polyethylene and polypropylene pulps are essentially the same as the starting polymers. Polypropylene requires a stabilizer, which is readily introduced in the manufacturing process. The effectiveness of the stabilizer is demonstrated by the observation that the brightness loss of paper containing 20 wt % polypropylene pulp is no more than a filled cellulose control under accelerated aging conditions simulating 25 yr.

Typical characteristics of standard commercial synthetic pulps are listed in Table 1.

Manufacture

The main processes used to prepare synthetic polyolefin pulps are solution flash spinning, emulsion flash spinning, melt extrusion/fibrillation, and shear precipitation. Solution flash spinning is outlined in Figure 6 (17–21). It consists of forming a true solution of polyolefin in a low boiling organic solvent at high temperature and pressure. The solution is passed into a specially designed spurting nozzle or spinneret, in which a small, controlled pressure drop is effected. Two liquid phases, one of which is polymer-rich and the other polymer-lean, result. This two-phase mixture exits through a small orifice at high shear into a chamber of low temperature and approximately atmospheric pressure to quickly and completely evaporate or flash the solvent in an almost explosive manner. Vaporization of solvent provides the energy to form the fibrous product, and the resulting cooling causes rapid crystallization of the polyolefin. The solvent is condensed and reused. With proper nozzle design and control of process variables, fine, discrete pulp fibers form at the last nozzle orifice. Early processes of this type gave continuous fibers which could be cut to appropriate lengths (22–23). The pulp fibers are conveyed into water, passed through refiners or deflakers, or both,

Table 1. Typical Characteristics of Pulpex^a Polyolefin Pulps

Property	Pulpex E-A (polyethylene)	Pulpex P-AD (polypropylene)
melting range, °C	130–135	160–165
specific area, m^2/g^b	10–15	5–10
poly(vinyl alcohol) content, wt %	1–1.5	0.6–0.8
drainage value, %^{c}	ca 120	ca 80
average length, mm	0.8–1.2	0.8–1.2
maximum length, mm	2.5	2.5
fines content^d, %	<3	<10
brightness, %^{e}	>94	>94
scattering coefficient	>0.15	>0.15
diameter, µm	10–20	20–40
polymer density, g/cm^3	0.965	0.91
pulp density, g/cm^3	ca 0.4	ca 0.2

^{a} Trademark of Lextar (Hercules, Inc. company)
^{b} To convert m^2/g to yd^2/oz, multiply by 29.5.
^{c} Drainage of a 70:30 blend of synthetic pulp and bleached kraft at 500 CSF; expressed as a percentage of the drainage of cellulose alone.
^{d} Percent of pulp passing through a 74-µm (200-mesh) screen.
^{e} MgO reference.

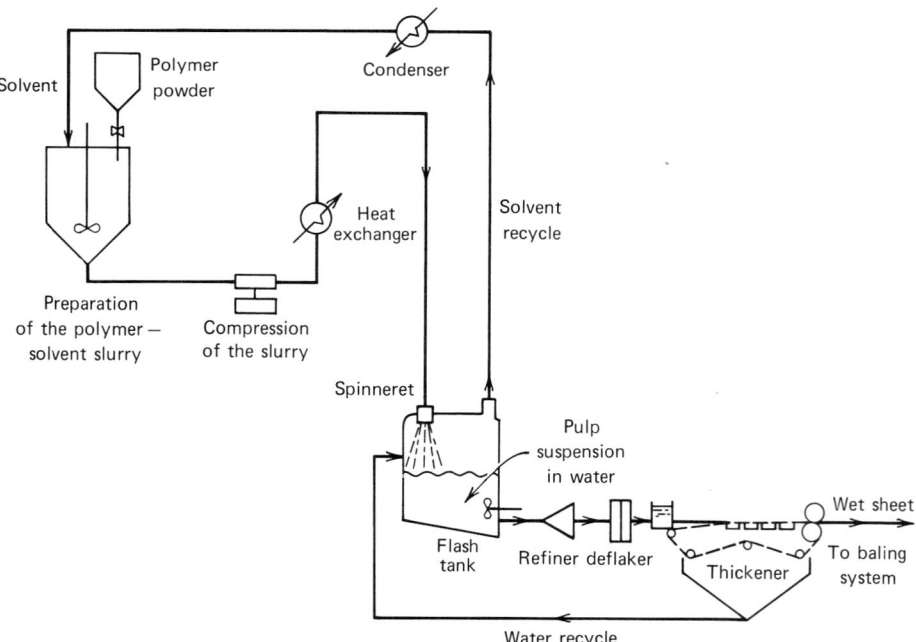

Figure 6. Flow sheet of flash spinning.

to control fiber length, if necessary, and then dewatered with conventional equipment. Wetting agents, eg, poly(vinyl alcohol), provide water dispersibility. Inorganic fillers (qv), eg, kaolin clay, talc, or calcium carbonate, also can be introduced into the polymer solution before flash spinning (24).

The emulsion flash-spinning method involves initial formation of an emulsion at high temperature and pressure from a polymer solution and water (25–27). The emulsion is passed through a small orifice at high shear into a chamber at low temperature and approximately atmospheric pressure so as partially to evaporate the solvent. Solvent evaporation is completed by refining the resulting fibrous gel in hot water at atmospheric pressure or under slight vacuum. The refining step, which typically involves multiple passes through a disk refiner, is an essential step in the process because it provides both fibrillation and adjustment of fiber length. A wetting agent, eg, poly(vinyl alcohol), is added before solvent removal and refining to effect good pulp dispersibility (16). Dewatering involves conventional technology. As is true for all polyolefin synthetic pulps, especially those based on polyethylene, very high pressures during wet pressing must be avoided to prevent formation of fiber bundles or fused knots.

The melt extrusion/fibrillation process begins with extrusion of a thermoplastic polymer (usually polyolefin) film followed by unidirectional molecular orientation by longitudinal stretching (28). Fibrillation then is effected by contacting the film over its width with many fine needles which move in the same direction as the film but 10–60 times faster. Commonly, inorganic fillers, such as barium sulfate, titanium dioxide, calcium carbonate, or clay, are incorporated into the polymer prior to extrusion to facilitate fibrillation and to provide water dispersibility. The fiber then must be cut to the desired length.

Several other processes for preparation of synthetic polyolefin pulps have been developed. Of these, the most important probably is the shear precipitation method, which is also called the fibrous gel process (29–32). It involves initial formation of a swollen polymer gel, for example, by direct polymerization of olefin in a suitable solvent, by cooling a high temperature, high pressure solution of preformed polymer or by partial evaporation of solvent under reduced pressure. Pulp fibrils are produced by subjecting the gel to very high shear forces, often with a disk refiner, in the presence of a wetting agent while heating to remove solvent. In another variant of shear precipitation, fibrillated polyolefin fibers are prepared by dispersing a solution of polyolefin in precipitant under high shear conditions (33). It also is possible to prepare pulps by melt-extruding a fiber with simultaneous fibrillation and cutting using high velocity hot air at the extrusion orifice; air at slow speeds fibrillates without cutting. Many types of polymers have been processed with this technology, but it is difficult to prevent formation of small amounts of nonfibrous polymer. Nascent polyolefin fibrils also have been formed directly during polymerization using Ziegler-Natta catalysts (34).

Modifications. In addition to poly(vinyl alcohol) and/or inorganic fillers for providing wettability of synthetic pulps, many materials and methods have been used. A wide variety of water-soluble polymers or even conventional surfactants give good pulp dispersion, provided that a reasonable amount of mechanical agitation is applied to an aqueous pulp slurry. However, poly(vinyl alcohol) is preferred, especially for polyethylene pulps, because of the superior stability of the resulting pulp slurry, ie, its resistance to floating. Cellulose at 2–10% based on synthetic pulp also is a useful dispersing agent (35). However, these treatments only provide superficial wettability rather than true compatibility with cellulose. This is reflected in the drop in strength of papers and boards that contain more than 5–10 wt % synthetic pulp with no subsequent thermal consolidation. Several types of pulp modifications improve blend sheet strengths and increase response to conventional paper strength additives. For example, a mixture of unmodified polyolefins and copolymers of α-olefins and acrylic acid or maleic anhydride has been subjected to a spurting process as an aqueous emulsion to effect polar groups in the resulting pulp (36). Chemical grafting of polar monomers, eg, acrylamide, acrylic acid, and maleic anhydride, also has been reported (37–39). The grafting can be carried out prior to the spurting process or onto preformed pulp fibers in an aqueous slurry. Ozone gas has been passed through a dilute aqueous slurry of polypropylene pulp (0.5–1.0% O_3 uptake based on fiber); this practice apparently generates carboxyl groups (40). All of the synthetic pulps containing carboxyl groups respond better toward paper wet-strength resins, eg, epichlorohydrin-modified aminopolyamides or polyamines, than do unmodified pulps. Special strength resins also have been designed for use with carboxylated pulps (see Papermaking additives). For example, mixtures of cationic wet-strength resins and anionic glyoxal-modified polyacrylamide resins are quite effective (41–42). It appears that a chemically bonded network forms between synthetic pulp fibers and cellulose as well as between the synthetic fibers.

Improved water dispersibility or bonding, or both, have been effected by precipitation of appropriate materials onto the surfaces of pulp fibers. Thus, anionic-cationic colloidal complexes have been formed in the presence of synthetic pulp, with preferred complexes being poly(ethylene-co-acrylic acid)/polyethyleneimine and carboxymethyl cellulose/melamine–formaldehyde resin (43). Another system involves alum precipitation of rosin salts onto pulp surfaces to render them hydrophilic and more receptive to starch, which is applied at the size press during manufacture of offset papers (44). Improved printability results.

Economic Aspects

Total manufacturing capacity for synthetic pulps was estimated in early 1982 to be ca 45,000 metric tons per year worldwide. A rough breakdown is shown in Table 2. The only full-size commercial unit is operated by Lextar in Deer Park, Texas. It came onstream early in 1981 and is capable of producing HDPE and PP pulps. Montedison and Shell produce developmental quantities of pulps containing substantial amounts of inorganic fillers.

Approximate prices (early 1981) of commercial synthetic pulps in the United States were: Pulpex E and P, $1.69/kg and $1.56/kg, respectively; Fybrel, ca $2.40/kg;

Table 2. Synthetic Pulp Production Facilities

Company	Location	Approximate size, t/yr	Polymer type	Trade name
Lextar[a]	United States	25,000	PE and PP	Pulpex E and P
Crown-Mitsui[b]	Japan	11,000	PE	Fybrel[c]
Montedison S.p.A.	Italy	6,000	PE[d]	Ferlosa
Lextar	Italy	5,000	PE and PP	Pulpex E and P
E. I. du Pont de Nemours & Co.	United States	na	Aramid	Kevlar pulp

[a] A Hercules, Inc. company.
[b] Joint venture of Crown Zellerbach and Mitsui Petrochemical.
[c] Called SWP in Japan and Hostapulp in Europe.
[d] Sometimes filled with kaolin clay or TiO_2.

Kevlar Pulp, $8.25/kg. The pulps are not competitive, on the basis of direct weight replacement, with even the higher priced cellulose fibers, eg, bleached softwood kraft, which is priced at ca $0.52/kg. However, because of the previously mentioned high bulk of the synthetic pulps, particularly of unfilled polypropylene, papers, boards, felts, etc, that are sold on an area basis from synthetic pulp can compete with conventional products. In some paper and board applications, the unusual properties conferred by synthetic pulps, eg, very high opacity and brightness, unusual porosity effects, embossability, dimensional stability, thermoformability, and outstanding wet properties after fusion, justify the higher price. The ban on asbestos in various applications creates needs for synthetic pulps at temperatures below their melting points. Although asbestos is much lower in price, replacement ratios as low as 1:10 synthetic:asbestos have been demonstrated so that competitive products are possible.

Although the pricing of synthetic pulps is on a dry basis, the pulps ordinarily are shipped as wet mats with solids levels of 40–60 wt %. Pulps from semicommercial plants have been provided as individually wrapped bales weighing 30–50 kg/bale. However, Lextar's commercial product weighs 454 kg, is shipped on a skid, and is overwrapped, with plastic film. For some applications, dry, voluminized, or fluffed synthetic pulps are required. Typically, these are prepared from wet lap by opening and dispersing the wet pulp fibers, flash-drying, collection, and mild compaction for shipment. Once dried at elevated temperatures, the pulps may not be redispersible in water, depending on the pulp. Prices for dry pulp are higher than for wet. However, if markets for dry pulp increase, dry pulp will probably be manufactured directly from polymer rather than going through the wet stage, so that dry and wet pulp prices could be about equal. A unique dry-pulp collection scheme has been patented (45).

In an era of increasing concern about hydrocarbon energy sources and petroleum feedstocks, the long-term economics of pulps made from plastics have been questioned. However, synthetic pulps are not designed to replace pulps from regenerable resources but rather to complement them. Also, the health hazards associated with asbestos require the use of substitute materials, and synthetic pulps are attractive candidates. Third, only ca 2% of world oil consumption is used for making plastic materials, so raw material sources for synthetic pulps seem assured (46). Lastly, energy consumption in plastics production, including the oil used as feedstock, is lower than that in production of conventional papers and boards (3,46).

Health and Safety Factors (Toxicology)

Synthetic polyolefin pulps that are rendered water-dispersible by treatment with poly(vinyl alcohol) comply with FDA Regulations 21 CFR 176.170, provided the extraction limitations are met, and 21 CFR 176.180. These regulations involve use of paper and paperboard components in contact with aqueous and fatty foods and with dry food, respectively. Both polyethylene and polypropylene in nonpulp form have been demonstrated to be practically nontoxic in animals and humans. The acute oral lethal dose for rats is greater than 500 mg/kg, which is the maximum feasible dose. Neither polymer is absorbed through the skin, and they do not irritate the skin nor are they skin sensitizers. Small fibrils of Pulpex P polypropylene pulp from Lextar caused only a typical foreign body reaction when injected directly into the lungs of rats. The rats were observed for the rest of their lives, and the Pulpex P caused no alteration or destruction of lung tissue. All particles were not promptly expelled, however,

428 PULP, SYNTHETIC

so it is recommended that atmospheric levels of Pulpex be kept below 5 mg/m^3 until additional exposure studies are carried out.

Uses

Commercially established applications for synthetic pulps include wallpaper, teabags, wet-lay nonwovens, flooring felts (qv), battery separators, filters, and textured compounds. Many other uses are being developed. The existing and potential applications can be divided into paper and nonpaper uses with the former applications further divided into those that do not require thermal treatment and those that do. In thermal treatment, or fusion or thermal consolidation, a formed sheet is heated to melt the synthetic fibers. Many of the nonpaper uses are for asbestos replacement and to control the rheology of various formulations. A representative, although by no means complete, list of applications is shown in Table 3.

Paper. In terms of paper-machine operability, there are no significant problems associated with the use of either polyethylene or polypropylene pulps. However, refining of synthetic pulps by themselves should be avoided by the papermaker, because certain types of refining equipment can form fiber bundles or knots. Although the bundles are not fused at this point, they are difficult to break apart and can lead to grease spots in the final sheet owing to subsequent fusion during calendering. High pressure can fuse fiber bundles even without heat, eg, in metal–metal calendering. Polyethylene pulps are more susceptible to balling than are polypropylene pulps. Difficulties usually are not encountered when blends of cellulose and synthetic pulps are refined; it appears that almost all of the energy is absorbed by cellulose. Unlike refining of cellulose, refining of synthetic pulps only shortens their length and no hydration or fibrillation occurs.

Drainage on the paper-machine wire generally is much faster for stocks containing

Table 3. Selected Synthetic Pulp Applications

Paper[a]		Nonpaper	
Unconsolidated	Consolidated	Rheology control	Miscellaneous
printing and writing	teabags	caulks, sealants[b]	dry-laid nonwovens
wallpaper[c]	wet-laid nonwovens	asphalt coatings[b]	vinyl tile[b]
sterilizable papers[c]	battery separators	joint cement	textured paints
flooring felts[b,c]	automotive board	spray cement[b]	textured compounds
filters[c]	drum lids and liners	adhesives	
coating base stock	labels		
	folding boxboard		
	corrugated boxes		
	luggage and shoe board		
	embossable papers		
	glassine, greaseproof		
	backing sheets		
	charts		
	release papers		

[a] Paper applications are arbitrarily defined here as those made on fairly conventional papermaking equipment.
[b] Synthetic pulp used for asbestos replacement.
[c] Sometimes also consolidated.

polypropylene pulps as well as some polyethylene pulps, depending on the grade, than it is for 100% cellulose. Easier drying has resulted, although it usually is more facile for polyethylene pulps. Care must be taken not to use temperatures that are greater than the melting point of polyethylene in the last dryer cans of the paper machine. Because of improved drainage and drying, higher productivity is possible on some machines, particularly those for higher refined cellulose furnishes eg, glassine. The response of synthetic pulp–cellulose blends to most wet-end sizing additives is not very different from that of 100% cellulose furnishes. Wet- and dry-strength resins also function normally as long as the synthetic pulp has not been modified to introduce carboxyl groups, in which case synergistic effects occur (41–42). Size press starch, however, does not uniformly cover polypropylene pulps unless special dispersion treatments are applied. Dyeing of mixed furnishes usually gives a lighter color shade than is ordinarily observed. Curiously, very intense colors develop upon heating of dyed papers containing polyolefin pulps. Because of the high surface area of synthetic pulps, excellent retention of pigments and inorganic fillers, ie, by wet-end addition, usually is obtained. Some care is required when rewinding paper containing synthetic pulps, especially polypropylene pulps, on equipment with stationary metal parts. For example, dust and needles of plastic can form at speeds greater than ca 500 m/min with polypropylene pulps wetted with poly(vinyl alcohol). Paper containing pulps that are modified to introduce carboxyl groups must be rewound more slowly. This problem can be rectified by rewinding at slower speeds, by using equipment with all-moving parts, by fabricating the stationary parts from high density polyethylene, or by wrapping the stationary metal parts with Teflon or polyethylene tape. Broke reworking of papers containing polyolefin synthetic pulps is not problematic as long as the pulp has not been melted but, once melting has occurred, broke reworkability depends heavily on synthetic pulp content. At synthetic pulp levels <10–15 wt %, equipment ordinarily used for wet-strength papers probably is sufficient but, at greater synthetic-pulp levels, special equipment that is capable of high shear is required.

Unconsolidated. In printing and writing papers, the properties of synthetic pulps, specifically polypropylene pulps, that can be used to advantage are high opacity, brightness, and bulk. Lighter weight papers can be made with caliper maintained even after calendering to high smoothness. Therefore, improved operating economics can result, provided that the paper can be sold on an area basis at the same price as heavier, filled cellulose sheets and that machine productivity on a weight basis is unchanged (40). Figure 7 shows the effect of polypropylene pulp on opacity, brightness, and smoothness. Deep and essential permanent embossability also can be attained. For example, embossable depths of 250 μm and 700 μm are obtained from simplex wallpapers containing 17 wt % and 25 wt % polyethylene pulp, respectively. Duplex and triplex wallpaper constructions sometimes are used; often one layer is rich in synthetic pulp and is treated thermally. Other useful properties of papers containing synthetic pulps are: heat sealability, porosity, smoothness, wet strength, thermoplasticity, thermoformability, chemical inertness, dielectric properties, dimensional stability, and barrier properties.

Although polyolefin pulps are dispersible in water, they do not absorb water so their dimensions are not affected by contact with liquid water or water vapor. As a result, sheets containing synthetic pulp tend not to change dimensions as relative humidity changes. This property seems to be affected only slightly by sheet calendering or fusion, as shown in Figure 8. Dimensionally stable papers are used for labels, charts, and wallpapers.

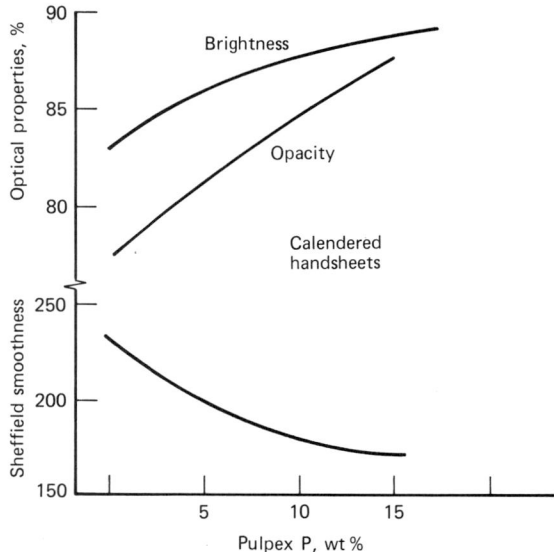

Figure 7. Effect of Pulpex P on optical properties and smoothness.

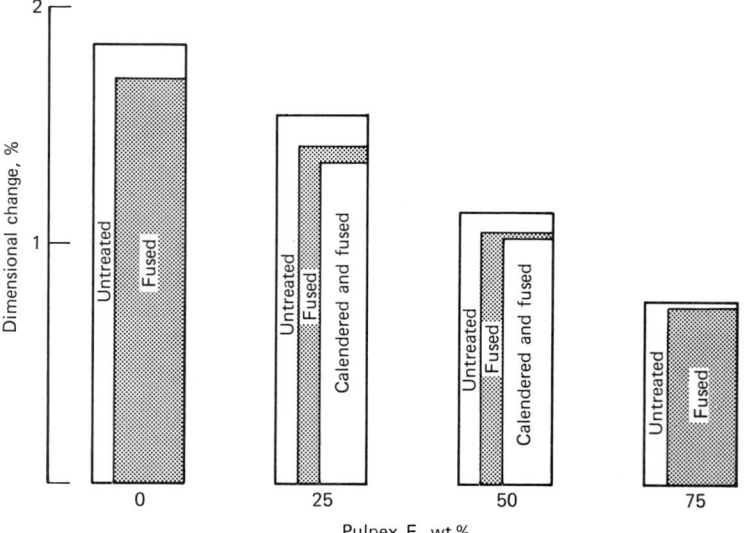

Figure 8. Effect of Pulpex E on dimensional stability.

Strengths of unfilled papers containing synthetic pulp that has not been thermally fused typically are less than those of corresponding 100% cellulose sheets. As illustrated in Figure 9, fold strength decreases more than tensile and burst strengths, although tear strength is essentially unaffected. Strength comparisons are much more favorable, especially in the z direction, for sheets containing unfused synthetic pulps as compared to highly filled, 100% cellulose sheets with comparable optical properties.

Specialty filters, eg, sterilizable papers, have been made from synthetic pulps because of their porous nature and very high surface area.

An application for synthetic pulps which is not truly in the form of paper but involves paper manufacturing equipment is the felt backing used in roll vinyl flooring.

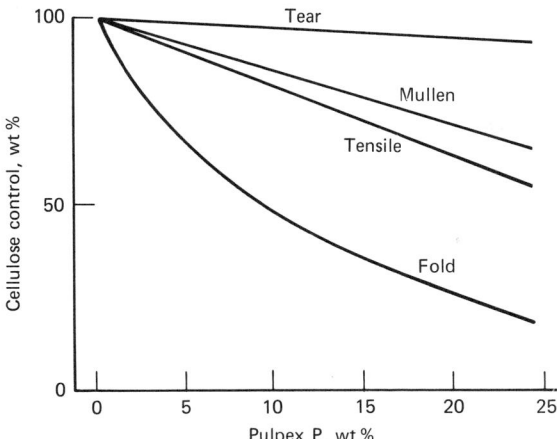

Figure 9. Effect of Pulpex P on x, y strength values in unfilled paper. Cellulose control, wt %, 67:33, bleached hardwood kraft:bleached softwood kraft, refined to 327 CSF. Experimental paper machine data.

Felts containing polypropylene pulp, glass, cellulose, large quantities of inorganic filler, latex, and sometimes bonding resins perform as well as the latex-bonded asbestos felts. Polypropylene pulp allows caliper to be maintained at reduced weight. Felt is sold on an area basis so that synthetic-based products can compete economically.

Consolidated. Polyolefin pulps in papers and boards can be fused by raising the temperature of the sheet above ca 130°C for polyethylene and above ca 165°C for polypropylene or by using a combination of pressure and lower temperatures. However, care must be taken since polyolefins may adhere to metal at temperatures close to their melting points. Many types of equipment can be used to fuse the synthetic pulps, eg, infrared heaters, hot-air tunnels, pull-through dryers, heated platens, heated molds, heated calenders, and embossing rolls. Exact conditions must be determined for the specific sheet being used. Detailed studies have been reported for certain compositions (47). It is easier and more energy-efficient to fuse boards containing polyethylene than those containing polypropylene, but several applications require the higher softening temperature of the latter.

Fusion has a dramatic effect on the physical properties of papers and boards containing polyolefin synthetic pulps, and the magnitude of the effect depends on synthetic pulp content. Fused composites at synthetic pulp levels above ca 50 wt % are regarded as a filled plastic made with paper equipment. Opacity decreases dramatically after fusion, as illustrated in Figure 10, and this is the basis for some unusual effects in decorative and embossed papers. If ca 70 wt % polyethylene pulp is present, resistance to moisture vapor can be achieved, eg, in folding boxboard, but lower synthetic pulp levels provide good resistance to penetration of aqueous and oily liquids, eg, in glassine and greaseproof papers. Heat sealability is possible for papers containing ca 40 wt % or more synthetic pulp, eg, in teabags and sterilizable papers. The high chemical resistance of synthetic pulps and their good dielectric properties have been utilized in battery separators (48).

Strength properties of papers and boards containing synthetic pulps are improved considerably by thermal consolidation. If fusion is carried out without use of pressure, the bulkiness imparted by polyolefin pulps can be maintained and strength increases considerably. Even higher tensile and burst strengths develop with sheet compaction (see Table 4).

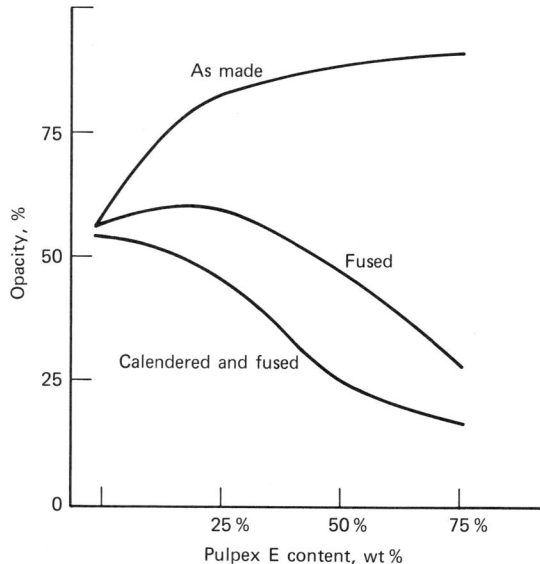

Figure 10. Effect of Pulpex on opacity. Experimental paper machine data.

Application of pressure at sheet temperatures above the polyolefin melting point is preferred to cold compaction before or after heating. Whether or not dry tensile and flexural properties of fused sheets containing polyolefin pulps are equal to or exceed those of sheets without synthetic pulps depends on the specifics of the system. For example, boards from an unfilled, well-refined, good quality virgin-cellulose pulp often have equal or better dry properties than those of fused boards containing polyolefin pulps. However, the dry properties of boards prepared from waste, unrefined, or highly filled cellulose can be improved by addition of polyolefin pulps followed by fusion and densification. In addition, wet tensile and flexural properties as well as water absorption are much better from fused boards containing polyolefin pulps, regardless of the type of cellulose, amount of refining, etc.

Table 4. Effect of Thermal Treatment of Boards Containing Pulpex E

Furnish	Board treatment	Basis weight, g/m²	Sp gr, g/cm³	Dry tensile strength, kN/m[a]	Wet tensile strength, kN/m[a]	Water absorption, wt %
cellulose[b]	none	244	0.6	12.5	0.1	190
80:20 cellulose:Pulpex E	none	250	0.6	5.1	0.2	184
	fused[c]	250	0.7	6.7	0.7	83
	fused and densified[d]	250	1.1	13.0	1.1	71
60:40 cellulose:Pulpex E	none	234	0.6	3.6	0.3	133
	fused	234	0.7	6.5	2.0	36
	fused and densified[d]	234	1.1	10.0	3.4	28

[a] To convert kN/m to lb/in., divide by 0.175.

[b] 50:50 bleached hardwood kraft:bleached softwood kraft; refined to 500 CSF. All boards were made on a pilot paper machine.

[c] At 150°C.

[d] Flat press at 150°C.

In contrast to all-cellulose boards, those that contain synthetic pulps can be thermoformed into various shapes. The depth of draw depends on several variables, eg, the amount of synthetic pulp, the nature of the cellulose used, the extent of refining, the kind and amount of inorganic fillers, and proper heating and forming equipment. Draws of 7–8 cm and more have been realized under optimum conditions, eg, use of 30–40 wt % synthetic pulp, unrefined or certain waste cellulose pulps, and good preheating equipment. Boards containing polyethylene pulp or polypropylene pulp are thermoformed equally well if the higher softening temperature of the latter pulp is taken into account.

The strength properties and thermoformability of wet-laid boards containing polyolefin pulps are useful in such products as automotive boards, drum lids and liners, rigid-when-wet corrugated boxes, luggage, shoeboard, and various backing sheets. Dry-laid boards also have been prepared from synthetic pulps and cellulose or wood fibers. The resulting materials are beginning to be competitive with wet-laid boards and plastics in automotive and furniture applications.

Mixtures of synthetic pulps and water-dispersible glass fibers can be formed into mats for subsequent fusion and, if desired, thermoforming. Fourdrinier machines, rotoformers, and special high dilution formers have been used. As is true for synthetic pulp–cellulose composites, the synthetic pulp content greatly determines properties. Thus, synthetic pulps can be used at a low level (5–15 wt %) to function as a binder or at higher levels (20–60 wt %) for deep or intricate thermoforming and to achieve properties approaching those of high performance plastics.

Nonpaper. *Rheology Control.* Use of synthetic pulps in rheology-control applications (ie, for control of flow properties) generally, although not always, is to replace asbestos. However, polyolefin synthetic pulps can be used only below their melting points. The properties provided by polyolefin synthetic pulps are viscosity control, slump resistance, prevention of crack propagation, pumpability, etc. In some applications, such as joint cements, spray cements, and emulsion asphalt coating, hydrophilic pulps are required. In others, eg, solvent cutback asphalt coatings, caulks, mastics, and adhesives, good compatibility with organic components is necessary to avoid separation or floating. For these applications, special pulp treatments have been devised and dry pulps usually are required. In many rheology-control uses, fiber lengths must be shorter than normal for maximum effectiveness. Even without reduction in fiber length, polyolefin synthetic pulps usually do not function as reinforcing fibers, ie, tensile and stiffness properties are not improved. However, higher impact strengths can be achieved.

Miscellaneous. In textured paints and compounds, polyolefin synthetic pulps serve as texturing agents as long as fiber length is not excessively short (see Paint). Improvements in hot strength during manufacture as well as dimensional stability result from synthetic pulps in vinyl floor tile. Synthetic pulps, particularly polyethylene, serve as bonding agents after thermal treatment of nonwoven webs formed both by air-lay and carding techniques. Despite the short fiber length of synthetic pulps compared with the length of staple fibers used in nonwovens, pulp retention during forming is very good. Use of 5–20 wt % synthetic pulp to bind rayon, polyester, polypropylene, or cellulose fluff pulp mats enables nonwoven manufacturers to use thermal bonding as opposed to latexes.

The largest use of aramid pulps is for asbestos replacement in gaskets for moderate and high temperature service, eg, engine head and manifold gaskets, protective apparel, and certain conveyer belts. Other possible applications are brake pads, brake linings, clutch faces, and felt furnace gaskets (1).

Acknowledgment. The author acknowledges the assistance of J. P. Pleska and G. Voituron (Solvay et Cie) in the preparation of this article.

BIBLIOGRAPHY

1. *Chem. Week*, 20 (Nov. 26, 1980).
2. R. A. Hentschel, *Nomex Aramid Papers, Properties and Uses*, TAPPI Paper Synthetics Conference Preprints, Pittsburgh, Pa., Oct. 6–8, 1975.
3. V. M. Wolpert, *Synthetic Polymers and the Paper Industry*, Miller-Freeman Publications, San Francisco, Calif., 1977.
4. Jpn. Kokai 75 154,318 (Dec. 12, 1975), S. Takada and co-workers (to Kanegafuchi Chemical Industry Company).
5. Jpn. Kokai 75 100,318 (Aug. 8, 1975), N. Nishioka and co-workers (to Mitsubishi Rayon Company, Ltd.).
6. U.S. Pat. 4,224,259 (Sept. 23, 1980), B. Sander, E. Bonitz, and K. Scherling (to BASF Aktiengesellschaft).
7. U. S. Pat. 3,963,821 (June 15, 1976), H. Takeda, T. Kobayashi, and K. Oka (to Toray Industries).
8. U.S. Pat. 3,950,473 (Apr. 13, 1976), E. Iwahori, M. Kurita, and M. Uno (to Chisso Corporation).
9. U.S. Pat. 4,260,565 (Apr. 7, 1981), F. D'Amico, G. Serbaki, and V. Foti (to Anic S.p.A.).
10. M. G. Halpern, *Synthetic Paper from Fibers and Film*, Noyes Data Corporation, Park Ridge, N.J., 1975.
11. U.S. Pat. 4,127,624 (Nov. 28, 1978), L. B. Keller and R. K. Jenkins (to Hughes Aircraft Co.).
12. U.S. Pat. 3,978,185 (Aug. 31, 1976), R. R. Buntin, J. P. Keller, and J. W. Harding (to Exxon Research and Engineering Company).
13. J. d'A. Clark, *Pulp Technology and Treatment for Paper*, Miller-Freeman Publications, Inc., San Francisco, Calif., 1978, pp. 496–97.
14. *TAPPI Standard T233-Su-64*, Technical Association of the Pulp and Paper Industry, Atlanta, Ga.
15. W. A. Kindler, *TAPPI* **58**(3), 103 (1975).
16. U.S. Pat. 3,848,027 (Nov. 12, 1974), D. L. Forbess, W. A. Kindler, and R. A. Marti (to Crown Zellerbach Corporation).
17. U.S. Pat. 4,010,229 (Mar. 1, 1977), J. P. Pleska and M. Marechal (to Solvay & Cie).
18. U.S. Pat. 4,025,593 (May 24, 1977), C. Raganato and G. Voituron (to Solvay & Cie).
19. U.S. Pat. 4,081,226 (Mar. 28, 1978), J. P. Pleska and M. Marechal (to Solvay & Cie).
20. U.S. Pat. 4,105,727 (Aug. 8, 1978), G. Zanella and L. Mancini (to Montedison S.p.A).
21. U.S. Pat. 4,211,737 (July 8, 1980), G. Di Drusco, D. Zaffagnini, and S. Biagio (to Montedison S.p.A.).
22. U.S. Pat. 3,081,519 (Mar. 19, 1963), H. Blades and J. R. White (to E. I. du Pont de Nemours & Co., Inc.).
23. U.S. Pat. 3,227,784 (Jan. 4, 1966), H. Blades and J. R. White (to E. I. du Pont de Nemours & Co., Inc.).
24. U.S. Pat. 4,129,629 (Dec. 12, 1978), W. Gordon (to Hoechst Aktiengesellschaft).
25. U.S. Pat. 3,987,139 (Oct. 19, 1976), J. Kozlowski and R. E. Howard (to Crown Zellerbach Corp.).
26. U.S. Pat. 4,054,625 (Oct. 18, 1977), J. H. Kozlowski, P. C. Litzinger, and F. J. Steffes (to Crown Zellerbach Corporation).
27. U.S. Pat. 4,167,548 (Sept. 11, 1979), G. Arduini, M. Ghirga, and G. Renda (to Societa Italiana Resine S.I.R. S.p.A.).
28. Brit. Pat. 3,891,540 (Jan. 12, 1978), (to Shell Internationale).
29. U.S. Pat. 3,891,610 (June 24, 1975), R. W. Fowells (to Crown Zellerbach Corporation).
30. U.S. Pat. 4,049,492 (Sept. 20, 1977), D. W. Lare (to Champion International Corporation).
31. U.S. Pat. 4,013,751 (Mar. 22, 1977), J. C. Davis, F. R. Galiano, and R. W. Hill (to Gulf Research and Development).
32. U.S. Pat. 3,882,095 (May 6, 1975), R. W. Fowells, R. A. Damon, and J. G. Coma (to Crown Zellerbach Corporation).

33. U.S. Pat. 3,743,272 (July 3, 1973), K. A. Nowotny and W. L. Shilling (to Crown Zellerbach Corporation).
34. P. Blais and R. St. John Manley, *Science* **153,** 539 (1966).
35. J. P. Bex, *New Developments in Synthetic Polyolefin Pulps*, 1976 International Synthetic Papers and Pulps Symposium, Mar. 31–Apr. 2, 1976.
36. Ger. Pat. 2,649,139 (Apr. 7, 1977), T. Fujita, H. Ishida, T. Kamaishi, and S. Otsu (to Toray Industries).
37. Ger. Pat. 2,649,139 (Nov. 3, 1975) G. Voituron and J. P. Pleska (to Solvay & Cie).
38. Belg. Pat. 824,531 (Jan. 21, 1974), P. Pirro (to Solvay & Cie).
39. U.S. Pat. 4,080,405 (Mar. 21, 1978), E. Agouri, R. Laputte, and J. Rideau (to ATO Chimie).
40. T. W. Rave, *Use of Polypropylene Synthetic Pulps in Printing and Writing Papers*, 1980 TAPPI International Symposium on Synthetic Pulps and Papers, Sept. 14–17, 1980.
41. U.S. Pat. 4,154,647 (May 15, 1979), T. W. Rave (to Hercules Incorporated).
42. U.S. Pat. 4,156,628 (May 29, 1979), T. W. Rave (to Hercules Incorporated).
43. U.S. Pat. 3,743,570 (July 3, 1973), C. Yang and W. A. Kindler (to Crown Zellerbach Corporation).
44. Eur. Pat. Appl. 14,534 (Aug. 20, 1980), R. W. Davison (to Hercules Incorporated).
45. U.S. Pat. 3,883,630 (May 13, 1975), C. Raganato (to Solvay & Cie).
46. V. M. Wolpert, *Review of Synthetic Pulps and Papers*, 1976 International Synthetic Papers and Pulps Symposium, Mar. 31–Apr. 2, 1976.
47. S. E. Back and L. J. Carlson, *Pap. Technol.* **16,** 309 (1975).
48. U.S. Pat. 4,216,281 (Aug. 5, 1980), D. O. O'Rell, N. I. Palmer, and V. H. Nguyen (to W. R. Grace & Co.).

TERENCE RAVE
Hercules, Incorporated

PUMPS.   See Supplement volume.

PURGATIVES.   See Gastrointestinal agents.

PURINES.   See Alkaloids; Chemotherapeutics, antimitotic

PYRAZOLES, PYRAZOLINES, PYRAZOLONES

Pyrazoles (1) are stable, five-membered, heterocyclic compounds with two nitrogen atoms in a 1,2 relationship. They can be represented by several tautomeric forms, of which two are the most common. Pyrazoles are very scarce in nature, whereas the isomeric imidazoles, which are characterized by two nitrogens in a 1,3 relationship, occur in most living systems. Pyrazolines (2) are dihydropyrazoles and are less stable than pyrazoles. Three tautomeric forms are known; 2-pyrazolines are the most common. Pyrazolones (3) or oxopyrazolines exist in three isomeric forms, and they also display keto-enol tautomerism. Of the three classes of compounds, pyrazolones have had the greatest commercial application, mainly as pharmaceuticals (qv) and dyes (see Dyes, application and evaluation). The pyrazolines are used mainly as textile brighteners, and the pyrazoles have pesticidal and pharmaceutical utility.

Pyrazoles

The pyrazole ring displays several tautomeric structures; the predominant ones are (4) and (5).

When the nitrogen atom carries no substituent, only one isomer has been isolated, but substitution on nitrogen forms two isomers, whereas removal of the substituent yields only one isomer. The two nitrogen atoms have different environments in the gas, solid, and solution states. A comprehensive review of tautomerism in pyrazoles and of the means of studying them is given in ref. 1. Nmr studies have shown that pyrazole [R = H] [288-13-1] consists of tautomers (4) and (5), which interconvert through the cyclic dimer (6). Because of very rapid interconversion, averaging occurs and the 3- and 5-hydrogens are magnetically equivalent. A number of excellent books on pyrazoles are available (2–4). The more recent reviews are brief (5–7).

Physical Properties. Pyrazoles are very stable compounds with high boiling points. Increasing substitution on the carbon atoms causes higher boiling points, but substitution on nitrogen lowers both melting and boiling points. For example, 3-methylpyrazole [1453-58-3] boils at 205°C, but 1-methylpyrazole [930-36-9] boils at 127°C

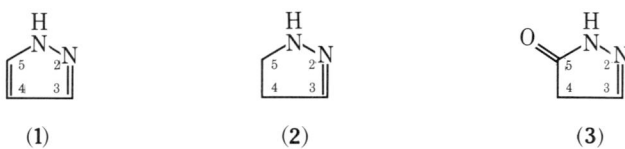

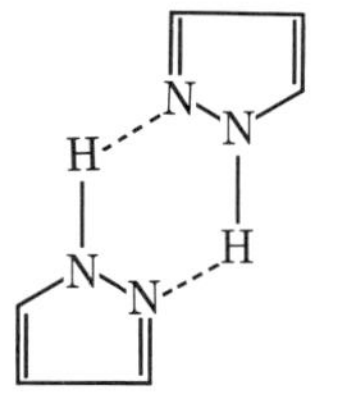

(6)

(8). These properties reflect an association of pyrazoles that have hydrogen on nitrogen.

Pyrazole and its lower homologues are soluble in water and most organic solvents. The specific gravities of pyrazoles are 0.89–1.02, and the refractive indexes are 1.46–1.48 (9). The low dipole moments of pyrazoles [8.0×10^{-30} C·m (2.4 D) for pyrazole] are usually attributed to strong association. Pyrazoles that have hydrogen on nitrogen are weakly acidic and weakly basic. They form salts with strong acids, and with metals by replacement of the hydrogen on the nitrogen atom (10). Although the pK_a of pyrazole is 2.53, N-substituted pyrazoles are not acidic.

The uv spectra of pyrazoles have been studied (11–12). Alkylpyrazoles have maxima at 210–225 nm (log ϵ 3.5–4.0), but arylpyrazoles absorb at 250–280 nm (log ϵ 3.4–4.2). Electronegative substituents give rise to a bathochromic shift of 25–40 nm. The ir spectra of pyrazoles reflect the expected bands (13). Absorption occurs at 3500–3100 cm^{-1} if NH is present and the C=N system gives rise to a strong band at 1590 cm^{-1}; weak bands at 1660 and 1550 cm^{-1} are attributed to C=C.

The ^{1}H nmr spectra of pyrazoles have been extensively reviewed (14). The method is most useful for the assignment of peaks for the hydrogens in the 3- and 5-positions in 1-substituted pyrazoles. Variation of the solvent from nonpolar to polar causes an appreciable shift for the 5-H but not the 3-H. The 4-H is normally at a higher field. The coupling constant between the 4- and 5-hydrogens is usually greater than between the 3- and 4-hydrogens. Detailed tables with chemical shifts for many pyrazole derivatives also appear in the above reference. Little has been published on ^{13}C nmr of pyrazoles (15–17). Carbon-13 nmr is a useful method for distinguishing isomers (18). Methylation of pyrazole causes an upfield shift of the carbon adjacent to the methylated nitrogen and a downfield shift to the carbon adjacent to the nonmethylated nitrogen. The ^{19}F nmr spectra of fluorine-substituted 1-methylpyrazoles have also been investigated (19).

Chemical Properties. Pyrazole is an aromatic molecule with six π electrons. In N-substituted pyrazoles, the 1-N has some cationic character and the 2-N retains two electrons and, therefore, has basic properties. The greater nuclear charges of the two nitrogen atoms reduce the charge density at positions 3 and 5, and this leads to a greater charge density at position 4 and, therefore, to that position being more prone to electrophilic attack. Conversely, nucleophilic attack is more likely to occur at the 5-position.

The halogenation of the silver salt of pyrazole has been studied (20). Stepwise halogenation can be performed; however, condensation of two or more pyrazole molecules also takes place. Chlorination with many reagents leads to the formation of 4-chloropyrazole [15878-00-4]. In acetic acid, chlorination of 3-methylpyrazole yields

3-(trichloromethyl)-4,5-dichloropyrazole [80294-32-2]. Pyrazoles that carry a phenyl group in the 1-position chlorinate only in the 4-position, but the chlorination of pyrazole itself leads to polymer formation (21).

Reaction of 1-hydroxypyrazoles with N-halosuccinimide gives the 4-halo derivatives and, with tert-butyl hypochlorite, the 4,4-dichloro-4H-pyrazole-1,1-oxides are obtained (22). 4-Chloro-4H-pyrazole derivatives (7) are prepared in a similar way from 3,5-diphenylpyrazole [1145-01-3]. They are easily converted to alkoxymethylpyrazoles (8) (23).

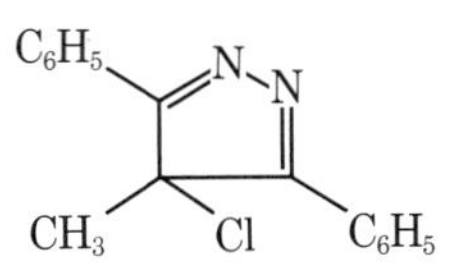

(7) [61355-01-9] 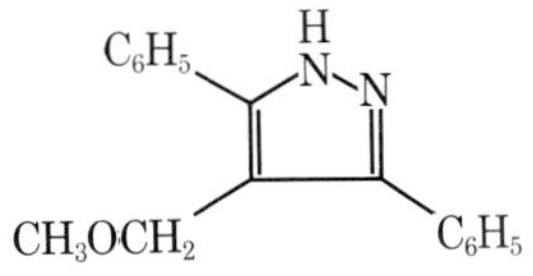   (8) [61355-04-2]

Bromination of pyrazole introduces one to three bromine atoms. With N-phenylpyrazoles, bromination also occurs in the phenyl ring. Tribromopyrazole [17635-44-8] can be obtained in almost quantitative yield by bromination in a sodium hydroxide solution (24). Iodination of pyrazoles is quite similar to bromination, with substitution occurring at all positions (25). Sulfonation of pyrazole occurs only under drastic conditions and yields amphoteric compounds. Ring or N-nitrosation is also difficult to effect, and 4-nitrosopyrazoles are best prepared from isonitroso-β-diketones (26). Nitropyrazoles can be obtained by direct synthesis or by nitration of pyrazole derivatives. Nitration usually occurs at the 4-position, but N-nitro compounds can also be prepared. These rearrange in acid to 4-nitropyrazoles or thermally to 3(5)-nitropyrazoles (27–28). Reduction of the basicity of the pyrazole ring enhances the nitration rate. Reduction with many reagents leads to formation of aminopyrazoles. In azopyrazoles, the azo group can be in the 3-, 4-, and 5-positions, as shown, for example, in (9).

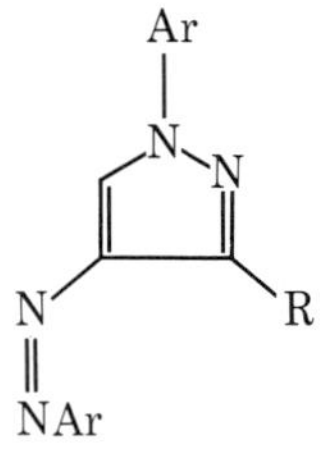

(9)

Other electrophilic reactions at the 4-position in pyrazoles are the formation of 4-formylpyrazoles by reaction with dimethylformamide and the Friedel-Crafts reaction, which is used to introduce acyl or aroyl groups. Replacement of halogen in the pyrazole ring by nucleophilic attack can only be achieved by activation at the 4-position. 1-Bromopyrazoles can be hydrogenolyzed in the presence of palladium on charcoal often in basic media. Lithium aluminum hydride removes a bromine atom from the 5-position (29). Treatment of tribromopyrazole with butyllithium, followed by carbon dioxide and thermal decarboxylation, furnishes 3,5-dibromopyrazole [67460-86-0] (30). Halogen atoms can also be removed by hydriodic acid and red phosphorus, zinc and hydrochloric acid, and sodium in alcohol. An aromatic nucleophilic substitution in the pyrazole ring also occurs during the reaction of 1,4-dinitropyrazoles (10) with secondary amines (31).

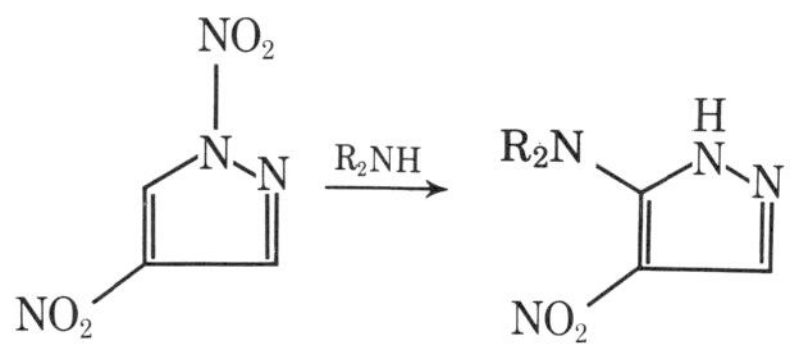

(10) [35852-77-8]

The pyrazole ring is as resistant to oxidation as it is to reduction. Alkyl side chains can be oxidized with permanganate or chromic acid to carboxylic acids, without destruction of the pyrazole ring. Only ozonolysis or electrolytic oxidation destroys the ring (32).

N-Alkylation of pyrazoles is best accomplished by sodium hydride in toluene or tetrahydrofuran. With unsymmetrically substituted pyrazoles, two products can be obtained; the ratio depends on the ring substituent and sometimes on the alkylating agent. Quaternization of N-substituted pyrazoles takes place at the 2-position. Pyrolysis causes reversion to the N-substituted pyrazole.

Other reactions on the unsubstituted nitrogen are acylations by acid chlorides (11; R = alkyl, aryl) and chloroformic esters (11, R = O-alkyl), which usually yield only one product, even with unsymmetrically substituted pyrazoles.

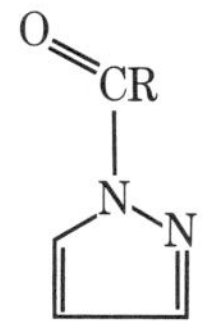

(11)

Isocyanates and carbamoyl halides yield carbamoylated pyrazoles [11; R = NH–alkyl or N(alkyl)$_2$]. Photolysis of (12) leads to migration of the $C_6H_5SO_2$ group from the 1- to the 4-position (33). 3,5-Disubstituted pyrazoles are photoisomerized to 2,4- and 2,5-disubstituted imidazoles in low yield (34). Irradiation of 1-p-nitrophenyl-3,5-disubstituted pyrazoles in the presence of acetate or cyanate ion leads to formylation in the 4-position (35).

Synthesis. There are over twenty methods for synthesizing pyrazole derivatives. However, only the most important are described below.

From β-Dicarbonyl Compounds and Hydrazine. β-Dicarbonyl compounds include any compound that has at least a contiguous 3-carbon fragment and functional derivatives that can give rise to a β-dicarbonyl system. Such systems are acetals, enamines, enols, enol ethers, etc. Hydrazine gives only one product with symmetrical and

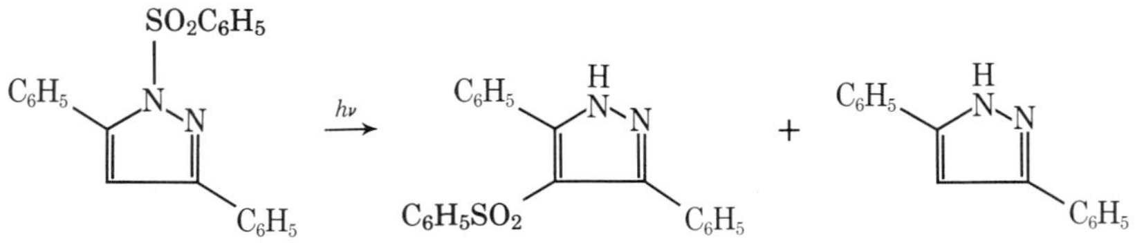

440 PYRAZOLES, PYRAZOLINES, PYRAZOLONES

(12) [52762-74-0] → [52762-75-1] + [1145-01-3]

unsymmetrical β-diketones, but substituted hydrazines give two products with unsymmetrical β-diketones.

$$RCCH_2CR' + R''NHNH_2 \longrightarrow$$

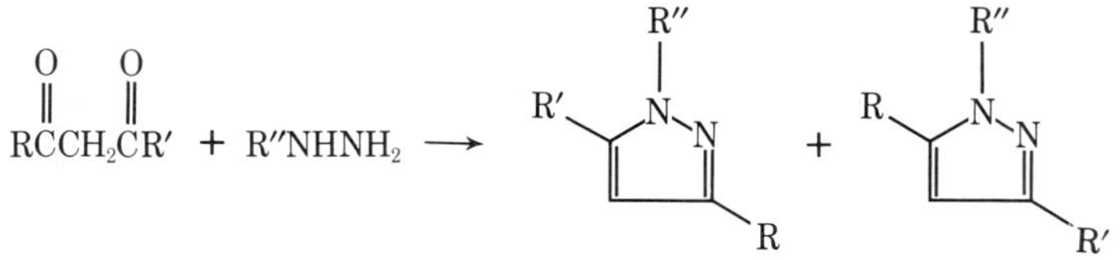

The reaction proceeds readily, although bulky substituents on the diketone can slow the rate of reaction considerably. Two substituents on the methylene give rise to 4H-pyrazoles.

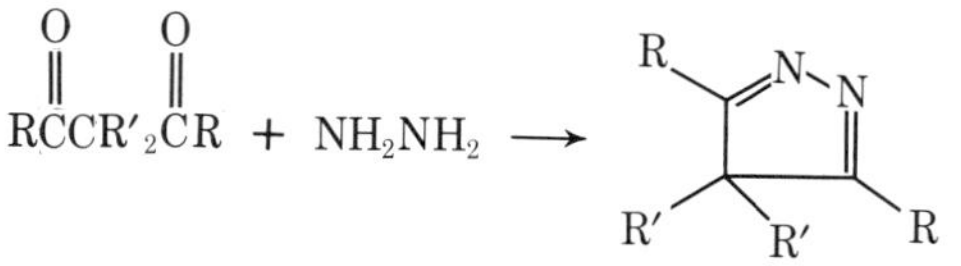

$$RCCR'_2CR + NH_2NH_2 \longrightarrow$$

The reaction is not limited to simple hydrazines. Sulfonylhydrazides, semicarbazides, aminoguanidines, and acyl hydrazines can all be employed in the synthesis of pyrazole derivatives. The N-acyl group can be removed easily by hydrolysis. β-Ketonitriles react with hydrazine to yield aminopyrazoles. A discussion of the reaction of hydrazines with 1,3-difunctional compounds is given in ref. 36.

From Unsaturated Carbonyl Compounds. α,β-Acetylenic carbonyl compounds react with substituted hydrazines to give a mixture of pyrazoles.

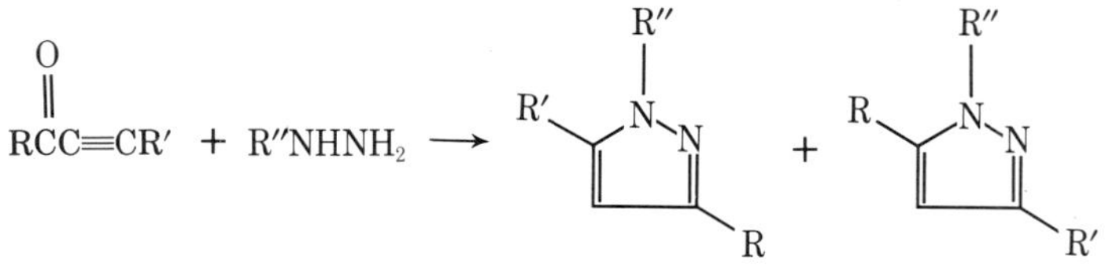

$$RCC \equiv CR' + R''NHNH_2 \longrightarrow$$

β-Chlorovinyl ketones react with phenylhydrazine to give only one isomer, which suggests that hydrazone (13) formation takes place first, followed by cyclization to 3-substituted pyrazoles (14).

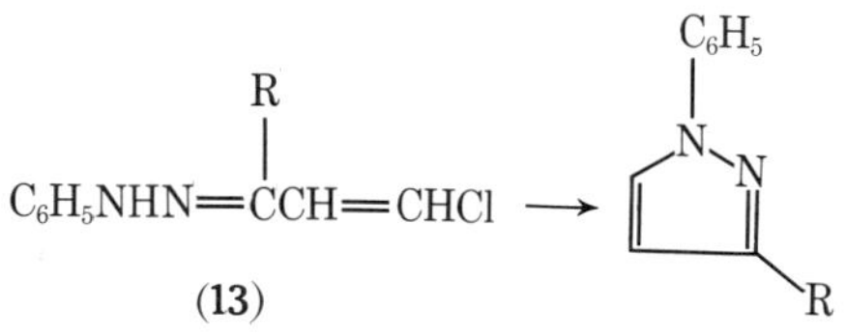

$$C_6H_5NHN=\overset{R}{C}CH=CHCl \longrightarrow \underset{(14)}{\text{[pyrazole with } C_6H_5 \text{ on N and R at position]}}$$

(13) → (14)

From Diazo Compounds. Pyrazoles can be synthesized by reaction of diazo derivatives with acetylenes. Nonsymmetrical acetylenes give a mixture of two products.

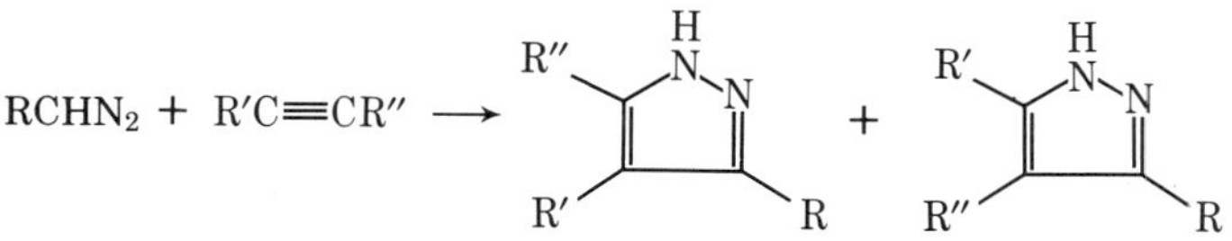

$$RCHN_2 + R'C\equiv CR'' \longrightarrow \text{[pyrazole isomer 1]} + \text{[pyrazole isomer 2]}$$

The reaction proceeds readily when the triple bond is activated by such groups as carbonyl or carboxyl; however, alkyl acetylenes also react under forcing conditions. The addition of diazomethane to α-acetylenic alcohols can also take place (37). The acetylene compounds can be replaced by olefins that include a good leaving group.

From Pyrazolines and Other Heterocyclic Systems. Pyrazoles can be obtained from pyrazolines by dehydrogenation, oxidation, and elimination. Dehydrogenation is the general method and can be effected with sulfur or selenium on alkyl, aryl, and N-substituted pyrazolines. Pyrazoles can also be obtained by elimination reactions, eg, loss of water or hydrogen halide from the pyrazolines. Depending on the substituent present, pyrazolines are oxidized to pyrazoles by the following reagents: bromine, lead tetraacetate, potassium permanganate, etc. Side chain groups can also be oxidized. 1,3,4-Oxadiazolium salts and imidazopyrimidines are converted to pyrazoles (38–39).

A novel synthesis of 3,4,5-trisubstituted pyrazoles (**16**) from 2,2-dioxoketene-*S,S*-diacetals (**15**) is described in ref. 40.

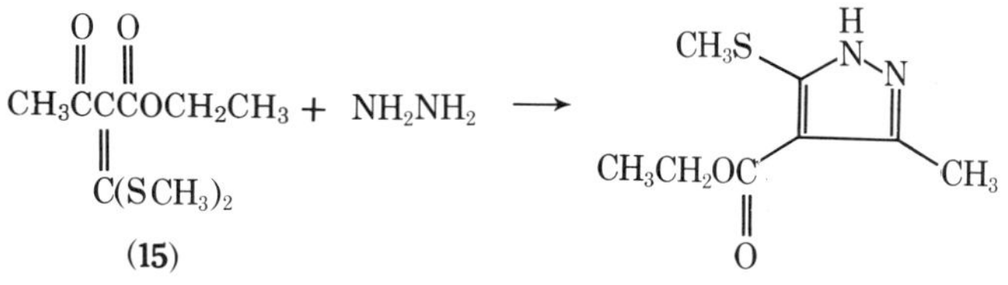

$$\underset{(15)}{CH_3\overset{O}{\overset{\|}{C}}\overset{O}{\overset{\|}{C}}COCH_2CH_3} + NH_2NH_2 \longrightarrow \underset{(16)\ [65551\text{-}58\text{-}8]}{\text{[pyrazole product]}}$$

with C(SCH$_3$)$_2$ substituent on (15); product (16) has CH$_3$S, CH$_3$CH$_2$OC(=O), and CH$_3$ substituents.

Health Factors. Pyrazoles that are substituted in the 4-position with alkyl or cycloalkyl groups strongly inhibit the activity of alcohol dehydrogenase in the human liver and may be used to combat the ill effects of alcohol abuse (41). They also are potentially useful in the treatment of methanol poisoning in human beings (42). A review of the complications and side effects of pyrazoles is given in ref. 43.

Uses. Although during the last decade a large number of pyrazole-related patents have been assigned, there are few registered products on the market. Some of the biological activities of pyrazoles are as follows: Compound (**17**) is useful for fruit abscission, and (**18**) and (**19**) have herbicidal activity (44–46) (see Herbicides).

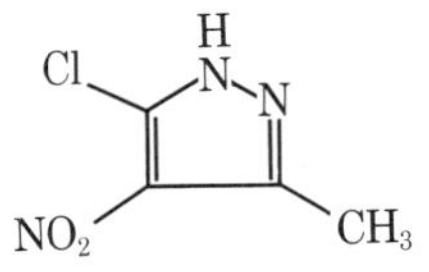

(**17**) [6814-58-0] 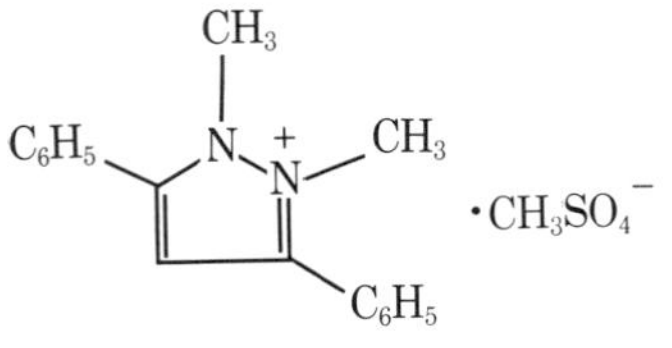   (**18**) [43222-48-6] 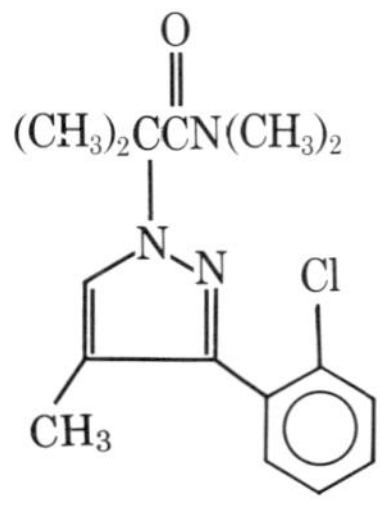   (**19**) [59845-24-8]

Other uses are as fungicides, antibacterials, and hypnotics and as dyes in the textile industry (see Antibacterial agents; Hypnotics, sedatives, anticonvulsants).

Pyrazolines

Pyrazolines are dihydropyrazoles, and three tautomeric forms are known: 1-pyrazoline [2721-43-9] (4,5-dihydro-3H-pyrazole) (**20**), 2-pyrazoline [109-98-8] (4,5-dihydro-1H-pyrazole) (**21**), and 3-pyrazoline [6569-23-9] (2,5-dihydro-1H-pyrazole) (**22**).

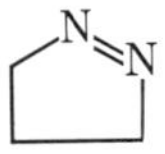

(**20**) 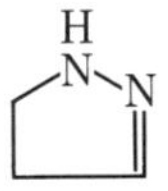   (**21**) 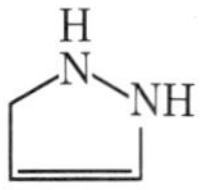   (**22**)

In cases where the individual structure is not fixed by substitution, only the most stable structure occurs and no tautomeric equilibrium exists. However, irreversible isomerization from one form to another can take place. Substitution on nitrogen increases the number of possible tautomers. Studies of the tautomerism of aminopyrazolines show that the imino (=NH) tautomer is present (47).

Physical Properties. Pyrazolines are colorless, high-boiling liquids or low-melting solids. Substitution on nitrogen lowers the melting point. Low molecular weight pyrazolines are water-soluble but, with an increase in molecular weight, they are more soluble in organic solvents. Pyrazolines are weakly basic and can be protonated depending on the position of the double bond. The uv spectra of simple pyrazolines show absorption at 240 nm. Introduction of a phenyl group gives rise to a peak at 280 nm. 3-Carboalkoxy-2-pyrazolines absorb at ca 292 nm. 2-Pyrazolines are strongly fluorescent. 3-Pyrazolines, which have a benzene ring in conjugation with the double bond, absorb at 229 and 289 nm. 1-Pyrazolines with aromatic substituents absorb at 327–329 nm. The ir spectra of unsubstituted 2-pyrazolines show absorption of the NH at 2.87–2.93 cm^{-1} and a strong C=N band at 6.37–6.39 cm^{-1} (48).

The ^{1}H nmr of the 3-pyrazolines and extensive tables of the chemical shifts of all protons are given in ref. 49. The ^{13}C nmr of pyrazolines can be used for the assignment of the position of the N-oxide function in tetra-substituted pyrazolines and in the assignment of (E) and (Z) isomers in 4-alkylidene-1-pyrazolines (50–51). Data on the mass spectral fragmentation of substituted 1,3-diphenyl-2-pyrazolines are also available (52).

Chemical Properties. A very facile reaction of pyrazolines is loss of nitrogen. Thus, 1-pyrazolines extrude nitrogen to yield mainly cyclopropanes; however, olefins may also be produced. Generally, *trans*-1-pyrazoline yields a mixture of cis and trans cyclopropanes. The ratio depends on the nature of substituents and on the use of photochemical or thermal decomposition.

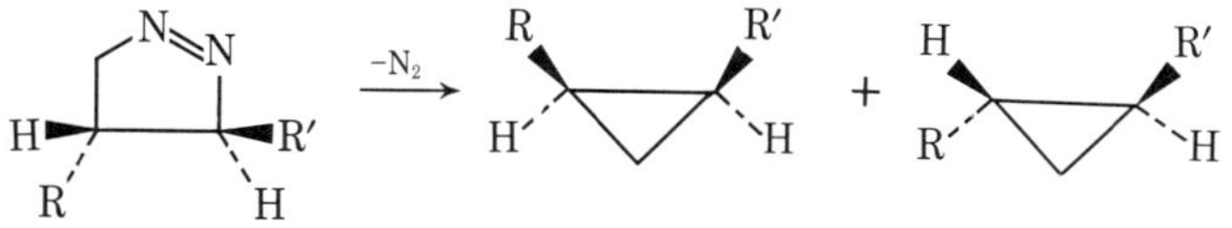

2-Pyrazolines do not lose nitrogen as easily as the 1-pyrazolines. Thermal isomerization from the 2-isomer to the 1-isomer does take place and is followed by loss of nitrogen. The stereochemistry of the resulting cyclopropanes is controlled by the intermediate 1-pyrazoline. There is some evidence that photochemical extrusion follows the same path. The stereochemistry of pyrazolines has been reviewed (53).

Pyrazolines can be oxidized to pyrazoles by means of a variety of reagents, eg, bromine or potassium permanganate; however, upon air oxidation, 1-pyrazoline undergoes ring cleavage to yield propylene and nitrogen. The selenium dioxide oxidation of a pyrazoline to a chalcone is described in ref. 54. Pyrazole formation is also possible by elimination reactions of pyrazolines. The primary reduction products from pyrazolines are pyrazolidines; however, ring cleavage can occur to yield amines.

Under alkaline conditions in an autoclave, pyrazolines produce good yields of β-amino acids (55). Several 1,3-diphenyl-2-pyrazolines dehydrogenate in the presence of light to 1,3-diphenylpyrazoles (56). 2-Pyrazolines that are unsubstituted at the 1-position can be easily acetylated, benzoylated, carbamoylated, alkylated, etc; rearrangement takes place on occasion. The anodic oxidation of N-aryl-2-pyrazolines has been studied (57). The products are pyrazoles and pyrazole–phenyl dimers.

Synthesis. The most common procedure for the synthesis of pyrazolines is the reaction of aliphatic or aromatic hydrazines with α,β-unsaturated carbonyl compounds.

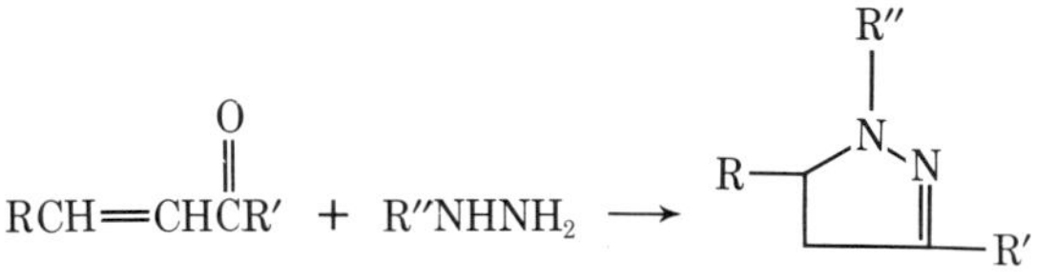

The reaction proceeds in an acidic or basic medium, with or without solvent, and at low or high temperatures. It proceeds through a hydrazone, although with aliphatic hydrazines its isolation before cyclization may be difficult. The hydrazine synthesis can also be applied to Mannich base, β-halo and β-quaternary ammonium aldehydes and ketones, and α-carbonyl derivatives of ethylene oxide and ethyleneimine. α,β-Unsaturated nitriles and hydrazine give 3-amino-2-pyrazolines.

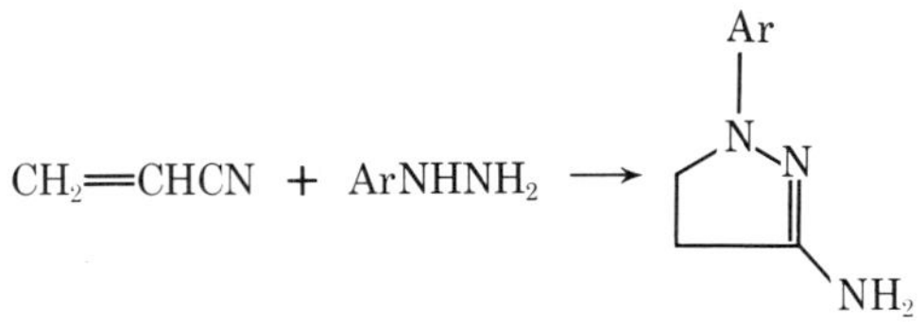

$$CH_2{=}CHCN + ArNHNH_2 \longrightarrow \underset{NH_2}{\underset{|}{\overset{Ar}{\underset{|}{N-N}}}}$$

The second most important method for the synthesis of pyrazolines is the cycloaddition of diazoalkanes with carbon–carbon double bonds. The chemistry of these reactions is described in ref. 58. Schematically, these [2 + 3] cycloaddition reactions are represented by letters, without definition of their character.

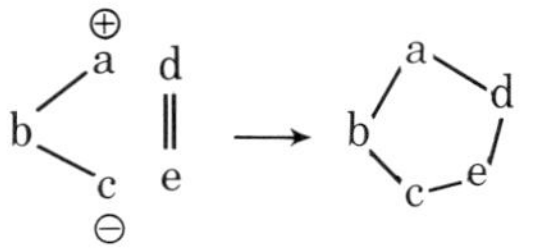

In the synthesis of 1-pyrazolines, the cycloaddition is cis and stereospecific, but the approach of the two reagents depends on the polarity of the olefin and the size of the substituent. The number of products is restricted by the regiospecificity of the reaction and the choice of the dipolarophile (59). Diazomethane is used mainly as the dipole, but numerous olefins have been used.

Reaction of diazomethane with an olefin possessing an electron-attracting substituent yields 2-pyrazolines. Again, the reaction is cis and regiospecific. These cycloaddition reactions are described in detail in refs. 59–60. Acylated 3-amino-2-pyrazolines (**24**) can be obtained by heating 1,2,4-oxadiazoles (**23**) that contain an aminoethyl substituent (61).

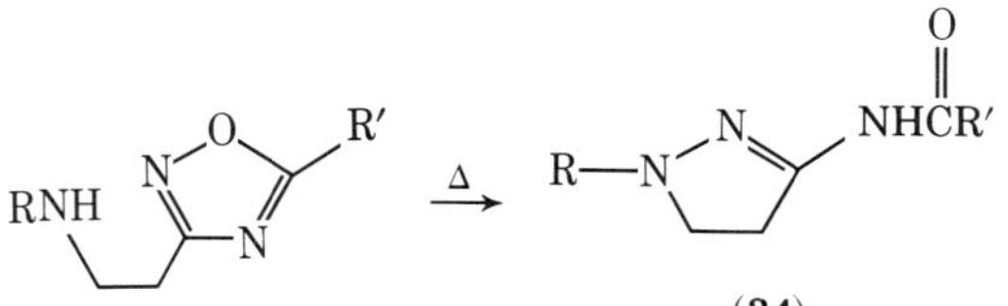

(**23**) → (**24**)

Uses. Use of pyrazolines as fluorescent optical whitening agents has been claimed (62) (see Brighteners, fluorescent). Compound (**25**) is representative of such agents.

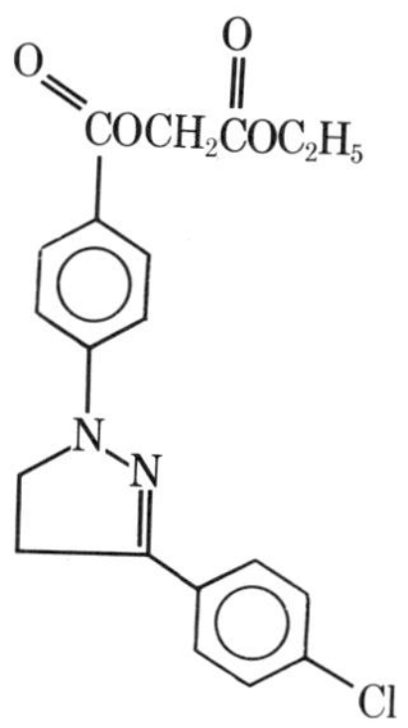

(25) [31874-76-7]

Pyrazolines, eg, derivative (26), also have insecticidal activity (63) (see Insect control technology).

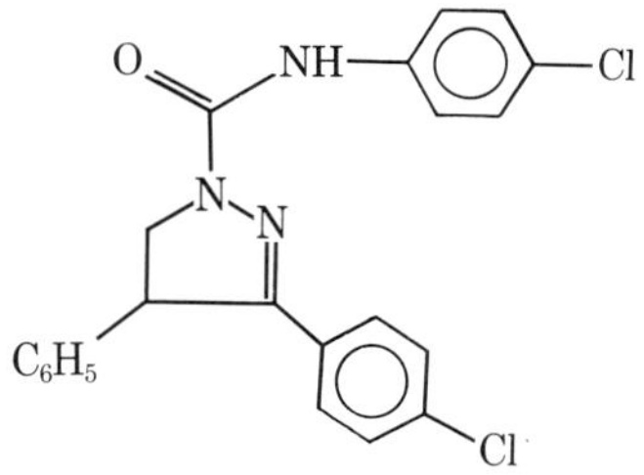

(26) [59074-27-0]

Some sulfonylurea pyrazolines also show hypoglycemic activity in rats.

Pyrazolones

Pyrazolones are oxygenated pyrazolines; the usual tautomeric forms are represented below.

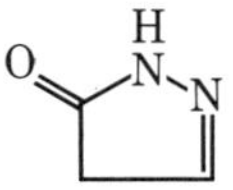

(27)
5-pyrazolone [137-44-0]
(2-pyrazolin-5-one)

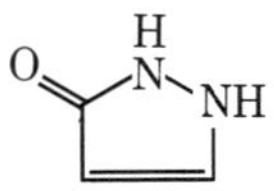

(28)
3-pyrazolone [137-45-1]
(3-pyrazolin-5-one)

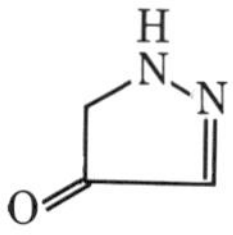

(29)
4-pyrazolone [27662-65-3]
(2-pyrazolin-4-one)

5-Pyrazolone and 3-pyrazolone are the most common tautomers. Pyrazolones are reviewed in ref. 64, which though published in 1964, is still relevant.

Structure. The tautomerism of pyrazolones is reviewed in ref. 65. The four principal techniques for the investigation of the phenomenon of tautomerism are basicity measurements, ^{1}H nmr, ir, and uv. The easiest way to discuss the tautomerism is to adopt Katritkzky's nomenclature of CH, OH, and NH for the three tautomeric structures of 1-substituted pyrazol-5-ones (65). By analogy, the three corresponding fixed forms are named CR, OR, and NR.

In ^{1}H nmr studies, the CH form can be distinguished by the 4-proton and, in ^{13}C nmr studies, by the carbon in the 4-position. In ir studies, the CH and CR forms exhibit a strong C=O band at 1700–1720 cm^{-1}, whereas uv studies show a maximum at ca 240–250 nm. Distinguishing between the OH- and NH-forms is difficult. If the 3- and 4-positions are unsubstituted, the coupling constant can be used for characterization. Carbon-13 nmr distinguishes between the OR and NR compounds. The properties of the tautomeric forms of 1-substituted pyrazol-5-ones are correlated in ref. 66. Solvent effects are important to the tautomeric composition of many 1-substituted pyrazol-5-ones (66). 1-Substituted pyrazol-5-ones exist mostly in the CH- or NH-forms.

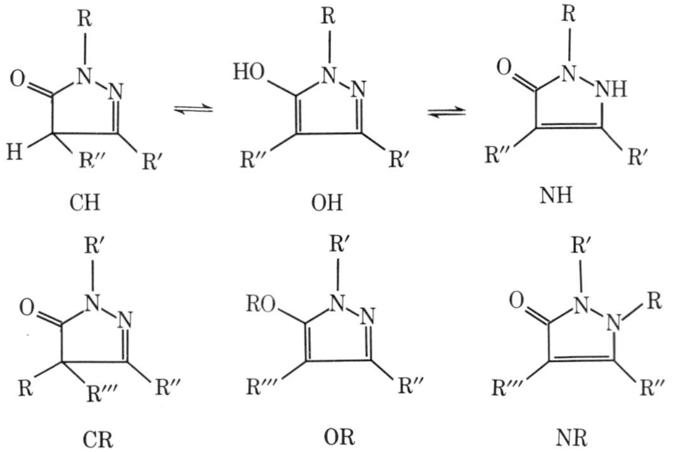

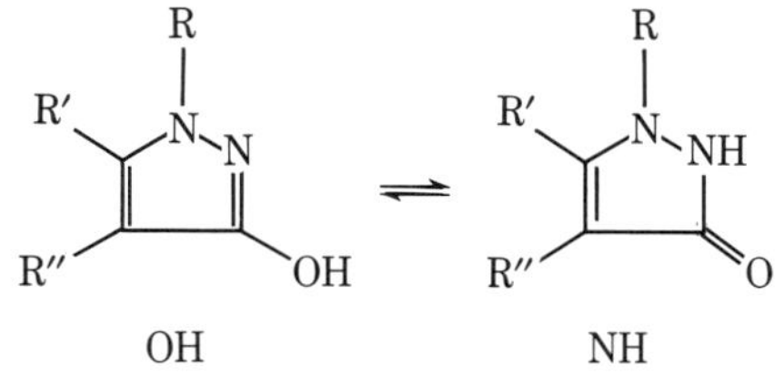

For 1-substituted pyrazol-3-ones only two tautomers can exist: the OH and the NH forms. Nuclear magnetic resonance is not very useful for determining the predominant tautomeric form, because only averaged signals are observed. In ir analysis, however, the OH form lacks C=O absorption and there is a band at 1560–1570 cm^{-1}, which is characteristic of the pyrazole. The NH form shows a C=O band at 1650–1670 cm^{-1}. As shown in uv spectra, 1-methyl-substituted pyrazolones absorb at 220 nm when in the OH-form and at 245 nm when in the NH-form; water is the solvent in both cases. Dipole moments and pK_a measurements can also be useful for determining the tautomeric ratios. Tables of tautomeric ratios for several analogues are available (67).

1-Substituted pyrazol-3-ones are usually in the OH-form and never in the CH-form. The N-unsubstituted pyrazol-3-ones can exist in many tautomeric forms and their analysis is quite difficult because of their insolubility. They exist mostly in the NH- or OH-form. Results of nmr studies of 3- and 5-pyrazolones are given in ref. 68.

Physical Properties. The pyrazolones are usually crystalline with poorly defined melting points. Generally, they are more soluble in polar solvents, including water. 5-Pyrazolones are both acidic and basic with pK_a values of 6.2–11.0. The 3-pyrazolones are more basic than the 5-pyrazolones, and the former can form salts, eg, hydrochlorides.

Chemical Properties. The 4-position in 5-pyrazolones is highly reactive and undergoes all of the typical reactions of an active methylene group. With ketones and aldehydes, a condensation product (**30**) and dimers (**31**) are formed; but with formaldehyde, a hydroxymethyl moiety is introduced into the 4-position.

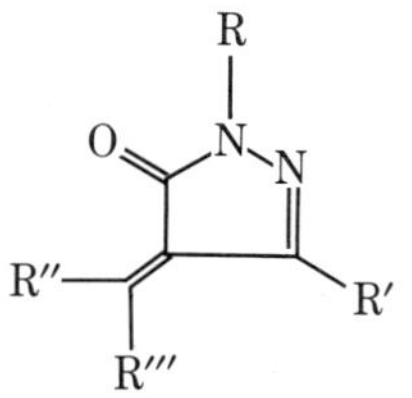

(30)

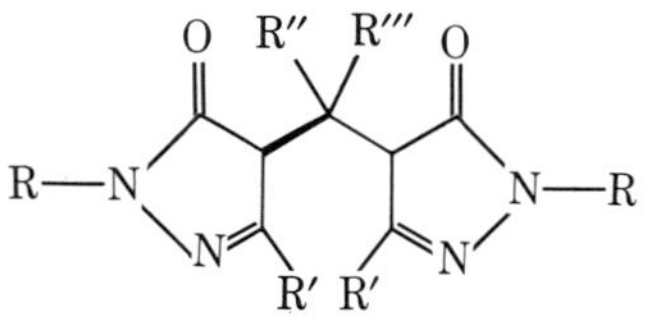

(31)

Compounds of type (**30**) undergo cycloaddition to give various heterocycles, eg, (**32**) (69).

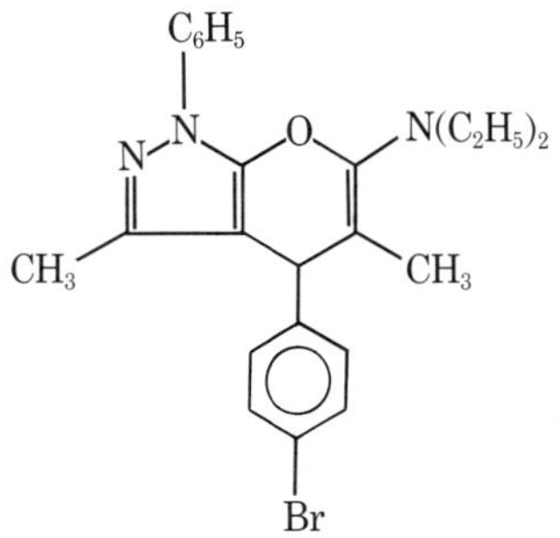

(32) [71327-19-9]

Pyrazolones undergo C-alkylation at the 4-position; O and N alkylation can also occur and lead to a mixture of products. Introduction of a halogen atom into the 4-position is also a facile reaction. The bromination of pyrazolones has been studied extensively (70). Acylation and carbamoylation at the 4-position are accomplished by the usual reagents. The Mannich reaction occurs in the 4-position, as does the introduction of the arylazo group. The latter reaction is used in the preparation of azo dyes (qv). The acylation of 1-arylpyrazol-5-ones usually gives 5-acyloxy pyrazoles in high yield.

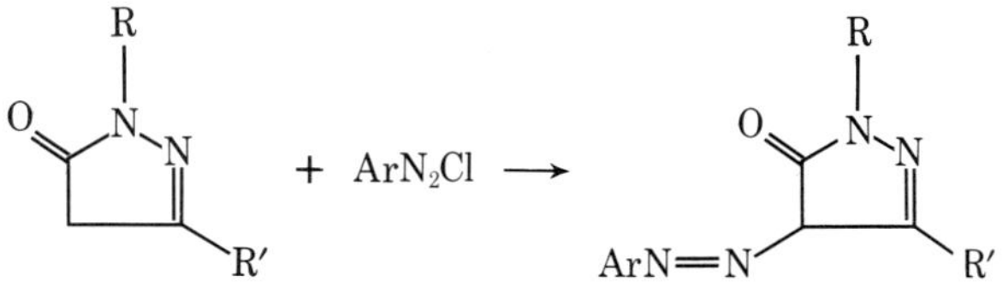

[Reaction scheme: R-substituted pyrazolin-5-one + ArN₂Cl → arylhydrazone product with ArN=N group at 4-position]

The alkylation of 2-pyrazolin-5-ones at 100°C yields 2-alkyl-3-pyrazolin-5-ones; this reaction is used in the synthesis of antipyrine. In the reaction with phosphorus oxychloride, the 5-one is replaced with a chlorine atom, and phosphorus pentasulfide introduces a sulfur atom. N-Acylation with the usual reagents is often accompanied by O- and C-acylation. Strong oxidizing agents rupture the pyrazolone ring, whereas catalytic reduction leaves the ring intact. Lithium aluminum hydride reduction yields 5-hydroxypyrazolines.

The chemical properties of 3-pyrazolones and 5-pyrazolones are similar. Both react by direct substitution at the 4-position in the following reactions: acylation, carbamoylation, halogenation, the Mannich reaction, nitrosation, and formylation. Reaction of 3-pyrazolones with aromatic aldehydes yields bispyrazolinones, just as for the 5-pyrazolones. With phosphorus oxychloride, the oxygen is replaced by chlorine. Alkylation of N-unsubstituted 3-pyrazolones leads to N-alkylation accompanied by some O-alkylation. The 3-pyrazolone ring is stable to normal catalytic reduction, but reduction under high pressure or hot acid treatment causes ring rupture.

Synthesis. The most common procedure for the synthesis of 5-pyrazolones is the reaction of almost any nonsubstituted or monosubstituted β-ketoester ($R''' = O$-alkyl) with almost any monosubstituted hydrazine.

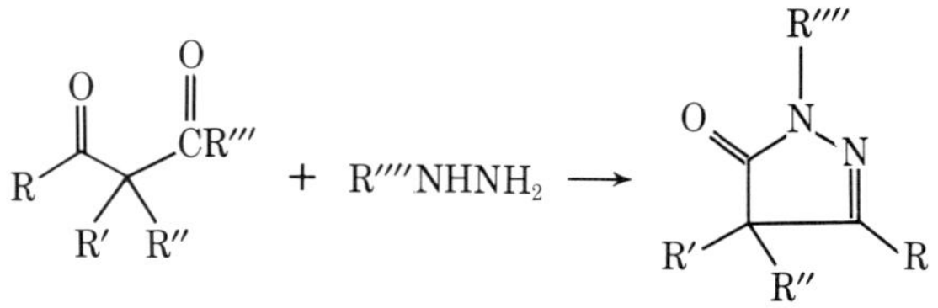

[Reaction scheme: β-ketoester R-CO-CR'R''-CR'''(=O) + R''''NHNH₂ → pyrazolone]

The ester group can be replaced by amides or anilides ($R''' = N$-alkyl or -aryl) and R'''' can be an acetyl group, which is eliminated during the cyclization.

$α,β$-Unsaturated acids, esters, or amides are also useful in pyrazolone synthesis [$R' = O$-alkyl or $N(\text{alkyl})_2$].

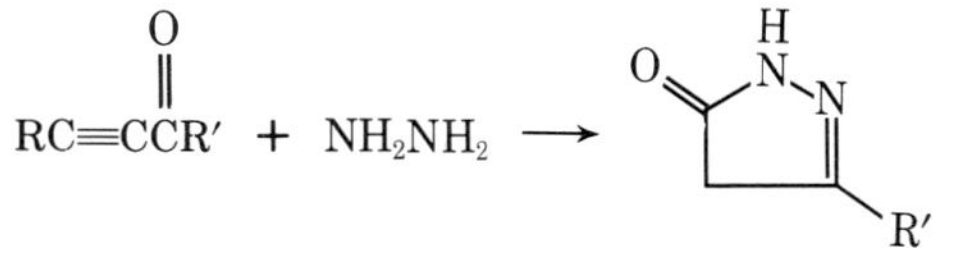

[Reaction scheme: RC≡CCR' (with C=O) + NH₂NH₂ → pyrazolone with NH]

A process for making 3-anilino-5-pyrazolones (**33**) was recently patented (R = aryl) (71).

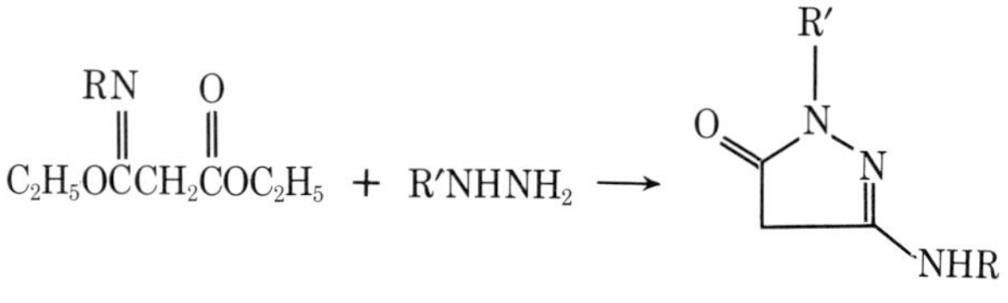

$$\text{C}_2\text{H}_5\text{OCCH}_2\text{COC}_2\text{H}_5 \text{ (with RN= and =O)} + R'\text{NHNH}_2 \rightarrow$$

[Product structure with R' on N, NHR at 3-position]

(**33**)

Various heterocyclic systems, eg, isooxazolines, pyrones, and recently 2,3-dihydrofurans, can be converted into 5-pyrazolones (72). Pyrazoles with the proper substituents in the 5-position, eg, an alkoxy group, can also be converted to pyrazolones.

3-Pyrazolones are synthesized mainly by the alkylation of 5-pyrazolones.

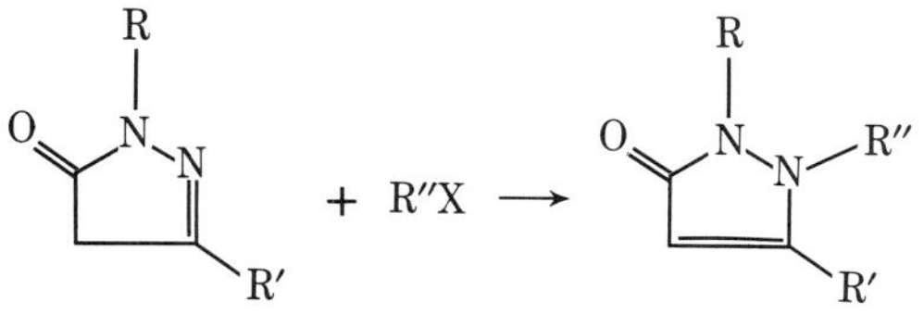

However, this reaction also gives O-alkylated products. In another procedure, the condensation of a β-ketoester with acetyl-2-phenylhydrazide gives a 2-phenyl product.

The reaction of substituted diethyl malonate with disubstituted hydrazine gives 3,5-pyrazolidinediones (**34**).

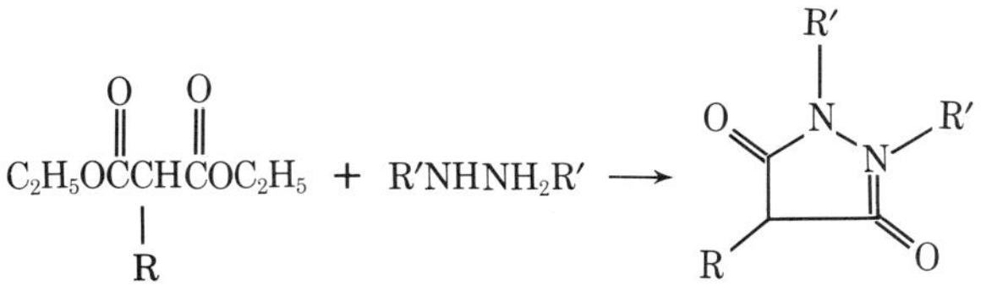

$$C_2H_5OCCHCOC_2H_5 + R'NHNH_2R' \longrightarrow$$

(**34**)

Health Factors. Pyrazolones are suspected of being carcinogenic because their ease of nitrosation can lead to the formation of dimethylnitrosamine, a potent carcinogen. Results from one study showed that the combination of aminopyrine–sodium nitrite is highly carcinogenic. Tumors occur very rapidly and consist mainly of angiosarcomas of the liver (73). The combination of the dipyrone–nitrite, however, is not carcinogenic nor is sodium nitrite alone. The toxicological consequences of nitrosation are different for drugs with very closely related structures, and each drug must be considered separately (73).

The reactivity of four frequently used pyrazolones with nitrite and their *in vitro* mutagenicity have been investigated (74). Positive results were obtained for Antipyrine, aminopyrine, and Sulpyrine; results for isopropylantipyrine were negative. Nitrosability of various pyrazolones have been reviewed, and detailed mechanisms for the formation of nitrosamines have been suggested (75) (see *N*-Nitrosamines).

Uses. The main uses of pyrazolones are in medicinals, dyes, and color photography (qv). Some of the important pyrazolone drugs are listed in Tables 1 and 2 (75). Their main uses are as antipyretics and analgesics, and for the relief of arthritis. In the United States, only three of these drugs are used extensively. Phenylbutazone purchases by drugstores and hospitals amounted to ca 19×10^6 in 1979, and the estimated sales for 1980 are 17.7×10^6. 1979 and 1980 sales of oxyphenbutazone were ca 6×10^6 and 5.5×10^6, respectively. 1979 and 1980 sulfinpyrazone sales were ca 7.5×10^6 and 10.8×10^6, respectively.

Phenylbutazone has been produced for many years by the reaction of malonate

Table 1. Medicinal 3-Pyrazolone Derivatives[a]

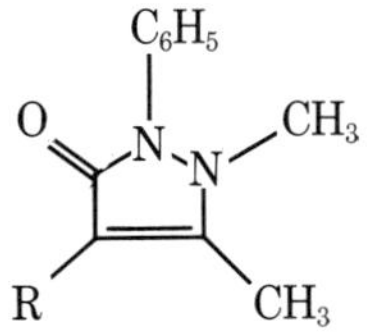

R	CAS Registry No.	Generic name	Trade name
H	[60-80-0]	phenazone	Antipyrine
$N(CH_3)_2$	[58-15-1]	aminophenazone or aminopyrine	Pyramidon
$CH(CH_3)_2$	[479-92-5]	propyphenazone	Saridon
$N(CH_3)CH_2SO_3Na$	[68-89-3]	metamizole or dipyrone	Novalgin
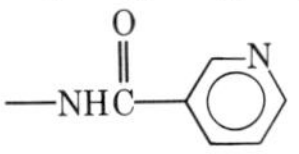 —NHC(O)—(pyridyl)	[2139-47-1]	nifenazone	Nicopyron

[a] Ref. 75.

Table 2. Medicinal Pyrazolidinediones

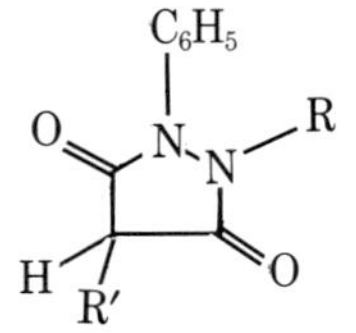

R	R'	CAS Registry No.	Generic name	Trade name
C_6H_5	$n\text{-}C_4H_9$	[50-33-9]	phenylbutazone	Butazolidin
$p\text{-}HOC_6H_4$	$n\text{-}C_4H_9$	[129-20-4]	oxyphenbutazone	Tanderil
C_6H_5	$C_6H_5SCH_2CH_2$	[57-96-5]	sulfinpyrazone	Anturane
H	C_6H_5	[3426-01-5]	phenopyrazone	Osadrin

esters with diphenylhydrazine, which yields (**34**) where R = H and R' = C_6H_5. Reaction with butyraldehyde followed by reduction yields phenylbutazone (**34**; R = C_4H_9, R' = C_6H_5). It is an excellent anti-inflammatory drug but it can produce severe side effects, such as gastrointestinal disorders and agranulocytosis. Some modified procedures for the synthesis of phenylbutazone have been described (76–77) (see Analgesics, antipyretics, and anti-inflammatory agents).

Pyrazolones are widely used as dyes. They can be represented by the general formula (**35**)

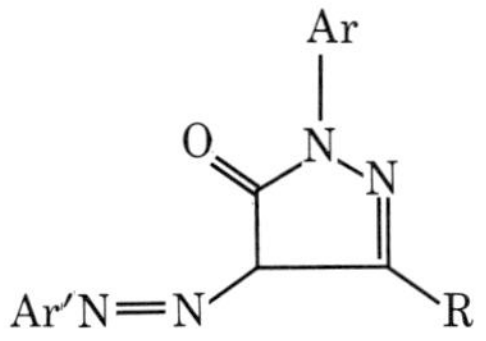

(**35**)

Table 3. U.S. Production and Sales of Some Pyrazolone Derivatives[a]

Product	Year	Production, t	Sales Quantity, t	Sales Unit value, $/kg
tartrazine (CI acid yellow 23, CI 19140) [1934-21-0]	1971	142	145	5.23
	1974	152	157	6.09
	1976	110	82.1	8.34
	1978	126	107	9.50
	1979	142	117	9.67
xylene light yellow 2G and 3G (CI acid yellow 17, CI 18965) [6359-98-4]	1971	258	229	5.07
	1974	83.5	81.6	5.58
	1976	80.3	68	6.76
	1978	83.9	64.4	10.87
	1979	42.2	51.7	9.36
pyrazolone red (CI pigment red 38, CI 21120) [6358-87-8]	1971	60.3	45	1.08
	1974			
	1976	64		
	1978			
	1979	74.8	64.9	18.78
benzidine orange (CI pigment orange 13, CI 21110) [3520-72-7]	1971	65.3	64.9	8.12
	1974	197	116	10.42
	1976	121	77.6	11.24
	1978	108	108	12.00
	1979	164	130	11.64

[a] Ref. 78.

where R is mostly methyl or carboxyl and Ar and Ar' are mainly benzenesulfonic acid derivatives. These pyrazolones can be either acid dyes, which are applied directly to the fabric, or complexed with metals and deposited on the fabric. Their great advantage is stability to light and wet treatments.

U.S. production and sales of some pyrazolone dyes are listed in Table 3 (78). Production figures for such large-volume pyrazolones as Developer Z [89-25-8], Pyrazolone T [118-47-8], and Pyrazolone G [89-36-1] were available from the U.S. International Trade Commission (78), but these figures have been made confidential and are no longer published.

Tartrazine (**35**; R = COONa; Ar = Ar' = 4-NaSO$_3$C$_6$H$_5$) is a food coloring agent and is prepared by the reaction of the pyrazolone with diazotized sulfanilic acid (see Colorants for foods, drugs, and cosmetics).

The use of pyrazolones in color photography (qv) is based on their reaction with phenylenediamines in the presence of oxidizing agents, which yields azomethine dyes (**36**) (79). A second use for pyrazolones in photography is as optical filter dyes in color films. Other uses for pyrazolones are as agricultural fungicides, reduction of the metal-ion content of aqueous solutions, separation of plutonium, and as diuretics (qv) (80–83) (see Fungicides, agricultural).

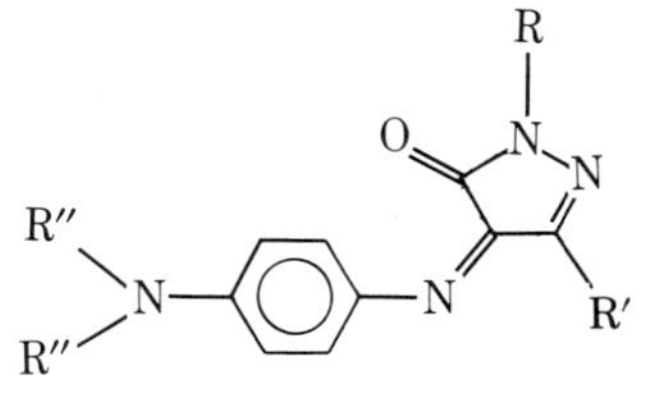

(**36**)

BIBLIOGRAPHY

"Pyrazoles, Pyrazolines, Pyrazolones" in *ECT* 2nd ed., Vol. 16, pp. 763–779, by Paul F. Wiley, The Upjohn Company.

1. J. Elguero, C. Marzin, A. R. Katritzky, and P. Linda in A. R. Katritzky and A. J. Boulton, eds., *The Tautomerism of Heterocycles, Advances in Heterocyclic Chemistry*, Supplement 1, Academic Press, Inc., New York, 1976, pp. 269–278.
2. T. L. Jacobs in R. C. Elderfield, ed., *Heterocyclic Compounds*, Vol. 5, John Wiley & Sons, Inc., New York, 1957, p. 45.
3. A. N. Kost and I. I. Grandberg in A. R. Katritzky and A. J. Boulton, eds., *Advances in Heterocyclic Chemistry*, Vol. 6, Academic Press, Inc., New York, 1966, pp. 347–429.
4. L. C. Behr, R. Fusco, and C. H. Jarboe in R. H. Wiley, ed., "Pyrazoles, Pyrazolines, Pyrazolidines, Indazoles and Condensed Rings," Vol. 22 of A. Weissberger, ed., *The Chemistry of Heterocyclic Compounds*, Wiley-Interscience, 1967.
5. F. H. Deis and D. Lester, *Biochem. Pharmacol. Ethanol*, 303 (1979).
6. G. S. Saharia and H. R. Sharma, *Rasayan Samiksha* **4,** 71 (1977).
7. R. E. Orth, *J. Pharm. Sci.* **57,** 537 (1968).
8. J. D. Loudon in E. H. Rodd, ed., *Chemistry of Carbon Compounds*, Vol. 4, Part A, Elsevier, North-Holland, Inc., Amsterdam, The Netherlands, 1957, p. 251.
9. Ref. 3, p. 352.
10. Ref. 3, p. 353.
11. D. DelMonte Casoni, A. Mangini, and R. Passerini, *Gazz. Chim. Ital.* **86,** 797 (1956).
12. I. I. Grandberg, *Zh. Obshch. Khim.* **33,** 518 (1963).
13. Ref. 3, p. 356.
14. T. J. Battermam, "NMR Spectra of Simple Heterocycles," in E. C. Taylor and A. Weissberger, eds., *General Heterocyclic Chemistry Series*, Wiley-Interscience, New York, 1973, p. 165.
15. F. J. Weigert and J. D. Roberts, *J. Am. Chem. Soc.* **90,** 3543 (1968).
16. R. G. Rees and M. J. Green, *J. Chem. Soc. B*, 387'(1968).
17. G. C. Levy, R. L. Lichter, and G. L. Nelson, *Carbon-13 Nuclear Magnetic Resonance Spectroscopy*, 2nd ed., John Wiley & Sons, Inc., New York, 1980, p. 119.
18. M. C. Agnew, V. J. Bauer, and R. C. Effland, *J. Heterocycl. Chem.* **17,** 1573 (1980).
19. F. Fabra, J. Vilarrasa, and J. Coll, *J. Heterocycl. Chem.* **15,** 1447 (1978).
20. H. Reimlinger and co-workers, *Chem. Ber.* **103,** 1942, 1949, 1954 (1970).
21. Ref. 3, p. 392.
22. J. F. Hansen and co-workers, *J. Org. Chem.* **45,** 76 (1980).
23. J. P. Freeman and J. F. Lorenc, *J. Org. Chem.* **42,** 177 (1977).
24. G. Kornis and E. Nidy, *Org. Prep. Proced. Int.* **5,** 141 (1973).
25. Ref. 3, p. 395.
26. G. L. McNew and N. K. Sundholm, *Phytopathology* **39,** 721 (1949).
27. R. Huttel and F. Buchele, *Chem. Ber.* **88,** 1586 (1955).
28. J. W. A. M. Janssen, H. J. Koeners, C. G. Kruse, and C. L. Habrakern, *J. Org. Chem.* **38,** 1777 (1973).
29. G. Kornis, unpublished observation.
30. R. Huttel and M. E. Schor, *Ann.* **625,** 55 (1959).
31. C. L. Habraken and E. K. Poels, *J. Org. Chem.* **42,** 2893 (1977).
32. Ref. 3, p. 427.
33. A. Lablache-Combier in O. Buchardt, ed., *Photochemistry of Heterocyclic Compounds*, John Wiley & Sons, Inc., New York, 1976, p. 191.
34. Ref. 33, p. 156.
35. P. Bouchet, G. Joncheray, R. Jacquier and J. Elguero, *Tetrahedron* **35,** 1331 (1979).
36. G. Coispeau and J. Elguero, *Bull. Soc. Chim. Fr.*, 2717 (1970).
37. J. Bastide and J. Lematre, *C.R. Acad. Sci. Ser. C.* **269,** 358 (1969).
38. G. V. Boyd and S. R. Dando, *J. Chem. Soc. C*, 225 (1971).
39. J. Clark and M. Curphey, *J. Chem. Soc. Chem. Commun.*, 184 (1974).
40. E. C. Taylor and W. R. Purdum, *Heterocycles* **6,** 1865 (1977).
41. B. R. Tolf and co-workers, *Acta. Chem. Scand. (Ser. B)* **33,** 483 (1979).
42. R. Blomstrand and co-workers, *Proc. Natl. Acad. Sci. U.S.A.* **76,** 3499 (1979).

43. J. Villiaumey and B. Larget-Piet, *Therapie* **31**, 299 (1976).
44. U.S. Pat. 4,025,530 (May 24, 1977), A. J. Crovetti and co-workers (to Abbott Laboratories).
45. Ger. Pat. 2,441,504 (Mar. 20, 1975), M. Garber, W. J. Stepek, and L. J. Ross (to American Cyanamid Co.).
46. Can. Pat. 1,053,231 (Apr. 24, 1979), M. W. Moon and G. Kornis (to Upjohn Co.).
47. J. L. Barascut, J. Elguero, and R. Jacquier, *Bull. Soc. Chim. Fr.*, 1571 (1970).
48. Ref. 4, p. 224.
49. Ref. 14, pp. 197–213.
50. E. Abushanab and co-workers, *J. Org. Chem.* **43**, 2017 (1978).
51. R. J. Crawford and co-workers, *Can. J. Chem.* **56**, 992 (1978).
52. D. Srzic, L. Klasinc, W. Seitz, and H. Guesten, *Croat. Chem. Acta* **53**, 33 (1980).
53. W. L. F. Armarego, "Stereochemistry of Heterocyclic Compounds" in E. C. Taylor and A. Weissberger eds., *General Heterocyclic Chemistry Series*, Wiley-Interscience, New York, 1977, pp. 103–112.
54. D. D. Berge and A. V. Kale, *Chem. Ind. London*, 662 (1979).
55. B. V. Ioffe and co-workers, *Dokl. Akad. Nauk USSR* **197**, 91 (1971).
56. N. A. Evans, D. E. Rivett, and J. F. K. Wilshire, *Aust. J. Chem.* **27**, 2267 (1974).
57. F. Pragst, *Z. Chem.* **14**, 236 (1974).
58. R. Huisgen, *Angew. Chem. Int. Ed. Engl.* **2**, 565, 633 (1963); R. Huisgen and co-workers, *Angew. Chem.* **92**, 198 (1980).
59. Ref. 53, p. 104.
60. Ref. 4, p. 200.
61. D. Korbonits, E. Bako, and K. Horvath, *J. Chem. Res. Synop.*, 64 (1979).
62. Brit. Pat. 1,209,631 (Oct. 21, 1970), G. H. Keats (to Imperial Chemical Industries).
63. A. C. Grosscurt, R. Vanhes, and K. Wellinga, *J. Agric. Food Chem.* **27**, 406 (1979).
64. R. H. Wiley and P. Wiley, "Pyrazolones, Pyrazolidines and Derivatives," Vol. 20 of A. Weissberger ed., *The Chemistry of Heterocyclic Compounds*, Interscience Publishers, a division of John Wiley & Sons, Inc., New York, 1964.
65. Ref. 1, pp. 313–352.
66. Ref. 1, pp. 326–328.
67. Ref. 1, p. 345.
68. Ref. 14, pp. 198–201.
69. G. Desimoni and co-workers, *J. Chem. Soc. Perkin Trans.* **1**, 856 (1979).
70. J. Elguero and co-workers, *Bull. Soc. Chim. Fr.*, 328 (1967).
71. U.S. Pat. 4,113,954 (Sept. 12, 1978), D. J. Tracy and W. F. Hoffstadt (to GAF Corp.).
72. C. Venturello and R. D'Aloisio, *Synthesis*, 283 (1979).
73. H. G. Berscheid and co-workers in F. T. Coulston and J. F. Dunne eds., *The Potential Carcinogenicity of Nitrosatable Drugs WHO Symposium*, Geneva, Switzerland, June 1978, Ablex Publishing Corporation, Norwood, N.J., 1980, p. 121.
74. M. Arisawa, M. Fujiu, Y. Suhara, and H. B. Marujama, *Mutat. Res.* **57**, 287 (1978).
75. H. Roth, *Pharm. Ztg.* **123**, 625 (1978).
76. Hung. Pat. 14,331 (Dec. 28, 1977), J. Odor and I. Szentpeteri (to Vegyterv); *Chem. Abstr.* **88**, P190,804 (1978).
77. Ger. Pat. 1,814,649 (June 25, 1970), K. Klemm and E. Langescheid (to BYK-Gulden Lomberg).
78. U.S. International Trade Commission, *Synthetic Organic Chemicals, United States Production and Sales 1974* Rept. No. 776, U.S. Government Printing Office, Washington, D.C., 1976; 1976 USITC Rept. No. 833, 1977; 1978 USITC Rept. No. 1001, 1979; 1979 USITC Rept. No. 1099, 1980.
79. Jpn. Pat. 76,112,341 (Oct. 4, 1976), T. Kojima and co-workers (to Konishiroku Photo Industry Co.).
80. Jpn. Pat. 76,86,130 (July 28, 1976), K. Shimada and co-workers (to Otsuka Chemical Drugs).
81. Europ. Patent 5859 (Dec. 12, 1979), P. M. Spaziante and co-workers (to DeNora, Oronzio).
82. M. K. Chmutova, P. N. Palei, and YU. A. Zolotov, *Zh. Anal. Khim.* **23**, 1476 (1968).
83. Ger. Pat. 2,363,138 (July 10, 1975), E. Moeller and co-workers (to Bayer A-G).

GABRIEL KORNIS
The Upjohn Company

PYRIDINE AND PYRIDINE DERIVATIVES

Pyridine (**1**) is the parent of a series of compounds that is important in medicinal, agricultural, and industrial chemistry. Although many polysubstituted pyridine compounds, like other heterocyclic compounds, are synthesized with their functional groups present from acyclic compounds, most derivatives are prepared by manipulation of pyridine and its simple homologues in a manner similar to the chemistry of the benzenoid aromatics. However, the simple pyridine compounds are prepared by cyclization of aliphatic raw materials.

Since pyridine has the symmetry of a monosubstituted benzene, there are three possible monosubstituted pyridine isomers, six compounds with two like substituents, etc (see Benzene). The three monomethylpyridines or picolines are 2- or α-picoline (**2**), 3- or β-picoline (**3**), and 4- or γ-picoline (**4**). Although pyridine and the picolines dominate the commercially important chemistry of pyridine derivatives, 2-methyl-5-ethylpyridine (**5**) (MEP or aldehyde collidine) also is important. Dimethylpyridines are called lutidines, and the 2,6- (**6**) and 3,5- (**7**) lutidines are readily available. The trivial name of trimethylpyridine is collidine, and the symmetrical 2,4,6-collidine (**8**) is the most common. Pyridine chemistry has been comprehensively reviewed (1–6).

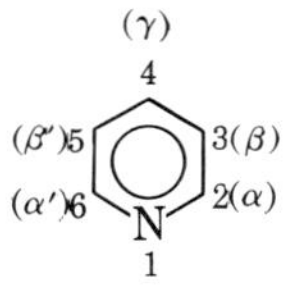

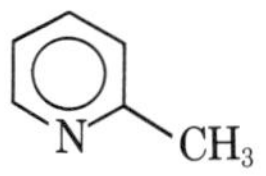

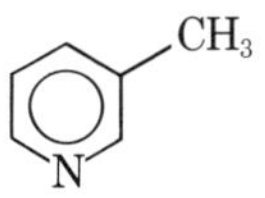

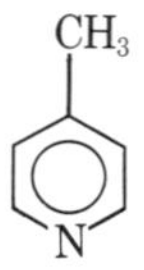

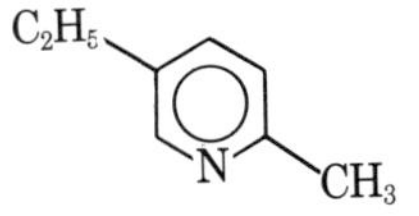

Although the volume of commercial pyridine chemicals is relatively large, the economic aspects resemble those of specialty chemicals more than those of commodities: commercial transactions occur with little publicity, trade secrets are carefully guarded, and patents proliferate, thus obscuring commercial processes.

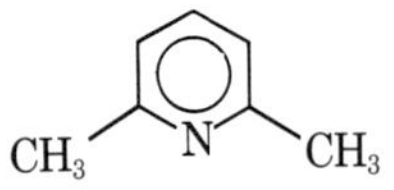

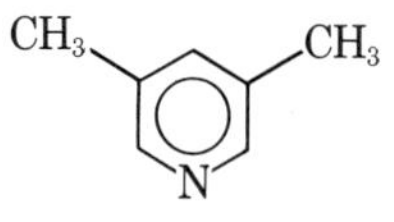

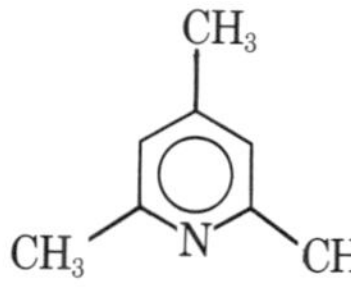

Properties

The physical properties of simple pyridine compounds (see Table 1) (7–12) are largely controlled by the presence of the basic electronegative nitrogen atom in the ring (13–14). Other properties of pyridine are listed in Table 2 (8).

Although the structure and molecular weight of pyridine are nearly identical with those of benzene, pyridine has a boiling point more than 35°C higher than that of benzene and it is miscible in all proportions with water. Boiling points of pyridines

Table 1. Properties of Pyridine Derivatives[a]

Compound	CAS Registry No.	Structure no.	Freezing point, °C	Boiling point, °C	Density at 20°C, g/cm³	pK_a thermodynamic in H$_2$O at 25°C[b]	Solubility in H$_2$O at 20°C, g/100 g[c]	Water azeotrope Bp, °C	Water azeotrope Wt % H$_2$O[d]
pyridine[e]	[110-86-1]	(1)	−41.6	115.3	0.9830	5.22	miscible	93.6	41.3
2-methylpyridine	[109-06-8]	(2)	−64	129.5	0.9462	5.96	miscible	93.5	48
3-methylpyridine	[108-99-6]	(3)	−18.3	143.9	0.957	5.63	miscible	96.7	63
4-methylpyridine	[108-89-4]	(4)	3.7	144.9	0.9558	5.98	miscible	97.4	63.5
2,3-dimethylpyridine	[583-61-9]		−15.5	161.5	0.9491	6.57	13.3[f]		
2,4-dimethylpyridine	[108-47-4]		−64	158.7	0.9325	6.63	miscible[g]		
2,5-dimethylpyridine	[589-93-5]		−15.7	157	0.9331	6.40	10.0[h]		
2,6-dimethylpyridine	[108-48-5]	(6)	−6.1	143.7	0.923	6.72	miscible	93.3	51.5
3,4-dimethylpyridine	[583-58-4]		−10.6	179.1	0.9534	6.46	5.2		
3,5-dimethylpyridine	[591-22-0]	(7)	−6.6	172.7	0.944	6.15	3.3		
2,4,6-trimethylpyridine	[108-75-8]	(8)	−44.5	170.4	0.913		3.6		
5-ethyl-2-methylpyridine (MEP)	[104-90-5]	(5)	−70.9	178.3	0.9208		1.2	98.4	72
2-vinylpyridine	[100-69-6]	(113)		110 (20 kPa[i])	0.9746	4.98	2.75	97	62.0
4-vinylpyridine	[100-43-6]	(114)		121 (20 kPa[i])	0.988		2.91	98	76.6
piperidine[j]	[110-89-4]	(26)	−11.0	106.3	0.8659 (15°C)	11.12	miscible	92.8	35

[a] Ref. 7 unless otherwise noted.
[b] Ref. 10.
[c] Refs. 7 and 11.
[d] Ref. 12.
[e] Refs. 8–9.
[f] Miscible below 16°C.
[g] Miscible below 23°C.
[h] Miscible below 13°C.
[i] To convert kPa to mm Hg, multiply by 7.5.
[j] Ref. 9.

Table 2. Properties of Pyridine[a]

Property	Value
enthalpy of fusion ΔH_{fus} (at -41.62°C), kJ/mol[b]	8.2785
enthalpy of vaporization ΔH_{vap}, kJ/mol[b]	
at 25°C	40.2
at 115.26°C	35.11
critical temperature T_c, °C	346.8
critical pressure P_c, MPa[c]	5.63
enthalpy of formation ΔH_f° (gas, at 25°C), kJ/mol[b]	140.37
Gibbs energy of formation ΔG_f° (gas, at 25°C), kJ/mol[b]	190.48
heat capacity C_p° (gas, at 25°C), J/(K·mol)[b]	78.23
surface tension γ (liquid, at 25°C), mN/m (= dyn/cm)	36.6
viscosity η (liquid, at 25°C), mPa·s (= cP)	0.878
dielectric constant ϵ (liquid, at 25°C)	13.5
thermal conductivity λ (liquid, at 25°C), W/(m·K)[d]	0.165

[a] Ref. 8.
[b] To convert J to cal, divide by 4.184.
[c] To convert MPa to atm, divide by 0.1013.
[d] To convert W/(m·K) to (Btu·ft)/(h·ft^2·F), divide by 1.73.

rise with substitution by alkyl groups in the usual manner, but the effect is attenuated when the substitution is in the 2-position. Thus, the boiling points of 3- and 4-picoline and 2,6-lutidine are within a very narrow range and cannot be separated practically by fractional distillation, although any or all are easily separated from the lower boiling 2-picoline.

Pyridine is a tertiary amine and all simple pyridine compounds are basic, but they are less basic in solution than typical aliphatic amines. The basicity of pyridine derivatives is increased by electron-donating substituents and decreased by electron-withdrawing substituents.

Most alkyl pyridines form azeotropes with water. Distillation of water azeotropes or other azeotropes with simple compounds frequently allows the separation of mixtures not separable by simple fractional distillation. Also, small amounts of water can be conveniently removed from pyridine or alkyl pyridines by distilling and discarding the low boiling water azeotrope. Additional purification methods are suggested in ref. 9.

Preparation and Manufacture

From Coal Tar. Originally, pyridine and its homologues, which are collectively called pyridine bases, were isolated from coal tar and coal gas, ie, from the volatile by-products in the pyrolysis of coal (see Coal; Dyes and dye intermediates). The noncondensable gas contains ammonia and most of the usable pyridine bases formed during coking. These basic compounds can be removed with sulfuric acid, and the gas fraction or town gas is used as fuel. Pyridine bases that are produced by neutralization of the acid solution, combined with bases produced by acid-base separation of the appropriate boiling-range material from fractionation of coal tar, can be further refined by fractional distillation. The compounds made available by this method include pyridine, 2-methylpyridine, and the β-picoline fraction, which includes 3- and 4-methylpyridine and 2,6-lutidine. Various methods have been used to separate the components of the latter mixture. Isolation of pyridine bases from coal tar has been practiced to only a very small extent in the United States since the introduction of synthetic processes.

Syntheses. The most important synthetic reactions for the manufacture of pyridine bases have been the reactions of aldehydes and ketones with ammonia. Although these reactions were explored first during the nineteenth century and later in more detail by Chichibabin in the first half of the twentieth century, synthetic pyridine bases have been commercially available only since the 1950s. Synthetic methods are described in refs. 15–17.

Reaction of acetaldehyde (qv), preferably in the form of its trimer paraldehyde (**9**), with aqueous ammonia in the liquid phase, takes place at 230°C and 5.6–20.8 MPa (800–3000 psig) with catalysis by ammonium salts, eg, ammonium acetate, to give MEP (**5**) (18).

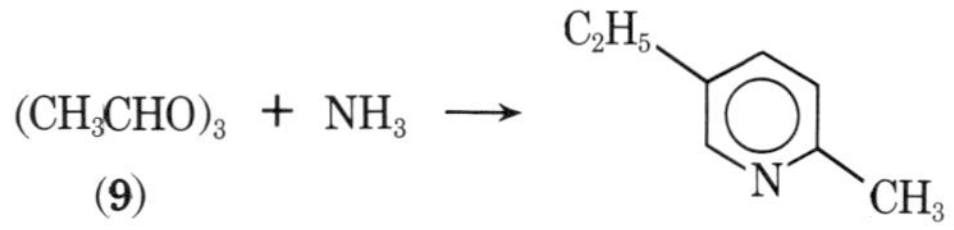

$$(CH_3CHO)_3 + NH_3 \rightarrow \underset{(5)}{\text{[2-methyl-5-ethylpyridine]}}$$
(9)

In the vapor phase, acetaldehyde reacts with ammonia in the presence of heterogeneous catalysts to give 2- and 4-methylpyridine in about equal proportions; acetylene reacts similarly. The reaction is carried out at 350–550°C, and the catalyst is usually alumina or silica–alumina which may bear a di- or trivalent metal as the oxide, halide, or phosphate.

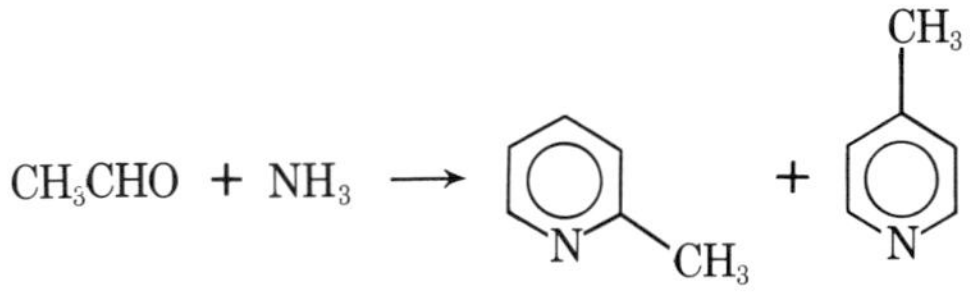

$$CH_3CHO + NH_3 \rightarrow \underset{(2)}{\text{[2-methylpyridine]}} + \underset{(4)}{\text{[4-methylpyridine]}}$$

Although the details of the reaction mechanism, particularly the function of the catalyst, are not understood, the reaction may be rationalized by a series of aldol and Michael reactions of imines and enamines (see Fig. 1) (15). Although the final step of picoline synthesis is expected to involve the loss of hydrogen, hydrogen is usually not observed in the gaseous products, and hydrogen transfer to other intermediates is suspected.

Under essentially the same vapor-phase reaction conditions, acetaldehyde, formaldehyde, and ammonia react to yield a mixture of pyridine and 3-methylpyridine; methanol is included in the feed for increased yield.

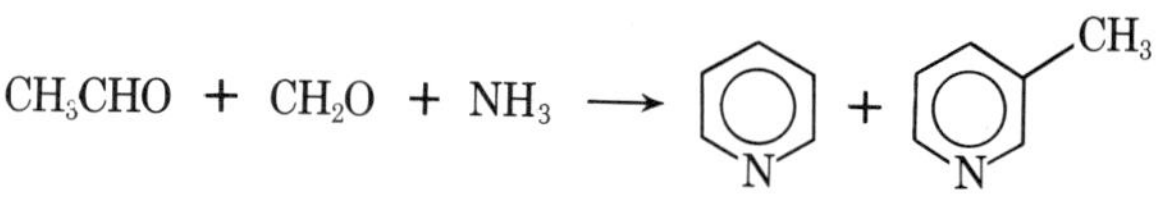

$$CH_3CHO + CH_2O + NH_3 \rightarrow \underset{(1)}{\text{[pyridine]}} + \underset{(3)}{\text{[3-methylpyridine]}}$$

Acrolein, some derivative of which may be an intermediate in the preceding reaction, under similar conditions yields mostly 3-methylpyridine and lesser amounts of pyridine (19).

PYRIDINE AND PYRIDINE DERIVATIVES

$$CH_2=CHCHO + NH_3 \longrightarrow \underset{(3)}{\text{[3-methylpyridine]}} + \underset{(1)}{\text{[pyridine]}}$$

Other aldehydes and ketones or mixtures thereof react similarly to give alkyl pyridines. Thus, acetone, formaldehyde, and ammonia give 2,6-lutidine under the same vapor-phase conditions, whereas acetone alone reacts with ammonia to give 2,4,6-collidine with apparent loss of methane.

A synthetic route that does not involve the carbonyl compound reaction with ammonia has recently been commercialized for the manufacture of 2-methylpyridine (20). Reaction of acrylonitrile with a large excess of acetone, catalyzed by a primary amine and a weak acid, occurs in the liquid phase at 180°C and 2.2 MPa (300 psig) to give predominantly the monocyanoethylation product (**10**). The latter compound on vapor-phase reaction over a palladium-containing catalyst in the presence of hydrogen appears to undergo reduction of the cyano group, cyclization, and dehydrogenation

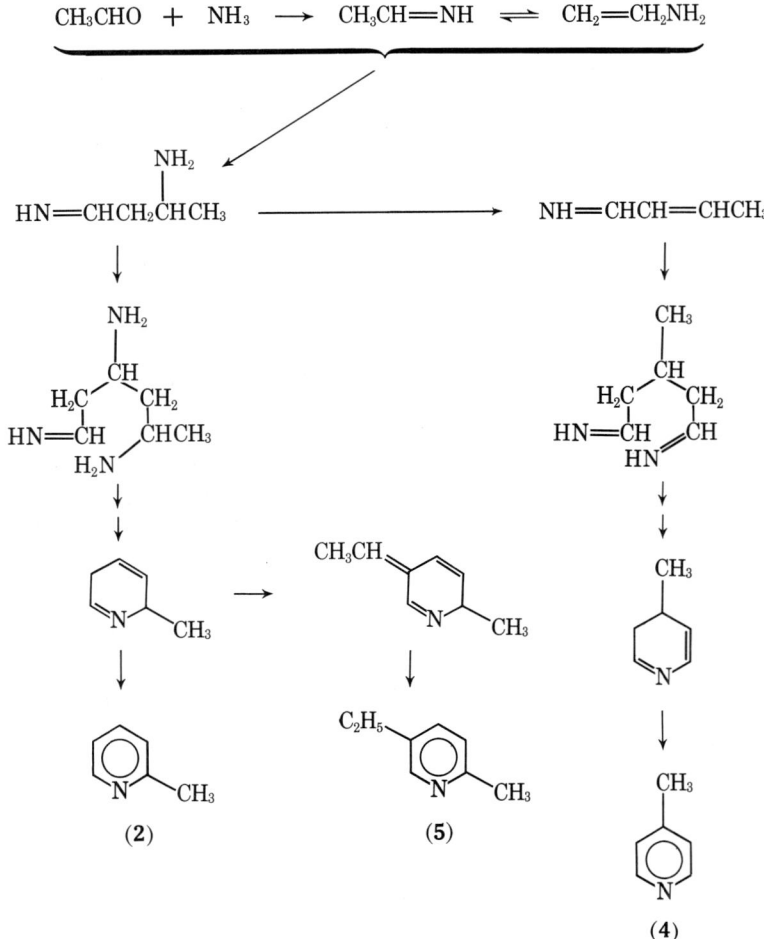

Figure 1. Reaction of acetaldehyde with ammonia.

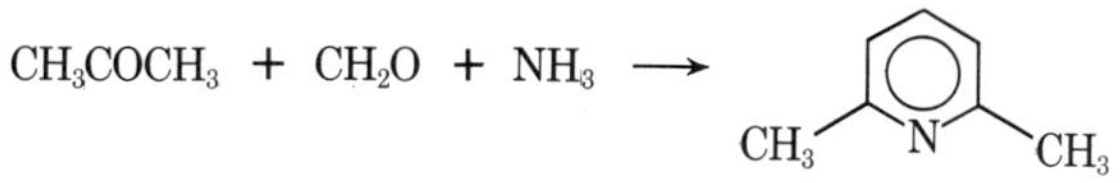

$$CH_3COCH_3 + CH_2O + NH_3 \longrightarrow \underset{(6)}{\text{[2,6-dimethylpyridine]}}$$

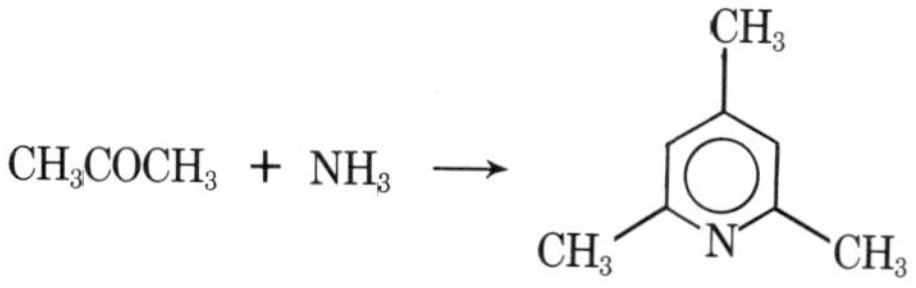

$$CH_3COCH_3 + NH_3 \longrightarrow \underset{(8)}{\text{[2,4,6-trimethylpyridine]}}$$

to give 2-methylpyridine. By similar routes, 2,3-lutidine can be prepared from 2-butanone, and quinoline [91-22-5] can be prepared from cyclohexanone.

$$CH_3COCH_3 + CH_2{=}CHCN \longrightarrow \underset{(10)}{CH_3COCH_2CH_2CH_2CN}$$

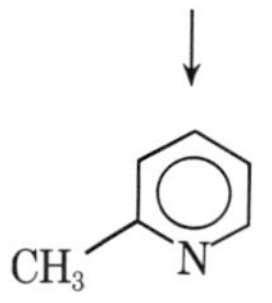

$$\downarrow$$

$$\underset{(2)}{\text{[2-methylpyridine]}}$$

Acetylene reacts with nitriles (11) at 120–180°C and 0.8–2.5 MPa (100–350 psig) to give 2-substituted pyridines (12) (21); the reaction is catalyzed by cobalt compounds. This reaction does not appear to have been commercialized.

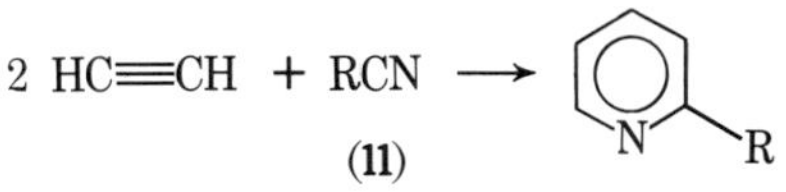

$$2\ HC{\equiv}CH + \underset{(11)}{RCN} \longrightarrow \underset{(12)}{\text{[2-R-pyridine]}}$$

The reactions described above are or could be suitable for the manufacture of pyridine and the simple alkyl pyridines; substitution and functional-group manipulation of these compounds yields end-use products. Many reactions can be used to give polysubstituted, functionalized pyridines directly; a few reactions of particular commercial utility are described below.

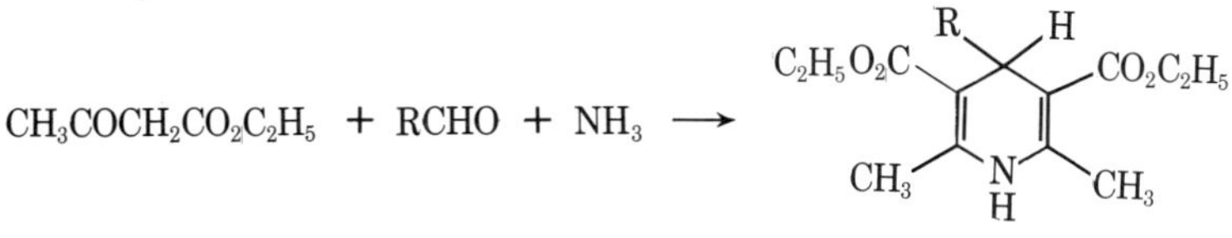

$$CH_3COCH_2CO_2C_2H_5 + RCHO + NH_3 \longrightarrow \underset{(13)}{\text{[dihydropyridine diester]}}$$

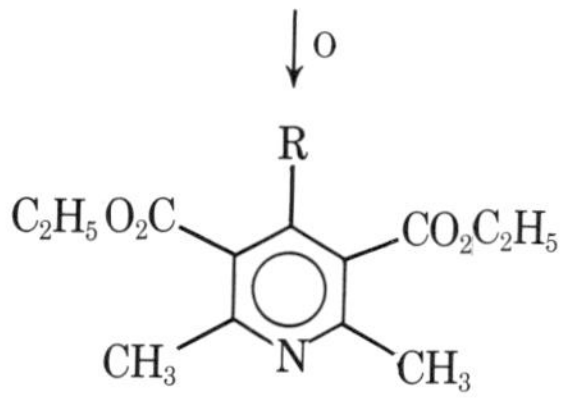

$$\downarrow \text{O}$$

$$\underset{(14)}{\text{[pyridine diester]}}$$

The Hantzsch reaction of a β-ketoester or related active methylene compound, an aldehyde, and ammonia or an aliphatic amine gives a 1,4-dihydropyridine (13) in high yield (22). Dehydrogenation or oxidation of the latter yields the corresponding pyridine derivative (14). A simple compound of type (13) where R = o-$O_2NC_6H_4$ is nifedipine [21829-25-4], a coronary vasodilator (23) (see Cardiovascular agents).

Typically, the Guareschi reaction is a cyclization of a β-ketoester and cyanoacetamide or a derivative, which is sometimes formed *in situ*. The resulting compounds (15) couple at the 5-position with diazonium salts to form dyes (24).

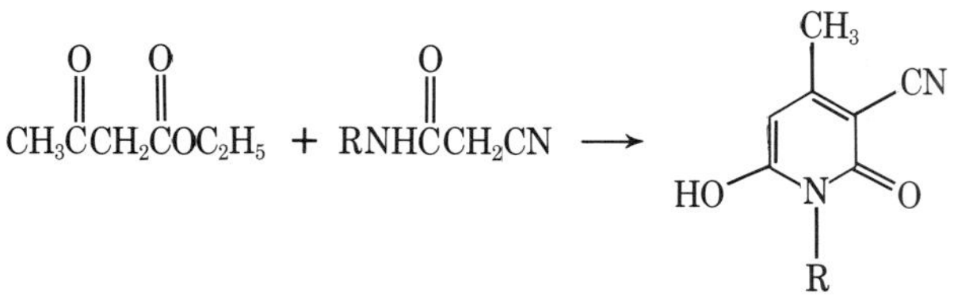

$$CH_3CCH_2COC_2H_5 + RNHCCH_2CN \longrightarrow (15)$$

Exhaustive formylation of 1,3-diarylacetone derivatives gives bis-enamines or -enolates of tricarbonyl compound (16); cyclization with amines yields 4-pyridinone derivatives (17). One such compound, fluridone [59756-60-4], [(17), R = C_6H_5, R' = m-$CF_3C_6H_4$, R" = CH_3], is an aquatic herbicide (25) (see Herbicides).

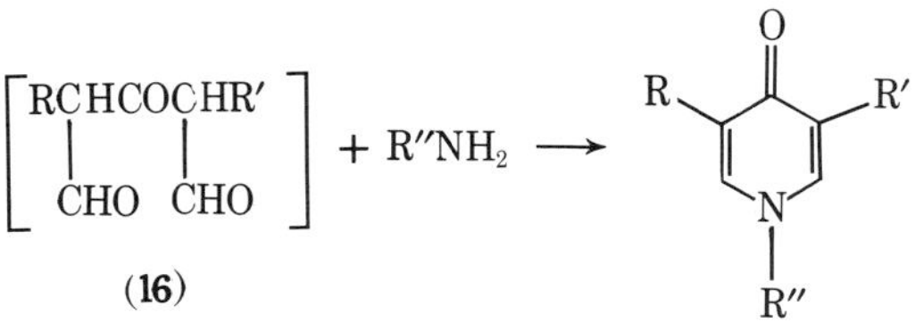

$$\begin{bmatrix} RCHCOCHR' \\ | \quad\quad | \\ CHO \quad CHO \end{bmatrix} + R''NH_2 \longrightarrow$$

(16)

(17)

Ring Transformations. Preparation of pyridine derivatives from other heterocyclic compounds is described in ref. 26. Oxazoles react with dienophiles to give pyridines after dehydration or other aromatization reactions (16,27). The commercially important example is the reaction of a 4-methyl-5-alkoxyoxazole (18) with butene-1,4-diol (19) or a derivative to yield pyridoxine [58-56-0] (20), which is a vitamin B_6 derivative. The utility of this synthesis is demonstrated by the importation of ca 204 metric tons of pyridoxine to the United States in 1979 at a price of ca $43/kg (28). By contrast, before the discovery of this synthesis, the sales price of the vitamin was ca $7600/kg in 1940 (see Vitamins, vitamin B_6).

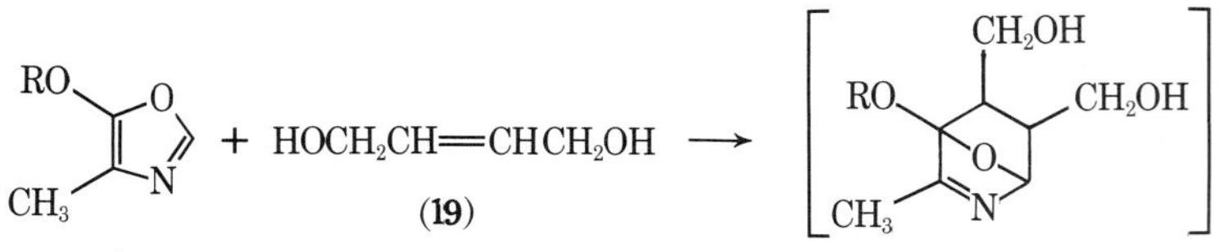

(18) + $HOCH_2CH=CHCH_2OH$ ⟶

(19)

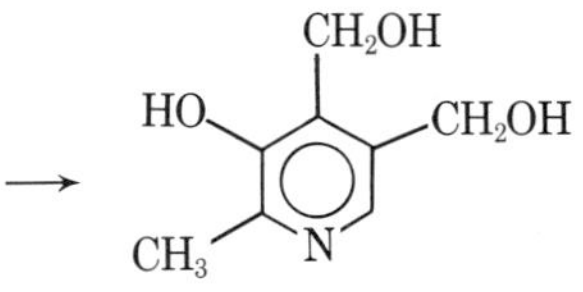

(20)

Pyrans and related compounds react with ammonia to give pyridines. A commercially useful example is the reaction of dehydroacetic acid (21) (derived from diketene) with ammonia to give 2,6-dimethyl-4-pyridinone [13603-44-6] (23); under milder conditions, 2,6-dimethyl-4-pyridinone-3-carboxylic acid [52403-25-5] (22) is formed. Chlorination of (23) gives clopidol [2971-90-6] (24), a coccidiostat (29–30) (see Chemotherapeutics, antiprotozoal).

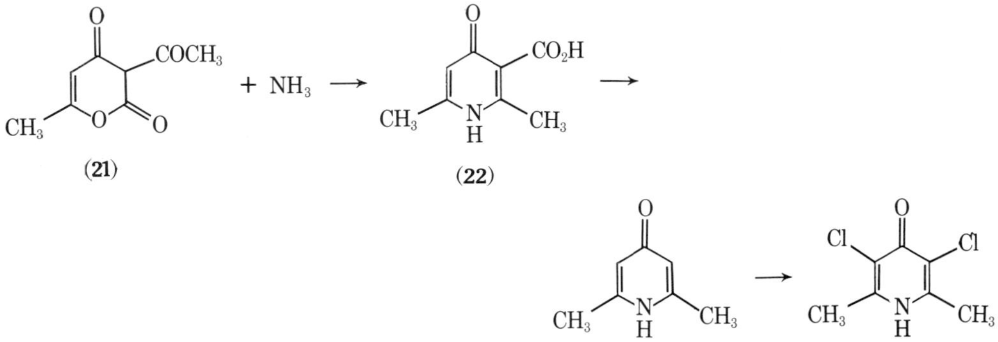

Ring expansion of compounds derived from furfural (see Furan derivatives) has several applications. Reaction of tetrahydrofurfuryl alcohol (25) with ammonia gives pyridine under dehydrogenating conditions, or piperidine (26) under reductive conditions.

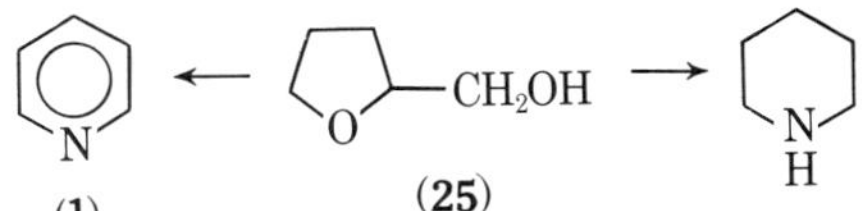

Furfurylamine (27) reacts with hydrogen peroxide and acid to give 3-hydroxypyridine [109-00-2] (28).

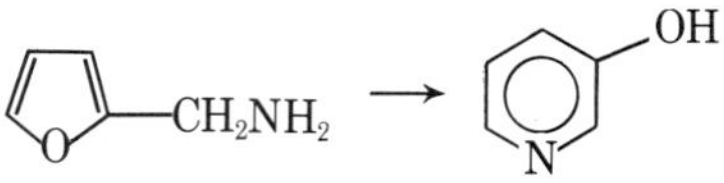

Production and Economics

In the United States, synthetic pyridine and alkyl pyridines are manufactured by Reilly Tar & Chemical Corporation in Indianapolis, Indiana, and by Nepera Chemical Co., Inc. (a subsidiary of Schering AG, Federal Republic of Germany) in Harriman, New York. Small amounts of coal tar-derived pyridine bases are manufactured by Koppers Co., Inc. in Pittsburgh, Pennsylvania. In Europe, Reilly Chemicals

SA (a division of Reilly Tar & Chemical Corporation) manufactures pyridine and alkyl pyridines in Belgium, and Lonza AG manufactures MEP in Switzerland. DSM (Stamicarbon B.V.) in the Netherlands manufactures 2-methylpyridine. Coal tar pyridine bases are available from many sources in Europe. Pyridine and alkyl pyridines are manufactured in Japan by Koei Chemical Co., Ltd. and by Daicel, Ltd. Coal tar bases are available from Nippon Steel. Downstream pyridine derivatives are manufactured by most of these companies and many others. Because of the small number of manufacturers, accurate production volumes and values of pyridine chemicals are not usually available. The U.S. market for pyridine has been estimated to be thousands of tons.

As with all petrochemicals, prices of pyridine chemicals are rising rapidly and reflect the inflation in cost of their raw materials. March 1981 list prices were synthetic pyridine (2°) $4.48/kg; nicotinic acid or nicotinamide (feed grade), $7.00/kg (31). Prices will no doubt rise in direct proportion to those of petroleum feedstocks.

Specifications, Analysis, Shipping, and Storage

Pyridine is generally available in two grades. Synthetic pyridine (2°) has a 2°C specification for the boiling-point range, which includes the normal boiling point of pyridine. Most 2° pyridine actually has a boiling-point range of less than 1°C and is greater than 99.8% pure by gas chromatographic analysis. ACS pyridine is a reagent grade and has the additional specifications listed in Table 3 (32) (see Fine chemicals).

Nicotinic acid and nicotinamide are available according to specifications of the USP for use as a vitamin supplement in humans (33) and as a feed grade for use in animal-feed supplements. Other industrial pyridine derivatives are generally sold as 98% or higher purity, with assay by gas chromatography, acid titration, or freezing point when applicable. Water content as determined by titration with the Karl Fischer reagent or by the Dean-Stark method is sometimes an important specification.

Pyridine compounds are shipped in drums, trucks, tank cars, and bulk containers in accordance with the appropriate regulations. Department of Transportation labeling requirements are shown in Table 4. Manufacturers recommend that 2- and 4-vinylpyridines be shipped and stored under refrigeration at less than 4°C to prevent polymerization.

Sales of piperidine in the United States are regulated by the Drug Enforcement Agency because of its use in the illicit synthesis of controlled drugs (37).

Table 3. Specifications of Pyridine, ACS Reagent Grade[a]

boiling range, °C	2.0 including 115.3 ± 0.1
solubility in water	to pass test there is no turbidity in a 10 wt % solution in 30 min
residue after evaporation, wt %	not more than 0.002
water, wt %	not more than 0.1
chloride, wt %	not more than 0.001
sulfate, wt %	not more than 0.001
ammonia, wt %	not more than 0.002
copper	to pass test, ie, ca 5 ppm
reducing substances	to pass test; 5 mL of pyridine does not entirely discharge color of 0.5 mL of 0.1 N $KMnO_4$ in 30 min

[a] Ref. 32.

Table 4. Health and Safety Factors and DOT Classification[a]

Compound	CAS Registry No.	Structure No.	Flash point[b] °C	Method[c]	DOT label	LD$_{50}$, rat (oral) mg/kg	LC$_{50}$, rat (inhl) ppm/hr	TLV, mg/m³	Minimum detectable odor[d], ppm
pyridine			20	TCC	red (flammable)	891	4000/4	15	0.012
2-methylpyridine			27	TCC	red (flammable)	790	4000/4		0.014
3-methylpyridine			38	TCC	none	400–800	8700/2[e]		
4-methylpyridine			39	TCC	none	1290	1000/4		
5-ethyl-2-methyl-pyridine			74	TCC	white (corrosive[f])	1540	1700/3.7[e]	2[g]	0.006
2-vinylpyridine			50	TCC	none	100–200		0.05[g]	<0.3
4-vinylpyridine			56	TCC	white (corrosive[f])	100–200			
nicotinic acid	[59-67-6]	(93)	over 150	TOC	none	5000[h]			
nicotinamide	[98-92-0]	(96)	over 150	TOC	none	3500			
piperidine			12	TCC	red (flammable)	400	4000/4		
2-aminopyridine	[504-29-0]	(46)	110	COC	Poison B	200		0.5 (ppm)	

[a] Refs. 34 and 35 unless noted otherwise.
[b] Ref. 7.
[c] TOC, Tag Open Cup; TCC, Tag Closed Cup; COC, Cleveland Open Cup.
[d] Ref. 36.
[e] LC$_{100}$.
[f] Causes skin irritation, but is not corrosive to metals.
[g] USSR.
[h] Mouse.

Health and Safety Factors (Toxicology)

Acute toxic responses to many pyridine derivatives have been measured in laboratory animals (see Table 4). All pyridine chemicals should be handled according to the best standards of laboratory and manufacturing safe practice (34). Some alkyl pyridines occur naturally in small quantities in foods, and some compounds are acceptable as food additives (38). Safety data for all commercial compounds are available from the manufacturers.

Pyridine was selected by the TSCA Inter-Agency Testing Committee for extensive testing of its long-term health effects; no testing rule has been made at this time (39). Pyridine has been selected by the National Cancer Institute for carcinogenesis assay; results will be available in mid-1982 (40). All mutagenicity tests, including the Ames test, on pyridine have given negative results.

In addition to the usual precautions for handling industrial chemicals, pyridine and the alkyl pyridines demand additional handling precautions to prevent the escape of vapors, because the compounds typically have intense odors that are detectable and disagreeable at very low concentrations. Contact of 2- or 4-vinylpyridine with the skin should be avoided, as temporary, painful skin burns frequently result. Some individuals may become especially sensitized to vinylpyridine, but adverse reaction may be prevented by the use of new rubber gloves, aprons, and face shields (34).

Nicotinic acid and nicotinamide, which therapeutically are nearly equivalent forms of vitamin B_3, exhibit low toxicity. Both forms of niacin cause relaxation of the capillaries and concomitant reduction in blood pressure. Inhalation or skin contact with nicotinic acid dusts, in particular, can cause objectionable skin flushing or faintness in many individuals (41) (see Vitamins, nicotinic acid.).

Pyridine and alkyl pyridines are excellent solvents for many materials, a property that must be taken into consideration when selecting O-rings, gaskets, and other sealants that are to be in contact with the liquids. Generally, only polytetrafluoroethylene, graphite, and asbestos-based gasket and O-ring materials are acceptable, whereas most rubbers are rapidly swollen or degraded by liquid alkyl pyridines.

Reactions and Uses of Derivatives

Reactions at Nitrogen. *Basicity and Acylation.* The basicity of simple pyridines (see Table 1), although less than that of aliphatic tertiary amines, characterizes pyridine and related compounds as useful acid acceptors in acylation reactions involving anhydrides or acid chlorides. Pyridine is typically used as a solvent in such laboratory-scale reactions. On an industrial scale, pyridine can frequently be replaced by a cross-linked polyvinylpyridine (see Vinyl polymers, *N*-vinyl monomers and polymers).

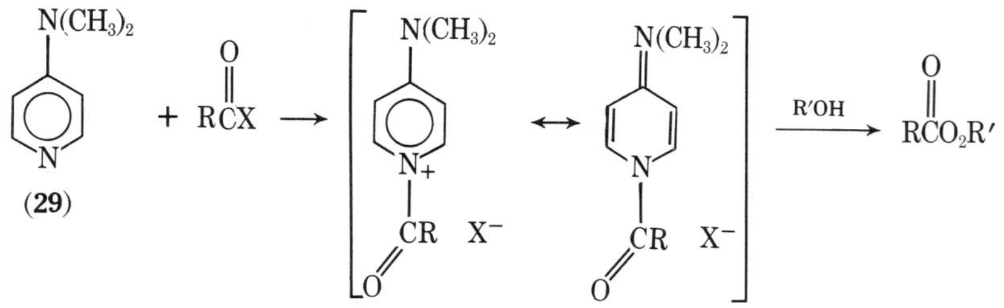

Additional acceleration of acylation reactions is observed when the nucleophilicity of the pyridine nitrogen is increased by the presence of an electron-donating substituent on the pyridine ring. The most useful compound exhibiting this property is 4-dimethylaminopyridine [*1122-58-3*] (DMAP) (29). The exceptional catalysis of acylations by DMAP is accounted for by its enhanced nucleophilicity as compared with that of pyridine, which causes the equilibrium formation of the N-acylpyridinium salt (**30**) to occur to a greater extent (42).

Quaternization. Pyridine and most of its derivatives react with alkylating agents, typically alkyl halides or sulfates, to produce quaternary (quat) salts (**31**) (43–45). The quaternary salt [(**31**) R = n-$C_{16}H_{33}$, X = Cl] formed from pyridine and cetyl chloride, 1-cetylpyridinium chloride [123-03-5], is an antiseptic (see also Disinfectants and antiseptics).

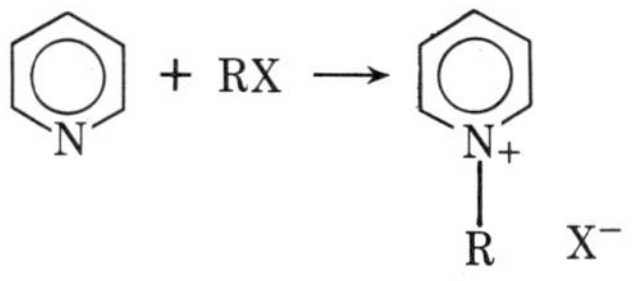

Quaternary salts of 2,2′- and 4,4′-bipyridyls are useful herbicides (qv). Reaction of pyridine with metallic sodium, followed by air oxidation and hydrolysis, yields 4,4′-bipyridyl [553-26-4] (**32**) (47). Reaction with a methylating agent gives paraquat [4685-14-7] (**33**), a contact herbicide and plant dessicant.

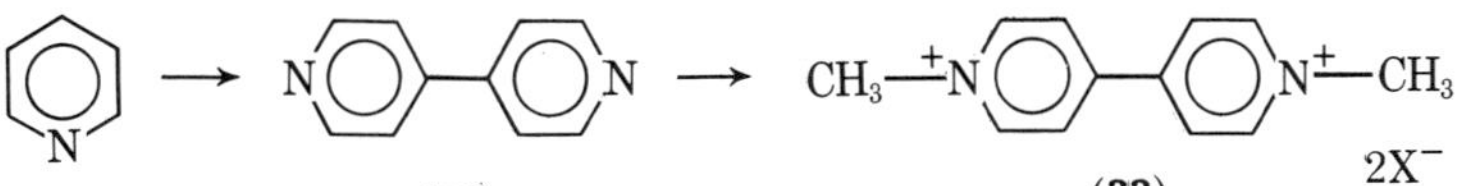

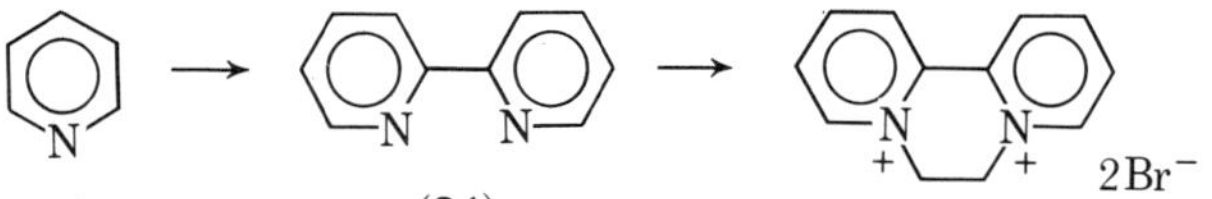

Similarly, dehydrogenation of pyridine with nickel or other metal catalysts gives 2,2′-bipyridyl [366-18-17] (**34**) (48), which on quaternization with ethylene bromide yields diquat [2764-72-9] (**35**), also a contact herbicide and plant dessicant.

Amprolium [121-25-5] (**36**), a coccidiostat, is a quaternary salt of 2-methylpyridine with a substituted pyrimidine ring (49).

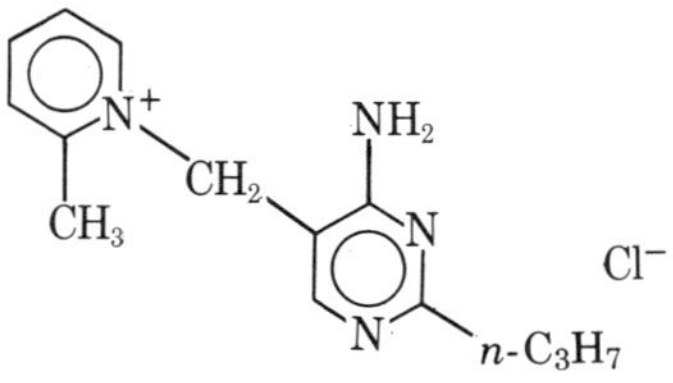

Quaternary hydroxides of pyridines (**37**) are in equilibrium with a small amount of the covalent pseudobase (**38**) because of the increased susceptibility of the pyridine ring to nucleophilic attack when quaternized (50–51). The intermediacy of pseudobases accounts for the easy oxidation of pyridinium salts to the corresponding pyridinones (**39**) by mild oxidizing agents, eg, potassium ferricyanide, or electrochemically. Quaternary hydroxides of pyridines that bear an α-hydrogen on a substituent in the 2- or 4-position are in equilibrium with a small amount of the corresponding anhydro base (**40**) or (**41**). The latter account for the enhanced reactivity of quaternary salts in condensations with aldehydes and other base-catalyzed reactions (see also Quaternary ammonium compounds).

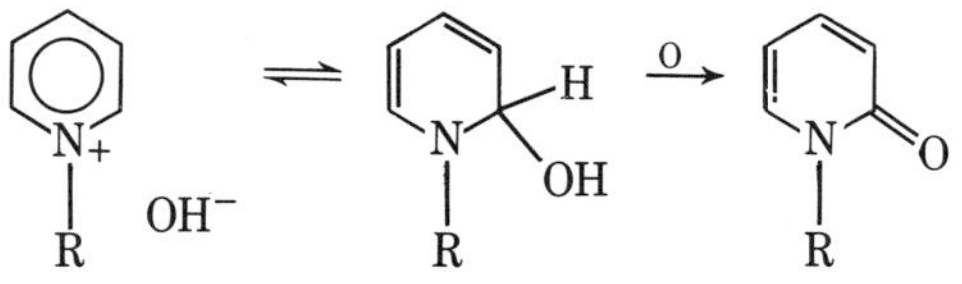

N-Oxidation. Most pyridine derivatives react with a peracid, usually peracetic acid, or hydrogen peroxide (qv) in the presence of a catalyst to give the corresponding N-oxides (**42**) (see Amine oxides). These N-oxides offer a useful increased reactivity in comparison with that of the corresponding pyridine bases (see below).

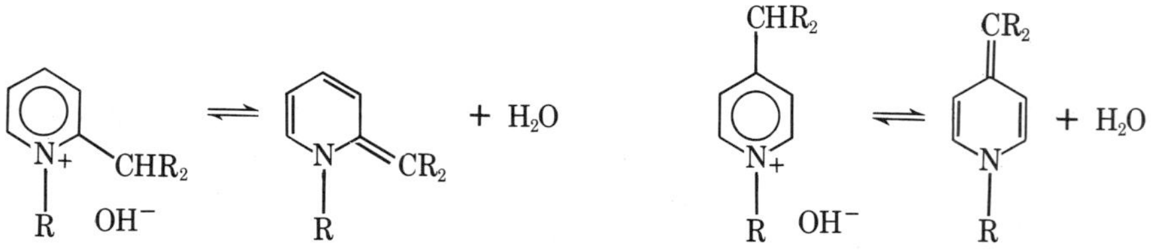

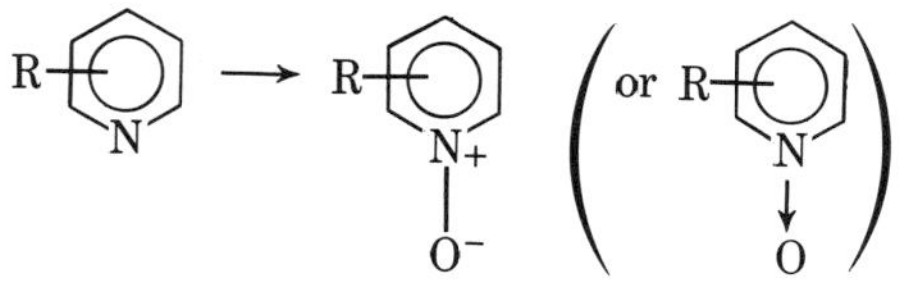

Reactions on the Nucleus. Electrophiles. Although much of benzene chemistry is concerned with the attack of electrophiles on the benzene nucleus, these reactions are generally not useful in the pyridine series for two reasons: First, the ring nitrogen withdraws electrons from the ring carbons, as rationalized by the resonance forms (**43**), (**44**), and (**45**); thus, the ring carbons in pyridine are about as unreactive toward electrophiles as those in nitrobenzene (**1**).

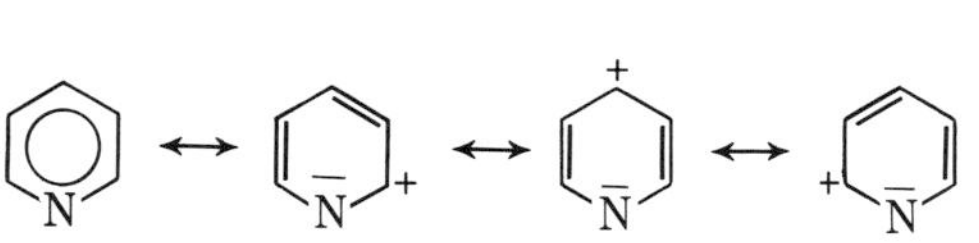

Second, most electrophiles react preferentially and reversibly with the basic pyridine nitrogen, which further reduces the reactivity of the ring carbons. Reactions of electrophiles are described in refs. 52 and 53.

Pyridine is nitrated only under drastic conditions, ie, in oleum at 300°C, to give 3-nitropyridine [2530-26-9] in very low yield. Sulfonation of pyridine with H_2SO_4 and $HgSO_4$ at 230°C gives pyridine-3-sulfonic acid [636-73-7] in good yield. Friedel-Crafts reactions of pyridine fail to give carbon-alkylated or -acylated products. Conversion of the basic pyridine nitrogen to the N-oxide function partly overcomes these effects, at least in the case of nitration.

Substitution of pyridine by an electron-donating group partly overcomes the reluctance of the ring to react with electrophiles. Thus, 2-aminopyridine (46) reacts easily with cold mixed nitric and sulfuric acid, first yielding N-nitro-2-pyridinamine [26482-54-2] (47). On warming of the reaction mixture, rearrangement occurs to give mostly 2-amino-5-nitropyridine [4214-76-0] (48) and a smaller amount of the 2,3-isomer [4214-75-9] (49) (54–56).

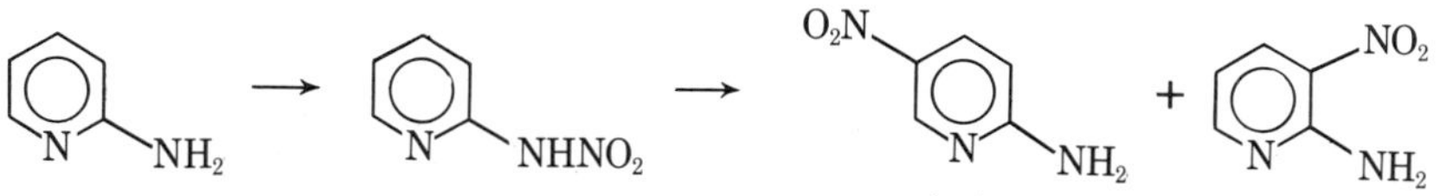

Chlorination of 2-aminopyridine occurs in acid to give predominantly the 5-chloro isomer [1072-98-6] (50), and subsequently, the 3,5-dichloro compound [4214-74-8] (51).

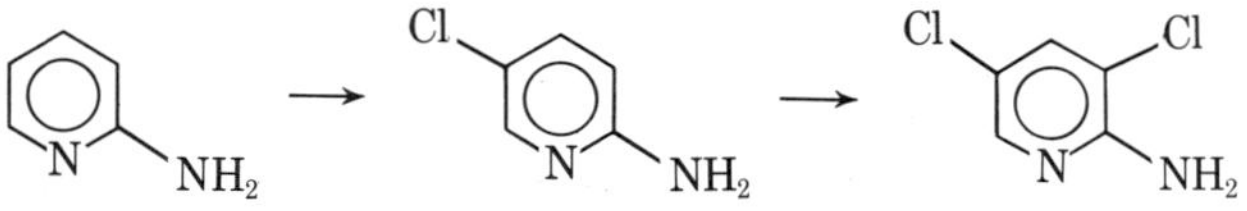

Related electrophilic reactions occur with 3-aminopyridine [462-08-8], 4-aminopyridine [504-24-5], 2-pyridinone [142-08-5], 4-pyridinone [108-96-3], and 3-hydroxypyridine [109-00-2] with typical ortho-para directing effects.

Amination. In contrast to its lack of useful reactivity in electrophilic substitution, some nucleophilic substitutions occur on pyridines in a synthetically useful fashion. Sodamide reacts with pyridine, typically in refluxing toluene or xylene (the Chichibabin amination) to give, after hydrolysis of the intermediate sodium salt (52), 2-aminopyridine (46) in high yield (57–60).

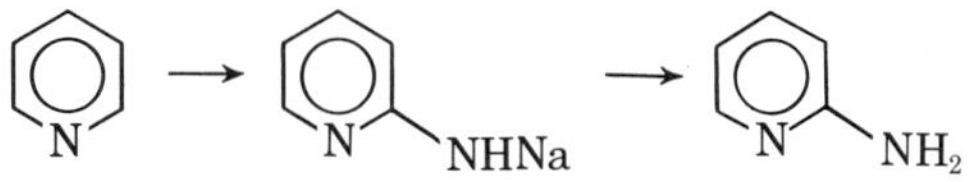

Similar reactions of 2- and 4-methylpyridine give 2-amino-6-methylpyridine [1824-81-3] (53) and 2-amino-4-methylpyridine [695-34-1] (54). Amination of 3-methylpyridine gives predominantly 2-amino-3-methylpyridine [1603-40-3] (55) and much smaller amounts of the 2,5-isomer [1603-41-4].

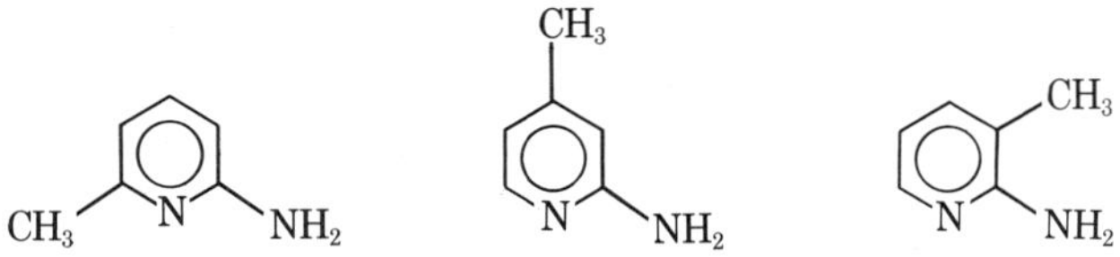

468 PYRIDINE AND PYRIDINE DERIVATIVES

Although many other simply substituted pyridines are aminated normally, compounds with large alkyl groups are coupled to give substituted 2,2′-bipyridyls under Chichibabin conditions. Thus, 4-(5-nonyl)pyridine [2961-47-9] (56) reacts with sodamide to give 4,4′-bis-(5-nonyl)-2,2′-bipyridyl [72230-93-4] (57) (61). Details of the mechanism of the Chichibabin amination, and the relevance of this coupling reaction, remain unclear.

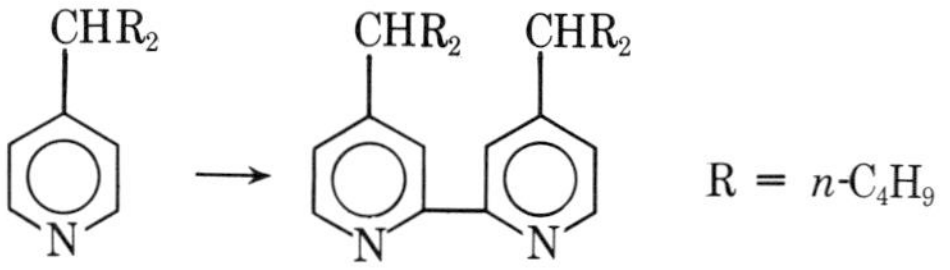

Amination of pyridine or 2-aminopyridine under more vigorous conditions gives 2,6-diaminopyridine [141-86-6] (58). Coupling with a benzenediazonium salt gives phenazopyridine [136-40-3] (59), which is an antiseptic (62).

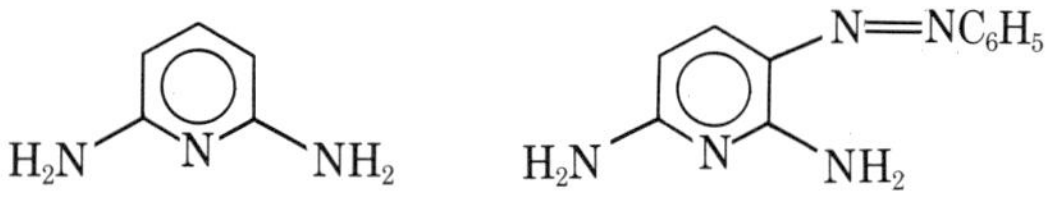

Aminopyridine. Reactions of aminopyridines are described in refs. 59 and 60. Reaction of 2-aminopyridine with N-acetylsulfanilyl chloride, followed by hydrolysis, gives sulfapyridine [144-83-2] (60) (see Antibacterial agents, synthetic).

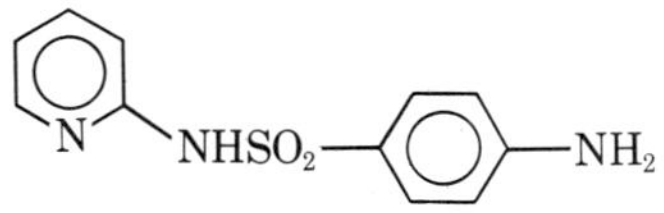

Aminopyridines generally react with alkylating agents at the ring nitrogen to give derivatives of (61). However, the metal salt (52), available from treatment of 2-aminopyridine with sodamide or as an intermediate in the Chichibabin amination, is alkylated on the exocyclic nitrogen. Such reactions can be used to produce common antihistamines of type (62) (see Histamine and histamine antagonists).

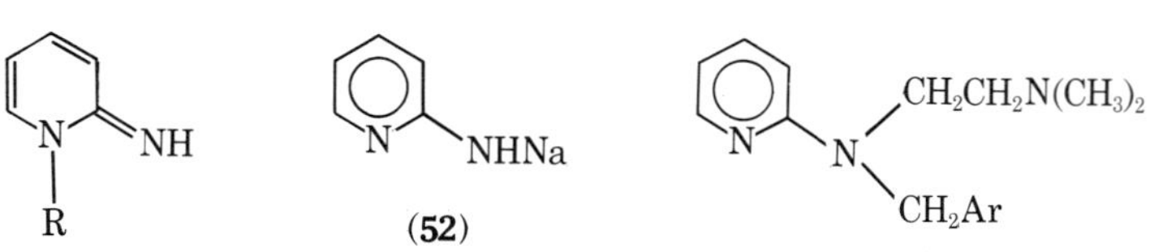

Reaction of 2-amino-6-methylpyridine with an alkoxymethylene–malonic ester gives the substituted aminomethylene compound [13250-95-8] (63). Cyclization gives the naphthyridinone [35482-56-5] (64), which can be converted to nalidixic acid [389-08-2] (65) (63) (see Antibacterial agents, synthetic).

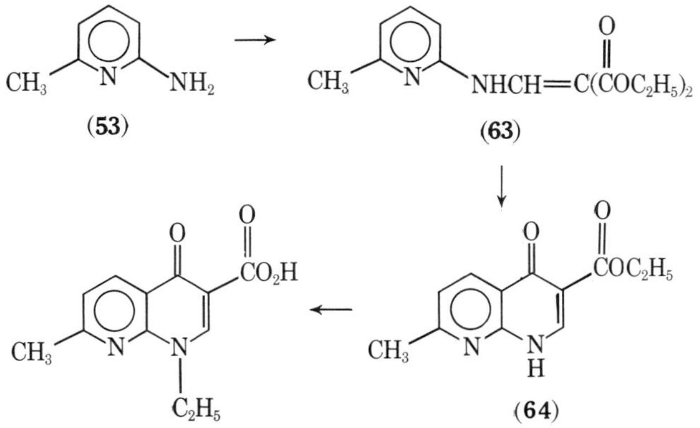

Amino groups on the pyridine ring can be diazotized with the usual reagents. However, the diazonium group in the 2- and 4-positions hydrolyzes so rapidly that special reaction conditions must be chosen to achieve acceptable yields of substitution products. On the other hand, 3-aminopyridine can be diazotized, and substitution reactions can be carried out in the usual manner (64). Thus, treatment of 2-aminopyridine with sodium nitrite and hydrochloric acid gives 2-pyridinone [142-08-5] (66), which also is named as its tautomer 2-hydroxypyridine [142-08-5] (67), in good yield. Diazotization of 4-aminopyridine occurs similarly to give 4-pyridinone [626-64-2] (68) (4-hydroxypyridine [626-64-2] (69)).

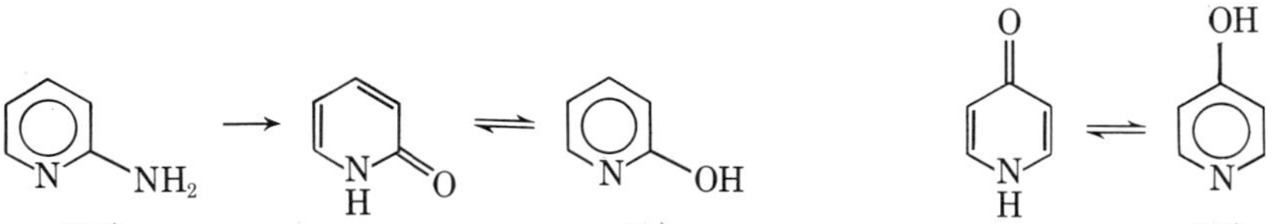

The subject of these tautomeric equilibria has been reviewed (65–67).

Diazotization of 2-aminopyridine in hydrobromic acid containing bromine gives 2-bromopyridine [109-04-6] (70).

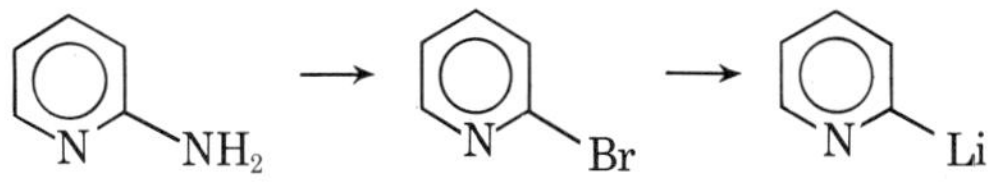

Metal–halogen exchange of (70) with butyllithium gives 2-pyrridyllithium [17624-36-1] (71), which reacts normally with electrophiles (68–69). A commercial reaction is that of (71) with the appropriate Mannich ketone to give alcohol [70708-28-0] (72) which, on dehydration, yields the antihistamine triprolidine [486-12-4] (73) (70).

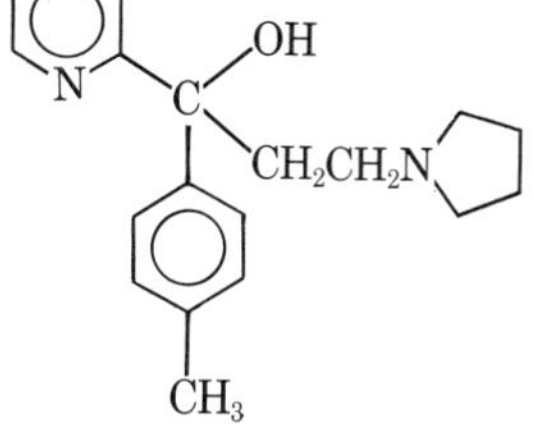

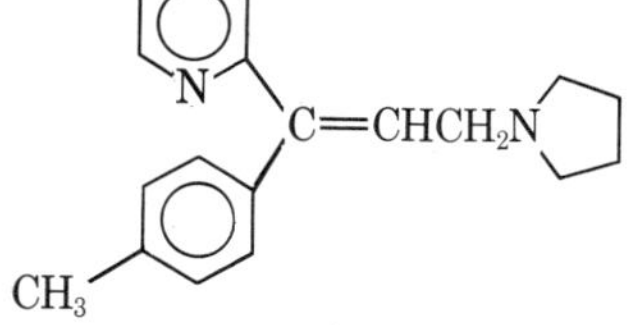

Chlorination.

Chlorination reactions are described in refs. 71–75. Direct chlorination of pyridine occurs in the vapor phase at over 300°C in the presence of a diluent to give 2-chloropyridine [109-09-1] (**74**); 2,6-dichloropyridine [2402-78-0] (**75**) is a by-product. Vigorous chlorination of pyridine or most of its alkyl derivatives eventually gives pentachloropyridine [2176-62-7] (**76**).

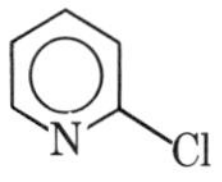

(74) 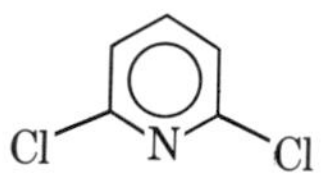   (75) 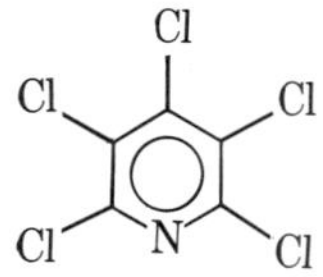   (76)

Chloride can be displaced from 2-chloropyridine by nucleophiles. Pyridylation of aryl acetonitrile derivatives using strong base is followed by further alkylation, hydrolysis, and decarboxylation to produce antihistamines, eg, pheniramine [86-21-5] [(**77**), Ar = C_6H_5]. Disopyramide [3737-09-5] (**78**), which is an antiarrhythmic, is produced similarly (76) (see Cardiovascular agents).

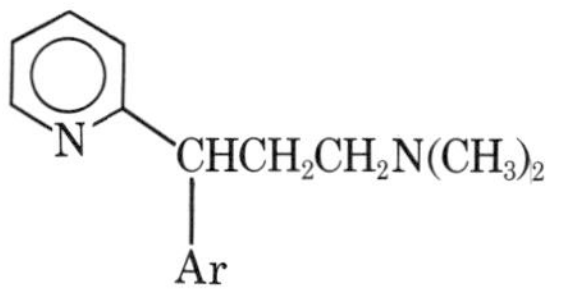

(77) 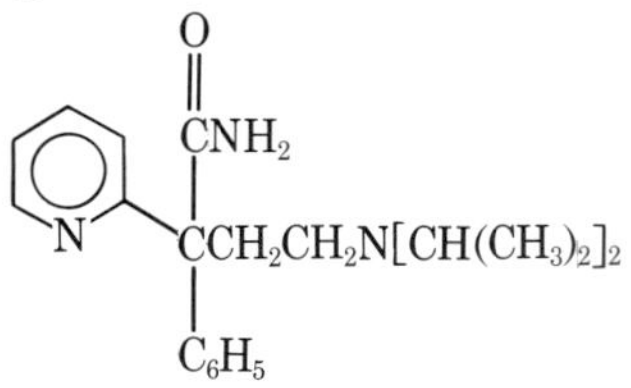   (78)

Reaction of 2-chloropyridine N-oxide [20295-64-1] (**79**) with sodium hydrosulfide gives pyrithione [1121-31-9] (**80**); the zinc salt [13463-41-7] is widely used as an antifungal agent (77) (see Cosmetics; Hair preparations).

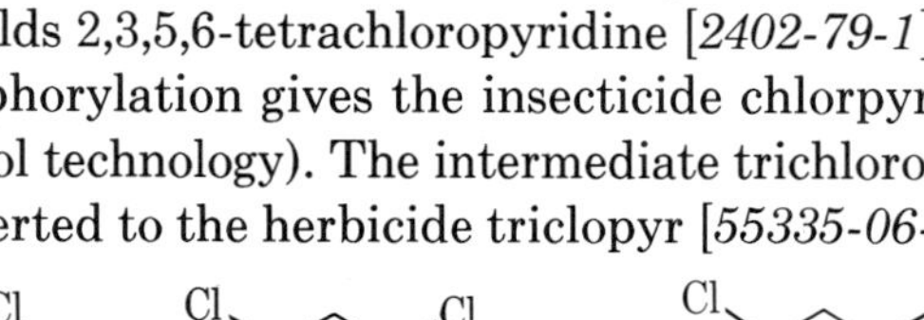

(79)   (80)

Pentachloropyridine is most reactive at the 4-position. Reduction with zinc, or electrochemically, yields 2,3,5,6-tetrachloropyridine [2402-79-1] (**81**) (78–79). Partial hydrolysis and phosphorylation gives the insecticide chlorpyrifos [2921-88-2] (**83**) (80) (see Insect control technology). The intermediate trichloropyridinol [6515-38-4] (**82**) can also be converted to the herbicide triclopyr [55335-06-3] (**84**) (81).

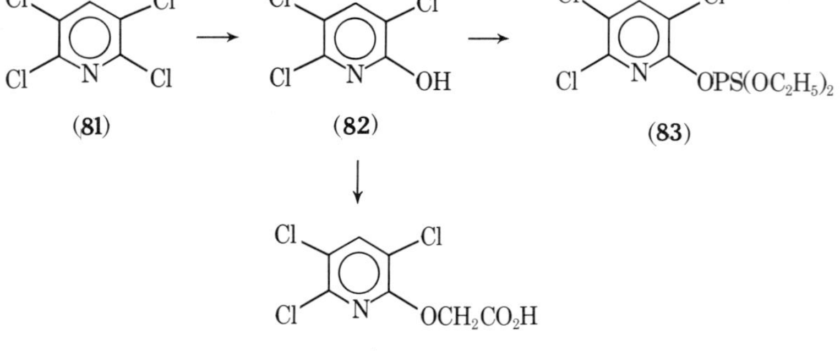

(81)   (82)   (83)

(84)

Exhaustive chlorination of 2-methylpyridine or derivatives, followed by displacement with ammonia and hydrolysis, gives the herbicide picloram [1918-02-1] (**85**) (82).

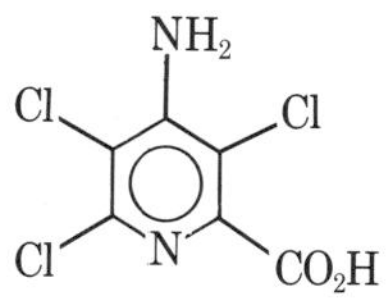

(**85**)

Radicals. Alkyl and aryl radicals react with pyridine to give substitution at all three positions (83). Thus, phenyl radicals that are generated from nitrobenzene, benzenediazonium chloride, or benzoyl peroxide react with pyridine to give a mixture of 2- [1008-89-5], 3- [1008-88-4], and 4-phenylpyridine [939-23-1]; 2-phenylpyridine predominates. In strong acid solution, nucleophilic radicals react preferentially at the 2 and 4 positions.

Pyridine or alkyl pyridines react with methanol or higher primary alcohols in the presence of nickel catalysts to give pyridines that are methylated in the 2 position. 3-Alkylpyridines react preferentially at the less hindered position to give 2-methyl-5-alkylpyridines (84). In the Wibaut-Arens reaction, alkyl pyridines (**86**) that are not substituted in the 4 position react with zinc, or iron, and acetic anhydride to give 4-ethylpyridines (**89**). Intermediates (**87**) and (**88**) have been suggested (85–86).

Reactions of Side Chains. Alkyl groups attached to the pyridine ring undergo many of the typical oxidation and substitution reactions of alkyl aromatic compounds (85–86). In addition, hydrogens on carbon atoms adjacent to the ring are particularly acidic, leading to additional kinds of reactions.

Halogenation. Chlorination of 2-methylpyridine in carbon tetrachloride solution in the presence of sodium carbonate gives 2-chloromethylpyridine [4377-33-7] (**90**). The compound is conveniently stored as the hydrochloride [6959-47-3] (87).

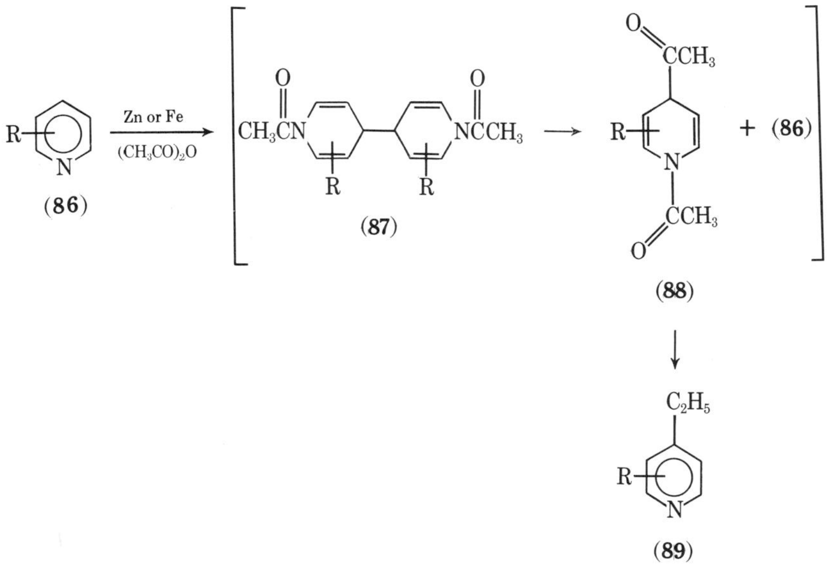

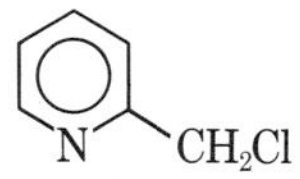

Chlorination under more vigorous conditions gives nitrapyrin [1929-82-4] (**91**), which is used in agriculture to prevent loss of ammonia fertilizers from soil as a result of metabolism by bacteria (88).

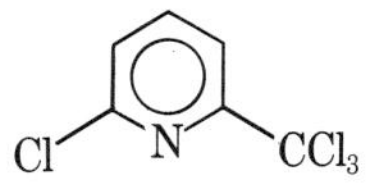

Oxidation. *Carboxylic Acids.* By the usual laboratory methods, oxidation of alkyl pyridines yields carboxylic acids (see Table 5) (89–90). Carboxy groups in the 2 or 4 positions are easily lost. Oxidation of MEP (**5**) with nitric acid gives nicotinic acid (**93**), following decarboxylation of the intermediate (**92**) under the reaction conditions (91).

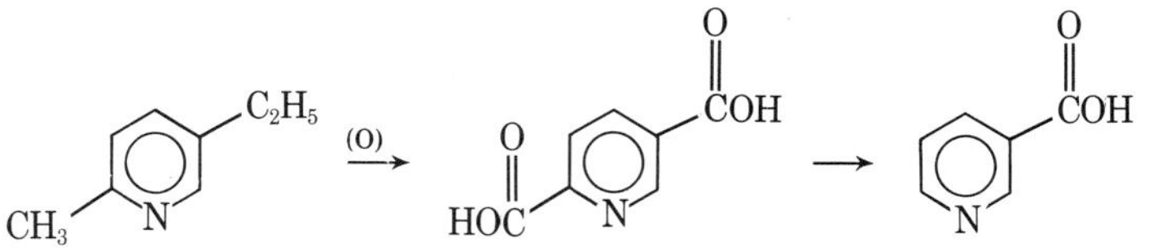

Nicotinic acid was first named because of its preparation by degradation of nicotine [54-11-5] (**94**); the name niacin was coined to avoid sanctioning the structural relationship of the vitamin to the alkaloid (see Alkaloids; Vitamins).

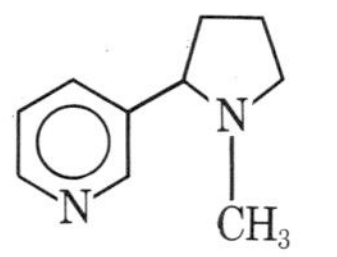

Ammoxidation of the methylpyridines occurs catalytically in the vapor phase in the presence of air and ammonia to give the cyanopyridines (92). Of these, 3-cyanopyridine [100-54-9] (**95**) is the most important commercially, since partial hydrolysis occurs readily with base catalysis to give nicotinamide (niacinamide) (**96**). Further hydrolysis to nicotinic acid can also be carried out (93).

Table 5. Pyridinecarboxylic Acids[a]

Common name	CAS Registry No.	Structure no.	Position of carboxy group(s)	Mp, °C	Solubility (at 20°C), g/100 g H_2O
picolinic acid	[98-98-6]		2-	134–136	>90
nicotinic acid (niacin)	[59-67-6]		3-	236	1.7
isonicotinic acid	[55-22-1]		4-	314–315	0.5
quinolinic acid	[89-00-9]		2,3-	190	0.5
lutidinic acid	[499-80-9]		2,4-	243; 248–250[b]	0.6[c]
isocinchomeronic acid	[100-26-5]	(**92**)	2,5-	254; 257[b]	
dipicolinic acid	[499-83-2]		2,6-	237	0.5[c]
cinchomeronic acid	[490-11-9]		3,4-	256; 266–268[b]	
dinicotinic acid	[499-81-0]		3,5-	320–323	

[a] Refs. 7, 89, and 90.
[b] Different melting points are reported in different references.
[c] At 25°C.

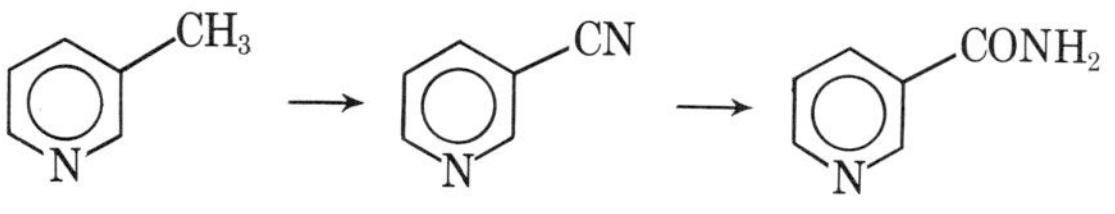

(3) → (95) → (96)

Nicotinic acid or a derivative is required by all living cells and occurs in all cells, as the amide, as part of several coenzymes involved in oxidation–reduction reactions. Although niacin can be synthesized from tryptophan in certain animal tissues, nicotinic acid or nicotinamide is essential in the diet of nonruminant animals; it may be beneficial in the diet of ruminants as well. Nicotinic acid or nicotinamide, both of which are sometimes called niacin, are manufactured in a feed grade that is routinely incorporated in prepared feeds for swine and poultry. Nicotinamide is more effective than the acid, as shown by chick growth-rate assay (94).

The availability of 3-cyanopyridine, nicotinamide, and nicotinic acid makes possible their common use as synthetic intermediates. Hydrogenation of (95) gives 3-picolylamine [3731-52-0] (97) which, upon diazotization, gives 3-pyridyl carbinol [100-55-0] (98). The latter is used as a peripheral vasodilator (95–97). The urea [53558-25-1] (99) is a rodenticide (98–99).

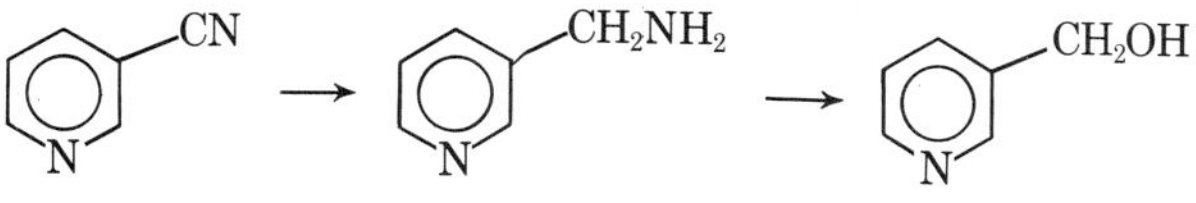

(95) → (97) → (98)

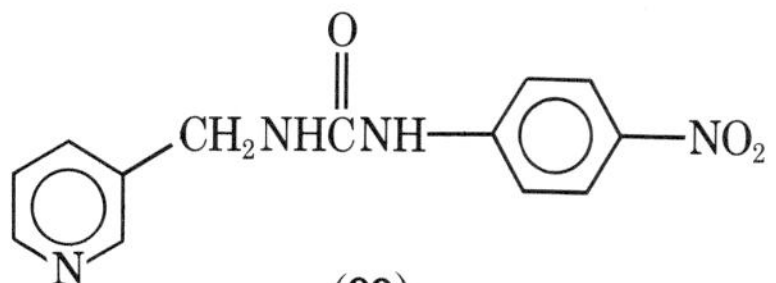

(99) 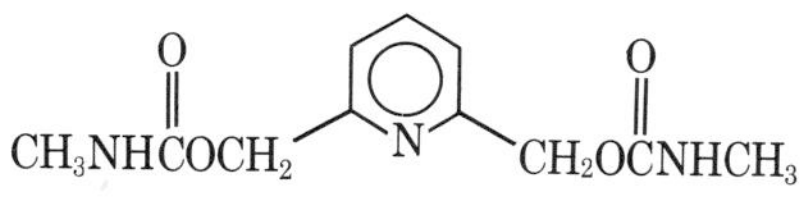   (100)

Several esters of nicotinic acid are used internally as vasodilators or externally as rubifacients. The bis(methylcarbamate) [1882-26-4] (100) of pyridine-2,6-dimethanol [1195-59-1], available from reduction of the dipicolinic ester, is used to lower serum cholesterol and lipid levels (100).

Hofmann reaction of nicotinamide yields 3-aminopyridine (101). Isonicotinamide [1453-82-3] reacts similarly to give 4-aminopyridine, which is used in agriculture as a repellent or poison for crop-destroying birds (102) (see Poisons, economic).

Reaction of an ester of isonicotinic acid with hydrazine yields isonicotinic hydrazide (isoniazid) [54-83-3] (101). Isoniazid was the first highly effective drug in the treatment of tuberculosis (103). However, the compound has been determined to be carcinogenic in laboratory animals, and its manufacture and use in the United States has declined (104).

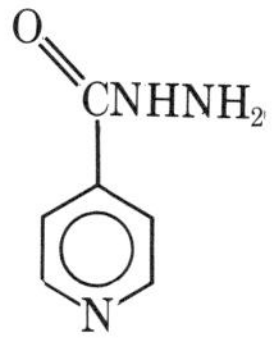

(101)

Aldehydes and Ketones. Catalytic air oxidation of the methylpyridines in the vapor phase with steam dilution gives 2- [1121-60-4], 3- [500-22-1], or 4-pyridinecarboxaldehyde [872-85-5]. Quaternary salts of oximes, particularly 2-pyridinealdoxime methochloride (PAM-2 chloride) [51-15-0] (**102**) and obidoxime chloride [114-90-9] (**103**) are used as antidotes for poisoning by organophosphate acetylcholinesterase inhibitors (see Cholinesterase inhibitors).

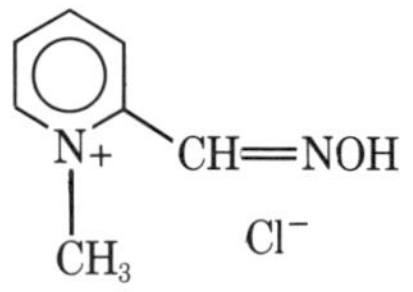

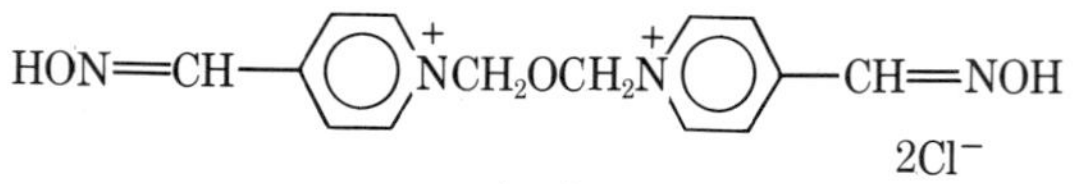

Catalytic air oxidation of ethylpyridines in the liquid phase yields the 2- [1122-62-9], 3- [350-03-8], and 4-acetylpyridines [1122-54-9] (**105**). These compounds are especially reactive in typical methyl ketone reactions, addition to the carbonyl and condensations at the methyl group, because of the electron-withdrawing effect of the pyridine ring. Aldehyde and ketone reactions are described in refs. 106 and 107.

Base-Catalyzed Reactions. The acidity of side-chain hydrogens in methylpyridines and related compounds is accounted for by the inductive electron-withdrawing effect of the electronegative nitrogen atom in the ring and, in the 2 and 4 positions, by resonance stabilization of the negative charge in the conjugate base (**104**)–(**107**) (108). Such resonance structures are not possible for anions that are adjacent to the 3-position; therefore anion-forming reactivity in 3-alkylpyridines is reduced.

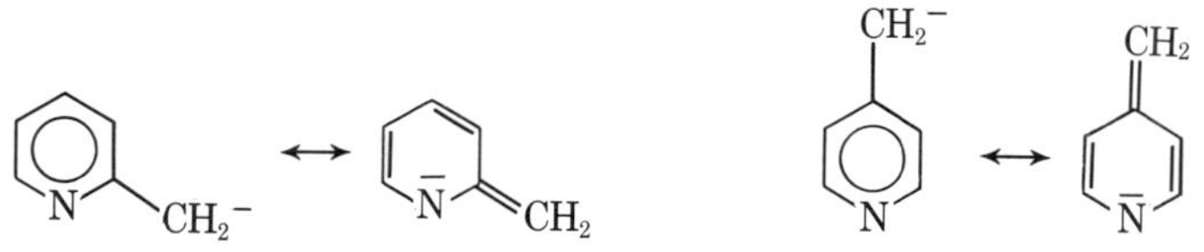

Alkylation or acylation of methyl or related groups in the 2, 3, or 4 position is easily accomplished by sequential addition of the alkyl pyridine and the alkylating or acylating agent to a solution or a suspension of a strong base. The strong base of choice is frequently sodamide in liquid ammonia. Typical alkylating agents include primary and secondary alkyl halides and epoxides. The acylating agent is usually an ester. Michael reactions in which 2- or 4-methylpyridine are donors are often carried out with metallic sodium as the catalyst. Compounds such as 4-phenylpropylpyridine [2057-49-0] (**108**) and 1,3-di(4-pyridyl)propane [17252-51-6] (**109**) can be prepared in this manner.

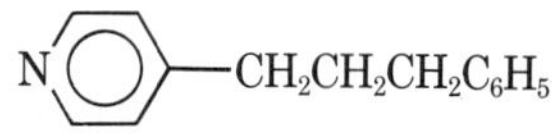

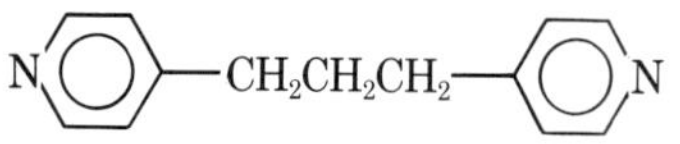

The acidity of methyl groups in 2- and 4-methylpyridine is sufficient to enable them to take part in aldol reactions. Condensations with aromatic aldehydes take place in refluxing acetic anhydride to give styrylpyridines or stilbazoles, eg, 2-(2-phenyl-vinyl)pyridine [714-08-9] (110).

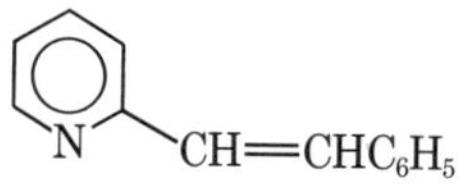

Quaternary salts can form anhydro bases as their conjugate bases, allowing aldol condensations to occur under milder conditions. Condensation of methylpyridine quaternary salts with aromatic aldehydes takes place rapidly under mild conditions with catalysis by amines to give quaternary derivatives of stilbazoles, eg, trans-1-methyl-2-(2-phenylvinyl)pyridinium iodide [1718-64-5] (111).

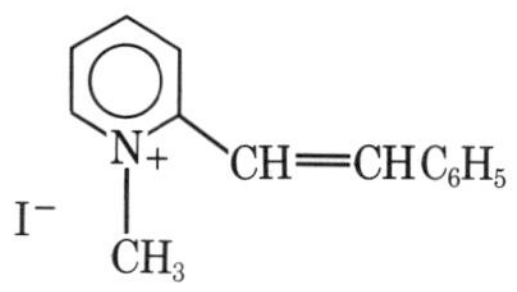

Condensation of 2- or 4-methylpyridine or MEP with formaldehyde occurs on heating, to yield initially the corresponding aldol, eg, 2-(2-hydroxyethyl)pyridine [103-74-2] (112). Dehydration by treatment with base yields 2- [100-69-6] (113) or 4-vinylpyridine [100-43-6] (114) or 5-ethyl-2-vinylpyridine [5408-74-2] (115), respectively. Dehydrogenation of MEP gives 2-methyl-5-vinylpyridine [140-76-1] (116), the remaining common vinylpyridine derivative.

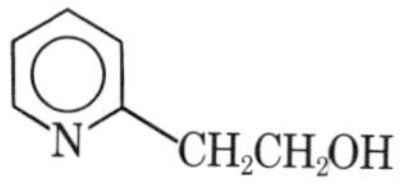

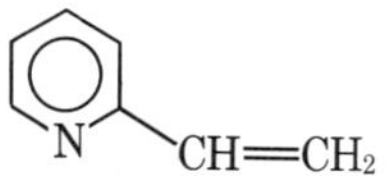

The first commercial utility for the vinylpyridines was the use of copolymers of 2-vinylpyridine as tire-cord binders. Typically, a tire cord (qv) is treated first with a resorcinol–formaldehyde polymer and then with a latex copolymer of styrene, butadiene, and 2-vinylpyridine. This treatment is commonly called RFL for resorcinol–formaldehyde–latex (109).

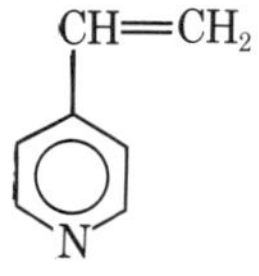

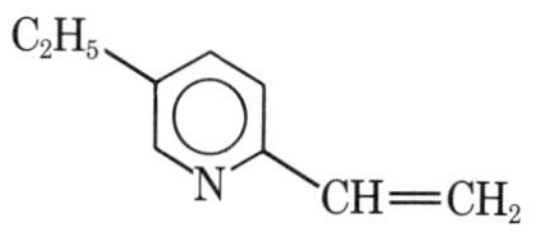

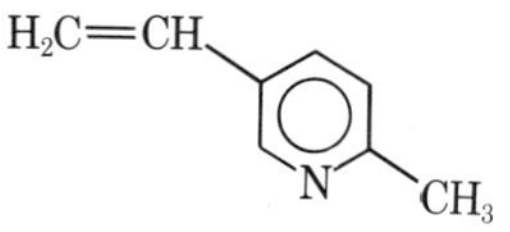

Homopolymers of vinylpyridines are easily prepared by means of anionic or free-radical initiators. Such polymers are usually soluble in many organic solvents and also in aqueous acid, since the basicity of the pyridine functionality remains unchanged. Polymers containing a cross-linking agent, eg, divinylbenzene, are solids that are insoluble in all solvents including acids. These insoluble polymers are convenient substitutes for pyridine and other monomeric amines in acylations and other acid-generating reactions, since they are easily removed from reaction mixtures by filtration or centrifugation, and their basic properties are easily regenerated by treatment with alkali. The polymers can also serve as replacements for pyridine in base-catalyzed condensation reactions (110).

Since a pyridine ring stabilizes a negative charge on an adjacent carbon in the 2- or 4-position, 2- and 4-vinylpyridines are reactive Michael acceptors. Methanol adds to 2-vinylpyridine to yield 2-methoxyethylpyridine [114-91-0] (117), which is a veterinary anthelmintic (111) (see Veterinary drugs). Addition of methylamine to 2-vinylpyridine gives 2-methylaminoethylpyridine [5638-76-6] (118); the hydrochloride [5579-84-0] is a vasodilator (112). Addition of pyridine to 2-vinylpyridine in the presence of hydrochloric acid gives 1-[2-(2-pyridinyl)ethyl]pyridinium chloride [41067-53-2] (119) (113). Vinylpyridines also react as dienophiles in the Diels-Alder reaction, and as dipolarophiles in 1,3-dipolar cycloadditions (see Vinyl polymers, N-vinyl monomers and polymers).

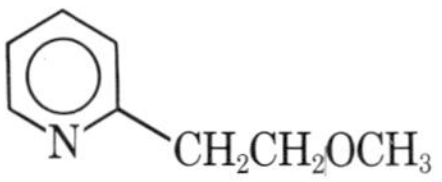

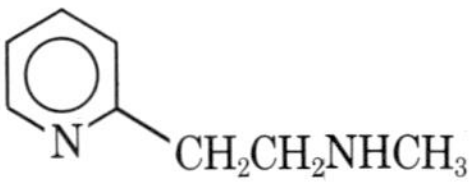

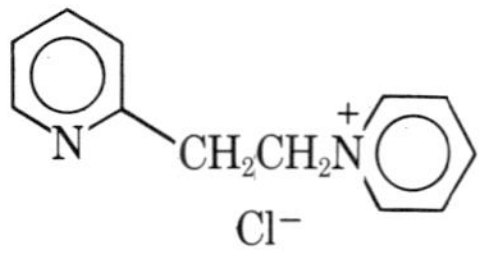

Reduction and Reduced Derivatives. Pyridine and most of its derivatives are easily hydrogenated at elevated temperatures and pressures with nickel, palladium, or ruthenium catalyst. Platinum is a very effective catalyst in acetic acid solution. Quaternary salts are more easily hydrogenated than free bases (114–118). Pyridine reduction can also be carried out electrochemically (119).

Piperidine (26) was first isolated from piperine [94-62-2] (120), which occurs in black pepper *Piper niger*. The relation of the names piperidine and pyridine by inclusion of the unaccented "pe" syllable is coincidental, but provides a consistent although cacophonous common nomenclature.

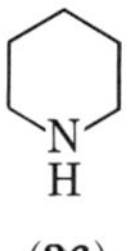

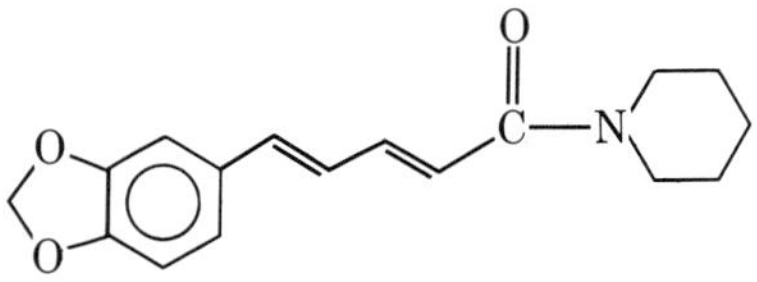

Methylpiperidines are pipecolines, dimethylpiperidines are lupetidines, and 2-, 3-, and 4-piperidinecarboxylic acids are pipecolic [535-75-1], nipecotic [498-95-3], and isonipecotic [498-94-2] acids, respectively. Extension to more highly substituted compounds is rare.

A major use of piperidine has been in the manufacture of the dithiuram tetrasulfide [120-54-7] (121), which is used as a vulcanization accelerator in rubber. Other uses include the formation and use of piperidine enamines and other compounds in specialty chemical manufacture.

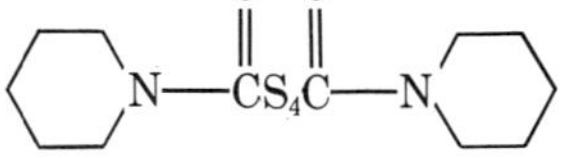

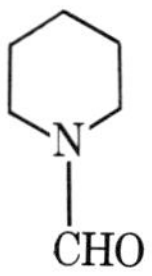

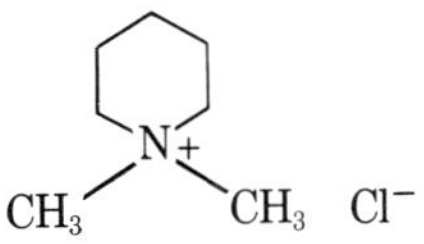

The dipolar aprotic solvent 1-formylpiperidine [2591-86-8] (122) is particularly useful because of its miscibility with water and organic solvents, its high boiling point (222°C) and extensive liquid range, and its lack of azeotropes. Mepiquat chloride [24307-26-4] (123) controls vegetative growth of cotton (120) (see Plant-growth substances). Several partly reduced pyridines are used in pharmaceuticals (13,121–122).

Reactions of N-Oxides. The N-oxide group alters the reactivity of the pyridine ring in a manner that is particularly useful, since the N-oxide group can usually be removed if desired or necessary (123–126). Resonance structures of pyridine N-oxide (124)–(128) illustrate electron-withdrawal from the 2 and 4 positions in the pyridine ring, as in pyridine itself, and the back-donation of π-electrons from the oxygen to the ring. Related resonance forms can be used to account for the increased reactivity of pyridine N-oxide derivatives in electrophilic and nucleophilic substitution, and in side-chain reactions. Pyridine N-oxides (pyridine-1-oxides) are well-characterized stable solids (see Table 6). Many can be distilled without decomposition, although distillation should be avoided if peroxides are or may be present.

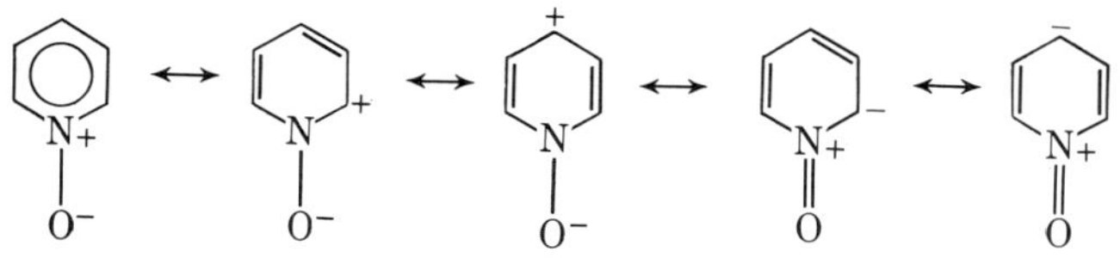

Nitration of pyridine N-oxide and of 2- and 3-methylpyridine N-oxides occurs in the 4-position under relatively mild conditions, ie, with mixed sulfuric and nitric acids at 100°C, and in good yields (127). The nitro group can be displaced by a variety of nucleophiles, or it can be reduced, with or without concomitant loss of the N-oxide group, as shown in Figure 2. Phosphorus(III) reagents generally deoxygenate pyridine N-oxide derivatives under mild conditions; phosphorus trichloride is most commonly used. Reactive inorganic chlorides without reducing properties generally introduce

Table 6. Commercially Available Pyridine-N-Oxides[a]

Compound	CAS Registry No.	Structure No.	Freezing point, °C	Solubility, g/100 g H_2O
pyridine N-oxide	[694-59-7]	(124)	67.0	>100
2-methylpyridine N-oxide	[931-19-1]	(136)	49.5	>100
3-methylpyridine N-oxide	[1003-73-2]	(130)	40.5	>100
4-methylpyridine N-oxide	[1003-67-4]		186.3	>100
2,6-dimethylpyridine N-oxide	[1073-23-0]		23.3	>100
3,5-dimethylpyridine N-oxide	[3718-65-8]		105.6	>100
nicotinic acid N-oxide	[2398-81-4]	(134)	260–261[b]	0.7
isonicotinic acid N-oxide	[13602-12-5]		270–271[b]	1.1

[a] Ref. 7.
[b] Melting point.

a chloride into the pyridine nucleus or into a side chain with loss of the N-oxide group:

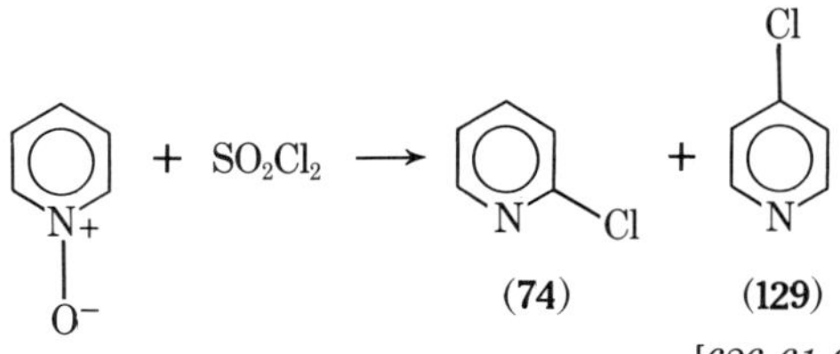

(124) + SO$_2$Cl$_2$ → (74) + (129)
[626-61-9]

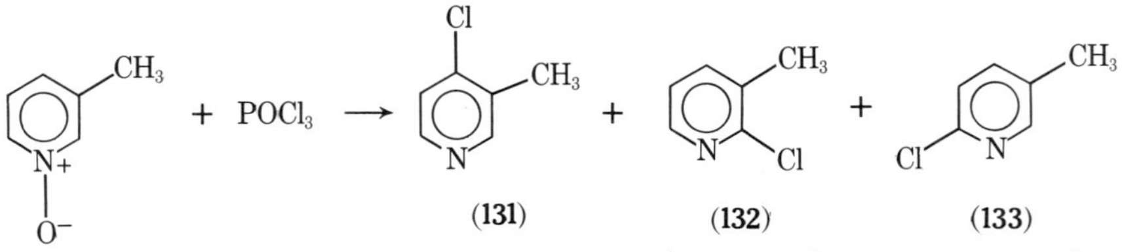

(130) + POCl$_3$ → (131) + (132) + (133)
[1681-36-3] [18368-16-8] [18368-64-4]

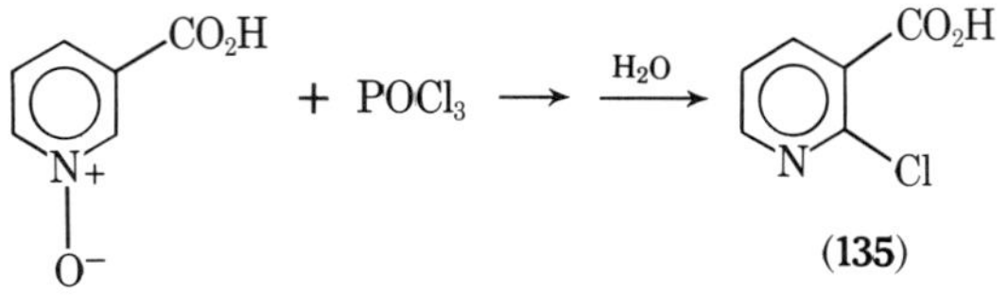

(134) + POCl$_3$ →^{H_2O} (135)
[2942-59-8]

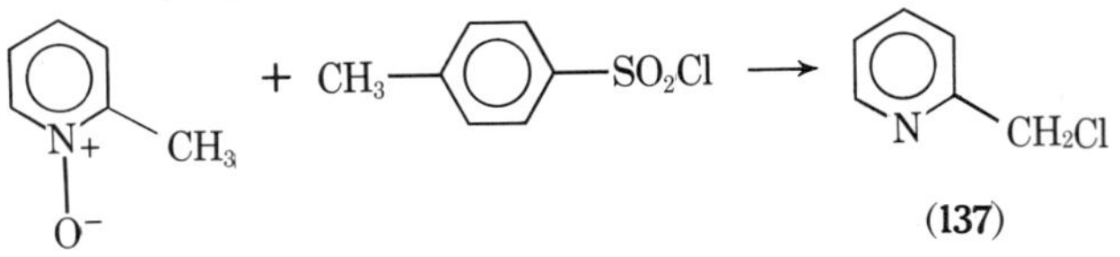

(136) + CH$_3$—C$_6$H$_4$—SO$_2$Cl → (137)
[4377-33-7]

Pyridine N-oxide derivatives can be alkylated at the oxygen by reactive alkylating agents, eg, methyl sulfate, to give N-alkoxypyridinium salts, eg, (146), which can be isolated when desirable. Nucleophilic attack by cyanide ion on (146) occurs to give 2- and 4-cyanopyridine derivatives, generally as a mixture of isomers.

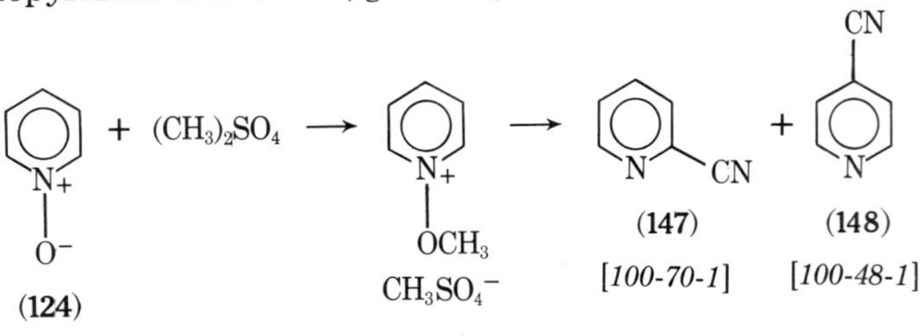

(124) + (CH$_3$)$_2$SO$_4$ → (146) → (147) + (148)
[51342-19-9] [100-70-1] [100-48-1]

Acylation at the oxygen atom of pyridine N-oxide derivatives gives N-acyloxy-pyridinium salts, which are usually not isolated. Heating pyridine-N-oxide or 3-methylpyridine N-oxide with acetic anhydride gives pyridones after hydrolysis. Similar reaction conditions with 2- or 4-methylpyridine N-oxides give the picolyl acetates as the main products.

Condensation of 2- or 4-methylpyridine N-oxide with aromatic aldehydes to give stilbazole N-oxides occurs with catalysis by alkali metal hydroxides or alkoxides. The

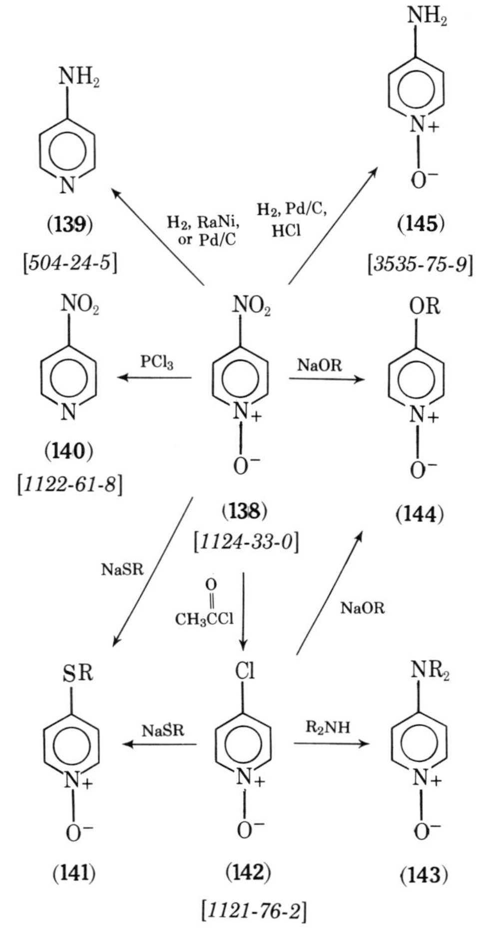

Figure 2. Reaction of 4-nitropyridine N-oxide.

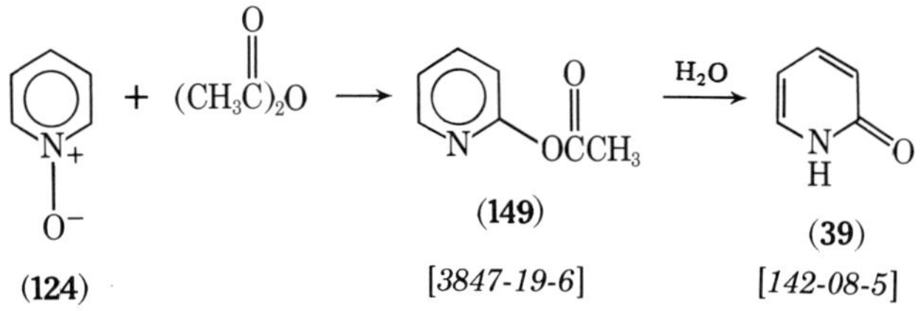

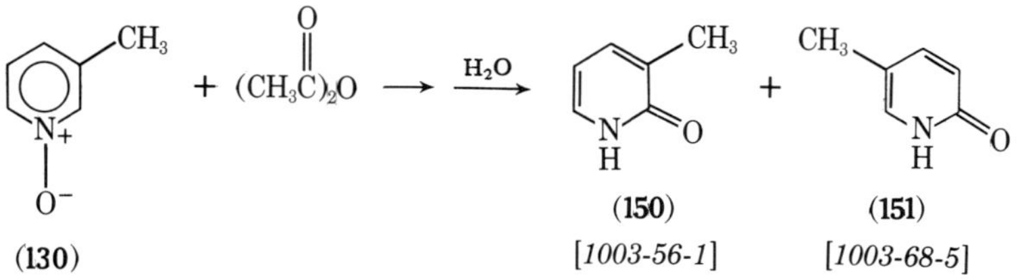

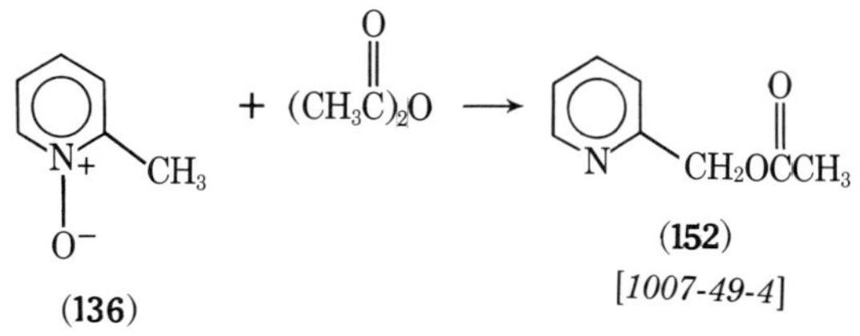

reaction conditions are intermediate in severity between those used for the methylpyridines and those used for the methylpyridinium salts, as expected.

$$\underset{\underset{O^-}{N+}}{\bigcirc}-CH_3 + ArCHO \longrightarrow \underset{\underset{O^-}{N+}}{\bigcirc}-CH=CHAr$$

(136)

BIBLIOGRAPHY

"Pyridine and Pyridine Bases" in *ECT* 1st ed., Vol. 11, pp. 278–293, by Harry S. Mosher, Stanford University; "Pyridine and Pyridine Derivatives" in *ECT* 2nd ed., Vol. 16, pp. 780–806, by R. A. Abramovitch, University of Alabama.

1. H. S. Mosher in R. C. Elderfield, ed., *Heterocyclic Compounds*, John Wiley & Sons, Inc., New York, Vol. 1, 1950, p. 397.
2. E. Klingsberg, ed., *Pyridine and Its Derivatives*, Interscience Publishers, Inc., New York, 1960.
3. R. A. Abramovitch, ed., *Pyridine and Its Derivatives*, Supplement, John Wiley & Sons, Inc., New York, 1974.
4. M. H. Palmer in S. Coffey, ed., *Rodd's Chemistry of Carbon Compounds*, 2nd ed., Vol. IV, Part F, Elsevier Scientific Publishing Co., Amsterdam, The Netherlands, 1976, pp. 1–26.
5. D. M. Smith in ref. 4, pp. 27–226.
6. H. Beschke, A. Kleeman, W. Clauss, W. Kurze, K. Mathes, and S. Habersang in *Ullmanns Encyklopaedie der Technischen Chemie*, 4th ed., Vol. 19, Verlag Chemie GmbH, Weinheim, 1980, pp. 591–617.
7. Values not reported elsewhere were obtained in the laboratories of Reilly Tar & Chemical Corporation, Indianapolis, Ind.
8. A. P. Kudchadker and S. A. Kudchadker, *Pyridine and Phenylpyridines*, API Publication 710, American Petroleum Institute, Washington, D.C., 1979.
9. J. A. Riddick and W. B. Bunger, *Organic Solvents*, 3rd ed., Vol. 2 of A. Weissberger, ed., *Techniques of Chemistry*, Wiley-Interscience, New York, 1970, pp. 438–439, 441–442.
10. R. J. L. Andon, J. D. Cox, and E. F. G. Herington, *Trans. Faraday Soc.* **50**, 918 (1954).
11. R. J. L. Andon and J. D. Cox, *J. Chem. Soc.*, 4601 (1952).
12. L. H. Horsley, *Advances in Chemistry Series 116*, American Chemical Society, Washington, D.C., 1973.
13. R. A. Barnes in ref. 2, Vol. 1, p. 2.
14. R. A. Abramovitch and G. M. Singer in ref. 3, Vol. 1, p. 1.
15. F. Brody and P. R. Ruby in ref. 2, Vol. 1, p. 99.
16. N. S. Boodman, J. O. Hawthorne, P. X. Masciantonio, and A. W. Simon in ref. 3, Vol. 1, p. 183.
17. I. Ya. Lazdin'sh and A. A. Avots, *Khim. Geterotsikl. Soedin.*, 1011 (1979).
18. R. L. Frank, F. J. Pilgrim, and E. F. Riener, *Org. Synth. Coll. Vol. IV*, 451 (1963).
19. H. Beschke and H. Friedrich, *Chem. Ztg.* **101**, 377 (1977).
20. *Chem. Mark. Rep.*, (Oct. 17, 1977); U.S. Pat. 3,780,082 (Dec. 18, 1973), J. M. Deumens and S. H. Groen (to Stamicarbon N.V.); Brit. Pat. 1,378,464 (Dec. 27, 1974), (to Stamicarbon B.V.); Brit. Pat. 1,304,155 (Jan. 24, 1973), (to Stamicarbon N.V.).
21. H. Boennemann, *Angew. Chem., Int. Ed. Engl.* **17**, 505 (1978).
22. U. Eisner and J. Kuthan, *Chem. Rev.* **72**, 1 (1972); F. Bossert, H. Meyer, and E. Wehinger, *Angew. Chem. Int. Ed. Engl.* **20**, 762 (1981).
23. U.S. Pat. 3,485,847 (Dec. 23, 1969), F. Bossert and W. Vater (to Bayer AG).
24. Brit. Pat. 1,256,095 and 1,256,340 (Dec. 8, 1971), A. H. Berrie and N. Hughes (to Imperial Chemical Industries Ltd.).
25. Ger. Pat. 2,537,753 (Mar. 11, 1976); H. M. Taylor (to Eli Lilly and Co.); J. W. Waldrep and H. M. Taylor, *J. Agr. Food Chem.* **24**, 1250 (1976).
26. H. C. van der Plas, *Ring Transformations of Heterocycles*, Academic Press, Inc., New York, 1973.
27. M. Ya. Karpeiskii and V. L. Florent'ev, *Usp. Khim.* **38**, 1244 (1969).
28. Estimated from import statistics from the U.S. Department of Commerce, Bureau of Census.

29. U.S. Pat. 3,206,358 (Sept. 14, 1965), G. T. Stevenson (to Dow Chemical Co.).
30. W. M. Reid and L. R. McDougald, *Feedstuffs* **53,** 27 (Jan. 12, 1981).
31. *Chem. Mark. Rep.*, (March 23, 1981).
32. *Reagent Chemicals*, American Chemical Society, Washington, D.C., 1974, p. 520.
33. *The United States Pharmacopeia XX*, The United States Pharmacopeial Convention Inc., Rockville, Md., 1980, pp. 548–549.
34. C. F. Reinhardt and M. R. Brittelli in G. D. Clayton and F. E. Clayton, eds., *Patty's Industrial Hygiene and Toxicology*, 3rd ed., Vol. 2A, John Wiley & Sons, Inc., New York, 1981, pp. 2688–2690, 2719–2745, 2751–2761.
35. R. J. Lewis, ed., *Registry of Toxic Effects of Chemical Substances*, U.S. Department of Health, Education and Welfare, U.S. Government Printing Office, Washington, D.C., 1978.
36. K. Verschueren, *Handbook of Environmental Data on Organic Chemicals*, Van Nostrand Reinhold Co., New York, 1977, pp. 95, 455–456, 537, 556–558, 635.
37. Title II of Public Law 95-633, *Fed. Regist.* **43,** 57922 (Dec. 11, 1978).
38. *Scientific Literature Review of Pyridine and Related Substances in Flavor Usage*, PB-296005, Food and Drug Administration, U.S. Department of Commerce, Washington, D.C., 1979.
39. *Fed. Regist.* **43,** 16684 (Apr. 19, 1978); 24907 (June 8, 1978).
40. *National Cancer Institute Carcinogenesis Bioassay*, agent number C 55301, 1978.
41. Ref. 34, p. 1836.
42. G. Hoefle, W. Steglich, and H. Vorbrueggen, *Angew. Chem. Int. Ed. Engl.* **17,** 569 (1978).
43. E. N. Shaw in ref. 2, Vol. 2, 1961, p. 1.
44. G. F. Duffin, *Adv. Heterocycl. Chem.* **3,** 1 (1964).
45. O. R. Rodig in ref. 3, Vol. 1, p. 309.
46. J. A. Zoltewicz and L. W. Deady, *Adv. Heterocycl. Chem.* **22,** 72 (1978).
47. M. A. E. Hodgson in *Modern Chemistry in Industry*, Society of Chemical Industry, London, UK, 1968, p. 49; L. A. Summers, *The Bipyridinium Herbicides*, Academic Press, Inc., New York, 1980.
48. G. M. Badger and W. H. F. Sasse, *Adv. Heterocycl. Chem.* **2,** 179 (1963).
49. E. F. Rogers and co-workers, *J. Am. Chem. Soc.* **82,** 2974 (1960).
50. D. Beke, *Adv. Heterocycl. Chem.* **1,** 167 (1963).
51. J. W. Bunting, *Adv. Heterocycl. Chem.* **25,** 1 (1979).
52. G. Illuminati, *Adv. Heterocycl. Chem.* **3,** 285 (1964).
53. R. A. Abramovitch and J. G. Saha, *Adv. Heterocycl. Chem.* **6,** 229 (1966).
54. L. W. Deady, M. R. Grimmett, and C. H. Potts, *Tetrahedron* **35,** 2895 (1979) and references cited therein.
55. R. H. Mizzoni in ref. 2, Vol. 2, 1961, p. 469.
56. R. H. Mizzoni in ref. 3, Vol. 3, p. 1.
57. M. T. Leffler, *Org. Reactions* **1,** 91 (1942).
58. A. F. Pozharskii, A. M. Simonov, and V. N. Doron'kin, *Russ. Chem. Rev.* **47,** 1042 (1978).
59. A. S. Tomcufcik and L. N. Starker in ref. 2, Vol. 3, 1962, p. 1.
60. C. S. Giam in ref. 3, Vol. 3, p. 41.
61. U.S. Pat. 4,177,349 (Dec. 4, 1979), C. K. McGill (to Reilly Tar & Chemical Corporation).
62. U.S. Pat. 1,680,108; 1,680,109; 1,680,111 (Aug. 7, 1928), I. Ostromuislenskii (to Pyridium Corporation).
63. U.S. Pat. 3,149,104 (Sept. 15, 1964), G. Y. Lesher and M. D. Gruett (to Sterling Drug Inc.).
64. R. N. Butler, *Chem. Rev.* **75,** 241 (1975).
65. H. Meislich in ref. 2, Vol. 3, 1962, p. 509.
66. H. Tieckelmann in ref. 3, Vol. 3, p. 597.
67. A. R. Katritzky and J. M. Lagowski, *Adv. Heterocycl. Chem.* **1,** 311, 339 (1963); P. Beak, *Acc. Chem. Res.* **10,** 186 (1977).
68. H. L. Yale in ref. 2, Vol. 2, 1961, p. 421.
69. H. L. Yale in ref. 3, Vol. 2, p. 489.
70. U.S. Pats. 2,712,020, 2,712,022, and 2,712,023 (June 28, 1955), D. W. Adamson (to Burroughs Wellcome & Co., Inc.).
71. H. E. Mertel in ref. 2, Vol. 2, 1961, p. 299.
72. J. J. Eisch, *Adv. Heterocycl. Chem.* **7,** 1 (1966).
73. B. Iddon and H. Suschitzky in H. Suschitzky, ed., *Polychloroaromatic Compounds*, Plenum Publishing Corp., New York, 1974, p. 197.
74. M. B. Green in ref. 73, p. 403.

75. M. M. Boudakian in ref. 3, Vol. 2, p. 407; U.S. Pat. 3,920,657 (Nov. 18, 1975), H. Beschke and W. A. Schuler (to Deutsche Gold- und Silber-Scheideanstalt vormals Roessler).
76. U.S. Pat. 3,225,054 (Dec. 21, 1965), J. W. Cusic and H. W. Sause (to G. D. Searle & Co.).
77. E. Shaw, J. Bernstein, K. Losee and W. A. Lott, *J. Am. Chem. Soc.* **72**, 4362 (1950); U.S. Pat. 2,745,826 (May 15, 1956), S. Semenoff and M. A. Dolliver (to Olin Mathieson Chemical Corp.).
78. U.S. Pat. 3,993,654 (Nov. 23, 1976), N. L. Dean, W. E. Embrey, and J. T. Marshall (to Dow Chemical Co.).
79. U.S. Pat. 3,694,332 (Sept. 26, 1972), V. D. Parker (to Dow Chemical Co.).
80. U.S. Pat. 3,244,586 (Apr. 5, 1966), R. H. Rigterink (to Dow Chemical Co.).
81. U.S. Pat. 3,862,952 (Jan. 28, 1975), L. D. Markley (to Dow Chemical Co.).
82. Belg. Pat. 628487 (Aug. 15, 1963), H. Johnston and M. S. Tomita (to Dow Chemical Co.).
83. F. Minisci and O. Porta, *Adv. Heterocycl. Chem.* **16**, 123 (1974).
84. C. V. DiGiovanna, P. J. Cislak, and G. N. Cislak in L. F. Albright and A. R. Goldsby, ed., *Industrial and Laboratory Alkylations*, ACS Symposium Series No. 55, American Chemical Society, Washington, D.C., 1977, p. 397.
85. L. E. Tenenbaum in ref. 2, Vol. 2, 1961, p. 155.
86. R. G. Micetich in ref. 3, Vol. 2, p. 263.
87. W. Mathes and H. Schuely, *Angew. Chem. Int. Ed. Engl.* **2**, 144 (1963).
88. Belg. Pat. 624,800 (May 14, 1963), H. Johnston, M. S. Tomita, F. H. Norton, and W. H. Taplin, III (to Dow Chemical Co.).
89. E. P. Oliveto in ref. 2, Vol. 3, 1962, p. 179.
90. P. I. Pollak and M. Windholz in ref. 3, Vol. 3, p. 257.
91. Ger. Pat. 2046556 (Apr. 22, 1971), A. Stocker, O. Marti, T. Pfammatter, G. Schreiner, and S. Brander (to Lonza Ltd.).
92. H. Beschke, H. Friedrich, H. Schaefer, and G. Schreyer, *Chem. Ztg.* **101**, 384 (1977), and references cited therein.
93. C. B. Rosas and G. B. Smith, *Chem. Eng. Sci.* **35**, 330 (1980).
94. D. H. Baker, J. T. Yen, A. H. Jensen, R. G. Teeter, E. N. Michel, and J. H. Burns, *Nutr. Rep. Int.* **14**, 115 (1976).
95. E. V. Brown in ref. 2, Vol. 4, p. 1.
96. E. V. Brown in ref. 3, Vol. 4, 1975, p. 1.
97. U.S. Pat. 2,547,048 (Apr. 3, 1951), R. Schlaepfer (to Hoffmann-La Roche).
98. Ger. Pat. 2,409,686 (Oct. 10, 1974), J. E. Ware, D. L. Peardon, and E. E. Kilbourn (to Rohm and Haas).
99. D. L. Peardon, *Pest. Contr.* **42**, 14 (1974).
100. Jpn. Pat. 22,185 (Dec. 24, 1966), I. Matsumoto (to Banyu).
101. C. F. H. Allen and C. N. Wolf, *Org. Synth. Coll. Vol. IV*, 45 (1963).
102. J. F. Besser, *Proc-Vertebr. Pest. Conf.* **7**, 11 (1976).
103. G. B. Kauffman, *J. Chem. Ed.* **55**, 448 (1978).
104. *IARC Monographs on the Evaluation of the Carcinogenic Risk of Chemicals to Man*, Vol. 4, International Agency for Research on Cancer, World Health Organization, Geneva, 1974, p. 159.
105. R. H. Mizzoni in ref. 2, Vol. 4, 1964, p. 123.
106. R. H. Mizzoni in ref. 3, Vol. 4, 1975, p. 115.
107. U.S. Pat. 3,979,400 (Sept. 7, 1976), W. H. Rieger and E. W. Crowe (to Reilly Tar & Chemical Corporation).
108. H. Pines and W. M. Stalick, *Base-Catalyzed Reactions of Hydrocarbons and Related Compounds*, Academic Press, Inc., New York, 1977, pp. 309–382.
109. D. B. Wootton, *Dev. Adhes.* **1**, 181 (1977).
110. E. E. Sowers, private communication.
111. U.S. Pat. 3,223,710 (Dec. 14, 1965), P. Arnall and N. Greenhalgh (to Imperial Chemical Industries Ltd. and Midland Tar Distillers Ltd.).
112. L. A. Walter, W. H. Hunt, and R. J. Fosbinder, *J. Am. Chem. Soc.* **63**, 2771 (1941).
113. U.S. Pat. 2,512,789 (June 27, 1950), F. E. Cislak and L. H. Sutherland (to Reilly Tar & Chemical Corporation).
114. H. S. Mosher in ref. 1, p. 617.
115. P. N. Rylander, *Catalytic Hydrogenation over Platinum Metals*, Academic Press, Inc., New York, 1967, pp. 375–384.

116. M. Freifelder, *Practical Catalytic Hydrogenation*, John Wiley & Sons, Inc., New York, 1971, pp. 582–600.
117. M. Freifelder, *Catalytic Hydrogenation in Organic Synthesis Procedures and Commentary*, John Wiley & Sons, Inc., New York, 1978, pp. 152–162.
118. P. Rylander, *Catalytic Hydrogenation in Organic Syntheses*, Academic Press, Inc., New York, 1979, pp. 213–219.
119. H. Lund, *Adv. Heterocycl. Chem.* **12,** 213 (1970).
120. Ger. Pat. 2,207,575 (Aug. 23, 1973), B. Zeeh, J. Jung, C. Rentzea, and K.-H. Koenig (to BASF); B. Zeeh, K.-H. Koenig, and J. Jung, *Kem.-Kemi*, 621 (1974).
121. R. E. Lyle in ref. 3, Vol. 1, p. 137.
122. R. T. Couts and A. F. Casy in ref. 3, Vol. 4, 1975, p. 445.
123. E. N. Shaw in ref. 2, Vol. 2, 1961, p. 97.
124. E. Ochiai, *Aromatic Amine Oxides*, Elsevier Scientific Publishing Co., Amsterdam, 1967.
125. A. R. Katritzky and J. M. Lagowski, *Chemistry of the Heterocyclic N-Oxides*, Academic Press, Inc., New York, 1971.
126. R. A. Abramovitch and E. M. Smith in ref. 3, Vol. 2, p. 1.
127. 3-Methyl-4-nitropyridine *N*-oxide has been reported to cause tumors in animals: M. Araki, C. Koga, and Y. Kawazoe, *Gann* **62,** 325 (1971); K. Takahashi, G.-F. Huang, M. Araki, and Y. Kawazoe, *Gann* **70,** 799 (1979).

GERALD L. GOE
Reilly Tar & Chemical Corporation

PYRIDOXINE, PYRIDOXAL, AND PYRIDOXAMINE. See Vitamins.

PYRITE, FeS_2. See Iron; Pigments, inorganic; Sulfur; Sulfuric acid.

PYROCATECHOL (1,2-BENZENEDIOL), $C_6H_4(OH)_2$. See Hydroquinone, pyrocatechol, and resorcinol.

PYROGALLOL (1,2,3-BENZENETRIOL), $C_6H_3(OH)_3$. See Polyhydroxybenzenes.

PYROMETALLURGY. See Extractive metallurgy.

PYROMETRIC CONES. See Ceramics.

PYROMETRY. See Temperature measurement.

PYROTECHNICS

Pyrotechnics is the art of using chemically generated light, heat, or sound for entertainment, convenience, or war (see also Chemiluminescence). Pyrotechnics, in terms of manufacturing fireworks, refers to the general discipline of civilian pyrotechnics which encompasses fireworks, matches (qv), and such devices as highway flares, gopher bombs, flashbulbs, automotive airbag inflators, thermitic welding kits, and items for theatrical effects. Military pyrotechnics include a wide range of devices for illumination, signaling, incineration, and gas generation. Military devices are characterized by more rugged construction and greater resistance to adverse environmental conditions with concomitant higher cost, reliability, and safety than are civilian pyrotechnics. In many ways, unlike its sister pyrochemical technologies of explosives and propellants, pyrotechnics is largely an empirical practical discipline with an incomplete theoretical foundation. The current status of pyrotechnics is descriptive, and rigorous application of the various scientific disciplines is taking place gradually (see Explosives and propellants).

Principles

The usefulness of pyrotechnic reactions derives from their being exothermic, self-sustaining, and self-contained. Commonly, chemical reactions require an input of energy or they occur by interaction with other substances from external sources; combustion in air is a typical example. Most pyrotechnic reactions occur independently of any external oxidizer, although in some instances the pyrotechnic effect is enhanced by interaction with the environment. Such characteristics of pyrotechnics are shared by explosives and propellants, which have also exothermic and self-propagating reactions so that in some cases, the distinction is derived solely from the application of the device and not from the mechanism of the reaction.

In many instances, the products of pyrotechnic reactions are solid so that the released energy is concentrated in a small volume. This consequent independence of most pyrotechnic reactions from environmental conditions assures performance with equal reliability at high pressures and in the high vacuum of space. Most pyrotechnic devices contain no moving parts and are small and lightweight. Compared with their mechanical analogues, pyrotechnic devices tend to be inexpensive and often are highly reliable. On the other hand, pyrotechnic devices function only once and do not lend themselves to reuse.

Pyrotechnics is based on the established principles of thermochemistry and thermophysics, which are extensions of the more general science of thermodynamics (qv). Unlike the other pyrochemical technologies, there has been little work on the kinetics of pyrotechnic reactions. Important expositions on the fundamentals of pyrotechnics have been published in Russian (1) and English (2–4). Thermochemical data are contained in general chemical handbooks and more specialized publications (5–9). Their use presupposes an understanding of chemical thermodynamics.

The heat output of a pyrotechnic reaction is the difference between the sum of all heats of formation of the products of the reaction and the sum of the heats of formation of the reactants. If the starting and final conditions of pressure and temperature are close to those of the initial surroundings, it is permissible to use the tabulated standard heats of formation. If the reaction products are condensed, an exothermic is a reliable indication of a tendency to self-propagate.

Chemical equilibrium depends on the total pressure of the system if the reaction products are totally or partly gaseous. This case is conveniently treated by calculation of the free energy change for the reaction from tabulated values of the standard enthalpy of formation $\Delta H°_{298}$ for reactants and products or from the standard heat of reaction ΔH_R; values are corrected by the tabulated entropy functions:

$$\Delta G_R = \Delta H_R - T\Delta S$$

where G, H, T, and S are the free energy, the heat of reaction, the temperature, and the entropy, respectively. This equation is used to calculate the adiabatic, ie, the maximum, reaction temperature at the condition $\Delta G_R = 0$.

As long as the pyrotechnic process is represented by a single chemical reaction, the preceding analytical methods are adequate. If, however, chemical processes take place at extreme temperatures and pressures, a large number of simultaneous equilibria may exist. As the number of possible reactions increases, so does the mathematical difficulty, because the simultaneous equilibrium constants cannot be obtained explicitly and iterative solutions of many simultaneous equations are required. Most calculations of chemical equilibria are performed with computers, which are programmed for the determination of the minimum free energy attainable at any given composition and temperature (10).

The mechanism of pyrotechnic reactions is controversial. The thermochemical views of Shidlovski and of Ellern have been most influential (1–2). These authorities hold that the tendency for oxidant–fuel pairs to react can best be correlated with the acid–base properties of the pairs and the heats of decomposition of the salts. Certain characteristics that are inherent in the systems are important to judging their pyrotechnic performance. These characteristics must be present in addition to a favorable heat of reaction, which merely determines the net energy output.

Magnesium and aluminum powders are the most common fuels in pyrotechnic mixtures. Magnesium reacts in an acidic environment, whereas aluminum is more affected in an alkaline environment. Mixtures that contain magnesium and alkali metal salts as oxidizers are stabilized with a few percent of an organic binder. Similar compositions containing aluminum are more stable. For example, illuminating flare and tracer compositions contain magnesium with barium or potassium nitrates in an organic binder, whereas the more sensitive photoflash compositions can be prepared with aluminum powder and potassium perchlorate.

Addition of sulfur to pyrotechnic mixtures, other than those containing magnesium, seems to improve their ignitability but to retard their reaction rate. Sulfur–magnesium mixtures are unstable. Particular care must be exercised in specifying only flour of sulfur, ie, ground and sieved crystalline material, and not the flowers of sulfur, ie, sublimed sulfur as commonly employed in agriculture. The latter is always contaminated with the hydrated oxides, ie, sulfurous and sulfuric acids, which make pyrotechnic mixtures unsafe.

Of particular concern in pyrotechnic formulations is the possibility of exchange reactions. Addition of ammonium salts to compositions containing magnesium and a metallic chlorate is particularly undesirable because such mixtures are hygroscopic. Moreover, the ammonium chlorate is an unstable salt and can react spontaneously at near ambient temperatures. Spontaneous ignition of chlorates upon mixing with sulfur or red phosphorus is especially likely and such compositions cannot be handled safely without special precautions to neutralize the acidity of the materials.

The ignition temperature of pyrotechnic mixtures is a complex function of the heat of formation and the rate of heat dissipation from the reaction zone. Mixtures containing chlorates are the easiest to ignite because the heats of formation of chlorates are close to zero. Metallic oxide–fuel mixtures (thermites) have high ignition temperatures unless special means are employed to incorporate a fusible or reactive constituent. The rate of reaction depends on the fineness of the powders and the availability of liquid or gaseous reaction products at the reaction temperature.

Not all pyrotechnic reactions depend on a gas phase. Alloying reactions can be extremely fast and energetic and yet occur entirely in the interface between solid reagents (11). Certain thermitic reactions, such as between iron powder and barium peroxide, are solid-phase reactions (3). The problem of catalysis in condensed-phase reactions has not been examined extensively. Modern semiconductor theory is believed to be applicable to the understanding of this problem (see Semiconductors). Crystal defects and impurity states are used to describe the cause of altered reactivity of pyrotechnic reagents (3). The reactivity of gasless systems is known to be enhanced by the presence of a liquid phase (12). Another view is that volatile impurities, such as hydrocarbons in charcoal, suboxides in boron, hydrides in zirconium, etc, are determining factors.

Civilian Pyrotechnics

Fireworks. Fireworks art is historically the art of using black powder as the principal ingredient in firecrackers. Modern displays are based on mortar-fired aerial shells whose effects derive from a combination of devices containing colored stars. The assembly of aerial fireworks is described in reference 13. Many important facets of manufacture are trade secrets. For example, the correct choice of certain metal salts or color producers is complex and difficult. Color-donor combinations which are effective and are devoid of the prohibited arsenic and mercury compounds are shown in Table 1. A timely review of the status of the current fireworks industry is given in reference 14. The complexity of the formulation of an aerial shell is described in reference 4. Fireworks are divided into "dangerous fireworks" whose manufacture, transport, and display is specially regulated, "exempt fireworks" which are of utilitarian nature, and "safe-and-sane" fireworks which are sold to the general public, some of which are described below.

Safe-and-Sane Fireworks. Waterfalls are made of a mixture of potassium perchlorate and aluminum. As in all pyrochemical formulations, careful attention must be paid to the distinction between chlorate and perchlorate. Chlorate mixtures are hazardous owing to their ease of initiation.

Roman candles are cardboard tubes containing a small quantity of black powder on the bottom with stars, delay mixtures, and black powder alternating along the length of the tube. The delay mixture consists of black powder diluted with charcoal.

Sparklers are made by dipping wires into a viscous mixture of an oxidizer with steel filings and pyroaluminum in an excess of dextrin or shellac. Variations of fountains are wheels, saxons, pinwheels, flower pots, and gerbs. All are composed of black powder whose action is reduced by the addition of excess potassium nitrate, sulfur,

Table 1. Color Donors for Aerial Stars

Color	Additives
red	strontium carbonate or strontium oxalate
yellow	sodium oxalate or cryolite
green	barium chlorate[a]
blue	cupric carbonate with ammonium perchlorate or with potassium chlorate and poly(vinyl chloride) or hexachlorobenzene[a]
silver	aluminum with ammonium perchlorate[b]
white	black powder and antimony or strontium and barium nitrates with magnesium in poly(vinyl chloride)
white glitter (twinkle)	aluminum, antimony, and black powder[b]
gold glitter	aluminum, antimony, black powder, sodium oxalate[b]
gold streamers	charcoal or lampblack

[a] One must be careful not to incorporate an ammonium salt into the same shell as a chlorate because, in the presence of moisture, ammonium chlorate may form through an exchange reaction. Ammonium chlorate may detonate on standing.

[b] For all formulations requiring the addition of aluminum, the choice of the appropriate grade of aluminum flake is critical. Descriptions of the various grades of metallic powders of interest in pyrotechnic applications is given in ref. 2.

or charcoal. The spark-forming ingredient may be pyroaluminum, cast-iron turnings, or coarse titanium.

A torpedo is a device that initiates on impact. A small quantity of a primary explosive mixed with gravel is enclosed in small bags. These are made to function by tossing them on the pavement. A railroad torpedo is a utilitarian device for signaling a track hazard to a moving locomotive. It typically contains the hazardous sulfur and potassium chlorate mixture stabilized and encased in a strong container.

A whistle is composed of a mixture of chlorates, perchlorates or nitrates with certain aromatic compounds, eg, trihydroxybenzoic acids, potassium benzoate, or sodium salicylate. The mixture burns with a shrill sound when burned in a slender tube.

Pharaoh's serpents are made of nitrated naphthol pitch which is compounded with a mild oxidizer, eg, picric acid. Upon ignition, the mixture burns progressively and forms large quantities of black ash.

Theatrical Effects. Most theatrical effects for motion pictures and television, such as smoke, fog, lightning, and flame, are not pyrotechnic in origin but are produced from oil dispersions, dry ice, and arc lamps (15). Erupting vulcanoes, however, have been simulated by dropping glycerol on potassium permanganate, or by igniting mixtures of ammonium dichromate, potassium nitrate, and dextrin, or by a gently prepared mixture of potassium chlorate and powdered sugar initiated with a drop of concentrated sulfuric acid. The ensuing reactions are frequently so violent that visual effects are obtained by filming the reactions in slow motion. Cannon and artillery fire is simulated by the use of special blanks. Blanks in muzzle-loading weapons contain small quantities of black powder which produce a better visual effect than do commercially produced blanks. Bullet effects or bullet hits are electrically initiated igniters which have been built up to various degrees of intensity from various amounts of primary explosive. Bullet holes as depicted in movies are not technically correct. Normally, when a bullet penetrates an object, it produces a small entrance hole nearly its own diameter, but as it tears its way through the material, it forces more and more substance ahead of it. Upon exit, large chunks of material tear out to form the characteristic exit crater. Usually, for the sake of visual effect in motion pictures, the entrance hole is created to resemble an exit crater, which is then filled with explosive, covered with plaster and then painted to conceal the presence of the hole (15).

A list of manufacturers and suppliers of special effects materials and equipment as well as references, articles, and public documents on this subject is given in reference 15. Pyrotechnics for theatrical display can only be prepared by specially licensed operators (16).

Model Rockets and Missiles. A new form of pyrotechnic device is the model rocket, which is marketed as a hobby or toy. It is sold as a kit with modular, preassembled, solid-propellant engines. These are constructed of nonmetallic casings, clay nozzles, and a compressed black-powder charge, which is ignited by an electric match. Unlike military solid-propellant systems, the burn-time is brief, ie, less than a second, compared with the ascent time of the missile, ie, several seconds, so that despite the end-burning character, a combustible confinement, ie, the paper tube, is adequate because of the rapid rate of thrust decay. The impulse and the thrust can be specified over moderately wide ranges for incorporation into single- or multiple-stage rockets. The concept of prepackaging has added a significant element of safety to what had been a dangerous sport because of the tendency of gas-forming pyrotechnic mixtures to explode rather than to burn progressively when compressed to a nonuniform density with voids and other structural defects. Solid-propellant engines are of necessity not reusable or refillable. Black powder, when compressed, burns in a similar manner to military solid propellants but with only 20–30% of their specific impulse because of the high solids content of the combustion products and the comparatively low chamber pressure (17). In contrast with model rockets, skyrockets are mortar-launched aerial shells with delay fuses; thus, no thrust is generated by the device itself.

Utilitarian Devices. Utilitarian devices, eg, highway flares, are made chiefly of strontium nitrate mixed with sawdust, wax, sulfur, and potassium perchlorate and contained in a waterproof cardboard tube. A small quantity of a safety-match composition also is incorporated and, upon ignition, the device burns for up to 30 minutes with a distinctive red flame.

Of some interest is the commercial photoflash bulb, which contains thin, shredded zirconium in a partial pressure of oxygen within a glass tube. For reasons of technology and safety, the oxygen pressure in flashbulbs must be as small as possible. During the combustion of zirconium in oxygen 17 $(lm \cdot s)/(cm^3 \cdot kPa)$ [1700 $(lm \cdot s)/(cm^3 \cdot atm)$] are emitted, whereas aluminum generates only 10 $(lm \cdot s)/(cm^3 \cdot kPa)$ [1000 $(lm \cdot s)/(cm^3 \cdot atm)$]. Therefore, even though, because of its higher atomic weight, zirconium emits less light per unit mass than aluminum does, zirconium is preferred because of its higher light intensity. The combustion temperature is limited by the stability of the oxide, zirconium dioxide being unstable above 4000 K, hafnium dioxide above 4500 K (18). Flashbulbs are ignited electrically or by means of a wire, which is coated with a friction-sensitive match composition. Upon activation of the shutter, the wire is pushed into the bulb where it rubs against an actuating surface (19).

Pyrotechnic devices are used in agriculture. A variant of the highway fuse, which contains perchlorates and sulfur, serves as gopher bombs. Firecrackerlike cartridges are set off periodically to discourage birds from frequenting fruit orchards.

For many years silver iodide or lead iodide was dispersed for cloud seeding. Two variations of dispersions have been employed: the iodides could be contained in a solid gas-generating mixture, eg, ammonium perchlorate in an organic binder, or the iodide was formed by the combustion of the lead or silver nitrates with ammonium iodide or iodoform. Alternatively, the iodates of silver or lead were used as oxidizers of metallic fuels (2).

Activation of the proposed airbags for automotive use was to have been based on the explosive combustion of metallic fuels with perchlorates.

Certain light-generating devices are based on chemiluminescence. Some cyclic compounds, eg, trichlorocarbobutoxyphenyl oxalate (TCCPO) are mixed with hydrogen peroxide in the presence of a fluorescer, eg, bisphenylethynylanthracene (BPEA) (20). These systems constitute Coolite light sticks (the Coolite Corporation) and Cyalume chemical light (American Cyanamid Corporation), which are marketed as emergency light sources or for special effects (see Chemiluminescence).

Economic Aspects. In the United States, in excess of $100 $\times$ 10^6 of fireworks is sold annually (14). The domestic production is approximately 30% of total sales and the remainder is imported chiefly from China and Japan. Domestic production has declined by ca 50% since 1967 and imports have increased dramatically. Principal U.S. manufacturers are Keystone Fireworks Manufacturing Co. in Dunbar, Pa.; New Jersey Fireworks Manufacturing Co. in Elkton, Md.; New York Pyrotechnic Products in Bellport, N.Y.; Red Devil Fireworks Co. in Anaheim, Calif.; Tri-State Manufacturing Co. in Loveland, Ohio; and Zambelli Internationale Fireworks Manufacturing Co. in New Castle, Pa. The most significant development of fireworks during the 1980s is expected to occur in the People's Republic of China, which is the world's largest producer.

Regulations. The control of commercial fireworks, except for the regulations governing their transport, is not uniform and is by individual states (21). Some Federal regulation of fireworks is provided by the Consumer Product Safety Commission, which restricts the amount of primary explosives of the safe-and-sane variety. Dangerous fireworks are fireworks that contain arsenic, boron, magnesium, mercury, zirconium, chlorates, or red phosphorus, except the latter two as contained in toy caps and party poppers. Other dangerous fireworks are firecrackers, sky rockets, Roman candles, torpedoes, chasers, and devices that involve an element of surprise for the user, eg, cigarette loads, trick matches, and trick noisemakers. All other devices are deemed safe-and-sane and are thought to provide a reasonable degree of safety and do not create a fire hazard when used according to the manufacturer's instructions. Use of exempt fireworks is limited to signaling, to commercial, agricultural, and hobby use, or to religious ceremonies. Great losses of property and life have occurred as the result of the private use and clandestine manufacture of safe-and-sane fireworks, which are subject to increasingly strict local control on county and municipal levels.

Separate regulations and licensing requirements exist for public display of dangerous fireworks. In addition to the appropriate licenses, permits and coverage by adequate public liability insurance are required. Separate licenses are issued by the Treasury Department's Bureau of Alcohol, Tobacco, and Firearms for the manufacture, wholesale, import, and export of dangerous fireworks and pyrotechnics. Separate regulations pertain to special effects, ie, pyrotechnic articles designed for television, theater, or motion pictures.

Surface and air transport is federally regulated and pyrotechnic devices are Class

C explosives (21). Specific regulations can be obtained from state and local fire marshalls.

Military Pyrotechnics

Uses of military pyrotechnics include the following: generation of light, eg, in flares, flash charges, and tracers; concealment or signaling by means of smoke; generation of heat, eg, in incendiaries; ignition and propagation of pyrotechnic reactions, eg, with delays; and sources of prime ignition, eg, in percussion primers and matches. Many aspects of military pyrotechnics have been reduced to quantitative analysis through application of physical and chemical principles.

Photometry. Quantities of interest in describing luminous devices are the intensity I (cd), the illuminance E (lm/m^2), and the light output C (lm/W). For a flare burning at height h and illuminating an area of radius r, E is given in terms of I as follows

$$E = \frac{Ih}{(r^2 + h^2)^{3/2}}$$

The light output in terms of the heat of the pyrotechnic reaction and the intensity and time t is given by

$$C = \frac{4\pi It}{Qm}$$

where Q is the heat of reaction, and m is the mass of pyrotechnic mixture. The radiance W/(sr·cm^2), the burn time t, and the intensity as function of the spectral distribution constitute a complete description of a flare.

Flares have a grey-body spectral distribution, although the output of a luminous flame has spectrally discrete components. Infrared flares depend on the high degree of reflectance of natural terrain at long wavelengths. Recent developments in electro-optical devices, such as image intensifiers and low level light television, for the purpose of detecting such images have enhanced performance in the red or near infrared spectral region. These pyrotechnic mixtures are obtained from alkali nitrates other than sodium nitrate (see Table 2).

If the visible spectral component is to be discarded, as in a decoy, the flare can be burned within a long cylindrical tube in which the inner walls are protected by an air stream. Another reaction system suitable for decoys utilizes the magnesium–

Table 2. Emitted Wavelengths of Infrared Flare Components[a]

Component	Wavelength emitted		
	0.76 μm	0.79 μm	0.8–0.9 μm
silicon	10	10	16.3
KNO$_3$	70		
CsNO$_3$			78.7
RbNO$_3$		60.8	
hexamethylenetetramine	16	23.2	
epoxy resin	4	6	5

[a] Ref. 18.

fluorocarbon reaction, which forms hot carbon particles as the source of radiant energy in the infrared regime. Conversely, magnesium/alkali nitrate flares maximize the brightness in the visible region, and this maximization is aided by the high vapor pressure of the magnesium whose combustion is an event of relatively long duration. Consequently, flares have high intensity [$(0.5-2) \times 10^6$ cd] and produce a useful ground illumination of ca 0.25 lm/m^2 for 1–5 minutes when suspended from a parachute.

The intensity of a flare is largely determined by its temperature, which depends on the stability of the reaction products. The higher the decomposition temperature of the oxide, the higher the flame temperature. Hydrogen, carbon, boron, silicon, and phosphorus form oxides that dissociate at comparatively low temperatures. A flare temperature greater than 3000 K is required to generate grey-body radiation, which is optimum for the spectral sensitivity of the human eye (0.4–0.74 μm). Whereas this is possible by the combustion of the transition elements, such as zirconium and titanium, in practice magnesium and aluminum are best in terms of heat output, cost, and transparency to visible radiation. Also, the human eye has the greatest sensitivity to the yellow sodium emission. Consequently, illuminating shells and flares contain 53–58 wt % magnesium powder, 36–40 wt % sodium nitrate, and 4–8 wt % alkyd–styrene resin. Thermodynamic data for flare reactions are given in reference 22.

Signal flares may also be compact stars, which are designed to burn with clearly distinguishable colors. The colors and the chemicals that produce them are red, $Sr(NO_3)_2$; yellow, $NaNO_3$; green, $Ba(NO_3)_2$ and copper powder; and white, $Ba(NO_3)_2$; the fuel is magnesium. They differ from illuminants in that the signal-light source must be discriminated from the background and from other signal-light sources.

Photoflash bombs are explosive devices that are used in night photography and that, for time intervals as short as 0.1 s, provide more than 3×10^9 cd. Flash charges are loosely packed, powdered mixtures of aluminum powder with potassium perchlorate and barium nitrate. Flash compositions are among the most hazardous in pyrotechnics and are processed by remote control in armored bays. Initiation is by explosive charge. These mixtures or flash charges are also useful for simulating explosions when packaged into smaller units as projectile ground-burst simulators (2).

Smoke Generators. Smoke generators are pyrotechnic devices for daytime obscuration and signaling. Concealment of troop movements and structures is not generally done pyrotechnically but through atomization of fog oil or release of titanium tetrachloride (FM smoke) or of sulfur trioxide. An exception is HC smoke, which is formed by the combustion of aluminum with hexachloroethane and zinc oxide. Signal smokes may be white or colored but not black, because black smoke is not sufficiently opaque to be distinguished against the background. White signal smoke is derived from red phosphorus or an HC mixture or from a combination thereof. Colored smokes derive their color from organic dyes which, because of their low combustion temperatures, evaporate and recondense. This process is accompanied by considerable loss as a result of partial combustion. The mixtures are cool-burning, since they are composed of potassium chlorate with added sulfur or sugar containing sodium bicarbonate to inhibit the chlorate. The organic dyes are substituted anthraquinones, which are carcinogenic and which, therefore, pose an additional problem on loading and disposal (see Anthraquinone derivatives). The visibility of smoke depends on the ability of the smoke particles to scatter light, ie, on the size of the smoke particles; the illumination of the smoke cloud; and the visual contrast against the background. Yellow and orange smokes are more reliably discriminated than are blue or green, particularly under hazy atmospheric conditions.

PYROTECHNICS

Tracer Munitions. Tracer bullets guide the direction of the fire, aid in range estimation, mark target impact, and act as incendiaries. Tracers can, through preselected tracer colors, serve for nighttime identification of the combatants and, by delay-train action, to self-destruct munitions. Rocket-assisted projectiles have also been proposed. Daylight smoke tracers have been developed which produce colored trails by dissemination of a dry powder, by the sublimation of organic dyes or by the combustion of phosphorus (white smoke) or of cadmium with sulfur (yellow smoke) (23).

Although unfused tracers require multiple hits to ignite flammable targets, such as fuel tanks, substitution of the magnesium in the body of incendiary rounds with zirconium and titanium improves one-round effectiveness. Red tracers are more visible under adverse atmospheric conditions; therefore, these are preferred although green tracers based on barium salts also are used. Of particular interest is the dim igniter, which serves as a delay composition to protect the gunner from exposure and from being blinded by the glare of the trace (see Table 3). Daylight visibility is enhanced by increasing the fraction of magnesium. The approximate intensity required for color visibility at 5000 m is shown in Table 4. Tracer and spotting munitions performance is described in reference 23.

Incendiary Devices. Incendiary devices are used to initiate destructive fires in a variety of targets. Small-arms incendiaries are used primarily for starting fires in aircraft fuels. Whereas they are highly effective against subsonic aircraft, such as

Table 3. Typical Compositions of Igniter and Tracer Formulations, wt %[a]

	Delay-action igniter	Dim igniter	Daylight (bright) igniter	Red tracer
strontium peroxide	90			
magnesium		6	15	28
delay-action igniter		94		
calcium resinate	10			4
barium peroxide			83	
zinc stearate			1	
strontium nitrate				40
strontium oxalate				8
potassium perchlorate				20
toluidine red (identifier)			1	

[a] Ref. 23.

Table 4. Approximate Intensity Required for Color Visibility at 5000 m, cd

Weather conditions	Relative intensities			
	Red	Amber	White	Green
Night				
clear	1.0	2.0	2.5	2.8
light rain	1.2	2.1	3.0	3.2
overcast and haze	3.2	4.1	3.1	5.9
heavy rain	8.9	33.5	132.0	567.0
Day				
clear	2,000	2,111	3,222	4,000
overcast and haze	4,778	7,556	11,111	10,000

helicopters, the problem of defeating supersonic aircraft by incendiary action alone is unsolved owing to the high flash point of jet fuels. Most small-arms incendiary compositions consist of a metallic fuel, eg, magnesium–aluminum alloy, and an oxidizer, eg, barium nitrate. In recent years, the use of pyrophoric fragment generators, eg, zirconium, titanium, misch metal, or depleted uranium as incendiary components of the projectiles has been the subject of intensive study (24). The reason some metals ignite and burn progressively when fractured is uncertain. Generally, these metals have higher densities than do their oxides, so that the oxides flake from the metal during combustion, thereby exposing fresh metallic surface to oxidizer attack. Below diameters of 20 mm, incendiary rounds are unfused and ignite by impact, whereas larger caliber ammunition is fused, partly because of the larger fraction of explosive charge which these rounds can carry.

Incendiary bombs may also contain pyrophoric metal components if the bombs are intended for use against flammable targets, such as fuel depots. Because it is usually difficult to ignite other targets with short-duration burns, incendiary bombs for deployment against structures rely either on Napalm or on thermitic reactions. Military thermites are composed of granular aluminum and coarse iron oxide scale; these are made easily ignitable by the admixture of a small amount of barium nitrate.

Ignition. An important function of pyrotechnic devices is to start a fire without the use of another fire, ie, by prime ignition. Many methods for the initiation of pyrochemical reactions are not pyrotechnic but rely on primary explosives. Others consist of materials and mixtures that ignite by the action of friction, shock, heat, electrical impulse, or by environmental action as from air, water, or chemical reagents. The selection of the best pyrotechnical technique for prime ignition depends on the desired functioning time and the availability of suitable external stimuli.

Priming compositions are either noncorrosive or corrosive formulations. Noncorrosive compositions are typical of commercial products for small arms usage. These are styphnate-based and, although exact formulations are proprietary, the PA-101 mixture listed in Table 5 is typical. These mixtures are quite sensitive but have low temperature stability. Corrosive mixtures contain no primary explosives and these are represented by the FA-70 mixture in Table 5. The corrosive effects result from

Table 5. Composition and Characteristics of Typical Percussion Primer Compositions[a]

Component and characteristic	FA-70	PA-101	G-11
lead styphnate, wt %		53	
barium nitrate, wt %		22	
potassium chlorate, wt %	53		51 ± 1%
antimony sulfide, wt %	17	10	26.0
calcium silicide, wt %			13.0
Tacot[b], wt %			10 ± 1%
lead thiocyanate, wt %	25		
trinitrotoluene (TNT), wt %	5		
aluminum powder, wt %		10	
tetracene, wt %		5	
max continuous temp tolerance, °C	70	95	200
firing sensitivity (drop height ± 5 σ) × wt, N·m	0.15	0.16	0.67

[a] Ref. 25.
[b] A tetranitro benzotriazolo benzotriazol (E. I. du Pont de Nemours & Co., Inc.).

the formation of soluble chloride combustion by-products, which originate from the chlorate oxidizer. Use of these igniter compositions is commonly limited to ca 70°C; above this temperature, dudding becomes an increasingly serious problem. Solar radiation on an aircraft cockpit can produce internal temperatures of 95°C. This finding gave rise to the development of the G-11 priming mixtures, which have achieved successful firing after 3000 hours storage at 200°C (25). The sensitivity of the G-11 mixtures is less than that of styphnate-based compositions and the former, therefore, require increased firing energy which is provided by detonating cords.

Air-reactive igniter materials include white phosphorus, phosphine and its derivatives, silane and its derivatives, and boranes especially if these materials contain a halide and are complexed. Metallic borohydrides and cacodyl (tetramethyldiarsenic) also are pyrophoric in air. The only application of air-reactive reactions is in the MK2 night marine-location marker, which relies on the action of seawater on a mixture of metallic phosphides and carbides. The resulting mixture of acetylene and phosphine rises to the surface and burns with a brilliant flame. Other applications have been the addition of triethylaluminum to liquid rocket propellants and to flame-thrower fuels to make them hypergolic, ie, ignite spontaneously (2,26).

Metals that are not normally pyrophoric can be made so. Cerium amalgams, thorium–silver alloys, and Raney nickel are spontaneously flammable when dry. Chromic anhydride reacts with ethyl alcohol in an incendiary manner (26). Under certain circumstances, finely divided lead, iron, and cobalt can be made pyrophoric (2).

Water-reactive materials include the alkali metals and their hydrides, alloys, carbides, and amides. Mixtures of ammonium salts, chlorates, and magnesium ignite when wet because of the exchange reaction that yields the unstable ammonium chlorate. An intimate mixture of sodium hydroxide and ferrosilicon glows with incandescence when moistened and aluminum powder mixed with iodine reacts violently when wet (26). Magnesium powder and silver nitrate ignite and burn with a blinding flash when water is dropped on the mixture as does a mixture of aluminum or magnesium with sodium peroxide.

Many systems can be made to ignite by the action of strong oxidizing agents. Glycerol, benzaldehyde, or dimethyl sulfoxide in contact with solid potassium permanganate burst into flame. Calcium permanganate is more reactive than potassium permanganate and ignites paper and cotton. Such systems can be made to explode if the action of the permanganate is augmented by concentrated sulfuric acid (2). Red phosphorus ignites on impulse or with friction when mixed with finely ground permanganate, perchlorates, and nitrates.

Spontaneous reactions are aided if the reactants are liquid and are well dispersed. If the reactants are solid, intimacy of contact is improved by shock or friction. Most important in this context are match compositions (qv).

The reliable initiation of pyrochemical reactions is of crucial importance. This is generally accomplished with mechanically or electrically initiated primers. There are special initiators, such as exploding bridgewires (EBW), which are high voltage devices in contact with PETN that are used in applications for which the use of primary explosives is too hazardous. Another safety device is the fluidic initiator, which relies on acoustic resonance energy to generate the heat for ignition (27). The choice of the igniter depends on the permissible ignition delay. Igniters containing primary explosives function in microseconds, whereas purely pyrotechnic igniters require milliseconds.

A special category of pyrotechnic system is the cartridge-actuated device (CAD) which is also called the propellant-actuated device (PAD) (18). The CAD produces mechanical movement for the closing of electrical switches, the opening or closing of valves, cutting of cables, or the performance of other mechanical work. They usually contain a small quantity of a gas-generating mixture which is ignited electrically.

Igniter compositions known as first fires are formulated for specific desired rise times, ignition sensitivities, and gas output. Sensitivity also relates to questions of safety, ie, response to discharge of static electricity, impact, friction, or elevated temperature, and compatibility, storability and hygroscopicity. Gas output determines the suitability of the formulation as a flame agent, as a propellant in squibs and PADs, and for use at high altitudes. Examples of igniter materials are mixtures of aluminum, boron, magnesium, silicon, titanium, or zirconium with ammonium or potassium perchlorate, barium or potassium nitrate, barium or lead chromate and cupric or lead oxides. Illustrative compositions are shown in Table 6. A traditional igniter is black powder, which is an efficient flame carrier because of the high fraction of hot solid particles in its combustion products.

Special igniter systems consist of compressed mixtures of magnesium–fluorocarbon, which are gas-forming, or of thin layers of palladium or platinum on aluminum foil or wire, which do not form gas. The latter derive their activity from the intense heat of alloying between the two elements, which is sufficient to melt the resulting intermetallic compound. Compositions that burn with little or no gaseous products tend to have reaction characteristics, eg, burn rate, that are highly dependent on the ambient temperature but that are not very dependent on pressure. The opposite generalization can be made for gas-forming mixtures, an exception being black powder

Table 6. Some First-Fire, Starter, and Igniter Compositions

Component	Composition, wt %					
	1	2	3	4	5	6
aluminum						13
boron				10		
charcoal						4
magnesium					25	
silicon	20	25				26
titanium		25				
zirconium			20			
zirconium hydride	15					
barium chromate						
barium nitrate	50			90	75	
iron oxide (Fe$_3$O$_4$)		25				
iron oxide (Fe$_2$O$_3$)		25				
iron oxide (scale)						22
lead oxide (PbO$_2$)			80			
lead oxide (Pb$_3$O$_4$)						35
potassium nitrate						
tetranitracarbazole	5					
Laminac binder	5^b	c	c	c	c	c

^{a} Ref. 22.
^{b} Laminac 99 wt %; Lupersol 1 wt %.
^{c} Most of these compositions can be used as a loose powder mixture or with binders, such as celluloid, nitrocellulose, or NC (nitrocellulose) lacquer.

which has good high altitude capability. Manufacturers of military ordnance components have prepared proprietary compounds, which can be tailored to meet specific requirements in temperature, shock, and electrical discharge tolerance. Manufacturers include Unidynamics/Phoenix, Inc. in Phoenix, Arizona, or Teledyne/McCormick-Selph in Hollister, California. A review of igniter compositions has been published (28).

Delay Elements. The time delay for the initiation of a pyrotechnic device varies from microseconds for systems initiated by a primary explosive or an EBW to many milliseconds if the initiation mechanism is thermal, as from a hot wire or electric match. Delay elements that are used are those for sequential functioning of, for example, pilot-ejection mechanisms, missile-separation devices, or grenades. Delay elements are self-contained pyrotechnic devices consisting of an initiator, a delay column, and an output terminal charge. Delays are either obturated or vented. An obturated element contains all of the reaction products and, therefore, it is not affected by ambient pressure. Obturation is an aid in protecting other components of the hardware from the effects of unwanted smoke and debris, but it also tends to increase the burn rate of the composition. So-called gasless compositions may nevertheless produce a pressure spike upon functioning because of the gaseous nature of the reaction products at the reaction temperature and, therefore, require obturation for proper functioning. Vented delays are used if large amounts of gaseous products must be disposed of and if the device is used where the effect of changes in ambient pressure can be ignored. Delay elements based on black powder are necessarily vented and their burn rates are affected by the rotational speed of the projectile as well as by the ambient pressure. Compositions and characteristics of modern delay compositions are listed in Table 7.

Health and Safety Factors, Storage and Handling

Safety concerns permeate all aspects of pyrotechnics. The number of fatalities and serious injuries experienced annually in the manufacture and handling of pyrotechnic devices far exceeds those resulting from any other pyrochemical activity. This may result from the lesser presumed damage potential because of the smaller quantities of materials involved or from a degree of unpredictability of the hazard. Death and injury are caused usually by burns and lung damage and, to a lesser extent, by shock and by impact of flying fragments. Often accidents are unreported, especially when the damage is limited to the test facilities. One must therefore assume that hazardous conditions are common in the pyrotechnics industry and that the best preventative is continual vigilance.

Except on an industrial manufacturing scale, remote operations are not practical and the protection of personnel should include shielding for the eyes, face, and hands as well as thorough protection for the hair and the use of antistatic clothing. Complete eye shields are desirable and safety glasses are the minimum form of eye protection. Rubber and plastic gloves may present a hazard when used for handling flammable materials because they may melt and stick to damaged skin. Suede leather gloves are preferable as these are wettable and washable (2). Water in copious quantities is a good fire-fighting agent except against burning reactive metals; the latter must be smothered with sand. Alkali metal fires can be extinguished with special carbonaceous powders. Chlorinated hydrocarbons must never be used because reactive metals burn in a halogen donor and in carbon dioxide.

Table 7. Survey of Delay Mixtures[a]

Type of mixture	Typical composition, component, wt %	Burn time, s/cm	Temperature coeff, %/K	Change in delay time with storage, s/(yr·cm)
manganese delay	Mn, 29 $PbCrO_4$, 26 $BaCrO_4$, 45	0.8–5.4[b]	0.17	0.02
"T-10"	B, 3–15 $BaCrO_4$, 97–85	0.23–0.32[b]	0.23–0.32	0.15–0.38
Zr–Ni delays	Ni–Zr, 26 $BaCrO_4$, 60 $KClO_4$, 14 CeO_2 up to 10 wt %	0.8 for 70% Zr–30% Ni 4.6 for 30% Zr–70% Ni 7.8 with CeO_2[c]	0.16	0.06
tungsten-Viton	W, 30 $BaCrO_4$, 55 $KClO_4$, 10 diatomaceous earth, 4 Viton, 1	ca 0.8–6.2[c]	0.1	not known, not suitable for storage in high humidity
tungsten delay	W, 30 $BaCrO_4$, 55 $KClO_4$, 10 diatomaceous earth, 5	0.04–16; up to 24 with CaF_2[c]	0.1	sensitive to moisture

[a] Ref. 18.
[b] Obturated.
[c] Vented.

The risk of electrostatic discharge can be minimized by grounding materials, containers, and personnel. Water must never be used in the blending of mixtures containing zinc, aluminum, or magnesium, nor should it be used with titanium or zirconium powder unless water is present in excess. Wet slurries, which form when fine zirconium powder settles, are particularly dangerous if they are "dug out" without excess water. Otherwise, no mixture should be blended when dry. The danger of dust explosion is present when fuels are finely divided. Of particular concern is the handling of zirconium powder (<10 µm in diameter), which may be pyrophoric in air. Black powder also must be handled in a humid atmosphere, since the static discharge from the operator can ignite it (17).

Pyrotechnic mixtures must be stored in metal or electrically conductive rubber containers. Caution must be taken not to blend chlorates with reducing agents, eg, metals, sugar, phosphorus, etc. Pyrotechnic compositions have an abundance of toxicological hazards. Organic dyes and chromic oxides, as used in smokes and delay compositions, are suspected of being carcinogenic. Heavy metal ions of barium, mercury, and lead can chronically impair liver and kidney functions. Antimony and arsenic have been restricted in articles of commerce for this reason. Primary explosives should not be handled by inexperienced personnel. Mixtures must not be ground, and plastic screw caps and glass-stoppered bottles must not be used.

A work plan should be prepared prior to the commencement of any pyrotechnic activity and should include descriptions of objectives and step-by-step procedures.

498   PYROTECHNICS

Advance planning for the possibility of an accident greatly minimizes the consequences of such an accident. A work plan aids in the supervision of inexperienced personnel, and if an accident should occur, the procedure can be replicated with a minimum of uncertainty.

BIBLIOGRAPHY

"Pyrotechnics" in *ECT* 1st ed., Vol. 11, "Military Pyrotechnics," pp. 322–332, by David Hart, Picatinny Arsenal; "Commercial Pyrotechnics," pp. 332–338, V. C. Allison, Aerial Products, Inc.; "Pyrotechnics" in *ECT* 2nd ed., Vol. 16, pp. 824–840, by Herbert Ellern, Pyrotechnic Consultant, formerly UMC Industries, Inc.

1. A. A. Shidlovski, *Principles of Pyrotechnics*, Mashinostroyeniye Press, Moscow, 1973; translation, Publ. No. AD A001 859, 1974.
2. H. Ellern, *Military and Civilian Pyrotechnics*, The Chemical Publishing Company, New York, 1968.
3. J. H. McLain, *Pyrotechnics*, The Franklin Institute Press, Philadelphia, Pa., 1980.
4. T. Shimizu, *Fireworks, the Art, Science, and Technique*, Maruzen Co., Ltd., Tokyo, Japan, 1981.
5. N. A. Lange, ed., *Handbook of Chemistry*, 12th ed., McGraw-Hill, New York, 1981.
6. *Handbook of Chemistry and Physics*, 62nd ed., CRC Publishing Co., Boca Raton, Fla., 1981.
7. *JANAF Thermochemical Tables*, NSRDS-NBS 37, Dow Chemical Company, Midland, Mich.; published as Catalogue No. C 13.48.37, U.S. Government Printing Office, Washington, D.C.
8. C. E. Wicks and F. E. Block, *Thermodynamic Properties of 65 Elements, Their Oxides, Halides, Carbides and Nitrides*, U.S. Bureau of Mines Bulletin 605, 1963.
9. D. D. Wagman and co-workers, *Selected Values of Chemical Thermodynamic Properties*, U.S. Department of Commerce, Bureau of Standards Technical Notes 270-3 and 270-4, 1968, 1969.
10. F. J. Zeleznik and S. Gordon, *Ind. Eng. Chem.* **60**(6), 25 (June 1968); S. Gordon and B. J. McBride, *Computer Program for the Calculation of Complex Chemical Equilibrium Compositions, Rocket Performance, Incident and Reflected Shocks and Chapman-Jouguet Detonations*, NASA SP-273, Lewis Research Center, published as NTIS N-71-37775, U.S. Department of Commerce, Springfield, Va., 1971.
11. A. P. Hardt and P. V. Phung, *Effect of Heat Loss on Delay Column Performance*, Proceedings of the Fifth International Pyrotechnic Seminar, The Denver Research Institute, Denver, Co., 1976, p. 223; A. P. Hardt, *Proceedings of the Eighth International Pyrotechnic Seminar*, The Illinois Institute of Technology, Chicago, Ill., 1982.
12. Yu. M. Maksimov, A. G. Merzhanov, A. T. Pak, and M. N. Kuchkin, *Phys. Combust. Explos.* (Russian) **17**(4), 51 (1981).
13. R. Lancaster, *Fireworks Principles and Practices*, The Chemical Publishing Company, New York, 1972.
14. J. A. Conkling, *Chem. Eng. News*, 24 (June 29, 1981).
15. F. P. Clark, *Special Effects in Motion Pictures*, Society of Motion Picture and Television Engineers, Inc., Scarsdale, N.Y., 1966.
16. California Administrative Code, Title 19, Sec. 992.6–992.10, 1981.
17. J. E. Rose and A. P. Hardt, *Proceedings of the 10th Symposium on Explosives and Pyrotechnics*, The Franklin Research Center, Philadelphia, Pa., Feb. 14–16, 1979, pp. 6a 1–10.
18. S. M. Kaye, ed., *Encyclopedia of Explosives and Related Items*, Vol. 7, ADA 019502, Vol. 8, ADA 057762, Vol. 9, ADA 097595, and Vol. 10, U.S. Army Research and Development Command, Dover, N.J., 1960–1982.
19. U.S. Pat. 3,535,063 (Oct. 20, 1980), L. F. Anderson, S. R. Bennet, and J. W. Shaffer (to Sylvania Electric Products, Inc.); U.S. Pat. 3,535,064 (Oct. 20, 1970), L. F. Anderson and P. Bader (to Sylvania Electric Products, Inc.); U.S. Pat. 3,674,411 (July 4, 1972), J. W. Shaffer (to Sylvania Electric Products, Inc.).
20. D. R. Maulding and co-workers, *Development of a Three Component Thixotropic Chemical Lighting System*, RDTR 183, AD 738 806, Naval Ammunition Depot, Crane, Ind., Oct. 1971.
21. *Code of Federal Regulations, Title 49, Sec. 173.100–173.114.*
22. *Engineering Design Handbook, Military Pyrotechnics Series, Part One, Theory and Application*, AMCP 706-185, Apr. 1967, pp. 6–38.

23. J. J. Caren and T. Stevenson, *Pyrotechnics for Small Arms Ammunition*, R-1968, Department of the Army, Frankford Arsenal, Philadelphia, Pa., July 1970.
24. W. W. Hillstrom, *Formation of Pyrophoric Fragments*, BRL MR 2306, AD 735447, Ballistics Research Laboratory, Aberdeen, Md., 1973.
25. A. P. Hardt and F. J. Valenta, *Proceedings of the Eleventh Symposium on Explosives and Pyrotechnics*, The Franklin Research Center, Philadelphia, Pa., Sept. 1981, p. 19-1.
26. W. Ripley, *Air and Water Reactive Materials*, RDTR 124. U.S. Naval Ammunition Depot, Crane, Ind., Aug. 8, 1968.
27. J. W. Morris and V. P. Marchese, *Flueric Cartridge Initiation Development*, ADPA Pyrotechnics and Explosives Omnibus, Los Alamos, N.M., Oct. 7–9, 1975.
28. W. E. Robertson, *Igniter Materials Handbook*, TR 690, AD 51001, Naval Ammunition Depot, Crane, Ind., Sept. 1969.

ALEXANDER P. HARDT
Lockheed Missiles and Space Company, Inc.

PYRROLE AND PYRROLE DERIVATIVES

Pyrrole [109-97-7], a five-membered, heterocyclic system, is a fundamental structural subunit of many of the most important biological molecules, eg, heme, chlorophyll, the bile pigments, certain amino acids, many alkaloids, and some enzymes. Early interest in the chemistry of pyrrole began with the discovery of indole (benzopyrrole) as the fundamental nucleus of indigo. Pyrrole was first obtained in 1834 from the destructive distillation of coal, bone, or proteins (1). It was characterized in 1858, and its composition was determined in 1870 (2–3). Ring positions in pyrrole are designated by number or Greek letter.

Physical Properties of Pyrrole

Pyrrole is a colorless, slightly hygroscopic liquid and, if fresh, it emits an odor like that of chloroform. However, it darkens on exposure to air and eventually produces a dark brown resin. It can be preserved by excluding air from the storage container, preferably by displacement with ammonia to prevent acid-catalyzed polymerization. Some physical properties of pyrrole are listed in Table 1. General properties of pyrroles are given in ref. 4.

Pyrrole has a planar, pentagonal (C_{2v}) structure and is aromatic in its reactions since it has an aromatic sextet of electrons. It is isoelectronic with the cyclopentadienyl

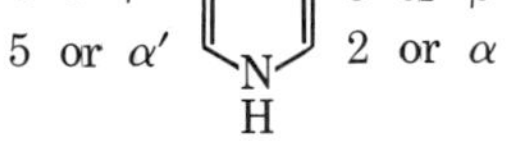

4 or β' 3 or β
5 or α' 2 or α

(1)
pyrrole

Table 1. Physical Properties of Pyrrole

Property	Value
melting point, °C	−18.5
boiling point, °C	130
critical temperature, °C	366
density, d_4^{20}, g/cm³	0.9698
refractive index, n_D^{20}	1.5085
dielectric constant (at 20°C), ϵ	8.00
flash point (closed-cup), °C	39

anion. The π-electrons are delocalized throughout the ring system; thus, pyrrole is best characterized as a resonance hybrid with contributing structures (1)–(5). These structures explain its lack of basicity, which is less than pyridine, its unexpectedly high acidity, and its pronounced aromatic character.

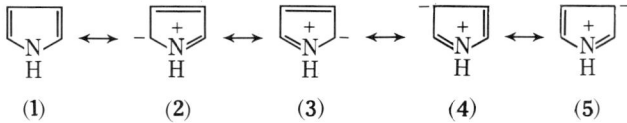

The resonance energy is ca 100 kJ/mol (24 kcal/mol) or about two-thirds that of benzene (5). Its resonance energy is intermediate between those of furan and thiophene; thiophene has the highest value.

The contributions from the canonical forms have been calculated, and the contributions from the equivalent polar structures (2) and (3) dominate those of (4) and (5) (6). Thus, electrophilic substitution is predicted to occur in the α-position, and this is proven experimentally in most cases. Nitrosation and selenocyanation occur at the β-position (7–8).

Many of the physical characteristics of pyrrole indicate at least partial association. In particular, the boiling point is 98°C higher than that of furan. It has been postulated that various associated dimeric and higher structures occur as results of hydrogen bonding (9–10).

Pyrrole is freely soluble in alcohol, benzene, and diethyl ether, but is only sparingly soluble in water and in aqueous alkalies. It dissolves with decomposition in dilute acids. Pyrroles with substituents in the β-position are usually less soluble in polar solvents than the corresponding α-substituted pyrroles. Pyrroles that have no substituent on nitrogen readily lose a proton to form the resonance-stabilized pyrrolyl anion, and alkali metals react with pyrrole in liquid ammonia to form metal salts. However, pyrrole (pK_a = ca 17.5) is a weaker acid than methanol (11). The acidity of the pyrrole hydrogen is greatly increased by electron-withdrawing groups, eg, the pK_a of 2,5-dinitropyrrole [32602-96-3] is 3.6 (12–13).

The dipole moment varies according to the solvent; it is ca 5.14×10^{-30} C·m (ca 1.54 D) in pure pyrrole and ca 6.0×10^{-30} C·m (ca 1.8 D) in a nonpolar solvent such as benzene or cyclohexane (14–15). In solvents to which it can hydrogen bond, the dipole moment may be much higher. The dipole is directed toward the ring from a positive nitrogen atom, whereas the saturated nonaromatic analogue pyrrolidine [123-75-1] has a dipole moment of 5.24×10^{-30} C·m (1.57 D) and is oppositely directed.

Pyrrole and its alkyl derivatives are π-electron rich and form colored charge-transfer complexes with acceptor molecules, eg, iodine, tetracyanoethylene, etc (16).

Infrared spectra of pyrrole and its derivatives have been described in detail (13). The $N-H$ absorption of nonassociated pyrroles varies predictably with the substituents and is related to the acidity of the pyrrole. It has been used as evidence for intermolecular association, which results from hydrogen bonding, between pyrrole units. Nuclear magnetic resonance studies have provided evidence for ring currents and for an estimate that the aromaticity of pyrrole is 59% that of benzene (17). However, these studies have been criticized (18). The effect of substituents upon the various chemical shifts has also been reported (19). Pyrrole gives an intense molecular ion in the mass spectrum. Smaller fragments which arise often result from ring-size reduction.

Syntheses of Pyrroles

Knorr. The Knorr reaction and its modifications are the most important and widely used methods for the synthesis of pyrroles (6).

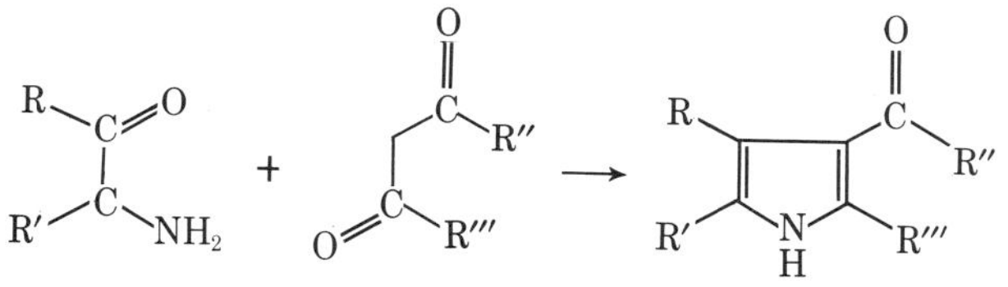

(6)

Since the α-aminoketone is subject to self-condensation, the condensation with a β-dicarbonyl derivative (8) is usually carried out by generating the α-aminoketone *in situ* through reduction of an oximino derivative (7); zinc in glacial acetic acid is used as the reductant (20). For example,

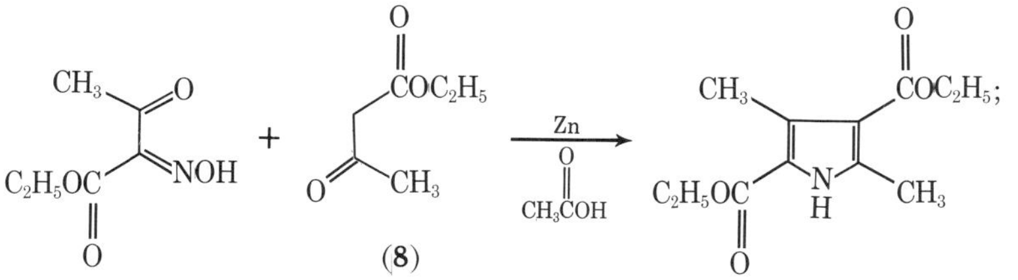

(7)   (8)   Knorr's pyrrole [2436-79-5]

Modifications include the use of β-ketoaldehydes as acetals, eg, (9), which leads to loss of the acetyl group (21):

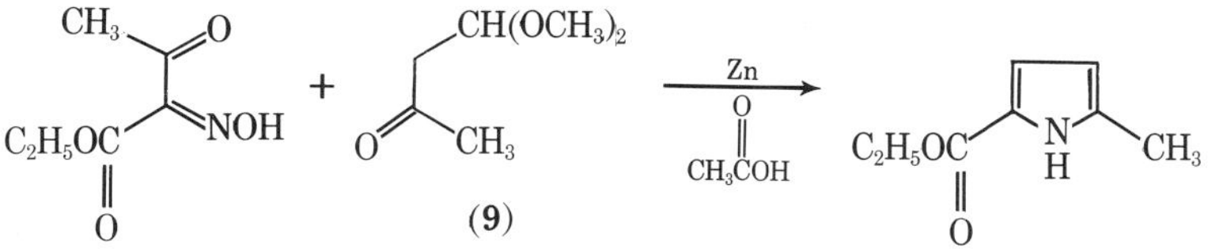

(7)   (9)   [3284-51-3]

The Knorr synthesis is not particularly sensitive to the nature of R and R''', ie, they may be alkyl, acyl, aryl, or carbalkoxy without significantly affecting the yield. Similarly, good yields are obtained if R' and R" are acyl or carbalkoxy, but poor yields are obtained if they are alkyl or aryl.

Hantzsch and Feist. The Hantzsch synthesis of pyrroles involves condensation of an α-haloketone (**10**) with a β-keto ester (**8**) in the presence of ammonia or an amine (22).

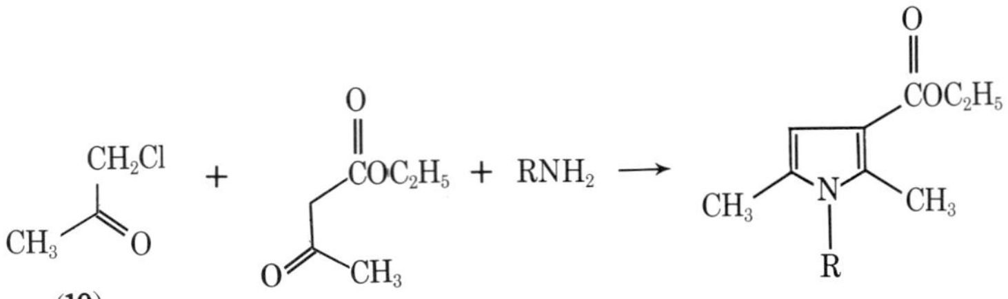

The Feist synthesis is similar to the Hantzsch method and involves condensation of acyloins, eg, (**11**) with aminocrotonic esters, eg, (**12**) in the presence of zinc chloride (23).

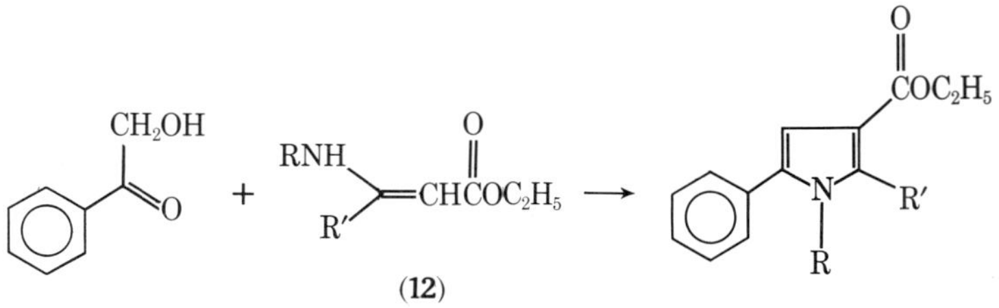

Paal-Knorr. The condensation of a 1,4-diketone, eg, (**13**) with ammonia or a primary amine generally gives good yields of pyrroles; many syntheses have been reported (24).

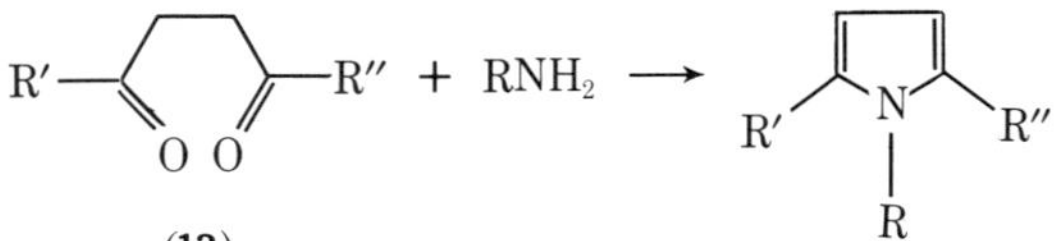

The lack of availability of the appropriate 1,4-diketone sometimes limits the usefulness of the reaction.

Miscellaneous. Acetylenic compounds have often been used as precursors to certain pyrroles. Thus, 2-butyne-1,4-diol (**14**) reacts with aniline in the presence of alumina to produce 1-phenylpyrrole (**15**) (25).

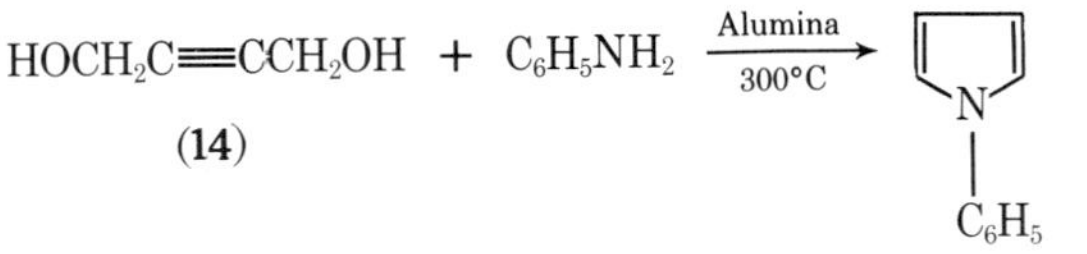

$$HOCH_2C{\equiv}CCH_2OH + C_6H_5NH_2 \xrightarrow[300°C]{Alumina}$$

(**14**)

(**15**)

[635-90-5]

Acetylenedicarboxylic esters, eg, (**16**) also react with phenylhydroxylamines, eg, (**17**) to give pyrroles (26).

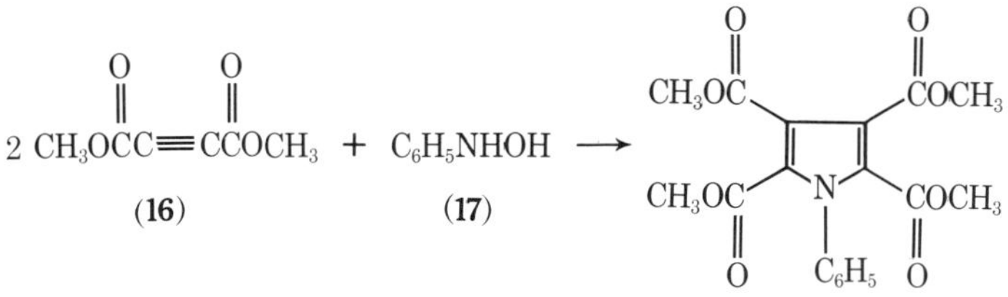

$$2\ CH_3OCC \equiv CCOCH_3 + C_6H_5NHOH \longrightarrow$$
(**16**) \qquad (**17**)

[*37802-39-4*]

α-Aminoketones, eg, (**18**) and acetylenic carbonyl compounds, eg, (**19**) cyclize and dehydrate to give pyrroles in high yields by a Michael-type addition (27).

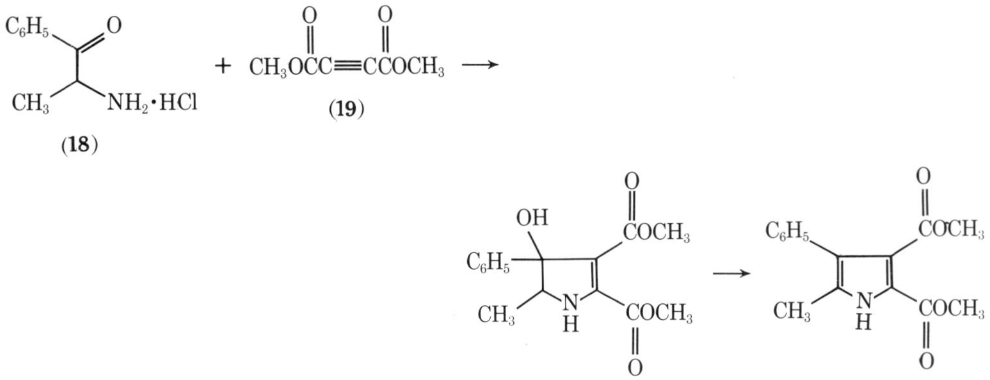

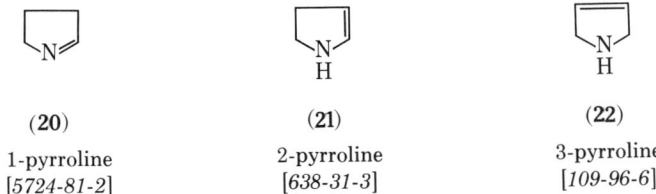

[*53252-73-6*]

Pyrrolines and Pyrrolidines

The pyrrolines or dihydropyrroles can exist in three isomeric forms:

(**20**)   (**21**)   (**22**)

1-pyrroline   2-pyrroline   3-pyrroline
[*5724-81-2*]   [*638-31-3*]   [*109-96-6*]

1-Pyrroline (**20**) (3,4-dihydro-2H-pyrrole) is an unstable material that resinifies upon exposure to air. 2-Pyrroline (**21**) (2,3-dihydro-1H-pyrrole) is even more unstable. Only 3-pyrroline (**22**) (2,5-dihydro-1H-pyrrole) is reasonably stable. It boils at 91°C and has a density (d_4^{28}) of 0.9097 g/cm^3 and a refractive index (n_D^{20}) of 1.4664. Pyrrolidine (**23**) or tetrahydropyrrole is a water-soluble, strong base with the usual properties of a secondary amine.

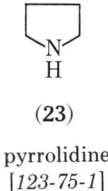

(23)

pyrrolidine
[123-75-1]

An important synthesis of pyrrolidines is the reaction of reduced furans with excess amine or ammonia over an alumina catalyst in the vapor phase at 400°C. However, if labile substituents are present on the tetrahydrofuran, pyrroles may form (28).

Pyrrolidines also can be obtained by reaction of 1,4-dihydroxyalkanes with amines in the presence of dehydrating agents at elevated temperatures or by reaction of primary amines with 1,4-dihalides. The dry distillation of 1,4-butanediamine dihydrochloride, eg, (24) also generates pyrrolidine.

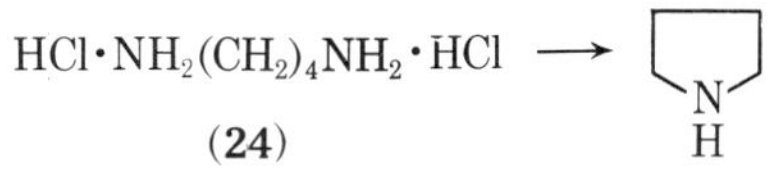

$$HCl \cdot NH_2(CH_2)_4NH_2 \cdot HCl \longrightarrow$$

(24)   (23)

Pyrroles can also be catalytically hydrogenated to pyrrolidines.

Reactions of Pyrroles

In keeping with its decidedly aromatic character, pyrrole is relatively difficult to hydrogenate, it does not ordinarily serve as a diene in Diels-Alder reactions, and it does not undergo typical olefin reactions. Electrophilic substitutions are the most characteristic reaction, and pyrrole has often been compared with phenol or aniline in its reactivity. Acids strong enough to form salts with pyrrole destroy the aromaticity and cause polymerization.

Although N-alkyl pyrroles are normally obtained by the Knorr synthesis, they also form from the reaction of the pyrrolyl metallates, ie, Na, K, and Tl, with alkyl halides, eg, compound (25).

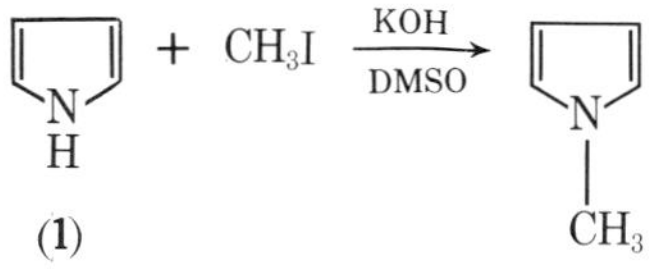

$$\underset{(1)}{\text{pyrrole}} + CH_3I \xrightarrow[DMSO]{KOH} \underset{(25)}{\text{N-CH}_3}$$

1-methylpyrrole
[96-54-8]

Alkylation of pyrroles at the other ring positions can be carried out under mild conditions with allylic or benzylic halides or under more stringent conditions (100–150°C) with CH_3I. However, unless most of the other ring positions are blocked, polyalkylation and polymerization tend to occur.

N-Acylation is readily carried out by reaction of the alkali-metal salts with the appropriate acid chloride. *C*-Acylation of pyrroles carrying negative substituents occurs in the presence of Friedel-Crafts catalysts. Pyrrole and alkyl pyrroles can be acylated noncatalytically with an acid chloride or an acid anhydride. The formation of trichloromethyl 2-pyrryl ketone [*35302-72-8*] (**27**, R = CCl_3) is a particularly useful procedure, since the ketonic product can be readily converted to the corresponding pyrrolecarboxylic acid or ester by treatment with aqueous base or alcoholic base, respectively (29).

The most generally useful method for acylation or formylation of pyrroles is the Vilsmeier-Haak reaction (30–31). The pyrrole is treated with the phosphoryl chloride complex (**26**) of an *N*,*N*-dialkylamide and the intermediate imine salt is hydrolyzed.

(**1**) (**26**) (**27**)

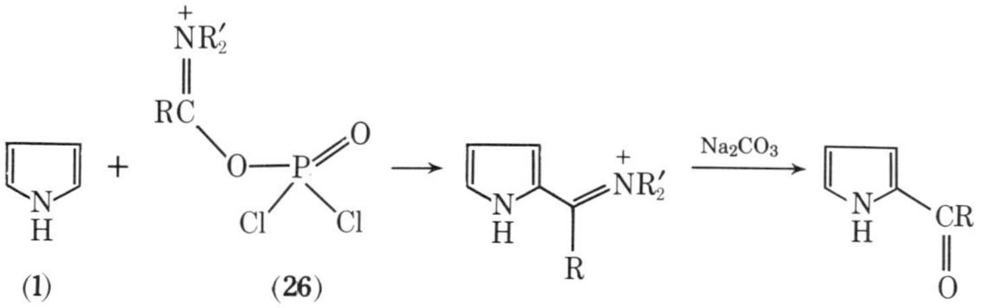

Nitration of pyrroles by the usual methods leads to extensive degradation. However, nitration can be achieved with an equimolar nitric acid–acetic anhydride mixture at low temperatures. In the case of pyrrole, the reaction leads predominantly to substitution at the β-position (32).

(**1**) 51% 3-nitropyrrole [*5930-94-9*] 13% 2-nitropyrrole [*5919-26-6*]

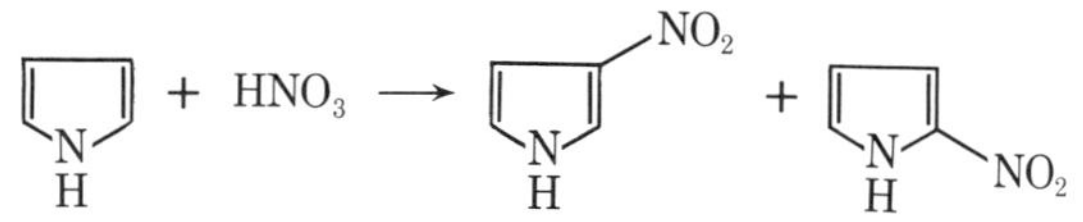

Halogenation reactions usually involve pyrroles with electronegative substituents. Mixtures are usually obtained and polysubstitution products, ie, tetrahalopyrroles, predominate. The monohalopyrroles are difficult to prepare and are not very stable in air or light.

The following rules pertain to electrophilic substitution in pyrroles (33):

1. An electron-withdrawing substituent in the α-position directs substitution into the β'- and α'-positions.

2. An electron-releasing substituent in the α-position directs substitution to the neighboring β-position or to the α'-position.

3. An electron-withdrawing substituent in the β-position leads to substitution in the α'-position.

4. An electron-releasing substituent in the β-position tends to direct substitution into the neighboring α-position.

Pyrrole can be reduced catalytically to pyrrolidine over a variety of metal catalysts, ie, Pt, Pd, Rh, and Ni. Of these, rhodium on alumina is one of the most active. Less active reducing agents have been used to produce the intermediate 3-pyrroline (**34**). The 2-pyrrolines are ordinarily obtained by ring-closure reactions. Nonaromatic pyrrolines can be reduced easily with H_2 to pyrrolidines.

Pyrrole oxidizes in air to red or black pigments of uncertain composition. More useful is the preparation of 2-oxo-Δ^3-pyrrolines, eg, (**29**), which is best carried out by oxidation of the appropriate pyrrole (**28**) with H_2O_2 in pyridine (**35**).

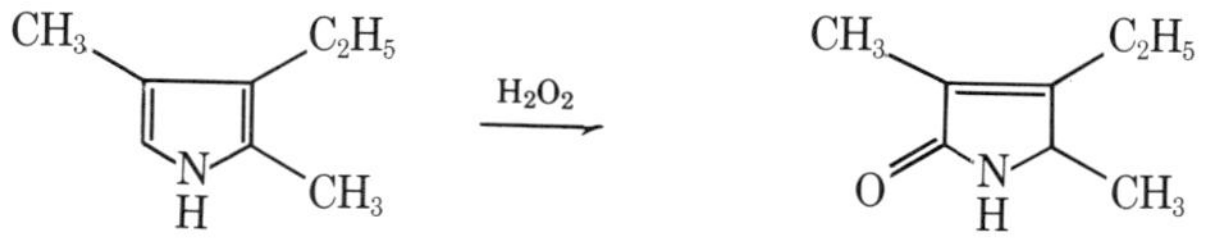

(**28**)
2,4-dimethyl-3-ethylpyrrole
[517-22-6]

(**29**)
3,5-dimethyl-4-ethyl-3-pyrrolin-2-one
[4030-24-4]

Perbenzoic acid oxidizes *N*-methylpyrrole to *N*-methylsuccinimide (**36**).

Ring openings of pyrrole commonly occur at the carbon–nitrogen bond. Treatment of pyrrole or 2,5-dimethylpyrrole [625-84-3] ((**30**), R = CH_3) with hydroxylamine leads to ring opening and formation of 2,5-dioximes (**31**) (**37**).

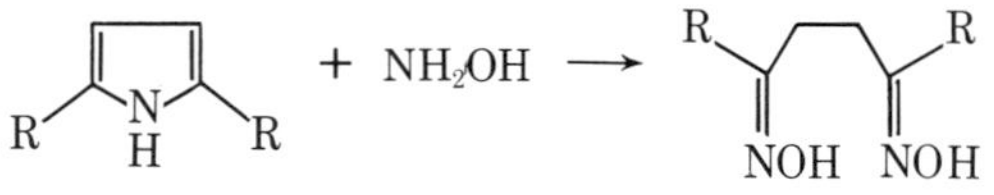

Reaction of pyrrole with carbenes yields enlarged ring systems as well as 2-formylpyrrole [1003-29-8] (**38**).

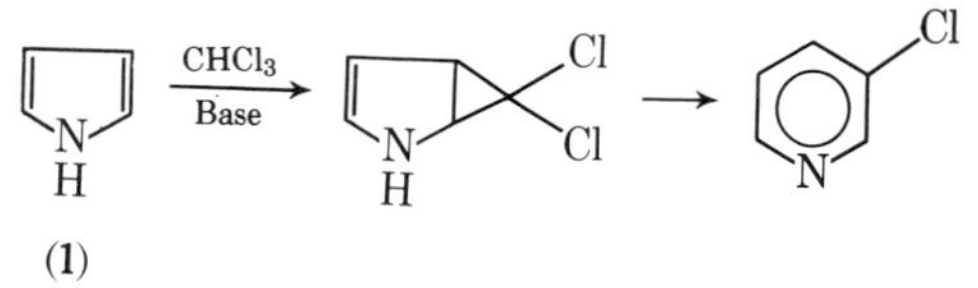

Analytical and Test Methods

In addition to the modern spectroscopic methods of detection and identification of pyrroles, there are several chemical tests. The classical Runge test with HCl yields pyrrole red, an amorphous polymer mixture. In addition, all pyrroles with a free α- or β-position or with groups, eg, ester, that can be converted to such pyrroles under acid conditions undergo the Ehrlich reaction with *p*-(dimethylamino)benzaldehyde (**32**) to give purple products.

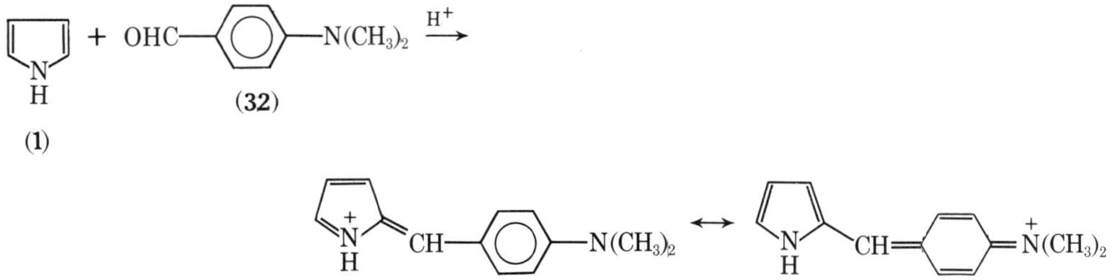

[Reaction scheme showing pyrrole (1) + OHC-C6H4-N(CH3)2 (32) with H+ giving resonance structures of the product cation]

Both pyrrole and indole react with selenium dioxide in the presence of nitric acid to give a deep violet solution. Very small quantities (ca 4×10^{-5} g) of pyrrole can be detected by this method.

Functional Derivatives

Hydroxypyrroles. Pyrroles with nitrogen-substituted side chains containing hydroxyl groups are best prepared by the Paal-Knorr cyclization. Pyrroles with hydroxyl groups on carbon side chains can be made by reduction of the appropriate carbonyl compound with hydrides, by Grignard synthesis, or by insertion of ethylene oxide or formaldehyde.

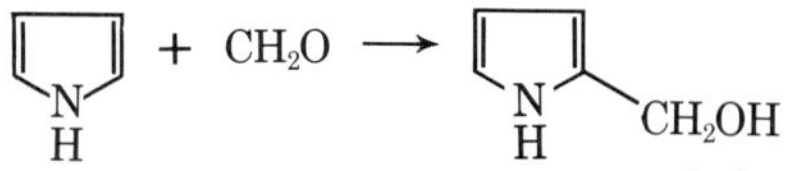

[Reaction: pyrrole (1) + CH2O → 2-hydroxymethylpyrrole [27472-36-2]]

The hydroxymethylpyrroles do not act as normal primary alcohols because of resonance stabilization of any carbonium ion formed.

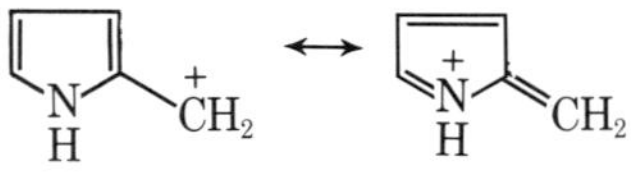

[Resonance structures of the pyrrolyl methyl cation]

The α-hydroxypyrroles, which exist primarily in the tautomeric pyrrolin-2-one form, can be synthesized either by oxidation of pyrroles that are unsubstituted in the α-position or by ring synthesis (33). β-Hydroxypyrroles also exist primarily in the keto form but do not display the ordinary reactions of ketones because of the contributions of the polar form. They can be readily O-alkylated and -acylated (39).

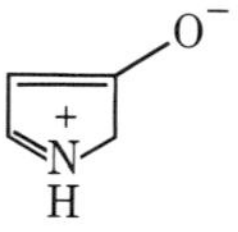

[Structure: polar resonance form of β-hydroxypyrrole with N+–H and C–O−]

Aldehydes and Ketones. Pyrrole aldehydes and ketones are somewhat less reactive than the corresponding benzenoid derivatives and are vinylogous amides. The aldehydes do not undergo Cannizzaro or Perkin reactions but condense with a variety of compounds that contain active methylene groups. They also react with unsubstituted pyrroles under acidic conditions to form dipyrrylmethenes (**33**).

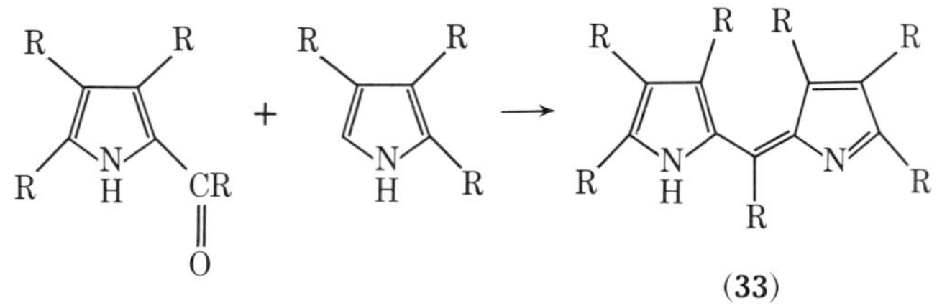

(33)

The aldehydes can be reduced to the methyl or carbinol structures. The ketones undergo normal carbonyl reactions.

Pyrrole Carboxylic Acids and Esters. The acids are considerably less stable than benzoic acid and often decarboxylate readily on heating. However, electron-withdrawing substituents tend to stabilize them toward decarboxylation. The pyrrole esters are important synthetically, since they stabilize the ring and may also act as protecting groups. Thus, the esters can be utilized synthetically and then hydrolyzed to the acid, which can be decarboxylated as described above. Often β-esters are hydrolyzed more easily than the α-esters.

Condensed Pyrroles. Pyrroles can be condensed to compounds containing two, three, or four pyrrole nuclei. These are important in synthetic routes to the tetrapyrrolic porphyrins, corroles, and bile pigments and to the tripyrrolic prodigiosins. The pyrrole nuclei are joined by either a one-carbon fragment or a direct pyrrole–pyrrole bond.

Bipyrroles. Although four different types of bipyrroles, ie, 1,1'-, 2,2'-, 2,3'-, and 3,3'-, are known, the most important is the 2,2'-type. The parent compound can be made by the Vilsmeier condensation of 2-pyrrolidinone [616-45-5] (**34**) (40).

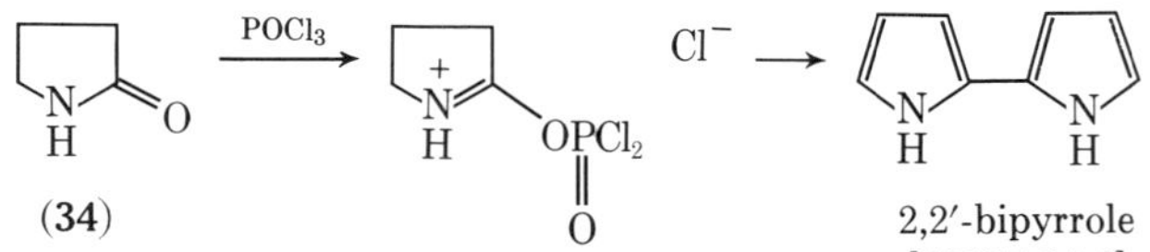

(34)   2,2'-bipyrrole [10087-64-6]

Other syntheses utilize α-bromo or -iodo compounds, which condense on heating with copper (41).

Dipyrrylmethanes. The most important dipyrrylmethanes are the 2,2'-derivatives. The parent compound is not very stable but electron-withdrawing substituents increase its stability considerably. Symmetrical dipyrrylmethanes, eg, (**35**) and (**36**) can be synthesized by acid-catalyzed self-condensation of α-halomethyl, α-acetylmethyl, or α-methoxymethylpyrroles (42).

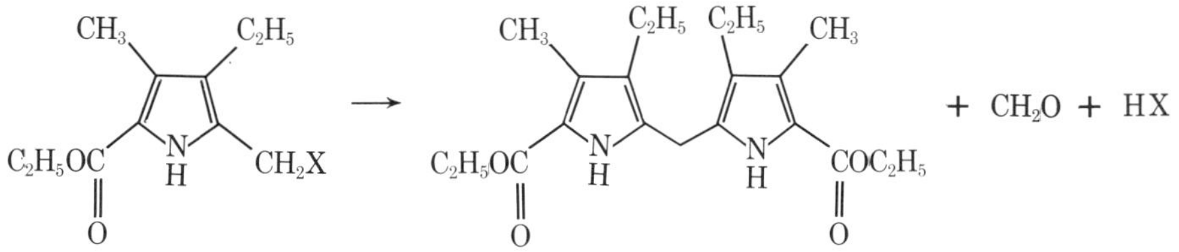

(35)
[6305-93-7]

Unsymmetrical dipyrrylmethanes are obtained through condensation of α-bromomethyl or α-hydroxymethylpyrroles with a pyrrole that has an open α- or β-position.

Dipyrrylmethenes. Oxidation of the dipyrrylmethanes by bromine or sulfuryl chloride yields dipyrrylmethenes, eg, (33) and (37).

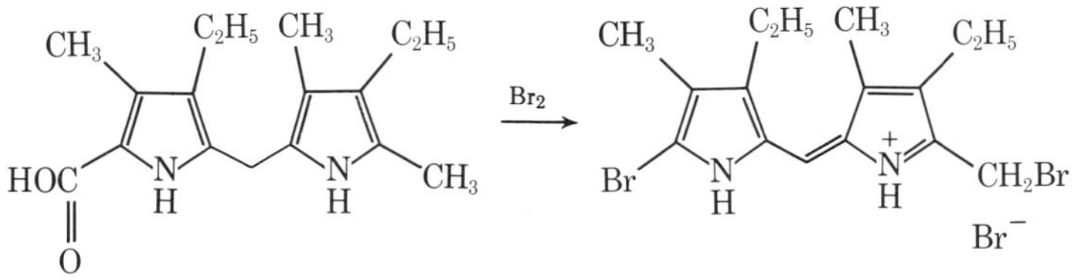

(36)　　　　　　(37)
[26030-65-9]　　[80294-31-1]

Polypyrroles. Highly stable, flexible films of polypyrrole are obtained by electrolytic oxidation of the appropriate pyrrole monomers (43). The films are not affected by air and can be heated to 250°C with little effect. It is believed that the pyrrole units remain intact and that linking is by the α-carbons. Copolymerization of pyrrole with N-methylpyrrole (25) yields compositions of varying electrical conductivity, depending upon the monomer ratio. Conductivities as high as $10^4/(\Omega\cdot m)$ have been reported (44).

Natural Products. Prodigiosins. The prodigiosins are antibacterial and antifungal orange-red pigments based upon the basic pyrryl–dipyrrylmethene unit (38).

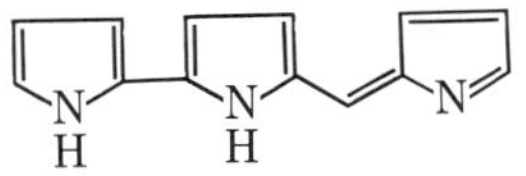

(38)
[22187-69-5]

Bile Pigments. The oxidative degradation of heme yields open-chain tetrapyrroles as waste products in humans and other higher animals. The yellow color of the skin in jaundice victims is caused by the presence of bilirubin (39).

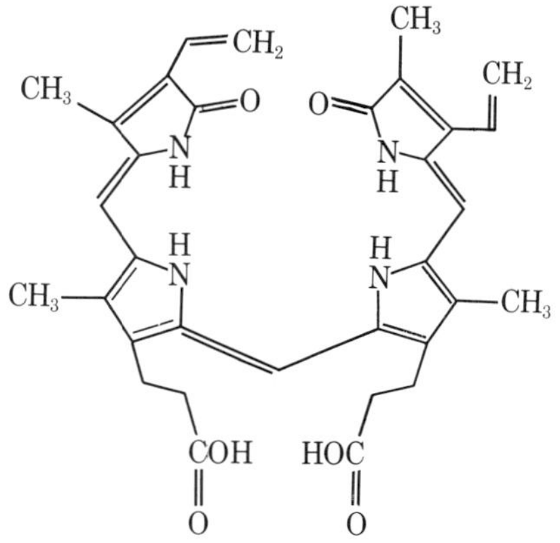

(39)
bilirubin [635-65-4]

Phthalocyanines. An important group of pigments that contains the pyrrole ring system is the blue or blue-green phthalocyanines (see Pigments, organic pigments).

Pyrrolidinone and Derivatives. *2-Pyrrolidinone.* 2-Pyrrolidinone (**34**) (2-pyrrolidone, butyrolactam or 2-Pyrol) was first reported in 1889 as produced by dehydration of 4-aminobutanoic acid (45). The synthesis used for commercial manufacture, ie, condensation of butyrolactone with ammonia, was first described in 1936 (46). Other synthetic routes include carbon monoxide insertion into allylamine (47–48), hydrolytic hydrogenation of succinonitrile (49–50), and hydrogenation of ammoniacal solutions of maleic or succinic acids (51–53). Properties of 2-pyrrolidinone are listed in Table 2. 2-Pyrrolidinone is completely miscible with water, lower alcohols, lower ketones, ether, ethyl acetate, chloroform, and benzene. It is soluble to ca 1 wt % in aliphatic hydrocarbons.

Table 2. Properties of 2-Pyrrolidinone

Property	Value
melting point, °C	25.6
boiling point, °C	
at 0.133 kPa^a	103
at 1.33 kPa^a	122
at 13.2 kPa^a	181
at 101.3 kPa^a	245
density (liquid), g/cm^3	
d_4^{25}	1.107
d_4^{50}	1.087
refractive index, n_D^{30}	1.4840
viscosity (at 25°C), mPa·s (= cP)	13.3
flash point (open-cup), °C	129.4

^{a} To convert kPa to mm Hg, multiply by 7.5.

Reactions. 2-Pyrrolidinone undergoes the reactions of a typical lactam, eg, ring opening, attack on the carbonyl group, and replacement of hydrogens alpha to the carbonyl group. Many of the reactions involve the hydrogen on the nitrogen atom. 2-Pyrrolidinone can be polymerized with anionic catalyst systems to polypyrrolidinone, (nylon-4 (**40**)), which is a high molecular weight linear polymer of potential interest as a textile fiber, film former, and molding compound (54–55) (see Polyamides; Textiles).

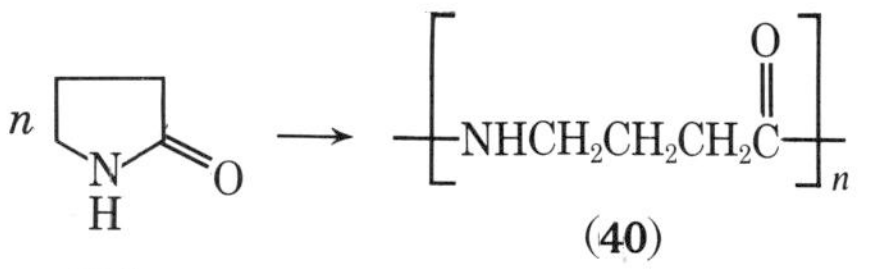

$$n \underset{\text{(34)}}{\text{[pyrrolidinone]}} \longrightarrow \underset{\text{(40)}}{[-\text{NHCH}_2\text{CH}_2\text{CH}_2\text{C(=O)}-]_n}$$

Strong acids or bases catalyze the hydrolysis of 2-pyrrolidinone to 4-aminobutanoic acid [γ-aminobutyric acid (GABA) (**41**)]. The latter is involved in the functioning of the brain and nervous system and is of considerable interest as a potential dietary supplement (56).

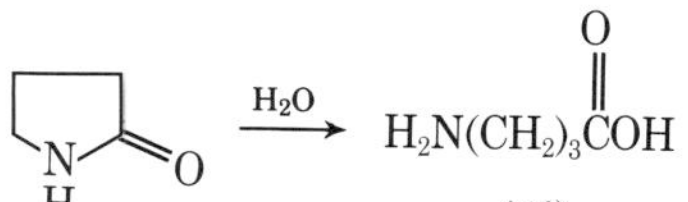

$$\underset{\text{(34)}}{\text{[pyrrolidinone]}} \xrightarrow{\text{H}_2\text{O}} \underset{\text{(41)}}{\text{H}_2\text{N(CH}_2)_3\text{COH}}$$

2-Pyrrolidinone forms alkali-metal salts by direct reaction with alkali metals or their alkoxides or with their hydroxides under conditions in which the water of reaction is removed. The mercury salt has been described, as have the *N*-bromo and *N*-chloro derivatives (57–58).

2-Pyrrolidinone can be alkylated by reaction with an alkyl halide or sulfate and an alkaline acid acceptor (59–60). This reaction can be advantageously carried out with a phase-transfer catalyst (61) (see Catalysis, phase-transfer).

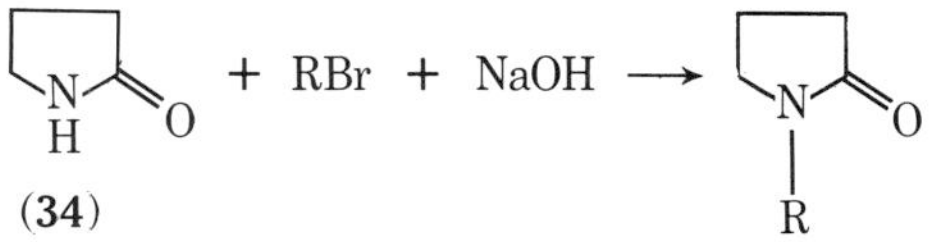

$$\underset{\text{(34)}}{\text{[pyrrolidinone]}} + \text{RBr} + \text{NaOH} \longrightarrow \underset{\underset{R}{|}}{\text{[N-alkyl pyrrolidinone]}}$$

Alkylation can also be accomplished with alcohols and either copper chromite or heterogenous acid catalysts (62–63).

Treatment of 2-pyrrolidinone with an acid anhydride or acyl halide results in *N*-acylation, eg, (**42**) (15).

The amide hydrogen readily adds across the carbonyl group of an aldehyde yielding *N*-hydroxyalkyl-substituted pyrrolidinones, eg, (**43**) (64). In the presence

(34)

(42) [932-17-2]

of secondary amines or alcohols, the hydroxyl groups are replaced eg, yielding (**44**) and (**45**) (65–66).

(34)

(43) [15438-71-8]

(34)

(44) [66297-50-5]

(34)

(45) [51576-82-0]

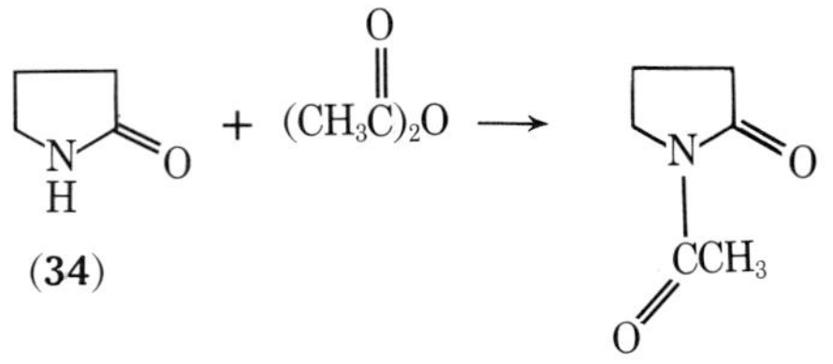

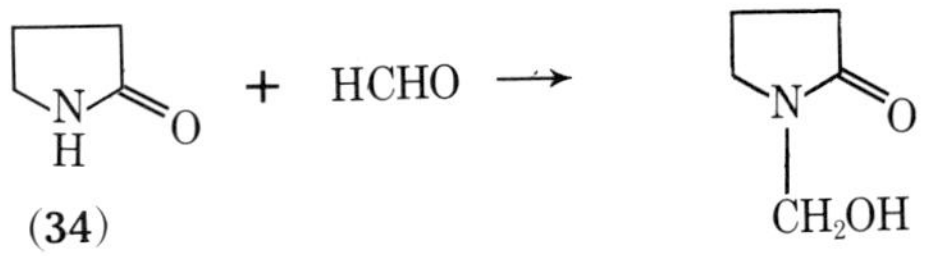

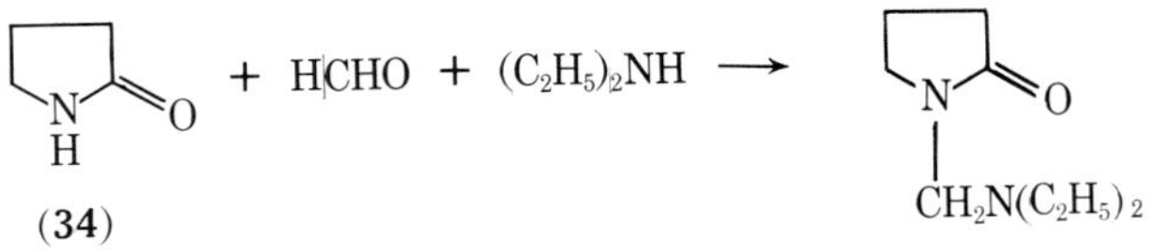

2-Pyrrolidinone is readily *N*-alkylated by styrene to give (**46**). Additional styrene alkylates the 3-position yielding (**47**) (67).

(34)

(46) [10135-23-6]

(47) [60549-73-7]

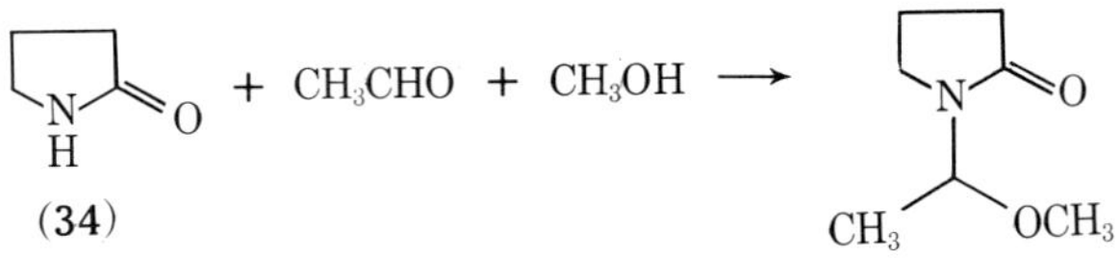

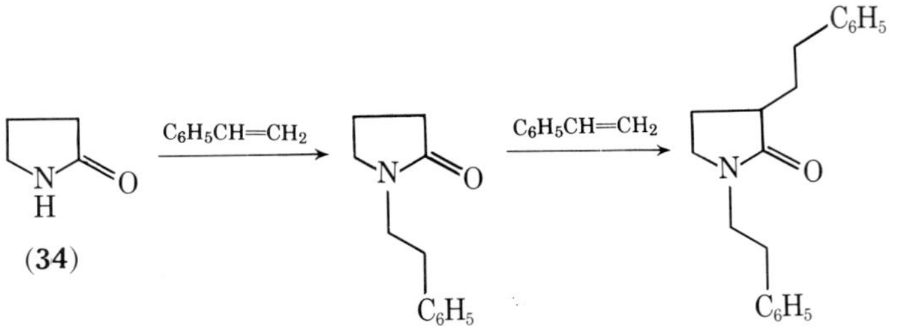

High temperature hydrogenation with a cobalt catalyst gives pyrrolidine (**23**) (68).

(34)

(23)

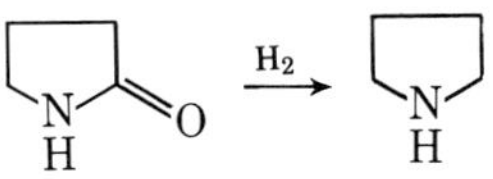

Under dehydrating conditions, 2-pyrrolidinone condenses with itself to form 1-(Δ1'-pyrrolin-2-yl)-2-pyrrolidinone (**48**) (69).

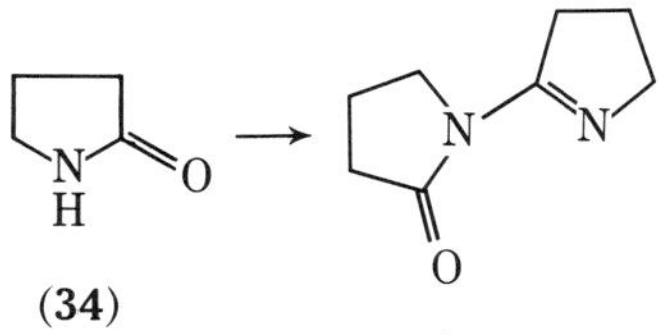

(**34**) → (**48**) [7060-52-8]

2-Pyrrolidinone can also condense with primary or secondary amines.

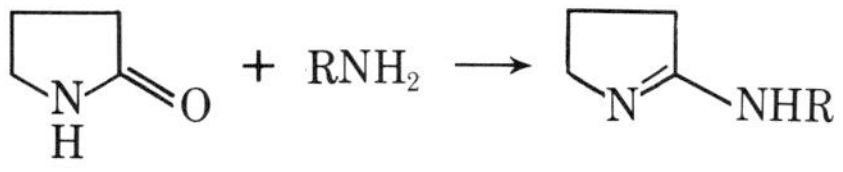

(**34**) + RNH$_2$ →

Under suitable conditions, O-alkylation rather than N-alkylation takes place, eg, (**49**) (70–72).

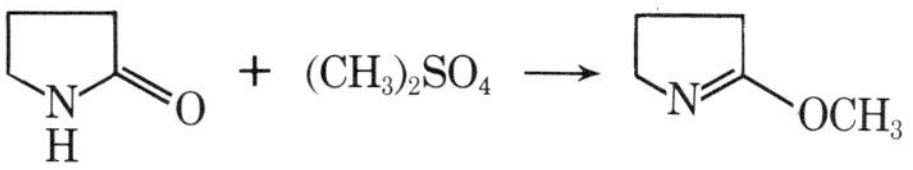

(**34**) + (CH$_3$)$_2$SO$_4$ → (**49**) [5264-35-7]

Manufacture. There are two main 2-pyrrolidinone producers. GAF has manufacturing facilities in Calvert City, Kentucky, and Texas City, Texas, and BASF manufactures it at Ludwigshaven, FRG. Both producers consume nearly all of their production in the manufacture of 1-vinyl-2-pyrrolidinone.

Butyrolactone (**50**) and a moderate excess of ammonia are passed through a reactor at ca 250°C and 8–9 MPa (80–90 atm). Yields of 90–95% have been reported (73). The reaction proceeds in two steps, but the intermediate 4-hydroxybutyramide (**51**) is not ordinarily isolated.

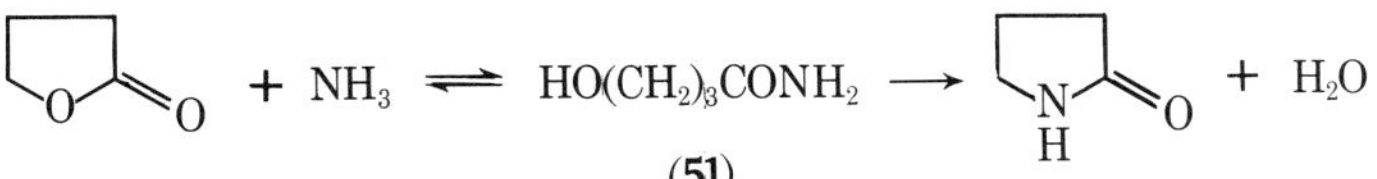

(**50**) + NH$_3$ ⇌ HO(CH$_2$)$_3$CONH$_2$ → (**34**) + H$_2$O
(**51**)

Shipment, Storage, and Price. 2-Pyrrolidinone is available in various-sized steel drums and in aluminum or stainless-steel tank cars and tank trailers. Because of its high freezing point, bulk shipments are made in tanks equipped with heating coils to facilitate unloading of the frozen material. Hot water as opposed to steam is recommended as the heating material to avoid product discoloration. Steel, stainless steel, and aluminum are satisfactory materials for storage containers. Because 2-pyrrolidinone is hygroscopic, it must be protected from atmospheric moisture. In February 1981, the U.S. price for tank car quantities was $3.53/kg.

Specifications and Analytical Methods. The purity of 2-pyrrolidinone is determined by gas chromatography and is specified as 98.5 wt % minimum. Maximum moisture content is specified as 0.5 wt %. Typical purities are much higher than specification.

Health and Safety Factors. Results of acute oral toxicity studies of 2-pyrrolidinone on white rats and guinea pigs showed the LD_{50} to be 6.5 mL/kg. Skin patch tests on 200 human subjects indicated that 2-pyrrolidinone is a skin irritant, but there was no indication of sensitizing action (74).

Uses. Because of the labile hydrogen on the nitrogen, 2-pyrrolidinone is not as good a solvent as 1-methylpyrrolidinone (**52**). Nevertheless, moderate amounts are sold as a solvent and as a plasticizer and coalescing agent for polymer emulsion coatings. There is also continuing interest in 2-pyrrolidinone as a monomer for polypyrrolidinone and as a source of 4-aminobutanoic acid. The main use of 2-pyrrolidinone is as an intermediate for the manufacture of 1-vinyl-2-pyrrolidinone (see Vinyl polymers, *N*-vinyl).

1-Methyl-2-pyrrolidinone. *N*-Methyl-2-pyrrolidinone (**52**) (NMP or methyl-2-pyrrolidone, M-Pyrol) was first reported in 1907 as prepared by alkylation of 2-pyrrolidinone with methyl iodide (75). The present commercial route, ie, condensation of butyrolactone with methylamine, was first described in 1936 (46). Other preparative routes include hydrogenation of succinonitrile in the presence of methylamine and hydrogenation of solutions of maleic or succinic acid and methylamine (76–77). Properties are listed in Table 3. 1-Methyl-2-pyrrolidinone is completely miscible with water, lower alcohols, lower ketones, ether, ethyl acetate, chloroform, and benzene. It is moderately soluble in aliphatic hydrocarbons and dissolves many organic and inorganic compounds.

Reactions. Although usually a stable and unreactive solvent, 1-methyl-2-pyrrolidinone can undergo a number of characteristic chemical reactions. In particular, these involve ring opening, attack of the carbonyl group, or replacement of hydrogens alpha to the carbonyl group. Although it is very resistant to hydrolysis under neutral conditions, with strong acids or bases 1-methyl-2-pyrrolidinone can be hydrolyzed

Table 3. Properties of 1-Methyl-2-Pyrrolidinone

Property	Value
freezing point, °C	−24.4
boiling point, °C	
at 0.133 kPa[a]	41
at 1.33 kPa[a]	79
at 13.3 kPa[a]	136
at 101.3 kPa[a]	202
density, g/cm^3, d_4^{25}	1.028
refractive index, n_D^{25}	1.4690
viscosity at 25°C, mPa·s (= cP)	1.65
flash point (open-cup), °C	95
solubility parameter, δ	11

[a] To convert kPa to mm Hg, multiply by 7.5.

to 4-methylaminobutyric acid (**53**).

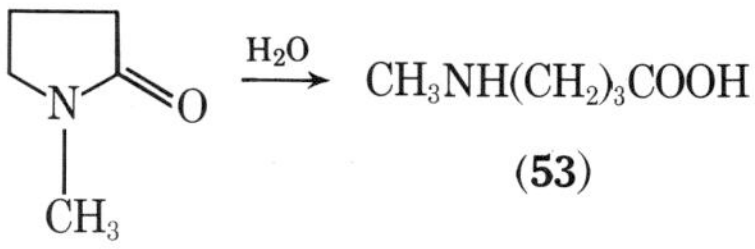

$$\underset{\underset{\text{CH}_3}{|}}{\text{pyrrolidinone}} \xrightarrow{\text{H}_2\text{O}} \text{CH}_3\text{NH}(\text{CH}_2)_3\text{COOH}$$
(**53**)

(**52**) [872-50-4]

Borohydride reduction under suitable conditions yields 1-methylpyrrolidine (**54**) (78).

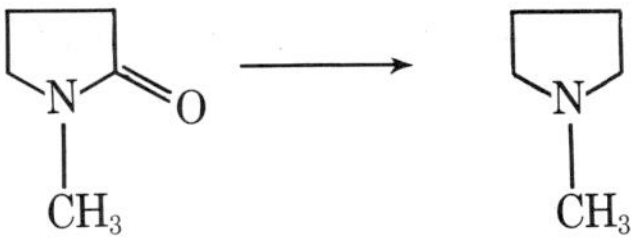

(**52**) → (**54**) [120-94-5]

1-Methyl-2-pyrrolidinone reacts with chlorinating agents, eg, COCl$_2$, SOCl$_2$, POCl$_3$, PCl$_5$, etc, to form the amide (79). The amide, in turn, reacts with a variety of substituents (80–82):

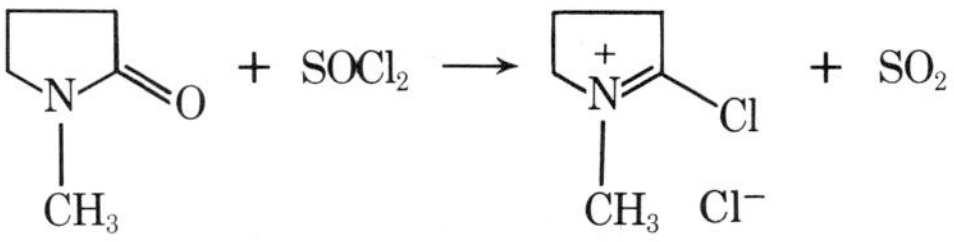

(**52**) + SOCl$_2$ → (**55**) [15862-82-5] + SO$_2$

(**55**) + RONa → [pyrrolidine with OR, OR substituents]

(**55**) + NH$_2$OH → (**56**) [35197-40-1]

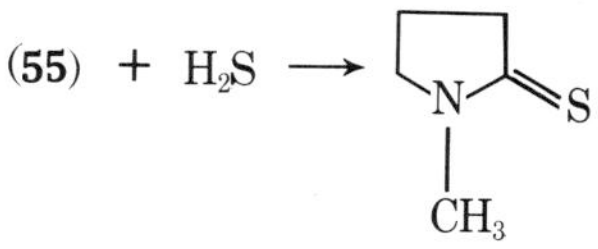

(**55**) + H$_2$S → (**57**) [10441-57-3]

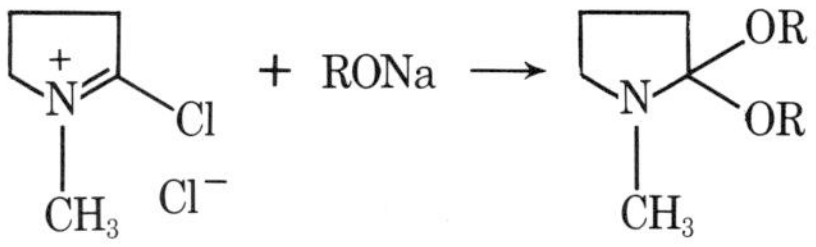

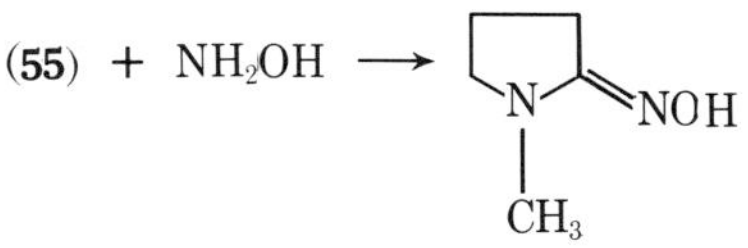

(**55**) + RNH$_2$ → [pyrrolidine with =NR]

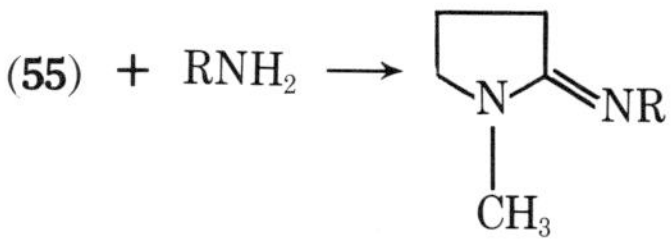

1-Methyl-2-thiopyrrolidinone (**57**) can also be prepared by reaction of 1-methyl-2-pyrrolidinone with sulfur or carbon disulfide at high temperatures and pressures (83–84).

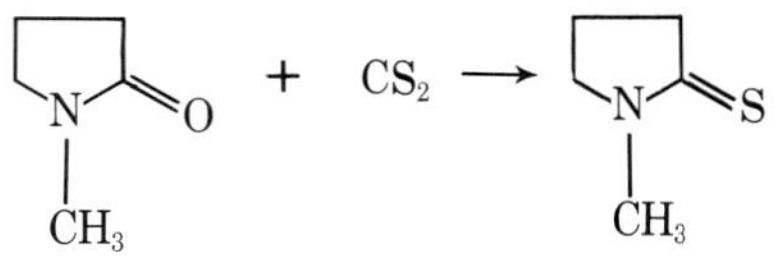

(**52**) + CS$_2$ → (**57**)

With anionic catalysts, one or both of the hydrogen atoms alpha to the carbonyl group can be added across styrene (85).

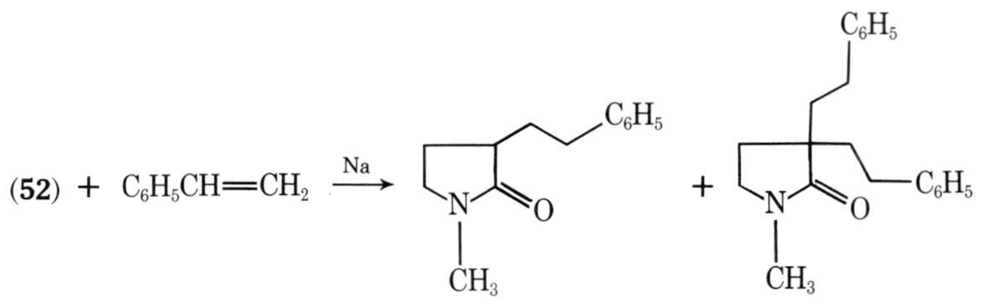

(**52**) + C$_6$H$_5$CH=CH$_2$ $\xrightarrow{\text{Na}}$ (**58**) [21053-47-4] + (**59**) [21053-48-5]

At low temperatures, 1-methyl-2-pyrrolidinone can be nitrated by amyl nitrate and potassium t-butoxide (86).

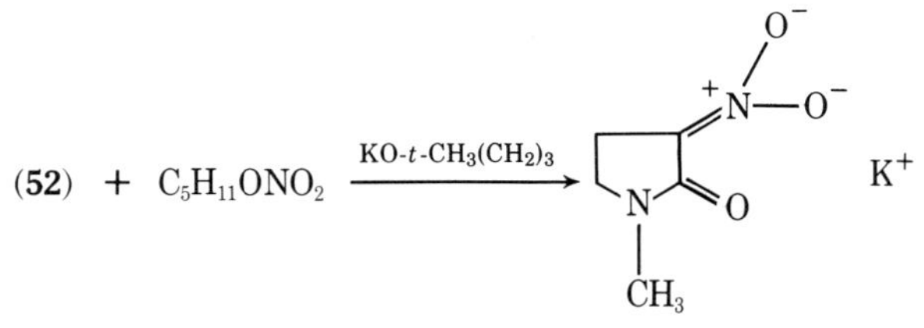

(**52**) + C$_5$H$_{11}$ONO$_2$ $\xrightarrow{\text{KO-}t\text{-CH}_3(\text{CH}_2)_3}$ (**60**) [80294-30-0] K$^+$

Oxalic esters alkylate 1-methyl-2-pyrrolidinone in the 3-position. The resulting ketoesters can be treated with aldehydes to give olefins (87).

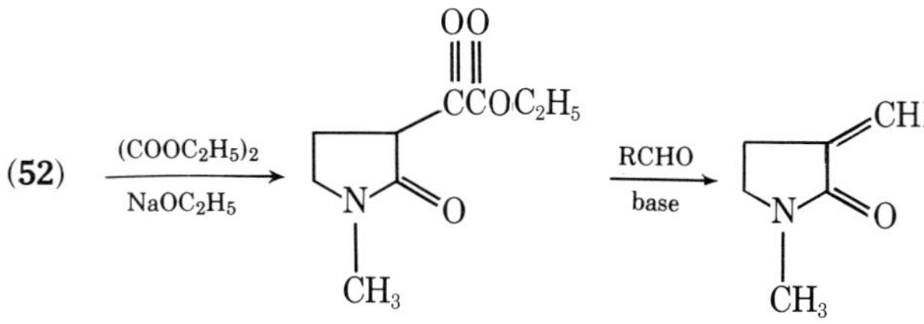

(**52**) $\xrightarrow[\text{NaOC}_2\text{H}_5]{(\text{COOC}_2\text{H}_5)_2}$ (**61**) [36821-26-8] $\xrightarrow[\text{base}]{\text{RCHO}}$

Manufacture. 1-Methyl-2-pyrrolidinone (**52**) is manufactured at the same sites and by essentially the same process as described for 2-pyrrolidinone (**34**). In general, capacity for the two materials involves the same equipment with only moderate differences in auxiliaries.

$$\text{(50)} + CH_3NH_2 \rightleftharpoons HO(CH_2)_3CONHCH_3 \rightarrow \text{(52)} + H_2O$$

Shipment, Storage, and Price. 1-Methyl-2-pyrrolidinone is available in tank cars or tank trailers as well as in drums. Shipping containers are normally of unlined steel. Rubber hose is unsuitable for handling; standard steel pipe or braided steel hose is acceptable. Ordinarily 1020 carbon steel (0550) is satisfactory as a storage material. Stainless-steel 304 and 316, nickel, and aluminum are also suitable. Methylpyrrolidone is hygroscopic and must be protected from atmospheric moisture. In February 1981, the U.S. price for tank car quantities was $2.43/kg.

Specifications and Analytical Methods. The purity of 1-methyl-2-pyrrolidinone is determined by gas chromatography and is specified as 99.5 wt % minimum. Maximum moisture content is specified as 0.05 wt % by ir spectroscopy.

Health and Safety Factors. 1-Methyl-2-pyrrolidinone is less toxic than many competitive dipolar aprotic solvents. The LD_{50} for white rats is 4.2 mL/kg. Although it does not appear to be a sensitizing agent, prolonged skin contact should be avoided. It is a severe eye irritant.

Uses. 1-Methyl-2-pyrrolidinone is a dipolar aprotic solvent. It has a high dielectric constant and cannot donate protons for hydrogen bonding. All of its commercial uses involve its strong and frequently selective solvency. In recent years it has replaced other solvents of poorer stability, higher vapor pressures, greater toxicities, or more facile skin penetration.

The largest use of NMP is in extraction of aromatics from lube oils. In this application, it has been replacing phenol and, to some extent, furfural. Other petrochemical uses involve separation and recovery of aromatics from mixed feedstocks; recovery and purification of acetylenes, olefins, and diolefins; removal of sulfur compounds from natural and refinery gases; and dehydration of natural gas.

Large amounts of NMP are consumed in the polymer industry as a medium for polymerization and as a solvent for finished polymers. Polymers that are soluble in NMP are poly(vinyl acetate), poly(vinyl fluoride), polystyrene, nylon and aromatic polyamides and polyimides (qv), polyesters (qv), acrylics, polycarbonates (qv), cellulose derivatives, and synthetic elastomers (see Vinyl polymers; Polyamides; Polyimides;

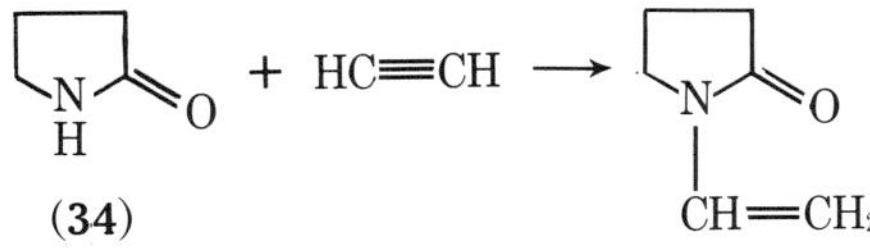

$$\text{(34)} + HC\equiv CH \rightarrow \text{(62)}\ [88\text{-}12\text{-}0]$$

Table 4. N-Substituted 2-Pyrrolidinones

N-Substituent	CAS Registry No.	Bp at 670 Pa (5 mm Hg), °C
ethyl	[2687-91-4]	85
isopropyl	[3772-26-7]	84
cyclohexyl	[6837-24-7]	138
fatty alkyls[a]		190–255
2-hydroxyethyl	[3445-11-2]	154
N-dimethylaminopropyl	[7375-15-7]	132

[a] Two blends offered; coco and tallow.

Cellulose; Cellulose derivatives; Elastomers, synthetic). 1-Methyl-2-pyrrolidinone is also useful for stripping potting resins, eg, epoxies.

1-Vinyl-2-pyrrolidinone. 1-Vinyl-2-pyrrolidinone (VP) (1-ethenyl-2-pyrrolidinone, N-vinyl-2-pyrrolidone, and V-Pyrol) is manufactured by GAF in the United States and by BASF in Germany by vinylation of 2-pyrrolidinone with acetylene (see Acetylenic-derived chemicals; Vinyl polymers, N-vinyl polymers).

Other Pyrrolidinones. A number of other N-substituted 2-pyrrolidinones have been promoted as developmental products. These materials offer different and sometimes unique solvency properties. All are prepared by reaction of butyrolactone with suitable primary amines. Principal examples are listed in Table 4.

BIBLIOGRAPHY

"Pyrrole and Pyrrole Derivatives" in *ECT* 1st ed., Vol. 11, pp. 339–353, by W. R. Vaughan, University of Michigan; "Pyrrole and Pyrrole Derivatives" in *ECT* 2nd ed., Vol. 16, pp. 841–858, by E. V. Hort and R. F. Smith, GAF Corporation.

1. R. Runge, *Ann. Physik* **31,** 67 (1834).
2. T. Anderson, *Ann.* **105,** 349 (1858).
3. A. Baeyer and H. Emmerling, *Ber.* **3,** 517 (1870).
4. *The Beilstein Handbook of Organic Chemistry*, Vols. 21–22, Beilstein Institute, Frankfurt/Main, FRG, 1978.
5. R. Alan Jones in A. R. Katrizky and A. J. Boulton, eds., *Advances in Heterocyclic Chemistry*, Vol. 11, Academic Press, Inc., New York, 1970, p. 386.
6. B. Bak, *Acta Chem. Scand.* **9,** 1355 (1955).
7. H. Fischer and H. Orth, *Die Chemie des Pyrrols*, Vol. 1, Akad, Verlagsges, Leipzig, FRG, 1934.
8. L-B. Agenas and B. Lindgren, *Arkiv Kemi* **28,** 145 (1968).
9. A. Marinangelli, *Ann. Chim. Rome* **44,** 211 (1954).
10. M. L. Josien, M. Paty, and P. Pineau, *J. Chem. Phys.* **24,** 126 (1956).
11. G. Yagil, *Tetrahedron* **23,** 2855 (1967).
12. A. Gossauer, *Die Chemie der Pyrrole*, Springer-Verlag, Berlin, FRG, 1974, p. 130.
13. Ref. 12, pp. 60–77.
14. L. Janelli and P. G. Orsini, *Gazz. Chim. Ital.* **89,** 1467 (1959).
15. H. Kofod, L. E. Sutton, and J. Jackson, *J. Chem. Soc.*, 1467 (1952).
16. A. Albert, *Heterocyclic Chemistry*, University of London, London, UK, 1959.
17. J. A. Elvidge and L. M. Jackman, *J. Chem. Soc.*, 859 (1961); J. Elvidge, *Chem. Comm.*, 160 (1965).
18. R. J. Abraham and W. A. Thomas, *J. Chem. Soc. Part B*, 127 (1966).
19. R. A. Jones, T. Mc. L. Spotswood, and P. Chenychit, *Tetrahedron* **23,** 4469 (1967).
20. L. Knorr and H. Lange, *Ber.* **35,** 2998 (1902); H. Fisher, *Org. Syn.* **15,** 17 (1935); **17,** 96 (1937).
21. H. Fisher and E. Fink, *Z. Physiol. Chem.* **280,** 123 (1944); **283,** 152 (1948).
22. A. Hantzsch, *Ber.* **23,** 1474 (1890).

23. M. Feist and E. Stenger, *Ber.* **35**, 1558 (1902).
24. H. S. Broadbent, W. S. Burnham, R. K. Olsen, and R. M. Shelley, *J. Heterocycl. Chem.* **5**, 757 (1968).
25. Yu. K. Yurev, I. K. Karobitsyma, R. D. Ben-Yarik, L. A. Savina, and P. A. Akeshin, *Vestn. Mosk. Univ.* **6**(2), *Ser. Fiz. Mat. Estestv. Nauk* (1), 37 (1951); *Chem. Abstr.* **47**, 124 (1953).
26. E. H. Huntress, T. E. Lesslie, and W. M. Hearon, *J. Am. Chem. Soc.* **78**, 419 (1956).
27. J. B. Hendrickson and R. Rees, *J. Am. Chem. Soc.* **83**, 1250 (1961).
28. Yu. K. Yurev and I. S. Levi, *Vestn. Mosk. Univ. Ser. Mat. Mekh. Astron. Fiz. Khim* **11**(2), 153 (1956); *Chem. Abstr.* **52**, 355 (1958).
29. S. Clementi and G. Marino, *Tetrahedron* **25**, 4599 (1969).
30. P. Rothemund, *J. Am. Chem. Soc.* **58**, 625 (1936).
31. W. C. Anthony, *J. Org. Chem.* **25**, 2049 (1960); P. A. Burbridge, G. L. Collier, A. H. Jackson, and G. W. Kenner, *J. Chem. Soc. Part B*, 930 (1967).
32. A. R. Cooksey, K. J. Morgan, and D. P. Morrey, *Tetrahedron* **26**, 5101 (1970).
33. A. Treibs and G. Fritz, *Ann.* **611**, 162 (1958).
34. A. H. Corwin in R. C. Elderfield, ed., *Heterocyclic Compounds*, Vol. 1, John Wiley & Sons Inc., New York, 1950, pp. 277–342.
35. J. H. Atkinson, R. S. Atkinson, and A. W. Johnson, *J. Chem. Soc.*, 5999 (1964).
36. I. Nabih and E. Helmy, *J. Pharm. Soc.* **56**, 649 (1967).
37. S. P. Findlay, *J. Org. Chem.* **21**, 644 (1956).
38. J. Hine and J. M. Van der Veen, *J. Am. Chem. Soc.* **81**, 6446 (1959).
39. R. Chong and P. S. Clezy, *Aust. J. Chem.* **20**, 935 (1967).
40. H. Rapaport and G. Castagnoli, Jr., *J. Am. Chem. Soc.* **84**, 2178 (1962); H. Rapaport and J. Bordmer, *J. Org. Chem.* **29**, 2727 (1964).
41. D. Dolphin, R. Grigg, A. W. Johnson, and J. Leng, *J. Chem. Soc.*, 1460 (1965).
42. A. W. Johnson, I. T. Kay, E. Markham, R. Price, and K. B. Shaw, *J. Chem. Soc.*, 3416 (1959); J. Ellis, A. H. Jackson, A. C. Vain, and G. W. Kenner, *J. Chem. Soc.*, 1935 (1964).
43. A. F. Diaz, K. K. Kanazawa, and G. P. Gardini, *J. Chem. Soc. Chem. Comm.*, 635 (1979).
44. K. K. Kanazawa, A. F. Diaz, R. H. Geiss, W. D. Gill, J. F. Kwak, J. A. Logan, J. F. Rabalt, and G. B. Street, *J. Chem. Soc. Chem. Comm*, 854 (1979).
45. S. Gabriel, *Ber.* **22**, 3335 (1889).
46. E. Späth and J. Lintner, *Ber.* **69**, 2727 (1936).
47. U.S. Pat. 4,110,340 (Aug. 29, 1978), J. F. Knifton (to Texaco Inc.).
48. U.S. Pat. 4,111,952 (Sept. 5, 1978), J. F. Knifton (to Texaco Inc.).
49. U.S. Pat. 4,036,836 (July 19, 1977), J. L. Green (to Standard Oil Co., Ohio).
50. U.S. Pat. 4,181,662 (Jan. 1, 1980), W. A. Sweeney (to Chevron Research Co.).
51. U.S. Pat. 3,812,148 (May 21, 1974), E. J. Hollstein and W. A. Butte, Jr. (to Sun Research and Development Co.).
52. U.S. Pat. 3,812,149 (May 21, 1974), E. J. Hollstein (to Sun Research and Development Co.).
53. U.S. Pat. 3,884,936 (May 20, 1975), E. J. Hollstein (to Sun Research and Development Co.).
54. U.S. Pat. 2,638,463 (May 12, 1953), W. O. Ney, Jr., W. R. Nummy, and C. E. Barnes (to GAF Corp.).
55. U.S. Pat. 2,739,959 (March 27, 1956), W. O. Ney, Jr., and M. Crowther (to Arnold, Hoffman and Co.).
56. Span. Pat. 398,322 (Sept. 16, 1974), S. A. Hebron; *Chem. Abstr.* **83**, 43755 (1975).
57. J. Tafel and M. Stern, *Ber.* **33**, 2228 (1900).
58. U.S. Pat. 3,850,920 (Nov. 26, 1974), W. E. Walles (to The Dow Chemical Co.).
59. T. Tafel and O. Wassmuth, *Ber.* **40**, 2835 (1907).
60. W. Gaffield, L. K. Keefer, and P. P. Roller, *Org. Prep. Proced. Int.* **9**(2), 49 (1977).
61. J. Palecek and J. Kuthan, *Z. Chem.* **17**(7), 260 (1977).
62. Jpn. Kokai 74 117,459 (Nov. 9, 1974), T. Ayusawa and S. Fukami (to Mitsubishi Petroleum Co.).
63. Jpn. Kokai 76 16,657 (Aug. 1, 1974), S. Enomoto and Y. Takahashi (to Asahi Chemical Industry Co.).
64. U.S. Pat. 3,073,843 (Jan. 15, 1963), S. R. Buc (to GAF Corp.).
65. Span. Pat. 447,346 (Oct. 16, 1977), A. Surroca and J. José (to Farma-Lepori S.A.); *Chem. Abstr.* **88**, 169962 (1978).
66. Jpn. Pat. 73 26,753 (Aug. 15, 1973), S. Asai and S. Nomura (to Kao Soap. Co.).
67. A. T. Malkhasyan, G. G. Sukiasyan, S. G. Matinyan, and G. T. Martirosyan, *Arm. Khim Zh.* **29**(5), 458 (1976); *Chem. Abstr.* **85**, 142966 (1976).

68. T. Kamiyama, M. Inoue, and S. Enomoto, *Yuki Gosei Kagaku Kyokaishi* **36**(1), 65 (1978).
69. G. Dannhardt, *Arch. Pharm. Weinheim, Ger.* **311**(4), 294 (1978).
70. V. G. Granik, A. M. Zhedkova, N. S. Kuryatov, V. P. Pakhomov, and R. G. Glushkov, *Khim. Geterotsikl. Soedin*, (11), 1132 (1973); *Chem. Abstr.* **80**, 82016 (1974).
71. U.S. Pat. 3,816,454 (June 11, 1974), Y.-H. Wu and W. G. Lobeck (to Mead Johnson and Co.).
72. U.S. Pat. 3,706,766 (Dec. 19, 1972), F. M. Hershenson (to G. D. Searle and Co.).
73. J. W. Copenhaver and M. H. Bigelow, *Acetylene and Carbon Monoxide Chemistry*, Reinhold, New York, 1949, pp. 163–164.
74. *2-Pyrrole Technical Bulletin 7543-120*, rev. 1, GAF Corp., New York.
75. J. Tafel and O. Wassmuth, *Ber.* **40**, 2839 (1907).
76. U.S. Pat. 4,152,331 (May 5, 1979), P. J. N. Meijer and L. H. Geurtz (to Stamicarbon BV).
77. U.S. Pat. 3,448,118 (June 3, 1964), G. Chickery, P. Benite, and P. Perras (to Rhone-Poulenc SA).
78. A. Basha, J. Orlando, and S. M. Weinreb, *Syn. Commun.* **7**(8), 549 (1977).
79. H. Eilingsfeld, M. Seefelder, and H. Weidinger, *Angew. Chem.* **78**, 836 (1960).
80. H. Eilingsfeld, M. Seefelder, and H. Weidinger, *Chem. Ber.* **96**, 2671 (1963).
81. U.S. Pat. 3,787,576 (Jan. 22, 1974), E. Enders and W. Stendel (to Bayer A.-G.).
82. T. Jen, H. Van Hoeven, W. Groves, R. A. McLean, and B. Loev, *J. Med. Chem.* **18**(1), 90 (1975).
83. U.S. Pat. 3,306,911 (Feb. 28, 1967), R. C. Doss (to Phillips Petroleum).
84. J. C. Meslin and G. Duguay, *Bull. Soc. Chim. Fr.* (7-8 Pt. 2), 1200 (1976).
85. G. G. Sukiasyan and A. T. Malkhasyan, *Tezisy Dokl. Molvdezhnaya Konf. Org. Sint. Bioorg. Khim.*, pp. 47–48, 1976; *Chem. Abstr.* **88**, 169867 (1978).
86. H. Feuer, L. R. Blecker, R. W. Jans Jr., and J. W. Frost, *J. Heterocycl. Chem.* **16**(3), 481 (1979).
87. G. M. Ksander, J. E. McMurray, and M. Johnson, *J. Org. Chem.* **42**(7), 1180 (1977).

General References

References 4, 5, and 34 are general references.
E. Baltazzi, *Chem. Rev.* **63**, 511 (1963).
J. M. Patterson, *Synthesis*, 281 (1976).
A. H. Jackson, *Compr. Org. Chem.* **4**, 275 (1979).
M-Pyrol, N-Methyl-2-Pyrrolidone Handbook, GAF Corp., 1972.

EUGENE V. HORT
L. R. ANDERSON
GAF Corporation

Q

QUATERNARY AMMONIUM COMPOUNDS

Quaternary ammonium compounds are usually tetrasubstituted ammonium salts. Originally, it was considered that the R groups were only hydrocarbon radicals attached to the nitrogen by a C–N bond, but now a large variety of substituents can be used, eg, nitrogen or oxygen atoms. The alkyl radicals may be substituted or unsubstituted, saturated or unsaturated, aliphatic or aromatic, or branched or normal chains. Also, there may be a great variety of substituents on the R group. In all cases, the nitrogen atom is pentavalent and is in the positively charged portion of the molecule. Thus, quaternary ammonium salts or hydroxides are cationic electrolytes. Originally, reaction products of tertiary amines with alkyl halides were thought to be complexes formed by simple addition of the two reagents. Later work showed their true structure and led to the elucidation of the various valence states of nitrogen derivatives. One of the first quaternary ammonium compounds prepared was ethyltrimethylammonium iodide [51-93-4] from trimethylamine and ethyl iodide (1). It can also be prepared from dimethylethylamine and methyl iodide. The product is the same regardless of which reactants are used. Later, the tetrahedral configuration was proved by a synthesis of an optically active quaternary ammonium salt in which there were four different substituents: methyl, alkyl, phenyl, and benzyl (1). However, the resolution of optically active quaternary ammonium salts is difficult because of the ease with which they undergo dissociation and racemization at room temperature (see Amines).

Nomenclature

The quaternary ammonium salts are usually named as substituted nitrogen compounds. The expressions pyridinium and quinolinium are used for the corresponding quaternaries, for example,

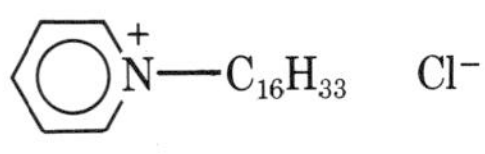

$$\text{C}_5\text{H}_5\overset{+}{\text{N}}-\text{C}_{16}\text{H}_{33}\quad \text{Cl}^-$$

is hexadecylpyridinium chloride [140-72-7]. The substituents may be cited either in alphabetical order and without prefixes (eg, di- and tri-) or in order of increasing complexity. However, the *Chemical Abstracts* system is based on alphabetical order, since the increasing-complexity rule can become exceedingly complicated and difficult to apply. Thus, the expressions methyl(ethyl)propyl(butyl)ammonium chloride and butyl(ethyl)methyl(propyl)ammonium chloride are both satisfactory.

Selected quaternary ammonium compounds and some economic data are listed in Table 1.

Properties

Physical. The structure of the quaternary ammonium compound (quaternary) determines the physical properties of the material. The lowest molecular weight quaternary, ie, tetramethylammonium chloride [75-57-0], is very soluble in water and insoluble in nonpolar solvents such as ether, benzene, and aliphatic compounds. As the molecular weight of the quaternary increases, its solubility in polar solvents, including water, decreases and its solubility in nonpolar solvents increases. The introduction of hydroxy-substituted groups in the quaternary structure increases the solubility of the compound in polar solvents. Specific data are given in references 2–4. For example, trimethyloctadecylammonium chloride dissolves in water only up to ca 27 wt %. Further increases in molecular weight of the alkyl units further decrease the solubility of the salt in water. Thus, dimethyldioctadecylammonium chloride [107-64-2] has very limited solubility in water. In this case, dispersions in water are possible and are the form in which most compounds of this type are used.

Quaternary ammonium compounds have indefinite melting points. Those that form crystalline solids decompose on heating, and the amount of water present and rate of heating determine the decomposition points. Some quaternaries form hydrates or other solvates, which in some instances can be crystallized.

Quaternary ammonium salts in which one of the alkyl groups contains twelve or more carbon atoms are often referred to as invert soaps, because the lipophilic portion of the molecule is cationic instead of anionic, as in sodium stearate. Figure 1 is an example of how this type of surfactant behaves in water (5). The diagram has been simplified by the omission of the curves representing an unstable polymorph. These salts are surface-active colloidal electrolytes that form aggregates in solution. The aggregates or micelles form as a result of the hydrophilic nature of the ammonium group and the hydrophobic nature of the hydrocarbon chain. The micelle is assumed to be made up of molecules with polar groups oriented toward the surrounding aqueous solution. The concentration at which micelle formation begins is the critical micelle

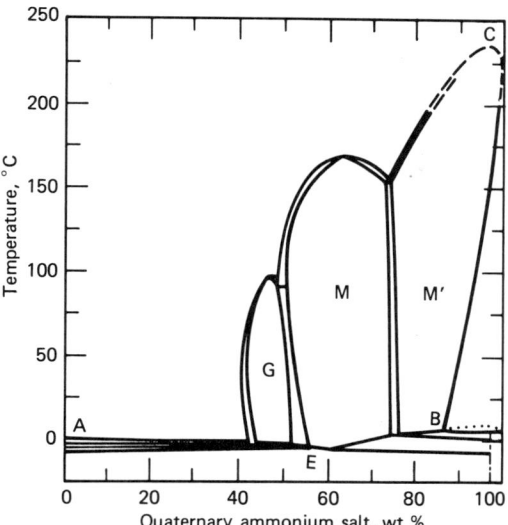

Figure 1. The system dodecyltrimethylammonium chloride–water: A, freezing point of water; E, eutectic of water plus hydrate of the quaternary ammonium salt; B, incongruent melting point of the hydrate; C, melting point of the anhydrous salt; S, region of isotropic liquid; G, region of isotropic gel; M and M′, liquid crystalline phases.

concentration and is a measure of the tendency for the specific compound to leave solution. This tendency is an important factor in surface activity. Most of the uses of quaternary ammonium compounds are based on the lipophilic portion of the molecule being positively charged. Most surfaces are negatively charged, and solutions or emulsions made from these cationic surface-active agents are readily adsorbed to the negatively charged surfaces, which causes reaction, breaking of emulsions, and protection of these surfaces (see Surfactants and detersive systems).

Biological. One of the most important uses for quaternary ammonium compounds depends on their biological activity. The most comprehensive screening of biocidal activity of nitrogen derivatives is given in ref. 6. Generally, optimum activity of completely aliphatic compounds is achieved if the higher aliphatic group contains a normal chain of 16–18 carbon atoms. The bactericidal activity of benzyl quaternary compounds is optimum if the higher aliphatic chain contains 14 carbon atoms. The anion has little influence except on solubility (see also Disinfectants and antiseptics).

The mechanism of the bactericidal action is closely related to the surface activity of the quaternary ammonium compound. Undoubtedly, interaction of the bactericidal agent with the cell wall interferes with the metabolic process of the organism, and this causes the inhibiting or killing action. Results from pharmacological and toxicological studies of certain higher aliphatic quaternary ammonium compounds indicate that these compounds are innocuous in concentrations required for germicidal effectiveness. In all cases, the biocidal quaternaries are sold as solutions in isopropanol–water or ethanol–water.

Quaternary ammonium compounds play an important part in biological functions. The vitamin B complex contains two components that have the quaternary nitrogen atom (see Vitamins, vitamin B). Vitamin B_1 [67-03-8] (thiamine) is a heterocyclic compound containing a thiozolium ring and is part of the enzyme carboxylase which

Table 1. Selected Quaternary Ammonium Compounds

Compound	CAS Registry No.	Trade name	Manufacturer	Form	Bulk price, $/kg
trimethyltallowammonium chloride	[8030-78-2]	Arquad T-50	Armak	liquid	1.87
		Adogen 471	Sherex		
trimethylsoyaammonium chloride	[61790-41-8]	Arquad S-50	Armak	liquid	2.07
		Adogen 415	Sherex		
trimethylcocoammonium chloride	[61789-18-2]	Arquad C-33	Armak	liquid	
			Sherex		
		Arquad C-50	Armak	liquid	2.44
		Adogen 461	Sherex		
dimethyldicocoammonium chloride	[61789-77-3]	Arquad 2C-75	Armak	liquid	3.10
		Adogen 462	Sherex		
dimethyldi(hydrogenated tallow)ammonium chloride	[61789-80-8]	Arquad 2HT-75	Armak	solid	1.85
		Adogen 442	Sherex		
trimethyldodecylammonium chloride	[112-00-5]	Arquad 12-33	Armak	liquid	3.48
		Arquad 12-50	Armak	liquid	
trimethyloctadecylammonium chloride	[112-03-8]	Arquad 18-50	Armak	liquid	2.64
trimethylhexadecylammonium chloride	[112-02-7]	Arquad 16-29	Armak	liquid	1.36
		Variquat 228	Sherex		
		Arquad 16-50	Armak	liquid	2.22
dimethylalkylbenzylammonium chloride R = n-$C_{12}H_{25}$	[139-07-1]	Arquad B-100	Lonza	liquid	3.12
n-$C_{14}H_{29}$	[139-08-2]		Armak		
n-$C_{16}H_{33}$	[122-18-9]		Sherex		
n-$C_{18}H_{37}$	[122-19-0]		Mason		
1:1 mixture of trimethyltallowammonium chloride and dimethyldicocoammonium chloride	[8030-78-2] [61789-77-3]	Arquad T-2C-50	Armak	liquid	3.00
N,N,N',N'-pentamethyl-N-tallow-1,3-propanediammonium dichloride	[68607-29-4]	Duoquad T-50	Armak	liquid	2.40
		Adogen 477	Sherex		
methylbis(2-hydroxyethyl)cocoammonium chloride	[58449-53-9]	Ethoquad C/12	Armak	liquid	3.23
		Variquat 638	Sherex		
methylpolyoxyethylene (15) cocoammonium chloride	[61791-10-4]	Ethoquad C/25	Armak	liquid	2.75
		Variquat K215	Sherex		

Name	CAS	Product	Company	Form	Value
methylbis(2-hydroxyethyl)oleylammonium chloride	[55957-12-5]	Ethoquad O/12	Armak	liquid	2.64
methylpolyoxyethylene (15) oleylammonium chloride	[37208-39-2]	Ethoquad O/25	Armak	paste	2.90
methylbis(2-hydroxyethyl)octadecylammonium chloride	[28724-32-5]	Ethoquad 18/12	Armak	liquid	3.09
methylpolyoxyethylene (15) octadecylammonium chloride	[28724-32-5]	Ethoquad 18/25	Armak	liquid	2.77
n-dodecyl (61%) tetradecyl (23%) dimethylbenzylammonium chloride	[139-07-1] [139-08-2]	Maquat LC-125	Mason	liquid	2.11
n-tetradecyl (60%) hexadecyl (30%) dimethylbenzylammonium chloride	[139-08-2] [122-18-9]	MC-1416	Mason	liquid	3.32
n-dodecyl (40%) tetradecyl (50%) dimethylbenzylammonium chloride	[139-07-1] [139-08-2]	MC-1412	Mason	liquid	3.32
n-dodecyl (61%) tetradecyl (23%) dimethyldichlorobenzylammonium chloride	[33377-87-6] [68568-47-8]	DLC-1214	Mason	liquid	5.68
n-octadecyldimethylbenzylammonium chloride	[122-19-0]	SC-18 Varisoft FDC	Mason Sherex		1.16
42% n-tetradecyl (40%) hexadecyl (60%) dimethylbenzylammonium chloride	[139-08-2] [122-18-9]	TC-76	Mason	liquid	2.00
8% dialkylmethylbenzylammonium chloride n-dodecyl (35%) tetradecyl (5%) hexadecyl (60%) dimethylbenzylammonium chloride	[139-07-1] [139-08-2] [122-18-9]	MQ-2525	Mason	liquid	3.32
n-dodecyl (20%) tetradecyl (50%) hexadecyl (30%) dimethylethylbenzylammonium chloride	[27479-28-3] [27479-29-4] [29656-52-8]				
methyl sulfate quaternary of ethoxylated tallow diethylenetriamine condensate	[68410-69-5]		Capital City Products Sherex	viscous liquid	1.21
methyl sulfate quaternary of propoxylated tallow diethylenetriamine condensate	[68413-04-5]		Capital City Products Sherex	liquid	1.21
1-(tallow amidoethylene)-2-nor (tallow alkyl)-2-imidazolinium, methyl sulfate quaternary	[68122-86-1]		Capital City Products Sherex	liquid	1.21

participates in carbohydrate metabolism. Choline [62-49-7] (qv), $(CH_3)_3\overset{+}{N}CH_2$-$CH_2OH^-OH$, trimethyl(2-hydroxyethyl)ammonium hydroxide, functions in fat metabolism in transmethylation reactions. Acetylcholine [60-31-1] is involved in the transmission of nerve impulses (see Cholinesterase inhibitors).

Curare action, muscarinic–nicotinic action, and ganglia-blocking action are three general types of physiological actions attributed to quaternary ammonium compounds. The active substance in curare is capable of producing muscular paralysis without affecting the central nervous system or the heart; thus, curare has been used to induce muscular relaxation during surgery. A large number of similar quaternary ammonium compounds have been synthesized in attempts to optimize properties. Direct stimulation of smooth muscles is called muscarinic action, and primary transient stimulation and secondary persistent depression of sympathetic and parasympathetic ganglia is called nicotinic action. Many quaternary ammonium compounds cause these actions in varying degrees.

Preparation

The methods of preparation of quaternary ammonium compounds are many and varied, depending on the structure of the final compound. The most convenient reaction is one in which the suitable tertiary amine reacts with an alkylating agent, which can be an alkyl ester.

$$\underset{\underset{R''}{|}}{\overset{\overset{R'}{|}}{R N}} + R'''X \longrightarrow \underset{\underset{R'''}{|}}{\overset{\overset{R'}{|+}}{R N R''}} + X^-$$

There are many variations in the final product because of the large number of diverse starting amines and alkylating agents. The main quaternary structures obtainable from industrially available alkylating agents are listed in Table 2.

The tertiary amines used commercially are derived from synthetic or natural raw materials. Although some of these amines are reported in ref. 7, the methods of

Table 2. Some Quaternary Structures Obtainable from Alkylating Agents

Tertiary amine	Alkylating agent	Final quaternary ammonium salt
R_3N	CH_3Cl	$R_3\overset{+}{N}CH_3\ Cl^-$
R_3N	$(CH_3)_2SO_4$	$R_3\overset{+}{N}CH_3\ CH_3SO_4^-$
R_3N	$(C_2H_5)_2SO_4$	$R_3\overset{+}{N}C_2H_5\ C_2H_5SO_4^-$
R_3N	⟨C₆H₅⟩—CH₂Cl	$R_3\overset{+}{N}CH_2$—⟨C₆H₅⟩ Cl^-

preparation of the most used tertiary amines are given below:

Dimethylalkylamines

$$RNH_2 + 2\ CH_2O + H_2 \xrightarrow{catalyst} RN(CH_3)_2 + 2\ H_2O$$

$$RCl + NH(CH_3)_2 \xrightarrow{NaOH} RN(CH_3)_2 + NaCl + H_2O$$

$$RBr + HN(CH_3)_2 \xrightarrow{NaOH} RN(CH_3)_2 + NaBr + H_2O$$

Diethoxylated monoalkylamines

$$RNH_2 + 2\ \underset{O}{CH_2\!-\!\!-\!CH_2} \longrightarrow RN(CH_2CH_2OH)_2$$

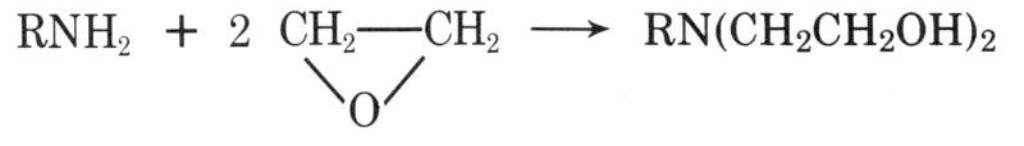

Methyldialkylamines

$$R_2NH + CH_2O + H_2 \xrightarrow{catalyst} R_2NCH_3$$

Ethoxylated dialkylamines

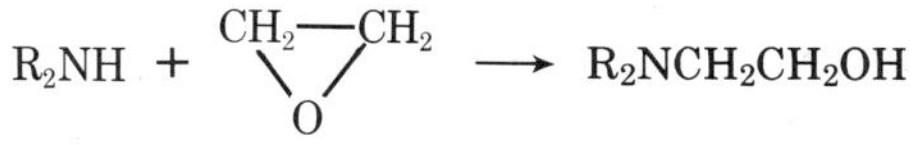

$$R_2NH + \underset{O}{CH_2\!-\!\!-\!CH_2} \longrightarrow R_2NCH_2CH_2OH$$

Propoxylated monoalkylamines

$$RNH_2 + 2\ \underset{O}{CH_3CH\!-\!\!-\!CH_2} \longrightarrow RN(CH_2CHOHCH_3)_2$$

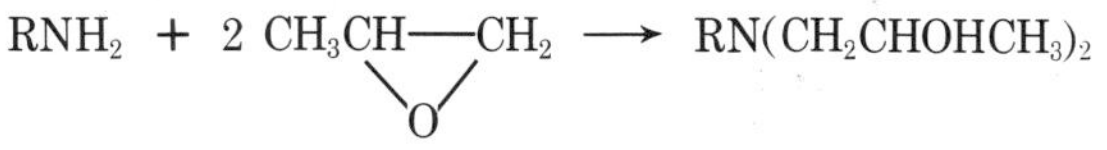

Ethoxylated fatty amidoamines

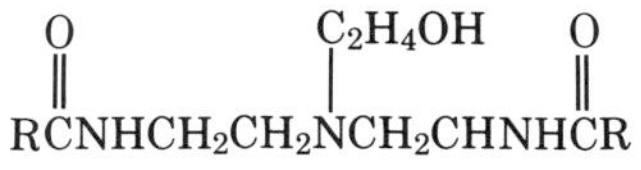

$$\underset{\underset{RCNHCH_2CH_2NCH_2CHNHCR}{}}{\overset{O\qquad\quad C_2H_4OH\quad O}{\|\qquad\qquad\quad|\qquad\quad\ \|}}$$

Imidazolines

$$2\ RCO_2H + H_2NCH_2CH_2NHCH_2CHNH_2 \xrightarrow{-2\ H_2O}$$

$$\underset{RCNHCH_2CH_2NHCH_2CH_2NHCR}{\overset{O\qquad\qquad\qquad\qquad O}{\|\qquad\qquad\qquad\qquad\ \|}} \xrightarrow{-H_2O} R\!-\!\!\!\left[\begin{array}{c}\text{imidazoline ring}\\ CH_2CH_2NHCR\\ \|\\ O\end{array}\right]$$

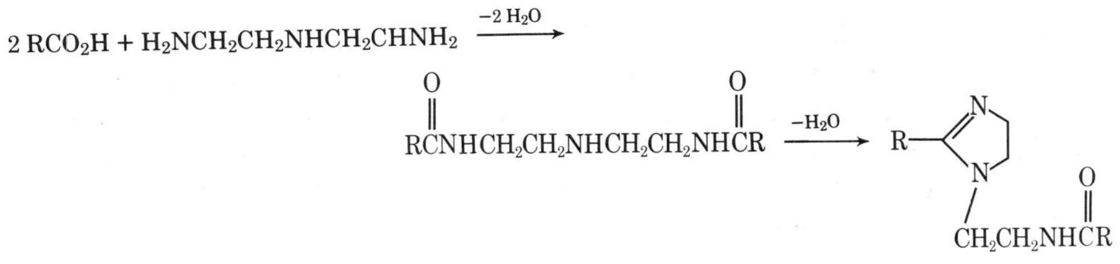

The preceding list of tertiary amines probably accounts for at least 95% of those marketed.

Primary or secondary amines can also be used in the preparation of quaternary ammonium compounds. In these cases, a basic material must be used to neutralize the acidic by-product of the reaction.

$$RNH_2 + 3\ CH_3Cl + 2\ NaOH \rightarrow R\overset{+}{N}(CH_3)_3\ Cl^- + 2\ NaCl + 2\ H_2O$$

$$R_2NH + 2\ CH_3Cl + NaOH \rightarrow R_2\overset{+}{N}(CH_3)_2\ Cl^- + NaCl + H_2O$$

Certain side reactions do occur. For example, some of the alkylating reagent reacts with the neutralizing agent; therefore, more than stoichiometric amounts of reactants are used. The difficulty in the manufacture of quaternary ammonium compounds from primary or secondary amines is the removal by filtration of salt from the finished product, which results in yield loss.

Quaternary ammonium compounds are usually prepared in stainless-steel or glass-lined equipment. The amine and solvent, eg, isopropyl alcohol, water, or both are loaded into the reactor and heated to the proper temperature (usually 80–100°C), and then the alkylating reagent is added. Quaternization of tertiary amines with alkyl halides is bimolecular. The rate of reaction is influenced by a number of factors, including basicity of the amine, steric effects, reactivity of the halide, and the polarity of the solvent. Polar solvents promote the reaction by stabilizing the ionic products. The following data illustrate the effect of solvent on the reaction between pyridine and ethyl iodide (7):

Solvent:	Benzene	Ethanol	Methanol	Acetone	Nitrobenzene
Relative rate:	1	1.4	2.5	12.8	25.0

In the case of methyl chloride, the system is under pressure. The reaction is exothermic and cooling is required. The resulting product, after reaction, is analyzed, the activity and pH is adjusted, and usually any insolubles are removed by filtration (8).

A unique type of quaternary ammonium compound can be prepared from the reaction of alkylene oxides with amine salts, in which the anion can be one of a variety of possibilities and can be chosen to be noncorrosive (9):

$$R\overset{+}{N}H_3\ X^- + 3\ \underset{O}{CH_2\!\!-\!\!CH_2} \rightarrow R\overset{+}{N}(CH_2CH_2OH)_3 X^-$$

Toxicity

Because of their biocidal, algicidal, and fungicidal properties, quaternary ammonium compounds are toxic to some sewage systems. However, the organisms in the sewage systems develop so that, unless the quaternary ammonium concentrations are too high, they can biodegrade most compounds. Seventeen methods of determining toxicity of quaternary compounds are given in ref. 10. The toxicities of selected compounds are listed in Table 3.

Uses

Quaternary ammonium compounds have scores of uses because of their affinity for negatively charged surfaces (11). Their single largest market is as fabric softeners, which accounts for 36,000 metric tons of 75% active material and is increasing at an annual rate of 4–5%. There are three types of commercial product. The original product is a 4–8 wt % dispersion of quaternary and is added to the rinse cycle of the washing process. The second product is a quaternary formulation applied to a nonwoven sheet

Table 3. Toxicity of Quaternary Ammonium Salts[a,b,c]

Compound	Acute oral toxicity, LD$_{50}$	Chronic oral toxicity, results	Skin, results	Irritation of eye mucous membrane, mg/kg body wt, results	Irritation of genital mucous membrane, mg/kg body wt, results
trimethyltallowammonium chloride	over 500 mg/kg over 600 mg/kg		1 wt % soln—skin irritation and tissue breakdown	1 wt % soln >30 0.1 wt % soln <30	
trimethylsoyaammonium chloride	800 mg/kg 300 mg/kg	4:100 dilution—no toxicity		10 wt % soln 14.3	10 wt % soln 0.66
trimethylcocoammonium chloride	1000 mg/kg 500 mg/kg			10 wt % soln 17.2	10 wt % soln 1.2
dimethyldicocoammonium chloride	1100 mg/kg 600 mg/kg	1:1000 dilution—no toxicity	2,5,10 wt % soln—mild irritation	2 wt % soln 16	2 wt % soln 0.75
dimethyldi(hydrogenated tallow)-ammonium chloride	7000 mg/kg	1 (g/kg body wt)/day threshold[d]	2,5,10 wt % soln for 21 days—mild irritation	10 wt % soln 11.7	10 wt % soln 0.43
trimethyldodecylammonium chloride	250–300 mg/kg			1 wt % soln 3.6 10 wt % soln 59.6	10 wt % soln 0.9 1 wt % soln 0.5
trimethyloctadecylammonium chloride	1000 mg/kg			10 wt % soln 11.9	10 wt % soln 1.8
trimethylhexadecylammonium chloride	250–300 mg/kg			1 wt % soln 3.6 10 wt % soln 47.5	10 wt % soln 1.08 1 wt % soln 0.0
dimethylalkylbenzylammonium chloride	0.300–0.350 ML/kg	acute dermal 2.98 ML/kg	0.1 wt % aqueous soln, nonirritating, nonsensitizing	0.1 wt % aqueous soln, nonirritating	

[a] Ref. 10.
[b] The technical information and suggestions for use made herein are based on Armak's research and experience and are believed to be reliable, but such information and suggestions do not constitute a warranty, and no patent liability can be assumed. Since Armak has no control over the conditions under which the product is transported, stored, handled, used or applied, the buyers must determine, by preliminary tests or otherwise, the suitability of the product for their purposes.
[c] Rats were used as test animals.
[d] In the determination of the chronic oral toxicity of Arquad 2HT, subacute dietary levels of 2800 ppm were administered orally to beagles and rats for 90 days. No abnormalities were observed with respect to body weights, food consumption, reactions, mortality, urine analyses, or hematologic, blood chemistry, or gross pathologic or histopathologic studies.

or a polyurethane foam, which is added with the wet clothes into the dryer. The formulation contains a transfer agent, usually a fatty-acid ester, which allows the quaternary to transfer from the substrate to the clothes. The latest innovation is the introduction of combined detergent, softener, and antistatic formulations, which allow the introduction of all ingredients in the wash cycle (see Antistatic agents). In all cases, the benefits to the users are fabric softening, antistatic properties, ease of ironing, and odor improvement, the latter because of the addition of perfumes to the formulation. The most widely used and most effective products are the dimethyl(dihydrogenated tallow)ammonium chlorides or methyl sulfates. The imidazolinium and amidoamine quaternaries are not as effective and cause fabric yellowing (see Drycleaning and laundering, laundering).

The second largest market for quaternary compounds is in the manufacture of organomodified clays. The main use for compounds of this type is in the addition of organomodified clay to drilling mud to improve the lubricity and rheology of the systems (see Petroleum, drilling fluids). Because of the rapid increase of well drilling and the greater depths of drilling, the use of the organoclays has grown rapidly. Current estimates are that ca 14,000 t of quaternaries is used annually in the organoclay market. Production of organoclays is by an ion-exchange reaction. The clay compound is dispersed in water and then combined with an equivalent amount of quaternary dispersion under thorough mixing conditions. The positively charged quaternary ions replace the positively charged inorganic ions on the clay surface. The organoclay product is then removed from the dispersion, dried, and ground to the proper particle size. There are three main quaternaries used for this reaction: dimethyldi(hydrogenated tallow)ammonium chloride, dimethyl(hydrogenated tallow)benzylammonium chloride, and methyldi(hydrogenated tallow)benzylammonium chloride. Besides drilling mud, the organoclays are used as thixotropic agents in paints, various coatings, grease additives, foundry additives, cosmetics (qv), resins, and printing inks. The properties desired in each specific system dictates the organoclay needed.

The third largest use for quaternaries is in the bactericidal or sanitizer market. The most popular types are those prepared from the reaction of benzyl chloride and a dimethylalkylamine, wherein the alkyl group is C_{12}–C_{16}. These compounds in high concentration are viscous liquids but are usually sold as aqueous solutions. A wide variety of these solutions is used as sanitizing agents for cleansing eating utensils and food-processing equipment and as cleaning compounds in restaurants, dairies, and hospitals. They have advantages over phenols and chlorine-containing disinfectants in that they are nonirritating and odor-free and have relatively long activity. Quaternary ammonium compounds containing three methyl groups and one long alkyl chain, eg, trimethyloctadecylammonium chloride, also exhibit excellent germicidal activity. The only drawback may be in their formulation. Compounds containing two methyl groups and two alkyl groups, such as those derived from coconut-oil fatty acids, are most effective against anaerobic bacteria, eg, those that occur in oil wells. These bacteria are sulfate reducers, and their growth frequently causes severe corrosion problems in oil wells as well as plugging of formations. The surface-active property of the quaternary ammonium compound also helps in removing oil from the sandstone formation.

The advent of pollution concern and the rising price of petroleum products have also increased the use of cationic emulsifiers in producing water-based asphalt emulsions for road building and maintenance. Compounds such as the trimethyl quaternaries are particularly useful for a noncorrosive cationic system. The cationic emulsions can be produced without any polluting solvents and can be designed to break on contact with the negatively charged aggregates used in road maintenance (12) (see Asphalt).

Another significant use for quaternary ammonium compounds is in hair treatment (see Hair preparations). Quaternaries have a high affinity for proteinaceous substrates, and this property makes them useful for hair treatment and imparts antistatic effects, increases wetting, improves both wet and dry combing, and improves feel and luster. The uses are widely diversified, and the formulations vary from one cosmetic manufacturer to another, depending on the qualities to be emphasized. The main quaternaries used are trimethylalkylammonium chloride, pentaethoxystearylammonium chloride [80462-94-8], dimethylstearylbenzylammonium chloride [122-19-0], and dimethyldialkylammonium chlorides.

BIBLIOGRAPHY

"Quaternary Ammonium Compounds" in *ECT* 2nd ed., Vol. 19, pp. 859–865, by Richard A. Reck, Armour Industrial Chemical Co.

1. W. J. Pope and S. J. Peachey, *J. Chem. Soc.*, 1127 (1899).
2. D. N. Eggenberger, F. K. Broome, R. A. Reck, and H. J. Harwood, *J. Am. Chem. Soc.* **72,** 4135 (1950).
3. A. W. Ralston, R. A. Reck, H. J. Harwood, and P. L. DuBrow, *J. Org. Chem. Soc.* **13**(2), 186 (1948).
4. R. A. Reck, H. J. Harwood, and A. W. Ralston, *J. Org. Chem.* **12,** 517 (1947).
5. F. K. Broome, C. W. Hoerr, and H. J. Harwood, *J. Am. Chem. Soc.* **73,** 3350 (1951).
6. H. J. Hueck, D. M. M. Adema, and J. R. Wiegmann, *Appl. Microbiol.* **14**(3), 308 (May 1966).
7. L. P. Hammett, *Physical Organic Chemistry: Reaction Rates, Equilibria, and Mechanisms*, 2nd ed., McGraw-Hill Book Co., Inc., New York, 1970, p. 215.
8. L. D. Metcalfe, R. J. Martin, and A. A. Schmitz, *J. Am. Oil Chem. Soc.* **43,** 355 (1966).
9. Brit. Pat. 1,283,730 (Mar. 3, 1971), R. A. Reck (to Armak Co.).
10. *Toxicity Data for Aliphatic Nitrogen Compounds*, Armak Co., Chicago, Ill., 1980.
11. *Armak Quaternary Ammonium Salts*, Armak Co., Chicago, Ill., 1980.
12. *Redicote Reference Manual*, Armak Co., Chicago, Ill., 1980.

RICHARD A. RECK
Armak Company

QUARTZ, SiO$_2$. See Silica.

QUENCHING OILS. See Petroleum products.

QUINHYDRONE. See Quinones.

QUININE. See Alkaloids.

QUINOLINE DYES. See Quinolines and isoquinolines.

QUINOLINES AND ISOQUINOLINES

Quinoline [91-22-5] (1) and isoquinoline [119-65-3] (2) are the two isomeric benzopyridines. They have the same relationship with pyridine that naphthalene has with benzene (see Pyridine and pyridine derivatives).

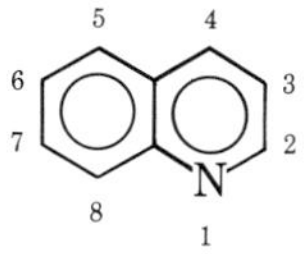

(1) 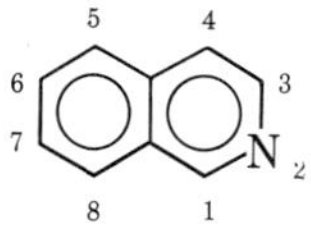   (2)

Quinoline

Quinoline, benzo[b]pyridine, C_9H_7N, is a heterocyclic base and is comparable to naphthalene with a methine group replaced by —N=. The determination of its structure is based on methods of synthesis and degradation. The benzene ring of quinoline is destroyed by strong oxidizing agents with the formation of 2,3-pyridine-dicarboxylic acid (3) (quinolinic acid).

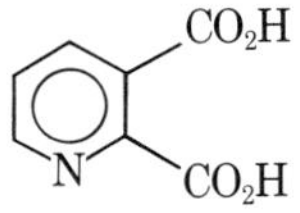

(3)
[89-00-9]

Physical Properties. Quinoline is a colorless, highly refractive liquid with a pungent odor. It is very hygroscopic, is more soluble in hot than in cold water, and distills in steam. As a weak tertiary base (basic ionization constant, 8.9×10^{-10}), quinoline dissolves in acids and forms characteristic salts, eg, the sparingly soluble dichromate [56549-24-7], $2C_9H_7N \cdot H_2Cr_2O_7$ (1). Quinoline is soluble in ethanol, ethyl ether, acetone, carbon disulfide, and other common organic solvents. The physical properties of quinoline and isoquinoline are listed in Table 1.

Data from heats of combustion show that there is considerable resonance in the quinoline system (4). The uv spectrum of quinoline in cyclohexane has been reported (5). The Raman spectrum showed the strongest lines at 521 cm^{-1}, 758 cm^{-1}, and 1372 cm^{-1} (6). The π-electron densities as shown in (4) and (5) have been calculated for quinoline and isoquinoline by the molecular-orbital method (2).

Reactions. Quinoline and quinoline derivatives exhibit the reactions common to benzene and pyridine. Electrophilic substitution occurs almost exclusively in the benzene ring. Nucleophilic substitution occurs in the pyridine ring. These substitutions are influenced by the ring nitrogen and the polarizability of different positions of the molecule by attacking reagents.

Table 1. Physical Properties of Quinoline[a] and Isoquinoline[b]

Property	Value	
	Quinoline	Isoquinoline
mp, °C	ca −15	6.4
bp, °C	237.63	243.25
ΔH_{vap}, kJ/mol[c]	46.4[d]	49.0
n_D	1.62928[e]	1.62078[f]
d^{30}, g/cm^3	1.08579	1.09101
K_a	8.9×10^{-10} [g]	2.5×10^{-9} [h]
viscosity (at 30°C), mPa·s (= cP)	2.997	3.2528
T_c	509	530

[a] Ref. 2.
[b] Ref. 3.
[c] To convert J to cal, divide by 4.184.
[d] Calculated.
[e] At 15°C.
[f] At 30°C.
[g] At 25°C.
[h] At 20°C.

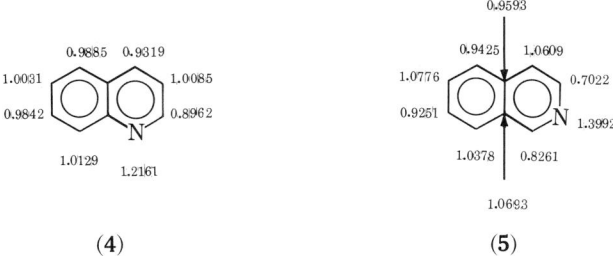

(4)   (5)

Nitration. In nitration of quinoline, the position at which the nitro group enters is determined by the conditions of the nitration. For example, a 52:48 wt % mixture of 5 [607-34-1]- and 8-nitroquinoline [607-35-2] is obtained by nitration of quinoline in concentrated sulfuric acid at 0°C. It has been shown that, in mixed acid, the nitronium ion attacks the protonated quinoline molecule (7). Under less acidic conditions of nitration, as when acetic anhydride is used with nitric acid or with dinitrogen tetroxide, the main product is 3-nitroquinoline [17576-53-3]; the 3-nitro compound is not obtained if the reaction conditions preclude formation of nitrous fumes (8). This orientation presumably results through the initial 1,2-addition of the nitrating reagent to the heterocyclic part of the compound. 7-Nitroquinoline [613-51-4] forms from the nitration of 1,2,3,4-tetrahydroquinoline [635-46-1] with mixed acid followed by dehydrogenation (9). In this case, substitution occurs in the carbocyclic part of the compound at a position meta to the protonated tetrahydroquinoline, which exhibits reactivity similar to that of the corresponding N-alkylaniline. Reaction of quinoline with tetranitratotitanium(IV) at RT gives mainly 3-nitroquinoline and, with tetranitratozirconium(IV) or tetranitratonitrosyliron(III), mainly 7-nitroquinoline. A 90% yield of 7-nitroquinoline was obtained from the tetranitratozirconium(IV) reaction (10).

The nitration of quinoline N-oxide [1613-37-2] with warm mixed acid gives 4-nitroquinoline 1-oxide [56-57-5], which is readily reduced to 4-nitroquinoline [3741-15-9] or 4-aminoquinoline [578-68-7], depending upon the reducing agent. A 65% yield of 4-nitroquinoline 1-oxide is obtained by oxidation of quinoline with H_2O_2–CH_3COOH to give the 1-oxide, which is nitrated by HNO_3–H_2SO_4 without purification (11).

Sulfonation. Sulfonation of quinoline at 220°C gives mainly 8-quinolinesulfonic acid [85-48-3] which, when heated to 300°C, rearranges to 6-quinolinesulfonic acid [65433-95-6]. Optimum conditions for sulfonation of quinoline, ie, two-hour reaction time at 140°C and a one-to-four quinoline-to-40 wt % oleum ratio, give an 80% yield of the 8-sulfonic acid (12). At a 1:3 quinoline:oleum ratio at 130°C, the yield is 64% (13). 8-Quinolinesulfonic acid hydrochloride [85-48-3] has been obtained in 82–93% yields by treatment of quinoline hydrochloride with excess chlorosulfonic acid at 100–120°C (14).

1,2-Addition. Quinoline reacts with allylmagnesium chloride in THF (tetrahydrofuran) in the absence of air to form 2-allyl-1,2-dihydroquinoline [55570-23-5] in 80% yield (15). This is a labile compound and isomerizes to 2-n-propylquinoline [1613-32-7] when heated to 170°C in an inert atmosphere, or it is converted to 2-propenylquinoline [57078-89-4] when exposed to air or oxidizing agents. The vinyl derivative reacts similarly (16). Heating the vinyl Grignard adduct with quinoline directly or heating the isolated dihydroquinoline causes isomerization to the corresponding 2-n-alkylquinoline or its magnesium salt. Similarly, the reaction of quinolines with organolithium reagents leads to the corresponding 2-substituted 1,2-dihydro-1-lithioquinolines which, upon hydrolysis and oxidation, provides 2-substituted quinolines (17). Acylation of quinoline with RCOCl (R = C_6H_5, CH_3, OC_2H_5) in acetonitrile at 0°C gives an intermediate which is phosphorylated with trimethyl phosphite and sodium iodide to give (**6a**), (**6b**), and (**6c**) in 65, 36, and 91% yields, respectively (18).

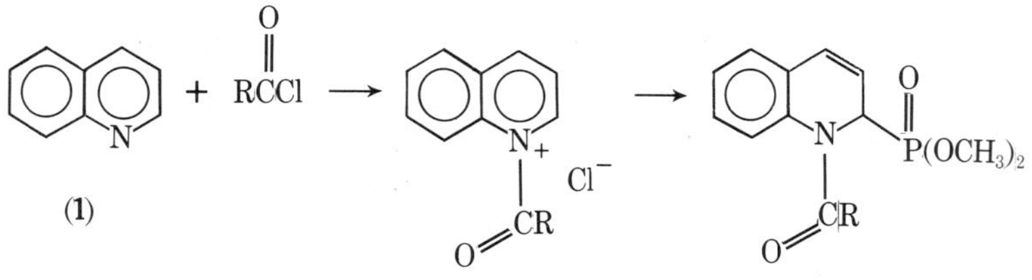

(**6a**) R = C_6H_5 [69664-77-3]
(**6b**) R = CH_3 [69664-78-4]
(**6c**) R = OC_2H_5 [69664-79-5]

The von Braun reaction involves reaction of cyanogen bromide with a tertiary amine. Quinoline reacts with cyanogen bromide in moist ether to give 1-cyano-2-hydroxy-1,2-dihydroquinoline and its ether (19). However, when the reaction occurs in methanol in the presence of sodium bicarbonate, a high yield of 1-cyano-2-methoxy-1,2-dihydroquinoline (**7**) is obtained (20). The latter is converted in high yield into 3-bromo-1-cyano-2,4-dimethoxy-1,2,3,4-tetrahydroquinoline (**8**) by treatment with bromine in methanol, followed by reaction with sodium carbonate (21). Subsequent reaction with concentrated hydrochloric acid in methanol affords 3-bromoquinoline [5332-24-1] quantitatively. Under similar conditions, the 4 [491-35-0]-, 5 [7661-55-4]-, 7 [612-60-2]-, and 8 [611-32-5]- methylquinolines give the corresponding 3-bromo derivatives in high yield.

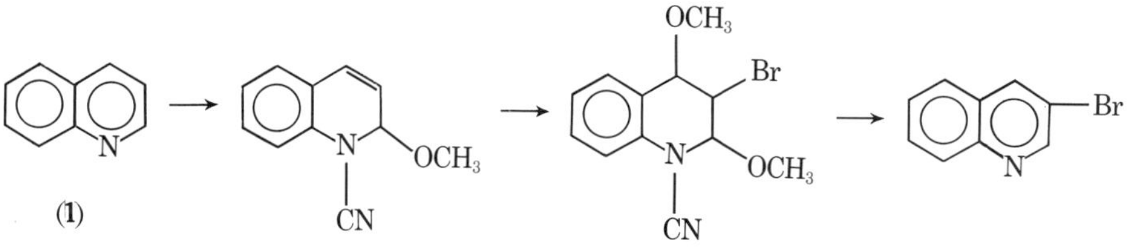

(1)

(7)
[880-95-5]

(8)
[66438-70-8]

Amination. Treatment of quinoline with barium amide in liquid ammonia affords an 80% yield of 2-aminoquinoline [580-22-3], which can also be obtained together with 4-aminoquinoline [578-68-7] by amination of quinoline with sodium or potassium amide in solvents, eg, xylene, toluene, or dimethylaniline (22). Amination of quinoline N-oxide with ammonium salts in the presence of p-toluenesulfonyl chloride gives 2-aminoquinoline (23). The 2-, 3 [580-17-6]- and 4-aminoquinolines form from the reaction of the corresponding chloro compounds with ammonia.

Halogenation. Halogenation is analogous to nitration of quinoline in that the 3-substitutions involve electrophilic attack on the neutral quinoline molecule and the 5- and 8-substitutions involve attack on the protonated molecule. 3-Bromoquinoline [5332-24-1] is conveniently prepared in 82% yield by heating the quinoline–bromine adduct in an inert solvent with pyridine as the hydrogen bromide acceptor (24). 3-Chloroquinoline [612-59-9] is prepared similarly. An apparent exception to the generalization above is the bromination of quinoline hydrochloride with excess bromine in nitrobenzene, which also gives 3-bromoquinoline (25–26).

The treatment of quinoline in concentrated sulfuric acid containing silver sulfate with one mole of bromine results in a mixture of 5 [165-18-3]- and 8-bromoquinoline [16567-18-3] in about equal quantities and some 5,8-dibromoquinoline (27). 5-Bromoquinoline is formed in 78% yield if a quinoline–aluminum chloride complex is heated with bromine (28). Further bromination in sulfuric acid or in the presence of aluminum chloride yields 5,8-dibromo [81278-86-6]-, 5,6,8-tribromo [81278-87-7]-, and 5,7,8 [81278-88-8]- tribromoquinolines. Bromination of quinoline at 450°C gives 2-bromoquinoline and presumably proceeds through a radical mechanism.

The addition of chlorine to quinoline or to 4-chloroquinoline [611-35-8] in carbon tetrachloride gives mainly 3,4,6,8-tetrachloroquinoline [25777-78-2] (29). Polychlorination of quinoline with sulfuryl chloride at 55–60°C gives a mixture of 3,4,5,6,7,8-hexachloroquinoline [42194-05-8] (57%), 3,4,6,8-tetrachloroquinoline (37%), and minor amounts of a product believed to be 2,3,4,6,7,8-hexachloroquinoline [37884-98-3] (30).

Oxidation. Oxidation of quinoline electrolytically or with alkaline permanganate, or boiling concentrated sulfuric acid in the presence of selenium dioxide or with ozone produces quinolinic acid (**3**), which can be thermally decarboxylated to nicotinic acid (3-pyridinecarboxylic acid) (**9**) (31). A 79% yield of nicotinic acid by oxidation of quinoline with nitric acid in sulfuric acid at 220–230°C has been reported (32). Hypochlorous acid converts quinoline into carbostyril [2-quinolinol or 2(1H)-quinolone] [59-31-4].

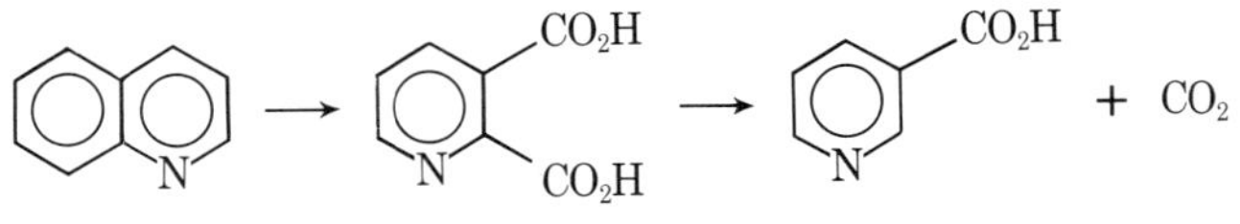

Peroxy acids, eg, peroxyacetic, monoperoxyphthalic, or peroxybenzoic acids, convert the tertiary nitrogen of quinoline to the N-oxide. This also can be accomplished by dissolving quinoline in acetic acid in the presence of 30% hydrogen peroxide and heating at 50°C for 24 hours. Hydroperoxides, eg, *tert*-butyl hydroperoxide, have also been used in the presence of catalysts to prepare N-oxides. Quinoline N-oxide is converted to 2 [612-62-4]- and 4 [611-35-8]- chloroquinolines in good yield with sulfuryl chloride; phosphorus oxychloride is preferable for the chlorination of substituted quinoline N-oxides. Deoxygenation of N-oxides can be accomplished by DMSO (33).

Quaternary Salts. Quinoline forms quaternary salts with alkyl halides, dimethyl sulfate, and a wide variety of aliphatic and aromatic acid halides, anhydrides, or sulfonyl chlorides. The reaction with acid chlorides, eg, benzoyl chloride and potassium cyanide, gives quinoline intermediates that are 1,2-adducts of quinoline known as Reissert compounds (34). Thus, treatment of quinoline with benzoyl chloride gives N-benzoylquinolinium chloride (10), which forms 1-benzoyl-2-cyano-1,2-dihydroquinoline (11) on reaction with aqueous potassium cyanide. The Reissert compound (11) affords either 2-carboxyquinoline (12) and benzaldehyde when treated with concentrated hydrochloric acid, or it yields 2-cyanoquinoline (13) when treated with phosphorus pentachloride. Alternatively, (13) can be prepared in excellent yield by mixing quinoline N-oxide with benzoyl chloride followed by treatment of the reaction mixture with alkali cyanide.

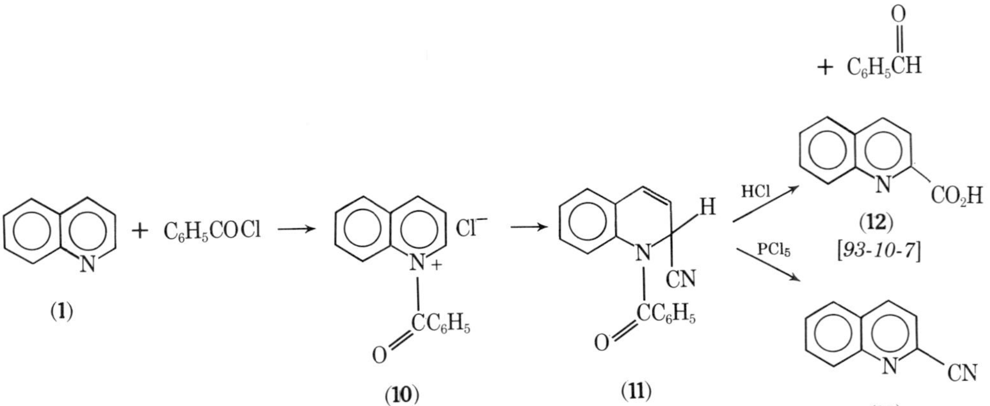

The reaction of N-alkylquinolinium salts, eg, (14), with alkali cyanides follows a different course in that the cyano group unexpectedly enters the 4-position and forms ultimately 4-cyanoquinoline (15).

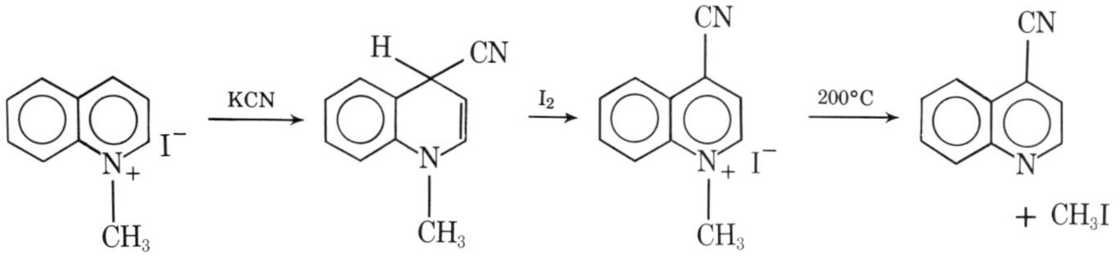

[Reaction scheme: (14) [3947-76-0] → (via KCN) [828-69-3] → (via I₂) [64275-22-5] → (200°C) (15) [23395-72-4] + CH₃I]

A similar reaction occurs with certain thiocarbonyl chlorides, eg, ethoxythiocarbonyl and phenoxythiocarbonyl chlorides, quinoline, and KCN in methylene chloride–water (35). The use of phase-transfer catalysts, eg, benzyltriethylammonium chloride, gives improved yields of Reissert compounds (36) (see Catalysis, phase-transfer).

Alkylation and Arylation. Photochemical alkylation of quinoline by irradiation of a solution in benzene containing acetic acid gives 2-methylquinoline [91-63-4] (20% yield), 4-methylquinoline [491-35-0] (10% yield), and 2,4-dimethylquinoline [1198-37-4] (5% yield) (37). Reductive alkylation, in which 4-alkyl-1,2,3,4-tetrahydroquinolines are formed with the expected products, has also been observed (38). A similar reductive alkylation occurs when benzyl radicals that are generated by oxidation of phenylacetic acid with $Na_2S_2O_8$, Ti^{3+}-$Na_2S_2O_8$, or Ag^+-$Na_2S_2O_8$ react reversibly with quinoline to give 2,4-dibenzyltetrahydroquinoline (16) and other products, including 2- and 4-benzylquinoline, dibenzyl, etc (39).

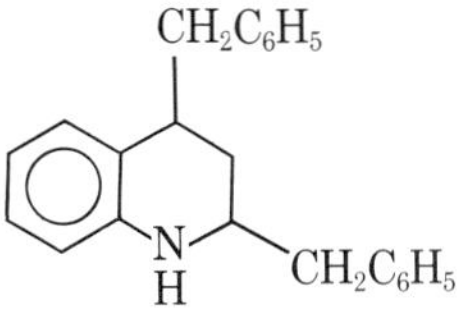

(16)
[69998-80-7]

The alkylation of quinolines in 60–83% yield with carboxylic acids in the presence of ammonium persulfate and silver nitrate has been patented (40). Quinoline is methylated by lead tetraacetate or t-butyl peroxide at positions 2-, 4-, 5- and 8- in nonacidic solutions; the reactivities are enhanced in acidic media. Total yield of methylquinolines, based on the amount of radical source used, is ca 35% (41). Generation of methyl radicals from DMSO with Fenton's reagent, ie, hydrogen peroxide and ferrous ion, in the presence of quinoline at 60°C produces 2-methyl- and 4-methylquinoline and an unknown compound, perhaps 5-methylquinoline, in a 1.5:1:2 ratio and a 45% total yield (42).

Hydroxyalkylation of quinoline or of its methyl derivatives has been effected with alcohols, eg, methanol, ethanol, or n-propanol, in the presence of an oxidizing agent, eg, ammonium persulfate (43). N-Substituted quinoline derivatives, eg, N-oxide, N-methyl, and N-amino, convert to the methyl compound substituted at C-2, C-4, or both by radical methylation and by radical hydroxymethylation (44). Alkyl-substituted derivatives at C-2 are selectively prepared by alkylation after formation of the N-oxide.

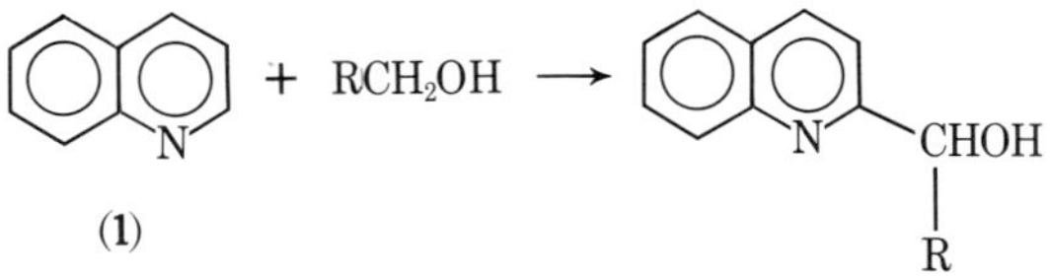

$$\text{(1)} + RCH_2OH \longrightarrow \text{quinoline-CHOH-R}$$

(R = H, CH$_3$, C$_2$H$_5$)

Catalytic methylation of quinoline with methanol over a Ni–NiO$_2$ catalyst at 295°C gives 2-methylquinoline (45). Similarly, catalytic methylation by CO–H$_2$ at 450–470°C and 6.89 MPa (68 atm) over 22% Cr$_2$O$_3$–78% ZnO, 60% CuO–40% ZnO, or ca 50% CuO–50% Cr$_2$O$_3$ has been reported (46). The reaction of quinoline N-oxide with isoxazole derivatives, in which the 4-position is free, has been carried out and the product hydrogenated in the presence of Raney Ni catalysts to give quinolines with side chains in the 2-position (47). Arylation of quinoline is produced by heating with a nitro-substituted compound, eg, nitrobenzene, at 400–1000°C, ca 1–100 MPa (10–1000 atm), and reaction times of 0.1 s–10 h (48).

Reduction. Catalytic hydrogenation of quinoline yields products in which the pyridine ring, the benzenoid ring, or both are reduced, depending upon the absence or presence and nature of the substituents as well as upon reaction conditions. Reduction of quinoline over Raney Ni at 70–100°C and 6.1–7.1 MPa (60–70 atm) affords 70% yields of 1,2,3,4-tetrahydroquinoline [635-46-1], whereas an increase in temperature to 210–270°C affords 52–62% yields of decahydroquinoline [2051-28-7] (49). Hydrogenation over an Adkins catalyst, ie, copper–chromium oxide, at 225–235°C gives 65–68% conversions to tetrahydroquinoline (50). A catalyst containing 30 wt % WS$_2$ on Al$_2$O$_3$ at 250–270°C and 7.8–8.8 MPa (77–87 atm) yields up to 85–90% conversion of quinoline to tetrahydroquinoline. However, hydrogenolysis occurs over MoS$_2$ at 360°C and ca 10 MPa (100 atm) to give 1,2,3,4- and 5,6,7,8-tetrahydroquinoline [10500-57-9], trans-decahydroquinoline [767-92-0], 2- and 3-methylindole, indene, and many other products (51). Catalytic hydrogenation over PtO$_2$ in strongly acidic solutions under mild conditions leads predominantly to reduction of the benzene ring (52). For example, reduction at RT and 0.34 MPa (3.4 atm) provides a 70% yield of 5,6,7,8-tetrahydroquinoline, which is further reduced with sodium–ethanol to give a 90% yield of trans-decahydroquinoline [767-92-0] (53).

Treatment of quinoline with formic acid followed by reaction with methyl formate affords N-formyltetrahydroquinoline (**17**) in overall yields of 88–95% (54). One-step reduction–alkylation of quinoline with sodium borohydride and carboxylic acids gives (**18**) in 17–79% yields (55). Reaction of quinoline with sodium borohydride–acetic acid in the presence of acetone gives predominantly 1,2,3,4-tetrahydro-1-isopropylquinoline [21863-25-2], whereas sodium cyanoborohydride in acetic acid reduces quinoline without alkylation to give the tetrahydroquinoline.

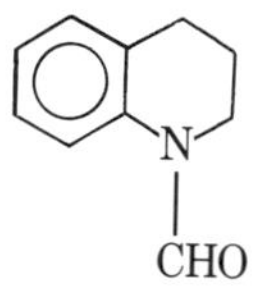

(17)
[2739-16-4]

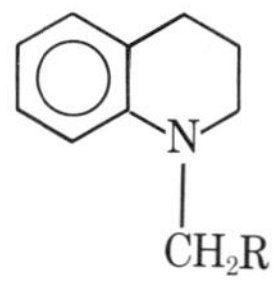

(18)

R = H [491-34-9]
R = CH$_3$ [16768-69-7]
R = C$_2$H$_5$ [6613-29-2]
R = CF$_3$ [57928-03-7]

Manufacture and Synthesis. *From Coal Tar.* A good but somewhat dated description of the occurrence, recovery, and separation of pyridine and quinoline bases from coal tar is given in ref. 56. Quinoline is isolated from the chemical oil fraction of coal-tar distillates. After removal of tar acids by a caustic extraction, the acid-free oil is distilled to afford the methylnaphthalene fraction (bp, 230–280°C). This fraction is washed with dilute sulfuric acid to remove the higher boiling tar bases. The tar bases are liberated from the sulfate salts with caustic and then are distilled. Properties of commercial refined quinoline are 90% min purity, distillation range of 2°C from 235 to 238°C, and sp gr at 15.5°C, 1.095. Typical composition (glc) of such a product is 92 wt % quinoline, 5 wt % isoquinoline, and 3 wt % others (57).

Isoquinoline, 2-methylquinoline, and 4-methylquinoline occur in the cruder fractions of tar bases that distill at 230–251°C. Studies of carefully controlled fractional distillation and refractionation of such crude tar bases boiling at 230–265°C were made, and small or trace amounts of all the possible isomeric monomethylquinolines in addition to quinoline and isoquinoline and isolated 2,8-dimethylquinoline [1463-17-8] as well as homologues of isoquinoline were present (58). Isoquinoline and quinoline homologues are not separated in a pure form but are marketed as refined QSR (quinoline still residue) bases. A refined QSR-bases product (bp, 235–255°C) has the following typical composition, as determined by glc: xylidenes (5 wt %), quinoline (40 wt %), isoquinoline (30 wt %), 2-quinaldine [91-63-4] (10 wt %), 8-quinaldine [611-32-5] (5 wt %), and indole (19 wt %) (59).

Separation of quinoline and isoquinoline mixtures can be effected by chromatography or complex formation. Chromatographic separation of the quinoline–isoquinoline fraction by complexing the components with metal salts on a sorbent, such as alumina saturated with cobalt chloride, liquid chromatographic separation with reversed-phase and adsorbent packings, and capillary gas chromatographic separation with deactivated glass columns have been reported (60–62).

A 90% yield of >95% pure isoquinoline is obtained from an isoquinoline fraction containing 24.3 wt % isoquinoline, 40.7 wt % quinoline, 19.6 wt % 2-methylquinoline, 4.9 wt % 8-methylquinoline, and 10.5 wt % other compounds by treatment with hydrochloric acid followed by addition of an alcoholic solution of cupric chloride (cupric chloride to isoquinoline mole ratio = 1:2). The complex that forms is purified by washing with methanol (63). Similarly, a 75–80% yield of 97.5 wt % 2-methylquinoline is obtained by treatment of a quinaldine fraction of bituminous coal tar containing 2-methylquinoline, isoquinoline, quinoline, and 4-methylquinoline with 30% aqueous urea and subsequent separation and decomposition of the clathrate. The remaining liquor is treated with hydrochloric acid and cupric chloride to yield 95% pure isoquinoline (64).

Quinoline and Quinoline Derivatives. *Skraup Synthesis.* The Skraup synthesis is a general reaction that can be used for the synthesis of many quinolines (65). It consists of heating a primary aromatic amine with glycerol, concentrated sulfuric acid, and an oxidizing agent, eg, the nitro compound corresponding to the aromatic amine that is used, arsenic acid, or ferric chloride. The reaction proceeds through dehydration of glycerol to acrolein, addition of acrolein to the amine to form β-anilinopropionaldehyde (**19**), cyclization to 1,2-dihydroquinoline (**20**) and oxidation to quinoline (**1**) in 84–91% yields. The preparation of quinoline in 75% yield by the Skraup method with the use of methanesulfonic acid as the dehydrating agent has been reported (66).

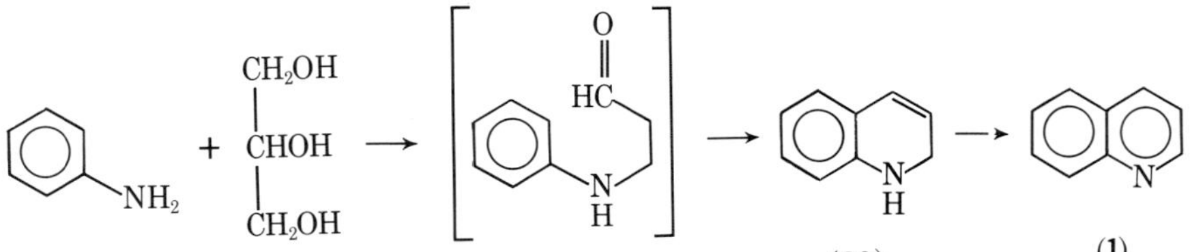

The Skraup method, which was originally designed for the synthesis of quinolines substituted in the benzene ring, has been extended to the synthesis of quinolines substituted in the heterocyclic ring. This is done by means of substituted acroleins instead of glycerol. For example, crotonaldehyde and α-methylacrolein yield 2- and 3-methylquinoline [612-58-8], respectively. Similarly, methyl vinyl ketone, 4-methoxybutanone, or 1,3,3-trimethoxybutane give 4-methylquinoline. A *meta*-substituted aromatic amine can give rise to two isomeric quinoline derivatives, depending on the orienting influence of the substituent. Aromatic diamines react twice in the Skraup reaction to form the corresponding phenanthrolines. The Skraup synthesis has been extended to the preparation of phosphorus-containing compounds. For example, reaction of dibutyl-3-aminophenylphosphine oxide with glycerol in sulfuric acid containing stannic chloride gives 56% of the quinoline derivative (**21**) (67).

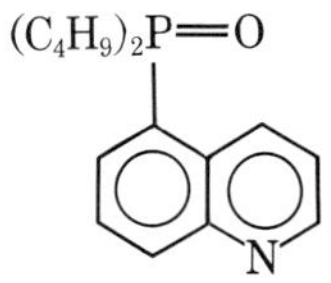

Döbner-von Miller Synthesis. The Döbner-von Miller synthesis is very closely related to the Skraup synthesis but is experimentally simpler and not nearly as violent (68). It consists of the reaction of one mole of an aromatic amine with two moles of acetaldehyde in the presence of hydrochloric acid or zinc chloride. The reaction does not involve an oxidant to form 2-methylquinoline (quinaldine). The reaction sequence consists of reaction of crotonaldehyde with an aromatic amine to form 3-anilinobutyraldehyde (**22**), which cyclizes to the aldol base (**23**), which reacts with another molecule of aniline to form (**24**). Dehydration of (**23**) or the loss of aniline from (**24**) results in 1,2-dihydroquinaldine (**25**). The latter is converted into quinaldine (**26**) by dehydrogenation. A number of mechanistic studies have been reported (69).

(**22**) (**23**) *[36191-13-6]*

(**24**) *[81045-46-7]* (**25**) *[42051-93-4]* (**26**) *[91-63-4]*

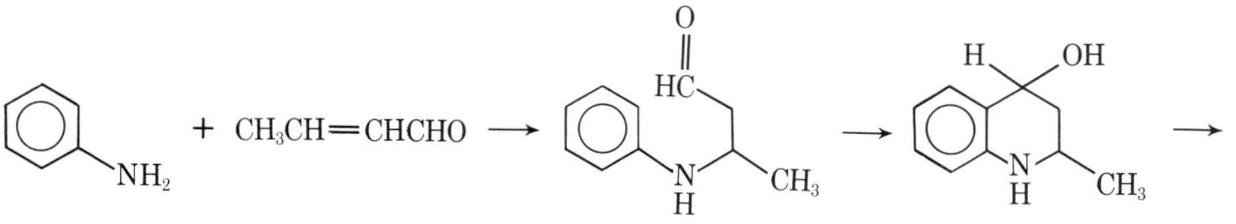

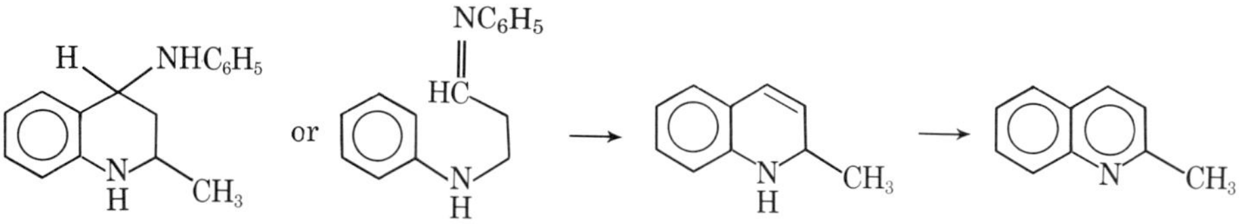

The Döbner-von Miller synthesis has been extended and improved. For example, an improved yield of 3-methylquinoline from the cyclocondensation of aniline with propionaldehyde and formaldehyde is obtained by the use of an ethanolic reaction medium and ferric chloride and zinc chloride as condensing agents. Trimethylquinolines are similarly prepared from 2,3-, 2,4-, or 2,5-dimethylaniline, and 2-isopropylquinoline [*17507-24-3*] is obtain from isovaleraldehyde (70). Another improvement is the addition of zinc chloride to the crude reaction mixture to give a precipitate of a 2:1 complex of the quinaldine–HCl and zinc chloride as an easily purified solid, from which pure quinaldine is recovered by treatment with aqueous base (71).

A variant of the Döbner-von Miller synthesis is that of Beyer, in which a mixture of an aldehyde, ketone, and aromatic amine is condensed under acid conditions to a 2,4-disubstituted quinoline. The Combes synthesis involves condensation of an aromatic amine with a 1,3-diketone to given an intermediate, which is cyclized under acid conditions to give, eg, 2,4-dimethylquinoline (**27**). Polyphosphoric acid is a convenient reagent (72).

(**27**)
[1198-37-4]

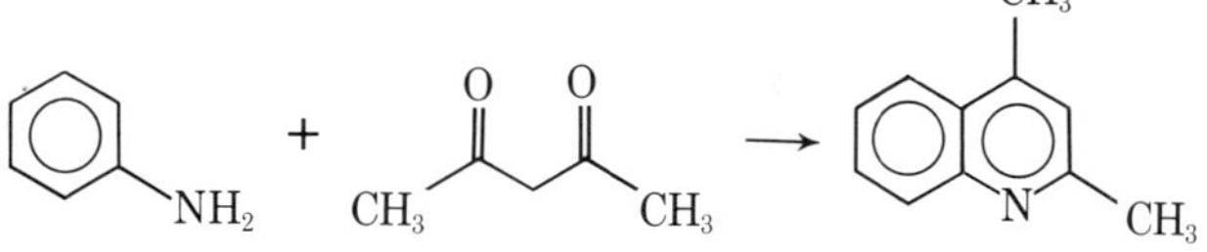

542 QUINOLINES AND ISOQUINOLINES

Conrad-Limpach-Knorr Synthesis. The Conrad-Limpach-Knorr reaction involves the condensation of β-keto esters with aromatic amines (73). It has been applied to the synthesis of many quinoline derivatives, eg, 2- and 4-hydroxyquinolines that are intermediates for chemotherapeutic agents (see Chemotherapeutics). Aniline condensed with ethyl acetoacetate (28) at below 100°C forms ethyl 3-anilinocrotonate (29), which can be cyclized to 4-hydroxy-2-methylquinoline (30) by adding it to a suitable material preheated to ca 250°C. If, however, aniline reacts with ethyl acetoacetate at ca 160°C or with diketene, acetoacetanilide (31) forms and can be cyclized to 2-hydroxy-4-methylquinoline (32) by treatment with concentrated sulfuric acid.

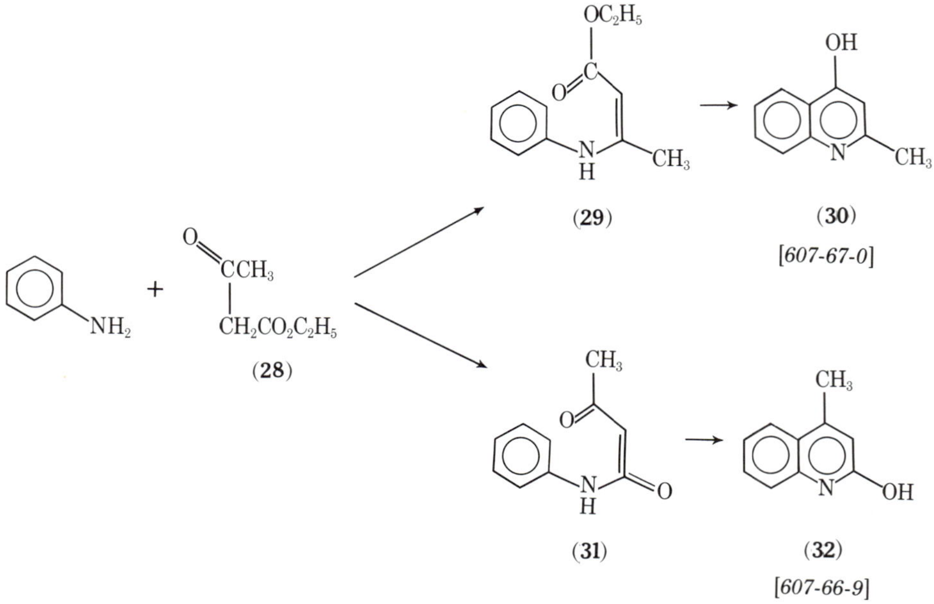

Pfitzinger Reaction. The Pfitzinger reaction involves the condensation of isatin (33) with carbonyl compounds, eg, aldehydes, ketones, acids, and esters (34), to quinoline-4-carboxylic acids (35) (74). The acids are easily decarboxylated to the corresponding quinolines.

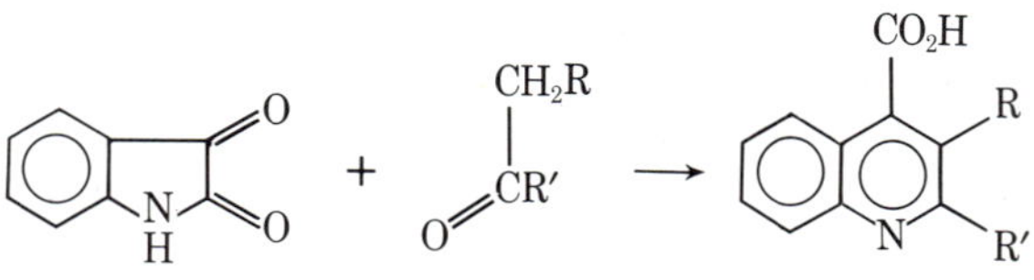

Cyclic ketones, such as cyclohexanone or α-tetralone, also have been utilized in the Pfitzinger reaction. Cyclohexanone treated with isatin yields (36), which is oxidized to quinoline-2,3,4-tricarboxylic acid (37) (75). Similarly, α-tetralone yields (38) (76).

The Pfitzinger reaction has also been utilized in the preparation of poly(quinolinecarboxylic acid)s and polyquinolines by polycondensation of bis(isatinic acids) with diketones in basic aqueous solutions of alcohol or pyridine, followed (in the case of the polyquinolines) by decarboxylation (77). The polymers [(**39**); X and Y are CH_2, O, or a direct bond] are obtained in high yield.

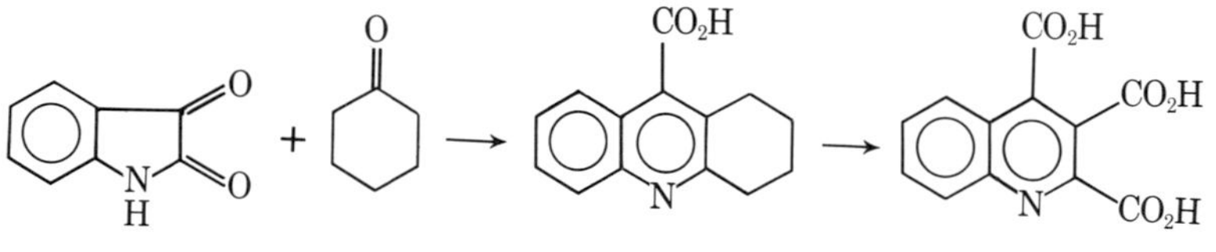

(**33**)   (**36**) [38186-54-8]   (**37**) [16880-83-4]

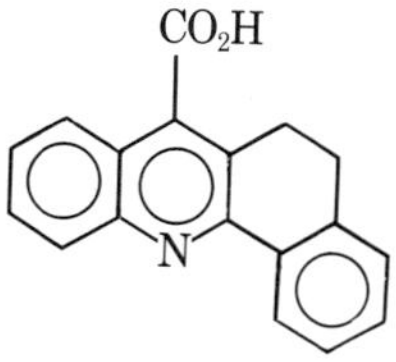

(**38**) [83-93-2] 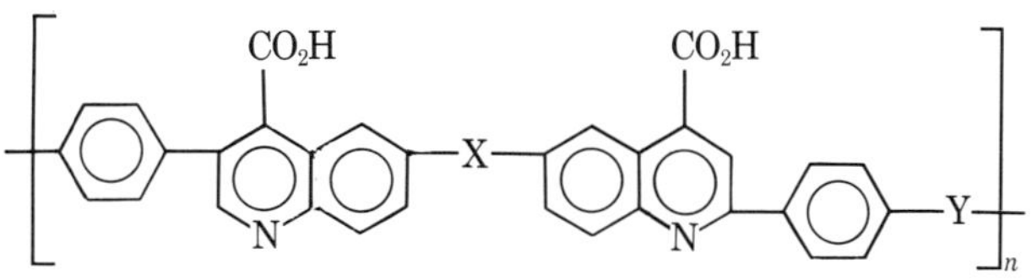   (**39**)

Friedländer, Camps, and von Niementowski Reactions. The Friedländer reaction involves the base-catalyzed condensation of an *o*-aminobenzaldehyde [(**40**), R = H] or *o*-aminophenyl ketone [(**40**), R = alkyl or aryl] with a second aldehyde [(**41**), R″ = H] or ketone [(**41**), R″ = alkyl or aryl] to form the quinoline (**42**). The main limitation is the difficulty in preparing the required *o*-aminoaldehydes or -ketones (**40**). However, the reaction has been successful for the preparation of aromatic polymers containing 2,6-quinoline units in the main chain by polymerization of 4,4′-diamino-3,3′-dibenzoyldiphenyl ether with bisketomethylene monomers (78).

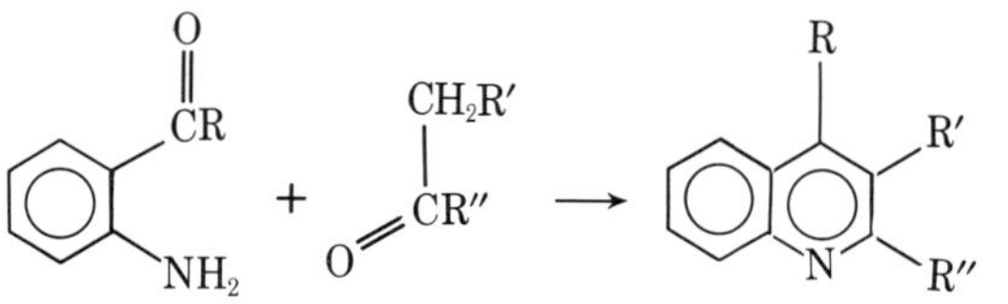

(**40**)   (**41**)   (**42**)

The Camps modification involves the base-catalyzed cyclization of an acylated *o*-aminoketone or *o*-aminoaldehyde to form substituted quinolones. Two possible modes of cyclization are possible with ketones such as (**43**).

In general, the von Niementowski variation, which involves the condensation of anthranilic acid derivatives and carbonyl compounds to quinolones, is less versatile than the original Friedländer synthesis. In some cases, improved yields are obtained when the acetal is used in place of the free ketone (79).

Miscellaneous. Quinoline has been prepared in 51% yield from an amine, eg, *N*-propylaniline and iodine by passage over calcium oxide at 300–500°C (80). Quinoline can be prepared in >90% yield from 2-(2-cyanoethyl)cyclohexanone (**44**) by passage

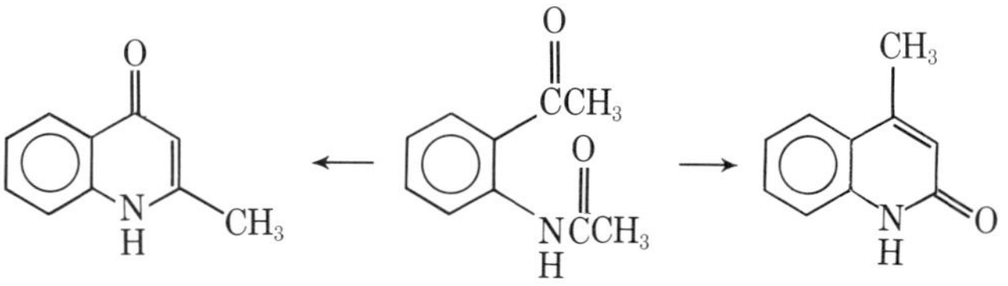

544   QUINOLINES AND ISOQUINOLINES

(30)   (43)   (32)

over a Pd–Al$_2$O$_3$ catalyst at 204–211°C to give a mixture of quinoline, decahydroquinoline, and 1,2,3,4- and 5,6,7,8-tetrahydroquinoline. The mixture is then passed over the same catalyst at 266–300°C to give 28–34% conversions to quinoline (81). Cinnamanilides (45) are converted to carbostyrils (46) in 67–84% yields by treatment with anhydrous aluminum chloride at 90°C in chlorobenzene (82). Conversion of o-nitrocinnamaldehyde, from o-nitrobenzaldehyde and acetaldehyde, by reduction with potassium tetracarbonylhydridoferrate under mild conditions gives quantitative yields of quinoline (83). A new synthesis of 2,3-cycloalkenopyridines [(47); $n = 3-13$; R = H, CH$_3$] involves the catalytic gas-phase reaction of cycloalkanols with unsaturated aldehydes or ketones and ammonia (84). For example, 5,6,7,8-tetrahydroquinoline is obtained in 76% yield from acrolein and cyclohexanone. Reaction of acetone with p-C$_2$H$_5$OC$_6$H$_4$NH$_2$ in the presence of aryl sulfonic acids affords 2,2,4-trimethyl-6-ethoxy-1,2-dihydroquinoline [91-53-2] in 90–93% yields and 98–99% purity (85).

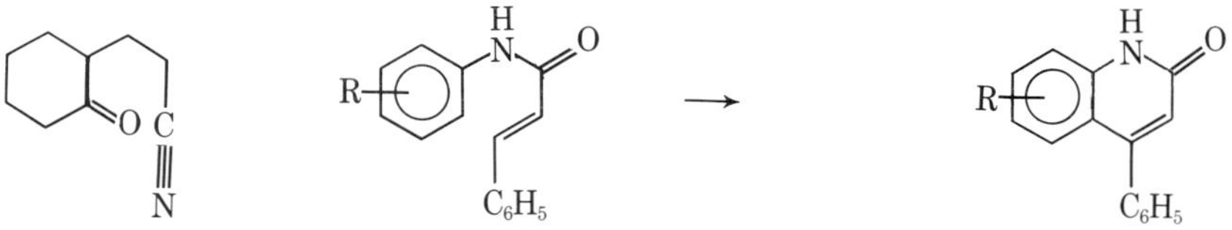

(44)   (45) R = H, 4-CH$_3$, 3-CH$_3$ (46) R = H [5855-57-3], 5-CH$_3$ [4053-33-2]
                                                6-CH$_3$ [4053-34-3], 7-CH$_3$ [4053-35-4]

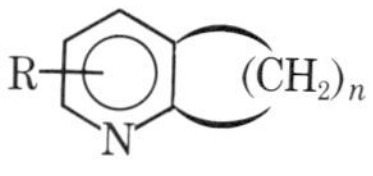

(47)

Economic Aspects. In 1978, the production and consumption of quinoline and quinoline still residue was ca 226.8 metric tons and ca 113.4 t, respectively. Prices of quinoline are given in Table 2. The production and sales of quinoline derivatives are given in Table 3 (see Derivatives).

Table 2. Prices of Quinoline

Year	Price, $/kg	
	Tank cars	Drums
December 1970	1.10	1.12
December 1975	1.10	1.12
December 1980	2.20	8.80
December 1981	6.05	7.15

Table 3. Production and Sales of Some Quinoline Derivatives in the United States[a]

Year	Chemical	Production, t	Sales Quantity, t	Sales Value, 10^3 $	Unit value, $/kg
1967	quinoline dyes	353	362[b]	2539	7.01
	8-quinolinol derivative (medicinal)	365	180	1785	9.92
	8-quinolinol copper salt (pesticide)	87	88[b]	472	5.36
1968	quinoline dyes	563	505	3603	7.13
	8-quinolinol derivative (medicinal)	421			
	8-quinolinol copper salt (pesticide)	61	127[b]	374	2.94
1969	quinoline dyes	732	643	4747	7.38
	8-quinolinol derivative (medicinal)	244	154	2049	13.31
1970	quinoline dyes	787	745	5370	7.21
	8-quinolinol derivative (medicinal)	171	84	930	11.07
	8-quinolinol copper salt (pesticide)	32	31	120	3.87
1971	quinoline dyes	1015	971	7574	7.80
	diiodohydroxyquin	10			
1972	quinoline dyes	1227	1033	7876	7.62
	8-quinolinol copper salt (pesticide)	18			
1973	8-oxyquinoline sulfate (medicinal)	3.6	4.5	44	9.78
1974	8-oxyquinoline sulfate (medicinal)	4.1	5.0	55	11.00

[a] Reports of U.S. Tariff Commission and United States International Trade Commission on synthetic organic chemicals.
[b] Quantity sold includes production from prior years.

Toxicology. Quinoline vapors are irritating to the eyes, nose, and throat and may cause headaches, dizziness, nausea, etc. The liquid is rapidly absorbed through the skin; it causes skin burns and eye irritation. Toxicity values are oral LD_{50} (rat), 330–460 mg/kg; dermal LD_{50} (rat), 540 mg/kg (86).

8-Quinolinol is highly toxic intraperitoneally and moderately toxic if taken orally. It is an experimental carcinogen and causes neoplasma of the uterus, rectum, brain, and bladder by oral, implantation, and intravenous routes. Acute toxicity data are oral LD_{50} (rat), 1299 mg/kg; ip LD_{50} (mice), 48 mg/kg.

Uses. *Antioxidants.* Most quinoline antioxidants are 1,2-dihydroquinoline derivatives. For example, 1,2-dihydro-6-decyl-2,2,4-trimethylquinoline [*81045-48-9*], 1,2-dihydro-6-ethoxy-2,2,4-trimethylquinoline [*91-53-2*], and 1,2-dihydro-2,2,4-trimethylquinoline have been produced commercially in the United States as antioxidants, antiozonants, and stabilizers in rubber processing (see Antioxidants and antiozonants) (87–88). Poly(1,2-dihydro-2,2,4-trimethylquinoline) [*26780-96-1*] has been used as an antioxidant in biscyclopentadienyl resins (89). The polymer has also been used in water curable ethylene–trimethoxyvinylsilane graft copolymers (90), in lubricant oils containing esters or mineral oils (91), and in heat-stable lubricants for synthetic fibers (92). 1,2-Dihydro-2,2,4-trimethylquinoline and its 6-ethoxy derivative are useful as food antioxidants (93). Similarly, santoquin [91-53-2] prevents the decomposition of carotenes and significantly decreases the losses of vitamins A and B_{12} and choline during storage of feed (94). Treatment of 1,2-dihydro-2,2,4-trimethylquinolines with aldehydes produces materials useful as food antioxidants (95). The cross-linking of polyethylene by peroxides and other monomers in the presence of 1,2-dihydro-6-ethoxy-2,2,4-trimethylquinoline [91-53-2] gives polymers with a chemically bonded antioxidant (96).

Corrosion Inhibitors. Quinoline and several of its derivatives have been recommended as corrosion inhibitors (see Corrosion and corrosion inhibitors). The corrosion of steel-reinforcing wire or rods embedded in concrete can be prevented by the addition of a composition containing quinoline or quinoline chromate to the cement mix (97). Quinoline protects 63:37 brass in 0.5 N acetic acid against corrosion (98). Surface treatment of metals with 8-hydroxyquinoline inhibits tarnishing and corrosion (99). Quinoline, 2-chloro [612-62-4]-, 4-amino-, 8-nitro-, or 8-hydroxyquinoline [148-24-3] have been added to ethylene glycol-type antifreezes as corrosion inhibitors (100).

Agricultural Chemicals. Quinolyl esters of N-substituted dithiocarbamic acids have been prepared from aliphatic primary and secondary amines by reaction with carbon disulfide to form the corresponding amine salt. The salt is added to 2-vinylquinoline [772-03-2] to form (48). The herbicidal activity of the product is comparable to that of 2,4-dichlorophenoxyacetic acid (2,4-D) (101). 7-Chloroquinoline N-oxide [22614-94-4] has been used as a herbicide against broad-leaved weeds in cereal cultures. 7-Chloroquinoline, when applied preemergence to prevent seed germination, completely controls certain weeds, with no phytotoxicity to cotton and wheat (102) (see Herbicides). Substituted 8-aminoquinolines have been used to prepare herbicides that possess phytotoxic activity, especially when applied preemergence (103). Decahydroquinaldine derivatives (49), phenoxyquinolines synthesized from, eg, 2-chloroquinoline and m-ethylphenol, and 2,3-disubstituted 4-methylquinolines also are herbicides (qv) (104–106).

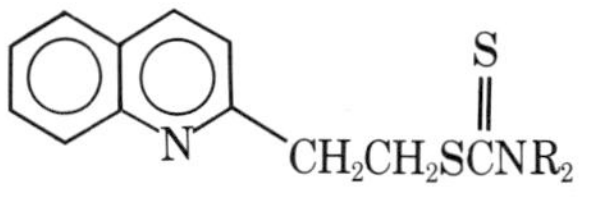

(48)

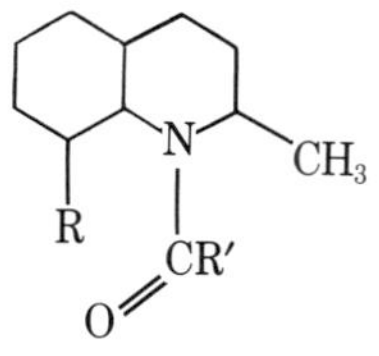

(49)

8-Quinolinecarboxylic acid derivatives and salts exhibit insecticidal properties (106) (see Insect control technology). Also, quinolyl carbamates are useful as insecticides and do not adversely affect useful insects and fish at normal concentrations (107). Coating cotton seeds with quinoline N-oxide prior to sowing increases the activity of enzymes, improves the quality of the cotton plants, and raises their yield (108) (see Plant-growth substances). Mixtures of 1-naphthol compounds with hydroxyquinoline compounds repel termites in low concentrations (<100 ppm) (109). The copper salt of 8-hydroxyquinoline has been manufactured in the United States and is an effective fungicide (see Fungicides).

Polymers. Quinoline and quinoline derivatives are either added to, or incorporated in, polymers to impart ion-exchange capability (see Ion exchange). For example, quinoline, quinaldine, or lepidine is added before or during removal of water in the manufacture of phenol–formaldehyde polymers (110). Similarly, polymers that have quinoline groups, which are anion exchangers, have been prepared by treatment of Amberlite IRA 402 or Amberlite IRA 900 with 2-methylquinoline to form a resin that has strongly basic exchange capacity (1.9 meq/g) and weakly basic exchange capacity (0.8 meq/g) (111).

Chelate sorbents that can form complexes with platinum-group metals have been synthesized from various polymer matrixes and various 8-mercaptoquinoline [491-33-8] derivatives (112) (see Chelating agents). Usually, quinoline derivatives, eg, hydroxy-substituted ones, are incorporated in resins that form by phenol–formaldehyde condensations (113–116). Upon treatment with methylquinolines, cross-linked polymers with chloromethylphenyl groups form anion exchangers with weak-base and strong-base exchange capabilities (117). Stannic chloride has been used as a catalyst to prepare polymers by condensation of bis(chloromethyl)benzene with quinoline (118).

Metallurgy. Quinoline and some of its derivatives are used in plating baths, extraction of metals from aqueous solutions, and separation of metals (see Electroplating; Metal surface treatments; Extraction).

The reaction product of an aldehyde with 2- or 4-methylquinoline when added to nickel-plating baths of the type containing nickel, chloride, and sulfate improves ductility, brightness, and leveling in the deposits (119). 8-Hydroxyquinaldine has been used as a component of soft-zinc electroplating baths (120). Smooth, ductile, uniformly lustrous to semibright or fully bright copper deposits for electroforming, rotogravure applications, or copper deposits of high throwing power for plating printed-circuit boards are obtained from aqueous acidic copper-plating baths containing quinoline or benzoquinoline [123-31-9] (121). An improved copper-electroplating bath includes a soluble organic leveling compound (0.007 g/L), which forms from the reaction of an epihalohydrin with nitrogen-containing compounds, eg, quinoline (122). The bath for bright tin electroplating of steel sheets contains stannous sulfate, phenolsulfonic acid, and 8-hydroxyquinoline as brightener (123).

The extraction of metal ions depends on the chelating ability of 8-hydroxyquinoline. Modification of the compound may give improved properties, eg, higher solubility in organic solvents (124–128). Nickel, cobalt, copper, and zinc are extracted from acid sulfate solutions by treatment with a chelating agent, eg, 8-hydroxyquinoline, in an immiscible solvent (126). Halogen-substituted 8-hydroxyquinolines in conjunction with oximes have been used to recover copper and zinc from aqueous solutions (127). Heavy metals, eg, mercury, cadmium, copper, lead, zinc, etc, at 0.2 mg/L are removed from aqueous solution by quinoline-8-carboxylic acid that is adsorbed on carbon, silica gel or silica–alumina gel (128).

Polymers containing 8-hydroxyquinoline are claimed to be selective adsorbents for the recovery of tungsten from alkaline brines (129). The separation of zinc from gallium and indium with quinaldic acid [93-10-7] in the presence of tartrate and citrate has been reported (130). Also, selective separation of lead and zinc from oxides ores involves treatment of the ore with a complexing agent, eg, quinaldic acid or 8-hydroxyquinoline, leaching the lead, and organic-phase extraction of zinc (131). Extractive separation of micro amounts of rhenium(VII) from ca a 100-fold excess of tungsten(VI); ca a ten-fold excess of copper(II), vanadium(V) and chromium(VI); and various amounts of molybdenum has been achieved by extraction with quinoline from 4 M sodium hydroxide solution and back-extraction with a 1:1 water–chloroform system (132). Extraction–separation of cobalt, copper, and nickel has been effected by 8-sulfamidoquinolines (133).

Catalysts. Rigid foams (qv) have been prepared by treatment of unsaturated dicarboxylic acids with diols and decarboxylating the adduct at 170°C in the presence of 2 wt % quinoline (134). Lithium complexes of quinoline are claimed to be effective catalysts for 1,4-addition polymerization of acrolein and methacrolein (135). Normal paraffins are selectively dehydrogenated to normal olefins in the presence of hydrogen and a catalyst consisting of platinum supported on sodium mordenite and by the addition of an organic base, eg, quinoline, to the paraffin stream (136). In the oxidation of $\geq C_{26}$ α-olefins by peracetic acid, partially esterified poly(phosphoric acid) salts, 8-hydroxyquinoline, or both are used as catalysts to give 1-alkene epoxides (137). Improved hydroformylation catalysts have been prepared and consist of $Co_2(CO)_8$ and 2-quinolone [59-31-4] (138) (see Catalysis).

Analytical Reagents. As with the metallurgical applications, analytical applications of quinoline derivatives rely on their chelating properties. The determination of numerous metals can be performed by gravimetric methods with 8-hydroxyquinoline (see Analytical methods). A summary of the determination of various metals with 8-hydroxyquinoline and substituted hydroxyquinolines has been reported (139) (see Fine chemicals). Other quinoline derivatives used in the determination and analysis of metals include 8-hydroxyquinoline substituted with hydrocarbon residues, aminoquinolines, and sulfur derivatives of quinoline (140–145).

Medicine. Quinoline derivatives are the parent substances of quinine and other plant alkaloids (qv) and of synthetic medicinals. The principal quinoline-derived drugs that are marketed in the United States are given in Table 4.

Quinoline Dyes. Quinoline dyes are chromophores containing a quinophthalone. Acid yellow 3 (**50**) (CI 47005) and Direct yellow 22 (direct yellow 3, CI 13925) are manufactured in small quantities by Sandoz in Switzerland. Solvent yellow 33 (**51**) (CI 47000) is manufactured by American Cyanamid Co. and Allied Corp.

Derivatives. Alkyl Quinolines. Alkyl quinolines are present in small amounts in tars from low and high temperature carbonization of coal and in coal-liquefaction products (147–148). Many alkyl derivatives of quinoline have been synthesized by the methods described previously, especially the Skraup and Döbner-von Miller syntheses. The preparation of 4-methylquinoline (lepidine), 4,6-dimethylquinoline [826-77-7] (55% yield), and 4,8-dimethylquinoline [13362-80-6] (65% yield) by the reaction of 4-(diethylamino)-2-butanone with aniline hydrochloride, p-toluidine, and o-toluidine, respectively, has been reported (149). Irradiation of a solution of quinoline and acetic acid in benzene has given 2-methylquinoline (20% yield), 4-methylquinoline (10% yield), and 2,4-dimethylquinoline (5% yield) (37). 2-Methylquinoline (97% purity) is obtained from a mixture of aniline, palladium chloride, triphenylphosphine, and ethylene which is autoclaved at 150°C (150). Treatment of vinylacetylene with mercuric chloride affords an intermediate which reacts with aniline or p-toluidine to yield 4-methylquinoline and 4,6-dimethylquinoline, respectively (151).

The 2- and 4-alkyl derivatives of quinoline, especially quinaldine and lepidine, are more reactive than the 3-, 5-, 6-, 7-, and 8-derivatives. Thus, quinaldine and lepidine condense with benzaldehyde to give $C_9H_6NCH=CHC_6H_5$ (2 [4945-26-0]- and 4

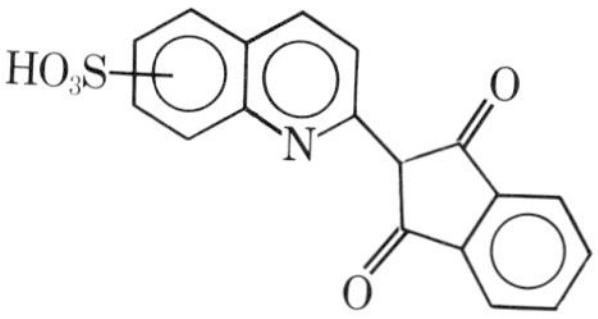

(50) [8004-92-0]

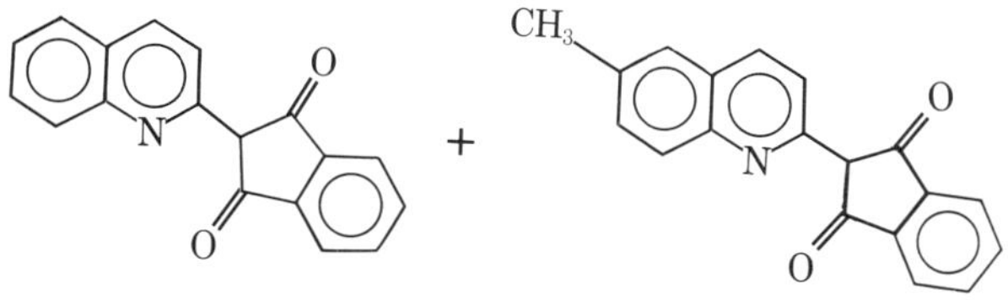

(51) [8003-22-3]
67% [83-08-9]
33% [6493-58-9]

[13362-63-5]-β-styrylquinolines) as well as the corresponding diquinaldine [81340-59-2] and dilepidine [81340-60-5] derivatives, eg, (C$_9$H$_6$NCH$_2$)$_2$CHC$_6$H$_5$. The reaction has been extended to heterocyclic aldehydes to give 55%, 78%, and 51% yields of (52), which is derived from the corresponding 2-, 3- or 4-pyridinecarboxaldehyde (the β(2 [13206-41-2]-, 3 [1586-51-2]-, and 4 [18633-00-6]-pyridinyl)-2-vinylquinolines), respectively (152).

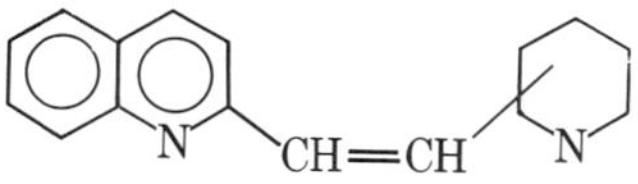

(52)

Treatment of o-phenylenediamine with 2-methylquinoline in the presence of sulfur at 150–210°C affords the benzimidazole (53) (153). Bimolecular dehydrogenation to 1,2-bis(2-quinoyl)ethane and other products occurs when quinaldine is heated with sulfur or with sulfuric acid (154).

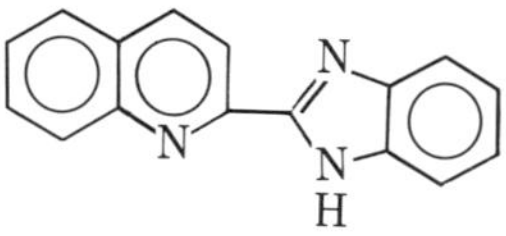

(53) [14044-48-5]

The methyl groups also undergo the typical reactions of aromatic derivatives. Chlorination of quinaldine in the presence of acetic anhydride or phosphorus pentoxide at 95°C gives 88% pure 2-(trichloromethyl)quinoline [4032-53-5], without resinification (155). The 2-, 3-, and 4-methylquinolines oxidize to the carboxylic acids by forming the palladium chloride complex followed by treatment with hydrogen peroxide (156).

Hydroquinolines. 1-Ethoxycyclohexene heated with R'CH=C(R)CHO at 210°C forms derivatives of 5,6-dihydro-6-ethoxy-5,6-tetramethylene-4H-pyran [(54); R = R' = H] which, upon treatment with hydroxylamine, gives 59–72% yields of 5,6,7,8-tetrahydroquinolines (55), which can be dehydrogenated to the corresponding quinolines (157).

Table 4. Quinoline-Derived Drugs Marketed in the United States[a]

Name	CAS Registry No.	Structure	Application	Brand name (manufacturer)
7-chloro-4-[(4-diethylamino-1-methylbutyl)-amino]quinoline	[54-05-7]	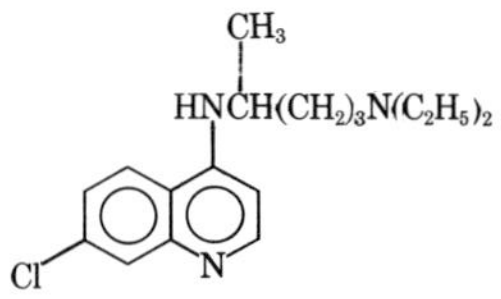	antirheumatic	Aralen (Winthrop)
2-butoxyquinoline-4-carboxylic acid, β-diethylaminoethylamide	[61-12-1]	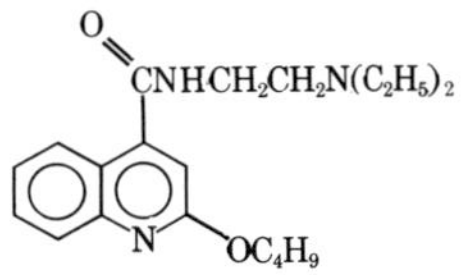	local anesthetic	Nupercaine (Ciba)
5-chloro-7-iodo-8-hydroxyquinoline	[130-26-7]	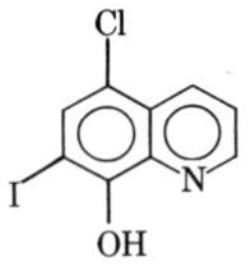	wound and intestinal antiseptic	Vioform (Ciba)
5,7-diiodo-8-hydroxyquinoline	[83-73-8]	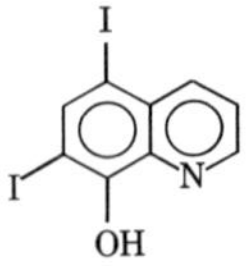	intestinal antiseptic	various names

Name	CAS Registry Number	Structure	Use	Trade name (manufacturer)
5,7-dichloro-8-hydroxyquinoline	[8067-69-4]	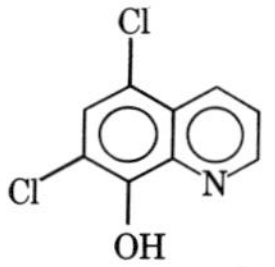	intestinal antiseptic, antidiarrheal	Quinolor (Squibb)
4-(4-[ethyl-(2-hydroxyethyl)amino]-1-methylbutylamino)-7-chloroquinoline	[118-42-3]	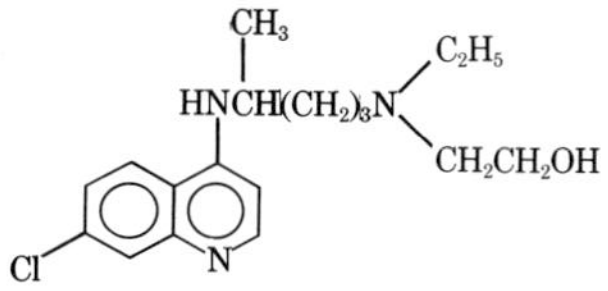	antirheumatic, antimalarial	Plaquenil (Winthrop)
1-ethyl-6,7-methylenedioxy-4-oxo-1,4-dihydroquinoline-3-carboxylic acid	[14698-29-4]	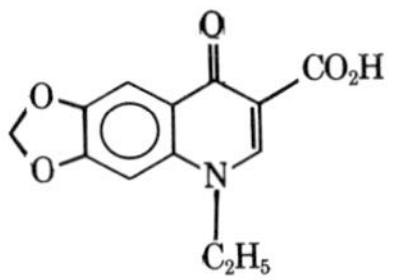	antibacterial	Utibid (Warner)
2-[2-(2,5-dimethyl-1-phenyl-3-pyrrolyl)vinyl]-6-dimethylamino-1-methylquinolinium pamoate	[3546-41-6]	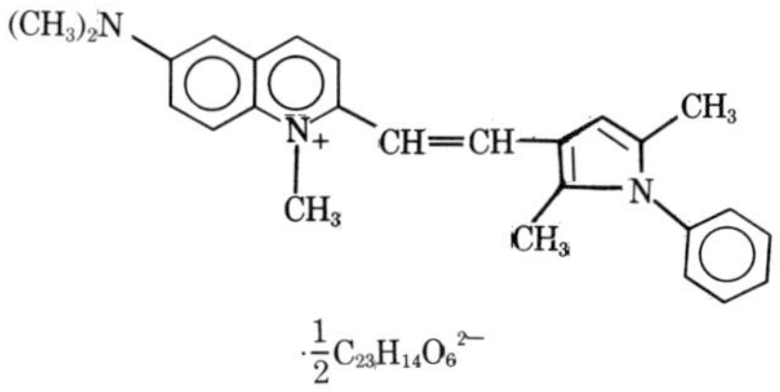	anthelmintic	Povan (Parke-Davis)

[a] Ref. 146.

(**54**)

(**55**)

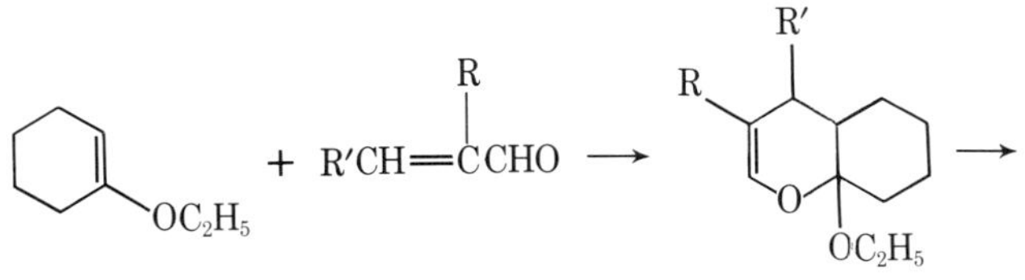

Similarly, the reduced pyran [(**54**); R = R′ = H] and $CH_3CH(OC_2H_5)_2$ yields (**56**), which yields (**57**) (R = H,C_2H_5) with ammonia on γ-alumina or (**57**) and (**58**) with Pt/Al_2O_3 (158). 5,6,7,8-Tetrahydroquinoline can also be prepared by heating *O*-allylcyclohexanone oxime (159). Alkyl 1,2-dihydroquinolines are prepared by reaction of acetone or acetone-derived products such as diacetone alcohol or mesityl oxide with aromatic amines (160).

(**56**) (**57**) (**58**)

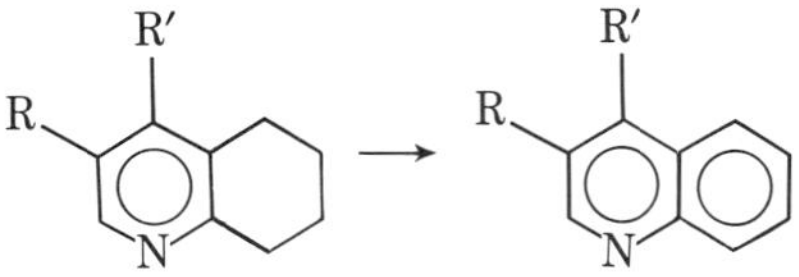

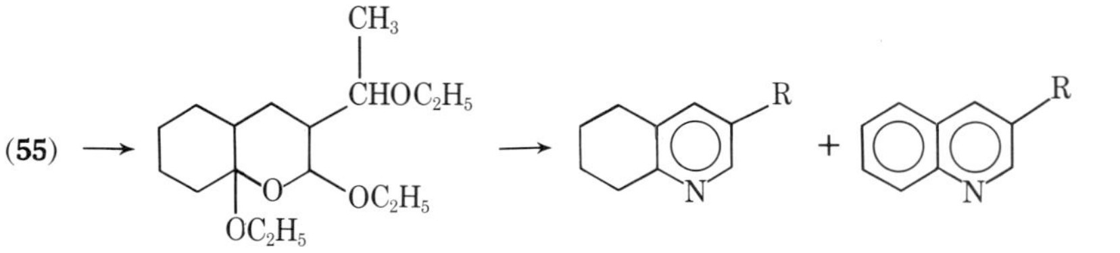

1,2,3,4-Tetrahydroquinoline [*635-46-1*] behaves like a monoalkylaniline. It forms an *N*-nitroso derivative, which rearranges readily to 6-nitro-1,2,3,4-tetrahydroquinoline [*14026-45-0*]. Mild oxidation with potassium permanganate in acetone forms 3,3′,4,4′-tetrahydro-1,1′(2*H*,2′*H*) biquinoline [*34555-59-4*] (**59**) which can be converted to the 6,6′-isomer (**60**) by a benzidine-type rearrangement. Thermolysis of 1,2,3,4-tetrahydroquinoline at 650–750°C in the presence of water vapor affords a 64% yield of indole (qv) (161).

The reactions of 5,6,7,8-tetrahydroquinoline have been reviewed (162–164).

(**59**) [*34555-59-4*]

(**60**) [*53899-16-4*]

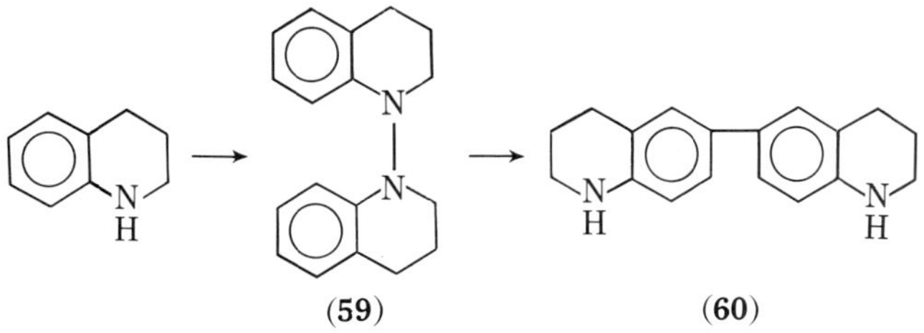

Hydroxyquinolines (Quinolinols). The preparation of the hydroxyquinolines involves a variety of reactions. 2-Hydroxyquinoline [2(1H)-quinolinone, carbostyril] [59-31-4] can be prepared in 70% yield by a modified Chichibabin reaction of quinoline in a KOH–NaOH melt at 240°C (165). Other preparations involve treatment of 3-hydroxyquinoline [580-18-7] with aqueous sodium hydroxide at 300°C to give an 88% yield of 2-hydroxyquinoline and treatment of quinoline with cupric sulfate in an autoclave to give 2(1H)-quinolinone in 25% yield based on consumed quinoline (166–167). A 93% yield of 7-chloro-4-hydroxyquinoline [86-99-7] has been claimed for the ring closure, hydrolysis, and decarboxylation of m-ClC$_6$H$_4$NHCH=C(CO$_2$C$_2$H$_5$)$_2$ in the presence of p-toluenesulfonic acid, phosphoric acid, or zinc oxide catalyst (168).

Both 5 [578-67-6]- and 8-hydroxyquinoline have been prepared in 79–90% yields by acid hydrolysis of the corresponding aminoquinoline in the presence of sulfuric acid, ammonium acid sulfate, or phosphoric acid at 180–235°C (169). 8-Hydroxyquinoline can also be prepared by sulfonation-fusion of quinoline, by hydrolysis of 8-chloroquinoline [611-33-6] (93% yield), or by a modified Skraup method with o-aminophenol (80% yield) (170–171).

Dihydroxyquinolines are obtained by synthetic methods and are formed *in vivo*. 2,4-Dihydroxyquinolines are prepared by heating 3,1-benzoxazin-4-ones with strong bases and in 81% yield by treatment of N-acetoacetylanthranilate, with aqueous base followed by acid (172–173). An enzymatic decarboxylation of 5-hydroxykynurenine to yield 4,5-dihydroxyquinoline through 5-hydroxykynurenamine has been reported (174).

Properties of 2- and 4-hydroxyquinolines are of both the enol (**61** and **63**) and keto (**63** and **64**) tautomers and, hence, they are also called 2(1H)- and 4(1H)-quinolinones. The other isomeric hydroxyquinolines do not exhibit keto–enol tautomerism.

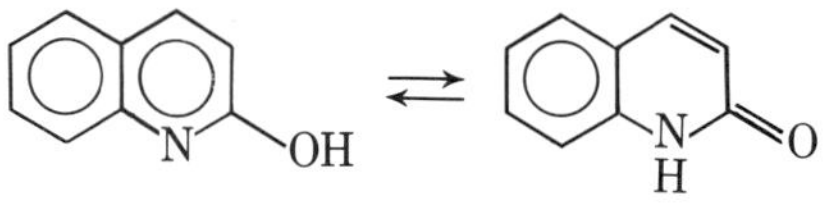

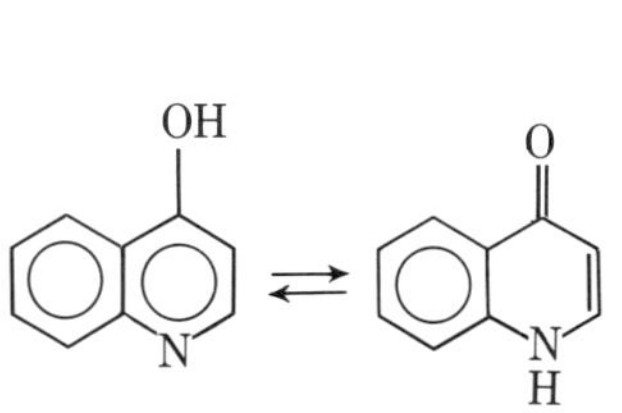

Halogenation of 8-hydroxyquinoline affords the 5,7-dihalo derivatives. Reported syntheses include 5,7-dichloro-8-hydroxyquinoline [733-76-2] (175), 5,7-dibromo-8-hydroxyquinoline [521-74-4] (176), and 5-dichloro-8-hydroxyquinoline [130-16-5] and 5-chloro-7-iodo-8-hydroxyquinoline [130-26-7] (177).

Haloquinolines. The 2 [*612-62-4*]- and 4 [*611-35-8*]-chloroquinolines are prepared from the corresponding hydroxyquinoline by the reaction of phosphorus oxychloride and phosphorus pentachloride. Reaction of substituted anilines with acrylic acid in the presence of hydroquinone gives the 3-anilinopropionic acid, which is treated with iodine and phosphorus oxychloride to give the corresponding substituted 4-chloroquinoline (178). Similarly, reaction of chloroanilines with β-propiolactone followed by treatment with polyphosphoric acid and then phosphorus oxychloride and iodine affords the respective 4,6 [*4203-18-3*]-, 4,7 [*86-98-6*]-, or 4,8-dichloroquinoline [*21617-12-9*] (179). Trichloroquinolines ((**65**); R = H [*40335-02-2*], CF$_3$) have been prepared in 55–70% yields by heating 3-RC$_6$H$_4$N=CClCCl=CCl$_2$ at 260–300°C (180).

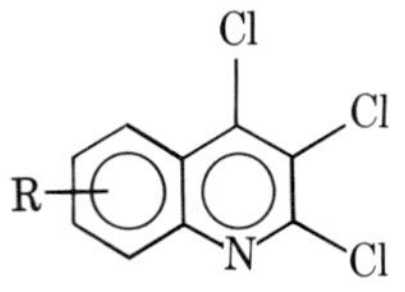

(**65**)

Aminoquinolines. Most of the isomeric aminoquinolines can be prepared from quinoline derivatives as discussed previously, eg, by reduction of nitro compounds, amination of quinoline, or ammonolysis of halogen compounds. 8-Aminoquinoline [*578-66-5*] has also been prepared in 72% yield by treating 8-hydroxyquinoline with ammonium sulfite (181). Reactions of aryl isocyanates with ynamines yield substituted 4-amino-2-quinolones and 2-amino-4-quinolones (182). The reaction of quinazoline 3-oxide (**66**) with malononitrile, phenylacetonitrile, or ethyl cyanoacetate affords the corresponding 3-substituted 2-aminoquinolines [(**67**); R = CN, C$_6$H$_5$, CO$_2$C$_2$H$_5$, respectively] (183).

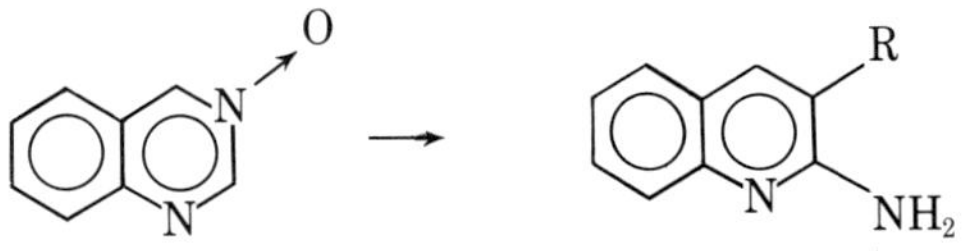

(**66**)

(**67**) R = CN [*31407-25-7*]
R = C$_6$H$_5$ [*36926-84-8*]
R = CO$_2$C$_2$H$_5$ [*36926-83-7*]

Aldehydes, Ketones, and Acids. 2 [*5470-96-2*]- and 4-Quinolinecarboxaldehyde [*4363-93-3*] can be prepared from the corresponding methyl derivatives by oxidation with selenium dioxide or by bromination to the dibromomethyl derivative followed by hydrolysis. The aldehydes undergo the usual reactions of aromatic aldehydes (184). Preparation of ketones is described in ref. 185.

Carboxylic acids are obtained by oxidation of methyl groups and hydrolysis of cyano groups. Substituted 1,4-naphthoquinones undergo ring expansion with hydrazoic acid to yield 1*H*-1-benzazepine derivatives, ie, (**68**) (R = alkyl). The hydroxy derivative (**68**) (R = H) gives kynurenic acid (**69**) upon treatment with base (186).

Biquinolines. Biquinolines and their derivatives can be obtained by application of standard quinoline syntheses to bifunctional molecules, ie, aromatic diamines, or

(68) (69)
[492-27-3]

by the application of the Ullmann synthesis. Biquinolines are also obtained by less familiar routes. For example, a mixture of quinoline, powdered aluminum, and mercuric chloride heated at 190–200°C and then refluxed with quinoline and nitrobenzene gives a 46% yield of 2,2′-biquinoline (**70**) (187). The interaction of quinoline 1-oxide and 7-methoxy-4-phenylquinoline 1-oxide with dimethylsulfonium methylide gives 2,2′-biquinoline (**70**) and 2,2′-bis(7-methoxy-4-phenylquinoline) (188). Thermolysis of *N*,*O*-di(4-quinolyl)hydroxylamine produces rearrangement to 3,3′-(4-amino-4′-hydroxy)biquinoline (**71**) (189). Heating silver salts of quinoline carboxylic acids at 100–500°C gives decarboxylation-dimerization products (190).

(**70**)
[*119-91-5*]

(**71**)
[*64372-81-2*]

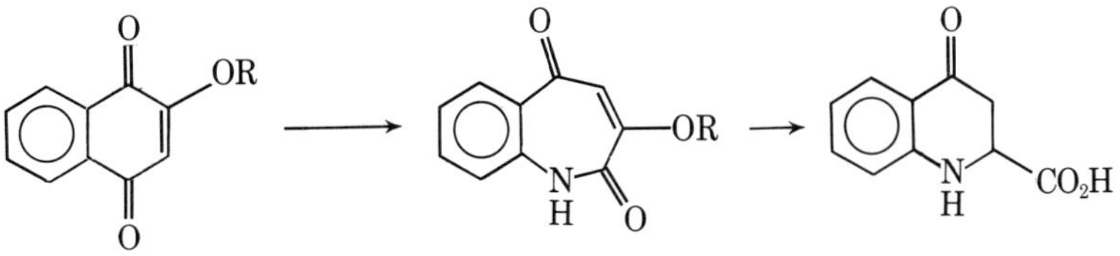

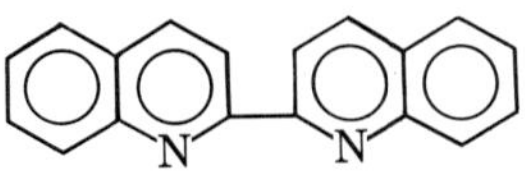

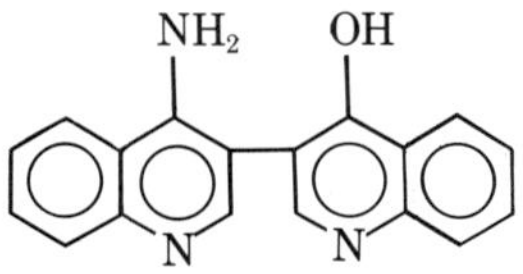

Benzoquinolines. A summary of the synthesis of benzoquinolines, some of which characterize certain alkaloids (qv), has been reported (191). Heating of 4-methoxy-1-naphthylamine with epichlorohydrin gives 34% 3-hydroxy-6-methoxy-1,2,3,4-tetrahydrobenzo[*h*]quinoline [*38419-41-9*], which can be aromatized to yield 24% 6-methoxybenzo(*h*)quinoline (**72**) (192). 2-(Chloromethyl)benz[*g*]indoline (**73**) heated with concentrated hydrochloric acid provides a 39% yield of benzo[*h*]quinoline (**74**) and a 27% yield of the 1,2,3,4-tetrahydroderivative (193).

(**72**)
[*38419-43-1*]

(**73**) (**74**)
[*230-27-3*]

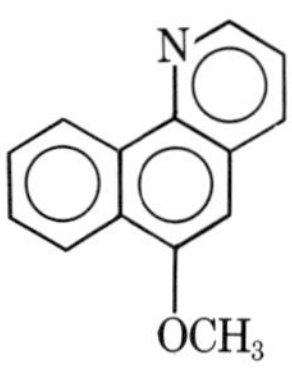

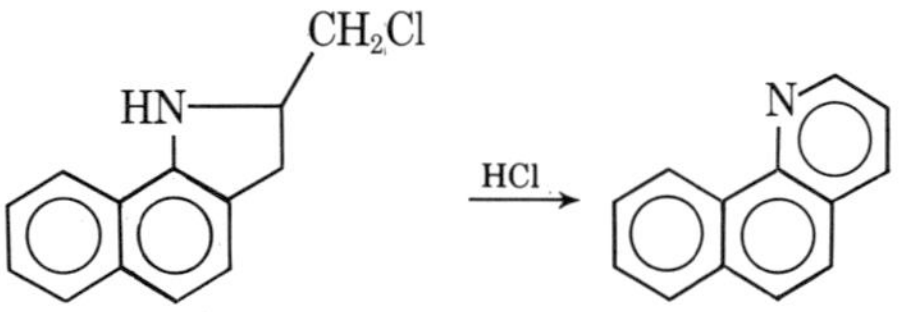

Mercaptoquinolines (Quinolinethiols). Mercaptoquinolines generally are prepared by diazotization of amines, by reduction of sulfides or sulfonyl chlorides, or by displacement reactions involving an active halogen. Preparations of various mercaptoquinolines are described in refs. 194–199.

Isoquinoline

Isoquinoline (2) (2-benzazine, leucoline) is a heterocyclic compound formed by the fusion of a benzene and a pyridine ring with nitrogen in the 2-position. The structure is assigned on the basis of synthesis and degradation studies. Isoquinoline is oxidized to phthalic acid (76) and cinchomeronic (3,4-pyridinedicarboxylic) acid (75) on treatment with alkaline permanganate.

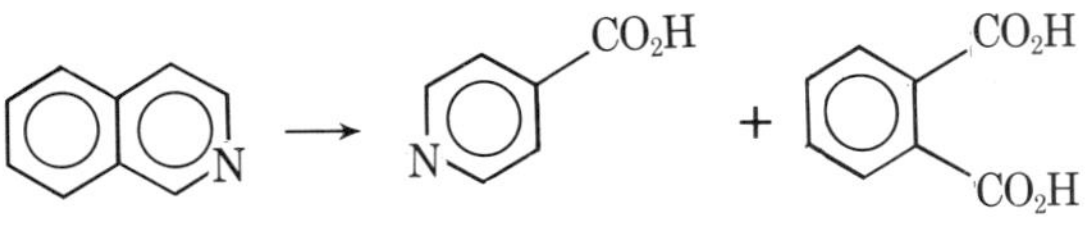

The isoquinoline nucleus occurs in many plant alkaloids, such as the cactus alkaloids, the opium bases, and the curare alkaloids. Isoquinoline also occurs in products from hydrogenation of coal (qv) and in coal tar. Isoquinoline is one of the main components of the QSR bases (63).

Physical Properties. Isoquinoline has an odor resembling that of benzaldehyde, is a stronger base than quinoline, and reacts vigorously with alkyl halides to form quaternary salts. Selected physical constants are given in Table 1. Isoquinoline is sparingly soluble in water and volatile in steam. It dissolves readily in ethanol, ethyl ether, and the common organic solvents.

A total assignment of the vibrational spectrum of isoquinoline has been reported (200) and compared with both quinoline and naphthalene. The uv absorption spectrum of isoquinoline closely resembles that of both quinoline and naphthalene with bands at 317, 266, and 217 nm (201). High resolution ^{1}H nmr in carbon tetrachloride (202) exhibited the following chemical shifts (δ, ppm): position 1 (9.114); 3 (8.448); 4 (7.501); 5 (7.697); 6 (7.563); 7 (7.582); and 8 (7.853).

Calculations of the π-electron densities in isoquinoline give rise to a wide range of results dependent on the values of parameters used and the approximations made; the most recent values are given in structure (5) (203).

Reactions. In general, isoquinoline undergoes electrophilic substitution reactions in the 5-position and nucleophilic reactions in the 1-position. Nitration of isoquinoline with mixed acid at 0°C affords a 90/10 mixture of 5-nitroisoquinoline [607-32-9] and 8-nitroisoquinoline [7473-12-3] and at 100°C the proportions are 85/15 (204). Sulfonation of isoquinoline results in a mixture of products with isoquinoline-5-sulfonic acid as the principal product.

Amination with sodamide in neutral solvents gives 1-aminoisoquinoline [1532-84-9], which may be converted to isocarbostyril [491-30-5], 1-hydroxyisoquinoline, by diazotization and decomposition. Isocarbostyril as well as 3-methylisocarbostyril can be prepared in 98 and 76% yields, respectively, by a modified Chichibabin reaction of isoquinoline or 3-methylisoquinoline in a KOH–NaOH or KOH–Na$_2$O melt (165).

Direct bromination of isoquinoline hydrochloride in an inert solvent, preferably nitrobenzene, with excess bromine affords an 81% yield of 4-bromoisoquinoline [1532-97-4] (25–26). However, bromination of an isoquinoline–aluminum chloride complex with gaseous bromine gives 5-bromoisoquinoline [34784-04-8] in 78% yield (28). Continued bromination leads to 5,8-dibromo [81045-39-8]- and 5,7,8-tribromoisoquinoline [81045-40-1].

As noted, oxidation of isoquinoline with alkaline permanganate or with hot concentrated sulfuric acid yields cinchomeronic acid (3,4-pyridinedicarboxylic acid) (**75**) and phthalic acid (**76**). However, the formation of phthalic acid (58% yield) as the main product from the reaction of ruthenium tetroxide with isoquinoline is a different result from those obtained by oxidation by other methods (205). Like quinoline, isoquinoline is oxidized by peracids to the N-oxide [1532-72-5]. The N-oxide has also been obtained from 2-(2,4-dinitrophenyl)isoquinolinium chloride [33107-14-1] by treatment with hydroxylamine hydrochloride followed by refluxing with concentrated hydrochloric acid (206).

Isoquinoline and substituted isoquinoline N-oxides undergo a variety of reactions. For example, trialkylboranes react with isoquinoline N-oxide to give the 1-alkyl derivatives (207). Isoquinoline N-oxide reacts with cyanogen bromide in ethanol to give ethyl N-(1- and 4-isoquinolyl) carbamates (**77** and **78**) and some by-products (208). An unusual reaction is that of isoquinoline N-oxide with methacrylonitrile in the presence of hydroquinone; products are 1-acetonylisoquinoline, 1-cyanoisoquinoline [1198-30-7], and isoquinoline (**2**) (209). Isoquinoline N-oxide undergoes direct acylamination with an N-benzoylanilinoisoquinoline chloride or the corresponding nitrilium salt to form 1-N-benzoylanilinoisoquinoline [53112-30-4] in 55% yield (210). The reaction of N-sulfinyl-p-toluenesulfonamide with isoquinoline N-oxide readily forms 1-(tosylamino)isoquinoline [25770-51-8], which affords 1-aminoisoquinoline upon hydrolysis (211).

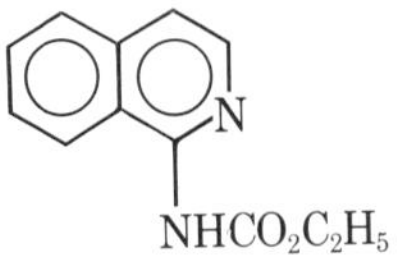

(**77**)
[36160-16-4]

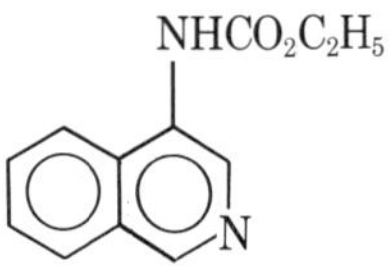

(**78**)
[55417-69-1]

Isoquinoline can be reduced to hydroisoquinolines. Isoquinoline in acetic acid solution containing mineral acid can be quantitatively hydrogenated over platinum to a mixture of cis- [2744-08-3] and trans-decahydroisoquinoline [2744-09-4] (212). Hydrogenation of isoquinoline over platinum oxide in strong acid under mild conditions leads to selective hydrogenation of the benzene ring to afford a 90% yield of 5,6,7,8-tetrahydroisoquinoline [36556-06-6] (52–53). Sodium hydride reduces isoquinoline in nearly quantitative yield in a dipolar aprotic solvent, eg, hexamethylphosphoric triamide, to the adduct (**79**) (213). The adduct (**79**) can be converted by acid anhydrides or chlorides to the *N*-acyl derivative (**80**), which can be converted to 4-substituted 1,2-dihydroisoquinolines. One-step reduction–alkylation with sodium borohydride–carboxylic acid gives (**81**) (R = H [*91-21-4*], CH$_3$ [*16768-69-7*], C$_2$H$_5$ [*57928-05-9*], CF$_3$ [*57928-06-0*]) in 21–79% yields (55). Sodium cyanoborohydride reduces isoquinoline without *N*-alkylation to give 1,2,3,4-tetrahydroisoquinoline. 1,2,3,4-Tetrahydroisoquinoline reacts with cyanamide to afford an 87% yield of 2-amidino-1,2,3,4-tetrahydroisoquinoline (**82**) (214).

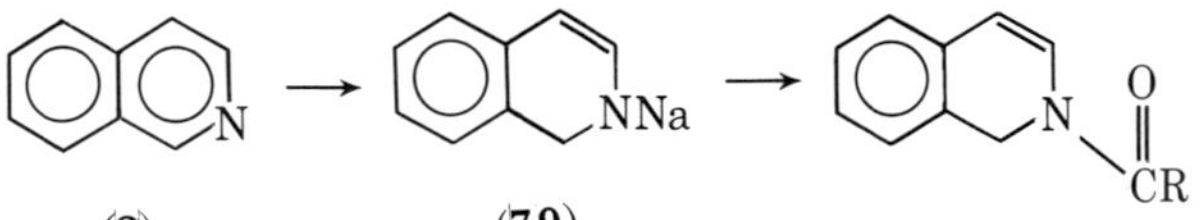

Isoquinoline, like quinoline, forms the Reissert compound when treated with benzoyl chloride and alkali cyanide (34). The use of phase-transfer catalysts improves the yields of Reissert reactions and has been applied to the preparation of phosphorus-containing Reissert compounds in 13–82% yields (36). The *N*-phenylsulfonyl Reissert compound ((**83**); R = C$_6$H$_5$SO$_2$, R' = H) has been converted to 1-cyanoisoquinoline ((**84**); R = CN) with sodium borohydride under very mild conditions (215). If R in (**83**) were benzoyl and R' were alkyl, reductive fission would occur to give the 1-alkyl isoquinoline (**84**). The synthesis of (±)-apomorphine from the Reissert compound derived from isoquinoline has been reported (216). Treatment of isoquinoline with sulfuryl chloride and potassium cyanide gives 4-chloro [*53491-80-8*]-, 1-cyano [*1198-30-7*]-, and 1-carbamoyl-3-cyanoisoquinoline [*81045-35-4*] in one step by means of Reissert compounds; their yields depend on reagent proportions (217).

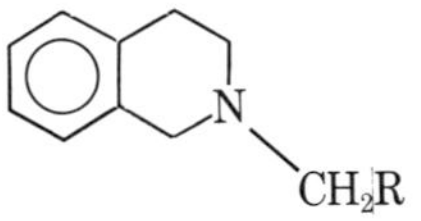

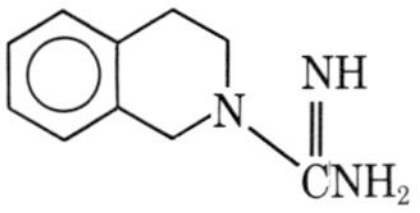

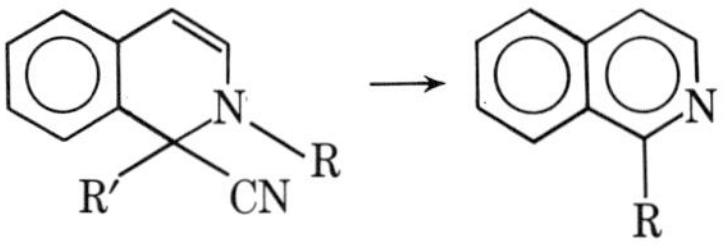

Photoreaction of isoquinoline with aliphatic carboxylic acids results in decarboxylation and formation of 1-alkyl isoquinolines (37–38). Similarly, homolytic alkylation of isoquinolines is effected by the silver-catalyzed oxidative decarboxylation of acids by peroxydisulfate to afford 1-alkyl isoquinolines in 84–100% yields (218). Heterocyclic cyclopentadiene, indene, and azulene derivatives have been prepared by reaction with isoquinolone in the presence of benzoyl chloride in benzene to afford the corresponding 2-benzoyl-1,2-dihydro-1-isoquinolyl derivative (219).

Synthesis of Isoquinoline and Isoquinoline Derivatives. Bischler-Napieralski Reaction.

The Bischler-Napieralski reaction consists of the cyclodehydration of N-acyl derivatives (86) of β-phenethylamines (85) to 3,4-dihydroisoquinolines (87) with Lewis acids, eg, phosphorus pentoxide, phosphoryl chloride, polyphosphoric acid, or zinc chloride in a dry inert solvent (220). The 3,4-dihydroisoquinoline (87) can be either dehydrogenated catalytically to the corresponding isoquinoline or reduced to the corresponding tetrahydroisoquinoline.

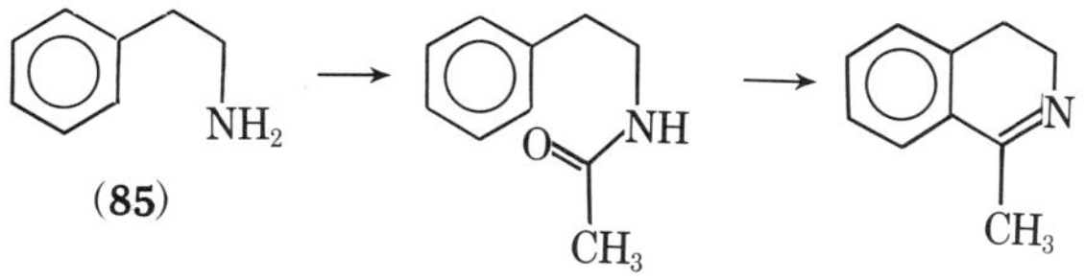

A modification of the Bischler-Napieralski reaction involves the use of isocyanate (88) or urethane (89) derivatives for the synthesis of 3,4-dihydroisoquinolines (90) (221). The isocyanate or urethane is treated with phosphoryl chloride followed by treatment with stannic chloride to give (90).

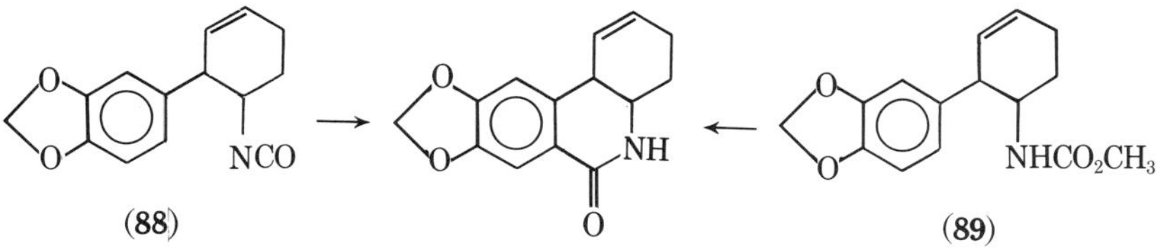

The Pictet-Gams modification of the Bischler-Napieralski reaction involves the cyclization of β-hydroxy- or β-methoxy-β-phenethylamides, which yields the isoquinoline derivative instead of the 3,4-dihydro compound. A further extension of this modification is based on a methoxyethylamine rather than a hydroxyethylamine.

Pictet-Spengler Synthesis. The Pictet-Spengler synthesis involves the condensation of β-phenethylamines with carbonyl compounds in the presence of an acidic catalyst to give 1,2,3,4-tetrahydroisoquinolines (222). 2-Arylethylamines (91) are heated with strong hydrochloric acid and a slight excess of an aldehyde to form the anil or Schiff base (92), which is cyclized with or without isolation to the tetrahydroisoquinoline (93). The enzyme-catalyzed Pictet-Spengler condensation has been used to prepare the enantiomers of hydroxy-substituted tetrahydroisoquinolines (223). Similarly, biogenetic synthesis of (S)-reticuline (94) [485-19-8] has been accomplished by an asymmetric Pictet-Spengler reaction (224).

Pomeranz-Fritsch Synthesis. The Pomeranz-Fritsch synthesis involves cyclization of benzalaminoacetals in the presence of acid to yield isoquinolines (225). A Schiff base (95) forms and then is cyclized to the isoquinoline (96) in the presence of acid. Although only moderate yields are obtained, the use of polyphosphoric acid for the cyclizing agent is successful in all cases, particularly for the preparation of 8-substituted isoquinolines (226).

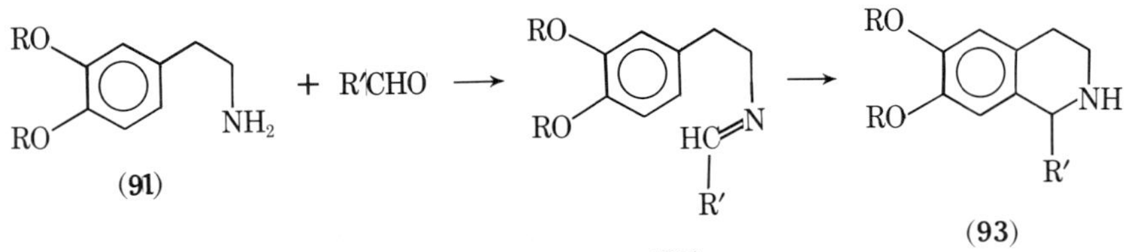

Bobbitt's modification (227) gives 1,2,3,4-tetrahydroisoquinoline derivatives in good yield. The Schiff base (95) is reduced to the amine (97), which is cyclized with 6 N hydrochloric acid. The resulting 1,2,3,4-tetrahydro-4-hydroxyisoquinoline (98) is hydrogenolyzed to afford the 1,2,3,4-tetrahydroisoquinoline (99). This modification is widely used because of the mild reaction conditions, simple procedure, and generally good yields.

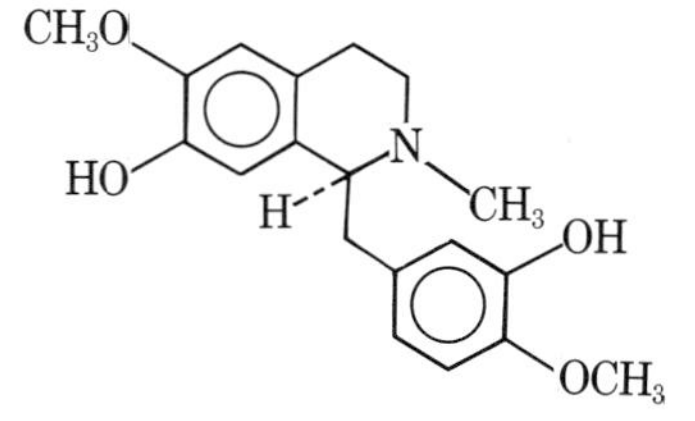

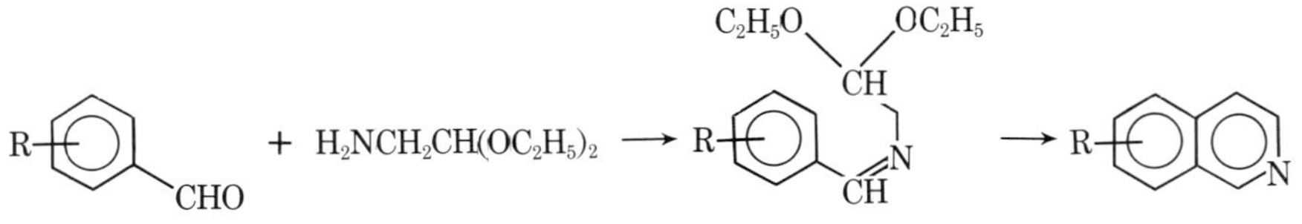

The Schlitter and Müller (228) variation involves the reaction of benzylamine (100) with glyoxal hemiacetal to form (101). Cyclization of (101) with sulfuric acid gives the same isoquinoline as that obtained from the Schiff base derived from the aromatic aldehyde and aminoacetal. This method is especially useful for the synthesis of 1-substituted isoquinolines.

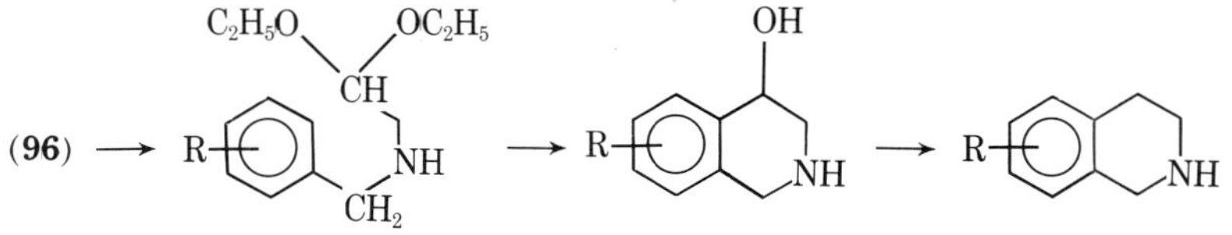

Alkyl Isoquinolines. 1-Methylisoquinoline [1721-93-3], 3-methylisoquinoline [1125-80-0], and 1,3-dimethylisoquinoline [1721-94-4] have been obtained in small amounts from coal tar. The methyls of 1- and 3-methylisoquinoline are more reactive than those of the other positions and oxidize readily with selenium dioxide to form the corresponding isoquinoline aldehydes (229). The aldehyde can also be obtained by selective halogenation to the dihalomethyl derivative followed by hydrolysis. The 1- and 3-methylisoquinolines also condense with benzaldehyde in the presence of zinc chloride or acetic anhydride to the corresponding 1- and 3-styrylisoquinolines. Carboxylic acids afford 1-alkylisoquinolines through the radicals formed by photolysis or oxidative decarboxylation.

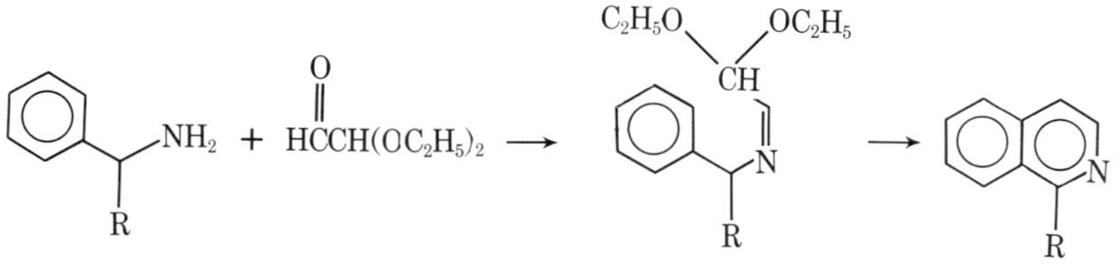

Hydroisoquinolines. The hydroisoquinolines are prepared by ring-closure reactions or by reduction of isoquinoline. Lithium aluminum hydride or sodium in liquid ammonia initially convert isoquinoline to 1,2-dihydroisoquinoline (230–231). 1,2,3,4-Tetrahydroisoquinolines are obtained by further reduction of the 1,2-dihydroisoquinolines, by reduction of isoquinolines with tin and hydrochloric acid or with sodium and alcohol, and catalytically over platinum in neutral solution. A 90% yield of 5,6,7,8-tetrahydroisoquinoline can be obtained over platinum oxide in strong acid (52–53). Isoquinoline in acetic acid solution reduces quantitatively over platinum to a mixture of cis- and trans-decahydroisoquinoline (212).

The hydroisoquinolines are more susceptible to ring cleavage than the isoquinolines. Ring cleavage occurs with nitrogen elimination when 3,4-dihydroisoquinolines (**102**) are heated with alkali and dimethyl sulfate to form o-acylstyrenes (**103**) in 41–98% yields (232). If there is no substituent in the 1-position, the product is an aldehyde (**103**; R = H). Ring opening without loss of nitrogen occurs if the 3,4-dihydroisoquinoline is heated with formaldehyde, formic acid, and diethylamine or sodium formate to form the o-substituted N,N-dimethylphenethylamine (233).

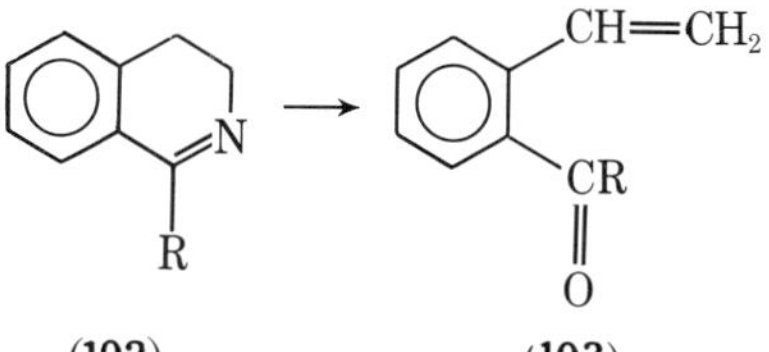

Hydroxyisoquinolines. Hydroxy groups on the benzene ring exhibit the usual phenol reactions. Thus, 5-, 6-, 7-, and 8-hydroxyisoquinolines undergo the Bücherer reaction to the corresponding aminoisoquinolines, the Mannich condensation, the azo coupling reaction, and nitrosation. 1-Hydroxyisoquinoline (isocarbostyril) gives both O-methyl and N-methyl derivatives on methylation; this indicates the existence of both tautomeric forms. The oxygens in isocarbostyrils, homophthalimides, and 4-hydroxyisocarbostyrils are removed by heating or distilling the compounds with zinc dust. Treatment of 1-isoquinolinol with phosphorus tribromide yields 1-bromoisoquinoline [1532-71-4] (234).

Haloisoquinolines. Generally, chloroisoquinolines are prepared from aminoisoquinolines by the Sandmeyer reaction or from isocarbostyrils by treatment with chlorides of phosphorus. The chloro substituent may also be present in the aromatic intermediates used in the synthesis, eg, in the preparation of 1,3-dichloroisoquinoline [7742-73-6] and 1,3,7-trichloroisoquinoline [21902-41-0] from benzyl cyanide and 4-chlorobenzyl cyanide (235). The addition of bromine to a slurry of isoquinoline hydrochloride in nitrobenzene affords 4-bromoisoquinoline [1532-97-4] in 70–80% yields (25). 1-Bromo-3-aminoisoquinoline [13130-79-5] has been obtained from 2-cyanobenzyl cyanide by treatment with dry hydrogen bromide followed by treatment with sodium bicarbonate (236). 1-Iodoisoquinoline [19658-77-6] forms as a result of heating 1-chloroisoquinoline [19493-44-8] with sodium iodide and 50 wt % hydriodic acid (237).

The haloisoquinolines undergo reactions typical of such compounds. 1-Nitroisoquinoline [19658-76-5] is obtained by heating 1-iodoisoquinoline at 100°C with sodium nitrite (237). Both 1- and 3-bromoisoquinoline are converted by potassium amide in liquid ammonia into the corresponding amine derivatives in excellent yields (238). In the case of 4-bromo- and 4-iodoisoquinolines, reduction and coupling reactions occur simultaneously and form isoquinoline, 4,4'-biisoquinoline, and 1-amino-4,4'-biisoquinoline (239). Reaction of 5-bromoisoquinoline with potassium amide in liquid ammonia at −33°C yields 6-aminoisoquinoline [23687-26-5] in 47% yield and 5-aminoisoquinoline [1125-60-6] in 21% yield. The latter is converted to 5-bromoisoquinoline [34784-04-8] by the Sandmeyer reaction (240).

Other. o-Disubstituted benzenes are utilized to synthesize isoquinoline. o-Cyanomethylbenzoic acid (**104**) when heated affords homophthalimide (**105**) or, when treated with sodium cyanide at 215°C, affords 3-(o-carboxybenzoyl)-4-cyano-1(2H)-isoquinolone (**106**) (241). Treatment of (**104**) by the Radziszewski method followed by dehydration gives a 90% yield of homophthalimide (242). 4-Acetylisochroman-1,3-diones, which are obtained by reaction of (**105**) with acetic anhydride, react with ammonia or methylamine to give 3-methyl- or 2,3-dimethyl-1(2H)-isoquinolone (**107**) and (**108**), respectively in good yields (243). Chloroisocoumarin (**109**) upon treatment with ammonia affords (**110**), which cyclizes to (**105**) when treated with base (244).

Thermal Curtius reaction of *trans*-cinnamoyl azides, which form from the reaction of the acid chloride with sodium azide, gives the corresponding *trans*-β-styryl isocyanates (**111**). The latter can be efficiently cyclized to the corresponding isocarbostyril (**112**) (245).

N,N-Dimethyl-β-phenethylamine is lithiated in the o-position and treated with dry paraformaldehyde to give N,N-dimethyl-o-hydroxymethyl-β-phenylethylamine, which cyclizes upon treatment with the methyl ester of chlorosulfonic acid (CH_3O-SO_2Cl) in pyridine and eventually yields isoquinoline (246). Benzylamine, upon reaction with bromoethanol followed by reaction with hydrogen bromide, yields the hydrobromide of the N-(2-bromoethyl) derivative, which is cyclized with aluminum chloride to give 1,2,3,4-tetrahydroisoquinoline (247).

(104)

(105) [88-97-1]

(106) [81045-38-7]

(107) [59816-89-6]

(108) [7114-78-5]

(109)

(110)

(111)

(112)

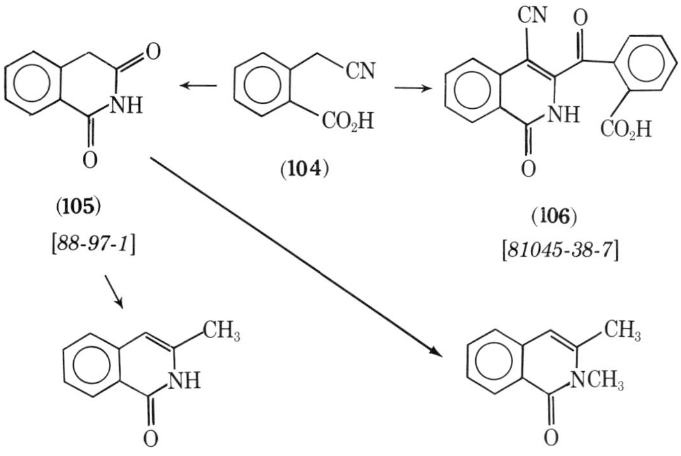

Toxicology

Isoquinoline vapors are irritating to the eyes, nose, and throat and may cause headaches, dizziness, nausea, etc. The liquid is rapidly absorbed through the skin and it causes skin burns and eye irritation. Toxicity data include oral LD_{50} (rat), 350 mg/kg; dermal LD_{50} (rabbit), 590 mg/kg (86). Isoquinoline is considered to be highly toxic if ingested and moderately toxic if absorbed through the skin.

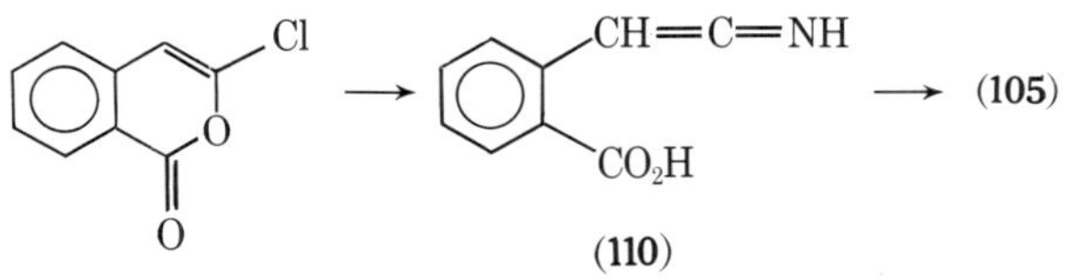

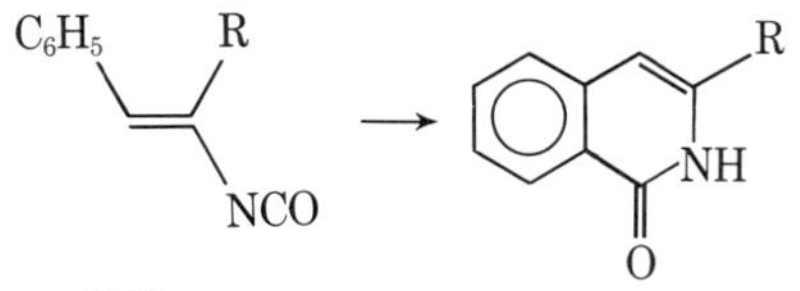

Uses. Isoquinoline and isoquinoline derivatives are useful as corrosion inhibitors, antioxidants, pesticides, and catalysts, and they are used in plating baths and miscellaneous applications, such as in photography, polymers, and azo dyes (qv). Numerous derivatives have been prepared and evaluated as pharmaceuticals. Isoquinoline is a main component in quinoline still residue bases, which are sold as corrosion inhibitors and acid inhibitors for pickling of iron and steel.

4-Aminoisoquinoline is a component of an ethylene glycol-based corrosion-in-

Table 5. Isoquinoline-Derived Drugs Marketed in the United States[a]

Name	CAS Registry No.	Structure	Application	Brand name (manufacturer)
2-acetoxy-9,10-dimethoxy-1,3,4,6,7,11b-hexahydro-2H-benzo[a]quinolizin-3-carboxylic acid diethylamide	[63-12-7]	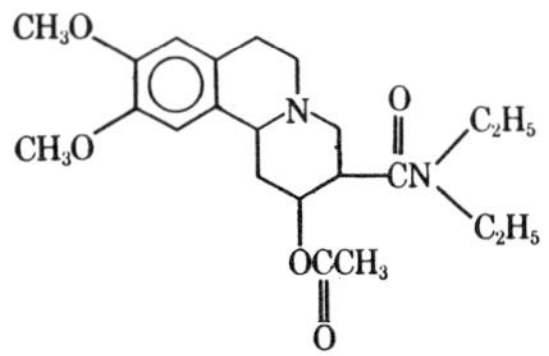	antiemetic, tranquilizer	Emete-Con (Roerig)
1,2,3,4-tetrahydroisoquinoline-2-carboxamidine	[1131-64-2]	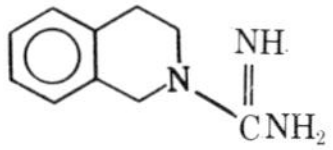	muscle relaxant	Metubine (Lilly)
1-(4-ethoxy-3-methoxybenzyl)-6,7-dimethoxy-3-methylisoquinoline	[147-27-3]	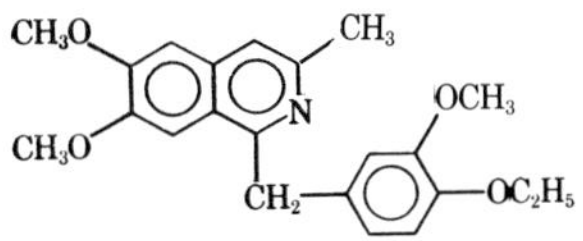	spasmolytic	Paveril (Lilly)
1-(3,4-diethoxybenzyl)-6,7-diethoxyisoquinoline	[486-47-5]	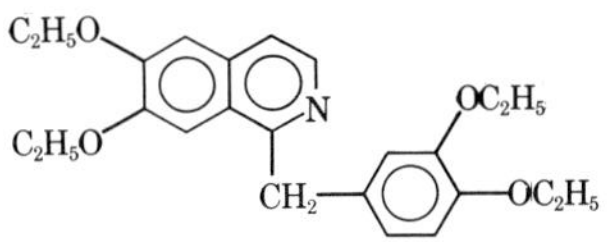	spasmolytic	Ethoquin (Ascher) Tensodin (Knoll)
papaverine	[58-74-2]	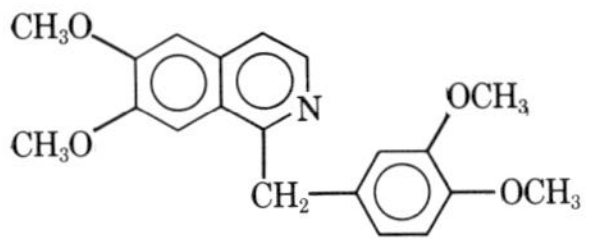	spasmolytic, vasodilator	numerous names
d-tubocurarine chloride	[57-94-3]	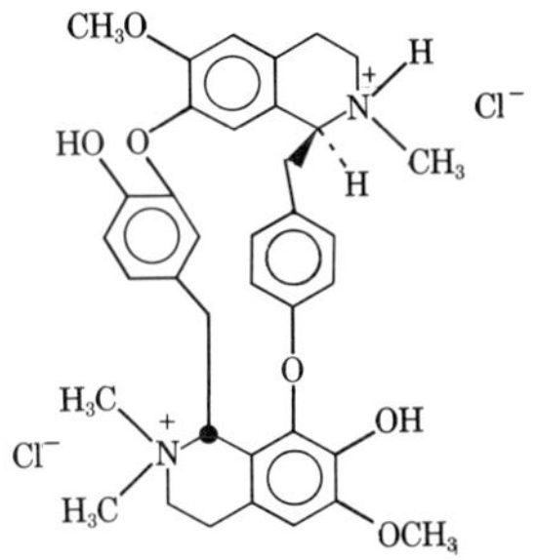	muscle relaxant	Tubocurarine Chloride (Abbott, Lilly, Squibb)

[a] Ref. 146.

hibiting antifreeze agent (248) (see Antifreezes). s-Triazolo[5,1-a]isoquinolines (113) and their derivatives are antioxidants, corrosion inhibitors, and acid acceptors, and they are prepared by cyclization of the amidines (114) obtained by heating 1-aminoisoquinoline with RCN (249). 1,2-Dihydroisoquinoline-1-phosphonates ((115); R = H [*39233-31-3*], CH_3 [*39233-30-2*]) are effective corrosion inhibitors for steel (250). Dihydro- (116) and tetrahydroisoquinolines (117) are used as antioxidants for fats and oils (251). The *N*-methyl derivative (118) also has antioxidant properties (252).

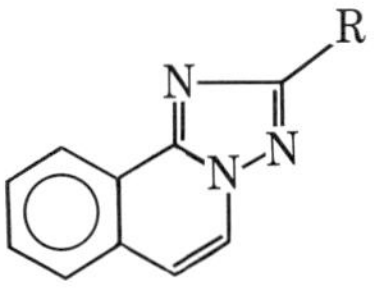

(113) 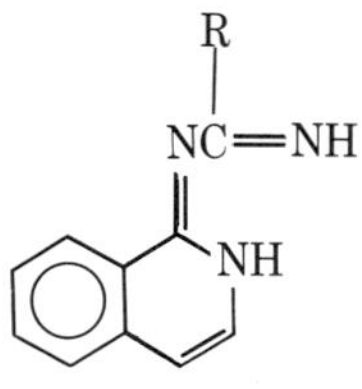   (114) 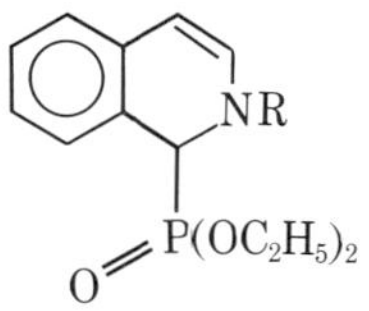   (115)

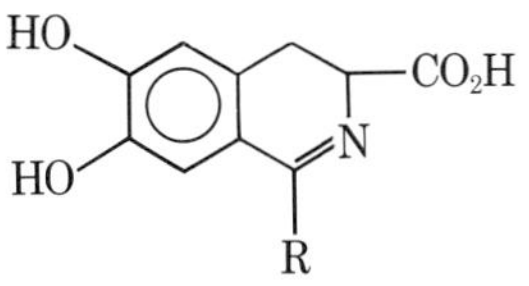

(116) 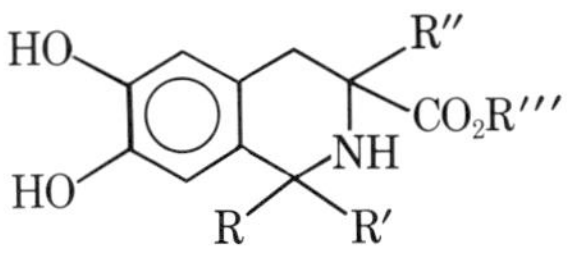   (117) 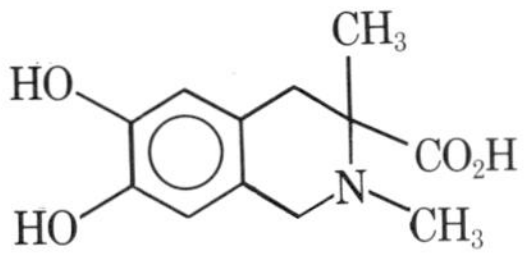   (118)
                [*62356-02-9*]

A number of isoquinoline derivatives are useful as pesticides, eg, (119) has fungicidal properties (253). Substituted isoquinolines are effective in controlling undesired vegetation, insects, acarina, or fungi (254–255). 2-Substituted-1,3(2*H*,4*H*)-isoquinolinediones [(120); X and Y = CH_3, CH_3O, or Cl] are plant-growth regulators (256) (see Plant-growth substances).

Isoquinoline is used as a catalyst in several applications (257–259). A number of miscellaneous applications of isoquinoline involve photography and dyes (260–263).

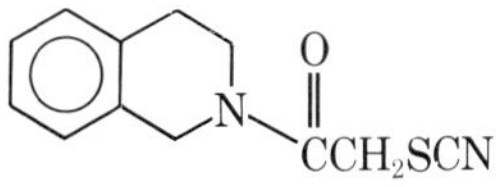

(119)
[*41910-26-3*]

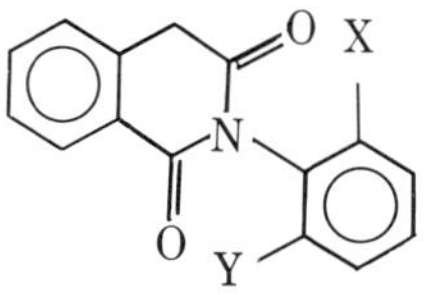

(120)

As with quinoline, isoquinoline derivatives are the parent substances of many plant alkaloids and of synthetic medicinal compounds. The principal drugs that contain the isoquinoline structure and that are marketed in the United States are listed in Table 5.

BIBLIOGRAPHY

"Quinoline and Isoquinoline" in *ECT* 1st ed., Vol. 11, pp. 389–401, by Alice G. Renfrew, Mellon Institute; "Quinoline and Isoquinoline" in *ECT* 2nd ed., Vol. 16, pp. 865–885, by Marshall Kulka, Uniroyal (1966) Ltd.

1. W. K. Miller, S. B. Knight and A. Roe, *J. Am. Chem. Soc.* **72,** 4763 (1950).
2. G. Jones, "Quinolines, Part I," in A. Weissberger and E. C. Taylor, eds., *The Chemistry of Heterocyclic Compounds,* Vol. 32, Wiley-Interscience, New York, 1977.
3. G. Grenthe, "Isoquinolines, Part One," in A. Weissberger and E. C. Taylor, eds., *The Chemistry of Heterocyclic Compounds,* Vol. 38, Wiley-Interscience, New York, 1981.
4. G. W. Wheland, *The Theory of Resonance,* John Wiley & Sons, Inc., New York, 1944, p. 70.
5. R. A. Friedel and M. Orchin, *Ultraviolet Spectra of Aromatic Compounds,* John Wiley & Sons, Inc., New York, 1951, pp. 270–271.
6. H. Luther and C. Reichel, *Z. Physik. Chem. (Leipzig)* **195,** 103 (1950).
7. M. W. Austin and J. H. Ridd, *J. Chem. Soc.,* 4204 (1963).
8. Dewar and Maitlis, *J. Chem. Soc.,* 2521 (1957).
9. M. Kulka and R. H. F. Manske, *Can. J. Chem.* **30,** 720 (1952).
10. R. G. Coombes and L. W. Russell, *J. Chem. Soc. Perkin Trans I,* 1751 (1974).
11. V. N. Konyukhov and M. K. Murshtein, *Izv. Vyssh. Uchebn. Zaved. Khim. Khim. Technol.* **18,** 1267 (1975); *Chem. Abstr.* **84,** 30833 (1976).
12. D. W. Rangnekar and S. V. Sunthankar, *Indian J. Technol.* **13,** 460 (1975).
13. A. R. Firth and R. D. Smith, *J. Appl. Chem. Biotechnol.* **28,** 857 (1978).
14. O. F. Sidorov, M. K. Murshtein, and L. F. Vasil'chenko, *Izv. Vyssh. Uchebn. Zaved. Khim. Khim. Tekhol.* **15,** 723 (1972); *Chem. Abstr.* **77,** 75107 (1972).
15. H. Gilman, J. Eisch, and T. S. Soddy, *J. Am. Chem. Soc.* **81,** 4000 (1959).
16. J. J. Eisch and D. R. Comfort, *J. Org. Chem.* **40,** 2288 (1975).
17. C. E. Crawforth, O. Meth-Cohn, and C. A. Russell, *J. Chem. Soc. Perkin Trans. I,* 2807 (1972) and references therein; D. I. C. Scopes and J. A. Joule, *J. Chem. Soc. Perkin Trans. I,* 2810 (1972).
18. K. Akiba, Y. Negishi, and N. Inamoto, *Synthesis,* 55 (1979).
19. H. A. Hageman, "The Von Braun Cyanogen Bromide Reaction," in R. Adams and co-workers, eds., *Organic Reactions,* Vol. 7, John Wiley & Sons, Inc., New York, 1953.
20. Y. Hamada and M. Sugiura, *Yakugaku Zasshi* **99,** 445 (1979).
21. *Ibid.,* **98,** 1 (1978).
22. F. W. Bergstrom, *Chem. Rev.* **35,** 77 (1944).
23. M. A. Solekhova and co-workers, *Khim. Geterotsikl. Soedin.,* 229 (1976); *Chem. Abstr.* **84,** 150472 (1976).
24. J. J. Eisch, *J. Org. Chem.* **27,** 1318 (1962).
25. T. J. Kress and S. M. Constantino, *J. Heterocycl. Chem.* **10,** 409 (1973).
26. Fr. Demande 2,207,100 (June 14, 1974), T. J. Kress (to Eli Lilly and Co.).
27. P. B. D. De la Mare, M. Kiamud-din, and J. H. Ridd, *Chem. Ind. London,* 1958, 361; *J. Chem. Soc.,* 561 (1960).
28. M. Gordon and D. E. Pearson, *J. Org. Chem.* **29,** 329 (1964).
29. T. Kato, H. Yamanaka, and T. Shimizu, *Yakugaku Zasshi* **93,** 73 (1973).
30. Ch. Churdaru and co-workers, *Synthesis,* 356 (1974).
31. U.S. Pat. 2,964,529 (Dec. 13, 1960), M. G. Sturrock and co-workers (to Koppers Co.); L. P. Yurkina and co-workers, *Chem. Abstr.* **69,** 35876 (1968).
32. J. Bialek, *Przem. Chem.* **46,** 526 (1967); *Chem. Abstr.* **68,** 114393 (1968).
33. M. E. C. Biffin, J. Miller, and D. B. Paul, *Tetrahedron Lett.,* 1015 (1969).
34. W. E. McEwen and R. L. Cobb, *Chem. Rev.* **55,** 511 (1955); R. L. Cobb and W. E. McEwen, *J. Am. Chem. Soc.* **77,** 5042 (1955); R. F. Collins and T. Henshall, *J. Chem. Soc.,* 1881 (1956); F. D. Popp and W. Blount, *Chem. Ind. London,* 550 (1961); F. D. Popp, *Adv. Heterocycl. Chem.* **9,** 1 (1968).
35. R. Hull, *J. Chem. Soc. C,* 1777 (1968).
36. T. Koizumi and co-workers, *Synthesis,* 497 (1977).

37. H. Nozaki and co-workers, *Tetrahedron Lett.*, 4259 (1967).
38. R. Noyori and co-workers, *Tetrahedron* **25,** 1125 (1969).
39. O. Porta and G. Sesana, *Tetrahedron Lett.*, 3571 (1978).
40. Ital. Pat. 906,418 (Feb. 1, 1972), F. Minisci and co-workers (to Montedison S.p.A.).
41. K. C. Bass and P. Nababsing, *J. Chem. Soc. C*, 2169 (1970).
42. B.-M. Bertilsson and co-workers, *Acta Chem. Scand.* **24,** 3590 (1970).
43. Ital. Pat. 918,225 (May 2, 1972), F. Minisci and co-workers (to Montedison S.p.A.).
44. H. Itokawa and co-workers, *Chem. Pharm. Bull.* **26,** 1015 (1978).
45. U.S. Pat. 3,428,641 (Feb. 18, 1969), R. C. Myerly and K. Weinberg (to Union Carbide Corp.).
46. U.S. Pat. 3,718,704 (Feb. 27, 1973), D. K. Chapman and co-workers (to Ashland Oil, Inc.); Fr. Demande 2,208,864 (June 28, 1974), D. K. Chapman and co-workers (to Ashland Oil, Inc.).
47. H. Yamanka, H. Egawa, and T. Sakamoto, *Chem. Pharm. Bull.* **26,** 2759 (1978).
48. U.S. Pat. 3,891,656 (June 24, 1975), E. K. Fields (to Standard Oil Co., Indiana).
49. R. C. Elderfield, *Heterocyclic Compounds*, Vol. 4, John Wiley & Sons, Inc., New York, 1952, p. 282.
50. J. Neiser, *Ropa Uhlie* **9,** 191 (1967).
51. S. Landa, Z. Kafka, V. Galik, and M. Safai, *Collect. Czech. Chem. Commun.* **34,** 3967 (1969).
52. J. Z. Ginos, *J. Org. Chem.* **40,** 1191 (1975).
53. E. W. Vierhapper and E. L. Eliel, *J. Am. Chem. Soc.* **96,** 2256 (1974).
54. Swiss Pat. 540,908 (Oct. 15, 1973), T. Voelker and K. Hering (to Lonza Ltd.).
55. G. W. Gribble and P. W. Heald, *Synthesis*, 650 (1975).
56. R. Ruzicka, *Phenole und Basen-Vorkommen und Gewinnung*, Akademie-Verlag, Berlin, FRG, 1958.
57. *Koppers Industrial Chemicals Technical Data Sheet*, *Refined Quinoline*, Pittsburgh, Pa., March 1980.
58. E. Jantzen, *Das fraktionierte Destillieren und das fraktionierte Verteilen als Methoden zur Trennung von Stoffgemischen*, Dechema Monographie 48, Vol. V, Verlag Chemie, Berlin, FRG, 1932, pp. 117–142.
59. *Koppers Industrial Chemicals Technical Data Sheet*, *Refined QSR Bases*, Pittsburgh, Pa., March 1980.
60. USSR Pat. 289,089 (Dec. 8, 1970), Yu. I. Chumakov and M. S. Alyab'eva (to Kiev Institute of Civil Aviation Engineering).
61. M. Dong and co-workers, *J. Chromatogr. Sci.* **15,** 32 (1977).
62. M. Novotny and co-workers, *Anal. Chem.* **52,** 401 (1980).
63. J. Vymetal and A. Kulhankova, *Chem. Prum.* **26,** 193 (1976); *Chem. Abstr.* **86,** 19301 (1977).
64. *Ibid.*, **29,** 308 (1979); *Chem. Abstr.* **91,** 195671 (1979).
65. R. H. F. Manske and M. Kulka, "The Skraup Synthesis of Quinolines," in R. Adams and co-workers, eds., *Organic Reactions*, Vol. 7, John Wiley & Sons, Inc., New York, 1953.
66. A. P. Sandy, Jr. and M. Lodolini, *Res. Discl.* **164,** 9 (1977).
67. V. V. Kormachev and co-workers, *Zh. Obshchei Khim.* **45,** 307 (1975).
68. I. T. Millar and H. D. Springall, *Sidgwick's Organic Chemistry of Nitrogen*, 3rd ed., Clarendon Press, Oxford, UK, 1966, Chapt. 24.
69. A. B. Turner, *Chem. Commun.*, 1659 (1960); G. A. Dauphinee and T. P. Forrest, *J. Chem. Soc. D*, 327 (1969); Y. Ogata, A. Kawasaki and S. Suyama, *J. Chem. Soc. B*, 805 (1969); O. Schindler and W. Michaelis, *Helv. Chim. Acta* **53,** 776 (1970).
70. P. A. Claret and A. G. Osborne, *J. Chem. Technol. Biotechnol.* **29,** 175 (1979).
71. C. M. Leir, *J. Org. Chem.* **42,** 911 (1977).
72. G. Saint-Ruf, A. De, and J. C. Perche, *Bull. Soc. Chim. Fr.*, 2514 (1973).
73. R. H. Reitsema, *Chem. Rev.* **43,** 43 (1968).
74. Ng. Ph. Buu-Hoi and R. Roger, *J. Chem. Soc.*, 106 (1948).
75. F. A. Al-Tai, G. Y. Sarkis, and J. M. Al-Janabi, *Bull. Coll. Sci.* **9,** 55 (1966).
76. F. A. Al-Tai, A. M. El-Abbady, and A. S. Al-Tai, *J. Chem. U.A.R.* **10,** 339 (1967).
77. I. Shopov, *Vysokomol. Soedin. Ser. B* **11,** 248 (1969).
78. S. O. Norris and J. K. Stille, *Macromolecules* **9,** 496 (1976).
79. R. C. Fuson and D. M. Burness, *J. Am. Chem. Soc.* **68,** 1270 (1946).
80. Brit. Pat. 1,184,242 (Mar. 11, 1970), W. H. Bell and R. A. C. Rennie (to Imperial Chemical Industries, Ltd.).
81. Ger. Pat. 2,459,095 (June 26, 1975), J. A. S. Thoma and P. A. M. J. Stijfs (to Stamicarbon B.V.).

82. T. Manimaran and co-workers, *Synthesis*, 739 (1975).
83. Y. Watanabe and co-workers, *Bull. Chem. Soc. Jpn.* **51,** 3397 (1978).
84. H. Beschke and H. Friedrich, *Chem. Stg.* **101,** 377 (1977).
85. Hung. Pat. 157,370 (Apr. 22, 1970), D. Ambrus and co-workers; *Chem. Abstr.* **73,** 45371 (1970); Span. Pat. 369,261 (May 16, 1971), C. M. Sarabia.
86. N. I. Sax, *Dangerous Properties of Industrial Materials*, 5th ed., Van-Nostrand Reinhold Co., New York, 1979.
87. Ger. Offen. 2,822,722 (Dec. 7, 1978), H. Nagasaki, T. Kojima, and Y. Shiro (to Sumitomo Chemical Co., Ltd.); Jpn. Kokai Tokkyo Koho 78,145,854 (Dec. 19, 1978).
88. Ger. Offen. 2,832,126 (Feb. 8, 1979), S. Sato and T. Watanabe (to Bridgestone Tire Co., Ltd.).
89. Brit. Pat 1,113,360 (May 15, 1968), (to Ciba-Geigy Ltd.); S. Afr. Pat. 67 04,722 (Jan. 19, 1968), H. Rembold and A. Renner (to Ciba-Geigy Ltd.).
90. Jpn. Kokai 76 148,739 (Dec. 21, 1976), T. Fujita and M. Okada (to Dainichi Nippon Cables, Ltd.).
91. Jpn. Kokai 78 51,206 (May 10, 1978), T. Uematsu and co-workers (to Maruzen Oil Co., Ltd.; Hitachi, Ltd.).
92. Ger. Offen. 2,833,600 (Feb. 22, 1979), K. Katabe and T. Hirota (to Kao Soap Co., Ltd.); Jpn. Kokai Tokkyo Koho 79 30,998 (Mar. 7, 1979), K. Katabe and T. Hirota (to Kao Soap Co., Ltd.).
93. Jpn. Pat. 71 36,625 (Oct. 27, 1971), T. Gono, E. Tanaka, and Z. Enomoto (to Yoshitomi Pharmaceutical Industries, Ltd.).
94. A. Pelevin, *Krmivastvi Sluzby* **12,** 224 (1976); *Chem. Abstr.* **86,** 138271 (1977).
95. Jpn. Kokai Tokkyo Koho 78 46,839 (Dec. 16, 1978), (to Material Vegyi KSZ).
96. Ger. Offen. 2,439,534 (Mar. 4, 1976), H. U. Voigt (to Kabel-und Metallwerke Gutehoffnungshuette A.-G.).
97. Brit. Pat. 1,153,178 (May 29, 1969), I. Norvick.
98. N. K. Patel, M. M. Patel, A. B. Patel, and J. C. Vora, *Labdev*, Part A **12-**A, 113 (1974).
99. Fr. Pat. 1,529,230 (June 14, 1968), I. G. Rose (to National Research Development Corporation).
100. Ger. Offen. 2,149,138 (Apr. 5, 1973), C. Rasp (to Bayer A.-G.); Jpn. Kokai 77 94,880 (Aug. 9, 1977), K. Mitamura and H. Yokota (to Nippon Oil, Ltd.).
101. G. Buchmann and O. Wolniak, *Pharmazie* **21,** 650 (1966).
102. Ger. Offen. 2,322,143 (Dec. 13, 1973), D. Cartwright and co-workers (to Imperial Chemical Industries, Ltd.).
103. G. Pagani and G. Caccialanza, *Farmaco. Ed. Sci.* **31,** 364 (1976).
104. Swiss Pat. 550,171-2 (June 14, 1974), E. Sturm and C. Vogel (to Agripat S.A.).
105. Jpn. Kokai 77 72,821 (June 17, 1977), T. Takematsu and co-workers (to Mitsubishi Petrochemicals Co., Ltd.).
106. Ger. Offen. 2,437,297 (Feb. 20, 1975), F. Gialdi and co-workers (to Montedison, S.p.A.).
107. Ger. Offen. 2,361,438 (June 26, 1975), A. Studeneer and co-workers (to Hoechst A.-G.).
108. A. S. Sadykov and co-workers, *Uzb. Biol. Zh.* **11,** 19 (1967); *Chem. Abstr.* **67,** 107563 (1967).
109. D. L. Lewis and co-workers, *J. Econ. Entomol.* **71,** 818 (1978).
110. Czech. Pat. 158,491 (June 15, 1975), J. Ciernik and D. Ambrus.
111. Jpn. Kokai Tokkyo Koho 79,110,292 (Aug. 29, 1979), H. Kanbara (to Tokyo Organic Chemical Industries, Ltd.).
112. G. V. Myasoedova and co-workers, *Org. Reagenty Anal. Khim. Tezisy Dokl. Vses. Konf. 4th*, **1,** 116 (1976); *Chem. Abstr.* **87,** 189883 (1977).
113. V. N. Tolmachev and N. N. Orlova, *Vysokomol. Soedin Ser. B* **11,** 284 (1969); *Chem. Abstr.* **71,** 30963 (1969).
114. F. Vernon and K. M. Nyo, *Sep. Sci. Technol.* **13,** 273 (1978).
115. F. Vernon and K. M. Nyo, *Anal. Chim. Acta* **93,** 203 (1977).
116. Ger. Offen. 2,407,306-7 (Aug. 28, 1975), K. Idel and co-workers (to Bayer A.-G.).
117. Jpn. Kokai Tokkyo Koho 79,139,989 (Oct. 30, 1979), H. Kanbara (to Tokyo Organic Chemical Industries, Ltd.).
118. N. Grassie and I. G. Meldrum, *Eur. Polym. J.* **4,** 571 (1968).
119. Fr. Pat. 1,497,474 (Oct. 13, 1967), P. Enthone.
120. Jpn. Kokai 76,126,936 (Nov. 5, 1976), S. Emori (to Seiken Chemical Co., Inc.).
121. U.S. Pat. 4,009,087 (Feb. 22, 1977), O. Kardos and co-workers (to M&T Chemicals, Inc.).
122. U.S. Pat. 4,038,161 (July 26, 1977), W. E. Eckles and T. W. Starinshak (to R. O. Hull and Co., Inc.).
123. Indian Pat. 140,766 (Dec. 18, 1976), J. Adhya; *Chem. Abstr.* **92,** 84920 (1980).

124. Ger. Offen. 2,443,743 (Mar. 27, 1975), J. A. Hartlage (to Ashland Oil, Inc.).
125. U.S. Pat. 4,045,441 (Aug. 30, 1977) and Ger. Offen. 2,710,491 (Sept. 15, 1977), H. J. Richards and B. C. Trivedi (to Ashland Oil, Inc.).
126. Fr. Demande 2,309,645 (Nov. 26, 1976), (to International Nickel Co. of Canada, Ltd.).
127. S. Afr. Pat. 75 00,412 (Dec. 8, 1975), P. L. Mattison (to General Mills Chemicals, Inc.).
128. Jpn. Kokai 73 55,558 (Aug. 4, 1973), N. Yokota and co-workers (to Osaka Soda Co., Ltd.).
129. U.S. Appl. 925,672 (Mar. 2, 1979), W. N. Marchant and P. T. Brooks (to United States Department of the Interior); P. B. Altringer and co-workers, *Min. Eng.* (*Littleton, Colo.*) **31,** 1220 (1979).
130. E. A. Ostroumov and A. V. Kulumbegashvii, *Zh. Anal. Khim.* **31,** 1338 (1976).
131. Fr. Demande 2,401,228 (Mar. 23, 1979), (to Consiglio Nazionale delle Ricerche).
132. D. Dorz and J. Dobrowolski, *Chem. Anal. Warsaw* **24,** 465 (1979).
133. U.S. Pat. 4,100,163 (July 11, 1978), M. J. Virnig (to General Mills Chemicals, Inc.).
134. U.S. Pat. 3,671,471 (June 20, 1972), S. E. Jamison (to Celanese Corp.).
135. Jpn. Pat. 15,621 (Aug. 29, 1967), R. Wakasa, S. Ishida, and Y. Kitahama (to Asahi Chemical Industry Co., Ltd.).
136. U.S. Pat. 4,000,210 (Dec. 28, 1976), E. E. Sensel and A. W. King (to Texaco, Inc.); Neth. Appl. 76 13,473 (June 6, 1978), (to Texaco Development Corp.).
137. Jpn. Kokai Tokkyo Koho 79 07,764 (Apr. 10, 1979), S. Nagato and co-workers (to Daicel Ltd.).
138. Ger. Offen. 2,901,347 (July 19, 1979), H. Kojima and co-workers, (to Daicel Ltd.).
139. I. M. Kolthoff and E. B. Sandell, *Textbook of Quantitative Inorganic Analysis*, 3rd ed., The Macmillan Company, New York, 1952, pp. 87–90.
140. Ger. Offen. 2,320,880 (Oct. 31, 1973), P. H. Cardwell and J. A. Olander (to Deepsea Ventures, Inc.).
141. Jpn. Kokai Tokkyo Koho 80 10,506 (Jan. 25, 1980), T. Tarui and M. Hashimoto (to Idenitsu Kosan Co., Ltd.).
142. K. Yamamoto and co-workers, *Bunseki Kagako* **16,** 229 (1967).
143. H. F. Schaeffer, *Microchem. J.* **14**(1), 90 (1969).
144. K. Watanabe and co-workers, *Bunseki Kagaku* **26,** 570 (1977).
145. A. Sturis and J. Bankovskis, *Org. Reagenty Anal. Khim. Tezisy Dokl. Vses. Konf.*, *4th*, **1,** 98 (1976); *Chem. Abstr.* **88,** 31500 (1978).
146. A. Kleemann, *Pharmazeutische Wirkstoffe—Syntheses·Patente·Anwendungen*, Georg Thieme Verlag, Stuttgart, FRG, 1978.
147. W. W. Pandler and M. Cheplen, *Fuel* **58,** 775 (1979).
148. F. K. Schweighardt and co-workers, *ACS Symp. Ser.* **71,** 240 (1978).
149. B. I. Ardashev and co-workers, *Khim. Geterotsikl. Soedin.*, 857 (1968); *Chem. Abstr.* **70,** 106351 (1969).
150. Jpn. Kokai 77 83,378 (July 12, 1977), T. Nakano and T. Morimoto (to Mitsui Petrochemical Industries, Inc.).
151. N. S. Kozlov and G. P. Korotyshova, *Tezisy. Dokl. Bsec. Konf. Khim. Atsetilena*, *5th*, 288 (1975); *Chem. Abstr.* **88,** 169924 (1978).
152. S. Biniecki and W. Modrzejewska, *Acta Pol. Pharm.* **24,** 561 (1967).
153. USSR Pat. 541,846 (Jan. 5, 1977), Yu. M. Yulitov and L. I. Kovaleva (to L. V. Pisarzhevskii Institute of Physical Chemistry, Dovetsk); *Chem. Abstr.* **86,** 171448 (1977).
154. S. Skidmore and E. Tidd, *J. Chem. Soc.*, 1098 (1961).
155. Jpn. Pat. 73 05,598 (Feb. 17, 1973), M. Fukushima and co-workers (to Chisso Corp.).
156. S. Paraskewas, *Synthesis*, 819 (1974).
157. Yu. I. Chumakov and N. B. Bulgakova, *Ukr. Khim. Zh.* **36,** 514 (1970); *Chem. Abstr.* **73,** 55950 (1970).
158. Yu. I. Chumakov and N. B. Bulgakova, *Khim. Geterotsikl. Soedin.* **7,** 1533 (1971); *Chem. Abstr.* **77,** 5307 (1972).
159. T. Kusumi, Ko Yoneda, and H. Kakisawa, *Synthesis*, 221 (1979).
160. Jpn. Kokai 77 116,478 (Sept. 29, 1977), K. Takimoto (to Mitsubishi Monsanto Chemical Co.).
161. Ger. Offen. 2,822,907 (Nov. 29, 1979), K. Handrick and G. Koelling (to Bergwerksverband G.m.b.H.).
162. H. Beschke, *Aldrichimica Acta* **11,** 13 (1978).
163. Brit. Pat. 1,059,701, 1,059,702 and 1,059,703 (Feb. 22, 1967), (to Allied Chemical Corp.).
164. Brit. Pat. 1,463,583 (Feb. 2, 1977), A. C. W. Curran and D. G. Hill (to Wyeth, John and Brother Ltd.); Brit. Pat. 1,463,584 (Feb. 2, 1977), A. C. W. Curran, D. G. Hill and Crossley (to Wyeth, John and Brother, Ltd.).

165. J. J. Wandewalle and co-workers, *Chem. Ber.* **108,** 3898 (1975).
166. K. M. Dyumaev and E. P. Popova, *Khim. Geterotsikl Soedin.*, 85 (1973); *Chem. Abstr.* **79,** 78567 (1973).
167. P. Tomasik and A. Woszczyk, *Tetrahedron Lett.*, 2193 (1977).
168. Hung. Teljes 6251 (May 28, 1973), A. Ujhidy and co-workers (to E. Gy. T. Gyogyszervegyeszeti Gyar); *Chem. Abstr.* **79,** 92026 (1973).
169. Ger. Pat. 2,352,976 (June 20, 1974), F. M. Covelli (to Koppers Co., Inc.).
170. Jpn. Pat. 72 37,436 (Sept. 20, 1972), I. Onishi and co-workers (to Yuki Gosei Kogyo Co., Ltd.).
171. Ger. Offen. 2,545,704 (Apr. 22, 1976), J. M. Cognion (to Ugine Kuhlmann).
172. Fr. Demande 2,009,119 (Jan. 30, 1970), (to Badische Anilin- und Soda-Fabrik A.G.).
173. Fr. Demande 2,009,125 (Jan. 30, 1970) and Ger. Offen. 1,924,362 (Nov. 19, 1970), H. J. Sturm and H. Goerth (to Badische Anilin- und Soda-Fabrik A.-G.).
174. T. Noguchi and co-workers, *J. Biochem. Tokyo* **67**(1), 113 (1970).
175. Czech. Pat. 180,910 (Sept. 15, 1979), J. Simek and M. Ulrich.
176. Ger. Offen. 2,515,476 (Oct. 21, 1976), H. Jenkner and G. Neisen (to Chemische Fabrik Kalk G.m.b.H.).
177. USSR Pat. 591,467 (Feb. 5, 1978), A. P. Zuev and co-workers (to Ordzhonikidze, S., All-Union Scientific Research Chemical-Pharmaceutical Institute); *Chem. Abstr.* **88,** 152449 (1978).
178. Ger. Offen. 2,843,781 (Apr. 12, 1979), J. Bulidon and C. Pavan (to Roussell-UCLAF).
179. Fr. Pat. 1,514,280 (Feb. 23, 1968), R. Joly, J. Warnant, and B. Goffinet (to Roussell-UCLAF).
180. Ger. Offen. 2,213,233 (Sept. 27, 1973), E. Degener and H. Holtschmidt (to Bayer A.-G.).
181. U.S. Pat. 3,312,708 (Apr. 4, 1967), C. J. Lind (to Allied Chemical Corp.).
182. M. E. Kuehen and P. J. Sheeran, *J. Org. Chem.* **33,** 4406 (1968).
183. T. Higashino, U. Nagano, and E. Hayashi, *Chem. Pharm. Bull.* **21,** 1943 (1973).
184. K. K. Hsu and S. F. Chang, *T'ai wan K'o Hsueh* **29,** 51 (1975); **31,** 130 (1977).
185. Ref. 38, p. 186.
186. B. R. Birchall and A. H. Rees, *Can. J. Chem.* **52,** 610 (1974); Brit. Pat. 1,340,334 (Dec. 12, 1973), A. H. Rees.
187. A. K. Sheinkman, V. A. Ivanov, and S. N. Baranov, *Metody Poluch. Khim. Reaktin. Prep.*, No. 23, 64 (1971); *Chem. Abstr.* **78,** 29588 (1973).
188. V. N. Gogte, K. M. More, and B. D. Tilak, *Indian J. Chem.* **12,** 1238 (1974).
189. H. Sawanishi and Y. Kamiya, *Chem. Pharm Bull.* **23,** 2949 (1975).
190. U.S. Pat. 4,174,447 (Nov. 13, 1979), E. K. Fields (to Standard Oil Co., Indiana).
191. Ref. 38, Chapt. 5.
192. S. Kutkevicius and R. Valite, *Khim. Geterotsikl. Soedin*, 1117 (1972); *Chem. Abstr.* **77,** 139764 (1972).
193. Ref. 192, p. 1121.
194. U.S. Pat. 3,509,159 (Apr. 28, 1970), D. Kealey.
195. S. Kubota and co-workers, *Yakugaku Zasshi* **93,** 354 (1973).
196. A. Sturis and co-workers, *Latv. PSR Zinat. Akad. Vestis. Kim. Ser.*, 476 (1966); *Chem. Abstr.* **66,** 94900 (1967).
197. A. Sturis and co-workers, *Latv. PSR Zinat. Akad. Vestis. Kim. Ser.*, 181 (1979); *Chem. Abstr.* **91,** 39289 (1979).
198. A. O. Fritton and F. Ridgway, *J. Med. Chem.* **13,** 1008 (1970).
199. USSR Pat. 553,245 (Apr. 5, 1977), J. Bankovskii and M. Cirule (to Institute of Inorganic Chemistry, Academy of Sciences, Latvian SSR).
200. S. C. Wait and J. C. McNerney, *J. Molec. Spectrosc.* **34**(1), 56 (1970).
201. S. F. Mason, "The Electronic Absorption Spectra of Heterocyclic Compounds," in A. R. Katritsky, ed., *Physical Methods in Heterocyclic Chemistry*, Vol. 2, Academic Press, Inc., New York, 1963, p. 24.
202. F. Balkau and M. L. Heffernan, *Aust. J. Chem.* **24,** 2311 (1971).
203. M. J. S. Dewar and O. Trinajstic, *J. Chem. Soc. A*, 1220 (1971).
204. Ref. 8; R. A. Robinson, *J. Am. Chem. Soc.* **69,** 1943 (1947); C. G. LeFevre and R. J. W. LeFevre, *J. Chem. Soc.*, 1470 (1935).
205. D. C. Ayres and A. M. M. Hossain, *Chem. Soc. Trans Perkin I*, 707 (1975).
206. Y. Tamura, N. Tsuijimoto, and M. Uchimura, *Chem. Pharm. Bull.* **19,** 143 (1971); Jpn. Pat. 72 36,740 (Sept. 14, 1972), N. Tsujimoto and T. Tamura.
207. T. Koduo, T. Nose, and M. Hamana, *Yakugaku Zasshi* **95,** 521 (1975).

208. M. Hamana and S. Kumadaki, *Yakugaku Zasshi* **95,** 87 (1975).
209. M. Hamana and co-workers, *Chem. Pharm. Bull.* **23,** 346 (1975).
210. R. A. Abramovitch, R. B. Rogers, and G. M. Singer, *J. Org. Chem.* **40,** 41 (1975).
211. T. Onaka, *Itsuu Kenkyusho Nempo*, 29 (1968).
212. Ref. 49, p. 385.
213. M. Natsume and co-workers, *Tetrahedron Lett.*, 2335 (1973).
214. Brit. Pat. 1,149,463 (Apr. 23, 1969), (to Hoffmann-LaRoche Inc.).
215. I. Saito, Y. Kikugawa, and S. Yamada, *Chem. Pharm. Bull.* **22,** 740 (1974).
216. U.S. Pat. 3,717,639 (Feb. 20, 1973), J. L. Neumeyer (to Arthur D. Little, Inc.).
217. G. W. Kirby, S. L. Tan, and B. C. Uff, *J. Chem. Soc. Perkin Trans. I*, 270 (1979).
218. F. Minisci and co-workers, *Tetrahedron* **27,** 3575 (1971).
219. USSR Pat. 434,076 (June 30, 1974), A. K. Sheinkman and G. V. Samoilenko (to Donetsk State University).
220. W. M. Whaley and T. R. Govindachari, "Preparation of 3,4-Dihydroisoquinolines and Related Compounds by the Bischler-Napieralski Reaction," in R. Adams and co-workers, eds., *Organic Reactions*, Vol. 6, John Wiley & Sons, Inc., New York, 1951, Chapt. 2.
221. Y. Tsuda and co-workers, *Heterocycles* **5,** 157 (1976).
222. Ref. 220, Chapt. 3.
223. S. Teitel and A. Brossi, *Lloydia* **37,** 196 (1974).
224. M. Konda, T. Shioiri, and S. Yamada, *Chem. Pharm. Bull.* **23,** 1063 (1975).
225. W. J. Gensler in ref. 220, Chapt. 4.
226. M. J. Bevis, *Tetrahedron* **25,** 1585 (1969).
227. J. M. Bobbitt and co-workers, *J. Org. Chem.* **30,** 2247, 2459 (1965); ref. 3, pp. 222–226.
228. E. Schlitter and J. Muller, *Helv. Chim. Acta.* **31,** 914 (1948).
229. A. Roe and C. E. Teague, *J. Am. Chem. Soc.* **73,** 687 (1951).
230. L. M. Jackman and D. I. Packham, *Chem. Ind. London*, 360 (1955); H. Schmid and P. Karrer, *Helv. Chim. Acta* **32,** 960 (1949).
231. W. Huckel and G. Graner, *Chem. Ber.* **90,** 2017 (1957).
232. W. J. Gensler and co-workers, *J. Am. Chem. Soc.* **78,** 1713 (1956).
233. J. Gardent, *Compt. Rend.* **243,** 1042 (1956).
234. J. L. Butler, F. L. Bayer, and M. Gordon, *Trans. Ky. Acad. Sci.* **38,** 15 (1977).
235. Jpn. Pat. 71 39,699 (Nov. 22, 1971), S. Komori and S. Uanagita.
236. U.S. Pat. 3,277,096 (Oct. 4, 1966), F. Johnson (to The Dow Chemical Co.).
237. B. Hayashi and co-workers, *Yakugaku Zasshi* **87,** 1342 (1967).
238. G. M. Sanders, M. Vandijk, and H. J. Den Hertog, *Recl. Trav. Chim. Pays Bas* **93,** 198 (1974).
239. Ref. 244, p. 273.
240. H. Poradowska and E. Huczkowska, *Synthesis*, 733 (1975).
241. G. Pangon and co-workers, *Bull. Soc. Chim. Fr.*, 1991 (1970).
242. S. Nanya and E. Maekawa, *Nagoya Kogyo Daigaku Gakuho* **20,** 161 (1968).
243. S. A. Foster, J. Leyshorn, and D. G. Saunders, *Indian J. Chem.* **10,** 1060 (1972).
244. S. Karnik and R. N. Usganokar, *Curr. Sci.* **43,** 43 (1974).
245. G. J. Mikol and J. H. Boyer, *J. Org. Chem.* **37,** 724 (1972).
246. N. S. Narasimhan and A. C. Ranade, *Chem. Ind. London*, (3) 120 (1967).
247. R. D. Topsom, L. W. Deady and N. Pirzada, *J. Chem. Soc. D*, 799 (1971).
248. Ger. Offen. 2,149,138 (Apr. 5, 1973), C. Rasp (to Bayer A.-G.).
249. U.S. Pat. 3,758,480 (Sept. 11, 1973), H. K. Reimlinger and J. J. M. Vanderwalle (to Mallinckrodt Chemical Works).
250. U.S. Pat. 3,809,694 (May 7, 1974) and U.S. Pat. 3,830,815 (Aug. 20, 1974), D. Redmore (to Petrolite Corp.).
251. Ger. Offen. 2,342,474 (Mar. 7, 1974) and Jpn. Kokai 74 39,584 (Apr. 13, 1974), S. Okumura and co-workers (to Ajinomoto Co., Inc.).
252. Jpn. Kokai 76,105,982 (Sept. 20, 1976), I. Chibata and co-workers (to Tanabe Seiyaku Co., Ltd.).
253. U.S. Pat. 3,818,010 (June 18, 1974), U.S. Pat. 3,869,471 (March 4, 1975), U.S. Pat. 3,917,843 (Nov. 4, 1975), S. B. Richter and A. A. Levin (to Velsicol Chemical Corp.); U.S. Pat. 3,929,830 (Dec. 30, 1975), S. B. Richter (to Velsicol Chemical Corp.).
254. Aust. Pat. 465,390 (Sept. 25, 1975), A. Serban (to ICI Australia Ltd.).
255. U.S. Pat. 3,930,837 (Jan. 6, 1976), A. Serban (to ICI Australia Ltd.).
256. U.S. Pat. 4,097,260 (June 27, 1978), J. J. D'Amico (to Monsanto Co.).

257. Pol. Pat. 54,191 (Oct. 31, 1967), M. Uhniat, L. Kubiczck and M. Nowakowska (to Instytut Ciezkiej Syntezy Organicznej); *Chem. Abstr.* **68,** 96641 (1968).
258. Jpn. Kokai 77 95,608 (Aug. 11, 1977), S. Nakamura and co-workers (to Sumitomo Chemical Co., Ltd.); Jpn. Kokai 77 93,711 (Aug. 6, 1977).
259. Ger. Offen. 2,630,086 (Jan. 12, 1978), R. Platz and co-workers (to BASF A.-G.).
260. Ger. Offen. 2,332,317 (Jan. 16, 1975), P. Bergthaller and co-workers (to Agfa-Gevaert).
261. Ger. Offen. 2,444,422 (Mar. 27, 1975), N. Furutachi and A. Arai (to Fuji Photo Film Co., Ltd.).
262. Ger. Offen. 2,347,756 (Apr. 24, 1975), E. Fleckenstein and R. Mohr (to Farbwerke Hoechst A.-G.).
263. Ger. Offen. 2,650,226 (May 11, 1978), D. Rose (to Henkel und Cie G.m.b.H.).

General References

References 2, 3, and 49 are general references.

SAMUEL N. HOLTER
Koppers Company, Inc.

QUINONES

The oldest and best known example of this colorful and reactive class of cyclic enones is 1,4-benzoquinone. The first synthesis of this compound in 1838 involved the oxidation of quinic acid (**1**) (1). The generic name quinone (**2**) is derived from this starting material.

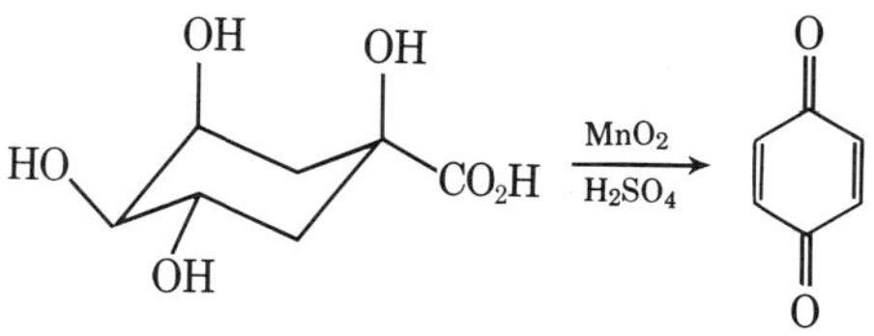

Oxidation continues to be the method of choice for the synthesis of quinones (2–4). The cross-conjugated system of two α,β-unsaturated carbonyl groups is closely related to the aromatic compound 1,4-benzenediol (hydroquinone) (**3**), and the two form a reversible redox couple.

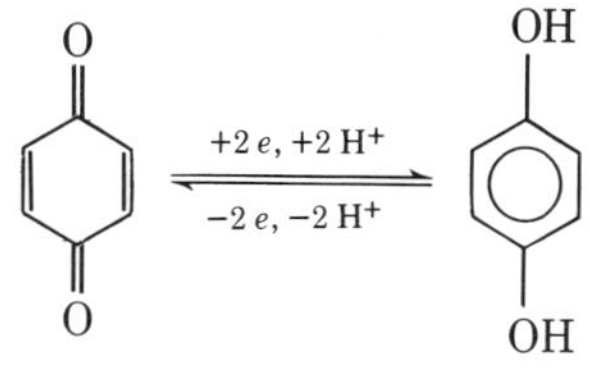

(structures of 2 ⇌ 3 with +2e, +2H⁺ / −2e, −2H⁺)

(2)   (3)

A similar relationship exists with 1,2-benzoquinone (4) and 1,2-benzenediol (catechol) (5).

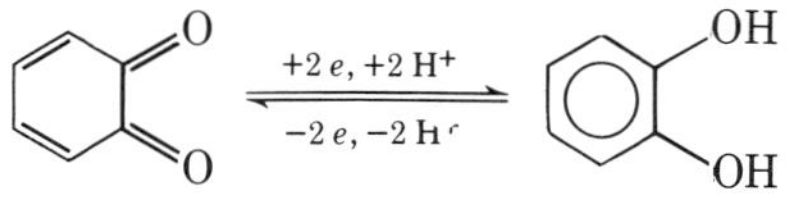

(4)   (5)

Both 1,2- and 1,4-dione types occur in many polynuclear hydrocarbons, eg, in naphthalene (qv), eg, (6) and (7), anthracene, eg, (8), (9), and (10), phenanthrene, eg, (11),

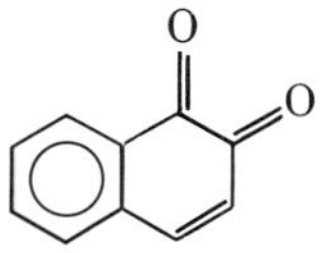

(6) 1,2-naphthoquinone
[524-42-5]

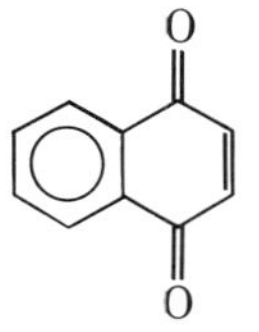

(7) 1,4-naphthoquinone
[130-15-4]

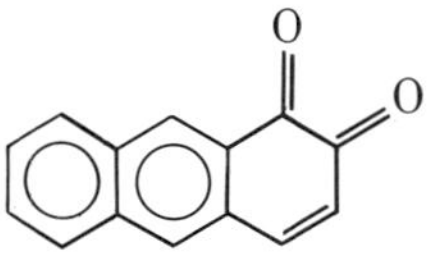

(8) 1,2-anthraquinone
[655-04-9]

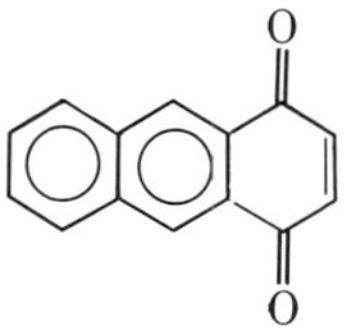

(9) 1,4-anthraquinone
[635-12-1]

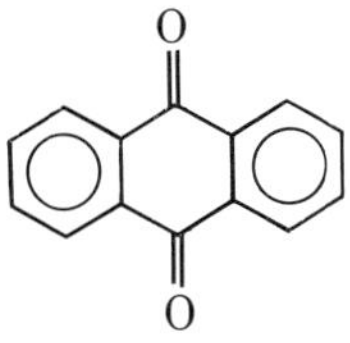

(10) 9,10-anthraquinone
[84-65-1]

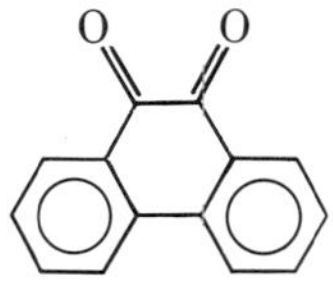

(11) 9,10-phenanthraquinone
[84-11-7]

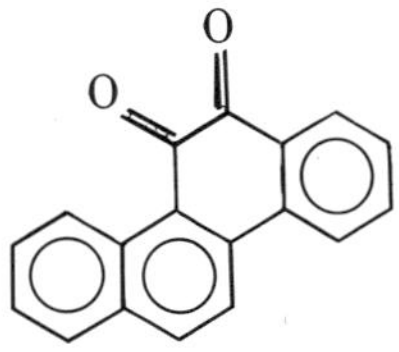

(12) 5,6-chrysenequinone (5,6-dihydrochrysene-5,6-dione)
[2051-10-7]

and chrysene, eg, (12). The carbonyl groups may be located in different rings, but they occupy positions corresponding to the 1,2- or 1,4-orientation of monocyclic quinones; eg, in naphthalene, eg, (13), biphenyl, eg, (14) and (15), and stilbene, eg, (16).

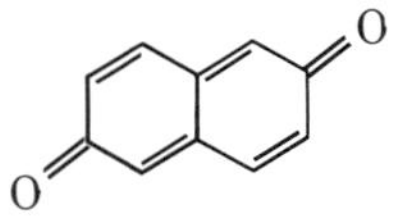

(13) 2,6-naphthoquinone
[613-20-7]

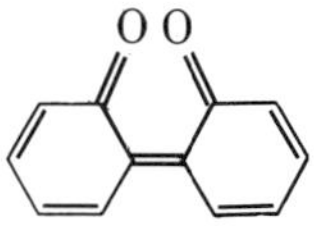

(14) 2,2'-diphenoquinone
[59869-78-2]

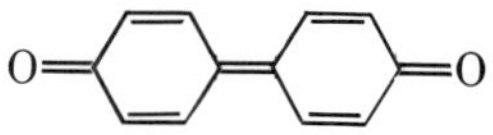

(15) 4,4'-diphenoquinone
[494-72-4]

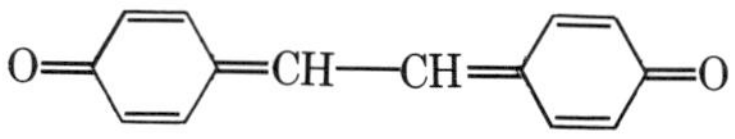

(16) 4,4'-stilbenequinone (stilbene-4,4'-dione)
[3457-53-2]

Some 1,2-diones continue to be described as nonbenzenoid quinones, eg, (17) acenaphthenequinone (acenaphthene 1,2-dione). Others, such as camphorquinone (18), are rarely used in current nomenclature.

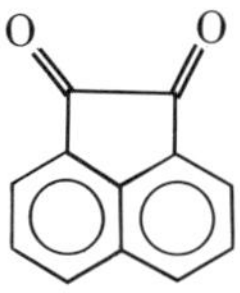

(17) acenaphthenequinone (acenaphthene-1,2-dione)
[82-86-0]

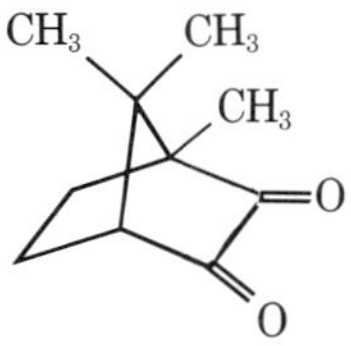

(18) camphorquinone (2,3-camphandione)
[465-29-2]

Chemical Abstracts has abandoned the use of "quinone" as a nomenclature unit. The examples in Table 1 illustrate the general principles of quinone nomenclature. The slightly older but widely accepted quinone names are used in this article, since they are preferable for the simple benzene and naphthalene derivatives.

Two outstanding physical properties of the simple quinones are their odor and color. The 1,4-benzo- and 1,4-naphthoquinones and many of their derivatives have high vapor pressures and pungent, irritating odors. Compounds of the former class are often synthesized by insects to repel predators. In general, the 1,2-quinones are vibrant in color, ranging from orange to red, whereas the 1,4-quinones are usually lighter, ie, yellow to orange.

The quinones have excellent redox properties; thus, they are important oxidants in laboratory and biological applications. The presence of an extensive array of conjugated systems, especially the α,β-unsaturated ketone arrangement, allows the quinones to participate in a variety of reactions. Characteristic of quinone chemistry are nucleophilic substitution; electrophilic, radical, and cycloaddition reactions; photochemistry; and normal and unusual carbonyl chemistry.

Table 1. Quinone Nomenclature

Common name	CAS Registry No.	Structure no.	Chemical Abstracts nomenclature	Synonyms
1,4-benzoquinone	[*106-51-4*]	(**2**)	2,5-cyclohexadiene-1,4-dione	*p*-benzoquinone, quinone
1,2-benzoquinone	[*583-63-1*]	(**4**)	3,5-cyclohexadiene-1,2-dione	*o*-benzoquinone
1,4-naphthoquinone	[*130-15-4*]	(**7**)	1,4-naphthalenedione	α-naphthoquinone
p-chloranil	[*118-75-2*]	(**19**)	2,3,5,6-tetrachloro-2,5-cyclohexadiene-1,4-dione	tetrachloro-*p*-benzoquinone
2,3-dichloro-5,6-dicyano-1,4-benzoquinone	[*84-58-2*]	(**20**)	4,5-dichloro-3,6-dioxo-1,4-cyclohexadiene-1,2-dicarbonitrile	DDQ
4,4′-diphenoquinone	[*492-72-4*]	(**15**)	[bi-2,5-cyclohexadien-1-ylidene]-4,4′-dione	

(**19**)

(**20**)

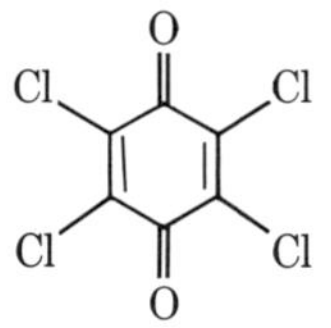

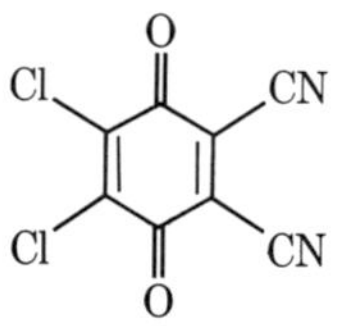

Physical Properties

Selected physical constants of various quinones are given in Table 2 (5). An additional and excellent compilation of data for many more complex quinones is given in ref. 6.

(**21**)

(**22**)

(**23**)

(**24**)

(**25**)

(**26**)

(**27**)

(**28**)

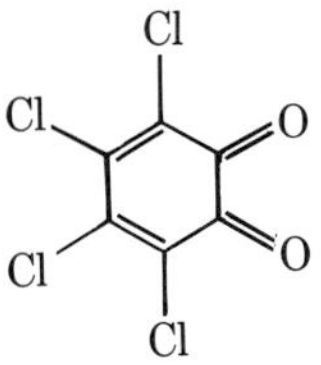

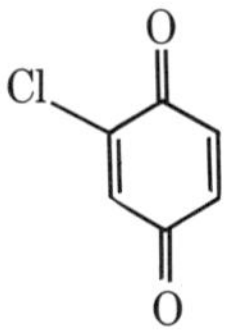

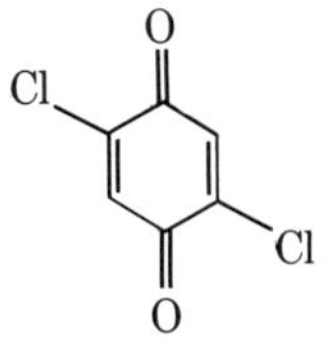

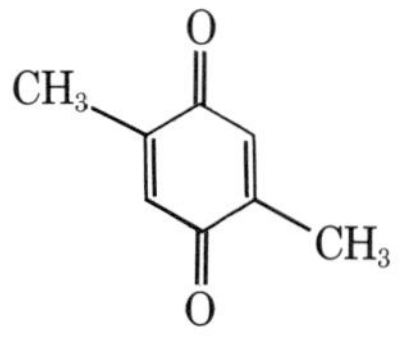

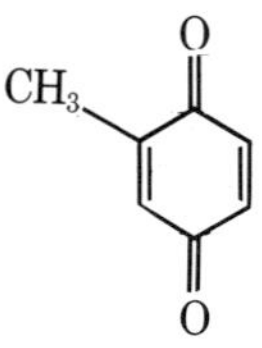

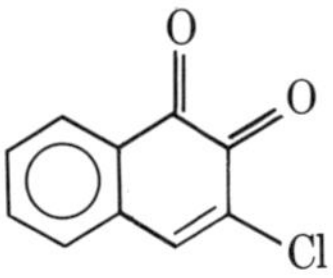

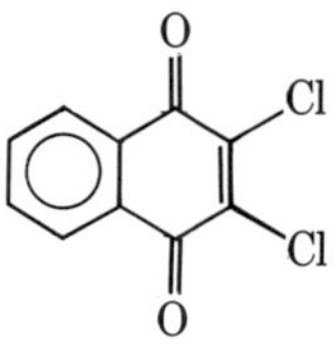

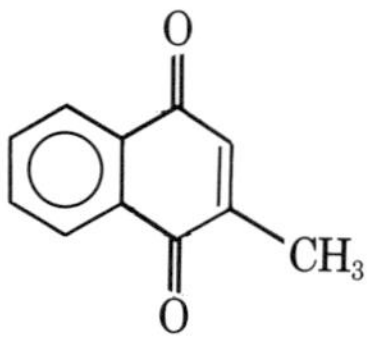

Table 2. Physical Properties of Selected Quinones[a]

Name	CAS Registry No.	Structure no.	Color	Melting point, °C	Crystalline form
1,2-benzoquinone		(4)	red	60–70 (dec)	plates or prisms
3,4,5,6-tetrachloro-1,2-benzoquinone	[2435-53-2]	(21)	orange-red	133, 122–127	
1,4-benzoquinone		(2)	yellow	113, 116	monoclinic prisms
2-chloro-1,4-benzoquinone	[695-99-8]	(22)	yellow-red	57	rhombic-hexagonal
2,5-dichloro-1,4-benzoquinone	[615-93-0]	(23)	pale yellow	161–162	monoclinic prisms
2,3-dichloro-5,6-dicyano-1,4-benzoquinone		(20)	bright yellow	201–203 (dec)	plates
2,5-dimethyl-1,4-benzoquinone	[137-18-8]	(24)	yellow	125	
2-methyl-1,4-benzoquinone	[553-97-9]	(25)	yellow	69	plates or needles
2,3,5,6-tetrachloro-1,4-benzoquinone		(19)	yellow	290, 294	monoclinic prisms
1,2-naphthoquinone	[524-42-5]	(6)	yellow-red, orange	145–147	needles
3-chloro-1,2-naphthoquinone	[18099-99-5]	(26)	red	172 (dec)	needles
1,4-naphthoquinone		(7)	bright yellow	125, 128.5	needles
2,3-dichloro-1,4-naphthoquinone	[117-80-6]	(27)	yellow	193, 195	needles
2-methyl-1,4-naphthoquinone	[58-27-5]	(28)	yellow	105–107	needles

[a] Ref. 5.
[b] In many cases, more than one set of data is given.

Solubility		Uv spectra Wavelength at max absorption (ϵ); solvent[b]	$-E_{1/2}$ 25°C, SCE,	
Sol	Insol		CH_3CN	$(C_2H_5)_4NClO_4$
ether, benzene	pentane	580 (30), 385 (1585); ether	0.31	0.90
			0.1	−0.71
alcohol, ether	water, pentane	243; CH_3OH	0.51	1.14
water, alcohol		251 (7740); CH_3OH	0.34	0.97
ether, chloroform	water, alcohol	271 (5710); CH_3OH	0.18	0.81
			−0.51	0.30
ether, alcohol	water, alcohol	293 (1770), 251 (11650), 220 (4080); CH_3OH	0.67	1.27
ether, alcohol	water	429 (19), 314 (589), 246 (13804); C_2H_5OH	0.58	1.12
ether	water, ligroin	364 (248), 286 (12600); CH_3OH	−0.01	0.71
water, alcohol	ligroin	398 (1800), 336 (2280), 248 (20400); CH_3OH	0.58	1.18
alcohol, benzene	water			
alcohol, benzene	water, ligroin	330 (3020), 250 (19953), 246 (20417); CH_3CN	0.71	1.25
benzene, chloroform	water, alcohol	337, 279, 252, 246; CH_3OH		
ether, benzene	water, alcohol	328, 264, 253, 249, 244; C_6H_{12}	0.77	1.28

Chemical Properties

The quinones in biological systems play a variety of important roles (6). In addition to their defense purposes, eg, in insects, the vitamin K family members, eg, (**29**)–(**30**), which are based on 2-methyl-1,4-naphthoquinone, are blood-clotting agents (see Vitamins). Two groups of substituted 1,4-benzoquinones are associated with photosynthetic and respiratory pathways; the plastoquinones, eg, (**31**), and the ubiquinones, eg, (**32**), are involved in these processes, but their specific roles are unclear.

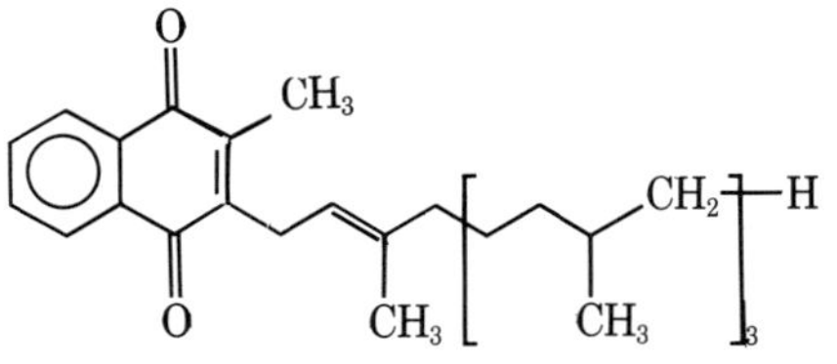

(**29**) vitamin K_1 [*11104-38-4*]

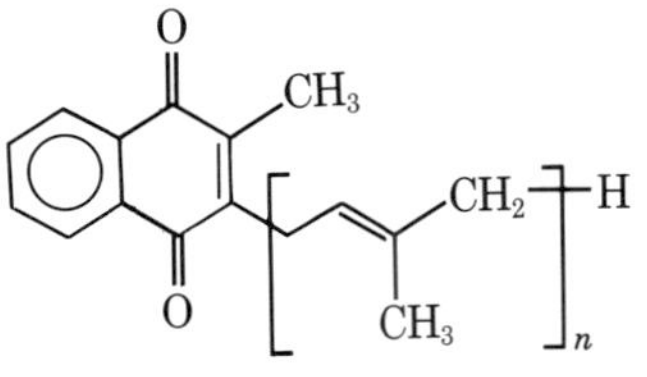

(**30**) vitamin K_2 [*11032-49-8*]

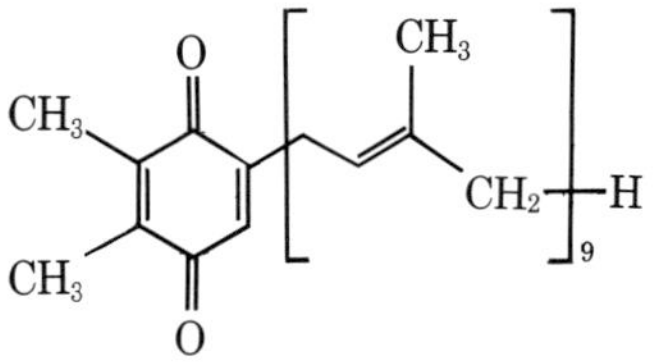

(**31**) plastoquinone [*4299-57-4*]

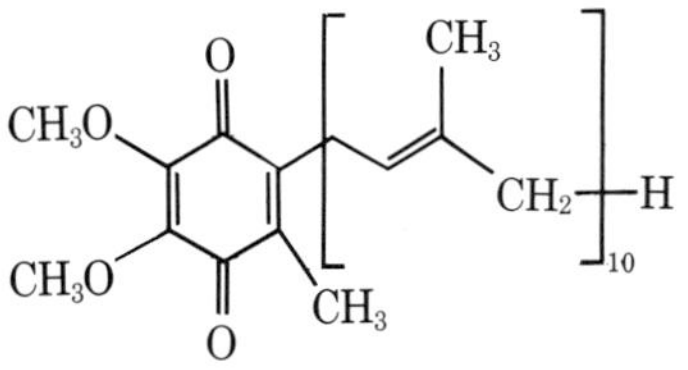

(**32**) ubiquinone [*1339-63-5*]

Quinones of various degrees of complexity have antibiotic, antimicrobial, and anticancer activity, eg, (**33**)–(**35**) (see Antibiotics; Chemotherapeutics, antimitotic).

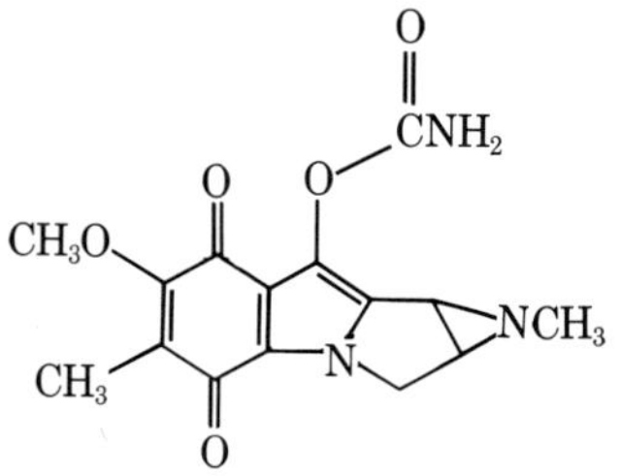

(**33**) aziridinomitosene
[*80954-63-8*]

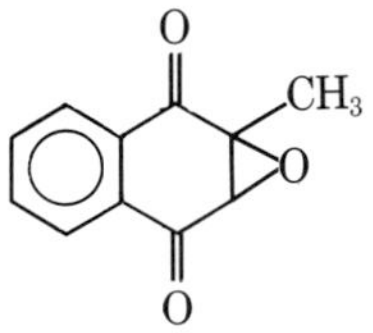

(**34**) [*61840-91-3*]

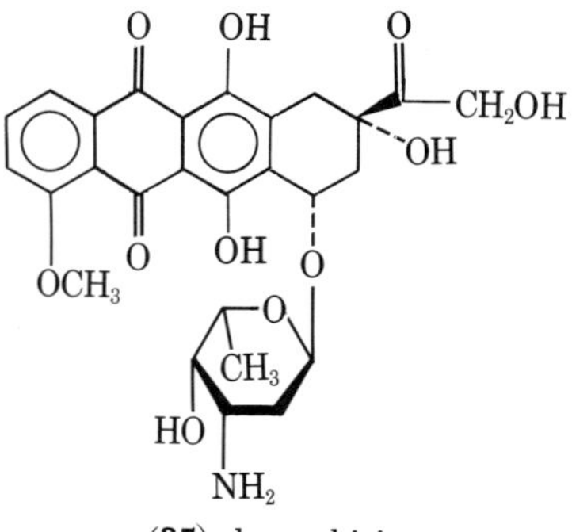

(**35**) doxorubicin
(adriamycin)
[23214-92-8]

Interest in the use of quinones as oxidants has not been restricted to biosynthetic and biochemical problems. Since 1960, the literature of quinone oxidation and dehydrogenation chemistry has grown rapidly. The most widely used oxidants have been 1,4-benzoquinone (**2**) (oxidation potential, 711 mV), 2,3,5,6-tetrachloro-1,4-benzoquinone (**19**) or chloranil (742 mV), and 2,3-dichloro-5,6-dicyano-1,4-benzoquinone (**20**) (DDQ) (ca 1000 mV). Low cost, availability, stability, and high oxidation potential characterize these reagents. Although the oxidation potential is of primary importance, it can be outweighed by other considerations; for example, 2,3,5,6-tetracyano-1,4-benzoquinone [4032-03-5] has been prepared, but its high reactivity and moisture sensitivity have greatly restricted its application. The 1,2-quinones tend to undergo Diels-Alder dimerization and only 3,4,5,6-tetrachloro-1,2-benzoquinone (**21**) has been used extensively. Some more complex quinones, which are readily available and have high oxidation potentials, have not been studied, ie, the diphenoquinones.

The oldest and an important synthetic use of quinones is as dehydrogenation agents, especially for aromatization (7).

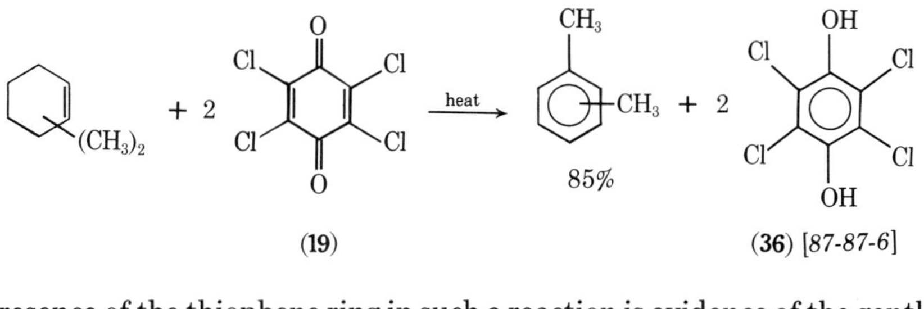

The presence of the thiophene ring in such a reaction is evidence of the gentle nature of quinonoid oxidants (8).

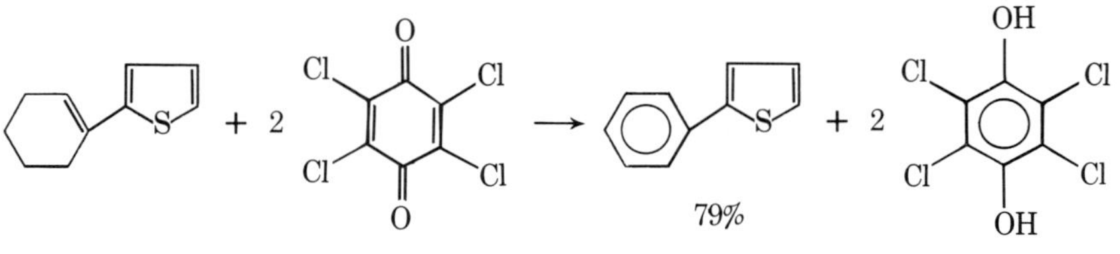

580   QUINONES

The use of 1,4-benzoquinone in combination with palladium(II) chloride converts terminal alkenes to alkyl methyl ketones in very high yield (9).

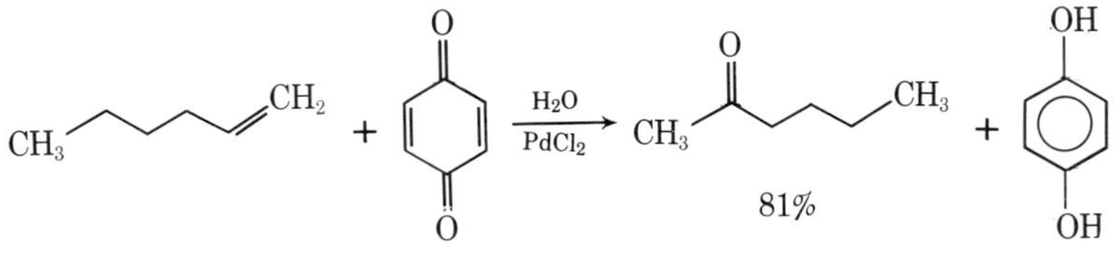

$$CH_3\text{-}CH_2\text{-}CH_2\text{-}CH=CH_2 + \text{(2)} \xrightarrow[PdCl_2]{H_2O} CH_3\text{-}CO\text{-}CH_2\text{-}CH_2\text{-}CH_2\text{-}CH_3 + \text{(3)}$$

81%

(2) = 1,4-benzoquinone; (3) = hydroquinone

The quinone appears to reoxidize the palladium.

Although saturated alcohols are sufficiently stable towards quinones to be used as solvents for these oxidation reactions, benzylic and allylic alcohols are often readily converted to the corresponding carbonyl compounds (10–11).

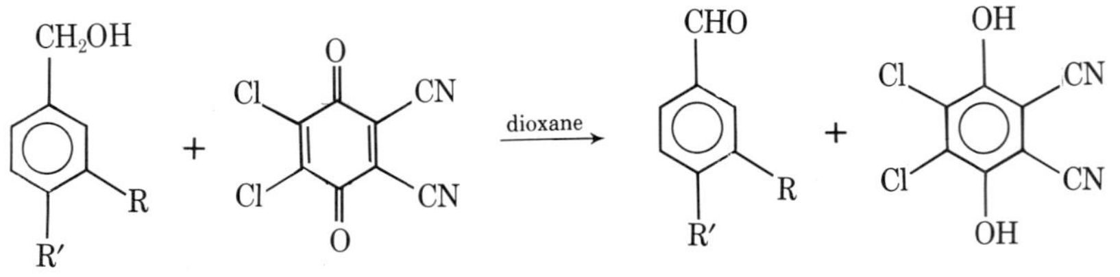

(20) + benzyl alcohol derivative → (37) + tetrasubstituted hydroquinone

		yield
R = H	R' = CH$_3$	93%
R = H	R' = C$_6$H$_5$SO$_2$	14%
R = OH	R' = H	83%
R = OH	R' = OCH$_3$	97%

Cinnamyl alcohol + (21) → cinnamaldehyde + (38)

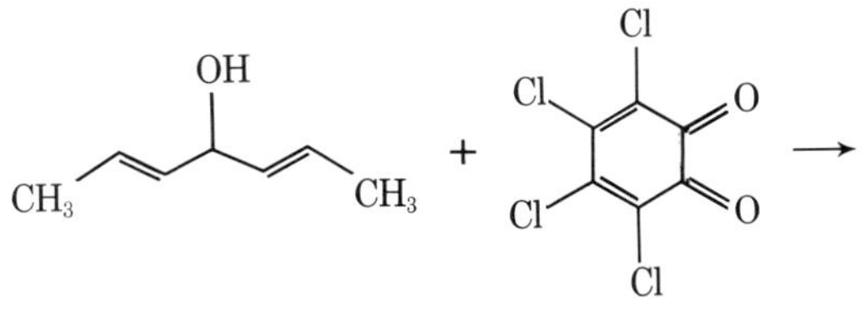

CH$_3$-CH=CH-CH(OH)-CH=CH-CH$_3$ + (21) →

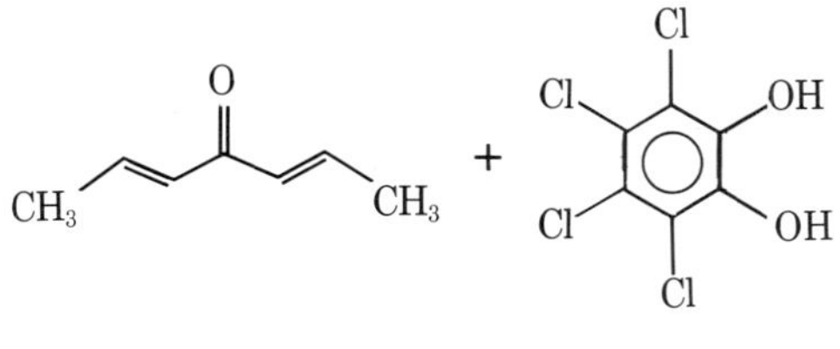

CH$_3$-CH=CH-CO-CH=CH-CH$_3$ + (38)

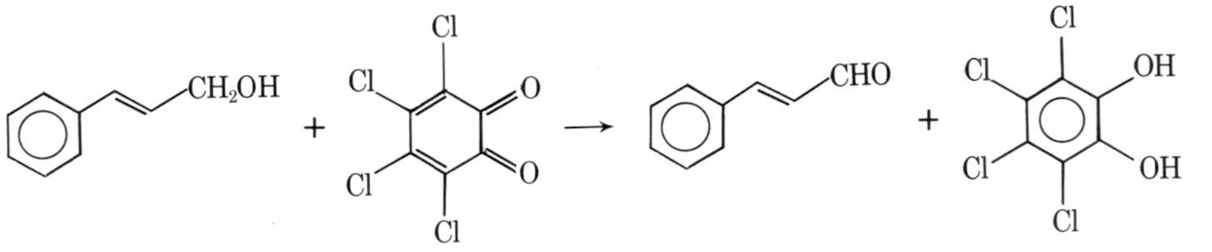

The low yield obtained for the phenylsulfonyl substituent is typical of electron-withdrawing groups.

Quinones are extensively used in the dehydrogenation of steroidal ketones (12–13). Such reactions are marked by high yields and high selectivity.

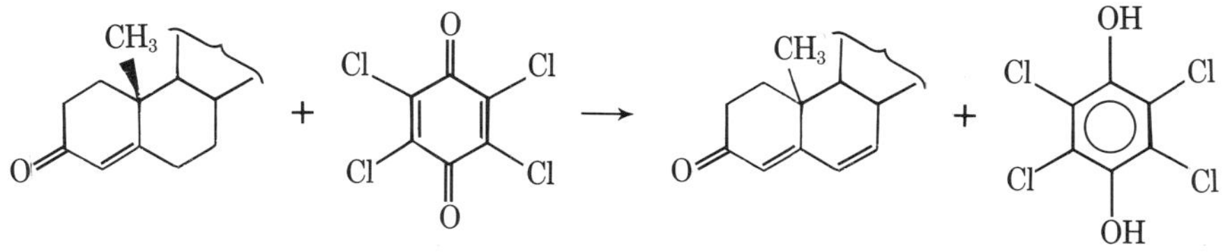

(19)   40–80%   (36)

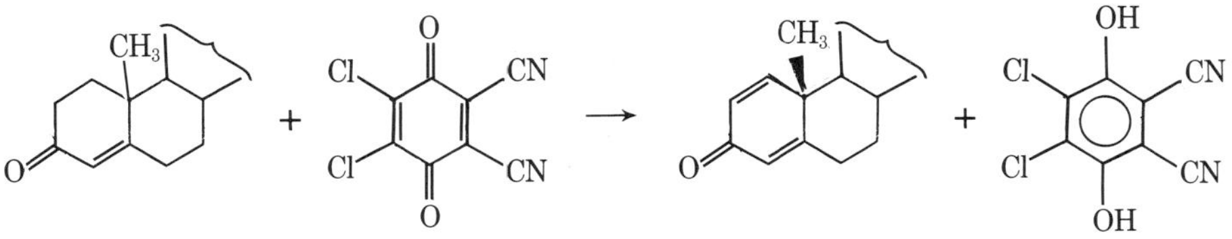

(20)   (37)

Generally, the results with nonsteroidal ketones are disappointing.

The powerful accelerating effect of electron-donating substituents in dehydrogenation reactions is illustrated in (20) and (37) (14–15). A spectacular example of selective dehydrogenation in the steroid ring system has been attributed to stereoelectronic effects (16). Several related steroids also show this chemistry. An extensive review containing many additional examples and a mechanistic discussion is available (17).

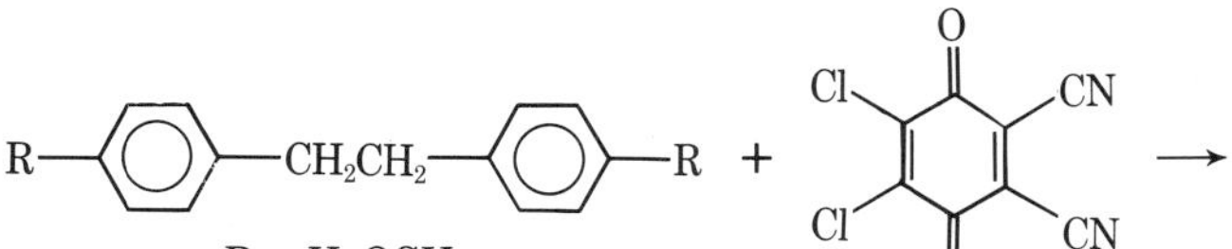

R = H, OCH$_3$

(20)

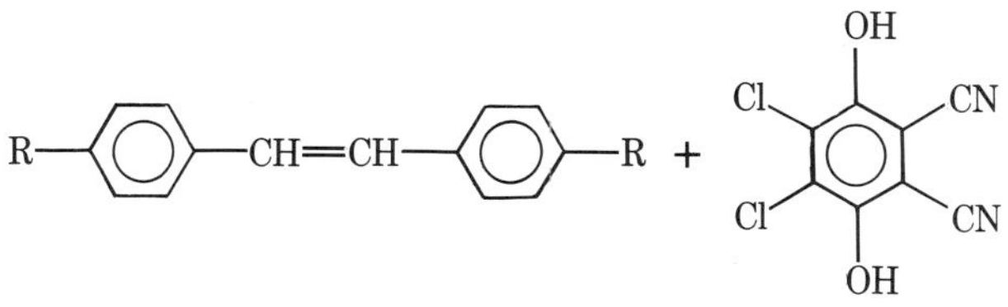

(37)

R = H   yield = 10%
R = OCH$_3$ yield = 85%

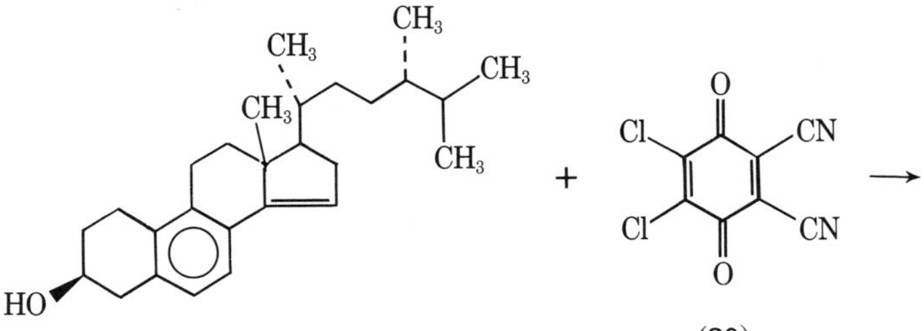

Other Reactions. Increased knowledge of the importance of quinones in photosynthesis has stimulated renewed interest in their photochemical behavior. Work has centered on the 1,4-quinones and the two reaction types most frequently observed, ie, [2 + 2] cycloaddition and hydrogen abstraction.

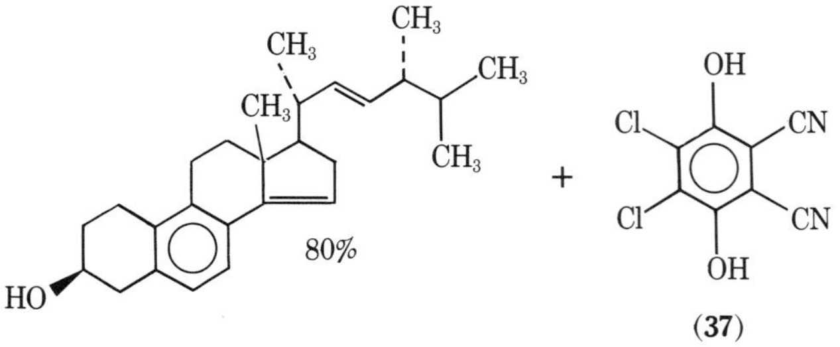

The products of [2 + 2] cycloaddition are usually of the cyclobutane-type (**39**) or spirooxetane-type (**40**). The product distribution appears to depend upon the radiation used for quinone excitation, the structure of the quinone, and the quinone: alkene ratio. In the example cited, 1,4-benzoquinone gives only the spirooxetane, whereas chloranil (**19**) gives both products in amounts related to the ratio of starting materials (18–19).

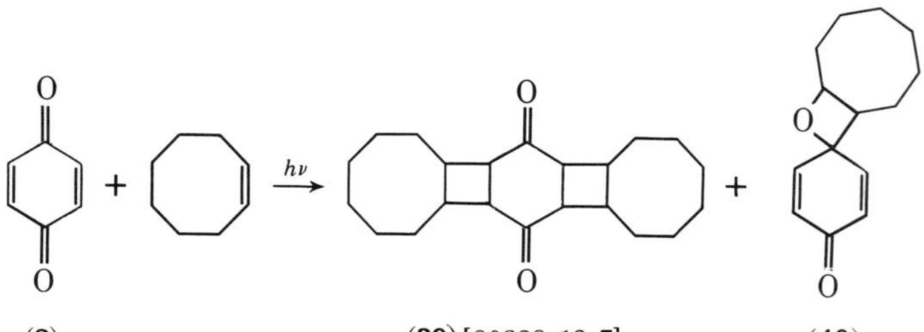

Hydrogen abstraction reactions have been carried out with a variety of substrates. In general, the reactions involve removal of hydrogen either directly as a hydrogen atom or indirectly by electron transfer followed by proton transfer. The products are derived from ground-state reactions; for example, chloranil probably reacts with cycloheptatrienyl radicals to produce the ether shown below (20).

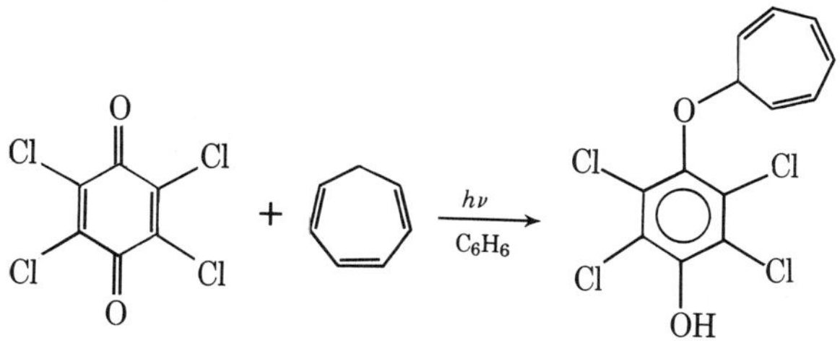

(19)

This chemistry is contrasted with the ground-state reaction in which 2,3-dichloro-5,6-dicyano-1,4-benzoquinone (20) produces tropylium quinolate (41) (21).

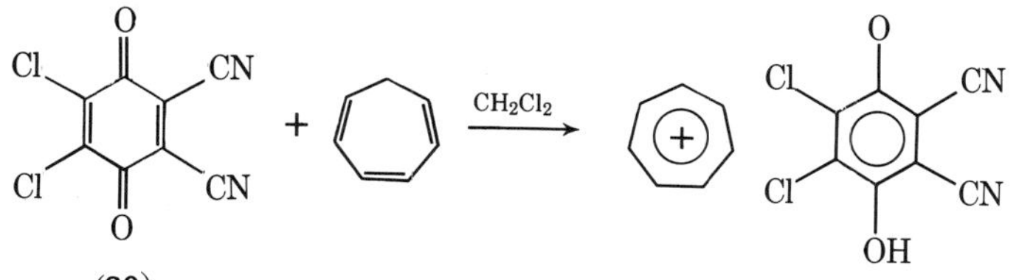

(20)   (41)
        91%

An interesting example involving an ether starting material is shown below (22). The initial product probably is oxidized by a second mole of quinone to give (42).

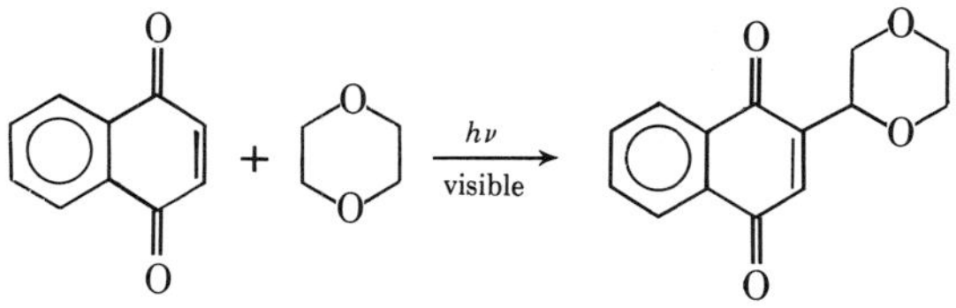

(7)   (42) [24161-37-3]

The synthesis of polycyclic hydrocarbons that are highly carcinogenic is effected by photocycloaddition, which results in aromatic ring formation, eg, (44) (23).

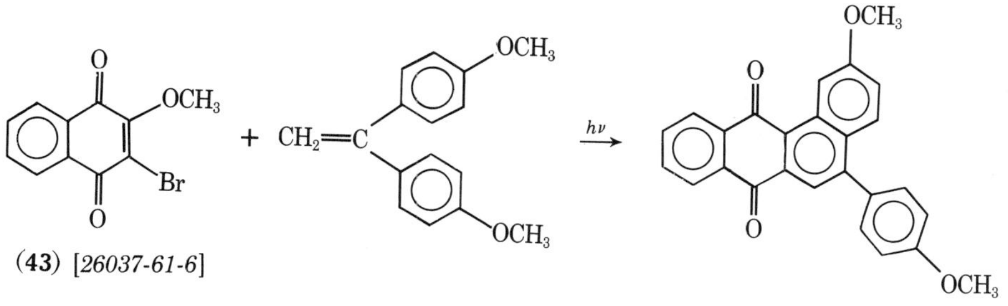

(43) [26037-61-6]

(44) [72735-91-2]

Although the yields are not always high and isomeric products are obtained with unsymmetrical 1,1-diarylethylenes, the reaction conditions are mild and show promise of control of regioselectivity.

The role of rose bengal and other dyes in the photodimerization of 2-acetyl-1,4-benzoquinone (45) involves electron transfer but not singlet oxygen (24).

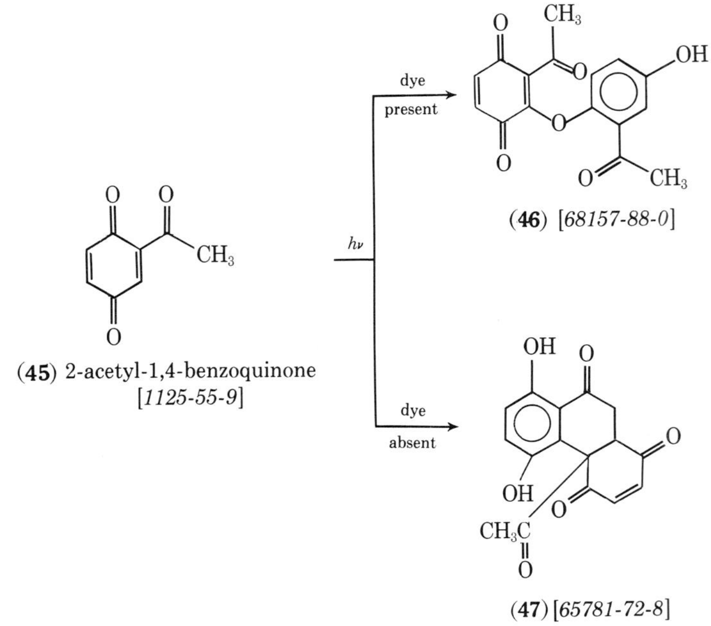

(45) 2-acetyl-1,4-benzoquinone [1125-55-9]

(46) [68157-88-0]

(47) [65781-72-8]

The addition of nucleophiles to quinones is often an acid-catalyzed, Michael-type, reductive process.

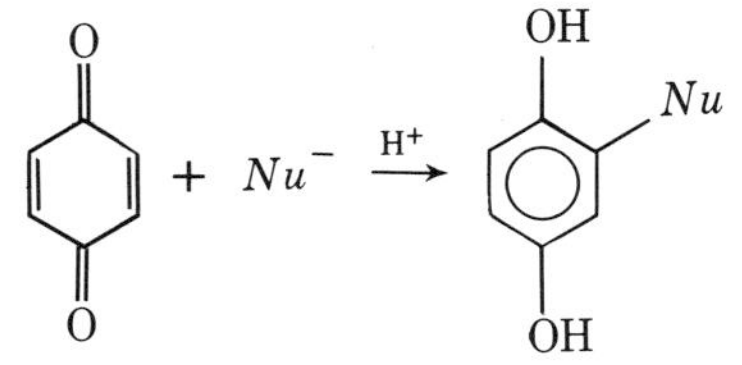

Michael employed the addition of benzenethiol to 1,4-benzoquinone (**2**) to develop a better understanding of valence in organic molecules (25).

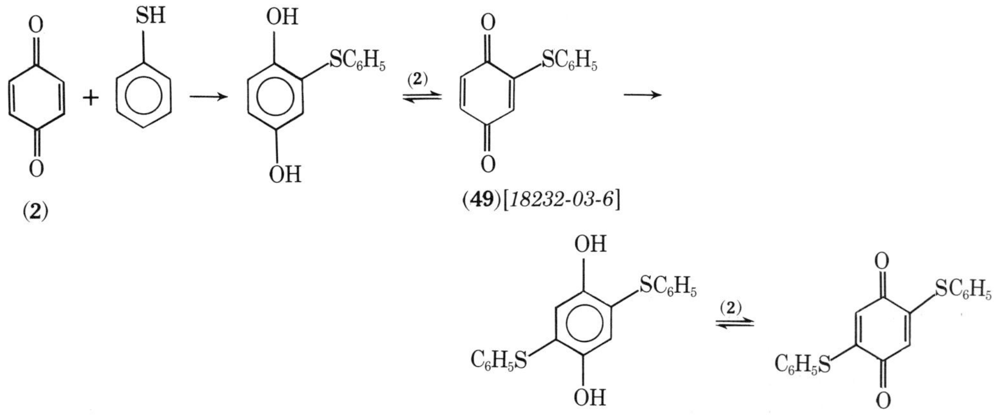

(49) [18232-03-6]

(50) [17058-53-6]

Such chemistry may be complicated by subsequent cross-oxidation and further addition. The main factor involved is the electronic influence of the substituent introduced. The phenyl sulfide linkage is electron donating and thereby makes the hydroquinone product susceptible to oxidation by the quinone starting material and to a second addition–oxidation sequence. The 2,5-bis(phenylthio)-1,4-benzoquinone (**50**) predominates for electronic reasons, but it is not the only isomer that is obtained. Similar chemistry is observed in the addition of organic nitrogen and oxygen nucleophiles as well as inorganic anions. Some notable exceptions, eg, (**51**)–(**53**), exist, although they are quite easily explained in terms of electronics (26–28).

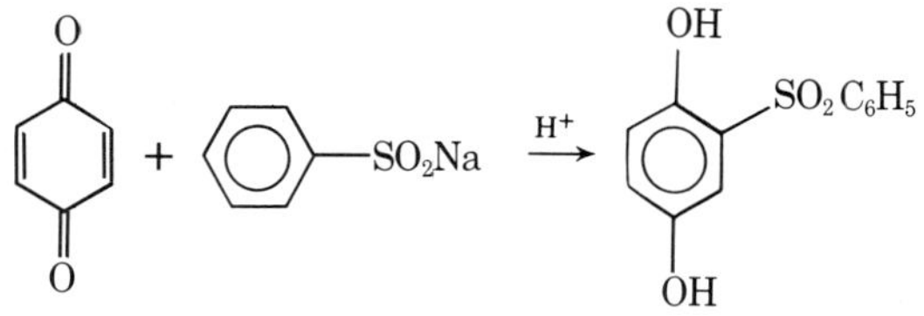

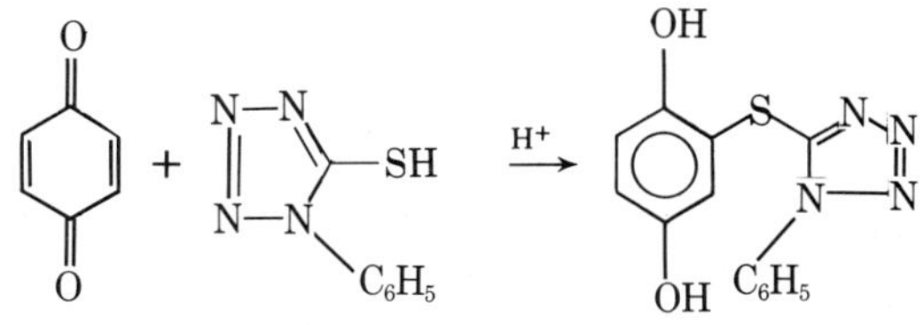

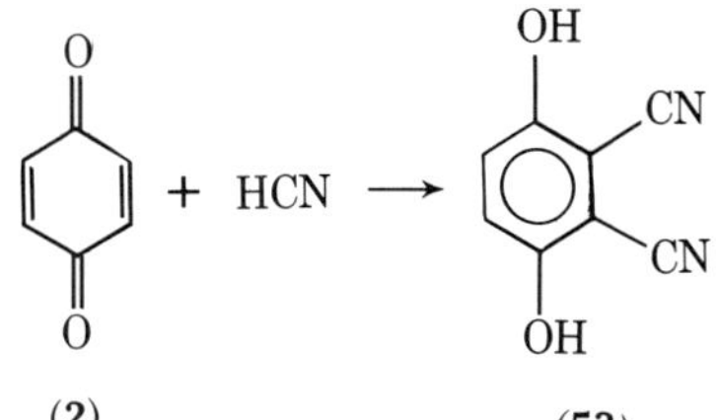

The product of each of these reactions contains an electron-withdrawing substituent, which makes further oxidation unfavorable. The introduction of the first cyano group provides a new conjugated system for further 1,4-addition, which leads to the only observed product, ie, 2,3-dicyano-1,4-benzenediol.

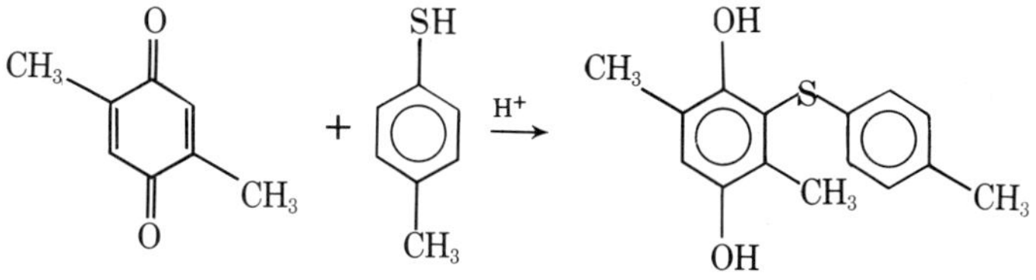

586   QUINONES

Steric considerations also are involved in determining the oxidation state of the product (26). When a methyl group must be adjacent to a newly created arylsulfide linkage, the combined effect of three electron-donating substituents prevents significant cross-oxidation.

It appears that the methyl group prevents the planarity required for efficient electron-donation by the sulfur atom's unshared electron pair.

An especially interesting case of oxygen addition to quinonoid systems involves acidic treatment with acetic anhydride, and it produces both addition and esterification. The Thiele acetylation has been used extensively for synthesis, structure proof, isolation, and purification (29).

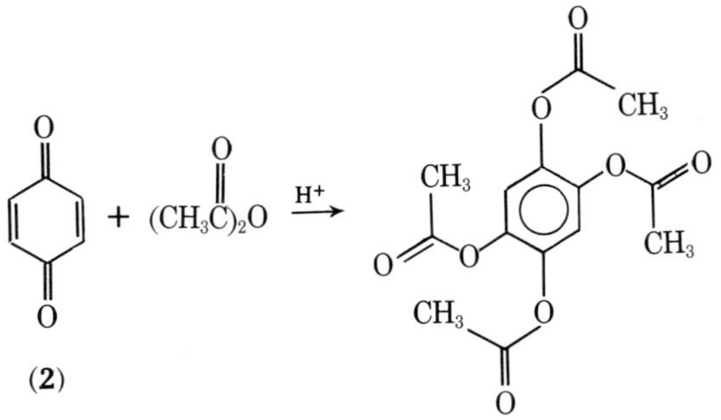

(2)

The kinetics and mechanism of the Thiele acetylation are described in ref. 30. Although the electrophilic acetylium ion is involved, the results of many studies of electronic effects show a definite relationship to nucleophilic-addition chemistry (31).

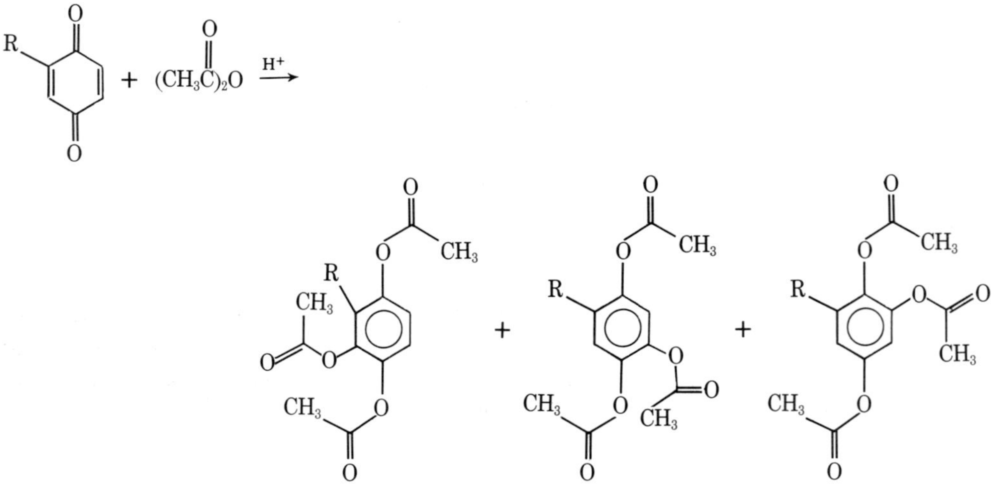

R = (45) COCH$_3$ yield = 92%
R = (25) CH$_3$ yield = 78%   yield = 15%
R = (54) C$_6$H$_4$-p-NO$_2$ yield = 18%   yield = 56%   yield = 19%
   [15394-91-9]
R = (55) C$_6$H$_5$ [363-03-1]   yield = 21%   yield = 52%

Electrophilic addition, eg, in the synthesis of aryl-substituted quinones with diazonium salts, is in marked contrast to the Thiele reaction in terms of product isomer distribution (32). The presence of significant amounts of all three isomeric products and the dependence on the aryl substituent are noteworthy.

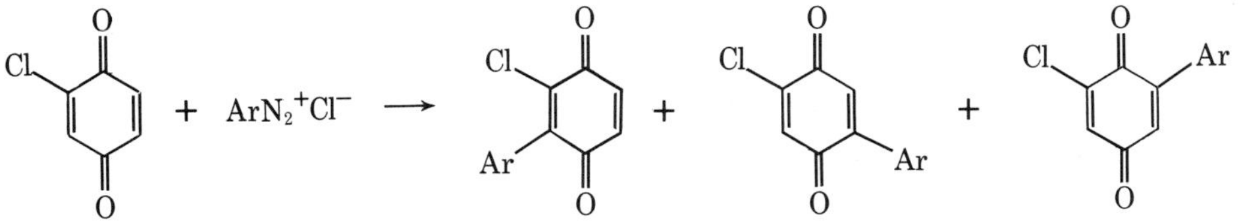

	(56)	(57)	(58)
Ar = C$_6$H$_5$	yield = 8% [80632-59-3]	75% [39171-11-4]	17% [80632-60-6]
Ar = C$_6$H$_4$-p-Cl	yield = 28% [80632-61-7]	35% [62120-48-3]	37% [80954-64-9]
Ar = C$_6$H$_4$-p-NO$_2$	yield 41% [80632-62-8]	19% [80632-63-9]	40%

The reaction of 2-methyl-1,4-benzoquinone with a wide variety of diazonium salts has been reported to produce only the 2,5-isomer (33).

The quinones undergo Michael addition with nitrogen nucleophiles in much the same fashion as observed for thiols (34). Recent interest in the synthesis of quinones that contain heterocyclic nitrogen prompted a detailed study of the addition of pyrrolidine to unsymmetrical benzoquinones (35). Regioselectivity was obtained in several cases. Information on nucleophilic-addition chemistry of quinones and various mechanistic rationalizations have been discussed, and molecular orbital calculations have been proposed as more definitive approaches for explanation and prediction (36).

Two synthetic methods avoid the problems of instability, sequential addition, and poor regiospecificity, which often occur in quinone addition reactions. For example, the 1,2-benzoquinones react in much the same fashion as the 1,4-benzoquinones, but the lower stability of the former makes their generation *in situ* advisable. A number of nascent quinone reactions, such as those shown below, have been reviewed (37).

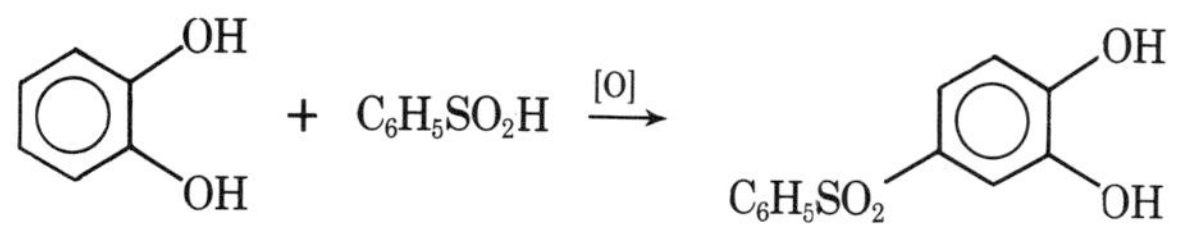

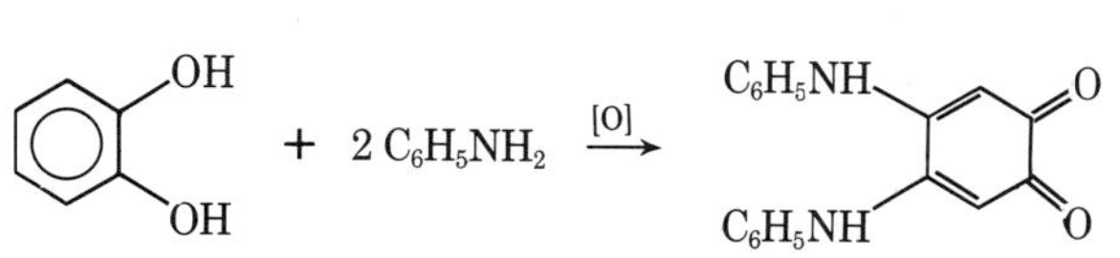

The quinone monoacetals show regiospecific addition of active methylene compounds (38).

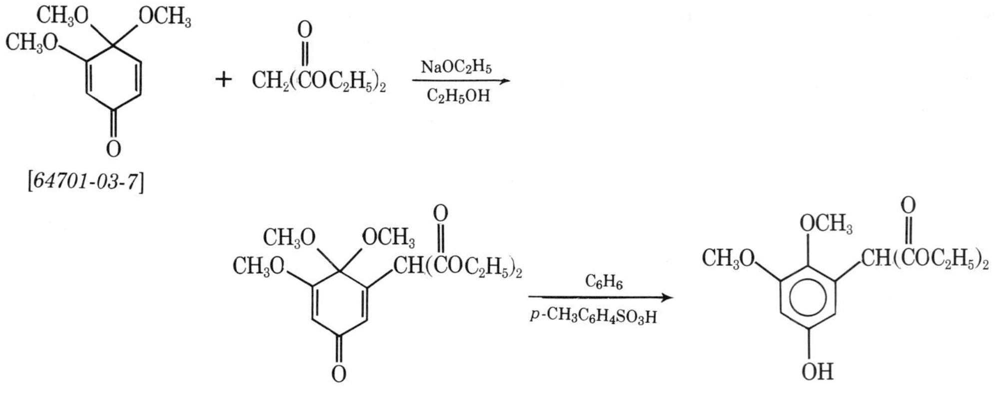

[64701-03-7]

83%   63%

The synthesis of optically active epoxy-1,4-naphthoquinones with the use of benzylquininium chloride as the chiral catalyst under phase-transfer conditions has been reported (39). The reaction is characteristic of the usual Michael-addition of hydroperoxide anion, and it yields enantiomeric excesses of up to 45%.

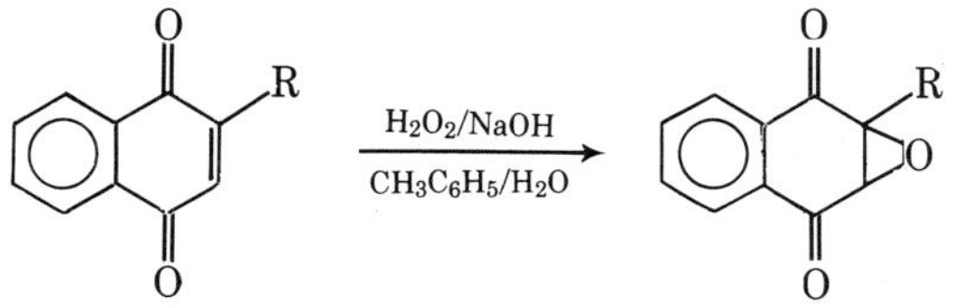

(28) R = —CH$_3$ yield = 70% (**34**) (−)
(**60**) R = —C$_6$H$_{11}$ [34987-31-0]   yield = 85% [73377-78-3] (+)
(**61**) R = C$_6$H$_5$ [2348-77-8]   yield = 92% [73377-82-9] (−)

Several reactions of quinones with radicals have been explored and, from synthetic and mechanistic viewpoints, alkylation with diacyl peroxides is the most interesting (40).

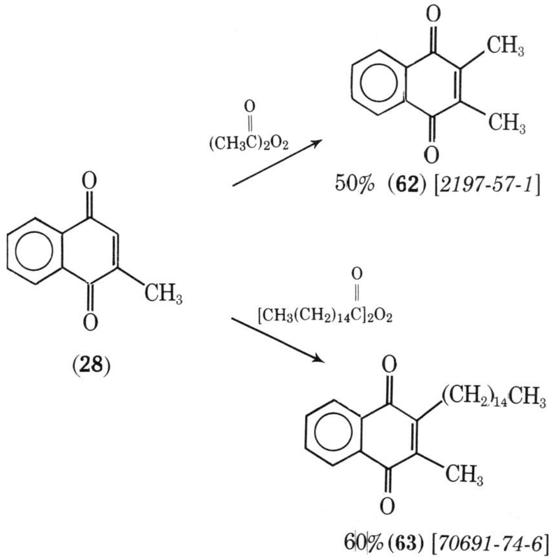

50% (**62**) [2197-57-1]

(**28**)

60% (**63**) [70691-74-6]

Although there are some limitations, the range of substituents that can be introduced in good yield is impressive. Substituents include alkyl chains ending with functional groups, eg, (**64**) (41).

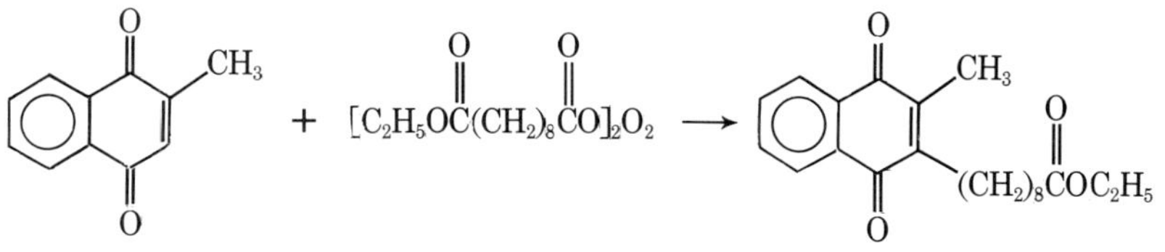

(**28**)   +   [C$_2$H$_5$OC(CH$_2$)$_8$CO]$_2$O$_2$ →   50% (**64**) [80632-67-3]

The importance of quinones with unsaturated side chains in respiratory, photosynthetic, blood-clotting, and oxidative phosphorylation processes has stimulated much research in synthetic methods. The application of allyl- or polyisoprenyltin reagents is shown below (42–44).

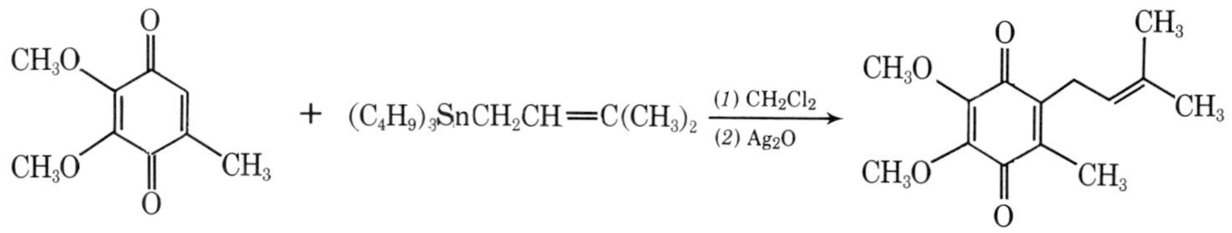

(**65**) [605-94-7]   +   (C$_4$H$_9$)$_3$SnCH$_2$CH=C(CH$_3$)$_2$ $\xrightarrow{\text{(1) CH}_2\text{Cl}_2}_{\text{(2) Ag}_2\text{O}}$ 75% (**66**) [727-81-1]

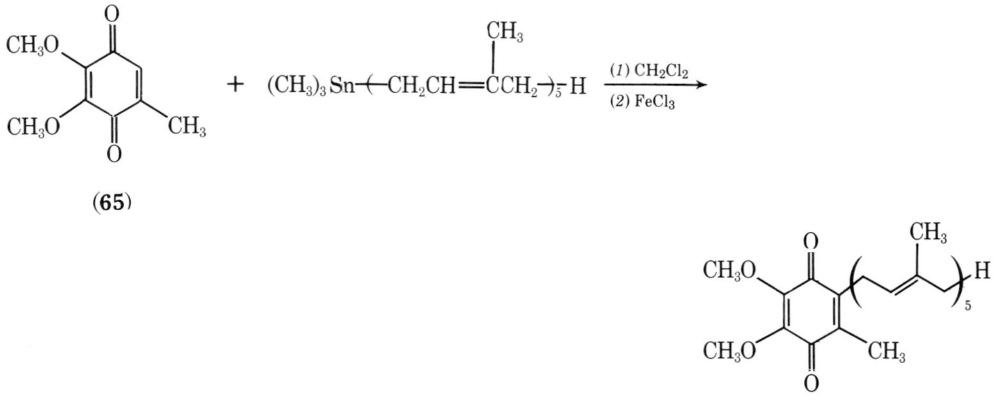

(**65**)   +   (CH$_3$)$_3$Sn(CH$_2$CH=CCH$_2$)$_5$H $\xrightarrow{\text{(1) CH}_2\text{Cl}_2}_{\text{(2) FeCl}_3}$

94% (**67**) [4370-61-0]

A selective method for preparing unsymmetrical 2-alkynyl-5-alkyl- or aryl-1,4-benzoquinones involves the sequential introduction of lithium salts (45). Since the lithium reagents are generated *in situ* and the intermediates are not isolated, the transformation may be applicable to the synthesis of various natural products, including some antibiotics.

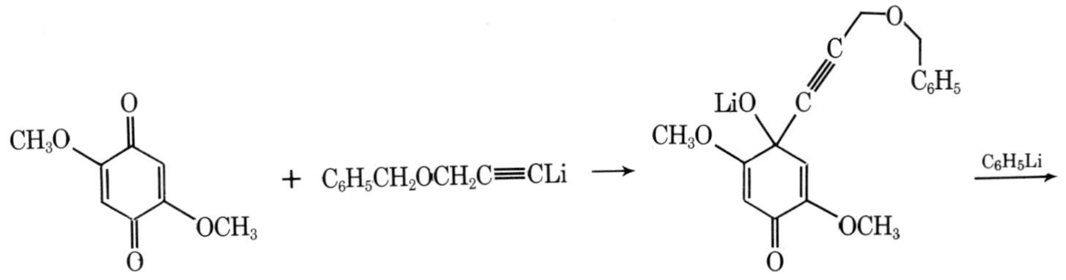

590 QUINONES

(**68**) [3117-03-1]

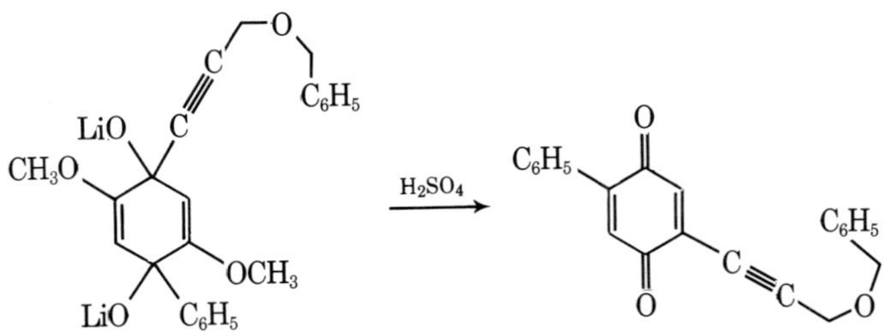

60% (**69**) [64080-65-5]

The quinone ketals undergo both metalation by lithium and addition to carbonyl compounds (46).

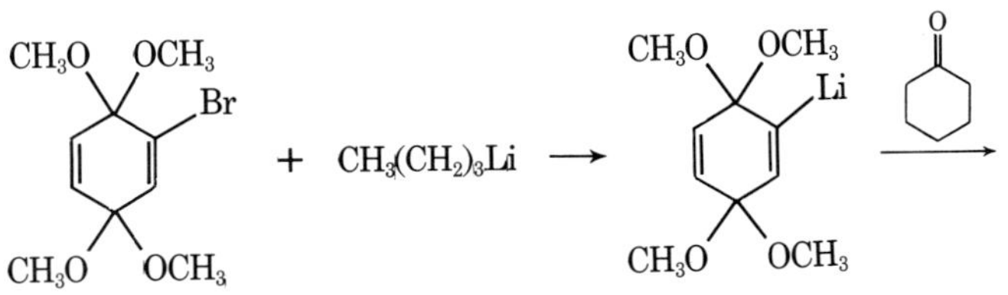

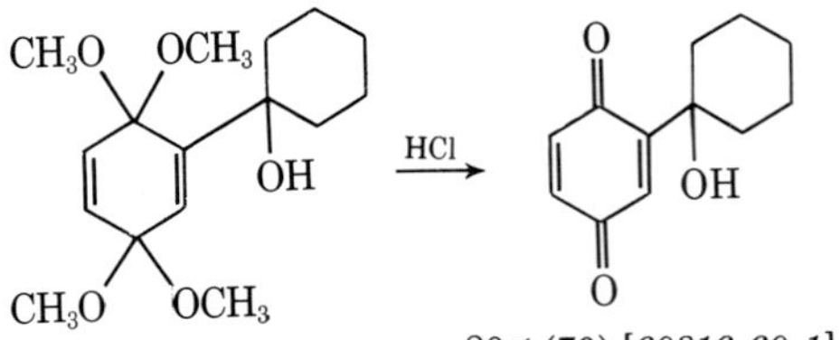

81%   80% (**70**) [60316-60-1]

Diels-Alder cycloaddition of quinonoid dienophiles has been reported for a wide range of dienes (47–49).

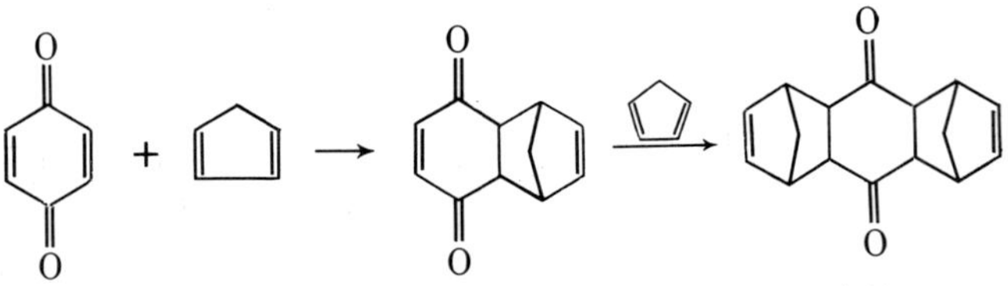

(**2**)   (**71**) [1200-89-1]   (**72**) [5439-22-5]

The analogous 1,3-cyclohexadiene adduct has been the subject of ^{13}C-nmr and x-ray studies, which have shown that it has the endo-anti-endo stereostructure (50). When unsymmetrical dienes are involved in this synthesis, high regioselectivity is observed (51).

(73) [3958-79-0] 97% (74) [80328-15-0]

(75) [4622-04-2] 58% (76) [80328-16-1]

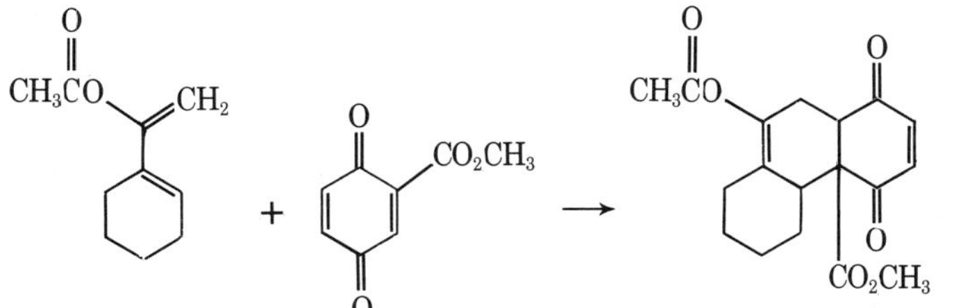

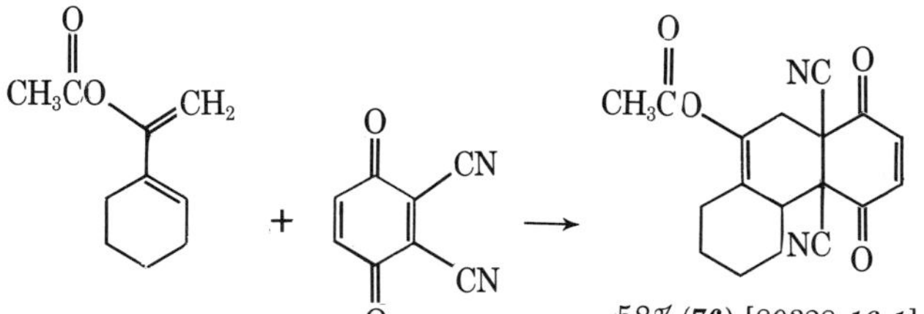

The importance of the electronic influence is obvious, but steric effects do play a role as in cross-oxidation.

(77) [77220-15-6]

30% (78) [80328-13-8] 20% (79) [80328-14-9]

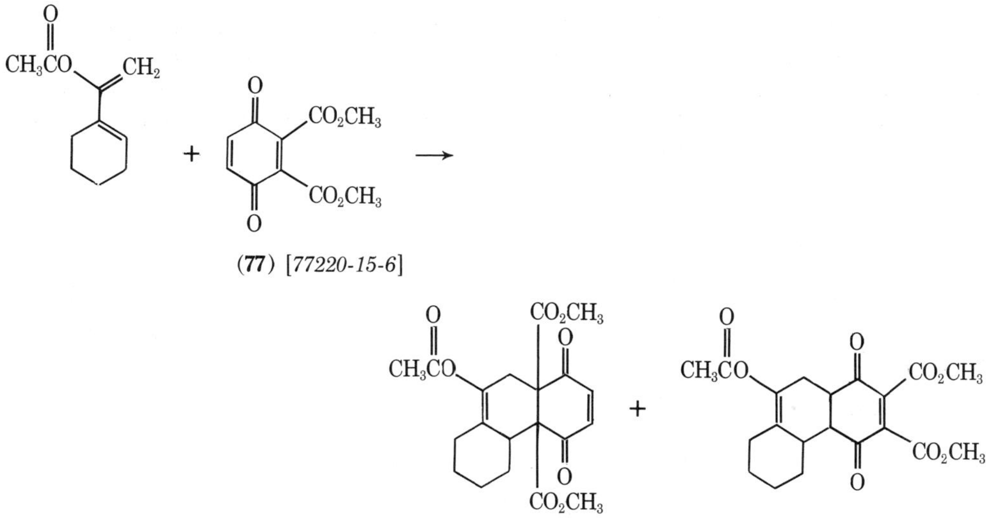

Interest continues in the application of the diene synthesis to complex natural products, eg, the aromatic portion of the rubradirin antibiotics (52).

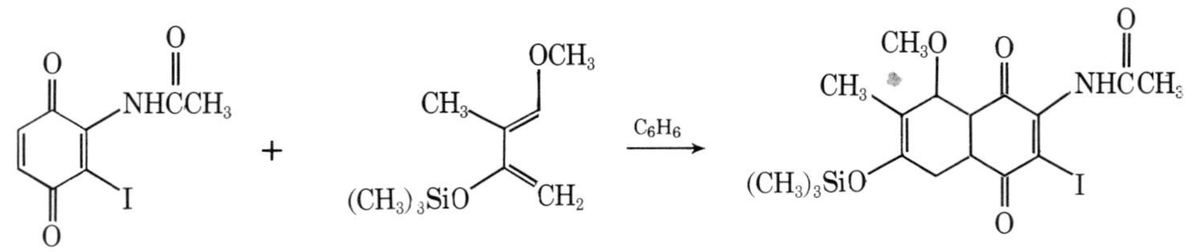

(80) [39015-18-8]   100% (81) [76160-18-4]

Sequential application of this method can yield naturally occuring anthraquinones, eg, macrosporin (83) (53).

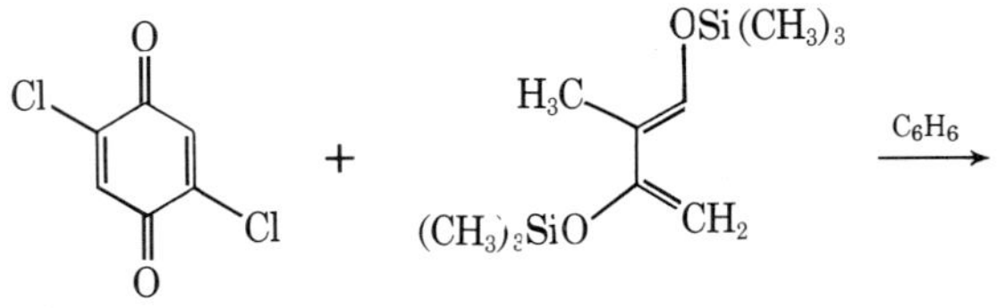

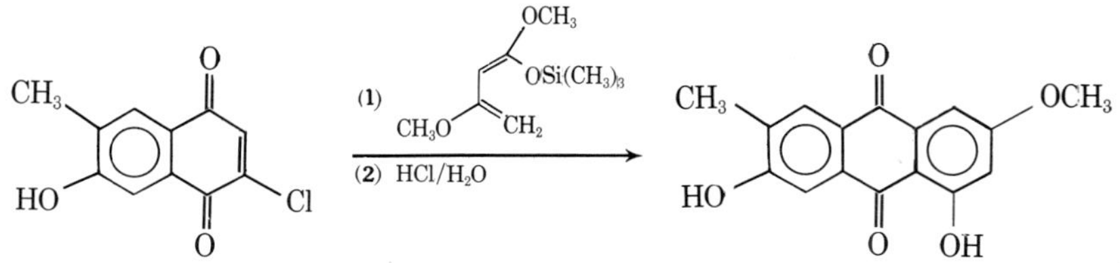

78% (82) [76665-67-3]   83% (83) [22225-67-8]

The problems associated with predicting regioselectivity in quinone Diels-Alder chemistry have been studied (54–55), and a mechanistic model based on frontier molecular-orbital theory has been proposed. In certain cases of poor regioselectivity, eg, 2-methoxy-5-methyl-1,4-benzoquinone (84) with alkyl substituted dienes, the use of Lewis acid catalysts is effective (56).

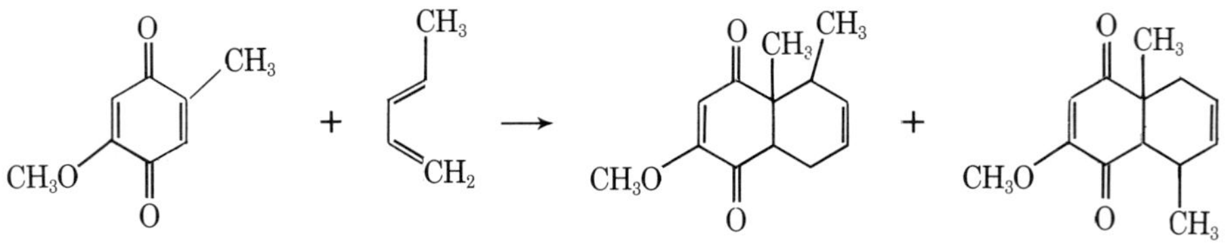

(84)   (85)   (86) [58822-91-6]

Conditions	Yield	Ratio of products		
thermal	80%	1	:	1
SnCl$_4$	>85%	1	:	20
BF$_4$	>85%	4	:	1

Especially sensitive quinones, eg, (87)–(89), can be generated *in situ* and the diene adduct is obtained in excellent yield (57).

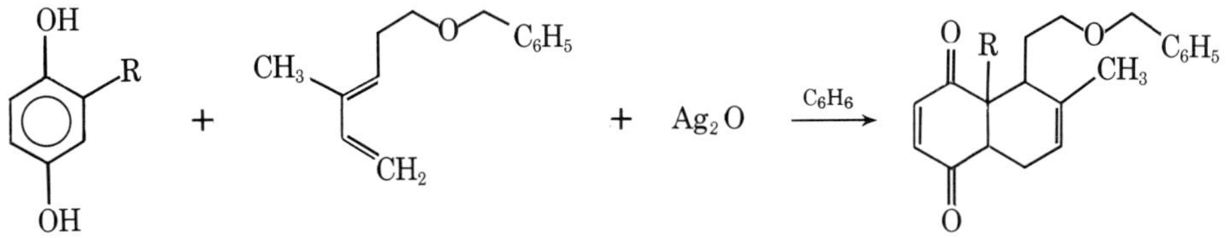

R = COCH$_3$ yield = 95% (**87**) [80867-99-1]

R = CHO   yield = 97% (**88**) [80866-99-5]

R = CCH$_3$ (with =O)   yield = 100% (**89**) [80866-98-4]

The potentially important 3,5,5-trialkoxy-1,2,4-trichlorocyclopentadienes react with 1,4-benzo- and 1,4-naphthoquinone (58).

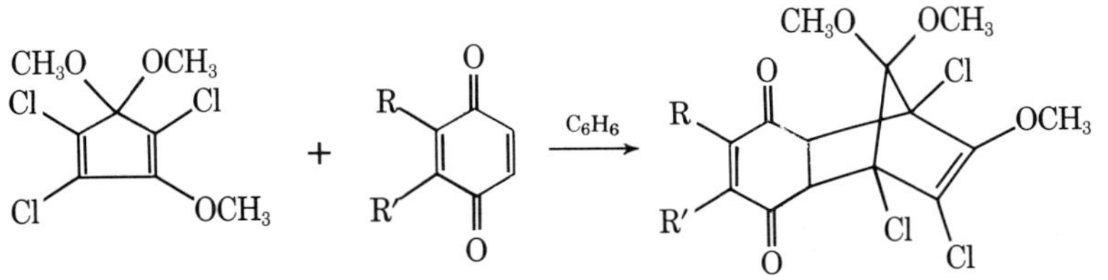

(**2**) R, R' = H, H   (**90**) [73286-38-1]   yield = 95%

(**7**) R, R' = (benzo)   (**91**) [73346-47-1]   yield = 60%

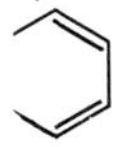

Reactions of 1,2-quinones are depicted below (59–61).

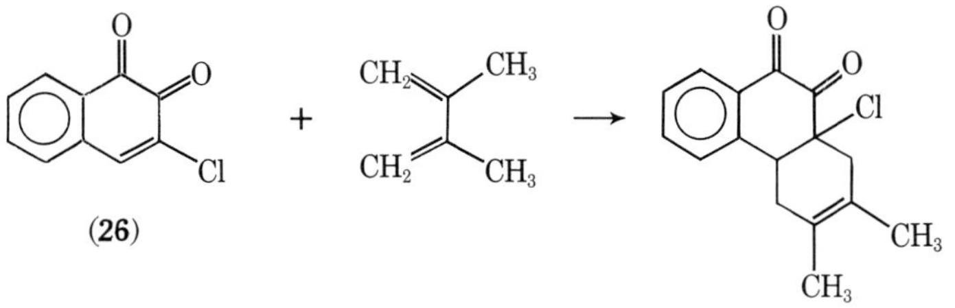

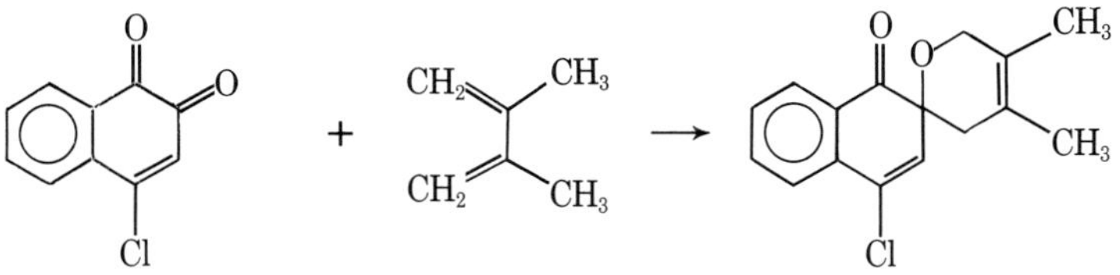

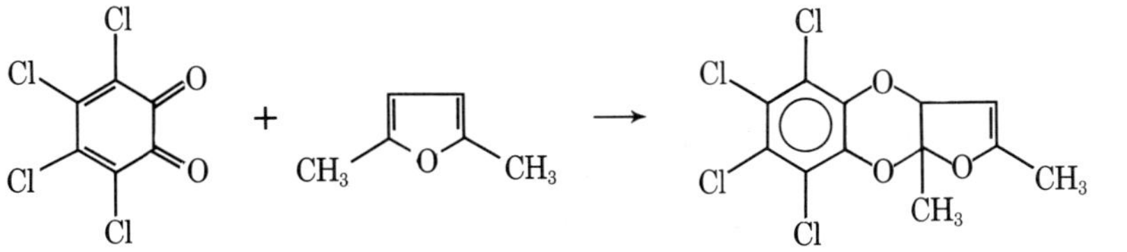

594 QUINONES

A wide variety of other substituents, eg, CH$_3$O, Br, CN, CH$_3$OC(O), and NO$_2$, were investigated in the system with (26) and (92), and the products have been described (62). The Diels-Alder reactions of 1,2-benzoquinones generated *in situ* have also been carried out (63).

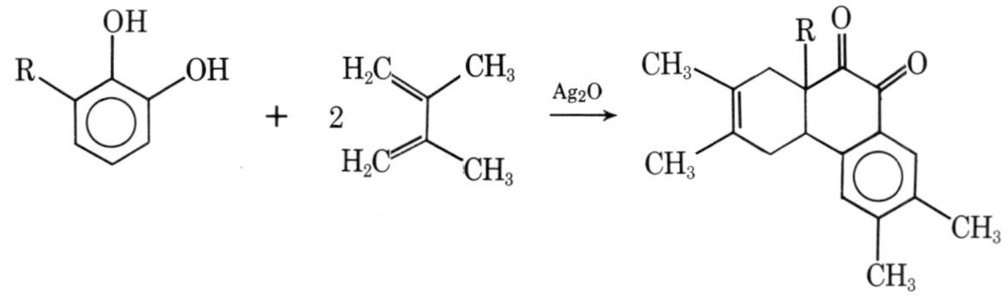

R = $\overset{O}{\overset{\|}{C}}CH_3$ yield = 51% [*67984-83-2*]

R = $\overset{O}{\overset{\|}{C}}C_6H_5$ yield = 65% [*67984-84-3*]

The formation of heterocyclic quinones through various modes of cycloaddition is an important synthetic technique, and reactions with diazo compounds or enamines have been especially significant (64–65).

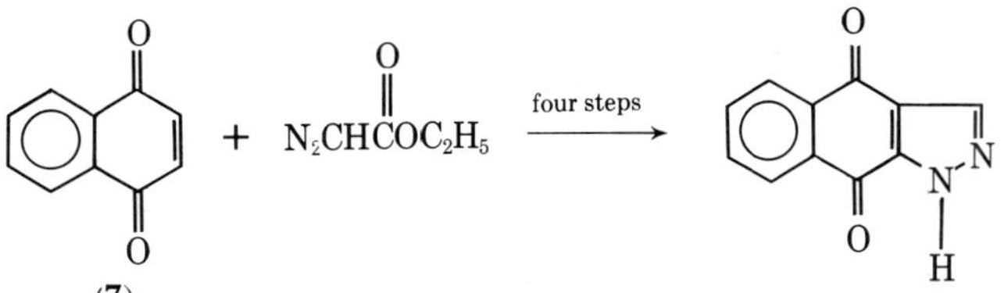

[*1015-97-0*]

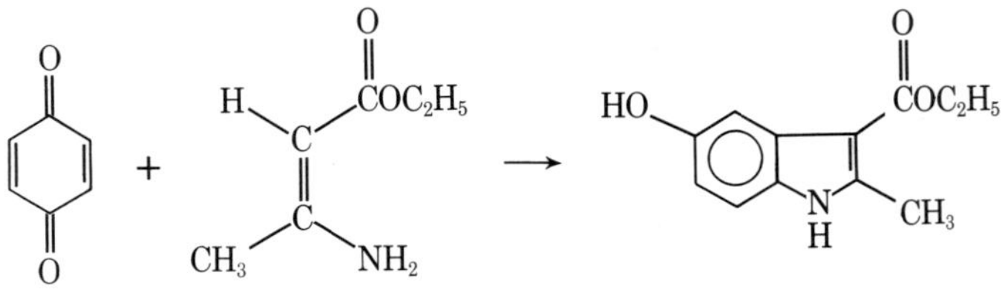

The addition of thioureas to 1,4-benzoquinones is an excellent preparation of 5-hydroxy-1,3-benzoxathiol-2-one (93) and a reasonable synthesis of 2-amino-6-hydroxybenzothiazole (94) (66–67). In the former case, monosubstituted quinones give good yields and high regioselectivity. The addition of thioacetamide or thiobenzamide to 1,4-benzoquinone produces modest yields of 2-substituted 3a-hydroxy-1,3-benzothiazol-6(3a*H*)-ones (95) (68).

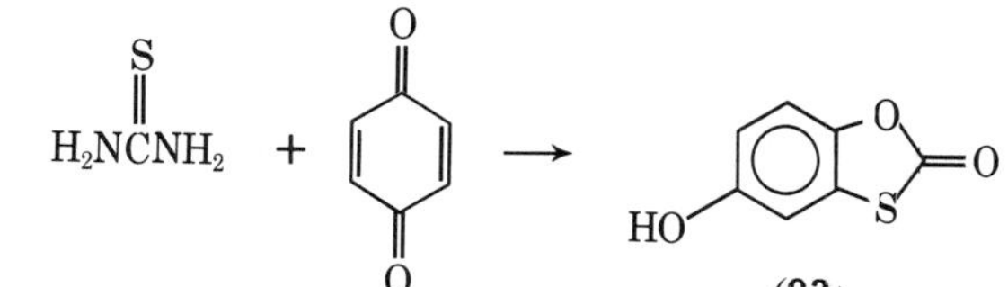

$$\text{H}_2\text{NCNH}_2 \ (\overset{\text{S}}{\|}) + \text{(2)} \longrightarrow \text{(93)} \quad 92\%$$

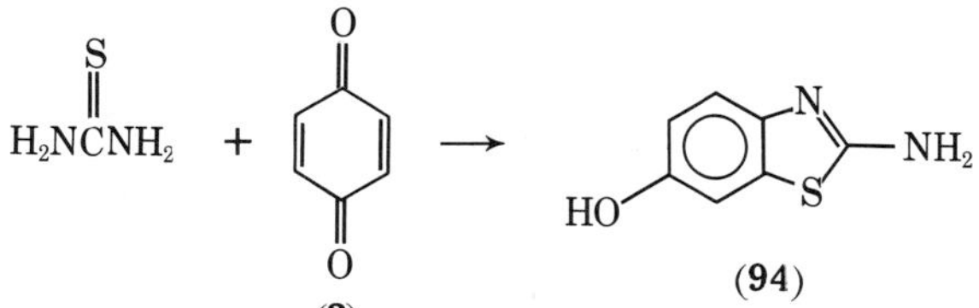

$$\text{H}_2\text{NCNH}_2 \ (\overset{\text{S}}{\|}) + \text{(2)} \longrightarrow \text{(94)} \quad 76\%$$

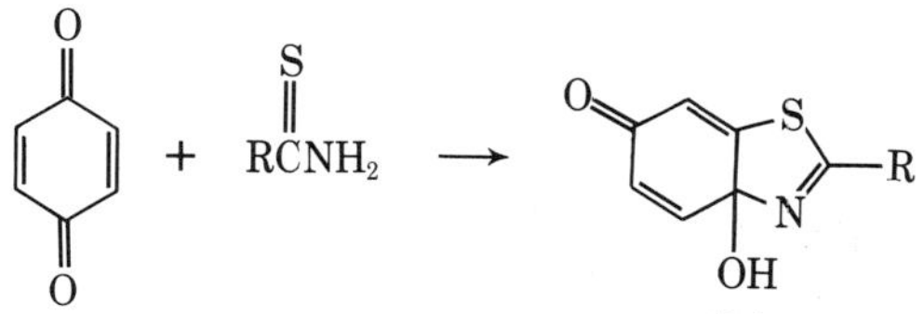

$$\text{(2)} + \text{RCNH}_2 \ (\overset{\text{S}}{\|}) \longrightarrow \text{(95)}$$

R = CH₃   yield = 27.5% [68001-47-8]
R = C₆H₅   yield = 51% [68001-54-7]

A useful synthesis of benzoxazoles was developed in an attempt to oxidize an antibiotic α-amino acid with 3,5-di-*tert*-butyl-1,2-benzoquinone (**96**) (69).

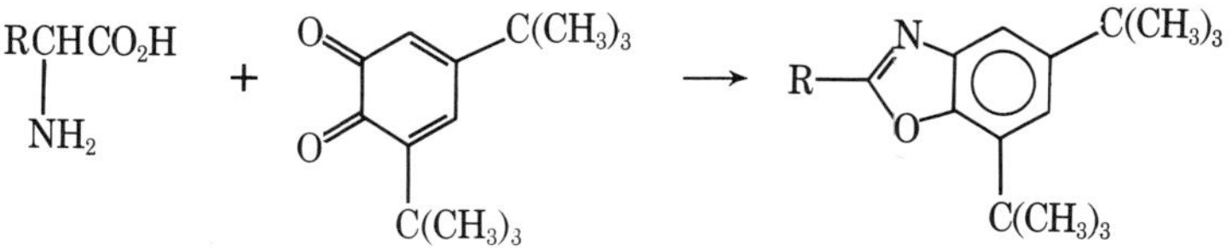

$$\underset{\text{NH}_2}{\text{RCHCO}_2\text{H}} + \text{(96)} \ [3383\text{-}21\text{-}9] \longrightarrow$$

R = CH₃   yield = 80%
R = CH₂C₆H₅   yield = 32%

A general method for the preparation of 2*H*-isoindol-4,7-diones involves 1,3-dipolar addition of oxazolium-5-oxides (sydnones) (**97**) to (**25**) (70).

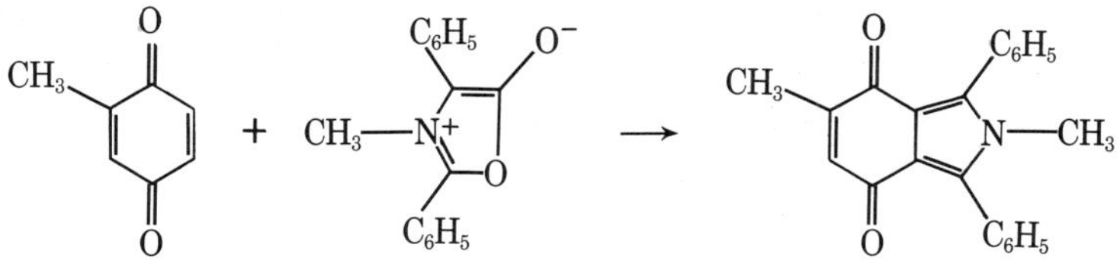

(**25**) + (**97**) → 37% [72726-02-4]

The formation of oximes is strongly influenced by ring substituents of the quinone. The mono and dioximes form if at least one position adjacent to the carbonyl group is unsubstituted (71).

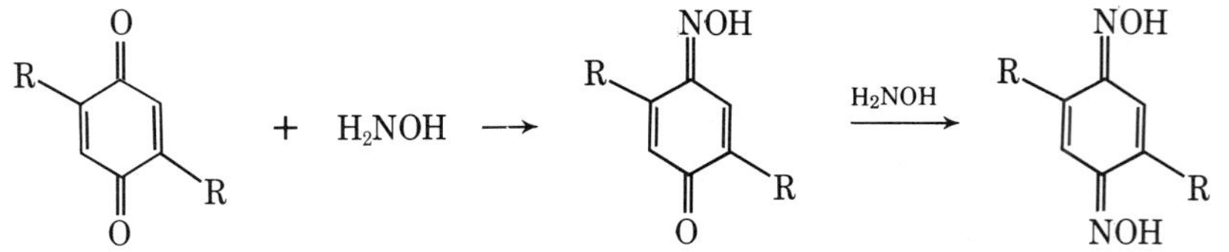

596 QUINONES

(2) R = H
(24) R = CH₃

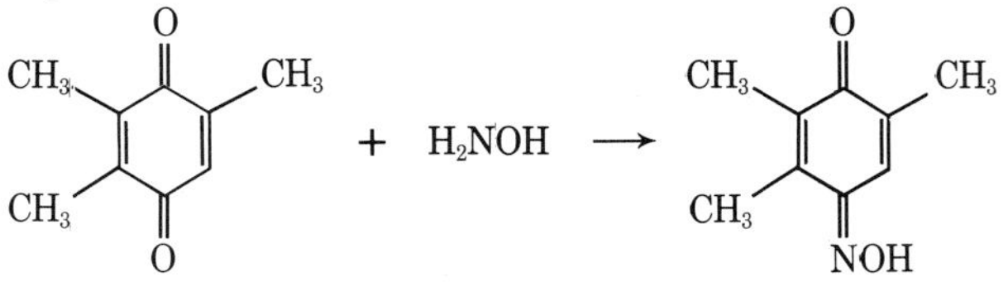

The monoximes, which are tautomers of 4-nitrosophenols, have been used in syntheses of quinones (72).

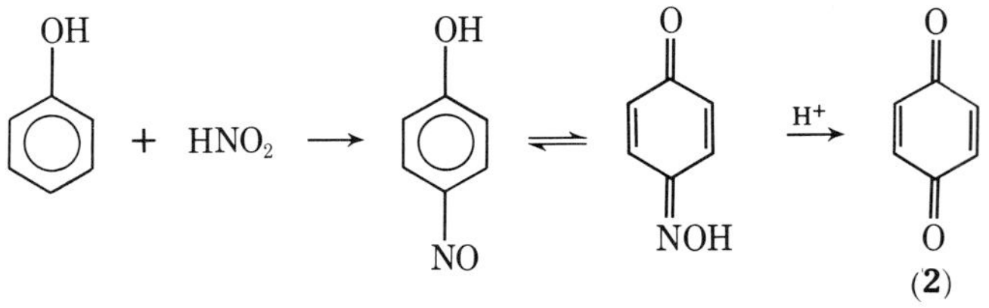

Other nitrogen carbonyl derivatives also exist in tautomeric equilibrium.

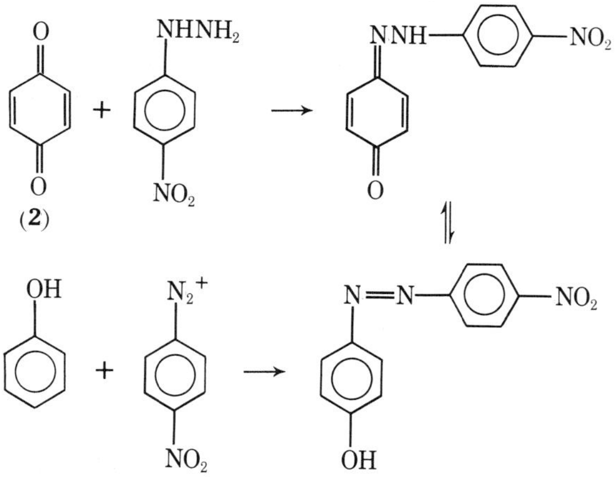

The relationship to diazonium coupling is used in structure determination (73). Unlike most simple carbonyl compounds, the quinones do not yield bisulfite addition products but undergo ring addition. Another significant carbonyl reaction is the addition of tertiary phosphites under anhydrous conditions (74).

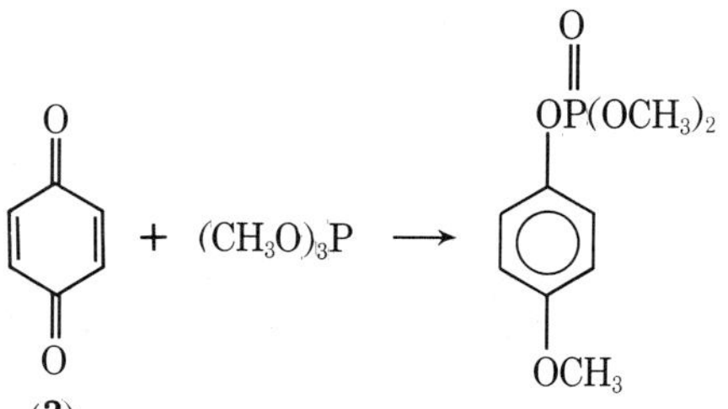

The ester product is easily hydrolyzed, and the reaction sequence provides an excellent synthesis of hydroquinone monoethers. The exact path of the reaction is uncertain.

A third main area of quinone chemistry involves the nucleophilic substitution of labile groups. Most of this chemistry has involved amines and is a complementary method to the addition sequence, but oxygen nucleophiles and inorganic anions have also been studied (75–77). A smaller number of studies have been done with sulfur nucleophiles (78).

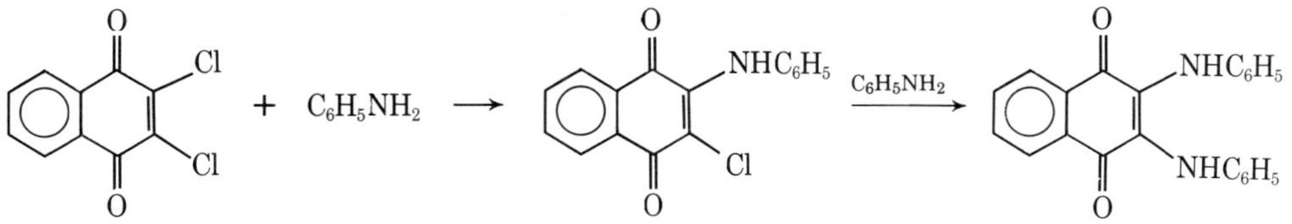

(**27**) (**98**) [*1090-16-0*] (**99**) [*80632-71-9*]

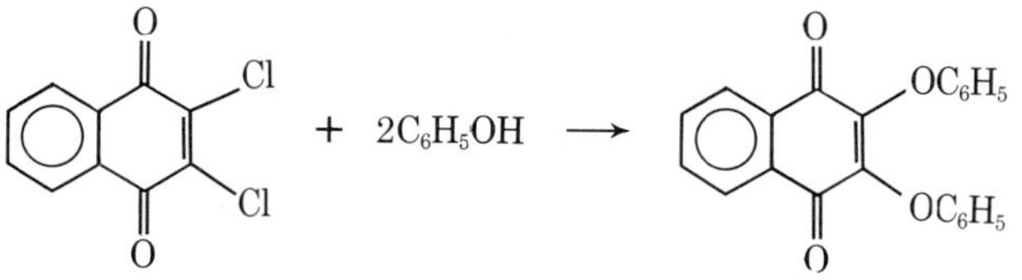

(**27**) (**100**) [*36626-24-1*]

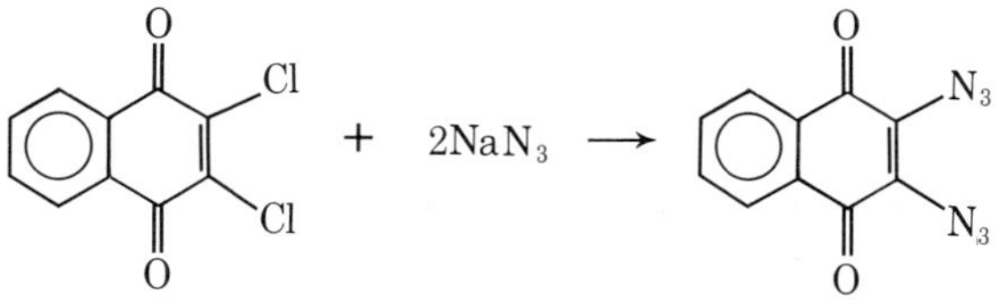

(**27**) (**101**) [*6427-95-8*]

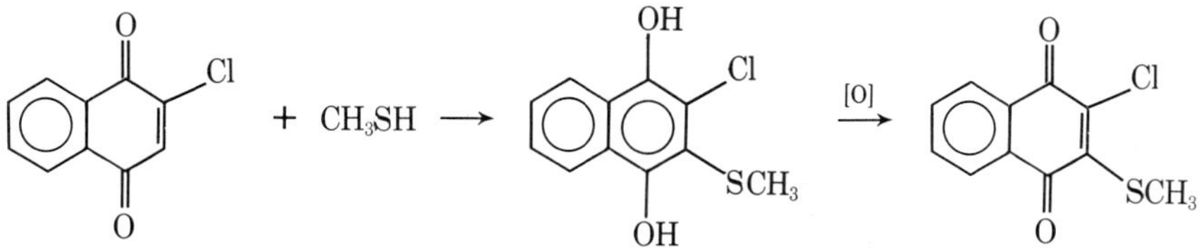

(**102**) (**103**) [*80632-72-0*]

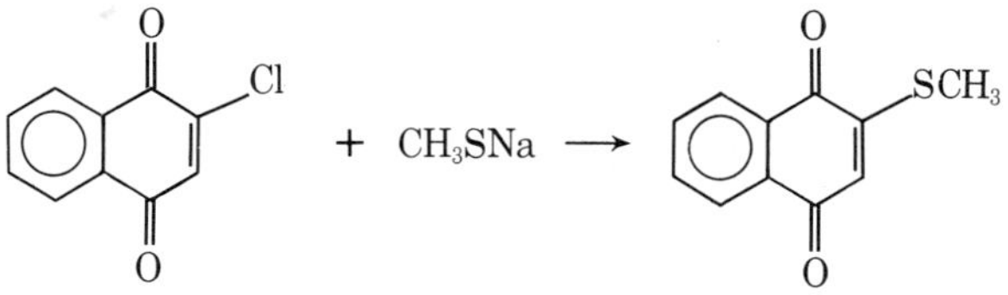

(**102**) (**104**) [*26037-60-5*]

The nitrogen substitution chemistry of 2,3-dichloro-1,4-naphthoquinone (**27**) is shown below (79). The product mixtures are complex and the yields generally are modest. The analogous reaction of 2,3-dibromo-1,4-naphthoquinone (**107**) has been used in the synthesis of mitomycin antibiotics (80).

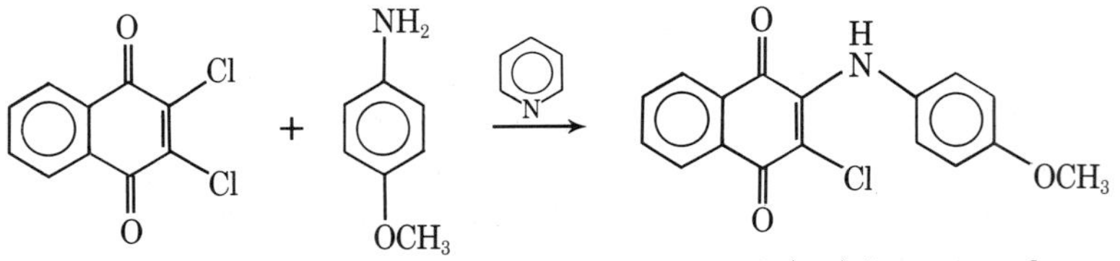

44% (**105**) [22359-32-6]

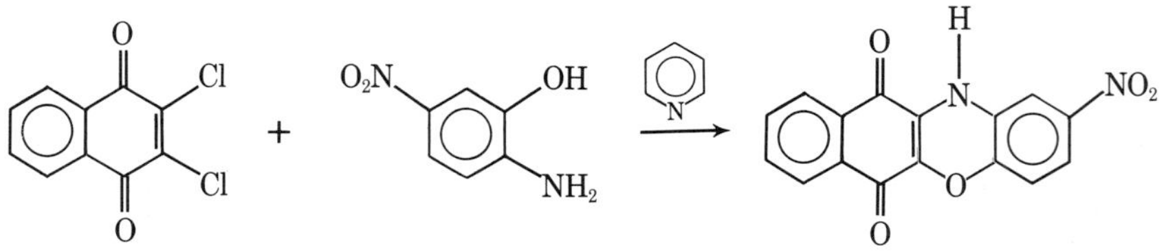

65.5% (**106**) [75197-88-5]

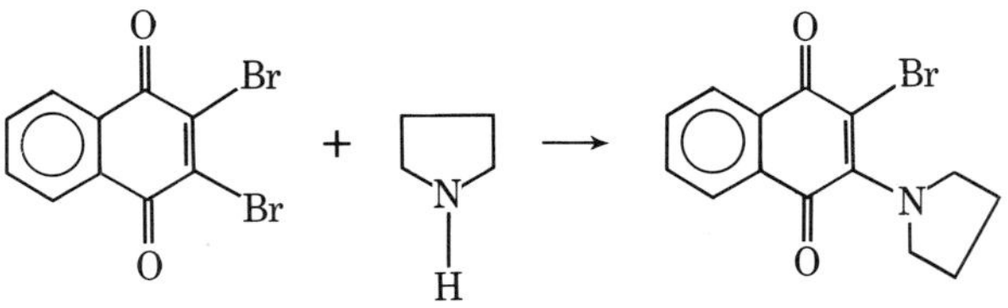

(**107**) [13243-65-7]   91% (**108**) [72866-63-8]

The addition and substitution chemistry of quinones has been reveiwed (81).

Syntheses

Syntheses of quinones often involve oxidation, since it is the only completely general method (82). Thus, in several instances, quinones are the reagents of choice for the preparation of other quinones. Oxidation has been especially useful with catechols and hydroquinones as starting materials (17). The preparative utility of these reactions depends largely on the relative oxidation potentials of the quinones (83–84).

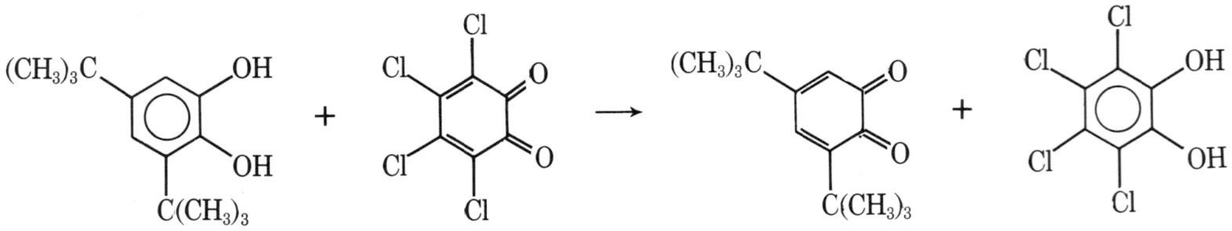

(**21**)   (**96**)   (**38**)

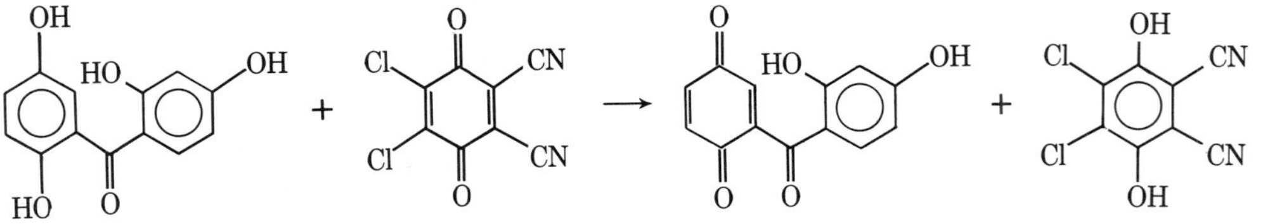

(**20**)   (**109**) [16850-72-9]   (**37**)

For the preparation of ten grams or less of a quinone, the oxidation of a phenol with Fremy's salt (Teuber reaction) is perhaps the method of choice (85–86). A wide range of phenols has been used, including some with 4-substituents.

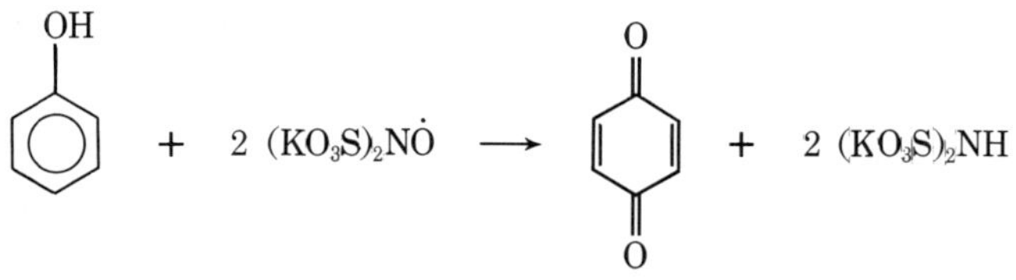

$$\text{PhOH} + 2\ (KO_3S)_2N\dot{O} \longrightarrow \text{benzoquinone} + 2\ (KO_3S)_2NH$$

(2)

The latter usually produce 1,2-benzoquinones, but the 4-chloro- and 4-*tert*-butyl groups can be eliminated to yield a 1,4-benzoquinone. The yield for simple phenols is frequently in excess of 70%, and some complex phenols show highly selective oxidation. With an occasional exception, substituents and side chains are not attacked by Fremy's salt (87–88).

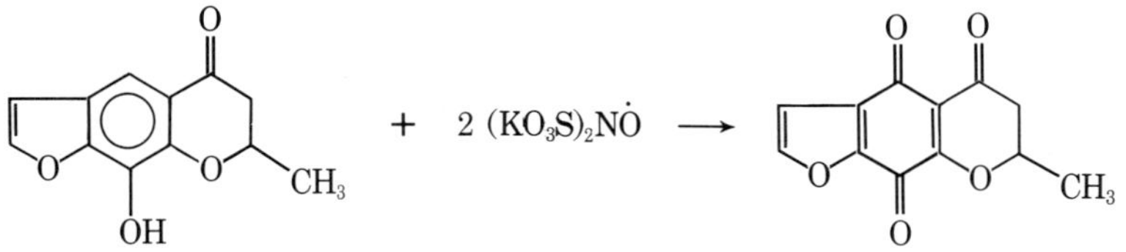

(**110**) [*481-71-0*]

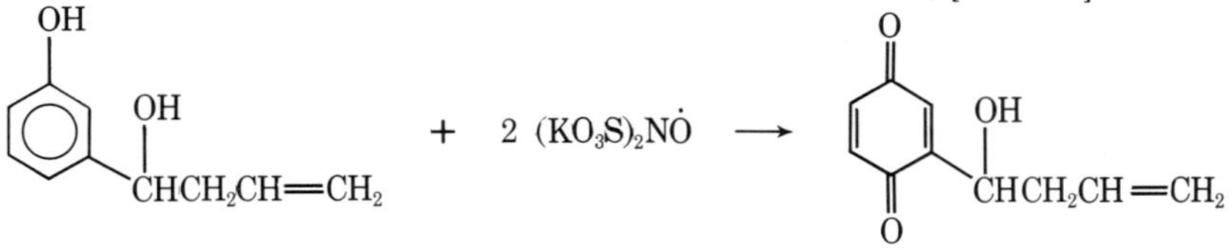

(**111**) [*32578-45-3*]

The extraordinary delicacy and selectivity that is possible is illustrated below; this reaction is achieved only with Fremy's salt (89). A wide variety of oxidants has been employed in the preparation of quinones from phenols.

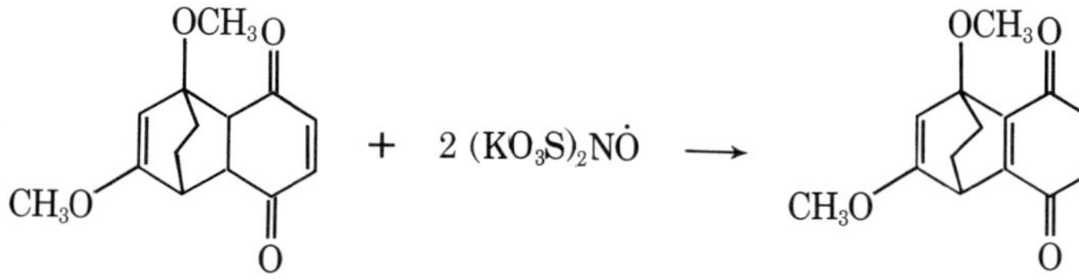

(**112**) [*58785-57-2*]   (**113**) [*58785-59-4*]

Of these reagents, chromic acid, ferric ion, or silver oxide demonstrate significant usefulness in the oxidation of hydroquinones. Thallium(III) trifluoroacetate converts 4-halo- or 4-*tert*-butylphenols to 1,4-benzoquinones in high yield (90).

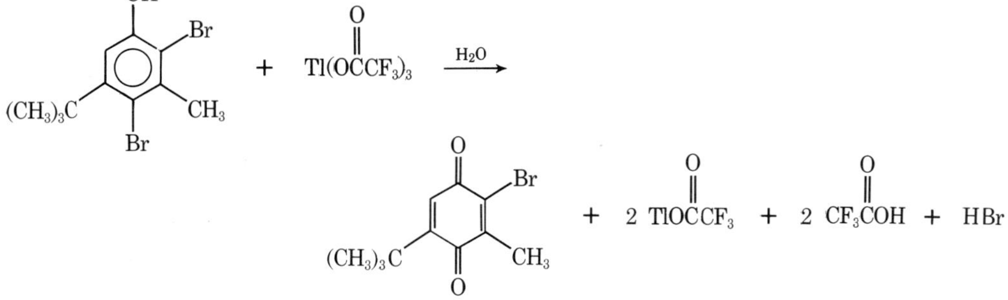

(**114**) [*25441-20-3*]

600 QUINONES

Thallium trinitrate oxidizes naphthols and hydroquinone monoethers to quinones or 4,4-dialkoxycyclohexa-2,5-dienones, respectively (4,91).

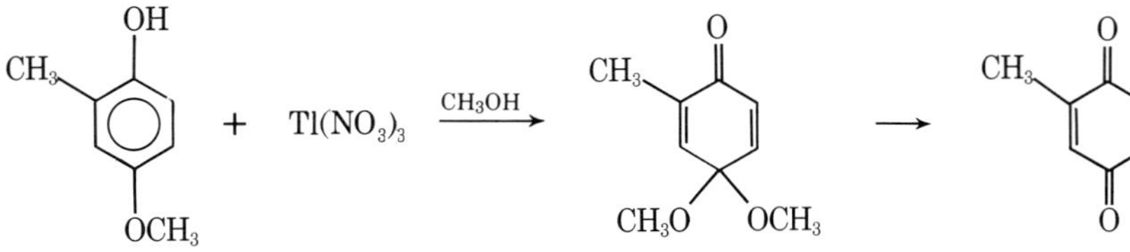

89% [57197-11-2]   (25)

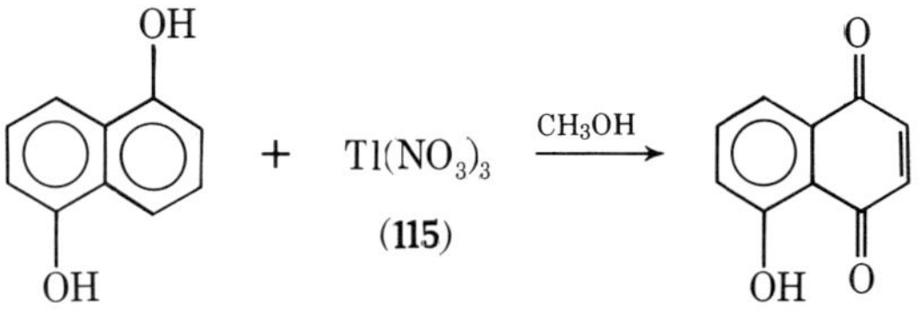

64% (115) [481-39-0]

The oxidation of 4-bromophenols to quinones can also be accomplished with periodic acid (92).

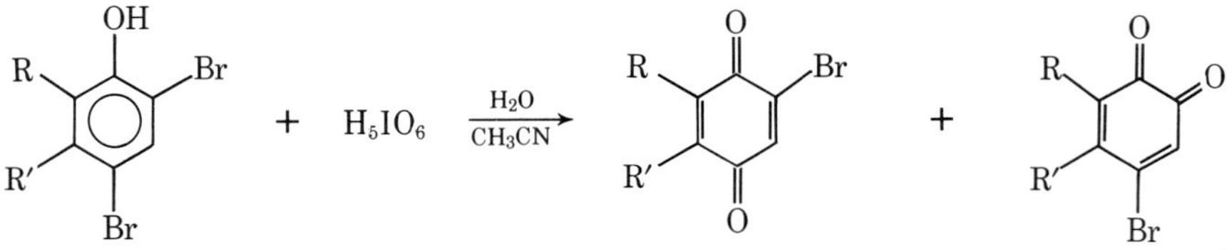

R, R' = CH=CH—CH=CH
R = Br, R' = H

(116) [2065-37-4] yield = 71%   (117) [18099-96-2]
(118) [19643-45-9] yield = 75%   yield = 20%

A detailed study of this reagent with sterically hindered phenols provides information about factors that influence the quinonoid product (93).

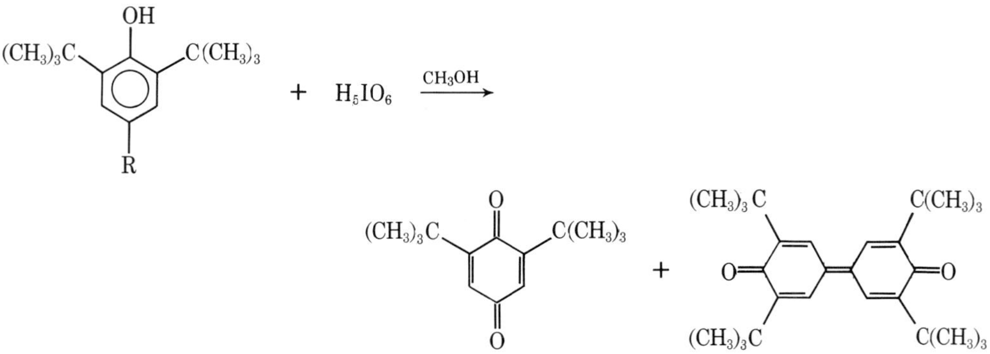

(119) [719-22-2]   (120) [2411-18-9]

R	yield (119)	yield (120)
R = H	7%	61%
R = Br	33%	24%
R = OCH$_3$	60%	0%

The anodic oxidation of hydroquinone ethers to quinone ketals yields synthetically useful intermediates that can be hydrolyzed to quinones at the desired stage of a sequence (46). Trimethylsilyl ethers are oxidized directly to quinones in nonaqueous media (3).

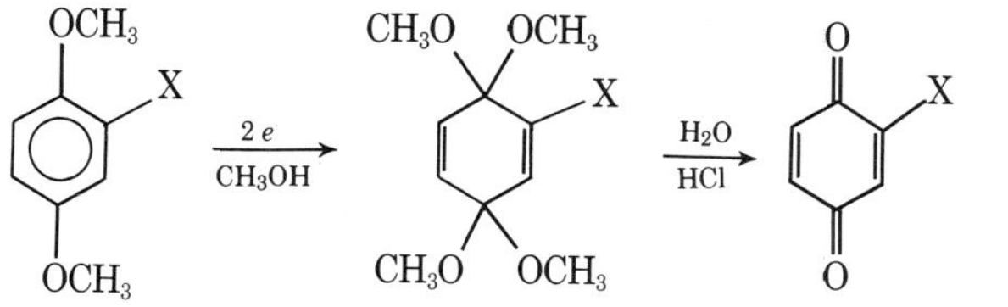

X = Cl   yield = 83%
X = Br   yield = 75%   (22)
                                 (121) [3958-82-5]

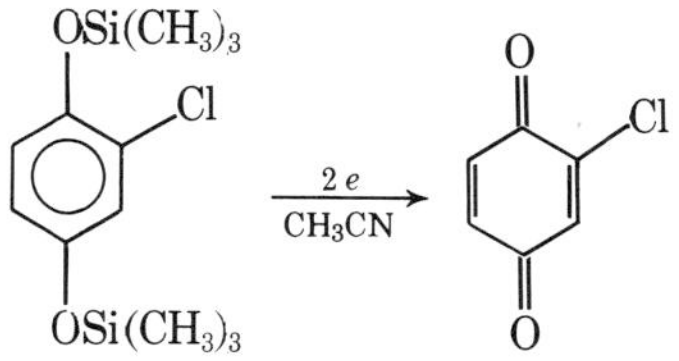

(22) 80%

The synthesis of 2-halohydroquinones or their disilyl ethers has been reported (94).

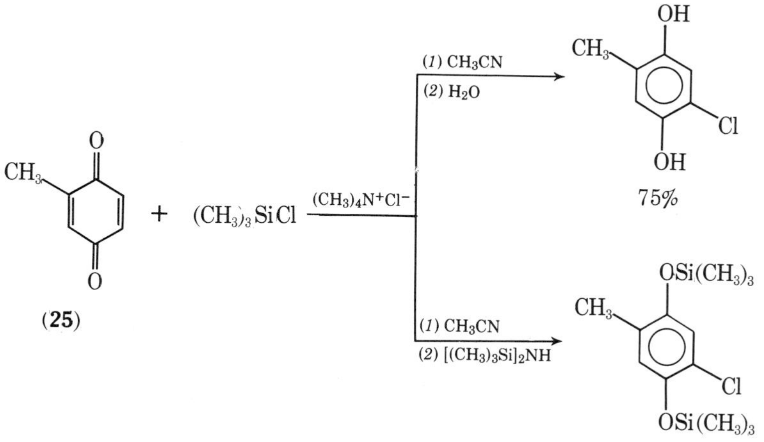

(25)

75%

77%

Derivatives of the natural product juglone are obtained in a single reaction involving halogenation and oxidation by N-bromosuccinimide (95).

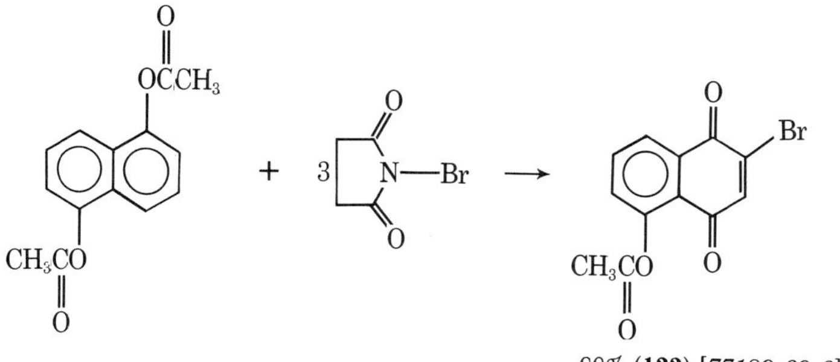

90% (122) [77189-69-6]

602   QUINONES

Several oxygenated naphthalenes produce high yields of regiospecific product.

The dimethyl ethers of hydroquinones and 1,4-naphthalenediols can be oxidized with silver(II) oxide or ceric ammonium nitrate.

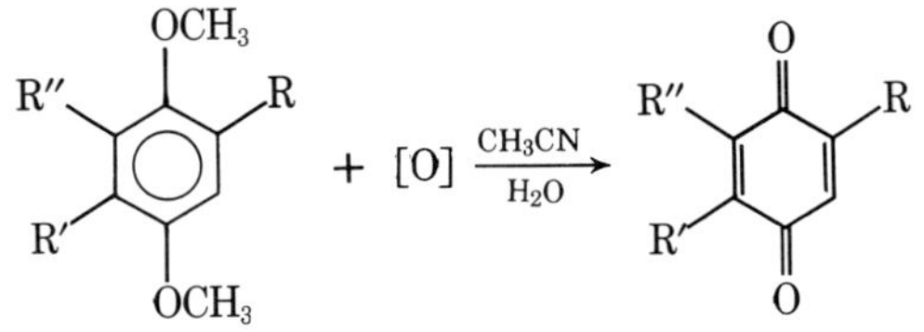

R = CH_3	R′,R″ = CH_3, H	[O] = AgO	yield = 91%
R = CH_3	R′,R″ = CH_3, H	[O] = $Ce(NH_4)_2(NO_3)_6$	yield = 99% (**24**)
R = i-Pr	R′,R″ = H, i-Pr	[O] = AgO	yield = 79%
R = i-Pr	R′,R″ = H, i-Pr	[O] = $Ce(NH_4)_2(NO_3)_6$	yield = 92% (**123**) [67902-52-7]
R = CH_3	R′,R″ = CH=CHCH=CH	[O] = AgO	yield = 84%
R = CH_3	R′,R″ = CH=CHCH=CH	[O] = $Ce(NH_4)_2(NO_3)_6$	yield = 99% (**28**)

The yields of quinones are excellent when pyridine- or pyrazinecarboxylic acids are used as catalysts (96).

Aqueous sodium hypochlorite with phase transfer has also produced efficient conversion of catechols and hydroquinones to 1,2- and 1,4-benzoquinones (97).

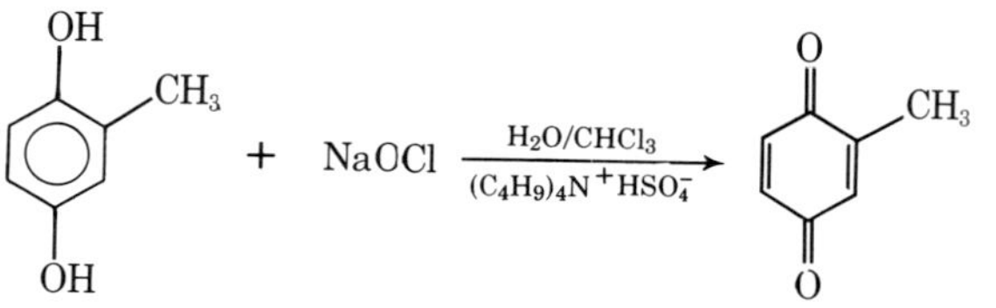

72% (**25**)

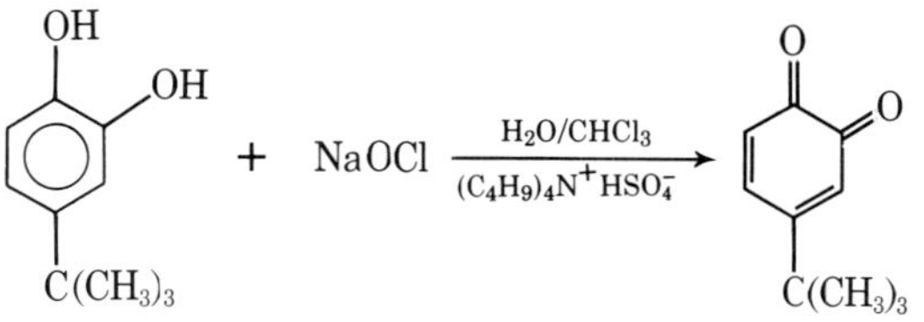

92% (**124**) [1129-21-1]

Manufacture

With the exceptions of 1,4-benzoquinone and 9,10-anthraquinone, quinones do not have a substantial market, but a few of them are commercially available (see Anthraquinone). Some 1981 prices of quinones are listed in Table 3 (98–99). Most of the compounds are prepared by the methods described earlier (see Syntheses). The few large-scale preparations involve oxidation of aniline, phenol, or aminonaphthols (eg, (**125**)).

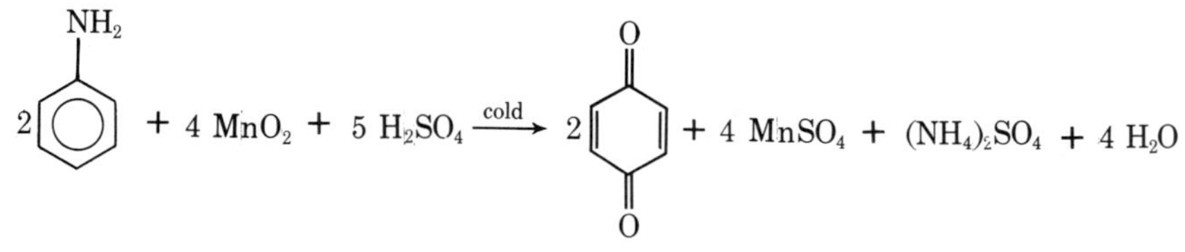

$$2\,\underset{}{C_6H_5NH_2} + 4\,MnO_2 + 5\,H_2SO_4 \xrightarrow{cold} 2\,\underset{(2)}{C_6H_4O_2} + 4\,MnSO_4 + (NH_4)_2SO_4 + 4\,H_2O$$

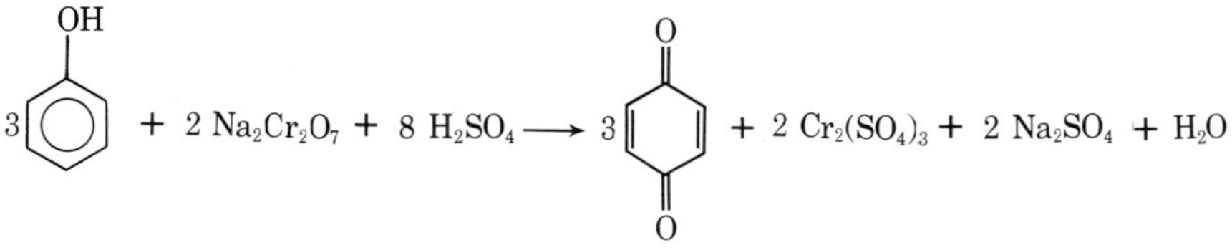

$$3\,\underset{}{C_6H_5OH} + 2\,Na_2Cr_2O_7 + 8\,H_2SO_4 \longrightarrow 3\,\underset{(2)}{C_6H_4O_2} + 2\,Cr_2(SO_4)_3 + 2\,Na_2SO_4 + H_2O$$

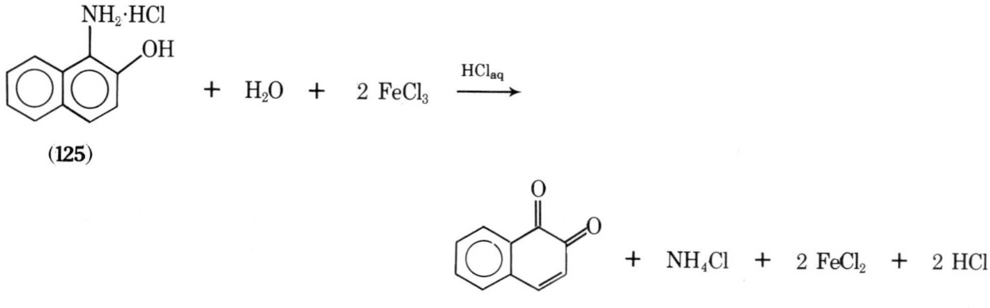

(125) 1-amino-2-naphthol·HCl + H_2O + 2 $FeCl_3$ $\xrightarrow{HCl_{aq}}$ 1,2-naphthoquinone (6) 93% + NH_4Cl + 2 $FeCl_2$ + 2 HCl

In the case of 1,4-benzoquinone, the product is steam-distilled, chilled, and obtained in high yield and purity. The direct oxidation of the appropriate unoxygenated hydrocarbon has been described for a large number of ring systems, but it is generally utilized only for the polynuclear quinones. A few representative uses of the quinones are given in Table 4.

Table 3. Commercially Available Quinones

Quinone	Structure no.	Price (1982), $ Eastman[a]	Aldrich[b]
1,4-benzoquinone	(2)	41.60/kg[c]	29.00/kg
1,2-naphthoquinone	(6)	43.55/25 g	54.00/25 g
1,4-naphthoquinone	(7)	30.80/500 g	34.40/500 g
2-methyl-1,4-benzoquinone	(25)	73.35/500 g	21.25/100 g
p-chloranil	(19)	38.80/kg	
o-chloranil	(21)	61.15/100 g	23.35/100 g
2,3-dichloro-5,6-dicyano-1,4-benzoquinone	(20)	54.85/25 g	110.00/100 g
2,3-dichloro-1,4-naphthoquinone	(27)	47.55/100 g	7.00/100 g

[a] Ref. 98.
[b] Ref. 99.
[c] $9.83 in 13, 607-kg lots.

604 QUINONES

Table 4. Uses of Some Quinones

Quinone	Structure no.	Use
1,4-benzoquinone	(2)	oxidant, amino acid determination
2-chloro-, 2,5-dichloro-, and 2,6-dichloro-1,4-benzoquinones	(22), (23), (126)	bactericides
2,3-dichloro-5,6-dicyano-1,4-benzoquinone	(20)	oxidation and dehydration agent
chloranils	(19), (21)	intermediate, oxidant
2-methyl- and 2,3-dimethyl-1,4-naphthoquinones	(28, 61)	vitamin K substitutes, antihemorrhagic agents
2,3-dichloro-1,4-naphthoquinone	(27)	intermediate, fungicide

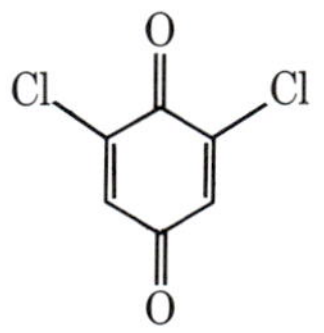

(126)

Health and Safety Factors

Because of the high vapor pressures of the simple quinones and their penetrating odor, adequate ventilation should be provided in areas where they are handled or stored. Quinone vapor can harm the eyes, and a limit of 0.1 ppm of 1,4-benzoquinone (2) in air has been recommended. The solid or solutions can cause severe local damage to the skin and mucous membranes. The higher quinones represent less of a problem because of their decreased volatility (100).

BIBLIOGRAPHY

"Quinones" in *ECT* 1st ed., Vol. 11, pp. 410–424, by J. R. Thirtle, Eastman Kodak Company; "Quinones" in *ECT* 2nd ed., Vol. 16, pp. 899–913, by J. R. Thirtle, Eastman Kodak Company.

1. A. Woskresensky, *Justus Liebigs Ann. Chem.* **27,** 257 (1838).
2. L. F. Fieser and M. Fieser, *Advanced Organic Chemistry*, Reinhold Publishing Corp., New York, 1961, pp. 845–875.
3. R. F. Stewart and L. L. Miller, *J. Am. Chem. Soc.* **102,** 4999 (1980).
4. D. J. Crouse, M. M. Wheeler, M. Goemann, P. S. Tobin, S. K. Basu, and D. M. S. Wheeler, *J. Org. Chem.* **46,** 1814 (1981).
5. R. C. Weast, ed., *Handbook of Chemistry and Physics*, 62nd ed., CRC Press, Boca Raton, Fla., 1981–1982; J. G. Grasselli and W. M. Ritchey, eds., *Atlas of Spectral Data and Physical Constants for Organic Compounds*, 2nd ed., CRC Press, Cleveland, Ohio, 1975; L. Meites and P. Zuman, eds., *CRC Handbook Series in Electrochemistry*, Vols. 1–4, CRC Press, Cleveland, Ohio, 1977, W. Palm Beach, Fla., 1978, and Boca Raton, Fla., 1980.
6. R. Bentley and I. M. Campbell in S. Patai, ed., *The Chemistry of the Quinonoid Compounds*, Wiley-Interscience, New York, pp. 683–736, 1974.
7. N. Dost and K. van Nes, *Rec. Trav. Chim. Pays-Bas* **70,** 403 (1951).
8. A. I. Kosak, R. J. F. Palchak, W. A. Steele, and C. M. Selwitz, *J. Am. Chem. Soc.* **76,** 4450 (1954).
9. U.S. Pat. 3,365,499 (Jan. 23, 1968), W. H. Clement and C. M. Selwitz (to Gulf Research Development Co.).
10. H.-D. Becker, A. Björk, and E. Adler, *J. Org. Chem.* **45,** 1596 (1980).
11. E. A. Braude, R. P. Linstead, and K. R. Wooldridge, *J. Chem. Soc.*, 3070 (1956).
12. E. J. Agnello and G. D. Laubach, *J. Am. Chem. Soc.* **82,** 4293 (1960).

13. D. Burn, D. N. Kirk, and V. Petrov, *Proc. Chem. Soc.*, 14 (1960).
14. N. Dost and K. van Nes, *Rec. Trav. Chim. Pays-Bas* **71**, 857 (1957).
15. J. W. A. Findlay and A. B. Turner, *Organic Syntheses*, Vol. 5, John Wiley & Sons, Inc., New York, 1973, pp. 428–431.
16. W. Brown and A. B. Turner, *J. Chem. Soc. C*, 2057 (1971).
17. H.-D. Becker in ref. 6, pp. 335–423.
18. D. Bryce-Smith, A. Gilbert, and M. G. Johnson, *J. Chem. Soc. C*, 383 (1967).
19. D. Bryce-Smith and A. Gilbert, *Tetrahedron Lett.*, 3471 (1964).
20. G. O. Schenck, E. K. von Gustorf, B. Kim, G. v. Bünau, and G. Pfundt, *Angew. Chem. Int. Ed. Engl.* **1**, 516 (1964).
21. D. H. Reid, M. Fraser, B. B. Molloy, H. A. S. Payne, and R. G. Sutherland, *Tetrahedron Lett.*, 530 (1961).
22. H. J. Pick, *Tetrahedron Lett.*, 1169 (1969).
23. K. Maruyama, T. Otsuki, and K. Mitsui, *J. Org. Chem.* **45**, 1424 (1980).
24. K. Maruyama and N. Narita, *J. Org. Chem.* **45**, 1421 (1980).
25. A. Michael, *J. Prakt. Chem.* **79**(2), 418 (1909).
26. E. R. Brown, K. T. Finley, and R. L. Reeves, *J. Org. Chem.* **36**, 2849 (1971).
27. H. S. Wilgus, E. Frauenglass, E. T. Jones, R. F. Porter, and J. W. Gates, Jr., *J. Org. Chem.* **29**, 594 (1964).
28. C. F. H. Allen and C. V. Wilson, *J. Am. Chem. Soc.* **63**, 1756 (1941).
29. J. Thiele, *Chem. Ber.* **31**, 1247 (1898).
30. H. A. E. Mackenzie and E. R. S. Winter, *Trans. Faraday Soc.* **44**, 159 (1948).
31. H. S. Wilgus and J. W. Gates, Jr., *Can. J. Chem.* **45**, 1975 (1967).
32. J. F. Bagli and P. L'Ecuyer, *Can. J. Chem.* **39**, 1037 (1961).
33. P. Brassard and P. L'Ecuyer, *Can. J. Chem.* **37**, 1505 (1959).
34. J. Kumanotani, F. Kogawa, A. Hikosaka, and K. Sugita, *Bull. Chem. Soc. Jpn.* **41**, 2118 (1968).
35. J. R. Luly and H. Rapoport, *J. Org. Chem.* **46**, 2745 (1981).
36. M. D. Rozeboom, I-M. Tegmo-Larsson, and K. N. Houk, *J. Org. Chem.* **46**, 2338 (1981).
37. W. H. Wanzlick in W. Foerst, ed., *Newer Methods of Preparative Organic Chemistry*, Vol. IV, Academic Press, Inc., New York, 1968, pp. 139–154.
38. K. A. Parker and S-K. Kang, *J. Org. Chem.* **45**, 1218 (1980).
39. H. Pluim and H. Wynberg, *J. Org. Chem.* **45**, 2498 (1980).
40. L. F. Fieser and A. E. Oxford, *J. Am. Chem. Soc.* **64**, 2060 (1942).
41. L. F. Fieser and R. B. Turner, *J. Am. Chem. Soc.* **69**, 2338 (1947).
42. K. Maruyama and Y. Naruta, *J. Org. Chem.* **43**, 3796 (1978).
43. Y. Naruta, *J. Org. Chem.* **45**, 4097 (1980).
44. Y. Naruta, *J. Am. Chem. Soc.* **102**, 3774 (1980).
45. H. W. Moore, Y-L. L. Sing, and R. S. Sidhu, *J. Org. Chem.* **45**, 5057 (1980).
46. M. J. Manning, P. W. Rynolds, and J. S. Swenton, *J. Am. Chem. Soc.* **98**, 5008 (1976).
47. W. Albrecht, *Justus Liebigs Ann. Chem.* **348**, 31 (1906).
48. O. Diels and K. Alder, *Justus Liebigs Ann. Chem.* **460**, 98 (1928).
49. O. Diels, K. Alder, G. Stein, P. Pries, and H. Winckler, *Chem. Ber.* **62**, 2337 (1929).
50. R. K. Hill, M. G. Newton, N. S. Pantaleo, and K. M. Collins, *J. Org. Chem.* **45**, 1593 (1980).
51. M. F. Ansell, B. W. Nash, and D. A. Wilson, *J. Chem. Soc.* 3006 (1963).
52. A. P. Kozikowski, K. Sugiyama, and J. P. Springer, *Tetrahedron Lett.*, 3257 (1980).
53. C. Brisson and P. Brassard, *J. Org. Chem.* **46**, 1810 (1981).
54. I-M. Tegmo-Larsson, M. D. Rozeboom, and K. N. Houk, *Tetrahedron Lett.*, 2043 (1981).
55. I-M. Tegmo-Larsson, M. D. Rozeboom, N. G. Rondan, and K. N. Houk, *Tetrahedron Lett.*, 2047 (1981).
56. J. S. Tou and W. Reusch, *J. Org. Chem.* **45**, 5012 (1980).
57. G. A. Kraus and M. J. Taschner, *J. Org. Chem.* **45**, 1174 (1980).
58. U. O'Connor and W. Rosen, *J. Org. Chem.* **45**, 1824 (1980).
59. M. F. Ansell and R. A. Murray, *Chem. Commun.*, 1583 (1968).
60. *Ibid.*, pp. 1111–1112.
61. W. M. Horspool, J. M. Tedder, and Z. U. Din, *J. Chem. Soc. C*, 1694 (1969).
62. M. F. Ansell and R. A. Murray, *J. Chem. Soc. C*, 1429 (1971).
63. R. Al-Hamdang and B. Ali, *Chem. Commun.*, 397 (1978).
64. L. F. Fieser and M. A. Peters, *J. Am. Chem. Soc.* **53**, 4080 (1931).
65. U. Kuckländer, *Tetrahedron Lett.*, 157, 2093 (1971).

66. P. T. S. Lau and M. Kestner, *J. Org. Chem.* **33,** 4426 (1968).
67. P. T. S. Lau and T. E. Gompf, *J. Org. Chem.* **35,** 4103 (1970).
68. V. Horak and W. B. Manning, *J. Org. Chem.* **44,** 120 (1979).
69. M. C. V. Zwan, F. W. Hartner, R. A. Reamer, and R. Tull, *J. Org. Chem.* **43,** 509 (1978).
70. J. A. Myers, L. D. Moore, Jr., W. L. Whitter, S. L. Council, R. M. Waldo, J. L. Lanier, and B. U. Omoji, *J. Org. Chem.* **45,** 1202 (1980).
71. F. Kehrmann, *Chem. Ber.* **21,** 3315 (1888).
72. H. H. Hodgson and F. H. Moore, *J. Chem. Soc.* **128,** 2036 (1926).
73. L. I. Smith and W. B. Irwin, *J. Am. Chem. Soc.* **63,** 1036 (1941).
74. F. Ramirez and S. Dershowitz, *J. Am. Chem. Soc.* **81,** 587 (1959).
75. A. Plagemann, *Chem. Ber.* **15,** 484 (1882).
76. F. Ullmann and M. Ettisch, *Chem. Ber.* **54,** 259 (1921).
77. F. Fries and P. Ochwat, *Chem. Ber.* **56,** 1291 (1923).
78. L. F. Fieser and R. H. Brown, *J. Am. Chem. Soc.* **71,** 3609 (1949).
79. N. L. Agarwal and W. Schäfer, *J. Org. Chem.* **45,** 2155, 5139 (1980).
80. S. N. Falling and H. Rapoport, *J. Org. Chem.* **45,** 1260 (1980).
81. K. T. Finley in ref. 6, pp. 877–1144.
82. R. H. Thomson in ref. 6, pp. 111–161.
83. M. F. Ansell, A. F. Gosden, V. J. Leslie, and R. A. Murray, *J. Chem. Soc. C*, 1401 (1971).
84. R. C. Ellis, W. B. Whalley, and K. Ball, *Chem. Commun.*, 803 (1967).
85. H.-J. Teuber and O. Glosauer, *Chem. Ber.* **98,** 2643 (1965).
86. H. Zimner, D. C. Lankin, and S. W. Horgan, *Chem. Rev.* **71,** 229 (1971).
87. O. Dann and H.-G. Zeller, *Chem. Ber.* **93,** 2829 (1960).
88. J. M. Bruce, D. Creed, and K. Dawes, *J. Chem. Soc. C*, 2244 (1971).
89. V. H. Powell, *Tetrahedron Lett.*, 3463 (1970).
90. A. McKillop, B. P. Swann, and E. C. Taylor, *Tetrahedron* **26,** 4031 (1970).
91. A. McKillop, D. H. Perry, M. Edwards, S. Antus, L. Farkas, M. Nógrádi, and E. C. Taylor, *J. Org. Chem.* **41,** 282 (1976).
92. P. T. Perumal and M. V. Bhatt, *Synthesis*, 205 (1979).
93. H.-D. Becker and K. Gustafsson, *J. Org. Chem.* **44,** 428 (1979).
94. L. L. Miller and R. F. Stewart, *J. Org. Chem.* **43,** 3078 (1978).
95. S. W. Heinzman and J. R. Grunwell, *Tetrahedron Lett.*, 4305 (1980).
96. L. Syper, K. Kloc, J. Mlochowski, and Z. Szulc, *Synthesis*, 521 (1979).
97. F. Ishii and K. Kishi, *Synthesis*, 706 (1980).
98. *Eastman Laboratory Chemicals*, Catalog No. 51, Rochester, N.Y., Jan. 1, 1981.
99. *Aldrich Catalog Handbook of Fine Chemicals 1981–1982*, Catalog 20, Milwaukee, Wisc., 1980.
100. N. I. Sax, ed., *Dangerous Properties of Industrial Materials*, 4th ed., D. Van Nostrand Co., New York, 1975, pp. 1074–1075.

General References

Reference 6 is a general reference.
J. S. Glasby, *Encyclopedia of Antibiotics*, 2nd ed., Wiley-Interscience, New York, 1979.
J. A. Kent, ed., *Riegel's Handbook of Industrial Chemistry*, 7th ed., D. Van Nostrand Co., New York, 1974.
R. A. Morton, ed., *Biochemistry of Quinones*, Academic Press, Inc., New York, 1965.
W. A. Remers, ed., *The Chemistry of Antitumor Antibiotics*, Wiley-Interscience, New York, 1979.
R. H. Thomson, *Naturally Occuring Quinones*, 2nd ed., Academic Press, Inc., New York, 1971.

K. THOMAS FINLEY
State University College at Brockport

R

R ACID, HOC$_{10}$H$_5$(SO$_3$H)$_2$. See Azo dyes; Naphthalene derivatives.

RACEMIC ACID, HOOCCHOHCHOHCOOH. See Tartaric acid.

RADIATION CURING

The interaction of electromagnetic radiation with organic substrates is of widespread interest and has broad commercial applications. The use of electromagnetic radiation to alter the physical and chemical nature of a material is sometimes termed radiation-curing technology. The following discussion of radiation curing concerns processes that involve interaction of electromagnetic radiation with organic substrates to develop cross-linked or solvent-insoluble network structures. For example, a preformed thermoplastic polymer that interacts directly with certain types of ionizing radiation from a given source of energy can develop into cross-linked or network structures with higher melting points, improved tensile strengths, and improved chemical resistance than the original thermoplastic polymer starting materials (see Fig. 1) (1–2). Similar types of radiation curing are used in organic coating technologies if the liquid coating composition contains reactive vinyl-unsaturated components (see Fig. 2) (3–4). In general, radiation-curing technology involves consideration of at least four main variables: type of radiation source; type of organic substrate to be irradiated; kinetics and mechanisms of radiation energy–organic substrate interactions; and final chemical, physical, and mechanical properties of network formation.

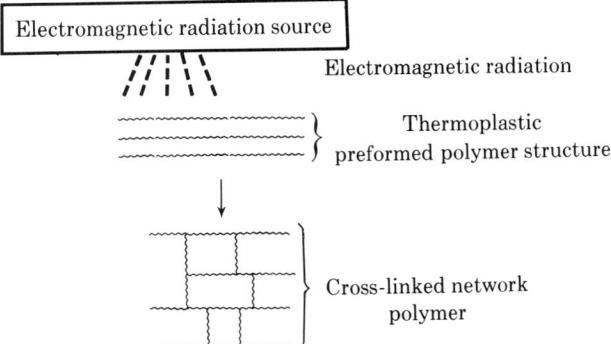

Figure 1. Interaction of electromagnetic radiation with a preformed thermoplastic polymer to develop cross-linked network polymer structures.

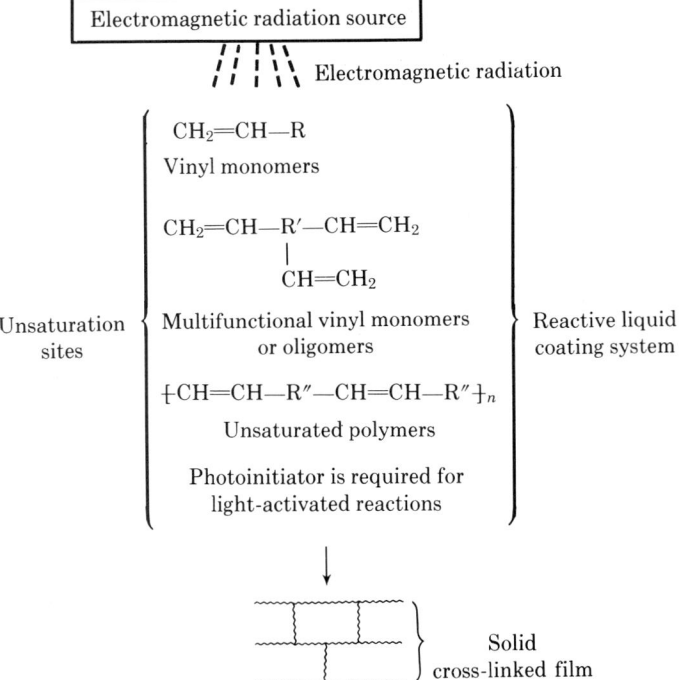

Figure 2. Interaction of electromagnetic radiation with a reactive liquid coating-system containing reactive vinyl-unsaturated components to develop solid cross-linked film.

Radiation and Electromagnetic Radiation Sources

Radiation curing, as applied to cross-linking of polymers or coating materials, involves the full spectrum of electromagnetic radiation energies to effect chemical reaction. These forms of radiation energy include ionizing radiation, ie, α, β, and γ rays from radioactive nuclei; x rays; high energy electrons; and nonionizing radiation such as are associated with uv, visible, ir, microwave, and radio frequency wavelengths of energy (see Table 1) (5).

Table 1. Electromagnetic Spectrum[a]

Types of radiation	Wavelengths, nm	Frequency, Hz	Energy, eV
gamma ray	10^{-4}–10^{-2}	10^{19}–10^{22}	10^5–10^8
electron beam	10^{-3}–10^{-1}	10^{18}–10^{21}	10^4–10^7
x-ray	10^{-2}–10	10^{16}–10^{19}	10^2–10^5
ultraviolet	10–400	10^{15}–10^{16}	5–10^2
visible	400–750	10^{15}	1–5
infrared	750–10^5	10^{12}–10^{14}	10^{-2}–1
microwave	>10^6	10^{11}–10^{12}	<10^{-2}
radio frequency	>10^6	<10^{11}	<10^{-2}

[a] Ref. 5.

Table 2. Radioactive Nuclei[a]

Isotopes	Half-life	Type of radiation
polonium-210	138 d	α,γ-rays
radium-226	1620 yr	α,γ-rays
cesium-137	30 yr	β,γ-rays
cobalt-60	5.27 yr	β,γ-rays
strontium-90	28 yr	β rays

[a] Ref. 6.

One of the most common sources of α, β, and γ rays is radioactive nuclei, such as those listed in Table 2 (6). The only main ionizing radiation sources with somewhat limited commercial polymer-coating curing applications is γ rays that are produced from either cobalt-60 (^{60}Co) or cesium-137 (^{137}Cs) radioactive nuclei. Detailed descriptions of these sources, ie, energies, cost of operation, shielding requirements, comparisons between ^{60}Co and ^{137}Cs efficiencies, reactor geometries, etc, are available in references 5–8 (see Radioisotopes).

X rays can be produced through deceleration of high speed electrons through the electric field of an atomic nucleus. Various types of accelerator equipment capable of producing x rays are listed in Table 3 (6,9). High voltage electron accelerators have a distinct advantage over γ-ray and certain types of x-ray processing equipment. High energy electrons produced by machine acceleration, in comparison with radioisotopes, can be applied easily to industrial processes for the following reasons: easy on–off switching capability; less shielding is required than with gamma radiation; accelerator beams are directional and less penetrating than gamma or x rays; and electron radiation provides high dose rates, ie, maximum penetration per unit density of material, and is well suited for on-line, high speed processing applications. Commercially available high or low energy electron-processing equipment include the dynamitron, dynacote,

Table 3. X-Ray Processing Equipment[a]

Accelerator	Energy (MeV)
x-ray machine	(5×10^{-2})–(3×10^{-1})
resonant transformer	10^{-1}–3.5
Van de Graaff accelerator	1–5
betatron	10–300
linear electron accelerator	3–630

[a] Refs. 6 and 9.

insulating-core transformers, linear accelerators, Van de Graaff accelerator, pelletron, laddertron, and planar cathodes (10). Manufacturers of high voltage electron-accelerator equipment are High Voltage Engineering Corporation, Burlington, Mass.; Radiation Dynamics, Inc., Westbury, N.Y.; Energy Science, Inc., Woburn, Mass.; and Radiation Polymer Company, Plainfield, Ill.

Electromagnetic radiation, ie, from photons in the uv and visible ranges can also produce chemical changes, but these energies do not cause direct ionization of organic substrates. Chemical reaction depends upon the ability of the organic substrate to absorb light energy and to undergo photophysical processes involving electronically excited states, which ultimately result in free-radical-intermediate formation.

Ultraviolet light sources are based on the mercury-vapor arc. The mercury is enclosed in a quartz tube and a potential is applied to electrodes at either end of the tube. The electrodes can be of mercury, iron, tungsten, or other metals and the pressure in a mercury-vapor lamp may be <101.3 kPa–>1.013 MPa (<1 atm–>10 atm). As the mercury pressure and lamp operating temperatures are increased, the radiation becomes more intense and the width of the emission lines increases (11).

Visible light sources can be obtained from high pressure mercury arcs by addition of rare gases or metal halides, which increase the number of emission lines in the 350–600 nm region of the spectrum. Fluorescent lamps, tungsten halide lamps, and visible lasers are also used for light-induced photochemical reactions as applied to the curing of polymers and to coating technologies (12).

Infrared radiation (λ = 0.7–400 μm) has been used to fabricate plastics and to cure coating systems for a wide variety of commercial applications. In these applications, either gas fuel or electricity generates the infrared radiation; the types of processing equipment are listed in Table 4 (13–14) (see Infrared technology).

The energy density of microwaves is proportional to the square of the electrical field intensity at a given point of reference. Microwaves can generate thermal energy through resistive losses in a conductor, magnetic losses in magnetic materials, and dielectric losses in materials with high dielectric constants (13). Microwave heating equipment consists of the following five main elements: a power supply, which converts 440 V, three-phase, 60 Hz to 10^3–(2×10^4) V dc; a high frequency generating system (magnetron or klystron tube circuits); a high frequency transmission system (microwave waveguides); a control system; and work application fixtures.

Radio-frequency energies (4–5 MHz, 1–100 W) initiate glow discharge, ie, plasma polymerization reactions for a wide variety of organic starting materials. These types of energetic gas-phase reactions produce very thin cross-linked films which have a broad range of useful physical properties (15–16). Typical plasma or glow-discharge monomer polymerization or polymer-modification reaction equipment consists of the following components: a radio-frequency power source, a standing-wave ratio bridge, a matching network, inductive coupling or capacitor electrodes, and a vacuum system (see Microwave technology; Plasma technology, Supplement Volume).

Table 4. Infrared Processing Equipment

radiant-tube burners	
surface-combination burners	gas-fired infrared-generating systems
direct-fired refractory burners	
short-wave emitter lamps	
radiant metallic rods	electrically powered infrared-generating system
ceramic, quartz, or glass tubes	

Radiation Energy–Organic Substrate Interaction: Mechanisms

High energy interaction with organic substrates produces excited states which undergo secondary reactions, eg, electron capture, charge neutralization, intermolecular and intramolecular energy-transfer processes, ion formation, and molecular dissociation to produce free-radical intermediate species. The resulting chemical reactions are caused by the excited species and the formation of reactive intermediates (6). Ionizing radiation can produce excited molecules and secondary reactions through the following direct interaction processes:

$$A:B \xrightarrow{\gamma \text{ ray, x ray, or high energy electrons}} (A:B)^* \text{ (excited state)}$$

$$(A:B)^* \rightsquigarrow (A \cdot B)^{\ddagger} \text{ (radical-ion)} + e \text{ (electron ejection)}$$
$$\text{intermediate}$$

$$(A \cdot B)^+ \rightsquigarrow A^+ + B \cdot$$
$$\text{(ion)} \quad \text{(free radical)}$$

$$(A:B)^* + C:D \rightarrow A:B + (C:D)^* \text{ (energy transfer)}$$

$$(A \cdot B)^+ + e \rightarrow (A:B)^* \text{ (electron capture)}$$

$$C:D + e \rightarrow (C:D)^- \text{ (electron capture)}$$

$$(A \cdot B)^+ + (C:D)^- \rightarrow (A:B)^* + (C:D)^* \text{ (charge neutralization)}$$

$$(A:B)^* \rightarrow A \cdot + B \cdot \text{ (molecular dissociation)}$$

In the case of photochemical reactions, light energy must be absorbed by the system so that excited states of the molecule can form and subsequently produce free-radical intermediates (17) (see Photochemical technology).

$$A:B \xrightarrow[\text{uv or visible light energies}]{h\nu \text{ (photons)}} (A:B)^* \text{ (excited states)}$$

$$(A:B)^* \rightarrow A \cdot + B \cdot \text{ (free-radical intermediates)}$$

The difference between ionizing radiation production of excited states and those produced photochemically is that an incident photon does not have sufficient energy to eject an electron completely from the molecule but only displaces it into a new orbital farther from the nucleus. Ionizing radiation produces the same types of excited states as photochemical processes, but ionizing radiation also produces higher excited states, ie, of more intrinsic energy, which can not be formed directly by absorption of light energies (6,18). Energies emitted from an infrared source are transmitted directly to the surface of an organic substrate where they are absorbed, reflected, and/or transmitted, thereby causing vibrational and rotational molecular processes, which are subsequentially converted into heat (13,19).

Microwave curing of organic substrates involves dielectric loss of energy, which results in heat formation. In an oscillating electric field, organic substrates with high dipole moments (ie, with high dielectric constants and high tan δ power factors) align, rotate, and realign, and these changes cause internal molecular friction and conversion of the electromagnetic energy into thermal energy. Rapidly oscillating electric fields cause a greater rate of conversion to heat than do lower frequency electromagnetic waves. Since microwave fields vary 1–10 GHz, the rate of electromagnetic energy conversion to heat energy is significant (13).

Microwave or radio frequencies above 1 MHz that are applied to a gas under low pressure produce high energy electrons, which can interact with organic substrates in the vapor and solid state to produce a wide variety of reactive intermediate species.

gas + S (organic substrate in the vapor phase or solid polymer)

$$\xrightarrow{\text{radio frequency}} \text{gas}^+ \text{ (ions)} + e \text{ (high energy electrons)} + S^- \text{ (anions)} + S^* \text{ (excited states)}$$

$$+ S^+ \text{ (ions)} + S\cdot \text{ (radicals)} + S^{\dagger} \text{ (ion radicals)}$$

These intermediates can combine or react with other substrates to form cross-linked polymer surfaces and cross-linked coatings or films (15–16,20).

Curing of Polymers with γ-Ray, X-Ray, and High Energy Electron Sources

Radiation curing of preformed polymers with ionizing-radiation processing equipment can result in two types of chemical change that are associated with cross-linking and degradation reaction mechanisms. Cross-linking reaction mechanisms on preformed polymer substrates usually involve removal of hydrogen atoms to form a macroradical intermediate. These macroradical intermediates can then couple to form a single molecule. This coupling results in an increase in the original average molecular weight of the starting polymer.

$$\text{polymer}-CH_2CH_2-\text{polymer} \xrightarrow[\text{(direct absorption)}]{\text{ionizing radiation}} \text{macroradicals} \begin{cases} \text{polymer}-\overset{\bullet}{C}HCH_2\sim\text{polymer} \\ \text{polymer}-\overset{\bullet}{C}HCH_2\sim\text{polymer} \end{cases} + H_2$$

$$\downarrow$$

$$\text{polymer}-CHCH_2-\text{polymer}$$
$$| \leftarrow \text{cross-link site}$$
$$\text{polymer}-CHCH_2-\text{polymer}$$

If irradiation continues, the original polymer substrate is transformed into one gigantic molecule of infinite molecular weight with lower solvent solubility, higher melting points, and improved physical properties over the original material (21). Enhancement of cross-linking can be facilitated through the use of multifunctional vinyl monomers or oligomers which copolymerize and propagate much more rapidly than in a direct coupling reaction to form greater amounts of gel or cross-linked material at lower dose rates and shorter reaction times (22).

$$\text{polymer} + (CH_2=CH)_n-R \xrightarrow{\text{ionizing radiation}} \text{rapid gel formation}$$

(multifunctional vinyl monomers or oligomers)

Radiation-induced degradation reactions are in direct opposition to cross-linking or curing processes, in that the average molecular weight of the preformed polymer decreases because of chain scission and without any subsequent recombination of its broken ends.

$$\text{polymer—polymer (high molecular weight)} \xrightarrow{\text{ionizing radiation}} \text{polymer (low molecular weight)} + \text{polymer (low molecular weight)}$$

In order for efficient radiation curing of a polymer to take place, these degradation processes must be minimized in favor of the desired cross-linking reaction (23).

Examples of typical radiation curable polymer systems, experimental conditions, and applications are listed in Table 5.

Curing of Coatings with Electron Beams, γ-Ray, X-Ray, and Planar Cathodes

In conventional gas-oven, thermal curing of coatings, a mixture of polymers, cross-linking oligomers, catalysts, additives, pigments, and fillers is dissolved or dispersed in organic or water-based solvents to form a coating system. The coating is applied to a substrate and the solvents are thermally removed. The coating cross-links into a three-dimensional network by an energy-rich chemistry, which requires a high degree of thermal energy to convert the polymers into those with useful commercial

Table 5. Ionizing Radiation Interactions with Polymeric Substrate

Polymer	Radiation source	Results	Refs.
polyethylene	^{60}Co	6% gel content at 12 kGy (1.2 Mrads)	22, 24
polyethylene plus 0.5 wt % ethylene glycol diacrylate	^{60}Co	30% gel content at 12 kGy (1.2 Mrads)	24
polyethylene plus 4 wt % allyl methacrylate	^{60}Co	improved heat stability and tensile strength over polyethylene without added cross-linking monomer	22, 24
polypropylene	^{60}Co	onset of gel formation required 500 kGy (50 Mrad) with large amounts of degradation products	22, 25
polypropylene plus allyl acrylate	^{60}Co	70% gel content at 50 kGy (5 Mrad)	22, 25
polyethylene	electron beam, ^{60}Co	10–800 kGy (1–80 Mrad)—6.4 mm (thick slabs); electrical applications	24
poly(vinyl chloride)	electron beam	electrical applications	26–27
cross-linked silicone rubber	electron beam	cable termination cover, other electrical applications	10, 28
polyester	electron beam	degradation and cross-linking reactions are correlated with chemical structure of the polyester	23
polyethylene, neoprene, and silicone rubbers	electron beam	heat shrinkable articles for film packaging and electrical connector applications	29–30
poly(vinylidene fluoride), polyimides, ethylene-alkyl acrylate copolymers, nylons and natural polymers	electron beam, ^{60}Co	wide range of applications	5, 10
polysulfones	x rays	photoresist technologies	31

properties. Much of the energy is absorbed by the substrate before heat reaches the polymers to initiate the curing chemistry (3–4,32).

High energy electron- and light-energy radiation-curable coatings generally are of multifunctional acrylic or methacrylic unsaturated polymers. They differ from conventional coatings in that the solvents for the polymers are high boiling, nonvolatile, and 100% coreactive with themselves and with all of the other organic components in the film. The curing process for these coatings is a free-radical chain reaction. Ionizing radiation from the processing equipment is absorbed directly in the coating where the free radicals form uniformly in depth. Since electron energies of only 100 eV or less are required to break chemical bonds and to ionize or to excite components of the coating system, the shower of scattered electrons produced in the coating leads to a uniform population of free radicals throughout the coating. These initiate the polymerization. This polymerization process results in a dry, three-dimensional cross-linked coating (Figure 3). In this process, most of the energy is absorbed into the coating and is not lost to the substrate, as is the case in thermal curing reactions (4). Neither cobalt-60 nor x-ray energy sources are used in current radiation curable coating systems.

Electron Beam. An electron-beam processing unit consists mainly of a power supply and an electron-beam acceleration tube. The power supply increases and rectifies line current and the accelerator tube generates and focuses the beam and controls the electron scanning. The beam is produced when high voltage energizes a tungsten filament thereby causing electrons to be produced at very high rates. These fast electrons are concentrated to form a high energy beam and are accelerated to full velocity inside the electron gun. Electromagnets on the sides of the accelerator tube allows deflection or scanning of the beam as with a television tube. Scanning widths and depths vary from 61–183 cm to 10–15 cm, respectively. The scanner opening is covered with a thin metal foil, usually titanium, that allows passage of electrons but maintains the high vacuum required for high free-path lengths. Characteristic power, current, and dose rates of accelerators are 200–500 kV, 25–200 mA, and 10–100 kGy/s (1–10 Mrad/s).

Electron processors have several disadvantages. The most severe of these is the large area which must be shielded, since any surface enclosing the electron accelerator scanner acts as a source of x rays generated by electrons which are scattered to the wall, and these emissions are along the entire length of the system. Another disadvantage is the large space requirement for housing the equipment.

Advantages of the electron-beam processor are its ability to penetrate very thick coatings. It is used to cross-link polymers, insulation, and wire-cable coverings (3,10) (see Insulation, electric—electric wire and cable coverings).

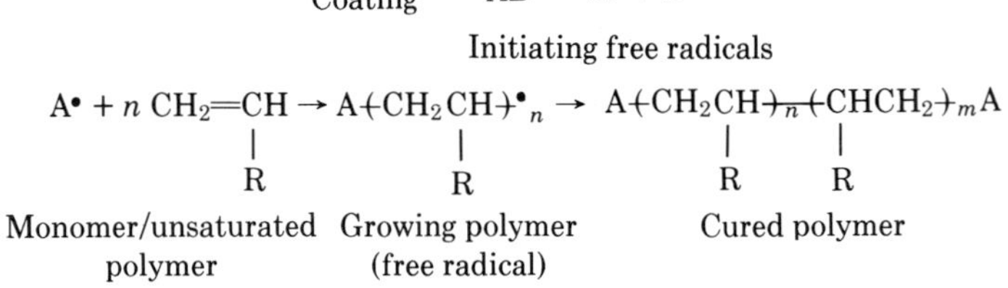

$$AB \underset{Coating}{\nearrow} \begin{matrix} AB^+ \\ \updownarrow e \\ AB\bullet \to A\bullet + B\bullet \end{matrix}$$

Initiating free radicals

$$A\bullet + n\ CH_2{=}CH \underset{R}{|} \to A{+}CH_2CH{+}\bullet_n \underset{R}{|} \to A{+}CH_2CH{+}_n{+}CHCH_2{+}_mA \underset{R\quad R}{|\quad |}$$

Monomer/unsaturated polymer   Growing polymer (free radical)   Cured polymer

Figure 3. Polymerization initiation and propagation by radiation-generated free radicals.

Electrocurtain. The Electrocurtain processor (Energy Sciences, Woburn, Mass.) is a high voltage (150 kV) electron tube that provides a continuous strip of energetic electrons from a linear filament or cathode, which is on the axis of symmetry of the system. The cylindrical electron gun shapes and processes the electron system in a grid-controlled structure. The stream is then accelerated across a vacuum gap to a metal window where it emerges directly into air and travels onto the product. The energetic electrons from the processor are absorbed directly in the coating, where they create the initiating free radicals uniformly in depth. In the liquid-phase systems, the polymerization process propagates until the activity of the growing chain is terminated. These energetic electrons can penetrate many different types of pigmented coatings and are capable of producing through-cure to the substrate–polymer coating interface. Both the electron beam and the Electrocurtain cure pigmented films, but the power of the Electrocurtain is substantially less: its maximum-curing-film thickness range is ca 0.025–0.36 mm.

In the Electrocurtain processor, the shielding is clad directly to the tube housing. Housing space is relatively small, since a shielded tube 25 cm in diameter replaces the 3-m high structure required for the scanned electron-beam apparatus. The electrocurtain has a more flexible geometry and can be adapted readily to many different types of curing applications (33).

Multiple Planar-Cathode Processors. The design criterion for this electron-accelerator system is a planar array of concentrated cathode-control grid elements. The modular cathode construction allows for broad-beam (250 cm wide) processing of materials with powers of 30 kGy (3 Mrad) at 300 m/min. The system also includes integrated shielding and high terminal voltages (34).

Coatings Ingredients. Ingredients of liquid radiation-curable coatings are analogues of components contained in conventional solvent-based thermal-curing coating systems. In conventional solvent-based coatings systems, a preformed polymer (usually 3,000–25,000 mol wt) is dissolved in an organic solvent (30–80% solids), and a cross-linking oligomer and various flow agents, catalysts, pigments, etc, are added. The coating is applied to a substrate by conventional methods, ie, spray, roll coating, flow coating, etc, and subsequently is cured in gas or ir thermal ovens. Curing of conventional solvent-based coatings involves solvent removal and thermal initiation of chemical reactions between the polymers and cross-linking oligomers that are involved in developing final film properties through three-dimensional network formation of cross-link sites (35).

Monomer, Oligomers, and Cross-linking Agents. The monomer in radiation-curable coatings is the analogue of the solvent in conventional paint. Although, like a solvent, it is a medium for all of the other ingredients and provides the necessary liquid physical properties and rheology, the monomer differs in that it enters the copolymerization and is not lost on cure. Most radiation-curable monomers (mol wt, ca 100–500) contain single unsaturation sites and are high boiling acrylate esters, although in some special coatings styrene is the monomer. Where styrene is used, usually most or all of the polymer-polyester unsaturation is fumarate rather than acrylic.

Cross-linking oligomers in conventional thermosetting coatings formulations are usually melamine resins, ie, involve acid, hydroxyl-transetherification cross-linking; amine–amide hardeners which involve oxirane ring opening; and blocked isocyanate prepolymers. Oligomers and cross-linking materials in radiation-curing systems are similar to single, vinyl monomers except they contain di-, tri-, or multifunctional unsaturation sites (mol wt, ca 200–1000). These multifunctional components cause polymer propagation reactions to grow into three-dimensional network structures of a cured film (Fig. 4).

Polymers. The molecular weights of polymers used in radiation-curable coating systems are ca 1,000–25,000 and the polymers usually contain acrylic, methacrylic, or fumaric vinyl unsaturation along or attached to the polymer backbone. The chemical class of the polymer backbone varies with coating requirements, and the different chemical classes give the same overall properties as a related thermal cure coating system (Table 6). Examples of radiation curable coating systems, experimental conditions, and their application are listed in Table 7 (36).

Curing with Ultraviolet, Visible, and Infrared Processing Equipment

Polymers. Upon direct absorption of uv or visible wavelengths of light, polymer substrates undergo chain scission and cross-linking. Cross-linking or curing of preformed polymeric materials, ie, of thermoplastics, can be markedly enhanced through use of special photosensitive molecules that are mixed into the polymer matrix or that chemically attach to the backbone of the polymer chains. These special photosensitive molecules absorb uv or visible light energies much more efficiently than the polymer; they rapidly form excited states which undergo photochemical reactions which in turn form reactive free-radical intermediates that effect polymer dimerization or cross-linking. These special photosensitive molecules, when compounded into the preformed polymer matrix, can undergo light-induced radical abstraction or insertion reactions which result in coupling of the polymer chains and in network formation (44) (see also Photoreactive polymers; Uv stabilizers).

Similar types of cross-linking reactions are observed for polymers to which photosensitive molecules are chemically attached.

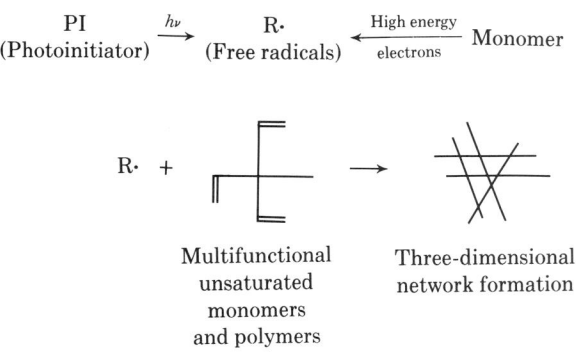

Figure 4. Radiation-induced cross-linking reactions.

Table 6. Performance Characteristics of Polymer Backbone Systems

System	Cost	Chemical resistance	Physical properties	Outdoor durability
epoxies	medium	excellent	very good	very poor
acrylics	low–medium	very good	good	very good
polyesters	low	fair–good	very good	very good
aliphatic polyurethanes	high	very good	excellent	very good
aromatic polyurethanes	high	very good	excellent	very poor

Table 7. Ionizing Radiation Curing of Coatings

Reactive coating formulation ingredients	Conditions	Refs.
Single-functional vinyl monomers		
2-ethylhexyl acrylate, styrene, *N*-vinyl-pyrrolidinone, vinyltoluene, lauryl methacrylate		36
Multifunctional vinyl monomers		
1,6-hexanediol diacrylate, tetraethylene glycol diacrylate, trimethylolpropane triacrylate, pentaerythritol triacrylate		36
Unsaturated polymers		
maleic–fumaric acid unsaturated polyesters, acrylic copolymers containing pendant vinyl unsaturation, epoxy acrylates, polyurethane acrylates		36
Coating composition		
65 wt % unsaturated polyester, 35 wt % vinyl monomer: 2-ethylhexyl acrylate or styrene	cured with 300 keV electrons at 200 kGy/min (20 Mrad/min)	37–38
acrylic copolymers containing pendant vinyl unsaturation (unsaturation levels, 0.5–1.75 mol of double bonds per 1000 mol wt) and 35–45 wt % of a vinyl monomer: 2-ethylhexyl acrylate or styrene	cured with a total dosage of 150 kGy (15 Mrad) electron beam; used to coat plastic substrates	39
acrylic monomers: acrylic unsaturated epoxy and acrylic unsaturated polyurethanes	electron-curtain curing; used in general coating applications	40
monomers: polyfunctional vinyl intermediates	metal, wood, and plastic finishing applications with electron beam processes	41–42
vinyl monomer	^{60}Co curing of monomer; vacuum impregnation of wood	43

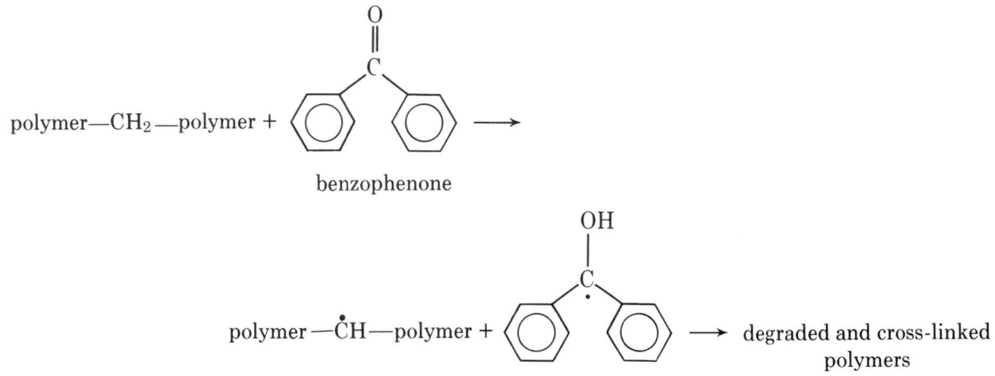

polymer—CH$_2$—polymer + benzophenone →

polymer—ĊH—polymer + [diphenyl·COH radical] → degraded and cross-linked polymers

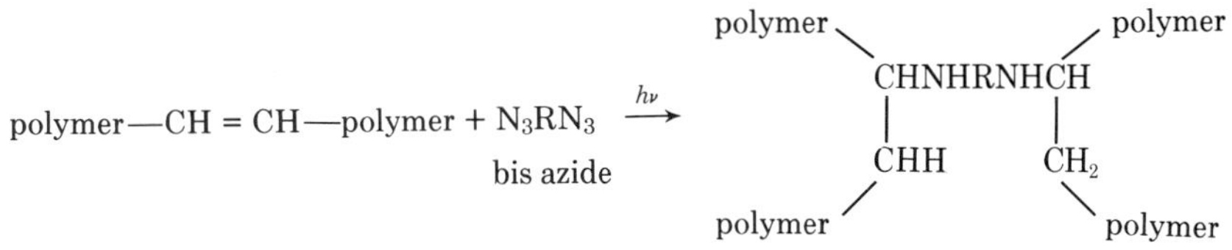

polymer—CH=CH—polymer + N$_3$RN$_3$ $\xrightarrow{h\nu}$
bis azide

$$\begin{array}{c}\text{polymer} \diagdown \quad \diagup \text{polymer} \\ \text{CHNHRNHCH} \\ | \qquad | \\ \text{CHH} \qquad \text{CH}_2 \\ \diagup \qquad \diagdown \\ \text{polymer} \qquad \text{polymer}\end{array}$$

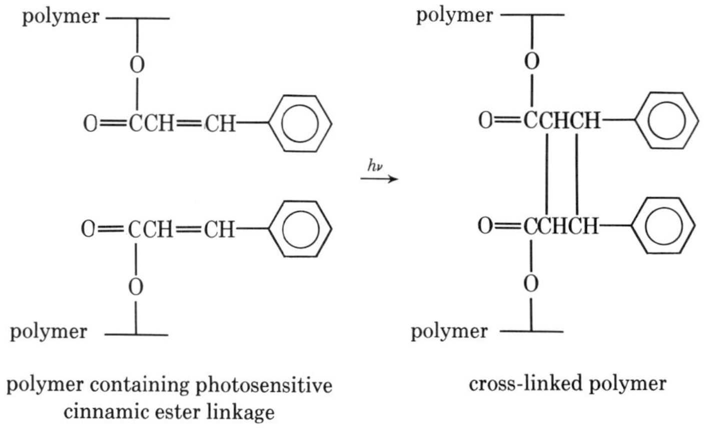

polymer containing photosensitive cinnamic ester linkage → cross-linked polymer

Radiation curing of polymers with uv and visible-light energies is used widely in photoimaging and photoresist technology (Table 8) (45). Infrared processing is involved with thermoforming or heat-bonding of thermoplastic polymeric materials. These polymer heat-forming or melting processes do not usually cure the polymer but only cause physical changes and maintain original polymer thermoplastic characteristics (13).

In order to cure, ie, form three-dimensional network structures through chemical changes, polymer systems with ir radiation, it is necessary to design a reactive functionality within the polymer structure so that coupling reactions can take place between polymer chains.

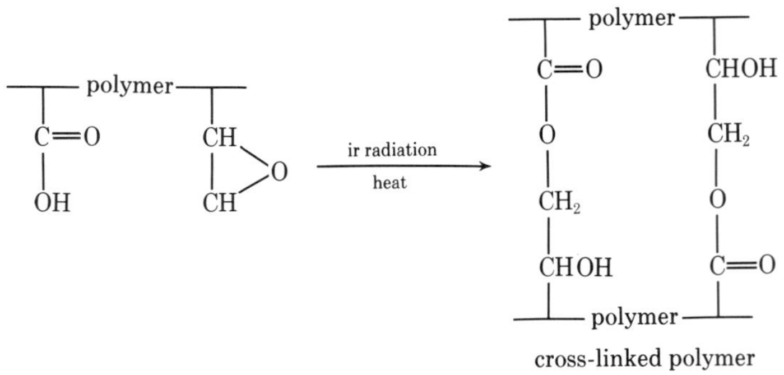

cross-linked polymer

Certain polymeric structures also can be blended with other coreactive polymers or multifunctional reactive oligomers that effect curing reactions when exposed to ir radiation. These coreactive polymers and cross-linking oligomers undergo condensation or addition, which cause network formation (see Table 9) (3–4,36).

Coatings. There are five characteristics of uv and visible-light-energy irradiation or photocuring of coatings. (1) The stable light source must be capable of producing uv and visible wavelengths of light, ie, near and far uv, 200–400 nm to visible, 400–700 nm, with sufficient power or intensity to be commercially feasible. (2) The photoinitiator must be capable of absorbing uv and visible-light radiation at appropriate

Table 8. Photocurable Polymer Systems

Polymers	Remarks	Refs.
poly(vinyl cinnamate)	uv- and visible light-induced photodimerization reactions; used in negative photoresist technologies	46
poly(vinyl cinnamylidene-acetate)	visible light-induced photodimerization reactions; 50% more efficient than poly(vinyl cinnamate)	47
polychalcones	photodimerization or addition reactions; used in negative photoresist technologies	48
polycoumarins	photodimerization or addition reactions; used in negative photoresist technologies	49
polystilbenes	photodimerization or addition reactions; used in negative photoresist technologies	50
cyclized rubber	cross-linked with bis-azide-nitrene insertion reactions	51
allylic esters	cross-linked with bis-azide-nitrene insertion reactions	52
thiolene polymers	unsaturated polymers (allyl or vinyl functionality) cross-linked with multifunctional mercaptans and photosensitizer molecules	53
phenolic polymers and acid functional acrylic resins	diazide photosensitizers for light-induced hydrophobic–hydrophilic reactions associated with positive photoresist technology	54

wavelengths of energy as emitted from the light source. (3) Active free radicals must be produced through the action of light absorption by the photochemically active photoinitiator. These free radicals initiate polymerization of unsaturated monomers, oligomers, and polymers. Catalysts are not required in high energy electron-curing processes. (4) Unsaturated, high boiling acrylic or methacrylic monomers, oligomers, cross-linking agents, and low molecular weight polymers comprise the fluid, low viscosity, light-curable coating system and are similar to the coatings materials used in high energy electron-curing processes. (5) After free-radical initiation of the reactive liquid coating, the monomers propagate into a fully cured, cross-linked solid coating or film (4,58).

Light Source. The light source normally used in commercial photocuring reactions is the medium pressure mercury-arc lamp with a quartz or Vicor envelope. These lamps may contain electrodes for electrical to light-energy conversion or may be electrodeless, in which case a radio-frequency wave causes mercury atom excitation and subsequent

Table 9. Thermally Curable Polymer Systems

Polymers	Remarks	References
epoxy polymers plus acid functional polymers or oligomers	powder coatings: poor outdoor weatherability but good chemical resistance	55
epoxy polymers plus polyfunctional amines	powder coatings: excellent corrosion resistance	56
hydroxyl-containing polymers and melamine oligomers	powder coatings: good outdoor weatherability	57
hydroxyl-containing polymers plus blocked isocyanates	powder coatings: good outdoor weatherability	56
carboxyl-containing polymers plus oxazolines	powder coatings: good outdoor weatherability	56

light emission. The normal power input levels are 40–160 W/cm arc length resulting in sharp peak outputs with ca 10-nm bandwidths. The main peaks are at 365, 404, 436, 546, and 578 nm with relative outputs (80 W/(cm arc) Hanovia lamp) of 2.4–5.9 W/(cm arc).

Many other types of uv and visible light sources can be used for photopolymerization reactions, eg, low pressure mercury arcs, flash lamps, fluorescent or germicidal lamps, and lasers (qv). A complete review of light sources used in photopolymerization reactions is given in references 11–12, and 59.

Light-wavelength output energies from the various sources are very small compared to electron-beam or Electrocurtain processors (2–5 eV to thousands of eV). The energy associated with 365 nm wavelengths of light is equivalent to 3.4 eV or 343 kJ/mol (82 kcal/mol) and is sufficient to cause very selective rearrangement and cleavage of aromatic carbonyl–alkyl carbon bonds (aromatic—C(O)—alkyl).

Light energy alone is not sufficient to cause direct, efficient monomer-initiation reactions. Commercial light-induced curing reactions require the use of a special photosensitive catalyst in the coatings formulation. These photosensitive catalysts or photoinitiators are an integral part of the formulation and the cost of a light-sensitive radiation-curable coating system. The type and amount of photoinitiator also influence the relative rate of cure and the final properties of the cured film or coating.

Photoinitiators. Many theories of photoinitiated polymerization reactions with different light-sensitive catalysts have been reviewed in references 60–62. There are, however, two general classes of photoinitiators: those that undergo direct photofragmentation upon exposure to uv or visible-light irradiation and produce active free-radical intermediates and those that undergo electron transfer followed by rearrangement into a free-radical species. The absorption bands of photosensitizers and photoinitiators should overlap the emission spectra of the various commercial light sources (14).

The alkyl ethers of benzoin undergo direct photofragmentation upon absorption of uv energy at ca 360 nm to produce two free-radical intermediates.

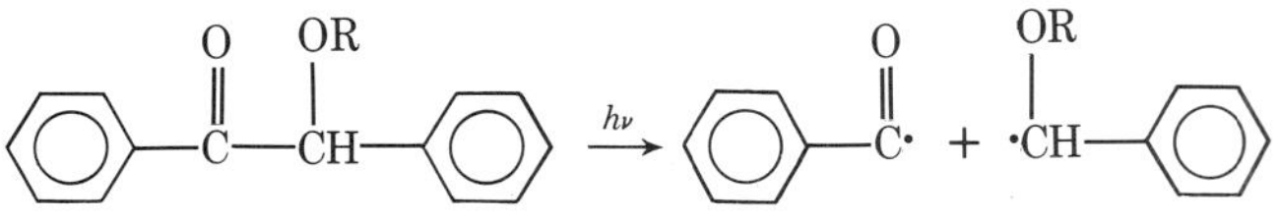

Other similar structures undergo the following photofragmentation rearrangement decomposition processes:

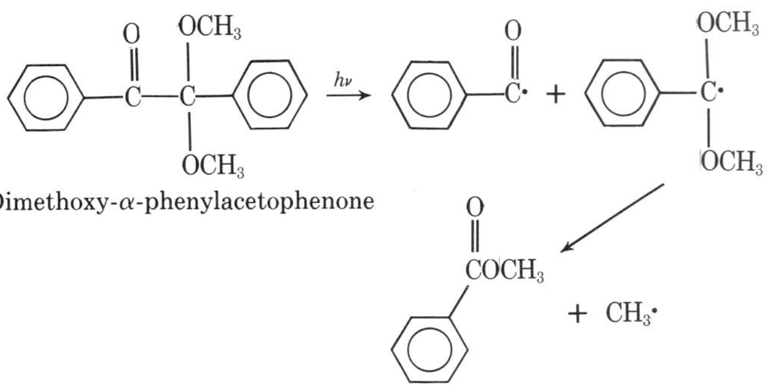

α,α-Dimethoxy-α-phenylacetophenone

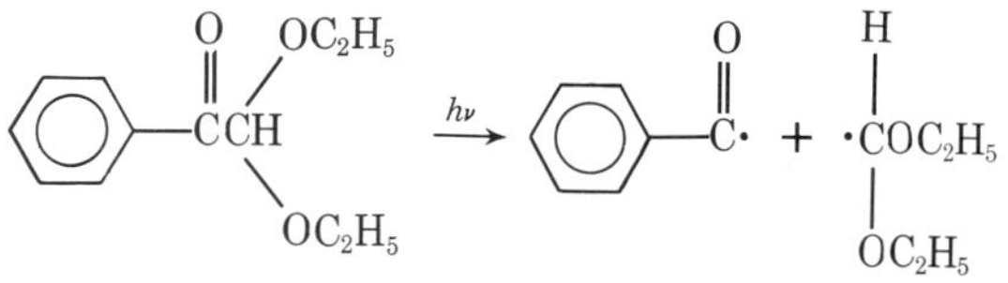

$$\underset{\alpha,\alpha\text{-Diethoxyacetophenone}}{\underset{\displaystyle OC_2H_5}{\underset{|}{\underset{\displaystyle -\overset{\displaystyle O}{\overset{\|}{C}}CH}{}}}-OC_2H_5} \xrightarrow{h\nu} \underset{}{C_6H_5-\overset{O}{\overset{\|}{C}}\cdot} + \cdot\underset{\displaystyle OC_2H_5}{\underset{|}{\overset{\displaystyle H}{\overset{|}{C}}OC_2H_5}}$$

α,α-Diethoxyacetophenone

The second type of photoinitiators are those that undergo electron transfer followed by proton transfer to give free-radical species:

$$(C_6H_5)_2C{=}O \xrightarrow{h\nu} (C_6H_5)_2C{=}O_{S_1} \xrightarrow{kst} (C_6H_5)_2{=}O_{T_1}$$

$$(C_6H_5)_2C{=}O_{T_1} + (C_2H_5)_3\ddot{N}[(C_6H_5)_2\overset{-}{\dot{C}}-\overset{+}{\ddot{O}}\quad\dot{N}(C_2H_5)_2]$$
$$\underset{\text{exciplex}}{|\atop HCHCH_3}$$

$$(C_6H_5)_2\dot{C}OH + CH_3CH\dot{C}H\ddot{N}(C_2H_5)_2$$

where kst is the rate constant for intersystem-crossing efficiency. Benzophenone in its ground state (S_0) undergoes absorption of uv energy (340–360 nm) and is excited to its singlet excited states (S_1) followed by intersystem crossing (kst ≃ 1) to its triplet excited state (T_1). From the triplet excited state, benzophenone forms an encounter complex, or exciplex, with the ground-state alkyl amine, which then undergoes electron transfer from the nitrogen to the excited carbonyl followed by proton transfer of a hydrogen atom on the carbon α to the nitrogen atom. This results in reduction of the benzophenone to form the semibenzpinacol radical and radical formation on the carbon α to the nitrogen atom. Both of these free-radical species can initiate or terminate polymerization of acrylic monomers. These free-radical species can also cross-couple or dimerize to form unreactive compounds. Photoinitiators or photosensitizers having absorption capabilities in the visible-light-energy range are based on dyes, quinones (qv), diketones, and heterocyclic chemical structures (60).

Coating Compositions. Light-induced, radiation-curable coating systems are similar to those for high energy, electron radiation-curing coatings. Examples of typical coating compositions and their applications for light-induced radiation-curing processes are given in reference 63.

Photoinduced Cationic Ring-Opening Polymerization Reactions. All of the previous discussions on radiation curing of materials were based on free-radical initiation, propagation and termination mechanisms. It should be noted, however, that it is also possible to photochemically release Lewis acids such as BF_3, PF_5, ArF_5, SbF_5 (by photodecomposition of aryl diazonium, iodonium, and sulfonium ion salts) which catalyze epoxide ring opening and subsequent curing reactions for use in photoresist and printed-circuit technologies (61).

Infrared Radiation Curing. Thermal curing of a conventional coating system requires solvent removal and chemical cross-linking of polymers or oligomers. Commercial processors emit short wavelengths (0.7 μm–2.0 μm), medium wavelengths (2 μm–4 μm), and long wavelengths (4 μm–1,000 μm) of infrared radiation. The shortwave ir is near the visible end of the spectrum and, therefore, it is higher in energy and more penetrating than either the medium or long wavelength emissions. The shortwave ir radiation is also reflective and can be focused for improved efficiency. A possible difficulty in using short wavelength ir processing equipment is its color selectivity characteristics. A dark-colored substrate or coating absorbs more thermal energy than a light-colored substrate; such color selectivity can lead to differential curing processes. Medium wavelengths of ir radiation are not substrate-color sensitive but they are more difficult to control in terms of reflection or focusing and are somewhat less efficient for certain coating cure requirements than short wavelength ir radiation (64).

Many conventional coatings systems containing polymer functionalities that cure through thermal processes involved with gas-fired oven technologies can also be cured or processed very efficiently with ir radiation energies (see Table 9).

Curing with Microwave or Radio-Frequency Processing Equipment

Polymers and Coatings. Polymer surfaces can be easily modified with microwave or r-f-energized glow-discharge techniques. The polymer surface cross-links or oxidizes depending upon the nature of the plasma atmosphere. Oxidizing (oxygen) and nonoxidizing (helium) plasmas can have a wide variety of effects on polymer surface wettability characteristics (65).

One particularly promising approach to the development of improved coatings for use as metal primers, in fiber treatment, in packaging materials, etc, involves the use of vacuum-plasma deposition techniques. Thin, ie, micrometers thick, pinhole-free, polymeric coatings can be developed from a variety of organic moieties, both monomeric and nonmonomeric. Depending on the applications, such coatings can be developed to provide corrosion protection, abrasion and chemical resistance, improved adhesion, barrier properties, etc. With the plasma technique, both capacitively coupled and inductively coupled r-f-energized vacuum chambers can be used for the deposition. In the capacitively coupled chamber, the substrate to be coated is attached to one of the electrodes. Monomer then is introduced into the chamber such that an operating pressure of ca 13 Pa (ca 0.1 mm Hg) is maintained. A glow discharge between the electrodes polymerizes a coating on the substrate. In the inductively coupled reactor, r-f energy is supplied by coils surrounding the chamber. The glow discharge, therefore, fills the entire chamber and coats any item placed in the chamber.

Plasma deposition has been used to prepare acrylic, silicone, and fluorocarbon coatings on a variety of substrates in thicknesses of up to ca 8 μm. Coating variables include applied power, substrate temperature, deposition time, and monomer pressure. The coatings are similar to those produced from the same monomers by conventional polymerization techniques. However, there are structural differences and the resulting coatings are highly cross-linked. All exhibit excellent chemical resistance and very good adhesion to the different substrates. Certain coatings have good optical clarity. Coating hardness appears to vary appreciably with monomer type and deposition conditions (15–16,65).

BIBLIOGRAPHY

"Radiochemical Technology" in *ECT* 2nd ed., Vol. 17, pp. 53–64, by William H. Beamer, The Dow Chemical Company.

1. A. Chapiro, *Radiation Chemistry of Polymeric Systems*, Interscience Publishers, a division of John Wiley & Sons, Inc., New York, 1962.
2. J. E. Wilson, *Radiation Chemistry of Monomers, Polymers, and Plastics*, Marcel Dekker, Inc., New York, 1974.
3. V. D. McGinniss, L. J. Nowacki, and S. V. Nablo, *ACS Symposium*, No. 107, 1979, pp. 51–70.
4. V. D. McGinniss, *National Symposium on Polymers in the Service of Man*, ACS, 16th State-of-the-Art Symposium, 1980, pp. 175–180.
5. F. A. Bovey in H. P. Mark, ed., *Polymer Devices*, Vol. 1, Interscience Publishers, New York, 1958.
6. J. W. T. Spinks and R. J. Woods, *An Introduction to Radiation Chemistry*, John Wiley & Sons, Inc., New York, 1964.
7. A. Danno, "Applications of Ionizing Radiation to Polymer Chemistry" in *Radiation Chemistry and Its Applications*, IAEA Tech. Rpt. Series No. 84, Panel Meeting, Vienna, Apr. 17–21, 1967, pp. 23–41.
8. S. Jefferson and co-workers, "Industrial Applications of Ionizing Radiation" in *Advances in Nuclear Science and Technology*, Vol. 4, Academic Press, New York, 1968, pp. 335–338.
9. G. W. Grodstein, *X-Ray Attenuation Coefficients from 10 keV to 100 MeV*, Circ. 583, U.S. National Bureau of Standards, Washington, D.C., 1957.
10. A. F. Readdy, Jr., *Application of Ionizing Radiations in Plastics and Polymer Technology*, Plastic Report R41, Plastics Technical Evaluation Center, Picatinny Arsenal, Dover, N.J., 1971.
11. W. E. Elanbass, *Light Sources*, Crane, Russak & Co., Inc., New York, 1972.
12. V. D. McGinniss, "Light Sources" in S. P. Pappas, ed., *UV Curing Science and Technology*, Technology Marketing Corporation, Stamford, Conn., 1978, pp. 96–132.
13. A. F. Readdy, Jr., *Plastics Fabrication by Ultraviolet, Infrared, Induction, Dielectric and Microwave Radiation Methods*, Plastic Report R43, Plastics Technical Evaluation Center, Picatinny Arsenal, Dover, N.J., 1972.
14. R. W. Pray, *A New Look at Infrared*, SME Publ. FC78-543, Association for Finishing Processes of SME, Dearborn, Mich., 1978.
15. J. R. Hollahan and A. T. Bell, *Techniques and Applications of Plasma Chemistry*, John Wiley & Sons, Inc., New York, 1974.
16. M. Shen and A. T. Bell, eds., *Plasma Polymerization*, ACS Symposium Series No. 180, American Chemical Society, Washington, D.C., 1979.
17. N. J. Turro, *Modern Molecular Photochemistry*, Benjamin/Cummings Publishing Co., Menlo Park, Calif., 1978.
18. R. O. Bolt and J. D. Carroll, *Radiation Effects on Organic Materials*, Academic Press, New York, 1963.
19. J. F. Kinstle, *Paint Varn. Prod.* **63**(6), 17, 1973.
20. L. F. Thompson and K. G. Mayhan, *J. Appl. Polym. Sci.* **16,** 2291 (1972).
21. G. Alder, *Science* **141,** 321 (1963).
22. G. Odian and B. S. Bernstein, *Nucleonics* **21,** 80 (1963).
23. G. F. D'Alelio, R. Haberli, and G. F. Pezdirtz, *Effect of Ionizing Radiation on a Series of Saturated Polyesters*, NASA-SP-58, National Aeronautics and Space Administration, Washington, D.C., 1964.
24. B. S. Bernstein and co-workers, *J. Appl. Polym. Sci.* **10,** 143 (1966).
25. B. S. Bernstein and J. Lee, *Ind. Eng. Chem. Prod. Res. Dev.* **6,** 211 (1967).
26. M. Izumi and co-workers, *Sumitomo Electr. Tech. Rev.*, 50 (March 1967).
27. V. L. Lanza and R. M. Halperin, *Proceedings, 13th Annual Symposium on Communication Wires and Cables*, Dec. 2–4, 1964.
28. "Self-Adhering Silica Rubber Tape" in *General Electric Company Bulletin*, Insulating Materials Data PD 1302.
29. V. L. Lanza and P. M. Cook, *9th Annual Symposium on Communication Wire and Cable*, Nov. 30, Dec. 1–2, 1960.
30. R. C. Becker, *Plast. World* **35,** 48 (Feb. 1977).
31. *Photopolymers: Principles, Processes and Materials*, Regional Technical Conference of the Society of Plastic Engineers, Inc., Mid-Hudsen Section, Ellenville, N.Y., Oct. 13–15, 1976.
32. A. Banov, *Paints and Coatings Handbook*, Structures Publishing Company, Farmington, Mich., 1978.
33. S. V. Nablo, *SME Technical Paper*, FC75-311, Society of Manufacturing Engineers, Dearborn, Mich., 1975.

34. W. J. Ramler, *J. Radiat. Curing* **1**(3), (Aug. 1974).
35. A. J. Chompff and S. Newman, *Polymer Networks*, Plenum Press, New York, 1971.
36. J. L. Gardon and J. W. Prane, eds., *Nonpolluting Coatings and Coating Processes*, Plenum Press, New York, 1973.
37. W. Burlant and J. H. Hinsch, *J. Polym. Sci.* **A2,** 2135 (1964).
38. *Ibid.*, **A3,** 3587 (1965).
39. S. S. Labana and E. O. McLaughlin, *J. Elastoplast.* **2,** 3 (1970).
40. S. V. Nablo and E. P. Tripp, *Radiat. Phys. Chem.* **9,** 325 (1977).
41. T. J. Miranda and T. F. Huemmer, *J. Paint Technol.* **41,** 118 (1969).
42. K. H. Morganstern, *SAE Engineering Congress*, Detroit, Ill., Jan. 10, 1967.
43. J. H. Frankfort, *Proceedings of Information Meeting on Irradiated Wood-Plastic Materials*, Report ORNL-11C-7, Chicago, Ill., Sept. 15, 1965.
44. S. S. Labana, ed., *Ultraviolet Light Induced Reactions in Polymers*, ACS Symposium Series No. 25, American Chemical Society, Washington, D.C., 1976.
45. E. D. Feit, "Photoresists: Photoformation of Relief Images in Polymeric Films" in S. P. Pappas, ed., *UV Curing: Science and Technology*, Technology Marketing Corporation, Stamford, Conn., 1978, pp. 229–256.
46. K. Nakamura, T. Sakata, and S. Kikuchi, *Bull. Chem. Soc. Jap.* **41,** 1765 (1968).
47. T. A. Shankoff and A. M. Trozzolo, *Photo. Sci. Eng.* **19,** 143 (1975).
48. M. Tsuda, *J. Polym. Sci.* **A2,** 2907 (1964).
49. G. A. Delzenne, "Photoresists" in N. Bikales, ed., *Encyclopedia of Polymer Science and Technology*, Supp. 1, Interscience Publishers, a division of John Wiley & Sons, Inc., New York, 1976.
50. F. A. Stuber and co-workers, *J. Appl. Polym. Sci.* **13,** 2217 (1969).
51. F. C. DeSchryver, N. Boens, and G. Smith, *J. Polym. Sci.* **A1,** 1939 (1970).
52. S. Shimizm and G. R. Bird, *J. Electrochem. Soc.* **124,** 1394 (1977).
53. W. S. DeForest, *Photo-resists: Materials and Processes*, McGraw-Hill Book Company, New York, 1975.
54. J. Kosar, *Light-Sensitive Systems*, John Wiley & Sons, Inc., New York, 1965, Chapt. 6, pp. 194–320.
55. U.S. Pat. 3,888,943 (June 10, 1975), S. S. Labana (to Ford Motor Corporation).
56. S. Gabriel, *J. Oil Col. Chem. Assoc.* **59,** 52 (1976).
57. N. J. H. Gulpen and A. J. Van DeWerff, *J. Paint Technol.* **47,** 81 (1975).
58. V. D. McGinniss and D. M. Dusek, *J. Paint Technol.* **46,** 23 (1974).
59. L. R. Koller, *Ultraviolet Radiation*, John Wiley & Sons, Inc., New York, 1965.
60. S. P. Pappas and V. D. McGinniss, "Photoinitiation of Radical Polymerization" in ref. 45, Chapt. 1, pp. 1–22.
61. V. D. McGinniss, *Photogr. Sci. Eng.* **23**(3), 124 (1979).
62. V. D. McGinniss, *J. Radiat. Curing* **2,** 3 (1975).
63. *Paint Varn. Prod.* **64**(8), 19 (1974).
64. W. C. Hankins, *J. Oil. Col. Chem. Assoc.* **60,** 300 (1977).
65. J. F. Kinstle, *J. Radiat. Curing* **2**(2), 15 (1975).

Vincent D. McGinniss
Battelle Columbus Laboratory

RADIOACTIVE DRUGS

Radioactive drugs are diagnostic or therapeutic agents by virtue of the physical properties of their constituent radionuclides. Thus, their utility is not based on any pharmacologic action. Most clinically used drugs of this class are diagnostic agents incorporating a gamma-emitting nuclide which, because of its physical or metabolic properties, localizes in a specific organ after intravenous injection. Images reflecting organ structure or function are then obtained by means of a scintillation camera that detects the distribution of ionizing radiation emitted by the radioactive drug. The principal isotope used in clinical diagnostic nuclear medicine is reactor-produced metastable technetium-99m [14133-76-7] (^{99m}Tc). It either is injected directly as sodium pertechnetate [23288-60-0] (NaTcO$_4$) or is added to instant kits consisting of nonradioactive carrier molecules, to which it spontaneously binds to form the final radiopharmaceutical product (see Kits). Other clinically important diagnostic isotopes are thallium-201 [15064-65-0] (^{201}Tl), gallium-67 [14119-09-6] (^{67}Ga), and iodine-123 [15715-08-9] (^{123}I). All are cyclotron-produced. A few radioactive drugs contain beta-emitting radionuclides and are used for therapeutic purposes. These agents localize in pathologic tissue and destroy it by ionizing radiation. The most important members of this group, ie, iodine-131 [10043-66-0] (^{131}I) and phosphorus-32 [14596-37-3] (^{32}P), are reactor-produced. *In vitro* radioimmunoassay reagents in widespread clinical laboratory use contain tracers, eg, ^{125}I [14158-31-7] and ^{57}Cr [36819-21-3], and are employed to measure minute quantities of hormones, drugs, and other materials in biological samples (see Radioactive tracers).

Diagnostic Radioactive Drugs for Imaging

Design of diagnostic radioactive drugs requires the combination of low toxicity, specific biodistribution, low radiation dose, and radionuclidic emissions compatible with currently available instrumentation. Since the sensitivity of scintillation cameras, also called gamma cameras, is great, only trace amounts of the radioactive drug need be administered, and the potential for toxicity is thus reduced. Nonetheless, acute and subacute animal toxicology testing with decayed, nonradioactive material is essential in the development of a useful radioactive drug. The utility of these entities depends entirely upon their ability to localize in specific tissues and thereby reflect anatomic and pathologic structure or function. Certain radionuclides are useful by virtue of their natural biodistribution, eg, ^{99m}TcO$_4^-$ to the thyroid, brain, and kidneys; ^{201}Tl to the heart; ^{131}I and ^{123}I to the thyroid; ^{67}Ga to certain tumors and to abscesses. For imaging of other structures or functions, carrier molecules are designed to bind usable radioisotopes and to carry them to specific sites or through specific metabolic or physiologic processes. For example, ^{99m}Tc combines with pyrophosphate (PYP) to form ^{99m}Tc-PYP, which localizes in the skeleton in general and in areas of hyperactive bone metabolism in particular. Biodistribution of this agent combines rapid, high concentration by target tissue and low uptake by surrounding tissue with rapid blood clearance and urinary excretion of the nonspecifically localized radioactive drug. As a consequence, the high target-to-background concentration of radioisotope needed to resolve the image is obtained. Rapid excretion also minimizes radiation dose to the patient.

Isotope selection is based on concerns for minimizing patient radiation exposure, maintaining useful biodistribution of carrier, compatibility with existing instrumentation, and stability sufficient for radiochemical processing, distribution, and use. Essentially pure gamma emission is necessary to maintain acceptable patient dosimetry. All widely used diagnostic radioactive drugs expose the patient to no more radiation than routine x-ray procedures. A useful radioisotope must yield an abundance of gamma rays that can be imaged with current cameras. Internal scatter and absorption of low energy gamma emission and sensitivity and resolution constraints imposed by the collimators, sodium iodide detectors, and electronics of current scintillation cameras limit useful isotopic emission from 50–60 keV to ca 350 keV.

Short radioisotopic half-lives are necessary to minimize patient exposure, but they cannot be so short as to preclude commercial processing and transport. Commercial production and distribution of a short-lived isotope are sometimes possible with the use of a generator system. Generators take advantage of parent–daughter nuclidic pairs in which a relatively long-lived parent spontaneously decays to a short-lived daughter that is useful for imaging. The differing chemistry of the parent may allow adsorption to a column support, from which the daughter may be periodically eluted at the site of clinical use. The molybdenum–technetium generator, which is in widespread use, is designed to take advantage of the affinity of ^{99}Mo ($t_{1/2}$ = 66 h) to an alumina column from which its decay product ^{99m}Tc can be periodically eluted with sterile saline as sodium pertechnetate (Na ^{99m}TcO$_4$) ($t_{1/2}$ = 6 h). The useful life of a generator system for pharmaceutical use is roughly two half-lives of the parent. In clinical use, such devices must yield a sterile and pyrogen-free eluate and should permit minimal elution of parent nuclide or column-support material.

If a carrier molecule is to be used, the radioisotope binding must be stable *in vitro* and *in vivo* and must not degrade the useful aspects of carrier biodistribution. Iodine-131 fatty acids, for example, are rapidly deiodinated by lysosomal deiodinases after injection, so that high background concentrations of free ^{131}I decrease target-to-background ratios, and the biodistribution of ^{131}I does not parallel that of the fatty acid. Double-label studies of new radiopharmaceutical entities are always necessary. Carrier biodistribution is typically followed by incorporation of a ^{14}C or ^{3}H label for comparison to biodistribution of the associated gamma emitter.

Reactor Produced. Nuclear-reactor production of radioisotopes is limited to neutron reactions, whereas cyclotrons can accelerate a variety of particles selectively to optimize a desired reaction (1) (see Nuclear reactors; Radioisotopes). Reactor-production processes of radiopharmaceutical interest are basically of two types: fission of a heavy nucleus upon interaction with a neutron to form two or more fission fragments of greater stability that are neutron-rich and that decay by beta emission and reactions of the type (n, γ) which, for a monoisotopic or highly enriched target, yield mainly a single, final product. Thermal-neutron bombardment produces high probabilities of nuclear interaction through the (n, f) reaction, but the product is not carrier-free and is of lower specific activity than are radionuclides produced by fission-product isolation (2). On the other hand, the high specific activity of fission products implies the need for careful and often complex radiochemical separation processes to yield products of high radionuclidic purity.

Reactor-produced molybdenum-99 (^{99}Mo) is used exclusively in commercial ^{99}Mo–^{99m}Tc generator systems. High specific activity fission ^{99}Mo can be produced by thermal-neutron fission of ^{235}U:

$$^{235}\text{U} \ (n, \text{fission}) \ ^{99}\text{Zr} \xrightarrow[\beta^-]{30 \text{ s}} \ ^{99}\text{Nb} \xrightarrow[\beta^-]{3 \text{ min}} \ ^{99}\text{Mo}$$

Thermal neutron activation of a MoO_3 target yields ^{99}Mo through the reaction

$$^{98}Mo\ (n, \gamma)\ ^{99}Mo$$

Molybdenum-99 has a 66-hour half-life and decays to ^{99m}Tc by β^- decay. ^{99m}Tc, with a half-life of six hours, decays to stable ^{99}Tc ($t_{1/2} = 2 \times 10^5$ yr) with the emission of a 140-keV gamma (3). For generator production, molybdenum is adsorbed onto a column of aluminum oxide as ammonium molybdate or sodium phosphomolybdate. It is then surrounded with specially designed lead shielding with access ports, which permits aseptic elution of ^{99m}Tc with sterile saline. Radiochemical purity of the eluate may be assessed with ascending paper chromatography or thin-layer chromatography and solvent systems such as 85 wt % methanol (4).

Sodium pertechnetate that is eluted from the generator may be administered intravenously and is useful for imaging the brain, kidney, thyroid, and salivary glands. Pertechnetate-99m is trapped in the thyroidlike iodide and is secreted by cells of the gastric mucosa. Pertechnetate-99m is also used to label carrier molecules radioisotopically to act as markers of organ anatomy, metabolism, or physiologic function.

Iodine-131 is also reactor-produced. It has a half-life of eight days and emits beta and gamma radiation. It can be recovered from ^{235}U at ca 3% yield after a four-week irradiation with a neutron flux of 2×10^{14} n/(cm^2·s). Radiochemical purification consists of dissolving the target in 4.5 M NaOH, removing precipitants, purging with a caustic scrub, converting Na ^{131}I to H ^{131}I with 4.5 M H_2SO_4, oxidizing to elemental iodine with air, and finally purifying the Na ^{131}I from NaOH–Na_2SO_3 by adsorption onto platinum. Iodine-131 can also be produced by bombarding a tellurium dioxide target with thermal neutrons. Heating the target material to 800°C after irradiation in a stream of nitrogen vaporizes the ^{131}I, which can then be collected by NaOH scrubbers (2).

Iodine-131 is available commercially as sodium iodide-131 [7790-26-3] and is used for diagnosis and treatment of thyroid disease. Physiologic thyroid function is evaluated by measuring the percent uptake of a few tens of GBq (microcuries) of administered ^{131}I. Imaging of the thyroid anatomy after ^{131}I administration is an important adjunct in the detection of thyroid nodules or of diffuse glandular enlargement. Much larger doses are given to destroy pathologic thyroid tissue or thyroid tumors and their metastases.

Iodine-125 is considered less suitable for therapeutic use than in the past because of its relatively long half-life of 60 days and the resulting less favorable dosimetry.

Xenon-133 [14932-42-4] (^{133}Xe) is a gas with a half-life of 5.3 days and is reactor-produced as a uranium fission by-product. It decays to stable ^{133}Cs with emission of an imageable 81 keV photon. Xenon-133 is used clinically to determine the regional distribution of pulmonary ventilation by imaging after inhalation. After inhalation, the biological half-life is essentially limited to a few minutes by minimal tissue absorption, and ^{133}Xe that does enter the blood tends to be exhaled after a single pass through the circulation. However, xenon is fat-soluble and may be retained longer by fatty tissue. It frequently is used clinically in conjunction with imaging of pulmonary blood flow in the evaluation of the ventilation–perfusion lung ratio. In the past, it was widely administered intravenously as a saline solution to measure cerebral blood flow.

Cyclotron Produced. Indium-111 (^{111}In) has a 2.8-day half-life, has imageable gamma emissions of 171 and 245 keV. Indium-111 chloride [50800-85-6] was used experimentally for tumor and bone-marrow imaging but yielded indifferent results (5). Currently, its most common experimental clinical use is in combination with oxine for the labeling of white blood cells for *in vivo* imaging.

By accelerating charged particles of specific energy to collide with targets, cyclotrons produce nuclides on the proton excess side of the region of stability in a proton–neutron configuration plot of the nuclides (6). In most commercial cyclotrons, protons bombard the targets, although deuterons, alpha and ^{3}He particles, are also used in research machines.

Although extremely short-lived positron-emitting isotopes, eg, ^{11}C ($t_{1/2}$ = 20 min), ^{13}N ($t_{1/2}$ = 10 min), and ^{15}O ($t_{1/2}$ = 122 s), have been produced for immediate on-site imaging use, isotopes in commercial production and widespread use, eg, ^{201}Tl, ^{67}Ga, and ^{123}I, have substantially longer decay times. Although none of the positron emitters has shown general clinical utility, ^{18}F-fluorodeoxyglucose has been extremely useful in research of central nervous system metabolism.

Thallium (^{201}Tl) has become the most widely used cyclotron-produced clinical radioisotope by virtue of its tendency to localize in the heart and thereby reflect regional blood flow. Thallium-201 decays by electron capture with a half-life of 73 h and emits low abundance 135- and 167-keV gamma rays and high abundance (94.5%) imageable mercury K-x-rays of 69–83 keV. Cyclotron production is by the 31-MeV proton bombardment of a 99.999% pure natural thallium target (7). Target ^{203}Tl is converted to ^{201}Pb by the nuclear reaction ^{203}Tl (p, $3n$) ^{201}Pb. Lead-201 has a half-life of 9.4 h and decays to ^{201}Tl. After irradiation, the thallium target is dissolved in concentrated nitric acid, evaporated to dryness, redissolved in 0.025 M EDTA, and is passed over an ion-exchange column to separate adherent thallium from eluted ^{203}Pb and ^{201}Pb. After standing, during which ^{201}Pb decays to ^{201}Tl, the eluate is passed over a second ion-exchange column to which ^{201}Tl adheres and from which the product can be isolated. Thallium-201 is available commercially as the chloride in sterile, nonpyrogenic isotonic saline solution, which is preserved with 0.9 wt % benzyl alcohol. It is essentially carrier-free, as it contains less than 0.25 wt % ^{203}Pb and less than 1.9 wt % ^{202}Tl.

The biodistribution of carrier-free ^{201}Tl is analogous to potassium. After being administered intravenously, it is rapidly cleared from the blood with maximum target-to-background heart ratios 10–20 min postinjection. Uptake by the kidneys, GI (gastrointestinal) tract, and actively exercising muscle also occurs in proportion to blood flow. The principal route of excretion is renal. It is injected intravenously for myocardial perfusion imaging as an adjunct in the diagnosis of coronary artery disease and myocardial infarction.

When the relatively low specific activity bone-seeker gallium-72 (^{72}Ga) was determined to be ineffective in treating bone cancer, cyclotron production of ^{67}Ga was undertaken in an effort to produce a carrier-free isotope of high specific activity. Gallium-67 is cyclotron-produced by bombardment of a ^{67}Zn-target with 21-MeV protons or of a ^{66}Zn-target with deuterons. It decays by electron capture to ^{67}Zn with a half-life of 78.3 h and emits four primary gamma rays of 93, 184, 300, and 393 keV. Because gallium chloride is a protein precipitant, the isotope is supplied commercially in sterile, pyrogen-free isotonic saline solution, which is preserved with 0.9 wt % benzyl alcohol, as ^{67}Ga citrate [*41183-64-6*] (pH 4.5–7.5). Carrier-free ^{67}Ga shows diverse biodistribution: it localizes in the lymphoid tissue of the nasopharynx, liver, spleen, and genitalia, as well as in the bones (8). In addition, the isotope is useful clinically by virtue of its tendency to localize in areas of abscess and inflammation and in certain specific tumors, eg, bronchogenic carcinoma, Hodgkins disease, and other lymphomas. ^{67}Ga citrate clears slowly from the body by renal and fecal excretion; 65% is retained seven days after administration. Because of its slow clearance from blood and background tissues, target-to-background ratios suitable for clinical imaging are only obtained at 24–72 h postinjection.

The third principal cyclotron-produced isotope in clinical use is iodine-123. It is clinically used in a manner analogous to ^{131}I but has the advantage of superior dosimetry by virtue of its 13.2-h half-life and its nonparticulate 159-keV gamma-ray emission. Commercially available ^{123}I is cyclotron produced by deuterium bombardment of a purified tellurium target according to the reaction ^{122}Te (d, n) ^{123}I. The currently available Na ^{123}I product contains >93.75 wt % ^{123}I, <3.2 wt % ^{130}I, <1.1 wt % ^{124}I, <0.8 wt % ^{131}I, and <1.1 wt % ^{126}I. Traces of tellurium and aluminum may also be present. Oak Ridge has described an alternative production process based on the ^{123}Te (p, n) ^{123}I reaction. Production involves use of 15-MeV protons to bombard an 80% isotopically enriched ^{123}Te target. The product of this reaction is ca 97% pure (2).

Kits. Radiopharmaceutical kits are composed of sterile, nonpyrogenic, nonradioactive carrier materials that are configured so that they can be labeled by the aseptic addition of a radioisotope. All kits in use in North America are designed primarily for use in conjunction with the molybdenum–technetium generator and for labeling with its product, ie, sodium pertechnetate (^{99m}TcO$_4^-$) in sterile saline. In general, tin is the catalyst in the technetium labeling process. Although the exact chemistry of ^{99m}Tc-labeling is unknown in all cases, it is clear that Sn(II) acts to reduce TcO$_4$ to the Tc(IV) oxidation state, where labeling occurs (9). The carrier ligand competes for binding of Tc(IV), which tends to form spontaneously insoluble and unreactive TcO$_2$. Technetium reduction for labeling has also been accomplished electrolytically with zirconium electrodes and by catalysis with iron and ascorbate (see also Medical diagnostic reagents).

Skeletal imaging is routinely performed with ^{99m}Tc-labeled phosphate and phosphonate compounds that localize in actively metabolizing bone (10). The phosphonates, in particular, have the advantage of resistance to *in vivo* enzymatic hydrolysis and produce images of higher target-to-background ratios than the phosphate compounds. Technetium-99m methylenediphosphonate [80908-09-4] (^{99m}Tc-MDP) is the clinical agent of choice, but ^{99m}Tc-ethylenehydroxydiphosphonate [65330-43-0] (^{99m}Tc-EHDP), ^{99m}Tc-hydroxymethylenediphosphonate [72945-61-0] (^{99m}Tc-HMDP), and ^{99m}Tc-pyrophosphate [80908-08-3] (^{99m}Tc-PYP) are also commercially available (11–13).

Technetium-99m methylenediphosphonate is available commercially as a sterile, nonpyrogenic lyophilized powder in vials suitable for reconstitution with Na ^{99m}TcO$_4$ to form ^{99m}Tc-MDP (11). Kits contain 10 mg ^{99m}Tc-MDP and 0.85 mg stannous chloride dihydrate whose pH has been adjusted to pH 7.0–7.5. Vial contents are stored under nitrogen to inhibit stannous oxidation. Approximately 50 wt % of the injected ^{99m}Tc–MDP localizes in the skeleton 2–3 h postinjection. Approximately 50 wt % of the administered dose is excreted by the kidneys in the first 24 h postinjection. Radiocalcium kinetic studies suggest that the diphosphonates act by adsorbing to hydroxyapatite in forming bone. Thus, areas of active mineral metabolism stimulated by fracture or by tumor metastases avidly concentrate the tracer.

Technetium-99m ethylenehydroxydiphosphonate is available commercially lyophilized in vials containing 5.9 mg ^{99m}Tc–EHDP and 0.16 mg stannous chloride (12); the solution can be reconstituted with Na ^{99m}TcO$_4$. Lyophilized ^{99m}Tc-PYP is available commercially in vials containing 10–25 mg ^{99m}Tc-PYP (11,13). Both drugs are used for skeletal imaging as described above. The latter also localizes in areas of tissue damage and is used for imaging regions of myocardial infarction. In addition, ^{99m}Tc-PYP is used for *in vivo* labeling of red blood cells for blood-pool imaging. This procedure involves the injection of ^{99m}Tc-PYP reconstituted with sterile saline, followed in 5–15 min by injection of Na ^{99m}TcO$_4$ in saline. Red-cell labeling with pertechnetate occurs spontaneously by an unknown mechanism mediated by tin.

Colloids are cleared from the circulation by elements of the reticuloendothelial system (RES), which is principally in the liver, spleen, and bone marrow. Available kits for preparation of a sulfur colloid contain 1 N HCl and thiosulfate. When mixed with ^{99m}TcO$_4$ and heated briefly to 100°C, thiosulfate converts to thiosulfuric acid, which decomposes to form elemental sulfur and sulfur dioxide labeled with ^{99m}Tc (14). The pH is adjusted with phosphate buffer (pH 7.4) after heating. Gelatin was once used to stabilize the colloid, but was associated with occasional adverse reactions and is no longer employed. The product is described in USP XX (15) (see Fine chemicals). Radiochemical purity can be assessed with ascending paper chromatography in 85 wt % methanol.

A preformed stannous-microaggregated human serum albumin colloid 0.5–2.0 μm in diameter has been used clinically on a trial basis for RES imaging (11). Unlike sulfur colloid, this material is biodegradable and does not require heating for ^{99m}Tc-labeling.

Small colloids (<50 μm in dia) prepared from microaggregated human serum albumin and from antimony sulfide have been studied for their ability to localize in regional lymphatic nodes after subcutaneous or intramuscular injection.

Technetium-99m-labeled macroaggregated human serum albumin (MAA) particles ca 20–70 μm in dia are used for imaging lung perfusion because they randomly distribute throughout the pulmonary circulatory bed and temporarily lodge in capillaries. In aggregate formation, human serum albumin in solution with tin is denatured by varying combinations of pH, heat, and agitation. Commercial, lyophilized MAA for labeling with TcO$_4$ is supplied in a kit containing 1–2 mg human serum albumin and 70–125 μg of stannous chloride dihydrate (11,13). Various stabilizers and preservatives, such as benzyl alcohol, sodium chloride, sodium acetate, and acetic acid, can be used. Since toxicity is directly related to the size and number of particles administered, aggregates greater than 150 μm in diameter are excluded, and vials contain $(5–10) \times 10^6$ particles. The pharmaceutical product is described in USP XX (15).

Renal imaging and brain imaging are most commonly performed with ^{99m}Tc-Gluceptate (glucoheptonate) or ^{99m}Tc-diethylenetriaminepentaacetate [80908-06-1] (^{99m}Tc-DTPA), which localize in areas of blood–brain barrier disruption and in the kidneys by a combination of glomerular filtration and tubular mechanisms. Lyophilized glucoheptonate is commercially available in vials containing 200 mg glucoheptonate sodium and 0.07 mg tin (11); it is suitable for labeling by addition of Na ^{99m}TcO$_4$. In humans, 25% of the injected dose is excreted in the urine during the first hour postinjection. Lyophilized DTPA is commercially available in vials containing 10 mg CaNa$_3$ ^{99m}Tc-DTPA and 0.50 mg SnCl$_2$.2H$_2$O (11). Unlike ^{99m}Tc-glucoheptonate, ^{99m}Tc-DTPA shows little or no binding to renal parenchyma. Approximately 4 wt % of injected ^{99m}Tc-DTPA binds to serum proteins.

N-[2,6-dimethylphenylcarbamoylmethyl]iminodiacetic acid [59160-29-1] (HIDA) is the prototype of a family of molecules that is readily cleared by the hepatobiliary system and is useful in imaging the liver and biliary tract and in assessing biliary function (16). A number of iminodiacetic acid (IDA) derivatives have been evaluated for clinical use in an effort to optimize hepatic uptake, biliary clearance rate, and renal clearance rates. All members of this family compete with bilirubin for anion transport into the bile; thus, biliary excretion tends to diminish and renal excretion to increase with increasing serum bilirubin levels. Diisopropyl-IDA [80908-07-2] (Disofenin) demonstrates highest biliary clearance and lowest urinary excretion, followed by the diethyl-IDA [6290-05-7], N-p-butylphenyl-IDA [66292-52-2], N-p-isopropylphenyl-IDA [66292-53-3], and dimethyl-IDA [6096-81-7] (17). Disofenin is largely cleared from the circulation by the normal healthy liver within five minutes postinjection and appears in the gall bladder and intestines by 60 min postinjection. Diisopropyl-IDA has been developed commercially for human use and is available lyophilized in kit form. The kit contains 20 mg diisopropyl-IDA and the kit product can be reconstituted with Na ^{99m}TcO$_4$. Similar commercial kits have been prepared for the other congeners (11).

Principal North American commercial radiopharmaceutical suppliers include New England Nuclear, Mallinckrodt, Medi-Physics, and E. R. Squibb. Total U.S. sales in 1980 were ca $\$90 \times 10^6$.

Radioimmunoassay

Radioimmunoassay (RIA) is the generic term for systems of quantitative *in vitro* measurement based on the principle of saturation analysis, displacement analysis, or competitive protein binding (18–19). Since they are extremely sensitive and specific, RIA techniques are used widely for the determination of drug and hormone levels in biological fluids. The development of this field stems from the observation that unlabeled insulin displaces ^{131}I-labeled insulin from insulin antibody *in vitro*. With the antibody concentration and radioiodinated antigen held constant, the binding of the label is quantitatively related to the amount of unlabeled antigen that is added. Thus, in the insulin RIA, known insulin standards are used initially to prepare a plot of fraction of bound ^{131}I-insulin against the concentration of insulin added. The amount of unlabeled insulin in a serum sample can subsequently be determined by measuring the fraction of bound labeled insulin in its presence from the standard curve.

In general, RIAs involve the separation of a labeled antigen of interest into bound and unbound fractions after its interaction with an antibody in the presence of an unknown quantity of unlabeled antigen. The ratio of bound-to-free labeled antigen is related to the concentration of unknown antigen in the system by comparison to a curve demonstrating binding in the presence of standards of known concentration (20).

Many modifications of the original assay technique have been described. Iodine-125 and ^{57}Co [13981-50-5] are the most commonly used radiotracers. Antibodies are the most widely used binding reagents, but some RIAs involve naturally occurring receptor molecules or nonimmunoglobulin binding proteins. Intrinsic factor for binding vitamin B$_{12}$ and folic acid-binding milk proteins are two examples in clinical use (see Vitamins). Binding of radioisotopically labeled antigen is the paradigm in these assays; however, fluorescence labels and enzyme labels are being used with increasing frequency. In addition, various techniques have evolved for separating bound from unbound labeled antigen, eg, precipitation in the presence of polyethylene glycol, second antibody precipitation, adsorption to dextran-coated charcoal, adsorption to antibody-coated tubes, ammonium sulfate precipitation, and molecular sieving (see Molecular sieves) (21).

The most commonly used RIAs include those for hepatitis, thyroxine, digoxin, cortisol, estrogens, and human chorionic gonadotrophins. There are approximately 45 commercially available kits for serum thyroxine determination. The largest manufacturers include Abbott Laboratories, Nuclear Medical Laboratory, and Clinical Assays. Total U.S. sales in 1980 were ca 160×10^6.

BIBLIOGRAPHY

"Radioactive Drugs and Tracers," in *ECT* 2nd ed., Vol. 17, pp. 1–8, by Seymour Hopfan, Memorial Hospital for Cancer and Allied Diseases.

1. J. K. Poggenburg, *Semin. Nucl. Med.* **4,** 229 (1974).
2. F. P. Castronovo, "Principles, Properties and Quality Control of Nuclear Medicine Agents," in D. Rollo, ed., *Nuclear Medicine Physics, Instrumentation and Agents*, C. V. Mosby Company, St. Louis, Mo., 1977, pp. 560–636.
3. P. V. Harper, K. A. Lathrop, F. Jimenez, and co-workers, *Radiology* **85,** 101 (1965).
4. Th. Miller and E. Steinnes, *Scand. J. Clin. Lab. Invest.* **28,** 213 (1971).
5. P. A. McIntyre, S. M. Larson, E. A. Eikman, and co-workers, *J. Nucl. Med.* **15,** 856 (1974).
6. R. S. Tilbury and J. S. Laughlin, *Semin. Nucl. Med.* **4,** 245 (1974).
7. E. Lebowitz, M. W. Greene, R. Fairchild, and co-workers, *J. Nucl. Med.* **16,** 151 (1975).
8. G. S. Johnston and A. E. Jones, *Atlas of Gallium-67 Scintigraphy*, Plenum Publishing Corp., New York, 1973.
9. I. Hambright, R. J. McRae, P. E. Volk, and co-workers, *J. Nucl. Med.* **16,** 478 (1975).
10. M. D. Francis, A. J. Tofe, J. J. Benedict, and J. A. Beran, *paper presented at Second International Symposium on Radiopharmaceuticals*, Society of Nuclear Medicine, New York, 1979, pp. 603–614.
11. Package insert, New England Nuclear Corp., No. Billerica, Mass.
12. Package insert, Proctor and Gamble, Cincinnati, Ohio.
13. Package insert, Mallinckrodt, Inc., St. Louis, Mo.
14. H. S. Stern, J. G. McAfee, and G. Subramanian, *J. Nucl. Med.* **7,** 665 (1966).
15. *USP XX–NF XV*, The United States Pharmacopeial Convention, Inc., Rockville, Md., 1980.
16. M. D. Loberg and A. T. Fields, *Int. J. Appl. Radiol. Isot.* **29,** 167 (1978).
17. H. S. Weissman, L. A. Sugarman, and L. M. Freeman in L. M. Freeman and H. H. S. Weissman, eds., *Nuclear Medicine Annual 1981*, Raven Press, New York, 1981, pp. 35–89.
18. J. H. Howanitz and P. J. Howanitz, "Radioimmunoassay and Related Techniques," in J. B. Henry, ed., *Clinical Diagnosis and Management*, 16th ed., W. B. Saunders Company, Philadelphia, Pa., 1979, pp. 385–401.
19. S. Goldsmith, *Semin. Nucl. Med.* **5,** 125 (1975).
20. A. Zettner and P. E. Duly, *Clin. Chem.* **20,** 5 (1974).
21. R. F. Schall and H. J. Tenosos, *Clin. Chem.* **27,** 1157 (1981).

ALLAN M. GREEN
IRWIN GRUVERMAN
New England Nuclear Corporation

RADIOACTIVE TRACERS

Radiochemical tracers, since their introduction in 1945, have become a basic analytical tool for scientists. By virtue of their radioactive emission, which is readily observed, they are detectable in matter and, with some noncritical exceptions, mimic completely the physicochemical properties of the product to be traced. A molecule or chemical is termed labeled, or radioactive, if a radioactive atom is substituted for a stable atom in that molecule. If the radioactive chemical is used to trace its movement in any medium, it is called a radioactive tracer.

Properties

Any radioactive element can be used as a radioactive tracer, eg, chromium-51 [14392-02-0], cobalt-60 [10198-40-0], tin-110 [15700-33-1], and mercury-203 [13982-78-0], but the preponderance of use is with carbon-14, hydrogen-3, sulfur-35, phosphorus-32, and iodine-125. By far the greater number of radioactive tracers produced are based on carbon-14 and hydrogen-3 since these atoms exist in almost all the known natural and synthetic chemical compounds. The properties of the principal radioactive elements are listed in Table 1. The isotopes are available for use as barium carbonate (^{14}C) [1882-53-7], tritium, phosphoric acid (^{32}P) [15364-02-0], sulfuric acid (^{35}S) [13770-01-9], and sodium iodide (^{125}I) [24359-64-6]. It is from these chemical forms that all radiotracer chemicals are synthesized.

Syntheses

Syntheses of radioactive tracers involve all of the classical biochemical and synthetic chemical reactions used in the synthesis of nonradioactive chemicals (2). There are, however, specialized techniques and considerations required for the safe handling of radioactive chemicals, strategic synthetic considerations in terms of their relatively high cost, and synthesis scale constraints governed by specific activity requirements.

Basic precursor materials, eg, carbon dioxide (^{14}C), benzene (^{14}C), methyl iodide

Table 1. Properties of the Principal Radioactive Elements[a]

Element	CAS Registry No.	Half-life, $t_{1/2}$	Specific activity, Bq/mmol[b]	meV, %	Decay product	Biological $t_{1/2}$
^{14}C	[14762-75-5]	5.73×10^3 yr	2.38×10^{10}	0.156 β^-, 100	^{14}N	10 d
^{3}H	[10028-17-8]	12.3 yr	1.1×10^{12}	0.018 β^-, 100	^{3}He	12 d
^{35}S	[15117-53-0]	86.7 d	5.2×10^{13}	0.17 β^-, 100	^{35}Cl	90 d
^{32}P	[14596-37-3]	14.3 d	3.6×10^{14}	1.70 β^-, 100	^{32}S	257 d
^{125}I	[14158-31-7]	60.2 d	7.86×10^{13}	0.03 β^-, 90; 0.027 x ray; 0.035 γ, 7	^{125}Te	138 d

[a] Ref. 1.
[b] To convert Bq to Ci, divide by 3.70×10^{10}.

(^{14}C), sodium acetate (^{14}C), sodium cyanide (^{14}C), etc, require vacuum-line handling in well-ventilated fume hoods. Tritium gas and methyl iodide (^{3}H), iodine, and tritium, which are the most difficult of the isotopes to contain, must be handled in specialized closed systems. Sodium sulfate (^{35}S) and sodium iodide (^{125}I) must be handled similarly in closed systems to avoid the liberation of volatile sulfur oxides (^{35}S) and iodine (^{125}I). Adequate shielding must be provided when handling phosphoric acid (^{32}P) to prevent an external radiation dose.

A multistep synthesis is strategically designed such that the labeled species is introduced as close to the last synthetic step as possible in order to minimize yield losses and cost. Use of indirect reaction sequences frequently maximizes the yield of the radioactive species at the expense of time and labor.

Radioactive tracers are most useful when the radioactive emissions are great relative to the mass of the compound, ie, if they have high specific activity. In order to study the metabolic pathway of the drug digitoxin in a mouse, the injectable mass dose of drug must be less than the lethal dose and yet be of sufficient radioactivity to be measured accurately in the various mouse organs. Radiation scale requirements dictate that micro or semimicro synthetic methods are used.

Synthetic chemical approaches to the preparation of carbon-14 labeled materials utilize a number of basic building blocks that are prepared from barium carbonate (^{14}C) (2). These are carbon dioxide (^{14}C), acetylene (^{14}C), benzene (^{14}C(U)) (U = uniformly labeled), sodium acetate (1- and 2-^{14}C), methyl iodide (^{14}C), methanol (^{14}C), sodium cyanide (^{14}C), and urea (^{14}C). Many complicated radiotracers are synthesized from these materials. Some examples are 8,11,14-eicosatrienoic acid (1-^{14}C) [3435-80-1] from carbon dioxide (^{14}C), phenylisothiocyanate (ring-^{14}C(U)) [77590-93-3] from acetylene (^{14}C), norepinephrine (7-^{14}C) [18155-53-8] from acetic acid (1-^{14}C), cholesterol (4-^{14}C) [1976-77-8] from methyl iodide (^{14}C), glucose (1-^{14}C) [4005-41-8] from sodium cyanide (^{14}C), and uracil (2-^{14}C) [626-07-3, 27017-27-2] from urea (^{14}C). All the syntheses of the basic radioactive building blocks are described in ref. 3.

Biosynthetic techniques are ideally suited for the synthesis of many radiolabeled compounds. Plants, eg, potato and tobacco, when grown in an exclusive atmosphere of radioactive carbon dioxide, utilize the labeled chemical as their sole source of carbon. After a suitable period of growth, almost every carbon atom in the plant is radioactive. Thus, plants can serve as an available source of labeled carbohydrates. Algae grown under similar conditions provide labeled amino acids, lipids, nucleotides, etc. Enzyme-catalyzed biochemical reactions are also used because of their low mass requirement and reaction specificity. Sugar transferases and phosphokinases are a few of the available synthetic enzymes that are used for the syntheses of complicated labeled nucleotides.

The introduction of tritium into molecules is most commonly achieved by reductive methods, including catalytic reduction by ditritium gas (^{3}H$_2$) of olefins, catalytic reductive replacement of halogen (Cl, Br, or I) by ^{3}H$_2$, and metal hydride (^{3}H) reduction of carbonyl compounds, ie, ketones and some esters, to tritium-labeled alcohols (4). The use of tritium-labeled building blocks, eg, methyl iodide (^{3}H) and acetic anhydride (^{3}H), is an alternative route to the preparation of high specific activity, tritium-labeled compounds.

Iodination of organic compounds with iodine-125 gives radiotracers that are in most cases modified forms of the compound being traced. The identity of behavior between the radiotracer and the nontagged parent substance must be assured before acceptance of any derived data is valid. In the case of thyroxine, which is a naturally occurring iodine-containing substance, iodination is achieved by exchange with NaI (^{125}I).

Noniodine-containing substances that are to be iodinated must have a moiety that can be iodinated directly, eg, phenol, imidazole, pyrimidine, etc, with such reagents as KI$_3$ (^{125}I), NaI (^{125}I)/lactoperoxidase, CII (^{125}I), etc. An alternative method for iodination is the use of the reactive Bolton-Hunter reagent [60285-92-9] (^{125}I-iodinated p-hydroxyphenylpropionic acid N-hydroxysuccinimide ester). Proteins are most readily labeled with iodine and in most cases their properties are unaffected by iodination. Phosphorus (^{32}P) and sulfur (^{35}S) are introduced mostly through the use of biosynthetic techniques acting upon phosphate (^{32}P) and sulfate (^{35}S). Such reagents as PCl$_5$ (^{32}P), POCl$_3$ (^{32}P), PCl$_3$ (^{32}P), P$_2$S$_5$ (^{32}P), H$_2$S (^{35}S), Na$_2$SO$_4$ (^{35}S), Na$_2$SO$_3$ (^{35}S), KSCN (^{35}S), and P$_2$S$_5$ (^{35}S) have been prepared and are available as synthetic precursors.

Detection and Quantification

The methods for detection and quantification of radiolabeled tracers are determined by the type of emission (β or γ) the tracer affords, the energy of the emission, and the efficiency of the system by which it is measured. Detection of radioactivity can be achieved in all cases with the Geiger counter. However, in the case of the weaker emitting isotopes, ie, ^{3}H, ^{14}C, and ^{35}S, large amounts of isotopes are required for detection of a signal. This is in most cases undesirable and impractical. Thus, more sensitive methods of detection and quantitation have been developed.

Liquid scintillation counting is by far the most common method of detection and quantitation of β-emission (5). This technique involves the conversion of the emitted β radiation into light by a solution of a mixture of fluors (the liquid scintillation cocktail), and the sensitive detection of this light by a pair of matched photomultiplier tubes in the dark chamber which is amplified, measured, and recorded with the liquid scintillation counter. Efficiencies of detection are typically 25–60% for tritium; >90% for ^{14}C, ^{35}S, and ^{32}P; and 60–70% for ^{125}I. A lesser-used technique for the detection and quantitation of β-emissions is Planchette counting. A film of the sample on a Planchette, which is a flat metal pan, is brought into proximity, but at a fixed distance, to a Geiger counter. The emissions are measured and recorded. Typical efficiencies are ^{14}C, ca 30–40%; ^{35}S, ca 30–40%; and ^{32}P, ca 50%.

The detection and quantitation of γ-emission from ^{125}I is accomplished by Well counting. A thallium-activated sodium iodide crystal, with a well or drilled hole which contains the sample, converts the emission to light. The light is amplified, counted, and recorded by a photomultiplier tube. The efficiency for ^{125}I is typically 70%.

The β-emission of ^{32}P is energetic enough in its passage through water to emit light (the Cherenkov effect). This can be measured by a photomultiplier tube with a typical efficiency of ca 40%.

The nonquantitative detection of radioactive emission often is required for special experimental conditions. Autoradiography, which is the exposure of photographic film to radioactive emissions, is a commonly used technique for locating radiotracers of thin-layer chromatographs, gel electrophoreses, tissue mounted on slides, and whole-body animal slices (6). After exposure to the radiolabeled emitters, black silver grains appear as the film develops. This technique is especially useful for tritium detection but is also used for ^{14}C, ^{35}S, ^{32}P, and ^{125}I.

Gas-flow counting is a method for detecting and quantitating radioisotopes on paper chromatography strips and thin-layer plates. Emissions are measured by interaction with an electrified wire in an inert gas atmosphere. All isotopes are detectable; however, tritium is detected at very low efficiency (<1%).

Other methods of sensitive detection of radiotracers have recently been developed. Fourier transform nmr has been used to detect ^{3}H (nuclear spin $\frac{1}{2}$), which has an efficiency of detection equivalent to that of ^{1}H. The development of field-desorption mass spectrometry (FDMS) as an analytical tool for radioisotopes has been reported (7). Although not limited to radioisotopes, FDMS can be used to detect samples in nanogram quantities and therefore is suitable for radiolabeled tracers.

Decomposition

The emission of ionizing radiation by a radioactive element results in transformation of the radioactive element, ie, decay, and leads to altered species which contaminate the labeled product (8). This physical decomposition is termed primary decomposition. Impurities arising from the decay process are generally minor compared to the chemical impurities arising from the destruction of molecules in the surrounding environment by the emitted energy of the radioactive element. This energy can cause free-radical formation in neighboring solvent molecules as well as in adjacent labeled molecular species, and can trigger a cascade of unwanted chemical reactions and products.

The chemical reactions initiated by the radiation, unlike decomposition arising from decay, can be minimized by storage at low temperature, the addition of suitable radical or electron scavengers, and storage in dilute solution. The initial avoidance of chemical impurities, which by themselves can generate additional impurities when subjected to ionizing emission, is obligatory in maximizing shelf-life.

Economic Aspects

The radioisotope industry is segmented into producers, processors, fabricators, and equipment manufacturers. The processers convert the radioisotopes made by the producers into radiochemicals, and this segment is referred to as the radiochemical or radioactive-tracer industry. The industry is comprised of full-line manufacturers and part-line manufacturers. The full-line manufacturers offer a complete line of radioactive tracers with all of the common isotopes used as tracers as well as a complete line of services, eg, custom syntheses and custom labeling. The full-line companies are New England Nuclear Corp. (Boston, Mass.), which was acquired by DuPont, and Amersham International (Amersham, UK). The part-line manufacturers specialize in one or several of the classes of radioactive tracer chemicals and one or more of the services. Part-line manufacturers are ICN (Cleveland, Ohio), Schwartz Radiochemicals (Orangeburg, N.Y.), CEA (Gif-sur-yvette, France), and Pathfinder (St. Louis, Mo.). Collectively the part-line manufacturers account for 5–10% of the radioactive chemicals sold in the industry.

The estimated world growth rate of radiotracer sales was 10% per year during 1970–1980. World sales in 1980 were ca 64×10^6. It is estimated that sales of radiochemicals in the United States during 1981 amounted to 35×10^6. Of that amount, ca 17×10^6 was sold by New England Nuclear and ca $(5-7) \times 10^6$ was sold by Amersham Corp. (Arlington Heights, Ill.), which distributes products imported from Amersham International. It is estimated that sales of radiochemicals for tracer use in 1985 will be 55×10^6 (9).

Health and Safety Factors

Allowable external radiation doses are described in reference 10. Depending on the quantities used and type of operation, the more energetic emissions of ^{32}P (β-ray) and ^{125}I (γ- and x rays) may require appropriate shielding to minimize personnel exposure. The energy of the β-rays of ^{35}S, ^{14}C, and ^{3}H are weak enough to require no shielding. All isotopes, however, present toxicity problems if the isotopes are ingested. Both tritium and ^{125}I are particularly problematic because of their volatility as H_2O (^{3}H) and $I_2(^{125}I)$. However, personal safety precautions are related to the relative quantities of radioactive materials handled. Basic laboratory procedures to be followed protect the user from oral ingestion, skin contact, and inhalation. The use of closed systems, well-ventilated work areas (ie, with hoods and glove boxes), disposable gloves, disposable lab coats, etc, and neat work habits provide a safe working environment. Personal monitoring may include urine and exhaled breath (CO_2) analysis, thyroid uptake (^{125}I) radioassay, use of personal monitors for ^{32}P detection, and contamination surveys with appropriate instruments. The choice of test and frequency of testing vary with quantities and use (11).

Protection of the environment from uncontrolled radioactive release is also a consideration in the use of radiotracers. Drain discharge into sewer systems is limited by the NRC. Similarly, airborne emission limits have been established by the NRC for nonrestricted areas. Limits of surface contamination must be established to provide a safe work place for users (12). The application of the as-low-as-reasonably-achievable (ALARA) principle to the above draws upon the creative talents of the user to regard the limits as nonapproachable barriers and not as tolerable maxima for discharge.

U.S. radiation protection limits are established by the NCR and are based upon the recommendations of the International Council for Radiation Protection (ICRP) (12). The National Academy of Sciences also sponsors a report from its advisory committee on the biological effects of ionizing radiations (13).

Uses

The detectability of minute quantities of radiolabeled tracers makes possible the determination of micro quantities of substances. The most effective use of the radiotracer has been in biomedical research. A radiolabeled, nonmetabolized tracer for glucose, ie, 2-deoxyglucose (1-^{3}H) [77590-94-4], is administered to a test animal to identify areas of brain activity, ie, of glucose metabolism, corresponding to particular external stimuli. An external stimulus is given and the animal is sacrificed. The brain is frozen, sectioned, and exposed to x-ray film (autoradiography), and the location of the radioactivity is noted. In this way it is possible to relate the areas and to produce a brain map.

Radiotracers have also been used extensively for the quantitative micro determination of blood serum levels of hormones, protein, neurotransmitters, and other physiologically important compounds (see Hormones; Proteins; Neuroregulators).

Radioimmunoassay, which involves the competition of a known quantity of radiolabeled tracer (usually ^{125}I or ^{3}H) with the unknown quantity of serum component for binding to a specific antibody that has been raised against the component to be determined, is used in the micro determination of physiologically active materials in live animals (14). The development of monoclonal antibodies will increase the effectiveness of this technique.

The determination of the presence of reverse transcriptase in virus-infected cells can be done with nucleotide triphosphates (^{32}P) or (^{3}H). Reverse transcriptase is an enzyme capable of synthesizing DNA from RNA and it is thought to play an important role in virus-mediated cell modification. It utilizes radiolabeled nucleotides with nonlabeled substrates to synthesize tagged DNA; the degree of radioactive incorporation reflects the reverse transcriptase level.

The use of adenosine triphosphate (γ-^{32}P) [2964-07-0] as a source of phosphate for end-labeling of DNA is an integral part of the Maxam-Gilbert DNA sequencing procedure. The end-radiolabeled DNA (^{32}P) is selectively cleaved by reagents that are specific for the rupture of selected chemical bonds between the various base sequences. Gel electrophoresis of the experiments followed by autoradiographic detection of the fragments permits determination of the sequence of the bases.

Radioactive tracers also are used in agriculture. A test field containing a food crop is sprayed with either an organic fertilizer, pesticide, or fungicide that is laced with the appropriate radioactive tracer. Run off, leaching, or contamination of the water table can then be determined by measuring radioactivity in local ponds or rivers. Possible incorporation of these chemicals into the plants also is easily determined as is longevity or breakdown in the soil (see also Soil chemistry of pesticides).

In the petroleum industry, the size of an underground oil deposit is determined by the injection of radiolabeled substances into a well head. The occurrence of radioactivity in the oil-water mixture, which is pumped out of adjoining wells, gives an indication of the pocket size of the oil deposit (see Petroleum).

BIBLIOGRAPHY

"Radioactive Drugs and Tracers" in *ECT* 2nd ed., Vol. 17, pp. 1–9, by Seymour Hopfan, Memorial Hospital for Cancer and Allied Diseases.

1. Y. Wang, *Handbook of Radioactive Nuclides*, The Chemical Rubber Company, Cleveland, Ohio, 1969, pp. 16–63.
2. J. R. Catch and Y. Cohen in *Radionuclides in Pharmacology*, Vol. 1, Pergamon Press, New York, 1971, pp. 97–130; ref. 1, pp. 339–379.
3. A. Murray and D. Williams, *Organic Synthesis with Isotopes*, Interscience Publishers, New York, 1958.
4. E. A. Evans, *Tritium and Its Compounds*, 2nd ed., John Wiley & Sons, Inc., New York, 1974.
5. Y. Kobayashi and D. V. Maudsley, *Biological Applications of Liquid Scintillation Counting*, Academic Press, New York, 1974.
6. A. Rogers, *Techniques of Autoradiography*, Elsevier Scientific Publishing Co., Amsterdam, 1979.
7. H. R. Schulten, *J. Mass Spectrosc.* **23** (1979) (entire issue on FDMS).
8. Ref. 1, pp. 274–338.
9. S. C. Stinson, *Chem. Eng. News*, 10 (June 8, 1981).
10. *Standards for Protection Against Radiation*, Code of Federal Regulations, Title 10, Chapt. 1, 10CFR20.
11. J. Shapiro, *Radiation Protection*, Harvard University Press, Cambridge, Mass., 1972.
12. *Health Physics Surveys for By-Product Material at N.R.C. Licensed Processing and Manufacturing Plants*, U.S. NRC Regulatory Guide 8.21, Office of Standards Development, October 1971.

13. *The Effects on Populations of Exposure To Low Levels of Ionizing Radiation*, National Academy Press, National Academy of Sciences, Washington, D.C., 1980
14. E. A. Evans and M. Muramatsu, *Radiotracer Techniques and Applications*, Marcel Dekker, Inc., New York, 1977.

General References

Reference 1 is also a general reference.
Ref. 2 is also a general reference which contains an excellent bibliography.
NCRP reports, obtained from NCRP Publications, P.O. Box 30175, Washington, D.C. 20014.
ICRP reports, obtained from Page Bros., Publishers, Norwich, UK.

ROBERT E. O'BRIEN
New England Nuclear Corporation

RADIOACTIVITY, NATURAL

Radioactivity is a widespread nuclear property exhibited by some isotopes of every chemical element (see Radioisotopes; Isotopes; Radioactive drugs; Radioactive tracers). This article is limited to those radionuclides and radioelements that are found in nature and to those phenomenological aspects and applications of radioactivity that are prominently associated with natural substances. The chemical properties of natural radioelements that are technologically important as elements are discussed in articles devoted to their general nonnuclear aspects (see Bismuth and bismuth alloys; Potassium; Lead; Thorium and thorium compounds; Uranium and uranium compounds; Actinides and transactinides). However, this article does include some general chemistry of elements that are encountered principally as natural radioelements and are not otherwise covered (polonium, astatine, radon, francium, and radium).

The spontaneous emission of penetrating energetic radiations by matter, now called radioactivity, was discovered accidentally in 1896 by Becquerel, who was seeking a connection between x rays, discovered the preceding year, and phosphorescence. The photographic effect of a uranium salt caused by penetrating radiation was found to be independent of prior illumination and to be associated with the uranium, regardless of its chemical or physical state. Chemical investigations by the Curies led to the discovery in 1898 that not only uranium but also other chemically distinct substances associated with it in its ores exhibited the phenomenon of radioactivity; the first of these were named polonium and radium. Marie Curie and Schmidt discovered independently in the same year that thorium is also naturally radioactive.

Others soon discovered a host of radioactive substances associated with uranium and thorium, including some that behaved chemically as inert gases, called emanations (now collectively called radon). Many of these, unlike uranium and thorium, have short lifetimes. They are now designated as *secondary* natural radionuclides, and they owe their existence to continual generation by *primary* natural radionuclides, ie, those whose survival through geologic history is due to their slow decay and long lifetimes.

Searches for radioactivity in other elements led to the discovery of β activity in potassium and rubidium in 1906. The natural α activity of platinum was discovered in 1921 and that of samarium in 1932. After the discovery of so-called artificial radioactivity in 1934, weak natural radioactivity was found in a number of other elements; its occurrence is explained by the fact that a small fraction of unstable nuclides have half-lives exceeding a few hundred million (10^8) years.

Following a prediction by Libby that cosmic-ray-induced nuclear reactions in the earth's atmosphere should produce ^{14}C and ^{3}H, he and his collaborators succeeded in detecting their natural occurrence. Carbon-14 and hydrogen-3 were the first of many *induced* natural radionuclides now known and provided the bases for the ^{14}C- and tritium-dating methods (see below). Naturally induced radioactivity is even more prevalent in meteorites and lunar rocks, which are not shielded from the cosmic radiation by atmospheres. They provide dating methods for the radiation-exposure history of cosmic materials, including the surface features of the moon.

A fourth class of natural radioactivity, ie, *extinct*, comprises nuclides whose lifetimes are too short for survival from their pre-solar-system production, yet long enough for them to have left observable effects in nature, especially in meteorites. In 1960 extinct ^{129}I was discovered through excess of its β-decay product ^{129}Xe in a meteorite. This was followed by the discovery in 1965 of extinct ^{244}Pu through heavy-xenon-isotope excess and fission-fragment radiation-damage tracks. The most important extinct natural radionuclide subsequently discovered is ^{26}Al. Extinct natural radionuclides provide the means for dating the earliest history of the solar system.

The ^{238}U, ^{235}U, and ^{232}Th Radioactive Series

Genetic Relationships. By 1903 Rutherford and Soddy had deduced that the radioactive emission of energetic radiations was an accompaniment of spontaneous transformation of an atom from one kind to another, generally of a different element. The α particles were deduced and promptly demonstrated to be helium ions, which explained the high abundance of helium in uranium and thorium minerals. The atomic weight of the transforming atom must then be reduced by about 4 units, the atomic weight of helium. The β particles were found to be high speed electrons, whereas the γ radiation was found to be electromagnetic in nature; in neither case was the atomic weight appreciably affected. Extensive studies of the chemical properties and genetic relationships of uranium and thorium descendents led Soddy in 1910 to recognize that, in several cases, two or more substances of different radioactive properties (radiations, half-lives) had identical chemical properties and hence belonged in the same place in the periodic table. In 1913, he introduced the concept of isotopes, ie, atomic species of the same element differing in atomic weight. Concurrently he formulated the displacement laws: α transformation results in a displacement two places downward in the periodic table, whereas β transformation results in a displacement upward by one place.

In 1911 Rutherford determined the nuclear structure of the atom. It followed that the nucleus is the determinant of chemical behavior and thus of elementary identity through nuclear charge and of atomic weight through nuclear mass; it also is the site of radioactive transformation. The concept of nuclear charge number and its identification with elementary atomic number Z was the result of Moseley's x-ray studies in 1913–1914. Mass-spectrographic studies in the 1920s permitted a distinction between mass number A and atomic weight. The discovery of the neutron in 1932 led to the concept of neutron number N, which is simply A minus Z. In terms of these quantities, the displacement laws assume their modern formulation:

α decay:	$\Delta Z = -2$	$\Delta A = -4$	$\Delta N = -2$
β^- decay:	$\Delta Z = +1$	$\Delta A = 0$	$\Delta N = -1$

The emission of γ radiation was recognized as an adjustment of energy content subsequent to an α or β transformation. In one case, two substances called UX_2 and UZ were found by Hahn to have not only the same atomic number (91) but also the same mass number (234); the former is transformed to the latter by an apparently radiationless process. Such pairs are now called nuclear isomers, and the transformation, which can be effected by γ emission, is called isomeric transition, IT. The displacement laws could thus be extended:

γ emission:	$\Delta Z = 0$	$\Delta A = 0$	$\Delta N = 0$
IT:	$\Delta Z = 0$	$\Delta A = 0$	$\Delta N = 0$

Radioactive disintegration by positron (β^+) emission was discovered along with the phenomenon of artificial radioactivity by Frédéric Joliot and Irène Curie in 1934. Four years later, the occurrence of electron capture, EC, was established. These processes follow identical displacement laws

β^+ decay:	$\Delta Z = -1$	$\Delta A = 0$	$\Delta N = +1$
EC:	$\Delta Z = -1$	$\Delta A = 0$	$\Delta N = +1$

and therefore often compete in a given nuclear transformation.

Since in a radioactive family connected by α and β transitions A can differ only by multiples of 4, there are four such families, designated by $4n+0$, $4n+1$, $4n+2$, and $4n+3$. Three of these comprise the prominent heavy natural radionuclides; each is headed by a long-lived α-emitting uranium or thorium isotope, each contains a number of shorter-lived α- or β-emitting nuclides, and each terminates in a stable Pb isotope. In each series, branching occurs at one or more points where an intermediate nuclide decays by either α or β disintegration, but in all cases the branches converge on the same end-product.

Figures 1, 2, and 3 show the members and their genetic relationships for the $4n+2$ series headed by ^{238}U, the $4n+3$ series headed by ^{235}U, and the $4n+0$ series headed by ^{232}Th, respectively. Each nuclide is designated both by its classical radiochemical symbol, eg, RaA = radium A, as well as its modern nuclear symbol, eg, ^{218}Po. Throughout this article, the former is occasionally given in parenthesis, eg, ^{210}Pb (RaD). Except where otherwise indicated, all nuclear parameters in this article are the preferred values in the 1978 *Table of Isotopes*, edited by C. M. Lederer and V. S. Shirley (see References for all citations).

The ^{238}U and ^{235}U series in natural uranium are linked by the relative proportions of those isotopes, which are nearly constant in most terrestrial sources: on an atomic basis, $^{238}U/^{235}U = 137.88$. This and certain other nuclear constants important in geochronology are taken from the 1976 agreement of the IUGS *Subcommittee on*

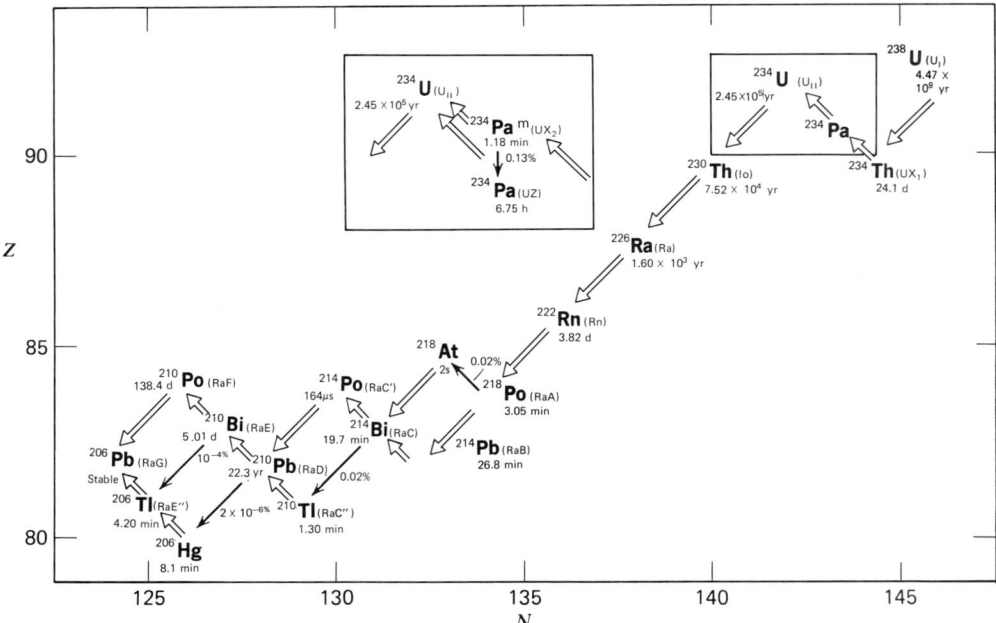

Figure 1. The ^{238}U ($4n + 2$) disintegration series. ∠ = α decay; ↖I= β decay; ↓ = IT, isomeric transition; ⇑ denotes main branch. The ^{230}Th half-life has been changed to the preferred value. Modified with permission from G. Friedlander, J. W. Kennedy, E. S. Maceas, and J. Miller, *Nuclear and Radiochemistry*.

Geochronology: Convention on the Use of Decay Constants in Geo- and Cosmochronology. The specific activities of ^{238}U and ^{235}U in the natural element are 1.2346×10^4 and 568.5 Bq/g [7.408×10^5 and 3.411×10^4 dpm (disintegrations per minute) per gram], respectively. As seen below, the ^{234}U specific activity in the natural element is ordinarily the same as that of ^{238}U. The specific activity of ^{232}Th, which comprises virtually all of the natural element from U-free sources, is 4.07×10^3 Bq/g [2.44×10^5 dpm/g].

Temporal Variations in Activity. In 1902, Rutherford and Soddy formulated the law of radioactive decay for an individual radioactive substance; its differential and integral forms are

$$-dN/dt = \lambda N; \quad N(t) = N(0)e^{-\lambda t} \tag{1}$$

where N = number of atoms; t = time; λ = disintegration constant; and $-dN/dt$ = disintegration rate = R = activity. In such exponential decay, the mean life is given by $\tau = 1/\lambda$ and the half-life by $t_{1/2} = (\ln 2)/\lambda = 0.69315/\lambda$.

For sequential (chain) transformations, the differential equation for the ith member of the series

$$dN_i/dt = \lambda_{i-1}N_{i-1} - \lambda_i N_i \tag{2}$$

when solved for the general case leads to the following integral equations:

$$N_1(t) = N_1(0)e^{-\lambda_1 t} \tag{3}$$

$$N_2(t) = N_1(0)\frac{\lambda_1}{\lambda_2 - \lambda_1}(e^{-\lambda_1 t} - e^{-\lambda_2 t}) + N_2(0)e^{-\lambda_2 t} \tag{4}$$

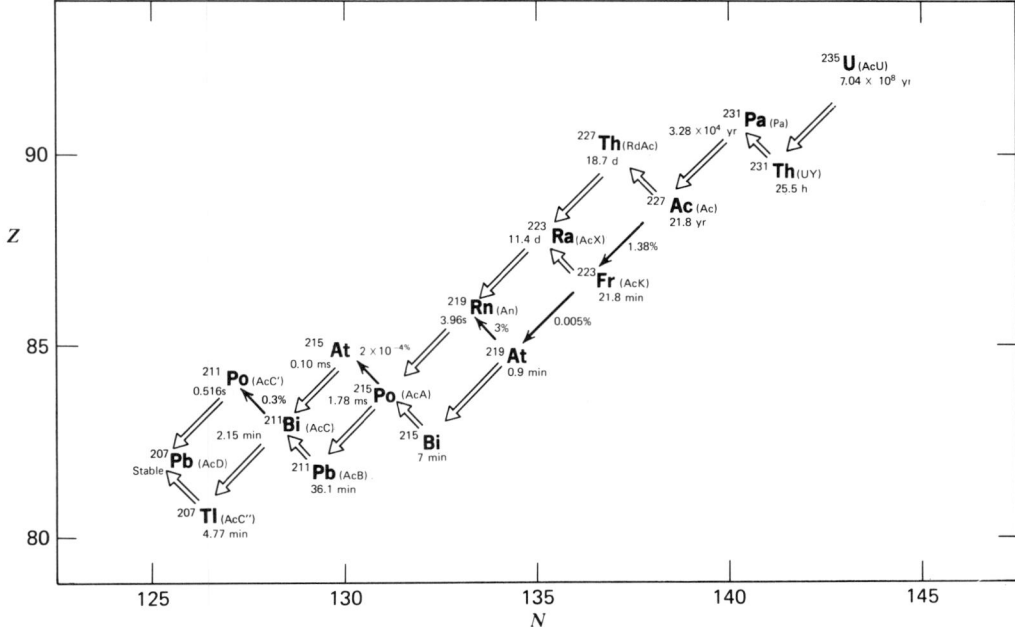

Figure 2. The ^{235}U ($4n + 3$) disintegration series. See Figure 1 legend for details.

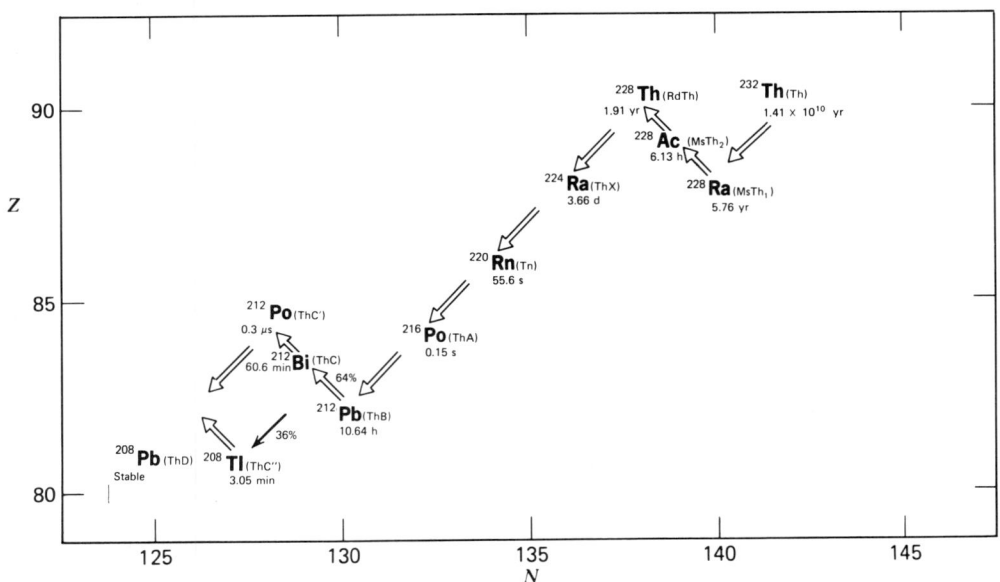

Figure 3. The ^{232}Th ($4n + 0$) disintegration series. See Figure 1 legend for details.

$$N_i(t) = N_1(0)P_{i,1} + N_2(0)P_{i,2} + \ldots + N_i(0)P_{i,i}$$

$$P_{i,j} = Q_{i,j,j}e^{-\lambda_j t} + Q_{i,j,j+1}e^{-\lambda_{j+1} t} + \ldots + Q_{i,j,i}e^{-\lambda_i t}$$

$$Q_{i,j,k} = \frac{(1)\lambda_j \lambda_{j+1} \ldots \lambda_{i-1}}{(1)(\lambda_2 - \lambda_k)(\lambda_{j+1} - \lambda_k) \ldots (\lambda_i - \lambda_k)} \qquad (5)$$

omitting $(\lambda_k - \lambda_k)$ from denominator

Where branching occurs, the appropriate branching fraction must be included as a factor for each branch member; for members produced by more than one branch, the numbers of atoms must be summed over the branches. For the stable series product, $\lambda_i = 0$.

If the series is left alone for a time such that $t_{1/2,1} \gg t \gg$ all $t_{1/2,>1}$, a steady state is reached in which all activities are equal (except for branching factors) to that of the parent, which may be regarded as constant:

$$R_1 = R_2 = R_3 = \ldots \tag{6}$$

Furthermore, since $R_i = \lambda_i N_i = N_i (\ln 2)/t_{1/2,i}$

$$N_1/t_{1/2,1} = N_2/t_{1/2,2} = N_3/t_{1/2,3} = \ldots \tag{7}$$

This state is referred to as secular equilibrium, and it occurs in old undisturbed uranium and thorium minerals, etc. For a two-membered chain in which the second member is somewhat shorter-lived than the first, the ratio of activities becomes constant after a time considerably longer than the shorter half-life:

$$R_2/R_1 = \lambda_2/(\lambda_2 - \lambda_1); \quad N_2/N_1 = \lambda_1/(\lambda_2 - \lambda_1) \tag{8}$$

This is called transient equilibrium.

More complicated cases can be solved by formulating and solving the appropriate differential equations; terms for production and destruction in nuclear reactions may be added. Simplifying approximations (neglect of small terms, etc) are often justifiable and useful.

Radioactive disintegrations occur randomly, and the disintegration constant λ is actually a statistical probability that an atom disintegrates per unit time; this theoretically justifies the Rutherford-Soddy decay law. It follows that, when the number of atoms is essentially constant, the standard deviation of the number of disintegrations in a given time is the square root of that number. The same is also true for the number of disintegrations actually counted (as with <100% counting yield), as it is for any count of independent random events of constant mean rate.

Radiations. For details of the disintegration schemes and radiations of the heavy natural radionuclides, the 1978 *Table of Isotopes* should be consulted. Nuclear energies are usually expressed in terms of the multiples of electron volt, keV and MeV.

The α-particle energies for ground-state-to-ground-state transitions encountered among the heavy natural radionuclides range from 4.01 MeV (^{232}Th) to 8.78 MeV (^{212}Po). In general they increase down the disintegration chains because of the special nuclear stability of $Z = 82$ (Pb) and $N = 126$ configurations. The lifetimes are inversely correlated with disintegration energies and directly with Z; both factors contribute to a pronounced general decrease in α half-lives down the chains. Because of the generally close spacing and large number of energy levels in heavy nuclei, many α emitters have complex spectra consisting of from 2 to many groups of particles of discrete energies.

In addition to α transitions originating from nuclear ground states, long-range α particles are emitted from excited states of ^{212}Po (intensity = 2.1×10^{-4} of the ground-state α particles, maximum energy = 10.54 MeV) and ^{214}Po (3.1×10^{-5}, 10.50 MeV) in competition with the normal γ de-excitation. These long-range α particles are temporally associated with the β decay of the respective precursor nuclides ^{212}Bi and ^{214}Bi.

The β spectra of most heavy β emitters are likewise complex. In addition to the overlapping continuous components of the β^- particles of nuclear origin, the negative-electron spectra include conversion electrons of discrete energies from low energy γ transitions, Auger electrons from atomic x-ray transitions, and pair-creation negatrons from high energy γ transitions; also present may be secondary electrons of instrumental origin. Additional complexity results from short lifetimes which, in many cases, requires that the spectra of several genetically related nuclides be observed together.

As with α emitters, the disintegration energies of β emitters generally increase down the disintegration chains, but less regularly. In spite of the smaller energy dependence and the opposite Z dependence, the shortest β lifetimes are found near the ends of the chains. Observed β^- end points range from 0.017 MeV (^{210}Pb) to 2.38 MeV (^{208}Tl).

The γ spectra are likewise often complex. In addition to true γ rays of nuclear origin, there are x rays of atomic origin and weak continuous inner-bremsstrahlung components. Secondary electromagnetic radiations include bremsstrahlung, x rays, and β^+–β^- annihilation photons of 0.511 MeV. In solid-state spectrometers, such as NaI(Tl) scintillators and Ge(Li) and intrinsic-Ge semiconductor detectors, there may be pseudolines caused by pile-up of two or more coincident γ rays, whereas with intense sources accidental-coincidence pile-up lines may occur; moreover, line shapes are modified by escape phenomena.

The highest-energy abundant γ radiation is that of ^{208}Tl in the ^{232}Th series with an energy of 2.6145 MeV. In ^{214}Bi in the ^{238}U series, γ energies up to at least 3.184 MeV occur in very low abundance, but the highest energy of any prominence in that series is 2.4478 MeV, also from ^{214}Bi.

The γ spectra of uranium minerals are dominated by the ^{238}U series; only the 0.1875-MeV ^{235}U γ photon of the ^{235}U series is quantitatively important, and minor amounts of ^{232}Th-series γ radiations are sometimes present. Likewise, thorium minerals show only the ^{232}Th-series spectrum, unless contaminated by significant amounts of uranium. In terrestrial igneous rocks, uranium, thorium, and potassium show roughly parallel variations with each other and with acidity (silica content), and thus inversely with density and depth of occurrence in the earth. The Th:U mass ratio is ca 3 or 4 in most crustal rocks, whereas the ^{232}Th:^{238}U specific-activity ratio is 0.33. Hence the ^{232}Th and ^{238}U series are usually represented with comparable intensities in the γ spectra of most rocks, soils, siliceous building materials, etc. In addition, the 1.4608-MeV γ ray of ^{40}K usually shows prominently in the spectra as well.

Portable γ-radiation detectors for radiometric geochemical prospecting for uranium, thorium, and potassium, and for borehole logging and aerial surveys of the near-surface distribution of these elements, take advantage of the fact that the strong 2.6145-MeV γ ray of ^{208}Tl in the ^{232}Th-series stands alone at the highest energy, and thus indicates thorium abundance. The 1.7645-MeV ^{214}Bi photon in the ^{238}U-series is similarly used to indicate uranium abundance, with minor interference from thorium. The 1.4608-MeV ^{40}K radiation indicates potassium abundance, with moderate interference from uranium and thorium. With the commonly used NaI(Tl) scintillation detectors, energy windows of 1.37–1.57 MeV, 1.66–1.86 MeV, and 2.41–2.81 MeV are used to accentuate the responses caused by potassium, uranium, and thorium, respectively. From the simultaneous counts in all three windows, the abundance of each element can be calculated. The coefficients needed are different from those derived from laboratory spectra because of absorption and scattering phenomena in natural environments. Concrete slabs of variable contents of the three elements are used for empirical calibration.

The ^{238}U-series, ^{232}Th-series, and ^{40}K γ radiations contribute comparably to the natural radiation environment under most ordinary circumstances. However, uranium and thorium are capable of enrichment to many times the typical norms and can then constitute radiation hazards. Ore-process residues in which ^{226}Ra and ^{228}Ra are concentrated are particularly hazardous. Furthermore, gaseous ^{222}Rn can emanate from uranium-rich and radium-rich materials and, with its short-lived decay products, can present special hazards. There is evidence that release of radon into wells and the near-surface atmosphere is correlated with seismic and preseismic activity, and radon monitoring is being tested for its potential in earthquake warning.

Other Primary Natural Radionuclides

All primary natural radionuclides whose radioactivity or transformation has been detected are listed in Table 1, along with decay-product particulars that are relevant to geochronology (see Radiometric Dating). Not included in Table 1 are several observationally stable nuclides known to be energetically β-labile, eg, ^{48}Ca, ^{50}V, ^{96}Zr, ^{123}Te; many even-Z/even-A nuclides that are stable against ordinary β^- or EC decay but energetically unstable to double-β^- or double-EC decay, where this has not been detected; and many β-stable heavy nuclides that are energetically α-labile but of extremely long lifetime. As techniques improve, ultimately some of these may be demonstrated to be radioactive.

Induced Natural Radioactivity

There are two main sources of nuclear reactions that occur in nature in materials accessible for observation: α particles and cosmic rays. The α particles emitted by heavy natural radionuclides, particularly by the short-lived end members, are sufficiently energetic to induce nuclear reactions in light elements; oxygen is the most abundant target. Although some induced radioactivity results directly, the secondary neutrons emitted are responsible for more, since they can interact with heavy nuclei. In addition, the spontaneous fission of ^{238}U yields neutrons directly. Uranium and thorium are the most abundant natural targets in n-rich regions, and the induced fission of ^{235}U enhances the neutron flux. Cosmic rays are protons and nuclei of helium and heavier elements accelerated in space or remote astrophysical sites to very high energies. These rays are capable of producing nuclear reactions in planetary atmospheres and exposed solid matter. Somewhat less energetic particles are ejected by the sun, particularly from flares. The direct interactions produce spallation-type reactions and liberate neutrons that can produce additional reactions, mainly by capture after thermalization. Accordingly, a considerable variety of induced natural radionuclides results (see Table 2).

Neutron capture by ^{238}U produces ^{239}Pu by the same pathway involving ^{239}U and ^{239}Np as in nuclear reactors. Fast neutrons interact with ^{238}U through the $(n,2n)$ reaction to produce ^{237}U, which transforms to ^{237}Np. Neutron capture in ^{232}Th produces ^{233}Th, which decays through ^{233}Pa to ^{233}U, the same overall process used in advanced converter or breeder reactors operating on the Th–U fuel cycle (see Nuclear reactors). Uranium-233 is also generated in uranium minerals by the decay of ^{233}Pa in equilib-

Table 1. Known Primary Natural Radionuclides

Nuclide	CAS Registry No.	Disintegration modes (%)	Half-life, yr	Isotopic abundance, %	Element specific activity, Bq/g[a]	Decay products (isotopic abundance, %)
^{235}U	[15117-96-1]	$7\alpha + 4\beta^-$	7.038×10^{8} [b]	0.720	5.685×10^{2}	^{207}Pb[c] (22.1)
		SF[d]	3.5×10^{17}		1.14×10^{-6}	FP[e,c]
^{40}K	[13966-00-2]	β^- (89.5)[b]	1.250×10^{9} [b]	0.01167[b]	28.3	^{40}Ca[c] (96.94)
		EC (10.5)[b]			3.32	^{40}Ar[c] (99.60)
		β^+ (0.0010)			3.2×10^{-4}	^{40}Ar[c] (99.60)
^{238}U	[24678-82-8]	$8\alpha + 6\beta^-$	4.468×10^{9} [b]	99.275	1.2346×10^{4}	^{206}Pb[c] (24.1)
		SF[d]	8.2×10^{15}		6.7×10^{-3}	FP[e,c]
^{232}Th	[7440-29-1]	$6\alpha + 4\beta^-$	1.401×10^{10} [b]	100[f]	4.07×10^{3}	^{208}Pb[c] (52.4)
		SF[d,g]	$>1 \times 10^{21}$		$<6 \times 10^{-8}$	FP[e]
^{176}Lu	[14452-47-2]	β^-	3.6×10^{10}	2.61	5.5	^{176}Hf[c] (5.2)
^{187}Re	[14391-29-8]	β^-	4×10^{10}	62.60	1.1×10^{3}	^{187}Os[c] (1.6)
^{87}Rb	[13982-13-3]	β^-	4.88×10^{10} [b]	27.83	8.8×10^{2}	^{87}Sr[c] (7.0)
^{147}Sm	[14392-33-7]	α	1.06×10^{11}	15.1	1.25×10^{2}	^{143}Nd[c] (12.2)
^{138}La	[15816-87-2]	EC (68)	1.1×10^{11}	0.089	0.52	^{138}Ba (71.7)
		β^- (32)			0.25	^{138}Ce (0.25)
^{190}Pt	[15735-68-9]	α	6×10^{11}	0.013	1.5×10^{-2}	^{186}Os (1.58)
^{152}Gd	[14867-54-0]	α	1.1×10^{14}	0.21	1.6×10^{-3}	^{148}Sm (11.3)
^{115}In	[14191-71-0]	β^-	5.1×10^{14}	95.7	0.22	^{115}Sn (0.38)
^{174}Hf	[14922-49-7]	α	2.0×10^{15}	0.16	6×10^{-5}	^{170}Yb (3.2)
^{144}Nd	[14834-76-5]	α	2.1×10^{15}	23.8	1.0×10^{-2}	^{140}Ce (88.5)
^{148}Sm	[14913-64-5]	α	8×10^{15}	11.3	1.2×10^{-3}	^{144}Nd (23.8)
^{113}Cd	[25284-77-9]	β^-	9×10^{15}	12.2	1.6×10^{-3}	^{113}In (4.3)
^{82}Se	[52788-46-2]	double-β^-	1.4×10^{20}	9.2	1.1×10^{-7}	^{82}Kr[c] (11.6)
^{130}Te	[14390-76-2]	double-β^-	2×10^{21}	34.5	2×10^{-8}	^{128}Xe[c] (4.1)
^{128}Te	[14390-75-1]	double-β^-	1.5×10^{24}	31.7	2.2×10^{-11}	^{130}Xe[c] (1.91)
all α emitters						^{4}He[c] (99.99986)

[a] Bq = disintegration per second.
[b] Value recommended for geo- and cosmochemistry by International Union of Geological Sciences, Subcommittee on Geochronology, 1977.
[c] Decay product has been observed with enhanced isotopic abundance.
[d] SF = spontaneous fission. Half-life is that which would result if SF were only disintegration mode.
[e] FP = fission products, radioactive and stable.
[f] In Th from natural sources containing U, some ^{230}U will also be present.
[g] Not observed.

Table 2. Observed Induced Natural Radionuclides

Nuclide	Half-life, d[a]	Observation[b]	Nuclide	Half-life, yr	Observation[b]
^{34m}Cl[c]	32.0 min	T	^{22}Na	2.60	T, M, L
^{38}Cl	37.3 min	T	^{55}Fe	2.7	M, L
^{39}Cl	56 min	T	^{60}Co	5.27	M, L
^{31}Si	2.62 h	T	^{3}H	12.33	T, M, L
^{38}S	2.8 h	T	^{44}Ti	47	M, L
^{24}Na	15.0 h	T	^{32}Si	108	T, M
^{28}Mg	21.0 h	T	^{39}Ar	269	T, M, L
^{198}Au	2.70	T	^{14}C	5370	T, M
^{199}Au	3.14	T	^{94}Nb	2.0×10^4	T
^{52}Mn	5.59	M	^{239}Pu[d]	2.41×10^4	T
^{32}P	14.3	T, M	^{59}Ni	7.5×10^4	T, M, L
^{48}V	16.0	M, L	^{233}U[d]	1.59×10^5	T
^{33}P	25.3	T, M	^{60}Fe	$\sim 2 \times 10^5$	M
^{51}Cr	27.7	M, L	^{81}Kr	2.1×10^5	M
^{37}Ar	35.0	T, M, L	^{36}Cl	3.0×10^5	T, M, L
^{59}Fe	44.6	M	^{26}Al	7.2×10^5	T, M, L
^{7}Be	53.3	T, M, L	^{10}Be	1.6×10^6	T, M, L
^{58}Co	70.8	M	^{237}Np[c]	2.14×10^6	T
^{56}Co	78.8	M, L	^{53}Mn	3.7×10^6	T, M, L
^{46}Sc	83.8	M, L	^{40}K	1.25×10^9	M
^{35}S	87.4	T			
^{45}Ca	165	M			
^{57}Co	271	M, L			
^{54}Mn	312	M, L			
^{49}V	330	M, L			

[a] Unless otherwise noted.
[b] T = terrestrial materials; M = meteorites; L = lunar materials.
[c] m = metastable.
[d] Also radioactive precursors and descendants.

rium with the decay of its parent, ^{237}Np, and together these nuclides head the (4 n + 1)-disintegration series ending with ^{209}Bi. Figure 4 gives the members, decay modes, and half-lives of this disintegration series. Although only the long-lived parents are listed in Table 2, their presence implies also that of their antecedents and descendants as described above.

Extinct Natural Radioactivity

Known and putative extinct natural radionuclides are listed in Table 3. Information relevant to their possible use as chronometers or indicators of cosmochemical processes is included (see also Space chemistry).

Radiochemistry of the Elements Thallium to Uranium

Several heavy natural radioelements, eg, Po, At, Rn, Fr, Ra, Ac, and Pa, have no stable or very long-lived isotopes, and hence are usually encountered in trace quantities. Furthermore, radioisotopes of most other radioelements are often generated by radioactive parents in the absence of their ordinary forms, and they are also unweighable. Thus, radiochemistry of these elements involves carrier-free techniques, in which

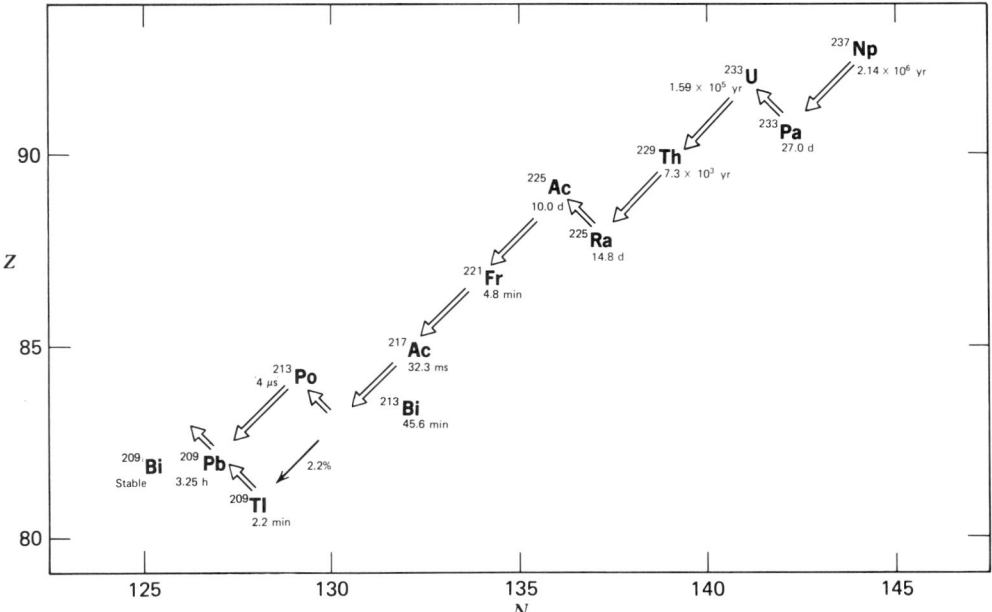

Figure 4. The ^{237}Np–^{233}U $(4n+1)$ series. See Figure 1 legend for details.

chemical separations are made by physical separations of macroscopic phases involving uneven distribution of the low concentration species between the phases. Phase pairs may be liquid–solid, liquid–gas, solid–gas, or liquid–liquid.

Where a stable or long-lived form of the radioelement exists, an isotopic carrier may be used. The fact that the radioisotope cannot be separated from the carrier by ordinary chemical means is often of no consequence.

Heavy natural radioelements are almost always detected by measuring the radioactivity. Time dependence, ie, analysis of growth–decay curves, differential absorption of α, β, and γ radiations, and energy spectroscopy are employed. For uranium and thorium, both radiometric and ordinary chemical analytical methods are available, and for thallium, lead, and bismuth, parallel chemical analyses of the elements in bulk are often useful. When highly concentrated in pure form, ^{210}Pb, ^{210}Po, ^{222}Rn, ^{226}Ra, ^{227}Ac, ^{230}Th, and ^{231}Pa are handled by ordinary chemical techniques, albiet usually on a microchemical scale.

The phenomenon of isotopy provides the basis for the use of radioactive indicators, now recognized as a special case of isotopic indicators or tracers.

Thallium. For the general chemistry of this element, see Thallium and thallium compounds.

All isotopes of thallium in the natural radioactive series are short-lived ($t_{1/2} = 1.3–4.8$ min) and, therefore, only rapid radiochemical techniques can be used for their preparation. The most readily prepared and useful is ^{208}Tl (Th″C; $t_{1/2} = 3.1$ min); it can be isolated from ^{212}Bi, ^{212}Pb, aged ^{228}Ra (mesothorium) and even thorium salts.

Pure ^{208}Tl can be obtained by α recoil from ^{212}Bi, the daughter of ^{212}Pb, prepared as an active deposit from ^{220}Rn. The recoil ions are best collected on a negatively charged plate in air. ^{210}Tl and ^{207}Tl can be similarly prepared from ^{222}Rn and ^{219}Rn, respectively.

Table 3. Known and Putative Extinct Natural Radionuclides

Nuclide	CAS Registry No.	Half-life, yr	Disintegration modes (%)	Decay products (isotopic abundance, %)
Established				
^{129}I	[15046-84-1]	16.0×10^6	β^-	^{129}Xe (26.4)
^{244}Pu	[14119-34-7]	81×10^6	$3\alpha + 2\beta^-$ (99.9)	^{232}Th (100)
			SF (0.123)	^{131}Xe (21.2), ^{132}Xe (26.9), ^{134}Xe (10.4), ^{136}Xe (8.9), ^{83}Kr (11.5), ^{84}Kr (57.0), ^{86}Kr (17.3), others
^{26}Al	[14682-66-7]	0.72×10^6	β^+, EC	^{26}Mg (11.01)
Claimed but not verified				
^{146}Sm	[14280-31-0]	103×10^6	α	^{142}Nd (27.2)
^{107}Pd	[17637-99-9]	ca 6.5×10^6	β^-	^{107}Ag (51.83)
"G" (SHE)[a]	[14119-34-7]	$>20 \times 10^6$ (?)	SF	
^{22}Na	[13966-32-0]	2.60	β^+, EC	^{22}Ne (9.22)
Suggested to account for certain observations				
^{41}Ca	[14092-95-6]	0.10×10^6	EC	^{41}K (6.73)
^{202}Pb	[15752-86-0]	ca 0.3×10^6	2 EC	^{202}Hg (29.8)
^{135}Cs	[15726-30-4]	3×10^6	β^-	^{135}Ba (6.59)
Suggested additional possibilities				
^{247}Cm	[15758-32-4]	15.6×10^6	$2\alpha + 3\beta^-$	^{235}U (0.720)
^{205}Pb	[14119-28-9]	15.1×10^6	EC	^{205}Tl (70.5)
^{60}Fe	[32020-21-6]	ca 0.2×10^6 (?)	$2\beta^-$	^{60}Ni (26.1)

[a] Superheavy element; see ^{244}Pu above for decay products.

Thallium(III) is strongly adsorbed on anion-exchange resins from concentrated HCl and can be eluted by dilute HCl saturated with SO_2 to reduce it to thallium(I). Thallium(III) can be extracted into ethers from strong HCl solutions and back-extracted into dilute HCl after reduction. These procedures are useful mainly for long-lived artificial thallium radioisotopes, of which ^{204}Tl ($t_{1/2}$ = 3.77 yr) is most important. The naturally occurring Tl radioisotopes are all energetic β^- emitters; β counting is the most suitable method of measurement.

Lead. For the general chemistry of this element, see Lead; Lead compounds.

Of the naturally occurring lead isotopes, ^{212}Pb (ThB) is the most useful as a tracer because of its convenient $t_{1/2}$ (10.6 h) and easily measured β radiations, especially when in equilibrium with its daughter ^{212}Bi (ThC; $t_{1/2}$ = 60.6 min). It is prepared from thorium or mesothorium MsTh$_1$ (^{228}Ra) sources, which occur conveniently in the active deposit from thoron (^{220}Rn). The long-lived ^{210}Pb (RaD; $t_{1/2}$ = 22.3 yr) is important as a tracer of longer duration and as a parent source of ^{210}Bi (RaE) and ^{210}Po (RaF) (see also Radioisotopes). Moreover, its natural occurrence has many interesting and important facets. In uranium minerals and ores, it is inevitably diluted by much ordinary and radiogenic stable lead, but it can be prepared isotopically pure from radium or from spent radon tubes.

In aqueous solution, Pb is usually encountered as Pb(II). As such, it is moderately adsorbed on anion-exchange resins from 1–2 M HCl and can be eluted slowly with the same medium or rapidly with very strong or very dilute HCl or water. If a solution of radium, polonium, bismuth, and lead in 2 M HCl is passed through an anion-exchange column, radium passes through promptly, and lead can be eluted with several column volumes of the acid. Bismuth can be eluted with concentrated HCl and polonium with concentrated HNO_3. Lead(IV) is strongly adsorbed from 12 M HCl and can be eluted by 12 M HCl saturated with SO_2.

A number of complex compounds of lead can be extracted from aqueous media into organic solvents. Lead and other metallic elements can be extracted selectively with dithizone by controlling the pH and adding complexing agents such as citrate and cyanide.

Separation of lead from many elements, final purification, and preparation of samples for radioassay can be achieved by anodic electrodeposition of PbO_2 from 5–9% HNO_3 with a potential of ca 2.5 V.

Lead-212 can be measured most easily by β counting, allowing several hours for ^{212}Bi to grow into secular equilibrium. The ^{210}Pb β particles are too weak (maximum energy 61 keV) for easy detection, but ^{210}Bi, a strong β emitter, grows in with a $t_{1/2}$ of 5.01 d. Over a period of several days, β counting yields a growth or decay curve that can be analyzed for the initial amounts of parent and daughter. The α emitter ^{210}Po ($t_{1/2}$ = 130.4 d) grows in more slowly, and analysis of the α growth curve permits calculation of the ^{210}Pb activity. Alternatively, after a suitable growth period, ^{210}Po may be separated from the purified ^{210}Pb solution and counted separately. The α counting takes longer than β counting, but sensitivity is higher; this may be essential for weak environmental samples. Mixed ^{210}Pb, ^{210}Bi, and ^{210}Po can be determined separately by liquid-scintillation counting simultaneously in three energy windows.

Bismuth. For the general chemistry of this element, see Bismuth and bismuth alloys; Bismuth compounds.

The most useful radioisotope of bismuth is ^{210}Bi, which can be separated easily from its longer-lived parent ^{210}Pb.

Bismuth(III) is strongly adsorbed on anion-exchange resins from HCl solutions. The distribution coefficient decreases with increasing acid concentration, and bismuth can be eluted from a column with several column volumes of 12 M HCl. It can be eluted more rapidly with 1 M H_2SO_4 or 0.25 M HNO_3.

Bismuth is strongly extracted from aqueous solution by dithizone in CCl_4 between pH 2 and 10. Since lead is not extracted appreciably below pH 3, extraction at pH 2 gives a good separation of bismuth from lead. 2-Thenoyltrifluoroacetone (TTA) in benzene extracts many heavy natural radioelements at varying pH. At pH 1, polonium, thorium, and protactinium are extracted. At pH 2, bismuth is extracted; thallium, lead, actinium, and radium remain in the aqueous phase.

Bismuth-210 is readily assayed by simple β counting. An absorber or counter window of suitable thickness absorbs the ^{210}Po α particles and ^{210}Pb β particles.

Polonium. Although Marie Curie and Debière were able to prepare and study polonium and some of its compounds in weighable amounts, most studies of the natural element were made by tracer techniques, for which ^{210}Po ($t_{1/2}$ = 138.4 d) is eminently suitable. With the advent of nuclear reactors, large quantities of ^{210}Po have been prepared by neutron irradiation of ^{209}Bi through ^{210}Bi ($t_{1/2}$ = 5 d) and have been used for studies of the pure element and its compounds. The α radiations of massive quantities of polonium create problems through ionization of the surrounding air, decomposition of containing solvents, and heating. Because of the considerable health hazard, special handling procedures are necessary.

In the periodic table, polonium is a member of Group VIA, and thus shares valence states and many chemical properties with tellurium. It also resembles its horizontal congeners, lead, bismuth, and astatine, since it is a predominantly metallic element forming many compounds of high volatility. The metal is easily deformed, melts at 252°C, and has a vapor pressure of 101.3 kPa (1 atm) at 962°C. It is prepared by cathodic electrodeposition or by thermal decomposition of PoS and purified by sublimation in an inert atmosphere, hydrogen, or vacuum. It forms compounds in which it has oxidation states of -2, $+2$, $+4$, and $+6$, with a suggestion of $+3$. In aqueous solution the $+4$ state is most stable, and many of the Po(IV) compounds are soluble. Insoluble compounds include $Po(OH)_2$, PoS, and PoI_4. Polonium(IV) forms many anionic complexes with a coordination number of six, such as $PoCl_6^{2-}$.

Trace levels of polonium can be precipitated from aqueous solution with the use of insoluble hydroxides or sulfides as carriers. Elementary tellurium precipitated from Po-containing solutions by $SnCl_2$ reduction in $<6\ M$ HCl carries polonium and separates it from bismuth, whereas tellurium precipitated from $>2\ M$ HCl by SO_2 reduction does not carry polonium.

Aged ^{210}Pb (RaD) contains ^{210}Bi (RaE) and ^{210}Po (RaF). In the preparation of the latter from RaDEF, advantage may be taken of the differing electrochemical behavior of the three elements. If a weakly acidic solution is exposed to metallic Ag, the polonium deposits, and the other two remain in solution. Nickel can be used to deposit the bismuth. If H_2 is bubbled over a platinum electrode, polonium and bismuth deposit together. Polonium can also be deposited on metallic bismuth; any radiobismuth and radiolead are left in solution. The separations can be made more cleanly with the use of an inert electrode such as platinum and control of its cathodic potential.

Polonium can be separated from lead and bismuth by extraction with isopropyl ether or 20% tributyl phosphate in dibutyl ether.

In the large-scale production of ^{210}Po from neutron-irradiated bismuth, advantage is taken of the much greater volatility of polonium at ca 900°C, which can also be used to separate ^{210}Po from its decay product, stable ^{206}Pb.

Polonium isotopes having all mass numbers from 193 to 218, plus metastable isomers of some, have been produced artificially. The α-emitting ^{208}Po ($t_{1/2}$ = 2.9 yr) is a yield-determining tracer for the quantitative assay of ^{210}Po in natural materials. Both are measured by α counting or spectrometry.

Astatine. Only rare and short-lived isotopes of astatine occur in nature. Astatine-218 ($t_{1/2}$ = ca 2 s), ^{219}At (0.9 min), and ^{215}At (0.10 ms) occur in minor branches of the ^{238}U, ^{235}U, and ^{235}U decay chains, respectively. Astatine-217 (32 ms) occurs in the main decay chain of ^{237}Np and ^{233}U, which are rare induced natural radionuclides found in uranium and thorium minerals, but is hardly detectable. The natural occurrences were discovered only after astatine had been discovered and chemically characterized in cyclotron bombardments of bismuth with α particles. The circumstances obviously preclude application of ordinary radiochemical techniques to naturally occurring astatine, except for ^{219}At derived from very strong ^{227}Ac preparations.

Many artificially produced astatine isotopes are known. The longest-lived are ^{210}At ($t_{1/2}$ = 8.3 h) and ^{211}At ($t_{1/2}$ = 7.21 h). They are generally produced together and are the most useful tracers for the element. Its chemistry is known only from tracer studies.

Astatine, which occurs below iodine in the halogen family of the periodic table, has some resemblances to that element. Oxidation states of $-1, 0, +1, +3,$ and $+5$ are known. Astatine, like its horizontal congeners Po, Bi, etc, also has some metallic characteristics. It is highly volatile and can be separated from irradiated bismuth by heating the latter to its melting point at 271°C. It is purified by further sublimation. Astatine-219 can be separated from ^{223}Fr as well as radium, bismuth, and lead by evaporating a solution on a platinum filament, heating the latter, and condensing the astatine on a cold Pt plate. Astatine can be distilled out of aqueous solutions, but less readily than iodine.

In the -1 state, astatine is carried by precipitating AgI. Elementary tellurium, precipitated from 3 M HCl by SO_2, carries astatine and gives excellent separation from polonium. At zero oxidation state, astatine is soluble in CCl_4, benzene, diisopropyl ether, etc, but extractions are unreliable, presumably because of variability of oxidation state and reactions with impurities, especially halides.

Astatine is distributed in living organisms in a manner similar to iodine, including a tendency to concentrate in thyroid tissue. Biological materials are prepared for At assay by wet-ashing with nitric and perchloric acid, followed by precipitating with tellurium as described above.

Astatine-219 is measured by its α particles. Astatine-210 and ^{211}At decay predominately by electron capture to Po isobars and can be measured conveniently by counting the ca 80–90-keV x rays of the daughter element. Product activities must be considered in all cases.

Radon. The term radon refers both to the element of atomic number 86 and the nuclide ^{222}Rn. A member of the ^{238}U decay series, ^{222}Rn owes its importance to its convenient $t_{1/2}$ (3.823 d), the powerful γ radiation associated with its short-lived descendants, and the long lifetime of its immediate antecedent, ^{226}Ra. Thoron (^{220}Rn) and actinon (^{219}Rn) in the ^{232}Th and ^{235}U series, respectively, are of considerably less practical importance because of their short lifetimes. In spite of the relatively short $t_{1/2}$ of ^{222}Rn, it has been possible to make physical, chemical, and spectrographic studies of the pure element. However, the results have been supplemented by a great deal of tracer work.

Radon is an inert or rare gas, below xenon in the periodic table, and resembles it most in physical and chemical behavior. It is less inert than xenon, melts at a higher temperature (-71 or $-113°C$), boils at a higher temperature ($-65°C$), adsorbs on surfaces more readily, and dissolves more readily in water and organic solvents. It shares with xenon the ability to form compounds with the highly electronegative F as well as clathrate compounds with some inorganic and organic crystals (see also Helium-group gases).

Radon is separated from gas streams by removal of reactive gases by chemical reaction or cold-trapping. It is then frozen in a liquid-N_2 trap or adsorbed on silica gel or activated charcoal. Adsorption on charcoal is partial at RT and virtually complete at dry-ice and lower temperatures. Xenon and other rare gases adsorbed at low temperatures can be pumped off at 0°C, whereas the radon is pumped off at 350–500°C.

Radon is generated in aqueous solutions by dissolved radium and may be removed by boiling or flushing with helium or any other gas; this procedure is called de-emanation.

Diffusive loss of radon from solids, called emanation, varies greatly. Finely divided, cryptocrystalline, amorphous solids may have high emanating power, particularly for the longer-lived ^{222}Rn. The same is true of heavy-metallic-element fatty acid salts, eg, barium stearate, which may contain some radium. Extensive studies of the structures of solids, aging of precipitates, and solid-state reactions have been made by incorporating tracer amounts of radium or radiothorium and determining the emanating power for the generated radon isotope. Radium adsorbed as sulfate or carbonate on Fe(OH)$_3$ or Al(OH)$_3$ gels which are dried by special procedures has >95% emanating power and is used in practical radon generators.

Small amounts of ^{222}Rn are measured by introducing it with a carrier gas into an ionization chamber and counting the α particles of radon and its active deposit. Large quantities are measured through the high energy γ radiation from the decay products. Short-lived Rn isotopes are measured with a flow-type ionization chamber and a carrier-gas stream. A great variety of detectors and techniques have been developed.

Francium. The longest-lived isotope of francium is ^{223}Fr ($t_{1/2}$ = 22 min), formed by minor branching of ^{227}Ac. No other isotope occurs in the natural radioactive series except for ^{221}Fr ($t_{1/2}$ = 4.8-min), which should be present in the rare induced ^{237}Np–^{233}U series.

The chemistry of francium, an alkali element, is rather simple and resembles that of cesium, which is the best carrier for francium. Thus, it coprecipitates with cesium as the perchlorate, chloroplatinate, cobaltinitrite, and silicotungstate. It is also precipitated without carrier by silicotungstic acid and other heteropolyacids in cold concentrated HCl. It remains in solution with most precipitations that carry down all other heavy natural radioelements.

Francium is adsorbed on cation-exchange resins and is thus separated from carrier silicotungstic acid. The francium is eluted with concentrated HCl.

In the preparation of ^{223}Fr from a parent ^{227}Ac solution, the highly active main-branch decay products of the latter, particularly ^{227}Th and ^{223}Ra, must first be removed by selective precipitation, anion exchange, or solvent extraction. Equilibrium between ^{223}Fr and its parent is achieved in a few hours. Francium-223 can then be separated from the latter by the silicotungstic-acid method already mentioned. Alternatively, actinium, thorium, and radium can be precipitated as carbonates with the use of lanthanum and barium as carriers; francium remains in solution.

Francium-223 is easily measured by β counting.

Radium. The name radium was initially given to the nuclide ^{226}Ra and is still used in that sense, even though it has been assigned to the entire element 88. Because of its relatively long $t_{1/2}$ (1600 yr), it occurs in substantial quantities in uranium ores (Ra:U = 3.4 × 10^{-7} by mass). Tracer methods are used to separate ^{224}Ra (ThX; $t_{1/2}$ = 3.66 d) in the thorium series; it can be isolated from an ordinary thorium salt, but best from a source of ^{228}Th (radiothorium, RdTh).

Radium is a homologue of barium, which it resembles closely as an alkaline-earth element. It melts at ca 700°C and boils at ca 1140°C. It occurs only as Ra(II) in aqueous solution, usually as Ra^{2+}, although weakly bound complexes with EDTA and other chelating ligands are formed. Insoluble compounds are the same as those of barium. The hydroxide is relatively soluble, but is appreciably carried by Fe(OH)$_3$ at high pH, especially in the presence of carbonate. Barium is the best carrier for analytical purposes. Lead is an excellent nonisotopic carrier and can be separated readily.

The separation of radium from barium was solved by the Curies with fractional crystallization. The system RaBr$_2$–BaBr$_2$–HBr—H$_2$O gives the most efficient separation and yielded most of the radium ever produced.

Today, however, cation-exchange-resin columns are used. After adsorption from weak HCl solution, elution with ammonium citrate at pH 7.8 or EDTA at pH 6.25 removes barium first and lighter alkaline earths as well as lanthanum and some other elements, then the Ra decay products (Pb, Bi, Po), and finally radium. When a strong HCl solution is passed through an anion-exchange resin, many nonradioactive impurities and uranium, bismuth, and polonium remain on the column, whereas radium as well as barium and many other impurities pass through.

In order to dissolve radium for analysis, $HF-H_2SO_4$ evaporation destroys silicates, whereas Zn fusion decomposes sulfates. Rapid ^{226}Ra assay, as in ore-process solutions, can be made by separating radium by suitable precipitation or anion-exchange procedures; the solution is evaporated, thus driving off Rn, and the Ra is determined by direct α counting. More often, however, radium is analyzed by de-emanating a solution with a carrier gas and counting the α particles of radon and its decay products in a gas ionization chamber or a scintillation counting bottle. Large quantities of radium are assayed by ionization measurements of the γ radiation, usually after waiting >30 d for ^{222}Rn and its decay products to come to equilibrium.

Initially, the unit of radioactive intensity, the curie (Ci), was defined as a disintegration (dis) rate equal to that of 1 g ^{226}Ra. It is now defined as exactly 3.7×10^{10} Bq (= dis/s) and can be applied to any radionuclide.

Actinium. For the chemistry of this element, see Actinides and transactinides.

By far the most important isotope of actinium is ^{227}Ac ($t_{1/2}$ = 21.77 yr), which occurs in the ^{235}U disintegration series. Until the advent of the nuclear reactors, its chemistry was known only from tracer studies. Milligram quantities are now prepared by neutron irradiation of ^{226}Ra through ^{227}Ra, which permits ordinary microchemical operations. Actinium-225 ($t_{1/2}$ = 10.1 d) belongs to the decay series of ^{237}Np–^{233}U, rare induced natural radionuclides in uranium and thorium ores; it has been detected in such ore concentrates. Synthetic ^{225}Ac provides an alternative tracer for actinium. It can be extracted as needed from ^{229}Th ($t_{1/2}$ = 7.3×10^3 yr), which has been separated from aged reactor-produced ^{233}U.

Because it occurs only in one nonzero oxidation state, Ac(III), actinium resembles the +3 lanthanides. Lanthanum is usually the carrier of choice. Actinium is separated by coprecipitation with LaF_3 from acidic solutions. It is also carried by lanthanum or thorium oxalate from very dilute mineral acids. Actinium forms complexes with many organic compounds and anions; its bonds are weaker than those of the lanthanides because of its larger ionic radius.

Actinium and thorium are adsorbed strongly from 2 M HNO_3 on a cation-exchange resin, whereas radium passes through with excess solvent. Actinium is eluted with 4 M HNO_3; thorium remains on the column. Actinium adsorbs more strongly than +3 lanthanides and can be separated chromatographically with ammonium citrate at pH 3–6. Actinium is not adsorbed on anion-exchange resins from HCl solutions, which permits separation from uranium, protactinium, polonium, bismuth, lead, thallium, and many nonradioactive elements.

Direct counting of the ^{227}Ac β particles is difficult; their maximum energy is 44 keV and most are considerably weaker; α-particles occur in 1.4% of its disintegrations, and α particles from its decay products (a total of five per disintegration) begin to grow in immediately, controlled by ^{227}Th ($t_{1/2}$ = 18.72 d) and ^{223}Ra ($t_{1/2}$ = 11.44 d). Hard β particles from ^{211}Pb and ^{207}Tl also grow in, somewhat slower at first. Either α or hard-β counting of thin deposits, which have been ignited to minimize emanating power for ^{219}Rn, over periods of several days or weeks is practical. Actinium-225 ($t_{1/2}$

= 10 d) can be measured by its α particles (with those of its descendents) or by the relatively hard β particles of ^{213}Bi and ^{209}Pb, which come to equilibrium within a day.

Thorium. For the chemistry of this element, see Thorium and thorium compounds.

Because all descendents of ^{232}Th have half-lives <6 yr, even geologically young Th-containing minerals and rocks contain the entire series in equilibrium. The same is nearly true of aged refined Th salts and metal. Purification removes all disintegration products except ^{228}Th (radiothorium, RdTh; $t_{1/2}$ = 1.913 yr). The subsequent β particles and γ rays grow in rapidly and then decay with the ^{228}Th half-life. Meanwhile, ^{228}Ra (mesothorium, MsTh$_1$; $t_{1/2}$ = 5.76 yr) grows in, replenishing ^{228}Th. The γ activity of the purified thorium passes through a minimum in 4.5 yr, and 90% equilibrium is restored in 22 yr. Purified ^{228}Th is a long-lived tracer and a source of its descendents. It is prepared by isolating ^{228}Ra from thorium. After several months or years, the generated ^{228}Th is purified. A more useful and readily available tracer is ^{234}Th (UX$_1$; $t_{1/2}$ = 24.10 d), which can be separated in radiochemically and isotopically pure form from a purified uranium salt or solution after several days or months of standing. Thorium-230 (ionium, Io; $t_{1/2}$ = 7.52 × 10^4 yr) can be isolated together with significant quantities of ^{232}Th from natural uranium sources or in pure form from large quantities of purified uranium after several years of growth. It is useful as a spike for quantitative analysis of thorium by mass spectrometry (isotope-dilution analysis), especially at very low concentrations in natural materials.

In aqueous solution and compounds, thorium is always encountered as Th(IV). In separation procedures, it is coprecipitated with Fe(OH)$_3$, LaF$_3$, Zr(IO$_3$)$_4$, or Zr peroxide; alternatively, the insoluble sulfates or sulfides are precipitated, and thorium is left in solution.

In the absence of complexing agents, Th(IV) behaves like Th^{4+} and is absorbed on cation-exchange resins more strongly than most other ions. Thorium is separated from most other elements on a small cation-exchange column followed by washing with 6 M HCl. Thorium is not adsorbed on anion-exchange resins at any HCl concentration.

Thorium-IV is notably complexed by 1,3-diketones, eg, TTA, which is most useful for solvent extraction from moderately strong acids. In a separation of ^{234}Th (UX$_1$) from U(VI), the latter is converted to a soluble uranyl carbonate complex by excess (NH$_4$)$_2$CO$_3$, raising the pH to 8.0–8.5; aqueous cupferron is then used to extract thorium with chloroform.

Pure ^{234}Th and ^{230}Th are radioassayed by β counting (with the ^{234}Pa isomers in equilibrium) and α counting, respectively. Thorium-228 is always present in ^{232}Th; the ratio depends on the history of the source. Mixtures of ^{228}Th and ^{227}Th are unscrambled and the isotopes determined individually with α or γ spectrometry.

The fission-track method of determining and mapping uranium in solids can be extended to thorium by first irradiating a specimen-detector assembly with slow neutrons to determine the uranium content or distribution; a fresh detector is then used with high energy protons or α particles to induce fission in uranium and thorium. The Th content or distribution is determined by difference.

Protactinium. This element was named after its longest-lived isotope, ^{231}Pa ($t_{1/2}$

= 3.28×10^4 yr), which occurs in the ^{235}U series. Like ^{226}Ra, it occurs in weighable quantities in uranium ores (Pa:U = 3.3×10^{-7} by mass) and has been used for both tracer and microchemical studies. Protactinium-234m (UX$_2$; $t_{1/2}$ = 1.17 min) is too short-lived for effective tracer use, but the ground-state ^{234}Pa (UZ; $t_{1/2}$ = 6.75 h), though occurring in much lower abundance, can be isolated in useful quantities from strong preparations of ^{234}Th. Protactinium-233 ($t_{1/2}$ = 27.0 d) prepared artificially is now the most readily available and useful tracer for the element.

Although protactinium is an actinide corresponding to Pr of the lanthanides, it actually behaves more like niobium and tantalum in group VB of the periodic table. It is known in oxidation state +3, which is very unstable; +4, which in solution is oxidized by O$_2$; and +5, in which it is usually encountered. It shares with niobium and tantalum the formation of acidic oxides and hydrates that are insoluble in solutions of ordinary acids. Slow and generally irreversible hydrolytic condensation, particularly at high concentrations (including tracer Pa in the presence of Nb or Ta) cause erratic behavior and often chemical intractability. Protactinium and its carriers are kept in solution in the presence of HF; Pa(V) forms stable fluoride complexes, including PaF$_7^{2-}$ and PaF$_8^{3-}$. It is also soluble in >3 M H$_2$SO$_4$.

Protactinium(V) is precipitated by ammonium hydroxide, though hydrolysis of the fluoro complex and complete precipitation may take hours. The hydroxide is soluble in citric, tartaric, and oxalic as well as hydrofluoric acids. Protactinium can be coprecipitated in acidic solution with several oxides and fluorides.

On anion-exchange resins, protactinium is strongly adsorbed from >4 M HCl, whereas Th, Ac, Ra, and Fr are not adsorbed. Lead and bismuth are eluted with 9 M HCl, and protactinium with 9 M HCl + 0.04 M HF; polonium and uranium remain on the column. In the absence of complexing agents, protactinium has a tendency to be adsorbed on cation-exchange resins, but in the presence of dilute HF it passes through.

Protactinium-231 is a useful analytical tracer for ^{233}Pa, which is an intermediate in the production of ^{233}U in breeder reactors. Protactinium-233 serves the same function for ^{231}Pa in studies of the geochemistry of the latter. Protactinium-230 is measured by α counting or spectrometry; ^{233}Pa and ^{234}Pa isomers, UX$_2$ and UZ, are measured by β counting.

Uranium. For the chemistry of this element, see Uranium and uranium compounds.

Each natural uranium isotope decays to a thorium isotope, which in several instances transforms to an isotope of protactinium. Thus, the radiochemistry of natural uranium involved only thorium and protactinium. With the discovery of artificial radioactivity and particularly of nuclear fission, that situation was dramatically changed.

Separations of thorium and protactinium from uranium have been mentioned already. When uranyl nitrate hexahydrate is dissolved in ethyl ether, a small amount of aqueous phase forms, and this contains the thorium and protactinium. With irradiated uranium, most of the fission products and transuranics are also concentrated in this solution.

Commercial uranium is now mostly depleted in ^{235}U. It may also contain some ^{236}U if it has been recovered from nuclear reactors.

The uranium isotopes that occur in nature can be measured by α spectrometry. However, for minute traces, mass spectrometry is more sensitive; it also yields more precise isotope ratios. For quantitative analysis of small amounts, as in rocks and meteorites, mass-spectrometric isotope dilution is commonly used. The usual spike is highly enriched ^{235}U. The techniques of isolation, purification, and preparation for mass spectrometry are very similar to radiochemical techniques (see Analytical methods).

A sensitive and versatile method for determining the content and distribution of uranium in solids is the fission-track method (see below). A uranium-poor track detector (muscovite mica, quartz, fused silica, Lexan plastic) is placed between a flat sample and a standard glass of known uranium content; the assembly is irradiated with slow neutrons. The detector is etched to render visible the fission tracks, which are counted on both sides. The bulk uranium content of the unknown is calculated from the ratio of track densities. The uranium distribution in a mineral, rock, meteorite, archeological object, etc, can be mapped by a modification of the technique.

Production of Natural Radioactive Substances

Radium. After the discovery of radioactivity, the chief motivation for locating, mining, and processing uranium ores was their radium (^{226}Ra) content. For a number of years, the most important source was pitchblende from the St. Joachimsthal mines operated by the Czechoslovakian government. Later, some radium was also produced elsewhere in Europe. Before World War I, extensive though relatively low grade carnotite deposits of the Uravan mineral belt in southwest Colorado and adjacent Utah were exploited by the Standard Chemical Corporation and processed in Denver, Pittsburgh, and Canonsburg, Pennsylvania. About the same time, the rich Katanga (Belgian Congo, now Zaire) pitchblende deposits were discovered. Low cost production from this source soon dominated the world market. In the 1930s, extensive pitchblende deposits at Great Bear Lake in the Northwest Territories of Canada were found, and the refinery of the Eldorado Gold Mines, Ltd. at Port Hope, Ontario, started producing radium in 1933. For a number of years, the Belgian and Canadian producers divided the market on a 60:40 basis by a cartel agreement. Production was greatly accelerated during World War II. After the development of nuclear reactors and the widespread production of artificial radionuclides with a variety of properties and low costs, commercial radium production virtually ceased. Today, most commercial radium is recycled from hospitals and industrial users who have changed to other radiation sources.

The commercial production of radium followed the procedure developed by Marie Curie. Typically, ground ore is dissolved in a mixture of nitric and sulfuric acids with some barium. The insoluble sulfates of lead, barium, and radium and the siliceous gangue are removed by filtration, and $PbSO_4$ and some silica are leached out with boiling NaOH or NaCl. The residue is converted to barium and radium carbonates in an autoclave with Na_2CO_3, filtered, dissolved in HCl, and filtered from the remaining silica. This procedure is repeated for purification, and the carbonates are dissolved in HBr and subjected to systematic fractional crystallization. In ten stages the radium is enriched from a few ppm to several parts per thousand. Accumulated batches are combined, repurified, and further fractionated until 90–99% $RaBr_2$ is obtained. The crystals are sealed in glass tubes for storage, shipment, and use. The nature and quality of the ores, recovery of other constituents, and local conditions and preferences govern variations in the procedure.

In the early days, radium was processed and handled without shielding, and with poor ventilation and little contamination control. Many scientists' and production workers' lives were shortened by cancer or other radiation effects.

Originally, radium was interchanged freely among scientific and medical investigators, but then became an extremely expensive commercial product. Table 4 gives typical prices for newly produced radium, which have increased in spite of availability as a uranium by-product and the development of artificial radionuclides. Early in the 1960s, surplus used radium became available at ca 5000 $/g, dropping to 2000 $/g by the end of the decade. Now it is frequently given away. A radium depository is operated by the EPA at its Eastern Environmental Radiation Facility in Montgomery, Alabama.

The chief world supplier at present is Amersham-Buchler, Braunschweig, FRG. In the United States, radium is supplied by the Radium Chemical Company, Woodside, New York, as distributor for Union Minière S.A. of Brussels. Calibrated radium standards can be obtained from the National Bureau of Standards, Washington, D.C.

Radon. Sealed radon (^{222}Rn) sources are preferred to radium itself for many uses for reasons of economy, logistics, and small physical size. Such preparations are made at central facilities that serve both medical and industrial users.

Typically, ca 1 g radium is kept in solution, or preferably as a highly emanating solid. Semiautomatic devices have been developed for extracting and purifying the radon and compressing it into long capillary tubes. Glass capillaries can be sectioned by a fine torch and metal ones by a crimping cutter to yield sources of desired strength. For medical applications, these are usually sealed into tubes of Ag, Au, or Pt to provide screens against the β radiations. For industrial use, they are encased in rugged aluminum or stainless-steel capsules.

Other Nuclides. Protactinium (^{231}Pa) and actinium (^{227}Ac) can be extracted from radium-process residues in which they are concentrated.

By-product lead contains large quantities of ^{210}Pb (RaD) at low specific activity (radiolead). Commercial lead seems to be free of this fraction. Highly concentrated ^{210}Pb is obtained by crushing or dissolving spent radon needles or radiographic sources and purifying the lead by radiochemical techniques. Either can be used for the extraction of polonium (^{210}Po).

The thorium extracted from high grade uranium minerals is predominantly ^{232}Th by mass but ionium (^{230}Th) by activity.

Mesothorium (^{228}Ra) separated from Th minerals and ores, principally monazite, is heavily diluted with barium and contaminated with ^{226}Ra. It can be enriched, though not completely separated from barium, by fractional crystallization. Pure ^{228}Ra can

Table 4. Typical Market Prices for Newly Produced Radium

Year	$/g	Year	$/g
1915	135,000	1946	19,000
1925	50,000	1951	16,000
1929	70,000	1960–1970	15,000
1937	25,000	1975	20,000
1943	14,000	1978	35,000
		1981	65,000

be isolated from a purified Th salt after several months or years of growth. Either can be used for subsequent isolation of radiothorium (^{228}Th) after several months or years of growth.

Radiometric Dating

Radiometric dating is based on the functional relationship between the rate of nuclear transformation, the time during which the transformation has occurred, and the extent of the transformation. If the rate and the extent can be determined, the time can be calculated.

Radioactive transformations occur at fixed rates expressed as disintegration constants (or half-lives). The occasional small variations observed in the laboratory are unimportant. In a few cases, the disintegration constants are uncertain by significant amounts, but this does not often limit the precision and accuracy of dates.

Radiometric geochronology provides information about geochemical processes and the nature of geologic events as well as dates and ages.

Methods Based on Primary Natural Radioactivity. In 1906 Rutherford, having noted that the amount of helium in uranium and thorium minerals depends on their geologic ages, used He:U ratios to estimate geologic time based on radioactivity. Shortly thereafter dating based on lead accumulation was used, and the ages of a number of uranium minerals were calculated from their Pb:U ratios. True geochronology, ie, the use of specific ages to date geologically significant events and to quantify the general geologic time scale, followed. In the 1920s, mass spectrometry was applied to lead, and it was determined that its principal isotopes had mass numbers 206, 207, and 208. Their respective ultimate ancestors were recognized to be ^{238}U, ^{235}U and ^{232}Th. The discovery of the nonradiogenic ^{204}Pb paved the way for modern mass-spectrometric age determinations by the U,Th–Pb methods.

The discovery that the β-active isotope of rubidium is ^{87}Rb launched the development of the Rb–Sr method of age determination. Furthermore, it was recognized that ^{40}K, in addition to its β^- emission, undergoes decay partly by electron capture to ^{40}Ar, which explains the high abundance of ^{40}Ar in the earth's atmosphere. After nearly pure ^{40}Ar was found in potassium minerals, the K–Ar dating method was developed. The Sm–Nd method, based on ^{147}Sm α transformation to ^{143}Nd, is a recent but important development.

In all these methods, the transforming radionuclide is usually available for measurement. A closed system, in which there is no change in the amounts of parent and daughter nuclides except for the transformation of one to the other, obeys the relationship:

$$N_d^r = N_p(e^{\lambda T} - 1); \quad T = \frac{1}{\lambda} \ln \left(\frac{N_d^r}{N_p} + 1 \right) \tag{9}$$

where T = derived age; N = number of atoms; p = parent; d = daughter; r = radiogenic (produced by radioactive transformation); and λ = disintegration constant. If some portion N_d^c of the daughter nuclide is nonradiogenic or common (c), and was presumably present in the system at its formation, that portion must be estimated and subtracted from the measured amount (N_d) to obtain the radiogenic moiety: $N_c^r = N_d - N_d^c$. As implied above, only ratios of numbers of atoms are important. Moreover, wherever the daughter has a stable or very long-lived isotope, it (or one of several) is chosen as a reference nuclide (s), and the experiments are designed to yield atomic ratios based on that nuclide. For the daughter element, the ratios are directly measured isotope ratios. Thus,

$$T = \frac{1}{\lambda} \ln \left[\frac{\frac{N_d}{N_s} - \frac{N_d^c}{N_s}}{\frac{N_p}{N_s}} + 1 \right] \quad (10)$$

The half-lives and disintegration modes of the primary natural radionuclides, their isotopic abundances in their natural elements, their ultimate decay products, and the mean isotopic abundances of the latter on earth are given in Table 1.

Rubidium–Strontium Method. Rubidium forms no minerals of its own. However, it is widespread and occurs in many igneous rocks at sufficient concentration for ^{87}Rb to produce measurable variations in the ^{87}Sr content of the Sr also present. Consequently, this is a highly successful geochronometric method. The conventional reference nuclide is ^{86}Sr.

For a single phase which has been closed for an interval T, equation 10 becomes

$$T = \frac{1}{\lambda} \ln \left[\frac{\frac{^{87}\text{Sr}}{^{86}\text{Sr}} - \left(\frac{^{87}\text{Sr}}{^{86}\text{Sr}}\right)_o}{\frac{^{87}\text{Rb}}{^{86}\text{Sr}}} + 1 \right] \quad (11)$$

The subscript o refers to the initial Sr incorporated into the phase at its formation. With two cogenetic phases (formed at the same time with the same initial Sr), there are two equations that can be solved simultaneously for the two unknowns, $(^{87}\text{Sr}/^{86}\text{Sr})_o$ and T.

More than two such phases yield a check on the closed-system assumption and better values for the unknowns. A plot is made of ^{87}Sr/^{86}Sr vs ^{87}Rb/^{86}Sr. If the points fit a straight line, called an isochron, of the form

$$\frac{^{87}\text{Sr}}{^{86}\text{Sr}} = \left(\frac{^{87}\text{Sr}}{^{86}\text{Sr}}\right)_o + (e^{\lambda T} - 1) \frac{^{87}\text{Rb}}{^{86}\text{Sr}} \quad (12)$$

the zero-intercept is $(^{87}\text{Sr}/^{86}\text{Sr})_o$ and the slope yields T. This isochron method is also used in other radiometric dating methods.

In the simplest case, the derived age T is the interval since formation, called formation age. However, the apparent age may be less than the formation age, corresponding to an isotopic reequilibration of the Sr in a rock because of solid-state diffusion, phase transformations, etc, that accompany the metamorphism of the rock, which resets the clock. In this case the derived age is called metamorphic age.

The Rb–Sr method has also been applied to meteorites and lunar samples. In cosmic bodies, metamorphic episodes that are causally related to impact have been recognized and dated.

In a large reservoir, such as the earth as a whole or a large segment of the mantle or crust, the ^{87}Sr:^{86}Sr ratio undergoes slow secular growth because of the low Rb:Sr ratios in undifferentiated matter. Many meteorites, presumably derived from a common solar nebular reservoir, yield Rb–Sr ages close to 4.6×10^9 yr, but the initial

Sr ratios indicate differences in times of formation of stony meteorites up to ca 40 × 10^6 yr. The lowest $(^{87}Sr/^{86}Sr)_o$ observed is 0.6988, regarded as close to that of primordial Sr in the solar system. Terrestrial rocks have initial ratios from ca 0.700 to 0.735.

Uranium, Thorium–Lead Methods. The natural radioactive transformations ^{238}U to ^{206}Pb, ^{235}U to ^{207}Pb, and ^{232}Th to ^{208}Pb are closely connected by the isotopy of the series products. The first two are intimately coupled by the isotopy of the parents, whereas the third is loosely connected to them by the considerable geochemical coherence of uranium and thorium. Fortunately, there is a nonradiogenic Pb isotope, ^{204}Pb, that can be used as a reference nuclide to which the abundances of the radiogenic Pb isotopes and the parent nuclides can be compared.

The disintegration constants of ^{238}U, ^{235}U, and ^{232}Th are designated λ_8, λ_5, and λ_2, respectively. For transformation of ^{238}U to ^{206}Pb, the closed-system equation 10 takes the specific form

$$T_6 = \frac{1}{\lambda_8} \left[\frac{\frac{^{206}Pb}{^{204}Pb} - \left(\frac{^{206}Pb}{^{204}Pb}\right)_o}{\frac{^{238}U}{^{204}Pb}} + 1 \right] \tag{13}$$

Transformations ^{235}U to ^{207}Pb and ^{232}Th to ^{208}Pb have parallel equations. Thus, three ages, T_6, T_7, and T_8, can be calculated for a sample for which the elemental abundances of lead, uranium, and thorium and the isotopic composition of the lead have been measured. It is necessary to incorporate isotope ratios for the initial Pb of the sample. These may be deducted from geologically associated or coeval common Pb samples, mainly galena, PbS, or from model compositions for common crustal Pb at the time of formation (see below).

A fourth though not independent age, the lead–lead age, is that value of T which satisfies the equation:

$$\frac{\frac{^{207}Pb}{^{204}Pb} - \left(\frac{^{207}Pb}{^{204}Pb}\right)_o}{\frac{^{206}Pb}{^{204}Pb} - \left(\frac{^{206}Pb}{^{205}Pb}\right)_o} = \frac{^{235}U}{^{238}U} \frac{e^{\lambda_5 T} - 1}{e^{\lambda_8 T} - 1} \tag{14}$$

This can be calculated from the isotopic composition of the lead only, without chemical analyses for uranium and lead.

When discordant calculated ages are obtained, a more complex sample history is indicated. Data on series of samples are often treated by a concordia plot, in which the abscissa is $^{207}Pb:^{235}U$ and the ordinate $^{206}Pb:^{238}U$. A curve called concordia, the locus of points

$$\frac{^{207}Pb}{^{235}U} = e^{\lambda_5 T} - 1; \quad \frac{^{206}Pb}{^{238}U} = e^{\lambda_8 T} - 1 \tag{15}$$

is drawn and marked at intervals of T. Each sample is plotted as a point at $^{207}Pb^r:^{235}U$, $^{208}Pb^r:^{238}U$, the common Pb-isotope ratios having been subtracted from the measured Pb-isotope ratios. Samples plotting on or close to a concordia plot are concordant with the age given by the T-scale of the curve. Discordant samples frequently plot inside (below and to the right) of the curve, and indicate relative loss of lead or gain of uranium; more likely the former, since radiogenic lead is crystallographically labile and geochemically mobile.

Related to the U,Th–Pb method is the common lead method. Most galenas are virtually devoid of uranium and thorium and, therefore, the Pb has the unmodified isotopic composition of the reservoir from which it was formed at the time of mineralization. As a result of the coexistence of uranium, thorium, and lead in the reservoir, ^{206}Pb:^{204}Pb, ^{207}Pb:^{204}Pb, and ^{208}Pb:^{204}Pb have increased monotonically through geologic time. By dating galenas through associations with datable minerals or stratigraphically, a functional relationship between those ratios and the time for average earth material has been derived and accommodated by a simple theoretical model. From this model, a curve can be drawn on a plot of ^{207}Pb:^{204}Pb versus ^{206}Pb:^{204}Pb and marked with T values, enabling dating of conformable leads and understanding of anomalous leads with multistage histories.

Helium Method. From the measured contents of uranium and thorium in a natural sample, He production can be calculated, considering that ^{238}U, ^{235}U, and ^{232}Th generate 8, 7, and 6 He atoms, respectively, per disintegration. Comparison with the measured He content then gives a nominal U,Th–He age, or simply helium age.

Because of diffusive loss of helium, such calculated ages are only lower limits to formation ages. For U- or Th-rich minerals, the ages are much too low, because radiation damage and substantial helium pressure cause appreciable He loss.

The method is little used in general geochronology. However, it has utility in meteorite studies. Cosmic rays generate ^{4}He in meteorites and also some ^{3}He. Hence, the cosmogenic ^{4}He component can be estimated from the ^{3}He and subtracted to yield the radiogenic ^{4}He when that is not too small.

Potassium–Argon Method. Potassium is a principal element in the earth's crust and is widely distributed in silicate minerals and rocks. Argon is largely excluded from solids at their formation and, therefore, the inherited ^{40}Ar is low. However, some ^{36}Ar and ^{38}Ar are always found along with ^{40}Ar, and it is necessary to determine the isotopic composition to distinguish atmospheric and radiogenic ^{40}Ar.

The simple age equation (see eq. 9) must be modified to reflect the branching decay. If λ_e is the disintegration constant for electron capture to produce ^{40}Ar, and λ_β is that for ^{40}Ca production by β^- emission, then $\lambda = \lambda_e + \lambda_\beta$ is the total disintegration constant determining ^{40}K half-life, and

$$\frac{^{40}\text{Ar}^r}{^{40}\text{K}} = \frac{\lambda_e}{\lambda}(e^{\lambda T} - 1); \quad T = \lambda^{-1}\ln\left[\frac{\lambda}{\lambda_e}\frac{^{40}\text{Ar}^r}{^{40}\text{K}} + 1\right] \tag{16}$$

Like helium, argon is lost from most rocks and minerals, particularly if they are old or have been exposed to high temperatures or weathering. Young unaltered basaltic rocks, basaltic glass, and volcanic ash retain large amounts of argon and can yield reliable data on formation ages. This information is valuable for dating ancient human habitations. Usually, however, Ar loss results in low apparent ages; the K–Ar date is assumed to correspond to the last metamorphism with cooling to the blocking temperature for Ar diffusion in the particular mineral. Different minerals in the same rock may exhibit different ages corresponding to different blocking temperatures, which allows deductions about the thermal history.

The K–Ar method in its various modifications has been applied to meteorites and lunar samples. The derived ages are often intermediate between the U,Th–Pb or Rb–Sr and the He ages in the same samples, particularly in shocked materials, and may date metamorphism caused by impacts that may have ejected meteoritic fragments from asteroidal bodies or produced lunar craters and breccias.

Samarium–Neodymium Method. Both ^{147}Sm and its α-decay product ^{143}Nd belong to rare-earth (lanthanide) elements (see Rare-earth elements). Because of their chemical and geochemical similarity, samarium and neodymium are much less fractionated in nature than other parent-daughter pairs useful for geochronometry. Moreover, since the half-life of ^{147}Sm is quite long, the transformation rate is slow. Nevertheless, modern high precision mass spectrometry can measure precisely the small variations in the ^{143}Nd content of neodymium found in nature, even when these elements are present only as trace elements in igneous rocks.

The isochron method, fitting an equation analogous to equation 12 with ^{144}Nd as the reference nuclide, is often used to determine the initial ratio $(^{143}\text{Nd}/^{144}\text{Nd})_0$ and age T of a suite of cogenetic rocks or minerals.

In general Sm–Nd ages are less affected by metamorphism than the other ages discussed. The main significance of this method is in the applicability of $(^{143}\text{Nd}/^{144}\text{Nd})_0$ as a geochemical and petrologic tracer of the source regions and magmas of igneous rocks.

Fission-Track Method. Spontaneous-fission fragments from ^{238}U in mineral grains produce radiation-damage tracks that can be made visible under the microscope in polished sections by chemical etching. Neutron irradiation of identical specimens produces induced-fission tracks from ^{235}U. The calibrated ratio of spontaneous-to-induced tracks in a mineral yields the age since it cooled through its annealing temperature.

This method can be applied to artificial glass less than one year old or to the entire range of geologic time. Applications have included dating of archeological artifacts, ancient human habitations, deep-sea-sediment cores, glassy skins of ocean-bottom basalts associated with sea-floor spreading, and a wide variety of geologic specimens, tektites, and meteorites. It is particularly valuable in the range of ca 40,000 to 10^6 yr, between the effective ranges of the radiocarbon and K–Ar dating methods.

Other Methods. *Potassium–Calcium Method.* In old minerals and rocks with high potassium and low calcium contents, radiogenic ^{40}Ca can be detected and measured. The K–Ca method is similar to the Rb–Sr method.

Rhenium–Osmium Method. Radiogenic ^{187}Os from ^{187}Re β decay can be detected in a few minerals, particularly molybdenite, MoS_2, and allows ages to be calculated. This method is more useful for iron meteorites than for terrestrial materials.

Lutetium–Hafnium Method. The rare-earth nuclide ^{176}Lu generates ^{176}Hf by β decay. The Lu:Hf ratios are usually low, but radiogenic ^{176}Hf has been measured and used for age calculations.

Methods Based on Secondary Natural Radioactivity. In old undisturbed geologic materials, the ^{238}U, ^{235}U, and ^{232}Th series are in secular equilibrium; consequently, the activity, ie, disintegration rate, of each member of a chain is equal to that of the parent except as reduced by branching. Processes such as partial melting, magmatic crystallization, groundwater leaching, weathering, biogenic mineralization, and sedimentation usually cause chemical differentiation and may disrupt the radioactive equilibrium. In derived matter that is subsequently left undisturbed, a nuclide that is enriched relative to a longer-lived antecedent decays, and one that is depleted grows, until new secular equilibrium is restored. Measurements of the pair often date the disturbing events on a time scale usually up to several half-lives of the shorter-lived nuclide.

Uranium-234. The generation of a ^{234}U atom from ^{238}U by α decay through ^{234}Th is a hot-atom process, in which the recoil from the α particle dislodges the atom from its position in the crystal lattice and disorganizes its local surroundings. All ^{234}U atoms ($t_{1/2}$ = 245,000 yr) in a uranium-bearing rock are replaced in ca 10^6 yr by labile ones that are much more mobile than the ^{238}U atoms and are more readily oxidized from U(IV) to UO_2^{2+}. They are readily leached by ground- and surface-waters. Such waters contain uranium with ^{234}U:^{238}U activity ratios enhanced by factors of up to ca 10. Weathered rocks and soils have ^{234}U:^{238}U generally less than unity. The ratio for uranium dissolved in ocean water is fairly constant at ca 1.14.

The ^{234}U:^{238}U ratio in groundwaters is a useful tracer for their origin, transport, and mixing. Water-derived minerals, coral, etc, can be dated by their residual excess ^{234}U.

Ionium and Protactinium. Thorium and protactinium are much less soluble in sea water than uranium, as reflected in oceanic residence times of ca 300, 300, and 500,000 yr, respectively. Consequently, ^{230}Th (ionium, Io) and ^{231}Pa carried into the oceans by rivers and generated within the oceans by decay of ^{234}U and ^{235}U are rather promptly carried to the bottom sediments and, in the topmost layers, have activities in excess of those of their parents. In an undisturbed sediment layer, the excess of Pa activity above that of ^{235}U decreases with its $t_{1/2}$ of 32,800 yr. The decay of excess Io is somewhat more complicated, since it approaches equilibrium with ^{234}U with its $t_{1/2}$ of 75,200 yr while ^{234}U returns to equilibrium with ^{238}U with its threefold longer lifetime.

In the ionium–thorium method, sediment-core samples are leached with acid to enrich authigenic minerals and adsorbed matter relative to detrital (rock) grains, and ^{232}Th is used for normalization. The Th-fraction is radiochemically purified, and the ^{230}Th:^{232}Th ratio is determined by spectrometric α counting. This ratio is plotted against sediment depth. In an ideal case, the initial ratio is given by the zero-depth intercept, and ^{238}U:^{232}Th is given by the leveling-off ratio at considerable depth. Then, after subtraction of the latter, the sedimentation rate is derived from the slope of the logarithm of the ratio versus depth. Usually the plots are nonideal and show effects of top-layer mixing, variable sedimentation rates, slumping, etc.

Marine carbonates tend to incorporate uranium and exclude thorium, and hence have an initial deficiency of ionium. By measuring the ^{230}Th:^{234}U ratio in a sample of coral, mollusk shell, or calcareous ooze, if one assumes it was zero at the time of deposition and the ^{234}U:^{238}U was initially 1.14, the age can be calculated from the theoretical growth curve.

The ^{231}Pa:^{235}U ratio in the acid leachings of a deep-sea sediment might serve as an age indicator, but would be complicated by the unequal leachability of protactinum and uranium. The protactinium–ionium method assumes that the ^{231}Pa:^{230}Th ratio is a more reliable indicator because both nuclides are derived from uranium and have similar sedimentation characteristics. Ideally, the initial ratio of excess ^{231}Pa to excess ^{230}Th should be (^{235}U:^{234}U) ($\lambda_{231}/\lambda_{230}$); with ^{234}U:^{238}U = 1.14 and ^{235}U:^{238}U = 0.0460, that is 0.093. The ratio should decrease with age with $\lambda_{eff} = \lambda_{231} - \lambda_{230}$, corresponding to an effective $t_{1/2}$ of 58,000 yr. An age can be calculated for an individual sediment sample by assuming the theoretical initial ratio and half-life, but it is more useful to plot the ratio vs depth and interpret as described above.

The most important application of the Io and Pa methods has been in dating sediment cores used for the determination of paleotemperatures and paleoclimates by oxygen-isotope measurements and microfossil analyses.

Radium-226 and Lead-210. In young volcanic rocks, ^{226}Ra ($t_{1/2}$ = 1600 yr) and ^{210}Pb (22.3 yr), like ^{230}Th, are enriched relative to ^{238}U, their common ancestor. The ratio of each to its immediate longer-lived progenitor is of chronological importance. The ^{226}Ra:^{230}Th ratios observed are slightly to greatly in excess of unity, indicating qualitatively that the magmatic radium enrichment occurred less than a few thousand years ago. ^{210}Pb:^{226}Ra ratios less than unity indicate lead depletion within the preceding century or two.

Lead-210 is formed in the atmosphere by the decay of ^{222}Rn, which is continually emanated by soils and discharged by natural gases and volcanos. It has a residence time in the troposphere of days or weeks and is removed mainly by rain and snow. Much of it enters soils, rivers, lakes, freshwater sediments, the oceans, and marine sediments; in arctic and mountainous regions it accumulates in glacial firn and ice.

In glacial snow and ice, ^{210}Pb decreases with depth by radioactive decay. The same is true in undisturbed lacustrine and marine sediments after subtraction of ^{226}Ra activity, which supports an equal ^{210}Pb activity at depth. Analysis of the activity-versus-depth curves allows deduction of the initial ^{210}Pb activities and the precipitation or sedimentation rates in the respective situations, as well as the occurrence of disturbances.

Lead paints also contain traces of ^{210}Pb from uranium present in the lead ore as well as smaller amounts of ^{226}Ra. By measuring these two nuclides and estimating the initial ^{210}Pb from recent paints of the same type, an age can be calculated from the remaining excess ^{210}Pb over ^{226}Ra. This method has been used to date or check the authenticity of works of art.

Methods Based on Induced Natural Radioactivity. As mentioned earlier, a host of induced natural radionuclides is found in the earth's atmosphere and surface materials, meteorites and lunar rocks, and uranium and thorium minerals. Any one abundant enough for reasonably precise measurement can be used for chronological purposes.

Carbon-14. Carbon-14 is the most abundant terrestrially induced natural radionuclide. The nascent ^{14}C atoms in the atmosphere rapidly become oxidized to CO_2 and, on a time scale of years, are thoroughly mixed into the atmospheric CO_2 pool. Consequently, before the industrial and atomic ages, a rather constant specific activity of ca 235 mBq/g C (14.1 dpm/g) prevailed. Atmospheric CO_2 exchanges with oceanic HCO_3^- on a time scale of decades and results in a specific activity of ca 227 mBq/g (13.6 dpm/g) in the well-mixed upper layer of the ocean. Deep-ocean waters average ca 205 mBq/g (12.3 dpm/g).

Atmospheric CO_2 with its ^{14}C component is taken up in reduced form by plants, from which it enters the animal food chains. Thus, all young living matter resembles the atmosphere in its specific ^{14}C activity. At death, living matter that has remained as a closed system with respect to carbon undergoes a steady decrease in its ^{14}C activity as the radionuclide decays with its $t_{1/2}$ of 5370 yr. A radiocarbon date is the time necessary for the ^{14}C activity to have fallen to its present level through exponential decay.

Carbon-14 dating presents a number of complications. The burning of fossil fuels devoid of ^{14}C, which became quantitatively important in the latter half of the 19th century, has not only caused a substantial increase in the atmospheric CO_2 content (ca 13% in the late 1970s), but also caused a measurable reduction in the atmospheric ^{14}C specific activity (ca 3% by 1950), ie, the Suess effect. Neutrons from nuclear bombs exploded in the atmosphere beginning in 1945 and especially in the 1960s have substantially increased the ^{14}C activity in the atmosphere and in all materials deriving carbon from it. Secular variations in the atmospheric ^{14}C specific activity with a quasi-periodicity of ca 100 yr occurred during the past few centuries; this is known as the de Vries effect. Variations on a time scale of millenia also occur; the level ca 7500 yr ago was ca 8% higher than that in the AD era. The causes of these variations may be variations in the cosmic-ray flux at the earth's surface and in the carbon mass in its terrestrial reservoirs. These variations in the ^{14}C-specific activity have been established by measurements of wood dated by historical records and dendrochronology (tree-ring matching and counting). Isotope fractionation in geochemical, biological, and laboratory processes changes the ^{14}C:^{12}C ratio to different extents for standards and samples of different types.

The 1850 ^{14}C level, before the burning of fossil fuel became important, is often used as the zero-age level. Because of ^{14}C produced by nuclear bombs, radiocarbon dating is limited to materials that incorporated carbon from the atmosphere before 1950, which is taken as the zero time for reporting radiocarbon ages; BP means before 1950. To compensate for variations in ^{14}C-specific activity, for samples less than 8000-yr old, the initial ^{14}C-specific activity is corrected by a calibration curve constructed from the historical and dendrochronology samples. To compensate for isotope fractionation, the ^{13}C:^{12}C ratio in each standard and unknown CO_2 sample to be counted is determined by mass spectrometry. Correction to a standard ^{13}C:^{12}C ratio is made by assuming that the carbon-isotope fractionations are linearly mass-dependent, ie, the ^{14}C varies relative to ^{12}C just twice as much as does ^{13}C.

Carbon-14 is usually measured by converting the carbon in the sample to CO_2, a hydrocarbon gas or liquid benzene, and counting the β emissions in a low-background system.

Isotope enrichment of the heavy carbon isotopes is sometimes employed. A recent development is the measurement of the ^{14}C/^{12}C ratio directly by using modified cyclotrons or electrostatic accelerators as mass spectrometers. This can greatly reduce required sample sizes, increase precision, and extend the range of radiocarbon dating.

Radiocarbon dating has revolutionized chronology in archeology, life sciences, and geology and has produced valuable nonchronological information in many scientific areas.

Tritium. Spallation reactions by cosmic-ray and solar-proton primary and secondary radiations produce ^{3}H (tritium, T) in the atmosphere (see Deuterium and tritium). In addition, significant amounts of tritium from the sun's surface are believed to be brought to the earth by solar wind and flare emissions. This tritium is rapidly incorporated into H_2O molecules and mixed into the water of the atmosphere, hydrosphere, and biosphere. However, the $t_{1/2}$, 12.33 yr is too short for mixing to be complete, and activity levels are much more variable than those of ^{14}C. Moreover, tritium injection into the atmosphere by nuclear-bomb tests, particularly thermonuclear bombs, has been much greater than that of ^{14}C. Tritium produced by nuclear bombs is redistributed in much the same way as its natural counterpart.

The abundance of tritium in hydrogen is conventionally expressed as the ratio of atomic ^{3}H:^{1}H in tritium units (TU): 1 TU = 1 ^{3}H atom per 10^{18} ^{1}H atoms, corresponding to a specific radioactivity in water of ca 119 mBq/kg (7.1 dpm/kg). In the prenuclear age, the tritium content of rainwater ranged from <1 to >100 TU, depending mainly on the time between evaporation from seawater and precipitation of the water mass. Continental river water and biosphere hydrogen typically contained several TU. The well-mixed surface layer of the oceans generally had a few tenths of a TU, and the deep-sea waters had undetectable levels. At the height of bomb testing in 1963, the tritium content of rainwater attained a maximum of ca 10^4 TU in the northern hemisphere and about a tenth as high in the southern hemisphere. The content of surface seawater increased to ca 5–20 TU. Bomb-produced tritium still exceeds natural tritium in the environment, and concern is expressed about reactor-produced tritium and accidental release of artificial tritium.

Although the variability of tritium concentration in natural water prevents precise dating such as is possible with radiocarbon, several dating applications had been proposed and tested before the advent of nuclear explosions. These included the dating of agricultural products including wines, of water in underground aquifers, and of glacial snow and ice. However, more important than dating were applications of tritium as an indicator of meteoric water (derived from atmospheric precipitation) versus juvenile water (derived from the earth's interior) in geysers, fumaroles, springs, and wells, and as a tracer for the transport of water within and between its various terrestrial reservoirs.

The measurement of natural tritium requires isotopic enrichment, which is usually done by electrolysis of water but sometimes by thermal diffusion of H_2. The parallel enrichment of deuterium (^2H) is measured and used to estimate the enrichment factor for tritium. Low level gas or liquid β counting is employed.

Tritium, like ^{14}C, can be detected at very low levels with the use of a cyclotron or electrostatic accelerator as a mass spectrometer. This may decrease sample requirements and increase precision, particularly for pre-bomb water samples.

Other Terrestrial Nuclides. Most reaction products of cosmic rays striking the earth are generated in the stratosphere, preponderantly at high geomagnetic latitudes. Sufficiently long-lived products are mixed into the troposphere, carried by adsorption on settling particles, transported laterally in both the stratosphere and troposphere, and precipitated in rain, snow, or dry fallout. Some become locked into glacial ice or adsorbed in soil, and some are redistributed by ground water. Most products arrive rapidly at the upper mixing layer of the sea, whence they are slowly mixed into the deeper waters. They are then carried into the sediments at rates depending on their solubilities and residence times.

Besides ^{14}C and tritium, the nuclides ^{10}Be ($t_{1/2} = 1.5 \times 10^6$ yr) and ^{26}Al ($t_{1/2} = 720{,}000$ yr) are particularly interesting. They have similar geochemical properties, including short oceanic residence times, and have been detected in glacial ice and deep-sea sediments. Beryllium-10 is produced from the abundant nitrogen and oxygen in the atmosphere at a flux of ca 1×10^{-2} atom/(s·cm^2) of the earth's surface; ^{26}Al is produced from the rarer target Ar at ca 1×10^{-4} atom/(s·cm^2). The corresponding activity fluxes are ca 5×10^{-9} and 10^{-10} Bq/(cm^2·yr), respectively, leading to typical concentrations of ca 5×10^{-7} and 10^{-8} Bq/L in glacial ice and 0.05 and 0.001 Bq/kg in dry ocean sediments (1 Bq = 60 dpm). The decay of either with depth can give the mean ice- or sediment-accumulation rate and the ages at various depths.

Use of ^{10}Be and ^{26}Al for such purposes is in the early stages of development because of the extreme difficulty of measuring the radioactivity. The Be carrier of ^{10}Be must be cyclically purified and counted in a low-level β^- counter to constant specific activity; ^{26}Al is determined by an elaborate low background γ–γ coincidence-spectrometer system. Here also, accelerator mass spectrometry is promising.

Silicon-32 ($t_{1/2}$ = 108 yr), another cosmogenic nuclide of promise, has been measured in ground waters, polar ice, and ocean sediments. In low Si materials, the energetic-β-emitting ^{32}Si daughter ^{32}P, with its characteristic $t_{1/2}$ of 14.3 d, can be counted as it grows into the purified Si carrier. With siliceous materials, ^{32}P can be extracted and counted separately.

Chlorine-36 ($t_{1/2}$ = 300,000 yr) can be measured by atom-counting accelerator mass spectrometry and offers an interesting possibility as a tracer and chronometer for water and ice of Pleistocene origin.

Nuclides in Meteorites and Lunar Samples. The greater variety of cosmogenic radionuclides occurring in meteorites and lunar-surface materials as compared to terrestrial sources is due to the higher abundance of heavy target elements and the direct exposure to the primary cosmic radiation. Cosmogenic reactions essentially cease and the radionuclides decay without replenishment after the meteorite or lunar sample arrives on earth. Thus, meteorite finds and museum specimens contain only relatively long-lived radionuclides.

In addition to radionuclides, cosmogenic stable nuclides of the rare-gas elements are readily observable because of the very low contents of their primordial forms.

If the conditions of exposure to cosmic radiation of a meteorite and the intensity of the radiation are essentially constant averaged over intervals comparable to the half-life of a radionuclide, the content of the latter builds up to a steady state in which its disintegration rate equals its formation rate. Thus, the specific activity of a radionuclide is a measure of the cosmic-ray flux over the past two or three half-lives of the nuclide. This flux can be evaluated if excitation functions, target compositions, and shielding modifications can be determined. By measuring nuclides of different half-lives, comparisons have been made of the cosmic-ray intensity over various past intervals from a few days to ca 10^9 yr. The cosmic-ray environment of the meteorites seems to have been within a factor of two of its present value for the past approximate 10^7 yr when variations within decades and less are averaged. Short-lived enhancements of solar particles from flares and variability in the galactic cosmic-ray flux and in the solar-particle intensity related to the 11-yr solar-activity cycle have been reflected by short-lived nuclides. The spatial intensity of cosmic radiation in the solar system has been investigated by measurements of freshly fallen meteorites of determined orbits and by statistical analyses of other meteorites.

Terrestrial ages of meteorite finds are indicated by absence or deficiency of radionuclides of comparable or shorter lifetime. Carbon-14 deficiencies indicate that some stone meteorites have been preserved on earth for ca 3,000 to >30,000 yr. Low ^{26}Al in some antarctic meteorites indicates terrestrial ages up to 700,000 yr.

The exposure age of a meteorite is the time since its meteoroid became a small object as a result of collisional fragmentation from a larger body in which it was presumably shielded from the cosmic radiation. The most useful method of determining exposure ages is by measuring the content of one or more stable cosmogenic rare-gas nuclides and dividing by the estimated production rates. Most such ages are calculated from pairs of nuclides, a stable nuclide and a radionuclide expected to have a similar production rate. Only the ratio of production rates needs to be determined.

Methods Based on Extinct Natural Radioactivity. Whereas secondary and induced natural radionuclides give chronological information on the most recent geologic and cosmological history, extinct natural radionuclides give information on the earliest history of the solar system and even earlier events.

Iodine-129. Since sensitive and precise rare-gas mass-spectrometric techniques are well established, excess ^{129}Xe attributable to the decay of now extinct ^{129}I ($t_{1/2}$ = 16.0 × 10^6 yr) is commonly observed in meteorites; in fact, its absence indicates an origin or outgassing of the particular sample some time after the formation of the primordial solid matter of the solar system. For the content of radiogenic ^{129}Xe to be meaningful, the content of ^{127}I, the only stable isotope of ^{129}I, in the same material must also be measured (see Iodine and iodine compounds). The sample is usually irradiated in a neutron flux, and thereby some of the ^{127}I is converted through ^{128}I to ^{128}Xe. The latter is liberated by stepwise heating, and the amount and isotopic composition of the xenon is determined in each fraction.

The largest $(^{129}\text{I}/^{127}\text{I})_o$ ratio thus observed in any meteorite is 1.6×10^{-4}, and many have ratios $>0.7 \times 10^{-4}$. Since ^{129}I and ^{127}I were probably produced at comparable rates in the nucleosynthesis of solar-system iodine, probably in supernovae, the formation of the host matter must have occurred no more than ca 10–12 times the ^{129}I $t_{1/2}$ after the last nucleosynthetic contribution, which may be taken as a formal definition of the origin of the solar system. This suggests a time scale of ca 10^8 yr for the evolution of the solar nebula to the formation of the earliest planetary bodies.

Different $(^{129}\text{I}/^{127}\text{I})_o$ ratios indicate different absolute times of formation or of cooling to below the Xe-diffusion-blocking temperature. By convention, the chondrite Bjurböle, for which $(^{129}\text{I}/^{127}\text{I})_o = 1.09 \times 10^{-4}$, is taken for a time reference (t_o). Relative to this, about a score of other meteorites, mostly chondrites, have Δt values ranging from ca −10 to ca +10 × 10^6 yr with a resolution of ca 10^6 yr.

Lunar rocks are virtually devoid of ^{129}I-derived ^{129}Xe, which reflects an early thermal regime in the moon's history, although at least one Apollo-14 breccia appears to have excess ^{129}Xe (see Space chemistry).

Excess ^{129}Xe has been observed in terrestrial well gases and probably identified in atmospheric xenon. It has been deduced that the earth did not retain xenon until ca 10^8 yr after the main chondrite-forming period.

Plutonium-244. Analysis of the same kind of meteoritic Xe samples has shown excesses of the heavier Xe isotopes ^{131}Xe, ^{132}Xe, ^{134}Xe, and ^{136}Xe in proportions identical with their fission yields from the spontaneous fission of ^{244}Pu ($t_{1/2}$ = 81 × 10^6 yr for α decay; 6.6 × 10^{10} for SF) (see Plutonium and plutonium compounds). Unfortunately, ^{244}Pu does not have a stable or very long-lived isotope with which the former amount of ^{244}Pu can be compared. Consequently, results are expressed as $(^{244}\text{Pu}/^{232}\text{Th})_o$ or $(^{244}\text{Pu}/^{238}\text{U})_o$.

Fission-track densities in some meteoritic minerals in excess of those attributable to ^{238}U spontaneous fission are ascribed to ^{244}Pu. This permits determination of times of cooling through the radiation-damage annealing temperatures and thereby deducing cooling rates.

In early-formed meteorites $(^{244}\text{Pu}/^{232}\text{Th})_o$ values from 0.006 to 0.015 are found. In lunar-surface materials, effects due to ^{244}Pu are rare, which reflects the complex and prolonged processes by which they were formed and conditioned. However, a few Apollo-14 breccias dated by other methods at ca 4 × 10^9 yr contain fissiogenic xenon attributable to ^{244}Pu.

Terrestrial well-gas and atmospheric xenon likewise seem to contain ^{244}Pu-derived components. The inferred outgassing chronology of the earth is consistent with that deduced from ^{129}Xe.

By considering the amounts of ^{129}I, ^{127}I, ^{244}Pu, ^{235}U, ^{238}U, and ^{232}Th that were simultaneously present in the early solar system together with theoretical relative production rates, it is possible to evaluate the parameters in simple models of the chronology of nucleosynthesis of the matter of the solar system. These nuclides originated by the same nucleosynthetic mechanism. The calculations indicate a quasi-exponentially decaying general nucleosynthetic activity for ca 5–10 $\times$ 10^9 yr before the formation of the solar system ca 4.7 $\times$ 10^9 yr ago and strongly suggest a terminal burst of nucleosynthesis from one or several local supernovae; an interval of ca 150 $\times$ 10^6 yr then elapsed before the formation of the earliest solids which are now accessible as meteorites.

Aluminum-26. Certain rare refractory-mineral inclusions in some carbonaceous chondrites contain excess ^{26}Mg correlated with Al:Mg ratios and hence almost certainly derived from the *in situ* decay of ^{26}Al ($t_{1/2}$ = 720,000 yr) (see Aluminum and aluminum alloys).

Although now-extinct ^{26}Al seems to have been present in some early-solar-system matter, there is considerable disagreement as to the time and locale of its transformation to ^{26}Mg. The half-life is much too short for ^{26}Al to fit into the nucleosynthesis planetology chronology derived from ^{129}I and ^{244}Pu (see above). According to one theory, the gas–dust cloud from which the solar system was formed was injected with ^{26}Al-containing material freshly nucleosynthesized from a local supernova. This theory also suggests that the supernova explosion may have been responsible for initiating the collapse of the interstellar medium to form the self-gravitating solar nebula. The lack of more general ^{26}Mg variations could be explained if most primordial Mg was deficient in that isotope. Another theory postulates that the time constraints of the first theory are too severe. It favors instead a scenario in which the ^{26}Al was incorporated into dust grains condensing in the expanding and cooling ejecta of the supernova in which it was synthesized; it then decayed locally to ^{26}Mg. These grains were subsequentially mixed into the interstellar medium and thence into the solar nebula. Some survived vaporization and homogenization to be incorporated into planetary matter.

Other Nuclides. Samarium-146 has a very favorable $t_{1/2}$ (103 $\times$ 10^6 yr), but ^{142}Nd anomalies resulting from its α decay would be very small because of the minor fractionations of samarium and neodymium (see Samarium–Neodymium Method above). Nevertheless, small ^{142}Nd:^{144}Nd variations correlated with Sm:Nd ratios seem to have been observed in minerals of some meteorites. However, the ^{142}Nd excesses are too small relative to their uncertainties for the early-solar-system existence of ^{146}Sm to be firmly established.

In several high-Ni and relatively high-Pd iron meteorites that are very low in Ag, the ^{107}Ag:^{109}Ag ratios are up to ca 160% above normal, almost certainly because of the former presence of ^{107}Pd ($t_{1/2}$ ca 6.5 $\times$ 10^6 yr). The implied (^{107}Pd/^{108}Pd)$_0$ is ca 2 $\times$ 10^{-5}. Since these meteorites are products of melting and solidification in the parent minor planets, the ^{107}Ag excesses could not have been brought into the solar nebula through interstellar dust grains, but would have to have been generated *in situ*.

The xenon in carbonaceous chondrites generally contains excesses of the fissiogenic isotopes ^{131}Xe, ^{132}Xe, ^{134}Xe, and ^{136}Xe with relative abundance different from their fission yields in actinide nuclides, but about those expected for fission of a much heavier nuclide. The chief carrier of this component (xenon-X) has been traced to rare grains of ferrichromite, spinel, and carbyne. It is believed that this xenon owes its origin to spontaneous fission of an isotope of a superheavy element (SHE) of volatile character, $Z \approx$ 113–115, and $t_{1/2}$ >20 $\times$ 10^6 yr. However, this attribution is disputed.

In certain meteorites, neon contains an isotopic component (neon-E) that is rich in ^{22}Ne, possibly pure ^{22}Ne generated by the EC + β^+ decay of ^{22}Na ($t_{1/2}$ = 2.60 yr). Because of its short lifetime, ^{22}Na would have to have been incorporated into solid grains very close to the site of its nucleosynthesis, as suggested for ^{26}Al (see above). If this origin is established, it would affirm this mode of manifestation of extinct natural radioactivity and predict that still other nuclides of $t_{1/2} > 1$ yr may exhibit such manifestations.

Calcium-41 ($t_{1/2}$ = 100,000 yr) is such a nuclide. Limited observations have suggested the occurrence of potassium with enhanced ^{41}K:^{39}K in the same types of refractory carbonaceous-chondrite inclusions that exhibit radiogenic ^{26}Mg from ^{26}Al. These inclusions have very high Ca:K ratios, and the EC decay of ^{41}Ca would produce ^{41}K.

Neutron-activation analysis of mercury in some meteorites has occasionally yielded abnormally low ^{197}Hg:^{203}Hg ratios, possibly reflecting low ^{196}Hg:^{202}Hg ratios in the meteoritic mercury. The ^{202}Hg in these samples may be enhanced by the EC decay through ^{202}Tl of now-extinct ^{202}Pb ($t_{1/2}$ = 300,000 yr).

High temperature carbonaceous-chondrite inclusions exhibit an occasional apparent deficit of ^{135}Ba associated with low Cs:Ba. A possible interpretation is the incorporation into the inclusions of barium with low ^{135}Ba and ^{135}Cs ($t_{1/2}$ = ca 3.0 × 10^6 yr), whose β^- decay eliminated the ^{135}Ba deficit in other phases.

It has generally been considered that ^{60}Fe is too short-lived for naturally observable effects. However, its $t_{1/2}$, deduced from estimated bombardment cross-sections as 200,000 yr, is very uncertain. It could generate ^{60}Ni enhancements.

Radiogenic Heat

Heat Generation by Radioactivity. The kinetic energy of α and β particles, ejected electrons and recoil nuclei, the electromagnetic energy of γ and x radiations, and the mass-equivalent energy of positrons and their annihilation partners from radioactive disintegrations in matter are absorbed and promptly converted to thermal energy of that matter. However, neutrinos, which are emitted in all β processes, are virtually unabsorbed and escape to space even from the interiors of large planets and ordinary stars. The radiogenic heat of α and IT (isomeric transition) disintegrations is the total disintegration energy, equivalent to the mass difference between disintegrator and ultimate product. For β disintegrators, the neutrino energy must be subtracted. For β^- and β^+ emission, the neutrino energy spectrum and mean energy can be deduced from the spectra of the electrons. For approximate purposes, the mean neutrino energy can be assumed to be two thirds of the maximum electron energy for allowed low Z β^- transitions, somewhat higher for high Z and highly forbidden transitions, and one-third for β^+ emission. For EC, the total transition energy must be reduced by the weighted mean x-ray-level excitation of the product.

The total disintegration and thermal energies of the most prominent radiogenic heat sources among the primary natural radionuclides and some extinct natural radionuclides postulated to have been present in early planetary matter are listed in Table 5. For each, the heat-production rate of its element is calculated for the present and at the time of its possible incorporation into planetary matter ca 4.6 × 10^9 yr ago. Typical potassium, thorium, and uranium contents of various terrestrial and meteoritic materials and their calculated specific heat-generation rates at the present time are given in Table 6.

Table 5. Radiogenic Heat in Natural Radionuclides and Their Elements

Nuclide	Half-life, 10^9 yr	Disintegration modes (%)	Energy, MeV Disintegration	Energy, MeV Thermal	Isotopic abundance Now	Isotopic abundance 4.6×10^9 yr ago	Elementary heat-production rate, W/g Now	Elementary heat-production rate, W/g 4.6×10^9 yr ago
Extant								
^{238}U	4.468	$8\alpha + 6\beta^-$	51.70	48.8	0.99275	0.718	9.65×10^{-8}	7.01×10^{-8}
^{235}U	0.7038	$7\alpha + 4\beta^-$	46.41	44.4	0.00720	0.239	4.04×10^{-9}	1.35×10^{-7}
U total							1.005×10^{-7}	2.05×10^{-7}
^{232}Th	14.01	$6\alpha + 4\beta^-$	42.66	40.4	1.00	1.00	2.63×10^{-8}	2.63×10^{-8}
^{40}K	1.250	β^- (89.5)	1.312	0.59	0.000117	0.00150		
		EC (10.5)	1.505	1.46				
				0.68				
^{40}K total							3.4×10^{-12}	4.4×10^{-11}
^{147}Sm	106	α	2.31	2.31	0.151	0.150	4.64×10^{-11}	4.61×10^{-11}
^{87}Rb	48.8	β^-	0.273	ca 0.03	0.2783	0.279	ca 4×10^{-12}	ca 5×10^{-12}
Extinct								
^{244}Pu	0.081	$3\alpha + 2\beta^-$ (99.9)	17.08	15.2	0	1.00^a	0	
		SF (0.123)	ca 200	ca 190				
^{244}Pu total				15.4				$1.7 \times 10^{-6\ b}$
^{146}Sm	0.103	α	2.54	2.54	0	0.01^a	0	ca $3 \times 10^{-9\ b}$
^{129}I	0.0160	β^-	0.192	ca 0.02	0	0.0001^a	0	ca $2 \times 10^{-12\ b}$
^{26}Al	0.00072	β^+, EC	4.005	3.26	0	0.00005^a	0	ca $2 \times 10^{-8\ b}$
^{60}Fe	0.0001^a	$2\beta^-$	3.018	2.72	0	$10^{-10\ a}$		ca $1 \times 10^{-13\ b}$
	0.001^a					$10^{-7\ a}$		ca $1 \times 10^{-11\ b}$
	0.01^a					$10^{-4\ a}$		ca $1 \times 10^{-9\ b}$

^{a} Values assumed for illustrative calculations only.
^{b} Calculated using assumed values indicated by ^{a}.

Table 6. Typical Contents of Radioelements and Their Heat-Production Rates in Some Rocks and Minerals

Material	Mass fraction			Specific radiogenic heat-production rate, W/g			
	K, %	Th, ppm	U, ppm	K	Th	U	Total
silicic igneous rocks[a]	4.0	20	4.7	1.4×10^{-13}	5.3×10^{-13}	4.7×10^{-13}	1.1×10^{-12}
granodiorites	2.6	9	2.6	8.8×10^{-14}	2.4×10^{-13}	2.6×10^{-13}	5.9×10^{-13}
intermediate rocks[b]	1.0	8	2.0	3.4×10^{-14}	2.1×10^{-13}	2.0×10^{-13}	4.5×10^{-13}
mafic rocks[c]	0.5	2.7	0.9	1.7×10^{-14}	7.1×10^{-14}	9.0×10^{-14}	1.8×10^{-13}
abyssal tholeiite	0.3	0.35	0.1	1.0×10^{-14}	9.2×10^{-15}	1.0×10^{-14}	2.9×10^{-14}
eclogite	0.1	0.4	0.1	3.4×10^{-15}	1.1×10^{-14}	1.0×10^{-14}	2.4×10^{-14}
peridotite	0.01	0.04	0.01	3.4×10^{-16}	1.1×10^{-15}	1.0×10^{-15}	2.4×10^{-15}
dunite	0.001	0.05	0.014	3.4×10^{-17}	1.3×10^{-15}	1.4×10^{-15}	2.8×10^{-15}
"pyrolite"[d]	0.025	0.07	0.02	8.5×10^{-16}	1.8×10^{-15}	2.0×10^{-15}	4.7×10^{-15}
chondrites[e]	0.088	0.057	0.017	3.0×10^{-15}	1.5×10^{-15}	1.7×10^{-15}	6.2×10^{-15}
meteoritic iron	ca 0	ca 0	ca 0	ca 0	ca 0	ca 0	ca 0

[a] Includes granites.
[b] Diorites and quartz diorites.
[c] Includes continental basalts.
[d] Hypothetical material, consisting mostly of the constituents of pyroxene and olivine, believed by some to constitute the bulk of the earth's mantle below the lithosphere. Assumed to contain 21% Si and Th:U:Si as in type-I carbonaceous chondrites.
[e] Si content of ordinary chondrites (18%) with K:Th:U:Si of type-I carbonaceous chondrites.

Radiogenic Heat in the Earth. It was early discovered that radium preparations evolve heat, and that radium is ubiquitous in common rocks, so that a thin rocky crust could account for the heat flow from the earth's interior. It is now recognized that most of the terrestrial heat flow results from the radioactive disintegration of ^{238}U, ^{235}U, ^{232}Th (with their decay products), and ^{40}K distributed in the mantle and crust.

The surface heat flux of the earth is usually determined as the product of the temperature gradient measured *in situ* in boreholes, wells, tunnels, and ocean-sediment probes and the thermal conductivity of the rock or sediment measured in the laboratory. A great number of such measurements taken at many continental and oceanic locations reveal considerable variability, in many cases correlating with local rock composition and geologic environment. The mean heat fluxes are estimated to be ca 53 mW/m^2 for the continental crust, ca 100 mW/m^2 for the ocean basins, and ca 81 mW/m^2 for the terrestrial average. In volcanically active regions fluxes as high as ca 200–400 mW/m^2 are encountered (see Geothermal energy).

Radiogenic Heat and Planetary Evolution. During most of the past lifetimes of the terrestrial planets and the Moon, the dominant internal heat sources have probably been uranium, thorium, and potassium. This radiogenic heat has contributed to internal differentiation and volcanic activity in those bodies. For smaller bodies, however, heat conduction to and radiation from the surfaces would prevent long-lived radioactivity from creating internal temperatures sufficiently high for melting. Yet some asteroids and meteorites show evidences of igneous differentiation early in the history of the solar system. It has been postulated that extinct natural radioactivity may have been responsible. Aluminum-26 has been most frequently mentioned, but its short half-life (0.72×10^6 yr) and the absence of ubiquitous ^{26}Mg anomalies makes its thermal importance questionable. If the half-life of ^{60}Fe (now poorly determined as ca 0.2×10^6 yr) is actually longer than ca 10^6 yr, then that nuclide could have been a significant source of heat in the early solar system.

Uses

Medical Radiology. Becquerel found that γ rays, like x rays, could produce absorption-contrast radiographs, but they never became competitive with the latter for medical diagnostic radiography (see X-ray technology). Early workers established that radioactive radiations were damaging to living tissue and could be lethal to micro- and macroorganisms. Therapeutic researches on radium were first undertaken around 1900, and the foundations of therapeutic radiology were laid a few years later. First radium and then radon were found to be highly effective in the treatment of both superficial abnormalities and deep-seated cancers, the latter because of the great penetrability of high energy γ rays as compared to x radiation available before the 1930s. The curative powers of radioactivity were for a while overestimated, and radioactive springs became popular as health resorts in the first half of the century. However, it is now recognized that ionizing radiations have only destructive effects on biological matter. Many early experimenters, radiologists, and patients died of radiation-induced cancer.

Radium and radon are used mainly because of their γ radiations. Sealed radium sources develop equilibrium levels of γ radiation in about a month. Radon sources develop their maximum γ intensity within hours, then decay with the 3.823-d $t_{1/2}$. Implanted sources are often prepared to deliver the desired doses over the radon lifetime and left in place indefinitely. A great variety of source holders have been developed for various external and internal applications. Radium Chemical Co. is the chief supplier of these in the United States.

In the 1930s, high energy electron accelerators were developed and used to produce x rays of γ-ray and higher energies for medical therapy. Later on, synthetic radionuclides, especially ^{60}Co, ^{137}Cs, and ^{192}Ir, became available and have largely supplanted natural radionuclides for therapeutic use (see Radioactive drugs).

Industrial Radiography. Radiography for technological and industrial purposes was dominated by x rays until it was demonstrated that high energy radium-series γ rays could radiograph metallic objects of great thickness.

As with medical sources, radon offers practical advantages over radium. More importantly, an active radon source can be made much smaller than the radium from which it was derived, yielding sharper radiographs. For this purpose, an activated-charcoal button is placed at the bottom of a tube, the radon is adsorbed on the carbon, and the tube is sealed. With this procedure ca 5×10^{10} Bq (1.5 Ci) Rn can be concentrated into a volume of ca 0.1 mm^3. However, today very little natural radioactivity is used for such purposes (see Nondestructive testing).

Luminescent Materials. An important use of concentrated α-emitting substances has been as radiation sources for phosphors—usually ZnS—for illuminating clock, watch, and instrument dials, for signs visible in the dark, and similar applications. Unfortunately, many workers in the early years of this industry, in the habit of moistening the paint brushes with saliva, suffered cancer of the mouth, often fatally. This usage increased during World War II, but subsequently α emitters have been almost completely replaced by artificial β emitters (see Luminescent materials, phosphors).

The chief α source was ^{226}Ra, but impure ^{224}Ra, ^{228}Th, ^{210}Po, and ^{210}(Pb,Bi,Po) were sometimes used. Typically, ca 1 mg radium or an activity-equivalent amount of mesothorium was mixed with 1 kg ZnS and small amounts of activator, binder, etc. Besides the hazards of manufacture, user hazards resulted from the γ radiation and radon leakage. Moreover, the concentrated α-radiation damage results in rapid decrease of brightness and limited lifetimes of devices, whose disposal presents additional radiation hazards.

The artificial β-emitting nuclides currently used for self-luminous paints and devices are ^{3}H, ^{90}(Sr,Y), ^{85}Kr, and ^{147}Pm; the latter is preferred for most applications.

Ionization Devices. Static electricity is a serious problem in many industrial operations. It produces electrical discharges with fire hazard, causes dust to adhere to optical surfaces and photographic negatives, damages semiconductor electronic components, etc. Ionization of the ambient air provides a conduction path for discharging static and preventing its buildup. Air can be ionized in a separate chamber or tube and blown over the devices or workspace. The preferred radiation source is ^{210}Po; ^{3}H is also used. The polonium is incorporated in ceramic microspheres bound to a tape with epoxy resin. Typically, a few GBq (tens of mCi) are used, and the source must be replaced annually. (The decay factor for a $t_{1/2}$ of 138 d in 365 d is 0.16.) Although it is a natural radionuclide, ^{210}Po is now manufactured artificially by irradiation of Bi-metal slugs in nuclear reactors.

Some smoke detectors use α-emitting ionization sources. Although radium and ^{210}Po have been used, γ radiation and radon escape make the former undesirable, and the sources of the latter have to be frequently replaced. Reactor-produced ^{241}Am is used almost exclusively in such commercial devices today.

Specialized Radiation Sources. Natural radioactive substances provided the only sources of high-energy projectiles and radiations for research into the structure of matter until the development of ion accelerators in 1932.

Most early neutron research employed sources made of polonium, radon, or radium intimately mixed with beryllium, eg, the work on slow neutrons by Fermi and collaborators in 1934–1935 and the discovery of nuclear fission by Hahn and Strassmann in 1939. Portable Ra–Be neutron sources were used extensively on the Manhattan Project in World War II, and some are still in use today. However, today ^{252}Cf, a synthetic spontaneously fissioning nuclide, is preferred.

A convenient laboratory α-energy standard can be prepared by placing a flat receptor into a closed vessel containing a highly emanating preparation of ^{224}Ra or ^{228}Th. In a few hours, the plate acquires the ^{220}Th (thoron) active deposit, in which ^{212}Pb (ThB) ($t_{1/2}$ = 10.64 d) supports ^{212}Bi (ThC) and ^{212}Po (ThC′) with intense α-particles from 6.051 to 8.784 MeV and weak ones from 5.302 to 10.543 MeV; this can be used for several days. Longer-lived monoenergetic (5.305 MeV) α sources are prepared from ^{210}Po.

Radium γ sources have been used in a variety of thickness and density gauges in metal, paper, plastics, and other industries, for measuring the density of coal slurries, etc. Although generally less effective than those employing artificial radionuclides, they do not have to be licensed in the United States by the NRC, and many are still in use.

Intense γ sources are used in well and borehole logging, in which the Compton-scattered γ intensities measured by a detector shielded from the source by a heavy-metal plug provide information about the density and other characteristics of the strata. Radium sources have been and still are used, but have been largely replaced by ^{137}Cs and ^{60}Co. Their monoenergetic (or nearly so) γ rays that pass through the absorber can be electronically discriminated against in favor of the lower energy scattered γ rays to give higher signal-to-background ratios than is possible with radium.

BIBLIOGRAPHY

"Radioactive Elements, Natural" in *ECT* 1st ed., Vol. 11, pp. 438–462, by Boris Pregel, Canadian Radium & Uranium Corporation; "Radioactive Elements, Natural," in *ECT* 2nd ed., Vol. 17, pp. 9–34, by Boris Pregel, Conrad Precision Industries, Inc.

General, Historical, and Nuclear References

L. Badash, *Radioactivity in America: Growth and Decay of a Science*, Johns Hopkins University Press, Baltimore, Md., 1979 (Period 1900–1920).
G. Friedlander, J. W. Kennedy, E. S. Macias, and J. M. Miller, *Nuclear and Radiochemistry*, 3rd ed., John Wiley & Sons, Inc., New York, 1981.
C. S. Garner, "Radioactivity Applied to the Discovery and Investigation of the Newer Elements" in A. C. Wahl and N. A. Bonner, eds., *Radioactivity Applied to Chemistry*, John Wiley & Sons, Inc., New York, 1951, Chapt. 7, pp. 179–243.
N. E. Holden, *Isotopic Composition of the Elements and Their Variation in Nature: A Preliminary Report*, U.S. Energy Research and Development Administration Report, **BNL-NCS-50605**, Brookhaven National Laboratory, Upton, N.Y., 1977.

E. K. Hyde, I. Perlman, and G. T. Seaborg, "Natural Radioactivity of the Heavy Elements" in *The Nuclear Properties of the Heavy Elements*, Vol. II, Prentice-Hall, Inc., Englewood, Cliffs, N.J., 1964, Chapt. 6, pp. 409–576.

G. E. M. Jancey, "The Early Years of Radioactivity," *Am. J. Phys.* **14,** 226 (1946) (Period 1896–1904).

C. M. Lederer and V. S. Shirley, eds., *Table of Isotopes*, 7th ed., John Wiley & Sons, Inc., New York, 1978.

F. Soddy, "Radioactivity," *Ann. Repts. Prog. Chem.* **1–17,** (1905–1921).

The ^{238}U, ^{235}U, and ^{232}Th Radioactive Series

J. A. S. Adams and P. Gasparini, *Gamma-Ray Spectrometry of Rocks*, Elsevier North-Holland, Inc., Amsterdam, The Netherlands, 1970.

J. A. S. Adams and W. M. Lowder, eds., *The Natural Radiation Environment*, University of Chicago Press, Chicago, Ill., 1964.

J. A. S. Adams, W. M. Lowder, and T. F. Gesell, eds., *The Natural Radiation Environment II*, Vols. 1 and 2, U.S. Energy Research and Development Administration, 1972 (National Technical Information Service, Springfield, Va. 22161).

J. G. Baird and S. Nargolwalla, *Nuclear Exploration Techniques*, Scintrex Limited, Concord, Ontario, Can., 1978.

T. F. Gesell and W. M. Lowder, eds., *Natural Radiation Environment III*, Vols. 1 and 2, Technical Information Center, U.S. DOE, Washington, D.C., 1980 (National Technical Information Service, Springfield, Va. 22161).

R. L. Grasty, "Uranium Measurement by Airborne Gamma-Ray Spectrometry," *Geophysics* **40,** 503 (1975).

T. P. Kohman, "Natural Radioactivity" in H. Blatz, ed., *Radiation Hygiene Handbook*, McGraw-Hill Inc., New York, 1959, Chapt. 6, pp. 2–32. (γ radiations; U, Th, and K minerals; radioactivity of rocks.)

W. Rubinson, "The Equations of Radioactive Transformation in a Neutron Flux," *J. Chem. Phys.* **17,** 542 (1949).

R. H. Steiger and E. Jäger, "Subcommittee on Geochronology: Convention on the Use of Decay Constants in Geo- and Cosmochronology," *Earth Plan. Sci. Lett.* **36,** 359 (1977).

K. K. Turekian, Y. Nozaki, and L. K. Benninger, "Geochemistry of Atmospheric Radon and Radon Products," *Ann. Rev. Earth Plan. Sci.* **5,** 227 (1977).

Induced Natural Radioactivity

P. E. Damon, J. C. Lerman, and A. Long, "Temporal Fluctuations of Atmospheric ^{14}C: Causal Factors and Implications," *Ann. Rev. Earth. Plan. Sci.* **6,** 457 (1979).

M. Honda and J. R. Arnold, "Effects of Cosmic Rays on Meteorites" in *Handbuch der Physik*, Vol. 46, Part 2, Springer-Verlag, Berlin, 1967, pp. 613–634.

T. P. Kohman and M. L. Bender, "Nuclide Production by Cosmic Rays in Meteorites and on the Moon" in B. S. P. Shen, ed., *High-Energy Nuclear Reactions in Astrophysics*, W. A. Benjamin, New York, 1967, Chapt. 7, pp. 169–245.

D. Lal and B. Peters, "Cosmic Ray Produced Radioactivity in the Earth" in *Handbook der Physik*, Vol. 46, Part 2, Springer-Verlag, Berlin, 1967, pp. 551–612.

R. C. Reedy and J. R. Arnold, "Interaction of Solar and Galactic Cosmic-Ray Particles With the Moon," *J. Geophys. Res.* **77,** 537 (1972).

J. P. Shedlovsky and co-workers, "Pattern of Bombardment-Produced Radionuclides in Rock 10017 and in Lunar Soil," in *Proc. Apollo 11 Lun. Sci. Conf.*, Pergamon Press Inc., New York, 1970, pp. 1503–1532.

Extinct Natural Radioactivity

E. Anders, "A Superheavy Element in Meteorites?" in W. O. Milligan, ed., *Cosmochemistry*, Robert O. Welch Foundation, Houston, Texas, 1978, pp. 246–263.

F. Begemann, "Isotopic Anomalies in Meteorites," *Rep. Prog. Phys.* **43,** 1309 (1980).

T. Kaiser, W. R. Kelly, and G. J. Wasserburg, "Isotopically Anomalous Silver in the Santa Clara and Piñon Iron Meteorites," *Geophys. Res. Lett.* **7,** 271 (1980).

T. Lee, D. A. Papanastassiou, and G. J. Wasserburg, "Aluminum-26 in the Early Solar System: Fossil or Fuel?," *Astrophys. J.* **211,** L107 (1977).

T. Lee, "New Isotopic Clues to Solar System Formation," *Rev. Geophys. Space Phys.* **17,** 1591 (1979).

G. W. Lugmair and K. Marti, "Sm-Nd-Pu Timepieces in the Angra dos Reis Meteorite," *Earth Plan. Sci. Lett.* **35,** 273 (1977).

J. H. Reynolds, "Isotope Cosmochemistry: The Rare Gas Story and Related Matters" in W. O. Milligan, ed., *Cosmochemistry*, Robert O. Welch Foundation, Houston, Texas, 1978, Chapt. 5, pp. 201–244.

Radiochemistry of the Elements Tl–U

The publications in the series *Radiochemistry of the Elements* prepared under the sponsorship of the U.S. National Academy of Science and National Research Council in the broader *Nuclear Science Series*, designated here as NAS-NRC Publ. NS-30. ., are available from, inter alia, the National Technical Information Service, U.S. Department of Commerce, Springfield, Va. 22161. The series is undergoing revision, and parts should become available during the early 1980s.

E. Anders, "Technetium and Astatine Chemistry," *Ann. Rev. Nucl. Sci.* **9,** 203 (1959).

E. H. Appelman, *The Radiochemistry of Astatine*, NAS-NRC Publ. NS-3012, 1960.

K. W. Bagnell, *Chemistry of the Rare Elements: Polonium-Actinium*, Butterworths, London, 1957. (Po, Fr, Rn, Ra, Ac).

D. Brown and A. G. Maddock, "The Analytical Chemistry of Protactinium" in H. A. Elson and D. C. Stewart, eds., *Progress in Nuclear Energy, Ser. IX: Analytical Chemistry*, Vol. 8, Part 1, Pergamon Press Inc., Oxford, UK, 1967, Chapt. 1, pp. 1–47.

R. E. Elson, *The Chemistry of Protactinium* in G. T. Seaborg and J. J. Katz, eds., *The Actinide Elements*, McGraw-Hill Inc., New York, 1954, Chapt. 5, pp. 103–129.

W. D. Fairman and J. Sedlet, "Direct Determination of Lead-210 by Liquid Scintillating Counting," *Anal. Chem.* **40,** 2004 (1968).

P. E. Figgins, *The Radiochemistry of Polonium*, NAS-NRC Publ. NS-3037, 1961.

R. L. Fleischer, P. B. Price, and R. M. Walker, "Element Mapping and Isotopic Analysis," in *Nuclear Tracks in Solids*, University of California Press, Berkeley, Calif., 1975, Chapt. 8, pp. 489–520.

W. M. Gibson, *The Radiochemistry of Lead*, NAS-NRC Publ. NS-3040, 1961.

J. E. Grindler, *The Radiochemistry of Uranium*, NAS-NRC Publ. NS-3050, 1962.

F. T. Hagemann, "The Chemistry of Actinium" in G. T. Seaborg and J. J. Katz, eds., *The Actinide Elements*, McGraw-Hill, Inc., New York, 1954, Chapt. 2, pp. 14–44.

O. Hahn, *Applied Radiochemistry*, Cornell University Press, Ithaca, N.Y., 1936.

E. K. Hyde, "Radiochemical Separations of the Actinide Elements" in G. T. Seaborg and J. J. Katz, eds., *The Actinide Elements*, McGraw-Hill Inc., New York, 1954, Chapt. 15, pp. 542–595 (Ac, Th, Pa, U).

E. K. Hyde, *The Radiochemistry of Francium*, NAS-NRC Publ. NS-3003, 1960.

E. K. Hyde, *The Radiochemistry of Thorium*, NAS-NRC Publ. NS-3004, 1960.

H. W. Kirby, *The Radiochemistry of Protactinium*, NAS-NRC Publ. NS-3016, 1959.

H. W. Kirby and M. L. Salutsky, *The Radiochemistry of Radium*, NAS-NRC Publ. NS-3057, 1964.

H. W. Kirby, "The Analytical Chemistry of Actinium" in H. A. Elson and D. C. Stewart, eds., *Progress in Nuclear Energy, Ser. IX: Analytical Chemistry*, Vol. 8, Part 1, Pergamon Press Inc., Oxford, UK, 1967, Chapt. 3, pp. 89–139.

F. F. Momyer, Jr., *The Radiochemistry of the Rare Gases*, NAS-NRC Publ. NS-3025, 1960 (Rn).

J. Sedlet, "Actinium, Astatine, Francium, Polonium, and Protactinium" in I. M. Kolthoff and P. J. Elving, eds., *Treatise on Analytical Chemistry*, Part II, Vol. 6, Wiley-Interscience, New York, 1964, pp. 435–610.

J. Sedlet, "Radon and Radium" in I. M. Kolthoff and P. J. Elving, eds., *Treatise on Analytical Chemistry*, Part II, Vol. 4, Wiley-Interscience, New York, 1966, pp. 219–366.

H. L. Smith, ed., *Collected Radiochemical Procedures*, U.S. Energy Research and Development Administration Report, **LA-1721,** 4th ed., 1976.

P. C. Stevenson and W. E. Nervik, *The Radiochemistry of the Rare Earths, Scandium, Yttrium, and Actinium*, NAS-NRC Publ. NS-3020, 1961.

A. C. Wahl, "Emanation Methods" in A. C. Wahl and N. A. Bonner, eds., *Radioactivity Applied to Chemistry*, John Wiley & Sons, Inc., New York, 1951, Chapt. 9, pp. 284–310 (Rn).

Production of Natural Radioactive Substances

J. A. T. Dawson, "Radon: Its Properties and Preparation for Industrial Radiography," *J. Sci. Inst.* **23**, 138 (1946).
O. Erbacher and H. Käding, "Mechanism of Precipitation of a Radium Preparation of High Emanating Power," *Z. Physik. Chem.* **A149**, 439 (1930); *Science Abst.* **34** (523), (1931).
W. A. Jennings and S. Russ, *Radon: Its Technique and Use*, John Murray, London, UK, 1948.
S. F. Johnstone, *Minerals for the Chemical and Allied Industries*, John Wiley & Sons, Inc., New York, 1954. (Production and early uses of mesothorium and radium, pp. 523–538, 597–612.)
M. Pochon, "Radium Recovery. Canada's Unique Chemical Industry," *Chem. Met. Eng.* **44**, 362 (1937).
J. E. Rose and R. W. Swain, "Remote-Controlled Radium Emanation Plant and Automatic Measuring Apparatus for Gold Implants," *J. Nat. Cancer Inst.* **10**, 605 (1949).
Various authors, *Minerals Yearbook*, Bureau of Mines, U.S. Department of the Interior, Washington, D.C., annual volumes. (Information on production, commerce, and uses of radium and occasionally other natural radioactive substances, under "Minor Metals" or other headings.)

Radiometric Dating

I. Almodóvar, *Thorium Isotopes Method for Dating Marine Sediments*, Thesis, Carnegie Institute of Technology, Pittsburgh, Pa., 1960; U.S. Atomic Energy Commission Report **NYO-8919**; and T. P. Kohman, "idem," *Preprints, Div. Petrol. Chem., Am. Chem. Soc.* **5**, 165 (1960) "Method for the Isolation of Thorium from Siliceous Materials, *Anal. Chim. Acta* **33**, 426 (1965).
E. Anders, "Meteorite Ages," *Rev. Mod. Phys.* **34**, 287 (1962); in B. Middlehurst and G. P. Kuiper, eds., *The Solar System, Vol. IV: The Moon, Meteorites, and Comets*, University of Chicago Press, Chicago, Ill., 1963, Chapt. 13, pp. 402–495.
G. B. Dalrymple and M. A. Lanphere, *Potassium-Argon Dating*, W. H. Freeman and Co., San Francisco, Calif., 1969.
B. R. Doe, *Lead Isotopes*, Springer-Verlag, New York, 1970.
D. Elmore and co-workers, "Analysis of ^{36}Cl in Environmental Water Samples Using an Electrostatic Accelerator," *Nature* **277**, 22, 246 (1979).
G. Faure and J. L. Powell, *Strontium Isotope Geology*, Springer-Verlag, New York, 1972.
G. Faure, *Principles of Isotope Geology*, John Wiley & Sons, Inc., New York, 1977.
R. L. Fleischer, P. B. Price, and R. M. Walker, "Earth and Space Sciences," *Nuclear Tracks in Solids*, University of California Press, Berkeley, Calif., 1975, Pt. II, pp. 157–431.
M. A. Forman and O. A. Schaeffer, "Cosmic Ray Intensity Over Long Time Scales," *Rev. Geophys. Space Phys.* **17**, 552 (1979).
W. A. Fowler, "Nuclear Cosmochronology" in W. O. Milligan, ed., *Cosmochemistry*, Robert O. Welch Foundation, Houston, 1978, Chapt. 3, pp. 61–93.
E. I. Hamilton, *Applied Geochronology*, Academic Press, Inc., London, UK, 1965.
E. I. Hamilton and R. M. Farquhar, eds., *Radiometric Dating for Geologists*, Wiley-Interscience, Chichester, UK, 1968.
B. Keisch, R. L. Feller, A. S. Levine, and R. R. Edwards, "Dating and Authentication of Works of Art By Measurement of Natural Alpha Emitters," *Science* **155**, 1238 (1967).
B. Keisch, "Dating Works of Art Through Their Natural Radioactivity: Improvements and Applications," *Science* **160**, 413 (1968).
T. Kirsten, "Time and the Solar System" in S. F. Dermott, ed., *The Origin of the Solar System*, John Wiley & Sons, Inc., Chichester, UK, 1978, pp. 267–346.
T. P. Kohman, "Chronology of Nucleosynthesis and Extinct Natural Radioactivity," *J. Chem. Educ.* **38**, 73 (1961).
T.-L. Ku, "The Uranium-Series Methods of Age Determination," *Ann. Rev. Earth Plan. Sci.* **4**, 347 (1976).
W. F. Libby, *Radiocarbon Dating*, 2nd ed., University of Chicago Press, Chicago, Ill., 1955.
R. A. Muller, "Radioisotope Dating With a Cyclotron," *Science* **196**, 489 (1977).
R. K. O'Nions, S. R. Carter, N. M. Evenson, and P. J. Hamilton, "Geochemical and Cosmochemical Applications of Nd Isotope Analysis," *Ann. Rev. Earth Plan. Sci.* **7**, 11 (1979).
G. M. Raisbeck, F. Yiou, M. Fruneau, and J. M. Loiseaux, "Beryllium 10 Mass Spectrometry With a Cyclotron," *Science* **202**, 215 (1978).
R. D. Russell and R. M. Farquhar, *Lead Isotopes in Geology*, Interscience Publishers, New York, 1960.

T. P. Sarma, *Dating of Marine Sediments by Ionium and Protactinium Methods*, Thesis, Carnegie Institute of Technology, Pittsburgh, Pa., 1964; U.S. Atomic Energy Commission Report, **NYO-8925**.
O. A. Schaeffer and J. Zähringer, eds., *Potassium Argon Dating*, Springer-Verlag, New York, 1966.

Radiogenic Heat

D. E. Alburger, "Heat Production in Potassium," *Phys. Rev.* **81,** 888 (1951).
S. P. Clark, Jr., Z. E. Peterman, and K. S. Heir, "Abundances of Uranium, Thorium, and Potassium," in S. P. Clark, Jr., ed., *Handbook of Physical Constants*, rev. ed., Geological Society of America, New Haven, Ct., 1966, Section 24, pp. 521–541.
G. F. Davies, "Review of Oceanic and Global Heat Flow Estimates," *Rev. Geophys. Space Phys.* **18,** 718 (1980).
A. M. Jessop, M. Hobart, and J. G. Sclater, *World-Wide Compilation of Heat-Flow Data*, Geothermal Series No. 5, Dept. of Energy, Mines, and Resources, Ontario, Can., 1976.
G. M. Jones, "The Thermal and Dynamic State of the Earth," *EOS* **62,** 609 (1981).
T. P. Kohman and M. S. Robison, *Iron-60 as a Possible Heat Source and Chronometer in the Early Solar System*, Lunar and Planetary Science XI, 564, 1980.
B. Mason, "Cosmochemistry: Meteorites," Chapt. B, Part 1 of M. Fleisher, ed., *Data of Geochemistry*, 6th ed., *U.S. Geological Survey Prof. Paper* **444-B-1,** U.S. Government Printing Office, Washington, D.C., 1977.
A. E. Ringwood, *Composition and Petrology of the Earth's Mantle*, McGraw-Hill, Inc., New York, 1975.
A. E. Ringwood, *Origin of the Earth and Moon*, Springer-Verlag, New York, 1979.
J. G. Sclater, C. Jaupert, and D. Galen, "The Heat Flow Through Oceanic Crust and the Heat Loss of the Earth," *Rev. Geophys. Space Phys.* **18,** 269 (1980).
D. J. Stevenson, "Models of the Earth's Core," *Science* **214,** 611 (1981).
M. N. Toksoz, A. T. Hsui, and D. H. Johnston, "Thermal Evolution of the Terrestrial Planets," *Moon and Planets* **18,** 281 (1978).
J. A. Wood, "Review of the Metallographic Cooling Rates of Meteorites and a New Model for the Planetesimals in Which They Formed" in T. Gehrels, ed., *Asteroids*, University of Arizona Press, Tucson, Ariz., 1979, pp. 849–891.

TRUMAN P. KOHMAN
Carnegie-Mellon University

RADIOCHEMICAL ANALYSIS AND TRACER APPLICATIONS. See Radioactive tracers.

RADIOCHEMICAL TECHNOLOGY. See Radiation curing.

RADIOGRAPHY. See X-ray technology.

RADIOIMMUNOASSAY. See Radioactive drugs.

RADIOISOTOPES

Radioactivity was discovered in 1896 by Becquerel following the discovery of x rays by Roentgen in the previous year. The penetrating radiation emitted was later shown to consist of the particulate α and β rays and the electromagnetic γ rays. By 1898 β rays were shown to be identical to electrons, and by 1909 α rays were shown to be identical to nuclei of the element helium.

In 1911 Soddy (1) established the existence of chemically identical atoms with different atomic weights, for which he proposed the name isotopes (qv).

Decay Constants, Half-Lives

Radioisotopes decay spontaneously by the emission of particle or electromagnetic radiation. In the context of this article, the term radioisotope can refer to either the ground state or an isomeric state of a nucleus. Each radioisotope, whether a ground state or an isomeric state, is characterized by a probability of decay per unit time, λ, also called the total radioactive decay constant. For radioisotopes that decay by more than one mode, a partial decay constant, λ_i, is associated with each mode; thus $\lambda = \Sigma_i \lambda_i$. An example of such a nuclear species is ^{40}K, which decays by β^- emission to ^{40}Ca ($\lambda_\beta = 0.486 \times 10^{-9}$ yr^{-1}), and by electron capture decay to ^{40}A ($\lambda_\epsilon = 0.0570 \times 10^{-9}$ yr^{-1}). All other partial decay constants are zero. The total decay constant for ^{40}K is thus $\lambda = \lambda_\beta + \lambda_\epsilon = 0.543 \times 10^{-9}$ yr^{-1}.

The decay constant λ determines the rate of decay of a nuclear species. Given an initially large number of atoms, N(0) at a time $t(0) = 0$, of a radioisotope with decay constant λ, the number of atoms N, present at a later time t, is given by the well-known exponential relation

$$N = N(0)e^{-\lambda t} \qquad (1)$$

The number of disintegrations per unit time at time t, called the activity, is given by the product λN. From equation 1, the activity at time t is related to the activity at time $t(0) = 0$ by

$$\frac{dN}{dt} = -\lambda N = -N(0)\lambda e^{-\lambda t} \qquad (2)$$

Thus both the number of radioactive atoms and the activity of these atoms decrease exponentially with time.

Concerning the decay constant, two fundamental assumptions are made: λ is the same for all atoms of a given nuclear species, and λ is independent of time, that is, the probability of decay of an atom is independent of the age of that atom. No experimental evidence is known that contradicts these assumptions. Without the first assumption, the concept of a decay constant would have no meaning. The second assumption is already implicit in the derivation of equation 1.

A further assumption usually made concerning λ is that it is unaffected by external conditions such as temperature, pressure, or chemical state. This assumption is valid to a high degree; however, recent experiments have shown some variation in λ for a few radioisotopes under certain conditions (2). The measured effects are slight and unimportant for most chemical or biological uses (3–4).

It should be emphasized that the preceding discussion refers to neutral atoms. Dramatic changes in λ can occur for atoms that are stripped, even partially, of their orbital electrons. For example, a radioisotope that decays by electron capture only would, if fully stripped, become stable. Similarly, for an isomeric transition, λ would become zero for that fraction of the decays that would normally proceed by internal conversion. A special case of β^- decay, where orbital stripping dramatically affects λ, is ^{187}Re with a decay energy to ^{187}Os of only 2.6 keV. The atomic binding energies of ^{187}Re and ^{187}Os are such that a completely stripped ^{187}Re nucleus would have a mass less than that of ^{187}Os and would thus be stable against β^- decay. Although these effects can in principle be significant, in practice they are unimportant.

Rather than the decay constant λ, a more commonly used characteristic of a radioisotope is the half-life $t_{1/2}$, defined as the time interval during which the number of atoms of that radioisotope decreases by one-half. From equation 1, with N = N(0)/2:

$$t_{1/2} = \ln 2/\lambda = 0.693/\lambda \qquad (3)$$

Whenever a decay mode is energetically allowed, then $\lambda \neq 0$ for that decay mode and the nuclide in question is, by definition, radioactive. However, the association of a unique half-life with each radioisotope leads to an empirical definition of nuclear stability based on the upper limit of measurability of the quantity $t_{1/2}$. The present limit is about 10^{14} to 10^{19} yr for single-process decays, depending upon the type of emission. Longer $t_{1/2}$ values have been measured for a few nuclides that can decay by double β^- or double ϵ emission (5–7). Each element from hydrogen to bismuth, with the exception of technetium (Z = 43) and promethium (Z = 61), has at least one stable isotope. All isotopes above bismuth are unstable.

Naturally Occurring Radioisotopes

Some radioisotopes occur in nature. They are classified according to their origin (see also Radioactivity, natural).

Primary. These radioisotopes have half-lives greater than about 10^9 yr, a value comparable to the age of the earth. They are assumed to have been present since the formation of the earth's crust (see Table 1).

Secondary. Radioisotopes in this group are decay products of the three chains beginning with the primary isotopes ^{235}U, ^{238}U, and ^{232}Th. Each chain proceeds by a series of α and β decays and ends in a stable isotope of lead (see Tables 2–4).

Induced. These radioisotopes have half-lives much shorter than the age of the earth, but their loss owing to decay is compensated for by their continual production by nuclear reactions occurring in nature. High-energy cosmic rays interacting with nuclei in the atmosphere can break up these nuclei into neutrons and residual fragments. Although some fragments may themselves be radioactive, the neutrons, n, give rise to most of the naturally occurring short-lived radioisotopes. The interaction of neutrons with nitrogen produce ^{3}H (tritium) and ^{14}C:

$$^{14}N + {}'n \begin{cases} \rightarrow {}^{12}C + {}^3H \\ \rightarrow {}^{14}C + {}^1H \end{cases}$$

whereas ^{239}Pu is formed by

Table 1. Naturally Occurring Radioisotopes

Radioisotope	Isotopic abundance, %	Half-life, yr[a]	Decay mode	Q-value, keV[a]	Daughter radioisotope
^{40}K	0.012	1.277×10^9 8	β^-	1311.6 5	^{40}Ca
			ϵ	1505.0 6	^{40}Ar
^{87}Rb	27.83	4.80×10^{10} 13	β^-	273.3 19	^{87}Sr
^{113}Cd	12.2	9.3×10^{15} 19	β^-	322 5	^{113}In
^{115}In	95.7	4.41×10^{14} 25	β^-	495 8	^{115}Sn
^{123}Te	0.89	$>10^{13}$	ϵ	52 23	^{123}Sb
^{138}La	0.089	1.35×10^{11} 16	β^-	1041 12	^{138}Ce
			ϵ	1749 5	^{138}Ba
^{144}Nd	23.8	2.4×10^{15} 3	α	1910.3 31	^{140}Ce
^{147}Sm	15.1	1.06×10^{11} 2	α	2310.5 15	^{140}Nd
^{148}Sm	11.3	8×10^{15} 2	α	1986.2 13	^{144}Nd
^{152}Gd	0.20	1.08×10^{14} 8	α	2206.2 34	^{148}Sm
^{174}Hf	0.16	2.0×10^{15} 4	α	2504 6	^{170}Yb
^{176}Lu	2.6	3.60×10^{10} 16	β^-	1186.5 22	^{176}Hf
^{180}Ta	0.012	$>10^{13}$	β^-	710 11	^{180}W
			ϵ	865 13	^{180}Hf
^{186}Os	1.58	2.0×10^{15} 11	α	2816.5 28	^{182}W
^{187}Re	62.60	5×10^{10}	β^-	2.64 4	^{187}Os
^{190}Pt	0.013	6×10^{11}	α	3243 20	^{186}Os

[a] Each one or two of the uncertainty digits (the numbers after the spaces) refers to the last one or two digits of the numerical value.

$$^{238}\text{U} + n \rightarrow {}^{239}\text{U} \xrightarrow{\beta^-} {}^{239}\text{Np} \xrightarrow{\beta^-} {}^{239}\text{Pu}$$

and ^{242}Pu and ^{244}Pu are formed by multiple neutron capture on, for example, ^{238}U and ^{239}Pu.

Artificially-Produced Radioisotopes

By far the largest group of radioisotopes are produced in the laboratory, either by charged-particle or neutron-induced reactions on targets of stable isotopes. Cobalt-60, for example, is commonly produced by the (n,γ) reaction on ^{59}Co (relative isotopic abundance = 100%). Cobalt-58 is produced by the (p,n) reaction on ^{56}Fe (relative isotopic abundance = 91.7%). Iodine-131 is a product of the neutron-induced fission of ^{235}U. Although nuclear reactions have also been carried out on a few radioactive targets, such as ^{210}Bi (isomeric state, $t_{1/2} = 3.0 \times 10^6$ yr) (8), these are specialized cases not suitable for commercial production. So far, more than 2100 radioisotopes have been artificially produced (see Actinides and transactinides).

The above methods are suitable for the production of radioisotopes that have half-lives long enough for production and use without significant loss of activity. Some short-lived radioisotopes can be obtained in radioisotope generators and separated, usually in an ion-exchange column, from the relatively long-lived parent. Preparation of ^{99}Tc ($t_{1/2} = 6.02$ h) from ^{99}Mo ($t_{1/2} = 66.0$ h) is routinely carried out in such a generator. A list of radioisotopes available from generators is given in ref. 9. For a discussion of radiochemical methods, see Analytical methods, Vol. 2, p. 64.

Table 2. ^{232}Th Decay Chain

Parent radioisotope	Half-life^a	Decay mode	Branching, %^{a}	Q-value, keV^a	Daughter radioisotope
^{252}Cf	2.638 yr 10	α	99.908 8	6217.1 5	^{248}Cm
		SF	0.092 8		
^{248}Cm	3.39×10^5 yr 3	α	91.74 3	5162 4	^{244}Pu
		SF	8.26 3		
^{244}Pu	8.26×10^7 yr 9	α	99.875 6	4665.5 10	^{240}U
		SF	0.125 6		
^{244}Cm	18.11 yr 2	α	100	5901.8 1	^{240}Pu
^{240}U	14.1 h 2	β^-	100	500 60	^{240}Np
^{240}Np	7.4 min 2	β^-	100	2090 60	^{240}Pu
^{240}Pu	6537 yr 10	α	100	5255.96 15	^{236}U
^{236}U	2.3145×10^7 yr 14	α	100	4569.8 21	^{232}Th
^{232}Th	1.405×10^{10} yr 6	α	100	4081 4	^{228}Ra
^{228}Ra^b	5.75 yr 3	β^-	100	45.6 10	^{228}Ac
^{228}Ac^c	6.13 h	β^-	100	2137 7	^{228}Th
^{228}Th^d	1.9131 yr 9	α	100	5520.26 25	^{224}Ra
^{224}Ra	3.66 d 4	α	100	5789.05 15	^{220}Rn
^{220}Rn^e	55.6 s 1	α	100	6404.88 11	^{216}Po
^{216}Po	0.145 s 2	α	100	6906.6 5	^{212}Pb
^{212}Pb (ThB)	10.64 h 1	β^-	100	572.8 36	^{212}Bi
^{212}Bi (ThC)	60.55 min 6	β^-	64.06 6	2246 4	^{212}Po
		α	35.94 6	6207.36 7	^{208}Tl
^{212}Po (ThC')	0.298 s 3	α	100	8953.58 9	^{208}Pb
^{208}Tl (ThC'')	3.053 min 4	β^-	100	4992 4	^{208}Pb
^{208}Pb (ThD)	stable				

^{a} Each one or two of the uncertainty digits (the number after the spaces) refers to the last one or two digits of the numerical value.
^{b} Mesothorium I.
^{c} Mesothorium II.
^{d} Radiothorium.
^{e} Thoron.

Radiation Processes

The decay of a nuclear species to a different species can take place by the emission of β particles (positively or negatively charged electrons, β^+, β^-) α particles, neutrinos, and, less commonly, neutrons and protons. Accompanying these emissions, depending upon the type of decay, can be γ rays (photons), x rays, Auger electrons, and bremsstrahlung. For the isomeric branch of an isomeric state decay, the parent and daughter states are in the same nucleus and therefore, of the decays mentioned, only γ rays, internal-converison electrons, x rays, and Auger electrons would be observed.

α-Particle Transitions. In α decay, an atom of atomic number Z and mass number A, emits an α particle (^{4}He nucleus, Z = 2, A = 4) and decays to an atom with atomic number Z − 2 and mass number A − 4, and two surplus electrons.

For α decay between ground states of parent and daughter atoms, the maximum energy available for the α particle is,

$$E = Q_\alpha - E_R \qquad (4)$$

Table 3. ^{235}U Decay Chain

Parent radioisotope	Half-life[a]	Decay mode	Branching, %[a]	Q-value, keV[a]	Daughter radioisotope
^{243}Am	7380 yr 40	α	100	5438.7 10	^{239}Np
^{239}Np	2.355 d 4	β^-	100	721.4 19	^{239}Pu
^{239}Pu	2.411 × 10^4 yr 3	α	100	5243.7 5	^{235}U
^{235}U	7.038 × 10^8 yr 5	α	100	4679.3 20	^{231}Th
^{231}Th	25.52 h 1	β^-	100	389.0 18	^{231}Pa
^{231}Pa	3.276 × 10^4 yr 11	α	100	5148.2 8	^{227}Ac
^{227}Ac	21.773 yr 3	β^-	98.62 4	43.7 20	^{227}Th
		α	1.38 4	5042.7 20	^{223}Fr
^{227}Th	18.718 d 5	α	100	6146.64 20	^{223}Ra
^{223}Fr	21.8 min 4	β^-	>99[b]	1147.6 28	^{223}Ra
^{223}Ra	11.434 d 2	α	100	5979.1 4	^{219}Rn
^{219}Rn	3.96 s 1	α	100	6946.32 30	^{215}Po
^{215}Po	1.780 × 10^{-3} s 4	α	>99[c]	7526.5 8	^{211}Pb
^{211}Pb	36.1 min 2	β^-	100	1373 6	^{211}Bi
^{211}Bi	2.14 min 2	α	99.725[d] 4	6751.3 6	^{207}Tl
^{207}Tl	4.77 min 2	β^-	100	1422 6	^{207}Pb
^{207}Pb	stable				

[a] Each one or two of the uncertainty digits (the numbers after the spaces) refers to the last one or two digits of the numerical value.
[b] % α = 0.006.
[c] % β^- = 0.00023 2.
[d] % β^- = 0.275 4.

where Q_α is the difference, in energy units, between the mass of the parent atom and the sum of the masses of the daughter atom and a helium atom. The recoil energy E_R of the daughter atom, if M and M_α are the masses of the daughter atom and the α particle, respectively, is given by

$$E_R = Q_\alpha \frac{M_\alpha}{M_\alpha + M} \qquad (5)$$

For decay to a particular energy level E_L in the daughter nucleus from an excited level E^* in the parent nucleus, the maximum energy available for the alpha particle is $Q_\alpha + E^* - E_L - E_R$.

β-Particle Transitions. *β^- Decay.* In β^- decay, an antineutrino ($\bar{\nu}$) and a negative electron (β^-) are emitted from the nucleus as a result of the process $n \to p + \beta^- + \bar{\nu}$. The decay increases the nuclear charge by one unit.

The energy released in a single β transition is divided between the β particle and the antineutrino in a statistical manner in such a way that, when a large number of transitions is considered, both the antineutrinos and β particles have energy distributions extending from zero up to some maximum value. For decay to a particular energy level, E_L, in the daughter nucleus, the maximum energy available is $E_{max} = Q^- + E^* - E_L$, where Q^- is the atomic mass difference, expressed in energy units, between ground states of parent and daughter nuclides, and E^* is the excitation energy of the decaying level in the parent nucleus. For decay from the ground state, $E^* = 0$; for an isomeric state, $E^* > 0$. This definition is also true for β^+ decay. The average energy of a β particle in this transition is given by

$$E_{av} = \int_0^{E_{max}} E N(E) dE \Big/ \int_0^{E_{max}} N(E) dE$$

Table 4. ^{238}U Decay Chain

Parent radioisotope	Half-life[a]	Decay mode	Branching, %[a]	Q-value, keV[a]	Daughter radioisotope
^{254}Fm	3.240 h 2	α	99.9408 2	7304 3	^{250}Cf
		SF	0.0592 2		
^{250}Cf	13.08 yr 9	α	99.923 3	6129.0 6	^{246}Cm
		SF	0.077 3		
^{246}Cm	4730 yr 100	α	99.97386 5	5476.0 26	^{242}Pu
		SF	0.02614 5		
^{242}Pu	3.763 × 10^5 yr 20	α	100	4983.1 12	^{238}U
^{242}Am	16.02 h 2	ε	17.3 3	747.7 16	^{242}Pu
		β$^-$	82.7 3	661.2 18	^{242}Cm
^{242}Cm	162.8 d 4	α	100	6215.76 13	^{238}Pu
^{238}U	4.468 × 10^9 yr 3	α	100[b]	4270 4	^{234}Th
^{238}Np	2.117 d 2	β$^-$	100	1291.9 11	^{238}Pu
^{238}Pu	87.74 yr 4	α	100	5593.27 20	^{238}U
^{234}Th	24.10 d 3	β$^-$	100	262.5 20	^{234}Pa[c]
^{234}Pa[c]	1.17 min 3	IT	0.13 1	ca 74	^{234}Pa
		β$^-$	99.87 1	ca 2281	^{234}U
^{234}Pa	6.70 h 5	β$^-$	100	2207 5	^{234}U
^{234}U	2.446 × 10^5 yr 7	α	100	4856.4 18	^{230}Th
^{230}Th	7.578 × 10^4 yr 30	α	100	4770.6 15	^{226}Ra
^{230}Pa	17.4 d 5	ε	90.5 6	1304.2 24	^{230}Th
		β$^-$	9.5 6	559 5	^{230}U
^{230}U	20.8 d	α	100	5992.8 20	^{226}Th
^{226}Ra	1600 yr 7	α	100	4870.79 30	^{222}Rn
^{226}Th	30.9 min	α	100	6451.7 20	^{222}Ra
^{222}Rn	3.8235 d 3	α	100	5590.50 30	^{218}Po
^{222}Ra	38.0 s 5	α	100	6676 4	^{218}Rn
^{218}Po (RaA)	3.05 min	α	99.980[d] 2	6114.88 10	^{214}Pb
^{218}Rn	35 × 10^{-3} s 5	α	100	7266.4 20	^{214}Po
^{214}Pb (RaB)	26.8 min	β$^-$	100	1024 12	^{214}Bi
^{214}Bi (RaC)	19.9 min 4	β$^-$	99.979[e] 1	3270 12	^{214}Po
^{214}Po (RaC′)	164.3 × 10^{-6} s 20	α	100	7833.73 6	^{210}Pb
^{210}Pb (RaD)	22.3 yr 2	β$^-$	100[f]	63.0 5	^{210}Bi
^{210}Bi (RaE)	5.013 d 5	β$^-$	99.99987[g] 1	1161.4 10	^{210}Po
^{210}Po (RaF)	138.378 d 7	α	100	5407.63 10	^{206}Pb
^{206}Tl		β$^-$	100	1526.3 19	^{206}Pb
^{206}Pb (RaG)	stable				

[a] Each one or two of the uncertainty digits (the numbers after the spaces) refers to the last one or two digits of the numerical value.
[b] % SF = 5.4 × 10^{-5} 8.
[c] Metastable state.
[d] % β$^-$ = 0.020 2.
[e] % α = 0.021 1.
[f] % α = 1.7 × 10^{-6}.
[g] % α = 0.00013 1.

where $N(E)$ is the number of β particles with energy between E and $E + dE$. The β particles from a particular transition can thus be characterized by the quantities E_{max} and E_{av}.

$β^+$ Decay. In $β^+$ decay, a neutrino ($ν$) and a positive electron ($β^+$) (positron) are emitted from the nucleus as a result of the process $p \rightarrow n + β^+ + ν$. This decay decreases the nuclear charge by one unit.

As in β^- decay, the β^+ particles emitted in a transition to a particular level in the daughter nucleus have a continuous distribution of energies with a definite $E_{\max}$ and E_{av}. For positron decay,

$$E_{\max} = Q^+ + E^* - E_L - 2mc^2$$

β^+ decay to a particular energy level, E_L, cannot occur unless $Q^+ + E^* - E_L > 2mc^2$ ($2mc^2 = 1022$ keV).

Every positron eventually combines with a negative electron in an interaction in which the total rest mass of the pair spontaneously converts into electromagnetic radiation called annihilation radiation (10). It always accompanies positron decay.

Electron-Capture Decay. In electron-capture decay ϵ, an atomic electron is captured by a nucleus and a neutrino is emitted as a result of the process $p + e \rightarrow n + \nu$. This decay decreases the nuclear charge by one unit and leaves the daughter nucleus with a vacancy in one of its atomic shells. K-shell electron capture, for example, refers to a capture process where the final-state vacancy is in the K shell.

Electron-capture always competes with β^+ decay, but can occur also when the transition energy is too small to allow β^+ emission, that is, when $0 < (Q^+ + E^* - E) < 1022$ keV.

γ-Ray Transitions. Electromagnetic radiation is emitted by a nucleus in a transition from a higher to a lower energy state. The energy of this γ ray, $E(\gamma)$, is equal to the energy difference between the two levels (except for a usually negligible amount of nuclear recoil energy

$$E_r \text{ (keV)} \approx 5.4 \times 10^{-7} E(\gamma)^2/A$$

where A is the mass number and $E(\gamma)$ is in keV).

Internal-Conversion-Electron Transitions (ce). An atomic electron from the cloud of electrons orbiting the nucleus can be emitted as an alternative to γ-ray emission in the transition of a nucleus from a higher to a lower energy state. In the internal-conversion process, the energy difference between the states is transferred directly to a bound atomic electron that is ejected from the atom. A letter hyphenated to ce refers to the shell from which the atomic electron is ejected. Thus, ce-K denotes K-shell conversion and ce-MNO denotes conversion in the M, N, and O shells combined. For a transition with energy $E(\gamma)$, the K-shell internal-conversion electron is emitted with energy $E(\text{ce-K}) = E(\gamma) - E_K$, where E_K is the K-shell binding energy.

For a particular transition, the ratio of the probability for emission of a K-conversion electron to that for emission of a γ ray, $I(\text{ce-K})/I(\gamma)$, is called the K-shell conversion coefficient for that transition, where I = probability of emission.

X-Ray and Auger-Electron Transitions. Whenever a vacancy is produced in an inner electron shell of an atom, the filling of this vacancy is accompanied by the emission of either an x ray (X) or Auger electron (e_A). Vacancies created by the filling of the initial vacancy produce, in turn, further x rays or Auger electrons. This cascade of radiations continues until all vacancies have been transferred to the outermost electron shell. Inner-shell vacancies are always produced in two types of nuclear decay, ie, electron capture and internal conversion. Other processes that lead to electron vacancies following a nuclear decay occur rarely.

The fluorescence yield for a particular atomic shell (ω_K, $\bar{\omega}_L$, etc) is defined as the probability that a vacancy in that shell results in the emission of an x ray. If n_K is the number of K-shell vacancies per transition, the number of K x rays is $n_K\omega_K$, and the number of K Auger electrons is $n_K(1 - \omega_K)$.

An x ray emitted as the result of the filling of a K-shell vacancy by an electron from a higher shell, eg, the Y shell, has an energy $E_K - E_Y$, where E_K and E_Y are the electron binding energies in the K and Y shells, respectively. This transition can be denoted by K − Y. In order of decreasing intensity, the most important transitions are $K_{\alpha 1} = K - L_3$, $K_{\alpha 2} = K - L_2$, $K_{\beta 1} = K - M_3$, $K_{\beta 2} = K - N_3$, $K_{\beta 3} = K - M_2$, $K_{\beta 4} = K - N_2$, and $K_{\beta 5} = K - M_4$.

Since the number of K x rays per disintegration is $n_K\omega_K$, the number of $K_{\alpha 1}$ x rays is $n_K\omega_K/(1 + K_\beta/K_\alpha)(1 + K_{\alpha 2}/K_{\alpha 1})$ with similar expressions for $K_{\alpha 2}$ and K_β. The number of K-shell vacancies per disintegration is the sum of the vacancies produced by K capture and those produced by internal conversion in the K shell. Thus, $n_K = \epsilon_K + I(\text{ce-K})$, where ϵ_K is the total number of K-capture events per disintegration, and $I(\text{ce-K})$ is the total number of K-shell internal-conversion electrons per disintegration.

As in the case of the K shell, many transitions contribute to the L x-ray spectrum. The calculation of the number of L x rays per disintegration, $n_L\omega_L$, is similar to that for K x rays except that, in addition to vacancies produced by direct L-shell capture and conversion, those created by the transfer of L-shell electrons to K-shell vacancies must be included. Thus,

$$n_L = \epsilon_L + I(\text{ce-L}) + n_{KL}n_K, \qquad (6)$$

where n_{KL} is the number of L-shell vacancies created per K-shell vacancy, and the other symbols have meanings analogous to those used above for the K shell.

The Auger process competes with the emission of x rays as a means of carrying off the energy released in the filling of an inner-shell vacancy by an electron from an outer shell. In the Auger process, the filling of an inner-shell vacancy is accompanied by the simultaneous ejection to the continuum of an outer-shell electron. The resulting atom is left with two vacancies. The Auger-electron yield per disintegration of a particular atomic shell is $n_K(1 - \omega_K)$, $n_L(1 - \overline{\omega}_L)$, etc. If the original vacancy is in the K shell and if this vacancy is filled by an electron from shell X with the ejection of an electron from shell Y, the transition is denoted by KXY. The energy of the ejected electron is $E_K - E_X - E'_Y$ where E_K and E_X are neutral atom K- and X-shell-binding energies, respectively, and E'_Y is the binding energy of a Y-shell electron in an atom containing a vacancy in the X shell. The most intense K-Auger transitions are of the type KLL.

Bremsstrahlung. In addition to any monoenergetic x rays or γ rays that may be present in radioactive decay, every β or electron-capture decay produces continuous electromagnetic radiation called bremsstrahlung; two processes contribute to this continuous spectrum:

External bremsstrahlung is produced by collisions between β particles or conversion electrons and the atoms of the material surrounding the radiating atoms. The intensity of the external bremsstrahlung depends on the atomic number, Z, of the surrounding material. An approximate value for the average energy of the external bremsstrahlung associated with a β group of maximum energy E_β (in keV) is given (10) for the thick-target approximation as

$$E_{\text{av}} \approx 1.4 \times 10^{-7}\, Z\, E_\beta^2 \text{ keV per } \beta \qquad (7)$$

Internal bremsstrahlung, originating within a decaying atom, is produced by the sudden change of nuclear charge that occurs in β^+, β^-, or ϵ decay. The average energy of the internal bremsstrahlung associated with a β group of maximum energy E_β (in keV) is given (11) for $E_\beta \gg mc^2$ (511 keV) as

$$E_{av} \approx 1.5 \times 10^{-3} E_\beta \log (0.004 E_\beta - 2.2) \text{ keV per } \beta \qquad (8)$$

In electron capture (11), the corresponding expression for a transition with energy E_ϵ (in keV) is

$$E_{av} \approx 1.5 \times 10^{-7} E_\epsilon^2 \text{ keV per capture} \qquad (9)$$

The low average energy in both external and internal bremsstrahlung is a result of the low probability of the process. The actual spectrum of this electromagnetic radiation extends in energy up to the E_{max} for the β transition giving rise to it or, in the case of electron capture, up to E_ϵ.

Spontaneous Fission (SF). This mode of decay has been established for several transuranic nuclides. In this process, a heavy nucleus decays into two lighter fragments with emission of several prompt neutrons. Each of the resulting fission fragments is neutron rich and undergoes several successive β^- decays before reaching a stable nucleus. In the case of ^{252}Cf, the two fragments are centered around A = 106 and A = 142 (12), and the width at one-tenth maximum of each peak is approximately 27 mass units (13). The fragment with A = 106 decays as follows:

$$^{106}\text{Nb } \beta^- \rightarrow {}^{106}\text{Mo } \beta^- \rightarrow {}^{106}\text{Tc } \beta^- \rightarrow {}^{106}\text{Ru } \beta^- \rightarrow {}^{106}\text{Rh } \beta^- \rightarrow {}^{106}\text{Pd (stable)}$$

whereas the fragment with A = 142 follows the chain:

$$^{142}\text{I } \beta^- \rightarrow {}^{142}\text{Xe } \beta^- \rightarrow {}^{142}\text{Cs } \beta^- \rightarrow {}^{142}\text{Ba } \beta^- \rightarrow {}^{142}\text{La } \beta^- \rightarrow {}^{142}\text{Ce (stable)}$$

Hence, the radiations from spontaneous fission include the fission fragments themselves, neutrons, β particles, and prompt and delayed gammas. The energy released in spontaneous-fission decay from these radiations is about 200 MeV. Thus, when compared to the typical α energies of about 5 MeV, spontaneous fission can be the most important mode of decay in terms of total energy released, even though the SF branch may be small.

In order to account for the radiations from a spontaneous-fission decay mode, the yield of each of the fission-product β^--decaying mass chains must be known. With these yields and a knowledge of the decay schemes and half-lives of each β^--decaying member in the chain, the individual γ, β^-, etc, energies and intensities can be computed as a function of time. In general this approach is difficult since the decay schemes of the very short-lived members of the chains are usually not known. If the average values for the delayed γ and β^- radiations are sufficient, then general expressions depending only on properties of the fissioning parent can be derived. Expressions for the intensities and average energies of the radiations accompanying spontaneous fission are given in Table 5 (14–15).

IT Decay. The decay of an isomeric state can proceed by any of the modes discussed in this section. The term IT decay, ie, isomeric transition decay, refers to those decay modes that can involve deexcitation of the isomeric state to states with lower energy in the same nucleus. Thus, there is no change in Z or A.

Other Processes. In addition to the single-process decay modes discussed above, double-beta decay ($\beta^-\beta^-$, $\epsilon\epsilon$, $\epsilon\beta^+$) occurs when a nuclide decays to an isobar (same mass number) with a charge number differing from that of the parent by ±2. This decay mode would be unlikely to compete with a single-decay mode and is of interest

Table 5. **Formulas for the Calculation of Intensities and Average Energies of Radiations Accompanying Spontaneous Fission**

Radiation type	Average energy, keV[a]	Intensity, % (per parent decay)[a]
neutrons	$750 + 650\,(\bar{\nu} + 1)^{1/2}$	$\bar{\nu} f_{sf}$
fission fragments	$139.6\,Z^2/A^{1/3} - 21975$	f_{sf}
prompt gammas[b]	884.7	$8.636\,f_{sf}$
delayed gammas[c,d]	957.8	$0.2102\,m^2 f_{sf}$
β^- particles[d]	$205.8\,m$	$m f_{sf}$

[a] Where $\bar{\nu}$ = average number of neutrons released per fission, f_{sf} = % decay branch for spontaneous fission, and $m = 5.98 - Z + 92\,A/236$.
[b] Emitted during deexcitation of the fission fragments (ca 10^{-3} s).
[c] Emitted at each step in the chain of successive β^- decays.
[d] Entries given are integrated values over the time interval 10^{-3} s to infinity.

only when the parent is stable with respect to the single-decay modes. For example, ^{128}Te is stable against β^- decay to ^{128}I, but has been observed to decay by double β^- emission directly to ^{128}Xe. There are 58 pairs of stable isobars that could lead to double-beta decay, but only in three cases has the decay been established, namely, ^{82}Se, ^{128}Te, and ^{130}Te.

Table of Radioisotope Decay Data

The data given in Table 6 contain the atomic and nuclear radiations that are emitted by all known radioisotopes with half-life greater than one hour and for which the decay schemes are sufficiently well known that intensities per decay can be obtained.

The basic nuclear decay data, as of August 1981, were assembled from the Evaluated Nuclear Structure Data File designed by the Nuclear Data Project, Oak Ridge National Laboratory, edited and maintained by the National Nuclear Data Center, Brookhaven National Laboratory, on behalf of the International Network for Nuclear Structure Evaluation. A summary of the file contents may be found in any issue of the Nuclear Data Sheets. Documentation can be obtained from the National Nuclear Data Center, Brookhaven National Laboratory, Upton, N.Y.

Based on the radioactive decay data sets in the Evaluated Nuclear Structure Data File and computerized tabulations of the relevant Z-dependent constants (ω_K, n_{KL}, etc) discussed earlier, the computer program, MEDLIST (16), calculates the energies and intensities of the atomic radiations (x-ray and Auger-electron transitions). The program then combines these radiations with the nuclear radiations, arranges them according to radiation type (ie, α, β, γ, ce, etc) and, within each type, arranges and numerically labels them in order of increasing energy.

The energy and intensity values given in Table 6 have been computed from the formulas in Table 1 and are given for spontaneously fissioning radioisotopes with % SF > 10^{-3} with $t_{1/2} > 1$ h (17).

Uncertainties in all experimental quantities, including the atomic constants, are consistently carried through the calculations. A 3% uncertainty is assigned to all theoretical conversion coefficients and is propagated, within MEDLIST, along with the experimental uncertainties. Each one or two uncertainty digits refers to the last one or two digits of the numerical value.

A variable low-intensity cutoff limit is built into MEDLIST. For Table 6 this

Table 6. Atomic and Nuclear Radiations Emitted by Selected Radioisotopes [a]

Isotope	CAS Registry No.	Half-life	Type of radiation	Energy, keV Max	Energy, keV Avg	Intensity, %
^{3}H (β^- decay)	[10028-17-8]	12.35 yr 1	β^-	18.600 20	5.680 10	100
^{7}Be (EC decay)	[13966-02-4]	53.3 d 1	γ 1		477.605 3	10.34 6
^{10}Be (β^- decay)	[14390-89-7]	1.6 × 10^6 yr 2	β^- 1	555.7 7	252.2 3	100
^{14}C (β^- decay)	[14762-75-5]	5730 yr 40	β^- 1	156.478 9	49.470 3	100
^{18}F (EC decay)	[13981-56-1]	109.77 min 5	Auger K		0.52	3.069 11
			β^+ 1	633.5 6	249.8 3	96.73 4
			max $\gamma\pm$			193.46
^{22}Na (β^+ decay)	[13966-32-0]	2.602 yr	Auger K		0.82	9.20 5
			β^+ 1	545.5 5	215.54 21	89.84 10
			1 weak β			0.06
			γ 1		1274.542 7	99.944 14
			max $\gamma\pm$			179.79
^{24}Na (β^- decay)	[13982-04-2]	15.020 h 7	β^- 1	1390.2 7	553.9 4	99.944 4
			3 weak βs			0.06
			γ 2		1368.633 6	100
			γ 3		2754.030 14	99.944 4
			3 weak γs			0.05
^{26}Al (β^+ decay)	[14682-66-7]	7.16 × 10^5 yr 32	Auger K		1.18	16.4 19
			β^+ 1	1174.0 5	543.8 23	81.80 20
			γ 1		1129.67 10	2.50 20
			γ 2		1808.65 7	99.76 4
			1 weak γ			0.24
			max $\gamma\pm$			163.60
^{28}Mg (β^- decay)	[15092-71-4]	20.90 h 3	Auger K		1.39	26 6
			ce-L-1		29.080 20	27 7
			ce-K-1		30.522 20	2.6 7
			β^- 1	211.8 20	65.2 7	4.60 20
			β^- 2	458.9 20	155.9 8	94.0 20
			β^- 3	859.6 20	319.3 9	1.40 20
			total β^-		154.0 9	100.0 21
			γ 1		30.640 20	66 4
			γ 2		400.690 20	36.6 10
			γ 5		941.45 3	38.3 10
			γ 8		1342.25 3	52.6 16
			γ 9		1372.89 6	4.70 20
			γ 10		1589.36 3	4.20 20
			5 weak γs			0.43
^{28}Al (β^- decay)	[14999-04-3]	2.240 min 1	β^- 1	2864.2 6	1247.2 3	100
			γ 1		1778.85 3	100
^{31}Si (β^- decay)	[14276-49-4]	157.3 min 3	β^- 1	1491.6 11	596.0 5	99.93 5
			1 weak β			0.07
			1 weak γ			0.07
^{32}Si (β^- decay)	[15092-72-5]	105 yr 18	β^- 1	213 7	64.722	100
^{32}P (β^- decay)	[14596-37-3]	14.26 d 4	β^- 1	1710.4 6	695.0 3	100
^{33}P (β^- decay)	[15749-66-3]	25.34 d 12	β^- 1	249.0 20	76.6 6	100
^{35}S (β^- decay)	[15117-53-0]	87.51 d 12	β^- 1	167.47 19	48.80 10	100
^{36}Cl (β^- decay)	[13981-43-6]	3.01 × 10^5 yr 2	β^- 1	709.5 3	251.31 13	98.10 10
^{38}S (β^- decay)	[15759-21-4]	170.3 min 7	β^- 1	184 12	54 4	1.60 20
			β^- 2	994 12	372 6	84.0 20
			β^- 3	1190 12	458 6	2.50 20
			β^- 4	2936 12	1291 6	12.0 20
			total β^-		479 8	100 3
			γ 1		1745.52 14	2.50 20
			γ 2		1941.90 20	84.0 20
			γ 3		2751.7 5	1.60 20
^{38}Cl (β^- decay)	[14158-34-0]	37.24 min 5	β^- 1	1107.2 9	420.4 1	32.5 6
			β^- 2	2749.6 9	1181.5 4	11.4 8
			β^- 3	4917.2 9	2244.1 4	56.0 5
			total β^-		1529.5 4	99.9 12
			γ 1		1642.42 2	32.5 6
			γ 2		2167.51 5	44.0 5
			1 weak γ			0.03
^{39}Ar (β^- decay)	[25729-41-3]	269 yr 3	β^- 1	565 5	218.8 21	100
^{40}K (EC decay)	[13966-00-2]	1.277 × 10^9 yr 8	Auger K		2.66	7.24 15
			γ 1		1460.75 6	10.70 20
^{40}K (β^- decay)	[13966-00-2]	1.277 × 10^9 yr 8	β^- 1	1311.6 5	588 6	89.30 20
^{41}Ar (β^- decay)	[14163-25-8]	1.827 h 7	β^- 1	1198.1 8	459.3 3	99.170 20
			2 weak βs			0.83
			γ 1		1293.64 4	99.160 20
			1 weak γ			0.05
^{42}Ar (β^- decay)	[31581-51-8]	32.9 yr 11	β^- 1	600 40	233 17	100
^{42}K (β^- decay)	[14378-21-3]	12.360 h 3	β^- 1	1996.4 16	822.3 8	17.5 5
			β^- 2	3521.1 16	1563.9 8	82.1 5
			total β^-		1429.8 9	100.0 7
			3 weak βs			0.44
			γ 6		1524.665 20	17.9 5
			7 weak γs			0.46

Table 6 (*continued*)

Isotope	CAS Registry No.	Half-life	Type of radiation	Energy, keV Max	Energy, keV Avg	Intensity, %
^{43}K (β^- decay)	[14903-02-7]	22.3 h 1	β^- 1	422 10	137 4	2.24 9
			β^- 2	827 10	298 4	92.1 13
			β^- 3	1224 10	469 5	3.7 4
			β^-	1817 10	762 5	1.3 3
			total β^-		309 5	100.1 14
			1 weak β			0.80
			γ 2		220.608 18	4.11 22
			γ 3		372.763 15	87.3 5
			γ 4		396.870 20	11.43 12
			γ 6		593.40 8	11.0 3
			γ 7		617.494 25	80.5 14
			γ 11		1021.79 18	1.88 8
			6 weak γs			1.11
^{43}Sc (EC decay)	[14276-61-0]	3.891 h 12	Auger L		0.3	30.3 10
			Auger K		3.3	15.0 5
			β^+ 1	825.3 19	344.2 9	57.2 16
			β^+ 2	1198.1 19	507.8 9	22.9 19
			total β^+		391.0 10	80.1 25
			x ray Kα_1		3.69168 4	1.72 18
			γ 1		372.81 5	75.0 20
			max $\gamma\pm$			160.20
^{44}Sc (β^+ decay)	[14391-94-7]	3.927 h 8	Auger L		0.3	8.56 16
			Auger K		3.3	4.22 9
			β^+ 1	1475.8 20	632.7 9	94.370 20
			γ 1		1157.002 11	99.884 15
			5 weak γs			1.03
			max $\gamma\pm$			188.74
^{44}Sc (EC decay)		58.6 h 1	Auger L		0.3	2.08 4
			Auger K		3.3	1.027 21
			γ 1		1001.82 4	1.370 10
			γ 2		1126.06 4	1.370 10
			γ 3		1157.002 11	1.370 10
^{44}Sc (IT decay)		58.6 h 1	Auger L		0.37	18.1 4
			Auger K		3.64	8.79 18
			ce-K-1		266.748 10	10.8493 11
			ce-L-1		270.741 10	1.0849 1
			x ray Kα_1		4.09060 20	1.21 11
			γ 1		271.241 10	86.3 3
^{44}Ti (EC decay)	[15749-33-4]	47.3 yr 12	Auger L		0.37	172 6
			Auger K		3.64	83 3
			ce-K-1		63.36 3	10.8 21
			ce-L-1		67.35 3	1.1 3
			ce-K-2		73.91 6	2.9 5
			x ray Kα_2		4.08610 20	5.8 6
			x ray Kα_1		4.09060 20	11.5 11
			x ray Kβ		4.46	2.26 21
			γ 1		67.85 3	87.7 15
			γ 2		78.40 6	94.7 15
			1 weak γ			0.10
^{45}Ca (β^- decay)	[13966-05-7]	163 d 1	β^- 1	256.9 10	77.2 4	100
^{45}Ti (EC decay)	[14392-00-8]	3.08 h 1	Auger L		0.37	22.7 5
			Auger K		3.64	11.02 24
			β^+ 1	1040.6 24	439.1 11	85.09 23
			2 weak βs			0.01
			x ray Kα_1		4.09060 20	1.52 13
			13 weak γs			0.34
			max $\gamma\pm$			170.21
^{46}Sc (β^- decay)	[13967-63-0]	83.83 d 2	β^- 1	357.3 8	112.0 3	99.9964 7
			γ 1		889.277 3	99.9840 10
			γ 2		1120.545 4	99.9870 10
^{47}Ca (β^- decay)	[14391-99-2]	4.536 d 2	β^- 1	691 3	241.2 10	82.0 20
			β^- 2	1988.3 25	817.2 12	18.0 20
			total β^-		344.9 13	100 3
			2 weak βs			0.11
			γ 2		489.23 10	6.7 3
			γ 5		807.86 10	6.9 3
			γ 6		1297.09 10	74.9 19
			4 weak γs			0.33
^{47}Sc (β^- decay)	[14391-96-9]	3.351 d 2	β^- 1	440.6 20	142.5 8	68.0 20
			β^- 2	600.0 20	203.8 8	32.0 20
			total β^-		162.1 9	100 3
			γ 1		159.381 15	68.0 20
^{48}Sc (β^- decay)	[14391-86-7]	43.7 h 1	β^- 1	482 6	157.9 23	9.85 18
			β^- 2	657 6	226.5 25	90.0 20
			total β^-		220 3	99.8 20
			γ 1		175.357 10	7.47 18
			γ 2		983.501 10	100.0 5

Table 6 (*continued*)

Isotope	CAS Registry No.	Half-life	Type of radiation	Energy, keV Max	Energy, keV Avg	Intensity, %
			γ 3		1037.496 10	97.5 21
			γ 4		1212.849 10	2.38 5
			γ 5		1312.087 10	100.0 21
^{48}V (EC decay)	[14331-97-6]	16.238 d 25	Auger L		0.42	72.5 21
			Auger K		4	34.3 10
			β^+ 1	697 3	291.4 13	49.4 5
			2 weak βs			0.19
			x ray Kα_2		4.50486 4	2.85 25
			x ray Kα_1		4.51084 4	5.6 5
			x ray Kβ		5	1.13 10
			γ 3		944.101 12	7.76 18
			γ 4		983.501 10	100.0 3
			γ 5		1312.087 10	97.5 20
			γ 7		2240.341 20	2.41 7
			6 weak γs			1.06
			max γ±			99.18
^{48}Cr (EC decay)	[14833-09-1]	22.96 h 3	Auger L		0.47	143.4 13
			Auger K		4.38	66.3 7
			β^+ 1	212 8	91 4	1.47 24
			x ray Kα_2		4.94464 4	6.54 21
			x ray Kα_1		4.95220 4	13.0 4
			x ray Kβ		5.43	2.60 9
			γ 1		112.440 20	99.0 14
			γ 2		308.330 20	100.0 10
			max γ±			2.94
^{49}V (EC decay)	[14392-01-9]	330 d 15	Auger L		0.42	147 4
			Auger K		4	69.7 16
			x ray Kα_2		4.50486 4	5.8 5
			x ray Kα_1		4.51084 4	11.5 10
			x ray Kβ		5	2.29 20
^{51}Cr (EC decay)	[14392-02-0]	27.704 d 4	Auger L		0.47	143.8 12
			Auger K		4.38	66.9 7
			x ray Kα_2		4.94464 4	6.60 21
			x ray Kα_1		4.95220 4	13.1 4
			x ray Kβ		5.43	2.62 9
			γ 1		320.0840 10	9.83 14
^{52}Mn (EC decay)	[14092-99-0]	5.591 d 3	Auger L		0.54	99.4 20
			Auger K		4.78	44.7 10
			β^+ 1	575.3 23	241.6 10	29.4 7
			x ray Kα_2		5.40551 5	5.20 18
			x ray Kα_1		5.41472 5	10.3 4
			x ray Kβ		6	2.06 8
			γ 8		744.214 11	90.0 19
			γ 9		848.13 3	3.32 8
			γ 11		935.520 20	94.5 20
			γ 13		1246.246 12	4.21 10
			γ 15		1333.615 16	5.07 11
			γ 16		1434.056 16	100
			15 weak γs			2.91
			max γ±			58.80
^{52}Mn (EC decay)		21.1 min 2	Auger L		0.54	2.37 5
			Auger K		4.78	1.068 24
			β^+ 1	2632.8 23	1173.8 11	96.5 20
			4 weak βs			0.21
			γ 4		1434.056 16	98.2 20
			13 weak γs			0.39
			max γ±			193.43
^{52}Mn (IT decay)		21.1 min 2	γ 1		377.738 11	1.68 4
^{52}Fe (β^+ decay)	[14093-04-0]	8.275 h 8	Auger L		0.6	62.6 23
			Auger K		5.19	27.4 12
			β^+ 1	804 12	340 6	56.0 13
			x ray Kα_2		5.88765 3	3.7 3
			x ray Kα_1		5.89875 3	7.3 6
			x ray Kβ		6.49	1.4 12
			γ 1		168.684 11	99.21 3
			max γ±			112.00
^{53}Mn (EC decay)	[14999-33-8]	3.7×10^6 yr 4	Auger L		0.54	142.0 12
			Auger K		4.78	63.9 7
			x ray Kα_2		5.40551 5	7.43 21
			x ray Kα_1		5.41472 5	14.7 4
			x ray Kβ		6	2.95 10
^{54}Mn (EC decay)	[13966-31-9]	312.5 d 5	Auger L		0.54	142.1 12
			Auger K		4.78	63.9 7
			x ray Kα_2		5.40551 5	7.43 21
			x ray Kα_1		5.41472 5	14.7 4
			x ray Kβ		6	2.95 10
			γ 1		834.843 6	99.9750 10

Table 6 (*continued*)

Isotope	CAS Registry No.	Half-life	Type of radiation	Energy, keV Max	Energy, keV Avg	Intensity, %
^{55}Fe (EC decay)	[14681-59-5]	2.7 yr 1	Auger L		0.6	139 4
			Auger K		5.19	60.7 22
			x ray Kα_2		5.88765 3	8.2 7
			x ray Kα_1		5.89875 3	16.3 12
			x ray Kβ		6.49	3.3 3
^{55}Co (β^+ decay)	[13982-25-7]	17.54 h 4	Auger L		0.67	33.3 15
			Auger K		5.62	14.1 7
			β^+ 1	1024.1 11	437.0 5	26.2 22
			β^+ 2	1116.0 11	477.6 5	3.0 8
			β^+ 3	1501.2 11	650.3 5	45 5
			total β^+		567.7 6	74 6
			3 weak βs			0.05
			x ray Kα_2		6.39084 3	2.23 12
			x ray Kα_1		6.40384 3	4.40 23
			γ 1		91.8 3	2.7 6
			γ 4		477.2 3	20.3 13
			γ 6		803.8 3	2.10 23
			γ 8		931.5 3	75 4
			γ 11		1316.7 3	7.1 7
			γ 12		1370.0 3	3.0 3
			γ 13		1408.7 3	16.5 12
			14 weak γs			4.49
			max $\gamma\pm$			148.49
^{56}Mn (β^- decay)	[14681-52-8]	2.5785 h 6	β^- 1	324.8 12	98.8 5	1.16 3
			β^- 2	734.6 12	254.8 5	14.6 4
			β^- 3	1037.0 12	381.5 6	27.8 8
			β^- 4	2847.7 12	1216.3 6	56.3 9
			total β^-		829.8 9	100.0 13
			3 weak βs			0.12
			γ 1		846.754 20	98.9 3
			γ 4		1810.72 4	27.2 8
			γ 5		2113.05 4	14.3 4
			7 weak γs			2.27
^{56}Co (β^+ decay)	[14093-03-9]	78.76 d 12	Auger L		0.67	108.8 15
			Auger K		5.62	46.1 8
			β^+ 1	423.4 19	179.7 8	1.07 4
			β^+ 2	1461.3 19	632.2 9	18.6 7
			total β^+		607.6 10	19.7 7
			x ray Kα_2		6.39084 3	7.27 21
			x ray Kα_1		6.40384 3	14.3 4
			x ray Kβ		7	2.89 10
			γ 7		846.764 6	99.930 6
			γ 9		977.42 4	1.40 3
			γ 11		1037.844 4	14.09 20
			γ 15		1175.099 8	2.26 6
			γ 17		1238.287 6	67.0 7
			γ 20		1360.206 6	4.29 4
			γ 24		1771.350 15	15.51 14
			γ 27		2015.179 11	3.03 5
			γ 28		2034.759 11	7.77 12
			γ 34		2598.460 10	16.74 22
			γ 36		3009.596 17	1.03 9
			γ 37		3201.954 14	3.02 21
			γ 38		3253.417 14	7.4 5
			γ 39		3272.998 14	1.73 12
			30 weak γs			5.04
			max $\gamma\pm$			39.34
^{56}Ni (EC decay)	[14932-64-0]	6.10 d 2	Auger L		0.75	135 5
			Auger K		6	55.4 25
			x ray Kα_2		6.91530 4	10.1 8
			x ray Kα_1		6.93032 4	19.9 15
			x ray Kβ		7.65	4.1 3
			γ 1		158.38 3	98.8 10
			γ 2		269.500 20	36.5 8
			γ 3		480.440 20	36.5 8
			γ 4		749.95 3	49.5 12
			γ 5		811.85 3	86.0 9
			γ 6		1561.80 5	14.0 6
^{57}Co (EC decay)	[13981-50-5]	270.9 d 6	Auger L		0.67	249 3
			Auger K		5.62	105.5 14
			ce-K-1		7.3007 10	69.6 4
			ce-L-1		13.5666 6	7.79 22
			ce-MNO-1		14.3198 10	1.15 7
			ce-K-2		114.9494 10	1.87 10
			ce-K-3		129.3623 11	1.40 12
			x ray Kα_2		6.39084 3	16.6 5
			x ray Kα_1		6.40384 3	32.8 8

Table 6 (*continued*)

Isotope	CAS Registry No.	Half-life	Type of radiation	Energy, keV Max	Energy, keV Avg	Intensity, %
^{57}Ni (β^+ decay)	[13981-99-2]	36.08 h 9	x ray Kβ		7	6.63 21
			γ 1		14.4127 4	9.54 13
			γ 2		122.0614 3	85.59 19
			γ 3		136.4743 5	10.61 18
			7 weak γs			0.19
			Auger L		0.75	80 4
			Auger K		6	32.7 17
			β^+ 1	716 7	304 3	5.0 4
			β^+ 2	843 7	359 3	34.1 9
			total β^+		346 3	40.4 10
			2 weak βs			1.27
			x ray Kα_2		6.91530 4	6.0 5
			x ray Kα_1		6.93032 4	11.8 9
			x ray Kβ		7.65	2.40 19
			γ 1		127.19 3	12.9 8
			γ 11		1377.59 4	77.9 19
			γ 14		1757.48 8	7.1 7
			γ 16		1919.43 8	14.7 10
			16 weak γs			0.75
			max $\gamma\pm$			80.74
^{58}Co (β^+ decay)	[13981-38-9]	70.80 d 8	Auger L		0.67	116.5 13
			Auger K		5.62	49.4 6
			β^+ 1	474.6 14	201.2 6	15.00 5
			x ray Kα_2		6.39084 3	7.78 21
			x ray Kα_1		6.40384 3	15.4 4
			x ray Kβ		7	3.10 10
			γ 1		810.757 18	99.4 3
			2 weak γs			1.19
			$\gamma\pm$			30.00
^{58}Co (IT decay)		9.15 h 10	Auger L		0.75	120 4
			Auger K		6	43.0 20
			ce-K-1		17.180 21	69.5 6
			ce-L-1		23.963 21	22.9 5
			ce-MNO-1		24.788 21	7.60 20
			x ray Kα_2		6.91530 4	7.8 6
			x ray Kα_1		6.93032 4	15.5 12
			x ray Kβ		7.65	3.15 24
			1 weak γ			0.04
^{59}Fe (β^- decay)	[14596-12-4]	44.529 d 7	β^- 1	130.8 20	35.8 6	1.27 3
			β^- 2	273.4 20	80.8 6	45.6 8
			β^- 3	465.8 20	149.3 7	52.8 12
			total β^-		117.4 8	99.5 15
			2 weak βs			0.26
			γ 1		142.648 4	1.00 3
			γ 2		192.344 6	3.00 7
			γ 5		1099.251 4	56.1 12
			γ 6		1291.596 7	43.6 8
			3 weak γs			0.35
^{59}Ni (EC decay)	[14336-70-0]	7.5 × 10^4 yr 13	Auger L		0.75	134 5
			Auger K		6	54.9 24
			x ray Kα_2		6.91530 4	10.0 8
			x ray Kα_1		6.93032 4	19.8 14
			x ray Kβ		7.65	4.0 3
^{60}Co (β^- decay)	[10198-40-0]	5.271 yr 1	β^- 1	317.81 11	95.80 10	99.920 20
			2 weak βs			0.08
			γ 3		1173.2384	99.900 20
			γ 4		1332.502 5	99.9824 5
			4 weak γs			0.02
^{61}Co (β^- decay)	[13981-83-4]	1.65 h	Auger L		0.84	15.5 7
			Auger K		6.54	6.1 4
			ce-K-1		59.082 10	10.5 4
			ce-L-1		66.407 10	1.09 4
			β 1	398 18	124 7	3.6 4
			β 2	1240 18	468 8	96.4 4
			total β^-		456 9	100.0 6
			x ray Kα_2		7.46089 4	1.29 10
			x ray Kα_1		7.47815 4	2.53 19
			γ 1		67.415 10	85.1 5
			γ 3		909.2 5	3.0 4
			1 weak γ			0.59
^{61}Cu (β^+ decay)	[15128-03-7]	3.408 h 10	Auger L		0.84	50 3
			Auger K		6.54	20.0 13
			β^+ 1	567.1 23	241.7 10	2.6 3
			β^+ 2	940.1 24	402.2 11	5.9 8
			β^+ 3	1155.7 24	497.2 11	2.9 9
			β^+ 4	1223.1 23	527.4 11	50.0 24
			total β^+		501.4 12	61 3
			1 weak β			0.04

Table 6 (continued)

Isotope	CAS Registry No.	Half-life	Type of radiation	Energy, keV Max	Energy, keV Avg	Intensity, %
			x ray Kα_2		7.46089 4	4.2 4
			x ray Kα_1		7.47815 4	8.2 7
			x ray Kβ		8.26	1.68 14
			γ 1		67.37 22	5.1 11
			γ 2		283.0 4	12.8 15
			γ 3		372.9 4	2.02 22
			γ 6		588.60 20	1.28 11
			γ 8		656.00 20	10.1 13
			γ 12		908.8 6	1.13 14
			γ 22		1185.7 6	4.0 8
			24 weak γs		11	2.06
			max $\gamma\pm$			122.88
^{62}Zn (EC decay)	[14833-23-9]	9.26 h 2	Auger L		0.92	139 7
			Auger K		7	53 3
			ce-K-1		31.87 6	14.7 14
			ce-L-1		39.75 6	1.56 15
			β^+ 1	605 11	259 5	8.4 12
			x ray Kα_2		8.027830 10	12.6 8
			x ray Kα_1		8.047780 10	24.8 14
			x ray Kβ		9	5.1 3
			γ 1		40.85 6	25.2 23
			γ 3		243.36 6	2.49 23
			γ 4		246.95 6	1.88 18
			γ 5		260.43 7	1.34 13
			γ 9		394.03 6	2.21 17
			γ 11		507.60 10	14.6 14
			γ 12		548.35 11	15.2 14
			γ 13		596.56 13	25.7 19
			21 weak γs			1.17
			max $\gamma\pm$			16.80
^{63}Ni (β^- decay)	[13981-37-8]	100.1 yr 20	β^- 1	65.87 20	17.13 5	100
^{64}Cu (β^+ decay)	[13981-25-4]	12.701 h 2	Auger L		0.84	59.2 21
			Auger K		6.54	23.4 12
			β^+ 1	652.9 8	278.1 4	17.90 18
			x ray Kα		7.46089 4	4.9 4
			x ray Kα_1		7.47815 4	9.7 7
			x ray Kβ		8.26	1.97 14
			1 weak γ			0.48
			max $\gamma\pm$			35.80
^{64}Cu (β^- decay)		12.701 h 2	β^- 1	578.2 15	190.2 6	37.1 4
^{65}Ni (β^- decay)	[14833-49-4]	2.520 h 2	β^- 1	654.7 16	220.7 7	28.1 10
			β^- 2	1021.0 16	371.8 7	9.8 5
			β^- 3	2136.5 16	875.5 8	60.7 14
			total β^-		632.1 11	100.0 18
			3 weak βs			1.38
			γ 2		366.27 3	4.61 20
			γ 8		1115.546 4	14.8 6
			γ 9		1481.84 5	23.5 8
			8 weak γs			1.46
^{65}Zn (EC decay)	[13982-39-3]	243.9 d 1	Auger L		0.92	126.7 18
			Auger K		7	48.3 8
			β^+ 1	329.9 11	143.0 5	1.460 20
			x ray Kα_1		8.027830 10	11.5 3
			x ray Kα_1		8.047780 10	22.6 5
			x ray Kβ		9	4.61 13
			γ 3		1115.546 4	50.75 10
			max $\gamma\pm$			2.92
^{66}Ni (β^- decay)	[15766-33-3]	54.6 h 9	β^- 1	233 19	67 7	100
^{66}Ga (β^+ decay)	[14119-08-5]	9.40 h 7	Auger L		0.99	56 3
			Auger K		7.53	20.6 13
			β^+ 1	361 3	157.0 13	1.00 10
			β^+ 2	924 3	397.1 13	4.10 20
			β^+ 3	4153 3	1904.1 14	49.9 10
			total β^+		1726.3 19	56.4 11
			4 weak βs			1.45
			x ray Kα_2		8.61578 5	5.6 4
			x ray Kα_1		8.63886 5	11.0 8
			x ray Kβ		9.57	2.25 16
			γ 8		833.56 10	6.12 15
			γ 15		1039.29 10	38.4 8
			γ 21		1333.20 20	1.25 3
			γ 30		1918.64 10	2.17 5
			γ 33		2190.00 20	5.76 15
			γ 37		2422.70 10	1.97 5
			γ 40		2752.10 20	23.5 6
			γ 46		3229.26 10	1.51 4
			γ 48		3381.32 10	1.43 4
			γ 54		3791.47 10	1.025 25

Table 6 (*continued*)

Isotope	CAS Registry No.	Half-life	Type of radiation	Energy, keV Max	Energy, keV Avg	Intensity, %
			γ 58		4086.36 10	1.15 4
			γ 59		4295.70 20	3.53 9
			γ 61		4806.59 15	1.49 5
			48 weak γs			7.72
			max γ±			112.90
^{66}Ge (β+ decay)	[15756-84-0]	2.27 h 5	Auger L		1	125 5
			Auger K		8	43.5 17
			ce-L-1		21.13 6	1.20 23
			ce-K-6		33.52 7	18.3 18
			ce-L-6		42.59 7	2.63 25
			ce-K-9		54.75 11	1.47 13
			β+ 1	543 13	234 6	2.70 20
			β+ 2	698 13	300 6	12.0 10
			β+ 3	1036 13	447 6	8.5 14
			total β+		346 7	24.3 18
			3 weak βs			1.10
			x ray Kα$_2$		9.22482 7	13.4 6
			x ray Kα$_1$		9.25174 7	26.3 10
			x ray Kβ		10.3	5.60 23
			γ 1		22.43 6	1.56 13
			γ 5		42.83 20	1.1 3
			γ 6		43.89 7	28.6 15
			γ 9		65.12 11	7.1 5
			γ 13		108.85 3	10.4 6
			γ 15		147.79 3	1.3 3
			γ 18		182.03 4	5.6 4
			γ 19		190.20 3	5.6 5
			γ 24		245.71 3	5.3 3
			γ 25		272.97 4	10.4 6
			γ 27		302.52 3	2.47 23
			γ 30		338.05 3	8.6 7
			γ 32		381.85 5	27.8 12
			γ 35		470.62 6	7.3 5
			γ 36		472.00 11	3.2 3
			γ 40		536.74 7	6.1 4
			γ 49		705.94 3	4.2 3
			75 weak γs			12.46
			max γ±			48.61
^{67}Ga (EC decay)	[14119-09-6]	78.26 h 3	Auger L		0.99	168 9
			Auger K		7.53	61 4
			ce-K-2		83.652 5	28.7 13
			ce-L-2		92.117 5	3.52 15
			x ray Kα$_2$		8.61578 5	16.8 12
			x ray Kα$_1$		8.63886 5	32.9 23
			x ray Kβ		9.57	6.7 5
			γ 1		91.266 5	3.07 10
			γ 2		93.311 5	38.3 12
			γ 3		184.577 10	20.9 6
			γ 4		208.951 10	2.37 7
			γ 5		300.219 10	16.8 4
			γ 6		393.529 10	4.70 14
			4 weak γs			0.28
^{68}Ga (EC decay)	[15757-14-9]	68.06 min 26	Auger L		0.99	13.8 7
			Auger K		7.53	5.1 3
			β+ 1	821.7 12	352.6 6	1.12 12
			β+ 2	1899.1 12	836.0 6	88.0 4
			total β+		829.9 6	89.1 5
			x ray Kα$_2$		8.61578 5	1.38 10
			x ray Kα$_1$		8.63886 5	2.71 18
			γ 3		1077.40 10	3.0 3
			8 weak γs			0.35
			max γ±			178.24
^{68}Ge (EC decay)	[15756-77-1]	288 d 6	Auger L		1	121.5 18
			Auger K		8	42.4 7
			x ray Kα$_2$		9.22482 7	13.1 3
			x ray Kα$_1$		9.25174 7	25.6 5
			x ray Kβ		10.3	5.46 14
^{69}Zn (β− decay)	[13982-23-5]	57 min 1	β− 1	905 3	320.9 13	100
^{69}Zn (IT decay)	[13982-23-5]	13.76 h 3	Auger L		0.99	6.3 4
			Auger K		7.53	2.28 15
			ce-K-1		428.975 18	4.38 14
			x ray Kα$_1$		8.63886 5	1.23 9
			γ 1		438.634 18	94.90 16
^{69}Ge (β+ decay)	[15034-49-8]	39.05 h 10	Auger L		1	80.9 19
			Auger K		8	28.7 8
			β+ 1	629.7 24	270.9 11	1.94 19
			β+ 2	1203.5 24	521.4 11	32.5 8

Table 6 (*continued*)

Isotope	CAS Registry No.	Half-life	Type of radiation	Energy, keV Max	Energy, keV Avg	Intensity, %
			total β+		505.1 12	34.7 9
			2 weak βs			0.21
			x ray Kα₂		9.2248 7	8.8 3
			x ray Kα₁		9.2514 7	17.3 5
			x ray Kβ		10.3	3.69 12
			γ 5		318.40 20	1.31 15
			γ 10		573.90 20	11.1 12
			γ 15		871.70 20	9.8 10
			γ 19		1106.40 20	25.7 14
			γ 23		1336.20 20	2.9 4
			32 weak γs			4.03
			max γ±			69.31
^{71}Zn (β− decay)	[14914-52-4]	3.94 h 5	β− 1	728 12	249 4	6.6 6
			β− 2	1481 12	572 5	89 4
			total β−		539 5	99 4
			11 weak βs			3.21
			γ 2		121.48 5	2.9 3
			γ 3		142.60 5	5.6 6
			γ 4		386.28 5	93 4
			γ 5		389.87 5	2.6 3
			γ 6		453.08 7	1.12 11
			γ 7		487.34 5	62 4
			γ 8		511.55 5	28.4 23
			γ 14		596.07 7	27.9 23
			γ 15		620.19 5	57 4
			γ 16		753.41 7	3.3 4
			γ 17		771.26 7	2.05 21
			γ 22		964.70 10	4.3 5
			γ 24		988.60 20	1.21 11
			γ 30		1107.40 20	2.0 3
			47 weak γs			8.17
^{71}Ge (EC decay)	[14374-81-3]	11.8 d 4	Auger L		1	121.9 19
			Auger K		8	42.9 7
			x ray Kα₂		9.22482 7	13.2 3
			x ray Kα₁		9.25174 7	25.9 5
			x ray Kβ		10.3	5.52 14
^{71}As	[16685-55-5]	64.8 h 7	Auger L		1.19	100 4
			Auger K		8.56	33.5 20
			ce-K-1		12.40 10	3.3 8
			ce-K-2		163.80 10	6.75 21
			β+ 1	816 4	352.0 18	28.5 5
			6 weak βs			0.55
			x ray Kα₂		9.85532 7	11.6 6
			x ray Kα₁		9.88642 7	22.7 12
			x ray Kβ		11	5.1 3
			γ 2		174.90 10	83.1 5
			γ 7		326.80 20	2.66 25
			γ 20		500.00 20	2.83 25
			γ 68		1095.70 20	4.2 5
			99 weak γs			7.27
			max γ±			58.11
^{72}Zn (β− decay)	[15743-55-2]	46.5 h 1	Auger L		1	134 25
			ce-K-1		6.0 3	92 19
			Auger K		8	47 10
			ce-L-1		15.1 3	10.0 21
			ce-M-1		16.2 3	1.5 3
			ce-K-6		92.43 10	1.35 8
			ce-K-8		134.33 10	1.57 6
			β− 1	249 6	72.0 20	15.0 20
			β− 2	296 6	87.5 21	85.0 20
			total β−		85.2 21	100 3
			x ray Kα₂		9.22482 7	15 3
			x ray Kα₁		9.25174 7	29 6
			x ray Kβ		10.3	6.1 12
			γ 1		16.4 3	8.3 17
			γ 4		79.40 20	1.74 10
			γ 5		88.70 10	2.16 10
			γ 6		102.80 10	2.32 10
			γ 7		112.10 10	2.07 10
			γ 8		144.70 10	83.0 20
			γ 9		191.50 20	9.4 3
			2 weak γs			1.41
^{72}Ga (β− decay)	[13982-22-4]	14.10 h 1	β− 1	650 3	217.1 12	15.00 20
			β− 2	667 3	223.7 12	21.7 3
			β− 3	956 3	341.9 13	27.7 5
			β− 4	1049 3	381.0 13	1.87 4
			β− 5	1477 3	569.1 14	8.90 15

Table 6 (*continued*)

Isotope	CAS Registry No.	Half-life	Type of radiation	Energy, keV Max	Energy, keV Avg	Intensity, %
			β^- 6	1927 3	774.2 14	2.99 12
			β^- 7	2528 3	1055.1 15	8.5 6
			β^- 8	3158 3	1354.5 15	10.3 8
			total β^-		497.7 18	100.2 12
			16 weak βs			3.26
			γ 18		600.95 3	5.54 11
			γ 19		629.96 4	24.8 5
			γ 24		786.44 8	3.20 6
			γ 25		810.20 9	2.01 4
			γ 26		834.03 3	95.63 7
			γ 29		894.25 10	9.88 17
			γ 34		970.55 6	1.104 17
			γ 39		1050.69 5	6.91 12
			γ 44		1230.86 7	1.454 20
			γ 46		1260.10 7	1.13 3
			γ 47		1276.76 7	1.565 17
			γ 50		1464.00 7	3.55 6
			γ 56		1596.68 8	4.24 9
			γ 63		1861.09 6	5.25 8
			γ 68		2109.50 9	1.042 20
			γ 69		2201.66 7	25.9 5
			γ 73		2490.98 7	7.68 23
			γ 74		2507.79 7	12.78 23
			71 weak γs			9.85
^{72}As (EC decay)	[15755-33-6]	26.0 h 1	Auger L		1.19	14.5 6
			Auger K		8.56	4.9 3
			β^+ 1	1865 7	822 4	5.81 18
			β^+ 2	2495 7	1115 4	64.1 15
			β^+ 3	3329 7	1526 4	16.3 17
			total β^+		1166 4	87.7 23
			8 weak βs			1.49
			x ray Kα_2		9.85532 7	1.69 9
			x ray Kα_1		9.88642 7	3.30 17
			γ 6		629.92 5	7.92 22
			γ 10		833.99 3	79.5 17
			γ 29		1464.00 6	1.11 3
			76 weak γs			8.70
			max $\gamma\pm$			175.40
^{72}Se (EC decay)	[14809-46-2]	8.40 d 8	Auger L		1.24	169 12
			Auger K		9.1	55 5
			ce-K-1		34.1 3	38.8 18
			ce-L-1		44.5 3	4.16 19
			x ray L		1.28	1.2 4
			x ray Kα_2		10.50800 10	21.0 15
			x ray Kα_1		10.54370 10	41 3
			x ray Kβ		11.7	9.5 7
			γ 1		46.0 3	58.0 20
^{73}Ga (β^- decay)	[15034-51-2]	4.87 h 3	Auger L		1.19	248 7
			ce-K-1		2.171 19	34.8 4
			Auger K		8.56	59 4
			ce-L-1		11.860 17	75.0 9
			ce-MNO-1		13.094 17	15.0 5
			ce-K-2		42.35 5	92.6 7
			ce-L-2		52.04 5	13.6 4
			ce-MNO-2		53.27 5	4.4388
			β^- 1	430 40	133 15	7.2 3
			β^- 2	1170 40	432 18	7.1 4
			β^- 3	1200 40	445 18	78.6 23
			β^- 4	1490 40	576 18	7.0 10
			total β^-		429 20	101 3
			6 weak βs			1.02
			x ray L		1.19	1.5 5
			x ray Kα_2		9.85532 7	20.3 11
			x ray Kα_1		9.88642 7	39.6 20
			x ray Kβ		11	8.9 5
			γ 2		53.45 5	10.2 4
			γ 6		297.32 5	79.8 24
			γ 7		325.70 7	11.2 4
			γ 15		739.42 5	4.2 3
			γ 16		767.80 10	1.44 9
			γ 18		1065.10 10	1.28 9
			12 weak γs			2.90
^{73}As (EC decay)	[15422-59-0]	80.30 d 6	Auger L		1.19	320 10
			ce-K-1		2.171 19	27.9 3
			Auger K		8.56	88 5
			ce-L-1		11.860 17	60.2 7
			ce-MNO-1		13.094 17	12.0 4

Table 6 (continued)

Isotope	CAS Registry No.	Half-life	Type of radiation	Energy, keV Max	Energy, keV Avg	Intensity, %
			ce-K-2		42.334 12	75.1 5
			ce-L-2		52.023 9	11.0 3
			ce-MNO-2		53.257 9	3.60 10
			x ray L		1.19	1.9 7
			x ray Kα_2		9.85532 7	30.3 16
			x ray Kα_1		9.88642 7	59 3
			x ray Kβ		11	13.3 7
			γ 2		53.437 9	10.3 3
			weak γ			0.09
^{73}Se (EC decay)	[15422-57-8]	7.15 h 8	Auger L		1.24	65 6
			Auger K		9.1	21.0 22
			ce-K-1		55.20 10	17 3
			ce-L-1		65.54 10	1.9 3
			ce-K-2		349.3 3	1.13 12
			β^+ 1	1290 10	562 5	65 7
			7 weak βs			0.64
			x ray Kα_2		10.50800 10	8.1 8
			x ray Kα_1		10.54370 10	15.7 15
			x ray Kβ		11.7	3.6 4
			γ 1		67.07 10	70 11
			γ 2		361.2 3	97 10
			61 weak γs			2.56
			max $\gamma\pm$			131.28
^{74}As (β^+ decay)	[14304-78-0]	17.76 d 2	Auger L		1.19	43.5 19
			Auger K		8.56	14.7 9
			β^+ 1	944.3 18	407.9 8	25.7 7
			β^+ 2	1540.1 18	700.9 9	3.4 8
			total β^+		442.1 9	29.1 11
			x ray Kα_2		9.85532 7	5.1 3
			x ray Kα_1		9.88642 7	9.9 6
			x ray Kβ		11	2.22 13
			γ 1		595.80 8	59.2 15
			10 weak γs			0.91
			max $\gamma\pm$			58.20
^{74}As (β^- decay)		17.76 d 2	β^- 1	718 3	242.9 11	15.4 3
			β^- 2	1353 3	530.9 12	18.8 16
			total β^-		400.8 14	34.2 17
			1 weak β			0.04
			γ 1		634.78 8	15.4 8
			2 weak γs			0.15
^{75}Ge (β^- decay)	[14687-40-2]	82.78 min 4	β^- 1	913 3	323.0 11	11.5 12
			β^- 2	1178 3	435.8 12	87.1 13
			total β^-		420.8 12	100.0 18
			5 weak βs			1.41
			γ 3		198.60 10	1.18 12
			γ 4		264.60 10	11.3 11
			9 weak γs			0.66
^{75}Se (EC decay)	[14265-71-5]	119.770 d 10	Auger L		1.24	130 7
			Auger K		9.1	42 3
			ce-K-1		12.7 8	5.3 10
			ce-L-1		23.1 8	1.2 3
			ce-K-4		84.8673 22	2.70 11
			ce-K-6		124.135 3	1.55 5
			x ray Kα_2		10.50800 10	16.1 10
			x ray Kα_1		10.54370 10	31.3 18
			x ray Kβ		11.7	7.3 5
			γ 2		66.060 7	1.14 3
			γ 4		96.7340 20	3.48 9
			γ 5		121.119 3	17.32 25
			γ 6		136.002 3	59.0 8
			γ 7		198.596 6	1.47 4
			γ 9		264.656 4	59.1 8
			γ 10		279.538 3	25.2 4
			γ 11		303.924 3	1.342 18
			γ 13		400.6570 20	11.56 15
			11 weak γs			0.09
^{75}Br (EC decay)	[14809-47-3]	97 min 2	Auger L		1.32	33.5 24
			Auger K		9.67	10.4 10
			β^+ 1	1093 20	475 9	1.00 11
			β^+ 2	1129 20	491 9	3.5 4
			β^+ 3	1324 20	578 9	3.4 3
			β^+ 4	1402 20	613 9	1.25 12
			β^+ 5	1560 20	685 9	4.7 7
			β^+ 6	1695 20	746 10	39 4
			β^+ 7	1701 20	749 10	15 4
			total β^+		700 10	71 6
			13 weak βs			3.35

Table 6 (*continued*)

Isotope	CAS Registry No.	Half-life	Type of radiation	Energy, keV Max	Energy, keV Avg	Intensity, %
			γ 1		6.5 4	53 5
			x ray Kα₂		11.18140 20	4.5 4
			x ray Kα₁		11.22240 20	8.8 7
			x ray Kβ		12.5	2.10 16
			γ 2		112.10 10	1.75 19
			γ 3		141.19 10	6.9 6
			γ 6		286.50 20	92.0 10
			γ 7		292.85 10	2.79 15
			γ 14		377.39 11	4.1 4
			γ 15		427.79 13	4.5 5
			γ 16		431.75 13	4.0 5
			γ 27		572.93 10	2.08 24
			γ 31		608.90 12	1.76 19
			γ 37		733.94 12	1.61 19
			γ 44		912.05 15	1.06 11
			γ 46		952.10 15	1.74 19
			43 weak γs			10.69
			max γ±			142.40
^{76}As (β^- decay)	[15575-20-9]	26.32 h 7	β^- 1	313.6 19	92.8 7	1.19 7
			β^- 2	540.1 19	173.8 8	1.89 12
			β^- 3	1181.2 19	436.4 9	2.02 12
			β^- 4	1752.8 19	691.7 9	7.6 5
			β^- 5	2409.8 19	996.5 9	34.6 15
			β^- 6	2968.9 19	1267.1 9	51.0 20
			total β^-		1064.1 11	100 3
			11 weak βs			1.72
			γ 7		559.10 5	44.7 19
			γ 8		563.23 8	1.17 6
			γ 12		657.03 5	6.1 4
			γ 31		1212.72 18	1.63 12
			γ 32		1216.02 7	3.84 25
			γ 33		1228.52 8	1.39 8
			45 weak γs			3.53
^{76}Br (β^+ decay)	[15765-38-5]	16.2 h 2	Auger L		1.32	49 3
			Auger K		9.67 7	15.2 13
			β^+ 1	774 15	336 7	1.34 9
			β^+ 2	864 15	375 7	5.9 4
			β^+ 3	983 15	427 7	5.1 3
			β^+ 4	1264 15	551 7	1.34 10
			β^+ 5	2245 15	999 7	1.0 3
			β^+ 6	2718 18	1221 7	4.1 12
			β^+ 7	2812 15	1265 7	1.83 16
			β^+ 8	3375 15	1532 8	27.6 14
			β^+ 9	3934 15	1800 8	6.0 11
			total β^+		1203 11	57.1 23
			9 weak βs			2.93
			x ray Kα₂		11.18140 20	6.6 4
			x ray Kα₁		11.2240 20	12.8 8
			x ray Kβ		12.5	3.06 19
			γ 3		472.91 6	1.89 9
			γ 5		559.11 5	72.3 15
			γ 6		563.22 5	2.82 16
			γ 10		657.00 5	15.5 9
			γ 31		1129.85 6	4.34 23
			γ 34		1213.10 10	1.16 8
			γ 35		1216.10 5	8.7 5
			γ 37		1228.65 6	2.02 10
			γ 41		1380.56 8	2.41 11
			γ 45		1471.14 7	2.31 11
			γ 58		1853.68 5	14.0 8
			γ 65		2096.78 8	1.28 8
			γ 66		2111.27 8	2.31 13
			γ 72		2391.29 6	4.5 3
			γ 76		2510.85 8	1.80 12
			γ 81		2792.72 6	5.3 3
			γ 83		2950.55 5	7.6 4
			γ 92		3603.99 8	1.58 12
			85 weak γs			21.25
			max γ±			114.27
^{76}Kr (EC decay)	[28522-17-0]	14.8 h 1	Auger L		1.4	142 9
			Auger K		10.2	42 4
			ce-K-5		32.03 20	19.2 22
			x ray L		1.48	1.4 5
			x ray Kα₂		11.87760 20	20.4 14
			x ray Kα₁		11.92420 20	40 3
			x ray Kβ		13.3	9.8 7
			γ 5		45.50 20	18.1 16
			γ 13		103.20 20	3.7 4

Table 6 (*continued*)

Isotope	CAS Registry No.	Half-life	Type of radiation	Energy, keV Max	Energy, keV Avg	Intensity, %
			γ 17		134.80 20	2.7 3
			γ 18		136.30 20	1.07 15
			γ 25		199.80 20	1.27 15
			γ 31		252.10 20	6.7 5
			γ 32		270.30 20	21.5 17
			γ 33		271.70 20	4.7 6
			γ 37		309.80 20	2.5 3
			γ 38		315.70 20	40 4
			γ 41		355.30 20	5.2 6
			γ 43		406.50 20	12.3 12
			γ 50		452.00 20	9.0 14
			γ 62		552.6 3	1.54 22
			γ 67		582.5 3	1.07 15
			69 weak γs			13.01
^{77}Ge (β^- decay)	[*14687-59-3*]	11.30 h 1	Auger L		1.24	3.70 22
			Auger K		9.1	1.20 10
			ce-K-7		199.164 19	1.84 8
			β^- 1	390 50	118 18	1.080 19
			β^- 2	400 50	122 18	2.137 22
			β^- 3	630 50	208 20	2.077 23
			β^- 4	740 50	252 21	7.6 4
			β^- 5	770 50	264 21	2.366 25
			β^- 6	1170 50	431 22	1.754 18
			β^- 7	1180 50	436 22	6.71 8
			β^- 8	1210 50	451 22	1.657 24
			β^- 9	1280 50	481 22	4.4 3
			β^- 10	1340 50	508 23	1.70 3
			β^- 11	1550 50	601 23	17.2 4
			β^- 12	2110 50	857 24	19.6 5
			β^- 13	2260 50	930 24	16.5 8
			β^- 14	2520 50	1052 24	5.4 7
			β^- 15	2740 50	1154 24	5 5
			total β^-		680 30	99 6
			19 weak βs			3.71
			γ 5		194.762 20	1.68 5
			γ 7		211.031 19	29.2 8
			γ 8		215.505 22	27.1 8
			γ 11		264.440 17	51.0 5
			γ 19		367.397 16	13.27 22
			γ 21		416.328 14	20.6 3
			γ 22		419.75 3	1.165 17
			γ 26		461.378 15	1.198 17
			γ 34		558.018 13	15.20 18
			γ 41		631.823 13	6.59 8
			γ 42		634.389 15	1.969 22
			γ 55		714.345 12	6.77 8
			γ 63		784.770 12	1.244 14
			γ 68		810.352 12	2.148 24
			γ 104		1085.188 13	5.72 7
			γ 114		1193.263 13	2.43 3
			γ 132		1368.4 5	3.2 4
			162 weak γs			28.32
^{77}As (β^- decay)	[*14687-61-7*]	38.83 h 5	β^- 1	445 9	139 4	1.55 4
			β^- 2	684 9	229 4	97.24 8
			total β^-		226 4	100.02 9
			5 weak βs			1.23
			γ 7		239.10 20	1.59 4
			12 weak γs			1.35
^{77}Br (EC decay)	[*15765-39-6*]	57.036 h 6	Auger L		1.32	116 7
			Auger K		9.67	36 3
			1 weak β			0.74
			x ray L		1.38	1.05 24
			x ray Kα_2		11.18140 20	15.5 9
			x ray Kα_1		11.22240 20	30.1 17
			x ray Kβ		12.5	7.2 5
			γ 3		87.59 7	1.45 4
			γ 8		161.83 8	1.14 3
			γ 12		200.40 7	1.25 6
			γ 14		238.98 7	23.9 5
			γ 16		249.77 7	3.08 10
			γ 19		281.65 7	2.37 7
			γ 20		297.23 8	4.30 22
			γ 21		303.76 9	1.22 4
			γ 32		439.47 6	1.62 5
			γ 34		484.57 7	1.03 4
			γ 37		520.69 6	23.2 6
			γ 41		574.64 8	1.23 4

Table 6 (*continued*)

Isotope	CAS Registry No.	Half-life	Type of radiation	Energy, keV Max	Energy, keV Avg	Intensity, %
			γ 42		578.91 7	3.06 10
			γ 43		585.48 7	1.62 5
			γ 48		755.35 7	1.72 5
			γ 51		817.79 6	2.15 7
			45 weak γs			4.88
^{77}Kr (EC decay)	[14983-72-3]	74.4 min 6	Auger L		1.4	33 4
			Auger K		10.2	9.5 15
			ce-K-3		92.4 3	6.2 5
			ce-L-3		104.1 3	1.59 12
			ce-K-4		116.23 10	2.78 9
			ce-K-5		132.93 10	4 3
			β^+ 1	1860 100	820 50	2.7 8
			β^+ 2	2000 100	890 50	38.6 19
			β^+ 3	2150 100	960 50	40.2 19
			β^+ 4	2170 100	970 50	1.0 10
			β^+ 5	2280 100	1020 50	1.8 18
			total β^+		920 50	86 4
			7 weak βs			1.37
			x ray Kα_2		11.87760 20	4.6 6
			x ray Kα_1		11.92420 20	8.9 12
			x ray Kβ		13.3	2.2 3
			γ 2		77.0 20	11.0 7
			γ 3		105.9 3	1.28 9
			γ 4		129.70 10	80.0 10
			γ 5		146.40 10	37.6 17
			γ 8		276.10 20	2.88 17
			γ 10		312.10 20	3.4 8
			29 weak γs			2.91
			max γ+			171.35
^{78}Ge (β^- decay)	[15756-83-9]	88 min 1	β^- 1	686 20	231 8	4.0 10
			β^- 2	703 20	237 8	96.0 10
			total β^-		237 8	100.0 15
			γ 1		277.3 3	96.0 10
			γ 2		293.9 5	4.0 8
^{78}As (β^- decay)	[15755-35-8]	90.7 min 2	β^- 1	790 70	270 30	1.41 18
			β^- 2	1000 70	360 30	3.3 5
			β^- 3	1060 70	380 30	1.66 24
			β^- 4	1150 70	420 30	2.5 3
			β^- 5	1450 70	560 40	6.5 8
			β^- 6	1610 70	630 40	16.8 20
			β^- 7	2290 70	940 40	1.0 3
			β^- 8	2980 70	1270 40	14.4 21
			β^- 9	3680 70	1600 40	17.7 24
			β^- 10	4290 70	1890 40	32 8
			total β^-		1280 50	100 9
			12 weak βs			2.70
			γ 4		354.30 20	1.9 3
			γ 11		545.40 10	2.9 4
			γ 13		613.90 10	54 6
			γ 18		694.90 10	17.3 22
			γ 22		828.00 10	8.6 11
			γ 28		888.70 10	2.1 3
			γ 35		1079.80 20	1.78 23
			γ 36		1145.00 20	1.78 23
			γ 40		1240.30 10	6.2 9
			γ 43		1308.70 10	12.4 15
			γ 45		1373.40 20	5.0 7
			γ 49		1530.00 10	2.6 4
			γ 52		1713.7 3	1.89 24
			γ 56		1835.80 20	1.51 20
			γ 61		1995.60 20	1.35 19
			γ 72		2680.80 20	1.7 3
			60 weak γs			18.24
^{79}Se (β^- decay)	[15758-45-9]	≤65000 yr	β^- 1	159 6	55.8 22	100
^{79}Kr (β^+ decay)	[15478-11-2]	35.04 h 10	Auger L		1.4	105 7
			Auger K		10.2	31 3
			β^+ 1	609 8	265 4	6.9 7
			3 weak βs			0.21
			x ray L		1.48	1.1 4
			x ray Kα_2		11.87760 20	14.9 11
			x ray Kα_1		11.92420 20	29.0 22
			x ray Kβ		13.3	7.2 6
			γ 2		135.99 10	1.00 10
			γ 5		217.02 10	2.40 10
			γ 6		261.26 10	12.7 4
			γ 8		299.51 10	1.57 7
			γ 9		306.31 15	2.60 10

Table 6 (continued)

Isotope	CAS Registry No.	Half-life	Type of radiation	Energy, keV Max	Energy, keV Avg	Intensity, %
			γ 11		389.00 10	1.52 7
			γ 12		397.56 10	9.5 3
			γ 19		606.07 10	8.10 20
			γ 25		832.04 10	1.26 6
			32 weak γs			3.62
			max $\gamma\pm$			14.21
^{80}Br (EC decay)	[14391-61-8]	17.68 min 2	Auger L		1.32	7.0 4
			Auger K		9.67	2.17 18
			β^+ 1	848.3 20	368.2 10	2.17 22
			x ray Kα_1		11.22240 20	1.82 10
			γ 1		665.80 20	1.15 12
^{80}Br (β^- decay)	[14391-61-8]	17.68 min 2	β^- 1	1389 11	526 5	6.2 6
			β^- 2	2006 11	805 5	84.9 7
			total β^-		783 5	91.6 10
			2 weak βs			0.50
			γ 1		616.3 5	7.24 10
			6 weak γs			0.59
^{80}Br (IT decay)	[14391-61-8]	4.42 h 1	Auger L		1.4	175 9
			Auger K		10.2	48 5
			ce-K-1		23.5783 21	53.6 20
			ce-L-1		35.2700 21	6.04 22
			ce-K-2		35.4 4	72 4
			ce-MNO-1		36.7955 21	1.27 3
			ce-L-2		47.1 4	22.4 10
			ce-MNO-2		48.6 4	5.05 16
			x ray L		1.48	1.8 6
			x ray Kα_2		11.87760 20	22.9 14
			x ray Kα_1		11.92420 20	44 3
			x ray Kβ		13.3	11.0 7
			γ 1		37.0520 20	39.0 8
			1 weak γ			0.33
^{81}Kr (EC decay)	[15678-91-8]	2.1×10^5 yr 2	Auger L		1.4	111 6
			Auger K		10.2	31 3
			x ray L		1.48	1.14
			x ray Kα_2		11.87760 20	15.1 9
			x ray Kα_1		11.92420 20	29.3 17
			x ray Kβ		13.3	7.2 5
			γ 1		275.990 11	2.0 20
^{82}Br (β^- decay)	[14686-69-2]	35.30 h 3	β^- 1	264.6 15	75.6 5	1.360 20
			β^- 2	444.3 15	137.6 5	98.6 9
			total β^-		136.8 5	100.0 9
			γ 6		221.45 3	2.26 7
			γ 12		554.320 20	70.8 7
			γ 14		606.30 10	1.17 9
			γ 15		619.070 20	43.5 5
			γ 17		698.330 20	28.5 3
			γ 19		776.49 3	83.6 9
			γ 20		827.81 3	24.04 25
			γ 24		1007.57 9	1.272 13
			γ 25		1043.97 3	27.2 3
			γ 30		1317.47 5	26.5 3
			γ 33		1474.82 8	16.32 17
			26 weak γs			4.38
^{82}Rb (β^+ decay)	[14391-63-0]	6.2 h 5	Auger L		1.5	85 5
			Auger K		10.8	24.1 22
			β^+ 1	810 40	352 14	19.4 23
			β^+ 2	1640 40	720 15	2.24 4
			total β^+		390 15	23.1 23
			6 weak βs			1.45
			x ray L		1.59	1.1 5
			x ray Kα_2		12.5980 20	12.8 8
			x ray Kα_1		12.6490 20	24.9 15
			x ray Kβ		14	6.3 4
			γ 4		183.2 5	1.73 25
			γ 5		221.45 3	1.73 25
			γ 11		554.320 20	63 5
			γ 12		606.30 10	1.57 25
			γ 13		619.070 20	37 3
			γ 14		698.330 20	24.0 17
			γ 15		776.49 3	82.6 9
			γ 16		827.81 3	20.6 17
			γ 18		1007.57 9	6.9 9
			γ 19		1043.97 3	33 3
			γ 21		1081.40 20	1.24 17
			γ 25		1317.47 5	25.6 17
			γ 27		1474.82 8	17.3 17
			γ 28		1650.30 10	1.32 17

Table 6 (*continued*)

Isotope	CAS Registry No.	Half-life	Type of radiation	Energy, keV Max	Energy, keV Avg	Intensity, %
^{83}Br (β^- decay)	[14687-62-8]	2.39 h 2	25 weak γs			6.15
			max $\gamma\pm$			46.18
			β^- 1	392 15	119 6	1.3 4
			β^- 2	921 15	324 7	98.6 4
			total β^-		321 7	100.0 4
			2 weak βs			0.11
			γ 6		529.5 3	1.3 4
			8 weak γs			0.12
^{83}Kr (IT decay)	[13965-98-5]	1.83 h 2	Auger L		1.5	165.2 17
			ce-L-1		7.479 10	77.70 20
			ce-M-1		9.112 10	12.70 10
			ce-NOP-1		9.376 10	4.220 10
			Auger K		10.8	8.6 8
			ce-K-2		17.834 20	24.30 10
			ce-L-2		30.239 20	61.70 10
			ce-M-2		31.872 20	10.40 10
			ce-NOP-2		32.136 20	3.450 10
			x ray L		1.59	2.2 9
			γ 1		9.400 10	5.42 16
			x ray Kα_2		12.5980 20	4.57 23
			x ray Kα_1		12.6490 20	8.9 5
			x ray Kβ		14	2.26 12
			1 weak γ			0.05
^{83}Rb (EC decay)	[17056-36-9]	86.2 d 1	Auger L		1.5	109 6
			Auger K		10.8	31 3
			x ray L		1.59	1.4 6
			γ 1		9.35 7	6.03 19
			x ray Kα_2		12.5980 20	16.4 10
			x ray Kα_1		12.6490 20	31.8 18
			x ray Kβ		14	8.1 5
			γ 5		520.35 10	46.1 15
			γ 6		529.54 10	30.0 10
			γ 7		552.50 10	16.3 5
			7 weak γs			1.12
^{83}Sr (β^+ decay)	[14809-51-9]	32.4 h 2	Auger L		1.68	82 4
			Auger K		11.4	22.0 12
			β^+ 1	803 8	351 4	1.4
			β^+ 2	1185 8	519 4	9.7
			β^+ 3	1227 8	537 4	12 3
			total β^+		506 5	24 3
			3 weak βs			0.92
			x ray L		1.69	1.2 5
			x ray Kα_2		13.33580 20	12.9 7
			x ray Kα_1		13.39530 20	25.0 13
			x ray Kβ		15	6.6 4
			γ 2		42.30 10	1.57 21
			γ 8		381.6 3	7.7 12
			γ 9		381.6 3	11.9 15
			γ 10		389.4 3	1.55 20
			γ 11		418.4 3	5.0 6
			γ 12		423.7 3	1.56 19
			γ 34		762.7 3	30 3
			γ 35		778.5 3	1.94 24
			γ 73		1147.7 3	1.23 14
			γ 74		1160.4 3	1.53 17
			γ 92		1562.3 4	1.96 22
			104 weak γs			11.49
			max $\gamma\pm$			48.63
^{84}Rb (β^+ decay)	[15765-86-3]	32.87 d 11	Auger L		1.5	75 4
			Auger K		10.8	21.3 19
			β^+ 1	777 3	338.5 13	12.4 9
			β^+ 2	1658 3	756.3 13	13.5 7
			total β^+		556.3 16	25.9 12
			x ray Kα_2		12.5980 20	11.3 6
			x ray Kα_1		12.6490 20	22.0 11
			x ray Kβ		14	5.6 3
			γ 1		881.46 20	67.9 12
			2 weak γs			1.16
			max $\gamma\pm$			51.72
^{84}Rb (β^- decay)		32.87 d 11	β^- 1	890 4	331.2 15	4.0 5
^{85}Kr (β^- decay)	[13983-27-2]	10.72 yr 2	β^- 1	687.0 20	251.4 8	99.563 10
			1 weak β			0.44
			1 weak γ			0.43
^{85}Kr (IT decay)		4.480 h 8	Auger L		1.5	7.8 5
			Auger K		10.8	2.14 21
			ce-K-1		290.53 3	6.04 25
			x ray Kα_2		12.5980 20	1.14 8

Table 6 (*continued*)

Isotope	CAS Registry No.	Half-life	Type of radiation	Energy, keV Max	Energy, keV Avg	Intensity, %
^{85}Kr (β^- decay)		4.480 h 8	x ray Kα_1		12.6490 20	2.20 14
			γ 1		304.86 3	13.8 4
			Auger L		1.68	3.86 15
			Auger K		11.4	1.05 5
			ce-K-1		135.97 3	3.16 12
			β^- 1	840.7 20	290.4 9	79.0 6
			x ray Kα_1		13.39530 20	1.19 5
			γ 1		151.17 3	75.4 18
^{85}Sr (EC decay)	[13967-73-2]	64.84 d 3	Auger L		1.68	108.2 23
			Auger K		11.4	29.1 9
			x ray L		1.69	1.6 6
			x ray Kα_2		13.33580 20	17.1 4
			x ray Kα_1		13.39530 20	33.0 7
			x ray Kβ		15	8.66 24
			γ 1		513.993 14	98.3 10
			1 weak γ			0.01
^{86}Rb (β^- decay)	[14932-53-7]	18.66 d 2	β^- 1	698.0 20	232.6 8	8.78 8
			β^- 2	1774.6 20	709.4 9	91.22 8
			total β^-		667.5 10	100.00 12
			γ 1		1076.60 10	8.78 8
^{86}Y (β^+ decay)	[14809-53-1]	14.74 h 2	Auger L		1.79	70 4
			Auger K		12	18.1 16
			β^+ 1	933 10	408 5	1.26 20
			β^+ 2	1066 10	467 5	2.1 5
			β^+ 3	1195 10	524 5	1.40 12
			β^+ 4	1254 10	550 5	12.4 5
			β^+ 5	1578 10	696 5	5.7 5
			β^+ 6	1769 10	783 5	1.7 10
			β^+ 7	2021 10	899 5	3.6 9
			β^+ 8	2397 10	1093 5	1.0 10
			β^+ 9	3174 10	1452 5	2.0 12
			total β^+		672 6	33.4 23
			14 weak β^+			2.23
			x ray L		1.8	1.1 4
			x ray Kα_2		14.09790 20	11.8 6
			x ray Kα_1		14.16500 20	22.7 11
			x ray Kβ		15.8	6.1 3
			γ 4		187.87 13	1.26 5
			γ 5		190.80 13	1.01 4
			γ 12		307.00 10	3.46 9
			γ 17		382.86 23	3.63 12
			γ 20		443.13 10	16.9 5
			γ 26		515.18 20	4.89 15
			γ 27		580.57 10	4.78 15
			γ 28		608.29 10	2.01 15
			γ 30		627.72 10	32.6 10
			γ 32		644.82	2.2 4
			γ 33		645.87	9.2 11
			γ 37		703.33 10	15.4 5
			γ 38		709.90 10	2.62 8
			γ 40		740.81 13	1.36 5
			γ 41		767.63 13	2.4 4
			γ 43		777.37 10	22.4 6
			γ 45		826.02 13	3.30 9
			γ 46		833.72	1.5 4
			γ 47		835.67	4.4 6
			γ 50		955.35 20	1.04 5
			γ 53		1024.04 10	3.79 17
			γ 54		1076.63 10	82.5 4
			γ 61		1153.05 10	30.5 10
			γ 63		1163.03 10	1.18 5
			γ 64		1253.11 10	1.53 5
			γ 70		1349.15 10	2.95 10
			γ 81		1801.70 10	1.65 5
			γ 82		1854.38 13	17.2 5
			γ 83		1920.72 13	20.8 7
			γ 92		2567.97 18	2.25 11
			γ 93		2610.11 20	1.24 8
			73 weak γs			17.58
			max $\gamma\pm$			66.78
^{86}Y (IT decay)		48 min 1	Auger L		2	82 3
			ce-L-1		7.83 10	78.9 5
			ce-M-1		9.81 10	15.4 24
			Auger K		12.7	1.4 3
			ce-K-2		191.06 20	4.8 8
			x ray L		2	1.8 6
			x ray Kα_1		14.95840 20	1.9 3

Table 6 (*continued*)

Isotope	CAS Registry No.	Half-life	Type of radiation	Energy, keV Max	Energy, keV Avg	Intensity, %
^{86}Zr (EC decay)	[15743-56-3]	16.5 h 1	γ 1		208.10 20	93.6 7
			Auger L		2	189 13
			ce-K-1		12.06 10	70 6
			Auger K		12.7	46 6
			ce-L-1		26.73 10	8.7 7
			ce-M-1		28.71 10	1.46 11
			ce-K-9		225.76 10	3.56 11
			x ray L		2	4.2 14
			x ray Kα_2		14.88290 20	33.0 23
			x ray Kα_1		14.95840 20	64 5
			x ray Kβ		16.7	17.5 13
			γ 1		29.10 10	21.6 15
			γ 9		242.80 10	95.80 10
			γ 10		612.00 10	5.7 3
			9 weak γs			1.19
^{87}Kr (β^- decay)	[14809-68-8]	76.3 min 5	β^- 1	928 5	326.4 21	4.4 4
			β^- 2	1334 5	500.4 22	9.4 7
			β^- 3	1475 5	562.6 23	5.5 3
			β^- 4	3044 5	1294.2 24	7.3 4
			β^- 5	3486 5	1502.0 24	41 3
			β^- 6	3889 5	1694.8 24	30.5 22
			total β^-		1328 3	100 4
			7 weak βs			2.12
			γ 2		402.578 20	50 4
			γ 6		673.83 8	1.89 11
			γ 9		845.44 4	7.3 4
			γ 15		1175.40 8	1.11 7
			γ 23		1740.52 8	2.04 11
			γ 25		2011.88 10	2.88 18
			γ 28		2554.80 20	9.2 7
			γ 29		2558.10 20	3.9 4
			26 weak γs			4.48
^{87}Rb (β^- decay)	[13982-13-3]	4.80×10^{10} yr 13	β^- 1	273.3 19	78.8 7	100
^{87}Sr (IT decay)		2.81 h 1	Auger L		1.79	18.4 11
			Auger K		12	4.6 5
			ce-K-1		372.30 5	15.0 5
			ce-L-1		386.18 5	2.20 7
			x ray Kα_2		14.0979 20	3.01 16
			x ray Kα_1		14.1650 20	5.8 3
			x ray Kβ		15.8	1.56 8
			γ 1		388.40 5	82.26 16
^{87}Y (EC decay)	[14274-68-1]	80.3 h 3	Auger L		1.79	105 6
			Auger K		12	27.0 23
			1 weak β			0.21
			x ray L		1.8	1.7 6
			x ray Kα_2		14.09790 20	17.5 8
			x ray Kα_1		14.16500 20	33.8 14
			x ray Kβ		15.8	9.1 4
			γ 2		484.90 5	92.2 10
^{87}Y (IT decay)		12.9 h 4	Auger L		2	20.5 12
			Auger K		12.7	4.9 6
			ce-K-1		364.0 7	17.0 6
			ce-L-1		378.6 7	2.57 8
			x ray Kα_2		14.88290 20	3.49 20
			x ray Kα_1		14.95840 20	6.7 4
			x ray Kβ		16.7	1.85 11
			γ 1		381.0 7	78.05 13
^{87}Zr (β^+ decay)	[15743-57-4]	104.0 min 5	Auger L		2	20.8 20
			Auger K		12.7	5.1 7
			β^+ 1	2180 80	970 40	79 4
			7 weak βs			1.20
			x ray Kα_2		14.88290 20	3.6 4
			x ray Kα_1		14.95840 20	7.0 7
			x ray Kβ		16.7	1.93 20
			γ 20		1227.0 10	3.03 10
			34 weak γs			4.73
			max γ±			160.39
^{88}Kr (β^- decay)	[14995-61-0]	2.84 h 2	Auger L		1.68	14.8 7
			Auger K		11.4	4.00 20
			ce-K-1		12.313 14	10.7 6
			ce-L-1		25.448 14	1.23 8
			ce-K-7		181.120 15	1.14 12
			β^- 1	365 17	109 6	2.65 16
			β^- 2	521 17	165 7	67 3
			β^- 3	681 17	227 7	9.1 5
			β^- 4	1198 17	441 8	1.92 11
			β^- 5	2051 17	825 8	1.31 25

Table 6 (*continued*)

Isotope	CAS Registry No.	Half-life	Type of radiation	Energy, keV Max	Energy, keV Avg	Intensity, %
			β^- 6	2717 17	1136 8	1.80 25
			β^- 7	2913 17	1233 8	14 4
			total β^-		358 13	100 5
			14 weak βs			2.52
			x ray Kα_2		13.33580 20	2.35 12
			x ray Kα_1		13.39530 20	4.54 21
			x ray Kβ		15	1.19 6
			γ 1		27.513 14	1.94 17
			γ 4		165.98 4	3.10 15
			γ 7		196.320 15	26.0 13
			γ 12		362.226 13	2.25 12
			γ 34		834.830 3	13.0 7
			γ 42		985.780 16	1.31 7
			γ 48		1141.33 6	1.28 7
			γ 54		1250.67 4	1.12 6
			γ 60		1369.50 20	1.48 9
			γ 63		1518.39 3	2.15 12
			γ 64		1529.77 3	10.9 6
			γ 74		2029.84 3	4.53 23
			γ 75		2035.411 18	3.74 21
			γ 77		2195.842 7	13.2 7
			γ 78		2231.772 21	3.39 17
			γ 82		2392.11 4	34.6 16
			70 weak γs			13.11
^{88}Rb (β^- decay)	[14928-36-0]	17.8 min 1	β^- 1	795 11	271 5	2.13 11
			β^- 2	2575 11	1068 6	13.3 7
			β^- 3	3473 11	1494 6	4.2 3
			β^- 4	5309 11	2370 6	77.4 14
			total β^-		2067 7	99.5 16
			9 weak βs			2.44
			γ 5		898.042 4	14.0 8
			γ 13		1836.063 13	21.4 13
			γ 18		2677.86 5	1.96 12
			24 weak γs			2.58
^{88}Y (EC decay)	[13982-36-0]	106.64 d 8	Auger L		1.79	105 6
			Auger K		12	27.1 23
			1 weak β			0.22
			x ray L		1.8	1.7 6
			x ray Kα_2		14.09790 20	17.6 8
			x ray Kα_1		14.16500 20	33.9 14
			x ray Kβ		15.8	9.1 4
			γ 2		898.042 4	93.4 7
			γ 4		1836.063 13	99.35 3
			4 weak γs			0.73
^{88}Zr (EC decay)	[14681-75-5]	83.4 d 3	Auger L		2	105 6
			Auger K		12.7	26 3
			ce-K-1		375.86 10	2.33 7
			x ray L		2	2.4 8
			x ray Kα_2		14.88290 20	18.2 9
			x ray Kα_1		14.95840 20	35.1 16
			x ray Kβ		16.7	9.7 5
			γ 1		392.90 10	97.24 9
^{89}Sr (β^- decay)	[14158-27-1]	50.5 d 1	β^- 1	1492 4	583.1 13	99.9910 10
^{89}Zr (β^+ decay)	[13981-27-6]	78.43 h 8	Auger L		2	79 4
			Auger K		12.7	19.5 21
			β^+ 1	900 3	394.9 13	22.64 19
			x ray L		2	1.8 6
			x ray Kα_2		14.88290 20	13.9 7
			x ray Kα_1		14.95840 20	26.8 12
			x ray Kβ		16.7	7.4 4
			γ 1		909.20 10	99.870 10
			4 weak γs			1.07
			max $\gamma\pm$			45.28
^{90}Sr (β^- decay)	[10098-97-2]	29.12 yr 24	β^- 1	546.0 20	195.8 8	100
^{90}Y (β^- decay)	[10098-91-6]	64.0 h 1	β^- 1	2284 4	934.8 18	99.984 3
			1 weak β			0.02
^{90}Y (IT decay)		3.19 h 1	Auger L		2	12.1 7
			Auger K		12.7	2.9 4
			ce-K-1		185.47 3	2.65 8
			ce-K-2		462.49 4	7.52 23
			ce-L-2		477.16 4	1.05 4
			x ray Kα_2		14.88290 20	2.09 11
			x ray Kα_1		14.95840 20	4.03 21
			x ray Kβ		16.7	1.11 6
			γ 1		202.51 3	96.58 18
			γ 2		479.53 4	90.71 7
			1 weak γ			0.36

Table 6 (*continued*)

Isotope	CAS Registry No.	Half-life	Type of radiation	Energy, keV Max	Energy, keV Avg	Intensity, %
^{90}Nb (β^+ decay)	[14681-65-3]	14.60 h 5	Auger L		2	46 4
			Auger K		13.4	10.8 16
			β^+ 1	1500 4	662.2 18	53 6
			x ray L		2	1.3 5
			x ray Kα_2		15.69090 20	8.5 8
			x ray Kα_1		15.77510 20	16.3 15
			x ray Kβ		17.7	4.6 5
			γ 1		132.59 3	4.5 7
			γ 2		141.149 20	70 5
			γ 4		371.01 15	1.5 4
			γ 12		890.60 20	1.73 12
			γ 15		1129.10 20	92.0 11
			γ 17		1270.60 20	1.45 13
			γ 20		1611.80 20	2.39 19
			γ 30		1913.30 20	1.30 23
			γ 34		2186.40 20	18.0 10
			γ 36		2319.20 20	82.0 9
			26 weak γs			8.67
			max $\gamma\pm$			106.20
^{91}Sr (β^- decay)	[14331-91-0]	9.52 h 6	β^- 1	477 4	148.8 15	1.48 11
			β^- 2	617 4	200.9 16	2.07 15
			β^- 3	1104 4	398.9 17	34.7 24
			β^- 4	1138 4	413.4 18	1.82 13
			β^- 5	1379 4	518.0 18	25.0 18
			β^- 6	2031 4	812.9 19	3.4 5
			β^- 7	2684 4	1121.2 19	29 5
			total β^-		646.6 22	99 6
			8 weak βs			1.58
			γ 4		274.70 20	1.03 8
			γ 14		620.10 10	1.77 13
			γ 17		652.3 3	3.0 3
			γ 18		652.90 20	8.0 7
			γ 21		749.80 10	23.6 17
			γ 29		925.80 20	3.8 3
			γ 32		1024.30 10	33.4 23
			42 weak γs			6.80
^{91}Y (β^- decay)	[14234-24-3]	58.51 d 6	β^- 1	1543.0 20	603.8 9	99.70 7
			1 weak β			0.30
			1 weak γ			0.30
^{92}Sr (β^- decay)	[14928-29-1]	2.71 h 1	β^- 1	550 30	174 12	96 12
			β^- 2	1930 30	777 14	3 12
			total β^-		193 13	99 17
			1 weak β			0.21
			γ 1		241.52 3	3.0 4
			γ 3		430.56 5	3.3 5
			γ 8		953.32 9	3.6 5
			γ 9		1142.30 10	2.9 4
			γ 10		1383.94 6	90 11
			5 weak γs			0.82
^{92}Y (β^- decay)	[15751-59-4]	3.54 h 1	β^- 1	1294 16	480 7	6.5 7
			β^- 2	2138 16	869 8	1.15 20
			β^- 3	2251 16	920 8	2.3 3
			β^- 4	2699 16	1123 8	3.5 10
			β^- 5	3634 16	1563 8	85.7 16
			total β^-		1446 9	100.0 21
			7 weak βs			0.82
			γ 1		448.50 10	2.3 3
			γ 3		561.10 10	2.4 3
			γ 4		844.30 10	1.25 14
			γ 6		934.50 10	13.9 16
			γ 9		1405.40 10	4.8 6
			15 weak γs			1.91
^{92}Nb (EC decay)	[13982-37-1]	3.5 × 10^7 yr 3	Auger L		2	99 5
			Auger K		13.4	23 3
			x ray L		2	2.8 10
			x ray Kα_2		15.69090 20	17.6 9
			x ray Kα_1		15.77510 20	33.8 15
			x ray Kβ		17.7	9.5 5
			γ 1		561.1 11	100
			γ 2		934.51 8	100
^{92}Nb (EC decay)		10.15 d 2	Auger L		2	100 6
			Auger K		13.4	23 3
			1 weak β			0.06
			x ray L		2	2.8 10
			x ray Kα_2		15.69090 20	18.3 9
			x ray Kα_1		15.77510 20	35.3 16
			x ray Kβ		17.7	9.9 5

Table 6 (*continued*)

Isotope	CAS Registry No.	Half-life	Type of radiation	Energy, keV Max	Energy, keV Avg	Intensity, %
^{93}Zr (β^- decay)	[15751-77-6]	1.53×10^6 yr 10	γ 1		912.60 20	1.78 10
			γ 2		934.44 10	99.12 4
			1 weak γ			0.88
			β^- 1	62.0 20	19.6 7	100
^{93}Nb (IT decay)	[7440-03-1]	13.6 yr 3	Auger L		2.15	79.3 12
			ce-K-1		11.4 3	14.5
			Auger K		14	3.7 5
			ce-L-1		27.7 3	66.5
			ce-M-1		29.9 3	14.3
			ce-NOP-1		30.3 3	4.72
			x ray L		2.17	2.5 9
			x ray Kα_2		16.52100 20	3.12 14
			x ray Kα_1		16.61510 20	6.0 3
			x ray Kβ		18.6	1.72 8
^{93}Mo (EC decay)	[14119-13-2]	3.5×10^3 yr 7	Auger L		2.15	98 5
			Auger K		14	21 3
			x ray L		2.17	3.0 11
			x ray Kα_2		16.52100 20	18.1 9
			x ray Kα_1		16.61510 20	34.8 16
			x ray Kβ		18.6	10.0 5
^{93}Mo (IT decay)		6.85 h 7	Auger L		2.27	38.6 21
			Auger K		14.8	7.2 10
			ce-K-2		243.1005 3	30.2 10
			ce-L-2		260.2345 3	8.7 3
			ce-M-2		262.5954 3	2.62 9
			x ray L		2.29	1.6 6
			x ray Kα_2		17.3743 14	6.7 4
			x ray Kα_1		17.47930 10	12.8 7
			x ray Kβ		19.6	3.77 21
			γ		263	58.1 8
			γ		684.7	99.78 5
			γ		1477.2 3	99.18 5
			2 weak γs			1.32
^{94}Nb (β^- decay)	[14681-63-1]	2.03×10^4 yr 16	β^- 1	471 3	145.6 10	100
			γ 1		702.645 6	100
			γ 2		871.119 4	100
^{94}Tc (EC decay)	[14809-55-3]	293 min 1	Auger L		2.27	86 5
			Auger K		14.8	18.3 25
			β^+	816 5	360.5 22	11.00 10
			2 weak βs			0.03
			x ray L		2.29	3.6 12
			x ray Kα_2		17.3743 14	17.1 8
			x ray Kα_1		17.47930 10	32.7 15
			x ray Kβ		19.6	9.6 5
			γ 2		449.1 3	3.3 3
			γ 3		532.10 10	2.4 3
			γ 4		702.64 7	99.8 6
			γ 5		742.30 20	1.20 20
			γ 6		849.70 7	97.7 6
			γ 7		871.03 7	100
			γ 8		916.12 10	7.6 4
			γ 9		1591.9 3	2.30 20
			2 weak γs			0.50
			max $\gamma^\pm$			22.06
^{95}Zr (β^- decay)	[13967-71-0]	63.98 d 6	β^- 1	365 4	108.8 13	55.0 3
			β^- 2	398 4	120.0 13	44.6 6
			total β^-		115.3 14	100.3 7
			1 weak β			0.70
			γ 2		724.199 5	44.5 6
			γ 3		756.74 4	55.0 3
^{95}Nb (β^- decay)	[13967-76-5]	35.15 d 3	β^- 1	159.7 5	43.33 15	100
			γ 1		765.83 4	100
^{95}Nb (IT decay)		86.6 h 8	Auger L		2.15	71 4
			Auger K		14	14.7 19
			ce-K-1		215.71 14	58.4 5
			ce-L-1		232.00 14	11.70 10
			ce-MNO-1		234.23 14	3.86 3
			x ray L		2.17	2.2 8
			x ray Kα_2		16.52100 20	12.6 6
			x ray Kα_1		16.61510 20	24.2 11
			x ray Kβ		18.6	6.9 4
			γ 1		234.70 14	26.1 6
^{96}Nb (β^- decay)	[15832-32-3]	23.35 h 5	β^- 1	746 4	249.7 16	2.3 3
			β^- 2	749 4	250.5 16	96.7 19
			total β^-		249.1 16	100.0 20
			3 weak βs			1.01
			γ 4		219.02 9	3.78 20

Table 6 (*continued*)

Isotope	CAS Registry No.	Half-life	Type of radiation	Energy, keV Max	Energy, keV Avg	Intensity, %
			γ 5		241.40 9	3.88 20
			γ 11		371.91 9	2.91 20
			γ 15		460.05 5	28.6 8
			γ 16		480.69 6	6.2 3
			γ 17		568.84 5	56.5 16
			γ 18		591.16 12	1.07 10
			γ 20		719.49 12	7.3 4
			γ 22		778.20 4	96.88 19
			γ 23		810.24 7	10.0 5
			γ 24		812.47 17	2.9 3
			γ 25		847.60 17	1.45 20
			γ 26		849.95 7	20.6 8
			γ 27		1091.33 5	48.7 16
			γ 29		1200.19 5	19.9 7
			γ 33		1497.71 7	2.91 20
			18 weak γs			5.47
^{96}Tc (EC decay)	[14808-44-7]	4.28 d 6	Auger L		2.27	96 7
			Auger K		14.8	20 3
			x ray L		2.29	4.0 14
			x ray Kα_2		17.3743 14	18.9 12
			x ray Kα_1		17.47930 10	36.1 22
			x ray Kβ		19.6	10.6 7
			γ 5		314.27 5	2.43 24
			γ 6		316.50 6	1.40 20
			γ 19		778.22 4	99.760 10
			γ 20		812.54 4	82 4
			γ 22		849.86 4	98 4
			γ 24		1091.30 4	1.10 8
			γ 25		1126.85 6	15.2 12
			22 weak γs			4.00 10
^{97}Zr (β^- decay)	[14928-30-4]	16.90 h 5	Auger L		2.15	1.76 12
			ce-K-17		724.37 10	1.65 5
			β^- 1	559 16	178 6	5.5 3
			β^- 2	901 16	312 6	2.10 20
			β^- 3	1414 16	532 7	4.1 5
			β^- 4	1922 16	760 8	86.0 7
			total β^-		700 9	99.9 10
			6 weak βs			2.24
			γ 4		254.15 20	1.25 14
			γ 9		355.39 10	2.27 24
			γ 11		507.63 10	5.1 6
			γ 13		602.41 20	1.39 14
			γ 17		743.36 10	92.7 9
			γ 22		1021.3 3	1.34 14
			γ 24		1147.95 10	2.6 3
			γ 26		1362.66 10	1.34 14
			γ 27		1750.46 10	1.34 14
			19 weak γs			5.86
^{97}Nb (β^- decay)	[18496-04-3]	72.1 min 7	β^- 1	910 16	315 6	1.10 10
			β^- 2	1276 16	470 7	98.30 10
			total β^-		467 7	99.99 15
			4 weak βs			0.59
			γ 5		657.92 10	98.34 11
			13 weak γs			0.91
^{97}Tc (EC decay)	[15759-35-0]	2.6 × 10^6 yr 4	Auger L		2.27	96 6
			Auger K		14.8	20 3
			x ray L		2.29	4.0 14
			x ray Kα_2		17.3743 14	18.9 9
			x ray Kα_1		17.47930 10	36.1 16
			x ray Kβ		19.6	10.6 5
^{97}Ru (EC decay)	[15758-35-7] 1	2.9 d 1	Auger L		2.17	97 6
			Auger K		15.5	20 3
			ce-K-5		194.64 4	2.94 8
			x ray L		2.42	4.6 16
			x ray Kα_2		18.2508 8	20.0 9
			x ray Kα_1		18.3671 8	38.3 17
			x ray Kβ		20.6	11.5 6
			γ 5		215.68 4	86.0 4
			γ 6		324.55 7	10.2 4
			17 weak γs			1.46
^{98}Tc (β^- decay)	[32025-58-4]	4.2 × 10^6 yr 3	β^- 1	397 22	119 8	100
			γ 1		652.41 5	100
			γ 2		745.35 5	100
^{99}Mo (β^- decay)	[14119-15-4]	66.0 h 2	Auger L		2.17	5.4 3
			Auger K		15.5	1.11 17
			ce-K-2		19.543 15	3.77 7
			β^- 1	436.1 10	133.0 4	16.55 7
			β^- 2	847.6 10	289.6 4	1.17 3

Table 6 (continued)

Isotope	CAS Registry No.	Half-life	Type of radiation	Energy, keV Max	Energy, keV Avg	Intensity, %
			β^- 3	1214.1 10	442.7 5	81.96 18
			total β^-		388.7 6	99.94 20
			4 weak βs			0.26
			x ray Kα_2		18.2508 8	1.12 6
			x ray Kα_1		18.3671 8	2.15 10
			γ 2		40.587 15	1.15 4
			γ 4		140.466 15	4.95 9
			γ 8		181.057 15	6.06 8
			γ 11		366.421 15	1.193 24
			γ 25		739.500 15	12.194 17
			γ 27		777.921 20	4.32 7
			26 weak γs			0.40
^{99}Tc (β^- decay)	[14133-76-7]	2.13 × 10^5 yr 5	β^- 1	293.5 19	84.6 7	100
^{99}Tc (IT decay)		6.02 h 3	ce-M-1		1.630 5	86.84 16
			ce-NOP-1		2.106 5	12.18 20
			Auger L		2.17	10.4 7
			Auger K		15.5	2.1 4
			ce-K-2		119.422 15	8.8 4
			ce-L-2		137.423 15	1.06 4
			x ray Kα_2		18.2508 8	2.12 13
			x ray Kα_1		18.3671 8	4.06 24
			x ray Kβ		20.6	1.22 8
			γ 2		140.466 15	88.97 24
			2 weak γs			0.02
^{101}Rh (EC decay)	[14378-53-1]	3.3 yr 3	Auger L		2.53	107 7
			Auger K		16.2	20 4
			ce-K-4		105.09 7	10.8 10
			ce-L-4		123.99 7	1.36 14
			ce-K-7		175.88 20	3.0 3
			x ray L		2.56	5.6 18
			x ray Kα_2		19.15040 20	22.3 13
			x ray Kα_1		19.27920 20	42.5 24
			x ray Kβ		21.7	13.0 8
			γ 4		127.21 7	73 6
			γ 7		198.00 20	71 6
			γ 12		325.00 20	13.4 16
			13 weak γs			2.77
^{101}Rh (EC decay)		4.34 d 1	Auger L		2.53	87 8
			Auger K		16.2	17 3
			ce-K-6		284.743 20	1.18 10
			x ray L		2.56	4.6 15
			x ray Kα_2		19.15040 20	18.3 16
			x ray Kα_1		19.27920 20	35 3
			x ray Kβ		21.7	10.7 10
			γ 6		306.860 20	86 7
			γ 13		545.06 5	4.0 3
			14 weak γs			1.96
^{101}Rh (IT decay)		4.34 d 1	Auger L		2.39	6.6 8
			Auger K		17	1.03 22
			ce-K-1		134.10 3	5.3 7
			ce-L-1		153.91 3	1.68 22
			x ray Kα_2		20.07370 20	1.24 17
			x ray Kα_1		20.21610 20	2.3 4
			1 weak γ			0.25
^{101}Pd (EC decay)	[15749-54-9]	8.47 h 6	ce-K-1		1.240 10	70 7
			Auger L		2.39	160 12
			Auger K		17	29 5
			ce-L-1		21.048 10	8.9 8
			ce-M-1		23.833 10	1.66 15
			β^+ 1	778 4	346.7 18	4.9 4
			1 weak β			0.19
			x ray L		2.7	8 3
			x ray Kα_2		20.07370 20	35.2 22
			x ray Kα_1		20.21610 20	67 5
			x ray Kβ		22.7	20.9 14
			γ 1		24.460	3.9 4
			γ 10		269.67 7	6.4 6
			γ 11		296.29 3	19.2 15
			γ 23		565.98 5	3.4 3
			γ 24		590.44 6	12.1 10
			γ 29		723.75 10	2.05 24
			γ 51		1202.04 6	1.52 14
			γ 53		1289.05 5	2.28 19
			59 weak γs			5.88
			max $\gamma\pm$			10.18
^{102}Rh (EC decay)	[15765-82-9]	2.9 yr	Auger L		2.53	99 6
			Auger K		16.2	19 3
			x ray L		2.56	5.2 16

Table 6 (*continued*)

Isotope	CAS Registry No.	Half-life	Type of radiation	Energy, keV Max	Energy, keV Avg	Intensity, %
			x ray Kα₂		19.15040 20	20.5 11
			x ray Kα₁		19.27920 20	39.0 19
			x ray Kβ		21.7	12.0 6
			γ 3		415.25 15	2.1 3
			γ 4		418.52 18	9.31 22
			γ 5		420.40 20	3.2 3
			γ 6		475.06 4	94 5
			γ 7		628.05 5	8.2 5
			γ 8		631.29 5	55.5 23
			γ 9		692.40 20	1.60 20
			γ 10		695.6 3	2.8 4
			γ 11		697.49 8	43.2 21
			γ 12		766.84 6	33.8 32
			γ 13		1046.59 7	33.8 21
			γ 14		1103.16 6	4.5 3
			γ 15		1112.84 7	18.8 11
			6 weak γs			1.75
¹⁰³Ru (β⁻ decay)	[13968-53-1]	39.35 d 5	Auger L		2.39	78 4
			ce-K-1		16.535 12	9.6 4
			Auger K		17	2.1 4
			ce-L-1		36.343 12	72 3
			ce-M-1		39.128 12	14.4 6
			ce-NOP-1		39.674 12	4.82 20
			β⁻ 1	113 4	29.8 10	6.5 4
			β⁻ 2	226 4	63.2 12	90.0 20
			β⁻ 3	763 4	254.7 15	3.50 20
			total β⁻		67.9 14	100.3 21
			2 weak βs			0.33
			x ray L		2.7	4.1 14
			x ray Kα₂		20.07370 20	2.47 14
			x ray Kα₁		20.21610 20	4.7 3
			x ray Kβ		22.7	1.46 9
			γ 14		497.080 13	89.5 10
			γ 18		610.330 20	5.64 19
			18 weak γs			1.98
¹⁰³Rh (IT decay)	[7440-16-6]	56.12 min 1	Auger L		2.39	77 4
			ce-K-1		16.535 12	9.5 6
			Auger K		17	1.8 4
			ce-L-1		36.343 12	71 4
			ce-M-1		39.128 12	14.4 9
			ce-MNO-1		39.128 12	4.8 4
			x ray L		2.7	4.0 13
			x ray Kα₂		20.07370 20	2.19 16
			x ray Kα₁		20.21610 20	4.2 3
			x ray Kβ		22.7	1.30 10
			1 weak γ			0.07
¹⁰³Pd (EC decay)	[14967-68-1]	16.96 d 2	Auger L		2.39	167 7
			ce-K-1		16.535 12	9.5 3
			Auger K		17	18 3
			ce-L-1		36.343 12	71.2 23
			ce-M-1		39.128 12	14.4 5
			x ray L		2.7	9 3
			x ray Kα₂		20.07370 20	22.1 9
			x ray Kα₁		20.21610 20	41.9 17
			x ray Kβ		22.7	13.1 6
			9 weak γs			0.10
¹⁰⁴Ag (β⁺ decay)	[15116-79-7]	69.2 min 10	Auger L		2.5	77 6
			Auger K		17.7	13.6 23
			β⁺ 1	960 30	428 14	5.0 6
			β⁺ 2	1050 30	465 14	2.9 4
			β⁺ 3	1150 30	509 14	4.0 4
			total β⁺		458 14	12.5 9
			9 weak βs			0.65
			x ray L		2.84	4.5 16
			x ray Kα₂		21.02010 20	17.5 10
			x ray Kα₁		21.17710 20	33.3 18
			x ray Kβ		23.8	10.6 6
			γ 4		263.20 20	1.0 5
			γ 5		289.70 20	1.21 19
			γ 6		362.30 20	1.3 3
			γ 7		444.50 20	1.7 3
			γ 8		479.20 20	1.02 19
			γ 10		555.80 20	92.8 21
			γ 12		623.20 20	2.5 5
			γ 14		740.50 20	7.2 10
			γ 15		758.70 20	6.4 9
			γ 16		767.60 20	65.9 23

Table 6 (*continued*)

Isotope	CAS Registry No.	Half-life	Type of radiation	Energy, keV Max	Energy, keV Avg	Intensity, %
			γ 17		785.70 20	9.6 14
			γ 19		839.70 20	1.4 3
			γ 20		857.90 20	10.4 14
			γ 21		863.0 3	6.9 10
			γ 25		908.0 3	4.5 6
			γ 26		923.3 5	7.0 10
			γ 27		925.9 5	12.5 15
			γ 28		941.6 3	25.1 24
			γ 31		1075.3 3	2.1 4
			γ 36		1265.2 3	4.3 7
			γ 40		1341.8 3	7.3 11
			γ 41		1451.2 4	1.11 19
			γ 44		1526.6 3	7.1 10
			γ 47		1600.2 4	1.02 19
			γ 48		1625.8 3	5.1 8
			γ 55		1781.8 4	3.2 6
			43 weak γs			16.38
			max γ±			25.10
^{105}Ru (β⁻ decay)	[14331-95-4]	4.44 h 2	Auger L		2.39	20.1 11
			Auger K		17	2.9 5
			ce-K-3		106.39 7	14.5 6
			ce-L-3		126.20 7	6.5 3
			ce-M-3		128.98 7	1.29 6
			β⁻ 1	541 4	170.1 14	1.61 9
			β⁻ 2	573 4	181.8 14	3.48 20
			β⁻ 3	948 4	329.4 16	3.90 20
			β⁻ 4	1112 4	397.6 16	18.6 6
			β⁻ 5	1132 4	406.0 16	16.6 5
			β⁻ 6	1193 4	432.2 16	47.3 6
			β⁻ 7	1448 4	542.9 17	3.3 7
			β⁻ 8	1788 4	694.4 17	2.5 9
			total β⁻		412.2 17	99.6 16
			16 weak βs			2.33
			x ray L		2.7	1.1 4
			x ray Kα₂		20.07370 20	3.43 20
			x ray Kα₁		20.21610 20	6.5 4
			x ray Kβ		22.7	2.03 12
			γ 3		129.61 7	5.60 16
			γ 5		149.10 7	1.74 15
			γ 11		262.83 10	6.49 16
			γ 14		316.44 15	11.0 4
			γ 15		326.14 10	1.05 12
			γ 20		350.18 10	1.00 12
			γ 22		393.36 10	3.73 7
			γ 24		413.53 10	2.22 19
			γ 25		469.37 10	17.3 6
			γ 29		499.3 3	2.03 24
			γ 44		656.21 10	2.03 24
			γ 45		676.36 8	15.5 5
			γ 48		724.21 8	46.7 5
			γ 56		875.85 15	2.47 10
			γ 60		969.44 10	2.08 8
			73 weak γs			8.38
^{105}Rh (β⁻ decay)	[14913-89-4]	36.36 h 6	β⁻ 1	248 3	69.9 10	19.7 5
			β⁻ 2	261 3	73.9 10	5.2 3
			β⁻ 3	567 3	179.4 11	75.1 6
			total β⁻		152.3 13	100.0 9
			2 weak βs			0.04
			γ 3		306.10 20	5.13 22
			γ 4		318.90 11	19.2 4
			3 weak γs			0.23
^{105}Ag (EC decay)	[14928-14-4]	41.29 d 7	Auger L		2.5	103 7
			Auger K		17.7	18 3
			ce-K-2		39.63 3	13.3 5
			ce-L-2		60.38 3	1.67 7
			x ray L		2.84	6.1 22
			x ray Kα₂		21.02010 20	23.5 10
			x ray Kα₁		21.17710 20	44.6 18
			x ray Kβ		23.8	14.2 6
			γ 2		63.98 3	11.15 24
			γ 16		280.44 8	31.0 4
			γ 22		319.16 6	4.41 6
			γ 25		331.51 7	4.10 5
			γ 26		344.52 6	41.6 4
			γ 32		392.64 6	1.99 3
			γ 40		443.37 7	10.77 14
			γ 52		617.85	1.190 17

Table 6 (*continued*)

Isotope	CAS Registry No.	Half-life	Type of radiation	Energy, keV Max	Energy, keV Avg	Intensity, %
			γ 55		644.55 6	10.07 13
			γ 58		650.72 5	2.50 4
			γ 66		807.46 6	1.14 3
			γ 73		1087.94 6	3.58 6
			62 weak γs			7.50
^{106}Ru (β^- decay)	[13967-48-1]	371.63 d 17	β^- 1	39.4 3	10.03 8	100
^{106}Rh (β^- decay)	[14234-34-5]	130 min 2	Auger L		2.5	1.29 10
			β^- 1	728 9	241 4	14.5 20
			β^- 2	924 9	319 4	85 3
			total β^-		308 4	100 4
			γ 2		221.80 10	6.5 8
			γ 3		228.6 3	2.1 3
			γ 5		328.3 4	1.22 23
			γ 7		390.8 4	3.6 5
			γ 8		406.00 10	11.8 14
			γ 10		429.40 10	13.5 16
			γ 12		450.80 10	25 3
			γ 14		511.70 10	87 10
			γ 16		601.2 3	3.0 4
			γ 17		616.10 10	20.5 24
			γ 18		645.80 20	2.8 4
			γ 19		680.6 3	1.91 24
			γ 20		703.10 20	4.5 7
			γ 21		717.20 10	29 4
			γ 22		748.50 10	19.7 23
			γ 23		793.80 20	5.7 7
			γ 24		804.60 20	13.2 16
			γ 25		808.40 20	7.6 9
			γ 26		825.00 10	13.8 17
			γ 27		848.00 20	1.65 19
			γ 28		848.00 20	2.00 23
			γ 30		1020.5 3	2.00 25
			γ 31		1046.70 10	31 4
			γ 33		1127.70 10	13.7 17
			γ 35		1200.50 10	11.6 15
			γ 36		1224.20 10	8.3 11
			γ 37		1395.50 10	2.9 5
			γ 38		1529.40 10	17.8 24
			γ 40		1573.90 20	6.8 9
			γ 41		1724.60 20	2.3 5
			γ 42		1840.60 20	1.9 5
			11 weak γs			6.49
^{106}Ag (EC decay)	[14333-39-2]	8.46 d 10	Auger L		2.5	90 6
			Auger K		17.7	16 3
			x ray L		2.84	5.3 19
			x ray Kα_2		21.02010 20	20.1 10
			x ray Kα_1		21.17710 20	38.1 19
			x ray Kβ		23.8	12.1 7
			γ 3		221.701 15	6.6 5
			γ 4		228.633 21	2.11 16
			γ 5		328.463 23	1.14 9
			γ 7		391.03 3	3.7 3
			γ 8		406.182 20	13.5 9
			γ 10		429.646 22	13.2 9
			γ 12		450.976 22	28.3 18
			γ 14		511.85 3	88 6
			γ 17		601.17 7	1.62 13
			γ 18		616.17 3	21.6 14
			γ 19		646.03 5	1.46 13
			γ 21		680.420 10	1.55 12
			γ 22		703.11 8	4.5 3
			γ 23		717.34 9	29.0 19
			γ 24		748.36 11	20.7 14
			γ 25		793.17 10	5.9 5
			γ 26		804.28 10	12.4 9
			γ 27		808.36 11	4.0 5
			γ 28		824.69 7	15.4 10
			γ 29		847.270 20	1.6 6
			γ 30		847.430 20	2.8 7
			γ 35		1019.72 15	1.05 17
			γ 36		1045.83 8	29.7 20
			γ 41		1128.02 7	11.8 9
			γ 45		1199.39 10	11.3 9
			γ 46		1222.88 12	7.0 6
			γ 48		1394.35 14	1.50 20
			γ 50		1527.65 19	16.4 17
			γ 52		1572.35 15	6.6 7
			γ 54		1722.76 18	1.41 20
			γ 57		1839.05 10	2.0 3

Table 6 (*continued*)

Isotope	CAS Registry No.	Half-life	Type of radiation	Energy, keV Max	Energy, keV Avg	Intensity, %
^{107}Pd (β^- decay)	[17637-99-9]	6.5 yr	28 weak γs			7.32
			β^- 1	33 3	9.3 10	100
^{107}Cd (EC decay)	[14709-52-5]	6.49 h	Auger L		2.6	165 9
			Auger K		18.5	22 4
			ce-K-2		67.59 5	43 4
			ce-L-2		89.29 5	41 4
			ce-M-2		92.38 5	8.3 7
			ce-NOP-2		93.00 5	1.49 13
			1 weak β			0.22
			x ray L		3	11 4
			x ray Kα_2		21.9903 3	30.6 14
			x ray Kα_1		22.16290 10	58.0 25
			x ray Kβ		24.9	18.7 9
			γ 2		93.10 5	4.6 4
			30 weak γs			0.38
^{108}Ag (EC decay)	[14391-65-2]	127 yr 21	Auger L		2.5	82 5
			Auger K		17.7	14.2 24
			x ray L		2.84	4.8 17
			x ray Kα_2		21.02010 20	18.3 8
			x ray Kα_1		21.17710 20	34.7 14
			x ray Kβ		23.8	11.0 5
			γ 1		433.936 4	90.3 6
			γ 2		614.281 4	90.8 6
			γ 3		722.929 4	90.9 6
^{108}Ag (IT decay)		127 yr 21	Auger L		2.6	8.2 5
			ce-L-1		26.57 6	6.5 5
			ce-MNO-1		29.66 6	2.16 16
			ce-K-2		53.69 5	1.84 14
			γ 2		79.20 5	6.8 5
^{109}Pd (β^- decay)	[14981-64-7]	13.427 h 14	Auger L		2.6	79.7 7
			Auger K		18.5	7.1 13
			ce-K-2		62.53 5	42 5
			ce-L-2		84.23 5	44 5
			ce-M-2		87.32 5	8.9 9
			ce-MNO-2		87.32 5	1.59 15
			β^- 1	1027.9 20	361.9	99.900 10
			12 weak βs			0.11
			x ray L		3	5.3 18
			x ray Kα_2		21.9903 3	9.9 11
			x ray Kα_1		22.16290 ·10	18.7 20
			x ray Kβ		24.9	6.0 7
			γ 2		88.04 5	3.6 4
			35 weak γs			0.13
^{109}Cd (EC decay)	[14109-32-1]	464 d 1	Auger L		2.6	166 8
			Auger K		18.5	21 3
			ce-K-1		62.5180 21	41.7 17
			ce-L-1		84.2262 21	44.0 18
			ce-M-1		87.3145 21	9.0 4
			ce-NOP-1		87.9368 21	1.59 7
			x ray L		3	11 4
			x ray Kα_2		21.9403 3	29.1 11
			x ray Kα_1		22.16290 10	55.1 19
			x ray Kβ		24.9	17.7 7
			γ 1		88.0320 20	3.61 10
^{109}In (β^+ decay)	[14833-35-3]	4.2 h 1	Auger L		2.72	82 5
			Auger K		19.3	13.1 24
			β^+	791 8	417 4	7.88 24
			2 weak βs			0.02
			x ray L		3.13	5.8 20
			x ray Kα_2		22.98410 20	19.6 8
			x ray Kα_1		23.17360 20	36.9 14
			x ray Kβ		26	12.1 5
			γ 2		74.8	2.205 15
			γ 3		84	2.940 20
			γ 5		203.50 20	73.5 5
			γ 8		288.4 8	1.76 15
			γ 11		347.5 3	2.20 15
			γ 13		426.2 3	4.23 15
			γ 23		613.6 4	2.50 23
			γ 24		619.0 5	1.76 22
			γ 26		623.5 4	6.0 3
			γ 28		649.8 4	3.01 23
			γ 29		652.9 4	1.91 15
			γ 38		822.5 5	1.40 8
			γ 46		949.1 5	1.54 22
			γ 50		1049.7 6	1.16 6
			γ 52		1149.1 6	4.3 4

Table 6 (continued)

Isotope	CAS Registry No.	Half-life	Type of radiation	Energy, keV Max	Energy, keV Avg	Intensity, %
			γ 54		1196.5 6	1.76 22
			γ 59		1419.2 7	1.32 15
			γ 64		1622.3 8	2.06 22
			49 weak γs			14.91
			max γ±			15.81
^{110}Ag (IT decay)	[14391-76-5]	249.9 d 1	Auger L		2.6	1.11 8
^{110}Ag (β^- decay)		249.9 d 1	β^- 1	83.9 19	21.5 6	67.5 9
			β^- 2	530.7 19	165.1 7	30.6 4
			total β^-		66.2 14	98.7 10
			5 weak βs			0.59
			γ 17		446.811 3	3.66 4
			γ 23		620.360 3	2.78 3
			γ 26		657.7620 20	94.7 10
			γ 28		677.6230 20	10.72 11
			γ 29		687.015 3	6.49 7
			γ 30		706.682 3	16.74 12
			γ 32		744.277 3	4.66 5
			γ 33		763.944 3	22.36 23
			γ 35		818.031 4	7.32 8
			γ 36		884.685 3	72.9 8
			γ 37		937.493 4	34.3 4
			γ 50		1384.300 4	24.35 25
			γ 52		1475.788 6	3.99 4
			γ 53		1505.040 5	13.11 14
			γ 54		1562.302 5	1.184 13
			46 weak γs			1.94
^{110}In (β^+ decay)	[14133-75-6]	4.9 h 1	Auger L		2.72	82 7
			Auger K		19.3	13 3
			2 weak βs			0.03
			x ray L		3.13	5.8 20
			x ray Kα_2		22.98410 20	19.6 15
			x ray Kα_1		23.1736 20	37 3
			x ray Kβ		26	12.1 10
			γ 3		461.10 10	2.23 21
			γ 4		461.80 13	4.7 5
			γ 6		560.32 11	1.84 12
			γ 7		581.93 9	8.4 4
			γ 8		584.21 9	6.4 3
			γ 9		626.24 7	1.45 11
			γ 10		641.68 5	25.6 10
			γ 12		657.750 10	97 4
			γ 13		677.6 4	4.5 5
			γ 15		707.400 20	29.1 14
			γ 16		708.12	1.65 20
			γ 17		744.26 3	1.94 12
			γ 18		759.87 6	3.10 14
			γ 22		818.016 12	2.23 12
			γ 23		844.667 13	3.20 14
			γ 25		884.667 13	92 4
			γ 26		901.53 5	1.94 12
			γ 27		937.478 13	67.5 25
			γ 28		997.16 4	10.4 4
			γ 32		1085.52 4	1.35 7
			γ 33		1117.360 20	4.17 17
			γ 40		1475.76 3	1.23 8
			29 weak γs			9.77
^{110}In (β^+ decay)		69.1 min 5	Auger L		2.72	33.0 24
			Auger K		19.3	5.2 10
			β^+ 1	2260 30	1015 14	61 4
			8 weak βs			0.69
			x ray L		3.13	2.3 8
			x ray Kα_2		22.98410 20	7.9 5
			x ray Kα_1		23.17360 20	14.8 9
			x ray Kβ		26	4.9 3
			γ 2		657.750 10	98 5
			γ 11		1125.75 5	1.02 5
			γ 29		2129.48 10	2.13 9
			γ 30		2211.49 10	1.76 7
			γ 31		2317.54 10	1.31 5
			45 weak γs			7.46
			max γ±			123.38
^{111}Pd (IT decay)	[14928-31-5]	5.5 h 1	Auger L		2.5	34 4
			Auger K		17.7	5.1 11
			ce-K-1		147.83 8	28 4
			ce-L-1		168.58 8	9.5 12
			ce-M-1		171.51 8	1.85 23
			x ray L		2.84	2.0 8

Table 6 (*continued*)

Isotope	CAS Registry No.	Half-life	Type of radiation	Energy, keV Max	Energy, keV Avg	Intensity, %
			x ray Kα_2		21.02010 20	6.6 9
			x ray Kα_1		21.17710 20	12.4 16
			x ray Kβ		23.8	4.0 5
			γ 1		172.18 8	34 4
^{111}Pd (β^- decay)		5.5 h 1	Auger L		2.6	24 3
			Auger K		18.5	2.3 5
			ce-K-1		34.26 4	4.5 7
			ce-K-2		44.93 7	8.5 18
			ce-L-1		55.96 4	12.0 18
			ce-M-1		59.05 4	2.5 4
			ce-L-2		66.63 7	1.17 25
			β^- 1	380 50	114 18	1.5 3
			β^- 2	470 50	143 18	1.20 20
			β^- 3	550 50	173 19	5.4 9
			β^- 4	590 50	188 19	1.20 20
			β^- 5	820 50	277 20	2.1 4
			β^- 6	1410 50	525 22	1.6 5
			β^- 7	1550 50	585 23	1.2 3
			β^- 8	2240 50	899 23	7 3
			total β^-		460 30	26 4
			9 weak βs			4.41
			x ray L		3	1.6 6
			x ray Kα_2		21.9903 3	3.2 5
			x ray Kα_1		22.16290 10	6.0 9
			x ray Kβ		24.9	1.9 3
			γ 2		70.44 7	8.3 17
			γ 9		289.80 10	1.04 19
			γ 14		391.30 10	5.4 9
			γ 15		413.5 3	1.8 5
			γ 16		415.5 3	1.6 5
			γ 21		454.50 20	1.12 25
			γ 25		525.60 10	1.30 24
			γ 29		575.00 10	3.2 6
			γ 34		632.80 20	3.6 6
			γ 38		694.20 10	2.0 3
			γ 46		762.20 10	1.26 21
			γ 47		797.80 10	1.02 16
			γ 69		1115.90 20	1.10 19
			γ 76		1282.50 20	1.04 19
			γ 81		1691.10 20	1.28 22
			73 weak γs			14.09
^{111}Ag (β^- decay)	[15760-04-0]	7.45 d 1	β^- 1	686 3	223.5 12	7.1 5
			β^- 2	783 3	278.9 12	1.00 20
			β^- 3	1028 3	360.4 13	92 5
			total β^-		349.8 13	100 5
			4 weak βs			0.05
			γ 2		245.422 6	1.24 9
			γ 4		342.118 7	6.7 4
			12 weak γs			0.25
^{111}In (EC decay)	[15750-15-9]	2.83 d 1	Auger L		2.72	100 8
			Auger K		19.3	16 3
			ce-K-2		144.57 3	8.4 5
			ce-L-2		167.26 3	1.06 7
			ce-K-3		218.64 4	5.0 3
			x ray L		3.13	7.1 24
			x ray Kα_2		22.98410 20	23.6 14
			x ray Kα_1		23.17360 20	45 3
			x ray Kβ		26	14.6 9
			γ 2		171.28 3	90 5
			γ 3		245.35 4	94 5
^{113}Cd (β^- decay)	[14336-66-4]	9.3 × 10^{15} yr 19	β^- 1	322 6	93.3 20	100
^{113}Cd (β^- decay)		14.1 yr 5	β^- 1	586 5	185 3	100
^{113}In (IT decay)	[14885-78-0]	1.658 h 1	Auger L		2.84	29.7 18
			Auger K		20	4.2 9
			ce-K-1		363.751 8	28.2 3
			ce-L-1		387.453 8	5.48 7
			ce-M-1		390.865 8	1.100 10
			x ray L		3.29	2.3 8
			x ray Kα_2		24.00200 20	6.8 3
			x ray Kα_1		24.20970 20	12.9 5
			x ray Kβ		27.3	4.27 18
			γ 1		391.691 8	64.90 20
^{113}Sn (EC decay)	[13966-06-8]	115.1 d 2	Auger L		2.84	85 6
			Auger K		20	12.8 25
			x ray L		3.29	6.7 23
			x ray Kα_2		24.00200 20	20.7 8
			x ray Kα_1		24.20970 20	39.0 14

Table 6 (*continued*)

Isotope	CAS Registry No.	Half-life	Type of radiation	Energy, keV Max	Energy, keV Avg	Intensity, %
			x ray Kβ		27.3	13.0 5
			γ 1		255.115 15	1.85 6
^{114}In (EC decay)	[13981-55-0]	49.51 d 1	Auger L		2.72	3.5 3
			x ray Kα$_1$		23.17360 20	1.56 13
			γ 1		558.43 3	4.4 5
			γ 2		725.24 3	4.3 5
^{114}In (IT decay)		49.51 d 1	Auger L		2.84	64 3
			Auger K		20	6.0 12
			ce-K-1		162.33 3	39.8 13
			ce-L-1		186.03 3	31.6 10
			ce-MNO-1		189.44 3	8.83 5
			x ray L		3.29	5.0 17
			x ray Kα$_2$		24.00200 20	9.6 5
			x ray Kα$_1$		24.20970 20	18.2 9
			x ray Kβ		27.3	6.0 3
			γ 1		190.27 3	15.41 8
^{115}Cd (β$^-$ decay)	[14336-68-6]	53.46 h 10	Auger L		2.84	4.4 3
			ce-K-1		7.63 6	3.88 16
			β$^-$ 1	583.5 20	184.6 8	33.1 7
			β$^-$ 2	619.0 20	197.8 8	3.3 3
			β$^-$ 3	850.5 20	287.5 8	1.16 3
			β$^-$ 4	1111.4 20	394.4 9	62.6 18
			total β$^-$		317.3 10	100.2 20
			3 weak βs			0.01
			x ray Kα$_1$		24.20970 20	1.83 10
			γ 4		260.896 3	1.94 8
			γ 10		492.351 4	8.0 4
			γ 11		527.901 7	27.4 12
			14 weak γs			1.27
^{115}In (β$^-$ decay)	[14191-71-0]	5.1 × 10^{14} yr 4	β$^-$ 1	495 8	152 3	100
^{115}In (IT decay)		4.486 h 4	Auger L		2.84	42 3
			Auger K		20	6.0 12
			ce-K-1		308.30 3	39.7 8
			ce-L-1		332.00 3	8.36 20
			ce-M-1		335.42 3	1.70 5
			x ray L		3.29	3.3 12
			x ray Kα$_2$		24.00200 20	9.6 4
			x ray Kα$_1$		24.20970 20	18.1 8
			x ray Kβ		27.3	6.0 3
			γ 1		336.241 25	45.8 5
^{115}In (β$^-$ decay)		4.486 h 4	β$^-$	831 8	279 4	5.0 7
			1 weak β			0.05
			1 weak γ			0.05
^{115}Sb (EC decay)	[17620-10-9]	32.1 mo	Auger L		3	57 4
			Auger K		21	8.2 17
			β$^+$ 1	1511 20	675 9	33.0 20
			2 weak βs			0.23
			x ray L		3.44	4.9 17
			x ray Kα$_2$		25.04400 20	14.2 8
			x ray Kα$_1$		25.27130 20	26.6 14
			x ray Kβ		28.5	9.0 5
			γ 6		489.3 7	1.3 3
			γ 7		497.31 8	98 5
			47 weak γs			3.27
			max γ±			66.46
^{117}Cd (β$^-$ decay)	[15139-70-5]	2.49 h 4	Auger L		2.84	7.3 6
			ce-K-2		61.790 10	5.4 4
			ce-L-2		85.492 10	1.72 13
			β$^-$ 1	216 14	60 5	6.2 7
			β$^-$ 2	356 14	105 5	1.9 3
			β$^-$ 3	418 14	126 5	1.6 3
			β$^-$ 4	464 14	141 5	1.50 20
			β$^-$ 5	506 14	156 5	2.2 3
			β$^-$ 6	531 14	165 5	8.2 9
			β$^-$ 7	636 14	204 6	32 4
			β$^-$ 8	815 14	274 6	3.4 5
			β$^-$ 9	1779 14	685 7	13.2 17
			β$^-$ 10	1868 14	726 7	2.6
			β$^-$ 11	1939 14	758 7	3.7 6
			β$^-$ 12	2213 14	882 7	21.0 20
			total β$^-$		429 11	101 5
			12 weak βs			3.85
			x ray Kα$_2$		24.00200 20	1.56 11
			x ray Kα$_1$		24.20970 20	2.94 21
			γ 2		89.730 10	3.26 22
			γ 10		220.92 3	1.17 9
			γ 12		273.349 18	27.9 7

Table 6 (*continued*)

Isotope	CAS Registry No.	Half-life	Type of radiation	Energy, keV Max	Energy, keV Avg	Intensity, %
			γ 20		344.459 10	17.9 6
			γ 26		434.190 17	9.8 5
			γ 47		831.80 3	2.26 11
			γ 52		880.710 17	3.96 22
			γ 53		945.67 3	1.53 10
			γ 67		1051.70 10	3.79 22
			γ 70		1116.60 5	1.03 7
			γ 73		1142.43 3	1.67 12
			γ 78		1247.89 4	1.20 7
			γ 80		1260.00 3	1.14 7
			γ 85		1303.27 3	18.4 6
			γ 89		1337.57 7	1.62 12
			γ 92		1408.72 3	1.28 7
			γ 100		1562.24 4	1.42 7
			γ 103		1576.62 3	11.2 4
			γ 111		1706.93 4	1.00 7
			γ 112		1723.06 3	2.01 10
^{117}In (IT decay)	[14914-66-0]	116.5 min 7	Auger L		2.84	23.6 17
			Auger K		20	3.3 17
			ce-K-1		287.362 13	22.0 11
			ce-L-1		311.064 13	4.82 23
			x ray L		3.29	1.9 7
			x ray Kα_2		24.00200 20	5.3 4
			x ray Kα_1		24.20970 20	10.0 6
			x ray Kβ		27.3	3.33 21
			γ 1		315.302 13	19.1 7
^{117}In (β^- decay)		116.5 min 7	Auger L		3	2.09
			ce-K-1		129.40 20	2.15 24
			β^- 1	1612 8	610 4	18.3 18
			β^- 2	1770 8	680 4	34.6 24
			total β^-		656 4	53 3
			2 weak βs			0.03
			γ 1		158.60 20	15.9 17
			4 weak γs			0.03
^{117}Sn (IT decay)	[13981-59-4]	13.61 d 4	Auger L		3	91 5
			Auger K		21	10.8 22
			ce-K-1		126.82 3	64.9 20
			ce-K-2		129.360 20	11.7 4
			ce-L-1		151.56 3	26.2 8
			ce-L-2		154.095 20	1.48 5
			ce-M-1		155.14 3	5.64 18
			ce-NOP-1		155.88 3	1.35 5
			x ray L		3.44	8 3
			x ray Kα_2		25.04400 20	18.8 9
			x ray Kα_1		25.27130 20	35.2 15
			x ray Kβ		28.5	11.9 6
			γ 1		156.02 3	2.113 12
			γ 2		158.560 20	86.4 4
^{117}Sb (EC decay)	[15755-18-7]	2.80 h 1	Auger L		3	94 6
			Auger K		21	14 3
			ce-K-1		129.362 15	11.7 4
			ce-L-1		154.097 15	1.46 5
			β^+ 1	564 18	258 8	1.7 3
			x ray L		3.44	8 3
			x ray Kα_2		25.04400 20	23.5 9
			x ray Kα_1		25.27130 20	44.1 16
			x ray Kβ		28.5	14.9 6
			γ 1		158.562 15	85.9 4
			11 weak γs			0.99
			max γ±			3.40
^{117}Te (β^+ decay)	[15758-20-0]	62 min 2	Auger L		3	62 4
			Auger K		21.8	8.5 19
			β^+ 1	1540 30	690 14	1.47 25
			β^+ 2	1750 30	783 14	22.8 10
			total β^+		757 15	25.5 11
			4 weak βs			1.21
			x ray L		3.6	5.7 20
			x ray Kα_2		26.11080 20	15.8 8
			x ray Kα_1		26.35910 20	29.7 15
			x ray Kβ		29.7	10.1 6
			γ 3		719.7 7	64.7 14
			γ 5		886.7 7	1.49 20
			γ 6		923.9 7	6.2 7
			γ 8		996.7 7	3.9 4
			γ 9		1090.7 7	6.9 8
			γ 16		1716.4 7	15.9 17
			γ 19		2300.0 7	11.2 12

Table 6 (*continued*)

Isotope	CAS Registry No.	Half-life	Type of radiation	Energy, keV Max	Energy, keV Avg	Intensity, %
			14 weak γs			5.89
			max γ±			50.97
^{119}Sn (IT decay)	[14314-35-3]	293.0 d 13	Auger L		3	137 6
			ce-L-1		19.405 8	67 3
			Auger K		21	4.5 10
			ce-M-1		22.986 8	13.0 5
			ce-NOP-1		23.733 8	4.33 17
			ce-K-2		36.460 10	32.2 11
			ce-L-2		61.195 10	52.3 17
			ce-M-2		64.776 10	12.4 4
			ce-NOP-2		65.523 10	3.10 10
			x ray L		3.44	12 4
			γ 1		23.870 8	16.1 4
			x ray Kα_2		25.04400 20	7.9 4
			x ray Kα_1		25.27130 20	14.8 7
			x ray Kβ		28.5	4.99 25
^{119}Sb (EC decay)	[14914-68-2]	38.1 h 2	Auger L		3	145 7
			ce-L-1		19.405 8	67 3
			Auger K		21	12.0 24
			ce-M-1		22.986 8	13.0 6
			x ray L		3.44	13 5
			γ 1		23.870 8	16.1 5
			x ray Kα_2		25.04400 20	20.8 8
			x ray Kα_1		25.27130 20	39.1 14
			x ray Kβ		28.5	13.2 5
^{119}Te (β^+ decay)	[14914-67-1]	16.05 h 5	Auger L		3	81 5
			Auger K		21.8	11.2 24
			β^+	628.0 20	285.9 9	1.90 4
			1 weak β			0.16
			x ray L		3.6	8 3
			x ray Kα_2		26.11080 20	20.8 8
			x ray Kα_1		26.35910 20	39.0 14
			x ray Kβ		29.7	13.3 5
			γ 6		644.01 4	84.5 5
			γ 9		699.85 6	10.1 6
			γ 22		1413.19 8	1.10 9
			γ 26		1749.65 8	3.9 3
			24 weak γs			3.11
			max γ±			4.12
^{119}Te (β^+ decay)		4.69 d 4	Auger L		3	87 6
			Auger K		21.8	12 3
			ce-K-2		123.10 3	3.26 17
			ce-K-8		240.04 4	1.011 17
			1 weak β			0.41
			x ray L		3.6	8 3
			x ray Kα_2		26.11080 20	22.1 11
			x ray Kα_1		26.35910 20	41.5 19
			x ray Kβ		29.7	14.2 7
			γ 2		153.59 3	67 3
			γ 3		164.34 5	1.31 6
			γ 8		270.53 4	28.2 5
			γ 17		912.60 5	6.30 10
			γ 19		942.21 6	5.13 8
			γ 23		976.37 7	2.735 21
			γ 24		979.29 7	3.03 8
			γ 25		1013.20 8	1.667 13
			γ 27		1048.44 6	3.21 6
			γ 29		1081.35 10	1.61 4
			γ 30		1095.75 10	2.25 4
			γ 32		1136.75 7	7.72 9
			γ 33		1212.73 7	66.7 5
			γ 37		1366.39 14	1.074 22
			γ 44		2089.57 12	4.72 7
			33 weak γs			4.67
^{120}Sb (EC decay)	[14391-68-5]	5.76 d 2	Auger L		3	83 5
			Auger K		21	11.6 24
			x ray L		3.44	7.2 25
			x ray Kα_2		25.04400 20	20.2 8
			x ray Kα_1		25.27130 20	37.9 13
			x ray Kβ		28.5	12.8 5
			γ 1		89.8 3	80 3
			γ 2		197.3 3	88 3
			γ 3		1023.3 4	99.0 20
			γ 4		1113.4 6	1.30 10
			γ 5		1171.7 3	100
^{121}Sn (β^- decay)	[14683-06-8]	27.06 h 4	β^-	386.6 25	114.5	100 9
^{121}Sn (IT decay)		55 yr 5	ce-L-1		1.83 8	48.9 13

Table 6 (*continued*)

Isotope	CAS Registry No.	Half-life	Type of radiation	Energy, keV Max	Energy, keV Avg	Intensity, %
^{121}Sn (β^- decay)		55 yr 5	Auger L		3	45.0 18
			ce-M-1		5.41 8	22.5 6
			x ray L		3.44	3.9
			Auger L		3	17.0 17
			ce-K-1		6.66 4	17.7 17
			Auger K		21.8	2.3 6
			ce-L-1		32.45 4	2.29 22
			β^-	356 3	120.7 9	22.4 20
			x ray L		3.6	1.6 6
			x ray Kα_2		26.11080 20	4.4 5
			x ray Kα_1		26.35910 20	8.2 9
			x ray Kβ		29.7	2.8 3
			γ 1		37.15 4	1.85 17
^{121}Te (EC decay)	[14304-79-1]	16.78 d 35	Auger L		3	84 5
			ce-K-1		6.647 10	1.12 6
			Auger K		21.8	11.6 25
			x ray L		3.6	8 3
			x ray Kα_2		26.11080 20	21.4 8
			x ray Kα_1		26.35910 20	40.2 14
			x ray Kβ		29.7	13.7 6
			γ 3		470.472 13	1.41 5
			γ 4		507.591 11	17.7 6
			γ 5		573.139 11	80.3 25
			2 weak γs			0.38
^{121}Te (EC decay)		154 d 7	Auger L		3	18.0 15
			ce-K-1		6.647 10	9.0 14
			Auger K		21.8	2.5 6
			ce-L-1		32.440 10	1.17 17
			x ray L		3.6	1.7 6
			x ray Kα_2		26.11080 20	4.6 4
			x ray Kα_1		26.35910 20	8.6 7
			x ray Kβ		29.7	2.94 24
			γ 8		1102.149 18	2.5 3
			9 weak γs			1.10
^{121}Te (IT decay)		154 d 7	Auger L		3.19	72 4
			Auger K		22.7	5.1 12
			ce-K-1		49.974 15	34.4 16
			ce-L-1		76.849 15	41.6 19
			ce-M-1		80.782 15	9.9 5
			ce-NOP-1		81.620 15	2.65 12
			ce-K-2		180.38 3	6.12 20
			x ray L		3.77	7.1 24
			x ray Kα_2		27.20170 20	10.1 6
			x ray Kα_1		27.47230 20	18.8 10
			x ray Kβ		31	6.5 4
			γ 2		212.19 3	81.4 11
			1 weak γ			0.05
^{121}I (EC decay)	[15755-17-6]	2.12 h 1	Auger L		3.19	77 11
			Auger K		22.7	10 3
			ce-K-3		180.38 3	6.3 7
			β^+	1136 4	509.2 18	12.7 23
			12 weak βs			0.48
			x ray L		3.77	8 3
			x ray Kα_2		27.20170 20	20 3
			x ray Kα_1		27.47230 20	38 6
			x ray Kβ		31	13.1 19
			γ 3		212.19 3	84 9
			γ 9		319.7 5	1.04 11
			γ 13		475.0 5	1.04 11
			γ 14		531.9 3	6.1 7
			γ 17		598.7 10	1.54 16
			62 weak γs			6.14
			max γ±			26.36
^{121}Sb (EC decay)	[14265-72-6]	2.70 d 1	Auger L		3	2.00 15
			1 weak γ			0.74
^{122}Sb (β^- decay)		2.70 d 1	β^- 1	724 4	236.6 15	4.64 13
			β^- 2	1417 4	522.3 17	66.6 3
			β^- 3	1981 4	772.1 17	26.40 20
			total β^-		576.1 18	97.7 4
			3 weak βs			0.04
			γ 1		564.24 4	70.04 22
			γ 3		692.65 4	3.82 13
			5 weak γs			0.85
^{123}Sn (β^- decay)	[14683-07-9]	129.2 d 4	β^- 1	1397 4	523.1 17	99.37 11
			5 weak βs			0.63
			9 weak γs			0.64
^{123}Te (IT decay)	[14304-80-4]	119.7 d 1	Auger L		3.19	91 5

Table 6 (*continued*)

Isotope	CAS Registry No.	Half-life	Type of radiation	Energy, keV Max	Energy, keV Avg	Intensity, %
			Auger K		22.7	7.2 17
			ce-K-1		56.65 3	44.2 25
			ce-L-1		83.52 3	46 3
			ce-M-1		87.45 3	10.8 6
			ce-NOP-1		88.29 3	2.87 16
			ce-K-2		127.19 3	13.73 6
			ce-L-2		154.06 3	1.791 9
			x ray L		3.77	9 3
			x ray Kα_2		27.20170 20	14.4 8
			x ray Kα_1		27.47230 20	26.9 15
			x ray Kβ		31	9.4 6
			γ 2		159.00 3	84.0 4
			2 weak γs			0.09
^{123}I (EC decay)	[15715-08-9]	13.2 h 1	Auger L		3.19	94 6
			Auger K		22.7	12 3
			ce-K-1		127.16 5	13.61 7
			ce-L-1		154.03 5	1.773 10
			x ray L		3.77	9 4
			x ray Kα_2		27.20170 20	24.7 9
			x ray Kα_1		27.47230 20	46.0 16
			x ray Kβ		31	16.0 6
			γ 1		158.97 5	83.3 4
			γ 26		528.96 5	1.39 4
			43 weak γs			1.73
^{123}Xe (β^+ decay)	[15700-10-4]	2.08 h 2	Auger L		3.3	82 7
			Auger K		23.6	10.0 25
			ce-K-3		115.73 20	16.2 18
			ce-L-3		143.71 20	4.0 5
			ce-K-4		144.93 20	1.94 23
			β^+ 1	1324 15	593 7	1.06 17
			β^+ 2	1476 15	661 7	3.8 5
			β^+ 3	1505 15	674 7	17.6 21
			total β^+		666 7	22.6 22
			7 weak βs			0.10
			x ray L		4	9 3
			x ray Kα_2		28.3172 4	21.3 15
			x ray Kα_1		28.6120 3	40 3
			x ray Kβ		32.3	13.9 10
			γ 3		148.90 20	49 5
			γ 4		178.10 20	14.9 17
			γ 6		330.20 20	8.6 10
			γ 25		899.6 4	2.4 4
			γ 40		1093.4 3	2.8 4
			γ 41		1113.1 3	1.57 22
			γ 68		1807.3 3	1.25 18
			103 weak γs			12.65
			max γ^+			45.13
^{124}Sb (β^- decay)	[14683-10-4]	60.20 d 3	β^- 1	212.2 19	56.7 6	8.80 20
			β^- 2	612.2 19	192.1 7	52.0 10
			β^- 3	866.6 19	290.0 8	3.63 7
			β^- 4	948.1 19	322.6 8	2.0 4
			β^- 5	1580.4 19	591.1 8	5.39 11
			β^- 6	1657.4 19	625.0 8	2.45 5
			β^- 7	2303.2 19	916.3 9	22.6 5
			total β^-		380.5 13	100.0 13
			14 weak βs			3.15
			γ 12		602.730 3	97.92 5
			γ 14		645.8550 20	7.21 22
			γ 16		709.31 5	1.42 5
			γ 17		713.781 5	2.39 12
			γ 18		722.786 4	11.26 16
			γ 26		968.201 4	1.83 5
			γ 29		1045.131 4	1.84 4
			γ 35		1325.512 6	1.41 3
			γ 37		1368.164 7	2.35 5
			γ 40		1436.563 7	1.02 3
			γ 47		1690.980 6	48.8 5
			γ 53		2090.942 8	5.58 10
			52 weak γs			6.23
^{124}I (β^+ decay)	[14158-30-6]	4.18 d 2	Auger L		3.19	64 4
			Auger K		22.7	8.4 19
			β^+ 1	1532 4	685.9 18	11.3 6
			β^+ 2	2135 4	973.6 18	11.3 6
			total β^+		823.7 19	22.9 9
			4 weak βs			0.30
			x ray L		3.77	6.322
			x ray Kα_2		27.20170 20	16.7 7

Table 6 (*continued*)

Isotope	CAS Registry No.	Half-life	Type of radiation	Energy, keV Max	Energy, keV Avg	Intensity, %
			x ray Kα_1		27.47230 20	31.1 13
			x ray Kβ		31	10.8 5
			γ 9		602.730 3	61 5
			γ 17		722.786 4	10.1 9
			γ 40		1325.512 6	1.45 12
			γ 43		1376.00 10	1.68 14
			γ 49		1509.49 4	3.01 25
			γ 55		1690.980 6	10.5 9
			72 weak γs			8.37
^{125}Te (IT decay)	[14390-73-9]	58 d 1	Auger L		3.19	153 8
			ce-K-1		3.65 3	80.10 20
			Auger K		22.7	16 4
			ce-L-1		30.52 3	10.50 10
			ce-M-1		34.45 3	2.090 10
			ce-K-2		77.456 20	51.70 10
			ce-L-2		104.331 20	37.20 10
			ce-M-2		108.264 20	8.560 10
			ce-NOP-2		109.102 20	2.240 10
			x ray L		3.77	15 5
			x ray Kα_2		27.20170 20	32.8 12
			x ray Kα_1		27.47230 20	61.2 21
			x ray Kβ		31	21.3 8
			γ 1		35.46 3	6.67 18
			1 weak γ			0.28
^{125}I (EC decay)	[14158-31-7]	60.14 d 11	Auger L		3.19	156 9
			ce-K-1		3.6781 6	80.0 6
			Auger K		22.7	20 5
			ce-L-1		30.5527 6	10.5 3
			ce-M-1		34.4859 6	2.20 20
			x ray L		3.77	15 6
			x ray Kα_2		27.20170 20	39.8 14
			x ray Kα_1		27.47230 20	74.2 25
			x ray Kβ		31	25.8 10
			γ 1		35.4919 5	6.67 22
^{126}I (EC decay)	[14158-32-8]	13.02 d 7	Auger L		3.19	43 3
			Auger K		22.7	5.7 13
			β^+ 1	1134 5	508.4 23	3.34 22
			x ray L		3.77	4.3 15
			x ray Kα_2		27.20170 20	11.3 7
			x ray Kα_1		27.47230 20	21.1 13
			x ray Kβ		31	7.3 5
			γ 1		666.331 12	33.1 25
			γ 3		753.819 13	4.24
			5 weak γs			0.304
^{126}I (β^- decay)		13.02 d 7	β^- 1	371 5	108.9 17	3.6 3
			β^- 2	862 5	289.7 20	32 3
			β^- 3	1251 5	449.5 22	8 3
			total β^-		304.1 22	44 5
			γ 1		388.633 11	34 3
			γ 2		491.243 11	
			1 weak γ			2.85 22
^{127}Sn (β^- decay)	[15690-89-8]	2.10 h 4	β^- 1	332.8	97	1.15 10
			β^- 2	353.8	104	1.25 15
			β^- 3	394.8	117	1.25 8
			β^- 4	437.8	132	2.69 18
			β^- 5	504.2	155	1.86 16
			β^- 6	561.5	176	4.5 3
			β^- 7	613.2	195	11.2 6
			β^- 8	615.1	196	1.55 16
			β^- 9	670.3	216	4.09 25
			β^- 10	699.3	227	6.2 4
			β^- 11	744.1	245	3.75 19
			β^- 12	752.7	248	1.10 16
			β^- 13	793.7	264	1.6 3
			β^- 14	841.6	283	5.5 8
			β^- 15	925.2	316	3.0 5
			β^- 16	1049.6	367	1.67 23
			β^- 17	1106.6	391	4.7 17
			β^- 18	1196.5	429	2.0 7
			β^- 19	1262.6	457	1.7 23
			β^- 20	1279.8	464	4.8 16
			β^- 21	1615.7	610	1.0 9
			β^- 22	2085.7	821	3 5
			β^- 23	2104.4	830	1 4
			β^- 24	3200	1337	22 8
			total β^-		521	103 11
			23 weak βs			10.2

Table 6 (*continued*)

Isotope	CAS Registry No.	Half-life	Type of radiation	Energy, keV Max	Energy, keV Avg	Intensity, %
			γ 12		119.7 4	2.16 23
			γ 16		152.5 4	1.33 16
			γ 19		169.2 4	2.01 19
			γ 24		184.7 4	1.10 23
			γ 41		262.5 4	2.31 23
			γ 42		266.2 4	2.12 23
			γ 46		284.3 4	2.7 3
			γ 47		292.9 4	1.25 12
			γ 58		390.5 4	1.25 12
			γ 61		407.1 4	1.51 16
			γ 64		438.2 4	6.1 8
			γ 70		490.9 4	5.3 4
			γ 71		493.2 4	3.1 4
			γ 72		500.7 4	1.51 16
			γ 73		509.0 4	1.4 3
			γ 80		545.4 4	2.27 23
			γ 84		583.3 4	3.2 3
			γ 85		592.3 4	2.01 19
			γ 97		805.9 4	8.2 9
			γ 98		823.1 4	10.6 23
			γ 99		824.7 4	6.1 12
			γ 101		859.5 4	8.0 8
			γ 106		916.5 4	1.17 12
			γ 109		979.2 4	7.2 16
			γ 112		997.9 4	1.93 19
			γ 113		1002.6 4	1.74 19
			γ 114		1036.1 4	1.97 19
			γ 117		1093.3 4	3.8 8
			γ 118		1095.6 4	19 4
			γ 119		1114.3 4	38 4
			γ 123		1160.4 4	2.4 5
			γ 136		1472.5 4	1.2 3
			γ 140		1584.3 4	1.78 19
			γ 143		1647.8 4	1.02 12
			γ 157		2003.4 5	5.3 6
			γ 166		2317.4 5	1.10 12
			γ 173		2584.9 5	1.55 16
			γ 174		2695.9 5	1.63 16
			139 weak γs			38.7
^{127}Sb (β^- decay)	[13968-50-8]	3.85 d 5	Auger L		3.19	3.8 4
			ce-K-1		29.29 10	3.5 3
			β^- 1	439 5	132.3 18	1.54 18
			β^- 2	504 5	155.0 18	5.40 16
			β^- 3	657 5	210.9 20	1.3 3
			β^- 4	795 5	264.1 20	4.56 16
			β^- 5	797 5	265.0 20	17.8 5
			β^- 6	895 5	303.9 21	35.6 5
			β^- 7	950 5	325.7 21	4.5 5
			β^- 8	1108 5	390.9 21	23.2 10
			β^- 9	1240 5	446.6 22	2.2 5
			β^- 10	1493 5	561.9 21	2.0 5
			total β^-		308.8 21	100.1 16
			5 weak βs			2.01
			x ray Kα_2		27.20170 20	1.03 9
			x ray Kα_1		27.47230 20	1.92 16
			γ 1		61.10 10	1.43 12
			γ 5		252.4 3	8.5 4
			γ 7		290.8 5	2.01 12
			γ 11		412.1 5	3.8 4
			γ 13		445.1 5	4.32 13
			γ 16		473.0 4	25.7 8
			γ 18		543.3 5	2.9 5
			γ 20		603.5 5	4.43 13
			γ 26		685.7 5	36.6 5
			γ 27		698.5 5	3.62 9
			γ 28		722.2 5	1.87 12
			γ 30		783.7 5	15.0 4
			25 weak γs			8.60
^{127}Te (β^- decay)	[13981-49-2]	9.35 h 7	β^- 1	276 5	78.2 16	1.190 20
			β^- 2	694 5	224.7 19	98.810 20
			total β^-		222.9 20	100.0 33
			3 weak βs			0.03
			9 weak γs			1.26
^{126}Te (IT decay)		109 d 2	Auger L		3.19	74 3
			Auger K		22.7	5.2 12
			ce-K-1		56.45 8	41.58 13
			ce-L-1		83.32 8	43.04 14

Table 6 (*continued*)

Isotope	CAS Registry No.	Half-life	Type of radiation	Energy, keV Max	Energy, keV Avg	Intensity, %
			ce-M-1		87.25 8	10.15 10
			ce-NOP-1		88.09 8	2.704 12
			x ray L		3.77	7.3 25
			x ray Kα_2		27.20170 20	10.4 4
			x ray Kα_1		27.47230 20	19.3 7
			x ray Kβ		31	6.71 25
			1 weak γ			0.09
^{127}Te (β^- decay)		109 d 2	β^- 1	725 5	253.0 19	2.40 20
			3 weak βs			0.02
			5 weak γs			0.01
^{127}Xe (EC decay)	[13994-19-9]	36.41 d 2	Auger L		3.3	96 6
			Auger K		23.6	12 3
			ce-K-1		24.431 20	4.28 18
			ce-K-2		112.05 3	1.54 7
			ce-K-3		138.93 3	3.65 16
			ce-K-4		169.67 3	6.63 7
			x ray L		4	10 4
			x ray Kα_2		28.3172 4	25.1 10
			x ray Kα_1		28.6120 3	46.7 17
			x ray Kβ		32.3	16.4 7
			γ 1		57.600 20	1.33 6
			γ 2		145.22 3	4.29 14
			γ 3		172.10 3	25.5 8
			γ 4		202.84 3	68.3 4
			γ 5		374.96 5	17.2 6
			1 weak γ			0.01
^{129}Te (β^- decay)	[14269-71-7]	69.6 min 2	Auger L		3.3	59 7
			ce-L-1		22.62 4	65 7
			ce-M-1		26.74 4	13.1 14
			ce-NOP-1		27.62 4	4.4 5
			β^- 1	1011 4	350.0 17	8.9 4
			β^- 2	1470 4	544.5 18	89.3 9
			total β^-		521.5 18	100.1 10
			8 weak βs			1.88
			x ray L		4	6.2 22
			γ 1		27.81 4	15.6 17
			γ 14		459.60 5	7.4 4
			γ 16		487.39 6	1.35 6
			44 weak γs			2.47
^{129}Te (IT decay)		33.6 d 1	Auger L		3.19	50 4
			Auger K		22.7	4.1 10
			ce-K-1		73.69 5	33 3
			ce-L-1		100.56 5	24.9 24
			ce-MNO-1		104.49 5	7.3 7
			x ray L		3.77	4.9 17
			x ray Kα_2		27.20170 20	8.1 9
			x ray Kα_1		27.47230 20	15.2 15
			x ray Kβ		31	5.3 6
			1 weak γ			0.15
^{129}Te (β^- decay)		33.6 d 1	β^- 1	908 4	308.3 16	3.1 4
			β^- 2	1604 4	602.8 18	31 6
			total β^-		567.2 18	35 6
			7 weak βs			0.95
			γ 19		695.88 6	3.1 6
			38 weak γs			1.12
^{129}I (β^- decay)	[15046-84-1]	1.57 × 10^7 yr 4	Auger L		3.43	74 4
			ce-K-1		5.02 3	79.10 20
			Auger K		24.6	8.8 16
			ce-L-1		34.13 3	10.6 3
			ce-M-1		38.44 3	2.10 10
			β^- 1	150 5	40 5	100
			x ray L		4.1	8.2 25
			x ray Kα_2		29.4580 10	20.0 6
			x ray Kα_1		29.7790 10	37.1 9
			x ray Kβ		33.6	13.2 4
			γ 1		39.58 3	7.50 20
^{129}Xe (IT decay)	[13965-99-6]	8.0 d 2	Auger L		3.43	147 8
			ce-K-1		5.02 3	79.10 20
			Auger K		24.6	16 3
			ce-L-1		34.13 3	10.6 3
			ce-M-1		38.44 3	2.10 10
			ce-K-2		162.00 3	63.90 10
			ce-L-2		191.11 3	24.4 7
			ce-M-2		195.42 3	5.50 20
			ce-NOP-2		196.35 3	1.50 10
			x ray L		4.1	16 5
			x ray Kα_2		29.4580 10	36.2 10

Table 6 (*continued*)

Isotope	CAS Registry No.	Half-life	Type of radiation	Energy, keV Max	Energy, keV Avg	Intensity, %
			x ray $K\alpha_1$		29.7790 10	67.1 17
			$K\beta$		33.6	23.9 7
			γ 1		39.58 3	7.50 20
			γ 2		196.56 3	4.70 20
^{129}Cs (EC decay)	[15047-05-9]	32.06 h 6	Auger L		3.43	110 9
			ce-K-1		5.020 15	31.4 22
			Auger K		24.6	13 3
			ce-L-1		34.128 15	4.2 3
			ce-MNO-1		38.439 15	1.02 7
			x ray L		4.1	12 4
			x ray $K\alpha_2$		29.4580 10	29.7 21
			x ray $K\alpha_1$		29.7790 10	55 4
			x ray $K\beta$		33.6	19.6 14
			γ 1		39.581 15	2.99 18
			γ 7		278.614 4	1.33 8
			γ 10		318.1800 20	2.46 14
			γ 14		371.9180 20	30.8 16
			γ 16		411.4900 20	22.5 12
			γ 21		548.945 8	3.42 18
			27 weak γs			2.85
^{130}I (β^- decay)	[14914-02-4]	12.36 h 1	β^- 1	624 10	198 4	46.0 10
			β^- 2	814 10	271 4	2.14 5
			β^- 3	1042 10	362 4	47.7 12
			β^- 4	1178 10	419 4	1.45 4
			total β^-		280 5	99.1 16
			11 weak βs			1.78
			γ 9		418.010 20	34.2 7
			γ 13		536.090 20	99.0 20
			γ 14		539.10 3	1.40 3
			γ 16		586.050 20	1.69 4
			γ 19		668.540 10	96.1 21
			γ 20		685.990 10	1.07 3
			γ 22		739.480 20	82.3 18
			γ 40		1157.470 10	11.31 25
			44 weak γs			6.66
^{131}Te (IT decay)	[14683-12-6]	30 h 2	Auger L		3.19	16.9 13
			Auger K		22.7	1.8 5
			ce-K-1		150.436 20	14.4 11
			ce-L-1		177.311 20	5.4 4
			ce-M-1		181.244 20	1.19 9
			x ray L		3.77	1.7 6
			x ray $K\alpha_2$		27.20170 20	3.6 3
			x ray $K\alpha_1$		27.47230 20	6.7 6
			x ray $K\beta$		31	2.32 19
			1 weak γ			0.85 2
^{131}Te (β^- decay)		30 h 2	Auger L		3.3	10.7 7
			Auger K		23.6	1.3 4
			ce-K-14		47.971 20	4.79 22
			ce-K-21		68.891 10	4.87 23
			ce-K-36		116.541 10	1.06 15
			β^- 1	264 6	74.3 19	1.10 20
			β^- 2	318 6	91.6 20	2.0 4
			β^- 3	421 6	125.9 21	2.6 3
			β^- 4	431 6	129.4 21	5.4 6
			β^- 5	452 6	136.5 21	37.1 19
			β^- 6	508 6	156.1 22	2.1 5*
			β^- 7	533 6	165.2 22	16.1 12
			β^- 8	545 6	169.3 22	1.30 20
			β^- 9	786 6	260.2 24	2.6 9
			β^- 10	2432 6	970 3	3.8 4
			total β^-		191 3	78 3
			17 weak βs			4.15
			x ray L		4	1.1 4
			x ray $K\alpha_2$		28.3172 4	2.87 14
			x ray $K\alpha_1$		28.6120 3	5.34 24
			x ray $K\beta$		32.3	1.87 9
			γ 14		81.140 20	4.06 14
			γ 21		102.060 10	7.9 3
			γ 36		149.710 10	5.1 7
			γ 48		200.630 20	7.54 25
			γ 58		240.930 10	7.58 24
			γ 65		278.560 20	1.78 7
			γ 75		334.270 10	9.6 3
			γ 85		364.98 10	1.20 16
			γ 95		452.30 4	1.5 4
			γ 96		462.92 5	1.82 7
			γ 107		586.30 3	1.97 10

Table 6 (*continued*)

Isotope	CAS Registry No.	Half-life	Type of radiation	Energy, keV Max	Energy, keV Avg	Intensity, %
			γ 113		665.05 3	4.33 15
			γ 118		713.10 4	1.43 16
			γ 120		744.20 4	1.59 6
			γ 122		773.67 3	38.1 12
			γ 124		782.49 4	7.8 3
			γ 125		793.75 3	13.8 5
			γ 127		882.78 4	6.11 20
			γ 131		852.21 3	20.6 8
			γ 136		910.00 3	3.29 13
			γ 137		920.62 5	1.20 9
			γ 149		1059.69 4	1.55 6
			γ 153		1125.46 4	11.4 4
			γ 156		1148.89 7	1.5 4
			γ 162		1206.60 4	9.7 4
			γ 179		1646.01 5	1.24 6
			γ 183		1887.70 7	1.35 6
			γ 187		2000.94 6	2.01 7
			162 weak γs			21.03
^{131}I (β^- decay)	[10043-66-0]	8.04 d 1	Auger L		3.43	5.1 3
			ce-K-1		45.622 10	3.54 13
			ce-K-14		329.919 11	1.54 6
			β^- 1	247.9 6	69.40 20	2.13 3
			β^- 2	333.8 6	96.60 20	7.36 10
			β^- 3	606.3 6	191.6 3	89.4 10
			total β^-		181.7 3	100.0 10
			3 weak βs			1.10
			x ray Kα_2		29.4580 10	1.37 5
			x ray Kα_1		29.7790 10	2.54 9
			γ 1		80.183 10	2.62 5
			γ 7		284.298 11	6.06 9
			γ 14		364.480 11	81.2 12
			γ 17		636.973 10	7.27 11
			γ 19		722.893 10	1.80 3
			14 weak γs			1.33
^{131}Xe (IT decay)	[14683-11-5]	11.9 d 1	Auger L		3.43	75 4
			Auger K		24.6	6.8 13
			ce-K-1		129.369 8	61.2 7
			ce-L-1		158.477 8	28.6 6
			ce-M-1		162.788 8	6.50 20
			ce-NOP-1		163.722 8	1.78 6
			x ray L		4.1	8 3
			x ray Kα_2		29.4580 10	15.5 5
			x ray Kα_1		29.7790 10	28.7 8
			x ray Kβ		33.6	10.2 4
			γ 1		163.930 8	1.96 6
^{131}Cs (EC decay)	[14914-76-2]	9.69 d 1	Auger L		3.43	79 4
			Auger K		24.6	9.3 17
			x ray L		4.1	9 3
			x ray Kα_2		29.4580 10	21.1 6
			x ray Kα_1		29.7790 10	39.1 10
			x ray Kβ		33.6	13.9 4
^{131}Ba (EC decay)	[14914-75-1]	11.8 d 2	Auger L		3.55	104 7
			Auger K		25.5	11.4 15
			ce-K-2		42.770 13	1.23 9
			ce-K-5		87.818 12	18.1 8
			ce-L-5		118.089 12	5.99 25
			ce-MNO-5		122.586 12	1.66 5
			ce-K-10		180.11 3	1.79 6
			x ray L		4.29	13 4
			x ray Kα_2		30.6251 3	27.7 17
			x ray Kα_1		30.9728 3	51 3
			x ray Kβ		35	18.4 12
			γ 5		123.803 12	29.1 9
			γ 7		133.607 14	2.19 9
			γ 10		216.09 3	19.9 4
			γ 11		239.63 3	2.41 8
			γ 13		249.44 3	2.81 10
			γ 18		373.25 3	13.3 15
			γ 19		404.04 3	1.29 9
			γ 25		486.48 4	1.89 21
			γ 26		496.28 3	44 4
			γ 30		585.02 3	1.23 9
			γ 31		620.05 3	1.57 9
			γ 44		1047.56 4	1.194 24
			36 weak γs			5.03
^{132}Te (β^- decay)	[14234-28-7]	78.2 h 8	Auger L		3.3	75 7
			ce-K-1		16.551 10	70 6

Table 6 (continued)

Isotope	CAS Registry No.	Half-life	Type of radiation	Energy, keV Max	Energy, keV Avg	Intensity, %
			Auger K		23.6	9.4 23
			ce-L-1		44.532 10	9.3 7
			ce-MNO-1		48.648 10	2.44 17
			ce-K-4		194.99 6	7.14 22
			ce-L-4		222.97 6	1.34 4
			β^- 1	215 4	59.4 12	100
			x ray L		4	8 3
			x ray Kα_2		28.3172 4	19.9 15
			x ray Kα_1		28.6120 3	37 3
			x ray Kβ		32.3	13.0 10
			γ 1		49.720 10	14.4 10
			γ 2		111.76 8	1.85 18
			γ 3		116.30 8	1.94 18
			γ 4		228.16 6	88.2 4
^{132}I (β^- decay)	[14683-16-0]	2.30 h	β^- 1	741 20	242 8	12.8 8
			β^- 2	740 20	242 8	1.81 10
			β^- 3	910 20	309 8	3.60 20
			β^- 4	967 20	331 9	8.1 4
			β^- 5	996 20	344 9	3.79 16
			β^- 6	1155 20	409 9	2.11 7
			β^- 7	1185 20	422 9	19.0 7
			β^- 8	1413 20	519 9	1.7 6
			β^- 9	1470 20	543 9	10.2 10
			β^- 10	1468 20	543 9	1.9 8
			β^- 11	1617 20	608 9	12.7 7
			β^- 12	2140 20	841 9	17.6 22
			total β^-		490 10	98 3
			22 weak βs			2.29
			γ 5		262.70 10	1.44 9
			γ 23		505.90 15	5.03 20
			γ 24		522.65 9	16.1 6
			γ 26		547.10 20	1.25 9
			γ 31		621.2 10	1.579 4
			γ 32		630.22 9	13.7 6
			γ 33		650.60 20	2.66 20
			γ 34		667.69 8	98.70 20
			γ 35		669.8 3	4.9 8
			γ 36		671.6 3	5.2 4
			γ 37		727	2.2 6
			γ 38		727.2 10	3.2 6
			γ 39		729.5 4	1.1 3
			γ 41		772.60 8	76.2 18
			γ 42		780.2 3	1.23 6
			γ 45		809.80 20	2.9 3
			γ 46		812.20 20	5.6 5
			γ 48		876.80 20	1.08 5
			γ 53		954.55 9	18.1 6
			γ 64		1136.03 12	2.96 20
			γ 66		1143.40 20	1.38 10
			γ 68		1173.20 20	1.09 10
			γ 72		1290.7 3	1.14 6
			γ 73		1295.3 3	1.97 10
			γ 77		1372.07 13	2.47 10
			γ 78		1398.57 10	7.1 3
			γ 80		1442.56 10	1.42 6
			γ 101		1921.08 12	1.18 9
			γ 103		2002.30 12	1.09 10
			91 weak γs			13.41
^{132}Cs (β^+ decay)	[15758-03-9]	6.475 d 10	Auger L		3.43	78 4
			Auger K		24.6	9.2 17
			1 weak β			0.30
			x ray L		4.1	9 3
			x ray Kα_2		29.4580 10	21.0 6
			x ray Kα_1		29.7790 10	39.0 10
			x ray Kβ		33.6	13.9 4
			γ 3		630.22 9	1.01 8
			γ 4		667.67 6	97.47 11
			7 weak γs			2.16
^{132}Cs (β^- decay)		6.475 d 10	β^- 1	810 30	267 11	1.57 12
			2 weak βs			0.43
			γ 1		464.55 6	1.87 14
			3 weak γs			0.43
^{132}La (β^+ decay)	[15066-93-0]	4.8 h 2	Auger L		3.67	49 3
			Auger K		26.4	5.3 14
			β^+ 1	1090 50	492 22	1.57 25
			β^+ 2	1280 50	578 23	1.59 19
			β^+ 3	2630 50	1186 23	10.77 24

Table 6 (continued)

Isotope	CAS Registry No.	Half-life	Type of radiation	Energy, keV Max	Energy, keV Avg	Intensity, %
			β^+ 4	3190 50	1449 24	14.01 17
			β^+ 5	3660 50	1660 23	9.14 15
			total β^+		1292 25	41.5 5
			16 weak βs			4.45
			x ray L		4.47	6.6 17
			x ray Kα_2		31.8171 3	13.8 5
			x ray Kα_1		32.1936 3	25.4 8
			x ray Kβ		36.4	9.2 4
			γ 1		19 3	1.15 6
			γ 6		464.55 6	77 7
			γ 8		479.47 3	2.23 21
			γ 10		515.78 9	5.1 6
			γ 11		540.360 20	7.8 8
			γ 13		567.14 3	15.9 15
			γ 17		663.07 3	9.2 8
			γ 25		899.32 3	4.7 4
			γ 33		1031.70 3	7.9 7
			γ 35		1046.56 3	3.5 3
			γ 43		1221.23 3	2.97 24
			γ 55		1533.66 4	1.49 8
			γ 59		1604.03 3	3.7 3
			γ 66		1909.91 4	9.1 8
			γ 71		2102.84 5	5.9 5
			γ 83		2754.73 5	1.62 13
			78 weak γs			19.12
			max γ^1			83.07
^{133}I (β^- decay)	[14834-67-4]	20.8 h 1	β^- 1	370 30	110 10	1.24 4
			β^- 2	460 30	140 11	3.74 6
			β^- 3	520 30	162 11	3.12 6
			β^- 4	880 30	299 12	4.15 11
			β^- 5	1020 30	352 13	1.81 5
			β^- 6	1230 30	441 13	83.5 18
			β^- 7	1530 30	573 13	1.07 5
			total β^-		407 14	100.0 18
			3 weak βs			1.35
			γ 15		510.530 11	1.81 5
			γ 18		529.872 11	86.3 18
			γ 27		706.578 13	1.49 4
			γ 31		856.278 12	1.23 4
			γ 32		875.329 11	4.47 10
			γ 39		1236.411 12	1.49 4
			γ 40		1298.223 11	2.33 6
			37 weak γs			4.61
^{133}Xe (β^- decay)	[14932-42-4]	5.245 d 6	Auger L		3.55	49.0 20
			Auger K		25.5	5.5 7
			ce-K-2		45.0124 4	52.0 3
			ce-L-2		75.2827 4	8.49 20
			ce-MNO-2		79.7799 4	2.3 3
			β^- 1	346.3	100.5 10	99.34 10
			2 weak βs			0.67
			x ray L		4.29	6.1 17
			x ray Kα_2		30.6251 3	13.3 3
			x ray Kα_1		30.9728 3	24.6 5
			x ray Kβ		35	8.84 20
			γ 2		81	37.1 4
			5 weak γs			0.29
^{133}Xe (IT decay)		2.19 d 1	Auger L		3.43	70.4
			Auger K		24.6	7.1 13
			ce-K-1		198.62 4	63.7 8
			ce-L-1		227.73 4	20.8 6
			ce-M-1		232.04 4	4.16 13
			ce-NOP-1		232.97 4	1.04 17
			x ray L		4.1	7.8 24
			x ray Kα_2		29.4580 10	16.1 5
			x ray Kα_1		29.7790 10	29.9 8
			x ray Kβ		33.6	10.6 4
			γ 1		233.18 4	10.3 3
^{133}Ba (EC decay)	[13981-41-4]	10.74 yr 5	Auger L		3.55	135 6
			ce-K-1		17.170 16	10.5 4
			Auger K		25.5	14.0 16
			ce-K-2		43.636 11	3.72 15
			ce-K-3		45.012 5	46.9 10
			ce-L-1		47.441 16	1.43 20
			ce-L-3		75.283 5	7.64 24
			ce-M-3		79.780 5	1.78 14
			ce-K-8		320.020 17	1.31 5
			x ray L		4.29	17 5

Table 6 (*continued*)

Isotope	CAS Registry No.	Half-life	Type of radiation	Energy, keV Max	Energy, keV Avg	Intensity, %
			x ray $K\alpha_2$		30.6251 3	34.0 8
			x ray $K\alpha_1$		30.9728 3	62.9 12
			x ray $K\beta$		35	22.6 6
			γ 1		53.155 16	2.17 4
			γ 2		79.621 11	2.66 8
			γ 3		80.997 5	33.5 5
			γ 6		276.397 12	7.09 13
			γ 7		302.851 15	18.40 20
			γ 8		356.005 17	62.1 7
			γ 9		383.851 15	8.91 10
			2 weak γs			1.08
^{133}Ba (IT decay)		38.9 h 1	Auger L		3.67	130 6
			ce-L-1		6.30 4	77.6 2
			ce-M-1		11.00 4	15.9 5
			ce-NOP-1		12.04 4	5.20 20
			Auger K		26.4	5.9 16
			ce-K-2		238.65 15	59.2 3
			ce-L-2		270.10 15	18.0 6
			ce-M-2		274.80 15	4.00 10
			ce-NOP-2		275.84 15	1.15 3
			x ray L		4.47	18 5
			γ 1		12.29 4	1.39 4
			x ray $K\alpha_2$		31.8171 3	15.2 5
			x ray $K\alpha_1$		32.1936 3	28.0 9
			x ray $K\beta$		36.4	10.2 4
			γ 2		276.09 15	18.0 5
^{134}Cs (β^- decay)	[13967-70-9]	2.062 yr 5	β^- 1	88.6 4	23.09 11	27.40 20
			β^- 2	415.2 4	123.43 14	2.47 5
			β^- 3	658.0 4	210.15 15	70.1 5
			total β^-		156.8 3	100.0 6
			2 weak βs			0.05
			γ 3		475.35 5	1.46 4
			γ 4		563.227 15	8.38 5
			γ 5		569.315 15	15.43 11
			γ 6		604.699 15	97.6 3
			γ 7		795.845 22	85.4 4
			γ 8		801.932 22	8.73 4
			γ 10		1167.94 3	1.80 3
			γ 11		1365.15 3	3.04 4
			3 weak γs			1.04
^{134}Cs (IT decay)		2.90 h 1	Auger L		3.55	133 5
			ce-L-1		5.5457 4	77.1 4
			ce-M-1		10.0429 4	15.77 13
			ce-NOP-1		11.0292 4	5.23 3
			Auger K		25.5	3.7 5
			ce-K-2		91.4354 4	34.7 3
			ce-L-2		121.7057 4	40.4 3
			ce-M-2		126.2029 4	9.01 6
			ce-NOP-2		127.1892 4	2.272 16
			x ray L		4.29	16 5
			x ray $K\alpha_2$		30.6251 3	8.95 20
			x ray $K\alpha_1$		30.9728 3	16.6 4
			x ray $K\beta$		35	5.95 15
			γ 2		127.502 2	12.9 8
			2 weak γs			0.96
^{135}I (β^- decay)	[14834-68-5]	6.61 h 1	β^- 1	300 30	86 10	1.03 5
			β^- 2	350 30	103 10	1.39 5
			β^- 3	460 30	138 11	4.73 15
			β^- 4	480 30	145 11	7.33 20
			β^- 5	620 30	196 12	1.57 7
			β^- 6	670 30	213 12	1.10 5
			β^- 7	740 30	243 12	7.9 3
			β^- 8	920 30	313 12	8.7 3
			β^- 9	1030 30	359 13	21.5 5
			β^- 10	1150 30	405 13	8.0 3
			β^- 11	1250 30	451 13	7.6 4
			β^- 12	1450 30	535 13	23.8 12
			β^- 13	1580 30	591 14	1.1 9
			β^- 14	2180 30	858 14	1.2 6
			total β^-		366 15	99.1 19
			10 weak βs			2.19
			γ 7		220.502 15	1.75 6
			γ 12		288.451 16	3.09 12
			γ 23		417.63 3	3.52 12
			γ 29		546.557 16	7.13 25
			γ 43		836.804 16	6.67 24
			γ 47		972.6	1.20 6

Table 6 (continued)

Isotope	CAS Registry No.	Half-life	Type of radiation	Energy, keV Max	Energy, keV Avg	Intensity, %
			γ 49		1038.760 21	7.9 3
			γ 51		1101.58 4	1.60 6
			γ 52		1124.00 4	3.61 12
			γ 53		1131.511 18	22.5 8
			γ 61		1260.409 17	28.6 10
			γ 70		1457.56 3	8.6 3
			γ 71		1502.79 4	1.07 5
			γ 73		1566.41 3	1.29 6
			γ 74		1678.03 3	9.5 4
			γ 75		1706.46 3	4.09 18
			γ 77		1791.20 3	7.70 25
			75 weak γs			13.10
^{135}Xe (β^- decay)	[14995-62-1]	9.09 h 1	Auger L		3.55	5.12 25
			ce-K-3		213.809 15	5.63 17
			β^- 1	550 9	171 4	3.12 10
			β^- 2	908 9	307 4	95.98 3
			total β^-		302 4	99.98 11
			weak βs			0.88
			x ray Kα_2		30.6251 3	1.43 6
			x ray Kα_1		30.9728 3	2.66 9
			γ 3		249.794 15	90.13 8
			γ 9		608.185 16	2.90 9
			11 weak γs			1.08
^{135}Cs (β^- decay)	[15726-30-4]	2.3×10^6 yr 3	β^- 1	205 5	56.3 15	100
^{135}Ba (IT decay)	[14698-58-9]	28.7 d 2	Auger L		3.67	63 4
			Auger K		26.4	5.9 16
			ce-K-1		230.797 10	59.9 7
			ce-L-1		262.249 10	18.7 5
			ce-M-1		266.945 10	4.19 12
			ce-NOP-1		267.985 10	1.20 3
			x ray L		4.47	8.62 2
			x ray Kα_2		31.8171 3	15.3 6
			x ray Kα_1		32.1936 3	28.3 10
			x ray Kβ		36.4	10.3 4
			γ 1		268.238 10	16.0 5
^{136}Cs (β^- decay)	[14234-29-8]	13.16 d	Auger L		3.67	20.6 13
			Auger K		26.4	1.9 5
			ce-K-1		29.47 5	7.3 7
			ce-K-2		48.85 5	1.85 11
			ce-L-1		60.92 5	1.04 9
			ce-K-4		115.78 5	2.38 9
			ce-K-5		126.45 5	5.16 20
			ce-L-5		157.90 5	4.11 16
			ce-K-14		303.13 5	1.21 4
			β^- 1	174.4 20	47.2 6	2.30 20
			β^- 2	341.0 20	98.8 7	94 5
			β^- 3	408.0 20	121.0 7	1.20 20
			β^- 4	681.5 20	219.0 8	1.80 20
			total β^-		100.0 8	100 5
			2 weak βs			0.31
			x ray L		4.47	2.8 8
			x ray Kα_2		31.8171 3	4.87 24
			x ray Kα_1		32.1936 3	9.0 5
			x ray Kβ		36.4	3.27 16
			γ 1		66.91 5	12.5 10
			γ 2		86.29 5	6.3 3
			γ 4		153.22 5	7.47 16
			γ 5		163.89 5	4.62 10
			γ 7		176.55 5	13.59 21
			γ 10		273.65 4	12.69 21
			γ 14		340.57 5	48.6 5
			γ 18		818.50 4	99.7 4
			γ 19		1048.07 7	79.7 9
			γ 20		1235.34 5	19.8 8
			13 weak γs			3.55
^{136}Pr (EC decay)	[22095-53-0]	131.1 min 1	Auger L		4	33.6 21
			Auger-K		28.4	3.4 10
			β^+ 1	1920 40	865 19	1.50 16
			β^+ 2	2010 40	905 19	2.14 15
			β^+ 3	2530 40	1140 19	7.0 3
			β^+ 4	2990 40	1353 19	39.9 11
			β^+ 5	3530 40	1604 19	4.42 8
			total β^+		1294 19	56.8 12
			18 weak βs			1.85
			x ray L		4.84	5.0 12
			x ray Kα_2		34.27890 20	9.8 4
			x ray Kα_1		34.71970 20	18.0 7

Table 6 (continued)

Isotope	CAS Registry No.	Half-life	Type of radiation	Energy, keV Max	Energy, keV Avg	Intensity, %
			x ray Kβ		39.3	6.7 3
			γ 3		460.80 20	7.7 7
			γ 5		539.80 20	52 5
			γ 6		552.20 20	76 5
			γ 9		761.3 4	1.5 3
			γ 14		1000.8 3	5.0 4
			γ 19		1092.0 5	18.5 16
			γ 27		1514.8 4	1.92 24
			γ 30		1602.8 3	3.9 4
			γ 49		2066.8 3	3.0 3
			87 weak γs			17.88
^{137}Cs (β$^-$ decay)	[10045-97-3]	30.0 yr 2	β$^-$ 1	511.5 9	173.5 3	94.6 3
			β$^-$ 2	1173.2 9	415.4 3	5.4 3
			total β$^-$		186.6 4	100.05
^{137}La (EC decay)	[14834-69-6]	6 × 10^4 yr 2	Auger L		3.67	77 5
			Auger K		26.4	8.3 22
			x ray L		4.47	10 3
			x ray Kα$_2$		31.8171 3	21.4 7
			x ray Kα$_1$		32.1936 3	39.6 12
			x ray Kβ		36.4	14.4 5
^{138}Xe (β$^-$ decay)	[15751-81-2]	14.17 mo	Auger L		3.55	50.0 24
			ce-M-1		3.63 5	31 5
			ce-NOP-1		4.62 5	10.5 15
			ce-L-2		5.14 5	52.0 19
			ce-M-2		9.63 5	10.4 4
			ce-NOP-2		10.62 5	2.55 9
			ce-K-5		117.77 3	1.61 25
			ce-K-8		222.33 5	1.80 10
			β$^-$ 1	490 80	150 30	3.07 13
			β$^-$ 2	570 80	180 30	9.5 4
			β$^-$ 3	800 80	270 40	32.7 13
			β$^-$ 4	2380 80	950 40	20.1 7
			β$^-$ 5	2420 80	970 40	13.4 5
			β$^-$ 6	2570 80	1040 40	5.6 21
			β$^-$ 7	2820 80	1150 40	5 5
			β$^-$ 8	2810 80	1150 40	9 6
			total β$^-$		660 60	100 9
			12 weak βs			2.10
			x ray L		4.29	6.2 17
			x ray Kα$_2$		30.6251 3	1.02 8
			x ray Kα$_1$		30.9728 3	1.89 13
			γ 5		153.75 3	5.95 25
			γ 7		242.56 5	3.50 14
			γ 8		258.31 5	31.5 13
			γ 14		396.43 5	6.3 3
			γ 15		401.36 5	2.17 13
			γ 17		434.49 5	20.3 9
			γ 63		1114.29 10	1.47 9
			γ 84		1768.26 13	16.7 7
			γ 88		1850.86 13	1.42 7
			γ 91		2004.75 14	5.35 23
			γ 92		2015.82 14	12.3 5
			γ 94		2079.17 14	1.44 7
			γ 95		2252.26 14	2.29 11
			88 weak γs			12.40
^{138}Cs (β$^-$ decay)	[15758-29-9]	32.2 mo	β$^-$ 1	2510 70	1010 40	1.59 7
			β$^-$ 2	2650 70	1070 40	8.70 23
			β$^-$ 3	2710 70	1100 40	1.67 8
			β$^-$ 4	2840 70	1160 40	43.0 7
			β$^-$ 5	2980 70	1220 40	7.90 20
			β$^-$ 6	3070 70	1270 40	13.0 3
			β$^-$ 7	3390 70	1410 40	13.7 7
			β$^-$ 8	3850 70	1630 40	4.6 18
			total β$^-$		1200 40	99.8 21
			24 weak βs			5.63
			γ 2		138.10 6	1.49 9
			γ 6		227.76 6	1.51 4
			γ 12		408.98 6	4.66 10
			γ 14		462.79 7	30.7 7
			γ 16		546.94 7	10.76 24
			γ 31		871.80 8	5.11 14
			γ 36		1009.78 8	29.8 7
			γ 39		1147.22 9	1.24 7
			γ 43		1343.59 9	1.14 6
			γ 47		1435.86 9	76.3 16
			γ 65		2218.00 10	15.2 4
			γ 71		2639.59 13	7.63 23

Table 6 (*continued*)

Isotope	CAS Registry No.	Half-life	Type of radiation	Energy, keV Max	Energy, keV Avg	Intensity, %
^{138}La (EC decay)	[15816-87-2]	1.35×10^{11} yr 16	74 weak γs			10.03
			Auger L		3.67	22.3 3
			Auger K		26.4	2.7 8
			x ray L		4.47	2.9 9
			x ray Kα_2		31.8171 3	7.0 9
			x ray Kα_1		32.1936 3	13.0 17
			x ray Kβ		36.4	4.7 7
			γ 1		1435.6 3	67.1 18
^{138}La (β^- decay)		1.35×10^6 yr 16	β^- 1	254 12	95 5	32.9 18
			γ 1		788.4 5	32.9 8
^{138}Pr (EC decay)	[15481-22-8]	2.1 h 1	Auger L		4	66 5
			Auger K		28.4	6.7 20
			ce-K-10		262.26 10	9.6 9
			β^+ 1	1650 30	742 12	23.0 20
			2 weak βs			0.48
			x ray L		4.84	9.9 24
			x ray Kα_2		34.27890 20	19.6 11
			x ray Kα_1		34.71970 20	36.0 20
			x ray Kβ		39.3	13.4 8
			γ 10		302.70 10	80 5
			γ 13		390.90 10	6.1 3
			γ 15		547.50 10	5.2 3
			γ 16		635.70 10	1.77 10
			γ 19		788.70 10	99.590 10
			γ 21		1037.80 10	97.6 20
			55 weak γs			6.64
^{139}Ba (β^- decay)	[14378-25-7]	82.8 min 2	β^- 1	2140 5	837.0 23	30 3
			β^- 2	2306 5	912.0 23	70 3
			total β^-		887.0 23	100 5
			15 weak βs			0.42
			γ 1		165.853 7	23.8 3
			27 weak γs			0.42
^{139}Ce (EC decay)	[13982-30-4]	137.66 d 2	Auger L		3.8	89 5
			Auger K		27.4	8.4 24
			ce-K-1		126.928 7	17.05 16
			ce-L-1		159.587 7	2.30 8
			x ray L		4.65	13 4
			x ray Kα_2		33.03410 20	23.1 8
			x ray Kα_1		33.44180 20	42.6 14
			x ray Kβ		37.8	15.6 6
			γ 1		165.853 7	79.94 13
^{140}Ba (β^- decay)	[14798-08-4]	12.746 d 10	Auger L		3.8	86 7
			ce-L-1		7.58 5	51 6
			ce-M-1		12.49 5	10.6 12
			ce-L-2		23.70 5	46 5
			ce-M-2		28.61 5	9.6 10
			ce-K-7		123.72 5	1.48 5
			β^-	454 10	136 4	24.42 22
			β^-	567 10	177 4	9.85 6
			β^-	872 10	292 4	3.81 4
			β^-	991 10	340 4	39 6
			β^-	1005 10	345 4	23 6
			total β^-		273 5	100 9
			x ray L		4.65	13 4
			γ 1		13.85 5	1.20 13
			γ 2		29.97 5	10.7 10
			γ 7		162.64 5	6.21 10
			γ 8		304.840 20	4.30 7
			γ 9		423.69 3	3.12 15
			γ 10		437.55 3	1.93 5
			γ 12		537.32 8	24.39 22
			5 weak γs			0.43
^{140}La (β^- decay)	[13981-28-7]	40.272 h 7	Auger L		4	1.98 13
			β^- 1	1238.2 21	440.9 9	11.1 3
			β^- 2	1243.9 21	443.3 9	5.70 10
			β^- 3	1295.7 21	465.1 9	5.60 10
			β^- 4	1347.7 21	487.2 9	44.8 4
			β^- 5	1411.8 21	514.5 9	5.10 20
			β^- 6	1676.5 21	629.3 9	20.8 5
			β^- 7	2163.5 20	846.0 9	4.8 6
			total β^-		525.8 9	99.8 10
			10 weak βs			1.90
			γ 10		328.768 12	20.74 19
			γ 12		432.53 3	2.99 4
			γ 15		487.029 19	45.9 4
			γ 17		751.83 8	4.4 4
			γ 18		815.80 9	23.64 18

Table 6 (continued)

Isotope	CAS Registry No.	Half-life	Type of radiation	Energy, keV Max	Energy, keV Avg	Intensity, %
			γ 19		867.82 14	5.59 5
			γ 20		919.63 15	2.68 3
			γ 21		925.24 9	7.05 8
			γ 26		1596.49 24	95.40 8
			γ 31		2521.7 5	3.43 8
			26 weak γs			3.65
^{140}Nd (EC decay)	[14952-28-4]	3.37 d 2	Auger L		4	73 4
			Auger K		29.4	7.0 21
			x ray L		5	12.0 18
			x ray Kα_2		35.55020 20	21.6 7
			x ray Kα_1		36.02630 20	39.5 12
			x ray Kβ		40.7	14.8 5
^{141}Ce (β^- decay)	[13967-74-3]	32.50 d 1	Auger L		4	16.4 9
			Auger K		29.4	1.6 5
			ce-K-1		103.450 3	18.9 6
			ce-L-1		138.606 3	2.62 9
			β^- 1	434.6 15	129.6 6	70.5 7
			β^- 2	580.0 15	180.7 6	29.5 7
			total β^-		144.7 7	100.0 10
			x ray L		5	2.7 4
			x ray Kα_2		35.55020 20	4.91 22
			x ray Kα_1		36.02630 20	9.0 4
			x ray Kβ		40.7	3.38 15
			γ 1		145.440 3	48.4 4
^{141}Nd (EC decay)	[14877-64-6]	2.49 h 3	Auger L		4	72 4
			Auger K		29.4	7.0 21
			β^+ 1	793 8	363 4	2.50 10
			x ray L		5	11.7 18
			x ray Kα_2		35.55020 20	21.4 7
			x ray Kα_1		36.02630 20	39.2 12
			x ray Kβ		40.7	14.7 5
			15 weak γs			2.01
^{143}Pr (β^- decay)	[14981-79-4]	13.58 d 3	β^- 1	935.3 19	315.6 8	100
^{143}Pm (EC decay)	[14834-72-1]	265 d 7	Auger L		4.23	72 4
			Auger K		30.5	6.6 20
			x ray L		5.23	12.7 18
			x ray Kα_2		36.8474 3	21.7 10
			x ray Kα_1		37.3610 3	39.5 18
			x ray kβ		42.3	15.0 7
			γ 1		741.98 4	38.5 24
^{144}Ce (β^- decay)	[14762-78-8]	284.9 d 2	Auger L		4	10.1 5
			ce-L-1		26.785 10	1.08 9
			ce-K-5		38.115 5	2.40 25
			ce-K-7		91.553 5	5.5 3
			β^- 1	184.7 20	50.2 6	20.1 8
			β^- 2	238.1 20	66.1 6	3.1 4
			β^- 3	318.2 20	91.1 7	77.0 9
			total β^-		82.1 7	100.2 13
			x ray L		5	1.64 25
			x ray Kα_2		35.55020 20	2.35 13
			x ray Kα_1		36.02630 20	4.30 22
			x ray Kβ		40.7	1.62 9
			γ 5		80.106 5	1.13 12
			γ 7		133.544 5	11.1 5
			5 weak γs			0.82
^{145}Pm (β^- decay)	[15706-44-2]	5.98 h 2	β^- 1	1057 10	364 4	1.050 20
			β^- 2	1805 10	683 5	97.6 10
			total β^-		677 5	99.7 10
			8 weak βs			1.03
^{145}Pm (EC decay)		17.7 yr 2	Auger L		4.23	80 6
			ce-K-1		23.63 10	1.86 11
			ce-K-2		28.83 10	5.7 3
			Auger K		30.5	6.6 21
			ce-L-1		60.07 10	2.68 15
			x ray L		5.23	14.0 22
			x ray Kα_2		36.8874 3	21.4 18
			x ray Kα_1		37.3610 3	39 4
			x ray Kβ		42.3	14.9 13
			γ 2		72.40 10	1.85 7
			1 weak γ			0.55
^{145}Sm (EC decay)	[15065-02-8]	340 d 3	Auger L		4.38	127 7
			ce-K-1		16.07 5	65.5 18
			Auger K		31.5	11 4
			ce-L-1		53.82 5	10.1 8
			ce-MNO-1		59.60 5	3.3 3
			x ray L		5.43	24 4
			x ray Kα_2		38.1712 5	39.0 14

Table 6 (*continued*)

Isotope	CAS Registry No.	Half-life	Type of radiation	Energy, keV Max	Energy, keV Avg	Intensity, %
^{147}Nd (β^- decay)	[*14269-74-0*]	10.98 d 1	x ray Kα_1		38.7247 5	70.8 23
			x ray Kβ		43.8	27.2 10
			γ 1		61.25 5	12.36 23
			Auger L		4.38	41.7 24
			Auger K		31.5	3.7 12
			ce-K-1		45.922 20	48.5 20
			ce-L-1		83.678 20	7.1 3
			ce-M-1		89.457 21	1.51 6
			β^- 1	209.9 9	57.6 3	2.20 20
			β^- 2	364.8 9	106.0 3	15.3 10
			β^- 3	804.7 9	264.0 4	81 4
			total β^-		233.3 5	100 5
			4 weak βs			1.47
			x ray L		5.43	7.9 11
			x ray Kα_2		38.1712	12.9 7
			x ray Kα_1		38.7247	23.5 12
			x ray Kβ		43.8	9.0 5
			γ 1		91.106 20	27.9 11
			γ 6		319.411 18	1.95 14
			γ 9		439.895 22	1.20 10
			γ 11		531.016 22	13.1 9
			11 weak γs			3.71
^{147}Pm (β^- decay)	[*14380-75-7*]	2.6234 yr 2	β^- 1	224.7 4	61.96 12	99.9940 10
^{147}Eu (EC decay)	[*14191-78-7*]	24 d 1	Auger L		4.53	93 7
			ce-K-1		29.32 5	2.3 5
			Auger K		32.6	8 3
			ce-K-2		74.42 5	19 3
			ce-L-2		113.51 5	3.3 5
			ce-K-4		150.52 5	4.1 5
			ce-L-4		189.61 5	1.25 14
			3 weak βs			0.36
			x ray L		5.64	19 3
			x ray Kα_2		39.5224 3	28.7 20
			x ray Kα_1		40.1181 3	52 4
			x ray Kβ		45.4	20.2 14
			γ 2		121.25 5	23 3
			γ 4		197.35 5	26 3
			γ 10		601.43 10	6.8 10
			γ 11		677.60 10	10.7 14
			γ 14		798.81 12	5.5 8
			γ 18		857.07 10	3.1 4
			γ 22		933.11 12	3.6 5
			γ 24		955.94 12	3.9 5
			γ 30		1077.16 12	6.4 9
			γ 36		1255.91 18	1.01 13
			39 weak γs			4.28
^{147}Gd (EC decay)	[*14952-31-9*]	38.1 h 1	Auger L		4.69	82 5
			Auger K		33.7	6.9 16
			ce-K-14		180.801 20	9.4 11
			ce-L-14		221.268 20	1.34 15
			ce-K-33		347.48 10	4.2 5
			2 weak βs			0.16
			x ray L		5.85	18.0 22
			x ray Kα_2		40.9019 3	26.4 13
			x ray Kα_1		41.5422 3	47.7 22
			x ray Kβ		47	18.7 9
			γ 12		216.90 10	1.24 21
			γ 14		229.320 20	61 7
			γ 16		240.64 5	1.48 17
			γ 18		261.10 10	1.90 21
			γ 24		309.96 10	4.0 5
			γ 25		318.60 10	2.10 23
			γ 30		346.3 3	2.05 22
			γ 31		370.00 10	16.5 18
			γ 33		396.00 10	33 4
			γ 42		484.90 10	2.9 4
			γ 52		559.07 10	6.2 7
			γ 60		610.43 10	1.53 20
			γ 61		619.00 10	3.5 4
			γ 62		625.18 10	4.5 6
			γ 63		632.35 10	1.64 18
			γ 76		755.01 10	1.99 22
			γ 77		765.81 10	10.9 13
			γ 78		775.9 10	1.06 11
			γ 79		776.33 10	4.2 5
			γ 80		778.04 5	4.8 6
			γ 81		782.60 20	1.15 13

Table 6 (*continued*)

Isotope	CAS Registry No.	Half-life	Type of radiation	Energy, keV Max	Energy, keV Avg	Intensity, %
			γ 89		861.70 10	1.68 10
			γ 93		893.50 10	7.8 9
			γ 97		929.01 7	19.4 21
			γ 108		1006.40 10	1.31 16
			γ 114		1069.35 10	6.9 9
			γ 119		1130.90 10	6.2 8
			γ 131		1235.70 10	1.11 14
			150 weak γs			19.32
^{148}Pm (β^- decay)	[14683-19-3]	5.37 d 1	β^- 1	999 9	340 4	35 6
			β^- 2	1914 9	729 4	10.0 20
			β^- 3	2464 9	975 4	54 5
			total β^-		726 5	99 8
			γ 2		550.10 20	23.3 20
			γ 4		611.1 3	1.12 12
			γ 6		914.90 20	12.5 12
			γ 7		1465.10 20	22.2 20
^{148}Pm (IT decay)		41.3 d 1	Auger L		4.38	5.5 6
			ce-K-2		30.52 10	3.0 6
			ce-L-1		54.07 10	3.5 5
			ce-M-1		59.85 10	1.00 14
			x ray L		5.43	1.04 18
			x ray Kα_1		38.7247 5	1.4 3
			γ 2		75.70 10	1.00 21
^{148}Pm (β^- decay)		41.3 d 1	Auger L		4.53	5.8 7
			ce-K-1		51.67 20	5.7 8
			β^- 1	408 9	120 3	56.0 20
			β^- 2	506 9	154 4	17.0 10
			β^- 3	696 9	222 4	22.0 10
			β^- 4	1007 9	343 4	1.1 6
			total β^-		152 4	96 3
			x ray L		5.64	1.19 20
			x ray Kα_2		39.5224 3	2.14 23
			x ray Kα_1		40.1181 3	3.9 4
			x ray Kβ		45.4	1.50 16
			γ 1		98.50 20	3.83 13
			γ 2		189.5 3	1.24 5
			γ 3		288.00 20	12.4 4
			γ 4		311.7 3	3.98 13
			γ 5		414.1 3	18.6 6
			γ 6		432.7 3	5.66 21
			γ 7		501.1 4	6.90 21
			γ 8		550.10 20	93.7 24
			γ 9		599.5 3	12.4 4
			γ 10		611.1 3	5.5 3
			γ 11		629.90 20	89.2 15
			γ 12		725.60 20	32.8 9
			γ 13		915.30 10	19.0 8
			γ 14		1013.7 3	20.4 7
^{148}Sm (α decay)	[14913-64-5]	8.0×10^{15} yr 2	α 1	1932.5 13		100
^{149}Nd (β^- decay)	[15749-81-2]	1.73 h 1	Auger L		4.38	34.2 21
			ce-K-2		13.699 20	1.42 8
			ce-L-1		22.57 3	4.8 5
			ce-M-1		28.35 3	1.10 11
			ce-K-4		29.15 3	3.7 6
			Auger K		31.5	2.6 9
			ce-K-8		51.823 15	2.22 20
			ce-L-4		66.90 3	1.7 4
			ce-K-10		69.137 14	17.1 19
			ce-L-10		106.893 14	2.6 3
			ce-K-23		166.123 8	4.3 3
			ce-K-27		195.034 8	2.18 21
			β^- 1	1034 4	354.7 17	19.5 13
			β^- 2	1151 4	402.4 17	24.1 18
			β^- 3	1292 4	461.1 17	3.8 5
			β^- 4	1419 4	514.7 17	19.1 16
			β^- 5	1478 4	539.8 18	26.5 21
			β^- 6	1500 4	549.5 18	2.9 13
			β^- 7	1575 4	581.5 18	ca 1
			total β^-		457.4 17	101 4
			9 weak βs			3.63
			x ray L		5.43	6.5 9
			x ray Kα_2		38.1712 5	9.1 6
			x ray Kα_1		38.7247 5	16.6 11
			x ray Kβ		43.8	6.4 5
			γ 2		58.883 20	1.52 7
			γ 4		74.33 3	1.26 20
			γ 8		97.007 15	1.53 13

Table 6 (continued)

Isotope	CAS Registry No.	Half-life	Type of radiation	Energy, keV Max	Energy, keV Avg	Intensity, %
			γ 10		114.321 14	18.8 20
			γ 16		155.876 10	6.1 6
			γ 19		188.640 9	1.99 24
			γ 21		198.928 9	1.46 16
			γ 22		208.148 10	2.9 4
			γ 23		211.307 8	27.3 18
			γ 27		240.218 8	4.0 4
			γ 28		245.699 8	1.04 11
			γ 32		267.692 9	6.1 6
			γ 33		270.165 8	10.7 11
			γ 42		326.556 11	4.7 5
			γ 45		349.233 10	1.48 16
			γ 53		423.554 10	9.4 10
			γ 55		443.550 12	1.50 15
			γ 64		540.510 10	7.7 8
			γ 65		556.43 5	1.2 5
			γ 72		654.831 14	7.3 8
			133 weak γs			12.36
^{149}Pm (β^- decay)	[15765-31-8]	53.08 h 5	β^- 1	786.5 20	256.6 8	3.1 3
			β^- 2	1072.4 20	369.6 9	96.7 3
			total β^-		365.2 9	100.1 5
			8 weak βs			0.35
			γ 9		285.90 5	2.8 3
			27 weak γs			0.36
^{149}Eu (EC decay)	[14907-89-2]	93.1 d 4	Auger L		4.53	70 4
			Auger K		32.6	6.0 19
			x ray L		5.64	14.3 18
			x ray Kα_2		39.5224 3	21.8 7
			x ray Kα_1		40.1181 3	39.6 11
			x ray Kβ		45.4	15.3 5
			γ 7		277.00 20	3.30 20
			γ 9		327.50 20	3.90 20
			12 weak γs			3.18 10
^{150}Eu (EC decay)	[15840-16-1]	12.62 h 10	Auger L		4.53	7.6 12
			1 weak β			0.64
			x ray L		5.64	1.5 3
			x ray Kα_2		39.5224 3	2.4 5
			x ray Kα_1		40.1181 3	4.3 8
			x ray Kβ		45.4	1.7 3
			γ 3		333.90 10	3.8 8
			γ 4		406.50 10	2.7 6
			19 weak γs			1.29
^{150}Eu (β^- decay)		12.62 h 10	β^- 1	1013 4	344.4 16	89.0 20
			1 weak β			0.02
^{150}Eu (EC decay)		34.2 yr 12	Auger L		4.53	72 4
			Auger K		32.6	6.1 20
			x ray L		5.64	14.7 19
			x ray Kα_2		39.5224 3	22.2 10
			x ray Kα_1		40.1181 3	40.3 18
			x ray Kβ		45.4	15.6 7
			γ 16		333.960 10	94 3
			γ 29		439.390 10	79 3
			γ 41		505.510 10	4.71 18
			γ 51		584.260 10	51.5 20
			γ 67		712.190 10	1.06 4
			γ 69		737.440 10	9.4 4
			γ 72		748.040 10	5.05 20
			γ 74		751.050 10	2.10 9
			γ 88		869.230 10	1.81 7
			γ 101		1049.020 20	5.24 20
			γ 108		1170.560 10	1.31 5
			γ 110		1197.080 10	1.11 6
			γ 112		1246.940 10	1.87 8
			γ 118		1343.740 10	2.54 10
			γ 125		1485.450 10	1.80 9
			118 weak γs			20.36
^{150}Tb (EC decay)	[15065-95-9]	3.27 h 10	Auger L		4.84	40.6 25
			Auger K		34.9	3.3 12
			β^+ 1	2190 30	998 14	1.60 18
			β^+ 2	2280 30	1028 14	2.0 4
			β^+ 3	2420 30	1092 14	1.0 4
			β^+ 4	2570 30	1163 14	2.7 4
			β^+ 5	3070 30	1390 14	17.5 18
			β^+ 6	3710 30	1686 14	10 5
			total β^+		1323 15	41 6
			15 weak βs			6.0
			x ray L		6	9.5 12

Table 6 (continued)

Isotope	CAS Registry No.	Half-life	Type of radiation	Energy, keV Max	Energy, keV Avg	Intensity, %
			x ray Kα$_2$		42.3089 3	13.3 7
			x ray Kα$_1$		42.9962 3	24.0 13
			x ray Kβ		48.7	9.5 5
			γ 4		496.20 20	15.1 11
			γ 8		565.9 4	1.15 16
			γ 9		569.2 4	2.5 3
			γ 11		638.20 20	72 4
			γ 12		650.4 4	4.1 5
			γ 16		792.3 4	4.7 6
			γ 18		820.9 4	1.44 17
			γ 19		880.3 4	3.2 4
			γ 21		949.9 4	1.08 13
			γ 22		954.5 4	1.22 16
			γ 23		1045.4 4	1.22 16
			γ 28		1291.5 4	1.66 17
			γ 30		1349.7 4	1.30 16
			γ 33		1430.4 4	2.4 3
			γ 36		1453.5 4	3.7 5
^{151}Sm (β$^-$ decay)	[15705-94-3]	90 yr 6	β$^-$ 1	76.1 6	19.68 16	99.12 6
			1 weak β			0.88
			1 weak γ			0.03
^{151}Gd (EC decay)	[14937-17-8]	120 d 20	Auger L		4.69	117 8
			ce-L-1		13.488 20	52 9
			ce-M-1		19.740 20	11.3 19
			Auger K		33.7	6.0 13
			ce-K-4		105.05 6	2.4 4
			ce-K-5		126.13 6	4.6 8
			x ray L		5.8 5	26 4
			γ 1		21.540 20	2.3 4
			x ray Kα$_2$		40.9019 3	23.1 6
			x ray Kα$_1$		41.5422 3	41.7 9
			x ray Kβ		47	16.3 4
			γ 4		153.57 6	5.1 8
			γ 5		174.65 6	2.4 4
			γ 10		243.22 6	4.6 8
			14 weak γs			1.35
^{151}Tb (EC decay)	[14998-51-7]	17.6 h 1	Auger L		4.84	67 4
			Auger K		34.9	5.4 19
			9 weak βs			1.26
			x ray L		6	15.7 18
			x ray Kα$_2$		42.3089 3	21.9 6
			x ray Kα$_1$		42.9962 3	39.5 10
			x ray Kβ		48.7	15.6 5
			γ 1		108.26 10	25 5
			γ 4		180.41 12	11.4 22
			γ 5		192.09 15	3.8 8
			γ 9		251.73 13	26 5
			γ 11		287.04 12	25 5
			γ 14		380.40 20	4.3 9
			γ 15		385.4 6	1.14 22
			γ 16		395.30 20	9.6 19
			γ 17		416.40 20	1.4 3
			γ 18		426.50 20	4.1 8
			γ 19		443.70 20	10.4 20
			γ 20		467.40 20	1.04 20
			γ 21		479.0 3	16 3
			γ 25		587.27 13	17 4
			γ 26		604.81 13	3.2 7
			γ 27		616.60 13	10.4 20
			γ 30		692.26 15	1.5 3
			γ 31		703.75 14	3.8 8
			γ 32		731.14 13	9.1 18
			γ 41		905.71 15	1.01 20
			79 weak γs			12.95
^{152}Eu (EC decay)	[14683-23-9]	13.33 yr 4	Auger L		4.53	74 4
			Auger K		32.6	5.7 19
			ce-K-1		74.9482 7	19.4 6
			ce-L-1		114.0456 7	10.6 4
			ce-M-1		120.0596 9	2.43 8
			2 weak βs			0.03
			x ray L		5.64	15.1 19
			x ray Kα$_2$		39.5224 3	21.0 7
			x ray Kα$_1$		40.1181 3	38.0 11
			x ray Kβ		45.4	14.7 5
			γ 1		121.7824 4	28.38 23
			γ 9		244.6989 10	7.51 7
			γ 29		444	2.80 3

Table 6 (*continued*)

Isotope	CAS Registry No.	Half-life	Type of radiation	Energy, keV Max	Energy, keV Avg	Intensity, %
			γ 63		867.388 8	4.21 5
			γ 70		964.13	14.49 6
			γ 75		1085.914 13	9.92 5
			γ 76		1112.116 17	13.56 6
			γ 79		1212.950 12	1.397 19
			γ 84		1408.011 14	20.85 9
			81 weak γs			7.59
^{152}Eu (β$^-$ decay)		13.33 yr 4	β$^-$ 1	176 4	47.5 10	1.820 20
			β$^-$ 2	385 4	112.5 11	2.410 20
			β$^-$ 3	696 4	221.8 13	13.80 10
			β$^-$ 4	1475 4	535.6 14	8.20 20
			total β$^-$		297.8 18	27.89 23
			8 weak βs			1.66
			γ 7		344.2810 20	26.58 19
			γ 10		411.115 5	2.233 13
			γ 26		778.910 10	12.96 7
			γ 32		1089.700 15	1.710 21
			γ 36		1299.124 12	1.626 21
			33 weak γs			2.98
^{152}Eu (EC decay)		9.32 h 1	Auger L		4.53	26 3
			Auger K		32.6	2.1 8
			ce-K-1		74.9482 7	4.9 8
			ce-L-1		114.0456 7	2.7 4
			x ray L		5.64	5.4 9
			x ray Kα_2		39.5224 3	7.7 10
			x ray Kα_1		40.1181 3	14.0 17
			x ray Kβ		45.4	5.4 7
			γ 1		121.7824 4	7.2 11
			γ 23		841.63 4	14.6 21
			γ 27		963.37 4	12.0 18
			36 weak γs			1.62
^{152}Eu (β$^-$ decay)		9.32 h 1	β$^-$ 1	550 4	168.9 12	1.60 20
			β$^-$ 2	1521 4	554.1 14	1.70 20
			β$^-$ 3	1865 4	704.1 15	68 4
			total β$^-$		687.4 16	71 4
			3 weak βs			0.16
			γ 5		344.2810 20	2.4 3
			21 weak γs			1.87
^{153}Sm (β$^-$ decay)	[15766-00-4]	46.7 h 1	Auger L		4.69	54 3
			ce-K-4		21.1544 5	23.5 14
			Auger K		33.7	4.5 10
			ce-K-10		54.6617 5	40.8 15
			ce-L-4		61.6214 5	3.83 22
			ce-L-10		95.1287 5	6.17 23
			ce-MNO-10		101.3807 6	1.78 4
			β$^-$ 1	644 4	203.0 15	34.7 16
			β$^-$ 2	714 4	228.8 15	43 3
			β$^-$ 3	817 4	267.9 16	20.9 17
			total β$^-$		227.6 16	99 4
			12 weak βs			0.86
			x ray L		5.85	11.9 14
			x ray Kα_2		40.9019 3	17.3 7
			x ray Kα_1		41.5422 3	31.3 12
			x ray Kβ		47	12.2 5
			γ 4		69.67340 20	5.25 25
			γ 10		103.1807 3	28.3 6
			63 weak γs			2.02
^{153}Gd (EC decay)	[14276-65-4]	242 d 1	Auger L		4.69	110 5
			ce-K-3		21.1544 5	10.8 7
			Auger K		33.7	9.0 20
			ce-K-7		48.9126 5	7.6 4
			ce-K-8		54.6617 5	30.4 14
			ce-L-3		61.6214 5	1.77 10
			ce-L-7		89.3796 5	1.13 5
			ce-L-8		95.1287 5	4.60 22
			ce-MNO-8		101.3807 6	1.33 5
			x ray L		5.85	24 3
			x ray Kα_2		40.9019 3	34.6 12
			x ray Kα_1		41.5422 3	62.5 20
			x ray Kβ		47	24.5 9
			γ 3		69.67340 20	2.42 12
			γ 7		97.4316 3	29.5 9
			γ 8		103.1807 3	21.1 8
			6 weak γs			0.99
^{155}Eu (β$^-$ decay)	[14391-16-3]	4.96 yr 1	ce-L-1		2.024 20	1.21 14
			Auger L		4.84	34 3
			ce-K-9		9.7709 21	8.5 9

Table 6 (continued)

Isotope	CAS Registry No.	Half-life	Type of radiation	Energy, keV Max	Energy, keV Avg	Intensity, %
			ce-L-3		10.40 4	14 3
			ce-M-3		16.90 4	3.1 7
			ce-NOP-3		18.40 4	1.03 20
			Auger K		34.9	1.6 6
			ce-K-11		36.3039 21	11.2 8
			ce-L-9		51.6344 21	1.71 18
			ce-K-12		55.069 3	4.4 4
			ce-L-11		78.1674 21	1.72 13
			β^- 1	129 3	34.0 9	2.2 3
			β^- 2	141 3	37.6 9	46 5
			β^- 3	160 3	42.9 9	26 4
			β^- 4	187 3	50.6 9	8.3 11
			β^- 5	247 3	68.5 9	18 5
			total β		45.4 10	101 9
			1 weak β			0.69
			x ray L		6	8.0 11
			x ray Kα_2		42.3089 3	6.5 4
			x ray Kα_1		42.9962 3	11.8 7
			γ 7		45.2980 20	1.28 10
			x ray Kβ		48.7	4.6 3
			γ 9		60.0100 20	1.14 12
			γ 11		86.5430 20	30.9 20
			γ 12		105.308 3	20.6 14
			9 weak γs			0.65
^{155}Tb (EC decay)	[14391-17-4]	5.32 d 6	Auger L		4.84	93 6
			ce-K-13		9.7709 21	5.5 13
			ce-L-2		10.40 4	12 5
			ce-M-2		16.90 4	2.8 10
			Auger K		34.9	6.6 22
			ce-K-16		36.3039 21	8.2 19
			ce-L-13		51.6344 21	1.11 25
			ce-K-20		55.069 3	3.9 9
			ce-L-16		78.1674 21	1.3 3
			x ray L		6	22 3
			x ray Kα_2		42.3089 3	26.4 10
			x ray Kα_1		42.9962 3	47.6 17
			γ 9		45.2980 20	1.2 3
			x ray Kβ		48.7	18.8 7
			γ 16		86.5430 20	23 5
			γ 20		105.308 3	18 4
			γ 25		148.65	1.9 5
			γ 29		161.32	2.1 5
			γ 31		163.3	3.3 8
			γ 33		180.14	6.5 15
			γ 49		262.45	4.3 10
			71 weak γs			8.56
^{156}Eu (β^- decay)	[14280-35-4]	15.19 d 6	Auger L		4.84	22.7 21
			ce-K-1		38.7246 25	14.0 17
			ce-L-1		80.5881 25	16.1 19
			ce-M-1		87.0829 25	3.8 5
			ce-NOP-1		88.5879 25	1.05 13
			β^- 1	183 9	50 3	7.59 18
			β^- 2	248 9	69 3	2.42 15
			β^- 3	266 9	75 3	11.3 6
			β^- 4	426 9	126 3	6.0 3
			β^- 5	487 9	147 4	32.1 10
			β^- 6	1087 9	374 4	2.37 21
			β^- 7	1211 9	425 4	5.1 9
			β^- 8	1285 9	456 4	4.5 5
			β^- 9	1404 9	505 4	1.44 12
			β^- 10	2453 9	966 4	27 6
			total β^-		395 9	98 7
			8 weak βs			1.44
			x ray L		6	5.3 8
			x ray Kα_2		42.3089 3	3.8 5
			x ray Kα_1		42.9962 3	6.8 9
			x ray Kβ		48.7	2.7 4
			γ 1		88.9637 24	8.9 11
			γ 23		599.47 5	2.26 18
			γ 25		646.29 5	7.0 5
			γ 28		723.47 5	5.9 3
			γ 33		811.77 5	10.2 5
			γ 40		867.01 8	1.38 14
			γ 46		944.35 7	1.37 11
			γ 48		960.50 8	1.59 13
			γ 58		1065.14 5	5.1 4
			γ 60		1079.16 5	4.8 5

Table 6 (continued)

Isotope	CAS Registry No.	Half-life	Type of radiation	Energy, keV Max	Energy, keV Avg	Intensity, %
			γ 65		1153.47 5	7.0 8
			γ 66		1154.09 5	5.2 5
			γ 72		1230.71 6	8.8 6
			γ 73		1242.427 24	6.6 5
			γ 75		1277.43 5	3.15 20
			γ 76		1366.41 5	1.72 11
			γ 80		1877.03 14	1.69 11
			γ 81		1937.68 10	2.10 15
			γ 83		1965.95 11	4.1 3
			γ 84		2026.61 10	3.47 23
			γ 86		2097.68 10	4.2 3
			γ 91		2180.91 11	2.39 16
			γ 92		2186.71 11	3.9 5
			γ 97		2269.90 12	1.10 7
			77 weak γs			10.45
^{157}Tb (EC decay)	[14391-18-5]	150 yr 30	Auger L		4.84	60 5
			x ray L		6	14.0 19
			x ray Kα_2		42.3089 3	2.7 14
			x ray Kα_1		42.9962 3	4.8 24
			x ray Kβ		48.7	1.9 10
^{157}Dy (EC decay)	[14981-97-6]	8.1 h 1	Auger L		5	71 4
			ce-K-1		8.82 7	2.1 9
			ce-K-2		31.01 4	1.7 3
			Auger K		36	5.5 20
			ce-K-7		274.16 20	1.09 4
			x ray L		6.27	17.7 20
			x ray Kα_2		43.7441 3	23.5 8
			x ray Kα_1		44.4816 3	42.2 13
			x ray Kβ		50.4	16.8 6
			γ 4		182.20 20	1.95 13
			γ 7		326.16 20	95.2 12
			24 weak γs			1.47
^{159}Gd (β^- decay)	[14041-42-0]	18.56 h 8	Auger L		5	17 4
			ce-K-1		6.004 10	21 6
			Auger K		36	1.3 6
			ce-L-1		49.292 10	3.5 10
			β^- 1	611.2 18	190.4 7	11 3
			β^- 2	916.7 18	305.7 7	26 7
			β^- 3	974.7 18	328.6 8	62 9
			5 weak βs			0.37
			x ray L		6.27	4.2 11
			x ray Kα_2		43.7441 3	5.6 15
			x ray Kα_1		44.4816	10 3
			x ray Kβ		50.4	4.0 11
			γ 1		58.000 10	2.3 6
			γ 11		363.56 3	11 3
			16 weak γs			0.70
^{159}Dy (EC decay)	[14280-34-3]	144.4 d 2	Auger L		5	83 4
			ce-K-2		6.004 10	20.1 12
			Auger K		36	6.3 22
			ce-L-2		49.292 10	3.40 20
			x ray L		6.27	20.8 23
			x ray Kα_2		43.7441 3	26.8 9
			x ray Kα_1		44.4816 3	48.1 16
			x ray Kβ		50.4	19.1 7
			γ 2		58.000 10	2.22 13
^{160}Tb (β^- decay)	[13981-29-8]	72.3 d 2	Auger L		5.16	39.8 20
			ce-K-1		32.9995 21	20.9 12
			Auger K		37.2	1.3 4
			ce-L-1		77.7422 21	31.3 18
			ce-M-1		84.7412 21	7.5 5
			ce-NOP-1		86.3717 21	2.05 12
			β^- 1	438.4 20	129.9 7	4.49 15
			β^- 2	478.7 20	143.6 7	9.9 4
			β^- 3	550.6 20	168.6 7	3.35 12
			β^- 4	572.6 20	176.4 8	46.5 16
			β^- 5	788.2 20	255.7 8	6.4 4
			β^- 6	871.2 20	287.5 8	26.5 12
			total β^-		206.9 9	99.0 21
			5 weak βs			1.84
			x ray L		6.5	10.6 11
			x ray Kα_2		45.2078 4	6.0 4
			x ray Kα_1		45.9984 4	10.7 6
			x ray Kβ		52	4.29 25
			γ 1		86.7880 20	13.2 7
			γ 4		197.035 7	5.15 19
			γ 5		215.648 8	3.95 15

Table 6 (*continued*)

Isotope	CAS Registry No.	Half-life	Type of radiation	Energy, keV Max	Energy, keV Avg	Intensity, %
			γ 10		289.575 4	26.9 11
			γ 15		392.500 15	1.34 5
			γ 19		765.28 4	2.00 8
			γ 21		879.362 14	29.5 9
			γ 22		962.302 19	9.8 4
			γ 23		966.155 16	25.0 9
			γ 24		1002.88 4	1.02 5
			γ 28		1115.12 3	1.53 6
			γ 29		1177.938 18	15.2 6
			γ 30		1199.89 3	2.32 9
			γ 32		1271.861 19	7.5 3
			γ 35		1312.14 4	2.92 11
			20 weak γs			3.00
^{165}Er (EC decay)	[14041-43-1]	10.36 h 4	Auger L		5.33	64 3
			Auger K		38.4	4.6 17
			x ray L		6.72	18.2 18
			x ray Kα_2		46.6997 4	21.5 6
			x ray Kα_1		47.5467 4	38.3 10
			x ray Kβ		53.9	15.5 5
^{166}Ho (β^- decay)	[13967-65-2]	26.80 h 2	Auger L		5.5	25.7 14
			ce-K-1		23.088 8	10.6 6
			ce-L-1		70.823 8	24.5 14
			ce-M-1		78.367 8	5.9 4
			ce-NOP-1		80.125 8	1.62 10
			β^- 1	1775.9 14	652.0 6	47.7 24
			β^- 2	1856.5 14	694.8 6	51.0 20
			total β^-		666.7 6	100 4
			4 weak βs			1.29
			x ray L		7	7.7 8
			x ray Kα_2		48.2211 4	2.86 18
			x ray Kα_1		49.1277 4	5.1 4
			x ray Kβ		55.7	2.07 14
			γ 1		80.574 8	6.2 3
			8 weak γs			1.31
^{166}Ho (β^- decay)		1.20 × 10^3 yr 18	Auger L		5.5	72 4
			ce-K-1		23.088 8	21.8 17
			Auger K		39.7	2.3 9
			ce-L-1		70.823 8	50 4
			ce-M-1		78.367 8	12.2 10
			ce-NOP-1		80.125 8	3.3 3
			ce-K-9		126.921 15	15.5 12
			ce-L-9		174.656 15	7.3 6
			ce-M-9		182.200 15	1.72 14
			ce-K-15		222.970 20	1.87 15
			β^- 1	34.0 25	8.6 7	18.8 12
			β^- 2	74.6 25	19.2 7	80 5
			total β^-		17.2 8	99 6
			x ray L		7	21.4 22
			x ray Kα_2		48.2211 4	11.1 7
			x ray Kα_1		49.1277 4	19.7 11
			x ray Kβ		55.7	8.0 5
			γ 1		80.574 8	12.7 9
			γ 9		184.407 15	75 6
			γ 12		215.88 3	2.65 25
			γ 14		259.716 20	1.12 9
			γ 15		280.456 20	30.4 22
			γ 16		300.744 20	3.8 3
			γ 18		365.739 25	2.57 19
			γ 19		410.941 25	11.8 9
			γ 20		451.524 25	3.12 22
			γ 21		464.83 4	1.25 11
			γ 22		529.81 3	10.4 8
			γ 23		571.00 3	5.9 5
			γ 25		611.52 7	1.42 11
			γ 28		670.51 4	5.9 5
			γ 29		691.21 5	1.56 12
			γ 30		711.69 4	60 5
			γ 32		752.27 4	13.4 10
			γ 33		778.82 4	3.37 24
			γ 34		810.31 4	64 5
			γ 35		830.56 4	10.8 8
			γ 37		950.94 6	3.10 22
			γ 41		1241.44 6	1.02 8
			22 weak γs			6.07
^{167}Tm (EC decay)	[14391-22-1]	9.24 d 2	Auger L		5.5	110 6
			Auger K		39.7	5.5 22
			ce-L-1		47.35 5	15.0 7

Table 6 (*continued*)

Isotope	CAS Registry No.	Half-life	Type of radiation	Energy, keV Max	Energy, keV Avg	Intensity, %
			ce-M-1		54.89 5	3.51 16
			ce-K-3		150.3145 5	20 3
			ce-L-3		198.0487 4	29 5
			ce-M-3		205.5935 6	7.1 11
			ce-NOP-3		207.3509 10	2.0 3
			x ray L		7	33 4
			x ray Kα_2		48.2211 4	27.1 11
			x ray Kα_1		49.1277 4	48.2 19
			x ray Kβ		55.7	19.7 9
			γ 1		57.10 5	3.55 12
			γ 3		207.8	41 7
			γ 8		531.5	1.60 6
			5 weak γs			0.03
^{168}Tm (EC decay)	[14900-13-1]	93.1 d 1	Auger L		5.5	113 9
			ce-K-4		22.3145 12	19.1 15
			Auger K		39.7	5.4 21
			ce-K-5		41.4985 21	1.56 11
			ce-K-6		41.8035 12	1.04 7
			ce-L-4		70.0487 11	45 4
			ce-M-4		77.5935 12	11.0 9
			ce-NOP-4		79.3509 15	3.01 23
			ce-K-10		126.796 3	3.40 24
			ce-K-11		140.735 3	2.25 16
			ce-L-10		174.530 3	1.60 12
			x ray L		7	34 4
			x ray Kα_2		48.2211 4	26.4 23
			x ray Kα_1		49.1277 4	47 4
			x ray Kβ		55.7	19.1 17
			γ 4		79.8000 10	11.0 8
			γ 6		99.2890 10	3.74 23
			γ 10		184.281 3	16.4 11
			γ 11		198.221 3	50 4
			γ 19		447.461 24	21.9 15
			γ 22		546.73 6	2.41 16
			γ 25		631.67 3	7.8 5
			γ 27		645.56 4	1.41 11
			γ 29		720.17 3	10.9 7
			γ 30		730.58 3	4.5 3
			γ 31		741.30 3	11.3 8
			γ 33		815.90 3	46 3
			γ 34		821.09 3	11.1 8
			γ 35		829.89 4	6.2 4
			γ 37		914.86 3	2.88 19
			γ 41		1277.33 6	1.62 12
			28 weak γs			3.05
^{169}Yb (EC decay)	[14269-78-4]	32.01 d 2	ce-K-3		3.729 7	38.0 25
			Auger L		5.67	162 10
			ce-M-1		6.094 8	76 3
			ce-NOP-1		7.929 8	18.4 7
			ce-L-2		10.63 5	10.4 7
			ce-M-2		18.44 5	2.31 16
			ce-K-4		34.223 7	8.0 6
			Auger K		40.9	10 4
			ce-K-5		50.387 7	35.0 18
			ce-L-3		53.003 7	6.7 5
			ce-K-7		58.797 7	1.35 7
			ce-M-3		60.812 7	1.51 10
			ce-K-8		71.130 7	6.3 4
			ce-L-4		83.497 7	1.38 9
			ce-L-5		99.661 7	5.64 23
			ce-M-5		107.470 7	1.26 5
			ce-L-7		108.071 7	1.39 7
			ce-K-10		117.820 7	10.7 6
			ce-L-8		120.404 7	5.4 3
			ce-M-8		128.213 7	1.30 7
			ce-K-11		138.563 7	13.2 7
			ce-L-10		167.094 7	1.92 10
			ce-L-11		187.837 7	2.15 11
			x ray L		7.18	51 7
			x ray Kα_2		49.7726 4	53.2 23
			x ray Kα_1		50.7416 4	94 4
			x ray Kβ		57.5	38.6 17
			γ 3		63.119 7	41.6 24
			γ 4		93.613 7	2.55 16
			γ 5		109.777 7	17.4 7
			γ 7		118.187 7	1.91 8
			γ 8		130.520 7	11.5 5

Table 6 (continued)

Isotope	CAS Registry No.	Half-life	Type of radiation	Energy, keV Max	Energy, keV Avg	Intensity, %
			γ 10		177.210 7	22.3 9
			γ 11		197.953 7	35.9 14
			γ 13		261.072 7	1.68 7
			γ 14		307.730 7	9.9 4
			27 weak γs			0.79
^{170}Tm (β$^-$ decay)	[13981-30-1]	128.6 d 3	Auger L		5.84	12.1 8
			ce-K-1		22.9228 6	4.7 3
			ce-L-1		73.7687 5	12.2 7
			ce-M-1		81.8570 5	3.00 18
			β$^-$ 1	883.7 10	290.5 4	24.0 10
			β$^-$ 2	968.0 10	323.1 4	76.0 10
			total β$^-$		315.3 4	100.0 15
			x ray L		7.42	4.0 6
			x ray Kα$_2$		51.3540 5	1.27 8
			x ray Kα$_1$		52.3889 5	2.25 14
			γ 1		84.2551 3	3.26 16
^{171}Er (β$^-$ decay)	[14391-45-8]	7.52 h 3	ce-L-2		2.269 8	6.4 10
			ce-M-1		2.718 6	70 4
			ce-NOP-1		4.553 6	23.3 14
			Auger L		5.67	47 3
			ce-M-2		10.078 8	1.42 22
			Auger K		40.9	2.6 10
			ce-K-4		52.231 4	40 3
			ce-K-5		57.266 6	1.72 10
			ce-K-6		64.627 4	5.8 4
			ce-L-4		101.505 4	6.5 5
			ce-L-5		106.540 6	1.82 11
			ce-M-4		109.314 4	1.45 10
			ce-L-6		113.901 4	5.4 4
			ce-M-6		121.710 4	1.32 8
			ce-K-16		248.901 18	1.06 6
			β$^-$ 1	577.3 12	177.4 5	2.17 11
			β$^-$ 2	1065.4 12	362.2 5	97 5
			β$^-$ 3	1485.3 12	534.7 5	2.30 20
			total β$^-$		359.5 5	103 5
			9 weak βs			1.24
			x ray L		7.18	14.7 20
			x ray Kα$_2$		49.7726 4	13.4 8
			x ray Kα$_1$		50.7416 4	23.6 14
			x ray Kβ		57.5	9.7 6
			γ 4		111.621 4	21.0 12
			γ 5		116.656 6	2.35 12
			γ 6		124.017 4	9.3 5
			γ 15		295.901 14	29.5 15
			γ 16		308.291 18	66 4
			60 weak γs			5.03
^{171}Tm (β$^-$ decay)	[14333-45-0]	1.92 yr 1	ce-K-1		5.386 7	1.02 20
			Auger L		5.84	1.17 17
			β$^-$ 1	30.0 10	7.6 3	2.1 4
			β$^-$ 2	96.7 10	25.2 3	97.9 4
			total β$^-$		24.8 3	100.0 6
			1 weak γ			0.15
^{171}Lu (EC decay)	[15752-27-9]	8.22 d 3	Auger L		5.84	61 4
			Auger K		42.2	4.0 16
			x ray L		7.42	20 3
			γ 1		9.15	31 3
			γ 2		19.384	89 15
			γ 3		27.126	2.4 6
			γ 4		46.516	3.2 5
			x ray Kα$_2$		51.3540 5	21.7 6
			x ray Kα$_1$		52.3889 5	38.3 9
			γ 5		55.679	6.2 9
			x ray Kβ		59.4	15.9 5
			γ 6		66.718	33 5
			γ 7		72.365	19 3
			γ 8		75.872	62 9
			γ 9		85.59	6.6 10
			γ 10		91.39	2.3 3
			γ 11		109.27	2.2 4
			γ 23		667.29	11.6 12
			γ 25		689.2	2.52 25
			γ 26		712.56	1.23 13
			γ 27		739.67	53 5
			γ 31		780.53	4.7 5
			γ 35		839.77	3.4 4
			γ 37		852.83	2.8 3
			31 weak γs			4.43

Table 6 (*continued*)

Isotope	CAS Registry No.	Half-life	Type of radiation	Energy, keV Max	Energy, keV Avg	Intensity, %
^{175}Yb (β^- decay)	[14041-44-2]	4.19 d 1	Auger L		6	3.1 4
			ce-K-1		50.489 4	3.7 5
			β^- 1	71.4 13	18.4 4	10.3 4
			β^- 2	353.9 13	101.6 5	3.30 20
			β^- 3	467.7 13	139.1 5	86.5 17
			total β^-		125.0 7	100.9 18
			1 weak β			0.80
			x ray L		7.66	1.10 18
			x ray Kα_2		52.9650 5	1.10 14
			x ray Kα_1		54.0698 5	1.93 24
			γ 1		113.803 4	1.91 25
			γ 5		282.517 14	3.1 4
			γ 6		396.322 20	6.5 8
			3 weak γs			0.53
^{177}Yb (β^- decay)	[14119-23-4]	1.9 h 1	Auger L		6	4.2 6
			ce-K-1		58.306 3	5.3 8
			ce-L-1		110.750 3	1.31 21
			β^- 1	152 7	40.4 20	3.9 5
			β^- 2	162 7	43.5 20	7.2 9
			β^- 3	243 7	67.2 22	1.28 15
			β^- 4	1104 7	376 3	2.2 5
			β^- 5	1243 7	432 3	13.0 24
			β^- 6	1271 7	444 3	9.0 14
			β^- 7	1393 7	494 3	62 5
			total β^-		419 5	100 6
			5 weak βs			1.38
			x ray L		7.66	1.5 3
			x ray Kα_2		52.9650 5	1.43 22
			x ray Kα_1		54.0698 5	2.5 4
			x ray Kβ		61.3	1.05 16
			γ 1		121.620 3	3.4 5
			γ 2		138.606 5	1.33 20
			γ 4		150.392 3	20 3
			γ 14		941.7 3	1.01 13
			γ 18		1080.1 3	5.5 7
			γ 24		1241.4 3	3.4 5
			19 weak γs			3.80
^{177}Lu (β^- decay)	[14265-75-9]	6.71 d 1	Auger L		6.18	8.9 6
			ce-K-2		47.601 3	5.2 4
			ce-L-2		101.681 3	7.0 5
			ce-M-2		110.351 3	1.74 12
			β^- 1	175.8 10	47.3 3	12.2 7
			β^- 2	384.1 10	111.3 4	9.1 12
			β^- 3	497.1 10	148.9 4	78.6 10
			1 weak β			0.05
			x ray L		7.9	3.3 4
			x ray Kα_2		54.6114 8	1.64 12
			x ray Kα_1		55.7902 8	2.87 20
			x ray Kβ		63.2	1.20 9
			γ 2		112.952 3	6.4 4
			γ 4		208.359 10	11.0 8
			4 weak γs			0.63
^{177}Lu (IT decay)		160.9 d 3	Auger L		6	24 5
			ce-K-1		52.52 4	1.3 5
			ce-K-2		58.306 3	9 4
			ce-K-3		83.851 5	3.3 12
			ce-L-1		104.96 4	13 5
			ce-K-4		108.554 8	2.8 11
			ce-L-2		110.750 3	2.3 8
			ce-M-1		113.34 4	3.5 13
			x ray L		7.66	8.3 18
			x ray Kα_2		52.9650 5	4.9 10
			x ray Kα_1		54.0698 5	8.6 18
			x ray Kβ		61.3	3.6 8
			γ 2		121.620 3	6.0 21
			γ 3		147.165 5	3.7 13
			γ 4		171.868 8	5.0 18
			γ 6		218.097 11	3.0 11
			γ 7		268.801 14	3.4 12
			γ 8		319.040 20	10 4
			γ 9		367.44 4	3.0 11
			γ 10		413.70 4	17 6
			2 weak γs			1.49
^{177}Lu (β^- decay)		160.9 d 3	Auger L		6.18	106 7
			ce-K-6		40.009 20	33 4
			Auger K		44.8	5.5 23
			ce-K-7		47.601 3	17.9 20

Table 6 (*continued*)

Isotope	CAS Registry No.	Half-life	Type of radiation	Energy, keV Max	Energy, keV Avg	Intensity, %
			ce-K-9		63.129 20	24 3
			ce-K-12		87.90 4	17.1 19
			ce-L-6		94.089 20	6.9 7
			ce-L-7		101.681 3	24 3
			ce-M-6		102.759 20	1.61 17
			ce-K-14		109.02 6	8.4 10
			ce-M-7		110.351 3	5.9 7
			ce-NOP-7		112.414 3	1.70 19
			ce-L-9		117.209 20	4.6 6
			ce-M-9		125.879 20	1.06 12
			ce-K-17		138.71 6	6.1 7
			ce-L-12		141.98 4	3.2 4
			ce-K-18		143.008 10	3.4 6
			ce-K-20		163.09 6	4.4 5
			ce-L-14		163.10 6	1.48 17
			ce-L-17		192.79 6	1.05 12
			ce-L-20		217.17 6	2.03 23
			β^- 1	151.9 10	40.5 3	79 5
			x ray L		7.9	39 5
			x ray $K\alpha_2$		54.6114 8	32.7 17
			γ 2		55.150 20	1.2 3
			x ray $K\alpha_1$		55.7902 8	57 3
			x ray $K\beta$		63.2	24.0 13
			γ 6		105.360 20	12.2 13
			γ 7		112.952 3	21.8 24
			γ 9		128.480 20	15.5 18
			γ 10		136.730 6	1.39 20
			γ 12		153.25 4	18.2 21
			γ 14		174.37 6	12.8 15
			γ 15		177.05 8	3.4 4
			γ 17		204.06 6	14.5 17
			γ 18		208.359 10	62 7
			γ 19		214.45 6	6.7 8
			γ 20		228.44 6	38 5
			γ 21		233.83 6	5.7 7
			γ 23		249.686 25	6.2 8
			γ 24		281.78 7	14.2 16
			γ 26		291.42 10	1.02 15
			γ 28		296.45 8	5.5 7
			γ 29		299.03 10	1.53 18
			γ 30		305.52 8	1.73 19
			γ 31		313.69 8	1.29 15
			γ 32		321.33 4	1.07 13
			γ 33		327.66 8	17.8 20
			γ 34		341.64 8	1.81 25
			γ 35		378.51 8	28 4
			γ 36		385.02 8	3.0 4
			γ 37		418.51 10	20.3 23
			γ 39		465.96 12	2.4 3
			12 weak γs			4.43
^{177}Ta (EC decay)	[15759-27-0]	56.6 h 1	Auger L		6.18	69 9
			Auger K		44.8	4.0 18
			ce-K-5		47.601 3	5.9 14
			ce-L-5		101.681 3	7.8 19
			ce-M-5		110.351 3	1.9 5
			x ray L		7.9	25 4
			x ray $K\alpha_2$		54.6114 8	24 4
			x ray $K\alpha_1$		55.7902 8	42 6
			x ray $K\beta$		63.2	17.5 25
			γ 5		112.952 3	7.2 18
			47 weak γs			2.00
^{179}Ta (EC decay)	[14391-27-6]	665 d 4	Auger L		6.18	58 8
			Auger K		44.8	2.7 12
			x ray L		7.9	21 4
			x ray $K\alpha_2$		54.6114 8	15.9 23
			x ray $K\alpha_1$		55.7902 8	28 4
			x ray $K\beta$		63.2	11.7 17
^{181}Hf (β^- decay)	[14900-21-1]	42.4 d 1	ce-M-1		1.19 10	2.64 14
			ce-NOP-1		3.33 10	1.36 7
			Auger L		6.35	40.3 24
			Auger K		46.2	1.5 7
			ce-K-3		65.604 20	21.4 13
			ce-K-4		68.834 20	8.5 6
			ce-K-5		69.44 4	2.5 4
			ce-L-3		121.338 20	25.0 15
			ce-L-4		124.568 20	1.71 12
			ce-M-3		130.312 20	6.2 4

Table 6 (*continued*)

Isotope	CAS Registry No.	Half-life	Type of radiation	Energy, keV Max	Energy, keV Avg	Intensity, %
			ce-NOP-3		132.454 20	1.81 11
			ce-K-8		414.58 20	1.54 10
			β^- 1	404 4	117.7 14	7 3
			β^- 2	408 4	118.9 14	93 3
			total β^-		118.8 14	100 5
			γ 1		3.90 10	4.00 20
			x ray L		8.15	15.7 18
			x ray Kα_2		56.2770 10	9.5 5
			x ray Kα_1		57.5320 10	16.5 8
			x ray Kβ		65.2	7.0 4
			γ 3		133.020 20	43.0 22
			γ 4		136.250 20	6.1 3
			γ 5		136.86 4	1.80 9
			γ 6		345.85 20	14.0 7
			γ 8		482.00 20	86 5
			4 weak γs			0.72
^{181}W (EC decay)	[15749-46-9]	121.2 d 3	ce-M-1		3.50 3	25 6
			ce-NOP-1		5.64 3	8.2 19
			Auger L		6.35	58 6
			Auger K		46.2	3.0 14
			x ray L		8.15	22 3
			x ray Kα_2		56.2770 10	19.0 20
			x ray Kα_1		57.5320 10	33 4
			x ray Kβ		65.2	14.0 15
			3 weak γs			1.11
^{182}Ta (β^- decay)	[13982-00-8]	115.0 d 2	Auger L		6.53	58 3
			ce-K-5		15.1558 5	15.8 10
			ce-K-6		30.5815 5	12.5 6
			ce-K-8		44.1473 5	5.1 4
			Auger K		45.7	1.6 8
			ce-L-3		53.6219 4	6.5 4
			ce-L-4		55.6502 4	6.5 4
			ce-M-3		62.9021 5	1.49 9
			ce-M-4		64.9304 5	1.48 10
			ce-L-5		72.5810 5	3.93 25
			ce-L-6		88.0067 5	32.4 14
			ce-M-6		97.286 9	8.1 4
			ce-NOP-6		99.5115 5	2.39 11
			ce-K-13		109.8698 6	1.63 12
			β^- 1	258 3	71.5 9	28.6 10
			β^- 2	324 3	91.8 10	2.70 20
			β^- 3	437 3	128.5 10	20 3
			β^- 4	480 3	142.8 11	2.3
			β^- 5	522 3	157.1 11	40 5
			β^- 6	590 3	180.6 11	5.0 20
			total β^-		126.2 11	100 7
			7 weak βs			1.44
			x ray L		8.4	25 3
			x ray Kα_2		57.9817 5	10.3 5
			x ray Kα_1		59.31820 10	17.9 7
			γ 3		65.72170 20	2.79 15
			x ray Kβ		67.2	7.6 4
			γ 4		67.75000 20	41.2 23
			γ 5		845.6808 3	2.65 15
			γ 6		100.1065 3	14.0 5
			γ 8		113.6723 4	1.92 11
			γ 11		152.4308 5	7.15 19
			γ 12		156.3874 5	2.72 9
			γ 13		179.3948 5	3.14 12
			γ 14		198.3530 6	1.54 8
			γ 15		222.1099 6	7.54 25
			γ 16		229.3220 9	3.63 13
			γ 17		264.0755 8	3.63 13
			γ 22		1001.68 7	2.09 11
			γ 25		1121.301 5	34.9 6
			γ 29		1189.050 5	16.4 4
			γ 30		1221.408 5	27.3 6
			γ 32		1231.016 5	11.55 25
			γ 33		1257.418 5	1.51 4
			γ 35		1289.156 5	1.41 4
			22 weak γs			5.69
^{183}Hf (β^- decay)	[15832-40-3]	64 min 1	ce-K-1		5.744 15	25 4
			Auger L		6.35	18 3
			Auger K		46.2	1.1 6
			ce-L-1		61.478 15	4.7 7
			ce-M-1		70.452 15	1.06 16
			β^- 1	466 19	138 7	3.2 3

Table 6 (*continued*)

Isotope	CAS Registry No.	Half-life	Type of radiation	Energy, keV Max	Energy, keV Avg	Intensity, %
			β^- 2	1152 19	394 8	68 6
			β^- 3	1550 19	557 8	27 3
			total β^-		430 9	98 7
			1 weak β			0.07
			x ray L		8.15	7.2 12
			x ray Kα_2		56.2770 10	7.0 11
			x ray Kα_1		57.5320 10	12.2 19
			x ray Kβ		65.2	5.2 8
			γ 1		73.160 15	38 6
			γ 7		315.860 20	1.22 18
			γ 9		397.860 20	2.9 5
			γ 10		459.070 20	27 4
			γ 15		783.73 3	65 7
			γ 19		1470.20 10	2.7 4
			15 weak γs			3.14
^{183}Ta (β^- decay)	[14683-36-4]	5.1 d 1	Auger L		6.53	116 7
			ce-K-4		13.3930 21	2.16 21
			ce-K-5		15.1860 21	8.5 4
			ce-L-1		28.8760 11	4.1 5
			ce-K-6		29.5540 21	6.0 5
			ce-K-7		32.409 3	1.21 9
			ce-K-8		32.956 3	3.8 4
			ce-L-2		34.3839 11	32 4
			ce-K-10		38.4060 21	34 3
			ce-K-11		40.201 3	1.83 12
			ce-L-3		40.4932 21	24 3
			ce-M-2		43.6641 11	7.3 9
			Auger K		45.7	3.8 17
			ce-M-3		49.7734 21	5.6 7
			ce-NOP-3		51.9980 21	1.69 19
			ce-L-5		72.6112 21	1.53 12
			ce-K-14		74.600 3	3.57 14
			ce-L-6		86.9792 21	16.0 13
			ce-L-8		90.381 3	1.25 13
			ce-K-16		91.817 5	9.0 5
			ce-K-17		92.794 4	4.53 20
			ce-L-10		95.8312 21	6.3 6
			ce-M-6		96.2594 21	4.0 4
			ce-NOP-6		98.4840 21	1.18 10
			ce-M-10		105.1114 21	1.44 14
			ce-L-16		149.242 5	1.51 7
			ce-K-26		176.536 5	8.53 17
			ce-L-26		233.961 5	1.35 5
			ce-K-31		284.468 8	1.33 5
			β^- 1	445 10	131 4	ca 4
			β^- 2	615 10	190 4	92 3
			β^- 3	656 10	204 4	1.7 9
			β^- 4	776 10	248 4	1.0 10
			total β^-		188 4	100 4
			x ray L		8.4	50 6
			γ 2		46.4837 10	4.9 6
			γ 3		52.5930 20	5.1 6
			x ray Kα_2		57.9817 5	24.5 11
			x ray Kα_1		59.31820 10	42.6 18
			x ray Kβ		67.2	18.1 8
			γ 5		84.7110 20	1.33 6
			γ 6		99.0790 20	6.6 5
			γ 10		107.9310 20	10.8 10
			γ 14		144.125 3	2.51 10
			γ 15		160.527 4	2.91 10
			γ 16		161.342 5	8.9 4
			γ 17		162.319 4	4.85 21
			γ 22		209.864 8	4.53 16
			γ 24		244.262 6	8.6 3
			γ 26		246.061 5	26.7 5
			γ 28		291.719 8	3.79 13
			γ 29		313.00 3	3.63 7
			γ 30		313.28 3	3.63 7
			γ 31		353.993 8	11.4 4
			17 weak γs			6.00
^{184}Hf (β^- decay)	[29687-28-3]	4.12 h 5	Auger L		6.35	177 20
			ce-L-1		29.72 20	77 10
			ce-L-2		32.22 20	40 4
			ce-L-3		36.22 20	87 22
			ce-M-1		38.69 20	17.3 22
			ce-M-2		41.19 20	9.0 12
			ce-M-3		45.19 20	22 6

Table 6 (*continued*)

Isotope	CAS Registry No.	Half-life	Type of radiation	Energy, keV Max	Energy, keV Avg	Intensity, %
			Auger K		46.2	1.0 5
			ce-K-4		71.68 20	21.4 19
			ce-L-4		127.42 20	22.8 21
			ce-M-4		136.39 20	5.7 5
			ce-NOP-4		138.53 20	1.65 15
			β^- 1	720 50	229 19	40 3
			β^- 2	890 50	290 20	16.0 20
			β^- 3	1110 50	378 20	41 12
			total β^-		302 21	97 13
			x ray L		8.15	69 10
			γ 1		41.40 20	9.9 12
			γ 2		43.90 20	6.1 8
			γ 3		47.90 20	1.2 3
			x ray Kα_2		56.2770 10	6.3 6
			x ray Kα_1		57.5320 10	10.9 10
			x ray Kβ		65.2	4.6 4
			γ 4		139.10 20	48 4
			γ 5		181.00 20	14.8 17
			γ 6		344.90 20	38 3
^{184}Ta (β^- decay)	[15701-21-0]	8.7 h 1	Auger L		6.53	70 4
			ce-K-4		21.75 4	4.3 3
			ce-K-5		41.667 20	17.8 10
			ce-L-1		43.23 4	1.6 4
			Auger K		45.7	1.3 6
			ce-L-2		51.60 4	35.0 13
			ce-M-2		60.88 4	8.8 4
			ce-NOP-2		63.10 4	2.54 13
			ce-L-4		79.17 4	1.73 13
			ce-K-9		91.75 6	2.9 8
			ce-L-5		99.092 20	34.5 19
			ce-M-5		108.372 20	8.7 5
			ce-NOP-5		110.597 20	2.55 14
			ce-K-17		183.32 4	4.0 3
			ce-L-17		240.75 4	1.84 14
			β^- 1	1120 30	380 11	15.0 6
			β^- 2	1170 30	399 11	81.3 20
			total β^-		394 11	100.2 21
			7 weak βs			3.86
			x ray L		8.4	30 4
			x ray Kα_2		57.9817 5	8.4 5
			x ray Kα_1		59.31820 10	14.6 7
			γ 2		63.70 4	1.77 9
			x ray Kβ		67.2	6.2 4
			γ 4		91.27 4	1.06 7
			γ 5		111.192 20	24.3 11
			γ 9		161.27 6	3.3 9
			γ 10		162	1.7 8
			γ 13		215.34 6	11.7 12
			γ 14		216.54 5	1.77 23
			γ 15		226.74 4	6.8 4
			γ 16		244.44 6	3.6 5
			γ 17		252.85 4	44 3
			γ 18		253	5.0 15
			γ 24		318.04 6	23.4 7
			γ 31		384.28 5	12.8 4
			γ 32		414.01 5	73.9 9
			γ 33		461.06 5	10.9 4
			γ 35		528.28 6	1.02 12
			γ 36		536.71 6	13.1 5
			γ 38		641.99 10	1.44 12
			γ 42		792.07 5	14.9 5
			γ 46		894.77 5	10.9 4
			γ 47		903.29 5	15.4 5
			γ 48		920.93 5	32.8 9
			γ 59		1110.12 8	2.29 8
			γ 62		1173.77 8	4.9 4
			44 weak γs			11.28
^{184}Re (EC decay)	[14983-46-1]	38.0 d 5	Auger L		6.53	81 5
			ce-K-1		41.682 7	12.5 6
			Auger K		45.7	4.0 18
			ce-L-1		99.107 7	24.3 12
			ce-M-1		108.387 7	6.1 3
			ce-NOP-1		110.612 7	1.80 9
			x ray L		8.4	35 4
			x ray Kα_2		57.9817 5	25.5 9
			x ray Kα_1		59.31820 10	44.3 15
			x ray Kβ		67.2	18.9 7

Table 6 (*continued*)

Isotope	CAS Registry No.	Half-life	Type of radiation	Energy, keV Max	Energy, keV Avg	Intensity, %
			γ 1		111.207 7	17.1 7
			γ 6		25.845 10	3.0 3
			γ 14		641.915 20	1.94 5
			γ 17		792.067 22	37.5 9
			γ 18		894.760 19	15.6 4
			γ 19		903.282 19	37.9 9
			23 weak γs			2.02
^{184}Re (EC decay)		165 d 5	Auger L		6.53	41.3 24
			ce-K-4		21.745 10	1.05 7
			ce-K-5		41.682 7	4.3 3
			ce-L-1		43.178 5	8.8 10
			Auger K		45.7	1.3 6
			ce-L-2		51.615 15	7.5 13
			ce-M-1		52.458 5	2.00 22
			ce-M-2		60.895 15	1.9 3
			ce-K-9		91.744 15	5.8 3
			ce-L-5		99.107 7	8.4 6
			ce-M-5		108.387 7	2.11 5
			ce-K-11		147.022 12	1.33 6
			ce-L-9		149.169 15	1.25 5
			x ray L		8.4	17.7 19
			γ 1		55.278 5	2.36 25
			x ray Kα_2		57.9817 5	8.5 3
			x ray Kα_1		59.31820 10	14.8 5
			x ray Kβ		67.2	6.29 21
			γ 5		111.207 7	5.9 4
			γ 9		161.269 15	6.64 22
			γ 10		215.326 12	2.84 12
			γ 11		216.547 12	9.6 4
			γ 12		226.748 10	1.51 6
			γ 14		252.845 10	10.9 5
			γ 16		318.008 10	5.88 19
			γ 18		384.250 12	3.20 11
			γ 20		536.674 15	3.37 11
			γ 24		792.067 22	3.77 14
			γ 26		894.760 19	2.81 13
			γ 27		903.282 19	3.82 14
			γ 28		920.933 21	8.3 3
			γ 32		1173.77 3	1.24 8
			18 weak γs			2.84
^{184}Re (IT decay)		165 d 5	Auger L		6.7	70 4
			ce-K-1		11.60 4	1.42 5
			ce-K-2		33.053 7	49.5 7
			Auger K		47	2.1 10
			ce-L-1		70.75 4	50.7 6
			ce-M-1		80.35 4	16.9 5
			ce-NOP-1		82.65 4	5.52 16
			ce-L-2		92.202 7	9.11 24
			ce-M-2		101.797 7	2.09 8
			x ray L		8.65	32 4
			x ray Kα_2		59.7179 6	14.1 4
			x ray Kα_1		61.1403 6	24.4 6
			x ray Kβ		69.3	10.4 3
			γ 2		104.729 7	13.3 4
^{185}W (β^- decay)	[14932-41-3]	75.1 d 3	β^- 1	432.5 10	126.8 4	100
			1 weak β			0.08
			1 weak γ			0.02
^{187}W (β^- decay)	[14983-48-3]	23.9 h 1	ce-K-7		0.384 10	9.1 6
			ce-M-1		4.2183 4	1.67 22
			Auger L		6.7	19.2 13
			Auger K		47	1.1 5
			ce-L-7		59.533 10	1.65 9
			ce-K-15		62.544 10	17.6 7
			ce-L-15		121.693 10	2.99 15
			β^- 1	540.0 18	163.1 7	4.33 20
			β^- 2	627.1 18	193.6 7	58.9 22
			β^- 3	687.4 18	215.2 7	1.56 7
			β^- 4	694.5 18	217.8 7	7.2 3
			β^- 5	1178.7 18	401.9 7	1.7 10
			β^- 6	1312.9 18	457.3 8	25.1 24
			total β^-		263.2 8	100 4
			11 weak βs			1.45
			x ray L		8.65	8.6 10
			x ray Kα_2		59.7179 6	7.7 4
			x ray Kα_1		61.1403 6	13.3 6
			x ray Kβ		69.3	5.69 24
			γ 7		72.060 10	11.9 5

Table 6 (*continued*)

Isotope	CAS Registry No.	Half-life	Type of radiation	Energy, keV Max	Energy, keV Avg	Intensity, %
			γ 15		134.220 10	9.4 4
			γ 30		479.530 10	23.4 10
			γ 34		551.550 10	5.45 22
			γ 41		618.370 10	6.7 3
			γ 42		625.520 10	1.17 5
			γ 47		685.810 10	29.3 12
			γ 52		772.880 20	4.42 18
			57 weak γs			2.36
^{188}W (β^- decay)	[24421-27-0]	69.4 d 5	β^- 1	349 3	99.7 10	99.00 9
			4 weak βs			1.00
			7 weak γs			0.75
^{188}Re (β^- decay)	[14378-26-8]	16.98 h 2	Auger L		6.88	6.5 4
			ce-K-1		81.169 20	4.9 3
			ce-L-1		142.072 20	5.5 3
			ce-M-1		151.991 20	1.40 8
			β^- 1	1486.7 9	527.5 4	1.60 14
			β^- 2	1964.7 9	728.6 4	25.1 12
			β^- 3	2119.7 9	795.1 4	71.6 13
			total β^-		764.4 4	100.0 18
			16 weak βs			1.72
			x ray L		9	3.1 4
			x ray Kα_2		61.4867 7	1.37 8
			x ray Kα_1		63.0005 7	2.36 14
			x ray Kβ		71.4	1.01 6
			γ 1		155.040 20	14.9 7
			γ 5		477.96 3	1.04 5
			γ 10		633.00 5	1.25 13
			39 weak γs			1.95
^{188}Ir (EC decay)	[15752-22-4]	41.5 h 5	Auger L		6.88	69 5
			Auger K		48.3	3.6 17
			ce-K-6		81.18 4	9.7 10
			ce-L-6		142.08 4	11.0 11
			ce-M-6		152.00 4	2.8 3
			2 weak βs			0.33
			x ray L		9	33 4
			x ray Kα_2		61.4867 7	25.3 11
			x ray Kα_1		63.0005 7	43.7 18
			x ray Kβ		71.4	18.7 9
			γ 6		155.05 4	30 3
			γ 17		322.91 4	1.62 15
			γ 34		477.99 4	14.7 6
			γ 48		633.02 10	18 3
			γ 49		634.91 15	5.0 8
			γ 55		672.50 5	1.44 12
			γ 71		824.34 8	1.03 10
			γ 73		829.42 6	5.1 5
			γ 92		1017.63 6	1.06 10
			γ 94		1096.54 6	1.46 14
			γ 100		1174.59 10	1.32 15
			γ 103		1209.77 6	7.0 7
			γ 117		1435.42 15	1.48 14
			γ 119		1452.28 15	1.06 10
			γ 120		1457.19 15	1.75 17
			γ 122		1465.24 15	1.35 15
			γ 127		1574.48 15	2.63 24
			γ 132		1704.9	1.04 16
			γ 133		1715.75 15	6.1 6
			γ 149		1944.08 20	3.9 4
			γ 154		2049.78 20	5.0 4
			γ 155		2059.65 20	7.0 6
			γ 157		2096.9 4	5.7 8
			γ 158		2099.1 4	4.8 6
			γ 164		2193.7 4	2.0 4
			γ 165		2214.59 20	18.7 16
			165 weak γs			23.31
^{188}Pt (EC decay)	[14922-70-4]	10.2 d 3	Auger L		7	93 6
			ce-K-5		22.26 5	1.86 20
			ce-L-1		28.56 5	5.9 7
			ce-M-1		38.81 5	1.37 17
			ce-L-2		41.43 5	14.9 17
			Auger K		49.6	4.0 19
			ce-M-2		51.68 5	3.8 5
			ce-NOP-2		54.16 5	1.13 13
			ce-K-7		64.24 10	4.6 3
			ce-K-8		111.48 10	16.9 11
			ce-K-9		118.94 10	14.6 11
			ce-L-8		174.17 10	2.77 17

Table 6 (continued)

Isotope	CAS Registry No.	Half-life	Type of radiation	Energy, keV Max	Energy, keV Avg	Intensity, %
			ce-L-9		181.63 10	2.38 17
			x ray L		9.18	46 5
			x ray Kα_2		63.2867 7	29.0 8
			x ray Kα_1		64.8956 7	49.9 13
			x ray Kβ		73.6	21.5 7
			γ 7		140.35 10	2.33 15
			γ 8		187.59 10	19.4 12
			γ 9		195.05 10	18.6 12
			γ 14		381.43 10	7.5 5
			γ 15		423.34 10	4.4 3
			γ 16		478.3 5	1.8 3
			10 weak γs			2.59
^{191}Os (β^- decay)	[14119-24-5]	15.4 d 1	Auger L		7	87 7
			ce-L-1		28.431 10	71 6
			ce-M-1		38.676 10	21.7 19
			Auger K		49.6	2.2 11
			ce-K-4		53.320 5	58 6
			ce-L-4		116.012 5	12.0 12
			ce-M-4		126.257 5	2.8 3
			β^- 1	139 3	37 1	100
			x ray L		9.18	43 5
			x ray Kα_2		63.2867 7	16.1 16
			x ray Kα_1		64.8956 7	28 3
			x ray Kβ		73.6	12.0 12
			γ 4		129.431 5	25.7 24
			3 weak γs			0.03
^{191}Os (IT decay)		13.10 h 5	ce-K-1		0.509 10	8.7 12
			Auger L		6.88	49 7
			ce-L-1		61.412 10	65 9
			ce-M-1		71.331 10	19 3
			ce-NOP-1		73.726 10	7.2 10
			x ray L		9	23 4
			x ray Kα_2		61.4867 7	2.4 4
			x ray Kα_1		63.0005 7	4.2 6
			x ray Kβ		71.4	1.8 3
			1 weak γ			0.07
^{192}Ir (EC decay)	[14694-69-0]	74.02 d 18	Auger L		6.88	3.13 19
			x ray L		9	1.47 15
			x ray Kα_2		61.4867 7	1.17 3
			x ray Kα_1		63.0005 7	2.02 5
			γ 3		205.7955 5	3.29 11
			γ 11		484.5779 13	3.16 8
			11 weak γs			2.13
^{192}Ir (β^- decay)		74.02 d 18	Auger L		7.4	7.5 5
			ce-K-2		217.5634 11	1.920 10
			ce-K-3		230.0621 11	1.786 25
			ce-K-4		238.1132 11	4.47 14
			ce-L-4		302.6281 9	1.95 6
			ce-K-6		389.6767 14	1.02 4
			β^- 1	256 4	70.8 12	5.59 8
			β^- 2	536 4	161.2 14	41.35 19
			β^- 3	672 4	208.9 15	48.0 6
			total β^-		179.8 16	95.0 7
			2 weak βs			0.10
			x ray L		9.44	4.0 5
			x ray Kα_2		65.1220 20	2.63 7
			x ray Kα_1		66.8320 20	4.52 10
			x ray Kβ		75.7	1.97 6
			γ 2		295.9582 8	28.96 12
			γ 3		308.4569 8	29.67 9
			γ 4		316.5080 8	82.84 9
			γ 6		468.0715 12	47.8 6
			γ 8		588.5851 16	4.52 8
			γ 10		604.4146 16	8.18 16
			γ 11		612.4657 16	5.33 10
			8 weak γs			1.24
^{192}Ir (IT decay)		241 yr 9	Auger L		7	50 3
			ce-L-1		148 5	74.5 23
			ce-MNO-1		158 5	24.6
			x ray L		9.18	24.8 24
			1 weak γ			0.13
^{193}Os (β^- decay)	[16057-77-5]	30.5 h 4	Auger L		7	20.0 20
			ce-K-7		30.882 10	2.69 17
			ce-L-3		59.593 7	17 3
			ce-K-9		62.781 7	8.1 5
			ce-M-3		69.838 7	4.1 7
			ce-NOP-3		72.322 7	1.25 20

Table 6 (continued)

Isotope	CAS Registry No.	Half-life	Type of radiation	Energy, keV Max	Energy, keV Avg	Intensity, %
			ce-L-9		125.473 7	1.55 10
			β^- 1	575 5	174.7 18	2.40 20
			β^- 2	672 5	208.9 18	7.9 4
			β^- 3	952 5	313.0 20	2.0 3
			β^- 4	993 5	328.8 20	12.5 9
			β^- 5	1059 5	354.4 20	20 4
			β^- 6	1132 5	383.1 20	53 5
			total β^-		345.2 20	100 7
			10 weak βs			2.56
			x ray L		9.18	9.8 13
			x ray Kα_2		63.2867 7	3.65 17
			x ray Kα_1		64.8956 7	6.3 3
			γ 3		73.012 7	3.2 5
			x ray Kβ		73.6	2.71 13
			γ 9		138.892 7	4.3 3
			γ 21		280.43 3	1.24 8
			γ 26		321.56 3	1.28 8
			γ 35		387.46 4	1.26 8
			γ 41		460.49 3	3.95 25
			γ 50		557.36 8	1.30 13
			60 weak γs			3.72
^{193}Ir (IT decay)	[13967-67-4]	10.60 d 11	Auger L		7	45.9 25
			ce-L-1		66.85 4	68.2 21
			ce-M-1		77.10 4	23.5 7
			ce-NOP-1		79.58 4	7.94 24
			x ray L		9.18	22.6 22
^{194}Os (β^- decay)	[15766-57-1]	6.0 yr 2	Auger L		7	16 4
			ce-L-1		29.7 3	23 5
			ce-M-1		39.9 3	5.3 12
			β^- 1	53.9 20	13.8 6	33 8
			β^- 2	97.0 20	25.2 6	67 8
			total β^-		21.4 7	100 12
			1 weak β			0.05
			x ray L		9.18	7.6 19
			γ 1		43.1 3	2.3 5
^{194}Ir (β^- decay)	[14158-35-1]	19.15 h 3	β^- 1	983.9 20	324.7 8	1.76 8
			β^- 2	1629.0 20	583.9 9	1.34 13
			β^- 3	1922.6 20	707.2 9	9.2 6
			β^- 4	2251.0 20	847.5 9	85.5 20
			total β^-		808.2 9	100.0 21
			19 weak βs			2.25
^{195}Pt (IT decay)	[14191-88-9]	4.02 d	Auger L		7.24	137 10
			ce-L-3		17.01 9	69 6
			ce-K-4		20.505 20	66 6
			ce-M-3		27.59 9	15.7 13
			ce-NOP-3		30.17 9	7.4 6
			Auger K		51	3.0 11
			ce-K-5		51.11 20	13.0 10
			ce-K-6		51.395 20	1.33 11
			ce-L-4		85.020 20	11.6 10
			ce-M-4		95.604 20	2.74 22
			ce-L-5		115.62 20	61 5
			ce-L-6		115.910 20	2.71 22
			ce-M-5		126.20 20	19.0 15
			ce-NOP-5		128.78 20	6.0 5
			x ray L		9.44	74 9
			γ 3		30.89 9	2.28 12
			x ray Kα_2		65.1220 20	22.4 16
			x ray Kα_1		66.8320 20	38 3
			x ray Kβ		75.7	16.7 12
			γ 4		98.900 20	11.4 9
			γ 6		129.790 20	2.83 21
			6 weak γs			0.21
^{195}Au (EC decay)	[14320-93-5]	133 d 2	Auger L		7.24	105 8
			ce-L-1		16.996 6	22.8 13
			ce-K-2		20.485 20	63 5
			ce-M-1		27.580 6	5.2 3
			ce-NOP-1		30.154 6	2.44 14
			Auger K		51	3.9 14
			ce-L-2		85.000 20	11.1 9
			ce-M-2		95.584 20	2.62 21
			x ray L		9.44	57 7
			x ray Kα_2		65.1220 20	29.0 18
			x ray Kα_1		66.8320 20	50 3
			x ray Kβ		75.7	21.7 13
			γ 2		98.880 20	10.9 9
			4 weak γs			1.59

Table 6 (*continued*)

Isotope	CAS Registry No.	Half-life	Type of radiation	Energy, keV Max	Energy, keV Avg	Intensity, %
^{196}Ir (β^- decay)	[14621-82-0]	52 s	β^- 1	1300 50	450 21	1.20 20
			β^- 2	2090 50	776 22	15 3
			β^- 3	2860 50	1083 22	2.2 6
			β^- 4	3220 50	1270 22	80 4
			total β^-		1175 23	99 5
			3 weak βs			0.80
			γ 1		332.8 3	4.3 4
			γ 2		355.4 3	18.9 12
			γ 3		446.6 3	4.5 5
			γ 4		779.4 3	10.4 8
			8 weak γs			3.01
^{196}Ir (β^- decay)		1.40 h 2	Auger L		7.24	42 5
			ce-K-1		24.91 20	11.4 15
			ce-L-1		89.42 20	45 6
			ce-M-1		100.00 20	11.5 15
			ce-NOP-1		102.58 20	3.5 5
			ce-K-3		277.51 20	3.81 21
			ce-K-4		315.11 20	1.16 6
			ce-L-3		342.02 20	1.41 8
			ce-K-6		368.71 20	2.26 11
			ce-K-7		443.01 20	1.60 7
			ce-K-10		568.91 20	1.00 6
			β^- 1	330 130	90 40	9.5 3
			β^- 2	470 130	140 50	4.7 3
			β^- 3	740 130	230 50	5.40 20
			β^- 4	1160 130	390 50	80 4
			2 weak βs			1.58
			x ray L		9.44	23 4
			x ray Kα_2		65.1220 20	5.9 5
			x ray Kα_1		66.8320 20	10.1 8
			x ray Kβ		75.7	4.4 4
			γ 1		103.30 20	16.3 20
			γ 2		340.7 4	1.54 20
			γ 3		355.90 20	94 5
			γ 4		393.50 20	97 4
			γ 5		420.9 3	2.50 13
			γ 6		447.10 20	94 4
			γ 7		521.40 20	96 3
			γ 9		633.5 3	1.10 6
			γ 10		647.30 20	91 4
			γ 11		693.90 20	4.2 4
			γ 13		727.30 20	2.59 13
			γ 15		835.60 20	6.3 3
			γ 24		1482.5 4	2.30 21
			11 weak γs			3.67
^{196}Au (EC decay)	[14914-16-0]	6.183 d 10	Auger L		7.24	51 4
			Auger K		51	2.9 10
			ce-K-2		254.64 5	1.17 5
			ce-K-3		277.34 5	3.52 12
			ce-L-3		341.85 5	1.30 5
			x ray L		9.44	27 4
			x ray Kα_2		65.1220 20	21.5 9
			x ray Kα_1		66.8320 20	36.9 14
			x ray Kβ		75.7	16.0 7
			γ 2		333.03 5	22.9 6
			γ 3		355.73 5	86.9 10
			13 weak γs			0.67
^{196}Au (β^- decay)		6.183 d 10	β^-	258 4	71.3 12	7.5 11
			γ 1		426.10 8	7.20 13
^{196}Au (IT decay)		9.7 h 1	ce-L-1		5.567 10	24.6 25
			Auger L		7.42	170 13
			ce-L-2		16.357 10	6.8 7
			ce-M-1		16.495 10	6.2 7
			ce-M-2		27.285 10	1.56 16
			Auger K		52.4	3.2 15
			ce-K-5		56.97 3	3.2 11
			ce-K-6		67.085 20	14.9 17
			ce-L-4		70.307 20	70 7
			ce-M-4		81.235 20	19.9 20
			ce-NOP-4		83.901 20	6.4 7
			ce-K-7		87.645 20	10.7 13
			ce-K-8		94.185 20	29 3
			ce-K-9		107.55 3	30 4
			ce-L-6		133.457 20	25 3
			ce-M-6		144.385 20	6.4 7
			ce-NOP-6		147.051 20	2.12 24
			ce-L-7		154.017 20	1.76 20

Table 6 (*continued*)

Isotope	CAS Registry No.	Half-life	Type of radiation	Energy, keV Max	Energy, keV Avg	Intensity, %
			ce-L-8		160.557 20	57 6
			ce-M-8		171.485 20	17.0 17
			ce-L-9		173.92 3	7.5 9
			ce-NOP-8		174.151 20	5.7 6
			ce-MNO-9		184.85 3	2.24 24
			x ray L		9.7	95 12
			x ray Kα_2		66.9895 8	24.6 15
			x ray Kα_1		68.8037 8	42.0 25
			x ray Kβ		78	18.4 12
			γ 5		137.69 3	1.3 5
			γ 6		147.810 20	42 5
			γ 7		168.370 20	7.6 9
			γ 9		188.27 3	37 4
			γ 10		285.49 7	4.3 6
			γ 11		316.19 5	2.9 4
			5 weak γs			0.96
^{196}Tl (EC decay)	[18724-77-1]	1.84 h 3	Auger L		7.6	49 5
			Auger K		53.8	2.3 14
			ce-K-3		342.60 20	2.3 3
			ce-L-11		739.2 5	11.5 20
			β^+ 1	2441.8 5	1102.18 23	1.12 24
			β^+ 2	3052.30 20	1375.81 9	10.9 14
			total β^+		1287.62 16	14.0 15
			7 weak βs			1.98
			x ray L		10	30 4
			x ray Kα_2		68.8950 20	19.2 11
			x ray Kα_1		70.8190 20	32.6 18
			x ray Kβ		80.3	14.6 18
			γ 2		354.5 5	1.25 24
			γ 3		425.70 20	84 10
			γ 6		610.5 5	11.9 17
			γ 7		635.2 5	9.8 14
			γ 8		705.0 10	1.34 24
			γ 9		713.6 10	1.25 24
			γ 11		754.0 5	1.44 25
			γ 12		778.4 5	1.15 23
			γ 19		964.6 10	3.6 6
			γ 22		1036.2 10	2.6 4
			γ 29		1289.1 15	1.15 23
			γ 31		1350.0 5	1.15 23
			γ 32		1389.0 5	2.5 4
			γ 35		1434.2 20	1.44 25
			γ 37		1495.8 5	8.2 13
			γ 39		1553.0 7	4.8 7
			γ 40		1586.7 10	2.3 4
			γ 41		1621.4 20	4.9 8
			γ 42		1696.7 20	3.0 5
			γ 43		1775.5 10	2.8 5
			γ 44		1844.9 20	1.9 3
			γ 46		2011.3 25	3.7 6
			γ 47		2049.2 20	1.15 23
			γ 48		2067.4 25	1.06 23
			γ 49		2102.1 25	1.15 23
			γ 50		2127.8 25	2.8 5
			γ 52		2212.0 20	3.4 6
			γ 53		2227.7 25	1.25 24
			γ 54		2392.7 20	1.7 4
			25 weak γs			10.07
^{197}Pt (β^- decay)	[15735-74-7]	18.3 h 3	Auger L		7.42	37 6
			ce-L-1		63.00 5	55 9
			ce-M-1		73.93 5	13.3 20
			ce-NOP-1		76.59 5	4.2 7
			ce-K-2		110.712 10	3.5 4
			β^- 1	450.2 6	132.13 20	8.2 8
			β^- 2	641.7 6	197.67 22	81 3
			β^- 3	719.0 6	225.32 22	11 3
			total β^-		195.23 22	100 5
			x ray L		9.7	21 4
			x ray Kα_1		68.8037 8	1.69 19
			γ 1		77.35 5	17.1 25
			γ 2		191.437 10	3.7 4
			1 weak γ			0.23
^{197}Pt (IT decay)		94.4 min 8	Auger L		7.24	90 8
			ce-L-1		39.07 5	72 8
			ce-M-1		49.65 5	18.4 19
			Auger K		51	1.8 7
			ce-NOP-1		52.23 5	5.6 6

Table 6 (continued)

Isotope	CAS Registry No.	Half-life	Type of radiation	Energy, keV Max	Energy, keV Avg	Intensity, %
			ce-K-2		268.11 20	49 5
			ce-L-2		332.62 20	27 3
			ce-M-2		343.20 20	7.3 8
			ce-NOP-2		345.78 20	2.3 3
			x ray L		9.44	48 7
			γ 1		52.95 5	1.07 12
			x ray Kα_1		65.1220 20	13.6 14
			x ray Kα_2		66.8320 20	23.3 24
			x ray Kβ		75.7	10.1 11
			γ 2		346.50 20	11.1 12
^{197}Pt (IT decay)		94.4 min	Auger L		7.24	90 8
			ce-L-1		39.07 5	72 8
			ce-M-1		49.65 5	18.4 19
			Auger K		51	1.8 7
			ce-NOP-1		52.23 5	5.6 6
			ce-K-2		268.11 20	49 5
			ce-L-2		332.62 20	27 3
			ce-M-2		343.20 20	7.3 8
			ce-NOP-2		345.78 20	2.3 3
			x ray L		9.44	48 7
			γ 1		52.95 5	1.07 12
			x ray Kα_2		65.1220 20	13.6 14
			x ray Kα_1		66.8320 20	23.3 24
			x ray Kβ		75.7	10.1 11
			γ 2		346.50 20	11.1 12
^{197}Pt (β^- decay)		94.4 min 8	Auger L		7.42	2.0 6
			ce-L-2		115.6472 4	2.3 8
			β^- 1	709.6 8	221.75 25	3.3 10
			x ray L		9.7	1.1 3
			γ 4		279	2.3 8
			4 weak γs			0.15
^{198}Au (β^- decay)	[10043-49-9]	2.696 d 2	Auger L		7.6	2.08 16
			ce-K-1		328.7021 14	2.87 9
			ce-L-1		396.9651 15	1.02 3
			β^- 1	286.1 10	79.8 3	1.30 10
			β^- 2	962.0 10	315.1 4	98.70 10
			total β^-		312.1 4	100.02 15
			1 weak β			0.03
			x ray L		10	1.27 15
			x ray Kα_1		70.8190 20	1.37 6
			γ 1		411.8044 11	95.50 10
			γ 2		675.8875 19	1.06 5
			1 weak γ			0.23
^{198}Tl (EC decay)	[15743-50-7]	5.3 h 5	Auger L		7.6	49 4
			Auger K		53.8	2.7 16
			14 weak βs			0.77
			x ray L		10	30 4
			x ray Kα_2		68.8950 20	22.3 6
			x ray Kα_1		70.8190 20	38.0 9
			x ray Kβ		80.3	16.7 5
			γ 12		411.8044 11	82 10
			γ 18		511.0 3	1.05 19
			γ 24		596.80 20	1.00 15
			γ 27		636.70 20	10.1 12
			γ 29		675.8875 19	10.9 10
			γ 35		759.6 3	1.45 19
			γ 39		798.7 3	1.07 13
			γ 51		1007.6 3	2.7 4
			γ 56		1087.690 3	2.4 4
			γ 63		1200.60 20	9.7 14
			γ 65		1219.2 3	1.08 14
			γ 69		1312.20 20	4.7 7
			γ 76		1420.6 3	8.0 12
			γ 77		1435.4 3	3.5 6
			γ 78		1447.0 3	4.3 6
			γ 82		1489.6 3	2.6 4
			γ 86		1593.60 20	2.1 3
			γ 91		1659.1 3	1.69 23
			γ 95		1720.8 3	2.8 4
			γ 100		1832.6 3	4.3 6
			γ 105		1899.3 3	2.2 3
			γ 109		2040.20 20	8.4 12
			γ 117		2190.5 3	2.7 4
			γ 135		2486.2 3	1.12 15
			139 weak γs			29.15
^{198}Tl (EC decay)		1.87 h 3	Auger L		7.6	25.8 20
			Auger K		53.8	1.4 9

Table 6 (continued)

Isotope	CAS Registry No.	Half-life	Type of radiation	Energy, keV Max	Energy, keV Avg	Intensity, %
			6 weak βs			0.93
			x ray L		10	15.8 18
			x ray Kα$_2$		68.8950 20	11.8 3
			x ray Kα$_1$		70.8190 20	20.0 5
			x ray Kβ		80.3	8.80 24
			γ 4		215.6 3	1.17 22
			γ 5		226.2 3	5.0 9
			γ 6		227.5 3	1.3 3
			γ 8		274.0 3	1.41 23
			γ 11		390.4 3	1.56 23
			γ 12		411.8044 11	53 7
			γ 14		423.3 4	1.00 17
			γ 15		441.8 3	2.0 4
			γ 16		489.6 3	4.2 6
			γ 17		519.2 3	3.3 5
			γ 21		587.20 20	49 4
			γ 23		636.70 20	53 7
			γ 26		767.3 3	1.04 17
			18 weak γs			7.9
^{198m}Tl (IT decay)		1.87 h 3	ce-L-1		7.75 10	2.7 4
			Auger L		7.78	30 4
			Auger K		55.2	1.1 5
			ce-K-2		174.1 3	1.4 3
			ce-K-3		175.4 3	20 5
			ce-K-4		197.27 20	11.1 18
			ce-L-3		245.6 3	19 4
			ce-M-3		257.2 3	5.4 12
			ce-NOP-3		260.1 3	1.8 4
			ce-L-4		267.45 20	1.9 3
			x ray L		10.3	20 3
			x ray Kα$_2$		70.8319 9	9.0 13
			x ray Kα$_1$		72.8715 9	15.2 22
			x ray Kβ		82.6	6.7 10
			γ 2		259.6 3	2.9 6
			γ 3		260.9 3	1.3 3
			γ 4		282.80 20	28 5
			1 weak γ			0.03
^{199}Au (β$^-$ decay)	[14391-11-8]	3.139 d 7	Auger L		7.6	20.9 15
			ce-L-1		34.986 7	2.66 11
			ce-K-2		75.273 7	10.5 5
			ce-K-3		125.099 7	5.5 3
			ce-L-2		143.536 7	17.1 5
			ce-M-2		154.813 7	4.5 4
			ce-NOP-2		157.575 7	1.34 6
			ce-L-3		193.362 7	1.02 4
			β$^-$ 1	244.8 20	67.3 6	18.9 6
			β$^-$ 2	294.6 20	82.4 6	66.4 10
			β$^-$ 3	453.0 20	132.9 6	14.7 5
			total β$^-$		87.0 7	100.0 13
			x ray L		10	12.8 14
			x ray Kα$_2$		68.8950 20	4.47 20
			x ray Kα$_1$		70.8190 20	7.6 4
			x ray Kβ		80.3	3.34 15
			γ 2		158.375 7	36.9 11
			γ 3		208.201 7	8.4 3
			1 weak γ			0.33
^{200}Au (β$^-$ decay)	[20091-45-6]	48.4 mo	β$^-$ 1	630 60	193 22	3.9 3
			β$^-$ 2	670 60	206 22	10.9 8
			β$^-$ 3	1890 60	690 30	4.7 7
			β$^-$ 4	2260 60	850 30	79.3
			total β$^-$		740 40	100 4
			11 weak βs			1.80
			γ 4		367.90 5	19 3
			γ 23		1225.41 10	10.7 16
			γ 25		1262.89 10	3.1 5
			34 weak γs			2.75
^{200}Tl (EC decay)	[15720-55-5]	26.1 h 1	Auger L		7.6	53 5
			Auger K		53.8	2.8 17
			ce-K-14		284.840 10	3.41 11
			ce-L-14		353.103 10	1.37 5
			2 weak βs			0.35
			x ray L		10	32 4
			x ray Kα$_2$		68.8950 20	23.3 8
			x ray Kα$_1$		70.8190 20	39.6 13
			x ray Kβ		80.3	17.4 7
			γ 14		367.942 10	87.2 4
			γ 32		579.28 9	13.8 7

Table 6 (continued)

Isotope	CAS Registry No.	Half-life	Type of radiation	Energy, keV Max	Energy, keV Avg	Intensity, %
			γ 36		628.63 13	1.00 8
			γ 37		661.35 9	2.28 13
			γ 40		701.56 14	1.29 10
			γ 45		787.1 3	1.03 18
			γ 46		828.32 10	10.8 7
			γ 49		886.15 10	2.02 13
			γ 60		1205.70 9	29.9 18
			γ 61		1225.50 9	3.36 21
			γ 64		1273.52 10	3.31 4
			γ 68		1362.9 3	3.4 4
			γ 70		1407.64 11	1.45 13
			γ 72		1514.90 10	4.0 3
			γ 76		1604.50 14	1.17 10
			81 weak γs			10.88
^{200}Pb (EC decay)	[16645-99-1]	21.5 h 4	Auger L		7.78	81 6
			ce-K-2		24.01 3	2.7 4
			Auger K		55.2	3.6 17
			ce-K-3		56.76 3	8.4 6
			ce-K-4		62.09 3	12.7 6
			ce-L-3		126.94 3	1.45 10
			ce-L-4		132.27 3	25.7 12
			ce-M-4		143.92 3	6.7 4
			ce-NOP-4		146.77 3	2.17 11
			ce-K-8		150.09 3	2.78 14
			ce-K-9		171.64 3	1.26 11
			ce-K-10		182.85 3	1.80 11
			x ray L		10.3	54 6
			x ray Kα_2		70.8319 9	30.6 10
			x ray Kα_1		72.8715 9	51.9 16
			x ray Kβ		82.6	22.9 8
			γ 3		142.29 3	3.16 19
			γ 4		147.62 3	37.7 14
			γ 8		235.62 3	4.30 17
			γ 9		257.17 3	4.46 17
			γ 10		268.38 3	3.96 20
			γ 11		289.11 10	1.1 4
			γ 12		289.94 10	1.7 4
			γ 17		450.53 5	3.33 8
			12 weak γs			2.57
^{201}Tl (EC decay)	[15064-65-0]	3.044 d 9	ce-NOP-1		0.78 4	38 22
			Auger L		7.6	76 6
			ce-L-2		15.76 3	11.4 6
			ce-L-3		17.35 3	9.1 5
			ce-MNO-2		27.04 3	3.63 16
			ce-MNO-3		28.63 3	2.85 12
			ce-K-4		52.24 4	7.5 4
			Auger K		53.8	3.3 20
			ce-K-6		84.33 7	15.5 6
			ce-L-4		120.50 4	1.27 7
			ce-L-6		152.59 7	2.62 9
			γ 1		158 4	38 22
			x ray L		10	47 6
			x ray Kα_2		68.8950 20	27.4 9
			x ray Kα_1		70.8190 20	46.6 14
			x ray Kβ		80.3	20.5 7
			γ 4		135.34 4	2.65 10
			γ 6		167.43 7	10.00 17
			3 weak γs			0.78
^{201}Pb (EC decay)	[17239-87-1]	9.4 h 2	Auger L		7.78	56 5
			Auger K		55.2	3.0 14
			ce-K-13		245.64 5	8.5 6
			ce-K-16		275.74 5	1.98 10
			ce-L-13		315.82 5	2.25 15
			1 weak β			0.04
			x ray L		10.3	37 5
			x ray Kα_2		70.8319 9	25.5 15
			x ray Kα_1		72.8715 9	43.2 25
			x ray Kβ		82.6	19.1 12
			γ 13		331.17 5	79 5
			γ 16		361.27 5	9.9 5
			γ 19		406.03 7	2.01 12
			γ 29		584.55 5	3.56 16
			γ 32		692.37 8	4.27 16
			γ 36		767.28 8	3.16 16
			γ 38		803.66 7	1.51 12
			γ 39		826.21 8	2.36 13
			γ 40		907.56 11	5.7 4

Table 6 (*continued*)

Isotope	CAS Registry No.	Half-life	Type of radiation	Energy, keV Max	Energy, keV Avg	Intensity, %
			γ 41		945.96 8	7.4 6
			γ 49		1069.95 8	1.14 11
			γ 51		1098.51 7	1.83 12
			γ 57		1238.76 7	1.18 8
			γ 58		1277.13 7	1.63 12
			65 weak γs			11.31
^{202}Tl (EC decay)	[15720-57-7]	12.23 d 2	Auger L		7.6	51 4
			Auger K		53.8	2.8 17
			ce-K-1		356.458 10	2.38 8
			x ray L		10	31 4
			x ray Kα_2		68.8950 20	22.8 7
			x ray Kα_1		70.8190 20	38.8 11
			x ray Kβ		80.3	17.1 6
			γ 1		439.560 10	91.4 2
			2 weak γs			1.03
^{202}Pb (EC decay)	[15752-86-0]	ca 3 × 10^5 yr	Auger L		7.78	35 9
			x ray L		10.3	24 6
^{202}Pb (EC decay)		3.62 h 3	Auger L		7.78	6.3 6
			ce-K-5		304.41 7	1.01 71
			x ray L		10.3	4.2 5
			x ray Kα_2		70.8319 9	2.84 22
			x ray Kα_1		72.8715 9	4.8 4
			x ray Kβ		82.6	2.13 17
			γ 5		389.94 7	6.2 6
			γ 6		459.72 7	8.6 7
			γ 7		490.47 7	9.1 7
			5 weak γs			2.63
^{202}Pb (IT decay)		3.62 h 3	Auger L		8	15 4
			ce-L-2		113.34 10	14 6
			ce-M-2		125.35 10	4.6 18
			ce-NOP-2		128.31 10	1.6 6
			ce-K-6		334.12 6	2.53 16
			ce-K-11		698.99 6	4.01 13
			ce-L-11		771.13 6	3.03 10
			x ray L		10.6	10.1 25
			x ray Kα_2		72.8042 9	2.14 8
			x ray Kα_1		74.9694 9	3.61 12
			x ray Kβ		84.9	1.60 6
			γ 6		442.12 6	84 5
			γ 9		657.49 6	31.8 15
			γ 11		786.99 6	48.9 3
			γ 13		960.70 15	90 8
			γ 14		1382.8 5	3.3 6
			9 weak γs			2.44
^{203}Hg (β$^-$ decay)	[13982-78-0]	46.60 d 2	Auger L		7.78	8.8 7
			ce-K-1		193.6663 14	13.37 21
			ce-L-1		263.8500 13	3.91 17
			ce-MNO-1		275.4926 13	1.30 17
			β$^-$ 1	212.2 20	57.7 6	100
			x ray L		10.3	5.9 7
			x ray Kα_2		70.8319 9	3.75 10
			x ray Kα_1		72.8715 9	6.36 16
			x ray Kβ		82.6	2.81 8
			γ 1		279.1967 12	81.5 8
^{203}Pb (EC decay)	[14687-25-3]	52.05 h 10	Auger L		7.78	56 5
			Auger K		55.2	3.0 14
			ce-K-1		193.659 5	13.14 21
			ce-L-1		263.842 5	3.89 6
			x ray L		10.3	38 4
			x ray Kα_2		70.8319 9	25.5 6
			x ray Kα_1		72.8715 9	43.2 9
			x ray Kβ		82.6	19.1 5
			γ 1		279.189 5	80.1 8
			γ 2		401.315 12	3.44 17
			1 weak γ			0.70
^{203}Bi (EC decay)	[24383-94-6]	11.76 h 5	Auger L		8	55 5
			ce-L-2		44.13 3	2.00 22
			Auger K		56.7	2.7 12
			ce-K-9		98.6 5	4.2 4
			ce-L-5		110.5992 5	1.79 11
			ce-K-15		176.2 5	2.3 6
			ce-K-83		737.2 5	3.2 3
			7 weak βs			0.21
			x ray L		10.6	38 5
			x ray Kα_2		72.8042 9	24.2 16
			x ray Kα_1		74.9694 9	41 3
			x ray Kβ		84.9	18.1 12

Table 6 (*continued*)

Isotope	CAS Registry No.	Half-life	Type of radiation	Energy, keV Max	Energy, keV Avg	Intensity, %
			γ 5		126.46	1.20 7
			γ 9		186.6 5	3.11 22
			γ 15		264.2 5	5.2 4
			γ 29		381.67	1.28 7
			γ 55		569.29	1.22 7
			γ 62		633.8 3	1.37 7
			γ 63		633.8	1.33 7
			γ 72		722.4 5	4.8 4
			γ 81		816.32	4.03 21
			γ 82		820.2 5	29.6 15
			γ 83		825.2 5	14.6 11
			γ 84		847.18	8.5 5
			γ 86		866.47	1.49 8
			γ 90		896.8 5	13.1 7
			γ 96		933.39	1.44 8
			γ 106		1033.73	8.8 5
			γ 125		1198.55	2.02 11
			γ 126		1203	1.54 8
			γ 132		1253.83	1.23 7
			γ 157		1506.7 5	3.7 3
			γ 159		1536.5 5	7.5 6
			γ 161		1552.55	1.48 8
			γ 167		1592.66	1.09 6
			γ 171		1679.6 5	8.8 7
			γ 173		1719.6 5	3.40 23
			γ 176		1748.4 8	1.89 10
			γ 185		1847.3 5	11.4 8
			γ 187		1888	1.94 10
			γ 188		1893.0 5	8.2 6
			γ 190		1928.16	1.12 6
			γ 198		2011.39	1.76 9
			194 weak γs			45.17
^{204}Tl (EC decay)	[13968-51-9]	3.78 yr 2	Auger L		7.6	1.23 9
^{204}Tl (β^- decay)		3.78 yr 2	β^- 1	763.40 20	243.93 7	97.45 5
^{204}Pb (IT decay)	[13966-26-2]	67.2 min 3	Auger L		8	7.3 6
			ce-K-2		286.74 10	3.50 15
			ce-L-2		358.88 10	1.52 7
			ce-K-6		823.74 15	5.3 4
			ce-L-6		895.88 15	3.07 22
			x ray L		10.6	5.1 6
			x ray Kα_2		72.8042 9	2.66 12
			x ray Kα_1		74.9694 9	4.50 19
			x ray Kβ		84.9	2.00 9
			γ 2		374.74 10	89 3
			γ 5		899.15 10	99 3
			γ 6		911.74 15	96 3
			4 weak γs			0.33
^{204}Bi (EC decay)	[14903-04-9]	11.22 h	Auger L		8	60 6
			ce-K-8		12.20 10	1.32 20
			ce-K-12		52.80 10	1.6 8
			Auger K		56.7	2.9 12
			ce-L-3		62.68 8	4.5 7
			ce-L-4		64.34 10	2.0 4
			ce-M-3		74.69 8	1.17 18
			ce-K-21		88.17 10	1.77 20
			ce-K-25		128.11 15	1.28 15
			ce-K-32		160.91 10	1.25 15
			ce-K-35		201.25 15	1.11 17
			ce-K-46		286.74 10	3.18 24
			ce-L-46		358.88 10	1.38 11
			6 weak βs			0.17
			γ 1		6.27 10	4
			x ray L		10.6	42 5
			γ 2		29	2.1 5
			x ray Kα_2		72.8042 9	25.4 18
			x ray Kα_1		74.9694 9	43 3
			x ray Kβ		84.9	19.1 14
			γ 21		176.17 10	1.11 12
			γ 25		216.11 15	1.43 16
			γ 27		219.46 15	2.30 24
			γ 32		248.91 10	2.06 24
			γ 35		289.25 15	2.8 4
			γ 46		374.74 10	81 6
			γ 50		421.55 10	1.08 12

Table 6 (*continued*)

Isotope	CAS Registry No.	Half-life	Type of radiation	Energy, keV Max	Energy, keV Avg	Intensity, %
			γ 53		440.34 15	2.5 4
			γ 65		532.72 10	1.35 16
			γ 79		661.55 15	2.6 4
			γ 83		670.70 10	10.6 10
			γ 87		709.13 15	1.43 24
			γ 88		710.13 15	1.43 24
			γ 93		753.78 15	1.06 12
			γ 96		791.16 10	3.2 4
			γ 101		834.10 15	1.03 16
			γ 104		899.15 10	98 8
			γ 105		911.74 15	13.5 16
			γ 106		911.96 15	11.1 16
			γ 107		918.26 15	10.8 8
			γ 117		983.98 10	58 7
			γ 118		990.34 15	1.11 16
			γ 123		1043.63 10	1.27 16
			γ 133		1111.27 10	1.43 16
			γ 144		1203.84 25	2.1 4
			γ 145		1211.74 10	3.1 4
			γ 150		1274.81 10	2.2 5
			γ 193		1703.32 10	1.98 24
			γ 198		1755.29 25	1.23 15
			γ 215		1896.31 25	1.35 16
			240 weak γs			57.96
^{205}Pb (EC decay)	[14119-28-9]	1.43×10^7 yr 15	Auger L		7.78	37 3
			x ray L		10.3	25 3
^{205}Bi (EC decay)	[14333-38-1]	15.31 d 4	Auger L		8	56 5
			ce-L-2		10.359 10	12.0 12
			ce-M-2		22.369 10	3.3 4
			Auger K		56.7	2.5 10
			1 weak β			0.11
			x ray L		10.6	39 4
			x ray Kα$_2$		72.8042 9	21.6 8
			x ray Kα$_1$		74.9694 9	36.4 12
			x ray Kβ		84.9	16.2 6
			γ 18		260.50 5	1.09 4
			γ 22		284.15 10	1.692 23
			γ 41		549.84 4	2.95 4
			γ 43		570.60 5	4.34 7
			γ 46		579.80 10	5.442 18
			γ 57		703.45 5	31.10 10
			γ 66		759.10 10	1.04 5
			γ 88		910.90 5	1.64 4
			γ 95		987.66 5	16.13 17
			γ 107		1043.75 5	7.51 10
			γ 114		1190.03 5	2.26 7
			γ 124		1351.52 5	1.06 4
			γ 135		1614.30 15	2.28 4
			γ 140		1764.30 10	32.5 7
			γ 141		1775.80 10	3.99 8
			γ 145		1861.70 10	6.17 10
			γ 146		1903.45 10	2.47 4
			133 weak γs			20.70
^{206}Bi (EC decay)	[15776-19-9]	6.243 d 3	Auger L		8	71 5
			Auger K		56.7	3.7 15
			ce-K-6		96.02 3	22.0 8
			ce-L-6		168.16 3	3.84 14
			ce-K-11		174.71 5	1.56 6
			ce-K-15		255.51 3	5.89 8
			ce-K-18		310.00 3	1.819 17
			ce-L-15		327.65 3	1.00 4
			ce-K-25		409.06 4	1.428 15
			ce-K-26		428.18 4	1.99 7
			ce-K-27		449.45 4	2.32 8
			ce-L-26		500.32 4	1.25 4
			x ray L		10.6	50 5
			x ray Kα$_2$		72.8042 9	32.4 7
			x ray Kα$_1$		74.9694 9	54.8 11
			x ray Kβ		84.9	24.3 6
			γ 6		184.02 3	15.8 4
			γ 11		262.71 5	3.02 5
			γ 15		343.51 3	23.4 3
			γ 18		398.00 3	10.74 10
			γ 25		497.06 4	15.31 15

Table 6 (*continued*)

Isotope	CAS Registry No.	Half-life	Type of radiation	Energy, keV		Intensity, %
				Max	Avg	
			γ 26		516.18 4	40.7 4
			γ 27		537.45 4	30.4 3
			γ 31		620.48 5	5.76 6
			γ 32		632.25 5	4.47 5
			γ 33		657.16 5	1.91 3
			γ 39		803.10 5	98.90 3
			γ 42		881.01 5	66.2 7
			γ 43		895.12 5	15.66 16
			γ 46		1018.63 8	7.60 8
			γ 50		1098.26 7	13.50 15
			γ 59		1405.01 8	1.434 25
			γ 66		1595.27 8	5.01 6
			γ 67		1718.70 7	31.8 4
			γ 69		1878.65 8	2.01 4
			57 weak γs			7.20
^{206}Po (EC decay)	[15735-86-1]	8.8 d 1	Auger L		8.15	100 11
			ce-L-4		43.520 18	67 11
			ce-M-4		55.909 18	18 3
			Auger K		58.2	3.1 16
			ce-NOP-4		58.970 18	5.9 10
			ce-K-22		195.88 3	10.6 14
			ce-K-24		221.03 3	1.47 20
			ce-K-27		247.92 4	5.4 7
			ce-L-22		270.02 3	1.84 24
			ce-K-36		420.83 6	2.2 3
			ce-K-37		431.94 6	1.38 18
			x ray L		10.8	72 9
			γ 4		59.908 18	1.23 20
			x ray Kα_2		74.8148 10	28.6 24
			x ray Kα_1		77.1079 10	48 4
			x ray Kβ		87.3	21.5 18
			γ 22		286.41 3	24 3
			γ 24		311.56 3	4.2 6
			γ 27		338.44 4	19.2 25
			γ 34		463.38 5	1.79 24
			γ 36		511.36 6	24 3
			γ 37		522.47 6	15.7 20
			γ 40		554.64 6	1.56 21
			γ 41		579.78 6	1.06 14
			γ 46		677.71 8	1.47 20
			γ 49		807.38 9	23 3
			γ 50		818.23 9	1.04 4
			γ 53		860.93 9	3.5 5
			γ 57		980.23 10	7.1 9
			γ 58		1007.15 10	3.1 4
			γ 60		1032.26 10	33 5
			51 weak γs			8.32
^{206}Po (α decay)		8.8 d 1	α_1	5223.4 15		5.45 5
^{207}Bi (EC decay)	[13982-38-2]	38 yr 3	Auger L		8	52 4
			Auger K		56.7	2.5 11
			ce-K-2		481.665 20	1.56 5
			ce-K-4		975.615 20	7.27 24
			ce-L-4		1047.759 20	1.80 6
			1 weak β			0.01
			x ray L		10.6	36 4
			x ray Kα_2		72.8042 9	21.9 6
			x ray Kα_1		74.9694 9	36.9 9
			x ray Kβ		84.9	16.4 5
			γ 2		569.670 20	97.8 5
			γ 4		1063.620 20	74.9 11
			γ 6		1770.23 4	6.85 20
			3 weak βs			0.29
^{207}Po (EC decay)	[15720-45-3]	350 min	Auger L		8.15	51 4
			ce-K-1		9.27 20	1.04 24
			Auger K		58.2	2.7 13
			ce-K-12		159.07 10	1.06 9
			ce-K-21		315.17 10	1.77 15
			ce-K-29		652.07 10	1.05 9
			ce-K-34		901.77 10	1.02 9
			2 weak βs			0.49
			x ray L		10.8	37 4
			x ray Kα_2		74.8148 10	24.6 6
			x ray Kα_1		77.1079 10	41.4 8
			x ray Kβ		87.3	18.5 5

Table 6 (continued)

Isotope	CAS Registry No.	Half-life	Type of radiation	Energy, keV Max	Energy, keV Avg	Intensity, %
			γ 9		222.10 10	1.26 7
			γ 12		249.60 10	1.62 13
			γ 13		297.20 10	1.01 8
			γ 16		345.20 20	2.00 16
			γ 17		369.50 10	1.93 17
			γ 21		405.70 20	10.1 8
			γ 25		629.80 10	1.48 17
			γ 27		687.60 10	2.03 20
			γ 29		742.60 10	29.2 22
			γ 32		911.80 10	18.0 14
			γ 33		947.90 10	1.08 11
			γ 34		992.30 10	60 5
			γ 36		1148.30 10	6.1 7
			γ 41		1372.40 20	1.39 14
			γ 52		2060.00 20	1.44 14
			37 weak γs			7.14
^{208}Bi (EC decay)	[14145-42-7]	3.68×10^5 yr 4	Auger L		8	45 4
			Auger K		56.7	1.4 6
			x ray L		10.6	31 3
			x ray Kα_2		72.8042 9	12.3 4
			x ray Kα_1		74.9694 9	20.7 5
			x ray Kβ		84.9	9.2 3
			γ 1		2610	100
^{208}Po (α decay)	[15735-87-2]	2.898 yr 2	α 1		5116.0 20	99.9982 2
^{209}Pb (β^- decay)	[14119-30-3]	3.253 h 14	β^- 1	644.6 12	197.6 5	100
^{209}Po (α decay)	[15735-81-6]	102 yr 5	α 1		4882 3	99.17 4
			3 weak αs			0.57
			2 weak γs			0.23
^{210}Pb (β^- decay)	[14255-04-0]	22.3 yr 2	Auger L		8.15	34 3
			ce-L-1		30.115 15	57.9 21
			ce-M-1		42.504 15	13.6 5
			ce-NOP-1		45.565 15	4.54 17
			β^- 1	16.5 5	4.14 13	80 3
			β^- 2	63.0 5	16.13 14	20 5
			total β^-		6.54 18	100 5
			x ray L		10.8	24.3 25
			γ 1		46.503 15	4.05 8
^{210}Bi (β^- decay)	[14331-79-4]	5.013 d 5	β^- 1	1161.4 10	389.0 4	100
^{210}Bi (α decay)		3.0×10^6 yr 1	Auger L		7.78	9.2 8
			ce-K-1		180.17 20	4.4 3
			ce-K-2		219.3 3	8.8 6
			ce-L-1		250.35 20	2.90 17
			ce-L-2		289.5 3	1.49 10
			α 1		4568 5	4.8 3
			α 2		4908 10	39.4 10
			α 3		4946 9	55.0 11
			4 weak αs			0.81
			x ray L		10.3	6.2 7
			x ray Kα_2		70.8319 9	3.80 20
			x ray Kα_1		72.8715 9	6.4 4
			x ray Kβ		82.6	2.84 15
			γ 1		265.70 20	51 3
			γ 2		304.8 3	27.5 16
			γ 8		649.8 10	2.86 16
			6 weak γs			2.20
^{210}Po (α decay)	[13981-52-7]	138.378 d 7	α 1		5304.51 7	100
^{211}At (EC decay)	[15755-39-2]	7.214 h 7	Auger L		8.33	26.0 20
			Auger K		59.7	1.3 7
			x ray L		11	19.7 20
			x ray Kα_2		76.862 5	12.7 3
			x ray Kα_1		79.290 5	21.3 4
			x ray Kβ		89.8	9.54 22
			1 weak γ			0.25
^{211}At (α decay)		7.214 h 7	α 1		5867.0 20	41.70 20
^{212}Pb (β^- decay)	[15092-94-1]	10.64 h 1	Auger L		8.15	20.9 17
			ce-K-1		24.650 7	3.42 16
			Auger K		58.2	1.1 6
			ce-K-4		148.1061 22	32.3 14
			ce-K-5		209.561 10	1.26 7
			ce-L-4		222.2445 21	5.59 23
			ce-M-4		234.6329 21	1.32 6
			β^- 1	158 4	41.9 11	5.09 17
			β^- 2	334 4	94.4 12	83.2 24
			β^- 3	573 4	172.7 13	11.8 24

Table 6 (*continued*)

Isotope	CAS Registry No.	Half-life	Type of radiation	Energy, keV Max	Energy, keV Avg	Intensity, %
			total β^-		101.0 13	100 4
			x ray L		10.8	15.1 16
			x ray Kα_2		74.8148 10	10.5 5
			x ray Kα_1		77.1079 10	17.6 7
			x ray Kβ		87.3	7.9 4
			γ 4		238.6320 20	43.6 13
			γ 5		300.087 10	3.34 11
			4 weak γs			0.67
^{212}Bi (A decay)	[14913-49-6]	60.55 min 6	Auger L		7.78	12.3 10
			ce-L-1		24.511 4	20.5 8
			ce-M-1		36.154 4	4.77 18
			ce-NOP-1		39.012 4	1.59 6
			α 1		6050.99 3	25.23 9
			α 2		6090.09 4	9.63 8
			6 weak αs			1.07
			x ray L		10.3	8.2 9
			γ 1		39.858 4	1.10 3
			12 weak γs			0.95
^{212}Bi (β^- decay)		60.55 min 6	β^- 1	625 4	190.6 14	1.89 5
			β^- 2	733 4	228.7 15	1.46 4
			β^- 3	1519 4	530.7 17	4.49 15
			β^- 4	2246 4	831.6 17	55.31 17
			total β^-		767.9 18	64.07 24
			3 weak βs			0.92
			γ 4		727.18 6	6.65 15
			γ 5		785.42 6	1.11 3
			γ 12		1620.735 10	1.51 5
			13 weak γs			1.59
^{222}Rn (A decay)	[14859-67-7]	3.8235 d 3	α 1		5489.7 3	99.920 10
			2 weak αs			0.08
			1 weak γ			0.08
^{223}Ra (A decay)	[15623-45-7]	11.434 d 2	ce-NOP-1		3.333 5	56.4
			ce-M-2		5.528 5	9.537
			Auger L		8.7	27 3
			ce-NOP-2		8.913 5	3.1
			ce-M-3		9.938 5	9.6
			ce-NOP-3		13.323 5	3.187
			ce-K-9		23.91 7	7.24 25
			ce-K-11		45.80 5	12.6 5
			ce-K-12		55.79 4	18.2 7
			ce-K-13		60.22 5	2.06 8
			Auger K		62.7	1.5 8
			ce-L-9		104.26 8	1.36 5
			ce-L-11		126.15 6	2.29 9
			ce-L-12		136.14 5	3.29 12
			ce-NOP-11		143.10 4	1.90 7
			ce-K-28		171.01 4	9.1 4
			ce-K-30		225.49 5	1.57 6
			ce-K-33		239.92 7	1.01 4
			ce-L-28		251.36 5	1.66 7
			α 1		5433.6 5	2.27 20
			α 2		5501.6 10	1.00 15
			α 3		5540.0 10	9.2 3
			α 4		5606.9 3	24.2 4
			α 5		5716.4 3	52.5 8
			α 6		5747.2 4	9.5 6
			21 weak αs			1.87
			x ray L		11.7	24 3
			x ray Kα_2		81.070 20	15.0 4
			x ray Kα_1		83.780 20	24.8 7
			x ray Kβ		94.9	11.3 4
			γ 9		122.31 6	1.190 20
			γ 11		144.20 4	3.26 7
			γ 12		154.19 3	5.59 10
			γ 28		269.41 3	13.6 3
			γ 30		323.89 4	3.90 9
			γ 33		338.32 6	2.78 7
			γ 47		444.94 5	1.27 6
			48 weak γs			2.78
^{224}Ra (α decay)	[13233-32-4]	3.66 d 4	α 1		5449	4.9 4
			α 2		5685.56 20	95.1 4
			3 weak αs			0.02
			γ 1		241.00 10	3.9 11
			3 weak γs			0.02

Table 6 (*continued*)

Isotope	CAS Registry No.	Half-life	Type of radiation	Energy, keV Max	Energy, keV Avg	Intensity, %
^{225}Ra (β^- decay)	[13981-53-8]	14.8 d 2	Auger L		9.28	14.2 25
			ce-L-1		20.2 10	29 4
			ce-M-1		35.0 10	7.2 11
			β^- 1	322 12	90 4	67.3
			β^- 2	362 12	103 4	32.7
			total β^-		94 4	100
			x ray L		12.7	15 3
			γ 1		40.0 10	29 4
^{225}Ac (α decay)	[14265-85-1]	10.0 d 1	α 1		5608 3	1.10 10
			α 2		5637 3	4.5 3
			α 3		5681 3	1.40 20
			α 4		5723 3	2.9 5
			α 5		5731 3	10.00 10
			α 6		5790.6 22	8.6 9
			α 7		5792.5 22	18.1 20
			α 8		5829.0 20	51.6 15
			33 weak αs			1.13
			γ 22		99.80 10	1.70 20
			54 weak γs			6.52
^{226}Ra (α decay)	[13982-63-3]	1600 yr 7	ce-L-1		167.94 6	1.20 3
			α 1		4601.9 5	5.55 5
			α 2		4784.50 25	94.45 5
			γ 1		185.99 4	3.28 3
^{227}Ac (α decay)	[14952-40-0]	21.773 yr 3	19 weak αs			1.37
^{227}Ac (β^- decay)			32 weak γs			0.65
			ce-M-1		4.12 10	33 7
			ce-NOP-1		7.97 10	10.8 23
			ce-M-2		10.02 10	6.7 15
			ce-NOP-2		13.87 10	2.2 5
			β^- 1	19.2 20	4.8 5	ca 10
			β^- 2	34.4 20	8.7 6	ca 35
			β^- 3	43.7 20	11.1 6	ca 55
			total β^-		9.6 7	100
			3 weak γs			0.06
^{227}Th (α decay)	[15623-47-9]	18.718 d 5	ce-L-5		1.03 10	1.2 5
			ce-M-2		3.18 20	2.6 6
			Auger L		9	41 9
			ce-L-9		10.67 3	41 15
			ce-L-10		12.383 11	15 5
			ce-M-9		25.09 3	11 4
			ce-M-10		26.798 11	4.1 13
			ce-NOP-9		28.70 3	3.5 13
			ce-L-20		29.06 10	2.3 6
			ce-NOP-10		30.412 11	1.3 5
			ce-L-23		30.96 10	4.5 10
			ce-L-30		42.27 10	5.7 19
			ce-M-23		45.38 10	1.10 23
			ce-M-30		56.69 10	1.5 5
			α 1		5668.0 15	2.06 12
			α 2		5693.0 16	1.50 10
			α 3		5700.8 16	3.63 20
			α 4		5709.0 16	8.2 3
			α 5		5713.2 16	4.89 20
			α 6		5757.06 15	20.3 10
			α 7		5807.5 15	1.270 20
			α 8		5866.6 15	2.42 10
			α 9		5959.7 15	3.00 15
			α 10		5977.92 10	2.34 10
			α 11		6008.8 15	2.90 15
			α 12		6038.21 15	24.5 10
			33 weak αs			2.25
			x ray L		12.3	41 9
			γ 23		50.20 10	8.5 18
			γ 43		79.77 6	2.1 5
			x ray Kα_2		85.430 10	1.61 11
			x ray Kα_1		88.470 10	2.66 18
			γ 45		94.00 6	1.4 3
			x ray Kβ		100	1.22 9
			γ 84		210.65 8	1.13 24
			γ 92		236.00 8	11.2 24
			γ 100		256.25 5	6.8 15
			γ 113		286.15 6	1.6 4
			γ 118		299.90 10	2.0 5
			γ 120		304.44 13	1.05 25

Table 6 (*continued*)

Isotope	CAS Registry No.	Half-life	Type of radiation	Energy, keV Max	Energy, keV Avg	Intensity, %
^{228}Ac (β^- decay)	[14331-83-0]	6.13 h	γ 130		329.82 10	2.8 6
			γ 131		334.40 14	1.00 23
			230 weak γs			9.33
			Auger L		9.48	38 6
			ce-L-1		37.29 6	60 9
			ce-M-1		52.58 6	16.3 24
			ce-NOP-1		56.43 6	6.0 9
			ce-K-9		74.9 8	5 4
			ce-L-2		78.98 8	4.2 18
			ce-M-2		94.27 8	1.0 5
			ce-L-3		108.6 3	7.5 25
			ce-M-3		123.9 3	2.1 7
			ce-L-9		164.1 8	1.2 8
			β^- 1	1950 7		3 3
			β^- 2	1809 7		1.0 10
			β^- 3	705 7		2.0 20
			β^- 4	413 7	118.5 23	1.6 3
			β^- 5	449 7	130.0 23	2.5 4
			β^- 6	454 7	131.7 23	1.6 3
			β^- 7	491 7	143.8 24	5.1 9
			β^- 8	499 7	146.3 24	1.3 3
			β^- 9	606 7	182.0 24	9 3
			β^- 10	910 7	290 3	1
			β^- 11	969 7	311 3	3.3 6
			β^- 12	983 7	317 3	7 5
			β^- 13	1014 7	328 3	6.7 12
			β^- 14	1115 7	366 3	3.7 9
			β^- 15	1168 7	386 3	32 7
			β^- 16	1741 7	611 3	13 4
			β^- 17	2079 7	748 3	9 6
			total β^-		357 3	107 13
			22 weak βs			4.50
			x ray L		13	41 7
			x ray Kα_2		89.9530 20	2.2 10
			x ray Kα_1		93.3500 20	3.6 16
			γ 2		99.45 8	1.4 6
			x ray Kβ		105	1.7 8
			γ 3		129.1 3	2.9 10
			γ 13		209.4 3	4.6 16
			γ 18		270.3 3	3.8 10
			γ 22		328.0 3	3.4 9
			γ 25		338.4 3	12 3
			γ 33		409.4 5	2.2 4
			γ 38		463.0 3	4.6 8
			γ 72		755.2 5	1.10 24
			γ 73		772.1 5	1.6 4
			γ 77		794.8 3	4.8 9
			γ 82		835.6 5	1.8 4
			γ 90		911.07 3	29 4
			γ 97		964.6 5	5.5 10
			γ 98		968.9 5	17 3
			γ 137		1459.2 5	1.04 22
			γ 140		1495.8 5	1.05 18
			γ 149		1587.9 4	3.7 8
			γ 152		1630.4 4	1.9 3
			155 weak γs			19.97
^{228}Th (α decay)	[14274-82-9]	1.9131 yr 9	Auger L		9	9 3
			ce-L-1		65.16 5	19 6
			ce-M-1		79.58 5	5.1 16
			ce-NOP-1		83.19 5	1.8 6
			α 1		5340.54 15	26.70 20
			α 2		5423.33 22	72.7
			3 weak αs			0.59
			x ray L		12.3	9 3
			γ 1		84.40 5	1.2 4
			4 weak γs			0.49
^{229}Th (α decay)	[15594-54-4]	7340 yr 160	α 1		4797.8 12	1.2
			α 2		4814.6 12	9.30 8
			α 3		4837	4.8
			α 4		4845.3 12	56.20 20
			α 5		4901.0 12	10.20 8
			α 6		4967.5 12	5.97 6
			α 7		4978.5 12	3.17 4
			α 8		5050	5.2

Table 6 (continued)

Isotope	CAS Registry No.	Half-life	Type of radiation	Energy, keV Max	Energy, keV Avg	Intensity, %
^{230}Th (A decay)	[14269-63-7]	7.58 × 10^4 yr 3	α 9		5052	1.6
			19 weak αs			2.09
			γ 1		11.10 10	13.3 13
			γ 5		31.30 20	4.1 4
			γ 15		86.44 5	3.1 3
			γ 17		124.50 10	1.22 12
			γ 23		137.03 6	1.63 16
			γ 27		148.30 20	1.39 14
			γ 31		156.48 4	1.12 11
			γ 39		193.63 6	4.6 5
			γ 41		210.97 10	3.3 4
			38 weak γs			5.82
			Auger L		9	8.5 9
			ce-L-1		48.49 3	17.0 6
			ce-M-1		62.91 3	4.60 14
			ce-NOP-1		66.52 3	1.66 5
			α 1		4621.0 15	23.40 10
			α 2		4687.5 15	76.3 3
			9 weak αs			0.43
			x ray L		12.3	8.5 9
			8 weak γs			0.45
^{230}Pa (EC decay)	[15766-10-6]	17.4 d 5	Auger L		9.48	54 7
			ce-L-1		32.73 5	40 5
			ce-M-1		48.02 5	10.9 13
			ce-NOP-1		51.87 5	4.0 5
			Auger K		69.2	1.5 11
			ce-L-3		100.43 5	1.17 21
			ce-K-26		334.10 5	1.09 11
			x ray L		13	59 7
			x ray Kα_2		89.9530 20	18.5 14
			x ray Kα_1		93.3500 20	30.2 22
			x ray Kβ		105	14.0 10
			γ 22		397.80 20	1.82 17
			γ 26		443.75 5	5.3 5
			γ 28		454.95 5	6.1 5
			γ 33		508.20 5	3.5 3
			γ 34		518.50 10	1.92 17
			γ 38		571.10 10	1.05 9
			γ 46		728.23 7	1.84 15
			γ 48		781.35 5	1.44 12
			γ 51		898.65 10	5.7 5
			γ 52		918.50 10	8.0 7
			γ 53		951.95 10	28.3 20
			γ 55		956.3 3	1.5 3
			γ 59		1009.60 20	1.05 9
			γ 60		1026.05 10	1.42 12
			47 weak γs			5.52
^{230}Pb (β^- decay)	[51634-84-5]	17.4 d 5	Auger L		9.89	3.0 7
			ce-L-1		29.99 5	6.8 12
			ce-M-1		46.20 5	1.9 4
			β^- 1	507 5	148.7 17	9.3 6
			2 weak βs			0.25
			x ray L		13.6	3.8 8
			6 weak γs			0.27
^{230}U (α decay)	[15743-51-8]	20.8 d	Auger L		9.48	11.2 17
			ce-L-1		51.73 4	23 3
			ce-M-1		67.02 4	6.4 8
			ce-NOP-1		70.87 4	2.3 3
			α 1		5817.7 7	32.00 20
			α 2		5888.5 7	67.4 4
			9 weak αs			0.65
			x ray L		13	12.1 18
			11 weak γs			0.93
^{231}Th (β^- decay)	[14932-40-2]	25.52 h 1	ce-L-5		4.535 20	49 5
			Auger L		9.68	120 140
			ce-M-3		11.8331 16	33 12
			ce-M-4		12.7031 16	100 170
			ce-M-5		20.273 20	12.7 12
			ce-L-8		37.465 20	55 4
			ce-M-8		53.203 20	15.2 11
			ce-NOP-8		57.183 20	5.6 4
			ce-L-13		60.135 20	6.6 6
			ce-L-14		61.005 20	2.5 5
			ce-L-15		63.105 20	11.5 10

Table 6 (*continued*)

Isotope	CAS Registry No.	Half-life	Type of radiation	Energy, keV Max	Energy, keV Avg	Intensity, %
			ce-M-13		75.873 20	1.65 14
			ce-M-15		78.843 20	3.7 8
			β^- 1	142.7 18	37.6 12	2.8 5
			β^- 2	206.5 18	55.7 12	12.8 11
			β^- 3	215.8 18	58.4 12	1.30 20
			β^- 4	287.7 18	79.6 12	12 7
			β^- 5	288.6 18	79.9 12	37 15
			β^- 6	305.8 18	85.1 13	35 20
			total β^-		77.0 13	100 30
			7 weak βs			0.41
			x ray L		13.3	140 160
			γ 5		25.640 20	14.8 13
			γ 15		84.210 20	6.5 6
			47 weak γs			5.96
^{231}Pa (α decay)	[14331-85-2]	3.276×10^4 yr 11	ce-L-12		5.70 7	14 3
			ce-L-14		7.520 21	33 11
			Auger L		9.28	44 8
			ce-L-15		10.11 3	19 5
			ce-M-5		11.50 10	1.7 5
			ce-M-7		13.898 5	30 7
			ce-NOP-7		17.631 5	9.3 21
			ce-L-20		18.36 3	8.5 20
			ce-M-12		20.54 6	3.2 7
			ce-M-14		22.358 12	8 3
			ce-L-25		24.32 3	1.8 5
			ce-M-15		24.948 21	4.8 12
			ce-NOP-14		26.091 12	2.4 9
			ce-NOP-15		28.681 21	1.5 4
			ce-L-28		32.90 3	1.5 4
			ce-M-20		33.198 2	2.1 5
			ce-L-31		37.35 4	4.3 10
			ce-L-33		43.83 4	3.3 8
			ce-M-31		52.19 3	1.2 3
			α 1		4680.0 20	1.5
			α 2		4712.0 20	1
			α 3		4736.0 20	8.4
			α 4		4851.0 20	1.4
			α 5		4933.0 ·20	3
			α 6		4950	22.8
			α 7		4984.0 20	1.4
			α 8		5011.0 20	25.4
			α 9		5028	20
			α 10		5030.5 20	2.5
			α 11		5057.3 20	11
			9 weak αs			0.67
			x ray L		12.7	46 8
			γ 14		27.360 10	9.3 19
			γ 57		283.67 6	1.6 4
			γ 59		300.08 6	2.3 5
			γ 60		302.67 6	2.3 5
			γ 61		302.67 6	2.3 5
			γ 66		330	1.3 4
			88 weak γs			3.22
^{232}Th (α decay)	[7440-29-1]	1.405×10^{10} yr 6	Auger L		9	6.6 7
			ce-L-1		39.8 10	13.1 4
			ce-M-1		54.2 10	3.54 11
			ce-NOP-1		57.8 10	1.27 4
			α 1		3953	23 3
			α 2		4010 5	77 3
			1 weak α			0.20
			x ray L		12.3	6.6 7
			2 weak γs			0.19
^{233}Pa (β^- decay)	[13981-14-1]	27.0 d 7	ce-L-2		6.78 5	17.1 22
			Auger L		9.89	39 6
			ce-M-1		11.7120 4	1.48 5
			ce-L-3		18.593 10	9 6
			ce-M-2		22.99 5	4.4 6
			ce-M-3		34.802 10	2.3 17
			ce-L-7		53.523 10	10.7 8
			ce-L-8		64.833 10	10.6 14
			ce-M-7		69.732 10	2.59 18
			ce-M-8		81.042 10	2.6 4
			ce-L-10		82.103 20	2.6 5
			ce-K-15		184.51 3	5.6 4

Table 6 (continued)

Isotope	CAS Registry No.	Half-life	Type of radiation	Energy, keV Max	Energy, keV Avg	Intensity, %
			ce-K-16		196.37 3	29.4 10
			ce-K-17		224.89 4	2.4 7
			ce-L-15		278.36 3	1.09 7
			ce-L-16		290.22 3	5.70 18
			ce-M-16		306.43 3	1.37 5
			β^- 1	156.5 24	41.5 7	24.2 17
			β^- 2	173.8 24	46.4 7	15.6 20
			β^- 3	231.8 24	63.1 7	29 4
			β^- 4	260.4 24	71.5 8	33 4
			β^- 5	572.3 24	170.4 8	4 4
			total β^-		62.4 8	106 8
			x ray L		13.6	49 7
			γ 7		75.280 10	1.26 8
			γ 8		86.590 10	1.89 24
			x ray Kα_2		94.6650 20	10.9 4
			x ray Kα_1		98.4390 20	17.7 7
			x ray Kβ		111	8.2 4
			γ 15		300.12 3	6.6 4
			γ 16		311.98 3	38.6 4
			γ 17		340.50 4	4.5 5
			γ 19		398.62 8	1.27 16
			γ 20		415.76	1.62 16
			13 weak γs			1.89
^{233}U (α decay)	[13968-55-3]	1.592×10^5 yr 2	α 1		4729	1.6
			α 2		4783.5 12	13.20 20
			α 3		4824.2 12	84.4 5
			29 weak αs			0.72
			149 weak γs			0.19
^{234}Th (β^- decay)	[15065-10-8]	24.10 d 3	ce-L-3		8.385 20	4.3 4
			Auger L		9.68	8.4 11
			ce-M-2		14.653 20	1.28 18
			ce-M-3		24.123 20	1.18 11
			ce-L-6		42.185 20	1.18 10
			ce-L-12		71.275 11	11.5 10
			ce-M-12		87.013 11	2.78 24
			ce-NOP-12		90.993 11	1.02 9
			β^- 1	75.8 20	19.5 6	2.7 4
			β^- 2	95.8 20	24.8 6	6.2 6
			β^- 3	96.2 20	24.9 6	18.6 15
			β^- 4	188.6 20	50.6 6	72.5 20
			total β^-		43.4 7	100 3
			x ray L		13.3	9.9 11
			γ 6		63.290 20	3.8 3
			γ 12		92.380 10	2.72 22
			γ 13		92.800 20	2.69 21
			16 weak γs			0.37
^{234}Pa (β^- decay)	[15100-28-4]	6.70 h 5	Auger L		9.89	60 11
			ce-L-1		12.54 4	5.0 6
			ce-L-2		21.64 5	75 13
			ce-K-15		24.69 20	2.2 4
			ce-M-1		28.75 4	1.36 15
			ce-K-18		37.09 10	1.23 17
			ce-M-2		37.85 5	21 4
			ce-L-9		77.94 10	42 7
			ce-M-9		94.15 10	11.6 18
			ce-L-15		118.54 20	1.36 22
			ce-L-18		130.94 10	7.9 11
			ce-M-18		147.15 10	2.2 3
			β^- 1	326 5	91.0 16	1.5 7
			β^- 2	396 5	112.7 16	1.16 17
			β^- 3	425 5	121.9 16	4.7 7
			β^- 4	445 5	128.6 17	2.6 7
			β^- 5	469 5	136.3 17	2.7 7
			β^- 6	484 5	141.0 17	29.0 25
			β^- 7	484 5	141.2 17	13.3 17
			β^- 8	514 5	150.8 17	5.8 9
			β^- 9	654 5	198.1 18	18.3 17
			β^- 10	711 5	217.6 18	ca 4.6
			β^- 11	932 5	296.6 19	1.2 6
			β^- 12	1013 5	316.8 17	1.7 18
			β^- 13	1114 5	364.0 19	8 5
			β^- 14	1138 5	372.9 19	ca 3.2
			β^- 15	1183 5	390.0 19	ca 1.3
			β^- 16	1183 5	390.0 19	ca 5

Table 6 (*continued*)

Isotope	CAS Registry No.	Half-life	Type of radiation	Energy, keV Max	Energy, keV Avg	Intensity, %
			β^- 17	1238 5	410.8 20	ca 7.1
			β^- 18	1244 5	413.3 20	ca 1.7
			total β^-		214.4 21	ca 113
			4 weak βs			1.49
			x ray L		13.6	77 12
			γ 5		63.00 20	2.7 4
			x ray Kα_2		94.6650 20	1.34 13
			x ray Kα_1		98.4390 20	2.16 21
			γ 9		99.70 10	4.0 6
			x ray K		111	1.01 10
			γ 12		131.20 20	16.6 18
			γ 18		152.70 10	5.6 8
			γ 23		186.00 20	1.7 3
			γ 26		199.70 20	2.5 3
			γ 31		226.4 4	4.9 6
			γ 32		227.20 20	4.6 5
			γ 34		248.90 20	2.3 4
			γ 41		293.7 3	3.2 5
			γ 49		369.8 4	2.4 4
			γ 50		372.4 4	1.08 21
			γ 56		458.8 3	1.24 16
			γ 65		506.8 5	1.3 3
			γ 66		513.7 5	1.08 21
			γ 74		565.9 10	1.2 3
			γ 75		568.7 5	2.5 3
			γ 76		569.5 5	8.9 10
			γ 77		574.0 10	1.66 18
			γ 95		664.8 10	1.1 4
			γ 96		666.7 6	1.3 4
			γ 97		669.9 5	1.2 4
			γ 100		692.7 5	1.2 5
			γ 101		699.0 5	3.8 5
			γ 102		706.1 3	2.6 6
			γ 106		733.0 5	7.1 11
			γ 108		742.81 3	2.0 7
			γ 110		755.6 10	1.2 6
			γ 118		786.27 3	1.2 4
			γ 119		793.6 10	1.24 14
			γ 120		796.3 5	3.2 6
			γ 122		805.8 5	2.7 6
			γ 126		819.6 6	2.2 5
			γ 128		826.3 6	3.3 8
			γ 129		831.6 8	4.6 8
			γ 134		876.4 8	3.3 17
			γ 135		880.5	3.3 4
			γ 136		880.51 4	7.5 8
			γ 137		883.24 4	10 4
			γ 138		899.0 5	3.4 8
			γ 141		925.0 10	2.4 3
			γ 142		926.0 8	9.1 20
			γ 143		927.1 8	7.5 19
			γ 144		946.00 3	10 5
			γ 145		949	6.6 8
			γ 148		978.8 10	1.2 6
			γ 149		980.5 5	1.66 18
			γ 150		980.5 5	2.5 3
			γ 151		984.0 10	1.6 6
			γ 170		1353.3 6	1.4 5
			γ 172		1394.1 5	2.5 8
			159 weak γs			44.47
^{234}U (α decay)	[13966-29-5]	2.454 × 10^5 yr 6	Auger L		9.48	9.7 14
			ce-L-1		32.73 5	20.1 18
			ce-M-1		48.02 5	5.5 5
			ce-NOP-1		51.87 5	2.02 18
			α 1		4723.7 20	27.5 15
			α 2		4775.8 20	72.5 20
			4 weak αs			0.24
			x ray L		13	10.5 14
			10 weak γs			0.16
^{235}U (α decay)	[15117-96-1]	7.038 × 10^8 yr 11	Auger L		9.48	42 6
			ce-MNO-1		14.4077 3	12.2 12
			ce-L-4		21.49 15	19.2 20
			ce-L-5		30.83 10	4 3
			ce-L-6		33.6279 5	2.08 13

Table 6 (*continued*)

Isotope	CAS Registry No.	Half-life	Type of radiation	Energy, keV Max	Energy, keV Avg	Intensity, %
			ce-K-17		34.109 20	1.68 22
			ce-MNO-4		36.78 15	4.8 4
			ce-MNO-5		46.12 10	1.5 12
			ce-L-8		52.23 20	4.2 5
			ce-MNO-8		67.52 20	1.54 16
			ce-K-24		76.064 5	4.9 5
			ce-K-27		92.469 20	2.1 3
			α 1		4218.5 20	5.5 8
			α 2		4326.4 20	4.4 5
			α 3		4366.8 20	17.0 20
			α 4		4397.8 20	54 3
			α 5		4417.1 20	2.10 20
			α 6		4558.4 20	4.2 3
			α 7		4599.7 20	5.0 5
			9 weak αs			2.55
			x ray L		13	46 6
			γ 1		19.59	61 6
			γ 6		54	2.6
			x ray Kα_2		89.9530 20	2.65 19
			x ray Kα_1		93.3500 20	4.3 3
			x ray Kβ		105	2.00 15
			γ 12		109.140 20	1.50 25
			γ 17		143.760 20	10.5 14
			γ 20		163.350 20	4.7 7
			γ 24		185.715 5	54 6
			γ 27		202.120 20	1.00 15
			γ 28		205.311 10	4.7 7
			47 weak γs			3.85
^{236}U (α decay)	[13982-70-2]	2.342×10^7 yr 4	α 1		4445 5	26 4
			α 2		4494 3	74 4
			1 weak α			0.26
			2 weak γs			0.02
^{237}U (β^- decay)	[14269-75-1]	6.75 d 1	ce-L-2		3.921 10	15.0 23
			ce-M-1		8.087 21	40 5
			Auger L		10	51 9
			ce-L-3		10.763 11	16 7
			ce-M-2		20.625 11	3.8 6
			ce-L-6		20.996 20	4.0 6
			ce-M-3		27.472 12	3.9 17
			ce-L-8		37.116 15	29 5
			ce-M-6		37.700 21	1.07 16
			ce-M-8		53.820 16	7.0 11
			Auger K		74.3	1.2 9
			ce-K-15		89.33 4	54 6
			ce-L-14		142.183 20	2.04 24
			ce-L-15		185.578 23	10.9 12
			ce-M-15		202.282 24	2.7 3
			β^- 1	187.0 11	50.1 4	3.0 5
			β^- 2	238.1 11	64.8 4	52 7
			β^- 3	251.9 11	68.8 4	44 7
			total β^-		66.0 4	100 10
			1 weak β			0.60
			x ray L		13.9	70 10
			γ 2		26.348 10	2.2 4
			γ 8		59.543 15	33 5
			γ 9		64.830 20	1.16 18
			x ray Kα_2		97.08 4	16.1 18
			x ray Kα_1		101.07 4	26 3
			x ray Kβ		114	12.2 14
			γ 14		164.610 20	1.83 21
			γ 15		208.005 23	21.7 24
			γ 20		332.36 4	1.20 14
			18 weak γs			1.85 6
^{237}Np (α decay)	[13994-20-2]	2.14×10^6 yr 1	ce-L-3		8.268 11	33 6
			Auger L		9.68	47 7
			ce-M-3		24.006 11	8.3 15
			ce-L-5		36.05 4	54 8
			ce-L-7		42.82 8	1.2 4
			ce-M-5		51.78 4	14.8 21
			ce-NOP-5		55.76 4	5.5 8
			ce-L-10		65.398 20	13.9 15
			ce-M-10		81.136 20	3.8 4
			α 1		4639.5 20	6.18 12
			α 2		4664.1 20	3.32 10

Table 6 (*continued*)

Isotope	CAS Registry No.	Half-life	Type of radiation	Energy, keV Max	Energy, keV Avg	Intensity, %
			α 3		4708.3 20	1.00 20
			α 4		4766.1 15	8 3
			α 5		4771.1 15	25 6
			α 6		4788.1 15	47 9
			α 7		4803.4 20	1.56
			α 8		4817.4 20	2.5 4
			α 9		4873.1 20	2.60 20
			12 weak αs			2.67
			x ray L		13.3	55 8
			γ 3		29.373 10	14.0 25
			γ 10		86.503 20	12.6 13
			46 weak γs			3.79
^{237}Pu (EC decay)	[15411-93-5]	45.3 d 2	ce-L-1		3.8732 9	1.59 11
			Auger L		10	38 6
			ce-L-2		10.7732 9	11.6 8
			ce-M-2		27.477 4	2.92 21
			ce-L-6		37.0732 9	2.79 17
			x ray L		13.9	52 6
			γ 6		59.5	3.25 17
			x ray Kα$_2$		97.08 4	12.7 10
			x ray Kα$_1$		101.07 4	20.6 16
			x ray Kβ		114	9.6 8
			10 weak γs			0.32
^{238}U (α decay)	[24678-82-8]	4.468 × 10^9 yr 5	Auger L		9.48	8.2 14
			ce-L-1		29.08 6	17.0 23
			ce-M-1		44.37 6	4.6 6
			ce-NOP-1		48.22 6	1.38 18
			α 1		4147 5	23 4
			α 2		4196 5	77 4
			1 weak α			0.23
			x ray L		13	8.9 15
			1 weak γ			0.07
^{238}Np (β$^-$ decay)	[15766-25-3]	2.117 d 2	Auger L		10.3	25 4
			ce-L-1		20.98 3	60 3
			ce-M-1		38.15 3	16.6 9
			ce-NOP-1		42.52 3	5.6 3
			ce-L-2		78.783 20	2.24 8
			β$^-$ 1	222.0 11	60.0 4	10.8 5
			β$^-$ 2	263.4 11	72.2 4	42.3 14
			β$^-$ 3	329.1 11	91.9 4	1.15 5
			β$^-$ 4	1247.8 11	412.4 5	44.8 15
			total β$^-$		223.0 9	100.3 22
			5 weak βs			1.28
			x ray L		14.3	37 4
			γ 19		923.980 20	2.48 14
			γ 26		984.450 20	23.8•6
			γ 27		1025.870 20	8.2 5
			γ 28		1028.540 20	17.3 10
			24 weak γs			3.63
^{238}Pu (α decay)	[13981-16-3]	87.74 yr 4	Auger L		9.89	9.1 13
			ce-L-2		21.723 10	20.7 8
			ce-M-2		37.932 10	5.69 22
			ce-NOP-2		42.039 10	1.88 8
			α 1		5456.5 4	28.3 6
			α 2		5499.21 20	71.6 6
			14 weak αs			0.10
			x ray L		13.6	11.6 14
			34 weak γs			0.05
^{239}Np (β$^-$ decay)	[13968-59-7]	2.355 d 4	ce-M-1		1.9171 14	6.6 6
			ce-NOP-1		6.2914 8	2.19 19
			Auger L		10.3	40 7
			ce-L-3		21.553 20	7.0 18
			ce-L-4		26.313 20	9.5 21
			ce-L-5		34.163 20	25 4
			ce-M-3		38.717 20	1.6 4
			ce-M-4		43.477 20	2.6 6
			ce-L-8		44.78 3	6.6 17
			ce-M-5		51.327 20	6.9 10
			ce-NOP-5		55.701 20	2.6 4
			ce-M-8		61.95 3	1.8 5
			ce-L-10		83.033 11	4.3 8
			ce-K-15		87.93 5	7.8 9
			ce-M-10		100.197 10	1.14 20
			ce-K-17		106.37 5	20.9 20

Table 6 (*continued*)

Isotope	CAS Registry No.	Half-life	Type of radiation	Energy, keV Max	Energy, keV Avg	Intensity, %
			ce-K-20		155.78 6	17.2 8
			ce-L-15		186.653 11	1.62 18
			ce-L-17		205.093 11	4.4 5
			ce-MNO-17		222.257 10	1.50 9
			ce-L-20		254.50 3	3.52 15
			ce-MNO-20		271.67 3	1.13 4
			β^- 1	209.6 19	57.0 10	1.80 20
			β^- 2	329.8 19	93.0 10	33.0 20
			β^- 3	391.3 19	112.0 10	7 3
			β^- 4	436.0 19	126.0 10	53 5
			β^- 5	664.2 19	201.0 10	2.0 10
			β^- 6	713.6 19	219.0 10	4.0 20
			total β^-		118.1 11	101 7
			4 weak βs			0.03
			x ray L		14.3	60 7
			x ray Kα_2		99.55 5	13.8 8
			x ray Kα_1		103.76 5	22.1 12
			γ 10		106.130 10	22.7 13
			x ray Kβ		117	10.4 6
			γ 15		209.750 10	3.24 25
			γ 17		228.190 10	10.7 7
			γ 20		277.60 3	14.1 4
			γ 22		315.88 4	1.59 11
			γ 23		334.30 5	2.03 18
			27 weak γs			3.62
^{239}Pu (α decay)	[15117-48-3]	2.411 × 10^4 yr 3	ce-M-2		7.3920 4	12.6 15
			Auger L		9.89	3.3 5
			ce-NOP-2		11.4992 4	5.4 7
			ce-L-4		16.93 3	2.64 12
			ce-L-8		29.86 3	4.78 21
			ce-MNO-4		33.14 3	1.93 7
			ce-MNO-8		46.07 3	1.73 6
			α 1		5105.1 5	10.6 13
			α 2		5143.2 5	15.10 20
			α 3		5155.9 5	73.2 7
			45 weak αs			0.26
			γ 1		0.073 5	100 10
			x ray L		13.6	4.2 5
			154 weak γs			0.05
^{240}U (β^- decay)	[15687-53-3]	14.1 h 2	Auger L		10	32 6
			ce-L-2		21.67 7	76 10
			ce-M-2		38.38 7	18.6 23
			ce-NOP-2		42.60 7	6.1 8
			β^- 1	360 20	102 16	100
			x ray L		13.9	44 7
			γ 2		44.10 7	1.69 20
			1 weak γ			0.11
^{240}Pu (α decay)	[14119-33-6]	6537 yr 10	Auger L		9.89	8.7 13
			ce-L-1		23.485 6	19.7 7
			ce-M-1		39.694 6	5.40 18
			ce-NOP-1		43.801 6	1.79 6
			α 1		5123	27.0 3
			α 2		5168.3	73.0 3
			5 weak αs			0.09
			x ray L		13.6	11.0 13
			13 weak γs			0.05
^{241}Pu (β^- decay)	[14119-32-5]	14.4 yr 2	β^- 1	20.81 20	5.23 5	100
^{241}Am (α decay)	[14596-10-2]	432.2 yr 5	ce-L-2		3.9182 14	16.1 9
			Auger L		10	31 5
			ce-L-5		10.768 11	16.8 16
			ce-M-2		20.622 4	4.08 21
			ce-L-9		20.993 20	9.1 14
			ce-M-5		27.472 12	4.2 4
			ce-L-14		37.1102 14	30.9 11
			ce-M-9		37.697 21	2.4 4
			ce-M-14		53.814 4	7.5 3
			α 1		5388.0 10	1.40 20
			α 2		5442.98 13	12.80 20
			α 3		5485.74 12	85.2 8
			27 weak αs			0.57
			x ray L		13.9	43 5
			γ 2		26.3450 10	2.40 10
			γ 14		59.5370 10	35.9 6
			144 weak γs			0.31

Table 6 (*continued*)

Isotope	CAS Registry No.	Half-life	Type of radiation	Energy, keV Max	Energy, keV Avg	Intensity, %
^{242}Pu (α decay)	[13982-10-0]	3.763 × 10^5 yr 20	Auger L		9.89	7.2 15
			ce-L-1		23.158 13	16.3 24
			ce-M-1		39.367 13	4.5 7
			ce-NOP-1		43.474 13	1.48 21
			α 1		4856.3 12	22.4 20
			α 2		4900.6 12	78 3
			2 weak αs			0.10
			x ray L		13.6	9.1 17
			3 weak γs			0.04
^{242}Am (EC decay)	[13981-54-9]	16.02 h 2	Auger L		10.3	8.4 13
			ce-L-1		21.42 6	7.7 3
			ce-M-1		38.59 6	2.15 8
			x ray L		14.3	12.7 13
			x ray Kα_2		99.55 5	3.66 9
			x ray Kα_1		103.76 5	5.89 13
			x ray Kβ		117	2.77 8
			1 weak γ			0.01
^{242}Am (β^- decay)		16.02 h 2	Auger L		10.7	11.9 20
			ce-L-1		17.67 11	33.17 12
			ce-M-1		35.86 11	9.30 4
			β^- 1	619.0 18	184.8 6	46 5
			β^- 2	661.2 18	199.0 7	37 4
			total		191.1 7	83 7
			x ray L		15	21.2 20
			1 weak γ			0.04
^{242}Cm (α decay)	[15510-73-3]	162.8 d 4	Auger L		10.3	7.6 12
			ce-L-1		20.98 3	19.0 4
			ce-M-1		38.15 3	5.25 15
			ce-NOP-1		42.52 3	1.74 7
			α 1		6069.63 12	25.9 5
			α 2		6112.93 8	74.1 5
			x ray L		14.3	11.4 12
			9 weak γs			0.04
^{243}Am (α decay)	[14993-75-0]	7380 yr 40	ce-L-1		8.67 15	9.2 10
			Auger L		10	12.1 19
			ce-L-3		21.10 15	4.8 5
			ce-M-1		25.38 15	2.23 24
			ce-M-3		37.81 15	1.21 11
			ce-L-6		52.24 15	13.9 8
			ce-M-6		68.95 15	3.42 19
			ce-NOP-6		73.17 15	1.21 7
			α 1		5181.0 10	1.10 20
			α 2		5233.5 10	10.730 10
			α 3		5275.4 10	87.8 5
			12 weak γs			0.29
			x ray L		13.9	16.7 19
			γ 3		43.53 15	5.5 5
			γ 6		74.67 15	66 3
			15 weak γs			1.11
^{244}Pu (α decay)	[14119-34-7]	8.26 × 10^7 yr 8	Auger L		9.89	6.2 9
			ce-L-1		23.2 10	14.2 6
			ce-M-1		39.5 10	3.90 19
			ce-NOP-1		43.6 10	1.28 7
			α 1		4546.0 10	19.4 8
			α 2		4589.0 10	80.6 8
			x ray L		13.6	7.9 10
			γ 1		45.0 10	19.4 8
^{244}Pu (SF decay)		8.26 × 10^7 yr 9	neutrons		1927	0.29
			fission fragments		175400	0.125
			prompt γ		885	1.08
			delayed γ		958	1.32
			β^- particles		1461	0.89
^{244}Cm (α decay)	[13981-15-2]	18.11 yr 2	Auger L		10.3	6.8 13
			ce-L-1		19.727 9	17.1 19
			ce-M-1		36.891 9	4.7 5
			ce-NOP-1		41.265 8	1.57 17
			α 1		5762.84 3	23.60 20
			α 2		5804.96 5	76.40 20
			5 weak αs			0.02
			x ray L		14.3	10.3 15
			17 weak γs			0.03
^{245}Cm (α decay)	[15621-76-8]	8500 yr 100	Auger L		10.3	ca 26
			ce-K-1		11.2 10	ca 45.3
			ce-K-2		52.2 10	ca 20.8

Table 6 (*continued*)

Isotope	CAS Registry No.	Half-life	Type of radiation	Energy, keV Max	Energy, keV Avg	Intensity, %
			Auger K		76	ca 1.3
			ce-L-1		109.9 10	ca 9.1
			ce-MNO-1		127.1 10	ca 3.04
			ce-L-2		150.9 10	ca 4.2
			ce-MNO-2		168.1 10	ca 1.39
			α 1		5303.8 10	5.00 10
			α 2		5362.0 7	93.2 5
			8 weak αs			1.85
			x ray L		14.3	ca 39
			x ray K$α_2$		99.55 5	ca 19.2
			x ray K$α_1$		103.76 5	ca 30.9
			x ray Kβ		117	ca 14.5
			γ 1		133.0 10	ca 4.9
			γ 2		174.0 10	ca 5
^{246}Cm (α decay)	[15757-90-1]	4730 yr 100	Auger L		10.3	6.1 10
			ce-L-1		21.448 10	15.3 8
			ce-M-1		38.612 10	4.26 21
			α 1		5343 3	21.0 10
			α 2		5386 3	79.0 10
			x ray L		14.3	9.2 11
			1 weak γ			0.03
^{246}Cm (SF decay)		4730 yr 100	neutrons		20 27	0.075
			fission fragments		183350	0.0261
			prompt γ		885	0.226
			delayed γ		958	0.190
			$β^-$ particles		1210	0.154
^{248}Cm (α decay)	[15758-33-5]	3.39 × 10^5 yr 3	Auger L		10.3	4.8 8
			ce-L-1		21.1 4	12.12 17
			ce-M-1		38.3 4	3.37 10
			ce-NOP-1		42.6 4	1.10 4
			α 1		5034.93 25	16.54 18
			α 2		5078.45 25	75.1 4
			2 weak αs			0.08
			x ray L		14.3	7.3 8
			γ 1		44.2 4	16.60 19
^{248}Cm (SF decay)		3.39 × 10^5 yr 3	neutrons		2073	25.9
			fission fragments		182800	8.26
			prompt γ		885	71.3
			delayed γ		958	77.0
			$β^-$ particles		1370	55.0
^{249}Bk ($β^-$ decay)	[14900-25-5]	320 d 6	$β^-$ 1	124.0 20	32.4 6	100
^{249}Cf (α decay)	[15237-97-5]	350.6 yr 21	Auger L		10.7	16 5
			ce-L-2		18.28 3	2.61 18
			ce-L-4		30.20 8	35 9
			ce-M-4		48.39 8	9.8 20
			ce-NOP-4		53.04 8	3.9 9
			ce-K-17		124.62 10	5.2 3
			ce-L-17		228.35 9	1.08 6
			ce-K-25		259.69 7	1.34 6
			α 1		5759.7 10	3.66
			α 2		5813.5 10	84.4
			α 3		5849.5 10	1
			α 4		5903.4 10	2.79
			α 5		5946.2 10	4
			α 6		6139.5 7	1.1
			α 7		6194.0 7	2.17
			26 weak αs			0.84
			x ray L		15	29 7
			x ray K$α_2$		104.61 5	2.10 10
			x ray K$α_1$		109.29 5	3.35 15
			x ray Kβ		123	1.59 8
			γ 17		252.88 8	2.73 11
			γ 22		333.44 5	15.5 5
			γ 25		387.95 5	66.0 20
			36 weak γs			2.21
^{250}Cf (α decay)	[13982-11-1]	13.08 yr 9	Auger L		10.7	4.0 8
			ce-L-1		18.32 4	11.0 9
			ce-M-1		36.516 19	3.1 3
			ce-NOP-1		41.167 21	1.03 9
			α 1		5989.1 6	15.1 12
			α 2		6030.8 6	84.5 12
			2 weak αs			0.31
			x ray L		15	7.0 9
			1 weak γ			0.01

Table 6 (*continued*)

Isotope	CAS Registry No.	Half-life	Type of radiation	Energy, keV Max	Energy, keV Avg	Intensity, %
^{250}Cf (SF decay)		13.08 yr 9	neutrons		2127	0.27
			fission fragments		190851	0.077
			prompt γ		885	0.67
			delayed γ		958	0.48
			β^- particles		1119	0.42
^{252}Cf (α decay)	[13981-17-4]	2.638 yr 10	Auger L		10.7	4.0 8
			ce-L-1		18.87 4	11.0 8
			ce-M-1		37.06 3	3.07 21
			α 1		6075.7 5	15.2 3
			α 2		6118.3 5	81.6 3
			3 weak αs			0.23
			x ray L		15	7.1 9
			3 weak γs			0.03
^{252}Cf (SF decay)		2.638 yr 10	neutrons		2164	11.5
			fission fragments		190286	3.09
			prompt γ		885	26.7
			delayed γ		958	25.1
			β^- particles		1279	19.2
^{253}Cf (β^- decay)	[15720-29-3]	17.81 d 8	β^- 1	289 10	79 3	100
^{253}Es (α decay)	[15840-02-5]	20.47 d 3	Auger L		10.9	2.0 5
			ce-L-3		16.52 6	4.6 5
			ce-M-3		35.24 6	1.16 13
			α 1		6591.90 23	5.9 5
			α 2		6594.01 23	1.4
			α 3		6633.03 23	91.0 10
			17 weak αs			2.07
			γ 1		8.80 10	3.0 3
			x ray L		15.3	3.8 6
			64 weak γs			0.14
^{254}Cf (SF decay)	[22095-76-7]	60.5 d 2	neutrons		2187	387.8
			fission fragments		189728	99.7
			prompt γ		885	861
			delayed γ		958	1026
			β^- particles		1440	698
^{254}Fm (α decay)	[15750-23-9]	3.240 h 2	Auger L		11.2	3.5 9
			ce-L-1		16 4	10.2 11
			ce-M-1		35 4	2.9 3
			α 1		7147	14.0 10
			α 2		7189 5	85.0 10
			2 weak αs			0.90
			x ray L		15.7	7.3 11
			3 weak γs			0.04
^{254}Fm (SF decay)		3.240 h 2	neutrons		2203	0.24
			fission fragments		198457	0.0592
			prompt γ		885	0.51
			delayed γ		958	0.31
			β^- particles		1028	0.30
^{257}Fm (α decay)	[15750-26-2]	100.5 d 2	Auger L		11.2	20 5
			ce-K-4		44.7 6	38 7
			ce-L-3		77.7 17	4.7 7
			Auger K		83.3	1.2 12
			ce-M-3		96.9 17	1.14 16
			ce-K-5		106.4 7	20 4
			ce-L-4		153.7 6	8.1 15
			ce-M-4		172.9 6	2.0 4
			ce-L-5		215.4 7	4.2 7
			ce-M-5		234.6 7	1.03 18
			α 1		6441 4	2.0 3
			α 2		6519.0 20	93.8 10
			α 3		6696 3	3.2 3
			x ray L		15.7	43 7
			x ray Kα_2		109.87 5	17.1 23
			x ray Kα_1		115.07 5	27 4
			x ray Kβ		129	13.0 18
			γ 4		179.7 6	6.4 12
			γ 5		241.4 7	7.5 13
			3 weak γs			0.75
^{275}Fm (SF decay)		100.5 d 2	neutrons		2199	0.83
			fission fragments		197596	0.210
			prompt γ		885	1.81
			delayed γ		958	1.66
			β^- particles		1263	1.29

[a] Each one or two of the uncertainty digits (the numbers after the spaces of columns 3, 5, 6, and 7) refers to the last one or two digits of the numerical value.

intensity limit is 1.0%. The listing of the radiation type is followed immediately by the number and the summed intensity of the radiations omitted because of the chosen cutoff limit.

The numerical labels on the γ radiations are based on the radiations contained in the original data set on which MEDLIST operates. These labels are maintained throughout the MEDLIST calculations and are carried into the output. In particular, when low-intensity radiations are omitted, the remaining radiations are not, in the γ-ray listing, relabeled. Thus the sequence γ 1, γ 2, γ 4, ..., may appear, indicating that γ 3, with energy between that of γ 2 and γ 4, has been omitted because its intensity is below the cutoff limit. However, a conversion line could still appear associated with γ 3, eg, ce-K-3, if the conversion coefficient is large enough such that the intensity of ce-K-3 exceeds the cutoff limit.

For those nuclei that decay by positron emission, MEDLIST prints a comment at the end of the γ-ray list, giving the maximum possible observable intensity of the annihilation radiation ($\gamma\pm$), calculated from $I(\gamma\pm) = 2 \Sigma_i I(\beta_i^+)$. The actual fraction of these events observed in practice depends upon source attenuation and the detector efficiency.

Table 6 contains data on the following atomic and nuclear radiations: Auger (K, L shells); x ray (L, K_{α_1}, K_{α_2}, K_β); β^-, β^+; α; γ; and ce (K, L, ... shells) emitted with intensity > 1.0 per 100 disintegrations of the parent nucleus (ie, >1.0%).

Auger electrons and x rays resulting from the filling of vacancies in the M, N, ... shells, although sometimes occurring with large intensities, are not included. These transitions have energies ranging from $\lesssim 2$ keV in ^{197}Hg to $\lesssim 0.6$ keV in ^{113}Sn decay. Their intensities, however, cannot, in general, be calculated since the contribution to the total number of, eg, M-shell vacancies due to initial K- or L-shell vacancies has not been measured. However, theoretical estimates of some of these quantities are available for limited Z regions. For ^{83m}Kr, based on theoretical values of the required atomic constants (18), the yield of M-shell Auger transitions (with energy ca 0.2 keV) is estimated to be ca 300%. For ^{125}I decay, where M-shell vacancies are produced both in electron-capture decay and the subsequent internal conversion process, the yield of M-shell Auger transitions is estimated to be ca 590%. The total energy emitted per decay by the low-energy Auger transitions (ca 0.5 keV) is ca 3 keV.

The intensity given for the total β entry is the sum of the intensities of the individual groups. For β^- emitters with no alternative mode of decay, this intensity must be exactly 100%. However, since the intensities of the individual β^- groups are usually determined from the γ-ray intensities, which in turn are usually determined by the requirement that the total ($\beta^- + \gamma$) feeding of the daughter ground state be 100%, the intensity of the total β^- contributing, as calculated from the individual β^- intensities, does not always add up to exactly 100%. This summed value could be replaced by 100%, but this has not been done in the tables.

The radioisotopes are listed in order of increasing mass number, A. For nuclei with the same A, the order is by increasing Z.

No separate entry is made in the γ-ray listing for the 511-keV annihilation radiation ($\gamma\pm$) accompanying positron decay. Instead, the intensity of this radiation, calculated from $I(\gamma\pm) = 2 \times \Sigma_i I(\beta_i^+)$, is given at the end of each data set. This intensity is presented separately, and as a maximum value, since the annihilation process takes place externally to the decaying atom in the medium surrounding the atom. The fraction of the maximum of importance in a given application therefore depends upon the extent of the region surrounding the decaying atom that is being considered. The intensity is calculated from the complete data listing and includes contributions from individual β^+ branches that may not be given in the table because of the adopted 1% intensity cutoff.

Each radioisotope entry in Table 6 is characterized by atomic mass number, chemical symbol, and half-life. The m that usually accompanies the mass number for decays that do not originate from the ground state, as in ^{99m}Tc and ^{110m}Ag, does not appear in the table.

Parent-Daughter Activity Ratios. A thorough discussion of the general case of activity relations among members of a radioactive decay chain is given in ref. 10. For the special case of a radioactive parent, p, with one radioactive daughter, d, fed in a fraction, f, of the parent decays, the ratio of daughter activity to that of the parent (both at time t) is given by

$$\frac{N_d \lambda_d}{N_p \lambda_p} = \frac{f \lambda_d}{\lambda_d - \lambda_p}[1 - e^{-(\lambda_d - \lambda_p)t}] \qquad (10)$$

where λ_p and λ_d are the decay constants ($\lambda = \ln 2/t_{1/2}$) of the parent and daughter nuclides, respectively. The activity of the daughter is taken to be zero at time $t = 0$ (ie, the parent source is initially pure).

$t_{1/2}, d < t_{1/2}, p$. Frequently, the daughter half-life is shorter than the parent half-life. In such cases, the ratio of daughter-to-parent activities first increases with time and then approaches a constant value. For time t large compared with $1/(\lambda_d - \lambda_p)$, equation (10) reduces to

$$\frac{f \lambda_d}{\lambda_d - \lambda_p} = \frac{f t_{1/2,p}}{t_{1/2,p} - t_{1/2,d}} \qquad (11)$$

$t_{1/2}, d > t_{1/2}, p$. In this case, the daughter-to-parent activity ratio increases continuously with time. An example is the chain

$$^{56}\text{Ni}(\epsilon) \rightarrow {}^{56}\text{Co}(\epsilon) \rightarrow {}^{56}\text{Fe (stable)}$$

for which $t_{1/2}, p = 6.10$ d 2 $t_{1/2}, d = 78.76$ d 12, and $f = 1$. From equation 10, the daughter-to-parent activity ratio as a function of time after preparation of an initially pure ^{56}Ni source is

$$(0.0840 \ 4)(e^{(0.1048 \ 4)t} - 1) \qquad (12)$$

When $t_{1/2}, d \gg t_{1/2}, p$ ($\lambda_p \gg \lambda_d$), and for $t \gg t_{1/2}, p$, equation 10 with the aid of equation 2 reduces to

$$N_d \lambda_d \approx f N_{op} \lambda_d e^{-\lambda_d t} \qquad (13)$$

The initial source of parent atoms, N_{op}, decaying exponentially as $e^{-\lambda_p t}$ has thus become a source of daughter atoms, N_{od} ($N_{od} \approx N_{op}$), decaying exponentially as $e^{-\lambda_d t}$.

Radioisotopes given in Table 6 that decay to radioactive daughters are listed in Table 7.

Table 7. Radioisotopes with Radioactive Daughters

Parent radioisotope	Half-life[a]	Decay mode	Branching, %[a]	Daughter radioisotope	Half-life[a]
^{28}Mg	20.90 h 3	β^-	100	^{28}Al	2.240 min 1
^{32}Si	105 yr 18	β^-	100	^{32}P	14.26 d 4
^{38}S	170.3 min 7	β^-	100	^{38}Cl	37.24 min 5
^{42}Ar	32.9 yr 11	β^-	100	^{42}K	12.360 h 3
^{44}Sc	58.6 h 1	IT	98.63 1	^{44}Sc	3.927 h 8
^{44}Ti	47.3 yr 12	ϵ	100	^{44}Sc	3.927 h 8
^{47}Ca	4.536 d 2	β^-	100	^{47}Sc	3.351 d 2
^{48}Cr	22.96 h 3	ϵ	100	^{48}V	16.238 d 25
^{52}Fe	8.275 h 8	ϵ	100	^{52}Mn	21.1 min 2
^{55}Co	17.54 h 4	ϵ	100	^{55}Fe	2.7 yr 1
^{56}Ni	6.10 d 2	ϵ	100	^{56}Co	78.76 d 12
^{57}Ni	36.08 h 9	ϵ	100	^{57}Co	270.9 d 6
^{58}Co	9.15 h 10	IT	100	^{58}Co	70.80 d 8
^{66}Ge	2.27 h 5	ϵ	100	^{66}Ga	9.40 h 7
^{68}Ge	288 d 6	ϵ	100	^{68}Ga	68.06 min 26
^{69}Zn	13.76 h 3	IT	100	^{69}Zn	57 min 1
^{71}As	64.8 h 7	ϵ	100	^{71}Ge	11.8 d 4
^{72}Zn	46.5 h 1	β^-	100	^{72}Ga	14.10 h 1
^{72}Se	8.40 d 8	ϵ	100	^{72}As	26.0 h 1
^{73}Se	7.15 h 8	ϵ	100	^{73}As	80.30 d 6
^{75}Br	97 min 2	ϵ	100	^{75}Se	119.770 d 10
^{76}Kr	14.8 h 1	ϵ	100	^{76}Br	16.2 h 2
^{77}Ge	11.30 h 1	β^-	100	^{77}As	38.83 h 5
^{77}Kr	74.4 min 6	ϵ	100	^{77}Br	57.036 h 6
^{78}Ge	88 min 1	β^-	100	^{78}As	90.7 min 2
^{80}Br	4.42 h 1	IT	100	^{80}Br	17.68 min 2
^{83}Br	2.39 h 2	β^-	99.975 8	^{83}Kr	1.83 h 2
^{83}Rb	86.2 d 1	ϵ	76.2 4	^{83}Kr	1.83 h 2
^{83}Sr	32.4 h 2	ϵ	100	^{83}Rb	86.2 d 1
^{85}Kr	4.480 h 8	IT	21.0 6	^{85}Kr	10.72 yr 2
^{86}Zr	16.5 h 1	ϵ	100	^{86}Y	14.74 h 2
^{87}Kr	76.3 min 5	β^-	100	^{87}Rb	4.80×10^{10} yr 13
^{87}Y	80.3 h 3	ϵ	100	^{87}Sr	2.81 h 1
^{87}Y	12.9 h 4	IT	98.43 16	^{87}Y	80.3 h 3
^{87}Zr	104.0 min 5	ϵ	99.70 6	^{87}Y	12.9 h 4
^{88}Kr	2.84 h 2	β^-	100	^{88}Rb	17.8 min 1
^{88}Zr	83.4 d 3	ϵ	100	^{88}Y	106.64 d 8
^{90}Sr	29.12 yr 24	β^-	100	^{90}Y	64.0 h 1
^{90}Y	3.19 h 1	IT	100	^{90}Y	64.0 h 1
^{91}Sr	9.52 h 6	β^-	59 5	^{91}Y	49.71 min 4
		β^-	41 5	^{91}Y	58.51 d 6
^{92}Sr	2.71 h 1	β^-	100	^{92}Y	3.54 h 1
^{93}Zr	1.53×10^6 yr 10	β^-	100	^{93}Nb	13.6 yr 3

Table 7 (*continued*)

Parent radioisotope	Half-life[a]	Decay mode	Branching, %[a]	Daughter radioisotope	Half-life[a]
^{93}Mo	3500 yr 700	ϵ	100	^{93}Nb	13.6 yr 3
^{93}Mo	6.85 h 7	IT	100	^{93}Mo	3500 yr 700
^{95}Zr	63.98 d 6	β^-	100	^{95}Nb	35.15 d 3
^{97}Zr	16.90 h 5	β^-	100	^{97}Nb	72.1 min 7
^{97}Ru	2.9 d 1	β^-	100	^{97}Tc	2.6×10^6 yr 4
^{99}Mo	66.0 h 2	β^-	100	^{99}Tc	6.02 h 3
^{99}Tc	6.02 h 2	IT	100	^{99}Tc	2.13×10^5 yr 5
^{101}Rh	4.34 d 1	IT	7.7 6	^{101}Rh	3.3 yr 3
^{101}Pd	8.47 h 6	ϵ	99.76 2	^{101}Rh	4.34 d 1
^{105}Ru	4.44 h 2	β^-	100	^{105}Rh	35.36 h 6
^{106}Ru	371.63 d 17	β^-	100	^{106}Rh	29.80 s
^{108}Ag	127 yr 21	IT	8.9 6	^{108}Ag	2.37 min 1
^{109}In	4.2 h 1	ϵ	100	^{109}Cd	464 d 1
^{110}Ag	249.9 d 1	IT	1.33 10	^{110}Ag	24.6 s 2
^{111}Pd	5.5 h 1	IT	73 3	^{111}Pd	23.4 min 2
		β^-	27 3	^{111}Ag	7.45 d 1
^{113}Sn	115.1 d 2	ϵ	100	^{113}In	1.658 h 1
^{114}In	49.51 d 1	IT	95.7 3	^{114}In	71.9 s 4
^{115}Cd	53.46 h 10	β^-	100	^{115}In	4.486 h 4
^{115}In	4.486 h 4	IT	95.0 7	^{115}In	5.1×10^{14} yr 4
^{117}Cd	2.49 h 4	β^-	91.7 4	^{117}In	116.5 min 7
		β^-	8.3 4	^{117}In	43.8 min 7
^{117}In	116.5 min 7	IT	47.1 15	^{117}In	43.8 min 7
^{117}Te	62 min 2	ϵ	100	^{117}Sb	2.80 h 1
^{119}Te	16.05 h 5	ϵ	100	^{119}Sb	38.1 h 2
^{119}Te	4.69 d 4	ϵ	100	^{119}Sb	38.1 h 2
^{121}Sn	55 yr 5	IT	77.6 20	^{121}Sn	27.06 h 4
^{121}Te	154 d 7	IT	88.6 11	^{121}Te	16.8 d 4
^{121}I	2.12 h 1	ϵ	100	^{121}Te	16.8 d 4
^{123}Xe	2.08 h 2	ϵ	100	^{123}I	13.2 h 1
^{127}Sn	2.10 h 4	β^-	100	^{127}Sb	3.85 d 5
^{127}Sb	3.85 d 5	β^-	86.1 8	^{127}Te	9.35 h 7
		β^-	13.9 8	^{127}Te	109 d 2
^{127}Te	109 d 2	IT	97.6 2	^{127}Te	9.35 h 7
^{129}Te	69.6 min 2	β^-	100	^{129}I	1.57×10^7 yr 4
^{129}Te	33.6 d 1	IT	65 6	^{129}Te	69.6 min 2
		β^-	35 6	^{129}I	1.57×10^7 yr 4
^{131}Te	30 h 2	IT	22.2 16	^{131}Te	25.0 min 1
		β^-	77.8 16	^{131}I	8.04 d 1
^{131}I	8.04 d 1	β^-	1.086 13	^{131}Xe	11.9 d 1
^{131}Ba	11.8 d 2	ϵ	100	^{131}Cs	9.69 d 1
^{133}I	20.8 h 1	β^-	2.883 23	^{133}Xe	2.19 d 1
		β^-	97.117 23	^{133}Xe	5.245 d 6

Table 7 (*continued*)

Parent radioisotope	Half-life[a]	Decay mode	Branching, %[a]	Daughter radioisotope	Half-life[a]
^{133}Xe	2.19 d 1	IT	100	^{133}Xe	5.245 d 6
^{133}Ba	38.9 h 1	IT	100	^{133}Ba	10.74 yr 5
^{134}Cs	2.90 h 1	IT	100	^{134}Cs	2.062 yr 5
^{135}I	6.61 h 1	β^-	15.8 5	^{135}Xe	15.29 min 3
			84.2 5	^{135}Xe	9.09 h 1
^{135}Xe	9.09 h 1	β^-	100	^{135}Cs	2.3×10^6 yr 3
^{137}Cs	30.0 yr 2	β^-	94.6 3	^{137}Ba	2.552 min 2
^{140}Ba	12.746 d 10	β^-	100	^{140}La	40.272 h 7
^{140}Nd	3.37 d 2	ϵ	100	^{140}Pr	3.39 min 1
^{144}Ce	284.9 d 2	β^-	100	^{144}Pr	17.28 min 5
^{145}Sm	340 d 3	ϵ	100	^{145}Pm	17.7 yr 4
^{147}Nd	10.98 d 1	β^-	100	^{147}Pm	2.6234 yr 2
^{147}Gd	38.1 h 1	ϵ	100	^{147}Eu	24 d 1
^{148}Pm	41.3 d 1	IT	4.6 5	^{148}Pm	5.37 d 1
^{149}Nd	1.73 h 1	β^-	100	^{149}Pm	53.08 h 5
^{151}Tb	17.6 h	ϵ	100	^{151}Gd	120 d 20
^{157}Dy	8.1 h 1	ϵ	100	^{157}Tb	150 yr 30
^{171}Er	7.52 h 3	β^-	100	^{171}Tm	1.92 yr 1
^{177}Yb	1.9 h 1	β^-	100	^{177}Lu	6.71 d 1
^{177}Lu	160.9 d 3	IT	21 5	^{177}Lu	6.71 d 1
^{183}Hf	64 min 1	β^-	100	^{183}Ta	5.1 d 1
^{184}Hf	4.12 h 5	β^-	100	^{184}Ta	8.7 h 1
^{184}Re	165 d 5	IT	74.7 6	^{184}Re	38.0 d 5
^{188}W	69.4 d 5	β^-	100	^{188}Re	16.98 h 2
^{188}Pt	10.2 d 3	ϵ	100	^{188}Ir	41.5 h 5
^{191}Os	13.10 h 5	IT	100	^{191}Os	15.4 d 1
^{192}Ir	241 yr 9	IT	100	^{192}Ir	74.02 d 18
^{194}Os	6.0 yr 2	β^-	100	^{194}Ir	19.15 h 3
^{196}Au	9.7 h 1	IT	100	^{196}Au	6.183 d 10
^{197}Pt	94.4 min 8	IT	96.7 10	^{197}Pt	18.3 h 3
^{198}Tl	1.87 h 3	IT	47 4	^{198}Tl	5.3 h 5
^{200}Pb	21.5 h 4	ϵ	100	^{200}Tl	26.1 h 1
^{201}Pb	9.4 h 2	ϵ	100	^{201}Tl	3.044 d 9
^{202}Pb	ca 3×10^5 yr	ϵ	100	^{202}Tl	12.23 d 2
^{202}Pb	3.62 h 3	ϵ	9.5 5	^{202}Tl	12.23 d 2
		IT	90.5 5	^{202}Pb	ca 3×10^5 yr
^{203}Bi	11.76 h 5	ϵ	100	^{203}Pb	52.05 h 10
^{205}Bi	15.31 d 4	ϵ	100	^{205}Pb	1.43×10^7 yr 15
^{206}Po	8.8 d 1	ϵ	94.55 5	^{206}Bi	6.243 d 3
^{207}Po	350 min 4	ϵ	100	^{207}Bi	38 yr 3
^{209}Po	102 yr 5	α	99.74 3	^{205}Pb	1.43×10^7 yr 15
^{211}At	7.214 h 7	ϵ	58.3 2	^{211}Po	0.516 s 3
		α	41.7 2	^{207}Bi	38 yr 3

[a] Each one or two of the uncertainty digits (the numbers after the spaces) refers to the last one or two digits of the numerical value.

Uses

In 1977, ERDA laboratories alone made 3310 shipments of 132 different radioisotopes with a total value of 10×10^6 to about 350 U.S. and 50 non-U.S. customers (19). The world market for radioactive isotopes was ca 38×10^7 in February 1982 (20). Any issue of *Energy Abstracts* or *Chemical Abstracts* reveals the wide variety of applications of these radioisotopes (see Radioactive drugs; Radioactive tracers). For example, superphosphates or nitrophosphates, labeled with ^{32}P for agronomical investigations, tritium-gas-filled light sources, and ^{85}Kr- and ^{147}Pm-activated self-luminous compounds. Cesium-137 irradiation is being studied for treatment of waste to be used both for land application and for animal feeding. Strontium-90 has application for unattended power generators using the heat of β emission (21) (see Power generation). An excellent detailed description of many of the industrial uses of radioisotopes is given in refs. 22–24 (see also Liquid-level measurement; Nondestructive testing; X-ray technology). A list of sources with complete address is given in the yearly "Buyers Guide" issue of *Nuclear News*.

Radioisotope Detectors. A thorough discussion of the various types of detectors and detector systems currently in common use is given in refs. 9, 22, and 25–28.

BIBLIOGRAPHY

"Isotopes" in *ECT* 1st ed., Vol. 8, pp. 89–104, by D. H. Templeton, University of California; "Radioisotopes" in *ECT* 1st ed., Second Supplement Volume, pp. 681–708, by Alan Beerbower, Esso Research and Engineering Co.; "Radioisotopes" in *ECT* 2nd ed., Vol. 17, pp. 64–130, by Alan Beerbower and Annelle E. von Rosenberg, Esso Research and Engineering, Co.

1. F. Soddy, *Chem. Soc. London, Annu. Rep. Prog. Chem.* **99,** 72 (1911).
2. H.-P. Hahn, H.-J. Born, and J. I. Kim, *Radiochim. Acta* **23,** 23 (1976).
3. H. W. Johlige, D. C. Aumann, and H.-J. Born, *Phys. Rev.* **C2,** 1616 (1970).
4. W. K. Hensley, W. A. Bassett, and J. R. Huizenga, *Science* **181,** 1164 (1973).
5. T. Kirsten and H. W. Muller, *Earth Planet. Sci. Lett.* **6,** 271 (1969).
6. B. Srinivasan, E. C. Alexander, and O. K. Manuel, *J. Inorg. Nucl. Chem.* **34,** 2381 (1972).
7. E. W. Hennecke, O. K. Manuel, and D. D. Sabu, *Phys. Rev.* **C11,** 1378 (1975).
8. K. A. Erb, W. D. Callender, and R. K. Sheline, *Phys. Rev.* **C20,** 2031 (1979).
9. W. R. Hendee, *Radioactive Isotopes in Biological Research*, John Wiley & Sons, Inc., New York, 1973.
10. R. D. Evans, *The Atomic Nucleus*, McGraw-Hill, Inc., New York, 1955.
11. H. F. Schopper, *Weak Interactions and Nuclear Beta Decay*, North-Holland, Amsterdam, 1966, pp. 76–84.
12. A. Prince, "Nuclear and Physical Properties of ^{252}Cf" in *Proceedings of a Symposium on Californium 252*, U.S. Atomic Energy Commission Report CONF-681032, 1969.
13. Earl K. Hyde, *Fission Phenomena*, Part 3 of *The Nuclear Properties of the Heavy Elements*, Prentice-Hall, Inc., N.J., 1964.
14. L. T. Dillman, *EDISTRA—A Computer Program to Obtain a Nuclear Decay Data Base for Radiation Dosimetry*, Oak Ridge National Laboratory Report ORNL/TM-6689, Oak Ridge, Tenn.
15. J. P. Unik and J. E. Gindler, *A Critical Review of the Energy Released in Nuclear Fission*, Argonne National Laboratory Report ANL-7748, Argonne, Ill.

16. W. B. Ewbank, Nuclear Data Project, and M. J. Kowalski, Great Lakes College Association (Oak Ridge Science Semester participant from Ohio Wesleyan University), MEDLIST.
17. W. B. Ewbank, Y. A. Ellis, and M. R. Schmorak, *Nucl. Data Sheets* **26,** 1 (1979).
18. W. Bambynek, B. Crasemann, R. W. Fink, H.-U. Freund, H. Mark, C. D. Swift, R. E. Price, and P. Venugopala Rao, *Rev. Mod. Phys.* **44,** 716 (1972).
19. J. L. Simmons, List of ERDA Radioisotope Customers with Summary of Radioisotope Shipments FY 1977, PNL-2572, July 1978.
20. *Chem. Week*, 15 (Feb. 24, 1982).
21. *Proc. Jpn. Conf. Radioisot.* (13), 265 (May 1978).
22. J. F. Cameron and C. G. Clayton, *Radioisotope Instruments*, Pergamon Press, New York, 1971.
23. L. H. Meyer, ed., *Applications of Radioisotopes*, Pt. 22 of *Nuclear Engineering*, American Institute of Chemical Engineers, New York, 1970.
24. *Science* **215,** 377 (1982).
25. G. F. Knoll, *Radiation Detectors and Measurements*, John Wiley & Sons, Inc., New York, 1979.
26. D. G. Miller, *Radioactivity and Radiation Detection*, Gordon and Breach, Science Publishers, Inc., New York, 1972.
27. P. Ouseph, *Introduction to Nuclear Radiation Detectors*, Plenum Press, New York, 1975.
28. W. B. Mann, R. L. Ayres, and S. B. Garfinkle, *Radioactivity and Its Measurement*, Pergamon Press, New York, 1980.

MURRAY J. MARTIN
Oak Ridge National Laboratory

RADIOPAQUES

Radiopaques are diagnostic agents that permit physicians to examine patients and diagnose abnormalities and impairment of organ functions (see also Medical diagnostic agents). By definition, radiopaques cause soft-tissue structures, such as the stomach, heart, and gallbladder, to become visible during x-ray examination by inhibiting the passage of x rays and producing a shadow of positive contrast (1). The term contrast media refers to radiopaques and also to materials that produce a shadow of negative contrast, that is, the part to be visualized is less dense than the surrounding tissue structures (1). Air is an example of a contrast medium with a negative shadow. The ideal contrast medium should permit rapid and adequate visualization of the organ under investigation, have no pharmacodynamic effect in the body, and be rapidly eliminated from the body. Almost all contrast media currently in use are radiopaques (see also Radioactive drugs).

X rays were discovered by Roentgen in 1895 and in a short time were used by physicians for examinations. The relationship between the atomic number of an element and its ability to absorb x rays is approximated by

$$\mu = k\lambda^3 Z^4 + 0.2$$

where μ is the absorption coefficient, k a constant, λ the wavelength of the x rays, Z the atomic number, and 0.2 the average coefficient of scattering (2). It can be seen from this equation that elements with high atomic number are much more effective for absorbing x rays than those of low atomic number (see also X-ray technology).

Iodine is the most useful element for producing radiopaques with satisfactory properties (see Iodine and iodine compounds). It can be incorporated into a large number of organic compounds in which it forms a strong covalent bond. Barium sulfate is currently the only widely used radiopaque that does not contain iodine. In the 1930s and 1940s colloidal thorium oxide [1314-20-1] was employed to visualize the liver, spleen, blood vessels, and spinal canal. It produced good diagnostic shadows but is radioactive (see Radioactivity, natural) and it could not be removed from the body. Numerous reports describe problems, including malignancies, after the employment of thorium oxide (3); therefore, its use was discontinued. Radiographic procedures, the organ or area involved, and the type of agent are given in Table 1.

Alimentary Tract

The alimentary tract was the first area of the body extensively investigated with radiopaques. For this purpose, barium sulfate [7727-43-7] was found to be more suitable (4) than other inorganic compounds such as salts of lead (5) and bismuth (6). Barium sulfate has a high degree of radiopacity and low cost. The low toxicity exhibited after oral administration is probably owing to lack of absorption and almost complete insolubility in aqueous media. Formulations are available that contain added ingredients to reduce agglomeration and maintain the suspension of barium sulfate. The two infrequent hazards in its application are the accidental entrance of barium sulfate into the peritoneal cavity and the danger of impaction if there is a partial obstruction.

Table 1. Some Radiographic Procedures

Procedure	Organ or area involved	Type of agent[a]
angiography	blood vessels	water-soluble
aortography	aorta	water-soluble
arthrography	joints	water-soluble
bronchography	bronchial tree	suspension or oil-type
cholangiography	bile ducts	cholangio-cholecystographic
cholecystography	gallbladder	cholecystographic
hepatography	liver	
hysterosalpingography	uterus and fallopian tubes	oil-type and water-soluble
lymphangiography	lymphatic ducts	oil-type and water-soluble
mammography	mammary glands	oil-type and water-soluble
myelography	subarachnoid space of the spinal canal	oil-type and water-soluble
sialography	salivary glands	oil-type
urography, pyelography	urinary tract	water-soluble
ventriculography	ventricules of the brain	air, water-soluble

[a] Concentrations and amounts vary greatly depending on patient, patient's age, agent, and other factors.

Sometimes aqueous solutions of the ionic organic iodine compounds employed in urography–angiography are used for visualization of the alimentary tract. These compounds are poorly absorbed from the gastrointestinal tract. The aqueous solutions will be readily absorbed and excreted if the material enters the peritoneal cavity and will not cause impaction. In addition, these solutions have a low viscosity which may produce better visualization of small lesions than barium sulfate. Disadvantages of these solutions are their high costs and their osmotic activity which dilutes the concentration and may cause poor visualization of the small intestine as well as diarrhea. The new low-osmolality myelographic–angiographic agents have less osmotic activity than the ionic iodoradiopaques, but they are very expensive and their use would not be practical for visualization of the alimentary tract except in special circumstances.

Urography and Angiography

Urography or pyelography is the visualization of the urinary tract by means of radiopaques. In excretion urography, the solution is given intravenously and the kidneys concentrate and excrete the radiopaque, whereas in retrograde urography, the solution is injected by catheter up the ureter into the pelvis of the kidney. Excretion urography is used more frequently and the drug may be administered by injection at a moderate rate or by drip infusion. The compounds developed for excretion urography are also employed for retrograde urography and angiography, which is the visualization of the blood vessels by means of radiopaques. Most of the urographic–angiographic agents are ionic, but recently nonionic agents were introduced for myelography and angiography. These nonionic media have lower osmolality and toxicity than the ionic media. Some urographic–angiographic agents and their iodine content, osmolality, and toxicity are given in Table 2.

Sodium iodide [7681-82-5] was the first agent reported for visualization of the urinary tract after intravenous injection (15), but because of irritation and systemic

Table 2. Some Urographic-Angiographic Agents

Name	CAS Registry Number	Structure	Iodine, %	LD$_{50}$ in mice, g I/kg	Osmolality, osmol/kg at 280 mg I/mL
sodium acetrizoate	[129-63-5]	(6) Na salt	65.8	5.5^a	
sodium diatrizoate	[737-31-5]	(7) Na salt	59.9	8.4^a	1.51^{b,c}
sodium iothalamate	[1225-20-3]	(10) Na salt	59.9	8.0^a	1.69^{b,c}
sodium metrizoate	[7225-61-8]	(13) Na salt	58.5	9.1^a	1.66^{b,c}
sodium iodamide	[10098-82-5]	(15) Na salt	58.5	9.0^a	1.79^{b,c,d}
metrizamide	[31112-62-6]	(17)	48.2	17.5^{e,f}	0.43^g
iopamidol	[60166-93-0]	(20)	49.0	21.8^h	0.47^i
iohexol	[66108-95-0]	(23)	46.4	23.4^j	0.62^g
sodium ioxaglate	[67992-58-9]	(26) Na salt	59.0	11.5^k	0.49^{i,l}

^{a} Calculated from data in ref. 7.
^{b} Ref. 8.
^{c} N-Methylglucamine (NMG) salt.
^{d} For a solution with 300 mg I/mL.
^{e} Ref. 9.
^{f} Male mice; the value for female mice was 18.7.
^{g} Ref. 10.
^{h} Ref. 11.
^{i} Ref. 12.
^{j} Ref. 13.
^{k} Ref. 14.
^{l} A 1:2 mixture of sodium NMG salts.

reactions, it was not widely employed. The first organic iodine compound used for urography was 5-iodo-1-methyl-2(1H)-pyridinone [60154-05-4] (1), but it was soon found that 5-iodo-2-oxo-1(2H)-pyridineacetic acid [80462-95-9] (2) was better suited (16). Within a short time a number of more effective urographic compounds were reported (17), including sodium methiodal [126-31-8] (3), sodium iodomethamate [519-26-6] (4), and iodopyracet [300-37-8] (5). Some of these were in clinical use for many years. The sodium salt of iodopyracet was first employed, but the diethanolamine salt was found to be more soluble and thus more useful (18).

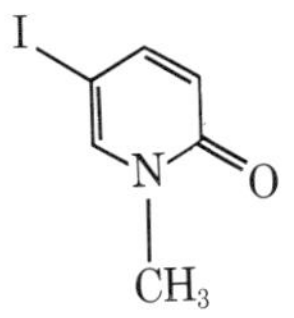

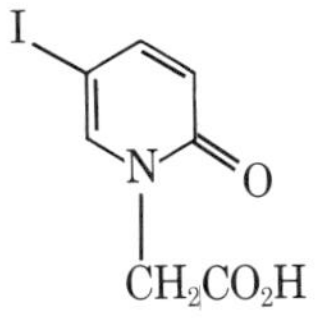

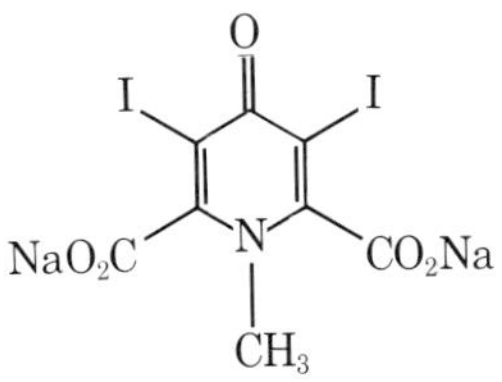

The introduction of acetrizoic acid [85-36-9] (6) was an important advance in radiopaques (19). This compound had a favorable toxicity compared with earlier media

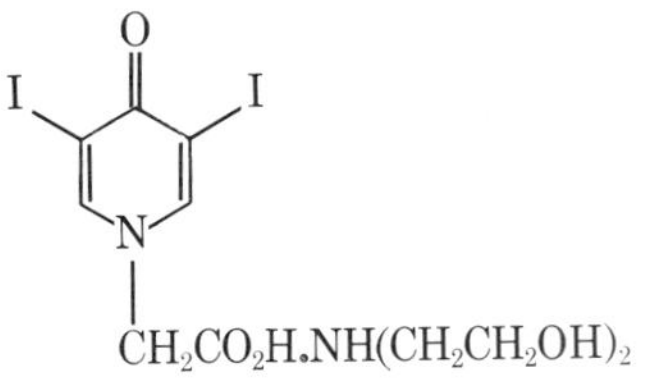

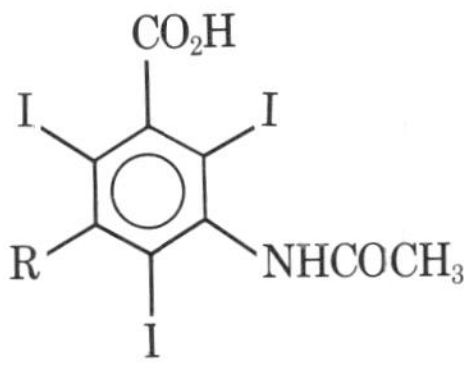

(6) R = H
(7) R = NHCOCH₃

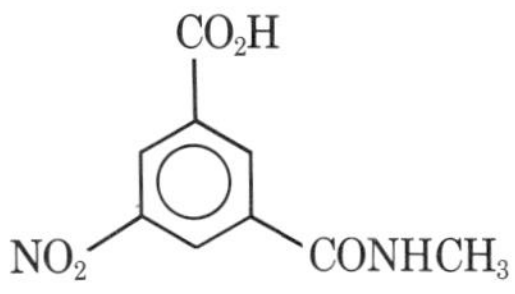

(8)

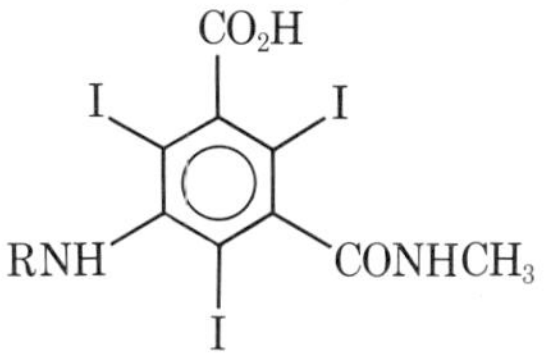

(9) R = H
(10) R = COCH₃

and was rapidly eliminated by the kidneys after intravenous administration of an aqueous solution of the sodium salt.

A number of less toxic analogues of acetrizoic acid are reported in which the 5-position hydrogen is replaced by an amide substituent. The first of these was diatrizoic acid [*117-96-4*] (**7**) (20–21). It is prepared by reduction of 3,5-dinitrobenzoic acid to 3,5-diaminobenzoic acid, which gives diatrizoic acid after iodination with potassium iododichloride and acetylation (21).

Iothalamic acid [*2276-90-6*] (**10**) is the second member of the group with a substituent in the 5-position. This acid is prepared from dimethyl 5-nitro-1,3-benzenedicarboxylate. Partial hydrolysis produces the half-ester which is converted to the acid (**8**) by treatment with aqueous methylamine. Reduction of the acid (**8**) followed by iodination of the reaction mixture gives the triiodo acid (**9**) which yields iothalamic acid (**10**) after acetylation (22).

Metrizoic acid [*1949-45-7*] (**13**) (23–24), another member of this group, is prepared by the *N*-methylation of diatrizoic acid (**7**) with dimethyl sulfate in aqueous sodium hydroxide. An alternative method is the *N*-methylation of the acetylamino group of the intermediate (**11**) in order to prepare the acid (**12**), from which metrizoic acid (**13**) is obtained by acetylation (23,25). The nonmethylated acid (**11**) is prepared from 3-amino-5-nitrobenzoic acid by acetylation of the amino group, reduction of the nitro group, and iodination (21). These two methods illustrate the interesting reaction of *N*-methylation of the (acetylamino)triiodophenyl group with dimethyl sulfate in aqueous alkali. Metrizoic acid (**13**) exhibits isomerism and selective acidification of an alkaline solution separates the two isomers.

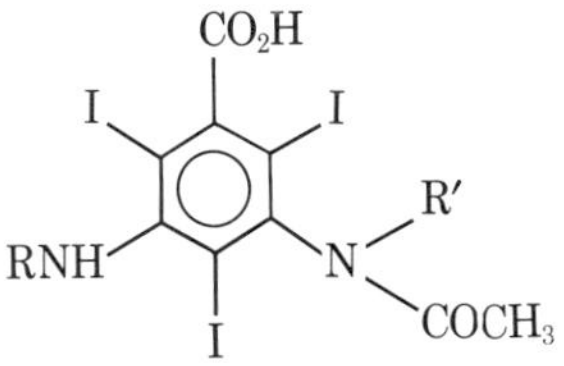

(11) R = H; R' = H
(12) R = H; R' = CH₃
(13) R = COCH₃; R' = CH₃

Analogues of metrizoic acid also show isomerism. It was postulated that the isomerism is caused by restricted rotation around the nitrogen–carbonyl carbon bond of the acetylmethylamino group. This conclusion was based upon nmr studies, a comparison of R_f values of the isomers with paper chromatography, and the observation of isomers with structures where no other type of isomers appeared possible (23). If this is correct, it is the first case where such isomers were isolated and not just observed spectrally (26–27).

Iodamide [440-58-4] (15) is a fourth member of the group with a substituent in the 5-position. The intermediate amine (14) is prepared from 4-chlorobenzoic acid by reaction with 2,2-dichloro-N-(hydroxymethyl)acetamide, hydrolysis, acetylation, nitration, and reduction. Iodamide (15) is obtained from the amine (14) by iodination followed by acetylation (28).

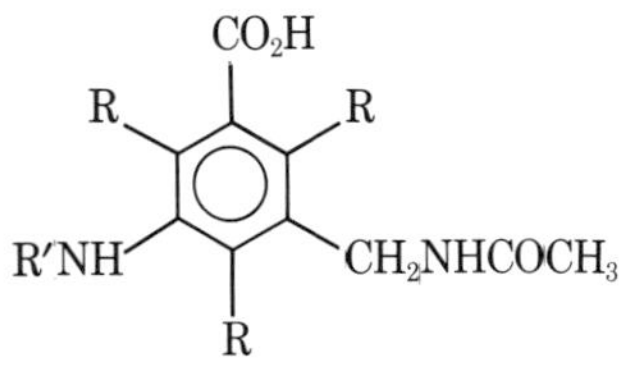

(14) R = H; R′ = H
(15) R = I; R′ = COCH$_3$

The acids (7), (10), (13), and (15) are all effective urographic agents with similar structure and toxicity. Their pK_a values are about 2.5. The presence of hydrophilic groups in preference to lipophilic groups may account for the rapid urinary excretion of these compounds. The sodium and N-methylglucamine (NMG) salts are the least toxic salts of these acids and generally the most water-soluble salts. The addition of small amounts of calcium or calcium and magnesium ions may improve the toxicity of an intravascularly injected ionic radiopaque solution (7,29–31). For excretion urography the concentration of the injected solution is not critical because the kidneys concentrate the radiopaque to a degree relatively independent of the concentration administered. Usually 25–60% wt/vol solutions (150–300 mg of organically bound iodine per mL) of the sodium or NMG salt or a mixture of the two is used for excretion urography. Solutions of similar concentration provide adequate radiopacity for retrograde urography. For angiography it is important to have a high concentration of the injected radiopaque because the blood rapidly dilutes the medium. Solutions of 50–90% wt/vol (280–480 mg I/mL) are employed for this purpose. These solutions have viscosities of about 3–19 mPa·s (= cP) at 37–38°C (32); automatic injectors are used sometimes to obtain the rapid administration required for adequate visualization. The osmolalities of these concentrated ionic media are about 1.5–3.6 osmol/kg water (8), whereas the osmolality of blood is about 0.3. The high osmolalities of the rapidly injected media may be the cause of many observed side effects (12,33).

It was suggested that water-soluble nonionic radiopaques may provide agents with lower osmolalities (33). On the basis of this idea, a series of compounds was prepared, some of which contain an amino sugar attached to a radiopaque nucleus (34). These compounds have lower osmolalities than the ionic media. Metrizamide (17), a successful myelographic and angiographic agent, is the preferred compound in this series. It can be prepared from compound (12) in three steps: treatment with thionyl chloride gives the benzoyl chloride, which is converted to the amide (16) by heating with acetic anhydride and sulfuric acid; reaction of the intermediate (16) with glucosamine hydrochloride in the presence of potassium carbonate gives metrizamide (17) (34).

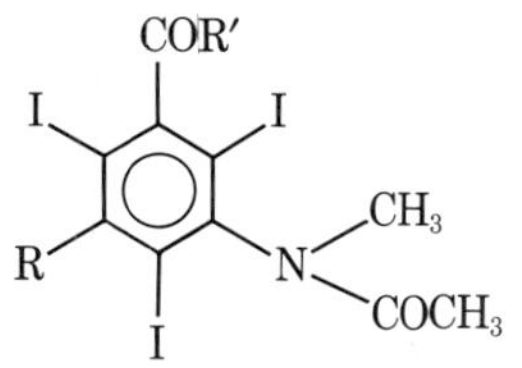

(12) R = NH$_2$; R' = OH
(16) R = NHCOCH$_3$; R' = Cl

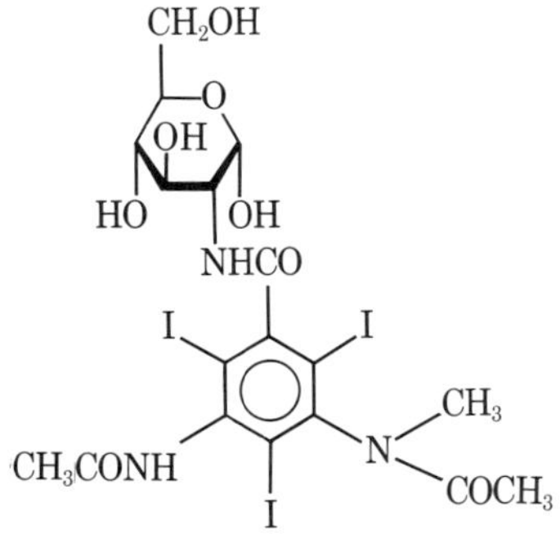

(17)

The ionic media (7), (10), (13), and (15) have one carboxyl group for three iodine atoms. Hence, aqueous solutions of the sodium and NMG salts of these media contain three iodine equivalents per two ions (particles). The nonionic agent metrizamide (17) is water-soluble without salt formation and its aqueous solutions have three iodine equivalents per one particle. If these radiopaques formed ideal solutions, the osmolality of metrizamide solutions should be about one-half that of the ionic media (7), (10), (13), and (15) for the same iodine concentration. In Table 2 the osmolalities of these media are shown for a concentration of 280 mg I/mL. Metrizamide was first introduced for myelography (see under Myelography). Animal studies and clinical angiographic studies of metrizamide give a favorable comparison with the ionic media (35–38). Metrizamide is marketed in Europe as an angiographic agent and is supplied as a lyophilized powder.

Iopamidol (20) (11) and iohexol (23) (13) are two additional nonionic agents that are under investigation and are expected to be effective myelographic and angiographic agents. In addition, a low-osmolality ionic agent, ioxaglic acid [59017-64-0] (26) (14), is in clinical use for angiography in Europe in the form of a solution of the sodium and NMG salts. Ioxaglic acid has one carboxyl group and two benzene rings, each of which contain three iodine atoms. Hence, when ioxaglic acid is in solution as the sodium or NMG salt, there are six equivalents of iodine per two ions. This is the same ratio of iodine per particle as the nonionic agents (17), (20), and (23), and the solutions of salts of ioxaglic acid have osmolalities similar to the nonionic agents.

Iopamidol (20) is prepared from 5-nitro-1,3-benzenedicarboxylic acid (18). Reduction of the acid (18) is followed by iodination, treatment with thionyl chloride, and conversion to the amide (19) by careful addition of L-2-acetyloxypropanoyl chloride.

792 RADIOPAQUES

Reaction of the amide (**19**) with 2-amino-1,3-propanediol and hydrolysis of the ester group with aqueous sodium hydroxide gives iopamidol (**20**) (11).

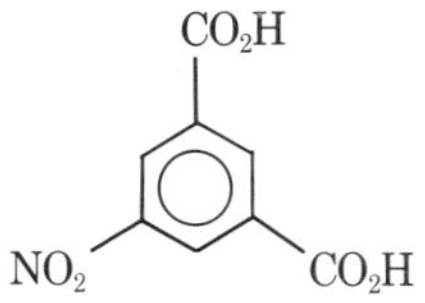

(**18**) 3,5-dinitrobenzoic acid derivative: NO₂, CO₂H, CO₂H on benzene ring

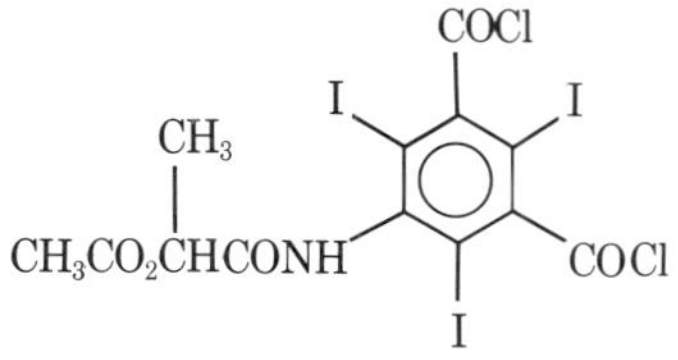

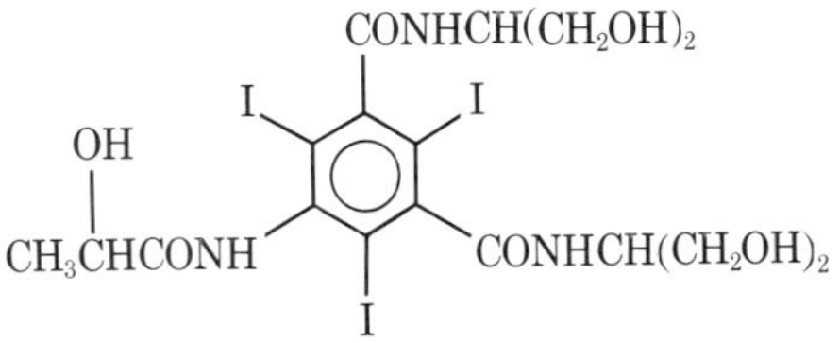

(**19**) CH₃CO₂CH(CH₃)CONH-C₆H(I)₃-(COCl)₂

(**20**) Iopamidol: CH₃CH(OH)CONH- and two -CONHCH(CH₂OH)₂ groups on triiodobenzene

The preparation of iohexol (**23**) starts with the acid (**18**) from which the dimethyl ester is prepared. This diester is converted to the bisamide (**21**) by refluxing with 3-amino-1,2-propanediol in methanol. Reduction followed by iodination gives the amine (**22**). Iohexol (**23**) is obtained from (**22**) by acetylation and then alkylation with 3-chloro-1,2-propanediol and sodium methoxide (13).

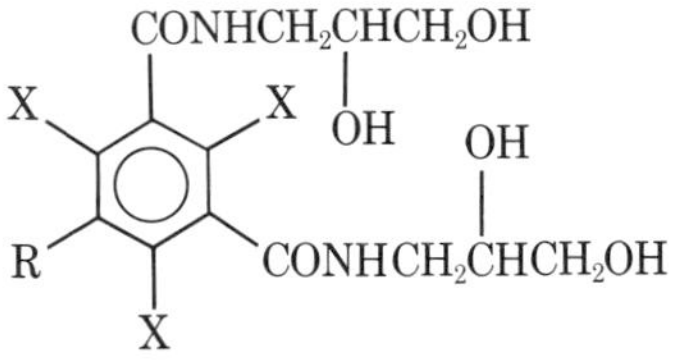

(**21**) R = NO₂; X = H
(**22**) R = NH₂; X = I

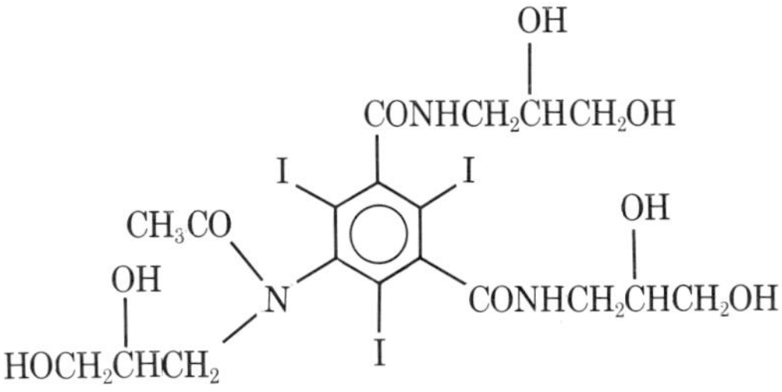

(**23**) Iohexol structure

Ioxaglic acid [*59017-72-0*] (**26**) is synthesized from the benzoyl chloride (**24**) and the acid (**25**) in the presence of triethylamine and dimethylacetamide (14). The intermediate benzoyl chloride (**24**) is made from iothalamic acid (**10**) by methylation in aqueous sodium hydroxide with dimethyl sulfate (39) and conversion to the acid chloride (**24**) with thionyl chloride (14). The benzoic acid (**25**) is prepared from the half ethyl ester (**27**) in five steps: treatment with aminoethanol gives an intermediate which is reduced with ammonium sulfide and iodinated to give (**28**) (40). Treatment of the acid (**28**) with more than two equivalents of phthalylglycine acid chloride gives a product that converts to the acid (**25**) by heating with aqueous hydrazine (14). The ionic agent ioxaglic acid is not used for myelography but it appears to produce fewer side effects than the ionic agents in angiographic studies (41–43).

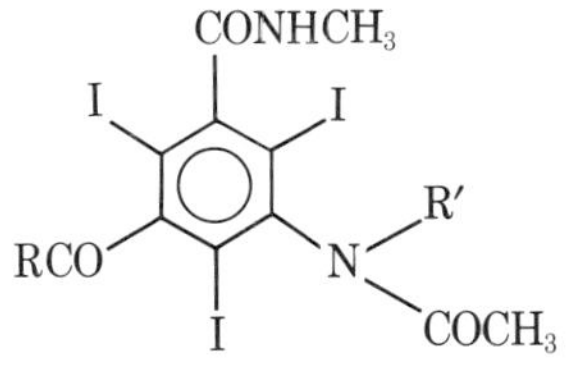

(**10**) R = OH; R' = H
(**24**) R = Cl; R' = CH_3

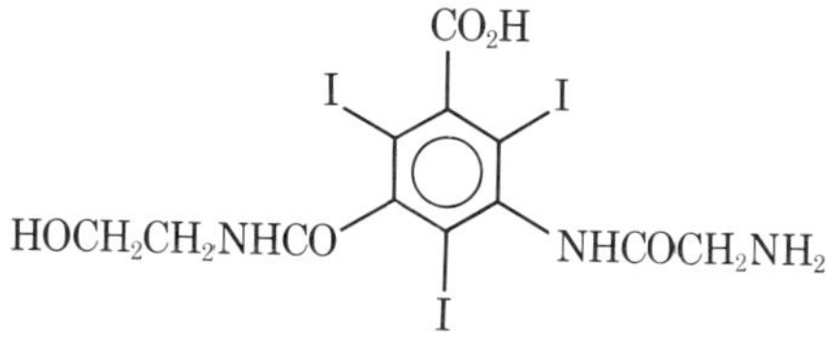

(**25**)

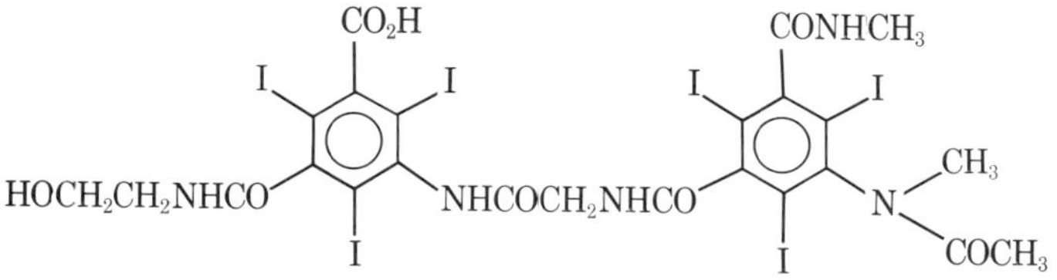

(**26**) [*59017-72-0*]

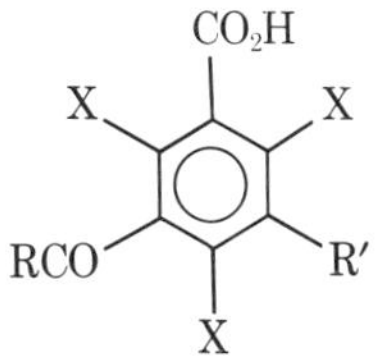

(**27**) R = OC_2H_5; R' = NO_2, X = H
(**28**) R = $NHCH_2CH_2OH$; R' = NH_2, X = I

The nonionic media (**17**), (**20**), and (**23**) and the ionic low-osmolality agent (**26**) are more expensive than the ionic media (**7**), (**10**), (**13**), and (**15**), but their lower osmolality and toxicity are distinct advantages for angiography and related procedures. It is anticipated that the nonionics and ioxaglic acid (**26**) will become the media of choice for angiography. However, their greater cost may limit their use to special urographic cases. The urographic–angiographic agents are excreted primarily in the urine and apparently no metabolic products have been observed.

Cholecystography

Oral Agents. Visualization of the gallbladder with radiopaques is called cholecystography. In this procedure the administered substance or its metabolite is excreted by the liver into the bile and collected in the gallbladder. After it was reported that chlorinated phenolphthaleins are excreted mainly in the bile (44), a series of brominated and iodinated phenolphthaleins was tested intravenously in animals as cholecystographic agents (45–46). Disodium tetraiodophenolphthalein (**29**) became the product of choice (47). It was used both orally and intravenously; the oral procedure was preferred, although vomiting and diarrhea occurred frequently. The development of iodoalphionic acid (**30**) (48–50) was an improvement. It was given orally as tablets, and the side effects were less severe.

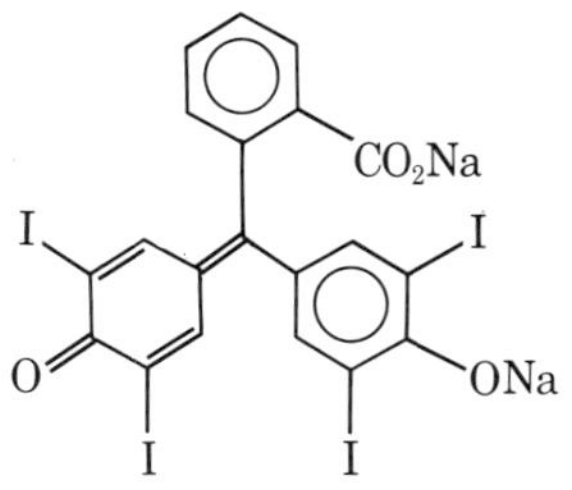

(**29**)

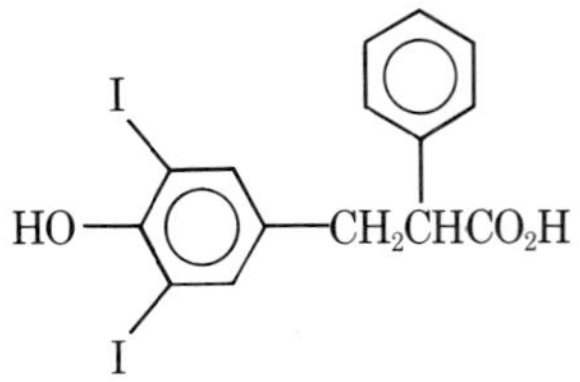

(**30**)

An important advance was the introduction of iopanoic acid (**32**), a compound widely used as an oral cholecystographic agent (51). It is prepared by the iodination of the intermediate (**31**), which is synthesized by reduction of the product from the Perkin reaction of 3-nitrobenzaldehyde and butanoic acid anhydride (51).

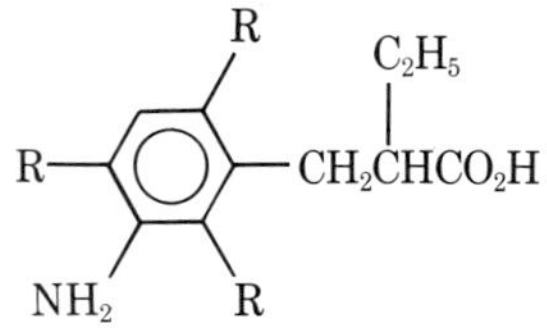

(**31**) R = H
(**32**) R = I

Many other aliphatic acids or salts that contain the 2,4,6-triiodophenyl group have been suggested as oral agents. Some of the more interesting examples are ipodoic acid [5587-89-3] (**33**) (52), sodium tyropanoate (**34**) (53), iocetamic acid (**35**) (54), iophenoxic acid (**36**) (55) and bunamiodyl (**37**) (56). All these compounds have a weak carboxylic acid group and a hydrocarbon portion that produces a suitable lipophilic–hydrophilic balance. Too much of the compound may be lost in the urine if the hydrocarbon portion is too small and the compound may be poorly absorbed from the intestinal tract if the hydrocarbon portion is too large (57–58). Although the oral cholecystographic agents are usually administered as solids, they may be considered to be water-soluble. Some are administered as water-soluble salts; others are capable of forming water-soluble salts. The agents are transported in the body bound primarily to albumin.

The final step in the preparation of ipodoic acid (**33**) is a reaction that apparently was first encountered in work with radiopaque-like structures (59). 3-Amino-2,4,6-triiodobenzenepropanoic acid, which can be prepared in a manner similar to iopanoic acid (**32**), and dimethylformamide combine in the presence of an acid chloride to give ipodoic acid (**33**) (59). It is sold as the sodium or calcium salt.

Sodium tyropanoate (**34**) is prepared from iopanoic acid (**32**) by heating with butanoic acid anhydride and sulfuric acid, followed by conversion to the sodium salt

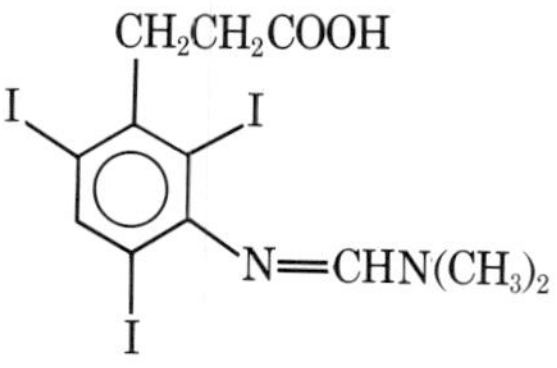

(33)

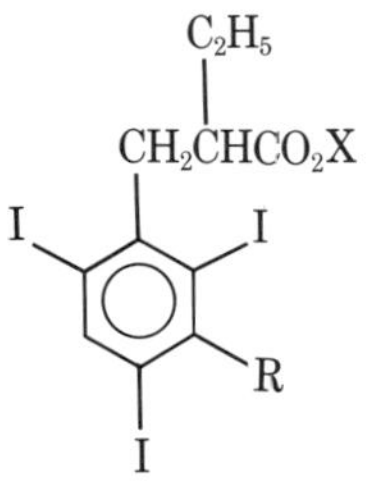

CH₂CH₂COOH, with N=CHN(CH₃)₂ substituent, triiodo phenyl

(34) R = NHCOCH₂CH₂CH₃; X = Na
(36) R = OH; X = H

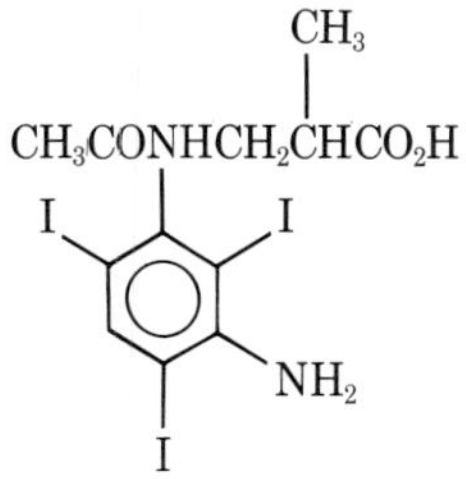

Structure with CH₂CHCO₂X, C₂H₅ side chain, triiodo, R substituent

(35)

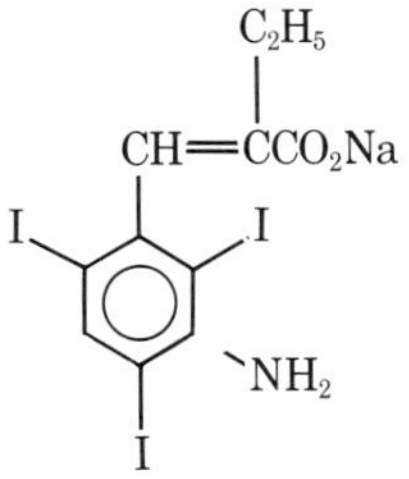

CH₃CONHCH₂CH(CH₃)CO₂H, triiodo phenyl with NH₂

(37)

CH=C(C₂H₅)CO₂Na, triiodo phenyl with NH₂

(53). Iocetamic acid (35) is obtained from the intermediate (38) by reduction, followed by iodination. The intermediate acid (38) was prepared by acetylation of the Michael addition product from 3-nitroaniline and methacrylic acid (54).

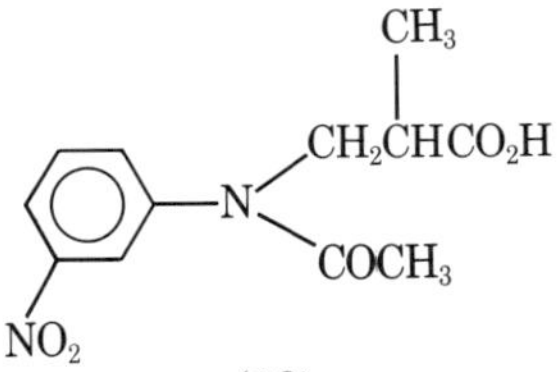

(38) 3-nitrophenyl-N(COCH₃)(CH₂CH(CH₃)CO₂H)

Iophenoxic acid (36) was removed from the market by the manufacturer after it was indicated that it interfered with the determination of protein-bound iodine (PBI) values which rendered this test of thyroid function in the patient useless for many years (60) (see Thyroid and antithyroid preparations). The other oral cholecystographic agents generally interfere with PBI values for a few months only. In addition, iophenoxic acid was not quite as effective as iopanoic acid (61–63). Bunamiodyl (37) was discontinued in the United States when its use was associated with a number of cases of renal shutdown (64–66). Some oral cholecystographic compounds are given in Table 3.

Table 3. Oral Cholecystographic Agents

Name	CAS Registry Number	Structure	Iodine, %
disodium tetraiodophenolphthalein	[2217-44-9]	(29)	54.1
iodoalphionic acid	[577-91-3]	(30)	51.4
iopanoic acid	[96-83-3]	(32)	66.7
sodium ipodate	[1221-56-3]	(33) Na salt	61.4
sodium tyropanoate	[7246-21-1]	(34)	57.4
iocetamic acid	[16034-77-8]	(35)	62.0
iophenoxic acid	[96-84-4]	(36)	66.6
bunamiodyl	[22613-45-2]	(37)	57.6

It appears that most oral cholecystographic agents are metabolized and formation of glucuronides occurs (67–70). The oral agents must be absorbed from the alimentary tract, and their metabolism in the body to a more polar form, such as glucuronic acids, may be required to prevent their absorption from the gallbladder. There is no evidence that deiodination leads to a principal metabolic product of any radiopaque.

Although the cholecystographic media are intended for visualization of the gallbladder, a large portion of many oral agents is eliminated through the kidney. A study of three oral agents indicated an average urinary excretion ranging from 37–50% after 4.5 days (71).

Sonography is used more and more in the diagnosis of liver, gallbladder, pancreas, and biliary-tree disease (72). It is a method for visualizing parts of the body with ultrasound without radiopaques (see Ultrasonics). In some cases sonography gives information similar to that obtained with radiopaques without subjecting the patient to drug and x-ray treatment. With some patients both cholecystography and sonography may be used.

Intravenous Cholangio-Cholecystographic Agents. Disodium tetraiodophenolphthalein (29) was originally used as an intravenous or oral cholecystographic agent, but because of its toxicity the oral route was preferred. Although the newer intravenous compounds are much less toxic, the oral procedure is still preferred in the United States because it is simpler and less hazardous. When the oral procedure does not produce adequate visualization of the gallbladder, intravenous administration is recommended (73). It frequently gives a better visualization of the bile ducts more rapidly than the oral procedure. The new intravenous agents are called cholangio-cholecystographic agents because they show the bile ducts as well as the gallbladder.

The first widely used intravenous medium was iodipamide [606-17-7] (39), which was prepared from 3-amino-2,4,6-triiodobenzoic acid and hexanedioyl dichloride (74). Currently, iodipamide is the only intravenous cholecystographic agent available in the United States and is marketed as the NMG salt in aqueous solution.

A number of analogues of iodipamide have been clinically tested as cholangio-cholecystographic agents, including ioglycamic acid [2618-25-9] (40) (74), iodoxamic acid [31127-82-9] (41) (75), iotroxic acid [51022-74-3] (42) (76), and iosulamide meglumine [63534-64-5] (43) (77). These compounds are synthesized by procedures similar to that for iodipamide. Ioglycamic acid (40) and iodoxamic acid (41) are marketed in Europe. These compounds are sometimes referred to as dimers of urographic–angiographic agents such as acetrizoic acid (6). The intravenous cholangio-cholecystographic agents are excreted mainly unchanged (67). The urinary excretion of iodipamide (39) was 13.8% in 48 h (78). Some urinary excretion of these media may be desirable since it permits the body to eliminate the drug if there is liver damage inhibiting biliary excretion.

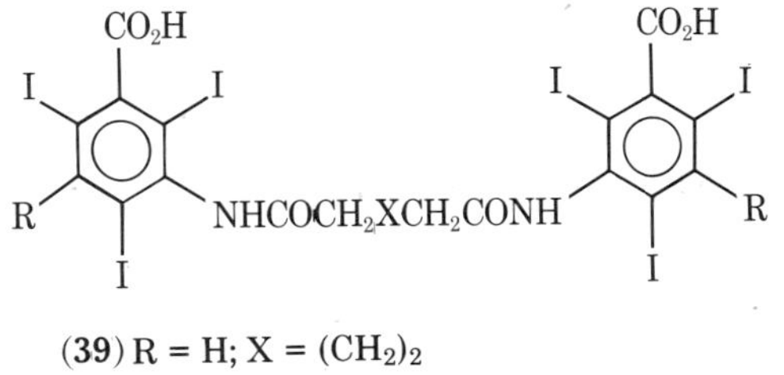

(39) R = H; X = $(CH_2)_2$
(40) R = H; X = O
(41) R = H; X = $(CH_2OCH_2)_4$
(42) R = H; X = $OCH_2CH_2OCH_2CH_2O$
(43) R = $N(C_2H_5)COCH_3$; X = $CH_2SO_2CH_2$

Myelography

Myelography is the visualization of the subarachnoid space of the spinal canal by means of a radiopaque or air (1). The use of air for myelography was described in 1919 (79). Although air is nontoxic, it disturbs the pressure balance of the subarachnoid space and causes severe side effects (80). Both water-soluble agents and oils not miscible with water are used. Iodinated poppyseed oil was the first radiopaque employed for myelography (81), but iophendylate [*99-79-6*] (**44**) became the preferred oil-type medium for this purpose (82). Both are absorbed slowly from the subarachnoid space and generally removed after the examination. Iophendylate is less viscous and has less tendency to form droplets than iodinated poppyseed oil (80). It is prepared from iodobenzene and ethyl undecenoate in the presence of aluminum chloride (82). The structure (**44**) is usually ascribed to iophendylate, but apparently it is a mixture. Examination of several samples of iophendylate by gas chromatography suggest the presence of four main components (>15%) and six or more lesser ones (1–3%) (83).

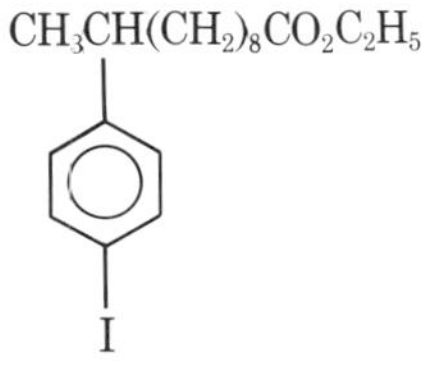

$$CH_3CH(CH_2)_8CO_2C_2H_5$$

(**44**)

Sodium methiodal (**3**) was the first water-soluble agent employed for myelography (84) and was widely used in Europe for this purpose. The procedure is restricted to the lumbar region of the spinal canal and was generally carried out under spinal anesthesia. Water-soluble media offer the advantage that the agent does not need to be removed from the patient after the procedure. These agents are excreted primarily in the urine after being absorbed from the cerebrospinal fluid.

Iothalamic acid (**10**) was introduced primarily as a urographic–angiographic agent, but in myelographic studies its NMG salt was found to be less toxic in the subarachnoid space than sodium methiodal. Furthermore, iothalamic acid can be introduced into the lumbar region without spinal anesthesia (80,85). The NMG salt of iocarmic acid [*10397-75-8*] (**45**) (86) was still less toxic than iothalamate in animal studies and was used in clinical investigations (87).

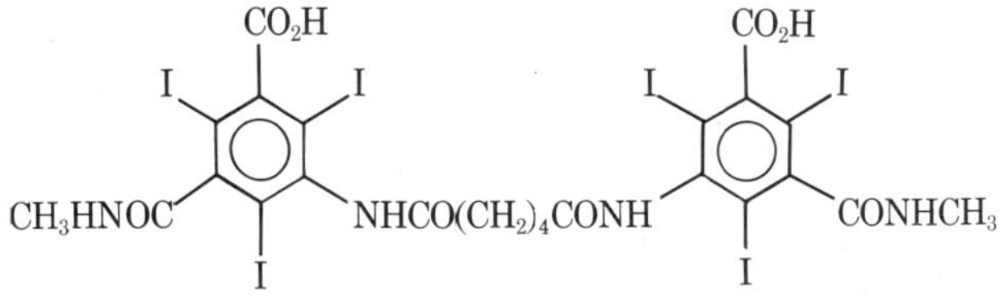

(**45**)

The nonionic agent metrizamide (**17**) improved the safety of myelography (88–89). In spite of its high cost, this agent has almost completely replaced the other water-soluble media for this purpose. The nonionic agents iopamidol (**20**) and iohexol (**23**) also show promise in this field.

Bronchography, Lymphangiography, and Other Radiopaque Procedures

Bronchography is the x-ray visualization of the bronchial tree with the help of a radiopaque. The first bronchographic agent was iodinated poppyseed oil (81). Suspension-type agents (90–91) and tantalum (92–93) have also been used. The difficulties with the procedure and the introduction of computed tomography (CT) have greatly diminished the application of bronchography. In CT, x-ray photographs are taken with the help of a computer. This method provides more information and detail than the usual x-ray procedure and is employed to investigate many parts of the body. Radiopaques are sometimes used in conjunction with CT scanning.

The visualization of the lymphatic ducts by radiopaques is called lymphangiography. Water-soluble agents and oils usually are used for this procedure. Iodetryl [8008-53-5], ethiodized oil, is the most commonly employed oil-type medium. The other radiographic procedures (see Table 1) are generally carried out with water-soluble and oil-type radiopaques.

Economic Aspects

Prices and suppliers information are given in Table 4.

Table 4. Prices and Suppliers

Radiopaque	Structure	Price, $[a]	Suppliers
barium sulfate, USP		7.00–8.30/kg	Baker Chemical Co., City Chemical Corp. Humco Laboratories
sodium diatrizoate, oral aqueous solutions of Na and NMG salts for urography and angiography	(7) Na salt	8.59 for 50 g in 120 mL 54.86 for 25 30 mL vials 38.93 for 10 50-mL vials	Winthrop Labs, Squibb
diatrizoic acid	(7)		Winthrop Labs, Squibb
iothalamic acid	(10)		Mallinckrodt
cholecystographic agents, oral		ca 30.00 for 25 3-g doses	
iopanoic acid	(32)		Winthrop Labs
calcium and sodium ipodate	(33) Ca and Na salts		Squibb
sodium tyropanoate	(34)		Winthrop Labs
iocetamic acid	(35)		Mallinckrodt
cholecystographic agent, iv			
iodipamide	(39)	185.52 for 25 20-mL vials	Squibb
iophendylate	(44)	72.44 for 6 6-mL ampules	Picker X-ray Corp. General Electric Westinghouse E. M. Parker Co.
metrizamide, lyophilized[b]	(17)	185.52 for 6 vials of 3.75 g	Winthrop Labs
iodetryl		13.98 for 2 10-mL ampules	Savage Laboratories

[a] All the given prices are taken from the Red Book.
[b] Made into a solution at time of use.

BIBLIOGRAPHY

"Radiopaques" in *ECT* 2nd ed., Vol. 17, pp. 130–142, by James Ackerman, Sterling-Winthrop Research Institute.

1. J. O. Hoppe in E. E. Campaigne and W. H. Hartung, eds., *Medicinal Chemistry*, Vol. 6, John Wiley & Sons, Inc., New York, 1963, pp. 290–349.
2. R. R. Newell in O. Glasser, ed., *Medical Physics*, Year Book Medical Publishers, Inc., Chicago, Ill., 1944, p. 1269.
3. R. L. Swarm in P. K. Knoefel, ed., *International Encyclopedia of Pharmacology and Therapeutics*, Vol. 76 (2 vols.), Pergamon Press, Oxford, 1971, pp. 431–441.
4. C. Bachem and H. Gunther, *Z. Roentgenk. Radiumforsch.* **12**, 369 (1910).
5. W. Becker, *Dtsch. Med. Wochenschr.* **22**, 202 (1896).
6. W. B. Cannon, *Am. J. Physiol.* **1**, 359 (1898).
7. J. O. Hoppe, L. P. Duprey, W. A. Borisenok, and J. G. Bird, *Angiology* **18**, 257 (1967).
8. Unpublished results, F. J. Rosenberg, and J. H. Ackerman, Sterling-Winthrop Research Institute, Rensselaer, N.Y. Data were presented at Contrast Media Toxicity International Symposium, Skytop, Pa., May 1970.
9. S. Salvesen, *Acta Radiol. Suppl.* **335**, 5 (1973).
10. J. Haavaldsen, *Acta Radiol. Suppl.* **362**, 9 (1980).
11. U.S. Pat. 4,001,323 (Jan. 4, 1977), E. Felder and D. E. Pitré (to Savac AG).
12. R. G. Grainger, *Brit. J. Radiol.* **53**, 739 (1980).
13. Brit. Pat. 1,548,594 (July 18, 1979), V. Nordal and H. Holtermann (to Nyegaard and Co. A/S).
14. U.S. Pat. 4,014,986 (Mar. 29, 1977), G. Tilly, M. J. C. Hardouin, and J. Lautrou (to Laboratories Andre Guerbet).
15. E. D. Osborne, C. G. Sutherland, A. J. Scholl, Jr., and L. G. Rowntree, *J. Am. Med. Assoc.* **80**, 368 (1923).
16. M. Swick, *J. Am. Med. Assoc.* **95**, 1403 (1930).
17. A. von Lichtenberg, *Brit. J. Urol.* **3**, 119 (1931).
18. T. D. Moore, *J. Urol.* **30**, 27 (1933).
19. V. H. Wallingford, H. G. Decker, and M. Kruty, *J. Am. Chem. Soc.* **74**, 4365 (1952).
20. H. Langecker, A. Harwart, and K. Junkmann, *Arch. Exp. Pathol. Pharmakol.* **222**, 584 (1954).
21. A. A. Larsen, C. Moore, J. Sprague, B. Cloke, J. Moss, and J. O. Hoppe, *J. Am. Chem. Soc.* **78**, 3210 (1956).
22. G. B. Hoey, R. D. Rands, G. DeLaMater, D. W. Chapman, and P. E. Wiegert, *J. Med. Chem.* **6**, 24 (1963).
23. Private communication from H. Holtermann, L. G. Haugen, J. Haavaldsen, V. Nordal, K. Wille, K. Tjonneland, J. Koutroulos, and N. Thorsdalen, Nyegaard and Co., Oslo, Norway, 1963.
24. E. R. Jolly and D. H. Baeder, *Pharmacologist* 203 (Fall 1964).
25. U.S. Pat. 3,178,473 (Apr. 13, 1965), H. Holtermann, L. G. Haugen, V. Nordal, and J. L. Haavaldsen (to Nyegaard and Co. A/S).
26. H. S. Gutowsky and C. H. Holm, *J. Chem. Phys.* **25**, 1228 (1956).
27. T. H. Siddall, III, *Tetrahedron Lett.* **1966**, 2027.
28. E. Felder, D. Pitré, and L. Fumagalli, *Helv. Chim. Acta* **48**, 259 (1965).
29. U.S. Pat. 3,175,952 (Mar. 30, 1965), J. G. Bird (to Sterling Drug, Inc.).
30. U.S. Pat. 3,325,370 (June 13, 1967), H. Holtermann and S. Salvesen (to Nyegaard and Co. A/S).
31. J. Fog, *Acta Radiol. Suppl.* **270**, 99 (1967).
32. H. W. Fischer, *Angiology* **16**, 759 (1965).
33. T. Almén, *J. Theor. Biol.* **24**, 216 (1969).
34. U.S. Pat. 3,701,771 (Oct. 31, 1972), T. Almén, J. Haavaldsen, and V. Nordal (to Nyegaard and Co. A/S).
35. T. Almén, *Acta Radiol. Suppl.* **355**, 419 (1977).
36. T. Almén, E. Boijsen, and S. E. Lindell, *Acta Radiol. Diagn.* **18**, 33 (1977).
37. I. Enge, S. Nitter-Hauge, E. Andrew, and K. Levorstad, *Radiology* **125**, 317 (1977).
38. V. Albrechtsson and C. G. Olsson, *Acta Radiol. Diagn.* **20**, 46 (1979).
39. Ger. Pat. 58,307 (Oct. 20, 1967), H. Cassebaum and K. Dierback.
40. Brit. Pat. 1,146,133 (Mar. 19, 1969), (to Laboratories Andre Guerbet).
41. V. Tillmann, R. Adler, and W. A. Fuchs, *Brit. J. Radiol.* **52**, 102 (1979).
42. M. Holm and J. Praestholm, *Brit. J. Radiol.* **52**, 169 (1979).
43. R.G. Grainger, *Brit. J. Radiol.* **52**, 781 (1979).
44. J. J. Abel and L. G. Rowntree, *J. Pharmacol. Exp. Ther.* **1**, 231 (1909).
45. E. A. Graham, *Am. J. Surg.* **12**, 330 (1931).

46. W. H. Cole, *Radiology* **76,** 354 (1961).
47. E. A. Graham and W. H. Cole, *J. Am. Med. Assoc.* **82,** 613 (1924).
48. M. Dohrn and P. Diedrick, *Dtsch. Med. Wochenschr.* **66,** 1133 (1940).
49. N. Kleiber, *Dtsch. Med. Wochenschr.* **66,** 1134 (1940).
50. K. Junkmann, *Klin. Wochenschr.* **20,** 125 (1941).
51. T. R. Lewis and S. Archer, *J. Am. Chem. Soc.* **71,** 3753 (1949).
52. H. Priewe and A. Poljak, *Chem. Ber.* **93,** 2347 (1960).
53. J. O. Hoppe, J. H. Ackerman, A. A. Larsen, and J. Moss, *J. Med. Chem.* **13,** 997 (1970).
54. J. A. Korver, *Rec. Trav. Chim. Pays-Bas* **87,** 308 (1968).
55. D. Papa, H. F. Ginsberg, I. Lederman, and V. DeCamp, *J. Am. Chem. Soc.* **75,** 1107 (1953).
56. M. Chao and P.-C. Hu, *Hua Hsueh Hsueh Pao* **23,** 361 (1957).
57. J. O. Hoppe and S. Archer, *Am. J. Roentgenol. Radium Ther. Nucl. Med.* **69,** 630 (1953).
58. B. S. Epstein, S. Natelson, and B. Kramer, *Am. J. Roentgenol. Radium Ther. Nucl. Med.* **56,** 201 (1946).
59. S. African Pat. Appl. 978/59 (March 15, 1959), H. Priewe and A. Poljak (to Schering, A.-G.).
60. E. B. Astwood, *Trans. Assoc. Am. Physicians* **70,** 183 (1957).
61. R. Shapiro, *Radiology* **60,** 687 (1953).
62. C. R. Weinberg, *Am. J. Roentgenol. Radium Ther. Nucl. Med.* **70,** 585 (1953).
63. J. C. Root and R. F. Lewis, *Radiology* **64,** 714 (1955).
64. J. A. Gunn, *J. Am. Med. Assoc.* **175,** 911 (1961).
65. J. G. Setter, J. F. Maher, and G. E. Schreiner, *J. Am. Med. Assoc.* **184,** 102 (1963).
66. J. E. Wennberg, R. Okun, E. J. Hinman, R. C. Northcutt, R. J. Griep, and W. G. Walker, *J. Am. Med. Assoc.* **186,** 461 (1963).
67. E. W. McChesney in ref. 3, p. 147.
68. E. W. McChesney and J. O. Hoppe, *Arch. Intern. Pharmacodyn. Ther.* **99,** 127 (1954).
69. *Ibid.*, **105,** 306 (1956).
70. E. W. McChesney, *Biochem. Pharmacol.* **13,** 1366 (1964).
71. E. W. McChesney and W. F. Banks, Jr., *Proc. Soc. Exp. Biol. Med.* **119,** 1027 (1965).
72. B. T. Burney, *Applied Radiology*, 9(5), 99 (1980).
73. R. E. Wise, *Radiol. Clin. North Am.* **4,** 521 (1966).
74. H. Priewe, R. Rutkowski, K. Pirner, and K. Junkmann, *Chem. Ber.* **87,** 651 (1954).
75. E. Felder, D. Pitré, L. Fumagalli, and E. Lorenzotti, *Farmaco Ed. Sci.* **28,** 912 (1973).
76. Ger. Offen. 2,405,652 (Aug. 21, 1975), H. Pfeiffer, U. Speck, and K. H. Kolb (to Schering A.-G.).
77. U.S. Pat. 3,732,293 (May 8, 1973), J. Ackerman (to Sterling Drug, Inc.).
78. H. Billion, W. Frommhold, K. Oeff, and W. Schutz, *Aerztl. Wochenschr.* **10,** 574 (1955).
79. W. E. Dandy, *Ann. Surg.* **70,** 397 (1919).
80. P. Amundsen in J. F. Sackett and C. M. Strother, eds., *New Techniques in Myelography*, Harper & Row, Publishers Inc., Hagerstown, Md., 1979, p. 2.
81. Sicard and Forestier, *Société Médicale des Hôpitaux de Paris* **46,** 463 (1922).
82. W. H. Strain, J. T. Platt, and S. L. Warren, *J. Am. Chem. Soc.* **64,** 1436 (1942).
83. Private communication from J. M. Lennon, Sterling-Winthrop Research Institute, 1967.
84. S. Arnell and F. Lidstrom, *Acta Radiol.* **12,** 287 (1931).
85. R. L. Campbell, J. A. Campbell, R. F. Heimburger, J. E. Kalsbeck, and J. Mealey, Jr., *Radiology* **82,** 286 (1964).
86. G. B. Hoey, R. D. Rands, P. E. Wiegert, D. W. Chapman, R. L. Zey, and G. B. DeLaMater, *J. Med. Chem.* **9,** 964 (1966).
87. R. Gonsette, *Clin. Radiol.* **22,** 44 (1971).
88. *Acta Radiol. Suppl.* **335,** (1973).
89. E. Lindgren, ed., *Acta Radiol. Suppl.* **355,** (1977).
90. D. Conway, *Brit. J. Radiol.* **25,** 573 (1952).
91. E. G. Tomich, B. Basil, and B. Davis, *Brit. J. Pharmacol.* **8,** 166 (1953).
92. J. A. Nadel, W. G. Wolfe, and P. D. Graf, *Invest. Radiol.* **3,** 229 (1968).
93. J. A. Nadel, W. G. Wolfe, P. D. Graf, J. E. Youker, N. Zamel, J. H. M. Austin, W. A. Hinchcliffe, R. H. Greenspan, and R. R. Wright, *New Engl. J. Med.* **283,** 281 (1970).

General References

Reference 1 is a general reference.
P. K. Knoefel, ed., *International Encyclopedia of Pharmacology and Therapeutics*, Vol. 76 (2 vols.), Pergamon Press, Oxford, 1971.
J. F. Sackett and C. M. Strothers, eds., *New Techniques in Myelography*, Harper & Row, Publishers Inc., Hagerstown, Md., 1979.
W. H. Strain, S. M. Rogoff, R. H. Greenlaw, R. M. Johnston, F. Huegin, and W. P. Berliner, *Med. Radiography Phot.* (*Suppl.*) **40**, (1964).

JAMES ACKERMAN
Sterling-Winthrop Research Institute

RADIOPROTECTIVE AGENTS

The discovery of x rays by Roentgen in 1895 and the isolation of the highly radioactive element radium by the Curies in 1896 were the two events that marked the beginning of the nuclear era. The ability of x rays to penetrate human tissue was appreciated immediately as a potential medical asset, and they were put to use enthusiastically. Early workers did notice an erythematous reaction when skin was exposed to the rays for lengthy periods of time. Later, lesions of the skin and cancer of the bone, primarily in the extremities, developed in some scientists, physicians, and patients receiving high doses of x rays. For centuries prior to this period, workers in the mines from which the Curies had obtained pitchblende ore for their radium isolation experienced a high incidence of lung cancer. Uranium miners in the Colorado Plateau were similarly afflicted.

The death, suffering, and permanent damage inflicted on many people by high levels of radiation incurred by the dropping of atomic bombs on Hiroshima and Nagasaki in 1945 stimulated research in several countries, notably the United States, the USSR, France, the United Kingdom, Canada, the Federal Republic of Germany, and Japan, to find the chemical agents that would minimize the effects of radiation.

In an early experiment performed under the Manhattan Project, it was discovered that an irradiated sulfur-containing enzyme could be reactivated by the addition of cysteine; this suggested that radioprotection of biological systems was possible. Several thousand compounds, mainly sulfur containing, have since been designed, synthesized, and tested in animals, mostly rodents. When administered in advance of irradiation, many of them show excellent ability to prolong the life of animals subjected to lethal radiation. In humans, some of the more promising radioprotective agents are being considered as adjuncts in cancer radiotherapy and chemotherapy.

A discussion of the biological effects of ionizing radiation is given in refs. 1–10. Reviews and compendia relating to chemical radiation protection also are available (11–22). The application of radioprotective agents to the treatment of cancer is described in ref. 23.

Biological Aspects of Radiation Damage

The Nature of Radiobiological Injury. The sequence of events leading from the initial absorption of radiation energy to the ultimate death of the organism is complex and incompletely understood. It is convenient to divide the process of radiation injury into events occurring at the molecular level, at the cellular level, and at the level of the organism. The following discussion is limited to the biological effects of high energy photon sources, ie, x and γ radiation of >200 kV_p (kilovolts peak). When high energy photons interact with molecules in biological systems, the initial relevant event is the production of Compton electrons, which interact with adjacent molecules to produce secondary electrons. It is these electrons that are primarily responsible for the initiation of radiation damage.

Molecular Level. The interaction of radiation with molecules in biological systems can be divided into direct action and indirect action. Direct action results when radiation energy is deposited in a target molecule and causes damage without the participation of mediators. In indirect action, radiation energy is deposited in the vicinity of a target molecule, and damage is transferred to the target by free radicals. Absorption of radiation by biological solvent molecules, followed by reaction of the radiolysis products of water with solute molecules, is the usual example of indirect action in biological systems. However, even in nonaqueous regions of such systems, there are few cases where secondary transfer of radiation energy, ie, indirect action, is not involved.

When a molecule in a biological system undergoes initial interaction with secondary electrons, it is either excited or ionized. When an electron is ejected from the molecule, a radical cation forms. Conversely, when such an electron is trapped, a radical anion forms (24). These primary ion-radical species then undergo transformations, eg, decarboxylation, deamination, and disulfide bond rupture, whereby deposited energy is transferred inter- or intramolecularly with the formation of semistable neutral bioradicals. The resulting molecular alterations may lead to the loss of the target molecule's ability to perform its particular biological role. These steps can be summarized:

$$RH \rightleftharpoons RH^+ + e \qquad (1)$$

$$RH + e \rightarrow RH^- \qquad (2)$$

$$RH^+ \rightarrow R\cdot + H^+ \qquad (3)$$

In oxygenated systems

$$R\cdot + O_2 \rightarrow RO_2\cdot \qquad (4)$$

In the presence of an endogenous or exogenous sulfhydryl

$$R\cdot + RSH \rightarrow RH + RS\cdot \qquad (5)$$

Such symbolic reactions suggest one way in which molecular oxygen can act as a radiosensitizer by preventing the reversal of the first reaction (see eq. 1), ie, fixing radiation damage by peroxy-radical formation (see eq. 4). Donation of a hydrogen atom (see eq. 5) by a thiol could result in instantaneous repair of the radical lesion formed in equations 1 and 3.

When a biological target molecule is considered in its *in vivo* environment, numerous factors may influence the mechanism of its damage by ionizing radiation. It may be located in a principally aqueous phase of a cell where its inactivation would be mediated by the processes of indirect action, in which the primary radiolysis products of water, ie, $H^{\cdot}$, $OH^{\cdot}$, and e_{aq}, are the damaging species. At the other extreme, it might be in a cellular region where other molecules like itself surround it, ie, in a nonaqueous environment approaching a solid-state system. Here, the processes of classical direct action and solid-state energy transfer could lead to its inactivation. These potential environmental extremes in the living cell encompass numerous variations which must be considered, especially when simplistic models are used to relate *in vitro* findings in well-defined model systems to findings in cell culture or *in vivo*.

With cautious interpretation, model systems for the study of molecular-level mechanisms of radiation damage can be extremely valuable. Reactions of water radicals with biomolecules are most frequency studied by pulse-radiolysis techniques (25–26), by spin trapping (27), or in esr studies in which frozen matrices are used to slow the interaction processes (28). One can determine which of the water radicals is most damaging to a particular macromolecule in dilute aqueous solution. Molecular-level investigations of single crystals or powders can lead to an understanding of the mediators of energy transfer in the solid state.

Cellular Level. Within the living cell, the dynamics of cell biochemistry and metabolism are superimposed on molecular-level interactions. Radiation inactivation of a single molecule can be transferred to other types of molecules through coupled reactions (29). If a number of molecules altered by radiation-induced rearrangements were involved in a common coupled reaction, there could result a marked enhancement of radiation damage suffered by dependent reactions. Radiation damage to DNA or RNA can result in the synthesis of inactive biological macromolecules, thereby, markedly amplifying the initial radiation damage. Disruption of subcellular organelles, such as chromosomes, microsomes, or mitochondria, can result in malfunction. Experimental evidence indicates that although both direct and indirect effects can contribute to the injury of mammalian cells, it is probably indirect action mediated by the hydroxyl radical that is most damaging to critical macromolecules of cells (30–31).

Identification of the critical site or macromolecular target whose damage is primarily responsible for cell injury has been the goal of extensive investigations. It is generally agreed that this target is in the cell nucleus (32) and that a principal target molecule in mammalian cells is DNA (33–34). Biological effects other than cell death, such as gene malfunction and chromosome aberration, are also related to DNA damage (35). Other important candidates for the cellular target include membranes of the cell, the nucleus, and the other cellular organelles (36–37) and sulfhydryl- and disulfide-containing biological molecules (38).

Radiation-induced cell death may be broadly separated into two classes: reproductive death, ie, where cell division is involved, and interphase death, ie, where cell death does not involve mitosis. Interphase death usually follows very high radiation doses [>200 Gy (>20 krad)] and may be related to impairment of membrane permeability, reduced ATP (adenosine triphosphate) synthesis, and nuclear disorganization, but the precise mechanism of cell death is unknown. Reproductive death of cells has been defined as the loss of reproductive integrity with subsequent loss of metabolic activity and cellular functions and may result from low radiation doses [<10 Gy (<1 krad)] (39). Death of the cell may take place on the first or second attempt at division and appears to be primarily the result of chromosome aberration (40).

Mammalian-cell radiosensitivity is not constant during the various stages of the cell cycle, but is generally greatest immediately before, during, and immediately after mitosis (40–41). When the radiation dose rate is lowered sufficiently, many of its damaging effects are considerably diminished, even though the same total radiation dose is given. Similar effects are observed in fractionated irradiation, in which the interval between irradiations is increased. Repair or recovery mechanisms have been proposed to explain such phenomena, and they generally consider repair of sublethal damage and circumvention of additive effects (42–45). Repair of sublethal damage to mammalian-cell DNA includes such processes as excision repair and postreplication repair (46). The nucleases involved in such repair processes must be considered as important radiation target molecules that have a significant role in modification of reproductive death. At low dose rates and when there is sufficient time between divided radiation doses, they may mediate recovery from sublethal or even lethal doses.

Organism Level. It is useful to categorize tissues in three classes, according to their rate of cell division (in order of increasing radiation sensitivity): steady-state populations, eg, adult nerve and muscle; expanding populations, eg, liver and kidney; and renewing populations, eg, bone marrow, thymus, spleen, and intestinal-crypt cells. Low doses of whole-body irradiation [<10 Gy (<1 krad)] damage stem cells of bone marrow and lymphoid tissue. Death from such radiation dose levels usually results from hemorrhaging, infection, or anemia; it follows 7–30 days after exposure and is termed hematopoietic death. At doses of 10–100 Gy (1–10 krad), intestinal-crypt cell renewal is inhibited; thus, before the elements of hematopoietic death are manifested, gastrointestinal death occurs 3–6 d after exposure. At that time, intestinal-crypt cells have been depleted with a fatal loss of electrolytes and water. If extremely high radiation doses are administered, ie, >200 Gy (>20 krads), function of the steady-state cells of the central nervous system is sufficiently altered, and CNS death follows within 24 h or even during irradiation, depending on the magnitude of the administered dose.

Evaluation of Potential Radioprotective Agents. Numerous preliminary questions must be answered before a potential antiradiation agent can be tested for efficacy. These include selection of the appropriate radiation type and dose level, the test organism, the meaningful end-point parameter, the drug dose and route of administration, the time separation of drug administration and irradiation, and the statistical method by which data are analyzed. Organism death is usually selected as the end point to judge antiradiation effectiveness, although numerous other radiobiological lesions have been used (47–51). Data derived from end points other than organism death are usually correlated with mammalian death. Because of the large number of animals required for the statistical analyses of screening results, the mouse is the most commonly used animal model.

The optimum dose of the potential radioprotective drug must be derived from toxicity studies with the same type of test animals. The drug dose that is lethal to half of a group of animals in 14 d ($LD_{50/14}$) is determined, and from this a maximum tolerated dose (MTD) is selected, usually as $1/2$–$2/3$ the $LD_{50/14}$. A radiation dose is selected that is just sufficient to kill all control animals in 30 d ($LD_{100/30}$). This dose is ca 5–10 Gy (500–1000 rads) for most mammals. Control animals receive equivalent volumes of the carrier or solvent and are exposed to irradiation simultaneously with the treated animals.

The optimal time between drug administration and irradiation must be determined empirically but is based on the suspected mechanism of protection involved and on the probable absorption time by the route of administration. In mice, this time is 15–30 min after ip (intraperitoneal) injection and 30–60 min after oral intubation. Each radioprotector has a characteristic time-interval optimum. For example, whereas 2-mercaptoethylamine [60-23-1] (MEA) protects irradiated human cells in culture maximally ten min after administration, the sulfhydryl form of the phosphorothioate WR 2721 [20537-88-6] protects optimally at ca 100 min; WR 2721 shows maximal protection after still longer times (52). Numerous additional factors affect the protective interval *in vivo* (53–54).

Although numerous indexes have been employed to describe the magnitude of chemical protection against radiation damage, the two most commonly reported are modifications of the percent survival and the dose-reduction factor (DRF) or dose-modification factor (DMF). The first is the percentage of irradiated animals that survive a particular dose of radiation for a certain time, usually 30 d, and is most useful in preliminary screening. For critical comparison studies, the DRF or DMF is employed. The dose-reduction or modification factor is the ratio of radiation doses administered to protected and control animals that produces the same biological effect in both animal groups. Again, the usual practice is to use organism death as the end point, but other parameters, including endogenous spleen-colony counts, have been employed (51). Thus,

$$\text{DRF} = \frac{\text{LD}_{50} \text{ protected}}{\text{LD}_{50} \text{ unprotected}}$$

where the LD_{50} may be determined at any selected time interval. When another end point of biological effect is selected, it is essential that an identical biological effect common to experimental and control animals can be identified.

Theories of Protection. *Molecular Level.* *Radical Scavenging.* To the extent that the radiolysis products of water play a role in the cause of cell injury in mammalian systems, the ability of radioprotective compounds to scavenge these mediators of the indirect effect must be considered a relevant mechanism in radioprotection. Moreover, since there are few examples in biological systems where radiation damage is mediated solely by the direct effect, it is axiomatic that scavenging of free radicals should at the very least be a contributing mechanism of radiation protection. However, even excellent OH· scavengers, eg, MEA or ethanol, which react with the hydroxyl radical at essentially diffusion-controlled rates, must be present *in vitro* at 10 mM and 1 M to assure minimal and good protection, respectively, of macromolecular function (30–31). On the other hand, if MEA is assumed to be distributed uniformly in the aqueous phase of the cell at a concentration of 3 mM, it gives excellent protection against radiation-induced cell death (30). This indicates that some phenomenon must be responsible for concentrating the protector at the site of critical target macromolecules by approximately a factor of 100 above that predicted from a uniform distribution of protector in cell water. Although this requirement is frequently overlooked, it has been the goal of hypotheses of radioprotection to develop mechanisms where this requirement is met so that radical scavenging and hydrogen-atom donation (see below) can be meaningful in the cellular milieu (55).

Model studies involving DNA as the target molecule indicate that the most effective radical-scavenging radioprotectors form some type of complex with DNA and are then able to scavenge radicals at this presumably critical site. For example, both cadaverine [462-94-2] $H_2N(CH_2)_5NH_2$, and WR 2721, $H_2NCH_2CH_2CH_2NHCH_2CH_2SPO_3H_2$ (or its *in vivo* sulfhydryl form) bind to DNA. However, presumably because WR 2721 can also scavenge OH· at a diffusion-controlled rate, it protects DNA from radiation damage, whereas cadaverine does not (56). DNA binding is not a universal requirement for radioprotection and does not correlate with radioprotective efficacy in the *N*-heterocyclic aminoethyl disulfides (57). Sulfur-containing radioprotective compounds are also excellent scavengers of hydrogen atoms and hydrated electrons, which are two other significant water radicals.

Hydrogen Transfer. A second fundamental phenomenon, which may account for reduction of radiation damage in biological systems, is indicated by equation 5. If the initial damage to the target consists effectively of loss of a hydrogen atom, then its restoration constitutes instantaneous repair and, thus, protection. Although H-atom transfer has only been observed in model systems (58–60), studies of mammalian cells in tissue culture have, by analysis of the shape of survival curves, resulted in the identification of two types of protection: competitive and restitutive (61). Existence of shoulders in such curves probably indicates protection involving competition for radiation products, ie, scavenging, whereas displacement of linear survival curves, which are observed after irradiation with high LET (linear energy transfer) particles, is thought to indicate repair of radiation lesions, perhaps by H-atom transfer. As in the case of radical scavenging, some mechanism must be invoked to account for localization of H-atom donors at critical sites.

The Mixed Disulfide Hypothesis. In one explanation of radioprotectant localization, it is proposed that the aminothiol protectors form temporary mixed disulfides with —SH and —SS— groups within cells (62–64). If the mixed disulfides were attacked by either direct- or indirect-radiation action, in at least half of such encounters radiolytic scission of the disulfide bond would restore the originally covered sulfhydryl moiety or allow restoration of a disulfide bond. Times for formation of mixed disulfides *in vivo* correspond well with times of observed optimal radiation protection, and good protectors exhibit the better propensities for mixed disulfide formation. The most protective sulfur compounds are probably able to induce polarization of the temporary mixed disulfide bond and, thereby, increase the probability that the S atom receiving the radiation insult would be that contributed by the protector to the mixed disulfide bond, which accounts for greater than 50% protection.

This hypothesis provides what is perhaps the most attractive postulate to account for localization of protectors at specific sites. However, since DNA has no —SH and —SS— moieties, it would seem that the mixed-disulfide hypothesis does not apply to protection of what has been considered the most important target macromolecule. Subsequent studies indicate that mixed-disulfide formation with nuclear proteins could, in part, account for the required enhanced concentration of potential radical scavengers and H-atom donors in the vicinity of DNA (65). Also, to the extent that cellular membranes constitute the target of ionizing radiation, the mixed-disulfide hypothesis becomes increasingly relevant (66).

Endogenous Nonprotein Sulfhydryl Compounds. A second hypothetical mechanism for localization of sulfur-containing radioprotective compounds at critical sites has been proposed (67–68). This hypothesis, like the former, involves the formation of temporary mixed disulfides between exogenous sulfhydryl radioprotective compounds and cellular disulfides; however, the subsequent focus of this concept is on the glutathione [70-18-8] or other nonprotein sulfhydryl (NPSH) species released when the temporary disulfide forms. The hypothesis suggests that it is the released endogenous sulfhydryl that scavenges water radicals. Thus, the NPSH hypothesis could account for protection of any cellular macromolecule, eg, protein, membrane constituent, or nucleic acid, exposed to increased concentration of released endogenous sulfhydryl. The fundamental observation on which the hypothesis is based is the correlation of increase in cellular sulfhydryl concentration induced by radioprotectors with their radioprotective efficacy. Compounds such as N-(2-mercaptopropionyl)-glycine [1953-02-2], with little ability to bind to DNA, protect DNA against single-strand breaks as effectively as MEA (69), which suggests that the shared ability to displace GSH (reduced glutathione) could be responsible. Enhanced protection by a combination of sulfur and nonsulfur protectants indicates cooperative effects in nonprotein thiol release (70). However, the postulated released endogenous scavenger, glutathione, is not a particularly effective radical scavenger (25–26). Moreover, selective oxidation of NPSH prior to irradiation does not enhance cell radiosensitivity (71). What differentiates the two mechanisms is the question of whether critical reactions of radiation damage occur primarily within the sphere of influence of the temporary mixed disulfide or in regions where only scavengers released from this zone are effective. It is probable that both hypotheses have relevance in chemical radioprotection.

Physiological-Biochemical Level. It is important that a mechanism of radioprotection not be sought at the molecular level when a far more critical mechanism of protection operates at a higher level. A reduction in body temperature is often associated with administration of various radioprotectants (72). A suggested mechanism for protection by hypothermia is that, during the period of lowered temperature, a reduced metabolic rate permits repair of crucial radiation damage before the demand of normal metabolism returns. Whereas such phenomena may be involved in protection by phenothiazines (73), hypothermia is usually considered only a side effect for the sulfur-containing radioprotectants.

Prevention of reactions involved in the oxygen effect (see eq. 4) is the apparent mechanism by which a number of compounds, particularly those related to histamine, are thought to be capable of protection. Drugs can induce hypoxia by either blocking hemoglobin function, increasing tissue oxygen utilization, or reducing local blood flow. Involvement of hypoxia in the mechanism of a compound's protection can be tested by irradiation under high oxygen pressure where the compound's hypoxic effects would be overwhelmed. Sulfhydryl radioprotective compounds maintain their efficacy under such conditions.

Administration of compounds that result in the release of interferon [9008-11-1] endogenously or of interferon itself has been reported to increase radioresistance of animals, if the effects of interferon were maximal at the time of irradiation (74–76). Although the mechanism of interferon protection is uncertain, it may involve an interruption of progression through the cell cycle at the most radioresistant stages. It may also induce the release of endogenous thiols (75).

Chemical Radioprotective Agents

Thiols. Cysteine. The vast majority of antiradiation agents are aminoalkyl thiols or derivatives thereof, the prototype of which is the sulfur-containing amino acid cysteine [4371-52-2] (1).

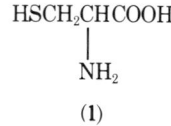

(1)

This compound protects 75–89% of rats subjected to 8 Gy (800 rad) if it is administered 5 min prior to x irradiation at 175–575 mg/kg (77). In this study, 19% of the irradiated control rats survived. Cysteine is equally effective if given up to one hour before irradiation. Mice given 1000 mg/kg of cysteine iv (intravenously) are protected to the extent of 50% from the effects of lethal radiation (78–79). Chromosome damage in irradiated human bone-marrow cells has been reduced 58% by cysteine (80).

A number of carboxylic esters of cysteine have been reported to give good protection, in terms of percent survival, to rats (81): cysteine methyl ester hydrochloride [2485-62-3], 70%; cysteine ethyl ester hydrochloride [3411-58-3], 55%; cysteine propyl ester hydrochloride [60654-26-4], 100%; cysteine isopropyl ester hydrochloride [73255-49-9], 40%; cysteine butyl ester hydrochloride [60654-27-5], 60%; cysteine isobutyl ester hydrochloride [81643-70-1], 100%; and cysteine isoamyl ester hydrochloride [81643-71-2], 70%. The oxidized form of cysteine, namely, cystine, and its diethyl ester impart no protection (77–78).

A reaction between cysteine and rutoside gives rutosidyl-2′-methylenecysteine, which is claimed to normalize serum-protein fractions in x-irradiated animals (82).

Interchanging the positions of the NH$_2$— and HS— groups of cysteine gives isocysteine, which has no radioprotective properties (83).

2-Mercaptoethylamine. The decarboxylated form of cysteine, namely 2-mercaptoethylamine (2) (MEA, cysteamine, 2-aminoethanethiol, mercamine, Becaptan) is an even more promising antiradiation agent than cysteine (84).

$$H_2NCH_2CH_2SH$$

(2)

2-Mercaptoethylamine, as the free base, is readily air-oxidized to its disulfide and probably exists in the zwitterionic form $H_3N^+CH_2CH_2S^-$ (see Sulfur compounds; Thiols). The latter concept is borne out by the unusual situation wherein the free base of MEA melts at a higher temperature (97–98.5°C) than its hydrochloride salt [156-67-0] (70–71°C). The hydrochloride salt of (2) is less susceptible to air-oxidation than the free base and is the form in which it is generally stored and administered.

Because of its structural simplicity, MEA hydrochloride is one of the most studied antiradiation agents. It is the compound that not only serves as a model for the design of other agents, but generally is also the standard by which the activity of other agents is judged. The compound confers greater protection to mice irradiated with a single 8-Gy (800-rad) dose than to the mice given four 2-Gy (200-rad) doses at intervals of 7 d (85). It offers protection against at least 3 repeated lethal exposures, provided they are at 30-day intervals (86). When administered in the drinking water of mice, MEA did not protect against chronic radiation (87). The compound protects the gastrointestinal tract and bone marrow of mice (88).

The antiradiation properties of MEA are optimized in mice if it is given 10 min prior to radiation (89), whereas in rats, best results are obtained 45 min before radiation (90). 2-Mercaptoethylamine protects mouse (91–92) and rat (93–94) spermatozoa. In the rat fetus, it prevents foot deformities and gait defects in the progeny if administered to mothers irradiated on the 14th day of pregnancy (95). Also, it reduces learning deficiency in surviving rats irradiated *in utero* (96).

2-Mercaptoethylamine has been prepared in a variety of salt forms, eg, the hydrobromide [*42954-15-4*], ascorbate [*16031-82-6*], nicotinate [*81643-72-3*], salicylate [*81643-73-4*], (97) and tartrate [*18594-39-3*] (98). The first three are more effective and less toxic than the hydrochloride [*156-67-0*] (99–100). Cytriphos (cysteamine adenosine triphosphate) shows a radioprotective effect in mice after both short-term and prolonged irradiation and is less toxic than MEA.HCl (101–103).

The radiation-prophylactic action of MEA has been ascribed to its ability to scavenge free radicals (30–31), to form mixed disulfides (104), to induce hypoxia (105–107), and to prevent cross-linking (108) and DNA breakdown induced by radiation (109).

Other Mercaptoalkylamines. The relative activities of the next higher homologues of MEA, namely 3-aminopropanethiol [*462-47-5*] (**3**) (3-mercaptopropylamine, MPA), 2-aminopropanethiol [*10229-29-5*] (**4**), and 1-amino-2-propanethiol [*598-36-7*] (**5**), are not particularly clear.

$$HSCH_2CH_2CH_2NH_2 \qquad\qquad CH_3\underset{\underset{NH_2}{|}}{C}HCH_2SH \qquad\qquad CH_3\underset{\underset{SH}{|}}{C}HCH_2NH_2$$

$$(3) \qquad\qquad\qquad (4) \qquad\qquad\qquad (5)$$

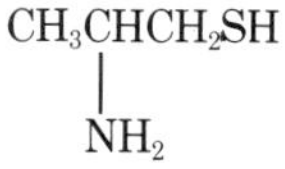

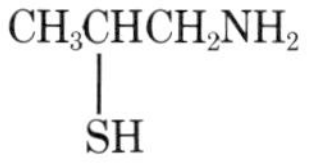

3-Aminopropanethiol is reported to be superior to MEA on a molar basis (78); however, it is also claimed that the same compound is totally ineffective (79,110). Compound (**4**) is reported to have about the same toxicity as MEA (111). Other investigations suggest that (**4**) and (**5**) have a greater prophylactic range than MEA (112). In general, there seems to be agreement that (**5**) is superior to its isomer (**4**). Whereas (**4**) offers protection to only 20% of irradiated mice at a dose of 175 mg/kg (79), (**5**) gives 60% protection at 300 mg/kg. In one study, (**5**) offered good protection (113). In another, 57–80% survival in mice and 65–75% in rats was obtained, depending on the administered dosage, but (**5**) is inactive when administered orally (114). In another report, (**5**) is judged to be superior to MEA as an antiradiation agent (115).

The placement of more than three carbon atoms between a thiol or a potential thiol and an amino group completely eliminates mammalian radioprotective properties (78,115–117). Other variations that seriously diminish or destroy activity include the alkyl-branched mercaptoethylamines (**6**) (118), *sec*-mercaptoalkylamine (**7**) (119), and 2-mercapto-2-phenethylamine [*934-14-5*] (**8**) (120).

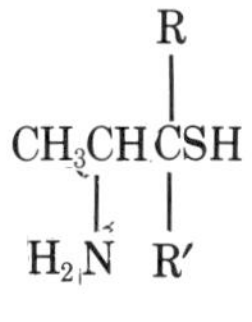

$$CH_3\underset{\underset{R'}{|}}{\underset{|}{C}H}\underset{\underset{}{|}}{\overset{\overset{R}{|}}{C}}SH \qquad\qquad R\underset{\underset{R'}{|}}{C}H\underset{\underset{}{|}}{C}HNH_2$$
$$H_2N\phantom{{}R'} \qquad\qquad SH$$

$$(6) \qquad\qquad\qquad (7)$$

Penicillamine [*52-67-5*] (**9**) long thought to be a radiosensitizer (121–122), is protective when administered 1 h before radiation (123).

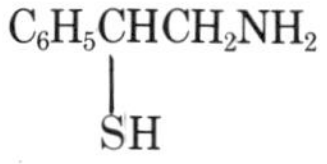

$$C_6H_5\underset{\underset{SH}{|}}{C}HCH_2NH_2$$

(8)

$$CH_3\underset{\underset{SH}{|}}{\overset{\overset{CH_3}{|}}{C}}CH_2NH_2$$

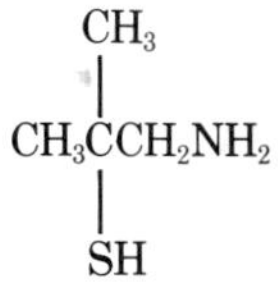

(9)

The placement of amino and thiol functions at adjacent positions in an alicyclic system, eg, DL-*trans*-2-aminocyclohexanethiol [20509-06-2] (10) (124–125) and *cis* and *trans*-cyclobutyl derivatives (11)–(13) (126), yields compounds with considerable antiradiation activity.

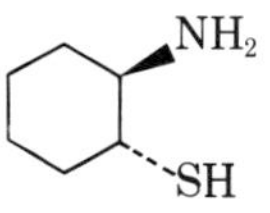

(10) 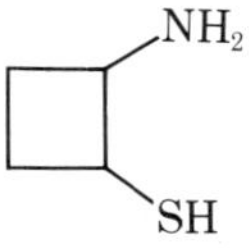 (11) (12) (13)

                cis [36455-65-9]   cis [81643-74-5]   cis [40830-54-4]

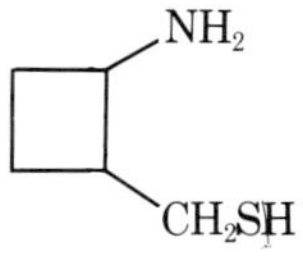

              trans [36455-66-0] trans [59276-24-3] trans [59273-23-2]

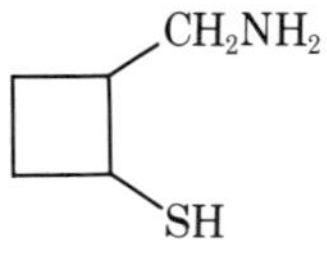

An extensive series of *N*-alkyl MEAs and 2-hydrazinoethanethiols was synthesized, but none is superior to MEA (127). The *N*-phenethyl- (128) and carboxymethyl-[cysteamine-*N*-acetic acid, (14)] derivatives of MEA have good antiradiation properties.

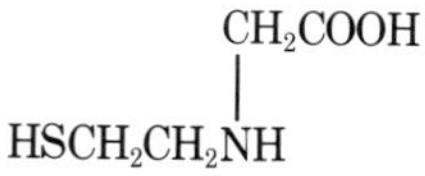

$$HSCH_2CH_2\underset{\underset{}{}}{\overset{\overset{CH_2COOH}{|}}{N}}H$$

(14)

cysteamine-*N*-acetic acid

The latter compound and its esters and salts are better tolerated and more effective than MEA (129–131).

N-Acetylation and *N,S*-diacetylation of MEA yields products with only slight activity (132), 2-Carboxyethyl- and 2-carbamidoethyl-*N*-substituted derivatives of MEA have excellent activity if given in high doses (133). 2-Mercaptoacetamidine [19412-52-3] (15) and its disulfide [44957-28-0], when given ip (intraperitoneal) or po (*per os* = oral) to mice prior to exposure to 9 Gy (900 rads) of x rays, increases their survival chances to 50% (134).

$$HSCH_2\overset{\overset{NH}{\|}}{C}NH_2$$

(15)

Aminoethyl sulfides (**16**) that form by S-alkylation of MEA tend to have little or no radioprotective activity (36,135–138).

$$H_2NCH_2CH_2SR$$
(**16**)

However, the increased stability of these derivatives has been reported, and S-methyl [*18542-42-2*], S-phenyl [*2014-75-7*], S-benzimidazoyl [*7673-88-3*], S-benzothiazoyl [*60372-30-7*], and S-furfuryl [*81643-73-6*] derivatives of MEA have antiradiation activity comparable to MEA (135). A caffeine derivative of MEA is reported to have good activity (99).

S-Acetyl MEA [*6197-31-5*] offers 60% protection to mice when given at 400 mg/kg, but S-benzoyl MEA is, for the most part, devoid of activity (139). The trithiocarbonate [*15547-18-9*] (**17**) protects 75% of irradiated mice at a dose of 350 mg/kg (139–141).

$$H_3\overset{+}{N}CH_2CH_2S\overset{\overset{S}{\|}}{C}S^-$$
(**17**)

An extensive series of MEA hemimercaptals derived from glycolic acid has been prepared. The most active member is (**18**) [*32641-24-0*], which protects mice against 8.5 Gy (850 rads) at a dose of one-half its LD_{50} (142).

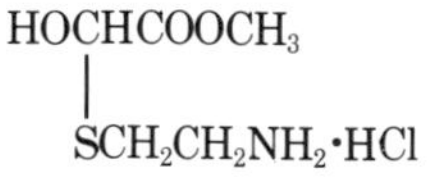

$$\underset{\underset{SCH_2CH_2NH_2 \cdot HCl}{|}}{HOCHCOOCH_3}$$
(**18**)

Glutathione [*70-18-8*] (**19**) is a tripeptide possessing a cysteine moiety and is reported by one group to give moderate radiation protection (143), whereas another group indicates that it is inactive (78).

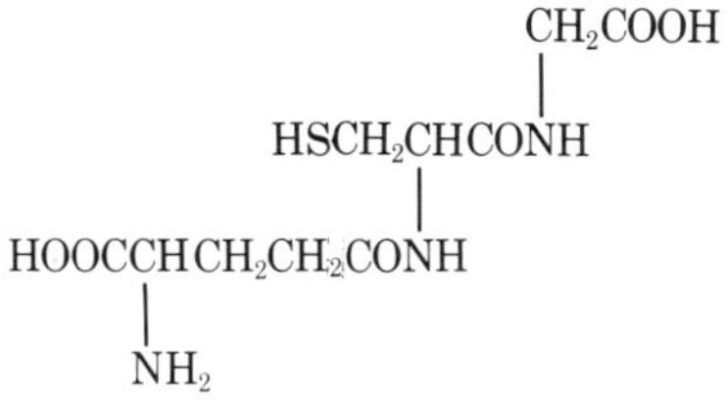

$$\underset{\underset{NH_2}{|}}{\underset{\underset{HOOCCHCH_2CH_2CONH}{|}}{\underset{\underset{HSCH_2CHCONH}{|}}{CH_2COOH}}}$$
(**19**)

Alkylthiols Lacking an Amino Group. One of the most interesting compounds developed as an antiradiation agent is sodium 2,3-dimercaptopropanesulfonate [*4076-02-2*] (**20**) (Unithiol), which was first synthesized in the USSR (144).

$$\underset{SH \quad SH}{\underset{|\quad\;\; |}{CH_2CHCH_2SO_3Na}}$$
(**20**)

It is more protective and less toxic (LD_{50}, 1400 mg/kg) in mice than MEA (99). It also is protective in rats and dogs. Unithiol is structurally related to the heavy-metal antidote, 2,3-dimercaptopropanol (BAL). In rodents, it is an efficient chelating agent which, when complexed, is eliminated from mammalian systems in water-soluble form. Unithiol has been studied in the treatment of poisoning by mercury (145–146), arsenic (147), antimony (148), gold and cadmium (149), and mixtures of metals, eg, mercury, nickel, copper, and cadmium (150).

Cleland's reagent [3483-12-3], which is also a dimercaptan, protects ca 30% of irradiated mice (151–152), whereas its cyclized, ie, disulfide, form [25902-99-2] protects 56% of irradiated mice (151–152).

Disulfides and Trisulfides. 2-Mercaptoethylamine can be oxidized to its disulfide, ie, bis(2-aminoethyl) disulfide [51-85-4] (cystamine). The free base is a water-soluble liquid; however, it is usually administered as a solution of its crystalline dihydrochloride salt [56-17-7] (**21**). The compound has lower acute toxicity, is as about as effective as MEA, and exhibits activity when administered orally to mice, rats, and guinea pigs (153–154).

$$(HCl \cdot H_2NCH_2CH_2S)_2$$
(**21**)

The compound affords 60% survival at a dose of 146 mg/kg to rats subjected to lethal radiation and also protects antibody production in rats (155–156). Cystamine is not effective when incorporated into the diet of mice (157). Cystamine at a dose of 60 mg/kg depresses the clinical signs of radiation sickness in dogs and accelerates their rate of recovery after exposure to 3 Gy (300 rads). The severity of leukopenia is also diminished (158). A clue to its mechanism of action may be in the reduction of cystamine to MEA *in vivo* during irradiation (159). Other beliefs are that cystamine protects DNA by complexing with it, thereby stabilizing the DNA helix (160), and that mobilization of endogenous catecholamines may be involved (161). Recently, a drug called resin-amine, ie, cystamine bound to Dowex 50 resin, is claimed to prolong the action of cystamine and to increase its radioprotective effect (162). Cystamine pyrophosphate [58480-03-8], when given ip or iv to rats, accumulates in the bone marrow to a greater extent and is more radioprotective than the dihydrochloride salt (163).

None of the *N*-substituted derivatives of cystamine, including a series of *N*-heterocyclic aminoethyl disulfides, exceeds the activity of the parent compound (164). The pharmacology of a few *N,N'*-dialkylated cystamines has been reported (131).

A limited number of aromatic mixed (unsymmetrical) disulfides, eg, *o*-(2-aminoethyldithio)benzoic acid hydrochloride [1204-52-0] (**22**), show possible activity (165–167), whereas those obtained from mercaptoterephthalic acid (**23**) fail entirely to protect irradiated animals (168).

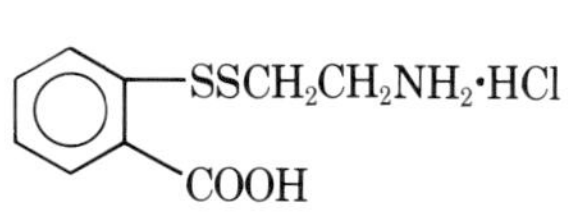

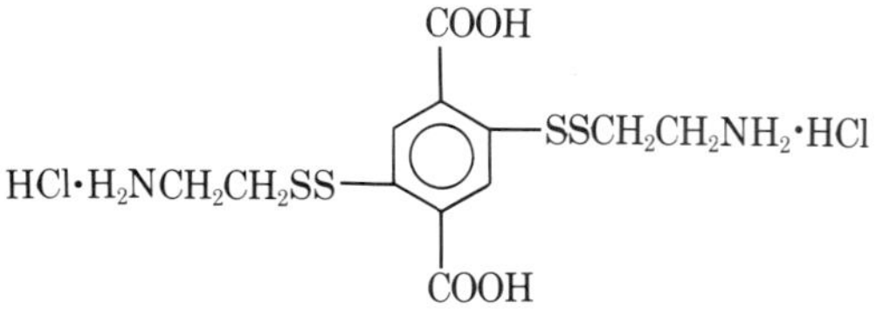

Of the totally aliphatic unsymmetrical disulfides of the type exemplified by structures

(24) (169) and (25) (170), the only active compound is (25), where $n = 3$ [*15386-71-7*].

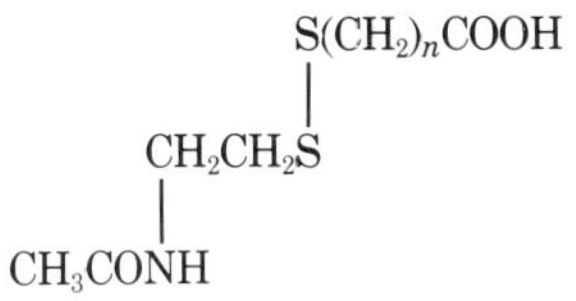

$$\begin{array}{c} \text{S(CH}_2)_n\text{COOH} \\ | \\ \text{CH}_2\text{CH}_2\text{S} \\ | \\ \text{CH}_3\text{CONH} \end{array} \quad n = 1\text{-}4$$

(24)

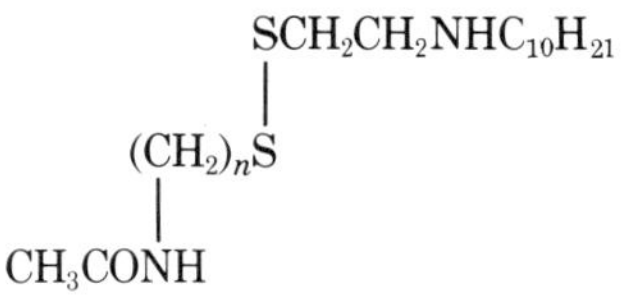

$$\begin{array}{c} \text{SCH}_2\text{CH}_2\text{NHC}_{10}\text{H}_{21} \\ | \\ (\text{CH}_2)_n\text{S} \\ | \\ \text{CH}_3\text{CONH} \end{array} \quad n = 2, 3$$

(25)

The mixed disulfides of type (26) give good protection to mice at moderately low doses (171).

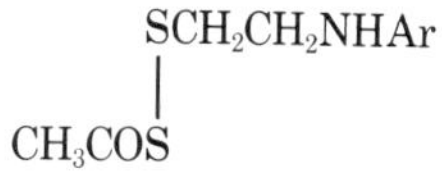

$$\begin{array}{c} \text{SCH}_2\text{CH}_2\text{NHAr} \\ | \\ \text{CH}_3\text{COS} \end{array}$$

(26)

Since unsymmetrical disulfides tend to disproportionate, especially under alkaline conditions, so as to give a mixture of symmetrical disulfides, it may be that under physiological conditions the combined effects of the symmetrical disulfides are observed.

$$2 \text{ RSSR}' \rightleftharpoons \text{RSSR} + \text{R}'\text{SSR}'$$

The cyclic disulfide, thioctic acid [*62-46-4*] (27) is reported to be toxic and nonprotective in mice (172). However, other investigators claim that, when given 10 min prior to irradiation [5.4 Gy (540 rads)], it protects liver, spleen, and kidneys somewhat better than MEA (173).

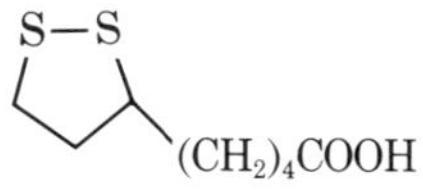

(27)

An interesting class of antiradiation compounds bearing both disulfide and butanesulfinate groups but lacking a basic amino moiety has been developed. The most active of the class are (28) [*19293-56-2*] and its disproportionation product (29) [*34915-82-7*], which is devoid of a nitrogen-containing functionality (174–175).

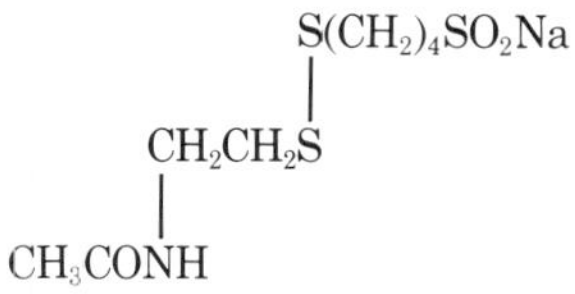

$$\begin{array}{c} \text{S(CH}_2)_4\text{SO}_2\text{Na} \\ | \\ \text{CH}_2\text{CH}_2\text{S} \\ | \\ \text{CH}_3\text{CONH} \end{array}$$

(28)

$$\begin{array}{c} \text{S(CH}_2)_4\text{SO}_2\text{Na} \\ | \\ \text{NaO}_2\text{S(CH}_2)_4\text{S} \end{array}$$

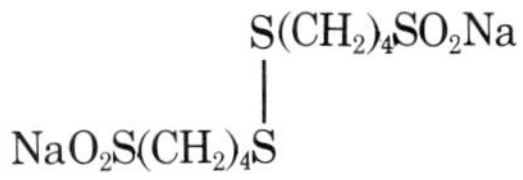

(29)

These compounds protect 93% and 73% of lethally irradiated mice at doses of 172 mg/kg and 200 mg/kg, respectively. The former also protects 100% of the mice if it is administered at a dose of 278 mg/kg po (175–176). The trisulfide corresponding to (29) [56527-86-7] protects 100% of irradiated mice when 300 mg/kg is given ip (175). At a dose of 37.5 mg/kg, the trisulfide protects 73–93% of irradiated mice when given ip (176). The main disadvantage of the sulfinates is their long-term instability and the difficulty with which pure samples are prepared.

Initially, the controlled oxidation of cystamine dihydrochloride yields 2-aminoethyl 2-aminoethanethiolsulfinate dihydrochloride (30) (177), and then the related thiolsulfonate [81643-76-7] (31) (178).

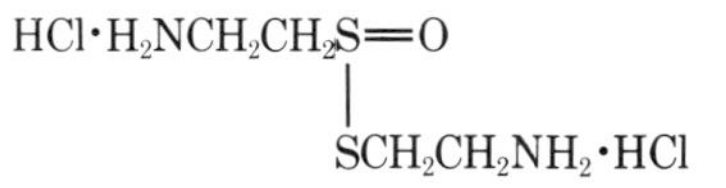

$$HCl \cdot H_2NCH_2CH_2S=O \qquad\qquad HCl \cdot H_2NCH_2CH_2\overset{\overset{O}{\|}}{S}=O$$
$$| \qquad\qquad\qquad\qquad\qquad\qquad |$$
$$SCH_2CH_2NH_2 \cdot HCl \qquad\qquad SCH_2CH_2NH_2 \cdot HCl$$
$$(30) \qquad\qquad\qquad\qquad\qquad (31)$$

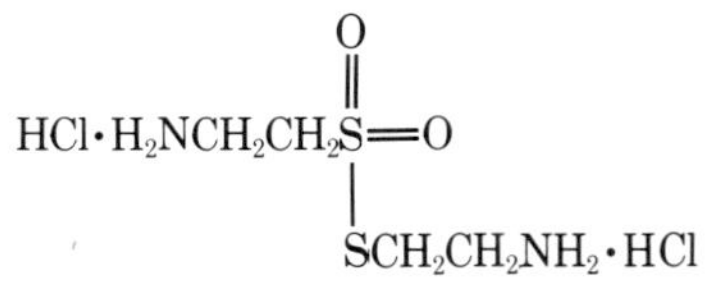

Whereas (30) is almost nonprotective (179), (31) is protective, as are its N-acetylated and N-decylated derivatives (166).

Organic Thiosulfates (Bunte Salts). In contrast to thiols, which are susceptible to air oxidation, organic thiosulfates (Bunte salts) are essentially unaffected by air. In addition, they can be solubilized by formation of their alkali salts. Furthermore, the latter react *in vitro* and *in vivo* with sulfhydryls to form mixed disulfides (180–181). A review on the chemistry and applications of Bunte salts is given in ref. 182.

$$RS^- + R'SSO_3^- \rightleftharpoons RSSR' + SO_3^{2-}$$

Bunte salts usually have lower acute toxicity than the corresponding thiols; however, their antiradiation properties tend to be inferior to the latter, especially when they are administered orally (79,183).

2-Aminoethanethiosulfuric acid [2937-53-3] (32) increases the survival of irradiated mice (184).

$$H_2NCH_2CH_2SSO_3H$$
$$(32)$$

When administered at a dose of 150 mg/kg ip to lethally irradiated mice, (32) protects 73% of them (79). The same compound is effective when administered orally to mice subjected to x irradiation of 6–8 Gy (600–800 rads) for more than 6 h (185). The material, like MEA, protects against chromosomal aberrations in the bone marrow of mice exposed to x rays (186). Also, if incubated in rat tissue homogenates, (32) reacts rapidly and nonenzymatically with a protein sulfhydryl group to form MEA, cystamine, protein-bound MEA disulfides, and sulfite ion (181).

3-Aminopropanethiosulfuric acid [13286-24-3] is a less effective antiradiation agent than either its lower homologue (32) or MEA and protects 40% of irradiated mice at a dose of 500 mg/kg (79,187–188). 1-Aminopropane-2-thiosulfuric acid [2403-34-1] (33) and its isomeric Bunte salt 2-aminopropane-1-thiosulfuric acid [2403-32-9] (34) have been tested, but only the latter shows radioprotective activity (80% survival at a dose of 350 mg/kg) (79).

$$H_2NCH_2CHCH_3 \atop |\atop SSO_3H$$
(33)

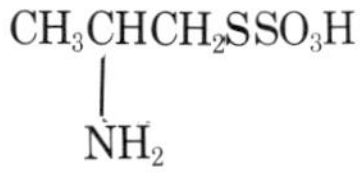

$$CH_3CHCH_2SSO_3H \atop |\atop NH_2$$
(34)

Increasing the number of methylene groups that separate the amino and thiosulfuric acid functions to 4 and above results in loss of antiradiation activity (79). Bunte salts derived from amino acids have shown limited radioprotective properties (189–190).

N-Alkylated-2-aminoethanethiosulfuric acids, of which over 100 have been synthesized, lose their water solubility and become waxlike as the chain length increases. High antiradiation activity is, nevertheless, observed from many of the poorly water-soluble compounds. 2-Methylaminoethanethiosulfuric acid [1000-68-6] possesses about the same activity as the parent compound in mice. When R ranges from ethyl to hexyl, derivatives (35) are nonprotective (191). However, activity returns if R = heptyl (2-heptylaminoethanethiosulfuric acid [1191-49-7]) and is maximized at R = decyl. The latter compound, decylaminoethanethiosulfuric acid [3752-51-0] (36) (WR 1607), protects 90% of irradiated mice at the low dose of 5 mg/kg (191–192).

$RNHCH_2CH_2SSO_3H$ $n\text{-}C_{10}H_{21}NHCH_2CH_2SSO_3H$
(35)   (36)

Placement of the amino function at the 2- or 3-positions of the decyl group results in the need for the greater dose of agent to maintain the same level of radioprotection. The incapacitation of rhesus monkeys that have received massive doses of radiation is prevented by prior administration of 10 mg/kg of (36) (193). The same compound also protects 50% of *Mucaca mulatta* monkeys against 8.5 Gy (850 rads) (194–195). Analogues above C_8 are not effective when administered orally.

Of a group of N-phenylalkylaminoethanesulfuric acids, the greatest activity is shown by N-(4-phenylbutyl)aminoethanethiosulfuric acid [23464-46-2] (37) (191). The presence of a *p*-methoxy group on the phenyl ring of (37) yields N-[4-(4-methoxyphenyl)butyl]aminoethanethiosulfuric acid [21208-80-0], which has improved activity (196).

$C_6H_5(CH_2)_4NHCH_2CH_2SSO_3H$
(37)

Cyclohexyl and cyclohexenyl moieties in place of the phenyl ring in (37) also yield active compounds, ie, N-4-(cyclohexyl)butylaminoethanethiosulfuric acid and N-4-(cyclohex-3-enyl)butylaminoethanethiosulfuric acid (197). An interesting terpenelike compound is (38) (198), which affords 100% survival in mice at a dose of 15 mg/kg.

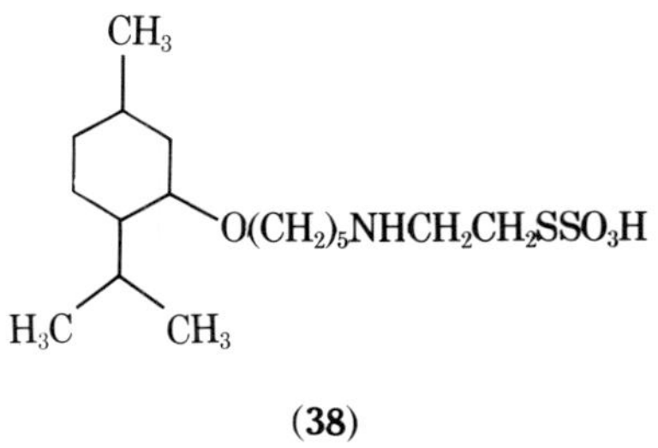

(38)

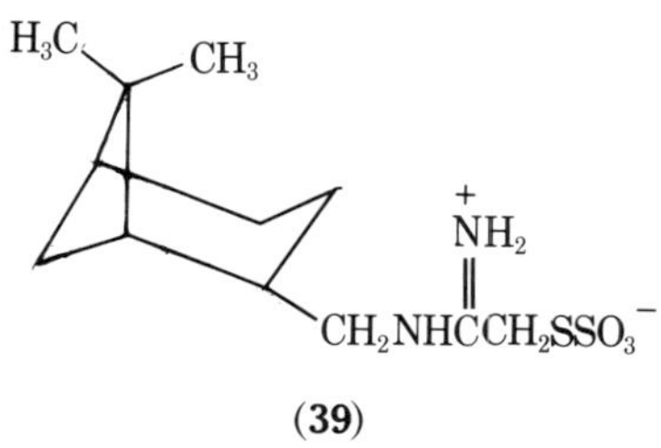

(39)

The bridging of two molecules of 2-aminoethanethiosulfuric acid by 2–6 methylene groups, as in [HO$_3$SSCH$_2$CH$_2$NH(CH$_2$)$_n$NHCH$_2$SSO$_3$H], eliminates antiradiation activity (199). A series of N-heterocyclic 2-aminoethanethiosulfuric acids is essentially inactive (164).

Many Bunte salts with amidino groups have been prepared (200–203). Among them are many alkyl, cycloalkyl, cycloalkylalkyl, and aralkyl compounds that show good activity (203). Some of the most effective possess terpenoid structures, eg, (**39**). The substitution of an amidino group for an amino group in a series of well-known antiradiation compounds does not significantly affect the radioprotective properties of the compound (204). A small number of α-amidrazonium thiosulfates (**40**) show poor activity (205).

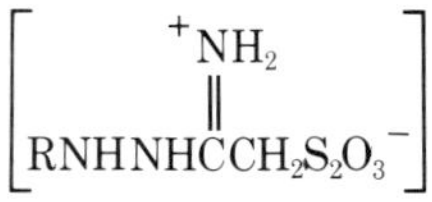

$$\left[\begin{array}{c} {}^+\text{NH}_2 \\ \| \\ \text{RNHNHCCH}_2\text{S}_2\text{O}_3{}^- \end{array} \right]$$

(**40**)

2-Guanidinoethanethiosulfuric acid [*7176-65-0*] (**41**) affords 80% survival in lethally irradiated mice when administered at a dose of 100 mg/kg (206–207).

$$\begin{array}{c} \text{NH} \\ \| \\ \text{H}_2\text{N CNHCH}_2\text{CH}_2\text{SSO}_3\text{H} \end{array}$$

(**41**)

Structure modification in the guanidino group or extension of the N–S distance beyond two methylene groups results in great loss of activity.

The Bunte salt related to cysteine, ie, S-sulfocysteine [*1637-71-4*] (**42**), is weakly protective (79,208).

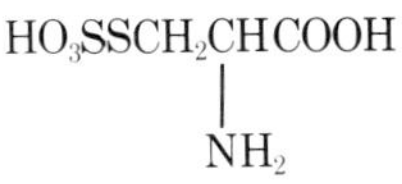

$$\begin{array}{c} \text{HO}_3\text{SSCH}_2\text{CHCOOH} \\ | \\ \text{NH}_2 \end{array}$$

(**42**)

The related sodium cysteinethiosulfate [*7381-67-1*] (**43**) is protective if given 5 min before irradiation at a dose of 250 mg/kg (209–211).

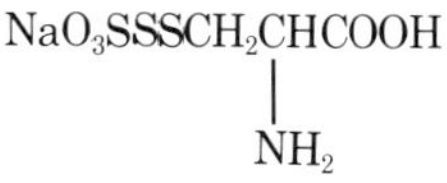

$$\begin{array}{c} \text{NaO}_3\text{SSSCH}_2\text{CHCOOH} \\ | \\ \text{NH}_2 \end{array}$$

(**43**)

Phosphorothioates. The most promising of the modified thiol groups to be incorporated into potential radioprotective agents is the phosphorothioate functionality. Compounds bearing this group do not undergo typical thiophilic displacements, to which disulfides or Bunte salts are subject, to give mixed disulfides. Phosphorothioates, however, are hydrolyzed very rapidly in the presence of acid to give the corresponding thiols and are enzymatically converted to thiol and orthophosphate by human erythrocytes, bovine brain, rat liver homogenates, and isolated acid or alkaline phosphatases (213–216).

Sodium 2-aminoethanephosphorothioate [3724-89-8] (**44**) (sodium 2-aminoethanethiol dihydrogen phosphate, WR 638, cystaphos) was first prepared and studied in 1959 (217).

$$H_2NCH_2CH_2SPO_3HNa$$
(**44**)

It has excellent radioprotective action [(>95% survival in lethally irradiated mice) (218)] and is superior to MEA when given orally (219). In rats, the compound exerted its maximum radioprotective action when given 60–90 min prior to irradiation (220). Its ability to protect the DNA molecule from the effects of γ radiation (221) and to work synergistically with 2-aminoethylisothiuronium bromide hydrobromide [151-16-6] (AET) in mice (222) has been demonstrated.

Many N-substituted derivatives of (**44**) are ineffective (218), the notable exceptions being the 1-adamantyl derivative and some related alicyclic compounds that have moderate activity (223). The homologue of (**44**) in which the backbone is extended to 3 carbon atoms is similarly inactive (218,224). Placement of a hydroxyl group on C-2 restores activity, but the resultant compound is inactive when administered orally (225). In general, placement of alkyl groups on either of the two methylenes separating the amine and phosphorothioate functions does not have detrimental effects (189–200,226–227). Linkage of 2 molecules of 2-aminoethanephosphorothioate by an N,N'-polymethylene chain gives good protection if either 3 or 4 methylene groups are present. Similarly constructed Bunte salts are inactive. Members of a series of S-2-ω-diaminoalkyl dihydrogen phosphorothioates of type (**45**) for the most part provide high antiradiation activity (89–100%) at moderately high dose levels (200–400 mg/kg) (228).

$$H_2N(CH_2)_nCHCH_2SPO_3H_2$$
$$|$$
$$NH_2$$
(**45**)

The compounds lack activity if administered orally. 2-Guanidinoethanephosphorothioate [54978-25-5] and 3-guanidinopropanephosphorothioate protect 97% and 80%, respectively, of irradiated mice (218). Many amidino-phosphorothioates have been synthesized. One of the most active is (**46**) [16886-54-7], which affords 100% protection to mice at a dose of 8 mg/kg (203).

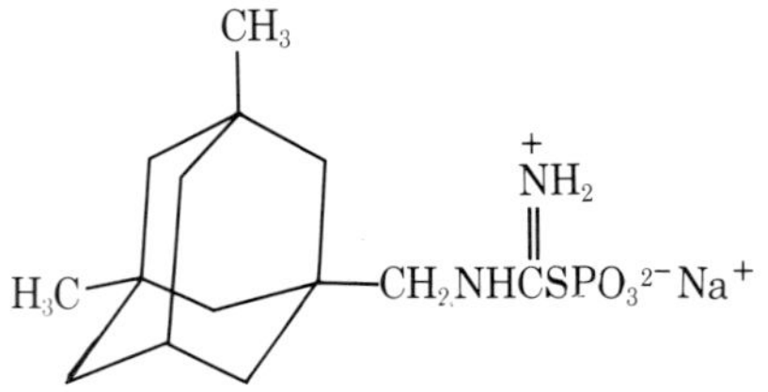

(46)

The inorganics (47) [16886-55-8] and (48) [16886-55-8] have reduction factors superior to MEA (226–227).

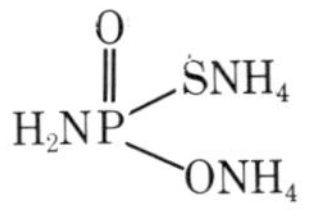

(47) 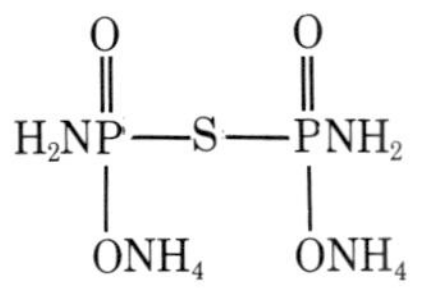 (48)

A particularly valuable series of antiradiation agents consists of 2-(ω-aminoalkylamino)ethyl- (49) and -propyl- (50) dihydrogen phosphorothioates (224).

$H_2N(CH_2)_n NHCH_2CH_2SPO_3H_2$ $H_2N(CH_2)_n NHCH_2CH_2CH_2SPO_3H_2$
(49)   (50)

High survivals and low toxicities characterize the former series when $n = 2$–6 and in the latter, when $n = 2,3$. The comparable Bunte salts are inactive. The excellent radioprotective ability of 2-(5-aminopentylamino)ethanephosphorothioate [20724-76-9] in mice exposed to x or neutron radiation has been verified (229); injury to hematopoietic organs and the gastrointestinal tract is reduced.

Probably the most effective of all antiradiation agents is 2-(3-aminopropylamino)ethanephosphorothioic acid [41510-53-6] (51) [WR 2721, amifostine (World Health Organization), gammaphos (USSR), YM-08310 (Japan)]. This compound protects mice, dogs, and rhesus monkeys against the effects of γ and x radiation.

$H_2NCH_2CH_2CH_2NHCH_2CH_2SPO_3H_2$
(51)

The compound protects 86% of irradiated mice at a dose of 300 mg/kg. WR 2721 promotes wound healing in irradiated rats and increases the resistance of the immune response to radiation injury (231). When administered topically, it does not provide skin protection in the mouse (232). The compound protects mouse intestine against fission neutrons (233) and x irradiation (234–235). WR 2721 is being considered for application in cancer radio- and chemotherapy.

Thioureas. Thiourea [62-56-6] was reported initially to be nonprotective (236), as was a group of S-alkyl thioureas (thiopseudoureas) (237), a series of α,ω-bis(thiopseudoureas) (132), and *N*- and *S*-substituted thioureas (238). In a more recent paper, thiourea, methylthiourea [*2986-19-8*], ethylenethiourea [*96-45-7*], methylthiopseudourea [*2986-19-8*], and ethylthiopseudourea [*2986-20-1*] are described as radioprotectors with low toxicity in x irradiated mice (239). Favorable results have been reported more recently with a series of α,ω-bis(thiopseudoureas), in which the methylene bridges are 2–5 carbon atoms in length (240). A series of phosphorus-containing derivatives of alkylthiopseudoureas, exemplified by *S*-ethylisothiuronium ethyl phosphite [*16400-82-1*], are active when given ip to rats prior to being exposed to γ irradiation [9 Gy (900 rads)] (241).

There has been intense interest in aminoalkylthiopseudoureas and, particularly, 2-aminoethylisothiuronium bromide hydrobromide (AET) [*56-10-0*] (**52**).

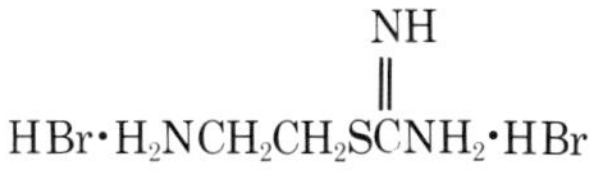

NH
HBr·H2NCH2CH2SCNH2·HBr

(**52**)

In aqueous solution, especially near neutrality, AET undergoes a rearrangement through an intermediate diaminothiazolidine to give 2-mercaptoethylguanidine [*1190-74-5*] (**53**) (MEG, 2-guanidinoethanethiol).

H2NCH2CH2
S
C=NH2+
NH2
HN
S
H2N
NH3+
HN
SH
C
NH2
NH2
-H+
HN
NH
SH
NH2
H+
N
S
+ NH4+
NH2

(**53**)

(**54**)

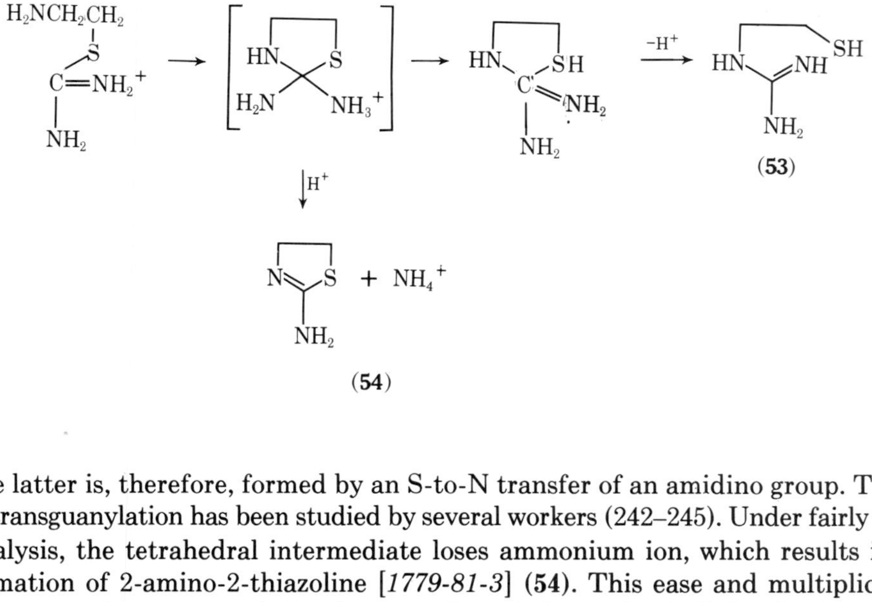

The latter is, therefore, formed by an S-to-N transfer of an amidino group. The intratransguanylation has been studied by several workers (242–245). Under fairly acidic catalysis, the tetrahedral intermediate loses ammonium ion, which results in the formation of 2-amino-2-thiazoline [*1779-81-3*] (**54**). This ease and multiplicity of conversions in aqueous solution has complicated the study of AET and related compounds.

2-Aminoethylisothiuronium bromide hydrobromide protects 88% of lethally irradiated mice at a dose of 250 mg/kg and is, thus, more effective than MEA on a molar basis (132). Other investigators have obtained excellent results with AET and have noted its lack of chronic toxicity (246–250). It appears to minimize functional and genetic damage to the reproductive system (251–254). The compound is poorly tolerated by dogs (255) but seems to be nonlethal at a dose of 125 mg/kg if administered by rapid iv injection (256). The protective dose (85–100 mg/kg) is only slightly below the lethal dose.

Numerous mixtures of AET with other compounds, eg, mexamine [66-83-1] (257–258), dimethyl sulfoxide [67-68-5] (DMSO) (259–260), barbital [57-44-3] (261), and cysteine (262), have been studied. Aminoethylisothiuronium adenosine triphosphate (Adeturon) protects the lymphocytes in human blood against chromosomal aberrations (263).

2-Mercaptoethylguanidine (**53**) is extremely difficult to isolate, however, its oxidized form, bis(2-guanidinoethyl) disulfide [1072-13-5] (**55**) (GED), is readily obtained in the pure state and is stable.

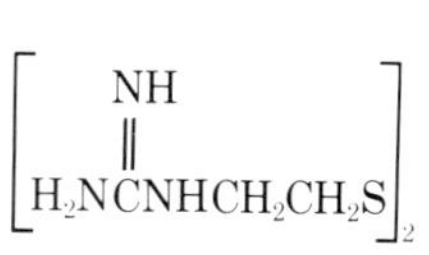

$$\left[\begin{array}{c} \text{NH} \\ \| \\ \text{H}_2\text{NCNHCH}_2\text{CH}_2\text{S} \end{array} \right]_2$$

(**55**)

The latter protects the bone marrow and gastrointestinal tract of mice (56,264–265), but is more slowly adsorbed than the corresponding thiol (266–267).

3-Aminopropylthiopseudourea bromide hydrobromide [7072-40-4] is superior to MEA on a molar basis but it is not as effective as AET (78). Its slower intratransguanylation has been investigated (243,268–269). 2-Aminobutylthiopseudourea dihydrobromide [33977-39-8] is also an active protector (270). Its optical resolution indicates that the D(−) isomer is about twice as effective as the L(+) isomer. A backbone greater than 3 carbon atoms in length eliminates radioprotective action in the AET series (271).

Thiazolines. The condensation of MEA or an *N*-substituted MEA with an aldehyde or ketone yields a thiazolidine (**56**). Numerous compounds of this type possess antiradiation activity, probably because of their ability to hydrolyze slowly *in vivo* to form their constituent aminoalkylthiols.

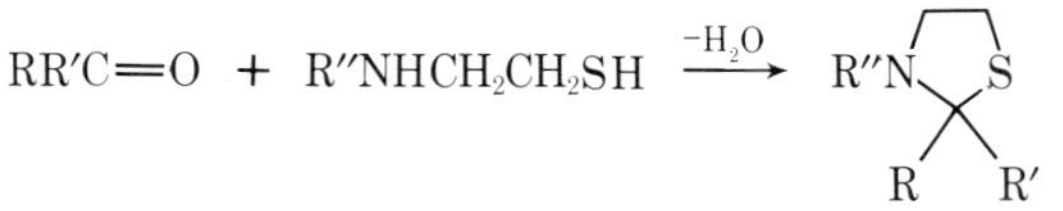

$$\text{RR'C}=\text{O} + \text{R''NHCH}_2\text{CH}_2\text{SH} \xrightarrow{-\text{H}_2\text{O}} \text{R''N} \diagup \text{S}$$
$$\diagdown \text{R} \quad \text{R'}$$

(**56**)

The correlation between radioprotection and the rate of hydrolysis as related to the substituents at the 2-positions of the heterocycle has been studied (272–273). The more active thiazolidines [eg, 2-propylthiazolidine [24050-10-0] (**57**), which effects 71% survival (187)], must be administered in larger doses to achieve the same order of protection as provided by the parent aminothiol.

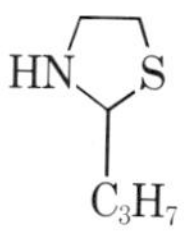

(57)

Numerous *N*-substituted thiazolidines, the alkyl chains of which have an oxy- or thiocycloalkyl, aryl, or heterocyclic group at the terminal position, have been made (198,274). One of the most active of the series is (58), which affords 93% survival to irradiated mice.

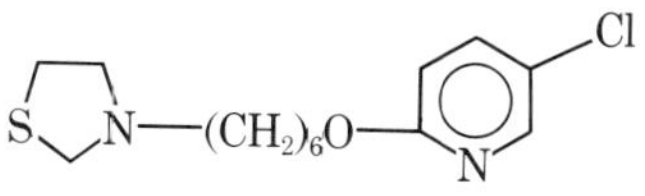

(58) 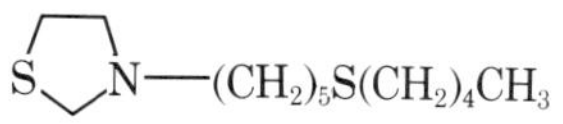   (59)

Compound (59) protects 92% if it is administered orally (275). The other active thiazolidines in animals have been correlated with their protection of irradiated human erythrocytes (276). A series of 2-phenylthiazolidines offers complete protection against one-half the LD_{50} radiation dose in mice (277–278).

2-Amino-2-thiazoline (54) effects 70% (261) and 35% (279) survival in mice, and its salts also are active (280). The 5-methyl- [10416-80-5] and 5-hydroxymethyl [35525-88-3] derivatives of (54) are active protectors of x-irradiated mice (281). The most effective of a series of 23 thiadiazoles is 2-amino-1,3,4-thiadiazole-5-thione hydrochloride [59909-21-6] (60), which affords 25–45% protection to x-irradiated mice at doses of 100–400 mg/kg (282).

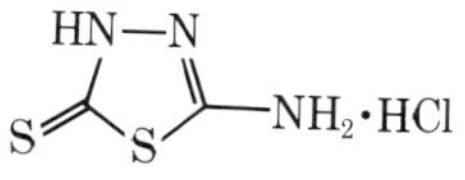

(60)

Selenium Compounds. Selenium analogues of the better known sulfur-containing antiradiation agents, eg, 2-aminoethaneselenol [21681-94-7], 2-aminoethaneselenosulfuric acid [2697-60-1] (79,283), and 2-aminoethylselenopseudourea [1704-04-7] (284), are toxic and nonprotective. However, in a more recent paper, it was reported that the latter compound gives significant protection to mice (285). Selenourea [630-10-4] protects rats subjected to 7.5 Gy (750 rads) of γ irradiation (286), whereas selenosemicarbazide [21198-79-8] [$H_2NNH(C=Se)NH_2$], at a dosage of 4 mg/kg subcutaneous, protects 50% of x-irradiated rats [6.01 Gy (601 rads)] (287). An extensive review of organoselenium compounds as potential medicinal agents is given in ref. 288.

Many inorganic selenium compounds, eg, sodium selenate [13410-01-0], minimize postirradiation effects of radiation in mammalian and enzyme systems (289–291).

Other Radioprotective Agents. Dimethyl sulfoxide in a dosage of 4500 mg/kg ip protects rats against 8 Gy (800 rads) (292), and is effective when applied topically to the animals' tails prior to irradiation (293–294).

Radioprotective antioxidants which have been claimed to be effective are gallic acid derivatives, eg, sodium gallate [2053-21-6] (295–297) and propyl gallate [121-79-9] (298). p-Aminoacetophenone [99-92-3] and especially p-aminopropiophenone [70-69-9] (**61**) (PAPP) have radioprotective action through their methemoglobin-inducing properties in rats and dogs (299).

(61)

The latter has been used in combination with MEA and AET because it probably acts by a different mechanism than MEA and AET (300–302). A study involving a series of 27 analogues of PAPP revealed that the only consistent structural feature for activity is a free amino or hydroxylamino group (303).

Mitotic supressive agents, eg, methyl trimethylcolchicinate [3476-50-4] (304) and colcemid [477-30-5] (305), either alone or in combination with MEA, have beneficial antiradiation properties.

Psychotropic drugs and tranquilizers, which are generally administered 3–4 h prior to irradiation, act beneficially by their hypothalmic and metabolism-depressing effects and probably are mediated through hypoxia. Compounds of varying degrees of activity are reserpine [50-55-5] (306–308); its N-oxide [474-48-6] (309); chloropromazine [50-53-3] (310–312); Sordinol [982-24-1], Mellerill [50-52-2], Truxal [113-59-7], and Fluanxol [2709-56-0] (314–315); Librium [438-41-5] (316); Valium [439-14-5] (317); Imipramine [50-49-7] (318–319); Trimipramine [739-71-9] (320); thiopental [76-75-5] (321); and phencyclidine [77-10-1] (322).

Two biogenic amines, serotonin [50-67-9] (**62**) (307,323–326) (5-hydroxytryptamine, 5-HT) and mexamine [66-83-1] (**63**) (327) (5-methoxytryptamine) are moderately active antiradiation agents in rodents when used alone.

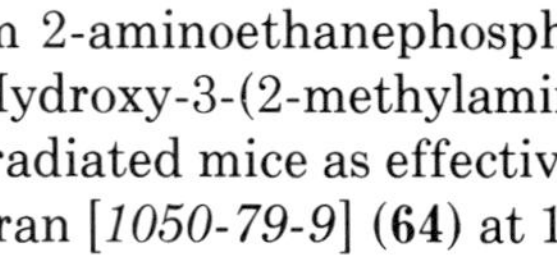

(62) 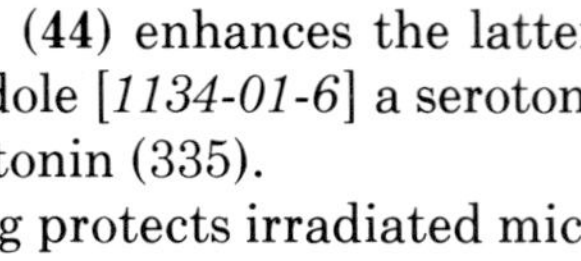   (63)

Serotonin has been studied in combination with MEA (**2**) (328) and (**44**) (329), and synergistic effects occur. Similarly, mexamine with MEA (330), cystamine (331–333), and sodium 2-aminoethanephosphorothioate (**44**) enhances the latter's protective effects. 5-Hydroxy-3-(2-methylaminoethyl)indole [1134-01-6] a serotonin derivative, protects irradiated mice as effectively as serotonin (335).

Luvatran [1050-79-9] (**64**) at 19–20 mg/kg protects irradiated mice (336).

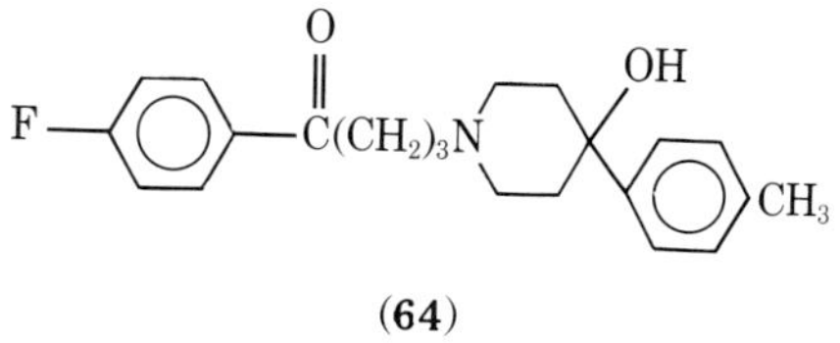

(64)

Certain interferon inducers, eg, tilorone [27591-69-1] and E. coli lipopolysaccharide, have radioprotective properties (337).

8-Mercaptocaffeine derivatives bearing β-aminoethyl- and β-hydroxyethyl groups have radioprotective activity in mice that is quantitatively similar to that of cystamine (338). Other compounds and groups of compounds purported to have antiradiation properties include N-(dimethylamino)ethylacridones (339), prodigiosan [82-89-3] (340), Vitamin C [50-81-7] (341), orotic acid [65-86-1] and its derivatives (342–344), and adenosine derivatives (345) (see Vitamins, vitamin C).

Sulfur-containing polymers have been developed which could, presumably, extend the period during which the radioprotective action is in effect. Success in x-irradiated mice has been claimed for vinylpyrrolidinone formaldehyde S-vinyl S-ethylmercaptal copolymer [35661-69-9], vinylpyrrolidinone N-methacrylhomocysteine thiolactone copolymer [34411-25-1], and vinylpyrrolidinone acrylic acid N-acryl-2-methylthiazolidine copolymer [34411-26-2] (346). Some polymeric dithiocarbamates show antiradiation activity in mice (37) (see also Polymers containing sulfur).

Potassium iodide or iodine is used to prevent thyroid damage in humans exposed to high levels of radioiodine (^{131}I), which is attached to macromolecules, eg, an antibody or fibrinogen, during cancer therapy or diagnosis (348–349). Here, the body, particularly the thyroid gland, is saturated with nonradioactive iodine given as Lugol's solution at a dose of about 250 mg iodide per day before, during, and after administration of the radioiodinated macromolecule. Any radioiodine released from the macromolecule during metabolism or by autoradiolysis is then excreted with the excess iodide, rather than sequestered in an iodine-requiring organ, which would result in radiation damage. Thus, iodine or potassium iodide serve as radioprotective agents and the latter is, in fact, now being distributed as a chemical prophylactic for use in the event of a nuclear accident involving ^{131}I release.

Additional Uses

Cancer Treatment. In clinical radiotherapy of malignant tumors, adjacent tissue is unavoidably damaged to some extent. It is, therefore, desirable to chemically protect normal tissue from radiation injury without affecting the radiosensitivity of the tumor. Sensitizers, eg, misonidazole [13551-87-6], aid in achieving some selectivity (350). Modest success in the protection of normal tissue has been achieved with the use of several phosphorothioate antiradiation agents. The compounds WR 638 (**44**) and WR 2721 (**51**) are apparently less absorbed by solid tumors than by the surrounding tissue (351–356). Whereas both types of tissue actively concentrate the radioprotective agents, the deficient vascularity of the tumor or lack of some concentration mechanism places it at a competitive disadvantage (353–354). The partition of AET and MEA between normal and tumor tissues has also been examined (357). The possible utilization of mixtures of radiation sensitizers and antiradiation agents in radiotherapy is being studied (23) (see also Chemotherapeutics, antimitotic).

Numerous antiradiation agents have been investigated for cancer chemotherapy. In one effort, MEA was used to treat 11 leukemia patients but no clear benefit was observed (358). In another study, MEA showed no antileukemic activity in mice (359). The toxic side effects of several antitumor drugs in mice and rats is reduced by AET (360). Administered to mice that were given ascites cells before irradiation, AET prolonged their life although it was taken up by normal and cancerous tissues equally (361). Greater selectivity was shown by MEG (362).

To the extent that water radicals are involved in the promotion phase of carcinogenesis, use of radioprotectants may be valuable in the design of cancer-prevention regimens.

Shock Therapy. Competitive inhibition of α-adrenergic receptors can be achieved through the use of the antiradiation agents 2-(5-aminopentylamino)ethanephosphorothioic acid [20724-76-9] (**65**) (WR 2823), its corresponding thiol [14653-79-3] (WR 1729), and its corresponding disulfide tetrahydrochloride (**66**) [31235-39-9] (WR 149,024). The order of their ability to act as α-adrenergic blockers is the opposite order of radioprotective efficacy: WR 149,024 > WR 1729 > WR 2823.

$$H_2N(CH_2)_5NHCH_2CH_2SPO_3H_2 \qquad [H_2N(CH_2)_5NHCH_2CH_2S]_2$$
$$(65) \qquad\qquad\qquad (66)$$

WR 2823 has shown potential usefulness in the treatment of hemorrhagic (363–364) and endotoxic shock (364–365). WR 149,024 effectively attenuated anaphylactic shock in mice (366) and aided in ameliorating the effects of hemorrhagic shock in dogs (367).

Space Flight. Travel through space not only subjects humans to the possibility of exposure to ionizing radiation but also to the stress of vibration and acceleration. Studies with rodents subjected to all three factors indicate that the added trauma does not substantially affect the course of the radiation sickness (368). The extrapolation of animal data to humans regarding the use of chemical antiradiation agents during space flight has been studied (369–370) (see Space chemistry). Mixtures containing MEA are of special interest in the USSR (371).

The additional effect of radiation upon hypokinesia (immobilization) has also been studied in the USSR. In general, the radioprotective action of such agents, eg, mexamine and cystamine, is reduced in animals subjected to both types of trauma (332).

BIBLIOGRAPHY

1. T. Alper, *Annu. Rev. Nucl. Sci.* **10**, 489 (1960).
2. V. P. Bond, T. M. Fleidner, and J. O. Archambeau, *Mammalian Radiation Lethality*, Academic Press, Inc., New York, 1965.
3. A. P. Casarett, *Radiation Biology*, Prentice-Hall, Inc., Englewood Cliffs, N.J., 1968.
4. H. Dertinger and H. Jung, *Molecular Radiation Biology*, Springer-Verlag, Berlin and New York, 1970.
5. E. Fahr, *Angew. Chem. Int. Ed. Engl.* **8**, 578 (1969).
6. A. M. Kuzin, *Radiation Biochemistry*, Daniel Davey, New York, 1964.
7. F. Hutchinson, *Cancer Res.* **26**, 2045 (1966).
8. S. Okada, ed., *Radiation Biochemistry*, Vol. 1, Academic Press, Inc., New York, 1969.
9. J. Schubert and R. E. Lapp, *Radiation: What it Is and How it Affects You*, The Viking Press, New York, 1957.
10. G. Silini, ed., *Radiation Research*, Elsevier-North-Holland Inc., Amsterdam, The Netherlands, 1967.
11. E. R. Atkinson, *Annu. Rep. Med. Chem.*, 327 (1968); 346 (1970).
12. Z. M. Bacq, *Chemical Protection Against Ionizing Radiation*, Charles C Thomas Publisher, Springfield, Ill., 1964.
13. V. S. Balabukha, ed., *Chemical Protection of the Body against Ionizing Radiation*, Macmillan, Inc., New York, 1964.
14. W. O. Foye, *Annu. Rep. Med. Chem.*, 324 (1966); 330 (1967).
15. W. O. Foye in W. E. Wolff, ed., *Burger's Medicinal Chemistry*, 4th ed., Part III, John Wiley & Sons, Inc., New York, 1980, p. 11.

16. P. C. Jocelyn, *Biochemistry of the SH Group*, Academic Press, Inc., New York, 1972, p. 323.
17. D. L. Klayman and E. S. Copeland in E. J. Ariëns, ed., *Drug Design*, Vol. 6, Academic Press, Inc., New York, 1975, p. 81.
18. R. R. Overman and S. J. Jackson, *Annu. Rev. Med.* **18,** 71 (1967).
19. H. M. Patt, *Progr. Radiat. Ther.* **1,** 115 (1958).
20. E. F. Romantsev, *Radiation and Chemical Protection*, Atomizdat, Moscow, 1968.
21. J. F. Thomson, *Radiation Protection in Mammals*, Van Nostrand-Reinhold, Princeton, N.J., 1962.
22. L. A. Tiunov, G. A. Vasil'ev, and V. P. Paribok, *Radioprotective Compounds*, Academy of Sciences, Moscow, USSR, 1961; L. A. Tiunov, G. A. Vasil'ev, and E. A. Wald'shtein, *Radioprotective Compounds*, Nauka, Moscow, 1964.
23. L. W. Brady, ed., *Radiation Sensitizers*, Masson Publishing, USA, Inc., New York, 1980.
24. H. C. Box, *Annu. Rev. Nucl. Sci.* **22,** 355 (1972).
25. G. E. Adams, G. S. McNaughton, and B. D. Michael in G. R. A. Johnson and G. Scholes, eds., *The Chemistry of Ionization and Excitation*, Taylor & Francis, London, UK, 1967, p. 281.
26. G. E. Adams, J. W. Boag, J. Curvant, and B. D. Michael in M. Ebert and co-workers, eds., *Pulse Radiolysis*, Academic Press, Inc., New York, 1965; p. 131.
27. P. Riesz and S. Rustgi, *Radiat. Phys. Chem.* **13,** 21 (1979).
28. M. D. Sevilla, J. B. D'Arcy, and K. M. Morehouse, *J. Phys. Chem.* **83,** 2893 (1979).
29. Ref. 8, p. 70.
30. T. Sanner and A. Pihl, *Radiat. Res.* **37,** 216 (1969).
31. Ref. 8, p. 78.
32. R. E. Zirkle, *Adv. Biol. Med. Phys.* **5,** 103 (1957).
33. Ref. 8, p. 103.
34. H. S. Kaplan and L. E. Moses, *Science* **145,** 21 (1964).
35. W. C. Dewey, B. A. Sedita, and R. M. Humphrey, *Int. J. Radiat. Biol.* **12,** 597 (1967).
36. Z. M. Bacq and P. Alexander, *Fundamentals of Radiobiology*, 2nd ed., Pergamon Press, Inc., Oxford, UK, 1961.
37. K. Fonk and A. W. Konings, *Br. J. Radiol.* **51,** 832 (1978).
38. L. Eldjarn and A. Pihl, *J. Biol. Chem.* **223,** 341 (1956).
39. Ref. 8, p. 230.
40. W. K. Sinclair, *Radiat. Res.* **33,** 620 (1968).
41. J. W. Harris, R. B. Painter, and G. M. Hahn, *Int. J. Radiat. Biol.* **15,** 289 (1969).
42. M. M. Elkind and H. Sutton, *Nature London* **184,** 1293 (1959).
43. M. M. Elkind, H. Sutton-Gilbert, W. B. Moses, and C. Kamper, *Nature London* **214,** 1088 (1967).
44. R. F. Kallman, *Nature London* **197,** 557 (1963).
45. M. M. Elkind and W. K. Sinclair in M. Ebert and A. Howard, eds., *Current Topics*, Vol. 1, Elsevier North-Holland Inc., Amsterdam, 1965, p. 165.
46. P. C. Hanawalt, P. H. Cooper, A. K. Ganesan, and C. A. Smith, *Ann Rev. Biochem.* **48,** 783 (1979).
47. D. R. A. Wharton and M. L. Wharton, *Radiat. Res.* **16,** 723 (1962).
48. A. J. Vergroesen, L. Budke, and O. Vos, *Int. J. Radiat. Biol.* **13,** 77 (1967).
49. P. Eker and A. Pihl, *Radiat. Res.* **21,** 165 (1964).
50. M. L. Beaumariage and P. van Caneghem, *Radiat. Res.* **33,** 74 (1968).
51. K. E. Kinnamon, L. L. Ketterling, H. F. Stampfli, and M. M. Grenan, *Proc. Soc. Exp. Biol. Med.* **164,** 370 (1980).
52. J. W. Purdie, *Radiat. Res.* **77,** 303 (1979).
53. J. M. Yuhas, *Experientia Suppl.* **27,** 63 (1977).
54. W. F. Ward, A. Shih-Hoellwarth, and P. M. Johnson, *Radiat. Res.* **81,** 131 (1980).
55. E. S. Copeland, *Photochem. Photobiol.* **28,** 839 (1978). Discussion following.
56. G. Kollman and B. Shapiro, *Radiat. Res.* **27,** 474 (1966).
57. W. O. Foye, M. M. Karkaria, and W. H. Parsons, *J. Pharm. Sci.* **69,** 84 (1980).
58. E. S. Copeland, T. Sanner, and A. Pihl, *Eur. J. Biochem.* **1,** 312 (1967).
59. T. Henriksen, *Scand. J. Clin. Lab. Invest. Suppl.* **22**(106), 7 (1968).
60. G. Adams in H. L. Morrison and M. Quintiliani, eds., *Radioprotection and Sensitization*, Taylor & Francis, London, 1970, p. 3.
61. R. P. Bird, *Radiat. Res.* **82,** 290 (1980).
62. L. Eldjarn and A. Pihl, *Progress in Radiobiology*, Oliver & Boyd, London, 1956, p. 249.

63. L. Eldjarn and A. Pihl, *Det Norske Videnskaps-Akademi i Oslo I. Mat. Naturv. Klasse. 1958*, No. 3, Gunderson, Oslo, Norway, 1959, p. 254.
64. L. Eldjarn, A. Pihl, and B. Shapiro, *Proc. Int. Conf. Peaceful Uses Atom. Energy* **11**, 335 (1956).
65. J. Sümegi, T. Sanner, and A. Pihl, *Biochim. Biophys. Acta* **262**, 145 (1972).
66. R. M. Sutherland, J. N. Stannard, and R. I. Weed, *Int. J. Radiat. Biol.* **12**, 551 (1967).
67. H. G. Modig, M. Edgren, and L. Révész, *Int. J. Radiat. Biol.* **22**, 257 (1971).
68. H. G. Modig and L. Révész, *Int. J. Radiat. Biol.* **13**, 469 (1967).
69. H. G. Modig, M. Edgren, and L. Révész, *Acta Radiol. Ther. Stockholm* **16**, 245 (1977).
70. A. Stoklasová, J. Kŕiźala, and M. Ledvina, *Strahlentherapie* **156**, 205 (1980).
71. J. W. Harris in H. L. Moroson and M. Quintiliani, eds., *Radioprotection and Sensitization*, Taylor & Francis, London, 1970, p. 189.
72. C. O. Criborn and C. Rónnbáck, *Acta Radiol. Oncol. Radiat. Phys. Biol.* **18**, 31 (1979).
73. E. H. Betz, D. J. Mewissen, and J. Closon, *Arch. Int. Pharmacodyn. Ther.* **121**, 134 (1959).
74. M. Tálas and E. Szolgay, *Arch. Virol.* **56**, 309 (1978).
75. M. Tálas, A. S. Novokhatsky, and L. Bátkai, *Acta Microbiol. Acad. Sci. Hung.* **26**, 213 (1979).
76. J. R. Ortaldo and J. L. McCoy, *Radiat. Res.* **81**, 262 (1980).
77. H. M. Patt, E. B. Tyree, R. L. Straube, and D. E. Smith, *Science* **110**, 213 (1949).
78. D. G. Doherty, W. T. Burnett, Jr., and R. Shapira, *Radiat. Res.* **7**, 13 (1957).
79. D. L. Klayman, M. M. Grenan, and D. P. Jacobus, *J. Med. Chem.* **12**, 510 (1969).
80. S. Romito, *Fracastoro* **62**, 571 (1969).
81. V. G. Yacovlev, *Khim. Zashch. Organizma Ioniz. Izluch.*, 14 (1960); *Chem. Abstr.* **56**, 1724 (1962).
82. Rom. Pat. 62,021 (Dec. 24, 1976), E. Grigorescu, I. Selmiciu, T. Baran, M. Lazar, and I. Vasiliu; *Chem. Abstr.* **89**, 100350 (1978).
83. I. V. Filippovich, E. E. Kolesnikov, T. N. Sheremet'evskaya, A. T. Tarasenko, and E. F. Romantsev, *Biochem. Pharmacol.* **22**, 815 (1973).
84. Z. M. Bacq, A. Herve, J. Lecompte, P. Fischer, J. Blavier, G. Dechamps, H. LeBihan, and P. Rayet, *Arch. Int. Physiol.* **59**, 442 (1951).
85. S. N. Aleksandrov and K. F. Galkovskaya, *Dokl. Akad. Nauk USSR* **150**, 665 (1963).
86. M. P. Dacquisto and S. M. Benson, *Nature London* **195**, 1116 (1962).
87. H. L. Andrews, D. C. Peterson, and D. P. Jacobus, *Radiat. Res.* **23**, 13 (1964).
88. J. R. Maisin and D. G. Doherty, *Radiat. Res.* **19**, 474 (1963).
89. Z. M. Bacq and M. L. Beaumariage, *Arch. Int. Pharmacodyn. Ther.* **153**, 457 (1965).
90. V. Smoliar, *C. R. Soc. Biol.* **156**, 1202 (1962).
91. A. M. Mandl, *Int. J. Radiat. Biol.* **1**, 131 (1959).
92. K. G. Lüning, H. Frölén, and A. Nelson, *Radiat. Res.* **14**, 813 (1961).
93. C. M. Starkie, *Int. J. Radiat. Biol.* **3**, 609 (1961).
94. B. A. Fedorov, *Med. Radiol.* **10**, 42 (1965); *Chem. Abstr.* **64**, 8613 (1966).
95. B. H. Ershoff, C. W. Steers, Jr., and L. Kruger, *Proc. Soc. Exp. Biol. Med.* **111**, 391 (1962).
96. J. M. Roberts, *Radiat. Res.* **49**, 311 (1972).
97. F. Yu. Rachinskii, N. M. Slavachevskaya, and D. V. Ioffe, *Zh. Obshch. Khim.* **28**, 2998 (1958).
98. M. C. Tikhomirova, V. G. Yakovlev, L. S. Isupova, and R. A. Klimova, *Radiobiologiya* **9**, 854 (1969).
99. S. J. Arbusov, *Pharmazie* **14**, 132 (1959).
100. L. I. Tank, *Tiolovye Soedin. Med., Tr. Nauch. Konf. Kiev 1957*, 260 (1959); *Chem. Abstr.* **55**, 796 (1961).
101. T. Pantev, N. Bokova, I. Nikolov, V. D. Rogozkin, M. F. Sbitneva, and K. S. Chertkov, *Acta Med. Bulg.* **2**, 50 (1974); *Chem. Abstr.* **88**, 32052b (1978).
102. I. Nikolov, T. Pantev, and N. Bokova, *Rentgenol. Radiol.* **12**, 126 (1973).
103. T. Pantev, N. Bokova, A. I. Nikolov, V. D. Rogozkin, M. F. Sbitneva, and K. S. Chertkov, *Radiobiologiya* **13**, 709 (1973).
104. A. Pihl and T. Sanner, *Radiat. Res.* **19**, 27 (1963).
105. E. F. Romantsev, *Med. Radiol.* **5**, 19 (1960); *Chem. Abstr.* **55**, 1908 (1961).
106. L. I. Korchak, *Radiobiologiya* **10**, 940 (1970).
107. R. L. Mundy and M. H. Heiffer, *Radiat. Res.* **13**, 381 (1960).
108. M. G. Ormerod and P. Alexander, *Radiat. Res.* **18**, 495 (1963).
109. J. H. Stuy, *Radiat. Res.* **14**, 56 (1961).
110. A. Kaluszyner, P. Czerniak, and E. D. Bergmann, *Radiat. Res.* **14**, 23 (1961).

111. A. Smailiene and N. M. Slavachevskaya, *Farmakol. Toksikol. Moscow* **33**, 271 (1970); *Chem. Abstr.* **73**, 64750 (1970).
112. D. W. Bekkum and H. T. M. Nieuwerkerk, *Int. J. Radiat. Biol.* **7**, 473 (1963).
113. G. R. Handrick, E. R. Atkinson, F. E. Granchelli, and R. J. Bruni, *J. Med. Chem.* **8**, 762 (1965).
114. V. A. Kozlov, *Radiobiologiya* **5**, 892 (1965).
115. L. I. Tank, *Med. Radiol.* **5**, 34 (1960); *Chem. Abstr.* **55**, 10697 (1961).
116. H. Langendorff and R. Koch, *Strahlentherapie* **99**, 567 (1956).
117. M. M. Grenan and E. S. Copeland, *Radiat. Res.* **47**, 387 (1971).
118. G. W. Stacy, B. F. Barnett, and P. L. Strong, *J. Org. Chem.* **30**, 592 (1965).
119. F. I. Carroll, J. D. White, and M. E. Wall, *J. Org. Chem.* **28**, 1236 (1963).
120. Khr. Tonchev, *Nauch. Tr. Vissh. Med. Inst. Sofia* **39**, 143 (1960); *Chem. Abstr.* **55**, 19007 (1961).
121. H. Langendorff, M. Langendorff, and R. Koch, *Strahlentherapie* **107**, 121 (1958).
122. B. Percarpio and J. J. Fischer, *Radiology* **131**, 791 (1979).
123. W. F. Ward, A. Shih-Hoellwarth, and P. M. Johnson, *Radiat. Res.* **81**, 131 (1980).
124. R. Katsuhara, *Nippon Igaku Hoshasen Gakkai Zasshi* **19**, 73 (1959).
125. Y. Yoshihara, *Nippon Igaku Hoshasen Gakkai Zasshi* **23**, 1 (1963).
126. R. W. Hart, R. E. Gibson, D. J. Chapman, A. P. Reuvers, B. K. Sinha, R. K. Griffith, and D. T. Witiak, *J. Med. Chem.* **18**, 323 (1975).
127. R. D. Westland, J. D. Holmes, B. Green, and J. R. Dice, *J. Med. Chem.* **11**, 824 (1968).
128. A. Ferris, O. L. Salerni, and B. A. Schutz, *J. Med. Chem.* **9**, 391 (1966).
129. E. Felder and F. Bonati, and S. Bianchi, *Experientia* **15**, 32 (1959).
130. Ger. Pat. 1,062,705 (Aug. 6, 1959), E. Felder and S. Bianchi; *Chem. Abstr.* **55**, 22241 (1961).
131. A. Smailene, *Farmakol. Tsent. Kholinolitikov Drugikh Neirotropnykh Stredstv.*, 308 (1969); *Chem. Abstr.* **73**, 118961 (1970).
132. D. G. Doherty and W. T. Burnett, *Proc. Soc. Exp. Biol. Med.* **89**, 312 (1955).
133. F. I. Carroll and M. E. Wall, *J. Pharm. Sci.* **59**, 1350 (1970).
134. N. I. Bicheikina, Z. I. Zhulanova, N. B. Pushkareva, B. A. Titov, E. F. Romanstev, T. V. Klimova, V. I. Stanko, V. Yu. Kovtun, V. G. Rashunskii, and E. A. Krasheninnikova, *Radiobiologiya* **19**, 235 (1979).
135. L. I. Tank, *Med. Radiol.* **6**, 76 (1961); *Chem. Abstr.* **56**, 7664 (1962).
136. R. Rugh and E. Grupp, *Atompraxis* **6**, 143 (1960).
137. H. Langendorff and R. Koch, *Strahlentherapie* **98**, 245 (1955).
138. A. S. Mozzhukhin, F. Yu. Rachinskii, and L. I. Tank, *Med. Radiol.* **5**, 78 (1960).
139. W. O. Foye, R. N. Duvall, and J. Mickles, *J. Pharm. Sci.* **51**, 168 (1962).
140. F. V. Morriss, J. J. Downs, B. A. Schutz, and A. F. Ferris, *J. Pharm. Sci.* **52**, 409 (1963).
141. W. O. Foye, J. R. Marshall, and J. Mickles, *J. Pharm. Sci.* **52**, 406 (1963).
142. E. J. Jezequel, H. Frossard, M. Fatome, R. Perles, and P. Poutrain, *C. R. Acad. Sci. Ser. D* **272**, 2826 (1971).
143. W. H. Chapman and E. P. Cronkite, *Proc. Soc. Exp. Biol. Med.* **75**, 318 (1950).
144. V. E. Petrunkin, *Ukr. Khim. Zhur.* **22**, 603 (1956).
145. B. Gabard, *Arch. Toxicol.* **35**, 15 (1974).
146. B. Gabard, *Acta Pharmacol. Toxicol.* **39**, 250 (1976).
147. W. Hauser and N. Weger, *7th International Congress of Pharmacology*, Paris, France, July 16–21, 1978, p. 764, Abstr. 2382.
148. M. T. Khayyal, I. H. Kemper, H. P. Bertram, and M. Renhof, *7th International Congress of Pharmacology*, Paris, France, July 16–21, 1978, p. 55, Abstr. 132.
149. J. Schubert and S. K. Derr, *Nature* **275**, 311 (1978).
150. G. C. Battistone and R. A. Miller, *J. Dent. Res.* (*IADR Suppl.*) **58**, 312, Abstr. 882 (1979).
151. C. Falconi, P. Scotto, and P. deFranciscis, *Boll. Soc. Ital. Biol. Sper.* **44**, 326 (1968).
152. C. Falconi, P. Scotto, and P. deFranciscis, *Experientia* **26**, 172 (1970).
153. Z. M. Bacq and P. Alexander, *Fundamentals of Radiobiology*, 2nd ed., Pergamon Press, Inc., Oxford, UK, 1961, p. 460.
154. Il. Belokonski, *Sov. Med.* **11**, 3 (1960); *Chem. Abstr.* **55**, 1908 (1961).
155. M. L. Beaumariage, *C. R. Soc. Biol.* **151**, 1788 (1957).
156. G. Booz, L. J. Simar, and E. H. Betz, *Int. J. Radiat. Biol.* **9**, 429 (1965).
157. Z. M. Bacq and P. van Caneghem, *Int. J. Radiat. Biol.* **10**, 595 (1966).
158. O. K. Makhalova, A. S. Mozzhukhin, and V. I. Bertash, *Radiobiologiya* **6**, 883 (1966).
159. A. V. Titov, D. A. Glubentsov, and V. V. Mordukhovich, *Radiobiologiya* **10**, 606 (1970).

160. E. Jellum, *Int. J. Radiat. Biol.* **10,** 577 (1966).
161. V. I. Kulinskii and L. P. Miller, *Radiobiologiya* **10,** 596 (1970).
162. T. Pantev, *Rentgenol. Radiol.* **41,** 44 (1975); *Chem. Abstr.* **83,** 141856 (1975).
163. I. I. Krasil'nikov, E. I. Levachevskaya, N. M. Slavachesvskaya, and L. G. Tarnopol'skaya, *Radiobiologiya* **15,** 902 (1975).
164. W. O. Foye, Y. H. Lowe, and J. J. Lanzillo, *J. Pharm. Sci.* **65,** 1247 (1976).
165. R. R. Crenshaw and L. Field, *J. Org. Chem.* **30,** 175 (1965).
166. L. Field and H. K. Kim, *J. Med. Chem.* **9,** 397 (1966).
167. L. Field, H. Härle, T. C. Owen, and A. Ferretti, *J. Org. Chem.* **29,** 1632 (1964).
168. L. Field and P. R. Engelhardt, *J. Org. Chem.* **35,** 3647 (1970).
169. Y. H. Khim and L. Field, *J. Org. Chem.* **37,** 2714 (1972).
170. L. Field, H. K. Kim, and M. Bellas, *J. Med. Chem.* **10,** 1166 (1967).
171. L. Field and J. D. Buckman, *J. Org. Chem.* **32,** 3467 (1967).
172. R. Goutier and M. L. Beaumariage, *Arch. Int. Pharmacodyn. Ther.* **126,** 341 (1960).
173. G. Pennisi and F. Candela, *Rass. Med. Sper.* **6,** 82 (1959).
174. L. Field and R. B. Barbee, *J. Org. Chem.* **34,** 1792 (1969).
175. L. Field and Y. H. Khim, *J. Med. Chem.* **15,** 312 (1972).
176. P. K. Srivastava, L. Field, and M. M. Grenan, *J. Med. Chem.* **18,** 798 (1975).
177. D. L. Klayman and G. W. A. Milne, *J. Org. Chem.* **31,** 2349 (1966).
178. L. Field, T. C. Owen, R. R. Crenshaw, and A. W. Bryan, *J. Am. Chem. Soc.* **83,** 4414 (1961).
179. M. M. Grenan, Walter Reed Army Institute of Research (personal communication).
180. D. L. Klayman, J. D. White, and T. R. Sweeney, *J. Org. Chem.* **29,** 3737 (1964).
181. J. J. Kelley, N. F. Hamilton, and O. M. Friedman, *Cancer Res.* **27,** 143 (1967).
182. D. L. Klayman and R. J. Shine, *Q. Rep. Sulfur Chem.* **3,** 189 (1968); H. Distler, *Angew. Chem. Int. Ed. Engl.* **6,** 544 (1967); R. Milligan and J. M. Swan, *Rev. Pure Appl. Chem.* **12,** 72 (1962).
183. Yu. B. Kudryashov, M. L. Kakushkina, S. M. Mechtieva, F. Yu. Rachinskii, G. V. Sumarukov, and O. F. Filenko, *Radiobiologiya* **6,** 272 (1966).
184. B. Holmberg and B. Sörbo, *Nature London* **183,** 832 (1959).
185. Y. Takagi, M. Shikita, and S. Akaboshi, *Chem. Pharm. Bull.* **20,** 1102 (1972).
186. B. Padeh, P. Czerniak, E. Akstein, and A. Kalusziner, *Harefuah* **77,** 393 (1969); *Chem. Abstr.* **72,** 107605 (1970).
187. A. Kaluszyner, P. Czerniak, and E. D. Bergmann, *Radiat. Res.* **14,** 23 (1961).
188. J. J. Kelley, K. A. Herrington, S. P. Ward, A. Meister, and O. M. Friedman, *Cancer Res.* **27,** 137 (1967).
189. H. Gershon and R. Rodin, *J. Med. Chem.* **8,** 864 (1965).
190. J. R. Piper, C. R. Stringfellow, Jr., and T. P. Johnston, *J. Med. Chem.* **9,** 911 (1966).
191. D. L. Klayman, M. M. Grenan, and D. P. Jacobus, *J. Med. Chem.* **13,** 251 (1970).
192. F. A. Hodge and M. S. Silverman, *U.S. Govt. Res. Dev. Rep.* **68,** 45 (1968).
193. J. C. Sharp, D. D. Kelly, and J. V. Brady, *Use Nonhuman Primates Drug Evaluation Symposium, 1967,* 1968, p. 338.
194. H. E. Hamilton, G. S. Melville, and E. J. Stork, *U.S. Clearinghouse Fed. Sci. Tech. Inform., AD Rep.* **AD 691409** (1968).
195. C. L. Turbyfill, R. M. Roudon, R. W. Young, and V. A. Kieffer, *U.S. Nat. Tech. Inform. Serv., AD Rep.* **745284** (1972).
196. R. D. Westland and J. L. Holmes, *J. Med. Chem.* **15,** 976 (1972).
197. R. D. Westland, J. L. Holmes, M. L. Mouk, D. D. Marsh, R. A. Cooley, Jr., and J. R. Dice, *J. Med. Chem.* **11,** 1190 (1968).
198. R. D. Westland, M. L. Mouk, J. L. Holmes, R. A. Cooley, Jr., J. S. Hong, and M. M. Grenan, *J. Med. Chem.* **15,** 968 (1972).
199. J. R. Piper, C. R. Stringfellow, Jr., and T. P. Johnston, *J. Med. Chem.* **9,** 563 (1966).
200. L. Bauer and B. K. Ghosh, *J. Org. Chem.* **30,** 4298 (1965).
201. A. P. Parulkar and L. Bauer, *J. Heterocycl. Chem.* **3,** 472 (1966).
202. J. M. Barton and L. Bauer, *Can. J. Chem.* **47,** 1233 (1969).
203. R. D. Westland, M. M. Merz, S. M. Alexander, L. S. Newton, L. Bauer, T. T. Conway, J. M. Barton, K. K. Khullar, P. B. Devdhar, and M. M. Grenan, *J. Med. Chem.* **15,** 1313 (1972).
204. V. G. Vladimirov, Ya. L. Kostyukovskii, N. M. Slavachevskaya, and Yu. E. Strelinkov, *Radiobiologiya* **14,** 415 (1974).
205. T. L. Fridinger and J. E. Robertson, *J. Med. Chem.* **12,** 1114 (1969).

206. A. Kaluszyner, *Bull. Res. Counc. Isr. Sect. A: Chem.* **9**, 35 (1960).
207. D. L. Klayman, M. M. Grenan, and D. P. Jacobus, *J. Med. Chem.* **12**, 723 (1969).
208. B. Hansen and B. Sörbo, *Acta Radiol.* **56**, 141 (1961).
209. T. W. Szczepkowski, T. Zebro, and J. Stachura, *Acta Biochim. Pol.* **11**, 235 (1964).
210. T. W. Szczepkowski, T. Zebro, and J. Stachura, *Acta Med. Pol.* **4**, 305 (1963).
211. T. Zebro, *Acta Med. Pol.* **7**, 83 (1966).
212. T. Zebro, *Folia Med. Cracov.* **8**, 149 (1966).
213. S. Åkerfeldt, *Acta Chem. Scand.* **14**, 1980 (1960).
214. *Ibid.*, p. 1019.
215. *Ibid.*, **15**, 1813 (1962).
216. K. A. Herrington, C. J. Small, A. Meister, and O. M. Friedman, *Cancer Res.* **27**, 148 (1967).
217. S. Åkerfeldt, *Acta Chem. Scand.* **13**, 1479 (1959).
218. S. Åkerfeldt, *Acta Radiol. Ther. Phys. Biol.* **1**, 465 (1963).
219. G. M. Airapetyan and P. G. Zherebchenko, *Radiobiologiya* **4**, 259 (1964).
220. G. M. Airapetyan, G. A. Anorova, L. N. Kublik, G. K. Otarova, M. I. Pekarskii, V. N. Stanko, L. Kh. Eidus, and S. P. Yarmonenko, *Med. Radiol.* **12**, 58 (1967); *Chem. Abstr.* **67**, 621 (1967).
221. S. V. Tselikova, M. Kh. Levitman, and N. V. Kondakova, *Biol. Nauchno Tekh. Prog. Tezisy Dokl. Vses. Konf. Molodykh Och. Spets.*, 343 (1974); *Chem. Abstr.* **85**, 137893t (1976).
222. T. N. Oganesova, V. G. Ovakimov, S. P. Yarmonenko, and G. M. Airapetyan, *Mater. Vses. Konf., "Farmakol. Prolivoluchevykh Prep.,"* 1st (1970); *Chem. Abstr.* **81**, 9707r (1974).
223. R. D. Elliot, J. R. Piper, C. R. Strongfellow, and J. P. Johnston, *J. Med. Chem.* **15**, 595 (1972).
224. J. R. Piper, C. R. Stringfellow, Jr., R. D. Elliott, and T. P. Johnston, *J. Med. Chem.* **12**, 236 (1969).
225. J. R. Piper, L. M. Rose, T. P. Johnston, and M. M. Grenan, *J. Med. Chem.* **18**, 803 (1975).
226. S. Åkerfeldt, C. Rönnbäch, and A. Nelson, *Radiat. Res.* **31**, 850 (1967).
227. S. Åkerfeldt, *Acta Chem. Scand.* **20**, 1783 (1966).
228. J. R. Piper, L. M. Rose, T. P. Johnston, and M. M. Grenan, *J. Med. Chem.* **22**, 631 (1979).
229. A. G. Sverdlov, A. V. Bagatyoev, N. G. Nikamorova, S. I. Timoshenko, and G. I. Kalmykova, *Radiobiologiya* **19**, 293 (1979).
230. L. R. Stromberg, M. M. McLaughlin, and R. M. Donati, *Proc. Soc. Exp. Biol. Med.* **129**, 140 (1968).
231. J. M. Yuhas, *Cell. Immunol.* **4**, 256 (1972).
232. R. O. Lowry and D. G. Baker, *Radiology* **105**, 425 (1972).
233. C. P. Sigdestad, A. M. Connor, and R. M. Scott, *Radiat. Res.* **65**, 430 (1976).
234. *Ibid.*, **62**, 267 (1975).
235. A. G. Sverdlov, S. A. Grachev, A. V. Bogatyrev, A. I. Shabarova, G. A. Bagiyan, and S. T. Timoshenko, *Radiobiologiya* **14**, 304 (1974).
236. R. H. Mole, J. St. L. Philpot, and G. R. V. Hodges, *Nature London* **166**, 515 (1950).
237. G. V. Andreenko, V. M. Fedoseev, and N. P. Sytina, *Vestn. Mosk. Univ. Ser. Biol. Pochvoved. Geol. Geogr.*, 39 (1958); *Chem. Abstr.* **54**, 17703 (1960).
238. V. D. Lyashenko, S. M. Pekkerman, L. F. Semenov, and I. B. Simon, *Tiolovye Soedin. Med. Tr. Nauch. Konf. 1957*, 276 (1959).
239. M. Shinoda, S. Ohta, Y. Takagi, and S. Akaboshi, *Yakugaku Zasshi* **94**, 1419 (1974).
240. N. A. Izmozhevov, N. E. Ainbindev, A. N. Filaretov, G. V. Girshik, Yu. V. Gilve, and E. I. Izmozhevova, *Radiobiologiya* **15**, 923 (1975).
241. V. V. Kozhevnikov, *Izv. Estestvennonauchn. Inst. Permsk. Gos. Univ.* **15**, 99 (1974); *Chem. Abstr.* **85**, 186592c (1976).
242. A. Hanaki, *Chem. Pharm. Bull.* **16**, 2023 (1968); **17**, 1146 (1969); **18**, 766 (1970); **19**, 326 (1971).
243. A. Hanaki, T. Hanaki, K. Oya, A. Andou, T. Hino, and S. Akaboshi, *Chem. Pharm. Bull.* **14**, 108 (1966).
244. T. Hino, K. Tana-ami, K. Yamada, and S. Akaboshi, *Chem. Pharm. Bull.* **14**, 1193, 1201 (1966).
245. M. D. Hallas, P. B. Reed, and R. B. Martin, *J. Chem. Soc. D*, 1506 (1971).
246. P. G. Zherebchenko and T. G. Zaitseva, *Radiobiologiya* **9**, 701 (1969).
247. H. Kakehi, H. Ichikawa, and M. Nakano, *Radioisotopes* **10**, 240 (1961).
248. V. J. Plzak, J. Doull, and D. G. Oldfield, *Radiat. Res.* **25**, 228 (1965).
249. S. Antoku and S. Sawada, *J. Radiat. Res.* **11**, 70 (1970).
250. A. Leonard and G. Mattelin, *C. R. Soc. Biol.* **164**, 907 (1970).
251. J. Maisin and J. Moutschen, *Exp. Cell Res.* **21**, 347 (1960).
252. N. F. Barakina and M. I. Yanushevskaya, *Radiobiologiya* **4**, 226 (1964).

253. U. H. Ehling and D. G. Doherty, *Proc. Soc. Exp. Biol. Med.* **110,** 493 (1962).
254. A. Leonard and J. R. Maisin, *Radiat. Res.* **23,** 53 (1964).
255. R. E. Benson, S. M. Michaelson, W. L. Downs, E. A. Maynard, J. K. Scott, H. C. Hodge, and J. W. Howland, *Radiat. Res.* **15,** 561 (1961).
256. J. R. Newsome, D. H. Knott, and R. R. Overman, *Radiat. Res.* **17,** 847 (1962).
257. S. P. Yarmonenko, *Zh. Obshch. Biol.* **26,** 501 (1965).
258. B. L. Sztanyik and V. Varteresz in H. L. Moroson and M. Quintiliani, eds., *Radioprotection and Sensitization*, Taylor & Francis, London, UK, 1970, p. 363.
259. M. J. Ashwood-Smith, *Int. J. Radiat. Biol.* **5,** 201 (1962).
260. E. Schroder and R. Huber, *Acta Biol. Med. Ger.* **9,** 62 (1962).
261. G. S. Melville, Jr. and T. P. Leffingwell, *U.S. Dep. Comm. Off. Tech. Serv.*, AD Rep. **AD 268549** (1961).
262. D. R. Anderson, *Diag. Treat. Acute Radiat. Inj. Proc. Sci. Meet. 1960*, 363 (1961).
263. M. Bulanova, M. Mileva, B. Ivanov, and T. Pantev, *Suvrem. Med.* **28,** 31 (1977); *Chem. Abstr.* **89,** 259d (1978).
264. E. E. Schwartz and B. Shapiro, *Radiat. Res.* **13,** 768 (1960).
265. B. Shapiro, E. E. Schwartz, and G. Kollman, *Radiat. Res.* **18,** 17 (1963).
266. *Ibid.*, **20,** 17 (1963).
267. G. Kollmann, B. Shapiro, and E. E. Schwartz, *NASA Doc. N63-23812*, Washington, D.C., 1963.
268. A. Hanaki, *Chem. Pharm. Bull.* **18,** 1653 (1970).
269. *Ibid.*, **19,** 1223 (1971).
270. D. G. Doherty and R. Shapira, *J. Org. Chem.* **28,** 1339 (1963).
271. R. Shapira, D. G. Doherty, and W. T. Burnett, Jr., *Radiat. Res.* **7,** 22 (1957).
272. R. Riemschneider and G.-A. Hoyer, *Z. Naturforsch. B* **18,** 25 (1963).
273. F. Yu. Rachinskii, M. S. Kushakovskii, B. V. Matveev, N. M. Slavachevskaya, and L. I. Tank, *Radiobiologiya* **4,** 266 (1964).
274. R. D. Westland, R. A. Cooley, Jr., J. L. Holmes, J. S. Hong, M. L. Lin, M. L. Zwiesler, and M. M. Grenan, *J. Med. Chem.* **16,** 319 (1973).
275. R. D. Westland, M. H. Lin, R. A. Cooley, Jr., M. L. Zwiesler, and M. M. Grenan, *J. Med. Chem.* **16,** 328 (1973).
276. M. L. Kakushkina, Yu. B. Kudryashov, F. Yu. Rachinskii, and N. G. Dmitrieva, *Radiobiologiya* **4,** 632 (1964).
277. R. Granger, H. Orzalesi, Y. Robbe, J. P. Chapet, A. Terol, M. Rendon, J. Bitoun, F. Simon, and J. P. Fernandez, *Chim. Therap.* **6,** 439 (1971).
278. A. Terol, J. P. Fernandez, Y. Robbe, J. P. Chapat, R. Granger, M. Fatome, L. Andrieux, and H. Sentenac-Roumanou, *Chim. Therap.* **13,** 149 (1978).
279. V. S. Shashkov and V. M. Fedoseev, *Med. Radiol.* **6,** 25 (1961); *Chem. Abstr.* **56,** 6334 (1962).
280. K. Taira, R. Katsuhara, and M. Kubota, *Boei Eisei* **12,** 99 (1965).
281. G. P. Bogatyrev, Ya. L. Lys, V. I. Mal'ko, E. N. Goncharenko, Yu. B. Kudryashov, and V. M. Fedoseev, *Radiobiologiya* **13,** 864 (1973).
282. G. N. Krutovskikh, A. M. Rusanov, G. F. Gornaeva, L. P. Vartanyan, and M. B. Kolesova, *Khim. Farm. Zh.* **11,** 48 (1977); *Chem. Abstr.* **87,** 62373d (1977).
283. D. L. Klayman, *J. Org. Chem.* **30,** 2454 (1965).
284. S.-H. Chu and H. G. Mautner, *J. Org. Chem.* **27,** 2899 (1962).
285. I. Kozak, L. Kronrad and Z. Dienstbier, *Strahlentherapie* **150,** 539 (1975).
286. R. Badrillo, E. Gattavecchia, M. Mattii, and M. Tamba, *Z. Naturforsch. Teil C* **29,** 647 (1974).
287. G. B. Abdullaev, M. A. Mekhtiev, R. N. Ragimov, G. V. Teplyakova, K. M. Akhmedli, S. M. Beibutov, R. A. Babaev, and N. B. Alieva, *Selen Biol. Mater. Nauchn. Konf.* 2nd 1975, **2,** 159 (1976); *Chem. Abstr.* **86,** 65636g (1977).
288. D. L. Klayman in D. L. Klayman and W. H. H. Günther, eds., *Organic Selenium Compounds: Their Chemistry and Biology*, John Wiley & Sons, Inc., New York, 1973, p. 727.
289. Z. M. Hollo and S. Zlatarov, *Naturwissenschaften* **47,** 328 (1960).
290. G. B. Abdullaev, G. G. Gasanov, M. A. Mekhtiev, A. I. Sadykhov, A. I. Dzhafarov, D. M. Mazanov, and R. A. Babaev, *Selen Biol. Mater. Nauch. Konf.*, 122, 182 (1974); *Chem. Abstr.* **84,** 69541d (1976).
291. R. Badiello, M. D. Solenghi, and M. Tamba, *Ital. J. Biochem.* **26,** 255 (1977).
292. B. Highman, J. R. Hansell, and D. C. White, *Radiat. Res.* **30,** 563 (1967).
293. W. S. Moos and S. E. Kim, *Experientia* **22,** 814 (1966).

294. S. E. Kim and W. S. Moos, *Health Phys.* **13**, 601 (1967).
295. A. A. Gorodetskii and V. A. Baraboi, *Fiziol. Zh. Akad. Nauk Ukr. RSR* **7**, 617 (1961); *Chem. Abstr.* **56**, 2688 (1962).
296. A. A. Gorodetskii and V. A. Baraboi, *Patog. Eksp. Profil. Ter. Luchevykh Porazhenii*, Sb. Stat., 228 (1964); *Chem. Abstr.* **66**, 8674 (1967).
297. V. A. Baraboi, *Fiziol. Zh. Akad. Nauk Ukr. RSR* **10**, 89 (1964); *Chem. Abstr.* **65**, 18967 (1966).
298. F. Pozza and G. Verza, *Bol. Soc. Ital. Biol. Sper.* **36**, 1120 (1960).
299. E. F. Romantsev and M. V. Tikhomirova, *Radiobiologiya* **3**, 126 (1963).
300. J. B. Storer, *Radiat. Res.* **47**, 537 (1971).
301. J. R. Newsome and R. R. Overman, *Radiat. Res.* **21**, 530 (1964).
302. L. T. Blouin and R. R. Overman, *Radiat. Res.* **16**, 699 (1962).
303. T. J. Fitzgerald, J. Doull, and F. G. DeFeo, *J. Med. Chem.* **17**, 900 (1974).
304. W. W. Smith, *Science* **127**, 340 (1958).
305. W. E. Rothe and M. M. Grenan, *Science* **133**, 888 (1961).
306. H. Langendorff, H.-J. Melching, and H.-A. Ladner, *Strahlentherapie* **108**, 57 (1959).
307. *Ibid.*, p. 251.
308. A. K. Rupke, G. Gold, R. Rugh, and S. G. Wang, *Radiat. Res.* **19**, 88 (1963).
309. T. J. Haley, A. M. Flesher, and L. Mavis, *Nature London* **192**, 1309 (1961).
310. S. Liébecq-Hutter, D. J. Mewissen, and Z. M. Bacq, *Arch. Int. Physiol. Biochim.* **66**, 546 (1958).
311. D. Jamieson and H. A. S. van den Brenk, *Radiobiol. Proc. Australas. Conf. 3rd 1960*, 142 (1961).
312. L. Dimitrov, *C. R. Acad. Bulg. Sci.* **15**, 231 (1962).
313. L. A. Dimitrov, *Izv. Inst. Fiziol. Bulgar. Akad. Nauk.* **6**, 253 (1963); *Chem. Abstr.* **61**, 16410 (1964).
314. A. Locker and P. Weish, *Experientia* **26**, 771 (1970).
315. A. Locker, P. Weish, and H. Krumpholz, *Oesterr. Studienges. Atomenerg. SGAE*, **SGAE SS-7** (1971).
316. A. Locker and H. Ellegast, *Experientia* **18**, 363 (1962).
317. *Ibid.*, **20**, 389 (1964).
318. Z. Uray and T. Holan, *Naturwissenschaften* **53**, 552 (1966).
319. E. A. Pora, A. D. Abraham, Z. Uray, and T. Holan, *Rev. Roum. Biol. Ser. Zool.* **14**, 203 (1969).
320. Z. Uray, M. Farcasanu, and M. Maniu, *Agressologie* **11**, 135 (1970).
321. A. Vacek and T. Tacev, *Int. J. Radiat. Biol.* **10**, 509 (1966).
322. J. H. Wilkins and J. H. Barnes, *Nature London* **195**, 1172 (1962).
323. Z. M. Bacq, M. L. Beaumariage, and D. Radivojevic, *Bull. Acad. Roy. Med. Belg.* **1**, 519 (1961).
324. H. A. S. van den Brenk and K. Elliott, *Nature London* **182**, 1506 (1958).
325. H. Langendorff, H.-J. Melching, and H.-A. Ladner, *Int. J. Radiat. Biol.* **1**, 24 (1959).
326. H. Miyata, H.-J. Melching, and H. Rösler, *Strahlentherapie* **113**, 193 (1960).
327. P. Dukor, *Strahlentherapie* **117**, 330 (1962).
328. H. Langendorff and U. Hagen, *Strahlentherapie* **117**, 321 (1962).
329. K. K. Zaitseva, P. G. Zherebchenko, and T. G. Zaitseva, *Radiobiologiya* **9**, 236 (1969).
330. I. G. Krasnykh, P. G. Zherebchenko, V. S. Murashova, N. N. Suvorov, and N. P. Sorokina, *Radiobiologiya* **2**, 298 (1962).
331. P. G. Zerebchenko and T. G. Zaitseva, *Mater. Vses. Konf. "Farmakol. Protivoluchevykh Prep.,"* 1st Conference, 1970; *Chem. Abstr.* **80**, 78374w (1974).
332. L. Ya. Kolemeeva, V. S. Shashkov, and B. B. Egorov, *Kosm. Biol. Aviakosmicheskaya Med.* **9**, 78 (1975); *Chem. Abstr.* **84**, 84169g (1976).
333. J. Kautska, J. Misutstova, and L. Novak, *Radiobiologiya* **16**, 83 (1976).
334. O. P. Ol'shevskaya and S. P. Yarmonenko, *Radiobiologiya* **9**, 58 (1969).
335. P. Dukor and R. Schuppli, *Experientia* **17**, 257 (1961).
336. Z. Uray and M. Farcasanu, *Stud. Biophys.* **27**, 81 (1971).
337. M. Talas and E. Szolgay, *Arch. Virol.* **56**, 309 (1978).
338. G. N. Krutovskikh, M. B. Kolesova, A. M. Rusanov, L. P. Vartanyan, and M. Q. Shagoyam, *Khim. Farm. Zh.* **9**, 21 (1975); *Chem. Abstr.* **83**, 108312f (1975).
339. D. Postescu and I. Mustea, *Stud. Biophys.* **43**, 233 (1974).
340. B. I. Monastyrskaya, G. A. Lavronov, A. G. Soerdlov, G. E. Vaisberg, and Z. V. Ermol'eva, *Antibiotiki Moscow* **19**, 1108 (1974); *Chem. Abstr.* **82**, 11894a (1975).
341. J. Manowska, *Folia Biol. Krakow* **24**, 97 (1976).

342. E. L. Izmozherova, D. R. Kolodenko, L. P. Shipitsina, and S. A. Kashlyunov, *Izuch. Biol. Deistviya Nov. Prod. Org. Sint. Prir. Soedin.*, 87 (1977); *Chem. Abstr.* **89**, 99933d (1978).
343. E. L. Izmozherova, T. D. Afonina, and A. S. Novichkova, *Izuch. Biol. Deistviya Nov. Prod. Org. Sint. Prir. Soedin.* 80 (1977); *Chem. Abstr.* **89**, 99932c (1978).
344. N. A. Izmozherov, T. D. Afonina, and S. N. Vaganova, *Izuch. Biol. Deistviya Nov. Prod. Org. Sint. Prir. Soedin.*, 93 (1977); *Chem. Abstr.* **89**, 99934e (1978).
345. H. Asakura, *Paper presented at the International Symposium on Biological Aspects of Radiation Protection*, 1969, p. 132.
346. H. Ringsdorf, A. G. Heisler, F. H. Mueller, E. H. Graul, and W. Ruether in ref. 345, p. 138.
347. V. S. Etlis, B. V. Dubovik, A. P. Sineokov, F. N. Shomina, and A. S. Shevchuk, *Khim. Farm. Zhur.* **10**, 33 (1976); *Chem. Abstr.* **85**, 78581g (1976).
348. W. F. Bale, I. L. Spar, and R. L. Goodland, *Cancer Res.* **20**, 1488 (1960).
349. S. E. Order, *Cancer Res.* **40**, 3001 (1980).
350. J. J. Clement, I. Wodinsky, R. K. Johnson, and D. M. Silveira in L. W. Brady, ed., *Radiation Sensitizers*, Masson Publishing USA Inc., New York, 1980, p. 232.
351. T. L. Phillips, L. J. Kane, and J. F. Utley, *Cancer* **32**, 528 (1973).
352. T. L. Phillips in ref. 350, p. 321.
353. J. M. Yuhas, J. M. Spellman, and F. Culo in ref. 350, p. 303.
354. J. M. Yuhas, *Cancer Res.* **40**, 1519 (1980).
355. M. M. Kligerman, M. T. Shaw, M. Slavik, and J. M. Yuhas in ref. 350, p. 426.
356. J. F. Utley and L. J. Kane in ref. 350, p. 516.
357. R. Huber and B. Teichmann, *Arch. Geschwulstforsch.* **37**, 213 (1971).
358. Z. M. Bacq, J. Bernard, H. Ramioul, and G. Deltour, *Bull. Acad. Roy. Med. Belg.* **17**, 460 (1952).
359. J. Bichel, *Paper presented at the 8th International Congress of Hematology*, 1960, Vol. 1, p. 473.
360. L. F. Larionov, G. N. Platonova, I. G. Spasskaya, and E. N. Tolkacheva, *Byull. Eksp. Biol. Med.* **53**, 68 (1962); *Chem. Abstr.* **57**, 15754 (1962).
361. J. R. Maisin, J. Hugon, and A. Leonard, *J. Belg. Radiol.* **47**, 871 (1974).
362. E. Schwartz, B. Shapiro, and G. Kollmann, *Cancer Res.* **24**, 90 (1964).
363. U.S. Pat. 3,629,410 (Dec. 21, 1971), M. H. Heiffer and D. P. Jacobus (to the U.S.A. as represented by the Secretary of the Army).
364. J. A. Vick, M. H. Heiffer, C. R. Roberts, R. W. Caldwell, and A. Nies, *Mil. Med.* **138**, 490 (1973).
365. S. Phillips and J. A. Vick, *Pharmacologist* **12**, 284 (1970).
366. G. E. Demaree, J. S. Frost, M. H. Heiffer, and W. E. Rothe, *Proc. Soc. Exp. Biol. Med.* **136**, 1332 (1971).
367. R. W. Caldwell and G. E. Demaree, *Fed. Proc. Fed. Amer. Soc. Exp. Biol.* **30**, 1745 (1971).
368. P. P. Saksonov, V. V. Antipov, N. N. Dubrov, V. S. Shashkov, V. A. Kozlov, V. S. Parshin, B. I. Davydov, B. C. Razgovorov, V. S. Morozov, and M. D. Nikitin, *Probl. Kosm. Biol.* **4**, 119 (1965); *Chem. Abstr.* **65**, 1010 (1966).
369. G. Schaefer and K. H. Weiner, *Deut. Luft. Raumfahrt Forschungsber.* **DLR-FB-65-40**, 20 (1965).
370. P. P. Saksonov, V. V. Antinov, B. I. Davydov, and N. N. Dobrov, *Kosm. Biol. Med.* **4**, 17 (1970); *Chem. Abstr.* **75**, 72183 (1971).
371. V. D. Rogozkin, S. A. Davydova, M. V. Tikhomirova, and K. S. Chertkov, *Kosm. Biol. Med.* **4**, 13 (1970); *Chem. Abstr.* **76**, 135691r (1972).

DANIEL L. KLAYMAN
Walter Reed Army Institute of Research

EDMUND S. COPELAND
National Institutes of Health

RADIUM. See Radioactivity, natural.

RADON. See Helium-group gases.

RARE-EARTH ELEMENTS

The rare earths comprise a group of 17 elements in the periodic table and have very similar properties in aqueous solutions. They are all metals in the elemental state and all form salts that are strong electrolytes when dissolved in water. They ionize in this medium to give triply charged ions and, because of the high charge on these ions, react strongly with water dipoles to form a tight sheath of water molecules about them. Other ions in aqueous solutions only contact this sheath. Thus, the properties of rare-earth cations in water are very similar.

The group consists of the following elements: scandium (^{21}Sc), yttrium (^{39}Y), lanthanum (^{57}La), all of which appear in the column III B of the periodic table, and cerium (^{58}Ce), praseodymium (^{59}Pr), neodymium (^{60}Nd), promethium (^{61}Pm), samarium (^{62}Sm), europium (^{63}Eu), gadolinium (^{64}Gd), terbium (^{65}Tb), dysprosium (^{66}Dy), holmium (^{67}Ho), erbium (^{68}Er), thulium (^{69}Tm), ytterbium (^{70}Yb), and lutetium (^{71}Lu). The elements with atomic numbers of 58–71 seem to have properties that are characteristic of lanthanum and share the same space in the original Mendeleev periodic table. They are usually placed at the bottom of the periodic table and are called the lanthanide rare earths or, more generally, the lanthanides (Ln). The term lanthanons is used when lanthanum is also included.

All the naturally occurring lanthanides were discovered and isolated before chemists understood why such an anomaly should appear at this point in the Mendeleev chart, and the existence of these elements played an important role in the understanding of atomic structure.

Quantum mechanics requires that each electron in an atom be designated by four quantum numbers, ie, n, l, m_l, and m_s, and in an atom no two electrons can have the same set of quantum numbers. The quantum number n determines the shell occupied by a given electron, and the outermost shell occupied by electrons determines the row in the periodic table where the element is located.

The shells are also split into a number of subshells, and the angular momentum quantum number l determines in which subshell the electron belongs. In any shell, l can have any value from 0 to $n-1$. The subshells with the lower l quantum numbers are more tightly attached to the atom than those with a higher quantum number. Since the l quantum numbers were first discovered in atomic spectra, it is common to use a different designation for the value of the l quantum numbers. An electron for which $l = 0$ is referred to as an s electron; for $l = 1$, a p electron; for $l = 2$, a d electron; and for $l = 3$, an f electron. Any electron can have a magnetic quantum number m_l equal to $+l$ or some whole number in the range $+l \ldots 0 \ldots -l$ and m_s can be either $+\frac{1}{2}$ or $-\frac{1}{2}$. Thus the electrons in the $4f$ subshell can have quantum numbers $n = 4$ $l = 3$ $m_l = 3 \ldots 0 \ldots -3$ $m_s = \pm\frac{1}{2}$. The subshell is half-filled with seven electrons and completely filled with fourteen electrons (see Table 1).

Electrons on the outside of the atom can interact with neighboring atoms in the liquid and solid state and the electrons which do so are called valence electrons. In the sixth row of the periodic table, the energy of binding of the $4f$ and $5d$ subshells lie close together and the environment can shift the energy so as to determine which electron enters the atom first. For element 57 (La) one $5d$ electron enters the atom and, because the $5d$ orbital reaches the outside of the atom, it is a valence electron. Thus La is trivalent.

Table 1. **How Shells and Subshells Fill with Electrons as Atomic Number of Element Increases**

	Shells	Maximum number of electrons in subshells	Row in periodic table	Last element in row
	$1s^2$	2	1	He
[He]	$2s^2 2p^6$	8	2	Ne
[Ne]	$3s^2 3p^6$	8	3	Ar
[Ar]	$3d^{10} 4s^2 4p^6$ 2 1 3^a	18	4	Kr
[Kr]	$4d^{10} 5s^2 5p^6$ 2 1 3	18	5	Xe
[Xe]	$4f^{14} 5d^{1} 5d^{2-10} 6s^2 6p^6$ 3 2 4 1 5	32	6	Rn

^{a} Order in which subshells are filled.

For element 58 (cerium), the $4f$ ($l = 3$) subshell starts filling and the subshell is complete at element 71 (lutetium) after fourteen electrons are added. The $4f$ subshell lies inside the atom so that the electrons in it are well shielded from the electric fields of neighboring atoms by the completed $5s^2 5p^6$ subshells. Therefore, they do not contribute to the valency shell. This behavior accounts for the placement of all lanthanides with lanthanum in the Mendeleev table (1).

As the atomic number increases across the rare-earth series, the increased charge on the nucleus acts on all the electrons and tends to pull them closer to the nucleus. This shrinkage usually is not so large that it changes the crystal structure of a given rare-earth compound when the next higher member of the series is substituted for the original rare earth. Such substitution, however, does cause a shrinkage in the crystal lattice constants. This shrinkage, or lanthanide contraction, has made the rare-earth series a powerful tool for checking theoretical property relationships between two substances. Since all of the lanthanons are usually trivalent, the bulk and chemical properties are very much alike when one lanthanide is substituted for another in a given compound or solution. The shift of the lattice parameters should cause the properties to shift slightly in a given direction across the series.

If a theoretical relation or mathematical equation relating properties does not predict this tendency when one rare earth is substituted for another, then it is probably an empirical, not a true, relationship.

The binding of electrons is a little tighter when the subshell is half-filled, or filled. The $4f$ and $5d$ subshells have about the same energy of binding. The neighboring atoms in a condensed, ie, solid or liquid, state can then slightly shift the energies of these two subshells with respect to each other so that in certain environments, a $4f$ electron can be promoted to the $5d$ or valence shell; for example, cerium can exist in acid solution in the 4-valent state. In other environments, eg, oxide crystals, praseodymium and terbium can also be stabilized in the 4-valent state. Conversely, in environments with different crystal fields, one can cause a valence electron to shift to the $4f$ subshell so that europium, ytterbium, and (sometimes with difficulty) samarium and thulium can exist in the divalent state. When a rare earth is stabilized in a different valence state, it becomes a simple matter to separate it from the other rare earths, since its chemical properties are distinctly different from those of the regular trivalent rare earths. In the gaseous state of the neutral rare-earth atom, where there are no neighboring atoms to exert a crystal field, the extra electron added as the atomic number increases tends to go to the $4f$ subshell instead of the $5d$ subshell. Thus, the basic state or ground state for most of the rare-earth neutral atoms is $4f^{n+1} 6s^2$, and it is only in those atoms where the $4f$ shell is empty, half full, or full that the ground state is $4f^n 5d\ 6s^2$. For La, $n = 0$; for Gd, $n = 7$; and for Lu, $n = 14$. Here n is not a quantum number but the number of electrons in the $4f$ subshell when the atom is in the $4f$ condensed state. Although the filling in of the $4f$ subshell plays a minor role in the chemical properties of the atoms, it is important to certain physical properties, such as atomic spectra and the magnetic properties of various elements and their compounds.

The early Greeks believed that all nature was made up of four elements: air, earth, fire, and water. An earth was defined as a material that could not be changed further by subjecting it to any heat source then available. Until the beginning of the 19th century, the Greek influence on chemistry was still strong, and oxides of elements such as calcium, aluminum, and magnesium were known as earths and thought to be elements. An early history of the rare earths from 1790 to 1920 can be found in ref. 2.

The first rare earth was discovered during the investigation of a rare mineral that is now called gadolinite. An impure new oxide was isolated by Gadolin in 1796 and called yttria. In 1803, another new rare earth was reported by several authors, and was named ceria.

In 1808, it was demonstrated by Sir Humphrey Davey that earths as a class were not elements, but compounds composed of oxygen and a metallic element. Later, a number of chemists verified the existence of ceria and yttria in gadolinite and in a wide variety of other rare minerals.

From 1839 to 1843, it was discovered that yttria and ceria were not the oxides of single elements but were mixtures of several rare earths. It was reported that if ceria and yttria were dissolved in strong acids and their resulting solutions subjected to a long series of fractional separations, two new rare earths could be separated from ceria; the oxides were called lanthana and didymia. The new elements isolated from yttria in the form of oxides were erbia and terbia, and the metallic elements are known as erbium and terbium.

After 1859, the spectrograph was introduced in the study of the individual rare earths. Almost all the rare earths, except scandium, yttrium, lanthanum, and lutetium, had narrow characteristic absorption bands in solutions which, contrary to those of other elements, did not shift much in frequency regardless of the surrounding atoms in various compounds. The uv-vis spectra of the rare earths were not understood until the 20th century.

The Russian chemist Dmitri Ivanovich Mendeleev first proposed the periodic table in 1869, and he found it necessary to insert a blank in the position now occupied by scandium. In 1871, he predicted that a new element should be found in the third column of the periodic table, and predicted what its properties would be. Scandium was discovered in 1879; the agreement of its properties with those predicted by Mendeleev helped to bring about general scientific acceptance of Mendeleev's periodic table. Interestingly enough, one of the greatest weaknesses of his table was that it provided no logical place for the lanthanides.

From 1843 to 1939 the rare earths continued to be separated by extensive fractionation processes and were studied by many scientists. Didymia was resolved into several oxides: samaria, praseodymia, neodymia, and europia. The terbia and erbia fractions were resolved into holmia, thulia, dysprosia, ytterbia, and lutetia.

Shortly after Auer von Welsbach isolated praseodymia and neodymia in 1885, he invented the Welsbach gas mantle and later produced lighter flints. The demand for rare earths which developed as a result of von Welsbach's invention resulted in a worldwide search for rare-earth materials.

It was not until 1913 that it was determined how many lanthanide elements existed (3). About seventy new rare earths had been reported over the years. Most of them were not well established but the claims were made on the basis of one fraction of the long fractionation process applied to the mixed rare earth which showed different color, molecular weight, or spectral lines. Moseley discovered that a plot of the square root of the frequency of certain x-ray lines of the various elements against atomic number gives a straight line. All of the rare earths except promethium were discovered before this work. Promethium does not have a stable isotope, so it does not appear in the earth's crust. Kilogram amounts of Pm_2O_3 [*12036-25-8*] have been separated from reactor wastes at Hanford. It turns out that there are 14 elements in the lanthanide series but element 61 is missing (4). All isotopes of Pm 61 are intensely radioactive, so that its salts and metal alloys must be handled by remote control (see Nuclear reactors, special engineering for radiochemical plants; Radioactivity, natural).

Occurrence

The rare earths are widely distributed in low concentrations throughout the earth's crust (5). They occur as mixtures in many massive rock formations, eg, basalts, granites, gneisses, shales, and silicate rocks, wherein they are present in amounts of 10–300 ppm. They also occur in ca 160 discrete minerals, most of which are rare, but in which the rare-earth content, expressed as R_2O_3 or Ln_2O_3 (REO), can be as high as 60% (REO is an abbreviation used in industry for R_2O_3 content. Ln_2O_3 is used when Y_2O_3 is excluded.) Approximately ten of these occur in sufficient quantities that they may furnish some REO to commerce, but more than 95% of the REO occurs in three minerals: monazite and bastnaesite for the light rare earths, and xenotime for yttrium and the heavy rare earths. Xenotime occurs mixed with monazite in alluvial deposits.

Rare earths do not occur in nature in the elemental (metallic) state and, except for scandium, they do not occur in minerals as individual rare-earth compounds. As a result of multifarious geological processes, no two samples of the same rock or mineral, if taken from different locations, have exactly the same composition of the mixed rare earths.

The relative abundances of some rare earths in rocks are powerful tools for the study of the formation of rock deposits. When the molten basic basalt rocks, which lie deep in the earth, come in contact with the more acidic silicate rocks, the silicates extract some of the rare earths from the basalts. The silicate rocks extract a higher percentage of the light rare earths than they do the heavy ones. By studying the relative abundances of these elements in various rocks, geochemists can tell whether they have been molten. Most of the rich rare-earth minerals were formed deep in the earth's crust. They precipitated from superheated solutions of seawater or molten rock that were subject to very high pressures. If the relative abundance of europium is abnormally low, the geochemist can deduce that, when the minerals were formed, the reaction took place in a reducing atmosphere. Conversely, if the cerium abundance is low, the reaction must have occurred in an oxidizing atmosphere. The relative abundances of

Table 2. Typical Abundance of Rare Earths in Some Rocks and Minerals

Element	Rank[a]	Kilauea basalt, ppm	Precambrian granites, ppm	North American shales, ppm	South Carolina monazite, wt % R_2O_3	California bastnaesite, wt % R_2O_3	Malaysia xenotime, wt % R_2O_3
Y	31		31.00	35.00	3.0	0.1	60.8
La	35	10.50	49.00	39.00	19.5	32.0	0.5
Ce	29	35.00	97.00	76.00	44.0	49.0	5.0
Pr	45	3.90	11.00	10.30	5.8	4.4	0.7
Nd	32	17.80	42.00	37.00	19.2	13.5	2.2
Pm							
Sm	42	4.20	7.20	7.00	4.0	0.5	1.9
Eu	57	1.31	1.25	2.00	0.17	0.1	0.2
Gd	43	4.70	5.80	6.10	2.0	0.3	4.0
Tb	59	0.66	0.94	1.30	0.2		1.0
Dy	50	3.00			1.3		8.7
Ho	56	0.64	1.22	1.40	0.1	Total 0.1	2.1
Er	54	1.69	3.20	4.00	0.5		5.4
Tm	65	0.21	0.53	0.58			0.9
Yb	53	1.11	3.50	3.40	0.2		6.2
Lu	60	0.20	0.52	0.60			0.4

[a] Ranks of 1–90 are relative measures of the abundance of each element in the earth's crust.

the rare earths in some typical rocks and the relative abundances in monazite and bastnaesite are listed in Table 2. As is generally true for all elements, the abundances of the even-atomic-numbered rare-earth elements are considerably greater than the adjacent odd-atomic-numbered elements. The rank column in the table represents the relative abundances of the various elements in the earth's crust. The rare earths also occur in the earth's mantle, in chrondritic meteorites, on the moon, and in most stars. In certain classes of stars, they are much more abundant than they are on earth and, in some stars, astronomers have observed the spectrum of promethium, which does not occur in the earth's crust because of its short half-life. The relative abundance of the rare earths in chrondritic, ie, stony or granular, meteorites is accepted to be the average distribution of these elements in the original solar system. The relative abundances of the rare earths, and particularly of their isotopes, are of critical interest to astrophysicists, since they can relate such information to the formation of the elements and to theories concerning the origin of the universe (see also Space chemistry).

Properties

Physical. The rare-earth metals form alloys with most metals in which rare-earth metal atoms can be present interstitially, in solid solutions, or as intermetallic compounds as a second phase (5–7). In rare-earth-rich alloys, other materials can frequently change the properties of the pure metal by drastically lowering or raising the melting point, eg, by 200–300°C in some cases. Alloying with other elements can make the rare earth either pyrophoric or corrosion-resistant. Heating the metals and alloys can cause precipitation of small single crystals, and the physical properties of these crystals vary

Table 3. Some Properties of the Rare Earths

	Scandium	Yttrium	Lanthanum	Cerium	Praseodymium	Neodymium	Promethium
CAS Registry Number	[7440-20-2]	[7440-65-5]	[7439-91-0]	[7440-45-1]	[7440-10-0]	[7440-00-8]	[7440-12-2]
atomic number	21	39	57	58	59	60	61
atomic weight	44.956	88.905	138.91	140.12	140.907	144.24	145
color	silvery	silvery	silvery	silvery	silvery	silvery	silvery
melting point, °C	1541	1522	918	798	931	1021	1042
boiling point, °C	2836	3338	3464	3433	3520	3074	3000 (estd)
density, g/cm^3	2.9890	4.4689	6.1453	6.770	6.770	7.007	
heat of fusion, kJ/mol[a]	14.10	11.43	6.201	5.179	6.912	7.134	
heat of sublimation (at 25°C), kJ/mol[a]	377.8	424.7	431.0	422.6	355.6	327.6	348 (estd)
conduction electrons	3	3	3	3, 3.1	3	3	3
crystal structure	hcp	hcp	hcp	dhcp	dhcp	dhcp	
radius of atom, nm	0.1640	0.1801	0.1879	0.1824	0.1828	0.1821	0.1811
Curie point, °C							
Néel point, °C				ca 13			
valence in aqueous solutions	3	3	3	3.4	3	3	3
color of oxide[b]	white	white	white	CeO_2, off-white	Pr_6O_{11}, black Pr_2O_3, green	blue	
color of aqueous solution	colorless	colorless	colorless	colorless		rose	yellow
ionic radius, nm	0.0732	0.0893	0.1061	0.1034	0.1013	0.0995	0.0979

[a] To convert kJ/mol to kcal/mol, divide by 4.184.
[b] R_2O_3, otherwise described.

greatly depending on the direction in the crystal in which they are measured, ie, the materials are frequently anisotropic. For the above reasons it is extremely important, when determining physical constants or for certain solid-state devices, that the materials be very pure and well-characterized. The amount of all impurities and the previous heat treatment of the sample should be known.

The arc and spark spectra of the individual rare-earths elements are exceedingly complex. Thousands of emission lines are observed, so that such spectra are difficult to decipher. In the case of the triply ionized rare-earth ions that exist in solids, the absorption spectra are much better understood. However, the crystal fields of the neighboring atoms remove the degeneracy of some states so that there are several levels where only one existed. Many of these crystal field levels occur very close to a basic level. As the solid is heated, a number of the low lying levels become occupied; thus, some physical properties of rare-earth metals are very sensitive to temperature (7).

The magnetic moments of the $4f$ electrons do not cancel out unless the shell is empty or is completely full, ie, most of the lanthanide metals exhibit magnetic properties. At room temperature and above, they are paramagnetic; however, if the temperature is lowered, the spin and orbital moments begin to line up in various ways and the metals become antiferromagnetic. As the temperature is lowered further, a number of different ways of combining these moments can occur. Finally, for some of them, all the moments line up and the metals become strongly ferromagnetic. This magnetism, however, is highly anisotropic and is important in some industrial applications (see Uses).

Some properties of the rare earths are listed in Table 3.

Samarium	Europium	Gadolinium	Terbium	Dysprosium	Holmium	Erbium	Thulium	Ytterbium	Lutetium
[7440-19-9]	[7440-53-1]	[7440-54-2]	[7440-27-9]	[7429-91-6]	[7440-60-0]	[7440-52-0]	[7440-30-4]	[7440-64-4]	[7439-94-3]
62	63	64	65	66	67	68	69	70	71
150.35	151.96	157.95	158.9254	162.50	164.930	167.26	168.934	173.04	174.97
silvery	silvery	silvery	silvery	silvery	silvery	silvery	silvery	silvery	silvery
1074	822	1313	1365	1412	1474	1529	1545	819	1663
1794	1429	3273	3230	2567	2700	2868	1950	1196	3402
7.520	5.234	7.9004	8.2294	8.5500	8.7947	9.066	9.3208	6.9654	9.8404
8.623	9.221	10.05	10.80	10.782	16.874	19.90	16.84	7.657	18.65
206.7	144.7	397.5	288.7	290.4	300.8	317.10	232.2	152.1	427.6
3	2	3	3	3	3	3	3	2	3
rhomb	bcc	hcp	hcp	hcp	hcp	hcp	hcp	fcc	hcp
0.1804	0.20418	0.18013	0.17833	0.17743	0.17661	0.17566	0.17462	0.19392	0.17349
		292.7	220	86	19	18	32		
15	90		230	178	133	84	56		
3	3, 2	3	3	3	3	3	3	3	3
cream	white, greenish tinge	white	Tb_4O_7, brown	yellowish white	yellowish white	pink	white, greenish tint	white	white
colorless	colorless	colorless	Tb_2O_3, colorless	yellow tint	yellow	pink	white, greenish tint	colorless	colorless
0.0964	0.0950	0.0938	0.0923	0.0908	0.0894	0.0881	0.0870	0.0858	0.850

Chemical. The chlorides, bromides, nitrates, bromates, and perchlorate salts are soluble in water and, when their aqueous solutions evaporate, they precipitate as hydrated crystalline salts. The acetates, iodates, and iodides are somewhat less soluble. The sulfates are sparingly soluble and are unique in that they have a negative solubility trend with increasing temperature. The oxides, sulfides, fluorides, carbonates, oxalates, and phosphates are insoluble in water. The oxalate, which is important in the recovery of highly pure Ln, can be calcined directly to the oxide.

Anhydrous rare-earth salts usually cannot be prepared by evaporating the water of crystallization. For example, when heated, $LnCl_3.6H_2O$ partially converts to the oxychloride (LnOCl); under the same condition, the nitrates convert to basic nitrates. The vapor pressure above $LnCl_3.6H_2O$ is made up of H_2O and HCl, so that to prepare the anhydrous salt it is necessary to suppress the loss of HCl. The usual practice is to pass a stream of argon over the cold hydrated salt and then slowly heat the hydrate while a stream of anhydrous HCl gas is passed over it. The rate of increase in temperature must be slow if no oxychloride is to form. Many binary anhydrous salts can be prepared by directly combining the metal with a more electronegative element. For example, $Ln + Cl_2$ gives $LnCl_3$ and Ln plus sulfur can give Ln_2S_3.

The lanthanides can form hydrides of any composition up to LnH_3. Small amounts of hydrogen dissolve interstitially but, with increasing amounts of hydrogen, a second phase (LnH_2) appears. These alloys are metallic in their properties and lose hydrogen at relatively high temperatures. If more hydrogen is added, a LnH_3 phase occurs, and such compounds have saltlike properties. Yttrium dihydride contains more hydrogen per cubic centimeter than liquid hydrogen or water. Yttrium has a low nuclear cross section and well above a red heat must be applied to obtain a hydrogen vapor pressure of 101.3 kPa (1 atm) over the alloy. Therefore, yttrium hydride makes an excellent moderator for nuclear reactors where weight is a prime consideration.

The rare-earth salts readily form double salts, eg, $Ln_2(SO_4)_3.Na_2SO_4.2H_2O$, and $Ln_2(SO_4)_3.MgSO_4.24H_2O$. These salts were sometimes used in the fractionation process. The early rare-earth chemists would ship rare-earth elements as their salt because the cost per gram of mixed salt containing one rare-earth element was much less than when the element was furnished in some more concentrated form. Salts, such as $Ln_2(SO_4)_3.2H_2O$, are important because of their limited solubility in water and their easy conversion to $Ln(OH)_3$.

The rare earths form many compounds with organic ligands; some of the compounds are water-soluble. Chelating agents (qv), which are organic molecules that engulf rare-earth ions and displace some of the water of hydration, form complexes with rare-earth ions that have a much wider spread in their formation constants than do those of ordinary mineral-acid salts. This makes such reagents very useful as complexing agents that can be used in the band-displacement cation-exchange process for separating rare earths and in some liquid–liquid extraction schemes (see Extraction, liquid–liquid extraction).

Many binary compounds, such as the oxides, nitrides, and carbides, can exist as nonstoichiometric compounds. They form crystals where some of the negative atoms are missing from the site they should occupy in the crystal structure. Therefore, the same crystal structure is present over an appreciable concentration range in phase diagrams.

Nuclear. The rare earths as a group are very rich in the total number of isotopes (qv) that can exist for each lanthanide (14–19 each). The elements with odd atomic numbers have one or two stable isotopes, but those with even atomic numbers may have four to seven. The radioactive isotopes can be produced in many ways so that they are either rich or lean with regard to neutrons (see Radioisotopes). Neutron-rich lanthanide isotopes occur in the fission of uranium or plutonium and these and others can be produced by neutron bombardment, by the radioactive decay of neighboring atoms, and by nuclear reactions in accelerators where the rare earths are bombarded with charged particles. A study of the various isotopes which can be formed, along with their decay modes and half-lives, are of intense interest to physicists trying to understand the nucleus of the atom. In the rare-earth region of the periodic table, many nuclei are not spherically symmetric and their nuclear spectra are complicated. Whenever a nuclear subshell is complete, either with neutrons or protons, the nucleus is very stable. The nuclear numbers corresponding to filled subshells are called magic numbers (see Actinides and transactinides), and one such magic number for neutrons, ie, 82, occurs in the rare-earth series.

Production and Processing

Monazite and bastnaesite are the minerals used commercially to supply most of the rare-earth chemicals. Monazite is a brown, dense phosphate mineral. There are extensive deposits of enriched monazite sands on beaches in many parts of the world, eg, along the southwest coast of India and the east coast of Brazil; in the uplands of Australia, South Africa, the USSR; and in the United States in Idaho, South Carolina, and Florida. These deposits are thought to be an artifact of ancient sea beds that were forced up in the process of mountain building. Such deposits are dredged, pulverized if necessary, and further enriched by flotation (qv) methods. Sometimes they are also subjected to cross-belt magnetic separation, since they are weakly magnetic. Frequently another phosphate mineral, xenotime, which is rich in yttrium and the heavy rare earths, accompanies the monazite. Xenotime can be either processed with monazite or separated from it by magnetic separation (qv), since it is more magnetic than monazite. A high yttrium content in monazite usually indicates that xenotime mineral is mixed with it, and the xenotime content can be as high as 10 wt %.

Monazite is about 50 wt % lanthanons. The R_2O_3 fraction usually occurs as a mixture of ca 50 wt % ceria, 20–30 wt % lanthana, 15–20 wt % neodymia, and 5–6 wt % praseodymia. All the other rare earths make up less than 5 wt %. Many rare-earth chemical companies buy monazite as their starting material, but bastnaesite is of greater importance in the United States.

There are extensive deposits of the fluorocarbonate mineral bastnaesite near Mountain Pass in southeastern California. Typical rock analyses show ca 50 wt % calcite, 25 wt % baryte, 15 wt % bastnaesite, and 10 wt % silica. The ore, after being crushed and ground, is upgraded by flotation methods to ca 60 wt % lanthanons (REO). It can be upgraded further by roasting and leaching with hydrochloric acid to give a 70 wt % content. A typical analysis of the lanthanons is CeO_2, 50 wt %; La_2O_3, 32 wt %; Pr_6O_{11}, 4 wt %; Nd_2O_3, 13 wt %; Sm_2O_3, 0.5 wt %; and Eu_2O_3, 0.1 wt %. Bastnaesite contains almost no heavy rare earths. Bastnaesite also occurs in parts of Africa and in China. The lanthanide content of xenotime is ca 60 wt % yttria and it contains much larger amounts of the heavier rare earths than does monazite.

Other sources of rare-earth materials are occasional residues from uranium and apatite mining and processing operations. Frequently the rare earths are obtained as by-products, even though their content in the ores is low.

The rare-earth minerals contain almost no scandium. Scandium is generally produced as a by-product of uranium (qv) mining. The scandium, which is present in amounts up to 5 ppm, is recovered from uranium processing solutions. Thortveitite is a scandium-rich mineral that exists in Norway but is very rare. It contains 34–42 wt % scandium oxide.

Beneficiated monazite or bastnaesite can be broken open by sulfuric acid or sodium hydroxide and the lanthanons made water-soluble. It is then possible to proceed by several methods to produce mixed rare-earth oxides, fluorides, nitrates, hydroxides, or carbonates. Each rare-earth processor has its own procedures, which differ somewhat from those of others, but all are based on simple straightforward chemical operations (9).

Fractional Crystallization. Although a substantial market for industrial use of lanthanum, cerium, and mixed rare earths developed after 1885, there were no industrial uses for any other lanthanide. Small amounts of the latter were mostly isolated and purified on a laboratory scale. From 1804 to 1945, the rare earths, except for cerium, were separated one from another by laborious fractional processes, eg, fractional crystallization, fractional precipitation, fractional decomposition, or fractional extraction (8–9). It required thousands of such operations, however, to obtain some of the heavy lanthanons.

Elution. The elution method was developed during World War II by scientists working for the Manhattan District, Corps of U.S. Army Engineers. When uranium or plutonium atoms undergo fission, they form intensely radioactive fission fragments as well as emit a great deal of energy. The fragments are usually emitted in pairs consisting of a lighter atom whose atomic number is approximately that of yttrium and a heavier atom whose atomic number is approximately that of lanthanum. It is the radioactivity of these neutron-rich unstable fission products that makes it necessary for nuclear scientists to carry out all operations (after fission has occurred) by remote control from behind thick radiation shields. It is also this same radioactivity that poses the many problems associated with ultimate nuclear waste disposal.

If atomic bombs were to be produced based on fissile plutonium, large-scale chemical processes had to be developed to dissolve the spent fuel elements from the Hanford reactors and to separate the plutonium produced from the uranium and a variety of fission products. In developing such processes, it was helpful in the early stages to work with nonradioactive mixtures which closely approximated the radioactive mixtures.

It was known that a number of elements heavier than plutonium were also formed from Pu by subsequent neutron capture and beta decay in the reactors. From atomic structure considerations, it was expected that this group, ie, Am, Cm, Bk, Cf, etc, would resemble a new rare-earth series. Nonradioactive rare earths, therefore, were needed as carriers and substitutes; however, not enough separated rare earths were available to satisfy all such needs. Therefore, a group was established at Ames, Iowa, to develop rapid methods for separating large quantities of rare earths. Another group was established at Oak Ridge, Tennessee, to study the separation of fission products and their radioactive characteristics. The Oak Ridge group developed an elution technique during 1942 that usually was successful in separating most of the fission products in tracer amounts. Sometimes a few milligrams of a nonradioactive salt of a particular fission element was added to act as a carrier to help separate the element. Separations were followed by radiochemistry techniques.

Synthetic organic cation-exchange resins became available commercially ca 1940. They were marketed as porous beads of various sizes and consisted of an open, water-permeated polymeric framework of organic materials, such as sulfonated polystyrene–divinylbenzene copolymer. The H^+ cations in such a resin readily exchange for other cations in solutions that percolate through a bed of the material. If an acid solution is passed through the resin, ultimately all the cation-exchange sites in the exchanger become occupied by hydronium ions and the resin then is said to be in the hydrogen state. Similarly, if sodium chloride or ammonium chloride solutions are passed through the resin, the resin is converted by mass action to the sodium or ammonium state. Mixed rare earths in aqueous solution have trivalent cations and these tripositive ions are absorbed much more tightly than any monovalent cation. Therefore, if a resin in the ammonium state is packed into a long column and a mixed rare-earth solution is percolated through it, the trivalent rare-earth cations are absorbed in a compact band at the top of the column. If an ammonium chloride solution then is poured continually into the top of the column, the rare-earth band can be eluted down the column and it spreads out as the resin bed slowly becomes a mixed ammonium–rare-earth bed. When the dilute rare-earth solution starts coming out the bottom of the column, the rare earths are not appreciably separated even though the original rare-earth absorption band has traveled many times its original length. If a separation is to occur, a complexing agent that has significant differences in its affinities for various lanthanons must be added to the eluant.

The procedure first used by the Ames group was to exploit a buffered solution (pH 2–4) of ammonium citrate (NH_4Cit) as the eluant. This solution contains H_2Cit^-, and $HCit^{2-}$ ions that complex the Ln^{3+} ions in solution. As the absorbed mixed rare-earth band is washed down the column, the various lanthanides travel at different rates; eg, Lu^{3+} travels faster than La^{3+}. The individual, originally overlapped, rare-earth bands start to separate, although for much of the column they continue to overlap considerably. As the component bands progress down the system, they lengthen and become much more dilute on the resin, since relatively more and more sites on the resin are occupied by ammonium ions. If the concentration of a given lanthanide is plotted against the volume of the eluant that passes through the column, bell-shaped elution curves are obtained. The further the original rare-earth band travels down a column before being collected, the flatter and wider the individual bell-shaped curves become.

If an artificial mixture of cerium and yttrium is eluted down a column, the two bands separate rather rapidly and a considerable portion of each element can be obtained in relatively pure form in a short time. However, with a normal mixture of rare earths, there are 15 elution bands overlapping each other. Therefore, the bands must travel a considerable distance before they are strung out sufficiently to render their overlap small enough that an appreciable amount of any pure lanthanide can be obtained. However, by this time the elution bands are so wide that the concentrations of the rare earths in the effluent solution are low. Because the differences in the complex formation constants are greater for adjacent light lanthanides than for neighboring pairs of lanthanides with higher atomic numbers, the light rare earths separate more rapidly than the heavier ones.

The Ames group succeeded in separating several hundred grams of each of the light rare-earth elements, which are the main rare earths in the fission products of uranium. They did not, however, have much success in separating bulk quantities of the heavy rare earths by using citrate as the eluant at pH 2–4 on NH_4^+ resin. They then tried elutions at higher pHs on H^+-cycle beds. The latter process, both in cost and effort, was a considerable improvement over the older fractionation methods for separating the light rare earths in quantities from 100 milligrams to one kilogram. The citrate elution processes for separating bulk rare earths have been entirely superseded by the band-displacement techniques with ammonium ethylenediaminetetraacetate (NH_4^+ EDTA) and ammonium hydroxyethylethylenediaminetriacetate (NH_4^+ HEEDTA) (see Chelating agents; Citric acid).

The early process involving the NH_4^+ resin was excellent for radioactive-tracer chemistry and for microanalytical analyses where radioactive tracers are used. When very small quantities are involved, the absorbed band at the top of a resin column can be very narrow and, even though the band must travel many times its original length, it need not travel very far down a column. With sufficient travel, the component bands pull entirely apart and, even though the recovered solutions are extremely dilute in rare earths, radioactive detectors are sensitive enough to follow what is happening (see Ion exchange).

Band Displacement. The band-displacement technique has many features in common with the band-elution process. A number of glass or metal columns are connected to each other head-to-tail and are loosely packed with an organic cation-exchange resin, such as sulfonated polystyrene–divinylbenzene. The resin is first converted to a special cation state, depending on what complexing agent is to be used in the eluant, ie, Cu^{2+} or Zn^{2+} with buffered ethylenediaminetetraacetic acid, EDTA, and H^+ with buffered hydroxyethylethylenediaminetriacetate, HEEDTA. This is accomplished by pouring through the column a solution of, eg, cupric sulfate, if a cupric state resin is desired. Next, a solution of mixed rare-earth chlorides (nitrates or sulfates) is poured in the top of a column or column series until the absorbed rare-earth band is of the desired length.

It was clear from the band-elution process that, if the rare-earth band could be prevented from growing as it moved down the column and thus assuring that most of the active sites would be occupied by rare-earth ions, a much more rapid and efficient separation of the lanthanides could be obtained. This was achieved by imposing a strong chemical constraint on the system at the beginning and the end of the adsorbed rare-earth band. Multidentate chelate complexes of the rare earths have a much wider spread between their formation constants than do those of simpler complex ions. The chelate ligands are anions of organic molecules, eg, ethylenediaminetetraacetic acid, which engulf the rare-earth cations, thus displacing much of the water of hydration which normally surrounds the rare-earth ion in aqueous solution.

A buffered solution of NH_4^+ EDTA (ca pH 8.4) is prepared for the eluant. The formation constants of all of the lanthanide EDTA complexes are rather large, ie, 10^{14}–10^{19}, so that as the eluant solution encounters rare-earth ions on the column, the EDTA anion complexes with any rare-earth cation it encounters, removes it from the resin, and deposits ammonium ions in its place. The displacement boundary is very sharp, so the rear edge of the mixed rare-earth band moves down the column at a rate depending on the concentration of EDTA being poured through the column. Similarly, at the front edge of the mixed rare-earth band, if the retaining bed contains cations that form a much tighter complex with the EDTA than any Ln^{3+} cations, those ions on the resin bed are picked up by the eluant and the rare-earth ions are redeposited on the resin. As a result, the band front moves at the same rate as the back edge of the band. Accordingly, the total rare-earth band remains approximately the same length no matter how far it is washed down the column. It does expand slightly at first as the resin comes to equilibrium with the eluate solution, which contains hydrogen and ammonium ions, but after that point the length of the band remains constant, no matter how far it is eluted. If the elution curve of a solution leaving the column is plotted, ie, the concentration of rare earth versus the volume of eluant, flat-topped elution curves with sharp boundaries are obtained. Once the rare-earth mixture is resolved, the flat-topped curves for individual elements follow one another head-to-tail.

The overlap of individual adjacent rare-earth bands is quite narrow and is independent of the length of the rare-earth band. If the original absorbed band is long enough, 98% of the rare-earth mixture can be resolved into individually pure components. If the samples are taken from the middle of the bands, purities can reach 0.999999. By the time the rare-earth band has moved $1/10$ of its length, all of the Lu is located near the front edge of the absorbed rare-earth band and, by the time the elution has proceeded approximately one band length, all of the rare earths are resolved into their individual bands. Thus, one can collect the individual components issuing from the bottom of the column, one after the other, in a purity higher than is characteristic of most commercial chemicals.

The narrow overlap portions must be recycled. However, it is feasible to separate a mixture of 15 rare earths into pure components (>99.99%) with an overall yield of up to 98% in one pass through the system. The band-displacement process is capable of being scaled up to multiton quantities because the diameter of the columns can be almost any desired size. In the Ames pilot plant, 76-cm diameter columns were used, but the system can be larger than this if necessary, and the absorbed band can be any length. The water, retaining ion, and complexant can be recovered and recycled.

In the Ames pilot plant, tons of yttrium and many kilograms of all the individual rare earths were separated and a considerable amount of this material was supplied to scientists in the U.S. In the late 1950s, a number of industrial companies began to produce pure rare earths by this method. As a result, pure rare earths can be obtained in large amounts at reasonable prices, and industrial uses for many individual rare earths other than cerium and lanthanum have developed, eg, Y, Sm, Eu, Gd.

Band displacement can be considered a countercurrent extraction system. Since all parts of the total rare-earth band move at the same rate down the column, a reference point can be chosen on the moving band, and the resin bed can be considered to be moving in one direction and a solution phase in the other. Then a theoretical plate height, which is very narrow, can be determined and within the rare-earth band there are a great many stages operating (see Extraction, liquid–liquid extraction).

Liquid–Liquid Extraction. As the industrial demand for individual rare earths increased, some of the rare-earth companies developed liquid–liquid extraction systems (10). An organic stream immiscible with water is flowed countercurrent to the aqueous stream containing the rare-earth mixture. The organic phase may absorb neutral rare-earth molecules by complexing with them, eg, as does tributyl phosphate, or it may contain a complexing molecule or anion added to it for the same purpose. Also, complexing ligands may be added to the aqueous phase so as to effect a synergistic equilibrium between the rare-earth ions and the complexant. If the various equilibria between the organic and aqueous phases have overall equilibrium constants sufficiently different for various rare earths, separation occurs. The principle involved is the same as in the band-displacement ion-exchange technique. If the element desired is one that has an additional valence other than 3+, eg, Eu^{2+} or Ce^{4+}, the element can be put into that other oxidation state to make the liquid–liquid extraction process much simpler.

If many tons of a particular element are needed, liquid–liquid extraction is preferred in that more concentrated solutions can be used and the flow rates are faster than with other methods. Countercurrent liquid extraction suffers the disadvantages that the plate height in a liquid system is much greater than in the ion-exchange resin system and that the process only makes one cut in the series at a time instead of the 15 that are feasible with cation exchange. If only one element in a particular part of the series is desired, only two liquid–liquid extractions are needed. The purity of the rare earths obtained by solvent extraction is generally not as great as of those for the band-displacement ion-exchange system, but a number of companies claim that they can produce the materials at 99.9% or even 99.99% purity. If extreme purity is desired, the rare-earths from the liquid–liquid extraction system can be processed in ion-exchange columns.

Rare-Earth Metals. It is easy to reduce anhydrous rare-earth halides to the metal by their reaction with more electropositive metals, such as calcium, lithium, sodium, or potassium. The main difficulty is obtaining a solid billet that is free of impurities. The rare-earth metals have a great affinity for oxygen, sulfur, nitrogen, carbon, silicon, boron, phosphorus, and hydrogen and remove these elements from most other metals or from water and carbon dioxide. They form very stable compounds with the above electronegative elements, and such compounds are usually appreciably soluble in molten rare-earth metals. Molten rare-earth metals tend to attack any crucible in which they are melted and thus introduce into the metal billet as impurities any elements they contact (5–6,9,11).

In the first reduction of a rare earth to its metallic state, the anhydrous chloride salt of ceria was reduced with metallic sodium or potassium (2). The yields were low (26%) and the metal existed as small nuggets surrounded by the alkali chloride slag. It is difficult to separate such nuggets from a slag since, if one attempts to leach the chlorides, the water reacts slowly with the metal and introduces oxygen. The early Ce metal was very impure, since it contained the several other rare earths later determined to be in ceria and contained considerable quantities of oxides, nitrides, carbides, as well as other metallic elements. The reaction was usually carried out in air. The molten rare earths severely attacked the crucibles.

In 1935, samples of the purest rare-earth chlorides available were reduced by potassium metal at relatively low temperatures in enclosed glass capsules. This process gave relatively pure metals as fine powders imbedded in potassium chloride. No attempt was made to separate them from the potassium chloride, since it did not interfere with the properties of interest, ie, crystal structure, density, and magnetic susceptibility. The potassium chloride acted as an inner standard in the x-ray determination of the metal crystal structure, and the magnetic susceptibility measurements could be corrected for the amount of potassium chloride present. Although the metals were not pure by modern standards, since all contained some potassium and rare-earth impurities, in 1935 the investigators were able to obtain lattice constants and densities within 1% of the most accurate modern values.

In 1875 the first successful preparation of rare-earth metals by an electrolytic process was reported. A few grams of cerium, lanthanum, and didymium (Di), which consists of a mixture of neodymium and praseodymium, were obtained in compact form by electrolyzing molten rare earth in molten chloride electrolytes covered with a layer of ammonium chloride. The metals were not very pure. The electrolytic technique was steadily improved over the years and by 1905 misch metal, La, Ce, Pr, Nd, and Sm had been made. By 1925, several improvements in the cell design had been made, and purer samples of La, Ce, and Nd were prepared with Y, although most of the latter deposited as a powder.

The electrolytic process suffers from much the same difficulties as the metallothermal method. It is difficult to find cell materials that are not attacked by molten rare earths and thereby introduce impurities into the ingot. It is also difficult to design cells that completely exclude atmospheric elements. The method works best for rare-earth metals with low melting points. For these elements, the cells can be kept hot enough that the rare earth collects at the bottom of the cell as a molten pool. Rare earths with higher melting points deposit as powders at the upper limit of the cell's operating temperature and it is difficult to separate these powders from the electrolyte without introducing additional impurities. At present, there are few electrolytes that can be heated much above 1000°C.

By 1931, an electrolytic cell, specially designed for producing much purer metals, was employed to produce metallic Ce which contained only small amounts of impurities. Later, this apparatus was used to produce other rare-earth metals, including Eu, Gd, and Y. By 1939, most of the rare-earth metals had been made in fair purity and a number of their properties, eg, magnetic behavior, melting point, density, crystal structure, and chemical reactivities, had been studied. All of these metals, however, contained small amounts of various metallic impurities and unknown amounts of nonmetallic impurities. Most of the impurities were not reported in the literature because analytical methods to determine them at the concentrations that were present had not been developed.

Ultrapure Rare-Earth Metals. Many of the properties of rare-earth metals are very sensitive to very small amounts of impurities. Even the presence of a few tenths of an atom percent of some elements can change the measured value of certain constants significantly. This is especially true for light-element contaminants, such as hydrogen, oxygen, and carbon, since these impurities are usually expressed in weight percent. For example, if Lu contains 300 ppm hydrogen by weight, it is a 5 atomic % alloy of H with Lu. Certain properties are also very sensitive to small amounts of other rare-earth metals, particularly when magnetic properties are studied at very low temperatures (5–6,11).

For many industrial uses, extreme purity is generally not required nor desired because the less pure metals can be produced at much lower costs. However, the presence of impurities can be critical in metal produced for research purposes, especially when experimental properties are being compared with theories that are applied to predict their values. Trace impurities also can be critical in many solid-state devices that take advantage of some particular property of the rare earth involved. The painstaking processes described below are, therefore, used in making research-grade metals. If less pure metal is satisfactory, many of the steps described below can be omitted and the process can be terminated at the point where the desired purity is obtained.

Properties across the rare-earth-metal series change much more drastically than they do for the salts or solutions of rare-earth salts. For example, the melting point of Lu expressed in °C is almost twice that of La. Metal boiling points vary considerably in various parts of the series. Therefore, no one process for producing a rare-earth metal is satisfactory for the entire series. Although the basic process for obtaining the metal is the same for most lanthanides, several of the finer details have been modified for each element.

One process is the metallothermal reduction of anhydrous rare-earth fluoride with calcium metal. If ultrapure metals (99.99% or 99.999% pure) are to be obtained, it is essential that both the fluoride salt and the calcium metal be extremely pure, particularly with regard to oxygen and carbon, since these elements invariably end up in the rare earth billet if they are present at the outset. When the anhydrous fluoride has been prepared by passing dry HF over a rare-earth oxide, the material always contains a small amount of oxyfluoride. It is necessary, therefore, to pass purified anhydrous HF over the molten fluoride in a crucible that does not contaminate the fluoride. In the laboratory, where only 200- or 300-g batches are being prepared, a platinum crucible is employed. If such precautions are taken, the final anhydrous fluoride should contain less than 10 ppm oxygen. Commercial calcium always contains appreciable amounts of carbon and oxygen and rapidly picks up additional amounts of such elements from the air. Therefore, the calcium metal is distilled twice in an atmosphere of high purity argon with tantalum containers, and the redistilled Ca is either sealed or handled exclusively in a dry box whose atmosphere is ultrapure argon gas. Approximately 10 mol % of calcium over the stoichiometric amount is used in a reduction. If the same apparatus is used for reducing several rare earths consecutively, it must be cleaned thoroughly between each reduction to prevent cross-contamination of the final billets. Modern crucibles are made from sheet tantalum and tungsten. Oxide crucibles made from BeO, MgO, or REO are attacked by the molten metal and an appreciable amount of oxygen is introduced into the rare-earth billet. Tantalum and tungsten are attacked very slightly below 1000°C; and, at higher temperatures, attack is slow. Dissolved tantalum or tungsten can be removed in a straightforward manner.

The 17 rare-earth elements can be divided into 5 groups. Group I is comprised of those elements with low melting points and high boiling points: La, Ce, Pr, and Nd. The reduction is carried out in a furnace in a highly purified helium or argon atmosphere and, after the reaction takes place, the products are held at a temperature slightly above the melting point of the rare-earth metal for sufficient time to allow the metal and slag to separate. The crucible is allowed to cool to room temperature and is transferred to a dry box with a highly purified argon or helium atmosphere, wherein the slag is removed. The solid rare-earth ingot obtained is placed in another tantalum crucible and returned to the furnace. The system is attached to a high vacuum pump, which is allowed to run until the pressure is less than 1×10^{-5} Pa ($<10^{-7}$ mm Hg). A 1×10^{-3}-Pa (10^{-5}-mm Hg) vacuum is not good enough, since it does not remove all the fluoride and oxide. The crucible is slowly heated to ca 1500°C, at which temperature all of the volatile impurities, eg, Ca metal and fluorides, are removed. During this process, however, the rare-earth metal dissolves a certain amount of tantalum from the crucible. The molten metal is cooled slowly to a temperature just above the melting point of the particular rare-earth metal and held there for a sufficient time so that most of the tantalum can reprecipitate on the side of the crucible or settle to the bottom of the ingot. The crucible is finally allowed to cool to room temperature and is removed to the dry box, where the tantalum crucible material is stripped from the ingot. The rare-earth ingot is machined to remove the surface layer and the bottom part, which contain most of the tantalum. If the procedure is followed carefully, one obtains a rare-earth billet with less than 0.01 wt % total impurities. For Group I metals the solubility of tantalum in rare earths at their melting points is less than 50 ppm. Tungsten is attacked much less at a given temperature than is tantalum. However, the fabrication of tungsten crucibles is difficult and their cost makes it desirable to use tantalum, even though billets of slightly higher purity are obtained if tungsten is the crucible material.

The Group II metals are those elements with high melting points and high boiling points: Gd, Tb, Sc, Y, and Lu. They are prepared in the same manner as the metals in Group I. Because of their high melting points, however, the tantalum content of a finished ingot may be as high as several hundred ppm. The presence of this tantalum can usually be observed in metallographic photographs of the metal as small dark spots scattered through the grains of the cooled metal. If the presence of this tantalum, which is present mostly as a second phase, would be problematic in any subsequent experiments, it can be removed. One way of doing this is to place an inverted tantalum crucible over the reaction vessel. This crucible is left outside the heating zone around the original crucible and, because of radiation losses, is 400–500°C colder than the other vessel. The metal is then slowly distilled into this condenser at a few hundred Pascals (a few mm Hg), so that the tantalum remains behind.

Group III consists of the metals with high melting points, low boiling points, and appreciable vapor pressures at their melting points: Dy, Ho, Er, and Sc. These elements cannot be held at their melting points for long because a considerable quantity of metal would be lost; therefore, the boiling process is omitted. Sublimation or volatilization of the metal into a condenser works well and, by the right choice of temperature in the still and an appropriate condenser gradient, both the volatile and inert impurities can be eliminated.

The metallothermal process does not work for the Group IV elements. They are metals that have multiple valences, or extremely high vapor pressures at their melting points, or both: Sm, Eu, Yb, and Tm. If the metallothermal process is used with Sm, Eu, or Yb, a mixture that contains a large amount of the divalent salt is obtained. Thulium has such a high vapor pressure at its melting point that most of the metal is lost even if melted for a very short time. These metals are prepared by a distillation process. The oxide of the rare earth is mixed with portions of pure La, Ce, or misch metal that has been heated to a sufficient temperature to remove any of its volatile components, and the Group IV rare earths are removed by distillation. All have low boiling points, so that the vapor pressure of the metal at a given temperature can be appreciable, whereas the vapor pressure for the reductant is negligible and this drives the reaction to completion: $Eu_2O_3 + 2\,La \rightarrow La_2O_3 + 2\,Eu\uparrow$. The condensed Sm, Eu, and Yb are placed in a crucible and melted into an ingot. However, the vapor pressure of thulium at the melting point is so high that most of it is lost if the metal is melted at atmospheric pressure. If a solid ingot is desired, the usual practice is to pack the condensed thulium-metal crystals reduced at a temperature well below the melting point in a tantalum tube and then reduce the diameter of the tantalum tube by swaging it to less than half its diameter. At the right temperature, this compacts the thulium metal into a solid ingot.

Group V consists of one element: Pm. This element would be described as a Group II element if it were not so intensely radioactive. However, because of its radioactivity, all procedures must be carried out by remote control in special installations built to handle intense radioactivity (4).

Economic Aspects

Many companies do not publish consumption information and there is wide variation in the cost of raw materials. Quoted prices depend very much on the quantities wanted by a given individual or company and on how many of the other rare earths in the processed minerals are marketable (12).

Approximately 15,000 metric tons of rare-earth-oxide equivalent [REO] was consumed in the U.S. in 1979. The rare-earth industry is growing, and the average increase over the past several years has been 5–15%. The two principal sources of light rare earths are bastnaesite and monazite. These account for 95% of the world's consumption. In 1979, the price of monazite mineral, depending on its rare-earth content, was $273–353 per metric ton.

Bastnaesite, which supplies ca 55% of the world's rare-earth needs, is marketed as a rare-earth concentrate, and 1979 prices were ca $2.20/kg of rare-earth oxide contained. The 1979 price of Malaysian xenotime, which is rich in yttrium and heavy rare earths, was ca $4.40–6.60/kg and misch metal was ca $9.30/kg.

Toxicity

The rare earths are considered only slightly toxic according to the Hodge-Sterner classification system and can be handled safely with ordinary care (13). The worker should not breathe the vapors or dust arising from any manipulation, wash any skin surface which has accumulated dust or solution, and be careful not to swallow any when pipetting. If a sliver of the metal should be lodged in an eye, the sliver should be removed immediately. Blindness can result if the rare-earth material were left in contact with the eye for a considerable period of time. The worker should be particularly careful not to breathe the dust of any ore containing the rare earths, since these ores frequently contain some poisonous materials, eg, compounds of beryllium, uranium, and thorium.

The toxicity of rare earths on humans has not been reported, but extensive tests of toxicity have been made on animals. In animal tests, the toxicity is very dependent on the manner by which the rare earths are introduced. If the rare earths are administered orally, only a small fraction of the rare earths are absorbed by the intestines. This probably accounts for the low toxicity of the rare earths when taken orally. When rare-earth vapors or dust are inhaled, they are somewhat more toxic but tend to remain in the lungs and are only slowly absorbed into the body. If injected subcutaneously, most of the injected material remains in place. The amount that is absorbed tends to collect in the liver, spleen, and kidneys.

By far the most toxic reactions are obtained if the rare earths are introduced by means of intraperitoneal or intravenous injections. The symptoms of extreme toxicity include writhing, atoxia, labored respiration, walking on the toes with arched back, and sedation. Examination of the animals show degeneration of the spleen and yellow atrophy. They also show central lobular necrosis of the liver. Hyperglycemia may occur as well as decreased blood pressure.

A recent review includes literature on toxicity and the use of radioactive rare-earth tracers in medical research. The review also includes a discussion of the use of radioactive rare earths in the treatment of cancer.

Uses

The use of rare earths in industry can best be described if they are listed under two different categories: Class 1 rare-earth mixtures take advantage of the general properties of the rare-earth elements, eg, the large size of the ion, the trivalent nature of the cations, and the great proclivity of the metals to form very stable compounds with electronegative elements, such as oxygen, carbon, and nitrogen (12). This class can be divided into two subclasses. Class 1A, which consists of mixed rare earths having the same relative abundance as in their minerals, and Class 1B, which is the same as Class 1A except that one or more elements, eg, Ce, La, or Eu, have been removed. The latter also includes individual rare-earth elements where a single element has been concentrated to comprise 50–95 wt % of the mixture. Class 1A accounts for ca 65% of the tonnage and ca 25% of the dollar value of sales. Class 1A and 1B rare earths account for 99% of the tonnage and 66% of sales.

The Class 2 applications take advantage of the individual properties of separated elements, which usually are better than 99% pure. They account for ca 1% of the total tonnage but 33% of the dollar value of all sales. The Class 2 properties depend on the following: the presence or absence of $4f$ electrons, the different valences (other than +3) of a few of the elements, and the nuclear properties.

The main applications involving the mixed rare earths are as gasoline-cracking catalysts, the starting material for making misch metal, the rare-earth silicides for various metallurgical uses, mixed rare-earth oxides as a polishing compound, and carbon arcs. Practically all petroleum-cracking units in the world use zeolite catalysts containing up to 5 wt % REO (see Molecular sieves). Misch metal is used in the iron and steel industries to remove sulfur, carbon, and other electronegative elements from iron and steel. These impurities combine with the rare earths so that they end up either in the slag or as segregated nodules in the metal, for example, nodular cast iron, which is much more malleable than normal cast iron. These two uses account for 60% of the worldwide consumption of rare-earth elements. A use for mixed rare-earth fluorides is in carbon-arc cores. Such carbon arcs produce the intensely white light used in movie projectors and searchlights.

Class 1B applications include the use of highly pure lanthanum oxide in the making of optical glass. It imparts to the glass a high refractive index, and almost all high quality cameras contain La glass in their lenses. La glass is also used in making glass fibers for optical purposes (see Fiber optics). Another use for La_2O_3 is to improve the temperature dependence and dielectric properties of barium titanate and strontium titanate (see Ferroelectrics).

Cerium dioxide is used almost exclusively as a polishing agent for polishing lenses, mirrors, TV-tube faceplates, photomasks, etc. It is cleaner, faster, and longer lasting than the older polishing agents and gives the glass surfaces a superior finish. Cerium dioxide is also added to the glass used in beverage bottle manufacture, since it removes the color that impurities of iron impart to the bottle glass. In some of the above applications, a mixture of rare earths with removed Ce is also used where the higher valence Ce might be problematic. To a minor extent, CeO_2 is used to slow the darkening of glasses when exposed to solar, electron, x-ray, or neutron radiation. It prevents the browning effect induced by the radiation. Mixtures of rare earths with the Ce removed are used to produce a special misch metal used as an additive to some nonferrous alloys.

Didymium [8006-73-3] (Di), a mixture composed of Pr–Nd and no Ce, is sold as a metal to be added to magnesium-based alloys to improve their high temperature strength and creep resistance. Di is used in welding and glass-blowers' goggles to protect the worker's eyes from the intense sodium light. Another main use of Pr concentrate is as the coloring agent in $ZrSiO_4$-based yellow ceramic tiles. Also, Pr has a great potential as $PrCo_5$ in the manufacture of permanent magnets. Although it is as good as $SmCo_5$ for this purpose, in practice $PrCo_5$ has not yet obtained wide acceptance (see Magnetic materials).

Although several inorganic lanthanide compounds exhibit pleasing colors, Nd oxide is the most important as a glass coloring agent; Ce, Pr, Sm, Er, and Ho oxides also are used. Use of Nd_2O_3 is extensive in the production of laser materials (see Lasers).

The main use for Sm concentrate is in the manufacture of $SmCo_5$ for producing the strongest permanent magnets available. Today, all the Sm that companies can obtain by processing bastnaesite and monazite (ca 200 t/yr) is used in the manufacture of $SmCo_5$ magnets. Sm is a by-product of Eu extraction for use in color-TV tubes. It can be separated from Gd by solvent extraction. Samarium in large quantities is a by-product from the manufacture of misch metal. When mixed rare-earth chlorides are reduced electrolytically to metal, Sm^{3+} is only reduced to Sm^{2+}, and it remains in the molten electrolyte with Eu^{2+}. This electrolyte is chemically processed to give a Sm–Eu concentrate. The amount of Sm obtained by this method is comparable to that obtained from processing bastnaesite and monazite in other ways. The Sm concentrate used to produce the magnets is ca 96% pure and the presence of small amounts of other rare earths does not greatly effect the magnetic properties.

Mixed rare-earth applications have supported a substantial rare-earth industry since the turn of the century. On the other hand, the applications of highly purified individual rare earths has only been important since the late 1950s. By far, the largest use of highly purified rare earths is in the manufacture of color-TV screens. If Eu is used as the activator in a Y oxide or orthovanadate screen, an intense and truer red fluorescence can be obtained, and this and the intense blue and green phosphors make it possible to achieve bright, lifelike color patterns.

For different reasons, both Eu and Y can be readily isolated. Eu can be put in the divalent state. Y, because it is not a true lanthanide, can be shifted with different complexing agents from the heavy rare-earth region to the light rare-earth region. By use of two liquid–liquid extractions, Y can be separated from the light rare-earth elements with the heavies and then have its position in the rare-earth sequence shifted so that it can be separated from the heavy rare-earth elements. The use of rare-earth oxides to coat fluorescent lamps is minor, but its market potential is great. Such phosphors can also be used in street lights where intense white light is desired.

Rare-earth phosphors are used in x-ray intensifying screens. X rays activate the phosphors, causing them to emit visible light, which strikes and exposes a photographic film. The advantage of such screens is decreasing the x-ray exposure to the patient by a factor of 2–4. The resulting film images are sharp.

Yttrium aluminum garnet (YAG) has been used for making artificial diamonds (see Gems, synthetic). Because of its high refractive index and hardness, it is difficult to distinguish between a YAG diamond and a real diamond. Better artificial diamonds can be made from zirconium oxide (YSZ) that has had its high temperature cubic form stabilized by yttrium oxide. YSZ has an even higher refractive index and is harder than YAG.

In the electronic and magnetic areas, the most important rare earth compounds from an industrial point of view are garnet-based materials, eg, yttrium iron garnet (YIG) and gadolinium gallium garnet (GGG). The YIG-based materials are used in a polycrystalline form in a variety of microwave devices. These include attenuators, circulators, isolators, phase shifters, power limiters, and switches. They are also used in microwave integrated circuits (see Microwave technology). GGG is used in bubble devices for memory storage (see Magnetic materials). A recent growth market for Y_2O_3 involves the stabilization of high temperature cubic form of ZrO_2 (YSZ). The latter has long been used as a ceramic material for thermal insulation. Because of the oxygen deficiencies in the material, its electrical resistivity varies linearly over several orders of oxygen pressure. Recently, this property has been taken advantage of and devices have been developed for measuring the oxygen partial pressure in automotive exhaust gases. It is used for measuring and regulating the carbon potential produced in CO–O–CO_2 atmospheres maintained in carburizing furnaces and to determine the oxygen contents in molten steel. In 1980, General Motors cars alone required 5×10^6 such sensors. Yttrium oxide has been employed with nickel-based superalloys, which are used in critical parts of gas turbine engines (see High temperature alloys). The presence of the dispersed oxides improves the high temperature heat resistance of the superalloys.

In nuclear research, the rare earths are used usually in the form of oxides. An application of Gd, because of its extremely large nuclear cross section, is as an absorber of neutrons for regulating the control level and critically of nuclear reactors (see Nuclear reactors). The nuclear poisons disintegrate as the reactivity of the reactor decreases. One effect balances the other and the reactivity remains nearly constant. Sm, Eu, and Dy also have large nuclear cross sections and sometimes are used in reactor control rods. Europium is particularly effective because its two naturally occurring isotopes have high neutron capture cross sections. Europium-151 transmutes to ^{152}Eu which transmutes to ^{153}Eu which transmutes to ^{154}Eu, ^{155}Eu, and finally ^{156}Eu by successive neutron capture; the latter decays to ^{156}Gd. On the average, each natural Eu atom absorbs 4 neutrons before losing its effectiveness.

The rare earths, La and Y, are added to a number of superalloys and other specialty alloys to help getter the trace elements present and to improve the high temperature oxidation and creep resistances of the alloys. Another recent application is the use of RNi_5 compounds as hydrogen-storage materials (see Hydrogen; Hydrogen energy).

The prices of the highly purified individual rare earths vary greatly depending on how much is needed, the purity required, and the element desired. Typical bids for oxides are $10/kg for La, Ce, and Nd, ca $75/kg for Y and ca $4000/kg for Lu.

BIBLIOGRAPHY

"Rare Earth Metals" in *ECT* 1st ed., Vol. 11, pp. 503–521, by H. E. Kremers, Lindsay Chemical Company; "Rare Earth Elements" in *ECT* 2nd ed., Vol. 17, pp. 143–168, by W. L. Silvernail and N. J. Goetzinger, American Potash & Chemical Corporation.

1. C. A. Hampel and G. G. Hawley, eds., *Encyclopedia of Chemistry*, 3rd ed., Van Nostrand Reinhold Company, New York, 1973.
2. J. W. Mellor, *A Comprehensive Treatise of Inorganic and Theoretical Chemistry*, Vol. 5, 1924.
3. H. G. J. Moseley, *Phil. Mag.* **26,** 1024 (1913).
4. E. J. Wheelwright, *Promethium Technology*, American Nuclear Society, Hinsdale, Ill., 1973.
5. K. A. Gschneidner and L. Eyring, eds., *Handbook on the Physics and Chemistry of the Rare Earths; Metals*, Vol. I, *Alloys*, Vol. II, *Non-metallic Compounds*, Vols. III and IV, North Holland Publishing Co., Amsterdam, 1979, Reviews by 40 experts in the field.
6. F. H. Speeding and A. H. Daane, eds., *The Rare Earths*, John Wiley & Sons, Inc., New York, 1961. A good review to 1960.
7. R. J. Elliott, ed., *Magnetic Properties of the Rare Earth Metals*, Plenum Press, London, 1972. Nine review articles on various fields of magnetic research on rare earth materials.
8. D. M. Yost, H. Russell, and C. S. Garner, *The Rare Earth Elements and Their Compounds*, John Wiley & Sons, Inc., New York, 1947.
9. F. Trombe, ed., *Elements des terres Rares; Scandium* and *Yttrium*, Vols. 1 and 2, Masson et Cie, 1960, A review of rare earth chemistry in French to 1958.
10. A. Marinsky and Y. Marcus, eds., *Ion-Exchange and Solvent Extraction*, Marcel Dekker, New York, 1974.
11. K. A. Gschneidner, *Rare Earth Alloys*, D. Van Nostrand Co., Inc., New York, 1961.
12. K. A. Gschneidner, ed., *Industrial Applications of Rare Earth Elements*, Symposium sponsored by the Division of Industrial And Engineering Chemistry, American Chemical Society, Special Issues Sales Department, Washington, D.C., Aug. 1981.
13. T. J. Haley in ref. 5, Vol. 4, Chapt. 40.

General References

T. Moeller, *Comprehensive Inorganic Chemistry*, Vol. 4, Pergamon Press, Elmsford, N.Y.
Gmelin Handbuch der Anorganische Chemie, System No. 39, Rare Earth Metals, 1978. A German review.
E. J. McCarthy, J. J. Rhyne, and H. Silber, eds., *The Rare Earths in Modern Science and Technology: Papers Presented at the Meeting*, Plenum Publishing Company, New York, 1980. The Proceedings of the Rare Earth Research Conference are published approximately every 18 months.
A partial list of books and review articles on rare earths published in the past 10 years can be obtained from K. A. Gschneidner, Director, The Rare Earth Information Center, Ames Laboratory of the USDOE, Ames, Iowa 50011. This center also answers question on the rare earths and maintains an up-to-date file on the new developments.

F. H. SPEDDING
Ames Laboratory
Iowa State University

RASCHIG RINGS. See Absorption; Distillation.

RAYON

The history of the inception and early development of rayon is well-documented (1–2). In the late 1950s and early 1960s rayon manufacturers, realizing that their regular rayon fibers lacked resistance to alkali, ie, regular rayon fibers exhibited loss of strength and lack of dimensional stability, developed high wet-modulus-type rayons (HWM) that largely overcame these deficiencies.

The first high performance (HP) rayons were marketed in the United States by American Viscose (Avril) and American Enka (Fiber 700) and, in the UK, by Courtaulds (Vincel). Other HWM fiber producers in Europe were Snia Viscosa (Koplon) and Lenzing (Hoch Modul 333). These rayons were prepared by the addition of various amines and polyglycols to viscose and of zinc to the acidic coagulation baths, and modification of the overall spinning conditions. During this period, the Japanese at the Tachikawa Research Institute took a somewhat different approach and used cellulose of very high degrees of polymerization (DP) and high CS_2 concentrations combined with low acid-bath concentrations (without zinc) and high stretching to produce so-called polynosic fibers (3). These fibers had high wet modulus and low caustic solubility. Fabrics produced from these polynosic fibers displayed the cover and crisp hand of cotton. As originally produced, polynosics had low elongation, low loop-and-knot strength, and poor abrasion resistance. Subsequently, they were prepared by some of the HWM-type technology, and today's fibers have good overall properties.

In the 1970s, further improvements in HP rayons were introduced to make these fibers more competitive with cotton. Among these were Avtex Fibers' Avril III, a multilobal fiber, and ITT Rayonier's Prima, a HWM crimped fiber (4–6). Courtaulds' Viloft is produced by adding sodium carbonate to the viscose to give a hollow fiber of higher water-holding capacity (7). Unfortunately, Viloft fibers and many of the other modified fibers are not high wet-modulus types. Fibers of higher water-holding capacity are American Enka's Absorbit and Avtex Fibers' Maxisorb, which are produced by addition of hydrophilic colloids, eg, carboxylated cellulose, methyl cellulose, or various polyacrylic acids, to the viscose prior to spinning (8–11).

During the past decade, the quest for flame-retardant fibers resulted in considerable work on viscose additives (12). Among the fire-retardant rayons are Avtex Fibers' PFR Fiber in the United States, Lenzing's Flamgard in Europe, and Daiwabo's HFG Fiber and Kanebo's Bell Flam in Japan. At present, the commercial demand for such products appears to be small (see Flame retardants in textiles).

As this work to improve the viscose process was going on, other significant developments relative to future rayon manufacture began to emerge in the mid- and late 1970s. In 1972, S. Kwolek at DuPont patented the use of anisotropic spinning wherein properly constructed polymers having the correct aspect ratio to form liquid-crystal solutions are spun to give fibers of exceptional strength (13). This patent discloses examples of nylon and describes how to make liquid-crystal solutions of cellulose acetate as well as how to spin these to obtain acetate fibers having up to three times the strength of the usual acetate fibers. It was not until 1980 that a liquid-crystal solution of cellulose was reported (14). However, no one has reported production of a rayon-type of fiber from a liquid-crystal cellulose solution (see Liquid crystals; Aramid fibers).

During the late 1970s, a new phase of rayon research emerged. By recognizing that the viscose system was almost 90 years old and that investment and pollution problems were becoming prohibitive in cost, Enka, Rayonier, Snia Viscosa, Rhone Poulenc, Courtaulds, and others began to examine the possibility of making solvent-spun rayons. At a major American Chemical Society symposium in 1977, Rayonier described its efforts to make solvent-spun high wet-modulus-type rayon fibers from cellulose solutions in dimethylformamide–nitrogen tetroxide (DMF/N_2O_4) and from dimethyl sulfoxide (DMSO)-paraformaldehyde solutions (15). In 1979, American Enka announced its Newcell process for making reconstituted cellulose fibers. This process is disclosed in patents that deal with solutions of cellulose in N-methylmorpholine N-oxide with the use of cellulose concentrations of ca 17–20 wt % (16). This process appears to have good potential for economical manufacture of reconstituted cellulose fibers of quality by a nonpolluting closed-loop method. In 1981, patents issued to Rayonier describe preparation of up to 16% cellulose solutions in dimethylacetamide (DMAc) or N-methylpyrrolidinone containing dissolved lithium chloride (17). These novel cellulose solutions were spun to give HWM-type reconstituted cellulose fibers. The Newcell and the LiCl–DMAc systems dissolve cellulose directly without forming derivatives; therefore, to call fibers from these systems rayons may be improper because all rayons have been produced by regenerating cellulose from a chemical derivative.

By far the main production of rayon today is by the viscose process; therefore, a large part of this review is dedicated to viscose technology. However, significant effort is being invested in search of other routes to make rayons more efficiently. These involve reexamination of the cuprammonium process and new solvent systems for cellulose.

Viscose Rayon

Most of the world's rayon is made by the viscose process. The term rayon designates a range of products that have widely varying properties, for the most part owing to the versatility of the viscose process. This versatility, which is the result of the many process steps and spinning changes that can be made, is a mixed blessing, since each stage of processing and spinning requires close attention to guarantee the desired product properties. The viscose process is a demanding process that requires continuous, year-long operation to prevent gelling of the system and to yield high quality products.

Process. The stages in the preparation of a satisfactory cellulose xanthate solution are given in Figure 1. A few of the critical aspects of each stage are reviewed for both batch and continuous processes.

Steeping. In order to have cellulose react with carbon disulfide to form the corresponding cellulose xanthate, cellulose must be converted to alkali cellulose (see Cellulose). Normally, this is done by placing cellulose-pulp sheets in a steeping press and filling the press with a closely controlled (±0.2 wt %) concentration of sodium hydroxide at a desired level of 18–20 wt %, depending on the type of cellulose used. With wood pulp, ca 18 wt % NaOH is used; cotton linters usually require higher NaOH concentrations of about 19.5–20% because linters are more difficult to penetrate than wood pulp. The rate of filling of the press is critical for obtaining an adequate alkali cellulose; wood pulp and cotton linters require different filling rates. The rate of filling

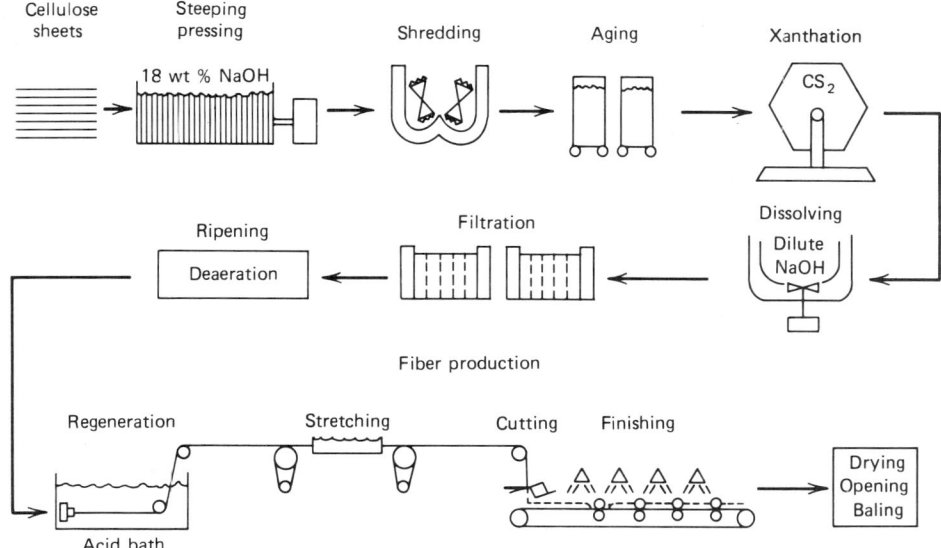

Figure 1. Viscose process.

should correspond as nearly as possible with the rate of sheet wetting by the NaOH. If too slow a fill rate is used, a wicking action is observed where NaOH of a more dilute concentration wicks and wets the sheet ahead of the rising caustic liquid level. Such dilute wicking caustic does not open the cellulose crystallites adequately and results in incomplete conversion of cellulose I structure to the desired cellulose II found in alkali cellulose. Conversely, if the level of concentrated NaOH solution rises too rapidly, air bubbles are trapped in unwetted regions, which causes untreated cellulose fibers to remain in the unwetted regions; these subsequently create severe filtration problems. Although most steeps are carried out at 25°C, careful experiments have shown that the sheet temperature at the advancing caustic level is actually about 28–32°C because of exothermic reactions. The criticality of a good steep to viscose dissolution cannot be overemphasized. If the initial exposure of cellulose to caustic solution in the first 5 min is not sufficient, the damage is done; soaking for hours will not improve the situation.

The swelling of cellulose by water decreases with increasing temperature; this inverse swelling behavior is more pronounced when NaOH is present. Maximum swelling occurs with cold caustic solution (18); under proper conditions, some of the low molecular weight cellulose may dissolve along with the hemicelluloses that normally are removed. The hemicellulose concentration in the steep caustic must be controlled since it affects ultimate fiber properties. For regular rayon, this concentration is 2%, whereas for HWM fibers it is ca 0.7%, and for tire-cord fibers ca 0.5%. Therefore, pure dissolving pulps must be used or the alkali must be dialyzed if low hemicellulose concentrations are desired. High hemicellulose concentrations lead to reduced physical properties and increased brittleness in the final product. The effect of temperature on swelling is further illustrated in continuous slurry-steeping operations where higher caustic concentrations and higher temperatures (≤45°C) are used to reduce swelling of dispersed 5% cellulose slurry to ensure that the resulting alkali cellulose has a reasonable pressed-weight ratio.

After steeping, the cellulose is pressed under high pressure to a pressed-weight ratio of 2.6–3.0 to give an alkali cellulose with ca 34 wt % cellulose, 15.3 wt % caustic, and the remainder as water. As much excess caustic as possible is removed, because any excess will react with CS_2 in later steps to give undesirable by-products. Many steeping presses are arranged so that the caustic squeezing during the last phases of pressing are shunted to a separate receiver, because these final squeezings are always high in hemicellulose concentrations. This fact was recognized by Sihtola, who proposed the use of double steeping (19). In the Sini process, about 18% caustic is used to make the alkali cellulose, which is treated with more dilute caustic of ca 12%. This decreases the overall excess alkali and reduces subsequent by-product formation while also removing extra hemicellulose; the latter occurs because the more dilute caustic swells the cellulose to a greater extent, thereby facilitating removal of more hemicellulose.

Shredding. In addition to converting the alkali cellulose to a crumblike material, shredding serves two additional functions: first, the squeezing action of the shredder blades distributes the caustic more uniformly in the cellulose. Most alkali cellulose sheets are wetter at the edges than in the centers and, when the caustic solution drains, the bottoms of the sheets have more alkali. Thus, shredding gives a more uniform distribution of caustic in the alkali-cellulose crumb. Second, the shredder heats the crumb to a temperature favorable for aging, usually about 30–32°C. Typical shredders turn at about 30–60 rpm and have a blade clearance of about 0.08 cm. Improper blade clearance leads to the formation of undesirable ball-like aggregates. Too much shredding can give a low alkali-cellulose crumb density; the resulting crumb takes up too much space in the churn during xanthation.

Aging. Aging is used to decrease and control the cellulose DP, ie, the alkali cellulose stands in covered containers in a temperature-controlled room. The alkali crumb should never become dry. In the continuous process, the alkali cellulose may be dried while conveyed in an air stream; therefore, humidified air must be recirculated to the conveying section. In the batch process, drying can occur during aging if the cans are not covered properly. Formation of excess sodium carbonate at this stage is equally undesirable, since a 1% increase in carbonate concentration reduces filterability by 10%. Normally, sufficient oxygen is present in the alkali cellulose crumb, ie, added air exposure is unnecessary. Although the main DP decrease occurs during aging, it begins immediately upon exposure to caustic in steeping and continues during shredding. DP loss occurs during xanthation, where typically 75–125 additional monomer units are lost.

If the aging conditions are altered, the effect on final pulp DP can be predicted by use of two equations. To estimate the effect of aging time on DP at constant temperature, the rate constant k is obtained from the DP in the operation

$$k = \frac{\left(\dfrac{10^4}{DP_t} - \dfrac{10^4}{DP_o}\right)}{t}$$

where k = reaction-rate constant; t = aging time in hours; DP_t = DP at end of aging; and DP_o = DP at start of aging. If this value of k is used, DP at a certain time can be calculated.

To estimate the change of aging rate with temperature, a linear approximation is sufficient because the range of temperatures is small; thus

$$\log k_2 = \log k_1 + 0.05 \, \Delta T(°C)$$

If these two equations are used judiciously, plant production cycles may be altered with minimum change in the product.

Xanthation. The four main types of xanthation units are the dry churn, the wet churn, the continuous belt xanthator (CBX), and a reciprocal screw of the Ko-Kneader-type called the Maurer unit. The latter two are essentially continuous xanthation units. Most viscose is made by the dry-churn method operating with addition of CS_2 at atmospheric pressure or at reduced pressure into evacuated churns containing alkali-cellulose crumb. The churn is rotated very slowly at ca 2.5 rev/min; thus, the crumb undergoes tumbling action and centrifuging action is avoided. A churn should be loaded to not more than 0.15 g/cm³ density with alkali crumb. Usually, about 32 wt % CS_2 is used, and the temperature is kept at ca 32–33°C, ie, below the CS_2 bp of 46°C. The exothermic reaction is usually completed in ca 75–90 min. Complete reaction of CS_2 results in a partial vacuum. A final high vacuum is applied to the churn to remove traces of CS_2 before the churn is opened and the xanthate crumb is dumped. If the churn was dry when the alkali cellulose was added, the xanthate emerges as a free-flowing crumb. If water condensed inside the churn before the alkali-cellulose crumb was added, large balls will be present. In commercial practice, an economic balance is reached between operational conditions and by-product formation.

Xanthation is a critical part of the viscose process; many reactions proceed simultaneously during this step. Pure cellulose xanthate is white; any yellow color is owing to by-product formation, usually in the form of trithiocarbonate. The lower the xanthation temperature and the better the alkali-cellulose preparation, the lighter the color of the xanthate crumb and the less the hydrogen sulfide (H_2S) evolution during spinning, because H_2S comes only from the yellow trithiocarbonate (see eqs. 1–5).

$$CS_2 + H_2O \longrightarrow HSC(=S)OH \longrightarrow H_2S + COS \quad (1)$$

$$COS + H_2O \longrightarrow HOC(=O)SH \longrightarrow H_2S + CO_2 \quad (2)$$

$$CO_2 + 2\ NaOH \longrightarrow Na_2CO_3 + H_2O \quad (3)$$

$$H_2S + CS_2 \longrightarrow HSC(=S)SH \xrightarrow{2\ NaOH} NaSC(=S)SNa + 2\ H_2O \quad (4)$$

$$NaSC(=S)SNa + H_2SO_4 \longrightarrow Na_2SO_4 + CS_2 + H_2S \quad (5)$$

In equations 1–4, sodium compounds such as Na_2S and NaOH or their respective ions could be used in place of H_2S and H_2O, but the reactions are depicted more simply as shown.

Analysis of xanthate crumb corresponds closely to the theoretical ratios of expected carbonate to trithiocarbonate. There is evidence to suggest that COS reaction with cellulose is much faster than CS_2 and may precede CS_2 addition to cellulose.

Typically, about 25 wt % of the CS_2 is consumed in forming by-product trithiocarbonate through the side reaction, which leaves 75 wt % to react with the cellulose to give

$$\text{Cell}-\underset{\underset{S}{\|}}{O}CS^-\ Na^+$$

However, these ratios can be changed significantly if improper steeping produces a poor alkali crumb or if xanthation temperatures are not controlled properly, in which cases far more by-product formation occurs.

The amount of cellulose xanthate actually formed often is reported as the extent of reaction, called the gamma (γ) number for xanthate formation. Thus, a γ number of 1.0 means that each anhydroglucose residue on the average has a 1.0 xanthate group. Correspondingly, a 0.5 γ indicates that every other anhydroglucose unit has a xanthate group attached to it. If 32 wt % CS_2 based on cellulose for xanthation is used and 75 wt % of this CS_2 gives cellulose xanthate, the xanthate crumb has a γ of ca 0.5. Xanthate sulfur was determined by iodine titration in the past (20); more rapid methods are available (21–22).

Cellulose has three hydroxyl groups: the C-2 and C-3 positions, which are secondary alcohols, and a C-6 position, which is a primary alcohol. During xanthation each of these hydroxyl groups reacts at a different rate mainly because of the nature of the primary or secondary alcohols, ie, its steric availability and the effect of adjacent hydroxyl groups. In freshly prepared xanthate crumb, more than half of the xanthate groups are on the 2 and 3 positions ($\gamma_{2,3}$) as compared to the 6 position (γ_6); the 2 and 3 positions are favored kinetically during xanthation even though reaction at the 6 position gives a thermodynamically more stable product. This is discussed in more detail under ripening.

Dissolving. Xanthate crumb is dumped into large, stirred tanks containing dilute caustic solution to dissolve the cellulose xanthate into a clear, honeylike, viscous dope known as viscose. The caustic and water in the dissolver are measured to give the desired cellulose:alkali ratios. Many dissolvers have standpipes that force the mixtures upwards in a cascading fashion while the moving blades propel the mixtures in a circular manner.

The most important factor in the dissolving cycle is the use of as cold a dilute caustic as possible. The initial dissolver temperature should be as low as possible, ie, preferably ca 5°C and definitely below 10°C. When the xanthate crumb is added, the temperature should not be permitted to rise above 10–12°C; it should be held <15–18°C during the 2–3 h required for dissolution to occur. Only during the last part of the dissolving cycle should the temperature be allowed to rise above 20°C to the temperature for ripening. Cellulose differs from almost all other polymers in that it and its derivatives dissolve more easily cold than hot. Faster stirring introduces more mechanical energy per unit time; this is converted to heat and warms the solution. Although fast stirring is not desirable, high shear does speed up dissolving. No one has successfully beaten cellulose into solution; it must be coaxed into solution with high shearing action at low temperatures. Any other approach produces too many undissolved fibers and gels that result in very poor filtrations.

Ripening. When the cellulose xanthate is first dissolved, it is not ready for spinning because it will not coagulate readily. The xanthate groups attached to the cellulose molecule are distributed over the three hydroxyl positions of the anhydroglucose monomer units; more xanthate groups are on the kinetically favored 2 and 3 positions ($\gamma_{2,3}$) than on the thermodynamically favored 6 position (γ_6). In essence, the xanthate groups in the 2 and 3 positions act as wedges to keep the cellulose chains from approaching one another; these must be removed or relocated to the 6 positions before closer chain packing can be achieved. This redistribution of xanthate groups is the main function of the ripening stage.

Under the alkaline conditions in the dissolved viscose, the 2- and 3-xanthate groups hydrolyze 15–20 times faster than the C-6 xanthate groups. The released xanthate groups can transxanthate with C-6 hydroxyl groups or react with the ever-present excess caustic to form the inorganic by-product sodium trithiocarbonate. The higher the ripening temperature, the more trithiocarbonate is formed. Therefore, ripening usually is done at rather low temperatures for long times and in vacuum to remove dissolved air.

During ripening, the xanthate concentration drops significantly. For example, a fresh viscose may have a 2.2% total sulfur and a concentration of about 1.6% xanthate sulfur (XS), which drops to about 0.9–1.1% XS after ripening. A significant fraction of the xanthate sulfur goes to by-product sulfur, mostly at the expense of the 2- and 3-xanthate groups. Several methods for measuring the readiness for spinning of ripened viscose have been developed. Since xanthate groups solubilize the cellulose, the number of such residual xanthate groups will have the greatest influence on viscose coagulability; a direct relationship between XS and ripening level exists. In the past, xanthate sulfur was determined by a tedious iodine titration.

Xanthate concentration alone does not determine coagulation rate; cellulose concentration, DP, caustic level, etc, influence coagulation rate. Therefore, the rayon industry adopted salt-index tests to measure coagulability. Two methods are used; for the sodium chloride index, saline solutions of varying concentrations are used to determine incipient coagulation when a single drop of viscose is added, whereas in the ammonium chloride or Hottenroth test, the viscose is diluted in a standard manner and titrated with 10% NH_4Cl to incipient precipitation.

A drop of fresh viscose may require about an 8% solution of NaCl to cause coagulation, whereas a 4% NaCl solution will coagulate the ripened viscose after 40 h, and 2% NaCl will work after 70 h. Similarly, it may be necessary to use 30 cm^3 of 10% NH_4Cl for fresh viscose and 10 cm^3 and 6 cm^3 after 40 h and 70 h, respectively. Each viscose plant uses its own standard methods, and great care must be taken when trying to compare salt indexes among various locations. Some use an acetic acid index. An approximate correlation of the three methods for a single viscose composition is given in Figure 2.

Spinning. Once the viscose is ripened to the proper xanthate level and the 2- and 3-xanthate wedges are removed so that proper rates of coagulation can be obtained, the viscose is ready for spinning. Chief goals in spinning are control of coagulation versus regeneration rates and the maximal use of the differences in these rates to obtain maximum responses to stretching. Stretching a soft gel does not give permanent alignment of molecules; when the gel hardens too much, further stretching is virtually impossible. In the spinning process, viscose is extruded into a bath containing both salt and acid. The salt usually is sodium sulfate at sufficiently high concentrations

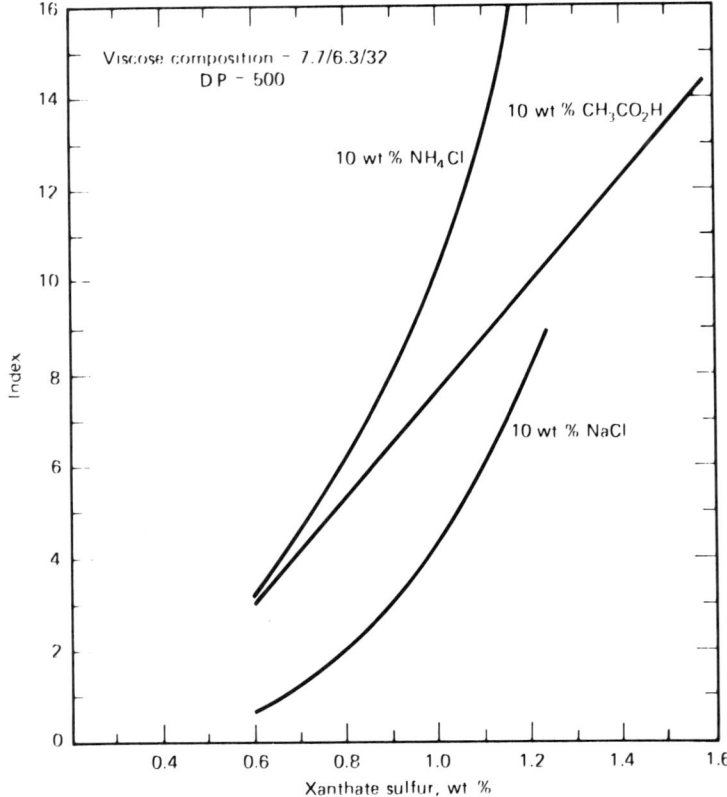

Figure 2. Maturity index comparison curves.

to ensure that the sulfate solution behaves as a dehydrating system that removes water from the entering 7–10%-cellulose xanthate solution. Sodium sulfate has an RT solubility of 280 g/L; normally, it is used at 120–200 g/L in the coagulation baths. If greater dehydration action is desired, ammonium sulfate is used as the salt in the coagulation bath; its solubility is 440 g/L at RT, and it can be used at 230–350 g/L as a coagulant. Sulfuric acid is present to regenerate cellulose from the coagulated viscose. The concentration of sulfuric acid relative to the concentration of salt is critical in controlling the extent of water removal from the extruded viscose (by the salt) before the system is rigidly fixed by the acid removing the solubilizing xanthate groups. The control of densification of the extruded viscose by removing water with salt and stretching the densified mass versus the rate of acid hardening is the gist of the spinning process. The denser the viscose, the higher the resulting wet modulus. A high wet-modulus fiber with minimal stretch has been made in this manner. The higher the density, the more effective the stretching in aligning cellulose molecules; the final, dry and wet tenacities of the product will also be higher. In practice, the interplay of simultaneous reactions and other processes leads to significant interactions among the various factors governing spinning.

A comparison of the stress-strain performance of rayon and other fibers is given in Figure 3; the cross-sectional shapes of several typical rayons are given in Figure 4.

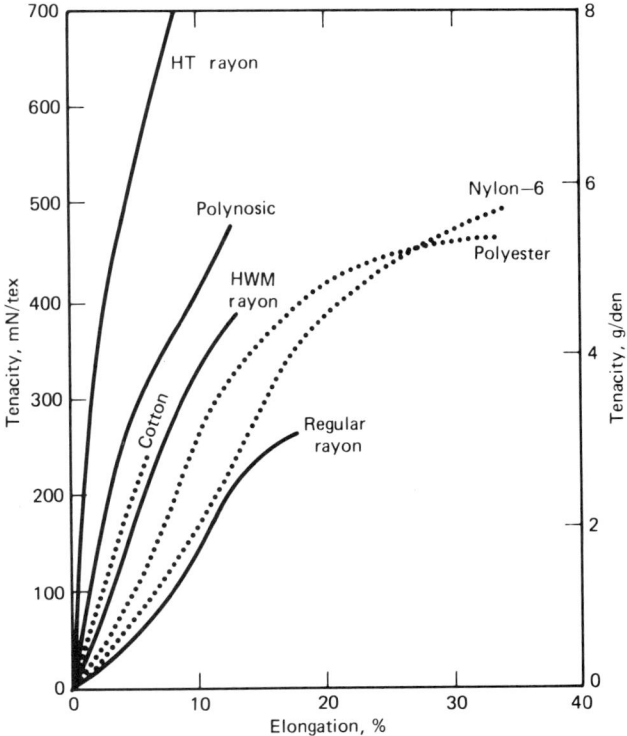

Figure 3. Physical properties of rayons vs other fibers (conditioned).

Spinning Additives. Rayon manufacturers became acutely aware of the need for controlling the complex interactions of coagulation, regeneration, and stretching when they first tried to prepare HWM fibers to improve rayon's competitive position relative to cotton. The simple approaches of using only Na_2SO_4 and H_2SO_4 were altered by the use of additives in both the viscose dope and the spin bath. The viscose ripeness level is an important factor in any study of additives. Several methods were developed to determine the rates of xanthate decomposition in the presence of various additives, eg, adding indicator-type dyes to the viscose and noting the distance from the spinneret where color changes occur; following the loss of the xanthate absorption peak at 303 nm when a cast viscose film is placed in a spin bath in a spectrophotometer cell; and dipping pH electrodes first in viscose and then in spin-bath liquor and measuring the pH changes with time. Hundreds of combinations of additives to the viscose and the spin bath were evaluated with these techniques. Controversy as to why certain combinations are effective exists; two main schools of thought seem to have developed—one believes that the modifying action takes place over the whole cross-section of the fiber, and the other feels that a surface layer of some type forms and controls further rates of transport into the fiber (23–24). Theories and mechanisms are in doubt, but it is certain that fiber properties are controlled by the use of additives. The best systems make use of some type of amine, eg, dimethylamine, and a polyetherglycol of ca 1500 mol wt added to the viscose as well as zinc in the coagulation bath (25). Proper combinations of these ingredients can increase regeneration times as much as 400–500% over unmodified systems.

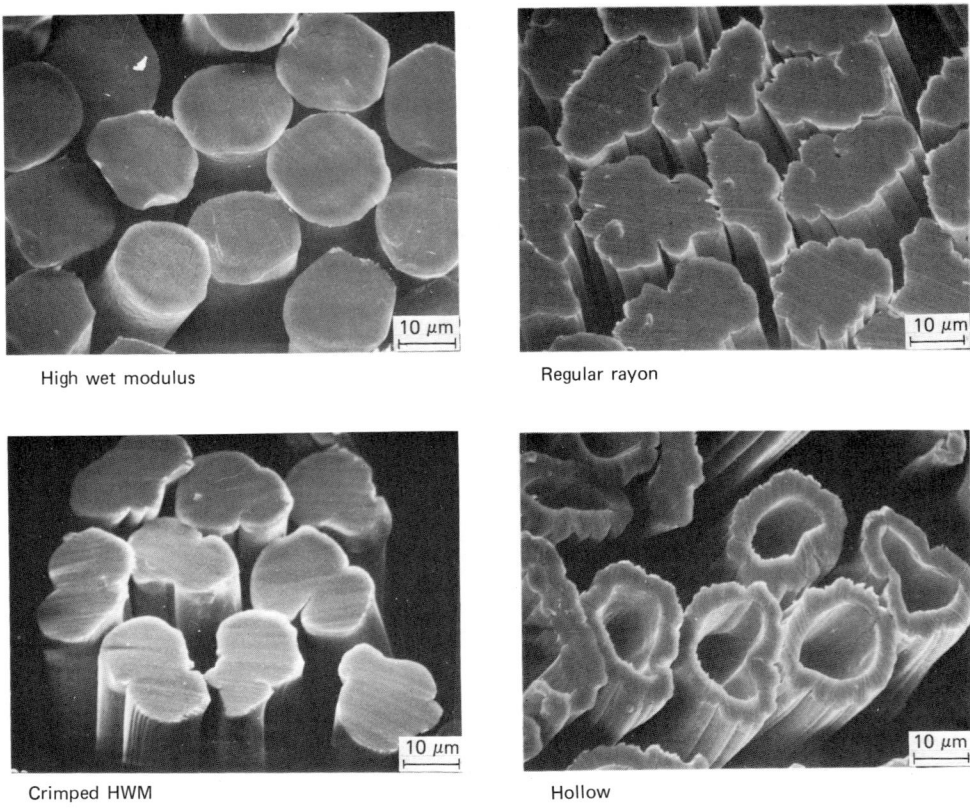
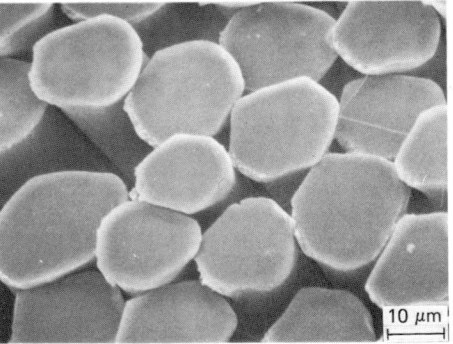

Figure 4. Cross-sections of typical rayons.

Stretching. An advantage of retarding regeneration is to make use of the densification of the extruded dope so as to obtain better molecular alignment on stretching. A fixed amount of stretching can occur; this is balanced between what is called jet stretch and godet stretch. Jet stretch is the ratio of the linear speed at the first godet roll to the average linear flow rate in the spinneret; it is expressed as follows:

$$\text{Jet stretch} = \text{linear speed of first godet} \cdot \left[\frac{\pi/4 \ (\text{hole dia})^2 \ (\text{no. of holes})}{\text{flow rate of dope}} \right]$$

Godet stretch is the stretching between the first and last godet rolls. The higher the godet stretch, the less residual elongation is left in the final fiber. Therefore, the manufacturer balances fiber strength and residual elongation.

Spinneret Design. There probably are as many individual spinneret designs as there are fiber producers; yet in all cases, certain basic design parameters are dictated by the diffusion-controlled nature of the wet-spinning process. Each spinneret hole has a defined length to diameter (L/D) ratio and a defined conical entrance angle (α) (see Fig. 5).

For the most part, spinnerets have been made from a gold–platinum alloy to provide long-term resistance to corrosion. However, less expensive materials are now under investigation.

Some rayon producers favor one large spinneret (ca 7.5-cm dia) with various radial or circular arrangements of holes. In this design, several thousand holes can be drilled, but the regenerant reaches the center holes with difficulty. Another design uses small units with ca 2000 holes in a 1.9-cm-dia spinneret. A number of these small units are arranged in a circular cluster of ca 10-cm dia. In this fashion, 30,000 or more filaments can be spun from a single cluster head; the spaces among the individual spinnerets allow more complete acid penetration to the center filaments.

Finishing. After the fibers are spun, cut, and washed, finishes are applied to give the proper frictional performance in subsequent textile processing. Slipperiness presents problems in spinning where physical properties of yarn are closely related to surface characteristics and friction, whereas stickiness results in improper fiber movement during processes such as carding and can lead to nep formation and fiber damage. Over the years, a wide variety of finishes have been proposed; one of the original and still one of the best is prepared from oleic and sulfonated oleic acids (see Textiles, finishes).

Modified Rayons. Control of the regeneration conditions coupled with specific additives gives fiber manufacturers the opportunity to produce a host of modified fibers (see Fibers, chemical). An excellent review is given in ref. 26.

High Wet Modulus. The production of HWM fibers by controlling regeneration kinetics through the use of additives was explained under the section on spinning. ISO permits any fiber with a wet strength of at least 220 mN/tex (2.5 g/den) and an elongation of less than 15% at this load to be called a Modal fiber. A wet modulus (tenacity at 5% stretch) of at least 45 mN/tex (0.5 g/den) is required for good performance.

Crimped Fibers. In the early 1970s, Rayonier patented a chemically crimped rayon trademarked Prima (5–6). The crimp is produced by selective control of the conditions in the primary and secondary regeneration baths wherein a gel-core fiber is allowed to leave the primary bath and is exposed to hot-acid conditions in a secondary bath. Under these conditions, significant tensions are established throughout the fiber, which causes permanent chemical crimping. These fibers also have high wet modulus and

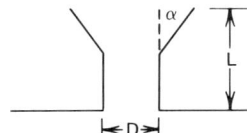

Figure 5. Cross-section of a spinneret hole.

give cover, hand, bulk, and working performance in finished fabrics superior to that of ordinary rayon. When produced in higher deniers, Prima blends with polyester closely resemble wool fabrics.

Hollow Fibers. Courtaulds has developed a hollow rayon fiber trademarked Viloft, (27). This fiber is much lighter than regular rayon and therefore gives significantly more coverage per unit weight. Moisture-absorbing capacity is about 50% greater than ordinary rayons; this makes Viloft a strong candidate for use in towels, sheets, pillow cases, and garments requiring improved absorbency. Although Viloft is not a high wet-modulus fiber, it reportedly behaves well if used up to 20 wt % in polyester or cotton blends. It is produced by adding sodium carbonate to the viscose and then controlling the regeneration conditions to capture the released CO_2 gases to give the hollow fiber. By increasing the carbonate content further, it is possible to produce a thinner-walled version of the hollow paper-making fiber, ie, PM fiber, which collapses and convolutes on drying. With further addition of carbonate, the fiber wall ruptures during regeneration and collapses from a superinflated state, ie, an SI fiber, to give a product resembling cotton in many respects (28) (see Hollow-fiber membranes).

Flame-Resistant Fibers. A large body of literature exists on the various efforts to make an acceptable flame-resistant rayon. Some commercial successes have been reported, eg, Avtex's PFR fiber (29). Most such fibers are based on the addition of complex phosphorus-containing compounds to viscose. For example, various products based on phosphonitrilic acid and complex phosphorus esters are commercially available at this time (see Flame retardants). Although these materials give definite improvements in flame resistance, much more work will be required before wholly acceptable products are available. A good review on the present state of the art is available (12).

Test Methods. A survey of tests is given in Table 1. In addition, ASTM and TAPPI standards should be consulted as well as the pulp suppliers.

Solvent-Spun Rayon

The viscose process is now 90 years old; current plant investment, labor, and operating and pollution costs make the construction of any new viscose plants unlikely in the United States. The present plants compete only because they have been written off for depreciation and new capital is not factored into the current selling prices. In a recent article in *Chemical Week*, an author summarizes the situation quite accurately by realizing the virtues of rayon as a fiber and simultaneously describing the many needs for modifying the overall viscose process (30). In spite of such candid realizations, some new HWM rayon capacity is being developed in Finland, Austria, and the USSR, although plant expansion may not be involved.

Several companies are exploring the possibility of using organic solvents with closed recycle–recovery loops to produce rayon-type fibers through solvent-spun processes because of the continued projected need for a fiber such as rayon. The first symposium on this subject was sponsored in 1977 by the American Chemical Society (15). In a subsequent publication, the past 130 years of cellulose-solvent technology was summarized, in which all cellulose solvents were categorized into four general classifications, and the considerations that would be required for a cellulose solvent to serve as a potential system for rayon manufacture were outlined (31). An expanded, companion review of the cellulose solvents and solubility is ref. 32. The American Enka

Table 1. Test Methods for Viscose Rayon

Test	Parameter measured	Significance
S_{18}[a]	hemicellulose content of starting pulp	high hemicellulose concentrations give brittle fibers
S_{10}[b]	hemicellulose plus low DP-ends of cellulose	low DP cellulose adversely affects fiber strength
DP[c]	av chain length of cellulose molecules	process control check and DP greatly influence fiber physical properties
plugging index	gels, fibers, etc, remaining undissolved	undissolved materials impair fiber quality; defines control efficiency
xanthate sulfur	xanthate groups on cellulose	xanthate concentration controls solubility and coagulability
total sulfur	organic plus inorganic sulfur	test for amount of CS_2 added and for inorganic by-product
salt index	coagulability by specified concentration of NaCl, NH_4Cl, or acetic acid	measures readiness for spinning
particle counts[d]	particle-size distribution in viscose	too many particles above 15 µm impair fiber properties and suggest problems in viscose preparation
neutralization point[e]	distance from spinneret at which viscose is neutralized by coagulating acid	spinning control; additives effectiveness
gel swell	water in fibers before drying; gel nature of fibers	effectiveness of regeneration coagulation stretch ratios; relates to ultimate fiber modulus
rewet swell	water retention or imbibition of dried and rewet fibers	relates to product field performance; also helps classify fiber type; HWM retains less water than regular rayon
$S_{6.5}$[f]	concentration of nonoriented cellulose left in product	predicts how much product will be lost on repeated laundering

[a] Soluble in 18% caustic.
[b] Soluble in 10% caustic.
[c] By cuen (copper ethylene diamine) viscosity.
[d] By Coulter, HIAC, or laser scanners.
[e] Diffusion value; film uv; electrode diffusion.
[f] Soluble in 6.5% caustic.

Company has issued reports about the Newcell process, which appears to be based on spinning solutions of up to 20 wt % cellulose dissolved in hot aqueous N-methylmorpholine N-oxide (16,33). This system appears to be amenable to closed-loop recovery–recycle and relatively rapid spinning speeds. Although most previously investigated solvent-spun-rayon systems were based upon the use of unstable, easily regenerated cellulose derivatives, the Newcell process involves no discernable derivative formation but rather uses what appears to be a solution of cellulose; this leads to important economic advantages for the system.

Although the solubility parameter (δ) of the solvent must be nearly the same as that of cellulose, many liquids with the proper solubility-parameter values do not dissolve cellulose. For example, DMSO and DMF each have solubility parameters in the cellulose range, but neither dissolves cellulose. A solubility parameter is the square root of the vaporization energy per molar volume. It can be used for solutions of nonpolar liquids and amorphous polymers in nonpolar liquids to estimate or explain heats of mixing. For polar, crystalline polymers, other factors become significant, eg, with crystalline polymers, the heat of fusion or melting energy must be provided before solubility-parameter considerations can apply. In natural polymers such as cellulose and proteins, hydrogen bonding may be sufficient to prevent melting phenomena from occurring. These factors can be summarized in a thermodynamic approach that considers crystalline forces, derivative formation, and ultimate mixing. The change in free energy in dissolving the polymer may be written as

$$\Delta G_{solution} = \Delta G_{fusion} + \Delta G_{derivatization} + \Delta G_{mixing}$$

For dissolution to occur, the sum of all the changes in free energies for processes on the right side of the equation must be negative. Also, $\Delta G_{derivatization}$ will be negative if spontaneous derivative formation occurs. The reacted or derivatized material would be a new dissolved species and, to the extent that it formed, would alter the corresponding free energies of fusion and mixing. The change in free energy of mixing is

$$\Delta G_{mixing} = \Delta H_{mix} - T\Delta S_{mix}$$

The entropy of mixing ΔS_{mix} usually is positive; therefore, ΔG_{mix} is negative if the heat in mixing ΔH_{mix} has a sufficiently small positive value. For mixing nonpolar liquids in which the interactions among unlike molecules are geometric means of interactions among like molecules, $\Delta H_{mix} = K(\delta_1 - \delta_2)^2$ where (δ_1) and (δ_2) are the solubility parameters of the solvent and solute, respectively. Therefore, if the solubility parameters for both materials are as nearly equal as possible, the heat of mixing will be a small positive number which will be less than the negative $(-T\Delta S_{mix})$ term to give a negative ΔG_{mix}. For cellulose, the additional term ΔG_{fusion}, which is the free energy of melting the polymer, must be considered. In nonpolar polymers, the crystal forces can be overcome by heating, and the resultant melts dissolve readily in solvents having the proper boiling points and solubility parameters, ie, polypropylene dissolves readily in hot decalin or tetralin. In cellulose and proteins, the energies of intermolecular bonding that result from dispersion forces, hydrogen bonding, and other dipole–dipole interactions are too great for melting to occur at temperatures below those of decomposition; therefore, these polymers usually char or decompose rather than melt. This positive ΔG_{fusion} is significant and some type of physical work input or chemical change is required to overcome the effects of the relatively large bonding forces. Alternatives are to make $\Delta G_{derivatization}$ and ΔG_{mixing} sufficiently negative to overcome this positive ΔG_{fusion}.

A new solvent system for cellulose has been developed that can give up to 16% solutions of 500 DP cellulose. This system, which does not cause cellulose degradation even on extended heating or air exposure, utilizes lithium chloride in dimethylacetamide or in N-methylpyrrolidinone to form solutions of cellulose that are stable in extended storage (17). This system permits total recycle and recovery of all components to give a nonpolluting process. The economics of the process appear to be competitive with a new viscose plant installation.

Cuprammonium Rayon

Essentially all of the rayon made today is produced by the viscose process; nevertheless, there are some uses where the cuprammonium process has retained an advantage because of specific performance factors. Fibers from cuprammonium rayon are significantly more supple than viscose fibers and are used where a very soft hand is desired. The use of films and hollow fibers from the cuprammonium process for making artificial kidneys is critically important in this branch of medicine (see Dialysis; Prosthetic and biomedical devices). At present, almost all artificial kidney units use membranes prepared from such films and fibers. These films and fibers exhibit superior clearance performance for urea, creatinine, and metabolites, possess better dewatering characteristics, and cause less blood clotting as compared to any synthetics or corresponding products from the viscose process.

An analysis of the cuprammonium process suggests that it might be developed into a closed-loop system with the capability of complete recovery and recycle. Some workers in the USSR believe that the cuprammonium process should be reevaluated as a nonpolluting process to meet present-day ecological requirements. The following brief description is based on an updated review of the cuprammonium process practiced at the former I.G. Farbenindustrie Dormagen plant (34).

Dissolving. The ability of selected cupric ion, ammonia and alkali mixtures to dissolve cellulose was originally reported in 1847. Such solutions, known as Schweitzer's reagent, are used not only for making cuprammonium rayon, but also for measuring dilute-solution viscosities that can be related to degrees of polymerization of cellulose.

In the early 1900s, reports of how and why cellulose should dissolve in mixtures of cellulose and ammonia were published. In modern terms, the solution is described by using Werner-complex theory, in which four of the six d^3sp^2 orbitals of cupric ion are filled by NH_3 molecules as ligands, which leaves two orbitals available for ligand donation by 2- and 3-hydroxyl groups of cellulose. This is illustrated in Figure 6. Usually, some sodium hydroxide is added to help in the interaction of the $Cu(NH_3)_4^{2+}$ ions with cellulose. For some uses of cuprammonium rayon, the copper is precipitated as copper hydroxide, and the supernatant liquid containing sulfate ions (which are considered deleterious to good dissolution of cellulose) is removed, and the cupric hydroxide is added to the cellulose and ammonia and sodium hydroxide solution.

The foregoing brief, theoretical discussion establishes definite minimum ratios for concentrations of reagents that must be used in dissolving the cellulose. The minimum molar ratio that can be employed is 1 cellulose:1 copper:4 NH_3. In actual practice, excess NH_3 approaching 7 moles is used and NaOH is added to aid swelling and dissolution. The mole and weight ratios used by the Dormagen plant are shown in Table 2. I.G. Farbenindustrie used 0.53 pt of Na_2SO_3 in order to inhibit oxidative degradation.

Several methods have been tried to improve dissolution. These include mechanical

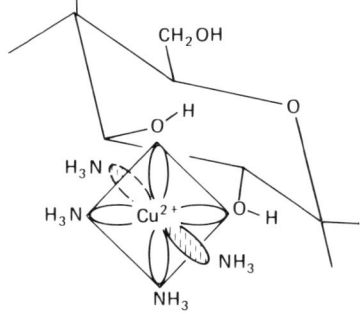

Figure 6. Orbital interactions in cuprammonium cellulose.

Table 2. Dormagen Plant Process Resulting in a 10% Cellulose Solution

Constituent	Mol ratio	Unit wt
cellulose	1.00	100
NH_3	7.52	79.1
copper	1.00	39.9
NaOH	0.655	16.2
Na_2SO_3[a]		0.53
water		764.27

[a] To inhibit oxidative degradation.

beating, ultrasonic dispersion, and use of air-dried slush pulp which, according to some investigators, dissolves faster than ordinary pulp.

The Dormagen plant used wood pulp containing only 89 wt % alpha cellulose for extensive year-by-year production of 40 metric tons per day of rayon staple and 13 t/d of filament. They employed both hot (>50°C) and cold (ca 20°C) purification processes. Pulps containing >99% alpha cellulose are available; therefore, wood pulp can be used in long-term commercial production through the cuprammonium process (see Pulp).

Cuprammonium rayon and the process compare favorably with viscose rayon and the viscose process. First, the DP of cuprammonium cellulose is 500–550 as compared to ca 350–400 for viscose rayon; second, the solutions are normally spun at 10% or more cellulose, which helps overall economics; third, the dissolution process for cuprammonium rayon is simpler than that for viscose; and fourth, the solutions are quite stable, a distinct advantage for the cuprammonium process.

Spinning. The spinning of cuprammonium yarns depends on the use of a funnel into which the coagulating–regenerating fluid flows and into which the dope is extruded.

As the flowing liquid travels down the funnel, its velocity increases and the entrained gelatinous cellulose fibers are stretched up to 400%. Subsequently, the fibers are washed as free as possible of occluded material, which is recycled, and the fibers enter a 5% H_2SO_4 bath where final removal of copper is achieved and where any remaining alkali or ammonia is converted to the corresponding sulfate.

The original stretch-spinning of cuprammonium rayon involved the use of reels. Filament on these reels had to be purified and rewound, which required considerable manpower and made the cuprammonium process noncompetitive relative to viscose. Since the mid 1940s, the cuprammonium process has been converted to a continuous process where yarn from 400–600 spinnerets is processed in line and wound directly onto a section beam. These section beams can be used directly for knitting, or several section beams can be combined to a larger warp beam for weaving. An excellent description of the continuous process is available in a patent issued to Beaunit Mills (35). A schematic presentation of both processes is given in Figure 7.

Fiber Properties. Cross-sectional views of fibers normally produced by the cuprammonium process are given in Figure 4; physical properties of cuprammonium and viscose rayons are given in Table 3. Conditioned and wet stress-strain properties of cuprammonium rayon and other rayons are compared in Figures 8 and 9. Cuprammonium rayon made to date is about equal to regular rayon in tensile properties and does not approach any of the improved viscose high wet modulus properties. If

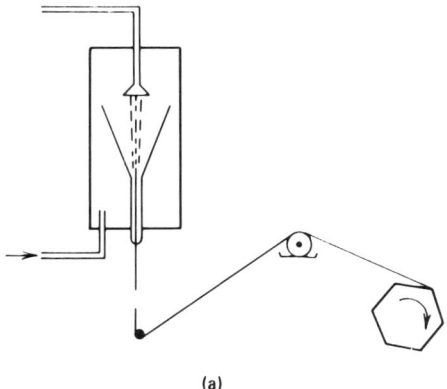

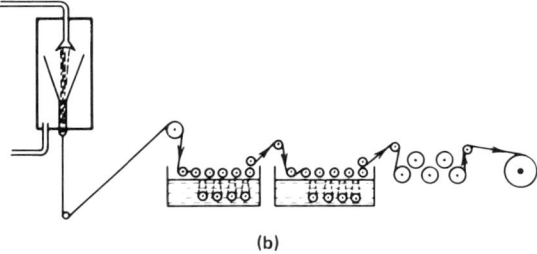

Figure 7. Cuprammonium process. (a) Reel spinning; (b) Continuous spinning.

this is the best that can be achieved in fiber properties, there is no hope for the cuprammonium process; to survive it must supply fibers that satisfy the varied commercial requirements. Recent patents suggest that improved fibers can be expected by spinning-process modifications.

Fiber properties are strongly dependent upon spinning conditions, which are related to recovery processes and conditions. Therefore, wide ranges of spinning conditions are found in the various references and patents. In the past, turbulent flow was regarded as harmful to fiber properties and the cause of filament sticking and breakage; consequently, many of the early patents were directed towards ways to cause smooth flow and prevent eddy currents. Later patents show how countercurrent flow is useful in preventing the deposit of copper precipitates that caused fiber breakage. Siphons were attached to the bottom of the funnels to help control flow and prevent precipitate buildup. Multicluster spinning into a partitioned funnel to increase production and various shapes for heads for multicluster spinning are reported. One enterprising inventor who used only water as the initial coagulation bath spun a highly ripened viscose dope in a cuprammonium unit and found no difficulty in obtaining yarns.

All patents agree that the spin coagulant liquid should be alkaline. The Dormagen plant used only 430 ppm of NH_3 (ca 0.025 M) in the spin water to help keep the funnels clean. Far more important than any of these factors is the comparison of flow rates in the spin funnel coupled with the new design of double-funnel spinning. Herein lies the possibility of achieving an improved wet modulus fiber through the cuprammonium process.

Table 3. Physical Properties of Commercial Rayons

Property	Cuprammonium	Regular	Viscose High tenacity LWM[a]	Viscose High tenacity IWM[a]	HWM	Polynosic, unmodified
stress-strain behavior tenacity, mN/tex[b]						
conditioned	150–200	106–238	265–503	441–574	397–883	300–485
wet	84–120	62–159	132–380	291–353	309–706	238–353
standard loop	88–230	88–132	203–318	88–221	62–132	62–106
standard knot	62–150	62–124	194–309	141–282	106–247	106–256
elongation at break, %						
conditioned[c]	7–23	15–30	9–26	14–18	5–10	6.5–10
wet	16–43	20–40	14–34	17–22	6–11	7–12
specific gravity	1.52–1.54	1.50–1.53	1.52	1.53	1.53	1.51
moisture regain[d]	12.5	13	13	11–13	11–12.5	11.5–12.5
birefringence, by refractive index	0.026	0.018	0.036	0.039–0.044	0.046–0.057	0.040–0.045
water retention, %	100	90–100	65–80	65–75	60–70	55–70
solubility in 6.5% NaOH at 20°C, %	20–30	20–30	30–40	15–20	1.0–5.0	4–6
fiber DP[e]	450–550	300–350	300–500	350–600	500–800	550–650
wet modulus, mN/tex[b]	9–18	16–18	18–26	44–53	71–177	71–177

[a] Low and intermediate wet modulus.
[b] To convert N/tex to g/den, multiply by 11.33.
[c] Conditioned at 21°C and 65% rh.
[d] After centrifuging at 1000 g (ASTM D 2402-65T); commercial standard is 11%.
[e] Tenacity at 5% stretch.

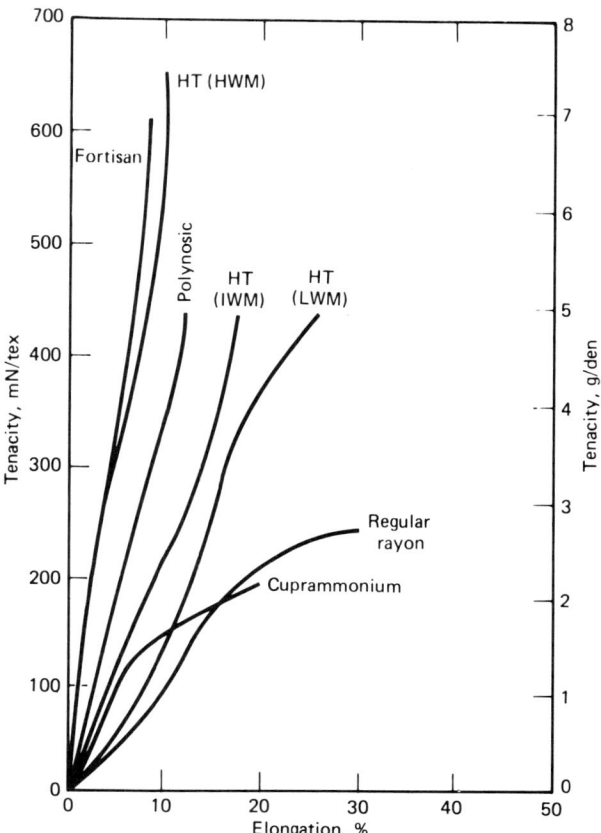

Figure 8. Conditioned stress-strain curves of rayons at 21°C and 65% rh.

At the Dormagen plant, 1500 holes of 0.16 tex (1.4 den) fiber at 90 m/min were spun. Spinning water was introduced at 15 L/min (ca 4 gal/min), which means that it takes ca 720 L of spin-bath flow to produce one kilogram of rayon. This translates to >650,000 L/t or 34.5 ML/d (9×10^6 gal/d) for the production of 53 t/d. Although this is in line with water consumption for viscose production, it is still very high, since all of this flow must be recirculated or treated for recovery and recycle. A city of 72,000 people uses ca 34 ML water each day.

Spin water is only part of the plant's water flow in fiber spinning. For example, a viscose plant producing 50,000 t/yr capacity of rayon reportedly requires ca 81.8 ML (21.6×10^6 gal) water usage per day. This usage, which is equivalent to the needs of a city of about 170,000 people, is broken down as follows:

12.5 ML (3.3×10^6 gal) for soft water to process and to boilers.

27.3 ML (7.2×10^6 gal) for filtered water for services.

42 ML (11.1×10^6 gal) for refrigeration, cooling, and spin-bath recovery condensers.

Cuprammonium rayon plants have similar needs. Lowered spin flow directly diminishes all supporting energy and flow requirements. Much of this spin flow is considered necessary to obtain sufficient stretching of filaments; such high flow rates lead to considerable dilution of copper and ammonia and consequently, increase recovery

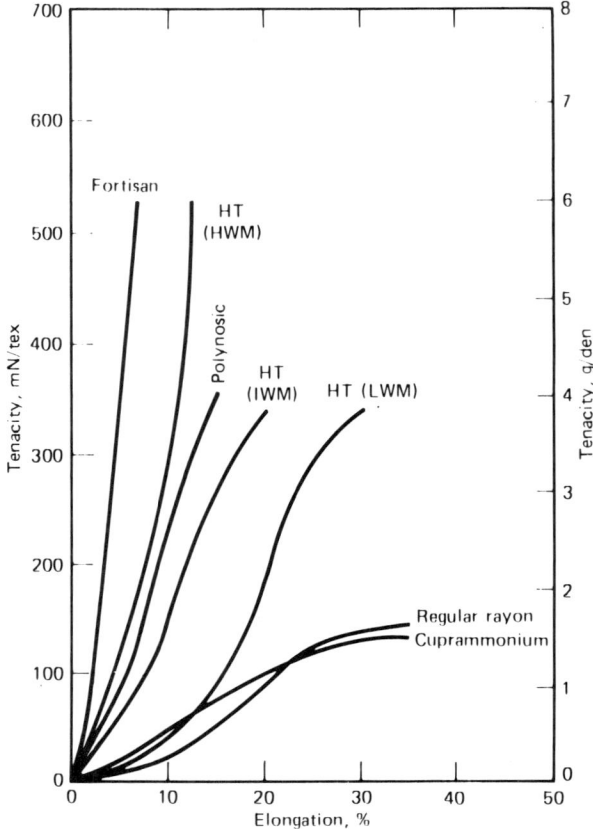

Figure 9. Wet stress-strain curves of rayons.

costs. The rayon industry realized these facts; most of the recent patents deal with improved methods for coagulating and stretching and the use of lower flow rates. Also, recent patents describe the use of various salts in the spin water to give coagulation followed by stretching, all of which should produce improved modulus fibers. For example, a British patent describes the addition of 0.8 mole $CaCl_2$ per liter spin water, which needs only 150% stretch to obtain 0.46 N/tex (5.2 g/den) conditioned tenacity and 0.39 N/tex (4.4 g/den) wet tenacity, ie, a high quality fiber (36). Bemberg SPA in Italy reports that precoagulated fibers are stretched 100% in the presence of 270 g/L $CaCl_2 \cdot 2H_2O$ at 150 m/min to give rayon of 0.38 N/tex (4.3 g/den) and 0.30 N/tex (3.4 g/den) conditioned and wet tenacity, respectively (37).

An improved spinning funnel is described by Bemberg (38). A double funnel is used in which two different liquid flow rates can be controlled separately. Coagulation takes place in the upper funnel at a flow of 100 cm³/min, whereas a faster flow of 500 cm³/min imparts a high stretch in the free air space between the funnels. This design improves overall recovery.

Apparently, manufacturers who are still making cuprammonium rayon realize the need for lowering flow rates and the need for making improved-modulus fibers. Issued patents indicate that they are making efforts that appear to be properly directed for making such improved fibers.

Recovery. Recovery data are most difficult to obtain from the literature. The data for copper and for ammonia recovery are reviewed separately.

Copper Recovery. The amount of copper removed from the fiber in the spinning bath is low. Only 30% of the copper appears in the blue water; because of the high spin-water flow rates, it is in dilute solution. At the Dormagen plant, this gave 80 mg/L in the spin-bath effluent for yarn and 180 mg/L in the spin-bath effluent from the staple operation. This effluent is passed through ion-exchange resins, and the effluent from these resins is circulated back to the spinning funnels.

The 70% copper remaining in the yarns is removed in the acid-wash operation and exists as a solution containing 6.5% H_2SO_4 and 1.0% copper. This acid solution is used to regenerate the ion-exchange beds and then is adjusted to 0.5% acid and 1.2% copper. It passes to the sludging operation where Na_2CO_3 is used to precipitate the copper at pH of 5.0 to give a basic copper sulfate sludge; this copper sludge is recirculated to the cellulose-dissolving area. The overall recovery efficiency is 95% copper even at these high dilutions.

A method for reclaiming copper from the cuprammonium waste solution by precipitating it first as CuS and then reclaiming $Cu(OH)_2$ by use of alkali and air oxidation has been reported (39). Such a method might be advantageous in effecting almost quantitative copper removal to meet present-day environmental standards. The CuS could be concentrated either by sludging or by frothing methods that are well established. The extreme insolubility of CuS makes such thorough cleansing of effluents possible. A patent describes the use of mixing the waste streams to obtain a pH of 11–12 followed by evaporation of liberated NH_3 and 95% recovery of copper (40). A final example of copper recovery is reported at 98% for the Kustanai plant of the USSR (41). Ref. 41 quotes data issued by the Japan Organo Co. Ltd., according to which up to 99.9% of the copper can be recycled with modern techniques.

The use of lower spin-bath flow rates would significantly enhance overall copper recovery. In any case, the technology is available to achieve essentially quantitative copper recovery even if sulfide precipitation or electrolysis of the final effluent has to be employed.

Ammonia Recovery. The ammonia recovery differs from the copper recovery in that a much higher percentage of the ammonia can be recovered from the spin-bath effluent and by washing prior to the final acid bath. The use of a short hot-air exposure chamber between the spin bath and the final acid to achieve more complete NH_3 removal has been suggested (42). The ammonia that enters the acid wash and is converted to ammonium salt is recovered when the acid-wash liquor is made alkaline to precipitate the copper. However, the problem again is that the relative concentration of the ammonia is low owing to the rather high flow rates; thus, large liquid volumes must be processed. No recovery process removes the small levels that remain. The low residual concentrations present in the large water volumes signify losses in material and money. The only way to recover NH_3 at low concentrations is to process at reduced pressure. This means additional energy costs and higher investment costs, since more distillation vessels are required in vacuum processing as compared to processing at atmospheric pressure. Thus, the Dormagen plant had to process 25 ML (6.6×10^6 gal) of ammonia water at 1.27% concentration at reduced pressure in order to recover 30 t of NH_3 from the 38 t of NH_3 originally used to dissolve 48 t of cellulose. This required six recovery towers, which were 3.6-m dia and 19.8-m high. This amounts to handling ca 585 L (154 gal) of NH_3 recovery liquor per kilogram of rayon produced.

The cuprammonium process may be made a closed-loop system producing reasonably good quality rayon fibers if overall economics for handling recovery and recycling of dilute effluent streams can be improved.

Ecological and Pollution Considerations

In most processes developed many years ago, little consideration was given to pollution and ecological impact; the older rayon-manufacturing processes are not exceptions. The cuprammonium and the viscose processes consume huge amounts of water for each kilogram of product; ca 420–750 L (110–200 gal) of water per kilogram of rayon is needed directly for processing and 8–10 times that amount of water must be handled to provide the plant supplementary-service facilities. When water effluents did not require special treatments, overall process economics were favorable. However, treating ca 835 L (220 gal) of water per kilogram of fiber imposes on such processes new penalties that cannot be overlooked. Similar considerations exist for air handling in present-day plants. Each of the three different methods for making rayon reviewed in this article has its own pollution and ecological problems.

Cuprammonium Process. About 40 kg copper and 80 kg ammonia are used for every 100 kg cellulose dissolved. Not only must this be recycled, but it must first be recovered from dilute solutions since ca 32,500 L (8600 gal) water are used to make this much fiber. The copper can be recovered with ion-exchange columns and by use of proper precipitation stages, but costs have become significant. Of more concern is the ammonia recovery in which dilute ammonia streams must be handled to give effective ammonia recovery. However, a group of rayon chemists in the USSR claim that the cuprammonium process is preferred to the viscose or solvent-spinning process from the recovery and overall economic viewpoints. Their data are needed for acceptance of such a position.

Viscose Process. The viscose process also requires rather large amounts of water per kilogram of rayon. In this case, the aqueous effluent contains large amounts of sodium sulfate, much of which is recovered and sold. However, excess salt must be released in the effluent from the plant. Viscose plants must handle the H_2S gas and CS_2 vapor. Removing small amounts of these vapors in a plant by sweeping ca $2.8 \times (10^3–10^5)$ m^3/min [$(10^5–10^7)$ cfm] is a difficult engineering problem. H_2S and CS_2 in air are scrubbed with cascading caustic (if the air flow rate is less than ca 2.8×10^4 m^3/min (10^6 cfm), or adsorbed on zeolites or activated carbon for the higher flow rates where scrubbing is impractical. The presence of H_2S in the exhaust stream complicates the CS_2 recovery on activated-carbon beds since the H_2S is oxidized to sulfurous and sulfuric acids; this reduces the carbon-bed adsorptive capacity.

Most processes involve removal of from 50 to 300 ppm of H_2S present as HS^- and S^{2-} from exit streams, with the use of some type of alkaline medium in conjunction with some type of iron catalyst. The overall method based on the Claus reaction is

$$H_2S + \tfrac{1}{2} O_2 \xrightarrow{Fe_2O_3} S + H_2O$$

In the Ferrox process, simple iron salts are used, whereas in the Cataban process, the iron is present as a soluble iron chelate complex. In both cases, the iron is returned to the ferric state by passing air through the solution. The Cataban process probably has some advantage over the Ferrox process in that it has a higher reaction rate; this

Table 4. Energy Used in Fiber Production from Naphtha[a]

Product	Ratio of energy of raw materials to energy of naphtha combustion	Energy to process raw material, GJ[b]/t			Total processing energy plus energy of raw materials, GJ[b]	Ratio of the energies of raw materials plus processing to energy of naphtha consumption
		Monomer production	Polymerization	Fiber spinning		
Filament yarns						
nylon-6	2.09	65.1	16.3	31.1	204	4.97
nylon-6,6	2.15	54.5	14.0	34.5	198	4.71
cellulose acetate	0.69	93.6		68.2	192	4.58
polyester	1.41	55.1	22.2	40.4	174.8	4.17
rayon		28.6		85.6	114	2.72
Staple fiber						
nylon-6,6	2.15	54.5	12.0	18.4	180	4.28
polyester	1.41	55.1	18.8	21.3	157	3.75
acrylic	1.55	42.5		46.3	157	3.75
polypropylene	1.42	5.8	14.2	14.9	97.5	2.33
rayon		26.3		44.1	70.7	1.69

[a] Ref. 43.
[b] To convert J to Btu, divide by 1054; to convert J/t to Btu/lb, multiply by 0.4302.

Table 5. Worldwide Rayon Production, 10^6 t

Year	Filament	Staple	Total
1965	1.040	1.905	2.945
1970	0.998	1.995	2.993
1975	0.816	1.814	2.630
1980	0.816	2.222	3.038

results in fewer undesirable sulfur-containing by-products. Processes that rely only on the use of alkaline scrubbing suffer from the use of large quantities of expensive caustic and from problems in disposing of the resulting alkaline sulfide liquors. Recent Federal regulations on effluent concentrations of Na_2SO_4, H_2S, and CS_2 place stringent restrictions on the viscose process.

Solvent-Spinning Process. In these cases, all recovery and recycle stages are considered to be closed loops with essentially none of the solvent being lost. This must be achieved because the solvent systems are too expensive to permit anything less than almost complete recycling. However, for solvent spinning, the biggest factors lie in the economics of recovery of sufficiently pure materials to allow for repeated recycle use. Small amounts of impurity can build up to cause some problems, but these should be resolvable. The discovery of a simple solvent that can be recycled easily will be the best approach to a long-range competitive rayon process.

Energy Requirements for Rayon vs Synthetics

As energy becomes more expensive, increasing emphasis is placed on the overall energy required to manufacture products. Energy considerations must include not only all aspect of fiber production but should also consider the energy involved in subsequent product performance. An approximate generalization of the production of an apparel garment is that it consumes between 7–8 kg of oil per kg of textiles. Much of this (1.5–2 kg oil/kg textile) is used in dyeing and finishing, whereas 20% is the energy content of the hydrocarbon starting materials, and the remainder is split among chemical processing, fiber spinning, and cloth production.

Since rayon is made from trees, no petroleum is used to make the original polymer; also, a significant part of the energy needed for separation and purification of cellulose is derived from pulping by-products as energy sources. These factors give rayon a favorable position relative to synthetic fibers with regard to total energy needed for fiber production, as demonstrated in Table 4 (43).

Economic Aspects

Over the last few years, the number of U.S. rayon plants in operation had decreased, whereas the number of rayon plants in the Far East and the USSR has increased. World rayon production since 1965 is given in Table 5.

Regardless of how rayon might be made in the future, whether by the viscose, cuprammonium, solvent-spun, or as yet some undiscovered process, fibers with the properties of rayon are and will be essential to the textile and related industries. A high demand for fibers with rayonlike properties will exist for many years to come. Rayon(s) or solvent-spun cellulose fibers will satisfy this demand.

BIBLIOGRAPHY

"Rayon and Acetate Fibers" in *ECT* 1st ed., Vol. 11, pp. 522–550, "Rayon," by Lionel A. Cox, American Viscose Corp., and "High-Tenacity Rayon," by P. M. Levin, E. I. du Pont de Nemours & Co., Inc.; "Rayon" in *ECT* 2nd ed., Vol. 17, pp. 168–209, by R. L. Mitchell and G. C. Daul, ITT Rayonier Inc.

1. G. E. Linton, *Natural and Man-Made Textile Fibers*, Duell Sloan & Pearce, a division of Meridith Press, New York, 1966.
2. R. L. Mitchell and G. C. Daul, "Rayon," in *Encyclopedia of Polymer Science and Technology*, Vol. 11, Interscience Publishers, a division of John Wiley & Sons, Inc., 1969, pp. 810–847.
3. U.S. Pat. 2,592,355 (Apr. 8, 1952), S. Tachikawa Sanjo and Higashiyama-Ku; U.S. Pat. 2,946,782 (July 26, 1960), S. Tachikawa, Higashiyama-Ku (to Tatsuji Tachikawa, heir).
4. I. H. Welch, *Am. Text. Rep. Bull. Edn.* **AT 8,** 49 (1978).
5. U.S. Pat. 3,632,468 (Jan. 4, 1972), and U.S. Pat. 3,793,136 (Feb. 19, 1974), G. C. Daul and F. B. Barch (to Rayonier, Inc.).
6. U.S. Pat. 3,720,743 (March 13, 1973), T. E. Muller and H. D. Stevens (to ITT Corp.).
7. J. S. Ward and R. Hill, *Text. Inst. Ind.* **7,** 5274 (1969).
8. U.S. Pats. 4,041,121 (Aug. 9, 1977); 4,063,558 (Dec. 20, 1977); 4,136,697 (Jan. 30, 1979), F. R. Smith (to Avtex Fibers, Inc.).
9. U.S. Pat. 4,066,584 (Jan. 3, 1978), T. C. Allen and D. B. Denning (to Akzona Inc.).
10. U.S. Pat. 4,104,214 (Aug. 1, 1978), A. W. Meierhoefer (to Akzona Inc.).
11. J. H. Welch and J. A. Combes, *Paper presented at the 8th Tech. Symp. of the International Nonwovens and Disposables Association*, 1980, p. 3.
12. N. A. Portnoy and G. C. Daul, *Paper presented at Nat. Tech. Conf. AATCC*, 1978, pp. 269–274.
13. U.S. Pat. 3,671,542 (June 20, 1972), S. Kwolek (to E. I. du Pont de Nemours & Co., Inc.).
14. H. Chanzy and A. Peguy, *J. Polym. Sci.* **18,** 1137 (1980); *Repr. 5th International Dissolving Pulps Conference*, Vienna, Austria, 1980.
15. A. F. Turbak, *Amer. Chem. Soc. Symp. Ser.* **58,** (1977).
16. U.S. Pat. 4,142,913 (March 6, 1979), C. C. McCorsley, III, and J. K. Varga (to Akzona Inc.); U.S. Pat. 4,144,080 (March 13, 1979), C. C. McCorsley, III (to Akzona Inc.); U.S. Pat. 4,145,532 (March 20, 1979), N. E. Franks and J. K. Varga (to Akzona Inc.).
17. U.S. Pat. 4,302,252 (Nov. 24, 1981), A. F. Turbak, A. El-Kafrawy, F. W. Snyder, Jr., and A. B. Auerbach (to ITT Corp.).
18. J. Marsh, *Mercerizing*, D. Van Nostrand Co., Inc., New York, 1942.
19. H. Sihtola and T. Rantanen, *Prepr. 4th International Dissolving Pulps Conference*, 1977, pp. 35–39.
20. W. H. Fock, *Kunstseide* **17,** 117 (1935).
21. D. Tunc, R. F. Bampton, and T. E. Muller, *Tappi* **52**(10), 1882 (1969).
22. M. Rahman, *Anal. Chem.* **43**(12), 1614 (1971).
23. D. K. Smith, *Text. Res. J.* **29,** 32 (1959).
24. F. R. Charles, *Can. Text. J.* **83**(16), 37 (1966).
25. U.S. Pat. 2,942,931 (June 28, 1960), R. L. Mitchell, J. W. Berry, and W. H. Wadman (to Rayonier, Inc.).
26. J. Dyer and G. C. Daul, *Ind. Eng. Chem. Prod. Res. Dev.* **20,** 222 (1981).
27. U.S. Pat. 3,626,045 (Dec. 7, 1971), C. Woodings (to Courtaulds Ltd.).
28. E. Attle, *Text. Mon.*, 24 (May 1977).
29. U.S. Pat. 4,040,483 (Aug. 9, 1977), B. R. Franko-Filipasc and J. F. Stuart (to FMC Corp.).
30. *Chem. Week*, 25 (July 29, 1981).
31. A. F. Turbak, R. B. Hammer, R. E. Davies, and H. L. Hergert, *Chemtech* **10,** 51 (Jan. 1980).
32. S. Hudson and J. Cuculo, *J. Macromol. Sci.* **C18**(1), 1 (1980).
33. R. Armstrong, J. K. Varga, and C. C. McCorsley, *Prepr. 5th International Dissolving Pulps Conf.*, Vienna, Austria, 1980.
34. *Synthetic Fiber Development in Germany*, compiled and edited by Leroy R. Smith, Textile Res. Inst., 10 East 40th St., New York 10016, 1946, based on the Technical Industrial Intelligence Committee Report on I.G. Farbenindustrie Plant, Dormagen, Germany, 1946.
35. U.S. Pat. 2,587,619 (March 4, 1952), H. Hoffman (to Beaunit Mills).
36. Brit. Pat. 815,189 (June 17, 1959) (to I. P. Bemberg Akt.-Ges.); H. Frind, *Kolloid Z.* **179,** 110 (1961).
37. U.S. Pat. 3,488,344 (Jan. 6, 1970), F. Biffi, E. Chesi, and O. Gallina (to Bemberg SpA).
38. U.S. Pat. 3,798,297 (March 19, 1974), O. Gallina (to Bemberg SpA).

39. Ger. Pat. 708,009 (June 5, 1941), G. T. Trout (to New Process Rayon, Inc.).
40. Ger. Pat. 1,006,147 (Apr. 11, 1957), P. Schubert (to I. P. Bemberg Akt. Ges.).
41. U. V. Grafov, G. S. Bykova and U. I. Malboroda, *Khim. Volokna* (3), 63 (May–June 1976).
42. U.S. Pat. 3,110,546 (Nov. 12, 1963), F. Hoelkeskamp (to J. P. Bemberg Akt.).
43. W. C. Firth, *Energy Balance of Man-Made Fiber Production*, Comité International de la Rayon et des Fibres Synthétiques, 1980.

JOHN LUNDBERG
ALBIN TURBAK
Georgia Institute of Technology

REACTION-PATH SYNTHESIS. See Supplement Volume.

REACTOR TECHNOLOGY

A reactor consists of the vessels used to produce desired products by chemical means and is the heart of a commercial processing plant. Its configurations, operating characteristics, and underlying engineering principles constitute reactor technology. Besides stoichiometry and kinetics, reactor technology includes requirements for introducing and removing reactants and products, supplying and withdrawing heat, accommodating phase changes and material transfers, assuring efficient contacting among reactants, and providing for catalyst replenishment or regeneration. These issues are taken into account when one translates reaction kinetics and bench-scale data into effective pilot plants (qv), scales to larger sized units, and ultimately designs and operates commercial plants (see Heat-exchange technology; Mass transfer; Simultaneous heat and mass transfer; Absorption, Distillation; Catalysis).

Many, but not all, reactor configurations are discussed in this review; a few are described to illustrate specific concepts. This article is not concerned with theoretical derivations, mathematical formulations, and design correlations. Topics not covered, but deserving attention, include reactor stability, process simulation and control (see Simulation and process design; Instrumentation and control), materials selection (see Materials reliability; Materials standards), catalyst manufacture, thermodynamics (qv), design of experiments (qv), and economic analysis (see Economic evaluation). Separations, which are unit operations whose technologies often are applicable to reactor technology, also are not included (see Separation systems synthesis; other unit-operations articles).

Reactor Types and Characteristics

Many reactor configurations and designs have evolved. The specific characteristics including size depend on the particular reactor's use as a laboratory reactor, pilot plant, or commercial unit. In each case, selection is based on the physical properties of the feed and products, ie, vapor, liquid, solid, or combinations; the characteristics of the chemical reactions, ie, reactant concentrations, reaction rates, operating conditions, and heat addition or removal; the nature of any catalyst used, ie, activity, life, and physical form; and the requirements for contacting reactants and removing products, ie, flow characteristics, transport phenomena, and mixing and separating mechanisms. These factors are interdependent and must be considered together. The requirements for contacting reactants and removing products are the paramount focus of reactor technology; the other factors usually are set by the original selection of the reacting system and intended levels of reactant conversion and product selectivity.

All reactors have in common selected characteristics of three basic reactor types: the well-stirred batch reactor, the continuous-flow stirred-tank reactor, and the tubular reactor (see Fig. 1). A reactor often may be represented by or modeled after one or a combination of the three types. The acceptability of a model depends upon the extent

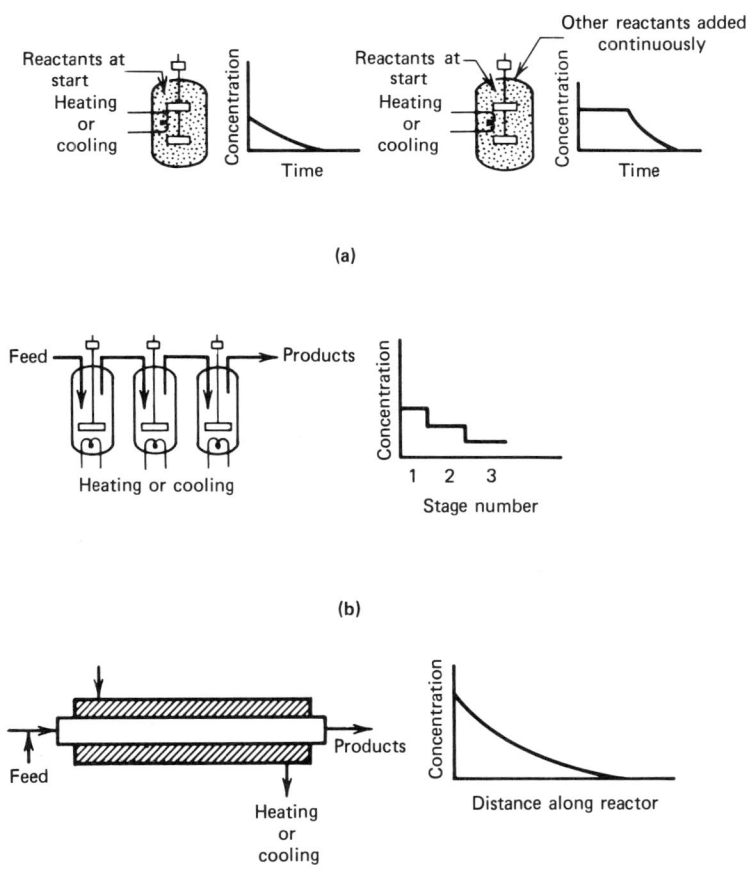

Figure 1. Three basic reactor types. (**a**) Batch. (**b**) Continuous-flow stirred-tank. (**c**) Tubular.

to which reactions, and thermal and transport processes, are satisfactorily depicted. Basic reaction engineering concepts used to develop and validate models are discussed in references 1–2.

Batch Reactor. A batch reactor is one in which a feed material is treated as a whole for a fixed period of time. Commercial reactors generally process continually rather than in single batches, because overall investment and operating costs of continuous processes usually are less. However, batch reactors may be preferred for small-scale production of high priced products. This is particularly true if a large number of sequential operations are employed to obtain high product yields, eg, a complex cycle of temperature–pressure–reactant additions. Batch reactors also may be justified when multiple, low volume products are produced in the same equipment or when the assurance of continuous flow is difficult, eg, when transporting highly viscous or sticky solids-laden liquids.

Batch reactors often are used to develop continuous processes because of their suitability and convenient use in laboratory experimentation. The data, except for very rapid reactions, can be well defined and used to predict performance of larger scale, continuous-flow reactors. Almost all batch reactors are well-stirred; thus, ideally, compositions are uniform throughout and residence times of all contained reactants are constant. In theory, this may also be achieved in a continuous reactor if the reactants flow at a single velocity as a plug. However, in practice, approximating plug-flow behavior is difficult, and nonuniform behavior, eg, variations in residence times of reactants, must be considered.

Because uniform residence times can be achieved with batch reactors, better yields and higher selectivities can be obtained than with continuous reactors. This advantage exists when undesired reaction products inhibit the reaction, side reactions are of lower order than the desired one, or the product is an unstable or reactive intermediate. These considerations must be taken into account when applying batch data to continuous processes. Similar considerations may be important when scaling from a smaller continuous unit if residence time distributions in the units differ. Thus, knowledge of the effects of internal flow characteristics on reactor design is essential to predicting reactor performance.

The semibatch reactor is a modification of the batch reactor and is characterized by the continuous addition of some reactants in addition to those initially placed in the reactor or by the continuous removal of one or more products. For example, gradual addition of chlorine to a stirred vessel containing benzene and catalyst results in higher yields of di- and trichlorobenzenes than if chlorine were included in the original batch. Similarly, thermal decompositions of organic liquids are enhanced by the continuous removal of gaseous products. Constant pressure can be maintained and chain-terminating reaction products can be removed from the system. In addition to better yields and selectivities, gradual addition or removal helps to control temperature when the net reaction is, respectively, exothermic or endothermic.

Continuous-Flow Stirred-Tank Reactor. If reactants and products are continuously added and withdrawn from a well-stirred vessel, the resultant reactor is the continuous-flow stirred-tank reactor (CSTR). In practice, mechanical or hydraulic agitation is required to achieve uniformity of composition and temperature. The CSTR is the idealized opposite of the well-stirred batch and tubular plug-flow reactors. Analysis of judicious combinations of these reactor types can be useful in quantitatively evaluating more complex gas-, liquid-, and solid-flow behaviors.

Since the compositions of mixtures leaving a CSTR are those within the reactor, the reaction driving forces, usually reactant concentrations, are necessarily low. Therefore, except for zero- and negative-order reactions, a CSTR requires the largest volume of the reactor types. However, the low driving force makes possible better control of rapid exothermic and endothermic reactions. When high conversions of reactants are needed, several CSTRs in series can be used. Equivalent results can be obtained by dividing a single vessel into compartments while minimizing back-mixing and short-circuiting. The larger the number of stages, the closer performance approaches that of a tubular plug-flow reactor.

Continuous-flow stirred-tank reactors in series are simpler and easier to design for isothermal operation than tubular reactors. Reactions with narrow operating temperature ranges or those requiring close control of reactant concentrations for optimum selectivity benefit from series arrangements. If heat-transfer requirements are severe, heating or cooling zones can be incorporated within or external to a stirred tank. For example, impellers or centrally mounted draft tubes circulate liquid upward and then downward through vertical heat-exchanger tubes. In a similar fashion, reactor contents can be recycled through external heat exchangers (see Heat-exchange technology).

Tubular Reactor. The tubular reactor is a vessel through which flow is continuous, usually at steady state, and configured so that process variables are functions of position within the reactor rather than of time. In the ideal tubular reactor, the fluids flow as if they were solid plugs or pistons, and reaction time is the same for all flowing material at any given tube cross section; hence, position is analogous to time in the well-stirred batch reactor. Tubular reactors also resemble batch reactors in providing initially high driving forces, which diminish as the reactions progress down the tubes.

In actual tubular reactors, flow can be laminar and, thus, greatly deviate from plug flow or, more likely, flow can be turbulent (see Fig. 2). Turbulent flow generally is preferred to laminar flow, because mixing and heat transfer when normal to flow are improved and less back-mixing is introduced in the direction of flow. For slow reactions and especially in small laboratory and pilot-plant reactors, establishing turbulent flow can result in inconveniently long reactors or may require unacceptably high feed rates. Consequently, compromises often are necessary but they may not be acceptable, depending upon consequences in the process development.

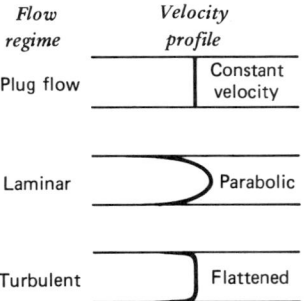

Figure 2. Flow characteristics for single-phase flows.

Multiphase Reactors

The presence of more than one phase, whether it is fixed or flowing, further compounds analyses of reactor performance and increases the multiplicity of reactor configurations. Gases, liquids, and solids each flow in characteristic fashions, either dispersed in other phases or separately. Flow patterns in these reactors are complex and the phases rarely exhibit plug-flow behavior. Some important reactor configurations are illustrated in Figures 3 and 4. The names presented here are often employed, but are not the only ones used.

A fixed-bed reactor is one that is packed with catalyst. If a single phase is flowing, the reactor can be analyzed as a tubular plug-flow reactor or, essentially, as plug flow modified for uniform axial diffusion. If both liquid and gas or vapor are injected downward through the catalyst bed or if substantial amounts of vapor are generated

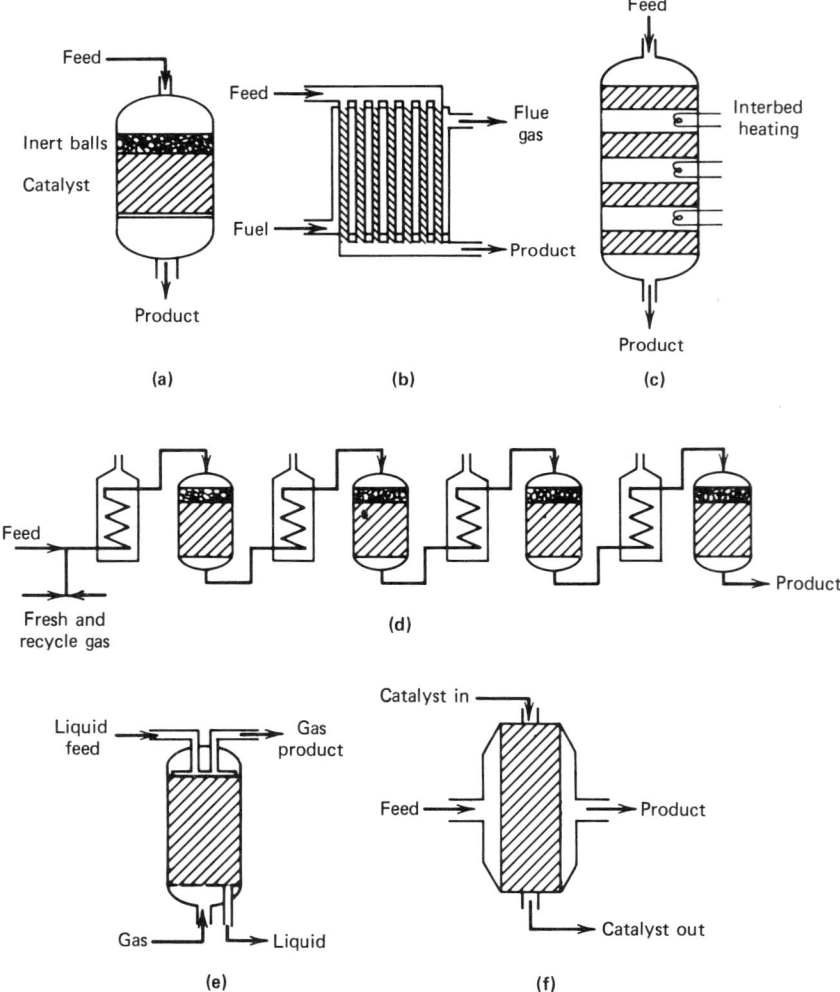

Figure 3. Multiple fixed-bed configurations. (**a**) Adiabatic fixed-bed reactor. (**b**) Tubular fixed beds. (**c**) Adiabatic fixed beds in series. (**d**) Trickle beds or pulsed columns in series. (**e**) Countercurrent packed-bed absorber. (**f**) Moving radial or panel fixed-bed reactor.

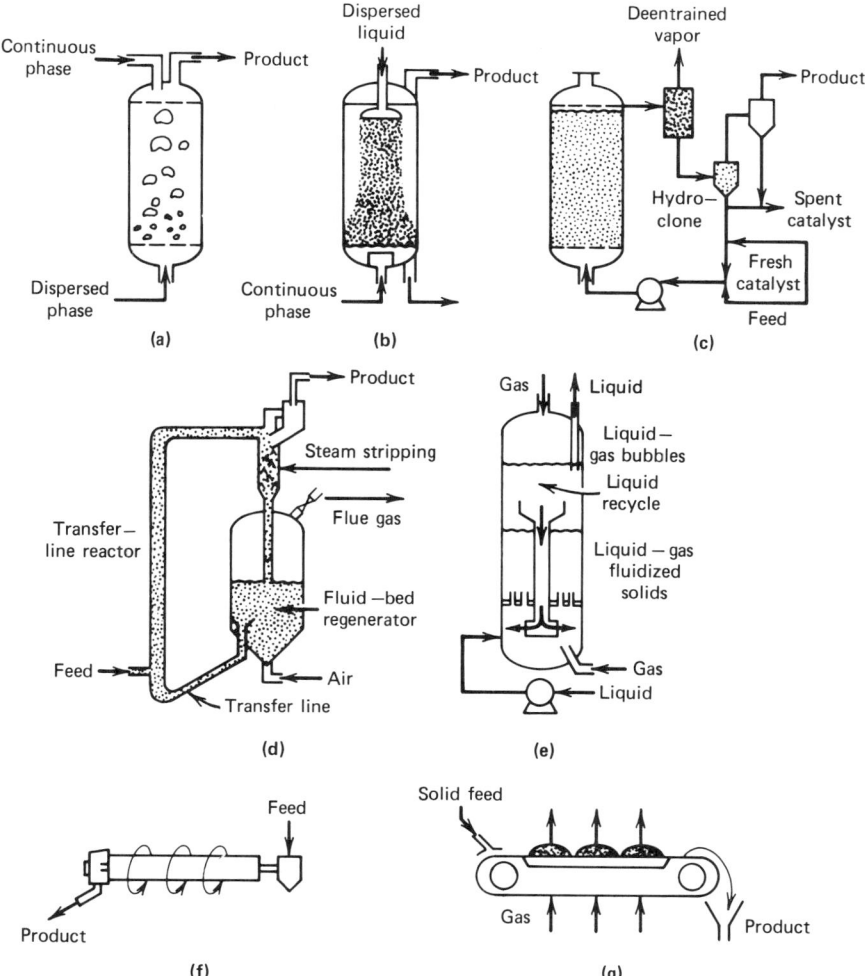

Figure 4. Multiphase fluid and fluid-solids reactors. (**a**) Bubble column. (**b**) Spray column. (**c**) Slurry reactor and auxiliaries. (**d**) Fluidization system. (**e**) Gas–liquid fluidized system. (**f**) Rotary kiln. (**g**) Traveling grate or belt dryer.

internally, the reactors are mixed-phase, downflow, fixed-bed reactors. If the liquid and gas rates are so low that the liquid flows as a continuous film over the catalyst, the reactors are called trickle beds. At higher total flow rates, particularly when the liquid is prone to foaming, the reactor is a pulsed column; this designation arises from the observation that the pressure drop within the catalyst bed cycles at a constant frequency. The pulsed column is not to be confused with the pulse reactor, which is a laboratory reactor used to obtain kinetic data on catalysts and into which a small pulse of reactant is introduced. At high liquid flow rates, gas becomes the dispersed phase and bubble flow develops, with flow characteristics being similar to those in countercurrent packed-column absorbers. At high gas rates, spray and slug flows can develop. Downflow is preferred because the reactors are more readily designed mechanically to hold a catalyst in place and are not prone to inadvertant excessive velocities, which upset the beds. Upflow is used less often but has the advantage of optimum contacting between gas, liquid, and catalyst over a wider range of conditions.

Moving beds are fixed-bed reactors in which spent catalyst or reactive solids are slowly removed from the bottom and fresh material is added at the top. A fixed bed that collects solid impurities present in the feed or produced in the early reaction stages is a guard bed. If catalyst deposits are periodically burned or otherwise removed, the operation is cyclic, and the catalyst remaining behind the combustion front is regenerated.

In bubble columns, gas bubbles upward through a slowly moving liquid. The bubbles, which rise in essentially plug flow, draw liquid in their wakes, thereby back-mixing the liquid with which they have come in contact. These reactors are used for slow reactions where contact is not critical. In spray columns, liquid as droplets descend through a fluid, which is usually a gas. These are used for reactions where high interfacial areas between phases are desirable. If beds of solids are lifted by either gas or liquid flows, the reactors are termed fluid beds because the suspended solids appear to behave as liquids. The process is usually referred to as fluidization (qv). The most common fluid bed is the gas-fluidized bed. With gas feeds, the excess gas over the minimum required for fluidization rises as discrete bubbles, through which the surrounding solids circulate. At higher gas rates, such beds lose their clearly defined surface, and the particles are fully suspended. Under different circumstances, these reactors are called dilute-phase, transport, fast fluidized, or entrainment reactors. In analogous conditions of no discrete bed interface for liquids, the reactors are called slurry reactors. In ebullating beds or gas–liquid fluidized systems (GLFSs), the solids are fluidized by liquid and gas.

In these multiphase systems, flows are generally cocurrent, or countercurrent. However, cross-flow also is used commercially. Solids in a moving bed can move downward under gravity while a gas passes horizontally through them. Such radial reactors or panel beds can be designed as thin annuli or slices, and they are used for rapid reactions to reduce stresses on the catalyst or to minimize pressure drops. Rotary kilns, belt dryers, and traveling grates are additional examples. Cross-flow reactors are not restricted to solids-containing systems. Venturis, in which atomized liquids are injected across the gas stream, are effective for fast reactions.

Reactors as Process Systems

The interacting effects of mixing, heat and mass transfer, and kinetics are essential elements in establishing reactors as process systems. The degree of back-mixing directly affects overall reactor size and catalyst requirements for reactions with greater than zero-order dependencies on limiting reactants. Back-mixing does not necessarily imply reduced conversions. Complete back-mixing can decrease catalyst deactivation by diluting catalyst poisons, if these are limiting constituents. Since total conversion is a function of the kinetic behavior of the entire reactor system, the overall effects of reactor back-mixing must be considered.

The full range of flows, ie, from plug to well-mixed, can be obtained by staging several reactor vessels or by introducing recycle around the entire system. Thus the back-mixing character of individual reactors is only consequential to their specific designs. Three or four well-mixed stages in series approach plug-flow behavior for a first-order reaction that is carried out to 70% conversion. On the other hand, one or more plug-flow reactors in series with a large recycle stream around the reactor system behaves as a well-mixed system. Thus, the degree of back-mixing sets criteria for system selection, but not necessarily for selecting the reactor type.

The impacts of heat generation and removal on kinetics similarly affect the criteria for selecting reactor systems. The problems of maximizing outputs from exothermic or endothermic reactions occur with reversible reactions, eg, ammonia synthesis or water–gas shift reactions; temperature selective reactions, eg, reforming or hydroformylation; or rapidly deactivating side reactions, eg, catalytic cracking of petroleum or polymerizations. Consider a reverse reaction in which its conversion rate increases more rapidly with temperature than that of a desired exothermic reaction. As the reaction progresses, the desired temperature for maximum conversion decreases. Consequently, a batch reactor with a heat exchanger or with small-diameter tubes jacketed in a coolant might be used. Actually, there are many options for limiting temperature rises or programming temperature distributions, so that even large-diameter fixed beds with their poor heat-transfer capabilities have been employed. Options include using interstage coolers, injection of a volatile liquid, eg, water, to cool by evaporation, recycling of unreacted feed or a portion of the product, and addition of inert diluent to the feed or interposed within catalyst beds. The advantage of complicating reactors by permitting large axial and radial temperature gradients is the lower cost of using these large, essentially adiabatic fixed-bed reactors.

Reactor Selection

Selection is often determined by economics, reliability, or availability of a proven system that is amenable to extension in a new service. For example, fixed beds and slurry reactors are favored at high pressures over fluid beds; fluidized systems are less likely to develop hot spots or be subject to temperature runaways; and downflow vapor-phase fixed-bed technology for desulfurization of naphthas and light gas oils has been extended to mixed-phase operation with higher boiling gas oils and, subsequently, to residua. Representative factors that can influence the selection process for catalytic reaction processes are described below.

Catalyst Size. Initial catalyst activity and catalyst life increase with decreasing size. Catalyst particles 20–40 μm dia can be used in fluidized reactors. Fixed beds require 3-mm or possibly 1.5-mm dia pellets; smaller sizes are precluded because of their high pressure drops and greater tendency to foul. Slurry and gas–liquid fluidized bed reactors (GLFS) can be operated with intermediate sized pellets. The higher activity and reduced rate of deactivation of a smaller catalyst may offset the larger reactor volumes required for bed expansion, which are inherent in these systems. If a single, fixed-bed reactor with 1.5-mm catalyst is considered as the basis of comparison, the increased activity advantage required for a multistaged GLFS bed with 0.8-mm catalyst to give a break-even reactor size is ca 20%; for a slurry containing 100 μm catalyst, ca 50% advantage is needed. For reactions controlled by diffusion of reactants into the pores of catalysts, even greater increases in activity may be achieved (3). Reactions often are limited by diffusion when liquid viscosity is high, reactant molecules are large with low diffusivities, high catalyst interfacial area exist in small pores, or reaction rates are fast; ie, reactions are diffusion-limited when the characteristic diffusion time is larger than that of the chemical reaction.

Catalyst Strength. A superior but physically weak catalyst may preclude the use of a fixed bed. Such a catalyst, however, might be suitable for one of the other reactor configurations, since catalyst strength can increase with decreasing size. For use in fixed beds, catalyst strengths of ca three kilograms per particle are desirable to prevent excessive breakage, particularly if pressure drop across the bed increases significantly with time. Lower strengths are required of catalysts that are used in GLFS reactors. For a slurry reactor, catalyst attrition in auxiliary equipment, eg, pumps and hydroclones, can be limiting. Catalyst replacement cost must be considered for catalyst lost by attrition and from the inherent inefficiencies in containing inventory within a system. Inventory maintenance is much more complex in fluidized systems, particularly when catalyst properties change with time in the reactor.

Catalyst Fouling and Deactivation. Fixed beds generally are favored if the catalyst deactivates in ca ≥ 3 mo. If plugging is more rapid than deactivation but fouling only occurs in the first bed of a train, fixed-bed reactors might still be used since this bed is accessible and can be designed as a moving or guard bed. Similarly, if only a first bed deactivates rapidly, this configuration may be preferred. If a catalyst deactivates rapidly or plugging is widespread, the fluidized systems are preferred because the catalyst can be rapidly removed from the reactor, regenerated in other vessels, and returned.

Scale-Up

The principle of scale-up is to select designs and operating conditions so that the effects of the main dependent variables are similar in the different sized units. The ultimate objective is ensuring comparable yields, product distributions, and time effects. Often the use of the smallest possible laboratory unit or pilot plant is necessary for practical or economic considerations. Where considerable uncertainties exist or large quantities of products are needed for market evaluations, demonstration units which are intermediate in size between pilot and commercial plants are used.

In the design and scale-up of reactors, key factors are the nature of the reaction sites, the specific reaction rates, and the mass- and heat-transport rates to and from the reaction sites. When reaction and transport rates are comparable, they affect each other and are said to be coupled. In these situations, increasing the reactor size alters mass- and heat-transport rates and, therefore, changes the apparent reaction rate. Effects on selectivity are also possible. In such situations, conversions are underestimated in the small reactor, but there is no general trend in selectivity. The effects of scale on selectivity must be determined by experimentation. If transport rates to and from the reaction sites are more than ca ten times greater than the specific reaction rate, the overall reaction rate will be uncoupled from the transport rates. Increasing reactor size then does not affect the apparent reaction rate.

The degree of coupling can be determined by evaluating radial or interfacial (in a multiphase reactor) concentration and temperature gradients within a reactor. If these gradients are ca 10% or less than the total concentration gradients of limiting reactants or inhibitor products, the reaction is uncoupled. At 10–90%, the reaction is coupled and, above 90%, the reaction is transport- or diffusion-controlled. Systematic experimentation is required to establish the reaction-induced gradients. Such work represents major efforts in studying the performance characteristics of heterogeneous and homogeneous catalysts and when determining the mechanisms of radical and ionic reactions.

For example, consider the vapor-phase and mixed-phase catalytic hydrodesulfurizations of a naphtha and a heavy gas oil with pelletized solid catalysts composed of cobalt molybdate on an alumina base. These pellets are highly porous with large internal surface areas. In the vapor-phase process, sulfur-laden naphtha molecules and hydrogen reach the reaction sites by diffusion into the pores from the pellet surface. After reaction on the interior pore surfaces, the desulfurized products and hydrogen sulfide reverse the process and diffuse from the catalyst pores. For mixed-phase catalytic hydrodesulfurization, the countercurrent diffusion of hydrogen and hydrogen sulfide between vapor and liquid phases also is required. Capillary forces in the pores keep the catalyst wet and the active sites separate from the vapor. The exothermic heats of reaction must also pass from the reaction sites to the reaction media or to a heat-exchange surface. Heat generated at the reaction sites is conducted from the catalytic surface and the adsorbed reaction liquids to the pellet surfaces and then into the reacting fluids within the reactor. In a tubular fixed-bed reactor, heat conduction to the wall through the catalyst and fluids also is operative. The rates of these steps must be determined and compared with intrinsic reaction rates to establish overall catalyst effectiveness.

The mass- and heat-transport rates can be estimated using phenomenological correlations describing the interrelated effects of fluid dynamic and geometric parameters on transport rate. Such generalized correlations have been developed from results obtained in a variety of systems. These are based on gradients being linear across the resistance paths, ie, the film theory, or being nonlinear because of insufficient time for reaching steady state, ie, the penetration theory. For example, in a vapor-phase, fixed-bed reactor, the rate of mass transport from flowing reactant into a catalyst pellet can be determined from the following correlation (4):

$$eJ_D = \frac{0.765}{Re^{0.82}} + \frac{0.365}{Re^{0.386}}$$

where e is the catalyst void volume, Re is the particle Reynolds number $D_p G/\mu$ (D_p is pellet diameter, G is mass flow rate, and μ is viscosity), and J_D is the mass transfer factor $\bar{k}\rho/G(S_c)^{2/3}$ ($\bar{k}$ is the mass transfer coefficient, ρ is gas density, and Sc is the Schmidt number $\mu/\rho D_v$ in which D_v is molecular diffusivity). Equivalent relationships are available for heat transport. Recommendations of mathematical techniques for analyzing mass- and heat-transport phenomenon in reacting systems are given in refs. 5–6. Because the correlations are highly dependent on and specific to the nature of the phases in the reactor, care is necessary to assure selecting only those correlations appropriate to the given system.

Dimensional analysis (qv) can be helpful in analyzing reactor performance and developing scale-up criteria. Seven dimensionless groups that are used in generalized rate equations for continuous flow reaction systems are listed in Table 1. Other dimensionless groups apply in specific situations (6–7). Compromising assumptions are often necessary, and their validation must be established experimentally or by analogy to previously studied systems.

Minimum Pilot-Plant Size

Minimum sizes of laboratory and pilot units often are set by operability factors not directly involving the reactor. These include feed and product-transfer line diameters, inventory control in feed and product separation systems, and preheat and temperature maintenance requirements. Most of these extraneous factors favor large units. Large commercial plants can be operated with high service factors for years,

Table 1. Dimensionless Groups in Chemical Reaction Systems

Name	Formula[a]	Proportional to:
Arrhenius Group	$\dfrac{\Delta E}{RT}$	$\dfrac{\text{activation energy}}{\text{potential energy of fluid}}$
Damköhler Group I	$\dfrac{rL}{UC}$	$\dfrac{\text{chemical reaction rate}}{\text{bulk mass flow rate}}$
Damköhler Group II	$\dfrac{rL^2}{D_v C}$	$\dfrac{\text{chemical reaction rate}}{\text{molecular diffusion rate}}$
Damköhler Group III	$\dfrac{QrL}{C_p \rho U \Delta T}$	$\dfrac{\text{heat liberated}}{\text{bulk transport of heat}}$
Damköhler Group IV	$\dfrac{QrL^2}{k \Delta T}$	$\dfrac{\text{heat liberated}}{\text{conductive heat transfer}}$
Reynolds number, Re	$\dfrac{LU\rho}{\mu}$	$\dfrac{\text{inertial force}}{\text{viscous force}}$
Thring's radiation group	$\dfrac{\rho C_p U}{\epsilon S T^3}$	$\dfrac{\text{bulk transport of heat}}{\text{radiant heat transfer}}$

[a] Nomenclature

ΔE = activation energy, J/mol
R = gas constant, J/(mol·K)
T = absolute temperature, K
r = reaction rate, mol/(cm^3·s)
L = characteristic length, m
U = fluid velocity, m/s
C = concentration, mol/cm^3
D_v = molecular diffusivity, cm^2/s
Q = heat generated, J/mol
C_p = specific heat, J/(g·K)
ρ = density, g/cm^3
ΔT = temperature rise, K
k = thermal conductivity, W/(m·K)
μ = viscosity, g/(m·s)
ϵ = emissivity of surface, dimensionless
S = Stefan-Boltzmann constant, W/(m^2·K^4)

whereas it is not unusual for pilot units to operate at sustained conditions for only days or even hours (see Pilot plants and microplants).

Small-diameter lines are prone to plugging and pressure drop buildup. Compensating with oversized lines, which reduce velocities, can be detrimental with mixed-phase flows. Reduced flow rates encourage solids deposition, impair heat transfer, and promote slug flow, and hence, cause coking and fouling. Extraneous reactions occurring in feed and product inventories confound any data analysis. In smaller reactors, greater portions of the total system inventory tend to be heated or cooled and so are more likely to have thermal and catalytic reactions outside the reactor confines.

Small reactors usually require additional heat inputs to match their higher heat losses per unit of reactor inventory, compared with larger commercial units. A large unit may require heat removal. The modes selected for redressing heat imbalances can affect results, particularly in systems of multiple reactions of different reaction orders, activation energies, or equilibria limits. Many different techniques are employed, and all represent substantial compromise. These include additional feed preheat or segmented heaters on the reactor. Alternatively, reactor variables, eg, temperature, pressure, catalyst activity, or feed compositions, can be modified to increase reaction rates of exothermic reactions. The resultant deviations may be substantial and may prevent the unit from operating as a fully integrated system and, consequently, they may be the factors setting minimum reactor size.

Requirements for mixing feed components or separating products may determine minimum pilot unit size. If reactants cannot be premixed before they are passed into the reactor, the effectiveness of the inlet distributor in mixing the reactants can markedly affect reactor performance. This is especially true for gases, multiple phases, or liquid streams of greatly different kinematic viscosities. Such a distributor in a commercial unit typically has a large number of injection points. Flow variations through these parallel paths can lead to poor flow distributions within a reactor; thus, decreased product yields and selectivities can result. In some circumstances, undesirable side products can foul portions of the distributor and further upset flow patterns. Where this is important, or where the possibilities and consequences are insufficiently understood and independent means cannot be employed to assure adequate distribution, the pilot plant must be sized to accommodate such a distributor. The distributor should have injection points around the reactor periphery and multiple injection points in its interior. Spacings should be comparable to those that are anticipated to be used commercially.

Additional definition of the operative mechanisms can obviate the need for the larger unit. It may be possible to assess limitations in a smaller unit that has only a few injection points on the distributor. The unit could be used to evaluate distributor designs that permit a wide range of acceptable operating conditions. However, if the acceptable range proves smaller than desired, the larger pilot unit would then be needed to establish acceptable performance.

High temperature gasification of high ash-content coals in a fluid bed is illustrative of such issues. In this process, coal is oxidized with oxygen, or air, and steam under reducing conditions to produce a high or low energy gas containing H_2 and CO. Unfortunately, the coal circulating within the bed near the distributor contacts oxygen under oxidizing conditions and then moves upward into the reducing zone. Some coal particles recirculate back into the oxidizing zone and the cycle repeats. Under these circumstances, coal particles may overheat, sinter, and foul the distributor (8). The successive contacting of coal with oxidizing and reducing atmospheres may alter the particle surfaces and consequently affect gasification kinetics. Thus, demonstrating nonfouling distributor operation in a unit that is sufficiently large to include these influences is essential to successful reactor performance.

Impact of Fluid Dynamics on Scale-Up

Matching of fluid flow characteristics within the reactor might determine the operative criteria. Ideally, the smaller reactor should act as a volume segment of the larger one. Flow distributions should not be markedly influenced by the wall, turbulent eddies should be approximately the same size, and bubbles in bubble columns and fluid beds should be the same size. Under these conditions, the residence time distributions of feeds, intermediates, and products should be comparable. In catalytic processes, catalyst particles should remain in the reactor for the same lengths of time (see Fluid mechanics).

In cases where a large reactor operates similarly to a CSTR, fluid dynamics sometimes can be established in a smaller reactor by external recycle of product. For example, the extent of solids back-mixing and liquid recirculation increases with reactor diameter in a GLFS. Consequently, if gas and liquid velocities are maintained constant when scaling and the same space velocities are used, then the smaller pilot unit should be of the same overall height and be tall and thin. The net result is that the large diameter reactor is well mixed and no temperature gradients occur even with a highly exothermic reaction. The smaller reactor approaches plug-flow behavior and exhibits a large temperature gradient. In this case, external recycle provides the same degree of back-mixing as is provided by internal circulation in the larger diameter reactor.

Trickle-bed reactors have complicated and as yet poorly defined fluid dynamic characteristics. Contacting between the catalyst and the dispersed liquid film and the film's resistance to gas transport into the catalyst, particularly with vapor generation within the catalyst, is not a simple function of liquid and gas velocities. Maximum contacting efficiency is attainable with high liquid mass velocities, ie, $1-5$ kg/(m$^2\cdot$s) or higher in all sized units (1,9), even in those only 8 catalyst particle diameters; however, $3-8$ kg/(m$^2\cdot$s) is a more preferable range of liquid mass velocities. These velocities require reactor diameters of 12 mm and 24 mm for 1.5-mm and 3-mm dia catalysts, respectively, as well as feed rates of up to 13 kg/h. For the handling of smaller quantities of reactants and to reduce the overall length of these pilot units, some process developments have been carried out at a tenth of these velocities, despite the lower product yields and differing responses to operating variables compared with the larger commercial units. However, more demanding process developments with trickle beds are best scaled in high mass velocity reactors.

The markedly different fluidization characteristics in different sized gas–fluid-bed units dictate process development that is based on a series of pilot and demonstration unit sizes. This is in marked contrast to fixed-bed process developments. Various fixed-bed hydrodesulfurization, hydrocracking, and catalytic reforming processes have been scaled based on data from pilot units with reactors only 13–44 mm dia; the commercial reactors have diameters of 2.5–9 m. In contrast, 300–600-mm dia vessels were used in low conversion, fluid catalytic cracking pilot units; 380-mm and 2.4-m dia were used for high conversion fluid-bed catalytic reforming; vessels 300-mm and 2.4-m dia were employed for a multiple fluid-bed iron ore reduction process development; and 75-mm and 2.9-m dia reactors were used for a high temperature coke gasification process. Capacity scale-up factors of only one or two orders of magnitude were employed in scaling to the 7.5–17-m dia vessels that are employed commercially in these fluid-bed processes. For comparison, capacity scale-up factors for the fixed-bed developments exceeded 50,000.

Application of Fluid Dynamics to Fluid-Bed Scale-Up

A difficulty in scaling up fluid-bed processes stems from fluid-bed behavior being as much a function of bed dimensions as of physical properties of the gas and solids. Thus, wall effects cannot be neglected and the various mechanisms that limit contacting must be taken into account. Scaling is possible if the effects of the important variables that influence the reactions are known and a means for predicting conditions can be specified.

Important to successful scaling is knowledge of the flow regime governing gas-solids contacting. The various flow regimes, ie, particulate fluidization, spouting, the grid produced and controlled bubbling, and free bubbling and slugging, obey different physical laws. Slugging or slug flow behavior exists when the rising bubbles approach the size of the column diameter. Fluid beds with bubbles that are only one third or one fourth of the reactor diameter are similarly affected by the wall and are governed by essentially the same equations as fully developed slugs of gas. Walls exert some effect on bubbles that are as small as a tenth of the reactor's diameter. The following analysis of one approach to achieving similitude is described in ref. 10.

The initial design of the pilot unit is critical. The fully developed large bubbles or slugs should be made to form as rapidly as possible above the inlet. Usually, the basic reaction conditions of feed composition, temperature, pressure, and catalyst activity are kept constant. Constant catalyst activity usually requires use of the same particle-size distribution and, therefore, constant minimum fluidization velocity. Generally, minimum fluidization velocity is much less than the superficial gas velocity, and mass transport from the bubble by diffusion may be less than by convective exchange between the bubble and the surrounding emulsion phase. These design fundamentals result in the following ratios, which should be held constant.

Effective space time. Conversion in a fluid bed is lower than in a fixed bed with equal gas flow rates and catalyst weights, because gas bubbles have short residence times in the bed and, at a given instant, only a small fraction of the gas circulating through the bubbles is in contact with the catalyst. Thus, the effective space time t_e is the product of bubble residence time in the bed t and the fraction of bubble-gas exchanged with the emulsion phase F.

$$t_e = \frac{0.61 \, H_0 U_0}{e_0 U \sqrt{gD}} \sim \frac{H_0}{U \sqrt{D}}$$

H_0 is the bed height at minimum fluidization velocity, which is roughly the settled bed height in a fixed bed; U_0 is the minimum superficial fluidization velocity; e_0 is the bed void fraction in the emulsion phase and at minimum fluidization velocity; U is the superficial gas velocity; g is the acceleration of gravity; and D is the reactor diameter.

Fraction of bubble-gas exchanged.

$$F = \frac{H_0 U_0}{k' U D} \sim \frac{H_0}{U D}$$

k' is the number of reactor diameters between slugs.

Bubble residence time. The time that a bubble remains in a bed is the bed height divided by the velocity at which bubbles are rising. This ratio is proportional to the bed height at minimum fluidization velocity and the rise velocity U_B of a single bubble injected into a quiescent bed. For slug flow and wall-controlled bubbles, $U_B = 0.35 \sqrt{gD}$.

$$t = \frac{H_0}{U_B} \sim \frac{H_0}{\sqrt{D}}$$

Relative bed expansion. The increase in bed height relative to its unexpanded bed height is proportional to superficial velocity divided by bubble velocity.

$$\frac{H - H_0}{H_0} = \frac{U}{U_B} \sim \frac{U}{\sqrt{D}}$$

Space time. The corresponding space time or the reciprocal of space velocity for a fixed-bed reactor is

$$t' = \frac{H_0}{U}$$

Therefore the effective space time of the gas in the fluid bed is proportional to that in a fixed bed by the factor $0.61\, U/\rho_0 \sqrt{gD}$. Space velocity is the dimensional group most commonly used in scaling catalytic reactors, and it is generally reported as the flow rate of reactants per unit weight of catalyst.

Length-to-diameter ratio. Geometry and bed expansion become important when entrance and exit effects are significant. Therefore, their constancy maintains these effects in proportion to scale, H_0/D.

All these conditions are not mutually compatible. With slow reactions and poor contacting, effective space time is important. For rapid reactions, the fraction of exchanged bubble gas is important. For a slow reaction with a large U_0, space time is of primary concern. In the case of noncatalytic reactions, bubble residence time may be the most significant factor.

In one study, ozone conversion in reactors with diameters of 50, 200, and 760 mm, bed heights of 0.9–3 m, and gas velocities of 6–24 cm/s, was studied (11). Conversions were calculated based on fixed-bed rate constants for the kinetics and effective space time and these were compared with measurements. As shown in Figure 5, there is good agreement among the various fluid-bed reactor sizes, although conversions are con-

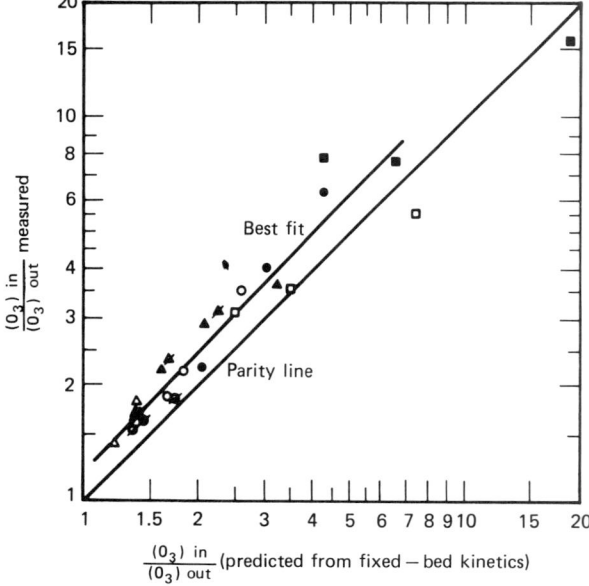

Figure 5. Measured and predicted fluid-bed conversions in different diameter units in terms of effective space time (10).

Approximate depth, m	Diameter, cm		
	5	20	76
3	■	●	▲
1.8	□	○	△
0.9		⊘	⩕

sistently higher than those predicted from the fixed-bed results obtained with a different batch of iron-scale catalyst. In the ozone conversion study, scale-up requires catalyst inventory to vary proportionally to reactor diameter and superficial velocity to vary proportionally to its square root.

Modeling Transient Processes

Reactor technology can also be applied to a time-dependent or transient chemical processes, eg, in the recovery of heavy crude oils by *in situ* combustion in massive unconsolidated sands (12) (see Petroleum, enhanced oil recovery). Underground burning offers promise for the economical recovery of crude reserves that are too viscous to be recovered by normal production methods, even though they are located in large, permeable, unconsolidated sand formations. Air is injected into the reservoir and a portion of the crude is burned and the temperature of the remainder is increased. The process involves multiphase fluid-flow, phase changes, gravity segregation, heat and mass transport, and chemical reactions. Scaling this unsteady-state process from field size down to the size and time scales related to laboratory equipment represents a feasible method for evaluating reservoir performance. The combustion front in this process is irregular because the air channels through the viscous crude and tends to rise because of gravity forces. The heat released into the unburned oil sand causes vaporization, thermal cracking, and viscosity reduction, all of which contribute to driving oil horizontally to the production well.

Six criteria, which include most but not all operative mechanisms, have been determined for scaling down the process:

Geometry. Constant well spacing-to-bed thickness ratio (L/D) is used to maintain comparable flow patterns.

Throughput. Constant UL, where U is superficial air rate, is used to maintain the same total air injection per unit volume of reservoir and similar convective heat and mass transport.

Diffusion. Constant t/L^2 maintains similarity for unsteady state heat and mass transfer; t is the time since air injection was initiated.

Dynamic and gravity forces. Scaling of the multiphase flow characteristics and the effects of gravity segregation requires a number of assumptions, including laminar liquid flow and comparable relative permeabilities and porosities of the sand strata as functions of the various fluid concentrations. This results in maintaining constant KL, where K is permeability as defined by Darcy's law for a single fluid.

Conduction heat loss. Oil-bearing sand is bounded top and bottom by impervious shale through which heat can pass. Thus, insulation, eg, concrete, with similar properties is used on a laboratory scale according to an insulation-to-bed thickness ratio (X/D).

Heat generation. The heat that is slowly generated away from the combustion front in a field operation is matched in the accelerated smaller model tests by raising their ambient temperatures. Physical properties are maintained approximately constant by distilling light components in the crude oil. The required temperature rise is determined from estimates of the additional heat-generation requirements of smaller models.

The practical consequences of these relationships are that the time since air injection is longer than that in the model by the field-model length ratio squared. Air and permeability flux must be greater in the model by this ratio of field and model dimensions. Higher permeability is obtained by using coarser sand.

This scaling concept was tested with two horizontal, sand-filled, cylindrical pressure vessels with internal concrete insulation. The units differed dimensionally by a factor of 6.9, and time was accelerated in the smaller unit by a factor of 48. Properties and operating conditions for the units and extrapolated values to a field operation are presented in Table 2. Five measures were used to establish the adequacies of the scaling relationships, percentages of oil recovered and injected air as functions of cumulative void volumes of the injected air, and the relative volumes and locations of burned and coked sand zones. Good agreement was obtained with these measures. In addition, published results from three field tests appeared reasonably consistent with these data. Following verification, a field element was modeled and several 4.8-yr field runs were simulated in one-hour tests. Recoveries of 60–70% were achieved with air requirements of 42–126 m^3/t crude oil produced (10,000–30,000 SCF/bbl).

Experimental Reactors

Data for the chemical kinetics that are operative in a process and the associated physical properties are best obtained with an experimental reactor that is different from the one that is to be used industrially or during process development. It is important that the reactor be similar to one of the basic types, have known flow patterns, operate isothermally, and provide data over wide ranges of variables. The range of the conditions to be studied must be sufficiently large for adequately defining the effects of variables, but the rate equations need not accurately reflect true reaction mechanisms. However, true rate equations can be useful to understand the process and to extrapolate results to conditions that have not been studied.

Stirred vessels are particularly useful for liquid and predominantly liquid multiphase systems. They are especially convenient for low pressure and modest temperature operations. Well-mixed batch reactors are preferred for slow reactions, since compositions can be accurately measured with time. The continuous-flow stirred-tank reactor (CSTR) is better for fast reactions, because conversion and composition at a given condition is well-defined. Interfacial areas in liquid–liquid and liquid–gas systems are not well-defined and may need to be established by some independent

Table 2. Model and Field Properties of Underground Burning[a]

Property	Small model	Large model	Field (extrapolated)
well spacing, m	0.26	1.3	183
sand thickness, m	0.09	0.6	61
sand permeability, μm^2 [b]	217	31.6	0.30
pressure, kPa[c]	2800	2800	2800
temperature, °C	173	97	38
air rate, $m^3/(m^2 \cdot s)$ [d]	146	21	0.2
exploitation time	0.5 h	24 h	27 yr

[a] Ref. 12.
[b] To convert μm^2 to darcy, multiply by 1.013.
[c] To convert kPa to psi, multiply by 0.145.
[d] STP.

means or by a different type reactor; liquid films or laminar liquid jets, because of their known interfacial areas, are useful for this purpose. Mechanical stirring rather than shaking or rocking should be used to produce well-defined mixing characteristics. Flat- or dish-bottomed, vertical cylindrical tanks containing four equally spaced baffles which are vertically mounted on the circumference are preferred. These baffles, with widths of one tenth of the vessel diameter, prevent swirling and vortexing and produce good mixing with either marine or flat-blade turbines. All experimental stirred reactors should be equipped with a variable-speed drive to ensure sufficient agitation.

The tubular reactor is excellent for obtaining data for fast thermal or catalytic reactions, especially for gaseous feeds. With sufficient volume or catalyst, high conversions, as would take place in a large scale unit, are obtained; conversion represents the integral value of reaction over the length of the tube. Short tubes or pancake-shaped beds are used as differential reactors to obtain instantaneous reaction rates. Instantaneous reaction rates can be computed directly, since composition changes can be treated as differential amounts. Initial reaction rates are obtained with a fresh feed. Reaction rates at higher conversions can be established either by using synthetic feeds, if the intermediates can be accurately specified, or by using a prereactor to supply feed of the correct concentration. In rare cases, recirculation around the differential reactor might be analyzed to follow the course of a reaction.

Maintenance of isothermal conditions requires special care. Temperature differences should be minimized and heat-transfer coefficients and surface areas maximized. Electric heaters, steam jackets, or molten salt baths are often used for such purposes. Separate heating or cooling circuits and controls are used with inlet and outlet lines to minimize end effects. Pressure or thermal transients can result in longer-lived transients in the individual catalyst pellets, because concentration and temperature gradients within catalyst pores adjust slowly. Even at steady-state, isothermal conditions, consideration must be given to the possible loss in catalyst activity resulting from gradients. The loss is usually calculated based on the effectiveness factor, which is the diffusion-limited reaction rate within catalyst pores divided by the reaction rate at catalyst surface conditions (13). For spherical catalysts and a first-order reaction, the effectiveness factor E can be related to the modulus ϕ:

$$E = \frac{3}{\phi^2}[\phi(\coth \phi) - 1]$$

$$\phi = R_p \sqrt{ka/D_e}$$

where R_p is the pellet radius, k is the first-order rate constant, a is the internal surface area, and D_e is the effective diffusivity. E should approximate unity. More complex formulations than the preceding equation have been developed for E. These are useful for reactors that are potentially subject to thermal instabilities, eg, hot spots and temperature runaways (1,14–15).

A partial list of recommended experimental reactor types is given in Table 3. The best selection depends upon a careful analysis of the anticipated reactions, the specific chemical and physical properties of the system, and appraisals of the objectives of the study (17).

Table 3. Experimental Reactors[a]

Reaction mixture	Experimental reactor	Remarks
gas	tubular	integral or differential
liquid	batch	slow reactions
	continuous-flow stirred-tank	moderately fast reactions
	tubular	fast reactions
solid	batch fluidized bed	inert gas or liquid for fluidization
gas–liquid	gas–liquid contactor[b]	slow reactions
	laminar film or jet	fast reactions
liquid–liquid	batch	good agitation
	continuous-flow stirred-tank	no settling of dispersed phase
	film	fast reactions
gas–solid	tubular	integral or differential with fixed bed
liquid–solid	batch	good dispersion of solids
	tubular	fixed or fluidized bed

[a] Ref. 16.
[b] Interfacial area must be known.

Model Reactions

As already discussed, independent measurements of interfacial areas are difficult to obtain in liquid–gas, liquid–liquid, and liquid–solid–gas systems. Correlations developed from studying nonreacting systems may be satisfactory. Comparisons of reaction rates in reactors of known small interfacial areas, eg, falling-film reactors, with the reaction rates in reactors of large but undefined areas, can provide an effective measure of such surface areas. Another approach is substitution of a model reaction, whose kinetics are well-established and where the physical and chemical properties of reactants are similar and the limiting mechanisms are comparable. The main advantage of employing a model reaction is the use of easily processed reactants, less severe operating conditions, and simpler equipment. Such an approach has been used in establishing the efficiency of catalyst utilization in mixed-phase, downflow, fixed-bed reactors operating as trickle beds (18). In this study, charcoal-catalyzed decomposition of hydrogen peroxide to water and oxygen was selected as the model reaction.

Complex Flow Behavior

The concepts of well-mixed and plug flow become inadequate when flow patterns deviate significantly from ideal behavior. Adsorption and subsequent desorption from within catalyst pellets in a trickle bed, solids recirculating downward to compensate for those entrained into rising bubbles in a fluid bed, and liquid held in the wake of a solid slurry exemplify back-mixing or bypassing in mixed-phase tubular reactors. Practical considerations, such as rapid mixing requirements, heating and reaction quenching demands, variations in flows through inlet distributors, the presence of internal baffles or structural obstructions, and inadequacies in locations of tank inlets and outlets, can effect substantial deviations.

Departures from ideal behaviors result in a range of residence times. Residence times can be coupled with reaction kinetics and used in assessing reactor performance.

The techniques for measuring residence times generally are used to define residence-time distributions (RTDs) of flowing and confined circulating elements. The RTD in a flow reactor can be determined by analyzing the effluent pattern resulting from a step function or pulsed change in an inlet flow property, eg, the concentration of one component or of an added tracer. Periodic functions, eg, sine waves, can also be used to measure time-averaged amplitude ratios and phase shifts. Thus, they provide inherently higher accuracy but require complex experimental technique. Comparable methods are applicable to analyze nonuniform batch reactors, but these require some means for introducing the tracer transient and for sampling or monitoring the resultant response.

The resulting flow responses are best illustrated graphically. Figure 6 shows the responses of outlet concentrations to step and pulse changes in inlet concentrations for various well-defined flow systems. For plug flow, an inert tracer is not observed until a time that is equal to the mean residence time in the vessel has elapsed. With complete mixing, a response to a step change shows immediately and rises progressively; a tracer also responds instantaneously but then diminishes. Intermediate distributions are obtained for plug flow with superimposed longitudinal mixing, as occurs with turbulent flow in an empty pipe or single-phase flow through a fixed-bed reactor. Laminar flow in pipes is also an intermediate distribution; a tracer is first observed in the effluent at half the mean residence time. When flow is restricted to part of the reactor vessel and stagnant fluid or recirculating eddies occupy the remainder, some tracer fluid does not appear in the effluent in a reasonable amount of time.

The residence-time distributions for these models can be expressed mathematically (1–2). Similarly, other RTDs can be predicted for other flow models or combinations of models and many of these are also available. However, response curves are not necessarily unique to specific flow patterns. Consequently, their validity depends on the degree to which the selected model adequately represents the effects of changing physical properties, operating variables, geometry, or size of the unit. On the other hand, these distributions can yield considerable insight in problem situations, when the responses do not meet expectations.

The concept of residence-time distribution is applicable to more than fluids. Varying catalyst deactivations in a reactor as results of loss of catalyst surface area by coking or other aging mechanisms can be evaluated with particle population-balance models. Other applications include monitoring particle breakup and agglomeration, drop coalescence, other mixing phenomena, and biochemical and crystal growth (1–2).

Perfect pulses and step functions are difficult experimental objectives. Pulses cannot be instantaneously injected nor step functions sharply outlined. Sinusoidal responses may take long times to reach steady states. Such analyses generally involve analyzing moments of both the transient inlet and outlet tracer signals. Difficulties exist in accurate determination of these concentrations and determination and extrapolation of concentrations in the long tail portion of a distribution where concentrations are low. An effective approach to reduce the contribution of the tail is to use the corresponding Laplace transform integrals. The ready availability of high speed computers permits multiple least-square curve fitting of the data using a range of values for the arbitrary constant s in the Laplace transform integrals (19).

Interpretation of the resultant response curves is accomplished in a manner that is somewhat analogous to determination of reaction orders. Combinations of plug flow

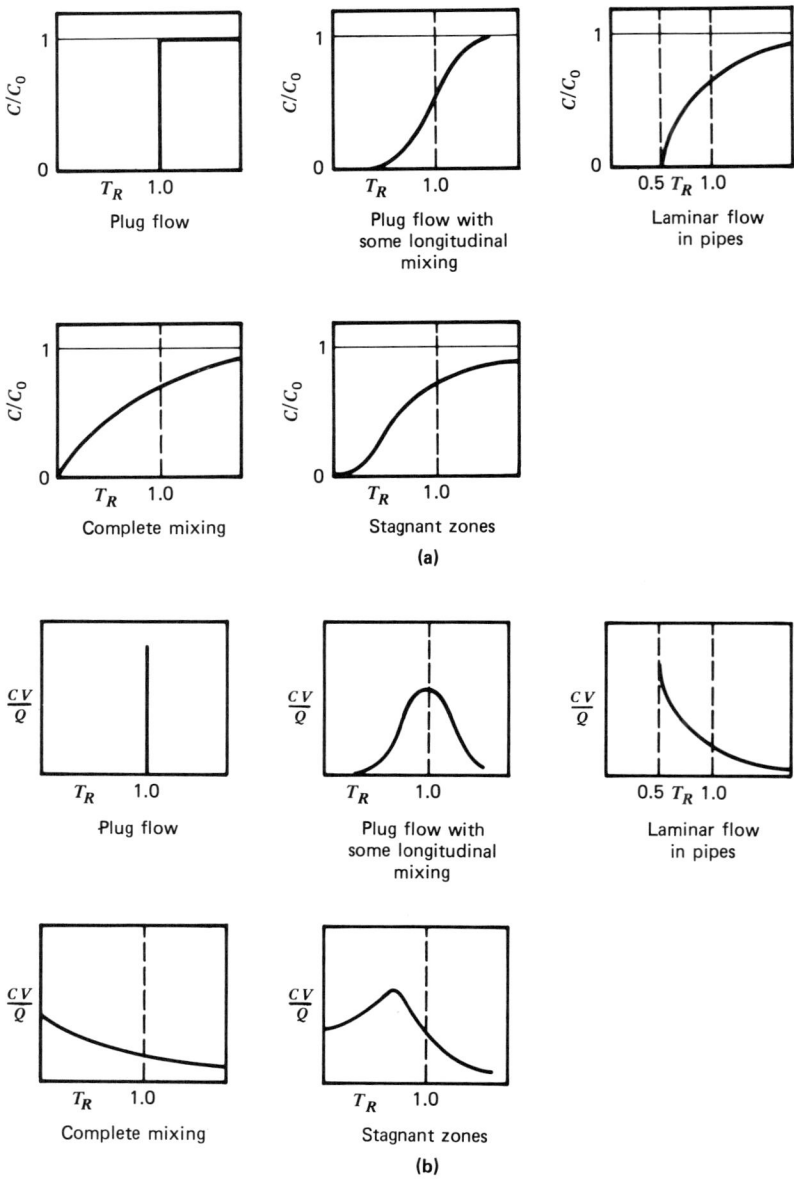

Figure 6. Characteristic residence time distributions. (a) Output responses to step changes. (b) Output responses to pulse inputs. $T_R = Ut/V$; U = velocity; V = reactor volume; t = time; C/C_0 = concentration relative to initial concentration; C = tracer concentration; Q = tracer volume.

and well-mixed reactors of various size fractions of the total reactor vessel are arranged in series and parallel combinations. The number of possible combinations can exceed the number of variables used to differentiate independently among them. Thus, reasonableness of a model is one of judgement. Once a model or combination of models is selected for further consideration, the quantitative expressions describing the models can be coupled with reaction kinetics into equations for predicting product yields. Many of the often used model combinations are described in refs. 1–2, 14, and 20–21.

In Figure 7(**a**) two plug-flow reactors are parallel and each is modified to account for longitudinal diffusion but their diffusivities differ. In Figure 7(**b**), the internal recycle gas-flow reactor comprises two countercurrent plug-flow reactors that are parallel with interconnecting distributed flows. In Figure 7(**c**), plug flow and well-mixed reactors are in series. Figure 7(**d**) is of a zero-intermixing model in which plug-flow reactors are parallel and a distribution of residence times duplicates that of a well-mixed reactor.

For zero- and first-order reactions, the combining sequence has no effect on predicted conversion. This is not the case for higher- or intermediate-order reactions. Residence-time distributions provide insight into the macromixed state of the system but provide no information about the mixing sequence, ie, the localized micromixed states within the reactor. The model combinations depicted in Figures 7(**c**) and 7(**d**) can describe reactors that are well macromixed but that are not micromixed or do not necessarily provide mixing sufficiently early in the process. In one case of a second-

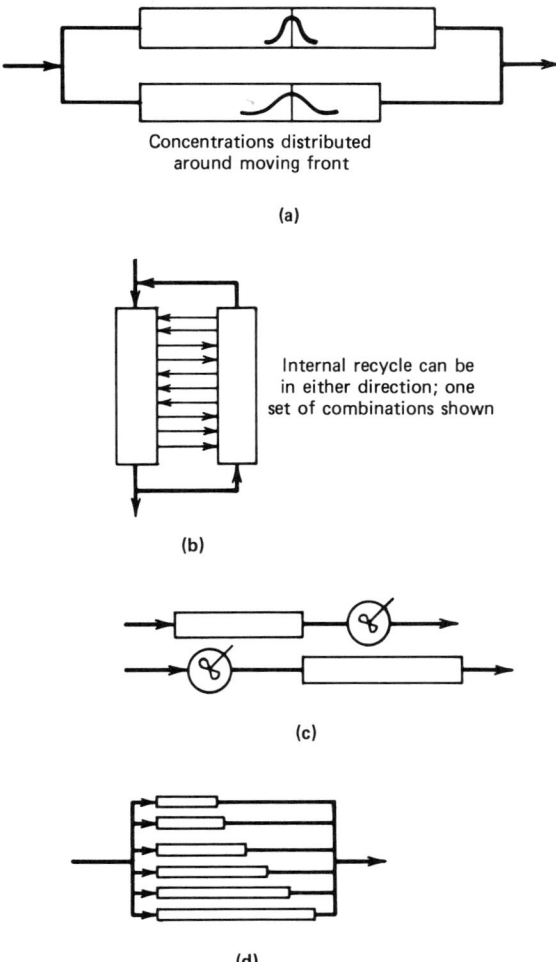

Figure 7. Combined flow reactor models. (**a**) Parallel flow reactors with longitudinal diffusion. (**b**) Internal recycle–cross-flow reactor. (**c**) Plug and well-mixed reactors in series. (**d**) Zero-intermixing model.

order reaction of premixed reactants and with well-stirred and plug-flow reactor volumes in a 4:1 ratio, the following conversions were calculated: plug flow followed by complete mixing, 75%; complete mixing followed by plug flow, 72%; and zero intermixing, 77%. These differences may represent either unacceptably high yield losses or the need to double the reactor volume (22). Despite these limitations, such models have been useful in developing, improving, and troubleshooting heterogeneous catalytic and noncatalytic thermal reactors, ion-exchange and chromatographic systems, polymerizations, and leaching and crystallization operations. More detailed discussions of micromixing and macromixing are given in refs. 23–24.

Cold-Flow Models

Use of tracers to obtain RTD characteristics followed by inferential analysis of flow patterns establishes conceptual but not actual detailed flow characteristics of a system. This limitation can be removed for reactions that are zero- or first-order or that can be so treated, eg, pseudofirst-order reactions, with the aid of cold-flow models. These are replicas of reactors or portions of reactors that operate at ambient conditions and are designed for detailed observation of flow patterns. The models are usually large so that little scaling is required. Similarly, fluids with known physical properties that are as similar as possible to those of the real system are used to minimize fluid dynamic scaling. Often the model is constructed in part of glass or plastic to permit photographing phenomena. The use of cold-flow models offers many advantages, including the potential for greater understanding, faster data collection because steady operation and accurate measurements are more easily obtained, and a greater range of geometries and flow conditions that can be studied. The development times should be shorter and research and development costs reduced as a result of adding this approach.

There are numerous applications of cold-flow models to all reactor types, eg, determining macro- and micromixing characteristics and contacting efficiencies; developing inlet distributor and quench hardware designs; selecting optimum feed-injection and product-removal locations; establishing the effects of axial- and radial-flow distributions; defining impeller and tank baffle configurations; determining mixing capabilities of jet nozzles; establishing limiting conditions of flow instabilities and extending operating ranges for acceptable performance; eliminating adverse flow distributions that promote erosion, attrition, misting, fouling, hot spotting, temperature runaway, etc; and providing generalized correlations for mixed-phase flow pressure drops and holdups. In one study, cold-flow modeling was used to eliminate large temperature gradients, improve catalyst efficiency, and increase throughput of a purportedly well-mixed commercial reactor (25).

A large cold-flow model requires large quantities of feed, possibly in excess of what is reasonable for a pilot-plant operation. One approach to reduce these requirements without sacrificing the scale of the experiment is to use a reactor slice. In effect, a two dimensional representation is employed. Such a unit was used to determine the effects of immersing large diameter tubes in a limestone fluid bed on fluidization performance and heat-transfer characteristics (26). This represented the first phase of the use of coal-fired fluid-bed technology for refinery and petrochemical plant process heating. The transparent test unit was 2.3 m wide by 3.7 m high by 0.3 m deep. Horizontal bundles of tubes up to 15 cm ID were used in experiments involving 2 t of limestone and ca 5 m³/s air at 170 kPa (10,000 SCF/min at 10 psig). Mixing characteristics were determined by photographs of color-tagged limestone as often as every second at three locations within the bed. Radial heat transfer from each tube was determined with electrically heated thin nichrome strips, which were imbedded in the tubes, and with specially constructed thermocouple probes. The conditions for acceptable and impaired solids movement and heat transport within the system were established as functions of geometric configurations and operating variables.

Numerical Experimentation

Numerical experimentation offers prospects for reducing dependencies on idealized models. Numerical experimentation involves combining modern quantitative representations of laminar and turbulent flows and new finite difference algorithms and other mathematics in order to solve coupled nonlinear differential equations and new advanced computer graphics in order to visualize results. The reduced cost of computation and the increased cost of experimentation makes numerical experimentation increasingly attractive.

Numerical experimentation has been applied principally in meteorology, nuclear energy and weapons development, aeronautics and space research, and petroleum reservoir engineering. Detailed quantitative predictions of free, wall, and recirculating flows, both turbulent and laminar, are possible even in representations of fluctuations and other transient behavior (27). The decay patterns of three-dimensional jets from rectangular orifices have been modeled according to a finite difference procedure, in which local turbulent energy and its dissipation are characterized by the k–ϵ model (where k is kinetic energy per unit mass of mean velocity field and ϵ is the dissipation rate) (27). Unsteady flows do not fundamentally alter turbulent structures or boundary layers. Flow separation that is caused by obstacles and other recirculating flows also have been simulated by the k–ϵ model. Improved characterizations of large eddies and small vortices is possible. Associated with these efforts has been the development of general steady-state and transient computer programs for carrying out finite element calculations. SOLA, which is a series of laminar flow codes, and TEACH-T and APACHE, which are, respectively, turbulent-flow and chemically reactive computer codes, are described in refs. 28–31.

Application of the mathematical formulations and techniques of numerical experimentation to reactor problems is recent and few validated examples have been published. Some data are available on the performance characteristics of industrial furnaces and gas turbines; such systems that operate with turbulent diffusion flames have been studied for simple two-dimensional geometries and selected conditions (32). Turbulent diffusion flames are produced when fuel and air are injected separately into the reactor. Second-order and infinitely fast reactions coupled with mixing have been analyzed with the k–ϵ model to describe the macromixing process. The performance with more complex multiple reactions of any order has been modeled by including micromixing through the addition of a coalescence–dispersion (c–d) model (33).

Axial and transverse velocity profiles and temperature distributions have been numerically determined for an epitaxial reactor that is a tubular reactor of rectangular cross section in which vapor deposition is used to produce high purity crystals and dielectric films and to treat material surfaces (34). Induction heating of a susceptor, eg, a carbon deposit that is placed on the bottom of the channel to 700–1200°C generates a vigorous but laminar, three-dimensional, free convective flow pattern. Finite difference equations to describe these patterns have been developed assuming constant physical properties. Calculated spiral flow patterns and temperature distributions show good agreement with experimental measurements in a reactor 2.6 × 4.3 × 170 cm (Fig. 8). Calculated reactant concentrations also agree with measurements for experiments in which the induction-heated carbon deposit is burned and the local concentrations of carbon dioxide are measured. Thickness and product purity, which are especially important in semiconductor manufacture, are functions of reactant concentration patterns.

Commercial Reactor Design and Use

Most reactors have evolved from concentrated efforts focused on one type of reactor. A few processes have emerged from parallel developments using markedly different reactor types. In most cases, the reactor selected for laboratory study has become the reactor type used commercially. Further development usually will favor extending this technology. Descriptions of some industrially important petrochemical processes and their reactors are available in several recent texts (35–37). Following are illustrative examples of reactor usages, classified according to reactor type (see also Petroleum).

Batch Reactors. The batch reactor is the preferred configuration for manufacturing plastic resins. Such reactors generally are 6–40-m³ (ca 200–1400-ft³), baffled tanks, in which there are blades or impellers which are connected from above by long shafts, and heat is transferred either through jacketed walls or by internal coils. Finger-shaped baffles near the top are used instead of full-length baffles. All resins, including polyesters, phenolics, alkyds, urea–formaldehydes, acrylics, and furans, can

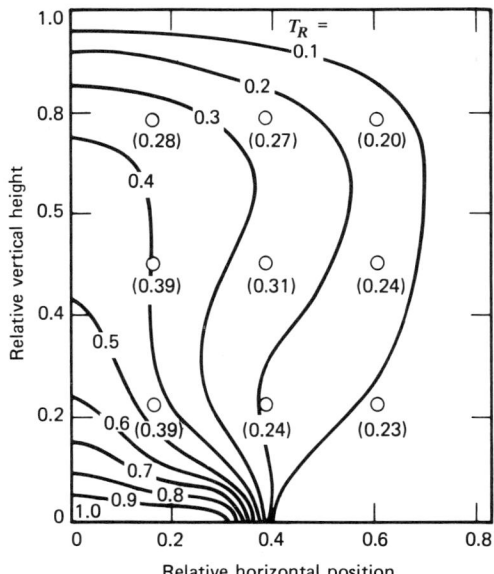

Figure 8. Calculated reduced temperature distributions for an epitaxial reactor cross section compared with experimental data (34). Gr (Grashof number) = 1.5×10^4; L/H = 0.827; X_A/H = 0.310; Pr (Prandtl number) = 0.69; Re = 12.8.

be produced in essentially the same way. Raw materials are held at temperatures of up to 275°C for ca 12 h until the polymerized liquid becomes sufficiently viscous. One plant has been designed to produce 200 different types and grades of synthetic resins in three 18-m^3 (ca 640-ft^3) reactors (38).

The cost savings of large-scale operations are possible for batch polymerizations. Worldwide production capacity of poly(vinyl chloride) (PVC) is greater than 15×10^6 t/yr, but no continuous process reactor has been developed for PVC production. However, commercial batch suspension-type PVC polymerization reactors of 200-m^3 (ca 7060-ft^3) volume, 5.5-m dia, and 10-m high have been developed (39).

Continuous-Flow Stirred-Tank Reactors. The synthesis of p-tolualdehyde (PTAL) from toluene and CO involves the use of CSTR equipment (40). p-Tolualdehyde is an intermediate in the manufacture of terephthalic acid. Hydrogen fluoride–boron trifluoride catalyzes the carbonylation of toluene to PTAL. In the commercial process, separate stirred tanks are used for each process step. Toluene and recycle HF and BF$_3$ come in contact in a CSTR to form a toluene–catalyst complex; the toluene complex reacts with CO to form the PTAL–catalyst complex in another CSTR. Once the complex decomposes to the product and the catalyst is regenerated, hydrated HF/BF$_3$ is processed in a separate vessel, because by-product water deactivates the catalyst and promotes corrosion. High yields (95% of theoretical), high product purity (99.3% PTAL), and low catalyst consumption are obtained.

The switch from the conventional cobalt-complex catalyst to a new rhodium-based catalyst represents a technical advance for producing aldehydes by olefin hydroformylation with H$_2$ and CO, ie, by the oxo process (qv) (41). A 200 metric ton per year CSTR pilot plant provided the scale-up data for the first commercial, 150,000 t/y plant.

The physical nature of the low density polyethylene is a direct function of reactor type (42). The autoclave reactor, which here is a CSTR, produces resin with long chain branches. However, a plug-flow tubular reactor can generate short chain branches of the same molecular weight. High pressures [100–300 MPa (10–30 atm)] and moderate temperatures (150–200°C) are used in both processes. Peroxides or oxygen initiators are added in the preheated ethylene feed in the autoclave reactors, but can be injected at multiple points along the long (10,000–100,000 tube diameters) tubular reactors. Because of these differences, the autoclave product is preferred for laminations and pipe moldings, and the tubular reactor is used to produce most films, blow and injection molding, and cable sheathing.

Thermal Tubular Reactors. Tubular reactors are used for the pyrolysis or thermal cracking of petroleum feeds to olefins, particularly ethylene. With the current uncertainty in feed supply, the trend is towards processing a wide range of feeds, including high molecular weight gas oils in single large units, ca 250 t/h (43). Reactor configurations strongly depend on feed composition and the degree of flexibility must be weighed against the cost. There are limits to the range of feeds that can be processed in tubular reactors. The residua are difficult to crack and thus must be processed in other reactor types, eg, fluid beds (44).

More than 8×10^6 t of hydrocarbons are converted each year to petrochemicals by low temperature, liquid-phase noncatalytic oxidation (see Feedstocks). Generally, conversions and selectivities to any given product are low. A few of the principle oxidations include butane to acetic acid and methyl ethyl ketone (MEK), p-xylene to terephthalic acid, cyclohexane to cyclohexanone and cyclohexanol, and n-alkanes to secondary alcohols. Oxidation proceeds by radical-chain reactions; therefore, continuous radical production is necessary. Since runaway branch chain reactions are possible, heat dissipation must be assured and oxygen concentrations must be controlled. These considerations often favor the use of a series of tubular reactors in plug flow with some back-mixing in each reactor to maintain sufficient radical concentrations to propagate the reactions. In butane oxidation, reaction rates are reduced as CO_2 and H_2O concentrations increase, so these products are best removed between reactor stages (45). In cyclohexane oxidation, selectivity to useful products decreases with conversion, but so does the ease of separation.

Bubble and Spray Columns. Bubble columns in series have been used to establish the same effective mix of plug-flow and back-mixing behavior required for liquid-phase oxidation of cyclohexane, as obtained with staged reactors in series. In one case, well-mixed behavior is established with both liquid and air recycle. The choice of this bubble column reactor was motivated by the need to minimize sticky by-products which accumulate on the walls (46). However, the high air rate also increases conversion by eliminating reaction water from the reactor. Thus, the choice of a reactor system need not always be based on compromise. Here, solutions to production and maintenance problems are complementary. Contrary to most bubble columns, the liquid in this reactor is intentionally well-mixed.

A spent sulfuric acid decomposition reactor illustrates the use of spray columns for processing. Thermal decomposition of spent acids is required as an intermediate step at temperatures sufficiently high to consume completely the organic contaminants by combustion. Decomposition requires temperatures above 1000°C. Concentrated acid can then be made from the sulfur oxides. Spent acid is sprayed into a vertical combustion chamber, where the energy required to heat and volatilize the feed and to support these endothermic reactions is supplied by complete combustion of fuel oil plus added sulfur, if additional acid is desired. High feed rates of up to 30 t/d of uniform spent-acid droplets are attained with a rotary atomizer. Decomposition rates of ca 400 t/d are possible in individual units (47).

Tubular Fixed-Bed Reactors. Bundles of downflow reactor tubes that are filled with catalyst and surrounded by heat-transfer media are tubular fixed-bed reactors. Such reactors are used most notably in steam reforming and phthalic anhydride manufacture. Steam reforming is the reaction of light hydrocarbons, preferably natural gas or naphthas, with steam over a nickel-supported catalyst to form synthesis gas, which is primarily H_2 and CO with some CO_2 and CH_4. Additional conversion to the primary products can be obtained by iron oxide-catalyzed water–gas shift reactions, but these are carried out in large-diameter, fixed-bed reactors rather than in small-diameter tubes (37). The physical arrangement of a multitubular steam reformer in a box-shaped furnace is described in ref. 1.

Approximately 45% of the world's phthalic anhydride production is by partial oxidation of o-xylene or naphthalene in tubular fixed-bed reactors. For the annual production of 31,000 t, ca 15,000 tubes of 25-mm dia would be used. Nitrate salts at 375–410°C are circulated from steam generators to maintain reaction temperatures. The resultant steam can be used for gas compression and distillation as one step in reducing process energy requirements (48) (see Phthalic acids).

For fixed-bed reactors containing rapidly deactivating catalysts, the scheduled changes in operating variables to accommodate activity loss can have a marked effect on run length. This is exemplified by acetylene hydrochlorination to produce vinyl chloride in tubular fixed-bed reactors. Steel reactors, of 2.4–3-m dia and 3–4.5 m high, with 150–500 tubes containing 6-mm $HgCl_2$-supported catalysts are used. Catalyst deactivation results from local overheating, which causes mercuric chloride to sublime from its carbon support. A variety of nearly optimal practices have been established, including changing both inlet and cooling-jacket temperatures to maintain conversion and decreasing mass flow rate (49).

Fixed-Bed Reactors. *Single-Phase Flow.* Fixed-bed reactors supplied with single-phase reactants are used extensively in the petrochemical industry for catalytic reforming, ammonia synthesis, various hydroprocesses, eg, hydrocracking and hydrodesulfurization, and oxidative dehydrogenation. The feeds in these processes are gases or vapors. The reactors generally are of large diameter, operate adiabatically, and often house multiple beds in individual pressure vessels. Bed geometries usually are determined by catalyst reactivity, so that the beds can be either tall and thin or short and squat. Variations of designs result from issues associated with the specific properties of the reactants or catalysts and differences in operating conditions. Either interbed cooling (or heating) or liquid quench additions are used to remove or supply reaction heat from the effluent of each bed. Multicomponent heterogeneous catalysts that require special activation treatments or unique handling requirements are used in most cases. Provisions must be made for restoring catalyst activity, either by replacement with fresh catalyst or by regeneration. Although reactants generally flow downward, air may be injected for catalyst regeneration at the bottom of the reactor.

Catalytic reforming is the primary refinery process for upgrading highly paraffinic, low octane gasoline-range feeds to higher octane products of about the same range of molecular weights. Because dehydrocyclization and dehydrogenation are principal reactions, catalytic reforming is a main source of aromatic chemicals, specifically benzene, toluene, and xylenes (see BTX processing). For the same reason, the process is an important net producer of hydrogen. Isomerization, which also improves gasoline octane number, and hydrogenation also occur. Process definition involves compromises among operating conditions (50). The balance among the variables is further modified if bimetallic catalysts, eg, Pt–Re, with their superior activity maintenance characteristics, rather than Pt-supported catalysts are used.

Consequent to these selection requirements, three basic reactor systems have evolved. The earliest involved three or four fixed beds in series and each was contained in a pressure vessel with reheat furnaces in between to compensate for the endothermic heat losses from dehydrogenation. Catalyst is regenerated after ca six months in this semiregenerative process. In cyclic reactors, there are four to five reactors in series with special valving arrangements, as shown in Figure 9, so that one reactor is regenerated while the others remain in service. These units are more flexible to process requirements and can produce high yields of high octane products. Slowly moving beds with radial gas flow are main features of the third reactor system. Essentially, continuous operation is achieved with steady catalyst addition to the top of three stacked reactors and subsequent removal from the bottom after about one to two months residence.

Reactor design for ammonia synthesis is subject to numerous constraints. Equilibrium favors high pressures [15–30 MPa (ca 150–300 atm)] and low temperatures (430–480°C) and kinetics favor the opposite conditions. The promoted iron catalyst tolerates a maximum temperature of 500°C. Cold synthesis gas is used to cool the

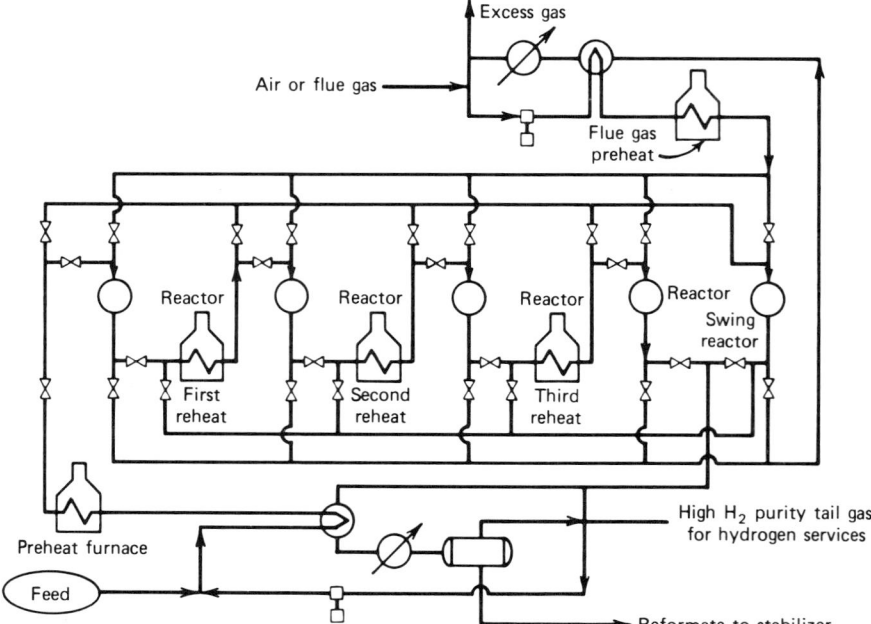

Figure 9. Cyclic regenerative catalytic reforming (36).

pressure vessel walls to prevent hydrogen embrittlement. Consequently, many ammonia converters have been designed and are in use (51). In one widely used design, several catalyst baskets are in series and cool synthesis gas is added between them. Radial reactors also have proven successful because smaller catalyst particles (1.5–3 mm) can be used without excessive pressure drop to increase catalyst utilization. Even horizontal converters containing three beds of catalysts have been designed (see Figure 10).

Hydrocracking, which is the undesired reaction in catalytic reforming, is used to crack and hydrogenate at high pressures [1–10 MPa (10–100 atm)] polycyclic aromatic feeds, eg, heavy gas oils, which are resistant to low pressure catalytic cracking. Hydrogen feed rate is the minimum required to suppress coke formation, since higher rates increase deleterious aromatic hydrogenations to naphthenes. Reaction temperatures are ca 200–400°C (52). Typically, four fixed beds are housed in a single pressure vessel, and hydrogen is injected between beds to control temperature. Temperature control is essential to prevent large temperature excursions. With the advent of new noble-metal catalysts, an additional set of fixed beds in a second pressure vessel as a second stage can be used to provide more flexibility in the relative amounts and qualities of gasoline and jet–diesel fuels produced (53).

The main impetus for such hydrotreating processes as hydrodesulfurization (HDS) was the availability of hydrogen from catalytic reforming and the deleterious effects of sulfur on downstream catalytic processing. Such processes minimize sulfur and nitrogen oxide emissions, which are generated from these products when they are burned as fuels. Relatively mild operating conditions [300–400°C and 1–7 MPa (ca 10–70 atm)] and simple adiabatic downflow fixed beds are used (54). Similar reactor configurations are used for dehydrogenation of n-butene to butadiene. Conventional dehydrogenation with steam is endothermic; the steam supplies heat and increases yields by reducing n-butene partial pressures. Oxidative dehydrogenation is the pri-

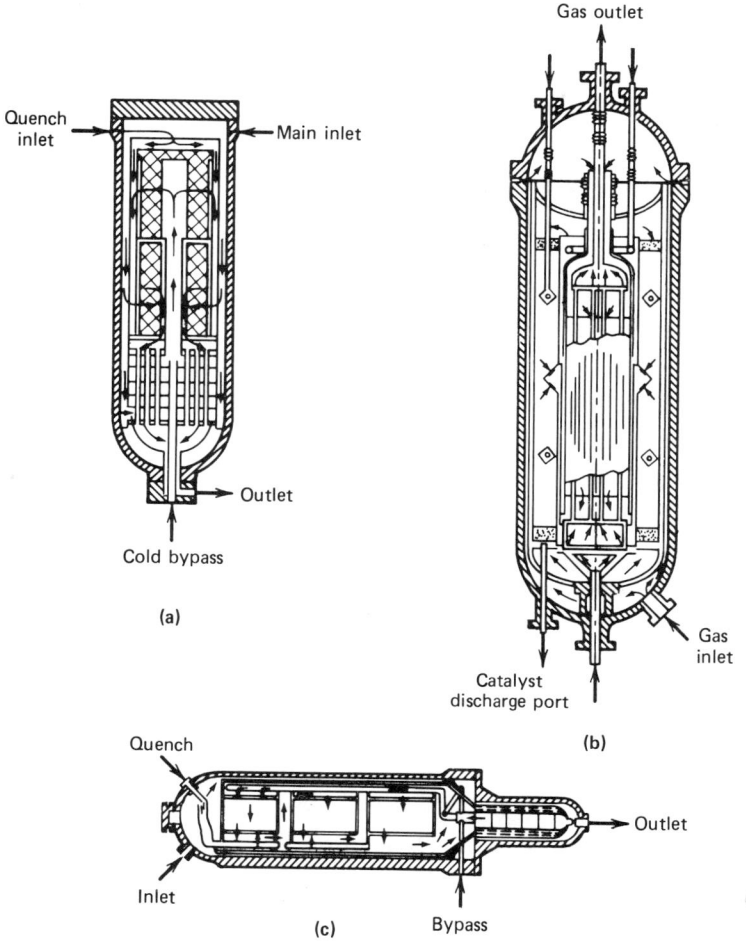

Figure 10. Three fixed-bed ammonia converter designs (1). (**a**) Radial H. Topsoe converter. (**b**) Upflow ICI reactor. (**c**) Horizontal multibed Kellog reactor.

mary butadiene process. It is an exothermic, autoregenerative, heterogeneous catalyzed reaction, in which the steam serves as a heat sink to moderate the temperature rise. Typical temperatures in the fixed bed are 350–600°C, and most of the 0.6-m deep beds operate at the higher temperatures (55).

Multiphase Flow. Flow regimes and contacting mechanisms in fixed-bed reactors that operate with mixtures of liquids and gases are totally different from those with single-phase feeds. Nevertheless, mixed-phase, downflow fixed-bed reactor designs are extensions of single-phase, fixed-bed hydroprocessing technology and outwardly resemble such reactors. The most generally used mixed-phase reactor is the trickle bed. Special distributors are used to uniformly feed the two phase mixtures (Fig. 11). Hydrodesulfurization, hydrocracking, hydrogenation, and oxidative dehydrogenation are carried out with high boiling feeds in such reactors. Pressures generally are higher than in their single-phase flow counterparts. Though some of these reactors may operate in the pulsed-flow regime, these reactors are called trickle beds because the same configurations are used. Further, reactor instrumentation may not be suited for recording the pulsation and, thus, the flow regime would not be noticed. Where required,

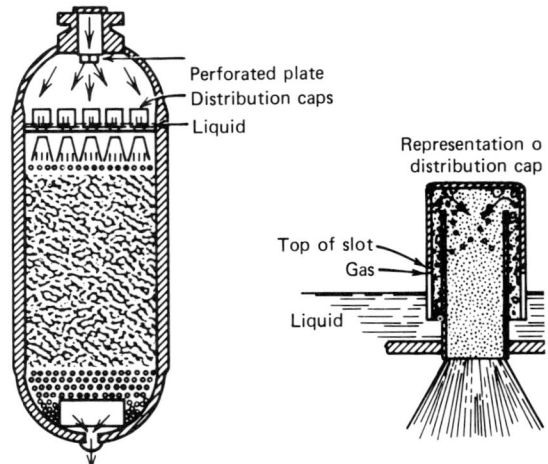

Figure 11. Mixed-phase, trickle-bed distributor used in the Unicracking-HDS process (56).

the temperature rise in trickle-bed reactors is controlled as with single-phase reactors. Generally, liquid quench systems are preferred.

High boiling feeds that have been successfully processed include virgin and cracked heavy gas oils and residua. Such feeds contain metals, which are hydrodemetalized and removed by deposition within the catalyst porous structures. Catalysts that are low in cost relative to noble metal supported catalysts slowly deactivate and are replaced after 1–3 years of use, depending upon the severity of service (56–57). Moving trickle beds are also used in these applications. Catalysts can be fed as an oil-slurry to the top and then flow downward through a series of beds. The conical bottom of each bed is designed to assure plug flow of catalyst (58).

Fluid Beds and Other Fluidization Reactors. The most well-known and one of the early applications of fluidization technology is catalytic cracking. In this process, gas oils are cracked at ca 500–525°C to produce gasoline and other light hydrocarbons, and fine silica–alumina-based catalysts averaging ca 50–60-μm dia are used. The reactors are either dense-phase fluid beds or dilute-phase transfer lines or combinations of the two fluidization regimes. In the various configurations developed for this process, deactivated catalyst, which flows downward through a steam-stripping zone to remove residual products, is air-lifted into a regenerator, where air at 600–700°C is used to burn the coke. Regenerated catalyst is recirculated to the reactor (59). The advent of high activity catalysts containing zeolites widened the range of design configurations, feed properties, and operating conditions. These units are large and can process 6000 t/d of feed. Most feature short residence-time, transfer-line reactor sections (Fig. 12).

Subsequent to this development, numerous other processes have been developed including fluid-bed coking of residua alone and combined with coke gasification (Fig. 13); catalytic reforming and catalytic oxidations of aromatics, eg, benzene, toluene, and naphthalene, to maleic and phthalic anhydrides; ammoxidation to acrylonitrile; and hydrogen chloride oxidation to chlorine. Process developments outside the petrochemical industry include iron-ore reduction and spent nuclear-fuel reprocessing. Thermal and catalytic coal gasifications, oil-shale retorting, flue-gas desulfurization, and ethanol dehydration represent new and renewed initiatives for fluidization technology. The latter process has been suggested for use in developing countries to

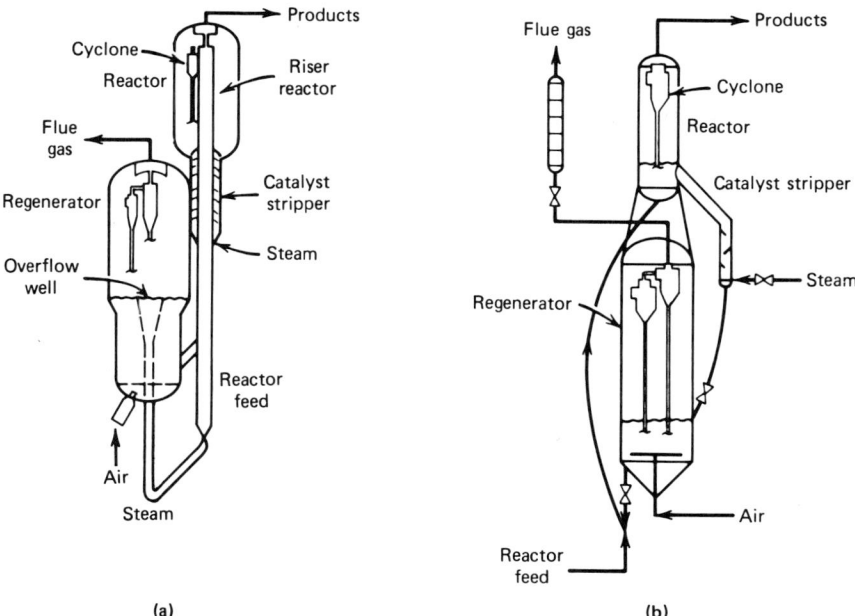

Figure 12. Two modern fluid-catalytic-cracking designs. (**a**) Exxon Flexicracker-riser configuration. (**b**) UOP stacked unit.

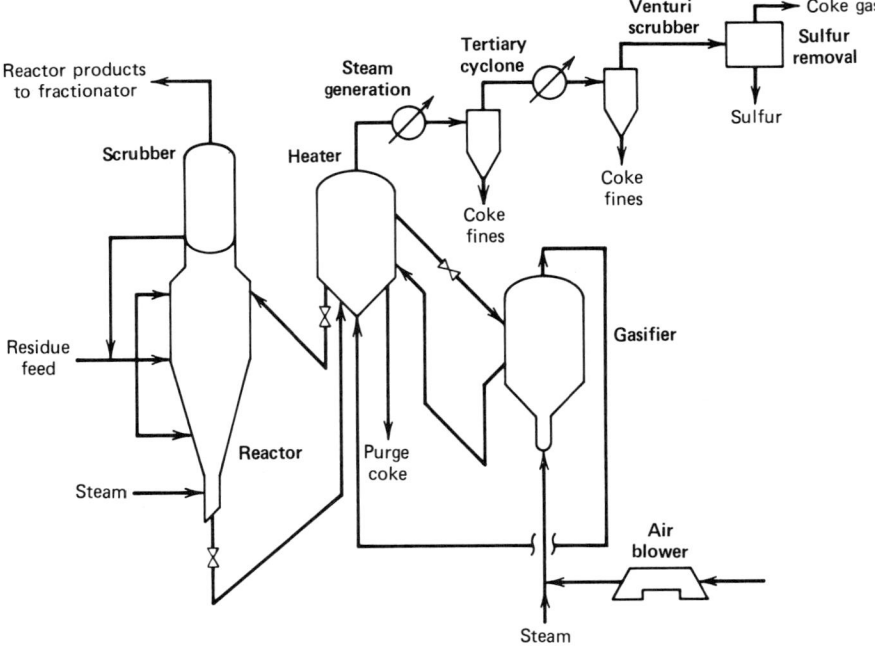

Figure 13. Simplified flow plan of a Flexicoking unit combining fluid coking and gasification in a three fluid-bed process.

reduce dependence on nonrenewable petroleum (60). Fundamental fluidization technology is described in refs. 61–62.

Logical extensions of the use of gas-fluidization technology are the gas-liquid,

912 REACTOR TECHNOLOGY

fluidized-bed (GLFS) reactors. These reactors offer the same advantages attributed to gas fluidization (63). A distinguishing feature of this fluidization regime is that liquid and gas fluidize the solids, because each alone may not be sufficient at the flow rates used. Such reactors can be used for catalytically hydrodesulfurizing and hydrodemetallizing residua fuels, upgrading heavy gas oils and residua for further processing, and converting nonpumpable heavy crudes and bitumens into transportable lighter crudes that are suitable for conventional processing. Liquid recycle is used to increase total flow rate to that desired for good fluidization quality (64). Catalyst activity is maintained by periodic addition and removal through lock hoppers (64–65).

Direct and indirect coal liquefactions are effected by gas–liquid fluidization. Indirect liquefaction involves coal conversion to synthesis gas which, with an inert hydrocarbon liquid, fluidizes a copper–zinc catalyst. The CO, CO_2, and H_2 mixture is converted to methanol at 230–250°C and 7.8 MPa (ca 77 atm) in this reactor. Catalyst activity increases as particle size decreases from 3 to 1 mm, which is the typical range for a gas–liquid fluidized-bed reactor (66). In one embodiment of direct liquefaction, crushed coal is contacted with molecular hydrogen and a hydrogen-donor diluent at ca 14–17 MPa (ca 140–170 atm) and 430–480°C. The dehydrogenated donor stream is regenerated in a conventional downflow, fixed-bed hydrotreater, and a heavy bottoms fraction from vacuum distillation is processed in an integrated fluid-bed coking and gasification unit (67). These new developments illustrate the merging of varied reactor technologies into one processing plant, as shown in Figure 14.

Miscellaneous. Descriptions of some industrially important petrochemical processes and the reactors associated with them are given in refs. 35–37. Additional reaction engineering texts are listed in refs. 68–73.

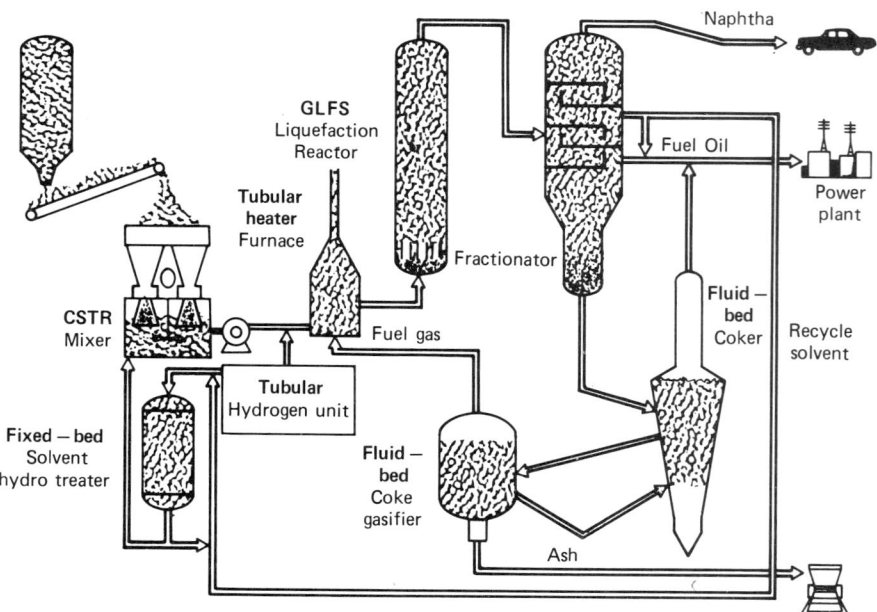

Figure 14. Schematic flow diagram of Exxon donor solvent (EDS) coal-liquefaction process.

BIBLIOGRAPHY

1. G. F. Froment and K. B. Bischoff, *Chemical Reactor Analysis and Design*, John Wiley & Sons, Inc., New York, 1979.
2. O. Levenspiel, *Chemical Reaction Engineering*, 2nd ed., John Wiley & Sons, Inc., New York, 1972.
3. P. B. Weisz and J. S. Hicks, *Chem. Eng. Sci.* **17,** 265 (1962).
4. P. N. Dwivedi and S. N. Upadhyay, *Ind. Eng. Chem. Process Des. Dev.* **16**(2), 157 (1977).
5. Y. T. Shah, *Gas–Liquid–Solid Reactor Design*, McGraw-Hill Book Company, New York, 1979.
6. R. B. Bird, W. E. Stewart, and E. N. Lightfoot, *Transport Phenomena*, John Wiley & Sons, Inc., New York, 1960, pp. 342, 646.
7. J. Beek in T. B. Drew, J. W. Hoopes, Jr., and T. Vermeulen, eds., *Advances in Chemical Engineering*, Academic Press, Inc., New York, 1962, pp. 203–271.
8. R. Lewis, J. P. Strackey, W. P. Haynes, R. R. Santore, and D. Dubis, *Proceedings of the Tenth Synthetic Pipeline Gas Symposium*, Chicago, Ill., Oct. 30–Nov. 1, 1978, pp. 207–219.
9. C. N. Satterfield, *Mass Transfer in Heterogeneous Catalysis*, The MIT Press, Cambridge, Mass., 1970.
10. J. M. Matsen and B. L. Tarmy, *Chem. Eng. Prog. Symp. Ser.* **66**(101), 1 (1970).
11. C. G. Frye, W. C. Lake, and N. C. Eckstrom, *AIChE J.* **4,** 403 (1958).
12. G. G. Binder, Jr., E. R. Elzinga, B. L. Tarmy, and B. T. Willman, *Proceedings of the 7th World Petroleum Congress*, Mexico City, Mexico, Vol. 3, 1967, p. 447.
13. Ref. 1, pp. 178–187.
14. H. Kramers and K. R. Westerterp, *Elements of Chemical Reactor Design and Operation*, Academic Press, Inc., New York, 1963, p. 228.
15. N. G. Karanth and R. Hughes, *Catal. Rev.* **9**(2), 169 (1974).
16. Ref. 14, p. 228.
17. V. W. Weekman, Jr., *AIChE J.* **20**(5), 833 (1974).
18. R. M. Koros in *Chemical Reaction Engineering 4th International/6th European Symposium*, Heidelberg, Federal Republic of Germany, Apr. 6–8, 1976, IX-372-381.
19. J. H. Seinfeld and L. Lapidus, *Mathematical Methods in Chemical Engineering*, Prentice Hall, Inc., Englewood Cliffs, N.J., 1974, pp. 372–382.
20. G. Astarita, *Mass Transfer With Chemical Reaction*, Elsevier Publishing Company, Amsterdam, The Netherlands, 1967.
21. C. Y. Wen and L. T. Fan, *Models for Flow Systems and Chemical Reactors*, Marcel Dekker, Inc., New York, 1975.
22. Ref. 1, p. 612; ref. 14, p. 84.
23. R. S. Brodkey, *Turbulence in Mixing Operations*, Academic Press, Inc., New York, 1975.
24. D. Hyman in ref. 7, pp. 119–202.
25. A. M. Goldstein, *Chem. Eng. Sci.* **28,** 1021 (1973).
26. D. C. Cherrington and L. P. Golan, *43rd Mid-Year Refining Meeting*, American Petroleum Institute, Toronto, Canada, May 9, 1978.
27. F. Durst, B. E. Launder, F. W. Schmidt, and J. H. Whitelaw, *Turbulent Shear Flows*, Vol. 1, Springer-Verlag, Berlin, 1979.
28. C. W. Hirt, B. D. Nichols, and N. C. Romero, *SOLA—A Numerical Solution Algorithm for Transient Fluid Flows*, Report LA-5852, Los Alamos Scientific Laboratory, N.M., Apr. 1978.
29. B. D. Nichols, C. W. Hirt, and R. S. Hotchkiss, *SOLA—VOF: A Solution Algorithm for Transient Fluid Flow with Multiphase Free Boundaries*, Report LA-8355, Los Alamos Scientific Laboratory, N.M., Aug. 1980.
30. A. D. Gosman and F. J. Ideriah, *TEACH-T: A General Computer Program for Two Dimensional Turbulent Recirculating Flows*, Imperial College Report, London, June 1976.
31. J. D. Ramshaw and J. K. Duchowicz, *APACHE: A Generalized Mesh Eulerian Computer Code for Multicomponent Chemically Reactive Fluid Flow*, Report LA-7427, Los Alamos Scientific Laboratory, Jan. 1979.
32. S. E. Elghobashi and W. M. Pun, *15th Symposium (International) on Combustion*, Tokyo, Japan, 1974, pp. 1353–1365.
33. G. K. Patterson, *70th Annual AIChE Meeting*, New York, Nov. 1977.
34. T. Hanzawa, Z. Sakauchi, U. Ito, K. Kato, and T. Tadaki, *J. Chem. Eng. Jpn.* **10**(4), 313 (1977).
35. A. V. Hahn, *The Petrochemical Industry—Markets and Economics*, McGraw-Hill Book Company, New York, 1970.

36. B. C. Gates, J. R. Katzer, and G. C. A. Schuit, *Chemistry of Catalytic Processes*, McGraw-Hill Book Company, New York, 1979.
37. C. N. Satterfield, *Heterogeneous Catalysis in Practice*, McGraw-Hill Book Company, New York, 1980.
38. *Can. Chem. Process.*, 21 (Apr. 1978).
39. S. Terwiesch, *Hydrocarbon Process.*, 117 (Nov. 1976).
40. S. Fujiyama and T. Kasahara, *Hydrocarbon Process.*, 147 (Nov. 1978).
41. *Chem. Eng.*, 110 (Dec. 5, 1977).
42. Y. Tomura, N. Nagashima, and K. Shirayama, *Hydrocarbon Process.*, 151 (Nov. 1978).
43. J. M. Dluzniewski and J. E. Wallace, *Oil Gas J.*, 64 (Oct. 16, 1978).
44. L. F. Hatch and S. Matar, *Hydrocarbon Process.*, 129 (Mar. 1978).
45. J. B. Saunby and B. W. Kiff, *Hydrocarbon Process.*, 247 (Nov. 1976).
46. J. Alagy, P. Trambouze, and H. Van Landeghem, *Advanced Chemical Series*, No. 133, Evanston, Ill., Aug. 27–29, 1974, pp. 644–645.
47. U. Sander and G. Daradimos, *Chem. Eng. Process.*, 57 (Sept. 1978).
48. A. Wiedmann and W. Gierer, *Chem. Eng.*, 62 (Jan. 29, 1979).
49. A. F. Ogunye and W. H. Ray, *Ind. Eng. Chem. Process Des. Dev.* **9**(4), 619 (1970).
50. Ref. 36, pp. 184–193; ref. 37, pp. 247–256.
51. Ref. 37, pp. 304–306.
52. Ref. 37, pp. 258–259.
53. A. Billon, J. P. Franck, J. P. Peries, E. Fehr, E. Gallei, and E. Lorenz, *Hydrocarbon Process.*, 122 (May 1978).
54. Ref. 37, pp. 259–265.
55. L. M. Welch, L. J. Croce, and H. F. Christmann, *Hydrocarbon Process.*, 131 (Nov. 1978).
56. R. L. Richardson, F. C. Riddick, and M. Ishikawa, *Oil Gas J.*, 80 (May 28, 1979).
57. *1978 Refining Process Handbook of Hydrocarbon Process.*, 121 (Sept. 1978).
58. W. C. Van Zull Langhout, C. Ouwerkerk, and K. M. A. Pronk, *Oil Gas J.*, 120 (Dec. 1, 1980).
59. Ref. 1, pp. 662–666.
60. U. Tsao and J. W. Reilly, *Hydrocarbon Process.*, 133 (Feb. 1978).
61. J. F. Davidson, D. Harrison, R. C. Darton, and R. D. LaNauze in L. Lapidus and N. Amundson, eds., *Chemical Reaction Theory: A Review*, Prentice Hall, Inc., Englewood Cliffs, N.J., 1977, pp. 583–685.
62. D. Kunii and O. Levenspiel, *Fluidization Engineering*, John Wiley & Sons, Inc., New York, 1969.
63. Ref. 5, p. 12.
64. R. P. Van Driesen, V. Caspers, A. R. Campbell, and G. Lunin, *Hydrocarbon Process.*, 107 (May 1979).
65. *Hydrocarbon Process.*, 133 (Sept. 1978).
66. M. B. Sherwin and M. E. Frank, *Hydrocarbon Process.*, 122 (Nov. 1976).
67. W. R. Epperly, K. W. Plumlee, and D. T. Wade, *Exxon Donor Solvent Coal Liquefaction Process: Development Program Status*, International Coal Show, Chicago, Ill., May 5–8, 1980.
68. J. Butt, *Reaction Kinetics and Reactor Design*, Prentice-Hall, Englewood Cliffs, N.J., 1980.
69. O. Levenspiel, *The Chemical Reactor Omnibook*, Oregon State University Bookstores, Corvallis, Ore., 1979.
70. C. D. Holland and R. G. Anthony, *Fundamentals of Chemical Reaction Engineering*, Prentice-Hall, Englewood Cliffs, N.J., 1979.
71. C. G. Hill, Jr., *An Introduction to Chemical Engineering Kinetics and Reactor Design*, John Wiley & Sons, Inc., New York, 1977.
72. J. J. Carberry, *Chemical and Catalytic Reaction Engineering*, McGraw-Hill, New York, 1976.
73. S. W. Churchill, *The Interpretation and Use of Rate Data: The Rate Concept*, McGraw-Hill, New York, 1974.

BARRY L. TARMY
Exxon Research and Engineering Company

RECORDING DISKS

The disk has been used for the storage of information for at least 100 yr (1). The recording of audio signals on disks as undulations representing sound waveforms began with the Berliner gramophone in the 1880s, in which a stylus, driven by a diaphragm at the throat of a horn aimed at a performer, scratched a clear track in lamp black coated on a glass disk. Design of the earliest disks exploited one of the fundamental advantages of disk-format storage: the registration of a large amount of basically one-dimensional information, eg, sound amplitude vs time, on a two-dimensional structure of unity aspect ratio and compact dimensions. This characteristic of disks makes them easy to manufacture by printing or molding, unlike cylinders, piano rolls, or tapes.

Berliner's gramophone record has enjoyed spectacular success as a distribution medium for sound. Despite order-of-magnitude improvements in playing time, dynamic range, and frequency response, the basic stylus-in-groove technology has persisted so well that quite old records play on modern equipment. The commercial benefit of this maintenance of compatibility has helped the record industry to grow to enormous size and the world population to have easy access to a huge resource of music and sound.

Magnetic Disks

The success of the disk format extends beyond the analogue audio disk. Since the 1950s, with the advent of the RAMAC product from IBM, magnetic disk devices have supplied computer systems with capacious data storage with rapid random access (2). The latter feature depends on a second fundamental advantage of the disk format, ie, convenient access to information by exploitation of two dimensions (radius and azimuth) for location of a segment of a one-dimensional signal. The use of two short motions to locate a block of information is advantageous in terms of time and mechanical simplicity over the use of one long motion, eg, searching a reel of tape.

Magnetic disk systems record data by writing patterns of magnetization on layers of ferromagnetic material which is coated on disks. This information can be read, erased, or overwritten; thus, the data storage is stable but alterable. In some magnetic disk systems, the disk is removable, and thereby provides a means of long-term or off-line storage or information interchange among systems. In the most recent high performance magnetic disk systems, the disk or disks, ie, the pack, is/are permanently mounted within the machines or drives that move them.

In less than thirty years, as computer systems of all sizes have become dependent on magnetic disks for rapid random access to large bodies of information, the worldwide annual sales volume of magnetic disk products has grown to a very large business with rapid growth; future growth is predicted to be at least as rapid. During the same time, storage capacities have increased and price per unit capacity has decreased by three orders of magnitude. The current rule of thumb calls for doubling of capacity every two to three years.

Video and PCM (Pulse Code Modulation) Audio Disks

During the late 1960s, consumer electronics and entertainment companies developed the concept of a record player for television; ie, a device connected to the antenna terminals of a standard television that plays inexpensive, prerecorded television records. In addition to the many magnetic tape and photographic film cassette formats, several disk formats, some of which read with a stylus and some with focused light spot, were developed.

In the first of the stylus systems (the TED (Telefunken-Decca) system from Teldec), a stylus and pressure pickup were designed to scan a groove (3). This system which was basically a miniaturization of a conventional audio playback system, suffered from poor performance and short stylus lifetime. In subsequent stylus systems, the stylus was one electrode of a variable capacitor and the vertically undulating floor of the track was the other electrode. In the first of these systems (the CED (capacitive electronic disk) system from RCA), a grooved record guided the stylus (4). A flat record is used in the more recent VHD (video high density) system for JVC (Japan Victor Company) and an electromechanical servo system guides the stylus. The laser optical systems read a flat record by means of an optical stylus in the form of a light spot produced by focusing the beam from a low powered laser (see Lasers). The spot is held in focus and on track by optoelectromechanical servos (5).

There has been much interest recently in the application of videodisk technology to audio recording by a PCM or digital encoding of the audio signal (6). Several systems of this sort, all of which are based on records which are incompatible with conventional phonograph equipment, are scheduled for introduction to the consumer market during the next few years. The disks in such systems reflect an emphasis on minimum cost, ie, manufacturing cost <$1/disk. The feasibility of this exemplifies the ease of manufacture afforded by the disk format; in comparison, the manufacturing cost for a consumer videotape cassette exceeds $10. In none of the disks is any provision made for user recording or erasure. The disks are typically embossed or molded from vinyl or acrylic plastic; the molds are made from master records by galvanic means derived from standard phonograph-record manufacturing techniques. The masters are cut either mechanically with a stylus or optically with a focused-light-exposing photoresist.

Optical Data Disks

Work on directly readable laser-written recording materials has been described (7–8). In the early 1970s, research efforts turned to materials and structures appropriate to recording at the submicrometer mark size, which is characteristic of optical video and data storage (9). Optical data disk systems have been developed to combine the rapid random access, local writing, and comprehensive error management and control characteristic of magnetic disk systems with the extremely high capacities characteristic of optical video disk systems. Optical data disk systems are expected to play an important role in the development and proliferation of computer and communication systems that are designed to handle text as well as numeric information (10).

The emphasis on direct readability for data storage is based on the desire to check the recorded data by reading and writing simultaneously; the desire to write and subsequently to read parts of a disk before writing other parts rules out gross chemical or physical development procedures affecting the whole disk. There are several additional generic requirements: the marking process should permit making contrast marks with minimal delivered energy density, the marks should be clean and repeatable, the marks should be permanent, and the frequency of bad or erroneous marks should be low.

Physical Properties

Audio master records are made by stylus-cutting the surface of appropriate master blanks (11). These blanks are made by spin-coating an acetate lacquer on both surfaces of a machined aluminum disk. The disks are available in standard diameters of 356, 301.6, and 254 mm, are 0.91 mm thick with an overall thickness with coatings of 1.3 mm, and are finished with a center hold 7.26 mm in diameter to fit the standard phonograph spindle (12).

The lacquer coating, typically 0.18 mm thick, must be very homogeneous in order that the walls of the groove cut by the stylus be smooth. The absolute requirement on this smoothness is set by the dynamic range expected of modern phonograph records; a groove wall roughness of ca 0.025 μm produces audible hiss. Heating the diamond cutting stylus helps in producing a smooth groove wall; a small resistance heating coil mounted on the stylus shank maintains the desired temperature. The thermoplastic property of the lacquer enables surface tension to smooth the resulting groove wall.

The cutting stylus differs from a playback stylus in that it has sharp edges and corners. In the plane of the disk, the stylus has an isosceles triangular cross section with the base facing forward. The resulting sharp-edged, flat surface digs material out of the lacquer coating. A tube connected to a vacuum pump sucks up the string of cut material (known as chip or swarf) to prevent fouling of the cutting process.

Magnetic Disks. Rigid magnetic disk platters look much like audio master blanks. They are aluminum disks that are spin-coated on both sides with layers of plastic (13). The plastic is a binder supporting fine ferromagnetic particles, called oxide, which are magnetized during the writing process. On some magnetic disk platters, continuous-film (plated) media are applied to the aluminum disk. Several standard disk formats are listed in Table 1 (14).

Table 1. Standard Dimensions of Rigid Magnetic Disk Platters

Disk diameter, mm	Thickness, mm	Center hole diameter, mm	ANSI standard
356.4[a]	1.91 or 1.27	168.3	X3.52-1976
210[b]	1.91	100	in process
200[b]	1.91	63.5	only ECMA[c] so far
130[d]	1.91	40.0	in process

[a] Actually 14.03 in., nominally 14 in.
[b] Nominally 8 in.
[c] Electronic Components Manufacturers' Association.
[d] Nominally 5¼ in.

Platters of the dimensions listed in Table 1 are used singly or stacked in packs of two to twelve platters. They are secured to hubs, which usually are cast aluminum, by mechanical clamping. The hubs in turn may be removable or may be permanently mounted to the spindles of the drives. There is a standard hub form for removable single platter disk cartridges that corresponds to the IBM 2315 product and is described by ANSI standard X3.52-1976. Removable packs meet the standards of individual drive manufacturers.

Rigid magnetic disk platters must meet stringent standards for smoothness and flatness and the coatings must be scrupulously clean. These strictures arise because the disks are written and read by heads which pass ("fly") over the disk surfaces at distances as small as 0.3 μm and are supported by a film of air that is viscously entrained by the rapidly moving (ca 100 km/h) surface of the spinning disk. If this film is disturbed even briefly, the head hits the disk. Such a head–media interference (crash) usually damages irreparably both head and disk. Avoidance of head–media interferences is the prime reason for the advent of fixed (ie, nonremovable) media drives.

In most current rigid disk drives, a servo keeps the head or heads on track despite disk eccentricity and vibration of the drive. The servo typically derives its guidance information from a pattern of servo tracks which exactly describe the desired data track format and which are written on one surface of one platter of the pack. An embedded servo pattern may be written on each active surface of the pack so as to guide without covering more than a small fraction of the useful surface area. Embedded servo patterns permit one-platter packs to be formatted efficiently and track densities not to be limited by platter-to-platter transfer accuracy of the head stack structure. Servo patterns are written on each pack by a servo writer, which is a spindle and moving arm assembly of extremely high stability and low mechanical error or runout. Servo patterns are determined by individual drive manufacturers.

Removable magnetic disk packs are available in various plastic cartridges or shrouds designed to protect the platters and surfaces from gross mechanical damage and contamination. Since these packages cannot provide a dust- or vapor-tight seal, the reliability of removable packs is affected by storage environment and operator habits. Usually some time is required for rigid disk packs to stabilize before use if they have been stored at low or high temperatures. Dropping a pack from desk height usually renders it nonfunctional.

Flexible or floppy magnetic disks or diskettes have oxide coatings similar to those on rigid disks, but the substrates are polyester. The disks are permanently sealed in plastic envelopes with access slots for the read/write heads and holes for reading sector and index marks. Standard floppy formats are listed in Table 2. The drives for floppy disks place the head or heads in rubbing contact with the oxide medium; therefore the flatness of the disk is not so important as for rigid disks. However, abrasiveness of the oxide layer directly affects head life, so coating smoothness and cleanliness remain important. Current floppy disks do not contain prewritten servo tracks.

Video and PCM Audio Disks. Master blanks for video or PCM audio disk recording are not standardized beyond the needs of each manufacturer. RCA typically masters its capacitive video disks by cutting an electroplated copper layer on a metal substrate with a stylus. This mastering is done through the use of a recently developed cutter which reciprocates at 6 MHz (4).

Capacitive disks can also be mastered by cutting with a laser, which is the preferred method for mastering of optical video and audio disks (16). A typical master

Table 2. Standard Dimensions of Flexible Magnetic Disks

Disk diameter, mm	Thickness, mm	Center hole diameter, mm	ANSI standard
200.2[a]	0.08	38.1	X3.73-1980
130.2[b]	0.08	28.6	X3.82-1980
88.9[c]			

[a] Actually 7.88 in., nominally 8 in.
[b] Actually 5.125 in., nominally 5¼ in.
[c] New from Sony; not yet standardized.

blank for laser cutting is a glass plate of 356-mm dia, 12.7-mm thick, and spin-coated on one side with a thin, eg, 150 nm, layer of positive-working photoresist. The stylus in this case is a spot of light focused from a laser beam and modulated so as to leave a trail of exposed and unexposed regions along a spiral trajectory on the resist surface. The exposed regions are removed in subsequent chemical development, leaving surface relief.

All master blanks for mass-replicated video and PCM audio disks are featureless; the master recorders have extreme mechanical accuracy and precision and, hence, guide themselves without help from the disk. The blanks must be clean and homogeneous enough to minimize errors. In video recorders, rates of less than one dropout (loss of signal, giving a black dot or line on the television screen) per television frame are standard. This corresponds to faults occupying ca 1 ppm of the surface area. PCM audio recording calls for similar defect control.

Optical Data Disks. There is increasing effort to develop disks capable of use in more than one brand of machine. Vacuum-deposited thin films of elemental metals and alloys have received the most investigation, but solvent-coated organics and combinations of the two are emerging as strong contenders, primarily because of their lower manufacturing costs.

Cleanliness and smoothness of the surface supporting the thin-film recording structure is crucially important to the achievement of low uncorrected error rates. After great efforts to rationalize grinding and polishing of glass substrates and clean molding of plastic substrates, it seems that a smoothing or subbing layer, which usually is a thin plastic layer that is cast, molded, rolled, or spun onto the main substrate, gives the best working surface (17).

For random-access data recording, it is necessary to preformat the disk with guidance information for the benefit of the radical tracking servo, so that data can be written in the prescribed location without the use of wasteful guard bands or overwriting previously written data. The straightforward but time-consuming approach of writing the servo pattern on the disk with a very stable recorder is available for any disk structure, but a guidance pattern that is molded into the subbing layer or the substrate, or that is printed onto the sensitive medium has been described (18).

Because light can be focused, it is possible to give the information-bearing layer physical protection with a transparent window or overcoating. This permits the physically accessible surface(s) of the disk to be a relatively large distance, ie, up to ca 1 mm, from the information-bearing surfaces(s), so that dust, scratches, and other environment-originating imperfections are far out of focus. Thus, their effect on writing and reading is minimized (19). Any optical recording disk standard must include the thickness and location of any such protective layer.

All emerging optical data disk formats include some sort of package, either in the form of a caddy, which holds the otherwise unprotected disk for storage, or a cartridge or cassette, all or part of which is loaded into the drive to carry the disk.

Chemical Properties and Manufacture

Lacquer recipes for audio master blanks are proprietary; however, most current lacquers are based on a nitrocellulose resin mixed with plasticizers, fillers, lubricants, and dye (20). This mixture, carried in a volatile solvent, is directed onto a spinning aluminum disk to produce a smooth, thin coating. The oxide media on most magnetic disks are layers of acicular gamma ferric oxide particles suspended in a polymer binder (21). The oxide particles are typically 100 nm long and 5–10 nm across; the binder is typically an epoxy resin. The layer is coated from solution, usually by spinning or sometimes by spraying. The finished layer, usually ca 1.3-μm thick on rigid disks and ca 4.6-μm thick on flexible disks, consists of ca 80 wt % epoxy and 20 wt % gamma ferric oxide (14). Plated media for magnetic disks historically have been electroplated layers of nickel–cobalt. However, thin continuous-film media have again been made to complement the demand for higher recording density. These newer media are usually vacuum-deposited and are sometimes single layers of cobalt alloyed with phosphorus, nickel, or rhenium, or both. Sometimes multilayer structures, eg, cobalt on chromium, are used (22). Layer thicknesses are always less than 1 μm. Plating is no longer the primary fabrication method for these media.

Master blanks for optical mastering of video and PCM audio records are typically spin-coated with positive photoresist, eg, Shipley AZ1350 (16). Disks for use by optical data recorders require sensitive, stable, light-responsive media (see Photoreactive polymers). Vacuum-deposited thin films of elemental metals and alloys, eg, rhodium (stable but very insensitive), bismuth (sensitive but very unstable), tellurium (sensitive and fairly stable), and alloys of arsenic, selenium, and germanium or tellurium (sensitive and stable) or both have been described (23). An issue concerning the vacuum deposition of heavy metals is their toxicity. However, appropriate traps in the vacuum pumping system collect vapor and particulate forms of the metals.

The high reflectivity of single-layer metal films results in rejection of most of the incident light and consequent low sensitivity. However, this problem is overcome by including the metal film in a multilayer, antireflection structure designed to couple light into the metal. Such structures include dielectric–metal bilayers, eg, TiO_2 over Te, metal–metal bilayers, eg, Bi over Ge (24), and metal–dielectric–metal trilayers, eg, Te or Ti over SiO_2 or polymethylmethacrylate (PMMA) over Al (25).

Organic and metal-organic materials have appeared as practical alternatives to metal films. The main reason for seeking alternatives to metal is the potential for reduced manufacturing cost when films are applied by spin or roll coating rather than by vacuum deposition. Eastman Kodak is developing a disk product with a dye-over-aluminum bilayer (26). Drexler Technology processes standard photographic silver halide emulsions to produce a dielectric with a metal crust. Workers at Xerox and elsewhere have produced optically dense dispersions of tiny iron particles in polymer binders.

Substrates considered for optical data disks include ground and polished glass, injection-molded plastic (sometimes vinyl but usually acrylic), cast and uv-cured plastic (usually acrylic), and polished aluminum as used in rigid magnetic disks (see Radiation curing). Investigations of molded and cast plastic substrates depend strongly on work done in the consumer video disk industry for replication of the consumer disk. Surface smoothing by spin-coated acrylic or other polymer layers is also being considered (17).

Economic Aspects

Worldwide volume of audio master blank sales are estimated to be $\$2 \times 10^6$/yr. Magnetic disk annual sales are more by approximately a factor of 100. Sales figures (1978) for rigid media worldwide were $\$30 \times 10^6$ (28), and for flexible media in the United States $\$40 \times 10^6$ (29).

The rigid-media market deals primarily with platters for fixed media drives. Except for replacements, this market is exactly proportional to that for new drives. The flexible-media market is far beyond that of companion drives because of media removability. Floppy disks are used as mass-storage media in many small computer and word-processing systems. This and the information interchange application originally considered by IBM when it introduced the original diskette guarantee that use of floppy disks is widespread.

Video and PCM audio master blanks are usually made by those who use them and so do not represent a significant market. The predicted value of the optical data disk market by 1985 is over $\$10^7$. These disks are to be sold to computer users and to users of office automation equipment.

BIBLIOGRAPHY

"Phonograph Record Compositions" in *ECT* 2nd ed., Vol. 15, pp. 225–232, by G. P. Humfeld, Radio Corporation of America.

1. *J. Audio Eng. Soc.* **25**(10/11), (1977).
2. R. White, *Sci. Am.* **243**(2), 138 (Aug. 1980).
3. D. Mennie, *IEEE Spectrum*, 34 (Aug. 1980).
4. *RCA Rev.* **39**(1), (1978).
5. L. Laub, *SMPTE J.* **85**(11), 881 (1976).
6. T. Doi, T. Otoh, and H. Ogawa, *J. Audio Eng. Soc.* **27**, 975 (1979).
7. D. Maydan, *Bell System Techn. J.* **50**, 1761 (1971).
8. Singer Librascope, product literature, Glendale, Calif., 1977.
9. A. Bell, *Nature* **287**, 583 (Oct. 16, 1980).
10. L. Laub, *Proceedings of COMPCON Spring '81*, IEEE, New York, 1981.
11. Ref. 1, p. 702.
12. Capitol Magnetics, product brochure, Hollywood, Calif., 1981.
13. R. Matick, *Computer Storage Systems and Technology*, John Wiley & Sons, Inc., New York, 1977, pp. 476–477.
14. Memorex, product literature, Santa Clara, Calif., 1981.
15. Memorex and Sony, product literature, New York, 1981.
16. J. Olijhoek and co-workers, *Proceedings of CLEOS*, Optical Society of America, Washington, D.C., 1980, p. 12.
17. A. Bell and co-workers, *IEEE Spectrum*, 27 (Aug. 1978).
18. Ref. 16, p. 22.
19. Ref. 16, p. 23.

20. Ref. 1, p. 783.
21. Ref. 13, pp. 480–481.
22. T. Chen, *Proceedings of the 1980 Meeting of the Materials Research Society*, K8.
23. Ref. 16, p. 24.
24. R. Willens, *Proceedings of the Symposium on Optical Storage Materials*, American Vacuum Society, 1980.
25. A. Bell, *IEEE J. Quantum Electron.* **QE-14**(7), 487 (July 1978).
26. D. Howe and co-workers, *Proceedings of Second SPSE Symposium on Optical Data Display, Processing and Storage*, Society of Photographic Scientists and Engineers, Springfield, Va., p. 97.
27. Ref. 26, p. 98.
28. *Worldwide Shipments of Rigid Disc Media*, IDC Report no. 1972, International Data Corporation, Waltham, Mass., April 1979.
29. *Market Outlook for Flexible Disc Media*, IDC Report no. 1992, International Data Corporation, Waltham, Mass., May 1979.

LEONARD LAUB
Vision Three, Inc.

RECREATIONAL SURFACES

For the purposes of this article, the term recreational surfaces means man-made surfaces that provide a durable area of consistent properties for recreational activities. The characteristics of the playing surface may be selected to match natural surfaces under ideal conditions, or may provide special characteristics not otherwise available. In all cases, the intent is to provide desirable and durable playing characteristics. Recreational surfaces are used for football, soccer, field hockey, cricket, baseball, tennis, track, jumping, golf, wrestling, and general purposes. Included in the latter category are indoor–outdoor carpets, patio surfaces, and similar materials designed for low maintenance in light recreational service.

A grasslike artificial surface was installed commercially for the first time in 1966 in the Houston Astrodome. Since that time, other commercial fabrics of various constructions have become available.

Resilient surfacing compositions for recreational use were introduced in tennis courts in the early 1950s (1). In 1958, the first all-weather resilient athletic track was installed at the University of Florida using a composition similar to that employed for tennis courts. The polyurethane running track used in the Olympic games at Mexico City in 1968 started a new era in this highly competitive sport, and the uniformity and resilience of the synthetic track contributed to establishment of new speed records. Similar tracks soon became standard throughout the world, probably contributing to the more frequent occurrence of the four-minute mile. In addition to the performance properties, the all-weather aspects of a synthetic track offered advantages for scheduled events and practice. The original system was a rubber and clay-filled polyurethane with polyurethane chips on the surface. A wide variety of systems followed, including polyurethane with sand finish for an all-weather skidproof surface, vinyl chips, rubber chips, so-called sandwich tracks consisting of a bound-rubber base with polyurethane surface, bound-rubber chips alone with a painted surface, etc. A similar smooth-surface product was used for basketball. For tennis, a sand finish proved acceptable, but the polyurethanes were not competitive in cost with coated asphalt and molded vinyls and polyolefins.

These grasslike and resilient installations require substantial amounts of synthetic materials. A typical football field covered with artificial turf requires approximately 15,000 kg fabric, 15–30,000 kg shock-absorbing underpad, and 5–10,000 kg adhesive and seaming materials. The artificial surface for a 0.40-km running track typically might require 50–70,000 kg composition. Proper coatings for smooth basketball courts, and striping and marking of turf tracks and courts require additional material.

Types

Recreational surfaces must provide certain performance characteristics at acceptable cost, with a reasonable lifetime, and with acceptable appearance. For classification, arbitrary but useful distinctions depending on the primary function of the surface may be made: a covering intended primarily to provide an attractive surface for private leisure activities, a surface designed for service in a specific sport, or a surface designed for a broad range of heavy-duty recreational activities, including professional athletics. Examples of these three categories are, respectively: patio surfaces, track surfaces, and artificial turf for outdoor sports.

Light-Duty Recreational Surfaces. Artificial surfaces intended for incidental recreational use are designed primarily to provide a practical, durable, and attractive surface, eg, for swimming-pool decks, patios, and landscaping. Material cost is a prime consideration. Many contemporary surfaces in this category utilize polypropylene ribbon and a tufted fabric construction (see below).

Specific Athletic Surfaces. Included here are running-tracks, tennis courts, golf tees and putting greens, and other applications designed for a particular sport or recreational use. Specific performance criteria are important.

In the case of a tennis surface, for example, friction and resilience characteristics affect footing and behavior of the tennis ball. The common asphalt tennis courts have been improved significantly by all-weather coatings with superior appearance and characteristics. Typical coatings are vinyl or acrylic compositions in various colors. Poured-in-place and preformed systems of polyurethane, vinyl, and rubber, although often used, are more expensive than coated asphalt or concrete. More recently, open molded mats that are easily installed as interlocking tiles have been used to construct new and repair old tennis courts. These mats are less expensive than the poured-in-place and preformed systems that require glueing. They are easily removed if necessary, and thus are easily and quickly repaired. In addition, the open structure allows rapid drainage after rain. Such molded systems are usually made of polyolefins, vinyls, or polyolefin–rubber blends, with appropriate pigmentation and stabilizers (see Olefin polymers; Vinyl polymers).

Characteristics and design criteria differ, eg, for a putting green vs a wrestling mat. Another specific surface is a warning area or track adjacent to a sports surface, eg, the area between the fence and the playing surface of baseball fields. This area must feel different from the main area to warn the player that he or she is approaching the fence. A football field may be surrounded by a full-size running track or, where the space is limited, by a warning track.

Multi-Purpose Recreational Surfaces. The performance demands for artificial surfaces in this category control the design. A good example is the playing surface for professional football in the United States. The shock absorbency of the system affects player safety and long-term performance under very heavy use. The grasslike fabrics used for these applications are made from various pile materials, including polypropylene, nylon-6,6, nylon-6, and polyester (see Polyamides; Polyesters). The fabric may be woven, knitted, or tufted. The critical underpad is derived from various materials, representing a compromise of performance properties. Because of the importance of safety and performance, fabric and installation costs are higher than the lighter duty surfaces.

Performance Characteristics

User-Related Properties. Shoe traction for light-duty consumer purposes requires modest footing and reasonable surface uniformity. The frictional characteristics are obviously of much greater importance in surfaces designed for athletic use. They can be significantly affected by pile density, pile height, and other aspects of fabric construction.

Measuring the coefficient of static friction between the playing surface and the shoe or other contact surfaces determines traction. To test traction for grasslike surfaces, the force required to initiate movement in a weighted sports shoe resting on the artificial turf is measured (2). The coefficient of static friction is defined as the force pull in a direction parallel to the playing surface divided by vertical force loading. The magnitude of the vertical force loading must be sufficient to approximate actual penetration, but is not otherwise critical. The shoe characteristics significantly affect the traction.

Typical static friction coefficients are given in Table 1. These data demonstrate that the absolute traction values for synthetic surfaces are satisfactory in comparison

Table 1. Typical Traction Characteristics of Surfaces[a]

Surface	Static-friction coefficient[b] Dry	Wet	Directionality index[c]
recreational surface			
tufted polypropylene	1.7–2.0	1.8–2.1	0.1–0.2
knitted nylon-6,6	1.8–2.0	1.6–1.8	0.10–0.25
tufted nylon-6	1.9–2.1		0.05–0.15
woven nylon-6	1.9–2.1		0.05–0.15
indoor-outdoor carpeting			
tufted polypropylene	0.4–1.5	0.4–1.5	
natural grass[d]	1.0–2.2	0.7–1.4	not applicable

[a] Ranges measured with appropriate sports shoes for the indicated surfaces.
[b] Defined as the average value measured in the four principal directions parallel to the fabric surface; two across the pile, one with, and one against the pile.
[c] Defined, from the same data under [b], as the average absolute deviation of the four traction values from the mean.
[d] The range is determined by the type and condition of grass.

with natural turf, provided that proper shoe surfaces are employed. Furthermore, synthetic surfaces, by virtue of their construction, are somewhat directional. This effect is evident in a measurement of shoe traction in various directions with respect to the turf-pile angle. Some traction characteristics are directly affected by the materials. Nylon pile fabrics, for example, show different traction characteristics under wet and dry conditions than polypropylene-based materials, since nylon exhibits higher moisture regain. Effects of artificial turf-fabric construction on shoe traction are given in Table 2.

For more specialized surfaces, shoe traction is equally critical. Gymnasium floors and running tracks must be uniform over their entire span. A balance of properties that gives good footing, traction and ball response is required. The surface must be tough and durable, provide long life, perform over a broad range of environmental conditions. A proper combination of energy and shock absorption enables the athlete to perform to maximum potential in relative comfort. For running tracks, resilience minimizes energy dissipation on the surface. A particular range is optimum for the track modulus of elasticity (3).

Spikes on track running shoes should be long enough to provide adequate traction with easy removal. Ideally, dull spikes afford traction by depression of the surface and gain additional energy for the runner as the resilient surface rebounds when the weight if lifted. As with turf, a specific shoe design for the surface is necessary.

Game-Related Properties. For some activities, like running and wrestling, the result of direct impact by the player is the only consideration. For others, like baseball or soccer, the system must also provide acceptable ball-to-surface contact properties. Important ball-response properties on the artificial surface are coefficients of restitution and friction, because these directly determine angle, speed, and spin of the ball.

The coefficient of restitution is defined as the ratio of the vertical components of the impact and rebound velocities resulting when a ball is dropped or thrown onto a playing surface. The velocities or related rebound heights are measured photographically. Criteria, such as ball inflation pressure, air temperature, and other details must be specified.

The coefficients of static friction between a sports ball and the playing surface are the ratios of the horizontal forces necessary to initiate sliding or rolling motion across the surface to the normal forces (wt) perpendicular to the surface. The sliding

Table 2. The Effect of Fabric Properties on Traction Characteristics

Fabric[a] construction, density	Description	Static-friction coefficient[c]	Directionality index[c]
standard	height = 1.27 cm	1.8–2.0	0.2
standard	increased curl[b]	1.8–2.0	0.15
high	height = 1.02 cm	1.8–2.2	0.15
standard	texturized[d]	1.8–2.2	0.02–0.05

[a] 55.5 tex (500 den) nylon-6,6 pile ribbon.
[b] Curl is an index of filament modification imparted to the ribbon during processing.
[c] See Table 1 for definitions.
[d] Refers to a fiber-modification process.

and rolling coefficients of dynamic friction are similarly defined in terms of the forces necessary to sustain uniform motion across the playing surface. These friction coefficients determine slip or retention of inertial effects present upon impact. In golf, for example, the driven ball may bounce forward after the first impact with the surface, then bounce backward after the second. In this particular example, the combination of velocities and friction create a slipping condition on the first bounce; on the second, the rotational back spin imparted to the ball when first hit is activated by sufficiently large friction. In soccer, on the other hand, the ball in play rarely slips because a coefficient of friction ≥ 0.4, which is almost always achieved, is sufficient to transfer momentum.

Some illustrative values for ball-response parameters in various sports are given in Table 3 (4). Artificial surfaces can be designed to match certain desirable game-response parameters of natural grass surface and provide these properties consistently. Response is confined to a narrower range and is less affected by weather. As a general rule, however, artificial turf surfaces tend to be somewhat livelier in ball response, velocity, and distance of roll, with coefficients of friction lower than those for natural grass.

Shock Absorption. Artificial playing surfaces for moderate to heavy use must provide a degree of shock absorbency for player comfort and safety. This requirement is achieved through incorporation of a resilient layer. For heavy recreational use, the shock-absorbing layer is a resilient underpad that provides a distinct layer between playing surface and substrate.

An ideal shock-absorbing medium, eg, for football, in the United States would provide a reasonable softness for player comfort in normal shoe contact combined with a high capacity for dissipation or distribution of kinetic energy involved in the impact of the player's fall. Various foamed elastomers are suitable for this purpose. The design

Table 3. Physical Parameters[a] for Ball Response from Sports Surfaces

Surface	Coefficient of restitution	Friction coefficients		
		Static	Sliding	Rolling[b]
soccer				
nylon turf-pad	0.7	0.4	0.3–0.4	
polyester turf-pad	0.7	0.4	0.3–0.4	
polypropylene turf-pad	0.7	0.5	0.5	
natural grass				
long	0.6	0.8	0.8	
soft	0.5	0.8	0.8	
short	0.7	0.8	0.8	
baseball				
nylon turf-pad	0.6	0.5–0.6	0.6	0.1–0.2
natural grass	0.5	1.0	0.8	0.2
golf				
nylon turf-pad	0.5	0.3–0.5		0.1–0.2
natural grass	0.4	0.5		0.1–0.2
tennis				
nylon turf-pad	0.7	0.4–0.6		0.1–0.2
natural grass	0.8	0.7		0.1–0.2

[a] Approximate values.
[b] Rolling resistance increases markedly with rolling velocity.

criterion is the ability to dissipate energy of motion by reducing the deceleration and through hysteresis losses in the material. A useful device for characterizing the required properties is a dynamic-mechanical impact tester (5). It employs an instrumental missile that is allowed to fall freely from a specific height onto the resilient surface. Suitable sensing components record electronically the force- and displacement-time profiles of the missile throughout the interval of first penetration and rebound from the playing surface. The plots shown in Figure 1 illustrate the acceleration and displacement between initial and final contact of the missile with the playing surface, both vs time. The profiles show that the deceleration forces increase as the missile penetrates into the surface, reach a maximum, and then decrease as the missile rebounds from the surface. The effectiveness of the shock-absorbing medium is indicated by the height of the maximum or, more accurately, the integrated profile throughout the duration of impact, calculated according to the severity index, $\int g^{5/2} dt$, where g = acceleration due to gravity, and t = time. The more effective the shock-absorbing material, the less sharply peaked is the g_{max} curve, and the broader and shallower the g_{max} profile. Effective performance would be achieved by a material displaying large hysteresis in which the impact of the falling weight is progressively absorbed without rebound. Clearly, since a useful system also must be reversible, this extreme example is not practical. Useful materials for shock absorbency have the ability to gradually dissipate the impact of the falling object, with a moderate conversion of the total kinetic energy to heat through hysteresis losses.

The shock-absorbing characteristics of underpad materials or resilient surfaces are functions of material selection, physical composition, thickness, and temperature. The sensitivity of performance to physical characteristics and temperature of the shock-absorbing medium is illustrated in Table 4 and in Figure 2. Clearly, thicker materials offer better shock absorbency. However, in practice, an excessively thick underpad may result in unsure footing and increased cost. The effects of temperature must be anticipated, since pad temperatures can easily range from below freezing to 65°C in the sun. The ideal system would provide a relatively flat g_{max} response over this temperature range. A compromise is a design for g_{max} peaks no greater than about 250 and a severity index below 1000 within a reasonable range of temperatures.

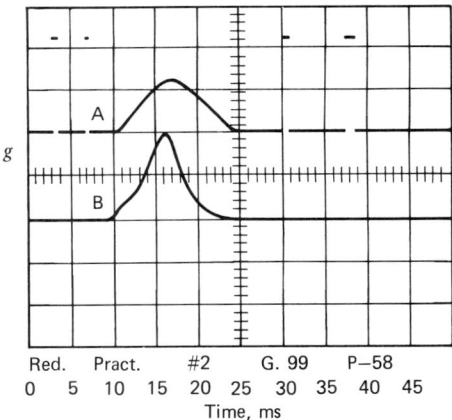

Figure 1. Deceleration and penetration curves from dynamic impact tester. Curve A is distance penetrated and curve B is deceleration plotted against time. Each square in curve A = 12.7 mm. Each square in curve B is equal to 50 g.

Table 4. Properties of Typical Underpad Materials

Property	Closed-cell foam, % closed cells		Poured elastomer
	75–85	90–95	
thickness, cm	1.6	1.6	1.0
tensile strength, kPa[a]	620	655	2700
density, kg/m³	96	256	1300–1400
g_{max} at °C			
21	85[b]	125[b]	105[c]
−12	105	150	
49	120	150	

[a] To convert kPa to psi, multiply by 0.145.
[b] 60-cm drop height of 9-kg flat-head missile (ASTM specifies 2 ft drop, 20 lb flat-head missile) (5).
[c] 22.8-cm (9-in.) drop height of hemispherical body (6).

Using the impact tester, values of g_{max} for grass-playing fields in late autumn range from about 75 for wet fields to 280 for frozen turf. The intermediate values observed depend upon soil type, moisture, condition, and other variables.

Durability. Grasslike surfaces intended for heavy-duty athletic use should have a service life of at least five years. The service life is approximately proportional to the amount of face ribbon available for wear. Pile density and height also affect the surface lifetime. In addition, different materials respond to abrasive wear to a greater or lesser extent. These effects cannot be measured except in simulated and controlled laboratory experiments, which do not necessarily reflect field conditions. Exposure to sunlight, uv light in particular, affects length of service. The Taber and Schiefer abrasion tests (7) evaluate fabrics and fabric constructions for potential wear properties. However, the data given in Table 5 indicate the unreliability of any specific accelerated wear test to predict longevity of fabrics, unless the tests are applied to closely related fabrics and for which actual wear-use data are available.

Other parameters provide indexes to surface durability including tuft bind (8), in which the force required to dislodge a surface element from its backing is measured; grab strength (9); and tensile strength. All are indications of strength and tear resistance. Artificial surfaces must be resistant to cigarette damage, vandalism, and the

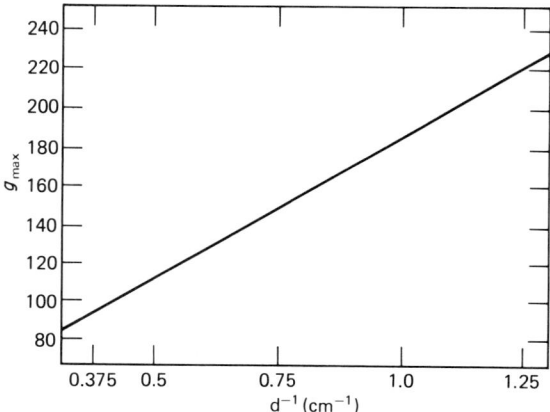

Figure 2. Shock absorption vs thickness for typical polyurethane resilient surface, measured according to Procedure B of ASTM F 355 using a 6.8-kg (ASTM specifies 15 lb) hemispherical missile dropped from a height of 30 cm (ASTM specifies 1 ft).

Table 5. Laboratory Simulations for Artificial Turf Wear

Surface	Effective pile loss, %	
	Taber method[a]	Schiefer method[b]
tufted polypropylene, 844 tex[c]	5	7
knitted nylon-6,6, 55.6 tex[c]	21	0.2
knitted polypropylene, 33.3 tex[c]		2.7

[a] ASTM D 1175, Rotary Platform, Double Head Method (5000 cycles).
[b] ASTM D 1175, Uniform Abrasion Method (5000 cycles).
[c] To convert tex to denier, divide by 0.1111.

like. A standard method for evaluating such resistance is the U.S. National Bureau of Standards Flooring Radiant Panel Test (10). In this test, a gas-fired panel maintains a heat flux, impinging on the sample to be tested, between 1.1 W/cm^2 at one end and 0.1 W/cm^2 at the other. The result of the burn is reported as the flux needed to sustain flame propagation in the sample. Higher values denote greater resistance to burning. The results obtained depend on both the material and surface construction. Polypropylene turf materials are characterized by critical radiant flux indexes that are considerably less than those for nylon, polyester, and acrylic polymers (11) (see also Flame retardants in textiles).

Materials and Components

The principal parts of a recreational surface system include the top surface material directly available for use and observation, backing materials that serve to hold together or reinforce the system, the fabric backing finish, the shock-absorbing underpad system, if any, and adhesives or other materials.

Surface Materials. Pile materials used in grasslike surfaces are selected from fiber-forming synthetic polymers, such as polyolefins, polyamides, polyesters, polyacrylates, vinyl polymers, and many others (see Fibers, elastomeric). These polymers exhibit good mechanical strength in the necessary direction. The materials listed in Table 6 are thermoplastic polymers that may be suitably pigmented before extrusion.

For the relatively smooth recreational surfaces of running tracks and tennis courts, polyurethanes, polyolefins, and other flexible, durable elastomers or composites are employed (see Table 7).

Backing Materials. Any fiber-forming polymer of reasonable strength may be used in backing materials, including polyamides, polyesters, and polypropylenes. The backing provides strength and offers a medium to which the pile fibers may be attached. The backing usually is not visible in the finished product, nor does its presence contribute much to the characteristics of the playing surface. However, it provides dimensional stability and prolongs service life. Some properties of fabric backing materials may be inferred from the data in Table 6; however, more highly drawn fiber equivalents of greater strength are employed for backing uses.

Backing Finish. Usually, the backing material is consolidated with a pile ribbon (see Textiles). In tufting, for example, the tufts are locked to the backing medium. In weaving and knitting, the finish seals and stabilizes the product. Backing materials

Table 6. Properties of Poured-in-Place Polyurethane Resilient Surfaces

Property	Test method	Desired range	Typical values[a]		
			A	B	C
impact resilience, %	ASTM D 2632	30–50	42	49	36
hardness, Shore A-2	ASTM D 2240	45–65	60	48	50
breaking strength, kPa[b]	ASTM D 412	>2,800	4,800	2,700	3,400
elongation to break, %	ASTM D 412	100–300	210	298	168
tear strength, N/m[c]	ASTM D 624	>11,000	17,200	15,200	11,200
10% compression, kPa[b]	ASTM D 575	590–860	720	500	520
compression recovery, %	ASTM D 395 (A)	>95	100	99.6	98.7

[a] Of laboratory samples prepared identically from different raw materials.
[b] To convert kPa to psi, multiply by 0.145.
[c] To convert N/m to lbf/in. (pli) sample thickness, divide by 171.1.

are usually applied as a coating that is subsequently heat-cured. For tufting, preferred choices are poly(vinyl acetate), poly(vinyl chloride), polyurethane resins, and various latex formulations. For knitted fabrics, poly(vinylidene chloride) and acrylics, or polystyrene–rubber latexes are used.

Underpads. Shock-absorbing underpad material is usually made of foamed elastomer, which provides good energy absorption at reasonable cost (see Foamed plastics). The foamed materials may be poly(vinyl chloride), polyethylene, polyurethanes, or combinations of these and other materials. Typical foam densities may range from 32–320 kg/m^3. Important criteria include a resistance to absorbing water, tensile strength and elongation, open-cell vs closed-cell construction, resistance to chemical attack, low cost, availability in continuous lengths, softness in energy-absorbing properties, and compression-set resistance (12). Water absorption is very important especially if the system is subjected to freezing temperatures. Generally, closed-cell materials are most resistant to water intrusion.

Some recreational surfaces, in particular the lighter-weight materials for patio surfaces, etc, either employ no shock-absorbing underpad system, or utilize a light coating that usually is joined directly to the turf during manufacture. This provides a certain amount of softness and grip adequate for the service intended. For running-track surfaces, on the other hand, the resiliency of the shock-absorbing medium and its relationship to athlete performance are obviously very important.

Adhesives and Joining Materials. Grasslike surfaces are employed over substantial areas, and lengths of rolls must be joined, glued, or sewn together. A variety of adhesives ranging from low cost poly(vinyl acetate) materials to cross-linked epoxy cements are available (see Adhesives).

Table 7. Typical Properties of Yarns Suitable for Pile Components of Artificial Surfaces

Property	Polypropylene	Poly(ethylene terephthalate)	Nylon-6,6
density, g/cm^3	0.91	1.38	1.14
melting point, °C	170	250	265
tenacity, N/tex[a]	0.22	0.18–0.35	0.31
elongation, %	25	30–100	33
moisture regain at 21°C and 65% rh, %	0.1	0.4	4

[a] To convert N/tex to g/den, multiply by 11.33.

Fabrication and Installation

Tufting. The tufting process is frequently employed in construction of grasslike surfaces (13). The techniques are essentially those developed for the carpet industry with economical high speed characteristics. In the tufting operation, pile yarn is inserted into one side of a woven or nonwoven fabric constituting the primary backing. Yarn is inserted by a series of needles, each creating a loop or tuft as the yarn penetrates the backing and forms the desired pattern on the other side. For artificial surfaces, the looped tufts that form in this process are cut to provide the desired individual blades in the playing surface. Cutting elements are incorporated in the tufting machine which sever the loops automatically in the process of forming the pile.

Depending upon the width of the fabric, a modern tufting machine may incorporate one to two thousand needles, simultaneously inserting the tufts along the fabric width. The needles may operate at speeds in excess of 500 strokes per minute, contributing to a highly efficient output of fabric yardage. The primary backing for tufted surfaces is usually a woven, synthetic filament fabric. After the tufts have been inserted, pile fiber and backing components are fused by application of the backing finish. The tufts are inserted in or looped through the backing fabric. Its quality determines the firmness of the attachment.

Knitting. The knitting process as applied to manufacture of artificial turf and related products provides a high strength, interlocked assembly of pile fibers and backing yarns. Pile yarn, stitch yarn, and stuffer yarn are assembled in the operation. The pile and stitch yarns run in the machine or warp direction, whereas the stuffer yarns interlock the wales, ie, rows, formed by the pile and stitch yarns, knotting the whole system together. Knitted fabrics typically possess high strength and high tuft bind.

A machine with approximately 1000 needles may be used to produce continuously a 5-m wide fabric. The assembly process is more complex than tufting. The pile yarn and stitch yarn are inserted into the knitting needle, whereas the stuffer yarn is interlocked with the others through a separate feed mechanism of the machine. As with tufting, the loops of pile fabric formed are slit, thus creating the desired individual blades.

The knitted fabric is subjected to a finishing operation in which a suitable backing material is applied to penetrate the yarn-contact points and stabilize the structure. This process usually is accompanied by a heat treatment that stabilizes the fabric and conditions the pile.

Weaving. As a general rule, weaving is slower than tufting or knitting. The process consists of a two or three-dimensional intermeshing of warp, pile, and fill yarns that may be of different types. In contrast to a knitted fabric, the yarns are not knotted together, but interwoven at right angles. The pile yarns are cut by a series of wires that are continuously assembled into and withdrawn through the fabric loops. A suitable finish further stabilizes the fabric.

Finishing. In each of the processes discussed above, the artificial turf fabric is subjected to a finishing operation in which a suitable adhesive is applied to the back side, thus bonding the components and stabilizing the material. The finish may be applied with a knife or brush, in foam or paste form, followed by a heating and drying stage. The temperatures also affect the pile-ribbon properties.

Underlay. An installed artificial turf system may or may not include components between the fabric and the subbase. As mentioned earlier, such components are not a requirement for light-duty use, but are essential in attaining the shock-absorbing properties required by heavy-duty surfaces. The foam underpads used in shock-absorbing systems are made by incorporating a chemical blowing agent into the foam latex or plastisol. Under the proper processing conditions, voids of controlled size and number are uniformly distributed throughout the foam material. Closed-cell foam structures are most desirable for outdoor use, because they resist water absorption.

Installation. In general, grass-like surfaces are glued down or otherwise affixed to a subbase. For light-duty purposes, it may suffice to tack the edges to the perimeter of the area to be covered. For heavy-duty systems, a solid subbase of asphalt or other material is first installed. The shock-absorbing underpad is glued to this surface, followed by glueing the turf on top of the pad. Other constructions with partial or complete glueing are also possible. However, it is essential to securely anchor the perimeter.

Additional fabric panels are bonded together by glueing onto a reinforcing tape, sewing, or some other technique. In Europe a secure stitch-seaming technique employing high-strength sewing yarns has an excellent durability record.

In another technique, common in Europe, the bonded turf–pad system is laid loosely onto permeable asphalt without glueing. This special subbase allows water to trickle through holes punched into the pad through the turf and to drain away laterally through the aggregate. Side-seam sewing is the preferred assembly technique.

For artificial surfaces in the athletic category, eg, running tracks, the installation techniques are different. A poured-in-place or interlocking-tile technique may be employed; the latter is used for tennis courts. Adequate provision for weathering and water drainage are essential. In general, the resilient surfaces are installed over a hard base that contains the necessary curbs to provide the proper finished level. Out-of-doors, asphalt is the most common base, and indoors, concrete. A poured-in-place polyurethane surface (14) is mixed on-site and cast from at least two components, an isocyanate and a filled polyol, of the polyether or polyester type. The latter usually contains an organic mercury catalyst (15) which provides a system with selective reaction toward organic hydroxyl groups, thus lowering the sensitivity toward moisture (see Urethane polymers). Amine-type catalysts, eg, Dabco 33 LV, may be used also. The isocyanate is of the polymeric type or a toluene diisocyanate or methylene bis-(phenyl isocyanate)-based prepolymer (see Isocyanates). Similar systems are used as binders for scrap-rubber granules. The surface properties can be varied by the type and amount of fillers and the size of the rubber granules.

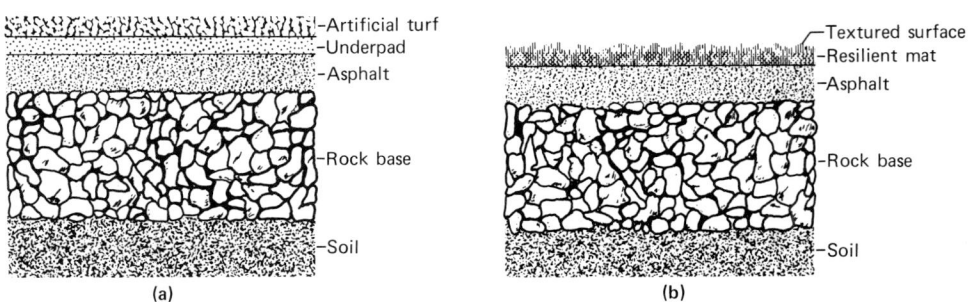

Figure 3. Cross-sections of typical artificial turf (**a**) and resilient track surfaces (**b**).

Table 8. Manufacturers and Trade Names for Artificial Turf Surfaces

Product	Manufacturer[a]	Description	Year
All-Pro Turf	All-Pro Turf	3.6-m width tufted fabric with polypropylene face yarn and synthetic backing yarns	current
AstroTurf Stadium Surface	Monsanto Company	4.5-m width fabric employing 55.5 tex (500 den) nylon-6,6 pile ribbon and high-strength polyester backing yarns	current
Clubturf	Clubturf, UK	woven polypropylene turf	current
Dunlop	Dunlop, UK	tufted nylon-6 fabric	current
Grand Turf 7000	Kureha Chemical Ind. Co., Ltd., Japan	a woven fabric with poly(vinylidene chloride) ribbon for the face pile and polyester warp and fill yarns	1977
Gräs	Fieldcrest/Karastan	woven fabric with textured nylon-6 face ribbon, synthetic yarns in backing	current
Grass Sport 500	Chevron	polypropylene turf–polyurethane pad system for sport use	current
Instant Turf	Instant Turf Industries, Inc.	polyolefin synthetic turf products for consumer and recreational use	current
Lancer	Lancer Enterprises, Inc.	light-duty recreational surfaces utilizing polypropylene or polyester face yarns, for consumer and marine applications	current
Playfield	Playfield Industries, Inc.	tufted product utilizing Chevron Polyloom II polypropylene yarn	current
Poligras	J. F. Adolff AG, FRG	knitted polypropylene pile fabric supplied with a bonded underpad	current
Poly-Turf	Sports Surfaces International Ltd., UK	nylon-6 tufted fabric installed with nitrile rubber–poly(vinyl chloride) pad	current
SuperTurf	SuperTurf, Inc.	tufted product utilizing polypropylene yarn and a synthetic backing	current
Tartan Turf	3M Company	tufted fabric with low-tex (den) nylon-6,6 filament	ca 1976
Marubeni/Toray GS-2	Mitsubishi Trading Company, Japan	tufted nylon-6 fabric utilizing woven polyester backing	current

[a] United States, unless otherwise indicated.

The mixed liquid is pumped into the area, where it levels and cures into the slab. It may be poured in two layers to eliminate imperfections in the base. The first layer may be a preformed rubber slab which is glued to the base, or a mixture of reground rubber and binders or rubber and polyurethane. A textured surface may be imparted to the second coat with sand or chips.

A permeable system may be strip-glued over a permeable asphalt base for drainage; this procedure is common in Europe. In some cases, colored binders impart the desired color, and a final protective coating of a urethane lacquer may be sprayed on the surface. Preformed slabs also may be used indoors, and covered by poured-in-place polyurethane. Whether solid or layered, indoor systems, particularly those used for basketball, require proper finish and maintenance coats to ensure adequate performance. The finish is selected according to the amount of use and the frequency of maintenance.

The general construction of artificial turf and resilient track surfaces is indicated in the cross-section drawings of Figure 3.

Table 9. Manufacturers[a] and Trade Names for Resilient Surfaces

Product	Manufacturer	Description	Year
AstroTurf Track Surface	Monsanto Company	poured-in-place filled polyurethane with smooth, sand, or chip surfaces	to 1977
Cal Track	California Products Corporation	highly porous rubber-urethane mat installed over subbase	current
Chevron 440 track	Chevron	poured-in-place filled polyurethane base covered with a layer of colored granules adhered to the base to form the running surface	current
Chevron 400 track	Chevron	breathable surface over a 0.95-cm base mat of ground recycled rubber with a urethane textured coating containing resilient granules	current
Mateflex	Mateflex by Mele	patented molded polypropylene rubber interlocking modules 33-cm square tiles; installed over hard base; permeable; for tennis courts	current
Mondo Sport Surfaces	Robbins, Inc.	calendered and vulcanized surface with a base of polychloroprene rubber, mineral aggregates, stabilizing agents, pigments; preformed rolls; adhered to base; for track, tennis	current
Recaflex	C. Voight Söhne, FRG	sandwich-type track surface with ground rubber underlay and polyurethane top surface	current
Rekortan	C. Voight Söhne, FRG	single-layer, cast, filled polyurethane trace surface	current
RoyalDek	Uniroyal, Inc.	resilient, shock-absorbing artificial tennis surface formed by molded tiles	current
Rubaturf track	Rubaturf Sport Surfaces, Inc.	rubber granulates from ground scrap tires combined with rubber latex fillers, antioxidants, and hardening agents poured-in-place	current
Sportan resilient surfacing	Sportan Surfaces, Inc.	liquid poured-in-place polyurethane; smooth texture or granular topping for gymnasiums, tennis, track, or other resilient recreational surfaces	current
Sport-Tred	Pandel Chemical, Inc.	prefabricated, solid cast vinyl flooring; factory coating available; glued to subbase	current
Swiss Flex	Swiss Flex	modular molded surface square (645 cm^2); permeable; installed over hard surface on tennis courts	current
Tartan resilient surfacing	3M Company	liquid poured-in-place polyurethane; smooth on granule topping for gymnasium or track	current
Tracklite	Tracklite Systems	resilient track surface	current

[a] United States, unless otherwise indicated.

Table 10. Current U.S. Manufacturers and Trade Names of Various Recreational Surfaces and Components

Product	Manufacturer	Description
Ensolite	Uniroyal, Inc.	gym, wrestling, and judo mats
Interlock rubber floor systems	Pawling Rubber	interlocking tiles for exercise and recreational areas
Port a Pit	Ampro Corporation	pole-vault landing surface
Pro-Pit	Vantel Corporation	pole-vault landing surface
Quest	A. E. Quest & Sons, Inc.	pole-vault and high jump pits
Resilite	Resilite Sports Products, Inc.	wrestling mats, wall mats
Supreme Court	AllWeather Surfaces, Inc.	roll-down sports surfaces

Paints and Striping. Outdoor running tracks, indoor basketball courts, field house surfaces, and stadium turf require line markings and appropriate decorations. These are painted on with two-component epoxy paints or water-based acrylic latex, depending on the permanence desired. In multiple-use stadiums, the football striping is removed when the area is converted to baseball, and vice versa. Multiple markings in different colors for football, field hockey, lacrosse, and soccer may be desired in community and school installations. Indoor paints usually are permanent, multiple color, and compatible with the surface, ie, deform with the resilient surface without cracking and accept any finish and maintenance coatings.

Economic Aspects

Manufacturers and trade names of commercial artificial surface products are given in Tables 8, 9, and 10.

Notice

Nothing contained herein should be construed as a recommendation to use any product, process, or apparatus in conflict with any patent.

BIBLIOGRAPHY

1. J. E. Nordale in N. M. Bikales, ed., *Encyclopedia of Polymer Science and Technology*, Vol. 15, Interscience Publishers, a division of John Wiley & Sons, Inc., New York, 1971, p. 490.
2. F. B. Roghelia and J. J. Burke, Monsanto Company Test Method, 1978.
3. T. A. McMahon and P. R. Greene, *J. Biomech.* **12**, 893 (1979); A. Chase, *Science* **81**, 90 (Apr. 1981).
4. J. J. Burke and F. B. Roghelia (1978), G. Raumann (1967), J. Vinicki (1966), Monsanto Company Report.
5. *ASTM F 355, Procedure A*, American Society for Testing and Materials, Philadelphia, Pa., 1978.
6. Ref. 1, p. 493.
7. *ASTM D 1175*, American Society for Testing and Materials, Philadelphia, Pa., 1980.
8. *ASTM D 1335*, American Society for Testing and Materials, Philadelphia, Pa., 1972.
9. *ASTM D 1682*, American Society for Testing and Materials, Philadelphia, Pa., 1975.
10. I. A. Benjamin and S. Davis, *Final Report No. NBSIR 78-1436*, National Bureau of Standards, U.S. Department of Commerce, Apr. 1978; T. Kashiwagi, *JFF/Consumer Product Flammability* **1**, 267 (1974).
11. I. A. Benjamin and C. H. Adams, *Fire J.*, 63 (Mar. 1976).
12. *ASTM D 3574*, 1977; *ASTM D 1667*, 1976; *ASTM D 624*, 1973; *ASTM D 2856*, 1976, American Society for Testing and Materials, Philadelphia, Pa.
13. Ref. 1, Vol. 13, 1970, p. 692.

14. J. H. Saunders and K. C. Frisch, *Polyurethanes, Chemistry, and Technology*, Vol. 1, John Wiley & Sons, Inc., New York, 1963; Vol. 2, 1964.
15. U.S. Pat. 3,583,945 (June 8, 1971), J. Robins (to Minnesota Mining and Manufacturing Company).

W. F. Hamner
T. A. Orofino
Monsanto Company

RECYCLING

Introduction, 936
Ferrous metals, 952
Glass, 963
Nonferrous metals, 967
Oil, 979
Paper, 986
Plastics, 993
Rubber, 1002

INTRODUCTION

Recycling is the recovery for reuse of the economic values of materials and energy from wastes that are usually destined for disposal. Although the term is applied most often to municipal wastes, it applies also to industrial or any other generated wastes or arisings. Many industrial waste streams, including most sources of industrial metal scraps, are never destined for disposal but for recycling (see Wastes, industrial).

Municipal solid waste comes from three main sources: construction and demolition, commercial and light industry, and households, ie, domestic waste. In addition, there are special categories, eg, pathogenic, hazardous, and radioactive wastes. Special wastes can sometimes be recycled, eg, in nuclear-waste processing (see Nuclear reactors, waste management).

Recycling is part of the total system of solid-waste management, which is shown diagrammatically in Figure 1. As a general rule, but subject to wide variations, the three main sources of municipal waste shown in Figure 1 each make up about one third of the total. Figure 1 illustrates that recycling of either materials or energy is an alternative to disposal but is only one segment of the total solid-waste management system (see also Fuels from waste). Litter, which is a misplaced solid waste, is not included in the figure. Litter is the result of slovenly behavior and, once collected, it becomes part of the management system. Compost also is not included in Figure 1. Although much has been written about conversion of the organic portion of wastes into a soil conditioner (not a fertilizer) (1), and despite the widespread practice in Europe and

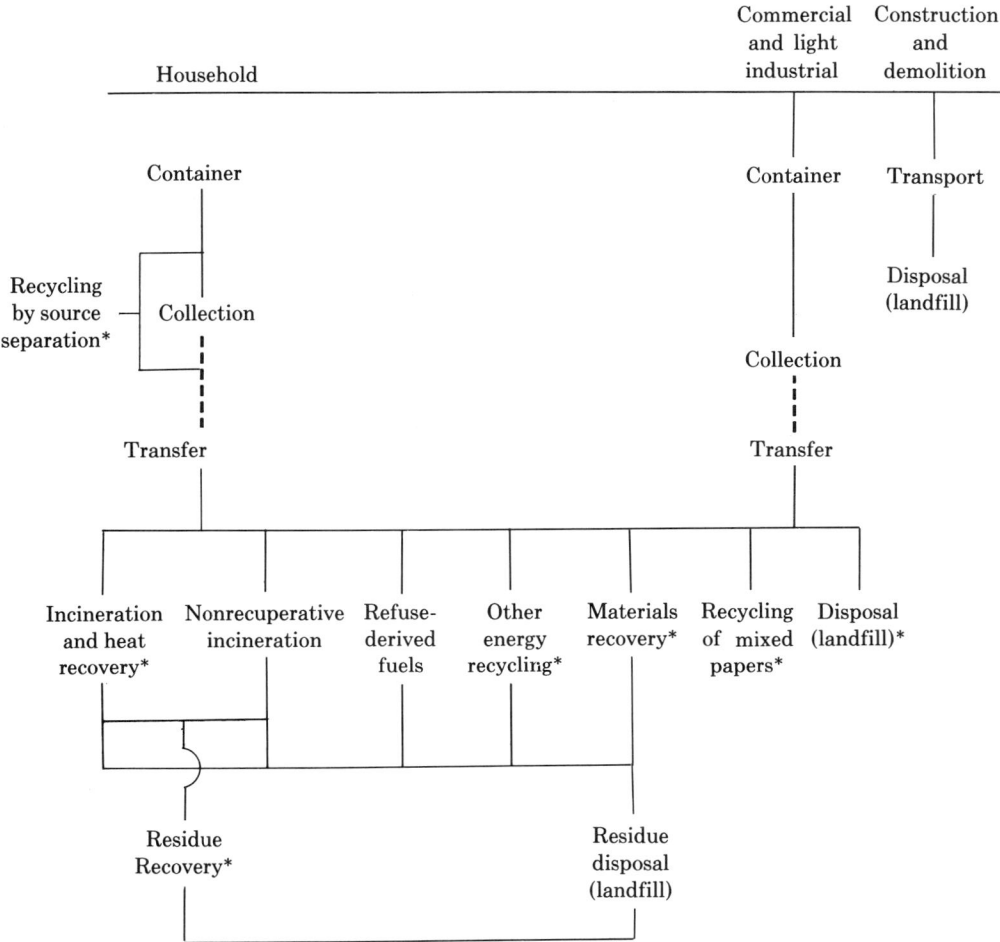

Figure 1. Municipal solid waste management system. Recycling options are indicated by an asterisk (*).

in parts of the Far East, composting is a well-documented economic failure in the United States, at least for large municipal plants (2).

Almost since the advent of municipal incinerators (qv) in the latter part of the nineteenth century, and with a revival of interest and research involving modern processing techniques, laboratories all over the world have studied residue-recovery processes (3). Commercial plants are nonexistent except for limited recovery of magnetic metals. One of many reasons for this is the difficulty of recovering products economically to specification after mixed waste has been exposed to the high temperatures in incinerators. An essential feature of recycling is that products must be recovered in a form and to a specification acceptable by users. Otherwise, it has no economic value and no use.

Methods

Early recycling activities, particularly in New York City, involved mechanical- and hand-separation methods and are described in ref. 4.

Scrap Processing. The scrap industry generally processes obsolete capital equipment, eg, railroad cars, automobiles, ships, etc, to produce metals or other valuable materials to users' specifications (5). In addition, the scrap industry processes metal trim and like wastes from manufacturing processes, surplus odd lots of plastics, or various sources of waste papers, etc, to forms that meet users' specifications. Specifications for these purposes have been published (6). The wastes that the scrap industry receives for processing are rarely destined for disposal; rather, their services are part of the total domestic manufacturing process. By far most of the recycling in the world is performed by the scrap industry.

In parallel, but often overlooked, is the small but growing industry that processes chemical wastes or contaminated chemicals to users' specifications. This is not to be confused with waste exchange by which providers and users of waste products are brought together, often without intermediate processing of the wastes. However, waste exchange is also a form of recycling.

The scrap industry performs essential services in terms of other forms of recycling. For example, a municipal-waste recovery plant may produce a dirty product, eg, contaminated used food cans, which can be upgraded by a local scrap processor to meet buyers' specifications.

Source Separation. Broadly, source separation can take three forms: householders separate recoverable materials and take these to a recycling center; householders separate recoverable materials, which are collected separately from the rest of the waste; and householders separate recoverable materials in a mixed form, eg, metals and glass, which is collected, and mechanically separated and processed to meet buyers' specifications. The economic success of any of the three methods depends greatly on the willing cooperation of the householders. With the exception of separate newspaper collection, participation of householders has generally been low and few programs have been economically self-sustaining. Experience is contrary to some optimistic projections which overstate the degree of participation and salability of the products (7).

The use of recycling centers has rarely been successful. The participants often must go out of their way to bring the separated materials to the central collection point. Also, the method relies on volunteer help to make certain that the contributed materials meet specifications, to arrange shipping and sale, and to keep the area surrounding the center clean. The centers usually cannot afford to replace the volunteers with paid help.

The second method of source separation includes a collection step, which is the most expensive element of solid-waste management since it requires additional use of trucks and labor. Again, only source separation of newspaper has been economical this way, even if the disposal cost, which is avoided, is considered. An exception to either method is the aluminum-beverage-can return program. However, such programs rely greatly on people collecting relatively large quantities of cans from central points, eg, bars and restaurants.

The first and second methods also require the householders' cooperation as quality-control agents for the separated materials, ie, the materials must be clean and prepared according to the customers' need. The householders or their surrogates bear a large responsibility to make certain the glass is clean of contaminants, that aluminum and steel cans are separated, and that newspaper is free of other grades of paper, string, and similar contaminants. Often, the surrogate may be a child who missorts. Also, these methods of source separation are subject to those who may intentionally perturbate the system.

The third method of source separation relies on the establishment of an intermediate processing facility. Generally, these are labor-intensive small operations for the separation (often by hand) of glass by color and of steel, aluminum, and whatever else is necessary to separate the crude mixture collected from households. These facilities have been established and operate on low budgets. New facilities have been designed to be more efficient and to meet worker safety requirements. However, they have not been built, almost without exception, because they are not economical without subsidization.

There are some arguments in favor of source separation. First, the participants realize that they are contributing directly to conservation and recovery ethics. As a result, the advocates are enthusiastic and often are vocal lobbyists. Second, source separation is often the only method of resource recovery suitable for small communities. Third, it is a way of establishing resource recovery quickly.

The general experience is that only the higher socioeconomic groups are likely to participate. Many working-class communities may not wish to participate and too little material is collected or sold to cover the costs of collection and other required activities. This is a principal reason why estimates of the likely success of such programs have been optimistically overstated. Nonetheless, there is still great interest in this approach, especially in some European nations (8).

Source separation and capital-intensive resource recovery are not competitive programs primarily because the former manages so little of the waste. Even the more successful source-separation programs leave most of the waste for conventional processing and disposal. Part of the controversy or competition has to do with newspaper, which can be used in the manufacture of certain paper products or as fuel. It is possible to compute the economic tradeoffs and thus identify the circumstances under which it is more advantageous to use paper one way as opposed to another.

Mechanical and Chemical Processing. Mechanical processing includes unit operations, eg, size reduction (qv), screening, air classification, and magnetic separation (qv), to reduce the waste to a homogeneous mixture suitable for materials handling and separation. After these unit operations, several others, eg, froth flotation (qv) or optical sorting for glass recovery, or eddy-current or heavy-media separation for aluminum recovery, may be used (see Gravity separation). Such unit operations also prepare homogeneous refuse-derived fuels (RDF) for burning or feedstocks for pyrolysis or biological fermentation. The application of unit operations to recovery and details of unit operations engineering are discussed in refs. 9 and 10.

The waste can be incinerated without prior processing to raise steam in an incinerator. This is perhaps the only method of fuel use or resource recovery that relies on a nonspecification fuel (11). As a result, the burning is not the easiest process to control.

Mixed Municipal Waste

Processing mixed municipal wastes includes wastes from household and light industrial and commercial sources. Construction and demolition wastes are discussed separately below. The processing of mixed municipal solid waste (MSW) is more detailed and involved than processing of most scrap sources. The mixture of materials is more complex and the content of valuable components, such as metals, is less. Further, the drive for resource recovery of MSW is coupled with the need for disposal. Therefore, the objectives of recovery include processing the complex mixture of organic materials in the MSW, which make up most of the total weight and volume.

Quantity. The quantity of MSW in the United States is ca $(90–100) \times 10^6$ metric tons annually in the standard metropolitan statistical areas (SMSAs). Other waste and hence other estimates are unimportant because it is only the waste that can be collected and assembled for recovery which is of interest here.

Various reports from the EPA have estimated the *per caput* generation of municipal waste in the United States. However, the estimates have been criticized as too large (12). For example, the estimates for 1971, 1973, 1974, and 1975 are 1.60, 1.70, 1.68, and 1.54 kg/(person·d), respectively (13). Although these estimates are for all sources of MSW, the actual amount of household waste in one medium size city (43,000 population) was 0.8 kg/(person·d) and was constant during 1971–1975 (14). A similar result was determined for a larger city (Milwaukee) as well as differences in amounts and composition resulting from differences in socioeconomic factors of communities (15). Other similar studies are described in refs. 14 and 15. The use of national averages in planning several municipal resource-recovery plants has resulted in large economic mistakes because of underestimates of the amount of waste to be received. It is important to use actual numbers for the community. Further, the *per caput* generation depends on the size of the community, especially in Europe (12), and likely in the United States.

Composition. The composition of MSW worldwide depends on the local use of paper, glass, and metal packaging; the less packaging material in the waste, the greater is the amount of food waste (16). The composition also depends on several demographic factors (15). The average composition of MSW in the United States is given in Table 1 (13). An average chemical analysis of the proximate and ultimate waste is given in Table 2 (17–18).

There are misconceptions about the sources of various chemical elements in waste, particularly those that are potential acid formers when the waste is incinerated or mechanically converted and used as a refuse-derived fuel. For example, it is often mistakenly stated that the source of chlorine in waste, hence a potential source of HCl emissions, is poly(vinyl chloride). The relative contents of selected, potentially acid-forming elements in the organic portion of a sample of waste collected from various households in one U.S. east-coast city is given in Table 3 (19). In this city, a chief source of chlorine in the waste is NaCl, probably from food waste.

The composition of waste is likely to vary by area according to local marketing and use habits. Such variations are not well documented in the United States but have been for parts of the United Kingdom (20). However, it is known that the quantity, and presumably the composition, vary considerably with the season (14). The composition of the waste is important in determining the amounts of recoverable materials and, therefore, potential revenues from recycling. The composition must be determined

Table 1. Amount of Postconsumer Residential and Commercial Solid Waste Generated[a] in 1975[b]

Material		Net waste disposed of, percent of total waste
paper		29.0
glass		10.4
metals		
ferrous	8.6	
aluminum	0.7	9.6
other nonferrous	0.3	
plastics		3.4
rubber and leather		2.6
textiles		1.6
wood		3.8
Total nonfood product waste		*60.4*
food waste		17.8
yard waste		20.2
miscellaneous inorganic waste		1.6

[a] As-generated weight basis is an assumed normal moisture content of material in its final use prior to discard; for example, paper at an air-dry 7 wt % moisture or glass and metals at 0 wt %. Total waste, including food and yard categories, is estimated at 26 wt % moisture.

[b] Ref. 13.

with recovery in mind. For example, a piece of copper wire is not recoverable either for the Cu or for the insulation, eg, plastic, which should not be included in either fraction. For fuel production, the composition should be considered in three categories: recoverable metals and glass, the inert or otherwise unusable fraction, and the fuel fraction, which consists of micellaneous organic matter, eg, papers, cardboard, plastics, and composites of such materials. The yield and composition of any refuse-derived fuel depends on the composition of the waste and on the methods of processing. The

Table 2. Analysis of MSW Composition, wt %[a]

Composition	Proximate analysis	Elemental analysis (magnetic metals removed)	
		As received, 19.7–31.3 wt % H_2O	Dry basis
moisture	19.7–31.3		
ash	9.4–26.8		
volatile	36.8–56.2		
fixed C	0.6–14.6		
C		23.45–33.47	48.7
H		3.38–4.72	
N		0.19–0.37	0.82
Cl		0.13–0.32	0.66
S		0.19–0.33	0.26
P			0.10
O		15.37–31.90	

[a] Refs. 17 and 18.

Table 3. Relative Contents of Selected Elements as Percentage of the Total[a]

Refuse category	Dry weight basis, organic portion only, wt %				
	C	N	S	P	Cl
textiles	7.29	43.35	25.64	13.38	5.55
wood	4.74	0.80	1.75	1.07	0.73
garden waste	8.65	18.18	5.53	16.63	3.66
rubber and leather	5.81	4.10	17.14	1.05	14.17
food waste	9.21	29.31	8.23	49.62	17.04
paper	54.39	1.80	40.19	17.28	22.98
plastics	9.91	2.46	1.52	0.97	35.87
total	*100.00*	*100.00*	*100.00*	*100.00*	*100.00*

[a] Ref. 19.

yield is higher if the extensive ash, ie, crushed glass, rock sand, etc, is not removed by screening. However, if these noncombustible materials are not removed, the quality of the fuel is lower.

Energy Savings from Recycling. The amount of the various materials potentially recoverable from municipal solid waste in the standard metropolitan statistical areas (SMSAs) has been estimated as (in 10^6 metric tons) magnetic metals, 5.8; Al, 0.4; other nonferrous metals, 0.2; and glass, 5.0. If these quantities of recovered materials are used as sources of raw materials to manufacture new materials, a considerable amount of energy [ca 0.4 EJ/yr (0.38×10^{15} Btu/yr)] would be conserved at the point of manufacture (21).

The organic fraction of the waste generated in the SMSAs in 1974 had an estimated higher heating value of 0.9 EJ (0.85×10^{15} Btu) (21). Similar estimates have been made for Europe (22). The organic fraction is the material that is suitable for conversion to fuels or feedstock for chemical or biological conversion. Such conversion and use of the material consumes energy. However, a resource-recovery plant producing a solid refuse-derived fuel and recovering metals and glass and processing ca 600 t/d would produce enough fuel to generate nine times more electrical power than it would consume, not including the energy savings from using the recovered materials (23).

The thermodynamic efficiency of various methods of resource recovery has been estimated, including the energy savings from using the recovered materials (24). The net energy efficiency, defined as $(E_{\text{out}} - E_{\text{in}})E_{\text{w}}$, where E_{w} is the heat content of the waste, varies in a ratio from -0.89 for nonrecuperative incineration to $+0.83$ for the use of shredded MSW (less the magnetic metals) as a fuel for cement manufacture. Some selected values are given in Table 4 (24). The efficiencies in Table 4 were computed by a consistent method and so provide some relative comparisons. Smaller differences among the various methods of resource recovery are reported elsewhere and take into account the gains or losses when the energy product from recovery is substituted for conventional fuels (25). The original references should be consulted as to how the thermodynamic boundaries are defined for the calculations. Nonetheless, the conclusion is that resource recovery produces much more energy than it consumes.

Table 4. Thermodynamic Efficiency of Recycling Processes[a]

Option	Efficiency, $(E_{out} - E_{in})/E_w$
incineration	
nonrecuperative	−0.089
steam generation	0.44
electricity generation	0.42
RDF	
supplementary fuel	0.62
pulverized for cement manufacture	0.78
direct firing	0.38
wet pulping	
incineration and electricity	0.16
wet RDF	0.52
fiber and wet RDF	0.25
dried RDF	0.36
pyrolysis	0.14–0.44
anaerobic digestion	0.28
materials recovery	0.02

[a] Ref. 24.

Preparation and Properties of Refuse-Derived Fuels

The conversion of MSW to refuse-derived fuel (RDF) in one form or another is a popular method of resource recovery. Some nine different RDFs have been defined, as listed in Table 5 (25). There are several ways to prepare RDF-3, which is perhaps the most popular form and is the feed used in the preparation of densified refuse-derived fuel (d-RDF). All forms of RDF are part of the broader set of waste-derived fuels (WDF), which includes various waste biomass, eg, from silvaculture or agriculture (see Fuels from biomass).

RDF-3 is intended for use as a supplement with coal for semisuspension or suspension firing or for use by itself in similar boilers. d-RDF is intended as a supplement with stoker coal or for use by itself in stoker boilers. The several methods and alternatives for producing RDF-3 or d-RDF have been described (27–28).

Because there is no single material called RDF and because the composition and, therefore, fuel properties, depend on the composition of the starting MSW and the methods of processing, it is impossible to give what might be an average set of fuel

Table 5. Definitions of Refuse-Derived Fuels[a]

RDF-1	wastes used as fuel in discarded form
RDF-2	wastes processed to coarse particle size with or without removal of magnetic metals
RDF-3	as MSW-derived shredded fuel which has been processed for the removal of metal, glass, and other entrained inorganic material. Generally, this material has a particle size such that 95 wt % passes through a 5-cm mesh screen.
RDF-4	combustible waste processed into powder form; 95 wt % passes through a 2.0-mm (10-mesh) screen
RDF-5	combustible waste compressed into pellets, slugettes, cubettes, or briquettes
RDF-6	combustible waste processed into gaseous fuel

[a] Ref. 26.

properties. Nonetheless, various analyses for RDF and d-RDF that are made in the UK are given in ref. 20. In addition, the properties of RDF made by a process which did not include sufficient screening to remove the extensive ash, ie, crushed glass, sand, etc, are listed in Table 6 (28). The effect of moisture and ash on RDF fuel properties has been examined as has the thermodynamic balance for possible drying of the fuel (23,29). It is unlikely that any RDF production process will be able to afford drying the material.

Resource Recovery Systems

Several developers of resource-recovery systems have built plants or otherwise described their system for recovery of materials and some form of RDF. The information about plants changes with time as new ones come on-stream or others are shut down or modified. Systems designs also tend to change with time. For information, the reader is directed to the list of general references.

Chemistry of Heat Recovery

The organic portion of MSW is likely to be used as a fuel or feedstock for chemical or biological conversion. Its utilization as a feedstock for incineration or as RDF has more chemical implications than merely burning the contained carbon. When unprocessed waste is burned in an incinerator, there is considerable corrosion (30). The key corrosive element in postulated corrosion schemes is chlorine. It has been suggested that HCl is released by the reaction:

$$2\ NaCl + SO_2 + \tfrac{1}{2} O_2 + H_2O \rightarrow Na_2SO_4 + 2\ HCl \qquad (1)$$

and that the corrosion of steel boiler tubes can continue by a combination of steps:

Table 6. Properties of Refuse-Derived Fuel[a]

Property	Sample 1 6 laboratories (ave)		Sample 2 6 laboratories (ave)		Sample 3 12 laboratories (ave)		
	$\bar{x}^b$	n^c	$\bar{x}^b$	n^c	$\bar{x}^b$	n^c	CV^d, %
calorific value[e], MJ/kg[f]	17.7	24	15.9	32	17.4	48	3.54
moisture, wt %	20.53	24	19.81	32	30.09	44	5.66
dry weight basis, wt %							
ash	15.70	28	25.28	32	22.17	48	11.7
S	0.34	28	0.24	32	0.48	48	17.4
C	46.33	32	40.02	32	42.60	44	3.14
H	6.17	32	5.37	32	5.84	44	4.98
N	0.76	32	0.64	28	0.77	44	9.21
Cl_t[g]	0.58	20	0.38	28	0.57	44	40.0
Cl_s[h]	0.35	20	0.24	24	0.27	42	25.1

[a] Ref. 28.
[b] $\bar{x}$ = mean of property.
[c] n = number of determinations.
[d] CV = coefficient of variation.
[e] Dry basis.
[f] To convert MJ/kg to Btu/lb, multiply by 430.4.
[g] Cl_t = total chlorine.
[h] Cl_s = determined as Cl^- after water extraction.

$$2\,HCl + \tfrac{1}{2}\,O_2 \rightarrow H_2O + Cl_2 \qquad (2)$$

which is postulated to be catalyzed by metal oxides, eg, Fe_2O_3 or PbO (30). The following reactions, which are not necessarily in order, occur:

$$Fe + Cl_2 \rightarrow FeCl_2 \qquad (3)$$

$$Fe + 2\,HCl \rightarrow FeCl_2 + H_2 \qquad (4)$$

It has been shown that (29):

$$4\,FeCl_2 + 3\,O_2 \rightarrow Fe_2O_3 + 4\,Cl_2 \qquad (5)$$

A key question is whether reaction 1 is possible. If so, reaction 2 need not maintain the cycle for corrosion by either HCl or Cl. In the absence of direct experimental evidence, the possibility of reaction 1 occurring was tested by calculating the free energy change $\Delta G°$ and enthalpy $\Delta H°$ of the reaction. The results are: $\Delta G° = -320$ kJ (-77 kcal) and $\Delta H° = -413$ kJ (-98.7 kcal); the equilibrium constant was calculated at boiler temperatures from these values. The values are large ($>10^6$) and it was concluded that reaction 1 is thermodynamically possible (31).

This result does not address the reports that less corrosion is observed when burning some form of RDF instead of unprocessed waste. As shown in Table 2, MSW contains as much as 0.66 wt % Cl (dry basis). Table 3 indicates that only 36% of this Cl can be accounted for by plastics, only a small portion of which can be poly(vinyl chloride) and the source of HCl on thermal decomposition. For reaction 1 to occur, there must be a source of NaCl. The total Cl content of RDF in one instance is reported to be 0.47 wt % with all but 0.10 wt % water-soluble, presumably NaCl (32). This report may be compared to similar results in Table 6.

In terms of the possible corrosion from RDF, the molar ratio of $NaCl/SO_2$ (presuming all of the sulfur is converted to SO_2) is approximately 2.5/1, so that reaction 1 would not consume all of the NaCl likely to be present when reaction 1 is driven to completion. The key is that it is unlikely that the processing of MSW to RDF can be accomplished in such a way as to control the content of NaCl, hence the corrosion mechanisms must be controlled by some other means. In one study, the ratio of RDF to coal was controlled so as to control the S/Cl ratio (33). Addition of S drives reaction 1 toward completion so that the metal chlorides in the combustion products are converted to sulfates in the flue-gas stream. It is believed that the HCl released in the gas stream does not react with the metal surface as it would by *in situ* formation beneath a deposit on the boiler tube (33).

The chemistry of slag deposits on boiler tubes from burning MSW or RDF has also been investigated (30). These are mostly sulfates, particularly Na_2SO_4, with some metal chlorides.

Accumulation of Impurities with Recycling

It is mistakenly believed that if products containing impurities are recycled several times, the impurities can accumulate without limit. This fallacy can be examined for the recycling of tinplate, eg, in the form of food cans, recovered from MSW mixed with other scrap steel. This method has been elaborated in ref. 33; the simplified example from ref. 31 is considered here.

Can scrap recovered from household waste has a Sn content c_0 weight fraction. If can scrap is charged to a steelmaking furnace at a weight fraction p, so that $p \times c_0$

= β, and is mixed with weight fraction q of other scrap of Sn content c'_0, and $q \times c'_0 = \gamma$ and $(q + p) = r$, so that $(1 - r) =$ weight fraction of added new hot metal of no Sn content, then the concentration of Sn for the nth recycle is

$$c_n = (\beta + \gamma) + (\beta + \gamma)r + (\beta + \gamma)r^2 + \ldots (\beta + \gamma)r^{n-1}$$

The limit of the series is:

$$\lim_{n \to \infty} c_n = \frac{(\beta + \gamma)}{(1 - r)}$$

This series has been evaluated for several combinations of scrap charge (34). For example, the Sn content of steel scrap from MSW is 0.17–0.48 wt %, with an average of ca 0.3 wt %, compared to other scrap grades used in the steel industry containing ca one-tenth this amount (34). Thus, $c_0 = 0.003$. If such scrap were repeatedly recycled at $p = 0.1$ (10 wt % charge), with other scrap at $q = 0.3$ (a high scrap charge) of Sn content $c'_0 = 0.0003$, then $c_n < 0.07$ wt %, which is a suitable concentration for several metallurgical applications. This approach to calculating the limiting value of impurity buildup is applicable to other products as well.

Production and Economic Aspects

The daily capacity of municipal resource-recovery plants, but not of modular incinerators, that are constructed in the United States and likely to be completed through 1981 was 27,110 t/d. The capacity of modular incinerators was 1416 t/d (35). However, these values do not reflect actual operating capacity because of various design faults, eg, not enough waste in the locality to fulfill the design capacity, or other problems. The capital value of the plants was 1.5×10^9 (35). The total worth of the industry is more because it includes industrial waste processing, the scrap industry, and other activities that cannot be estimated.

The growth of municipal resource recovery in the United States decreased in the early 1980s because of high interest rates and perhaps the unlikely expectation by municipal authorities of Federal aid. In 1972, the Federal government through the EPA invested in demonstration plants. In retrospect, this exercise slowed technological development because it did not always permit technical innovation or corrections of faulty designs. Future Federal investment might similarly freeze technology by attempting to avoid risk or innovation. Various ways have been described to make resource recovery economical (36).

Despite fluctuation in the industry, the rate of implementation of new facilities has been good. One prediction of this rate assumes that, because municipalities seek to follow the lead of their counterparts in other localities, the rate of implementation of new plants is proportional to the number of plants in existence at a given time (21,37).

Economics. The construction and operating costs of resource-recovery plants change rapidly. There are a few exceptions, notably the plants in Ames, Iowa, and New Orleans, La. (38). Costs of the Ames plant include operating and maintenance costs over a long running period. The New Orleans plant reports describe the development of the plant from the feasibility study through construction and shakedown.

The costs and economic risk of constructing and operating resource-recovery plants of different designs have been analyzed extensively (39). The method of analysis was the same for all types of plants so that the reported costs are relative figures of merit. Some of the results are given in Table 7. One method has been structured to compute the tipping fee necessary for a given return (40). Another approach is based on the indifference value (at which an indifferent operator of resource recovery decides whether to use resource recovery or competitive disposal) of the fuel fraction that must be received in order for resource recovery to be competitive with landfill in a community (41). This method was elaborated to introduce the sensitivity of the indifference value to the uncertainties in the estimates of capital costs, operating costs, or materials revenues (42). Application of this method helps a community to determine whether or not they can sell the fuel product for a price that is competitive with alternative fuels and still have an economical resource-recovery plant. It may be worth examining cost figures for RDF plants in the UK, although such figures are quickly outdated (20).

Specifications

The first step in planning resource recovery is to obtain the sales agreements for the products expected to be produced and the accompanying specifications. It is the required specifications that determine the technology to be employed. The bases for many specifications for products recovered from wastes have been defined (43). Their development and their evolution and need are described in refs. 44 and 45. Several

Table 7. Unit Cost of Recycling and Disposal Options for Municipal Solid Waste[a]

Option	Unit cost, $/t
landfill	6.96
with truck transfer	17.25
with shredding, two shifts	16.11
with baling	26.1
incineration	
nonrecuperative	38.2
steam generation	41.0
electricity generation	43.4
composting	41.0
RDF	
supplementary fuel	27.2
direct firing, suspension	52.9
wet pulping	
incineration and electricity	60.3
wet RDF	28.7
fiber and wet RDF	36.2
dried RDF	41.5
pyrolysis	53.7–90.4
anaerobic digestion	71.3
additional materials recovery	4.06–5.02

[a] Unit costs (June, 1978) are given gross of any revenues from sale of recovered products. The unit costs were calculated by a consistent method and should be used as relative figures of merit. Interpretation and additional costs are given in the original work (in June, 1978, one pound sterling was equivalent to ca U.S. $1.84) (39).

specifications, particularly for recovered glass, steel, and aluminum, and some standard test methods for RDF, have been published (46).

Environmental and Health Factors

Two factors of principal concern are air emissions from burning solid waste or refuse-derived fuels and the environmental health of workers in resource-recovery plants. The first factor has been addressed for both the United States and the UK (20,42). The overall conclusions, including those from several unpublished experimental trials using d-RDF–coal mixtures, are that the particulate and other priority pollutant emissions do not increase, either beyond legal limits or those from coal alone, until an RDF–coal weight ratio of $\leq 60/40$ is used. Exceptions may be for Cl and S; Cl can increase with increasing amounts of RDF more than coal, assuming the coal has a low Cl content, and S is likely to decrease when RDF is burned mixed with most coals. The S content of RDF is ca 0.2 wt % at the lower end of concentrations determined for western subbituminous coals. Heavy-metal emissions do not increase beyond levels believed to be undesirable (there are no National Ambient Air Quality Standards for most metals) when burning RDF. However, some effects depend on the volatility and melting points of the metalloid compounds released, which can lead to relative enrichment of certain metals in the emissions (47). Modern pollution-control equipment must be used.

There have been several reports of physiochemical and microbiological analyses of dusts in resource recovery plants, and most of these have been reviewed (48). The conclusions from a study of a pilot plant where the dusts were particularly heavy are described in ref. 48. In the studies cited in ref. 48, there was no reported indication of any severe health hazards resulting from working in a resource-recovery plant, provided normal hygiene precautions are taken.

Special Topics

The recovery and reuse of materials from construction and demolition wastes have been described (49). Thermodynamic data of many wastes for incineration are given in ref. 50. A description of the effects of contaminants, ie, tin, aluminum, and organic materials, on the reuse of magnetic metals recovered from MSW, and of contaminants in newspaper collected separately and commingled with other waste are described in refs. 51 and 52, respectively. Much has been written about the approach to resource recovery in Europe as contrasted to the practices in the United States (12,20,53). Resource recovery in Japan is described in ref. 54. The composition and properties of many industrial wastes and their disposal and recovery are discussed extensively in ref. 55.

In summary, there must be markets before waste products can have value and be reused. Methods of finding markets are described in ref. 48. Uncertainties in planning and implementing recycling schemes are reviewed in refs. 11, 21, and 35.

BIBLIOGRAPHY

1. R. T. Haug, *Compost Engineering: Principles and Practice*, Ann Arbor Science Publishers, Inc., Ann Arbor, Mich., 1980.

2. G. E. Stone, C. Wiles, and C. Clemons, *Composting at Johnson City*, Final Report on Joint USEPA-TVA Composting Project, Vols. 1 and 2, U.S. Environmental Protection Agency, Cincinnati, Ohio, 1975; Report EPA/530/SW-31r.2; *Guidelines for Local Government on Solid Waste Management*, U.S. Environmental Protection Agency, Report SW-17c, Washington, D.C., 1971.
3. *Solid Wastes* **66,** 536 (1976) (reprinted from *My Magazine* (July 1924)); *Engineering* **126,** 577, 641, 710 (1928).
4. R. Fenton, *Resource Recovery and Conservation* 1(2), 167 (1975); E. L. Armstrong, M. C. Robinson, and S. Hoy, eds., *History of Public Works in the United States, 1776–1876*, American Public Works Association, Chicago, Ill., 1976; S. Hoy and M. C. Robinson, *Recovering the Past: A Handbook of Community Recycling Programs 1890–1945*, Public Works Historical Society, Chicago, Ill., 1979.
5. C. Lipsett, *Industrial Wastes and Salvage*, Atlas Publishing Company, New York, 1963.
6. *Standard Classification for Nonferrous Scrap Metals*, National Association of Recycling Industries, New York, 1973; *Accepted Specifications for Selected Remelting Grades of Steel Scrap*, Institute of Scrap Iron and Steel, Washington, D.C., 1973.
7. *Materials and Energy from Municipal Waste*, Office of Technology Assessment, Washington, D.C., 1979.
8. *Resource Recovery and Conservation* **5**(1), 1 (1980).
9. H. Alter in ref. 8, p. 73.
10. P. A. Vesilind and A. E. Rimer, *Unit Operations in Resource Recovery Engineering*, Prentice-Hall, Inc., Englewood Cliffs, N.J., 1981.
11. C. O. Velzey, *Resource Recovery and Conservation* 4(1), 83 (1979); N. J. Weinstein and N. J. Toro, *Thermal Processing of Municipal Solid Waste for Resource and Energy Recovery*, Ann Arbor Science Publishers, Inc., Ann Arbor, 1976.
12. H. Alter and J. J. Dunn, Jr., *Solid Waste Conversion to Energy, Current European and U.S. Practice*, Marcel Dekker, Inc., New York, 1980.
13. *Resource Recovery and Waste Reduction: Fourth Report to Congress*, Report SW-600, U.S. Environmental Protection Agency, Washington, D.C., 1977.
14. J. C. Even, P. Arberg, J. R. Parker, and H. Alter, *Resources and Conservation* 6(3/4), 187 (1981).
15. W. L. Rathje and B. Thompson, *The Milwaukee Garbage Project*, The Solid Waste Council of the Paper Industry, American Paper Institute, Washington, D.C., 1981.
16. H. Alter in ref. 8, p. 39.
17. *Resource Recovery and Waste Reduction: Third Report to Congress*, Report SW-161, U.S. Environmental Protection Agency, Washington, D.C., 1975.
18. R. A. Lowe, *Energy Recovery from Waste*, Report SW-36d.ii, U.S. Environmental Protection Agency, Washington, D.C., 1975.
19. H. Alter, G. Ingle, and E. R. Kaiser, *Solid Wastes Manage. (England)* 64(12), 706 (1974).
20. A. Porteous, *Refuse Derived Fuels*, Applied Science Publishers, London, 1981.
21. H. Alter, *Environ. Conservation* 4(1), 11 (1977).
22. H.-C. Bailly and C. Tayart de Borms, *Material Flows in the Post Consumer Waste Stream of the EEC*, Graham & Trotman, London, 1977; Europool, *Secondary Materials in Domestic Refuse as Energy Sources*, Graham & Trotman, London, 1977.
23. H. P. Sheng and H. Alter, *Resource Recovery and Conservation* 1(1), 85 (1975).
24. D. C. Wilson, *Resource Recovery and Conservation* 4(2), 161 (1979).
25. H. Alter in R. A. Matula, ed., *Energy Recovery from Solid Waste: Looking Through a Dark Furnace Slowly, Present Status and Research Needs in Energy Recovery from Wastes*, American Society of Mechanical Engineers, New York, 1976, pp. 9–22.
26. ASTM Standards E 775, E 776, E 777, E 778, E 790, E 791 and others in preparation by subcommittee E 38.01, American Society for Testing and Materials, Philadelphia, Pa., 1981.
27. Ref. 12, Chapt. 5.
28. H. Alter and J. A. Campbell in J. L. Jones and S. B. Radding, eds., *The Preparation and Properties of Densified Refuse-Derived Fuel, Thermal Conversion of Solid Wastes and Biomass*, American Chemical Society, Washington, D.C., 1980, pp. 127–42.
29. Ref. 12, Chapt. 8.
30. H. H. Krause, D. A. Vaughn, and P. D. Miller, *ASME J. Eng. Power* **95**(1), 45 (1973); **96**(3), 216 (1974); H. H. Krause, D. A. Vaughn, and W. K. Boyd, *ASME J. Eng. Power* **97**(3), 448 (1975); **98**(3), 369 (1976); H. H. Krause, D. A. Vaughn, P. A. Cover, P. W. Boyd, and D. A. Oberacker, *ASME J. Eng. Power* **99**(3), 449 (1977); H. H. Krause, D. A. Vaughn, P. A. Cover, and W. K. Boyd, *ASME J. Eng. Power* **101**(4), 592 (1979).

31. H. Alter, *Resource Recovery from a Chemical Viewpoint*, *Materials and National Policy*, American Chemical Society, Washington, D.C., 1978, pp. 35–44.
32. K. Klumb, *Resource Recovery and Conservation* **1**(3), 225 (1976).
33. D. A. Vaughn, H. H. Krause, and W. K. Boyd, *Mater. Perform.* **14**, 16 (1975).
34. E. J. Duckett, *Resource Recovery and Conservation* **2**(4), 301 (1976).
35. *NCRR Bulletin* **10**(3), 57 (1980).
36. H. Alter, *Phoenix Quarterly* **12**(3), 6 (1980); H. Alter, *Proceedings, Seventh Mineral Waste Symposium*, U.S. Bureau of Mines and IIT Research Institute, Chicago, Ill., 1980, pp. 53–58.
37. H. Alter, *Resources and Conservation* **7**, 327 (1981).
38. *New Orleans Resource Recovery Facility*, National Center for Resource Recovery, Inc., Washington, D.C., 1972 (engineering feasibility), 1976 (design and cost), 1980 (shakedown); J. C. Even, Jr., S. K. Adams, P. Gheresus, A. W. Joensen, J. L. Hall, D. E. Fiscus, and C. A. Romine, *Evaluation of the Ames Solid Waste Recovery System: Part I, Summary of Environmental Emissions: Equipment Facilities, and Economic Evaluations*, U.S. Environmental Protection Agency, Cincinnati, Ohio, 1977; J. C. Even, Jr., C. Kosolchargen, and A. W. Joensen, *Resource Recovery and Conservation* **5**(3), 239 (1980).
39. D. C. Wilson, *Resource Recovery and Conservation* **4**(3), 261 (1979).
40. J. G. Abert, H. Alter, and J. F. Bernheisel, *Science* **183**, 1052 (1974).
41. H. W. Gershman, *Proceedings, 1976 National Waste Processing Conference*, American Society of Mechanical Engineers, New York, pp. 1–12 (1976).
42. Ref. 12, Chapt. 7.
43. H. Alter and E. Horowitz, eds., *Resource Recovery and Utilization*, STP 592, American Society for Testing and Materials, Philadelphia, Pa., 1975.
44. H. Alter, *Conservation Recycling* **2**(1), 71 (1978).
45. H. Alter, *ASTM Standardization News* **7**(11), 11 (1979).
46. *Annual Book of ASTM Standards*, Part 41, American Society for Testing and Materials, Philadelphia, Pa., 1981.
47. R. R. Greenberg, W. H. Zoller, and G. E. Gordon, *Environ. Sci. Technol.* **12**, 566 (1978); R. R. Greenberg, G. E. Gordon, W. H. Zoller, R. B. Neuendorf, K. J. Yost, and W. H. Jacko, *Environ. Sci. Technol.* **12**, 1329 (1978).
48. E. J. Duckett, J. Wagner, R. Welker, B. Rogers, and V. Usdin, *Am. Ind. Hyg. Assoc. J.* **41**(12), 908 (1980).
49. D. G. Wilson, *Resource Recovery and Conservation* **1**(2), 129 (1975); Environmental Resources Ltd., *Demolition Waste*, The Construction Press, London, 1980.
50. E. S. Domalski, W. H. Evans, and T. L. Jobe, Jr., *Thermodynamic Data for Waste Incineration*, Report NBSIR 78-1479, National Bureau of Standards, Washington, D.C., 1978.
51. E. J. Duckett, *Contaminants of Magnetic Metals Recovered from Municipal Solid Waste*, Report NSF/RA-770244, National Science Foundation, Washington, D.C., 1977.
52. H. Alter, K. L. Woodruff, A. Fookson, and B. Rogers, *Resource Recovery and Conservation* **2**(1), 79 (1976).
53. D. J. De Renzo, ed., *European Technology for Obtaining Energy from Solid Waste*, Noyes Data Corporation, Park Ridge, Ill., 1978; W. Palz and P. Chartier, eds., *Energy from Biomass in Europe*, Applied Science Publishers, London, 1980.
54. A. G. Buekens, *Resource Recovery and Conservation* **3**(3), 275 (1978); S. Gotoh, *Recent Developments in Resource Recovery from Municipal Solid Waste in Japan*, paper presented at the Second International Symposium on Materials and Energy Recovery from Refuse, Antwerp, Belgium, 1981.
55. C. L. Mantell, *Solid Wastes: Origin, Collection, Processing, and Disposal*, John Wiley & Sons, Inc., New York, 1975.

General References

J. G. Abert and H. Alter in K. V. Sarkanen and D. Tillman, eds., *Municipal Waste* in *Progress in Biomass Conversion*, Vol. 1, Academic Press, Inc., New York, 1979, pp. 145–213.

H. Alter and E. Horowitz, eds., *Resource Recovery and Utilization*, STP 592, American Society for Testing and Materials, Philadelphia, Pa., 1975.

H. Alter and J. J. Dunn, Jr., *Solid Waste Conversion to Energy, Current European and U.S. Practice*, Marcel Dekker, Inc., New York, 1980.

A. F. M. Barton, *Resource Recovery and Recycling*, John Wiley & Sons, Inc., New York, 1979.

T. C. Frankiewicz, *Energy from Waste*, Ann Arbor Science Publishers, Inc., Ann Arbor, Mich., 1980.

J. L. Jones and S. B. Radding, *Solid Wastes and Residues, Conversion by Advanced Thermal Processes*, American Chemical Society, Washington, D.C., 1978.

J. L. Jones and S. B. Radding, eds., *Thermal Conversion of Solid Wastes and Biomass*, American Chemical Society, Washington, D.C., 1980.

A. Porteous, *Refuse Derived Fuels*, Applied Science Publisher, London, 1981.

N. J. Sell, *Industrial Polution Control, Issues and Techniques*, Van Nostrand-Reinhold Company, New York, 1981.

P. A. Vesilind and A. E. Rimer, *Unit Operations in Resource Recovery Engineering*, Prentice-Hall, Inc., Englewood, Cliffs, N.J., 1981.

Proceedings, National Solid Waste Processing Conferences, American Society of Mechanical Engineers, New York, 1976, 1978, 1980, etc.

Proceedings, First, 2nd, 3rd Recycling World Congresses, Exhibitions for Industry, Oxted, Surrey, UK, 1978, 1979, and 1980.

Mineral Waste Utilization Symposia, Proceedings, U.S. Bureau of Mines and IIT Research Institute, Chicago, Ill. Seven biannual symposia have been held through 1980.

K. J. Thomé-Kozmiensky, *Recycling Berlin '79. International Recycling Congress*, E. Freitag-Verlag für Umwelttechnik, Berlin, Germany, 1979.

H. Alter, ed., *Resources Conservation* (formerly *Resource Recovery Conservation*), Elsevier Scientific Publishing Company, 1000 AH, Amsterdam, The Netherlands.

M. Henstock, ed., *Conservation Recycling*, Pergamon Press, Oxford, UK, OX3 OBW.

C. B. Kenahan, R. S. Kaplan, J. T. Dunham, and D. G. Linnehan, *Bureau of Mines Research Programs on Recycling and Disposal of Minerals-, Metal-, and Energy-Based Wastes*, Bureau of Mines Information Circular 8595, U.S. Department of the Interior, Washington, D.C., 1973.

M. J. Spendlove, *Bureau of Mines Research on Resource Recovery*, Bureau of Mines Information Circular 8750, U.S. Department of Interior, Washington, D.C., 1977.

HARVEY ALTER
Chamber of Commerce of the United States

FERROUS METALS

The following discussion is limited to iron and steel scrap, ie, ferrous scrap. Virgin or primary metals are produced from ores. Scrap or secondary metals are derived from primary metals or previously fabricated metal products. This scrap can be utilized in place of virgin metal or scrap from one source can replace scrap from another source in manufacturing a new product.

Ferrous scrap is consumed by industries other than iron-and-steel producers, iron foundries, and ferroalloy producers. The detinning and copper-precipitation industries use ferrous scrap. In terms of annual scrap consumption, however, iron and steel producers are the dominant force in the ferrous scrap market, as suggested in Table 1 (1–4).

In the iron-and-steel industry, the first efforts to make use of scrap iron or ferrous scrap in the production of iron and steel were closely associated with the development in the 1850s of the first modern steelmaking process, ie, the acid-Bessemer process (5). From the 1850s on, substantial amounts of scrap iron and steel were accumulated from the expanding iron and steelmaking industries, fabricating operations, and worn-out or obsolete iron and steel products from railroads and other industries. This ferrous scrap, readily available and inexpensive, did not have a market because of the absence of a steel-refining process that could consume substantial quantities of scrap (see Iron; Steel).

In the 1860s, the invention of the regenerative-heating principle for steelmaking, the open-hearth process, permitted heating of solid pig iron, iron ore, and scrap to high temperatures for refining (5). The solid charge or feedstock for the open hearth could be scrap; thus, the open-hearth furnace has a considerably greater capacity to consume ferrous scrap than the acid-lined Bessemer furnace. Growth of the open-hearth process was rapid and, by 1908, exceeded Bessemer steel production (6).

There have been two other modern steelmaking developments that have had different impacts on the ferrous-scrap industry. First, in 1878, Siemens built the first electric-arc furnace for making steel (7). Advantages of this process were low capital costs per ton of capacity and a charge or feedstock that was usually 100% solid iron or scrap; therefore, molten pig iron from a blast furnace was not needed. In the years following World War II, numerous large-capacity electric-arc furnaces up to 360 t in size were installed and many were used to produce the common carbon-steel grades (8). The second development resulted in the rebirth of the pneumatic Bessemer-type

Table 1. Total Ferrous-Scrap Consumption by Industry

Industry	Consumption, 10^6 metric tons	Year	Ref.
iron and steel production	66	1978	1
iron foundries	13.6	1979	2
ferroalloy	0.45	1973	3
detinning	0.64[a]	1974	4
copper precipitation	0.45	1974	4

[a] Based on 4.5 kg of tin per metric ton of tinplate scrap.

process. Although early developers of pneumatic steelmaking recognized the value of blowing oxygen instead of air to oxidize impurities and to adjust the steel chemistry, the technology for producing large quantities of low cost oxygen was not available until after World War II. Thus, the inability of air, which contains ca 80 wt % nitrogen, to refine steel to sufficiently low nitrogen concentrations and the difficulty of removing phosphorus in the air-blown, acid-lined furnace resulted by 1950 in the decrease of Bessemer steel production to less than 5% of the U.S. total production.

A steelmaking process known as the basic-oxygen process or LD process was developed in Austria between 1949 and 1952. Top-blown oxygen instead of air was used as the oxidizing or fuel agent in a basic-lined Bessemer-type furnace, and less scrap per furnace charge was used than in the open-hearth furnace. Steel low in phosphorus and nitrogen could be produced more rapidly and at lower costs than in the open-hearth furnace; the capital costs of these basic-oxygen furnaces (BOF) were lower than new open-hearth construction (6,9). By 1970, the production of BOF steel surpassed open-hearth steel and became the dominant steelmaking process in the United States. The development of modern steelmaking processes is described in refs. 1, 5, 6, 8, and 9 (see Steel).

Iron foundries originally used hot metal from blast furnaces or solidified pig iron as the primary charge in the cupola iron-making furnace (10). However, by the early 1950s, ferrous scrap had replaced ca 40% of the cupola charge. Furthermore, during the early 1960s and concurrent with the growth in electric-arc steelmaking capacity, iron foundries added electric-arc furnaces for the production of cast iron (3). By 1973, electric-arc furnaces produced 23% of the cast iron, and cupola-furnace production declined to 73% (11). Total production of ferrous castings in 1979, including gray cast iron, ductile cast iron, and malleable cast iron, totaled ca 15.5×10^6 metric tons with gray iron representing 72% or ca 11×10^6 t (2). Over 90% of the furnace charge in the gray cast-iron industry is ferrous scrap (see Furnaces).

The copper-precipitation industry uses ferrous scrap as a precipitating agent in processing low grade copper ores and mine tailings. The copper-bearing material is dissolved in sulfuric acid to produce copper sulfate. Light-gauge ferrous scrap is added to the copper sulfate solution, which causes the copper to be chemically displaced by the iron and to form a copper precipitate. The principal sources of the light-gauge ferrous scrap is the ferrous residue from the solid-waste incinerators and detinned ferrous scrap from the detinning industry. Detinners process only tinplate scrap, usually from can manufacturers and steel tinplate mills, in order to reclaim the tin and to generate a high quality, tin-free ferrous scrap as a by-product that can be consumed directly by the iron-and-steel and copper-precipitation industries. The secondary-metals-industry designation for this detinned can scrap is no. 1 bundles. In the detinning process, the tin coating is dissolved in hot sodium hydroxide. The tin is then electrolytically precipitated from the sodium stannate solution.

Sources of Ferrous Scrap

Traditionally, ferrous scrap has been called either home scrap, prompt industrial scrap, or old scrap. Home scrap or revert scrap is generated during the production of steel or cast iron and is always recycled. Manufacturing industries produce prompt industrial scrap during the fabrication of various industrial, commercial, or consumer steel products. Prompt industrial scrap or new scrap is also widely recycled because its chemical and physical characteristics can be well documented. However, in some cases, additional processing, eg, detinning or compacting, must be carried out to process the scrap in a form suitable for recycling. The availability of prompt industrial scrap is directly related to the level of industrial economic activity. Prompt-industrial-scrap producers usually cannot allow it to accumulate because of storage requirements and costs of inventory control. Thus, it is rapidly available at current prices to the steelmaker or ferrous-scrap industry.

All other ferrous scrap is included in the third category, called old scrap, obsolete scrap, or postconsumer scrap. This category includes all goods or products in which the iron content can at least in theory be recovered and recycled. The main types of old scrap recycled in the United States are railroad, machinery, and automotive, whereas in foreign countries with low labor costs, shipbreaking is a principal source of obsolete scrap. Ferrous scrap recovered from municipal solid waste (MSW), sometimes called MSW magnetics or municipal ferrous scrap (MFS), is a new nontraditional source of obsolete scrap. In terms of quality, desirability, and cost, home scrap generally ranks highest; prompt industrial scrap, second; and old scrap, including MFS, last. This lower ranking of old scrap is a direct consequence of its greater heterogeneity in chemical and physical characteristics, which makes it more costly to process.

Since 1974, $(7-10) \times 10^6$ automobiles and light trucks were deregistered or discarded annually and represented potential generation of up to 11×10^6 t/yr of obsolete ferrous scrap (10–13). In 1974, ca 89% of the ferrous content from discarded automobiles was recycled, and this accounted for ca 33% of all recycled obsolete ferrous scrap. Recent estimates suggest that $(9-11) \times 10^6$ t/yr of ferrous scrap is discarded into municipal solid waste; however, <2%, ie, less than 450,000 t, has been recovered in any year. In 1979, ca 180,000 t of municipal ferrous scrap was recovered (14). The composition of MSW exhibits regional, seasonal, and other variations as described in refs. 15–17. Approximately 80% of MSW (including ca 20% moisture) is combustible. Ferrous scrap accounts for ca 85% of the noncombustible MSW. Beverage and food cans, excluding all-aluminum cans, make up 50–90% of the ferrous fraction in municipal solid waste.

Role of Ferrous Scrap in Steelmaking

Modern steelmaking processes can be divided into two categories based on materials flow. One category is the blast furnace and the open-hearth and BOF refining furnaces, and the second category is the electric refining furnaces. In an integrated steel mill where iron ore is converted into finished steel products, the ore is chemically reduced with coke and limestone in a blast furnace to molten pig iron or hot metal. The hot metal is combined with ferrous scrap in an open-hearth furnace or a BOF and is refined until the desired grade of steel is produced. The open-hearth furnace has the greatest flexibility in its consumption of scrap because a portion of the heat needed to melt the scrap is supplied by an external fuel. The furnace can process 100% hot metal, 100% scrap, or solid pig iron, or any combination thereof. Usually 40–60% scrap is charged with hot metal. However, the importance of the open-hearth process is rapidly declining and so flexibility in scrap usage is diminishing. Scrap usage in the BOF is limited because the oxidation of carbon and silicon is the only source of heat for melting the scrap. Without preheating, the maximum amount of scrap per charge consumed in the BOF is ca 33%.

The electric-arc furnace, like the open-hearth furnace, refines a material charge with any ratio of hot metal to solid metal but typically operates with a solid charge of almost all scrap. The rapid increase in electric-furnace capacity in the last two decades has been paced by the growth of minimills, small capacity steel mills with only electric-arc furnaces based on 100% scrap, and expansion of electric-arc-furnace capacity by integrated steel mills that produce carbon steels. Growth in electric-furnace capacity has led to increased demand for ferrous scrap.

As indicated in Table 2, the basic oxygen furnaces in 1978 produced ca 61% of domestic steel, electric-arc furnaces ca 23%, and open-hearth furnaces ca 16% (1). However, ca 49% of the total ferrous scrap consumed was in electric-arc furnaces, 34% in the BOF, and almost 17% in open-hearth furnaces. Over 52% of the total steel production was from ferrous scrap. By 1988, the steel industry forecasts the decline of the open-hearth furnace process; thus, the BOF is expected to account for ca 68% of the total steel production and electric-arc furnaces to account for the remaining 32%. In terms of ferrous-scrap consumption, however, electric-arc furnaces will use 65% of the total scrap and the BOF only 35%. The increase in scrap consumption is expected to occur primarily through growth in electric-arc furnace capacity. Moreover, as open-hearth capacity decreases, additional hot-metal capacity will be available to supply the BOF. Thus, BOF scrap usage as a percent of the charge may fall slightly.

The transition from open-hearth dominance to BOF dominance from 1956 to 1970 was accompanied by an increase in the cost of molten pig iron and a decrease in the cost of ferrous scrap because the reduced scrap needs in the BOF were more easily met by the supply of home scrap and prompt industrial scrap (18). The availability of lower cost scrap was a main contributing factor to almost doubling electric-furnace steel production, especially for carbon-steel grades, from 1956 to 1970.

Since 1970, changes in the scrap market occurred that have led to some recent anomalies in the domestic supply–demand relationship for ferrous scrap. Typically, rising steel production increases the demand and cost of scrap and vice versa when steel production falls. The supply of prompt industrial scrap, however, is not strongly related to scrap prices but is closely related to the level of steel consumption. The supply of obsolete scrap, on the other hand, is more closely tied to scrap prices. In a study from 1950 to 1973, it was concluded that, over the short term, the supply of obsolete scrap was directly proportional to its price (19). The price of ferrous scrap is highly volatile. Contributing to this instability in the short term are several characteristics of typical scrap-purchase agreements (11). Ferrous scrap usually is pur-

Table 2. Steel Production and Scrap Consumption in the United States, 10^6 t [a]

Year	Basic-oxygen furnace (BOF)		Electric-arc furnace		Open-hearth furnace	
	Scrap consumed	Steel produced	Scrap consumed	Steel produced	Scrap consumed	Steel produced
1978	23	77	32	30	11	20
1988[b]	26	97	49	46		

[a] Ref. 1.
[b] Projected.

chased on the basis of 30-d or 60-d delivery contracts; thus, short-term changes in steel production directly affect the demand for scrap. Moreover, the basis for acceptance in terms of quality control is often inconsistent, which enables scrap buyers to hedge against short-term decreases in price at the time of delivery by rejecting the scrap shipment on the grounds of its quality. The weekly composite price per metric ton for the largest tonnage grade of ferrous scrap reflects these price fluctuations, as shown in Figure 1 (20). The maximum price change as a percent of the annual low price was 56% for 1977, 23% for 1978, 48% for 1979, 57% for 1980, and 38% for 1981 (20). Long-term effects on the price of scrap include international demand and technology changes that affect both scrap sellers, eg, auto shredders, and scrap buyers, eg, for electric furnaces.

Excess scrap accumulated during periods of low steel production is consumed during high demand years. However, during two periods in 1979 and 1980, scrap prices rose substantially as domestic scrap demand decreased. One factor contributing to the unusual behavior was the increased export demand for scrap. From 1978 to 1980, ca 9×10^6 t/yr of scrap was exported, whereas the average for the previous ten years was ca 7.5×10^6 t/yr or ca 20% less. Another explanation, one favored by steel-industry consumers of scrap, is that the demand for obsolete scrap probably exceeds and will continue to exceed the supply that is economically available (21). This demand is attributed to increases in the number of minimills, expansion of electric-furnace capacity by domestic integrated steel producers, and increased electric-arc-furnace capacity by foreign steelmakers, which raises the number of foreign countries buying U.S. scrap. However, according to the scrap industry, the continued high demand and high price for scrap is expected in the long run to result in an increased recovery of obsolete scrap, and the present domestic inventory of obsolete scrap is more than sufficient to meet the future needs of both the domestic and the export scrap markets. The potential market for new sources of obsolete scrap, including MFS, includes domestic and growing export markets.

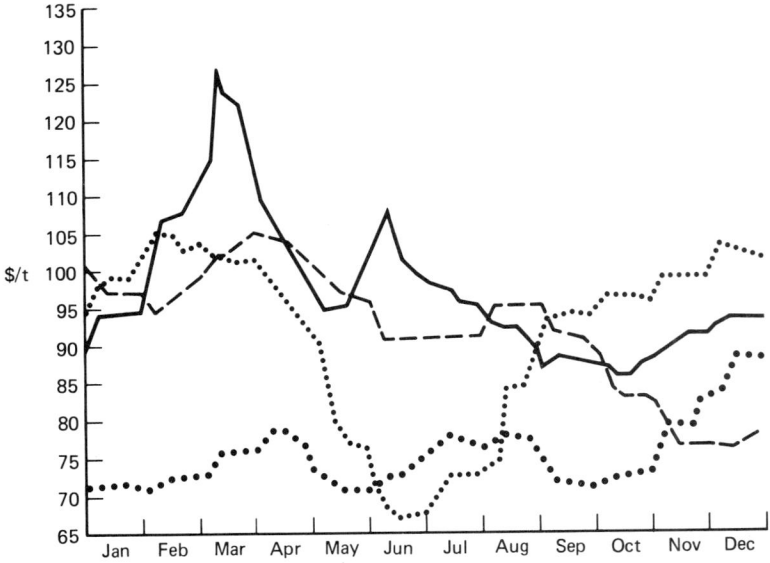

Figure 1. Weekly composite price, no. 1 heavy-melting steel scrap, based on Pittsburgh, Chicago, and Philadelphia. •••• = 1978; —— = 1979; ····· = 1980; --- = 1981 (20).

Recycling

Automobiles. The recycling of ferrous scrap from automobiles has two basic steps: dismantling the car to recover used parts of value and processing the stripped car hulk to recover ferrous scrap. Traditional recycling, which was done mostly by hand, is described in ref. 22. By the late 1950s, the development of the BOF steelmaking process with its reduced scrap requirements began to depress scrap prices and led to the buildup of a substantial backlog of unrecycled automobiles in most regions of the country. During the early 1960s, large fragmentizing machines called shredders were developed to upgrade auto scrap and to reduce labor and processing costs. These shredders could reduce a car hulk in minutes to 15-cm or smaller-sized pieces. This shredded scrap was usually further processed by magnetic separation and air classification to remove the contaminants and resulted in improved scrap quality. Shredding is now the dominant automobile-recycling process, and there are ca 200 automobile shredders in the United States (12).

Municipal Solid Waste (MSW). The development in the 1960s of a national interest in recovery of various materials, including ferrous scrap from municipal solid waste, or resource recovery, was a response to increasing solid-waste disposal problems (16). Resource recovery offered a way to achieve a significant reduction in the volume of MSW. Solid waste can be separated into glass, ferrous, aluminum, and the organic or combustible fractions called front-end separation. The sale of the materials would partially offset separation costs and disposal of the unsalable residue. Recycling of ferrous scrap from municipal solid waste, however, cannot be a main materials supplier because the quantities of ferrous scrap cannot provide enough material to satisfy annual consumption. However, municipal ferrous scrap could supply up to 10% of annual scrap consumption in the United States.

A more recent incentive for recycling is the emphasis on energy conservation and the development of alternative energy sources. The combustible fraction in municipal solid waste can be used as either a supplementary boiler fuel or as the only fuel in mass-burning facilities for the generation of steam and electricity. Ferrous scrap and other noncombustible materials can be removed from the solid waste as the waste is processed into a form suitable for a fuel or, if the unseparated solid waste is used as the fuel, the ferrous scrap can be separated and recovered from the boiler ash.

Finally, economic value and ease of recovery can be strong motivating factors for recycling a particular material. Many metals are recycled because of their high economic value. Some of these metals have physical properties, eg, high density or magnetism, that permit mechanical separation; others have chemical properties, eg, ease of selective dissolution, that facilitate chemical separation. The cost of separation and recovery by mechanical systems is usually lower than for chemical systems (16). Moreover, ferrous materials are among the easiest to recover because iron and most steels, excluding certain grades of stainless steel, are ferromagnetic and can be routinely separated from nonmagnetic materials by means of standard industrial drum- or belt-magnetic separators with either permanent magnets or electromagnets (23).

Several systems have been developed for processing municipal solid waste in order to separate and recover the materials of interest. The principal unit operations for separating municipal ferrous scrap include incineration or mass burning of the solid waste followed by magnetic separation of the ferrous residue from the ash; front-end separation or shredding the solid waste followed by magnetic separation, air classification, or both; and wet pulping (24). Most facilities in operation use either of the first two approaches, although a single system combining shredding and the separation of the noncombustible MSW followed by mass burning to generate steam and electricity combines the best technical features of both systems (15). Since each unit-operation sequence affects the chemical and physical properties of the recovered ferrous scrap differently, the relationship between processing, contamination, and market varies.

Impediments to Increased Use of Municipal Ferrous Scrap

Market consumption of municipal ferrous scrap (MFS) has decreased since 1976 (25). By the end of 1979, there were 56 municipal resource-recovery systems in operation, but only 27 were separating and recovering ferrous scrap (14). There are institutional and technical obstacles to increased consumption of municipal ferrous scrap. Many institutional barriers to recycling MSW have been identified (11,26–27). The obstacles are artificial or arbitrary and almost all contain significant questions of public policy. The technical barriers are primarily of two types: lack of national standards and specifications for the materials recovered from municipal solid waste and technological obstacles resulting both from the impact of municipal ferrous scrap on processes based on the scrap and the properties and quality of products made with the material. Lack of national material specifications and standard test methods for recovered materials place the producer of recovered products, eg, a resource-recovery facility, at a disadvantage with regard to identifying which of the potentially recoverable materials has real markets (16).

The primary technical obstacle that limits demand for scrap, especially municipal ferrous scrap, is the presence of certain chemical impurities or residual elements, primarily tin, copper, and aluminum, in amounts in excess of that in either home or prompt industrial scrap or in primary metal produced from virgin ore. In MFS, tin originates primarily from tin-coated food and beverage cans, and aluminum contamination arises from the aluminum tops of bimetal beverage cans. Copper is contributed by electrical components, eg, motors and wire, although increased copper contamination occurs in the metallic residue of incinerators or mass-burning plants as a result of vapor deposition of copper onto the residue (17).

Municipal ferrous scrap as produced by a single magnetic-separation operation almost always exhibits a high level of contamination not only by tramp elements but also by organic materials attached to the magnetic material. Although municipal ferrous scrap in this form can be used in each of the five main scrap-consuming industries, these levels of contamination severely limit the amount of scrap that can be used. For example, such MFS is limited in the iron-and-steel industry (16,28) and iron foundries (17) to ca 10% of the respective furnace-scrap charges.

Markets

Iron and Steel Production. Municipal ferrous scrap can be charged to either a blast furnace that is used to produce hot metal or to BOF or electric-arc steelmaking furnaces. Charging MFS as part of the electric furnace or BOF charge without preheating merely displaces another type of scrap and does not result in increased scrap consumption. Scrap usage would increase only if more steelmaking furnaces are built.

True recycling can occur if municipal ferrous scrap is charged to the blast furnace, because the scrap would displace iron ore. Increased scrap consumption would result even if overall iron and steel production did not increase. The iron content in 7 kg of ore could be replaced by ca 5 kg of scrap (29–30). However, the tramp elements tin and copper are not removed in either the blast furnace or the steel-refining furnace. Although quantitative upper limits on copper and tin content for particular grades of steel are not always known, copper and tin increase susceptibility to surface melting or hot shortness during hot-rolling operations and embrittlement at high temperatures and decrease tensile ductility and toughness; all of these are undesirable effects. Tramp-element concentrations in front-end separated scrap and incinerator residue are listed in Table 3.

The aluminum content of MFS varies widely, depending on the fraction of bimetallic cans present, but an upper limit of 2% can be estimated based on available data (31). When this scrap is charged in a blast furnace, the aluminum exothermically reduces some of the iron oxide from the ore charge to form alumina. The combination of increased temperature and alumina decreases the life of the blast-furnace lining. The use of this scrap in a steelmaking furnace creates somewhat different problems. In the latter, the aluminum reacts with the iron oxide in the slag and alters the slag chemistry. As the iron oxide content of the slag decreases, phosphorus reversion in the slag occurs and the phosphorus concentration in the steel increases, thereby decreasing the formability of the steel.

Processing the waste can reduce the detrimental effects of these tramp elements. Shredding the solid waste followed by air classification greatly reduces organic contamination because it removes most of the light paper and plastics (23). If the ferrous fraction is then magnetically separated, the copper level is greatly reduced (17). Subsequent finer shredding and magnetic separation can reduce the aluminum by up to one third, because many of the aluminum can tops become separated from the can bodies (34). Alternatively, heating the scrap to 315°C oxidizes the aluminum and reduces the aluminum concentration by 70% (35). At this point, the tin concentration can be reduced to 0.03–0.1 wt % by chemical detinning and, thus, a material that is similar in quality to no. 1 bundles is produced (36–38).

Iron Foundries. The tramp elements tin and aluminum and, to a lesser extent, copper also affect the properties and structure of the various cast irons, including gray cast iron, ductile cast iron, and malleable cast iron (28,39). The aluminum concentration is sufficient to cause casting problems. Excessive slag formation can occur as the aluminum is oxidized to alumina but, more importantly, the presence of as little as 0.02 wt % aluminum increases the susceptibility to pinhole formation or porosity

Table 3. Tramp-Element Concentrations in Municipal Ferrous Scrap, wt %

Tramp element	Typical tolerance limit, max[a]	Municipal ferrous scrap (MFS)	
		Front-end separated	Incinerator residue[b]
tin	0.03	0.5[b]	0.18
copper	0.01	0.21[c]	0.62

[a] Ref. 31.
[b] Ref. 32.
[c] Ref. 33.

as a result of hydrogen absorption from the air or from moisture in the mold. In cast iron, tin levels above 0.04 wt % completely stabilize the pearlitic microstructure and thus change the properties. For ferritic ductile-iron grades, the maximum tin concentration should be lower than 0.04 wt %; for unannealed malleable cast iron, the upper limit is ca 0.02 wt % (32). Copper up to 0.5 wt % reduces the ductility of ductile cast iron. Trial-cupola heats containing 10% of the charge as undetinned magnetically separated municipal ferrous scrap, produces satisfactory gray cast-iron automotive castings (39). Although cast irons are more tolerant of some tramp elements than steel, processing operations similar to those identified for upgrading MFS for iron and steel production are needed to lower the tin and aluminum concentrations before significantly higher proportions of this scrap can be routinely used in foundries.

Ferroalloy Production. Ferroalloys, used as additions in producing alloy steels, are made primarily in electric-arc furnaces. Municipal ferrous scrap has been used as part of the furnace charge for producing ferroalloys, even though alloy steels often have greater limitations on tramp-element concentrations than carbon steels. Incinerator scrap is preferred because all of the organic contaminants are eliminated, even though significant tin and copper impurities from the scrap ultimately occur in the ferroalloy product. However, since the ferroalloy is typically only a small addition to the steelmaking furnace, the dilution of the tin and copper results in acceptable concentrations in the steel product (32).

Detinning and Copper Precipitation. The detinning industry produces the only domestic supply of tin, the source of which is clean tinplate scrap. Municipal ferrous scrap with its high percentage of tin-coated food and beverage cans is another potential tin source. However, bimetallic cans with aluminum tops and possibly organic contaminants pose severe problems in the commercial detinning process. The sodium hydroxide solution used to dissolve the tin reacts with the aluminum to form sodium aluminate, thus consuming the detinning reagent. The sodium aluminate further causes an increased loss of solution owing to increased solution viscosity when the detinned material is removed from the bath and reduces the efficiency of the electrodeposition process, in which the tin is recovered from the bath (28,16). Although aluminum contamination can be considered an economic problem as well as a technical obstacle, the impact is continuous with no minimum threshold and a magnitude directly proportional to the aluminum content (35). The organic contaminants, including textiles and food residue, are retained in the sodium hydroxide solution and thus interfere with the optimum operation of the bath. Additional processing of municipal ferrous scrap, eg, a subsequent shredding and magnetic separation or a chemical pretreatment, is often necessary to reduce the aluminum and organic contamination to acceptable concentrations.

The primary requirement for municipal ferrous-scrap use in the copper-precipitation industry is a high surface area to volume ratio and a low level of organic contamination. Although detinned municipal ferrous scrap is more chemically reactive in the precipitation reaction than incinerated scrap, both types of scrap are essentially free of organics and are used to recover copper. In 1974, almost 15% of the domestic copper production was produced by the precipitation process (4). Further expansion of this market is uncertain because of a trend towards the use of ion-exchange (qv) techniques to recover the copper (40).

Standards

The development and operation of resource-recovery systems has been hindered by the absence of widely accepted national standards and specifications applicable to materials recovered from municipal solid waste (16,41–42). The first codification of ferrous-scrap specifications occurred in 1926 (43). These specifications and their modern successors are origin specifications, which describe the source of the material and the limits of the major contaminants. Origin specifications are most successful when the materials are derived from established processes (42). The principal focus has been on industrial scrap and certain special categories of obsolete scrap, eg, railroad products and automobiles, in which the scrap can be characterized according to origin specifications.

There is a strong need for quantitative specifications for materials recovered from MSW. Although the value of recovered materials generally increases with purity and homogeneity, the key to the usefulness of these materials and, thus, their potential for increased markets is the ability to ensure a reproducible quality (16).

Recognition of this problem led to the adoption of two national consensus standards for municipal ferrous scrap: ASTM E 701-80 and ASTM E 702-79. The specification document defines the chemical and physical requirements for municipal ferrous scrap in the five market areas of copper precipitation, iron foundries, iron and steel production, detinning, and ferroalloys. Requirements for chemical composition, metallic yield, cleanliness as measured by combustible content, and bulk density are included in ASTM E 702-79.

Outlook

Present trends in steelmaking technology and ferrous-scrap supply could result in increased demand for scrap, including municipal ferrous scrap (10,44). The reduced size of automobiles will eventually reduce the supply of automotive scrap, and changes in materials used in automobiles may lower the resulting scrap quality. Anticipated growth in continuous-casting capacity, which has lower scrap generation than ingot casting, and improved electric-furnace productivity will reduce the quantity of home scrap. Efforts by some steel fabricators to recycle their in-house scrap will reduce the availability of prompt industrial scrap. The continued growth of scrap-based electric-arc-furnace capacity combined with the increase of overseas exports of scrap will also contribute to the need for additional sources of scrap.

BIBLIOGRAPHY

1. *Steel at the Crossroads: The American Steel Industry in the 1980s*, American Iron and Steel Institute, Washington, D.C., 1980.
2. *1980 U.S. Industrial Outlook*, U.S. Department of Commerce, Industry and Trade Administration, Washington, D.C., 1980.
3. *J. Met.* **26**, 33 (1974).
4. *Progress Report on Recycling*, Tin Mill Products Producers, American Iron and Steel Institute, Washington, D.C., 1974.
5. L. F. Reinartz, *History of Iron and Steelmaking in the United States*, Metallurgical Society, American Institute of Mining, Metallurgical, and Petroleum Engineers, New York, 1961, pp. 73–76.
6. A. B. Wilder in ref. 5, pp. 61–72.
7. D. A. Fischer, *The Epic of Steel*, Harper & Row, Publishers Inc., New York, 1963, p. 290.

8. S. B. Casey in ref. 5, pp. 92–95.
9. H. E. McGannon, ed., *The Pneumatic Steelmaking Processes, The Making, Shaping, and Treating of Steel*, 9th ed., United States Steel Corporation, Pittsburgh, Pa., 1971, pp. 486–497.
10. M. B. Bever, *Conserv. Recycling* **1,** 55 (1976).
11. J. A. Commins, V. A. Hathaway, E. F. Palermo, B. M. Sattin, and M. A. Timothy, *U.S. Bureau of Mines Report NTIS PB 271*, 1977, p. 814.
12. R. F. Testin, *J. Met.* **33**(9), 21 (1981).
13. W. L. Swager, *paper presented at 1981 Annual Meeting of the American Institute of Mining, Metallurgical, and Petroleum Engineers*, Chicago, Ill, Feb. 23, 1981.
14. *Summary Report of Solid Waste Processing Facilities*, Committee of Tin Mill Products Producers, American Iron and Steel Institute, Washington, D.C., 1978.
15. J. G. Abert, *paper presented at International Recycling Congress*, Berlin, FRG, 1979, pp. 18–27.
16. *Mineral Resources and the Environment, Supplementary Report: Resource Recovery from Municipal Solid Wastes*, NAS, Washington, D.C., 1975.
17. E. J. Ostrowski, *Trans. Am. Foundrymen's Soc.*, 111 (1977).
18. D. J. Carney, *J. Met.* **26,** 41 (1974).
19. H. B. Jensen, *Electric Furnace Proceedings*, Vol. 30, Metallurgical Society of the American Institute of Mining, Metallurgical and Petroleum Engineers, Chicago, Ill., 1973.
20. *Am. Met. Mark./Metalworking News* **88,** 43 (Oct. 13, 1980); **90,** 28 (Feb. 22, 1982).
21. H. B. Jensen, *paper presented at ASM Congress*, Cleveland, Ohio, Oct. 1980.
22. M. B. Bever, *paper presented at the 7th Mineral Waste Utilization Symposium*, Chicago, Ill., 1980, pp. 174–183.
23. H. Alter and K. Woodruff, *U.S. EPA Rep. EPA/530/SW-559*, 1977.
24. N. T. Neff, *U.S. EPA Rep. NTIS PB 213*, 1972, p. 646.
25. *Solid Waste Syst.* **1**(1), 10 (1978).
26. J. G. Abert, *ASTM* **STP 592,** 31 (1975).
27. J. N. Humber in ref. 26, pp. 40–52.
28. R. S. Kaplan in ref. 26, pp. 91–105.
29. E. J. Ostrowski, *Proc. Ironmaking Conf., Am. Inst. Mining, Metallurgical, Petrol. Eng.*, Philadelphia, Pa., Vol. 30, 1971, pp. 115–124.
30. E. J. Ostrowski, *paper presented at Annual Mining Symposium of Minnesota Section*, American Institute of Mining, Metallurgical, and Petroleum Engineers, Duluth, Minn., Jan. 16–18, 1974.
31. H. V. Makar and co-workers, *U.S. Bureau of Mines Report RI 8037*, 1975.
32. E. J. Duckett, *Resource Recov. Conserv.* **2,** 301 (1976–1977).
33. E. J. Ostrowski, National Steel, Pittsburgh, Pa., 1975, private communication.
34. E. J. Ostrowski, *Iron Steel Eng.* **48,** 65 (July 1971).
35. E. J. Duckett, *National Center for Resource Recovery Report NSF/RA-770244*, Washington, D.C., 1977.
36. W. L. Hunter, *U.S. Bureau of Mines Report 8147*, 1976.
37. P. M. Sullivan and H. V. Makar, *paper presented at the Fifth Mineral Waste Utilization Symposium*, U.S. Bureau of Mines and IIT Research Institute, Washington, D.C., 1976.
38. *Metalworking News, Steelmaking Today Suppl.* **88,** 6A (Sept. 8, 1980).
39. R. C. Helmink, G. F. Ruff, and J. F. Wallace, *Trans. Am. Foundrymen's Soc.* **82,** 525 (1974).
40. E. Meschter, *Metalworking News* **86,** 28 (Aug. 14, 1978).
41. C. G. Interrante, *ASTM* **STP 592,** 146 (1975).
42. H. Alter, *Am. Soc. Test. Mater. Standard. News*, 11 (Nov. 1979).
43. *Specifications for Iron and Steel Scrap*, Institute of Scrap Iron and Steel, Washington, D.C., 1971, p. 1.
44. D. Prizinsky, *Metalworking News* **88,** (Feb. 25, 1980).

JAMES EARLY
National Bureau of Standards

GLASS

Glass (qv) recycling involves the recovery of used glass (postconsumer glass) and reusing it as a raw material in useful products. The main repository for recovered glass is glass containers. Other applications, eg, fiber glass and aggregate for road building and construction products, are not significant because only very small quantities are used on an experimental basis. Recycled glass that is in bulk form and is suitable for melting is called cullet. Waste glass, ie, off-quality material and scrap from the manufacture of glass products, may also be called cullet but is not considered recycled because it has not been used by the consumer.

There are two broad types of recovered postconsumer glass: source-separated glass and resource-recovery glass. Source-separated glass is separated usually in the home from other wastes. Approximately 10% of municipal refuse is glass most of which is in the form of discarded containers (1). Segregated glass is generally handled in two ways. It may be collected from individuals or may be delivered by individuals to a regional recycling center. Resource-recovered glass is separated by automated methods from municipal solid waste at a central location.

Inducements for recovery and reuse of postconsumer glass are twofold. First it permits cost reductions at the container manufacturing plants. Recovered cullet is lower in cost than virgin raw materials. Because of its lower melting temperature, less energy is required to melt cullet than raw materials. In addition, use of cullet may reduce particulate emissions sufficiently to obviate the use of expensive air-pollution-abatement equipment (2). The other inducement is cost reductions at central resource-recovery plants. The primary incentive to operating a resource-recovery plant is to convert the combustible portion of the refuse into a valuable energy product. However, revenue from the sale of recovered glass can contribute to a reduction in the overall cost of plant operation. In addition, the quantity of material which must be hauled away and used as landfill is reduced and results in cost savings.

Properties of Recovered Glass

Source-segregated glass is normally in the form of unbroken containers, which have a bulk density of 350–435 kg/m^3 (22–27 lb/ft^3). Glass recovered at a central resource-recovery plant is generally sized at less than 6 mm and has a density of 1300–1500 kg/m^3 (81–94 lb/ft^3). The color distribution of the glass in postconsumer municipal solid waste is ca 65% flint (colorless), 20% amber, and 15% green. A predominate proportion of the glass in soda-lime bottle glass with a composition of 66–75 wt % SiO_2, 1–7 wt % Al_2O_3, 9–13 wt % CaO and MgO, and 12–16 wt % Na_2O.

Processing

Usually, multiple operations are required to control or eliminate contaminants, which are inherently in postconsumer glass. Organic matter, metals, and refractory materials must each be reduced to a small fraction of a percent, generally with no particle more than 6 mm in size.

Source Separation. Processing is usually necessary at some central point before the cullet can be used in the manufacture of glass containers. This intermediate processing may in its simplest form consist of handsorting to segregate bottles by color and to remove nonglass containers (3). The bottles may be crushed to increase their bulk density for shipment and to facilitate removal of ferrous metal and other contaminates. In areas where bottle bills (government-imposed deposits on nonreturnable containers) have been adopted, crushing prevents another cycle of deposit redemption. An intermediate processing operation designed to process glass with high contaminant levels might consist of manual sorting, magnetic separation, crushing, and screening.

Typically, the intermediate processors of source-segregated glass are dealers in scrap, large beverage distributors, and processors organized specifically to beneficiate and market used glass. However, one container manufacturer (Glass Containers Corp. in Dayville, Ohio) operates its own upgrading plant where all foreign cullet is inspected at the plant. Contaminated shipments, which otherwise would be rejected, are sent to the on-site decontamination plant, which can process cullet at ca thirteen metric tons per hour. Treatment consists of manual sorting, magnetic separation, and two stages of crushing and screening. Rejected contaminants include steel cans and other ferrous metals; aluminum caps, neck rings, and pull tabs; wood; stone; and ceramics.

Resource Recovery. Several large solid-waste resource-recovery plants are equipped with glass-recovery modules. The glass-recovery system is generally combined with and inseparable from nonferrous-metal recovery equipment. Two basic final purification methods are employed by resource-recovery plants; the froth flotation process and the optical sorting process. Both approaches are, as yet, unproven.

The froth flotation process is generally applied to a glass-rich fraction separated from dry shredded, solid waste and produces a mixed colored product. The important preparation steps are magnetic separation of ferrous metal, separation of a heavy, glass-rich product by screening and air classification, removal of aluminum with so-called aluminum magnets, separation of residual organics and heavy metals by jigging and grinding. The aluminum magnet operates on the principle that when metals pass through an electromagnetic field, eddy currents are generated in each piece. These currents have a magnetic moment, which is phased to repel the moment of the applied field. This force is sufficient to cause the metal to be thrown away from the nonmetallic particles (see Magnetic separation). The jig has an action that produces a loose vibrating bed of solids in a liquid medium. The solids separate into layers of different apparent specific gravities. Organics move to the top and are skimmed off. Heavy metals, eg, lead, zinc, and copper, form the bottom layer and are drawn off. A glass-rich middle layer is removed separately.

Froth flotation is a minerals-processing technique which has been adapted for use in waste-glass recovery (see Flotation). The procedure was first applied to glass cleaning by the U.S. Bureau of Mines (4). Treating an aqueous mixture of finely ground (≤ 850 μm) glass and mineral particles with a cationic fatty amine, which is selectively absorbed by the glass, causes the glass particles to float as a froth. The tailing or sediment is composed of sand, ceramics, and other nonglass particles. The froth flotation process is used at plants in New Orleans, Louisiana, and Monroe County, New York.

In the optical sorting process, the glass-rich feed stock is prepared according to concepts developed by Sortex Company of North America and The Black Clawson Company (5–7). The process has been applied on a large scale (10,000 metric tons per week of refuse) plant in Hempstead, Long Island, New York. It consists of wet milling whole refuse to <25-mm size; separation of a heavy glass-rich product from the lighter organic material in a liquid cyclone; screening the heavy fraction to produce a product >6-mm size; using heavy media to remove heavy organics; separation of a mixture of glass, stones, and ceramics from metal by jigging; drying; high tension electrostatic separation of residual metal; and optical sorting.

Separation by heavy media is based on a sink–float principle. The solid feed is mixed with a liquid slurry of magnetite in water with an effective specific gravity of 1.6–2.0. Bone, wood, and plastic float from the heavier metal, glass, stones, and ceramics, which sink. Separation of metal by electrostatic high tension takes place in a sorter designed to impart static electric charge to the feed mixture, which is spread on a grounded rotating roll. The metal quickly loses its charge to ground and is lifted off the roll by high tension electric forces.

Optical sorters scan each particle of the feedstock, and opaque particles are removed in transparency sorters. When a light beam is broken, an air jet knocks the opaque particle out of the mixed colored feed stream. Color sorting is achieved by comparing each particle to a standard and rejecting off-color material.

Economic Aspects

Neither source separation nor the centralized resource-recovery processing techniques separate all of the glass present in solid waste. A recovery rate of ca 50% for source-separated refuse is very high (8). An average operation yields less than 25% of the glass present. Low recovery results from low public participation caused by indifference to or ignorance of the program. Recovery from central resource facilities is generally 40–50%. Glass is lost with the fuel fraction and losses also are inherent in the recovery process.

Although there are large divergences between grades, companies, and localities, the price paid for cullet by the user is ca 80–95% of the cost of virgin raw materials. This coupled with approximately 15% energy savings which results from the low melting point of cullet and the lower pollution-abatement-equipment operating costs provide the glass-bottle manufacturer with a cost advantage if recycled material is used. It is doubtful that there is much economic incentive to the householder or community in source separation of glass. Their motivation is mainly to extend landfill life and to preserve natural resources. The cost to the resource-recovery plant operator of processing and delivering cullet is ca $20–30/t (January, 1980). The price paid by glass-container manufacturers generally is less than $50/t. Maximum net revenue to the processor is not over $20/t. However, if glass is not sold, the processor must bear the cost of disposal, ie, hauling and landfilling.

Mixed colored glass can only be used in amber and green-glass bottle furnaces and in proportions of no greater than ca 5% for amber and 20% for green. Therefore, the locations and capacities of the colored-glass furnaces are important when considering markets for froth-floated glass. Clean, optically color-sorted material has an advantage over froth-floated glass since it has the potential of being used interchangeably with cullet produced in-house or by hand sorting.

Specifications

In order to be suitable for reuse in melted glass containers, glass cullet must meet rigid specifications. In general, glass particle size must be less than 50 mm and organic materials, metals, and refractory materials must each be a small fraction of a percent. Large particles of glass may contribute to materials handling problems. Organic materials can result in the formation of bubbles or seeds in the melted glass. Iron oxide can cause a green or amber tint in flint glass. Particles of metals may form stones in the finished product. Aluminum is particularly troublesome, since at glass melting temperatures aluminum may reduce silica to silicon, which is refractory. Refractory materials, eg, brick and ceramics are normal components of solid waste. Particles of these materials may produce stones in melted glass products, particularly if they are greater than ca 400 μm in size.

ASTM has formed a Subcommittee (E 38.05) for development of standard testing methods and specifications for recycled glass. Testing standards (ASTM E 688-79) and standard specifications (ASTM E 708-79) for waste glass as a raw material for the manufacture of glass containers have been developed by ASTM for particulate glass cullet material smaller than 6 mm. Standards for glass cullet particulate material over 6 mm in size are being developed.

BIBLIOGRAPHY

1. *NCRR Bulletin*, Vol. 3, No. 2, National Center for Resource Recovery, Inc., Washington, D.C., 1973.
2. Pollution Control Section, *Bus. Week* (March 31, 1975).
3. D. B. Weiss, *NCRR Bulletin*, Vol. 9, No. 3, National Center for Resource Recovery, Inc., Washington, D.C., 1979.
4. J. H. Heginbotham, *Proceedings of the Sixth Mineral Waste Utilization Symposium*, Illinois Institute of Technology, Chicago, Ill., 1978.
5. U.S. Pat. 3,650,396 (March 21, 1972), R. M. Gillespie and H. R. Rhys (to Sortex Company of North America, Inc.).
6. U.S. Pat. 3,788,568 (Jan. 29, 1974), P. G. Marsh (to Black Clawson Fiberclaim, Inc.).
7. U.S. Pat. 3,945,575 (March 23, 1978), P. G. Marsh (to Black Clawson Fiberclaim, Inc.).
8. *Multimaterial Source Separation in Marblehead and Somerville, Massachusetts, Composition of Source-Separation Materials and Refuse*, EPA Report SW 823, EPA, Washington, D.C., 1979.

PAUL MARSH
Marsh-Eco-Service Co., Inc.

NONFERROUS METALS

This article concerns the recycling of nonferrous metals in solid wastes, specifically nonmagnetic automobile shredder residues and municipal refuse. The quantities of nonferrous metals becoming obsolete each year and the quantity of these materials that have been recycled are reported in ref. 1. The most widely used nonferrous metals are aluminum, copper, copper-base alloys, and zinc. Because of their high cost and current rate of recycling or limited use in products that would be discarded, or both, other nonferrous metals rarely occur in solid waste streams. Prompt, old, or obsolescent scrap is recycled at a high rate, since most other nonferrous metals are much more valuable than aluminum, copper, copper alloys, and zinc.

New or prompt industrial scrap is generated from wrought or cast products as they are processed by fabricators into consumer or industrial products. Old scrap is retrieved from postconsumer wastes, eg, automobiles, beverage cans, etc. Old scrap also includes obsolescent products, eg, transmission cable, aircraft, motor vehicles, etc.

In 1969, of ca 1.2×10^6 metric tons of aluminum that became obsolete, only 159,000 t or 13% was recycled. A similar analysis of obsolete copper and zinc showed recycling rates of 43% and 14%, respectively. The low recycling rate for zinc is attributed to its predominant use in corrosion protection, a dissipative use. In contrast, aluminum scrap is dispersed but can be concentrated as a step in the process of recycling automobile-shredder residues and municipal refuse. Estimates of availability and recycling rates for all three metals, based on the original markets for each, are reported in ref. 1.

Approximately 125×10^6 t of mixed municipal refuse was generated in 1975 (2). Substantial increases have occurred in the aluminum recycling rate, ie, from 11% in 1975. Aluminum amounts to two thirds or more of the typical 1% nonferrous metal content of refuse (3–4). The remaining one third consists of varying quantities of heavy nonferrous metals, including copper-base alloys, zinc, small quantities of stainless steel, tin, lead in solders, and only traces of other metals. The aluminum scrap in municipal refuse consists largely of discarded packaging products, most of which are aluminum beverage cans. In a study of thirteen municipalities, the aluminum concentration in municipal refuse ranged from 0.6% in Erie County, Pennsylvania, to 1.1% in Washington, D.C. (5). The heavy (>2.7 g/cm^3), nonferrous metal fraction of municipal refuse in these cities was 0.1–0.3%.

Although some of the resource recovery facilities that have operated since 1970 have included aluminum, or nonferrous-metal recovery operations, or both, actual metal recovery has been virtually nil. Results from analyses of samplings received from a number of laboratory and pilot-scale operations and from several full-scale facilities are listed in Table 1 (4).

Since expanded scrap usage must come from old or postconsumer scrap, sources of additional old scrap can be identified from raw-material marketing studies and product-obsolescence estimates. Aluminum-marketing data for 1979, as listed in Table 2, show that the metal supplied to the transportation, consumer durables, and containers and packaging markets have the greatest potential for expanded recycling activity. Estimates for these markets in 1985 are shown in Table 3.

Table 1. Chemical Analyses of Aluminum Recovered from Municipal Refuse, Wt % [a,b]

	Si	Fe	Cu	Mn	Mg	Cr	Ni	Zn	Ti	Pb	Bi	Sn	Assay, % recovery
Unseparated metals[c]													
low	0.16	0.04	0.40	0.30	0.03	0.02	0.05	1.30	0.03	0.05	0.02	0.06	75.9
high	3.00	1.00	8.80	0.80	0.20	0.10	0.90	23.00	>0.10	1.20	>0.10	1.50	93.3
Arithmetic													
median	0.60	0.50	5.10	0.65	0.10	0.05	0.07	3.30	0.05	0.30	0.05	>0.10	84.7
avg	0.91	0.57	4.20	0.57	0.09	0.05	0.20	5.90	0.06	0.37	0.05	0.46	83.9
Separated metals[d]													
low	0.10	0.40	0.10	0.40	<0.05	<0.05	<0.03	0.04	0.03	<0.03	<0.03	<0.05	76.8
high	3.20	1.00	1.80	0.90	0.80	0.05	0.10	4.00	0.05	0.05	0.05	0.10	94.4
Arithmetic													
median	0.20	0.50	0.40	0.74	0.10	0.05	0.05	0.10	0.05	0.05	0.05	0.05	87.8
avg	0.48	0.57	0.47	0.70	0.16	0.05	0.05	0.62	0.05	0.05	0.05	0.06	86.4
Incinerated[e,f]													
low	0.55	0.50	0.15	0.08	<0.05	<0.05	<0.05	0.15	<0.05	0.08	<0.05	<0.05	69.9
high	4.10	1.30	3.00	0.60	0.10	0.07	1.10	4.20	0.08	0.37	0.08	0.05	95.6
Arithmetic													
median	1.20	0.85	1.10	0.50	0.08	0.06	0.07	0.50	0.05	0.15	0.05	0.05	86.1
avg	1.60	0.81	1.20	0.45	0.07	0.06	0.07	1.20	0.06	0.12	0.06	0.05	84.1
Wet pulp, heavy media and jig[g]													
low	0.14	0.40	0.20	0.60	0.05	<0.05	<0.05	0.05	<0.05	0.02	<0.01	0.05	59.6
high	0.70	1.10	1.40	1.00	0.70	0.10	0.05	1.50	0.05	0.37	0.05	0.08	88.3
Arithmetic													
median	0.15	0.50	0.45	0.70	0.30	0.05	<0.05	0.20	<0.05	0.13	<0.05	0.05	81.6
avg	0.22	0.56	0.58	0.73	0.25	<0.06	<0.05	0.43	<0.05	0.13	<0.05	0.07	78.6

[a] Ref. 4.
[b] The lines identified in all samples as low and high are the lowest and highest incidences of that element in the samples tested from that category of samples.
[c] Nonmagnetic nonferrous metals from refuse processed in front-end systems. Analyses are of 10 samplings from 7 cities.
[d] Aluminum fraction recovered by means of dense-media equipment on nonmagnetic metals and concentrates from front-end systems. Analyses are of 21 samplings from 7 cities. Four cities are represented in the separated and unseparated analyses.
[e] Recovered from incinerator ash with dense-media equipment. Analyses represent 10 samplings from 5 cities.
[f] Three samples with no dense-media separations show almost identical analyses except for: Cu, 1.3–20.0 wt %; Zn, 2.2–7 wt %; and Pb, 0.05–0.55 wt %. Three separate samplings from an incinerator in one city.
[g] Sample material from Black Clawson in Franklin, Ohio. Separation of the heavy fraction was conducted by Black Clawson with a 2.0-g/cm^3 dense media followed by use of a mineral jig.

Table 2. Estimated Maximum Old Aluminum Scrap Availability: 1979, 10^3 t[a]

Market	Total potential	Not recovered	Recovered
building/construction	4	2	2
transportation	721	432	289
consumer durables	385	272	113
electrical	131	108	23
machinery/equipment	147	99	48
containers/packaging	1100	898	202
other	190	140	50
Total	*2678*	*1951*	*727*

[a] Ref. 6.

The transportation market, which is made up primarily of manufacturers of automobiles, aircraft, railroad equipment, ships, and trucks, provides a potentially large supply of scrap metals at the end of a given product's life cycle. Approximately 2–4% of an automobile's curb weight and auto-shredder residues are nonferrous metals (7–8). A typical composition of the nonferrous metal fraction is 30 wt % aluminum, 60 wt % zinc, and 10 wt % copper, brass, stainless steel, solders, and other metals. The current downsizing of automobiles by the American automobile industry and substitution of lightweight materials, eg, aluminum and plastics for steel and cast iron, will change the composition of residues at the time of disposal.

At the end of the typical 7–10-yr life cycle of an automobile, the hulk is usually delivered to an auto dismantler and the reusable or repairable parts are removed.

The stripped hulk is shipped to ca 200 U.S. automobile shredders. Ferrous scrap enters the commercial scrap market and the nonmagnetic residues are sold to ca 20 U.S. processors of such materials. Recovery of nonmagnetic metals is based on dense-media separation of aluminum and recovering heavy nonferrous metal residue. Heavy nonferrous metals are separated after aluminum separation by handsorting red and yellow metals as well as metals that are identifiable on the basis of shape, eg, thin strips of stainless-steel moldings. Residues from sorting can be processed in a sweat furnace to recover zinc which can be purified by vacuum distillation. Copper-base metals are generally shipped to copper refiners, where they are processed in blast furnaces or open-hearth furnaces and then are electrolytically refined.

Table 3. Estimated Maximum Old Aluminum Scrap Availability: 1985, 10^3 t[a]

Market	Total potential	Not recovered	Recovered
building/construction	143	105	38
transportation	868	519	349
consumer durables	395	280	116
electrical	201	165	36
machinery/equipment	210	141	69
containers/packaging	1468	932	536
other	94	72	21
Total	*3379*	*2214*	*1165*

[a] Ref. 6.

Sources of Nonferrous Metal Scrap

In 1979, scrap amounted to ca 23% of the supply to the aluminum industry. Of this, "old" scrap increased from ca 25% of scrap sources in the 1950s to ca 35% in 1979. "New" scrap made up the balance of the scrap supply. In 1979, old scrap use was equivalent to ca 8% of industry shipments.

Projections of the container and packaging market through 1986 show that more than 800,000 t of additional scrap will be available than in 1979 from these short-life-cycle products. It also is projected that for 1985 aluminum beverage cans will represent ca 75% of the aluminum tonnage in this market (6). Some products containing aluminum, eg, architectural building panels, electrical conductors, industrial equipment, etc, have long life cycles and, for the most part, do not appear to be significant contributors to typical postconsumer scrap sources.

The other principal aluminum market, ie, transportation, also projects significant increases in aluminum use by 1985. Use of aluminum in automobiles is increasing, with ca 59 kg being used in the average 1981 U.S. automobile; this is almost a 70% increase from the 1967 value. Aluminum industry projections are for 91 kg/automobile by 1985 which, in addition to significant fuel savings, will result in the displacement of zinc, copper, cast iron, and to a lesser extent, steel (6).

The lead–acid battery is generally removed from every dismantled automobile for reasons of safety as well as to recover readily marketable lead. Recycled lead amounts to ca 60% of lead-industry production, most of which comes from automobile batteries. Copper radiators are also stripped from automobiles, as they represent high quality material; typically 8 kg of Cu is recycled from one radiator.

Both stripped and unstripped automobiles are shredded and the shredded products are magnetically separated.

The various product components of principal appliances, eg, refrigerators, ranges, air conditioners, dishwashers, dryers, etc, and their values are reported in refs. 1, 3–4. Estimated life cycles, annual sales, and projected scrappage rates of these products are reported in refs. 1 and 3. A significant number of these products are shredded in automobile-shredder plants and are recycled in the existing network of dense-media plants.

Recycling by Consumers. Recycling of aluminum scrap from solid wastes involves separate, complementary approaches. Recycling by consumers consists of the purchasing of clean, consumer-type aluminum scrap delivered to a recycling location. The complementary approach is the recovery of aluminum from municipal refuse streams. Most of the scrap aluminum products generated in the home originate in the containers and packaging market and consist of aluminum beverage cans, aluminum cookware, frozen-food trays, foil, etc. Other discards include lawn chairs, storm windows and doors, siding, etc. This material is relatively uncontaminated and, in the case of cans, is remelted for production of new cans.

At the outset of consumer aluminum-recycling pilot programs in 1968, the purchase price of this material was $0.18/kg and has increased over 180% to the 1981 level of $0.51/kg. The consumer price index increased by ca 160% during the same period, from 104.2 in 1968 to 269 in June 1981. Estimated industry receipts of consumer-type scrap amounted to over 450,000 metric tons excluding exports in 1981, up more than 65% over 1980 receipts. Recycling by consumers is carried out at over 2500 locations and in every state in the United States. In 1981, the industry recycled about one of every two aluminum cans produced that year, leaving the remainder in the solid-waste stream.

From 1970 to 1979, the recovery of aluminum from old scrap increased 200% (179,000–557,000 t), and recovery from new or prompt industrial scrap increased 45% (728,000–1,055,000 t). This increase in old scrap recovery is almost entirely the result of beverage-can recycling. Wrought products predominate in municipal refuse and cast alloys predominate in automobile-shredder scrap. The wrought alloys in municipal refuse are generally compatible, ie, a mixture of these alloys can be put into a processing furnace, melted, and cast into new ingot with only minor additions normally required.

Recovery Systems

Automobile Scrap. With the advent of more restrictive steel-making scrap specifications during the 1960s, automobiles were shredded and magnetically separated to produce a ferrous product of acceptable quality; residues were landfilled. As the secondary metal processors realized the value of the residues, the residues were handpicked to recover copper or brass or both, and were then processed in sweat furnaces to recover zinc and aluminum. In the late 1960s, the handpicking operation was replaced in some plants with the use of dense-media equipment.

Municipal Refuse. In the early 1970s, as municipal refuse recycling facilities were being designed and constructed, unit operations similar to those used by automobile shredders were incorporated. The systems most often used for refuse processing were of four types: those that produced shredded, magnetically separated ferrous metals and air-separated light materials, eg, paper, plastic, and fabric, and heavy materials, eg, metals, glass, and rock; those that incinerated the incoming refuse and separated the metals from the ash; those that composted the organic fraction of refuse; or those that wet-pulped the refuse in machinery similar to paper-pulping equipment, with recovery of glass and metals following the pulping operation. Variations or combinations of the above were also part of the early activities in resource recovery.

Since there is no currently available nonferrous metal recovery equipment analogous to the magnet used for ferrous metals, nonferrous metals are often by-products or residues from other unit operations. If in automobile-shredder residues and municipal refuse processing facilities, heavy nonferrous metals are not recovered as by-products of aluminum separation processes, they remain as contaminants in aluminum (4). The composition of the resulting, heavy, nonferrous metal by-product is related to the efficiencies of prior aluminum separation stages, since in most separation processes some cross contamination occurs.

Front-end processing includes shredding, magnetic separation, and air classification; screening of the heavy product from the air classifier; and concentration or recovery of the metals. Eddy-current separation or dense-media separation can be used to recover nonferrous metals (see Magnetic separation; Gravity concentration).

Processing after receipt at a dense-media separation facility includes screening and possibly some additional shredding. Water elutriation and dense media are used for separation of the aluminum from the heavy nonferrous metals. Dense-media processing is described extensively in refs. 9–11. The active element in the dense-media equipment used can consist of either a conical vessel, a rotating drum with internal lifters, an oscillating blade, or a cyclone separator. In each, the water–media slurry is magnetite or a mixture of magnetite and ferrosilicon at densities up to 3.0 g/cm^3. These media are removed from the rinse water with a magnet and must be demagnetized prior to reuse. The U.S. Bureau of Mines is using barite ($BaSO_4$) in a dense-medium test program (12). This nonmagnetic medium is recovered for reuse in a settling operation.

Preliminary data from some front-end separation systems show that aluminum or a mixed nonferrous metal product can usually be recovered and separated into the equivalent of ASTM grades 1 or 2 (4) (see Specifications). In one other location, the recovered aluminum met grade 1 or grade 2 specifications (4) without additional processing. In June 1981, grades 1 and 2 were valued at over $0.70/kg. Grades 3 and 4 are valued at 20–30% less than grade 1, depending on the market. Grades 5 and 6 have about equal value as grades 3 and 4.

Back-end separation, ie, the recovery of materials following thermal or biological processing of wastes, potentially can become a principal source of nonferrous metals because of the trend toward increased use of mass-burning incinerators (qv) with waste-heat recovery. Current European practice is to charge the unprocessed refuse directly into the incinerator.

The U.S. Bureau of Mines has developed a process involving magnetic separation, crushing, and screening to allow recovery of ferrous metals, nonferrous metals, and glass from the ash (13). This concentrate can be further separated by dense-media processing. One of the disadvantages of refuse incineration is that the lighter gauge aluminum foil and some containers may oxidize during incineration. In addition, the ferrous metal can become contaminated with lead, tin, and copper, the presence of which restricts the markets for the material.

There are substantial differences between incinerators (14). In general, any material with a melting point less than that of copper (1084°C) should melt in an incinerator. If the ash–combustibles bed on the grates becomes hot enough to melt metals, these metals can fall through the holes in the grates. Material that does not fall through remains molten as it progresses through the incinerator and entrains ash, burned glass, or other metals. The materials that pass through the grates and those that come off the end of the grate are deposited into an ash sluice. The nonferrous metal content of incinerator ash is ca 1–2 wt % of the ash. In consideration that the raw refuse charged into an incinerator typically contains 1 wt % nonferrous metals and is reduced to ca 25% its original weight, it appears that ca 50 wt % of the nonferrous metals is lost.

Unit Operations. There are several facilities in operation that involve a rotary screen or trommel for processing raw refuse. Depending on the screen sizes used, a trommel can remove fine dirt and grit, glass, and wet putrescible material from the refuse and can recover a concentrate of metal containers, eg, beverage cans. The oversize trommel product, consisting mostly of paper and plastic, contains only a relatively small amount of the metallic and glass contaminants that are problematic in combustion systems. Since the most widely used size (355-mL or 12-oz) beverage can is ca 6.3-cm dia and 12-cm high, an appropriate hole size in the trommel screens generally permits passage of a large percentage of the beverage cans as a concentrate. Further screening can separate most of the broken glass, rock, and fine putrescible material from the metal cans. The successive concentrating steps facilitate eddy-current separation, handpicking, or the shredding of a can-rich fraction prior to shipment to a dense-media system. Most of the heavy nonferrous metals, including coins, remain in the fine screen fraction (<5 cm) from a trommel. Because of the low metal concentration in this fraction, little reported data are available on composition or recovery techniques.

In addition to the use of trommels, disk screening has been successfully used in several facilities to remove glass, metals, grit, and dirt from combustible-rich fractions. The disk screen consists of a series of shafts with disks of predetermined contours mounted at intervals along the shafts. The centerline-to-centerline distance between the shafts and the lateral distance between the disks determine the screen opening. Reported throughput capacity and screening efficiency are high. In comparison, flat-deck screens tend to become obstructed in the separation of paper, plastic, metals, and textile wastes.

There are several commercial eddy-current separators (15–16). The basic eddy-current mechanism involves the use of either an electromagnetic or a permanent magnetic field to induce eddy currents in the conductors that enter the magnetic field. The eddy currents are then affected by the magnetic field and repel the conductor from the field area. Measured thrust data in the fields are less than 20 g and are related to the a-c frequency and the distance from the primary magnetic field to the conductor (17). A main operating problem is entrainment with contaminating materials. If the size of the entraining article, eg, a piece of cloth or paper, is too great for the force present in that eddy-current field to overcome, the conductor is not removed from the process stream. For most efficient operation, each discrete particle should be exposed to the eddy-current field, where it can be acted on and can react according to whether or not it is a conductor. Practically, this is not feasible because of the quantity of material to be processed. However, any method that provides an enriched concentrate of material to an eddy-current separation system improves its efficiency. Eddy-current separators are sensitive to the geometry of the conductor in the field, and some machines are more sensitive than others. Some are unable to remove materials smaller than 0.75 cm in diameter readily from the field whereas others can process material as small as 0.15 cm in diameter (18). The eddy-current separator cited in ref. 17 has been used for automobile-shredder scrap to separate aluminum from glass and rubber contaminants. It has also been used to separate a mixture of aluminum and heavy nonferrous metals from glass and rock contaminants. The mixed nonferrous metal recovered must be further separated to provide a commercial grade of aluminum. Such separation involves dense-media systems.

Electrostatic separation has been used in some operations, eg, by the U.S. Bureau of Mines, College Park, Maryland, and by Black-Clawson, Hempstead, Long Island, New York, to separate conductors from nonconductors (5,19). However, these separators are sensitive to density, moisture, and surface contaminants, including magnetite, which change nonconductors into conductors and can cause a false separation.

Other density-separation systems are based on water elutriation or a rising-current classifier (11,20). Their main purpose is to cause the contaminants to float from metals prior to further separation. These units can remove contaminants with densities of as much as 1.6–1.7 g/cm^3.

A mineral jig works best on material that is smaller than 5 cm in every dimension. Most nonferrous metals in municipal refuse are of light-gauge material. Proper shredding and screening provide other dimensional requirements for successful jigging. The mineral jig can produce a concentrate of nonferrous metals and, if properly operated, can produce a heavy nonferrous metal concentrate and a light, ie, aluminum-rich, nonferrous-metal concentrate. These concentrates are amenable to separation in either dense-media or eddy-current separation equipment.

Ferrofluid separation, which involves the use of a dense medium of colloidal magnetite in kerosene, was developed in the early 1970s but was not adopted industrially (21). However, the recovery and processing of many nonferrous metals in commercial use are reviewed in ref. 22.

Institutional and Technical Barriers

The institutional barriers to increased recycling of nonferrous metals include:

1. Lack of clear ownership of refuse, particularly municipal refuse (23). However, this is not a barrier to the recycling of automobile scrap and other scraps.
2. Inability of some local governments to sign long-term contracts for the sale of material from a municipally owned recycling facility.
3. Availability of capital for construction of high cost resource-recovery facilities.
4. Inability of local governments to enter into areas of unproved technology because of lack of technical expertise necessary to design, build, and operate large-scale recovery facilities.
5. Lack of a data base and statistical reporting structure for the materials recovered from municipal refuse.

The technical barriers include the following:

1. In both municipal refuse and in automobile-shredder scrap, nonferrous metals are not the main constituents.
2. Aluminum and other nonferrous metals have very few unique physical characteristics that facilitate ready separation.
3. The aluminum- and other nonferrous-metal producers produce alloys that are sometimes incompatible when mixed and remelted.
4. The traditional municipal-refuse and scrap-processing industries, eg, automobile shredders, often do not have the technical expertise to initiate and monitor the recycling technologies.

In addition to the impediments listed, many resource-recovery projects have had problems recovering the principal constituent of solid wastes, ie, refuse-derived fuel, and have not seriously addressed recovery of nonferrous metals. The new generation of advanced technology systems will allow relatively high quality fuel to be produced and, therefore, allow system operators to concentrate on recovery of the nonferrous metals. Automobile shredding and municipal refuse processing are hampered by the wearing out of equipment, excessive dust, and incidences of explosions caused by discarded volatile or explosive materials in shredders. These factors inhibit recycling efforts in terms of safety and cost.

Economic Aspects

Facility capital costs are as little as 2.7×10^6 for the provision of refuse-derived fuel in Madison, Wisconsin, to over 100×10^6 for incineration facilities with steam- or electric-generating capacity. In 1977 the capital cost of the Madison facility, which had a capacity of 450 t/d, was approximately $6,000/(d·net ton) of capacity (24). In contrast, the 1980 capital cost of the Pinellas County, Florida, electric-generating facility, which has a 2,000-t/d capacity, was $80,000/(d·net ton) of capacity (25).

The projected growth in aluminum consumption, particularly in the consumer sector, indicates that more and more aluminum products will be recovered through separation programs by consumers. However, the quantity of material entering the solid-waste stream can provide the incentive for installation of aluminum-recovery subsystems in resource-recovery facilities. Typically, ca 5 kg of aluminum and 0.75 kg of other nonferrous-related metals are present in one metric ton of refuse. The potential fuel value of the combustible fraction (ca 70%) of refuse, based on coal at ca $2/GJ (ca $1.9/10^6 Btu) is ca $18/t. Recovery of the nonferrous metals can add $3–4/$t_{recovered}$ to the revenues of the system.

The production of primary aluminum entails substantial capital costs for the mining, transportation, and refining of bauxite, which is used in alumina recovery. For electrolytic reduction of alumina and production of ingot, 1981 capital costs were ca $3.00–3.50 per annual kilogram of new primary reduction capacity. The capital investment for aluminum-recycling facilities with the same production capacity are an order of magnitude lower. The principal long-term risks in primary production are that the supply of foreign bauxite could be cut off, power costs increased, or power availability reduced.

Energy Savings. Energy saving is a main incentive for the recycling of aluminum and, to a lesser but significant extent, other nonferrous metals (22). Approximately 220 MJ/kg (ca 95,000 Btu/lb) are necessary to produce primary ingot. Approximately 10 MJ/kg (ca 4300 Btu/lb) are needed to convert scrap into molten metal comparable to that recovered in primary production facilities. Current (1981) energy prices are $(2–8)/GJ [ca $(2.1–8.4)/10^6 Btu]; thus, depending on the fuel used, substantial savings on energy costs are potentially available.

Resolution of existing problems in finding long-term markets for a fuel or sale of thermal energy, ie, steam or electricity, from municipal refuse appear to be the most

Table 4. Chemical Requirements: Municipal Aluminum Scrap [a]

Element [b]	Composition, max wt % allowable					
	Grade 1	Grade 2	Grade 3	Grade 4	Grade 5	Grade 6
silicon	0.30	0.30	0.50	1.00	9.00	9.00
iron	0.60	0.70	1.00	1.00	0.80	1.00
copper	0.25	0.40	1.00	2.00	3.00	4.00
manganese	1.25	1.50	1.50	1.50	0.60	0.80
magnesium	2.00	2.00	2.00	2.00	2.00	2.00
chromium	0.05	0.10	0.30	0.30	0.30	0.30
nickel	0.04	0.04	0.30	0.30	0.30	0.30
zinc	0.25	0.25	1.00	2.00	1.00	3.00
lead	0.02	0.04	0.30	0.50	0.10	0.25
tin	0.02	0.04	0.30	0.30	0.10	0.25
bismuth	0.02	0.04	0.30	0.30	0.10	0.25
titanium	0.05	0.05	0.05	0.05	0.10	0.25
others (each)	0.04	0.05	0.05	0.08	0.10	0.10
others (total)	0.12	0.15	0.15	0.20	0.30	0.30
aluminum	balance	balance	balance	balance	balance	balance

[a] ASTM E 753-80.
[b] By agreement between the purchaser and the seller, analysis may be required and limits may be established for elements or compounds specified in this table.

Table 5. Aluminum Analysis: Automobile Scrap[a]

Processor and date of analysis	Aluminum recovery after remelting, %	Contained elements, wt %									
		Si	Fe	Cu	Mn	Mg	Cr	Ni	Zn	Ti	Pb	Bi	Sn
Virginia, Sept. 25, 1972	83.2	8.2	0.9	2.8	0.2	0.05	<0.05	0.1	0.8	<0.05			
		6.5	0.7	3.2	0.2	<0.05	<0.05	0.1	0.7	<0.05			
California, Sept. 25, 1972	92	0.3	0.9	1.1	0.2	0.4	<0.05	0.1	0.1	<0.05	<0.05	<0.05	
Michigan, Feb. 13, 1974	95.8–97.0	4.0–5.0	0.7	1.4	0.3	0.1	<0.1	<0.1	0.6	<0.1	<0.1	<0.1	
Wisconsin I[b], March 29, 1974		9.0–10.0	0.8–0.9	3.0	0.1–0.2	0.05–0.1	<0.05	0.05–0.1	1.0–2.0	<0.05	0.06	<0.05	<0.05
Wisconsin II[c], March 29, 1974		9.0–10.0	0.8	2.6	0.1–0.2	0.35–0.40	<0.05	0.05–0.1	1.0–2.0	<0.05	0.06	<0.05	<0.05
Illinois[c], March 29, 1974		9.0	0.6–0.7	3.0	0.1–0.2	0.3	<0.05	0.05–0.1	7.0–8.0	<0.05	0.2	<0.05	<0.05
South Carolina[c] Jan. 1, 1974		5.5	0.6	2.5	0.3	0.4	<0.05	0.2	3.5	0.08	<0.1	<0.05	<0.05
		4.5	0.5	1.5	0.3	0.5	<0.05	0.2	3.5	<0.05	<0.05	<0.05	<0.05
		5.5	0.5	1.5	0.3	0.5	<0.05	0.2	5.5	<0.05	<0.05	<0.05	<0.05

[a] Ref. 29.
[b] Direct remelt after dense-media separation.
[c] Sweat furnace at 816°C.

powerful incentives for resource-recovery implementation. Additionally, several projects currently underway in the United States are expected to demonstrate that the recovery of aluminum and other nonferrous metals can be separate and economically favorable operations.

Specifications

Specifications, eg, NF-80, identify metal uses and enable the scrap processor to recover scrap of an acceptable quality for the ultimate consumer (26). The chemical requirements of the six grades of municipal aluminum scrap, as listed in ASTM E 753-80, are listed in Table 4. The chemical analysis limits of aluminum alloys commonly used in consumer products are given in refs. 27–28.

The aluminum alloys that usually are in automobile-shredder residues are largely casting alloys (grades 5 and 6). The assay recovery and chemical analyses of aluminum recovered from automobile scrap are listed in Table 5. The analysis of aluminum from shredded automobiles depends on the air classifier efficiency after shredding, how much aluminum is removed during dismantling, and the model of the automobile.

In 1981, an ASTM subcommittee approved the first draft standard for mixed nonferrous metals. The proposed standard established categories based on total nonferrous metal content. Identification of the heterogeneous mixtures is related to the type of processing used for recovery, product size, moisture content, and the amount of aluminum present.

BIBLIOGRAPHY

1. *A Study to Identify Opportunities for Increased Solid Waste Utilization*, Vols. II–III, and V, Battelle Columbus Laboratories, June 1972.
2. *Fourth Report to Congress–Resource Recovery and Waste Reductions* (SW-600), U.S. Environmental Protection Agency, Washington, D.C., 1977.
3. G. F. Bourcier, K. H. Dale, and R. F. Testin, *Proceedings of the Third Mineral Waste Utilization Symposium*, U.S. Bureau of Mines and ITT Research Institute, Chicago, Ill., 1972, pp. 345–352.
4. G. F. Bourcier and K. H. Dale, *Proceedings of the Sixth Mineral Waste Utilization Symposium*, U.S. Bureau of Mines and ITT Research Institute, Chicago, Ill., 1978, pp. 178–187.
5. R. S. DeCesare, F. J. Palumbo, and P. M. Sullivan, *U.S. Bureau of Mines Report of Investigations RI 8429*, Washington, D.C., 1980.
6. R. F. Testin, *Proceedings of the AIME Annual Meeting*, Chicago, Ill., Feb. 24, 1981.
7. L. J. Frosiland and co-workers, *Bureau of Mines Report of Investigation, RI 8049*, Washington, D.C., 1975.
8. E. G. Valdez, *Proceedings of the Fifth Mineral Waste Utilization Symposium*, U.S. Bureau of Mines and ITT Research Institute, Chicago, Ill., 1976, pp. 345–352.
9. K. C. Dean and co-workers in ref. 3, pp. 213–221.
10. E. L. Michaels and co-workers, *Trans. Soc. Min. Eng. AIME* **258**, 34 (1975).
11. H. H. Dreissen and A. T. Basten in ref. 8, pp. 377–385.
12. J. Sterner, U.S. Bureau of Mines, private communication, Dec. 1981.
13. J. J. Henn, *U.S. Bureau of Mines Information Circular IC8691*, Washington, D.C., 1975.
14. *Refuse Fired Systems in Europe: An Evaluation of Design Practices—An Executive Summary Report SW 771*, U.S. Environmental Protection Agency, Washington, D.C., 1979.
15. E. Schloeman and D. B. Spencer, *Resource Recovery Conserv.* **1**, 151 (1975).
16. J. A. Campbell, *Transactions of the AIME Annual Meeting*, Dallas, Texas, 1974.
17. U.S. Pat. 4,137,156 (Jan. 30, 1979), B. Morey and S. Rudy (to Occidental Petroleum Corp.).
18. B. Morey and co-workers in ref. 16.
19. N. Reuth, *Mech. Eng.*, 24 (Dec. 1977); G. E. Easterbrook, *Waste Age*, 50 (April 1978).

20. V. R. Degner, *Proceedings of the Fourth Mineral Waste Utilization Symposium*, U.S. Bureau of Mines and ITT Research Institute, Chicago, Ill., pp. 63–70.
21. G. W. Reimers and co-workers in ref. 8, pp. 371–376.
22. C. L. Kusik and C. B. Kenahan, *U.S. Bureau of Mines Information Circular IC 8781*, Washington, D.C., 1978.
23. B. N. Sattin in ref. 4, p. 370.
24. *1980 National Waste Processing Conference*, Solid Waste Processing Division, ASME, New York, pp. 411–425.
25. D. F. Acenbrack, *Public Works* (Nov. 1980), *paper presented at the ASME National Solid-Waste Processing Conference*, 1980.
26. *Standard Classifications for Nonferrous Scrap Metals*, NARI Circular NF-80, National Association of Recycling Industries, New York, July 1, 1980.
27. *Registration Record of Aluminum Association Designations and Chemical Composition Limits for Wrought and Wrought Aluminum Alloys*, The Aluminum Association, Inc., Washington, D.C., July 1, 1981.
28. *Registration Record of Aluminum Association Alloy Designations and Chemical Composition Limits for Aluminum Alloys in the Form of Castings and Ingots*, The Aluminum Association, Washington, D.C., June 1981.
29. G. F. Bourcier, *paper presented at the Eighth Society Die Casting Engineers International Exposition and Congress*, Detroit, Mich., March 17–20, 1975.

General References

1980 National Waste Processing Conference, Solid Waste Processing Division, ASME Proceedings, Library of Congress Cat. Card No. 70-124402.
Proceedings of the International Conference on European Waste-to-Energy Technology, ANL/CNSV-TM-14, U.S. Department of Energy, Washington, D.C., Oct. 1980.
Resource Recovery Plant Implementation: Guides for Municipal Officials, U.S. Environmental Protection Agency, Washington, D.C., 1976, 8 parts, SW-157.1-SW157.8.
First through *Seventh Mineral Waste Utilization Symposia*, U.S. Bureau of Mines and the Illinois Institute of Technology Research Institute, Chicago, Ill., 1968–1980.
Aluminum Recycling Casebook, The Aluminum Association, Inc., Washington, D.C., Oct. 1981.
U.S. Bureau of Mines, Washington, D.C., annual reports.
M. B. Bever, *The Impact of Materials and Design Changes on the Recycling of Automobiles*, Massachusetts Institute of Technology, Cambridge, Mass., 1980.
M. B. Bever, *Review of Scrap Metal Recovery*, Massachusetts Institute of Technology, Cambridge, Mass., 1980.
M. B. Bever, "The Recycling of Metals-II Nonferrous Metals," *Conservation Recycling* **1**, 137 (1976).
M. B. Bever, "Systems Aspects of Materials Recycling," *Conservation Recycling* **2**, 1 (1978).
M. B. Bever, "The Dissipative Uses of Lead," *Proceedings of the Council of Economics*, AIME, New York, 1976, pp. 17–26.
N. L. Drobny, H. E. Hull, and R. F. Testin, *Recovery and Utilization of Municipal Solid Waste*, Report SW-10c, U.S. Environmental Protection Agency, Washington, D.C., 1971.
Resource Recovery from Municipal Solid Waste, National Center for Resource and Recovery, Inc., 1974, 182 pp., Library of Congress Cat. Card No. 73-9765.
H. Alter and E. Horowitz, *Resource Recovery and Utilization, National Materials Conservation Symposium, 1974*, American Society for Testing and Materials, Philadelphia, Pa., 1975.

Periodicals

Aluminum Association Statistical Review, The Aluminum Association, Inc., Washington, D.C., updated annually.
Am. Met. Mark./Metalwork. News, Fairchild Publishing Co., New York, daily.
Solid Wastes Management, Communication Channels, Inc., Atlanta, Ga., monthly.
Waste Age, National Solid Waste Management Association, Inc., Washington, D.C., monthly.
Scrap Age, Three Sons Publishing Co., Niles, Ill., monthly.
Met. Prog., American Society for Metals, Metals Park, Ohio, monthly.

Resource Recovery and Conservation, Vols. 1–7, Elsevier Publishing Co., Amsterdam, The Netherlands.
Metal Statistics, 73rd annual edition, Fairchild Publications, Inc., New York, 1980.
Conservation and Recycling, Vol. 1, Pergamon Press, London, 1976–1982.

GILBERT BOURCIER
Reynolds Metals Co.

OIL

The term oil includes a variety of liquid or easily liquefiable, unctuous, combustible substances that are soluble in ether but not in water and that leave a greasy stain on paper or cloth (1). These substances can include animal, vegetable, and synthetic oils, but usually the word oil refers to a mineral oil produced from petroleum (qv). An oil that has been used or contaminated, or both, but not consumed, can often be recycled to regain a useful material regardless of its origin. In the past, used or contaminated oil was usually considered a waste and was disposed of in a variety of ways including incineration, road oiling, landspreading, and dumping on the ground and into storm sewers. However, with the current petroleum shortages, increased prices for most kinds of oils and oil products, and increased concern about pollution, there is strong interest in developing ways to conserve the valuable energy and resource content of these products.

The following definitions are useful to the discussion of oil recycling (2).

Used Oil. Oil whose characteristics have changed since being originally manufactured. It may be suitable for further use and is economically recyclable. It includes used lubricating oils of all types as well as dirty or contaminated fuel or other oils that can be economically recycled.

Waste Oil. Oil whose characteristics have changed markedly since being originally manufactured, has become unsuitable for further use, and is not economically recyclable.

Oil Recycling. The generic term for processing used oil to regain useful material.

Oil Reclaiming or Oil Laundering. The application of cleaning methods to used oil primarily to remove insoluble contaminants, thus making the oil suitable for further use. The methods may include settling, heating, dehydration, filtration, and centrifuging. The product is reclaimed oil.

Oil Re-refining. The application of petroleum refining processes to used lubricating oil to produce clean, high quality lubricating base oil, which is oil basestock before its formulation with additives. Re-refining many include distillation, and hydrotreating, or treatments with acid, caustic, solvent, clay, other chemicals, or both. The product is re-refined lubricating base oil.

For the purposes of this article, only the recycling of used petroleum oils will be considered further.

Oil recycling dates to ca 1920. At that time lubricating oils contained few or no additives, ie, chemical compounds added to oils to improve lubrication characteristics, eg, wear, oxidation, and corrosion (see Lubrication and lubricants). Recycling these oils required only limited processing involving heating to remove volatile components, settling to separate water, dirt, and sludge, and centrifuging or filtering to remove most of the remaining insoluble contaminants. With this limited processing, lubricating oils could be recycled to essentially original oil quality. The types of lubricating oils in the marketplace have expanded and many contain high levels of additives, which can make recycling a more difficult task.

In 1980, new lubricating oil sales in the United States were estimated at 9.5×10^6 m^3/yr (2.5×10^9 gal/yr). Of this, automotive lubricants represented 4.5×10^6 m^3 (1.2×10^9 gal) and industrial lubricants represented the remainder, or ca 5×10^6 m^3 (1.3×10^9 gal). These industrial lubricants include the following types: hydraulic, quenching, cutting, metal-working, electrical, and process oils. From the 9.5×10^6 m^3 of lubricating oils sold, ca 4.8×10^6 m^3 (1.3×10^9 gal) of collectable used oil are generated annually in the United States. It is estimated that, of the used oil collected each year, ca 7% is re-refined into motor oil, 17% is reclaimed into industrial oil, ca 50% is burned as a fuel oil or is used in road oiling and dust control, often with little processing to remove contaminants, with the use of the remaining 25% unidentified.

There are ca 30–50 oil recyclers in the United States. Of these, only 10–20 are re-refiners. The individual companies involved in oil recycling are small, averaging only 10–15 employees (3). These recyclers fall into three categories: processors who process used oil for reuse as a fuel, often only removing free-standing water and larger pieces of dirt or rust; reclaimers; and re-refiners. In many cases, an oil recycler may be involved with more than one of these categories and at times all three.

Technically, recycled lubricating oil products appear to be potentially suitable for all uses, given proper clean-up and additive treatment, since apparently the basic hydrocarbon structure is not significantly altered during use. Although nontechnical factors, such as economics and availability, determine whether all potential uses for recycled oil can be made available on a realistic basis, the national goals of energy conservation, resource conservation, pollution control, and improved international balance of payments certainly suggest that a strong effort be made in pursuing the goal of effective oil recycling in the United States.

Characteristics of Used Oils

Used petroleum oils to be recycled can be obtained from a variety of sources. These sources include automotive garages and service stations, truck and taxi fleets, military installations, individuals, industrial plants and manufacturing facilities of all types, and wastewater treatment plants. The main types of used petroleum oils that are being recycled are motor oils and hydraulic and industrial oils. The additives and contaminants typically in these oils can cause both performance-related and environment-related problems (4–5). As much as 90% of the easily recoverable used oil may have been used as fuel with little or no processing to remove contaminants (3). Used oils may often be comingled with each other and with water, solvents and other chemicals before being colle ed for recycling.

Chemical analyses have been performed on used oils and waste oils, primarily for inorganic constituents. These analyses are described in references, 3–4 and 6–15 and are summarized in Table 1. The data show that used oils contain some contaminants, which often should be removed during the recycling process to protect the environment. Many organic contaminants contained in such oils have not been analyzed and thus are not included in Table 1. In particular, poly(chlorinated biphenyls)s (PCBs) are often not determined in used or waste oils, yet sometimes occur in waste oils and are strictly regulated at concentrations as low as 500 ppm (0.05 wt %) in some oils and to 50 ppm (0.005 wt %) for certain other applications (see Chlorocarbons and chlorohydrocarbons). These regulations on PCBs are established by the Toxic Substances Control Act (17) with additional regulations based on hazardous waste provisions in the Resource Conservation and Recovery Act (18).

Technology

Used Oil Recycled as Burner Fuel. Combustion of used oil as burner fuel has often been condemned because it destroys a valuable resource and it can cause substantial environmental pollution through widely dispersed distribution of metal oxides and stable organic contaminants. Further, these contaminants may cause scaling of heat-transfer surfaces and fouling of burners and fuel-transfer lines if not handled or processed properly. However, under certain conditions and with suitable precautions, the recycling of used oil to recover the energy as heat is a valid recycling option. The conditions and precautions are described in ref. 4.

Table 1. Summary of Reported Used Oil Analyses[a]

Property or test	Motor oils	Industrial oils[b]
viscosity (at 40°C), SUs	87–837	143–330
API gravity (at 15.6°C)	19.1–31.3	25.7–26.2
specific gravity (15.6°C/15.6°C)	0.9396–0.8692	0.9002–0.8972
water, vol %	0.2–33.8[c]	0.1–4.6
bottom settlings and water, vol %	0.1–42	
benzene insolubles, wt %	0.56–3.33	
gasoline dilution, vol %	2.0–9.7	
flash point, °C	79–220	157–179
heating value, MJ/kg[d]	31.56–44.88	40.12–41.84
ash, sulfated, wt %	0.03–6.43	3.2–5.9[e]
carbon residue, wt %	1.82–4.43	
fatty oils, wt %		0–60
chlorine, wt %	0.17–0.47	<0.1–0.83
sulfur, wt %	0.17–1.09	0.54–1.03
zinc, ppm	260–1,787	
calcium, ppm	211–2,291	
barium, ppm	9–3,906	
phosphorus, ppm	319–1,550	
lead, ppm	85–21,676	
aluminum, ppm	<0.5–758	
iron, ppm	97–2,401	

[a] Data taken from references 3–4 and 6–15, did not provide data on all tests listed; therefore, data may be inconsistent between different tests.
[b] Very limited data were available for used industrial oils; see ref. 16.
[c] One sample had a water content of 46.5 wt % but is considered an outlier.
[d] To convert MJ/kg to Btu/lb, multiply by 430.4.
[e] Values for the industrial oils were stated to be for the regular not the sulfated ash.

Processing techniques for the recycling of used oil into fuel include pretreatment of the used oil to remove all or most of the contaminants which cause the environmental or operational concerns, or use of specialized facilities with acceptable environmental control, ie, electrostatic precipitators, Venturi scrubbers, or a fabric-filter baghouse (4). Lower level pretreatment options include settling or centrifugation, or both, to remove coarse solids and free water. Improved cleaning occurs with the application of heat and demulsifiers to remove water, volatiles, and most suspended solids. However, these methods do not significantly reduce the soluble contaminants or submicrometer-sized particulates, which contribute to ash formation. Higher level treatments of the used oil can be applied to clean the oil further, but they are not normally employed for economic reasons.

Reclaiming. Used oils can be reclaimed within the user facility or can be sent outside the facility to a commercial reclaimer. The primary difference between these two options is usually the level of treatment available. Often the in-plant reclaiming facility is limited to gravity purification or settling, filtration, centrifuging, and heating to remove volatiles and water. A commercial reclaimer can usually perform all of the above plus provide clay treatment, chemical treatment (acid–caustic), demulsification, distillation, and reformulation with additives if necessary. A general description of such processes may include any or all of the following: (*1*) removal of solid particles by settling, centrifuging, or filtering; (*2*) neutralization of acidic components with clay or alkalies, and removal of resulting soaps by washing; (*3*) heating–distillation to remove volatile solvents, gasoline, and water; (*4*) clay contacting to remove oxygenated components and spent additives or for decolorization; (*5*) aeration and use of biocides to reduce bacterial levels; and (*6*) replenishment of additives.

The product of a reclaiming procedure is a reclaimed oil, which often meets original oil specifications for such uses as gear lubricants, cutting and grinding oils, metal-rolling lubricants, etc. Perhaps the most important requirement for effective reclaiming is segregation of the used oils according to types. In general, a mixture of high quality oil and low quality oil can be reclaimed only as a low quality oil or as a fuel. Thus, the lack of segregation at the source may carry a high economic penalty.

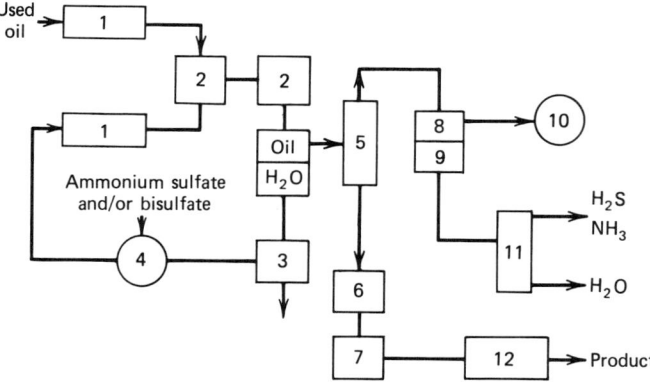

Figure 1. Flow chart of the Phillips PROP process (Phillips Petroleum, Bartlesville, Okla.) for refining used oil (19). 1, preheater; 2, reactor; 3, filter; 4, mixing; 5, flash; 6, adsorbent (clay); 7, filter; 8, gasoline; 9, water; 10, storage; 11, stripper; 12, hydrotreater.

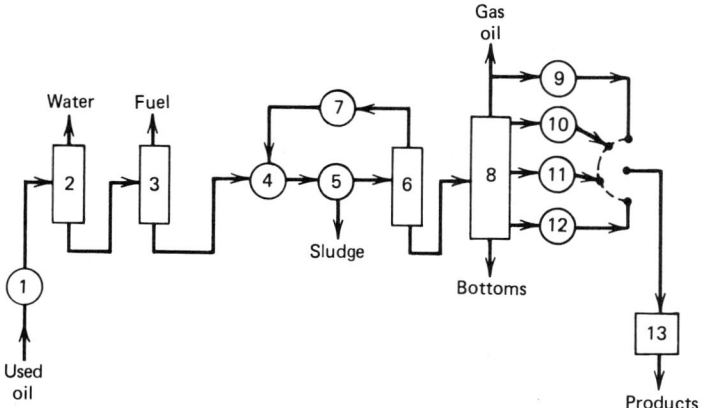

Figure 2. Flow diagram of BETC solvent-distillation re-refining process (Bartlesville Energy Technology Center, U.S. DOE) (19). SN = solvent neutral. 1, preheater; 2, atmospheric distillation; 3, vacuum distillation; 4, mixer-solvents/oil; 5, centrifuge or settlers; 6, stripper (vacuum); 7, solvent storage; 8, fractional distillation; 9, 80 SUs SN distillate; 10, 150 SUs SN distillate; 11, 250 SUs SN distillate; 12, >400 SUs distillate; 13, hydrofinishing or clay contactor.

Re-Refining. Petroleum refining processes are employed for used lubricating oil to produce clean, high quality, lubricating base oil. These processes often include a pretreatment to reduce the impurity content by one or more of the following methods: application of heat, filtration, acid, caustic, solvents, or other chemicals. The pretreatment is usually followed by one of the following refining processes: vacuum distillation with clay or hydrogen finishing, acid–clay treatment with plate-and-frame filtration, chemical pretreatment followed by hydrotreating, solvent extraction with clay or hydrogen finishing, and extensive treatment with only clay; the latter is limited to highly segregated used oils.

Four modern re-refining processes are described in ref. 19 and are shown schematically in Figures 1–4. These processes have been selected for inclusion because of the substantial differences between them, and no implications as to the suitability or quality of the processes or the products is expressed or implied. A principal concern

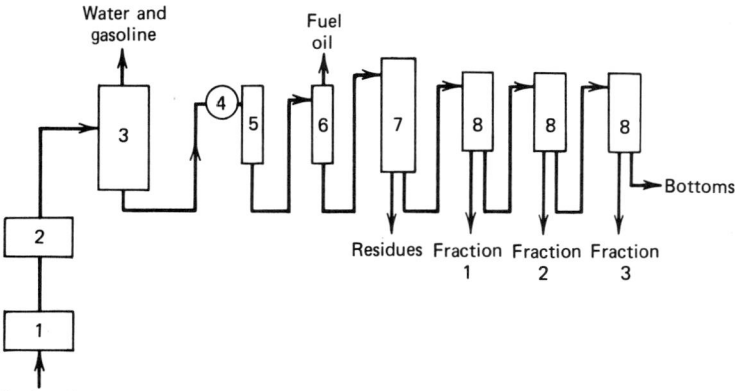

Figure 3. Flow chart of the Recyclōn process (Leyhold-Heraeus/Adolf Schmids Erban/Degussa, Federal Republic of Germany) for the re-refining of used oils (19). 1, filter; 2, preheater; 3, dehydration-gasoline separation; 4, sodium emulsion; 5, mixer; 6, flasher; 7, short-path distillation total evaporation; 8, wiped-film evaporators.

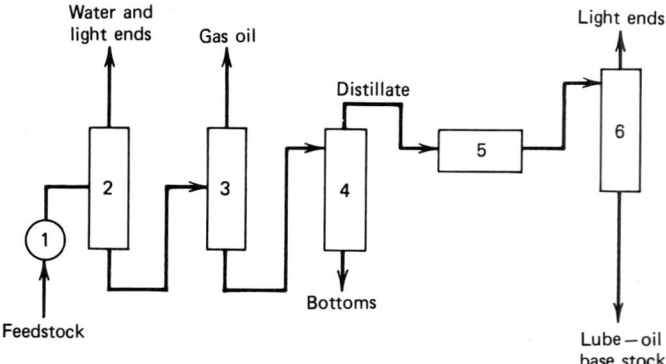

Figure 4. KTI process (Kinetics Technology International/Gulf S and T, The Netherlands). 1, preheaters; 2, atmospheric distillation dehydration; 3, vacuum distillation gas oil; 4, distillation; 5, hydrofinishing; 6, fractional distillation.

with many of these new re-refining technologies is the lack of technical data and long-term experience establishing the quality of the re-refined base-oil product. This subject is discussed in great detail in several recent publications (14–15,20). Additional information on a variety of re-refining and recycling processes is given in ref. 21.

Specifications

One important reason that recycled oil products have not met with widespread consumer acceptance is the lack of appropriate specifications and standards. As the result of recent legislation (18,22–23), both the Federal specification for burner fuel (VV-F-815d) and the military specification for administrative-vehicle engine crankcase oil (MIL-L-46152B) have been revised to allow the use of recycled materials (16,20). The technical requirements for both specifications, ie, test procedures and limits, have also been modified in order to assure adequate quality control for these materials. In addition to the above, there are research efforts at several government laboratories (eg, Recycled Oil Program, National Bureau of Standards; Mobility Equipment Research and Development Command (MERADCOM), U.S. Army/Fort Belvoir) involved in the development of test procedures and standards to allow more effective use of recycled oil products, particularly lubricants.

BIBLIOGRAPHY

1. *Webster's New Collegiate Dictionary*, G. & C. Merriam Co., Springfield, Mass., 1979.
2. DRAFT definitions currently under discussion by Technical Division P, Committee D-2, American Society for Testing and Materials, Philadelphia, Pa., 1981.
3. J. W. Swain, Jr., *Assessment of Industrial Hazardous Waste Management Practices for the Petroleum Re-refining Industry (SIC 2992)*, Rept. No. EPA-ISW-144c, Environmental Protection Agency, Washington, D.C., 1977.
4. S. Chansky and co-workers, *Waste Automotive Lubricating Oil—Reuse as a Fuel*, Rept. No. EPA-600/5-74-032, Environmental Protection Agency, Washington, D.C., 1974.
5. Recon Systems, Inc., *Used Oil Burned as a Fuel*, Rept. No. SW-892, Environmental Protection Agency, Washington, D.C., 1980.
6. J. W. Swain, Jr., *Recycled Oils for Fuel Uses*, unpublished report to the National Bureau of Standards, Jan. 1978.

7. Report to the Congress: *Waste Oil Study*, Environmental Protection Agency, Washington, D.C., April, 1974.
8. *Environ. Sci. and Technol.* **6,** 25 (Jan. 1972).
9. E. E. Berry, L. P. MacDonald, and D. J. Skinner, *Experimental Burning of Waste Oil as a Fuel in Cement Manufacture*, Environment/Canada Rept. No. EPS 4-WP-75-1, Ottawa, Canada, June 1975.
10. *Energy From Used Lubricating Oils*, API Pub. No. 1588, American Petroleum Institute, Washington, D.C., Oct. 1975.
11. P. Cukor, *Economic and Environmental Assessment of Used Oil Recovery and Disposal for Ontario*, report by Teknekron, under contract to the Ontario Energy Management Program, March 1976.
12. N. Suprenant, J. Sahagian, and R. Hall, *Waste Oil Recovery and Refuse Program—Residue Management*, GCA/Technol. Rept. No. GCA-TR-75-3-G, prepared for Maryland Environmental Services, by GCA/Tech., Bedford, Mass., April 1975.
13. M. L. Whismann, J. W. Goetzinger, and F. O. Cotton, Waste Lubricating Oil Research, series of research reports by the Bartlesville Energy Research Center, DOE, Bartlesville, Okla., 1974–1977.
14. D. A. Becker, ed., *Measurements and Standards for Recycled Oil—I*, NBS Special Pub. 488, National Bureau of Standards, Washington, D.C., Aug. 1977.
15. D. A. Becker, ed., *Measurements and Standards for Recycled Oil—II*, NBS Special Pub. 556, National Bureau of Standards, Washington, D.C., Sept. 1979.
16. E. A. Frame and T. C. Bowen, Jr., *U.S. Army/Environmental Protection Agency Re-refined Engine Oil Program*, AFLRL Rept. No. 98, U.S. Army MERADCOM, Ft. Belvoir, Va., 1978.
17. *Toxic Substance Control Act*, Public Law 94-469, Oct. 11, 1976; 15 USC 2601 *et seq.*
18. *Resource Conservation and Recovery Act*, Public Law 94-580, Oct. 21, 1976; 42 USC 6901 *et seq.*
19. M. L. Whisman, *Lubr. Eng.* **35,** 249 (1979).
20. D. A. Becker, ed., *Joint Conference on Measurements and Standards for Recycled Oil/Systems Performance and Durability*, NBS Special Pub. 584, National Bureau of Standards, Washington, D.C., Nov. 1980.
21. L. Y. Hess, *Reprocessing and Disposal of Waste Petroleum Oils*, Noyes Data Corp., Park Ridge, N.J., 1979.
22. *Energy Policy and Conservation Act*, Public Law 94-163, Dec. 22, 1975; 42 USC 6901 *et seq.*
23. *Used Oil Recycling Act of 1980*, Public Law 96-463.

DONALD A. BECKER
National Bureau of Standards

PAPER

Recycling of paper (qv) is a commercial practice involving the collection and processing of wastepaper for secondary fiber supply to the paper industry. In 1979, the world production of paper and paperboard was 171×10^6 metric tons, and recycled paper furnished 25% of this fiber supply (1). The use of recycled fiber will increase to help meet world paper demand, which is forecast to grow 3.5%/yr through 1990 (2). The cost and availability of energy, raw material, water, manpower, and money provide new economic incentives for paper recycling. New technology has been developed to control wastepaper contaminants, and modern machines can economically produce competitive, quality paper products from recycled paper.

The history of paper recycling follows that of early papermaking (3–4). It is assumed that paper recycling occurred as spoiled handmade sheets were reprocessed and formed again. The hand method of papermaking continued until the invention of the first paper machine in France in 1799. The latter was improved and developed in England with the advent of a continuous papermaking machine. During this time, methods of deinking recycled printed papers were developed. In 1809, the cylinder machine was patented and later was developed to produce multi-ply paperboard.

Recycled paper is a generic term for wastepaper that is first generated as scrap material during the original manufacture and conversion of paper and board products. No wastepaper grade is consciously produced as a raw material; the generation of wastepaper is a consequence of use. After consumer use, these products are discarded as part of the solid-waste stream. These sources of wastepaper are mostly available for recovery in concentrated population centers of developed countries. The available supply of recycled paper is a function of per capita paper consumption. The United States led the world in 1979 with 289 kg per person. This compares to an average of 128 kg in western Europe and a total world average of 40 kg per person (1). As a commodity, recycled paper follows the market forces of supply and demand. Wastepaper supply is inelastic in the short term, ie, an increase in demand does not automatically lead to an increase in supply. During periods of extraordinary demand, eg, 1973, wastepaper prices rose rapidly because of short supply. World market virgin-pulp price is a principal factor in the value–price scale of wastepaper grades.

Recycled fibers differ from virgin pulp and paper fibers in physical and mechanical properties because of the presence of contaminants and the changes within the fiber that result from drying, aging, and the recycling process (see Pulp). A recycled fiber has less strength value, lower specific volume, and often a greater mean specific surface area than its original virgin-fiber counterpart (5). The fiber lumen is dewatered and collapsed, which produces a flattened structure. The S-1 and S-2 layers are unraveled and separated at the fiber ends and often the fiber is split, broken, and shortened in average length. Its formation and interfiber bonding characteristics equal or exceed those of virgin fibers. Because of its fractured, expanded structure, recycled fibers offer excellent bulking to the formed sheet. The degradation of individual fiber structure during original virgin-pulp processing followed by repulping causes the recycled-fiber sheet to have generally lower strength and rigidity (6–7).

The recycled-fiber industry has developed distinct grades of wastepaper called paper stocks, which are defined by fiber type and degree of contamination. Amounts of pulp, brightness, and strength describe its general recycle use. The value of wastepaper is directly proportional to its economic substitution for virgin fiber. Paper-stock grades, eg, office waste, mixed paper, and old corrugated containers, have the lowest values because they are highly contaminated, are low in brightness, and require substantial processing. Other grades that have not been extensively converted receive a higher price. Converting adds contaminants in the form of coatings, films, inks, and adhesives. Uncirculated newspapers, ledgers, new corrugated cuttings, and publication trimmings are examples of unconverted, less contaminated paper stocks. These grades are suitable for deinking and production of newsprint, tissue, toweling, and printing papers. Old corrugated and new corrugated clippings contain high percentages of strong, unbleached, softwood kraft fiber, and they are used extensively in container boards and recycled boxboard. Direct pulp substitutes are the highest valued paper-stock grades, are uncoated, clean, and high in brightness, and are made of strong chemical-pulp fibers. They require little or no processing for contaminant removal. White envelope cuttings, unprinted ledgers, and blank newsprint are examples of direct pulp substitutes. The average 1980 selling prices of common paper-stock grades in the midwest United States are listed in Table 1 (8).

The wastepaper industry has established itself in most of the major cities and industrial areas of the world to serve the recycled-fiber market. Individuals, civic groups, companies, and local governments collect and deliver wastepaper. Waste is brought to the plant by automobile, truck, or railcar in bundles, boxes, and loose form. Modern plants have convenient layouts for weighing, dumping, and processing received stock. The seller is paid an amount at the time of delivery based on the purchased weight and current scale price per grade. Separate piles of news, corrugated, and mixed waste are accumulated and baled. Sorting of mixed waste into separate grades may be done by hand from a moving conveyor, and the grades are separated into bins or boxes for baling. Although this method can produce high value paper grades, it is labor-intensive and has been generally discontinued in the United States. The industry has concentrated on separating specific grades directly at the source. Paper-stock

Table 1. 1980 Average Market Prices for Common Paper-Stock Grades in the Midwest United States; Fob, Shipping Point [a]

Paper-stock grade	Baled price from supplier, $/t
white envelope cuttings	347.00
manila tab cards	308.50
manifold white ledger	187.50
printed bleached kraft	220.50
newsblanks	143.50
computer printouts	215.00
colored ledger	154.50
double-lined kraft	92.50
regular corrugated	49.50
regular news	60.50
shavings	66.00
boxboard cuttings	44.00
mixed papers	27.50

[a] Ref. 8.

dealers promote this source-separation by offering higher purchase prices. The application of local ordinances that require the separation of newspapers from home refuse is an effective method of source-separation for increased paper recycling.

With rising costs and demand, paper-stock suppliers have increased their productivity with new machinery developments. Methods of conveying, shredding, and baling have improved substantially. Old upstroke single-unit balers have been replaced by automatic continuous balers with production rates of 13–18 metric tons per hour (1). Using automatic conveyors, powerful hydraulic rams, and continuous strapping systems, a single operator can bale 100 t of news grade in 8 h. Large standard-size bales are produced at uniform weight (700–900 kg) for efficient handling and economical transportation by trucks, train, or ship.

In addition to conventional wastepaper-recovery practices, new technology is being applied to reclaim fiber from municipal refuse or garbage. There are reports of dismal performance records for some existing municipal reclamation systems, but increasing refuse-disposal costs combined with improved process technology will make such systems more effective. The disposal of refuse as landfill is expensive and unfavored because of limited space and possible public health and environmental hazards. New processing methods provide economic incentive to recover glass, metal, and paper from refuse. Large commercial operations that exist in the United States, Europe, and Japan process 800–1600 t/d of municipal refuse. These large resource-recovery plants are supported by local transfer stations that receive and compact the garbage for processing. The Americology system (American Can Co.) hammers the refuse to small pieces and classifies them by density with air-flotation separation of paper, plastics, and other light material (9). Ferrous and nonferrous metals and glass are sorted and sold in the reclaimed-material market. The recovered paper is purchased by local paper-stock dealers. The remaining refuse is formed into dense dry bales and sold as solid fuel. New developments by Aenco Inc., Delaware, have replaced the typical shredding step with air-classification to avoid frequent shredder explosions and excessive equipment wear (10). This air-classification method permits separation of news from corrugated grades. In Rome, Italy, a refuse-recovery plant separates the paper from plastics and cleans, screens, and disperses contaminants in wet form by means of the Maule Kneader-Beater system (11). The final press-dried product is sold commercially as recycled fiber. Another method of processing refuse paper is Fibre-Flow by Ahlstrom, in Finland, which employs a long, large-diameter rotating drum with various-sized holes. Paper and plastics are gradually processed inside the drum at high consistency. Washing and screening sections separate the fiber from contaminants, with very little attrition and energy input. The recovered fiber can be pressed, flash dried, baled, and sold commercially as a fiber resource (10). It has been concluded that recovered wastepaper has greater economic value as a raw material than as a fuel source for incineration (12).

Paper and board mills that make use of recycled fiber require special equipment and facilities, as illustrated in Figure 1. The grade being produced determines the basic wastepaper furnish and the degree of required stock cleaning. Recent developments in pulping, cleaning, screening, and refining have permitted the increased use of wastepaper in higher quality paper grades. Nonfibrous contaminants, eg, dirt, glass, sand, and paper clips, are removed by selective screening or centrifugal cleaning at various stock consistencies. Fine debris is removed by pressure screens with small perforated openings or fine flots. Low density contaminants, eg, plastic, styrofoam,

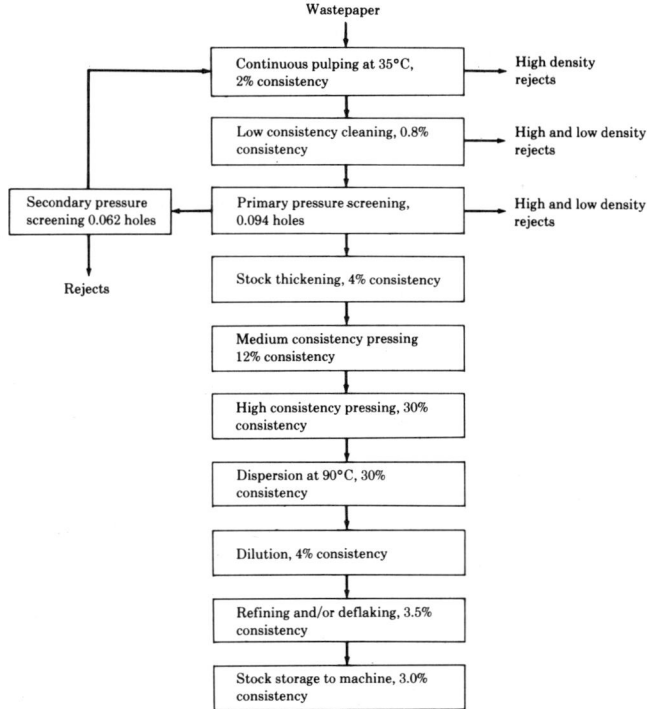

Figure 1. Typical process flow sheet of recycled fiber-stock preparation.

and coating films, are concentrated in stock-flow vortexes, which are generated in special screens and reverse-flow cleaners. Chemical contaminants are the most difficult to remove. These include pressure-sensitive adhesives, inks, colors, special coatings, and internal additives. For acceptable stock quality, chemical contaminants must be thoroughly dispersed as small particles or must be removed. Deinking methods involve various chemical reactions to reduce color, disperse inks, and dissipate chemical additives. Figure 2 shows the recycled-fiber deinking and bleaching process used by Abitibi Provincial Paper Ltd. (13). Mechanical dispersion systems operate at various temperatures and pressures on high consistency stock, and rotors disperse or separate adhesives and waxes. The details of these basic-process operations are described in the literature (14). A good overview on contaminant problems is given in ref. 15. The pulping developments in the recycled-fiber field and the advantages of low attrition pulping in handling friable contaminants are discussed in ref. 16. Screening and cleaning methods for processing wastepaper are described in refs. 17–20. Modern methods of deinking in North America have been reported (21–22). The energy usage in wastepaper-stock preparation is given in ref. 23. The recent developments in refining and deflaking are reported in ref. 24.

Properly prepared recycled fiber can be successfully formed on modern, high speed paper and board machines. The development of multi-ply formers with the latest stock-delivery systems has made it possible for the recycled-paper industry to compete more effectively with virgin-fiber products. A complete history and description of multi-ply web forming is given in ref. 25. New developments in press-water removal involving softer roll covers and special drilling patterns have significantly increased the production efficiency and quality of recycled paperboard (26–28).

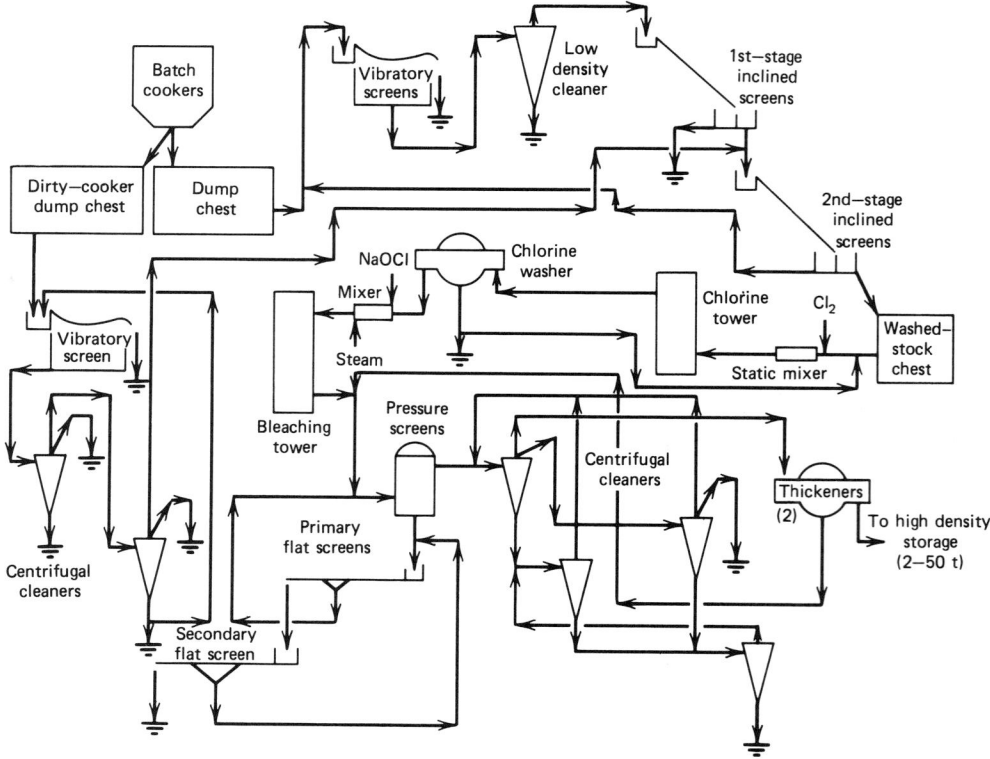

Figure 2. Schematic diagram of the deinking–bleaching process for recycled fibers used in the production of fine papers.

Recycled fiber-based grades of linerboard, newsprint, boxboard, printing papers, and tissue compete effectively in the market at competitive prices with acceptable product function and quality. Deinked, bleached, recycled fiber from colored ledger and printed bleached kraft has strength properties similar to those of bleached, softwood sulfite pulp (13).

Production

The world capacity and production of pulp, paper, and paperboard in 1979 is given in Table 2. Approximately 41×10^6 t of recycled paper was used to achieve this production. All main regions of the world have significantly less pulp capacity compared to paper-and-board production capability. This fact describes the growing opportunity

Table 2. 1979 Pulp, Paper, and Board Capacity and Production, 10^3 t [a]

Region	Capacity pulp	Paper and board	Production pulp	Paper and board
North America	71,224	77,386	64,819	72,370
Europe	47,183	67,097	40,210	58,907
Asia/Australia	18,248	31,239	17,779	30,864
Latin America	5,520	7,812	4,761	7,269
Africa	1,707	1,809	1,370	1,605
Total	*143,882*	*185,343*	*128,939*	*171,015*

[a] Ref. 1.

for paper recycling. Differences in geography, natural resources, population, and standards of living create a complex world-market demand for paper products. However, use of secondary, recycled fiber as a raw-material resource serves this market on the basis of consumer acceptance and cost effectiveness.

The United States is the world's largest consumer and supplier of recycled paper. In 1979, it consumed 12.4×10^6 t of recycled wastepaper and exported another 1.9×10^6 t (1). Total wastepaper recovery was 22% of 64.3×10^6 t of apparent paper and board consumption. It is estimated that the paperboard segment of the world industry used 72% of all recycled paper (29). In 1980, the United States produced 27.8×10^6 t of paperboard of which 5.7×10^6 t was made from recycled fiber (30). Since 1969, wastepaper recovery has been 18–22%, which is down from a high of 35% in 1944 (31). By 1983, wastepaper usage is projected to be 15.9×10^6 t, or 23.2% of total consumption (32). In another projection, wastepaper demand is expected to increase to 23.1×10^6 t in 1985 and 28.1×10^6 t in 1990 at respective recovery rates of 26.1% and 27.4% (33). Most of the increase is in news and corrugated grades. The former is used in deinking mills that manufacture newsprint, and the corrugated grade is used increasingly in linerboard and corrugating medium. These recovery rates for the United States are low compared to those for the United Kingdom and the Federal Republic of Germany, which had respective consumption rates of 52% and 45% in 1979. Recycled paper furnished 38% of all paper and board production in western Europe and 42% in Asia in 1979 (34).

Economic Aspects

The economic future of the recycled-paper industry in the United States was analyzed in 1975 (35). This detailed analysis of the demand and supply for secondary fiber in the United States paper and paperboard industries includes economic-model analysis of deinked newsprint and recycled-linerboard operations compared to their conventional virgin-fiber pulp and paper facilities. Capital investment, energy usage, manufacturing costs, profitability, and product quality are compared. The unit investment required for repulping secondary fiber is lower than for pulping virgin fiber, and the minimum economic size of a secondary-fiber pulping facility was also smaller. Similarly, the total energy consumption, water usage, and pollution liability is less. Paperboard manufacturing with recycled fibers requires ca $(12–14) \times 10^6$ kJ [$(11–13) \times 10^6$ Btu] of total energy and less than 12,000 liters of water per metric ton (1.44 gal/lb) of net salable production (36). The return on investment for a secondary-fiber facility is approximately the same for an integrated virgin-pulp and paper-manufacturing operation.

BIBLIOGRAPHY

1. M. Bayliss and L. Haas, *Pulp Pap.* **54**(8), 64 (1980).
2. J. E. Huber, *Kline Guide To The Paper Industry*, 4th ed., Charles H. Kline & Co., Fairfield, N.J., 1980.
3. D. Hunter, *Papermaking*, 2nd ed., Alfred A. Knopf, Inc., New York, 1947.
4. D. C. Smith, *History of Papermaking In The United States (1691–1969)*, Lockwood Publishing Co., New York, 1970.
5. H. Takahashi, H. Suzuki, and K. Endoh, *Tappi* **62**(7), (1979).
6. I. B. Sachs and T. A. Kuster, *Tappi* **63**(10), (1980).

7. D. H. Page and R. S. Seth, *Tappi* **63**(6), 113 (1980).
8. *Official Board Markets*, January–December 1981, 20 N. Wacker Dr., Chicago, Ill., 60606.
9. Americology Division, American Can Co., 1313 West Mount Veron Ave., Milwaukee, Wisc.
10. A. R. Nollet, E. T. Sherwin, and A. R. Baldwin, *Proceedings of The Technical Association of the Pulp and Paper Industry, 1980 Annual Meeting*, Tappi Press, Atlanta, Ga., 1980.
11. Ing. S. Maule & C. S. p. A., 10095 Grugliasco, Casilla Postale 32, Torino, Italy.
12. N. Lipschutz, *Paper Trade J.* **162**(11), 57 (1978).
13. D. Duncan, *Tappi* **62**(7), 31 (1979).
14. R. G. MacDonald and J. N. Franklin, *Pulp and Paper Manufacture*, 2nd ed., Vol. 3; *Papermaking and Paperboard Making*, McGraw-Hill Book Co., New York, 1970.
15. *Paper Recycling—The Impact of Contaminates, 1973–1985*, American Paper Institute, New York; W. E. Franklin, *Mid-West Research Institute & Franklin Associates*, October 1975.
16. R. Q. Koffinke, *Tappi* **63**(11), 51 (1980).
17. W. E. Crossland, *Recycle Fiber—Proceedings of the Technical Association of the Pulp and Paper Industry, 1980 Annual Meeting*, Tappi Press, Atlanta, Ga., 1980.
18. B. H. Mathews, *Tappi* **55**(6), 695 (1972).
19. R. L. Morgan in ref. 18, p. 697.
20. R. J. Getman and D. E. Lehman, *Tappi* **60**(7), 137 (1977).
21. H. M. Cody, *Pulp Pap.* **52**(11), 123 (1978).
22. H. E. Ortner, *Tappi* **63**(10), 83 (1980).
23. G. Miller in ref. 22, p. 87.
24. V. W. Cancilla, *Refining In The Pulp, Paper, and Board Industry*, Beloit Corporation, Massachusetts, 1975.
25. B. Attwood, *Proceedings of the Technical Association of the Pulp and Paper Industry, 1980 Annual Meeting*, Tappi Press, Atlanta, Ga., 1980.
26. A. J. Killeen, *Pap. Trade J.* **163**(11), 39 (1979).
27. W. C. Bliesner, *Pulp Pap.* **52**(9), 75 (1978).
28. S. Lord, *Pulp Pap.* **54**(7), 77 (1980).
29. J. P. Lucey and L. William, *Tappi* **61**(6), (1978).
30. *American Paper Institute*, Fourdrinier Kraft-Paperboard Group, Monthly Production Statistics, Dec. 1980.
31. P. D. Van Derveer and K. E. Lowe, *Proc. Pulp Pap. Seminar*, Chicago, Ill., May 1974, pp. 19–23.
32. *Paper, Paperboard, Woodpulp Capacity 1979–1982*, American Paper Institute, New York, 1980.
33. *Estimated Paper Industry Usage of Wastepaper For Recycling By Major Grade Category 1972–1990*, Midwest Research Institute, Mo., 1974.
34. W. Williams, *Paper* **192**(6), 32 (1980).
35. *Analysis of Demand and Supply for Secondary Fiber in the United States Paper and Paperboard Industry*, Arthur D. Little, Inc., Acorn Park, Mass., contract study for Resource Recovery Division, EPA, Washington, D.C., March 1975.
36. *1980 Energy Survey*, Boxboard Research and Development Association, Michigan.

JAMES H. ROBINS
JAMES R. GRANT
Mytec, Inc.

PLASTICS

Plastics are common materials in everyday life and, because of their low density, they are more apparent in wastes than might be indicated by statistics of nationwide or worldwide production. As a result, it is often mistakenly believed that plastics represent a serious waste disposal problem. Judging from extensive reports (1), this does not appear to be the case, including if substantial amounts of chlorinated plastics are disposed of with mixed municipal solid waste in municipal incinerators (2).

Plastics are made from hydrocarbon feedstocks for the most part; therefore, they should be recycled as a means of energy conservation (see Plastics processing). Although most plastics can easily be included in waste-derived fuels, more energy conservation may be achieved by recycling as plastic materials. In Table 1, the energy equivalent values of some common plastics are compared to the combustion enthalpies when these plastics are burned as fuels (3). Despite the energy advantages of plastics recycling, little has been accomplished compared to the amount of waste generated (see Fuels from waste).

Many industrial waste-plastic streams are recycled, although some are difficult or impossible to recycle because of inherent limitations in the nature of plastics and because the plastics are often intimately combined with other nonplastic materials. Because of fabrication and service requirements, there are formidable technological difficulties in separating the plastics in forms and to specifications that are suitable for reuse.

Availability of Waste Plastic

In 1974, ca 4.5×10^6 metric tons of plastic waste was discarded in the United States, ca 3.6×10^6 t in the European Economic Community, and ca 1.8×10^6 t in Japan (4). Waste was presumably from all sources: manufacturing, fabrication, industrial, and residential. For example, in the United States discarded plastic from industry accounts for ca 15–20% of the total (4).

Table 1. Energy Equivalent Value and Enthalpy of Some Common Plastics[a]

Plastic	Energy equivalent, MJ/kg[b]	Combustion enthalpy, MJ/kg[b]
polyethylene	69–72	43
polypropylene	73	44
polystyrene (clear)	80	40
polystyrene (high impact)	81	40
poly(acrylonitrile-co-butadiene-co-styrene) (ABS)	84	87
poly(vinyl chloride)	53	18
nylon-6,6 or -6	154–156	29
poly(ethylene terephthalate) (PET)	84	31
polycarbonate	107	29

[a] Ref. 3.
[b] To convert MJ/kg to Btu/lb, multiply by 430.4.

The use of plastics in many applications conserves energy compared to the use of alternative materials. For example, Table 2 indicates the uses of some plastics and the cumulative energy savings over various periods associated with those uses (5). However, the properties that make plastics uniquely suitable for some applications are the same as those that restrict the ability to recycle them.

Origin of Plastic Wastes

Home scrap is the waste and off-specification materials produced as by-products during polymer manufacture and compounding. Much of this is mixed, reformulated, or downgraded in specification and product application as a form of recycling.

Prompt scrap comes from fabrication and conversion operations as off-specification products and trim, eg, cutting sprues and gates. It often is recycled by the fabricator as regrind by reducing it in size, sometimes by pelletizing and further compounding, eg, by adding more heat stabilizer, and by mixing it with virgin material. The specifications for plastic items may include the permissible content of regrind. Its use is a form of recycling.

Some trim and other manufacturing waste may be disposed of because it is not economical to reprocess. Special extruders and mixers are available and are claimed to be suitable for processing such waste. However, the amounts of waste produced may not justify the investment.

Obsolete scrap, ie, postconsumer wastes, is usually destined for disposal and, because these are often in the form of final products or composites, they present the greatest challenge for separation and preparation for recycling. Typical postconsumer plastic wastes occur in mixed municipal waste, source-separated items such as plastic beverage containers, used agricultural mulch, or scrapping of major items such as obsolete automobiles.

Some mixed municipal waste is recycled as one form or another of refuse-derived fuel. The small amount of plastic in such waste, ca 3 wt %, adds to the calorific value of the fuel. This waste is not destined for disposal and the plastics contained in it cannot be considered as part of the waste-plastic stream available for recycling.

Table 2. Cumulative Energy Savings Associated With Plastics Used for Selected Purposes Through 1977 or 1978[a]

Use	Cumulative energy savings, PJ[b]	Years included in the analysis
soft-drink containers	12	5
pipe and fittings	1723	7
blow-molded bottles	155	15
refrigerators and freezers	21	3
household fans	15	2
building insulation[c]	40	3
automobile light-weighting	110	11

[a] Ref. 5.

[b] To convert PJ to 10^{12} Btu, divide by 1.054.

[c] Average annual net energy savings from 30 yr of use of plastic insulation produced in the 3-yr period.

Properties

Broadly, plastics are of two types: thermoplastic and thermosetting. Thermoplastic materials can be reheated and reformed, often several times. Thermosetting materials cannot; the initial heating and fabrication cause permanent chemical changes and subsequent reheating can cause degradation. Most plastics used either in durable goods or single-service items are thermoplastic. Thermosetting plastics are used mostly in durable goods and in much lower quantity than thermoplastics. Thermosetting plastic waste arises during manufacture and when obsolete items or capital goods, eg, automobiles or buildings, are scrapped. Thermoplastic waste is predominantly from packaging, and thermoplastic manufacturing waste is often recycled.

A commercial plastic item is likely to consist of a base resin or polymer mixed with adjuvant ingredients. Molded or extruded items are likely to contain such additives as heat stabilizers, plasticizers, or fillers. Many plastic items contain colorants and sometimes the plastic is foamed. Frequently, plastics are fabricated as part of a composite material or item. For example, plastic films may be made by lamination, coating, or coextrusion. Composite films may consist of mixtures of plastics or combinations of plastics with paper or metal foil.

By the nature of polymer manufacture, the base resins are often quite clean. The compounding of these resins to plastics for manufacture is well controlled so that the properties of the products for fabrication are usually narrowly controlled and tailored for a specific use or fabrication technique. For example, a material intended for extrusion as film is not well-suited for molding. In the recycling of any material, the users' specifications must be met and, in turn, the specifications determine the separation and recycling technology to be used. When plastics are separated from a mixture of materials, especially if they are separated as a mixture of different kinds of plastics, the final properties must be predictable within a consistent and useful range.

A characteristic of plastics is that generally they cannot be used in mixture. Each generic type has a range of mechanical properties when fabricated. When mixed with another, the mixture has the properties of neither but often is weak and cheesy. There have been attempts to overcome this by a number of methods, eg, by adding a compatibilizer to achieve a consistent set of properties. It is best to attempt to separate mixtures according to generic types, but that is not always possible. There are several reviews of the recycling of plastics (6–7). For a general discussion, see ref. 8.

A number of plastic-recycling programs have failed because they did not meet the four essential criteria necessary for successful recycling: a continuous source of suitable scrap, technology for recycling, applications and markets for the products based on the wastes, and an economical enterprise.

Processing

Mixed municipal wastes from domestic, commercial, and light industrial sources in the United States contain less than 4 wt % plastics (9); there are higher concentrations in other countries (10). Some of this plastic is in the form of composite packaging materials and, as such, cannot be considered suitable for separation or recovery. Municipal-waste-separation processes in several European countries reportedly recover a plastic-rich fraction of usable plastics (11). There are similar reports from Japan (12).

The methods used to separate the plastics differ considerably. In a commercial plant in Doncaster, UK, a paper and plastic mixture derived by screening municipal solid waste is separated by sensing the pieces of plastic film and blowing them off a conveyor belt; the sensing is by reflection of a laser beam. In two commercial waste-processing plants built by the Swedish firm AB Svenska Fläktfabriken in Stockholm, Sweden, and in Wijster, The Netherlands, a mixture of paper and plastics from screening and air classification is heated, the plastic pieces shrink, and they are separated from paper in a cyclone device. In the commercial Sorain-Cecchini waste-processing plant in Perugia, Italy, the paper and plastics are separated by proprietary classifiers. These plants and processes are described in refs. 10 and 11. The separated plastic from most of these plants is sold.

In addition, there are several experimental or pilot processes for the separation of plastics from mixed municipal solid waste. A French development is the separation of plastic from paper on a slanted, wet conveyor belt by differences in adhesion of the various pieces (11). In a Japanese method, mixtures of paper and plastics are wetted and the wet paper is pulped through sieve holes, leaving behind a plastics-rich fraction (12). A similar process has been reported in France (11). An experimental Dutch system wets a paper and plastics mixture prior to air classification so that the low density dry plastic can be blown from the wet paper (11).

Research results have been reported for recovering plastics from municipal waste, after concentrating a mixture of paper and plastics, by electrodynamic techniques. The mixture is passed through a device that imposes an electrostatic charge, causing the plastic to separate on a grounded drum (13–14). The method appears to be effective if the paper contains more than a certain critical amount of moisture and if it is not wet. Similar processes have been investigated in the United States and in the UK (13–14).

The processes described above yield a mixture of several types of plastic. One separation method involves the float-sink of the particles in liquids of differing densities (15); however, the method may not take into account fillers and foams. A somewhat more advanced method was developed for sources of industrial plastic waste or sources from the processing of obsolete plastic industrial items, and the technique is similar to froth flotation of minerals (16).

Froth Flotation Separation. A method similar to froth flotation of minerals has been described for the separation of plastics mixtures from various industrial streams, eg, the plastic residues from chopping scrap wire and cable for recovery of the metal conductor or the residue base films after recovery of silver from obsolete x-ray plates (16) (see Flotation). Such mixtures of waste plastics are chopped to pieces ca 0.5 cm in diameter and then passed through a series of froth-flotation cells containing proprietary aqueous mixtures of surfactants. At the same time, air is bubbled through the water solutions to form a froth. Separation is believed to occur because of differences in densities of the plastic pieces and their wettability in the different baths (17). The degree of separation is high and the properties of the separated plastics indicate that they can be used on a commercial scale (16).

The froth-flotation separation was applied in the United States to a mixture of hand-sorted plastic items from municipal solid waste (18). In Japan, a mixture was separated by froth flotation into a residue of dirt and three plastics fractions: polyolefins, polystyrene contaminated with unidentified materials, and poly(vinyl chloride) contaminated with pieces of rubber. Of these, only the mixture of polyolefins had physical properties of any interest or commercial value when molded. The polyolefin mixture is likely to have different properties from one batch to another of municipal waste, as the relative proportions of the various polyolefin plastics may change. Presumably, this does not apply to the processing of industrial waste streams. The method seems more suitable for industrial waste streams of plastics. A similar method has been described in the FRG (19).

Mechanical Methods. Several special extruders and similar devices have been developed to process mixtures of waste plastics for recycling (19). One such machine is the Mitsubishi Reverzer, which is intended to mix intimately a molten mixture of plastics so as to achieve acceptable final physical properties. Several criteria for using this machine have been defined (20). Principally, the feed should not be widely dispersed mixtures of plastics. The mixture must be subjected for a short time to a high rate of shear at high temperature to achieve intensive mixing at a viscosity much lower than that of the normal plastic melt. After homogenization, the melt should be transformed directly into the finished product to avoid an additional processing step. For efficient production, the raw material for recycling must have a roughly constant composition. Other considerations for production have been described (20).

Another device for thermally mixing and extruding mixed plastics from various sources has been developed in Belgium (21). The residence time for the processed plastic is very short so as to allow even the less thermally stable materials, eg, poly(vinyl chloride), to be processed.

A unique machine for the extrusion of mixtures of plastic wastes seems to be the Klobbie (22). This machine, offered by Rehsif S.A. (Switzerland), is an adiabatic extruder that homogenizes and plasticizes the feedstock and intrudes the molten mixture into an externally cooled mold. The possible products are stakes, fence posts, etc, and the device accepts industrial wastes containing up to 15 wt % of nonthermoplastic materials. Sometimes ca 5 wt % paper is added as a reinforcing agent. The content of poly(vinyl chloride) in the feed is kept below 15%. The machine's function can be likened more to extrusion–injection molding than to continuous extrusion.

Agricultural Mulch. A method of recycling waste agricultural film, used as mulch, has been developed by the Plastic Waste Management Institute and Horai Iron and Steel Works (23). The film is shredded and then passed through a water-wash to remove sand and other foreign matter. The washed film is shredded again and another water-wash tank of different design is used to separate polyethylene from poly(vinyl chloride). The plastics are mechanically dehydrated and dried. The various water effluents are cleaned. No properties of the final products or costs have been published.

Household Wastes. The mixed plastic wastes from household source separation in Japan have been used to encapsulate the residues from incinerators so as to reduce possible leaching of heavy metals (24). An additional benefit of the separation is that it reduces the volume of the plastics waste. The plastic wastes and incinerator residue are prepared in narrow size ranges, mixed, and heated in a rotating kiln to form plastic-coated pellets.

Compost. A pilot process for separating plastic film from compost derived from municipal solid waste has been described and operates in Vienna, Austria (25). When household refuse that is collected in plastic bags is composted, the plastic is left behind more or less intact. The plastic film is removed by screening, passed through a series of water washes, dried, and granulated. It is estimated from pilot tests carried out at near full scale with regard to plant parameters that the granulated material could be produced at 55–70% of the price of new material.

Plastic Beverage Bottles. Plastic beverage bottles consist of a body made from poly(ethylene terephthalate) (PET) and sometimes with a foot or cup of medium density polyethylene. Bottles returned for recycling are likely to include an aluminum cap and an adhesively bonded paper or metal foil label. These various components must be separated before the plastics can be reused. The general method of separation is to chop the bottle into small pieces and to use float–sink techniques, air separation, or both. Many of the methods are proprietary but there is no reason why the proper combination of air or liquid-density separations could not achieve usable polyethylene and PET (26).

The PET is not likely to be suitable for reuse as a beverage container because of possible contamination which could interfere with blow molding or with the cleanliness required for a food package. Apparently, some of the recovered PET is used for molded objects, some for making low grade fiber, eg, fiberfill, some in forming new thermosetting plastics, and some in chemical conversions to unsaturated polyester molding resins (26–27). A likely chemical conversion step would be reaction with a glycol to form a liquid polyester and subsequent incorporation of a difunctionality for later reaction to a thermosetting resin. The largest use for recycled PET is spinning for use as fiberfill batting, eg, for use as an interliner in clothing or for upholstery.

Obsolete Telephone Equipment. Obsolete equipment, eg, handsets, is reground and remolded to parts, trays, and similar objects (28). This type of recycling downgrades the use of the plastic to a less demanding application.

Scrap Automobile Upholstery. The polyurethane foam commonly used in automobile upholstery is a thermosetting resin (see Urethane polymers). There are several reports of attempts to recycle either prompt or obsolete scrap foam by chemical conversion to useful monomers; two of the principal methods are hydrolysis and alcoholysis. Hydrolysis can be achieved by several methods, eg, in an extruder (8,29). A recent report of alcoholysis addresses the types of products recovered, which depend on the particular polyester or polyethers used as starting materials in making the foam (30). None of the processes appear to have been used commercially in 1981.

Scrap Nylon and Polyacrylonitrile Fiber. A report from the People's Republic of China describes the chemical conversion of scrap nylon-6 and polyacrylonitrile fiber or textile waste (31). The nylon-6 (polycaprolactam) is thermally cracked to form caprolactam and resultant impurities must be removed later. The scrap polyacrylonitrile is saponified with NaOH, acidified to form poly(acrylic acid), and separated by extraction with alcohol. The same report addresses recovery of terephthalic acid from PET by alkaline hydrolysis, decolorization with activated carbon, and acidification (see Acrylonitrile polymers; Polyamides).

Other Chemical Conversions. Commercial and experimental chemical conversion of plastics are of interest. A few polymers, notably poly(methyl methacrylate), unzip to regenerate monomer when heated. This method has been used on a commercial scale to produce monomer from turnings and other sources of suitable prompt scrap (see Methacrylic polymers). Polytetrafluoroethylene also generates monomer when heated in a vacuum (see Fluorine compounds, organic). Many more common plastics do not.

Waste plastics can be recycled to a variety of chemical products by pyrolysis (32); the plastic waste might be pyrolyzed to a fuel gas or liquid. There have been proposals and some experimental data for acetylation, hydrogenation, and etherification (33). The thermodynamics and kinetics of several possible reactions of polyolefins, polystyrene, and poly(vinyl chloride) have been calculated (34) (see Olefin polymers; Styrene plastics; Vinyl Polymers). Experimental data have been published for the metal-catalyzed oxidation of polyethylene with oxygen or ozone. The reported products are similar to sugars and conceivably could be used as feed for biological conversion or as a chemical intermediate (35). Pyrolysis of certain waste plastics, particularly polyethylene, forms olefinic linkages. The pyrolysis can be accomplished in the presence of a reactive molecule. An example is the pyrolysis of polyethylene in the presence of maleic anhydride to form copolymers containing carboxyl groups. The pyrolysis of poly(vinyl chloride) yields hydrochloric acid. Waste plastic of this type can be a source of HCl.

Economic Aspects

Cost calculations of the various processes described here are rough estimates at best, because few operating plants have been built beyond the pilot stage. The full-scale operating plants in Europe for the processing of mixed municipal waste and separation of plastics are not good measures of the cost of recycling plastics. The approximate cost of processing mixed plastics to useful products may be within the range reported in two papers describing widely different processing methods (21,27). In the first, an industrial plant was designed to recycle 2400 t/yr of plastic wastes obtained by source separation from households in Liege, Belgium; 25% of the plant was to be paid for by the European Economic Community. The mixed plastics would be processed through the F.N. Plastifer for extrusion to useful shapes. Few details are given but the total investment is estimated at ca $700,000 with ca $200,000 for buildings. The plant was intended to employ 13 people working in two shifts. The economics of the plant are given in Table 3.

The other published estimates of the cost of processing collected plastic waste is for the conversion of returned PET beverage containers to unsaturated molding resins. The chopped PET is heated with propylene glycol to effect an alcoholysis; a separate step is the reaction with maleic anhydride (27). The resin produced is dissolved with 45 wt % styrene prior to use. Processing costs for energy, labor, overhead, etc, were estimated to be ca $325/h for an 18,000-kg reactor system. Table 4 shows the

Table 3. Approximate Costs Associated With Processing Source-Separated Plastics Through Extrusion[a]

Costs	$/kg[b]
wages	0.066
electricity	0.037
gas	0.011
water	0.011
depreciation	0.040
financial costs	0.033
research and maintenance	0.008
Total	0.206

[a] Ref. 21.
[b] Original costs were reported in Belgian francs, converted as 1 BF = $0.02.

dependence of the finished cost on the price paid for the starting material, ie, PET. However, all prices shown for the starting material are considerably above the prices being paid for such plastic waste in 1981. The prices shown in Table 4 presumably include costs of preparation needed to achieve the required cleanliness.

Table 4. Calculated Costs for PET-Modified Unsaturated Polyesters With 45 wt % Styrene[a,b]

Price paid for recovered PET, $/kg	Finished cost, $/kg
0.55	0.875
0.66	0.906
0.77	0.939

[a] Ref. 27.
[b] Propylene glycol, $0.77/kg; styrene, $0.79/kg; maleic anhydride, $0.99/kg; processing costs, $325/h.

Uses

Once separated from a waste stream, the thermoplastic material must be heated to be reformed. Thermoplastics contain heat stabilizers so that the initial processing and subsequent forming does not cause excessive degradation or decomposition. Some plastics are more thermally labile than others and so require addition of more heat stabilizer prior to recycling. Chemical stability after subsequent reforming of separated plastics is discussed in ref. 36.

There have been proposals of various new applications for recycled plastics (7,37–38). Few, if any, have been commercialized, usually because the user has not recognized the chemical nature of plastics as it affects recycling. As an example, old polyethylene milk bottles have been chopped and fed into a molding machine or extruder. When the plastic was heated, it gave off unbearable noxious odors, probably from the burning of the milk residues. The key to widespread use of separated waste plastics is to meet specifications for common uses and not be dependent on a few customers for specialty applications. Previous recycling in the form of novelty items failed when people were no longer satisfied with the novelty (37).

Outlook

Few of the methods described in this article are commercially practiced. Plastics recovery from mixed wastes present formidable problems, eg, the mixture of plastics and many other materials that result from automobile shredding (39). However, automobile shredding produces probably the largest stream of obsolete plastics and this situation will continue as manufacturers take advantage of the high strength-to-weight ratio of plastics and use more in future automobile models.

Various methods have been proposed for the separation of waste-plastics mixtures from shredded automobiles and are based on liquid elutriators, which rely on density differences (39). Little is known as to whether even well-separated plastics from this source retain their mechanical properties after many years of use.

A recent proposal is to reuse plastics in mixtures by fabrication involving foaming and sintering (40). It is difficult to consider seriously complex processes for relatively low cost scrap plastics, which must compete with relatively low cost virgin plastics. The economics of recycling will also favor the use of relatively clean or homogeneous streams of prompt or obsolete scrap plastics. Many of these are recycled. Any future technology also must be capable of consistently producing products to specification.

BIBLIOGRAPHY

1. A. J. Warner, C. H. Parker, and B. Baum, *Solid Waste Management of Plastics*, Manufacturing Chemists Association, Washington, D.C., 1970; B. Baum and C. H. Parker, *Plastics Waste Management*, Manufacturing Chemists Association, Washington, D.C., 1974.
2. E. R. Kaiser and A. A. Carotti, *Municipal Incineration of Refuse with 2% and 4% Additions of Four Plastics*, Society of the Plastics Industry, New York, 1971.
3. H. Kindler and A. Nikles, *Kunststoffe* **70**(12), 802 (1980).
4. J. Milgrom, *2nd Recycling World Congress*, Manila, Philippines, March, 1979, Exhibitions for Industry, Ltd. Oxted, England.
5. Franklin Associates, Ltd., *Energy Savings Associated with Selected Uses of Plastics*, Society of the Plastics Industry, New York, 1979.
6. W. A. Mack in *Disposal of Plastics with Minimum Environmental Impact*, STP 533, American Society for Testing and Materials, Philadelphia, Pa., 1973, pp. 1–16.
7. R. J. Sperber and S. L. Rosen, *Polym. Plast. Technol. Eng.* **3**(2), 215 (1974).
8. J. Milgrom, *Conservation Recycling* **3**(3/4), 327 (1979).
9. *Resource Recovery and Waste Reduction: Fourth Report to Congress*, Report SW-600, U.S. Environmental Protection Agency, Washington, D.C., 1977.
10. H. Alter and J. J. Dunn, Jr., *Solid Waste Conversion to Energy, Current European and U.S. Practice*, Marcel Dekker, New York, 1980, Chapt. 3.
11. *Ibid.*, Chapt. 4.
12. I. Ito and Y. Hirayama, *Resource Recovery Conservation* **1**(1), 45 (1975).
13. M. Grubbs and K. H. Ivey, "Recovering Plastics from Urban Refuse by Electrodynamic Techniques" *Bureau of Mines Technical Progress Report 50*, U.S. Department of Interior, 1972.
14. M. J. Pearce and T. J. Hickey, *Resource Recovery Conservation* **3**(2), 179 (1978).
15. J. L. Hulman, J. B. Stephenson, and J. W. Jensen, "Processing the Plastics from Urban Refuse" in ref. 13.
16. K. Saitoh, I. Nagana, and S. Izumi, *Resource Recovery Conservation* **2**(2), 127 (1976).
17. H. Alter, *J. Adhes.* **9**, 135 (1978).
18. H. Alter, unpublished, 1964.
19. A. Bahr in K. J. Thomé-Kozmiensky, ed., *Recycling Berlin '79*, Vol. 2, E. Freitag-Verlag für Umwelttechnik, Berlin, Germany, 1979, pp. 1202–1207.
20. H. Verity Smith, *First Recycling World Congress*, Basel, Switzerland, 1978, Exhibitions for Industry, Ltd., Oxted, England.
21. G. Micheels and P. Adriaenssens, *3rd Recycling World Congress*, Basel, Switzerland, 1980, Exhibitions for Industry, Ltd., Oxted, UK.
22. H. Verity Smith in ref. 19, pp. 1199–1201.
23. T. Ohnuma in ref. 21.
24. Y. Nabeshima, H. Kajimoto, and S. Tomizawa in ref. 21.
25. C. Hemmer in ref. 21.
26. *Mod. Plast. Int.* **10**(4), 64 (1980).
27. R. Calendine, M. Palmer, and P. von Bramer, *Mod. Plast.* 64 (May 1980).
28. L. A. Squitieri in *Disposal of Plastics with Minimum Environmental Impact*, STP 533, American Society for Testing and Materials, Philadelphia, Pa., 1973, pp. 40–49.
29. E. Gruber in ref. 19, pp. 1208–1212.
30. B. Prajsnar in ref. 19, pp. 1213–1218.
31. Xu Weiye in ref. 21.
32. W. Kaminsky, *Resource Recovery Conservation* **5**(3), 205 (1980).
33. J. F. Barbour, R. R. Groner, and V. H. Freed, *The Chemical Conversion of Solid Wastes to Useful*

Products, Report EPA 670/2-74-027, U.S. Environmental Protection Agency, Cincinnati, Ohio, 1974, 167 pp.
34. M. E. Banks, W. D. Lusk, and R. S. Ottinger, *New Chemical Concepts for Utilization of Waste Plastics*, Report SW-16c, U.S. Environmental Protection Agency, Washington, D.C., 1971, 129 pp.
35. H. Alter, *Ind. Eng. Chem.* **52,** 121 (1960).
36. G. Scott, *Resource Recovery Conservation* 1(4), 381 (1976).
37. International Research & Technology, *Recycling Plastics, A Survey and Assessment of Research and Technology*, Society of the Plastics Industry, New York, 1973, 56 pp.; J. L. Holman, J. B. Stephenson, and M. J. Adam, "Recycling of Plastics from Urban and Industrial Refuse" in *Bureau of Mines Report of Investigations 7955*, U.S. Department of Interior, 1972.
38. J. R. Lawrence in ref. 28, pp. 50–62.
39. J. H. Bilbrey, Jr., J. W. Sterner, and E. G. Valdez in ref. 20.
40. G. Menges and E. Haberstroh in ref. 19, pp. 1193–1197.

HARVEY ALTER
Chamber of Commerce of the United States

RUBBER

A technique for vulcanizing rubber into a useful material was developed in the 1800s, and a rubber-reclaiming process based on steam pressure to devulcanize natural rubber was developed in 1858 (1). Today, high cost silicone rubber polymers are reclaimed in much the same way, although most synthetic polymers require complicated reclaiming techniques. Rubber in asphalt (qv), scrap rubber as fuel, rubber pyrolysis, rubber reuse, eg, tire splitting and as reefs and crash barriers, etc, and other studies were conducted as rubber-recycling technology advanced. However, the discovery of plastics and oil-extended rubbers has led to a reduction in the use of rubber as a reclaimed material except for more costly polymers, eg, silicones and fluorocarbons. It is uneconomical to pyrolyze rubber scrap because of uncompetitive costs and few product markets, especially in the United States. In most instances, it is more expensive to prepare and burn scrap rubber than to burn natural gas, fuel oil, or coal. Higher fuel costs and petroleum scarcity in Europe and other parts of the eastern hemisphere make rubber reuse as a fuel source more economical there than in the United States. Rubber reuse will become more prevalent in the United States as fuel and petroleum-derivative costs for polymers increase. Approximately 67% of the scrap rubber, primarily as tires, is used as landfill (2). Owners of landfills charge up to $3.00 per tire, and disposal costs at municipal landfills are ca $0.10–0.20 per tire (3–4). As the economics become more favorable, rubber reuse will gain more recognition in resource-conservation and -recovery activities (5–6).

Fuel Source

The use of scrap rubber for fuel is one of the best alternatives for reusing rubber as natural-gas and fuel-oil costs increase. Whole tires and 2.5-cm tire chips are the most economical fuels because of shredding-cost savings. Tires that contain more than 90% organic materials have a heat value of ca 32.6 MJ/kg (ca 14,000 Btu/lb). Coal varies from 18.6 to 27.9 MJ/kg (ca 8000–12,000 Btu/lb). A cyclonic, rotary-hearth, whole-tire-fired boiler was operated by Goodyear Tire and Rubber Co. from 1975 to 1977 and was designed to burn 1400 kg of automobile tires per hour, thus generating 11,300 kg of steam per hour (5,7). In the Lucas furnace, tires are conveyed into an airtight vestibule and then onto the outer rim of a rotating hearth. The vestibule prevents flashback fires and limits furnace air leaks. An air-velocity head of 5.1 cm provides the necessary turbulence for combustion (8). Residues from the burning tires form a char that increases combustible heat loss and tends to clog furnace grates, but the carbon-black content also reduces slagging problems. Improper combustion at the ash-removal area of the furnace may prevent burning of the carbon black. The Lucas-Goodyear tire-burning furnace was shut down because of mechanical problems and failure to comply with Michigan's air-pollution emission standards (9).

Shredded tire chips have been burned successfully in stoker-fired boilers. Uniroyal fired a 15% mixture of tire chips with coal and both General Motors and B.F. Goodrich burned a 10% tire-chip mixture with coal (10–12). Tire-grinding size-reduction problems and supplier delivery costs have stymied cofiring projects based on tire and coal fuel. The Lucas furnace was developed to burn tires without size reduction. Scrap-tire transportation can cost $0.04/kg, not including grinding costs, and thus limits tire-fired boiler facilities to areas with ample scrap-tire supplies, eg, large cities or tire manufacturers. The cost of burning one metric ton of tires per hour in a tire incinerator was ca $0.20–$0.40 per tire in 1974, which escalates to $0.35–0.70 per tire today, based on 10% straight-line inflation (13).

The Japanese are using whole scrap tires to fuel portland-cement kilns. The process was developed by the Bridgestone Tire Company and has grown substantially since 1980. The Saitama plant of Nihon Co. Ltd. burns 140,000 tires per month as well as oil to produce 287,000 t of cement per month. Use of tires as fuel is considered a waste-tire treatment system (14).

Pyrolysis

Scrap-tire pyrolysis has been the subject of several research studies by rubber, oil, and carbon-black interests throughout the world (see Rubber compounding; Petroleum; Carbon, carbon black). The Tosco II process pyrolysis-research study was conceived to develop process equipment and to maximize carbon-black production and quality (15–17). The Tosco II process is shown schematically in Figure 1 (5). Chopped tires are fed into a rotary drum with hot ceramic balls at 480–549°C in a reducing atmosphere. The rubber pyrolyzes and forms a solid residue, an oil vapor, and off-gases. Condensed oil separates in a fractionator, and the gas is used for fuel to heat the ceramic balls. A trommel screen separates the fine carbon black from the ceramic balls. The carbon is pelletized after steel, fiber glass, and other contaminants have been removed. The off-gas is basically a combination of ethylene, propylene, and butylene. The oil, which contains about 1% sulfur, can be substituted directly for fuel

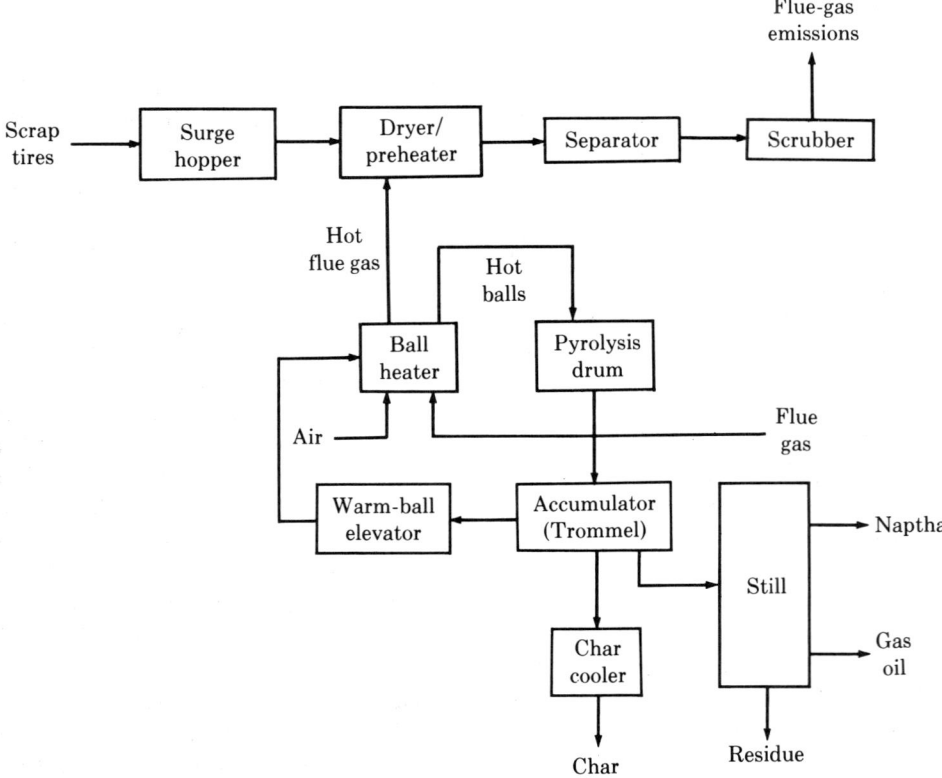

Figure 1. Tosco II process.

oil. In the Tosco II process, higher temperatures produce more gas and less liquid. The pilot-plant process was designed to handle 15 t of tires per day, and generally one ton of tires produced 0.5–0.6 m^3 (3–4 bbl) of oil, 1270–1540 kg of carbon black, 190–220 kg of steel, and 154–176 kg of fiber glass. The Tosco II project has been completed, but further work has halted partly because the carbon-black industry in the United States already has excess capacity. Also, the types of blacks and other related contents in the residue from tire pyrolysis, eg, ZnO, possibly other inorganic materials, and glass fiber from old tires, are unpredictable. However, carbon blacks are carefully chosen to achieve specific properties in compounded rubbers. Thus, the diverse and unpredictable residue mixture is useful only as a low grade filler for mechanical goods and is uncompetitive as a carbon-black source.

Foster Wheeler has two pyrolysis plants operating in the Federal Republic of Germany, and other European countries and Japan are pyrolyzing scrap rubber. Tyrolysis is building a 136-t/d pyrolysis plant in the United Kingdom (18), and the company will use vertical cross-flow reactors (19). Intennco in Houston has recently shut down a 45-t/d scrap-tire pyrolysis pilot plant. In the Intennco process, two reactors in series pyrolyzed shredded tires with indirect heat at ca 540°C in a reducing atmosphere. The process yielded ca 0.5 cubic meters (ca 3 bbl) of aromatic oil (30–40% light oil) per ton of scrap rubber. Contaminants in the carbon-black char were removed by several confidential processing steps (20). In the United States, tire pyrolysis has been studied as a method for recovering carbon black; however, production of carbon black from petroleum oils is less expensive, easier to control, and better in quality.

Whole tires have been pyrolyzed in a semifluidized-bed reactor (21). The tires are pyrolyzed on a tilting grate and the grate tilts to discharge the steel belt and bead wire. This test unit has potential because whole tires are pyrolyzed, thus eliminating rubber-grinding costs.

Nippon Zeon also had a tire-pyrolysis pilot plant. The company estimated that the break-even cost was $0.25 per tire (22–23). A recent study indicates that pyrolysis of tires and other polymers should be considered as a means for disposing of scrap within environmental constraints. However, a plant processing 90,000 t/yr of scrap could be profitable, based on reclaimed-product sales (24).

Other techniques include oxidative, steam-atmosphere (25), and molten-salt (26) pyrolyses. Rubber pyrolysis is an exothermic reaction in a partial-air atmosphere. The reaction rate and ratio of char black and oil products are controlled by the oxygen-flow rate. Rubber pyrolysis in a steam atmosphere results in a cleaner char with a greater surface area than char pyrolyzed in an inert atmosphere. The steam-pyrolyzed char impairs the physical properties of cured rubber. Because of the greater surface area, this black could be used for activated carbon, but production costs are prohibitive. Molten-salt baths produce pyrolyzed char and oil products. The product characteristics and quantities vary, depending upon the salt used. Recovery of the carbon black from the molten salt also is a problem.

Depolymerized Scrap Rubber

In the depolymerized-scrap-rubber (DSR) experimental process, ground scrap-rubber tires are used to produce a carbon-black dispersion in oil (27). Initially, aromatic oils are blended with the tire crumb, and the mixture is heated at 250–275°C in an autoclave for 12–24 h. The oil acts as a heat-transfer medium and swelling agent, and the heat and oil cause the rubber to depolymerize. As more DSR is produced and rubber is added, less aromatic oil is needed, and eventually virtually 100% of the oil is replaced by DSR. The DSR reduces thermal oxidation of polymers and increases the tack of uncured rubber (28–29). Depolymerized scrap rubber has a heat value of 40 MJ/kg (17,200 Btu/lb) and has been blended with No. 2 fuel oil as a fuel extender (30).

In Asphalt

The United States generates ca 220×10^6 scrap tires per year. Approximately 58 t of reclaimed rubber and 16,000 t of crumb rubber were produced in 1977 (31). Approximately 4500 t of reclaimed and crumb rubber were used in asphalt–rubber compounds in 1980, which is less than 5% of the recycled rubber produced.

There are several methods for mixing and applying asphalt rubber to roadways. One conventional method is to mix the rubber and asphalt at ca 175–220°C for 1–2 h. The hot rubber asphalt is applied to the roadway and is covered with a layer of stone chips to form a chip seal. The rubber crumb is usually scrap tires that are ground into particles that are less than 2 mm in diameter. Besides chip seals, rubber asphalt is used for waterproofing membranes, crack-and-joint sealers, hot-mix binders, and roofing materials (see Waterproofing and water repellancy; Sealants). The rubber improves asphalt ductility and increases the temperature at which the asphalt softens. The asphalt and aggregate adhesive bond is stronger, and long-term asphalt durability increases. Production of asphalt rubber has increased from 450 t in 1970 to 30,000 t in 1980 (32). About 2.2 t/km of rubber is used for typical asphalt-rubber application. If it is assumed that rubber asphalt contains ca 25% rubber and 75% asphalt, potential demand for scrap rubber is ca 45,000 t/yr or ca 2% of the amount available. In 1980, the United States used ca 4,500 t of rubber in rubber asphalt.

Reclaiming

Three basic processes are used in rubber reclaiming: digester, heater or pan, and the Reclaimator processes (33–34). Goodyear has recently devulcanized and reclaimed scrap rubber with microwaves (35). Tires are most commonly reclaimed by digesting. Two-roll mills or other grinding devices reduce whole tires to relatively uniform particle size. Fiber is mechanically separated from the rubber with hammer mills, blown into collectors, and baled. Metal chlorides might also be used to reduce tire fiber chemically during digesting. Reclaiming oils and processing aids are blended with the crumb rubber in ribbon blenders or similar mixers and are transferred to a digester, which is a steam-pressurized tank equipped with horizontal mixing paddles. The blend is mixed continuously at steam pressures of 1.03–1.70 MPa (10–17 atm) for 4–6 h. The pressurized digester batch is forced into a blowdown tank and is washed and dried. Compounding ingredients, ie, carbon black, clay, etc, are added to modify and maintain certain physical and chemical properties of the rubber for specific applications. Metal and other contaminants are strained from the digested rubber by extruders. High friction refining mills smooth the digested rubber into sheets. The number of mill passes vary, depending on the desired smoothness and physical properties needed in the final product. The reclaim is then baled, extruded into pellets, or made into slabs for shipment.

Butyl and natural-rubber tubes and other fiber-free scrap rubber are reclaimed by means of the heater or pan process. Brass tube fittings and other metal are removed from the scrap. The scrap is mechanically ground, mixed with reclaiming agents, loaded into pans or devulcanizing boats, and autoclaved at steam pressures of 1.03–1.40 MPa (10–14 atm) for 3–8 h. The reclaim is refiner-milled, extruded, and finish-milled much the same as in the digester process.

The Reclaimator, a high pressure extruder, devulcanizes fiber-free rubber continuously with reclaiming oils and other materials. High pressure and shear between the rubber mixture and the extruder-barrel walls effectively reclaim the rubber mixture. Devulcanizing occurs at 175–205°C in 1–3 min. The Lancaster-Banbury method applies high temperature, pressure, and shear to the rubber in a batch process that is otherwise similar to the Reclaimator. Another high pressure process devulcanizes scrap rubber and reclaiming agents at 5.5–6.9 MPa (54–68 atm) for ca 5 min. The reclaim product is milled, baled, or pelletized as in other processes.

The reclaiming oils and chemicals are complex wood and petroleum derivatives that swell the rubber and provide access for oxidizing the rubber bonds with heat, pressure, chemicals, and mechanical shearing. Approximately 2–4 pt of oil are used per 100 pt of scrap rubber. Some examples of reclaiming oils include monocyclic and mixed terpenes, ie, pine-tar products, saturated polymerized petroleum hydrocarbons; aryl disulfides in petroleum oil; cycloparaffinic hydrocarbons; and alkyl aryl polyether alcohols.

Natural-rubber (NR) scrap is reclaimed with solvent naphthas, terpenes, dipentenes, and resins that swell, tackify, and aid in bond cleavage (see Rubber, natural). Tires, mainly synthetic styrene–butadiene rubber (SBR), are less susceptible to oxidation. Aryl disulfides, phenyl disulfides, high molecular weight mercaptans and other sulfur-containing chemicals are used to swell and lubricate the rubber bonds and to devulcanize the rubber by chemical oxidation and bond cleavage. The reclaiming oils control the reclaimed-rubber chemical and physical properties. Oils soften the rubber and increase the elongation. Reinforcing agents and fillers (qv), eg, carbon black, aluminum silicate, and calcium carbonate, also control the reclaimed-rubber properties and aid processing. Nonreinforcing fillers reduce the tack of devulcanized rubber and make the reclaimed rubber easier to handle, and they abraid vulcanized particles during finish milling so as to make the texture smoother. Reinforcing fillers increase the tensile strength of the reclaimed material.

Reclaimed rubber can be used by itself in adhesives (qv) and solid-rubber-tire compounds, or it can be blended with virgin rubber. Compounding is based on rubber hydrocarbon content (RHC). To make a compound with 100 pt RHC, 70 pt of virgin rubber and 60 pt of reclaim that is 50% RHC are needed. Rubber compounds with reclaim require the same fillers, reinforcing agents, and plasticizers as the virgin compound. However, reclaim already contains some zinc oxide and sulfur; usually 0.5 pt of zinc oxide and 2.5 pt of sulfur are used per 100 pt of reclaim.

Tires, natural-rubber tubes, and butyl tubes are the main sources of scrap and reclaim. Specialty reclaims are made from scrap silicone, chloroprene (CR), nitrile–butadiene (NBR), and ethylene–propylene–butadiene-*ter*-polymer (EPDM) rubber scraps. Tires, hoses, belts, molded and extruded goods, and asphalt products consume ca 80% of the reclaim manufactured. Reclaimed rubber is used in the tire carcass and sidewall compounds. Reclaimed natural rubber is used in cements, dispersions, and pressure-sensitive tape, and butyl reclaim is used in tubes and inner liners for tires.

Until the 1960s, reclaimed rubber was a principal raw material in many molded and extruded rubber goods, ie, tires, rubber mats, and hard-rubber battery cases. With the advent of vinyls and other plastics and less expensive oil-extended synthetic polymers, reclaimed-rubber sales stabilized and then began to decrease. In 1973, oil embargoes and the resulting increased energy costs caused the energy-intensive rubber-reclaiming process costs to rise sufficiently to match virgin-polymer costs. Increased radial-tire production also led to the need for more crack-resistant rubber compounds than could be provided by reclaimed-rubber compounds.

Tire Retreading

About 36×10^6 automobile tires and 13×10^6 truck, bus, and off-the-road tires were retreaded in 1974 (36), and it is doubtful that retreading has increased since then with the increase in radial-tire sales. Retreading has been the most cost-effective alternative to recycling rubber. However, worn retreaded tires usually are discarded in a landfill. Approximately 50% of discarded tires are retreadable and only about half of these are retreaded, since the remaining tires are not inspected before they are discarded.

Grinding

Rubber-reclaiming devulcanization, pyrolysis, rubber in asphalt, and other recycling processes generally use ground scrap-rubber tires. The tires are mechanically ground sometimes by cryogenic or solvent swelling techniques. In one process, a polar solvent is used to swell the rubber and then to reduce particle size by shearing (37). In the rubber-reclaiming industry, ground tire-crumb rubber is commonly referred to as rubber reclaim, even though the rubber has not been devulcanized.

Generally, tires are mechanically ground with a two-roll, grooved rubber mill. The two-mill rolls turn at a ratio of ca 1:3, thus providing the shearing action necessary to rip the tire apart. The rubber chunks are screened and the larger material is recycled until the desired size is reached. Bead wire is removed by hand or with magnets. For most applications, ie, devulcanizing, pyrolyzing, asphalt, crumb-rubber particles smaller than 1.19 mm (16 mesh) are desired. Several milling steps are required to reduce the rubber to this size. Tire fiber is removed in intermediate operations with hammer mills, reel beaters, and air tables that blow a steady stream of air across the rubber, thus separating the fiber. The fiber is baled or is sold as landfill. Because tire grinding is very energy-intensive, tire slitters have been used to cut the tire initially. Steel-belted tires usually are not mechanically ground because the steel contaminates the rubber.

Cryogenics (qv) in conjunction with mechanical action has been used to make crumb rubber. Nitrogen cools the rubber below the glass transition temperature, and the brittle rubber is pulverized in a grinding mill. A small cryogenic system can be installed at the site to recycle process scrap rubber crumb into the compound mixing process. Cryogenic grinding of different rubber types, related costs, and rubber-compound properties are well documented (37–41). Cryogenic-grinding costs are $0.20/kg–$0.40/kg, depending on the desired particle size and the type of rubber. Generally, harder rubber tends to grind more easily. Ground-rubber scrap can be devulcanized, pyrolyzed, or recycled directly into the rubber compound. Ground rubber also has been added to plastics (42).

Other Uses

Scrap whole tires have been used for artificial fishing reefs, oyster beds, and as a floating breakwater. Goodyear has installed more than 2000 fishing reefs made from old tires. One of the largest reefs is made of 3×10^6 tires and stretches ca 2.4 km off Ft. Lauderdale, Florida (43). Tires have also been used as impact absorbers around highway and bridge abutments (44). Tires are also used in playgrounds, flower planters, and shoe soles. The tire-splitting industry cuts tires into pieces for gaskets, shims, dock bumpers, shock absorbers, blasting mats, etc. Rubber is sliced from tire tread to make strips for floor mats. Approximately 3×10^6 of the estimated 200×10^6 scrap tires generated each year are reused as reefs or by tire splitters.

Economic Aspects

Roughly 75% of the discarded tires are disposed of in landfills; 20% are retreaded; and 5% are reclaimed, burned for fuel, split, etc. Tire-disposal costs are $0.10–$3.00 per tire. Cost for incineration without heat recovery is $0.35–$0.70 per tire. Dis-

Table 1. Comparative Fuel Costs

Source	$/GJ[a]
coal (at $44/t)[b]	1.58
oil (No. 6 fuel oil at $0.195/L[c])[b]	4.66
natural gas (at $70/10^3 m^3 [d])[b]	1.89
rubber	
whole tire (at $0.06/kg)	2.03
ground (at $0.20/kg)	6.77

[a] To convert $/GJ to $/10^6 Btu, multiply by 1.054.
[b] Ref. 45.
[c] To convert L to gal, divide by 3.785.
[d] To convert m^3 to ft^3, multiply by 35.31.

carded-tire transportation can cost $0.04/kg, and size reduction can cost $0.20/kg–$0.60/kg.

Probably the best alternative for rubber recycling on a large scale would be as fuel, although the fuel-cost increments for natural gas and coal are lower than scrap rubber. Costs of various fuels are compared in Table 1. As fuel costs escalate, there will be more incentive to use scrap tires for fuel.

BIBLIOGRAPHY

1. G. Crane, R. A. Elefritz, E. L. Kay, and J. R. Layman, *Rubber Chem. Technol.* **51,** 577 (1978).
2. R. H. Snyder, V. R. Vincent, and F. C. Querry, *paper presented at the National Tire Disposal Symposium*, June 1977, Washington, D.C.
3. Personal communication with A-1 Refuse Service, Crow and Sons Sanitary Landfill and Pit, Estes' Service Co., and West Side Sanitary Landfill, Sept. 3, 1981.
4. Personal communication with the cities of Fort Worth, Arlington, and Dallas, Texas, Sept. 3, 1981.
5. L. L. Gaines and A. M. Wolsky, *Energy Conservation Through Alternative Uses ANL/CNSV-5*, Argonne National Laboratory, Dec. 1979.
6. W. J. Markiewicz and M. J. Granksy, *Solid Waste Management Series SW-22*, U.S. Department of Health, Education, and Welfare, PB203619, 1972.
7. E. R. Moats, *Resource Recov. Conserv.* 1(3), 315 (1976).
8. M. Weintraub, A. A. Orning, and C. H. Schwartz, *Experimental Studies of Incineration in a Cylindrical Combustion Chamber*, U.S. Bureau of Mines RI 6908, U.S. Department of the Interior, Washington, D.C., 1967.
9. Anonymous; and personal conversation with D. Kennedy, Dyna Electronics Corp., McLean, Va., Dec. 1981 and Jan. 1982.
10. R. Niles, Uniroyal, Inc., Oxford, Ct., personal communication, June 1981.
11. F. Querry, *Paper presented at National Tire Disposal Symposium*, June 1977, Washington, D.C.
12. R. H. Taggart, *Shredded Tires as Auxillary Fuel*, General Motors Report, March, 1975 and personal communication with Warren Underwood (GM), Jan. 1982.
13. C. C. Humpstone, E. Ayres, S. G. Keahey, and T. Schell, *Internat. Res. Technol.* **PB234602** (1974).
14. *A Unique and Effective Waste Tire Treatment Process*, Japan External Trade Organization (Jetro), Jan. 1980, pp. 32–35.
15. B. L. Schulman and P. A. White, *Pyrolysis of Scrap Tires Using the Tosco II Process—Progress Report*, rept. ACS Symp. Ser. 76 (1978).
16. C. E. Haberman, *paper presented at Rubber Division ACS Symposium*, May, 1977, Chicago, Ill.
17. *Chem. Eng.* **52** (Aug. 2, 1976).
18. Personal communication with Paul White, Tosco Co., Los Angeles, Calif., Dec. 21, 1981.
19. R. Fletcher and H. T. Wilson, *Resource Recov. Conserv.* **5**(4), 333 (1981).
20. J. E. Lunde, private conversation on July 22, 1981.

21. W. Kaminsky, *Resource Recov. Conserv.* **5**(3), 205 (1980); W. Kaminsky, W. Menzel, and H. Sinn, *Conserv. Recycl.* **1**(1), 91 (1976); W. Kaminsky and H. Sinn, *Kunstoffe* **68**(5), 14 (1978); H. Sinn, W. Kaminsky, and J. Janning, *Angew. Chem. Int. Ed. Engl.* **15**, 660 (1976).
22. Personal communication with Dr. M. Matsuo, Nippon Zeon of America, New York, Dec. 21, 1981.
23. Y. Saeki and G. Suzuki, *Rubber Age* **108**, 33 (Feb. 1976).
24. G. P. Bracker, *Conserv. Recycl.* **4**(3), 161 (1981).
25. J. A. Beckman, G. Crane, R. A. Elefritz, E. L. Kay, and J. R. Laman, *paper presented at the National Tire Disposal Symposium*, June 14–15, 1977, Washington, D.C.
26. J. W. Larsen and B. Chang, *Rubber Chem. Technol.* **49**, 1120 (1976).
27. G. Crane and E. L. Kay, *paper presented at Rubber Division, ACS Symposium*, Oct. 1974, Philadelphia, Pa.
28. G. Crane, J. W. Fieldhouse, and E. L. Kay, *Rubber Chem. Technol.* **48**, 62 (1975).
29. G. Crane, E. L. Kay, and L. B. Wakefield, *J. Elastomers Plast.* **7**, 372 (1975).
30. G. Crane and E. L. Kay, *Rubber Chem. Technol.* **48**, 50 (1975).
31. *Industrial Recovered Materials Utilization Targets for the Rubber Industry*, prepared for U.S. DOE, Assistant Secretary for Conservation and Solar Energy, Office of Industrial Programs, by Hittman Associates, Inc., 1980.
32. *Data Collection and Analysis Pertinent to the PA's Development of Guidelines for Procurement of Highway Construction Products Containing Recovered Materials*, Draft, Vol. 1, Issues and Technical Summary, EPA Contract **68-01-6014**, by Franklin Associates, Ltd., and Valley Forge Laboratory, Inc., July 6, 1981.
33. J. P. Paul, *Chemtech* **9**, 104 (Feb. 1979).
34. D. S. le Beau, *Rubber Chem. Technol.* **40**, 217 (1967).
35. *Chem. Week*, 35 (May 23, 1979).
36. W. J. Sears, *paper presented at EPA conference*, Apr. 2, 1975. Copies available from the Rubber Manufacturers Association, 1901 Pennsylvania Avenue, NW, Washington, D.C., 20006.
37. W. O. Murtland, *Elastomerics* **110**, 26 (Mar. 1978).
38. L. J. Ricci, *Chem. Eng.*, (July 4, 1977).
39. W. O. Murtland, *Elastomerics* **109**(12), 39 (1977).
40. D. J. Zolin, N. B. Frable, and J. F. Gintilcore, *paper presented at the 112th Meeting of ACS Rubber Division*, Cleveland, Ohio, Oct. 4–7, 1977.
41. M. C. Kazarnowicz, E. C. Osmundson, J. F. Boyle, and R. W. Savage in ref. 40.
42. D. Tuchman and S. L. Rosen in ref. 40.
43. *Rubber World* **67** (Oct. 1978).
44. E. R. Moats, *paper presented at the National Tire Disposal Symposium*, June 14–15, 1977, Washington, D.C.
45. U.S. DOE, Energy Information Administration, National Energy Information Center, Washington, D.C., December 21, 1981.

General References

Refs. 7, 19, and 21 are general references.
W. J. Search and T. E. Ctvrtnicek, "Resource Recovery Systems for Nonrecappable Rubber Tires," *Resource Recov. Conserv.* **2**(2), 159 (1976).

JOHN PAUL
Pedco Environmental, Inc.

RECYCLING, WATER. See Water, reuse.

REFRACTION. See Analytical methods.